Greenland
(Kalatdlit-Nunat)
Alaska
Yukon
N.W.T.
Nunavut
[Labrador]
B.C.
Alta.
Sask.
Man.
Nfld.
Que.
St.Pierre
and Miquelon
P.E.I.
Ont.
N.B.
N.S.
Wash.
Mont.
N.Dak.
Minn.
Vt.
Maine
N.H.
Mass.
Oreg.
Idaho
Wyo.
S.Dak.
Wis.
Mich.
N.Y.
R.I.
Conn.
Iowa
Pa.
N.J.
Nev.
Utah
Nebr.
Ill.
Ind.
Ohio
W.Va.
Del.
Md.
Colo.
Kans.
Mo.
Ky.
Va.
Calif.
N.C.
Tenn.
Okla.
Ark.
S.C.
Ariz.
N.Mex.
Ala.
Ga.
Miss.
Tex.
La.
Fla.
Mexico

Flora of North America

Contributors to Volume 26

James D. Ackerman
Geraldine A. Allen
Loran C. Anderson
George W. Argus
John T. Atwood
Richard M. Bateman
Rudolf W. Becking
John Beckner
Norlyn L. Bodkin
David J. Bogler
Paul Martin Brown
Frederick W. Case Jr.
Paul M. Catling
Mark W. Chase
Anita F. Cholewa
Karen H. Clary
Ronald A. Coleman
Robert W. Cruden
Robert L. Dressler
Germán Carnevali Fernández-Concha
P. L. Fiedler
Raymond O. Flagg
Walter S. Flory†
John V. Freudenstein
Dawn Frame
Peter Goldblatt
Douglas H. Goldman
Edward W. Greenwood†
Katharine B. Gregg
Eric Hágsater
Roger L. Hammer
Douglass M. Henderson†
Norlan C. Henderson
Alan Herndon
William J. Hess
Tim Hogan
Wendy C. Hodgson
Walter C. Holmes
Michael A. Homoya
Charles N. Horn
T. D. Jacobsen
J. A. Jernstedt
Jacquelyn A. Kallunki
Shoichi Kawano
James V. LaFrankie
Deborah Q. Lewis
C. A. Luer
Lawrence K. Magrath
Elizabeth McClintock
Dale W. McNeal Jr.
J. McNeill
Susan J. Meades
Max E. Medley
Loyal A. Mehrhoff
Randall Morgan
Bryan Ness
John G. Packer
J. Chris Pires
Tom A. Ranker
Lauren Raz
James L. Reveal
R. Laurie Robbins
Kenneth R. Robertson
Gustavo A. Romero-González
Fayla Schwartz
Aaron D. Shaw
H. L. Sherman
Charles J. Sheviak
Mark W. Skinner
Gerald L. Smith
Gerald B. Straley†
Victoria I. Sullivan
Frederick H. Utech
Susan Verhoek
R. David Whetstone
Heather L. White
R. K. Zebell

Editors for Volume 26

George W. Argus,
Taxon Editor for Orchidales

Kanchi Gandhi,
Nomenclatural Editor

Peter Goldblatt,
Co-editor for Iridaceae

William J. Hess,
Co-editor for Agavaceae

Robert W. Kiger,
Lead Editor, Bibliographic Editor,
and Taxon Editor for Liliales

John L. Strother,
Reviewing Editor
and Co-editor for Liliaceae

Frederick H. Utech,
Co-editor for Liliaceae

James L. Zarucchi,
Managing Editor

Volume Composition

Pat Harris
Douglas R. Harrison
Kay Yatskievych
Claire A. Hemingway

Iris brevicaulis

Flora of North America

North of Mexico

Edited by FLORA OF NORTH AMERICA EDITORIAL COMMITTEE

VOLUME 26

Magnoliophyta: Liliidae: Liliales and Orchidales

LILIALES (Lily order)
ORCHIDALES (Orchid order)

NEW YORK OXFORD • OXFORD UNIVERSITY PRESS • 2002

Oxford University Press

Oxford New York
Auckland Bangkok Buenos Aires Cape Town Chennai
Dar es Salaam Delhi Hong Kong Istanbul Karachi Kolkata
Kuala Lumpur Madrid Melbourne Mexico City Mumbai Nairobi
São Paulo Shanghai Singapore Taipei Tokyo Toronto

Published by Oxford University Press, Inc.
198 Madison Avenue, New York, New York 10016
www.oup-usa.com

Library of Congress Cataloging-in-Publication Data
(Revised for volume 26)
Flora of North America north of Mexico
edited by Flora of North America Editorial Committee.
Includes bibliographical references and indexes.
Contents: v. 1. Introduction—v. 2. Pteridophytes and gymnosperms—
v. 3. Magnoliophyta: Magnoliidae and Hamamelidae—
v. 22. Magnoliophyta: Alismatidae, Arecidae, Commelinidae (in part), and Zingiberidae—
v. 26. Magnoliophyta: Liliidae: Liliales and Orchidales

ISBN 0-19-515208-5 (v. 26)
1. Botany—North America.
2. Botany—United States.
3. Botany—Canada.
I. Flora of North America Editorial Committee.
QK110.F55 2002 581.97 92-30459

9 8 7 6 5 4 3 2 1
Printed in the United States of America
on acid-free paper

We dedicate this volume to

Gerald B. Straley

1945–1997

botanist, colleague, and friend

FLORA OF NORTH AMERICA EDITORIAL COMMITTEE

Contents

FOUNDING MEMBER INSTITUTIONS

Flora of North America Association

Arnold Arboretum
Jamaica Plain, Massachusetts

Biosystematics Research Institute
Ottawa, Ontario

Canadian Museum of Nature
(now part of Eastern Cereal and Oilseed Research Centre)
Ottawa, Ontario

Carnegie Museum of Natural History
Pittsburgh, Pennsylvania

Field Museum of Natural History
Chicago, Illinois

Fish and Wildlife Service
United States Department of the Interior
Washington, D.C.

Harvard University Herbaria
Cambridge, Massachusetts

Hunt Institute for Botanical Documentation
Carnegie Mellon University
Pittsburgh, Pennsylvania

Jacksonville State University
Jacksonville, Alabama

Jardin Botanique de Montréal
Montréal, Québec

Kansas State University
Manhattan, Kansas

Missouri Botanical Garden
St. Louis, Missouri

New Mexico State University
Las Cruces, New Mexico

New York State Museum
Albany, New York

Northern Kentucky University
Highland Heights, Kentucky

The New York Botanical Garden
Bronx, New York

The University of British Columbia
Vancouver, British Columbia

The University of Texas
Austin, Texas

Université de Montréal
Montréal, Québec

University of Alaska
Fairbanks, Alaska

University of Alberta
Edmonton, Alberta

University of California
Berkeley, California

University of California
Davis, California

University of Idaho
Moscow, Idaho

University of Illinois
Urbana-Champaign, Illinois

University of Iowa
Iowa City, Iowa

University of Kansas
Lawrence, Kansas

University of Michigan
Ann Arbor, Michigan

University of Oklahoma
Norman, Oklahoma

University of Ottawa
Ottawa, Ontario

University of Southwestern Louisiana
Lafayette, Louisiana

University of Western Ontario
London, Ontario

University of Wyoming
Laramie, Wyoming

Utah State University
Logan, Utah

For their support of the preparation of this volume, we gratefully acknowledge and thank:

National Science Foundation
Chanticleer Foundation
The David and Lucile Packard Foundation
National Fish and Wildlife Foundation
The William and Flora Hewlett Foundation
The Andrew W. Mellon Foundation
The Fairweather Foundation
ESRI

Project Staff — past and present
involved with the preparation of this volume

Barbara Alongi, *Illustrator*
Cheryl Casey, *Technical Editor and Acting Map Editor*
Gwen R. Ericson, *Technical Editor*
Lisa Ferrugia, *Editorial Coordinator*
Tricia L. Frye, *Map Editor*
Bee F. Gunn, *Illustrator*
Pat Harris, *Editorial Assistant and Compositor*
Douglas R. Harrison, *Editorial Assistant and Compositor*
Claire A. Hemingway, *Technical Editor and Compositor*
Helen K. Jeude, *Senior Technical Editor*
Daniel Murray, *Mapping Assistant*
John Myers, *Illustrator*
Gina Otterson, *Departmental Secretary*
Elizabeth Polen, *Technical Editor*
Mary Ann Schmidt, *Technical Editor*
Anne Keats Smith, *Map Editor*
Judith Unger, *Project Coordinator*
Yevonn Wilson-Ramsey, *Art Director and Illustrator*
Kay Yatskievych, *Production Coordinator and Compositor*
Eleanor Zeller†, *Clerk*

Contributors

James D. Ackerman
Department of Biology
University of Puerto Rico
San Juan, Puerto Rico

Geraldine A. Allen
Department of Biology
University of Victoria
Victoria, British Columbia

Loran C. Anderson
Department of Biological Science
Florida State University
Tallahassee, Florida

George W. Argus
Merrickville, Ontario

John T. Atwood
Marie Selby Botanical Gardens
Sarasota, Florida
and *Missouri Botanical Garden*
St. Louis, Missouri

Richard M. Bateman
Department of Botany
Natural History Museum
London, England

Rudolf W. Becking
College of Natural Resources and Sciences
Humboldt State University
Arcata, California

John Beckner
Marie Selby Botanical Gardens
Sarasota, Florida

Norlyn L. Bodkin
Harrisonburg, Virginia

David J. Bogler
Missouri Botanical Garden
St. Louis, Missouri

Paul Martin Brown
Native Orchids of North America Journal
Ocala, Florida (winter)
Acton, Maine (summer)

Frederick W. Case Jr.
Saginaw, Michigan

Paul M. Catling
Metcalfe, Ontario

Mark W. Chase
Molecular Systematics Section
Royal Botanic Gardens, Kew
Richmond, Surrey, England

Anita F. Cholewa
University of Minnesota
St. Paul, Minnesota

Karen H. Clary
Austin, Texas

Ronald A. Coleman
Tucson, Arizona

Robert W. Cruden
University of Iowa
Iowa City, Iowa

Robert L. Dressler
Micanopy, Florida

Germán Carnevali Fernández-Concha
Centro de Investigacion Cientifica de Yucatán
Mérida, Yucatán, Mexico

P. L. Fiedler
Piedmont, California

Raymond O. Flagg
Carolina Biological Supply Company
Burlington, North Carolina

Walter S. Flory†
Wake Forest University
Winston-Salem, North Carolina

Dawn Frame
University Herbarium
University of California, Berkeley
Berkeley, California

John V. Freudenstein
Department of Evolution, Ecology, and Organismal Biology
Herbarium
Ohio State University
Columbus, Ohio

Peter Goldblatt
Missouri Botanical Garden
St. Louis, Missouri

Douglas H. Goldman
Plant Biology Department
L. H. Bailey Hortorium
Cornell University
Ithaca, New York

Edward W. Greenwood†
Herbario AMO
Mexico City, Mexico

Katharine B. Gregg
Department of Biology
West Virginia Wesleyan College
Buckhannon, West Virginia

Eric Hágsater
Herbario AMO
Mexico City, Mexico

Roger L. Hammer
Miami-Dade Parks Department
Homestead, Florida

Douglass M. Henderson†
University of Idaho Herbarium
Moscow, Idaho

Norlan C. Henderson
Kansas City, Missouri

Alan Herndon
Tradewinds Tropicals
Perrine, Florida

William J. Hess
Morton Arboretum
Lisle, Illinois

Wendy C. Hodgson
Desert Botanical Garden
Phoenix, Arizona

Tim Hogan
University of Colorado Herbarium
Boulder, Colorado

Walter C. Holmes
Department of Biology
Baylor University
Waco, Texas

Michael A. Homoya
Indiana Department of Natural Resources
Indianapolis, Indiana

Charles N. Horn
Department of Biology
Newberry College
Newberry, South Carolina

T. D. Jacobsen
Hunt Institute for Botanical Documentation
Carnegie Mellon University
Pittsburgh, Pennsylvania

J. A. Jernstedt
Agronomy and Range Science
University of California
Davis, California

Jacquelyn A. Kallunki
The New York Botanical Garden
Bronx, New York

Shoichi Kawano
Department of Botany
Kyoto University
Kyoto, Japan

James V. LaFrankie
Smithsonian Tropical Research Institute
Balboa, Ancon, Republic of Panama

Deborah Q. Lewis
Department of Botany
Iowa State University
Ames, Iowa

C. A. Luer
Missouri Botanical Garden
St. Louis, Missouri

Lawrence K. Magrath
University of Science and Arts of Oklahoma
Chickasha, Oklahoma

Elizabeth McClintock
San Francisco, California

Dale W. McNeal Jr.
Department of Biological Sciences
University of the Pacific
Stockton, California

J. McNeill
Royal Botanic Garden
Edinburgh, Scotland
and *Royal Ontario Museum*
Toronto, Ontario

Susan J. Meades
Department of Biology
Algoma University College
Sault Ste. Marie, Ontario

Max E. Medley
Department of Biology
University of Louisville
Louisville, Kentucky

Loyal A. Mehrhoff
National Park Service
Fort Collins, Colorado

Randall Morgan
Soquel, California

Bryan Ness
Department of Biology
Pacific Union College
Angwin, California

John G. Packer
University of Alberta
Edmonton, Alberta

J. Chris Pires
Department of Agronomy
University of Wisconsin
Madison, Wisconsin

Tom A. Ranker
University Museum Herbarium
University of Colorado
Boulder, Colorado

Lauren Raz
Institute of Systematic Botany
The New York Botanical Garden
Bronx, New York

James L. Reveal
Montrose, Colorado

R. Laurie Robbins
Department of Biological Sciences
Emporia State University
Emporia, Kansas

Kenneth R. Robertson
Center for Biodiversity
Illinois Natural History Survey
Champaign, Illinois

Gustavo A. Romero-González
Harvard University Herbaria
Cambridge, Massachusetts

Fayla Schwartz
Department of Biology
Everett Community College
Everett, Washington

Aaron D. Shaw
Stockton, California

H. L. Sherman
Mississippi University for Women
Columbus, Mississippi

Charles J. Sheviak
Biological Survey
New York State Museum
Albany, New York

Mark W. Skinner
United States Department of Agriculture
Natural Resources Conservation Service
National Plant Data Center
Baton Rouge, Louisiana

Gerald L. Smith
Department of Biology
High Point University
High Point, North Carolina

Gerald B. Straley†
The Botanical Garden
University of British Columbia
Vancouver, British Columbia

Victoria I. Sullivan
New Iberia, Louisiana

Frederick H. Utech
Hunt Institute for Botanical Documentation
Carnegie Mellon University
Pittsburgh, Pennsylvania

Susan Verhoek
Lebanon Valley College
Annville, Pennsylvania

R. David Whetstone
Jacksonville State University
Jacksonville, Alabama

Heather L. White
Department of Biology
Baylor University
Waco, Texas

R. K. Zebell
Natural Areas Program
Recreation and Park Department
San Francisco, California

Taxonomic Reviewers

James D. Ackerman
University of Puerto Rico
San Juan, Puerto Rico

John T. Atwood
Marie Selby Botanical Gardens
Sarasota, Florida
and *Missouri Botanical Garden*
St. Louis, Missouri

Kerry A. Barringer
Brooklyn Botanic Garden
Brooklyn, New York

Paul M. Catling
Metcalfe, Ontario

Mark W. Chase
Royal Botanic Gardens, Kew
Richmond, Surrey, England

Eric A. Christenson
Sarasota, Florida

Phillip J. Cribb
Royal Botanic Gardens, Kew
Richmond, Surrey, England

Robert W. Cruden
University of Iowa
Iowa City, Iowa

Lawrence J. Davenport
Samford University
Birmingham, Alabama

Robert L. Dressler
Micanopy, Florida

Susan Farmer
University of Tennessee
Knoxville, Tennessee

John V. Freudenstein
Ohio State University Herbarium
Columbus, Ohio

Peter Goldblatt
Missouri Botanical Garden
St. Louis, Missouri

Edward W. Greenwood†
Herbario AMO
Mexico City, Mexico

Eric Hágsater
Herbario AMO
Mexico City, Mexico

James W. Hardin
Raleigh, North Carolina

Robert R. Haynes
University of Alabama
Tuscaloosa, Alabama

Norlan C. Henderson
Kansas City, Missouri

Wesley Higgins
Cape Coral, Florida

Mark Hooten
Fort Myers, Florida

Kathleen L. Hornberger
Widener University
Chester, Pennsylvania

Ken Inoue
Shinshu University
Matsumoto, Japan

David Jarrell
Mary Washington College
Fredericksburg, Virginia

Robert Kral
Vanderbilt University Herbarium
Botanical Research Institute of Texas
Fort Worth, Texas

Joseph E. Laférriere
Woonasquatucket Valley Botanical Foundation
Providence, Rhode Island

C. A. Luer
Missouri Botanical Garden
St. Louis, Missouri

Hiltje Maas
State University of Utrecht
Utrecht, The Netherlands

Paul J. M. Maas
State University of Utrecht
Utrecht, The Netherlands

Lawrence K. Magrath
University of Science and Arts of Oklahoma Herbarium
Chickasha, Oklahoma

Leonid I. Malyshev
Siberian Central Botanical Garden
Novosibirsk, Russia

James R. Massey
University of North Carolina
Chapel Hill, North Carolina

Brian F. Mathew
Royal Botanic Gardens, Kew
Richmond, Surrey, England

Dale W. McNeal Jr.
University of the Pacific
Stockton, California

Alan W. Meerow
United States Department of Agriculture
Agricultural Research Service
Subtropical Horticulture Research Station
Miami, Florida

Bryan Ness
Pacific Union College
Angwin, California

Jany Renz†
Basel, Switzerland

James L. Reveal
Montrose, Colorado

R. Laurie Robbins
Emporia State University
Emporia, Kansas

Gustavo A. Romero-González
Harvard University Herbaria
Cambridge, Massachusetts

Liz Slauson
Desert Botanical Garden
Phoenix, Arizona

Gerald L. Smith
High Point University
High Point, North Carolina

Victoria Sosa
Instituto de Ecologia, A. C.
Xalapa, Veracruz, Mexico

Victor G. Soukup
University of Cincinnati
Cincinnati, Ohio

John W. Thieret
Northern Kentucky University
Highland Heights, Kentucky

Thomas K. Todsen
New Mexico State University
Las Cruces, New Mexico

Norris H. Williams
University of Florida
Gainesville, Florida

Wendy B. Zomlefer
University of Georgia
Athens, Georgia

Regional Reviewers

ALASKA

Alan Batten
University of Alaska Museum
Fairbanks, Alaska

Robert Lipkin
Alaska Natural Heritage Program
University of Alaska
Anchorage, Alaska

Mary Stensvold
Forest Service, Alaska Region
United States Department of Agriculture
Sitka, Alaska

PACIFIC NORTHWEST

Geraldine A. Allen
University of Victoria
Victoria, British Columbia

Edward R. Alverson
The Nature Conservancy
Eugene, Oregon

Adolf Ceska
British Columbia Conservation Data Centre
Victoria, British Columbia

Kenton L. Chambers
Oregon State University
Corvallis, Oregon

George W. Douglas
British Columbia Conservation Data Centre
Victoria, British Columbia

Richard R. Old
XID Services, Inc.
Pullman, Washington

Jim Pojar
British Columbia Forest Service
Smithers, British Columbia

Scott Sundberg
Oregon State University
Corvallis, Oregon

Peter F. Zika
Seattle, Washington

SOUTHWESTERN UNITED STATES

Roxanne L. Bittman
California Department of Fish and Game
Sacramento, California

Steve Boyd
Rancho Santa Ana Botanic Garden
Claremont, California

Barbara J. Ertter
University of California
Berkeley, California

H. David Hammond
Flagstaff, Arizona

G. Frederic Hrusa
Courtland, California

Charles T. Mason Jr.
University of Arizona
Tucson, Arizona

James Morefield
Nevada Natural Heritage Program
Carson City, Nevada

Donald J. Pinkava
Arizona State University
Tempe, Arizona

WESTERN CANADA

William T. Cody
Agriculture and Agri-food Canada
Crop Protection Program
Central Experimental Farm
Ottawa, Ontario

Bruce A. Ford
University of Manitoba
Winnepeg, Manitoba

Vernon L. Harms
University of Saskatoon
Saskatoon, Saskatchewan

ROCKY MOUNTAINS

Bonnie Heidel
Wyoming Natural Diversity Database
University of Wyoming

Tim Hogan
University of Colorado Museum
Boulder, Colorado

B. E. Nelson
Rocky Mountain Herbarium
Department of Botany
University of Wyoming

J. Stephen Shelly
U.S.D.A. Forest Service, Northern Region
Missoula, Montana

NORTH CENTRAL UNITED STATES

William T. Barker
North Dakota State University
Fargo, North Dakota

Anita F. Cholewa
University of Minnesota
St. Paul, Minnesota

Theodore S. Cochrane
University of Wisconsin
Madison, Wisconsin

Patricia Folley
Noble, Oklahoma

Neil A. Harriman
University of Wisconsin
Oshkosh, Wisconsin

Robert B. Kaul
University of Nebraska
Lincoln, Nebraska

Gary E. Larson
South Dakota State University
Brookings, South Dakota

Deborah Q. Lewis
Iowa State University
Ames, Iowa

Ronald L. McGregor
University of Kansas
Lawrence, Kansas

Lawrence R. Stritch
Midewin National Tallgrass Prairie
Wilmington, Illinois

Connie Taylor
Southeastern Oklahoma State University
Durant, Oklahoma

George Yatskievych
Missouri Botanical Garden
St. Louis, Missouri

SOUTH CENTRAL UNITED STATES

David E. Lemke
Southwest Texas State University
San Marcos, Texas

Jackie M. Poole
Texas Parks and Wildlife Department
Austin, Texas

Robert C. Sivinski
New Mexico Energy, Mineral, and Natural Resources Department
Santa Fe, New Mexico

EASTERN CANADA

Jacques Cayouette
Agriculture and Agri-Food Canada
Ottawa, Ontario

William J. Crins
Ontario Ministry of Natural Resources
Peterborough, Ontario

Stuart G. Hay
Herbier Marie-Victorin
Université de Montréal
Montréal, Québec

Harold Hinds†
University of New Brunswick
Fredrickton, New Brunswick

John K. Morton
University of Waterloo
Waterloo, Ontario

Michael J. Oldham
Natural Heritage Information Centre
Peterborough, Ontario

Anton A. Reznicek
University of Michigan
Ann Arbor, Michigan

Claude Roy
Université Laval
Québec, Québec

Marian Zinck
Nova Scotia Museum of Natural History
Halifax, Nova Scotia

GREENLAND

Bent Fredskild
University of Copenhagen
Copenhagen, Denmark

Geoffrey Halliday
University of Lancaster
Lancaster, England

NORTHEASTERN UNITED STATES

Ray Angelo
New England Botanical Club
Cambridge, Massachusetts

Steven E. Clemants
Brooklyn Botanic Garden
Brooklyn, New York

Tom S. Cooperrider
Kent State University
Kent, Ohio

Arthur Haines
New England Wild Flower Society
Framingham, Massachusetts

Les Mehrhoff
University of Connecticut
Storrs, Connecticut

Edward G. Voss
University of Michigan
Ann Arbor, Michigan

SOUTHEASTERN UNITED STATES

Raymond Cranfill
University of California
Berkeley, California

John B. Nelson
University of South Carolina
Columbia, South Carolina

Bruce A. Sorrie
Longleaf Ecological
Whispering Pines, North Carolina

R. Dale Thomas
University of Louisiana at Monroe
Monroe, Louisiana

Lowell E. Urbatsch
Louisiana State University
Baton Rouge, Louisiana

Alan S. Weakley
Association for Biodiversity Information/The Nature Conservancy
Durham, North Carolina

Thomas F. Wieboldt
Virginia Polytechnic Institute and State University
Blacksburg, Virginia

B. Eugene Wofford
University of Tennessee
Knoxville, Tennessee

FLORIDA

Loran C. Anderson
Florida State University
Tallahassee, Florida

Edwin L. Bridges
Botanical and Ecological Consultant
Bremerton, Washington

Bruce F. Hansen
University of South Florida
Tampa, Florida

Walter S. Judd
University of Florida
Gainesville, Florida

Preface

In its makeup, the Flora of North America Editorial Committee has changed somewhat over the years since the project's inception but has been remarkably stable overall. Since 2000, when Volume 22 was published, Richard Rabeler has joined the Committee as Taxon Editor, and Craig Freeman has taken on responsibilities as Taxon Editor in addition to his role as Regional Coordinator. Also since then, Kanchi Gandhi has assumed the duties of Nomenclatural Editor, replacing John McNeill, who continues to serve in the capacity of Nomenclatural Advisor.

Editorial processing for this volume of the *Flora* was coordinated by the project's center at the Hunt Institute for Botanical Documentation. The maps were prepared by the center at The Arboretum at Flagstaff / Northern Arizona University. Pre-press production, including typesetting and layout, was done by the center at the Missouri Botanical Garden. As well, the St. Louis center provided coordination for all aspects of planning and executing the illustrations included in this volume.

In addition to her duties as the project's Art Director, Yevonn Wilson-Ramsey prepared the illustrations for numerous taxa: Iridaceae (including the color frontispiece), Liliaceae (except the eight genera credited to others below), and *Cypripedium* in the Orchidaceae. Other illustrations were prepared by John Myers, for Burmanniaceae, Dioscoreaceae, Haemodoraceae, and Pontederiaceae; by Barbara Alongi, for Smilacaceae, Orchidaceae except *Cypripedium,* and *Aletris, Amianthium, Chamaelirium, Tulipa,* and *Uvularia* in the Liliaceae; by Bee F. Gunn, for Agavaceae, Aloaceae, and Stemonaceae; and by Kimberly J. Martin, a student intern at the Missouri Botanical Garden, for *Crinum, Galanthus,* and *Hippeastrum* in the Liliaceae.

We remain deeply grateful to the very many people who have helped to create and sustain the *Flora.*

Introduction

Scope of the Work

Flora of North America North of Mexico is a synoptic account of the plants of North America north of Mexico: the continental United States of America (including the Florida Keys and Aleutian Islands), Canada, Greenland (Kalâtdlit-Nunât), and St. Pierre and Miquelon. The *Flora* is intended to serve both as a means of identifying plants within the region and as a systematic conspectus of the North American flora.

The *Flora* will be published in 30 volumes. Volume 1 contains background information that is useful for understanding patterns in the flora. Volume 2 contains treatments of ferns and gymnosperms. Families in volumes 3–26, the angiosperms, are arranged according to the classification system of A. Cronquist (1981). Bryophytes will be covered in volumes 27–29. Volume 30 will contain the cumulative bibliography and index.

The first two volumes were published in 1993, Volume 3 in 1997, and Volume 22, the first of five covering the monocotyledons, appeared in 2000. The correct bibliographic citation for the *Flora* is: Flora of North America Editorial Committee, eds. 1993+. *Flora of North America North of Mexico*. 5+ vols. New York and Oxford.

Volume 26 contains treatments of 11 families, 177 genera, and 908 species. For additional statistics, please refer to Table 1.

Contents · General

The *Flora* includes accepted names, selected synonyms, literature citations, identification keys, descriptions, chromosome numbers, phenological information, summaries of habitats and geographic ranges, and other biological observations. Economic uses, weed status, and conservation status are provided from specified sources. Each volume contains a bibliography and an index to the taxa included in that volume. The treatments, written and reviewed by experts from throughout the systematic botanical community, are based on original observations of herbarium specimens and, whenever possible, on living plants. These observations are supplemented by critical reviews of the literature.

Table 1. *Statistics for Volume 26 of Flora of North America.*

Family	Total Genera	Total Species	Endemic Genera	Endemic Species	Introduced Genera	Introduced Species	Conservation Taxa
Pontederiaceae	4	11	0	0	2	3	0
Haemodoraceae	2	2	1	1	0	0	0
Liliaceae	70	478	19	373	16	54	76
Iridaceae	16	92	0	61	9	19	10
Aloaceae	1	2	0	0	1	2	0
Agavaceae	9	84	0	38	2	6	10
Stemonaceae	1	1	0	1	0	0	0
Smilacaceae	1	20	0	16	0	0	1
Dioscoreaceae	1	6	0	2	0	4	0
Burmanniaceae	3	5	0	1	0	0	1
Orchidaceae	69[a]	207[b]	3	96	1	6	18
Totals	177[a]	908[b]	23	589	31	94	116

Italic = introduced

[a]does not include *Spathoglottis* [see footnote on page 491]

[b]does not include *Spathoglottis plicata* [see footnote on page 491]

Basic Concepts

Our goal is to make the *Flora* as clear, concise, and informative as practicable so that it can be an important resource for both botanists and nonbotanists. To this end, we are attempting to be consistent in style and content from the first volume to the last. Readers may assume that a term has the same meaning each time it appears and that, within groups, descriptions may be compared directly with one another. Any departures from consistent usage will be explicitly noted in the treatments (see also References).

Treatments are intended to reflect current knowledge of taxa throughout their ranges worldwide, and classifications are therefore based on all available evidence. Where notable differences of opinion about the classification of a group occur, appropriate references are mentioned in the discussion of the group.

Documentation and arguments supporting significantly revised classifications are published separately in botanical journals before publication of the pertinent volume of the *Flora*. Similarly, all new names and new combinations are published elsewhere prior to their use in the *Flora*. No nomenclatural innovations will be published intentionally in the *Flora*.

Taxa treated in full include extant and recently extinct native species, hybrids that are well established (or frequent), and waifs or cultivated plants that are found frequently outside cultivation and give the appearance of being naturalized. Taxa mentioned only in discussions include waifs or naturalized plants now known only from isolated old records and some nonnative, economically important or extensively cultivated plants, particularly when they are relatives of native species. Excluded names and taxa are listed at the ends of appropriate sections, e.g., species at the end of genus, genera at the end of family.

Treatments are intended to be succinct and diagnostic but adequately descriptive. Characters and character states used in the keys are repeated in the descriptions. Descriptions of related taxa at the same rank are directly comparable.

With few exceptions, taxa are presented in taxonomic sequence. If an author is unable to produce a classification, the taxa are arranged alphabetically, and the reasons are given in the discussion.

Treatments of hybrids follow that of one of the putative parents. Hybrid complexes are treated at the ends of their genera, after the descriptions of species.

We have attempted to keep terminology as simple as accuracy permits. Common English equivalents usually have been used in place of Latin or Latinized terms or other specialized terminology, whenever the correct meaning could be conveyed in approximately the same space, e.g., "pitted" rather than "foveolate," but "striate" rather than "with fine longitudinal lines." See *Categorical Glossary for the Flora of North America Project* (R. W. Kiger and D. M. Porter 2001; also available online at http://huntbot.andrew.cmu.edu) for standard definitions of generally used terms. Very specialized terms are defined, and sometimes illustrated, in the relevant family or generic treatments.

References

Authoritative general reference works used for style are *The Chicago Manual of Style*, ed. 14 (University of Chicago Press 1993); *Webster's New Geographical Dictionary* (Merriam-Webster 1988); and *The Random House Dictionary of the English Language*, ed. 2, unabridged (S. B. Flexner and L. C. Hauck 1987). *B-P-H/S. Botanico-Periodicum-Huntianum/Supplementum* (G. D. R. Bridson and E. R. Smith 1991) has been used for abbreviations of serial titles, and *Taxonomic Literature*, ed. 2 (F. A. Stafleu and R. S. Cowan 1976–1988) and its supplements by F. A. Stafleu and E. A. Mennega (1992+) have been used for abbreviations of book titles.

Graphic Elements

All genera, and approximately one out of three species, are illustrated. The illustrations may be of typical or of unusual species, or they may show diagnostic traits or complex structures. Most illustrations have been drawn from herbarium specimens selected by the authors. In some cases living material or photographs have been used. Data on specimens that were used and parts that were illustrated have been recorded. This information, together with the archivally preserved original drawings, is deposited in the Missouri Botanical Garden Library and is available for scholarly study.

Specific Information in Treatments

Keys

Dichotomous keys are included for all ranks below family if two or more taxa are treated. For dioecious species, keys are designed for use with either staminate or pistillate plants. Keys are designed also to facilitate identification of taxa that flower before leaves appear. More than one key may be given, and for some groups tabular comparisons may be presented in addition to keys.

Nomenclatural Information

Basionyms of accepted names, with author and bibliographic citations, are listed first in synonymy, followed by any other synonyms in common recent use, listed in alphabetical order, without bibliographic citations.

Vernacular names in common use are given in the appropriate language. In general, such names have not been coined for use in the *Flora.* Those preferred by governmental or conservation agencies are listed if known.

The last names of authors of taxonomic names have been spelled out. The conventions of *Authors of Plant Names* (R. K. Brummitt and C. E. Powell 1992) have been used as a guide for including first initials to discriminate individuals who share surnames. Two exceptions are "Alph. Wood" instead of "A. W. Wood" (Brummitt, pers. comm.) and "Gerald L. Smith" instead of "G. Lom. Smith."

If only one infraspecific taxon within a species occurs in the flora area, nomenclatural information (literature citation, basionym with literature citation, relevant other synonyms) is given for the species, as is information on the number of infraspecific taxa in the species and their distribution worldwide, if known. A description and detailed distributional information are given only for the infraspecific taxon.

Descriptions

Character states common to all taxa are noted in the description of the taxon at the next higher rank. For example, if corolla color is yellow for all species treated within a genus, that character state is given in the generic description. Characters used in keys are repeated in the descriptions. Characteristics are given as they occur in plants from the flora area. Characteristics that occur only in plants from outside the flora area may be given within square brackets, or instead may be noted in the discussion following the description. In families with one genus and one or more species, the family description is given as usual, the genus description is condensed, and the species are described as usual.

In reading descriptions, the reader may assume, unless otherwise noted, that: the plants are green, photosynthetic, and reproductively mature; woody plants are perennial; stems are erect; roots are fibrous; leaves are simple and petiolate; flowers are bisexual, radially symmetric, and pediceled; perianth parts are hypogynous, distinct, and free; and ovaries are superior. Because measurements and elevations are almost always approximate, modifiers such as "about," "circa," or "±" are usually omitted.

Unless otherwise noted, dimensions are length × width. If only one dimension is given, it is length or height. All measurements are given in metric units. Measurements usually are based on dried specimens but these should not differ significantly from the measurements actually found in fresh or living material.

Chromosome numbers generally are given only if published, documented counts are available from North American material or from an adjacent region. No new counts are published intentionally in the *Flora.* Chromosome counts from nonsporophyte tissue have been converted to the $2n$ form. The base number ($x =$) is given for each genus. This represents the lowest known haploid count for the genus unless evidence is available that the base number differs.

Flowering time and often fruiting time are given by season, sometimes qualified by early, mid, or late, or by months. Elevations over 50 m generally are rounded to the nearest 100 m; those 50 m and under are rounded to the nearest 10 m. Mean sea level is shown as 0 m, with the understanding that this is approximate. Elevation often is omitted from herbarium specimen labels, particularly for collections made where the topography is not remarkable, and therefore precise elevation is sometimes not known for a given taxon.

The term "introduced" is defined broadly to refer to plants that were released deliberately or accidentally into the flora and that now exist as wild plants in areas in which they were not recorded as native in the past. The distribution of non-native plants is often poorly documented and may be ephemeral.

If a taxon is globally rare or if its continued existence is threatened in some way, the words "of conservation concern" appear before the statements of elevation and geographic range. Criteria for taxa of conservation concern are based on The Nature Conservancy's designations of global rank (G-rank), G1 and G2:

G1 Critically imperiled globally because of extreme rarity (5 or fewer occurrences or fewer than 1000 individuals or acres) or because of some factor(s) making it especially vulnerable to extinction.

G2 Imperiled globally because of rarity (5–20 occurrences or fewer than 3000 individuals or acres) or because of some factor(s) making it very vulnerable to extinction throughout its range.

Range maps are given for each species or infraspecific taxon. The Nunavut boundary on the maps has been provided by the GeoAccess Division, Canada Centre for Remote Sensing, Earth Science. The maps are generalized and, in order to represent the probable range of a taxon, parts of states or provinces may be shaded even though documentation of occurrence there may be lacking. Occurrences in states or provinces listed in distribution statements are documented by specimens. We have assumed that details such as "northeastern Florida" are apparent on the map; consequently, directional qualifiers are not given in the list of territories, provinces, and states. Authors are expected to have seen at least one specimen documenting each state record and have been urged to examine as many specimens as possible from throughout the range of each taxon. Additional information about distribution may be given in the discussion.

Distributions are stated in the following order: Greenland; St. Pierre and Miquelon; Canada (provinces and territories in alphabetic order); United States (states in alphabetic order); Mexico (11 northern states may be listed specifically, in alphabetic order); West Indies; Bermuda; Central America (Belize, Costa Rica, El Salvador, Guatemala, Honduras, Nicaragua, Panama); South America; Europe, or Eurasia; Asia (including Indonesia); Africa; Pacific Islands; Australia; Antarctica.

Discussion

The discussion section may include information on taxonomic problems, distributional and ecological details, interesting biological phenomena, economic uses, and toxicity (see "Caution," p. xxvi).

Selected References

Major references used in preparation of a treatment or containing critical information about a taxon are cited following the discussion. These, and other works that are referred to in discussion or elsewhere, are included in Literature Cited at the end of the volume.

CAUTION

The Flora of North America Editorial Committee **does not encourage, recommend, promote, or endorse** any of the folk remedies, culinary practices, or various utilizations of any plant described within these volumes. Information about medicinal practices and/or ingestion of plants, or of any part or preparation thereof, has been included only for historical background and as a matter of interest. Under no circumstances should the information contained in these volumes be used in connection with medical treatment. Readers are strongly cautioned to remember that many plants in the flora are toxic or can cause unpleasant or adverse reactions if used or encountered carelessly.

Key to boxed codes following accepted names:

- [C] of conservation concern
- [E] endemic to the flora area
- [F] illustrated
- [I] introduced to the flora area
- [W1] weedy according to D. T. Patterson et al. (1989)
- [W2] weedy according to R. H. Callihan et al. (1995)

Flora of North America

Phylogeny and Classification of the Monocotyledons: An Update

James L. Reveal
J. Chris Pires

Introduction

Given the rapid and continuing changes in monocotyledonous systematics, we present here an update on the phylogenetic relationships and classification of all monocots. We review A. Cronquist's (1981, 1988) system, adopted by FNA, in comparison with the more recent systems proposed by A. L. Takhtajan (1997), W. S. Judd et al. (1999), R. F. Thorne (2000, pers. comm.), A. B. Doweld (2001), P. F. Stevens (http://www.mobot.org/MOBOT/research/APweb/, hereafter cited as "P. F. Stevens www") and the Angiosperm Phylogeny Group (1998, hereafter cited as "APG"; in prep., hereafter cited as "APG II"). In deciding to adopt the Cronquist system of classification for flowering plants, the FNA Editorial Committee sought to select a stable system that would be subject to only minor modification. Cronquist's only changes between 1981 and 1988 were the addition of a few families, most of them newly proposed (A. Cronquist 1983, 1993; J. L. Reveal 1993e). The Editorial Committee assumed that Cronquist's system would have long-term stability well into the 21st century, much like the successful plant classification scheme proposed by H. G. A. Engler and K. Prantl (see H. G. A. Engler 1897) that served systematic botany from the 1890s until the 1950s (A. Cronquist 1957).

Arthur Cronquist (1919–1992) knew that his system would have challenges. His good friends Rolf M. T. Dahlgren (1932–1986), Armen L. Takhtajan (1910–), and Robert F. Thorne (1920–) were already proposing major treatments of the flowering plants. Dahlgren had published several significant papers (R. M. T. Dahlgren 1975, 1977, 1980, 1983; R. M. T. Dahlgren et al. 1981), especially on the monocotyledonous angiosperms (R. M. T. Dahlgren and H. T. Clifford 1982; R. M. T. Dahlgren et al. 1985) and had begun work on the whole of the angiosperms (R. M. T. Dahlgren and K. Bremer 1985; G. Dahlgren 1989, 1989b) by the time of his death in 1987. While there was a fairly high degree of stability in Dahlgren's systems, this was not the case for those generated by A. L. Takhtajan (1959, 1966, 1969, 1970, 1980, 1983, 1986, 1987, 1997) and R. F. Thorne (1968, 1973, 1974, 1976, 1977, 1981, 1983, 1992, 1992b, 2000, pers. comm.; see also http://www.inform.umd.edu/PBIO/pb250/thorne.html and http://www.inform.umd.edu/PBIO/fam/thorne99mono.html). Both of these authors would significantly alter their arrangements as to both placement and circumscription of families. Unlike Cronquist, Takhtajan and Thorne were willing to apply new information to their understanding of the phylogeny of the flowering plants, and to modify their systems accordingly. Cronquist had hoped to create a system that was stable, morphologically useful, and structurally consistent. However, he could not have anticipated in 1981 that his system would form the basis for thousands of experiments testing his hypotheses on relationships. The advent of molecular phylogenetic studies bought into serious question any system based solely on the traditional methodologies of data gathering.

The molecular revolution challenging the Cronquist system began seriously with a publication by M. W. Chase et al. in 1993. The amount of molecular systematic data generated since then has been enormous and impossible to ignore (for reviews, see D. J. Crawford 2000; P. S. Soltis and D. E. Soltis 2001; K. J. Sytsma and W. J. Hahn 1996, 2000). When morphological data were added to molecular data, the phylogenetic results were generally more robust than with either data set alone (P. K. Endress et al. 2000). This research brought into question the definition of Cronquist's subclasses, as well as the number and arrangement of his orders and families. Chase et al. also questioned the conclusions of R. M. T. Dahlgren, A. L. Takhtajan, and R. F. Thorne. In short order, Takhtajan (1997) and Thorne (2000) responded to the findings in that work and the host of similar ones that soon followed (M. W. Chase et al. 1995, 1995b, 2000; M. R. Duvall et al. 1993b; W. J. Kress 1990, 1995; H. P. Linder and E. A. Kellogg 1995; H. Loconte and D. W. Stevenson 1990; O. I. Nandi et al. 1998; D. L. Nickrent and D. E. Soltis 1995; Z. K. Shinwari et al. 1994; J. F. Smith et al. 1993; D. E. Soltis et al. 1997, 1997b, 2000; D. W. Stevenson and H. Loconte 1995). Both adopted some (but by no means all) of these findings, and neither relied solely upon the cladistic or molecular findings to define a taxon. The same may be said for K. Kubitzki et al. (1990+, vols. 3, 4), in their treatment of the monocotyledonous families (see http://www.inform.umd.edu/PBIO/fam/kubitzkimonocot.html). In Thorne's latest revision (pers. comm.) his arrangement of taxa is even closer to that proposed by APG.

W. S. Judd et al. (1999) made the first attempt to summarize this new molecular information (mainly APG), in a systematic botany textbook concentrating on the vascular plants of the United States, and P. F. Stevens (www) has presented an online revision of the system in Judd et al. based on APG II for all flowering plants. In contrast, the latest edition of D. W. Woodland's textbook (2000) retained its Cronquist format, although this may change in future editions (Woodland, pers. comm.). A. B. Doweld (2001) proposed a system similar to Takhtajan's (1997), but recognized even more subclasses, superorders, orders, and families, and clearly attempted to take into account the most recent information on angiosperm phylogeny.

Research on higher flowering-plant phylogeny has expanded from the efforts of a few individuals to those of a large number of people working in teams, seeking consensus. The latest such team to emerge is the Angiosperm Phylogeny Group, which published a major overview of the flowering plants (APG) in 1998 and as of this writing had a revision (APG II) in final preparation. APG recognized 40 orders with 462 families, 86 of which were not classified among the recognized orders. D. E. Soltis et al. (2000) produced a summation based on three genes and accounted for 46 orders and 317 families, identifying several potential orders and/or families to which they did not assign names. Although Soltis et al. made no effort to account for all families mentioned by APG, some consolidation is evident from their implied synonymy. APG II accounts for 45 orders and 458 families. Of these families, 41 are not assigned to orders and an additional two are unplaced within the system. However, within the monocots, the five families not assigned to orders in APG were reduced to four in M. W. Chase et al. (2000), and now only two (Petrosaviaceae and Dasypogonaceae) in APG II. See Table 1 for a list of all monocot names at and above the rank of family that are in current use, with their authorships; for their bibliographic citations, see http://matrix.nal.usda.gov:8080/star/supragenericname.html; http://www.inform.umd.edu/PBIO/fam/revfam.html.

Although the power of phylogenetic methods and large data sets underlying the APG system are impressive, controversies in classification still remain. Most pertinent here are differences in principles of circumscription, such as whether or not to recognize paraphyletic groups (R. K. Brummitt and M. S. M. Sosef 1998). Both Takhtajan and Thorne maintained that paraphyletic groups are "the inevitable result of evolution." Contrary to J. V. Freudenstein's (1998) argument, some systematists believe that paraphyletic groups are inevitably expressed in any system of classification (R. K. Brummitt and M. S. M. Sosef 1998; M. S. M. Sosef 1997). Otherwise, following Freudenstein's argument to its logical extreme, one might regard the flowering plants as a minor offshoot of the algal genus *Coleochaete*, a group that in turn is only a slightly more sophisticated compilation of bacteria (K. G. Karol et al. 2001). Still, most systematists strive to use monophyletic groups in studies of comparative biology. To our knowledge, all of the families and orders recognized here are monophyletic, as they are in APG and APG II.

In this review, we have two goals. The first is to summarize the phylogenetic relationships of the monocots and basal angiosperms (figs. 1–4) and compare them to the traditional supraordinal categories of Cronquist and others. The second is to review the orders and families of the monocots, focusing on the relationships among families, especially those that occur in North America. We take a conservative approach in depicting only those phylogenetic relationships that received 75% or higher statistical support in a variety of studies that employed three or more data sets. We also chose to represent polytomies where different data sets from different studies showed strong support but were in conflict (e.g., Mayacaceae). The most recent and primary sources of information for large-scale monocot phylogenies are M. W. Chase et al. (2000), M. F. Fay et al. (2000), Qiu Y. L. et al. (2000), P. J. Rudall et al. (2000, 2000b), V. Savolainen et al. (2000), D. E. Soltis et al. (2000), and D. W. Stevenson et al. (2000). In addition, many other papers cited in the sections that follow tempered our understanding of the phylogeny of particular groups of the monocots. We also acknowledge information from a number of individuals who were willing to share unpublished work with us. These phylogenies dramatically illustrate two things: (1) the progress systematists have made in our understanding of monocot evolution since FNA's adoption of the Cronquist system; and (2) the many areas of uncertainty that remain in monocot phylogenetics and higher-level classification.

The Class Liliopsida: Rooting the Monocots within the Angiosperms

A. Cronquist (1981) subdivided the angiosperms (Magnoliophyta) into two classes, the Magnoliopsida and the Liliopsida. What is obvious now is that the dicotyledons, as traditionally defined (A. P. de Candolle [1817]1818–1821, vol. 1), do not form a natural group and must be divided into several groups (fig. 1). A final answer to the question of whether the monocots should retain the formal name Liliopsida can come about only by the resolution of their root among the traditional dicotyledonous angiosperms. Our representation of the phylogeny of the basal angiosperms in relationship to the monocotyledons (fig. 1) is based on several multiple-gene studies (T. J. Barkman et al. 2000; S. W. Graham and R. G. Olmstead 2000; Qiu Y. L. et al. 1999, 2000; D. E. Soltis et al. 2000; M. Zanis et al., pers. comm.), though critical information is still needed for morphological characters (W. E. Friedman and S. K. Floyd 2001). The three basal angiospermous groups may be tentatively termed the Amborellales, Nymphaeales, and Austrobaileyales. Recent phylogenetic analyses place Amborellales or Nymphaeales or both as the sister group to all remaining angiosperms. While we have conservatively placed Amborellales and Nymphaeales as a polytomy at the base, there is increasing evidence that Amborellaceae alone are the sister group to all the remaining angiosperms (M. Zanis et al., pers. comm.).

The critically unresolved polytomy is the one involving the relationships of the monocots to the Chloranthales, Ceratophyllales, and magnolids (Magnoliales and their immediate relatives), plus the remaining dicots (informally called the "eudicots"). Until this phylogeny is better understood, assignment of rank to the monocots is dubious and our use of the name Liliopsida is merely a matter of convenience. What is important is that the dicotyledonous angiosperms are not monophyletic and that the monocots are not derived from the Piperales (W. C. Burger 1977). Historically, the dicotyledonous Nymphaeales were thought to be the sister group to the putatively basal monocot group Alismatidae, since both share morphological characters adapted for the aquatic habit (A. Arber 1920, 1925; H. Hallier 1905; J. Hutchinson 1926–1934, vol. 2; E. Sargent 1903, 1904; see also A. Cronquist 1988; G. L. Stebbins 1974; A. L. Takhtajan 1997). However, molecular analyses show that the Alismatidae are not the basal-most monocots and Nymphaeales are not the sister group to monocots (D. H. Les and E. L. Schneider 1995). In any case, the monocots form a distinct, monophyletic group morphologically and molecularly, and neither their recognition nor their overall circumscription is in doubt (see K. Kubitzki et al. 1990+, vols. 3, 4; W. S. Judd et al. 1999; P. J. Rudall et al. 1995; K. L. Wilson and D. A. Morrison 2000).

Traditional Monocot Subclasses: The Main Lineages within the Monocots

A. Cronquist (1988) accepted five subclasses in the Liliopsida: Alismatidae, Arecidae, Commelinidae, Zingiberidae, and Liliidae. A. L. Takhtajan (1997) recognized six by proposing the Triurididae and Aridae and submerging Cronquist's Zingiberidae in his Commelinidae. R. F. Thorne (2000, pers. comm.) recognized only three subclasses: Alismatidae, Liliidae, and Commelinidae. A. B. Doweld (2001) accepted the six subclasses suggested by Takhtajan and added a seventh, Juncidae. A separation of the lower monocots, or what D. E. Soltis et al. (2000) termed the Acorales and Alismatales (including Arales, Potamogetonales, and Tofieldiales), from the more advanced lilioid and commelinid monocots, is justified morphologically. R. M. T. Dahlgren and H. T. Clifford (1982) and R. M. T. Dahlgren et al. (1985) suggested long ago a far closer relationship between Alismatidae and Aridae than was acknowledged by Cronquist (1981), who assigned Aridae to Arecidae. Takhtajan's separation of the Arecidae, Alismatidae, Triurididae, and Aridae resolved the inconsistency in Cronquist's scheme, but such a solution now is unfeasible given the positioning of the alismatoid monocots (Alismatidae/Aridae) as a clade that is well isolated from the palms (Arecidae/Arecales) in Commelinidae.

Recognizing subclasses within the monocots seems premature at this time. If one accepts the monocots as a taxon at the rank of class, then the acoroid members ought to be recognized at the rank of subclass, as they are sister to the remainder of the monocotyledonous angiosperms (fig. 2). Continued recognition of subclasses Alismatidae (including Aridae) and Liliidae (including Triurididae) presents no problem. There is a difficulty within Liliidae, however, since the commelinids are not sister to the remaining lilioid monocots (fig. 3) but rather are probably sister to a single order (e.g., the Asparagales). Thus, assigning even the rank of superorder to the commelinids would mean that the remaining lilioid monocots would be in a single paraphyletic superorder or would have to be divided into monophyletic superorders that would be essentially synonymous with orders. If the latter were attempted, and the scheme depicted in Figure 3 were adopted, only the names

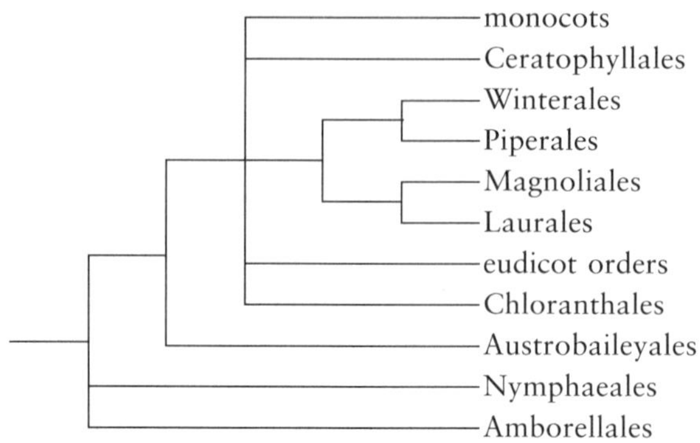

FIGURE 1. Rooting of the monocotyledonous angiosperms.

Petrosavianae, Dioscoreanae, Pandananae, and Lilianae would be available, as a name based on *Narthecium* has yet to be proposed. Still, if Commelinanae are sister only to the Asparagales, then all of the commelinids would fall into a single order, Commelinales. As we will show, much remains to be learned about these higher-level relationships, and naming weakly supported clades cannot be justified regardless of whether a ranked or rankless classification system is employed.

Summary of the Orders and Families of the Monocots

We will discuss the orders and families of the monocots by arranging them into three informal groups: (1) acoroids and alismatids (fig. 2); (2) lilioid monocots (fig. 3); and (3) commelinids (fig. 4). This follows the current consensus as to the circumscriptions of families and orders, while restricting the supraordinal classification to informal groupings (APG; APG II; M. W. Chase et al. 2000; D. E. Soltis et al. 2000; P. F. Stevens www). To facilitate comparisons between Cronquist and some of the other current systems, we have adopted, for the most part, the linear sequence of orders and families suggested by APG, Chase et al., Soltis et al., Stevens, and APG II. In a few cases we found the realignments of APG II controversial and difficult to defend. Thus, we deviate slightly from these APG-affiliated systems with respect to monocot classification: we recognize two additional families and three orders within APG's Alismatales; we recognize Nartheciales as separate from APG's Dioscoreales; and we recognize Petrosaviales and Dasypogonales (unplaced in APG) at the ordinal level.

I. The Acoroids and Alismatids

The phylogenetic relationships among the acoroids and alismatids shown in Figure 2 are well supported by several studies (M. W. Chase et al. 2000; S. Fuse and M. N. Tamura 2000; D. H. Les et al. 1997; D. E. Soltis et al. 2000; N. Tanaka et al. 1997). We recognize five orders in the two basal clades of the monocots: Acorales, Arales, Alismatales, Potamogetonales, and Tofieldiales. This differs slightly from APG, APG II, Chase et al., and P. F. Stevens (www), who recognize only two orders (Acorales and Alismatales), but we believe that splitting the single APG order Alismatales into four monophyletic and morphologically well-defined orders is justified (fig. 2). Morphology does not yet support the monophyly of a broadly defined Alismatales without including Acorales (M. Buzgo 2001, pers. comm.), and some of the molecular data sets also indicate that APG Alismatales are paraphyletic. For the families that we assign to two orders (Alismatales and Potamogetonales), A. Cronquist (1981, 1988) recognized three (Alismatales, Hydrocharitales, and Najadales), which was reasonable given that he emphasized floral evolution. A. L. Takhtajan (1997) did not justify his linear arrangement of Butomales, Hydrocharitales, Najadales, Alismatales, Aponogetonales, Juncaginales (including Scheuchzeriales), Potamogetonales, Posidoniales, Cymodoceales (including Zannichelliaceae), and Zosterales. He assigned these to a single superorder (Alismatanae), while A. B. Doweld (2001) distributed them between two (Alismatanae and Zosteranae). The sequences given by APG, Chase et al., and APG II are alphabetical. Our linear sequence is strongly influenced by those given by D. H. Les and R. R. Haynes (1995), R. F. Thorne (2000, pers. comm.), Fuse and Tamura, and Stevens. W. S. Judd et al. (1999) outlined a general sequence for their Alismatales, noting that there were two lineages; these correspond to our Alismatales and Potamogetonales. Aside from our recognition of Najadaceae and Zannichelliaceae, the number of families and their circumscriptions given here are similar to

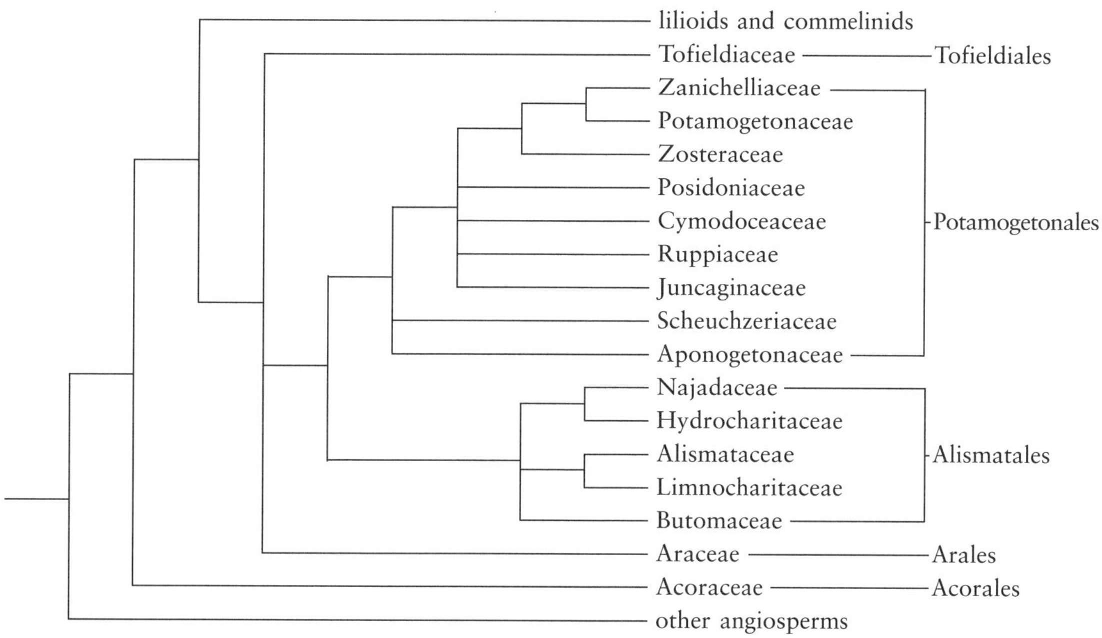

FIGURE 2. Phylogenetic relationships of the acoroid and alismatid monocots.

those proposed by APG, APG II, Chase et al., Thorne, and Stevens.

The changes in our understanding of monocot phylogeny over the last two decades are elegantly reflected in just one small group alone. Portions of what A. Cronquist (1981, 1988) distributed among three subclasses (Alismatidae, Arecidae, and Liliidae) have now been assigned to a single order (Alismatales) by P. F. Stevens (www). This pattern is reflected throughout the angiosperms as understood by Cronquist, and long-assumed relationships must now be put aside to better comprehend the evolutionary history of these plants.

1. Acorales

According to most of the recent molecular analyses, the monogeneric Acoraceae are basal among the monocotyledonous angiosperms (M. W. Chase et al. 1993, 1995b, 2000; but see D. W. Stevenson et al. 2000; Qiu Y. L. et al. 2000). Morphologically *Acorus* resembles Araceae (M. Buzgo 2001), and the genus was traditionally included therein (A. Cronquist 1981; R. M. T. Dahlgren et al. 1985). R. F. Thorne (1983) was opposed to this placement of *Acorus*; M. H. Grayum (1987) championed its removal from Araceae, and Cronquist (1988), G. Dahlgren (1989), and A. L. Takhtajan (1997) eventually accepted recognition of Acoraceae, as did J. Bogner and S. J. Mayo (1998) and P. F. Stevens (www). The only name currently available at a higher rank is Acoranae (table 1).

2. Arales

The Arales comprise a single family, Araceae (including Lemnaceae). A. Cronquist (1981) recognized both families and assigned these, along with the Arecaceae, to his subclass Arecidae. The arums and palms are unrelated morphologically, as demonstrated by R. M. T. Dahlgren and H. T. Clifford (1982; see also R. M. T. Dahlgren et al. 1985), a point supported by M. H. Grayum (1990) and acknowledged by A. L. Takhtajan (1997). Now well supported by molecular (APG; M. W. Chase et al. 2000; J. C. French et al. 1995; W. S. Judd et al. 1999; D. E. Soltis et al. 2000) and morphological (M. Buzgo 2001) data, the Arales are positioned in the same clade as Alismatales/Potamogetonales and Tofieldiales (fig. 2). P. F. Stevens (www) and APG II reduced Arales to synonymy under a broadly defined Alismatales, a concept rejected by R. F. Thorne (pers. comm.), who maintains each in a separate superorder. Should there be a need for higher taxonomic ranks, Aranae and Aridae are available (table 1).

The Araceae are diverse (S. J. Mayo et al. 1997) and may be divided into a number of relatively distinct subfamilies (J. Bogner and D. H. Nicolson 1991; M. H. Grayum 1990; S. J. Mayo et al. 1997, 1998; A. L.

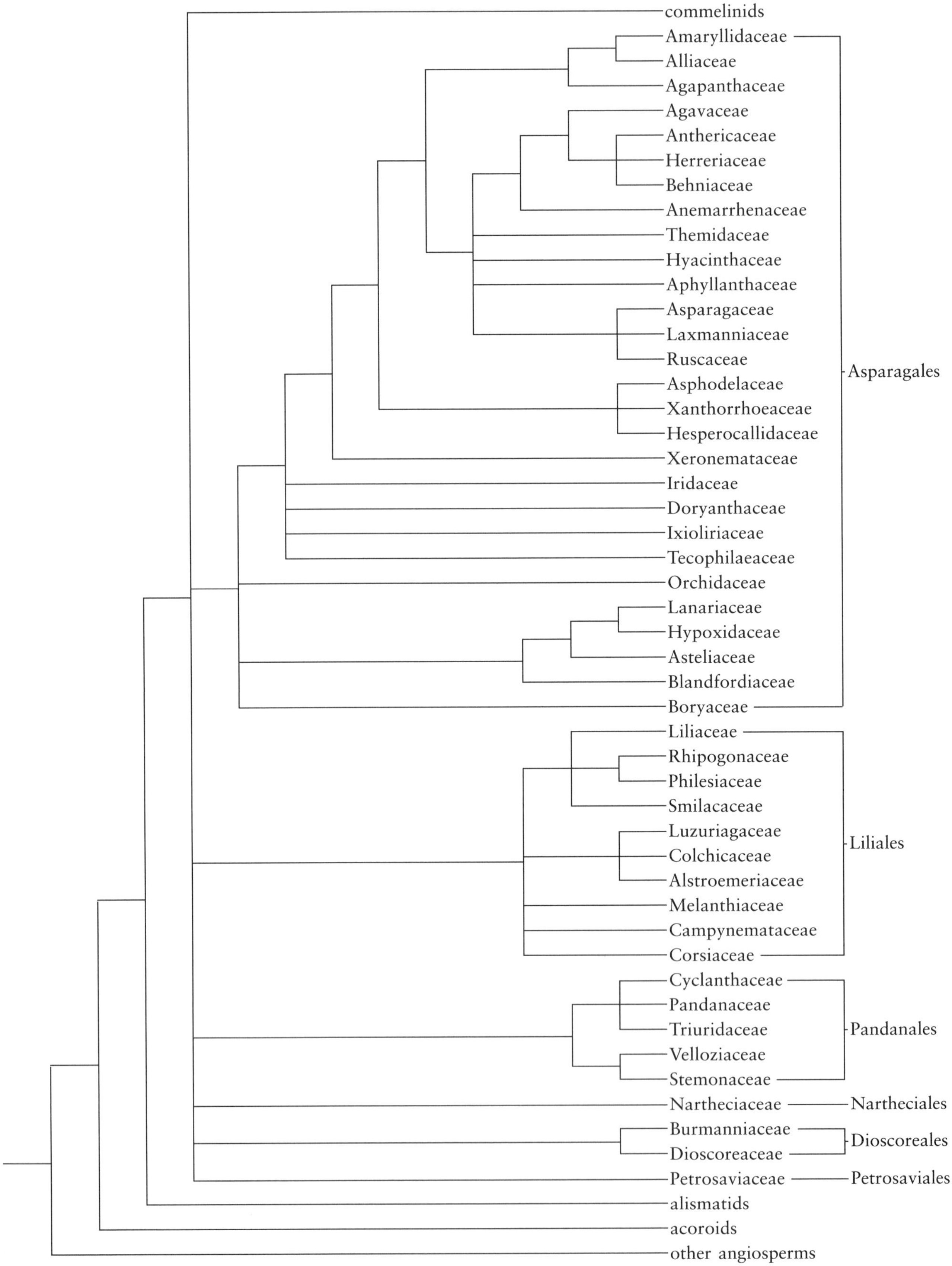

FIGURE 3. Phylogenetic relationships of the lilioid monocots.

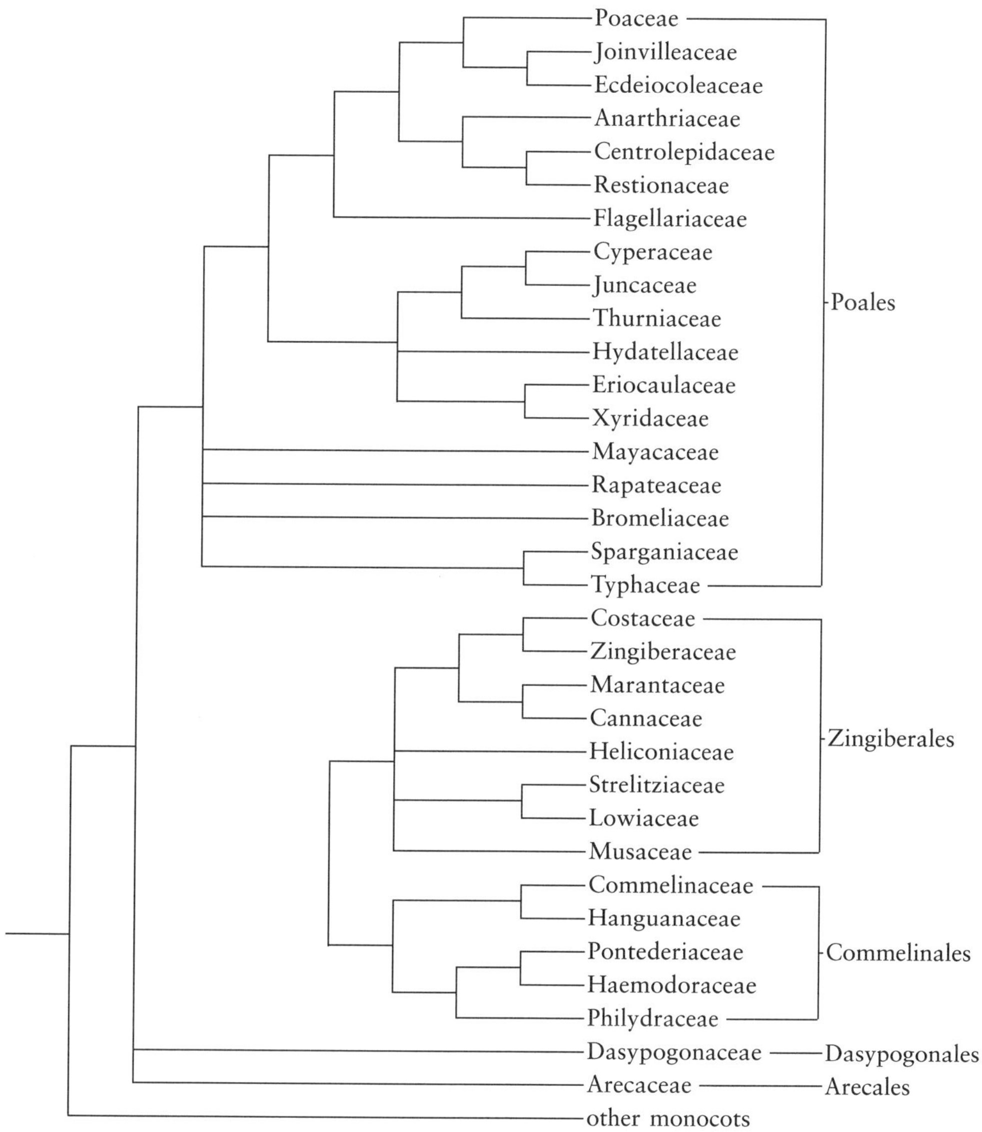

FIGURE 4. Phylogenetic relationships of the commelinid monocots.

Takhtajan 1997; R. F. Thorne 2000; P. F. Stevens www). Takhtajan and A. B. Doweld (2001) removed Pistiaceae from Araceae, but this distinction cannot be justified (W. S. Judd et al. 1999).

Lemnaceae (E. Landolt 1986, 1998; D. H. Les et al. 1997b; R. F. Thorne 2000), the duckweeds, are a specialized group nested within Araceae (M. W. Chase et al. 1993, 2000). Either they are recognized as the Lemnoideae (including Wolffioideae) or they are assigned to the Lemneae in the Aroideae (J. C. French et al. 1995; R. A. Stockey et al. 1997). Morphologically the duckweeds are easily recognized and are not confused with any other segment of Araceae.

3. Alismatales

The Alismatales are defined here to include five families: Butomaceae, Limnocharitaceae, Alismataceae, Hydrocharitaceae, and Najadaceae (D. H. Les et al. 1997), the five basal families of the Alismatanae as circumscribed by A. L. Takhtajan (1997) and the whole of that superorder according to A. B. Doweld (2001).

Butomaceae, Limnocharitaceae, and Alismataceae all have hypogynous flowers with distinct or only basally fused carpels and a differentiated perianth. If one defines Hydrocharitaceae narrowly (sensu A. Cronquist 1988; W. S. Judd et al. 1999; A. L. Takhtajan 1997), then the family can be incorporated easily into Alismatales, since the two groups share basically the same features, the primary difference being that the flowers of Hydrocharitaceae are epigynous rather than hypogynous. M. W. Chase et al. (2000), D. E. Soltis et al. (2000), P. F. Stevens (www), and R. F. Thorne (2000, pers. comm.) have suggested that *Najas* should be included within Hydrocharitaceae. Les (pers. comm.) remains dubious. Accordingly, we follow Takhtajan and retain Hydrocharitaceae and Najadaceae, positioning them well away from Aponogetonaceae (see P. F. Stevens www). Should higher taxonomic ranks be required, Alismatanae and Alismatidae are available (table 1).

Hydrocharitaceae are the most diverse family in the complex. D. J. Mabberley (1997) and R. F. Thorne (2000) recognized three subfamilies (Hydrocharitoideae, Thalassioideae, and Halophiloideae); A. L. Takhtajan (1997) considered each of these to be an independent family, although N. Tanaka et al. (1997) have shown that such separation is unjustified at any suprageneric rank. A. Cronquist (1981, 1988) adopted the broad circumscription without mentioning subfamilies. There is no evidence to support the placement of the Hydrocharitaceae (sans *Najas*) in the Arales as proposed by K. Kubitzki et al. (1990+, vol. 3).

Najadaceae is considered synonymous with Hydrocharitaceae by APG, APG II, and R. F. Thorne (2000, pers. comm.), and P. F. Stevens (www) informally assigns *Najas* to its own subfamily. Unlike Hydrocharitaceae, Najadaceae lacks a perianth, and the ovary is unilocular and apparently superior rather than compound and inferior.

4. Potamogetonales

The Potamogetonales are defined here to encompass nine mostly aquatic families: Aponogetonaceae, Scheuchzeriaceae, Juncaginaceae, Posidoniaceae, Ruppiaceae, Cymodoceaceae, Zosteraceae, Zannichelliaceae, and Potamogetonaceae (fig. 2). These constitute the superorder Zosteranae according to A. B. Doweld (2001). They are freshwater or marine aquatic herbs with inconspicuous flowers in which the perianth is either undifferentiated or absent. All but Aponogetonaceae have orthotropous ovules. Except for the assignment of Najadaceae to Alismatales, we retain a traditional view of the families based on their overall morphology (C. D. K. Cook 1998; P. A. Cox and C. J. Humphries 1993; W. S. Judd et al. 1999; D. W. Stevenson and H. Loconte 1995; see also D. H. Les et al. 1997; J. Kuo and A. J. McComb 1998, 1998b, 1998c; N. Tanaka et al. 1997). Aside from the repositioning of Najadaceae, the linear sequence suggested here is essentially the same as the one given by A. Cronquist (1981, 1988).

Aponogetonaceae are the basal member of the Potamogetonales. In that family as well as in Scheuchzeriaceae, the ovules are typically several per carpel, unlike the other families in the order.

Juncaginaceae, as circumscribed by A. Cronquist (1981, 1988), include four or five genera, which A. L. Takhtajan (1997) assigned to three families: Juncaginaceae (including *Triglochin*), Lilaeaceae (*Lilaea*), and Maundiaceae. R. F. Thorne (2000) recognized two subfamilies, regarding *Maundia* a member of Juncaginoideae.

Ruppiaceae, Potamogetonaceae, and Zannichelliaceae are considered closely related and are sometimes all assigned to Potamogetonaceae (e.g., I. L. Wiggins 1980), though they have also been variously combined with other taxa and given different names. P. F. A. Ascherson (1889) included within his circumscription of Potamogetonaceae all of the Potamogetonineae except Najadaceae (Ruppiaceae, Potamogetonaceae, Zannichelliaceae, Cymodoceaceae, Posidoniaceae, and Zosteraceae). J. K. Small (1903) combined all of the fresh- and brackish-water families into a single family, Zannichelliaceae. B. L. Robinson and M. L. Fernald (1908) called the same group Najadaceae, but later Fernald (1950) excluded Najadaceae from the freshwater families and called the result Zosteraceae. Occasionally only Potamogetonaceae and Ruppiaceae are combined. In North American floras, there is no consistent circumscription or even name. A. Cronquist (1981, 1988) avoided the nomenclatural morass and opted to maintain all of the families. This option was largely adopted by W. S. Judd et al. (1999). M. W. Chase et al. (2000) and P. F. Stevens (www), however, joined Zannichelliaceae and Potamogetonaceae, although, as Stevens pointed out, the former family "is rather weakly embedded within Potamogetonaceae." Here, like Cronquist as well as A. L. Takhtajan (1997), D. J. Mabberley (1997), K. Kubitzki et al. (1990+, vol. 4), Judd et al., R. F. Thorne (2000, pers. comm.), and A. B. Doweld (2001), we retain each of the families.

A. L. Takhtajan (1997) suggested that Zannichelliaceae are closely related to Cymodoceaceae. However, based on floral morphology and molecular evidence, a relationship with Potamogetonaceae is more likely (J. Kuo and A. J. McComb 1998c). Importantly, D. H. Les et al. (1997) showed that inclusion of the Ruppiaceae into Potamogetonaceae can not be supported.

5. Tofieldiales

Placing Tofieldiales with the alismatids is supported by molecular evidence (M. W. Chase et al. 2000; S. Fuse and M. N. Tamura 2000), while the traditional association with Nartheciales is not.

Tofieldiaceae, the only member of the order (W. B. Zomlefer 1999), includes four genera (W. B. Zomlefer 1997c), or five when *Pleea* is considered distinct from *Tofieldia*.

A. L. Takhtajan (1997) placed Tofieldiaceae (and Nartheciaceae) in the Melanthiales (Liliidae), but assigned Petrosaviaceae to the Petrosaviales in his Triurididae. In this way he differed from R. M. T. Dahlgren et al. (1985) and R. F. Thorne (1992, 1992b), who retained members of all three families in the Melanthiales. A. Cronquist (1981) considered *Tofieldia* and *Narthecium* members of Liliaceae, while assigning Petrosaviaceae to his Triuridales, in the Alismatidae. S. Fuse and M. N. Tamura (2000) considered Nartheciaceae to be related to Melanthiales but placed Petrosaviaceae in a separate clade basal to Melanthiales and the remainder of the lilioid and commelinid monocotyledons. The available evidence suggests that Petrosaviaceae should be considered basal within the lilioid monocots, along with Dioscoreales, Nartheciales, and Pandanales (fig. 3), but that, contrary to Thorne (2000, pers. comm.) and following M. W. Chase et al. (2000) and Fuse and Tamura, the Tofieldiales should be retained with their alismatid relatives.

II. The Lilioid Monocots

The phylogenetic relationships shown in Figure 3 are based on several recent studies (H.-D. Behnke et al. 2000; L. R. Caddick et al. 2002; M. W. Chase et al. 2000; M. F. Fay et al. 2000; S. Fuse and M. N. Tamura 2000; K. Hayashi and S. Kawano 2000; A. W. Meerow et al. 2000; P. J. Rudall et al. 2000, 2000b; V. Savolainen et al. 2000; D. E. Soltis et al. 2000; J. Yamashita and M. N. Tamura 2000). A number of clades are well supported, and these correspond for the most part to the orders in the APG system (APG; APG II; M. W. Chase et al. 2000; P. F. Stevens www). We have been conservative in depicting these relationships by showing a polytomy among most of them. However, there is emerging evidence that the commelinids are sister to the Asparagales (T. J. Givnish, pers. comm.; S. W. Graham, pers. comm.) and this is the relationship seen in several published studies, although with low statistical support (M. W. Chase et al. 2000; V. Savolainen et al. 2000; D. E. Soltis et al. 2000). Until these relationships are resolved, it would be premature to assign a name to these clades (e.g., commelinids + Asparagales), let alone assign a supraordinal Linnaean rank to the monophyletic commelinids or the remaining paraphyletic grade of lilioid monocots.

We recognize six orders of lilioid monocots in the following sequence: Petrosaviales, Dioscoreales (including Burmanniales), Nartheciales, Pandanales (including Cyclanthales, Triuridales, and Velloziales), Liliales (including Smilacales), and Asparagales (including Orchidales; see L. R. Caddick et al. 2002; M. W. Chase et al. 2000; W. S. Judd et al. 1999; V. Savolainen et al. 2000; D. E. Soltis et al. 2000; P. F. Stevens www). This differs slightly from APG and Chase et al., both of which leave Petrosaviaceae unplaced as to order and include Nartheciaceae within Dioscoreales.

All of this differs fundamentally from what A. Cronquist (1981, 1988) assigned to his Liliidae and where he placed that taxon. He restricted the subclass to two orders, Liliales and Orchidales, and positioned it at the end of Liliopsida. All evidence currently available is contrary to Cronquist's view. The phylogenies of M. W. Chase et al. (2000) and D. E. Soltis et al. (2000) have Dioscoreales followed by the Pandanales and then the Liliales/Asparagales. V. Savolainen et al. (2000) did the same. The first group of petaloid monocot families in Chase et al. consists of Nartheciaceae, Burmanniaceae, Taccaceae, Dioscoreaceae, and Petrosaviaceae. L. R. Caddick et al. (2002) have Petrosaviaceae as sister to a polytomy of Pandanales/Dioscoreales, Liliales, and Asparagales. D. W. Stevenson and H. Loconte (1995) considered the Dioscoreales, Taccales, and Stemonales to be the basal group to the remainder of the monocots. As a consequence, most recent classification systems have changed accordingly, with K. Kubitzki et al. (1990+, vol. 3), APG, A. L. Takhtajan (1997), R. F. Thorne (2000, pers. comm.), and A. B. Doweld (2001) departing radically from the Cronquist arrangement, although each differs as to the number of orders and families recognized, and the placement of each.

1. Petrosaviales

One of the more significant differences between what A. Cronquist (1981) proposed and what is suggested here is the inclusion of Petrosaviales among the lilioid monocots rather than Alismatidae. M. W. Chase et al. (2000) found Petrosaviaceae to be within the lilioid monocot grade, but they considered the family "unplaced" as to order since it has no clear relationships among the lilioids. Both Cronquist and R. F. Thorne (2000, pers. comm.) assigned the family to the Triuridales, but the evidence is clear that the two are unrelated. A. B. Doweld (2001) established Petrosavianae and placed both Nartheciales and Petrosaviales in that superorder, but this relationship is not supported by any recent phylogenetic analyses. We prefer to recognize Petrosaviales as a monofamilial order rather than leave it unplaced.

A. Cronquist (1981, 1988) restricted Petrosaviaceae to a single genus, unlike M. N. Tamura (1998), who included the genera of Petrosaviaceae and Tofieldiaceae within his Nartheciaceae. *Japonolirion*, a Japanese endemic and type of the family Japonoliriaceae recognized by A. L. Takhtajan (1997) and accepted by A. B. Doweld (2001), is now placed firmly in Petrosaviaceae based on the molecular evidence reported by M. W. Chase et al. (2000) and S. Fuse and M. N. Tamura (2000). R. F. Thorne (2000) questionably assigned the family to the Tofieldiaceae.

2. Dioscoreales

L. R. Caddick et al. (2002b) and APG II define the Dioscoreales to include three families: Nartheciaceae, Burmanniaceae, and Dioscoreaceae. Those accustomed to A. Cronquist's (1981, 1988) system will be surprised to find Burmanniaceae associated with Dioscoreales. Burmanniaceae traditionally have been considered related to the orchids and positioned either in the same order (Cronquist) or in a closely associated order, Burmanniales (A. L. Takhtajan 1997). A cladistic analysis by D. W. Stevenson and H. Loconte (1995) initially supported the traditional alignment, but the molecular data reported by M. W. Chase et al. (1995b) suggested another position, in an expanded Dioscoreales. In an interesting compromise, R. F. Thorne's (2000, pers. comm.) linear sequence is Liliales, Taccales (the same as our Dioscoreales), Burmanniales, and then Orchidales (or Asparagales sensu APG; APG II; M. W. Chase et al. 2000). The findings of S. Fuse and M. N. Tamura (2000) and Caddick et al. (2002) firmly support the association of Burmanniaceae (including Thismiaceae) with Dioscoreaceae (including Avetraceae, Stenomeridaceae, Taccaceae, and Trichopodaceae). Only P. F. Stevens (www), who kept Taccaceae and Trichopodaceae distinct from Dioscoreaceae, and A. B. Doweld (2001) have adopted the view expressed here, although the latter recognized more entities and assigned his three orders and eight families to Dioscoreanae.

As defined by A. Cronquist (1981), Dioscoreaceae included both Stenomeridaceae and Trichopodaceae but excluded Taccaceae, a family he positioned after Stemonaceae and Smilacaceae. He assigned six genera to the yam family. R. F. Thorne (2000) accepted only three genera, whereas both A. L. Takhtajan (1997) and W. S. Judd et al. (1999) recognized five. L. R. Caddick et al. (2002b) accepted four: *Dioscorea*, *Stenomeris*, *Tacca*, and *Trichopus*. They reduced *Avetra* to synonymy under *Trichopus*, whereas *Borderea*, *Epipetrum*, *Nanarepenta*, *Rajania*, *Tamus*, and *Testudinaria* were all placed in *Dioscorea*.

3. Nartheciales

The assignment of the Nartheciaceae to Dioscoreales by L. R. Caddick et al. (2002, 2002b) and by M. W. Chase et al. (2000) is not well supported, and we indicate a polytomy in our phylogenetic diagram (fig. 3). Nartheciaceae also do not possess simultaneous microsporogenesis as do the remaining Dioscoreales. As a result, we suggest that the family be considered the only member of a distinct order. Regardless, Nartheciaceae cannot be assigned to Liliaceae as was done by A. Cronquist (1981, 1988), or allied with the acoroids as done by R. F. Thorne (2000, pers. comm.), for in fact the family is well removed from that clade.

4. Pandanales

The union of Pandanales, Cyclanthales, Triuridales, and Velloziales into a single order is now strongly supported (fig. 3) by morphological (H.-D. Behnke et al. 2000) and molecular (L. R. Caddick et al. 2002; M. W. Chase et al. 2000; V. Savolainen et al. 2000; D. E. Soltis et al. 2000) data. This surprising group raises many questions about classification systems based on gross morphology, since only R. M. T. Dahlgren et al. (1985) ever suggested that Pandanaceae-Cyclanthaceae were not related to the palms (Arecaceae are now placed with the commelinids; see fig. 4). A. B. Doweld (2001) assigned Pandananae, Triuridanae, Dioscoreanae, and Petrosavianae to the Commelinidae, along with Pontederianae and Commelinanae. Only a single genus assigned to the Pandanales (*Croomia*) is native to the flora, although numerous members of each family frequently are cultivated out-of-doors in the warmer areas of the continental United States.

Stemonaceae, retained in the Taccales (or Dioscoreales) by R. F. Thorne (pers. comm.), were circumscribed by A. Cronquist (1981) to include three genera. A. L. Takhtajan (1997) and A. B. Doweld (2001) separated Croomiaceae for two of those genera, *Stichoneuron* and the native American genus *Croomia*. Pentastemonaceae (F. Bouman and N. Devente 1992; B. E. E. Duyfjes 1991, 1992) is considered here to be a synonym of Stemonaceae, following the findings of P. F. Stevens (www), APG II, and L. R. Caddick et al. (2002), although it was accepted by Cronquist (J. L. Reveal 1993e) and by Thorne (2000, pers. comm.). Currently, Stemonaceae are thought to be sister to Velloziaceae (M. W. Chase et al. 2000; D. E. Soltis et al. 2000; but see Caddick et al. 2002), but further taxon sampling is needed to confirm this relationship.

Velloziaceae are a surprising addition to the Pandanales. The family was placed by A. Cronquist (1981, 1988) between Iridaceae and Aloaceae in his Liliales, even though it had long been regarded as unrelated to either, especially given the evidence from detailed work by E. S. Ayensu (1973, 1974) and L. B. Smith and E. S. Ayensu (1974, 1976). The general assumption was that Velloziales were a distinct order in Bromelianae (R. M. T. Dahlgren et al. 1985; A. L. Takhtajan 1987, 1997; R. F. Thorne 1992, 1992b), so it was somewhat amazing when molecular data suggested another home for the family (APG II; M. W. Chase et al. 1995b, 2000; D. E. Soltis et al. 2000; P. F. Stevens www), a view now supported morphologically as well (APG; L. R. Caddick et al. 2002). The family (including Acanthochlamydaceae) is more or less equally divided, in numbers of species, into two subfamilies, Vellozioideae and Xerophytoideae (including Barbacenioideae).

Triuridaceae are firmly assigned to Pandanales by APG II, L. R. Caddick et al. (2002), and P. F. Stevens (www), although the family was unassigned previously by APG and M. W. Chase et al. (2000). The phylogenetic relationships among Triuridaceae, Pandanaceae, and Cyclanthaceae remain unresolved, although there is some evidence for a sister relationship between Triuridaceae and Pandanaceae (fig. 3; Chase et al. 2000; P. J. Rudall, pers. comm.). A. L. Takhtajan (1997) considered Triuridaceae to be the only member of Triurididae, a subclass that he positioned between Alismatidae and Aridae, a view maintained by R. F. Thorne (2000, pers. comm.). Takhtajan divided the family into two tribes, Sciaphileae with five genera, and Triurideae with four. A. Cronquist (1981, 1988) placed Triuridaceae at the end of his Alismatidae in the Triuridales. A. B. Doweld (2001) referred the family to Triuridanae and positioned it between Pandananae and Dioscoreanae. The Mexican genus *Lacandonia*, the type of the Lacandoniaceae, is variously recognized (R. F. Thorne 2000) or submerged in *Triuris* (Takhtajan 1997). The floral morphology is unique in that the gynoecium is positioned outside the androecium (A. Igersheim et al. 2001).

Pandanaceae are divided into two subfamilies, Pandanoideae and Freycinetioideae. Cyclanthaceae also are divided into two subfamilies, Cyclanthoideae and Carludovicoideae (R. Eriksson 1994). The circumscription of Cyclanthaceae is not in doubt, nor is its relationship to Pandanaceae questioned (H.-D. Behnke et al. 2000; L. R. Caddick et al. 2002; M. W. Chase et al. 2000; V. Savolainen et al. 2000; D. E. Soltis et al. 2000). R. F. Thorne (1992, 1992b) suggested that the two should be placed in different superorders, but today the evidence suggests that they can be combined into a single higher taxon, as Thorne (2000, pers. comm.) now does.

5. Liliales

The heart of the matter regarding the classification of Liliopsida, insofar as this volume is concerned, is the circumscription of Liliaceae. As A. Cronquist (1993) noted, he "would be happy enough to divide this group [Liliaceae] into several families, if I could find a reasonable way to do it, but I have not found the way." Earlier he (Cronquist 1981) had written in a more hopeful vein that "studies now underway... will permit the reorganization of the Liliaceae into a series of smaller and more homogeneous families." Clearly the former comment, published a few months after his death, indicated a belief that the efforts had been unsuccessful. He objected to the attempt by R. M. T. Dahlgren et al. (1985) and dismissed those of A. L. Takhtajan (1983, 1987), commenting dryly that we "still await a comprehensive reorganization of the lilies into several families more comparable to other recognized families of angiosperms" (Cronquist 1993). There is no question that the evidence available today strongly supports the wholesale dismemberment of the family as he circumscribed it. It seems clear from his 1993 comments that he objected strenuously to the concept of several orders for what he felt was "a single rather amorphous family." Even reducing the number of orders to two as suggested by R. F. Thorne (2000), M. W. Chase et al. (2000), and V. Savolainen et al. (2000)— Liliales and Asparagales (or Orchidales)—would have bothered Cronquist. It is unlikely that he would have been thrilled with the seed coat features used to differentiate Liliales from Asparagales (H. Huber 1969; W. B. Zomlefer 1994; W. S. Judd et al. 1999). Still, all molecular analyses also indicate that Liliales and Asparagales are separate clades.

Our arrangement of Liliales (fig. 3) is based on the studies of R. Neyland (2002), T. B. Patterson and T. J. Givnish (2002), P. J. Rudall et al. (2000b, pers. comm.), A. Vinnersten and K. Bremer (2001), and W. B. Zomlefer et al. (2001). As outlined by M. W. Chase et al. (2000), Rudall et al., P. F. Stevens (www), and APG II, the Liliales include such recently accepted orders as Alstroemeriales, Campynematales, Colchicales, Melanthiales, Smilacales, and Trilliales (see also P. Goldblatt 1995). The families of the order fall into five groups, three of which are monofamilial: Corsiaceae, Campynemataceae, and Melanthiaceae. The other two groups are Colchicaceae together with Alstroemeriaceae and Luzuriagaceae, and Liliaceae together with Smilacaceae, Philesiaceae, and Rhipogonaceae. The circumscriptions of these families here often are markedly different from those outlined by A. Cronquist (1981, 1988), A. L. Takhtajan (1997), R. F. Thorne (2000), and, to a lesser degree, from those in A. B. Doweld (2001) and in Thorne's (pers. comm.) latest revision.

Corsiaceae are the most surprising addition to Liliales. Recent studies (R. Neyland 2002; P. J. Rudall and A. Eastman 2002; P. J. Rudall pers. comm.) have provided ample evidence for this association, and the family is so placed in APG II. A. Cronquist (1981) accepted the family and positioned it—as was done traditionally—near Burmanniaceae in his Orchidales. Although P. F. Stevens (www) considers the family "unplaced," its position in Liliales rather than Pandanales now seems certain. Corsiaceae may include three genera, since recent studies are in conflict as to whether *Arachnitis* belongs to Burmanniaceae or Corsiaceae (Rudall and Eastman 2002; Rudall, pers. comm.).

Campynemataceae are firmly associated with the Liliales. A. Cronquist (1981) included the family in Liliaceae (J. L. Reveal 1993e), R. F. Thorne (2000) placed it next to his Melanthiaceae, and A. L. Takhtajan (1997) assigned it to his Colchicales near Uvulariaceae. A. B. Doweld (2001) recently established an order for Campynemataceae and placed it in his Lilianae (see K. Hayashi et al. 1998) rather than his Melanthianae, to which the molecular data now suggest a closer relationship.

Melanthiaceae here include Trilliaceae, as recently suggested by M. W. Chase et al. (2000) and supported by W. B. Zomlefer et al. (2001). However, S. Fuse and M. N. Tamura (2000) retained Trilliaceae (see also S. Kazempour Osaloo and S. Kawano 1999). The two families are markedly different morphologically and are never confused, and yet segregation of Trilliaceae creates a paraphyletic Melanthiaceae. For monophyletic groupings, Melanthiaceae should be one family divided into five tribes (Melanthieae, Xerophylleae, Heloniadeae, Chinonographideae, and Parideae [including Trillieae]), or alternatively each of those taxa should be a segregate family (W. B. Zomlefer 1997b; W. B Zomlefer et al. 2001). Recognition of the three additional families accepted by A. L. Takhtajan (1997)—Xerophyllaceae, Heloniadaceae, and Chionographidaceae—is not supported. As noted previously, Tofieldiaceae (including Japonoliriaceae) and Nartheciaceae are also removed from Takhtajan's circumscription of the Melanthiales, and this disassociation is well supported by molecular and anatomical data (Fuse and Tamura; W. B. Zomlefer 1997, 1997b, 1997c; W. B. Zomlefer et al. 2001).

Within Melanthiaceae, *Melanthium* is doubtfully distinct from *Veratrum* (W. B. Zomlefer 1997b), and *Anticlea* and *Toxicoscordion* should be recognized as distinct from *Zigadenus*, leaving that genus monospecific (W. B. Zomlefer et al. 2001; W. B. Zomlefer and W. S. Judd 2002).

Alstroemeriaceae are markedly distinct and accepted as a family by all modern workers except A. Cronquist (1981, 1988), who retained them in Liliaceae. A relationship with Colchicaceae is well supported by morphological (R. M. T. Dahlgren and H. T. Clifford 1982; R. M. T. Dahlgren et al. 1985) and molecular (APG; M. W. Chase et al. 2000; D. E. Soltis et al. 2000) data.

Colchicaceae are here defined to include three subfamilies: Colchicoideae, Wurmbeoideae, and Uvularioideae (K. Kubitzki et al. 1990+, vol. 3; R. F. Thorne 2000). While the traditional definition of the first two is maintained here (A. L. Takhtajan 1997), the circumscription of the latter is markedly different from that proposed by W. S. Judd et al. (1999) and Thorne. Here, based on the findings of M. W. Chase et al. (2000), P. J. Rudall et al. (2000), and T. B. Patterson and T. J. Givnish (2002), we remove *Tricyrtis*, *Scoliopus*, *Prosartes*, *Streptopus*, and *Clintonia* and assign them to Liliaceae (defined here to include Calochortaceae). All of these genera were included in Liliaceae by A. Cronquist (1981, 1988). Five of the species traditionally assigned to *Disporum* (e.g., H. A. Gleason and A. Cronquist 1991) are now referred to *Prosartes* (Liliaceae). It should be noted that the Australian genus *Burchardia* (the type of Burchardiaceae) is retained in Colchicaceae and belongs to Uvularioideae.

Luzuriagaceae have been difficult to circumscribe, and as a result their relationship to other families always has been dubious. R. M. T. Dahlgren et al. (1985) proposed a close relationship with Philesiaceae and Rhipogonaceae, and A. L. Takhtajan (1997) presented a workable key to these families. A. Cronquist (1981, 1988) considered Luzuriagaceae (and other families) synonymous with Smilacaceae. However, a new circumscription (two genera, *Luzuriaga* and *Drymophila*; R. F. Thorne 2000) makes it more akin to Alstroemeriaceae and Colchicaceae than to either of the families suggested in 1985. *Behnia* (Behniaceae), *Geitonoplesium* (the type of Geitonoplesiaceae = Hemerocallidaceae), and *Eustrephus* (the type of Eustrephaceae = Laxmanniaceae) are now assigned to Asparagales.

A. Cronquist (1981) included Philesiaceae, Rhipogonaceae, Petermanniaceae, and Luzuriagaceae in a single, broadly defined Smilacaceae, a close alliance that is akin to Liliaceae (A. Vinnersten and K. Bremer 2001). A. L. Takhtajan (1997) and A. B. Doweld (2001) placed all five, as separate families, in Smilacales, but Doweld assigned the order to Melanthianae rather than Lilianae. Smilacaceae are defined here to include only *Smilax* and *Heterosmilax*. Philesiaceae (including Lapageriaceae) includes two monospecific genera restricted to Chile, and Rhipogonaceae (often misspelled "Ripogonaceae") comprises a single genus found primarily in Australia, New Zealand, and New Guinea. Among recent authors, only W. S. Judd (1998) and W. S. Judd et al. (1999) retain *Rhipogonum* in Smilacaceae.

Classification of the core Liliales is unsettled. At family rank, M. F. Fay and M. W. Chase (2000), M. W. Chase et al. (2000), P. F. Stevens (www), and APG II recognized only Liliaceae. W. S. Judd et al. (1999), K. Kubitzki et al. (1990+, vols. 3, 4), T. B. Patterson (1998), T. B.

Patterson and T. J. Givnish (2002), and R. F. Thorne (2000, pers. comm.) recognized Calochortaceae separate from Liliaceae. A. L. Takhtajan (1997) recognized Medeolaceae, Scoliopaceae, and Tricyrtidaceae as well. As defined by Patterson and Givnish, Calochortaceae include five genera: *Calochortus*, *Tricyrtis*, *Prosartes*, *Scoliopus*, and *Streptopus*. *Prosartes* now includes several North American species recently transferred from *Disporum* (Colchicaceae), which now includes only species from eastern Asia. Takhtajan assigned *Prosartes* and *Scoliopus* to Uvulariaceae. D. J. Mabberley (1997) accepted Calochortaceae, assigned *Tricyrtis* to Liliaceae and *Scoliopus* to Trilliaceae, and placed *Prosartes* (as *Disporum*) and *Streptopus* in Convallariaceae (= Ruscaceae). If Calochortaceae is accepted and defined to include Tricyrtidaceae, the latter, a conserved name, is the correct one at the moment. A proposal has been drafted to conserve Calochortaceae so as to retain this commonly used name.

Liliaceae sensu M. N. Tamura (1998c) consists of two subfamilies restricted to the Northern Hemisphere: Medeoloideae, comprising *Medeola* and *Clintonia*; and Lilioideae, comprising two tribes, Lilieae (consisting of *Lilium*, *Fritillaria*, *Nomocharis*, *Cardiocrinum*, and *Notholirion*) and Tulipeae (consisting of *Tulipa*, *Erythronium*, *Gagea*, and *Lloydia*). Others (A. L. Takhtajan 1997; A. B. Doweld 2001) have segregated the Medeoloideae as Medeolaceae.

6. Asparagales

Liliales are generally distinguished from Asparagales by having nectaries typically at the base of the tepals or filaments, extrorse anthers, and petals that are often spotted (W. B. Zomlefer 1994). Unfortunately, none of these distinctions is consistent. Still, the vast majority of Asparagales have nectaries on the ovary and few ever have spotted petals. Within Asparagales there is a single developmental feature relating to microsporogenesis that distinguishes two major groups (R. M. T. Dahlgren and H. T. Clifford 1982; R. M. T. Dahlgren et al. 1985; W. S. Judd et al. 1999; P. J. Rudall et al. 1997; W. B. Zomlefer 1998). The first group, informally known as the "lower asparagoids," is paraphyletic, and the second, called the "higher asparagoids," is monophyletic. Obviously the microsporogenesis condition is not a feature that can be easily observed. Nonetheless, it is consistent and corresponds with molecular evidence that also distinguishes the two groups within the order. As a whole, Asparagales are one of the more intensely studied groups of monocots, and recent works (APG; M. W. Chase et al. 1995, 1995b, 2000, 2001; M. F. Fay et al. 2000; A. W. Meerow et al. 1999, 2000; P. J. Rudall 1994; P. J. Rudall et al. 1997; T. T. Sauza-Chies et al. 1997; V. Savolainen et al. 2000; D. E. Soltis et al. 2000) are in agreement as to circumscription of the order.

The arrangement of Asparagales shown in Figure 3 is based on M. W. Chase et al. (2000, 2001), M. F. Fay et al. (2000), V. Savolainen et al. (2000), D. E. Soltis et al. (2000), R. F. Thorne (2000, calling the taxon Orchidales), P. F. Stevens (www), and APG II. Integrity of the order as a whole is not strongly supported statistically in these studies, but unpublished data (T. J. Givnish, pers. comm.; S. W. Graham, pers. comm.) strongly support its monophyly. If the group of families Boryaceae through Orchidaceae is excluded, the remaining group (Tecophilaeaceae through Amaryllidaceae) is well supported even in the published studies. Our current conception of Asparagales is close to that first proposed by R. M. T. Dahlgren and H. T. Clifford (1982) and R. M. T. Dahlgren et al. (1985), with the notable difference that Iridaceae and Orchidaceae are now included in Asparagales.

A. Cronquist (1981, 1988) assigned all the genera in our Asparagales to his Liliales, and distributed them among many fewer families: Liliaceae, Iridaceae, Aloaceae, Agavaceae, and Xanthorrhoeaceae. Today, the circumscriptions of only the last two have any similarity to the ones that Cronquist drew for those families two decades ago. Cronquist's Orchidales also have been dismantled. Geosiridaceae are now included in Iridaceae, Burmanniaceae have been moved to Dioscoreales, and Corsiaceae to Liliales. What now appear to be close to the orchids are Boryaceae, Blandfordiaceae, Asteliaceae, Hypoxidaceae, and Lanariaceae, all of which were subsumed in Cronquist's Liliaceae. A. L. Takhtajan (1997) distributed the non-orchid and non-iris members of our Asparagales primarily among three orders: Amaryllidales, Asparagales, and Xanthorrhoeales. A. B. Doweld (2001) retained Takhtajan's three orders and in addition adopted Agavales, Asphodelales, and Asteliales.

A. The "lower asparagoids"

The 14 families of the "lower asparagoids" include: Boryaceae, an astelioid clade of four families (Blandfordiaceae, Asteliaceae, Hypoxidaceae, and Lanariaceae); Orchidaceae; another four families unresolved in a polytomy (Tecophilaeaceae, Ixioliriaceae, Doryanthaceae, and Iridaceae); Xeronemataceae; and an aloe clade (Hemerocallidaceae, Xanthorrhoeaceae, and Asphodelaceae). At the base of the order, the phylogenetic relationships among Boryaceae, the astelioid clade, and Orchidaceae are uncertain (fig. 3). In the analyses of M. F. Fay et al. (2000) and S. W. Graham (pers. comm.), Boryaceae are sister to the astelioid clade, but with little statistical support. Within the astelioid clade, relationships found by molecular data are confirmed by morphological synapomorphies, with branched hairs and mucilage canals characterizing the subordinate clade Asteliaceae, Hypoxidaceae, and Lanariaceae (P. J. Rudall et al. 1998b; A. Kocyan and P. K. Endress 2001). Orchidaceae sometimes have been included in the astelioid clade (M. W. Chase et al. 2000; D. E. Soltis et al. 2000), but with little statistical support. We have chosen a conservative approach in depicting a polytomy among the well-supported astelioid lineage, Boryaceae, and Orchidaceae.

Boryaceae include two Australian genera, *Borya* and *Alania* (M. W. Chase et al. 1996), which were often associated with Anthericaceae (L. Watson and M. J. Dallwitz, http://biodiversity.uno.edu/delta/angio/index.htm). They are well removed from the Anthericaceae of Asparagales (APG; M. W. Chase et al. 2000; D. E. Soltis et al. 2000). K. Kubitzki et al. (1990+, vol. 3), R. F. Thorne (2000), P. F. Stevens (www), and A. B. Doweld (2001) have all recognized the family.

Until recently, the position of the monogeneric Blandfordiaceae, endemic to Australia, was unsettled. P. J. Rudall et al. (1998b) placed the family firmly in Asparagales, but were uncertain about its location within the order. K. Kubitzki et al. (1990+, vol. 3) and R. F. Thorne (2000) placed it next to Asteliaceae. A. L. Takhtajan (1997) retained it near Tecophilaeaceae, as the findings of Rudall et al. were not known to him. A. Cronquist (1988) considered Blandfordiaceae a synonym of Liliaceae.

A. Cronquist (1981, 1988) assigned the whole of Asteliaceae to Liliaceae. Until the recent molecular work of M. W. Chase et al. (1995b, 2000), the family had been associated with Asparagales (A. L. Takhtajan 1997) or Burmanniales (R. M. T. Dahlgren et al. 1985). L. Watson and M. J. Dallwitz (http://biodiversity.uno.edu/delta/angio/index.htm) retained the genus *Cordyline* in Asteliaceae, but here we assign it to Laxmanniaceae, following P. J. Rudall et al. (1998b).

Among the lower asparagoids, Hypoxidaceae, which A. Cronquist (1981, 1988) included in his Liliaceae, share an inferior ovary with the distantly related Doryanthaceae (A. Kocyan and P. K. Endress 2001; P. J. Rudall 2002). The position of Hypoxidaceae near Orchidaceae is firmly established (APG; APG II; M. W. Chase et al. 2000; W. S. Judd et al. 1999; V. Savolainen et al. 2000; D. E. Soltis et al. 2000; P. F. Stevens www; A. L. Takhtajan 1997), although A. B. Doweld (2001) assigned the family to his Iridanae rather than his Orchidanae.

Lanariaceae are now removed from their traditional place near Tecophilaeaceae (R. M. T. Dahlgren et al. 1985; A. L. Takhtajan 1997) in the Tecophilaeales to a position much closer to Hypoxidaceae in the Asparagales (P. J. Rudall et al. 1998b). Curiously, D. J. Mabberley (1997) assigned the single genus, *Lanaria*, to the commelinid family Haemodoraceae. A. Cronquist (J. L. Reveal 1993) placed it in the Liliaceae.

R. M. T. Dahlgren et al. (1985; see also G. Dahlgren 1989) suggested that the orchid family should be fragmented into three or perhaps four families (Apostasiaceae, Neuwiediaceae, Cypripediaceae, and Orchidaceae proper). D. W. Stevenson and H. Loconte (1995) adopted this view, but the concept has had little support save from A. B. Doweld (2001). We retain a broadly defined Orchidaceae, within which the apostasioid and neuwiedioid genera are decidedly less specialized than the cypripedioid ones, and these in turn are significantly less specialized than the true orchids. The extent to which a clear distinction can be made between the apostasioid and neuwiedioid groups now seems less well supported than suggested initially by R. M. T. Dahlgren et al. and later by P. Burns-Balogh and V. A. Funk (1986). If one were to divide the orchids into separate families, these two groups should be united in one family, as done by A. B. Doweld (2001).

Orchidaceae are customarily divided into subfamilies, tribes, and subtribes (P. Burns-Balogh and V. A. Funk 1986; K. M. Cameron et al. 1999; R. L. Dressler 1981, 1993; J. V. Freudenstein and M. W. Chase 2001; J. V. Freudenstein and F. N. Rasmussen 1999; D. L. Szlachetko 1995). There now appears to be at least some consensus for recognizing five basically monophyletic groups that correspond to the currently recognized subfamilies Apostasioideae (including Neuwiedioideae; see A. Kocyan and P. K. Endress 2001b), Cypripedioideae, Vanilloideae, Orchidoideae (including Ophrydoideae, Spiranthoideae, and Tropidioideae), and Epidendroideae (including Neottioideae, Thelymitroideae, and Vandoideae). The most recent major changes reflected in this scheme have been to recognize a distinct subfamily based upon the tribe Vanilleae, while placing Spiranthoideae into a single clade containing members of Orchidoideae and sister to Diurideae (K. M. Cameron et al. 1999). Freudenstein and Chase found molecular phylogenetic patterns consistent with recognition of either four or five subfamilies, and with separating the vanilloid and epidendroid groups. However, the relationships among the numerous tribes and subtribes is still largely unresolved (J. V. Freudenstein and F. N. Rasmussen 1999).

Recent molecular systematic studies have shown that several orchid taxa are not monophyletic, including Diseae (E. J. P. Douzery et al. 1999), Diurideae (P. J. Kores et al. 2001), Arethuseae sensu Dressler and their component subtribes (D. H. Goldman et al. 2001), and *Coelogyne* (B. Gravendeel et al. 2001; S. E. C. Sierra et al. 2000). Conversely, W. M. Whitten et al. (2000) found Maxillarieae and their subtribes and genera to agree largely with molecular phylogenies, and A. M. Pridgeon et al. (2001) found that Pleurothallidinae are basically monophyletic, confirming the results of a previous morphological study (R. Neyland et al. 1995). These molecular studies within Orchidaceae have prompted new morphological studies of character evolution within the family (e.g., W. L. Stern and W. S. Judd 2000, 2001) and the lower asparagoids (A. Kocyan and P. K. Endress 2001; P. J. Rudall 2002).

The remaining families of the Asparagales (Tecophilaeaceae through Amaryllidaceae) are strongly supported as a monophyletic group. However, the basal portion of this clade (Tecophilaeaceae, Ixioliriaceae, Doryanthaceae, and Iridaceae) is an unresolved polytomy. M. F. Fay et al. (2000) suggested a sister relationship for Tecophilaeaceae and Ixioliriaceae, with Doryanthaceae and Iridaceae successively sister to the remaining families of Asparagales. Clear resolution of this polytomy will be significant because representatives of Doryanthaceae, Iridaceae, Xeronemataceae, and the higher asparagoids generally have been found to lack *Arabidopsis*-type telomere repeat sequences (S. P. Adams et al. 2001). Regardless of how it is resolved, Xeronemataceae are strongly supported as sister to the higher asparagoids (M. W. Chase et al. 2000; M. F. Fay et al. 2000; S. W. Graham, pers. comm.). A. Cronquist (1981) included Tecophilaeaceae and Ixioliriaceae in his Liliaceae, and Doryanthaceae in his Agavaceae. A. L. Takhtajan (1997) recognized Tecophilaeales and placed Tecophilaeaceae and Ixioliriaceae therein. He placed Doryanthaceae in his Asparagales and considered the family related to Phormiaceae (= Hemerocallidaceae).

Tecophilaeaceae can be divided into two subfamilies, the Wallerioideae being somewhat isolated from the Tecophilaeoideae (R. M. T. Dahlgren and A. E. van Wyk 1988), to such an extent that A. L. Takhtajan proposed a separate Walleriaceae.

The single, Asian genus of the Ixioliriaceae usually has been assigned to the Amaryllidaceae, but the family was recognized separately by A. L. Takhtajan (1997) and positioned close to Tecophilaeaceae.

Doryanthaceae, consisting of a single genus restricted to Australia, usually have been placed near what is here termed Hemerocallidaceae.

While the circumscription of Iridaceae is not in doubt on either morphological or molecular grounds (P. Goldblatt 1990; M. W. Chase et al. 1995, 2000; C. J. Prychid and P. J. Rudall 1999; G. Reeves et al. 2001; P. J. Rudall 1994; T. T. Sauza-Chies et al. 1997; V. Savolainen et al. 2000; D. E. Soltis et al. 2000), the family's placement within the overall system of petaloid monocots today differs from that of the past. A. Cronquist (1981, 1988) and A. L. Takhtajan (1997) adopted the traditional association of the lilies and irises, and the cladistic analysis of morphological features by D. W. Stevenson and H. Loconte (1995) supported this arrangement. Nonetheless, molecular data consistently have suggested a closer association with the asparagoid lilies than with the true lilies. Analysis of a combined data set of *rbc*L gene sequences and morphological traits (M. W. Chase et al. 1995b) initially indicated this relationship, and that finding has remained tenable. R. F. Thorne's (2000) and all other current systems except A. B. Doweld's (2001) place the iris family in Asparagales.

Within Iridaceae, the Iridoideae are divided into several tribes, notably Sisyrinchieae (*Libertia*, *Olsynium*, and *Sisyrinchium*), Irideae (*Iris* and *Belamcanda*), and Tigridieae (*Herbertia*, *Nemastylis*, and *Calydorea*). The Crocoideae (including Ixioideae) also are divided into several tribes. In a molecular systematic study, G. Reeves et al. (2001) found that all subfamilies of Iridaceae were resolved as monophyletic with the exception of Nivenioideae. However, the paraphyly of Nivenioideae was not well supported statistically. Reeves et al. found that the achlorophyllous genus *Geosiris* (sometimes referred to Geosiridaceae or Burmanniaceae) fell within the nivenioid group. They also found most of the tribes of Iridaceae to be monophyletic, with *Isophysis* sister to the rest of the family.

Studies by M. W. Chase et al. (1995b) and P. J. Rudall et al. (1997) had indicated that the monogeneric Xeronemataceae, restricted to New Caledonia and New Zealand, are isolated within the lower asparagoid monocots, and with the restructuring of Anthericaceae by M. W. Chase et al. (1996), the distinctiveness of the family became obvious. M. W. Chase et al. (2001) and M. F. Fay et al. (2000) found Xeronemataceae to be basal to the remainder of the Asparagales.

Hemerocallidaceae, Asphodelaceae, and Xanthorrhoeaceae form a small clade that P. F. Stevens (www) characterized by the presence of anthraquinones, styloids, and nonphotosynthetic cotyledons. A. Cronquist (1981) assigned the types of the first two to his Liliaceae and positioned the third, which he accepted as a family, near Agavaceae. However, he also assigned some genera in what we define as Asphodelaceae to his Agavaceae, whereas some that he placed in Agavaceae now belong to Hemerocallidaceae. At the time, Cronquist also placed the genus *Geitonoplesium* in Smilacaceae, its traditional home.

A. L. Takhtajan (1997) placed Asphodelaceae—circumscribed to include Xeronemataceae—in Asparagales. However, he assigned Xanthorrhoeaceae to Xanthorrhoeales. Hemerocallidaceae, as he defined the family, were monogeneric, whereas here, based on molecular evidence (M. W. Chase et al. 2000, 2001), we include also the genera that Takhtajan placed in Phormiaceae and Dianellaceae, as well as the Geitonoplesiaceae (J. G. Conran 1994), an odd taxon that Takhtajan included in Luzuriagaceae (Smilacales). K. Kubitzki et al. (1990+, vol. 3) recognized

Hemerocallidaceae (including Dianellaceae, Geitonoplesiaceae, and Phormiaceae) and Xanthorrhoeaceae.

Hemerocallidaceae are defined here as outlined by W. B. Zomlefer (1999), M. W. Chase et al. (2000), and M. F. Fay et al. (2000). This family of 19 genera probably should be divided into three subfamilies. The first would be a monogeneric Hemerocallidoideae. The second subfamily would include the genera formerly associated with Phormiaceae and Dianellaceae as recognized by A. L. Takhtajan (1997), the Geitonoplesiaceae as recognized by J. G. Conran (1994), and the strange genus *Simethis*, which G. Dahlgren (1989) suggested might represent a new family. The third potential subfamily of Hemerocallidaceae is an Old World taxon typified by the genus *Johnsonia*, which recently has been recognized at family rank (M. W. Chase et al. 1996; K. Kubitzki et al. 1990+, vol. 3; R. F. Thorne 2000, pers. comm.). In the past this group was generally recognized at tribal rank, as Johnsonieae (A. L. Takhtajan 1997). As circumscribed by Kubitzki and by Thorne, the Johnsoniaceae included some of the genera typically assigned to other tribes in Anthericaceae, notably *Caesia* and *Corynotheca*, and some genera previously placed in Johnsonieae were moved to Laxmanniaceae (e.g., *Chamaescilla*, *Laxmannia*, *Sowerbaea*, and *Thysanotus*).

According to R. F. Thorne (2000), the Xanthorrhoeaceae are monogeneric and restricted to Australia. This circumscription differs significantly from that outlined by A. Cronquist (1981, 1988) and D. J. Mabberley (1997), both of whom included also the genera here placed in the commelinid family Dasypogonaceae and the asparagalian Laxmanniaceae. All these taxa, along with Baxteriaceae (here subsumed in Dasypogonaceae), made up the Xanthorrhoeales of A. L. Takhtajan (1997). Cronquist (1981) did suggest the possibility that the genera that he included in the family did "not all properly belong together."

The circumscription of Asphodelaceae is a matter of some debate. A. L. Takhtajan (1997) and A. B. Doweld (2001) considered Asphodelaceae and Aloaceae distinct families, while D. J. Mabberley (1997), K. Kubitzki et al. (1990+, vol. 3), W. S. Judd et al. (1999), M. W. Chase et al. (2000), D. E. Soltis et al. (2000), and R. F. Thorne (2000, pers. comm.) united the two. We adopt the latter approach. A. Cronquist (1981, 1988) included the type of Asphodelaceae in his Liliaceae. Most of the genera that Cronquist assigned to his Aloaceae (often incorrectly spelled "Aloeaceae") are now assigned to Asphodelaceae (W. S. Judd 1997). Thorne (2000) divided Asphodelaceae into two subfamilies, Asphodeloideae and Alooideae. The data presented by M. W. Chase et al. (2000b) indicate that the Alooideae are monophyletic, but this leaves a paraphyletic Asphodeloideae because *Bulbine*/*Jodrellia* are sister to Alooideae. Because the characteristics used to separate Alooideae as a group, such as secondary growth and bimodal karyotypes, are found also in some members of Asphodeloideae, particularly *Bulbine* and *Jodrellia*, Alooideae can be difficult to distinguish.

B. The "higher asparagoids"

R. F. Thorne (2000) divided the 14 families of the "higher asparagoids" (Hemerocallidaceae through Amaryllidaceae, see fig. 3) into "Asparagineae" and "Amaryllidineae," implying that these were groups at the rank of suborder (to our knowledge neither name is available). His division of the higher asparagoids into two clades is supported by molecular data (M. W. Chase et al. 2000; M. F. Fay et al. 2000; A. W. Meerow et al. 2000). Interestingly, APG II proposes to recognize these two clades at the rank of family, as Asparagaceae and Alliaceae. However, APG II uses a bracketing system to semisynonymize the previously recognized APG families. Adding to the potential confusion is that a "superconservation proposal" is being prepared to mandate the use of Amaryllidaceae (1805) over Alliaceae (1797) (A. W. Meerow, pers. comm.). Here we recognize the two main clades within the higher asparagoids but maintain the original APG families.

The first clade within the higher asparagoids consists of 11 families (Ruscaceae through Agavaceae), which tend to have scapose inflorescences (the main exception being Themidaceae) and rhizomes, though the rootstocks may vary. The second clade consists of three families (Alliaceae, Agapanthaceae, and Amaryllidaceae) characterized by the presence of bulbous stems and umbellate inflorescences. The families of Thorne's "Asparagineae" (or the "Asparagaceae sensu lato" of APG II) can be divided into three groups. The first is a well-supported clade consisting of Ruscaceae, Laxmanniaceae, and Asparagaceae. The second group (Aphyllanthaceae, Hyacinthaceae, and Themidaceae) is currently shown as a polytomy. However, unpublished molecular data indicate a relationship among these three families (S. W. Graham, pers. comm.; J. C. Pires). The third group is a strongly supported clade that includes Anemarrhenaceae, Behniaceae, Herreriaceae, Anthericaceae, and Agavaceae.

As broadly defined here, the Ruscaceae include 26 genera, which recently have been subjected to intense study (APG; D. J. Bogler and B. B. Simpson 1995, 1996; M. W. Chase et al. 1996, 2000; M. F. Fay et al. 2000; A. W. Meerow et al. 1999; D. E. Soltis et al. 2000; J. Yamashita and M. N. Tamura 2000). The most recent dispositions of these genera have been either all to a single family (P. J. Rudall et al. 2000), or distributed among a series of closely related families (R. F. Thorne 2000, pers. comm.). Most of the above-cited papers used the name

Convallariaceae instead of Ruscaceae, but the latter is a conserved name and has priority when the two are merged. In the following discussion, we use tribal names to indicate the subunits within Ruscaceae s. lat., instead of the informal group names used by Rudall et al. ("ophiopogoneae," "nolinoids," "dracaenoids," "polygonatae," "ruscoids," and "convallarieae"). Yamashita and Tamura showed that all of these are well-defined groups. Each corresponds with a family name — Ophiopogonaceae, Nolinaceae, Dracaenaceae (including Sansevieriaceae), Polygonataceae, Ruscaceae, and Convallariaceae (including Aspidistraceae) —and A. L. Takhtajan (1997), Thorne, and A. B. Doweld (2001) have recognized some of these families. Thorne arranged the families as Dracaenaceae, Nolinaceae, Ruscaceae, and Convallariaceae (including Ophiopogonaceae and Polygonataceae). While some of Takhtajan's families included genera now excluded from the entire mix, Thorne's latest generic alignments conform closely to the informal groupings suggested by Rudall et al.

Eriospermeae are variously included or excluded from a broadly defined Ruscaceae. W. S. Judd et al. (1999) excluded them, and P. J. Rudall et al. (2000) were unable to assign *Eriospermum* firmly. A. L. Takhtajan (1997) recognized the taxon at family rank but assigned it to his Tecophilaeales. R. F. Thorne (2000) also recognized it as a family and placed it after both Ruscaceae and Convallariaceae. APG II assigned the taxon to the basal position within Ruscaceae, and we are following this suggestion since it is well supported by M. F. Fay et al. (2000).

Ophiopogoneae are basal to the remainder of Ruscaceae s. lat. (P. J. Rudall et al. 2000). R. F. Thorne (2000) included them in his Convallariaceae along with the genus *Comospermum*, the latter a decidedly odd addition given that its most recent home was with Anthericaceae (M. W. Chase et al. 1996; A. L. Takhtajan 1997; see P. F. Stevens www). While the final disposition of *Comospermum* awaits additional study, Ophiopogoneae may be recognized by their dehiscent capsules but berrylike seeds. Representatives of this taxon are not at all woody and have well-developed leaves. A. Cronquist (1981, 1988) included them (*Ophiopogon* and *Liriope*) in Liliaceae (H. G. Bedell and J. L. Reveal 1982; J. L. Reveal 1993, http://www.inform.umd.edu/PBIO/usda/usdaindex.html). The Asian genus *Peliosanthes* also has been associated with Ophiopogoneae (Liang S. Y. and Dai L. K. 1992; A. L. Takhtajan 1997; J. Yamashita and M. N. Tamura 2000; but see T. Nakai 1943 and P. J. Rudall et al. 2000). Unlike the flowers of other members of Ophiopogoneae, those of *Peliosanthes* are bright blue.

Nolineae are defined here as a small taxon of four genera, the same circumscription adopted by A. L. Takhtajan (1997) and R. F. Thorne (2000) for the family Nolinaceae (D. J. Bogler and B. B. Simpson 1995, 1996). The plants are generally arborescent, with terminal rosettes of well-developed leaves. The fruits are non-fleshy and mostly dry, few-seeded, winged capsules or samaras probably derived from berries (D. J. Bogler, pers. comm.). A. Cronquist (1981, 1988) included this group within Agavaceae.

Dracaeneae were included in Agavaceae by A. Cronquist (1981, 1988), who recognized both *Dracaena* and *Sansevieria*, the latter sometimes submerged, along with *Pleomele*, in the former. A. L. Takhtajan (1997) recognized the group at family rank but defined it to include *Cordyline* (Laxmanniaceae, fide P. J. Rudall et al. 1998b). Plants of this tribe have slightly woody stems, well-developed leaves, and fleshy berries.

The genera of Ruscaceae s. lat. that are the most difficult to place are those in the Polygonatae and the Convallarieae (including Aspidistreae). Both are distinct morphologically, yet the Rusceae are clearly derived from the Convallarieae, and molecular data indicate no obvious differences among the groups. When P. J. Rudall et al. (2000) combined the available data, they found the Polygonatae to be sister to both the Convallarieae and Rusceae. Clearly this was troublesome also for K. Kubitzki et al. (1990+, vol. 3) and R. F. Thorne (2000), both of whom decided to combine the Polygonatae and Convallarieae in a single family, the Convallariaceae, along with Ophiopogoneae.

Polygonateae include four (or five) genera: *Disporopsis*, *Heteropolygonatum*, *Polygonatum*, and *Maianthemum* (including *Smilacina*), which A. Cronquist (1981, 1988) referred to his Liliaceae. They are herbs that arise from creeping rhizomes, with well-developed leaves dispersed along unbranched, elongated aerial stems, and have distinct or slightly connate tepals. The fruits are indehiscent berries, as also in Convallarieae (P. F. Stevens www).

Convallarieae (including Aspidistreae) include eight genera. All are herbs with creeping rhizomes and well-developed leaves. The aerial stems are usually branched. The tepals, like those in Polygonateae but unlike those of Ophiopogoneae, lack nectaries (P. F. Stevens www; A. L. Takhtajan 1997). A. Cronquist (1981) assigned all of these genera to his Liliaceae.

Rusceae include the three Old World genera traditionally assigned to Ruscaceae s. str.: *Ruscus*, *Semele*, and *Danae* (A. L. Takhtajan 1997; R. F. Thorne 2000). They may be distinguished quickly from immediate relatives by the reduced, scalelike leaves bearing phylloclades in their axils. In this case the phylloclades function as leaves. The plants have woody stems, their flowers are small and inconspicuous, and the fruits are red berries (P. F. Stevens www; A. L. Takhtajan 1997). A. Cronquist (1981) assigned these genera to his Liliaceae.

Laxmanniaceae (including Lomandraceae and Eustrephaceae) are a family of 14 genera (P. F. Stevens www; R. F. Thorne 2000) found primarily in Australia and also in India, Madagascar, and South America. A. Cronquist (1981, 1988) considered *Laxmannia* to be a member of the Liliaceae, whereas he placed *Lomandra* in Xanthorrhoeaceae and *Eustrephus* in Smilacaceae.

Asparagaceae consist of two genera, *Asparagus* and *Hemiphylacus*. The former is a large and complex taxon that is occasionally fragmented into smaller genera. Traditionally, *Hemiphylacus* has been assigned to the Asphodelaceae (A. L. Takhtajan 1997; W. S. Judd 1997). Its transfer to Asparagaceae is supported by the molecular evidence (M. F. Fay et al. 2000), but there is some uncertainty on morphological grounds (P. J. Rudall et al. 2000, see figs. 1, 2). A. Cronquist (1981, 1988) included the Asparagaceae in his Liliaceae.

Although the relationships among Aphyllanthaceae, Hyacinthaceae, and Themidaceae remain somewhat uncertain (M. F. Fay et al. 2000; M. Pfosser and F. Speta 1999; J. C. Pires et al. 2001), unpublished data confirm that they are indeed related (S. W. Graham, pers. comm.; J. C. Pires).

The single, Mediterranean genus of Aphyllanthaceae has been associated previously with Eriocaulaceae, Xanthorrhoeaceae, Liliaceae, and Anthericaceae. Its assignment to a segregate family is now generally accepted (APG; A. B. Doweld 2001; K. Kubitzki et al. 1990+, vol. 3; P. F. Stevens www; R. F. Thorne 2000), but whether it should be situated nearer Anthericaceae or assigned more basally as we have done here remains to be settled.

Hyacinthaceae include 40 or more genera (P. F. Stevens www; R. F. Thorne 2000) distributed among several subfamilies and tribes (A. L. Takhtajan 1997; see also K. Kubitzki et al. 1990+, vol. 3). Most arise from a bulb and have a racemose or spikelike inflorescence. Although Takhtajan included the tribe Chlorogaleae (*Camassia*, *Chlorogalum*, *Hastingsia*, and *Schoenolirion*) within his circumscription of this family, he noted that D. J. Bogler and B. B. Simpson (1995) had demonstrated that those genera were much closer to Agavaceae. Accordingly, Thorne (pers. comm.) will propose to recognize a new family for the Chlorogaleae. Excluding the Chlorogaleae, M. Pfosser and F. Speta (1999) found four major lineages within Hyacinthaceae: the monotypic Oziroeoideae; Urgineoideae; Ornithogaloideae, including the Ornithogaleae and Dipeadieae; and Hyacinthoideae, consisting of Massonieae and Hyacintheae.

Themidaceae was resurrected for *Brodiaea* and its relatives when M. F. Fay and M. W. Chase (1996) found that those genera are not related to Alliaceae, but instead are allied with Hyacinthaceae and related families. The family consists of two tribes, Milleae and what traditionally has been called "Brodiaeeae," a name inadmissible for use in Themidaceae. J. C. Pires et al. (2001) found that there are four lineages within the family, Milleae being embedded in a paraphyletic "Brodiaeeae" (J. C. Pires 2000). Plants of Themidaceae arise from a corm, typically have an umbellate inflorescence, and lack the onionlike odor of Alliaceae. All 12 genera of the family were included in the Liliaceae by A. Cronquist (1981, 1988).

All recent phylogenetic analyses find a monophyletic group that includes Anemarrhenaceae, Behniaceae, Herreriaceae, Anthericaceae, and Agavaceae (M. W. Chase et al. 2000; M. F. Fay et al. 2000). The newly proposed families Anemarrhenaceae and Behniaceae (J. G. Conran et al. 1997) seem to be related (P. F. Stevens www). R. F. Thorne (2000) positioned the two side by side. K. Kubitzki et al. (1990+, vol. 3), on the other hand, separated them, placing Anemarrhenaceae near Xanthorrhoeaceae and associating Behniaceae with Ruscaceae. A. L. Takhtajan (1997) retained *Anemarrhena* in Anthericaceae but referred *Behnia* to Luzuriagaceae, which he placed in Smilacales. A. Cronquist (1981) had included *Behnia* in Smilacaceae. P. J. Rudall et al. (2000b) placed Anemarrhenaceae near Agavaceae. A. B. Doweld (2001) arranged the families essentially as we have here, but separated Agavaceae from Hostaceae and Hesperocallidaceae.

Behniaceae, Herreriaceae, and Anthericaceae together form a small clade within the larger group of families (M. F. Fay et al. 2000). M. W. Chase et al. (1996) proposed a circumscription of Anthericaceae considerably narrower than A. L. Takhtajan's (1997). They suggested that the majority of the 30 genera in the broader circumscription should be assigned to Laxmanniaceae (including Lomandraceae), with additional genera going to Johnsoniaceae. According to R. F. Thorne (2000), there are now eight genera in Anthericaceae, fourteen in Laxmanniaceae, and eight in Johnsoniaceae. A. Cronquist (1981, 1988) had assigned these genera variously. *Lomandra*, for example, was placed in his Xanthorrhoeaceae, while Anthericaceae were assigned to his Liliaceae, along with *Laxmannia* and *Johnsonia* (J. L. Reveal 1993, http://www.inform.umd.edu/PBIO/usda/usdaindex.html). More recent evidence (APG II) calls for a slight realignment. Johnsoniaceae are now placed in synonymy under Asphodelaceae, along with Aloaceae. Laxmanniaceae (including Lomandraceae and Eustrephaceae) are positioned near Asparagaceae.

The circumscription of Agavaceae as given by A. Cronquist (1981, 1988) has been abandoned for some time. M. W. Chase et al. (1993), D. J. Bogler and B. B. Simpson (1995, 1996), APG, W. S. Judd et al. (1999), M. W. Chase et al. (2000), D. E. Soltis et al. (2000), P. J. Rudall et al. (2000b), and P. F. Stevens (www) have moved essentially half of the genera to Ruscaceae. A. L.

Takhtajan (1997), K. Kubitzki et al. (1990+, vols. 3, 4), R. F. Thorne (2000), and A. B. Doweld (2001) assigned many of the same genera to either Nolinaceae or Dracaenaceae. By restricting Agavaceae to just two subfamilies, Agavoideae and Yuccoideae, the resulting family of 10 genera is elegantly defined and easy to characterize (Bogler and Simpson; see also Takhtajan). However, there is some question as to the accuracy of this arrangement. In a broad sense, Agavaceae includes Hesperocallidaceae and Hostaceae, and all three have a bimodal karyotype. Herreriaceae also have a bimodal karyotype, although not as pronounced as that seen in Agavaceae. For this reason among others, Takhtajan assigned Herreriaceae to Asparagales next to Anthericaceae, whereas he placed Agavaceae, Hesperocallidaceae, and Hostaceae in Amaryllidales. Thorne removed Herreriaceae from their traditional spot near Anthericaceae (K. Kubitzki et al., vol. 3) and grouped the four families together.

R. F. Thorne (2000) combined Hesperocallidaceae and Hostaceae into a single family, following R. M. T. Dahlgren et al. (1985, under the illegitimate name Funkiaceae). A. L. Takhtajan (1997), K. Kubitzki et al. (1990+, vol. 3), and A. B. Doweld (2001) maintained the two monogeneric families. Plants of both arise from a short rhizome and have tubular flowers, those of *Hosta* occasionally being slightly zygomorphic. Their pronounced bimodal karyotypes, along with compelling molecular data (APG; APG II; M. W. Chase et al. 2000; J. C. Pires, unpublished), clearly place them within Agavaceae. A. Cronquist (1981, 1988) had included both genera in Liliaceae.

The relationship of the tribe Chlorogaleae to Agavaceae is problematic. This taxon, composed of *Camassia*, *Chlorogalum*, *Hastingsia*, and *Schoenolirion*, traditionally has been associated with those members of the lilioid monocots now recognized as Hyacinthaceae, as by A. L. Takhtajan (1997) and R. F. Thorne (2000). A. Cronquist (1981, 1988) had retained them in his expanded Liliaceae. Molecular data (APG II; D. J. Bogler and B. B. Simpson 1995, 1996; M. W. Chase et al. 2000; M. Pfosser and F. Speta 1999) firmly align Chlorogaleae with Agavaceae. P. F. Stevens (www) makes clear, however, that there are readily distinguishable differences among Agavoideae, Yuccoideae, Hostaceae, Hesperocallidaceae, and Chlorogaleae, and so the problem is one of rank rather than position; as noted above, Thorne (pers. comm.) will propose family rank for Chlorogaleae. Here all are assigned to Agavaceae, since recognizing Hostaceae, Hesperocallidaceae, and/or Chlorogaleae at the rank of family seems to us premature given the available evidence.

The Agapanthaceae, Alliaceae, and Amaryllidaceae are a monophyletic group (M. F. Fay et al. 2000; A. W. Meerow et al. 1999) corresponding to R. F. Thorne's (2000) "Amaryllidineae" and to APG II's "Alliaceae sensu lato" (M. W. Chase et al. 2000; M. F. Fay et al. 2000; A. W. Meerow et al.). Plants of these families have bulbous stems (except for Agapanthaceae and a few African Amaryllidaceae) and umbellate inflorescences (A. Meerow, pers. comm.). Themidaceae also have umbellate inflorescences, but corms rather than bulbs.

Agapanthaceae are a monogeneric African family. A. L. Takhtajan (1997) retained *Agapanthus* within his circumscription of Alliaceae (as Agapanthoideae), noting that it differs from the rest of that family in having branched rhizomes with fleshy roots rather than tunicate bulbs. M. F. Fay and M. W. Chase (1996) recommended adoption of Agapanthaceae and suggested that the family is closer to Amaryllidaceae than to Alliaceae. This was supported by A. W. Meerow et al. (1999) and accepted by K. Kubitzki et al. (1990+, vols. 3, 4), R. F. Thorne (2000), P. F. Stevens (www), and A. B. Doweld (2001).

Alliaceae are composed of 13 genera and perhaps 600 species, according to R. F. Thorne (pers. comm.). Within Alliaceae as defined here, A. L. Takhtajan (1997) recognized two subfamilies, Allioideae and Gilliesioideae, to which we add Tulbaghioideae, following M. F. Fay and M. W. Chase (1996) and Thorne. Most members of the family have an onion- or garlic-like odor. The inflorescences are typically umbellate and have a bract that subtends a cluster of flowers rather than an individual flower. All but *Tulbaghia* arise from bulbs, which led A. B. Doweld (2001) to segregate that genus in the Tulbaghiaceae. A. W. Meerow (pers. comm.) is of the opinion that the South American Gilliesioideae are not well circumscribed and that the genera likely have been too narrowly delimited.

Amaryllidaceae are a well-defined monophyletic family that is complex taxonomically and has been divided into numerous tribes (A. L. Takhtajan 1997; K. Kubitzki et al. 1990+, vol. 3). A satisfactory proper resolution of the tribal classification awaits more study, however (M. Ito et al. 1999; A. W. Meerow et al. 1999, 2000). Detailed studies have been done on Amaryllideae, both morphological (D. A. Snijman and H. P. Linder 1996) and molecular (A. W. Meerow and D. A. Snijman 2001), the latter resulting in a revised classification of that tribe.

III. The Commelinid Monocots

The monophyly of the commelinid monocots is supported by both molecular (M. W. Chase et al. 1993, 1995, 2000; T. J. Givnish et al. 1999; D. E. Soltis et al. 2000) and morphological (R. M. T. Dahlgren and F. N. Rasmussen 1983) data. Morphological features that characterize the commelinids include UV fluorescent compounds in their cell walls, *Strelitzia*-type epicuticular wax, and starchy endosperm (W. S. Judd et al. 1999; P. F. Stevens www). The palms (Arecaceae), however, lack starchy endosperm and may be the sister group to the remaining commelinids. (We follow APG II in referring to this clade by the informal name "commelinids," rather than "commelinoids" as by APG, to avoid confusion with the subfamily Commelinoideae. Judd et al. referred to it as Commelinanae.)

The phylogenetic relationships among the commelinids shown in Figure 4 are based on M. W. Chase et al. (2000), H. P. Linder et al. (2000), and D. E. Soltis et al. (2000) as well as unpublished data (J. I. Davis, pers. comm.; T. J. Givnish pers. comm.; S. W. Graham, pers. comm.). A number of clades are well supported, and for the most part they correspond to orders in the APG system (APG; APG II; M. W. Chase et al. 2000; P. F. Stevens www). We have been conservative in depicting these relationships by showing a polytomy among most of these orders, with the notable exception of the sister relationship between the Commelinales and Zingiberales. We recognize five orders of commelinids: Arecales, Dasypogonales, Commelinales (including Hanguanales), Zingiberales, and Poales. This differs slightly from APG, Chase et al., and APG II, in which Dasypogonaceae were unplaced as to order.

A. Cronquist (1978b, 1981, 1988) recognized the subclasses Arecidae, Commelinidae, and Zingiberidae. A. L. Takhtajan (1980, 1983, 1986, 1997) included the families of Zingiberidae in his Commelinidae, as did R. F. Thorne (1999, 2000). Recent works (M. W. Chase et al. 1995b, 2000; APG; W. S. Judd et al. 1999; V. Savolainen et al. 2000; D. E. Soltis et al. 2000; P. F. Stevens www; APG II) clearly show that recognition of both the Liliidae (lilioid monocots) and Commelinidae (commelinid monocots) renders Liliidae paraphyletic since portions of Liliidae (e.g., Asparagales) are sister to the commelinids. Because the relationship between the commelinids and the lilioid monocots is uncertain, it is unwise to assign any higher Linnaean rank to the commelinids at this time.

A. Cronquist (1981, 1988) defined Arecidae to encompass four orders: Arecales, Cyclanthales, Pandanales, and Arales. The latter three orders are placed here outside the commelinids. Araceae are assigned to Alismatidae, while Cyclanthales and Pandanales are positioned among the lilioid monocots. A. L. Takhtajan (1997) separated Arales from Arecidae and established Aridae, which he positioned near Alismatidae. He recognized four superorders in his Aridae: Aranae, Cyclanthanae, Pandananae, and Typhanae. Of these, only the Typhanae (= Poales) here remain among the commelinid monocots. As for Arecidae proper, Takhtajan restricted the taxon to Arecaceae.

The core elements of the commelinids have remained constant. Aside from the showy petaloid members of Commelinales, Zingiberales, and Poales (Bromeliaceae, Rapateaceae, and Xyridaceae), all the remaining taxa are grasslike plants with typically dry and chaffy perianth parts. A. Cronquist (1981, 1988) put Xyridaceae into Commelinales, although it is now aligned with Eriocaulaceae. Otherwise, Cronquist divided the graminoid commelinids into six orders: Eriocaulales, Restionales, Juncales, Cyperales (including Poales), Hydatellales, and Typhales. A. L. Takhtajan (1997) included the same series of families, although his ordinal circumscriptions were narrower. Significantly, Takhtajan had an expanded view of the commelinids that included the whole of Cronquist's Zingiberidae as well as Philydraceae, Pontederiaceae, and Haemodoraceae. Cronquist had included these last three families in his Liliales; here they are placed within the commelinids, in Zingiberales and Commelinales. R. M. T. Dahlgren et al. (1985), G. Dahlgren (1989), and R. F. Thorne (1976, 1981, 1992, 1992b, 2000; see also http://www.inform.umd.edu/PBIO/fam/thorne99mono.html) circumscribed the core commelinids much the same as did Cronquist and Takhtajan. Among these systems there were slight differences in the numbers of orders and families, and in the linear sequences, but all associated essentially the same taxa in a small number of related superorders. Our scheme here does not vary greatly from those in terms of membership, but we recognize far fewer orders, reflecting the views outlined by M. W. Chase et al. (2000), P. F. Stevens (www), and APG II, who divided the commelinids into four orders: Arecales, Poales, Commelinales, and Zingiberales. This arrangement is similar to that of APG but differs considerably from the one suggested by W. S. Judd et al. (1999), who divided their Commelinanae into eight orders: Arecales, Bromeliales, Poales, Juncales, Typhales, Commelinales, Philydrales, and Zingiberales. T. J. Givnish et al. (1999) recognized those eight orders as well as Dasypogonales and Eriocaulales. Judd et al. noted that the apparent monophyly of some of these orders varies according to whether morphological or molecular data are analyzed.

1. Arecales

The Arecales are among the most distinctive monocots (W. J. Baker et al. 1999). The circumscription of Arecaceae to include all genera in the order is no longer much debated, although some, mainly in the field of paleobotany, still recognize Nypaceae separately. The palms are mostly unbranched trees or shrubs with plicate leaves and unilocular ovaries (each locule containing one ovule) that become drupes. Among the monocots, palms have unusually varied pollen, with 17 aperture types (M. M. Harley and W. J. Baker 2001).

Arecaceae generally have been divided into six subfamilies, with tribes and subtribes (C. B. Asmussen et al.

2000; W. J. Baker and J. Dransfield 2000; N. W. Uhl and J. Dransfield 1987; see also J. Dransfield and N. W. Uhl 1998; D. J. Mabberley 1997; N. W. Uhl et al. 1995; A. L. Takhtajan 1997; R. F. Thorne 2000, pers. comm.). Each of the subfamilies appears to be well defined, but not all seem to be monophyletic (Asmussen et al.); they include Nypoideae (*Nypa* only), Calamoideae (including *Calamus* and *Lepidocaryum*), Coryphoideae (including *Phoenix*), Phytelephoideae (including *Phytelephus*), Ceroxyloideae (including *Pseudophoenix*, *Ravenea*, and *Chamaedorea*), and Arecoideae (including *Caryota*, *Iriartea*, *Podococcus*, *Cocos*, *Areca*, *Roystonea*, *Geonoma*, etc.). The linear order of the subfamilies varies somewhat among the various systems, but for the most part not their circumscriptions (see also C. B. Asmussen and M. W. Chase 2001; W. J. Baker et al. 2000; W. J. Hahn 2002; W. J. Hahn et al. 1994; C. E. Lewis and J. J. Doyle 2001).

2. Dasypogonales

Dasypogonales include a single, Australian family with two subfamilies and four genera (R. F. Thorne 2000; see also K. Kubitzki et al. 1990+, vols. 3, 4). APG, M. W. Chase et al. (2000), and APG II left the group unplaced as an order, while T. J. Givnish et al. (1999) recognized it as sister to Arecales. Dasypogonaceae are thought to be closely related to the palms (M. W. Chase et al. 1993; P. J. Rudall and M. W. Chase 1996; T. J. Givnish et al. 1999; R. F. Thorne 2000), and a sister relationship has been strongly supported in recent multi-gene analyses (T. J. Givnish, pers. comm.). The dispositions here differ from those of A. Cronquist (1981, 1988) and D. J. Mabberley (1997), both of whom synonymized Dasypogonaceae under the lilioid family Xanthorrhoeaceae. A. L. Takhtajan (1997) defined Dasypogonaceae narrowly and kept it, along with Baxteriaceae and Calectasiaceae, in Xanthorrhoeales. A. B. Doweld (2001) formally established Dasypogonales and included therein Baxteriaceae, Calectasiaceae, and Dasypogonaceae.

3. Commelinales

The Commelinales, as outlined by T. J. Givnish et al. (1999), M. W. Chase et al. (2000), and APG II, include Philydraceae, Haemodoraceae, Pontederiaceae, Hanguanaceae, and Commelinaceae. The association of these five families in a single order differs from past dispositions based on evidence from traditional methodologies. Givnish et al. showed that A. Cronquist's (1981) Commelinales were unnatural and encompassed elements of three different clades of commelinids (Eriocaulaceae/Xyridaceae, Rapateaceae/Mayacaceae, and Commelinaceae). Cronquist had retained Philydraceae, Pontederiaceae, and Haemodoraceae as a group at the beginning of his Liliales. This grouping was recognized by R. M. T. Dahlgren et al. (1985) based on cladistic analyses done prior to molecular studies. Hanguanaceae also were included in the Liliales, but this family was well removed from the other three. A. L. Takhtajan (1997) retained Hanguanales in his Liliidae but associated Philydrales, Pontederiales, and Haemodorales in the superorder Pontederianae of his Commelinidae. Both Cronquist and R. F. Thorne (2000) positioned these four families well away from the Commelinaceae themselves. Clearly, from a traditional point of view, Philydraceae, Haemodoraceae, and Pontederiaceae form a single group that is well isolated from both Hanguanaceae and Commelinaceae. Recent morphological studies (e.g., of styloid-type calcium oxalate crystals; see C. J. Prychid and P. J. Rudall 1999) continue to unite Philydraceae, Pontederiaceae, and Haemodoraceae. W. S. Judd et al. (1999) divided this array of families into two orders, Philydrales and Commelinales, and circumscribed the former to include Philydraceae, Haemodoraceae, and Pontederiaceae; however, in the next edition of their textbook, Philydraceae will be included in Commelinales (Judd, pers. comm.). The association of the Philydrales and Commelinales is well supported (APG; APG II; M. W. Chase et al. 1993, 1995b, 2000; M. R. Duvall et al. 1993b; H. P. Linder and E. A. Kellogg 1995; D. E. Soltis et al. 2000), and this point was acknowledged by Takhtajan and Thorne. Our representation of the phylogeny of the Commelinales is derived from M. W. Chase et al. (2000), T. J. Givnish et al. (1999), D. E. Soltis et al. (2000), and unpublished studies by T. J. Givnish (pers. comm.), S. W. Graham (pers. comm.), and J. I. Davis (pers. comm.). Recent evidence (S. W. Graham, pers. comm.) indicates sister relationships between Commelinaceae and Hanguanaceae, and between Pontederiaceae and Haemodoraceae (fig. 4).

Haemodoraceae usually are divided into two subfamilies (M. G. Simpson 1998; S. D. Hopper et al. 1999): Haemodoroideae with eight genera, and Conostylidioideae with five or six, and often divided into tribes (A. L. Takhtajan 1997). Takhtajan considered each of these subfamilies to represent a distinct family.

R. F. Thorne (2000) placed the monogeneric Asian and Australian Hanguanaceae (as Hanguanales) at the end of his Arecanae, which he placed next to Hydatellanae and Commelinanae, but now he (pers. comm.) positions the family next to Commelinaceae within Commelinales. K. Kubitzki et al. (1990+, vol. 3) put the family in the Bromeliales. A. L. Takhtajan (1997) recognized Hanguanales and located it next to his Xanthorrhoeales, which contained some members now assigned to Dasypogonales. Based on morphological characters, P. J. Rudall et al. (1999) placed Hanguanaceae as sister to the other families of their Zingiberales. However, based on molecular data, T. J. Givnish et al. (1999) showed that it is sister to Commelinaceae, and M. W. Chase et al. (2000) assigned it to the Commelinales. More recent molecular data also support a strong relationship between Hanguanaceae and Commelinaceae (S. W. Graham, pers. comm.). All analyses indicate that Hanguanaceae are well removed from Arecaceae, as maintained by R. F. Thorne (2000) and A. B. Doweld (2001).

A. L. Takhtajan (1997) and R. F. Thorne (2000, pers. comm.) divided the Commelinaceae into two subfamilies, Commelinoideae with 39 genera and Cartonematoideae with only two. The family is characterized by the presence of closed, distinct leaf sheaths and seeds with distinctive conical caps. The stamens often are rather elaborate, with anther connectives and occasionally moniliform staminal hairs (P. F. Stevens www; A. L. Takhtajan 1997). T. M. Evans et al. (2000, 2000b) found incongruent morphological and molecular phylogenies when analyzing relationships within Commelinaceae, although much of the incongruence appears related to rampant sexual selection on the androecium. From most analyses to date, it appears that *Cartonema* is basal to the rest of the family.

4. Zingiberales

The Zingiberales (or Cannales) have long been recognized as a natural group of plants (A. Cronquist 1981; R. M. T. Dahlgren et al. 1985; W. S. Judd et al 1999; W. J. Kress 1990; W. J. Kress et al. 2001; L. M. Prince and W. J. Kress, http://www.els.net/; W. B. Zomlefer 1994). Distinctive morphological features include isomorphic root hair cells, presence of intracellular silica bodies, epigynous flowers, pollen grains with distinctive features, nuclear endosperm development, and arillate seeds. Cronquist (1978b, 1981, 1988) regarded the groups that A. L. Takhtajan and R. F. Thorne recognized as Zingiberanae and Bromelianae to be closely related and proposed that they form a single subclass, Zingiberidae. The evidence currently available suggests that this is not the case, Zingiberales being sister to Commelinales (see APG; APG II; M. W. Chase et al., 1995b, 2000; A. B. Doweld 2001; W. S. Judd et al. 1999; D. E. Soltis et al. 2000; P. F. Stevens www; A. L. Takhtajan 1997; R. F. Thorne 2000).

The Zingiberales include eight families and about 95 genera (L. M. Prince and W. J. Kress, http://www.els.net/). Our representation of the phylogeny of the order follows that suggested by the recent molecular and morphological analyses of W. J. Kress et al. (2001): a "ginger" group and a "banana" group (fig. 4). The four ginger families form a monophyletic clade of two lineages: Zingiberaceae as sister to Costaceae, and Cannaceae as sister to Marantaceae. The four banana families are unresolved in a paraphyletic assemblage, but Strelitziaceae are supported as sister to Lowiaceae. Zingiberales have been divided variously into orders (A. B. Doweld 2001; A. L. Takhtajan 1997) and suborders (W. J. Kress, http://www.nmnh.si.edu/botany/projects/zingib.html; R. F. Thorne 2000). Takhtajan combined Strelitziaceae, Musaceae, and Heliconiaceae in a single order, Musales, while Kress (1990), Kress et al., and Thorne suggested that each family is the only member of its own suborder. M. W. Chase et al. (2000) positioned Strelitziaceae closest to Lowiaceae, and Kress et al. assign the two families to a single suborder, Strelitziineae.

Musaceae are the basal family and consist of three genera (L. M. Prince and W. J. Kress www). Kress (1995) and Kress et al. (2001) assign the family to Musineae.

Lowiaceae include a single, Asian and Pacific genus, *Orchidantha*, a name slightly earlier than *Lowia*. The family was considered to be rather isolated within the Zingiberales (S. Carlquist and E. L. Schneider 1998; W. J. Kress 1990), but the evidence presented by M. W. Chase et al. (2000) and Kress et al. (2001) suggests otherwise. A. L. Takhtajan (1997) considered the family to be ordinally distinct from the remaining taxa of Zingiberanae.

Strelitziaceae are large to medium-sized arborescent plants from Africa and South America with more or less zygomorphic flowers that are subtended by showy bracts (A. Cronquist 1981; A. L. Takhtajan 1997). The family includes three genera: *Ravenala*, *Strelitzia*, and *Phenakospermum*. This family and Lowiaceae together seem to be distinguishable as Strelitziineae.

Heliconiaceae include the single large genus *Heliconia*. A. L. Takhtajan (1997), R. F. Thorne (2000), W. J. Kress (www) and Kress et al. (2001) assigned the family to the suborder Heliconiineae.

Cannaceae are a monogeneric family (L. M. Prince and W. J. Kress www; N. Tanaka 2001). The spirally arranged leaves and warty capsules are distinctive features.

Marantaceae include some 31 genera (L. M. Prince and W. J. Kress www). A. L. Takhtajan (1997) divided the family into two tribes, Phrynieae and Maranteae, but a recent molecular systematic analysis (L. Andersson and M. W. Chase 2001) found no evidence to support this division.

Zingiberaceae are the largest family of the order, with 51 genera (L. M. Prince and W. J. Kress www; P. F. Stevens www). A. L. Takhtajan (1997) divided the family into four tribes: Hedychieae, Zingibereae, Alpinieae, and Globbeae.

Costaceae comprise four genera (P. F. Stevens www; R. F. Thorne 2000) found mainly in Central and South America. Recent phylogenetic analyses suggest that the generic boundaries need to be altered, since *Costus* is broadly paraphyletic (C. D. Specht et al. 2001; C. D.

Specht, pers. comm.). Most modern workers accept the distinctiveness of the family (though a contrary view was expressed by D. J. Mabberley 1997).

5. Poales

While available evidence strongly supports the Poales as a monophyletic group, relationships within the order are still uncertain. Here we follow M. W. Chase et al. (2000), D. E. Soltis et al. (2000), P. F. Stevens (www), and APG II in assigning all of the graminoid commelinids to a single order. This makes Poales a large and heterogeneous order of 18 families, and several recent systematists have argued for splitting them into three or more orders (A. B. Doweld 2001; T. J. Givnish et al. 1999; W. S. Judd et al. 1999; R. F. Thorne 2000, pers. comm.). However, the next edition of Judd et al. will recognize the larger Poales (Judd, pers. comm.).

Our representation of the phylogenetic relationships among the Poales in Figure 4 is based on several published (M. W. Chase et al. 2000; T. J. Givnish et al. 1999; H. P. Linder et al. 2000; D. E. Soltis et al. 2000) and unpublished (J. I. Davis, pers. comm.; T. J. Givnish, pers. comm.; S. W. Graham, pers. comm.) studies. According to these sources, the base of the Poales is best represented as a polytomy composed of four lineages: Typhaceae/Sparganiaceae, Bromeliaceae, Rapateaceae, and Mayacaceae (fig. 4). The remaining families fall into two clades, one containing Cyperaceae and the other containing Poaceae. The bulk of these commelinids form a phalanx of well-established families that have long been associated. However, A. Cronquist (1981, 1988) assigned Bromeliaceae to his Zingiberidae.

The position of the monogeneric Typhaceae within the monocotyledons has been greatly disputed. Traditionally, the family was positioned near Pandanales (e.g., see A. B. Rendle 1930–38, vol. 1), but R. M. T. Dahlgren et al. (1985) suggested it belongs near Pontederiales. A. Cronquist (1981, 1988) retained it in his Commelinidae rather than in Arecidae (Pandanales) or Liliidae (Pontederiales), listing Typhales last among his series of orders. D. W. Stevenson and H. Loconte (1995) supported Cronquist's disposition in that subclass, but placed Typhales as the basal unit within the subclass. M. W. Chase et al. (1995b) immediately altered this and placed Typhales well within the commelinids and nearest to Hydatellaceae, basal to the rest of the graminoid commelinids. This position was retained by Chase et al. (2000) and D. E. Soltis et al. (2000) and is reflected in our linear arrangement here.

Bromeliaceae include 58 genera (L. B. Smith and W. Till 1998; see J. R. Grant and G. Zijlstra 1998) and traditionally have been divided into three or four subfamilies: "Brocchinioideae," Tillandsioideae, Bromelioideae, and Pitcairnioideae (A. L. Takhtajan 1997; R. F. Thorne 2000). The "Brocchinioideae" (the name is in current use but not yet validly published) are a monogeneric subfamily that has the only "tank-like" leaves among the lower bromeliads; such tanks recur in the Tillandsioideae and Bromelioideae. Tillandsioideae comprise some 10 genera and are apparently monophyletic, being distinct morphologically (R. M. T. Dahlgren et al. 1985; A. J. Gilmartin and G. K. Brown 1987) as well as molecularly (M. W. Chase et al. 1993, 1995, 1995b; R. Horres et al.; H. P. Linder and E. A. Kellogg 1995; T. A. Ranker et al. 1990; R. G. Terry et al. 1997). Bromelioideae, with 32 genera, also appear to be distinct and monophyletic (W. S. Judd et al. 1999).

The monophyly of the Pitcairnioideae as traditionally defined (L. B. Smith and R. J. Downs 1974; G. S. Varadarajan and A. J. Gilmartin 1988) has been questioned (D. M. Crayn et al. 2000; R. Horres et al. 2000; R. G. Terry et al. 1997, 1997b). In the literature, this conflict has yet to be fully resolved; however, unpublished data (T. J. Givnish, pers. comm.) that include information on almost all pitcairnioid genera indicate that the group is strongly paraphyletic, with both Tillandsioideae and Bromelioideae emerging from within it. R. F. Thorne (2000) considered the subfamily to include 15 genera (see also H. Robinson and D. C. Taylor 1999).

Although a relationship between Bromeliaceae and Rapateaceae has been found in several phylogenetic analyses (M. W. Chase et al 1995; H. P. Linder and E. A. Kellogg 1995; T. J. Givnish et al. 1999), this result needs to be corroborated further.

Rapateaceae comprise 16 or 17 genera (D. J. Mabberley 1997; A. L. Takhtajan 1997; R. F. Thorne 2000). The family typically is divided into two subfamilies, but the lowland Rapateoideae are paraphyletic since they include the largely montane Saxofridericioideae (T. J. Givnish et al. 2000). The possession of nectaries by members of the Schoenocephalieae became known only after A. Cronquist's death; absence of nectaries was one characteristic that he had used to unite Rapateaceae with Mayacaceae, Xyridaceae, and Commelinaceae in his Commelinales. Placement of Rapateaceae is unsettled, since morphological and molecular studies find alternative relationships (M. W. Chase et al. 2000; T. J. Givnish et al. 1999). A. B. Doweld (2001), for example, has placed the family in its own order and superorder, and positioned it between his Typhanae and Juncanae.

Mayacaceae are a monogeneric family (D. W. Stevenson 1998; A. L. Takhtajan 1997; R. F. Thorne 2000) that A. Cronquist (1981, 1988) placed in his Commelinales next to Xyridaceae. As with Rapateaceae, conflicting morphological and molecular data (M. W. Chase et al. 2000; J. I. Davis, pers. comm.; T. J. Givnish et al. 1999; S. W. Graham, pers. comm.) render placement of Mayacaceae uncertain, and various phylogenetic analyses have placed the family with Bromeliaceae,

Rapateaceae, Eriocaulaceae, or Xyridaceae. For example, A. B. Doweld (2001) defined his Typhanae to include two orders, Mayacales and Typhales.

A. Cronquist (1981, 1988) placed Xyridaceae in Commelinales and Eriocaulaceae in Eriocaulales. D. W. Stevenson and H. Loconte (1995) challenged Cronquist's recognition of Eriocaulales and submerged that taxon in Commelinales. A. L. Takhtajan (1997) accepted both Xyridales and Eriocaulales. The two families form a branch in a larger clade dominated by Cyperaceae. A. B. Doweld (2001) termed this branch Juncanae.

Xyridaceae include five genera (R. F. Thorne 2000) and are divided into two subfamilies, with Xyridoideae being monogeneric.

Eriocaulaceae comprise nine or ten genera and have long been considered distinct and monophyletic (A. M. Giuletti et al. 2000; H. P. Linder and E. A. Kellogg 1995; D. W. Stevenson and H. Loconte 1995; J. A. Thorsch and V. I. Cheadle 1996).

Hydatellaceae include only two genera, *Hydatella* and *Trithuria*, from Australia, New Zealand, and the Pacific. M. W. Chase et al. (2000) removed the family from the Poales as suggested by APG, but it is being restored to that order in APG II. A. Cronquist (1981, 1988) had assigned it to a spot near the end of his Commelinidae, remarking that its affinities were uncertain. A. L. Takhtajan (1997) assigned it to Hydatellanae, noting that it probably is related to his Commelinanae. The minute flowers of Hydatellaceae lack both a perianth and bracteoles. While graminoid in habit, the plants typically are found in aquatic environments and seem to be highly specialized (D. A. Cooke 1987; U. Hamann 1976, 1998; U. Hamann et al. 1979; T. D. Vyshenskaya 1985). Recent evidence (D. W. Stevenson et al. 2000; APG II) suggests that the family belongs to a clade including Xyridaceae and Cyperaceae, but its exact placement therein remains uncertain. A. B. Doweld (2001), however, assigned it to Hydatellanae and positioned that superorder next to his Juncanae.

Thurniaceae, in a strict sense, are monogeneric, from the New World tropics. These are coarse herbs with upright rhizomes quite unlike the creeping, sympodial rhizomes of Juncaceae in the narrow sense adopted here. Following a detailed review of the embryology, S. L. Munro and H. P. Linder (1997, 1998) established Prioniaceae for *Prionium*, a pachycaulous genus that had long been included in Juncaceae (e.g., A. Cronquist 1981). R. F. Thorne (2000, pers. comm.) continues to accept the family, even though T. J. Givnish et al. (1999) showed that *Prionium* is sister to *Thurnia* at the base of Cyperaceae-Juncaceae, and M. W. Chase et al. (2000) and APG II considered Prioniaceae a synonym of Thurniaceae. The latter view has been followed by A. B. Doweld (2001) and herein.

The relationships between the rushes and the sedges and between those and the grasses have always been a contentious problem. The general trend has been to assign each to an order, Juncales, Cyperales, and Poales, in that linear sequence (e.g., J. Hutchinson 1959, 1969; M. Melchior 1964, vol. 2). However, A. Cronquist (1981, 1988) adopted only Juncales and Cyperales, assigning the grasses to the latter. K. Kubitzki et al. (1990+, vols. 3, 4) accepted the same two orders but assigned Cyperaceae to Juncales. While the traditional circumscriptions of Cyperales and Poales have been limited mainly to Cyperaceae and Poaceae respectively, circumscriptions of Juncales have varied, with some authors adding Restionaceae and their relatives (J. Hutchinson) or segregating those as the Restionales (A. Cronquist). A. L. Takhtajan (1997) assigned Juncales and Cyperales to his Juncanae, whereas in Poanae he accepted Flagellariales, Restionales, Centrolepidales, and Poales. R. F. Thorne (2000) had basically the same arrangement, except that he reduced Cyperales to synonymy under Juncales and assigned all of the orders in Takhtajan's Poanae to Poales. The evidence presented by G. M. Plunkett et al. (1995), D. Simpson (1995), T. J. Givnish et al. (1999), and M. W. Chase et al. (2000) supports Thorne's arrangement.

Juncaceae include six genera (H. Balslev 1998), the largest being *Juncus* and *Luzula*. Unlike those of Cyperaceae, the perianth segments of Juncaceae are in two whorls of three, and while sometimes reduced to bristles or even entirely absent, they are never as modified as those seen in Cyperaceae (A. Cronquist 1981; A. L. Takhtajan 1997).

Cyperaceae are a huge family in numbers of both genera and species. A. L. Takhtajan (1997) suggested that there are 125 genera and some 5300 species; R. F. Thorne's (2000) figures are somewhat lower, with 104 genera and 4945 species; and P. Goetghebeur (1998) cited 104 genera and 5010 species. D. J. Mabberley (1997) was even more conservative, suggesting 98 genera and 4350 species, whereas the numbers given by W. S. Judd et al. (1999) are 122 genera and 4500 species. Whichever numbers are correct, Cyperaceae are among the most speciose of the monocotyledonous families (third largest after Orchidaceae and Poaceae). There is now general agreement that Juncaceae and Cyperaceae are more closely related to each other than either is to Poaceae. This relationship is supported by both morphological and molecular data (A. M. Muasya et al. 1998, 2000; Yen A. C. and R. G. Olmstead 2000). The genus *Oxychloe*, traditionally assigned to Juncaceae, is here considered a member of Cyperaceae based on early *rbc*L data (G. M. Plunkett et al. 1995; D. Simpson 1995). Thorne placed this genus in the subfamily Cyperoideae.

Cyperaceae have been divided variously into three (A. L. Takhtajan 1997), four (P. Goetghebeur 1998; R. F. Thorne 2000), or five (R. M. T. Dahlgren et al. 1985) subfamilies. The circumscriptions of the subfamilies are not firmly established in the literature, so it is difficult to compare systems. For example, Takhtajan, Goetghebeur, and Thorne recognized Mapanioideae, but D. J. Mabberley (1997) placed *Mapania* in Rhynchosporoideae, a taxon that Takhtajan and Goetghebeur considered a synonym of Cyperoideae and that Thorne assigned to Sclerioideae. Goetghebeur also accepted Sclerioideae but restricted the subfamily to just four tribes. The subfamily Caricoideae seem to be fairly well defined, with four or five genera all assigned to a single tribe (Yen A. C. and R. G. Olmstead 2000b). Even so, in phylogenetic analyses, A. M. Muasya et al. (2000) found that current subfamilial classifications are generally not supported, and in his latest revision Thorne (pers. comm.) recognizes no subfamilies in Cyperaceae. Furthermore, several tribes and genera are polyphyletic (A. M. Muasya et al. 1998, 2000, 2000b, 2001; E. H. Roalson and E. A. Friar (2000); E. H. Roalson et al. (2001); J. R. Starr et al. (1999); Yen A. C. and R. G. Olmstead 2000, 2000b). Clearly, even the most recent arrangements of genera in subfamilies and tribes must be questioned.

The final group in the Poales is a fascinating complex of seven families (fig. 4). Our representation of the phylogeny of these families is based on several works, including M. W. Chase et al. (2000), H. P. Linder and E. A. Kellogg (1995), B. G. Briggs et al. (2000), H. P. Linder (2000), H. P. Linder et al. (2000), the Grass Phylogeny Working Group (http://www.virtualherbarium.org/grass/gpwg/default.htm), D. E. Soltis et al. (2000), and APG II. The relationships among the various taxa are gradually falling into place: Flagellariaceae are sister to the remaining families, which are resolved into two clades, one of which is Anarthriaceae, sister to Centrolepidaceae/Restionaceae, and the other is made up of Ecdeiocoleaceae/Joinvilleaceae, sister to Poaceae. Those families not evaluated by Chase et al. are placed here according to their morphological similarities (R. M. T. Dahlgren et al. 1985; A. B. Doweld 2001; K. Kubitzki et al. 1990+, vol. 4; P. F. Stevens www; A. L. Takhtajan 1997; R. F. Thorne 2000) and the evidence presented by Briggs et al. and Linder et al.

A. L. Takhtajan (1997) divided his Poanae into four orders: Flagellariales, Restionales, Centrolepidales, and Poales. A. B. Doweld (2001) retained all but Centrolepidales, which he synonymized under Restionales. K. Kubitzki et al. (1990+, vol. 4), APG, M. W. Chase et al. (2000), R. F. Thorne (2000), P. F. Stevens (www), and APG II reduced all four orders to Poales.

The monogeneric Flagellariaceae are Old World tropical lianas with perianths that are somewhat petaloid (P. F. Stevens www). As such, they are markedly distinct from related groups, and for this reason A. L. Takhtajan (1997) placed the family in its own order.

Joinvilleaceae, Restionaceae, Ecdeiocoleaceae, and Anarthriaceae were assigned to Restionales by A. L. Takhtajan (1997). All have a graminoid habit, unlike the Flagellariaceae. A. Cronquist (1981) had assigned these, plus Flagellariaceae and Centrolepidaceae, to a broader Restionales. R. F. Thorne (2000) considered the former four families, plus the newly proposed Hopkinsiaceae and Lyginiaceae (B. G. Briggs and L. A. S. Johnson 2000), to form a separate suborder. All six of these families have small, membranous perianth parts (P. F. Stevens www); most have been described only recently (B. G. Briggs 2000). Recognition of Lyginiaceae and Hopkinsiaceae is not accepted by APG II because the families are monogeneric.

Restionaceae are complex taxonomically, with convergent evolution between the African and Australian taxa (B. G. Briggs et al. 2000; P. K. Eldenas and H. P. Linder 2000; E. Dorrat Haaksma and H. P. Linder 2000; J. B. Harborne et al. 2000; H. P. Linder et al. 2000; K. A. Meney and J. S. Pate 1999; K. A. Meney 2000). The family includes approximately 53 genera (R. F. Thorne 2000) of often xeromorphic, tufted or rhizomatous herbs with solid or hollow stems. The number of perianth parts is highly variable (0, 3, 4, or 6) and the fruit is a capsule or a small nut (P. F. Stevens www; A. L. Takhtajan 1997).

Centrolepidaceae, from southeast Asia, Australia, New Zealand, the western Pacific, South America, and the Falkland Islands, include three genera (R. F. Thorne 2000) of small, tufted herbs with unisexual flowers arranged in synanthia. The perianth is absent and the flowers are minute (P. F. Stevens www; A. L. Takhtajan 1997). Takhtajan segregated the family in its own order.

Joinvilleaceae (two species) are tall, reedlike herbs with hollow stems, two whorls of scarious or chartaceous perianth parts, and drupaceous fruits (P. F. Stevens www; A. L. Takhtajan 1997).

Poaceae are one of the largest modern plant families, following only Asteraceae, Orchidaceae, and Fabaceae (s. lat.) in size, with ca. 660–680 genera and approximately 10,000 species displaying tremendous morphological, ecological, and genetic diversity. Subfamilial classifications of Poaceae have differed markedly. J. Hutchinson (1973) accepted two subfamilies, whereas J. A. Caro (1982) recognized thirteen. A. Cronquist (1981) recognized three (Bambusoideae, Panicoideae, and Pooideae) but indicated that more might be appropriate. A. L. Takhtajan (1997) recognized only Arundinoideae (including Anomochlooideae, Bambusoideae, and Pharoideae) and Pooideae (including Aristidoideae, Centothecoideae, Chloridoideae, Ehrhartoideae, Oryzoideae, and Panicoideae). In so doing, he basically followed N. N. Tzvelev (1989). Recently, with the acquisition of large amounts of molecular data, many aspects of the classifications by W. D. Clayton and S. A. Renvoize (1986) and L. Watson and M. J. Dallwitz (1992) have been supported by phylogenetic studies (e.g., N. P. Barker et al. 1995; P. Catalan et al. 1997; L. G. Clark et al. 1995, 2000; M. R. Duvall and B. R. Morton 1996; M. R. Duvall et al. 2001; B. S. Gaut et al. 1999; Hilu K. W. and L. A. Alice 2001; Hilu K. W. et al. 1999; C. Hsiao et al. 1998, 1999; S. A. Kelchner and L. G. Clark 1997; E. A. Kellogg 1998, 2000, 2001; E. A. Kellogg and H. P. Linder 1995; Liang H. P. and Hilu K. W. 1996; R. J. Mason-Gamer and E. A. Kellogg 1996; S. Mathews and R. A. Sharrock 1996; S. Mathews et al. 2000; B. R. Morton et al. 1996; G. Petersen and O. Seberg 1997; R. J. Soreng and J. I. Davis 1998; R. Spangler et al. 1999; Zhang W. P. 2000), although the circumscriptions of Bambusoideae and Arundinoideae have been modified. This research has culminated in a large collaborative effort that has yielded a comprehensive phylogeny and a new subfamilial classification (Grass Phylogeny Working Group, http://www.virtualherbarium.org/grass/gpwg/default.htm).

Acknowledgments

A review of this sort requires the willing assistance of many individuals. Foremost have been Mark W. Chase and Paula J. Rudall (Kew), and Peter F. Stevens (Missouri Botanical Garden). The following have provided other useful information, much of it unpublished: Mary E. Barkworth, David J. Bogler, Kåre Bremer, Barbara G. Briggs, Richard K. Brummitt, Matyas Buzgo, Linda R. Caddick, Jerrold I. Davis, Alexander B. Doweld, John Faden, Kanchi Gandhi, Thomas J. Givnish, Peter Goldblatt, Sean W. Graham, Jason R. Grant, Walter S. Judd, Elizabeth A. Kellogg, Robert W. Kiger, W. John Kress, Donald H. Les, H. P. Linder, John McNeill, Marc McPhereson, Alan Meerow, Linda M. Prince, Qiu Y. L., Eric H. Roalson, Douglas E. Soltis, Rob J. Soreng, Chelsea D. Specht, Dennis W. Stevenson, Robert F. Thorne, Frederick H. Utech, Michael Zanis, James L. Zarucchi, and Wendy B. Zomlefer. Our thanks to all.

The taxonomic opinions and any errors in this discussion belong solely to the authors, who hope that they have not misinterpreted the findings of hundreds of botanists who have done so much over the last two decades to clarify our understanding of the monocotyledonous angiosperms.

Table 1. *Summary of Familial and Suprafamilial Names and Dispositions*

Here we list names of monocot taxa at and above family rank, with their dispositions in A. Cronquist (1988) vis-à-vis those suggested by more recent evidence. We have generally followed APG II, except that we recognize five additional orders (Arales, Potamogetonales, Tofieldiales, Nartheciales, and Dasypogonales), and accept two additional families (Najadaceae and Zannichelliaceae). For names at the rank of class, subclass, and superorder used by Cronquist, A. L. Takhtajan (1997), R. F. Thorne (2000), A. B. Doweld (2001), and others, we have not suggested dispositions since, as already noted, assignments of taxa at these supraordinal ranks is premature at present.

Readers wishing to keep up with the latest bibliographic information for suprageneric names will find updates provided by the Index Nominum Supragenericorum Plantarum Vascularium Project (http://matrix.nal.usda.gov:8080/star/supragenericname.html; http://www.inform.umd.edu/PBIO/fam/revfam.html).

The St. Louis Code (W. Greuter et al. 2000) removed 1789 as the informal starting date for conserved angiosperm family names, instead permitting the conservation of names published from 1753 onward. Papers on this subject will appear in future issues of *Taxon*, and a proposal—already with broad international support—will be submitted to the Vienna Botanical Congress in 2005 to formally establish 4 August 1789 as the starting date for all angiosperm family names. The family nomenclature here assumes that the proposal will pass, and only names published on or after 4 August 1789 are considered valid. Other than some minor questions of priority, no name in current use is affected.

In this listing we make no recommendations for disposition of supraordinal taxa. The suggested dispositions for families and orders are exactly as in APG II except that: we recognize two additional families and three orders within APG's Alismatales; we recognize Nartheciales as separate from APG's Dioscoreales; and we recognize Petrosaviales and Dasypogonales (unplaced in APG II) at ordinal rank.

Name of Taxon	*A. Cronquist (1988)*	*Suggested Disposition*
Acanthochlamydaceae P. C. Kao	Velloziaceae	Velloziaceae
Acoraceae Martynov	Acoraceae	Acoraceae
Acorales Reveal	Arales	Acorales
Acoranae Reveal	N/A	none
Agapanthaceae F. Voigt	Liliaceae	Agapanthaceae

Agavaceae Dumortier, nom. cons.	Agavaceae	Agavaceae
Agavales Hutchinson	Liliales	Asparagales
Alismataceae Ventenat, nom. cons.	Alismataceae	Alismataceae
Alismatales Dumortier	Alismatales	Alismatales
Alismatanae Takhtajan	N/A	none
Alismatidae Takhtajan	Alismatidae	none
Alliaceae Batsch ex Borkhausen, nom. cons.	Liliaceae	Alliaceae
Alliales Traub	Liliales	Asparagales
Aloaceae Batsch	Aloaceae	Asphodelaceae
Alstroemeriaceae Dumortier, nom. cons.	Liliaceae	Alstroemeriaceae
Alstroemeriales Hutchinson	Liliales	Liliales
Amaryllidaceae Jaume Saint-Hilaire, nom. cons.	Liliaceae	Amaryllidaceae
Amaryllidales Bromhead	Liliales	Asparagales
Anarthriaceae D. F. Cutler & Airy Shaw	Liliaceae	Anarthriaceae
Anemarrhenaceae Conran, M. W. Chase & Rudall	Liliaceae	Anemarrhenaceae
Anomochloaceae Nakai	Poaceae	Poaceae
Anthericaceae J. Agardh	Liliaceae	Anthericaceae
Aphyllanthaceae Burnett	Liliaceae	Asparagales
Aponogetonaceae J. Agardh, nom. cons.	Aponogetonaceae	Aponogetonaceae
Aponogetonales Hutchinson	Najadales	Potamogetonales
Apostasiaceae Lindley, nom. cons.	Orchidaceae	Orchidaceae
Araceae Jussieu, nom. cons.	Araceae	Araceae
Arales Dumortier	Arales	Arales
Aranae Thorne ex Reveal	N/A	none
Arecaceae Schultz Schultzenstein, nom. cons.	Arecaceae	Arecaceae
Arecales Bromhead	Arecales	Arecales
Arecanae Takhtajan	N/A	none
Arecidae Takhtajan	Arecidae	none
Aridae Takhtajan	Arecidae	none
Asparagaceae Jussieu, nom. cons.	Liliaceae	Asparagaceae
Asparagales Bromhead	Liliales	Asparagales
Asphodelaceae Jussieu	Liliaceae	Asphodelaceae
Asphodelales Doweld	Liliales	Asparagales
Aspidistraceae Endlicher	Liliaceae	Ruscaceae
Asteliaceae Dumortier	Liliaceae	Asteliaceae
Asteliales Dumortier	Liliales	Asparagales
Avetraceae Takhtajan	Dioscoreaceae	Dioscoreaceae
Bambusaceae Burnett	Poaceae	Poaceae

Baxteriaceae Takhtajan	Xanthorrhoeaceae	Dasypogonaceae
Behniaceae Conran, M. W. Chase & Rudall	Smilacaceae	Behniaceae
Blandfordiaceae R. Dahlgren & Clifford	Liliaceae	Blandfordiaceae
Boryaceae (Baker) M. W. Chase, Rudall & Conran	Liliaceae	Boryaceae
Bromeliaceae Jussieu, nom. cons.	Bromeliaceae	Bromeliaceae
Bromeliales Dumortier	Bromeliales	Poales
Bromelianae R. Dahlgren ex Reveal	N/A	none
Burchardiaceae Takhtajan	Liliaceae	Colchicaceae
Burmanniaceae Blume, nom. cons.	Burmanniaceae	Burmanniaceae
Burmanniales Heintze	Orchidaceae	Dioscoreales
Butomaceae Mirbel, nom. cons.	Butomaceae	Butomaceae
Butomales Hutchinson	Alismatales	Alismatales
Butomanae Takhtajan ex Reveal	N/A	none
Calectasiaceae Endlicher	Xanthorrhoeaceae	Dasypogonaceae
Calochortaceae Dumortier, nom. cons. prop.	Liliaceae	Liliaceae
Campynemataceae Dumortier	Liliaceae	Campynemataceae
Campynematales Doweld	Liliales	Liliales
Cannaceae Jussieu, nom. cons.	Cannaceae	Cannaceae
Cannales Dumortier	Zingiberales	Zingiberales
Cartonemataceae Pichon, nom. cons.	Commelinaceae	Commelinaceae
Centrolepidaceae Endlicher, nom. cons.	Centrolepidaceae	Centrolepidaceae
Centrolepidales R. Dahlgren ex Takhtajan	Restionales	Poales
Chionographidaceae Takhtajan	Liliaceae	Melanthiaceae
Colchicaceae Candolle, nom. cons.	Liliaceae	Colchicaceae
Colchicales Dumortier	Liliales	Liliales
Commelinaceae Mirbel, nom. cons.	Commelinaceae	Commelinaceae
Commelinales Dumortier	Commelinales	Commelinales
Commelinanae Takhtajan	N/A	none
Commelinidae Takhtajan	Commelinidae	none
Conantheraceae (Endlicher ex Meisner) Pfeiffer	Liliaceae	Tecophilaeaceae
Conostylidaceae (Bentham) Takhtajan	Haemodoraceae	Haemodoraceae
Convallariaceae Horaninow	Liliaceae	Ruscaceae
Corsiaceae Beccari, nom. cons.	Corsiaceae	Corsiaceae
Costaceae Nakai	Costaceae	Costaceae
Croomiaceae Nakai	Stemonaceae	Stemonaceae
Cyanellaceae Salisbury	Liliaceae	Tecophilaeaceae
Cyclanthaceae Poiteau ex A. Richard	Cyclanthaceae	Cyclanthaceae
Cyclanthales J. H. Schaffner	Cyclanthales	Pandanales

Cyclanthanae Thorne ex Reveal	N/A	none
Cymodoceaceae N. Taylor, nom. cons.	Cymodoceaceae	Cymodoceaceae
Cymodoceales Nakai	Najadales	Potamogetonales
Cyperaceae Jussieu, nom. cons.	Cyperaceae	Cyperaceae
Cyperales Hutchinson	Cyperales	Poales
Cypripediaceae Lindley	Orchidaceae	Orchidaceae
Dasypogonaceae Dumortier	Xanthorrhoeaceae	Dasypogonaceae
Dayspogonales Doweld	Liliales	Dasypogonales
Dianellaceae Salisbury	Liliaceae	Hemerocallidaceae
Dioscoreaceae R. Brown, nom. cons.	Dioscoreaceae	Dioscoreaceae
Dioscoreales J. D. Hooker	Liliales	Dioscoreales
Dioscoreanae (R. Brown) Takhtajan ex Reveal & Doweld	N/A	none
Doryanthaceae R. Dahlgren & Clifford	Agavaceae	Doryanthaceae
Dracaenaceae Salisbury, nom. cons.	Agavaceae	Ruscaceae
Ecdeiocoleaceae D. F. Cutler & Airy Shaw	Restionaceae	Ecdeiocoleaceae
Eriocaulaceae Martynov, nom. cons.	Eriocaulaceae	Eriocaulaceae
Eriocaulales Nakai	Eriocaulales	Poales
Eriospermaceae Endlicher	Liliaceae	Ruscaceae
Eustrephaceae Chupov	Smilicaceae	Laxmanniaceae
Flagellariaceae Dumortier, nom. cons.	Flagellariaceae	Flagellariaceae
Flagellariales (Meisner) Takhtajan ex Reveal & Doweld	Restionales	Poales
Geitonoplesiaceae R. Dahlgren ex Conran	Smilicaceae	Hemerocallidaceae
Geosiridaceae Jonker, nom. cons.	Geosiridaceae	Iridaceae
Gramineae Jussieu, nom. cons. et nom. alt.	Poaceae	Poaceae
Haemodoraceae R. Brown, nom. cons.	Haemodoraceae	Haemodoraceae
Haemodorales Hutchinson	Liliales	Commelinales
Halophilaceae J. Agardh	Hydrocharitaceae	Hydrocharitaceae
Hanguanaceae Airy Shaw	Hanguanaceae	Hanguanaceae
Hanguanales R. Dahlgren ex Reveal	Liliales	Commelinales
Heliconiaceae Nakai	Heliconiaceae	Heliconiaceae
Heloniadaceae J. Agardh	Liliaceae	Melanthiaceae
Hemerocallidaceae R. Brown	Liliaceae	Hemerocallidaceae
Herreriaceae Endlicher	Liliaceae	Herreriaceae
Hesperocallidaceae Traub	Liliaceae	Agavaceae
Hopkinsiaceae B. G. Briggs & L. A. S. Johnson	Restionaceae	Anarthriaceae
Hostaceae B. Mathew	Liliaceae	Agavaceae
Hyacinthaceae Batsch ex Borkhausen	Liliaceae	Hyacinthaceae
Hydatellaceae U. Hamann	Hydatellaceae	Hydatellaceae

Hydatellales (U. Hamann) Cronquist ex Reveal & Doweld	Hydatellales	Poales
Hydatellanae Takhtajan ex Reveal	N/A	none
Hydrocharitaceae Jussieu, nom. cons.	Hydrocharitaceae	Hydrocharitaceae
Hydrocharitales Dumortier	Hydrocharitales	Alismatales
Hypoxidaceae R. Brown, nom. cons.	Liliaceae	Hypoxidaceae
Hypoxidales (Bernhardi) Takhtajan ex Reveal & Doweld	Liliales	Asparagales
Iridaceae Jussieu, nom. cons.	Iridaceae	Iridaceae
Iridales Rafinesque	Liliales	Asparagales
Iridanae Doweld	N/A	none
Isophysidaceae (Hutchinson) F. A. Barkley	Iridaceae	Iridaceae
Ixioliriaceae Nakai	Liliaceae	Ixioliriaceae
Japonoliriaceae Takhtajan	Liliaceae	Petrosaviaceae
Johnsoniaceae Lotsy	Liliaceae	Hemerocallidaceae
Joinvilleaceae Tomlinson & A. C. Smith	Joinvilleaceae	Joinvilleaceae
Juncaceae Jussieu, nom. cons.	Juncaceae	Juncaceae
Juncaginaceae Richard, nom. cons.	Juncaginaceae	Juncaginaceae
Juncaginales Hutchinson	Najadales	Potamogetonales
Juncales Dumortier	Juncales	Poales
Juncanae Takhtajan	N/A	none
Juncidae Doweld	Commelinidae	none
Lacandoniaceae E. Martínes & Ramos	Triuridaceae	Triuridaceae
Lanariaceae H. Huber ex R. Dahlgren & A. E. van Wyk	Liliaceae	Lanariaceae
Lapageriaceae Kunth	Smilacaceae	Philesiaceae
Laxmanniaceae Bubani	Liliaceae	Laxmanniaceae
Lemnaceae Martynov, nom. cons.	Lemnaceae	Araceae
Lilaeaceae Dumortier, nom. cons.	Juncaginaceae	Juncaginaceae
Liliaceae Jussieu, nom. cons.	Liliaceae	Liliaceae
Liliales Perleb	Liliales	Liliales
Lilianae Takhtajan	N/A	none
Liliidae J. H. Schaffner	Liliidae	none
Liliopsida Batsch	Liliopsida	none
Limnocharitaceae Takhtajan ex Cronquist	Limnocharitaceae	Limnocharitaceae
Lomandraceae Lotsy	Xanthorrhoeaceae	Laxmanniaceae
Lophiolaceae Nakai	Haemodoraceae	Nartheciaceae
Lowiaceae Ridley	Lowiaceae	Lowiaceae
Lowiales Takhtajan ex Reveal & Doweld	Zingiberales	Zingiberales
Luzuriagaceae Lotsy	Smilacaceae	Luzuriagaceae
Lyginiaceae B. G. Briggs & L. A. S. Johnson	Restionaceae	Anarthriaceae

Marantaceae R. Brown, nom. cons.	Marantaceae	Marantaceae
Maundiaceae Nakai	Juncaginaceae	Juncaginaceae
Mayacaceae Kunth, nom. cons.	Mayacaceae	Mayacaceae
Mayacales Nakai	Commelinales	Poales
Medeolaceae (S. Watson) Takhtajan	Liliaceae	Liliaceae
Melanthiaceae Batsch, nom. cons.	Liliaceae	Melanthiaceae
Melanthiales R. Dahlgren ex Reveal	Liliales	Liliales
Melanthianae Doweld	N/A	none
Milulaceae Traub	Liliaceae	Alliaceae
Musaceae Jussieu, nom. cons.	Musaceae	Musaceae
Musales Reveal	Zingiberales	Zingiberales
Najadaceae Jussieu, nom. cons.	Najadaceae	Najadaceae
Najadales Dumortier	Najadales	Alismatales
Najadanae Takhtajan ex Reveal	N/A	none
Nartheciaceae Fries ex Bjurzon	Liliaceae	Nartheciaceae
Nartheciales Reveal & Zomlefer	Liliales	Nartheciales
Neuwiediaceae (Burns-Balogh & V. A. Funk) R. Dahlgren ex Reveal & Hoogland	Orchidaceae	Orchidaceae
Nolinaceae Nakai	Nolinaceae	Ruscaceae
Nypaceae Brongniart ex Le Maout & Decaisne	Arecaceae	Arecaceae
Ophiopogonaceae Endlicher	Liliaceae	Ruscaceae
Orchidaceae Jussieu, nom. cons.	Orchidaceae	Orchidaceae
Orchidales Rafinesque	Orchidales	Asparagales
Orchidanae Doweld	N/A	none
Palmae Jussieu, nom. cons. et nom. alt.	Arecaceae	Arecaceae
Pandanaceae R. Brown, nom. cons.	Pandanaceae	Pandanaceae
Pandanales Lindley	Pandanales	Pandanales
Pandananae Thorne ex Reveal	N/A	none
Peliosanthaceae Salisbury	Liliaceae	Ruscaceae
Pentastemonaceae Duyfjes	Stemonaceae	Stemonaceae
Petermanniaceae Hutchinson, nom. cons.	Smilacaceae	Colchicaceae
Petrosaviaceae Hutchinson, nom. cons.	Petrosaviaceae	Petrosaviaceae
Petrosaviales Takhtajan	Triuridales	Petrosaviales
Petrosavianae Doweld	N/A	none
Philesiaceae Dumortier, nom. cons.	Smilacaceae	Philesiaceae
Philydraceae Link, nom. cons.	Philydraceae	Philydraceae
Philydrales Dumortier	Liliales	Commelinales
Phormiaceae J. Agardh	Agavaceae	Hemerocallidaceae

Pistiaceae Richard ex C. Agardh	Araceae	Araceae
Poaceae (R. Brown) Barnhart, nom. cons.	Poaceae	Poaceae
Poales Small	Cyperales	Poales
Poanae Takhtajan ex Reveal & Doweld	N/A	none
Polygonataceae Salisbury	Liliaceae	Ruscaceae
Pontederiaceae Kunth, nom. cons.	Pontederiaceae	Pontederiaceae
Pontederiales J. D. Hooker	Liliales	Commelinales
Pontederianae Takhtajan ex Reveal	N/A	none
Posidoniaceae Hutchinson, nom. cons.	Posidoniaceae	Posidoniaceae
Posidoniales Nakai	Najadales	Potamogetonales
Potamogetonaceae Reichenbach, nom. cons.	Potamogetonaceae	Potamogetonaceae
Potamogetonales Dumortier	Najadales	Potamogetonales
Prioniaceae S. L. Munro & H. P. Linder	Juncaceae	Thurniaceae
Rapateaceae Dumortier, nom. cons.	Rapateaceae	Rapateaceae
Rapateales (Meisner) Colella ex Reveal & Doweld	Commelinales	Poales
Rapateanae Doweld	N/A	none
Restionaceae R. Brown, nom. cons.	Restionaceae	Restionaceae
Restionales J. D. Hooker	Restionales	Poales
Rhipogonaceae Conran & Clifford	Smilacaceae	Rhipogonaceae
Ruppiaceae Horaninow, nom. cons.	Ruppiaceae	Ruppiaceae
Ruppiales Nakai	Najadales	Potamogetonales
Ruscaceae Sprengel, nom. cons.	Liliaceae	Ruscaceae
Sansevieriaceae Nakai	Agavaceae	Ruscaceae
Scheuchzeriaceae F. Rudolphi, nom. cons.	Scheuchzeriaceae	Scheuchzeriaceae
Scheuchzeriales B. Boivin	Najadales	Potamogetonales
Scoliopaceae Takhtajan	Liliaceae	Liliaceae
Smilacaceae Ventenat, nom. cons.	Smilacaceae	Smilacaceae
Smilacales Lindley	Liliales	Liliales
Sparganiaceae Hanin, nom. cons.	Sparganiaceae	Sparganiaceae
Stemonaceae Carrel, nom. cons.	Stemonaceae	Stemonaceae
Stemonales Takhtajan ex Doweld	Liliales	Pandanales
Stenomeridaceae J. Agardh, nom. cons.	Dioscoreaceae	Dioscoreaceae
Strelitziaceae Hutchinson, nom. cons.	Strelitziaceae	Strelitziaceae
Streptochaetaceae Nakai	Poaceae	Poaceae
Taccaceae Berchtold & J. Presl, nom. cons.	Taccaceae	Dioscoreaceae
Taccales Dumortier	Dioscoreales	Dioscoreales
Tecophilaeaceae Leybold, nom. cons.	Liliaceae	Tecophilaeaceae
Tecophilaeales Traub ex Reveal	Liliales	Asparagales

Thalassiaceae Nakai	Hydrocharitaceae	Hydrocharitaceae
Themidaceae Salisbury	Liliaceae	Themidaceae
Thismiaceae J. Agardh, nom. cons.	Burmanniaceae	Burmanniaceae
Thurniaceae Engler, nom. cons.	Thurniaceae	Thurniaceae
Tofieldiaceae Takhtajan	Liliaceae	Tofieldiaceae
Tofieldiales Reveal & Zomlefer	Liliales	Tofieldiales
Trichopodaceae Hutchinson, nom. cons.	Dioscoreaceae	Dioscoreaceae
Tricyrtidaceae Takhtajan, nom. cons.	Liliaceae	Liliaceae
Trilliaceae Chevallier, nom. cons.	Liliaceae	Melanthiaceae
Trilliales Takhtajan	Liliales	Liliales
Triuridaceae Gardner, nom. cons.	Triuridaceae	Triuridaceae
Triuridales J. D. Hooker	Triuridales	Pandanales
Triuridanae Thorne ex Reveal	N/A	none
Triurididae Takhtajan ex Reveal	Liliidae	none
Typhaceae Jussieu, nom. cons.	Typhaceae	Typhaceae
Typhales Dumortier	Typhales	Poales
Typhanae Thorne ex Reveal	N/A	none
Uvulariaceae A. Gray ex Kunth, nom. cons.	Liliaceae	Melanthiaceae
Vallisneriaceae Link	Hydrocharitaceae	Hydrocharitaceae
Vallisneriales Nakai	Hydrocharitales	Alismatales
Velloziaceae Hooker	Velloziaceae	Velloziaceae
Velloziales R. Dahlgren ex Reveal	Liliales	Pandanales
Walleriaceae (R. Dahlgren) Takhtajan, nom. cons.	Liliaceae	Tecophilaeaceae
Xanthorrhoeaceae Dumortier, nom. cons.	Xanthorrhoeaceae	Xanthorrhoeaceae
Xanthorrhoeales Takhtajan ex Reveal & Doweld	Liliales	Asparagales
Xeronemataceae M. W. Chase, Rudall & M. F. Fay	Liliaceae	Xeronemataceae
Xerophyllaceae Takhtajan	Liliaceae	Melanthiaceae
Xyridaceae C. Agardh, nom. cons.	Xyridaceae	Xyridaceae
Xyridales Lindley	Commelinales	Poales
Zannichelliaceae Dumortier, nom. cons.	Zannichelliaceae	Zannichelliaceae
Zingiberaceae Martynov, nom. cons.	Zingiberaceae	Zingiberaceae
Zingiberales Grisebach	Zingiberales	Zingiberales
Zingiberanae Takhtajan ex Reveal	N/A	none
Zingiberidae Cronquist	Zingiberidae	none
Zosteraceae Dumortier, nom. cons.	Zosteraceae	Zosteraceae
Zosterales Nakai	Najadales	Potamogetonales
Zosteranae Doweld	N/A	none

Table 2. *Suggested Familial and Ordinal Dispositions of Select Genera in Volume 26*

Genus	Disposition
223. LILIACEAE	
1. *Harperocallis*	Tofieldiaceae – Tofieldiales
2. *Pleea*	Tofieldiaceae – Tofieldiales
3. *Tofieldia*	Tofieldiaceae – Tofieldiales
4. *Triantha*	Tofieldiaceae – Tofieldiales
5. *Aletris*	Nartheciaceae – Nartheciales
6. *Narthecium*	Nartheciaceae – Nartheciales
7. *Chamaelirium*	Melanthiaceae – Liliales
8. *Helonias*	Melanthiaceae – Liliales
9. *Xerophyllum*	Melanthiaceae – Liliales
10. *Veratrum*	Melanthiaceae – Liliales
11. *Melanthium*	Melanthiaceae – Liliales
12. *Schoenocaulon*	Melanthiaceae – Liliales
13. *Zigadenus*	Melanthiaceae – Liliales
14. *Stenanthium*	Melanthiaceae – Liliales
15. *Amianthium*	Melanthiaceae – Liliales
16. *Trillium*	Melanthiaceae – Liliales
17. *Scoliopus*	Liliaceae – Liliales
18. *Calochortus*	Liliaceae – Liliales
19. *Prosartes*	Liliaceae – Liliales
20. *Streptopus*	Liliaceae – Liliales
21. *Uvularia*	Melianthiaceae – Liliales
22. *Medeola*	Liliaceae – Liliales
23. *Clintonia*	Liliaceae – Liliales
24. *Erythronium*	Liliaceae – Liliales
25. *Fritillaria*	Liliaceae – Liliales
26. *Lilium*	Liliaceae – Liliales
27. *Lloydia*	Liliaceae – Liliales
28. *Tulipa*	Liliaceae – Liliales
29. *Alstroemeria*	Alstroemeriaceae – Liliales
30. *Hypoxis*	Hypoxidaceae – Asparagales
31. *Odontostomum*	Tecophilaeaceae – Asparagales
32. *Convallaria*	Ruscaceae – Asparagales
33. *Maianthemum*	Ruscaceae – Asparagales
34. *Polygonatum*	Ruscaceae – Asparagales
35. *Asparagus*	Asparagaceae – Asparagales
36. *Echeandia*	Anthericaceae – Asparagales
37. *Eremocrinum*	Anthericaceae – Asparagales
38. *Leucocrinum*	Anthericaceae – Asparagales
39. *Asphodelus*	Asphodelaceae – Asparagales
40. *Hemerocallis*	Hemerocallidaceae – Asparagales
41. *Dianella*	Hemerocallidaceae – Asparagales
42. *Hesperocallis*	Agavaceae – Asparagales
43. *Hosta*	Agavaceae – Asparagales
44. *Allium*	Alliaceae – Asparagales
45. *Nothoscordum*	Alliaceae – Asparagales
46. *Amaryllis*	Amaryllidaceae – Asparagales
47. *Crinum*	Amaryllidaceae – Asparagales
48. *Galanthus*	Amaryllidaceae – Asparagales
49. *Habranthus*	Amaryllidaceae – Asparagales
50. *Hymenocallis*	Amaryllidaceae – Asparagales
51. *Leucojum*	Amaryllidaceae – Asparagales
52. *Narcissus*	Amaryllidaceae – Asparagales
53. *Zephyranthes*	Amaryllidaceae – Asparagales
54. *Camassia*	Agavaceae – Asparagales
55. *Chlorogalum*	Agavaceae – Asparagales
56. *Hastingsia*	Agavaceae – Asparagales
57. *Schoenolirion*	Agavaceae – Asparagales
58. *Chionodoxa*	Hyacinthaceae – Asparagales
59. *Hyacinthoides*	Hyacinthaceae – Asparagales
60. *Muscari*	Hyacinthaceae – Asparagales
61. *Ornithogalum*	Hyacinthaceae – Asparagales
62. *Scilla*	Hyacinthaceae – Asparagales
63. *Brodiaea*	Themidaceae – Asparagales
64. *Dichelostemma*	Themidaceae – Asparagales
65. *Triteleiopsis*	Themidaceae – Asparagales
66. *Androstephium*	Themidaceae – Asparagales
67. *Muilla*	Themidaceae – Asparagales
68. *Bloomeria*	Themidaceae – Asparagales
69. *Triteleia*	Themidaceae – Asparagales
70. *Milla*	Themidaceae – Asparagales
224. ALOACEAE	
1. *Aloe*	Asphodelaceae – Asparagales
225. AGAVACEAE	
1. *Agave*	Agavaceae – Asparagales
2. *Nolina*	Ruscaceae – Asparagales
3. *Dasylirion*	Ruscaceae – Asparagales
4. *Yucca*	Agavaceae – Asparagales
5. *Hesperoyucca*	Agavaceae – Asparagales
6. *Hesperaloe*	Agavaceae – Asparagales
7. *Sansevieria*	Ruscaceae – Asparagales
8. *Furcraea*	Agavaceae – Asparagales
9. *Manfreda*	Agavaceae – Asparagales
226. STEMONACEAE	
1. *Croomia*	Stemonaceae – Pandanales
227. SMILACACEAE	
1. *Smilax*	Smilacaceae – Liliales

220. PONTEDERIACEAE
• Pickerel-weed Family

Charles N. Horn

Herbs, annual or perennial, aquatic, rooted in mud or free-floating. **Stems** of 2 types, vegetative and flowering; vegetative stems indeterminate, bearing many leaves, glabrous; flowering stems determinate, usually emersed, bearing single leaf, spathe, and terminal inflorescence. **Leaves** of 2 types, sessile and petiolate; stipule marcescent, apex truncate, associated with petiolate leaves only; sessile leaves submersed or rarely emersed, forming basal rosette or alternate on elongate stem, blade linear to occasionally oblanceolate, base sheathing, margins entire; petiolate leaves floating or emersed, blade cordate, reniform, or ovate. **Inflorescences** paniculate, spicate, umbellate, or 1-flowered; spathes bractlike, folded or clasping, commonly with acute to caudate extension. **Flowers** sessile, 3-merous; perianth with tepals connate proximally, yellow, blue, mauve, or white, tube tubular or funnelform, limb 6-lobed; stamens 3 or 6, adnate to perianth; pistils 3-locular, 1 or 3 developing to maturity; placentation parietal or basal; ovules 1–many. **Fruits** capsular or utriculate. **Seeds** smooth or with longitudinal wings.

Genera 6, species ca. 30 (4 genera, 11 species in the flora; 2 genera, 3 species introduced): nearly worldwide, primarily of tropical and subtropical regions.

SELECTED REFERENCES Alexander, E. J. 1937. Pontederiaceae. In: N. L. Britton et al., eds. 1905+. North American Flora.... 47+ vols. New York. Vol. 19, pp. 51–60. Barrett, S. C. H. and D. E. Seaman. 1980. The weed flora of California rice fields. Aquatic Bot. 9: 351–376. Eckenwalder, J. E. and S. C. H. Barrett. 1986. Phylogenetic systematics of Pontederiaceae. Syst. Bot. 11: 373–391. Kaul, R. B. and C. N. Horn. 1986. Pontederiaceae. In: Great Plains Flora Association. 1986. Flora of the Great Plains. Lawrence, Kans. Pp. 1239–1241. Rosatti, T. J. 1987. The genera of Pontederiaceae in the southeastern United States. J. Arnold Arbor. 68: 35–71. Solms-Laubach, H. 1883. Pontederiaceae. In: A. L. P. P. de Candolle and C. de Candolle, eds. 1878–1896. Monographiae Phanerogamarum.... 9 vols. Paris. Vol. 4, pp. 501–535.

1. Inflorescences commonly with more than 50 flowers; fruits utriculate, 1-seeded . . . . 4. *Pontederia*, p. 45
1. Inflorescences with fewer than 30 flowers; fruits capsular, 10–200-seeded.
    2. Tepals connate only basally . . . . 1. *Monochoria*, p. 38
    2. Tepals connate 1/2 or more of their length.
        3. Petiolate leaf blade coriaceous; perianth limb lobes ovate, longer than 2 cm; stamens 6 . . . . 2. *Eichhornia*, p. 38
        3. Petiolate leaf blade membranous or absent; perianth limb lobes linear to oblanceolate, shorter than 2 cm; stamens 3 . . . . 3. *Heteranthera*, p. 41

1. MONOCHORIA C. Presl, Reliq. Haenk. 1: 27. 1827 [I]

Carigola Rafinesque; *Gomphima* Rafinesque; *Limnostachys* F. Mueller

Herbs, annual or facultatively perennial, rooting in mud. **Vegetative stems** submersed with elongate internodes, or emersed and short. **Flowering stems** submersed and extending to water surface, or emersed. **Sessile leaves** forming basal rosette. **Petiolate leaves** floating or emersed; blade cordate to hastate, apex acute to acuminate. **Inflorescences** umbellate or paniculate, up to 30-flowered, elongating in 1–several days; spathes folded, sometimes with ovate to caudate extension. **Flowers:** open 1 day only; perianth with tepals connate basally, blue or white, limb lobes distinct, narrowly ovate to lanceolate, glabrous, apex acute; stamens 6, 1 with anther 2 times larger than others; filaments winged, glabrous; anthers yellow or blue, ovoid; ovary incompletely 3-locular; ovules many; style 1. **Fruits** capsular, ovoid. **Seeds** 10–200, ovoid, testa with longitudinal wings.

Species ca. 4 (1 in the flora): introduced; Africa, Asia, Australia.

This is a poorly understood genus in need of detailed work.

SELECTED REFERENCES Baker, C. A. 1951. Pontederiaceae. In: C. G. G. J. van Steenis, ed. 1950+. Flora Malesiana.... Series I. Spermatophyta. 11+ vols. in parts. Djakarta and Leiden. Vol. 4, part 3, pp. 255–261. Cook, C. D. K. 1989. Taxonomic revision of *Monochoria* (Pontederiaceae). In: K. Tan, ed. 1989. The Davis & Hedge Festschrift.. Plant Taxonomy, Phytogeography, and Related Subjects. Edinburgh. Pp. 149–184. Kaul, M. K. 1986. Weed Flora of Kashmir Valley. Jodhpur.

1. Monochoria vaginalis (Burman f.) C. Presl, Reliq. Haenk. 1: 128. 1827 [F] [I] [W1] [W2]

Pontederia vaginalis Burman f., Fl. Indica, 80. 1768; *Monochoria plantaginea* (Roxburgh) Kunth; *Monochoria vaginalis* var. *pauciflora* (Blume) Merrill

Flowering stems 8–16.5 cm, distal internode 3–15 mm. **Sessile leaves** up to 5 cm, blade linear, apex acuminate. **Petiolate leaves:** stipule 3.5–5.7 cm; petiole 7–28 cm, rigid; blade cordate, 4–8 × 1.8–5.2 cm. **Inflorescences** paniculate, 3–8-flowered, all flowers opening same day; spathes 1.9–4 cm, sometimes with lanceolate to ovate extension. **Perianth:** limb lobes blue or white, narrowly ovate, 10–14 mm; larger stamen with hook, 4–6 mm, shorter stamens 3–4 mm; filaments linear, glabrous; style glabrous. **Seeds** 10–11-winged, 0.8–1.2 × 0.4–0.6 mm.

Flowering (Jul–)Sep–Oct. Rice fields, ephemeral ponds; 0–200 m; introduced; Calif.; Asia (India, China, Japan, and Malayan Islands).

By the 1970s, *Monochoria vaginalis* had become part of the rice-field flora of California (S. C. H. Barrett and D. E. Seaman 1980). C. D. K. Cook (1989) described this species as characteristic of rice fields; however, no records are known from the rice fields of Missouri to Louisiana.

As is typical of many aquatic annuals, plant size, leaf shape, and flower number are highly variable in relation to the amount of water (C. D. K. Cook 1989). Specimens from California typically have three to eight flowers, while descriptions of those from other parts of the world (H. Aston 1977; C. D. K. Cook 1989) suggest that the inflorescences of this species may have up to 20 flowers.

2. EICHHORNIA Kunth, Eichhornia, 3. 1842, name conserved • Water-hyacinth, jacinthe d'eau [for Johann A. F. Eichhorn, 1779–1856, Prussian statesman] [I]

Piaropus Rafinesque, name rejected

Herbs, annual or perennial, rooting in mud or free-floating. **Vegetative stems** submersed and growing to surface, or emersed and short. **Flowering stems** emersed. **Sessile leaves** forming basal rosette or alternate on elongate vegetative stem; apex acuminate. **Petiolate leaves** floating or

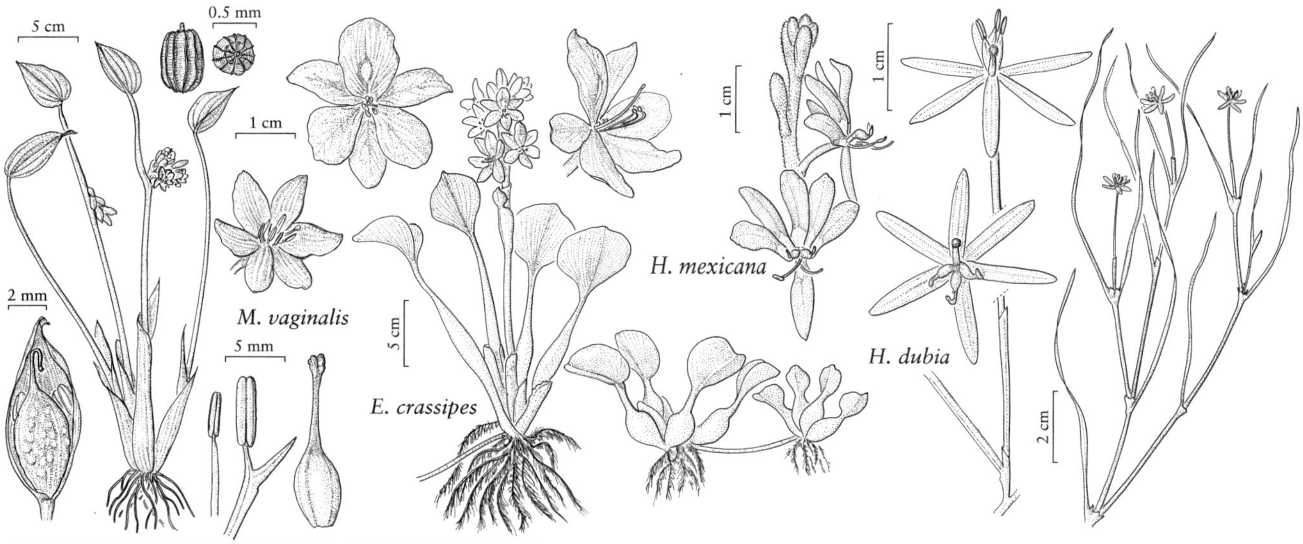

MONOCHORIA ◦ EICHHORNIA ◦ HETERANTHERA

emersed; blade cordate to ovate or round, coriaceous, apex obtuse. **Inflorescences** spicate, up to 30-flowered, elongating over several days; spathes folded, with acuminate to caudate extension. **Flowers** open 1 day only; perianth with tepals connate for more than 1/2 its length, blue, mauve, or rarely white, funnelform, limb lobes obovate to oblanceolate, longer than 2 cm, glandular-pubescent, apex obtuse to acute; stamens 6, distal 3 shorter than proximal 3; filaments curved upward near apex, purple, glandular-pubescent; anthers yellow, rounded to oblong; ovary incompletely 3-locular; ovules many; style 3-lobed. **Fruits** capsular, elongate. **Seeds** 10–200, ovoid, testa with longitudinal wings.

Species 7 (2 in the flora): introduced; Neotropics and Africa.

Eichhornia paniculata (Sprengel) Solms-Laubach [syn. *Piaropus paniculatus* (Sprengel) Small] was reported by J. K. Small (1913) as occurring in drainage ditches of peninsular Florida. The collection was made in 1907 in Manatee County (*J. K. Small s.n.*, NY). This species can be recognized by its growing rooted in mud and by the presence of cordate leaves and an elongate flowering stem terminated by a panicule of 20–100 flowers.

Eichhornia diversifolia (Vahl) Urban, native to Puerto Rico, is not known to occur in the flora area.

SELECTED REFERENCE Barrett, S. C. H. 1988. Evolution of breeding systems in *Eichhornia* (Pontederiaceae): A review. Ann. Missouri Bot. Gard. 75: 741–760.

1. Plants typically free-floating; sessile leaves produced in rosette; petioles commonly inflated; perianth limb lobe margins entire . . . . 1. *Eichhornia crassipes*
1. Plants typically rooted in mud; sessile leaves alternate on elongate stem; inflated petioles absent; perianth limb lobe margins erose . . . . 2. *Eichhornia azurea*

1. Eichhornia crassipes (Martius) Solms in A. L. P. P. de Candolle and C. de Candolle, Monogr. Phan. 4: 527. 1883 F I W1 W2

Pontederia crassipes Martius, Nov. Gen. Sp. Pl. 1: 9, plate 4. 1823; *Eichhornia speciosa* Kunth; *Piaropus crassipes* (Martius) Rafinesque

Plants perennial, typically free-floating. **Vegetative stems** condensed, except when branching. **Flowering stems** erect, bending over after flowering, to 25 cm, distal internode less than 4 cm. **Sessile leaves** in basal rosette. **Petiolate leaves** floating or emersed; stipule 2.5–14 cm, apex truncate; petiole at least slightly swollen, 3.5–33 cm; blade ovate to round, 2.5–11 × 3.5–9.5 cm. **Spikes** 4–15-flowered; spathes obovate, 4–11 cm; peduncle 5–12.5 cm, glabrous. **Flowers** opening individually within 2 hours after sunrise, wilting by night; perianth blue or mauve-blue, limb lobes obovate, 16–37 mm, margins entire, central distal lobe with dark blotch in center and yellow spot within blotch; proximal stamens 20–35 mm, distal 14–19 mm; anthers 1.7–2.1 mm; style 3-lobed. **Seeds** 11–14-winged, 1.1–2.1 × 0.6–0.9 mm. $2n = 32$.

Flowering early spring–late fall. Ponds, ditches, canals, calm waters of rivers; 0–300 m; introduced; Ala., Ariz., Calif., Fla., Ga., La., Miss., N.C., S.C., Tex., Va.; worldwide, tropics and subtropics.

This is probably the most aggressive aquatic weed ever known in the tropics (T. D. Center and N. R. Spencer 1981; R. E. Fitzsimons and R. H. Vallejos 1986; W. T. Penfound and T. T. Earle 1948; H. J. Webber 1897) and has been documented to cause extensive physiological changes in the aquatic environment (G. R. Ultsch 1973). It is on the USDA/APHIS noxious weeds list and the Florida Department of Natural Resources prohibited list (D. C. Schmitz 1990). Extensive research has gone into pesticide and biological control (R. G. Baer and P. C. Quimby Jr. 1981; T. D. Center and T. K. Van 1989; J. C. Joyce and W. T. Haller 1985; D. C. Schmitz 1990), resulting in its decline in Florida. Plants were probably first introduced into the U.S. at the Louisiana exposition in 1884 (W. T. Penfound and T. T. Earle 1948), and the species is now known in all states where freezing temperatures are minimal. Research results from the 1970s suggest that the species may have an economic importance by removing toxic substances from water (F. Chigbo et al. 1982).

The inflated petiole has long been a characteristic used to separate *Eichhornia crassipes* from other species of the genus. Yet on older plants under crowded conditions, the petiole shows little or no swelling (W. T. Penfound and T. T. Earle 1948). This has resulted in several misidentifications of specimens as *E. azurea*.

Inflorescences typically elongate overnight, and all flowers open the same day (W. T. Penfound and T. T. Earle 1948). However, some robust specimens have inflorescences in which the proximal flowers open the first day and the distal 2–4 flowers open on the subsequent day. Individual flowers open shortly after dawn and wilt by nightfall. By the next morning, the flowering stem has bent such that the developing fruits are typically submersed, allowing seed development to occur under water (W. T. Penfound and T. T. Earle 1948).

S. C. H. Barrett (1977, 1979, 1980, 1980b, 1988, 1989) has studied the reproductive biology in depth. The water hyacinth is a tristylous species; however, the only naturally occuring populations with all three morphs are in northeastern Brazil. This led S. C. H. Barrett and I. W. Forno (1982) to conclude that *Eichornia crassipes* is native to Amazonia. All other studied populations in the world have only one or two of the floral morphs. In some populations, a breakdown in the tristylous condition was observed whereby the stigma was found to be adjacent to anthers in some flowers (semihomostylous condition), resulting in some seed production (S. C. H. Barrett 1979). In general, seed production in temperate populations was found to be half that of tropical populations, mainly due to differences in levels of insect visitation (S. C. H. Barrett 1980b).

2. Eichhornia azurea (Swartz) Kunth, Eichhornia, 4. 1842 • Rooted water-hyacinth I W1 W2

Pontederia azurea Swartz, Prodr., 57. 1788

Plants perennial, typically rooted in mud. **Vegetative stems** elongate, developing to and growing at water surface. **Flowering stems** erect, 8–12 cm, glabrous, distal internode 2–10 cm. **Sessile leaves** submersed, alternate on elongate stem. **Petiolate leaves** emersed; stipule 7–13 cm, apex truncate; petiole never inflated, 11–25 cm; blade round, 7–16 × 2.3–16 cm. **Spikes** 7–50-flowered, individual flowers open 1 day only; spathes obovate, 3–6 cm; peduncle 1.9–15 cm, pubescent with orange hairs. **Perianth** blue or white, limb lobes obovate, 13–25 mm, margins erose, central distal lobe dark blue at base with yellow distal spot; proximal stamens 15–29 mm, distal 6–20 mm; anthers 1.2–2.3 mm; style 3-lobed. **Seeds** 10–13-winged, 1–1.8 × 0.8–1.1 mm.

Flowering (probably) Jun–Oct. Coastal rivers, lakes; 0–100 m; introduced; Fla.; Central America to South America (n Argentina).

Eichhornia azurea is on the USDA/APHIS noxious weeds list (D. C. Schmitz 1990) and is known only as a waif from several locations in Florida (R. G. Westbrooks 1990). It has been reported from Texas (L. H. Shinners 1962b), but this record was based on a misidentified specimen (*Jones 1160*, SMU) that actually is *E. crassipes*. All known populations in Florida have been eradicated by the Florida Department of Natural Resources. R. G. Westbrooks specifically noted a population growing in a residential estate pool in Palm Beach that was eradicated in June 1988. Another collection from Columbia County (*N. Coile 5793*, NBYC) was also eradicated.

The breeding system of *Eichornia azurea*, a tristylous species, is more like that of *Pontederia* species than that of *E. crassipes* (S. C. H. Barrett 1978). Populations typically have a mix of all three floral morphs, allowing for outcrossing.

The above description is based on South American material.

3. HETERANTHERA Ruiz & Pavón, Fl. Peruv. Prodr., 9, plate 2. 1794, name conserved • Mud plantain [Greek *heteros*, different, and *antheros*, anther]

Eurystemon E. J. Alexander; *Heterandra* Palisot de Beauvois; *Schollera* Schreber; *Zosterella* Small

Herbs, annual or perennial, rooting in mud. **Vegetative stems** submersed and growing to surface, or emersed and procumbent or short. **Flowering stems** submersed and growing to water surface, or emersed. **Sessile leaves** submersed, forming basal rosette, or emersed on elongate stem; blade linear or sometimes oblanceolate, apex acuminate to obtuse. **Petiolate leaves** floating or emersed; blade reniform, cordate, or oblong, membranous, apex acute to obtuse. **Inflorescences** spicate or solitary, 1–30-flowered, elongating over 1–several days; spathes folded or clasping, with acute to caudate apex, or rarely expanded leaflike extension. **Flowers:** individuals open 1 day only; perianth with tepals connate 1/2 or more of its length, yellow, blue-mauve, mauve, or white, tubular or salverform, limb lobes linear to oblanceolate or narrowly elliptic, shorter than 2 cm, apex obtuse to acuminate; stamens 3, unequal, 2 lateral ones usually shorter; filaments yellow or purple, inflated in some species, glabrous, glandular-pubescent, or pilose; anthers yellow or purple, rounded, oblong, or sagittate; ovary incompletely 3-locular; ovules 10–many; style 3-lobed. **Fruits** capsular, elongate. **Seeds** 10–200, ovoid, testa with longitudinal wings.

Species 12 (7 in the flora): Western Hemisphere, Africa.

In the flora area, *Heteranthera peduncularis* Bentham is known only from one collection in Arizona (*C. Mohr s.n.*, US), probably from about 1900. The center of the distribution of this species is in the highlands of Mexico. It is closely related to *Heteranthera multiflora* and *H. reniformis*, and may be distinguished by distinctly cordate leaves, mauve perianth limb lobes, and glabrous filaments.

SELECTED REFERENCE Horn, C. N. 1988. Developmental heterophylly in the genus *Heteranthera* (Pontederiaceae). Aquatic Bot. 31: 197–209.

1. Sessile leaf blade linear; inflorescences 1-flowered; stamens equal; anthers coiling . . . . . 1. *Heteranthera dubia*
1. Sessile leaf blade linear to sometimes oblanceolate, or absent; inflorescences 1–24-flowered; stamens unequal; anthers not coiling.
    2. Inflorescences 1-flowered; perianth limbs essentially actinomorphic.
        3. Vegetative stems commonly elongating unless plant emersed from early age; petiolate leaf blade round to oblong, base cordate to truncate; distal central perianth limb lobe with lateral flanges . . . . . 5. *Heteranthera rotundifolia*
        3. Vegetative stems elongating only on plants in over 5 cm of water; petiolate leaf blade oblong to ovate, base truncate to cuneate; distal central perianth limb lobe without lateral flanges . . . . . 6. *Heteranthera limosa*

2. Inflorescences 2–24-flowered; perianth limbs zygomorphic.
   4. Petiolate leaves not produced; emersed peduncle glandular-pubescent; inflorescences elongating over several days . . . . 4. *Heteranthera mexicana*
   4. Petiolate leaves present on mature plants; emersed peduncle glabrous; inflorescences elongating in 1–2 days.
      5. Spikes shorter than spathes, terminal flower sometimes extending past spathe apex; filament hairs white; spathes equal to or barely longer than subtending internode of flowering stem . . . . 2. *Heteranthera reniformis*
      5. Spikes longer than spathes, except on cleistogamous inflorescences; filament hairs purple; spathes 3 or more times longer than subtending internode of flowering stem . . . . 3. *Heteranthera multiflora*

1. Heteranthera dubia (Jacquin) MacMillan, Metasp. Minnesota Valley, 138. 1892 • Hétéranthère litigieuse F W1 W2

Commelina dubia Jacquin, Observ. Bot. 3: 9. 1768; *Heteranthera graminea* (Michaux) Vahl; *Heteranthera liebmannii* (Buchneau) Shinners; *Schollera graminea* (Michaux) Rafinesque; *Zosterella dubia* (Jacquin) Small; *Z. longituba* E. J. Alexander

Plants perennial. **Vegetative stems** submersed and growing to surface, or emersed and contracted, procumbent. **Flowering stems** less than 5 mm, no visible distal internode. **Sessile leaves** alternate or appearing opposite at nodes with inflorescences, blade linear, thin; submersed leaves 3–10 cm × 1–5 mm, emersed leaves 2–4 cm × 1–3 mm. **Petiolate leaves and stipules** not produced. **Inflorescences** 1-flowered; spathes 12–45 mm, glabrous; peduncle absent. **Flowers** opening within 2 hours after dawn, wilting by dusk; perianth yellow, salverform, tube 1.5–11 cm, limb lobes equal, of uniform color, linear, 4–11 mm; stamens equal, 2–10 mm; filaments centrally inflated, glabrous; anthers coiling after anthesis, by early afternoon; style glabrous or scarcely pubescent. **Seeds** 7–30, 12–22-winged, 1–1.6 × 0.5–1.3 mm. $2n = 30$.

Flowering Apr–Nov. Streams, rivers, ponds, lakes; 0–1500 m; B.C., Man., N.B., Ont., Que.; Ala., Ariz., Ark., Calif., Conn., Del., D.C., Fla., Ill., Ind., Iowa, Kans., Ky., La., Md., Mass., Mich., Minn., Mo., Nebr., N.H., N.J., N.Y., N.C., N.Dak., Ohio, Okla., Oreg., Pa., S.Dak., Tenn., Tex., Vt., Va., Wash., W.Va., Wis.; Mexico; West Indies (Cuba).

Extensive vegetative variation has been noted in the leaf morphology of *Heteranthera dubia* (C. N. Horn 1983). Plants in ponds and lakes tend to have longer, narrower leaves than plants from streams and rivers with a unidirectional current. Laboratory growth experiments revealed at least some genetic component to the differences (C. N. Horn 1981).

Plants are most extensive in their growth in alkaline waters (J. B. Moyle 1945), and, if plants are in a great enough density, they can cause an increase in water pH (E. B. Hollingsworth 1966). Growth in some lakes, streams, and canals has resulted in the species being considered a weed (E. B. Hollingsworth 1966).

Some authors (J. K. Small and J. J. Carter 1913; E. J. Alexander 1937) have transferred this species to the segregate genus *Zosterella*, based on the equal stamens (vs. nonequal in *Heteranthera* s.s.) and narrow, erect anthers (vs. broad and versatile). Small and Carter first described *Zosterella* citing only the above differences. Alexander later reported several more distinguishing characteristics: anthers coiling after anthesis, spathes sessile, and seeds relatively few. Recent molecular work by S. W. Graham et al. (1998) using chloroplast DNA suggests that *Heteranthera dubia* is clearly nested within the genus *Heteranthera* and is most closely related to *H. seubertiana* Solms, a tropical species.

Variation in perianth size has been a basis for separate recognition of *Heteranthera liebmannii*, or its equivalent, *Zosterella longituba* (E. J. Alexander 1937; D. S. Correll and H. B. Correll 1972). Alexander used differences in the lengths of perianth tube and lobes, and length of the perianth tube relative to that of the spathe to distinguish that segregate species. Study has revealed three causes of variation in perianth tube length: plants from populations farther north have a shorter perianth tube (herbarium study); flowers that develop under water have minimal perianth tube elongation, and the perianth remains completely within the spathe (J. W. Thieret 1972b; R. B. Wylie 1917); and lower overnight air temperatures result in less perianth tube elongation (controlled experiments). The specific epithet *liebmannii* was first applied to this entity by F. Buchneau (1872), under *Potamogeton*, and the type material consists of a series of specimens from Pital, Mexico, having only cleistogamous flowers. Later, the true generic identity of the specimens was realized. E. J. Alexander (1937), convinced that they represent a species distinct from those in Texas and Mexico, proposed the name *Zosterella longituba* for that entity, and placed *Potamogeton liebmannii* in synonymy. Since Alexander should have used the epithet *liebmannii*, which was not blocked in *Zosterella*, the name *Z. longituba* is illegitimate. An additional complication is that the types of *P. liebmannii* are morphologically ambiguous, since they include only cleistogamous flowers. Hence it is impossible, with current knowledge, to determine conclusively which of the two forms of *Heteranthera dubia* (as I recognize them) is represented by those types.

Populations are known to be infected by the fungal parasite *Membranosorus heterantherae* Ostenfeld & Peterson (H. S. Forest et al. 1986). The subsequent growths may be of two forms depending on whether they develop above the substrate on stems (nodal galls) or underground on roots (root galls).

2. Heteranthera reniformis Ruiz & Pavón, Fl. Peruv. 1: 43. 1798 W1 W2

Heterandra reniformis Ruiz & Pavón Palisot de Beauvois

Plants annual or facultatively perennial. **Vegetative stems** submersed with elongate internodes, or emersed and procumbent. **Flowering stems** 1–9 cm, distal internode 0.5–4 cm. **Sessile leaves** forming basal rosette, blade linear to oblanceolate, thin, 2.4–3.7 cm × 3–8 mm. **Petiolate leaves** floating or emersed; stipule 1–5 cm; petiole 2–13 cm; blade reniform, 1–4 × 1–5 cm, length equal to or less than width, apex obtuse. **Inflorescences** spicate, 2–8-flowered, elongating in 1 day, usually shorter than spathes, terminal flower sometimes extending beyond spathe apex; spathes 0.8–5.5 cm, glabrous; peduncle 0.5–4.2 cm, glabrous. **Flowers** opening ca. 3 hours after sunrise, wilting by early afternoon; perianth white, salverform, tube 5–10 mm, limbs zygomorphic, lobes narrowly elliptic, 3–6.5 mm, distal central lobe with yellow or green region at base, sometimes with distal brown spot; stamens unequal, lateral stamens 0.9–2.2 mm, filaments linear, pubescent with white multicellular hairs toward apex; central stamen 2.2–4.7 mm, filament sparsely pubescent with multicellular hairs; style pubescent with multicellular hairs. **Seeds** 8–14-winged, 0.5–0.9 × 0.3–0.5 mm. $2n$ = 48.

Flowering Jun–Oct. Roadside ditches, edges of streams and ponds, freshwater tidal mudflats; 0–2600 m; Ala., Conn., Del., Fla., Ga., Ill., Ind., Ky., La., Md., Miss., Mo., N.J., N.Y., N.C., Ohio, Pa., S.C., Tenn., Tex., Va., W.Va.; Mexico; throughout Central America; scattered in South America (Argentina, Brazil, Paraguay); naturalized in Italy.

3. Heteranthera multiflora (Grisebach) C. N. Horn, Phytologia 59: 290. 1986

Heteranthera reniformis Ruiz & Pavón var. *multiflora* Grisebach, Abh. Königl. Ges. Wiss. Göttingen 24: 323. 1879

Plants annual. **Vegetative stems** submersed with elongate internodes, or emersed and procumbent. **Flowering stems** 1–4 cm, distal internode less than 1 cm. **Sessile leaves** alternate; blade linear to oblanceolate, thin, 37–70 × 4–6 mm, margins entire. **Petiolate leaves** floating or emersed; stipule 3–5 cm; petiole 5–15 cm; blade cordate, 2–5 × 2–5 cm, length usually greater than width, apex obtuse. **Inflorescences** spicate, 3–16-flowered, elongating in 1–2 days, longer than spathes (except on cleistogamous inflorescences); spathes 2–7 cm, glabrous; peduncle 4–11 cm, glabrous when emersed. **Flowers** opening ca. 2 hours after dawn, wilting early afternoon; perianth mauve or white, salverform, tube 3–12 mm, limbs zygomorphic, lobes narrowly elliptic, 3.3–5.9 mm, distal central lobe dark purple in basal $1/3$, 2 yellow spots within dark area; stamens unequal, lateral stamens 1.3–2.4 mm, filaments linear, densely pubescent with purple multicellular hairs toward apex; central stamen 2.6–4 mm, filament sparsely pubescent with multicellular hairs; style pubescent with multicellular hairs. **Seeds** 9–12-winged, 0.6–0.9 × 0.4–0.5 mm. $2n$ = 32.

Flowering Jul–Nov. Roadside ditches, pond edges; 0–1800 m; Ark., Del., Ill., Iowa, Kans., La., Md., Miss., Mo., Nebr., N.J., N.C., Okla., Pa., Tenn., Tex., Va.; South America (Argentina, Brazil, Paraguay, Venezuela).

This species has in the past been incorrectly identified as either *Heteranthera reniformis* (J. A. Steyermark 1963) or *H. peduncularis* (R. B. Kaul and C. N. Horn 1986). However, detailed study of development, floral morphology, and chromosome number revealed distinct differences (C. N. Horn 1985, 1988).

Three distinct genetic forms of this species exist, two in the flora range. Populations along the Atlantic coast have white flowers and inflorescences with most of the flowers within the spathe. Populations in the Great Plains have blue flowers and inflorescences with most flowers typically beyond the spathe. These populations have not been, nor should they be, recognized at any taxonomic level.

4. Heteranthera mexicana S. Watson, Proc. Amer. Acad. Arts 18: 166. 1883 [F]

Eurystemon mexicanum (S. Watson) E. J. Alexander

Plants annual. **Vegetative stems** submersed with elongate internodes, or emersed and short. **Flowering stems** 6–15 cm, distal internode 4–8.5 cm. **Sessile leaves** submersed and thin, or emersed and thickened, 5–15 cm × 1–6 mm. **Petiolate leaves and stipules** not produced. **Inflorescences** spicate, 7–24-flowered, elongating over several days; spathes 1.2–4 cm; peduncle 4–9 cm, glandular-pubescent when emersed. **Flowers** 2–10 per inflorescence opening each day; perianth mauve, salverform, tube 2–6 mm, limbs zygomorphic, lobes narrowly elliptic, 8–13 mm, proximal lobe shorter; stamens unequal, lateral stamens 2.7–7 mm, filaments obliquely inflated at center, bent upward apically, glabrous; central stamen 5.9–10 mm, filament linear; style glabrous. **Seeds** 12–15-winged, 0.7–1 × 0.6–0.9 mm, outer integument composed of additional surface layer of tetragonal to hexagonal isodiametric cells.

Flowering Jun–Dec. Ephemeral pools; 0–400 m; Tex.; Mexico (Coahuila, Nuevo León, Tamaulipas).

Heteranthera mexicana is very rare and should be considered for national endangered or threatened status. Populations are located in swales and ditches in an area that is subject to irregular rainfall. In several attempts to locate populations during the mid-1980s, the swales were discovered to be dry; yet in 1987, a wet year in south Texas, a population of several thousand individuals was found. This suggests that seeds are adapted to survive at least several years in dry soil.

5. Heteranthera rotundifolia (Kunth) Grisebach, Cat. Pl. Cub., 252. 1866

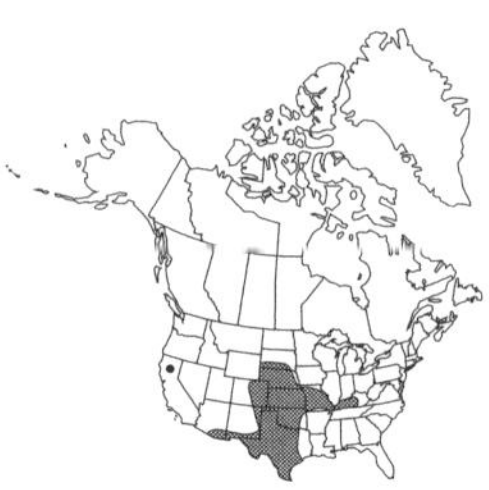

Heteranthera limosa (Swartz) Willdenow var. [ß] *rotundifolia* Kunth, Enum. Pl. 4: 122. 1843

Plants annual. **Vegetative stems** submersed with elongate internodes, or if emersed at early age, then procumbent. **Flowering stems** 2–12 cm, distal internode 1–6 cm. **Sessile leaves** submersed, forming basal rosette; blade linear to oblanceolate, thickened, 2.4–5 cm × 2–4 mm. **Petiolate leaves** floating or emersed; stipule 1–5 cm; petiole 3–11 cm; blade round to oblong, 1–5 cm × 5–25 mm, length equaling (or, on young plants, greater than) width, base cordate to truncate, apex obtuse. **Inflorescences** 1-flowered; spathes 1–2.8 mm, glabrous. **Flowers** opening within 1 hour after dawn, wilting by midday; perianth blue or white, salverform, tube 11–29 mm, limbs essentially actinomorphic, lobes narrowly elliptic, 5.2–18.2 mm, proximal lobe longer, distal central lobe with lateral flanges toward base; stamens unequal, lateral stamens 2.8–8 mm, central stamen 3.9–8.5 mm; filaments linear, curved downward apically, glandular-pubescent; style glabrous. **Seeds** 8–15-winged, 0.5–1 × 0.3–0.6 mm. $2n = 14$.

Flowering Jun–Oct. Shallow water or exposed mud of ephemeral pools, roadside ditches, small ponds; 0–2800 m; Ariz., Calif., Colo., Kans., Ky., Mo., Nebr., N.Mex., Okla., S.Dak., Tex.; West Indies; Central America; South America (except Chile, French Guiana, Guyana, Peru, Suriname); naturalized in Italy.

Heteranthera rotundifolia is closely related to *H. limosa* and has been considered conspecific with it by many, including D. S. Correll and H. B. Correll (1972), who noted differences in stem development but did not recognize them taxonomically. Close study of reproduction and seedling development (C. N. Horn 1988) has revealed other differences, besides yet others in flower morphology (cf. the descriptions). Seedlings of *H. rotundifolia* have flexible petioles and commonly develop elongate stems, while those of *H. limosa* have rigid petioles, even in deep water, and rarely develop elongate stems.

6. Heteranthera limosa (Swartz) Willdenow, Ges. Naturf. Freunde Berlin Neue Schriften 3: 439. 1801 [W1] [W2]

Pontederia limosa Swartz, Prodr., 57. 1788; *Leptanthus ovalis* Michaux

Plants annual. **Vegetative stems** submersed with elongate internodes only on plants in over 5 cm of water, or emersed and short. **Flowering stems** 2–24 cm, distal internode 1–11 cm. **Sessile leaves** forming basal rosette, blade linear to oblanceolate, thin or sometimes thickened, 3.1–6 cm × 3–5 mm. **Petiolate leaves** emersed; stipule 1–6 cm; petiole 2–13 cm; blade oblong to ovate, 1–5 cm × 4–33 mm, length equaling or greater than width; base truncate to cuneate, apex acute. **Inflorescences** 1-flowered; spathes 0.9–4.5 cm, glabrous. **Flowers** opening within 1 hour after dawn, wilting by midday; perianth blue or white, salverform, tube 15–44 mm, limbs essentially actinomorphic, lobes equal, narrowly elliptic, 5.2–26.3 mm, distal lobes yellow toward base; stamens unequal, lateral stamens 2.3–7.8 mm, central stamen 3.3–7.2 mm; filaments linear, glandular-pubescent; style glabrous. **Seeds** 9–14-winged, 0.5–0.8 × 0.2–0.6 mm. $2n = 14$.

Flowering May–Nov. Shallow water or emersed at pond edges and in roadside ditches; 0–2000 m; Ariz., Ark.,

Calif., Colo., Fla., Ill., Iowa, Kans., Ky., La., Miss., Mo., Nebr., N.Mex., Okla., S.Dak., Tenn., Tex.; Mexico; West Indies; Central America; South America (Argentina, Bolivia, Brazil, Colombia, Ecuador, Paraguay, Venezuela).

Personal observation of *Heteranthera limosa* suggests that it is best adapted for growth in water less than 5 cm deep and is a poor competitor with the closely related *H. rotundifolia* in deeper waters.

4. PONTEDERIA Linnaeus, Sp. Pl. 1: 288. 1753; Gen. Pl. ed. 5, 140. 1754 • Pickerelweed [for Giulio Pontedera, 1688–1757, Italian botanist]

Narukila Adanson; *Umsema* Rafinesque; *Unisema* Rafinesque

Herbs, annual or perennial, rooted in mud. **Vegetative stems** submersed and growing to surface, or rhizomatous. **Flowering stems** submersed and growing to surface, or emersed and slightly constricted just below first node. **Sessile leaves** forming basal rosette, blade thin. **Petiolate leaves** floating or emersed; blade cordate to reniform, apex obtuse to acuminate. **Inflorescences** spicate, commonly 50 or more-flowered, elongating over many days; spathes folded, apex acute to acuminate; peduncle glandular-pubescent or pilose. **Flowers** open 1 day only; perianth mauve, blue, or white, funnelform, limb lobes oblong to oblanceolate, glandular-pubescent or pilose, apex obtuse to acuminate; stamens 6, distal 3 shorter than proximal 3; filaments purple, glandular-pubescent; anthers yellow, ovoid to oblong; ovary 3-locular, only 1 developing to maturity; ovule 1; style 1 or more. **Fruits** utriculate, ovoid, with spinulose, dentate or smooth longitudinal ridges. **Seed** 1, ovoid, smooth.

Species 6 (1 in the flora): Western Hemisphere.

SELECTED REFERENCES Fernald, M. L. 1925. *Pontederia* versus *Unisema*. Rhodora 27: 76–81. Horn, C. N. and R. R. Haynes. 1987. Typification of *Pontederia* and its Linnaean species names (Pontederiaceae). Taxon 36: 621–625. Lowden, R. M. 1973. Revision of the genus *Pontederia* L. Rhodora 75: 426–487. Whigham, D. F. and R. L. Simpson. 1982. Germination and dormancy studies of *Pontederia cordata* L. Bull. Torrey Bot. Club 109: 524–528.

1. Pontederia cordata Linnaeus, Sp. Pl. 1: 208. 1753 • Pontédérie cordée, langue du boeuf F W1 W2

Narukila cordata (Linnaeus) Nieuwland; *Pontederia angustifolia* Pursh; *P. cordata* var. *lanceolata* (Nuttall) Grisebach; *P. cordata* var. *lancifolia* (Muhlenberg) Torrey; *P. lanceolata* Nuttall; *P. lancifolia* Muhlenberg; *Unisema cordata* (Linnaeus) Farwell

Plants perennial, rooted in mud. **Vegetative stems** contracted, rhizomatous. **Flowering stems** erect, to 120 cm. **Sessile leaves:** blade linear. **Petiolate leaves** emersed; stipule 7–29 cm; petiole distinctly constricted just below blade, to 60 cm; blade lanceolate to cordate, 6–22 × 0.7–12 cm. **Spikes** with up to several hundred flowers, 2–15 cm; spathes 5–17 cm. **Perianth** mauve, tube 3–9 mm, limb lobes oblanceolate, 5–8 mm, distal central lobe with 2-lobed yellow spot; proximal stamens 7–13 mm, distal 1.5–6.3 mm; style 3-lobed. **Utricles** with dentate ridges, 4–6 × 2–3 mm.

Flowering Mar–Nov in Florida; flowering season shorter farther north. Pond and lake margins; 0–500 m; N.B., N.S., Ont., P.E.I., Que.; Ala., Ark., Conn., Del., D.C., Fla., Ga., Ill., Ind., Iowa, Kans., Ky., La., Maine, Md., Mass., Mich., Minn., Miss., Mo., Nebr., N.H., N.J., N.Y., N.C., Ohio, Okla., Pa., R.I., S.C., S.Dak., Tenn., Tex., Vt., Va., W.Va., Wis.; Mexico; Central America (Belize); South America.

Pontederia cordata has a large number of synonyms, at the levels of species, variety, and form (see R. M. Lowden 1973 for most names). Almost all these names are based on variations in leaf shape, but extensive morphological variation has been observed within single populations and even in individual plants. Study is needed to determine the cause of the extreme leaf base forms of cuneate on lanceolate blades and cordate on ovate blades. Variation has also been observed in peduncle pubescence. A velutinous peduncle and ovate leaf blade with slightly cuneate base is consistently found among some South American populations, hence recognition there of var. *ovalis* (Martius) Solms.

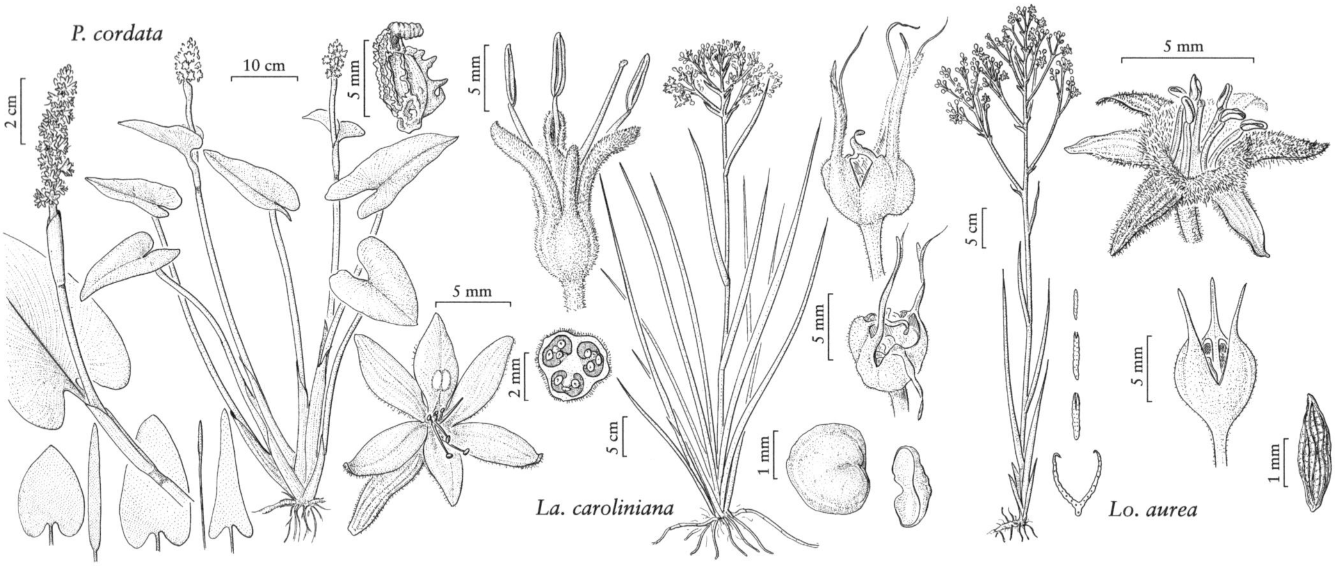

PONTEDERIA ◦ LACHNANTHES ◦ LOPHIOLA

The reproductive biology of *Pontederia cordata* has been well studied (R. Ornduff 1966; S. D. Price and S. C. H. Barrett 1982, 1984). It is a tristylous species, and most populations contain all three morphs (S. D. Price and S. C. H. Barrett 1982). At least some degree of self-incompatibility exists, being strongest with the short-style morphs and weakest with the midstyle morphs (R. Ornduff 1966).

221. HAEMODORACEAE R. Brown
• Bloodwort Family

Kenneth R. Robertson

Herbs, perennial or rarely annual, rhizomatous, often stoloniferous; sap often red or orange. **Stems** simple. **Leaves** mostly basal, distichous, irislike; blade linear, sheathing proximally, folded lengthwise distally, folded portions becoming fused. **Inflorescences** terminal, cymose or corymbose, many-branched, many-flowered, loose or condensed. **Flowers:** tepals persistent, 6 in 2 whorls of 3, distinct above level of insertion on ovary, ± equal, aestivation valvate; stamens 3 or 6; filaments elongate; anthers basifixed, 2-locular, often basally sagittate, dehiscence introrse by longitudinal slits; pistil 1, compound, 3-carpellate; ovary inferior or half-inferior to nearly superior, 3-locular with axile placentae, or 1-locular distally with parietal placentae and 3-locular proximally with axile placentae; ovules 1, 3–6, or numerous in each locule, arranged in vertical rows or around margin of peltate placenta, orthotropous or anatropous; style 1, undivided, elongate; stigma 1, terminal, very small. **Fruits** capsular, dehiscence loculicidal. **Seeds** discoid and peltate, or elongate and variously curved; endosperm present.

Genera ca. 14, species ca. 80 (2 genera, 2 species in the flora): e North America, Mexico, Central America, n South America, s Africa, New Guinea, Australia.

The two genera of Haemodoraceae found in North America are very different from each other. *Lachnanthes* belongs to tribe Haemodoreae and is morphologically similar to *Dilatris* P. J. Bergius of South Africa and *Haemodorum* Smith of Oceania. The relationships of *Lophiola* are more obscure and are discussed more fully in the references cited below and under that genus.

SELECTED REFERENCES Geerinck, D. 1969. Genera des Haemodoraceae et des Hypoxidaceae. Bull. Jard. Bot. Natl. Belg. 39: 47–82. Hopper, S. D., M. F. Fay, M. Rossetto, and M. W. Chase. 1999. A molecular phylogenetic analysis of the bloodroot and kangaroo paw family, Haemodoraceae: Taxonomic, biogeographic and conservation implications. Bot. J. Linn. Soc. 131: 285–299. Maas, P. J. M. and H. Maas. 1993. Haemodoraceae. In: Organization for Flora Neotropica. 1968+. Flora Neotropica. 75+ nos. New York. No. 61. Robertson, K. R. 1976. The genera of Haemodoraceae in the southeastern United States. J. Arnold Arbor. 57: 205–216. Simpson, M. G. 1990. Phylogeny and classification of the Haemodoraceae. Ann. Missouri Bot. Gard. 77: 722–784. Simpson, M. G. 1993. Septal nectary anatomy and phylogeny of the Haemodoraceae. Syst. Bot. 18: 593–613. Zomlefer, W. B. 1999. Advances in angiosperm systematics: Examples from the Liliales and Asparagales. Bull. Torrey Bot. Club 126: 58–62.

1. Stamens 3, longer than tepals; ovary inferior; tepals glabrous adaxially; seeds discoid, peltate, few per locule . . . . 1. *Lachnanthes*, p. 48
1. Stamens 6, shorter than tepals; ovary half-inferior to nearly superior; tepals with adaxial crest of yellow trichomes; seeds elongate, variously curved, many per locule . . . . 2. *Lophiola*, p. 48

1. LACHNANTHES Elliott, Sketch Bot. S. Carolina 1: 47. 1816, name conserved • Redroot [Greek *lachne*, wool, and *anthos*, flower, in reference to pubescent flowers]

Rhizomes reddish, roots fibrous. **Stems** erect; sap red. **Leaves** mostly basal, some cauline, reduced; blade linear-ensiform. **Inflorescences:** bracts conspicuous, subtending each flower. **Flowers:** tepals glabrous adaxially; stamens 3, longer than tepals; ovary inferior, 3-locular; ovules ca. 6 per locule, arranged around margin of peltate, stipitate, axile placenta. **Capsules** 3-lobed. **Seeds** discoid, peltate, few per locule.

Species 1: North America, West Indies.

1. Lachnanthes caroliniana (Lamarck) Dandy, J. Bot. 70: 329. 1932 [F] [W1] [W2]

Dilatris caroliniana Lamarck in J. Lamarck and J. Poiret, Tabl. Encycl. 1: 127. 1791 (as caroliana); *Heritiera tinctorum* J. F. Gmelin; *Lachnanthes tinctoria* (J. F. Gmelin) Elliott

Stems 3–10.5 dm, distally whitish villous-tomentose when young, becoming tawny-hoary. **Leaves:** basal leaves 15–45 × 0.5–2.1 cm proximally, shorter distally; cauline leaves shorter than basal, decreasing in size distally, becoming bracts in inflorescence. **Inflorescences** initially rounded, becoming open and corymbose after anthesis, each branch resembling a helicoid cyme. **Flowers:** tepals erect-spreading at anthesis, pale yellow, 7–9 × 1–1.5 mm, densely pubescent abaxially, persisting and incurving to form beak around fruit; stamens spreading at anthesis; filaments 8–10 mm; anthers yellow; style persistent, slightly exceeding anthers, 10–13 mm. **Capsules** globose or oblate, 3–5 mm diam. **Seeds** reddish brown, 2.5–3 mm diam., faintly wrinkled. $2n = 48$.

Flowering mid–late summer. Wet, acid, often sandy soil of bogs, swamps, ditches, low areas in savannas and pinelands, margins of hammocks and pocosins; 0–600 m; N.S.; Ala., Conn., Del., Fla., Ga., La., Md., Mass., Miss., N.J., N.Y., N.C., R.I., S.C., Tenn., Va.; West Indies (Cuba).

Lachnanthes caroliniana is sometimes a serious weed in commercial cranberry bogs.

2. LOPHIOLA Ker Gawler, Bot. Mag. 39: plate 1596. 1813 • Golden-crest [Greek *lophia*, mane or crest, in reference to pubescence on adaxial sides of tepals] [E]

Rhizomes brown to whitish, stoloniferous. **Stems** erect. **Leaves** basal and cauline, blade linear. **Inflorescences:** bracts present. **Flowers:** tepals with adaxial crest of moniliform trichomes; stamens 6, shorter than tepals; ovary half-inferior to nearly superior, 1-locular distally with 3 parietal placentae and 3-locular proximally with axile placentae; ovules numerous in each locule, arranged in several vertical rows. **Capsules** ± globose or ovoid. **Seeds** thin, elongate, variously curved, many per locule; seed coat reticulate; hilum basal.

Species 1: North America.

M. L. Fernald (1921) recognized three species of *Lophiola*: *L. aurea* in the southeastern United States, *L. americana* in New Jersey and Delaware, and *L. septentrionalis* in Nova Scotia. Pending more detailed studies, only one species is recognized here. While the genus is retained in the Haemodoraceae in this treatment, recent research indicates that *Lophiola* is quite different from other Haemodoraceae and may be best placed in Liliaceae subfam. Melanthioideae (J. D. Ambrose 1985), Tecophilaeaceae (R. M. T. Dahlgren et al. 1985), or Nartheciaceae (W. B. Zomlefer 1997).

SELECTED REFERENCES Ambrose, J. D. 1985. *Lophiola*, familial affinity with the Liliaceae. Taxon 34: 140–150. Fernald, M. L. 1921. The Gray Herbarium expedition to Nova Scotia, 1920. Rhodora 23: 153–171, 223–245. Zavada, M., Zu X. L., and J. M. Edwards. 1983. On the taxonomic status of *Lophiola aurea* Ker-Gawler. Rhodora 85: 73–81. Zomlefer, W. B. 1997. The genera of Nartheciaceae in the southeastern United States. Harvard Pap. Bot. 2: 195–211.

1. Lophiola aurea Ker Gawler, Bot. Mag. 39: plate 1596. 1813 E F

Lophiola americana (Pursh) Alph. Wood; *L. septentrionalis* Fernald

Stems 3.5–8.5 dm, distal portions woolly. **Leaves:** basal leaves smaller than proximal cauline leaves; cauline leaves erect, linear; proximal leaf blade 8–40 cm × 3–16 mm, decreasing in size distally, glabrous. **Inflorescences** corymbose, open. **Flowers:** tepals reflexed at anthesis, yellow, triangular, 3–4 × 1–2 mm, densely woolly abaxially, apex brownish to maroon, adaxial crest bright yellow; stamens erect-spreading at anthesis, 1.5–3 mm; style persistent, ± equaling stamens, stigmatic area slightly 3-lobed. **Capsules** enclosed by tepals before dehiscence, 3–4 mm diam.; beak prominent. **Seeds** whitish, 1–1.5 × 0.5 mm; seed coat finely open-reticulate.

Flowering early summer. Wet, acid, often sandy soil of bogs, pocosins, low, wet areas in savannas and pine barrens, roadside ditches; 0–100 m; N.S.; Ala., Del., Fla., Ga., La., Miss., N.J., N.C.

222. LILIACEAE Jussieu
• Lily Family

Frederick H. Utech

Herbs or shrubs, sometimes **vines** in *Asparagus*, perennial, mostly geophytic, scapose or caulescent, sometimes woody, from elongate, sometimes tuberous rhizomes, or from scaly or tunicate bulbs, or from solid corms. **Leaves** only rarely persistent, simple, basal and/or cauline, alternate, opposite, or whorled, herbaceous (scalelike in *Asparagus*), sometimes sheathing; blade typically narrow and parallel-veined, occasionally broad and/or reticulate-veined. **Inflorescences** racemose, spicate, paniculate, cymose, umbellate, or with flowers single or paired in leaf axils; bracts 1–several, sometimes involucrate or sheathing, or bracts absent. **Flowers** usually bisexual, sometimes bisexual and unisexual, or unisexual only, usually pedicellate, occasionally sessile; perianth actinomorphic or zygomorphic, often very showy; tepals 6, distinct or less often connate proximally forming tube that may also bear a corona, usually petaloid and ± equal in 2 whorls of 3, or those of outer whorl narrower, greener, more sepaloid; tepal nectaries often present; stamens 6, rarely 3 or 4, sometimes 3 fertile and 3 staminodial, free or adnate to perianth; filaments slender to dilated, occasionally connate-coroniform and/or with bases dilated to form wings; anthers basifixed with latrorse dehiscence or dorsifixed, versatile, and with introse or extrorse dehiscence, cordate to linear; ovary superior to inferior, (2–)3(–4)-locular, often with septal nectaries, ovules usually several or many per locule; styles 1 or 3(–4); stigmas several and distinct or 1 and capitate. **Fruits** capsular and loculicidal or septicidal, membranaceous to leathery, or baccate, or dry and indehiscent. **Seeds** 1–many, often flat and wind-distributed, sometimes thicker and with fleshy elaiosomes. x = 3–27+.

Genera ca. 280, species ca. 4200 (70 genera, 478 species in the flora; 16 genera, 54 species introduced): nearly worldwide, primarily tropical regions, widely cultivated.

There is no question that the evidence available today (Angiosperm Phylogeny Group 1998; P. J. Rudall et al. 1995; K. L. Wilson and D. A. Morrison 2000) strongly supports extensive dismemberment of A. Cronquist's (1981, 1988, 1993) very broadly circumscribed Liliaceae. No fewer than 30 segregate families have been recognized, though there is not universal acceptance of all of them, and in some cases their ordinal associations are not yet settled. For the genera of Cronquist's Liliaceae that are present in the flora, Table 1 summarizes their dispositions among the maximum number of segregate families in recent use. Whether ranked as families or otherwise, these sets of genera represent identifiable lineages that have been variously grouped in recent classifications of the monocots (A. L. Takhtajan 1997; K. Kubitzki et al. 1990+, vol. 3; W. S. Judd et al. 1999; R. F. Thorne 2000; A. B. Doweld 2001). Comments on these assemblages appear below, as well as within the various generic discussions and in the introductory chapter on monocot classification by J. L. Reveal and J. C. Pires at the front of this volume.

Traditionally, *Pleea, Triantha, Isidrogalvia* Ruíz & Pavón (5 species, South America), *Tofieldia*, and *Harperocallis* have been included in the tribe Tofieldieae Horaninow within a polyphyletic Melanthiaceae or Liliaceae sensu lato. However, the vastly different morphologies, anatomies, and cytologies of this lineage (J. D. Ambrose 1975, 1980; M. Takahashi and S. Kawano 1989; R. W. Cruden 1991; P. Goldblatt 1995; M. N. Tamura 1995, 1998b; W. B. Zomlefer 1997c) support its recognition as a separate family, Tofieldiaceae (A. L. Takhtajan 1994b, 1997), in a monotypic order, Tofieldiales (J. L. Reveal and W. B. Zomlefer 1998).

The segregate family Nartheciaceae includes three genera that are present in the flora: *Aletris*, *Narthecium*, and *Lophiola* Ker Gawler (treated under Haemodoraceae herein see p. 47).

Many botanists now consider *Trillium* and the closely related genera *Daiswa* Rafinesque, *Paris* Linnaeus, *Kinugasa* Tatenaki ex Suto, and *Trillidium* Kunth (when recognized separately) to constitute the separate family Trilliaceae (R. Y. Berg 1962b; S. Kazempour Osaloo, F. H. Utech, M. Ohara and S. Kawano 1999; S. Kazempour Osaloo and S. Kawano 1999; W. B. Zomlefer 1996). Others (M. W. Chase et al. 2000; W. B. Zomlefer et al. 2001) have defined the Melanthiaceae to include these genera, though the two groups have markedly different morphologies and karyologies.

Most recently, *Uvularia* has been associated not as before with the Melanthiaceae (J. D. Ambrose 1975, 1980; W. B. Zomlefer 1997b) or the Uvulariaceae of R. M. T. Dahlgren et al. (1985) or the Convallariaceae of A. L. Takhtajan (1980, 1997), but with the east Asian *Disporum* Salisbury in an expanded Colchicaceae (B. Nordenstam 1998; K. Hayashi et al. 1998).

As defined by A. Cronquist (1981), the Liliaceae contained "about 280 genera and nearly 4000 species." In a much more restricted, recent sense, the family was considered to include just 11 genera and perhaps 545 species (R. F. Thorne 2000). Thorne recognized two subfamilies, of which the Medeoloideae (*Medeola* and *Clintonia*) have sometimes been segregated as the Medeolaceae (A. L. Takhtajan 1997; A. B. Doweld 2001).

T. B. Patterson (1998), K. Kubitzki et al. (1990+, vol. 3), W. S. Judd et al. (1999), R. F. Thorne (2000), and T. B. Patterson and T. J. Givnish (1998) have recognized Calochortaceae separate from Liliaceae. As circumscribed by Patterson and Givnish, the family includes *Calochortus*, *Prosartes*, *Scoliopus*, *Streptopus*, and *Tricyrtis* Wallich (not in the flora). A. L. Takhtajan (1997) distributed these genera among three segregate families: Calochortaceae, Scoliopaceae, and Tricyrtidaceae.

Hesperocallis is currently treated as the sole representative of the segregate family Hesperocallidaceae (H. P. Traub 1972; A. L. Takhtajan 1997). Karyologically and embryologically, *Hesperocallis* is nearest to *Hosta* (Hostaceae) and the Agavaceae (M. S. Cave 1948, 1970), and even though their base chromosome numbers are different [$x = 24$ in *Hesperocallis* and $x = 30$ in *Hosta* and Agavaceae (T. W. Whitaker 1934; D. Satô 1935; S. Sen 1975, F. Maekawa and K. Kaneko 1968; M. N. Tamura 1995)], they share a strongly bimodal karyotype. As well, the pollen grains of *Hosta plantaginea* and *Hesperocallis* have similar unibaculate muri (A. Alvarez and E. Köhler 1987). *Hosta* has long been associated with *Hemerocallis* and *Leucocrinum* in the liliaceous tribe Hemerocallideae. However, recent molecular and morphological evidence (M. W. Chase et al. 1996; P. J. Rudall and D. F. Cutler 1995) supports separating these genera—*Hemerocallis* in the Hemerocallidaceae, *Leucocrinum* in the Anthericaceae (J. G. Conran 1998), and *Hosta* in a monotypic Hostaceae (K. Kubitzki 1998b; A. L. Takhtajan 1997; W. B. Zomlefer 1998).

Table 1. *Dispositions of FNA Genera of Cronquist's Liliaceae among Maximum Number of Segregate Families.*

Tofieldiaceae Takhtajan
- *Harperocallis* (p. 58)
- *Pleea* (p. 59)
- *Tofieldia* (p. 60)
- *Triantha* (p. 61)

Nartheciaceae Fries ex Bjurzon
- *Aletris* (p. 64)
- *Narthecium* (p. 66)

Chionographidaceae Takhtajan
- *Chamaelirium* (p. 68)

Heloniadaceae J. Agardh
- *Helonias* (p. 69)

Xerophyllaceae Takhtajan
- *Xerophyllum* (p. 71)

Melanthiaceae Batsch
- *Amianthium* (p. 89)
- *Melanthium* (p. 77)
- *Schoenocaulon* (p. 79)
- *Stenanthium* (p. 88)
- *Veratrum* (p. 72)
- *Zigadenus* (p. 81)

Trilliaceae Chevallier
- *Trillium* (p. 90)

Colchicaceae de Candolle
- *Uvularia* (p. 147)

Medeolaceae Takhtajan
- *Clintonia* (p. 151)
- *Medeola* (p. 150)

Liliaceae Jussieu
- *Erythronium* (p. 153)
- *Fritillaria* (p. 164)
- *Lilium* (p. 172)
- *Lloydia* (p. 198)
- *Tulipa* (p. 199)

Tricyrtidaceae Takhtajan
- *Prosartes* (p. 142)
- *Streptopus* (p. 145)

Scoliopaceae Takhtajan
- *Scoliopus* (p. 118)

Calochortaceae Dumortier
- *Calochortus* (p. 119)

Alstroemeriaceae Dumortier
- *Alstroemeria* (p. 200)

Hypoxidaceae R. Brown
- *Hypoxis* (p. 201)

Tecophilaeaceae Leybold
- *Odontostomum* (p. 204)

Convallariaceae Horaninow
- *Convallaria* (p. 205)
- *Maianthemum* (p. 206)
- *Polygonatum* (p. 210)

Asparagaceae Jussieu
- *Asparagus* (p. 213)

Anthericaceae J. Agardh
- *Echeandia* (p. 215)
- *Eremocrinum* (p. 216)
- *Leucocrinum* (p. 217)

Asphodelaceae Jussieu
- *Asphodelus* (p. 218)

Hemerocallidaceae R. Brown
- *Dianella* (p. 221)
- *Hemerocallis* (p. 219)

Hesperocallidaceae Traub
- *Hesperocallis* (p. 221)

Hostaceae B. Mathew
- *Hosta* (p. 222)

Alliaceae Batsch ex Borkhausen
- *Allium* (p. 224)
- *Nothoscordum* (p. 276)

Amaryllidaceae J. Saint-Hilaire
- *Crinum* (p. 278)
- *Galanthus* (p. 280)
- *Habranthus* (p. 281)
- *Hippeastrum* (p. 282)
- *Hymenocallis* (p. 283)
- *Leucojum* (p. 293)
- *Narcissus* (p. 294)
- *Zephyranthes* (p. 296)

Hyacinthaceae R. Brown
- *Camassia* (p. 303)
- *Chionodoxa* (p. 314)
- *Chlorogalum* (p. 307)
- *Hastingsia* (p. 310)
- *Hyacinthoides* (p. 315)
- *Muscari* (p. 316)
- *Ornithogalum* (p. 318)
- *Schoenolirion* (p. 312)
- *Scilla* (p. 320)

Themidaceae Salisbury
- *Androstephium* (p. 333)
- *Bloomeria* (p. 336)
- *Brodiaea* (p. 321)
- *Dichelostemma* (p. 328)
- *Milla* (p. 346)
- *Muilla* (p. 334)
- *Triteleia* (p. 338)
- *Triteleiopsis* (p. 332)

In the past, several taxonomic affinities have been suggested for *Androstephium*, *Bloomeria*, *Brodiaea*, *Dichelostemma*, *Milla*, *Muilla*, *Triteleia*, and *Triteleiopsis* (Liliaceae, Amaryllidaceae, Alliaceae), but most recently they have been placed in the resurrected family Themidaceae based on molecular and anatomical evidence (M. F. Fay and M. W. Chase 1996; J. C. Pires 2000; J. C. Pires et al. 2001).

The Liliaceae include numerous important ornamentals such as *Amaryllis*, *Hemerocallis*, *Hosta*, *Lilium*, *Narcissus*, and *Tulipa*. The family is a dominant component in the temperate spring flora, which includes both native and introduced species. Many of the introductions, or cultivars derived from them, are from ecologically equivalent, temperate zones and their naturalization potential is high. *Asparagus* and *Allium* have edible species of major economic importance, while numerous other genera (e.g., *Convallaria*, *Ornithogalum*, *Veratrum*, *Zigadenus*) are highly toxic due to the presence of various alkaloids and cardenolides (G. E. Burrows and R. J. Tyrl 2001).

Several horticultural exotics that have been reported as escaped in the flora (J. T. Kartesz and C. A. Meacham 1999) are not clearly naturalized and are not treated herein. They include: *Colchicum autumnale* Linnaeus, *Gagea fistulosa* Ker Gawler, *G. villosa* (M. Bieberstein) Duby, *Gloriosa superba* Linnaeus, *Kniphofia uvaria* (Linnaeus) Oken, *Liriope muscari* (Decaisne) L. H. Bailey, *L. spicatum* Loureiro, *Lycoris radiata* (L'Heritier) Herbert, *L. squamigera* Maximowicz, *Ophiopogon jaburan* (Siebold) Loddiges, *Sternbergia lutea* (Linnaeus) Ker Gawler ex Sprengel, and *Tricyrtis hirta* (Thunberg) Hooker.

SELECTED REFERENCES Ambrose, J. D. 1975. Comparative Anatomy and Morphology of the Melanthioideae (Liliaceae). Ph.D. dissertation. Cornell University. Ambrose, J. D. 1980. A re-evaluation of the Melanthioideae (Liliaceae) using numerical analyses. In: C. D. Brickell et al., eds. 1980. Petaloid Monocotyledons: Horticultural and Botanical Research. London and New York. Pp. 65–81. Badawi, A. and Z. Elwan. 1986. A taxonomic study of Liliaceae sensu lato: I. Numerical analysis. Phytologia 60: 201–213. Badawi, A. and Z. Elwan. 1986b. A taxonomic study of Liliaceae sensu lato: II. Evaluation of Engler's subfamilies. Phytologia 60: 214–221. Barnard, C. 1960. Floral histogenesis in the monocotyledons. IV. The Liliaceae. Austral. J. Bot. 8: 213–225. Bayer, E. 1998. Alstroemeriaceae. In: K. Kubitzki et al., eds. 1990+. The Families and Genera of Vascular Plants. 4+ vols. Berlin, etc. Vol. 3, pp. 79–83. Chase, M. W., P. J. Rudall, and J. G. Conran. 1996. New circumscriptions and a new family of asparagoid lilies: Genera formerly included in Anthericaceae. Kew Bull. 51: 667–680. Cheadle, V. I. and H. Kosakai. 1971. Vessels in Liliaceae. Phytomorphology 21: 320–333. Clifford, H. T., R. J. F. Henderson, and J. G. Conran. 1998. Hemerocallidaceae. In: K. Kubitzki et al., eds. 1990+. The Families and Genera of Vascular Plants. 4+ vols. Berlin, etc. Vol. 3, pp. 245–253. Conran, J. G. 1998. Anthericaceae. In: K. Kubitzki et al., eds. 1990+. The Families and Genera of Vascular Plants. 4+ vols. Berlin, etc. Vol. 3, pp. 114–121. Conran, J. G. and M. N. Tamura. 1998. Convallariaceae. In: K. Kubitzki et al., eds. 1990+. The Families and Genera of Vascular Plants. 4+ vols. Berlin, etc. Vol. 3, pp. 186–198. Fay, M. F. and M. W. Chase. 1996. Resurrection of Themidaceae for the *Brodiaea* alliance, and recircumscription of Alliaceae, Amaryllidaceae and Agapanthoideae. Taxon 45: 441–451. Fay, M. F. and M. W. Chase. 2000. Modern concepts of Liliaceae with a focus on the relationships of *Fritillaria*. Bot. Mag., [n. s.] 17: 146–149. Fuse, S. and M. N. Tamura. 2000. A phylogenetic analysis of the plastid *mat*K gene with emphasis on Melanthiaceae sensu lato. Pl. Biol. 2: 415–427. Gates, R. R. 1918. A systematic study of the North American Melanthaceae from a genetic standpoint. J. Linn. Soc., Bot. 44: 131–172. Ito, M. et al. 1999. Phylogenetic relationships of Amaryllidaceae based on *mat*K sequence data. J. Pl. Res. 112: 207–216. Johnson, R. G. 1969. A Taxonomic and Floristic Study of the Liliaceae and Allied Families in the Southeastern United States. Ph.D. dissertation. West Virginia University. Judd, W. S. 1997. The Asphodelaceae in the southeastern United States. Harvard Pap. Bot. 2: 109–123. Judd, W. S. 2001. The Asparagaceae in the southeastern United States. Harvard Pap. Bot. 6: 223–244. Kato, H. et al. 1995b. Molecular systematics of the Trilliaceae sensu lato as inferred from *rbc*L sequence data. Molec. Phylogen. Evol. 4: 184–193. Komar, G. A. 1983. Morphology of Liliaceae ovules. Bot. Zhurn. (Moscow & Leningrad) 68: 417–427. Kosenko, V. N. 1987. Pollen morphology in Tofieldieae, Narthecieae, Melanthieae, Xerophylleae (Melanthiaceae). Bot. Zhurn. (Moscow & Leningrad) 72: 1318–1330. Kosenko, V. N. 1988. Pollen morphology in Chionographideae, Uvularieae, Tricyrtideae, Scoliopeae, Anguillarieae, Iphigenieae, Glorioseae, Colchiceae (Melanthiaceae). Bot. Zhurn. (Moscow & Leningrad) 73: 172–185. Kosenko, V. N. 1991. Palynomorphology of the family Liliaceae s. str. Bot. Zhurn. (Moscow & Leningrad) 76: 1696–1710. Kubitzki, K. 1998b. Hostaceae. In: K. Kubitzki et al., eds. 1990+. The Families and Genera of Vascular Plants. 4+ vols. Berlin, etc. Vol. 3, pp. 256–260. Kubitzki, K. and P. J. Rudall. 1998. Asparagaceae. In: K. Kubitzki et al., eds. 1990+. The Families and Genera of Vascular Plants. 4+ vols. Berlin, etc. Vol. 3, pp. 125–129. Kupchan, M. S., J. H. Zimmermann, and A. Afonso. 1961. The alkaloids and taxonomy of *Veratrum* and related genera. Lloydia 24: 1–26. Meerow, A. W. and D. A. Snijman. 1998. Amaryllidaceae. In: K. Kubitzki et al., eds. 1990+. The Families and Genera of Vascular Plants. 4+ vols. Berlin, etc. Vol. 3, pp. 83–110. Nair, P. K. K. and M. Sharma. 1965. Pollen morphology of Liliaceae. J. Palynol. 1: 38–61. Nordenstam, B. 1998. Colchicaceae. In: K. Kubitzki et al., eds. 1990+. The Families and Genera of Vascular Plants. 4+ vols. Berlin, etc. Vol. 3, pp. 175–185. Oganezowa, G. G. 1982. On the anatomical structure of seed and fruit of some Liliaceae in connection with the systematics of the family. II. Scilloideae. Bot. Zhurn. (Moscow & Leningrad) 67: 729–742. Oganezowa, G. G. 1984. Morphological and anatomical specific features of seed and fruit in some representatives of the subfamily

Melanthioideae (Liliaceae) in relation with their systematics and phylogeny. Bot. Zhurn. (Moscow & Leningrad) 69: 772–781. Oganezowa, G. G. 1986. Morphological and anatomical characters of seed and fruit in some members of the subfamily Allioideae (Liliaceae) in relation to their systematics and phylogeny. Bot. Zhurn. (Moscow & Leningrad) 71: 300–310. Patterson, T. B. 1998. Phylogeny, Biogeography, and Evolutionary Trends in the Core Liliales and *Calochortus* (Calochortaceae): Insights from DNA Sequenced Data. Ph.D. dissertation. University of Wisconsin. Pfosser, M. and F. Speta. 1999. Phylogenetics of Hyacinthaceae based on plastid DNA sequences. Ann. Missouri Bot. Gard. 86: 652–875. Pires, J. C. 2000. Biosystematics and Molecular Phylogenetics of *Brodiaea* (Themidaceae) and Related Lilioid Monocots. Ph.D. dissertation. University of Wisconsin. Pires, J. C. et al. 2001. Molecular and morphological phylogenetic analyses of Themidaceae (Asparagales). Kew Bull. 56: 601–626. Rahn, K. 1998. Alliaceae. In: K. Kubitzki et al., eds. 1990+. The Families and Genera of Vascular Plants. 4+ vols. Berlin, etc. Vol. 3, pp. 70–78. Rahn, K. 1998b. Themidaceae. In: K. Kubitzki et al., eds. 1990+. The Families and Genera of Vascular Plants. 4+ vols. Berlin, etc. Vol. 3, pp. 436–441. Rudall, P. J. et al. 2000. Systematics of Ruscaceae/Convallariaceae: A combined morphological and molecular investigation. Bot. J. Linn. Soc. 134: 73–92. Savile, D. B. O. 1961. Some fungal parasites of Liliaceae. Mycologia 53: 31–52. Schulze, W. 1975. Beiträge zur Taxonomie der Liliifloren: I. Asphodelaceae. Wiss. Z. Friedrich-Schiller-Univ. Jena, Math.-Naturwiss. Reihe 24: 403–415. Schulze, W. 1975b. Beiträge zur Taxonomie der Liliifloren: II. Colchicaceae. Wiss. Z. Friedrich-Schiller-Univ. Jena, Math.-Naturwiss. Reihe 24: 417–428. Schulze, W. 1978. Beiträge zur Taxonomie der Liliifloren: III. Alstroemeriaceae. Wiss. Z. Friedrich-Schiller-Univ. Jena, Math.-Naturwiss. Reihe 27: 79–85. Schulze, W. 1978b. Beiträge zur Taxonomie der Liliifloren: IV. Melanthiaceae. Wiss. Z. Friedrich-Schiller-Univ. Jena, Math.-Naturwiss. Reihe 27: 87–95. Schulze, W. 1980. Beiträge zur Taxonomie der Liliifloren: V. Alliaceae. Wiss. Z. Friedrich-Schiller-Univ. Jena, Math.-Naturwiss. Reihe 29: 595–606. Schulze, W. 1980b. Beiträge zur Taxonomie der Liliifloren: VI. Der Umfang der Liliaceae. Wiss. Z. Friedrich-Schiller-Univ. Jena, Math.-Naturwiss. Reihe 29: 607–636. Schulze, W. 1982. Beiträge zur Taxonomie der Liliifloren: X. Asparagaceae. Wiss. Z. Friedrich-Schiller-Univ. Jena, Math.-Naturwiss. Reihe 31: 309–330. Shinwari, Z. K. et al. 1994b. Phylogenetic relationships among genera in the Liliaceae–Asparagoideae–Polygonatae s.l. inferred from *rbc*L gene sequence data. Pl. Syst. Evol. 192: 263–277. Smith, G. F. and B. E. van Wyk. 1998. Asphodelaceae. In: K. Kubitzki et al., eds. 1990+. The Families and Genera of Vascular Plants. 4+ vols. Berlin, etc. Vol. 3, pp. 130–140. Speta, F. 1998. Hyacinthaceae. In: K. Kubitzki et al., eds. 1990+. The Families and Genera of Vascular Plants. 4+ vols. Berlin, etc. Vol. 3, pp. 261–285. Sterling, C. 1977. Comparative morphology of the carpel in the Liliaceae: Uvularieae. Bot. J. Linn. Soc. 74: 345–354. Sterling, C. 1979. Comparative morphology of the carpel in the Liliaceae: Tofieldieae. Bot. J. Linn. Soc. 79: 321–332. Sterling, C. 1980. Comparative morphology of the carpel in the Liliaceae: Helonieae. Bot. J. Linn. Soc. 80: 341–356. Sterling, C. 1982. Comparative morphology of the carpel in the Liliaceae: Veratreae. Bot. J. Linn. Soc. 84: 57–77. Takahashi, M. and S. Kawano. 1989. Pollen morphology of the Melanthiaceae and its systematic implications. Ann. Missouri Bot. Gard. 76: 863–876. Tamura, M. N. 1998. Melanthiaceae. In: K. Kubitzki et al., eds. 1990+. The Families and Genera of Vascular Plants. 4+ vols. Berlin, etc. Vol. 3, pp. 369–380. Tamura, M. N. 1998b. Nartheciaceae. In: K. Kubitzki et al., eds. 1990+. The Families and Genera of Vascular Plants. 4+ vols. Berlin, etc. Vol. 3, pp. 381–392. Tamura, M. N. 1998c. Liliaceae. In: K. Kubitzki et al., eds. 1990+. The Families and Genera of Vascular Plants. 4+ vols. Berlin, etc. Vol. 3, pp. 343–353. Tamura, M. N. 1998d. Calochortaceae. In: K. Kubitzki et al., eds. 1990+. The Families and Genera of Vascular Plants. 4+ vols. Berlin, etc. Vol. 3, pp. 165–172. Tamura, M. N. 1998e. Trilliaceae. In: K. Kubitzki et al., eds. 1990+. The Families and Genera of Vascular Plants. 4+ vols. Berlin, etc. Vol. 3, pp. 444–452. Watson, S. 1879. Contributions to American botany. I. Revision of the North American Liliaceae. Proc. Amer. Acad. Arts 14: 213–288. Williams, C. A. 1975. Biosystematics of the Monocotyledoneae—Flavonoid patterns in leaves of the Liliaceae. Biochem. Syst. & Ecol. 3: 229–244. Yamashita, J. and M. N. Tamura. 2000. Molecular phylogeny of the Convallariaceae (Asparagales). In: K. L. Wilson and D. A. Morrison, eds. 2000. Monocots: Systematics and Evolution. Melbourne. Pp. 387–400. Zomlefer, W. B. 1996. The Trilliaceae in the southeastern United States. Harvard Pap. Bot. 1(9): 91–120. Zomlefer, W. B. 1997. The genera of Nartheciaceae in the southeastern United States. Harvard Pap. Bot. 2: 195–211. Zomlefer, W. B. 1997b. The genera of Melanthiaceae in the southeastern United States. Harvard Pap. Bot. 2: 133–177. Zomlefer, W. B. 1997c. The genera of Tofieldiaceae in the southeastern United States. Harvard Pap. Bot. 2: 179–194. Zomlefer, W. B. 1998. The genera of Hemerocallidaceae in the southeastern United States. Harvard Pap. Bot. 3: 113–145. Zomlefer, W. B., N. H. Williams, W. M. Whitten, and W. S. Judd. 2001. Generic circumscription and relationships in the tribe Melanthieae (Liliales, Melanthiaceae), with emphasis on *Zigadenus*: Evidence from ITS and *trn*L-F sequence data. Amer. J. Bot. 88: 1657–1669.

1. True leaves small, scalelike, their function taken over by lanceolate or ovate, filiform, or flattened cladophylls that are borne singly or in clusters in leaf axils . . . . 35. *Asparagus*, p. 213
1. True leaves (bracts in *Trillium*) present, herbaceous, cladophylls absent.
  2. Ovary subterranean . . . . 38. *Leucocrinum*, p. 217
  2. Ovary above ground.
    3. Ovary inferior (or partly inferior in *Zigadenus*).
      4. Inflorescences racemose, corymbose, or paniculate . . . . 13. *Zigadenus*, p. 81
      4. Inflorescences 1-flowered or umbellate.
        5. Flowers with corona.
          6. Corona funnelform or rotate . . . . 50. *Hymenocallis*, p. 283
          6. Corona cupular to trumpetlike.
            7. Perianth yellow and/or white; corona tubular and separate from filaments; stigma 3-lobed . . . . 52. *Narcissus*, p. 294

7. Perianth reddish to salmon, each outer tepal with whitish adaxial midstripe; corona minute, reduced to small crown; stigma capitate or slightly 3-lobed . . . . 49. *Hippeastrum*, p. 282

5. Flowers without corona.
  8. Plants from tubers, rhizomes, or corms.
    9. Rootstocks tuberous; plants with leafy stems; flowers slightly zygomorphic; tepals red, orange, purple, green or white, frequently spotted . . . . 29. *Alstroemeria*, p. 200
    9. Rootstocks rhizomatous or cormlike; plants scapose, with short, subterranean stem; flowers actinomorphic; tepals often greenish abaxially, yellow adaxially, ± pilose . . . . 30. *Hypoxis*, p. 201
  8. Plants from bulbs.
    10. Inflorescences few- to many-flowered; style capitate; capsules prominently beaked . . . . 46. *Crinum*, p. 278
    10. Inflorescences 1–7-flowered; style filiform or clavate; capsules globose or ± subglobose.
      11. Flowers declinate or ± erect; perianth 2–16 cm.
        12. Stamens of two lengths . . . . 53. *Zephyranthes*, p. 296
        12. Stamens of four lengths . . . . 48. *Habranthus*, p. 281
      11. Flowers nodding; perianth 2.5 cm or shorter.
        13. Inflorescences 2–7-flowered; scape hollow; tepals equal . . . . 51. *Leucojum*, p. 293
        13. Inflorescences 1-flowered; scape solid; tepals unequal . . . . 47. *Galanthus*, p. 280

[3. Shifted to left margin.—Ed.]

3. Ovary superior (only partly superior in *Aletris*, *Amianthium*, *Melanthium, Stenanthium*, and *Veratrum*).
  14. Inflorescences umbellate (or sometimes racemose in *Dichelostemma*), always subtended by spathes or bracts.
    15. Tepals distinct or connate basally into very short perianth tube.
      16. Plants smelling of onion . . . . 44. *Allium*, p. 224
      16. Plants not smelling of onion.
        17. Filaments sometimes dilated at base into cup, cup sometimes with basal filament appendages; pedicel articulate . . . . 68. *Bloomeria*, p. 336
        17. Filaments sometimes overlapping or connate but without appendages; pedicel not articulate . . . . 67. *Muilla*, p. 334
    15. Tepals distinctly connate proximally into perianth tube.
      18. Bracts 4; flowers sessile but appearing pedicellate because of long, slender perianth tube . . . . 70. *Milla*, p. 346
      18. Bracts 2 or more; flowers pedicellate.
        19. Fertile stamens 3 (or 6 but 3 much smaller).
          20. Leaves rounded abaxially; scape rigid; inflorescences open; staminodia usually 3, alternating with 3 fertile stamens; corona absent; stigma 3-lobed, lobes distinctly spreading and recurved . . . . 63. *Brodiaea*, p. 321
          20. Leaves keeled abaxially; scape curved to twining; inflorescences usually dense; perianth appendages arising from intersection of perianth tube and lobes, forming corona; stigma weakly 3-lobed . . . . 64. *Dichelostemma*, p. 328
        19. Fertile stamens 6.
          21. Filaments connate into tube with 2-fid appendages forming crown . . . . 66. *Androstephium*, p. 333
          21. Filaments distinct, or crown not formed by 2-fid lobes.
            22. Plants from bulbs with membranous outer coats . . . . . . 45. *Nothoscordum*, p. 276
            22. Plants from fibrous-coated corms.
              23. Scape 1–5 mm diam.; stigma weakly 3-lobed . . . . 69. *Triteleia*, p. 338
              23. Scape 7–15 mm diam.; stigma not lobed . . . . 65. *Triteleiopsis*, p. 332

[14. Shifted to left margin.—Ed.]

14. Inflorescences various or 1-flowered, sometimes bracteate, never spathaceous.
  24. Styles 3 (rarely 4), distinct, or stigmas 3, distinct (sometimes connate in *Trillium*), borne directly on ovary.
    25. Leaves (bracts in *Trillium*) in 1 or 2 whorls on flowering stem, or 2, opposite, never long and grasslike.
      26. Flowers solitary, terminal . . . . 16. *Trillium*, p. 90
      26. Flowers in umbellate arrays.
        27. Leaves 2 at base of each flowering stem; stamens 3 . . . . 17. *Scoliopus* (in part), p. 118
        27. Leaves in 2 whorls on each flowering stem; stamens 6 . . . . 22. *Medeola*, p. 150
    25. Leaves several to many, all basal or borne on flowering stem, if 2 and opposite then grasslike, at least 8 times longer than broad.
      28. Some part of plant obviously hairy or covered with scales (except glabrous in *Amianthium*).
        29. Tepals conspicuously clawed . . . . 11. *Melanthium* (in part), p. 77
        29. Tepals not clawed.
          30. Leaf blades broadly ovate to elliptic with many conspicuous veins, narrowed towards stalklike, sheathing bases . . . . 10. *Veratrum*, p. 72
          30. Leaf blades narrow, without conspicuous veins, not narrowing towards sheathing bases.
            31. Perigonal nectaries absent . . . . 15. *Amianthium*, p. 89
            31. Perigonal nectaries not well developed to prominent.
              32. Seeds broadly winged . . . . 11. *Melanthium* (in part), p. 77
              32. Seeds irregularly compressed or angled . . . . 12. *Schoenocaulon*, p. 79
      28. No part of plant obviously hairy or covered with scales.
        33. Plants from bulbs and reduced rhizomes . . . . 14. *Stenanthium*, p. 88
        33. Plants from rhizomes or fleshy roots.
          34. Anther dehiscence extrorse; capsules loculicidal.
            35. Plants from rhizome terminating in bulb, roots thick and cordlike; leaves rigid, filiform; inflorescences bracteate . . . . 9. *Xerophyllum*, p. 71
            35. Plants from rhizomes with fibrous roots; leaves flexible, broad; inflorescences ebracteate.
              36. Flowers dioecious, occasionally polygamodioecious; tepals white to greenish white, drying to yellow; gynoecium weakly syncarpous (carpels coherent); style arising from ovary apex; seeds with winglike arils . . . . 7. *Chamaelirium*, p. 68
              36. Flowers bisexual only; tepals purplish pink; gynoecium syncarpous; style sunken into ovary apex; seeds caudate at both ends . . . . 8. *Helonias*, p. 69
          34. Anther dehiscence introrse; capsules septicidal.
            37. Inflorescences 1-flowered; braceteoles of epicalyx distinct; anthers with appendages; ovary and fruits conspicuously and densely tuberculate; seeds yellowish . . . . 1. *Harperocallis*, p. 58
            37. Inflorescences racemose or thyrsate; bracteoles of epicalyx connate; anthers without appendages; ovary and fruits glabrous or with minute, scattered tubercles; seeds reddish brown to brown.
              38. Anthers 9 . . . . 2. *Pleea*, p. 59
              38. Anthers 6.
                39. Stems smooth; seeds without appendages . . . . 3. *Tofieldia*, p. 60
                39. Stems glandular-pubescent; seeds with appendages . . . . . 4. *Triantha*, p. 61
  24. Style 1, sometimes 3-branched or -lobed at apex, or absent and stigmas connate, borne directly on ovary.
    40. Tepals connate basally for more than $^1/_{10}$ total length.
      41. Flowering stems leafy.
        42. Leaves distinctly petiolate; racemes 1-sided . . . . 32. *Convallaria*, p. 205

42. Leaves clasping, subsessile, or short-petiolate; racemes not 1-sided . . . . 34. *Polygonatum*, p. 210

41. Flowering stems leafless, leaves all or mostly basal.

43. Plants from rhizomes with fleshy or fibrous roots.

44. Leaves distinctly petiolate . . . . 43. *Hosta*, p. 222

44. Leaves sessile.

45. Inflorescences racemose; style 3-branched at apex . . . . 5. *Aletris*, p. 64

45. Inflorescences 1-flowered or in terminal helicoid cymes; style unbranched at apex . . . . 40. *Hemerocallis*, p. 219

43. Plants from bulbs.

46. Inflorescences dense, racemose . . . . 60. *Muscari*, p. 316

46. Inflorescences open, racemose or paniculate.

47. Perianth 4.5–6 cm, funnelform . . . . 42. *Hesperocallis*, p. 221

47. Perianth 2 cm or shorter, lobes spreading or reflexed.

48. Plants 10–20 cm; tepals blue or blue and white; anther dehiscence introrse . . . . 58. *Chionodoxa*, p. 314

48. Plants 12–55 cm; tepals creamy white to yellowish; anthers dehiscing through pores . . . . 31. *Odontostomum*, p. 204

[40. Shifted to left margin.—Ed.]

40. Tepals distinct or connate basally for less than 1/10 total length.

49. Plants from rhizomes with fleshy or fibrous roots (except from corms in *Echeandia*).

50. Leaves basal.

51. Fruits baccate . . . . 23. *Clintonia*, p. 151

51. Fruits capsular.

52. Stamens 3; tepals dimorphic . . . . 17. *Scoliopus* (in part), p. 118

52. Stamens 6; tepals equal.

53. Ovules 1 or 2 per locule . . . . 39. *Asphodelus*, p. 218

53. Ovules 4–8+ per locule.

54. Anthers dorsifixed near base . . . . 36. *Echeandia*, p. 215

54. Anthers basifixed . . . . 37. *Eremocrinum*, p. 216

50. Leaves cauline (or basal in *Dianella*).

55. Fruits capsular.

56. Inflorescences racemose; tepals spreading; filaments pubescent . . . . . . 6. *Narthecium*, p. 66

56. Inflorescences 1-flowered; tepals imbricate; filaments glabrous . . . . 21. *Uvularia*, p. 147

55. Fruits baccate.

57. Anthers opening by pores . . . . 41. *Dianella*, p. 221

57. Anthers opening by slits.

58. Flowering stem with only 2 or 3 leaves along its length or with several leaves concentrated towards base. . . . . 33. *Maianthemum* (in part), p. 206

58. Flowering stem leafy over most of its length below inflorescence.

59. Flowers solitary or in pairs; peduncle adnate to stem, arising opposite next leaf axial, free portion of pedicel geniculate . . . . 20. *Streptopus*, p. 145

59. Flowers and pedicel not as above, peduncle absent.

60. Stems simple; tepals neither swollen nor slightly inflated above base; ovary with septal nectaries . . . . 33. *Maianthemum* (in part), p. 206

60. Stems branched distally; tepals weakly gibbous proximally; ovary lacking septal nectaries . . . . 19. *Prosartes*, p. 142

49. Plants from bulbs or corms.

61. Plants caulescent or scapose; inflorescences usually racemose, never paniculate; perianth never blue (except in a few species of *Calochortus*).

62. Tepals without nectaries at base . . . . 28. *Tulipa*, p. 199

62. Tepals with nectaries at base.

63. Style absent, stigmas borne directly on ovary . . . . 18. *Calochortus*, p. 119

63. Style present.

64. Inflorescences racemose, umbellate, or paniculate.

65. Tepals twisting together after flowering; pedicel articulate . . . . 55. *Chlorogalum*, p. 307
65. Tepals not twisting together after flowering; pedicel not articulate . . . . 26. *Lilium*, p. 172
64. Inflorescences racemose only.
66. Tepals reflexed . . . . 24. *Erythronium*, p. 153
66. Tepals not reflexed, or reflexed only at apex.
67. Tepals in 2 distinct whorls; bulb tunics usually absent, thin and white if present . . . . 25. *Fritillaria*, p. 164
67. Tepals not in 2 distinct whorls; bulb tunics always brown . . . . 27. *Lloydia*, p. 198
[61. Shifted to left margin.—Ed.]
61. Plants scapose, never caulescent; inflorescences varying, sometimes paniculate; perianth often blue.
68. Perianth white, each tepal with wide green abaxial stripe . . . . 61. *Ornithogalum*, p. 318
68. Perianth variously colored.
69. Ovules 2 per locule; tepals 3–12 mm.
70. Pedicel 6–15(–30) mm; filaments equal, 1–2 mm . . . . 57. *Schoenolirion*, p. 312
70. Pedicel 2–3 mm; filaments dimorphic, 4–8 mm . . . . 56. *Hastingsia*, p. 310
69. Ovules 1–12 per locule; tepals (6–)12–40 mm.
71. Tepals each 3–4-veined . . . . 54. *Camassia*, p. 303
71. Tepals each 1-veined.
72. Bracts 2 per flower . . . . 59. *Hyacinthoides*, p. 315
72. Bracts 1 per flower, or absent . . . . 62. *Scilla*, p. 320

1. HARPEROCALLIS McDaniel, J. Arnold Arbor. 49: 36, figs. 1, 2. 1968 • Harper's beauty [for Roland MacMillan Harper, 1878–1966, southeastern American botanist, and Greek *kallos*, beautiful, alluding to the attractive flower] [C] [E]

Frederick H. Utech

Loran C. Anderson

Herbs, perennial, from slender, nodose rhizomes with scarious scale leaves. **Stems** erect, simple, bracteate; bracts 2–5, much reduced, widely spaced, chaffy. **Leaves** basal, equitant, simple, sheathing; blade unifacial, striate, stiff; sheath margin scarious. **Inflorescences** terminal, erect, 1-flowered, bracteolate; bracteoles (2–)3(–4), in epicalyx, distinct, deltate-rhombic, small. **Flowers**: tepals persistent, 6, spreading, erect in fruit, distinct, oblanceolate, apex acute; stamens 6, hypogynous; filaments distinct; anthers basifixed, linear-lanceolate, with short appendages, introrse; gynoecium 3(–6)-carpellate, distally apocarpous; ovary superior, sessile, tuberculate; intercarpellary nectary present; styles 3(–6), slightly recurved; stigmas capitate. **Fruits** capsular, ellipsoid, prominently tuberculate, subtended by persistent tepals, dehiscence septicidal, then adaxially loculicidal. **Seeds** pale yellow, narrowly fusiform, straight or curved, short-appendaged at both ends. $x = 15, 16$.

Species 1: Florida.

Discovered in 1968, *Harperocallis* is the only genus endemic to the Apalachicola River region of Florida's western panhandle, where numerous endemic species occur (D. B. Ward [1979]). Carpel numbers often greater than 3, to 6, are frequently encountered (J. D. Ambrose 1975, 1980; L. C. Anderson and G. L. Smith 1994). The capsule is primarily septicidal, similar to those in *Tofieldia*, *Triantha*, and *Pleea* (J. D. Ambrose 1975, 1980; F. H. Utech 1993), and not simply loculicidal as originally described (S. McDaniel 1968; R. K. Godfrey [1979]).

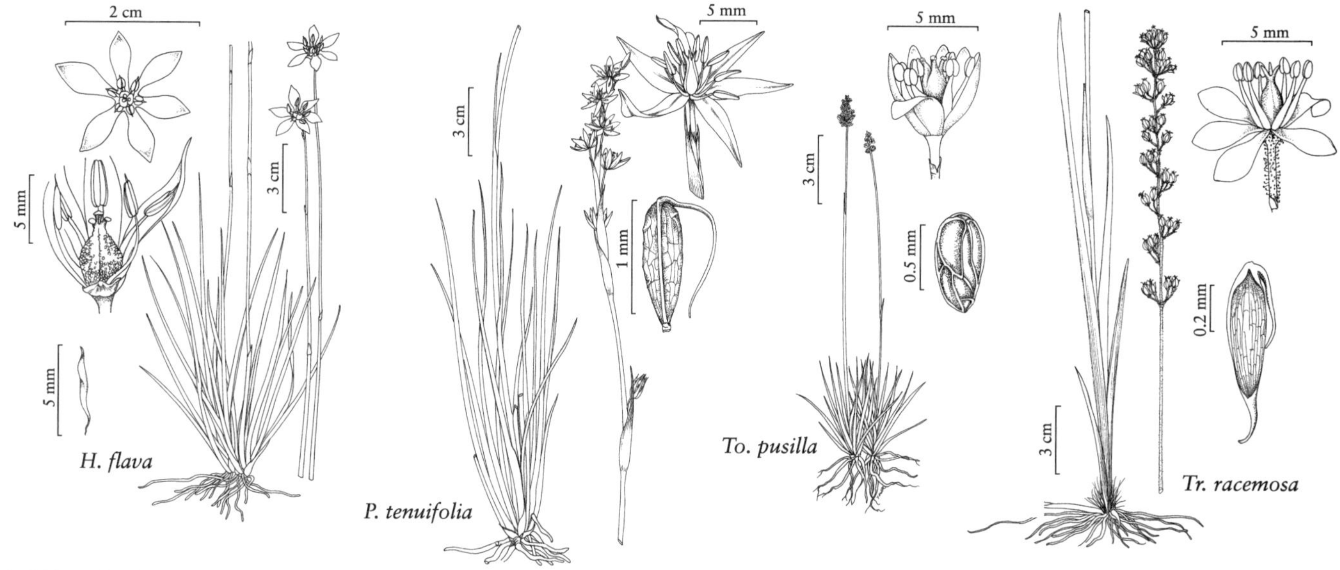

HARPEROCALLIS ◦ PLEEA ◦ TOFIELDIA ◦ TRIANTHA

SELECTED REFERENCES Godfrey, R. K. [1979.] Endangered: Harper's beauty. In: D. B. Ward, ed. [1979.] Rare and Endangered Biota of Florida. Vol. 5. Plants. Gainesville. Pp. 32–33. McDaniel, S. 1968. *Harperocallis*, a new genus of the Liliaceae from Florida. J. Arnold Arbor. 49: 35–40. Utech, F. H. 1993. Floral vascular anatomy of *Harperocallis flava* McDaniel and its systematic relationship within the Tofieldieae (Liliaceae). [Abstract.] In: International Botanical Congress. 1993. Abstracts. XV International Botanical Congress. Congress Center of Pacifico Yokohama, Japan, August 28 to September 3, 1993. [Tokyo.] P. 207.

1. **Harperocallis flava** McDaniel, J. Arnold Arbor. 49: 38, figs. 1, 2. 1968 [C] [E] [F]

Plants 22–55 cm, glabrous. **Leaf blades** linear, 0.2–0.3 × 5–21.5 cm, margins entire. **Flowers:** tepals yellow adaxially and greenish abaxially, becoming green to coppery, golden brown abaxially and purple-margined adaxially after anthesis,10–12-veined; outer tepal (10–)12–15 × (2.5–)3–3.5 mm; inner tepal (8–)9–12 × (2–)2.5–3.5 mm; stamens 8–9 mm; anthers 2.5–3.5 mm; styles 0.5–1 mm. **Capsules** 8–10 mm. **Seeds** 2–3 mm.

Flowering late spring. Open ti-ti and pineland swamps to wet roadside ditches; of conservation concern; 0–10 m; Fla.

Highly localized, *Harperocallis flava* occurs on gentle slopes, seepage savannas between pinelands, and cypress swamps to precariously open roadside depressions (R. K. Godfrey [1979]; R. Kral 1983; S. W. Leonard and W. W. Baker 1983; D. A. Wood 1996). It has distinctive showy, yellow, solitary flowers, and tuberculate ovaries and fruits. Genetic uniformity among individuals and populations of this species, in which 22 loci were found to be monomorphic, suggests low variability and serious conservation concern (M. J. W. Godt et al. 1997).

2. PLEEA Michaux, Fl. Bor.-Amer. 1: 247. 1803 • Rush-featherling [for Auguste Plée, 1787–1825, French traveller in the New World] [E]

John G. Packer

Herbs, perennial, rhizomatous, glabrous. **Leaves** 2-ranked, equitant, mostly basal; blade linear. **Inflorescences** terminal, racemose, open, bracteate, bracteolate; bracteoles connate in epicalyx. **Flowers** arising singly; tepals persistent, 6, in 2 somewhat dissimiliar series, distinct; stamens 9(–10), 2 opposite each outer tepal, 1 opposite each inner; filaments dilated basally, flattened; anthers versatile, 2-locular, introrse, without appendages; ovary superior, stipitate, apocarpous basally, glabrous; intercarpellary nectary present; styles 3. **Fruits** capsular, ovoid to broadly ellipsoid, glabrous, dehiscence septicidal, then adaxially loculicidal. **Seeds** appendaged. x =15.

Species 1: North America.

F. H. Utech (1978, 1979) clearly demonstrated the relationship of *Pleea* to *Tofieldia* sensu lato and reassigned the only species of the former to the latter. Morphologically, though, *P. tenuifolia* is a very distinctive species and, while it shares characteristics with both *Tofieldia* and *Triantha*, it is not in any way an intermediate.

SELECTED REFERENCES Utech, F. H. 1978. Floral vascular anatomy of *Pleea tenuifolia* Michx. (Liliaceae-Tofieldieae) and its reassignment to *Tofieldia*. Ann. Carnegie Mus. 47: 423–454. Utech, F. H. 1979. Karyotype analysis, palynology, and external seed morphology of *Tofieldia tenuifolia* (Michx.) Utech (Liliaceae-Tofieldieae). Ann. Carnegie Mus. 48: 171–184.

1. Pleea tenuifolia Michaux, Fl. Bor.-Amer. 1: 248, plate 25. 1803 [E] [F]

Tofieldia tenuifolia (Michaux) Utech

Stems scapelike, with 0–4 leaves progressively reduced distally, 30–85 cm. **Leaves** to 50 cm × 5 mm, with short sheathing base and much longer blade arising abaxially. **Inflorescences** 3–8-flowered, 5–20 cm; bracts sheathing, 2–7 cm, apex aristate; bracteoles 3, distinct to base or connate to 3/4 their length, 3.5–8 mm, forming sheathing involucre ca. mid pedicel. **Flowers:** tepals white to yellowish, 9–17 × 2–4 mm, outer series slightly longer and wider; filaments to 5.5 mm; anthers to 4 mm; ovary narrowly conical, ca. 5 mm; styles 1–1.8 mm; pedicel 1–3.5 cm. **Capsules** 5.5–9 mm, coriaceous. **Seeds** deep reddish brown, 1–1.5 mm, with appendage 1–2 times longer than seed at one end and sometimes a much shorter one at opposite end. $2n$ = 30.

Flowering late summer–autumn. Savannas, pocosins, marshy ground; 0–100 m; Ala., Fla., N.C., S.C.

3. TOFIELDIA Hudson, Fl. Angl. ed. 2, 157. 1778 • False asphodel [for Thomas Tofield, 1730–1779, English botanist]

John G. Packer

Herbs, perennial, rhizomatous, glabrous. **Stems** scapelike, sometimes with 1–3 leaves near base, smooth. **Leaves** equitant, 2-ranked; blade linear. **Inflorescences** terminal, racemose, open to dense, sometimes spikelike, bracteate, sometimes bracteolate, elongating in fruit; bracteoles connate in epicalyx. **Flowers** inserted singly; tepals persistent, 6, in 2 somewhat dissimilar series, distinct; stamens 6; filaments dilated basally, strongly flattened (except *T. coccinea*); anthers basifixed, 2-locular, introrse, without appendages; ovary superior, stipitate, apocarpous basally, glabrous; intercarpellary nectary present; styles 3. **Fruits** capsular, broadly ellipsoid to globose or obovoid, chartaceous, glabrous, dehiscence septicidal, then adaxially loculicidal. **Seeds** reddish brown, without appendages. x = 15.

Species 7 or 8 (3 in the flora): North America, Eurasia.

Following the example of J. G. Baker (1879) and R. R. Gates (1918), this treatment of *Tofieldia* excludes the species of sect. *Triantha*, which is recognized instead as a distinct genus; see J. G. Packer (1993). *Tofieldia* sometimes has been treated very broadly, to include not only *Triantha* but also *Pleea* (W. B. Zomlefer 1997c).

1. Margins of bracts 3-lobed; bracteoles absent . . . . 3. *Tofieldia pusilla*
1. Margins of bracts ± entire; bracteoles 3-lobed.
    2. Stems 2–20 cm; inflorescences 0.5–2.5 cm; tepals 2–2.5 mm . . . . 1. *Tofieldia coccinea*
    2. Stems 25–90 cm; inflorescences 3–25 cm; tepals 3–4.5 mm . . . . 2. *Tofieldia glabra*

1. Tofieldia coccinea Richardson in J. Franklin et al., Narr. Journey Polar Sea, 736. 1823 • Northern false asphodel, tofieldie écarlate

Stems 2–20 cm. **Leaf blades** to 7.5 cm × 3.5 mm. **Inflorescences** open to dense, sometimes spikelike, 5–30-flowered, 0.5–2.5 cm; proximal bracts conspicuous, ovate, margins entire, distal bracts reduced or absent; bracteoles forming epicalyx, deeply 3-lobed, lobes often appearing distinct, ovate, sometimes weakly 3-fid with 2 small lateral lobes. **Flowers:** tepals greenish, tinged pinkish cream to deep crimson, 2–2.5 mm, inner tepals slightly longer than outer; stamens ± equaling or slightly longer than tepals; filaments not strongly flattened or dilated basally; ovary ellipsoid; styles 0.3–0.8 mm; pedicel 0–3 mm, often becoming arcuate in fruit. **Capsules** broadly obovoid to globose, 2–2.3 mm. **Seeds** ca. 1 mm. $2n = 30$.

Flowering summer. Moist to dry open areas, generally calcareous; 0–1500 m; Greenland; B.C., N.W.T., Nunavut, Que., Yukon; Alaska; n Asia.

See under 3. *Tofieldia pusilla* regarding hybridization with that species.

2. Tofieldia glabra Nuttall, Gen. N. Amer. Pl. 1: 235. 1818 • White featherling, white false asphodel [E]

Stems 25–90 cm. **Leaf blades** to 30 cm × 6 mm. **Inflorescences** open, 20–100-flowered, 3–25 cm; bracts navicular, margins entire; bracteoles forming epicalyx, 3-lobed, lobes acute. **Flowers:** tepals white, 3–4.5 mm, inner tepals longer, wider than outer; stamens 3–4 mm; ovary narrowly ovoid; styles 0.4–0.8 mm; pedicel 1.5–4 mm. **Capsules** broadly ellipsoid, 3–4 mm. **Seeds** 1–1.3 mm.

Flowering late summer–fall. Savannas, pocosins, marshy ground, moist sandy soil; 0–200 m; N.C., S.C.

3. Tofieldia pusilla (Michaux) Persoon, Syn. Pl. 1: 399. 1805 • Scotch false asphodel [F]

Narthecium pusillum Michaux, Fl. Bor.-Amer. 1: 209. 1803

Stems 1.5–30 cm. **Leaf blades** to 8 cm × 3 mm. **Inflorescences** open, 3–35-flowered, 0.3–5 cm; bracts whitish, margins deeply 3-lobed; bracteoles absent. **Flowers:** tepals white to yellowish green, 1.5–3 mm, inner tepals slightly longer than outer; stamens ± equaling tepals; ovary ellipsoid; styles 0.3–0.4 mm; pedicel 1–3 mm. **Capsules** subglobose, 2.5–3 mm. **Seeds** 0.6–0.8 mm. $2n = 30$.

Flowering summer. Moist, open areas, often calcareous; 0–2300 m; Greenland; Alta., B.C., Man., Nfld. and Labr., N.W.T., Nunavut, Ont., Que., Sask., Yukon; Alaska, Mich., Minn., Mont.; Eurasia.

What appear to be hybrids between *Tofieldia pusilla* and *T. coccinea* occasionally occur in Greenland and the Northwest Territories. They are characterized by a leafy bract at or near the middle of the stem and enlarged, crimson bracts below the pedicels. In some suspected hybrids, the bracts are absent, and fused bracteoles, truncate or lobed, sheathe the pedicels well below the flowers.

4. TRIANTHA (Nuttall) Baker, J. Linn. Soc., Bot. 17: 490. 1879 • [Greek *tri-*, three-, and *anthos*, flower, alluding to the flowers aggregated in threes]

John G. Packer

Tofieldia Hudson sect. *Triantha* Nuttall, Gen. N. Amer. Pl. 1: 235. 1818

Herbs, perennial, rhizomatous. **Stems** glandular and/or glandular-pubescent. **Leaves** mostly basal with 0–3 towards stem base, 2-ranked, equitant; blade linear. **Inflorescences** terminal, racemose, open or dense and spikelike, elongating in fruit, bracteate, bracteolate; bracteoles connate in epicalyx. **Flowers** borne in clusters of 2–7; tepals persistent, 6, in 2 somewhat dissimiliar series, distinct; stamens 6; filaments strongly flattened, dilated basally; anthers basifixed, 2-locular, introrse, without appendages; ovary superior, stipitate, apocarpous basally, glabrous; intercarpellary nectary present; styles 3. **Fruits** capsular, ovoid to broadly ellipsoid or cylindrical, glabrous, dehiscence septicidal, then adaxially loculicidal. **Seeds** appendaged. $x = 15$.

Species 4 (3 in the flora): North America, Japan.

Recognition here of the genus *Triantha* follows J. G. Baker (1879) and R. R. Gates (1918); see J. G. Packer (1993). R. W. Cruden (1991) provided cladistic evidence supporting this segregation from *Tofieldia*.

In the absence of any clear understanding of evolutionary relationships within *Triantha*, the species are here listed alphabetically.

SELECTED REFERENCES Hitchcock, C. L. 1944. The *Tofieldia glutinosa* complex of western North America. Amer. Midl. Naturalist 31: 487–498. Packer, J. G. 1993. Two new combinations in *Triantha* (Liliaceae). Novon 3: 278–279.

1. Styles connate basally into column $1/4$–$2/3$ their length; capsules hard, ± equaling or slightly longer than tepals and ± enclosed by them . . . . 3. *Triantha racemosa*
1. Styles distinct to base, not forming column; capsules chartaceous and easily ruptured, considerably longer than tepals and not enclosed by them.
   2. Seed coat present; inflorescence sometimes forming globose head (subsp. *occidentalis*); glandular hairs on stem below inflorescence sometimes uniformly 4–6 times longer than wide (subsp. *montana*); w Canada, nw United States . . . . 2. *Triantha occidentalis*
   2. Seed coat absent; inflorescence never forming globose head; glandular hairs on stem below inflorescence never uniformly 4–6 times longer than wide; widespread in Canada, also in Alaska, some north, northeastern states, and Oregon . . . . 1. *Triantha glutinosa*

1. Triantha glutinosa (Michaux) Baker, J. Linn. Soc., Bot. 17: 490. 1879 • Sticky false asphodel, tofieldie glutineuse [E]

Narthecium glutinosum Michaux, Fl. Bor.-Amer. 1: 210. 1803; *Tofieldia glutinosa* (Michaux) Persoon; *T. racemosa* (Walter) Britton, Sterns & Poggenburg var. *glutinosa* (Michaux) H. E. Ahles

Stems leafless, or with 1–3 leaves toward base, 5–50 cm, glandular below inflorescence, glands dome-shaped to conical, $1/2$–2 times longer than wide, rarely glandular-pubescent with coarse cylindrical hairs. **Leaf blades** to 30 cm × 6 mm. **Inflorescences** forming cylindric-ovate spikelike heads, 3–30-flowered, sometimes interrupted or open, 1–6.5 cm, glandular-pubescent; bracts subtending pedicel in cluster; bracteoles forming ascending, truncate, or shallowly 3-lobed involucre around pedicel, lobes spreading, occasionally cleft from proximal $1/3$ to base, narrow, usually glabrous, margins usually entire or nearly so, apex rounded to acute. **Flowers** borne in clusters of 3, proximal sometimes remote; perianth white or yellowish; tepals 3.8–5 mm, inner series slightly longer; stamens 2.8–4 mm; ovary ellipsoid, tapering gradually to style base; styles distinct, 0.6–1.5 mm; pedicel 1–9 mm. **Capsules** ovoid to broadly ellipsoid, 3.5–7.5 mm, clearly longer than tepals and not enclosed by them, chartaceous, easily ruptured. **Seeds** reddish brown, ca. 1 mm; appendages 2, one at each end of seed, one contorted, 1–4 times longer than seed, other at opposite end, much shorter; coat absent.

Flowering summer. Marshes, wet meadows, calcareous soil; 0–2100 m; St. Pierre and Miquelon; Alta., B.C., Man., N.B., Nfld. and Labr., N.W.T., N.S., Ont., Que., Sask., Yukon; Alaska, Ill., Ind., Maine, Mich., Minn., N.H., N.Y., N.C., N.Dak., Ohio, Oreg., Tenn., Vt., Va., W.Va., Wis.

Hybrids between *Triantha glutinosa* and *T. racemosa* occur in Burlington County, New Jersey; see note under *T. racemosa*. On the Queen Charlotte Islands in British Columbia, there appears to be some evidence of hybridization between *T. glutinosa* and the two subspecies of *T. occidentalis* that occur there.

2. Triantha occidentalis (S. Watson) R. R. Gates, J. Linn. Soc., Bot. 44: 137. 1918 [E]

Tofieldia occidentalis S. Watson, Proc. Amer. Acad. Arts 14: 283. 1879; *T. glutinosa* (Michaux) Persoon var. *occidentalis* (S. Watson) C. L. Hitchcock

Stems leafless, or with 1–3 leaves towards base, 10–80 cm, variously glandular-hairy or only glandular below inflorescence, glands uniformly 4–6 times longer than wide. **Leaf blades** to 50 cm × 8 mm. **Inflorescences** forming globose or cylindric-ovoid, spikelike heads, 3–45-flowered, sometimes interrupted or open, 1–8 cm, glandular-pubescent; bracts subtending pedicel in cluster; bracteoles shallowly and symmetrically 3-lobed to cleft from proximal 1/3 to base, lobes rounded to acute, often markedly unequal. **Flowers** usually borne in clusters of 3, proximal sometimes remote; perianth white or yellowish; tepals 3–7 mm, inner series somewhat longer and narrower; stamens 3–6 mm; ovary ellipsoid, tapering gradually to style base; styles distinct, 0.6–3 mm; pedicel 1–12 mm. **Capsules** ovoid to broadly ellipsoid, 4–9 mm, clearly longer than tepals and not enclosed by them, chartaceous, easily ruptured. **Seeds** reddish brown, ca. 1 mm; appendages 1 or 2 with one at each end, rarely absent; coat white, inflated, reticulate.

Subspecies 3 in the flora: w Canada, nw United States.

The subspecies of *Triantha occidentalis* recognized here are for the most part readily distinguishable from one another. Only in the area of southwest Oregon where subsp. *occidentalis* and subsp. *brevistyla* make contact might it be said that some intergradation occurs, as was previously observed by C. L. Hitchcock (1944). It should also be noted that some specimens of subsp. *occidentalis* from Del Norte County in northern California and the adjacent Josephine County in Oregon are not entirely typical, being very robust with large, more elongate inflorescences.

1. Seed coat not strongly inflated; seeds 3–4 times longer than wide; stems coarsely glandular-pubescent with cylindrical hairs below inflorescence, hairs 4–6 times longer than wide, or glands absent . . . . 2b. *Triantha occidentalis* subsp. *montana*
1. Seed coat strongly inflated; seeds usually 1–2(–3) times longer than wide; stems glandular-pubescent below inflorescence, with glands 1/2–2 times longer than wide, or pubescent with cylindrical hairs 2–4 times longer than wide, sometimes glands and hairs intermixed.
   2. Styles 0.6–1.5 mm; inflorescences usually cylindrical-ovoid; Oregon to Alaska, occasionally Idaho and Alberta . . . . 2a. *Triantha occidentalis* subsp. *brevistyla*
   2. Styles 1.3–3 mm; inflorescences usually globose; California, Oregon . . . . 2c. *Triantha occidentalis* subsp. *occidentalis*

2a. Triantha occidentalis (S. Watson) R. R. Gates subsp. **brevistyla** (C. L. Hitchcock) Packer, Novon 3: 279. 1993 [E]

Tofieldia glutinosa (Michaux) Persoon subsp. *brevistyla* C. L. Hitchcock, Amer. Midl. Naturalist 31: 497, figs. 5, 6. 1944; *T. glutinosa* subsp. *absona* C. L. Hitchcock; *T. glutinosa* var. *absona* (C. L. Hitchcock) R. J. Davis; *T. glutinosa* var. *brevistyla* (C. L. Hitchcock) C. L. Hitchcock

Stems 10–60 cm, glandular below inflorescence, glands dome-shaped to conical, 1/2–2 times longer than wide, often intermixed with cylindrical hairs 2–4 times longer than wide; sometimes only cylindrical hairs present. **Inflorescences** spikelike, 3–21-flowered, forming cylindrical-ovoid heads, sometimes interrupted or open, 1–6 cm; bracteoles rarely glandular. **Flowers:** tepals 3.5–6 mm; stamens 3.5–5 mm; styles 0.6–1.5 mm; pedicel 2–9 mm. **Capsules** 5–9 mm. **Seeds** often lobed, 1–2 times longer than wide; appendage usually 1, sometimes absent, to 2 times longer than seed; coat strongly inflated. $2n = 30$.

Flowering summer. Wet meadows, marshes; 0–2400 m; Alta., B.C.; Alaska, Idaho, Oreg., Wash.

With the exception of the retuse tepals, the characteristics by which C. L. Hitchcock (1944) distinguished subsp. *absona* from the other subspecies that he recognized under *Tofieldia glutinosa* are commonly found in subsp. *brevistyla*, and the former is treated here as a disjunct Idaho population of the latter.

2b. Triantha occidentalis (S. Watson) R. R. Gates subsp. **montana** (C. L. Hitchcock) Packer, Novon 3: 279. 1993 [E]

Tofieldia glutinosa (Michaux) Persoon subsp. *montana* C. L. Hitchcock, Amer. Midl. Naturalist 31: 496, fig. 4. 1944; *T. glutinosa* var. *montana* (C. L. Hitchcock) R. J. Davis

Stems 5–55 cm, coarsely glandular-pubescent below inflorescence, hairs usually cylindrical, 4–6 times longer than wide, or stems occasionally clavate and glands absent. **Inflorescences** spikelike, 3–24-flowered, forming cylindrical-ovate heads, sometimes interrupted or open, 1–4.5 cm; bracteoles often glandular. **Flowers:** tepals 3–5 mm; stamens 3–4.2 mm; styles 0.7–1.5 mm; pedicel 2–12 mm. **Capsules** 4–7.5 mm. **Seeds** 3–4 times longer than wide; appendages usually 2, one at each end, one often highly contorted, to 4 times longer than seed, the other short; coat not strongly inflated. $2n = 30$.

Flowering summer. Wet meadows, marshes; 0–2700 m; Alta., B.C.; Alaska, Idaho, Mont., Wyo.

2c. Triantha occidentalis (S. Watson) R. R. Gates subsp. **occidentalis** [E]

Stems 10–80 cm, glandular below inflorescence with glands dome-shaped to conical, 1/2–2 times longer than wide, or pubescent with cylindrical hairs 2–4 times longer than wide, rarely both glandular and pubescent. **Inflorescences** usually forming globose heads ca. as long as wide, less often spikelike and forming cylindrical-ovoid heads, 3–45-flowered, sometimes interrupted or open, 1–8 cm; bracteoles usually glandular. **Flowers:** tepals 3–7 mm; stamens 3.5–6 mm; styles 1.3–3 mm; pedicel 1–12 mm. **Capsules** 4–8 mm. **Seeds** often lobed, 1–2(–3) times longer than wide; appendages occasionally absent, usually 1, to 3 times longer than seed, rarely also with short appendage at opposite end; coat strongly inflated. $2n = 30$.

Flowering summer. Moist meadows, marshes; 0–3300 m; Calif., Oreg.

3. Triantha racemosa (Walter) Small, Fl. S.E. U.S., 249, 1328. 1903 • Coastal false asphodel [E] [F]

Melanthium racemosum Walter, Fl. Carol., 126. 1788; *Tofieldia racemosa* (Walter) Britton, Sterns & Poggenburg

Stems leafless, or with 1–3 leaves towards base, 20–70 cm, coarsely glandular-pubescent below inflorescence. **Leaf blades** to 35 cm × 6 mm. **Inflorescences** racemose, 15–80-flowered, usually open, 3–22 cm; bracts subtending pedicel in cluster; bracteoles shallowly 3-lobed or cleft from proximal 1/3 to base, lobes acute or rounded, usually glandular. **Flowers** borne in clusters of (2–)3(–7), proximal sometimes remote; perianth white, drying orange; tepals 2.5–5 mm, inner series narrower, longer; stamens 2.5–4.5 mm; ovary ellipsoid to cylindrical, usually tapering abruptly to style base, forming rounded shoulder; styles connate basally into column 1/4–2/3 their length, 1–1.8 mm; pedicel 2–12 mm. **Capsules** ovoid to subglobose, 3–5 mm, ± equaling or slightly longer than tepals and ± enclosed by them, hard. **Seeds** reddish brown, ca. 1 mm; appendages 1 or 2 with one at one end ca. 1/2 to equaling seed, one at opposite end often much shorter; coat absent.

Flowering summer. Boggy areas, pine barrens, savannas; 0–400 m; Ala., Del., D.C., Fla., Ga., La., Md., Miss., Tex., Va.

Specimens of *Triantha* collected in Burlington County, New Jersey, have generally been determined as *T. racemosa*. However, the population is variable, and while there are aspects in some of these plants suggestive of *T. racemosa* that would explain the determinations, others are closer morphologically to *T. glutinosa*. This is seen in the typical "glutinosa" glands, the narrower, less-spreading bracteoles that usually lack glandular hairs, the more-ascending pedicels, the shorter internodes in most inflorescences, and the chartaceous capsules. I suppose that in the past the two species had overlapping ranges in this northeastern region where they no longer are found, and that some hybridization between *T. glutinosa* and *T. racemosa* has occurred. The Burlington County population is a surviving disjunct remnant with attributes of both species. The specimens have been annotated *T. glutinosa* × *T. racemosa*, and New Jersey is omitted from the list of states in which *T. racemosa* is found.

5. ALETRIS Linnaeus, Sp. Pl. 1: 319. 1753; Gen. Pl. ed. 5, 149. 1754 • Colic-root, stargrass [Greek *aletris*, a female slave who ground corn, alluding to the mealy texture of the perianths]

Victoria I. Sullivan

Herbs, perennial, scapose, rhizomatous. **Leaves** in dense basal rosettes, clasping erect branches; blade narrowly linear to lanceolate, oblanceolate, linear-elliptic, or elliptic, flat, leathery, distal margins fused to form subulate tips. **Scape** 2–10 dm. **Inflorescences** racemose. **Flowers** each subtended by 2 subulate, unequal bracts, short-pedicellate; perianth white, yellow, or golden orange, cylindrical, campanulate, or obovoid, abaxial surfaces rough; tepals 6, connate basally; stamens 6, included; filaments adnate to perianth; anthers oblong-lanceolate, longer than filaments; ovary half inferior with proximal portions of perianth adnate at maturity; style 3-

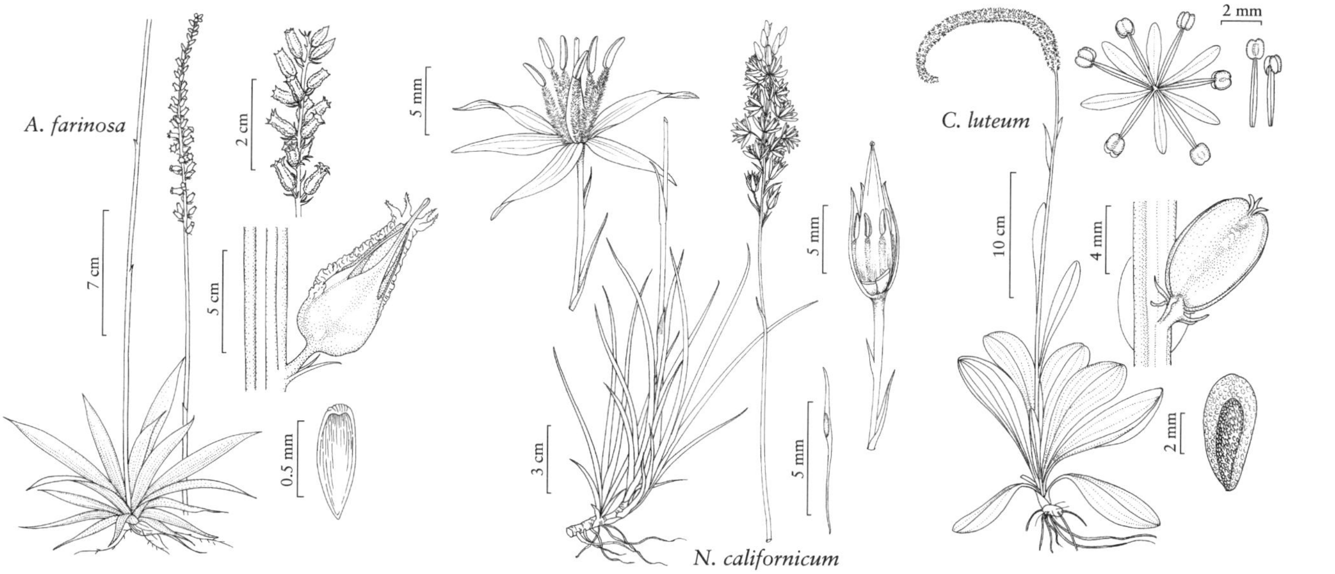

ALETRIS ∘ NARTHECIUM ∘ CHAMAELIRIUM

branched at apex. **Fruits** capsular, 3-locular, beaked. **Seeds** amber, deeply sulcate, ellipsoid to ovoid, 0.5–0.8 mm, lustrous.

Species ca. 25 (5 in the flora): North America, West Indies (Bahamas), eastern Asia.

Some species of *Aletris* (e.g., *A. lutea* and *A. obovata*) are quickly eliminated unless habitats are occasionally burned or otherwise kept clear of undergrowth.

SELECTED REFERENCE Sullivan, V. I. 1973. Biosystematics of *Aletris lutea* Small, *Aletris obovata* Nash and natural hybrids. Brittonia 25: 294–303.

1. Perianth usually wholly yellow to golden yellow, rarely white.
    2. Perianth cylindrical, 9–12 mm, more than 2.5 times as long as broad, lobes spreading . . . . . 4. *Aletris lutea*
    2. Perianth campanulate, 6–7 mm, 2 times or less as long as broad, lobes erect . . . . . 5. *Aletris aurea*
1. Perianth white to creamy white, lobes sometimes tipped with orange or pinkish orange.
    3. Perianth campanulate or obovoid, lobes turned slightly inward . . . . . 3. *Aletris obovata*
    3. Perianth cylindrical, lobes spreading.
        4. Leaves dull grayish green, 0.6–1 cm wide; beaks of fruits gradually tapering from body to tip . . . . . 1. *Aletris bracteata*
        4. Leaves bright yellowish green, 0.5–2.6 cm wide; beaks of fruits abruptly narrowed distally . . . . . 2. *Aletris farinosa*

1. Aletris bracteata Northrop, Mem. Torrey Bot. Club 12: 27, plate 1. 1902 [C]

Leaves 5–17 × 0.6–1 cm; blade dull grayish green, linear-lanceolate. **Scape** 5–6 dm. **Flowers:** perianth white, cylindrical, 6–8 mm, lobes spreading. **Fruits:** beaks gradually tapering distally.

Flowering year round. Everglades, pinelands; 0 m; of conservation concern; Fla.; West Indies (Bahamas: Abaco and Andros Islands).

2. Aletris farinosa Linnaeus, Sp. Pl. 1: 319. 1753 • Unicorn root, white stargrass [E] [F]

Leaves 4–20 × 0.5–2.6 cm; blade bright yellowish green, linear-lanceolate. **Scape** 4–10 dm. **Flowers:** perianth white to creamy white, cylindrical, 7–10 mm, lobes spreading. **Fruits:** beaks abruptly narrowed distally.

Flowering late spring–mid summer. Open sites of various habitats, moist bogs, dry to mesic prairies, and dry, upland woods and thickets; 0–2000+ m; Ont.; Ala., Ark., Conn., Del., D.C., Fla., Ga., Ill., Ind., Ky., La., Maine, Md., Mass., Mich., Miss., N.H., N.J., N.Y., N.C., Okla., Pa., R.I., S.C., Tenn., Tex., Va., W.Va., Wis.

3. Aletris obovata Nash ex Small in J. K. Small, Fl. S.E. U.S., 286. 1903 • White colic-root [E]

Leaves 3–13 × 1–2.8 cm; blade bright green, elliptic to lanceolate. **Scape** 4–7 dm. **Flowers:** perianth white to creamy white, sometimes tipped with orange or pinkish orange, campanulate or obovoid, 7–11 mm, lobes turned slightly inward (see note under 4. *Aletris lutea*). **Fruits:** beaks abruptly narrowed distally. $2n = 26$.

Flowering late spring–mid summer. Moist pine flatwoods and savannas, roadside ditches; 0–100 m; Ala., Fla., Ga., Miss., S.C.

4. Aletris lutea Small, Bull. New York Bot. Gard. 1: 278. 1896 • Yellow colic-root [E]

Leaves 7–18 × 1–2.6 cm; blade bright green, linear to lanceolate. **Scape** 5–10 dm. **Flowers:** perianth usually yellow to golden yellow, rarely white, cylindrical, 9–12 mm, more than 2.5 times as long as broad, lobes spreading. **Fruits:** beaks gradually narrowed distally. $2n = 26$.

Flowering spring. Seasonally wet to flooded pine flatwoods, savannas, bogs; 0 m; Ala., Fla., Ga., La., Miss.

A white-flowered form of *Aletris lutea* was described as forma *albiflora* E. T. Browne. Hybrids of *A. lutea* × *obovata* occur commonly in intermixed populations and have pale yellow corollas of intermediate shape (V. I. Sullivan 1973). These hybrids have been named as *A.* ×*tottenii* E. T. Browne.

5. Aletris aurea Walter, Fl. Carol., 121. 1788 [E]

Leaves 2–8 × 1–2 cm; blade bright green, elliptic to lanceolate. **Scape** 2–8 dm. **Flowers:** perianth golden yellow, campanulate, 6–7 mm, 2 times or less as long as broad, lobes erect. **Fruits:** beaks gradually narrowed distally. $2n = 26$.

Flowering late spring–mid summer. Dry to wet pine flatwoods, savannas, and bogs; 0–100+ m; Ala., Ark., Fla., Ga., La., Md., Miss., N.C., Okla., S.C., Tex., Va.

6. NARTHECIUM Hudson, Fl. Angl., 127. 1762, name conserved • Bog-asphodel [Greek *Narthex*, rod, alluding to appearance of stems]

Frederick H. Utech

Abama Adanson

Herbs, perennial, from slender, creeping rhizomes; roots fibrous. **Stems** simple, glabrous. **Leaves** mostly basal, astride rhizomes, equitant, overlapping; blade narrowly linear, apex acute; cauline leaves 3–6, reduced. **Inflorescences** terminal, racemose, simple. **Flowers** greenish yellow to yellow, glabrous; tepals persistent, 6, spreading, erect in age, distinct, weakly 3-veined, glandless, linear-lanceolate, margins scarious; stamens 6; filaments white, very densely pubescent; anthers dorsifixed, versatile, 2-locular, linear, introrse; ovary superior, 3-locular, lobes fused into long-pointed stylar beak, septal glands absent; stigma minutely 3-lobed; pedicels bracteolate proximally, usually with small bracteole near middle. **Fruits** capsular, weakly 3-lobed, oblong-lanceolate, thin walled, dehiscence loculicidal. **Seeds** pale yellow, filiform-ellipsoid, narrowly bristle-tailed at both ends. $x = 13$.

Species 4–8 (2 in the flora): primarily north-temperate regions.

The widely disjunct species of *Narthecium* (1–5 in Europe, 1 in eastern Asia, and 2 in North America) are morphologically and ecologically quite similar (A. A. Stabursvik 1959; W. B. Zomlefer 1997).

Narthecium has sometimes been included in the segregate family Melanthiaceae; based on morphology, anatomy, and karyology, it is best placed in its own family, Nartheciaceae (J. D. Ambrose 1975, 1980; P. Goldblatt 1995; W. B. Zomlefer 1997; A. L. Takhtajan 1997; M. N. Tamura 1998b), and order, Nartheciales (J. L. Reveal and W. B. Zomlefer 1998). The Nartheciales, considered basal to the Liliaceae in the broad sense and to the monocotyledons, includes only Nartheciaceae, with five genera: *Aletris* Linnaeus (14 species, eastern Asia and North America), *Metanarthecium* Maximowicz (1, eastern Asia), *Lophiola* Ker Gawler (2, eastern North America), and *Nietneria* Klotzsch ex Bentham (1, northern South America). In this flora, *Lophiola* is treated with the Haemodoraceae; see page 47.

Flowers of the common European species *Narthecium ossifragum* (Linnaeus) Hudson, which are fragrant, homogamous, and produce no nectar, are known to be pollinated by raindrops. The conspicuous filament hairs, common within the genus, trap pollen, and self-pollination occurs when raindrops accumulate around the filament hairs, allowing pollen to float to the stigma (O. Hagerup 1950; M. C. F. Proctor and P. F. Yeo 1972; M. C. F. Proctor et al. 1996). Such pollination probably occurs also in the North American species.

Steroidal saponins of *Narthecium* are thought to cause liver disease (icterus) in nonpigmented sheep via a deadly photosensitivity (A. A. Stabursvik 1959; R. Hegnauer 1962+, vol. 2, pp. 284–286).

SELECTED REFERENCE Stabursvik, A. 1959. A Phytochemical Study of *Narthecium ossifragum* (L.) Huds., with Additional Chapters on the Botany and the Veterinary History of the Plant. Trondheim.

1. Stems 2.5–4.5 dm; flowering racemes compact, 3–6 cm; pedicels 4–5 mm; tepals 4–7 mm; anthers yellow; e United States . . . . 1. *Narthecium americanum*
1. Stems 3.5–6 dm; flowering racemes open, 8–15 cm; pedicels 8–15 mm; tepals 7–10 mm; anthers brick red; w United States . . . . 2. *Narthecium californicum*

1. Narthecium americanum Ker Gawler, Bot. Mag. 37: plate 1505. 1812 • Yellow asphodel C E

Abama americana (Ker Gawler) Morong; *A. montana* Small; *Narthecium ossifragum* (Linnaeus) Hudson var. *americanum* (Ker Gawler) A. Gray

Stems 2.5–4.5 dm. **Leaf blade** 1–2 dm × 0.7–2.5 mm. **Racemes** compact, 3–6 cm. **Tepals** 4–7 × 0.5–1.5 mm; stamens 2–3.5 mm; anthers yellow; pedicel 4–5 mm. **Capsules** 10–14 mm including beaks. **Seeds** 1–1.5 mm, 7–8 mm including bristle-tails.

Flowering mid–late summer. Boggy pine-barrens, pocosins, and wet savannas; of conservation concern; 0–50 m; Del., Md., N.J., N.C., S.C.

Narthecium americanum is listed as rare or endangered in North Carolina (J. L. Amoroso 1997), South Carolina (D. A. Rayner et al. 1979), and New Jersey (D. B. Snyder and V. E. Vivian 1981), and it has been extirpated from Delaware (A. O. Tucker et al. 1979).

2. Narthecium californicum Baker, J. Linn. Soc., Bot. 15: 351. 1876 • California bog-asphodel E F

Abama californica (Baker) A. Heller; *A. occidentalis* (A. Gray) A. Heller; *Narthecium ossifragum* (Linnaeus) Hudson var. *occidentale* A. Gray

Stems 3.5–6 dm. **Leaf blade** 1–3 dm × 3–5 mm. **Racemes** open, 8–15 cm. **Tepals** 7–10 × 1.5–2.5 mm; stamens 3–4 mm; anthers brick red; pedicel 8–15 mm. **Capsules** 8–12 mm including beaks. **Seeds** 1.5–2 mm, 6–8 mm including bristle-tails. $2n = 26$.

Flowering mid–late summer. Springs, seeps, and bogs in mountain meadows; 700–2800 m; Calif., Oreg.

Narthecium californicum can be locally common in the northern Coast Ranges and Sierra Nevada of California, as well as in southwestern Oregon.

7. CHAMAELIRIUM Willdenow, Ges. Naturf. Freunde Berlin Mag. Neuesten Entdeck. Gesammten Naturk. 2: 18. 1808 • [Greek *chamae*, on the ground, and *lirion*, white lily] [E]

Frederick H. Utech

Herbs, perennial, from stout, nodose rhizomes; roots contractile, fibrous, fleshy. **Stems** persistent in pistillate plants, erect to nodding, simple, hollow, glabrous. **Leaves** persistent, evergreen, crowded in basal rosettes, reduced distally; blades glabrous, margins entire to minutely undulate; basal blades petiolate, spatulate to oblanceolate; cauline blades sessile, linear to lanceolate. **Inflorescences** terminal, ebracteate; in staminate plants racemose, rarely spiciform, 7–15 cm, apex nodding; in pistillate plants racemose or spiciform, to 35 cm in fruit. **Flowers** unisexual, staminate and pistillate on different plants, occasionally bisexual, ebracteate, weakly syncarpous; tepals persistent, 6, spreading or ascending, distinct, white to greenish white, becoming yellow, 1-veined, narrowly linear-spatulate, nectaries absent; stamens 6; filaments distinct, flattened, unequal; anthers basifixed, 2-locular, oblong-oblanceolate, extrorse; ovary superior, 3-locular, deeply 3-lobed; septal glands absent; styles persistent, 3, recurved, distinct, linear-clavate; stigmas papillate along adaxial surface. **Fruits** capsular, erect, 3-locular, dehiscence loculicidal. **Seeds** 2–4 per locule, elliptic to linear-oblong, with broad, winglike aril. $x = 12$.

Species 1: e North America.

Chamaelirium and the east Asian *Chionographis* Maximowicz (H. Hara 1968) form a classic Arcto-Tertiary, disjunct generic pair whose distinctive nature among the melanthoid genera (J. D. Ambrose 1975, 1980) has lent support to their recognition as a separate tribe, Chionographideae Nakai, within the Melanthiaceae sensu stricto (M. N. Tamura 1998; W. B. Zomlefer 1997b), or as the separate family Chionographidaceae (A. L. Takhtajan 1994, 1997). Both genera have distinctive tetraporate pollen grains (M. Takahashi and S. Kawano 1989), both are based on $x = 12$ (H. Hara and S. Kurosawa 1962), and they share other karyological similarities (N. Y. Tanaka and N. Tanaka 1985).

Floral sexuality is exceptionally diversified in the *Chamaelirium–Chionographis* lineage. *Chamaelirium* is dioecious or occasionally polygamodioecious, while *Chionographis* is often monoclinous or andromonoecious, sometimes gynodioecious, and rarely androdioecious (M. Maki 1992, 1992b, 1993; M. Maki and M. Masusa 1993, 1994).

Chamaelirium luteum has been the subject of numerous studies on dioecism associated with dimorphism. Staminate and pistillate plants differ markedly in morphology, life history, and spatial distribution. Mature pistillate plants are more robust than staminate plants, and there are proportionally more staminate than pistillate plants in a given population (T. R. Meagher 1978, 1980, 1981, 1982, 1984; T. R. Meagher and J. Antonovics 1982, 1982b; T. R. Meagher and E. Thompson 1987).

SELECTED REFERENCE Silliman, F. E. 1957. *Chamaelirium luteum* (L.) Gray: A Biological Study. Ph.D. dissertation. University of North Carolina.

1. Chamaelirium luteum (Linnaeus) A. Gray, Manual, 503. 1848 • Blazing-star, devil's-bit, fairy-wand, rattlesnake-root [E] [F]

Veratrum luteum Linnaeus, Sp. Pl. 2: 1044. 1753; *Chamaelirium carolinianum* Willdenow; *C. obovale* Small

Staminate plants 5–20-leaved, 1.5–3.5 dm. **Pistillate plants** 15–50-leaved, 3–6 dm, 1.5 m in fruit. **Leaf blades** spatulate to oblanceolate, 5–20 × 1.5–6 cm, apex obtuse, tapering proximally to broad petiole, 4–6 cm; distal blades oblanceolate to linear, 3–8 × 1–1.5 cm. **Staminate flowers** divergent, white; tepals 3–4 mm; filaments dimorphic, the outer longer; anthers white, 0.5 mm; pistils absent; pedicel 2–5 mm. **Pistillate flowers** ascending, white; tepals 2–3 mm; staminodes present; ovary elliptic to obovate; styles 1.5–2 mm; stigmas sessile. **Capsules** ovoid-oblong, 7–14 × 5–6 mm. **Seeds** reddish brown, 1.8–2 mm; 5–6 mm (including winglike aril).

Flowering late spring–summer. Moist meadows, thickets, rich wooded slopes, and coves; 0–1100 m; Ont.; Ala., Ark., Conn., Del., D.C., Fla., Ga., Ill., Ind., Ky., La., Md., Mass., Mich., Miss., N.J., N.Y., N.C., Ohio, Pa., S.C., Tenn., Va., W.Va.

Chamaelirium is rare in southern Ontario (J. H. Soper 1962; D. J. White et al. 1982) and quite local in several eastern states. The flowers turn yellow on drying, hence the specific name. The roots, called "starwort" or "unicorn root," are used medicinally.

8. HELONIAS Linnaeus, Sp. Pl. 1: 342. 1753; Gen. Pl. ed. 5, 159. 1754 • [Greek *helos*, marsh, alluding to the habitat] [E]

Frederick H. Utech

Herbs, perennial, subscapose, glabrous, from stout rhizomes; roots contractile, fibrous. **Stems** erect, simple, hollow, glabrous. **Leaves** evergreen, in basal rosettes, reduced distally to bractlike leaves; blade simple, oblong-spatulate to oblanceolate, margins entire, apex acute, glabrous. **Inflorescences** terminal, racemose, dense, ebracteate, pedicellate. **Flowers** spreading, fragrant, funnel-shaped; tepals persistent, 6, distinct to barely connate basally, purplish pink, becoming green, spatulate to oblong, nectary proximal, adaxial, weakly sulcate; stamens persistent, 6, equaling tepals; filaments distinct, inner 3 proximally adnate to ovary; anthers dorsifixed, versatile, 1-locular, extrorse, pollen sacs apically confluent; ovary superior, 3-locular proximally, 1-locular distally; septal nectaries absent; styles 3, depressed apically into ovary apex, ascending to arching, distinct, sessile; stigmas not papillate along adaxial surface. **Fruits** capsular, deeply 3-lobed, papery, dehiscence loculicidal. **Seeds** 16 per locule, linear-fusiform, caudate at both ends. $x = 17$.

Species 1: e North America.

The close relationships among *Helonias* and the east Asian *Heloniopsis* A. Gray and *Ypsilandra* Franchet have been noted repeatedly. They exhibit many similarities in morphology and anatomy (F. Buxbaum 1925, 1927; W. Schulze 1978b; C. Sterling 1980; F. H. Utech 1978b; F. H. Utech and S. Kawano 1981; N. Tanaka 1997, 1997b, 1997c, 1997d, 1997e, 1998), palynology (M. Takahashi and S. Kawano 1989), ecology (S. Kawano and J. Masuda 1980; H. Takahashi 1988), and karyology (F. H. Utech 1980). Based on these similarities, Tanaka merged the five species of *Heloniopsis* and the three species of *Ypsilandra* into *Helonias*. However, molecular evidence indicates that, although the genera are related, they are distinct (S. Fuse and M. N. Tamura 2000). This group has been recognized as a separate tribe, Heloniadeae Fries, within the Melanthiaceae sensu stricto (M. N. Tamura 1998; W. B. Zomlefer 1997b), or placed in a separate family, Heloniadaceae J. Agardh (A. L. Takhtajan 1997).

Helonias in eastern North America and the *Ypsilandra–Heloniopsis* lineage in eastern Asia comprise a paired Arcto-Tertiary element in which *Ypsilandra* (western China to the Himalayas)

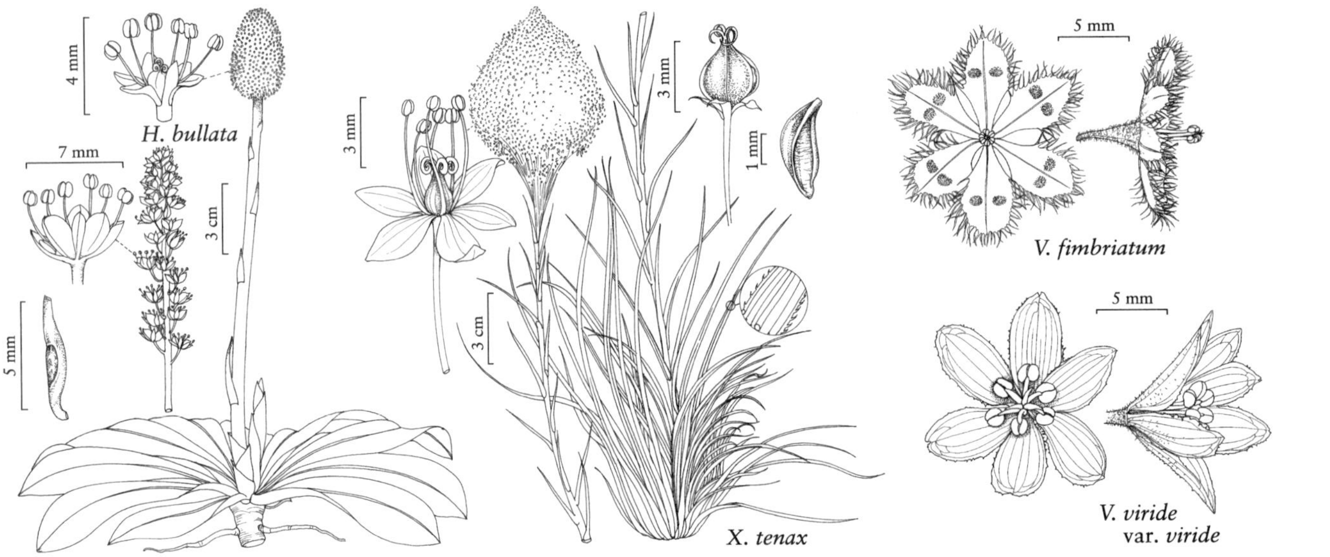

HELONIAS ◦ XEROPHYLLUM ◦ VERATRUM

and *Heloniopis* (Taiwan, Korea to Japan) have differentiated from each other in eastern Asia (N. Tanaka 1997e).

SELECTED REFERENCES Sutter, R. D. 1984. The status of *Helonias bullata* L. (Liliaceae) in the southern Appalachians. Castanea 49: 9–16. Tanaka, N. 1997. Taxonomic significance of some floral characters in *Helonias* and *Ypsilandra* (Liliaceae). J. Jap. Bot. 72: 110–116. Tanaka, N. 1997b. Evolutionary significance of the variation of the floral structure of *Heloniopsis*. J. Jap. Bot. 72: 131–138. Tanaka, N. 1997c. Phylogenetic and taxonomic studies on *Helonias, Ypsilandra* and *Heloniopsis*. I. Comparison of character states (1). J. Jap. Bot. 72: 221–228. Tanaka, N. 1997d. Phylogenetic and taxonomic studies on *Helonias, Ypsilandra* and *Heloniopsis*. I. Comparison of character states (2). J. Jap. Bot. 72: 286–292. Tanaka, N. 1997e. Phylogenetic and taxonomic studies on *Helonias, Ypsilandra* and *Heloniopsis*. II. Evolution and geographical distribution. J. Jap. Bot. 72: 329–336. Tanaka, N. 1998. Phylogenetic and taxonomic studies on *Helonias, Ypsilandra* and *Heloniopsis*. III. Taxonomic revision. J. Jap. Bot. 73: 102–115. Utech, F. H. 1978b. Vascular floral anatomy of *Helonias bullata* (Liliaceae–Helonieae), with a comparison to the Asian *Heloniopsis orientalis*. Ann. Carnegie Mus. 47: 169–191.

1. Helonias bullata Linnaeus, Sp. Pl. 1: 342. 1753 • Swamp-pink [E] [F]

Plants 1–3 dm in flower, 6 dm in fruit. **Leaves** dark green, 9–35 × 1.5–4 cm; bractlike leaves broadly triangular, 1–2 cm. **Racemes** 30–70-flowered, ovoid in flower, 2.5–10 cm, elongate in fruit, 10 17.5 cm. **Tepals** 4–9 mm; filaments filiform, 5–6 mm; anthers blue, 0.75–1 mm; styles 1.4–2.5 mm; pedicel 5–8 mm. **Capsules** obcordate, 3–8 × 8–10 mm. **Seeds** whitish brown, 4–6 mm. 2*n* = 34.

Flowering late spring–early summer. Swamps, bogs, and pocosins of the Blue Ridge Mountains and northern coastal plain; 0–1100 m; Del., Ga., Md., N.J., N.Y., N.C., S.C., Va.

Helonias bullata is threatened in several eastern coastal states (R. D. Sutter 1984). Linnaeus cited Pennsylvania as the type locality, but probably his material came from the New Jersey pine barrens, and all Pennsylvania populations are introduced (E. T. Wherry et al. 1979; F. H. Utech 1980). This species has horticultural potential in wet gardens.

9. XEROPHYLLUM Michaux, Fl. Bor.-Amer. 1: 210. 1803 • [Greek *xeros*, dry, and *phyllon*, leaf, alluding to the sclerified foliage] E

Frederick H. Utech

Herbs, perennial, from stout, tuberous, woody rhizomes ending in weakly developed, tunicate bulbs; offshoots not flowering for several years, then dying after fruiting; roots cordlike, thick. **Stems** erect, simple, robust. **Leaves** persistent, mostly basal, spiral, arching, dilated proximally, reduced distally; blade filiform-linear, keeled, rigid, margins serrulate, apex wiry pointed, densely tufted. **Inflorescences** at first corymbose, becoming elongate, terminal-racemose, densely flowered, bracteate. **Flowers** erect or spreading, semirotate; perianth hypogynous; tepals persistent, 6, spreading, distinct, creamy white, 5–7-veined, oblong or obovate, equal to subequal, claws absent, tepal nectaries absent; stamens 6, weakly epitepalous, divergent, distinct, 3–4 mm; filaments distinct, subulate, dilated proximally; anthers dorsifixed, versatile, 2-locular (not confluent), extrorse; gynoecium weakly syncarpous; ovary superior, 3-locular; septal nectaries absent; styles persistent, 3, recurved, distinct; stigmas papillate along adaxial apex; pedicel spreading or erect, 3–5 cm. **Fruits** capsular, 3-lobed, globular to ovoid, dehiscence loculicidal. **Seeds** 2–4 per locule, greenish brown, wingless, oblong-fusiform, 3-angled, 3–4 mm, glossy. $x = 15$.

Species 2 (2 in the flora): e, w North America.

Xerophyllum, an endemic genus with one species in eastern and one in western North America, is distinct from other melanthoid genera with which it has long been associated (J. D. Ambrose 1975, 1980; P. Goldblatt 1995; C. Sterling 1980; M. Takahashi and S. Kawano 1989; F. H. Utech 1978c) and has recently been segregated in the tribe Xerophylleae S. Watson within the Melanthiaceae sensu stricto (M. N. Tamura 1998; W. B. Zomlefer 1997b), or in the monotypic family Xerophyllaceae (A. L. Takhtajan 1994, 1997).

Xerophyllum tenax and *X. asphodeloides* are both cultivated for their showy, white racemes. Any given offshoot plant may not flower for several years, but ultimately an unbranched, strongly bracteate, flowering stalk is produced which, along with the surrounding leaves, dies after fruiting.

SELECTED REFERENCE Utech, F. H. 1978c. Comparison of the vascular floral anatomy of *Xerophyllum asphodeloides* (L.) Nutt. and *X. tenax* (Pursh) Nutt. (Liliaceae–Melanthioideae). Ann. Carnegie Mus. 47: 147–167.

1. Stems (0.8–)1.2–1.8(–2) m; leaf blade 2–8(–10) dm × 2–4(–6) mm; racemes 5–7 dm; w Cordilleras . . . . 1. *Xerophyllum tenax*
1. Stems (0.5–)0.8–1(–1.5) m; leaf blade 3–5 dm × 1–2.5 mm; racemes 1.5–3 dm; e mountains and coastal pine barrens . . . . 2. *Xerophyllum asphodeloides*

1. Xerophyllum tenax (Pursh) Nuttall, Gen. N. Amer. Pl. 1: 235. 1818 • Bear-grass, squaw-grass, elk-grass, Indian-basket-grass [E] [F]

Helonias tenax Pursh, Fl. Amer. Sept. 1: 243, plate 9. 1814; *Xerophyllum douglasii* S. Watson

Stems (0.8–)1.2–1.8(–2) m. **Leaf blade** 2–8(–10) dm × 2–4(–6) mm. **Racemes** 5–7 dm. **Tepals** oblong to lanceolate, 6–9 × 2–3 mm; styles 4 mm. **Capsules** 5–7 mm. **Seeds** 4 per locule. $2n = 30$.

Flowering spring–early summer. Open coniferous woods, dry ridges, rocky slopes, and clearings; 0–2300 m; Alta., B.C.; Calif., Idaho, Mont., Oreg., Wash., Wyo.

Xerophyllum tenax, variable in plant and flower size, is without evident geographic races (S. M. Maule 1959). Similar in most respects to *X. asphodeloides* but more robust, *X. tenax* typically has twice the number of flowers (F. H. Utech 1978c). The leaf fibers were used by native tribes for garments and decorative, watertight baskets. The bulbous rhizomes were roasted for several days before being eaten.

SELECTED REFERENCE Maule, S. M. 1959. *Xerophyllum tenax,* squawgrass, its geographic distribution and its behavior on Mount Rainier, Washington. Madroño 15: 39–48.

2. Xerophyllum asphodeloides (Linnaeus) Nuttall, Gen. N. Amer. Pl. 1: 235. 1818 • Turkey-beard, mountain-asphodel [E]

Helonias asphodeloides Linnaeus, Sp. Pl. ed. 2, 1: 485. 1762

Stems (0.5–)0.8–1(–1.5) m. **Leaf blade** 3–5 dm × 1–2.5 mm. **Racemes** 1.5–3 dm. **Tepals** oblong to ovate, 5–8 × 2–3 mm; styles 1.5–2 mm. **Capsules** 4–5.5 mm. **Seeds** 2 per locule. $2n = 30$.

Flowering spring–early summer. Pine barrens and dry mountain woods; 0–1400 m; Ala., Del., Ga., Ky., Md., N.J., N.C., S.C., Tenn., Va., W.Va.

Xerophyllum asphodeloides is disjunct from the xeric oak-hickory forests of the southern Blue Ridge Mountains to the dry sandy soils of the pine barrens of New Jersey (L. Artz 1951; W. B. Zomlefer 1997b). Xerophia, a unique alkaloid, has been reported from this species (W. Procter 1839).

10. VERATRUM Linnaeus, Sp. Pl. 2: 1044. 1753; Gen. Pl. ed. 5: 468. 1754 • False hellebore, skunk-cabbage, corn-lily, vérâtre, varaire [Latin *vere,* true, and *ater,* black, alluding to black rhizomes found in some species]

Dale W. McNeal Jr.

Aaron D. Shaw

Acelidanthus Trautvetter & C. A. Meyer; *Evonyxis* Rafinesque; *Leimanthium* Willdenow

Herbs, perennial, from short, thick, vertical rhizomes and swollen basal bulbs; roots contractile, fleshy. **Stems** erect, simple, hollow, robust, basally thickened, leafy. **Leaves** alternate, simple, strongly veined, narrowly to broadly orbicular, often plicate, reduced upwards, narrowed into closed, tubular, often overlapping sheaths. **Inflorescences** compound-racemose or large terminal-paniculate, bracteate, pubescent. **Flowers** bisexual or some proximal ones staminate; perianth perigynous, rotate to ± campanulate, with hypanthium fused to base of ovary; tepals persistent, 6, spreading to erect, distinct to weakly connate basally, off-white, green, yellow, or purple, petaloid, not clawed, equal to subequal, margins entire or erose-fimbriate; glands 1–2, basal, on adaxial surface, V-shaped over midrib or marginal pair; stamens 6, perigynous; filaments distinct, filiform; anthers persistent, 1-locular with confluent thecae, cordate-reniform, dehiscence apical/valvate; ovary superior or partly inferior, 3-locular; styles 3, erect to divergent, distinct, short; stigmas elongate, extending down inner style surfaces. **Fruits** capsular, deeply 3-lobed, style beaks persistent, dehiscence septicidal. **Seeds** ellipsoid to fusiform, flat, broadly winged (± globose with wings reduced to basal ridges in *V. fimbriatum*). $x = 8$.

Species 25–30 (5 in the flora): Northern Hemisphere.

The number of *Veratrum* species depends on the taxonomic treatment of four wide-ranging species complexes, three of which are entirely or mainly Eurasian (*V. album*, *V. nigrum* Linnaeus, and *V. maackii* Regel) and one North American (*V. viride*) (O. Loesener 1926, 1927, 1928; B. Mathew 1989; J. H. Zimmerman 1958). The Asian species have been variously circumscribed (Chen S. C. and H. Takahashi 2000; H. Kato et al. 1996; T. Nakai 1937, 1937b; N. S. Lee 1985, 1985b). Generic size further depends on inclusion or not of the closely related *Melanthium* (J. D. Ambrose 1975, 1980; P. Goldblatt 1995; M. N. Tamura 1998; W. B. Zomlefer 1997b), here treated as a separate genus.

The medicinal/poisonous properties of *Veratrum* involve a complex of cerveratrum and jerveratrum alkaloids, some with proven hypotensive properties and others that are highly toxic to humans and livestock (S. M. Kupchan et al. 1961; A. Osol et al. 1960; I. W. Southon and J. Buckingham 1989). Veraloid, a standard mixture of the most hypotensive *Veratrum* alkaloids, was widely prescribed until late in the nineteenth century, when emetic side effects greatly curtailed its use. The types and concentrations of alkaloids vary with the species, plant part, and season (C. A. Taylor 1956, 1956b).

Several *Veratrum* species are occasionally cultivated surrounded by low-growing plants for their architectural appeal in mesic or bog gardens. Century-old plants grow from the long-lived rhizome-bulb.

SELECTED REFERENCES Heller, A. A. 1905. The western veratrums. Muhlenbergia 1: 119–125. Leinfellner, W. 1961. Zur Kenntnis des Monokotyledonen—Perigons. III. Die Perigonblatter einiger weiterer Melanthioideen (*Melanthium, Zygadenus, Anticlea, Toxicoscordion, Veratrum* und *Kreysigia*). Oesterr. Bot. Z. 108: 108–210. Loesener, O. 1926. Studien über die Gattung *Veratrum* und ihre Verbreitung. Verh. Bot. Vereins Prov. Brandenburg 68: 105–166. Loesener, O. 1927. Übersicht über die Arten der Gattung *Veratrum*. Teil I. Repert. Spec. Nov. Regni Veg. 24: 61–72. Loesener, O. 1928. Übersicht über die Arten der Gattung *Veratrum*. Schluss. Repert. Spec. Nov. Regni Veg. 25: 1–10. Mathew, B. 1989. A review of *Veratrum*. Plantsman 11: 35–61. Taylor, C. A. 1956. Culture of false hellebore. Econ. Bot. 10: 155–165. Zimmerman, J. H. 1958. A Monograph of *Veratrum*. Ph.D. dissertation. University of Wisconsin.

1. Ovary and fruit densely tomentose; tepal margins (at least inner) irregularly and shallowly fringed . . . . 1. *Veratrum insolitum*
1. Ovary and fruit glabrous or sparsely pubescent with few ± straight hairs; tepal margins entire to deeply fringed.
    2. Tepal margins deeply fimbriate; seeds ± globose, wingless; coastal n California . . . . 2. *Veratrum fimbriatum*
    2. Tepal margins entire to finely erose-dentate; seeds flattened, winged; widespread.
        3. Tepals green to yellowish green; inflorescence branches drooping or, if spreading to ascending; from east of 85th meridian . . . . 3. *Veratrum viride*
        3. Tepals whitish to yellowish green; inflorescence branches spreading to ascending; west of 104th meridian.
            4. Stems 1–2.5 m, inflorescences freely branching; south of 50th parallel . . . . 4. *Veratrum californicum*
            4. Stems 0.3–1 m, inflorescences spiciform-racemose or few-branched proximally; Seward Peninsula, adjacent Alaska, w Aleutian Islands . . . . 5. *Veratrum album*

1. Veratrum insolitum Jepson, Fl. Calif. 1: 266. 1921 • Siskiyou false hellebore [E]

Stems 1–1.5 m, grayish-hairy distally. **Leaf blades** elliptic, distal lanceolate, 10–35 × 7–21 cm, reduced distally, grayish-hairy, at least when young. **Inflorescences** paniculate, with ascending branches, 20–50 cm, grayish-hairy; bracts ovate to lanceolate, shorter than flowers. **Tepals** creamy white to yellow, broadly ovate, not clawed, 6–15 mm, margins strongly erose-ciliate or inner shallowly fimbriate; gland 1, basal, green, V-shaped, elliptic; ovary densely tomentose; pedicel 6–15 mm. **Capsules** oblong-ovoid, 2–3 cm, densely tomentose. **Seeds** flat, broadly winged, 10–15 mm. $2n = 32$.

Flowering summer–fall. Stream banks, moist meadows, mixed-evergreen forest margins; 0–1500 m; Calif., Oreg., Wash.

The only other whitish-flowered species in North America, *Veratrum californicum,* which might be confused with *V. insolitum*, has entire, unfringed tepals, and only slightly hairy ovaries.

2. Veratrum fimbriatum A. Gray, Proc. Amer. Acad. Arts 7: 391. 1868 • Fringed false hellebore [E] [F]

Stems 1–2 m, densely tomentose distally. **Leaves** elliptic-lanceolate, reduced distally, 20–50 × 10–25 cm, glabrous or sparsely hairy. **Inflorescences** paniculate, with spreading to ascending branches, 30–50 cm, densely tomentose; bracts ovate to ovate-lanceolate, shorter than flowers. **Tepals** white, rhomboid-ovate to ovate, not clawed, 6–9 mm, margins ± deeply and irregularly fimbriate; glands 2, basal, ± mid tepal, yellow, elliptic; ovary glabrous; pedicel 6–12 mm. **Capsules** oblong-ovoid, 8–10 mm, glabrous. **Seeds** ± globose, wingless, 5–7 mm. $2n = 32$.

Flowering summer–early fall. Wet meadows in coastal scrub; 0–100 m; Calif.

This showy *Veratrum* species differs from others in its fragrant flowers with deeply fringed tepals, wingless seeds, and auxiliary bulbs in the lower leaf sheaths, and is restricted as a paleo-endemic to a 96-km coastal stretch of Sonoma and Mendocino counties (C. A. Taylor 1956, 1956b).

3. Veratrum viride Aiton, Hort. Kew. 3: 422. 1789 • American false hellebore, green false hellebore, Indian poke, vérâtre verti [E] [F] [W1] [W2]

Stems 0.5–2 m, nearly glabrous to densely tomentose. **Leaves** ovate to elliptic, reduced distally, to lanceolate, 15–35 × 8–20 cm, glabrous to densely hairy, especially on abaxial surface. **Inflorescences** paniculate, with ascending to spreading or distinctly drooping (particularly in w North America) branches, 30–70 cm, tomentose; bracts lanceolate, shorter than flowers. **Tepals** deep green to yellowish, lanceolate to oblong-elliptic, narrowed to short, broad claw, 5–12 mm, tomentose adaxially, margins of both whorls or at least inner obviously erose-serrulate; gland 1, basal, dark green or yellowish green, V-shaped; ovary glabrous; pedicel 2–10 mm. **Capsules** oblong-ovoid, 2–3 cm, glabrous. **Seeds** flat, broadly winged, 8–10 mm.

Varieties 2 (2 in the flora): North America.

Veratrum viride consists of two clearly related, disjunct populations, one in eastern and one in western North America. These were clearly separated by continental glaciation and have subsequently evolved in isolation. Nonetheless they show many critical features in common, and individual plants from either region occasionally show one or more features common to plants in the other. The two populations have been variously classified as separate species, varieties, or subspecies, or as a single taxon. We have chosen to recognize two distinctive, if subtle, varieties.

1. Inflorescence branches ascending to spreading, only rarely drooping; flowers spreading to rarely erect; east of 85th meridian . . . . 3a. *Veratrum viride* var. *viride*
1. Inflorescence branches spreading to commonly drooping; flowers erect; west of 110th meridian . . . . 3b. *Veratrum viride* var. *eschscholzianum*

3a. Veratrum viride Aiton var. **viride** [E] [F]

Stems glabrous proximally, ± tomentose distally. **Leaves** 15–25 × 10–18 cm. **Inflorescences** with branches ascending to spreading, only rarely drooping. **Flowers** spreading to rarely erect; tepals deep green, 6–10 mm. $2n = 32$.

Flowering early summer–fall. Moist clearings, shaded woodlands; 0–1600 m; N.B., Nfld. and Labr., Que.; Conn., Del., Ga., Maine, Md., Mass., N.H., N.J., N.Y., N.C., Ohio, Pa., R.I., S.C., Tenn., Vt., Va., W.Va.

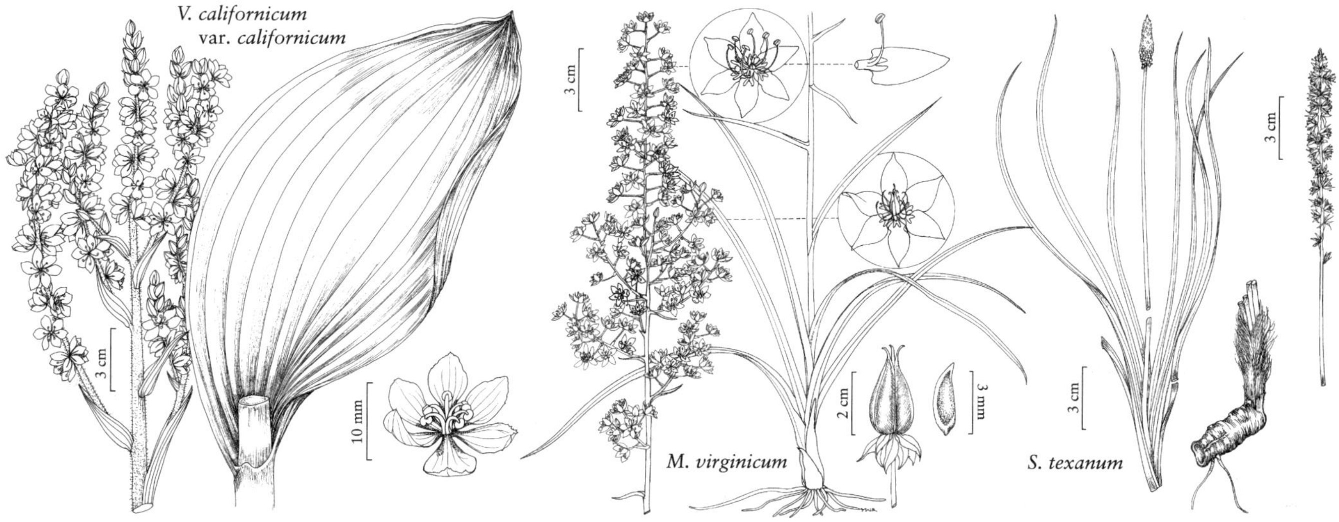

VERATRUM ◦ MELANTHIUM ◦ SCHOENOCAULON

Eastern Native Americans (Cherokee and Iroquois) used *Veratrum viride* var. *viride* as an antirheumatic and analgesic as well as a cold, skin, and orthopedic aid (D. E. Moerman 1986). Colonial settlers soaked corn seeds in an infusion of the plant to kill marauding birds (J. U. Lloyd 1897). This variety is considered a pasture weed in areas around Quebec and in the New England states (C. A. Taylor 1956).

3b. Veratrum viride Aiton var. **eschscholzianum** (Roemer & Schultes) Breitung, Canad. Field-Naturalist 71: 49. 1957 (as eschscholtzii) [E]

Veratrum lobelianum Bernhardi var. [ß] *eschscholzianum* Roemer & Schultes in J. J. Roemer et al., Syst. Veg. 7(2): 1555. 1830; *V. eschscholzianum* A. Gray (as eschscholtzii); *V. escholtzianum* Loesener; *V. eschscholtzianum* Rydberg ex A. Heller; *V. viride* Aiton subsp. *eschscholzii* (A. Gray) Á. Löve & D. Löve; *V. viride* var. *escholtzianoides* Loesener

Stems nearly glabrous proximally to densely tomentose distally. **Leaves** 15–30 × 10–18 cm. **Inflorescences** with branches spreading to commonly drooping. **Flowers** erect; tepals deep green to yellowish green, 5–12 mm. $2n = 32$.

Flowering summer–fall. Moist meadows, openings in coniferous forests; 0–2500 m; Alta., B.C., N.W.T., Yukon; Alaska, Calif., Idaho, Mont., Oreg., Wash.

Western Native Americans (Bella Colla, Cowlitz, Kwakiutl, Okanagan, Quinault, Salishan, Shuswap, and Thompson) used *Veratrum viride* var. *eschscholzianum* as an analgesic, antirheumatic, emetic, laxative, and poison, as well as a cold, blood, heart, orthopedic, and skin aid (D. E. Moerman 1986). Native Americans from northern British Columbia and the Yukon Territory consumed young plants as herbage (A. E. Porsild 1951; G. A. Mulligan and D. B. Munro 1987).

4. Veratrum californicum Durand, J. Acad. Nat. Sci. Philadelphia, ser. 2, 3: 103. 1855 • Californian false hellebore [E] [F] [W1] [W2]

Stems 1–2.5 m, ± glabrous proximally, tomentose distally. **Leaves** ovate, distalmost lanceolate to lance-linear, 20–40 × 15–25 cm, reduced distally, tomentose-ciliate, curly-hairy abaxially, glabrous or veins sparsely short-hairy adaxially. **Inflorescences** dense-paniculate, with spreading to stiffly erect branches to near tip or distal 1/3–1/2 unbranched, 30–70 cm, tomentose; bracts ovate-elliptic to lanceolate, shorter than to obviously longer than flowers. **Tepals** creamy white, greenish basally, lanceolate to elliptic or oblong-ovate, not or very slightly clawed, 8–17 mm, margins entire to denticulate, glabrous to abaxially tomentose; gland 1, basal, green, V-shaped; ovary glabrous or with few hairs; pedicel 1–6 mm. **Capsules** narrowly ovoid, 2–3 cm, glabrous. **Seeds** flat, winged, 10–12 mm.

Varieties 2 (2 in the flora): w North America.

Different geographic elements of *Veratrum californicum* have been described as separate species or varieties. The variation seems to be clinal, with most variants not consistent in their appearance or distribution. We have recognized two varieties that appear to be fairly consistent in their distributions and characteristics.

Western Native Americans (Blackfeet, Paiute, Shoshone, Thompson, and Washoe) used this species as an antirheumatic, poison, contraceptive, and emetic, as well as a skin, respiratory, blood, cold, snake bite, throat, and toothache aid (D. E. Moerman 1986).

1. Panicle branched more than 2/3 length of tip; bracts in unbranched portion ovate-elliptic, seldom exceeding flowers .... 4a. *Veratrum californicum* var. *californicum*
1. Panicle unbranched in distal 1/3–1/2; bracts in proximal unbranched portion lanceolate, frequently 2–3 times longer than flowers .... 4b. *Veratrum californicum* var. *caudatum*

4a. Veratrum californicum Durand var. **californicum** [E] [F]

Veratrum eschscholtzii A. Gray var. *watsonii* Baker; *V. jonesii* A. Heller; *V. speciosum* Rydberg; *V. tenuipetalum* A. Heller

Panicle usually branched nearly to tip, ± loosely flowered; bracts in unbranched portion ovate-elliptic, mostly less than 1 cm, shorter than flowers. **Tepals** elliptic to oblong-ovate, 8–15 mm. $2n = 32$.

Flowering summer–early fall. Moist to wet, marshy sites; 1000–3500 m; Ariz., Calif., Colo., Idaho, Nev., N.Mex., Oreg., Utah, Wash., Wyo.

Smaller-flowered plants from Colorado and New Mexico have been called *Veratrum tenuipetalum*; except for flower size, which is at the lower end of the species range, these plants are congruent with *V. californicum* var. *californicum*.

N. Chiariello et al. (1980), using the snow-bank-emergent characteristics of *Veratrum californicum,* documented that the plants differing in open versus closed bud morphologies also have different leaf areas, internal temperatures, and rates of early expansion growth. These differences in life-history characteristics are inferentially similar to those of other alpine and northern *Veratrum* species.

4b. Veratrum californicum Durand var. **caudatum** (A. Heller) C. L. Hitchcock in C. L. Hitchcock et al., Vasc. Pl. Pacif. N.W. 1: 809. 1969 [E]

Veratrum caudatum A. Heller, Bull. Torrey Bot. Club 26: 588. 1899

Panicle unbranched in distal 1/3–1/2, densely flowered; bracts in proximal unbranched portion lanceolate, 2–3 times longer than flowers, 2–3 cm. **Tepals** lanceolate, 12–17 mm.

Flowering summer–early fall. Moist to wet marshy meadows and coniferous forest openings; 0–1000 m; Calif., Idaho, Oreg., Wash.

Veratrum californicum var. *caudatum* occurs mostly west of the Cascade Mountains in the Pacific Northwest.

5. Veratrum album Linnaeus, Sp. Pl. 2: 1044. 1753 • White false hellebore

Varieties 2 (1 in the flora): nw Alaska, Eurasia.

This is the only *Veratrum* species to occur in Europe, Asia, and North America, and it is probably more widespread in northwestern Alaska than reported. *Veratrum album* is highly variable in size, flower color, pedicel length, and hairiness, and it has an extensive synonymy in Eurasia. Within the species at large, a polyploid series that includes 4*x*, 9*x*, 10*x*, and 12*x* accompanies the complex variability among constituent populations (M. N. Tamura 1998).

5a. Veratrum album Linnaeus var. **oxysepalum** (Turczaninow) Miyabe & Kudô in K. Miyabe and T. Miyake, Fl. Saghalin, 484. 1915

Veratrum oxysepalum Turczaninow, Bull. Soc. Imp. Naturalistes Moscou 13: 79. 1840

Stems 0.3–1 m, glabrous proximally, villous in inflorescence. **Leaves** basal and proximal cauline ovate to elliptic, 10–25 × 5–18 cm, reduced distally, glabrous to sparsely villous at least on veins. **Inflorescences** spiciform-racemose or with few, short, ascending branches proximally, 0–10 cm, sparsely villous; bracts lanceolate, 1/2 as long as to equaling flowers, sparsely villous, ciliate. **Tepals** yellowish green, lanceolate to lance-ovate, not clawed, 7–12 mm, margins denticulate, sparsely villous at least in proximal 1/2; gland 1, basal, green to yellowish green, V-shaped; ovary subglabrous. **Capsules** elliptic to oblong, 1–1.5 cm, subglabrous. **Seeds** flat, broadly winged, 5–7 mm. $2n = 64$.

Flowering early summer–late fall. Wet, marshy soils in mesic tundra and *Salix* thickets; 0–1000 m; Alaska; ne Russia, n Japan (Hokkaido).

11. MELANTHIUM Linnaeus, Sp. Pl. 1: 339. 1753; Gen. Pl. ed. 5, 157. 1754 • Bunch-flower [Greek *melas*, black, and *anthos*, flower, alluding to the black perianth in some species] [E]

Norlyn L. Bodkin

Frederick H. Utech

Herbs, perennial, subscapose, from erect, tunicate bulbs with short, vertical rhizomes; fibrils persistent, brown, stiff; roots contractile, fleshy. **Stems** erect, simple, leafy, hollow, basally thickened. **Leaves** basal, simple, reduced upwards, bases sheathing; blade linear, oblanceolate, elliptic to obovate, rarely plicate, glabrous, basally canaliculate, apex tapered to acute; sheaths tubular, closed. **Inflorescences** terminal, compound-racemose to open-paniculate, flexible, axes floccose; bracts clasping, linear to subulate. **Flowers** of distal racemes bisexual, progressively smaller distally, proximal flowers staminate; tepals persistent, 6, spreading, distinct to weakly connate basally, slightly adnate to ovary base, petaloid, gradually to abruptly narrowed into claws or claws absent, subequal, margin entire or undulate, apex acute to obtuse; perigonal nectaries obscure to very prominent, 2 per tepal, adaxially basal on either side of midvein; stamens 6, adnate to tepal bases; filaments strongly incurving; anthers basifixed, 1-locular, cordate-reniform; pollen sacs fugacious, confluent, dehiscence apical/valvate; ovary superior to partly inferior, nearly distinct, 3-locular; sepal nectaries absent; styles persistent, 3, spreading to recurved, turning inwards with age, distinct; stigmas minute; pedicel pubescent. **Fruits** capsular, deeply 3-lobed, dehiscence septicidal, then adaxially loculicidal; styles persistent, 3, beaklike. **Seeds** pale yellow to tan, broadly winged, flat, elliptic to lanceolate. $x = 8$.

Species 4 (4 in the flora): e North America.

Melanthium, often wholly or partly included in the closely related *Veratrum* (J. D. Ambrose 1975, 1980; S. M. Kupchan et al. 1961; J. H. Zimmerman 1958; W. B. Zomlefer 1997b; W. B. Zomlefer et al. 2001), is treated here as distinct (N. L. Bodkin 1978; B. Mathew 1989; M. N. Tamura 1998; S. Fuse and M. N. Tamura 2000). The eastern North American species of *Melanthium* differ from *Veratrum* primarily in their open inflorescences, which appear delicate and flexible; glabrous leaves; clawed tepals with two distinct, nonmarginal glands on either side of the midrib; and incurving, fugacious stamens adnate to the tepals (N. L. Bodkin 1978).

Resolution of the problematical *Melanthium–Veratrum* species series will require full molecular and phylogenetic analysis of the group, especially of the narrow-leaved Asian *Veratrum*, several of which approximate *Melanthium*, such as *V. mengtzeanum* Loesener, *V. micranthum* Wang & Tang, *V. stenophyllum* Diels, and *V. taliense* Loesener f. (B. Mathew 1989).

In eastern North America, an ecological replacement gradient occurs within the genus. *Melanthium virginicum* occurs mostly in wet habitats at lower elevations, while the Appalachian *M. latifolium* and the Ozark *M. woodii* commonly grow on mesic slopes, and *M. parviflorum* is most frequently found on or near mountain crests.

Species of *Melanthium* should be considered poisonous, since complex alkaloids derived from steroidal precursors, similar to those in *Veratrum*, are present (S. M. Kupchan et al. 1961; R. Hegnauer 1962+, vol. 2; R. M. T. Dahlgren et al. 1985).

SELECTED REFERENCES Bodkin, N. L. 1978. A Revision of North American *Melanthium* L. (Liliaceae). Ph.D. dissertation. University of Maryland. Leinfellner, W. 1961. Zur Kenntnis des Monokotyledonen — Perigons. III. Die Perigonblatter einiger weiterer Melanthioideen (*Melanthium, Zygadenus, Anticlea, Toxicoscordion, Veratrum* und *Kreysigia*). Osterr. Bot. Z. 108: 108–210. Zimmerman, J. H. 1958. A Monograph of *Veratrum*. Ph.D. dissertation. University of Wisconsin.

1. Leaves linear to narrowly oblanceolate; tepal bases clawed or abruptly attenuate.
    2. Tepals ovate to obovate-oblong, apex acute, margins entire, abruptly auriculate to hastate basally; stamens inserted at or above middle of claw . . . . 1. *Melanthium virginicum*
    2. Tepals orbiculate to rhombic, apex shortly acuminate, margins strongly undulate, abruptly attenuate basally; stamens inserted at or below middle of claw . . . . 2. *Melanthium latifolium*
1. Leaves broadly oblanceolate to obovate; tepal bases very gradually attenuate.
    3. Tepals rhombic-oblanceolate, greenish yellow to olive green; tepal glands obscure; ovary glabrous; filament bases barely 1/3 as wide as tepals at insertion point . . . . 3. *Melanthium parviflorum*
    3. Tepals obovate-oblanceolate, purplish maroon to chocolate brown; tepal glands dark purple to black, obcuneate; ovary finely tomentose; filament bases 1/2 or more as wide as tepals at insertion point . . . . 4. *Melanthium woodii*

1. Melanthium virginicum Linnaeus, Sp. Pl. 1: 339. 1753 • Virginia bunch-flower [E] [F]

Melanthium dispersum Small; *Veratrum virginicum* (Linnaeus) Aiton f.

Rhizomes 0.5–2 × 0.5–1.8 cm; bulbs 0.6–2.5 cm. **Stems** 1/4–1/3 floriferous, 0.6–2 m. **Leaf blades** linear, long-attenuate, 30–80 × 0.6–3.2 cm, apex acute. **Inflorescences** 1.3–8 dm; terminal raceme 1–2.6 dm; secondary racemes spreading to ascending, 0.2–1.1(–2.2) dm; tertiary racemes infrequent; bracts obovate to subulate, green to brown, 2–4(–6) mm, densely floccose abaxially and marginally. **Tepals** greenish yellow, aging dark reddish purple, ovate to obovate-oblong, 5.5–13 × 2.2–6 mm, base abruptly auriculate to hastate, clawed, margins entire, apex acute, claws 0.5–1.2(–2.5) mm; glands yellowish green, oblong to oblong-ovate, nectariferous; stamens 6.5–8.5 mm; filaments inserted at or above middle of claw, 0.7–2.5 mm from ovary base, not basally dilated, equaling tepal width at insertion point; anthers 0.9 mm; ovary ovoid, glabrous or pubescent; styles 1.5–3.5 mm; pedicel spreading to ascending, 5–20 mm. **Capsules** elliptic-ovoid, 10–18 × 6–10 mm, glabrous or glabrate. **Seeds** 5–8 × 2.5–4 mm (including wings). $2n = 16$.

Flowering late spring–summer. Bogs, marshes, wet woods, savannas, meadows, along railroads; 10–800 m; Ala., Ark., Del., D.C., Fla., Ga., Ill., Ind., Iowa, Kans., Ky., La., Md., Miss., Mo., N.J., N.Y., N.C., Ohio, Okla., Pa., S.C., Tenn., Tex., Va., W.Va.

The clawed tepals of *Melanthium virginicum* distinguish it from *M. parviflorum* and *M. woodii*, which have tepals with gradually attenuate bases.

2. Melanthium latifolium Desrousseaux in J. Lamarck et al., Encycl. 4: 25. 1797 • Crisped bunch-flower, slender bunch-flower [E]

Melanthium racemosum Michaux

Rhizomes 0.5–1.5 × 0.5–1 cm; bulbs 0.6–1.7 cm. **Stems** 1/4–1/2 floriferous, 0.5–1.6 m. **Leaf blades** narrowly oblanceolate, 25–55 × 1–7.2 cm, apex acute. **Inflorescences** 1.7–7 dm; terminal racemes 0.4–3.9 dm; secondary racemes spreading to ascending, 0.7–1.8(–2) dm; tertiary racemes common; bracts ovate to subulate, green to purplish brown, 2–6 mm, densely floccose proximally and marginally. **Tepals** greenish white to light yellow, aging green, orbiculate to rhombic, 3–8 × 2–4.8 mm, base abruptly attenuate, clawed, margins strongly undulate, basal 2/5 linear, involute, apex shortly acuminate; claws 0.3–6.9 mm; glands yellow, oblong to oblong-ovate, nectariferous; stamens 4–5.5 mm; filaments inserted at or below middle of claw, 0.4–1 mm from ovary base, not basally dilated, equaling tepal width at insertion point; anthers 0.7 mm; ovary ovoid, glabrous or pubescent; styles 1.7–3 mm; pedicel spreading to ascending, 7–17 mm. **Capsules** ellipsoid-oblong, 12–19 × 7–12 mm, glabrous or glabrate. **Seeds** 7.5–9 × 4–4.5 mm (including wings). $2n = 16$.

Flowering summer. Mesic to dry, rocky, wooded slopes; 300–1700 m; Ark., Conn., Del., D.C., Ga., Md., N.J., N.Y., N.C., Pa., S.C., Tenn., Va., W.Va.

Traditionally, *Melanthium latifolium* has been known by the misapplied name *M. hybridum* Walter (N. L. Bodkin 1979).

3. **Melanthium parviflorum** (Michaux) S. Watson, Proc. Amer. Acad. Arts 14: 276. 1879 • Appalachian bunch-flower [E]

Veratrum parviflorum Michaux, Fl. Bor.-Amer. 2: 250. 1803

Rhizomes 0.6–3 × 0.5–1.2 cm; bulbs 1–1.6 cm. **Stems** $^{2}/_{3}$–$^{4}/_{5}$ floriferous, 0.5–1.5 m. **Leaf blades** broadly oblanceolate to obovate, slightly plicate, 17–35 × 4.5–12.5 cm, apex obtuse to submucronate. **Inflorescences** 2.5–10 dm; terminal raceme 1–4.4 dm; secondary racemes spreading to occasionally perpendicular, 0.6–2.2(–2.4) dm; tertiary racemes rare, sometimes replacing first flowers on lower secondaries; bracts subulate, dark green, 2–4 mm, densely floccose abaxially, basally, and marginally. **Tepals** greenish yellow to olive green, aging to light green, narrowly rhombic to oblanceolate, 5–9 × 1.3–3 mm, base gradually attenuate to a narrow stalk, not clawed, margins entire, apex acute to acuminate; glands obscure, whitened in age, not nectariferous; stamens 3–4.5 mm; filaments inserted on tepal bases, 0.4 mm from ovary base, slightly dilated basally, barely $^{1}/_{3}$ as wide as tepals at insertion point; anthers 0.5 mm; ovary ovoid, glabrous; styles 1–1.5 mm; pedicel spreading to nearly perpendicular, 5.5–11 mm. **Capsules** elliptic-oblong, 10–18 × 7–10 mm, glabrous. **Seeds** 7.5–10 × 3.5–4.5 mm (including wings). $2n$ = 16.

Flowering summer. Rich, moist, deciduous woods on mountain slopes and crests; 800–2000 m; Ala., Ga., Ky., N.C., S.C., Tenn., Va., W.Va.

4. **Melanthium woodii** (Robbins ex Alph. Wood) Bodkin, Novon 8: 332. 1998 • Ozark bunch-flower, Wood's bunch-flower [E]

Veratrum woodii Robbins ex Alph. Wood, Class-book Bot. ed. 2, 557. 1847; *V. intermedium* Chapman

Rhizomes 1.5–2 × 0.5–1 cm; bulbs 1–1.6 cm. **Stems** $^{1}/_{2}$–$^{2}/_{3}$ floriferous, 0.7–1.5(–2) m. **Leaf blades** elliptic to oblanceolate, 20–50 × 3–10 cm, apex acute. **Inflorescences** 3–6 dm; terminal raceme 1–4 dm; secondary racemes spreading to ascending, 0.6–1.2(–2.2) dm; tertiary racemes rare; bracts lanceolate to subulate, brownish green, 2–6(–9) mm, floccose abaxially, proximally, and marginally. **Tepals** purplish maroon to chocolate brown adaxially, green abaxially, aging to dark green, obovate-oblanceolate, 5–10 × 1.8–4 mm, base gradually attenuate, not clawed, margins entire, apex obtuse; glands dark purple to nearly black, obcuneate, not nectariferous; stamens 4.5–6.5 mm; filaments adnate to tepal bases, 0.3 mm from ovary base, dilated basally, $^{1}/_{2}$ or more as wide as tepals at insertion point; anthers 0.7 mm; ovary conic-ovoid, finely tomentose, becoming glabrate; styles 1.6–4 mm; pedicel spreading to ascending, 2–11 mm. **Capsules** elliptic-oblong, 18–25 × 10–15 mm, sparsely pubescent. **Seeds** 8–13 × 4–6 mm (including wings). $2n$ = 16.

Flowering summer. Rich, moist, deciduous forests; 10–800 m; Ark., Fla., Ga., Ill., Ind., Iowa, Ky., Mo., N.C., Ohio, Okla., Tenn.

The purplish maroon to chocolate-brown tepals and the tomentose young ovaries are distinguishing characteristics of *Melanthium woodii*. In parts of Illinois and the Ozark regions of Missouri and Arkansas, this species sometimes is locally common, but often in the leafy sterile condition. Flowering is sporadic and appears to be promoted by fire (J. E. Ebinger 1993).

12. SCHOENOCAULON A. Gray, Ann. Lyceum Nat. Hist. New York 4: 127. 1837 • [Greek *schoenos*, rush, and *caulos*, stem]

Dawn Frame

Asagraea Lindley; *Sabadilla* Brandt & Ratzeburg; *Skoinolon* Rafinesque

Herbs, perennial, scapose, from tunicate bulb-rhizomes. **Leaves** narrow, grasslike, emerging conduplicate from neck of fibrous tunic. **Inflorescences** racemose, simple; racemes spikelike, flowers subtended by a persistent, membranous bract; distalmost bracts empty or concealing rudimentary flowers. **Flowers:** proximalmost bisexual, distalmost functionally staminate; tepals withering-persistent, 6, distinct, equal, leathery to slightly petaloid, yellowish green, green, or greenish purple, not clawed, linear-lanceolate to ovate, margins entire to crenulate or minutely denticulate; hyaline basal auricle sometimes present; perigonal nectaries not well developed; stamens 6, distinct; filaments withering-persistent, subulate or rarely distally dilated, 1.5–2.5 times tepal length; anthers basifixed, narrowly sagittate before dehiscence, explanate after, extrose; ovary superior, 3-locular, locules barely cohesive to connate; styles persistent, 3, distinct. **Fruits** capsular, 1–3-locular, dehiscence septicidal. **Seeds** 1–8 per locule, ellipsoid to ovoid or conic, irregularly compressed or angled by mutual pressure. $x = 8$.

Species 24 (3 in the flora): Western Hemisphere, especially Mexico.

Schoenocaulon has rarely been collected in the United States in recent times; all three species are possibly threatened. The genus is a source of cerveratrum alkaloids.

SELECTED REFERENCES Brinker, R. 1942. Monograph of *Schoenocaulon*. Ann. Missouri Bot. Gard. 29: 287–309. Frame, D. 1989. *Schoenocaulon*. In: R. McVaugh and W. R. Anderson. 1974+. Flora Novo-Galiciana: A Descriptive Account of the Vascular Plants of Western Mexico. 8+ volumes. Ann Arbor. Vol. 15, pp. 263–275. Frame, D. 2001. Chromosome studies in *Schoenocaulon* (Liliaceae: Melanthieae) a relict genus. Anales Inst. Biol. Univ. Nac. Auton. Mexico, Bot. 72: 123–129. Frame, D., A. Espejo, and A. R. López-Ferrari. 1999. A conspectus of Mexican Melanthiaceae including a description of new taxa of *Schoenocaulon* and *Zigadenus*. Acta. Bot. Mex. 48: 27–50.

1. Racemes loose, flowers well separated along axis; Florida . . . . 1. *Schoenocaulon dubium*
1. Racemes condensed, flowers closely packed along axis; New Mexico, Texas.
    2. Tepal margins crenulate to minutely denticulate . . . . 2. *Schoenocaulon ghiesbreghtii*
    2. Tepal margins entire . . . . 3. *Schoenocaulon texanum*

1. Schoenocaulon dubium (Michaux) Small, Fl. S.E. U.S., 250. 1903 E

Helonias dubia Michaux, Fl. Bor.-Amer. 1: 213. 1803; *Sabadilla dubia* (Michaux) Kuntze; *Schoenocaulon gracile* A. Gray

Bulbs ellipsoid to ovoid, 2–4.2 × 1.2–3.5 cm diam.; rhizome generally unnoticeable. **Leaves** 4–7, 14–75 cm × 1.5–3.5 mm, shorter than raceme. **Scape** 55–87 cm. **Racemes** loose, 55–220-flowered, 11.5–35 cm, flowers well separated along axis. **Flowers:** tepals green, elliptic to narrowly ovate or obpandurate, 2–3.2 mm, fleshy, margins entire; auricles sometimes present; filaments subulate, 1.5 times tepal length. **Capsules** 2–3-locular, ovoid, 7.5–11 × 3.2–5.5 mm. **Seeds** 1–3 per locule, 4–6.3 × 1.3–1.8 mm. $2n = 16$.

Flowering spring–early summer. Sandy, moderately to well-drained soils in dry pineland and sandhills; 0–100 m; Fla.

2. Schoenocaulon ghiesbreghtii Greenman, Proc. Amer. Acad. Arts 43: 20. 1907 • Green-lily

Sabadilla drummondii Kuntze; *Schoenocaulon yucatanese* Brinker

Bulbs narrowly ovoid to globose, 2.5–6 × 2–5.5 cm diam.; rhizome short, usually appressed to side of bulb. **Leaves** 6–11, 20–75 cm × 2.8–11 mm, usually shorter than or equaling raceme. **Scape** 21–93 cm. **Racemes** condensed, 45–180-flowered (–300-flowered, if cultivated), 5–23 cm, flowers closely packed along axis. **Flowers:** tepals linear, ovate, or obovate, 2.3–4.7 mm, slightly fleshy, margins crenulate to minutely denticulate; auricles absent; filaments often distally dilated, 2–2.5 times tepal length. **Capsules** 2–3-locular, ellipsoid to plumply ovoid, 8.5–13.5 × 4–7 mm. **Seeds** 1–4 per locule, 3–5 × 1.5–2 mm. $2n = 16$.

Flowering late summer–autumn. Locally abundant in sandy to moderately rich soils; 0–100 m; Tex.; Mexico.

In the United States, *Schoenocaulon ghiesbreghtii* grows at or near sea level and almost exclusively on the Reyosa gravel formation of the Texas Gulf prairie. In the late 1930s and early 1940s, it was used in a series of transplantation experiments conducted by the Texas Agricultural Experiment Stations. This has led to some confusion with regard to its natural distribution.

3. Schoenocaulon texanum Scheele, Linnaea 25: 262. 1852 [F]

Bulbs ellipsoid to plumply ovoid, 2–4.5 × 1.2–4 cm diam., usually arranged either in a circle or semicircle around a common rhizome, or one above the other on a vertical rhizome. **Leaves** 5–8, 15–60 cm × 2–6 mm, shorter than raceme to just surpassing it, rarely longer. **Scape** 13.5–55 cm. **Racemes** condensed, 30–250-flowered, 4–32 cm, flowers closely packed along axis. **Flowers:** tepals linear to cuneate proximally and broadly subulate distally, 2.3–3.8 mm, fleshy, margins entire or rarely irregularly crenate; auricles distinct or absent; filaments subulate, 2 times tepal length. **Capsules** 1–3-locular, ellipsoid to plumply ovoid, 0.8–1.9 cm × 5–12 mm. **Seeds** 2–4(–8) per locule, 3–6.7 × 1.6–2.2 mm. 2*n* = 16.

Flowering spring–early summer, sometimes later following unseasonable rainfall. Occasional in dry, rocky, limestone soils in chaparrals, usually growing under scrubby vegetation; 0–2200 m; N.Mex., Tex.; Mexico (Chihuahua, Coahuila, Durango, San Luis Potosí).

The morphology of *Schoenocaulon texanum* is peculiarly dependent upon annual rainfall. Plants of small stature but average bulb size can be readily distinguished as the products of one or several seasons of lower than normal rainfall. The hallmarks of such plants are congested racemes, few flowers, these often appearing withered and brown, and ovaries that do not mature or have perhaps only one small locule ripening to produce a few seeds.

13. ZIGADENUS Michaux, Fl. Bor.-Amer. 1: 213, plate 22. 1803 • Death camas, zigadène [Greek *zygos,* yoke, and *aden,* gland, referring to the pair of glands on each tepal of the type species]

Fayla C. Schwartz

Anticlea Kunth; *Chitonia* Salisbury; *Cyanotris* Rafinesque; *Oceanoros* Small; *Toxicoscordion* Rydberg; *Tracyanthus* Small

Herbs, perennial, subscapose, bulbous or rhizomatous, all parts containing toxic alkaloids; roots contractile. **Leaves** usually fewer than 10, mostly basal, alternate, simple, distal ones reduced and grading into floral bracts, sometimes completely sheathing at the base; blade with venation parallel, linear, glabrous, margins entire. **Inflorescences** terminal, racemose, corymbose, or paniculate, 5–125-flowered, glabrous. **Flowers** usually bisexual, but often staminate on branches in panicles, protandrous, pedicellate; perianth hypogynous or partly perigynous, white or cream colored to greenish or red tinged, actinomorphic, 0.5–2 cm diam.; tepals 6, often connate basally, each bearing 1 or 2, sometimes obscure to absent, yellow to green, adaxially basal nectary glands; stamens 6, anthers versatile, reniform, dehiscence transverse, extrorse or rarely introrse (*Z. glaberrimus*); ovary partly inferior; pistil 1, with 3 styles, tapering from ovary; stigmas 3, barely larger than diameter of style, rudimentary in staminate flowers; pedicel ascending to spreading, bracteate, bracts 2, green to red and foliaceous or white and membranous. **Fruits** capsular, with 3 ellipsoid, beaked sections; dehiscence septicidal, then adaxially loculicidal. **Seeds** 2–20 per capsule, elongate, twisted and flattened at ends. x = 8, 11, 26.

Species ca. 18–22 (14 in the flora): North America including Mexico, Central America, e Asia (Japan, e Siberia).

The generic name was spelled *Zigadenus* by Michaux, but subsequently was often rendered as "Zygadenus," presumably to reflect the Greek word roots more clearly. However, the original spelling has priority over that orthographic variant (W. B. Zomlefer 1997b).

As broadly circumscribed here, *Zigadenus* is polyphyletic (W. B. Zomlefer 1997b; W. B. Zomlefer et al. 2001) and this complex has had a complicated taxonomic and nomenclatural history (R. R. Gates 1918; O. S. Walsh 1940; S. J. Preece 1956; F. C. Schwartz 1994; W. B. Zomlefer 1997b). Numerous, variously defined segregates have been recognized, including *Anticlea*, *Chitonia*, *Cyanotris*, *Oceanoros*, *Toxicoscordion*, and *Tracyanthus*. *Zigadenus* shows morphological similarities to both *Stenanthium* (including *Stenanthella* Rydberg) and *Amianthium*.

The common name "death camas" is derived from the illness or death resulting from ingesting bulbs of *Zigadenus* that were mistaken for the edible ones of *Camassia*, an unrelated plant. Many alkaloids found in *Veratrum* have been isolated from most species of *Zigadenus* (W. B. Zomlefer 1997b; G. E. Burrows and R. J. Tyrl 2001), including zygadenine and zygacine (T. J. Gilbertson 1973; J. B. Harborne and H. Baxter 1993). All fresh parts of the plant (leaves, bulbs, flowers) are toxic. Dried parts (especially seeds and capsules) are even more potent, presumably because the alkaloids are more concentrated. Ingestion of *Zigadenus* by humans may result in severe illness and occasionally death even for adults. Livestock (sheep and cattle) poisoning is a serious problem in some rangeland areas of the western United States (C. D. Marsh and A. B. Clawson 1922; V. J. Tepidino 1982). Native Americans used these plants for their analgesic, antirheumatic, and emetic properties, as dermatological and orthopedic aids, and as a snake-bite remedy (D. E. Moerman 1986).

SELECTED REFERENCES Frame, D. M. 1999. A conspectus of Mexican Melanthiaceae including a description of new taxa of *Schoenocaulon* and *Zigadenus*. Acta Bot. Mex. 48: 27–50. Hess, W. J. and R. C. Sivinski. 1995. A new species of *Zigadenus* (Liliaceae) from New Mexico, with additional comments on the section *Anticlea*. Sida 16: 389–400. Preece, S. J. 1956. A Cytotaxonomic Study of the Genus *Zigadenus* (Liliaceae). Ph.D. dissertation. State College of Washington [Washington State University]. Schwartz, F. C. 1994. Molecular Systematics of *Zigadenus* Section *Chitonia* (Liliaceae). Ph.D. dissertation. University of Washington. Walsh, O. S. 1940. A Systematic Study of the Genus *Zigadenus* Michx. Ph.D. dissertation. University of California, Berkeley.

1. Plants rhizomatous; tepals with 2 glands .... 1. *Zigadenus glaberrimus*
1. Plants bulbose; tepals with 1, sometimes obscure, gland.
    2. Perianth partly perigynous, tepal gland obcordate.
        3. Perianth rotate to rotate-campanulate; pedicel erect or ascending at anthesis; tepals white or cream colored to greenish.
            4. Proximal stem with persistent leaf bases; tepals white .... 12. *Zigadenus vaginatus*
            4. Proximal stem without persistent leaf bases; tepals cream colored to greenish .... 11. *Zigadenus elegans*
        3. Perianth campanulate; pedicel recurved at anthesis; tepals greenish, sometimes tinged with purple.
            5. Tepals 4–6 mm .... 13. *Zigadenus virescens*
            5. Tepals 12–16 mm .... 14. *Zigadenus mogollonensis*
    2. Perianth hypogynous, tepal gland obovate or obscure.
        6. Tepal gland obscure; tepals narrowed at base but not sharply contracted .... 2. *Zigadenus densus*
        6. Tepal gland prominent; inner and sometimes outer tepals strongly contracted at base into claw.
            7. Bracts less than 1/2 length of corresponding pedicel; filaments thin and curved; capsules rounded, ca. 1/2 as wide as long .... 9. *Zigadenus brevibracteatus*
            7. Bracts 1/3–3 times length of corresponding pedicel; filaments thickened proximally and straight; capsules elongate, usually less than 1/2 as wide as long.
                8. Bulbs often clumped; outer tepals often persistently reflexed below mature capsule; Sierra Nevada foothills .... 8. *Zigadenus exaltatus*
                8. Bulbs not clumped; outer tepals not persistently reflexed below mature capsule; elsewhere than Sierra Nevada foothills.
                    9. Inflorescences broad; perianth rotate .... 7. *Zigadenus fremontii*
                    9. Inflorescences narrow, often pyramidal in early anthesis; perianth ± campanulate.

10. Filaments shorter than tepals . . . . 5. *Zigadenus fontanus*
10. Filaments equaling or longer than tepals.
    11. Distal margins of tepal gland obscure; outer tepals obtuse at apex . . . . 10. *Zigadenus nuttallii*
    11. Distal margins of tepal gland usually evident and often toothed; outer tepals often acute at apex.
        12. Proximal branches of inflorescences diverging from main axis at 60°–90° when in fruit . . . . 4. *Zigadenus micranthus*
        12. Proximal branches of inflorescences diverging from main axis at 10°–60° when in fruit.
            13. Inflorescences paniculate; outer tepals usually not clawed . . . . 6. *Zigadenus paniculatus*
            13. Inflorescences racemose, or paniculate with only 1 or 2 branches; outer tepals often clawed . . . . 3. *Zigadenus venenosus*

1. Zigadenus glaberrimus Michaux, Fl. Bor.-Amer. 1: 214, plate 22. 1803 • Sandbog death camas [E] [F]

Plants 6–12 dm, from thick, twisted rhizomes. **Leaves:** blade sometimes becoming reddish brown with age, 19–42 cm × 5–20 mm, glaucous abaxially. **Inflorescences** loosely paniculate, 30–75-flowered, with 2–6 ascending branches, pyramidal, 10–60 × 5–12 cm. **Flowers:** perianth hypogynous, 20–30 mm diam.; tepals persistent in fruit, white to cream colored, ovate to lanceolate, 9–15 × 3–5 mm, somewhat contracted at base; glands 2, elliptic, margins thickened; pedicel bracts ovate, 5–15 mm. **Capsules** conic, 10–15 mm. $2n = 52$.

Flowering mid Jul–Sep. Pine bogs, savannas, sandy pinelands; 0–100 m; Ala., Fla., Ga., Miss., N.C., S.C., Va.

2. Zigadenus densus (Desrousseaux) Fernald, Rhodora 42: 254. 1940 • Osceola's plume, black snakeroot, crow poison, pine-barren death camas [E] [F]

Melanthium densum Desrousseaux in J. Lamarck et al., Encycl. 4: 26. 1796; *Amianthium angustifolium* (Michaux) A. Gray; *A. texanum* (Bush) R. R. Gates; *Helonias angustifolia* Michaux; *Oceanoros leimanthoides* (A. Gray) Small; *Tracyanthus angustifolius* (Michaux) Small; *T. angustifolius* var. *texanus* Bush; *T. texanus* (Bush) Small; *Zigadenus angustifolius* (Michaux) S. Watson; *Z. leimanthoides* A. Gray

Plants 4–20 dm, from bulbs; bulbs not clumped, tunicate, ovoid, 0.5–2 × 1–2 cm. **Leaves:** proximal blades 10–50 cm × 2–12 mm. **Inflorescences** racemose or paniculate, 40–100-flowered, cylindrical, 4–15 × 2.5–5 cm. **Flowers:** perianth hypogynous, 5–10 mm diam.; tepals persistent in fruit, cream colored to greenish, ovate to elliptic, 3–5 mm, narrowed but not sharply contracted basally; gland 1, obscure; pedicel 1–2 cm, bracts often tinged with red, lanceolate, 2–12 mm. **Capsules** narrowly conic, 10–20 mm.

Flowering mid Mar–Jul. Pine bogs, flatlands; 0–1800 m; Ala., Del., Fla., Ga., Ky., La., Miss., N.J., N.Y., N.C., S.C., Tenn., Tex., Va., W.Va.

As treated here, *Zigadenus densus* includes *Z. leimanthoides* (W. W. McDearman 1984). These two entities are distinct in some parts of the range, but intergrade in size, inflorescence morphology, and flowering time in the Gulf coast region. The size of the plant (determined by age and environment) seems to affect inflorecence morphology, smaller plants bearing simple racemes and larger plants developing compound ones. Plants in more mountainous habitats develop paniculate inflorescences. Recognition of the two taxa at the varietal level may eventually be desirable.

3. Zigadenus venenosus S. Watson, Proc. Amer. Acad. Arts 14: 279. 1879 (as Zygadenus) [E] [W1] [W2]

Toxicoscordion venenosum (S. Watson) Rydberg

Plants 2–7 dm, from bulbs; bulbs not clumped, tunicate, ovoid, 10–30 × 8–20 mm. **Leaves:** proximal blades 12–50 cm × 2–10 mm. **Inflorescences** racemose or paniculate with 1 or 2 basal branches, 10–50-flowered, terminal raceme pyramidal in anthesis, 2–20 × 2–5 cm, branches (if any) $^1/_{10}$–$^1/_3$ length of entire inflorescence, diverging from main axis at 10°–60° when in fruit. **Flowers:** perianth hypogynous, campanulate, 5–10 mm diam.; tepals persistent in fruit, cream colored, 2–5 × 1–3 mm, outer often clawed, apex usually acute; gland 1, obovate, distal margins evident to obscure, irregular; filaments straight, usually equaling tepals, occasionally longer, thickened proximally; pedicel usually ascending in fruit, occasionally perpendicular to stem, 3–25 mm, bracts usually green, sometimes white, 5–25 mm. **Capsules** 8–20 × 4–7 mm. $2n = 22$.

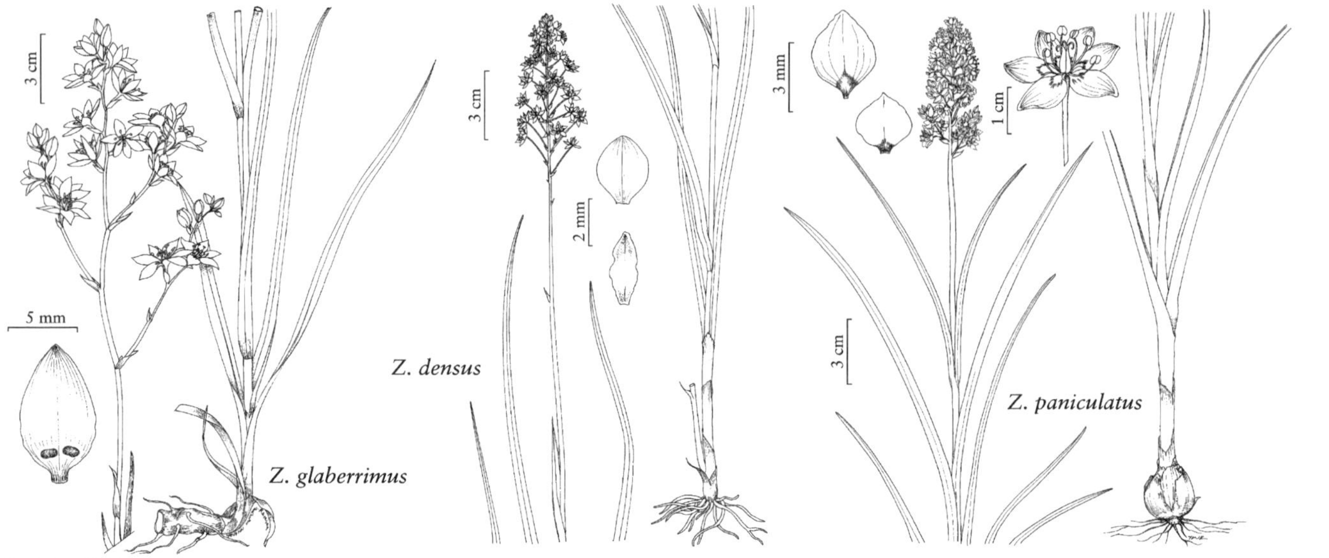

ZIGADENUS

Varieties 2 (2 in the flora): w North America.

Zigadenus venenosus is a smaller, more slender plant than either *Z. paniculatus* or *Z. fontanus*. It can be distinguished from *Z. micranthus* by its ascending pedicels. It has a much broader distribution than either *Z. micranthus* or *Z. fontanus*, and does not occur on serpentine substrates.

1. Outer tepals usually clawed ca. 5 mm; inflorescences mostly racemose, occasionally paniculate with 1 basal branch . . . . 3a. *Zigadenus venenosus* var. *venenosus*
1. Outer tepals not clawed, or clawed less than 5 mm; inflorescences mostly paniculate with 1–2 basal branches, occasionally racemose . . . . 3b. *Zigadenus venenosus* var. *gramineus*

3a. Zigadenus venenosus S. Watson var. **venenosus** • Meadow death camas E W1 W2

Toxicoscordion arenicola A. Heller; *T. salinum* (A. Nelson) R. R. Gates; *Zigadenus diegoensis* Davidson; *Z. salinus* A. Nelson; *Z. venenosus* var. *ambiguus* M. E. Jones

Inflorescences mostly racemose, occasionally paniculate with 1 basal branch, 2–20 × 2–5 cm. **Flowers:** outer tepals usually clawed to 5 mm. **Capsules** 8–12 × 4–5 mm.

Flowering Apr–Jul. Coastal areas, well-drained grasslands; 0–2500 m; B.C.; Calif., Idaho, Nev., Oreg., Utah, Wash.

The inflorescences of most plants in each population of var. *venenosus* are racemose, but those of a few individuals are paniculate, with a single short basal branch.

3b. Zigadenus venenosus S. Watson var. **gramineus** (Rydberg) O. S. Walsh ex C. L. Hitchcock in C. L. Hitchcock et al., Vasc. Pl. Pacif. N.W. 1: 815. 1969 • Grassy death camas E W1 W2

Zigadenus gramineus Rydberg, Bull. Torrey Bot. Club 27: 535. 1900 (as Zygadenus); *Toxicoscordion acutum* (Rydberg) Rydberg; *T. falcatum* (Rydberg) Rydberg; *T. gramineum* (Rydberg) Rydberg; *T. intermedium* (Rydberg) Rydberg; *Zigadenus acutus* Rydberg; *Z. falcatus* Rydberg; *Z. intermedius* Rydberg

Inflorescences mostly paniculate with 1–2 basal branches, occasionally racemose, 2–15 × 2–5 cm. **Flowers:** outer tepals not clawed, or clawed less than 5 mm. **Capsules** 9–20 × 4–7 mm.

Flowering Apr–Jul. Inland grasslands, open pine woods; 500–1300 m; Alta., B.C., Sask.; Colo., Idaho, Mont., Nebr., N.Dak., Oreg., S.Dak., Wash., Wyo.

Variety *gramineus* is most easily distinguished from var. *venenosus* by its paniculate inflorescence, although some individuals in each population are racemose. It occurs well inland, east of the Cascade and Rocky mountains and at high elevations within those ranges.

4. Zigadenus micranthus Eastwood, Bull. Torrey Bot. Club 30: 483. 1903 (as Zygadenus) • Small-flower death camas [E]

Toxicoscordion micranthum (Eastwood) A. Heller; *Zigadenus venenosus* S. Watson var. *micranthus* (Eastwood) Jepson

Plants 1–7 dm, from bulbs; bulbs not clumped, tunicate, ovoid, 10–35 × 8–30 mm. **Leaves:** proximal blades 12–45 cm × 2–10 mm. **Inflorescences** usually paniculate with 1–5 branches, occasionally racemose, 10–60-flowered, narrow, terminal raceme often pyramidal in early anthesis, 3–40 × 3–7 cm, proximal branches 2–13 cm, 1/10–1/2 length of entire inflorescence, diverging from main axis at 60°–90° when in fruit. **Flowers:** perianth hypogynous, campanulate, 5–12 mm diam.; tepals persistent in fruit, cream colored, 3–8 × 1–4 mm, outer sometimes with basal claw to 5 mm, usually with acute apex; gland 1, obovate, distal margins evident, dentate; filaments straight, ± equaling tepals or rarely longer, thickened proximally; pedicel usually perpendicular to stem in flower and in fruit, 12–40 mm, bracts white and membranous to green and foliaceous, narrowly lanceolate, 5–40 mm. **Capsules** 6–18 × 2–7 mm. $2n$ = 22.

Flowering Apr–Jul. Serpentine hillsides; 0–1000 m; Calif., Oreg.

5. Zigadenus fontanus Eastwood, Leafl. W. Bot. 2: 41. 1937 (as Zygadenus) • Small-flower death camas [E]

Zigadenus micranthus var. *fontanus* (Eastwood) O. S. Walsh ex McNeal

Plants 3–9 dm, from bulbs; bulbs not clumped, tunicate, ovoid, 12–40 × 12–40 mm. **Leaves:** proximal blades 18–80 cm × 3–20 mm. **Inflorescences** usually paniculate, 10–100-flowered, with 1–15 branches, narrow, appearing pyramidal in early anthesis, terminal raceme 5–60 × 3–7 cm, proximal branches 1/3–1/2 length of entire inflorescence, extending upwards from main axis of inflorescence at 10°–60° angle. **Flowers:** perianth hypogynous, campanulate, 5–15 mm diam.; tepals persistent in fruit, cream colored, 4–12 × 1–4 mm, outer usually lacking claw; gland 1, obovate, distal margins evident, serrate to dentate; filaments straight, shorter than tepals, thickened proximally; pedicel horizontally spreading, usually perpendicular to stem in fruit, 15–45 mm, bracts green, 5–35 mm. **Capsules** narrowly conic, 9–22 × 3–6 mm. $2n$ = 22.

Flowering late Mar–Aug. Serpentine marshes; 0–500 m; Calif.

6. Zigadenus paniculatus (Nuttall) S. Watson, Botany (Fortieth Parallel), 343. 1871 (as Zygadenus) • Foothill death camas, sand-corn [E] [F] [W1]

Helonias paniculata Nuttall, J. Acad. Nat. Sci. Philadelphia 7: 57. 1834; *Toxicoscordion paniculatum* (Nuttall) Rydberg

Plants 2–7 dm, from bulbs; bulbs not clumped, tunicate, ovoid, 18–40 × 12–35 mm. **Leaves:** proximal blades 15–35 cm × 3–15 mm. **Inflorescences** paniculate, 10–80-flowered, narrow, terminal raceme pyramidal in anthesis, 2–30 × 2–4 cm, proximal branches usually less than 6 cm, 1/10–1/2 length of entire inflorescence, diverging from main axis at 10°–60° when in fruit. **Flowers:** perianth hypogynous, campanulate, 5–10 mm diam.; tepals persistent in fruit, cream colored, ovate, 2–5 × 1–4 mm, outer usually not clawed, apex usually acute; gland 1, obovate, distal margins evident to obscure, irregular; filaments straight, usually equaling tepals, occasionally longer, thickened proximally; pedicel ascending to perpendicular to stem in fruit, 3–25 mm, bracts green, lanceolate, 5–15 mm. **Capsules** 5–20 × 3–8 mm. $2n$ = 22.

Flowering Apr–Jun. Sagebrush desert, dry open inland grasslands; 1300–2600 m; Ariz., Calif., Colo., Idaho, Mont., Nev., N.Mex., Oreg., Utah, Wash., Wyo.

7. Zigadenus fremontii (Torrey) Torrey ex S. Watson, Botany (Fortieth Parallel), 343. 1871 (as Zygadenus) • Fremont's death camas [E] [W2]

Anticlea fremontii Torrey, Pacif. Railr. Rep., 4(5): 144. 1856; *Toxicoscordion fremontii* (Torrey) Rydberg; *T. fremontii* var. *minor* R. R. Gates; *Zigadenus fremontii* var. *inezianus* Jepson; *Z. fremontii* var. *salsus* Jepson

Plants 1–8 dm, from bulbs; bulbs not clumped, tunicate, ovoid, 10–40 × 5–40 mm. **Leaves:** proximal blades 5–50 cm × 2–30 mm. **Inflorescences** racemose or paniculate, 2–80-flowered, broad, proximal branches 1/3–1/2 as long as entire inflorescence, spreading at 60°–90° angle. **Flowers:** perianth hypogynous, rotate, 15–25 mm diam.; tepals persistent in fruit, cream colored, 5–15 × 2–6 mm, outer very short clawed, inner distinctly clawed 2 mm; gland 1, obovate, distal margins evident, dentate; filaments straight, shorter than tepals, thickened proximally; pedicel ascending in flower to perpendicular to stem in fruit, 10–50 mm, bracts green, somewhat falcate, 5–50 mm. **Capsules** 8–30 × 4–11 mm. $2n$ = 22.

Flowering Feb–Jun. Coastal and inland, central valley grassland chaparral, mixed evergreen forests; 0–1000 m; Calif., Oreg; Mexico (Baja California).

8. Zigadenus exaltatus Eastwood, Bot. Gaz. 41: 283. 1906 (as Zygadenus) • Giant death camas [E]

Toxicoscordion exaltatum (Eastwood) A. Heller

Plants 3–10 dm, from bulbs; bulbs often clumped, tunicate, ovoid, 20–60 × 18–50 mm. **Leaves:** proximal blades 18–60 cm × 3–25 mm. **Inflorescences** paniculate, 10–80-flowered, with 2–10 branches, 3–40 × 2–8 cm. **Flowers:** perianth hypogynous, 12–20 mm diam.; tepals persistent in fruit, cream colored, elliptic to ovate, 5–10 × 2–4 mm, outer not clawed, often persistently reflexed below mature capsule; gland 1, obovate, distal margins evident, dentate; filaments straight, shorter than tepals, thickened proximally; pedicel ascending in flower, perpendicular to stem in fruit, 10–35 mm, bracts green and foliaceous to white and membranous, 3–25 mm. **Capsules** 9–20 × 4–9 mm. $2n = 22$.

Flowering Apr–Jun. Sierra Nevada foothill woodlands, yellow pine forests; 0–1200 m; Calif.

9. Zigadenus brevibracteatus (M. E. Jones) H. M. Hall, Univ. Calif. Publ. Bot. 6: 165. 1915 (as Zygadenus) • Desert death camas [E]

Zigadenus fremontii (Torrey) Torrey ex S. Watson var. *brevibracteatus* M. E. Jones, Contr. W. Bot. 12: 78. 1908 (as Zygadenus); *Toxicoscordion brevibracteatus* (M. E. Jones) R. R. Gates

Plants 3–6 dm, from bulbs; bulbs not clumped, tunicate, ovoid, 18–40 × 9–35 mm. **Leaves:** proximal blades 10–30 cm × 3–8 mm. **Inflorescences** usually paniculate, 20–80-flowered, with 1–8 branches, terminal raceme 10–45 × 3–8 cm. **Flowers:** perianth hypogynous, 10–20 mm diam.; tepals persistent in fruit, cream colored, elliptic, 4–7 × 2–5 mm, outer not clawed; gland 1, obovate, distal margins evident; filaments thin and curved, shorter than tepals; pedicel usually perpendicular to stem in fruit, 12–40 mm, bracts green, 2–8 mm. **Capsules** 8–16 × 5–10 mm. $2n = 22$.

Flowering Apr–Jun. Creosote bush scrub, Joshua tree woodlands, sandy areas of the Mojave and Colorado deserts; 600–1500 m; Calif.

10. Zigadenus nuttallii (A. Gray) S. Watson, Botany (Fortieth Parallel), 343. 1871 • Nuttall's death camas [E] [W2]

Amianthium nuttallii A. Gray, Ann. Lyceum Nat. Hist. New York 4: 123. 1837; *Toxicoscordion nuttallii* (A. Gray) Rydberg; *T. texense* Rydberg; *Zigadenus texensis* (Rydberg) J. F. Macbride

Plants 3–7.5 dm, from bulbs; bulbs not clumped, tunicate, ovoid, 15–40 × 9–35 mm. **Leaves:** proximal blades 15–45 cm × 3–15 mm. **Inflorescences** usually paniculate, 20–60-flowered, with 1–8 branches, narrow, terminal raceme pyramidal in early anthesis, 3–30 × 3–7 cm, proximal branches 1/10–1/2 length of entire inflorescence, ascending at 10°–60° angle. **Flowers:** perianth hypogynous, campanulate, 10–15 mm diam.; tepals persistent in fruit, cream colored, ovate, 3–8 × 1–4 mm, outer unclawed or rarely clawed to 5 mm, apex usually obtuse; gland 1, obovate, distal margins obscure, thin; filaments straight, usually equaling tepals, occasionally longer, thickened proximally; pedicel usually ascending in fruit, occasionally perpendicular to stem, 10–35 mm, bracts green, somewhat falcate, 3–20 mm. **Capsules** 8–16 × 3–8 mm. $2n = 32$.

Flowering Mar–Jun. Tallgrass prairie, calcareous glades, rocky hillsides; 500–1200 m; Ark., Kans., Mo., Okla., Tex.

11. Zigadenus elegans Pursh, Fl. Amer. Sept. 1: 241. 1814 • Mountain death camas; zigadène élégant [F]

Anticlea alpina A. Heller; *A. chlorantha* (Richardson) Rydberg; *A. coloradensis* (Rydberg) Rydberg; *A. elegans* (Pursh) Rydberg; *A. glauca* (Nuttall) Kunth; *A. gracilenta* (Greene) R. R. Gates; *A. longa* (Greene) A. Heller; *A. mohinorensis* (Greenman) R. R. Gates; *Melanthium glaucum* Nuttall; *Zigadenus alpinus* Blankinship; *Z. chloranthus* Richardson; *Z. coloradensis* Rydberg; *Z. dilatatus* Greene; *Z. elegans* subsp. *glaucus* (Nuttall) Hultén; *Z. elegans* var. *glaucus* (Nuttall) Preece ex Cronquist; *Z. glaucus* (Nuttall) Nuttall; *Z. gracilentus* Greene; *Z. longus* Greene; *Z. mohinorensis* Greenman; *Z. washakie* A. Nelson

Plants 2–8 dm, from bulbs; bulbs not clumped, tunicate, narrowly ovoid, 1.5–3 × 1–2 cm. **Stems** without persistent leaf bases. **Leaves:** proximal blades 10–30 cm × 3–15 mm. **Inflorescences** loosely racemose to paniculate, 10–50-flowered, with 1–4 branches, 0.9–2.5 dm × 3–6 cm. **Flowers:** perianth perigynous, rotate to rotate-campanulate, 15–20 mm diam.; tepals persistent in fruit, cream colored to greenish, ovate, 7–12 × 4–5 mm, somewhat narrowed at base; gland 1, obcordate; pedicel erect

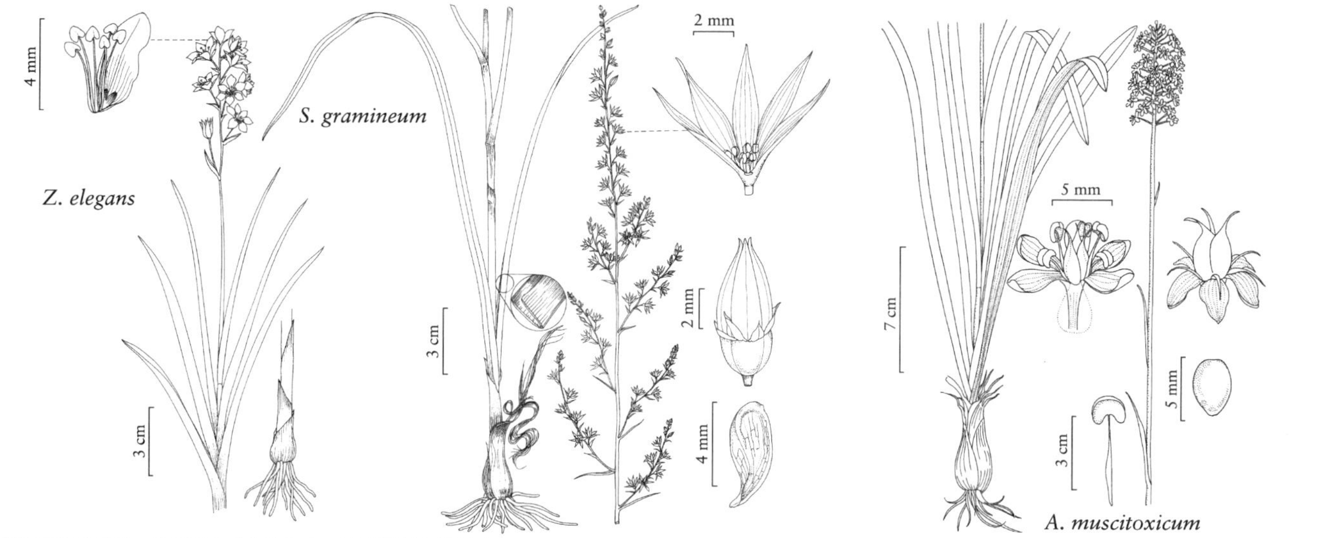

ZIGADENUS ◦ STENANTHIUM ◦ AMIANTHIUM

at anthesis, 1–2.3 cm, bracts often tinged with purple or pink, ovate, 5–20 mm. **Capsules** narrowly conic, 10–20 mm. 2*n* = 32.

Flowering Jun–Aug. Moist grasslands, river and lake shores, bogs in coniferous forests; 0–3600 m; Alta., B.C., Man., N.W.T., Ont., Que., Sask., Yukon; Alaska, Ariz., Colo., Idaho, Ill., Ind., Iowa, Md., Mich., Minn., Mo., Mont., Nebr., Nev., N.Mex., N.Y., N.C., N.Dak., Ohio, Oreg., Pa., S.Dak., Tenn., Tex., Utah, Vt., Va., Wash., W.Va., Wis., Wyo.; Mexico.

Zigadenus elegans has been treated previously as comprising two varieties, or two subspecies (W. B. Zomlefer 1997b). The western var. or subsp. *elegans* tends to be a smaller plant with a raceme or a 1–2-branched panicle and glabrous, sometimes glaucous leaves and stems; while the eastern var. or subsp. *glaucus* tends to be a larger plant with a paniculate inflorescence and glaucous leaves and stems. Because there is considerable evidence of intergradation between the two entities toward the middle of the range, including overlapping flowering times, they have not been formally distinguished here.

12. Zigadenus vaginatus (Rydberg) J. F. Macbride, Contr. Gray Herb. 53: 4. 1918 • Sheathed death camas [C] [E]

Anticlea vaginata Rydberg, Bull. Torrey Bot. Club 39: 108. 1912

Plants 4–6 dm, from bulbs; bulbs not clumped, tunicate, narrowly ovoid, 3–6 × 1.5–3 cm. **Stems** with persistent leaf bases proximally. **Leaves:** proximal blades 25–40 cm × 5–12 mm. **Inflorescences** paniculate, 15–30-flowered, with 1–4 branches, 1.5–3 dm × 3–6 cm, proximal branches 7–8 cm. **Flowers:** perianth perigynous, rotate to rotate-campanulate, 10–15 mm diam.; tepals white, ovate, 3–6 × 3–4 mm, somewhat narrowed at base, apex broadly rounded; gland 1, obcordate, thick; pedicel ascending at anthesis, 1–1.5 cm, bracts often streaked with purple, ovate, 3–6 mm. **Capsules** narrowly conic, 7–8 mm. 2*n* = 32.

Flowering Aug. Wet sandstone ledges of canyons, "hanging garden" community; of conservation concern; 1100–1900 m; Colo., Utah.

13. Zigadenus virescens (Kunth) J. F. Macbride, Contr. Gray Herb. 53: 4. 1918 • Green death camas

Helonias virescens Kunth in A. von Humboldt et al., Nov. Gen. Sp. 1: 267. 1816; *Anticlea mexicana* Kunth; *A. virescens* (Kunth) Rydberg; *Zigadenus mexicanus* (Kunth) Hemsley; *Z. porrifolius* Greene; *Z. virescens* var. *porrifolius* (Greene) O. S. Walsh ex Espejo & López-Ferrari

Plants 3.5–8.5 dm, from bulbs; bulbs not clumped, tunicate, narrowly ovoid, 3–5 × 1–1.5 cm. **Leaves:** proximal blades 25–45 cm × 3–20 mm. **Inflorescences** racemose to paniculate, 15–30-flowered, with 1–4 branches. **Flowers** nodding; perianth perigynous, campanulate, 4–8 mm diam.; tepals greenish, sometimes tinged with purple, ovate, 4–6 × 1.5–3 mm, somewhat narrowed at base, apex narrowly obtuse; gland 1, obcordate, thick; pedicel recurved at anthesis, 1.5–2 cm, bracts often tinged with purple or pink, ovate, 5–20 mm. **Capsules** narrowly conic, 10–20 mm. 2*n* = 32.

Flowering Jul–Aug. Montane coniferous forests; 1000–3200 m; Ariz., Colo., N.Mex.; s Mexico.

14. Zigadenus mogollonensis W. J. Hess & Sivinski, Sida 16: 390, fig. 1. 1995 • Mogollon death camas [E]

Plants 4.5–8.5 dm, from bulbs; bulbs not clumped, tunicate, narrowly ovoid. **Leaves:** proximal blades 25–45 cm × 10–20 mm. **Inflorescences** racemose, occasionally paniculate, 10–30-flowered, with 1–3 branches. **Flowers** nodding; perianth perigynous, campanulate, 8–20 mm diam.; tepals greenish, often tinged with purple, ovate to elliptic, 12–16 × 7–9 mm, somewhat narrowed at base; gland 1, obcordate, thick; pedicel recurved at anthesis, erect in fruit, bracts often tinged with purple or red, longer than pedicel, apex acute. **Capsules** 12–15 mm. $2n = 32$.

Flowering Aug–Sep. Mixed coniferous and spruce-fir forests of Mogollon Mountains; 2500–3200 m; N.Mex.

Zigadenus mogollonensis seems morphologically most similar to 13. *Z. virescens* and *Z. volcanicus* Bentham, a species from Guatemala.

14. STENANTHIUM (A. Gray) Kunth, Enum. Pl. 4: 189. 1843, name conserved • [Greek *stenos*, narrow, and *anthos*, flower, alluding to the narrow tepals]

Frederick H. Utech

Veratrum Linnaeus subg. *Stenanthium* A. Gray, Ann. Lyceum Nat. Hist. New York 4: 119. 1837; *Stenanthella* Rydberg

Herbs, perennial, from tunicate bulbs and reduced rhizomes; roots fibrous. **Stems** simple, mostly with 2–3 reduced bracts, glabrous. **Leaves** mostly basal, reduced distally, spiral, arching downward, sheathing proximally; blade narrowly linear to oblanceolate, strap-shaped, glabrous, apex acuminate or obtuse. **Inflorescences** terminal, racemose or paniculate, open, bracteate, glabrous. **Flowers:** proximalmost usually bisexual, distalmost staminate; perianth tubular-campanulate or rotate; tepals persistent, 6, distinct or weakly connate basally, glandular or not, oblong-lanceolate or lanceolate, equal to subequal, apex acute to acuminate; claws absent; tepal glands, when present, 2-lobed, adaxial; stamens 6, distinct or connate basally, equal or subequal; filaments lanceolate; anthers basifixed, 1-locular, obcordate-reniform; pollen sacs apically confluent, extrorse, opening into peltate disc; ovary superior to half-inferior, 3-locular proximally, 1-locular distal to ovules; septal nectaries absent; styles persistent, 3, spreading to recurved, distinct; stigmas minute. **Fruits** capsular, deeply 3-lobed, membranous, slenderly 3-beaked, dehiscence septicidal, then adaxially loculicidal. **Seeds** 3–4 per locule, brown to brownish black, narrowly oblong or ellipsoid to lanceoloid, flat, angled to winged. $x = 8, 10$.

Species 4 (2 in the flora): North America (including Mexico); e Asia.

The species of *Stenanthium* other than *S. gramineum* have at times been transferred to the segregate genus *Stenanthella* based on their racemose inflorescences, darker, campanulate flowers, and apically recurved tepals, leaving a monotypic *Stenanthium* with paniculate inflorescences, lighter colored, rotate flowers, and spreading tepals (P. A. Rydberg 1900). These differences are not sufficiently constant, however, to warrant the generic distinction (F. H. Utech 1987, 1987b). Among the melanthioid genera, *Stenanthium* shows greatest morphological similarities to *Zigadenus* Michaux (J. D. Ambrose 1975, 1980; S. J. Preece 1956; W. B. Zomlefer 1997b; W. B. Zomlefer et al. 2001).

Stenathium gramineum and *S. occidentale* are occasionally cultivated as garden ornamentals.

SELECTED REFERENCES Fernald, M. L. 1946c. *Stenanthium* in the eastern United States. Rhodora 48: 148–152. Utech, F. H. 1987. Biosystematic studies in *Stenanthium* (Liliaceae: Veratreae). I. Floral morphology, floral vascular anatomy, geography and taxonomy of *S. occidentale* A. Gray. Ann. Carnegie Mus. 56: 113–135. Utech, F. H. 1987b. Biosystematic studies in *Stenanthium* (Liliaceae: Veratreae). II. Floral morphology, floral vascular anatomy, geography and taxonomy of the Mexican *S. frigidum* (Schlecht. & Cham.) Kunth. Ann. Carnegie Mus. 56: 197–212.

1. Stems 1.5–4.5(–6) dm; leaves 15–30 cm; perianth tubular-campanulate; tepals greenish to brownish purple, 10–20 mm, recurved distally, apex gradually acuminate; w North America . . . . 1. *Stenanthium occidentale*
1. Stems 5–20 dm; leaves 20–70 cm; perianth rotate; tepals white to greenish yellow, 4–10 mm, not recurved distally, apex acute to narrowly acuminate; e North America . . . . 2. *Stenanthium gramineum*

1. Stenanthium occidentale A. Gray, Proc. Amer. Acad. Arts 8: 405. 1873 • Bronze-bells, mountain-bells [E]

Stenanthella occidentalis (A. Gray) Rydberg; *Stenanthium rhombipetalum* Suksdorf

Bulbs ovoid, 2–4 cm. **Stems** 1.5–4.5(–6) dm. **Leaves** several, mostly proximal; blade linear to oblanceolate, weakly keeled, 15–30 × 0.6–2.5 cm, apex acute. **Inflorescences** racemose and 3–6-flowered or paniculate and to 25-flowered; bracts lanceolate, 6–10 mm. **Flowers** drooping; perianth narrowly tubular-campanulate; tepals recurved distally, greenish to brownish purple, oblong-lanceolate, 10–20 mm, apex gradually acuminate; tepal glands present; stamens 7–8 mm; filaments subequal; pedicel ascending to erect, 1–3 cm. **Capsules** lanceoloid, 12–16 mm including slender style beaks, apex gradually acuminate. **Seeds** brownish black, narrowly oblong, 3–4 mm. 2*n* = 16.

Flowering late spring–mid summer. Wet cliffs, rocky crevices, montane meadows, mossy scree; 0–2500 m; Alta., B.C.; Calif., Idaho, Mont., Oreg., Wash.

Stenanthium occidentale is similar to a circum-northern Pacific and Sakhalin Island endemic, *S. sachalinense* F. Schmidt, which may be conspecific (S. M. Kupchan et al. 1961; F. H. Utech 1987).

2. Stenanthium gramineum (Ker Gawler) Morong, Mem. Torrey Bot. Club 5: 110. 1894 • Feather-bells, feather-fleece [E] [F]

Helonias graminea Ker Gawler, Bot. Mag. 39: plate 1599. 1813; *Stenanthium gramineum* var. *micranthum* Fernald; *S. gramineum* var. *robustum* (S. Watson) Fernald; *S. robustum* S. Watson

Bulbs slender, ellipsoid, 3–8 cm. **Stems** 5–20 dm. **Leaves** numerous, ascending; blade narrowly linear, channeled, 20–70 × 0.5–3 cm, apex acuminate to obtuse. **Inflorescences** freely branching, terminal panicles of many compound racemes; terminal racemes longer than lateral; bracts linear, 2 mm. **Flowers** erect; perianth rotate; tepals not recurved distally, white to greenish yellow, narrowly lanceolate, 4–10 × 1–2 mm, apex acute to narrowly acuminate; tepal glands absent; stamens 1–4 mm; filaments equal; pedicel divergent, 4–6 mm. **Capsules** oblong, cylindric to ovoid, 6–10 mm including divergent style beaks. **Seeds** brown, ellipsoid to lanceoloid, 5–8(–10) mm. 2*n* = 20.

Flowering mid–late summer. Moist woods, thickets, and meadows; 0–1200 m; Ala., Ark., D.C., Fla., Ga., Ill., Ind., Ky., La., Md., Mich., Mo., N.C., Ohio, Pa., S.C., Tenn., Tex., Va., W.Va.

Stenanthium gramineum shares molecular affinities with the *Zigadenus densus* (Desrousseaux) Fernald and *Z. leimanthoides* A. Gray species pair (W. B. Zomlefer et al. 2001).

The varieties of *Stenanthium gramineum* recognized by M. L. Fernald, but not here, are indistinct and sympatric (R. R. Gates 1918; R. G. Johnson 1969). The disjunct upper Michigan localities represent locally established waifs (E. G. Voss 1972–1985, vol. 1).

15. AMIANTHIUM A. Gray, Ann. Lyceum Nat. Hist. New York 4: 121. 1837, name conserved • Fly-poison, crow-poison, St. Elmo's-feather, staggergrass [merged Greek *amiantos*, unsoiled, and *anthos*, flower, alluding to the glandless tepals] [E]

Frederick H. Utech

Chrosperma Rafinesque 1825, name rejected

Herbs, perennial, from bulbs; bulbs tunicate, ovoid-oblong, ca. 5 × 2.5 cm; roots contractile. **Stems** erect, simple. **Leaves** mostly basal, reduced distally, spiral, arching downward, sheathing proximally; blade linear-elongate, strap-shaped, glabrous, apex obtuse to abruptly acute. **Inflorescences** terminal, racemose, dense, bracteate, glabrous. **Flowers** bisexual or both bisexual and unisexual flowers borne on the same or on different plants, cup-shaped to widely spreading, pedicellate; tepals persistent, 6, distinct, white, turning reddish green, claws absent, oblong-obovate, equal to subequal, apex acute; tepal glands absent; stamens 6, perigynous, distinct; filaments subulate, apex acute; anthers basifixed, 1-locular, cordate-ovate, extrorse, opening into peltate disc (apical/valvate dehiscence); pollen sacs apically confluent; ovary superior to partly inferior, 3-locular, diverging; septal nectaries absent; styles persistent, 3, widely diverging, distinct; stigmas terminal, minute. **Fruits** capsular, deeply 3-lobed, thin-walled, dehiscence septicidal, then adaxially loculicidal. **Seeds** 1–2 per locule, brownish to purplish black, wingless, ellipsoid, lustrous; testa fleshy. $x = 8$.

Species 1: e North America.

Amianthium is frequently included in a broadly circumscribed *Zigadenus* (J. D. Ambrose 1975, 1980; S. M. Kupchan et al. 1961; S. J. Preece 1956; W. B. Zomlefer 1997b). The absence of tepal nectaries or glands is a major generic characteristic of this Appalachian–Ozark taxon (F. H. Utech 1986). Narrow-leaved specimens of *Amianthium* are often confused with *Zigadenus densus*.

A self-incompatible, pollination-breeding system involving primarily beetles occurs in *Amianthium* (M. E. Palmer et al. 1989; A. M. Redmon et al. 1989; J. Travis 1984). An associated tepal color shift, as in *Zigadenus densus*, may represent possible pollinator signals (J. Travis 1984).

Two unique, toxic alkaloids, jervine and amianthine, found in *Amianthium* roots and leaves (N. Neuss 1953; S. M. Kupchan et al. 1961; R. F. Raffauf 1970; I. W. Southon and J. Buckingham 1989; G. E. Burrows and R. J. Tyrl 2001), caused deaths in cattle and sheep. Native Cherokee used the plant as a dermatological cure for itch and as a crow poison (D. E. Moerman 1986). Root extracts mixed with molasses or honey have been used as a housefly insecticide (C. D. Marsh et al. 1918).

SELECTED REFERENCE Utech, F. H. 1986. Floral morphology and vascular anatomy of *Amianthium muscaetoxicum* (Walter) A. Gray (Liliaceae–Veratreae) with notes on distribution and taxonomy. Ann. Carnegie Mus. 55: 481–504.

1. **Amianthium muscitoxicum** (Walter) A. Gray, Ann. Lyceum Nat. Hist. New York 4: 122. 1837 (as muscaetoxicum) [E] [F]

Melanthium muscitoxicum Walter, Fl. Carol., 125. 1788 (as muscaetoxicum); *Chrosperma muscitoxicum* (Walter) Kuntze; *Zigadenus muscitoxicus* (Walter) Regel

Stems 3–12(–14) dm. **Leaves** 10–45(–60) × 0.5–2.8 cm. **Inflorescences** densely ovoid in flower, cylindric in fruit, 3–13 × 2–3 cm; bracts ovate, 1–2 mm. **Tepals** 5–7 mm; pedicel (7–)10–17(–20) mm. **Capsules** 5–7 mm. **Seeds** 4–5 × 1.5–2 mm. $2n = 32$.

Flowering late spring–summer. Dry or mesic wooded slopes, meadows, savannas; 0–1500 m; Ala., Ark., Del., D.C., Fla., Ga., Ky., La., Md., Miss., Mo., N.J., N.Y., N.C., Okla., Pa., S.C., Tenn., Va., W.Va.

16. TRILLIUM Linnaeus, Sp. Pl. 1: 339. 1753; Gen. Pl. ed. 5, 158. 1754 • Trille [Latin, *trilix*, triple, alluding to the flowers having parts in threes]

Frederick W. Case Jr.

Herbs, perennial, scapose, rhizomatous; rhizomes subterranean, horizontal or semierect, monopodial, unbranched or weakly branched, elongated to thick and fleshy, compressed-shortened, distal end tapered to point or praemorse, apex bearing large terminal bud, numerous cataphylls, and contractile, adventitious, ringed roots. **Scapes** arising from terminal bud or axil of adjacent cataphylls, aerial (subterranean in *T. petiolatum*), erect, straight (decumbent, S-shaped in *T. decumbens* and *T. reliquum*). **Bracts** on mature plants in whorl of 3, uniformly green or mottled, foliaceous, petiolate or sessile, venation palmate-reticulate, with 3–5 major veins, ovate or obovate to elliptical. **Inflorescences** terminal, 1-flowered. **Flowers:** some totally to partially syncarpous, pedicellate or sessile; sepals persistent, 3, distinct, green, maroon, or with maroon markings, foliaceous, oblong, ovate, or lanceolate, alternating with bracts; petals shriveling after anthesis, typically 3, erect, spreading, or recurved, distinct, red, purple, pink, white, yellow, green, or combination of these, ovate or obovate to linear, sometimes clawed; stamens 6, alternating in 2 whorls of 3, erect, incurved, or divergent; filaments mostly short, basally expanded; anthers 2-locular, ± equaling or longer than filaments, dehiscence extrorse, latrorse, or introrse; connectives flat between (or in some species extending beyond) anther sacs; ovary superior, proximal portion 3-locular, 3- or 6-lobed, some axile, some parietal or a combination of both, distal portion forming stigmas; stigmas often persistent, 3, spreading, twisted, or erect, sometimes connate, sessile or with very short style, linear to subulate. **Fruits** capsular or baccate, fleshy with obscure sutures, not or rarely dehiscent along sutures, each shed as unit through abscission of thin-walled cells at base. **Seeds** many, elliptic, 2–4 mm, bearing white or yellowish, large, oily, myrmecochorous elaiosome (aril). $x = 5$. $2n = 10$ in all American species recorded.

Species 43 (38 in the flora): North America, Asia.

Trillium is traditionally divided into two subgenera, which overlap in some characters. Botanists consider subg. *Trillium* to be the more primitive group, because the genera considered closest to *Trillium* all have pedicellate flowers, as does subg. *Trillium*, except for one variety of *T. pusillum* that bears sessile to only barely pedicillate flowers.

There is not yet agreement among taxonomists about the relationships among the species within the genus. Certain ones, such as *Trillium recurvatum* and *T. lancifolium*, clearly are closely related, but the interrelationships of many others are not obvious. In the absence of definitive studies, the species are listed here alphabetically within subgenera.

Several species of *Trillium* contain sapogenins that have been used medicinally as astringents, coagulants, expectorants, and uterine stimulants, hence the common names birthwort and Indian balm, and *T. erectum*, *T. grandiflorum*, and possibly other species have been utilized commercially as beth root (W. B. Zomlefer 1996). Fruits, seeds, and rhizomes of trilliums are generally considered to be poisonous. Deer, however, feed voraciously on *T. grandiflorum* plants, especially in early spring.

Gardeners and wildflower enthusiasts, over most of the temperate zones of the world, consider trilliums to be among the most beautiful of wildflowers.

Species of *Trillium* exhibit few and obscure structural differences, making key construction difficult (J. D. Freeman 1975). The key below, adapted with permission from F. W. Case and R. B. Case (1997), works best with fresh, live material. Many species hybridize, especially within the *T. erectum* group. Hybrids and the occasional anomalous, odd-sized, or diseased individuals are not keyed here.

Morphologically, *Trillium* plants produce no true leaves or stems above ground. The horizontal rhizome produces dry, scalelike leaves (cataphylls). The above-ground plant is a flowering scape, and technically the leaflike structures are bracts subtending the flower. Origin aside, the bracts function in photosynthesis and have external and internal structure similar to that of a leaf, and many authors refer to them as leaves.

SELECTED REFERENCES Berg, R. Y. 1958. Seed dispersal, morphology, and phylogeny of *Trillium*. Skr. Norske Vidensk.-Akad. Oslo, Mat.-Natkurvidensk. Kl. 1958(1): 1–36. Case, F. W. and G. L. Burrows. 1962. The genus *Trillium* in Michigan: Some problems of distribution and taxonomy. Pap. Michigan Acad. Sci. 47: 180–200. Case, F. W. and R. B. Case. 1997. Trilliums. Portland. Gates, R. R. 1917b. A systematic study of the North American genus *Trillium*, its variability and its relation to *Paris* and *Medeola*. Ann. Missouri Bot. Gard. 4: 43–92. Kato, H. et al. 1995. Evolutionary biology of *Trillium* and related genera

(Trilliaceae). 1. Restriction site mapping and variation of chloroplast DNA and its systematic implications. Pl. Spec. Biol. 10: 7–30. Kazempour Osaloo, S., F. H. Utech, M. Ohara, and S. Kawano. 1999. Molecular systematics of Trilliaceae I. Phylogenetic analyses of *Trillium* using *mat*K gene sequences. J. Pl. Res. 112: 35–49. Samejima, K. and J. Samejima. 1987. *Trillium* Genus Illustrated. Sapporo.

1. Flower pedicellate; bract blades usually not mottled.
    2. Stigmas ± connate at least basally, or closely grouped and uniform, of uniform thickness throughout.
        3. Scapes usually 0.4–2 dm at anthesis.
            4. Bract blade bluish green, apex round-obtuse; scapes 6-gonal in cross section . . . . . 6. *Trillium nivale*
            4. Bract blade green, often with maroon undertones early, apex acuminate, blunt-acute, or rounded; scapes round in cross section.
                5. Pedicel recurved below bracts at anthesis, or rarely leaning to ± horizontal distal to bracts; petals strongly arcuate-recurved from base, white, pink, or rose . . . . 1. *Trillium catesbaei* (in part)
                5. Pedicel ± erect at anthesis; petals erect-spreading to widely spreading, not recurved or recurved only in distal 1/2, white or white with madder purple flecks.
                    6. Bracts strongly petiolate, blade apex long-acuminate, weakly to strongly cordate basally, glossy, bluish green, sometimes with lighter silvery green pattern on or near main veins; sepals shorter and narrower than petals, apex rounded-apiculate; petals erect-spreading, ovate-cordate to ± orbicular, widest at or beyond middle, round-tapered distal to middle to apiculate-acuminate tip, margins mostly without undulations, white, usually sparsely to very heavily flecked with madder-purple, color not changing with age; Siskiyou Mountains of n California and Oregon . . . . 10. *Trillium rivale* (in part)
                    6. Bracts very short-petiolate, subsessile, or sessile, blade apex obtuse, not glossy, green with maroon undertones; sepals about as large and prominent as petals, apex round to blunt; petals spreading-ascending, weakly recurved in distal 1/2, oblong to narrowly lanceolate, widest in proximal 1/3, margins strongly undulate, white, aging to deep rose abaxially; e U.S. . . . . 9. *Trillium pusillum* (in part)
        3. Scapes usually 2.0–4.5 dm at anthesis.
            7. Stigmas connate into style 2–6 mm; bract blades elliptic-ovate, ovate-lanceolate to lanceolate, usually 1.5–8 cm wide.
                8. Pedicel erect or only slightly leaning; petals erect to widely spreading, linear to narrowly elliptic, white at anthesis, darkening with age to light to medium deep pink, with inverted V-shaped, white base; bracts horizontal to drooping distally, blade lanceolate to lanceolate-ovate; stamens straight . . . . 8. *Trillium persistens*
                8. Pedicel leaning to almost horizontal or clearly recurved and declined; petals arcuate-recurved, oblong-lanceolate, white, pink, or rose, darkening to pink or nonfading with age; bracts and their margins often slightly raised, exposing flowers at anthesis, blade elliptic-ovate; stamens recurved . . . . 1. *Trillium catesbaei* (in part)
            7. Stigmas barely connate basally, rarely completely distinct; bract blade ovate to rhombic, usually 8–20 cm wide.
                9. Petals with distinct, dark red, inverted V-shaped mark basally; bracts strongly petiolate; ovary 3-lobed or -angled . . . . 14. *Trillium undulatum*
                9. Petals uniformly white or pink, or with faint pink blush, no distinct, V-shaped marking; bracts subsessile, sessile, or short-petiolate; ovary 6-lobed or -angled.
                    10. Petals erect basally, flaring and recurving above middle, margins undulate-wavy, overlapping basally, rolled, forming tube that conceals ovary when viewed from above, obovate to lanceolate, rarely suborbicular,

4–7.5 × 2–4 cm, white or rarely pink, aging to pink-purple; e North America . . . . 5. *Trillium grandiflorum*

10. Petals erect-ascending, widely spreading in distal 1/2, distal margins usually flat to weakly undulate to undulate, scarcely touching basally to overlapping and forming tube, not concealing ovary when viewed from above, linear to obovate, 1.5–7 × 1–4 cm, white, often with pink or blush markings, aging to deep rosy pink, purple, or dark red; w North America . . . . 7. *Trillium ovatum*

2. Stigmas distinct, often basally thickened or subulate.
11. Stigmas short, 2–4 mm, uniformly thick, recurved and appearing ± lobed adaxially, at least in dried specimens; bracts strongly petiolate, blade ovate-lanceolate, mostly less than 6 cm wide, weakly to strongly cordate basally . . . . 10. *Trillium rivale* (in part)
11. Stigmas of various lengths, basally thickened, gradually tapered, erect distally, reflexed, or coiled, not lobed adaxially; bracts sessile to subsessile, blade elliptic, obovate, or rhombic, mostly more than 6 cm wide, gradually tapered, rounded or weakly cuneate basally.
12. Anthers ± equaling filaments, pale lavender-pink or -gray; petals oblong-lanceolate, recurved-reflexed from just above base, extending behind plane of sepal bases for more than 1/2 their length, white or rarely pale pink, thin-textured, adaxial veins inconspicuous . . . . 2. *Trillium cernuum*
12. Anthers longer or shorter than, rarely equaling filaments, white, creamy white, yellow, purple, or maroon; petals ovate to obovate, spreading, flat, or, if recurved, in distal 1/2 only, red, maroon purple, rose, cream, or white, more rarely pale yellow, rarely 2-colored, heavy- or fleshy-textured, adaxial veins conspicuous, appearing engraved.
13. Ovary greenish white or white, occasionally with reddish purple stains at apex.
14. Pedicel strongly recurved beneath bracts; anthers thin, dark purple; petals ovate-elliptic, white, rarely 2-colored rose and white, apex acuminate; stigmas shorter than ovary; rhizome tapered to point distally . . . . . . 11. *Trillium rugelii* (in part)
14. Pedicel erect, declined, or recurved; anthers thick, creamy white or yellow; petals ovate-lanceolate to broadly ovate, creamy white, not 2-colored, apex acute; stigmas ± equaling ovary; rhizome praemorse . . . . 4. *Trillium flexipes*
13. Ovary dark purple or maroon.
15. Pedicel recurved or declined; flower held proximal to or below bracts.
16. Pedicel declined, ± straight; stamens 1.5–2 times taller than small, ± globose to conical-pyramidal ovary; anthers maroon to grayish purple; petals 3–6.5 × 4–6 cm, usually maroon to brownish red, ovate-suborbicular, spreading to weakly recurved, longer than sepals . . . . 15. *Trillium vaseyi*
16. Pedicel recurved beneath bracts; stamens ± equaling or not more than 1/3 longer than prominent, flask-shaped ovary; anthers dark purple; petals 2.5–5 × 0.8–3.5 cm, usually white or 2-colored, distally white, ovate to broadly ovate, recurved, ± equaling sepals . . . . 11. *Trillium rugelii* (in part)
15. Pedicel erect, or leaning from base, not recurved or declined; flower erect or dangling but not distinctly below bracts.
17. Perianth open, usually flat; petals carried in ± same plane as sepals, or ascending slightly, spreading, not recurved, lanceolate to ovate, to 2+ times width of sepals; sepals flat to somewhat sulcate apically; flower odor fetid, of wet dog . . . . 3. *Trillium erectum*
17. Perianth gaping, strongly 3-dimensional; petals erect-spreading or recurved in distal 1/2, ovate to ovate-orbicular, wider than sepals; sepal apex sulcate; flower odor musty or fruity.
18. Petals creamy white, 1.5 times sepal length, 4–7 × 1.5–4 cm, not recurved to weakly so at tip; pedicel not strongly reflexed, flowers facing ± upward, odor sweet, applelike; sepal apex mildly sulcate . . . . 12. *Trillium simile*
18. Petals dark reddish maroon to purplish, less than 1.5 times sepal

length, 1.8–5 × 1–3 cm, often recurved in distal 1/2; pedicel reflexed at tip; flowers facing outward at ± right angle to pedicel, odor funguslike; sepal apex strongly sulcate . . . . 13. *Trillium sulcatum*

1. Flower sessile, bract blades usually mottled (except in *T. pusillum*).
 19. Stigmas connate basally into short style, threadlike, tips widely spreading; petals spreading from base with distal 1/2 flat or weakly ascending, exposing stamens and ovary, of short duration, margins undulate . . . . 9. *Trillium pusillum* (in part)
 19. Stigmas distinct, erect, subulate with tips divergent-recurved, or linear and usually slightly thickened basally with tips recoiled; petals mostly erect from base, ± connivent, often at least partially concealing stamens and ovary, or erect-spreading, tips incurving slightly, long-lasting, sometimes spiral-twisted, margins entire.
  20. Scapes decumbent or semidecumbent, or subterranean; bracts resting near or on ground, or with petioles arising from scape apex at or near ground surface.
   21. Scapes ± erect, largely subterranean; bracts long-petiolate, blade not mottled; individual bracts *Plantago*-like; w United States . . . . 31. *Trillium petiolatum*
   21. Scapes decumbent or semidecumbent, S-shaped, aerial; bracts sessile or only short-petiolate, blade heavily mottled; se United States
    22. Anther dehiscence extrorse; scapes distinctly puberulent especially just below bracts; petals linear-lanceolate or oblanceolate, 4–8 cm or more; filaments 2–5 mm . . . . 21. *Trillium decumbens*
    22. Anther dehiscence introrse; scapes glabrous; petals narrowly elliptic-oblanceolate, 2.5–5.5 cm; filaments 1–2 mm . . . . 33. *Trillium reliquum*
  20. Scapes erect, straight; bracts held well above ground, or bract tips touching ground only in early anthesis.
   23. Sepals strongly reflexed at bases, often below bracts and parallel to scape, or adpressed to bracts; anther connectives strongly curved inward; rhizomes horizontal, elongated, slender, very brittle, white.
    24. Petals lanceolate-ovate, base attenuate to weakly clawed, ca. 2 times longer than wide; bracts petiolate . . . . 32. *Trillium recurvatum*
    24. Petals linear to narrowly spatulate, clawed, at least 4 times longer than wide; bracts sessile . . . . 26. *Trillium lancifolium*
   23. Sepals not strongly reflexed, spreading or somewhat erect; anther connectives straight; rhizomes horizontal or erect, thick or somewhat compressed-thickened and superficially bulblike, not brittle, brownish.
    25. Petals white, rarely tinged pale pink or purplish near bases.
     26. Anther dehiscence introrse; ovary purple . . . . 18. *Trillium chloropetalum* (in part)
     26. Anther dehiscence latrorse; ovary green or greenish white, rarely purple at base.
      27. Scapes usually 2.2–5.8 dm; petals oblanceolate to obovate, 4.8–8 cm, white to creamy white, rarely very pale pink or pinkish basally . . . . 16. *Trillium albidum*
      27. Scapes usually 1.7–3 dm; petals linear to linear-lanceolate, 2.2–4.5 cm, white, rarely purplish basally . . . . 30. *Trillium parviflorum*
    25. Petals maroon, reddish brown, bronze, yellow, or green.
     28. Petals yellow, pale yellow, greenish, or bicolored green with purplish base.
      29. Petal bases not clawed.
       30. Petals oblanceolate to lanceolate, apex long-acuminate; stamens 11–18 mm; anther connective extending 0.5 mm or less beyond sacs; stamens slightly longer than or ± equaling pistil; flower odor strongly of lemon; se United States . . . . 28. *Trillium luteum*
       30. Petals oblanceolate to obovate, apex variably acute to almost truncate; stamens 17–26 mm; anther connective extending 1–1.5 mm beyond sacs; stamens ca. 2 times as long as pistil; flower odor spicy, of rose; California . . . . . 18. *Trillium chloropetalum* (in part)
      29. Petal bases weakly to strongly clawed.

31. Petals widely spatulate, pale sulfur-yellow fading to ± cream, with broadly cuneate basal claw, claw green, sometimes obscurely marked with purple; apex apiculate, nipplelike in at least 1 petal . . . . 22. *Trillium discolor*
31. Petals narrowly spatulate to linear-spatulate, green, bicolored, green with purplish base, rarely all purple, obscurely to clearly narrow-clawed basally, claw purple or green; apex round-acute to obtuse, lacking nipple.
    32. Bract (under magnification) with numerous stomates across entire adaxial surface; pistil 2/3 stamens; anther dehiscence introrse-latrorse; petals green or yellowish green distal to weak purplish claw (very rarely purplish green throughout); flower odor of rotten fruit . . . . 37. *Trillium viride*
    32. Bract adaxially without stomates or with stomates near apex only; pistil 1/2 stamens; anther dehiscence latrorse; petals green or yellowish green distal to distinct purplish claw (rarely, petal purple throughout); flower odor spicy or musty . . . . 38. *Trillium viridescens*

[28. Shifted to left margin.—Ed.]

28. Petals usually maroon, reddish brown, pink, greenish maroon, purplish bronze, white, or brown throughout (except in pigment-free mutant individuals with petals pale green or yellowish green); occasionally yellow in *T. cuneatum, T. chloropetalum, T. gracile, T. maculatum,* and *T. foetidissimum,* occasionally 2-colored in *T. ludovicianum.*
    33. Anther dehiscence extrorse; stamens not obscured by petals; petals horizontal, with 1–2 spiral twists . . . . 35. *Trillium stamineum*
    33. Anther dehiscence introrse or latrorse; stamens at least partially obscured by petals; petals erect to divergent and slightly spreading with tips ± incurved, not spirally twisted.
        34. Anther dehiscence latrorse, occasionally introrse basally with age.
            35. Anther connectives extended 1–2 mm beyond sacs; margins of distal 1/3 of bract blade straight to apex.
                36. Scape 1.5 times longer than bracts; bract tips at anthesis often touching ground; petals 3–4 times longer than wide . . . . 36. *Trillium underwoodii*
                36. Scape 2.5–3 times longer than bracts; bract tips at anthesis held well away from ground; petals ca. 2 times longer than wide . . . . 20. *Trillium decipiens*
            35. Anther connectives extended scarcely if at all to 0.5 mm beyond sacs; margins of distal 1/3 of bract blade convex-curved to apex.
                37. Petals 0.9–2.7 cm wide, oblanceolate to elliptic-obovate, narrowed to cuneate but not clawed base, erect . . . . 19. *Trillium cuneatum*
                37. Petals 0.4–0.8 cm wide, linear-oblanceolate, weakly clawed basally, erect-spreading . . . . 27. *Trillium ludovicianum*
        34. Anther dehiscence introrse.
            38. Stamens ca. 2 times as long as pistil.
                39. Petals oblanceolate to obovate, 6.5–10 × 1.5–2.5 cm; anther connective extended 1–1.5 mm beyond sacs; California . . . . 18. *Trillium chloropetalum* (in part)
                39. Petals linear-elliptic to oblanceolate, 2–4 × 0.3–0.8 cm; anther connective extended 0.1–1 mm beyond sacs; e Texas and w Louisiana . . . . 24. *Trillium gracile*
            38. Stamens equaling, less than, or very slightly longer than pistil.
                40. Anther connective not extended beyond sac, or, if extended, rarely to 0.5–1 mm.
                    41. Scape 1.4–4 dm; bracts 7–15 × 4.6–6.7 cm; petals narrowly oblanceolate-spatulate to linear-spatulate, spreading-erect, becoming connivent-erect with age, 4–7 × 0.7–1.7 cm, clear dark red-maroon, reddish purple, or rarely yellow above purplish base or clear sulfur yellow, distal margins upraised from about widest point to near petal base; stigmas 2–4 mm; se United States. . . . . 29. *Trillium maculatum*

41. Scape 2.5–5.5 dm; bracts 11–18 × 12–17 cm; petals oblanceolate, spreading-erect, 5.5–11 × 2–3.5 cm, glossy dark maroon-red or purple, margins ± flat, widest at or below middle, cuneate basally; stigmas 6–8 mm; n California and extreme sw Oregon .... 25. *Trillium kurabayashii*

[40. Shifted to left margin.—Ed.]

40. Anther connective extended 1 mm or more beyond sac.

42. Anther connective extended 2–5+ mm beyond sac .... 34. *Trillium sessile*

42. Anther connective extended ca. 1–1.5 mm beyond sac.

43. Scapes 2.5–6 dm; petals linear, erect, 5–10 cm, 8–10 times longer than wide, glossy, dark purple to red-purple, inrolling with age; bract blades 10–22 × 8.7–15 cm, obscurely and sparsely or not at all mottled or spotted darker greenish brown, subsessile; c, s California .... 17. *Trillium angustipetalum*

43. Scapes 0.8–2.8 dm; petals narrowly elliptic to linear-lanceolate, weakly incurved, 2–5 × 0.3–0.5 cm, not glossy, pinkish purple, light to dark reddish or brownish purple, rarely yellow, not inrolling with age; bract blades 6.7–12 × 3.8–6 cm, strongly mottled, sessile; e Louisiana, sw Mississippi .... 23. *Trillium foetidissimum*

16a. TRILLIUM Linnaeus subg. TRILLIUM [E]

Bract blades not mottled (except sometimes in *T. ovatum*). **Flower** pedicellate except subsessile or sessile in *T. pusillum* var. *virginianum*. **Petals** wide-spreading, white, reddish purple, or cream, sometimes with patterns of another color.

Species 15 (15 in the flora): North America.

SELECTED REFERENCES Barksdale, L. 1938. The pedicellate species of *Trillium* found in the southern Appalachians. J. Elisha Mitchell Sci. Soc. 54: 271–296. Ihara, H. and K. Ihara. 1978. A biosystematic study on the pedicellate-flowered species of the North American *Trillium*. (1) Geographical distribution of major groups and their gynoecium norms. J. Geobot. 25: 139–172.

1. Trillium catesbaei Elliott, Sketch Bot. S. Carolina 1: 429. 1817 • Catesby's trillium, bashful trillium [E] [F]

Trillium affine Rendle; *T. nervosum* Elliott; *T. stylosum* Nuttall

Rhizomes short, tapered to a point distally. **Scapes** 1–2, round in cross section, 2–4.5 dm, slender, glabrous. **Bracts** usually raised somewhat, exposing flower; blade green, often with underlying purplish maroon in well-lighted plants, major veins on adaxial surface deeply engraved, elliptic-ovate, 6.5–15 × 4–8 cm, including elongated, petiolelike base, proximal 1/3 of leaf gradually tapered to base, margins often slightly raised, apex blunt-acute to acuminate. **Flower** opening at or recurved below bracts, rarely erect; sepals falcate-recurved, green or streaked with purple, linear-lanceolate, 20–45 × 7–8 mm, margins ± flat, apex acuminate; petals falcate-recurved distally, white, pink, or rose, darkening to pink or nonfading with age, veins visible but not appearing engraved, elliptic-oblong to oblong-lanceolate, 3.5–5 × 1–2+ cm, thin-textured, bases somewhat imbricated, forming funnel-like tube, margins undulate, apex acuminate; stamens prominent, recurved, 16–25 mm; filaments white, slightly longer and narrower than anthers; anthers recurving, bright yellow, 5–14 mm, thick, dehiscence introrse; ovary inconspicuous, white, angular-ovoid, deeply 6-angled, 4–12 × 3–7 mm, slender, base narrow; style 2–6 mm; stigmas prominent, strongly curled (circinate) to erect, connate basally, pale green or white, not lobed, 4–10 mm, uniformly thin; pedicel deflexed or recurved, rarely erect, angled, 2–4(–5) cm. **Fruits** baccate, greenish or whitish, ovoid-globose, angles less prominent when mature, 1–1.5 cm diam., pulpy but not juicy.

Flowering spring–summer (late Mar–early Jun). Acid soils, open dry or rich woods, laurel and rhododendron thickets, cove forests; 50–800 m; Ala., Ga., N.C., S.C., Tenn.

Elliott, author of *Trillium catesbaei*, apparently recognized one of its forms separately as *T. nervosum*. This name is still mentioned in British horticultural and botanical literature, but it is rarely found in U.S. literature. North American botanists treat *T. nervosum* as a synonym of *T. catesbaei*.

Plants of *Trillium catesbaei* from deep coves in South Carolina tend to have petals that are deeper pink and longer and wider than those of many other populations.

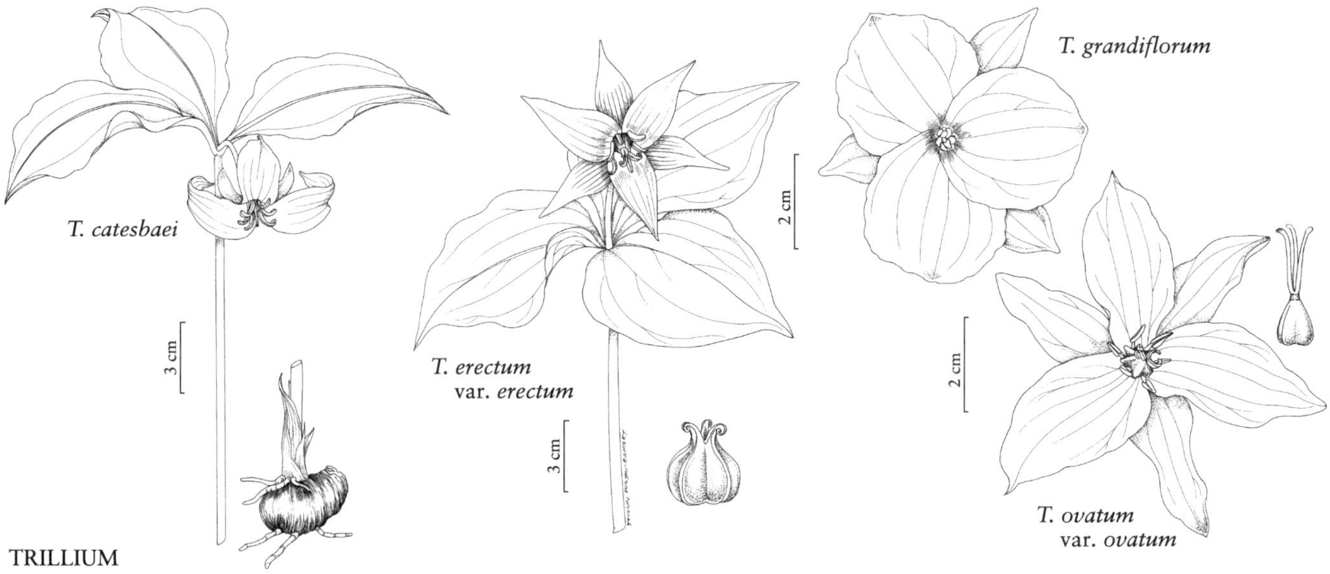

TRILLIUM

2. Trillium cernuum Linnaeus, Sp. Pl. 1: 339. 1753
• Nodding trillium, trille penché [E]

Trillium cernuum var. *macranthum* Eames & Wiegand

Rhizomes short, thick. **Scapes** 1–2(–3), round in cross section, 1.5–4+ dm, slender, glabrous. **Bracts** often overlapping, sessile or with a barely noticeable, petiolelike base, umbrellalike; blade bright green without red tones, broadly rhombic-ovate to suborbicular, 5–15 × 6–15+ cm, base attenuate, apex acuminate. **Flower** usually hidden beneath bracts, nodding, odorless; sepals spreading, green, lanceolate-ovate, 9–30 mm, slightly shorter than to equaling petals, margins slightly raised, apex acuminate; petals usually strongly recurved from above base, extending behind plane of sepal bases for more than 1/2 their length, white or rarely pale pink, adaxial veins not conspicuous, oblong-lanceolate, 1.5–2.5 × 0.9–1.5 cm, thin-textured, margins entire, apex acuminate; stamens ± straight, 6–15 mm, shorter than pistil, slender; filaments white, ± equaling anthers, slender; anthers straight, pale lavender-pink or -gray, 2–6.5 mm, dehiscence introrse to latrorse; ovary prominent, white to pinkish, pyramidal, strongly 6-angled, 3–12 × 3–10 mm, widest above basal attachment; stigmas erect, recurved, distinct, white, not lobed adaxially, 3–8 mm, widest at base, fleshy, basally thickened, gradually tapered; pedicel strongly recurved or declined below or angled between bracts, 1.5–3 cm. **Fruits** dark red, with fruity fragrance, ovoid, to 3 cm diam., fleshy, juicy. $2n = 10$.

Flowering spring–summer (Apr–Jul). Rich, mostly deciduous forest southward, mixed deciduous-coniferous forests, swamps, moist coniferous forests northward; 30–600 m; St. Pierre and Miquelon; Man., N.B., Nfld. and Labr. (Nfld.), N.S., Ont., P.E.I., Que., Sask.; Conn., Del., D.C., Ill., Ind., Iowa, Maine, Md., Mass., Mich., Minn., N.H., N.J., N.Y., N.Dak., Ohio, Pa., R.I., S.Dak., Vt., Va., W.Va., Wis.

Previous authors commonly recognized var. *cernuum* and var. *macranthum*, based primarily upon size differences. Plants attributed to var. *cernuum* are found from Delaware and eastern Pennsylvania northward to Newfoundland, while those attributed to var. *macranthum* are found mainly farther inland into the Midwest. Although there is a tendency for the eastern seaboard plants to be somewhat smaller and more delicate, and the midwestern and far northern plants to be more robust, there is much variation, largely dependent on soil nutrients. There are regional size trends, but based on my observations of this species in Newfoundland, Michigan, Wisconsin, and Minnesota, I do not believe that the two varieties can be maintained.

In the Gray Herbarium, there is a collection by Richardson labeled "Mackenzie River," which has been cited by H. M. Raup (1947) and others. W. J. Hooker ([1829–]1833–1840) reported *Trillium cernuum* "from Saskatchewan to Mackenzie River." Raup stated that "it is the only evidence for the occurrence of...*Trillium* in the entire Mackenzie Basin." Other writers have simply quoted that statement. In view of the relatively great disjunction from the known Saskatchewan stations and in the absence of any other supporting specimens from that area, I believe that there might be locality error on the Richardson sheet. However, it is not beyond possibility that *T. cernuum* could occur there. This apparent disjunct station is not mapped here.

3. Trillium erectum Linnaeus, Sp. Pl. 1: 340. 1753
• Wake robin, red trillium, stinking benjamin, stinking willie, birthwort [E] [F]

Rhizomes short, thick, praemorse. **Scapes** 1–2, often with numerous offsets forming heavy clumps, round in cross section, 1.5–6 dm, ± robust, glabrous. **Bracts** sessile; blade bright green, lacking dark pigmentation, major veins prominent, broadly rhombic to ovate-rhombic, 5–20 × 5–20 cm, about as broad as long, widest near middle, base attenuate, apex acuminate. **Flower** erect, ascending, or proximal to but above bracts, odor fetid, like a wet dog; perianth open, flat; sepals flat to sulcate apically, green, often streaked or overlain with maroon, occasionally entirely dark maroon, lanceolate-acuminate, 10–50 mm, equaling petals, ± 1/2 petal width, texture leafy, margins entire, apex acuminate; petals spreading, carried in same plane as sepals or ascending slightly, dark reddish brown, maroon, purple, or white, sometimes pale yellow, major adaxial veins prominent and appearing somewhat engraved, usually flat, lanceolate, ovate-lanceolate, or occasionally ovate, 1.5–5 × 1–3 cm, 2 times sepal width, widest near base, heavy-textured, apex acuminate; stamens erect to slightly recurved, 5–15 mm; filaments white, pinkish, or dark purple, ± equaling anthers, but variable within local populations, thin; anthers erect or weakly recurving, dark maroon, grayish maroon, or yellowish, strongly yellow when pollen exposed, 5–12 mm, dehiscence introrse; ovary dark purple to maroon, even in white-flowered forms, ovoid, elliptic to globose, 6-angled, angles forming very low ridges when fruit is ripe, 5–10 mm, broadly attached at base; stigmas recurved, distinct, dark purple, not lobed adaxially, subulate, short, 3–7 mm, ca. 1/2 or less length of ovary at anthesis, fleshy; pedicel straight, erect, or somewhat declined but not strongly recurved below bracts, 1–10+ cm. **Fruits** dark maroon, weakly fragrant of fruit, ± globose to slightly pyramidal, 1–1.6 × 1–1.5 cm, juicy. $2n = 10$.

Varieties 2 (2 in the flora): e North America.

SELECTED REFERENCE Case, F. W. and R. B. Case. 1993. *Trillium erectum* and its hybrids. Bull. Amer. Rock Gard. Soc. 51: 162–168.

1. Petals usually red, maroon, or dark purple .... 3a. *Trillium erectum* var. *erectum*
1. Petals white .... 3b. *Trillium erectum* var. *album*

3a. Trillium erectum Linnaeus var. **erectum** [E] [F]

Petals usually dark reddish maroon to purplish, fading purple with age, sometimes cream, yellow, pale green, pink, or white (see discussion).

Flowering spring (early–late Apr and May). Cool, rich, moist, neutral to acidic soils of upland deciduous forests, mixed deciduous-coniferous forests, coniferous swamp borders; 200–700 m; N.B., N.S., Ont., Que.; Conn., Del., Ga., Ill., Ind., Ky., Maine, Md., Mass., Mich., N.H., N.J., N.Y., N.C., Ohio, Pa., R.I., S.C., Tenn., Vt., Va., W.Va.

Variety *erectum* is more common than var. *album* over much of the species range, especially above 450 m; at lower elevations in the south it is largely replaced by var. *album*. Plants with yellow, pale green, pink, or white petals occasionally occur throughout the range.

3b. Trillium erectum Linnaeus var. **album** (Michaux) Pursh, Fl. Amer. Sept. 1: 245. 1814 [E]

Trillium rhomboideum Michaux var. *album* Michaux, Fl. Bor.-Amer. 1: 215. 1803

Petals white.

Flowering spring (late Mar–early May). Upland deciduous forested slopes, stream banks, wooded roadsides at lower elevations; 200–700 m; N.B., Ont., Que.; Conn., Del., Ga., Ill., Ind., Ky., Maine, Md., Mass., Mich., N.H., N.J., N.Y., N.C., Ohio, Pa., R.I., S.C., Tenn., Vt., Va., W.Va.

Variety *album* is occasional throughout the range stated above and is almost the only variety of *Trillium erectum* found at elevations less than 450 m in western North Carolina and areas surrounding the Great Smoky Mountains, but occasional white-flowered plants can occur in any population.

Variety *album* differs from var. *erectum* mainly just in petal color, and probably should be reduced to the rank of forma. Many other colors occur sporadically in otherwise typically maroon to purple populations, most frequently a soft pale yellow, but also white with a pink overlay, or paler forms of reddish maroon. Only some of these other forms seem to have been named. This is probably to the good, since some are forms in the true sense but others may represent interspecific hybrids (F. W. Case and G. L. Burrows 1962). Reports of *T. erectum* from west of the mapped distribution in Ohio, Michigan, and Illinois may represent *T. flexipes* forma *walpolei*, introgressive hybrids between outlying colonies of *T. erectum* and various other species (F. W. Case and G. L. Burrows 1962; F. W. Case and R. B. Case 1997), or escapes from gardens.

4. Trillium flexipes Rafinesque, Autik. Bot., 133. 1840 • Bent trillium, declined trillium [E]

Trillium declinatum Gleason; *T. gleasonii* Fernald

Rhizomes somewhat erect, thick, praemorse. **Scapes** 1–several from single rhizome, round in cross section, 2–5 dm, glabrous. **Bracts** sessile; blade medium green without red or maroon undertones, rhombic, 7–25 × 7–25 cm, frequently wider than long, base attenuate from just above middle, apex acuminate. **Flower** flexed ca. 90° on summit of pedicel to face outward rather than upward, or variously carried on strongly declined pedicels; sepals weakly recurved, green, lanceolate, 14–45 mm, shorter than or barely equaling petals, margins entire, apex acuminate; petals flat or recurved in distal $^{1}/_{2}$, creamy white, not 2-colored, veins conspicuously engraved, ovate-lanceolate to broadly ovate, 2–5 × 1–4 cm, texture heavy, margins entire, apex acute; stamens large, ± erect, 9–23 mm; filaments white, less than $^{1}/_{2}$ anther length, thin; anthers straight or very slightly recurving, creamy white or yellow, 5–18 mm, thick, dehiscence ± introrse; ovary white, ovoid to flask-shaped, strongly 6-angled, 5–16 × 4–12 mm, widely attached basally; stigmas recurved, distinct but closely grouped, white, not lobed adaxially, linear-subulate, short, thick, 4–13 mm, ± equaling ovary, fleshy; pedicel stiffly erect, carried horizontally, or declined beneath bracts, rarely recurved, 4–12 cm. **Fruits** baccate, rosy red to purplish, fragrant of ripe fruit, ovoid to somewhat pyramidal at summit, strongly angled, 2–3.5 × 1–3 cm, very juicy at maturity. 2*n* = 10.

Flowering spring (Apr–early Jun). Rich wooded slopes, floodplains in deciduous forests, especially over limestone; 100–600 m; Ont.; Ala., Del., Ill., Ind., Iowa, Ky., Mich., Minn., Mo., N.Y., Ohio, Pa., S.Dak., Tenn., Va., W.Va., Wis.

Although there are no named varieties of *Trillium flexipes*, many forms exist and at least one has been named, forma *walpolei* (Farwell) Fernald. F. W. Case and G. L. Burrows (1962) mapped the occurrence of forma *walpolei* for Michigan and found it to occur only along the contact zone between *T. flexipes* and *T. erectum*. F. W. Case and R. B. Case (1993) crossed typical *T. erectum* and *T. flexipes* to produce identical color variations as occur in these wild, mixed-species populations. I consider this form to be a hybrid expression. Some of the hybrids have petals colored proximally and white distally, superficially resembling *T. undulatum*. Dried specimens, when hastily examined, resemble *T. undulatum* superficially and probably account for reports of that species from locations near Ann Arbor, Michigan, and various places in Indiana and Illinois, all of which are well out of the range of *T. undulatum*.

5. Trillium grandiflorum (Michaux) Salisbury, Parad. Lond. 1: plate 1. 1805 • White trillium, great white trillium, white wake-robin, trille grandiflore [E] [F]

Trillium rhomboideum Michaux var. *grandiflorum* Michaux, Fl. Bor.-Amer. 1: 216. 1803; *T. erythrocarpum* Curtis 1805, not Michaux 1803

Rhizomes short, thick, praemorse. **Scapes** (1–)2–3(–many), round in cross section, 1.5–3+ dm, thick, glabrous. **Bracts** sessile or subsessile (occasionally weakly cuneate basally); blade dark green with maroon overtones early, ovate-rhombic, 12–20 × 8–15 cm, apex acuminate. **Flower** outfacing, erect, odorless; sepals spreading, flat, green, very rarely streaked with maroon-purple, lanceolate, 20–55 × 12–23 mm, margins entire, apex acuminate or acute; petals erect basally, recurving somewhat above middle to produce strongly funnelform corolla, obscuring ovary and base of style, white or rarely pink, without V-shaped or other markings, fading to dull pinkish purple, veins of adaxial surface conspicuous but not appearing engraved, shape variable, lanceolate to oblong, obovate, or, rarely, suborbicular, sides often parallel, 4–7.5 × 2–4 cm, thin-textured, base abruptly attenuate, margins overlapping basally, rolled, undulate-wavy in distal $^{1}/_{2}$, apex ± acuminate; stamens straight or barely recurved, 9–27 mm; filaments white, much shorter than anthers, relatively thin; anthers recurving slightly, pale yellow, strongly yellow when pollen exposed, long, 5–16 mm, slender, dehiscence introrse; ovary inconspicuous, pale green or white, ovoid, 6-angled, 8–18 mm, basal attachment narrower than ovary width; style barely united for 0.5–2 mm or merely closely grouped and separate; stigmas erect, becoming spreading, weakly connate basally, pale green-white, uniformly linear, 3–18 mm, equaling or exceeding ovary, slender; pedicel erect-ascending to strongly erect, 2–8+ cm. **Fruits** pale green, odorless, ± globose, obscurely 6-angled, 1.2–1.6 × 0.8–1.4 cm, mealy, moist (not juicy). 2*n* = 10.

Late spring–early summer (Apr–Jun). Rich deciduous or mixed coniferous-deciduous upland woods, floodplains, roadsides; 20–700 m; N.S., Ont., Que.; Conn., Del., Ga., Ill., Ind., Iowa, Ky., Maine, Md., Mass., Mich., Minn., N.H., N.J., N.Y., N.C., Ohio, Pa., S.C., Tenn., Vt., Va., W.Va., Wis.

Most variants of *Trillium grandiflorum* have green stripes or markings on the petals, many with numerous (4–30+) extra petals and/or bracts, and, often, much-deformed, monstrous characteristics. G. R. Hooper et al. (1971) showed that mycoplasmic organisms were present in all such forms examined, and were absent from normal plants. Most such forms should not be named taxonomically but, unfortunately, many have been. Nearly all of those that I examined represented stages in the development of the mycoplasma infection. *Trillium grandiflorum*, unlike most trilliums, produces many-petaled "double" forms. Forma *roseum* Farwell, opening a striking clear salmon-pink, occurs very rarely throughout the range, but is frequent in mixed or pure colonies along the Blue Ridge Mountains of Virginia.

SELECTED REFERENCE Patrick, T. S. 1973. Observations on the Life History of *Trillium grandiflorum* (Michaux) Salisbury. M.A. thesis. Cornell University.

6. Trillium nivale Riddell, Syn. Fl. West. States, 93. 1835 • Snow trillium, dwarf white trillium [E]

Rhizomes short, praemorse. **Scapes** usually 1, 6-gonal in cross section, 0.3–0.5 dm at onset of anthesis, expanding to 0.45–0.8 dm, slender, glabrous. **Bracts** distinctly petiolate; blade bluish green, elliptic-ovate to ovate, 1.5–4.5 cm × 0.7–34 mm, base abruptly rounded to petiole, apex round-obtuse. **Flower** erect, fragrance sweet; sepals spreading, flat, bluish green, lanceolate, 10–32 × 2–7 mm, shorter than petals, margins entire, apex obtuse; petals very showy, recurved to erect-spreading, white, veins not engraved on adaxial surface, ovate-elliptic to oblong, 1.5–3.5+ × 0.8–1.5 cm, thin- to firm-textured, margins entire to slightly wavy, apex obtuse-acuminate; stamens straight, 5–18 mm; filaments white, slightly shorter than anthers, slender; anthers straight, pale yellow, 2.5–11 mm, thin, dehiscence introrse; ovary greenish white, ± globular, obtusely 3-angled; style with tips spreading, elongate, 0.5–1.5 mm, slender; stigmas prominent, recurved or curled, connate basally, white, linear, threadlike, 4–12+ mm; pedicel erect in anthesis, rapidly recurving below leaves after pollination, 5–20 cm. **Fruits** greenish white, odorless, globose-ovate, 0.6–1 × 0.5–0.8 cm, pulpy, not juicy. $2n = 10$.

Flowering spring (early Mar–early Apr). Forested, limestone-derived soils, alkaline glacial drift or loess, creeping soils at head of ledges, talus of cliff bases, crevices in limestone cliffs, gravelly deposits on higher floodplain riverbanks; 100–300 m; Ill., Ind., Iowa, Ky., Md., Mich., Minn., Mo., Nebr., Ohio, Pa., S.Dak., W.Va., Wis.

Trillium nivale occurs primarily at the southern edge of Pleistocene glaciation and shuns humus, leaf deposits, and much plant competition.

7. Trillium ovatum Pursh, Fl. Amer. Sept. 1: 245. 1814 • Western white trillium, white trillium [E] [F]

Rhizomes semierect to horizontal, short, stout, praemorse. **Scapes** 1–2, round, 2–5 dm, ± slender, glabrous. **Bracts** sessile, subsessile, or short-petiolate; blade medium green, sometimes blotched and mottled, main veins prominent, ovate-rhombic, 7–12 × 5–20 cm, continuing to expand during anthesis, base rounded, apex acuminate. **Flower** erect or nodding, odorless; sepals spreading to horizontal, green, lanceolate to oblong-lanceolate, 15–50 × 6–20 mm, margins entire, apex acute; petals erect-ascending, usually wide-spreading from base, exposing entire pistil, white or with pink or blush markings, lacking V-shaped markings, fading to rosy pink, purple, or dark red, veins not deeply engraved, ± linear to widely obovate, 1.5–7 ×1–4 cm, widest at or above middle, thin-textured, margins flat to undulate, apex acuminate; stamens prominent, slightly recurved-spreading to straight, 10–18 mm; filaments white, shorter than anthers, slender; anthers yellow, 4–16 mm, slender, dehiscence latrorse-introrse; ovary green or white, ovoid, 6-angled, 5–12 mm, attachment ± 3/4 ovary width; stigmas recurved, barely connate basally, greenish white or white, linear, not lobed adaxially, 6–10 mm, uniformly thin; pedicel erect to leaning, 2–6 cm. **Fruits** baccate, green or white, ± odorless, broadly ovoid, obscurely winged, 1.2–2.8 × 0.7–1.9 cm, pulpy-moist. $2n = 10$.

Varieties 2 (2 in the flora): w North America.

1. Bracts sessile; petals lanceolate to obovate, 1.5–7 × 1–4 cm . . . . 7a. *Trillium ovatum* var. *ovatum*
1. Bracts distinctly short-petiolate; petals linear to linear-lanceolate, 0.5–2.4 × 0.2–0.6 cm . . . . 7b. *Trillium ovatum* var. *oettingeri*

7a. Trillium ovatum Pursh var. **ovatum** [E] [F]

Trillium ovatum var. *stenosepalum* R. R. Gates

Bracts sessile. **Flower** erect; petals lanceolate to obovate, 1.5–7 × 1–4 cm; pedicel erect.

Flowering late winter–spring (late Feb–Apr). Coniferous and mixed coniferous-deciduous forests, alder thickets, shrubby areas; 60–1600 m; Alta., B.C.; Calif., Colo., Idaho, Mont., Oreg., Wash., Wyo.

Two forms of *Trillium ovatum* should be noted. Forma *hibbersonii* T. M. C. Taylor & Szczawinski, a dwarf, grows on sea cliffs on Vancouver Island, British Columbia. Plants range from 3 to 10 cm tall, with all parts proportionally diminished, and petals opening clear pink, fading to white. T. M. C. Taylor and A. F. Szczawinski (1974) stated that this form occurs occasionally throughout the range of the species. Forma *maculosum* Case & R. B. Case, with bracts blotched or spotted with dark maroon or maroon-green markings on the adaxial (but not the abaxial) surface in the manner of the sessile trilliums, and with moderate-sized flowers, is locally abundant in Mendocino County near Gualala. This is the only reported instance of maculate leaves in a pedicellate *Trillium*.

7b. Trillium ovatum Pursh var. **oettingeri** (Munz & Thorne) Case, Taxon 50: 201. 2001 • Salmon Mountains wake-robin [E]

Trillium ovatum subsp. *oettingeri* Munz & Thorne, Aliso 8: 15, fig. 1. 1973

Bracts distinctly short-petiolate. **Flower** nodding; petals linear to linear-lanceolate, 0.5–2.4 × 0.2–0.6 cm; pedicel ± leaning to almost horizontal.

Flowering spring–summer (May–Jul). Coniferous or mixed montane forest; 1200–2000 m; Calif.

8. Trillium persistens W. H. Duncan, Rhodora 73: 244. 1971 [C] [E]

Rhizomes horizontal to erect, short, praemorse. **Scapes** 1–2, round in cross section, 2–3 dm, slender, glabrous. **Bracts** horizontal to drooping distally, sessile; blade 3–5-veined, ovate-lanceolate to lanceolate, 3–8.5+ × 1.5–3.5 cm, adaxial surface faintly glossy, rarely with ± 2 mm, winged, petiolelike base, apex acuminate. **Flower** opening above bracts; sepals spreading, green, elliptic to narrowly ovate, 11–22 × 5–6 mm, thin-textured, margins entire, apex acute; petals erect proximally, spreading distally, white, fading to deep pink with inverted V-shaped basal portion remaining white, veins not engraved, linear-elliptic to occasionally linear, 2–3.5 × 0.5–1 cm, thin-textured, margins undulate at least in distal portion, apex acute; stamens prominent, erect to divergent, straight, 9–14 mm; filaments ± equaling anthers; anthers straight, yellow or white, dehiscence introrse; connective barely longer than anther sacs; ovary white or greenish white, obovate, very sharply 6-angled, 2.5–6 mm; style 2–6 mm; stigmas erect, slightly divergent at tip, delicate, not lobed, shortly connate basally, uniformly thin; pedicel erect or slightly leaning, 1–3 cm, 1/4–1/2 bract length at anthesis. **Fruits** baccate, greenish white, 6-angled, pulpy, not juicy.

Flowering spring (early Mar–mid Apr). Humus soils in mixed deciduous-pine woodlands, along stream flats and at edges of *Rhododendron* thickets, occasionally in open *Vaccinium*-filled clearings; of conservation concern; 50 m; Ga., S.C.

Listed as a U.S. endangered species, *Trillium persistens* has appeared as such on a postage stamp. The species is very rare and known only from an approximately four-square-mile area at the head of Tallulah Gorge in Georgia and South Carolina.

9. Trillium pusillum Michaux, Fl. Bor.-Amer. 1: 215. 1803 • Dwarf trillium, least trillium [E]

Trillium pumilum Pursh

Rhizomes horizontal, branching, thin. **Scapes** 1–2, round in cross section, 0.7–2 dm, slender, becoming taller and more robust after flowering, glabrous. **Bracts** very short-petiolate, subsessile or sessile; blade dark green with maroon undertones when young, not mottled, 3–5 major veins from base, oblong to lanceolate-obtuse, 2.5–8+ × 1–3 cm, not glossy, apex obtuse. **Flower** above bracts, erect, odorless to faintly sweet, pedicellate or sessile; sepals conspicuous, spreading to same plane as petals, dark green with maroon undertones when young, oblong-lanceolate, 15–30 × 5–10 mm, margins entire, apex obtuse to strongly rounded; petals of short duration, spreading-ascending, exposing stamens and ovary, weakly recurved in distal 1/2, white, aging to deep rosy pink abaxially, veins not engraved but major petal veins clearly visible, oblong to narrowly lanceolate, 1.5–3 × 0.5–1.5 cm, thin-textured, widest above base, margins strongly undulate, quite variable in petal width and degree of undulation between individuals and populations, apex obtuse to weakly acute; stamens erect-spreading, 8–10 mm; filaments pinkish purple to white, ± equaling or slightly shorter than anthers, slender; anthers ± straight, pale lavender or yellow, 3–8(–10) mm, thicker than filaments, dehiscence introrse; connectives not extended beyond anther sacs; ovary conspicuous, white, ovoid, obscurely 6-angled, 2.5–8 mm, attachment narrower than ovary; stigmas confluent with style, greenish white to white, distally 3-lobed, lobes linear (threadlike), long-spreading, 3–12 mm, uniformly thin and threadlike; pedicel stiffly erect to leaning, 0.5–2 cm, or absent to much reduced. **Fruits** white or pale greenish, ovate, 1–1.5 cm, pulpy, moist but not juicy. $2n = 10$.

Varieties 2 (2 in the flora): se, sc United States.

Trillium pusillum comprises widely disjunct, regional populations, each varying somewhat from the others and variable within itself as well. Some of these populations have been named as varieties or separate species. In the wild, the plants that have been recognized as var. *ozarkanum* generally grow taller than others and are said to have bracts with five major veins instead of three. Plants attributed to var. *texanum*, on the other hand, are generally smaller in all parts, with narrower petals, and often revert to a single bract when not flowering. Only var. *virginianum* is easily distinguished at sight by its "sessile" flower. This variety has received extensive study. P. R. Cabe (1995), in a morphological study including statistical analysis, found variation within and between populations. He felt that his results were inconclusive, and also that some of the variation might be environmentally induced. The variation that he found did not correlate with a geographic pattern, and he suggested treating all Virginia populations as var. *virginianum*, or simply *T. pusillum*. In a later study, P. R. Cabe and C. R. Werth (1995), using isozyme evidence, obtained like results, and suggested treating all Virginia populations as a single variety pending further investigation. Until there has been such study, of the Virginia plants as well as the rest of the *T. pusillum* complex, I choose to retain the fairly distinctive and more or less traditionally known var. *virginianum*, and include all other populations in a broadly circumscribed var. *pusillum*.

SELECTED REFERENCES Cabe, P. R. 1995. The *Trillium pusillum* Michaux (Liliaceae) complex in Virginia. I. Morphological investigations. Castanea 60: 1–14. Cabe, P. R. and C. R. Werth. 1995. The *Trillium pusillum* Michaux (Liliaceae) complex in Virginia. II. Isozyme evidence. Castanea 60: 15–29.

1. Flower pedicillate; pedicel usually 0.5–2 cm . . . . 9a. *Trillium pusillum* var. *pusillum*
1. Flower sessile or subsessile; pedicel, if present, 0.1–0.3 cm . . . . 9b. *Trillium pusillum* var. *virginianum*

9a. Trillium pusillum Michaux var. **pusillum** [E]

Trillium ozarkanum E. J. Palmer & Steyermark; *T. pusillum* var. *monticulum* Bodkin & Reveal; *T. pusillum* var. *ozarkanum* (E. J. Palmer & Steyermark) Steyermark; *T. pusillum* var. *texanum* (Buckley) Reveal & C. R. Broome; *T. texanum* Buckley

Flower pedicellate; pedicel usually 0.5–2 cm.

Flowering spring (Mar–early May). Low, swampy woodlands, sometimes sphagnous, mixed deciduous forests, often associated with cherty rock or decomposing shale in beech and oak forests; 0–600 m; Ala., Ark., Ky., La., Miss., Mo., N.C., Okla., S.C., Tenn., Tex.

9b. Trillium pusillum Michaux var. **virginianum** Fernald, Rhodora 45: 397, plate 773, figs. 1, 2. 1943 [E]

Trillium virginianum (Fernald) C. F. Reed

Flower sessile or subsessile; pedicel, if present, 0.1–0.3 cm.

Flowering spring (Mar–early May). Acidic soils in low, swampy woodlands along streams, red maple (*Acer rubrum*) swamps, very wet in spring, plants often grouped on hummocks, possibly with sphagnum moss; 0–300 m; Md., N.C., Va., W.Va.

The flowers of var. *virginianum* are usually slightly smaller than those of var. *pusillum* and otherwise very similar. The name *virginianum* indicates the main area of occurrence, and this is the only variety found in the tidewater area, where it is frequent around Williamsburg. It can also be found in the Delmarva Peninsula.

10. Trillium rivale S. Watson, Proc. Amer. Acad. Arts 20: 378. 1885 • Brook wake-robin [E] [F]

Rhizomes horizontal, short, praemorse, fleshy. **Scapes** 1–2, round in cross section, 0.4–1.5 dm, slender, glabrous. **Bracts** distinctly petiolate; blade bluish green, occasionally with silvery green pattern along major veins, ovate-lanceolate, 1.5–8 × 0.8–6 cm, somewhat thick-textured, glossy, weakly to strongly cordate basally, apex long-acuminate; petiole (0.5–)1–3(–6) cm. **Flower** erect, rapidly recurving below bracts after pollination, ± odorless; sepals conspicuous, spreading, green, weakly sulcate, oblong or tapered apically, usually shorter and narrower than petals, 10–23 × 4–8 mm, margins entire, apex rounded-apiculate; petals erect-spreading, not recurved or only in distal 1/2, white (rarely pale pink), usually with dark madder-purple flecks, rarely almost entirely covered with madder-purple, color not changing with age, veins inconspicuous, not engraved, ovate-cordate to ± orbicular, tapered bluntly apically above middle, 1.3–2.8 × 1–2.4 cm, ± thin-textured, base rounded, cuneate, margins entire, mostly without undulations, apex ± apiculate-acuminate; stamens erect, 5–12 mm; filaments white, ± equaling anthers; anthers straight, yellow, 3–6 mm, dehiscence ± extrorse; connectives not extended beyond anther sacs; ovary creamy white, ovoid, weakly 6-angled but angles in close pairs, giving superficial impression of 3-angled ovary, short, 3–9 × 4 mm, somewhat obscured by anthers broadly attached; stigmas recurved and appearing ± lobed adaxially, at least in dried specimens, connate basally to distinct and closely

TRILLIUM

grouped, white or cream, 2–4 mm, uniformly thick, fleshy; pedicel strongly erect at anthesis, strongly recurving and elongating below bracts after pollination, 2.5–11 cm. **Fruits** greenish white, odorless, globose, 0.9–1.2 × 0.8–1 cm, pulpy, not juicy. $2n = 10$.

Flowering spring (early Mar–early Jun); withering in summer. Open, grassy hillsides, manzanita shrubbery, pine groves, clearings in sequoia groves, roadside banks, ledges, gravelly talus; 10–1400 m; Calif., Oreg.

Trillium rivale is confined to the Klamath and Siskiyou mountains of southwestern Oregon and California. It has a strong affinity for banks along streams and adjacent to trickles in *Darlingtonia* bogs, mainly on serpentine soils. After flowering, the entire plant may enlarge and become more robust and turgid with very glossy leaves.

11. Trillium rugelii Rendle, J. Bot. 39: 331, plate 426, fig. B 1901 (as rugeli) • Southern nodding trillium [E]

Rhizomes short, thick, tapered to point distally. **Scapes** 1–3, round in cross section, 1.5–4 dm, robust, glabrous. **Bracts** sessile to subsessile; blade bright green, veins not engraved, rhombic, broader than long, 6–15 × 6–16 cm, not glossy, base attenuate, apex acuminate. **Flower** strongly recurved below bracts; sepals shorter than petals and somewhat obscured by them, spreading, green, very rarely streaked with red, lanceolate-elliptic, 15–40 × 7–17 mm, margins entire, apex acuminate; petals recurving in distal $^1/_2$ in most forms (forms from Alabama with petals slightly narrower and strongly recurved just above base), white, rarely 2-colored with base rose or dark purple and white, adaxial veins conspicuous, broadly ovate-elliptic, 2.5–5 × 0.8–3.5 cm, heavy-textured, margins entire, without undulations, apex abruptly acuminate; stamens ± straight, 8–18 mm; filaments deep purple, 2–8 mm, slender; anthers ± straight, dark purple, 12–16 mm, thin, dehiscence introrse; connectives not extended beyond anther sacs; ovary prominent, purple-streaked or maroon distally or throughout, flask-shaped, 6-angled, 14–17 mm, widely attached at base; stigmas recurved, distinct, dark purple, not lobed adaxially, subulate, 2–5 mm, fleshy; pedicel strongly recurved beneath bracts, 1–7+ cm. **Fruits** baccate, dark reddish purple, fragrance faintly of fruit, ovoid to orbicular, 1.7 × 2 cm, fleshy, juicy. $2n = 10$.

Flowering mid spring (mid Apr–May). Deciduous forest hillsides and coves, inner piedmont, mostly in alluvial soils along stream banks and flats; 200–700 m; Ala., Ga., N.C., S.C., Tenn.

The Alabama plants of *Trillium rugelii* differ from those in the Great Smoky Mountains region in having usually smaller bracts, narrower ovate petals strongly recurved from the base, and strongly fragrant flowers with the odor of old-fashioned garden roses. The anther color and structure are the same as in the mountain form. Hybrids between *T. vaseyi* and *T. rugelii* occur frequently.

Trillium rugelii has been much confused with *T. cernuum* in past floras, and it is the taxon upon which most reports of *T. cernuum* south of Virginia are based.

12. Trillium simile Gleason, Bull. Torrey Bot. Club 33: 391. 1906 • Sweet white trillium, confusing trillium E

Trillium vaseyi Harbison var. *simile* (Gleason) Barksdale

Rhizomes forming clumps, stout, praemorse. **Scapes** 1–many, round in cross section, 3–6 dm, stout, glabrous. **Bracts** sessile to subsessile; blade green, major veins prominent, rhombic, 10–18 × 10–20 cm, not glossy, tapered basally, apex short-acuminate. **Flower** above bracts, erect to mostly leaning, odor faintly sweet, applelike; perianth gaping, strongly 3-dimensional; sepals spreading, green, flat, oblanceolate-lanceolate, 20–40 × 6–15 mm, margins entire, apex mildly sulcate; petals spreading-ascending, not recurved to weakly so at tip, creamy white, flat, adaxial veins faintly engraved, ovate to ovate-orbicular, 4–7+ × 1.5–4 cm, 1.5 times as long as sepals, heavy-textured, base rounded, margins entire, apex acuminate; stamens erect to weakly recurved, 7–20 mm; filaments purple or brownish, shorter than anthers, slender; anthers weakly recurved, yellow with brown undertones, 6–20 mm, longer than ovary, slender, dehiscence introrse; connectives purple-brown, not extending beyond anther sacs; ovary dark purplish black, pyramidal at anthesis, very strongly 6-angled, 7–12 mm, widely attached basally; stigmas short, mildly recurved, distinct, purple or yellow, not lobed adaxially, 2–5.5 mm, fleshy, basally widened to gradually tapered; pedicel ± erect to mostly leaning, 4–9 cm. **Fruits** baccate, dark purplish black, odorless, orbicular, 1–1.5 cm diam., fleshy, not juicy.

Flowering mid spring (Apr–May). Rich coves of mature forests, edges of *Rhododendron* thickets and at edges of forests, in moist humus soil; 500–700 m; Ga., N.C., Tenn.

A recent study of *mat*K gene sequencing (S. Kazempour Osaloo et al. 1999) placed *Trillium simile* with the *T. grandiflorum* group and *T. catesbaei*, which has united styles and white-angled ovaries, not with the *T. erectum* group. In my opinion, this may reflect an error in labeling samples, for *T. simile* has the dark, round ovary with three, separate, subulate stigmas and other characteristics of the *T. erectum* alliance. Also, it hybridizes with *T. erectum* and other species of that alliance. L. Barksdale (1938) described a complex of forms that he considered to be the result of such hybridization. I have seen such complexes near Maryville, Tennessee, where *T. simile* and *T. erectum* forma *album* occur together with a full range of intergrades between the two. *Trillium catesbaei* and the species related to *T. grandiflorum* do not hybridize with any species, and all have slightly to clearly fused, linear styles.

Clearly *Trillium simile* is closely related to *T. vaseyi* and *T. erectum*, but it seems to be a distinct species, though somewhat difficult to identify when not in its most robust condition.

J. K. Small (1933) reported *Trillium simile* to be deliciously fragrant, a quality I have not noticed in my plants.

13. Trillium sulcatum T. S. Patrick, Brittonia 36: 27, figs. 1–4. 1984 • Southern red trillium, Barksdale trillium E

Rhizomes horizontal–slightly erect, thick, praemorse. **Scapes** 1–4+, round in cross section, 3–7 dm, 2–2.5 times as long as bract, stout, glabrous. **Bracts** subsessile; blade obovate to broadly elliptic, 13–20 × 8–22 cm, not glossy, base attenuate, apex acuminate. **Flower** facing outward at right angle to pedicel, above bracts, odor faintly musty, like fresh fungus; perianth gaping, strongly 3-dimensional; sepals spreading, usually streaked or entirely purplish maroon, occasionally entirely green except on margins, elliptic, 15–38 × 9–15 mm, margins entire, apex strongly sulcate-acuminate; petals carried somewhat forward to recurved-spreading in distal 1/2, usually dark reddish maroon to purplish, rarely cream to soft yellow, pink, white, or 2-colored, heavy-veined, ovate to broadly ovate-overlapping, 1.8–5 × 1–3 cm, heavy-textured, apex acuminate, rarely somewhat sulcate; stamens erect, ± equaling or slightly longer than ovary, 15–18 mm; filaments purple to white, 3–5 mm, slender; anthers straight, purplish to yellow, 5–12 mm, thick, dehiscence introrse; connectives equaling or ± shorter than anther sacs; ovary usually dark purple, globose to flask-shaped, 6-angled, 14–18 mm, broadly attached basally; stigmas prominent, recurved, distinct, purple, not lobed adaxially, basally thickened, gradually tapered, 2–5 mm, fleshy; pedicel usually stiffly erect (rarely almost horizontal above bracts), straight, reflexed ± 90° at tip, 6–11 cm. **Fruits** red, fragrance of fresh mushrooms, ± globose to pyramidal, 1.5–3 × 1.2–2.8 cm, juicy to pulpy. $2n = 10$.

Flowering mid spring (Apr–May). Rich mesic woodlands, especially moist north- or east-facing slopes, wooded ledges and stream banks on neutral to slightly acid soil; 300–400 m; Ala., Ga., Ky., N.C., Tenn., Va., W.Va.

Trillium sulcatum is present mainly on the Cumberland Plateau and is absent from the Great Smoky Mountains and the southern Blue Ridge Mountains.

T. S. Patrick (1984) described *Trillium sulcatum* flowers as "relatively small and turned downward." This is true of plants from the type locality; however, in most plants of the Cumberland Plateau, the flowers are quite large and flattened, and the petals are recurved distally and face outward.

SELECTED REFERENCE Patrick, T. S. 1984. *Trillium sulcatum* (Liliaceae), a new species of the southern Appalachians. Brittonia 36: 26–36.

14. Trillium undulatum Willdenow, Ges. Naturf. Freunde Berlin Neue Schriften 3: 422. 1801 • Painted trillium, painted lady, trille ondulé [E]

Trillium erythrocarpum Michaux

Rhizomes horizontal, short, stout, ± praemorse. **Scapes** 1–3, dark greenish maroon, round in cross section, 1.1–4 dm, glabrous. **Bracts** strongly petiolate; blade dark green over maroon, main veins prominent, ovate to long-acuminate, 12–18 × 8–20 cm, ± glossy, base above petiole rounded, apex acuminate; petiole 4–17 mm. **Flower** opening above bracts, erect, odor unknown; sepals conspicuous, spreading, wine red to dull maroon green, rarely white-striated, lanceolate-acuminate, 13–37 × 4–10 mm, margins entire, apex acuminate; petals spreading, white, with distinct, inverted, V-shaped, dark-red mark basally, the red radiating outward along major veins, or white and lacking red marks, veins not engraved, oblong-ovate, occasionally lanceolate, usually widest just above middle, 2–5 × 1–2 cm, base tapering very gradually to attachment, margins not undulate basally, undulate apically, apex somewhat rapidly acuminate; stamens straight, 8–12 mm; filaments pink or white, equaling or longer than anthers, slender; anthers erect or slightly spreading, white or pink, 2–7 mm, thin, dehiscence extrorse; connectives pink, equaling anther sac; ovary fully exposed, white, pink tipped, obtusely 3-angled, becoming obscurely angled-rotund in cross section, 3–10 mm, broadly attached; stigmas ascending, barely connate basally, then strongly recurved, white, ± linear, 3–10 mm, uniformly thin; pedicel erect, 2–5 cm. **Fruits** quickly deciduous upon ripening, scarlet, obscurely 3-angled to cylindrical, 1–2 cm, fleshy, juicy. $2n = 10$.

Flowering late spring–early summer (late Apr–Jun). Deep acid humus of pine, spruce, hemlock, oak, mixed deciduous-coniferous woods, rhododendron or mountain laurel thickets in s range, almost any acidic forested situation n and ne, often around old *Pinus strobus* stumps in white birch-red maple-white pine second-growth woods in n range; prefers deep shade except at high elevations in range; not usually found on limestone-derived or basic soils unless in very deep acid humus; 10–1800 m; N.B., N.S., Ont., P.E.I., Que.; Conn., Ga., Ky., Maine, Md., Mass., Mich., N.H., N.J., N.Y., N.C., Ohio, Pa., R.I., S.C., Tenn., Vt., Va., W.Va.

Several forms of *Trillium undulatum* have been described. One, forma *enotatum* T. S. Patrick, lacks the usual red, V-shaped petal markings. Other named forms have green petals, extra leaves or petals, and deformities suggesting a mycoplasma infection as in *Trillium grandiflorum*.

This difficult-to-cultivate species has been reported from Wisconsin, but it cannot be accepted as native there.

15. Trillium vaseyi Harbison, Biltmore Bot. Stud. 1: 24. 1901 • Sweet beth, sweet trillium, Vasey's trillium [E]

Trillium erectum Linnaeus var. *vaseyi* (Harbison) H. E. Ahles

Rhizomes short, stout, praemorse. **Scapes** 1–2 per rhizome terminus, round in cross section, 3–6.5 dm, ± slender to stout, glabrous. **Bracts** sessile to subsessile; blade rhombic, 10–20 × 6–20 cm, often wider than long, not glossy, base attenuate, apex acuminate. **Flower** carried beneath bracts, odor faintly sweet; sepals spreading, pale green, lanceolate-acuminate, 25–50 × 6–18 mm, margins involute, apex acuminate; petals spreading to recurved distally, adaxially crimson, maroon-red, or brownish red, abaxially paler, grayish pink or rose, veins engraved, ovate-suborbicular, 3–6.5 × 4–6 cm, somewhat fleshy, base rounded and overlapping, margins entire, apex acute; stamens conspicuous, ± erect to weakly recurved, 15–25 mm, longer than pistil at anthesis; filaments grayish purple to blackish purple, longer than anthers, 5–12 mm, slender; anthers weakly recurving, grayish purple to maroon, slender, dehiscence introrse; connectives purple, barely equaling anther sacs; ovary small, maroon or dark reddish purple, ± globose to conical-pyramidal, 6-ridged, 3–12 mm, basal attachment less than ovary width; stigmas erect, recurved, distinct, gray-purple, not lobed adaxially, basally widened, ± linear distally, 2.5–6.5 mm, fleshy; pedicel horizontal to declined-drooping, ± straight, 2–13 cm. **Fruits** dark reddish maroon, ovoid, obtusely angled, relatively small, 1–1.4 × 1–2 cm, pulpy. $2n = 10$.

Flowering mid–late spring (late Apr–early Jun). Rich woods, often on steep slopes, ravines, stream banks, and deep, wind-sheltered, moist coves; 300–700 m; Ala., Ga., N.C., S.C., Tenn.

Although sometimes submerged within *Trillium erectum*, *T. vaseyi* has a later blooming season, a nodding flower of much larger size, a sweet fragrance, and cove habitat unlike that of *T. erectum*. In my garden hybridization experiments, *T. vaseyi* hybrids have different color patterns than *T. erectum* hybrids. *Trillium vaseyi* is clearly a distinct species. It frequently hybridizes with *T. rugelii*.

16b. TRILLIUM Linnaeus subg. PHYLLANTHERUM Rafinesque, J. Phys. Chim. Hist. Nat. Arts 91: 72. 1820 [E]

Bract blades green, variously mottled in darker green or bronze, except not mottled in *T. petiolatum*. **Flower** sessile. **Petals** ± erect, except spreading in *T. stamineum*, mostly dark purple, maroon, yellow, green, or white, sometimes bicolored (dark basally).

Species 23 (23 in the flora): nw, e United States.

SELECTED REFERENCE Freeman, J. D. 1975. Revision of *Trillium* subgenus *Phyllantherum* (Liliaceae). Brittonia 27: 1–62.

16. Trillium albidum J. D. Freeman, Brittonia 27: 48, fig. 11. 1975 • White toadshade [E]

Rhizomes horizontal, ± erect, brown, superficially bulblike, short, thick, not brittle. **Scapes** often several from same rhizome terminus, round in cross section, 2.2–5.8 dm, stout. **Bracts** held well above ground, sessile; blade weakly mottled with scattered, darker green spots, mottling often fading later in season, broadly ovate, 10–20 × 12–15 cm, not glossy, base rounded, apex obtuse. **Flower** borne directly upon bracts, erect, fragrance roselike; sepals conspicuous, spreading, displayed above bracts, pale green, lanceolate, 30–65 × 12–15 mm, margins entire, apex acute; petals long-lasting, usually erect or slightly spreading, ± connivent, ± concealing stamens and ovary basally, white to creamy white, rarely soft pinkish rose near base, veins prominent but not engraved, not spirally twisted, obovate, more lanceolate in young plants, 4.8–8 × 2.2–3 cm, widest at or just above middle, ± thick-textured, base cuneate, margins entire, ± flat, apex rounded to acute; stamens erect, greenish white, 15–25 mm; filaments whitish green, 3–4 mm, slender; anthers erect, straight, yellow, 11–20 mm, dehiscence latrorse; connectives straight, extended ca. 1 mm beyond anther sacs, rounded; ovary green or occasionally purple, ovoid, rounded 6-gonal, 6–11 mm; stigmas erect to spreading, distinct, sessile, subulate, 4–7 mm, thin distally. **Fruits** green or purplish green, fragrance unknown, ovoid to globose, pulpy, juicy.

Flowering spring (mid Mar–late Apr). Moist slopes in rich mixed deciduous-coniferous forests or coniferous stands, brushy thickets on flats, open fields, pastures, and fencerows, dense second-growth coniferous forests, floodplains along streams and larger rivers; 100–200 m; Calif., Oreg.

17. Trillium angustipetalum (Torrey) J. D. Freeman, Brittonia 27: 55. 1975 • Narrow-petaled trillium [E] [F]

Trillium sessile Linnaeus var. *angustipetalum* Torrey, Pacif. Railr. Rep. 4(5): 151. 1857; *T. giganteum* (Hooker & Arnott) A. Heller var. *angustipetalum* (Torrey) R. R. Gates

Plants tall, very robust. **Rhizomes** erect, brownish, thick, somewhat compressed-thickened, praemorse, not brittle. **Scapes** 1–2, round in cross section, 2.5–6 dm. **Bracts** held well above ground, spreading horizontally, subsessile; blade very sparsely mottled with dark greenish brown or rarely all green, mottling becoming obscure with age, broadly ovate, 10–22 × 8.7–15 cm, not glossy, often narrowed to falsely petiolate, very short, and narrowly cuneate base 10–20 mm, apex obtuse. **Flower** erect, odor spicy-musty, musty, or fetid; sepals conspicuous, spreading, often resting on bracts, maroon to green, linear to oblong-lanceolate, 35–47 × 8–10 mm, margins flat, entire, apex acute; petals long-lasting, erect, ± connivent, ± concealing stamens and ovary and partially obscuring stamens, dark purple to red-purple, not spirally twisted, veins obscure, linear, 5–10 × 0.7–1.4 cm, 8–10 times longer than wide, glossy, thick-textured, base linear, margins entire, at first flat but inrolling with age, apex variously acute-obtuse; stamens erect, 12–22 mm; filaments dark maroon, 2–4 mm, slender, widest at base; anthers erect, straight, purple, 12–18 mm, dehiscence introrse; connectives purple, slightly extended 1–1.5 mm beyond anther sacs; ovary dark, ovoid-ellipsoid, 6-angled toward apex, 7.5–12 mm; stigmas erect, divergent-recurved, distinct, purple, sessile, awl-shaped, thickly subulate, 5 mm, thick, fleshy. **Fruits** dark purple, fragrance unknown, subglobose, 6-angled, almost winged, fleshy. $2n = 10$.

Flowering spring–early summer (Mar–Jun). Big-tree (*Sequoiadendron*) groves and other mixed coniferous-deciduous flatwoods, slightly damper depressions under maples and deciduous shrubs; coastal mountains, oak (*Quercus*) groves in ravines and otherwise quite arid, almost treeless chaparral, wooded canyon slopes, dense woods near streams; 30–200 m; Calif.

Trillium angustipetalum occurs in the Sierra Nevada from Fresno County north to Placer County (J. D. Freeman 1975). It is disjunct in the coastal mountains and hills of Santa Barbara and San Luis Obispo counties.

B. D. Ness (1993) listed *Trillium kurabayashii* as a synonym of *T. angustipetalum*. In bract orientation, color, and texture, and in petal shape, the two are quite different and certainly not the same species. Cytologist Masataka Kurabayashi found chromosomal differences between the two species (reported by J. D. Freeman 1975).

18. Trillium chloropetalum (Torrey) Howell, Fl. N.W. Amer., 661. 1902 • Giant trillium [E]

Trillium sessile Linnaeus var. *chloropetalum* Torrey, Pacif. Railr. Rep. 4(5): 151. 1857; *T. giganteum* (Hooker & Arnott) A. Heller var. *chloropetalum* (Torrey) R. R. Gates

Rhizomes ± erect, brownish, somewhat compressed-thickened, superficially bulblike, praemorse, not brittle. **Scapes** 1–3, green, round in cross section, 2–6.5 dm, robust. **Bracts** held well above ground, sessile (narrowing of bract blade may give bract subsessile appearance); blade densely to weakly mottled in dark brownish green, mottling becoming more obscure to absent as bract matures, broadly ovate, 7–17.6 × 7.4–17.7 cm, not glossy, apex obtuse-rounded. **Flower** erect, odor roselike, spicy; sepals spreading-ascending above bracts, green, lanceolate, 35–65 × 7–12 mm, margins entire, flat, apex obtusely rounded; petals long-lasting, erect, connivent, ± concealing stamens and ovary, yellow, bronze, maroon, brown, deep purple, reddish brown, pink, dark purplish red, purplish bronze, rarely greenish white, not spirally twisted, veins not engraved, oblanceolate to obovate, 6.5–10 × 1.5–2.5 cm, thick-textured, base cuneate, margins entire, apex variably acute to almost truncate, erose; stamens erect, purplish, 17–26 mm; filaments purple, ca. 4 mm, widest at base, much shorter than anther sacs; anthers erect, straight, ± purple-brown, 13–22 mm, dehiscence introrse; connectives purple, straight, extended ca. 1–1.5 mm beyond anther sacs; ovary purple, ovoid, 6-angled, 6–12 mm; stigmas small, divergent or erect, distinct, purple, subulate, 4–8 mm, not fleshy. **Fruits** red-purple, fragrance not reported, ovoid, obscurely 6-angled, 2.5–3 cm, pulpy, juicy. $2n = 10$.

Varieties 2 (2 in the flora): California.

J. D. Freeman (1975) considered that *Trillium chloropetalum* differs from *T. albidum* in having introrse (not latrorse) anther sacs, and that the purple pigments present on anther and ovary tissue here are absent in *T. albidum*. In some places, hybridization between the two certainly has occurred, and a complete range of intergrades exists.

This species merits further study. The following varieties are only weakly differentiated and perhaps ought to be dropped.

1. Petal tissue always with yellow pigments, often present with other pigments; petals yellow, greenish yellow, greenish purple, or bronze-green or brown .... 18a. *Trillium chloropetalum* var. *chloropetalum*
1. Petal tissue lacking yellow pigments; petals purple, reddish purple, garnet red, pink, or greenish white .... 18b. *Trillium chloropetalum* var. *giganteum*

18a. Trillium chloropetalum (Torrey) Howell var. **chloropetalum** [E]

Petals yellow, yellowish green, greenish purple, to deep purple, bronze-green or brown, never white, yellow pigment always present even when masked by purple.

Flowering late winter–spring (late Feb–early or mid Apr). Variable habitat: edges, openings in and along ravines in redwood forest, thickets and tangles of cut-over forest, stream flats in deciduous woods, chaparral brushland, open wooded sloped on dunes, brushy islands in surprisingly open grasslands, but usually where soil remains moist most of the season; 0–2000 m; Calif.

Trillium chloropetalum var. *chloropetalum* is found in San Mateo and Santa Clara counties and in the counties along the east side of San Francisco Bay, north to Napa and Lake counties in California. It is more frequent east and south of the San Francisco Bay region.

J. D. Freeman (1975) determined chemically that all plants of var. *chloropetalum* have yellow pigment present in petals, though it may be masked by purple. My own quite limited field observations seem to indicate that plants of this variety are somewhat shorter at flowering than those of var. *giganteum*, and that the bracts of var. *chloropetalum* show a slightly metallic luster at that stage of development.

18b. Trillium chloropetalum (Torrey) Howell var. **giganteum** (Hooker & Arnott) Munz, Aliso 4: 88. 1958 [E]

Trillium sessile Linnaeus var. *giganteum* Hooker & Arnott, Bot. Beechey Voy., 402. 1840

Petals greenish white to deep red or purple, lacking any yellow pigments.

Flowering late winter–spring (late Feb–early or mid Apr). Variable habitat: edges, openings in and along ravines in redwood forest, thickets and tangles of cut-over forest, stream flats in deciduous woods, chaparral brushland, open wooded slopes on dunes, brushy islands in surprisingly open grasslands, but usually where soil remains moist most of the season; 0–600 m; Calif.

Trillium chloropetalum var. *giganteum* occurs in and around the San Francisco Bay region.

The epithet "giganteum" is appropriate, considering this variety's large size. In the populations that I have observed, it was more robust and generally larger than var. *chloropetalum*. Narrow-petaled forms of var. *giganteum* occasionally occur in mixed populations with the wider-petaled ones, and in a few places almost exclusively. J. D. Freeman (1975) suggested that these could represent the "undiluted" var. *giganteum*, free from influences of hybridization, as the type specimen has narrow petals.

19. Trillium cuneatum Rafinesque, Autik. Bot., 133. 1840 • Whip-poor-will flower, cuneate trillium, large toadshade, purple toadshade, bloody butcher, sweet betsy [E]

Rhizomes horizontal, brownish, short, thick, praemorse, not brittle. **Scapes** 1–5, 1.6–4.5 dm, smooth to rough near bract attachment. **Bracts** held well above ground, sessile; blade green to purplish green, weakly to strongly mottled, mottling becoming obscure with age, ovate, ovate-elliptic, occasionally ovate-orbicular with margins overlapping, 7–18.5 × 7–13 cm, usually widest below middle, not glossy, base ± rounded, margins of distal 1/3 convex-curved to apex, apex acuminate to acute. **Flower** borne upon bracts, erect, odor usually pleasant, faint, spicy, reminiscent of odor of bruised sweetshrub (*Calycanthus*) leaves, occasionally musty or unpleasant; sepals widely spreading, variably green, purple-streaked to all purple, oblong-lanceolate, 27–60 × 7–13 mm, margins entire, apex rounded to acute; petals long-lasting, erect, ± connivent, ± concealing stamens and ovary, maroon, maroon-purple, brownish purple, bronze, greenish purple, clear green, yellowish green, pale lemon yellow, or 2-colored, yellow distally with purple base, in occasional clones, flowers open or quickly fade to bright copper bronze, the particular color pattern is consistent from year to year, fading to browner tones with age except in yellow or green forms, not spirally twisted, shape quite variable across range, elliptic-obovate to oblanceolate, 4–7 × 0.9–2.7 cm, thick-textured, with widest portion at or above middle, narrowed to usually cuneate basally, not clawed, margins flat, entire, apex acute, rounded-acute to obtuse; stamens erect, straight, brownish purple-green, 11–18 mm; filaments brownish purple, 1.5–3.5 mm, widest at base; anthers erect, straight, brownish gray, 7–14 mm, dehiscence latrorse or occasionally introrse; connectives straight, scarcely (0.5 mm or less) if at all extended beyond anther sacs; ovary maroon, ovoid to vase-shaped, weakly 6-angled or -ridged when mature, 12–15 mm; stigmas erect, slightly diverging to spreading, distinct, purplish gray, linear-subulate to thickly subulate, 4–15 mm, fleshy. **Fruits** green or with purple streaks, ovoid, very obscurely angled or angles no longer apparent, 2 × 1–1.5 cm, mealy or pulpy, fleshy, not juicy. $2n$ = 10.

Flower late winter–mid spring (early Mar–mid Apr). Rich, mostly upland woods, especially limestone soils, also less calcareous sites, occasionally found in old fields, ditches, or coal-mine tailings; 50–400 m; Ala., Ga., Ky., Miss., N.C., S.C., Tenn.

Trillium cuneatum has escaped locally and become established in Michigan and other states. It is most frequent on the Ordovician limestone-derived soils of southern Kentucky and Tennessee, and is perhaps the most vigorous and certainly the largest of the eastern sessile trilliums. Numerous, mostly unnamed color forms occur. Plants from the lower piedmont of Georgia, Alabama, and Mississippi, considered by most botanists to be of this species, have smaller, narrower petals than specimens from northeastern Alabama northward to Kentucky, and they are therefore sometimes difficult to place with certainty.

20. Trillium decipiens J. D. Freeman, Brittonia 27: 17, fig. 3. 1975 • Deceiving trillium [E]

Rhizomes horizontal, brownish, thick, praemorse, not brittle. **Scapes** 1–3, green or bronze-green, round in cross section, 1.7–4.4 dm, stout, glabrous. **Bracts** held horizontally, not drooping, tips at anthesis held well above ground, sessile; blade usually very strongly marked with at least 3 shades of dark green, bronze green, and purplish green, often with light central strip, mottling becoming obscure with age, ovate-lanceolate to lanceolate, widest at ca. 1/3 of length from basal attachment, tapered very gradually to tip, 8–17+ × 4.9–8.5 cm, rounded basally, margins of distal 1/3 straight, apex acute. **Flower** faintly ill-scented; sepals divergent-ascending, streaked with green to maroon, lanceolate to ovate-lanceolate, 36–68 × 12–21 mm, margins entire, flat or slightly raised adaxially, apex acute; petals long-lasting, erect, ± connivent, ± partially concealing stamens and ovary, maroon-purple, brownish purple to brown, greenish streaked to green, rarely yellow, becoming brown, or occasionally bright copper-bronze with age, not spirally twisted, veins not engraved, obovate to oblanceolate, large in proportion to leaf size compared to many species, 5–9 × 1–2 cm, 2+ times longer than wide, widest at or just above middle, thick-textured, margins entire, flat, apex acute, obtuse, or rounded; stamens erect or incurving, 12–24 mm; filaments yellow, 2–3 mm; anthers erect, straight, rarely arcuate, yellow, 10–15 mm, dehiscence latrorse; connectives straight, projecting 1–2 mm beyond anther sacs; ovary dark red, brown, or gray, ellipsoid, strongly 6-angled, 6–13 mm; stigmas basally erect, tips recoiled upon ovary, distinct, green, white, or purple, linear, short, 3–12 mm, slightly thickened basally, not fleshy. **Fruits** baccate, dark green to purple, odor not reported, ellipsoid, strongly grooved and ridged, pulpy or mealy. $2n = 10$.

Flowering winter–mid spring (late Jan–early Apr). Rich woods and bluffs in mixed deciduous forests of oak, red maple, beech, elm, and others; also thinner upland oak woods, in depressions and in ravines, low sandy-alluvial slopes to local rivers; 50–100 m; Ala., Fla., Ga.

21. Trillium decumbens Harbison, Biltmore Bot. Stud. 1: 158. 1902 • Decumbent trillium [E]

Rhizomes short, thick. **Scapes** 1–2, decumbent, usually by an S-shaped curvature, round in cross section, 0.5–2 dm, somewhat stout, slightly expanded toward bracts, densely puberulent, especially just below bracts and on bases of main bract veins beneath. **Bracts** appearing early in season, in good condition for only a few weeks after anthesis, dying back to semipersistent bases early, resting on ground surface, sessile; blade strongly mottled in shades of green and bronze and with silvery overlay, mottling becoming obscure with age, ovate to suborbicular, 4–12 × 3.3–7 cm, apex acute to rounded. **Flower** erect, odor unreported; sepals divergent, green- or maroon-streaked, lanceolate-ovate, 22–48 × 9–14 mm, margins entire, flat, apex acute; petals long-lasting, rigidly erect, ± connivent, ± concealing stamens and ovary, dark maroon-purple when fresh, fading especially distally to dull reddish brown, greenish brown, or with creamy yellow tones, very rarely pale lemon yellow forms occur, twisted (but not spiraled), linear-lanceolate or oblanceolate, 4–8+ × 0.7–1 cm, thick-textured, margins entire, apex acute; stamens erect, straight, dark purple, 10–25 mm; filaments dark purple, 2–5 mm; anthers straight, dark purple, 9–20 mm, dehiscence extrorse; connectives broad, extending to 3.5–4 mm beyond anther sacs; ovary dark purple, oval, 6-angled, 5–10 mm; stigmas erect, divergent-recurved, distinct, sessile, pink or white, subulate, 4.5–8 mm. **Fruits** baccate, dark purple, broadly ovoid to subglobose, crownlike, strongly ridged, 1–1.5 × 1 cm, pulpy but not juicy, present and enlarging on naked scapes until early autumn. $2n = 10$.

Flowering late winter–spring (mid Mar–Apr). Thin, open rocky wooded slopes, mature deciduous woodlands, rocky talus and disintegrating shale, flats (floodplains) of small streams and adjacent slopes near river entrance; 50–200 m; Ala., Ga., Tenn.

Trillium decumbens occurs in the Ridge and Valley and Cumberland Plateau physiographic provinces. It has a large, deep, horizontal rhizome, the growing point of which always faces downslope.

22. Trillium discolor T. Wray ex Hooker, Bot. Mag. 58: plate 3097. 1831 • Pale yellow trillium, small yellow toadshade [E]

Rhizomes horizontal, brownish, short or somewhat compressed-thickened, bulblike, praemorse, not brittle. **Scapes** 1–2, 1–2.2 dm, slender, essentially glabrous. **Bracts** held well above ground, sessile; blade mottled dark green over lighter green, mottling fading or blurring with time, ovate, elliptic to almost circular, 6–13 × 4–7 cm, apex acute to acuminate. **Flower** erect, odor faintly pleasant, resembling sweetshrub (*Calycanthus* sp.); sepals displayed above bracts, spreading, green, oblong-lanceolate, shorter than petals, 20–30 × 7–19 mm, margins entire, apex acute; petals long-lasting, erect, ± connivent, ± concealing stamens and ovary to spreading distally, soft pale sulfur-yellow, fading to ± cream, sometimes slightly spirally twisted, widely spatulate, clawed basally, 2.2–5 × 0.9–1.5 cm, thick-textured, margins entire, often upraised adaxially, at least 1 petal with apex strongly apiculate, nipplelike, others with apex apiculate-acuminate; claw green, sometimes obscurely marked or mottled with purple, broadly cuneate; stamens incurved-erect, 8–15 mm; filaments purple, 1–2.5 mm; anthers erect, straight or arcuate, brown, 6.5–14 mm, dehiscence introrse-latrorse; connectives straight, extending 1–2 mm beyond anthers; ovary purplish, ovoid, 6-ridged, 2.5–8.5 mm; stigmas divergent to erect, distinct, purplish white, subulate, short, 2–6 mm, ± fleshy. **Fruits** baccate, greenish white, subglobose, weakly 6-winged, 1–5 × 1 cm, pulpy or mealy, not juicy. $2n = 10$.

Flowering spring (mid Apr–early May). Rather acidic to clearly circumneutral or basic soils of mixed deciduous forests, often under oaks, near *Rhododendron* thickets, or near tangles of *Leucothoë* on moist stream banks, slopes near streams; 100–200 m; Ga., N.C., S.C.

Trillium discolor occurs only in the upper drainage of the Savannah River, but it is locally frequent within its limited range.

23. Trillium foetidissimum J. D. Freeman, Brittonia 27: 31, fig. 7. 1975 • Stinking trillium, fetid trillium [E]

Rhizomes horizontal, brownish, thick, short, praemorse, not brittle. **Scapes** 1–2, green to maroon, round in cross section, 0.8–2.8 dm, papillose basally. **Bracts** often carried quite horizontally, well above ground, sessile; blade light green or bronze-green, strongly mottled in dark green with central light green stripe, mottling becoming obscure with age but less so than in most species, elliptic-ovate, rarely ± orbicular, 6.7–12 × 3.8–6 cm, not glossy, base evenly tapered to broad attachment, apex obtuse-acute. **Flower** borne directly on bracts, odor of putrid meat, especially when in strong sunlight; sepals displayed above bracts, carried almost horizontally, green or green streaked with dark maroon, lanceolate, 16–40 × 4–6 mm, thick-textured, margins entire, apex acute; petals long-lasting, erect, very gradually incurved from base to apex, ± connivent, ± concealing stamens and ovary, pinkish purple, light to reddish purple, brownish purple, rarely yellow, fading to brownish tones with age, not spirally twisted, not inrolling with age, veins not engraved, narrowly elliptic to linear-lanceolate, 2–5 × 0.3–0.5 cm, thick-textured, not glossy, margins entire, flat, acute at apex; stamens relatively prominent, erect, 9–25 mm; filaments dark maroon, 3–6 mm, dilated basally; anthers straight, dark maroon-black, 8–15 mm, dehiscence introrse; connectives straight, extended 1–1.5 mm beyond anther sacs; ovary red-purple, ovoid, hexagonal in cross section, 5–12 mm, broadly attached; stigmas erect, divergent-recurved, distinct, dark purple, subulate, nearly as long as ovary, fleshy. **Fruits** purplish brown, ovoid, 6-angled at least apically, fleshy.

Flowering late winter–early spring (early Mar [rarely Feb]–early Apr). River bluffs, ravines, floodplains, low ground, rich woods, road shoulders, silts, sandy-alluvium, loess soils, drier upland oak and pine woods; 40–50 m; La., Miss.

Trillium foetidissimum seems tolerant of a wide range of soil moistures and types, from low, swampy woods to high, dry bluffs and ravine slopes. This is the only *Trillium* known to occur within its Louisiana range (J. D. Freeman 1975). Freeman considered it to be closely related to *T. sessile*.

24. Trillium gracile J. D. Freeman, Sida 3: 289. 1969 • Slender trillium, graceful trillium [E]

Rhizomes horizontal, brownish, thick, short, praemorse, not brittle. **Scapes** 1–3, round in cross section, 1.6–3.5 dm, slender, glabrous. **Bracts** held well above ground, sessile; blade mottled with darker green blotches, mottling becoming obscure with age, elliptic-ovate to obovate, 6–8.5 × 2.6–4 cm, base rounded, apex obtuse or rounded, rarely acute. **Flower** odor musty or funguslike; sepals displayed above bracts, bases at right angle to scape axis, widely spreading with recurved tips, dark purple on adaxial surface, lanceolate to oblong, 20–25 × 4–5 mm, margins entire, apex obtuse; petals long-lasting, erect, weakly connivent, at least partially obscuring stamens, dark purple or maroon, rarely yellow, not spirally twisted, linear-elliptic to oblanceolate, 2–4 × 0.3–0.8 cm, ± thick-textured, margins entire, flat, apex acute; stamens erect, 12–16.5 mm; filaments purple, 2–3 mm, slender; anthers erect, straight, purple to yellow, 10–15 mm, slender, dehiscence introrse; connectives straight, extending 0.1–1 mm beyond anther sacs; ovary purple, ovoid, 3-angled, 4–11 mm; stigmas erect, spreading-recurved, distinct, purple to whitish, sessile, subulate, 2–4 mm, fleshy, thickened basally. **Fruits** dark greenish purple, fragrance not reported, ovoid, swollen enough to conceal its 3-angled nature, 1 cm, pulpy, moist.

Flowering spring (early–mid Apr). Mature pine and hardwood forests, banks and ridges of dissected streambeds, rather dense shade, low sandy flatwoods; 0–10 m; La., Tex.

25. Trillium kurabayashii J. D. Freeman, Brittonia 27: 56, fig. 12. 1975 [E]

Rhizomes horizontal, brownish, thick, praemorse, not brittle. **Scapes** often 2 from single terminal bud, round in cross section, 2.5–5.5 dm, stout, glabrous. **Bracts** held well above ground, sessile; blade bright green (in early anthesis rather succulent in appearance), usually well-marked with lighter and darker green spots, occasionally obscurely or scarcely mottled, mottling becoming obscure with age, ovate to broadly ovate, 11–18 × 12–17 cm, somewhat glossy, apex acuminate. **Flower** odor spicy in fresh flowers, becoming fetid in older flowers; sepals displayed above bracts, spreading, divergent, green, purple-streaked, lanceolate, 40–75 × 10–14 mm, margins entire, apex acute to round-acute; petals long-lasting, conspicuous, spreading-erect to erect, tips incurving slightly, ± connivent, partially to completely concealing stamens and ovary, glossy dark maroon-red or purple, not spirally twisted, flat, oblanceolate, 5.5–11 × 2–3.5 cm, widest at or below middle, glossy, thick-textured, cuneate basally, margins ± flat, entire, apex round-acute; stamens erect, straight, 15–26 mm, slightly concealing ovary; filaments dark purple, 2–4 mm; anthers dark maroon, 13–24 mm, dehiscence introrse; connectives straight, barely extended beyond anther sacs; ovary inconspicuous, purple, ovoid, round to 6-angled, 8–15 mm; stigmas erect, distinct, subulate, 6–8 mm, fleshy, thickened basally. **Fruits** dark reddish purple, ovoid to ellipsoid or weakly angled, 20–50 mm, fleshy.

Flowering spring (late Mar–early May). Rich, moist conifer-hardwood forest, slopes, especially lower slopes, predominantly deciduous flat woods along streams, edges of *Sequoia* groves, and alder, vine maple, and fern thickets along streams, especially older, higher flood terraces, not the lowest and wettest; at higher elevations, both in forests and in open grassy meadows with scattered oak trees; 20–500+ m; Calif., Oreg.

26. Trillium lancifolium Rafinesque, Autik. Bot., 132. 1840 • Lance-leaved trillium [E]

Rhizome horizontal, white, very slender-elongated, brittle; internodes elongated. **Scapes** 1–2, round in cross section, 1.5–3.2 dm, ca. 2.5–3 times longer than bracts, slender, glabrous. **Bracts** often downturned but leaves held well away from ground, sessile; blade mottled darker green, mottling becoming obscure in age, lanceolate to narrowly lanceolate-elliptic, 5–8.3 × 2–3.3 cm, not glossy, apex blunt or acute. **Flower** erect, no odor reported; sepals recurved basally and declining to ± same plane of and alternating with leaves, green, lanceolate, 13–20 × 5–7 mm, margins entire, apex acute; petals long-lasting, erect, ± connivent, not fully concealing stamens and ovary, maroon-red, purple, greenish tan, or 2-colored, claw dark reddish maroon basally, often twisted, linear to narrowly spatulate, 2.8–6.6 × 20–40 cm, widest above middle, thick-textured, basally clawed, margins entire, apex acute, claw to ± $^1/_2$ as long as expanded limb; stamens incurved, 13–21 mm; filaments purple, slender; anthers weakly to strongly incurved, purple, 4–6 mm, ± slender, dehiscence introrse; connectives weakly to strongly incurved, purple, extending 1 mm beyond anthers; ovary dark purple, ovoid-rhomboid, 6-angled, 6–7 mm; stigmas erect, somewhat divergent-recurved, distinct, purple, nearly linear, obscurely subulate, 3–4 mm, weakly fleshy. **Fruits** baccate, purplish, odorless, 6-angled, prolonged angle folds making fruit appear almost winged, 0.7–1.2 cm, pulpy. $2n = 10$.

Flowering later winter–spring (Feb–early May). Alluvial soils, floodplains, rocky upland woodlands, brushy thickets, canebrakes, heavy shade, or thin, open woods; 20–200 m; Ala., Fla., Ga., S.C., Tenn.

Trillium lancifolium occurs mostly in small, regionally disjunct populations.

27. Trillium ludovicianum Harbison, Biltmore Bot. Stud. 1: 23. 1901 • Louisiana trillium [E]

Rhizomes ± horizontal, brownish, short, thick, praemorse, not brittle. **Scapes** 1–3, round in cross section, 1.4–2.6 dm, ± slender, glabrous. **Bracts** held well above ground, sessile; blade strongly mottled in dark and bronzy green, often with central light strip, mottling becoming somewhat obscure with age, lanceolate-ovate, 5.3–9.5 × 2.3–5 cm, not glossy, margins of distal $^1/_3$ convex-curved to apex, apex rounded-acute. **Flower** erect, odor of carrion; sepals displayed above bracts, spreading, green, lanceolate-oblanceolate, 19–35 × 2.7–4 mm, margins entire, apex rounded or acute to sometimes weakly reflexed; petals long-lasting, faintly introrsely curved-erect-spreading, weakly connivent, ± concealing stamens and ovary, dark maroon-brown, purplish, or dull greenish, or 2-colored, basal portion purple, distal portion grayish green, not spirally twisted, oblanceolate-linear, 3.5–5.5 × 0.4–0.8 cm, thick-textured, thickened and weakly clawed basally, margins entire, apex acute; stamens erect, straight, 10–18 mm; filaments olive-orange, 2–3 mm, widened basally; anthers erect, straight, olive to orange, 7–20 mm, slender, dehiscence latrorse; connectives olive to orange, straight, scarcely extended beyond anther sac; ovary purple, ovoid, 6-angled, 8–9 mm; stigmas erect, with spreading or coiled tips, distinct, pale purple, subulate, 3–6 mm, ± fleshy. **Fruits** dark purplish green, little or no odor, ovoid, 6-angled, pulpy.

Flowering late winter–early spring (early Mar–Apr). Low flatwoods, floodplains along streams, steep ravine slopes leading to floodplains, mixed pine-beech woods; 50–500 m; La., Miss.

The range of *Trillium ludovicianum* is near to that of *T. cuneatum* in Mississippi, and the two appear to intergrade.

28. Trillium luteum (Muhlenberg) Harbison, Biltmore Bot. Stud. 1: 21. 1901 • Yellow trillium, yellow toadshade, wax trillium [E]

Trillium sessile Linnaeus var. *luteum* Muhlenberg, Cat. Pl. Amer. Sept., 38. 1813; *T. viride* L. C. Beck var. *luteum* (Muhlenberg) Gleason

Rhizomes horizontal, brownish, short, thick, praemorse, not brittle. **Scapes** 1–4, round in cross section, 1.4–4 dm, stout, glabrous. **Bracts** held well above ground, sessile; blade mottled, mottling becoming more obscure as growth matures, shape variable, ovate-elliptic to ± orbicular-acuminate, 6.5–17 × 6.5–9.8 cm, not glossy, apex acuminate to long-acuminate. **Flower** erect, odor strongly lemony; sepals displayed above bracts, widely spreading, green, oblong-lanceolate, lanceolate, or oblong-elliptic, 22–47 × 8–9 mm, margins entire, apex distinctly rounded-acute; petals long-lasting, erect, ± connivent, ± concealing stamens and ovary, greenish yellow to lemon yellow, not spirally twisted, not clawed, elliptic-lanceolate, lanceolate, or oblanceolate, 3.4–6.6 × 1–2.1 cm, thick-textured, widest near base, margins entire, tapering to long-acuminate apex; stamens erect, 11–18 mm; filaments greenish white, 1.5–2 mm, basally dilated; anthers erect, straight, yellow, 9–16 mm, dehiscence introrse (or appearing lateral); connectives green, straight, broad, extending only slightly (to 0.5 mm) or not at all beyond anther sacs; ovary pale green, ovoid to globose, 6-angled, 5–8 mm; stigmas erect, barely spreading, distinct, greenish white, sessile, linear to subulate, 3–4 mm, fleshy. **Fruits** green to greenish white, occasionally with dark streaks, odor not reported, ovoid, 6-angled, 3.5 × 2 cm, pulpy. $2n = 10$.

Flowering spring (Apr–May). Deciduous forests, thin open woods, rocky stream banks and flats, clearings and openings, old fields, rich mature forest on calcareous substrate; 200–400 m; Ga., Ky., N.C., Tenn.

Botanists have been confused by *Trillium luteum* for a long time. Some, such as A. E. Radford et al. (1968), appear to regard it as a form of *T. cuneatum*, while others confuse it with *T. viride*, a more western species. Early botanists confused *T. luteum* with the occasional individual or very local larger population of pallid color forms of other species. *Trillium cuneatum* rather frequently produces green, yellowish green, or pale lemon yellow forms (but with a cuneate larger and wider petal) that mimic *T. luteum*. These forms, when growing with *T. luteum*, hybridize, leading to so many intergrades that many plants cannot be placed in either species with any confidence. For these reasons, almost no work older than J. D. Freeman's (1975) can be used reliably to plot distribution of *T. luteum*.

In some parts of their ranges, *Trillium luteum* and *T. cuneatum* do not grow together. *Trillium luteum* is abundant in eastern Tennessee, extending a short distance into North Carolina and Georgia. *Trillium cuneatum* is rare or absent from this region except along the Little Tennessee River (J. D. Freeman 1975).

Trillium luteum has escaped and become established along the Grand and Saint Joseph rivers and elsewhere in Michigan, near Kingston and Grimsby, Ontario (J. K. Morton pers. comm.), and probably elsewhere.

29. Trillium maculatum Rafinesque, Med. Fl. 2: 103. 1830 • Spotted trillium [E] [F]

Rhizomes horizontal, brownish, short, thick, praemorse, not brittle. **Scapes** 1–3, round in cross section, 1.4–4 dm, slender to stout, glabrous. **Bracts** held well above ground, tips only touching ground in early anthesis, sessile; blade dark green, mottling darker, mottled to varying degrees, mottling becoming obscure with age, broadly ovate-elliptic to elliptic, 7–15 × 4.6–6.7 cm, not glossy, apex rounded or barely acuminate. **Flower** erect, odor faintly spicy and bananalike; sepals displayed above bracts, spreading almost to horizontal, often suffused or streaked with purple-maroon, lanceolate-linear, 22–50 × 5–7 mm, margins entire, apex rounded-acute, recurving slightly; petals long-lasting, spreading-erect, ± connivent, ± concealing stamens and ovary or sometimes slightly spreading early, only partially obscuring stamens, becoming more erect and even touching in the manner of *Trillium cuneatum* when older, clear dark red-maroon, or reddish purple, without muddy or brown tones of many purple sessile species, or, rarely, purple basally and yellow toward apex, or clear sulfur yellow, not spirally twisted, narrowly oblanceolate-spatulate to linear-spatulate, widest above middle, narrowing to apex, 4–7 × 0.7–1.7 cm, thick-textured, margins entire, slightly involute in proximal 1/2, apex acute to round-acute; stamens erect, 12–20 mm; filaments dark purple, 2–3 mm, widest at base, much shorter than anther sacs; anthers erect, straight, brownish purple, 10–16 mm, dehiscence introrse on broad connective; connectives brown-purple, straight, essentially not extended beyond anther sacs; ovary dark purple, ovoid, weakly 3-angled to smooth (rarely very obscurely 6-angled), angles often obscured in large, turgid ovary, 8–11 mm; stigmas erect, divergent-recurved, distinct, purplish, subulate, 2–4 mm, fleshy. **Fruits** dark purplish green, odor not reported, ovoid, obscurely 3- to 6-angled, bearing persistent stigma, 1 × 2 cm, pulpy.

Flowering mid winter–early spring (early Feb–early Apr). Rich mesic forests, particularly banks and bluffs of rivers, floodplains, often where quite brushy, rich soils, calcareous soils, alluvium; 0–500 m; Ala., Fla., Ga., S.C.

In their account of the Carolinas flora, A. E. Radford et al. (1968) did not treat *Trillium maculatum* or give any explanation of why they omitted this species, which is fairly widely distributed in southeastern South Carolina and morphologically is one of the more distinct sessile trilliums.

Many color forms occur and have been named. In forma *luteum* J. D. Freeman, for example, all floral organs lack purple pigment, and the petals are clear, soft yellow, whereas forma *simulans* J. D. Freeman has yellow petals with purple bases, and purple stamens and carpels. These are illustrated in color in F. W. Case and R. B. Case (1997).

30. **Trillium parviflorum** V. G. Soukup, Brittonia 32: 330, fig. 1. 1980 • Small-flowered trillium [E]

Rhizomes horizontal to ± erect, brownish, thick, praemorse, not brittle. **Scapes** 1–3, round in cross section, 1.7–3 dm, slender to robust, glabrous. **Bracts** held well above ground, sessile; blade green with widely scattered mottling, mottling becoming obscure with age, ovate to broadly ovate, 6.5–16 × 5–8 cm, not glossy, apex obtuse. **Flower** erect, odor spicy, of cloves; sepals displayed above bracts, spreading, divergent, green, lanceolate, 16–25 × 4–8 mm, margins entire, flat, apex variously obtuse to rounded; petals long-lasting, erect to erect-spreading, ± connivent, ± concealing stamens and ovary, white, occasionally purple stained basally, sometimes weakly spirally twisted, linear to linear-lanceolate, 2.2–4.5 × 0.4–1 cm, thin-textured, base occasionally cuneate, margins entire, apex obtuse; stamens erect, 10–15.5 mm; filaments white, greenish white, or purple stained, 1–3 mm, much shorter than anthers, slender; anthers erect, straight, greenish white, 9–11+ mm, ± slender, dehiscence latrorse; connective greenish, straight, barely extended (to 0.4 mm) beyond anther sacs; ovary green or green and purple basally, ovoid, obscurely 6-gonal, 4–8 mm; stigmas erect, ± divergent, distinct, green, outer surface purple, subulate, 3–4.5 mm, ± fleshy, thickened. **Fruits** dark reddish purple or maroon, fragrance not reported, subglobose, ± 1 cm, ± juicy. $2n = 10$.

Flowering spring (late Mar–early May). Mature fir (*Abies*), spruce (*Picea*), and hardwood forests in rich humus with mosses, open, somewhat grassy large groves of old oaks, with considerable underbrush and rather few herbaceous companions, in tangled, wet stream-bank alders (*Alnus* sp.), grasses, open clay hillside soils among shrubs; 20–60 m; Oreg., Wash.

Trillium parviflorum varies from very short, slender, small-bracted plants to tall, broad, umbrella-bracted giants. Regardless of plant or bract sizes, flower and petal sizes are remarkably constant, all plants having small, linear-lanceolate petals. This is not usually the case with *T. albidum*, the species with which this plant is most likely to be confused. In *T. albidum* also the plants can be enormous, but when they are, the petals are very long, broad, and conspicuously obovate-diamond-shaped. In large clonal clumps of *T. albidum*, the larger and more mature offsets show the typical petal shape, while the smaller and presumably youngest offsets sometimes produce smaller, narrower petals, more like those of *T. parviflorum*.

Some western botanists, more experienced with local populations than I, do not consider *Trillium parviflorum* to be distinct from *T. albidum*. They point out that since there is an extensive region of apparent intergradation (as indeed there is, well supported by herbarium vouchers), there exists a morphological cline from the long- and wide-petaled *T. albidum* to the narrower- and generally shorter-petaled *T. parviflorum*, and that *T. parviflorum*, therefore, should not be considered a separate species but rather a subspecies or a variety. Since no one to date has treated *T. parviflorum* at the subspecific or varietal level, and since in my own limited experience it does appear as a distinct species in Washington north of the Columbia River, I include it here as treated by Soukup. I have seen populations of considerable variation north of Corvallis, Oregon, and agree that there is much overlap with *T. albidum*. Obviously there is need for a much more extensive study of this situation.

A factor that exacerbates this problem (and many others in *Trillium*), is that nutrition, age, and even favorable position in the habitat can greatly influence plant and floral organ sizes. In many species, including *T. albidum*, when a single vigorous clonal clump produces many offsets, the oldest offsets may have flowers with very large petals, sepals, ovary, etc., while the younger offsets may have organs only half the size. In most sessile trilliums particularly, population averages are often more useful than isolated individual measurements, a difficult situation, indeed.

31. **Trillium petiolatum** Pursh, Fl. Amer. Sept. 1: 244. 1814 • Long-petioled trillium, purple trillium, round-leaved trillium [E] [F]

Rhizomes ± erect, often very deep, praemorse. **Scapes** typically 1, vertical but mostly subterranean, round in cross section, 0.4–1.7 dm, robust, glabrous. **Bracts** just at or slightly above soil, long-petiolate; blade medium green, not mottled, ovate to elliptic, 7–14 × 5.5–10.2 cm, not glossy, apex obtuse or

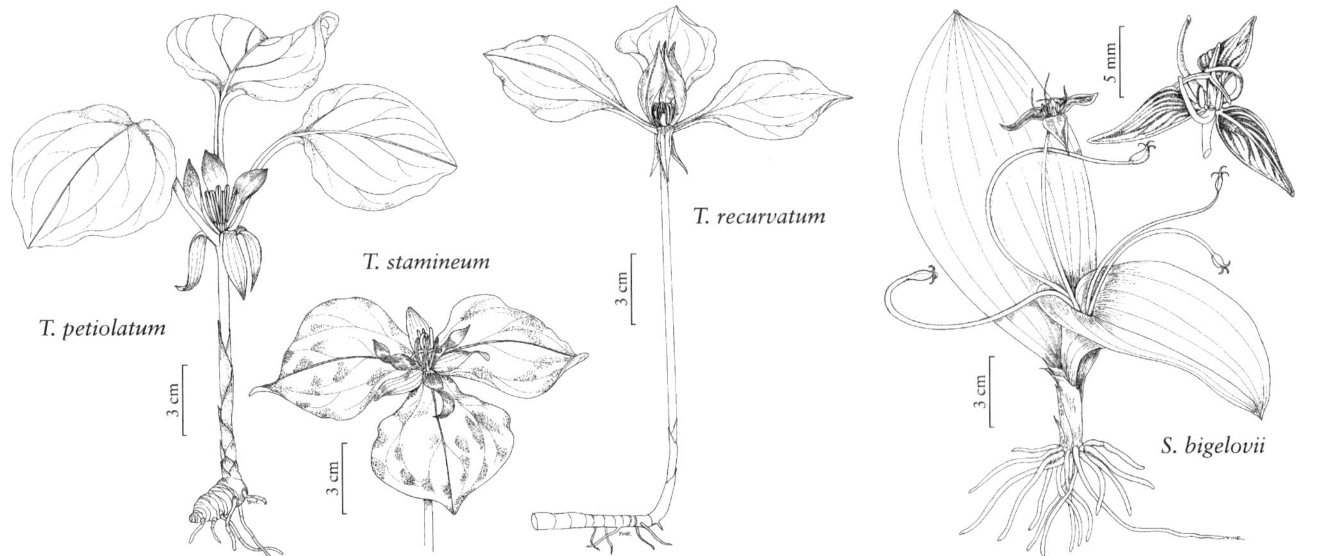

TRILLIUM ∘ SCOLIOPUS

rounded; petiole arising from scape apex at or near ground surface, 5–12 cm; bract and petiole strongly resembling leaves of *Plantago*. **Flower** in axil of bracts at or near ground level, ± erect, odor unknown; sepals erect to widely spreading, often weakly recurved near middle, green, oblong-elliptic to oblanceolate, 22–47 × 7–10 mm, margins entire, apex acute; petals long-lasting, ± connivent, ± concealing stamens and ovary, or erect-spreading early then ± connivent, erect to incurved, light maroon-red, purple, or greenish to yellowish, not spirally twisted, flat, linear-lanceolate, 3–5.5 × 0.4–1 cm, thick-textured, margins entire, apex acute; stamens erect, (15–)22–30 mm; filaments olive, (3–)5–7 mm, slender; anthers straight, brown or olive, 16–20 mm, slender, dehiscence latrose; connectives yellow or orange, ± not extending beyond anther sacs; ovary white, greenish, purplish distally, ovoid, sharply angled, 4–9 mm; stigmas erect, divergent, distinct, purple or olive, linear-subulate, 7–20 mm, apex somewhat recurved. **Fruits** fragrance not reported, ovoid, strongly angled/winged, ca. 1 cm, pulpy, moist. $2n = 10$.

Flowering spring (early Apr–late May). Lower rocky hillsides just above stream flats, under brush, edges of coniferous and deciduous forests, open grassy glades, river-flats, wet, seasonally swampy ground and edges of sloughs; 400–1400 m; Idaho, Oreg., Wash.

32. Trillium recurvatum L. C. Beck, Amer. J. Sci. Arts 11: 178. 1826 • Prairie trillium, toadshade, bloody noses [E] [F]

Rhizomes horizontal, white, slender, elongated, brittle. **Scapes** typically 1(–3), round in cross section, 1.5–4.8 dm, slender to robust, glabrous. **Bracts** held well above ground, strongly petiolate; blade at first strongly mottled in darker green or bronze, mottling fading with seasonal expansion after anthesis, rarely all green, ovate, elliptic, or lanceolate, 6–18 × 2.5–6.5 cm, not glossy, apex acuminate; petiole ca. 1/5 bract length. **Flower** erect, fragrance not reported; sepals strongly recurved basally and held against scape by turgor pressure, green, sometimes purple-streaked, ovate-lanceolate, 18–35 × 6–18 mm, margins entire, apex acute; petals long-lasting, erect, ± connivent, ± concealing stamens and ovary, dark maroon purple to clear yellow, occasionally 2-colored with purple and yellow, not spirally twisted, lanceolate to ovate, 1.8–4.8 × 0.9–2 cm, thick-textured, base attenuate to weakly clawed, margins entire, apex acute; stamens incurved, 10–15 mm; filaments erect, dark purple, 4–6 mm, ± slender; anthers strongly incurved above filaments, dark purple, 5–16 mm, ± thick, dehiscence introrse; connectives strongly incurved inward, dark purple, projecting about 1 mm beyond anther sacs; ovary greenish with ± purple stains distally, transversely rhombic to angular-ovate, somewhat 6-angled or -winged, 7–10 mm, ± equaling filament height; stigmas erect, divergent-recurved, distinct, ± linear, 4–6 mm, slightly thickened basally. **Fruits** green to white-

and purple-streaked, odorless, rhomboid-ovoid, 6-angled, almost winged, ca. 1 cm diam., pulpy. $2n = 10$.

Flowering spring (late Mar–late May). Rich clayey floodplain soils, plants often temporarily inundated while in flower; rich moist woods and bluffs, limestone-derived soils; 100–200 m; Ala., Ark., Ill., Ind., Iowa, Ky., La., Mich., Miss., Mo., Ohio, Tenn., Tex., Wis.

Trillium recurvatum has several named color forms, most notably forma *shayi* E. J. Palmer & Steyermark with clear yellow petals, and one foliose anomaly (possibly caused by mycoplasma).

33. Trillium reliquum J. D. Freeman, Brittonia 27: 21, fig. 5. 1975 • Relict trillium [C] [E]

Rhizomes short, stout, praemorse. **Scapes** 1–2, semidecumbent, decumbent, or weakly erect (especially in cultivation), S-shaped, round in cross section, 0.6–1.8 dm, glabrous. **Bracts** resting on or near ground surface, horizontally spreading, sessile; blade strongly mottled on each side of central light green stripe in shades of light green, dark green, bronze green, and dark purple, mottling becoming obscure with age, ovate to elliptic, rounded-tapered ± equally from base to tip from widest point, 5–12 × 6–10 cm, apex rounded or weakly acute. **Flower** borne directly on bracts, odor of putrid meat; sepals divergent, somewhat recurved, green, maroon streaked, lanceolate, 17–42 × 5–9.5 mm, margins entire, flat, apex rounded-acute; petals long-lasting, erect, ± connivent, ± concealing stamens and ovary, dark brownish maroon, greenish purple, or streaked with yellow, usually not spirally twisted, narrowly elliptic-lanceolate to oblanceolate, 2.5–5.5 × 0.6–1 cm, thick-textured, margins entire, ± flat, apex acute; stamens erect, incurved, 12–20 mm; filaments ± straight, reddish brown, 1–2 mm, slender; anthers ± straight, dark purple, 4–20 mm, ± thick, dehiscence introrse; connectives brown-purple, extending 1–2.5 mm beyond anther sacs, apex acute; ovary green-purple, ovoid, 6-angled, 5–10 mm; stigmas erect, divergent-recurved, distinct, linear, 2–4 mm, uniformly thin. **Fruits** baccate, dark maroon-purple, fragrance unreported, ovoid, 6-winged or -angled apically, 0.7–1 cm, pulpy, moist.

Flowering late winter–spring (early Mar–Apr). Rich mixed deciduous forested slopes, bluffs, stream-flats, lower slopes at edge of small stream floodplains; of conservation concern; 50–100 m; Ga., S.C.

Trillium reliquum recently has been reported from Alabama; I have not seen specimens from there. Otherwise, the species occurs in Clay, Early, and Richmond counties, Georgia, and in Aiken County, South Carolina. It is currently listed as an endangered species in the United States.

34. Trillium sessile Linnaeus, Sp. Pl. 1: 340. 1753 • Sessile trillium, toad trillium, toadshade [E]

Rhizomes horizontal, brownish, thick, praemorse, fleshy. **Scapes** 1–3, round in cross section, 0.8–2.5 dm, slender to stout, glabrous. **Bracts** held well above ground, sessile; blade green to bluish green, strongly to sparsely mottled, mottling becoming obscure with age, oval to suborbicular, 4–10 × 2–8 cm, base broadly attached, apex rounded-acuminate to bluntly parallel sided-acuminate (rounded basally to its broad attachment). **Flower** erect, odor pungent, spicy; sepals displayed above bracts, spreading, green, variously streaked with maroon, lanceolate-oblanceolate, 9–35 × 4–8 mm, margins entire, apex rounded-acuminate; petals long-lasting, erect, ± connivent, ± concealing stamens and ovary, maroon, brownish maroon, green, or yellowish green, not spirally twisted, oblanceolate to elliptic, occasionally almost orbicular, 1.7–3.5 × 0.7–2 cm, thick-textured, narrowed near basal attachment (but not truly clawed), margins entire, apex gradually rounded-tapered to acute; stamens straight, 10–23 mm; filaments red-purple, 2–5 mm, dilated basally; anthers erect, straight, gray-purple, 9–16 mm, thick, dehiscence introrse; connectives purplish brown, straight, projecting 2–5+ mm beyond anther sacs; ovary greenish white basally, purple distally, ovoid to globose, 6-angled, pyramidally narrowed to stigmas, 4–8.5 mm; stigmas erect, divergent-recurved, distinct, purple, subulate, 1–5 mm, ± fleshy. **Fruits** baccate, dark greenish purple, odorless, subglobose, 6-angled, angles somewhat winglike, pulpy, not juicy. $2n = 10$.

Flowering spring (Mar–early May). Rich woodlands, limestone districts, calcareous soils, floodplains, riverbanks, clayey alluvium, less fertile soils, high, dry limestone woods, persists under light pasturing, in fencerows and brushy areas after lumbering; 100–300 m; Ala., Ark., Ill., Ind., Kans., Ky., Md., Mich., Mo., N.Y., N.C., Ohio, Okla., Pa., Tenn., Va., W.Va.

Trillium sessile is rather uniform throughout its range, with few color forms.

35. Trillium stamineum Harbison, Biltmore Bot. Stud. 1: 23. 1901 • Twisted trillium [E] [F]

Rhizomes horizontal, brownish, short, thick, praemorse, not brittle. **Scapes** 1–3, round in cross section, 1.5–3 dm, slender to stout, pilose-pubescent, rarely glabrous. **Bracts** held well above ground, sessile; blade light silvery or bluish green with strong to faint mottling in darker colors, mottling becoming obscure with age, ovate-lanceolate to broadly ovate, 6.3–7.6 × 3.3–5 cm, larger bracts abaxially pilose-pubescent, margins entire, apex acute. **Flower** erect, odor strong, of carrion; sepals displayed above bracts, spreading to ± horizontal position, green, purple markings adaxially, lanceolate-elliptic, 17–40 mm, margins entire, purple, apex acuminate; petals long-lasting, spreading and carried in ± horizontal position unlike any other sessile trillium, very deep maroon to blackish red, rarely yellow, purple-streaked, with 1–2 spiral twists, narrowly elliptic to lanceolate or linear, narrow, 1.5–3.8 × 0.3–0.6 cm, rarely broader, thick-textured, margins entire, apex sharply acute to rounded, tips incurving slightly; stamens fully exposed, somewhat stiffly erect, dark purple, 16–24 mm, thick; filaments dark purple, 2–4 mm, basally dilated; anthers erect, straight, dark purple, 13–18 mm, thick, dehiscence extrorse; connectives dark purple, straight, coarse, flat, ± not extended beyond anther sacs; ovary dark purple, oval, 6-angled, 5–7 mm; stigmas erect, widely spreading, often strongly recurved or recoiled, distinct, purple, linear, 4–10 mm, slightly thickened basally. **Fruits** baccate, purple, odorless, ovoid, strongly 6-angled, sometimes winged, 2 × 1–1.5 cm, pulpy, moist. $2n = 10$.

Flowering spring (late Mar–mid May). Dry, upland woods of deciduous trees, deciduous forest mixed with pines, soil on limestone outcroppings, mesic woods, sandy flats along medium streams, steep wooded slopes, banks of rivers; 50–200 m; Ala., Miss., Tenn.

36. Trillium underwoodii Small, Bull. Torrey Bot. Club 24: 172. 1897 • Underwood's trillium [E]

Rhizomes horizontal, brownish, short, thick, praemorse, not brittle. **Scapes** 1–2, round in cross section, 0.8-2 dm, slender to stout, glabrous. **Bracts** usually drooping, often touching ground in early anthesis, sessile; blade pale silvery green, strongly mottled in 3 or more shades of dark green and bronze, sometimes maroon, mottling becoming obscure with age, ovate-lanceolate to obovate, 6.5–12 × 5–7.8 cm, margins of distal 1/3 straight from rounded base to apex, apex acuminate. **Flower** erect, odor fetid, especially when first open; sepals displayed above bracts, spreading, green or purple on adaxial surface, lanceolate, 4.5–5 × 8–12 mm, margins entire, flat, apex rounded-acute; petals long-lasting, erect-spreading, ± connivent, ± concealing stamens and ovary or at least partially obscuring stamens, tips ± incurved, dark maroon-red, or purplish red, brownish maroon, or yellowish green, not spirally twisted, oblanceolate to elliptic, 3–5.5+ × 1–1.5 cm, usually 3–4 times longer than wide, thick-textured, margins entire, apex acute; stamens ± erect to weakly incurved, 14–17 mm; filaments purple, 1–2 mm; anthers 8–15 mm, thick, dehiscence latrorse; connectives brown-purple, straight, extended 1–2 mm beyond anther sacs, apex somewhat acute; ovary purple, ellipsoid, 6-angled, 6.3–11 mm; stigmas erect, abruptly recurved upon ovary, distinct, purplish, linear, 1.5–5 mm, slightly thickened basally, fleshy. **Fruits** baccate, purple-black to dull greenish maroon, odorless, ovoid to obovoid, 6-angled (-ridged), 0.7–1 cm diam., pulpy, not juicy.

Flowering mid winter–spring (late Feb–mid Apr). Rich to dryish deciduous forests of mature or second-growth timber, dominated by oaks or with beech-oaks, occasionally with scattered pines present, flat ground along streams where soil can be fairly moist, rich clay or sand; 100–200 m; Ala., Fla., Ga.

Trillium underwoodii can cause considerable confusion for the beginning student of the genus. In some ways it closely resembles a small-statured *T. cuneatum*. As *T. underwoodii* continues to expand after flowering, it may become somewhat taller and less "ground hugging," and it then appears similar to a smaller plant of *T. decipiens*, which grows in parts of the same range, especially when its bracts do not touch the ground at anthesis. It also has vague similarities to *T. reliquum*, with which it grows sympatrically in parts of its range. Care should be taken to observe the details of floral structure accurately.

Distribution of *Trillium underwoodii* is much more restricted than that given by J. K. Small (1933), who indicated a range extending from Alabama to North Carolina and Arkansas.

37. Trillium viride L. C. Beck, Amer. J. Sci. Arts 11: 178. 1826 • Green trillium [E]

Rhizomes horizontal, brownish, short, slender to stout, praemorse, not brittle. **Scapes** 1–3, round in cross section, 2.3–3.4+ dm, ± slender, smooth to scabrous below bracts. **Bracts** held well above ground, sessile; blade green to bluish green, weakly mottled, rarely unmottled, mottling becoming obscure with age, adaxial surface covered with numerous stomata easily visible under a lens, occasionally visible as tiny white dots to the naked eye, narrowly to broadly elliptic, 8–20.5 × 5–8 cm, not glossy, apex blunt to rounded-acute. **Flower** erect, odor of decayed fruit; sepals displayed above bracts, widely spreading, distal 1/2 often weakly declined, green, lanceolate-acute, 28–60 × 7–9 mm, margins entire, apex acute; petals long-lasting, erect to widely spreading, ± connivent, ± concealing stamens and ovary, green, yellowish green, yellow, or variously purple streaked throughout, especially basally (rarely all purplish green), not spirally twisted, narrowly spatulate to linear-spatulate, 3.5–6.8 × 0.5–1.5 cm, widest above middle, thick-textured, base narrowed and thickened, ± forming claw, margins entire, apex obtuse, lacking nipple; stamens relatively straight-erect to slightly incurved, 15–25 mm; filaments brown-olive, 3–5.5 mm, dilated basally; anthers erect, straight, olive-brown or purplish, 8–21 mm, ± slender, dehiscence introrse-latrorse; connectives olive, straight, very slightly (± 0.5 mm) extended beyond sacs; ovary greenish basally, olive-purplish distally, ellipsoid to ovoid, deeply angled, 5–12 mm; stigmas spreading-erect, recurved, distinct, sessile, olive or purplish, subulate, 5–10 mm, somewhat fleshy. **Fruits** greenish white, odorless, ovoid, angled, 1–1.5 cm, somewhat pulpy, not juicy. $2n = 10$.

Flowering spring (late Apr–May). Rich woods, bluffs, rocky hillsides with some limestone outcrops, woodsy, humusy soil over fairly stiff, clayey substrate; 100–200 m; Ill., Mo.

Trillium viride has long been confused with *T. luteum* of the Smoky Mountain region, though it is far more slender and delicate than *T. luteum*. Because the pallid forms and so-called albinos of many other sessile trilliums described in written accounts sound much like the description of this plant, early authors ascribed far too wide a range to *T. viride*. The plant is closely limited to the woodlands of the counties adjacent to the Missouri River in northeastern Missouri and southern Illinois.

Although the ranges of *Trillium viride* and the very similar *T. viridescens* do not overlap, both species grow in Missouri, and authors earlier than J. D. Freeman (1975) frequently combined the two as a single species. If all other means of identification fail, the prevalence of numerous stomates on the adaxial bract surface, seen under magnification, will always distinguish *T. viride*.

38. Trillium viridescens Nuttall, Trans. Amer. Philos. Soc., n. s. 5: 155. 1837 • Ozark green trillium, Ozark trillium [C] [E]

Rhizomes horizontal, brownish, short, thick, praemorse, not brittle. **Scapes** 1–3, round in cross section, 2–5 dm, ± stout, glabrous to scabrous. **Bracts** touching ground in early anthesis or not at all, sessile; blade dark green, obscurely marked with few–many darker blotches (very rarely unmottled), mottling becoming obscure with age, few or no stomates adaxially, ovate-elliptic to broadly so, 8.5–14 × 6.8–9 cm, apex acuminate. **Flower** erect, odor, if present, spicy or musty; sepals displayed above bracts, widely spreading, green or variously purple marked, lanceolate, 38–60 × 5–12 mm, margins entire, apex acute; petals long-lasting, erect, ± connivent, ± concealing stamens and ovary, purplish black on claw, greenish to yellow-green distally, sometimes all dark purple, occasionally slightly twisted, linear to narrowly spatulate, 4–8 × 0.8–1.2 cm, thick-textured, base slightly thickened and clawed, margins entire, apex rounded, lacking nipple; stamens erect, connivent (clustered together, leaning upon each other), 16–25 mm; filaments olive or purplish brown, 2.5–5 mm, very slender, widened basally; anthers erect, straight or slightly incurved, olive-brown, 13–20 mm, ± slender, dehiscence latrorse; connectives brownish, barely extending beyond anther sacs; ovary pale greenish white basally, purplish distally, ovoid, 6-angled, 5.5–10 mm; stigmas erect, divergent-recurved, distinct, purplish abaxially, sessile, almost linear, 6–10 mm, ± equaling ovary, ± not fleshy, very slightly widened basally. **Fruits** dark purplish green or green, odor unreported, ovoid, obscurely angled, with remains of persistent stigma, 0.7–1.5 cm, pulpy, not juicy at time of separation from receptacle.

Flowering spring (early Apr–mid May). Deciduous forests, usually quite rich, on banks, bluffs, talus slopes, floodplain alluvium, with cane (*Arundinaria*) or on sloping banks just above normal flood levels in heavy, clayey soils, with common spring ephemerals; of conservation concern; 100–400 m; Ark., Kans., Mo., Okla., Tex.

J. D. Freeman (1975) considered that *Trillium viridescens* intergrades with *T. gracile* in northeastern Texas and stated that these putative intergrades produce purple petals, but purple-petaled forms also occur in Arkansas, far from the influence of *T. gracile*.

17. SCOLIOPUS Torrey, Pacif. Railr. Rep. 4(5): 145, plate 22. 1857 • Slink pod, fetid adder's-tongue, slink-lily [Greek *skolios*, crooked, and *-pous*, footed, alluding to the tortuous, recurved pedicels] [E]

Frederick H. Utech

Herbs, perennial, with short, knotty rhizomes; roots contractile. **Stems** subterranean, simple, vertical, short. **Leaves** 2–3(–4), petiolate or subsessile, sheathing at base; blade dark green distally, paler proximally, sometimes mottled with purple, elliptic to oblong, apex obtuse. **Inflorescences** umbellate, with several axillary fascicles of elongate, twisting pedicels. **Flowers** showy, odor unpleasant; perianth of 2 dissimilar whorls of tepals, outer spreading or recurving, distinct, petaloid, ovate to lanceolate to oblanceolate, with oblong gland basally, inner erect, distinct, linear, converging over pistil; stamens 3, opposite outer tepals; anthers versatile, 2-locular, oblong, extrorse; ovary superior, 1-locular, strongly 3-angled, placentation parietal; style erect, short; stigmas persistent, 3, spreading to recurved, linear, inner surface deeply channelled, ovules 20–40, in 2 rows on each placenta. **Fruits** capsular, brownish purple, oblong-lanceolate, strongly 3-angled, thin-walled, opening irregularly by decay, beaked by persistent style and stigmas. **Seeds** slightly curved, oblong, eliaosome present. $x = 7, 8$.

Species 2 (2 in the flora): w North America.

SELECTED REFERENCES Berg, R. Y. 1959. Seed dispersal, morphology and taxonomic position of *Scoliopus*, Liliaceae. Skr. Norske Vidensk.-Akad. Oslo, Mat.-Naturvidensk. Kl. 1959(4): 1–56. Cave, M. S. 1966. The chromosomes of *Scoliopus* (Liliaceae). Madroño 18: 211–213. Utech, F. H. 1979b. Floral vascular anatomy of *Scoliopus bigelovii* Torrey (Liliaceae-Parideae = Trilliaceae) and tribal note. Ann. Carnegie Mus. 48: 43–71. Utech, F. H. 1992. Biology of *Scoliopus* (Liliaceae). I. Systematics and phytogeography. Ann. Missouri Bot. Gard. 79: 126–142.

1. Leaves 14–24 cm; pedicel 11–22 cm; outer tepals 12.5–19 mm; style 4.5–6 mm . . . . 1. *Scoliopus bigelovii*
1. Leaves 8–14 cm; pedicel 6–10 cm; outer tepals 6.5–10 mm; style 2–2.5 mm . . . . 2. *Scoliopus hallii*

1. Scoliopus bigelovii Torrey, Pacif. Railr. Rep. 4(5): 145, plate 22. 1857 • Brownies, California fetid adder's-tongue [E] [F]

Stems 6–20 cm. **Leaves** 2–3(–4), petiolate; blade often mottled with purple, elliptic to oblong, 14–24 × 4.5–8 cm. **Flowers** 3–12; outer tepals mottled with green and purple, ovate-lanceolate, 12.5–19 mm; inner tepals linear-subulate, 11.5–19 × 0.5 mm; filaments 2.3–4.3 mm; anthers 2–2.8 mm; style 4.5–6 mm; stigmas strongly recurved; pedicel strongly recurved in fruit, 11–22 cm. **Capsules** 22–30 × 6–9 mm, including beak. $2n = 16$.

Flowering late winter–early spring. Redwood and coastal coniferous forests, mossy mountain stream banks, shaded slopes; 0–500 m; Calif.

R. Y. Berg (1959) demonstrated for *Scoliopus bigelovii* (and his findings also apply to *S. hallii*) that a single flower functions as three independent pollination units, a feature usually associated with *Iris* (F. H. Utech 1992).

2. Scoliopus hallii S. Watson, Proc. Amer. Acad. Arts 14: 272. 1879 • Oregon fetid adder's-tongue [E]

Stems very short, 2.5–10 mm. **Leaves** 2(–3), subsessile; blade rarely mottled with purple, oblong-elliptic, 8–14 × 2.5–5 cm. **Flowers** 1–5; outer tepals yellowish green with purple lines, lanceolate to oblanceolate, 6.5–10 mm; inner tepals linear-spatulate, 6–9 × 0.3 mm; filaments 2–4 mm; anthers 1.9–2.3 mm; style 2–2.5 mm; stigmas spreading to recurved; pedicel very slender, elongating to 6–10 cm. **Capsules** 16–18.5 × 5.5–7.5 mm, including beak. $2n = 14$.

Flowering late winter–early spring. Coniferous woods, mossy mountain stream banks; 0–800 m; Oreg.

18. CALOCHORTUS Pursh, Fl. Amer. Sept. 1: 240. 1814 • [Greek *kalos*, beautiful, and *chortos*, grass]

P. L. Fiedler

R. K. Zebell

Herbs, perennial, sometimes from bulbs; bulb coat membranous or fibrous-reticulate. **Stems** scapelike or leafy, simple or branched, glabrous, often glaucous; bulblets sometimes borne in leaf axils. **Leaves** sessile; basal persistent or withering by flowering, solitary, blade base sometimes attenuate and petiolelike; cauline 0–several, sometimes proximalmost appearing as basal, reduced. **Inflorescences** monochasiate or ± umbellate, 1–many-flowered, bracteate. **Flowers:** perianth globose to broadly campanulate; sepals 3, distinct, ovate to lanceolate, usually petaloid and glabrous; petals 3, distinct, usually longer and broader than sepals, sometimes clawed, usually hairy adaxially, bearing adaxial gland near base, often spotted to ± patterned; filaments widened at base; anthers usually basifixed or pseudobasifixed, linear to oblong; ovary superior; style absent; stigmas 3. **Fruits** capsular, 3-locular, 3-angled or -winged, linear, oblong, or globular, dehiscence septicidal. **Seeds** many, in 2 rows per locule, irregular or flat, coat usually hexagonally reticulate.

Species ca. 70 (56 in the flora): w North America, Central America.

Many species of *Calochortus* are in cultivation. Bulbs of some species were eaten by Native Americans. Petal shape, gland shape, and pubescence are exceedingly variable in many taxa but often are important in identifying species.

As part of a phylogenetic analysis of relationships within the Liliales, T. B. Patterson (1998) studied variation in noncoding cpDNA sequences throughout *Calochortus*. He resolved seven main clades that are geographically and ecologically strongly cohesive, including the mariposas of dry grasslands and semideserts, the star-tulips of wet meadows, the cat's ears of montane woodlands, and the fairy lanterns of closed forests. Patterson suggested that this pattern of relationships is the result of highly localized speciation within the genus.

SELECTED REFERENCES Buxbaum, F. 1958. Der morphologische Typus und die systematische Stellung der Gattung *Calochortus*. Beitr. Biol. Pflanzen 34: 405–452. Ness, B. D. 1989. Seed morphology and taxonomic relationships in *Calochortus* (Liliaceae). Syst. Bot. 14: 495–505. Ownbey, M. 1940. A monograph of the genus *Calochortus*. Ann. Missouri Bot. Gard. 27: 371–560. Ownbey, M. 1969. *Calochortus*. Univ. Wash. Publ. Bot. 17: 765–779.

1. Flowers nodding; perianth closed, globular to oblong.
    2. Petals white, pink, or rose.
        3. Sepals spreading; proximal petal gland membranes ± equal to petal width; petals rose . . . . 1. *Calochortus amoenus*
        3. Sepals appressed to petals; proximal petal gland membranes 1/3–2/3 petal width; petals usually white to pink . . . . 2. *Calochortus albus*
    2. Petals yellow.
        4. Stems usually not branching distal to base, 1–10 dm; petals with adaxial surface moderately hairy; leaves and stems conspicuously glaucous . . . . 5. *Calochortus raichei*
        4. Stems usually branching distal to base, 1–5 dm; petals with adaxial surface glabrous or sparsely hairy; leaves and stems not or slightly glaucous.
            5. Leaves and stems slightly glaucous; petals deep yellow, adaxial surface glabrous or with a few hairs near gland . . . . 4. *Calochortus amabilis*
            5. Leaves and stems not glaucous; petals pale yellow, adaxial surface glabrous or sparsely hairy . . . . 3. *Calochortus pulchellus*
1. Flowers erect, occasionally spreading; perianth open, usually campanulate.
    6. Capsules nodding; perianth usually not conspicuously marked.
        7. Adaxial petal surface glabrous or sparsely hairy.

8. Cauline leaves usually absent .... 17. *Calochortus minimus*
8. Cauline leaves usually present.
  9. Stems shorter than 5 (10) cm, bearing bulblets near base .... 15. *Calochortus uniflorus*
  9. Stems 8–25 cm, not bearing bulblets.
    10. Petals white to pale pink; c California .... 14. *Calochortus umbellatus*
    10. Petals bright lavender; s Oregon .... 16. *Calochortus indecorus*

7. Adaxial petal surfaces densely hairy.
  11. Petals deep yellow .... 6. *Calochortus monophyllus*
  11. Petals light green, white, pink, purple, or blue.
    12. Petal margins slightly ciliate or fringed, entire adaxial surface hairy or papillose (except *Calochortus elegans*, hairy only distal to gland).
      13. Stems usually branching, with bractlike cauline leaf............ 7. *Calochortus tolmiei*
      13. Stems usually simple, often scapelike.
        14. Petal glands nearly round .... 13. *Calochortus apiculatus*
        14. Petal glands lunate or transversely oblong.
          15. Adaxial leaf surface hairy or papillose; Oregon.
            16. Petal glands deeply depressed, surrounded by yellow hairs, reddish striations from petal base to gland .... 11. *Calochortus coxii*
            16. Petal glands only slightly depressed, hairs distal to gland not yellow, often surrounded by lime green coloration and purple striations .... 24. *Calochortus umpquaensis*
          15. Adaxial leaf surface glabrous; widespread.
            17. Distal petal gland membranes absent; sepals without basal glandular blotch; nw United States .... 10. *Calochortus elegans* (in part)
            17. Distal petal gland membranes present; sepals with basal glandular blotch; high elevations in Cascade Mountains .... 12. *Calochortus subalpinus*
    12. Petal margins conspicuously ciliate or fringed, adaxial surface hairy only distal to gland.
      18. Petals pink to light purplish .... 26. *Calochortus persistens*
      18. Petals light green, white, or light blue.
        19. Petals light green to white, usually with purple crescent distal to gland .... 10. *Calochortus elegans* (in part)
        19. Petals white to light blue, purple crescent absent.
          20. Petals obovate, apex acuminate, margins ciliate to apex .... 8. *Calochortus coeruleus*
          20. Petals lanceolate, apex acute, ciliate only on margins ... 9. *Calochortus westonii*

[6. Shifted to left margin. —Ed.]

6. Capsules erect; perianth usually conspicuously marked.
  21. Basal leaves usually persistent at flowering.
    22. Petals yellow or yellow-green.
      23. Petals deep yellow, streaked red-brown proximally .... 40. *Calochortus luteus*
      23. Petals light yellow-green, flecked and streaked purple-brown ...... 27. *Calochortus tiburonensis*
    22. Petals white, cream, pink, lavender, or purple, not flecked or streaked purple-brown.
      24. Petals with longitudinal green stripe on adaxial surface ..... 47. *Calochortus excavatus* (in part)
      24. Petals without longitudinal green stripe on adaxial surface.
        25. Adaxial petal surface with purple crescent or blotch.
          26. Petals pink or lavender.
            27. Proximal stem internodes very short; stems with bulblet in axil of cauline leaf at or below ground surface; petals light pink to ± lavender .... 19. *Calochortus longebarbatus*
            27. Proximal stem internodes elongate, cauline leaf ca. halfway up stem; stems rarely with bulblet; petals lavender .... 20. *Calochortus nitidus*
          26. Petals whitish or purplish.
            28. Petals obovate, cuneate at base, with median red-purple blotch .... 21. *Calochortus eurycarpus*

28. Petals ovate to lanceolate, conspicuously clawed, with median red-purple crescent . . . . 25. *Calochortus lyallii*

25. Adaxial petal surface without purple crescent or blotch.

29. Adaxial leaf surface pubescent; petals yellowish white, with purplish hairs toward base; gland with distally branching hairs . . . . 23. *Calochortus howellii*

29. Adaxial leaf surface glabrous; petals pale lavender or purplish, without purple hairs at base; gland with ciliate membranes.

30. Abaxial petal surface with dark purple crescent; glands deeply depressed; dry soils and slopes . . . . 22. *Calochortus greenei*

30. Abaxial petal surface without dark purple crescent; glands not deeply depressed; moist meadows and lake margins . . . . 18. *Calochortus nudus*

[21. Shifted to left margin.—Ed.]

21. Basal leaves usually withering by flowering.

31. Petal glands ± glabrous, surrounded by dense, obscuring hairs; bulb coat fibrous-reticulate.

32. Petals pale pink to rose . . . . 54. *Calochortus plummerae*

32. Petals cream, yellow to deep orange, purplish, or red-brown.

33. Petals broadly cuneate or obovate, apex usually without hairy tufts . . . . . 55. *Calochortus weedii*

33. Petals oblong-ovate, tipped purple-brown, apex with hairy tufts . . . . 56. *Calochortus obispoensis*

31. Petal glands hairy; bulb coat, when present, membranous.

34. Petal gland ± round, depressed, sometimes deeply so, surrounded by conspicuously fringed membrane.

35. Petals yellow or orange to red; stems twisted . . . . 50. *Calochortus kennedyi*

35. Petals yellow to gold, white, or lavender, usually tinged with lilac; stems not twisted.

36. Abaxial petal surface without longitudinal green stripe.

37. Petals lemon yellow to gold; sepals usually with red to brown blotch at base.

38. Perianth cup-shaped; stems flexuous; petal glands covered with branching hairs and surrounded by club-shaped hairs . . . . . 51. *Calochortus clavatus*

38. Perianth campanulate; stems not flexuous; petal glands covered with slender, unbranched hairs and surrounded by a few long hairs . . . . 49. *Calochortus concolor*

37. Petals white with reddish brown or purple band or blotch distal to gland, or lemon yellow with maroon blotch distal to gland.

39. Petals white, tinged with lilac or infrequently magenta; e Nevada and n Arizona north through Utah to e Idaho . . . . 43. *Calochortus nuttallii*

39. Petals lemon yellow; n Arizona to se Utah . . . . 48. *Calochortus aureus*

36. Abaxial petal surface with longitudinal green stripe.

40. Petals lavender; bracts 3–8 cm; anthers red-brown . . . . 47. *Calochortus excavatus* (in part)

40. Petals white, white tinged with lilac, or purplish; bracts 2–5 cm; anthers yellow, maroon, blue, purple, or red.

41. Sepals without dark purple or red blotch near base; petals cuneate to obovate . . . . 46. *Calochortus invenustus*

41. Sepals with dark purple or red blotch near base; petals narrowly obovate.

42. Petals with red or purple crescent distal to gland, glands surrounded by yellow border . . . . 44. *Calochortus bruneaunis*

42. Petals without red or purple crescent distal to gland, glands surrounded by red or purple border . . . . 45. *Calochortus panamintensis*

34. Petal glands not simultaneously round, depressed, and surrounded by fringed membrane (but sometimes one of these).

43. Plants rarely bulbose.

44. Petals with conspicuous purple striping, never spotted . . . . 33. *Calochortus striatus*

44. Petals spotted, not conspicuously striped.

45. Stems twining or straggling; each perianth member with transverse yellow bands . . . . 29. *Calochortus flexuosus*
45. Stems ± straight; perianth members without transverse yellow bands.
46. Petals lavender to deep purple, usually with purple blotch at base . . . . 32. *Calochortus splendens*
46. Petals white, usually flushed pink, with red-brown blotch distal to gland . . . . 30. *Calochortus dunnii*

[43. Shifted to left margin.—Ed.]

43. Plants usually bulbose.
47. Petal glands ± ovate, longitudinally oblong, or ± round.
48. Anthers sagittate . . . . 41. *Calochortus leichtlinii*
48. Anthers not sagittate.
49. Capsules narrow-oblong, rounded in cross section; stems ± flexuous . . . . 28. *Calochortus catalinae*
49. Capsules linear, angled in cross section; stems straight.
50. Sepals and petals ca. 4–5 cm; petal gland hairs slender, not branching . . . . 34. *Calochortus monanthus*
50. Sepals and petals ca. 2–3 cm; petal gland hairs thick, distally knobbed . . . . 31. *Calochortus palmeri*
47. Petal glands ± arched, transversely oblong, lunate, doubly lunate, ± triangular, or ± square.
51. Petal glands triangular-sagittate . . . . 42. *Calochortus macrocarpus*
51. Petal glands ± arched, transversely oblong, lunate, doubly lunate, or ± square.
52. Petal glands ± square.
53. Petal glands in lower middle of petal, not surrounded by red blotch . . . . 35. *Calochortus venustus*
53. Petal glands nearly adjacent to petal base, surrounded by red blotch . . . . 36. *Calochortus simulans*
52. Petal glands arched, transversely oblong, ± lunate, or doubly lunate.
54. Petal glands ± doubly lunate . . . . 39. *Calochortus vestae*
54. Petal glands ± lunate to transversely oblong.
55. Capsules lanceoloid; coastal . . . . 37. *Calochortus argillosus*
55. Capsules linear or linear-oblong; inland.
56. Petal glands round to lunate . . . . 52. *Calochortus ambiguus*
56. Petal glands transversely oblong, chevron-shaped, or weakly lunate, never round.
57. Petal glands strongly chevron-shaped to weakly lunate; petals with median brown or purple blotch surrounded by bright yellow zone; California . . . . 38. *Calochortus superbus*
57. Petal glands transversely oblong; petals with transverse purple band distal to gland, and purple blotch on claw; Rocky Mountains, se Arizona, w New Mexico . . . . 53. *Calochortus gunnisonii*

1. Calochortus amoenus Greene, Pittonia 2: 71. 1890 [E]

Stems branching, 2–5 dm. **Leaves:** basal persistent, 2–5 dm; blade linear, flat. **Inflorescences** 2–many-flowered. **Flowers** nodding; perianth closed at apex, globular; sepals spreading, lanceolate, 10–15 mm, glabrous, apex acute; petals rose, elliptic-obovate, 16–25 mm, margins sparsely ciliate and hairy; glands ± depressed, with 4–5 ciliate membranes, proximal membrane ± equaling petal width; filaments equaling anthers; anthers oblong, apex obtuse. **Capsules** nodding, narrowly winged, oblong, 2–3 cm, apex obtuse or acute. **Seeds** dark brown, irregular. $2n = 20$.

Flowering mid spring–mid summer. Grassy slopes in partial shade; 500–1500 m; Calif.

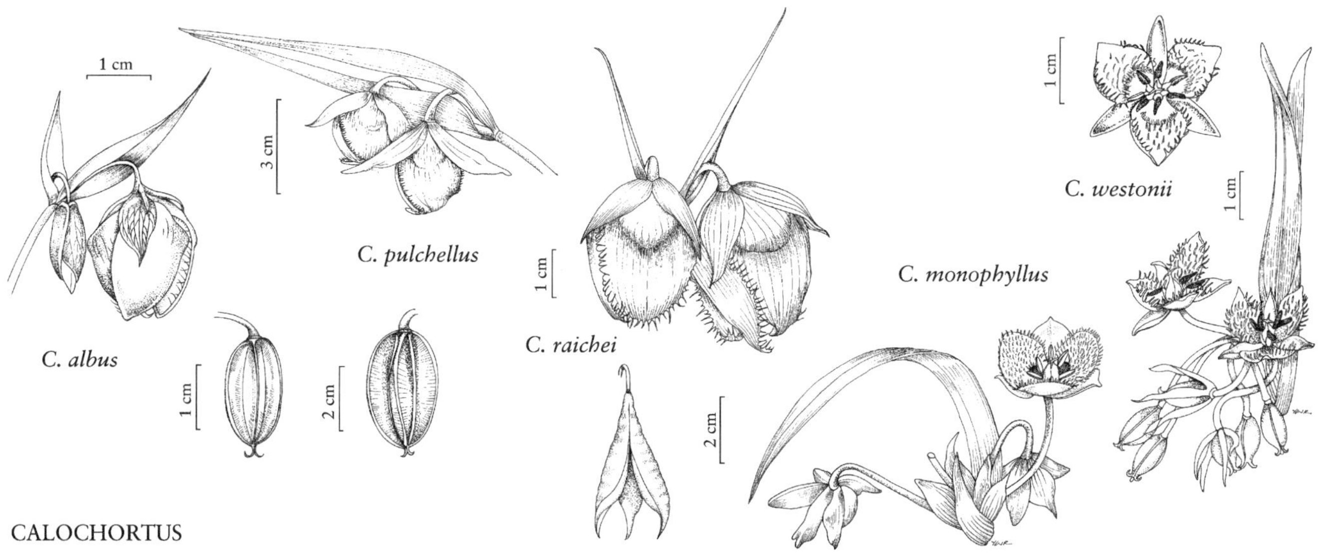

CALOCHORTUS

2. Calochortus albus (Bentham) Douglas ex Bentham, Botanist 2: no. 98. 1838 • White globe-lily, fairy lantern [E] [F]

Cyclobothra alba Bentham, Trans. Hort. Soc. London, ser. 2, 1: 413, plate 14, fig. 3. 1835; *Calochortus albus* var. *rubellus* Greene

Stems usually branching, 20–80 cm. **Leaves:** basal persistent, 3–7 dm; blade linear, flat. **Inflorescences** 2–many-flowered. **Flowers** nodding; perianth ± closed at apex, ± oblong; sepals appressed to petals, ovate to lanceolate, 10–15 mm, glabrous, apex acuminate; petals usually white to pink, ± elliptic, 20–25 mm, sparsely ciliate, hairs distal to gland slender; glands ± depressed, proximal ciliate membrane $1/3$–$2/3$ petal width; filaments equaling anthers; anthers oblong, apex obtuse or acute. **Capsules** nodding, winged, oblong, 2–4 cm, apex obtuse or acute. **Seeds** dark brown, irregular. $2n = 20$.

Flowering mid spring–mid summer. Shaded places in open woods and brush; 0–2000 m; Calif.

3. Calochortus pulchellus (Bentham) Alph. Wood, Proc. Acad. Nat. Sci. Philadelphia 20: 168. 1868 • Mount Diablo fairy lantern [C] [E] [F]

Cyclobothra pulchella Bentham, Trans. Hort. Soc. London, ser. 2, 1: 412, plate 14, fig. 1. 1835

Stems usually branching distal to base, 1–3 dm, not glaucous. **Leaves:** basal persistent, 1–4 dm; blade green, lanceolate, tapering at base, flat, not glaucous. **Inflorescences** 1–5-flowered. **Flowers** nodding; perianth closed at apex, ± globular; sepals spreading, yellow-green, ovate to lanceolate, 2–3 cm, glabrous, apex acuminate; petals pale yellow, lanceolate, 3–5 cm, conspicuously fringed with short, thick hairs, adaxial surface glabrous or sparsely hairy, apex obtuse; glands deeply depressed, bordered distally by long, slender hairs; filaments 6–8 mm; anthers oblong, 3–5 mm, apex obtuse or acute. **Capsules** nodding, winged, ellipsoid-oblong, 2–4 cm. **Seeds** dark brown, irregular. $2n = 20$.

Flowering late spring–early summer. Open places in woods, chaparral, typically on serpentine; of conservation concern; 200–800 m; Calif.

Calochortus pulchellus is known from Mt. Diablo, Contra Costa County.

4. Calochortus amabilis Purdy, Proc. Calif. Acad. Sci., ser. 3, 2: 119, plate 19, fig. 1. 1901 • Diogenes' lantern [E]

Stems usually branching distal to base, 1–5 dm, slightly glaucous. **Leaves:** basal persistent, 20–50 cm; blade linear, flat, slightly glaucous. **Inflorescences** 2–many-flowered. **Flowers** nodding; perianth closed at apex, globular; sepals spreading, ovate to lanceolate, 1.5–2 cm; petals deep yellow, with brown abaxial blotch, widely lanceolate, 1.6–2 cm, densely ciliate throughout, adaxial surface glabrous or with a few hairs near gland; glands deeply depressed, bordered distally by band of hairs; filaments longer than anthers; anthers oblong, apex obtuse or acute. **Capsules** nodding, winged, oblong, 2–3 cm, apex obtuse or acute. **Seeds** dark brown, irregular. $2n = 20$.

Flowering mid spring–mid summer. Shaded places in open woods or brush; 100–1000 m; Calif.

5. Calochortus raichei Farwig & V. Girard, Herbertia 43(1): 3, figs. 1, 2. 1987 • Cedars fairy lantern [C] [E] [F]

Stems usually not branching distal to base, 1–10 dm, conspicuously glaucous. **Leaves:** basal 3–4 dm; blade linear, attenuate, channeled, conspicuously glaucous. **Inflorescences** 1–2-flowered; bracts paired, opposite. **Flowers** nodding; perianth ± closed at apex, ± oblong; sepals appressed to petals, grayish mauve to pale green, broadly lanceolate, 1–2.5 cm, glabrous, apex acuminate; petals pale yellow, obovate, 2–4.5 cm, adaxial surface moderately hairy from gland to apex, margins moderately ciliate to apex, apex acuminate to acute; glands depressed, with long, thin, distal hairs; filaments 7–9 mm; anthers ± oblong, 6 mm, apex obtuse. **Capsules** nodding, ridged, not winged, oblong, 2–4 cm, apex acute. **Seeds** dark brown, irregular.

Flowering summer. Exposed serpentine soil in open woodlands; of conservation concern; less than 300 m; Calif.

Calochortus raichei is known only from the headwaters area of Big Austin Creek, Sonoma County. It is threatened by mining activities.

6. Calochortus monophyllus (Lindley) Lemaire, Fl. Serres Jard. Eur. 5: 430b. 1849 • Yellow star-tulip [E] [F]

Cyclobothra monophylla Lindley, J. Hort. Soc. London 4: 81. 1849

Stems slender, branching or not, 0.8–2 dm. **Leaves:** basal persistent, 1–3 dm; blade linear, flat. **Inflorescences** 1–6-flowered. **Flowers** ± erect at anthesis; perianth open, campanulate; sepals oblong-lanceolate, 16–20 mm, glabrous, apex acuminate; petals deep yellow, often with red-brown blotch distal to base, narrowly obovate, 12–20 mm, ciliate, adaxial surface densely branching-hairy; glands bordered proximally by ciliate membrane, distally by short hairs; anthers short-tipped. **Capsules** nodding, winged, oblong, 1–2 cm. **Seeds** dark brown, irregular. $2n = 20$.

Flowering spring. Wooded slopes, clay-loam soils; 400–1200 m; Calif., Oreg.

7. Calochortus tolmiei Hooker & Arnott, Bot. Beechey Voy., 398. 1840 • Pussy ears [E]

Stems ± slender, usually branching, 1–4 dm. **Leaves:** basal persistent, 1–4 dm; blade lanceolate, tapering at base, flat; cauline bractlike. **Inflorescences** ± umbellate, 1–several-flowered. **Flowers** erect or spreading; perianth open, campanulate; sepals 10–15 mm; petals white to pinkish or purplish, obovate, 12–25 mm, adaxial surface densely hairy, margins ciliate; glands bordered proximally by ciliate membrane, distally by short hairs; anthers lanceolate, apex acute to apiculate. **Capsules** nodding, winged, oblong, 2–3 cm. **Seeds** deep purple, dark brown in age, irregular. $2n = 20$.

Flowering mid spring–mid summer. Open grassy slopes or in woods, often on poor, dry soil; 0–2000 m; Calif., Oreg., Wash.

8. Calochortus coeruleus (Kellogg) S. Watson, Proc. Amer. Acad. Arts 14: 263. 1879 (as caeruleus) • Beavertail-grass [E]

Cyclobothra coerulea Kellogg, Proc. Calif. Acad. Sci. 2: 4. 1863; *Calochortus coeruleus* var. *fimbriatus* Ownbey; *C. maweanus* Leichtlin ex Baker

Stems not branching, 0.3–2 dm. **Leaves:** basal persistent, 1–2 dm; blade linear, flat. **Inflorescences** 1–10-flowered. **Flowers** erect to spreading; perianth open, campanulate; sepals oblong-lanceolate, ca. 10 mm, glabrous, apex acuminate; petals light blue, obovate, 8–12 mm, adaxial surface hairy only distal to gland, margins ciliate to apex, apex acuminate; glands slightly depressed, bordered proximally by ciliate membrane, distally by short hairs; anthers oblong, apex acute to acuminate. **Capsules** nodding, not winged, 1–2 cm, apex acute. **Seeds** irregular. $2n = 20$.

Flowering late spring–mid summer. Open gravelly places in woods; 600–2500 m; Calif.

Calochortus coeruleus appears to intergrade with *C. tolmiei* to some extent, especially in petal ciliation (K. L. Chambers, pers. comm.), and it might better be treated as a variety of the latter, more widespread species.

9. Calochortus westonii Eastwood, Proc. Calif. Acad. Sci., ser. 4, 20: 136. 1931 • Shirley Meadows star-tulip, mariposa-lily [C] [E] [F]

Calochortus coeruleus (Kellogg) S. Watson var. *westonii* (Eastwood) Ownbey

Stems slender and delicate, not branching, 3–15 cm. **Leaves:** basal persistent, 1–2 dm; blade linear, flat. **Inflorescences** 1–12-flowered. **Flowers** erect to spreading; perianth open, campanulate; sepals ca. 10 mm; petals white to light blue, lanceolate, 8–12 mm, ciliate only on margins, adaxial surface hairy only distal to gland, apex acute; glands slightly depressed, bordered proximally by ciliate membrane, distally by short hairs; anther apex acute to acuminate. **Capsules** nodding, angled, 1–2 cm, apex acute. **Seeds** irregular.

Flowering late spring. Open places in woods, meadows; of conservation concern; 1500–2000 m; Calif.

Calochortus westonii is known from the Greenhorn Mountains, Kern and Tulare counties.

10. Calochortus elegans Pursh, Fl. Amer. Sept. 1: 240. 1814 • Cat's ear [E] [F]

Stems ± slender, usually not branching, often scapelike, 0.5–2 dm. **Leaves:** basal persistent, 1–2 dm; blade linear, flat, glabrous adaxially. **Inflorescences** 1–7-flowered. **Flowers** erect to spreading; perianth open, campanulate; sepals oblong-lanceolate, 12–16 mm, usually glabrous, adaxial surface minutely papillose; petals light green to white, usually with purple crescent distal to gland, ± oblanceolate or broader, 10–15 mm, adaxial surface hairy only distal to gland, margins conspicuously ciliate; glands lunate, usually slightly depressed, bordered proximally by ± ciliate membrane, distally by short hairs; anthers lanceolate, apex apiculate. **Capsules** nodding, winged, ellipsoid-globular, 1–2 cm, apex obtuse or acute. **Seeds** light brown, irregular. $2n = 20$.

Varieties 3 (3 in the flora): nw United States.

1. Petal gland membranes erose to crenate . . . . 10c. *Calochortus elegans* var. *nanus*
1. Petal gland membranes deeply fringed.
   2. Petal glands broadly transverse-oblong, extending nearly across lower portion of petal, depressed, strongly arched distally; petals densely bearded . . . . . . 10a. *Calochortus elegans* var. *elegans*
   2. Petal glands shortly transverse-oblong, not depressed, straight or only slightly arched; petals only moderately bearded . . . . 10b. *Calochortus elegans* var. *selwayensis*

10a. Calochortus elegans Pursh var. **elegans** [E]

Petals densely bearded; glands broadly transverse-oblong, extending nearly across lower portion of petal, depressed, strongly arched distally, membranes deeply fringed.

Flowering late spring–mid summer. Grassy hillsides, open woods; 1500–2500 m; Idaho, Mont., Oreg., Wash.

10b. Calochortus elegans Pursh var. **selwayensis** (H. St. John) Ownbey, Ann. Missouri Bot. Gard. 27: 419. 1940 [E]

Calochortus selwayensis H. St. John, Proc. Biol. Soc. Wash. 41: 192. 1928.

Petals only moderately bearded; glands shortly transverse-oblong, not depressed, straight or only slightly arched, membranes deeply fringed.

Flowering late spring–mid summer. Grassy hillsides, open woods; 1500–2500 m; Idaho, Mont.

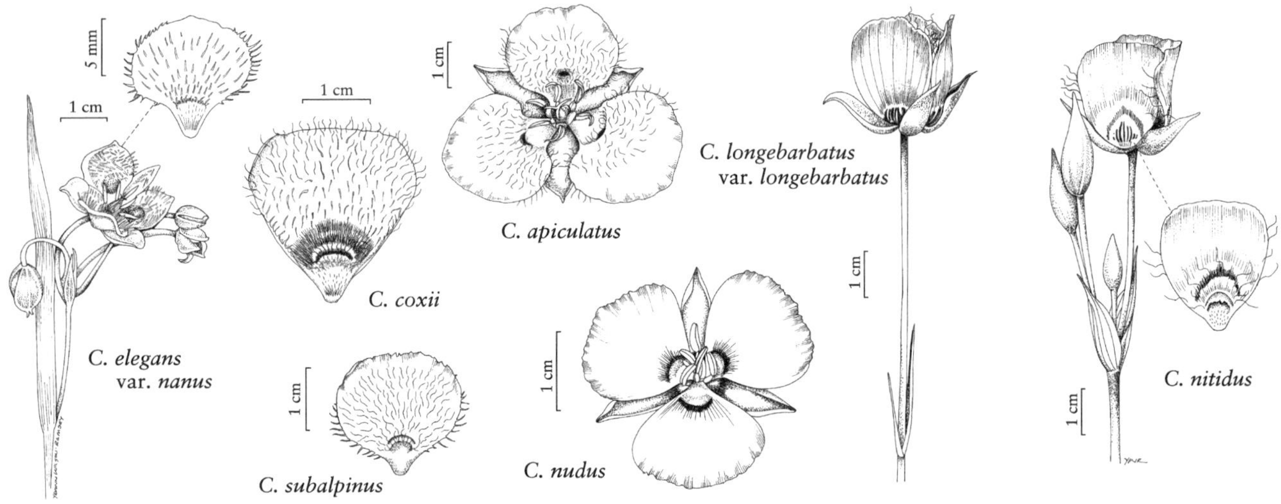

CALOCHORTUS

10c. Calochortus elegans Pursh var. **nanus** Alph. Wood, Proc. Acad. Nat. Sci. Philadelphia 20: 168. 1868 [E] [F]

Calochortus elegans var. *oreophilis* Ownbey

Petals more conspicuously fringed than other varieties, hairs usually sparser, shorter and thicker; gland membranes erose to crenate.

Flowering late spring–mid summer. Grassy hillsides, open woods; 1500–2500 m; Calif., Oreg.

11. Calochortus coxii M. R. Godfrey & Callahan, Phytologia 65: 216, fig.1g–k. 1988 • Cox's mariposa-lily [C] [E] [F]

Stems usually not branching, straight to flexuous, often scapelike, 15–25 cm. **Leaves:** basal ± erect, to 3 dm × 3–7 mm; blade with adaxial surface densely hairy, abaxial surface glabrous, shiny. **Inflorescences** erect, 1–7-flowered; bracts 1–several, 2.5–3 cm. **Flowers** erect; perianth open, campanulate; sepals ovate-acuminate, 20 × 8 mm; petals white, with reddish striations from base to gland and broad lavender chevron just distal to gland margins, broadly obovate, 2.5 cm, adaxial surface densely hairy, margins slightly ciliate; glands transversely oblong-lunate, deeply depressed, green at adaxial base, $^1/_2$ to nearly equaling petal claw width, surrounded by yellow hairs that grade to white at petal apex, covered with membranous scales, scales covered with very small, translucent, rodlike hairs; filaments 7 mm; anthers reddish brown, 3–7 mm, apex apiculate. **Capsules** nodding, 3-winged, ellipsoid-elongate, 3–4 cm. **Seeds** light beige, surface rough.

Flowering mid summer. North-facing open grassy slopes or woods, on serpentine; of conservation concern; 200–1000 m; Oreg.

Calochortus coxii is endemic to Douglas County from near the Umpqua River to Myrtle Creek Mountain.

12. Calochortus subalpinus Piper, Contr. U.S. Natl. Herb. 11: 195. 1906 [E] [F]

Calochortus lobbii (Baker) Purdy

Stems usually not branching, straight or flexuous, often scapelike, 0.5–3 dm. **Leaves:** basal 1–3 dm × 2–15 mm, usually equaling or exceeding stem length; blade flat, adaxial surface glabrous. **Inflorescences** subumbellate, 1–5-flowered; bracts 2–several, lanceolate to linear, unequal, 1–5 cm, apex acuminate; peduncle slender, becoming stouter, deflexed in fruit. **Flowers** erect or spreading; perianth open, campanulate; sepals typically with purple glandular blotch near base, oblong-lanceolate, shorter than petals, adaxial surface minutely hairy, apex acute to acuminate; petals yellowish white, sometimes lavender-tinged, frequently with narrow purple crescent distal to gland, broadly obovate, cuneate, moderately bearded nearly to apex, adaxial surface densely hairy, margins fringed, apex obtuse or acute; glands transversely oblong, arched upward, ± deeply depressed, bordered proximally by narrow, ascending, deeply fringed membrane, distally by narrow, crenate membranes, gland surface with rather long, slender hairs toward distal portion; anthers lanceolate, apex long-apiculate. **Capsules** nodding, 3-winged, ellipsoid, apex usually acute. **Seeds** pale yellow. 2*n* = 20.

Flowering summer. Open forest in loose volcanic soils; 1000–2200 m; Oreg., Wash.

13. Calochortus apiculatus Baker, J. Linn. Soc., Bot. 14: 305. 1874 [E] [F]

Stems usually not branching, straight, scapelike, stout, 1–3 dm. **Leaves:** basal 1–3 dm × 5–15 mm, usually shorter than stem. **Inflorescences** subumbellate, 1–5-flowered; bracts 2–several, lanceolate to linear, unequal, 1–5 cm, apex acuminate; peduncle stout, becoming strongly deflexed in fruit. **Flowers** erect or spreading; perianth yellowish white, sometimes streaked with purple, open, campanulate; sepals shorter than petals; petals oblong-lanceolate, clawed, adaxial surface densely hairy, margins fringed laterally and moderately bearded on proximal 1/2 distal to claw with slender, flexuous hairs, apex acute or obtuse; glands short, nearly round, depressed, typically bordered proximally by dark-colored, deeply fringed membrane, adaxial surface covered with short, thick hairs; filaments shorter than or equaling anthers; anthers lanceolate, apex long-apiculate. **Capsules** nodding, 3-winged, ellipsoid, apex acute. **Seeds** light brown, irregular. 2*n* = 20.

Flowering summer. Dry, rocky slopes in open, coniferous woods; 600–2400 m; Alta., B.C.; Idaho, Mont., Wash.

14. Calochortus umbellatus Alph. Wood, Proc. Acad. Nat. Sci. Philadelphia 20: 168. 1868 • Oakland star-tulip [E]

Stems usually branching, 8–25 cm, not bearing bulblets. **Leaves:** basal persistent, 2–4 dm; blade lanceolate, flat; cauline usually present. **Inflorescences** subumbellate, 3–12-flowered. **Flowers** ± erect; perianth open, campanulate; sepals 10–14 mm; petals white to pale pink, usually with purple blotch near gland, widely obovate to cuneate, 12–18 mm, adaxial surface not ciliate, ± glabrous, margins irregularly dentate distally; glands basally convex, covered by broad, ciliate membrane, bordered distally by short hairs; filaments longer than anthers; anthers oblong, apex acute or obtuse. **Capsules** nodding, 3-winged, 1–2 cm. **Seeds** dark brown, irregular. 2*n* = 20.

Flowering early spring–early summer. Openings in chaparral or woods, often on serpentine; 100–700 m; Calif.

Calochortus umbellatus grows around San Francisco Bay. It is reported also from Lake County.

15. Calochortus uniflorus Hooker & Arnott, Bot. Beechey Voy., 398, plate 94. 1840 • Large-flowered star-tulip [E]

Stems usually not branching, shorter than 5(–10) cm, bearing bulblets near base. **Leaves:** basal persistent, 1–4 dm; blade lanceolate, flat; cauline usually present. **Inflorescences** subumbellate, 1–5-flowered; bracts 2–4, linear, attenuate. **Flowers** erect; perianth open, campanulate; sepals elliptic-lanceolate, 12–16 mm, glabrous, apex acuminate; petals ± white to pink, often with purple blotch distal to gland, broadly obovate, cuneate, 15–28 mm, adaxial surface glabrous or sparsely hairy, margins irregularly dentate distally; glands covered with ciliate membranes, bordered distally by short hairs; filaments longer than anthers; anthers oblong, apex obtuse or acute. **Capsules** nodding, narrowly 3-winged, 1–3 cm. **Seeds** dark brown, irregular. 2*n* = 20, 40.

Flowering late spring–mid summer. Moist meadows; 0–200 m; Calif., Oreg.

16. Calochortus indecorus Ownbey & M. Peck, Leafl. W. Bot. 7: 191. 1954 [E]

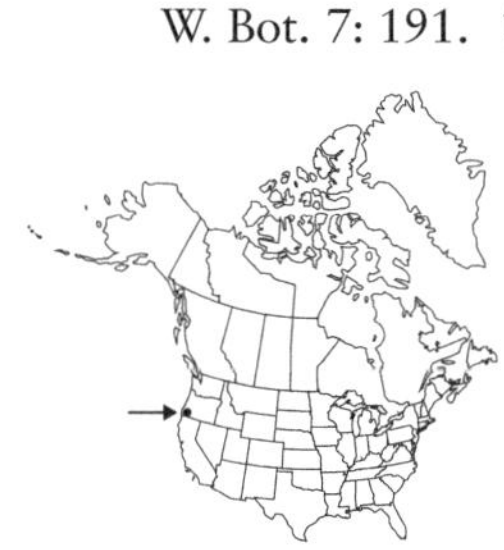

Stems stout, not branching, 16–22 cm, not bearing bulblets. **Leaves:** basal 2–25 dm × 8–12 mm; cauline usually present. **Inflorescences** 2–6 flowered. **Flowers** ± erect; perianth open, campanulate; sepals broadly lanceolate, somewhat shorter than petals; petals bright lavender, broadly obovate, adaxial surface glabrous or with short, purple hairs distal to gland, apex margins erose; glands slightly depressed, surrounded proximally by minutely denticulate membrane; filaments longer than anthers; anthers oblong, apex acute to obtuse. **Capsules** nodding, narrowly 3-winged, ovoid, 1.5–2 cm. **Seeds** unknown.

Flowering late spring. Oreg.

Calochortus indecorus was collected from the western slope of Sexton Mountain, northeastern Josephine County. Endemic to that area, the taxon is now presumed extinct.

17. Calochortus minimus Ownbey, Ann. Missouri Bot. Gard. 27: 430. 1940 [E]

Stems not branching, less than 1 dm. **Leaves:** basal persistent, 1–2 dm; blade linear, flat; cauline usually absent. **Inflorescences** 1–10-flowered. **Flowers** ± erect; perianth open, campanulate; sepals lanceolate, 8–10 mm, glabrous, apex acuminate; petals white, obovate, 10–14 mm, not ciliate, adaxial surface glabrous or sometimes with few hairs near gland; glands bordered proximally by ± ciliate membrane; filaments equaling anthers; anthers linear-oblong, apex acute. **Capsules** nodding, 3-winged, ellipsoid, 1–2 cm. **Seeds** irregular. $2n = 20$.

Flowering late spring–mid summer. Moist grassy areas in open woods, along lake margins; 1200–3000 m; Calif.

Derivatives of hybridization between *Calochortus minimus* and *C. nudus* (pink- to lavender-tinged petals with rounded tips) occur in the northern high Sierra Nevada (B. D. Ness et al. 1990).

18. Calochortus nudus S. Watson, Proc. Amer. Acad. Arts 14: 263. 1879 • Shasta star-tulip [E] [F]

Calochortus nudus var. *shastensis* (Purdy) Jepson; *C. shastensis* Purdy

Stems usually not branching, 1–3 dm. **Leaves:** basal persistent, 0.5–2 dm; blade lanceolate, tapering at base, flat, glabrous adaxially. **Inflorescences** 1–several-flowered. **Flowers** erect; perianth open, campanulate; sepals lanceolate, 10–12 mm, glabrous, apex acuminate; petals pale lavender, widely obovate, 14–16 mm, adaxial surface ± glabrous, not ciliate; glands not deeply depressed, bordered proximally by wide, ciliate membranes; anthers linear-oblong, apex obtuse or acute. **Capsules** erect, 3-winged, 1–2 cm. **Seeds** light brown, irregular. $2n = 20$.

Flowering late spring–mid summer. Moist grassy areas, meadows, lake and bog margins; 1200–2500 m; Calif., Oreg.

Calochortus nudus hybridizes with *C. minimus*.

19. Calochortus longebarbatus S. Watson, Proc. Amer. Acad. Arts 17: 381. 1882 • Long-haired star-tulip [E] [F]

Stems usually not branching, 1–3 dm, proximal internodes very short; bulblet in axil of cauline leaf at or below ground surface. **Leaves** single; basal persistent, 2–3 dm; blade linear. **Inflorescences** 1–4-flowered. **Flowers** erect; perianth open, campanulate; sepals ovate-lanceolate, 15–20 mm, apex acuminate; petals light pink to ± lavender, with red-purple adaxial crescent distal to gland, widely obovate-cuneate, 20–30 mm, not ciliate, with a few long hairs; glands bordered proximally by ciliate membrane, distally by short hairs; filaments twice length of anthers; anther apex obtuse to abruptly short-tipped. **Capsules** erect, 3-winged, ellipsoid to ± globular, 2–3 cm. **Seeds** light brown, irregular.

Recent collections of *Calochortus longebarbatus* from Oregon exhibit intergradation in the characters that heretofore have been thought to distinguish the following two varieties (K. L. Chambers, pers. comm.), and their continued recognition may prove unwarranted.

Varieties 2 (2 in the flora): w United States.

1. Perianth obconic-campanulate, petals not sharply bent at gland; basal leaves not reddish at base .... 19a. *Calochortus longebarbatus* var. *longebarbatus*
1. Perianth broadly campanulate, petals sharply bent at gland; basal leaves reddish at base .... 19b. *Calochortus longebarbatus* var. *peckii*

19a. Calochortus longebarbatus S. Watson var. **longebarbatus** [E] [F]

Basal leaves base typically not reddish. **Perianth** obconic-campanulate; petals not sharply bent at gland. $2n = 20$.

Flowering late spring–mid summer. Vernal meadows, heavy clay soil; 1200–1900 m; Calif., Oreg., Wash.

Calochortus longebarbatus var. *longebarbatus* is threatened by grazing.

19b. Calochortus longebarbatus S. Watson var. **peckii** Ownbey, Bot. Gaz. 104: 556. 1943 [E]

Basal leaves base typically reddish. **Perianth** broadly campanulate; petals sharply bent at gland. $2n$ = 30.

Flowering late spring–mid summer. Grassy margins of wet meadows and under pines; 1600–1800 m; Oreg.

Calochortus longebarbatus var. *peckii* is a triploid that bears only sterile anthers and reproduces only vegetatively through division of the bulb. It is found in the Ochoco Mountains.

20. Calochortus nitidus Douglas, Trans. Hort. Soc. London 7: 277, plate 9A. 1828 [E] [F]

Stems not branching, straight, 2–4 dm, proximal internodes elongate; bulblets rare. **Leaves:** basal usually persistent, 1–3 dm × 10–25 mm; blade flat, tapering toward both ends, becoming involute; cauline ca. halfway up stem. **Inflorescences** subumbellate, 1–4-flowered; bracts 2 or more, narrowly lanceolate to linear, long-attenuate, unequal, 2–10 cm. **Flowers** erect; perianth open, campanulate; sepals ovate to lanceolate, shorter than petals, glabrous, apex acuminate; petals lavender, with deep purple adaxial crescent distal to gland, obovate, cuneate, ± ciliate laterally, sparingly covered with long, flexuous hairs distal to gland; glands ± triangular-lunate, slightly depressed, bordered proximally by narrow, deeply fringed membrane, covered with short, thick hairs, hairs and membrane fringe densely covered with long papillae; filaments longer than anthers; anthers oblong, 6–10 mm, apex obtuse. **Capsules** erect, 3-winged, ellipsoid to nearly globular. **Seeds** light beige. $2n$ = 40.

Flowering summer. Low meadows along creeks; 700–900 m; Idaho, Oreg., Wash.

21. Calochortus eurycarpus S. Watson, Botany (Fortieth Parallel), 348. 1871 [E]

Calochortus nitidus Douglas var. *eurycarpus* L. F. Henderson

Stems not branching, straight, 1–5 dm. **Leaves:** basal persistent, subtended by inflorescence, 1–3 dm × 5–25 mm; blade flat, tapering toward both ends, becoming involute. **Inflorescences** subumbellate, 1–5-flowered; bracts 2 or more, narrowly lanceolate to linear, long-attenuate, unequal, 1–5 cm. **Flowers** erect; perianth open, campanulate; sepals ovate to lanceolate, usually much shorter than petals, glabrous, apex acute to acuminate; petals creamy white to lavender, with conspicuous median red-purple adaxial blotch, obovate, invested near gland with few long, flexuous hairs, base cuneate, apex rounded or acute; glands triangular-lunate, slightly depressed, bordered proximally by comparatively narrow, deeply fringed membrane, distally often by narrower, crenate membranes, enclosed surface densely covered with long yellowish hairs, which, with membrane fringe, are often inconspicuously papillose; filaments slightly longer than anthers; anthers oblong, apex obtuse. **Capsules** erect, 3-winged, ellipsoid-oblong. **Seeds** light beige. $2n$ = 20.

Flowering summer. Grasslands and open coniferous forests; 700–2700 m; Idaho, Mont., Nev., Oreg., Wash., Wyo.

22. Calochortus greenei S. Watson, Proc. Amer. Acad. Arts 14: 264. 1879 • Greene's mariposa-lily [C] [E] [F]

Stems usually branching, 1–3 dm. **Leaves:** basal persistent, ca. 2 dm; blade glabrous adaxially. **Inflorescences** 1–5-flowered. **Flowers** erect; perianth open, campanulate; sepals 25–30 mm; petals purplish, with darker purple abaxial crescent distal to gland, 30–40 mm, not ciliate at base, ± hairy; glands deeply depressed, bordered proximally by wide ciliate membrane, distally by short hairs. **Capsules** erect, 3-winged, ellipsoid, 20–25 mm. **Seeds** irregular.

Flowering mid summer. Shrubby hillsides, open woodlands, dry soils and slopes; of conservation concern; 700–1100 m; Calif., Oreg.

23. Calochortus howellii S. Watson, Proc. Amer. Acad. Arts 23: 266. 1888 [E] [F]

Stems slender, not branching or rarely branching in axil of cauline leaf, straight, 2–4 dm. **Leaves:** basal persistent, 20–35 cm; blade gradually tapering distally, adaxially pubescent. **Inflorescences** subumbellate, 1–3-flowered; bracts 2, subopposite, linear to narrowly lanceolate, attenuate, 1–3 cm. **Flowers** erect; perianth open, campanulate; sepals shorter than petals, ovate-acuminate, glabrous; petals yellowish white, with purplish hairs toward base, broadly obovate, cuneate, inconspicuously fringed later-

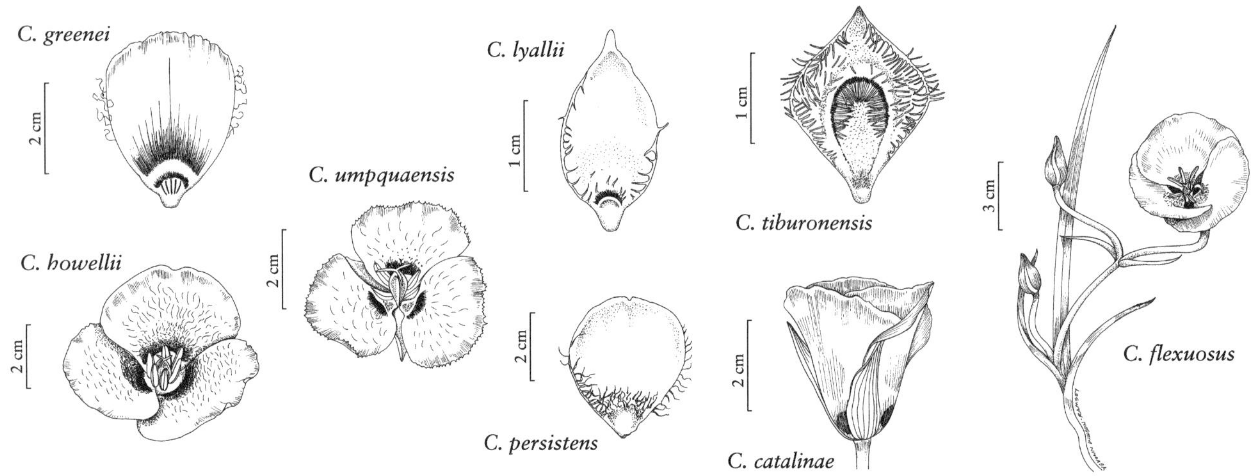

CALOCHORTUS

ally, sparingly short-hairy adaxially, apex rounded; glands transversely oblong, slightly depressed, densely covered with short, distally branching hairs, the bases of which unite at lower margin to form inconspicuous, discontinuous membranes; filaments shorter than anthers; anthers oblong-lanceolate, apex acute to short-apiculate. **Capsules** erect, 3-winged, ellipsoid, ca. 2 cm, base and apex acute. **Seeds** light-colored, irregular. 2*n* = 20.

Flowering late spring–mid summer. Dry, rocky, serpentine soils; 300–500 m; Oreg.

Calochortus howellii is known only from the Illinois River valley, Josephine County.

24. Calochortus umpquaensis Fredricks, Syst. Bot. 14: 12, figs. 1, 2, 3f–j, 4, 5. 1989 • Umpqua mariposa-lily [C] [E] [F]

Stems not branching, straight, often scapelike, 2–3 dm, glabrous or glaucous. **Leaves:** basal solitary, clasping; blade narrowly lanceolate, hairy, adaxially hispid, abaxially glabrous, sometimes glaucous. **Inflorescences** 1–several-flowered; bracts 2, subopposite, narrowly lanceolate. **Flowers** erect; perianth open, campanulate; sepals lanceolate-acuminate, ca. 2 cm; petals white to cream, with dark purple-black, pentagonal to lunate blotch, broadly oblong to obovate, 3.5 cm, bearded, adaxial surface typically minutely papillose, margins erose; glands transversely oblong-lunate, slightly depressed, with 0.7–1.4 mm-wide band of short dendritic hairs distally, hairs surrounded by lime-green coloration and purple striations; anthers lanceolate, apex acuminate. **Capsules** nodding, 3–5.4 cm. **Seeds** 2.8–3.5 mm, with inflated bulbous crest and hollow lateral ridge. 2*n* = 20.

Flowering late spring–mid summer. Grassland-forest ecotones in serpentine-derived soils; of conservation concern; 300–500 m; Oreg.

Calochortus umpquaensis is known only from Watson and Ace Williams mountains on both sides of the Little River, Douglas County.

25. Calochortus lyallii Baker, J. Linn. Soc., Bot. 14: 305. 1874 [E] [F]

Stems not branching or with branch in axil of cauline leaf, straight, 1–5 dm. **Leaves:** basal solitary, persistent, 1–3 dm × 2–20 mm, usually not exceeding inflorescence; blade tapering toward both ends. **Inflorescences** subumbellate, 1–9-flowered; bracts 2–several, lanceolate to linear, attenuate, unequal, 1–5 cm; peduncle slender, becoming stiffly erect in fruit. **Flowers** erect or spreading; perianth open, campanulate; sepals white or purplish-tinged, usually with purple crescent, lanceolate, usually equaling petals, often adaxially papillose, apex acuminate to attenuate; petals white or purplish-tinged, usually with median, red-purple, adaxial crescent distal to gland, ovate to lanceolate, abruptly contracted at base into short, conspicuous claw, usually conspicuously fringed with long, slender hairs, adaxial surface ± papillose, bearing a few long hairs distal to claw, apex acute to acuminate; glands transversely oblong, depressed, arched upward, bordered proximally by narrower, crenate membrane, with short, thick hairs toward distal margin of enclosed surface, both fringe of lower membrane and hairs densely long-papillose; filaments basally dilated, ± equaling anthers; anthers oblong-lanceolate, apex acute to short-apiculate. **Capsules** erect, 3-winged, ellipsoid, base and apex acute. **Seeds** irregular. $2n = 20$.

Flowering late spring–mid summer. Dry slopes, open coniferous forests; 300–1500 m; B.C.; Wash.

26. Calochortus persistens Ownbey, Ann. Missouri Bot. Gard. 27: 448, plate 38, figs. 1, 2. 1940 • Siskiyou mariposa-lily [C] [E] [F]

Stems not branching, ca. 1 dm. **Leaves:** basal persistent, ca. 2 dm; blade lanceolate, tapering at base, flat. **Inflorescences** subumbellate, 2-flowered. **Flowers** erect; perianth open, campanulate; sepals smaller than petals; petals pink to light purplish, obovate, 35–40 mm, adaxial surface hairy only distal to gland, margins yellow-ciliate; glands bordered proximally by wide, ciliate membrane, distally by short hairs; anthers lanceolate, apex apiculate. **Capsules** nodding, winged, ellipsoid, ca. 1 cm, apex acute. **Seeds** irregular. $2n = 20$.

Flowering late spring–mid summer. Open areas in rocky soils; of conservation concern; 1000–1500 m; Calif.

Calochortus persistens is restricted to the Klamath and Cascade ranges in Siskiyou County.

27. Calochortus tiburonensis A. J. Hill, Madroño 22: 100, fig. 1. 1973 • Tiburon mariposa-lily [C] [E] [F]

Stems usually branching, 1–6 dm. **Leaves:** basal persistent, 1–7 dm; blade lanceolate, flat. **Inflorescences** 2–several-flowered. **Flowers** erect; perianth open, campanulate; petals light yellow-green, flecked and streaked purplish brown, oblanceolate, ciliate to near apex, hairy; glands depressed, bordered proximally by ciliate membrane, distally by 2 or more rows of short hairs; anthers short-tipped. **Capsules** erect, angled, apex acute. **Seeds** dark brown, irregular. $2n = 20$.

Flowering late spring–early summer. Serpentine grasslands; of conservation concern; 0–200 m; Calif.

Calochortus tiburonensis is known only from Ring Mountain, Marin County.

28. Calochortus catalinae S. Watson, Proc. Amer. Acad. Arts 14: 268. 1879 • Catalina mariposa-lily [E] [F]

Mariposa catalinae (S. Watson) Hoover

Plants usually bulbose; bulb coat membranous. **Stems** usually branching distally, ± flexuous, 2–6 dm. **Leaves:** basal withering, 1–3 dm; blade linear. **Inflorescences** 1–few-flowered; bracts opposite pedicel, 2–10 cm. **Flowers** erect; perianth open, broadly campanulate; sepals with purple blotch near base, lanceolate, 2–3 cm, apex acuminate; petals white tinged with lilac, with purple blotch near base, usually obovate, ± concave, 2–5 cm, glabrous except few hairs near base, apex obtuse; glands oblong, not depressed, densely hairy, hairs branching; filaments 8–10 mm; anthers lilac, oblong, 4–5 mm. **Capsules** erect, narrow-oblong, rounded in cross section, 2–5 cm, apex obtuse. **Seeds** light yellow, translucent, flat. $2n = 14$.

Flowering spring–early summer. Heavy soil, open grassy slopes, openings in brush; less than 700 m; Calif.

Calochortus catalinae is increasingly threatened by habitat loss.

29. Calochortus flexuosus S. Watson, Amer. Naturalist 7: 303. 1873 [E] [F]

Plants rarely bulbose; bulb coat, when present, membranous. **Stems** twining or straggling over ground, branching, 1–2 dm, usually sinuous. **Leaves:** basal, withering; blade linear, attenuate. **Inflorescences** 1–4-flowered; bracts 1–3 cm. **Flowers** erect; perianth open, campanulate, each member with transverse yellow band; sepals lanceolate to lanceolate-ovate, 2–3 cm; petals white with lilac tinge, darker veined, especially distally, sometimes with purple blotch distal to gland, obovate to cuneate, 3–4 cm, with sparse or dense, short or long hairs near gland, apex rounded; glands transverse-lunate or wider, densely short-hairy; filaments 6–10 mm; anthers oblong, 5–7 mm, apex obtuse or acute. **Capsules** erect, lanceoloid, angled, stout, 3–4 cm. **Seeds** light beige, flat. $2n = 14$.

Flowering spring. Dry stony slopes, desert hills, mesas, creosote bush scrub, sagebrush scrub; 500–2000 m; Ariz., Calif., Colo., Nev., N.Mex., Utah.

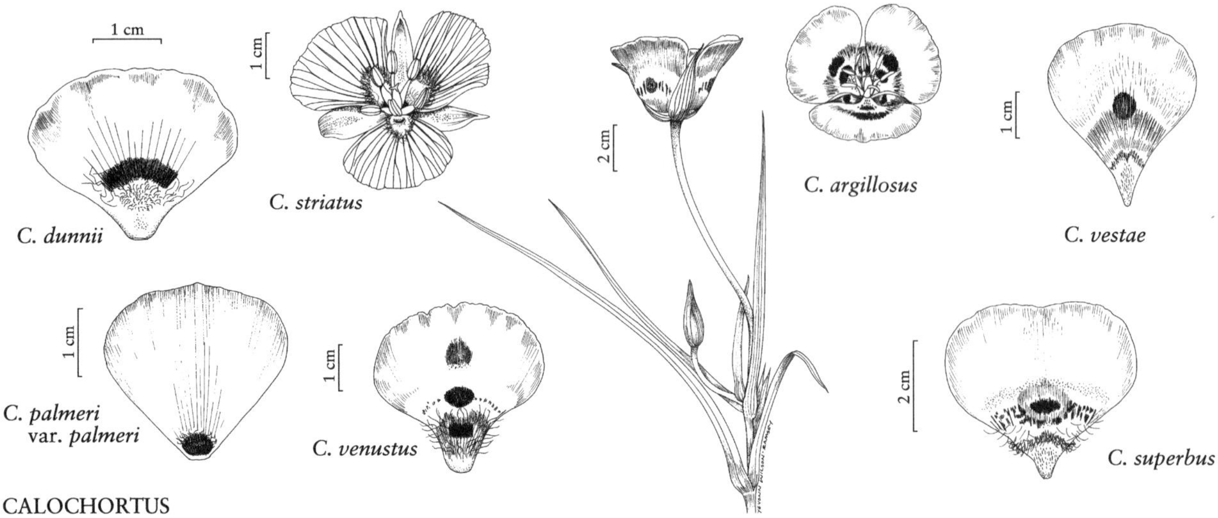

CALOCHORTUS

30. Calochortus dunnii Purdy, Proc. Calif. Acad. Sci., ser. 3, 2: 147, plate 19, figs. 14, 17. 1901 • Dunn's mariposa-lily [C] [F]

Calochortus palmeri S. Watson var. *dunnii* (Purdy) Jepson & Ames

Plants rarely bulbose; bulb coat, when present, membranous. **Stems** slender, usually branching, ± straight, 2–6 dm. **Leaves:** basal withering, 1–2 dm; blade linear-attenuate, channeled. **Inflorescences** 2–6-flowered; bracts 1–2 cm. **Flowers** erect; perianth open, campanulate; sepals lanceolate-ovate, 1–2 cm; petals white, usually flushed pink, with red-brown blotch distal to gland, obovate to cuneate, 2–3 cm, usually rounded distally, with yellowish hairs near gland; glands rounded, not depressed, covered with linear yellow hairs; filaments 5–6 mm; anthers white, oblong, 4–5 mm, apex acute. **Capsules** erect, linear, angled in cross section, 2–3 cm. $2n = 14$.

Flowering mid summer. Dry stony ridges in chaparral and pine forest; of conservation concern; 1500–1700 m; Calif.; Mexico (Baja California).

31. Calochortus palmeri S. Watson, Proc. Amer. Acad. Arts 14: 266. 1879 [C] [E] [F]

Plants usually bulbose; bulb coat membranous. **Stems** usually branching, straight, 3–6 dm; bearing bulblets or not. **Leaves:** basal withering, 1–2 dm; blade linear-attenuate. **Inflorescences** monochasiate, 1–4-flowered; bracts 1–2 cm. **Flowers** erect; perianth open, broadly campanulate; sepals usually with brownish blotch near base, lanceolate, 3 cm, apex acuminate; petals white to lavender, sometimes with brownish blotch distal to gland, obovate-cuneate, rounded distally, 2–3 cm, usually with yellow hairs near gland, apex acute to acuminate; glands many, rounded, not depressed, short, thick, distally knobbed, sometimes with yellowish or purplish hairs, otherwise glabrous; filaments 7–8 mm; anthers white, oblong, 5–7 mm, apex acute. **Capsules** erect, linear, angled in cross section, ca. 5 cm. $2n = 14$.

Varieties 2 (2 in the flora): California.

1. Petal glands yellow-hairy; stems bulbeliferous at base; bracts alternate . . . . 31a. *Calochortus palmeri* var. *palmeri*
1. Petal glands glabrous or purple-hairy; stems not bulbeliferous; bracts opposite . . . . 31b. *Calochortus palmeri* var. *munzii*

31a. Calochortus palmeri S. Watson var. **palmeri** • Palmer's mariposa-lily [C] [E] [F]

Calochortus palmeri var. *paludicola* (Davidson) Jepson & Ames; *C. paludicola* Davidson

Stems bearing bulbels at base. **Inflorescence bracts** alternate. **Petal glands** yellow-hairy, hairs straight.

Flowering late spring–mid summer. Meadows, vernally moist places in pine forest and chaparral; of conservation concern; 1200–2200 m; Calif.

31b. Calochortus palmeri S. Watson var. **munzii** Ownbey, Aliso 4: 88. 1958 • Munz's mariposa-lily [C] [E]

Stems not bearing bulbels. **Inflorescence bracts** opposite. **Petal glands** glabrous or purple-hairy, or absent.

Flowering late spring–mid summer. Pine forest; of conservation concern; 1200–2200 m; Calif.

Calochortus palmeri var. *munzii* is found in the San Jacinto Mountains, Riverside County.

32. Calochortus splendens Douglas ex Bentham, Trans. Hort. Soc. London, ser. 2, 1: 411, plate 15, fig.1. 1835

Calochortus davidsonianus Abrams; *Mariposa splendens* (Douglas ex Bentham) Hoover

Plants rarely bulbose; bulb coat, when present, membranous. **Stems** branching, ± straight, 2–6 dm. **Leaves:** basal withering, 1–2 dm; blade linear, attenuate. **Inflorescences** monochasiate, 1–4-flowered; bracts 2–5 cm. **Flowers** erect; perianth open, campanulate, narrow at base; sepals deep lilac, often with purple blotch, lanceolate-ovate to acuminate, 2–3 cm; petals lavender to deep purple, usually with purple blotch at base, obovate to cuneate, rounded, 3–5 cm, with scattered, white and/or purple hairs proximal to middle, sometimes with yellow hairs distal to gland, margins dentate distally; glands ± square, not depressed, usually with many branching hairs; filaments 7–8 mm; anthers purple or white, 5–7 mm, apex obtuse to abruptly pointed. **Capsules** erect, linear, angled, 5–7 cm. **Seeds** light-colored, flat. $2n = 14$.

Flowering late spring–mid summer. Dry slopes in heavy or granitic soil in chaparral, valley grassland, pine forest; less than 2800 m; Calif.; Mexico (Baja California).

33. Calochortus striatus Parish, Bull. S. Calif. Acad. Sci. 1: 122. 1902 • Alkali mariposa-lily [C] [E] [F]

Plants not bulbose. **Stems** 1–5 cm. **Leaves:** basal withering, 1–2 dm; blade linear. **Inflorescences** subumbellate, 1–5-flowered; bracts linear, 1–3 cm. **Flowers** erect; perianth open, campanulate, narrow at base; sepals lanceolate, 1–2 cm, apex acuminate; petals white to lavender, with conspicuous purple striping, obovate to cuneate, rounded, 2–3 cm, sparsely hairy around gland, margins irregularly dentate distally; glands oblong, not depressed, covered with linear hairs; filaments 5–7 mm; anthers oblong, 4–6 mm. **Capsules** erect, linear, angled, 4–5 cm. **Seeds** light-colored, flat.

Flowering mid spring–early summer. Alkaline meadows, springy places in creosote bush scrub; of conservation concern; 300–1400 m; Calif., Nev.

34. Calochortus monanthus Ownbey, Ann. Missouri Bot. Gard. 27: 465, plate 39, figs. 3, 4. 1940 • Single-flowered mariposa-lily [E]

Plants bulbose; bulb coat membranous. **Stems** not branching, straight. **Leaves:** basal withering; blade linear-attenuate. **Inflorescences** 1-flowered, long-peduncled; bracts opposite. **Flowers** erect; perianth open, turbinate to campanulate; sepals lanceolate, attenuate, ca. 4 cm; petals pinkish, with chevron-shaped, dark red blotch distal to gland, obovate, cuneate, rounded, ca. 5 cm, with few flexible hairs near gland, margins irregularly dentate distally; glands oblong, not depressed, densely covered with slender, unbranched hairs; filaments lanceolate-linear, shorter than anthers; anthers short-tipped. **Capsules** erect, linear, angled. **Seeds** unknown.

Flowering late spring–mid summer. Vernal meadows; 800 m; Calif.

Calochortus monanthus is known only from a single collection from meadow along the Shasta River, near Yreka, Siskiyou County, by E. L. Greene in June, 1876. It is presumed extinct.

35. Calochortus venustus Douglas ex Bentham, Trans. Hort. Soc. London, ser. 2, 1: 412, plate 15, fig. 3. 1835 E F

Mariposa venusta (Douglas ex Bentham) Hoover

Plants bulbose; bulb coat membranous. **Stems** usually branching, 1–6 dm. **Leaves:** basal withering, 1–2 dm; blade linear. **Inflorescences** subumbellate, 1–6(–10)-flowered; bracts 2–8 cm. **Flowers** erect; perianth open, campanulate; sepals lanceolate, 3–5 cm, apex recurved; petals white to brightly or deeply yellow, purple, or dark red, with dark red median blotch, often with second paler distal blotch, obovate or cuneate to clawed, rounded, 3–5 cm, with scattered proximal hairs, apex retuse to obtuse; glands in lower middle of petal, ± square, sometimes ovate, not depressed, covered with short, orange to red-purple hairs; filaments usually equaling anthers; anthers linear-lanceolate to oblong. **Capsules** erect, linear, angled, 5–6 cm. **Seeds** light beige, ± flat, 4–6 mm. $2n = 14$.

Flowering mid spring–mid summer. Light sandy soil, often in decomposed granite in valley grassland, foothill woodland, and pine forest; 300–2700 m; Calif.

Flower color and pattern are very showy and exceedingly variable in *Calochortus venustus*.

36. Calochortus simulans (Hoover) Munz, Suppl. Calif. Fl., 174. 1968 • San Luis Obispo mariposa-lily E

Mariposa simulans Hoover, Leafl. W. Bot. 4: 4. 1944

Plants bulbose; bulb coat membranous. **Stems** usually branching, 1–6 dm. **Leaves:** basal withering, 1–2 dm; blade linear. **Inflorescences** subumbellate, 1–3-flowered; bracts 2–8 cm. **Flowers** erect; perianth open, campanulate; sepals with dark red proximal blotch surrounding gland and extending to base, lanceolate, 2–3 cm, apex recurved; petals white or yellowish, often pink abaxially, with dark red blotch surrounding gland, often with small red spot immediately above gland, obovate to cuneate, 3–5 cm, with scattered hairs proximally; glands suprabasal, ± square, not depressed, covered with short hairs; filaments usually equaling anthers; anthers linear-lanceolate to oblong. **Capsules** erect, linear, angled, 5–6 cm. **Seeds** light beige, flat, 5–6 mm. $2n = 14$.

Flowering late spring–mid summer. Sand (often granitic), grasslands to pine forests; 0–1100 m; Calif.

Calochortus simulans is restricted to central San Luis Obispo County.

37. Calochortus argillosus (Hoover) Zebell & P. L. Fiedler, Madroño 39: 306. 1992 E F

Mariposa argillosa Hoover, Leafl. W. Bot. 4: 3. 1944

Plants bulbose; bulb coat membranous. **Stems** not branching, 4–6 dm. **Leaves:** basal withering, 2–3 dm; blade linear. **Inflorescences** subumbellate, 1–4-flowers; bracts 2–8 cm. **Flowers** erect; perianth open, campanulate; sepals lanceolate, 2–4 cm, apex acute to acuminate; petals white to purple, with median red blotch often surrounded by pale yellow, obovate to cuneate, rounded, 2–4 cm, with a few short hairs near gland, apex retuse to obtuse; glands arched, transversely oblong to weakly lunate or ± rectangular, not depressed, densely covered with short hairs; filaments 7–9 mm; anthers lanceolate-linear to oblong, 8–10 mm, apex acute to obtuse. **Capsules** erect, narrowly lanceoloid, stout, 4–6 cm. **Seeds** light beige, lanceolate, flat, 5–6 mm. $2n = 14$.

Flowering late spring–mid summer. Hard clay soil of volcanic or metamorphic rock; 0–400 m; Calif.

38. Calochortus superbus Purdy ex J. T. Howell, Leafl. W. Bot. 1: 11. 1932 E F

Mariposa superba (Purdy ex J. T. Howell) Hoover

Plants bulbose; bulb coat membranous. **Stems** 4–6 dm. **Leaves:** basal withering, 2–3 dm; blade linear. **Inflorescences** subumbellate, 1–3-flowered; bracts 2–8 cm. **Flowers** erect; perianth open, campanulate, each member with median brown or purple blotch surrounded by bright yellow zone; sepals lanceolate, attenuate, 2–4 cm; petals white to yellowish or lavender, usually streaked with purple toward base, obovate to cuneate, rounded, 2–4 cm, with a few short hairs near gland, apex retuse to obtuse; glands linear, strongly chevron-shaped to weakly lunate, not depressed, densely covered with short hairs; filaments 7–9 mm; anthers lanceolate-linear to oblong, 8–10 mm, apex acute to obtuse. **Capsules** erect, linear, angled, 5–6 cm, apex acute. $2n = 12, 14, 20, 26$.

Flowering late spring–mid summer. Open slopes, dry meadows or wooded places in valley grassland, foothill woodland, pine forest; 0–1700 m; Calif.

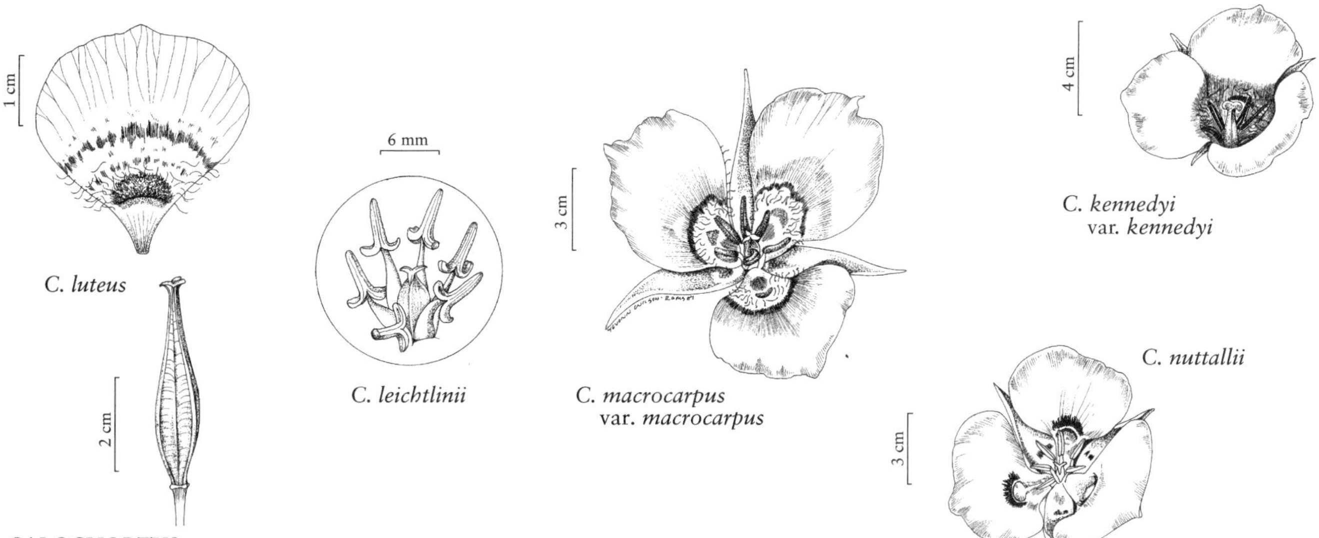

CALOCHORTUS

39. Calochortus vestae Purdy, Gard. Chron., ser. 3, 18: 14. 1895 (as vesta) E F

Plants strongly bulbose; bulb coat membranous. **Stems** usually branching, straight, 3–5 dm. **Leaves:** basal withering, 1–2 dm; blade linear. **Inflorescences** subumbellate, 1–3-flowered; bracts 2–5 cm. **Flowers** erect; perianth open, campanulate; sepals linear-lanceolate, attenuate, 2–3 cm; petals white to purplish, streaked red to purple proximal to gland, with median red-brown blotch surrounded by pale yellow zone, cuneate to obovate, 3–4 cm, with a few short hairs near gland, apex rounded; glands ± doubly lunate, not depressed, densely covered with short hairs; filaments equaling anthers; anthers oblong-linear, apex obtuse or acute. **Capsules** erect, linear, angled, 4–7 cm, apex acute. **Seeds** light beige, flat, 4–6 mm. $2n$ = 28.

Flowering late spring–mid summer. Heavy clay soils in mixed evergreen and pine forests; 500–900 m; Calif.

40. Calochortus luteus Douglas ex Lindley, Edwards's Bot. Reg. 19: plate 1567. 1833 E F

Mariposa lutea (Douglas ex Lindley) Hoover

Stems slender, 2–5 dm. **Leaves:** basal withering, 1–2 dm; blade linear. **Inflorescences** subumbellate, 1–4-flowered; bracts 1–8 cm. **Flowers** erect; perianth open, campanulate; sepals lanceolate-oblong, attenuate, 2–3 cm; petals deep yellow, usually streaked red-brown proximally, often with median red-brown blotch, cuneate to obovate, 2–4 cm, with a few slender hairs near gland; glands ± lunate to oblong, not depressed, covered with short, matted hairs; filaments 7–9 mm; anthers linear-oblong, 4–6 mm, apex obtuse or acute. **Capsules** erect, lanceoloid-linear, angled, 3–6 cm. **Seeds** light beige, flat. $2n$ = 14, 28.

Flowering late spring–mid summer. Heavy soils in grasslands, open woodlands, mixed evergreen forests; 0–700 m; Calif.

Coastal plants of *Calochortus luteus* are mostly triploid, while those of the interior are mostly diploid. Occasionally this species hybridizes with *C. superbus*.

41. Calochortus leichtlinii Hooker f., Bot. Mag. 96: plate 5862. 1870 E F

Mariposa leichtlinii (Hooker f.) Hoover

Plants bulbose; bulb coat membranous. **Stems** not branching, 2–6 dm. **Leaves:** basal withering, 1–1.5 dm; blade linear. **Inflorescences** subumbellate, 1–5-flowered; bracts paired. **Flowers** erect; perianth open, campanulate; sepals lanceolate-ovate, 1–2 cm; petals white to smoky blue, often tinged pink, with red to black blotch distal to gland, cuneate to obovate, 1–4 cm, with a few short hairs near gland, apex rounded; glands ± ovate, slightly depressed, with dense, short hairs; filaments 5 mm; anthers linear-oblong, 5–7 mm, base sagittate. **Capsules** erect, lanceoloid-linear, 3–6 cm. **Seeds** light beige, flat, inflated. $2n$ = 14, 28.

Flowering late spring–late summer. Open gravelly places, montane coniferous forests, open chaparral; 1300–4000 m; Calif., Nev., Oreg.

42. Calochortus macrocarpus Douglas, Trans. Hort. Soc. London 7: 276, plate 8. 1828 [E] [F]

Mariposa macrocarpa (Douglas) Hoover

Plants usually bulbose; bulb coat membranous. **Stems** usually not branching, stout, 2–5 dm. **Leaves:** basal withering, 0.5–1 dm; blade linear. **Inflorescences** subumbellate, 1–3-flowered; bracts 3–5 cm. **Flowers** erect; perianth open, campanulate; sepals lanceolate, 4–5 cm, apex acuminate; petals purple, with median green stripe on abaxial surface, obovate, 4–6 cm, bearded distal to gland, apex acuminate; glands triangular-sagittate, slightly depressed, densely covered with slender hairs; filaments 8–9 mm; anthers lanceolate-linear, ca. 1 cm, apex obtuse. **Capsules** erect, lanceoloid-linear, angled, 4–5 cm, apex acuminate. **Seeds** flat, inflated. $2n = 14$.

Varieties 2 (2 in the flora): w North America.

1. Petals purple, with or without dark purple band distal to gland . . . . 42a. *Calochortus macrocarpus* var. *macrocarpus*
1. Petals white, with reddish purple crescent distal to gland . . . 42b. *Calochortus macrocarpus* var. *maculosus*

42a. Calochortus macrocarpus Douglas var. **macrocarpus** [E] [F]

Petals purple, with or without dark purple band distal to gland. **Capsules** thick in cross section.

Flowering late spring–mid summer. Dry plains, slopes, sagebrush scrub, pine forests, usually in volcanic soil; 300–2700 m; B.C.; Calif., Idaho, Mont., Nev., Oreg., Wash.

42b. Calochortus macrocarpus Douglas var. **maculosus** (A. Nelson & J. F. Macbride) A. Nelson & J. F. Macbride, Contr. Gray Herb. 56: 14. 1918 [C] [E]

Calochortus maculosus A. Nelson & J. F. Macbride, Bot. Gaz. 56: 471. 1913

Petals white or nearly so, with conspicuous reddish purple crescent distal to gland. **Capsules** narrow in cross section.

Flowering late spring–mid summer. Dry plains, rocky slopes, sagebrush scrub, pine forests, usually in volcanic soil; of conservation concern; 300–2700 m; Idaho, Oreg., Wash.

43. Calochortus nuttallii Torrey in H. Stansbury, Exped. Great Salt Lake, 397. 1852 • Sego-lily [E] [F]

Plants usually bulbose; bulb coat, when present, membranous. **Stems** usually not branching or twisted, straight, 1.5–4.5 dm. **Leaves:** blade linear, becoming involute; basal withering. **Inflorescences** subumbellate, 1–4-flowered; bracts congested, unequal. **Flowers** erect; perianth open, campanulate; sepals marked similar to petals, usually shorter, lanceolate, glabrous, apex acuminate; petals white, tinged with lilac or infrequently magenta, yellow at base, with reddish brown or purple band or blotch distal to gland, broadly obovate, cuneate, sparsely invested near gland with slender hairs, apex usually short-acuminate; glands round, depressed, surrounded by conspicuously fringed membrane, densely covered with short, unbranched or distally branching hairs; filaments ca. equaling anthers; anthers yellowish or pinkish, oblong, apex obtuse. **Capsules** erect, linear-lanceoloid, 3-angled, apex acuminate. **Seeds** flat. $2n = 16$.

Flowering late spring–late summer. Dry soils; 700–3300 m; Ariz., Colo., Idaho, Mont., Nebr., Nev., N.Mex., N.Dak., S.Dak., Utah, Wyo.

44. Calochortus bruneaunis A. Nelson & J. F. Macbride, Bot. Gaz. 55: 372. 1913 [E]

Calochortus nuttallii Torrey var. *bruneaunis* (A. Nelson & J. F. Macbride) Ownbey

Plants usually bulbose; bulb coat, when present, membranous. **Stems** usually not branching or twisted, 1–4 dm. **Leaves:** basal withering, 1–2 dm; blade linear. **Inflorescences** 1–4-flowered; bracts 2–4 cm. **Flower** erect; perianth open, campanulate; sepals with dark red or purple blotch near base, lanceolate, 1–4 cm, usually glabrous, apex acuminate; petals white tinged with lilac, with longitudinal median green stripe on abaxial surface and dark red or purple crescent distal to gland, narrowly obovate, 2–4 cm, ± glabrous; glands surrounded by yellow border, round, depressed, surrounded by conspicuously fringed membrane, densely covered with short, unbranched or distally branching hairs; filaments 5–6 mm; anthers yellow, blue, or maroon, oblong, 5–7 mm, apex obtuse. **Capsules** erect, linear-lanceoloid, angled, 3–7 cm, apex acuminate. **Seeds** yellow, flat. $2n = 14$.

Flowering late spring–summer. Dry brushy, grassy slopes, flats, pinyon-juniper woodlands; 900–3000 m; Calif., Idaho, Mont., Nev., Oreg., Utah.

45. Calochortus panamintensis (Ownbey) Reveal in A. Cronquist et al., Intermount. Fl. 6: 501. 1977 • Panamint mariposa-lily [E]

Calochortus nuttallii Torrey var. *panamintensis* Ownbey, Ann. Missouri Bot. Gard. 27: 493. 1940

Plants usually bulbose; bulb coat, when present, membranous. **Stems** usually not branching or twisted, 4–6 dm. **Leaves:** basal withering, 1–2 dm; blade linear. **Inflorescences** 1–4-flowered; bracts 2–4 cm. **Flowers** erect; perianth open, campanulate; sepals with dark purple or red blotch near base, lanceolate, 1–4 cm, apex acuminate; petals white tinged with lilac, with longitudinal median green stripe on abaxial surface, not spotted, narrowly obovate, 2–4 cm, ± glabrous; glands surrounded by red or purple border, round, depressed, surrounded by conspicuously fringed membrane, densely covered with short, unbranched or distally branching hairs; filaments ca. 6 mm; anthers bluish, oblong, 5–7 mm, ± glabrous, apex obtuse. **Capsules** erect, tan to brown, linear-lanceoloid, angled, ca. 7 cm, apex acuminate. **Seeds** light yellow, flat. $2n = 14$.

Flowering early–mid summer. Dry rocky slopes in pinyon-juniper woodland; 2300–3200 m; Calif., Nev.

Calochortus panamintensis is restricted to the Panamint Mountains, Inyo County.

46. Calochortus invenustus Greene, Pittonia 2: 71. 1890 [E]

Plants usually bulbose; bulb coat, when present, membranous. **Stems** slender, usually not branching or twisted, 2–5 dm. **Leaves:** basal withering, 1–2 dm; blade linear. **Inflorescences** subumbellate, 1–5-flowered; bracts 2–5 cm. **Flowers** erect; perianth open, campanulate; sepals lanceolate-ovate, 2–3 cm, apex acuminate; petals white or dull lavender to purplish, with longitudinal median green stripe on adaxial surface and sometimes purplish blotch proximal to gland, cuneate to obovate, 2–4 cm, with a few short hairs near gland, apex obtuse to apiculate; glands ± round, slightly depressed, small, surrounded by conspicuously fringed membrane, densely covered with short, distally branching hairs; filaments 6–7 mm; anthers purplish or yellowish, oblong, 7–8 mm, apex obtuse. **Capsules** erect, lanceoloid-linear, angled, 5–7 cm, apex acute. **Seeds** flat. $2n = 14$.

Flowering late spring–late summer. Dry soil, usually granitic, usually in montane coniferous forests; 1500–3000 m; Calif.

47. Calochortus excavatus Greene, Pittonia 2: 71. 1890 • Inyo County star-tulip [E]

Stems slender, not branching, 1–3 dm. **Leaves:** basal usually persistent, 1–2 dm; blade linear. **Inflorescences** subumbellate, 1–4-flowered; bracts paired, 3–8 cm. **Flowers** erect; perianth open, campanulate; sepals lanceolate, 2–3 cm, apex acuminate; petals lavender, with longitudinal median abaxial green stripe, with or without blotches, broadly cuneate to obovate, 3–4 cm, with a few short hairs near gland; glands round, depressed, surrounded by conspicuously fringed membrane, densely covered with short, distally branching hairs; filaments 6–8 mm; anthers red-brown, oblong, 7–10 mm, apex obtuse. **Capsules** erect, brown, lanceoloid-linear, 3-angled, 2–3 cm. **Seeds** light beige, translucent, flat.

Flowering spring. Grassy meadows in shadscale scrub; 1200–2000 m; Calif.

Calochortus excavatus is found only in the eastern Sierra Nevada in Mono and Inyo counties. It is threatened by groundwater depletion.

48. Calochortus aureus S. Watson, Amer. Naturalist 7: 303. 1873 [E]

Calochortus nuttallii Torrey var. *aureus* (S. Watson) Ownbey

Plants usually bulbose; bulb coat, when present, membranous. **Stems** usually not branching or twisted, 1–3 dm. **Leaves:** basal withering, 1–2 dm; blade linear. **Inflorescences** erect, subumbellate, 1–4-flowered; bracts unequal at base of inflorescence, 1–6 cm. **Flowers** erect; perianth open, campanulate; sepals with maroon blotch distal to gland, broadly lanceolate, 1–4 cm, apex acuminate; petals lemon yellow, with maroon blotch distal to gland, broadly obovate, cuneate, 3–3.5 cm, ± glabrous or with a few lemon-yellow hairs near gland; glands round, depressed, surrounded by conspicuously fringed membranes, densely covered with short, unbranched or distally branching hairs; filaments ca. equaling anthers; anthers yellowish cream, oblong, apex obtuse. **Capsules** erect, light tan to light brown, linear-lanceoloid, 3-angled, ca. 2.5–5 cm, apex acuminate. **Seeds** light yellow, flat. $2n = 32$.

Flowering late spring–mid summer. Dry sandy–clayey places; 1200–2000 m; Ariz., N.Mex., Utah.

49. Calochortus concolor (Baker) Purdy, Proc. Calif. Acad. Sci., ser. 3, 2: 135. 1901 • Goldenbowl mariposa-lily

Calochortus luteus Douglas ex Lindley var. *concolor* Baker, Garden (London 1871–1927) 48: 440, plate 1043. 1895

Plants usually bulbose; bulb coat, when present, membranous. **Stems** usually stout, sparingly branching, not flexuous or twisted, 3–6 dm. **Leaves:** basal withering, 1–2 dm; blade linear, glaucous. **Inflorescences** subumbellate, 1–4-flowered; bracts opposite pedicel, 4–8 cm. **Flowers** erect; perianth open, campanulate; sepals usually with dark red blotch near base, lanceolate-ovate, 2–3 cm; petals yellow, often tinged purple in drying, cuneate to obovate, 3–5 cm; glands usually ± round, depressed, small, surrounded by conspicuously fringed membrane and a few long, yellow hairs, densely covered with slender, unbranched hairs; filaments 9–10 mm; anthers yellowish, oblong, 8–10 mm, apex obtuse. **Capsules** erect, lanceoloid-linear, angled, 5–8 cm, apex acuminate. **Seeds** flat. $2n$ = ca. 14.

Flowering late spring–mid summer. Dry slopes in chaparral and pine forest, frequently on decomposed granite; 700–2500 m; Calif.; Mexico (Baja California).

50. Calochortus kennedyi Porter, Bot. Gaz. 2: 79. 1877 [F]

Plants usually bulbose; bulb coat, when present, membranous. **Stems** usually not branching but sometimes twisted, 1–2(–5) dm. **Leaves:** blade linear-attenuate; basal withering, 1–2 dm; blade channeled, glaucous. **Inflorescences** subumbellate, 1–6-flowered; bracts 2–4 cm, bases dilated. **Flowers** erect; perianth open, campanulate, each member often with brown-purple blotch near base; sepals ovate, 2–3 cm, apex acute; petals orange to red or ± yellow, cuneate to obovate, 3–5 cm, with a few slightly enlarged hairs near gland; glands round, depressed, surrounded by conspicuously fringed membrane, densely covered with unbranched or cleft hairs; filaments 4–5 mm; anthers purplish, lanceolate, 5–8 mm. **Capsules** erect, longitudinally striped, lanceoloid-linear, angled, 4–6 cm, apex acuminate. **Seeds** flat. $2n$ =16.

Varieties 2 (2 in the flora): sw United States, nw Mexico.

1. Petals orange to red . . . . 50a. *Calochortus kennedyi* var. *kennedyi*
1. Petals ± yellow . . . . 50b. *Calochortus kennedyi* var. *munzii*

50a. Calochortus kennedyi Porter var. **kennedyi** [F]

Petals orange to red.

Flowering mid spring–mid summer. Heavy soil in open or brushy flats, or dry rocky slopes in creosote bush scrub, pinyon juniper woodland; 700–2200 m; Ariz., Calif., Nev.; Mexico (Sonora).

50b. Calochortus kennedyi Porter var. **munzii** Jepson, Man. Fl. Pl. Calif., 236. 1923

Petals ± yellow.

Flowering mid spring–mid summer. Heavy soil in open or brushy flats, dry rocky slopes in creosote-bush scrub, pinyon juniper woodland, in desert mountains; 700–2200 m; Ariz., Calif.; Mexico (Sonora).

51. Calochortus clavatus S. Watson, Proc. Amer. Acad. Arts 14: 265. 1879 [E] [F]

Mariposa clavata (S. Watson) Hoover

Plants usually bulbose; bulb coat, when present, membranous. **Stems** slender or coarse, stout, branching or not, strongly flexuous, 3–10 dm. **Leaves:** basal withering, 1–2 dm; blade linear. **Inflorescences** subumbellate, 1–6-flowered; bracts 4–8 cm, bases dilated. **Flowers** erect; perianth open, cup-shaped; sepals usually with red-brown blotch at base, lanceolate-ovate, 2–4 cm, apex acute; petals lemon to golden yellow, with transverse line distal to gland, broadly cuneate to obovate, 3–5 cm; glands round, deeply depressed, densely covered with short hairs with branching, coralline tips, surrounded by conspicuously fringed membrane and club-shaped hairs; filaments ca. 10 mm; anthers purplish brown, oblong, 4–10 mm. **Capsules** erect, lanceoloid-linear, angled, 6–9 cm, apex acuminate. **Seeds** light yellow, translucent, flat. $2n$ = 16.

Varieties 5 (5 in the flora): California.

1. Petals shorter than sepals; petal glands very deeply depressed . . . . 51e. *Calochortus clavatus* var. *avius*
1. Petals longer than sepals; petal glands shallowly to moderately depressed.
    2. Stems straight, slender, usually 3 dm or shorter.
        3. Petals 3–4 cm, sparsely bearded; leaves not recurved . . . . 51a. *Calochortus clavatus* var. *gracilis*
        3. Petals 4–5 cm, bearded; leaves strongly recurved . . . . 51d. *Calochortus clavatus* var. *recurvifolius*

2. Stems flexuous, coarse, usually 5–10 dm.
    4. Petals deep yellow, hairs very knobby; anthers deep purple . . . . 51b. *Calochortus clavatus* var. *clavatus*
    4. Petals light yellow, hairs not very knobby; anthers yellow to medium purple . . . . 51c. *Calochortus clavatus* var. *pallidus*

51a. Calochortus clavatus S. Watson var. **gracilis** Ownbey, Ann. Missouri Bot. Gard. 27: 504. 1940 • Slender mariposa-lily [E]

Stems slender, straight, 2–3 dm. **Leaves** not recurved. **Flowers:** petals with reddish brown line distal to gland, 3–4 cm, longer than sepals, sparsely bearded; glands shallowly to moderately depressed, small; anthers 4–7 mm.

Flowering late spring–mid summer. Shaded canyons; 0–1300 m; Calif.

Calochortus clavatus var. *gracilis* is restricted to the San Gabriel Mountains.

51b. Calochortus clavatus S. Watson var. **clavatus** [E]

Stems coarse, flexuous, 5–10 dm. **Leaves** not recurved. **Flowers:** petals deep yellow, longer than sepals, hairs very knobby; glands shallowly to moderately depressed; anthers deep purple.

Flowering late spring–mid summer. Usually on serpentine soils; 0–1300 m; Calif.

Calochortus clavatus var. *clavatus* is restricted to the southern outer Coast Range.

51c. Calochortus clavatus S. Watson var. **pallidus** (Hoover) P. L. Fiedler & Zebell, Madroño 42: 406. 1995 [E]

Mariposa clavata Hoover var. *pallida* Hoover, Leafl. W. Bot. 10: 126. 1964

Stems coarse, flexuous, ca. 5–10 dm. **Leaves** not recurved. **Flowers:** petals light yellow, longer than sepals, hairs not very knobby; glands shallowly to moderately depressed; anthers yellow to pale or medium purple.

Flowering late spring–mid summer. Dry, rocky slopes, chaparral, often on serpentine; 0–1300 m; Calif.

Calochortus clavatus var. *pallidus* is known from the Great Valley and western Transverse Ranges.

51d. Calochortus clavatus S. Watson var. **recurvifolius** (Hoover) P. L. Fiedler & Zebell, Madroño 42: 406. 1995 • Arroyo de la Cruz mariposa-lily [E] [F]

Mariposa clavata Hoover var. *recurvifolia* Hoover, Leafl. W. Bot. 10: 126. 1964

Stems slender, straight, 9–12 cm. **Leaves** strongly recurved. **Flowers:** petals 4–5 cm, longer than sepals, bearded; glands shallowly to moderately depressed.

Flowering late spring–mid summer. Ocean bluffs; 0–1300 m; Calif.

Calochortus clavatus var. *recurvifolius* is restricted to the southern outer Coast Range north of Arroyo de la Cruz, San Luis Obispo County.

51e. Calochortus clavatus S. Watson var. **avius** Jepson, Man. Fl. Pl. Calif., 235. 1923 • Pleasant Valley mariposa-lily [E]

Stems coarse, flexuous, 5–10 dm. **Leaves** not recurved. **Flowers:** petals shorter than sepals; glands very deeply depressed, more so than species.

Flowering late spring–mid summer. Dry, rocky slopes, chaparral, often on serpentine; 0–1300 m; Calif.

Calochortus clavatus var. *avius* is found in the northern Sierra Nevada foothills in Eldorado and Amador counties.

52. Calochortus ambiguus (M. E. Jones) Ownbey, Ann. Missouri Bot. Gard. 27: 505. 1940 [E]

Calochortus watsonii M. E. Jones var. *ambiguus* M. E. Jones, Contr. W. Bot. 14: 26. 1912

Plants often bulbose; bulb coat membranous. **Stems** usually branching, straight. **Leaves:** basal withering; blade linear-attenuate. **Inflorescences** subumbellate, 1–4-flowered. **Flowers** erect; perianth pinkish or bluish gray, infrequently with longitudinal gray stripe, open, campanulate; sepals invested near gland, with yellowish, distally enlarged or branching, gland-tipped hairs, base sometimes marked with dull purple, apex acuminate; petals with median blotch usually surrounded by yellow zone, ± equaling sepals, base sometimes marked with dull purple; glands round to lunate, depressed, surrounded with narrow, deeply fringed membrane, densely covered with short, distally branching

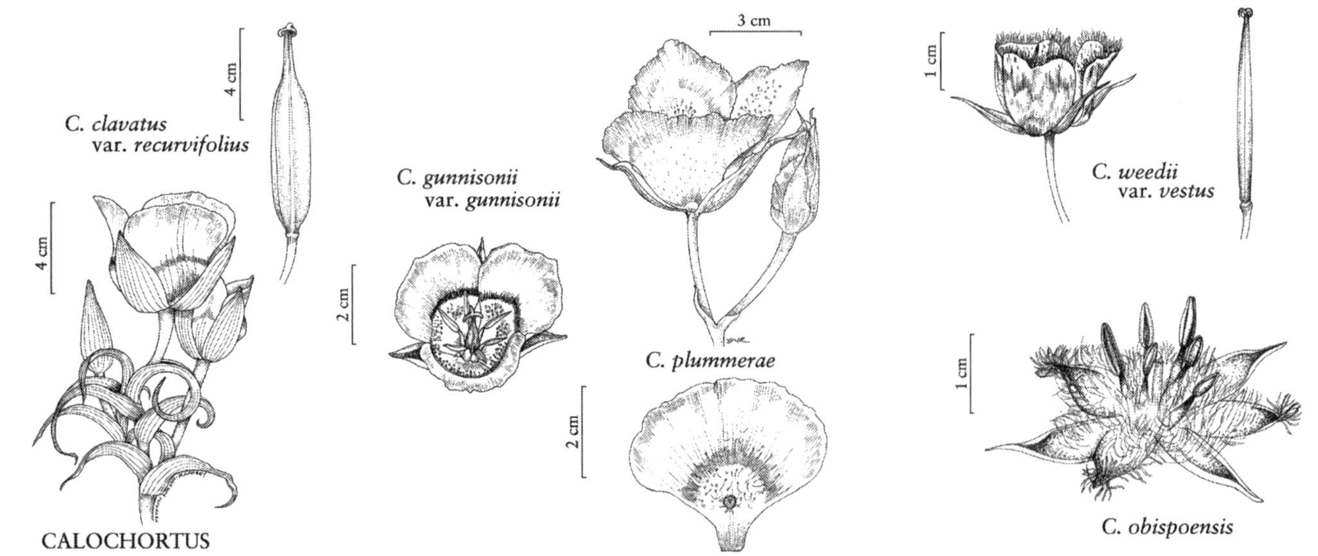

CALOCHORTUS

hairs; filaments shorter than anthers; anthers oblong-lanceolate, apex obtuse to acute. **Capsules** erect, linear-oblong, 3-angled, apex acute. **Seeds** flat. $2n = 18$.

Flowering late spring–summer. Dry soils; 1100–2700 m; Ariz., N.Mex., Utah.

53. Calochortus gunnisonii S. Watson, Botany (Fortieth Parallel), 348. 1871 (as gunnisoni) [E] [F]

Plants usually bulbose; bulb coat, when present, membranous. **Stems** not branching, straight, 2.4–5.5 dm. **Leaves:** basal withering, 18–35 cm; blade linear. **Inflorescences** subumbellate, 1–3-flowered. **Flowers** erect; perianth open, campanulate; sepals marked similar to petals, lanceolate, usually much shorter, glabrous, apex acute; petals white to purple, greenish adaxially, clawed, often with narrow, transverse purple band distal to gland and purple blotch on claw, obovate, cuneate, usually obtuse and rounded distally; glands transversely oblong, not depressed, densely bearded with distally branching hairs, outermost of which somewhat connate at base to form discontinuous, deeply fringed membranes; filaments shorter than anthers; anthers lanceolate, apex acute to apiculate. **Capsules** erect, linear-oblong, 3-angled, 3–6 cm, apex acute. **Seeds** flat, inflated. $2n = 18$.

Varieties 2 (2 in the flora): wc North America.

1. Petals white to purple, occasionally yellow; w United States .... 53a. *Calochortus gunnisonii* var. *gunnisonii*
1. Petals pale yellow; New Mexico only .... 53b. *Calochortus gunnisonii* var. *perpulcher*

53a. Calochortus gunnisonii S. Watson var. **gunnisonii** [E] [F]

Petals white to purple, occasionally yellow.

Flowering late spring–mid summer. Hills and mountains; 1200–3300 m; Ariz., Colo., Mont., Nebr., N.Mex., S.Dak., Utah, Wyo.

53b. Calochortus gunnisonii S. Watson var. **perpulcher** Cockerell, Bot. Gaz. 29: 281. 1900 [E]

Petals pale yellow.

Flowering late spring–mid summer. Hills and mountains; 2700–3000 m; N.Mex.

Calochortus gunnisonii var. *perpulcher* is found in Mora and San Miguel counties.

54. Calochortus plummerae Greene, Pittonia 2: 70. 1890 • Plummer's mariposa-lily [E] [F]

Plants usually bulbose; bulb coat fibrous-reticulate. **Stems** slender, usually branching, 3–6 dm. **Leaves:** basal withering; blade linear to lanceolate, 2–4 dm. **Inflorescences** 2–6-flowered; bracts resembling distal cauline leaves. **Flowers** erect; perianth open, broadly campanulate; sepals lanceolate, 3–5 cm, glabrous or with a few hairs at base, apex long-tapering; petals pale pink to rose, drying purplish, broadly cuneate to obovate, 3–4 cm, glabrous distally or nearly so, rarely fringed, with conspicuous, long, yellow hairs in broad median band, margins dentate; glands round, slightly depressed, ± glabrous, bordered by ring of dense, obscuring, orange hairs; filaments 9–11 mm, ± equaling anthers; anthers lanceolate-linear, apex acute to somewhat short-tipped. **Capsules** erect, linear, angled, 4–8 cm, apex acute. **Seeds** light beige. $2n = 18$.

Flowering late spring–mid summer. Dry rocky slopes, often in brush, chaparral, pine forest; 0–1700 m; Calif.

55. Calochortus weedii Alph. Wood, Proc. Acad. Nat. Sci. Philadelphia 20: 169. 1868 [F]

Plants usually bulbose; bulb coat fibrous-reticulate. **Stems** slender, usually branching, 3–9 dm. **Leaves:** basal withering, 2–4 dm; blade linear to lanceolate. **Inflorescences** 2–6-flowered; bracts resembling distal cauline leaves. **Flowers** erect; perianth open, broadly campanulate; sepals ovate to lanceolate, attenuate, 2–3 cm; petals orange-yellow, flecked, broadly cuneate or obovate, ca. 2–3 cm, adaxially bearded with long yellow hairs, margins often red-brown, dentate or fringed; glands round, slightly depressed, ± glabrous, surrounded by ring of long, dense, obscuring yellow hairs; filaments 8–12 mm; anthers lanceolate-oblong, apex acute. **Capsules** erect, linear, angled, 4–5 cm, apex acute. **Seeds** light beige. $2n = 18$.

Varieties 4 (3 in the flora): California, Mexico (Baja California).

1. Petals deep yellow .... 55b. *Calochortus weedii* var. *weedii*
1. Petals purplish or cream to red-brown.
   2. Petals purplish, apex rounded; anther apex rounded .. 55a. *Calochortus weedii* var. *intermedius*
   2. Petals usually cream or red-brown, never purplish, apex squarish; anther apex abruptly apiculate .... 55c. *Calochortus weedii* var. *vestus*

55a. Calochortus weedii Alph. Wood var. **intermedius** Ownbey, Ann. Missouri Bot. Gard. 27: 519. 1940 [E]

Flowers: petals purplish, conspicuously fringed, adaxial surface with long dark or yellow hairs, apex rounded; anther apex rounded.

Flowering late spring–mid summer. Dry, often heavy or rocky soil, in chaparral; less than 700 m; Calif.

Calochortus weedii var. *intermedius* is restricted to Orange County.

55b. Calochortus weedii Alph. Wood var. **weedii**

Flowers: petals deep yellow, ovate-lanceolate, often fringed with red-brown hairs; anther apex acute.

Flowering mid summer. Dry, rocky, open slopes; less than 1900 m; Calif.; Mexico (Baja California).

55c. Calochortus weedii Alph. Wood var. **vestus** Purdy, Proc. Calif. Acad. Sci., ser. 3, 2: 133. 1901 [E] [F]

Flowers: petals usually pale cream to red-brown, fringed with brown hairs, apex squarish; anther apex abruptly apiculate.

Flowering mid summer. Dry slopes in openings in coastal woodland and chaparral; 0–900 m; Calif.

56. Calochortus obispoensis Lemmon, Bot. Gaz. 11: 180. 1886 • San Luis mariposa-lily [C] [E] [F]

Plants usually bulbose; bulb coat fibrous-reticulate. **Stems** slender, branching, straight, 3–6 dm. **Leaves** basal withering, 2–3 dm; blade linear-lanceolate. **Inflorescences** 2–6-flowered. **Flowers** erect; perianth opening flat, campanulate; sepals often reflexed, lanceolate, 1–3 cm, glabrous, apex long-tapered; petals yellow to deep orange, purple-brown at tip, oblong-ovate, 1–2 cm, conspicuously bearded, margins fringed or dentate, apex fringed with hairy tufts; glands round, slightly depressed, glabrous, surrounded and obscured by ring of slender, dense hairs united at base; filaments 7–8 mm; anthers oblong, apex acute. **Capsules** erect, linear, 3-angled, 3–4 cm, apex acute. **Seeds** light yellow, translucent. $2n = 18$.

Flowering late spring–mid summer. Dry stony hills, canyons, edge of chaparral, serpentine; of conservation concern; 100–500 m; Calif.

Calochortus obispoensis is restricted to San Luis Obispo County.

19. PROSARTES D. Don, Proc. Linn. Soc. London 1: 48. 1839 • [Greek *prosarto*, to append, alluding to the pendulous ovules in the type species] [E]

Frederick H. Utech

Disporum Salisbury sect. *Prosartes* (D. Don) Q. Jones

Herbs, from slender, knotty rhizomes with fibrous roots, generally pubescent throughout. **Stems** branched distally, with 2–5 papery bracts sheathing proximally. **Leaves** sessile or subsessile; blade broadly ovate to oblanceolate, veinlets forming loose reticulum. **Inflorescences** strictly terminal. **Flowers** 1–4(–7) in a cluster, nodding, pedicellate; tepals deciduous, distinct, weakly gibbous proximally; stamens hypogynous, basally adnate to tepals; filaments filiform to basally dilated; anthers linear-oblong, extrorse; ovary superior, sessile, 3-locular, narrowly ellipsoid to obovoid, ovules 2–6 per locule, pendulous or horizontal; style included or exserted, filiform; stigma not lobed or weakly 3-lobed; pedicel slender. **Fruits** baccate, straw-colored to red, ± fleshy. **Seeds** light yellow to orangish brown, ellipsoid to oblong, smooth. $x = 6, 8, 9, 11$.

Species 5 (5 in the flora): North America.

The American species have long been treated as section *Prosartes* of the otherwise Asian genus *Disporum* (H. Hara 1988; Q. Jones 1951). However, cytological, morphological, and molecular evidence indicates a degree of difference that justifies generic status for this group (M. N. Tamura et al. 1992; Z. K. Shinwari et al. 1994; T. Fukuhara and Z. K. Shinwari 1994). Within *Prosartes* there are two disjunct, east-west pairs: *P. lanuginosa* and *P. hookeri,* and *P. trachycarpa* and *P. maculata* (F. H. Utech et al. 1995; C. E. Wood Jr. 1970).

SELECTED REFERENCES Jones, Q. 1951. A cytotaxonomic study of the genus *Disporum* in North America. Contr. Gray Herb. 173: 1–39. Shinwari, Z. K. et al. 1994. Recognition of the New World *Disporum* section *Prosartes* as *Prosartes* (Liliaceae) based on the sequence of the *rbc*L gene. Taxon 43: 353–366. Utech, F. H., Z. K. Shinwari, and S. Kawano. 1995. Biosystematic studies in *Disporum* (Liliaceae-Asparagoideae-Polygonateae). VI. Recognition of the North American section *Prosartes* as an autonomous genus. Mem. Fac. Sci. Kyoto Univ., Ser. Biol. 16: 1–41.

1. Ovary broadly ovoid to obovoid or obconic to obpyriform, becoming 3-lobed after anthesis, papillose, sometimes also with ascending, stellate glandular hairs; ovules horizontal; leaf margins moderately pubescent, hairs scattered, flattened.
    2. Leaf apex acute, with 7–9 prominent veins; tepals lacking purple spots and basal claws, 8–15 mm; ovary finely papillose; berries reddish orange to bright red; cordilleran and interior North America . . . . . 4. *Prosartes trachycarpa*
    2. Leaf apex acuminate, with (3–)5 prominent veins; tepals spotted with purple, narrowly clawed basally, 15–25 mm; ovary papillose with stellate glandular hairs; berries pale straw-colored; e North America . . . . . 5. *Prosartes maculata*
1. Ovary ellipsoid or ovoid to obovoid, glabrous or somewhat pubescent; ovules pendulous; leaf margins glabrous or short-pubescent.
    3. Stems freely branched; leaf margins glabrous or sparsely pubescent; perianth cylindrical, truncate basally; tepals 15–28 mm; berries 3–6-seeded; coastal w North America . . . . . 3. *Prosartes smithii*
    3. Stems sparingly branched; leaf margins short-pubescent, hairs pointed sharply forward; perianth turbinate, narrowed basally; tepals 8–22 mm; berries 1–4(–6)-seeded; e, w North America.
        4. Tepals greenish yellow, linear-lanceolate, apex long-acuminate; ovary glabrous; stigma weakly 3-lobed; abaxial leaf surface lanulose, densely so on veins; e North America . . . . . 1. *Prosartes lanuginosa*
        4. Tepals creamy white, oblanceolate, apex short-acuminate; ovary glabrous or somewhat pubescent; stigma unlobed or rarely 3-lobed; abaxial leaf surface scabrous or puberulent, especially on veins; w North America . . . . . 2. *Prosartes hookeri*

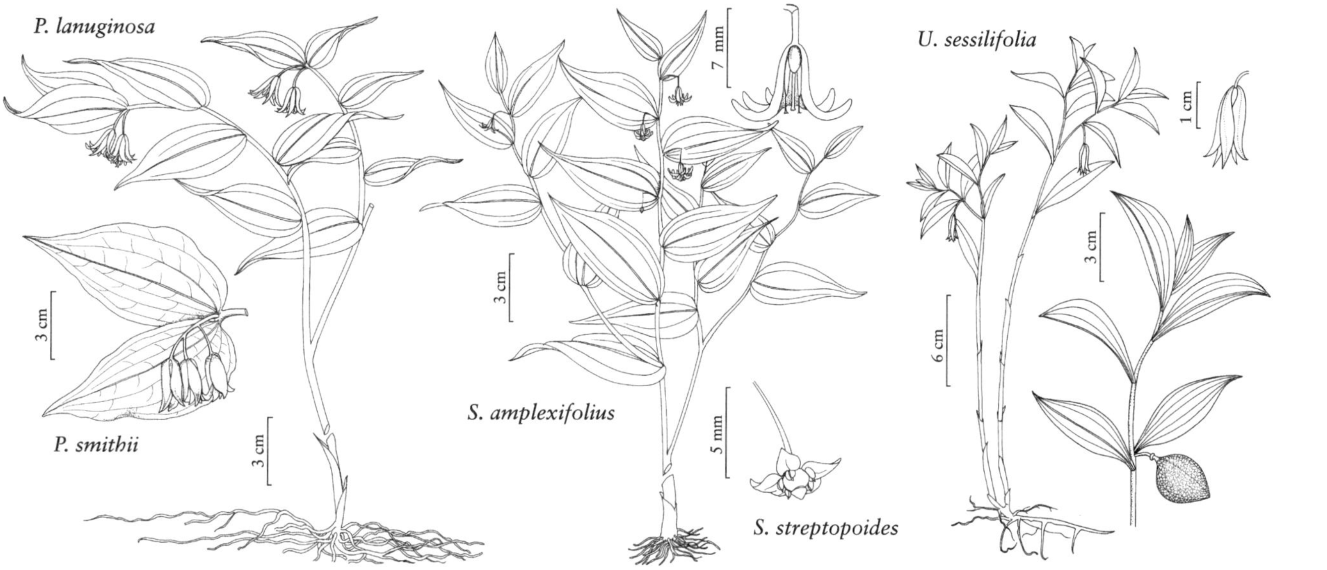

PROSARTES ◦ STREPTOPUS ◦ UVULARIA

1. Prosartes lanuginosa (Michaux) D. Don, Proc. Linn. Soc. London 1: 48. 1839 • Yellow mandarin, hairy mandarin [E] [F]

Streptopus lanuginosus Michaux, Fl. Bor.-Amer. 1: 201. 1803; *Disporum lanuginosum* (Michaux) G. Nicholson

Plants 4–7.5(–10) dm, finely pubescent. **Stems** sparingly branched. **Leaves** 5–15 × 2–5 cm; blade ovate to oblanceolate, rounded basally, adaxial surface glabrate or veins moderately pubescent, abaxial surface lanulose, densely so on veins, margins short-pubescent, hairs uniform, pointed sharply forward, apex narrowly acuminate. **Flowers** 1–3; perianth turbinate, narrowed basally; tepals spreading, greenish white to greenish yellow, linear-lanceolate to linear-elliptic, 15–22 mm, apex long-acuminate; stamens included, 8–15 mm; filaments narrowly dilated basally, 6–12 mm; anthers 2–4 mm, glabrous; ovary narrowly ellipsoid, glabrous, ovules 2 per locule, pendulous; style stout, 8–12 mm, glabrous; stigma weakly 3-lobed. **Berries** bright red, 1–2(–4)-seeded, globose, 10–15 mm, glabrous. **Seeds** 5–6 mm. 2*n* = 18.

Flowering late spring–early summer. Rich moist, deciduous woods, coves; 200–1600 m; Ont.; Ala., Ark., Ga., Ky., Md., N.Y., N.C., Ohio, Pa., S.C., Tenn., Va., W.Va.

Recent collections have extended the known range of *Prosartes lanuginosa* from the unglaciated Appalachians to the other ancient, eastern North American upland, the Ozark Mountains of Arkansas.

2. Prosartes hookeri Torrey, Pacif. Railr. Rep. 4(5): 144. 1857 • Fairy-bells [E]

Disporum hookeri (Torrey) G. Nicholson; *D. hookeri* var. *oblongifolium* (S. Watson) Britton; *D. hookeri* var. *oreganum* (S. Watson) Q. Jones; *D. hookeri* var. *trachyandrum* (Torrey) Q. Jones; *D. oreganum* (S. Watson) W. T. Miller; *D. parvifolium* (S. Watson) Britton; *D. trachyandrum* (Torrey) Britton; *Prosartes oregana* S. Watson; *P. parvifolia* S. Watson

Plants 3–10 dm, roughly pubescent. **Stems** sparingly branched. **Leaves** 3–9(–14) × 1.5–5 cm; blade ovate to lanceolate, rounded to obliquely subcordate, adaxial surface glabrate or sparsely pubescent, abaxial surface scabrous or puberulent especially on veins, margins minutely pubescent to scabrous, hairs pointed sharply forward, apex long-acuminate to acute. **Flowers** 2–3(–4); perianth turbinate, narrowed basally; tepals spreading, dull or creamy white to greenish white, oblanceolate, 8–18 mm, apex short-acuminate; stamens included or exserted, 8–15 mm; filaments filiform, 6–15 mm; anthers 3–5 mm, glabrous or minutely sharp-bristled; ovary ellipsoid, glabrous or somewhat pubescent, ovules 2 per locule, pendulous; styles glabrous or basal 1/2 pubescent; stigma not lobed or rarely weakly 3-lobed. **Berries** bright red, 4(–6)-seeded, ovoid to obovoid, beaked with persistent style bases, 8–16 mm. **Seeds** 5–6 mm. 2*n* = 18.

Flowering early spring–mid summer. Damp, shady forests and thickets; 100–2000 m; Alta., B.C.; Calif., Idaho, Mich., Mont., Oreg., Wash.

The three geographical races recognized as varieties by Q. Jones (1951), i.e., var. *hookeri* (coastal California northwards), var. *oreganum* (northern California north and northwestward), and var. *trachyandrum* (Sierra Nevada northwest to Oregon), are highly variable and intergrade with respect to their purportedly diagnostic differences in pubescence on the anthers, ovaries, and styles, and in the degree of stamen exsertion. This is especially true in their overlapping ranges in northern California and southern Oregon. Whereas the extremes may be distinctive, overall the varieties so intergrade that they are not here recognized.

From this same area, where the Coastal, Sierra, and Cascade ranges meet, the purported, sterile hybrid *Prosartes parvifolia* was first reported. The few older collections attributed to *P. parvifolia* and recent dwarf ones of *P. hookeri* from this area are similar, especially those from serpentine substrata, and the known variation in *P. hookeri* unquestionably encompasses the morphology described for *P. parvifolia*.

The recently discovered population of *Prosartes hookeri* in the Porcupine Mountains of upper Michigan (E. G. Voss 1972–1985, vol. 1) is a noteworthy disjunction for this otherwise western species.

3. **Prosartes smithii** (Hooker) Utech, Shinwari & Kawano, Taxon 43: 364. 1994 • Large-flowered fairy-bells, fairy-lantern [E] [F]

Uvularia smithii Hooker, Fl. Bor.-Amer. 2: 174, plate 189. 1838; *Disporum smithii* (Hooker) Piper

Plants 3–10 dm, crisp-pubescent or glabrate. **Stems** freely branched. **Leaves** 4–15 × 2–5 cm; blade mostly ovate to oblong-lanceolate, rounded to obliquely subcordate basally, abaxial surface glabrous, adaxial surface glabrate, margins glabrous or sparsely pubescent, hairs long, twisted, apex acute to abruptly acuminate. **Flowers** 2–5(–7); perianth cylindrical, truncate basally; tepals muddy to creamy white, oblong-lanceolate, flaring slightly at apex, 15–28 mm; stamens included, 10–23 mm, equal to or slightly shorter than style; filaments filiform or slightly dilated basally; anthers 4–5 mm; ovary ovoid to obovoid, becoming weakly 3-lobed after anthesis, glabrous, ovules 2–4 per locule, pendulous; style 10–23 mm, pubescent; stigma 3-lobed, lobes 0.5–1 mm. **Berries** yellowish orange to dark red, 3–6-seeded, ovoid to oblong, beaked with persistent style bases, 12–16 mm, smooth. **Seeds** 4 mm. $2n = 16$.

Flowering late spring–early summer. Moist shady forests, coastal mountains; 0–200 m; B.C.; Calif., Oreg., Wash.

4. **Prosartes trachycarpa** S. Watson, Botany (Fortieth Parallel), 344. 1871 • Rough-fruited fairy-bells, rough-fruited mandarin [E]

Disporum trachycarpum (S. Watson) Bentham & Hooker f.; *D. trachycarpum* var. *subglabrum* E. H. Kelso

Plants 3–8 dm, crisp-pubescent, glabrate or glabrous with age. **Stems** sparingly branched. **Leaves** 4–12 × 2–5 cm; blade ovate to oblong-lanceolate, subcordate to oblique basally, abaxial surface and margins moderately pubescent, hairs scattered, flattened, apex acute to short-acuminate, with 7–9 prominent veins. **Flowers** 1–2(–3); perianth narrowly campanulate; tepals creamy to greenish white, narrowly oblanceolate, 8–15 mm; stamens mostly exserted; filaments filiform, 10–15 mm; anthers 3–4(–5) mm; ovary broadly ovoid to obovoid, becoming 3-lobed and obpyriform after anthesis, finely papillose, ovules 2–6 per locule, horizontal; style 0.9–1.2 cm, glabrous or pubescent; stigma unlobed or 3-lobed. **Berries** reddish orange to bright red, 6–12(–18)-seeded, depressed-globose, 12–18 mm, strongly papillose. **Seeds** 4–5 mm. $2n = 22$.

Flowering late spring–summer. Rich, shady deciduous forests, aspen groves to open coniferous forests; 300–2500 m; Alta., B.C., Man., Ont., Sask.; Ariz., Colo., Idaho, Mich., Minn., Mont., Nebr., N.Mex., N.Dak., Oreg., S.Dak., Utah, Wash., Wyo.

The recently discovered populations of *Prosartes trachycarpa* in Isle Royale National Park, Michigan (E. J. Judziewicz et al. 1997), are noteworthy disjunctions for this otherwise western species.

5. **Prosartes maculata** (Buckley) A. Gray, Amer. J. Sci. Arts 47: 201. 1844 • Nodding mandarin [E]

Streptopus maculatus Buckley, Amer. J. Sci. Arts 45: 170. 1843; *Disporum cahnae* Farwell; *D. maculatum* (Buckley) Britton; *D. schaffneri* Moldenke

Plants 3–8 dm, glabrescent with age. **Stems** sparingly branched. **Leaves** 4–15 × 2–4 cm; blade ovate to lanceolate, rounded to subcordate basally, abaxial surface veins and margins pubescent, hairs scattered, flattened, apex sharply acuminate, with (3–)5 prominent veins. **Flowers** 1–3(–4); perianth broadly campanulate; tepals creamy white spotted with purple, ovate-lanceolate, abruptly narrowly clawed basally, 15–25 mm, apex acuminate; stamens exserted; filaments filiform, 15–20 mm; anthers 3–6 mm; ovary obconic to obpyriform, becoming 3-lobed after anthesis, papillose with ascending, stellate glandular hairs, ovules 2–4 per locule, horizontal; style 1.4–2.3 cm, glabrous; stigma 3-lobed. **Berries** pale straw-colored, 3-lobed, 6–10-seeded, 10–15 mm, papillose with stellate hairs arising from papillae; lobes subglobose. **Seeds** 4 mm. $2n = 12$.

Flowering early–late spring. Rich, moist, deciduous woods, slopes and ravines; 100–800 m; Ala., Ga., Ky., Mich., N.C., Ohio, Tenn., Va., W.Va.

This rare old-forest obligate is readily distinguished by its purple-spotted, clawed tepals and its three-lobed, straw-colored, hairy-papillose fruit (R. G. Johnson 1968).

20. STREPTOPUS Michaux, Fl. Bor.-Amer. 1: 200. 1803 • Twisted-stalk [Greek *streptos*, twisted, and *-pous*, -footed, alluding to the bent or twisted peduncles]

Frederick H. Utech

Herbs, perennial, rhizomatous. **Stems** simple to highly branched. **Leaves** numerous, sessile; blade elliptic to ovate, base rounded to cordate-clasping, apex acute to acuminate. **Inflorescences** 1–2-flowered, peduncle slender, adnate to stem for 1 internode and arising opposite next leaf axil, junction with pedicel abrupt or not, entire structure recurved. **Flowers** small, borne beneath leaves; perianth rotate or campanulate; tepals deciduous, erect to spreading or recurved, distinct, white to greenish yellow to rose, oblanceolate to oblong; stamens hypogynous; filaments short, broad, flat; anthers basifixed, apex minutely apiculate or with tapering, setose points, extrorse; ovary superior, 3-locular; style slender to bulbous; stigma unlobed or 3-lobed; pedicel geniculate. **Fruits** baccate, orangish to dark red, ellipsoid to globose. **Seeds** pale yellow, elongate, grooved longitudinally. $x = 8$.

Species 7 (3 in the flora): n temperate North America and Eurasia.

Streptopus ×*oreopolus* Fernald (as species) is a sterile hybrid ($2n = 24$) between *S. amplexifolius* and *S. lanceolatus*, and is found in subalpine woods and meadows in glaciated areas of Newfoundland, eastern Quebec, western Ontario, Maine, and New Hampshire (D. Löve and H. Harries 1963; C. Gervais 1979). Its characteristics are intermediate between those of the parent species, except that the leaves are distinctly ciliate-denticulate, the tepals are roseate to deep purple overall, and the berries are deep red.

SELECTED REFERENCE Fassett, N. C. 1935. Notes from the herbarium of the University of Wisconsin—XII. A study of *Streptopus*. Rhodora 37: 88–113.

1. Stems simple, 1–2 dm; perianth rotate; tepals 3–5 mm; anther apex minutely apiculate; style bulbous-conical, 0.5 mm . . . . 3. *Streptopus streptopoides*
1. Stems often branched, 1.5–12 dm; perianth campanulate; tepals 6–15 mm; anther apex long-pointed; style slender or stout, 3.5–5 mm.
   2. Stems simple or occasionally branched, 1.5–4(–8) dm; leaf base rounded; pedicel rarely geniculate at junction with peduncle, pubescent; tepals slightly spreading, streaked or spotted with pale rose to reddish purple; anthers 2 mm, apex 2-pointed; stigma weakly 3-lobed . . . . 1. *Streptopus lanceolatus*
   2. Stems freely branched, 5–12 dm; leaf base cordate-clasping; pedicel sharply geniculate at junction with peduncle, glabrous; tepals spreading, recurved at tips, whitish to greenish yellow; anthers 3–3.5 mm, apex 1-pointed; stigma fused from base to tip, unlobed . . . . 2. *Streptopus amplexifolius*

1. Streptopus lanceolatus (Aiton) Reveal, Phytologia 74: 187. 1993 E

Uvularia lanceolata Aiton, Hort. Kew. 1: 434. 1789; *Streptopus curvipes* Vail; *S. lanceolatus* var. *curvipes* (Vail) Reveal; *S. lanceolatus* var. *longipes* (Fernald) Reveal; *S. lanceolatus* var. *roseus* (Michaux) Reveal; *S. longipes* Fernald; *S. roseus* Michaux; *S. roseus* subsp. *curvipes* (Vail) Hultén; *S. roseus* var. *curvipes* (Vail) Fassett; *S. roseus* var. *longipes* (Fernald) Fassett; *S. roseus* var. *perspectus* Fassett

Plants from slender, sometimes matted rhizomes with variable internode lengths. **Stems** simple or occasionally branched, 1.5–4(–8) dm, glabrous, nodes sparsely pubescent-fringed. **Leaves** 5–10 × 2–4.5 cm; blade ovate-elliptic, base rounded, margins finely glandular-pubescent, apex acute to acuminate; peduncle: junction with pedicel not abrupt, entire structure (5–)9–20 mm, glands absent or obscure. **Flowers** 1(–2) per axil; perianth campanulate; tepals slightly spreading but not recurved, white or greenish yellow, streaked or spotted with pale rose to reddish purple, narrowly oblong-lanceolate, 6–10 mm; stamens 5–8 mm; anthers 2 mm, apex 2-pointed; style slender, 3.5–4 mm; stigma weakly 3-lobed; combined peduncles and pedicels recurved, (5–)9–20 mm, glands absent or obscure; pedicel rarely geniculate at junction with peduncle, coarsely pubescent. **Berries** reddish purple to red, subglobose, 6–9 mm. **Seeds** 2.5–3 mm. $2n = 16, 48$.

Flowering late spring–mid summer. Rich moist coniferous and deciduous woods; 50–2000 m; St. Pierre and Miquelon; Alta., B.C., Man., N.B., Nfld. and Labr., N.S., Ont., P.E.I., Que.; Alaska, Conn., Iowa, Ky., Maine, Md., Mass., Mich., Minn., Mont., N.H., N.J., N.Y., N.C., Ohio, Oreg., Pa., R.I., Tenn., Vt., Va., Wash., W.Va., Wis.

Streptopus lanceolatus has replaced the long-used name *S. roseus*, based on the recent lectotypification (J. L. Reveal 1993d) of Aiton's *Uvularia lanceolata*. This widespread North American species has been divided into four intergrading varieties or races (N. C. Fassett 1935) based on variation in rhizome internode lengths and density of leaf-margin ciliation. These include var. *roseus* in the southern Appalachians, var. *longipes* in the western Great Lakes region, var. *lanceolatus* (= var. *perspectus* Fassett) in the northeast, and var. *curvipes* in the west.

2. Streptopus amplexifolius (Linnaeus) de Candolle in J. Lamarck and A. P. de Candolle, Fl. Franç. ed. 3, 3: 174. 1805 F

Uvularia amplexifolia Linnaeus, Sp. Pl. 1: 304. 1753; *Streptopus amplexifolius* subsp. *americanus* (Schultes) Á. Löve & D. Löve; *S. amplexifolius* var. *americanus* Schultes; *S. amplexifolius* var. *chalazatus* Fassett; *S. amplexifolius* var. *denticulatus* Fassett; *S. amplexifolius* var. *grandiflorus* Fassett; *S. fassettii* Á. Löve & D. Löve; *Tortipes amplexifolius* (Linnaeus) Small

Plants from thick rhizomes. **Stems** freely branched, stout, 5–12 dm, often with reddish hairs basally, nodes not fringed. **Leaves** 5–15 × 2.5–6 cm; blade ovate-oblong to oblong-lanceolate, base cordate-clasping, apex acuminate; peduncle: junction with pedicel abrupt, with glands indicating transition, entire structure 2–5 cm, glands short-stalked. **Flowers** 1–2 per axil; perianth campanulate; tepals spreading, recurved at tips, white to greenish yellow, narrowly oblong-lanceolate, 9–15 mm; stamens unequal, outer shorter, 0.8–1 mm, inner 2–3 mm; anthers lanceolate, 3–3.5 mm, apex 1-pointed; style stout, 4–5 mm; stigma fused from base to tip, unlobed; combined peduncles and pedicels recurved, 2–5 cm, short-stalked glands indicating transition; pedicel sharply geniculate, glabrous. **Berries** whitish green maturing to yellowish orange or red, ellipsoid, 10–12 mm. **Seeds** 2.5–3 mm. $2n = 16, 32$.

Flowering late spring–mid summer. Rich moist coniferous and deciduous woods; 0–2800 m; Greenland; St. Pierre and Miquelon; Alta., B.C., Man., N.B., Nfld. and Labr., N.W.T., N.S, Ont., P.E.I., Que., Sask., Yukon; Alaska, Ariz., Calif., Colo., Conn., Idaho, Maine, Mass., Mich., Minn., Mont., N.H., N.Mex., N.Y., N.C., Oreg., Pa., S.Dak., Tenn., Utah, Vt., Va., Wash., W.Va., Wis., Wyo.; Eurasia.

Several poorly defined races described by N. C. Fassett (1935) as varieties based chiefly on minute difference in leaf-margin serration are not here recognized.

3. Streptopus streptopoides (Ledebour) Frye & Rigg, N.W. Fl., 109. 1912 F

Smilacina streptopoides Ledebour, Fl. Ross. 4: 128. 1853; *Kruhsea streptopoides* (Ledebour) Kearney; *Streptopus brevipes* Baker; *S. streptopoides* subsp. *brevipes* (Baker) Calder & Roy L. Taylor; *S. streptopoides* var. *brevipes* (Baker) Fassett

Plants from slender, creeping rhizome. **Stems** simple, 1–2 dm, glabrous, nodes coarsely fringed-ciliate. **Leaves** 3–5.5 × 1.5–2 cm; blade ovate-lanceolate to oblong-lanceolate, base rounded, apex acute to acuminate; peduncle bearing 1 flower per axil, 5–10 mm including pedicel, entire structure slender. **Flowers** 1–5; perianth rotate; tepals recurved at tips, wine-colored with yellowish green apex, oblong-lanceolate, 3–5 mm, apex acuminate; stamens 1.5–2 mm; anthers 1 mm, apex minutely apiculate; style bulbous-conical, 0.5 mm; stigma unlobed. **Berries** orangish red, globose, 5–6 mm. **Seeds** 2.5 mm. $2n = 16, 54$.

Flowering summer. Dense, damp coniferous forests; 0–1600 m; Alta., B.C.; Alaska, Idaho, Oreg., Wash.; e Asia.

The North American plants, usually referred to var. *brevipes* (N. C. Fassett 1935), differ from the typical Asiatic ones in lacking leaf-margin ciliation (F. H. Utech and S. Kawano 1975, 1976b; H. Takahashi 1976).

21. UVULARIA Linnaeus, Sp. Pl. 1: 304. 1753; Gen. Pl. ed. 5, 144. 1754 • Bellwort, merrybells [Latin *uvula*, alluding to the flowers hanging like that organ, and to formerly supposed efficacy in treating diseases of it] [E]

Frederick H. Utech

Shoichi Kawano

Oakesia S. Watson; *Oakesiella* Small

Herbs, caulescent, glabrous or pubescent, from short or elongate rhizomes bearing several fibrous or thickened roots. **Stems** simple or 1-branched, excluding flower-bearing branches, strongly angled or rounded, with sheathing, papery bracts proximally. **Leaves** alternate, sessile or perfoliate; blade oblong-linear to oblong-ovate, membranaceous to leathery. **Inflorescences** 1 per branch, terminal but appearing axillary; peduncles pendent. **Flowers:** perianth narrowly campanulate; tepals promptly deciduous, imbricate, distinct, linear to narrowly oblong, apex obtuse or acute, nectariferous; stamens distinct to weakly connate at perianth base; filaments dimorphic, glabrous; anthers linear-oblong, extrorse; connectives present; ovary superior, 3-locular, sessile or stipitate, rounded to sharply triangular; style 1, 3-lobed; stigmas 3, lobed, outwardly arching at maturity. **Fruits** capsular, greenish to yellowish brown, sessile or stipitate, leathery or submembranaceous, tardily dehiscent, dehiscence loculicidal. **Seeds** 1–3 per locule, brownish red, globose to ovoid; arils various. $x = 6, 7$.

Species 5 (5 in the flora): e North America.

Uvularia, a genus of attractive, spring-blooming plants, has been divided into sect. *Oakesiella* (Small) Wilbur with sessile leaves and sect. *Uvularia* with perfoliate leaves. Differences in other morphological (R. L. Wilbur 1963) and cytological (F. H. Utech 1978d) characters support recognition of these infrageneric groups.

D. K. Wijesinghe and D. F. Whigham (1997, 2001) compared the different, below-ground, vegetative morphologies of three *Uvularia* species and their relationship to their differing clonal populations and dynamics.

SELECTED REFERENCES Anderson, E. and T. W. Whitaker. 1934. Speciation in *Uvularia*. J. Arnold Arbor. 15: 28–42. Dietz, R. A. 1952. Variation in the perfoliate uvularias. Ann. Missouri Bot. Gard. 39: 219–247. Hayashi, K. et al. 1998. Molecular systematics of the genus *Uvularia* and selected Liliales based upon *mat*K and *rbc*L gene sequence data. Pl. Spec. Biol. 13: 129–146. Kawano, S. and H. H. Iltis. 1964. Cytotaxonomic and geographic notes on *Uvularia* (Liliaceae). Bull. Torrey Bot. Club 91: 13–23. Soper, J. H. 1952. Phytogeographic studies in Ontario I. The genus *Uvularia* in southern Ontario. Rhodora 54: 58–67. Wijesinghe, D. K. and D. F. Whigham. 2001. Nutrient foraging in woodland herbs: A comparison of three species of *Uvularia* (Liliaceae) with contrasting belowground morphologies. Amer. J. Bot. 88: 1071–1079. Wilbur, R. L. 1963. A revision of the North American genus *Uvularia* (Liliaceae). Rhodora 65: 158–188.

1. Stems rounded; leaf blades perfoliate, margins smooth; capsules obovoid to truncate; tepals smooth or densely papillose adaxially.
    2. Leaves pubescent (rarely glabrous) on abaxial veins; leaves below lowest branch typically 1; tepals smooth adaxially; capsule with 2 truncate beaks per lobe . . . . 4. *Uvularia grandiflora*
    2. Leaves glaucous, smooth abaxially; leaves below lowest branch (2–)3–4; tepals orange-papillose adaxially; capsule with 2 attenuate beaks per lobe . . . . 5. *Uvularia perfoliata*
1. Stems angled distally; leaf blades sessile, margins minutely papillose-denticulate; capsules generally ellipsoid; tepals smooth adaxially.
    3. Rhizomes short, 0.5–1 cm, bearing numerous, clustered, fleshy roots; stem nodes and abaxial leaf veins puberulent . . . . 1. *Uvularia puberula*
    3. Rhizomes elongate, 10–15 cm, bearing scattered, fibrous roots; stem nodes and abaxial leaf surfaces glabrous.
        4. Peduncles ebracteate; tepal apex rounded to acute; ovary and capsule stipitate . . . . 2. *Uvularia sessilifolia*
        4. Peduncles bearing 1, ovate, leafy bract; tepal apex acuminate; ovary and capsule sessile or subsessile . . . . 3. *Uvularia floridana*

1. Uvularia puberula Michaux, Fl. Bor.-Amer. 1: 199. 1803 • Mountain bellwort [E]

Oakesia puberula (Michaux) S. Watson; *Oakesiella puberula* (Michaux) Small; *Uvularia pudica* (Walter) Fernald

Rhizomes short, 0.5–1 cm, bearing numerous, clustered, fleshy roots; stolons absent. **Stems** 1–several, 1-branched, angled and puberulent distally, especially at nodes, 1–4.5 dm, bearing 1 leaf below lowest branch. **Leaf blades** sessile, broadly oblong-elliptic, 4–8(–8.5) × 1.5–3.5(–4) cm, puberulent on abaxial veins, margins minutely papillose, apex acute to acuminate. **Flowers** 1–3 per stem; peduncles 0.5–2 cm, ebracteate; tepals greenish to pale yellow, 10–25 × 2–4 mm, smooth adaxially, apex rounded; stamens 6.5–17 mm; anthers 5–12 mm; connectives 0.6–0.8 mm; ovary sessile or subsessile, sharply triangular; style 8–14 mm; stigma lobes 4–6 mm. **Capsules** sharply 3-winged, broadly ellipsoid, 1.5–3.7 × 1–2 cm, not beaked. **Seeds** 3–3.5 mm; arils crested. $2n = 14$.

Flowering spring–early summer. Moist to dry, open woods; 0–900 m; Ga., Md., N.J., N.Y., N.C., Pa., S.C., Tenn., Va., W.Va.

2. Uvularia sessilifolia Linnaeus, Sp. Pl. 1: 305. 1753 • Sessile-leaved bellwort, wild-oats, straw-lily, uvulaire à feuilles sessiles [E] [F] [W1]

Oakesia sessilifolia (Linnaeus) S. Watson; *Oakesiella sessilifolia* (Linnaeus) Small

Rhizomes elongate, slender, 10–15 cm, bearing scattered, fibrous roots; stolons present. **Stem** 1, 1-branched, angled distally, 1–4.5 dm, nodes glabrous, bearing 1–2 leaves below lowest branch. **Leaf blades** sessile, narrowly to broadly elliptic, 4–6.5(–8) × (1.3–)1.6–2.8(–4) cm, glabrous abaxially, margins minutely denticulate, apex acute to acuminate. **Flowers** 1 per stem; peduncles 0.6–1.6 cm, ebracteate; tepals pale straw-yellow, 13–25 × 2–4.5 mm, smooth abaxially, apex rounded to acute; stamens 8–15 mm; anthers 5–12 mm; connectives 0.3–0.5 mm; ovary stipitate, sharply triangular, flowering stipe 0.8–1.2 mm; style 10–15 mm; stigma lobes 1–2 mm. **Capsules** sharply 3-winged, stipitate, ellipsoid, 2–3 × 1–1.6 cm, not beaked; fruiting stipe 2–5 mm. **Seeds** 3–5 mm; arils crested. $2n = 14$.

Flowering spring–early summer. Moist hardwood coves, alluvial bottomlands, thickets and xeric woods northwards; 0–1000 m; Man., N.B., N.S., Ont., Que.; Ala., Ark., Conn., Del., D.C., Fla., Ga., Ill., Ind., Iowa, Ky., La., Maine, Md., Mass., Mich., Minn., Miss., Mo., N.H., N.J., N.Y., N.C., N.Dak., Ohio, Okla., Pa., R.I., S.C., Tenn., Tex., Vt., Va., W.Va., Wis.

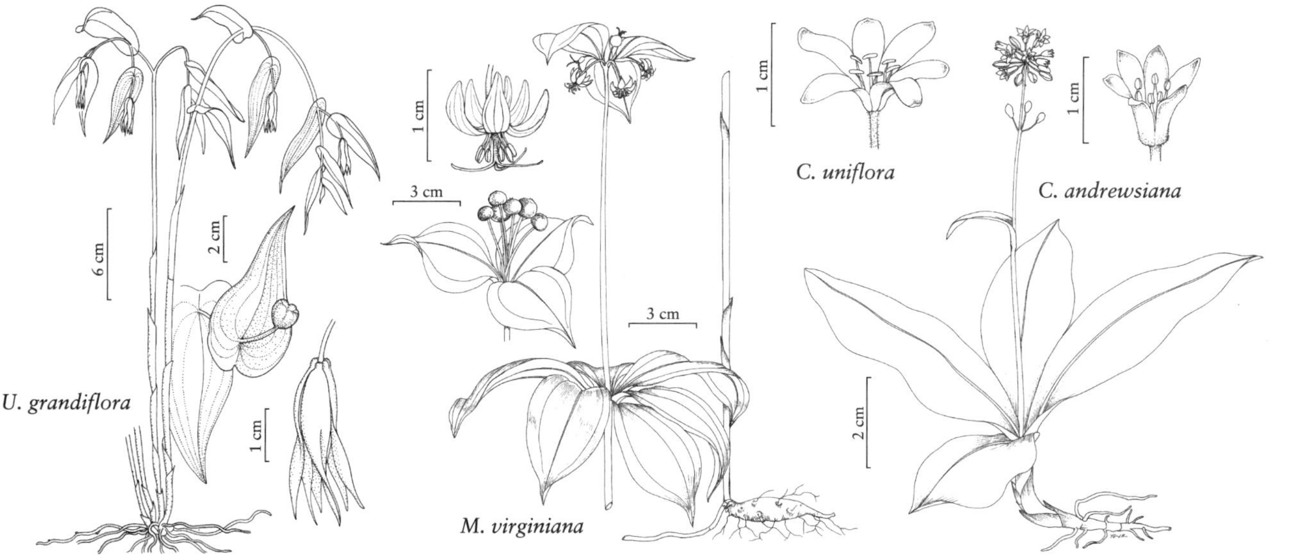

UVULARIA ◦ MEDEOLA ◦ CLINTONIA

3. Uvularia floridana Chapman, Fl. South. U.S., 487. 1860 • Florida bellwort [E]

Oakesia floridana (Chapman) J. F. Macbride; *Oakesiella floridana* (Chapman) Small

Rhizomes elongate, 10–15 cm, fleshy, bearing scattered, fibrous roots; stolons present. **Stem** 1, simple, angled distally, 2.5–4 dm, nodes glabrous, bearing 1–2 leaves below lowest branch. **Leaf blades** sessile, narrowly to broadly elliptic, 4–7(–8.5) × 1.5–3 cm, glabrous abaxially, margins minutely papillose, apex rounded to acute. **Flowers** 1 per stem; peduncles 0.2–0.8 cm, bearing 1, ovate, leafy bract, 0.7–2.5 cm; tepals pale whitish yellow, 20–30 × 3–4 mm, smooth adaxially, apex acuminate; stamens 10–15 mm; anthers 5–10 mm; connectives 0.5–2 mm; ovary sessile or subsessile, sharply triangular; style 10–15 mm; stigma lobes 3–5 mm. **Capsules** sessile or subsessile, sharply 3-winged, generally ellipsoid, 2–3.7 × 1–2 cm, apical attenuate beak 4–7 mm. **Seeds** 3–5 mm; arils crested. 2*n* = 12.

Flowering early spring. Rich hardwood forest and alluvial bottomlands; 0–100 m; Ala., Fla., Ga., Miss., S.C.

The 2*n* = 12 aneuploid chromosome report for *Uvularia floridana* from Florida's panhandle (F. H. Utech 1978d) represents a reduction from the typical 2*n* = 14 for the genus (S. Kawano and H. H. Iltis 1964).

4. Uvularia grandiflora Smith, Exot. Bot. 1: 99, plate 51. 1804 • Large-flowered bellwort, uvulaire grandiflore [E] [F]

Rhizomes short, ca. 1 cm, bearing clustered, fleshy roots; stolons absent. **Stems** 1–several, 1-branched, rounded, 2–7.5 dm, glabrous, typically bearing 1 leaf below lowest branch. **Leaf blades** perfoliate, elliptic to ovate-oblong, 6–13.5 × 2–6.5 cm, white-pubescent (rarely glabrous) on abaxial veins, margins smooth, apex acute. **Flowers** 1–3(–4) per stem; peduncles 1–2.5 cm, bearing 1 perfoliate bract; tepals golden yellow, 25–50 × 3–10 mm, smooth adaxially, apex acuminate; stamens 10–25 mm; anthers (8–)12–15(–20) mm; connectives 0.3–0.7 mm; ovary sessile, obovoid; style 8–12 mm; stigma lobes 2–5 mm. **Capsules** obovoid to obpyramidal, 3-lobed, 1–1.5 × 1–2 cm, 2 truncate beaks per lobe, lobes rounded. **Seeds** 2.5–4.5 mm; arils membranous. 2*n* = 14.

Flowering spring–early summer. Rich moist woods, calcareous to neutral soils; 0–1100 m; Man., Ont., Que.; Ala., Ark., Conn., Ga., Ill., Ind., Iowa, Kans., Ky., Md., Mass., Mich., Minn., Miss., Mo., N.H., N.Y., N.C., N.Dak., Ohio, Okla., Pa., S.Dak., Tenn., Vt., Va., W.Va, Wis.

5. Uvularia perfoliata Linnaeus, Sp. Pl. 1: 304. 1753
• Perfoliate bellwort [E]

Rhizomes short, ca. 1 cm, bearing numerous, clustered, fleshy roots and 1–2 slender, elongate, white stolons to 1.5 dm. **Stems** 1–several, simple or 1-branched, rounded, 1.5–5 dm, glaucous, bearing (2–)3–4 leaves below lowest branch. **Leaf blades** perfoliate, ovate to ovate-oblong, 4–20 (–12.5) × 1.5–4(–6) cm, glaucous and smooth abaxially, margins smooth, apex acute to short-acuminate. **Flowers** 1 per stem; peduncles 1–2 cm, bearing 1 perfoliate bract; tepals straw-yellow, (15–)20–35 × 3–5 mm, orange-papillose adaxially, apex acute; stamens 10–15 mm; anthers 6–10 mm; connectives 1 mm; ovary sessile, obovoid; style 8–10 mm; stigma lobes 3–5 mm. **Capsules** obovoid-truncate, 3-lobed, 0.7–1.3 × 1–1.6 cm, 2 attenuate beaks per lobe. **Seeds** 2.5–4.5 mm; arils membranaceous. $2n = 14$.

Flowering spring–early summer. Deciduous woods and upland thickets, acid to neutral soils; 0–1000 m; Ont.; Ala., Ark., Conn., Del., D.C., Fla., Ga., Ky., La., Maine, Md., Mass., Miss., N.H., N.J., N.Y., N.C., Ohio, Pa., R.I., S.C., Tenn., Tex., Vt., Va., W.Va.

Detailed life-history studies of *Uvularia perfoliata* (D. F. Whigham 1974; D. K. Wijesinghe and D. F. Whigham 1997, 2001; H. Kudoh et al. 1999) have established the genetic basis of its clonal structure.

22. MEDEOLA Linnaeus, Sp. Pl. 1: 339. 1753; Gen. Pl. ed. 5, 158. 1754 • Indian cucumber-root, concombre sauvage, jarnotte [for *Medea*, mythical sorceress] [E]

Frederick H. Utech

Herbs, perennial, from tapered, white, tuberlike, horizontal rhizomes. **Stems** simple, slender. **Leaves** in 2 whorls; proximal blades oblong-oblanceolate, base attenuate, margins entire, apex acuminate; distal blades ovate, rounded at base, apex acuminate. **Inflorescences** terminal, umbellate, (2–)3–9-flowered, sessile. **Flowers:** tepals caducous, 6, similar, recurved, distinct; stamens 6; anthers versatile, oblong, extrorse; ovary superior, 3-locular; styles 3, recurved, distinct to base, often purple; pedicel declined or spreading in flower, erect in fruit. **Fruits** baccate, dark purple to black, globose. **Seeds** few, shiny brown, subglobose. $x = 7$.

Species 1: e North America.

Medeola was removed from its long association with *Trillium* and *Paris* (tribe Parideae) when A. L. Takhtajan (1987, 1997) created the monotypic Medeolaceae, placing it next to a strictly defined Liliaceae. M. N. Tamura (1998c) included *Medeola* and *Clintonia* in the tribe Medeoloideae within a narrowly defined Liliaceae. The association of *Medeola* and *Clintonia* is supported by molecular analysis (K. Hayashi et al. 1998, 2001; T. B. Patterson and T. J. Givnish 1998) as well as morphology (R. Y. Berg 1962; M. Takahashi 1984; F. H. Utech 1978e), embryology (R. Y. Berg 1962b), and cytology (R. M. Stewart and R. Bamford 1942; M. N. Tamura 1995).

SELECTED REFERENCES Bell, A. D. 1974. Rhizome organization in relation to vegetative spread in *Medeola virginiana*. J. Arnold Arbor. 55: 458–468. Berg, R. Y. 1962. Morphology and taxonomic position of *Medeola*, Liliaceae. Skr. Norske Vidensk.-Akad. Oslo, Mat.-Naturvidensk. Kl., n. s. 3: 1–56. Berg, R. Y. 1962b. Contribution to the comparative embryology of the Liliaceae: *Scoliopus, Trillium, Paris* and *Medeola*. Skr. Norske Vidensk.-Akad. Oslo, Mat.-Naturvidensk. Kl., n. s. 4: 1–64. Utech, F. H. 1978e. Floral vascular anatomy of *Medeola virginiana* L. (Liliaceae-Parideae = Trilliaceae) and tribal note. Ann. Carnegie Mus. 47: 13–28.

1. Medeola virginiana Linnaeus, Sp. Pl. 1: 339. 1753 E F

Rhizomes 3–8 × 1 cm, decaying every year. **Stems** 2–9 dm, white-woolly when young, 1–3 deciduous bracts below proximal leaf whorl. **Leaf blades** sessile or short-petiolate, becoming purple-tinged basally in fruit, 5–9(–12) in proximal whorl, 6–16 × 1.5–5 cm, 3(–5) in distal whorl, 2.5–5 × 1.5–4 cm. **Tepals** yellowish green, 6–10 mm. **Fruits** (5–)8–10(–14) mm diam. **Seeds** 3 mm. $2n = 14$.

Flowering late spring. Moist slopes, mesic woods; 0–1600 m; N.B., N.S., Ont., P.E.I., Que.; Ala., Conn., Del., D.C., Fla., Ga., Ill., Ind., Ky., La., Maine, Md., Mass., Mich., Miss., Mo., N.H., N.J., N.Y., N.C., Ohio, Pa., R.I., S.C., Tenn., Vt., Va., W.Va., Wis.

The rhizome of *Medeola virginiana* has the taste and odor of cucumber and is edible. The whorled leaves look like those of the orchid *Isotria verticillata*. The Iroquois of eastern North America used *M. virginiana* as an anticonvulsive, pediatric aid (D. E. Moerman 1986).

23. CLINTONIA Rafinesque, Amer. Monthly Mag. & Crit. Rev. 2: 266. 1818 • Clintonia [for De Witt Clinton (1769–1828), statesman and several-times governor of New York]

Frederick H. Utech

Xeniatrum Salisbury

Herbs, perennial, rhizomatous, 1.5–8 dm; rhizomes bearing thin, fibrous roots. **Leaves:** basal 2–6, arising from rhizome crown, sessile, sheathing, hyaline, chaffy in age; cauline petiolate, blade with central vein prominent basally, obovate to oblanceolate, weakly villous-pubescent, base attenuate to cuneate, margins entire, apex acute to abruptly short-acuminate, often mucronate. **Inflorescences** terminal, in short racemes or umbellike clusters, 1–45-flowered, bracteate; bracts foliaceous or linear. **Flowers:** tepals caducous, 6, distinct, obovate to narrowly oblanceolate, nectaries present, sometimes obscure; stamens inserted at perianth base; filaments dilated, basally pubescent; anthers oblong-obovate to oblong-linear, semiversatile, extrorse; ovary superior, 2-locular [3-locular], ovoid to subcylindrical, glabrous, ovules 2–10 per locule; style columnar, compressed laterally; stigmas weakly 2-lobed [3-lobed]; pedicels unequal, elongate and ascending with age. **Fruits** baccate, metallic blue to black, 4–30-seeded, ellipsoid to ovoid, smooth. **Seeds** shiny brown, round abaxially, angled with 2 or 3 faces. $x = 7$.

Species 5 (4 in the flora): temperate North America, e Asia.

Eastern and western Native Americans used *Clintonia* as an eye and heart medicine, as well as a dermatological and gynecological aid (D. E. Moerman 1986).

The North American *Clintonia* species have bicarpellate ovaries by reduction (F. H. Utech 1973) and reticulate pollen, in contrast to the typical liliaceous tricarpellate ovaries and gemmate pollen that occur in the eastern Asian *C. udensis* (M. Takahashi and K. Sohma 1982; V. N. Kosenko 1991). The genus has a unique type of megasporogenesis, the "*Clintonia*-type" (A. N. Pahuja and V. Kumar 1971; F. H. Smith 1943; R. W. Smith 1911; R. I. Walker 1944), in which there is a selective elimination of genomes (heterotypic vs. homotypic divisions) before double fertilization, resulting in genetically identical diploid embryo and diploid endosperm nuclei. Such derived megasporogenesis is equivalent to female self-fertilization (pseudogamy).

Counts of $2n = 28$ are known in all *Clintonia* species except for a recently discovered sympatric diploid cytotype ($2n = 14$) in *C. udensis* (Li S. F. and Chang Z. Y. 1996; Li S. F. et al. 1996). It appears therefore that the *Clintonia* karyotype has an ancient allopolyploid origin and that the generic base number should be viewed as $x = 7$ and not $x = 14$. Based on these karyological data, an early eastern Asian and North American divergence with speciation and polyploidization occurring in North America is postulated.

M. N. Tamura (1998c) included *Medeola* and *Clintonia* in the tribe Medeoloideae within a narrowly defined Liliaceae, and molecular analysis supports the association of those two genera (K. Hayashi et al. 1998, 2001; M. F. Fay and M. W. Chase 2000).

SELECTED REFERENCES Hayashi, K., S. Yoshida, F. H. Utech, and S. Kawano. 2001. Molecular systematics in the genus *Clintonia* and related taxa based on *rbc*L and *mat*K gene sequence data. Pl. Spec. Biol. 16: 119–137. Utech, F. H. 1973. A Biosystematic Study of the Genus *Clintonia* Raf. (Liliaceae: Polygonatae). Ph.D. thesis. Washington University. Utech, F. H. 1975. Biosystematic studies in *Clintonia* (Liliaceae-Polygonatae) III. Cytogeography, chromosome numbers and chromosomal morphology of the North American species of *Clintonia*. Cytologia 40: 765–786.

1. Inflorescences 1–8(–10)-flowered, in short, terminal racemes.
   2. Plants 1.5–2.5 dm; inflorescences 1- or 2-flowered; tepals creamy white, 18–25 mm; anthers 3.5–5.5 mm; w North America . . . . 1. *Clintonia uniflora*
   2. Plants 2–5 dm; inflorescences 3–8(–10)-flowered, in short, terminal racemes; tepals yellow to yellowish green, 12–16 mm; anthers 2–3.5 mm; e North America . . . . 2. *Clintonia borealis*
1. Inflorescences 10–45-flowered, in terminal and/or lateral, umbel-like clusters.
   3. Inflorescences 20–45-flowered, lateral clusters usually present; tepals claret red, narrowly oblong to oblanceolate, 10–18 mm; berries blue to bluish black, 10–30-seeded; California, Oregon . . . . 3. *Clintonia andrewsiana*
   3. Inflorescences 10–25(–30)-flowered, in terminal clusters only; tepals white, often spotted purplish brown or green distally, ovoid-obovate, 5.5–8 mm; berries black, 2–4-seeded; e North America . . . . 4. *Clintonia umbellulata*

1. Clintonia uniflora (Menzies ex Schultes) Kunth, Enum. Pl. 5: 159. 1850 • Queen's-cup, bride's-bonnet [E] [F]

Smilacina borealis Ker Gawler var. *uniflora* Menzies ex Schultes in J. J. Roemer et al., Syst. Veg. 7(1): 307. 1829; *S. uniflora* (Menzies ex Schultes) Hooker

Plants 1.5–2.5 dm; rhizomes elongate. **Cauline leaves** 2–3; blade pale green adaxially, oblanceolate to obovate, 8–20 × 2.5–6.5 cm. **Inflorescences** 1(–2)-flowered, erect; bract 1, narrowly linear. **Flowers:** tepals creamy white, broadly obovate to oblanceolate, 18–25 × 5–7 mm; filaments 11–18 mm; anthers oblong-linear, 3.5–5.5 mm. **Berries** deep lustrous blue, subglobose to pyriform, 10–18-seeded, 6–12 mm. **Seeds** 3–4 mm. $2n = 28$.

Flowering late May–Jul. Montane coniferous forests; 0–2600 m; Alta., B.C.; Alaska, Calif., Idaho, Mont., Oreg., Wash.

2. Clintonia borealis (Aiton) Rafinesque, Atlantic J. 1: 120. 1832 • Blue bead-lily, yellow clintonia, corn-lily [E] [W1]

Dracaena borealis Aiton, Hort. Kew. 1: 454. 1789

Plants 2–5 dm; rhizomes thin, spreading. **Cauline leaves** 2–4; blade dark glossy green adaxially, oblong to elliptic obovate, 15–30 × 5–10 cm. **Inflorescences** in short, terminal racemes, 3–8(–10)-flowered; bracts 1(–3), narrowly foliaceous. **Flowers** nodding; tepals yellow to yellowish green, narrowly oblong, 12–16 × 3.5–4.5 mm; filaments 12–17.5 mm; anthers oblong, 2–3.5 mm. **Berries** ultramarine blue, ovoid, 8–16-seeded, 8–12 mm. **Seeds** 3–5 mm. $2n = 28, 32$.

Flowering early May–early Jul. Rich coniferous, mixed, or deciduous woods; 0–1600 m; St. Pierre and Miquélon; Man., N.B., Nfld. and Labr., N.S., Ont., P.E.I., Que.; Conn., Ga., Ill., Ind., Maine, Md., Mass., Mich., Minn., N.H., N.J., N.Y., N.C., Ohio, Pa., R.I., Tenn., Vt., Va., W.Va., Wis.

Throughout the range of *Clintonia borealis* a stabilized, $2n = 32$ cytotype has been reported (F. H. Utech and L. B. Thien 1973; B. M. Kapoor 1973).

3. Clintonia andrewsiana Torrey, Pacif. Railr. Rep. 4(5): 150. 1857 • Red clintonia [E] [F]

Plants 3–8 dm; rhizomes short, thick. **Cauline leaves** 4–6; blade dark green, oblanceolate to broadly elliptic, 20–35 × 5–15 cm. **Inflorescences** in terminal, umbel-like clusters, 1–3(–5) lateral clusters usually present, 20–45-flowered; bracts 1–3(–5), foliaceous. **Flowers:** tepals deep claret red, narrowly oblong to oblanceolate, 10–18 × 3–3.7 mm; filaments 7–10 mm; anthers oblong, 1.8–2.8 mm. **Berries** blue to bluish black, ovoid, 10–30-seeded, 8–12 mm. **Seeds** 3–4 mm. $2n = 28$.

Flowering late Apr–late Jun. Moist, coastal redwood forests; 0–400 m; Calif., Oreg.

4. Clintonia umbellulata (Michaux) Morong, Mem. Torrey Bot. Club 5: 114. 1894 • White clintonia, speckled wood-lily [E]

Convallaria umbellulata Michaux, Fl. Bor.-Amer. 1: 202. 1803; *Clintonia allegheniensis* Harned; *Xeniatrum umbellulatum* (Michaux) Small

Plants 2.7–6 dm; rhizomes short, thick. **Cauline leaves** 3–4; blade dark green, oblong to obovate-elliptic, 18–30 × 4.5–8 cm. **Inflorescences** in terminal clusters, 10–25(–30)-flowered; bract 1, foliaceous. **Flowers:** tepals white to greenish white, often spotted purplish brown or green distally, ovoid-obovate, 5.5–8 × 2.7–4 mm; filaments 5.5–7 mm; anthers short-oblong, 3.5–5.5 mm. **Berries** black, globose to ellipsoid, 2–4-seeded, 6–8 mm. **Seeds** 3–4 mm. $2n$ = 28.

Flowering early May–late Jun. Rich cove hardwood forests; 200–1000 m; Ga., Ky., Md., N.Y., N.C., Ohio, Pa., S.C., Tenn., Va., W.Va.

Clintonia umbellulata and *C. borealis* are allopatric in the southern Appalachians (P. P. Gunther 1972). A putative hybrid between the two, *C. allegheniensis* Harned (J. E. Harned 1931) is not recognized here and is regarded as a mere fruit-color variant of *C. umbellulata* (F. H. Utech 1973). Isozyme evidence supports this conclusion (M. T. Blain 1997).

24. ERYTHRONIUM Linnaeus, Sp. Pl. 1: 305. 1753; Gen. Pl. ed. 5, 145. 1754 • Trout-lily, fawn-lily, dog's-tooth-violet, érythrone [Greek *erythros*, red, alluding to the pink to purple flowers of *Erythronium dens-canis*]

Geraldine A. Allen

Kenneth R. Robertson

Herbs, perennial, scapose, from ovate to elongate bulbs, sometimes with small, beadlike segments of short, persistent rhizome attached; several species producing additional bulbs as sessile bulbels or at the ends of slender stolons or vertical droppers, these species typically flowering more sparingly than those without extensive vegetative reproduction. **Leaves** 2 (1 in nonflowering plants), basal, ± petiolate; blade green or mottled with purple, brown, or white, lanceolate to ovate (wider if solitary), flat to folded, 6–60 cm, glaucous in a few species, glabrous, base narrowed gradually or abruptly to petiole, margins entire or sometimes wavy. **Scape** green or sometimes reddish, typically elongating in fruit. **Inflorescences** terminal, racemose, 1–10-flowered. **Flowers** showy, usually nodding, sometimes held laterally or erect; tepals 6 (as few as 4 in *E. propullans*), spreading to reflexed, distinct, similar, white, yellow, pink, or violet, often with basal zone of yellow or other colors, lanceolate to ovate, inner tepals auriculate at base in many species, auricles appressed to ovary and forming sac- or pocketlike hollows on adaxial surfaces; stamens 6; filaments generally slender; ovary superior; style 1, abruptly attached to ovary (or forming a beak in *E. rostratum*); stigma unlobed or 3-lobed, lobes recurved to erect. **Fruits** capsular, erect, obovoid to oblong, apex rounded, truncate, or umbilicate (beaked in *E. rostratum*), dehiscence loculicidal. **Seeds** brown, ± angular, ± ovoid. x = 11, 12.

Species ca. 27 (23 in the flora): all North American, except for the Eurasian *Erythronium dens-canis* Linnaeus and its segregates *E. caucasicum* Woronow, *E. japonicum* Decaisne, and *E. sibiricum* (Fischer & C. A. Meyer) Krylov; north temperate in forest and montane meadow habitats, one species (*E. mesochoreum*) in prairies.

Erythronium is a well-marked and distinctive genus closely related to *Tulipa*. In North America, *Erythronium* consists of distinct eastern and western groups, the former clearly having an affinity with species of the Old World. Most of the species have attractive and showy flowers, and several are well suited for naturalizing in woodland gardens.

Pressed specimens tend to fade, so leaf and flower colors and markings should be recorded on specimen labels at the time of collection. The orientation of flowers and fruits should also be recorded, and it is useful to collect bulbs of both flowering and nonflowering plants when this will not damage a population. Bulbs of *Erythronium* species are often more than 10 cm deep. Collectors should press flowers so that the shape of the style, stigmas, filaments, and petal bases can be observed later.

Species of western North America (Rocky Mountains westward) and eastern North America (Great Plains eastward) seem to form two discrete evolutionary groups, and this geographic distinction is the first character used in the taxonomic key. Most western species can at least occasionally produce plants with multiple-flowered scapes, while eastern species always have a single flower. Bulbs of western species vary from ovoid to slender and elongate, while all eastern species have ovoid bulbs. Fruits of western species vary from obovoid to narrowly oblong, while all eastern species have obovoid fruits. In species with mottled leaves, the pattern of mottling in eastern species is a more or less random dappling (resembling that of the Eurasian *Erythronium dens-canis*), while in western species it takes the form of elongate lateral streaks or veining, often more or less symmetrical on either side of the midline. Several eastern species, but only a single western species (*E. multiscapideum*), propagate vegetatively by stolons originating largely from nonflowering bulbs. (The term "stolon" is used here to refer to the slender, white, elongate, underground structures found in *Erythronium*.) In *E. propullans*, one stolon per plant is produced from the flowering scape below ground level. In species with stolons, populations often contain relatively few flowering plants. In the following descriptions, measurements of bulbs and leaves refer to flowering plants.

The base chromosome number in *Erythronium* is $x = 12$, except for the white-flowered species of eastern North America, *E. albidum*, *E. mesochoreum*, and *E. propullans*, which have $x = 11$. Both diploid and tetraploid species occur with each base chromosome number ($2n$ = 22, 24, 44, and 48).

Although the pollination biology of most species of *Erythronium* is not well known, that of the bumblebee-pollinated *E. grandiflorum* has been the subject of extensive study (e.g., J. D. Thomson and D. A. Stratton 1985; J. D. Thomson and B. A. Thomson 1989). The oligolectic bee *Andrena erythronii* is associated with *Erythronium* in northeastern North America, although *A. erythronii* visits other species of plants, and other insects also visit *Erythronium* (P. Bernhardt 1977; W. E. LaBerge 1987).

SELECTED REFERENCES Applegate, E. I. 1935. The genus *Erythronium*: A taxonomic and distributional study of the western North American species. Madroño 3: 58–113. Mathew, B. 1992. A taxonomic and horticultural review of *Erythronium* L. (Liliaceae). Bot. J. Linn. Soc. 109: 453–471. Parks, C. R. and J. W. Hardin. 1963. Yellow erythroniums of the eastern United States. Brittonia 15: 245–259. Robertson, K. R. 1966. The genus *Erythronium* (Liliaceae) in Kansas. Ann. Missouri Bot. Gard. 53: 197–204. Shevock, J. R., J. A. Bartel, and G. A. Allen. 1990. Distribution, ecology and taxonomy of *Erythronium* (Liliaceae) in the Sierra Nevada of California. Madroño 37: 261–273.

1. Plants of e North America, from the Great Plains eastward.
   2. Tepals yellow.
      3. Capsule apex long-beaked; style persistent, forming beak on capsule; inner tepals with conspicuous, well-developed auricles . . . . 19. *Erythronium rostratum*
      3. Capsule apex rounded, truncate, apiculate, umbilicate, or indented; style deciduous or base forming a small apiculum; auricles of inner tepals small or absent.
         4. Capsule apex rounded, truncate, or apiculate, capsules held erect or at least off ground; style swollen distally or ± terete; stigma lobes erect or recurved; inner tepals with small auricles; stolons 1–3 per bulb, mostly on 1-leaved plants . . . . 18. *Erythronium americanum*
         4. Capsule apex indented, umbilicate, or rarely rounded, capsules ± resting on ground on reclining peduncles; style ± terete; stigma lobes spreading; tepals without auricles; stolons absent, or 1 per bulb of 1-leaved plants . . . . . 20. *Erythronium umbilicatum*

2. Tepals white to pink or lavender.
    5. Flowering plants reproducing vegetatively by stolons produced halfway up stem; tepals 8–15 mm; se Minnesota . . . . 23. *Erythronium propullans*
    5. Flowering plants reproducing vegetatively by droppers or offshoots or by stolons arising from bulbs; tepals 15–40 mm; widespread.
        6. Tepals strongly reflexed at anthesis; leaf blade mottled, ± flat; capsules held erect; usually mesic bottomlands . . . . 21. *Erythronium albidum*
        6. Tepals spreading at anthesis; leaf blade usually not mottled, conduplicate; capsules resting on ground; prairies, glades, dry, open woods . . . . . . 22. *Erythronium mesochoreum*

1. Plants of w North America, from the Rocky Mountains westward.
    7. Leaf blade distinctly irregularly mottled with irregular streaks of brown or white; tepals violet-pink to white or creamy white, often yellow at base but never throughout.
        8. Filaments flattened, wider than 2 mm; stigma lobes 2–6 mm.
            9. Tepals white to creamy white at anthesis, sometimes pinkish in age . . . . 6. *Erythronium oregonum*
            9. Tepals uniformly clear violet-pink at anthesis . . . . 5. *Erythronium revolutum*
        8. Filaments linear, less than 0.8 mm wide; stigma unlobed or with lobes shorter than 4 mm.
            10. Tepals violet to pink, dark purple at base . . . . 15. *Erythronium hendersonii*
            10. Tepals ± white with yellow base.
                11. Style 5–10 mm; anthers yellow, white, or cream (sometimes pink, reddish, or brownish red in forms of *E. citrinum*).
                    12. Anthers yellow; style often bent; tepals bright yellow at base . . . . 16. *Erythronium helenae*
                    12. Anthers white to cream; style straight; tepals usually pale yellow at base . . . . 14. *Erythronium citrinum*
                11. Style 10–15 mm; anthers white to cream.
                    13. Scape (when flowers more than 1) branched well above leaves; bulbels sessile or absent; flowering individuals generally abundant in populations . . . . 13. *Erythronium californicum*
                    13. Scape (when flowers more than 1) branched near ground level; bulbels produced at ends of long slender stolons; flowering individuals generally uncommon in populations (most plants 1-leaved and vegetative) . . . . 17. *Erythronium multiscapideum*
    7. Leaf blade uniformly green (faintly mottled with brown or white in *E. elegans* and *E. quinaultense*); tepals yellow or white, sometimes pinkish in age.
        14. Tepals yellow.
            15. Style and filaments yellow; tepals 15–28 mm; stigma unlobed or with very short, rounded lobes shorter than 1 mm; inflorescences 1–10-flowered . . . . 11. *Erythronium pluriflorum*
            15. Style and filaments ± white; tepals 20–35 mm; stigma unlobed or with slender, recurved lobes (1–)2–4 mm; inflorescences 1–4-flowered.
                16. Inflorescences usually 1-flowered; style 10–15 mm; stigma unlobed or with lobes (1–)2–4 mm; tepals (in live specimens) often with narrow, paler zone at base . . . . 1. *Erythronium grandiflorum* (in part)
                16. Inflorescences usually 1–4-flowered; style 8–10 mm; stigma unlobed or with lobes shorter than 1 mm; tepals uniformly yellow . . . . . . 12. *Erythronium tuolumnense*
        14. Tepals white to creamy white with yellow zone at base, sometimes tinged pink.
            17. Stigma ± unlobed, or with lobes shorter than 1 mm.
                18. Tepals 10–20 mm, auricles absent . . . . 8. *Erythronium purpurascens*
                18. Tepals 20–45 mm, inner auriculate.
                    19. Tepals less than 1/3 yellow; filaments white; leaves 6–17 cm; inflorescences 1–3-flowered; sw Oregon, nw California . . . . 7. *Erythronium klamathense*
                    19. Tepals 1/2–2/3 bright yellow proximally; filaments white or yellow; leaves 10–35 cm; inflorescences 1–8-flowered; Sierra Nevada of California.
                        20. Filaments white; anthers yellow . . . . 9. *Erythronium pusaterii*
                        20. Filaments yellow; anthers cream . . . . 10. *Erythronium taylorii*

[17. Shifted to left margin.—Ed.]

17. Stigma with slender, usually recurved lobes 1–5 mm.
    21. Tepals white to creamy white; filaments linear, less than 0.8 mm wide.
        22. Tepals narrowly ovate, 20–35 mm, length at least 4 times width; style 15 mm or shorter; leaf blade narrowed gradually to petiole; nw Idaho, w Montana, e Washington . . . . 1. *Erythronium grandiflorum* (in part)
        22. Tepals broadly ovate to broadly lanceolate, 25–45 mm, length (at least of the inner) less than 4 times width; style 13–25 mm; leaf blade narrowed ± abruptly to petiole; sw British Columbia, w Oregon, w Washington . . . . 2. *Erythronium montanum*
    21. Tepals tinged with pink; filaments linear to lanceolate, at least 0.8 mm wide.
        23. Inner tepals ± white, outer ± white and often strongly marked with pink, especially abaxially and along midline; Coast Ranges, Oregon . . . . 3. *Erythronium elegans*
        23. All tepals white proximally, shading to pink at outer margins and tips; Olympic Mountains, Washington . . . . 4. *Erythronium quinaultense*

1. Erythronium grandiflorum Pursh, Fl. Amer. Sept. 1: 231. 1814 • Glacier-lily [E] [F]

Bulbs slender, 30–50 mm. **Leaves** 5–20 cm; blade green, lanceolate, ± glaucous, base gradually narrowed to petiole, margins ± wavy. **Scape** 5–30 cm. **Inflorescences** usually 1-flowered, sometimes up to 5-flowered. **Flowers:** tepals recurved, bright yellow with (in live specimens) narrow paler zone at base, or white to creamy white with yellow base, narrowly ovate, 20–35 mm, length at least 4 times width, inner usually auriculate at base; stamens 11–18 mm; filaments white, ± slender, linear, less than 0.8 mm wide; anthers cream, yellow, red, or purplish red; pollen yellow or red; style white, 10–15 mm; stigma unlobed or with slender, recurved lobes (1–)2–4 mm. **Capsules** oblong to narrowly obovoid, 2–5 cm. $2n = 24$.

Subspecies 2 (2 in the flora): w North America.

This beautiful species is often very abundant in mountain meadows of western North America, especially in the Rocky Mountains, where it may form spectacular displays. It is often difficult to grow in cultivation outside its preferred habitats. Bulbs of this species were a staple food for native North American peoples of several tribes, and were eaten in large quantities and also traded. Within the typical subspecies, both anthers and pollen vary considerably in color; plants with pale anthers have been called var. *pallidum*, and forms with yellow anthers, var. *chrysandrum*. Plants possessing very short stigma lobes and lacking auricles on the tepals have been recognized as var. *nudipetalum*, but they do not appear sufficiently distinct to warrant taxonomic recognition and may simply represent depauperate forms.

1. Tepals white to creamy white, with yellow zone at base; anthers cream to yellow . . . . 1a. *Erythronium grandiflorum* subsp. *candidum*
1. Tepals bright yellow, with narrow paler zone at base; anthers cream, yellow, red, or purplish red . . . . 1b. *Erythronium grandiflorum* subsp. *grandiflorum*

1a. Erythronium grandiflorum Pursh subsp. **candidum** Piper in C. V. Piper and R. K. Beattie, Fl. S.E. Washington, 61. 1914 [E]

E. grandiflorum var. *candidum* (Piper) Abrams; *E. grandiflorum* var. *idahoense* (H. St. John & G. N. Jones) R. J. Davis; *E. idahoense* H. St. John & G. N. Jones

Flowers: tepals white to creamy white, with yellow zone at base; anthers cream to yellow.

Flowering late spring (Apr–Jun). Grasslands and forest openings; 500–1500 m; Idaho, Mont., Wash.

1b. Erythronium grandiflorum Pursh subsp. **grandiflorum** [E]

Erythronium giganteum Lindley; *E. grandiflorum* var. *chrysandrum* (Piper) Abrams; *E. grandiflorum* var. *nudipetalum* (Applegate) C. L. Hitchcock; *E. grandiflorum* var. *pallidum* H. St. John; *E. nudipetalum* Applegate (as nudopetalum); *E. parviflorum* (S. Watson) Goodding

Flowers: tepals bright yellow, with narrow pale zone at base; anthers cream, yellow, red, or purplish red.

Flowering spring–summer (Apr–Jul, depending on elevation). Widespread from lowlands to the subalpine, in open woodlands, grasslands, and subalpine meadows (where it is sometimes extremely abundant); 200–3100 m; Alta., B.C.; Calif., Colo., Idaho, Mont., N.Mex., Oreg., Utah, Wash., Wyo.

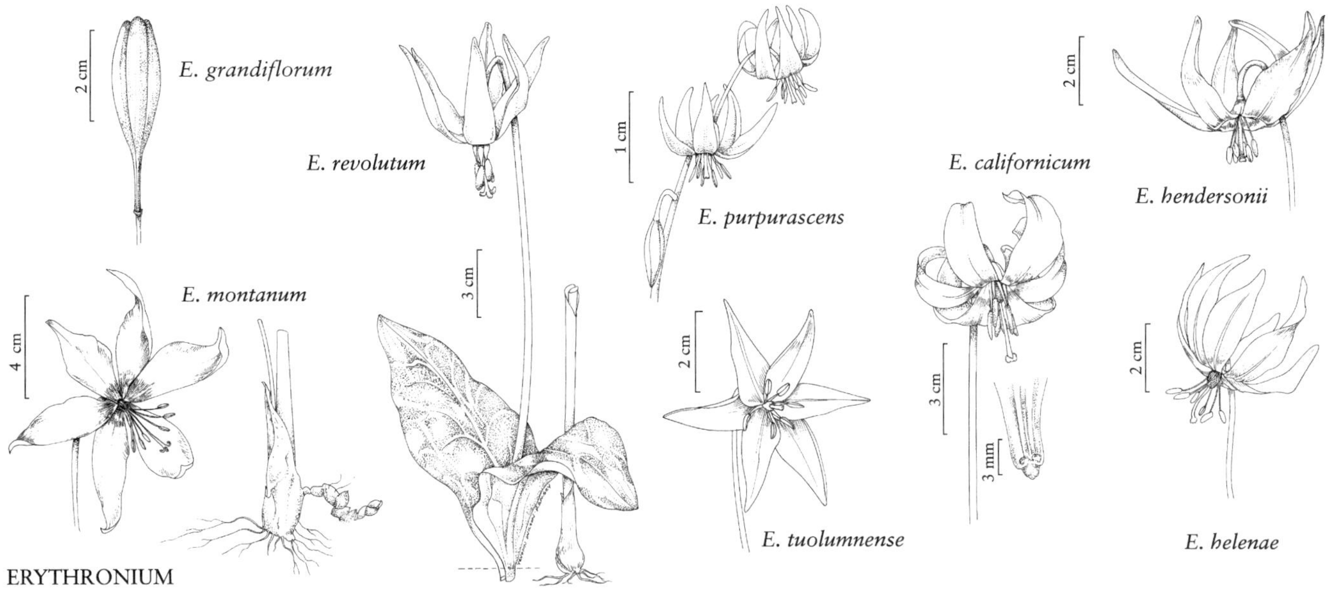

ERYTHRONIUM

2. Erythronium montanum S. Watson, Proc. Amer. Acad. Arts 26: 130. 1891 • Avalanche-lily [E] [F]

Bulbs narrowly ovoid, 25–60 mm. **Leaves** 10–20 cm; blade green, ovate to broadly lanceolate, base ± abruptly narrowed to petiole, margins wavy. **Scape** 12–35 cm. **Inflorescences** 1–3-flowered. **Flowers:** tepals white to creamy white with bright yellow zone at base, broadly ovate to broadly lanceolate, 25–45 mm, inner wider than outer, auriculate at base, length less than 4 times width; stamens 12–24 mm; filaments white, linear, slender, less than 0.8 mm wide; anthers bright yellow; style white, 13–25 mm; stigma with slender, usually recurved lobes 1–5 mm. **Capsules** oblong, 3–6 cm. $2n = 24$.

Flowering summer, usually soon after snowmelt (Jun–Aug). Montane and subalpine meadows, open coniferous forests; (300–)800–2000 m; B.C.; Oreg., Wash.

This species occurs in the Coast Ranges of southern British Columbia, and disjunctly to southern Vancouver Island, the Olympic Peninsula, and Cascade Mountains from Mount Rainier National Park in Washington to central Oregon.

3. Erythronium elegans P. C. Hammond & K. L. Chambers, Madroño 32: 49, fig. 1. 1985 • Elegant fawn-lily [C] [E]

Bulbs slender, 30–50 mm. **Leaves** 7–20 cm; blade green or faintly mottled with brown or white, narrowly ovate, margins often wavy. **Scape** 10–30 cm. **Inflorescences** 1–2(–4)-flowered. **Flowers:** tepals: inner ± white, outer ± white and tinged (often strongly) with pink, especially abaxially and along midline, becoming more generally pinkish with age, both inner and outer with yellow band at base, lanceolate to narrowly elliptic, 20–40 mm, abaxial surfaces and outer tepals often darker, inner auriculate at base; stamens 13–22 mm; filaments white, flattened, slightly widened, linear to lanceolate, 0.8–2 mm wide; anthers yellow; style white, 10–20 mm; stigma with slender, usually recurved lobes 2–4 mm. **Capsules** obovoid to oblong, 2–5 cm. $2n = 48$.

Flowering late spring (May–Jun). Meadows and open coniferous forests; of conservation concern; 800–1000 m; Oreg.

This species is endemic to the Coast Ranges of western Oregon.

4. Erythronium quinaultense G. A. Allen, Syst. Bot. 26: 269, fig. 3. 2001 • Quinault fawn-lily [E]

Bulbs narrowly ovoid, 35–75 mm. **Leaves** 12–20 cm; blade green or faintly mottled with white or brown, lanceolate to ovate, margins ± wavy. **Scape** 12–25 cm. **Inflorescences** 1–3-flowered. **Flowers:** tepals white proximally, shading to pink at outer margins, darkest toward tips, with yellow band at base, lanceolate to narrowly ovate, 30–50 mm, inner with small auricles at base; stamens 12–24 mm; filaments white, flattened, slightly widened, linear to lanceolate, 1–2 mm wide; anthers yellow; style white, 10–18 mm; stigma with slender, usually recurved lobes 1–5 mm. **Capsules** oblong to obovoid, 3–6 cm. $2n = 48$.

Flowering late spring (May). Openings and rocky ledges in coniferous forests; 500–900 m; Wash.

Erythronium quinaultense is a tetraploid species apparently derived from hybridization between *E. montanum* and *E. revolutum*. It is known only from the southwestern Olympic Peninsula.

5. Erythronium revolutum Smith in A. Rees, Cycl. 13: Erythronium no. 3. 1809 • Pink fawn-lily [E] [F]

Bulbs narrowly ovoid, 35–50 mm, sometimes producing sessile offsets. **Leaves** 10–25 mm; blade distinctly mottled with irregular streaks of brown or white, broadly lanceolate to ovate, margins entire to ± wavy. **Scape** 15–40 cm. **Inflorescences** 1–3-flowered. **Flowers:** tepals uniformly clear violet-pink at anthesis, with yellow banding at base, lanceolate to narrowly elliptic, 25–40 mm, inner with small auricles at base; stamens ± appressed to style, 12–22 mm; filaments white to pink (darkening with age), flattened, ± lanceolate, 2–3 mm wide; anthers bright yellow; style white to pink, 12–18 mm; stigma with slender recurved lobes 4–6 mm. **Capsules** oblong to obovoid, 3–6 cm.

Flowering early spring (Mar–Apr). Shaded stream banks, river terraces, wet places in forests; 0–600(–1000) m; B.C.; Calif., Oreg., Wash., generally within 100 km of Pacific Coast.

6. Erythronium oregonum Applegate, Madroño 3: 99. 1935 • Oregon fawn-lily [E]

Erythronium giganteum Lindley subsp. *leucandrum* Applegate; *E. oregonum* subsp. *leucandrum* (Applegate) Applegate

Bulbs narrowly ovoid, 25–60 mm, sometimes producing sessile offsets. **Leaves** 12–25 cm; blade distinctly mottled with irregular streaks of brown or white, ovate to broadly lanceolate, margins wavy. **Scape** ± reddish, 15–40 cm. **Inflorescences** 1–3-flowered. **Flowers:** tepals white to creamy white with yellow base at anthesis, sometimes pinkish in age, sometimes with red lines or bands, elliptic to narrowly ovate, 25–40 mm, inner with small auricles at base; stamens 12–25 mm; filaments white, flattened, ± lanceolate, 2–3 mm wide; anthers cream to yellow; style white, 12–18 mm; stigma with recurved lobes 3–6 mm. **Capsules** oblong to narrowly obovoid, 3–5 cm. $2n = 24$.

Flowering spring (Mar–May). Open coniferous forests, rocky outcrops, oak woodlands, meadows; 0–500 m; B.C.; Calif., Oreg., Wash.

Forms from the southern part of the range with cream-white tepals and pale anthers have been described as subsp. *leucandrum*. This species is closely related to *E. revolutum* and occasionally hybridizes with it where their ranges meet. In addition, *E. citrinum* and *E. hendersonii* are reported to hybridize with *E. oregonum* in the southern part of its range.

7. Erythronium klamathense Applegate, Contr. Dudley Herb. 1: 151. 1930 • Klamath fawn-lily [E]

Bulbs slender, 25–40 mm. **Leaves** 6–17 cm; blade green, lanceolate to narrowly elliptic, ± folded along midvein, margins entire to wavy. **Scape** 6–20 cm. **Inflorescences** 1–3-flowered. **Flowers:** tepals $^{2}/_{3}$ or more white, with yellow zone at base, ± pinkish in age, broadly lanceolate, 20–35 mm, inner with small auricles at base; stamens 8–14 mm; filaments white, slender; anthers ± yellow; style white, 4–9 mm; stigma ± unlobed. **Capsules** narrowly obovoid, 2–5 cm. $2n = 24$.

Flowering late spring to summer (Apr–Jun). Montane meadows, openings in coniferous forests; 1200–1900 m; Calif., Oreg.

8. Erythronium purpurascens S. Watson, Proc. Amer. Acad. Arts 12: 277. 1877 • Sierra Nevada fawn-lily E F

Bulbs slender, 25–40 mm. **Leaves** 6–15 cm; blade green, lanceolate to narrowly ovate, margins ± wavy. **Scape** 7–20 cm. **Inflorescences** 1–6-flowered. **Flowers:** tepals white, bright yellow on proximal $^1/_3$, pinkish purple in age, lanceolate, 10–20 mm, not auriculate at base; stamens 8–12 mm; filaments yellow, slender; anthers cream to yellow; style yellow, 4–5 mm; stigma ± unlobed. **Capsules** obovoid, 2–4 cm.

Flowering summer, soon after snowmelt (Jun–Aug). Open coniferous forests, meadows, rocky places; 1500–2700 m; Calif.

9. Erythronium pusaterii (Munz & J. T. Howell) Shevock, Bartel & G. A. Allen, Madroño 37: 264. 1991 • Kaweah Lakes fawn-lily C E

Erythronium grandiflorum Pursh subsp. *pusaterii* Munz & J. T. Howell, Leafl. W. Bot. 10: 104. 1964

Bulbs narrowly ovoid, 40–60 mm. **Leaves** 10–35 cm; blade green, lanceolate, margins ± wavy. **Scape** 12–40 cm. **Inflorescences** 1–8-flowered. **Flowers:** tepals white, proximal $^1/_2$–$^2/_3$ bright yellow, becoming pinkish in age, lanceolate, 25–45 mm, inner auriculate at base; stamens 8–15 mm; filaments ± white, slender; anthers yellow; style ± white, 7–10 mm; stigma ± unlobed, or with lobes shorter than 1 mm. **Capsules** obovoid, 2–4 cm.

Flowering late spring (Apr–May). Meadows, open forests, rocky ledges; of conservation concern; 2100–2500 m; Calif.

Erythronium pusaterii is known only from Tulare County in the southern Sierra Nevada.

10. Erythronium taylorii Shevock & G. A. Allen, Madroño 44: 360, fig. 1. 1998 • Taylor's fawn-lily E

Bulbs narrowly ovoid, 40–70 mm, often forming sessile offsets. **Leaves** 18–35 cm; blade green, elliptic to oblanceolate, margins wavy. **Scape** 25–40 cm. **Inflorescences** 1–4(–8)-flowered. **Flowers:** tepals white, proximal $^1/_2$–$^2/_3$ bright yellow, becoming pinkish in age, lanceolate, 25–45 mm, inner auriculate at base; stamens 10–16 mm; filaments yellow, slender; anthers cream colored; style white to cream, 9–11 mm; stigma ± unlobed or with lobes shorter than 1 mm. **Capsules** obovoid, 2–4 cm.

Flowering spring (Apr–May). Forest openings, rocky ledges; 1300–1400 m; Calif.

Erythronium taylorii is known only from Tuolumne County in the central Sierra Nevada.

11. Erythronium pluriflorum Shevock, Bartel & G. A. Allen, Madroño 37: 268, fig. 3. 1991 • Golden fawn-lily C E

Bulbs ± ovoid, 40–75 mm. **Leaves** 7–30 cm; blade green, oblanceolate to elliptic, margins ± wavy. **Scape** 8–35 cm. **Inflorescences** 1–10-flowered. **Flowers:** tepals yellow, bronze in age, lanceolate, 15–28 mm, not auriculate at base; stamens 8–12 mm; filaments yellow, slender; anthers yellow; style yellow, 6–8 mm; stigma unlobed or with very short, rounded lobes shorter than 1 mm. **Capsules** obovoid, 2–4 cm.

Flowering summer (Jun–Jul). Open montane coniferous forests; of conservation concern; 2300–2600 m; Calif.

Erythronium pluriflorum is known only from Madera County.

12. Erythronium tuolumnense Applegate, Contr. Dudley Herb. 1: 153. 1930 • Tuolumne fawn-lily C E F

Bulbs ovoid, 50–100 mm, readily forming sessile offsets. **Leaves** 15–35 cm; blade green, elliptic to ovate, margins entire to ± wavy. **Scape** ± reddish, 15–35 cm. **Inflorescences** usually 1–4-flowered. **Flowers:** tepals yellow, narrowly ovate, 20–35 mm, inner with small auricles at base; stamens 12–16 mm; filaments white to creamy white, slightly widened, 0.4–0.6 mm wide at base; anthers yellow; style ± white, 8–10 mm; stigma ± unlobed or with lobes shorter than 1 mm. **Capsules** obovoid, 2–4 cm.

Flowering early spring (Mar). Open woods, shady canyons; of conservation concern; 600–1000 m; Calif.

13. Erythronium californicum Purdy, Fl. & Sylva 2: 253. 1904 • California fawn-lily E F

Bulbs ovoid, 35–60 mm, sometimes producing sessile offsets. **Leaves** 7–18 cm; blade distinctly mottled with irregular streaks of brown or white, oblong to narrowly ovate, margins usually wavy. **Scape** ± reddish, branched well above leaves when flowers more than 1, 10–30 cm. **Inflorescences** 1–3-flowered. **Flowers:** flowering individuals generally abundant in populations; tepals white to cream, base yellow and often banded with brown or red, ± narrowly ovate, 25–40 mm, inner with small auricles at base; stamens 12–25 mm; filaments ± white, linear, slender, less than 0.8 mm wide; anthers white to cream; style ± white, 10–14 mm; stigma unlobed or with lobes less than 2 mm. **Capsules** obovoid, 3–6 cm. $2n = 24$.

Flowering spring (Mar–Apr). Dry woods, openings, cliffs; 0–1900 m; Calif.

Forms intermediate with *Erythronium citrinum* and *E. multiscapideum* are sometimes seen where ranges overlap.

14. Erythronium citrinum S. Watson, Proc. Amer. Acad. Arts 22: 480. 1887 • Pale fawn-lily E

Erythronium citrinum var. *roderickii* Shevock & G. A. Allen; *E. howellii* S. Watson

Bulbs slender, 40–50 mm, sometimes producing sessile offsets. **Leaves** 9–15 cm; blade mottled with irregular streaks of brown or white, lanceolate to narrowly ovate, margins ± wavy. **Scape** 12–35 cm. **Inflorescences** 1–3-flowered. **Flowers:** tepals ± white, often pinkish, usually pale yellow at base, lanceolate to narrowly elliptic, 25–45 mm, inner with or without small auricles at base; stamens 11–17 mm; filaments linear, white or pinkish, slender, less than 0.8 mm wide; anthers white, cream, pink, reddish, or brownish red; style straight, white or pink, 6–10 mm; stigma unlobed or with lobes shorter than 1 mm. **Capsules** obovoid, 2–5 cm. $2n = 24$.

Flowering spring (Mar–May). Dry woods, brushy slopes, common on serpentine substrates; 100–1300 (–1800) m; Calif., Oreg.

Plants lacking auricles on inner tepals are sometimes segregated as *Erythronium howellii*, Howell's fawn-lily, but they do not appear to differ from typical *E. citrinum* in any other characters. *Erythronium citrinum* intergrades with *E. californicum* and *E. hendersonii*, occasional populations or individuals displaying intermediate or recombined characteristics. Such plants from the upper Scott River drainage in Trinity County, California, which may be the result of introgression with *E. hendersonii*, have been recognized as variety *roderickii*.

15. Erythronium hendersonii S. Watson, Proc. Amer. Acad. Arts 22: 479. 1887 (as hendersoni) • Henderson's fawn-lily E F

Bulbs slender, 40–55 mm. **Leaves** 10–25 cm; blade mottled with irregular streaks of brown or white, oblong to ovate, margins entire to ± wavy. **Scape** reddish, 12–30 cm. **Inflorescences** 1–4-flowered. **Flowers:** tepals violet to pink, darker at tips, dark purple at base, broadly lanceolate, 18–35 mm, inner with small auricles at base; stamens 10–14 mm; filaments violet-purple, linear, slender, less than 0.8 mm wide; anthers pale brown to purple; style violet, 6–8 mm; stigma unlobed or with lobes shorter than 1 mm. **Capsules** obovoid, 2–4 cm. $2n = 24$.

Flowering spring (Mar–Apr). Dry woods and forest openings; 300–1600 m; Calif., Oreg.

16. Erythronium helenae Applegate, Contr. Dudley Herb. 1: 188. 1933 • Mount St. Helena fawn-lily E F

Bulbs ovoid, 30–55 mm, sometimes producing sessile bulbels. **Leaves** 7–20 cm; blade mottled with irregular streaks of brown or white, broadly lanceolate to ovate, margins ± wavy. **Scape** 12–30 cm. **Inflorescences** 1–3-flowered. **Flowers** fragrant; tepals ± white, bright yellow at base, pinkish in age, lanceolate to ovate, 25–40 mm, inner with small auricles at base; stamens 8–13 mm; filaments ± yellow, linear, ± slender, less than 0.8 mm wide; anthers yellow; style ± white, often bent to one side, 5–8 mm; stigma unlobed or with lobes shorter than 1 mm. **Capsules** obovoid, 2–4 cm. $2n = 24$.

Flowering spring (Mar–Apr). Dry woods or scrub, on serpentines; 500–1200 m; Calif. (vicinity of Mount St. Helena).

17. Erythronium multiscapideum (Kellogg) A. Nelson & Kennedy, Muhlenbergia 3: 137. 1908 (as multiscapoidea) • Sierra foothills fawn-lily [E]

Fritillaria multiscapidea Kellogg, Proc. Calif. Acad. Sci. 1: 46. 1855

Bulbs ovoid, 20–50 mm, producing bulbels (usually 1–3 per parent bulb) at ends of long, slender stolons. **Leaves** 4–16 cm; blade mottled with irregular streaks of brown or white, ± lanceolate, margins entire to wavy. **Scape** 8–23 cm, branching just above leaves near ground level when flowers more than 1. **Inflorescences** 1–4-flowered. **Flowers:** flowering individuals generally uncommon in populations, most plants 1-leaved and vegetative; tepals white to cream with yellow base, broadly lanceolate to elliptic, 16–40 mm, inner with small auricles at base; stamens 10–15 mm; filaments white, linear, slender, less than 0.8 mm wide; anthers white to cream; style white, 10–13 mm; stigma unlobed or with recurved lobes 1–4 mm. **Capsules** obovoid, 2–5 cm. $2n = 24$.

Flowering spring (Mar–Apr). Open woods, brushy slopes, sometimes on serpentines; 400–1000 m; Calif.

Erythronium multiscapideum is unusual among western species (and resembles some eastern species) in its tendency to reproduce vegetatively through the production of bulbels at the ends of stolons. It is similar in many respects to *E. californicum* and sometimes intergrades with it, resulting in occasional populations with the bulb characteristics of one species and the inflorescence branching pattern of the other.

18. Erythronium americanum Ker Gawler, Bot. Mag. 28: plate 1113. 1808 • Yellow trout-lily, érythrone d'Amérique [E] [F]

Bulbs ovoid, 15–28 mm; stolons 1–3, common, mostly on 1-leaved, nonflowering plants. **Leaves** 8–23 cm; blade green, irregularly mottled, elliptic-lanceolate to ovate or elliptic, ± flat, glaucous, margins entire. **Scape** 10–18 cm. **Inflorescences** 1-flowered. **Flowers:** tepals yellow, sometimes tinged light to dark purple-red abaxially, sometimes with reddish dots adaxially, strongly reflexed at anthesis, lanceolate, 20–33 mm, inner with small auricles; stamens 9–15 mm; filaments yellow, lanceolate; anthers yellow, chestnut brown, or lavender; pollen yellow or brown; style deciduous or base forming small apiculum, greenish yellow, 5–11 mm, swollen distally or ± terete; stigma lobes erect or recurved, 1.5 mm. **Capsules** held erect or at least off ground at maturity, obovoid, 12–15 mm, apex rounded, truncate, or apiculate. $2n = 48$.

Varieties 2 (2 in the flora): e North America.

Erythronium americanum is a very common and widespread species, particularly in northeastern North America, becoming less frequent towards the southern and western limits of its range. Nonflowering plants far outnumber flowering ones in most populations because of their extensive stolon production. Plants with brown anthers have been called forma *castaneum* L. B. Smith.

1. Capsule apex rounded, truncate, or short-apiculate; stigma lobes erect, not grooved; widespread in ne United States . . . . 19a. *Erythronium americanum* subsp. *americanum*
1. Capsule apex distinctly apiculate; stigma lobes recurved, grooved distally; n Alabama, Georgia, ne Mississippi, Tennessee . . . . 19b. *Erythronium americanum* subsp. *harperi*

18a. Erythronium americanum Ker Gawler subsp. **americanum** [E] [F]

Style swollen distally; stigma lobes erect, not grooved. **Capsules:** apex rounded, truncate, or short-apiculate. $2n = 48$.

Flowering spring. Usually deciduous wet-mesic to dry-mesic woods of bottomlands and uplands, slopes of ravines, lower ledges of bluffs, often near waterways, often in loamy soils; 0–900 m; N.B., Nfld. and Labr., N.S., Ont., Que.; Conn., Del., Ga., Ill., Ind., Iowa, Ky., Maine, Md., Mass., Mich., Minn., Mo., N.H., N.J., N.Y., N.C., Ohio, Pa., R.I., S.C., Tenn., Vt., Va., W.Va., Wis.

18b. Erythronium americanum Ker Gawler subsp. **harperi** (W. Wolf) C. R. Parks & Hardin, Brittonia 15: 252. 1963 • Harper's trout-lily [E]

Erythronium harperi W. Wolf, Castanea 6: 24. 1941

Style ± terete or slightly swollen distally; stigma lobes recurved, distinctly grooved distally. **Capsules:** apex distinctly apiculate. $2n = 48$.

Flowering spring. Wooded slopes, ravines; 0–300 m; Ala., Ga., Miss., Tenn.

C. R. Parks and J. W. Hardin (1963) indicated that subsp. *harperi* may represent merely a southern race of *E. americanum*, but that the transition seems to be more abrupt than expected in a cline.

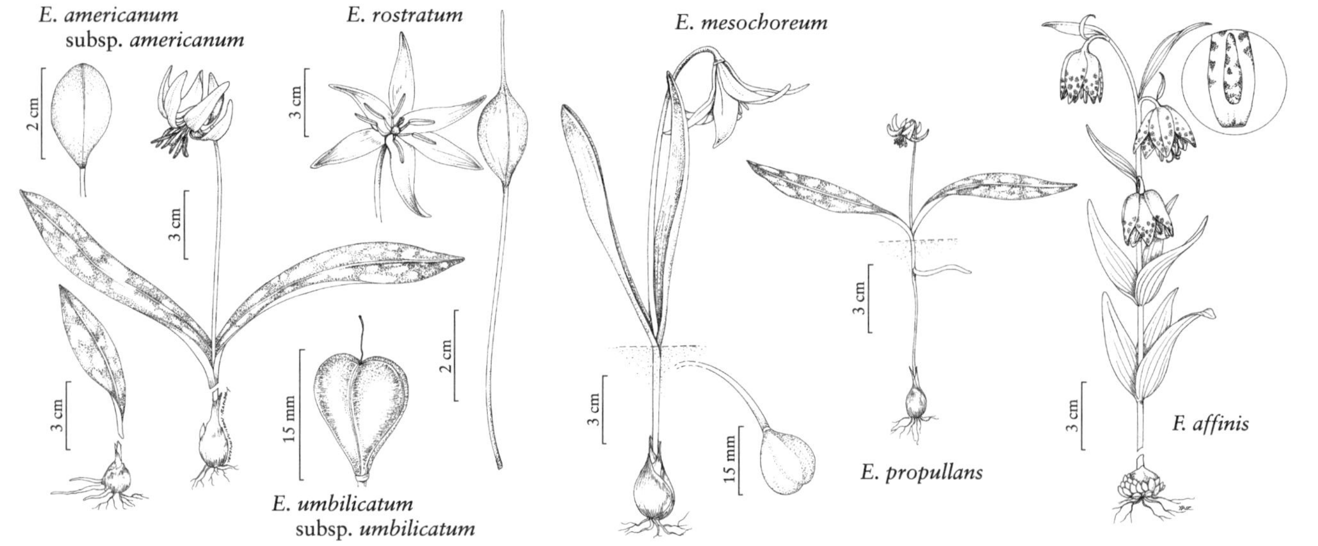

ERYTHRONIUM ° FRITILLARIA

19. Erythronium rostratum W. Wolf, Castanea 6: 24, plate 1. 1941 • Beaked trout-lily [E] [F]

Bulbs ovoid, 10–20 mm; stolons 1–3, common, mostly on 1-leaved, nonflowering plants. **Leaves** 5–18 cm; blade green, irregularly mottled, elliptic-lanceolate to ovate or elliptic, ± flat, not glaucous, margins entire. **Scape** 3–10 cm. **Inflorescences** 1-flowered. **Flowers** held erect at anthesis; tepals yellow, tinged red-purple or orange abaxially, mostly spreading at anthesis, lanceolate, 20–34 mm, inner with conspicuous, well-developed auricles almost encircling opposing filaments; stamens 13–17 mm; filaments yellow, lanceolate; anthers yellow; pollen yellow; style persistent, greenish yellow, swollen distally into stigma, tapering gradually proximally to ovary, forming beak on capsule, 8–11 mm; stigma lobes erect, short, 1 mm. **Capsules** held erect at maturity on upward-curving peduncle, ellipsoid to ellipsoid-obovoid, 15–25 mm (excluding beak), apex long-beaked. $2n = 24$.

Flowering spring. Mesic woods, often in flood plains and along waterways, also on shaded lower ledges of bluffs; 0–500 m; Ala., Ark., Kans., Ky., La., Miss., Mo., Ohio, Okla., Tenn., Tex.

Erythronium rostratum is found mostly on either side of the Mississippi River embayment, but rarely within it. It often forms extensive colonies where nonflowering, 1-leaved plants far outnumber flowering ones. This is the only species of *Erythronium* with erect rather than nodding flowers. Some plants at the western limits of the range are triploids (B. L. Carr 1986).

20. Erythronium umbilicatum C. R. Parks & Hardin, Brittonia 15: 252. 1963 • Dimpled trout-lily [E] [F]

Bulbs ovoid, 10–25 mm; stolons absent, or 1 per bulb on 1-leaved plants. **Leaves** 5–17 cm; blade green, irregularly mottled, elliptic to ovate-lanceolate, ± flat, not glaucous, margins entire. **Scape** 4–18 cm. **Inflorescences** 1-flowered. **Flowers:** tepals strongly reflexed at anthesis, yellow, sometimes with brownish adaxial spots, variously tinged brown-purple abaxially, lanceolate, 13–30 mm, auricles absent; stamens 9–18 mm; filaments yellow, lanceolate; anthers brown, purple, or infrequently yellow; pollen brown, purple, or infrequently yellow; ovary apex indented; style deciduous or forming small apiculum, ± terete, not yellow, 8–24 mm; stigma lobes spreading, 1.2–1.7 mm. **Capsules** ± resting on ground due to reclining peduncle, obovoid, 10–22 mm, apex indented, umbilicate, or rarely rounded. $2n = 24$.

Varieties 2 (2 in the flora): s and e United States.

1. Stolons absent; West Virginia to Alabama, Georgia, and n Florida . . . .
   . . . . 20a. *Erythronium umbilicatum* subsp. *umbilicatum*
1. Stolons present, 1 per bulb; North Carolina, Tennessee . . . .
   . . . . 20b. *Erythronium umbilicatum* subsp. *monostolum*

20a. Erythronium umbilicatum C. R. Parks & Hardin subsp. **umbilicatum** [E] [F]

Stolons absent. $2n = 24$.

Flowering spring. Woods of flood plains and adjacent slopes along creeks, streams, small rivers; 0–900 m; Ala., Fla., Ga., Ky., Md., N.C., S.C., Tenn., Va., W.Va.

Since subsp. *umbilicatum* does not produce stolons, most plants in a population produce flowers, and seedlings may be frequent. Although many *Erythronium* species have considerable variation in size at flowering, *E. umbilicatum* subsp. *umbilicatum* is especially variable in this regard.

20b. Erythronium umbilicatum C. R. Parks & Hardin subsp. **monostolum** C. R. Parks & Hardin, Brittonia 15: 254. 1963 [E]

Stolon 1 per bulb. $2n = 24$.

Flowering spring. Rich woods and slopes of high coves under either hardwoods or conifers, sometimes in open, grassy areas on peaks; 1400–1600 m; N.C., Tenn.

Within populations, nonflowering, 1-leaved plants far outnumber flowering, 2-leaved plants. Some populations form a nearly continuous ground cover.

21. Erythronium albidum Nuttall, Gen. N. Amer. Pl. 1: 223. 1818 • White trout-lily [E]

Bulbs ovoid, 15–30 mm; stolons 1–3, mostly on 1-leaved, nonflowering plants; flowering plants reproducing vegetatively by offshoots or droppers. **Leaves** 8–22 cm; blade green, irregularly mottled, elliptic-lanceolate to ovate-lanceolate or elliptic, ± flat, glaucous, margins entire. **Scape** 7–20 cm. **Inflorescences** 1-flowered. **Flowers:** tepals strongly reflexed at anthesis, white, tinged pink, blue, or lavender abaxially, with yellow adaxial spot at base, lanceolate, 22–40 mm, auricles absent; stamens 10–20 mm; filaments yellow, lanceolate; anthers yellow; pollen yellow; style white, 15–25 mm; stigma lobes recurving, 1.5 mm. **Capsules** held erect at maturity, obovoid, 10–22 mm, apex rounded to faintly apiculate or umbilicate. $2n = 44$.

Flowering spring. Mesic bottomlands, upland forests, woodlands, clay and silt bottomlands, floodplain forests; 0–300 m; Ont.; Ala., Ark., Ga., Ill., Ind., Iowa, Kans., Ky., La., Md., Mich., Minn., Miss., Mo., Nebr., N.J., N.Y., Ohio, Okla., Pa., S.Dak., Tenn., Tex., Va., W.Va., Wis.

Erythronium albidum often forms extensive colonies in which nonflowering, 1-leaved plants far outnumber flowering, 2-leaved ones. It is very widespread in eastern North America, more common in the central states than *E. americanum* and often occurs in slightly drier sites.

22. Erythronium mesochoreum Knerr, Midland College Monthly 2: 5. 1891 (as "messachoreum," corrected by Knerr, Trans. Kansas Acad. Sci. 13: 20. 1891) • Prairie trout-lily [E] [F]

Erythronium albidum Nuttall var. *coloratum* Sterns; *E. albidum* var. *mesochoreum* (Knerr) Rickett

Bulbs ovoid to ± globose, 10–25 mm; stolons absent; flowering plants reproducing vegetatively by droppers or offshoots. **Leaves** 5–14 cm; blade green, usually not mottled, elliptic-lanceolate to ovate-lanceolate, conduplicate, glaucous, base and margins sometimes purple-brown, margins even. **Scape** 5–15 cm. **Inflorescences** 1-flowered. **Flowers:** tepals spreading at anthesis, white, tinged pink, blue, or lavender abaxially, with yellow adaxial spot at base, lanceolate, 15–30 mm, auricles absent; stamens 8–15 mm; filaments yellow, lanceolate; anthers yellow; pollen yellow; style white, 7–10 mm; stigma lobes recurving, 1–1.5 mm. **Capsules** resting on ground at maturity due to reclining peduncle, obovoid, 10–15 mm, apex rounded to faintly apiculate or umbilicate. $2n = 22$.

Flowering spring. Prairies, glades, dry, open woods, occasionally pastures; 100–700 m; Ark., Ill., Iowa, Kans., Mo., Nebr., Okla., Tex.

Because stolons are absent in *Erythronium mesochoreum*, most plants in a population produce flowers; nonflowering plants may sometimes appear after flowering plants have flowered. This species is well adapted to prairie fires. The seeds have elaiosomes, and ants may act as dispersal agents (Great Plains Flora Association 1986). C. C. Deam et al. (1941) reported *E. mesochoreum* from Indiana, but an examination of the specimens cited, now at IND, showed that these are *E. albidum*. E. L. Braun (1967) reported that many populations of *E. albidum* in southwestern Ohio indicate introgression with "the Ozarkean var. *mesochoreum*...." However, personal observation by K. R. Robertson of several Ohio populations in flower confirmed that these plants are true *E. albidum*, although they occur in prairies and have more or less conduplicate leaves.

23. Erythronium propullans A. Gray, Amer. Naturalist 5: 300, fig. 74. 1871 • Minnesota dwarf trout-lily C E F

Bulbs ovoid, 10–25 mm; stolon 1 in flowering plants, arising from halfway up stem, 1–3 from bulbs of 1-leaved, nonflowering plants. **Leaves** 4–13 cm; blade green, irregularly mottled, elliptic-lanceolate to ovate-lanceolate or elliptic, ± flat, glaucous, margins entire. **Scape** 3.9–12 cm. **Inflorescences** 1-flowered. **Flowers:** tepals 4–6, strongly reflexed at anthesis, pale pink to white, darker abaxially, lanceolate, 8–15 mm, auricles absent; stamens 2–6, 6–8 mm; filaments white, lanceolate; anthers yellow; pollen yellow; style white, 6–10 mm; stigma ± unlobed. **Capsules** very rarely produced; when present, may be result of hybridization with *Erythronium albidum*.

Flowering spring. Mesic floodplain woods; of conservation concern; 300 m; Minn.

Erythronium propullans is known only from Goodhue and Rice counties. It often forms extensive colonies in which flowering plants are sometimes more abundant than nonflowering, 1-leaved ones, and sometimes the reverse. It grows mixed with *E. albidum* (J. A. Banks 1980), and putative hybrids between them have been reported (T. Morley 1988). Flowers frequently have fewer than six tepals and stamens (C. O. Rosendahl 1919), and may occasionally have only two carpels.

25. FRITILLARIA Linnaeus, Sp. Pl. 1: 303. 1753; Gen. Pl. ed. 5, 144. 1754 • Fritillary [Latin, *fritillus*, checkered, alluding to the markings on the tepals of many species]

Bryan Ness

Herbs, perennial, bulbose; bulbs with 1–several large fleshy scales and 0–many small scales (often called rice-grain bulblets). **Stem** 1, erect, simple, absent in nonflowering individuals. **Leaves** alternate or whorled proximally in some species, sessile; blade linear to ± ovate; nonflowering individuals with single elliptical, ovate, or obovate "bulb-leaf." **Inflorescences** loosely racemose, bracteate; bracts leaflike. **Flowers** 1–many, usually nodding, 3-merous; perianth hypogynous, campanulate or cupulate; tepals 6, in 2 similar whorls, distinct, nectaries present on all tepals, but better developed on inner ones; stamens 6, included; anthers adnate to filaments near middle; ovary superior, ± sessile; style unbranched or 3-branched. **Fruits** capsular, 3-locular, 6-angled or -winged, thin-walled, ± rounded, dehiscence loculicidal. **Seeds** many, in 2 rows per locule, yellowish to brownish, flat. x = 12, 13, often with chromosome fragments.

Species ca. 100 (20 in the flora): Northern Hemisphere.

A number of *Fritillaria* species are grown as ornamentals, although North American species can be difficult to grow and many have not been tried. Most require well-drained soil, full sun, and no summer watering. Native Americans used the bulbs of various species as food, typically roasted, sometimes dried for later use.

SELECTED REFERENCES Beck, C. H. 1951. Fritillaries: A Gardener's Introduction to the Genus *Fritillaria*. London. Beetle, D. F. 1944. A monograph of the North American species of *Fritillaria*. Madroño 7: 133–159. Farrens, B. M. 1947. A Taxonomic Study of the North American Species of *Fritillaria*. M.A. thesis. Stanford University. Marchant, C. J. 1981. *Fritillaria* in British Columbia. Davidsonia 12: 19–25. Rix, E. M. and D. Rast. 1975. Nectar sugars and subgeneric classification in *Fritillaria*. Biochem. Syst. & Ecol. 2: 207–209. Santana, D. O. 1984. Morphological and Anatomical Observations on the Bulbs and Tepals of *Fritillaria* (Liliaceae) Section *Liliorhiza* (Kellogg) Watson, Their Taxonomic Implications with a Synopsis and the Reproductive Biology of the Section. Ph.D. dissertation. University of California, Davis. Stapf, O. 1934. *Lilium, Notholirion* and *Fritillaria*. Bull. Misc. Inform. Kew 1934: 94–96. Turner, N. J. and H. V. Kuhnlein. 1983. Camas (*Camassia* spp.) and riceroot (*Fritillaria* spp.): Two liliaceous "root" foods of the Northwest Coast Indians. Ecol. Food Nutr. 13: 199–219. Turrill, W. B. 1950. Character combinations and distribution in the genus *Fritillaria* and allied genera. Evolution 4: 1–6. Turrill, W. B. and J. R. Sealy. 1980. Studies in the genus *Fritillaria* (Liliaceae). Hooker's Icon. Pl. 39.

1. Style unbranched or rarely with branches shorter than 1.5 mm; tepals never scarlet.
   2. Tepals shorter than 2 cm or, if longer, then yellow to orange.
      3. Leaves whorled proximally; tepals pinkish to purplish; s California . . . . 5. *Fritillaria brandegeei*

3. Leaves subopposite to scattered; tepals yellow to orange; n California to British Columbia, e to Wyoming . . . . 16. *Fritillaria pudica*
2. Tepals 2 cm or longer, white to pink or pinkish purple.
4. Tepals pinkish purple, obovate, apex rounded to acute, not recurved; flowers not noticeably fragrant; n California . . . . 15. *Fritillaria pluriflora*
4. Tepals white to pink, oblanceolate, apex acute to apiculate, usually recurved; flowers fragrant; s California . . . . 19. *Fritillaria striata*

1. Style obviously branched, branches longer than 1.5 mm; tepals sometimes scarlet.
5. Tepals red or scarlet, sometimes to maroon or purplish, clearly checkered or mottled, apex usually recurved, sometimes only spreading.
6. Perianth slender; tepal apex usually strongly recurved, nectaries ¼ tepal length or less; style branches ± erect; n California, s Oregon, w Nevada . . . . 18. *Fritillaria recurva*
6. Perianth broadly campanulate; tepal apex spreading, not recurved, nectaries ½ tepal length; style branches widely spreading; s Oregon . . . . 9. *Fritillaria gentneri*
5. Tepals sometimes mottled or rarely scarlet, but never both, apex not recurved, only rarely slightly recurved.
7. Leaves more than 10 or in whorls of 2–6(–9) per node proximally, blade linear to narrowly to broadly lanceolate to rarely ovate.
8. Flowers ± erect, occasionally nodding; distal leaves usually ca. ⅓–½ length of proximalmost leaf; leaves often longer than inflorescence; only in California . . . . 14. *Fritillaria pinetorum* (in part)
8. Flowers nodding or occasionally spreading; distal leaves usually ± equaling proximalmost leaf; leaves usually shorter than inflorescence; more widespread.
9. Tepals purplish brown, mottled yellow or white; leaves 2–3 per node proximally; 1000–3200 m, especially inland mountains . . . . . . 3. *Fritillaria atropurpurea* (in part)
9. Tepals greenish white, greenish yellow to red, purplish, or nearly black, mottled or not; leaves usually more than 3 per node proximally; 0–1800 m, especially in coastal mountains.
10. Tepals dull greenish yellow, dark-dotted, nectaries widely elliptic to ± diamond-shaped, paler than tepals; s California . . . . 13. *Fritillaria ojaiensis*
10. Tepals greenish white, pale green or greenish yellow to red, purplish, or almost black, sometimes mottled, nectaries lanceolate or linear, variously colored; s California to Alaska.
11. Tepals pale green to almost black, not mottled, nectaries ca. ½ tepal length, green; small bulb scales 0–4; s California . . . . 20. *Fritillaria viridea*
11. Tepals greenish white, pale greenish yellow to red, greenish brown to purplish, or pale yellowish green, sometimes mottled, nectaries variously colored, less than ½ tepal length (nearly equaling tepal length in *F. camschatcensis*, but very narrow and obscure; to ⅔ tepal length in *F. affinis*); small bulb scales usually 10 or more, rarely fewer; c California to Alaska.
12. Tepals usually 2 cm or longer, often clearly purple- or yellow-mottled.
13. Tepals clearly purple- or yellow-mottled and small bulb scales 20 or fewer, or tepals not mottled and bulb scales 50 or more; nectaries lanceolate, to ⅔ tepal length; flower odor not unpleasant . . . . 1. *Fritillaria affinis*
13. Tepals dark greenish brown to brownish purple; small bulb scales 30 or more; nectaries obscure, linear, ± equaling tepal length; flower odor unpleasant . . . . 6. *Fritillaria camschatcensis*
12. Tepals usually shorter than 2 cm, mottling absent or faint.
14. Style branches barely recurved; tepals greenish yellow to red, apex usually flared to slightly recurved, nectaries green, gold, or yellow . . . . 7. *Fritillaria eastwoodiae*

14. Style branches strongly recurved; tepals purplish to greenish white, apex not flared or recurved, nectaries greenish white, dotted purple . . . . 12. *Fritillaria micrantha*

[7. Shifted to left margin.—Ed.]

7. Leaves 10 or fewer and/or alternate, blade sometimes sickle-shaped.
   15. Tepals clearly mottled; small bulb scales usually 10 or more, rarely fewer.
      16. Leaf blade sickle-shaped; flowers erect; San Francisco Bay region and s coastal mountains of California . . . . 8. *Fritillaria falcata*
      16. Leaf blade usually not sickle-shaped; flowers usually horizontal or nodding, sometimes ± erect; not in San Francisco Bay region and s coastal mountains of California.
         17. Leaves 2–10, blade ovate, a few sometimes sickle-shaped; tepals white, spotted purple; small bulb scales 0–3 . . . . 17. *Fritillaria purdyi*
         17. Leaves usually 4 or more, blade linear to lanceolate; tepals purplish brown, mottled greenish yellow, yellow, or white; small bulb scales 45–50.
            18. Flowers nodding; distal leaves slightly less than or equaling proximalmost leaf, leaves usually shorter than inflorescence . . . . 3. *Fritillaria atropurpurea* (in part)
            18. Flowers ± erect; distal leaves usually 1/3–1/2 length of proximalmost leaf; leaves longer than inflorescence . . . . 14. *Fritillaria pinetorum* (in part)
   15. Tepals not clearly mottled; small bulb scales fewer than 10.
      19. Leaves 2–4, blade sickle-shaped . . . . 10. *Fritillaria glauca*
      19. Leaves usually more than 4, blade not sickle-shaped.
         20. Nectaries obscure, forming narrow band 1/2–2/3 tepal length; tepals white, striped green; flowers odorless or faintly fragrant; coastal . . . . 11. *Fritillaria liliacea*
         20. Nectaries prominent, forming narrow band 2/3 to equaling tepal length; tepals brown, purplish brown, or greenish purple at least adaxially; flowers odorless or with unpleasant odor; inland.
            21. Tepals greenish white or yellow abaxially, purplish brown adaxially; flower odor definitely unpleasant; usually in clay depressions . . . . 2. *Fritillaria agrestis*
            21. Tepals dark brown to greenish purple or yellowish green; flowers odorless or sometimes with unpleasant odor; usually on hillsides and mesas . . . . . 4. *Fritillaria biflora*

1. Fritillaria affinis (Schultes & Schultes f.) Sealy, Hooker's Icon. Pl. 39: 239. 1980 • Checker-lily E F

Lilium affine Schultes & Schultes f. in J. J. Roemer et al., Syst. Veg. 7(1): 400. 1829; *Fritillaria lanceolata* Pursh; *F. lanceolata* var. *gracilis* S. Watson; *F. lanceolata* var. *tristulis* A. L. Grant; *F. multiflora* Kellogg; *F. mutica* Lindley; *F. parviflora* Torrey; *F. phaeanthera* Purdy

Bulb scales: large 2–5; small 2–20 when tepals mottled, 50 or more when tepals unmottled. **Stem** 1–12 dm. **Leaves** in 1–4 whorls of 2–8 per node proximally, alternate distally, 4–16 cm, usually shorter than inflorescence; blade linear-lanceolate to ovate; distal leaves usually equaling proximalmost leaf. **Flowers** nodding, odor not unpleasant; tepals brownish purple to pale yellowish green, clearly mottled yellow or purple, or unmottled, oblong to ovate, 1–4 cm, apex not recurved; nectaries yellow, dotted purple, lanceolate, 1/2–2/3 tepal length; style obviously branched for 1/2 its length, branches longer than 1.5 mm. **Capsules** widely winged. $2n$ = 24, 36, 48.

Flowering Mar–Jun. Oak or pine scrub, grasslands; 0–1800 m; B.C.; Calif., Idaho, Mont., Oreg., Wash.

Fritillaria affinis has one of the broadest geographical distributions of all the North American species of the genus. It is also highly variable, which has resulted in the naming of several supposedly distinct species as well as some infraspecific taxa, all but one of which are in fact only poorly differentiated, and all of which are treated here as synonyms. Among the latter, *F. lanceolata* var. *tristulis* may actually merit formal recognition as a variety, but the new combination under *F. affinis* remains to be made. This entity is restricted to coastal grassland in Marin County, California, and has a perianth that is scarcely if at all mottled, and more than 50 small bulb scales.

Fritillaria affinis has long been known by the name *F. lanceolata*, which is illegitimate because when Pursh described it, he cited *Lilium camschatcense* (= *F. camschatcensis*) as a synonym but did not adopt that epithet. Actually, his synonymic reference was based on a misidentification, even though he stated that an illustration of *L. camschatcense* from a specimen in Pallas' herbarium was "an excellent figure" of his *F. lanceolata*.

Fritillaria affinis has been known to hybridize with *F. recurva*.

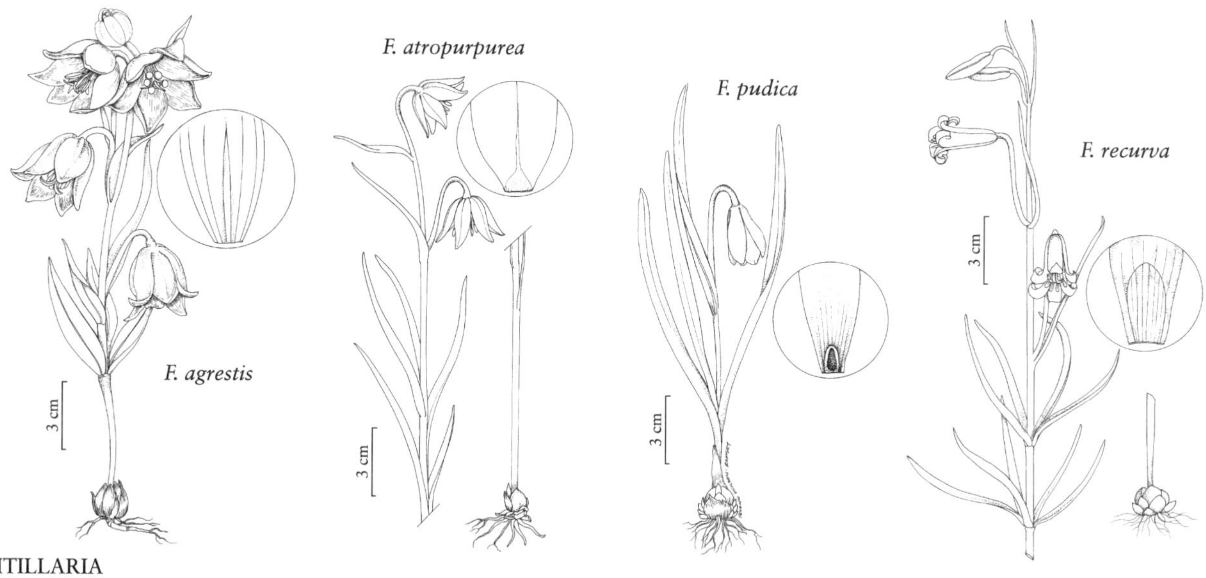

FRITILLARIA

2. Fritillaria agrestis Greene, Erythea 3: 67. 1895 • Stinkbells [E] [F]

Fritillaria biflora Lindley var. *agrestis* Greene

Bulb scales: large 2–9; small 0–2. **Stem** 3–6 dm. **Leaves** 5–12, alternate, crowded near proximal center of stem, 5–15 cm; blade linear to lanceolate-oblong. **Flowers** nodding, odor definitely unpleasant; tepals greenish to purplish brown adaxially, white or yellow abaxially, ovate, 1.8–3.5 cm, apex not recurved; nectaries prominent, green, narrowly linear, forming narrow band $^{2}/_{3}$ to equaling tepal length; style obviously branched for $^{1}/_{2}$ its length, branches longer than 1.5 mm. **Capsules** angled. $2n = 24$.

Flowering Mar–Apr. Depressions with clay or other heavy soils; 0–500 m; Calif.

This species is uncommon.

3. Fritillaria atropurpurea Nuttall, J. Acad. Nat. Sci. Philadelphia 7: 54. 1834 [E] [F]

Fritillaria adamantina M. Peck; *F. gracillima* Smiley; *F. linearis* J. M. Coulter & Fisher

Bulb scales: large 2–5; small 45–50. **Stem** 1–6 dm. **Leaves** in whorls of 2–3 leaves per node proximally, 4–12 cm, usually shorter than inflorescence; blade linear to lanceolate; distal leaves usually slightly less than or equaling proximalmost leaf. **Flowers** nodding; perianth widely open; tepals purplish brown, clearly mottled yellow or white, oblong to ± diamond-shaped, 1–2.5 cm, margins flat to revolute, apex not recurved; nectaries obscure, covering most of tepals, yellow with dark reddish dots, elliptic; style obviously branched for more than $^{1}/_{2}$ its length; branches longer than 1.5 mm. **Capsules** acutely angled.

Flowering May–Jul. Leaf mold under trees and shrubs; 1000–3200 m.; Ariz., Calif., Colo., Idaho, Mont., Nebr., Nev., N.Mex., N.Dak., Oreg., S.Dak., Utah, Wyo.

This species has the widest geographic distribution in the genus. It is easily confused with *Fritillaria pinetorum* as there is no one consistent, distinguishing trait. These two species are most easily identified in the field, but herbarium specimens can be successfully identified using a combination of flower posture and cauline leaf traits. In *F. pinetorum* the flowers are almost always erect, although some of the flowers may be ± spreading. In *F. atropurpurea* the flowers are nodding, with some flowers sometimes ± spreading. The habit of the flowers can be easily confused in pressed specimens, so careful attention to cauline leaf traits works best for them. *Fritillaria pinetorum* has long proximalmost cauline leaves that surpass the inflorescence, whereas the proximalmost cauline leaves in *F. atropurpurea* almost never exceed the inflorescence. For accurate identifications, it is best to look at several individuals from a location.

The Lakota tribe used the whole plant of this species as a cancer cure. Plants were pulverized into a salve and applied to scrofulous swellings.

4. Fritillaria biflora Lindley, Edwards's Bot. Reg. 20: under plate 1663. 1834 • Chocolate-lily, mission bells [E]

Bulb scales: large 2–8; small 0–4. **Stem** 1–4.5 dm. **Leaves** 3–7, alternate, often ± crowded just above ground level, 5–19 cm; blade linear or oblong to narrowly ovate. **Flowers** nodding, odorless or sometimes with unpleasant odor; tepals dark brown to greenish purple or yellowish green, narrowly ovoid, 1.8–4 cm, apex not recurved; nectaries prominent, purplish to greenish, narrowly linear, forming narrow band 2/3 to equaling tepal length; style obviously branched for 1/2–2/3 its length, branches longer than 1.5 mm. **Capsules** angled.

Varieties 2 (2 in the flora): California.

This species is sometimes confused with *Fritillaria agrestis*, which grows in heavier soil and has flowers with an unpleasant odor. *Fritillaria roderickii* is still recognized by some botanists, but it appears to be little more than a localized flower-color variant of *F. biflora* var. *biflora*, and is valued by horticulturists for this reason.

1. Leaf blade widely lanceolate to oblanceolate . . . . 4a. *Fritillaria biflora* var. *biflora*
1. Leaf blade linear to narrowly lanceolate . . . . 4b. *Fritillaria biflora* var. *ineziana*

4a. Fritillaria biflora Lindley var. **biflora** [E]

Fritillaria roderickii W. Knight

Leaves 8–40 mm wide; blade widely lanceolate to oblanceolate. **Flowers** 1.8–4 cm, odor not unpleasant. $2n = 24$.

Flowering Mar–Apr. Grassy slopes, mesas, serpentine barrens; 0–1200 m; Calif.

4b. Fritillaria biflora Lindley var. **ineziana** Jepson, Fl. Calif. 1: 306. 1921 [E]

Leaves 3–6 mm wide; blade linear to narrowly lanceolate. **Flowers** 1.5–2 cm, odor unpleasant.

Flowering Mar. Probably on serpentine; 0–200 m; Calif.

This variety is a narrow endemic found only in the Hillsborough area of San Mateo County. It is considered rare and endangered in California.

5. Fritillaria brandegeei Eastwood, Bull. Torrey Bot. Club 30: 484. 1903 (as brandegei) • Greenhorn fritillary [C] [E]

Bulb scales: large 8–12; small 60–200 or more. **Stem** 4–10 dm. **Leaves** in 1–2 whorls of 4–8 leaves per node proximally, alternate distally, 4–11 cm; blade lanceolate. **Flowers** nodding; tepals pinkish to purplish, oblong-lanceolate, 1.2–2 cm; nectaries green with reddish margins, lanceolate, 1/3 tepal length; style unbranched. **Capsules** winged. $2n = 24$.

Flowering Apr. Granitic soils, open forests; of conservation concern; 1500–2100 m; Calif.

This species is rare and under pressure from cattle grazing.

6. Fritillaria camschatcensis (Linnaeus) Ker Gawler, Bot. Mag. 30: under plate 1216. 1809

Lilium camschatcense Linnaeus, Sp. Pl. 1: 303. 1753

Bulb scales: large 6–15; small 30–190. **Stem** 2–6 dm. **Leaves** in 1–3 whorls of 5–9 leaves per node proximally, alternate distally, 4–10 cm, usually shorter than inflorescence; blade narrowly to broadly lanceolate; distal leaves usually ± equaling proximalmost leaf. **Flowers** spreading to nodding, odor unpleasant; tepals dark greenish brown to brownish purple, sometimes streaked or spotted with yellow, oblong-elliptic to elliptic-obovate, 2–3 cm, apex not recurved; nectaries obscure, same color as tepals, linear, ± equaling tepal length; style obviously branched for 2/3 its length, branches longer than 1.5 mm. **Capsules** cylindric-ovoid. $2n = 24, 36$.

Flowering May–Jul. Moist areas from near tideflats to mountain meadows; 0–1000 m; B.C.; Alaska, Oreg., Wash.; Asia (Japan, Russia).

Coastal Native Americans used bulbs of this species for food. Often the bulbs were dried and later added to other foods, especially soups and fish dishes.

SELECTED REFERENCE Matsura, H. and H. Toyokuni. 1963. A karyological and taxonomical study of *Fritillaria camschatcensis*. Sci. Rep. Tohoku Imp. Univ., Ser. 4, Biol. 29: 239–245.

7. Fritillaria eastwoodiae R. M. MacFarlane, Madroño 25: 95. 1978 • Butte County fritillary [E]

Fritillaria phaeanthera Eastwood, Leafl. W. Bot. 1: 55. 1933, not Purdy 1932

Bulb scales: large 2–5; small 10–60. **Stem** 2–8 dm. **Leaves** in 1–2 whorls of 3–5 leaves per node proximally, alternate distally, 5–10 cm, usually shorter than inflorescence; blade linear to narrowly lanceolate, ± glaucous; distal leaves usually ± equaling proximalmost leaf. **Flowers** nodding; tepals pale greenish yellow to red, narrowly elliptic, 1–1.7 cm, apex usually flared to slightly recurved; nectaries green, gold, or yellow, lanceolate, less than 1/3 tepal length; style obviously branched for less than 1/2 its length, branches barely recurved, longer than 1.5 mm. **Capsules** angled. $2n$ = 24, 34, 36.

Flowering Apr–May. Dry benches and slopes, sometimes on serpentine, in chaparral or beneath conifers; 500–1500 m; Calif., Oreg.

This species is highly variable and shows evidence of either being of hybrid origin between *Fritillaria recurva* and *F. micrantha*, or, if it is of separate origin, appearing to hybridize easily with those two species. In the northern part of its range, *F. eastwoodiae* intergrades with *F. recurva*, whereas in the southern part of its range it intergrades with *F. micrantha*. Occasional individuals throughout the range display traits of both *F. recurva* and *F. micrantha*.

SELECTED REFERENCE McFarlane, R. M. 1978. On the taxonomic status of *Fritillaria phaeanthera* Eastw. (Liliaceae). Madroño 25: 93–100.

8. Fritillaria falcata (Jepson) D. E. Beetle, Madroño 7: 148. 1944 (as folcata) • Talus fritillary [C] [E]

Fritillaria atropurpurea Nuttall var. *falcata* Jepson, Fl. Calif. 1: 309. 1921

Bulb scales: large 2–4; small 8–32. **Stem** 0.7–2 dm. **Leaves** 2–6, alternate, ± fleshy near stem, 3.5–8.5 cm; blade broadly linear, folded, sickle-shaped. **Flowers** erect; tepals clearly mottled rusty brown and yellow adaxially, greenish abaxially, obovate, 1.5–2.2 cm, apex not recurved; nectaries obscure, yellowish green, spotted brownish, broadly lanceolate to ± diamond-shaped, 1/2–2/3 tepal length; style obviously branched for ca. 2/3 its length, branches longer than 1.5 mm. **Capsules** acutely angled. $2n = 24$.

Flowering Mar–Apr. Serpentine talus; of conservation concern; 300–1200 m; Calif.

This is a rare, relatively restricted endemic found only in the San Francisco Bay region and slightly northward.

9. Fritillaria gentneri Gilkey, Madroño 11: 138, figs. 6–10. 1951 [C] [E]

Bulb scales: large several; small numerous. **Stem** 5–7 dm. **Leaves** in 1–3 whorls of 3–5 leaves per node proximally, alternate distally, 7–15 cm; blade broadly linear to lanceolate. **Flowers** spreading to nodding; perianth broadly campanulate; tepals red to purple, clearly mottled with yellow, 3.5–4 cm, apex spreading, not recurved; nectaries linear, 1/2 tepal length; style obviously branched for 1/3–1/2 its length, branches widely spreading, longer than 1.5 mm. **Capsules** winged.

Flowering Apr–Jun. Dry woodlands; of conservation concern; 300–1500 m; Oreg.

Fritillaria gentneri is a restricted endemic closely allied with *F. recurva* and *F. affinis*, with which it can be confused. Some evidence suggests that it may represent a hybrid between those two species. More study is needed to determine whether it should remain recognized as a distinct species.

10. Fritillaria glauca Greene, Erythea 1: 153. 1893 [E]

Bulb scales: large 3–9; small 1–9. **Stem** 0.8–2 dm. **Leaves** 2–4, alternate, 3.5–9 cm; blade lanceolate-oblong, sickle-shaped, glaucous. **Flowers** nodding; tepals purplish or greenish marked with yellow, lanceolate-oblong, 1.5–2.5 cm, apex not recurved; nectaries green with maroon dots, broadly lanceolate, less than 1/2 tepal length; style obviously branched for 1/2 its length, branches longer than 1.5 mm. **Capsules** broadly winged. $2n = 24$.

Flowering Jun–Jul. Serpentine talus slopes; 600–2100 m; Calif., Oreg.

11. Fritillaria liliacea Lindley, Edwards's Bot. Reg. 20: under plate 1663. 1834 • Fragrant fritillary [C] [E]

Bulb scales: large 2–7; small 1–2. **Stem** 1–3.5 dm. **Leaves** 2–20, alternate, 3.5–12 cm; blade linear to ovate. **Flowers** nodding, odorless or faintly fragrant; tepals white, striped green, 1–1.6 cm, apex not recurved; nectaries obscure, purplish to greenish, narrowly linear, forming narrow band 1/2–2/3 tepal length; style obviously branched for 1/2 its length, branches longer than 1.5 mm. **Capsules** obtusely angled. $2n = 24$.

Flowering Feb–Apr. Heavy soil, open hills and fields near coast; of conservation concern; 0–200 m; Calif.

Fritillaria liliacea is sometimes confused with *F. agrestis*, which has flowers with an unpleasant odor. This species is in cultivation, and is considered rare and endangered in California.

12. Fritillaria micrantha A. Heller, Muhlenbergia 6: 83. 1910 • Brown bells [E]

Fritillaria parviflora Torrey, Pacif. Railr. Rep. 4(5): 146. 1857, not Martius 1838

Bulb scales: large 1–4; small 12–60. **Stem** 4–9 dm. **Leaves** in 1–3 whorls of 4–6 leaves per node proximally, alternate distally, 4.5–15 cm, usually shorter than inflorescence; blade linear to narrowly lanceolate; distal leaves usually ± equaling proximalmost leaf. **Flowers** nodding; tepals purplish to greenish white, sometimes faintly mottled, 1–2 cm, apex not recurved; nectaries greenish white, dotted purple, narrowly lanceolate, more than $1/3$ but less than $1/2$ tepal length; style obviously branched for $1/3$–$2/3$ its length, branches strongly recurved, longer than 1.5 mm. **Capsules** broadly winged. $2n = 24$.

Flowering Apr–Jun. Dry benches and slopes to forest; 300–1800 m; Calif.

13. Fritillaria ojaiensis Davidson, Bull. S. Calif. Acad. Sci. 21: 41, plate 2. 1922 • Ojai fritillary [C] [E]

Bulb scales: large 3–5; small 1–3. **Stem** 4–7 dm. **Leaves** in 1–3 whorls of 3–5 leaves per node proximally, alternate or opposite distally, 13 cm, usually shorter than inflorescence; blade linear to narrowly lanceolate; distal leaves usually ± equaling proximalmost leaf. **Flowers** nodding; tepals dull greenish yellow with scattered to profuse dark dots, broadly lanceolate, 1.5–3 cm, apex not recurved; nectaries prominent or obscure, paler in color than rest of tepals, widely elliptic to ± diamond-shaped, $1/3$ tepal length; style obviously branched for $1/2$–$2/3$ its length, branches longer than 1.5 mm. **Capsules** winged. $2n = 14$.

Flowering Feb–Apr. Rocky slopes, river basins; of conservation concern; 300–500 m; Calif.

Fritillaria ojaiensis is considered rare and endangered in California. It is closely related to *F. affinis*, and is sometimes considered synonymous with it.

14. Fritillaria pinetorum Davidson, Muhlenbergia 4: 67. 1908 [E]

Fritillaria atropurpurea Nuttall var. *pinetorum* (Davidson) I. M. Johnston

Bulb scales: large 2–5; small 45–50. **Stem** 1–4 dm, ± glaucous. **Leaves** 4–20, in whorls of 2–3 leaves per node proximally, alternate distally, 5–15 cm, often equaling or larger than inflorescence; blade linear, glaucous; distal leaves usually ca. $1/3$–$1/2$ length of proximalmost leaf. **Flowers** ± erect, occasionally nodding; tepals purplish, clearly mottled greenish yellow, 1.4–1.9 cm, apex not recurved; nectaries obscure, yellow, dotted brown, widely ovate, $2/3$ tepal length; style obviously branched to near base, branches longer than 1.5 mm. **Capsules** angled. $2n = 26$.

Flowering May–Jun. Shaded or partially shaded granitic slopes; 1800–3200 m; Calif.

Fritillaria pinetorum is easily confused with *F. atropurpurea.*

15. Fritillaria pluriflora Torrey ex Bentham, Pl. Hartw., 338. 1857 • Adobe-lily [C] [E]

Bulb scales: large 1–12; small 0–2. **Stem** 1.5–4.5 dm. **Leaves** 3–10, alternate, clustered near ground, 6–15 cm; blade elliptic to obovate-oblong. **Flowers** nodding, not noticeably fragrant; tepals pinkish purple, obovate, 2–3.5 cm, apex rounded to acute, not recurved; nectaries lavender, narrowly linear, $2/3$ tepal length; style unbranched. **Capsules** obtusely angled. $2n = 24$.

Flowering Mar–Apr. Adobe soil of interior foothills; of conservation concern; 0–500 m; Calif.

Fritillaria pluriflora is considered rare and endangered in California and is in cultivation.

SELECTED REFERENCE Witzman, J. A. 1991. The Biology of *Fritillaria pluriflora* (Liliaceae): A Rare Endemic of the California Flora. M.S. thesis. California State University, Chico.

16. Fritillaria pudica (Pursh) Sprengel, Syst. Veg. 2: 64. 1825 • Yellow fritillary [E] [F]

Lilium pudicum Pursh, Fl. Amer. Sept. 1: 228, plate 8. 1814

Bulb scales: large 4–5; small 85–125. **Stem** 0.7–3 dm. **Leaves** 2–8, subopposite to scattered, 3–20 cm; blade linear to lanceolate. **Flowers** nodding; tepals yellow to orange, some lined brown, aging to brick red, 0.8–2.2 cm; nectaries near base of tepals, green, elliptic to round; style unbranched. **Capsules** angled. $2n = 24, 26$.

Flowering Mar–Jun. Grassy, shrubby, or wooded slopes; 0–2100 m; B.C.; Calif., Idaho, Mont., Nev., Oreg., Utah, Wash., Wyo.

Fritillaria pudica is highly variable and has one of the widest distributions of all the North American species of the genus. It was commonly used as food by Native Americans. The small bulbs were often eaten raw, and the larger ones were either dried or cooked in various ways. The Okanogan-Colville tribe used the appearance of *F. pudica* flowers as a sign that spring had arrived, and the Shuswap tribe used them in bouquets.

17. Fritillaria purdyi Eastwood, Bull. Torrey Bot. Club 29: 75, plate 6. 1902 • Purdy's fritillary [E]

Bulb scales: large 2–8; small 0–3. **Stem** 1–4 dm. **Leaves** 2–10, alternate, ± crowded near ground, 2.5–10 cm; blade ovate, occasionally a few sickle-shaped. **Flowers** horizontal or nodding; tepals white, with purple spots or lines and pink shading, 1.5–3 cm, apex often slightly recurved; nectaries obscure, colored like tepals, broadly linear, ± equaling tepal length; style obviously branched for 1/2 its length, branches longer than 1.5 mm. **Capsules** acutely angled. $2n = 24$.

Flowering Apr–Jun. Dry ridges, generally on serpentine; 400–2100 m; Calif., Oreg.

18. Fritillaria recurva Bentham, Pl. Hartw., 340. 1857 • Scarlet fritillary [E] [F]

Fritillaria recurva var. *coccinea* Greene

Bulb scales: large 6; small 20–30. **Stem** 3–9 dm. **Leaves** in 1–3 whorls of 2–5 leaves per node proximally, alternate distally, 3–15 cm; blade linear to narrowly lanceolate. **Flowers** nodding; perianth slender; tepals scarlet, checkered with yellow adaxially and purple abaxially, 1.5–3.7 cm, apex usually strongly recurved; nectaries yellow, narrowly lanceolate, 1/4 tepal length or less; style obviously branched for 1/4–1/2 its length, branches ± erect, longer than 1.5 mm. **Capsules** winged. $2n = 24, 36$.

Flowering Mar–Jun. Dry hillsides in shrubland or woodlands; 300–2200 m; Calif., Nev., Oreg.

Variety *coccinea* has been recognized by a number of authors, but examination of numerous specimens indicates no consistent distinction. In some cases, different plants from the same population appear to align with different varieties.

The Shasta and Yana tribes used *Fritillaria recurva* bulbs as food.

19. Fritillaria striata Eastwood, Proc. Calif. Acad. Sci., ser. 4, 20: 136. 1931 • Striped adobe-lily [C] [E]

Bulb scales: large 2–7; small 0–1. **Stem** 2.5–3.8 dm. **Leaves** 3–10, alternate, 6–7 cm; blade oblong-ovate, ± glaucous. **Flowers** nodding, fragrant; tepals white to pink, often striped red, oblanceolate, 2–3.5 cm, apex acute to apiculate, usually recurved; nectaries at base of tepals, green, outlined with lavender, linear; style barely branched, branches shorter than 1.5 mm. **Capsules** angled. $2n = 24$.

Flowering Mar–May. Adobe soil; of conservation concern; 0–1000 m; Calif.

Fritillaria striata is considered rare and endangered in California, the greatest threat at present being loss of habitat due to cattle grazing.

20. Fritillaria viridea Kellogg, Proc. Calif. Acad. Sci. 2: 9. 1863 • San Benito fritillary [E]

Bulb scales: large 3–5; small 0–4. **Stem** 3–6.5 dm. **Leaves** in 1–2 whorls of 3–4 leaves per node proximally, alternate distally, 4–10 cm, usually shorter than inflorescence; blade narrowly lanceolate; distal leaves usually ± equaling proximalmost leaf. **Flowers** nodding; tepals pale green to almost black, not mottled, lanceolate, 0.9–1.8 cm, apex not recurved; nectaries green, lanceolate, ca. 1/2 tepal length; style obviously branched for 1/2 its length, branches longer than 1.5 mm. **Capsules** winged.

Flowering Apr–May. Serpentine slopes; 200–1500 m; Calif.

Fritillaria viridea is considered rare and endangered in California. It is closely related to *F. affinis* and considered synonymous with it by some authors.

26. LILIUM Linnaeus, Sp. Pl. 1: 302. 1753; Gen. Pl. ed. 5, 143. 1754 • Lily [Greek *lirion*, white lily]

Mark W. Skinner

Herbs, perennial, bulbose. **Bulbs** whitish, rarely yellowish or purplish, often stained brown, erect and ovoid (hereafter "ovoid"), irregular and chunky ("chunky"), slanted in ground and ± elongate ("subrhizomatous"), or horizontally elongate ("rhizomatous"), sometimes branching if rhizomatous, rarely if not, 1.4–11.7 × 1.3–19 cm, 0.1–3 times taller than long, annual growth usually obscure; scales (modified leaves) numerous, fleshy and starchy, usually densely covering rhizomes, rarely bearing leaf blades known as basal leaves or their abscission scars, often notched or segmented, longest 0.8–11.9 cm; roots on each bulb either contractile and concentrically wrinkled and thick (to 5 mm), or for nutrition and thinner, fibrous. **Stems** erect, green, sometimes purple, rarely glaucous, to 3.1 m, ± glabrous, often with adventitious stem roots above bulb. **Buds** usually rounded in cross section, sometimes ± triangular. **Leaves** numerous, usually ± evenly distributed along stem, rarely concentrated proximally, scattered or more commonly in 1–12(–24) whorls with some scattered at stem base and apex, 3–20(–40) leaves per whorl, sessile, drooping at tips to ascending, 1.7–29 × 0.2–5.6 cm, 1.6–34 times longer than wide; blade green and somewhat lighter abaxially, rarely paler, linear, lanceolate, elliptic, or obovate, sometimes oblanceolate, especially in proximal leaves, often somewhat lanceolate in distal leaves, margins entire, undulate or not, usually glabrous and smooth or occasionally slightly papillose, sometimes roughened abaxially by ± deltoid epidermal spicules, apex acute to obtuse or rarely acuminate; principal veins usually 3, usually glabrous and smooth abaxially, sometimes with ± deltoid epidermal spicules, rarely impressed adaxially. **Inflorescences** maturing acropetally, terminal, racemose or umbellate (in small plants), usually open, bracteate, 1–25 (–45)-flowered; bracts usually 1–2 per flower, often with one lanceolate and very wide and the other linear or filiferous. **Flowers** pendent, nodding, horizontal, ascending, or erect, radially or slightly bilaterally symmetric, fragrant or not; perianth campanulate, funnelform, or with sepals and petals strongly reflexed in form of a "Turk's-cap"; sepals and petals usually differentiated, sometimes indistinctly so, recurved or reflexed, distinct, orange, red, yellow, pink, or white, usually with adaxial magenta or maroon spots concentrated in proximal $^{1}/_{2}$–$^{2}/_{3}$, ± lanceolate and narrowed or rarely clawed, glabrous (pubescent strip at base in *L. lancifolium*), nectaries present on each but often more developed on sepals, basal, green, usually hidden but occasionally exposed and forming visible green star at adaxial base of perianth; sepals 3, occasionally ridged abaxially, 3.1–12 × 0.6–2.6 cm, apex usually acute; petals 3, ridged abaxially, with 2 adaxial longitudinal median rounded ridges, 3–11.2 × 0.6–3.4 cm, apex usually acute, often more widely than sepal apex; stamens 6, opposite sepals and petals, distinct, included to strongly exserted; filaments ± parallel to style or spreading, diverging to 31° from flower axis, color variable but usually pale green or nearly translucent; anthers versatile, color variable, usually purplish, becoming darker, oblong, 0.3–2.6 cm; pollen cream, yellow, peach, tan, orange, rust, or brown, usually becoming lighter; pistil compound, 3-lobed, 3-locular, oblong, 2.1–10.5 cm; ovary superior, 0.8–3.5 cm, axile placentas 6, ovules as many as seeds, a few developing without embryos; style initially parallel to flower axis, usually elongating and curving toward periphery, usually pale green, round in cross section; stigma 3-lobed, hollow in older flowers; pedicel not articulate, 0.8–32 cm. **Fruits** erect, green maturing to brown, capsular, 3-valved, not strongly winged, ± oblong-obovate, 1.5–7.7 × 0.8–3.3 cm, 1.1–4.8 times longer than wide, base constricted, dehiscence loculicidal. **Seeds** 67–330, light brown with darker ovate embryo in center, 6-ranked, flattened into 60° wedge, verrucose. $x = 12$.

Species ca. 100 (22 in the flora): temperate Northern Hemisphere south to mountains of Asian tropics.

Perennial herbs of unsurpassed beauty and great commercial significance, the true lilies have been celebrated since time immemorial. *Lilium candidum* is the Madonna lily of biblical importance; *L. longiflorum* is the Easter lily. In China, lilies have been cultivated, eaten, and used medicinally for at least 2000 years (S. G. Haw 1986). Today, lilies are one of the mainstays of the worldwide horticultural bulb trade and many species are available. Usually these "species lilies" are susceptible to various diseases and cultural problems that render them difficult to maintain in gardens. Thousands of hybrids have been developed and about 6000 are registered, the best of which combine the delicate features of their wild relatives with disease resistance and hardiness. By and large, these hybrids predominate among the lilies grown in home gardens in North America, though the exotic *L. lancifolium* (tiger lily; China) is also widely grown, as to a lesser degree are several native species including *L. pardalinum*, *L. canadense*, and *L. superbum*. The other North American lilies rarely thrive in gardens, especially outside their native ranges, and most have suffered from excessive collecting. None should be removed from the wild.

The true lilies are distributed circumboreally, particularly in mountainous regions, and reach their southern limit in the tropical mountains of the Philippines and India. Eastern Asia and North America are the centers of highest worldwide diversity, with about 60 and 21 species respectively. The closest relatives are found in eastern Asia, where *Lilium* originated, in the genera *Fritillaria, Nomocharis* Franchet, *Notholirion* Wallich ex Boissier, and *Cardiocrinum* (Endicher) Lindley (H. D. Woodcock and W. T. Stearn 1950); many species now placed in these related genera have resided in *Lilium* at one time or another. The boundary between *Lilium* and *Nomocharis* has been particularly fluid taxonomically, and some recent molecular classifications (M. F. Fay and M. W. Chase 2000) include *Nomocharis* in *Lilium*. Of the closest relatives, only *Fritillaria* occurs in North America. It is distinguished from *Lilium* by smaller stature and shorter perianth parts (to ca. 4 cm), and by a small bulb usually consisting of a few large scales and numerous rice-grain-sized offset scales.

The 21 species of *Lilium* that are native to North America are derived from Asian stock, but since a suitable infrageneric classification for our species has not been developed, none is presented here. Nevertheless, there are several groups of evolutionary and taxonomic significance. *Lilium philadelphicum* and *L. catesbaei* are the only two North American lilies with erect flowers, highly clawed perianth parts, and spots confined to well-defined nectar guides. These two species almost certainly represent a single introduction from Asia (K. Hayashi and S. Kawano 2000; T. Nishikawa et al. 1999), the rest of the taxa probably another (M. W. Skinner 1988). The 12 species of the Pacific Coast fall into two groups. Species 4–9 are plants of usually dry ground with ± ovoid or oblique bulbs; this group is basal to those remaining, and is not monophyletic. Species 10–15 are plants of moist or wet places with rhizomatous bulbs that frequently branch; these apparently originated from within the dry-ground group (M. W. Skinner 1988). Species 16–22, the pendent-flowered eastern lilies, appear to have their closest relationship with *L. pardalinum* in the west, but within themselves form a coherent group of two rather well-defined alliances. Species 16–19 are characterized by southern and eastern distributions, buds that are triangular in cross section, and sepals with two abaxial longitudinal ridges. Species 20–22 occupy more northern distributions, and have in common roughened leaves due to spicules on the leaf margins and abaxially on the veins, usually yellowish bulbs, and red styles. If species 16–19 and 20–22 are considered separately, complementary allopatric or parapatric distributions—usually with hybridization in zones of contact—are the norm within all of the groups except for the one that contains species 4–9.

Most lilies are largely self-incompatible, and cross-pollination is required for seed set (J. S. Davis 1958). The flowers are weakly protandrous: pollen is offered upon anthesis, and then over time the pistil elongates, the style generally curves toward the periphery (particularly in pendent-flowered species), and the stigma enlarges. Nectar is offered especially from the sepal nectaries to reward insect pollinators, and also hummingbirds in North America. Phylogenetic analysis (M. W. Skinner 1988) suggests that the original pendent lilies to colonize North America were butterfly-pollinated, and other pollination syndromes involving moths and hummingbirds evolved in situ via pollinator mediation.

Identification of *Lilium* herbarium specimens is often difficult since bulbs are rarely collected and crucial features such as flower color, shape, and orientation are altered in pressed material. Leaf characters including shape, margin, and vestiture are sometimes diagnostic, and flower and anther size are frequently useful, but it is often necessary to resort to comparison of distributions during specimen identification. The keys and descriptions presented here are based on live plants, and anther measurements were taken following dehiscence. Descriptions of the flowers refer to the adaxial surface unless otherwise noted. A whorl or partial whorl consists of three or more leaves arranged cyclically on the stem, and bulb dimensions are given as height × length along the horizontal axis.

Regardless of their shape, the bulbs of the North American species are actually radially asymmetrical, slowly growing, scaly rhizomes. In some (e.g., *L. kelloggii* and *L. rubescens*), lateral growth is very slow and the bulb remains upright and more or less ovoid. In others, such as *L. philadelphicum,* the growing point appears to rotate around the underground stem, and the bulb remains compact but irregular. In *L. washingtonianum* the bulb is somewhat horizontally extended and slanted in the ground with the growing end deeper; this condition is termed subrhizomatous. In the majority of our species, however, the rhizome extends horizontally and the flowering stem moves incrementally each year; these bulbs are termed rhizomatous. Only in some of these rhizomatous lilies, for example *L. pardalinum* and *L. superbum*, do the bulbs branch often and evenly to produce two new plants. Other bulb types branch only rarely and irregularly. In the more extended bulb forms it is sometimes possible to count the partially visible annual stem scars to derive plant age, but this is usually an underestimate since the dying end of the bulb continually deteriorates.

Several species of Asian or European origin are sporadically naturalized following escape from cultivation, but none strays far or is widespread or common enough to be considered a pest. *Lilium bulbiferum* Linnaeus (orange lily; Europe) is tall, with linear alternate leaves and 1–3 (–20) erect, widely campanulate orange flowers; the perianth parts are recurved, clawed, and conspicuously papillose adaxially at the base. It is somewhat similar to native *L. philadelphicum* but for its papillose perianth and axillary bulbils, though the latter are not always present. It is occasionally found outside cultivation in temperate northeastern North America (Quebec, Ontario, and New Brunswick) and sparingly in the intermountain west in Utah (J. T. Kartesz and C. A. Meacham 1999). *Lilium candidum* Linnaeus (Mediterranean) is tall (1–2 m), with many lanceolate, ascending leaves and pure white, campanulate flowers of modest size (perianth parts to 8 cm) that are fragrant and horizontal to ascending on the stem; it is naturalized in two counties in Pennsylvania (A. F. Rhoads and W. M. Klein 1993). *Lilium martagon* Linnaeus (martagon lily; Eurasia) shares with many of the native species both whorled leaves and Turk's-cap flowers, but the latter are small (perianth parts to 3.5 cm), very numerous, and purple or sometimes white. It is weakly naturalized in Michigan near Ann Arbor, and perhaps in Quebec (H. J. Scoggan 1978–1979, part 2). Several similar Asian lilies with scattered leaves and long, funnelform white flowers are reportedly naturalized in the southeastern United States; careful examination is required to identify them correctly.

Lilium longiflorum Thunberg (Easter lily; Japan) is the common white lily of the florist trade; it is the most widely cultivated and commercially important lily. Plants are short (to 1 m), the leaves lanceolate and to 18 × 1.5 cm, and the pure white flower is fragrant and larger than in any of our native species; the perianth parts are 13–18 cm and glabrous at the base adaxially. It has been recorded from Utah (S. L. Welsh et al. 1993) and Florida. *Lilium formosanum* Wallace (Formosa lily; Formosa) is tall, with numerous long (to 20 cm), linear leaves of dark, lustrous green. The 1–2(–10) horizontal flowers are delicately fragrant, white, and usually suffused with wine-purple abaxially; the perianth parts are 13–20 cm long and basally papillose adaxially. The Formosa lily is evidently infrequently naturalized and has been reported from Louisiana and Florida, although most if not all of these records—like the two records of *L. regale* E. H. Wilson from Alabama (R. Kral 1981)—represent the next species. *Lilium philippinense* Baker (Philippine lily; Philippines) is similar to *L. formosanum* but with a thinner, longer floral tube. The delicately fragrant flower is white, occasionally streaked with green and red basally, and is the largest in the genus; the perianth parts are 18–25 cm and basally papillose. It is reported from one location in Kentucky (E. T. Browne Jr. and R. Athey 1992) and is becoming well established in parts of Florida, especially near Tallahassee.

Native Americans relied on the bulbs of many species of *Lilium* for food and medicine, and preparation varied widely (D. E. Moerman 1986). From *L. canadense* bulbs the Cherokee made flour and then bread for use during famine. In British Columbia, the Thompson Indians mixed *L. columbianum* bulbs with salmon roe and this was boiled and eaten as a favorite dish. In Saskatchewan, the Cree dried the bulb scales of *L. philadelphicum* as a snack. Otherwise, the bulbs were steamed, dried into cakes for winter use, baked in an earth oven, or made into soup. Lilies were equally versatile as medicine, and the mashed bulbs were variously employed in the treatment of spider bites, cuts and bruises, fever, coughs, consumption, stomach ailments, and rheumatism. Contemporary medical use seems to be largely limited to *L. lancifolium*, the bulbs of which are used to treat a variety of internal discomforts, including those associated with menstruation and menopause.

SELECTED REFERENCES Adams, R. M. and W. J. Dress. 1982. Nodding *Lilium* species of eastern North America (Liliaceae). Baileya 21: 165–188. Elwes, H. J. 1880. A Monograph of the Genus *Lilium*. London. Hayashi, K. and S. Kawano. 2000. Molecular systematics of *Lilium* and allied genera (Liliaceae): Phylogenetic relationships among *Lilium* and related genera based on *rbc*L and *mat*K gene sequence data. Pl. Spec. Biol. 15: 73–93. Nishikawa, T. et al. 1999. A molecular phylogeny of *Lilium* in the internal transcribed spacer region of nuclear ribosomal DNA. J. Molec. Evol. 49: 238–249. Skinner, M. W. 1988. Comparative Pollination Ecology and Floral Evolution in Pacific Coast *Lilium*. Ph.D. dissertation. Harvard University. Stewart, R. N. 1947. The morphology of somatic chromosomes in *Lilium*. Amer. J. Bot. 34: 9–27. Synge, P. M. 1980. Lilies: A Revision of Elwes' Monograph of the Genus *Lilium* and Its Supplements. New York. Tamura, M. N. 1998c. Liliaceae. In: K. Kubitzki et al., eds. 1990+. The Families and Genera of Vascular Plants. 4+ vols. Berlin etc. Vol. 3, pp. 343–353. Woodcock, H. D. and W. T. Stearn. 1950. Lilies of the World: Their Cultivation and Classification. London.

1. Flowers nodding to erect; perianth widely campanulate or funnelform; sepals and petals mostly recurved but not reflexed (except reflexed or rolled in *L. maritimum*; weakly reflexed in *L. parvum*).
    2. Sepals and petals distinctly clawed; inflorescences rarely more than 3-flowered; flowers erect.
        3. Sepal and petal apices very narrowly acute, sepals 8.2–12 cm; leaves scattered; se United States . . . . 2. *Lilium catesbaei*
        3. Sepal and petal apices obtuse to acute, sepals 4.9–8.2 cm; leaves whorled, or if scattered, then with at least 1 distal whorl, or infrequently entirely scattered; widespread . . . . 3. *Lilium philadelphicum*
    2. Sepals and petals not distinctly clawed; inflorescences occasionally 1-flowered, but usually 2–many-flowered; flowers nodding to erect (if erect, at least some horizontal to ascending on each plant).

4. Sepals and petals white, often aging purple or pink; flowers ± horizontal to erect.
   5. Sepals 4.3–6.6 cm; flowers ascending to erect; bulb scales always unsegmented; nw California . . . . 6. *Lilium rubescens*
   5. Sepals (6.1–)6.7–11.3 cm; flowers ± horizontal; bulb scales often notched but not fully segmented, or segmented; mountains of n California, Oregon . . . . 5. *Lilium washingtonianum*
4. Sepals and petals yellow, orange, red, or magenta, rarely pinkish, and if so, not darkening significantly with age; flowers nodding to ascending.
   6. Sepals and petals bright yellow; sepals 7.7–10.7 cm; mountains of s Arizona, s California . . . . 15. *Lilium parryi*
   6. Sepals and petals red, orange, magenta, or pink, rarely pale yellow; sepals 3.1–5.6 cm; West Coast or Appalachians.
      7. Leaves 1.8–7.1 cm, blade noticeably glaucous, margins nearly always undulate; bulbs ± ovoid, thus a species of dry ground; bulb scales unsegmented, longest always 3 cm or longer; flowers nodding to horizontal; sepals 3.1–4.7 cm; Klamath Mountains of Oregon and California . . . . 8. *Lilium bolanderi*
      7. Leaves 3.6–17.7 cm, blade not noticeably glaucous, margins never undulate; bulbs rhizomatous, thus species of moist or wet places; bulb scales variable, usually at least a few segmented, longest shorter than 3 cm (rare exceptions in large plants); flowers nodding to ascending; sepals 3.2–5.6 cm; w United States or Appalachians.
         8. Leaf veins and margins noticeably roughened abaxially with ± deltoid epidermal spicules; rhizomes with scaleless sections between annual bulbs; flowers nodding; sepals 3.2–5.6 cm; style red; s Appalachians . . . . 22. *Lilium grayi*
         8. Leaf veins and margins ± smooth abaxially; rhizomes continuously scaly; flowers nodding to ascending; sepals 3.2–5 cm; style green; n California.
            9. Flowers nodding or rarely horizontal; sepals and petals reflexed or rolled 1/2–4/5 along length from base, red or red-orange; sepals 3.4–5 cm; capsules 2.4–4.1 cm; bulb scales usually unsegmented, a few 2-segmented; extreme coastal n California . . . . 11. *Lilium maritimum*
            9. Flowers ± horizontal to ascending; sepals and petals somewhat recurved but not strongly reflexed, orange or yellowish proximally, darker (orange, red-orange, red, or rarely pinkish) on distal 2/5, occasionally uniformly light orange or rarely yellow; sepals 3.2–4.2 cm; capsules 1.6–2.7 cm; bulb scales (1–)2–3(–4)-segmented; Sierra Nevada . . . . 13. *Lilium parvum*

1. Flowers pendent (to nodding in *L. columbianum*); perianth Turk's-cap-shaped (campanulate in *L. canadense*); sepals and petals reflexed (somewhat recurved in *L. canadense*).
   10. Leaves scattered; axillary bulbils present on distal leaves; introduced, often persisting near dwellings; c and e North America . . . . 1. *Lilium lancifolium*
   10. Leaves whorled, though distal and proximal may be scattered; axillary bulbils entirely absent from leaves; endemic; geographically scattered.
       11. Sepals and petals pink or less often white, with median longitudinal yellow stripe, often aging deep pink; sepals 3.4–7.2 cm; mountains of nw California and sw Oregon . . . . 7. *Lilium kelloggii*
       11. Sepals and petals yellow, orange, or red; sepals 3.4–10.4 cm; geographically scattered.
           12. Leaf blade noticeably pale abaxially, texture noticeably fleshy, oblanceolate, occasionally obovate; flowers strongly fragrant; s and se United States . . . . 19. *Lilium michauxii*
           12. Leaf blade lighter green but not pale abaxially, texture not noticeably fleshy, shape variable; flowers odorless or mildly fragrant; geographically scattered.
               13. Bulbs subrhizomatous to ± ovoid, thus species of drier ground; longest bulb scales always longer than 3 cm; sepals and petals orange or yellow.

14. Pistil 2.4–3.7 cm; sepals 3.4–7.1 cm; anthers 0.5–1.3 cm; capsules 1.1–2 cm wide; n California to Montana, British Columbia . . . . 9. *Lilium columbianum*

14. Pistil 4.6–7.1 cm; sepals 5.2–9.8 cm; anthers 1.1–1.9 cm; capsules 1.8–3.3 cm wide; Sierra Nevada and s California . . . . 4. *Lilium humboldtii*

[13. Shifted to left margin.—Ed.]

13. Bulbs rhizomatous, thus species of moist, wet, or low places; longest bulb scales shorter than 3 cm (with rare exceptions in large plants); sepals and petals yellow, yellow-orange, or reddish.

15. Rhizomes continuously scaly; w of Rocky Mountains.

16. Flowers mildly fragrant; sepals and petals yellow or yellow-orange; sepals 4.3–5.7 cm; anthers 0.3–0.6 cm; pistil 2.6–3.4 cm; capsules 1.5–2.9 cm; Sierra Nevada above 2200 m . . . . 12. *Lilium kelleyanum*

16. Flowers almost never fragrant; sepals and petals ± two-toned with green, yellow, or orange proximally, darker orange to red on distal 1/5–2/3 (except ± uniformly orange or yellow-orange in *L. pardalinum* subsp. *wigginsii*); sepals 3.5–10.4 cm; anthers 0.5–2.2 cm; pistil 3–7.5 cm; capsules 2.1–5.7 cm; California and Oregon below 2000 m.

17. Sepals and petals green to yellow proximally, red to maroon (to orangish) distally, conspicuously green abaxially on proximal 2/5–1/2; sepals 4.4–8.1 cm; filaments ± parallel to style, diverging only 4°–12° from axis; rhizomes never branched; extreme coastal n California and s Oregon . . . . 10. *Lilium occidentale*

17. Sepals and petals yellow to orange proximally, orange to red distally (except subsp. *wigginsii*), conspicuously green abaxially only on proximal ± 1/5; sepals 3.5–10.4 cm; filaments spreading from style, diverging 7°–22° from axis; rhizomes usually branched; California and s Oregon . . . . 14. *Lilium pardalinum*

15. Rhizomes with scaleless sections between annual bulbs; e of Rocky Mountains.

18. Style red, at least distally, or brightly colored like sepals and petals; sepals not ridged abaxially; buds rounded in cross section; leaf veins and margins noticeably roughened abaxially with ± deltoid epidermal spicules; bulbs usually yellowish.

19. Sepals and petals reflexed; filaments parallel to style at first, then widely spreading, diverging 13°–23° from axis; stamens moderately exserted; midwestern North America . . . . 20. *Lilium michiganense*

19. Sepals and petals somewhat recurved; filaments ± parallel to style, barely spreading, diverging 4°–6° from axis; stamens barely exserted; e North America . . . . 21. *Lilium canadense*

18. Style pale green; sepals with 2 parallel, often faint abaxial ridges; buds ± triangular in cross section; leaf veins and margins ± smooth abaxially (except often rough in *Lilium iridollae*); bulbs whitish.

20. Rhizomes 9.6–18 cm, with 3–4 annual bulbs, scaleless sections between bulbs 2.7–5.4 cm; bulb-scale leaves or their abscission scars present; inflorescences 1–4-flowered; sepals and petals yellow-orange or yellow; coastal plain in Alabama and w Florida . . . . 17. *Lilium iridollae*

20. Rhizomes 5.2–10.2 cm, with 2(–3) annual bulbs, scaleless sections between bulbs 0.3–3.8(–4.6) cm; bulb-scale leaves or their abscission scars absent; inflorescences 1–22-flowered; sepals and petals orange or reddish; s and e United States.

21. Stems 1.2–2.8 m; leaves usually ± evenly distributed along stem, usually horizontal and drooping at tips, in 6–24 whorls of 3 or more, 7.1–26.1 cm, 4–18 times longer than wide; inflorescences 1–22-flowered; sepals and petals usually red-orange; s and e United States . . . . 16. *Lilium superbum*

21. Stems 0.6–1.6 m; leaves often concentrated proximally, usually ascending, in 0–12 whorls of 3 or more, 2.3–10.3(–12.2) cm, 1.6–7.6(–10.3) times longer than wide; inflorescences 1–7-flowered; sepals and petals variable, usually red-orange or dusky red, often somewhat pale; sandhills of Virginia and the Carolinas . . . . 18. *Lilium pyrophilum*

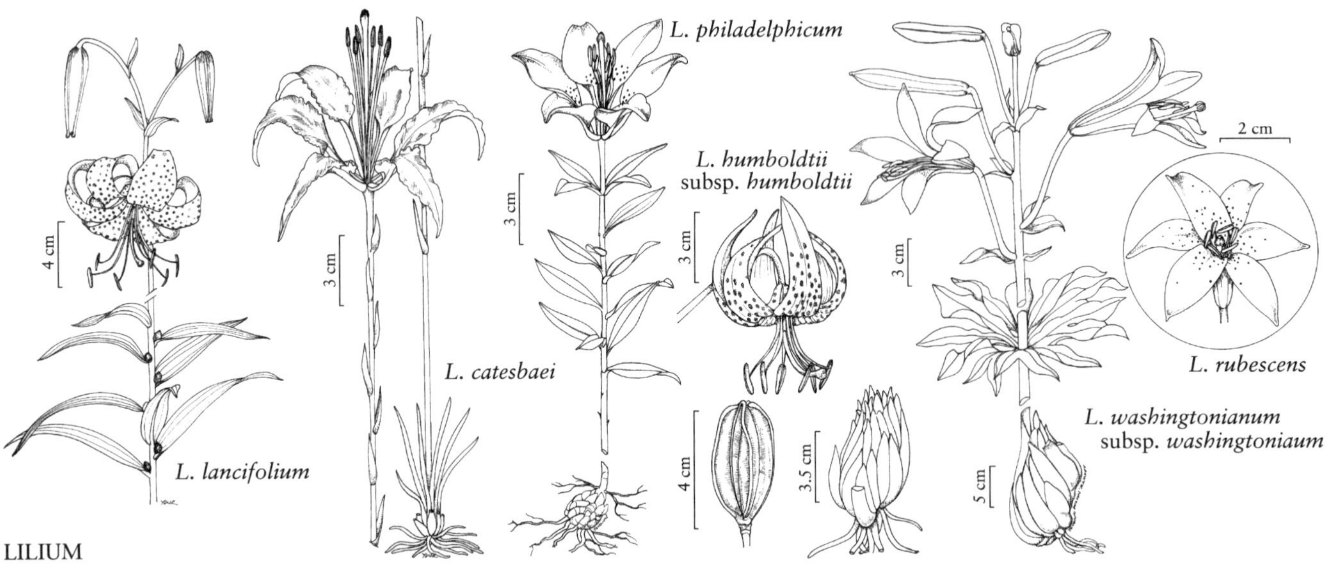

LILIUM

1. Lilium lancifolium Thunberg, Trans. Linn. Soc. London 2: 333. 1794 • Tiger lily, lis tigré [F] [I] [W1]

Lilium tigrinum Ker Gawler

Bulbs widely ovoid, 3.5(–8) × 4–8 cm, ca. 0.7–0.8 times taller than wide; scales broad, unsegmented, longest ca. 3–4 cm; stem roots usually present. **Stems** white-lanate, purplish, to 2 m. **Buds** usually flat-sided, somewhat triangular in cross section. **Leaves** scattered, horizontal and drooping at tips, 10–15(–18) × ca. 1.5 cm, ca. 3–6(–10) times longer than wide, distal bearing 1(–3) dark purple axillary bulbils; blade lanceolate, often narrowly so, margins not undulate, papillose, apex white-lanate, narrowly acute (rounded in distal leaves). **Inflorescences** racemose, 3–6(–25)-flowered. **Flowers** pendent, not fragrant; perianth Turk's-cap-shaped; sepals and petals reflexed ca. 1/5–1/4 along length from base, orange with many purple-brown spots, not distinctly clawed, 7–10 × 1–2 cm, adaxial base bearing pubescent strip; sepals not ridged abaxially; stamens strongly exserted; filaments very widely spreading, diverging ca. 25° from axis; anthers purplish, ca. 2 cm; pollen rust; pistil 6–9 cm; pedicel sometimes dichotomous, stout and relatively short, to 10 cm. **Capsules** 3–4 cm. **Seeds** not counted. $2n$ = 24, 36.

Flowering late summer (mid Jul–early Sep). Roadsides, railroad banks, near dwellings; 0–ca. 1000 m; introduced; N.B., N.S., Ont., Que.; Ala., Conn., Del., D.C., Ill., Ind., Iowa, Ky., La., Maine, Md., Mass., Mich., Minn., Mo., Nebr., N.H., N.J., N.Y., N.C., N.Dak., Ohio, Pa., R.I., Vt., Va., W.Va., Wis.; Asia (China).

Throughout most of modern botanical history this Chinese lily has been known as *Lilium tigrinum*, but recent nomenclatural reassessment affirms that Thunberg's description, published sixteen years earlier than Ker Gawler's, applies to this species. Though many North America species are known vernacularly as tiger lilies, the name is properly applied only to this one. Along with *L. candidum*, it is considered to be among the earliest domesticated lilies (H. D. Woodcock and W. T. Stearn 1950), no doubt because it is handsome, easy to grow, and the bulbs are edible and substantial. It is widely planted in North America, usually as a sterile triploid that is best propagated from the bulbils.

Perhaps the hardiest garden lily, *Lilium lancifolium* is a widespread but sporadic garden escape, and roadside lilies near habitation in eastern and northeastern North America are often this species. Despite its general use in gardens, it seems to be naturalized only in the better-watered eastern portion of the continent.

In North America, the tiger lily is rather easily diagnosed by its truly lanceolate and widely sessile alternating leaves that bear dark bulbils on the upper stem. The mature buds are usually high-shouldered and taper rather evenly to a flattish apex with three greenish, terminal, rounded processes. No native lily consistently displays any of these features.

2. Lilium catesbaei Walter, Fl. Carol., 123. 1788 • Pine lily [E] [F]

Lilium catesbaei subsp. *asprellum* Wherry; *L. catesbaei* var. *longii* Fernald

Bulbs ovoid, 1.6–2.5 × 1.3–2.4 cm, 0.8–1.6 times taller than long, 2 years' growth evident, newer bearing prominent basal leaves, older with abscission scars; scales few, loose, unsegmented or 2-segmented, longest 1–1.8 cm; stem roots usually present, often numerous. **Stems** to 0.9 m. **Buds** rounded in cross section. **Leaves** scattered, ascending, distal appressed, 1.8–8.2 × 0.2–1.2 cm, 3.1–10.5 times longer than wide; blade narrowly elliptic, sometimes linear or slightly oblanceolate, margins not undulate, apex acute, acuminate especially in distal leaves; veins and margins ± smooth abaxially. **Inflorescences** occasionally umbellate, 1(–3)-flowered. **Flowers** erect, not fragrant; perianth widely campanulate; sepals and petals recurved 2/5–1/2 along length from base, crimson or sometimes pink, distinctly clawed, apex very narrowly acute, nectar guides above claws yellow to pale yellow and spotted maroon or magenta, ± equal; sepals not ridged abaxially, 8.2–12 × 1.2–1.9 cm; petals at proximal widest point much wider than sepals, 7.6–11.1 × 1.8–3.4 cm; stamens moderately exserted; filaments ± parallel to style, barely spreading, diverging 0°–12° from axis, often purple at base; anthers variously colored tan-orange, brown, peachy magenta, or pale greenish, 0.4–1.6 cm; pollen burnt orange or dark tan; pistil 7.6–10.5 cm; ovary 1.4–3.5 cm; style pale green, sometimes darker distally; pedicel 1.8–9.5 cm. **Capsules** often ridged along valve margins, 2.2–5.3 × 0.8–1.6 cm, 1.7–3.8 times longer than wide. **Seeds** not counted. $2n$ = 24.

Flowering late summer–fall (late Jun–Oct) in most of range, sporadically spring and fall in peninsular Florida. Wet pine flatwoods and savannas, especially in pitcher plant (*Sarracenia*) bogs with *Sphagnum*; 0–200 m; Ala., Fla., Ga., La., Miss., N.C., S.C., Va.

The pine lily's flower is the largest of any North American lily and one of the largest among our native monocots. In small plants it dwarfs and sometimes topples the slender stem. Leaves are small and relatively few and the bulb is petite, and thus resource limitation in smaller plants undoubtedly contributes to the wide range of fruit sizes within populations. In other North American members of the genus, small plants produce one or a few capsules, but typically these approach normal size.

Lilium catesbaei subsp. *asprellum* Wherry and *L. catesbaei* var. *longii* Fernald have been proposed to account for individuals with leaves concentrated toward the middle of the stem or somewhat wide and lacking basal leaves, respectively. These variants are not emphasized here since both are based primarily on vegetative features that vary greatly in most lilies. Isotypes of var. *longii* are unremarkable, though with somewhat wide leaves, and the broadly overlapping distribution of this variety with nominate populations (A. E. Radford et al. 1968) strongly suggests that such differences are primarily environmentally induced. Variety *longii* was described from Virginia, and Fernald's observation that these northern plants lack basal leaves—which I have not investigated in the field—is unsurprising in those colder climates, and best considered in terms of the normal variation within a fairly wide-ranging species.

Although it is not yet rare, widespread alteration of native longleaf pine (*Pinus palustris* Miller) and slash pine (*P. elliottii* Engelmann) savanna, especially conversion to even-age pine plantations, is making steady inroads on populations of this most beautiful lily. It is adapted to frequent fires, and their suppression may contribute to this decline.

Lilium catesbaei is pollinated primarily by the palamedes swallowtail [*Papilio palamedes* (Drury), family Papilionidae], the only swallowtail that is widely endemic to this lily's coastal plain habitat. Spicebush swallowtails (*P. troilus* Linnaeus) visit the pine lily less frequently, and their smaller size suggests that they are less effective pollinators than the larger palamedes.

3. Lilium philadelphicum Linnaeus, Sp. Pl. ed. 2, 1: 435. 1762 • Wood lily, lis de Philadelphie [E] [F]

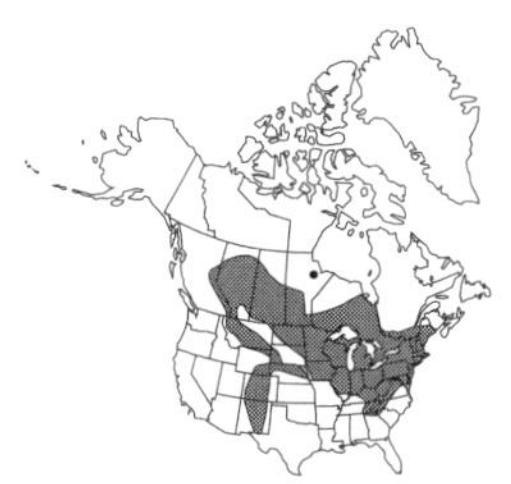

Lilium andinum Nuttall; *L. montanum* A. Nelson; *L. philadelphicum* var. *andinum* (Nuttall) Ker Gawler; *L. philadelphicum* var. *montanum* (A. Nelson) Wherry; *L. umbellatum* Pursh

Bulbs chunky, 1.6–2.9 × 2.2–4.7 cm, 0.6–1.1 times taller than long, 2(–4) years' growth visible; scales 1–2-segmented, longest 1.2–2.2 cm; stem roots present or absent. **Stems** to 1.2 m, glaucous. **Buds** rounded in cross section. **Leaves** scattered, or mostly scattered with at least 1 distal whorl, or in 1–5 whorls or partial whorls, 3–11 leaves per whorl, ± horizontal and drooping at tips, or ascending in sun, 2.9–10.2 × 0.3–2.3 cm, 3.5–18 times longer than wide; blade narrowly elliptic, sometimes linear, elliptic, or oblanceolate, margins not undulate, apex acute or barely acuminate; veins and margins ± smooth abaxially. **Inflorescences** umbellate, 1–3(–6)-flowered. **Flowers** erect, not fragrant; perianth widely campanulate; sepals and petals somewhat recurved 1/4–2/5 along length from base, red-orange or red-magenta, sometimes pale orange, pure red, or rarely yellow, distinctly clawed, apex often widely acute, rarely obtuse, nectar guides above claws yellow to orange and spotted maroon, more pronounced on sepals; sepals not ridged abaxially, 4.9–8.2 × 1.6–2.6 cm; petals 4.5–7.7 × 2–3.2 cm; stamens strongly exserted; filaments ± parallel to style, barely spreading, diverging 0°–8° from axis, ± same color as sepals and petals; anthers dull maroon, 0.5–1.2 cm; pollen variously colored dark orange, brown, brown-yellow, or yellow; pistil 5–8 cm; ovary 1.3–3.2 cm; style ± same color as sepals and petals; pedicel 2.5–10.5 cm. **Capsules** 2.2–7.7 × 1–1.8 cm, 3–4.8 times longer than wide. **Seeds** not counted. $2n = 24$.

Flowering late spring–summer (late May–Aug), earliest in s Appalachians, latest in n Rocky Mountains. Tallgrass prairies, open woods, thickets, roadsides, powerlines, e balds, barrens, dunes, and heathlands, w mountain meadows; 0–2700 m; Alta., B.C., Man., Ont., Que., Sask.; Colo., Conn., Del., D.C., Ga., Ill., Ind., Iowa, Ky., Maine, Md., Mass., Mich., Minn., Mo., Mont., Nebr., N.H., N.J., N.Mex., N.Y., N.C., N.Dak., Ohio, Pa., R.I., S.C., S.Dak., Tenn., Tex., Vt., Va., W.Va., Wis., Wyo.

Lilium philadelphicum is the widest ranging of our true lilies. Rather common in high meadows of the mountain west and some intact tallgrass prairies of the Great Plains and adjacent corn belt, it is decidedly rare to the east in lower midwestern prairies of the United States and in the southern Appalachians, where it is protected by several states. It has declined rapidly in the northeastern United States as prairies disappear and white-tailed deer [*Odocoileus virginianus* (Zimmermann), family Cervidae] continue to increase in number. In many places in the eastern U.S., the most reliable habitats are powerline right of ways that are maintained by brush-clearing.

The northernmost populations are introduced along the railway from Gillam to Churchill on Hudson Bay (H. J. Scoggan 1978–1979, part 2), though the wood lily is native and widespread farther south in Manitoba. Reports from Arkansas cannot be verified, and an 1869 D. E. Palmer specimen with the locality given as Arizona was probably collected in New Mexico or Colorado. Surprisingly, there are no known records for the area extending from southwestern Iowa to southeastern Alberta.

Lilium philadelphicum is highly variable in stature, leaf size and arrangement, flower color, and fruit length, and it is this variation that has accounted for the proliferation of names—only a few of which are cited above—applied to this taxon. Of these, only one remains in general use. Variety *andinum* has come to include usually western plants of smaller stature that have long capsules (4–8 cm vs. 2.5–3.5 cm in var. *philadelphicum*) and are characterized by scattered leaves usually topped by a single whorl (E. T. Wherry 1946). The typical habitat for these plants is the low grassy vegetation found in tall- and midgrass prairies and mountain meadows. The usually accepted break between this entity and var. *philadelphicum* runs north to south along the eastern border of Ohio (E. L. Braun 1967), and thus northeastern and Appalachian plants are normally assigned to the nominate variety. These are more often plants of open woods or thickets, though they do occur in low vegetation, including Appalachian balds and eastern prairies. Field studies show that these specimens tend to be large, with a mean stem height of 81 cm, compared to 48 cm in western plants. Leaf arrangement consists of 2–5 whorls of 3 or more leaves (averaging 3.8 whorls) as opposed to the more western plants typically assigned to var. *andinum*, which display 0–5 whorls, averaging 1.3 whorls per plant. Leaves are also longer and wider in these eastern and Appalachian plants and average 6.9 cm long (3.8–10.2 cm) by 1.3 cm wide (0.7–2.3 cm), compared with 5.1 cm long (2.9–7.7 cm) by 0.6 cm wide (0.3–1.4 cm) in plants sampled from western Ohio and Colorado. Other individuals disrupt this pattern. Plants from Nantucket, Massachusetts (presumably of low heath or grassland), fall within the range of variation of mountain plants from Colorado. These dwarf individuals have small flowers and short (ca. 4 cm), very narrow (ca. 0.4 cm) leaves, albeit mostly in whorls. Equally significant are massive plants of moist woods in Colorado with fully whorled, long (to 9.2 cm), and rather wide leaves (to 1.3 cm). Other Colorado plants are vegetatively indistinguishable from certain Connecticut material.

Somewhat surprisingly, given its relatively modest stature, the wood lily has the longest capsules of any *Lilium* in North America. The largest-fruited individuals seen come from the Appalachians, not from populations otherwise assignable to the purportedly long-fruited var. *andinum*. Fruit lengths average larger in these robust Kentucky plants (mean 6.2 cm, range 5.1–7.7) than in western Ohio plants (mean 4.3 cm, range 3–5.8). Plants from Vermont that are clearly assignable to var. *philadelphicum* on the basis of leaf size and arrangement have capsules of comparable size to those in western Ohio (mean 4.1 cm, range 3.1–5.3).

Therefore, it appears that local environment governs vegetative and fruit morphology to a great degree in *Lilium philadelphicum*, and in many cases overwhelms the presumed effect of genotype and the broad geographical patterns. Western plants from typically eastern habitats (moist woodlands) resemble eastern plants, and eastern plants from typically more western habitats (low prairie or heath) resemble western plants. The characters invoked to support the var. *andinum*—especially capsule length—vary continuously and somewhat independently. The status of var. *andinum* is unsettled and it is not accorded formal recognition here.

Lilium philadelphicum is pollinated by large swallowtail butterflies, in the west by the pale swallowtail (*Papilio eurymedon* Lucas, family Papilionidae) and western tiger swallowtail (*P. rutulus* Lucas), in the east by the eastern tiger swallowtail (*P. glaucus* Linnaeus; E. M. Barrows 1979) and undoubtedly most of the other species resident in its wide range. Hummingbirds occasionally visit the flowers but are unlikely to be equally effective pollinators due to a flower morphology that forces butterfly forewings to contact reproductive structures but probably allows birds ready nectar access without such contact.

The wood lily is the floral emblem of Saskatchewan.

The Cree, Meskwaki, and Blackfoot used *Lilium philadelphicum* bulbs as food, while other tribes used bulbs medicinally and in witchcraft (D. E. Moerman 1986). The Malecite mixed the roots with those of *Rubus* species and staghorn sumac (*Rhus typhina* Linnaeus) to treat coughs and fevers. The Chippewa made a poultice that was applied to dog bites and caused the dog's fangs to drop out. The Iroquois made a decoction of the whole plant to shed the placenta after childbirth, women used a decoction of the roots as a wash if the husband was unfaithful, and the whole plant was used as a romantic aid: if sun-dried plants twisted together, they signified a wife's infidelity.

4. Lilium humboldtii Roezl & Leichtlin ex Duchartre, J. Soc. Centr. Hort. France, sér. 2, 5: 43. 1871 • Humboldt lily [E] [F]

Bulbs often somewhat purplish, variable, subrhizomatous to ± ovoid, 3.3–11.7 × 2.4–14.4 cm, 0.4–2.6(–3) times taller than long; scales unsegmented or notched, longest 3.5–11.2 cm; stem roots present or absent. **Stems** to 3.1 m. **Buds** rounded in cross section. **Leaves** in 2–8 whorls or partial whorls, 3–16(–28) leaves per whorl, usually ascending, 4.6–14.5 × 0.8–3.6 cm, 2.9–7.2 times longer than wide; blade usually ± oblanceolate, though often weakly so, rarely elliptic or lanceolate, margins usually undulate, apex acute; veins and margins ± smooth abaxially. **Inflorescences** racemose, 1–33(–40)-flowered. **Flowers** pendent, not fragrant; perianth Turk's-cap-shaped; sepals and petals reflexed 1/5 along length from base, orange or yellow, with prominent red or magenta spots, not distinctly clawed; sepals not ridged abaxially, 5.2–9.8 × 1.4–2.4 cm; petals 5.6–9.6 × 1.5–2.9 cm; stamens strongly exserted; filaments parallel along most of length, then very widely spreading, diverging 17°–31° from axis; anthers purple, 1.1–1.9 cm; pollen rust, tan, or peach, becoming yellow or tan-yellow; pistil 4.6–7.1 cm; ovary 1.2–2.6 cm; style green, often pale; pedicel 7.8–21.2 cm. **Capsules** longitudinally keeled, 2.5–5.4 × 1.8–3.3 cm, 1.1–2.3 times longer than wide. **Seeds** not counted.

Subspecies 2 (2 in the flora): California.

Lilium humboldtii is declining throughout its range due to habitat destruction, primarily for housing. Forming large scattered colonies at foothill elevations under ponderosa pines or in oak canyons and chaparral, these massive plants with towering inflorescences and large flowers are quite striking. With *Lilium pardalinum* and *L. parryi*, they were used in the early part of the century to produce the Bellingham hybrid lilies, development of which continued with the Bellmaid hybrids. Though not as popular as various Asiatic hybrids, these are still in use.

A. M. Kellogg was aware as early as 1859 that the tall, orange-flowered, dry-land lily from the foothills of the Sierra Nevada was distinctive, and he presented this opinion to the California Academy of Sciences about that time (A. M. Kellogg 1872). By the time he had published a description attached to the name *Lilium bloomerianum*, P. E. S. Duchartre had described this same plant from material provided by B. Roezl and grown by M. Leichtlin, and named it after the German explorer and botanist Alexander von Humboldt. Carl Purdy and others then misapplied Kellogg's name in various combinations to the southern California expression recognized here as *Lilium humboldtii* subsp. *ocellatum*.

The Humboldt lily is pollinated primarily by large butterflies, especially western tiger swallowtails (*Papilio rutulus* Lucas, family Papilionidae) and pale swallowtails (*P. eurymedon* Lucas), and to a lesser extent by the monarch butterfly [*Danaus plexippus* (Linnaeus), family Nymphalidae].

1. Sepals and petals orange, spots magenta, without lighter red margins; bulbs off white, occasionally flecked with purple, scales always unsegmented; foothills of Sierra Nevada . . . . 4a. *Lilium humboldtii* subsp. *humboldtii*
1. Sepals and petals yellow or light orange, spots red or magenta, aureolated with lighter red margins; bulbs often purplish, scales usually notched; s California . . . . 4b. *Lilium humboldtii* subsp. *ocellatum*

4a. Lilium humboldtii Roezl & Leichtlin ex Duchartre subsp. **humboldtii** [E] [F]

Lilium bloomerianum Kellogg; *L. canadense* Linnaeus var. *humboldtii* Baker; *L. canadense* var. *puberulum* Torrey

Bulbs off white, occasionally flecked with purple; scales unsegmented; stem roots absent. **Racemes** 1–33(–40)-flowered. **Flowers:** sepals and petals orange, with magenta spots distributed distally or more proximally; pollen rust, rust-brown, rust-orange, occasionally to warm tan, becoming yellowish.

Flowering summer (mid Jun–early Aug). Ponderosa pine (*Pinus ponderosa* P. & C. Douglas) forest and chaparral openings; 600–1100 m; Calif.

Subspecies *humboldtii* is distributed from Tehama County south to Calaveras County; reports from farther south are erroneous. An anomalous population documented by A. A. Heller early in the twentieth century as occurring near the Sacramento River at valley elevations (ca. 100 m) in Butte County is unconfirmed and unlikely to be relocated.

4b. Lilium humboldtii Roezl & Leichtlin ex Duchartre subsp. **ocellatum** (Kellogg) Thorne, Aliso 9: 195. 1978 [E]

Lilium bloomerianum Kellogg var. *ocellatum* Kellogg, Proc. Calif. Acad. Sci. 5: 88. 1873; *L. fairchildii* M. E. Jones; *L. humboldtii* var. *magnificum* Purdy; *L. humboldtii* var. *ocellatum* (Kellogg) Elwes; *L. ocellatum* (Kellogg) Beane

Bulbs often purplish, especially apically; scales notched, segmented with 2–5 poorly defined segments, or occasionally unsegmented; stem roots frequently present. **Racemes** 1–25-flowered. **Flowers:** sepals and petals yellow or light orange, with large red or magenta spots aureolated with light red margins, spots bigger and their margins wider and lighter toward apex; pollen tan or peach, becoming yellow or tan-yellow, occasionally tan-orangish or rust. $2n = 24$.

Flowering late spring–summer (mid May–Jul). Oak canyons, chaparral; 200–1800 m; Calif.

Subspecies *ocellatum* is rather similar to subsp. *humboldtii*, but the yellowish sepals and petals with widely margined spots, lighter-colored pollen, and purplish bulb with notched scales are distinctive. In addition to the mainland, it occurs on the larger northern Channel Islands, where it is the only native lily (C. F. Smith 1976).

5. Lilium washingtonianum Kellogg, Hesperian (San Francisco) 3: 340. 1859 • Washington lily [E] [F]

Bulbs subrhizomatous to ± ovoid, 2.7–10 × 4.4–13.4 cm, 0.2–0.9(–1.4) times taller than long; scales unsegmented or notched with 2(–3) poorly defined segments, sometimes 2-segmented, longest 3.3–11.9 cm; stem roots absent. **Stems** to 2 m, often glaucous. **Buds** rounded in cross section. **Leaves** in 1–8 whorls or partial whorls, 3–16 leaves per whorl, horizontal and drooping at tips to ascending, occasionally nearly clasping stem, 3.7–12.3 × 0.9–4.7 cm, 2–6.5 times longer than wide; blade oblanceolate, sometimes obovate, rarely elliptic, margins undulate or not, apex acute, often widely so; veins and margins ± smooth abaxially. **Inflorescences** racemose, 1–33-flowered. **Flowers** opening before dusk, ± horizontal, slightly bilaterally symmetric, strongly fragrant; perianth ± funnelform; sepals and petals recurved $\frac{2}{3}$–$\frac{3}{4}$ along length from base and scarcely overlapping to form tube, lower usually less recurved than upper and forming landing platform, white, often aging deep pink or lavender, often with short yellowish stripe extending from basal median nectaries, often with fine magenta spots concentrated basally, not distinctly clawed; sepals sometimes purplish abaxially, not ridged abaxially, (6.1–)6.7–11.3 × 0.9–1.7 cm; petals noticeably wider than sepals, oblanceolate and often very wide distally, (6.1–)6.6–11.2 × 1.1–2.4 cm, apex widely acute, obtuse, or rounded; stamens barely exserted; filaments barely spreading, diverging 2°–8° from axis; anthers off white or cream, becoming pale pink or yellow, 0.8–1.5 cm; pollen yellow or cream; pistil 7.5–10.4 cm; ovary 1.7–3.4 cm; style pale green; pedicel 4.8–15 cm. **Capsules** often with 6 longitudinal ridges, 2.7–5.8 × 1.6–2.9 cm, 1.3–2.3 times longer than wide. **Seeds** 123–231.

Subspecies 2 (2 in the flora): California, Oregon.

1. Sepals and petals aging light pink or remaining white; bulbs sometimes (36%) without notched or segmented scales; sepals 8–11.3 cm; Sierra Nevada and s Cascades of California . . . . 5a. *Lilium washingtonianum* subsp. *washingtonianum*
1. Sepals and petals aging deep pink or lavender; bulbs usually (88%) with some notched or segmented scales; sepals (6.1–)6.7–9.5 cm; mountains of n California and Oregon . . . . 5b. *Lilium washingtonianum* subsp. *purpurascens*

5a. Lilium washingtonianum Kellogg subsp. **washingtonianum** [E] [F]

Lilium washingtonianum Kellogg var. *minus* Purdy

Bulbs subrhizomatous, 2.7–5.6 × 5.7–13.4 cm, 0.2–0.8 times taller than long; scales usually unsegmented, some notched with 2(–3) poorly defined segments, longest 3.5–7.7 cm. **Stems** to 2 m. **Leaves** in 2–5 whorls, 3–16 leaves per whorl or partial whorls, horizontal and drooping at tips to ascending, 5.4–10.9 × 1.3–2.8 cm, 2.9–5.4 times longer than wide; margins undulate or not. **Flowers:** sepals and petals recurved $\frac{2}{3}$–$\frac{3}{4}$ along length from base, white, rarely aging pink, usually with short yellowish stripe extending from basal median nectaries; sepals rarely purplish adaxially, 8–11.3 × 1.1–1.7 cm; petals 7.9–11.2 × 1.4–2.4 cm; anthers off white/cream, becoming pale pink or dirty yellow; pollen very pale yellow or cream. **Capsules** occasionally with 6 longitudinal ridges, 2.7–4.3 × 1.6–2.7 cm, 1.4–2.2 times longer than wide. **Seeds** 123–216. $2n$ = 24.

Flowering summer (mid Jun–early Aug). Mixed conifer forests, gaps, clearcuts, roadsides; 1300–1500 m; Calif.

Subspecies *washingtonianum* occurs primarily on the western slope of the Sierra Nevada and extends north to the southern Cascades at Mount Lassen and Mount Shasta. Early literature records this subspecies from southern California (Cuyamaca Mountains, San Diego County), but these reports probably represent *Lilium parryi*. *Lilium washingtonianum* var. *minus* (Shasta lily) was described by Purdy from Mt. Shasta primarily on the basis of reportedly smaller bulbs, but as this character is age-dependent, the variety is not recognized here. Plants from this area may be morphologically intermediate between the subspecies.

The Washington lily is pollinated by sphinx moths (*Sphinx* spp., family Sphingidae). Nectar secretion and scent production are timed to their arrival at dusk. The open nature of the flower and copious nectar secretion attract many diurnal visitors, including a variety of butterflies, bees, and flies, but these are not significant pollinators.

5b. Lilium washingtonianum Kellogg subsp. **purpurascens** (Stearn) M. W. Skinner, Novon 12: 258. 2002 • Cascade lily [E]

Lilium washingtonianum var. *purpurascens* Stearn, Gard. Chron., ser. 3, 124: 13. 1948; *L. purpureum* Purdy; *L. washingtonianum* var. *purpureum* (Purdy) Purdy 1919, not W. Bull ex Baker 1874

Bulbs variable, subrhizomatous to ± ovoid, 3–10 × 4.4–11.7 cm, 0.3–0.9(–1.4) times taller than long; scales variable, usually notched with 2(–3) poorly defined segments, some clearly 2-segmented or unsegmented, longest 3.3–11.9 cm. **Stems** to 1.7 m. **Leaves** in 1–8 whorls or partial whorls, 3–15 leaves per whorl, usually ascending and occasionally nearly clasping stem, rarely horizontal and drooping at tips, 3.7–12.3 × 0.9–4.7 cm, 2–6.5 times longer than wide; margins undulate. **Flowers:** sepals and petals recurved $^{2}/_{3}$ along length from base, mostly or entirely white but aging deep pink or lavender, sometimes with short yellowish stripe extending from basal median nectaries; sepals usually if obscurely purplish adaxially, (6.1–)6.7–9.5 × 0.9–1.6 cm; petals (6.1–)6.6–9.5 × 1.1–1.9 cm; anthers cream, sometimes spotted minutely with magenta on abaxial surface, becoming yellow; pollen pale yellow to occasionally yellow. **Capsules** usually with 6 longitudinal ridges, 2.8–5.8 × 1.6–2.9 cm, 1.3–2.3 times longer than wide. **Seeds** 144–231. $2n$ = 24.

Flowering summer (mid Jun–mid Aug). Forest openings, chaparral, burned clearcuts, roadsides; 300–1700 m; Calif., Oreg.

Subspecies *purpurascens* replaces the Sierran subsp. *washingtonianum* near Mount Shasta in Siskiyou County, California, and extends west through the Klamath Mountains and north through the Cascade ranges to Mount Hood in Clackamas County, Oregon. In addition to the characters mentioned in the key, subsp. *purpurascens* is also distinguished from subsp. *washingtonianum* by more compact bulbs with longer scales.

6. Lilium rubescens S. Watson, Proc. Amer. Acad. Arts 14: 256. 1879 • Redwood lily, chaparral lily [E] [F]

Lilium washingtonianum Kellogg var. *purpureum* W. Bull ex Baker

Bulbs ± ovoid, 3.9–9.4 × 2.5–7.1 cm, 1.1–2.6 times taller than long; scales unsegmented, longest 4–8.9 cm; stem roots absent. **Stems** to 2 m, often glaucous. **Buds** rounded in cross section. **Leaves** in 3–5 (–more) whorls or partial whorls, 3–15 leaves per whorl, usually ± ascending, sometimes horizontal and drooping at tips, 3.2–12.3 × 1–2.8 cm, 2–6.4 times longer than wide; blade oblanceolate, sometimes obovate, rarely elliptic, margins usually undulate, apex acute, often widely so, or obtuse; veins and margins ± smooth abaxially. **Inflorescences** racemose, often with flowers in loosely defined whorls, 1–40-flowered. **Flowers** ascending to erect, fragrant; perianth funnelform; sepals and petals recurved $^{1}/_{2}$–$^{2}/_{3}$ along length from base, white, aging pink-purple, with fine magenta spots concentrated proximally or less often generally distributed, often reddish or purplish abaxially, not distinctly clawed; sepals not ridged abaxially, 4.3–6.6 × 0.6–1.4 cm; petals noticeably wider than sepals, oblanceolate and often very wide proximally, 4.2–6.4 × 0.9–1.9 cm, apex widely acute, obtuse, or rounded; stamens included; filaments barely spreading, diverging 3°–12° from axis; anthers pale yellow, 0.4–0.8 cm; pollen yellow; pistil 2.7–3.8 cm; ovary 1.2–1.8 cm; style pale green; pedicel 1.4–9.5 cm. **Capsules** usually with 6 longitudinal ridges, 2–3.7 × 1.7–2.7 cm, 1.1–1.7 times longer than wide. **Seeds** 67–189. $2n$ = 24.

Flowering summer (late May–early Aug). Dry soils in chaparral, gaps in redwood [*Sequoia sempervirens* (D. Don) Endlicher] or mixed evergreen forests; 0–1500 m; Calif.

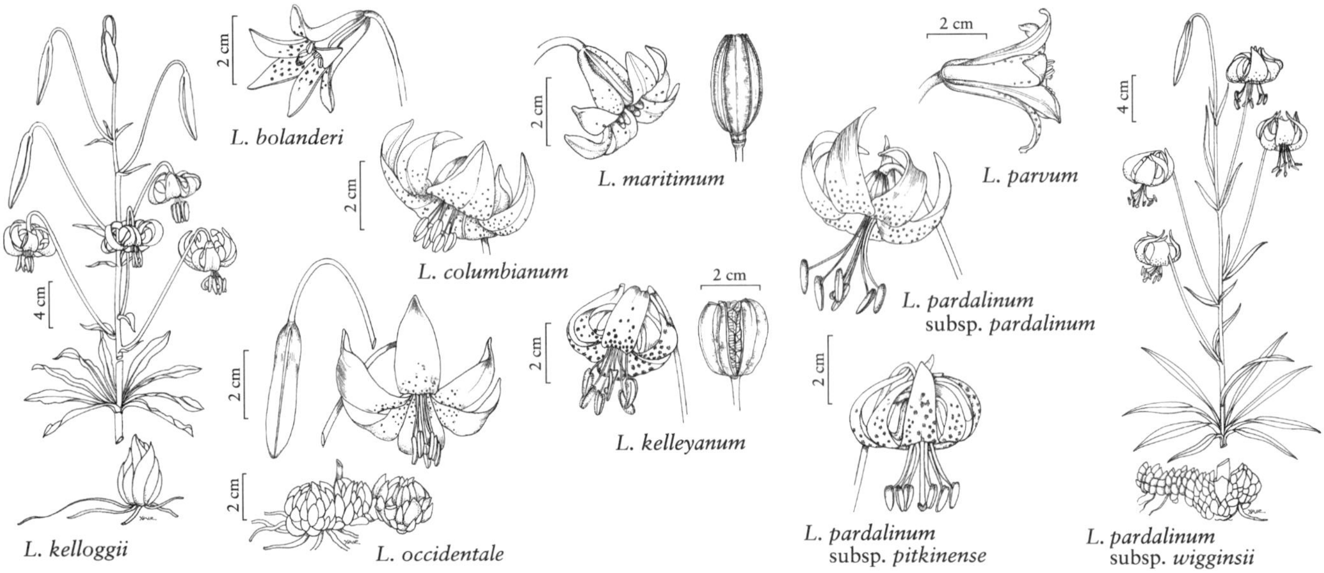

LILIUM

Lilium rubescens occurs from Santa Cruz County north through the Coast Ranges to Del Norte County, and is declining in the southern part of its range due to urbanization and over-collecting.

This is the only species in the genus in which the chromosome complement is known to consist of one pair of long metacentric chromosomes and eleven pairs of acrocentric chromosomes; all others possess two long metacentric pairs and ten shorter acrocentric pairs (M. W. Skinner 1988).

Various bee species visit the flowers during the day to collect nectar and pollen and may be the most significant pollinators. Pale swallowtails (*Papilio eurymedon* Lucas, family Papilionidae) also visit this lily.

The Karok used *Lilium rubescens* decoratively in bouquets (D. E. Moerman 1986).

7. **Lilium kelloggii** Purdy, Garden (London 1871–1927) 59: 331. 1901 • Kellogg's lily [E] [F]

Bulbs ± ovoid, 3.6–7.5 × 2–5.2 cm, 0.9–2.4(–2.7) times taller than long; scales unsegmented, longest 3.1–6.4 cm; stem roots absent. **Stems** to 2.1 m. **Buds** rounded in cross section. **Leaves** in 2–7 whorls or partial whorls (scattered and ± proximal in dry, exposed sites), 3–22(–40) leaves per whorl, drooping at tips to ascending, 6–16 × 0.9–4.4 cm, 2.6–8.5 times longer than wide; blade elliptic, usually narrowly so, often weakly oblanceolate, rarely obovate, margins undulate or not, apex acute; veins and margins ± smooth abaxially. **Inflorescences** racemose, 1–27-flowered. **Flowers** pendent, fragrant; perianth Turk's-cap-shaped; sepals and petals reflexed 1/3 along length from base, pink, less often white, and if so usually aging pink, usually with copious maroon spots and proximal median longitudinal yellow stripe extending from basal nectaries, not distinctly clawed; sepals darker pink abaxially and usually distally, not ridged abaxially, 3.4–7.2 × 0.9–1.7 cm; petals 3.2–7.1 × 0.9–1.9 cm; stamens moderately exserted; filaments ± widely spreading, diverging 10°–22° from axis; anthers pale red-orange or magenta, 0.5–1.4 cm; pollen orange; pistil 2.9–4.2 cm; ovary 1.5–3 cm; style green, often pale; pedicel 8.8–23.8 cm. **Capsules** 2.9–5.7 × 1–1.6 cm, 2.2–3.7 times longer than wide. **Seeds** 177–309. $2n = 24$.

Flowering summer (mid Jun–early Aug). Gaps and roadsides in redwood [*Sequoia sempervirens* (D. Don) Endlicher] or mixed evergreen forests, gaps and edges in chaparral; 200–1300 m; Calif., Oreg.

Kellogg's lily occurs from Humboldt County, California, north to just over the Oregon border. Like *Lilium bolanderi* and *L. pardalinum* subspp. *vollmeri* and *wigginsii*, it is endemic to the Klamath Mountains.

Occasionally plants occur with one or more whorls coalesced into tufts with as many as 40 leaves. Plants in open chaparral are usually of small stature with compact inflorescences and the ascending leaves are undulate on the margins. Plants of forest gaps are larger, with open inflorescences and flat, horizontal leaves.

Lilium kelloggii is pollinated by pale swallowtail butterflies (*Papilio eurymedon* Lucas, family Papilionidae) and western tiger swallowtails (*P. rutulus* Lucas).

8. Lilium bolanderi S. Watson, Proc. Amer. Acad. Arts 20: 377. 1885 • Bolander's lily [E] [F]

Lilium howellii I. M. Johnston

Bulbs ± ovoid, 3.5–7.9 × 2.6–5.3 cm, 0.9–2.1 times taller than long; scales unsegmented, longest 3–5.7 cm; stem roots absent. **Stems** to 1.1 m, glaucous. **Buds** rounded in cross section. **Leaves** in 1–5(–6) whorls or partial whorls, 3–19 leaves per whorl, ascending and often cupping stem, 1.8–7.1 × 0.7–2.8 cm, 2.2–4.8 times longer than wide; blade ± obovate, oblanceolate, or occasionally elliptic, noticeably glaucous, margins nearly always undulate, apex widely acute; veins and margins ± smooth abaxially. **Inflorescences** usually umbellate in small plants, in large plants racemose or in 2 whorls, 1–9-flowered. **Flowers** nodding to horizontal, not fragrant; perianth ± campanulate or funnelform; sepals and petals somewhat recurved 3/5–4/5 along length from base, red or magenta, occasionally salmon pink or pale yellow, with maroon spots, often yellowish on proximal 1/3–1/2, not distinctly clawed; sepals not ridged abaxially, 3.1–4.7 × 0.7–1.2 cm; petals 3–4.5 × 0.7–1.1 cm; stamens included; filaments barely spreading, diverging 0°–12° from axis; anthers reddish or magenta, 0.3–0.8 cm; pollen rust, orange, or yellow; pistil 2.1–3.5 cm; ovary 1–2.1 cm; style green, rarely reddish purple; pedicel 0.8–14.2 cm. **Capsules** 2–4.1 × 1.2–2.1 cm, 1.4–3 times longer than wide. **Seeds** 90–210. $2n = 24$.

Flowering summer (mid Jun–mid Aug). Dry serpentine soils in chaparral, gaps in open mixed conifer or Douglas-fir [*Pseudotsuga menziesii* (Mirbel) Franco] forests, associated with bear-grass [*Xerophyllum tenax* (Pursh) Nuttall]; 100–1500 m; Calif., Oreg.

I. M. Johnston (1923) noted that *Lilium bolanderi* Watson was based on a mixed collection including *L. kelloggii*, and he argued that Watson intended the name to apply primarily to the latter species. Thus he proposed the name *L. howellii* for this diminutive, red-flowered, serpentine endemic. A. D. Cotton (1936) correctly concluded that Watson's description applied primarily to the specimens here called *L. bolanderi*, and this view is now widely accepted.

Lilium bolanderi hybridizes with *L. rubescens*, *L. washingtonianum* subsp. *purpurascens*, and subspecies of *L. pardalinum*.

Bolander's lily is primarily pollinated by Allen's and rufous hummingbirds (*Selasphorus* spp., family Trochilidae).

9. Lilium columbianum Leichtlin ex Duchartre, J. Soc. Centr. Hort. France, sér. 2, 5: 98. 1871 • Columbia lily [E] [F]

Lilium canadense Linnaeus var. *parviflorum* Hooker; *L. lucidum* Kellogg; *L. parviflorum* (Hooker) Holzinger

Bulbs variable, subrhizomatous to ± ovoid, 3.1–8.1 × 3.6–10.6 cm, 0.3–1.4 times taller than long; scales 2–3(–5)-segmented, some unsegmented, longest 3.1–7.3 cm; stem roots absent. **Stems** to 1.7 m, glaucous or not. **Buds** rounded in cross section. **Leaves** in 1–9 whorls or partial whorls, 3–25 leaves per whorl, usually ascending, 1.7–15.7 × 0.4–4.7 cm, 2.4–7.4 times longer than wide; blade weakly oblanceolate to obovate or ± elliptic, margins usually somewhat undulate, apex acute; veins and margins ± smooth abaxially. **Inflorescences** racemose, 1–25(–45)-flowered. **Flowers** pendent to nodding, not fragrant; perianth Turk's-cap-shaped; sepals and petals reflexed 2/5–1/2 along length from base, yellow or orange to occasionally red, with copious maroon spots, often darker and especially reddish abaxially, not distinctly clawed; sepals not ridged abaxially, appearing wide for their length, 3.4–7.1 × 0.8–1.9 cm; petals 3.5–6.9 × 0.8–1.9 cm; stamens barely to moderately exserted; filaments moderately spreading, diverging 10°–20° from axis; anthers pale yellow to yellow, 0.5–1.3 cm; pollen orange or yellow; pistil 2.4–3.7 cm; ovary 1.1–2.2 cm; style green; pedicel 2.8–20.2 cm. **Capsules** 2.2–5.4 × 1.1–2 cm, 1.7–3.3 times longer than wide. **Seeds** 135–330. $2n = 24$.

Flowering summer (early May–early Aug). Coastal scrub and prairies, meadows, conifer or mixed forests, clearings, roadsides; 0–1800 m; B.C.; Calif., Idaho, Mont., Oreg., Wash.

The author citations often seen for this species derive from Baker (1874), who published the name as *Lilium columbianum* "Hanson in hort., Leichtlin"; this authority is given by various later writers as Hanson, or Baker, or Hanson ex Baker. However, Ducharte's (1871) recapitulation of a letter from M. Leichtlin is apparently the first confirmed and valid publication of *L. columbianum*, and hence that citation is used here.

This widespread lily is rather variable. In California plants the stamens are considerably less exserted than those of plants found farther north. *Lilium columbianum* may intergrade with *L. kelloggii* along Highway 199 at the border between California and Oregon; these plants are slightly fragrant, the stamens moderately exserted, and the bulb scales unsegmented. *Lilium columbianum* hybridizes with *L. pardalinum* subspp. *wigginsii* and *vollmeri*, and extensively with *L. occidentale* in Oregon.

Lilium columbianum is pollinated primarily by rufous hummingbirds [*Selasphorus rufus* (J. F. Gmelin), family

Trochilidae] and to a lesser extent by large butterflies, including the pale swallowtail (*Papilio eurymedon* Lucas, family Papilionidae).

Native Americans used *Lilium columbianum* bulbs as a food or peppery condiment, sometimes mixed with meat or salmon roe. For many, it was a staple food (J. Pojar and A. MacKinnon 1994).

10. Lilium occidentale Purdy, Erythea 5: 103. 1897 • Western lily, Eureka lily [C] [E] [F]

Bulbs rhizomatous, sometimes irregular, unbranched, continuously scaly, 1.8–4.3 × 2.1–9.9 cm, 0.3–0.7 times taller than long; scales 1–2(–3)-segmented, longest 0.9–2.5 cm; stem roots absent. **Stems** to 2.2 m. **Buds** rounded in cross section. **Leaves** scattered or in 1–9 whorls or partial whorls, 3–19 leaves per whorl, horizontal and drooping at tips to ascending, 4.1–26.3 × 0.4–3 cm, 4–29 times longer than wide; blade elliptic to linear, occasionally narrowly lanceolate, sometimes narrowly or weakly oblanceolate, margins not undulate, apex narrowly acute; veins and margins ± smooth abaxially. **Inflorescences** racemose, 1–35-flowered. **Flowers** pendent, not fragrant; perianth Turk's-cap-shaped; sepals and petals reflexed 2/5 along length from base, yellow, greenish yellow, or green proximally (and usually forming a green star), sometimes with an intercalary band of orange or yellow, then red-orange to red to maroon or very rarely orange on distal 1/2–2/3, maroon, dark purple, or rarely blackish spots concentrated proximally and rarely extending into distal reddish zone, conspicuously green abaxially on proximal 2/5–1/2, not distinctly clawed; sepals not ridged abaxially, 4.4–8.1 × 0.8–1.8 cm; petals 4.3–8 × 0.8–1.6 cm; stamens moderately exserted; filaments ± parallel to style, barely spreading, diverging 4°–12° from axis; anthers dull red, dull purple, or magenta, 0.5–1.4 cm; pollen rust or red-orange, occasionally orange or orange-brown to red-brown; pistil 3–5.5 cm; ovary 1–2 cm; style green, sometimes purple distally; pedicel 5–29.7 cm. **Capsules** 2.1–5.4 × 1–1.9 cm, 1.4–3.1 times longer than wide. **Seeds** 141–207. $2n = 24$.

Flowering summer (Jun–Aug). Coastal scrub, coastal prairies, gaps in coastal coniferous forests especially with Sitka spruce [*Picea sitchensis* (Bongard) Carrière], *Sphagnum* bogs and marshes in northern part of range; of conservation concern; 0–100 m; Calif., Oreg.

In the late nineteenth century, botanist and bulb merchant Carl Purdy employed Native Americans to collect bulbs of lilies and other desirable plants in the area around Eureka, California, and *Lilium occidentale* was among the species he was brought. After Purdy described *Lilium occidentale* in 1897, the plant came into great demand. He later (C. Purdy 1937) lamented that almost no plants were available for cultivation. Populations fluctuated in the middle of the twentieth century and frequently declined dramatically from overcollecting following publicity that the plants had recovered and flowered. Much habitat was subsequently lost to ecological succession and land conversion. Fortunately, several large populations have recently been discovered near Crescent City, California, and completion of a recovery plan (Fish and Wildlife Service 1998) bodes well for the future of this state and federally protected plant.

Lilium occidentale occurs in scattered populations from Humboldt Bay, California, to Coos Bay, Oregon, always within six kilometers of the ocean. Plants near Humboldt Bay occur in coastal prairies and scrub, whereas many Oregon populations are found in sphagnum bogs. Typical Oregon and extreme northern California plants in the vicinity of Crescent City (Del Norte County) have an irregular bulb that occasionally is nearly erect (0.3–0.7 times taller than long versus 0.3–0.5 in Humboldt Bay plants). In northern plants the leaves are often scattered—though they are whorled in larger plants—and extremely long and thin [4.2–26.3 × 0.4–2.7 cm (4.6–28.6 times longer than wide) versus blade 4.1–18.9 × 0.7–3 cm (4–17.4 times longer than wide)] in Humboldt Bay plants. Flowers of Oregon plants are marginally smaller, have somewhat narrower sepals (4.4–7.6 × 0.8–1.5 cm) than Humboldt Bay plants (4.4–8.1 × 1–1.8 cm), and typically appear somewhat later. These differences have provoked discussion about the merits of subdividing this species, but many of the differences appear to be related to habitat rather than genotype. For example, plants from Shores Acres State Park in Coos County, Oregon, occur in a coastal bluff setting that is similar to habitats near Humboldt Bay, and these lilies share many of the same vegetative and floral characteristics of Humboldt Bay plants. Subspecific differentiation appears to be unwarranted.

Lilium occidentale is distinguished from similar species such as *L. pardalinum* by the closely parallel stamens and adaxial green star formed by the conspicuously green sepal and petal bases, a characteristic also discernible in the abaxial surface of the buds. This green star is present in nearly all flowers when first opened, but changes to yellow or yellow-orange as the flowers age. The green star of certain eastern taxa (e.g., *L. superbum*) is also diagnostic for them but is primarily formed by glossy green adaxial nectaries and is not so prominent in bud.

Frequent hybrids with *Lilium columbianum* in Oregon occur primarily where human disturbance has introduced that species into boggy areas via drier open pathways such as powerline right of ways and roads.

With the likely exception of *Lilium canadense, L. occidentale* is the only North American pendent lily that is pollinated almost exclusively by hummingbirds (in

California primarily the Allen's hummingbird [*Selasphorus sassin* (Lesson), family Trochilidae]). Its pendent red flower is transitional between butterfly-pollinated forms like *L. pardalinum* and the classical bird-adapted flowers of *L. maritimum* and *L. bolanderi* that are more tubular and horizontal.

The Karok ate *Lilium occidentale* bulbs after baking them in an earth oven (D. E. Moerman 1986).

11. Lilium maritimum Kellogg, Proc. Calif. Acad. Sci. 6: 140. 1875 • Coast lily [C] [E] [F]

Bulbs rhizomatous, often irregular, unbranched, continuously scaly, 1.8–4.1 × 3–7.4 cm, 0.3–0.9 times taller than long; scales usually unsegmented, a few 2-segmented, longest 1.5–3.8 cm; stem roots absent. **Stems** to 1.2(–2.3) m, plants on coastal bluffs shorter than 0.25 m. **Buds** rounded in cross section. **Leaves** sometimes concentrated proximally, scattered or usually in 1–3 whorls or partial whorls, 3–17 leaves per whorl, horizontal and drooping at tips to ascending, 3.6–17.7 × 0.3–4.7 cm, 2.4–18 times longer than wide; blade elliptic, sometimes narrowly so, rarely weakly oblanceolate or obovate, margins not undulate, apex acute or narrowly so; veins and margins ± smooth abaxially. **Inflorescences** racemose, 1–13-flowered. **Flowers** nodding, rarely horizontal, not fragrant; perianth campanulate; sepals and petals reflexed or rolled 1/2–4/5 along length from base, red or red-orange, with maroon spots concentrated proximally along longitudinal median and surrounded by light orange or occasionally yellow-green, not distinctly clawed; sepals not ridged abaxially, 3.4–5 × 0.9–1.3 cm; petals 3.5–5 × 0.9–1.3 cm; stamens included; filaments somewhat spreading, diverging 3°–14° from axis; anthers light magenta, 0.4–1.2 cm; pollen orange; pistil 2.2–3.2 cm; ovary 1–1.8 cm; style green; pedicel 6.3–32 cm. **Capsules** 2.4–4.1 × 1.2–2 cm, 1.5–2.4 times longer than wide. **Seeds** 120–237. $2n = 24$.

Flowering summer (late May–Jul). Coastal prairies, north coastal scrub, sundew (*Drosera* spp.) bogs, gaps in closed-cone pine forests; of conservation concern; 0–200 m; Calif.

Lilium maritimum is known in California from below San Francisco to about Westport in Mendocino County, and invariably grows within a few miles of the ocean. The coast lily has suffered grave declines in the past century and has been extirpated from the southern part of its range. The recent rediscovery of a population in Marin County at Point Reyes National Seashore is heartening, but development of coastal prairie farther north continues to be a threat.

The combination of nodding habit and campanulate perianth with rolled sepal and petal tips is unique among the North American lilies. Hybrids with *Lilium pardalinum* are known; these can be clonal and their flowers are morphologically intermediate between the parents.

Lilium maritimum is pollinated sequentially, first by Allen's [*Selasphorus sassin* (Lesson), family Trochilidae] and Anna's hummingbirds [*Calypte anna* (Lesson)], and later in the blooming cycle by several different bumblebees (*Bombus* Latrielle, family Apidae).

12. Lilium kelleyanum Lemmon, Sierra Club Bull. 4: 300. 1903 • Kelley's lily [E] [F]

Lilium fresnense (Eastwood) Eastwood; *L. inyoense* Eastwood; *L. nevadense* Eastwood var. *fresnense* Eastwood; *L. nevadense* var. *monense* Eastwood; *L. pardalinum* Kellogg var. *parviflorum* Eastwood

Bulbs rhizomatous, unbranched, continuously scaly, 1.4–3.1 × 3.8–8.2 cm, 0.3–0.6 times taller than long; scales 2–3-segmented, a few unsegmented, longest 0.8–2.7 cm; stem roots absent. **Stems** to 2.2 m. **Buds** rounded in cross section. **Leaves** scattered or usually in 1–4 whorls or partial whorls, 3–10 leaves per whorl, horizontal and drooping at tips, 7.8–15.8 × 1.1–5.3 cm, 3.2–10.8 times longer than wide; blade usually elliptic, occasionally weakly lanceolate or oblanceolate, margins not undulate, apex narrowly acute; veins and margins ± smooth abaxially. **Inflorescences** racemose, 1–15(–25)-flowered. **Flowers** pendent, mildly fragrant; perianth Turk's-cap-shaped; sepals and petals reflexed 2/5 along length from base, yellow or yellow-orange, usually with copious maroon spots, not distinctly clawed; sepals not ridged abaxially, 4.3–5.7 × 0.9–1.5 cm; petals 4.2–5.7 × 0.9–1.5 cm; stamens moderately exserted; filaments moderately spreading, diverging 9°–19° from axis; anthers magenta or dull red, 0.3–0.6 cm; pollen pale orange, rust, orange, or tan-orange; pistil 2.6–3.4 cm; ovary 0.8–1.4 cm; style green; pedicel 8.5–20.7 cm. **Capsules** 1.5–2.9 × 1–1.9 cm, 1.1–2 times longer than wide. **Seeds** 96–169. $2n = 24$.

Flowering summer (Jul–Aug). Hillside seeps, wet thickets, along streams in coniferous forests; 2200–2900 m; Calif.

Lilium kelleyanum has in common with *L. parvum* a short, relatively wide fruit and small flowers, and divides with that species the montane and subalpine Sierra Nevada on both sides of the divide from Lake Tahoe south. The two species are parapatrically distributed with an intervening zone of introgression; intergrades occur, among other places, in the June Lake region about ten miles south of Mono Lake in Mono County and in adjacent Madera County. The two species appear to be sister taxa (M. W. Skinner 1988), which argues against subsuming *L. kelleyanum* as a subspecies of *L. pardalinum*, as has been suggested by several students of the group. Despite the close affinity of *L. kelleyanum* to *L. parvum*, taxonomic custom within the genus dictates that they be retained as separate species since floral features are well differentiated.

Lilium kelleyanum is pollinated by swallowtail butterflies (J. S. Davis 1956).

13. Lilium parvum Kellogg, Hesperian (San Francisco) 8: 163. 1862 • Alpine lily, Sierra tiger lily [E] [F]

Lilium canadense Linnaeus var. *parvum* (Kellogg) Hooker; *L. parvum* var. *crocatum* Stearn

Bulbs rhizomatous, unbranched, continuously scaly, 1.4–3.5 × 3.3–9.2 cm, 0.3–0.5 times taller than long; scales (1–)2–3(–4)-segmented, longest 1.1–3.4 cm; stem roots absent. **Stems** to 1.7 m. **Buds** rounded in cross section. **Leaves** in 2–5 whorls or partial whorls, 3–13 leaves per whorl, ± horizontal and drooping at tips or ascending in sun, 4–15.1 × 0.5–4.4 cm, 2.3–10.8 times longer than wide; blade ± elliptic, margins not undulate, apex acute, often narrowly so; veins and margins ± smooth abaxially. **Inflorescences** racemose, 1–26(–41)-flowered. **Flowers** ± horizontal to ascending, sometimes slightly bilaterally symmetric, not fragrant; perianth ± funnelform; sepals and petals somewhat recurved 3/5–2/3 along length from base, not strongly reflexed, lower often less recurved than upper and forming landing platform, orange or yellowish proximally, darker (light orange, orange, red-orange, red, or rarely ± pinkish) on distal 2/5, occasionally uniformly light orange or rarely yellow, spotted dark purple-brown or maroon, often pale abaxially, not distinctly clawed; sepals not ridged abaxially, noticeably wider than petals, 3.2–4.2 × 0.8–1.3 cm; petals 3.2–4.2 × 0.6–1.2 cm; stamens moderately exserted; filaments moderately spreading, diverging 8°–17° from axis; anthers pale yellow, orangish, or magenta, 0.3–0.8 cm; pollen yellowish to red-orange; pistil 2.3–3.7 cm; ovary 0.8–1.4 cm; style pale green; pedicel 3.5–15.5 cm. **Capsules** 1.6–2.7 × 1.2–1.8 cm, 1.2–1.9 times longer than wide. **Seeds** 90–132. $2n = 24$.

Flowering summer (mid Jun–Aug). Wet meadows, willow (*Salix* spp.) thickets, and streams in coniferous forests; 1400–2900 m; Calif., Nev.

At the lower elevations within its range, *Lilium parvum* tends to have clear yellow-orange flowers—a form that has been recognized as *L. parvum* var. *crocatum*—but perianth parts are redder apically at higher elevations in the Sierra Nevada. The pink-flowered form that is apparently localized at lower elevations in El Dorado County, California, is referred to informally as the ditch lily (O. H. Ballantyne 1983).

Lilium parvum hybridizes sporadically with *L. pardalinum* subsp. *pardalinum* at middle elevations (ca. 1200–1500 m) in the Sierra Nevada, resulting in variable swarms of plants with flowers that are intermediate in orientation, size, and perianth recurvature. Hybrids with *Lilium kelleyanum* are discussed under that species.

Lilium parvum is a pollination generalist visited by several species of hummingbirds (family Trochilidae), western tiger swallowtails (*Papilio rutulus* Lucas, family Papilionidae), pale swallowtails (*P. eurymedon* Lucas), and various bees (mostly family Apidae).

14. Lilium pardalinum Kellogg, Hesperian (San Francisco) 3: 300. 1859 [E] [F]

Bulbs rhizomatous, usually branching, continuously scaly, 1.4–5.1 × 3.9–19 cm, 0.2–0.6 times taller than long; scales sometimes unsegmented but always some 2–4-segmented on each bulb, longest 1–3.3 cm; stem roots absent. **Stems** to 2.8 m, strongly clonal and thus forming dense colonies, to weakly clonal and forming small colonies or clumps. **Buds** rounded in cross section. **Leaves** usually ± evenly distributed along stem, rarely concentrated proximally, scattered or in 1–6 whorls or partial whorls, 3–19 leaves per whorl, horizontal and drooping at tips to ascending, 4.9–26.5 × 0.3–5.6 cm, 3–34 times longer than wide; blade usually ± elliptic, wide or narrow, margins usually not undulate, apex acute, often narrowly so; veins and margins ± smooth abaxially. **Inflorescences** racemose, 1–28(–35)-flowered. **Flowers** pendent, usually not fragrant; perianth Turk's-cap-shaped; sepals and petals reflexed 1/4–1/3 along length from base, yellow, yellow-orange, or orange proximally, darker orange to red-orange to red on distal 1/5–3/5 (entirely orange or yellow-orange in subsp. *wigginsii*), with maroon spots concentrated proximally and always surrounded by yellow or orange if extending into distal reddish zone, conspicuously green abaxially on proximal ± 1/5, not distinctly clawed; sepals not ridged abaxially, 3.5–10.4 × 0.9–2.2 cm; petals 3.4–10.2 × 0.9–2.5 cm; stamens moderately to strongly exserted; filaments moderately to widely spreading, diverging 7°–22° from axis; anthers ± magenta or sometimes orange, orange-pink, or pale yellow, 0.5–2.2 cm; pollen red-brown, red-orange, brown-orange, rust, orange, or yellow; pistil 3.1–7.5 cm; ovary 1–2.2 cm; style green, often pale, rarely sordid; pedicel 6–32 cm. **Capsules** 2.2–5.7 × 1.2–2.1 cm, 1.5–3.7 times longer than wide. **Seeds** 123–264.

Subspecies 5 (5 in the flora): California, Oregon.

The subspecies of *Lilium pardalinum* display a classic pattern of discrete geographical ranges with intervening zones of introgression, and no two occur sympatrically without intermixing. Plants in the hybrid zones are confusing in appearance and cannot be assigned to subspecies. However, each subspecies is fairly well marked within its core distribution. With the exception of subsp. *pitkinense*, the subspecies of *L. pardalinum* can be common plants in the proper habitats within their rather narrow distributions.

Leaf size and shape are quite variable in *Lilium pardalinum* subspecies and often clearly dependent on environment. In populations that typically have narrow, ascending leaves, shaded plants often have wide, horizontal leaves. This hampers taxonomic separation as well as identification, especially of herbarium specimens. Further field study is desirable.

Lilium pardalinum is primarily pollinated by western tiger swallowtails (*Papilio rutulus* Lucas, family Papilionidae) and pale swallowtails (*P. eurymedon* Lucas); several species of hummingbirds (family Trochilidae) are also important visitors, especially when butterflies are rare.

The Atsugewi, Karok, and Yana ate *Lilium pardalinum* bulbs steamed or baked in an earth oven (D. E. Moerman 1986).

1. Sepals and petals uniformly yellow or yellow-orange; sepals 3.5–7.1 cm; anthers pale yellow, 0.5–1.3 cm; pollen yellow or orange; pistil 3.1–4.3 cm; capsules 2.3–4.2 cm; n California, s Oregon .... 14e. *Lilium pardalinum* subsp. *wigginsii*
1. Sepals and petals ± 2-toned, with yellow or orange proximally, distal $^{1}/_{5}$–$^{3}/_{5}$ darker orange to red; sepals 3.7–10.4 cm; anthers magenta, occasionally purple or orange, 0.5–2.2 cm; pollen yellow to rust; pistil 3.3–7.5 cm; capsules 2.2–5.7 cm; California, s Oregon.
   2. Sepals (5.9–)6.6–10.4 cm; anthers 1.1–2.2 cm; capsules 2.9–5.7 cm; leaves 3–12 times longer than wide, blade ± elliptic; stems strongly clonal, forming large colonies; California .. .. 14a. *Lilium pardalinum* subsp. *pardalinum*
   2. Sepals 3.7–8.3 cm; anthers 0.5–1.8 cm; capsules 2.2–4.8 cm; leaves 3–34 times longer than wide, blade elliptic to linear; stems weakly to moderately clonal, sometimes forming small colonies; n California, s Oregon.
      3. Leaves 7.3–34 times longer than wide, often concentrated proximally, often ascending, sometimes horizontal, blade ± linear; sepals (4.9–)5.3–8.3 cm; anthers 0.6–1.8 cm; pollen usually dark orange; extreme nw California, adjacent s Oregon .... 14d. *Lilium pardalinum* subsp. *vollmeri*
      3. Leaves 3–17 times longer than wide, ± evenly distributed along stem, ± ascending or horizontal, blade ± elliptic; sepals 3.7–7.6 cm; anthers 0.5–1.4 cm; pollen yellow, orange, or red- or brown-orange; n California, s Oregon.
         4. Pollen red- or brown-orange; anthers magenta; bulb scales usually 2-segmented; n Coast Ranges near Sebastopol, California .... 14b. *Lilium pardalinum* subsp. *pitkinense*
         4. Pollen usually yellow or bright orange; anthers orange to magenta; bulb scales (1–)2–4-segmented; ne California, adjacent s Oregon .... 14c. *Lilium pardalinum* subsp. *shastense*

14a. Lilium pardalinum Kellogg subsp. **pardalinum** • Leopard lily [E] [F]

Lilium harrisianum Beane & Vollmer; *L. pardalinum* var. *angustifolium* Kellogg; *L. roezlii* Regel

Bulbs branching dichotomously, 2–5.1 × 5.1–19 cm, 0.2–0.5 times taller than long, annual growth evident, generally forming discrete rounded module; scales usually 2-segmented but often unsegmented or 3-segmented, longest 1–2.7 cm. **Stems** to 2.8 m, strongly clonal, forming dense colonies. **Leaves** ± evenly distributed along stem, in 1–5 whorls or partial whorls, 3–18 leaves per whorl, usually horizontal and drooping at the tips, sometimes ± ascending, 8.2–24.5 × 0.9–5.6 cm, 3–12 times longer than wide; blade usually elliptic, sometimes ± oblanceolate, rarely lanceolate or obovate, margins usually straight but occasionally ± undulate. **Racemes** 1–28-flowered. **Flowers** rarely fragrant; sepals and petals reflexed $^{1}/_{4}$ along length from base, yellow, yellow-orange, or orange proximally, usually darker red-orange to red or sometimes pale orange on distal $^{1}/_{5}$–$^{1}/_{4}$; sepals (5.9–)6.6–10.4 × 1–2.1 cm; petals 5.9–10.2 × 1.1–2.5 cm; stamens strongly exserted; filaments often parallel over much of length, then spreading widely, diverging 14°–22° from axis; anthers usually purple, sometimes pale magenta or magenta, rarely yellowish, 1.1–2.2 cm; pollen orange-brown, rust, red-orange, or deep orange; pistil 5.6–7.5 cm; ovary 1.4–2.2 cm; pedicel 8.4–28 cm. **Capsules** 2.9–5.7 × 1.3–1.9 cm, 2–3.7 times longer than wide. **Seeds** 147–264. $2n$ = 24.

Flowering summer (Jun–late Aug). Streams and rivers, marshes, seeps, ditches, wet thickets, along coast; 0–1600 m; Calif.

Quite variable in flower size, color, and leaf size and shape, subsp. *pardalinum* is the easiest of the western lilies to cultivate. In gardens and in the wild, rhizomes branch regularly to produce clones of stems, and the resulting colonies often become large and the soil matted with bulbs.

Along the length of the Sierra Nevada this subspecies is replaced at higher elevations by congeners: to the north by subsp. *shastense*, then by *L. parvum*, and finally by *L. kelleyanum* in the high Sierra to the south. In the

Klamath Mountains it gives way to *L. pardalinum* subsp. *vollmeri* in Del Norte County, subsp. *wigginsii* in northern Humboldt County, and subsp. *shastense* in northeastern Trinity County. The range disjunction between the Coast Ranges in Santa Barbara County and the peninsular ranges in San Diego County probably exists because *L. parryi* replaces *L. pardalinum* subsp. *pardalinum* in the intervening mountains. Due to its poor condition, a collection from the Tehachapi Mountains—presumably from Kern County where the plant is otherwise unknown—cannot be confirmed as this taxon.

14b. Lilium pardalinum Kellogg subsp. **pitkinense** (Beane & Vollmer) M. W. Skinner, Novon 12: 255. 2002 • Pitkin Marsh lily [C] [E] [F]

Lilium pitkinense Beane & Vollmer, Contr. Dudley Herb. 4: 356, plate 19. 1955

Bulbs branching but less often than in subsp. *pardalinum*, otherwise similar; scales usually 2-segmented. **Stems** to 2 m, moderately clonal and usually forming small colonies. **Leaves** ± evenly distributed along stem, in 2–5 whorls or partial whorls, 3–14 leaves per whorl, ± ascending or horizontal and drooping at the tips, 6.8–22.3 × 1.2–3.6 cm, 4–13 times longer than wide; blade elliptic, sometimes narrowly so, or weakly oblanceolate, margins not undulate. **Racemes** 1–11-flowered. **Flowers** not fragrant; sepals and petals reflexed 1/3 along length from base, yellow-orange or orange proximally, darker orange-red to sometimes reddish on distal 1/3–1/2; sepals 4.9–7.1 × 1–1.7 cm; petals 4.9–7.1 × 1.2–2 cm; stamens moderately exserted; filaments moderately spreading, diverging 9°–18° from axis; anthers magenta, 0.6–1.1 cm; pollen red- or brown-orange; pistil 3.4–4.6 cm; ovary 1.2–1.9 cm; pedicel 8.5–26.5 cm. **Capsules** not measured. **Seeds** not counted.

Flowering summer (Jun–Jul). Marshes and adjacent scrub beneath valley oak (*Quercus lobata* Née); of conservation concern; 0–100 m; Calif.

The Pitkin Marsh lily is a local isolate of subsp. *pardalinum* with small flowers, short stamens, and brighter anthers known only from the vicinity of Sebastopol in Sonoma County, California. Celebrated and once quite common in its limited range, it is now protected by the state of California due to dramatic declines from land clearing for agriculture and housing, hydrological disruption, and depredations by lily fanciers and flower pickers.

The expression of subsp. *pitkinense* most faithful to the type is extremely rare and found only in Pitkin Marsh. Plants from nearby marshes are somewhat more variable, and some clones begin to approach subsp. *pardalinum* morphologically; hybrids or plants virtually indistinguishable from the nominate subspecies occur a few miles from Pitkin Marsh. Subspecies *pitkinense* can be distinguished with some difficulty from subsp. *shastense* by its darker pollen.

14c. Lilium pardalinum Kellogg subsp. **shastense** (Eastwood) M. W. Skinner, Novon 12: 255. 2002 [E]

Lilium nevadense Eastwood var. *shastense* Eastwood, Leafl. W. Bot. 1: 42. 1933; *L. nevadense* Eastwood; *L. parviflorum* (Hooker) Holzinger var. *luteum* (Purdy) Purdy; *L. parvum* Kellogg var. *luteum* Purdy; *L. shastense* (Eastwood) Beane

Bulbs branching less often and less regularly than subsp. *pardalinum*, 2.1–4.4 × 4.4–10.3 cm, 0.4–0.6 times taller than long; scales (1–)2–4-segmented, longest 1.7–3.3 cm. **Stems** to 2.1 m, weakly clonal and not forming large colonies. **Leaves** ± evenly distributed along stem, often scattered in young plants or in 1–4 whorls or partial whorls, 3–13 leaves per whorl, horizontal and drooping at tips or ascending, 5–20.5 × 0.6–3.7 cm, 3–17 times longer than wide; blade elliptic, sometimes narrowly so, occasionally weakly oblanceolate, margins undulate or not. **Racemes** 1–25(–35)-flowered. **Flowers** not fragrant; sepals and petals reflexed 1/3 along length from base, yellow-orange or orange proximally, darker orange to red-orange to red on distal 1/3–1/2; sepals 3.7–7.6 × 0.9–2.1 cm; petals 3.7–7.6 × 0.9–2.2 cm; stamens moderately exserted; filaments moderately spreading, diverging 7°–19° from axis; anthers orange, orange-pink, or ± magenta, 0.5–1.4 cm; pollen yellow to bright orange, occasionally red-orange; pistil 3.3–4.4 cm; ovary 1.2–1.8 cm; pedicel 6.3–27 cm. **Capsules** 2.2–4.3 × 1.2–1.9 cm, 1.6–2.9 times longer than wide. **Seeds** 132–177. $2n = 24$.

Flowering summer (Jul–mid Aug). Wet meadows, along streams in mixed conifer forests; 1100–1800 m; Calif., Oreg.

Subspecies *shastense* occurs in the Sierra Nevada from Plumas and Butte counties north to the region surrounding Mount Shasta, west to eastern Trinity County and to western Siskiyou County, where it intergrades with subsp. *wigginsii*. Plants from near Klamath Falls, Oregon, appear to be this subspecies, though live material from this locality has not been studied. The southern limit of subsp. *shastense* may extend as far as Placer County at moderately high elevations (ca. 1800 m) in the Sierra Nevada.

14d. Lilium pardalinum Kellogg subsp. **vollmeri** (Eastwood) M. W. Skinner, Novon 12: 257. 2002 • Vollmer's lily [E]

Lilium vollmeri Eastwood, Leafl. W. Bot. 5: 121. 1948

Bulbs branching less often and less regularly than subsp. *pardalinum*, 1.4–3.6 × 4.6–12.8 cm, 0.2–0.5 times taller than long; scales 1–2-segmented, longest 1–2.6 cm. **Stems** to 1.7 m, weakly clonal and not forming large colonies. **Leaves** often concentrated proximally, scattered, especially in small plants, or in 1–6 whorls or partial whorls, 3–15 leaves per whorl, often ascending, sometimes horizontal and drooping at the tips, 4.9–26.5 × 0.3–2.4 cm, 7.3–34 times longer than wide; blade linear to narrowly elliptic, sometimes lance-linear, especially in distal leaves, or weakly oblanceolate, especially in proximal leaves, margins not undulate. **Racemes** 1–13-flowered. **Flowers** not fragrant; sepals and petals reflexed $^{1}/_{4}$–$^{1}/_{3}$ along length from base, sometimes uniformly orange, usually yellow-orange or orange proximally, darker red-orange to red or crimson on distal $^{2}/_{5}$–$^{3}/_{5}$; sepals (4.9–)5.3–8.3 × 1–2.2 cm; petals 4.8–8 × 1–2.1 cm; stamens moderately exserted; filaments widely spreading, diverging 12°–22° from axis; anthers magenta or purple, 0.6–1.8 cm; pollen dark orange, sometimes rust-orange; pistil 3.5–5.3 cm; ovary 1.4–2.2 cm; pedicel 9–32 cm. **Capsules** 2.5–4.8 × 1.2–2 cm, 1.5–3.2 times longer than wide. **Seeds** 132–207. $2n = 24$.

Flowering summer (Jul–mid Aug). Bogs with California pitcher plant (*Darlingtonia californica* Torrey), hillside springs, streams; 100–1200 m; Calif., Oreg.

Subspecies *vollmeri* is narrowly endemic to the Siskiyou Mountains serpentine in extreme northwestern California and adjacent Oregon. A collection from near Wimer in Jackson County, Oregon, evidently represents the northernmost extent of this taxon. To the south it intergrades with subsp. *pardalinum*, but it can usually be told by its somewhat smaller and often redder flowers, and its narrowly elliptic or linear leaves that are often concentrated proximally on the stem. Northern populations in Curry County, Oregon, and those in the shade are rather similar to subsp. *pardalinum*, though the plants are usually less clonal. In the eastern part of its range, for example near Grayback Mountain in Josephine County, Oregon, and near Sanger Peak in Del Norte County, California, it intergrades extensively with subsp. *wigginsii*, producing swarms of individuals that vary in leaf arrangement and shape, and flower and anther coloration.

14e. Lilium pardalinum Kellogg subsp. **wigginsii** (Beane & Vollmer) M. W. Skinner, Novon 12: 258. 2002 • Wiggins' lily [E] [F]

Lilium wigginsii Beane & Vollmer, Contr. Dudley Herb. 4: 355, plate 18. 1955

Bulbs branching occasionally, often irregularly, 1.6–3 × 3.9–11.8 cm, 0.2–0.5 times taller than long; scales 2–4-segmented, longest 1–2 cm. **Stems** to 1.7 m, weakly clonal and not forming large colonies. **Leaves** ± evenly distributed along middle of stem, occasionally scattered, especially in young plants, or in 1–6 whorls or partial whorls, 3–19 leaves per whorl, usually horizontal and drooping at the tips, sometimes ascending, 6–24.6 × 0.5–3.5 cm, 4–22 times longer than wide; blade elliptic to narrowly linear, usually lanceolate in distal leaves, margins not undulate. **Racemes** 1–15-flowered. **Flowers** not fragrant; sepals and petals reflexed $^{1}/_{3}$ along length from base, usually uniformly orange or yellow-orange, rarely light orange or light red-orange on distal $^{1}/_{3}$–$^{2}/_{5}$ and lighter proximally; sepals appearing wide for their length, 3.5–7.1 × 1–2 cm; petals 3.4–7.1 × 0.9–2 cm; stamens moderately exserted, often malformed and/or shrunken; filaments moderately spreading, diverging 10°–18° from axis; anthers pale yellow, 0.5–1.3 cm, pollen yellow or orange; pistil 3.1–4.3 cm; ovary 1–2.2 cm; pedicel 6–29.5 cm. **Capsules** 2.3–4.2 × 1.2–2.1 cm, 1.6–2.6 times longer than wide. **Seeds** 123–198. $2n = 24$.

Flowering summer (Jul–Aug). Wet thickets and meadows among conifers; 800–2000 m; Calif., Oreg.

Subspecies *wigginsii* is a Klamath Mountain endemic that occurs widely along the county boundary between Del Norte and Siskiyou counties, California, and east through southeastern Josephine County, Oregon, to Mount Ashland in Jackson County (O. H. Ballantyne 1983). It intergrades with subsp. *shastense* in the Marble Mountains of Siskiyou County. The solid orange flowers and yellow anthers are distinctive. Genetic instability in this subspecies is frequently expressed as malformed flowers with shrunken or missing reproductive structures.

15. Lilium parryi S. Watson, Proc. Davenport Acad. Nat. Sci. 2: 188. 1878 • Lemon lily [E] [F]

Lilium parryi var. *kessleri* A. Davidson

Bulbs rhizomatous, unbranched, continuously scaly, 1.5–4.7 × 3.5–11 cm, 0.2–0.6 times taller than long; scales (1–)2(–4)-segmented, longest 0.9–3.7 cm; stem roots absent. **Stems** to 1.9 m. **Buds** rounded in cross section. **Leaves**

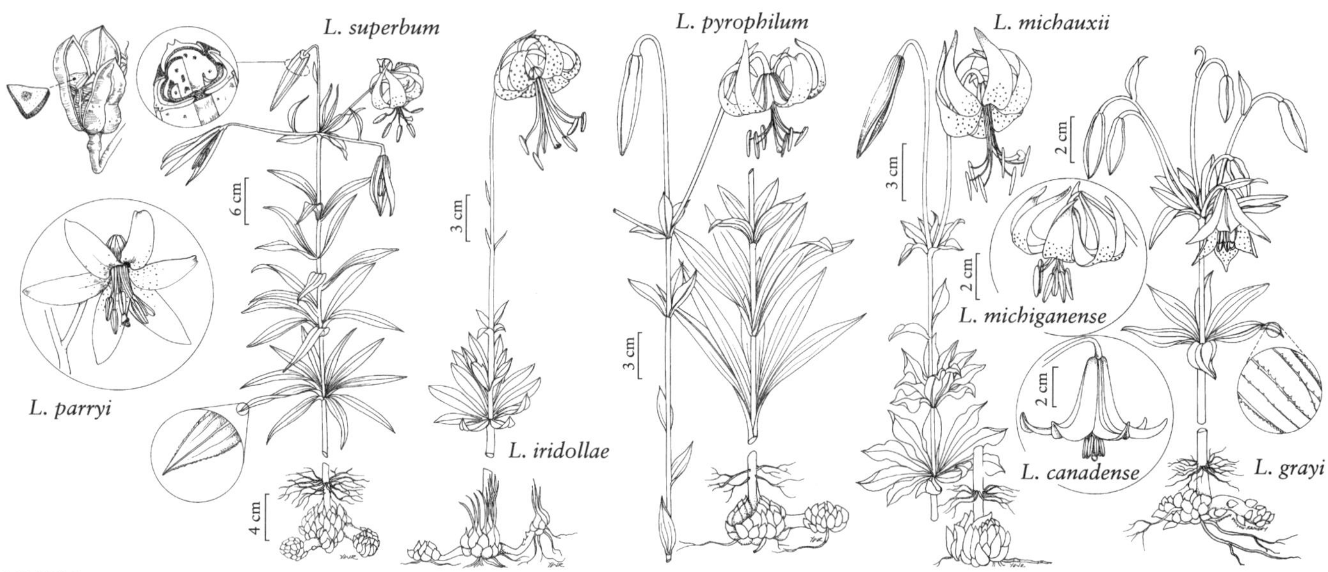

LILIUM

occasionally scattered in young plants or in 1–5 whorls or partial whorls, 3–18 leaves per whorl, ± horizontal and drooping at tips or ascending, 7.8–29 × 0.5–4.9 cm, 2.6–29 times longer than wide; blade elliptic to narrowly linear, or ± obovate, often lanceolate in distal leaves, margins not undulate, apex acute, often narrowly so; veins and margins ± smooth abaxially. **Inflorescences** racemose, 1–31-flowered. **Flowers** opening before dusk, horizontal or somewhat nodding, slightly bilaterally symmetric, strongly fragrant; perianth funnelform; sepals and petals recurved 3/5 along length from base, lower less recurved than upper and forming landing platform, bright yellow with sparse, usually minute maroon spots, not distinctly clawed; sepals not ridged abaxially, oblanceolate, 7.7–10.7 × 1.1–1.7 cm; petals noticeably wider than sepals, often very wide distally, 7.8–10.6 × 1.1–2.1 cm, apex widely acute or sometimes obtuse; stamens barely exserted; filaments barely spreading, diverging at 5°–12°; anthers pale magenta-brown, 0.8–1.4 cm; pollen rust-orange or orange-brown; pistil 5.3–9.3 cm; ovary 1.6–2.9 cm; style green, often pale; pedicel 2–17.5 cm. **Capsules** 3.9–5.9 × 1.1–1.7 cm, 2.5–4.4 times longer than wide. **Seeds** 141–303. $2n = 24$.

Flowering summer (late May–early Sep). Meadows, streams, and willow (*Salix* spp.) thickets in mixed conifer forests; 1300–2600 m; Ariz., Calif.

Plants in Arizona start to bloom in May (T. H. Kearney and R. H. Peebles 1960), while the majority of California plants flower in July and August.

Plants from the San Gabriel Mountains of California sometimes have wider leaves and have been given status as *Lilium parryi* var. *kessleri*, but this variation is due primarily to the rather shaded habitat of many of these populations. No significant vegetative discontinuity can be recognized across the range of this species, so no varieties are recognized here.

Lilium parryi probably arose from an ancestor in common with *L. pardalinum* (M. W. Skinner 1988), and subsequently diverged to become pollinated by various hawkmoths (family Sphingidae). The flowers are remarkably similar in form and function to those of *L. washingtonianum*, which is also moth-pollinated, but this resemblance is due to evolutionary convergence.

16. Lilium superbum Linnaeus, Sp. Pl. ed. 2, 1: 434. 1762 • Turk's-cap lily [E] [F]

L. canadense Linnaeus subsp. *superbum* (Linnaeus) Baker; *L. fortunofulgidum* Roane & J. N. Henry; *L. gazarubrum* Roane & J. N. Henry; *L. mary-henryae* Roane & J. N. Henry

Bulbs rhizomatous, often branching dichotomously at 120° from main axis, 2.4–4.3 × 6–10.2 cm, 0.2–0.6 times taller than long, 2(–3) years' growth evident as annual bulbs, the scaleless sections between these 0.6–3.8(–4.6) cm; scales 1–2-segmented (if 2-segmented, often only on inner scales), longest 1.2–3.9 cm; stem roots present or absent. **Stems** 1.2–2.8 m. **Buds** ± triangular in cross section. **Leaves** usually ± evenly distributed along stem, in 6–24 whorls or partial whorls, 3–20 leaves per whorl, usually ± horizontal and drooping at tips, distal leaves ascending in sun, 7.1–26.1 × 0.7–2.7 cm, 4–18 times longer than wide; blade narrowly elliptic, sometimes extremely so, occasionally barely oblanceolate, margins not undulate, apex acute, acuminate in distal leaves; veins and margins ± smooth abaxially. **Inflorescences** racemose, 1–22-flowered. **Flowers** pendent, not fragrant; perianth Turk's-cap-shaped; sepals and petals reflexed less than 1/5 along length from base, yellow or sometimes yellow-orange proximally, red-orange or sometimes red, red-purple, orange, or yellow barely suffused with red distally, spotted magenta, not distinctly clawed, nectaries exposed, forming visible green star; sepals with 2 parallel, often faint abaxial ridges, 6.8–10.5 × 1.1–2.1 cm; petals 7–10.2 × 1.4–2.6 cm; stamens strongly exserted; filaments parallel along much or most of length, then widely spreading, diverging (7°–)11°–30° from axis; anthers magenta, occasionally purple or dull purple, 1.4–2.6 cm; pollen rust; pistil 4.7–8.2 cm; ovary 1.5–3.4 cm; style pale green, often spotted purple; pedicel 7.4–19.1 cm. **Capsules** 2.9–6.2 × 1.7–2.5 cm, 1.7–3.3 times longer than wide. **Seeds** not counted. $2n = 24$.

Flowering summer (Jul–early Aug). Gaps and openings in rich woods, swamp edges and bottoms, streamsides, moist meadows and thickets, balds, pine barrens, roadsides; 0–1600 m; Ala., Ark., Conn., Del., D.C., Fla., Ga., Ill., Ind., Ky., La., Md., Mass., Miss., Mo., N.H., N.J., N.Y., N.C., Ohio, Pa., R.I., S.C., Tenn., Va., W.Va.

The largest *Lilium* east of the Rocky Mountains and the commonest over most of its range, the Turk's-cap lily is a familiar sight throughout much of the southern Appalachians and along the northern Atlantic coastal plain. Flowers are rather variable; those in the north are sometimes floridly colored with dark purple bases on the perianth parts. The single report from Vermont (F. C. Seymour 1993) remains unconfirmed; however, recent collections from Washington Parish in Louisiana, Perry County in Missouri, and Bamberg County in the coastal plain of South Carolina extend the range of this species.

Lilium superbum shares distinctive features with *L. michauxii*, *L. pyrophilum*, and *L. iridollae* that indicate a close relationship and are diagnostically useful, namely paired ridges on the backs of the sepals and buds that are triangular in cross section. None has the red style characteristic of the other eastern pendent lily clade (*L. michiganense*, *L. canadense*, and *L. grayi*) centered to the northwest, and only *L. iridollae* sometimes has the leaf margins and veins roughened abaxially as is characteristic of *L. canadense* and its close relatives.

Common inheritance of a suite of unique, derived features and peripheral allopatric distributions suggest that *Lilium pyrophilum* and *L. iridollae* may be geographical isolates of broadly distributed *L. superbum* ancestral stock; both restricted species occur in specialized wetland habitats. Geography aside, the three are unequivocally distinguished only by various combinations of leaf and bulb characteristics, flower color, habitat, and blooming time. However, the overall degree of separation is comparable to that between many other species in the genus. *Lilium superbum* blooms the earliest, is the largest, and has the most numerous and largest flowers, and the long, narrow leaves in many whorls are distinctive. *Lilium michauxii* overlaps in range with these three species but is easily separated morphologically and ecologically.

Lilium superbum is pollinated primarily by the swallowtail butterflies that are common within its range, among them the spicebush (*Papilio troilus* Linnaeus, family Papilionidae), pipevine (*Battus philenor* Linnaeus), and eastern tiger (*Papilio glaucus* Linnaeus). Great spangled fritillaries [*Speyeria cybele* (Fabricius), family Nymphalidae] also visit the Turk's-cap lily (R. M. Adams and W. J. Dress 1982).

17. Lilium iridollae M. G. Henry, Bartonia 24: 2. 1947 • Panhandle lily C E F

Bulbs rhizomatous, often branching, 1.4–2.9 × 9.6–18 cm, 0.1–0.3 times taller than long, 3–4 years' growth evident as annual bulbs, older growth often with perpendicular thin branches leading to small new bulbs and always with noticeable abscission scars, youngest 3 years' growth bearing basal leaves, scaleless sections between annual bulbs 2.7–5.4 cm; scales unsegmented, longest 1–2 cm; stem roots present or absent. **Stems** to 2 m. **Buds** ± triangular in cross section. **Leaves** in 1–5 whorls or partial whorls, 3–10 leaves per whorl, ascending, sometimes only slightly, or occasionally ± horizontal and drooping slightly at tips, 3.1–15.7 × 0.5–3.6 cm, 2.1–9.3 times longer than wide; blade weakly to sometimes narrowly oblanceolate, occasionally obovate, elliptic, or narrowly elliptic, margins not undulate, apex acute, acuminate in distal leaves; central vein in particular impressed adaxially, veins and margins often somewhat roughened abaxially with tiny ± deltoid epidermal spicules. **Inflorescences** umbellate or rarely racemose, 1–4-flowered. **Flowers** pendent, not fragrant; perianth Turk's-cap-shaped; sepals and petals reflexed less than 1/5 along length from base, yellow-orange or yellow, sometimes orange-yellow, often suffused with red distally, spotted maroon, often nearly to apex, not distinctly clawed, nectaries exposed, forming visible green star; sepals with 2 parallel, often faint abaxial ridges, 6.2–10.2 × 1.1–1.8 cm; petals 6.1–10 × 1.2–2.2 cm; stamens strongly exserted; filaments parallel along much of length, then widely spreading, diverging 14°–31° from axis; anthers magenta, 1–1.6 cm; pollen rust or sometimes rust-orange; pistil 4.7–6.8 cm; ovary 1.5–2.5 cm; style very pale green, often spotted purple near apex; pedicel 10–23 cm. **Capsules** 2.5–5 ×

1.7–3 cm, 1.2–2.2 times longer than wide. **Seeds** not counted.

Flowering summer (late Jul–late Aug). Streamsides, bogs, and seeps in wet pine woodlands, hardwood baygall scrub, wet roadside ditches, associated with pitcher plants (*Sarracenia* spp.); of conservation concern; 0–100 m; Ala., Fla.

As its vernacular name indicates, the panhandle lily is restricted to a small area in the western Florida panhandle and adjacent Alabama, where it has been much reduced in number by land-clearing for pasture and housing. It is now under consideration for federal protection.

Lilium iridollae is allopatrically distributed with its close relative *L. superbum*, though the latter occurs within 40 miles of the northernmost populations of the panhandle lily. Blooming times are staggered, *L. iridollae* typically starting to bloom as *L. superbum* is finishing. Diagnostic features include the usually brighter flower color, less numerous, shorter, noticeably though subtly oblanceolate leaves in five or fewer whorls (*L. superbum* has six or more), and few flowers, as well as very long rhizomes with basal leaves, as much as four years' visible growth, and extended scaleless sections between the annual growth bulbs.

Lilium iridollae is pollinated by all the larger swallowtail butterflies within its limited range, especially the spicebush (*Papilio troilus* Linnaeus, family Papilionidae), eastern tiger (*Papilio glaucus*, Linnaeus), and palamedes [*Papilio palamedes* (Drury)].

18. Lilium pyrophilum M. W. Skinner & Sorrie, Novon 12: 94, figs. 1, 2. 2002 • Sandhills lily [E] [F]

Bulbs rhizomatous, occasionally branching dichotomously at 120° from main axis, 2.4–2.8 × 5.2–8.6 cm, 0.3–0.5 times taller than long, 2–3 years' growth evident as annual bulbs, scaleless sections between these 0.3–2.5 cm; scales unsegmented, longest 1.1–1.9 cm; stem roots absent or sometimes present. **Stems** 0.6–1.6 m. **Buds** ± triangular in cross section. **Leaves** often concentrated proximally, occasionally scattered, usually in 1–12 whorls or partial whorls, 3–11(–15) leaves per whorl, barely to moderately ascending proximally, drooping at tips or not, ascending and appressed distally, 2.3–10.3(–12.2) × 0.8–2.4 cm, 1.6–7.6(–10.3) times longer than wide; blade narrowly elliptic, margins not undulate, apex acute, barely acuminate on distal leaves; veins and margins ± smooth abaxially. **Inflorescences** racemose, 1–7-flowered. **Flowers** pendent, not fragrant; perianth Turk's-cap-shaped; sepals and petals reflexed 1/5–1/4 along length from base, yellow (occasionally pale yellow or yellow-orange) proximally, red-orange or dusky red (occasionally magenta, pinkish, pale orange, or red) distally, spotted magenta, not distinctly clawed, nectaries exposed, forming visible green star (or triangle, when only sepal nectaries are exposed); sepals with 2 parallel, often faint abaxial ridges, 6.7–8.9 × 1.1–1.7 cm; petals 6.3–8.7 × 1.5–2.2 cm; stamens strongly exserted; filaments parallel at first, then widely spreading, diverging 12°–28° from axis; anthers magenta or sometimes purple, 1.1–1.8 cm; pollen rust; pistil 3.4–6.4 cm; ovary 1.5–2.8 cm; style pale green, often spotted purple; pedicel 6.8–16.5 cm. **Capsules** 2.8–4.7 × 1.5–1.9 cm, 1.7–2.8 times longer than wide. **Seeds** not counted.

Flowering summer (late Jul–mid Aug). Streamhead pocosins, sandhill seeps and streamsides, drainages in maintained powerlines; 0–200 m; N.C., S.C., Va.

Lilium pyrophilum is narrowly endemic to the Sandhills region of southern Virginia, North Carolina, and northern South Carolina. The most ecologically intact populations occur on military bases in the Carolinas that are carefully managed to ensure the frequent fires that promote this fire-dependent lily; many of these fires are initiated by exploding ordnance.

Its close relative, *Lilium superbum*, occurs to the west in the upper Piedmont, and is not known from the Sandhills proper. Prior to the description of *L. pyrophilum*, specimens now assignable to it were masquerading in herbaria as other species, primarily *L. michauxii*. A. E. Radford et al. (1968) mentioned robust *L. michauxii* bog plants that resembled *L. superbum*; these are considered here as *L. pyrophilum*.

Discriminating between *Lilium superbum* and larger specimens of *L. pyrophilum* may require reference to several characters. The shorter, blunter, more ascending leaves of *L. pyrophilum* average fewer per whorl and are grouped in fewer whorls, but overlap with *L. superbum* is extensive. In *L. pyrophilum* the whorls with the largest number of leaves are often clustered nearer the ground, and distally the stem bears scattered, appressed leaves, whereas the taller *L. superbum* displays whorling more or less evenly along the stem, and the few scattered leaves above the whorls are not appressed. *Lilium pyrophilum* blooms somewhat later than *L. superbum*, though some overlap occurs. The floral tube of *L. pyrophilum* is longer because the perianth parts are reflexed farther down their lengths, but this difference is subtle.

Hybrids between *Lilium michauxii* and *L. pyrophilum* are known and often display the wide petals characteristic of the former species.

The manner and frequency of visits by ruby-throated hummingbirds [*Archilochus colubris* (Linnaeus), family Trochilidae] suggest that this species is involved in pollination of the Sandhills lily. The palamedes swallowtail [*Papilio palamedes* (Drury), family Papilionidae] also visits and pollinates this species.

19. Lilium michauxii Poiret in J. Lamarck et al., Encycl., suppl. 3: 457. 1814 • Carolina lily [E] [F]

Lilium canadense Linnaeus var. *carolinianum* (Michaux) Baker; *L. carolinianum* Michaux 1803, not Bosc ex Lamarck 1792; *L. superbum* Linnaeus var. *carolinianum* (Michaux) Chapman

Bulbs chunky, 2–3.5 × 2.5–4.4 cm, 0.6–0.9 times taller than long, 2 years' growth evident; scales ± loose, unsegmented or rarely 1–2-segmented, longest 1.5–2.6 cm; stem roots present or rarely absent, often numerous. **Stems** to 1 m. **Buds** ± triangular in cross section. **Leaves** in 2–4(–5) whorls or partial whorls, 3–10(–14) leaves per whorl, ± horizontal or somewhat ascending, especially distal leaves, 3.6–11.1 × 1.5–3.8 cm, 1.8–5.1 times longer than wide; blade noticeably pale abaxially, oblanceolate, sometimes slightly or narrowly so, or occasionally obovate, texture noticeably fleshy, margins barely to strongly undulate, apex acuminate or sometimes acute, especially in distal leaves; veins and margins ± smooth abaxially. **Inflorescences** umbellate, 1–4-flowered. **Flowers** pendent, sweetly and strongly fragrant; perianth Turk's-cap-shaped; sepals and petals reflexed 1/4–1/3 along length from base, pale green then orange proximally, burnt orange-red distally, usually with numerous fine, dark maroon spots, dull abaxially, not distinctly clawed; sepals with 2 parallel, often faint abaxial ridges, 5.7–9.7 × 1.4–2.2 cm; petals 5.7–9.2 × 1.8–2.9 cm; stamens strongly exserted; filaments parallel along most of length, then widely spreading, diverging 14°–25° from axis; anthers purple, 1–2.2 cm; pollen rust or sometimes rust-brown; pistil 5–7.6 cm; ovary 1.4–2.5 cm; style very pale, often spotted purple; pedicel 9.3–17.4 cm. **Capsules** often longitudinally winged, 2.4–5.6 × 1.4–2.3 cm, 1.5–3.7 times longer than wide. **Seeds** not counted. $2n = 24$.

Flowering summer (Jul–mid Aug). Roadsides, pine-oak or open rich woods, bluffs; 0–1600 m; Ala., Fla., Ga., La., Miss., N.C., S.C., Tenn., Tex., Va., W.Va.

This is the only fragrant native lily east of the Rocky Mountains. The Carolina lily occurs in well-drained soils, while the three closely related congeners with which it is sympatric, *Lilium superbum*, *L. iridollae*, and *L. pyrophilum*, occur in either wetlands or moist woods. Where it occurs with *L. superbum*, as it often does along roadsides, *L. superbum* grows in the adjacent depressions and *L. michauxii* on the bank. Among its close relatives, its small stature, broad petals, and consistently oblanceolate leaves with undulate margins and acuminate tips are diagnostic. In these features it rather resembles several of the western taxa, especially smaller plants of *L. rubescens*; vegetative or fruiting herbarium material of the two can be confused. In *L. michauxii* the leaves are noticeably fleshy and pale abaxially, relatively few in number, and, perhaps because whorls are also few (commonly 2–4), the leaves seem concentrated near the middle of the stem.

This species is less common in the southwestern portion of its range, where it is sometimes encountered in vegetative form in shaded settings that preclude rapid growth and flowering.

Like the other southeastern pendent-flowered lilies, the Carolina lily is pollinated by large swallowtail butterflies including the eastern tiger (*Papilio glaucus* Linnaeus, family Papilionidae).

20. Lilium michiganense Farwell, Bull. Torrey Bot. Club 42: 353. 1915 • Michigan lily [E] [F]

Lilium canadense Linnaeus subsp. *michiganense* (Farwell) B. Boivin & Cody; *L. canadense* var. *umbelliferum* (Farwell) B. Boivin; *L. michiganense* var. *umbelliferum* Farwell; *L. michiganense* var. *uniflorum* Farwell

Bulbs usually yellowish, rhizomatous, unbranched, 1.6–5.8 × 4.9–14.1 cm, 0.3–0.5 times taller than long, 2 years' growth evident as annual bulbs, scaleless sections between these 2.6–6.2 cm; scales unsegmented, longest 1–3 cm; stem roots present or absent. **Stems** to 1.9 m. **Buds** rounded in cross section. **Leaves** in 4–12 whorls or partial whorls, 3–13 leaves per whorl, ± horizontal or ascending in sun, drooping at tips, 4.6–15.3 × 0.6–2.3 cm, 3.5–13.7 times longer than wide; blade narrowly elliptic, occasionally linear or slightly lanceolate, margins not undulate, apex acute, acuminate in distal leaves; principal and some secondary veins impressed adaxially, veins and margins noticeably roughened abaxially with tiny ± deltoid epidermal spicules, especially on proximal leaves. **Inflorescences** racemose, 1–11-flowered. **Flowers** ± pendent, not fragrant; perianth Turk's-cap-shaped; sepals and petals reflexed 1/4–2/5 along length from base, yellow-orange or sometimes orange-yellow or orange proximally, red-orange distally, with maroon, often large spots, red-orange or occasionally red or orange-red abaxially, not distinctly clawed; sepals not ridged abaxially, 5.5–9.3 × 1.2–2 cm; petals 5.3–9.1 × 1.5–2.2 cm; stamens moderately exserted; filaments parallel at first, then ± widely spreading, diverging 13°–23° from axis, pale yellow-green; anthers magenta or occasionally pink-magenta, 0.6–1.3 cm; pollen orange-rust, sometimes orange, rust, or rust-brown; pistil 3.4–6.5 cm; ovary 1.5–2.9 cm; style red entirely or only distally; pedicel 11–22 cm. **Capsules** 2.8–5 × 1.5–2.6 cm, 1.4–2.8 times longer than wide. **Seeds** not counted. $2n = 24$.

Flowering summer (mid Jun–Jul). Tallgrass prairies, streamsides, swamps and bottoms, moist woodland edges, lakeshores, ditches along roads and railways, often calciphilic; 100–600 m; Ont.; Ark., Ill., Ind., Iowa, Kans., Ky., Mich., Minn., Mo., Nebr., N.Y., Ohio, Okla., S.Dak., Tenn., Wis.

B. Boivin and W. J. Cody (1956) proposed uniting *Lilium michiganense* and *L. superbum* as subspecies of *L. canadense* on the basis of overall similarity, though it is now well accepted that *L. superbum* does not belong there. There can be little doubt as to the close relationship between *L. michiganense* and *L. canadense*, however, and vegetatively the two are often indistinguishable. Hybrid intermediates occur across a wide band where the distributions meet in central Ohio and northwestern New York (R. M. Adams and W. J. Dress 1982). It would not be unreasonable to include *L. grayi* and treat them as subspecies, but floral differences among the three are comparable to those between other species in the genus.

Farwell's proposed varieties *uniflorum* and *umbelliferum* were described from young plants with single flowers and umbellate inflorescences respectively, but young plants with these characteristics are found throughout the range of this species.

Plants examined from east-central Tennessee (e.g., Wayne and Coffee counties) that were previously referred to *Lilium michiganense* are *L. superbum* in some cases, in others *L. canadense* perhaps introgressed with *L. michiganense*.

The Michigan lily often co-occurs in tallgrass prairies with *Lilium philadelphicum*; here as everywhere it usually blooms much later. However, it flowers earlier than *L. canadense* where their ranges are contiguous in Ohio (E. L. Braun 1967).

Lilium michiganense is pollinated primarily by swallowtail butterflies; in the southern part of its range these include the pipevine [*Battus philenor* (Linnaeus), family Papilionidae]. Great spangled fritillaries [*Speyeria cybele* (Fabricius), family Nymphalidae] also visit this species and carry its pollen, though it is unlikely that this brushfooted butterfly is a major pollinator.

21. Lilium canadense Linnaeus, Sp. Pl. 1: 303. 1753 • Canada lily, lis du Canada [E] [F]

Lilium canadense var. *coccineum* Pursh; *L. canadense* subsp. *editorum* (Fernald) Wherry; *L. canadense* var. *editorum* Fernald; *L. canadense* var. *flavum* Pursh; *L. canadense* var. *rubrum* hort. ex T. Moore

Bulbs usually yellowish, rhizomatous, unbranched, 1.8–4.5 × 4.2–11.7 cm, 0.3–0.8 times taller than long, 2(–3) years' growth evident as annual bulbs, scaleless sections between these 0.7–5.3 cm; scales 1–2-segmented, longest 0.9–2.8 cm; stem roots present, often very many. **Stems** to 1.8 m. **Buds** rounded in cross section. **Leaves** in 6–10 whorls or partial whorls, 3–12 leaves per whorl, ± horizontal, occasionally slightly ascending, drooping at tips, 4–17.3 × 1–3.6 cm, 2.5–10 times longer than wide; blade narrowly elliptic, occasionally elliptic or slightly lanceolate, margins not undulate, apex acute, often acuminate in distal leaves; principal veins impressed adaxially, veins and margins very noticeably roughened abaxially with small ± deltoid epidermal spicules. **Inflorescences** racemose, 1–17-flowered. **Flowers** pendent, not fragrant; perianth ± campanulate; sepals and petals somewhat recurved $^1/_2$–$^3/_4$ along length from base, adaxial surface dirty yellow proximally and giving way to red dusting on tips, red or pale red abaxially, or orange adaxially and yellow-orange abaxially, or both surfaces solid yellow, spotted maroon, not distinctly clawed; sepals not ridged abaxially, 5.4–8.5 × 1.2–1.7 cm; petals 5.3–8 × 1.2–2 cm; stamens barely exserted; filaments ± parallel to style, barely spreading, diverging only 4°–6° from axis, ± same color as sepals and petals; anthers dull magenta or darker, 0.6–1.3 cm; pollen rust, sometimes light brown, rust-, tan-, or orange-brown; pistil 4.2–6.4 cm; ovary 1.5–2.8 cm; style ± same color as sepals and petals; pedicel 5–23.5 cm. **Capsules** 3–5.2 × 1.5–2.3 cm, 1.5–2.5 times longer than wide. **Seeds** not counted. $2n$ = 24.

Flowering summer (Jun–early Aug). Wet meadows, moist rich woods especially edges, streamsides and river alluvia, bogs, marshes, swamps, along wet roadsides and railroads; 0–1000(–1400) m; N.B., N.S., Ont., Que.; Ala., Conn., Del., D.C., Ga., Ky., Maine, Md., Mass., N.H., N.J., N.Y., N.C., Ohio, Pa., R.I., S.C., Tenn., Vt., Va., W.Va.

M. L. Fernald (1943e) proposed the variety *editorum* to account for the more montane plants with wide leaves (2–5 times longer than wide vs. 5–10 in var. *canadense*), red flowers with slender and elongate tubes with perianth parts arching near or above the middle but not recurved, and narrow petals (0.8–1.3 cm in dried material). Others, including E. T. Wherry (1946) and C. A. Best (1962), sought to characterize the variation better at the subspecific level, and placed more emphasis on ecological differences. In practice, most botanists who recognize var. *editorum* (e.g., R. M. Adams and W. J. Dress 1982) rely on flower color to designate the varieties, since other characters emerge as quite variable.

Field observations do not strongly support infraspecific splitting of *Lilium canadense*. Flower color varies widely, and various color forms—usually yellow and orange—are found within single populations in Massachusetts and elsewhere. As interpreted by Adams and Dress, the distributions of the proposed varieties overlap widely, and morphological evidence also offers little support. Leaves 2–10 times longer than wide occur within a sample of plants from Ohio and Alabama that is clearly referable to subsp. *editorum* in the sense of Adams and Dress, and in these plants the floral tube is wider than that of Massachusetts plants assignable to the nominate variety. Petal widths (fresh material) are 1.2–2 cm. In short, the increasingly refined attempts of the last 60 years to suitably characterize variation in this species suggest that is quite difficult or impossible to do so.

Though no specimens were seen, a report of *Lilium canadense* from Ashley County in extreme southeastern Arkansas is quite likely to represent *L. superbum*.

Field observations across the range of the species indicate that the Canada lily is pollinated primarily by ruby-throated hummingbirds [*Archilochus colubris* (Linnaeus), family Trochilidae].

Native Americans used *Lilium canadense* medicinally to treat irregular menstruation, stomach disorders, rheumatism, and snake bites. The Cherokee prepared a decoction of boiled rhizomes to fatten children (D. E. Moerman 1986).

22. Lilium grayi S. Watson, Proc. Amer. Acad. Arts 14: 256. 1879 • Gray's lily [E] [F]

Bulbs often yellowish, rhizomatous, unbranched, 2.2–2.6 × 3.8–5 cm, 0.5–0.6 times taller than long, 2 years' growth evident as annual bulbs, scaleless sections between these 1.2–2.5 cm; scales 1–2-segmented, longest 0.9–2.2 cm; stem roots present. **Stems** to 1.3 m. **Buds** rounded in cross section. **Leaves** in 3–5 whorls or partial whorls, 3–12 leaves per whorl, ± horizontal to occasionally slightly ascending, drooping at tips, 4.1–12.7 × 1.5–3.6 cm, 1.9–5 times longer than wide; blade elliptic, occasionally narrowly so or barely lanceolate, margins not undulate, apex acute, usually barely acuminate in distal leaves; principal veins impressed adaxially, veins and margins noticeably roughened abaxially with tiny ± deltoid epidermal spicules, especially apically and on proximal leaves. **Inflorescences** racemose, 1–9(–16)-flowered. **Flowers** nodding, not fragrant; perianth campanulate; sepals and petals barely recurved 2/3–9/10 along length from base, yellow-orange proximally, pale red distally, spotted maroon, pale red or sometimes red-orange abaxially, not distinctly clawed; sepals not ridged abaxially, 3.2–5.6 × 1.3–2 cm; petals 3.1–5.5 × 1.2–2 cm; stamens included; filaments ± parallel to style, barely spreading, diverging 3°–9° from axis, red; anthers magenta, 0.4–1.2 cm; pollen brown-rust; pistil 2.4–3.8 cm; ovary 0.8–1.7 cm; style red; pedicel 2.6–6.5 cm. **Capsules** 2.1–3.7 × 1.5–2.1 cm, 1.5–2.1 times longer than wide. **Seeds** not counted. $2n = 24$.

Flowering summer (late Jun–mid Jul). Grassy balds, openings in red spruce (*Picea rubens* Sargent)–Fraser fir (*Abies fraseri* (Pursh) Poiret) forests, moist hardwood bogs, seeps, and meadows at lower elevations; 1200–1900 m; N.C., Tenn., Va.

The narrowly endemic Gray's lily blooms predictably on or about July 4 in the balds and forest openings of the Roan Mountain massif shared by North Carolina and Tennessee. In its unadulterated form it also occupies the higher elevations of the Blue Ridge Mountains, including Grandfather Mountain in North Carolina and Mount Rogers and Whitetop Mountain in Virginia. A few populations occur at lower elevations (below 900 m) in streamside meadows along the Blue Ridge Parkway in northern North Carolina (Alleghany County), but in similar settings farther north in Virginia introgression with *L. canadense* occurs.

Lilium ×*pseudograyi* Grove (as species) is a name given to frequent hybrids between *L. grayi* and *L. canadense* that are scattered at somewhat lower elevations (usually 700–1000 m) in the southern Appalachians. The generally small stature of these hybrids is misleading and encourages the label of bona fide *L. grayi*, but in most respects they are intermediate. Sepal lengths of 4.8–6.2 cm and floral tube lengths of 3.2–4 cm predominate, and these are between the ranges of the two parent species. The freshwater wetland or moist hardwood habitat of these hybrids also reveals the contribution of *L. canadense* to their genome.

J. K. Small (1933) made reference to depredations by lily enthusiasts who sought Gray's lily because of its supposed rarity, and this continues today, though to a lesser degree. Of greater threat, perhaps, is succession on the high grassy balds that gradually shades and crowds the plants; like most lilies, this one requires open conditions for vigor and reproduction.

Although fritillaries (*Speyeria* spp., family Nymphalidae) pilfer nectar from flowers of Gray's lily, ruby-throated hummingbirds [*Archilochus colubris* (Linnaeus), family Trochilidae] are its only reliable pollinator. This red, tubular-flowered lily represents the zenith of pollinator-mediated evolution in the eastern true lilies, and is a high-elevation derivative of the ancestral stock that also produced *Lilium canadense*. The level of floral convergence with independently derived western *Lilium* species such as *L. bolanderi* and *L. maritimum* is remarkable and must be due to selection pressures exerted by hummingbirds during the floral evolution of these species.

27. LLOYDIA Salisbury ex Reichenbach, Fl. Germ. Excurs., 102. 1830, name conserved • Alp-lily [for Edward Lloyd (Lhwyd in Welsh), 1660–1709, curator of the Oxford Museum, who first found *Lloydia serotina* in the mountains of Wales]

James L. Reveal

Frederick H. Utech

Herbs, perennial, caulescent, glabrous, from short rhizomes with persistent leaf sheaths simulating brown, papery, tunicate bulbs. **Leaves** 2–8, reduced and alternate distally; blade linear or lanceolate. **Inflorescences** terminal, racemose, 1–2[–several]-flowered, loose, leafy-bracteate. **Flowers** slightly fragrant; perianth white with greenish to purplish veins, or yellow; tepals marcescent, 6, spreading, distinct, equal, transversely corrugate, with small, nectariferous gland above base; stamens 6, inserted at base of tepals; filaments slender; anthers basifixed, ellipsoid, latrorse; ovary superior, 3-locular; style persistent, distinct, 3-fid, short, lobes short. **Fruit** capsular, obovoid to globose, dehiscence loculicidal. **Seeds** black, 3-angled. $x = 12$.

Species 10–12 (1 in the flora): w North America, Eurasia.

1. Lloydia serotina (Linnaeus) Salisbury ex Reichenbach, Fl. Germ. Excurs., 102. 1830 [F]

Bulbocodium serotinum Linnaeus, Sp. Pl. 1: 294. 1753

Plants slender, 0.5–1.5(–1.9) dm. **Leaves** slightly dimorphic; basal 2, blade linear, 4–10(–20) cm × 0.8–1 mm; cauline 2–4, blade lanceolate, 1–5 × 0.1–0.2 cm. **Flowers** 1–2; tepals 0.9–1.2(–1.4) cm; stamens erect; filaments 3–8 mm; anthers 0.6–1(–1.2) mm. **Capsules** 6–8 mm.

Varieties 2 (2 in the flora): w North America, Eurasia.

1. Perianth white with greenish to purplish veins .... 1a. *Lloydia serotina* var. *serotina*
1. Perianth yellow with yellowish to greenish veins .... 1b. *Lloydia serotina* var. *flava*

1a. Lloydia serotina (Linnaeus) Salisbury ex Reichenbach var. **serotina** • Common alpine-lily [F]

Perianth white with greenish to purplish veins. $2n = 24, 36, 48$.

Flowering summer (Jun–Sep). Gravelly to rocky slopes and cliff faces or rock crevices, grassy tundra; 400–4400 m; B.C., N.W.T., Yukon; Alaska, Colo., Idaho, Mont., Nev., N.Mex., Oreg., Utah, Wash., Wyo.; Eurasia.

Variety *serotina*, the typical, arctic-alpine variety (T. Irmisch 1863; B. Jones and C. Gliddon 1999), is wide-ranging throughout the Northern Hemisphere (N. Woodhead 1951).

1b. Lloydia serotina (Linnaeus) Salisbury ex Reichenbach var. **flava** (Calder & Roy L. Taylor) B. Boivin, Naturaliste Canad. 94: 525. 1967 • Golden alpine-lily [E]

Lloydia serotina subsp. *flava* Calder & Roy L. Taylor, Canad. J. Bot. 43: 1392. 1965

Perianth yellow with yellowish to greenish veins. $2n = 24$.

Flowering summer (Jun–Sep). Rocky slopes and exposed habitats of subalpine zones; 100–1000 m; B.C.

Variety *flava* from the Queen Charlotte Islands (Graham and Moresby islands) differs from the typical variety in floral color and is more robust with well-developed leaves that nearly always exceed the flowering or fruiting scapes (J. A. Calder and R. L. Taylor 1965).

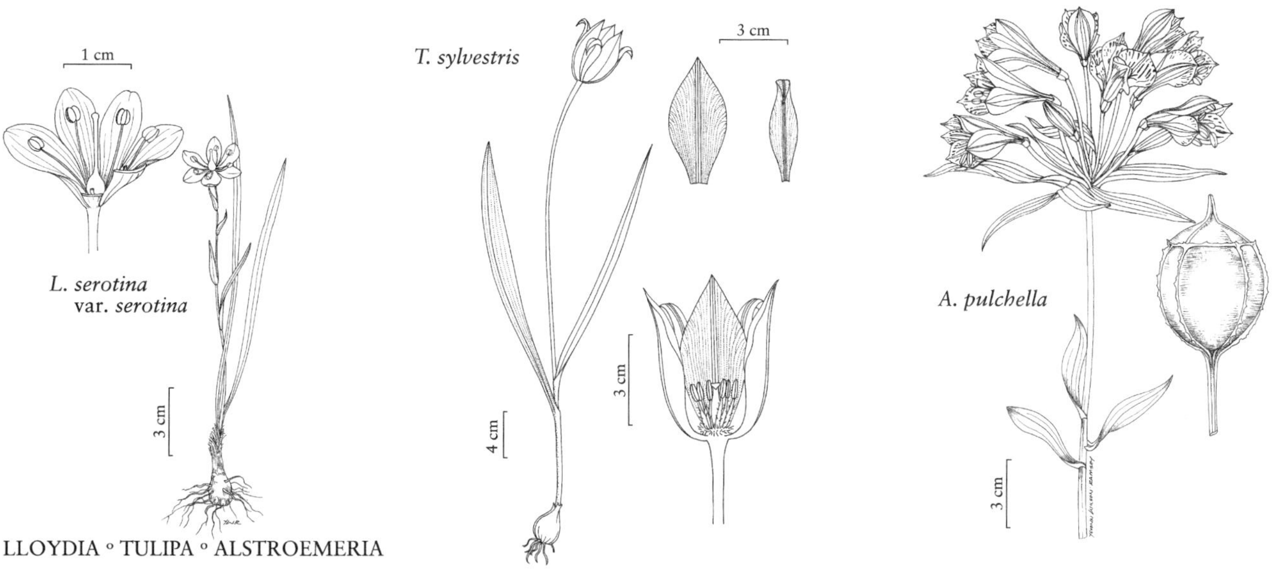

LLOYDIA ◦ TULIPA ◦ ALSTROEMERIA

28. TULIPA Linnaeus, Sp. Pl. 1: 305. 1753; Gen. Pl. ed. 5, 145. 1754 • Tulip, tulipe, tulipán [Persian *thoulyban* or Turkish *tulbend*, turban, alluding to the shape of the just-opening perianth] [I]

Gerald B. Straley†

Frederick H. Utech

Herbs, perennial, scapose or subscapose, bulbose; bulbs often stoloniferous, tunicate, papery to coriaceous; tunics variously hairy or glabrous adaxially. **Leaves** 2–6(–12), cauline, alternate, reduced distally; blade linear to oblong, somewhat fleshy. **Inflorescences** 1(–4)-flowered, bracts usually absent. **Flowers:** perianth campanulate to cup-shaped; tepals caducous, 6, distinct, often blotched near base, petaloid, ± equal; nectaries absent; stamens 6, distinct; filaments shorter than tepals, basally dilated; anthers basifixed, linear to narrowly elliptic, introrse; ovary superior, 3-locular; style very short or absent; stigma prominently 3-lobed. **Fruits** capsular, ellipsoid to subglobose, 3-angled, leathery, dehiscence loculicidal. **Seeds** many, in 2 rows per locule, flat. $x = 12$.

Species ca. 150 (1 in the flora): introduced; North America, temperate Eurasia (especially c, w Asia), n Africa, cultivated worldwide.

The common garden tulip (*Tulipa gesneriana* Linnaeus) and a number of other species (*T. bakeri* Hall, *T. clusiana* de Candolle, *T. fosteriana* Hoog ex W. Irving, *T. kaufmanniana* Regel, *T. tarda* Stapf), as well as a vast array of complex hybrid cultivars, are commonly planted for their spring flowers. Over 3500 names applied to tulips are currently listed (J. van Scheepen 1996). While some of these species or cultivars may persist for a short time, they rarely become truly naturalized in the flora. Taxonomic difficulties abound in *Tulipa* due to their long-established cultivation, hybridization, and selection.

Viral infection of tulips results in odd, yet often attractive, colored streaks in the flowers. In the early 1600s these variants, called "broken" tulips, became prized in the Netherlands, widely sought, and worth considerable money. The ensuing "tulipomania" lead to widespread trading, speculation, and then, as with most similar fads, a sudden market collapse in 1637 (W. Blunt 1950; M. Dash 1999; F. A. Stafleu 1963).

SELECTED REFERENCES Botschantzeva, Z. P. 1982. Tulips: Taxonomy, Morphology, Cytology, Phytogeography and Physiology, transl. and ed. H. Q. Varekamp. Rotterdam. Hall, A. D. 1940. The Genus *Tulipa*. London. Pavord, A. 1999. The Tulip. New York, London, and Bloomsbury.

1. **Tulipa sylvestris** Linnaeus, Sp. Pl. 1: 305. 1753 • Wild tulip [F] [I]

Plants (10–)30–45 cm; bulbs often stoloniferous, ovoid to obovate, 1.8–4(–5) × 1–2.5 cm; tunics brownish black, papery, apex with rusty, straight hairs adaxially. **Leaves** 2–3(–4), widely spaced, more or less declinate, channeled; blade dark green, linear to linear-lanceolate, keeled, glabrous; proximal cauline blade 25–35 × 1–1.5(–2) cm. **Flowers** 1(–2), nodding in bud; tepals 6(–8), golden yellow, midrib abaxially tinged with green; outer tepals narrowly elliptical to subrhombic, 30–40(–65) × 8–15(–25) mm, apex acuminate; inner tepals elliptic-oblanceolate, 30–40(–60) × 15–20(–25) mm, apex shortly acuminate; filaments flat, 8–14 mm, pubescent; anthers orange, 4–9 mm; ovary bright green, 10–12 mm. **Capsules** 15–30 × 14–16 mm. $2n = 36$.

Flowering late spring. Roadsides, fields, waste places; 0–200 m; introduced; Md., Mass., Pa.; Europe, n Africa, naturalized c, sw Asia, Siberia.

The exact native range of *Tulipa sylvestris* in the Old World is unknown. Stoloniferous, offset droppers account for much of the vegetative spread of this triploid species.

29. ALSTROEMERIA Linnaeus, Pl. Alströmeria, 8. 1762 • Lily-of-the-Incas, Peruvian-lily [for Clas Alströmer, 1736–1794, Swedish naturalist and pupil of Linnaeus] [I]

Walter C. Holmes

Herbs, perennial, from fascicles of fusiform tubers. **Stems** mostly simple; fertile stems to 1 m or more; sterile stems shorter, more leafy. **Leaves** alternate; petiole often twisted so as to invert leaf; blade parallel-veined, linear to ovate, margins entire. **Inflorescences** terminal, umbellate [or 1-flowered]. **Flowers** slightly zygomorphic; tepals 6, distinct, red, orange, purple, green, or white, frequently spotted, to 5 cm; stamens 6, inserted on perianth base, declinate, usually unequal; ovary inferior; style slender; stigma 3-lobed, filiform. **Fruits** capsular, 3-valved, dehiscence loculicidal.

Species ca. 60 (1 in the flora): introduced; North America, South America.

SELECTED REFERENCES Aker, S. and W. Healy. 1990. The phytogeography of the genus *Alstroemeria*. Herbertia 46: 76–87. Uphoff, J. C. T. 1952. A review of the genus *Alstroemeria*. Pl. Life 8: 37–53.

1. **Alstroemeria pulchella** Linnaeus f., Suppl. Pl., 206. 1782 [F] [I]

Alstroemeria brasiliensis Sprengel; *A. psittacina* Lehmann

Plants erect, leafy, weak-stemmed. **Leaves** 1.5–3 × 6–11 cm; blade narrowly lanceolate to elliptic-spatulate on fertile shoots, broader on sterile shoots, glabrous, base cuneate, apex acute to rounded; petiole 2–4 cm. **Umbels** 3–8-flowered, bracteate. **Flowers:** tepals red to wine, spotted purplish brown, greenish at apex, narrowly spatulate, 3.5–4.5 cm, apex and margins of base slightly puberulent; pedicel 1.5–3 cm. **Capsules** globose, 1.3 × 1 cm, bluntish terminal portion with several spinose acuminations, each valve narrowly winged adaxially.

Flowering mid spring–early summer. Loose or sandy soils in shade to partially sunny, moist, disturbed sites; 0–100 m; introduced; Ala., Fla., Ga., La., Miss.; Brazil.

Alstroemeria pulchella is used as an outdoor ornamental in the warmer parts of the United States, where it has escaped and apparently become naturalized.

30. HYPOXIS Linnaeus, Syst. Nat. ed. 10, 2: 986, 1366. 1759 • Star-grass [Greek *hypo*, under, and *oxys*, sharp, referring to the pointed bases of the ovaries]

Alan Herndon

Herbs, perennial, scapose, glabrous or sparsely to densely pubescent, often pilose, pubescence including at least some irregularly stellate trichomes, rhizomatous or cormose. **Stems** subterranean, usually vertical, fleshy. **Leaves** grasslike; blade linear to setaceous. **Scape** usually shorter than leaves. **Inflorescences** depauperate racemes or umbels, borne singly in leaf axils, bracteate. **Flowers:** tepals 6, spreading, distinct, often greenish abaxially, yellow adaxially, outer usually ± pilose abaxially; anthers 6, spreading, shortly connate at bases; ovary inferior, usually densely pubescent to pilose, sometimes glabrate; style erect. **Fruits** capsular, crowned by persistent flower parts throughout maturation. **Seeds** (5–)10–50 per capsule, ± globose, hilum and micropyle prominent, surfaces sharply to bluntly muricate or with rounded pebbling, sometimes with iridescent, membranous coat.

Species 100 (7 in the flora): mainly Southern Hemisphere.

Most species of *Hypoxis* are in southern Africa. The key to flowering plants presented below will work for most specimens; some specimens have intermediate characteristics and are impossible to identify without seeds. When mature seeds are available, identifications should be confirmed by examination of the seed characters. Relationships among the species in the flora are uncertain and will remain so until a better understanding of the primarily African *H. angustifolia* complex is reached. Therefore, species are treated here in alphabetical order.

SELECTED REFERENCE Britt, R. F. 1967. A revision of the Genus *Hypoxis* in the United States and Canada. Ph.D. dissertation. University of North Carolina.

1. Leaves glabrous or nearly so, or with scattered trichomes near base.
    2. Leaves 0.3–1 mm wide, hard, stiff; floral bracts 1–7(–12) mm; ovary densely pubescent; seeds with low, rounded sculpturing . . . . . 3. *Hypoxis juncea* (in part)
    2. Leaves 1–15 mm wide, soft, flexible; floral bracts 1–20(–80) mm; ovary glabrate or sparsely to densely pubescent; seeds muricate.
        3. Pedicel usually 2+ times longer than bracts; tepals much longer than ovary; ovary obconic, densely pubescent . . . . . 2. *Hypoxis hirsuta* (in part)
        3. Pedicel usually shorter than bracts; tepals usually 2 or less times longer than ovary; ovary cylindric, glabrate or sparsely pubescent . . . . . 1. *Hypoxis curtissii*
1. Leaves sparsely to densely pubescent, at least near base.
    4. Pedicel usually 2+ times longer than bracts.
        5. Leaves 1–15 mm wide; seeds coarsely muricate . . . . . 2. *Hypoxis hirsuta* (in part)
        5. Leaves 0.3–1 mm wide; seeds with low, rounded sculpturing . . . . . 3. *Hypoxis juncea* (in part)
    4. Pedicel usually less than 2 times as long as bracts.
        6. Pedicel usually longer than bracts; tepals longer than pedicel; anthers 1.5–3.5 mm; (old leaf bases persisting as bristles) . . . . . 5. *Hypoxis rigida*
        6. Pedicel shorter to longer than bracts; tepals shorter to longer than pedicel; anthers 0.6–2.2 mm.
            7. Tepals 1.5–2 times as long as ovary . . . . . 6. *Hypoxis sessilis*
            7. Tepals less than 1.5 times as long as ovary.
                8. Scape topped by 2 opposite bracts . . . . . 4. *Hypoxis mexicana*
                8. Scape topped by 1 bract or, if inflorescences 2- or more-flowered, bracts not opposite . . . . . 7. *Hypoxis wrightii*

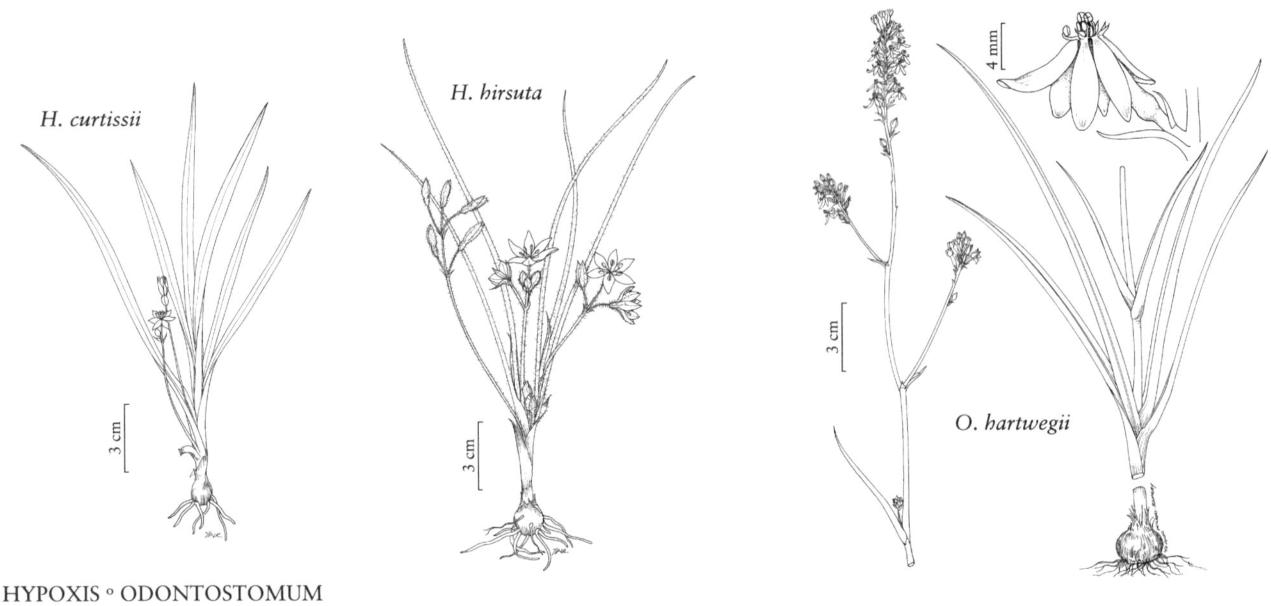

HYPOXIS ◦ ODONTOSTOMUM

1. **Hypoxis curtissii** Rose in J. K. Small, Fl. S.E. U.S., 287, 1329. 1903 [E] [F]

Hypoxis erecta Linnaeus var. *leptocarpa* Engelmann & A. Gray; *H. hirsuta* (Linnaeus) Coville var. *leptocarpa* (Engelmann & A. Gray) Brackett; *H. leptocarpa* (Engelmann & A. Gray) Small

Leaves 3–12 mm wide, soft, flexible, glabrous or nearly so. **Scape** (4–)5–13(–27) cm. **Inflorescences** racemose, 1–3(–7)-flowered; proximal 2 flowers, when present, not paired; bracts (3–)5–20(–80) mm. **Flowers:** tepals 4–8(–11) × 0. 5–2.1 mm, 2 or less times longer than pedicel, usually 2 or less times longer than ovary; anthers 1–1.8 mm; ovary cylindric, (2–)3–6(–7) × 1–3 mm, glabrate or sparsely pubescent; pedicel (2–)5–12 (–20) mm, usually shorter than bracts. **Seeds** black, lustrous, 1.2–1.6 mm, coarsely muricate.

Flowering year round. Riverbanks, floodplains; 0–100 m; Ala., Fla., Ga., La., Miss., N.C., S.C., Tex., Va.

When not in flower, the rosettes of *Hypoxis curtissii* can be easily mistaken for *Cyperus,* which grows in the same habitats.

2. **Hypoxis hirsuta** (Linnaeus) Coville, Bull. Torrey Bot. Club 5: 118. 1894 [E] [F]

Ornithogalum hirsutum Linnaeus, Sp. Pl. 1: 306. 1753; *Hypoxis carolinensis* Michaux; *H. erecta* Linnaeus; *H. graminea* Pursh; *H. grandis* Pollard; *H. micrantha* Pollard; *H. pallida* Salisbury; *H. villosa* Rafinesque

Leaves (1–)2–5(–15) mm wide, soft, flexible, essentially glabrous or sparsely to densely pubescent. **Scape** (4–)5–17(–25) cm. **Inflorescences** umbellate, 1–5(–7)-flowered; proximal 2 flowers, when present, always paired; bracts (1–)2–10 (–17) mm. **Flowers:** tepals (3–)6–10(–17) × 2–4 (–5) mm, shorter than pedicel, much longer than ovary; anthers 1.5–3.5(–4) mm; ovary obconic, (1–)2–4 × (1–)2–4 mm, densely pubescent; pedicel (6–)9–25(–30) mm, usually 2+ times longer than bracts. **Seeds** black, lustrous, (0.9–)1–1.5(–1.7) mm, coarsely muricate.

Flowering spring–summer. Wet to dry woodlands and prairies; 0–1700 m; Man., Ont., Sask.; Ala., Ark., Colo., Conn., Del., D.C., Ga., Ill., Ind., Iowa, Kans., Ky., La., Maine, Md., Mass., Mich., Minn., Miss., Mo., Nebr., N.H., N.J., N.Y., N.C., N.Dak., Ohio, Okla., Pa., R.I., S.C., S.Dak., Tenn., Tex., Vt., Va., W.Va., Wis.

Very rarely, specimens of *Hypoxis hirsuta* have cylindrical ovaries and/or bracts nearly equaling the pedicels. These specimens are recognized by the umbelliform inflorescence typical of *H. hirsuta*. The condition appears to be pathological and is associated with a lack of seed development.

3. **Hypoxis juncea** Smith, Spic. Bot., 15, plate 16. 1792 [E]

Hypoxis filifolia Elliott

Leaves 0.3–1 mm wide, hard, stiff, nearly glabrous or with scattered trichomes near base. **Scape** (5–)6–16(–19) cm. **Inflorescences** racemose, 1–2(–3)-flowered; proximal 2 flowers, when present, not paired; bracts 1–7(–12) mm. **Flowers:** tepals 6–13(–15) × 2–4 mm, usually shorter than pedicel, longer than ovary; anthers (1.3–)1.8–3.3 mm; ovary obconic, (2–)3–5(–7) × 1.5–2.5 mm, densely pubescent; pedicel (5–)10–26(–28) mm, usually 2+ times longer than bracts. **Seeds** black, lustrous, 0.7–1 mm, with low, rounded sculpturing.

Flowering year round, most commonly in spring. Seasonally flooded pinelands; 0–100 m; Ala., Fla., Ga., N.C., S.C.

4. **Hypoxis mexicana** Schultes & Schultes f. in J. J. Roemer et al., Syst. Veg. 7(2): 761. 1830

Leaves (1–)2–4(–9) mm wide, soft, stiff, densely pubescent. **Scape** 4–12(–16) cm, topped by 2 opposite bracts. **Inflorescences** racemose, 1–3-flowered; proximal 2 flowers, when present, always paired; bracts (3–)4–8(–11) mm. **Flowers:** tepals (3–)4–7(–8) × 1.3–2.5 mm, shorter to longer than pedicel, ± equaling ovary; anthers 1–2 mm; ovary oblanceoloid, 3–6(–9) × 1–2 mm, densely pubescent; pedicel 1–6 mm, shorter than bracts. **Seeds** light brown, dull, 1–1.4(–1.5) mm, muricate.

Flowering summer. Pine-oak woodlands; 2000–2500 m; Ariz.; Mexico.

Hypoxis mexicana is common in Mexico; it is rarely collected in Arizona.

5. **Hypoxis rigida** Chapman, Fl. South. U.S. ed. 2 repr. 1892, 696. 1892 [E]

Leaves (8–)10–20(–26) mm wide, soft, stiff, sparsely to densely pubescent, old leaf bases persisting as bristles. **Scape** (4–)6–13(–18) cm. **Inflorescences** racemose, 1–2(–4)-flowered; proximal 2 flowers, when present, not paired; bracts (2–)3–5(–6) mm. **Flowers:** tepals 5–8(–10) × 1.4–2.5 mm, longer than pedicel, longer than ovary; anthers 1.5–3.5 mm; ovary oblanceoloid, 3–5(–7) × 2–3 mm, densely pilose; pedicel (2–)3–8(–15) mm, usually less than 2 times as long as bracts. **Seeds** black, lustrous, 1–1.3 mm, with rounded pebbling.

Flowering spring–summer. Seasonally flooded pinelands and bogs; 0–100 m; Ala., Fla., Ga., La., Miss., S.C., Tex.

6. **Hypoxis sessilis** Linnaeus, Sp. Pl. ed. 2, 1: 439. 1762 [E]

Hypoxis erecta Linnaeus var. *aestivalis* Engelmann & A. Gray; *H. longii* Fernald

Leaves 0. 9–5 mm wide, soft, usually stiff, sparsely to densely pubescent. **Scape** (3–)4–9(–14) cm. **Inflorescences** racemose, 1(–2)-flowered; proximal 2 flowers, when present, not paired; bracts (2–)3–7(–8) mm. **Flowers:** tepals (6–)7–10(–13) × 0.9–2.4 mm (chasmogamous flowers), shorter to longer than pedicel, 1.5–2 times as long as ovary; anthers 1.1–2.2 mm; ovary oblanceoloid, (3–)4–5(–6) × 1.5–2.5 mm, densely pilose; pedicel (2–)3–7(–8) mm, equaling or slightly shorter than bracts. **Seeds** black beneath loose, iridescent, membranous coat, (1.3–)1.4–2 mm, with rounded pebbling.

Flowering spring–summer. Mesic to moist pinelands; 0–200 m; Ala., Ark., Fla., Ga., La., Miss., N.C., Okla., S.C., Tex., Va.

Although there is a strong tendency for summer flowering in *Hypoxis sessilis*, in the southern part of its range it will flower at any time of year in response to fire or other disturbances.

7. **Hypoxis wrightii** (Baker) Brackett, Rhodora 25: 140. 1923

Hypoxis juncea Smith var. *wrightii* Baker, J. Linn. Soc., Bot. 17: 106. 1878; *H. humilis* Tharp

Leaves (0.5–)0.7–2.1(–2.6) mm wide, soft, usually stiff, sparsely to densely pubescent to pilose. **Scape** (20–)35–85(–150) mm, topped by 1 bract or, if flowers 2 or more, bracts more than 1, not opposite. **Inflorescences** racemose, 1–2(–3)-flowered; proximal 2 flowers, when present, not paired; bracts (2–)3–8(–15) mm. **Flowers:** tepals 4–8(–10) × (1.4–)2–3(–3.5) mm, shorter to longer than pedicel, less than 1.5 times as long as ovary; anthers (0.6–)0.9–1.8(–2) mm; ovary oblanceoloid, (2–)3–6(–8) × 1.5–2.5 mm, usually densely pilose; pedicel (1–)3–12(–21) mm, usually longer than bracts. **Seeds** dark brown, dull, 0.9–1.1(–1.3) mm, minutely muricate.

Flowering spring. Pinelands, seasonally flooded prairies; 0–100 m; Ala., Fla., Ga., La., Miss., N.C., S.C., Tex.; West Indies.

In the southern part of its range, *Hypoxis wrightii* can be found flowering shortly after fire at any time of year. The name *H. micrantha* Pollard was misapplied to *H. wrightii* by A. E. Brackett (1923) and subsequent authors.

31. ODONTOSTOMUM Torrey, Pacif. Railr. Rep. 4(5): 150. 1857 • [Greek *odontos*, tooth, and *stoma*, mouth, alluding to the erect, subulate filaments at the flower throat] E

Mark W. Skinner

Herbs, ± scapose, bulbose; bulbs deep seated, cormose, fibrously tunicate. **Stems** solitary, ± erect, flexuous. **Leaves** few, mostly basal, sheathing; blade linear, obscurely scabrous. **Inflorescences** maturing acropetally, usually paniculate (racemose in small plants), many-flowered; floral bracts usually 1, minute, with adaxial bractlet near middle. **Flowers** ascending; perianth tardily deciduous; tepals 6, connate basally into tube, limb lobes sharply reflexed, strongly veined; stamens 6, opposite tepals and inserted on throat, alternating with 6 short staminodia, the abaxial 1 opposite a tepal standing alone and opposing 5 aggregate adaxial tepals; filaments narrowly subulate; anthers basifixed, dehiscent through apical pores; pistil 1, compound, 3-lobed, 3-locular; ovary superior, globose, ovules 2 per locule, ascending, only 1 developing; style deciduous, 1, filiform, slightly 3-fid; stigmas 3; pedicels shorter than flowers. **Fruits** capsular, apex loculicidal, valvate; valves 3, obovoid. **Seeds** 3, obovoid; testa dark brown, close, thin, slightly rugose. $x = 10$.

Species 1: California.

R. M. T. Dahlgren et al. (1985) placed *Odotostomum* in the family Tecophilaeaceae, where it fits securely if one is recognizing the many possible segregate families of Liliaceae. With some reservation, these authors put superficially similar Californian genera such as *Chlorogalum* and *Schoenolirion* in Hyacinthaceae, which suggests that the relationship of these genera to *Odontostomum* is somewhat distant.

1. **Odontostomum hartwegii** Torrey, Pacif. Railr. Rep. 4(5): 150. 1857 E F

Stems to 5.5 dm. **Bulbs** subglobose, slightly broader than tall, 1–3 cm wide. **Leaves** basal and cauline, tapering gradually to attenuate apex; cauline leaves smaller than basal, gradually reduced to bracts subtending panicle branches; blade (6.5–)10–25 × (0.2–)0.5–0.8 cm. **Inflorescences** 0–4-branched, each branch (1–)5–40-flowered; bracts subulate, ± equaling pedicels. **Flowers:** perianth creamy white to yellowish; perianth tube 4–6 mm, limb lobes ca. as long; outer 3 tepals 5–7-veined, ± lanceolate, apex acute; inner 3 tepals 3–5-veined, ± oblanceolate, apex obtuse; filaments 1–2 mm; staminodes 0.5–1 mm; pedicels 3–5 mm. **Capsules** ca. 4 mm wide.

Flowering spring (mid Apr–mid Jun). Grassland, savanna with blue oak (*Quercus douglasii* Hooker & Arnott) or digger pine (*Pinus sabiniana* Douglas ex D. Don), usually in well-drained, rocky clay soils (often serpentine), but sometimes on vernal pool margins; 0–600 m; Calif.

Odontostomum hartwegii ranges from Shasta County south to Mariposa County along the Sierra foothills, and locally in the Coast Ranges in Tehama and Napa counties. It is more frequent in the northern portion of this range, from Butte to Shasta counties.

32. CONVALLARIA Linnaeus, Sp. Pl. 1: 314. 1753; Gen. Pl. ed. 5, 383. 1754 • Lily-of-the-valley [Latin *convallis*, valley]

Frederick H. Utech

Herbs, perennial, scapose, from slender, branched rhizomes with upright buds (pips); forming colonies. **Leaves** 2–3, basal, erect, sheathing, petiolate; petiole erect, slender; blade oblong to widely elliptic, glabrous, apex acute to acuminate. **Scape** arising basally. **Inflorescences** racemose, 5–15-flowered, bracteate; raceme 1-sided, 1–3 dm; bracts subtending pedicel. **Flowers** deciduous, nodding, very fragrant; perianth globose-campanulate; tepals almost completely connate, distinct lobes outwardly recurved; stamens 6, included, attached at perianth tube base; anthers oblong to lanceolate, introrse; ovary superior, 3-locular, syncarpous, ovules 4–8 per locule; septal grooves absent; style included, connate, straight; stigmas weakly 3-lobed to capitate; pedicel recurved, filiform. **Fruits** baccate, orangish red, globose, pulpy. **Seeds** yellow to light brown, subglobose to angled, smooth. $x = 19$.

Species 1: n temperate North America and Eurasia.

As described here, *Convallaria* may be viewed as monotypic with three (or more) isolated infraspecific taxa, or else as three morphologically similar but isolated species (B. E. Streveler 1966). Our native taxon, *Convallaria majalis* var. *montana,* is more closely related to that in eastern Asia, var. *keiskei* (Miquel) Makino (= *C. keiskei* Miquel), than to that of western Europe, var. *majalis* (M. L. Fernald 1928c, 1944c). While the genus is well known for its sweet fragrance, potential poisonings are a concern due to the more than 38 cardenolide glycosides (G. E. Burrows and R. J. Tyrl 2001), including convallatoxin, found throughout these ornamentals (W. H. Lewis and M. P. F. Elvin-Lewis 1977).

SELECTED REFERENCES Streveler, B. E. 1966. A Taxonomic Study of the Genus *Convallaria* (Liliaceae). M.S. thesis. University of Wisconsin. Utech, F. H. and S. Kawano. 1976c. Floral vascular anatomy of *Convallaria majalis* L. and *C. keiskei* Miq. (Liliaceae-Convallarieae). Bot. Mag. (Tokyo) 89: 173–182.

1. Convallaria majalis Linnaeus, Sp. Pl. 1: 314. 1753 [F]

Plants 1–3 dm. **Leaves** 15–50 × 2.5–13 cm; petiole 4–30 cm; blade 9–26 cm, strongly or weakly veined. **Inflorescence bracts** scarious, 4–20 mm. **Flowers:** tepals 6, white, midribs prominent, white or green, lobes 1–3 mm, with microscopic glands; pedicel 7–11 mm. **Berries** 1–9, 1–5-seeded, 6–12 mm. **Seeds** 3–4 mm. $2n = 38$.

Varieties 3 (2 in the flora): deciduous forests of e North America, w Europe, e Asia.

1. Plants forming dense colonies; leaf blades green until frost; bracts 4–10 mm, shorter than pedicel; tepal midribs white . . . . 1a. *Convallaria majalis* var. *majalis*
1. Plants scattered or forming small groups; leaf blades yellowing in late summer; bracts 10–20 mm, equaling or longer than pedicel; tepal midribs green . . . . 1b. *Convallaria majalis* var. *montana*

1a. Convallaria majalis Linnaeus var. majalis • European lily-of-the-valley, muguet [I]

Plants forming dense colonies. **Leaves** 15–40 × 2.5–5 cm; petiole 8–24 cm; blade green until frost, veins and veinlets relatively faint in transmitted light, 9–20 cm. **Scape** 1–2.3 dm. **Inflorescences** 1/2 length of to equaling proximalmost sheathing leaves; bracts broadly lanceolate, 4–10 mm, shorter than pedicel. **Flowers:** tepal midribs white; pedicel 7–12 mm. $2n = 38$.

Flowering early–late spring. Disturbed, open areas; 0–1500 m; introduced; N.B., Nfld. and Labr., N.S., Ont., P.E.I., Que., Sask.; Ala., Ark., Conn., D.C., Ill., Ind., Iowa, Kans., Ky., Maine, Md., Mass., Mich., Minn., Mo., N.H., N.J., N.Y., N.C., Ohio, Pa., R.I., Tenn., Utah, Vt., Va., Wash., W.Va., Wis.; expected elsewhere; Eurasia.

Variety *majalis* is widely cultivated and naturalized throughout most of eastern North America and elsewhere

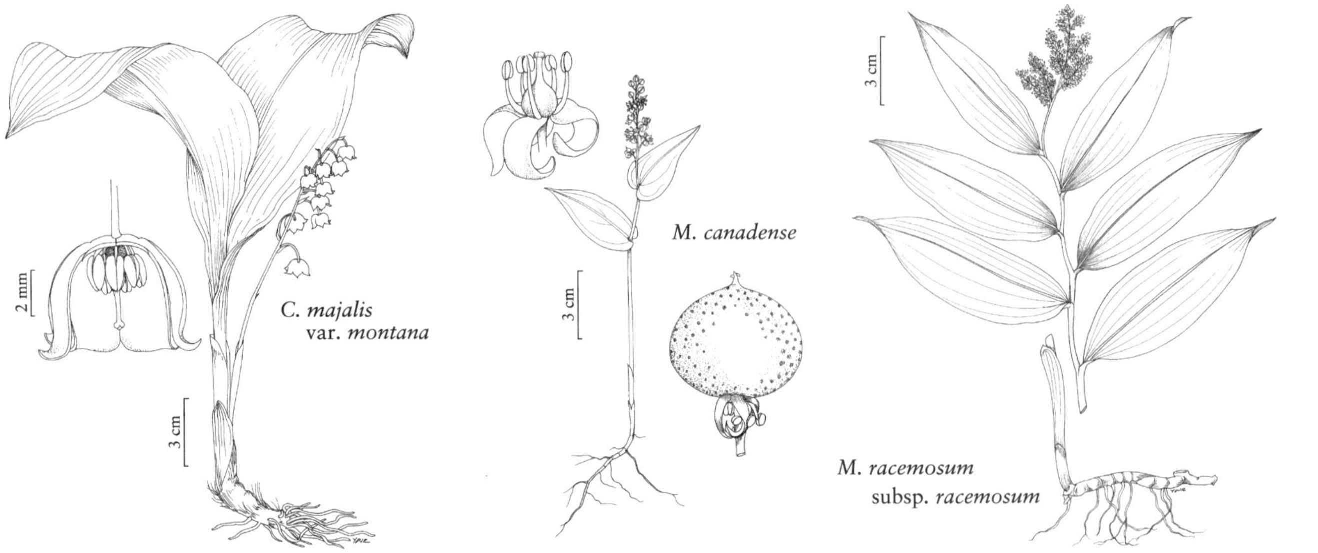

CONVALLARIA ∘ MAIANTHEMUM

near cemeteries, gardens, roadsides, and thickets. Rhizomatous spread results in highly dense, ground-covering carpets. Pink and double-flowered horticultural forms are known.

1b. Convallaria majalis Linnaeus var. **montana** (Rafinesque) H. E. Ahles, J. Elisha Mitchell Sci. Soc. 80: 172. 1964 • American lily-of the-valley [E] [F]

Convallaria montana Rafinesque, Autik. Bot., 66. 1840; *C. majuscula* Greene

Plants scattered or forming small groups. **Leaves** 20–50 × 5–13 cm; petiole 10–30 cm; blade yellowing in late summer, veins and veinlets visible in transmitted light, 10–26 cm. **Scape** 1.5–3 dm. **Inflorescences** to 1/2 length of proximalmost leaves; bracts lanceolate, 10–20 mm, equaling or longer than pedicel. **Flowers:** tepal midribs green; pedicel 7–11 mm. $2n = 38$.

Flowering mid–late spring. Open acid montane slopes, sandy woods, coves; 800–1700 m; Ga., Ky., N.C., S.C., Tenn., Va., W.Va.

33. MAIANTHEMUM F. H. Wiggers, Prim. Fl. Holsat., 14. 1780, name conserved • [Latin *Maius*, May, and Greek *anthemon*, flower]

James V. LaFrankie

Sigillaria Rafinesque; *Smilacina* Desfontaines

Herbs, perennial, terrestrial or aquatic, 1–12.5 dm, from rhizomes. **Rhizomes** persistent, sympodial, spreading and filiform, or densely clumped, cylindrical, and fleshy. **Stems** simple, arching or erect. **Leaves** 2–15, cauline, distichous, clasping or short-petiolate; blade usually ovate, glabrous or weakly pubescent, base rounded or cordiform, margins flat or undulate, denticulate or entire, apex acute or caudate. **Inflorescences** terminally paniculate or racemose, 5–250-flowered. **Flowers** 3-merous (6 tepals, 6 stamens) or, by reduction, 2-merous (4 tepals, 4 stamens); perianth spreading; tepals distinct, white, ovate or triangular, equal, 0.5–5 mm; stamens inserted at tepal base; anthers 4-locular, dehiscence introrse; ovary superior, 2–3-carpellate, septal walls with nectariferous canals; style shorter than 1.5 mm; stigma 2–3-lobed, less than 1 mm wide; pedicel subtended by 1 or more bracts. **Fruits** baccate, variously mottled when immature, bright red at maturity, usually lobed, 4–12 mm wide, pulp thin. **Seeds** 1–12, globose, 3–6 mm diam.; testa pale brown, thin; endosperm scaly. $x = 18$.

Species 30 (5 in the flora): North America, Central America, n Europe, e Asia to the Himalayas.

Outside the flora area, *Maianthemum* may be epiphytic, with foliage stems pendent or erect; two species are dioecious; tepals may be fused, the perianth campanulate; and tepals and anthers may be pink, green, or violet.

The two species of *Maianthemum* that bear 4-tepaled flowers, *M. canadense* and *M. dilatatum*, are very similar to one another and to the boreal Eurasian species *M. bifolium* (Linnaeus) F. W. Schmidt, and the three have sometimes been considered varieties of a single species. In most recent floristic works, the genus *Maianthemum* has been limited to these three species because of their reduced flowers, whereas the remaining species, which bear 6-tepaled flowers more typical of the Liliaceae, have been placed in the genus *Smilacina*. J. V. LaFrankie (1986) placed all the species in a single genus.

The monophyly of *Maianthemum* sensu lato is supported by morphological consistencies including baccate fruits that are marked when immature, a consistent and unique karyotype (S. Kawano and H. H. Iltis 1966; S. Kawano et al. 1967; N. H. Valentine and H. M. Hassan 1971; M. N. Tamura 1995), and molecular analysis (P. J. Rudall et al. 2000; J. Yamashita and M. N. Tamura 2000). The 2-merous floral condition in *Maianthemum* sensu stricto is the result of anatomical reduction from the 3-merous state (F. H. Utech and S. Kawano 1976).

SELECTED REFERENCES Kawano, S., M. Ihara, and M. Suzuki. 1968. Biosystematic studies on *Maianthemum* (Liliaceae-Polygonateae [sic]). II. Geography and ecological life history. Jap. J. Bot. 20: 35–65. Kawano, S., M. Ihara, and M. Suzuki. 1968b. Biosystematic studies on *Maianthemum* (Liliaceae-Polygonateae [sic]). IV. Variation in gross morphology of *M. kamtschaticum*. Bot. Mag. (Tokyo) 81: 473–490. Kawano, S., M. Ihara, M. Suzuki, and H. H. Iltis. 1967. Biosystematic studies on *Maianthemum* (Liliaceae). I. Somatic chromosome number and morphology. Bot. Mag. (Tokyo) 80: 345–352. Kawano, S. and H. H. Iltis. 1966. Cytotaxonomy of the genus *Smilacina*. II. Chromosome morphology and evolutionary consideration of the New World species. Cytologia 31: 12–28. Kawano, S. and M. Suzuki. 1971. Biosystematic studies on *Maianthemum* (Liliaceae-Polygonateae [sic]). VI. Variation in gross morphology of *M. bifolium* and *M. canadense* with special reference to their taxonomic status. Bot. Mag. (Tokyo) 84: 349–361. Kawano, S., M. Suzuki and S. Kojima. 1971. Biosystematic studies on *Maianthemum* (Liliaceae-Polygonateae [sic]). V. Variation in gross morphology, karyology and ecology of North American populations of *M. dilatatum* sensu lato. Bot. Mag. (Tokyo) 84: 299–318. LaFrankie, J. V. 1984. Anatomy of stem abcission in the genus *Smilacina* (Liliaceae). J. Arnold Arbor. 65: 563–570. LaFrankie, J. V. 1986. Transfer of the species of *Smilacina* Desf. to *Maianthemum* Wiggers (Liliaceae). Taxon 35: 584–589. LaFrankie, J. V. 1986b. Morphology and taxonomy of the New World species of *Maianthemum* (Liliaceae). J. Arnold Arbor. 67: 371–439. Takahashi, M. and K. Sohma. 1983. Pollen morphology of the genus *Smilacina* (Liliaceae). Sci. Rep. Tohoku Imp. Univ., Ser. 4, Biol. 38: 191–218. Utech, F. H. and S. Kawano. 1976. Biosystematic studies on *Maianthemum* (Liliaceae). VIII. Floral anatomy of *M. dilatatum*, *M. bifolium*, *M. canadense*. Bot. Mag. (Tokyo) 89: 145–157. Valentine, D. H. and H. M. Hassan. 1971. Cytotaxonomy of the genus *Maianthemum*. J. Indian Bot. Soc. 50: 437–446.

1. Rhizomes 8–14 mm wide; inflorescences paniculate, branches well developed; tepals inconspicuous, 0.5–1 mm . . . . 3. *Maianthemum racemosum*
1. Rhizomes 1–4.5 mm wide; inflorescences racemose, simple or complex, flowers 1–4 per node; tepals conspicuous, longer than 1 mm.
    2. Rhizomes 15–60 cm × 3–4.5 mm, roots scattered; leaves 8–11 on fertile shoots; tepals 6; immature fruits green striped with black . . . . 4. *Maianthemum stellatum*
    2. Rhizomes 1–30 cm × 1–2 mm, roots restricted to nodes; leaves 2–4 on fertile shoots; tepals 4 or 6; immature fruits green mottled or spotted with red.
        3. Leaf blade base tapered; racemes simple; tepals 6 . . . . 5. *Maianthemum trifolium*
        3. Leaf blade base with narrow or broad sinus; racemes complex, with 1–4 flowers per node; tepals 4.

4. Plants 10–25 cm; proximal leaves sessile, blade ovate, base with narrow sinus; distal leaf blade cordate; petiole 1–7 mm; flowers (1–)2(–3) per node . 1. *Maianthemum canadense*
4. Plants 20–45 cm; proximal leaves short-petiolate, blade triangular to cordate, base with broad, open sinus; distal leaf blade deeply cordate; petiole 7–10 cm; flowers (1–)3(–4) per node . . . . 2. *Maiathemum dilatatum*

1. **Maianthemum canadense** Desfontaines, Ann. Mus. Natl. Hist. Nat. 9: 54. 1807 • False lily-of-the-valley, two-leaved Solomon's-seal, Canadian may-lily, maïanthème du Canada [E] [F] [W1]

Maianthemum canadense var. *interius* Fernald; *Smilacina canadensis* (Desfontaines) Pursh; Rafinesque; *Unifolium canadense* (Desfontaines) Greene

Plants terrestrial, 10–25 cm. **Rhizomes** sympodial, proliferatively branching, units 2–30 cm × 1–1.5 mm, roots restricted to nodes. **Stems** erect, 1–1.8 dm × 1–2.5 mm. **Leaves** solitary on sterile shoots, 2–3 on fertile shoots; blade 4.5–7(–9) × 3–4.5(–5.5) cm, apex acute or short-caudate; proximal leaves sessile, blade ovate, base with narrow sinus; distal leaves petiolate, blade cordate, petiole 1–7 mm. **Inflorescences** racemose, complex, 12–25-flowered. **Flowers** (1–)2(–3) per node, 2-merous; tepals conspicuous, 1.5–2 × 0.8–1 mm; filaments 1–1.5 mm; anthers 0.2–0.4 mm; ovary globose, 0.8–1 mm wide; style 0.5–0.8 mm; stigma distinctly 2-lobed; pedicel 3–7 × 0.2–0.5 mm. **Berries** green mottled red when young, maturing to deep translucent red, globose, 4–6 mm diam. **Seeds** 1–2, globose, 3 mm. $2n = 36, 54, 72$.

Flowering early spring. Deciduous and coniferous forests, persisting in forest remnants and parks; 0–1800 m; St. Pierre and Miquelon; Alta., B.C., Man., N.B., Nfld. and Labr., N.W.T., N.S., Ont., P.E.I., Que., Sask., Yukon; Conn., Del., D.C., Ga., Ill., Ind., Iowa, Ky., Maine, Md., Mass., Mich., Minn., Mont., Nebr., N.H., N.J., N.Y., N.C., N.Dak., Ohio, Pa., R.I., S.C., S.Dak., Tenn., Vt., Va., W.Va., Wis., Wyo.

Pubescent specimens of *Maianthemum canadense* in the western half of the range with consistently larger leaves have been treated as var. *interius*.

2. **Maianthemum dilatatum** (Alph. Wood) A. Nelson & J. F. Macbride, Bot. Gaz. 61: 30. 1916 • May-lily, false lily-of-the-valley

Maianthemum bifolium (Linnaeus) F. W. Schmidt var. *dilatatum* Alph. Wood, Proc. Acad. Nat. Sci. Philadelphia 20: 174. 1868 (as Majanthemum); *M. bifolium* var. *kamtschaticum* (J. F. Gmelin) Jepson; *M. kamtschaticum* (J. F. Gmelin) Nakai; *Unifolium dilatatum* (Alph. Wood) Greene; *U. kamtschaticum* (J. F. Gmelin) Gorman

Plants terrestrial, 20–45 cm. **Rhizomes** sympodial, proliferatively branching, units 8–20 cm × 1–1.5 mm, roots restricted to nodes. **Stems** erect, 1.5–3.5 dm × 2–4 mm. **Leaves** solitary on sterile shoots, 2–3 on fertile shoots, petiolate; blade cordate, 6–10 × 5–8 cm; base lobed, with deep sinus; apex sharply acute; proximal leaves short-petiolate, blade triangular to cordate, petiole 4–7 cm; distal leaves petiolate, blade deeply cordate, petiole 7–10 cm. **Inflorescences** racemose, complex, 15–40-flowered. **Flowers** (1–)3(–4) per node, 2-merous; tepals conspicuous, 2–3.2 × 1.5 mm; filaments 1.5 mm; anthers 0.2–0.4 mm; ovary globose, 0.8–1 mm wide; style 0.4–0.5 mm; stigma distinctly 2-lobed; pedicel 3–5 × 0.2–0.4 mm. **Berries** green mottled with red when young, maturing to deep translucent red, globose, 4–6 mm diam. **Seeds** 1–2, globose, 2–3 mm. $2n = 36$.

Flowering early spring. Abundant in coniferous and deciduous forests, especially in forest margins; 0–800 m; B.C., Yukon; Alaska, Calif., Idaho, Oreg., Wash.; Asia (Kamtchatka peninsula in e Russia to Japan).

Variation in the gross morphology, karyology, and ecology of the North American populations has been documented (S. Kawano et al. 1971) and compared with that of disjunct populations in Japan (S. Kawano et al. 1968b).

3. **Maianthemum racemosum** (Linnaeus) Link, Enum. Hort. Berol. Alt. 1: 343. 1821 • Large false Solomon's-seal, false spikenard, Solomon's-plume [F]

Convallaria racemosa Linnaeus, Sp. Pl. 1: 315. 1753; *Smilacina racemosa* (Linnaeus) Desfontaines; *Unifolium racemosum* (Linnaeus) Britton

Plants terrestrial, to 12.5 cm. **Rhizomes** sympodial, cylindrical, units 30–40 cm × 8–14 mm, sometimes multiplied, roots scattered. **Stems** erect or arching, 7.5–12.5 dm × 7–9 mm. **Leaves** 7–12, sessile and clasping, or petiolate; blade elliptic to ovate, 9–17 × 5–8 cm; base rounded or tapered; apex acute or caudate. **Inflorescences** paniculate, 70–250-flowered, branches well developed, pyramidal. **Flowers** 3-merous; tepals inconspicuous, 0.5–1 × 0.5 mm; filaments 1 × 0.5 mm; anthers 0.5–1 mm; ovary globose, 1 mm wide; style 0.1–0.3 mm; stigma obscure; pedicel 0.5–1 × 0.5 mm. **Berries** green with copper spots when young, maturing to deep translucent red, globose or 3-lobed, 4–6 mm wide. **Seeds** 1–4, globose, 2.5–4 mm. $2n = 36, 72, 144$.

Subspecies 2 (2 in the flora): North America, n Mexico.

Maianthemum racemosum is sometimes cultivated. It was described as apomictic (A. L. Gorham 1953) but is much in need of cytogeographic and reproductive biological studies.

1. Stems erect; leaves sessile, clasping, blade base rounded; apex of third leaf below inflorescence acute, shorter than 2 mm; w North America .... 3a. *Maianthemum racemosum* subsp. *amplexicaule*
1. Stems arching; leaves petiolate, blade base tapered; apex of third leaf below inflorescence caudate, 12–25 mm; e North America .... 3b. *Maianthemum racemosum* subsp. *racemosum*

3a. Maianthemum racemosum (Linnaeus) Link subsp. **amplexicaule** (Nuttall) LaFrankie, J. Arnold Arbor. 67: 418. 1986

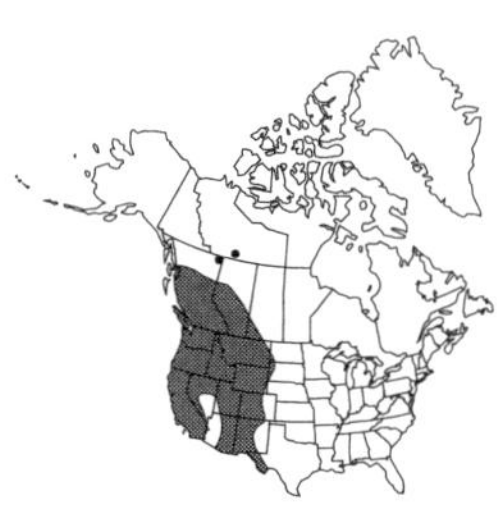

Smilacina amplexicaulis Nuttall, J. Acad. Nat. Sci. Philadephia 7: 58. 1834; *Unifolium amplexicaule* (Nuttall) Greene; *Vagnera amplexicaulis* (Nuttall) Greene; *V. pallescens* Greene

Stems erect. **Leaves** sessile, clasping; blade base rounded; apex of third leaf below inflorescence acute, shorter than 2 mm.

Flowering mid spring; fruiting early summer–autumn. Deciduous woodlands, persisting along shaded roadsides and urban and suburban wooded tracts; 0–700 m; Alta., B.C., N.W.T., Sask.; Alaska, Ariz., Calif., Colo., Idaho, Mont., Nev., N.Mex., N.Dak., Oreg., S.Dak., Tex., Utah, Wash., Wyo.; n Mexico.

3b. Maianthemum racemosum (Linnaeus) Link subsp. **racemosum** [E] [F]

Vagnera australis Rydberg; *V. retusa* Rafinesque

Stems arching. **Leaves** petiolate; blade base tapered; apex of third leaf below inflorescence caudate, 12–25 mm; petiole 2–6 mm.

Flowering mid spring; fruiting early summer–autumn. Deciduous woodlands, persisting along shaded roadsides and urban and suburban wooded tracts; 0–800 m; Man., N.B., Nfld. and Labr. (Nfld.), N.S., Ont., P.E.I., Que.; Ala., Ark., Conn., Del., D.C., Fla., Ga., Ill., Ind., Iowa, Kans., Ky., La., Maine, Md., Mass., Mich., Minn., Miss., Mo., Nebr., N.H., N.J., N.Y., N.C., N.Dak., Ohio, Okla., Pa., R.I., S.C., S.Dak., Tenn., Tex., Vt., Va., W.Va., Wis.

Subspecies *racemosum* is distinguished from subsp. *amplexicaule* by its erect habit, clasping leaves, and short-acute leaf apex; these differences can be seen in the two color photographs in H. W. Rickett ([1966–1973], vol. 1(1), plate 6; vol. 5(1), plate 8).

4. Maianthemum stellatum (Linnaeus) Link, Enum. Hort. Berol. Alt. 1: 343. 1821 • Little false Solomon's-seal, starflower Solomon's-seal, smilacine étoilée

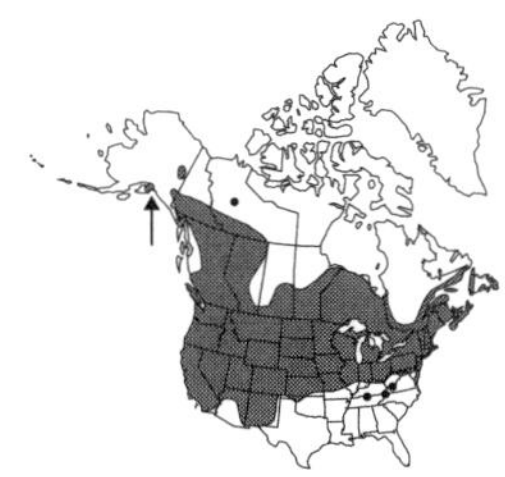

Convallaria stellata Linnaeus, Sp. Pl. 1: 316. 1753; *Smilacina liliacea* (Greene) Wynd; *S. stellata* (Linnaeus) Desfontaines; *Unifolium liliaceum* Greene; *U. sessilifolium* (Nuttall ex Baker) Greene; *U. stellatum* (Linnaeus) Greene; *Vagnera liliacea* (Greene) Rydberg; *V. sessilifolia* (Nuttall ex Baker) Greene; *V. stellata* (Linnaeus) Morong

Plants terrestrial, 15–45 cm. **Rhizomes** sympodial, proliferatively spreading, narrow-cylindrical, units 15–60 cm × 3–4.5 mm, roots scattered over surfaces. **Stems** erect, 2.5–5 dm × 2–3.5 mm. **Leaves** 8–11, sessile, clasping; blade ovate-elliptic to lanceolate, 5–6 × 2.5–3.5 cm; base rounded; apex acute. **Inflorescences** racemose, simple, 6–15-flowered. **Flowers** 1 per node, 3-merous; tepals conspicuous, 4–5 × 1.5–2 mm; filaments 1.2–1.5 mm; anthers 0.6–0.8 mm; ovary globose to cylindrical, ca. 1 mm wide; style 1.3–1.8 mm; stigma obscurely 3-lobed; pedicel 6–12 × 0.8–1 mm. **Berries** green with black stripes along median carpel vein when young, maturing to red, globose, 4–6 mm diam. **Seeds** 1–6, globose, 2.5–3 mm. $2n = 36$.

Flowering mid or late spring. Sand dunes, marginal woodlands, oak openings; 0–3200 m; Alta., B.C., Man., N.B., Nfld. and Labr. (Nfld.), N.W.T., N.S., Ont., P.E.I., Que., Sask., Yukon; Alaska, Ariz., Calif., Colo., Conn., Del., D.C., Idaho, Ill., Ind., Iowa, Kans., Ky., Maine, Md., Mass., Mich., Minn., Mo., Mont., Nebr., Nev., N.H., N.J., N.Mex., N.Y., N.Dak., Ohio, Okla., Oreg., Pa., R.I., S.Dak., Tenn., Utah, Vt., Va., Wash., W.Va., Wis., Wyo.; disjunct in nw Mexico.

5. Maianthemum trifolium (Linnaeus) Sloboda, Rostlinnictví, 192. 1852 • Three-leaf Solomon's-seal, three-leaf Solomon's-plume, smilacine trifoliée

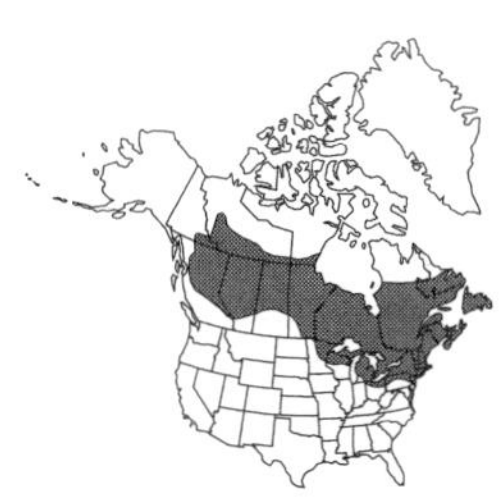

Convallaria trifolia Linnaeus, Sp. Pl. 1: 316. 1753; *Smilacina trifolia* (Linnaeus) Desfontaines; *Unifolium trifolium* (Linnaeus) Greene; *Vagnera trifolia* (Linnaeus) Morong

Plants aquatic, 4–15 cm. **Rhizomes** sympodial, proliferatively spreading, filiform, units 2–30 cm × 1–2 mm, roots restricted to nodes. **Stems** erect, 1–2.5 dm × 2–3 mm. **Leaves** 2–4 on fertile shoots, sessile; blade elliptic, 5–12 × 2.5–4 cm; base narrowly tapering; apex acute or acuminate. **Inflorescences** racemose, simple, 5–15-flowered. **Flowers** 3-merous; tepals conspicuous, 2.5–4 × 1.5–2 mm; filaments 1.5–2 mm; anthers 0.5–0.7 mm; ovary globose or cylindrical, 2–2.5 × 1.5–2.4 mm; style 0.5–1.5 mm; stigma distinctively 3-lobed; pedicel 1–3 × 0.5 mm. **Berries** green with fine red spots when young, maturing to red, globose, 4–6 mm diam. **Seeds** 1–3, globose, 2 mm. $2n = 36$.

Flowering May–Jun, fruits retained through Sep. Often forming dense clonal patches in sphagnum bogs, muskegs, and wet forests; 0–1000 m; St. Pierre and Miquelon; Alta., B.C., Man., N.B., Nfld. and Labr., N.W.T., N.S., Nunavut, Ont., P.E.I., Que., Sask., Yukon; Conn., Maine, Mass., Mich., Minn., N.H., N.J., N.Y., Ohio, Pa., R.I., Vt., Wis.; Asia (Siberia).

34. POLYGONATUM Miller, Gard. Dict. Abr. ed. 4, vol. 3. 1754 • Solomon's seal, sceau-de-Salomon [Greek *poly-*, many, and *gony*, knee, in reference to the jointed rhizome]

Frederick H. Utech

Herbs, perennial, from long, knotty, creeping, occasionally branched rhizomes; leaf scars conspicuous; roots fibrous to fleshy. **Stems** simple, leafy, erect to arching, glabrous or pubescent. **Leaves** simple, alternate [opposite or verticillate], dispersed along stem, short-petiolate, subsessile, or clasping; blade ovate or lanceolate to linear, margins entire. **Inflorescences** racemose, with 1–10(–15) flowers on axillary peduncles, except for distal- and proximal-most axils; peduncle glabrous [pubescent]. **Flowers** nodding or pendulous; tepals 6, persistent, marcescent, connate basally into cylindrical tube, distinct, tips short, valvate; stamens 6, included, adnate to perianth tube; filament surfaces various; anthers basifixed [dorsifixed], dehiscence introrse; ovary superior, 3-locular, ovoid to globose; septal nectaries present; style slender, included; stigma capitate to obscurely 3-lobed; pedicel articulate, glabrous. **Fruits** baccate, dark blue to black [red], globose, glaucous, pulpy. **Seeds** yellowish tan to brownish olive, globose or subglobose, 3–4.5 mm. $x = 10$ [9–15].

Species 57 (3 in the flora): temperate regions, North America, Eurasia, especially China and Japan.

Polygonatum is a member of the Polygonatae, a tribe now assigned to the Ruscaceae sensu lato (P. J. Rudall et al. 2000) or a more narrowly defined Convallariaceae (J. G. Conran and M. N. Tamura 1998; J. Yamashita and M. N. Tamura 2000). It is a complex genus with some species that are highly variable and wide-ranging, and others that are not. The flora provides examples of both. Polyploids arising from different base numbers and hybridization contribute to generic variation. The North American species all have $x = 10$, in contrast to variable base numbers among the Eurasian species (S. Kawano and H. H. Iltis 1963; E. Therman 1950, 1953, 1953b; M. N. Tamura 1993, 1995).

In *Polygonatum,* filament morphology and level of tube insertion provide useful distinctions at the species level (M. N. Tamura 1993). Bee vibration, associated with the pendulous perianth tubes, which include the anthers and style, is an essential feature of the pollination system (A. S. Corbet et al. 1988).

Young shoots and leaves can be boiled and eaten like asparagus, while the rhizomes reportedly have been dried and ground into flour or boiled and eaten like potatoes (L. Peterson 1978). Numerous eastern North American tribes, including the Cherokee, Chippewa, Iroquois, and Menominee, used plants of the genus as an analgesic or stimulant, or as a dermatological, cathartic, gynecological, and gastrointestinal aid (D. E. Moerman 1986). The berries contain an anthraquinone that causes vomiting and diarrhea.

The arching stems and fine foliage of *Polygonatum* add architectural interest and focus to woodland shade gardens. In addition to the native species, the Eurasian *P. odoratum* (Miller) Druce and the hybrid, *P.* ×*hybridum* Brügger (*P. multiflorum* × *P. odoratum*), are commonly cultivated. The European *P. multiflorum* (Linnaeus) Allioni has been reported as locally naturalized in eastern Ontario and southern Quebec (B. Boivin 1967c).

SELECTED REFERENCES Abramova, L. I. 1975. On the taxonomical structure of the genus *Polygonatum* Mill. Bot. Zhurn. (Moscow & Leningrad) 60: 490–497. Bush, B. F. 1927. The species of *Polygonatum*. Amer. Midl. Naturalist 10: 385–400. Gates, R. R. 1917. A revision of the genus *Polygonatum* in North America. Bull. Torrey Bot. Club 44: 117–126. Ownbey, R. P. 1944. The liliaceous genus *Polygonatum* in North America. Ann. Missouri Bot. Gard. 31: 373–413. Tamura, M. N., A. E. Schwarzbach, S. Kruse, and R. Reski. 1997. Biosystematics studies on the genus *Polygonatum* (Convallariaceae) IV. Molecular phylogenetic analysis based on restriction site mapping of the chloroplast gene *trn*K. Feddes Repert. 108: 159–168.

1. Leaves glabrous on both surfaces; axillary peduncle 2–10(–15)-flowered . . . . 1. *Polygonatum biflorum*
1. Leaves minutely hairy or pilose only on abaxial veins; axillary peduncle 1–3(–5)-flowered.
    2. Perianth yellowish green; filaments densely warty; e North America . . . . 2. *Polygonatum pubescens*
    2. Perianth whitish, tipped with green; filaments glabrous or sparsely downy; New England . . . . 3. *Polygonatum latifolium*

1. Polygonatum biflorum (Walter) Elliott, Sketch Bot. S. Carolina 1: 393. 1817 • Giant Solomon's seal [W2]

Convallaria biflora Walter, Fl. Carol., 122. 1788; *Polygonatum biflorum* var. *commutatum* (Schultes f.) Morong; *P. biflorum* var. *melleum* (Farwell) R. P. Ownbey; *P. cobrense* (Wooton & Standley) R. R. Gates; *P. commutatum* (Schultes f.) A. Dietrich; *P. giganteum* A. Dietrich; *P. melleum* Farwell

Rhizomes deeply set, 1–3 cm thick. **Stems** erect to arching, 5–20 dm; sheathing bract usually absent. **Leaves** sessile to clasping, (5–)9–20(–25) × 3–9(–13) cm; blade narrowly lanceolate to broadly elliptic, glabrous; prominent veins 1–19. **Inflorescences** in most leaf axils except distal 2–3 and proximal 2–7; peduncle flattened, not strongly reflexed, axillary 2–10(–15)-flowered, 6–9 cm in fruit. **Flowers:** perianth whitish to greenish yellow, tube (13–)17–22 mm, distinct tips gently spreading, 4–6.5 mm; stamens inserted near middle of perianth tube; filaments glabrous, sometimes minutely warty; pedicel 1–4 cm in fruit. **Berries** 8–12 mm. $2n = 20, 40$.

Flowering early–late spring. Dry to moist deciduous woods, in sandy soil, roadsides, railway embankments, old fields; 0–3000 m; Man., Ont., Que., Sask.; Ala., Ariz., Ark., Conn., Del., D.C., Fla., Ga., Ill., Ind., Iowa, Kans., Ky., La., Maine, Md., Mass., Mich., Minn., Miss., Mo., Nebr., N.H., N.J., N.Mex., N.Y., N.C., N.Dak., Ohio, Okla., Pa., R.I., S.C., S.Dak., Tenn., Tex., Vt., Va., W.Va., Wis., Wyo.; ne Mexico.

As a polyploid complex, *Polygonatum biflorum* is variable as to plant size, flower number, and ecology. Tetraploids, which have been recognized as var. *commutatum*, are usually larger and coarser than diploids, var. *biflorum*, with more flowers per peduncle and more major veins per leaf. They have also been called *P. commutatum*, *P. giganteum*, and *P. canaliculatum* (Mühlenberg) Pursh (misapplied). Some diploids, however, are fully as large as any tetraploids. Range-wide separation of species has been problematic (R. P. Ownbey 1944; E. G. Voss 1972–1985; J. A. Steyermark 1963; G. Yatskievych 1999+), while local distinction may be possible (M. L. Fernald 1944b; M. W. Richardson and D. Ugent 1974). Detailed molecular and enzymatic analysis at the population level is needed range-wide, in conjunction with karyological and morphological study.

Local populations of *Polygonatum biflorum* with honey-yellow flowers from southeastern Michigan and adjacent Ontario (E. G. Voss 1972–1985; R. P. Ownbey 1944) have been recognized as var. *melleum*. Disjunct populations in the trans-Pecos Texas and the mountains of Arizona and New Mexico have been called *P. cobrense*.

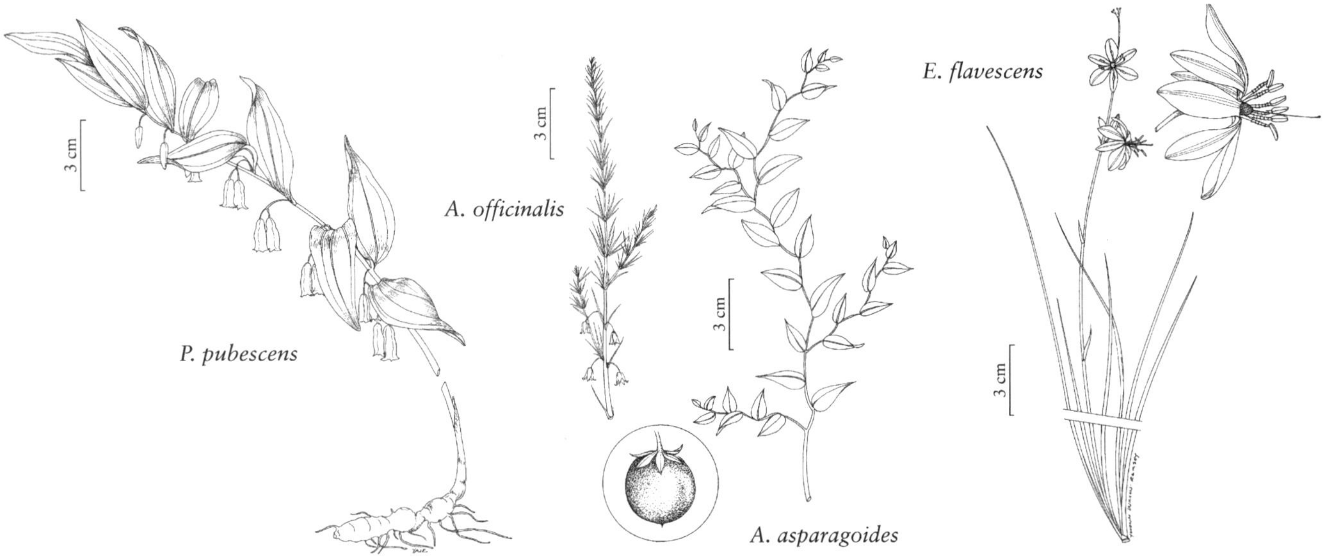

POLYGONATUM ◦ ASPARAGUS ◦ ECHEANDIA

2. **Polygonatum pubescens** (Willdenow) Pursh, Fl. Amer. Sept. 1: 234. 1814 • Hairy Solomon's seal, sceau-de-Salomon pubescent [E] [F]

Convallaria pubescens Willdenow, Hort. Berol. 1: 45. 1803

Rhizomes shallowly set, 1–1.8 cm thick. **Stems** erect, 5–9(–11) dm; sheathing bract cauline, papery, caducous. **Leaves** subsessile or short-petiolate, 4–15 × 2–5.5(–7.5) cm; blade elliptic-lanceolate to broadly ovate, glabrous adaxially, minutely hairy or pilose on abaxial veins; prominent veins 3–9. **Inflorescences** in most leaf axils except distalmost and proximal 2–4; peduncle sharply reflexed, axillary 1–3(–5)-flowered, to 2 cm in fruit. **Flowers:** perianth yellowish green, tube 10–13(–15) mm, distinct tips 2–3 mm; stamens inserted high in perianth tube; filaments densely warty; pedicel to 1.3 cm in fruit. **Berries** 6–9 mm. $2n = 20$.

Flowering early–late spring. Rich moist wooded slopes and coves; 0–1100 m; N.B., N.S., Ont., Que.; Conn., Del., D.C., Ga., Ill., Ind., Iowa, Ky., Maine, Md., Mass., Mich., Minn., N.H., N.J., N.Y., N.C., Ohio, Pa., R.I., Tenn., Vt., Va., W.Va., Wis.

3. **Polygonatum latifolium** (Jacquin) Desfontaines, Ann. Mus. Natl. Hist. Nat. 9: 50. 1807 [I]

Convallaria latifolia Jacquin, Fl. Austriac. 3: 18, plate 232. 1775; *Polygonatum hirtum* (Poiret) Pursh

Rhizomes shallowly set, 1–2 cm thick. **Stems** arching, angled, 1–12 dm, sparsely hirsute distally; sheathing bract cauline, papery, caducous. **Leaves** short-petiolate, 7–12.5 × 3–7 cm; blade elliptic-lanceolate to ovate, glabrous adaxially, minutely hairy or pilose on abaxial veins, apex abruptly acuminate; prominent veins 3–9. **Inflorescences** in most leaf axils except distalmost and proximal 2–4; peduncle reflexed, axillary 2–3(–4)-flowered. **Flowers:** perianth whitish, tipped with green, tube 10–18 mm, distinct tips 3 mm; stamens inserted near middle of perianth tube; filaments glabrous or sparsely downy. **Berries** 6–9 mm. $2n = 20$.

Flowering early–late spring. Roadside thickets; 0–200 m; introduced; Mass., N.H., N.Y., Vt.; Europe.

This European species is locally naturalized in New England (R. Angelo and D. E. Boufford 2000; D. W. Magee and H. E. Ahles 1999).

35. ASPARAGUS Linnaeus, Sp. Pl. 1: 313. 1753; Gen. Pl. ed. 5, 147. 1754 • Asparagus-fern, asperge, espárrago [Greek *asparasso*, to rip, alluding to the spiny leaves of some species] [I]

Gerald B. Straley†

Frederick H. Utech

Herbs, shrubs, or vines, perennial, from rhizomes, usually with fusiform tubers, often with fernlike appearance. **Stems** photosynthetic, erect, spreading or climbing, branched; cladophylls solitary or fasciculate, in nodes of reduced, scarious leaves. **Leaves** small, scale-like, membranous, or sometimes spiny with hardened base, subtending cladophylls. **Inflorescences** axillary or terminal, racemose, or umbellate, paired or solitary; racemes short. **Flowers** bisexual or unisexual; perianth greenish, white, or yellowish, campanulate to rotate; tepals 6, distinct or shortly connate basally, equal; stamens 6, distinct, equal; anthers versatile, 2-locular, dehiscence introrse; ovary superior, 3-locular, septal nectaries present; style 3-branched distally; pedicel with conspicuous joint. **Fruits** baccate, red or purplish black, globose, often with tepals persisting at base. **Seeds** 1–6, black, globose to angular. $x = 10$.

Species 170–300 (4 in the flora): introduced; Europe, Asia, Africa; some widely introduced, expected elsewhere.

Asparagus is a moderately important horticultural genus, with one species commonly grown for its edible young shoots and a number of others grown ornamentally. The genus is treated here in a broad sense (K. Kubitzki and P. J. Rudall 1998; W. S. Judd 2001) and encompasses species that have been segregated in several genera (A. L. Takhtajan 1997; S. T. Malcomber and Sebsebe D. 1993). Embryological features (P. J. Rudall et al. 1998) and DNA-based analyses (M. W. Chase et al. 1996; M. F. Fay et al. 2000; P. J. Rudall et al. 1997) support the monophyly of *Asparagus* and the Asparagaceae.

Asparagus virgatus Baker has been collected once as a garden escape in East Baton Rouge Parish, Louisiana, but it is probably not truly naturalized in the flora. *Asparagus falcatus* Linnaeus occasionally persists after cultivation in the Miami–Dade County area of southern Florida (W. S. Judd 2001).

SELECTED REFERENCES Cooney-Sovetts, C. and R. Sattler. 1986. Phylloclade development in the Asparagaceae: An example of homoeosis. Bot. J. Linn. Soc. 94: 327–371. Jessop, J. P. 1966. The genus *Asparagus* in southern Africa. Bothalia 9: 31–96. Malcomber, S. T. and Sebsebe D. 1993. The status of *Protasparagus* and *Myrsiphyllum* in the Asparagaceae. Kew Bull. 48: 63–78.

1. Plants erect; some flowers unisexual . . . . 1. *Asparagus officinalis*
1. Plants scrambling, twining, or arching; all flowers bisexual.
    2. Cladophylls solitary at each node, broadly lanceolate to ovate, with 20–24 veins . . . . 2. *Asparagus asparagoides*
    2. Cladophylls in fascicles of 3–20 per node, linear or filiform, with single vein.
        3. Cladophylls filiform; pedicels 1–3 mm; inflorescences 1–4-flowered terminal umbels; berries purplish black . . . . 3. *Asparagus setaceus*
        3. Cladophylls flattened; pedicels 5–8 mm; inflorescences 5–9(–17)-flowered axillary racemes; berries red . . . . 4. *Asparagus aethiopicus*

1. Asparagus officinalis Linnaeus, Sp. Pl. 1: 313. 1753 • Garden asparagus, common asparagus [F] [I] [W2]

Herbs, erect, 1–2.5 m; rhizomes fibrous. **Stems** annual, densely branched distally; branches finely dissected, ascending to perpendicular, unarmed; cladophylls in clusters of (2–)4–15(–25) per node, filiform, straight or curved, 1–3 cm. **Leaves** scalelike, 3–4 mm; blade lanceolate, base hardened. **Inflorescences** in axillary racemes, 1–3-flowered. **Flowers** some unisexual; perianth campanulate, yellow or yellowish green; tepals connate 1–2 mm, greenish white, 3–8 × 1–2 mm; pedicel 8–12 mm, jointed at or above middle. **Berries** red, 6–10 mm. **Seeds** 2–4. $2n$ = 20, 40.

Flowering summer. Fields, fencerows, roadsides, disturbed areas; 0–2500 m; introduced; St. Pierre and Miquelon; Alta., B.C., Man., N.B., Nfld. and Labr., N.S., Ont., P.E.I., Que., Sask.; Ala., Ariz., Ark., Calif., Colo., Conn., Del., D.C., Fla., Ga., Idaho, Ill., Ind., Iowa, Kans., Ky., La., Maine, Md., Mass., Mich., Minn., Miss., Mo., Mont., Nebr., Nev., N.H., N.J., N.Mex., N.Y., N.C., N.Dak., Ohio, Okla., Oreg., Pa., R.I., S.C., S.Dak., Tenn., Tex., Utah, Vt., Va., Wash., W.Va., Wis., Wyo.; Europe; Asia; n Africa; naturalized in temperate regions worldwide.

Eaten as a vegetable, *Asparagus officinalis* has been widely cultivated for its young shoots since ancient Greek times. The species is naturalized in many temperate climates. Mature asparagus has caused poisoning in cattle (J. M. Kingsbury 1964). Young plants can cause dermatitis, and the red berries are suspected of poisoning humans (E. M. Schmutz and L. B. Hamilton 1979). The species is dioecious (J. E. Lazarte and B. F. Palser 1979), and homomorphic sex chromosomes have been identified (H. Loptien 1979).

2. Asparagus asparagoides (Linnaeus) Druce, Rep. Bot. Exch. Club Soc. Brit. Isles 3: 414. 1914, not W. Wight 1909, invalid name • Smilax asparagus [F] [I]

Medeola asparagoides Linnaeus, Sp. Pl. 1: 339. 1753

Vines, erect to deflexed, branching, climbing to 1.5 m; roots tuberous. **Stems** slender, to 3 m, wiry, smooth to slightly ridged; cladophylls solitary at each node, broadly lanceolate to ovate, 15–35 × 5–15(–20) mm, leathery, with 20–24 parallel veins. **Leaves** membranous, 1–2 mm; blade broadly lanceolate. **Inflorescences** in axillary umbels, 1–3(–4)-flowered. **Flowers** bisexual; perianth campanulate; tepals white, with green median stripe abaxially, 5–7 × 1–1.5 mm; pedicel 5–8 mm, with hard, raised joint 1–3 mm above base. **Berries** red, 6–8 mm. **Seeds** 1–4(–6). $2n$ = 20.

Flowering spring–summer. Waste places, open forests, roadsides; 0–100 m; introduced; Calif.; s Africa.

Asparagus asparagoides is cultivated as an ornamental and has naturalized locally in southern Europe.

3. Asparagus setaceus (Kunth) Jessop, Bothalia 9: 51. 1966 • Climbing asparagus-fern, lace-fern [I]

Asparagopsis setacea Kunth, Abh. Königl. Akad. Wiss. Berlin 1842: 82. 1842; *Asparagus plumosus* Baker

Vines, woody, scrambling or climbing to 5 m; roots fibrous. **Stems** to 4 m, wiry, smooth, branches planate; cladophylls in fascicles of (5–)8–20 per node, filiform, 4–10 × 0.5 mm, ± rigid, with single vein. **Leaves** membranous, 1–2 mm; blade forming short spine with reflexed apex, base hardened. **Inflorescences** terminally umbellate, 1–4-flowered. **Flowers** bisexual, nodding; perianth spreading, campanulate; tepals white, 3–4 × 1–1.5 mm; pedicel 1–3 mm, jointed at or just above base. **Berries** purplish black, 4–5 mm. **Seeds** 1–3. $2n$ = 20.

Flowering spring–summer. Waste places, abandoned gardens; 0–100 m; introduced; Calif., Fla.; s, e Africa.

4. Asparagus aethiopicus Linnaeus, Mant. Pl., 63. 1767 • Sprenger asparagus-fern, emerald-fern [I]

Asparagus sprengeri Regel

Shrubs, sprawling or scrambling; roots fibrous and tuberous. **Stems** to 2 m, wiry, branches numerous, finely ridged; cladophylls solitary or in fascicles of 3 or more per node, linear, flattened, straight or curved, 8–22 × 2 mm, with single prominent vein. **Leaves** membranous, 1–2 mm; blade diamond-shaped, attached in middle, tapering to slender apex. **Inflorescences** in axillary racemes, 5–9(–17)-flowered. **Flowers** bisexual; perianth rotate to campanulate; tepals white to pale pink, 3–4 × 1.5–2 mm; pedicel 5–8 mm, jointed 2–3 mm above base. **Berries** red, 5–8 mm. **Seed** 1. $2n$ = 40, 60.

Flowering spring–summer. Waste places, abandoned gardens; 0–100 m; introduced; Fla.; s Africa.

The name *Asparagus densiflorus* (Kunth) Jessop (*Asparagopsis densiflora* Kunth) has been misapplied to this species (P. S. Green 1986; W. S. Judd 2001). *Asparagus aethiopicus* cv. 'Sprengeri' is the most common of the asparagus-fern cultivars.

36. ECHEANDIA Ortega, Nov. Pl. Descr. Dec., 135, plate 18. 1800 • [For Pedro Gregorio Echeandía, 1746–1817, Spanish botanist in Zaragosa]

Robert W. Cruden

Herbs, perennial, scapose, from corms with enlarged storage roots. **Leaves** basal and cauline; blade 2-faced, very narrowly linear to narrowly oblong or elliptic, base surrounded by fibrous leaf bases from previous year. **Inflorescences** racemose or paniculate, (1–)2–4 flowers per node. **Flowers** bisexual or pseudohermaphroditic-staminate, ± erect to pendulous; tepals strongly reflexed to spreading, distinct, yellow [orange, white, or cream], 3-veined, ± elliptic, equal; stamens 6; filaments linear to clavate, bearing scales at right angles to axis of filament [scaleless]; anthers connate into cone or distinct, dorsifixed near base, yellow; ovary superior, ellipsoid, ovules 8+ per locule; style 1. **Fruits** capsular, broadly to narrowly oblong, dehiscence loculicidal. **Seeds** black, irregularly compressed and folded, colliculose. $x = 8$.

Species ca. 80 (3 in the flora): sw United States to nw Argentina, s Bolivia, and s Peru.

More than 60 species of *Echeandia* occur in North America south of the United States. Two of those reach their northern distributional limits in the southwestern United States. The third species included here is known only from the southern tip of Texas.

SELECTED REFERENCES Cruden, R. W. 1994. *Echeandia.* In: G. Davidse et al., eds. 1994+. Flora Mesoamericana. 2+ vols. Mexico City. Vol. 6, pp. 27–30. Cruden, R. W. 1999. A new subgenus and fifteen new species of *Echeandia* (Anthericaceae) from Mexico and the United States. Novon 9: 325–338. Cruden, R. W. and R. McVaugh. 1989. *Echeandia.* In: R. McVaugh and W. R. Anderson, eds. 1974+. Flora Novo-Galiciana: A Descriptive Account of the Vascular Plants of Western Mexico. 8+ vols. Ann Arbor. Vol. 15, pp. 178–197.

1. Anthers connate, 4 mm or longer . . . . 1. *Echeandia texensis*
1. Anthers distinct, shorter than 4 mm.
    2. Filaments inserted in open pits; anthers versatile; most storage roots 1–2(–4) cm from corms . . . . 2. *Echeandia flavescens*
    2. Filaments inserted in covered pits; anthers not versatile, usually twisted and/or reflexed, thus appearing versatile; most storage roots 2–8 cm from corms or thickened through out . . . . 3. *Echeandia chandleri*

1. Echeandia texensis Cruden, Novon 9: 331, fig. 1. 1999 [E]

Storage roots enlarged 2–6 cm from corms, 1–3 cm. **Leaves:** basal 4–9, 33–60 cm × 10–20 mm, blade linear to elliptic, margins denticulate; cauline 4–5, blade long-acuminate, the proximal (4.3–)10–20 cm, frequently exceeding next node. **Inflorescence** 1, 0–3(–12)-branched, 55–105 cm, glabrous. **Flowers** nodding; tepals broadly elliptic, 12–19 mm, outer 2–3 mm wide, inner 5–8 mm wide; filaments inserted in covered pits, narrowly clavate, 5–9 mm; anthers connate, forming cone 4–5.5 mm, cone weakly to strongly tapered, 2–3 mm wide, apex broadly and deeply lobed, 1–1.5 mm; ovary 2–3.5 mm. **Capsules** oblong, 9–13 × 4–6 mm.

Flowering Apr, Jun, Nov. Clay dunes, llanos, open places; 0 m; Tex.

Echeandia texensis is distinguished from *E. reflexa*, a common species of eastern Mexico, by the short, broad anther cone, which has a wide, broadly and deeply lobed apex. The anther cone of *E. reflexa* is longer and narrower, and the apex is narrower and minutely lobed.

2. Echeandia flavescens (Schultes & Schultes f.) Cruden, Sida 9: 146. 1981 [F]

Anthericum flavescens Schultes & Schultes f. in J. J. Roemer et al., Syst. Veg. 7: 1692. 1830; *A. stenocarpum* Baker; *A. torreyi* Baker; *A. torreyi* var. *arizonicum* Poellnitz; *A. torreyi* var. *lanceolatum* Poellnitz; *A. torreyi* var. *neomexicanum* Poellnitz; *Echeandia leptophylla* Bentham; *Phalangium flavescens* Kunth

Storage roots enlarged 1–2(–4) cm from corms, (1.5–)2–5 cm. **Leaves:** basal 3–15, 8–40 cm × 0.5–9 mm, blade very narrowly linear to narrowly linear, margins denticulate, ciliate; cauline 0–3, (0.5–)1–9(–13) cm, blade long-acuminate in most specimens, the proximal not exceeding next node. **Inflorescences** 1–2, 0–1(–3)-branched, 20–60(–84) cm, wholly glabrous or scabrescent toward base. **Flowers** facing upward or away from scape; tepals elliptic, (7.5–)9–15 mm, outer 2–4 mm wide, inner 4–8.5 mm wide; filaments inserted in open pits, narrowly cylindric to narrowly clavate, 4–8.5 mm; anthers distinct, versatile, 1.5–3.5 mm; ovary 2–5 mm. **Capsules** broadly oblong to oblong, 7–16 × 3–6 mm. 2*n* = 16, 32, 48.

Flowering (late May–) mid Jul–early Oct. Desert grasslands, pinyon-juniper woodlands, juniper-oak-pine woodlands, openings in yellow-pine and pine-fir-spruce forests; 1500–2900 m; Ariz., N.Mex., Tex.; Mexico.

Echeandia flavescens is a highly variable polyploid complex. Most specimens from the flora fall into one of three groups: slender plants with leaves 1–4 mm wide and capsules 7–11 mm; slender plants with leaves 1–4 mm wide and capsules 12–15 mm; and more robust plants with leaves 4–9 mm wide and capsules 12–16 mm. The first plants are diploids and the last two are hexaploids and octoploids, respectively. The robust morph was described as *Anthericum stenocarpum* and has been collected at scattered locations throughout the distributional range of the species. Because the correlation between morphology and ploidy level breaks down across the distributional range, no infraspecific taxa are recognized.

3. **Echeandia chandleri** (Greenman & C. H. Thompson) M. C. Johnston, Vasc. Pl. Texas ed. 2, 18. 1990 • Lila de los llanos

Anthericum chandleri Greenman & C. H. Thompson, Ann. Missouri Bot. Gard. 1: 405. 1914

Storage roots enlarged 2–8 cm from corms, 2–5 cm, or thickened throughout. **Leaves:** basal (2–)5–12, 15–50 cm × 4–12(–16) mm, blade narrowly linear to narrowly elliptic, margins entire, often short-ciliate; cauline 1–5, 6.5–27 cm, blade long-acuminate, the proximal usually exceeding next node. **Inflorescence** 1, 0–4-branched, 18–76 cm, glabrous or scabrescent. **Flowers** facing outward or somewhat declinate; tepals elliptic, 10–20 mm, outer 2.5–4.5 mm wide, inner 4.5–8 mm wide; filaments inserted in covered pits, narrowly cylindric, (4–)5–9.5 mm; anthers distinct, usually not versatile, usually twisted and/or reflexed, thus appearing versatile, 2–3 mm; ovary 2–5 mm. **Capsules** broadly oblong to oblong, 8–14 × 5–7 mm.

Flowering Apr, May, Aug–Dec. Coastal plains, piedmont; 0–1000 m; Tex.; Mexico (San Luis Potosí, Tamaulipas).

37. EREMOCRINUM M. E. Jones, Zoë 4: 53. 1893 • Sand lily [Greek *eremos*, desert, and *krinon*, lily] E

James L. Reveal

Frederick H. Utech

Herbs, perennial, caulescent, glabrous, from short, vertical rhizomes with fleshy roots. **Leaves** 8–12, basal, tufted, each tuft surrounded basally by fibrous, persistent sheaths; blade linear. **Inflorescences** terminal, racemose, spikelike, elongate, dense, bracteate. **Flowers** rather showy, not obviously fragrant; perianth persistent, white to greenish white, with 3 greenish to brown stripes, these confluent terminally; tepals 6, shortly connate proximally, spreading distally, oblong, equal; perianth tube stipelike; stamens 6, hypogynous, 1-seriate; filaments linear, flattened, ca. $^1/_2$ length of tepals; anthers basifixed, becoming strongly incurved, introrse; ovary superior, 3-locular, ovules few per locule; style filiform, elongate; stigma discoid; pedicel short, slender, subtended by scarious bract. **Fruits** capsular, 3-lobed, globose, dehiscence loculicidal. **Seeds** 4–8(–12), black, angled.

Species 1: s Utah and n Arizona.

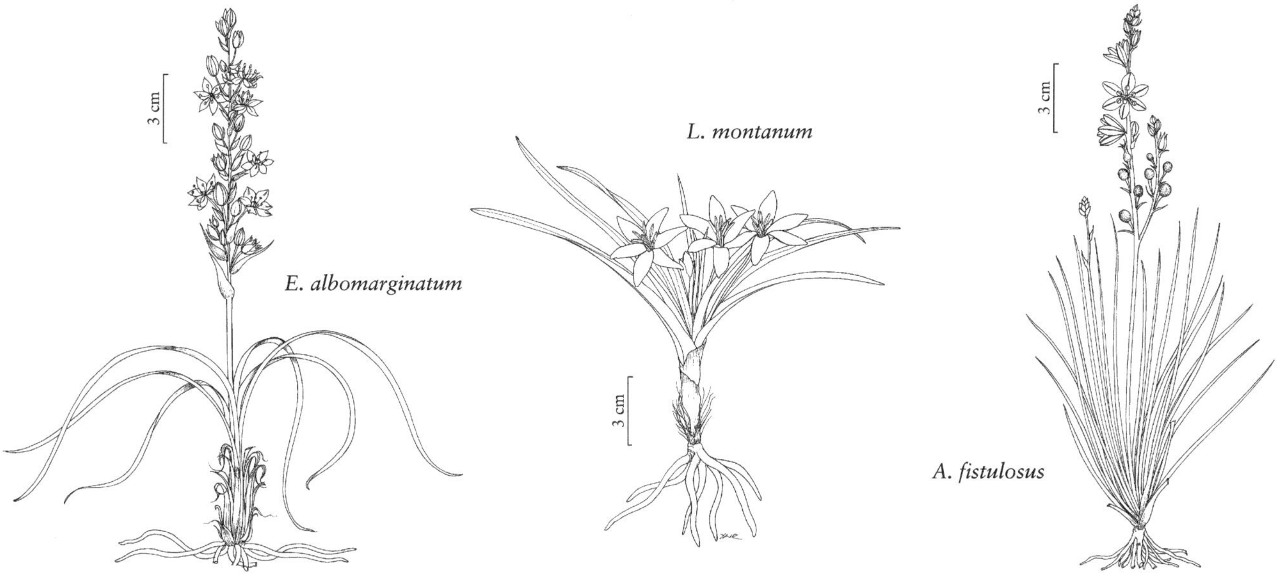

EREMOCRINUM ◦ LEUCOCRINUM ◦ ASPHODELUS

1. **Eremocrinum albomarginatum** (M. E. Jones) M. E. Jones, Zoë 4: 53. 1893 • Sand lily, lonely lily E F

Hesperanthes albomarginata M. E. Jones, Zoë 2: 251. 1891

Plants 1.5–3.5(–4) dm. **Leaves** erect to spreading, 10–40 × 0.1–0.3 cm; sheaths (4–)5–8(–10) × 0.5–0.9 cm. **Inflorescences:** bracts thin, membranous and whitish or sometimes chartaceous and greenish, triangular to awl-shaped, 0.3–2 cm. **Flowers** numerous; perianth 1–2 cm, tube 0.4–0.8(–1) cm; anthers 2.5–3 mm; style 3–4 mm; pedicel erect, peglike, 1 mm, concealed by bracts. **Capsules** 0.4–0.6 cm. **Seeds** 2–2.5 mm.

Flowering spring (Apr–Jun). Sandy flats, dunes; 1300–1800 m; Ariz., Utah.

38. LEUCOCRINUM Nuttall ex A. Gray, Ann. Lyceum Nat. Hist. New York 4: 108, 110. 1837 • Sand-lily, star-lily [Greek *leucos*, white, and *krinon*, lily] E

James L. Reveal

Frederick H. Utech

Herbs, perennial, acaulescent, glabrous, from short, deeply buried, fleshy roots. **Leaves** few, tufted, each tuft surrounded basally by membranous sheaths; blade linear. **Inflorescences** umbellike, central clusters at ground level. **Flowers** rather showy, fragrant; tepals 6, connate below middle, white, narrowly oblong, equal; perianth tube long, slender; limb lobes spreading; stamens 6, inserted near apex of perianth tube; filaments filiform, shorter than perianth lobes; anthers subversatile, often strongly curved or coiled after dehiscence, dehiscence introrse; ovary subterranean, 3-locular, ovoid, septal nectaries present; style filiform, elongate, 3-fid, lobes short; pedicel articulate, arising directly from rootstock, ebracteate. **Fruits** capsular, subterranean, obovoid, 3-angled, dehiscence loculicidal. **Seeds** black, angled. x = 11, 13, 14.

Species 1: w United States.

Molecular evidence (A. W. Meerow et al. 1999; M. F. Fay et al. 2000) indicates that *Leucocrinum* is most closely related to *Echeandia*.

SELECTED REFERENCE Ornduff, R. and M. S. Cave. 1975. Geography of pollen and chromosomal heteromorphism in *Leucocrinum montanum* (Liliaceae). Madroño 23: 65–67.

1. **Leucocrinum montanum** Nuttall ex A. Gray, Ann. Lyceum Nat. Hist. New York 4: 110. 1837 [E] [F]

Plants 5–10 cm. **Leaves** spreading; blade 10–20 cm × 2–8 mm; sheath 3–8 cm × 5–8 mm, distalmost occasionally fibrous. **Flowers:** perianth 5–10(–12) cm, tube (4–)5–8(–10) cm; tepals 2–2.5 cm × 3–7 mm; anthers 4–6 mm; pedicel slender, 0.5–3 cm. **Capsules** 5–8 mm. **Seeds** 3–4 mm. $2n$ = 22, 26, 28, ca. 50.

Flowering spring (Mar–Jun). Scrub flats, short-grass prairie, sagebrush deserts to open montane forests, sandy to rocky areas; 800–2400 m; Ariz., Calif., Colo., Idaho, Mont., Nebr., Nev., N.Mex., N.Dak., Oreg., S.Dak., Utah, Wyo.

Leucocrinum montanum exhibits an unusual chromosomal and pollen heteromorphism (M. S. Cave 1970; R. Ornduff and M. S. Cave 1975). Populations from the Rocky Mountain region, Utah, and central and eastern Nevada shed their pollen in monads and have x = 14, while populations from extreme western Nevada, California, and Oregon all shed tetrads and have x = 13. The occurrence of intraspecific dimorphism in pollen shedding is extremely rare.

The showy flowers of *Leucocrinum montanum* (L. S. Hannibal 1976; H. Rickabaugh 1975) with their long, white floral tubes are reportedly fragrant (V. A. Matthews 1986), and the subterranean capsules are more or less sessile on the rootstocks. Native Americans have eaten the roots (G. Kunkel 1984), and the Paiute and Shoshone tribes used the plant as a dermatological aid (D. E. Moerman 1986).

39. ASPHODELUS Linnaeus, Sp. Pl. 1: 309. 1753; Gen. Pl. ed. 5, 146. 1754 • Asphodel [Greek *asphodelos*, flower of Hades and the dead] [I]

Gerald B. Straley†

Frederick H. Utech

Herbs, annual or perennial [or biennial], scapose, from swollen rhizomes. **Leaves** numerous, basal; blade linear, cylindrical [or flat], base membranous, sheathing, margins entire. **Scape** hollow [or solid]. **Inflorescences** racemose or paniculate, many-flowered, bracteate; bracts persistent, narrowly lanceolate, scarious. **Flowers:** tepals 6, erect to spreading, distinct or barely connate basally, equal, each with single prominent vein; stamens 6, distinct, subequal [or equal], shorter than tepals; filaments expanded at base; anthers dorsifixed; ovary 3-locular, ovules 1 or 2 per locule; septal nectaries present; style 1; stigma weakly 3-lobed; pedicel articulate. **Fruits** capsular, globose, hard, dehiscence loculicidal. **Seeds** 3 or 6, black, angled [or winged]. x = 13, 14.

Species 12 (1 in the flora): introduced; sw Europe, n Africa, sw Asia (India); widely introduced elsewhere.

Sometimes placed in the segregate family Asphodelaceae, *Asphodelus* occurs mainly throughout the Mediterranean region, the Middle and Near East, and central Asia. *Asphodelus ramosus* Linnaeus has been collected once in the flora from ballast in New Jersey in 1879.

SELECTED REFERENCE Spellenberg, R., R., et al. 1986. Additions to the flora of New Mexico. Sida 11: 455–470.

1. **Asphodelus fistulosus** Linnaeus, Sp. Pl. 1: 309. 1753 • Onion-weed [F] [I]

Plants annuals or short-lived perennials; root crowns thickened with many fibrous roots. **Stems** branched or unbranched, 20–70 cm. **Leaves** 5–35 cm × 2–4 mm; blade cylindrical or slightly flattened adaxially, hollow, glabrous except on margins. **Inflorescences** simple racemes to open panicles with ascending branches, 15–70 cm; bracts white, 4–7 mm. **Flowers** diurnal, closing in evening and during cloudy weather; tepals white to pale pink, oblong, obtuse, 5–12 × 2–3 mm, vein dark pink or brown; longest stamens equaling style. **Capsules** 5–7 mm, transversely wrinkled. **Seeds** 3–4 mm, rugose. $2n$ = 28, 56.

Flowering late winter–spring. Roadsides, fields, waste places; 0–1400 m; introduced; Calif., N.Mex., Tex.; sw

Europe; sw Asia; n Africa; widely naturalized (Australia, Mexico, New Zealand); expected elsewhere.

Asphodelus fistulosus has become a noxious weed in California and in other places with Mediterranean climates worldwide.

40. HEMEROCALLIS Linnaeus, Sp. Pl. 1: 324. 1753; Gen. Pl. ed. 5, 151. 1754 • Daylily, hémérocalle [Greek *hemeros*, day, and *kallos*, beauty, alluding to the showy flowers, which bloom and wilt in one day] [I]

Gerald B. Straley†

Frederick H. Utech

Herbs, perennial, scapose, clump-forming, rhizomatous, from fibrous or fleshy contractile roots often enlarged at ends; rhizomes spreading. **Leaves** many, basal, sessile, 2-ranked, bases sheathing; blade long-linear, keeled, apex acuminate. **Inflorescences** 2, in terminal helicoid cyme, or solitary. **Flowers** mostly diurnal and ephemeral, slightly irregular, showy; tepals 6, connate basally into short, funnelform to campanulate tube, distinct parts imbricate, spreading, inner broader than outer; stamens 6, adnate to throat of perianth tube; filaments curved upward, distinct, unequal; anthers dorsifixed, 2-locular, linear-oblong, dehiscence introrse; ovary superior, green, 3-locular, conic, septal nectaries present; style curved upwards; stigma indistinctly 3-lobed or capitate. **Fruits** capsular, leathery, dehiscence loculicidal. **Seeds** rarely produced (sterile) or many. $x = 11$.

Species 15–30 (2 in the flora): introduced; temperate zones worldwide; temperate e Asia.

Hemerocallis is important economically as medicinal, poisonous, edible, and/or horticultural plants, which have been in Chinese culture for thousands of years (W. Erhardt 1992). Hemerocallin, a root neurotoxin, can be both poisonous and useful medicinally as an analgesic, diuretic, arsenic-poisoning antidote, and treatment for schistosomiasis (J. A. Duke and E. S. Ayensu 1985; W. Erhardt 1992; Hu S. Y. 1968). In Asia, flowers (buds and perianths), shoots, and tuberous roots (following suitable preparation) are important foods (G. Kunkel 1984). Daylilies are among the most popular North American garden plants. Registered cultivars of *Hemerocallis* now exceed 38,000, including more than 13,000 named clones of *H. fulva* (G. Grosvenor 1999; R. M. Kitchingman 1985; R. W. Munson Jr. 1989; W. B. Zomlefer 1998).

Hemerocallis has been included in a broadly circumscribed segregate family Hemerocallidaceae with 13–18 genera mainly from the Southern Hemisphere, especially Australia (W. B. Zomlefer 1998; H. T. Clifford et al. 1998), or placed alone in a monotypic Hemerocallidaceae (A. L. Takhtajan 1997).

The dwarf, yellow-flowered *Hemerocallis minor* P. Miller, grass-leaf daylily, has been reported as a local escape in Oregon.

SELECTED REFERENCES Erhardt, W. 1992. *Hemerocallis*: Daylilies. Portland. Hu S. Y. 1968. The species of *Hemerocallis*. Amer. Hort. Mag. 47: 86–111. Stout, A. B. 1934. Daylilies. New York. [Reprinted 1986, London.]

1. Plants 7–15 dm; flowers not fragrant; tepals tawny orange, reticulate-veined; inner tepal margins wavy; capsules not or only rarely developing . . . . 1. *Hemerocallis fulva*
1. Plants 5–10 dm; flowers fragrant; tepals lemon yellow, parallel-veined; inner tepal margins smooth; capsules fully developing . . . . 2. *Hemerocallis lilioasphodelus*

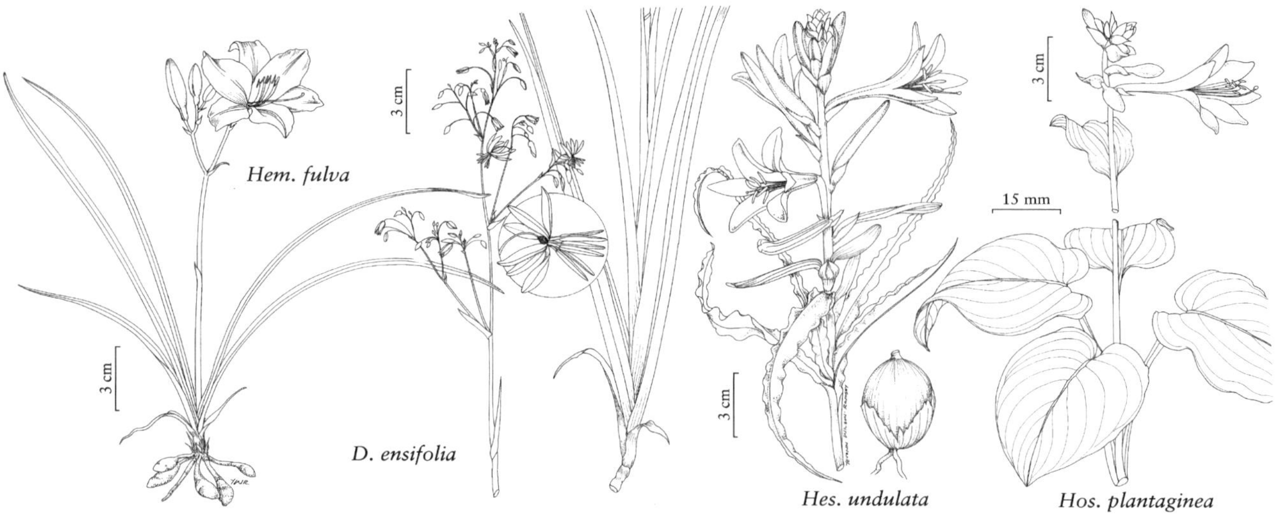

HEMEROCALLIS ◦ DIANELLA ◦ HESPEROCALLIS ◦ HOSTA

1. **Hemerocallis fulva** (Linnaeus) Linnaeus, Sp. Pl. ed. 2, 1: 462. 1762 • Orange daylily, tawny daylily, hémérocalle fauve, lis d'un jour [F] [I]

Hemerocallis lilioasphodelus Linnaeus var. *fulvus* Linnaeus, Sp. Pl. 1: 324. 1753

Plants 7–15 dm; main roots fleshy. **Leaf blade** yellowish green, 7–10 dm × (1–)2.5–3 cm. **Scape** branched, 10–20-flowered, taller than foliage. **Flowers** diurnal, not fragrant; perianth tube, widely funnelform, 2–3 cm; tepals yellow basally with darker tawny orange zones and stripes, veins reticulate; outer tepals 7–8 × 1.8–2.2 cm, margins smooth; inner tepals 7.5–8.5 cm × 3–3.5 cm, margins wavy; filaments 4.5–6.5 cm; anthers 5–7 mm; ovary 8–10 mm; style white to pale orange, 9–10 cm; pedicel 3–6 mm. **Capsules** not or rarely developing. **Seeds** rarely produced. $2n = 33$.

Flowering late spring–early summer. Roadsides, waste places, homesteads, open forests, stream banks; 0–1000 m; introduced; N.B., N.S., Ont., P.E.I., Que.; Ala., Ark., Conn., Del., D.C., Fla., Ga., Idaho, Ill., Ind., Iowa, Kans., Ky., La., Maine, Md., Mass., Mich., Minn., Miss., Mo., Mont., Nebr., N.H., N.J., N.Y., N.C., Ohio, Pa., R.I., S.C., S.Dak., Tenn., Tex., Utah, Vt., Va., Wash., W.Va., Wis.; e Asia (China, Japan); naturalized Eurasia; expected elsewhere.

Following an earlier European introduction from Asia, *Hemerocallis fulva* was brought to North America in the seventeenth century. This commonly cultivated daylily, the wild type, is distinguished as cultivar 'Europa' Stout and is a self-sterile triploid producing no seed. Essentially, it is a large, complex clone. Plants persist from cultivation or have arisen from root or rhizome fragments, which are capable of plant regeneration. Cultivar 'Kwanso' Regel, another ancient garden selection, persists in many areas along with the wild type and has fully doubled flowers. In eastern Asia, both diploids and triploids occur in the *H. fulva* complex and have been the basis for extensive breeding and tetraploid cultivar selection (A. B. Stout 1934).

2. **Hemerocallis lilioasphodelus** Linnaeus, Sp. Pl. 1: 324. 1753 • Yellow daylily, lemon daylily [I]

Hemerocallis flava (Linnaeus) Linnaeus

Plants 5–10 dm; roots enlarged, fibrous. **Leaf blade** dark green, 5–6.5 dm × 0.8–1.5 cm. **Scape** closely branched distally, 8–12-flowered, taller than foliage. **Flowers** often remaining open into night, fragrance strongly sweet, lemony; perianth tube shortly funnelform, 1.5–2.5 cm; tepals uniformly pale to bright lemon yellow, veins parallel; outer tepals 5–7 × 1–1.3 cm, margins smooth; inner tepals 5–7.5 × 1–2 cm, margins smooth; filaments 3–3.5 cm; anthers 2–3 mm; ovary 5–6 mm; style white to yellow, 7–8 cm; pedicel 2–4 mm. **Capsules** fully developed, oblong-elliptic, (2–)3–4 × (1–)1.5–2 cm. **Seeds** black, round or angular by compression, 3–5 mm, shiny. $2n = 22$.

Flowering summer. Roadsides, waste places, open woods; 0–500 m; introduced; N.B., Ont., Que.; Ark., Conn., Ill., Ind., Ky., Maine, Md., Mass., Mich., Minn., Miss., N.J., N.Y., N.C., Ohio, Okla., Pa., R.I., Texas, Vt., Va., W.Va., Wis.; e Asia; naturalized Europe.

Hemerocallis lilioasphodelus was an early introduction to Europe, where it naturalized, and then to North America (W. J. Dress 1955; Hu S. Y. 1968; W. B. Zomlefer 1998). This diploid species escapes only sporadically, unlike the more aggressive *H. fulva*, with true naturalization frequently questioned (W. B. Zomlefer 1998).

41. DIANELLA Lamarck ex Jussieu, Gen. Pl., 41. 1789 • Flax-lily [Latin *Diana*, Roman sylvan goddess, and *-ella*, diminutive suffix, alluding to the forest habitat and small stature] [I]

Frederick H. Utech

Herbs, perennial, evergreen, scapose, mat-forming, from stout, creeping, scaly rhizomes with fibrous, tuberous roots. **Leaves** basal, radical, distichous, crowded, sheathing, firm to subcoriaceous; blade linear to broadly ensiform, basal sheaths connate into short tubes, margins entire or serrulate, apex obtuse. **Scape** elongate, 0.5–1 m. **Inflorescences** paniculate, loosely branching with short, terminal racemes, bracteate; bracts small. **Flowers** nodding to ascending; tepals 6, persistent, withering, distinct, subequal, narrowly oblong to ovate, 3–7-veined; stamens 6, distinct; filaments barely adnate to tepal bases, thickened distally; anthers basifixed, dehiscence extrorse, opening by terminal pores that become longitudinal slits; ovary superior, 3-locular, septal nectaries present; style filiform; stigma minute, capitate; pedicel slender, articulate distally. **Fruits** rather long-persistent, baccate, blue to bluish purple, ovoid-globose. **Seeds** black, lustrous, ovate, somewhat flattened. $x = 8$.

Species 12–20 (1 in the flora): introduced; tropical Africa, Asia, Australia, and Polynesia.

SELECTED REFERENCES Schlittler, J. 1940. Monographie der Liliaceengattung *Dianella* Lam. Mitt. Bot. Mus. Univ. Zürich 163: 1–283. Schlittler, J. 1948. Unsere gegenwäertige Kenntnis über die Liliaceengattung *Dianella* in Malesien. Blumea 6: 200–228. Schlittler, J. 1957. Die Verbreitung der Liliaceengattung *Dianella* Lam. in Zusammenhang mit der Organdifferenzierung und der Arealbildung. Mitt. Bot. Mus. Univ. Zürich 207: 1–38.

1. **Dianella ensifolia** (Linnaeus) de Candolle in P. J. Redouté, Liliac. 1(1): plate 1. 1802 • Cerulean flax-lily, umbrella dracaena [F] [I]

Dracaena ensifolia Linnaeus, Mant. Pl. 1: 63. 1767

Plants glabrous. **Leaves** 50–80 × 1.5–2(–3) cm; blade lustrous, margins finely serrate, apex obtuse, basal sheath margins purplish; midvein impressed adaxially, keeled abaxially, keel finely serrate. **Inflorescences:** panicles 10–15 cm; flowers distal; bracts membranous. **Flowers:** tepals spreading, sky to light blue, yellow basally, 5-veined, 6 mm; filaments geniculate near middle; pedicel 7–15 mm. **Berries** 4–5 mm. $2n = 16, 32$.

Flowering late spring. Open, wet pinelands, near seashores; 0 m; introduced; Fla.; se Africa; Asia (s China, Formosa, India, Japan); Malaysia; Philippines; widely naturalized elsewhere.

In its native range, *Dianella ensifolia* occurs in a broad ecological range, from seashores, grasslands, and open scrubs to alpine and open lowland forests between sea level and 1600 m. In Florida, this naturalized ornamental probably spreads by its extensive rhizome systems and colorful fruits (W. B. Zomlefer 1998).

42. HESPEROCALLIS A. Gray, Proc. Amer. Acad. Arts 7: 390. 1867 • Desert-lily [Greek *hesperos*, western, and *kallos*, beauty] [E]

Frederick H. Utech

Herbs, perennial, scapose, from deep-set, tunicate bulbs with fleshy roots. **Leaves** mostly basal, reduced distally; blade bluish green, keeled, linear, margins white, strongly undulate. **Scape** simple to rarely branched, stout. **Inflorescences** terminal, racemose, open, bracteate, elongate; bracts withering-persistent, conspicuous, ovate, scarious. **Flowers** fragrant; perianth funnelform, 4.5–6 cm; tepals 6, withering-persistent, connate below middle into tube, limb lobes spreading, white adaxially, with bluish green midstripes abaxially, 5–7-veined, obovate-oblanceolate; stamens fused to perianth tube; filaments filiform; anthers dorsifixed, versatile, linear; ovary superior, 3-locular, sessile, oblong; style persistent, white, slender, equaling tepals; stigma capitate to slightly 3-lobed; pedicel jointed at apex. **Fruits** capsular, subglobose, 3-lobed, dehiscence loculicidal. **Seeds** numerous, jet black, flat. $x = 24$.

Species 1: sw North America.

SELECTED REFERENCES Cave, M. S. 1948. Sporogenesis and embryo-sac development of *Hesperocallis* and *Leucocrinum* in relation to their systematic position. Amer. J. Bot. 35: 343–349. Kubitzki, K. 1998. *Hesperocallis*. In: K. Kubitzki et al., eds. 1990+. The Families and Genera of Vascular Plants. 4+ vols. Berlin, etc. Vol. 3, p. 260. McNeal, D. W. Jr. 1993. *Hesperocallis*. In: J. C. Hickman, ed. 1993. The Jepson Manual. Higher Plants of California. Berkeley, Los Angeles, and London. P. 1198.

1. Hesperocallis undulata A. Gray, Proc. Amer. Acad. Arts 7: 391. 1867 [E] [F]

Plants 3–18 dm; bulbs ovoid, 4–6 cm. **Leaves** basal; blade 2–5 dm × 8–15 mm. **Racemes** 4–18-flowered, 1–3 dm; bracts 1–1.5 cm. **Flowers:** perianth tube 1.5–2 cm; limb lobes 3–4 cm × 6–10 mm; filaments 2–2.5 cm; anthers golden, 7 mm; pedicel 1 cm. **Capsules** 12–16 mm. **Seeds** 5 mm. $2n = 48$.

Flowering late Feb–May, following infrequent rains. Dry, sandy flats to rocky hills of creosote bush scrub in Mojave and Sonoran deserts; 300–800 m; Ariz., Calif., Nev.

The large, snowy white flowers of *Hesperocallis undulata* make it one of the showiest desert species, with consequent horticultural use. Native Americans used the bulbs for food (D. E. Moerman 1986), and the early Spanish colonists called the bulbs *ajo*, due to the garlic flavor.

43. HOSTA Trattinnick, Arch. Gewächsk. 1: 55, plate 89. 1811, name conserved • Plantain-lily, funkia [for Nicolaus Thomas Host, 1761–1834, Austrian botanist and physician to Emperor Frances II] [I]

Frederick H. Utech

Funkia Sprengel, *Niobe* Salisbury

Herbs, perennial, scapose, forming dome-shaped clumps, from rhizomes; rhizomes short, branching, sometimes stoloniferous, leaf scars prominent; roots fleshy. **Leaves** numerous, basal, spiral, distinctly petiolate; petiole sulcate, terete, sometimes ridged; blade light to dark green, often variegated, cordate to orbiculate to lanceolate, smooth to puckered, margins entire, slightly undulate [flat or crisped]; veins campylodromous, conspicuous, usually sunken adaxially, prominent abaxially. **Scape** usually surpassing leaves. **Inflorescences** simple, terminal, racemose, usually subsecund, elongate, subtended proximally by 1 or more sterile bracts, each flower usually bracteate. **Flowers:** perianth tubular to campanulate or urceolate-cylindric [funnelform]; tepals 6, similar, connate proximally into wide-throated tube, white, bluish purple, or purplish violet with darker markings or lines, lobes spreading, sometimes recurved, longer than perianth tube; stamens 6, inserted at base of perianth tube or ovary apex, exceeding tepals; filaments declinate; anthers dorsifixed in connective pits, dehiscence introrse; ovary superior, sessile, 3-locular, oblong, septal nectaries present; style filiform, exceeding stamens; stigma minute, capitate or 3-lobed; pedicel short. **Fruits** capsular, pendent at maturity, angled, elongate or triangular, dehiscence loculicidal. **Seeds** numerous, black, flattened, winged. $x = 30$.

Species ca. 40 (3 in the flora): introduced; temperate e Asia, especially Japan; cultivated worldwide.

Hosta species delimitation has been problematic due to a long history of cultivation, hybridization, and selection, particularly in Japan, from the eighth century onwards (W. G. Schmid

1991). *Hosta* nomenclature is further complicated because many names are based on types of garden origin or sports originating among wild populations. Earlier taxonomic treatments were largely based on materials cultivated regionally in Japan (N. Fujita 1976), North America (L. H. Bailey 1930), Korea (M. G. Chung 1990; M. G. Chung and J. W. Kim 1991), and Europe (N. Hylander 1954). *Hosta* can be considered to comprise as few as 23–26 species (F. Maekawa and K. Kaneko 1968; N. Fujita 1976), or 40 or more if a stricter species concept is applied (A. Huxley et al. 1992; F. Maekawa 1940; W. G. Schmid 1991).

Well over 1000 cultivars have been recorded with the International Registration Authority. Primarily used in temperate shade gardens, these cultivars feature various combinations of leaf size, shape, color, variegation, and texture (P. Aden 1988; D. Grenfell 1996, 1998; N. Hylander 1954; K. Kubitzki 1998b; W. G. Schmid 1991). While *Hosta* is mainly of ornamental importance economically, the leaves of some species are cooked and eaten in Korea and Japan, thus depleting local populations.

Funkia, a later generic name proposed by Sprengel for these plants, is an illegitimate later homonym of *Funckia* Wildenow, and the family name Funkiaceae based upon it is therefore invalid (B. Mathew 1988). However, "funkia," from the vernacular Japanese *fukurin fu,* long ago passed into many European languages as another common name for *Hosta.*

SELECTED REFERENCES Chung, M. G. and S. B. Jones. 1989. Pollen morphology of *Hosta* Tratt. and related genera. Bull. Torrey Bot. Club 116: 31–44. Currie, H. E. 1992. A Biosystematic Study of the Genus *Hosta* Tratt. (Liliaceae / Hostaceae) in Eastern Asia. Ph.D. dissertation. University of Georgia. Hylander, N. 1960. The genus *Hosta*. J. Roy. Hort. Soc. 85: 356–369. Kirkpatrick, N. S. 1993. A Chloroplast DNA Restriction Mapping Study of the Genus *Hosta* (Liliaceae). Ph.D. dissertation. Miami University. Maekawa, F. 1940. The genus *Hosta*. J. Fac. Sci. Univ. Tokyo, Sect. 3, Bot. 5: 317–425. Maekawa, F. and K. Kaneko. 1968. Evolution of karyotype in *Hosta* (Liliaceae). J. Jap. Bot. 43: 132–140. Schmid, W. G. 1991. The Genus *Hosta*. Portland. Stearn, W. T. 1931. The hostas and funkias: A revision of the plantain lilies. Gard. Chron., ser. 3, 90: 27, 47–49, 88–89, 110.

1. Leaf blades light yellowish green; flowers fragrant, to 13 cm, perianth long-tubular; tepals white . . . . 1. *Hosta plantaginea*
1. Leaf blades green; flowers not fragrant, 4–5.5 cm, perianth tubular-campanulate or urceolate-cylindric; tepals bluish purple or purplish violet.
    2. Scape 40–50 cm; leaf blades lanceolate to ovate-lanceolate, 10–17 × 5–7.5 cm, with 5–6 lateral vein pairs; tepals purplish violet; anthers purple . . . . 2. *Hosta lancifolia*
    2. Scape 80–95 cm; leaf blades broadly ovate to cordate, 20–30 × 15–20 cm, with 7–9 lateral vein pairs; tepals bluish purple; anthers spotted purple . . . . 3. *Hosta ventricosa*

1. Hosta plantaginea (Lamarck) Ascherson, Bot. Zeitung (Berlin) 21: 53. 1863 • August-lily, fragrant plantain-lily, white plantain-lily [F] [I] [W1]

Hemerocallis plantaginea Lamarck in J. Lamarck et al., Encycl. 3: 103. 1789

Plants forming open clumps 60–90 × 45 cm; rhizomes not stoloniferous. **Leaves:** petiole erect, spreading, light green, to 25 cm; blade light yellowish green, orbiculate to cordate, 20–28 × 15–24 cm, base decurrent, apex acuminate; veins in 6–9 lateral pairs. **Scape** 60–80 cm. **Inflorescences:** racemes bright green, 10–15-flowered, stout, solid, to 25 cm; floral bracts large, greenish white, ovate-lanceolate, navicular; sterile bracts leafy, sparse. **Flowers** to 13 cm, fragrant; perianth long-tubular; tepals white, lobes sharply recurved; anthers yellow. **Capsules** large, elongate, apex pointed. $2n = 60$.

Flowering late summer (August). Disturbed open areas; 0–100 m; introduced; Pa., expected elsewhere; se China.

Hosta plantaginea is widely cultivated for its fragrant white flowers, which open in late afternoon or evening. This showy species was the first *Hosta* cultivated in Europe, starting in 1784, and then in the New World. Following its introduction from China, it was planted by the thousands in public gardens and became known as Parisian funkia or hosta.

2. Hosta lancifolia Engler in H. G. A. Engler and K. Prantl, Nat. Pflanzenfam. 2[II,5]: 40, plates 17, 18. 1887 • Narrowleaf plantain-lily [I]

Plants forming open clumps 35–50 × 30 cm; rhizomes sometime stoloniferous. **Leaves:** petiole erect, green, dotted purple at base, 17–25 cm; blade deep olive green, lanceolate to ovate-lanceolate, 10–17 × 5–7.5 cm, apex narrowly acuminate; veins in 5–6 lateral pairs. **Scape** 40–50 cm. **Inflorescences:** racemes 10–20-flowered, lax, slender, 17–20 cm; floral bracts short, glossy green, narrowly navicular; sterile bracts 3–5, large, leafy. **Flowers** 4–4.5 cm, not fragrant; perianth tubular-campanulate; tepals purplish violet, lobes spreading, recurved; anthers purple. **Capsules** rarely developing (pod sterile). $2n = 60$.

Flowering late summer–early fall (September). Disturbed open areas; 0–300 m; introduced; Conn., Ill., Md., N.J., N.Y., Ohio, Pa., expected elsewhere; Japan (of garden origin); cultivated worldwide.

Hosta lancifolia produces an excellent ground cover with deep olive green leaves and vigorous vegetative growth. Since this species is of ancient horticultural origin and does not occur in the wild, it has been reduced to cultivar status in horticultural nomenclature as *Hosta* 'Lancifolia' (W. G. Schmid 1991).

3. **Hosta ventricosa** (Salisbury) Stearn, Gard. Chron., ser. 3, 90: 27. 1931 • Blue plantain-lily [I]

Bryocles ventricosa Salisbury, Trans. Hort. Soc. London 1: 335. 1812; *Niobe caerulea* (Andrews) Nash

Plants forming clumps 80–90 × 50 cm; rhizomes compact. **Leaves:** petiole spreading horizontally, light green with purple streaks at base, deeply grooved, 18–25 cm; blade lustrous dark green, broadly ovate to cordate, 20–30 × 15–20 cm, apex acuminate; veins in 7–9 lateral pairs. **Scape** 80–95 cm. **Inflorescences:** racemes stiffly erect, flushed red at base, 20–30-flowered, to 1 m; floral bracts broad, pale green, white at base; sterile bract 1, leafy, occurring at midpoint. **Flowers** 4–5.5 cm, not fragrant; perianth urceolate-cylindric; tepals bluish purple, lobes not recurved; anthers spotted purple. **Capsules** short, triangular, stubby, apex blunt. $2n = 120$.

Flowering summer (July). Disturbed open areas; 0–500 m; introduced; Conn., Del., Ky., Md., Mass., Mich., N.J., N.Y., N.C., Ohio, Pa., R.I., Vt., Va., W.Va., expected elsewhere; China; cultivated worldwide.

Hosta ventricosa, a natural tetraploid, undergoes pseudogamous apomixis and therefore breeds true, but is of no use as a seed parent in hybridizing. It can, however, act as a pollen parent. Like *H. plantaginea*, this species was an early introduction from China.

44. ALLIUM Linnaeus, Sp. Pl. 1: 294. 1753; Gen. Pl. ed. 5, 143. 1754 • Onion [Latin, classical name for garlic]

Dale W. McNeal Jr.

T. D. Jacobsen

Herbs, perennial, scapose, from tunicate bulbs, with onion odor and taste. **Bulbs** solitary or clustered, dividing at base, or on rhizomes, reforming annually; outer coats generally brown or gray, smooth, fibrous, or with cellular reticulation (generally important in identification); inner coats membranous. **Leaves** generally withering from tip by anthesis, usually persistent, 1–12, basal; blade usually linear, terete, channeled, or flat (carinate in *A. sativum*, *A. praecox*, *A. tuberosum*, *A. rotundum*, *A. neapolitanum*, *A. triquetrum*, *A. unifolium*, and *A. lacunosum*), straight or ± falcate (coiled or circinate in *A. nevadense* and *A. atrorubens*), broader in *A. victorialis* and *A. tricoccum*, not petiolate (except in *A. tricoccum* and *A. victorialis*). **Scape** usually persistent, terete or flattened. **Inflorescences** umbellate, flowering centripetally (centrifugally in *A. schoenoprasum*), sometimes replaced totally or partially by bulbils, subtended by spathe bracts; bracts conspicuous, ± fused, usually 3+-veined, equaling pedicel except in some introduced species, membranous. **Flowers** erect (pendent in *A. triquetrum*); tepals 6, in 2 similar whorls, ± distinct, petallike, usually becoming becoming dry and persisting; stamens 6, epipetalous; filaments in all but 1 native species broad at base, fused into ring (some introduced species and *A. victorialis* appendaged), linear, generally glabrous (*A. rotundum* and *A. hoffmanii* papillose to ciliate proximally); anthers and pollen variously colored; ovary superior, 3-lobed, sometimes crested with processes, 3-locular, usually 2 ovules per locule (6–8 in *A. nigrum*), crest processes 3 or 6, smooth except in *A. haematochiton*, *A. sharsmithiae*, and *A. lacunosum*; style 1; stigma capitate to ± 3-lobed; pedicel erect or spreading (lax in *A. triquetrum*). **Fruits** capsular, dehiscence loculicidal. **Seeds** black, obovoid, finely cellular-reticulate, cells smooth or minutely roughened, with 1–8 papillae, without caruncle except in *A. triquetrum*. $x = 7, 8, 9$.

Species 550–700 (96 in the flora): mostly Northern Hemisphere.

As with many other genera in the Liliaceae sensu lato, *Allium* has been segregated into a separate family, Alliaceae, by most recent authors (R. Dahlgren et al. 1985; K. Rahn 1998; A. L. Takhtajan 1997; R. F. Thorne 1992).

Old World species of *Allium* are generally placed in subgenera and sections. Attempts to treat New World species similarly have gained less acceptance. H. P. Traub (1972) recognized subg. *Amerallium*, encompassing all of the x = 7 North American members of the genus. P. Hanelt (1992) placed the Old World x = 7 species also in that subgenus, which includes all but three of the North American species (*A. schoenoprasum, A. tricoccum,* and *A. victorialis*), which have x = 8. These latter species have been placed in subg. *Rhizirideum* (P. Hanelt 1992). Resolution of the problematic subgeneric and sectional relationships among Old and New World species will require much more extensive molecular and phylogenetic analysis of the genus.

The characters used in the construction of the following key are the best that have been discovered for this purpose. Many of these are subject to sometimes rather wide variation. It is anticipated, therefore, that it can be used successfully only with considerable understanding of the natural units involved, their distribution, and comparison with descriptions and accurately named specimens.

The *Allium* bulb consists of a series of leaf bases, some with blades, others without, surrounding an apical meristem. The leaf bases are replaced annually as the meristem forms a new (renewal) bulb. Often the meristem will branch, resulting in two or more bulbs being formed from a single parent bulb (increase bulbs). In addition several types of rhizomes are formed in *Allium*. In some cases (e.g., *A. validum*) bulbs form atop a thick, iris-like rhizome that produces new bulbs in succeeding years. In other cases (e.g., *A. bolanderi, A. campanulatum,* and *A. unifolium*) each bulb produces one or more rhizomes that in turn produce a terminal renewal or increase bulb. There is variation even in this type of rhizome. In some species (e.g., *A. bolanderi* and *A. unifolium*) the bulb producing the rhizomes disappears during the development of the rhizomes and new bulbs except for the roots, which remain active until the new bulbs mature and become dormant. In others (e.g., *A. campanulatum*) the bulb produces a number of very short rhizomes around the roots, each of which develops a very small terminal bulbel. At the same time, the parent bulb remains intact and produces a renewal bulb, or may divide to produce two or more large increase bulbs. As a result, specimens with this type of rhizome will have one or more large bulbs with a number of much smaller bulbels among the roots.

The underground parts of *Allium* are often critical for successful identification, and every effort should always be made to collect them. The "dirty" brown or gray coats surrounding the bulbs are the only source for the cellular-reticulation patterns referred to in the key. These should be carefully collected and preserved as part of any specimen.

Several Old World species (*Allium ampeloprasum* Linnaeus, *A. cepa* Linnaeus, *A. oleraceum* Linnaeus, *A. sativum* Linnaeus, *A. nigrum* Linnaeus), grown as foodstuffs or ornamentals, may be encountered and are certainly represented in North American herbaria. For this reason, these species have been included in the key. Some do not reproduce by seeds and probably should not be considered as truly naturalized in our flora, although they may persist for long periods at or near places where they have been planted. Additionally, over 52 species contained in this treatment are considered choice garden plants and are readily available through catalogues and garden centers. Locally, these species and possibly others grown as garden plants may escape and become established. It is inevitable, therefore, that numerous specimens of these escapees will be represented in North American herbaria and may muddy the boundaries of what is regarded to be the natural distribution. We have attempted in all cases to map what we consider to be the natural ranges of these species, based on the specimens examined over the years. Material falling outside these ranges may include these escapees and should be suspect.

SELECTED REFERENCES Cronquist, A. and M. Ownbey. 1977. *Allium.* In: A. Cronquist et al. 1972+. Intermountain Flora. Vascular Plants of the Intermountain West, U.S.A. 5+ vols. New York and London. Vol. 6, pp. 508–522. New York. Davies, D. 1992. Alliums, the Ornamental Onions. Portland. Denison, S. S. and D. W. McNeal Jr. 1989. A reevaluation of the *Allium sanbornii* (Alliaceae) complex. Madroño 36: 122–130. Gregory, M. et al. 1998. Nomenclator *Alliorum*: *Allium* Names and Synonyms, a World Guide. Kew. Hanelt, P. 1992. Infrageneric grouping of *Allium*—The Gatersleban approach. In: P. Hanelt et al., eds. 1992. The Genus *Allium*: Taxonomic Problems and Genetic Resources.... Gatersleben. Pp. 107–123. Jacobsen, T. D. 1978. A Comparative Study of Three Alliances of the Genus *Allium*. Ph.D. thesis. Washington State University. Mathew, B. 1996. A Review of *Allium* sect. *Allium*. Kew. McNeal, D. W. Jr. 1970. Comparative Studies of the *Allium acuminatum* Alliance. Ph.D. thesis. Washington State University. McNeal, D. W. Jr. 1982. Taxonomy of the *Allium lacunosum* complex (Liliaceae). Madroño 29: 79–86. McNeal, D. W. Jr. 1992. A revision of the *Allium fimbriatum* (Alliaceae) complex. Aliso 13: 411–426. McNeal, D. W. Jr. 1992b. A reappraisal of *Allium cristatum* (Alliaceae) and its allies. Madroño 39: 83–89. Mingrone, L. V. 1968. A Comparative Study of the *Allium falcifolium* Alliance. Ph.D. thesis. Washington State University. Mortola, W. R. and D. W. McNeal Jr. 1985. Taxonomy of the *Allium tribracteatum* (Alliaceae) complex. Aliso 11: 27–35. Ownbey, M. 1947. The genus *Allium* in Arizona. Res. Stud. State Coll. Wash. 15: 211–232. Ownbey, M. 1950. The genus *Allium* in Idaho. Res. Stud. State Coll. Wash. 18: 3–39. Ownbey, M. 1950b. The genus *Allium* in Texas. Res. Stud. State Coll. Wash. 18: 181–222. Ownbey, M. 1969b. *Allium.* In: C. L. Hitchcock et al. 1955–1969. Vascular Plants of the Pacific Northwest. 5 vols. Seattle. Vol. 1, pp. 739–760. Ownbey, M. and H. C. Aase. 1955. Cytotaxonomic studies in *Allium*. I. The *Allium canadense* alliance. Res. Stud. State Coll. Wash., suppl. 1: 1–106. Peterson, P. M., C. R. Annable, and L. H. Rieseberg. 1988. Systematic relationships and nomenclatural changes in the *Allium douglasii* complex (Alliaceae). Syst. Bot. 13: 207–214.

1. Leaf blade flat, 15–90 mm wide, (tapering to base or distinctly petiolate).
   2. Leaves ephemeral, usually absent at anthesis; e North America . . . . 1. *Allium tricoccum*
   2. Leaves present at anthesis; Attu and Unalaska islands, Alaska . . . . 2. *Allium victorialis*
1. Leaf blade flat, channeled, or ± terete, never more than 30 mm wide, (never petiolate).
   3. Flowering pedicels mostly or completely replaced by bulbils.
      4. Outer bulb coats persisting as fibrous reticulum; leaf sheaths not extending more than 1/4 scape; spathe bract beakless or beak much shorter than base.
         5. Ovary, when present, crestless; spathe bracts 3–7-veined; east of 103rd meridian . . . . 3. *Allium canadense* (in part)
         5. Ovary, when present, obscurely crested with 6, low, central processes; spathe bracts 1-veined; west of 105th meridian . . . . 4. *Allium geyeri* (in part)
      4. Outer bulb coats membranous, if with fibers these not forming reticulum; leaf sheaths extending to midscape or above; spathe bract with beak equaling or longer than base.
         6. Spathe bract 1, caducous.
            7. Bulbs 1–2 cm diam.; leaf blade 2–4 mm diam., cylindric or filiform, not carinate, hollow below middle . . . . 5. *Allium vineale*
            7. Bulbs (1.5–)3–8 cm diam.; leaf blade 5–20 mm wide, flat, carinate, solid . . . . 6. *Allium sativum* var. *sativum*
         6. Spathe bracts 2–5, persistent.
            8. Spathe bracts 2–5, 4–9-veined, beak to 20 cm . . . . 7. *Allium oleraceum*
            8. Spathe bracts 3–5, 2–3-veined, beak to 10 cm . . . . 8. *Allium ampeloprasum* (in part)
   3. Flowering pedicels floriferous, bulbils almost unknown.
      9. Outer bulb coats persisting as fibrous reticulum.
         10. Ovary usually crestless; if obscurely crested, with 3 or 6 processes; east of 103rd meridian.
            11. Spathe bracts usually 1-veined.
               12. Spaces between bulb coat fibers filled in proximal 1/2 bulb; tepals white, pink, or red, rarely greenish yellow; central plains from n Mexico to Nebraska . . . . 9. *Allium drummondii*
               12. Spaces between bulb coat fibers open; tepals yellow; w Texas . . . . 10. *Allium coryi*
            11. Spathe bracts 3–7-veined.
               13. Umbel compact; pedicels much shorter than flowers . . . . 11. *Allium schoenoprasum* (in part)

13. Umbel loose; pedicels longer than flowers.
    14. Flowers substellate to urceolate-campanulate, ultimately withering somewhat and exposing capsule; reticula of bulbs finely or only moderately coarsely meshed.
        15. Bulbs 1–3, narrowly cylindric, attached to ± horizontal primary rhizome, often missing or not visible on herbarium specimens; leaf blade carinate; cells of seed coat smooth, shiny; occasional introduction . . . . 12. *Allium tuberosum*
        15. Bulbs 1–4+, ovoid, not attached to rhizome; leaf blades not carinate, channeled; cells of seed coat each with minute, central papilla; native east of 103rd meridian . . . . 3. *Allium canadense* (in part)
    14. Flowers urceolate, permanently investing capsule; reticula of bulbs usually very coarsely meshed.
        16. Flowering bulbs with cluster of stalked, basal bulbels; cells of innermost bulb coats contorted, with sinuous walls; extreme s Texas . . . . 13. *Allium runyonii*
        16. Flowering bulbs without basal bulbels; cells of innermost bulb coats vertically elongate, without sinuous walls; w Texas and e New Mexico to c South Dakota . . . . 14. *Allium perdulce*

[10. Shifted to left margin.—Ed.]

10. Ovary usually crested with 3 or 6 processes; if crestless, from west of 105th meridian.
    17. Ovary and capsule conspicuously crested with 6 contorted or horizontally spreading, ± lateral processes; tepals widely spreading to reflexed, se United States.
        18. Spathe bracts usually 5–7-veined; ovary crests conspicuously contorted; tepals spreading to reflexed . . . . 15. *Allium cuthbertii*
        18. Spathe bracts 1-veined; ovary crests flattened, horizontally spreading, not contorted; tepals widely spreading . . . . 16. *Allium speculae*
    17. Ovary crested with 6 ± erect, often obscure central processes; tepals erect to widely spreading; w North America.
        19. Leaves 3+ per scape; cells of seed coat each with minute, central papilla.
            20. Bulbs often short-rhizomatous basally; spathe bracts 3–5-veined; ovary conspicuously crested with 6 flattened, lacerate central processes; tepals spreading or reflexed, withering in fruit, not investing capsule . . . . 17. *Allium plummerae*
            20. Bulbs not short-rhizomatous; spathe bracts usually 1-veined; ovary obscurely crested with 6 rounded central processes; tepals erect, not withering in fruit, permanently investing capsule.
                21. Leaf blade flat, ± falcate, usually 3–6 mm wide; Box Elder County, Utah . . . . 18. *Allium passeyi* (in part)
                21. Leaf blade channeled, ± straight, usually less than 5 mm wide; widespread, n Great Plains and w North America . . . . 4. *Allium geyeri* (in part)
        19. Leaves usually 2 per scape; cells of seed coat ± smooth, with or without central papillae.
            22. Spathe bracts 3–5-veined; tepals becoming papery in fruit, midrib scarcely thickened, not investing capsule; ovary usually conspicuously crested with 6 flattened central processes, often to 2 mm . . . . 19. *Allium macropetalum*
            22. Spathe bracts 1-veined; tepals becoming callous-keeled, permanently investing capsule; ovary inconspicuously crested with 6 rounded central processes, to 1 mm.
                23. Leaf blade flat, ± falcate, usually 3–6 mm wide; cells of seed coat with minute central papilla; Box Elder County, Utah . . . . 18. *Allium passeyi* (in part)
                23. Leaf blade semiterete, channeled, ± straight, usually 1–3(–5) mm wide; cells of seed coat smooth; n Great Plains and w North America . . . . 20. *Allium textile*

9\. Outer bulb coats membranous to chartaceous, with or without distinct cellular markings (reticulation); without fibers or with some parallel fibers.
- 24\. Scape fistulose, 3–25 mm diam., not flattened and winged; leaves 2–10, blade flat and solid, or fistulose.
  - 25\. Leaf blade flat, solid.
    - 26\. Leaves not or scarcely sheathing base of scape . . . . 21. *Allium nigrum*
    - 26\. Leaves sheathing 1/3–1/2 scape . . . . 8. *Allium ampeloprasum* (in part)
  - 25\. Leaf blade fistulose.
    - 27\. Bulbs 1–3, to 10 cm diam., ± globose, not rhizomatous; leaf blade semicircular in cross section; occasional escape from cultivation . . . . 22. *Allium cepa*
    - 27\. Bulbs 1–2, 5 cm diam., cylindric, clustered on short rhizome (this often missing or not visible on herbarium specimens); leaf blade circular in cross section; native or introduced.
      - 28\. Flowers 8–18 mm; tepals lilac to pale purple; native or introduced . . . . 11. *Allium schoenoprasum* (in part)
      - 28\. Flowers 6–9 mm; tepals pale yellowish white; introduced . . . . 23. *Allium fistulosum*
- 24\. Scape solid, exceeding 5 mm wide only if flattened and winged; leaves 1–several, leaf blade solid.
  - 29\. Leaves (3–)5–40 mm wide, basal sheaths extending 1/3–1/2 scape.
    - 30\. Filaments unappendaged; leaf blade terete to semiterete; bulbels, if present, light brown . . . . 24. *Allium paniculatum*
    - 30\. Inner filaments appendaged with prominent tooth on each side of anther; leaf blade flat, channeled; bulbels very dark purple . . . . 25. *Allium rotundum*
  - 29\. Leaves 1–25 mm wide, basal sheaths never extending much above soil level.
    - 31\. Bulbs oblong, elongate, or ovoid, clustered on stout, primary rhizome, or short-rhizomatous; bulb coats membranous or chartaceous, finely striate with narrow, vertically elongate cells.
      - 32\. Bulbs on stout, iris-like rhizome; ovary crestless.
        - 33\. Tepals elliptic, apex obtuse; stamens ± equaling tepals; ec Arizona and adjacent New Mexico, and Santa Catalina Mountains, s Arizona . . . . 26. *Allium gooddingii*
        - 33\. Tepals narrowly lanceolate to lanceolate, apex acuminate; stamens much shorter than tepals or definitely exserted; widespread in w North America, not occurring in Arizona.
          - 34\. Stamens and style exserted; stigma capitate; Cascades and Sierras e to ne Nevada, e Oregon, w Idaho . . . . 27. *Allium validum*
          - 34\. Stamens and style ca. 1/2 tepals; stigma 3-lobed; Rocky Mountains from c Montana and ne Idaho to Wyoming, ne Utah, Colorado, and New Mexico . . . . 28. *Allium brevistylum*
      - 32\. Bulbs short-rhizomatous at base, rhizome not stout and iris-like; ovary strongly crested with 6 processes.
        - 35\. Stamens and styles included; outer bulb coats ± reddish brown, inner coats deep red to white; ovary crested with 6 short, rounded, densely papillose processes . . . . 29. *Allium haematochiton*
        - 35\. Stamens and styles exserted; outer bulb coats gray or brown, inner coats white to pink or reddish; ovary crested with 6 flattened, ± triangular processes, margins entire or toothed.
          - 36\. Flowers campanulate; tepals ± erect; scape nodding . . . . 30. *Allium cernuum*
          - 36\. Flowers stellate; tepals spreading; scape erect, or, if nodding at anthesis, becoming erect . . . . 31. *Allium stellatum*
    - 31\. Bulbs ovoid to subglobose, not clustered on stout, primary rhizome; rhizomes, if present, secondary, arising from bulbs, ± slender, terminated by new bulbs; bulb coats without reticulation or with ± isodiametric or transversely elongate cells that are sometimes intricately contorted.

[37. Shifted to left margin.—Ed.]

37. Leaf 1 per scape; leaf blade terete; ovary prominently crested with 6 ± triangular processes.
  38. Stigma unlobed or minutely 3-lobed, lobes ± stout, erect or spreading.
    39. Scape 18–60 cm; flowers 5–9 mm; tepals unequal, inner whorl 1/4–1/3 longer than outer, margins entire or irregular to erose; stamens exserted . . . . 32. *Allium sanbornii* (in part)
    39. Scape less than 25 cm; flowers 7–20 mm; tepals ± equal, margins entire; stamens included.
      40. Outer bulb coat reticulate with ± elongate, contorted meshes . . . . 33. *Allium nevadense*
      40. Outer bulb coat lacking reticulation, or meshes very indistinct, square or polygonal.
        41. Pedicels slender, longer than flowers; flowers 8–12 mm . . . . 34. *Allium atrorubens*
        41. Pedicels stout, generally shorter than flowers; flowers 12–20 mm.
          42. Tepals lanceolate to lance-linear, apex acute; lacking stalked, basal increase bulbs; rocky, sandy desert slopes, s California to w Arizona . . . . 35. *Allium parishii*
          42. Tepals lance-linear to lanceolate, apex long-acuminate; with 1–2 stalked basal increase bulbs; alpine ridges and talus, s California mountains . . . . 36. *Allium monticola*
  38. Stigma distinctly 3-lobed, lobes often slender and recurved.
    43. Stamens equaling tepals or exserted.
      44. Tepals unequal, inner 1/3–1/2 longer than outer . . . . 31. *Allium sanbornii* (in part)
      44. Tepals ± equal . . . . 37. *Allium howellii*
    43. Stamens included.
      45. Tepal (at least inner whorl) margins denticulate to erose.
        46. Scape 25–40 cm . . . . 38. *Allium jepsonii*
        46. Scape 5–20 cm.
          47. Outer bulb coats reddish brown; tepals erect, ± straight at tip; inner whorl margins denticulate . . . . 39. *Allium denticulatum*
          47. Outer bulb coats brown to gray; tepals erect, ± spreading-reflexed at tip; inner whorl margins denticulate to erose . . . . 40. *Allium abramsii*
      45. Tepal margins all ± entire.
        48. Margins of ovarian crest processes entire or notched at tip, outer margins sometimes irregular but never dentate or laciniate.
          49. Flowers 10–18 mm; tepals maroon or deep reddish purple.
            50. Tepals deep reddish purple, all reflexed at tip; Mount Hamilton Range, c California . . . . 41. *Allium sharsmithiae*
            50. Tepals maroon, outer curled back at tip, inner reflexed; Spanish Needle Peak, s Sierra Nevada, and Horse Canyon, Tehachapi Mountains, California . . . . 42. *Allium shevockii*
          49. Flowers 6–9 mm; tepals white to pink, darkening in age.
            51. Inflorescence loose; pedicels flexuous in fruit; tepals lanceolate to lance-ovate, apex acuminate . . . . 43. *Allium parryi*
            51. Inflorescence compact; pedicels straight; tepals ovate to nearly round, apex obtuse (rarely acute) to shallowly emarginate . . . 44. *Allium munzii*
        48. Margins of ovarian crest processes dentate to laciniate.
          52. Tepals deep reddish purple, erect, usually conspicuously recurved at tip . . . . 45. *Allium fimbriatum* (in part)
          52. Tepals white or flushed to pale lavender with darker midveins, spreading or erect, not conspicuously recurved at tip.
            53. Flowers usually 6–12 mm . . . . 45. *Allium fimbriatum* (in part)
            53. Flowers usually 6–8(–10) mm.
              54. Scape 25–50 cm; tepals spreading from base; serpentine soil, Rawhide Hill and Red Hills, foothills of Sierra Nevada, c California . . . . 46. *Allium tuolumnense*
              54. Scape 7–20(–30) cm; tepals erect; serpentine clay soils, s Coast Ranges and w Transverse Ranges, California . . . 47. *Allium diabolense*

37. Leaves usually 2 or more, if 1, blade flattened or broadly channeled; ovary crestless or variously crested.
55. Bulbs generally with numerous increase bulbs, these much smaller than parent bulb, enclosed by bulb coats, in basal cluster or on threadlike rhizomes to 10 cm.
56. Ovary crestless or obscurely crested with 3 low central processes.
57. Larger bulbs each with cluster of bulbels surrounding roots; s Texas . . . . 48. *Allium elmendorfii*
57. Larger bulbs each with cluster of small, basal bulbels on one side; ne Oregon and wc Idaho . . . . 49. *Allium madidum*
56. Ovary prominently crested with 6 triangular central processes, margins finely papillose or denticulate.
58. Leaves usually beginning to wither from tip by anthesis; tepals rigid (not papery), ± shiny in fruit, strongly involute at tip, carinate . . . . 50. *Allium campanulatum*
58. Leaves usually green at anthesis; tepals papery (not rigid and shiny) in fruit, not strongly involute, not carinate.
59. Tepals ovate to elliptic, apex acute to acuminate; foothills of Sierra Nevada, n, c California . . . . 51. *Allium membranaceum*
59. Tepals lanceolate, apex acuminate; Sierra Nevada, California, and intermountain region n to Oregon, Idaho . . . . 52. *Allium bisceptrum*
55. Increase bulbs absent or 1–4, ± equaling parent bulbs, enclosed by parental bulb coats, never appearing as basal cluster, not rhizomatous or rhizomes 2+ mm thick (not threadlike).
60. Leaf blade channeled to subterete, if flat, not falcate.
61. Bulb coats lacking reticulation or reticulum delicate, very obscure under hand lens.
62. Bulbs ovoid to subglobose; rhizomes absent, renewal bulbs formed within coats of parent bulb; native or introduced.
63. Scape terete throughout, 1–3 mm diam.; leaf blade 1–3 mm wide; native to w Texas to se Arizona . . . . 53. *Allium kunthii* (in part)
63. Scape triquetrous, 2-edged or slightly winged proximally, if terete only proximally so, 1–10 mm wide; introduced in California and Oregon near the Pacific coast.
64. Umbel erect, ± hemispheric; flowers ± erect; tepals broadly elliptic, apex obtuse . . . . 54. *Allium neapolitanum*
64. Umbel lax, ± 1-sided; flowers pendent; tepals lanceolate, apex acute . . . . 55. *Allium triquetrum*
62. Bulbs oblique or oblique-ovoid, renewal bulbs borne terminally on rhizomes outside coats of parent bulbs; native.
65. Rhizomes conspicuous, 2 cm or more, including renewal bulbs.
66. Rhizomes smooth, parent bulb disappearing by anthesis except for still-functional roots and bulb coat; leaf blade broadly concave-convex or ± flattened, carinate; tepals obovate to ovate, apex acute to obtuse or emarginate; Coast Ranges, California, Oregon . . . . 56. *Allium unifolium*
66. Rhizomes scaly, sometimes absent, often missing in herbarium specimens, parent bulb persisting after anthesis; leaf blade flat, not carinate; tepals lanceolate to oblong, apex acute to acuminate; trans-Pecos Texas to se Arizona . . . . 57. *Allium rhizomatum*
65. Rhizomes inconspicuous, 2 cm or less, including renewal bulb.
67. Tepals erect, red-purple, rarely pure white, at least inner tepal margins serrulate; nw California, sw Oregon . . . . 58. *Allium bolanderi*
67. Tepals ± spreading, white to pale pink, margins entire; w Texas to se Arizona . . . . 53. *Allium kunthii* (in part)
61. Bulb coats obviously reticulate with prominent meshes under hand lens.

68. Cells of outer bulb coat square or polygonal.
69. Ovary with 6 prominent, flat, ± triangular crest processes . . . . 59. *Allium bigelovii*
69. Ovary with 3 or 6 minute, rounded crest processes, or crest obscure.
70. Flowers 4–9 mm; tepals erect or spreading from base, margins entire . . . . 60. *Allium lacunosum*
70. Flowers 8–16 mm; tepals spreading at tip, inner tepal margins denticulate.
71. Bulb forming 1–3 renewal bulbs borne terminally on rhizomes outside coats of parent bulb; parent bulb disappearing by anthesis except for still-functional roots and shriveled bulb coats; near Weller Butte, Blue Mountains, se Washington . . . . 61. *Allium dictuon*
71. Bulbs not forming rhizomes, renewal bulbs formed within coats of parent bulb; widespread w of Rocky Mountains. . . . . 62. *Allium acuminatum*
68. Cells of bulb coat transversely elongate, V-shaped, arranged in ± vertical rows, forming herringbone pattern, or ± contorted.
72. Cells of bulb coat in wavy, transverse rows, forming indistinct herringbone pattern or ± contorted; tepals spreading, ± equal.
73. Scape (3–)5–15(–17) cm; umbel persistent; tepals erect, not connivent over capsule in fruit . . . . 63. *Allium hickmanii*
73. Scape 15–60 cm; umbel shattering, each flower with its pedicel falling as unit; tepals connivent over capsule in fruit.
74. Ovary crested with 6 ± rectangular lateral processes; umbel compact; pedicel 0.7–2 times perianth . . . . 64. *Allium amplectens*
74. Ovary crestless or crested with 3 minute, 2-lobed central processes; umbel loose; pedicel 1.5–4 times perianth.
75. Leaf blade to 10 mm wide, channeled or flattened, carinate; inner bulb coats white; tepals becoming papery (not hyaline) after anthesis . . . . 65. *Allium praecox*
75. Leaf blade 1–3 mm wide, channeled or subterete, not carinate; inner bulb coats light yellow or white; tepals becoming hyaline (not papery) after anthesis . . . . 66. *Allium hyalinum*
72. Cells of bulb coat in sharply serrate, transverse rows, forming distinct herringbone pattern; tepals erect, inner shorter, narrower.
76. Tepals connivent over capsule in fruit, not rigid; umbel shattering in fruit, each flower with its pedicel falling as a unit . . . . 67. *Allium serra*
76. Tepals not connivent over capsule, rigid in fruit; umbel persistent.
77. Leaves 3–6, blade arcuate to tortuous; umbel compact; pedicels 5–20 mm; sea cliffs, n, c California . . . . 68. *Allium dichlamydeum*
77. Leaves 2–3, blade straight to arcuate; umbel loose; pedicels 10–40 mm; not on sea cliffs, California Floristic Province, extending south in coastal ranges.
78. Inner tepal margins denticulate, crisped . . . . 69. *Allium crispum*
78. Inner tepal margins entire to denticulate, never crisped . . . . 70. *Allium peninsulare*
60. Leaf blade flat or broadly channeled, if flat, ± falcate.
79. Scape and leaves persisting after seeds mature or on pressing, or only tardily deciduous.
80. Stamens much shorter than tepals.
81. Bulb coat cellular-reticulate with elongate, ± obscure, intricately contorted cells (resembling *Allium madidum*, but never with cluster of basal bulbels) . . . . 71. *Allium fibrillum*
81. Bulb coat cellular-reticulate with ± narrowly hexagonal, transversely elongate cells . . . . 72. *Allium brandegeei*
80. Stamens equaling tepals or exserted.

82. Scape expanded proximal to inflorescence; leaf blade (2–)5–8 mm wide . . . . 73. *Allium columbianum*
82. Scape thickest immediately proximal to inflorescence; leaf blade 1–5(–15) mm wide.
83. Scape constricted just proximal to inflorescence, then expanded; leaf blade 1–3(–5) mm wide . . . . 74. *Allium constrictum*
83. Scape not expanded proximal to inflorescence; leaf blade 2–5(–15) mm wide.
84. Leaf blade usually more than 5 mm wide, flat; umbel 25–50-flowered; spathe bracts 3 . . . . 75. *Allium douglasii*
84. Leaf blade 2–3 mm wide, flat to channeled; umbel 10–30-flowered; spathe bracts 2.
85. Bulb coat with quadrate to polygonal reticulations; leaf blade ± equaling scape . . . . 76. *Allium nevii*
85. Bulb coat without reticulations or with 2–3 rows of ± quadrate cells just distal to roots; leaf blade exceeding scape . . . 77. *Allium macrum*

79. Scape and leaves forming abcission layer at soil surface and deciduous when seeds mature, also frequently breaking at soil surface after pressing.
86. Outer bulb coats cellular-reticulate throughout (often obscurely so in *A. aaseae* and *A. simillimum*).
87. Bulb coats obscurely cellular-reticulate with ± contorted cells; tepal margins denticulate to erose.
88. Tepals white with greenish or reddish veins, sometimes flushed pink; anthers purple or mottled purple and white; pollen white or gray . . . . 78. *Allium simillimum*
88. Tepals bright pink, rarely white; anthers yellow; pollen yellow . . . . 79. *Allium aaseae*
87. Bulb coats ± prominently cellular-reticulate; tepal margins entire.
89. Bulb coats reticulate, cells irregularly arranged, ± polygonal, rectangular, or transversely elongate, ± curved.
90. Cells of bulb coat irregularly arranged, ± transversely elongate, curved; Tuolumne County, c California . . . . 80. *Allium tribracteatum*
90. Cells of bulb coat irregularly arranged or in ± regular vertical rows, polygonal or ± rectangular; Sierra Nevada, California, and Nevada . . . . 81. *Allium obtusum*
89. Bulb coats reticulate, cells arranged in ± regular vertical rows, narrowly hexagonal to rectangular, transversely elongate.
91. Tepals linear-lanceolate . . . . 82. *Allium anceps*
91. Tepals oblanceolate to ovate.
92. Scape 3–10 cm; pedicel ± equaling perianth . . . . 83. *Allium punctum*
92. Scape 15–20 cm; pedicel 2–3 times perianth . . . . 84. *Allium lemmonii*
86. Outer bulb coats not cellular-reticulate or with 2–3 rows of cells just distal to roots.
93. Scape terete or ± compressed, not winged.
94. Stamens well included.
95. Leaf blade strongly falcate; umbel mostly 5–10-flowered . . . . 85. *Allium parvum* (in part)
95. Leaf blade linear or weakly falcate; umbel 20–30-flowered . . . . 86. *Allium cratericola*
94. Stamens ± equaling tepals or exserted.
96. Leaves 2 per scape.
97. Leaf blade ± equaling to 2 times scape; wc Idaho . . . . 87. *Allium tolmiei* (in part)
97. Leaf blade much longer than scape; c Sierra Nevada, California . . . . 88. *Allium yosemitense*
96. Leaf 1 per scape.

98. Filaments papillose proximally . . . . 89. *Allium hoffmanii*
98. Filaments smooth proximally . . . . 90. *Allium burlewii*
93. Scape flattened, 2-edged or usually winged distally.
99. Bulbs oblique or oblique-ovoid, renewal bulbs borne terminally on rhizomes outside coats of parent bulb; parent bulb disappearing by anthesis except for still-functional roots and shriveled bulb coat.
100. Pedicel ± equaling perianth; ovary obscurely 3-crested; barren, bald summits w of Cascade Mountains from Vancouver Island to sw Oregon, also at Jefferson Park, Oregon, and in Wenatchee Mountains, c Washington . . . . 91. *Allium crenulatum*
100. Pedicel 2–3 times perianth; ovary prominently 6-crested; mountains and scablands e of Cascade Mountains, Oregon . . . . 87. *Allium tolmiei* (in part)
99. Bulbs ovoid to subglobose, rhizomes absent, renewal bulbs formed within coats of parent bulb; parent bulbs persistent.
101. Tepals narrowly lanceolate, apex long-acuminate; stamens exserted . . . . 92. *Allium platycaule*
101. Tepals lanceolate to ovate or elliptic, apex obtuse to acuminate; stamens included.
102. Flowers 9–15 mm; tepal apex long-acuminate, inner margins usually denticulate . . . . 93. *Allium falcifolium*
102. Flowers 6–10(–12) mm; tepal apex obtuse to acute, or ± involute in age and appearing acuminate, inner margins denticulate or not.
103. Inner bulb coats usually pink or red; inner tepal margins sometimes ± denticulate; Siskiyou Mountains of nw California and sw Oregon . . . . . 94. *Allium siskiyouense*
103. Inner bulb coats white; inner tepal margins entire; w United States, e of Sierra–Cascade axis.
104. Tepals becoming rigid (not papery), carinate in fruit.
105. Tepals lanceolate, apex acute to acuminate, ± erect in fruit, involute at tip; ovary obscurely to prominently crested with 3 or 6 processes . . . . 86. *Allium tolmiei* (in part)
105. Tepals elliptic-oblong, apex obtuse, not involute at tip, connivent over ovary in fruit; ovary crestless or obscurely crested . . . . 95. *Allium scilloides*
104. Tepals becoming papery (not rigid), not carinate in fruit.
106. Ovary distinctly crested with 3 or 6 low processes; sand and gravel deposits, along Columbia River from Ferry County, ne Washington, to mouth of John Day River, nc Oregon . . . . 96. *Allium robinsonii*
106. Ovary obscurely crested with 3 low, rounded processes; rocky, clay slopes and talus, e Oregon, Idaho, to c California, n Nevada, nw Utah . . . . 85. *Allium parvum* (in part)

1. Allium tricoccum Solander in W. Aiton, Hort. Kew 1: 428. 1789 • Ramp [E]

Bulbs 2–6, usually borne on short rhizome, ovoid-conic, 1.5–6 × 1.5–3 cm; outer coats enclosing 1 or more bulbs, brownish to grayish, reticulate, cells finely fibrous; inner coat white, obscurely cellular, cells irregularly arranged. **Leaves** ephemeral, usually absent at anthesis, 2–3, basal; blade solid, flat, elliptic to elliptic-lanceolate, (15–)20–30(–40) cm × 15–90 mm, tapering to long, slender petiole, margins entire. **Scape** persistent, solitary, flexuous distally, terete, 10–40 cm × 2–4 mm. **Umbel** persistent, erect, loose, (6–)30–50-flowered, obconic to ± hemispheric, bulbils unknown; spathe bracts persistent, 2, ± 3-veined, lanceolate to lance-ovate, ± equal, apex acute, beakless. **Flowers** campanulate, 4–7 mm; tepals erect, white to cream or yellowish, oblong to ovate, ± equal, not withering in fruit, margins entire, apex obtuse; stamens ± equaling tepals; anthers white to light yellow; pollen white; ovary crestless; style included, linear, shorter than stamens; stigma capitate, scarcely thickened, unlobed; pedicel 10–20 mm. **Seed coat** shining; cells smooth.

Varieties 2 (2 in the flora): e North America.

SELECTED REFERENCE Jones, A. G. 1979. A study of wild leek, and the recognition of *Allium burdickii* (Liliaceae). Syst. Bot. 4: 29–43.

1. Bulbs mostly 4–6 cm; leaf blades (3–)5–9 cm wide; spathe bracts 2–3 cm; umbel (15–)30–50-flowered . . . . 1a. *Allium tricoccum* var. *tricoccum*
1. Bulbs 2–4(–5) cm; leaf blades (1.5–)2–4(–4.5) cm wide; spathe bracts 1–2 cm; umbel (6–)12–18 (–24)-flowered . . . . . 1b. *Allium tricoccum* var. *burdickii*

1a. Allium tricoccum Solander var. **tricoccum** [E]

Allium pictum Moldenke; *A. triflorum* Rafinesque; *Ophioscorodon tricoccum* (Solander) Wallroth; *Validallium tricoccum* (Solander) Small

Bulbs mostly 4–6 × 1.5–3 cm. **Leaf blade** (3–)5–9 cm wide, evidently petiolate, sheaths and petioles commonly anthocyanous. **Scape** (12–)25–35 cm. **Umbel** (15–)30–50-flowered; spathe bracts 2–3 cm. 2*n* = 16.

Flowering Jul–Aug. Moist ground in rich woods, depressions, streamside bluffs, and colluvial slopes; 0–1400 m; N.B., N.S., Ont., Que.; Ala., Conn., Del., D.C., Ga., Ill., Ind., Iowa, Ky., Maine, Md., Mass., Mich., Minn., Mo., N.H., N.J., N.Y., N.C., N.Dak., Ohio, Pa., R.I., S.C., S.Dak., Tenn., Vt., Va., W.Va., Wis.

1b. Allium tricoccum Solander var. **burdickii** Hanes, Rhodora 55: 243. 1953 [E]

Allium burdickii (Hanes) A. G. Jones

Bulbs mostly 2–4(–5) × 1–1.5 cm. **Leaf blade** (1.5–)2–4(–4.5) cm wide, scarcely petiolate or less strongly so than in var. *tricoccum*, sheaths and petioles not anthocyanous. **Scape** (10–)13–16(–18) cm. **Umbel** (6–)12–18 (–24)-flowered; spathe bracts 1–2 cm. 2*n* = 16.

Flowering early Jun. Dry soil in upland woods; 0–1800 m; Ont., Que.; Ala., Ga., Ill., Ind., Iowa, Ky., Maine, Mich., Minn., Mo., N.H., N.J., N.Y., N.C., N.Dak., Ohio, Pa., S.Dak. Tenn., Vt., Va., W.Va., Wis.

Allium tricoccum var. *burdickii* is apparently rare in the eastern part of its range and in the Ozark Plateau region.

With its overlapping characteristics and distribution, *Allium tricoccum* var. *burdickii* is doubtfully separate from var. *tricoccum*.

2. Allium victorialis Linnaeus, Sp. Pl. 1: 295. 1753

Allium victorialis subsp. *platyphyllum* Hultén

Bulbs 1–several, attached to ± oblique rhizome, cylindric-conic, 5–6 × 1–2 cm; outer coats enclosing 1 or more bulbs, light brown to grayish brown, reticulate, cells fine-meshed, open, ± fibrous; inner coat white to light brown, not appearing cellular, 10–18-veined, veins prominent, parallel. **Leaves** persistent, green at anthesis, 2–3, sheathing proximal $^{1}/_{3}$–$^{1}/_{2}$ scape; blade solid, flat, narrowly lanceolate to broadly elliptic, 10–20 cm × 15–90 mm, base evidently petiolate, margins entire; petiole 1–20 cm. **Scape** persistent, solitary, erect, terete distally, 2-edged proximally, 30–70 cm × 2–3 mm. **Umbel** persistent, erect, loose, 20–35-flowered, hemispheric, bulbels unknown; spathe bracts persistent, 1–2, 3–6-veined, lanceolate to ovate, ± equal, apex obtuse, nearly beakless. **Flowers** stellate, 4–5 mm; tepals spreading, white or greenish white, outer lanceolate, inner narrowly ovate, ± equal, reflexed in age, withering in fruit, margins entire, apex acute or obtuse; stamens exserted; anthers yellow; pollen yellow; ovary crestless; style exserted, linear, equaling stamens or longer; stigma capitate, scarcely thickened; pedicel 10–30 mm. **Seed coat** not known.

Flowering Jul–Aug. Rocky areas; 0–30 m; Alaska; e Asia.

Allium victorialis is found in North America only on Attu Island, where it is reported to be native, and on Unalaska Island, where it is reported to be introduced from Attu Island. Within this species, various subspecies and varieties have been described, but, as circum-

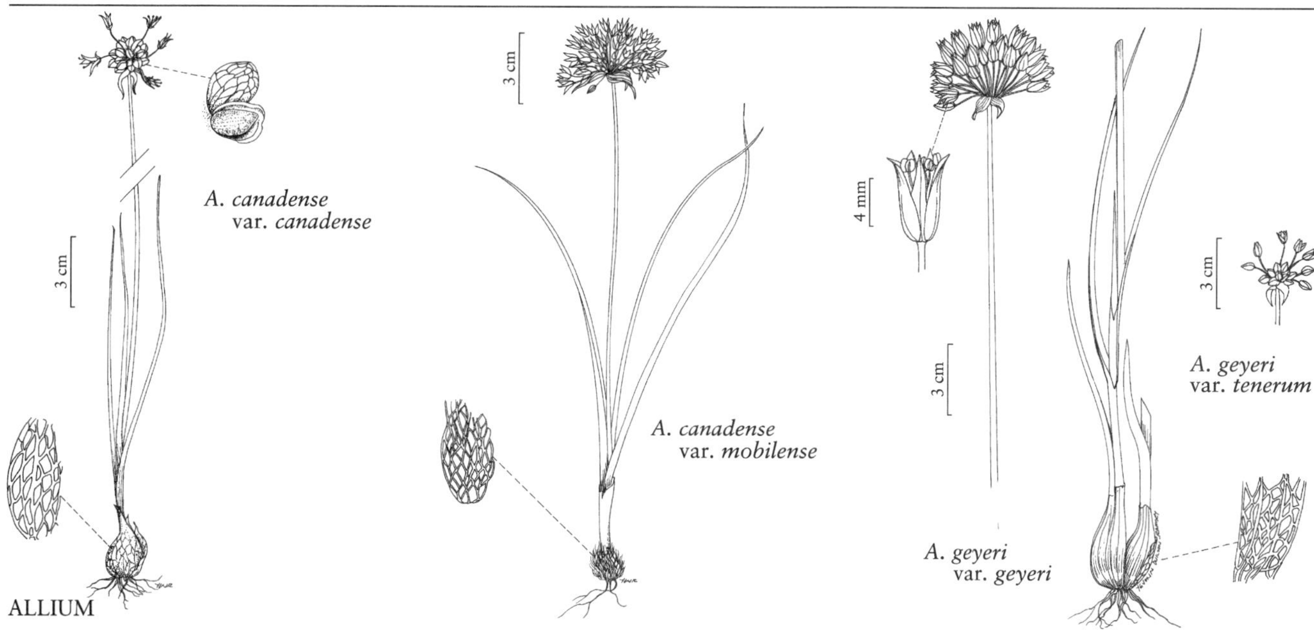

scribed, they are not clearly distinct. The North American plants have been known as subsp. *platyphyllum*, but until the variation within this widespread species can be worked out along natural lines, recognition of infraspecific taxa seems pointless.

3. Allium canadense Linnaeus, Sp. Pl. 2: 1195. 1753 [E] [F] [W1] [W2]

Bulbs 1–4+, without rhizome, with or without basal bulbels, often clustered, ovoid, 1–2.5 × 0.6–3 cm; outer coats enclosing 1 or more bulbs, brownish or grayish, reticulate, cells fine-meshed, open, fibrous; inner coats whitish, cells vertically elongate, sometimes contorted, walls straight or ± sinuous. **Leaves** persistent, green at anthesis, 2–6, basally sheathing, sheaths extending less than 1/4 scape; blade solid, flat, channeled, not carinate, 20–50 cm × 1–7 mm, margins entire or denticulate, apex acute to obtuse. **Scape** persistent, usually solitary, erect, terete, 10–60 cm × 1–5 mm. **Umbel** persistent, erect, loose, 0–60-flowered, hemispheric to globose, bulbils unknown or flowering pedicels replaced at least in part by bulbils; spathe bracts persistent, 3–4, 3–7-veined, ovate to lanceolate, ± equal, apex acuminate, beakless. **Flowers** urceolate-campanulate, 4–8 mm; tepals erect or spreading, white to pink or lavender, lanceolate to elliptic, ± equal, withering in fruit and exposing capsule, midribs somewhat thickened, margins entire, apex obtuse to acute; stamens included; anthers yellow; pollen yellow; ovary, when present, crestless; style linear, ± equaling stamens; stigma capitate, unlobed or obscurely 3-lobed; pedicel 8–70 mm. **Seed coat** shining; cells each with minute, central papilla.

Varieties 6 (6 in the flora): North America.

1. Flowering pedicels mostly or entirely replaced by bulbils, rarely producing capsules or seeds .... 3a. *Allium canadense* var. *canadense*
1. Flowering pedicels all floriferous, bulbils almost unknown, producing capsules and seeds.
    2. Tepals usually white .... 3f. *Allium canadense* var. *fraseri*
    2. Tepals pinkish to lilac.
        3. Pedicels filiform; scape 10–30(–50) cm .... 3b. *Allium canadense* var. *mobilense*
        3. Pedicels stouter, not filiform; scape 10–60 cm.
            4. Umbels 5–25-flowered; pedicels 3–4 times flowers; tepals thick .... 3e. *Allium canadense* var. *ecristatum*
            4. Umbels usually more than 25-flowered; pedicels 2–5 times flowers; tepals thin.
                5. Flowers fragrant; scape 15–30 cm; nc Texas to s Oklahoma .... 3c. *Allium canadense* var. *hyacinthoides*
                5. Flowers odorless; scape 20–60 cm; n Oklahoma to n Arkansas, northward .... 3d. *Allium canadense* var. *lavandulare*

3a. Allium canadense Linnaeus var. **canadense** [E] [F]

Allium canadense var. *ovoideum* Farwell; *A. canadense* var. *robustum* Farwell; *A. continuum* Small; *A. mutabile* Michaux

Bulbs without basal bulbels, 1.5–2.5 × 1–2 cm, inner coat cells obscure, regular or nearly so. **Leaves** 3–6; blade 1–5 mm wide, margins usually entire. **Scape** solitary, 15–50

cm. **Umbel** usually 0–10-flowered, rarely producing capsules or seeds, pedicels replaced, all or in part, by ovoid bulbils, some of which may, in turn, bear secondary umbels, bulbils sessile or stalk less than 0.5 mm; spathe bracts usually 3, lanceolate to ovate. **Flowers** fragrant, 4–7 mm; tepals spreading, white or pink, thin; flowering pedicel, if present, filiform, ± equal, 10–70 mm. $2n$ = 21, 28.

Flowering Apr–Jul. Roadsides, meadows, woods, and fields; 0–700 m; N.B., Ont., Que.; Ala., Ark., Conn., Del., D.C., Fla., Ga., Ill., Ind., Iowa, Kans., Ky., La., Maine, Md., Mass., Mich., Minn., Miss., Mo., Nebr., N.H., N.J., N.Y., N.C., Ohio, Okla., Pa., R.I., S.C., S.Dak., Tenn., Tex., Vt., Va., W.Va., Wis.

3b. Allium canadense Linnaeus var. **mobilense** (Regel) Ownbey, Res. Stud. State Coll. Wash. 18: 188. 1951 [E] [F]

Allium mobilense Regel, Trudy Imp. S.-Peterburgsk. Bot. Sada 3: 121. 1875; *A. arenicola* Small; *A. microscordion* Small; *A. mutabile* Michaux; *A. zenobiae* Cory

Bulbs sometimes with 1–2 basal bulbels, 1–1.5 × 0.6–1.2 cm; inner coat cells sometimes contorted or with sinuous walls. **Leaves** 2–5; blade 1–2 mm wide, margins usually entire. **Scape** solitary, 10–30(–50) cm. **Umbel** 15–50-flowered, producing capsules and seeds, bulbils almost unknown; spathe bracts usually 3, lanceolate. **Flowers** fragrant, 4–6(–7) mm; tepals spreading, pink (rarely white), thin; pedicel usually filiform, subequal, 8–22 mm. $2n$ = 14, 28.

Flowering Apr–May. Usually in sandy or rocky soils, rarely on limestone or in clay, woods and prairies; 0–300 m; Ala., Ark., Fla., Ga., La., Miss., Mo., Okla., S.C., Tex.

Allium canadense var. *mobilense* extends on the coastal plain perhaps to the southern border of North Carolina.

3c. Allium canadense Linnaeus var. **hyacinthoides** (Bush) Ownbey, Res. Stud. State Coll. Wash. 18: 191. 1951 [E]

Allium hyacinthoides Bush, Rep. (Annual) Missouri Bot. Gard. 17: 119. 1906

Bulbs without basal bulbels, 1.5–2.5 × 2–3 cm; inner coat cells obscure, regular or nearly so. **Leaves** 3–6; blade 2–7 mm wide, margins entire or denticulate. **Scape** solitary, 15–30 cm. **Umbel** 25–60-flowered, producing capsules and seeds, bulbils almost unknown; spathe bracts 3–4, lanceolate. **Flowers** fragrant, 5–7 mm; tepals erect, pink, thin; pedicel not filiform, subequal, 15–20 mm. $2n$ = 14.

Flowering late Mar–Apr. On calcareous prairies or more infrequently in sandy soils; 100–400 m; Okla., Tex.

3d. Allium canadense Linnaeus var. **lavendulare** (J. M. Bates) Ownbey & Aase, Res. Stud. State Coll. Wash., suppl. 1: 51. 1955 [E]

Allium lavendulare J. M. Bates, Amer. Bot. (Binghamton) 22: 58. 1916

Bulbs without basal bulbels, 1–2 × 1–2.5 cm; inner coat cells obscure, regular or nearly so. **Leaves** usually 3–6; blade 1–6 mm wide, margins entire. **Scape** solitary, 20–60 cm. **Umbel** 25–60-flowered, producing capsules and seeds, bulbils almost unknown; spathe bracts 3, ovate to lanceolate. **Flowers** odorless, 5–8 mm; tepals spreading, lavender-pink, rarely white, thin; pedicel linear, ± stout, becoming subequal, 15–20 mm. $2n$ = 14, 28.

Flowering May–Jul. Generally in calcareous soils; 100–600 m; Ark., Ill., Kans., Mo., Nebr., Okla., S.Dak.

3e. Allium canadense Linnaeus var. **ecristatum** Ownbey, Res. Stud. State Coll. Wash. 18: 193. 1951 [E]

Bulbs without basal bulbels, 1.5–2 × 2–2.5 cm; inner coat cells somewhat contorted. **Leaves** 3–6; blade 1–5 mm wide, margins entire or denticulate. **Scape** solitary, 10–30 cm. **Umbel** 5–25-flowered, producing capsules and seeds, bulbils almost unknown; spathe bracts 3, ovate. **Flowers** fragrant, 5–7 mm; tepals erect, deep pink, ± thick; pedicel linear, stout, subequal, 15–22 mm. $2n$ = 28.

Flowering Mar–early Apr. Known only from c coastal plain where it grows on prairies; 0–200 m; Tex.

Allium canadense var. *ecristatum* is endemic to Texas and was first published as a nomen nudum under *A. reticulatum*.

3f. Allium canadense Linnaeus var. **fraseri** Ownbey, Res. Stud. State Coll. Wash. 18: 195. 1951 [E]

Allium fraseri (Ownbey) Shinners; *A. lavendulare* J. M. Bates var. *fraseri* (Ownbey) Shinners

Bulbs without basal bulbels, 1.8–2.5 × 1.2–2 cm; inner coat cells obscure, ± regular. **Leaves** 3–6; blade 1–5 mm wide, margins entire or denticulate. **Scape** usually solitary, 20–50 cm. **Umbel** 10–50-flowered, producing capsules and seeds, bulbils almost unknown; spathe bracts usually 3, ovate to lanceolate. **Flowers** fragrant, 4–7 mm; tepals spreading, white (rarely pink), thin; pedicel linear, stout, becoming subequal, 10–20 mm. $2n$ = 14, 21.

Flowering Apr–Jun. Generally in sandy or rocky soils, woods or plains; 50–1100 m; Kans., Nebr., Okla., S.Dak., Tex.

4. Allium geyeri S. Watson, Proc. Amer. Acad. Arts 14: 227. 1879 E F W2

Bulbs 2–10+, not rhizomatous, ovoid or more elongate, 1–2.5 × 0.8–2 cm; outer coats enclosing 1 or more bulbs, gray or brown, reticulate, cells rather coarse-meshed, open, fibrous; inner coats whitish, cells vertically elongate and regular or obscure. **Leaves** persistent, usually green at anthesis, usually 3–5, sheathing less than 1/4 scape; blade solid, ± straight, flat, channeled, (6–)12–30 cm × 1–3(–5) mm, margins entire or denticulate. **Scape** persistent, solitary, erect, terete or somewhat 2-angled, 10–50 cm × 1–3 mm. **Umbel** persistent, erect, compact, 10–25-flowered, hemispheric to globose, not producing bulbils, or 0–5-flowered, largely replaced by ovoid, acuminate bulbils; spathe bracts persistent, 2–3, mostly 1-veined, ovate to lanceolate, ± equal, apex acuminate, beakless. **Flowers** urceolate-campanulate, (4–)6–8(–10) mm; tepals erect or spreading, pink to white, ovate to lanceolate, ± equal, not withering in fruit and permanently investing fruit, or withering if fruit not produced, midribs papillose, becoming callous-keeled, margins often obscurely toothed, apex obtuse to acuminate; stamens included; anthers yellow; pollen yellow; ovary when present, inconspicuously crested; processes 6, central, low, distinct or connate in pairs across septa, ± erect, rounded, margins entire, becoming variously developed or obsolete in fruit; style linear, ± equaling stamens; stigma capitate, unlobed or obscurely lobed; pedicel becoming rigid and stiffly spreading in fruit, 8–13 mm. **Seed coat** shining; cells each with minute, central papilla.

Varieties 2 (2 in the flora): w North America.

1. Umbel fully floriferous, not producing bulbils, ovaries mostly all producing seeds . . . . 4a. *Allium geyeri* var. *geyeri*
1. Umbel not fully floriferous, producing 0–5, mostly sterile flowers, flowers mostly replaced by ovoid, acuminate bulbils . . . . 4b. *Allium geyeri* var. *tenerum*

4a. Allium geyeri S. Watson var. **geyeri** E F

Allium dictyotum Greene; *A. funiculosum* A. Nelson; *A. pikeanum* Rydberg

Umbel 10–25-flowered, not producing bulbils. **Flowers:** tepals white to pink; pedicel 8–13 mm. **Seeds** commonly produced. $2n$ = 14, 28.

Flowering May–Sep. Moist, open slopes, meadows, or stream banks in mountains; 200–4000 m; Alta., Sask.; Ariz., Colo., Idaho, Mont., Nev., N.Mex., S.Dak., Tex., Utah, Wash., Wyo.

4b. Allium geyeri S. Watson var. **tenerum** M. E. Jones, Contr. W. Bot. 10: 28, fig. 55. 1902 E F

Allium arenicola Osterhout 1900, not Small 1900; *A. fibrosum* Rydberg 1897, not Regel 1887; *A. geyeri* var. *graniferum* Henderson; *A. rubrum* Osterhout; *A. rydbergii* J. F. Macbride; *A. sabulicola* Osterhout

Umbel 0–5-flowered, flowers mostly replaced by ovoid, acuminate bulbils; bulbils sessile or stalk less than 1 mm. **Flowers:** tepals white to pink; flowering pedicel, if present, 8–13 mm. **Seeds** rarely produced. $2n$ = 28, 35, 42.

Flowering May–Sep. Meadows and along streams; 0–3100 m; Alta., B.C.; Ariz., Colo., Idaho, Mont., Nev., N.Mex., Oreg., Utah, Wash., Wyo.

5. Allium vineale Linnaeus, Sp. Pl. 1: 299. 1753 I W1 W2

Bulbs 5–20, clustered, stipitate, hard-shelled, asymmetric, ovoid, 1–2 × 1–2 cm; outer coats enclosing bulbs, brownish to yellowish, membranous, vertically striate, splitting into parallel strips and fibers, cells arranged in ± wavy rows, vertical; inner coats white to light brown, cells obscure, vertically elongate. **Leaves** persistent, green at anthesis, 2–4, sheathing at least proximal 1/2 scape; blade hollow below middle, terete, cylindric or filiform, not carinate, 20–60 cm × 2–4 mm, margins entire. **Scape** persistent, solitary, erect, terete, 30–120 cm × 1.5–4 mm. **Umbel**

persistent, erect, ± compact, 0–50-flowered, subglobose to ovoid or hemispheric, flowering pedicels all or in part replaced by bulbils; bulbils sessile, basally narrowed, 4–6 × 2–3 mm; spathe bract caducous, 1, 2–several-veined, ovate, apex caudate, beaked, beak ± equaling or longer than base. **Flowers** campanulate, 3–4 mm; tepals erect, greenish to purple, elliptic-lanceolate, ± equal, withering in fruit, margins entire, apex obtuse; stamens exserted, outer 3 filaments without appendages, inner 3 filaments with 2 prominent lateral appendages; anthers purple; pollen white; ovary crestless; style exserted, linear, ± equaling stamen; stigma capitate, scarcely thickened, unlobed; flowering pedicel 10–20 mm. **Seed coat** shining; cells smooth. $2n = 32, 40$.

Flowering Jun–Aug. Disturbed areas often adjacent to agricultural lands; 0–700 m; introduced; Ont., Que.; Ala., Ark., Calif., Conn., Del., D.C., Ga., Ill., Ind. Iowa, Kans., Ky., La., Maine, Md., Mass., Mich., Miss., Mo., N.J., N.Y., N.C., Ohio, Okla., Pa., R.I., S.C., Tenn., Va., W.Va.; Europe.

Allium vineale is also expected to be found in Wisconsin and Texas; specimens were not seen. It is a noxious weed, apparently introduced from Europe in colonial times. The small, wheat-sized bulbils frequently contaminated wheat grown in infested areas. Bread made from such wheat was garlic-flavored, and cows grazing in infested pastures produce garlic-flavored milk.

6. **Allium sativum** Linnaeus, Sp. Pl. 1: 296. 1753 • Garlic [I]

Varieties 2 (1 in the flora): introduced; cultivated widely in Europe, Asia, and North America.

6a. **Allium sativum** Linnaeus var. **sativum** [I]

Bulbs comprising 5–15+ bulbels (cloves), depressed-ovoid, 3–6 × (1.5–)3–8 cm; outer coats enclosing bulbs, white to light brown, membranous, without reticulation; inner coats white, not appearing cellular. **Leaves** withering from tip by anthesis, 6–12, sheathing ± 1/2 scape; blade solid, flat, carinate, 20–100 cm × 5–20 mm, margins and keel scabrous, apex acute. **Scape** persistent, solitary, erect, terete, 25–100+ cm × 4–10 mm. **Umbel** persistent, erect, compact, 0–20-flowered, globose, with bulbils, flowers usually aborted in bud; spathe bract caducous, 1, 3-veined, ovate, apex beaked, beak much longer than base. **Flowers,** when present, ± campanulate, 3–5 mm; tepals erect, greenish white to pink, outer lanceolate, inner ovate-lanceolate, ± equal, withering without developing, margins entire, apex acute; ovary crestless; pedicel 10–20 mm. **Seed coat** not known; fruits usually abortive or without seeds.

Flowering Jun–Jul. Roadsides and disturbed areas; introduced; widely cultivated.

Allium sativum var. *sativum* is quite variable. It is possibly derived from *A. longicuspis* from central Asia. It generally reproduces by bulbels (cloves) as it produces few, if any, seeds. Prior to anthesis the scape of *A. sativum* var. *ophioscorodon* (Link) Doll is coiled into 1–2 wide loops. Both varieties are cultivated in North America and are usually found along roadsides and disturbed areas wherever garlic is cultivated. The bulbils make accidental introduction very easy.

7. **Allium oleraceum** Linnaeus, Sp. Pl. 1: 299. 1753 [I]

Bulbs 1 or more, not attached to rhizome, ovoid, 1.2–2 × 1–1.5 cm; outer coats enclosing bulbs, brown to grayish brown, fibrous, fibers close, ± parallel; inner coats white to light brown, not cellular. **Leaves** withering from tip by anthesis, 2–4, sheathing proximal 1/2+ scape; blade fistulose proximally, solid distally, terete, linear to filiform, prominently ribbed proximally, channeled distally, 1.5–2.5 cm × 0.5–5 mm, margins and veins usually scabrid with minute teeth, apex acute. **Scape** persistent, solitary, erect, terete, 25–100 cm × 4–8 mm. **Umbel** persistent, erect, compact to ± loose, 0–40-flowered, subglobose, with few to many bulbils or with bulbils only; spathe bracts persistent, 2, 4–9-veined, lanceolate, unequal, apex acuminate into beak, beak long, slender, to 20 cm, ± equaling or longer than base. **Flowers** usually aborting before capsules mature, if present, campanulate, 6–8 mm; tepals erect, whitish or pinkish to purple, outer narrowly obovate, inner ± elliptic, unequal, margins entire, apex obtuse; stamens included; anthers yellow to reddish; pollen yellow; ovary crestless; style linear, equaling stamens; stigma capitate, unlobed; pedicel 15–60 mm. **Seed coat** unknown; capsules only rarely produced.

Flowering late Jul–Aug. Roadsides and other disturbed ground; introduced; Europe.

Allium oleraceum is reported from New England, where it is sometimes found on roadsides and other disturbed ground. It persists and is spread easily by the bulbils.

8. **Allium ampeloprasum** Linnaeus, Sp. Pl. 1: 294. 1753 • Wild leek [I]

Bulbs 1–3+, variable, some with poorly developed bulbs, others ovoid with 1–2 large bulbs and several yellowish to light brown bulbels at base, 0.4–1(–3) × 0.4–1(–1.5) cm; outer coat enclosing 1 or more bulbs, yellowish, membranous; inner coats white to light brown, cells not evident, fibers ± parallel, few. **Leaves** withering from tips by anthesis, 6–9, sheathing 1/3–1/2 scape; blade solid, flat, channeled, 1–5 cm × 2–20(–30) mm, margins scabrid. **Scape** persistent, solitary, erect, fistulose, terete, 45–180 cm × 3–7 mm. **Umbel** persistent, erect, compact, to 500-flowered, few-flowered in variants with bulbils, globose; spathe bracts persistent, 3–5, 2–3-veined, lanceolate, ± equal, apex abruptly narrowed to beak, beak to 10 cm. **Flowers** urceolate, 4–5.5 mm; tepals erect, white, pink, or dark red, unequal, becoming papery and investing cap-

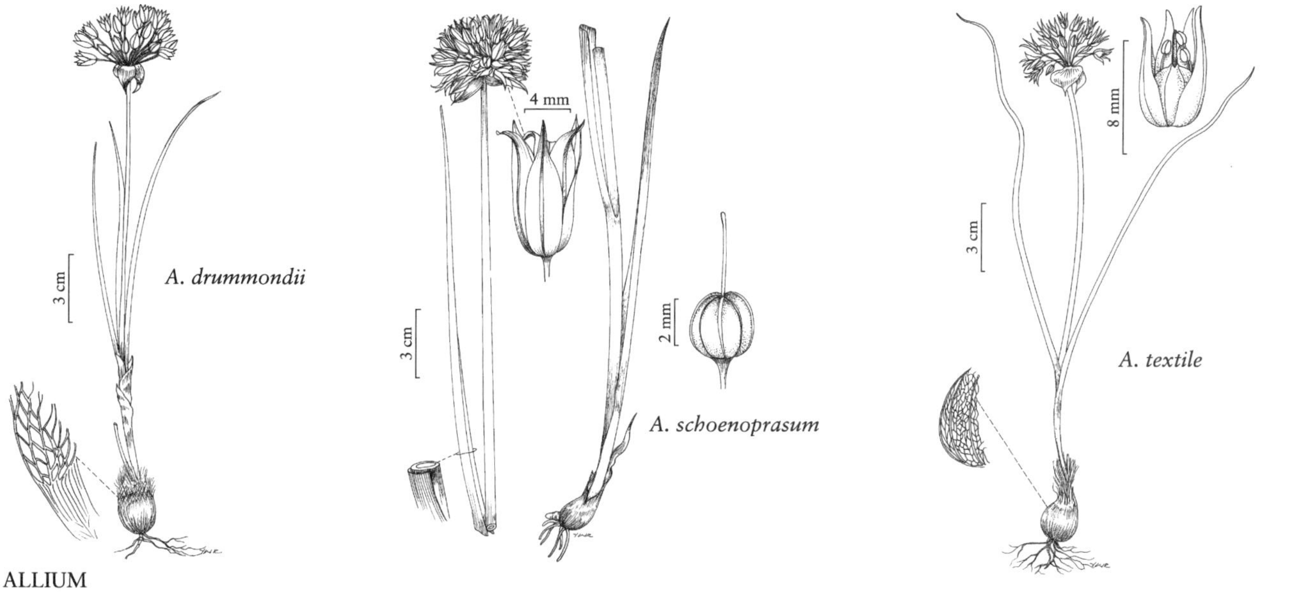

ALLIUM

sule in fruit; outer tepal oblong-lanceolate, margins entire, apex obtuse, sometimes mucronate; inner tepal narrowly ovate to spatulate, margins entire, apex obtuse; stamens equaling perianth or exserted; outer filaments simple, inner with 2 prominent lateral teeth that exceed anther-bearing portion, glabrous; anthers yellow or purple; pollen yellow; ovary crestless; style linear, equaling stamens; stigma capitate, scarcely thickened, unlobed; pedicel 15–50 mm. **Seed coat** not known.

Flowering Apr–Jul. Roadsides and other disturbed areas; 0–100 m; introduced; Europe; Asia; n Africa.

Normally the umbel of *Allium ampeloprasum* has no bulbils, but there are some variants with a few flowers that produce bulbils. The species has been reported as established in New England and adjacent Canada and can be found along roadsides and in other disturbed areas. It is probably conspecific with *A. porrum* Linnaeus, the leek of commerce. *Allium porrum* can be distinguished from *A. ampeloprasum* based on its unique bulb morphology and chemistry from centuries of cultivation and selection.

9. Allium drummondii Regel, Trudy Imp. S.-Peterburgsk. Bot. Sada 3: 112. 1875 [F]

Allium helleri Small; *A. nuttallii* S. Watson; *A. reticulatum* Fraser ex G. Don var. *nuttallii* (S. Watson) M. E. Jones

Bulbs 1–5, without basal bulbels, ovoid, 1–1.8 × 0.7–1.5 cm; outer coats enclosing 1 or more bulbs, brown, reticulate, cells fine-meshed, mostly closed in proximal 1/2 of bulb, fibrous; inner coats whitish or brownish, cells intricately contorted, walls usually not sinuous. **Leaves** persistent, green at anthesis, 2–5, sheathing; blade solid, flat, channeled, 10–30 cm × 1–3(–5) mm, margins entire. **Scape** persistent, solitary, erect, terete, 10–30 cm × 1–3 mm. **Umbel** persistent, erect, compact to ± loose, usually 10–25-flowered, hemispheric-globose, rarely replaced by bulbils; spathe bracts persistent, 2–3, 1-veined, ovate, ± equal, apex acuminate. **Flowers** campanulate to ± stellate, 6–9 mm; tepals spreading, white, pink, or red, rarely greenish yellow, ovate to lanceolate, ± equal, becoming papery and rigid in fruit, margins entire, apex obtuse or acute, midribs somewhat thickened; stamens included; anthers yellow; pollen light yellow; ovary crestless; style linear, equaling stamens; stigma capitate, unlobed or obscurely lobed; pedicel 5–20 mm. **Seed coat** shining; cells each usually with minute, central papilla. $2n = 14, 28$.

Flowering Mar–Jun. Plains, hills, and prairies, particularly in limestone soils; 0–1600 m; Ark., Kans., Neb., N.Mex., Okla., Tex.; Mexico.

10. Allium coryi M. E. Jones, Contr. W. Bot. 17: 21. 1930 [E]

Bulbs 1–5, without basal bulbels, ovoid, 1–1.8 × 0.7–1.5 cm; outer coats enclosing 1 or more bulbs, brown, reticulate, cells fine-meshed, open, fibrous; inner coats whitish or brownish, cells intricately contorted, walls not sinuous. **Leaves** persistent, green at anthesis, 3–5, sheathing; blade solid, flat, channeled, 10–30 cm × 1–3 mm, margins entire. **Scape** persistent, solitary, erect, ± terete, 10–30 cm × 1–3 mm. **Umbel** persistent, erect, compact or ± loose, usually 10–25-flowered, hemispheric-globose, bulbils unknown; spathe bracts persistent, 2–3, 1-veined, ovate, ± equal, apex acuminate. **Flowers** campanulate to ± stellate, 6–9 mm; tepals spreading, bright yellow, sometimes tinged with red, fading with age and sometimes upon drying, ovate to lanceolate, ± equal, becoming papery and rigid in fruit, margins entire, apex obtuse or acute, midribs somewhat thickened; stamens included; anthers yellow; pollen yellow; ovary crestless or rarely crested; processes 6, central, low, sometimes distinct or connate in pairs across septa, rounded, margins entire; style linear, ± equaling stamens; stigma capitate, unlobed or obscurely lobed; pedicel 5–20 mm. **Seed coat** shining; cells smooth. $2n = 14$.

Flowering Apr–May. Rocky slopes and plains, mountains; 800–1400 m; Tex.

Allium coryi is known only from western Texas.

11. Allium schoenoprasum Linnaeus, Sp. Pl. 1: 301. 1753 • Chive [F]

Allium schoenoprasum var. *laurentianum* Fernald; *A. schoenoprasum* var. *sibiricum* (Linnaeus) Hartman; *A. sibiricum* Linnaeus

Bulbs 1 or more, clustered, short-rhizomatous at base, cylindric, elongate, 0.5–0.9 × 1.5–2 cm; outer coats enclosing bulbs, grayish or brownish, persisting as fibrous reticulum but often appearing membranous as outer coats are lost during collecting, cells minutely striate; inner coats whitish or pinkish, cells closely parallel, elongate. **Leaves** persistent, green at anthesis, usually 2, distalmost usually ensheathing $^1/_3$–$^1/_2$ scape; blade hollow, terete, fistulose, 20–60 cm × 2–7 mm. **Scape** persistent, 2–12+, clustered, erect, terete, fistulose, 20–50 cm × 3–5 mm. **Umbel** persistent, erect, compact, 30–50-flowered, ± subglobose, bulbils unknown; spathe bracts persistent, 2, 3–7-veined, lanceolate to broadly ovate, ± equal, apex short-acuminate. **Flowers** campanulate, 8–12 mm; tepals erect, pale purple to deep lilac, drying pink, particularly on midrib, or white, elliptic to lanceolate, ± equal, becoming papery in fruit, permanently investing capsule, margins entire, apex acute to acuminate, tips ± recurved, midribs not thickened; stamens included; anthers purple; pollen white; ovary crestless; style linear, equaling stamens; stigma capitate, unlobed or obscurely lobed; pedicel 2–6 mm. **Seed coat** shining; cells minutely roughened, not pustuliferous. $2n = 16$.

Flowering Jun–Aug. Wet meadows, rocky or gravelly stream banks and lake shores, circumboreal; 0–3500 m; Alta., B.C., Man., N.B., Nfld. and Labr., N.W.T., N.S., Nunavut, Ont., P.E.I., Que., Sask., Yukon; Alaska, Colo., Conn., Idaho, Maine, Mass., Mich., Minn., Mont., N.H., N.J., N.Y., Ohio, Oreg., Pa., R.I., Vt., Wash., Wis., Wyo.; Siberia.

Allium schoenoprasum is native in North America, but it is also cultivated and has widely escaped. It is an extremely polymorphic species, and throughout its range both large and small races occur. These plants have been known as *A. sibiricum*, *A. schoenoprasum* var. *sibiricum*, or *A. schoenoprasum* var. *laurentianum*, and many, largely unsuccessful, attempts have been made to distinguish the varieties. Until the variation can be worked out along natural lines, if any, instead of unstable features such as plant size, and color and shape of the tepals, recognition of these varieties is unsound. Because we are unable to separate native populations from many of the escaped ones, we cannot reliably map the native distribution of this taxon in the flora.

SELECTED REFERENCE Tardiff, B. and P. Morisset. 1990. Clinal morphological variation of *Allium schoenoprasum* in eastern North America. Taxon 39: 417–429.

12. Allium tuberosum Rottler ex Sprengel, Syst. Veg. 2: 38. 1825 • Chinese chive, oriental garlic [I]

Bulbs 1–3, borne on stout, ± horizontal rhizome, elongate, cylindric or conic, 0.5–1.5 × 0.7–2 cm; outer coats enclosing 1 or more bulbs, brown, reticulate, cells ± fine-meshed, open, fibrous; inner coats white, cells closely parallel, elongate. **Leaves** withering from tip by anthesis, 2–5, sheathing scape to ± soil level; blade solid, flat, carinate abaxially, 20–40 cm × 2–6 mm, margins entire. **Scape** persistent, solitary, erect, terete, 30–50 cm × 1–3 mm. **Umbel** persistent, erect, loose, 20–50-flowered, hemispheric-globose, bulbils unknown; spathe bracts persistent, 1–3, 3–7-veined, lance-ovate, shorter than pedicel, apex acuminate. **Flowers** substellate, 4–9 mm; tepals spreading, white with green or brownish midveins, lanceolate to elliptic, ± equal, withering and exposing capsule, margins entire, apex obtuse or acute; stamens included; anthers purple; pollen white; ovary crestless; style linear, ± equaling stamens; stigma capitate, unlobed; pedicel 10–30 mm. **Seed coat** shining; cells smooth, irregularly shaped, with ± sinuous walls.

Flowering Jul–Aug. Roadsides and other disturbed ground; introduced; Iowa, Nebr., Wis.; se Asia.

Allium tuberosum is cultivated in China, Siberia, and North America, and is reported to be established in New England. It may escape anywhere the species is cultivated.

13. Allium runyonii Ownbey, Res. Stud. State Coll. Wash. 18: 198, fig. 4. 1951 [E]

Bulbs 1–3, bearing cluster of short-stalked, basal bulbels, ovoid, 1.2–2 × 1.2–2 cm; outer coats enclosing 1 or more bulbs, brownish, reticulate, cells very coarse-meshed, open, fibrous; inner coats whitish, cells contorted, walls very sinuous. **Leaves** green at anthesis, persistent, 3–6, basally sheathing; blade solid, flat, channeled, 10–40 cm × 1–4 mm, margins entire. **Scape** persistent, often 2 or more successively produced from single bulb, erect, ± terete, 10–45 cm × 1–4 mm. **Umbel** persistent, erect, loose, usually 10–25-flowered, hemispheric-globose, bulbils unknown; spathe bracts persistent, usually 3, 3–7-veined, ovate, ± equal, apex acuminate. **Flowers** urceolate, 5–7 mm; tepals erect, white with pinkish midribs, aging to pink, lanceolate, ± equal, becoming callous-keeled and permanently investing capsule, margins entire, apex obtuse or even emarginate to acute; stamens ± equaling tepals; anthers yellow; pollen yellow; ovary crestless; style linear, equaling stamens; stigma capitate, unlobed or obscurely lobed; pedicel 8–28 mm, elongating in fruit. **Seed coat** shining; cells smooth. $2n = 14$.

Flowering Mar–Apr. Sandy soils, Rio Grande plains; 10–200 m; Tex.

Allium runyonii is known only from extreme southern Texas.

14. Allium perdulce S. V. Fraser, Trans. Kansas Acad. Sci. 42: 124. 1940 [E]

Bulbs 2–20+, without basal bulbels, 1–2.5 × 1.2–2.8 cm; outer coats enclosing 1 or more bulbs, dark brown, reticulate, cells coarse-meshed, open, fibrous; inner coats whitish, cells regular, vertically elongate, walls not sinuous. **Leaves** persistent, green at anthesis, 3–5, sheathing; blade solid, flat, channeled, 8–30 cm × 1–2(–3) mm, margins entire. **Scape** persistent, solitary, erect, ± terete, 10–20 cm × 1–3 mm. **Umbel** persistent, erect, loose, 5–25-flowered, hemispheric-globose, bulbils unknown; spathe bracts persistent, 2–3, 3–7-veined, ovate, ± equal, apex acuminate. **Flowers** urceolate, 7–10 mm; tepals erect, white or pale pink with deep pink midribs to deep rose, lanceolate, ± equal, becoming callous-keeled and permanently investing capsule, margins entire, apex obtuse or acute; stamens included; anthers yellow or purple; pollen yellow; ovary crestless; style linear, ± equaling stamens; stigma capitate, unlobed or obscurely lobed; pedicel 3–17 mm, ± 2 times perianth at anthesis, elongating in fruit. **Seed coat** dull or shining; cells minutely roughened.

Varieties 2 (2 in the flora): c North America.

1. Tepals deep rose, aging to purple; n Texas and e New Mexico to se South Dakota and adjacent Iowa .... 14a. *Allium perdulce* var. *perdulce*
1. Tepals white to light pink, aging pink; trans-Pecos Texas .... 14b. *Allium perdulce* var. *sperryi*

14a. Allium perdulce S. V. Fraser var. **perdulce** [E]

Bulbs 1–2 × 1.2–1.5 cm. **Leaf blade** 1–2 mm wide. **Tepals** deep rose-colored, aging to purple. **Pedicel** 3–15 mm. $2n = 14$.

Flowering Apr–Jun. Usually sandy soils, plains; 300–1300 m; Iowa, Kans., Nebr., N.Mex., Okla., S.Dak., Tex.

14b. Allium perdulce S. V. Fraser var. **sperryi** Ownbey, Res. Stud. State Coll. Wash. 18: 202. 1951 [E]

Bulbs 1.5–2.5 × 1.6–2.8 cm. **Leaf blade** 1–3 mm wide. **Tepals** white or pale pink with deep pink midveins, aging to pink. **Pedicel** 10–17 mm. $2n = 28$.

Flowering Mar–Apr. Rocky margins entire, apex obtuse or acute, slopes and plains; 700–1500 m; Tex.

Allium perdulce var. *sperryi* is a well-marked, polyploid, geographic variant. Although both varieties of the species reach the same extremes of size, var. *sperryi* is, on average, considerably taller and stouter than var. *perdulce*. It also tends to have a larger number of flowers and longer pedicels. Distributions of the two varieties do not overlap.

15. Allium cuthbertii Small, Fl. S.E. U.S., 264, 1328. 1903 [E]

Bulbs usually solitary, without basal bulbels, ovoid, 1–1.8 × 1–1.2 cm; outer coats enclosing single bulb, grayish, reticulate, cells fine-meshed, open, fibrous; inner coats whitish, cells vertically elongate, regular. **Leaves** persistent, withering by anthesis, 2, sheathing; blade solid, flat, linear, 12–35 cm × 3–6 mm, margins entire or denticulate. **Scape** persistent, solitary, erect, terete or ± 3–4-angled, 20–40 cm × 1–3.5 mm. **Umbel** persistent, erect, loose, 10–25-flowered, hemispheric-globose, bulbils unknown; spathe bracts persistent, 2–3, 5–7-veined, lanceolate, ± equal, apex acuminate. **Flowers** ± stellate, 7–9 mm; tepals spreading to reflexed, white to pinkish or purple, lanceolate, ± equal, remaining thin and becoming strongly reflexed in fruit, margins entire, apex acute; stamens included; anthers yellow; pollen yellow; ovary conspicuously crested; processes 6, central, irregularly contorted; style linear, equaling stamens; stigma capitate, unlobed or obscurely 3-lobed; pedicel 14–45 mm. **Seed coat** shining; cells each with obscure, central papilla. $2n = 14$.

Flowering May–early Jun. On granitic "flat-rocks" of Piedmont and in sand on coastal plains; 0–300 m; Ala., Fla., Ga., N.C., S.C.

16. Allium speculae Ownbey & Aase, Rhodora 61: 70. 1959 C E

Bulbs 1–5+, not rhizomatous, ovoid, 1–5 × 1–2 cm; outer coats usually enclosing single bulb, grayish or brownish, reticulate, cells very fine-meshed, open; inner coats whitish, cells vertically elongate, regular. **Leaves** persistent, green at anthesis, 3+, sheathing; blade solid, flat, channeled, 10–30 cm × 1–2 mm, margins entire. **Scape** persistent, solitary (sometimes 2 in cultivated plants), erect, terete, 20–30 cm × 1–3 mm. **Umbel** persistent, erect, compact, 10–15-flowered, hemispheric-globose, bulbels unknown; spathe bracts persistent, 3, usually 1-veined, lanceolate, ± equal, apex long-acuminate. **Flowers** ± campanulate, 5–6 mm; tepals widely spreading, pinkish, elliptic, ± equal, remaining thin and withering over ovary, margins entire, apex obtuse, involute at tip; stamens included; anthers light yellow; pollen white to light yellow; ovary crested; processes 6, central, horizontally spreading, flattened, triangular, margins entire; style linear, equaling stamens; stigma capitate, obscurely 3-lobed; pedicel 9–20 mm. **Seed coat** shining; cells each with minute, central papilla. $2n = 14$.

Flowering Jun. On gneiss "flat-rocks" and sandy soil of Piedmont; of conservation concern; 300 m; Ala., Ga.

17. Allium plummerae S. Watson, Proc. Amer. Acad. Arts 18: 195. 1883

Bulbs usually 1 of a cluster, sometimes short-rhizomatous basally, elongate, to 5 × 1–1.5 cm; outer coats enclosing 1 or more bulbs, gray or brown, reticulate, cells very coarse-meshed, open, fibrous; inner coats whitish, cells vertically elongate and regular or obscure. **Leaves** persistent, green at anthesis, 3–6, sheathing; blade solid, flat, broadly channeled, 25–50 cm × 3–7 mm, margins entire or minutely denticulate. **Scape** persistent, solitary, erect, terete or ± 4-angled, 30–50 cm × 1–3.5 mm. **Umbel** persistent, erect, loose, 10–25-flowered, hemispheric-globose, bulbils unknown; spathe bracts persistent, 2–3, 3–5-veined, lanceolate, ± equal, apex long-acuminate. **Flowers** ± stellate, 5–10 mm; tepals spreading or reflexed, white or pink, lanceolate, ± equal, withering in fruit and exposing capsules, margins entire, apex acute, midrib becoming noticeably thickened; stamens ± equaling tepals to exserted; anthers purple; pollen light yellow; ovary conspicuously crested; processes 6, central, usually connate in pairs across septa, ± erect, flattened, to 2 mm, margins lacerate, usually well developed in fruit; style linear, equaling stamens; stigma capitate, unlobed or obscurely lobed; pedicel 15–25 mm, elongating in fruit. **Seed coat** shining; cells each with minute, central papilla. $2n = 28$.

Flowering Jun–Sep. Rocky slopes, stream banks, and marshy ground, mountains; 1600–2800 m; Ariz.; Mexico.

18. Allium passeyi N. H. Holmgren & A. H. Holmgren, Brittonia 26: 309, figs. 1, 4. 1974 C E

Bulbs 2–3+, not rhizomatous, ovoid, 1–2 × 1.2–2 cm; outer coats enclosing 1 or more bulbs, light brown, reticulate, fibrous; inner coats whitish to light brown, cells vertically elongate and regular or obscure. **Leaves** persistent, green at anthesis, 2–3, sheathing; blade solid, flat, ± falcate, 10–20 cm × 3–6 mm, margins entire. **Scape** persistent, solitary, erect, terete or ± winged-angled, mostly 10–20 cm × 1–2.5 mm. **Umbel** persistent, erect, compact, 8–27-flowered, hemispheric-globose, bulbils unknown; spathe bracts persistent, 2–4, mostly 1-veined, ovate, ± equal, apex acuminate to caudate. **Flowers** campanulate, 7–9 mm; tepals erect, light pink, narrowly lanceolate to lance-ovate, becoming callous-keeled, persistent and permanently investing capsule, margins entire, apex acuminate, inner shorter and narrower than outer, outer strongly keeled and gibbous at base; stamens included; anthers yellow; pollen yellow; ovary obscurely crested; processes 6, central, ± erect, rounded, minute, to 1 mm, margins entire; style linear, equaling stamens; stigma capitate, scarcely thickened, unlobed; pedicel 10–18 mm. **Seed coat** shining; cells smooth, each with minute, central papilla.

Flowering Jun. Shallow, stony, lithosolic soil over dolomitic limestone, hilltops; of conservation concern; 1400–1600 m; Utah.

Allium passeyi is known only from Box Elder County.

19. Allium macropetalum Rydberg, Bull. Torrey Bot. Club 31: 401. 1904 [E]

Allium deserticola (M. E. Jones) Wooton & Standley; *A. reticulatum* Fraser ex G. Don var. *deserticola* M. E. Jones.

Bulbs 1–5, without basal bulbels, ovoid, 1.5–2.5 × 1–2 cm; outer coats enclosing 1 or more bulbs, brown, reticulate, cells usually coarse-meshed, open, fibrous; inner coats whitish, cells vertically elongate and regular or obscure. **Leaves** persistent, green at anthesis, 2, sheathing; blade solid, channeled, semiterete, 8–20 cm × 1–3 mm, margins entire. **Scape** persistent, solitary, erect, terete or ± 4-angled, 5–20 cm × 1–4 mm. **Umbel** persistent, erect, compact to loose, 10–20-flowered, hemispheric to globose, bulbils unknown; spathe bracts persistent, 2–3, 3–5-veined, ovate to lanceolate, ± equal, apex acute to short-acuminate. **Flowers** campanulate, 8–12 mm; tepals spreading, pink with deeper pink or reddish midveins, lanceolate, ± equal, becoming papery in fruit, not investing capsule, margins entire, apex obtuse to acuminate, midrib scarcely thickened; stamens included; anthers yellow or purple; pollen yellow; ovary usually conspicuously crested; processes 6, central, usually connate in pairs across septa, ± erect, flattened, triangular, to 2 mm, margins entire, mostly well developed in fruit; style linear, equaling stamens; stigma capitate, unlobed or obscurely lobed; pedicel 5–20 mm. **Seed coat** shining; cells smooth, without central papillae. $2n = 14$.

Flowering late Mar–Jun. Desert plains and hills; 300–2500 m; Ariz., Colo., N.Mex., Utah, Tex.

20. Allium textile A. Nelson & J. F. Macbride, Bot. Gaz. 56: 470. 1913 [E] [F] [W2]

Allium aridum Rydberg; *A. reticulatum* Fraser ex G. Don 1827, not J. Presl & C. Presl 1817; *A. reticulatum* Fraser ex G. Don var. *playanum* M. E. Jones

Bulbs 1–3+, not rhizomatous, without basal bulbels, ovoid, 1.2–2.5 × 1–2 cm; outer coats enclosing 1 or more bulbs, gray or brown, reticulate, cells fine-meshed, open, fibrous; inner coats whitish, cells vertically elongate and regular or obscure. **Leaves** persistent, green at anthesis, 2, sheathing; blade solid, ± straight, channeled, semiterete, 10–40 cm × 1–3(–5) mm, margins entire or denticulate. **Scape** persistent, solitary, erect, ± terete, 5–30(–40) cm × 1–3 mm. **Umbel** persistent, erect, compact to ± loose, 15–30-flowered, hemispheric, bulbils unknown; spathe bracts persistent, 3, usually 1-veined, ovate, ± equal, apex acuminate. **Flowers** urceolate to campanulate, 5–7 mm; tepals erect, white or rarely pink, with red or reddish brown midribs; outer whorl broadly ovate to lanceolate, unequal, becoming callous-keeled and permanently investing capsule, margins often obscurely toothed apically, apex obtuse to acuminate; inner whorl narrower, margins entire, apex distinctly spreading; stamens included; anthers yellow; pollen yellow; ovary ± conspicuously crested; processes 6, central, distinct or connate in pairs across septa, ± erect, rounded, to 1 mm, margins entire, becoming variously developed or obsolete in fruit; style linear, equaling filaments; stigma capitate, unlobed or obscurely lobed; pedicel 5–20 mm. **Seed coat** shining; cells ± smooth, without central papillae. $2n = 14, 28$.

Flowering May–Jun. Dry plains and hills; 300–2400 m; Alta., Man., Sask.; Colo., Idaho, Iowa, Kans., Minn., Mont., Nebr., Nev., N.Mex., N.Dak., S.Dak., Utah, Wash., Wyo.

21. Allium nigrum Linnaeus, Sp. Pl. ed. 2, 1: 430. 1762 • Black garlic, Homer's garlic [I]

Bulbs 3–4, not clustered on stout, primary rhizomes, ovoid, asymmetric, 2.5–3 × 2–5 cm; outer coats enclosing 1 or more bulbs, light yellow to light brown, thin, papery, chartaceous, cells narrow, vertically elongate, without fibers; inner coats white, cells narrow, vertically elongate. **Leaves** usually persistent, 3–6, scarcely ensheathing scape even basally, sheaths not extending much above soil surface; blade solid, flat, broadly channeled, 10–60 cm × 10–25 mm, margins entire. **Scape** usually persistent, solitary, erect, solid, terete, fistulose, 60–100 cm × 5–10 mm. **Umbel** persistent, erect, compact, 75–150-flowered, hemispheric-globose, bulbils unknown; spathe bracts persistent, 2–4, 11+-veined, ovate, ± equal, apex acuminate. **Flowers** stellate to subcampanulate, 6–9 mm; tepals ± spreading, white with green midvein, narrowly oblong, ± equal, becoming ± rigid and reflexed in fruiting, margins entire, apex obtuse; stamens included; anthers yellow or purple, aborting without producing pollen; ovary crestless; style linear, ± equaling tepals; stigma capitate, scarcely thickened, obscurely 3-lobed; pedicel 25–40 mm. **Seed coat** not known.

Flowering Jun. Disturbed roadsides; 100–200 m; introduced; Oreg.; native to Mediterranean; cultivated in Europe

Native to the Mediterranean and long-cultivated in Europe, *Allium nigrum* is reported to be well established near Perrydale, Oregon, and may be expected elsewhere in North America. This species is remarkable in its lack of an onion/garlic odor. It is also unusual in having more than two ovules in each locule, a characteristic of sect. *Melanocrommyum* Webb & Berthelot but otherwise unusual in the genus.

22. Allium cepa Linnaeus, Sp. Pl. 1: 301. 1753 • Cultivated onion

Bulbs 1–3, not rhizomatous, mostly depressed-globose, varying in size from cultivar to cultivar, 5–8 × 3–10 cm; outer coats enclosing 1 or more bulbs, yellowish brown, red, or white, membranous, without reticulation; inner coats white to pink, cells obscure to quadrate. **Leaves** persistent, 4–10, sheathing proximal 1/6–1/4 scape; blade fistulose, usually ± semicircular in cross section, 10–50 cm × 4–20 mm. **Scape** persistent, solitary, erect, fistulose, inflated below middle, 30–100 cm × 3–20 mm. **Umbel** persistent, erect, compact, to 500-flowered, globose, bulbils occasionally found; spathe bracts caducous, 2–3, 3–4-veined, ovate, ± equal, apex acute to acuminate. **Flowers** stellate to campanulate to urceolate, 3–7 mm; tepals erect to ± spreading, white to pink with greenish midveins, withering in fruit, margins entire, apex obtuse or acute, outer ovate, inner oblong; stamens exserted; anthers white; pollen white; ovary crestless; style linear, ± equaling stamens; stigma capitate, unlobed; pedicel 10–50 mm. **Seed coat** not known.

Flowering Jun–Aug. Disturbed sites adjacent to areas where cultivated; 0–500 m; Ark., Calif., Kans., La., Mont., Oreg., Tex., Wash.; cultivated in Europe, Asia.

The onion of commerce, *Allium cepa* is widely cultivated as a biennial in North America, Europe, and Asia. It is unknown in the wild and is probably derived from *A. oschanini* of central Asia. The cultivated form is often polyploid ($2n$ = 16, 32, 54) and possibly of hybrid origin. It exists in numerous cultivars, a few of which form large bulbils in the umbel.

23. Allium fistulosum Linnaeus, Sp. Pl. 1: 301. 1753 • Welsh onion, Japanese bunching onion [I]

Bulbs 2–12+, borne on short rhizome, cylindric, 2–5 × 1–2.5 cm; outer coats enclosing 1 or more bulbs, white to light brown, membranous, without reticulation; inner coats white, cells obscure, quadrate. **Leaves** persistent, 2–6, sheathing lower 1/4–1/3 of scape; blade terete, fistulose, 10–40 cm × 10–25 mm. **Scape** persistent, solitary, erect, fistulose, inflated in middle, tapering to umbel, (12–)15–70 cm × 8–25 mm. **Umbel** persistent, erect, compact, 50–100-flowered, globose to ovoid, bulbils unknown; spathe bracts persistent, 1–2, 1–3-veined, ovate, ± equal, apex acute. **Flowers** narrowly campanulate to urceolate, 6–9 mm; tepals erect, yellowish white, withering in fruit, margins entire, apex acute, outer lanceolate, inner narrowly ovate, unequal; stamens long-exserted; anthers white to yellow; pollen white; ovary crestless; style linear, equaling stamens; stigma capitate, obscurely 3-lobed; pedicel 10–30 mm. **Seed coat** shining; cells 4–6-angled, ± rectangular.

Flowering Jul–Aug. Disturbed areas; introduced; N.W.T.; Alaska; cultivated in Europe, Asia.

Allium fistulosum is cultivated in Europe and Asia. It is reported to have escaped in Alaska and is established near the north end of Great Slave Lake. The species is to be expected elsewhere in Canada and the northern United States.

24. Allium paniculatum Linnaeus, Syst. Nat., ed. 10, 2: 978. 1759 [I] [W2]

Varieties 4 (1 in the flora): introduced; Mediterranean Basin.

24a. Allium paniculatum Linnaeus var. **paniculatum** [I] [W2]

Bulbs 1–8, not clustered on stout primary rhizome, increase bulbs absent or ± equaling parent bulbs, never appearing as basal cluster, rhizomes lacking, ovoid, 1–1.5 × 1–1.5 cm; outer coats enclosing renewal bulbs, dark brown, membranous, very obscurely reticulate, cells rectangular, vertically elongate, without fibers, meshes delicate; inner coats white to light brown, cells rectangular, vertically elongate. **Leaves** persistent, withering from tip at anthesis, 3–5, sheathing proximal 1/3–1/2 scape; blade solid, terete to semiterete, channeled, 5–35 cm × 1–5 mm, margins entire. **Scape** persistent, solitary, erect, solid, terete, 30–70 cm × 1–3(–4) mm. **Umbel** persistent, erect, loose, 25–100-flowered, globose to ovate, bulbils unknown; spathe bracts persistent, 2, 4–9-veined or veins not visible, ovate to narrowly lanceolate, unequal, longer bract 5–14 cm, greatly exceeding pedicels, apex contracted into appendage, appendage terete, 5–14 cm. **Flowers** campanulate, 5–7 mm; tepals erect, white to lilac pink, ovate, ± equal, becoming papery and withering in fruit, margins entire, apex obtuse; stamens included to slightly exserted; anthers yellow; pollen yellow; ovary crestless; style exserted, linear; stigma capitate, scarcely thickened; pedicel unequal, 10–45 mm. **Seed coat** not known.

Flowering Jun–Jul. Disturbed areas; 0–50 m; introduced; Calif.; Europe.

Allium paniculatum var. *paniculatum* is naturalized around San Francisco Bay and elsewhere in northern California. It is native to Europe and is potentially a noxious weed.

25. Allium rotundum Linnaeus, Sp. Pl. ed. 2, 1: 423. 1762 [I]

Bulbs 1–50+, ovoid, 0.2–1.5 × 0.2–1 cm; outer coats surrounding numerous, very dark purple bulbels, usually white or greenish, membranous, sometimes breaking into ± parallel, fibrous strips; inner coats white, cells obscure, quadrate. **Leaves** persistent, 2–5, sheathing 1/3–1/2 scape; blade solid, flat or channeled, carinate, 15–40 cm × 2–5(–7) mm, margins entire. **Scape** persistent, solitary, erect,

± solid, 25–90 cm × 1–4 mm. **Umbel** persistent, erect, very dense, 80–200-flowered, globose, bulbils unknown; spathe bracts caducous, 2, 7–12-veined, lanceolate, ± equal, apex narrowed to beak, beak to 2 cm. **Flowers** campanulate, 4–7 mm; tepals erect, purple, or inner with broad whitish margins and purple midvein, oblong to ovate, ± equal, becoming keeled and ± rigid in fruit, margins entire, apex acute to obtuse; stamens included; filaments: outer unlobed, inner with 2 prominent teeth on either side of anther, teeth 2–3 times anther-bearing portion, usually papillose to ciliate; anthers yellow or purple; pollen light yellow to white; ovary crestless; style linear, ± equaling anther-bearing portion of stamens; stigma capitate, unlobed; pedicels 5–30 mm, outer much shorter than inner. **Seed coat** dull; cells each with central papilla.

Flowering May–Jul. Disturbed fields and roadsides; introduced; Mich.; c, s Europe.

Allium rotundum is native to central and southern Europe. It is reported as established in southern Michigan, and is to be expected elsewhere in the Great Lakes region.

26. Allium gooddingii Ownbey, Res. Stud. State Coll. Wash. 15: 222. 1947 E

Bulbs 1–3, clustered on thick, iris-like rhizome, elongate, 2–3 × 0.5–1 cm; outer coats enclosing single bulb, brownish, membranous, minutely striate, cells in regular vertical rows, elongate, not fibrous-reticulate, fibers persistent, parallel, few, coarse; inner coats whitish or pinkish, minutely striate, cells in vertical rows, elongate. **Leaves** persistent, green at anthesis, 3–6, basally sheathing, sheaths not extending much above soil level; blade solid, flat, 8–25 cm × 4–8 mm, margins entire. **Scape** persistent, solitary, erect, solid, flattened, narrowly winged distally, 34–45 cm × 1–3 mm. **Umbel** persistent, erect, loose, 18–23-flowered, conic, bulbils unknown; spathe bracts persistent, 2, 3–5-veined, narrowly lanceolate, ± equal, apex acute. **Flowers** campanulate, 8–10 mm; tepals erect, pink, elliptic, ± equal, withering in fruit, margins entire, apex obtuse, midribs not thickened; stamens ± equaling tepals; anthers white or purple; pollen white; ovary crestless; style linear, equaling stamens; stigma capitate, unlobed; pedicel 15–20 mm, elongating in fruit. **Seed coat** dull or shining; cells each with minute, central papilla. $2n = 14$.

Flowering Jun–Sep. Steep, rocky slopes; 2400–2900 m; Ariz., N.Mex.

Allium gooddingii is known only from the mountains of east-central Arizona and adjacent New Mexico, and the Santa Catalina Mountains of southern Arizona.

27. Allium validum S. Watson, Botany (Fortieth Parallel), 350. 1871 E

Bulbs 2–20+, clustered on thick, iris-like rhizome, elongate, 2–5 × 1–2.5 cm; outer coats enclosing 1 or more bulbs, brownish, membranous, minutely striate, cells in regular vertical rows, elongate, not fibrous-reticulate, fibers persistent, parallel, few, coarse; inner coats reddish purple or whitish, minutely striate, cells in regular vertical rows, elongate. **Leaves** persistent, green at anthesis, 3–6, sheathing basally, sheaths not extending much above soil surface; blade solid, flat, 20–70(–80) cm × 4–15 mm, margins entire. **Scape** persistent, solitary, erect, solid, flattened and narrowly winged distally, 30–70 cm × 2–7 mm. **Umbel** persistent, erect, compact, 15–30-flowered, hemispheric, bulbils unknown; spathe bracts persistent, 2, 3–5-veined, broadly ovate, ± equal, apex acute. **Flowers** ± campanulate, 8–10 mm; tepals erect to ± spreading, pink, narrowly lanceolate, ± equal, withering in fruit, margins entire, apex acuminate, midribs scarcely thickened; stamens exerted; anthers yellow or purple; pollen yellow; ovary crestless; style exserted, linear, longer than stamens; stigma capitate, unlobed; pedicel 10–15 mm, elongating and becoming stout in fruit. **Seed coat** dull; cells smooth or minutely roughened. $2n = 28, 56$.

Flowering Jun–Aug. Swampy meadows in mountains; 1500–2900 m; B.C.; Calif., Idaho, Nev., Oreg., Wash.

Allium validum is a Cascade-Sierran species extending east to northeastern Nevada, eastern Oregon, and western Idaho.

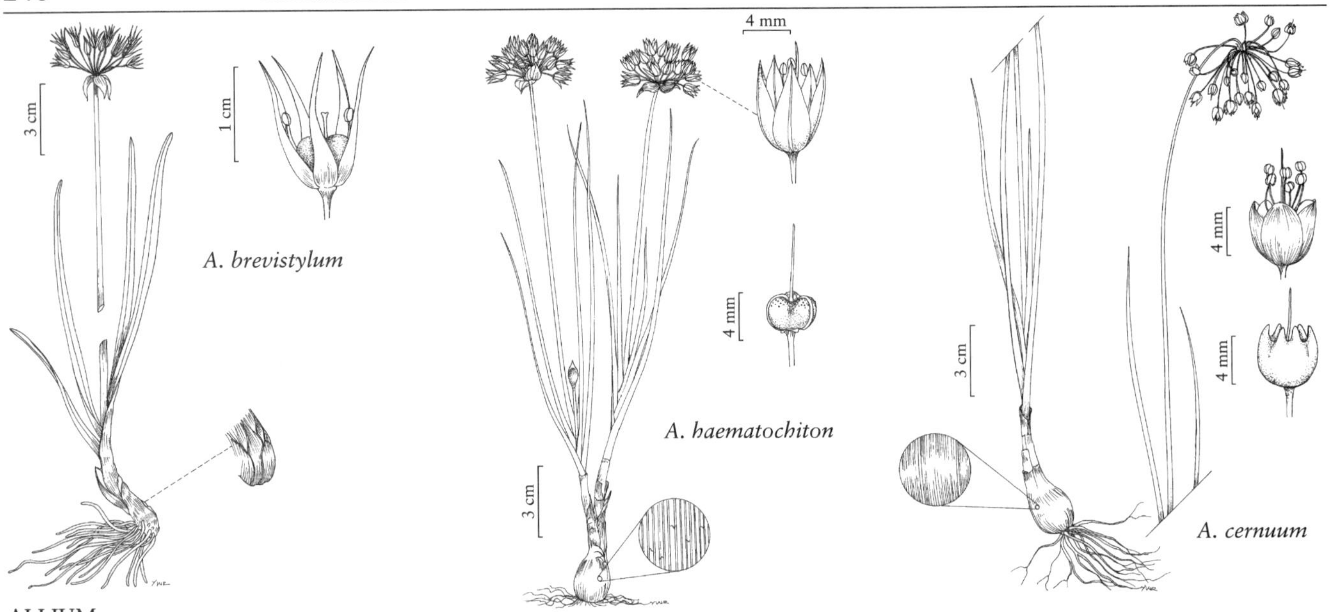

ALLIUM

28. Allium brevistylum S. Watson, Botany (Fortieth Parallel), 350. 1871 [E] [F]

Bulbs 2–4, terminating thick, iris-like rhizome, elongate, 2–3 × 0.6–1 cm; outer coats enclosing single bulb, grayish or brownish, membranous, minutely striate, cells in parallel vertical rows, narrow, elongate, not fibrous-reticulate, fibers persistent, parallel, few, coarse; inner coats whitish, cells narrowly vertically elongate. **Leaves** persistent, green at anthesis, 2–5, basally sheathing, sheaths not extending much above soil level; blade solid, flat, 10–40 cm × 2–8 mm, margins entire. **Scape** persistent, solitary, erect, solid, flattened and narrowly winged distally, 20–60 cm × 1.5–4 mm. **Umbel** persistent, erect, loose, 7–20-flowered, subhemispheric, bulbils unknown; spathe bracts persistent, 2, 3–5-veined, ovate, ± equal, apex acute. **Flowers** narrowly urceolate, 10–13 mm; tepals erect, pink, lanceolate, ± equal, withering in fruit, margins entire, apex acuminate, midribs somewhat thickened; stamens included, ca. 1/2 as long as tepals; anthers yellow; pollen light yellow; ovary crestless; style linear, equaling stamens; stigma capitate, distinctly 3-lobed; pedicel 8–35 mm, elongating and becoming stout and curved in fruit. **Seed coat** dull or shining; cells each with minute, central papilla, or obscurely and minutely roughened. $2n = 14$.

Flowering Jun–Aug. Swampy meadows and along streams, rarely on wooded slopes; 2200–3400 m; Colo., Idaho, Mont., N.Mex., Utah, Wyo.

Allium brevistylum is found only in the Rocky Mountains.

29. Allium haematochiton S. Watson, Proc. Amer. Acad. Arts 14: 227. 1879 [F]

Allium marvinii Davidson

Bulbs 2–3+, usually clustered, short-rhizomatous at base, rhizome not stout or iris-like, oblong, 2–3 × 1–1.5 cm; outer coats enclosing single bulb, ± reddish brown, membranous, striations fine, cells narrow, elongate, vertical; inner coats deep red to white, finely striate, cells narrow, vertically elongate. **Leaves** persistent, green at anthesis, 3–6, basally sheathing, sheaths not extending much above soil level; blade solid, terete, 3-angled, 15–30 cm × 1–4 mm, margins entire. **Scape** persistent, 1–2, clustered, erect, solid, slightly compressed, 10–40 cm × 1–3 mm. **Umbel** persistent, erect, compact, 10–30-flowered, hemispheric, bulbils unknown; spathe bracts persistent, 2–4, 2–3-veined, ovate-lanceolate, ± equal, apex obtuse. **Flowers** campanulate, 6–8 mm; tepals erect, white to rose, with darker midvein, broadly ovate to lance-ovate, ± equal, becoming rigid and permanently investing capsule, margins entire, apex acute; stamens included; anthers white; pollen white to dark gray; ovary crested; processes 6, rounded, short, densely papillose; style linear, equaling stamens; stigma capitate, scarcely thickened, unlobed; pedicel 1–2 cm. **Seed coat** dull; cells minutely roughened. $2n = 14$.

Flowering Mar–May. Dry slopes and ridges of clay or stony soil; 50–800 m; Calif.; Mexico (Baja California).

30. Allium cernuum Roth, Arch. Bot. (Leipzig) 1: 40. 1798 [F]

Allium allegheniense Small; *A. oxyphilum* Wherry; *A. recurvatum* Rydberg

Bulbs 2–5+, clustered, often short-rhizomatous at base, rhizome not stout or iris-like, oblong, elongate, 1–3 × 0.8–1.5 cm; outer coats enclosing 1 or more bulbs, grayish or brownish, membranous, minutely striate, cells in regular vertical rows, narrowly elongate, fibers persistent, parallel, few; inner coats white to pink or reddish, cells in regular vertical rows, narrowly elongate. **Leaves** persistent, green at anthesis, 3–5, basally sheathing, sheaths not extending much above soil level; blade solid, flat, channeled to broadly V-shaped in cross section, 10–25 cm × 1–6 mm, margins entire or denticulate. **Scape** persistent, sometimes 2 or more produced successively from single bulb, usually clustered, nodding, solid, terete or ridged, particularly distally, sometimes flattened and narrowly winged, abruptly recurved near apex, 10–50 cm × 1–3 mm. **Umbel** persistent, cernuous, loose, 8–35-flowered, hemispheric, bulbils unknown; spathe bracts persistent, 2, 3-veined, lanceolate, ± equal, apex acuminate, beakless. **Flowers** campanulate, 4–6 mm; tepals ± erect, pink or white, elliptic-ovate, ± equal, withering in fruit, margins ± entire, apex ± obtuse, at least outer tepals strongly incurved, midribs not thickened; stamens exserted; anthers yellow; pollen yellow; ovary conspicuously crested; processes 6, flattened, ± triangular, margins entire or toothed; style exserted, linear, ± equaling stamens; stigma capitate, scarcely thickened, unlobed; pedicel 6–25 mm, becoming stouter in fruit, elongating and bending abruptly upward from near point of attachment. **Seed coat** dull or shining; cells smooth, minutely roughened, or each with minute, central papilla. $2n = 14$.

Flowering Jul–Oct. Widely distributed on moist soils in mountainous and cool regions; 600–3500 m; Alta., B.C., Ont., Sask.; Ala., Ariz., Ark., Colo., D.C., Ga., Idaho, Ill., Ind., Iowa, Ky., Md., Mich., Minn., Mo., Mont., Nebr., N.Mex., N.Y., N.C., Ohio, Oreg., Pa., S.C., S.Dak., Tenn., Tex., Utah, Va., Wash., W.Va., Wis., Wyo.; Mexico.

Allium cernuum is the most widespread North American species of the genus. It is closely related to *A. stellatum*, and the character commonly used to differentiate them has been umbel orientation. In both species, the inflorescence is nodding in bud, but in *A. stellatum* it usually becomes erect by anthesis. In *A. cernuum* the peduncle remains permanently recurved near the apex, although the inflorescence may sometimes become erect overall, or nearly so. While this character is helpful in identification, an almost exclusive reliance on it (even by one of the present authors in his youth) has obscured other clearer distinctions between the species and has confused their geographic ranges. More reliable characters for differentiating these species are bulb shape (elongate in *A. cernuum*, ovoid in *A. stellatum*) and perianth shape (campanulate in *A. cernuum*, stellate in *A. stellatum*). Unfortunately, perianth shape is often difficult to see in herbarium specimens.

31. Allium stellatum Ker Gawler, Bot. Mag. 38: plate 1576. 1813 [E]

Bulbs 1–5+, usually clustered, often short-rhizomatous at base, rhizome not stout or iris-like, ovoid, 2–4 × 1–2.5 cm; outer coats enclosing 1 or more bulbs, grayish or brownish, membranous, fibers parallel, few, or sometimes reticulate, cells obscure, finely meshed; inner coats whitish to pinkish, cells obscure, ± quadrate. **Leaves** persistent, green at anthesis, 3–5, basally sheathing, sheaths never extending much above soil level; blade solid, flat, channeled, 14–35 cm × 1–5 mm, margins ± entire. **Scape** persistent, solitary, erect, or, if nodding at anthesis, becoming erect, solid, terete or ± 4-angled, particularly distally, 20–50 cm × 1–3.5 mm. **Umbel** persistent, nodding, becoming erect, ± loose, 9–40-flowered, hemispheric, bulbils unknown; spathe bracts persistent, 2, 2–4-veined, lanceolate, ± equal, apex acuminate. **Flowers** stellate, 5–8 mm; tepals spreading, deep pink, elliptic-lanceolate, ± equal, withering in fruit, margins entire, apex acute, midribs not thickened; stamens exserted; anthers yellow; pollen yellow; ovary conspicuously crested; processes 6, distinct, flattened, ± triangular, margins entire or toothed; style exserted, linear, ± equaling stamens; stigma capitate, scarcely thickened, unlobed; pedicel 0–20 mm. **Seed coat** dull; cells minutely roughened. $2n = 14$.

Flowering Jul–Oct. Often on calcareous soils; 300–2200 m; Man., Ont., Sask.; Ark., Ill., Iowa, Kans., Mich., Minn., Mo., Nebr., N.Dak., Okla., S.Dak., Tenn., Tex., Wis.

32. Allium sanbornii Alph. Wood, Proc. Acad. Nat. Sci. Philadelphia 20: 171. 1868 [E]

Bulbs 1–3, not clustered on stout, primary rhizome, ovoid, 1–2.5(–3) × 1.2–2 cm; outer coats enclosing 1 or more bulbs, dark reddish brown, chartaceous, lacking cellular reticulation or cells arranged in 2–3 rows distal to roots, ± quadrate, without fibers; inner coats light brown or white, cells obscurely quadrate. **Leaves** persistent, withering from tip by anthesis, 1, basally sheathing, sheath never extending much above soil level; blade solid, terete, 30–45 cm × 2–4 mm. **Scape** persistent, solitary, erect, ± solid, terete, 18–60 cm × 2–3 mm. **Umbel** persistent, erect, compact, 18–190-flowered, hemispheric to globose, bulbils unknown; spathe bracts persistent, 4, 3-veined, ovate, ± equal, apex long-acuminate. **Flowers** campanulate, 5–9 mm; tepals erect, white to pink with darker midveins, unequal, becoming papery in fruit, margins entire or irregular to erose, apex acute or acuminate to long-acuminate; outer tepals lanceolate to ovate, reflexed at tip; inner 1/4–1/3 longer than outer, ovate to broadly ovate; stamens exserted; anthers yellow or purple; pollen yellow or white; ovary crested; processes 6, prominent, erect, ± triangular, margins entire; style linear, equaling stamens; stigma capitate, obscurely to distinctly 3-lobed, lobes erect or spreading, ± stout; pedicel 8–22 mm. **Seed coat** dull; cells minutely roughened.

Varieties 2 (2 in the flora): California and Oregon.

1. Stigma obscurely 3-lobed; tepal apex acute; anthers mostly yellow; flowering late Jun–Aug . . . . 32a. *Allium sanbornii* var. *sanbornii*
1. Stigma distinctly 3-lobed; tepal apex acuminate to long-acuminate; anthers mostly purple; flowering late Jun–Jul . . . . . . 32b. *Allium sanbornii* var. *congdonii*

32a. Allium sanbornii Alph. Wood var. sanbornii [E]

Bulbs 1–2 × 1.3–2 cm. **Scape** 18–54 cm. **Flowers:** tepals: outer 3–5 mm, inner 4.5–6.5 mm, margins entire or ± irregular, apex acute; anthers mostly yellow; stigma obscurely 3-lobed; pedicel 10–22 mm. $2n = 14$.

Flowering late Jun–Aug. Heavy serpentine clay; 700–1400 m; Calif., Oreg.

32b. Allium sanbornii Alph. Wood var. congdonii Jepson, Fl. Calif. 1: 275. 1921 [E]

Allium intactum Jepson

Bulbs 1.5–2.5(–3) × 1.2–2 cm. **Scape** 24–60 cm. **Flowers:** tepals: outer 4.5–6 mm, inner 6–7 mm, margins ± erose, apex acuminate to long-acuminate; anthers mostly purple; stigma distinctly 3-lobed; pedicel 8–22 mm. $2n = 14$.

Flowering Jun–Jul. Heavy serpentine clay; 400–700 m; Calif.

Allium sanbornii var. *congdonii* is known only from the foothills of the Sierra Nevada.

33. Allium nevadense S. Watson, Botany (Fortieth Parallel), 351, plate 38, figs. 1–3. 1871 [E] [F]

Allium nevadense var. *macropetalum* M. Peck

Bulbs 1–3, not clustered on stout, primary rhizome, with 0–2 stalked, basal bulbels, ovoid, 1–2 × 0.8–1.5 cm; outer coats enclosing 1 or more bulbs, brown, membranous, usually prominently cellular-reticulate, cells ± transversely elongate, intricately contorted, without fibers; inner coats white or pinkish, cells elongate, intricately contorted. **Leaves** persistent, 1, basally sheathing, sheath not extending much above soil surface; blade solid, terete, 10–20 cm × 1–3 mm. **Scape** persistent, solitary, erect, ± solid, terete, 5–15 cm × 1–2.5 mm. **Umbel** persistent, erect, compact, 5–25-flowered, hemispheric, bulbils unknown; spathe bracts persistent, 2–3, 3–7-veined, ovate to lanceolate, ± equal, apex acuminate. **Flowers** stellate, 7–12 mm; tepals spreading, white or pinkish white with dark pink midveins, lanceolate to ovate, ± equal, becoming papery in fruit, margins entire, apex acute to acuminate; stamens included; anthers purple; pollen yellow; ovary crested; processes 6, prominent, ± triangular, margins entire to notched or shallowly toothed; style linear, equaling stamens; stigma capitate, scarcely thickened, unlobed; pedicel 8–16 mm. **Seed coat** dull; cells minutely roughened. $2n = 14, 28$.

Flowering Apr–Jun. Sandy, rocky, or occasionally clay soils on desert plains and hillsides; 1400–1700 m; Ariz., Calif., Colo., Idaho, Nev., Oreg., Utah.

34. Allium atrorubens S. Watson, Botany (Fortieth Parallel), 352, plate 38, figs. 4, 5. 1871 [E]

Bulbs 1–3, not clustered on primary rhizome, with 0–2 stalked basal bulbels, ovoid to globose, 1–1.6 × 0.8–1.4 cm; outer coats enclosing 1 or more bulbs, brown, membranous, lacking cellular reticulation or cells arranged in only 2–3 rows distal to roots, ± quadrate, without fibers; inner coats pink or white, cells obscure, quadrate. **Leaves** persistent, 1, basally sheathing, sheath not extending much above soil level; blade solid, terete, 8–22 cm × 1–2.5 mm. **Scape** persistent, solitary, erect, solid, terete, 5–17 cm × 1–3 mm. **Umbel** persistent, erect, ± compact, 5–50-flowered, hemispheric to globose, bulbils unknown; spathe bracts persistent, (2–)3, 3–6-veined, ovate to lanceolate,

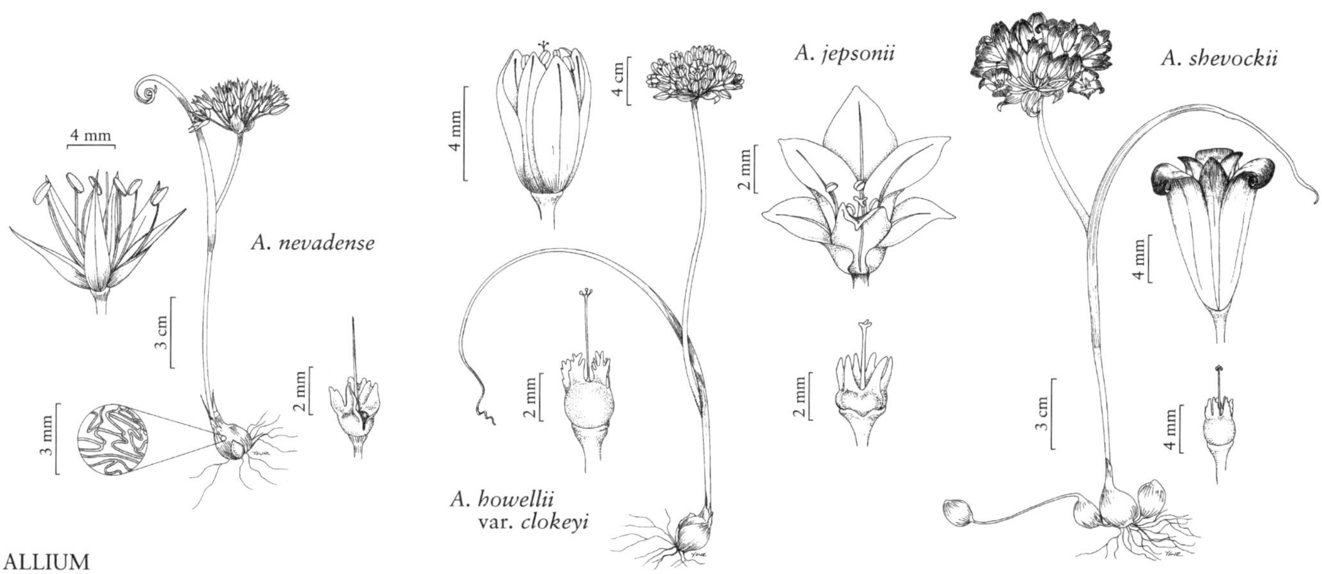

ALLIUM

± equal, apex acuminate. **Flowers** campanulate, 8–12 mm; tepals erect, pale pink to deep reddish purple, rarely white, lanceolate to ovate, ± equal or inner longer and narrower than outer, becoming rigid and ± keeled in fruit, margins entire, apex acute to acuminate; stamens included; anthers purple; pollen white or light yellow; ovary crested; processes 6, prominent, ± triangular, margins entire to notched or shallowly toothed; style linear, equaling stamens; stigma obscurely capitate, scarcely thickened, unlobed; pedicel 10–22 mm. **Seed coat** dull or shining; cells minutely roughened.

Varieties 2 (2 in the flora): intermountain w United States.

1. Tepals deep reddish purple (rarely pure white), lanceolate, apex margins becoming involute, apex thus appearing long-acuminate . . . . 34a. *Allium atrorubens* var. *atrorubens*
1. Tepals pale pink with darker midveins, lance-ovate to ovate, apex margins flat, not becoming involute . . . . 34b. *Allium atrorubens* var. *cristatum*

34a. Allium atrorubens S. Watson var. **atrorubens** [E]

Tepals deep reddish purple or rarely pure white, narrowly lanceolate to lanceolate, apex margins becoming involute, apex thus appearing long-acuminate. $2n = 14$.

Flowering May–Jun. Sandy, rocky, or clay soils on desert slopes; 1400–2100 m; Ariz., Calif., Nev., Utah.

34b. Allium atrorubens S. Watson var. **cristatum** (S. Watson) McNeal, Madroño 39: 86. 1992 [E]

Allium cristatum S. Watson, Proc. Amer. Acad. Arts 14: 232. 1879; *A. atrorubens* subsp. *inyonis* (M. E. Jones) Traub; *A. atrorubens* var. inyonis (M. E. Jones) Ownbey; *A. decipiens* M. E. Jones; *A. inyonis* M. E. Jones; *A. nevadense* subsp. *cristatum* (S. Watson) Traub; *A. nevadense* S. Watson var. *cristatum* (S. Watson) Ownbey

Tepals pale pink with darker pink midveins, lance-ovate to ovate, apex margins flat, not becoming involute. $2n = 14$.

Flowering late Apr–Jun. Sandy, rocky, or clay soils on desert slopes and mountains; 1200–2100 m; Ariz., Calif., Nev., Utah.

35. Allium parishii S. Watson, Proc. Amer. Acad. Arts 17: 380. 1882 [E]

Bulbs 1–2, not clustered on primary rhizome, without stalked basal increase bulbs, ovoid, 1–1.5 × 0.8–1.3 cm; outer coats enclosing bulbs, brown to reddish brown, membranous, lacking cellular reticulation or cells arranged in only 2–3 rows distal to roots, ± quadrate, without fibers; inner coats pinkish to reddish brown, cells obscure, ± rectangular, vertically elongate. **Leaves** persistent, withering from tip by anthesis, 1, basally sheathing, sheath not extending much above soil level; blade solid, terete, 5–30 cm × 1–3 mm. **Scape** persistent, solitary, erect, solid, terete, 5–25 cm × 1–3 mm. **Umbel** persistent, erect, compact, 6–15-flowered, hemispheric, bulbils unknown; spathe bracts persistent, 2–3, 5–7-veined, ovate, ± equal, apex acuminate. **Flowers** narrowly campanulate, 12–18 mm; tepals erect, spreading at tips, pale pink with darker midveins, lanceolate to lance-linear, ± equal, inner slightly shorter and narrower than outer, margins entire, apex acute, becoming rigid in fruit; stamens included; anthers yellow; pollen yellow; ovary crested; processes 6, prominent, ± triangular, margins entire to irregularly denticulate; style linear, equaling stamens; stigma capitate, scarcely thickened, minutely 3-lobed, lobes stout, ± spreading; pedicel 5–15 mm. **Seed coat** dull or shining; cells minutely roughened. $2n = 14$.

Flowering Apr–May. Rocky, sandy desert slopes; 900–1400 m; Ariz., Calif.

36. Allium monticola Davidson, Bull. S. Calif. Acad. Sci. 20: 51. 1921 E

Allium monticola var. *keckii* (Munz) Ownbey & Aase; *A. parishii* S. Watson var. *keckii* Munz; *A. peirsonii* Jepson

Bulbs 1–3+, generally with 1–2 stalked, basal bulbels, not clustered on stout, primary rhizome, ovoid, 1–2.2 × 1.2–2 cm; outer bulb coats enclosing 1 or more bulbs, gray-brown, membranous, lacking cellular reticulation or cells arranged in only 2–3 rows distal to roots, ± quadrate, without fibers; inner coats white to pinkish, cells obscurely quadrate. **Leaves** persistent, withering just at tip by anthesis, 1, basally sheathing, sheath not extending much above soil surface; blade solid, terete, 9–35 cm × 2–4 mm. **Scape** persistent, solitary, erect, solid, terete, 6–25 cm × 2–4 mm. **Umbel** persistent, erect, compact, 8–25-flowered, conic, bulbils unknown; spathe bracts persistent, 2–3, 5–7-veined, broadly ovate, ± equal, apex acuminate. **Flowers** urceolate, 12–19 mm; tepals erect, rose-purple distally, often white proximally, lance-linear to lance-ovate, ± equal, becoming rigid in fruit, margins entire, apex long-acuminate, flaring at tip; stamens included; anthers yellow; pollen yellow; ovary crested; processes 6, ± prominent, ± triangular to ± linear, margins entire; style linear, ± equaling stamens; stigma capitate, scarcely thickened, unlobed; pedicel 5–12 mm. **Seed coat** dull; cells minutely roughened. $2n = 14$.

Flowering late May–Jul. Loose rock and talus slopes, alpine ridges and talus; 1400–3200 m; Calif.

Allium monticola is known only from the southern California mountains.

37. Allium howellii Eastwood, Leafl. W. Bot. 2: 109. 1938 E F

Bulbs usually solitary, not clustered on stout, primary rhizome, ovoid to ± globose, 0.9–1.7 × 0.8–2 cm; outer coats enclosing bulb, dark reddish brown, membranous, lacking cellular reticulation or cells arranged in only 2–3 rows distal to roots, ± quadrate, without fibers; inner coats light brown to white, cells obscure, quadrate. **Leaves** persistent, withering from tip by anthesis, 1, basally sheathing, sheath not extending much above soil surface; blade solid, terete, 20–55 cm × 1–4 mm. **Scape** persistent, solitary, erect, solid, terete, 15–60 cm × 1–3.5 mm. **Umbel** persistent, erect, compact, 15–100-flowered, hemispheric, bulbils unknown; spathe bracts persistent, 2–4, 5–8-veined, lanceolate to broadly ovate, ± equal, apex mucronate to acuminate or long-acuminate. **Flowers** stellate to ± campanulate, 5–8 mm; tepals spreading to ± erect, white to pink or pale lavender with darker midveins, ovate to oblong, ± equal, becoming papery in fruit, margins entire, apex obtuse; stamens ± equaling tepals to long-exserted; anthers yellow or purple; pollen yellow; ovary crested; processes 6, prominent, ± triangular, margins ± irregularly denticulate to laciniate; style linear, equaling stamens or longer; stigma capitate, 3-lobed, lobes slender, recurved; pedicel 6–25 mm. **Seed coat** dull or shining; cells minutely roughened.

Varieties 3 (3 in the flora): California.

1. Scape 15–25(–35) cm; umbels generally 15–30 (–50)-flowered; tepals usually white to pink or lavender; ovary crests purple in fresh specimens .... 37a. *Allium howellii* var. *howellii*
1. Scape 20–60 cm; umbels generally 25–100-flowered; tepals usually white or pale pink; ovary crests white or greenish in fresh specimens.
   2. Stamens exserted, 2–4 mm .... 37b. *Allium howellii* var. *sanbenitense*
   2. Stamens ± equaling tepals............... .... 37c. *Allium howellii* var. *clokeyi*

37a. Allium howellii Eastwood var. **howellii** E

Scape 15–25(–35) cm. **Umbel** 15–30(–50)-flowered. **Flowers:** tepals white to usually pink or lavender; stamens equaling tepals or short-exserted; ovary crests purple in fresh specimens. $2n = 14$.

Flowering Mar–Apr. Clay soils, including serpentine, on open or grassy slopes; 200–900 m; Calif.

37b. Allium howellii Eastwood var. **sanbenitense** (Traub) Ownbey & Aase in P. A. Munz, Calif. Fl., 1376. 1959 [E]

Allium sanbenitense Traub, Herbertia 12: 68. 1945; *A. howellii* subsp. *sanbenitense* (Traub) Traub & Ownbey; *A. robustum* Eastwood 1938, not Karelin & Kirilov 1841

Scape 25–60 cm. **Umbel** (25–)50–90-flowered. **Flowers:** tepals white or pale pink; stamens exserted, 2–4 mm; ovary crests white or greenish in fresh specimens. $2n = 14$.

Flowering Apr–May. Grassy slopes in heavy clay soils including serpentine; 300–1000 m; Calif.

Allium howellii var. *sanbenitense* is known only from the inner central Coast Range.

37c. Allium howellii Eastwood var. **clokeyi** Ownbey & Aase ex Traub, Pl. Life 28: 63. 1972 [E] [F]

Scape 20–40 cm. **Umbel** 30–100-flowered. **Flowers:** tepals white or with pinkish midveins; stamens ± equaling tepals; ovary crests white or greenish in fresh specimens. $2n = 14$.

Flowering May–Jun. Heavy clay soils, including serpentine; 1300–1900 m; Calif.

Allium howellii var. *clokeyi* is known only from the Mount Piños region and San Bernardino Mountains.

38. Allium jepsonii (Ownbey & Aase ex Traub) S. S. Denison & McNeal, Madroño 36: 127. 1989 [C] [E] [F]

Allium sanbornii Alph. Wood var. *jepsonii* Ownbey & Aase ex Traub, Pl. Life 28: 63. 1972

Bulbs 1–2, not clustered on stout, primary rhizome, ovoid to subglobose, 1.5–2.5 × 1–2 cm; outer coats enclosing single bulb, gray-brown, membranous, lacking cellular reticulation or cells arranged in only 2–3 rows distal to roots, ± quadrate, without fibers; inner coats light brown or white, cells obscure, quadrate. **Leaves** persistent, withering from tip by anthesis, 1, basally sheathing, sheath not extending much above soil surface; blade solid, terete, 22–42 cm × 1–4 mm. **Scape** persistent, solitary, erect, solid, terete, 25–37 cm × 1–3 mm. **Umbel** persistent, erect, ± compact, 20–60-flowered, hemispheric, bulbils unknown; spathe bracts persistent, 3–4, 4–5-veined, ovate, ± equal, apex long-acuminate. **Flowers** campanulate, 7–8.5 mm; tepals erect, white, flushed with pink near dark pink midveins, ovate-elliptic, ± equal, becoming papery in fruit, margins erose, apex acute to apiculate, outer tepals with reflexed tips; stamens included; anthers yellow; pollen yellow; ovary crested; processes 6, prominent, ± triangular, margins erose; style linear, equaling stamens; stigma capitate, 3-lobed, lobes slender, recurved; pedicel 7–20 mm. **Seed coat** dull or shining; cells minutely roughened. $2n = 14$.

Flowering late May–early Jul. Clay soils, including serpentine; of conservation concern; 300–600 m; Calif.

Allium jepsonii is disjunct in the foothills of the Sierra Nevada.

39. Allium denticulatum (Ownbey & Aase ex Traub) McNeal, Aliso 13: 414. 1992 [E]

Allium fimbriatum S. Watson var. *denticulatum* Ownbey & Aase ex Traub, Pl. Life 28: 64. 1972

Bulbs 1–2, not clustered on stout, primary rhizome, ovoid to ± globose, 1–1.4 × 0.1–0.14 cm; outer coats enclosing single bulb, reddish brown, membranous, usually lacking cellular reticulation or cells arranged in only 2–3 rows distal to roots, ± quadrate, without fibers; inner coats pale brown to white, cells obscure, quadrate. **Leaves** persistent, withering from tip by anthesis, 1, basally sheathing, sheath not extending much above soil surface; blade solid, terete, 12–30 cm × 1–3 mm. **Scape** persistent, solitary, erect, solid, terete, 5–15 cm × 1–2 mm. **Umbel** persistent, erect, compact, 5–30-flowered, hemispheric, bulbils unknown; spathe bracts persistent, 2–3, 4–6-veined, lance-ovate to ovate, ± equal, apex acuminate. **Flowers** urceolate, 9–17 mm; tepals erect, purple, lanceolate to lance-ovate, unequal, becoming rigid and ± keeled in fruit, margins minutely denticulate at least distally, apex acute, ± straight at tip, inner slightly shorter and narrower than outer; stamens included; anthers yellow; pollen yellow; ovary crested; processes 6, prominent, ± triangular, margins entire to notched or irregularly papillose; style linear, equaling stamens; stigma capitate, 3-lobed, lobes slender, recurved; pedicel 5–25 mm. **Seed coat** dull or shining; cells minutely roughened. $2n = 14$.

Flowering mid Apr–mid Jul. Sandy, rocky soil, Techachapi Mountains and desert slopes, western Mojave Desert, California; 900–1600 m; Calif.

40. Allium abramsii (Ownbey & Aase ex Traub) McNeal, Aliso 13: 417. 1992 [E]

Allium fimbriatum S. Watson var. *abramsii* Ownbey & Aase ex Traub, Pl. Life 28: 64. 1972

Bulbs 1–3, not clustered on stout, primary rhizome, ovoid to globose, 1–1.5 × 1–1.5 cm; outer coats enclosing 1 or more bulbs, gray to dark brown, membranous, lacking cellular reticulation or cells arranged in only 2–3 rows distal to roots, ± quadrate, without fibers; inner coats white, cells obscure, quadrate. **Leaves** persistent, withering from tip by anthesis, 1, basally sheathing, sheath not extending much above soil surface; blade solid, terete, 12–30 cm × 1–3.5 mm. **Scape** persistent, solitary, erect, solid, terete, 6–15 cm × 1–3 mm. **Umbel** persistent, erect, compact, 10–45-flowered, hemispheric, bulbils unknown; spathe bracts persistent, 2–3, 3–5-veined, lanceolate, ± equal, apex acuminate. **Flowers** campanulate to urceolate, 8–15 mm; tepals erect, rose-purple, linear to lanceolate, unequal, becoming rigid and ± keeled in fruit, apex acute, ± spreading-reflexed at tip, inner tepals slightly shorter and wider than outer, outer margins denticulate on distal 1/4, inner margins denticulate to erose and crisped throughout; stamens included; anthers yellow; pollen yellow; ovary crested; processes 6, prominent, ± triangular, margins entire; style linear, equaling stamens; stigma capitate, 3-lobed, lobes slender, recurved; pedicel 4–12 mm. **Seed coat** dull or shining; cells minutely roughened. $2n = 14$.

Flowering May–Jul. Granite sands in open coniferous forest; 1400–2000 m; Calif.

Allium abramsii is known only from the central Sierra Nevada.

41. Allium sharsmithiae (Ownbey & Aase ex Traub) McNeal, Aliso 13: 417. 1992 (as sharsmithae) [C][E]

Allium fimbriatum S. Watson var. *sharsmithiae* Ownbey & Aase ex Traub, Pl. Life 28: 64. 1972 (as sharsmithae)

Bulbs usually solitary, not clustered on stout, primary rhizome, ovoid to globose, 1–1.8 × 1–1.5 cm; outer coats enclosing bulb, reddish brown, membranous, lacking cellular reticulation or cells arranged in only 2–3 rows distal to roots, ± quadrate, without fibers; inner coats pale brown, cells obscure, quadrate. **Leaves** persistent, withering from tip by anthesis, 1, basally sheathing, sheath not extending much above soil surface; blade solid, terete, 15–25 cm × 1–4 mm. **Scape** persistent, solitary, erect, solid, terete, 4–17 cm × 1–2.5 mm. **Umbel** persistent, erect, compact, 5–50-flowered, hemispheric, bulbils unknown; spathe bracts persistent, 2–3, 3–8-veined, lanceolate to lance-ovate, ± equal, apex long-acuminate to setaceous. **Flowers** urceolate, 10–18 mm; tepals erect, deep reddish purple, linear-lanceolate to lance-ovate, ± equal, becoming rigid in fruit, margins entire, apex acute to obtuse, recurved-spreading at tips; stamens included; anthers yellow; pollen yellow; ovary crested; processes 6, prominent, ± triangular, margins entire, surfaces usually papillate; style linear, equaling stamens; stigma capitate, 3-lobed, lobes slender, recurved; pedicel 6–19 mm. **Seed coat** dull or shining; cells minutely roughened. $2n = 14$.

Flowering Apr–May. Serpentine clay soil on talus slopes; of conservation concern; 500–1100 m; Calif.

Allium sharsmithiae is known only from the Mount Hamilton Range.

42. Allium shevockii McNeal, Madroño 34: 150, fig. 1. 1987 [C][E][F]

Bulbs 2–10+, forming threadlike rhizomes, rhizomes 3–10 cm, terminated by bulbels and/or 1–2 basal bulbels that in turn may produce threadlike rhizomes, ± globose, 1–1.5 × 1–1.5 cm; outer coats enclosing single bulb, brown, membranous, lacking cellular reticulation or cells arranged in only 2–3 rows distal to roots, ± quadrate, without fibers; inner coats light yellow, turning reddish on drying, cells obscure, ± quadrate. **Leaves** persistent, withering from tip by anthesis, 1, basally sheathing, sheath not extending much above soil surface; blade solid, terete, 20–40 cm × 2–4 mm. **Scape** persistent, solitary, erect, solid, terete, 10–20+ cm × 1–5 mm. **Umbel** persistent, erect, compact, 12–30-flowered, globose, bulbils unknown; spathe bracts persistent, usually 3, 5–7-veined, lanceolate, ± equal, apex apiculate. **Flowers** campanulate, 12–14 mm; tepals spreading, unequal; outer tepals maroon, oblanceolate, margins irregularly shallow-toothed, apex acute to mucronate, reflexed-curled; inner tepals maroon distally, white proximally, ovate, withering in fruit, curled back at tip, margins entire, apex acute, reflexed; stamens included; anthers yellow; pollen yellow; ovary crested; processes 6, prominent, narrowly triangular, margins entire, apex emarginate; style linear, equaling stamens; stigma capitate, 3-lobed, lobes slender, recurved; pedicel 10–16 mm. **Seed coat** shining; cells minutely roughened. $2n = 14$.

Flowering Jun–Jul. Soil pockets on schist outcrops; of conservation concern; 2200–2400 m; Calif.

Allium shevockii is known only from the upper slopes of Spanish Needle Peak in the southern Sierra Nevada and from Horse Canyon, Tehachapi Mountains.

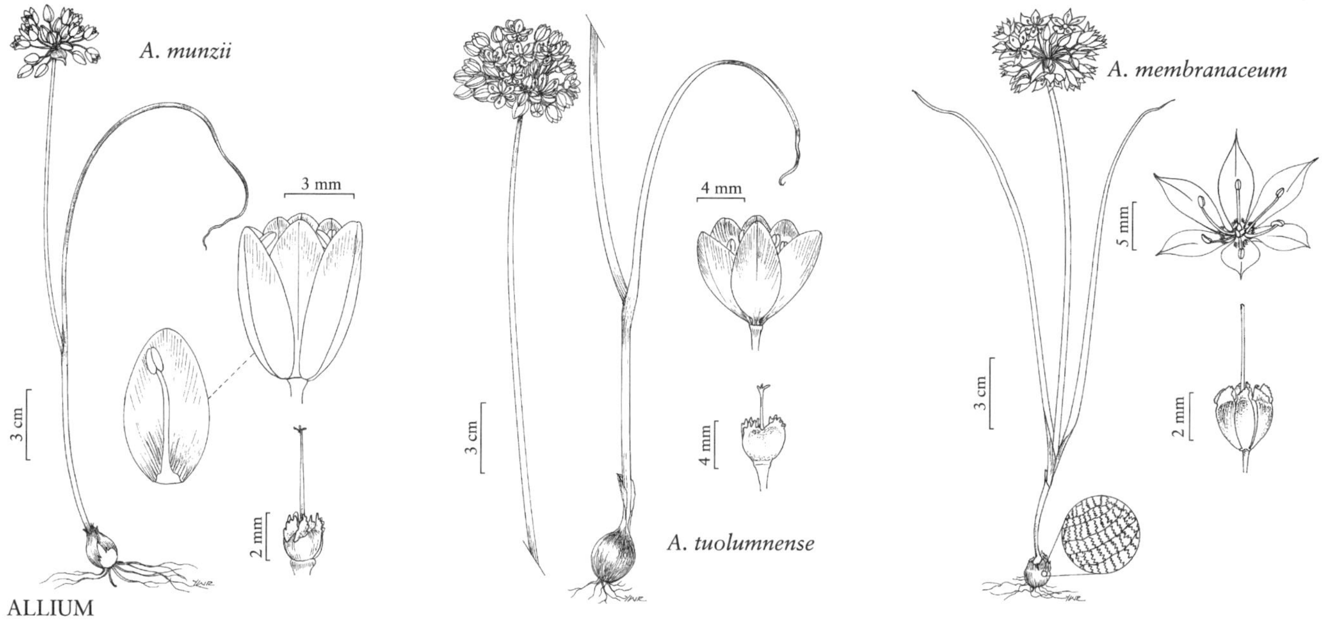

43. Allium parryi S. Watson, Proc. Amer. Acad. Arts 14: 231. 1879

Allium fimbriatum S. Watson var. *parryi* (S. Watson) Ownbey ex Traub; *A. kessleri* Davidson

Bulbs 1–3, not clustered on stout, primary rhizome, ovoid to globose, 0.8–1.4 × 0.8–1.5 cm; outer coats enclosing 1 or more bulbs, brown to reddish brown, chartaceous, lacking cellular reticulation or cells arranged in only 2–3 rows distal to roots, ± quadrate, without fibers; inner coats light brown, cells obscure, quadrate. **Leaves** persistent, withering from tip by anthesis, 1, basally sheathing, sheath not extending much above soil surface; blade solid, terete, 12–28 cm × 1–3 mm. **Scape** persistent, solitary, erect, solid, terete, 5–25 cm × 1–3 mm. **Umbel** persistent, erect, loose, 8–50-flowered, hemispheric, bulbels unknown; spathe bracts persistent, 2–3, 4–6-veined, ovate, ± equal, apex setaceous. **Flowers** campanulate, 6–9 mm; tepals erect, white to pink, becoming deep pink to red after anthesis, lanceolate to lance-ovate, ± equal, becoming ± rigid in fruit, margins entire, apex acute to acuminate, flaring distally; stamens included; anthers yellow; pollen yellow; ovary crested; processes 6, prominent, ± triangular, margins ± irregularly toothed, apex emarginate; style linear, equaling stamens or shorter; stigma capitate, 3-lobed, lobes slender, recurved; pedicel 6–20 mm, flexuous in fruit. **Seed coat** shining; cells minutely roughened. $2n$ = 14.

Flowering May–Jul. Clay or sandy soils on dry slopes; 900–2200 m; Calif.; Mexico (Baja California).

44. Allium munzii (Ownbey & Aase ex Traub) McNeal, Aliso 13: 415. 1992 [C] [E] [F]

Allium fimbriatum S. Watson var. *munzii* Ownbey & Aase ex Traub, Pl. Life 28: 64. 1972

Bulbs usually solitary, not clustered on stout, primary rhizome, ovoid, 1–1.5 × 0.8–1.4 cm; outer coats enclosing single bulb, reddish brown, membranous, lacking cellular reticulation or cells arranged in only 2–3 rows proximal to roots, ± quadrate, without fibers; inner coats pale brown to white or pink, cells obscure, quadrate. **Leaves** persistent, withering from tip by anthesis, 1, basally sheathing, sheath not extending much above soil surface; blade solid, terete, 15–25 cm × 0.5–1.5 mm. **Scape** persistent, solitary, erect, solid, terete, 15–35 cm × 1–2.5 mm. **Umbel** persistent, erect, compact, 10–35-flowered, hemispheric, bulbils unknown; spathe bracts persistent, 2–3, 5–7-veined, lanceolate to ovate, ± equal, apex abruptly acuminate. **Flowers** campanulate, 6–8 mm; tepals erect, white with greenish midveins, becoming deep pink to red in fruit, ovate to nearly round, ± equal, becoming papery in fruit, margins entire to finely denticulate, apex obtuse (rarely acute), shallowly emarginate; stamens included; anthers yellow; pollen yellow; ovary crested; processes 6, prominent, ± triangular, margins entire, apex emarginate; style linear, equaling stamens; stigma capitate, 3-lobed, lobes slender, recurved; pedicel spreading, straight, 7–12 mm, becoming rigid and ± reflexed in fruit. **Seed coat** dull; cells minutely roughened. $2n$ = 14.

Flowering Apr–mid May. Clay soil on grassy slopes; of conservation concern; 400–900 m; Calif.

45. Allium fimbriatum S. Watson, Proc. Amer. Acad. Arts 14: 232. 1879

Bulbs 1–3, not clustered on stout, primary rhizome, ovoid to ± globose, 1–1.7 × 0.8–1.7 cm; outer coats enclosing 1 or more bulbs, reddish brown, membranous, lacking cellular reticulation or cells arranged in only 2–3 rows proximal to roots, ± quadrate, without fibers; inner coats pale brown to white, cells obscure, quadrate. **Leaves** persistent, withering from tip by anthesis, 1, basally sheathing, sheath not extending much above soil surface; blade solid, terete, 12–50 cm × 1–4 mm. **Scape** persistent, solitary, erect, solid, terete, 10–35 cm × 0.5–3.5 mm. **Umbel** persistent, erect, compact, 6–75-flowered, hemispheric, bulbils unknown; spathe bracts persistent, 2–3, 4–7-veined, lanceolate to ovate, ± equal, apex attenuate to setaceous. **Flowers** urceolate to campanulate, 6–12 mm; tepals erect, dark reddish purple to pale lavender, or white, lanceolate to ovate, ± equal, becoming ± rigid to papery in fruit, margins entire or denticulate with few minute teeth near tip, apex acute to acuminate, recurved-spreading or not at tip; stamens included; anthers yellow; pollen yellow; ovary usually crested (rarely crestless); processes 6, prominent, ± triangular, margins denticulate to laciniate; style linear, equaling stamens; stigma capitate, 3-lobed, lobes slender, recurved; pedicel 6–20 mm. **Seed coat** dull; cells minutely roughened.

Varieties 3 (3 in the flora): California, Mexico (Baja California).

1. Tepals dark reddish purple, apex usually spreading or recurved at tip . . . . 45a. *Allium fimbriatum* var. *fimbriatum*
1. Tepals white to pale lavender with darker midvein, apex not spreading or recurved at tip.
   2. Scape 10–25 cm × 1–2 mm; Mojave Desert, se California . . 45b. *Allium fimbriatum* var. *mohavense*
   2. Scape 10–35 cm × 1.5–3.5 mm; known only from serpentine soils, n Coast Ranges, California . . . . 45c. *Allium fimbriatum* var. *purdyi*

45a. Allium fimbriatum S. Watson var. **fimbriatum**

Allium anserinum Jepson; *A. fimbriatum* var. *aboriginum* Jepson

Scape 10–20 cm × 0.5–1 mm. **Umbel** 6–35-flowered. **Flowers:** tepals dark reddish purple, apex ± spreading or recurved at tip; ovary crest process margins denticulate to laciniate. $2n = 14$.

Flowering Apr–Jun. Open clay (including serpentine) slopes and flats; 300–2700 m; Calif.; Mexico (Baja California).

45b. Allium fimbriatum S. Watson var. **mohavense** Jepson, Fl. Calif. 1: 273. 1921 [E]

Allium fimbriatum subsp. *mohavense* (Jepson) Traub & Ownbey; *A. mohavense* (Jepson) Tidestrom

Scape 10–25 cm × 1–2 mm. **Umbel** 12–60-flowered. **Flowers:** tepals white to pink or pale lavender with darker midvein, apex not spreading or recurved at tip; ovarian crest process margins laciniate, usually with additional protuberances on ovary surface at crest bases. $2n = 14$.

Flowering Apr–May. Sandy or rocky soils on open slopes and flats; 800–1400 m; Calif.

Populations of *Allium fimbriatum* var. *mohavense* from the north base of the San Bernardino Mountains intergrade with the typical variety and are placed here provisionally.

45c. Allium fimbriatum S. Watson var. **purdyi** (Eastwood) Ownbey ex McNeal, Aliso 13: 423. 1992 [E]

Allium purdyi Eastwood, Leafl. W. Bot. 2: 110. 1938

Scape 10–35 cm × 1.5–3.5 mm. **Umbel** 20–75-flowered. **Flowers:** tepals white to pale lavender with darker midvein, apex not spreading or recurved at tip; ovarian crest process margins denticulate to laciniate (crests rarely absent). $2n = 14$.

Flowering late Apr–Jun. Serpentine clay; 300–600 m; Calif.

Allium fimbriatum var. *purdyi* is known only from the vicinity of Clear Lake.

46. Allium tuolumnense (Ownbey & Aase ex Traub) S. S. Denison & McNeal, Madroño 36: 128. 1989 [C] [E] [F]

Allium sanbornii Alph. Wood var. *tuolumnense* Ownbey & Aase ex Traub, Pl. Life 28: 63. 1972

Bulbs usually solitary, not clustered on stout, primary rhizomes, ovoid, 1.3–2 × 1.4–2 cm; outer coats enclosing single bulb, dark reddish brown, membranous, lacking cellular reticulation or cells arranged in only 2–3 rows distal to roots, ± quadrate, without fibers; inner coats light brown, cells obscure, quadrate. **Leaves** persistent, withering from tip by anthesis, 1, basally sheathing, sheath not extending much above soil surface; blade solid, terete, 30–55 cm × 2–4 mm. **Scape** persistent, solitary, erect, solid, terete, 25–50 cm × 2–4 mm. **Umbel** persistent, erect, compact, 20–60-flowered, hemispheric, bulbils unknown; spathe bracts persistent, usually 3, 7–8-veined, ovate, ± equal, apex attenuate. **Flowers** saucer-shaped, 6–8 mm; tepals spreading from base, white or flushed with pink, broadly ovate, ± equal, becoming papery in fruit, margins entire, apex obtuse to nearly round, not recurved at tip; stamens included; anthers yellow; pollen yellow; ovary crested; processes 6, prominent, ± triangular, margins laciniate; style linear, equaling stamens; stigma capitate, 3-lobed, lobes slender, recurved; pedicel 7–20 mm. **Seed coat** dull; cells minutely roughened. $2n = 14$.

Flowering Mar–May. Serpentine soil on open hillsides; of conservation concern; 400–600 m; Calif.

Allium tuolumnense is known only from the foothills of the central Sierra Nevada, Rawhide Hill and Red Hills.

47. Allium diabolense (Ownbey & Aase ex Traub) McNeal, Aliso 13: 425. 1992 [E]

Allium fimbriatum S. Watson var. *diabolense* Ownbey & Aase ex Traub, Pl. Life 28: 64. 1972

Bulbs 1–3, not clustered on stout, primary rhizomes, ovoid to ± globose, 1–1.6 × 0.9–1.6 cm; outer coats enclosing 1 or more bulbs, reddish brown, membranous, lacking cellular reticulation or cells arranged in 2–3 rows proximal to roots, ± quadrate, without fibers; inner coats pale brown to white, cells obscure, quadrate. **Leaves** persistent, withering from apex by anthesis, 1, basally sheathing, sheath not extending much above soil surface; blade solid, terete, 10–30(–40) cm × 1–3 mm. **Scape** persistent, solitary, erect, solid, terete, 7–20(–30) cm × 1–3 mm. **Umbel** persistent, erect, compact to ± loose, 10–50-flowered, hemispheric, bulbils unknown; spathe bracts persistent, 2–3, 8–10-veined, lanceolate, ± equal, apex acuminate to long-acuminate. **Flowers** campanulate, 6–10 mm; tepals erect, white or tinged pink with dark red midveins, lanceolate to ovate, ± equal, becoming papery in fruit, margins entire, apex obtuse to acute or apiculate, not conspicuously recurved at tip; stamens included; anthers yellow; pollen yellow; ovary crested; processes 6, prominent, ± triangular, margins erose to ± laciniate; style linear, equaling stamens; stigma capitate, 3-lobed, lobes slender, recurved; pedicel 7–20 mm. **Seed coat** dull; cells minutely roughened. $2n = 14$.

Flowering mid Apr–Jun. Serpentine clay soils; 500–1500 m; Calif.

Allium diabolense is known only from the southern Coast Ranges and western Transverse Ranges.

48. Allium elmendorfii M. E. Jones ex Ownbey, Res. Stud. State Coll. Wash. 18: 218, fig. 8. 1951 [C] [E]

Bulbs 1–3, larger bulbs each bearing 3–10 basal bulbels surrounding roots, not clustered on stout, primary rhizome, ovoid, 1–1.5 × 0.8–1.5 cm; outer coats enclosing 1 or more bulbs, thin, membranous, lacking reticulation, without persistent fibers; inner coats whitish, cells vertically elongate or contorted, walls very sinuous. **Leaves** persistent, green at anthesis, 3–6, basally sheathing, sheaths not extending much above soil surface; blades solid, flat, channeled, 15–40 cm × 1–3 mm, margins entire. **Scape** persistent, 1–3, clustered, erect, solid, terete, 15–40 cm × 1–4 mm. **Umbel** persistent, erect, loose, 10–30-flowered, hemispheric, bulbils unknown; spathe bracts persistent, 3–4, 4–5-veined, ovate, ± equal, apex acuminate. **Flowers** ± stellate to campanulate, 5–6 mm; tepals erect or ± spreading, ± flexuous, white or pinkish, lanceolate, ± equal, papery and withering away from fruit, margins entire or nearly so, apex obtuse, midribs thickened; stamens included; anthers yellow; pollen yellow; ovary crestless; style linear, equaling stamens; stigma capitate, unlobed to distinctly lobed; pedicel 1–2.5 mm, elongating in fruit. **Seed coat** shining; cells smooth. $2n = 14$.

Flowering Mar–early Apr. Sandy soil; of conservation concern; 100–200 m; Tex.

Allium elmendorfii is known only from the Carrizo sands of eastern Bexar, Frio, Wilson, and Atacosa counties.

49. Allium madidum S. Watson, Proc. Amer. Acad. Arts 14: 228. 1879 [E]

Bulbs 1–3, larger bulbs each usually with cluster of 10–30 easily detached bulbels to one side of base, not clustered on stout, primary rhizome, globose to ovoid, 1–1.6 × 0.8–1.5 cm; outer coats enclosing 1 or more bulbs, brownish, membranous, without fibers; inner coats white to pink; inner and outer coats frequently without obvious reticulations, these, when present, contorted. **Leaves** persistent, green at anthesis, 2, basally sheathing, sheaths not extending much above soil surface; blade solid, flat or channeled, 10–25 cm × 1–4 mm, margins entire. **Scape** persistent, solitary, erect, solid, ± terete, sometimes ridged, 10–20 cm × 1–2 mm. **Umbel** persistent, erect, compact, 10–20-flowered, hemispheric, bulbils unknown; spathe bracts persistent, 2, 5–7-veined, ovate, ± equal, apex acuminate. **Flowers** campanulate, 6–10 mm; tepals erect or ± spreading, white with prominent green or pink midveins, lanceolate, ± equal, becoming papery in fruit, margins entire, apex obtuse to acuminate, becoming involute at tip; stamens included; anthers white or yellow; pollen yellow; ovary crestless or obscurely crested; processes 3, low, central, rounded, minute; style linear, equaling stamens; stigma capitate, unlobed or slightly lobed; pedicel 5–12 mm. **Seed coat** dull; cells ± smooth. 2*n* = 28, 42.

Flowering May–Jul. Seasonally wet meadows; 1100–2000 m; Idaho, Oreg.

Allium madidum is found in the Blue Mountains, Oregon and in Idaho near Payette Lake and New Meadows.

50. Allium campanulatum S. Watson, Proc. Amer. Acad. Arts 14: 231. 1879 [E]

Allium austinae M. E. Jones; *A. bidwelliae* S. Watson; *A. campanulatum* var. *bidwelliae* (S. Watson) Jepson

Bulbs 1–3, producing cluster of stalked basal bulbels distal to roots or filiform rhizomes to 10 cm and terminated by bulbels, ovoid, 1–2 × 0.6–1.4 cm; outer coats enclosing 1 or more bulbs, brown to gray, membranous, cellular-reticulate, cells ± quadrate, walls very sinuous, without fibers; inner coats pink to white, cells obscure, vertically elongate. **Leaves** persistent, withering from tip by anthesis, 2, basally sheathing, sheaths not extending much above soil surface; blade solid, flat, distinctly concave-convex, 8–40 cm × 1–5 mm, margins entire. **Scape** persistent, solitary or clustered, 1–3, erect, solid, terete, 10–30 cm × 1–5 mm. **Umbel** persistent, erect, loose, 10–50-flowered, ± globose, bulbils unknown; spathe bracts persistent, 2–3, 3–5-veined, ovate to lanceolate, ± equal, apex acuminate. **Flowers** stellate, 5–8 mm; tepals spreading, rose to purple (rarely white) with darker purple crescent adaxially basally, lanceolate to ovate, ± equal, carinate in fruit, becoming erect, ± shiny, rigid, margins entire, apex acuminate, strongly involute at tip; stamens included; anthers purple; pollen yellow; ovary crested; processes 6, prominent, low, central, triangular, margins minutely denticulate; style linear, ± equaling stamens; stigma capitate, scarcely thickened, unlobed; pedicel 10–20 mm, becoming flexuous and mostly strongly deflexed in fruit. **Seed coat** shining; cells each with minute, central papilla. 2*n* = 14, 28.

Flowering late May–Aug. Generally sandy soils on open or shaded slopes; 600–2600 m; Calif., Nev., Oreg.

51. Allium membranaceum Ownbey ex Traub, Pl. Life 28: 63. 1972 [E] [F]

Bulbs 1–3, producing bulbels in basal cluster around roots, ovoid, 1–1.6 × 0.9–1.4 cm; outer coats enclosing 1 or more bulbs, brown, membranous, obscurely cellular-reticulate, cells quadrate, walls very sinuous, without fibers; inner coats white or pink, cells obscure, quadrate. **Leaves** persistent, green at anthesis, 2–3, basally sheathing, sheaths not extending much above soil surface; blade solid, flat, 15–40 cm × 3–5 mm, margins entire. **Scape** persistent, solitary, erect, solid, terete, 15–40 cm × 1–3 mm. **Umbel** persistent, erect, loose, 15–35-flowered, ± globose, bulbils unknown; spathe bracts persistent, 2, 4–6-veined, lanceolate, ± equal, apex acuminate, setaceous. **Flowers** stellate, 7–12 mm; tepals spreading, pink or less commonly white, ovate to elliptic, ± equal, becoming papery in fruit, not carinate, margins entire, apex acute to acuminate, not involute; stamens included; anthers yellow; pollen yellow; ovary crested; processes 6, prominent, central, triangular, short, margins minutely denticulate; style linear, equaling stamens; stigma capitate, scarcely thickened, unlobed; pedicel 10–20 mm. **Seed coat** dull; cells each with minute, central papilla. 2*n* = 14.

Flowering May–Jun. Wooded, shady slopes; 200–1400 m; Calif.

Allium membranaceum is found in the foothills of the northern and central Sierra Nevada.

52. Allium bisceptrum S. Watson, Botany (Fortieth Parallel), 351, plate 37, figs. 1–3. 1871 [E]

Allium bisceptrum var. *palmeri* (S. Watson) A. Cronquist; *A. bisceptrum* var. *utahense* M. E. Jones; *A. palmeri* S. Watson

Bulbs 1–7+, commonly producing either cluster of stalked, basal bulbels or filiform rhizomes to 1 dm, terminated by bulbels, rhizomes generally lost when specimens are collected, ovoid, 1–2 × 0.6–1.8 cm; outer coats enclosing 1 or more bulbs, light brown to gray, membranous, obscurely cellular-reticulate, cells rectangular, walls minutely sinuous, vertical, varying to irregular, all sinuous, without fibers; inner coats white to pink, cells obscure, quadrate. **Leaves** persistent, green at anthesis, 2–5, basally sheathing, sheaths not extending much above soil surface; blade solid, flat, broadly channeled, 8–30 cm × 1–13 mm, margins entire. **Scape** persistent, solitary or clustered 1–3, erect, solid, terete, 10–30(–40) cm × 1–5 mm. **Umbel** persistent, erect, loose, 15–40-flowered, globose, bulbils unknown; spathe bracts persistent, 2, 3–4-veined, ovate to lanceolate, ± equal, apex acuminate. **Flowers** stellate, 7–10 mm; tepals spreading, lilac to white, lanceolate, ± equal, becoming papery in fruit, not carinate, margins entire, apex acuminate, not involute; stamens included; anthers purple; pollen yellow; ovary conspicuously crested; processes 6, central, distinct, flattened, triangular, margins papillose-denticulate; style included, linear, ± equaling stamens; stigma capitate, unlobed; pedicel 10–20 mm, often becoming flexuous and deflexed in fruit. **Seed coat** shining; cells each with minute, central papilla. $2n$ = 14, 28.

Flowering May–Jul. Meadows and aspen groves, less commonly on open slopes in mountains; 1100–3000 m; Ariz., Calif., Idaho, Nev., N.Mex., Oreg., Utah.

Two varieties of *Allium bisceptrum* have been recognized. Where their ranges overlap (central Utah and central Nevada), the putative varieties cannot be distinguished confidently except by chromosome number.

53. Allium kunthii G. Don, Mem. Wern. Nat. Hist. Soc. 6: 82. 1827

Allium scaposum Bentham; *Schoenoprasum lineare* Kunth, not *A. lineare* Linnaeus

Bulbs 1–4+, rhizomes, if present, secondary, inconspicuous, 2 cm or less including renewal bulb, ± thick, terminated by new bulb, parent bulbs disappearing by anthesis except for still-functional roots and bulb coat, not basally clustered, ovoid, 1–2 × 0.8–1.5 cm; outer coats enclosing renewal bulbs or not, grayish or brownish, with or without obscure, delicate, cellular markings, sometimes striate, membranous, cells elongate, in regular vertical rows, without fibers; inner bulb coats whitish or pinkish, cells obscure, ± quadrate or rectangular and vertically elongate. **Leaves** persistent, green at anthesis, 2–5, basally sheathing, sheaths not extended much above soil surface; blade solid, flat, channeled, 10–21 cm × 1–3 mm, margins and veins sometimes denticulate. **Scape** persistent, solitary, occasionally 2 or more produced successively from single bulb, erect, solid, terete, 15–30 cm × 1–3 mm. **Umbel** persistent, erect, loose, 5–20-flowered, conic, bulbils unknown; spathe bracts persistent, 2, 3–5-veined, lanceolate, apex acuminate. **Flowers** stellate to campanulate, 4–8 mm; tepals ± spreading, white or pale pink (particularly on midribs), lanceolate, ± equal, becoming papery and withering in fruit, margins entire, apex acute to acuminate; stamens included; anthers yellow or purple; pollen yellow; ovary crestless; style linear, equaling stamens; stigma capitate, unlobed; pedicel unequal, 10–20 mm. **Seed coat** dull; cells ± smooth. $2n$ = 14.

Flowering Jul–Sep. Dry, rocky hills and mountains, usually in limestone soils; 700–3000 m; Ariz., N.Mex., Tex.; Mexico.

Allium longifolium (Kunth) Sprengel (1825), based on *Schoenoprasum longifolium* Kunth (1816), may be the same as *A. kunthii*; the type material is inadequate for definite determination.

54. Allium neapolitanum Cirillo, Pl. Rar. Neapol. 1: 13. 1788 [I] [W2]

Bulbs 1–10+, not clustered on stout primary rhizome, rhizomes absent, bulbs absent or ± equaling parent bulbs, never appearing as basal cluster, subglobose, 1–2 × 1–2 cm; outer coats enclosing renewal bulbs, brown, cellular-reticulate, membranous, reticulum delicate, cells ± quadrate, without fibers; inner coats white, cells obscure, ± rectangular, vertically elongate. **Leaves** persistent, green at anthesis, 2–3, sheathing proximal 1/5–1/4 scape; blade solid, flat, not falcate, carinate, 15–50 cm × 5–20 mm, margins entire to denticulate. **Scape** persistent, solitary, erect, solid, ± triquetrous, 2-edged or slightly winged proximally, terete distally, 20–60 cm × 2–7 mm. **Umbel** persistent, erect, loose, 10–25-flowered, ± hemispheric, bulbils unknown; spathe bract persistent, 1, 8–9-veined, ovate, ± equal, equaling pedicel, apex acute. **Flowers** ± erect, saucer-shaped, 7–12 mm; tepals spreading, white, broadly elliptic, ± equal, becoming membranous and connivent over capsule, margins entire, apex obtuse; stamens included; anthers yellow; pollen yellow; ovary crestless; style linear, ± equaling stamens; stigma capitate, scarcely thickened, unlobed; pedicel 15–35 mm. **Seed coat** not known.

Flowering Mar–Apr. Disturbed sites; 0–100 m; introduced; Calif.; s Europe.

Allium neapolitanum is a garden escape, introduced from southern Europe.

55. Allium triquetrum Linnaeus, Sp. Pl. 1: 300. 1753 [I]

Bulbs 5–20+, not clustered on stout primary rhizome, rhizomes absent, increase bulbs absent or ± equaling parent bulbs, never appearing as basal cluster, ovoid, 1–2 × 1–2 cm; outer coats enclosing renewal bulbs, ± translucent, yellow-brown, obscurely cellular-reticulate, thin, membranous, meshes delicate, cells vertically elongate, contorted, without fibers; inner coats white, cells obscure, quadrate, or not visible. **Leaves** persistent, green at anthesis, 2–3, basally sheathing, sheaths not extending much above soil surface; blade solid, flat, not falcate, ± carinate, 15–50 cm × 3–15 mm, margins entire. **Scape** persistent, clustered, 1–4, erect, solid, sharply 3-angled, 10–40 cm × 1–10 mm. **Umbel** persistent, lax, loose, 3–15-flowered, ± 1-sided, bulbils unknown; spathe bracts persistent, 1–2, 3-veined, narrowly lanceolate, ± equal, apex acute. **Flowers** becoming pendent, campanulate, 10–18 mm; tepals erect to spreading, white with prominent green midrib, lanceolate, ± equal, becoming membranous in fruit, margins entire, apex acute; stamens included; anthers yellow; pollen yellow; ovary crestless; style linear, ± equaling stamens; stigma capitate, scarcely thickened, unlobed; pedicel 15–25 mm. **Seeds** appendaged with evident caruncle. **Seed coat** dull; cells minutely roughened. $2n = 18$.

Flowering Mar–Apr. Disturbed sites; 0–100 m; introduced; Calif.; sw Europe.

Allium triquetrum is a garden escape, introduced from southwestern Europe, and is potentially a noxious weed.

56. Allium unifolium Kellogg, Proc. Calif. Acad. Sci. 2: 112, fig. 35. 1863 [E]

Allium grandisceptrum Davidson; *A. unifolium* var. *lacteum* Greene

Bulbs solitary, replaced annually by new bulbs borne terminally on secondary rhizome; rhizomes 1–3, conspicuous, to 5 cm, smooth; parent bulbs disappearing by anthesis except for still-functional roots and bulb coat, ovoid to oblique-ovoid, 1–2 × 0.8–1.5 cm; outer coats not enclosing bulbs, pale brown, delicately cellular-reticulate, membranous, cells ± rectangular, without fibers; inner coats white, cells obscure, ± transversely elongate, contorted. **Leaves** persistent, green or withering from tip at anthesis, 2–3, basally sheathing, sheaths not extending much above soil surface; blade solid, flattened, sometimes carinate abaxially, ± falcate, 18–50 cm × 4–10 mm, margins entire. **Scape** persistent, solitary, erect, solid, terete, 20–80 cm × 2–7 mm. **Umbel** persistent, erect, loose, 15–35-flowered, hemispheric, bulbils unknown; spathe bracts persistent, 2, 6–8-veined, lance-ovate to broadly ovate, ± equal, apex acuminate. **Flowers** stellate, 11–15 mm; tepals spreading, bright pink or rarely white, obovate to ovate, unequal, becoming papery and connivent over capsule, margins entire, apex acute to obtuse or emarginate, inner shorter and narrower than outer; stamens included; anthers yellow or purple; pollen yellow or gray; ovary crestless, 3-grooved, with thickened ridge on either side of groove; style linear, equaling stamens; stigma capitate, scarcely thickened, unlobed or obscurely 3-lobed; pedicel 15–40 mm. **Seed coat** dull; cells minutely roughened. $2n = 14$.

Flowering May–Jun. Moist, clay soils, including serpentine, usually along streams; 0–1100 m; Calif., Oreg.

The long, relatively thick rhizomes that develop annually from the bulbs are very characteristic of *Allium unifolium* and almost unique in North America. Only *A. glandulosum* Link & Otto and *A. rhizomatum* Wooton & Standley have similar rhizomes, but these species are not closely related to *A. unifolium*.

Allium unifolium is known only from the Coast Ranges.

57. Allium rhizomatum Wooton & Standley, Contr. U.S. Natl. Herb. 16: 114. 1913

Bulbs solitary, not basally clustered, replaced annually by new bulbs borne terminally on rhizome; rhizomes 1–3, conspicuous, slender, 2–3 cm, scaly; parent bulbs persisting, often not collected, oblique-ovoid, 1–2.5 × 1 cm; outer coats enclosing parent bulbs, grayish, lacking cellular reticulation, membranous, without fibers; inner coats white or hyaline, cells obscure, ± quadrate. **Leaves** persistent, green at anthesis, 2–3, basally sheathing, sheaths not extending much above soil surface; blade solid, flat, not falcate, 20–35 cm × 2–3 mm, margins entire. **Scape** solitary, erect, solid, terete, 20–30 cm × 1–3 mm. **Umbel** persistent, erect, loose, 5–15(–22)-flowered, globose to hemispheric, bulbils unknown; spathe bracts persistent, 2, 3-veined, ovate to lance-ovate, ± equal, apex acute. **Flowers** stellate, 6–9 mm; tepals erect, pink with purplish or pinkish midveins, oblong to lanceolate, slightly carinate basally, ± equal, becoming papery in fruit, margins entire, apex acute to acuminate; stamens included; anthers yellow or pink; pollen yellow or white; ovary crestless, 3-grooved with thickened ridge on either side of groove; style linear, shorter than stamens; stigma capitate, scarcely thickened, unlobed; pedicel 10–20(–50) mm. **Seed coat** shining; cells smooth. $2n = 28$.

Flowering Jul–Sep. Dry, usually grassy areas; 1200–2200 m; Ariz., N.Mex., Tex.; Mexico.

Allium rhizomatum has often been included within the Mexican species *A. glandulosum*. This reduction is unwarranted. The perianth of *A. rhizomatum* is pale, with the color mainly confined to the midribs. Additionally, the species can be distinguished by its 3-lobed, apically 3-grooved ovary and lack of sepal glands. *Allium glandulosum* has a red perianth, an apically rounded ovary, and sepal glands. The nectar produced from these glands does not show in herbarium specimens.

58. Allium bolanderi S. Watson, Proc. Amer. Acad. Arts 14: 229. 1879 [E]

Bulbs 1–6+, not basally clustered, replaced annually by new bulbs borne terminally on rhizomes; rhizomes 1–3, inconspicuous, slender, less than 2 cm including renewal bulb; parent bulbs disappearing by anthesis except for still-functional roots and bulb coat, oblique-ovoid to ± oblong, 0.7–1.4 × 0.5–1.2 cm; outer coats not enclosing bulbs, brown to gray-brown, ± obscurely reticulate, membranous, reticulum delicate, cells transversely elongate, V-shaped or ± wavy, forming obscure herringbone pattern, without fibers; inner coats white, cells obscure, quadrate to ± rectangular, often contorted. **Leaves** persistent, withering from tip at anthesis, 2–3, basally sheathing, sheaths not extending much above soil surface; blade solid, subterete to ± channeled, 9–30 cm × 1–2 mm, margins entire. **Scape** persistent, solitary, erect, solid, terete, 10–35 cm × 1–3 mm. **Umbel** persistent, erect, compact to loose, 10–20-flowered, hemispheric, bulbils unknown; spathe bracts persistent, 2, 4–6-veined, lanceolate to lance-ovate, ± equal, apex acuminate. **Flowers** conic to campanulate, 7–14 mm; tepals erect, reddish purple, rarely white, narrowly lanceolate to lance-ovate, ± equal, becoming rigid and carinate in fruit, margins finely denticulate (inner tepal more prominently so), apex acute to obtuse, becoming involute at tip and appearing acuminate; stamens included; anthers yellow; pollen yellow; ovary crested; processes 3, central, 2-lobed, minute, margins entire; style linear, equaling stamens; stigma capitate, obscurely 3-lobed, scarcely thickened; pedicel 10–20 mm. **Seed coat** dull; cells minutely roughened.

Varieties 2 (2 in the flora): California, Oregon.

The varieties of *Allium bolanderi* are extremely difficult to distinguish in the absence of bulb material. Care must be taken to obtain good bulb material whenever collecting *Allium*, but it is extremely critical in this species.

1. Bulbs oblique-ovoid; bulb coat breaking with serrate edges; flowers 7–12 mm; tepals lance-ovate . . . . 58a. *Allium bolanderi* var. *bolanderi*
1. Bulbs oblique-oblong, irregular, suggesting small tubers; bulb coat breaking with irregular edge; flowers 9–14 mm; tepals narrowly lanceolate . . . . 58b. *Allium bolanderi* var. *mirabile*

58a. Allium bolanderi S. Watson var. **bolanderi** [E]

Allium bolanderi var. *stenanthum* (Drew) Jepson; *A. stenanthum* Drew

Bulbs oblique-ovoid, 0.7–1.2 cm; outer coats brown to gray-brown, breaking with serrate edges, cells of reticulum V-shaped. **Flowers** 7–12 mm; tepals lance-ovate, apex acute to obtuse. $2n = 14$.

Flowering late May–early Aug. Clay soils, including serpentine; 50–1000 m; Calif., Oreg.

58b. Allium bolanderi S. Watson var. **mirabile** (L. F. Henderson) McNeal, Phytologia 73: 307. 1992 [E]

Allium mirabile L. F. Henderson, Rhodora 32: 22. 1930

Bulbs oblique-oblong, ± irregularly shaped, suggesting small tuber, 0.9–1.4 cm; outer coats pale brown, breaking with irregular edges, cells of reticulum wavy. **Flowers** 9–14 mm; tepals narrowly lanceolate, apex acute. $2n = 14$.

Flowering late May–early Jul. Clay soils, including serpentine; 50–1000 m; Calif., Oreg.

59. Allium bigelovii S. Watson, Botany (Fortieth Parallel), 487, plate 38, figs. 8, 9. 1871 [E]

Bulbs usually solitary, not clustered on stout primary rhizome, ± globose, 1–1.5 × 1.2–1.5 cm; outer coats enclosing single bulb, dark brown, prominently reticulate, membranous, cells irregularly arranged, vertically elongate, rectangular to ± contorted, without fibers; inner coats white, cells obscure, quadrate. **Leaves** persistent, green at anthesis, 2, basally sheathing, sheaths not extending much above soil surface; blade solid, subterete to channeled, 16–21 cm × 2–4 mm, margins entire. **Scape** persistent, solitary, erect, solid, terete, 5–12 cm × 1–4 mm. **Umbel** persistent, erect, loose to ± compact, 10–25-flowered, hemispheric, bulbils unknown; spathe bracts persistent, 2, 2–11-veined, lance-ovate to ovate, ± equal, apex acute to acuminate. **Flowers** campanulate, (8–)10–14 mm; tepals erect, pink to reddish at tip and along midvein, otherwise white, lanceolate, ± equal, becoming papery and ± rigid in fruit, margins entire, apex acute; stamens included; anthers purple; pollen yellow; ovary crested; processes 6, prominent, flat, triangular, margins entire to coarsely toothed; style linear, equaling stamens; stigma capitate, scarcely thickened, obscurely 3-lobed; pedicel 10–15 mm. **Seed coat** shining; cells smooth. $2n = 14$.

Flowering late Mar–May. Open, rocky, gravelly slopes; 500–1700 m; Ariz., N.Mex.

Allium bigelovii is an anomalous species that does not appear to be closely related to any other from North America. With its prominent ovarian crests, a relationship with the group of species around *A. fimbriatum*, *A. nevadense*, and *A. sanbornii* is suggested. *Allium bigelovii* differs from this group, however, in having two leaves and a seed coat with smooth cells. In addition its bulb-coat reticulation is unlike that of any other North American species.

60. Allium lacunosum S. Watson, Proc. Amer. Acad. Arts 14: 231. 1879 E

Bulbs 1–3(–5), not clustered on stout primary rhizome, ovoid, 1–2 × 1–2 cm; outer coats enclosing 1 or more bulbs, brown, prominently cellular-reticulate, membranous, accumulating and investing bulb with thick cover, cells square, polygonal, or ± transversely elongate, walls thick, sinuous, without fibers; inner coats white to light brown, cells ± obscure, quadrate. **Leaves** persistent, withering from tip at anthesis, 2–3, basally sheathing, sheaths not extending much above soil surface; blade solid, flattened, subterete or ± channeled and carinate, ± falcate, 10–30 cm × 0.5–3 mm, margins entire. **Scape** persistent, solitary, erect, solid, terete, 10–35 cm × 1–4 mm. **Umbel** persistent, erect, compact or loose, 5–45-flowered, bulbils unknown, umbel shattering after seeds mature, each flower deciduous with its pedicel; spathe bracts persistent, 2–3, 4–6-veined, lance-ovate to ovate, ± equal, apex obtuse or acuminate. **Flowers** campanulate, 4–9 mm; tepals erect or spreading from base, white to deep pink with darker midveins, lance-ovate to ovate, ± equal, margins entire, apex obtuse, acute, or short-acuminate; stamens included; anthers yellow; pollen yellow; ovary 3-grooved with ridge on either side of each groove, ridges prolonged into crest, ridges and crest densely covered with minute, rounded papillae; processes 3, central, 2-lobed, minute; style linear, equaling stamens; stigma capitate, scarcely thickened, obscurely 3-lobed; pedicel 4–30 mm. **Seed coat** dull or shining; cells minutely roughened.

Varieties 4 (4 in the flora): California.

1. Umbel compact; pedicel 0.75–1.5 times flower; scape 10–25 cm.
   2. Scape 10–20 cm; leaf blade 1–2 times scape; flowers (6–)7–9 mm; Coast Ranges of c California and Channel Islands . . . . 60a. *Allium lacunosum* var. *lacunosum*
   2. Scape 15–25 cm; leaf blade equaling or shorter than scape; flowers 6–7(–8) mm; Sierra Nevada, California . . . . 60b. *Allium lacunosum* var. *kernensis*
1. Umbel loose; pedicel equaling–3.5 times flower; scape 15–35 cm.
   3. Leaf blade subterete, 1 mm or less wide, straight; spathe bracts 3, apex obtuse or acute; flowers 4–6 mm . . . . 60c. *Allium lacunosum* var. *micranthum*
   3. Leaf blade ± carinate, to 3 mm wide, ± falcate; spathe bracts 2, apex acuminate; flowers 6–8 mm . . . . 60d. *Allium lacunosum* var. *davisiae*

60a. Allium lacunosum S. Watson var. **lacunosum** E

Leaf blade subterete to ± terete, arcuate, 1–2 times scape, 1–2 mm wide. **Scape** 10–20 cm. **Umbel** compact, 5–25-flowered; spathe bracts 2, apex acuminate. **Flowers** (6–)7–9 mm; tepals white with deep pink or purple midveins; pedicel 5–15 mm. $2n = 14$.

Flowering Apr–Jun. Apparently restricted to serpentine clays on ridges and peaks; 300–1000 m; Calif.

Allium lacunosum var. *lacunosum* is known only from the central and south Coast Ranges, and on Santa Rosa and Santa Cruz islands.

60b. Allium lacunosum S. Watson var. **kernensis** McNeal & Ownbey, Madroño 29: 83. 1982 E

Leaf blade ± terete, straight, equaling or shorter than scape, 0.5–1 mm wide. **Scape** 15–25 cm. **Umbel** compact, 10–45-flowered; spathe bracts 2, apex acuminate. **Flowers** 6–7(–8) mm; tepals white or pale pink, often with green or deep pink midveins; pedicel 5–14 mm. $2n = 14$.

Flowering Apr–May. Rocky, sandy, or clay soils; 800–1300 m; Calif.

Allium lacunosum var. *kernensis* is known only from the southern Sierra Nevada.

60c. Allium lacunosum S. Watson var. **micranthum** Eastwood, Leafl. W. Bot. 2: 101. 1938 [E]

Leaf blade subterete, straight, equaling or shorter than scape, 1 mm wide or less. **Scape** 15–30 cm. **Umbel** loose, 10–30-flowered; spathe bracts 3, apex obtuse or acute. **Flowers** 4–6 mm; tepals white to pale pink with deep pink or purple midribs; pedicel 4–15 mm. $2n = 14$.

Flowering Apr–Jun. Rocky clay, including serpentine soils; 300–600 m; Calif.

Allium lacunosum var. *micranthum* is known only from the inner south Coast and Transverse ranges.

60d. Allium lacunosum S. Watson var. **davisiae** (M. E. Jones) McNeal & Ownbey, Madroño 29: 85. 1982 [E]

Allium davisiae M. E. Jones, Contr. W. Bot. 12: 78. 1908; *A. pseudobulbiferum* Davidson

Leaf blade channeled, carinate, ± falcate, equaling scape, 1–3 mm wide. **Scape** 15–35+ cm. **Umbel** loose, 10–35-flowered; spathe bracts 2, apex acuminate. **Flowers** 6–8 mm; tepals white to pale pink with deep pink or purple midribs; pedicel 4–21 mm. $2n = 14$.

Flowering Apr–May. Rocky, sandy, or clay soils; 600–2100 m; Calif.

Allium lacunosum var. *davisiae* is known only from the western Mojave Desert to the southern Central Valley.

61. Allium dictuon H. St. John, Proc. Biol. Soc. Wash. 50: 3, fig. 1. 1937 [C] [E] [F]

Bulbs usually solitary, forming rhizomes; rhizomes 1–2, terminated by 1–3 new bulbs, short, slender; parent bulb disappearing by anthesis except for still-functional roots and bulb coats, not basally clustered, oblique-ovoid, 1–1.5 × 0.9–1.5 cm; outer coats not enclosing bulbs, pale brown, prominently cellular-reticulate, membranous, cells polygonal to oblong, walls thick, obscurely sinuous, without fibers; inner coats white, cells ± transversely elongate, intricately contorted. **Leaves** persistent, withering from tip by anthesis, 2–3, basally sheathing, sheaths not extending much above soil surface; blade solid, subterete to ± channeled, 10–28 cm × 1–2 mm, margins entire. **Scape** persistent, solitary, erect, solid, terete, 20–40 cm × 1–3 mm. **Umbel** persistent, erect, loose, 10–25-flowered, hemispheric, bulbils unknown; spathe bracts persistent, 2, 3-veined, lanceolate to lance-ovate, ± equal, apex acuminate. **Flowers** campanulate, 11–16 mm; tepals erect, pink to rose-purple, lanceolate, unequal, outer tepal longer and wider than inner, becoming rigid and keeled in fruit, margins finely denticulate (inner tepal more prominently so), apex acute to narrowly obtuse, becoming involute and appearing long-acuminate, flaring toward tips; stamens included; anthers yellow; pollen yellow; ovary crested; processes 3, central, 2-lobed, rounded, minute, margins entire; style linear, equaling stamens; stigma capitate, scarcely thickened, obscurely 3-lobed; pedicel 10–25 mm. **Seed coat** shining; cells minutely roughened. $2n = 14$.

Flowering Jun–Jul. Rocky, sandy, basaltic soil; of conservation concern; 1500 m; Wash.

Allium dictuon is known only from the vicinity of Weller's Butte, Blue Mountains. It differs from *A. acuminatum* by its rhizomatous habit, in which it resembles *A. bolanderi*, and in the cellular pattern on the inner bulb coats.

62. Allium acuminatum Hooker, Fl. Bor.-Amer. 2: 184, plate 196. 1838 [E] [F]

Allium acuminatum var. *cuspidatum* Fernald; *A. cuspidatum* (Fernald) Rydberg

Bulbs 1–12+, not basally clustered, not forming rhizomes, ovoid to ± globose, 0.8–1.6 × 0.9–1.6 cm; outer coats enclosing 1 or more renewal bulbs, ± yellow-brown, prominently cellular-reticulate, membranous, cells square or polygonal, walls thick, obscurely sinuous, without fibers; inner coats white, cells obscure, ± quadrate. **Leaves** persistent, withering from tip by anthesis, 2–3, basally sheathing, sheaths not extending much above soil surface; blade solid, subterete or ± channeled, 7–30 cm × 1–3 mm, margins entire. **Scape** persistent, solitary, erect, solid, terete, 10–35 cm × 1–3 mm. **Umbel** persistent, erect, loose, 10–40-flowered, hemispheric, bulbils unknown; spathe bracts persistent, 2, 3–7-veined, lanceolate to ovate, ± equal, apex acuminate. **Flowers** campanulate, 8–15 mm; tepals erect, pink to rose-purple, or white, lanceolate to lance-ovate, unequal, becoming rigid and keeled in fruit, margins finely denticulate (inner tepal more prominently so), apex acuminate, outer tepal longer and wider than inner, spreading to recurved at tip, inner tepal with strongly recurved tips; stamens included; anthers yellow; pollen yellow; ovary crested; processes 3, central, 2-lobed, rounded, minute, margins entire; style linear, equaling

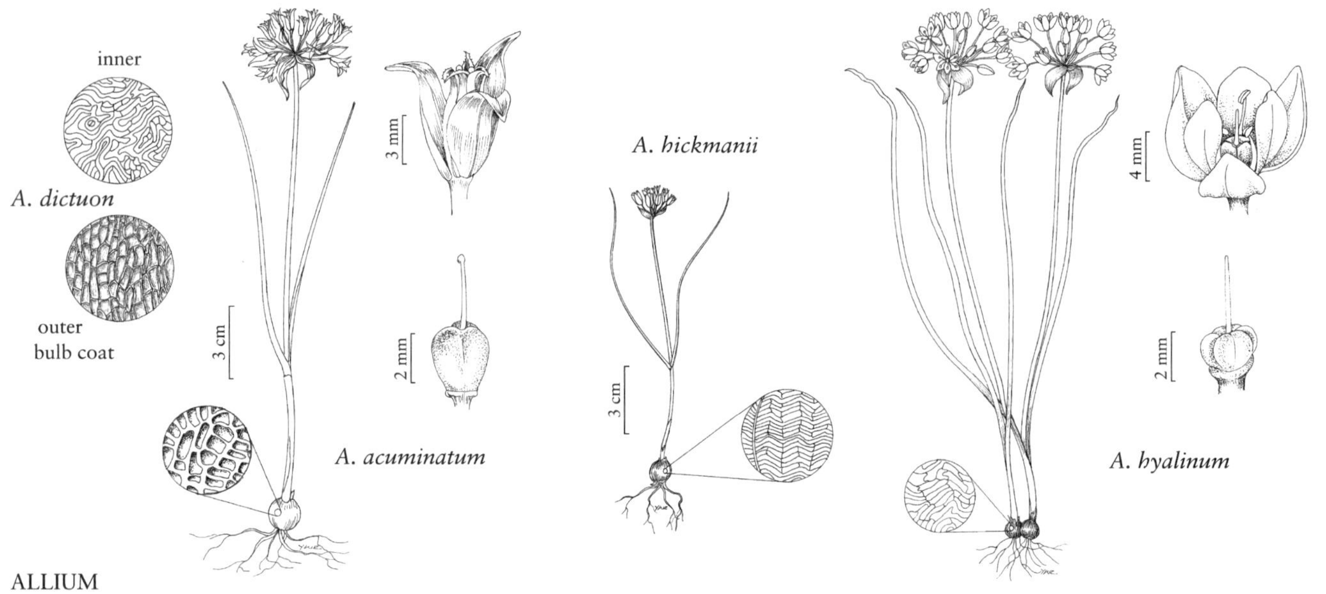

ALLIUM

stamens; stigma capitate, scarcely thickened, obscurely 3-lobed; pedicel 6–25 mm. **Seed coat** dull or shining; cells minutely roughened. $2n = 14$.

Flowering Apr–Jul. Dry slopes and plains; 100–1500 m; B.C.; Ariz., Calif., Colo., Idaho, Nev., Oreg., Utah, Wash., Wyo.

63. Allium hickmanii Eastwood, Bull. Torrey Bot. Club 30: 483. 1903 (as hickmani) [C] [E] [F]

Bulbs 1–5+, not clustered on stout primary rhizome, increase bulbs absent or ± equaling parent bulbs, never appearing as basal cluster, ovoid to ± globose, 0.8–1.2 × 0.7–1.2 cm; outer coats enclosing 1 or more bulbs, gray to gray-brown, prominently cellular-reticulate, membranous, cells arranged in ± vertical rows, forming irregular herringbone pattern, transversely elongate, wavy, V-shaped, without fibers; inner coats white, cells obscure, ± quadrate. **Leaves** persistent, withering from tip by anthesis, 2, basally sheathing, sheaths not extending much above soil level; blades solid, subterete, 3–20 cm × 0.5–1 mm, margins entire. **Scape** persistent, solitary, erect, solid, terete, (3–)5–15 (–17) cm × 0.5–2 mm. **Umbel** persistent, erect, compact, 4–15-flowered, hemispheric, bulbils unknown; spathe bracts persistent, 2, 3–5-veined, broadly ovate, ± equal, apex mucronate. **Flowers** campanulate, 5–7 mm; tepals erect, white to pale pink, lanceolate to lance-ovate, ± equal, becoming papery in fruit and remaining erect, margins entire, apex acute or obtuse, becoming ± involute at tip and appearing short-acuminate; stamens included; anthers white; pollen white; ovary crested or crestless; processes 3, central, rounded, very minute, margins entire; style linear, equaling stamens; stigma capitate, scarcely thickened, unlobed; pedicel 4–12 mm. **Seed coat** dull; cells minutely roughened. $2n = 14$.

Flowering Mar–May. Grassy, wooded slopes and flats; of conservation concern; 0–50 m; Calif.

Allium hickmanii is a rare endemic of the Monterey Peninsula and Arroyo de la Cruz.

64. Allium amplectens Torrey, Pacif. Railr. Rep. 4(5): 148. 1857 [E] [W2]

Allium acuminatum Hooker var. *gracile* Alph. Wood; *A. attenuifolium* Kellogg; *A. attenuifolium* var. *monospermum* (Jepson) Jepson; *A. monospermum* Jepson; *A. occidentale* A. Gray; *A. serratum* S. Watson

Bulbs 1–15+, increase bulbs absent or ± equaling parent bulbs, never appearing as basal cluster, not clustered on stout primary rhizome, ovoid to ± globose, 0.6–1.5 × 0.6–1.3 cm; outer coats enclosing 1 or more bulbs, brown, prominently cellular-reticulate, membranous, cells in ± vertical rows, forming irregular herringbone pattern, transversely elongate, V-shaped, without fibers; inner coats usually dark red, sometimes white to pink, cells obscure, quadrate. **Leaves** persistent, withering from tip at anthesis, 2–4, basally sheathing, sheaths not extending much above soil surface; blade solid, subterete or ± channeled, 10–36 cm × 0.5–2 mm, margins entire. **Scape** persistent, solitary, erect, solid, terete, 15–50 cm × 3–5 mm. **Umbel** shattering after seeds mature, each flower deciduous with its pedicel as a unit, erect, compact, 10–50-flowered, hemispheric, bulbels unknown; spathe bracts persistent, 2–3, 6–13-veined, ovate, ± equal, apex short-acuminate. **Flowers** stellate, 5–9 mm; tepals spreading at anthesis, white to pink, lanceolate, ± equal, becoming papery and connivent over capsule, margins entire, apex acute; stamens included; anthers yellow or purple; pollen yellow; ovary crested; processes 6, lateral, ± prominent, ± rectangular, margins entire; style linear, equaling stamens; stigma capitate, scarcely thickened, unlobed; pedicel 4–16 mm. **Seed coat** dull; cells minutely roughened. $2n = 14, 21, 28$.

Flowering Apr–Jul. Clay soils, including serpentine, dry slopes, and open plains; 0–1800 m; B.C.; Calif., Oreg., Wash.

All three chromosome races of *Allium amplectens* are widespread. The triploids are achiasmatic, causing a breakdown in the first meiotic division. This is followed by a normal second division resulting in pollen dyads that are, presumably, nonfunctional; seeds are produced by apomixis. The diploids and tetraploids produce normal pollen, in tetrads, that appears to be functional.

65. Allium praecox Brandegee, Zoë 5: 228. 1906

Allium hyalinum Curran var. *praecox* (Brandegee) Jepson

Bulbs 1–3, increase bulbs absent or ± equaling parent bulbs, never appearing as basal cluster, not clustered on stout primary rhizome, ovoid to globose, 1–1.8 × 1–1.7 cm; outer coats enclosing 1 or more bulbs, gray-brown to brown, prominently cellular-reticulate, membranous, cells arranged in ± vertical rows, forming irregular herringbone pattern, transversely elongate, V-shaped, without fibers; inner coats white, cells obscure, ± transversely elongate, contorted. **Leaves** persistent, withering from tip by anthesis, 2–3, basally sheathing, sheaths not extending much above soil surface; blade solid, flat, broadly channeled, carinate, 20–75 cm × 5–10 mm, margins entire. **Scape** persistent, solitary, erect, solid, terete, 20–60 cm × 2–4 mm. **Umbel** shattering after seeds mature, each flower deciduous with its pedicel as a unit, erect, loose, 5–40-flowered, hemispheric, bulbils unknown; spathe bracts persistent, 2, 5–7-veined, lanceolate to lance-ovate, ± equal, apex acuminate. **Flowers** stellate, 6–13 mm; tepals spreading at anthesis, white to pale pink with purple midveins, lance-ovate to ovate, ± equal, becoming papery and connivent over capsule, margins entire, apex acute, obtuse, or emarginate; stamens included; anthers purple and yellow; pollen yellow; ovary crested; processes 3, central, 2-lobed, minute, margins entire; style linear, equaling stamens; stigma capitate, scarcely thickened, unlobed; pedicel 15–40 mm. **Seed coat** shining; cells minutely roughened. $2n = 14$.

Flowering Mar–May. Clay soil on shaded, grassy slopes near coast; 50–800 m; Calif.; Mexico (Baja California).

66. Allium hyalinum Curran, Bull. Calif. Acad. Sci. 1: 155. 1885 [E] [F]

Bulbs generally 2–20+, increase bulbs absent or ± equaling parent bulbs, never appearing as basal cluster, not clustered on stout primary rhizome, ovoid to ± globose, 0.5–1.2 × 0.5–1.2 cm; outer coats enclosing 1 or more bulbs, gray-brown to brown, prominently cellular-reticulate, membranous, cells arranged in ± vertical rows, forming irregular herringbone pattern, transversely elongate, ± wavy, V-shaped, without fibers; inner coats yellow or white, cells obscure, arranged in vertical rows, forming a herringbone pattern or contorted, V-shaped. **Leaves** persistent, withering from tip by anthesis, 2–3, basally sheathing, sheaths not extending much above soil surface; blade solid, subterete or ± channeled, not carinate, 7–40 cm × 1–3 mm, margins entire. **Scape** persistent, solitary, erect, solid, terete, 15–45 cm × 2–4 mm. **Umbel** shattering after seeds mature, each flower deciduous with its pedicel as a unit, erect, loose, 5–25-flowered, hemispheric, bulbils unknown; spathe bracts persistent, 2, 3–5-veined, lanceolate to lance-ovate, ± equal, apex acuminate. **Flowers** stellate, 6–10 mm; tepals spreading at anthesis, white to pink, lance-ovate to broadly ovate, ± equal, becoming hyaline and connivent over capsule, margins entire, apex obtuse to rounded; stamens included; anthers yellow or purple; pollen yellow; ovary crestless; style linear, equaling stamens; stigma capitate, scarcely thickened, unlobed; pedicel 10–35 mm. **Seed coat** dull; cells minutely roughened. $2n = 14$.

Flowering Mar–May. Heavy, clay soils on shaded slopes; 50–1500 m; Calif.

Allium hyalinum is known from the Sierra Nevada foothills and inner south Coast Range.

67. Allium serra McNeal & Ownbey, Madroño 24: 25, fig. 1. 1977 [E]

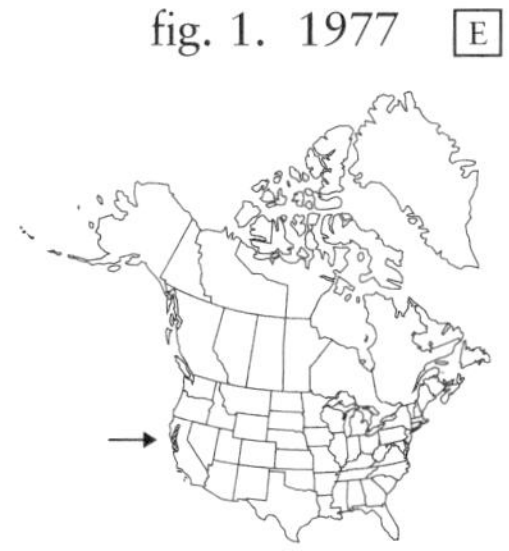

Bulbs 1–3, increase bulbs absent and ± equaling parent bulbs, never appearing as basal cluster, not clustered on stout primary rhizome, ovoid to ± globose, 0.8–1.2 × 0.8–1.2 cm; outer coats enclosing 1 or more bulbs, brown, prominently cellular-reticulate, membranous, cells arranged in ±

vertical rows, forming distinct herringbone pattern, transversely elongate, V-shaped, without fibers; inner coats white, cells obscure, arranged in vertical rows, forming distinct herringbone pattern or ± quadrate, V-shaped. **Leaves** persistent, withering from tips at anthesis, 2–3, basally sheathing, sheaths not extending much above soil surface; blade solid, subterete, channeled, 13–35 cm × 1–4 mm, margins entire. **Scape** persistent, solitary, erect, solid, terete, 1.5–4 cm × 2–3 mm. **Umbel** shattering after seeds mature, each flower deciduous with its pedicel as a unit, erect, compact, 10–40-flowered, hemispheric to globose, bulbils unknown; spathe bracts persistent, 2–3, 5–6-veined, lance-ovate to ovate, ± equal, apex acuminate. **Flowers** urceolate to campanulate, 8–11 mm; tepals erect, pink, lanceolate to lance-ovate, unequal, inner shorter and narrower than outer, becoming papery and connivent over capsule, margins entire, apex acute; stamens included; anthers yellow; pollen yellow; ovary crested; processes 3, central, 2-lobed, minute, margins entire; style linear, equaling stamens; stigma capitate, scarcely thickened, 3-lobed; pedicel 7–15 mm. **Seed coat** dull; cells minutely roughened. $2n = 14$.

Flowering Apr–May. Heavy, clay soils, including serpentine; 30–1200 m; Calif.

Allium serra is known only from the central and north Coast Ranges.

68. Allium dichlamydeum Greene, Pittonia 1: 166. 1888 [E]

Bulbs 1–3, increase bulbs absent or ± equaling parent bulbs, never appearing as basal cluster, not clustered on stout primary rhizome, ovoid to ± globose, 10–15 mm; outer coats enclosing 1 or more bulbs, brown, prominently cellular-reticulate, membranous, cells arranged in ± vertical rows, forming distinct herringbone pattern, transversely elongate, V-shaped, without fibers; inner coats white, cells obscure, arranged in vertical rows, forming distinct herringbone pattern, transversely elongate, V-shaped. **Leaves** persistent, withering from tip by anthesis, 3–6, basally sheathing, sheaths not extending much above soil surface; blade solid, subterete or ± channeled, arcuate to ± tortuous, 7–25 cm × 1–3 mm, margins entire. **Scape** persistent, solitary, erect, solid, terete, 10–30 cm × 2–3 mm. **Umbel** persistent, erect, compact, 5–30-flowered, conic to hemispheric, bulbils unknown; spathe bracts persistent, 2, 4–6-veined, broadly ovate, ± equal, apex abruptly acuminate. **Flowers** campanulate, 9–12 mm; tepals ± erect, deep reddish purple, ovate, unequal, inner shorter and narrower than outer, becoming rigid and ± carinate in fruit, margins entire or inner minutely denticulate, apex acute to obtuse, spreading to recurved at tips; stamens included; anthers yellow; pollen yellow; ovary crested; processes 3, central, 2-lobed, minute, margins entire; style linear, equaling stamens; stigma capitate, scarcely thickened, unlobed or minutely 3-lobed; pedicel 5–20 mm. **Seed coat** dull or shining; cells minutely roughened. $2n = 14$.

Flowering May–Jul. Dry, clay soils on or near sea cliffs; 10–50 m; Calif.

69. Allium crispum Greene, Pittonia 1: 166. 1888 [E] [F]

Allium peninsulare Lemmon ex Greene var. *crispum* (Greene) Jepson

Bulbs 1–3, increase bulbs absent or ± equaling parent bulbs, never appearing as basal cluster, not clustered on stout primary rhizome, ovoid to globose, 0.9–1.5 × 0.9–1.5 cm; outer coats enclosing 1 or more bulbs, brown or gray, prominently cellular-reticulate, membranous, cells arranged in ± vertical rows, forming distinct herringbone pattern, transversely elongate, V-shaped, without fibers; inner coats white, cells obscure, forming distinct herringbone pattern or ± quadrate, transversely elongate, V-shaped. **Leaves** persistent, withering from tip by anthesis, 2–3, basally sheathing, sheaths not extending much above soil surface; blade solid, subterete or ± channeled, straight to arcuate, 13–30 cm × 1–3 mm, margins entire. **Scape** persistent, solitary, erect, solid, terete, 15–35 cm × 1–3 mm. **Umbel** persistent, erect, loose, 10–40-flowered, hemispheric, bulbils unknown; spathe bracts persistent, 2, 4–6-veined, lance-ovate to ovate, ± equal, apex acuminate. **Flowers** campanulate, 8–13 mm; tepals erect, rose-purple, lance-ovate to broadly ovate, unequal, outer longer and wider than inner, becoming rigid and keeled in fruit, margins of outer tepal entire, inner denticulate and distinctly crisped, apex acuminate, recurved-spreading at tips; stamens included; anthers yellow; pollen yellow; ovary crested; processes 3, central, 2-lobed, minute, margins entire; style linear, equaling stamens; stigma capitate, 3-lobed; pedicel 10–35 mm. **Seed coat** dull; cells minutely roughened. $2n = 14$.

Flowering late Mar–Jun. Clay soils, including serpentine; 100–800 m; Calif.

Allium crispum is known only from the inner south Coast Ranges.

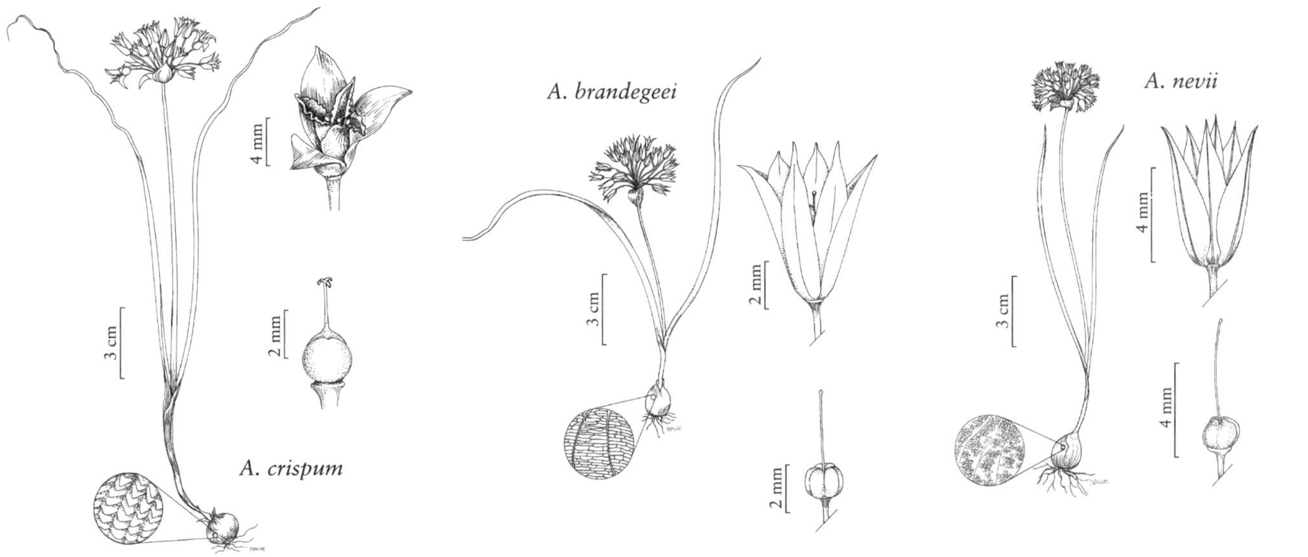

ALLIUM

70. Allium peninsulare Lemmon ex Greene, Pittonia 1: 165. 1888

Bulbs 1–5+, increase bulbs absent or ± equaling parent bulbs, never appearing as basal cluster, not clustered on stout primary rhizome, ovoid to ± globose, 0.6–1 cm × 6–10 mm; outer coats enclosing 1 or more bulbs, brown to gray-brown, prominently cellular-reticulate, membranous, cells arranged in ± vertical rows, forming distinct herringbone pattern, transversely elongate, V-shaped, without fibers; inner coats white, cells not visible or arranged in vertical rows, forming distinct herringbone pattern, ± transversely elongate. **Leaves** persistent, withering from tip by anthesis, 2–3, basally sheathing, sheaths not extending much above soil surface; blade solid, subterete or ± channeled, straight to arcuate, 8–30 cm × 1–3 mm, margins entire. **Scape** persistent, solitary, erect, solid, terete, 12–45 cm × 1–3 mm. **Umbel** persistent, erect, loose, 5–35-flowered, hemispheric to conic, bulbils unknown; spathe bracts persistent, 2, 3–6-veined, lance-ovate to ovate, ± equal, apex acuminate. **Flowers** campanulate, 8–15 mm; tepals erect, reddish purple, lanceolate to elliptic, unequal, outer tepals longer and wider than inner, becoming rigid and ± carinate in fruit, margins entire or inner minutely denticulate, never crisped, apex acute or short-acuminate, inner tepals spreading at tip, outer ± recurved; stamens included; anthers yellow; pollen yellow; ovary crested; processes 3, central, 2-lobed, minute, margins entire; style linear, equaling stamens; stigma capitate, unlobed to distinctly 3-lobed; pedicel 8–40 mm. **Seed coat** dull; cells minutely roughened.

Varieties 2 (2 in the flora): w United States; Mexico (Baja California).

1. Leaf blade straight or nearly so; flowers 10–15 mm; stigma capitate, distinctly 3-lobed . . . . 70a. *Allium peninsulare* var. *peninsulare*
1. Leaf blade arcuate; flowers 8–12 mm; stigma capitate, scarcely thickened, unlobed or minutely 3-lobed . . . . . 70b. *Allium peninsulare* var. *franciscanum*

70a. Allium peninsulare Lemmon ex Greene var. **peninsulare**

Allium montigenum Davidson

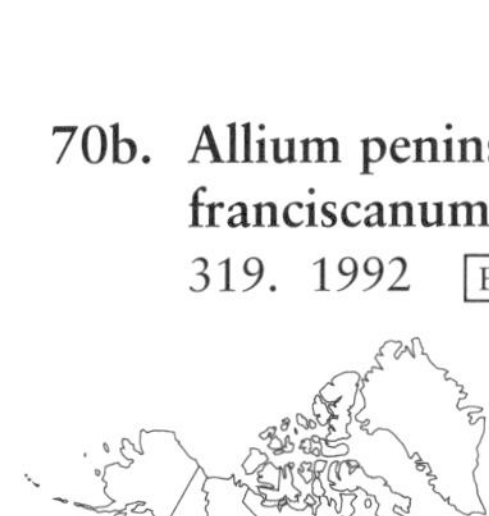

Leaf blade straight or nearly so. **Flowers** 10–15 mm; stigma distinctly 3-lobed; pedicel 10–40 mm. $2n = 14$.

Flowering Apr–May. Clay soils, including serpentine; 300–1100 m; Calif., Oreg.; Mexico (Baja California).

70b. Allium peninsulare Lemmon ex Greene var. **franciscanum** McNeal & Ownbey, Phytologia 73: 319. 1992 [E]

Leaf blade ± arcuate. **Flowers** 8–12 mm; stigma scarcely thickened, unlobed or minutely 3-lobed; pedicel 8–20 mm. $2n = 14$.

Flowering May–Jun. Clay soil on dry hillsides; 50–300 m; Calif.

Allium peninsulare var. *franciscanum* is found around San Francisco Bay to the north and west and does not overlap var. *peninsulare* in its distribution. It is morphologically intermediate between the typical variety and *A. dichlamydeum,* which it likewise does not overlap in distribution.

71. Allium fibrillum M. E. Jones ex Abrams in L. Abrams and R. S. Ferris, Ill. Fl. Pacific States 1: 393. 1923 [E]

Allium collinum Douglas ex S. Watson 1889, not Gussone 1831

Bulbs 1–5+, without cluster of basal bulbels, not clustered on stout primary rhizome, ± globose, 0.5–1.2 × 0.5–1 cm; outer coats enclosing 1 or more bulbs, brownish, membranous, reticulate; cells very fine, often obscure, narrow, ± transversely elongate, intricately contorted, without fibers; inner coats white to pink, cells obscure, transversely elongate, intricately contorted. **Leaves** usually persistent, withering from tip by anthesis, 2, basally sheathing, sheaths not extending much above soil surface; blade solid, flat, ± falcate, linear, channeled, 7–24 cm × 0.5–3 mm, margins entire. **Scape** persistent, solitary, erect, solid, terete or slightly flattened to very narrowly winged, 3–15 cm × 0.5–2 mm. **Umbel** persistent, erect, compact, 10–20-flowered, hemispheric to globose, bulbils unknown; spathe bracts persistent, 2, 4–6-veined, ovate, ± equal, apex acuminate. **Flowers** campanulate, 5–8 mm; tepals erect, white with prominent greenish or pink midribs, lanceolate, ± equal, becoming papery in fruit, margins entire, apex obtuse to acuminate, ± involute at tip; stamens included; anthers yellow; pollen yellow; ovary crestless or obscurely crested; processes 3, low, rounded, margins entire; style linear, equaling stamens; stigma capitate, unlobed; pedicel 3–10 mm. **Seed coat** dull or shining; cells smooth. $2n = 14$.

Flowering May–Jul. Moist, shallow soils; 300–2600 m; Idaho, Mont., Oreg., Wash.

Allium fibrillum closely resembles *A. madidum* but does not have a cluster of basal bulbels.

72. Allium brandegeei S. Watson, Proc. Amer. Acad. Arts 17: 380. 1882 (as brandegei) [E] [F]

Allium diehlii (M. E. Jones) M. E. Jones; *A. minimum* M. E. Jones; *A. tribracteatum* Torrey var. *diehlii* M. E. Jones

Bulbs 1–5, not clustered on stout, primary rhizome, ovoid to globose, 0.7–1.5 × 0.6–1.4 cm; outer coats enclosing 1 or more bulbs, brown to grayish brown, membranous, usually distinctly cellular-reticulate, cells isodiametric or ± narrowly hexagonal, transversely elongate, without fibers; inner coats red to purple or white, cells obscure, quadrate to ± transversely elongate. **Leaves** usually persistent, green at anthesis, 2, basally sheathing, sheaths not extending much above soil surface; blade solid, flat or channeled, ± falcate, 8–27 cm × 1–3 mm, margins entire. **Scape** persistent, solitary, erect, solid, terete, 3–10(–20) cm × 1–3 mm. **Umbel** persistent, erect, compact, 8–25-flowered, hemispheric, bulbils unknown; spathe bracts persistent, 2, 7–10-veined, ovate, ± equal, apex acuminate. **Flowers** campanulate, 5–8 mm; tepals erect, white with prominent green to purplish midvein, lanceolate to elliptic, ± equal, becoming involute at tip and rigid in fruit, margins entire, apex obtuse to acuminate; stamens included; anthers yellow; pollen yellow or white; ovary crestless or obscurely crested; processes 3, low, rounded, central, margins entire; style linear, equaling stamens; stigma capitate, scarcely thickened, unlobed; pedicel 5–15 mm. **Seed coat** shining; cells smooth. $2n = 14$.

Flowering late Apr–Jul. Sandy, rocky soil; 1200–3300 m; Colo., Idaho, Mont., Nev., Oreg., Utah, Wyo.

73. Allium columbianum (Ownbey & Mingrone) P. M. Peterson, Annable & Rieseberg, Syst. Bot. 13: 211. 1988 [E]

Allium douglasii Hooker var. *columbianum* Ownbey & Mingrone in C. L. Hitchcock et al., Vasc. Pl. Pacif. N.W. 1: 747. 1969

Bulbs 1–5+, not clustered on stout, primary rhizome, ovoid, 1–1.5 × 0.8–1.4 cm; outer bulb coats brownish or grayish, enclosing 1 or more bulbs, membranous, lacking cellular reticulation or cells arranged in only 2–3 rows distal to roots, ± quadrate, without fibers; inner coats white to pink, cells obscure, ± quadrate. **Leaves** usually persistent, green at anthesis, 2, basally sheathing, sheaths not extending much above soil surface; blade solid, flat, broadly channeled, ± falcate, 10–35 cm × (2–)5–8 mm, margins entire. **Scape** persistent, solitary, erect, solid, terete, enlarged proximal to inflorescence, (10–)20–30(–40) cm × 1.5–4 mm. **Umbel** persistent, erect, compact, 25–50-flowered, hemispheric to globose, bulbils unknown; spathe bracts persistent, 3, 5–7-veined, ovate, ± equal, apex acute. **Flowers** ± stellate, (6–)7–8(–10) mm; tepals spreading, light pink to purple with prominent green midribs, narrowly lanceolate, ± equal, becoming papery in fruit, margins entire, apex acuminate; stamens equaling or longer than tepals; anthers blue-gray; pollen light blue to gray; ovary crested; processes 6, 2 per lobe, low, rounded, margins entire; style exserted, linear; stigma capitate, scarcely thickened, unlobed; pedicel 10–20 mm. **Seed coat** shining; cells smooth. $2n = 14$.

Flowering May–Jul. Winter-wet, shallow soils on rock outcrops, or wet meadows; 300–1100 m; Idaho, Mont., Wash.

74. Allium constrictum (Ownbey & Mingrone) P. M. Peterson, Annable & Rieseberg, Syst. Bot. 13: 211. 1988 [C] [E]

Allium douglasii Hooker var. *constrictum* Ownbey & Mingrone in C. L. Hitchcock et al., Vasc. Pl. Pacif. N.W. 1: 747. 1969

Bulbs 1–5+, not clustered on stout, primary rhizome, ovoid, 1–1.5 × 0.8–1.4 cm; outer coats enclosing 1 or more bulbs, brownish, membranous, lacking cellular reticulation or cells arranged in only 2–3 rows distal to roots, ± quadrate, without fibers; inner coats white, cells obscure, quadrate. **Leaves** usually persistent, green at anthesis, 2, basally sheathing, sheaths not extending much above soil surface; blade solid, flat, broadly channeled, falcate, 10–35 cm × 1–3(–5) mm, margins entire. **Scape** persistent, solitary, erect, solid, terete, enlarged apically but distinctly constricted just proximal to inflorescence, 15–20 cm × 1.5–4 mm. **Umbel** persistent, erect, compact, 15–30-flowered, hemispheric to globose, bulbils unknown; spathe bracts persistent, 3, 5–7-veined, ovate, ± equal, apex acute. **Flowers** ± stellate, 7–8 mm; tepals spreading, light pink to rose with prominent green or reddish midribs, narrowly lanceolate, ± equal, becoming papery in fruit, margins entire, apex acuminate; stamens exserted; anthers blue-gray; pollen light blue to gray; ovary crested; processes 6, 2 per lobe, low, rounded, margins entire; style exserted, linear; stigma capitate, unlobed; pedicel 10–20 mm. **Seed coat** shining; cells smooth. $2n = 14$.

Flowering May–Jul. Shallow, lithosolic soils, often dry and sandy; of conservation concern; 300–500 m; Wash.

Allium constrictum is known only from Douglas, Grant, and Lincoln counties.

75. Allium douglasii Hooker, Fl. Bor.-Amer. 2: 184, plate 197. 1838 [E]

Allium hendersonii B. L. Robinson & Seaton

Bulbs 1–4, not clustered on stout, primary rhizomes, ovoid, 1.2–3 × 1–2 cm; outer coats enclosing 1 or more bulbs, light brown, membranous, lacking cellular reticulation, or cells arranged in only 2–3 rows distal to roots, ± quadrate, without fibers; inner coats white, sometimes pink, cells obscure, quadrate or linear. **Leaves** usually persistent, green at anthesis, 2, basally sheathing, sheaths not extending much above soil surface; blade solid, flat, falcate, 9–28 cm × (2–)5–15 mm, margins entire. **Scape** persistent, solitary, erect, solid, terete, not expanded proximal to inflorescence, (10–)20–30(–40) cm × 1–4 mm. **Umbel** persistent, erect, compact, 25–50-flowered, hemispheric to globose, bulbils unknown; spathe bracts persistent, 3, 4–6-veined, ovate, ± equal, apex acute. **Flowers** ± stellate, (6–)7–8(–10) mm; tepals spreading, light pink to purple with prominent green midribs, narrowly lanceolate, ± equal, becoming papery in fruit, margins entire, apex acuminate; stamens equaling tepals or exserted; anthers blue-gray; pollen white to light gray; ovary crested; processes 6, 2 per lobe, low, rounded, margins entire; style exserted, linear; stigma capitate, unlobed; pedicel 15–30 mm. **Seed coat** shining; cells smooth. $2n = 14$.

Flowering May–Jul. Winter-wet, shallow soils on rock outcrops; 400–1300 m; Oreg., Wash.

76. Allium nevii S. Watson, Proc. Amer. Acad. Arts 14: 231. 1879 [E] [F]

Allium douglasii Hooker var. *nevii* (S. Watson) Ownbey & Mingrone

Bulbs 1–5+, not clustered on stout, primary rhizome, ovoid, 1–2 × 1–2 cm; outer coats enclosing 1 or more bulbs, light brown, membranous, ± reticulate, cells quadrate to polygonal, vertically oblong, without fibers; inner coats white or pink, cells obscure, quadrate. **Leaves** usually persistent, withering from tip at anthesis, 2, basally sheathing, sheaths not extending much above soil surface; blade solid, flat or ± channeled, falcate, 12–25 cm × 2–3 mm, margins entire. **Scape** persistent, 1–3, erect, solid, terete or somewhat flattened, 2-edged, not expanded proximal to inflorescence, 15–30 cm × 1–3 mm. **Umbel** persistent, erect, compact, 10–30-flowered, hemispheric, bulbils unknown; spathe bracts persistent, 2, 5–6-veined, ovate, ± equal, apex acute. **Flowers** campanulate to substellate, 6–8 mm; tepals spreading, rose-colored, lanceolate, ± equal, becoming papery in fruit, margins entire, apex acuminate; stamens ± equaling tepals, or exserted; anthers blue-gray; pollen light blue or gray; ovary distinctly crested; processes 6, 2 per lobe, low, rounded, margins entire; style exserted, linear; stigma capitate, unlobed; pedicel 8–12 mm. **Seed coat** dull; cells smooth. $2n = 14$.

Flowering May–Jul. Heavy, rocky soils, wet meadows, or along streams; 30–1900 m; Oreg., Wash.

77. Allium macrum S. Watson, Proc. Amer. Acad. Arts 14: 233. 1879 [E]

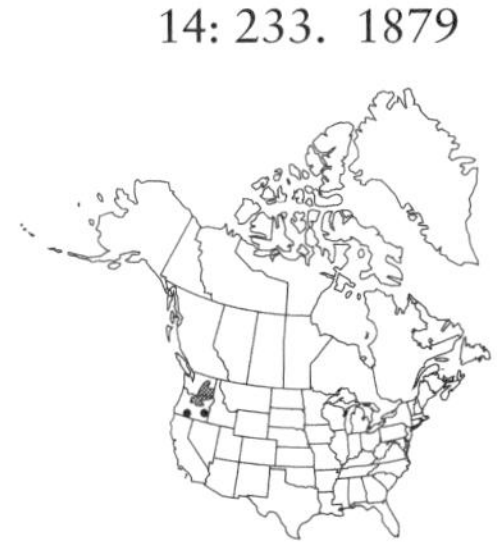

Allium equicaeleste H. St. John

Bulbs 1–3, not clustered on stout, primary rhizome, globose to ovate, 0.6–2 × 0.5–1.5 cm; outer coats enclosing 1 or more bulbs, brownish, membranous, lacking cellular reticulation or cells arranged in only 2–3 rows distal to roots, ± quadrate, without fibers; inner

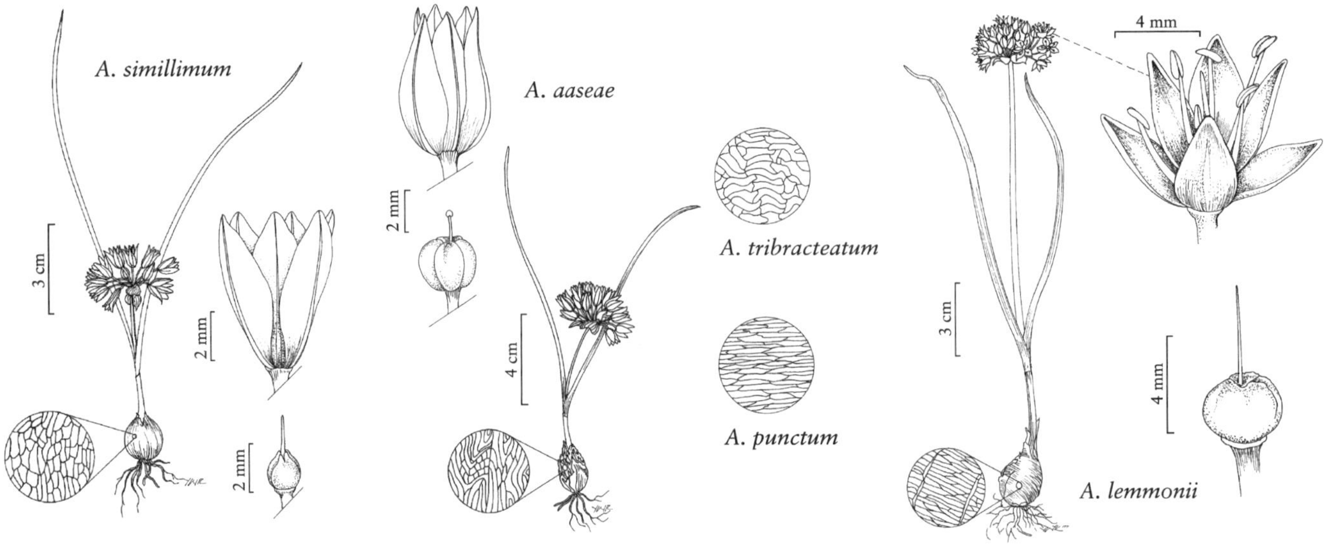

ALLIUM

coats white, lacking cellular reticulation. **Leaves** usually persistent, withering from tip at anthesis, 2, basally sheathing, sheaths not extending much above soil surface; blade solid, flat or channeled, ± falcate, 12–38 cm × 2–3 mm, margins entire. **Scape** persistent, solitary, erect, solid, ± terete, not expanded proximal to inflorescence, 9–32 cm × 0.5–4 mm. **Umbel** persistent, erect, loose, 10–20-flowered, globose, bulbils unknown; spathe bracts persistent, 2, 3–7-veined, lanceolate to lance-ovate, ± equal, apex abruptly acute. **Flowers** campanulate, 5–7 mm; tepals spreading, white or pinkish with prominent red midveins, narrowly lanceolate, ± equal, becoming papery in fruit, margins entire, apex acuminate; stamens equaling tepals or exserted; anthers yellow; pollen yellow; ovary crested; processes 3, thick, obtuse, margins entire; style linear, equaling stamens; stigma capitate, scarcely thickened, unlobed; pedicel 5–20 mm. **Seed coat** dull; cells ± smooth. $2n = 14$.

Flowering Apr–May. Barren, gravelly soils; 100–1400 m; Oreg., Wash.

78. Allium simillimum L. F. Henderson, Bull. Torrey Bot. Club 27: 355. 1900 [E] [F]

Bulbs usually solitary, not clustered on stout, primary rhizome, ovoid, 0.8–1.6 × 0.6–1.4 cm; outer coats enclosing single bulb, brownish, membranous, with ± obscure reticulations, cells isodiametric to transversely elongate, often contorted, without fibers; inner coats white to pink, cells obscure, quadrate. **Leaves** usually dedicuous with scape, green at anthesis, 2, basally sheathing, sheaths not extended much above soil surface; blade solid, flat to ± terete, linear, broadly channeled, 4–22 cm × 1–2 mm, margins entire. **Scape** usually forming abcission layer and deciduous with leaves after seeds mature, frequently breaking at this level after pressing, solitary, erect, solid, slightly flattened, very narrowly winged or not, 1–5 cm × 0.5–2 mm. **Umbel** persistent, erect, compact, 5–15-flowered, hemispheric, bulbils unknown; spathe bracts persistent, 2, 6–9-veined, ovate, ± equal, apex obtuse or acute. **Flowers** campanulate, 5–9 mm; tepals erect, white with green or reddish midribs or rarely pink, oblong to lanceolate, ± equal, becoming papery in fruit, margins obscurely to distinctly denticulate-serrulate, apex obtuse to acuminate; stamens included; anthers purple or mottled purple and white; pollen white or gray; ovary crested; processes 3, low, rounded, margins entire; style linear, equaling stamens; stigma capitate, scarcely thickened, unlobed; pedicel 2–5 mm. **Seed coat** dull or shining; cells smooth. $2n = 14$.

Flowering Apr–Jun. Coarse, sandy soils of granitic, calcareous, or basaltic origins; 1800–3400 m; Idaho, Mont.

79. Allium aaseae Ownbey, Res. Stud. State Coll. Wash. 18: 19, fig. 18. 1950 [E] [F]

Bulbs 2–10+, not clustered on stout, primary rhizome, ovoid, 1–2 × 0.8–1.5 cm; outer coats enclosing 1 or more bulbs, brownish, membranous, with or without obscure reticulations, cells isodiametric or transversely elongate, ± contorted, without fibers; inner coats white, pink, or red, cells obscure,

± quadrate. **Leaves** usually deciduous with scape, green at anthesis, 2, basally sheathing, sheaths not extending much above soil surface; blade solid, flat to ± terete, channeled, 7–25 cm × 1–4 mm, margins entire or minutely denticulate. **Scape** usually forming abcission layer and deciduous with leaves after seeds mature, frequently breaking at this level after pressing, solitary, erect, solid, slightly flattened, with or without narrow, sometimes crenulate-denticulate wings, 5–12 cm × 1–2 mm. **Umbel** persistent, erect, compact, 5–25-flowered, hemispheric, bulbils unknown; spathe bracts persistent, 2–3, 8–11-veined, ovate, ± equal, apex obtuse to acuminate. **Flowers** campanulate, 6–9 mm; tepals erect, bright pink fading to white with dark midribs, rarely white, oblong to lanceolate, ± equal, becoming papery in fruit, margins obscurely to distinctly serrulate-denticulate, apex obtuse to acuminate; stamens included; anthers yellow; pollen yellow; ovary crestless; style included, linear; stigma capitate, scarcely thickened, unlobed; pedicel 3–10 mm. **Seed coat** shining; cells smooth. $2n = 14$.

Flowering Apr–May. Coarse, sandy soil and gravelly river benches; 800–1100 m; Idaho.

Allium aaseae is known only from Ada and southern Gem counties and Rebecca Sand Hill, Washington County, and is considered of conservation concern by the Idaho Department of Fish and Game.

80. Allium tribracteatum Torrey, Pacif. Railr. Rep. 4(5): 148. 1857 C E F

Bulbs generally 2–5+, not clustered on stout primary rhizome, ovoid, 1–2 × 0.8–1.8 cm; outer coats enclosing 1 or more bulbs, brownish, membranous, prominently cellular-reticulate, cells irregularly arranged, ± transversely elongate, curved, without fibers; inner coats white, cells obscure, ± transversely elongate, irregularly arranged, curved. **Leaves** usually deciduous with scape, withering from tip at anthesis, 2, basally sheathing, sheaths not extending much above soil surface; blade solid, flat or broadly channeled, ± falcate, 12–20 cm × 1–3.5 mm, margins entire. **Scape** usually forming abcission layer and deciduous with leaves after seeds mature, frequently breaking at this level after pressing, solitary, erect, solid, terete, 2–7 cm × 0.5–1.5 mm. **Umbel** persistent, erect, compact, 10–30-flowered, hemispheric, bulbils unknown; spathe bracts persistent, (2–)3(–4), 4–5-veined, ovate, ± equal, apex abruptly acuminate to apiculate. **Flowers** campanulate, 6.5–8 mm; tepals erect, white to pink with dark pink or purplish midvein, lanceolate to ± elliptic, ± equal, becoming papery in fruit, margins entire, apex acute; stamens included; anthers purple; pollen gray; ovary crested; processes 3, acute, ± lateral, margins entire; style linear, shorter than stamens; stigma capitate, scarcely thickened, unlobed; pedicel 6–9 mm. **Seed coat** dull; cells ± smooth. $2n = 14$.

Flowering Mar.–May. Rocky, volcanic soils; of conservation concern; 1300–3000 m; Calif.

Allium tribracteatum is known only from the Sierra Nevada in Tuolumne County.

81. Allium obtusum Lemmon, Pittonia 2: 69. 1890 E

Bulbs 1–5, not clustered on stout primary rhizome, ovoid, 1–2 × 0.8–1.8 cm; outer coats enclosing 1 or more bulbs, brown, membranous, cellular-reticulate, cells quadrate, polygonal, or ± rectangular, ± transversely elongate, without fibers; inner coats white, reticulation absent or cells obscure, ± quadrate. **Leaves** usually deciduous with scape, withering from tip at anthesis, 1–2, basally sheathing, sheaths not extending much above soil surface; blade solid, flat or broadly channeled, ± falcate, 5–22 cm × 0.5–14 mm, margins entire. **Scape** usually forming abcission layer and deciduous with leaves after seeds mature, frequently breaking at this level after pressing, solitary, erect, solid, terete, 2–17 cm × 0.5–2 mm. **Umbel** persistent, erect, compact, 6–65-flowered, hemispheric, bulbils unknown; spathe bracts persistent, 2–3(–5), 5-8-veined, ovate, ± equal, apex abruptly acuminate to apiculate. **Flowers** campanulate, 4–12 mm; tepals erect, white or pink with dark purplish midveins, lanceolate to oblong-elliptic, ± equal, becoming papery in fruit, margins entire, apex obtuse to acute; stamens included; anthers yellow or purple; pollen yellow; ovary crested; processes 3, obscure to ± prominent, rounded, central, margins entire; style linear, equaling stamens or longer; stigma capitate, scarcely thickened, obscurely 3-lobed; pedicel 2–14(–20) mm. **Seed coat** dull; cells ± smooth.

Varieties 2 (2 in the flora): California, Nevada.

Allium obtusum is known only from the Sierra Nevada.

1. Leaf blade 0.5–4 mm wide; umbels 6–20-flowered; tepals white with prominent, dark midveins, oblong-elliptic, apex obtuse .... 81a. *Allium obtusum* var. *obtusum*
1. Leaf blade 2–14 mm wide; umbels 10–65-flowered; tepals pink with darker midveins, lanceolate, apex acute ..... 81b. *Allium obtusum* var. *conspicuum*

81a. Allium obtusum Lemmon var. **obtusum** [E]

Allium ambiguum M. E. Jones

Leaves 1–2; blade 5–18 cm × 0.5–4 mm. **Scape** 2–6.5 cm. **Umbel** 6–20-flowered. **Flowers** 4–7(–9) mm; tepals white with prominent, dark purplish midveins, oblong-elliptic, apex obtuse; pedicel 2–7(–11) mm. $2n = 14$.

Flowering May–Jun. Granitic sands; 1500–3500 m; Calif., Nev.

81b. Allium obtusum Lemmon var. **conspicuum** Mortola & McNeal, Aliso 11: 33, fig. 7. 1985 [E]

Leaves 1; blade 9–22 cm × 2–14 mm. **Scape** 2.5–17 cm. **Umbel** 10–65-flowered. **Flowers** 4–11.5 mm; tepals pink with dark purple or greenish purple midveins, lanceolate, apex acute; pedicel 3–14(–20) mm. $2n = 14$.

Flowering May–Jun. Granitic sands; 800–3000 m; Calif.

82. Allium anceps Kellogg, Proc. Calif. Acad. Sci. 2: 109. 1863 [E]

Bulbs 1–5, not clustered on stout primary rhizome, ovoid, 1.5–2 × 1.2–2 cm; outer coats enclosing 1 or more bulbs, brown to yellow-brown, membranous, ± prominently cellular-reticulate, cells in regular vertical rows, narrowly rectangular, transversely elongate, without fibers; inner coats white, cells obscure, transversely elongate. **Leaves** usually deciduous with scape, withering from tip at anthesis, 2, basally sheathing, sheaths not extending much above soil surface; blade solid, flat, falcate, 7–26 cm × 4–6 mm, margins entire. **Scape** usually forming abcission layer and deciduous with leaves after seeds mature, frequently breaking at this level after pressing, solitary, erect, solid, flattened, narrowly winged, 10–15 cm × 1–3 mm. **Umbel** persistent, erect, ± compact, 15–35-flowered, hemispheric, bulbils unknown; spathe bracts persistent, 2, 10–13-veined, ovate, ± equal, apex acute. **Flowers** stellate, 8–12 mm; tepals spreading, light pink with diffuse greenish midveins, linear-lanceolate, ± equal, papery, withering in fruit, margins entire, apex acute; stamens ± equaling tepals; anthers yellow; pollen yellow; ovary crested; processes 6, central, rounded, minute, margins entire; style linear, 0.5 times stamens; stigma capitate, scarcely thickened, unlobed; pedicel 15–30 mm. **Seed coats** dull; cells smooth. $2n = 14$.

Flowering Apr–May. Heavy, generally barren, clay soils; 1200–1600 m; Calif., Idaho, Nev., Oreg.

83. Allium punctum L. F. Henderson, Rhodora 32: 23. 1930 [E] [F]

Bulbs 1–3, not clustered on stout primary rhizome, ovoid to subglobose, 1–2 × 0.9–1.7 cm; outer coats enclosing 1 or more bulbs, brown to yellow-brown or gray, membranous, ± prominently cellular-reticulate, cells in ± regular vertical rows, narrowly rectangular, transversely elongate, without fibers; inner coats white, cells obscure, narrowly rectangular, transversely elongate, or quadrate. **Leaves** usually deciduous with scape, green or withering only at tip at anthesis, 2, basally sheathing, sheaths not extending much above soil surface; blade solid, flat, falcate, 9–18 cm × 2–3 mm, margins entire. **Scape** usually forming abcission layer and deciduous with leaves after seeds mature, frequently breaking at this level after pressing, solitary, erect, solid, ± flattened, not winged, 3–10 cm × 1–2 mm. **Umbel** persistent, erect, ± compact, 6–20-flowered, hemispheric, bulbils unknown; spathe bracts persistent, 2, 6–9-veined, lanceolate to ovate, ± equal, apex acute to abruptly acuminate. **Flowers** narrowly campanulate, 6–13 mm; tepals erect, white to pinkish with narrow to very broad, purple midvein, oblong-lanceolate, ± equal, becoming papery, investing capsule in fruit, margins entire, apex obtuse; stamens included; anthers light yellow; pollen yellow; ovary crested; processes 3, central, low, rounded, broad, margins entire; style linear, equaling stamens; stigma capitate, scarcely thickened, unlobed; pedicel 5–11 mm. **Seed coat** dull; cells smooth. $2n = 14$.

Flowering May–Jun. Sandy, rocky, or clay soils on open slopes and flats; 1300–1600 m; Calif., Nev., Oreg.

84. Allium lemmonii S. Watson, Proc. Amer. Acad. Arts 14: 234. 1879 [E] [F]

Allium incisum A. Nelson & J. F. Macbride; *A. scissum* A. Nelson & J. F. Macbride

Bulbs 1–5+, not clustered on stout primary rhizome, ovoid, 1.5–2.2 × 1–2 cm; outer coats enclosing 1 or more bulbs, brown, membranous, ± prominently cellular-reticulate, cells in ± regular vertical rows, narrowly rectangular, transversely elongate, without fibers; inner coats white to light brown, cells ± narrowly rectangular, transversely elongate, or quadrate. **Leaves** usually deciduous with scape, green or withering only at tip at anthesis, 2, basally sheathing, sheaths not extending much above soil surface; blade solid, flat, falcate, 8–30 cm × 3–5 mm, margins entire. **Scape** usually forming abcission layer and deciduous with leaves after seeds mature, frequently breaking at this level after pressing,

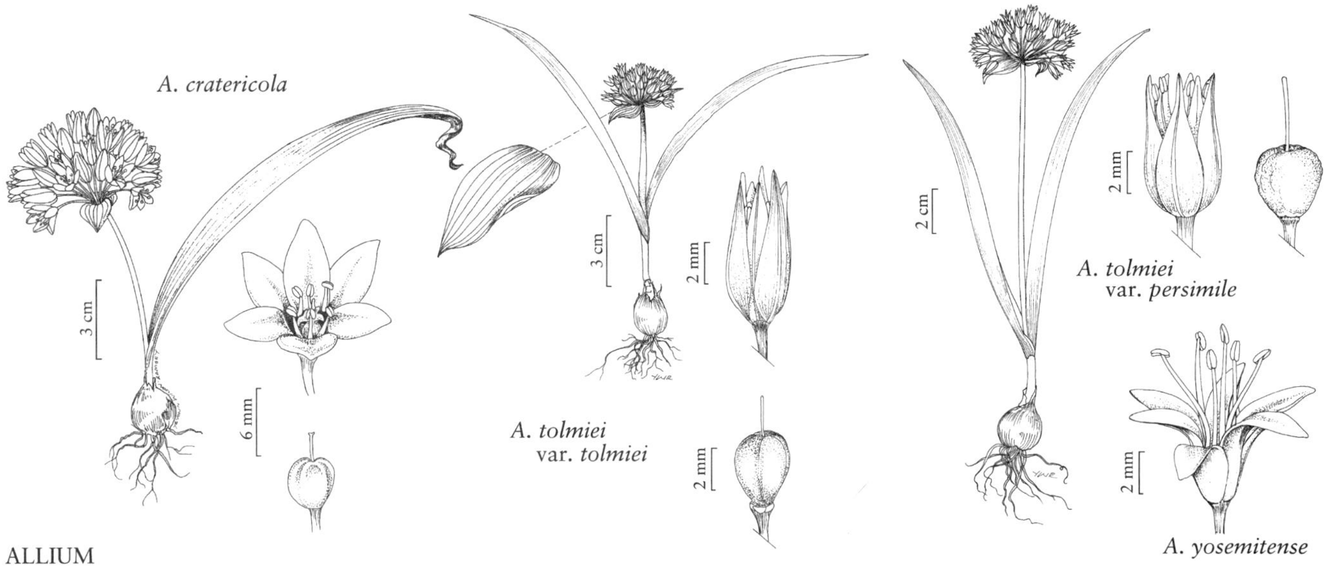

ALLIUM

solitary, erect, solid, flattened, narrowly winged, 15–20 cm × 1–4 mm. **Umbel** persistent, erect, compact to ± loose, 10–40-flowered, hemispheric, bulbils unknown; spathe bracts persistent, 2–3, 8–10-veined, broadly lanceolate to ovate, ± equal, apex long-acuminate. **Flowers** campanulate, 6–9 mm; tepals erect, pink to whitish, lance-ovate, ± equal, becoming ± rigid in fruit, margins entire, apex acute to acuminate; stamens ± equaling tepals; anthers light purple to yellow; pollen yellow; ovary obscurely crested; processes 6, low, central, 2 per lobe, margins entire; style linear, equaling stamens; stigma capitate, scarcely thickened, unlobed; pedicel 8–16 mm. **Seed coat** dull; cells smooth. $2n = 14$.

Flowering May–Jun. Drying, clay soils; 1200–1900 m; Calif., Idaho, Nev., Oreg., Utah.

85. Allium parvum Kellogg, Proc. Calif. Acad. Sci. 3: 54, fig. 13. 1863 [E]

Allium modocense Jepson; *A. pleianthum* S. Watson var. *particolor* M. E. Jones; *A. tribracteatum* Torrey var. *andersonii* S. Watson; *A. tribracteatum* var. *parvum* (Kellogg) Jepson

Bulbs 1–5+, not clustered on stout primary rhizome, rhizomes absent, renewal bulbs formed within coats of parent bulb, ovoid to ± globose, 1–2 × 0.7–2 cm; outer coats enclosing 1 or more bulbs, gray to grayish brown, membranous, lacking cellular reticulation or cells arranged in only 2–3 rows distal to roots, ± quadrate, without fibers; inner coats white or pinkish, cells obscure, ± quadrate. **Leaves** usually deciduous with scape, withering from tip at anthesis, 2, basally sheathing, sheaths not extending much above soil surface; blade solid, flat, ± falcate, 8–15 cm × 2–8 mm, margins entire. **Scape** usually forming abcission layer and deciduous with leaves after seeds mature, frequently breaking at this level after pressing, solitary, erect, solid, flattened, frequently ± winged distally, or, in smaller specimens, ± terete, 3–12 cm × 0.5–2 mm. **Umbel** persistent, erect, compact, 5–30-flowered, hemispheric to conic, bulbils unknown; spathe bracts persistent, 2, 12–14-veined, ovate, ± equal, apex acute to acuminate. **Flowers** narrowly campanulate, 6–9 mm; tepals erect, ± spreading toward tips, white or tinged with pink, with broad, dark, reddish purple or brown midvein, oblong to elliptic, unequal, outer longer, wider than inner, becoming papery and investing capsule, not carinate in fruit, margins entire, apex obtuse to ± acute; stamens included; anthers purple or yellow; pollen yellow; ovary obscurely crested; processes 3, central, low, rounded, margins entire; style linear, ± equaling tepals; stigma capitate, scarcely thickened, unlobed or obscurely 3-lobed; pedicel 3–12 mm. **Seed coat** dull; cells smooth. $2n = 14$.

Flowering late Apr–Jun. Rocky, clay slopes and talus; 1200–2800 m; Calif., Idaho, Nev., Oreg., Utah.

86. Allium cratericola Eastwood, Leafl. W. Bot. 1: 132. 1934 [E] [F]

Allium parvum Kellogg var. *brucae* M. E. Jones; *A. parvum* var. *jacintense* Munz

Bulbs 1–3, not clustered on stout primary rhizome, ovoid, 1.5–2.5 × 1–2 cm; outer coats enclosing 1 or more bulbs, brown or gray, membranous, lacking cellular reticulation or cells arranged in only 2–3

rows distal to roots, ± quadrate, without fibers; inner coats white, cells obscure, ± quadrate, or not visible. **Leaves** usually deciduous with scape, withering from tip at anthesis, 1–2, basally sheathing, sheaths not extending much above soil surface; blade solid, straight or weakly falcate, flat or broadly channeled, 10–30 cm × 1–21 mm, margins entire. **Scape** usually forming abcission layer and deciduous with leaves after seeds mature, frequently breaking at this level after pressing, solitary, erect, solid, terete, 2–12 cm × 1–3 mm. **Umbel** persistent, erect, compact, 20–30-flowered, hemispheric, bulbils unknown; spathe bracts persistent, 2–4(–6), 10–16-veined, ovate, ± equal, apex acuminate. **Flowers** campanulate, 7–14 mm; tepals erect, white or pink to purplish with dark greenish brown or purple midveins, lance-oblong, elliptic, or ± oblanceolate, ± equal, becoming papery and investing fruit, margins entire, apex obtuse; stamens included; anthers yellow; pollen yellow; ovary crested; processes 3, central, rounded, minute, margins entire; style linear, equaling stamens; stigma capitate, scarcely thickened, unlobed; pedicel 5–18 mm. **Seed coat** dull; cells ± smooth. $2n$ = 14, 28.

Flowering Mar–Jun. Serpentine, volcanic, and granitic soil; 300–1800 m; Calif.

Populations of *Allium cratericola* from southern California are 2-leaved, while those from the north are either 1- or 2-leaved or sometimes a mixture of both forms.

87. Allium tolmiei Baker, Bot. Mag. 32: under plate 6227. 1876 [E] [F]

Bulbs 1–10+, not clustered on stout primary rhizome, not forming rhizomes, renewal bulbs formed within coats of parent bulb, or borne terminally on rhizomes outside coats of parent bulb, parent bulb disappearing by anthesis except for still-functional roots and shriveled bulb coat, ovoid or oblique, 0.6–2 × 1–2.5 cm; outer coats enclosing 1 or more bulbs, brown, membranous, lacking cellular reticulation or cells arranged in only 2–3 rows distal to roots, ± quadrate, without fibers; inner coats white to pink, cells obscure, quadrate, or not visible. **Leaves** forming abcission layer at soil surface and deciduous or persistent with scape after seeds mature, also frequently breaking at soil surface after pressing, withering at tip at anthesis, 2, basally sheathing, sheaths not extending much above soil surface; blade solid, flat, ± falcate, 10–40 cm × 1–10 mm, margins entire. **Scape** deciduous or persistent, solitary, erect, solid, compressed to strongly flattened and winged or not, 5–40 cm × 1–5 mm. **Umbel** persistent, erect, compact to ± loose, 10–50-flowered, globose to hemispheric, bulbils unknown; spathe bracts persistent, 2–3, 8–17-veined, ovate to lanceolate, ± equal, apex acuminate. **Flowers** campanulate to ± stellate, 6–12 mm; tepals ± erect, almost white to pink, with prominent, darker midribs, lanceolate, ± equal, becoming rigid, spreading, carinate in fruit, margins entire, apex acute to acuminate, involute at tip; stamens included to ± equaling tepals or exserted; anthers purple to yellow; pollen yellow; ovary crestless or obscurely to prominently crested; processes 3 or 6, central, low, rounded to ± prominent, triangular, margins entire; style linear, ± equaling stamens; stigma capitate, unlobed or obscurely lobed; pedicel 5–25 mm. **Seed coat** dull; cells smooth.

Varieties 2 (2 in the flora): w United States.

1. Bulbs ovoid, not rhizomatous to oblique, short-rhizomatous; scape 5–28 cm; leaf blade 1.25–2 times scape; stamen $^{1}/_{2}$–$^{3}/_{4}$ tepals; widespread . . . . 87a. *Allium tolmiei* var. *tolmiei*
1. Bulbs ovoid, never rhizomatous; scape 10–40 cm; leaf blade shorter than to ± equaling scape; stamen ± equaling tepals or exserted; known only from s end of Seven Devils Mountains, w Idaho . . . . 87b. *Allium tolmiei* var. *persimile*

87a. Allium tolmiei Baker var. **tolmiei** [E] [F]

Allium anceps Kellogg var. *aberrans* M. E. Jones; *A. cusickii* S. Watson; *A. douglasii* var. *tolmiei* (Baker) Traub; *A. idahoense* Traub; *A. platyphyllum* Tidestrom; *A. pleianthum* S. Watson; *A. tolmiei* var. *platyphyllum* (Tidestrom) Ownbey

Bulbs ovoid, not rhizomatous or oblique, forming 1–2 short rhizomes terminated by new bulbs. **Leaves** forming abcission layer at soil level, deciduous with scape after seeds mature or frequently breaking at that level after pressing; blade 10–35 cm. **Scape** deciduous, strongly flattened, winged, 5–28 cm. **Umbel** 10–40-flowered; spathe bracts 2. **Flowers** 6–12 mm; tepals erect to ± spreading, apex acute to acuminate; stamens included, $^{1}/_{2}$–$^{3}/_{4}$ tepals; ovary crestless or obscurely to prominently crested; processes 3 or 6; pedicel 5–25 mm. $2n$ = 14, 28.

Flowering Apr–Jul. Rocky, gravelly, or clay soils, mountains and scablands; 400–2800 m; Calif., Idaho, Oreg., Wash.

87b. Allium tolmiei Baker var. **persimile** Ownbey, Res. Stud. State Coll. Wash. 18: 29, fig. 13. 1950 [E] [F]

Bulbs never rhizomatous, ovoid. **Leaves** usually persistent; blade 10–40 cm. **Scape** usually persistent, not strongly flattened or winged, 10–40 cm. **Umbel** 10–50-flowered; spathe bracts 2–3. **Flowers** 8–10 mm; tepals erect, apex acute; stamens ± equaling tepals or exserted; ovary crestless; pedicel 8–20 mm. $2n$ = 42.

Flowering Apr–Jul. Rocky, gravelly, or clay soils; 700–800 m; Idaho.

Allium tolmiei var. *persimile* is known only from the southern end of the Seven Devils Mountains.

88. Allium yosemitense Eastwood, Leafl. W. Bot. 1: 132. 1934 [C] [E] [F]

Bulbs 1–12+, not basally clustered on stout primary rhizome, ovoid, 2–3 × 1.5–2 cm; outer coats enclosing 1 or more bulbs, brown, membranous, lacking cellular reticulation or cells arranged in only 2–3 rows distal to roots, ± quadrate, without fibers; inner coats white, cells very obscurely quadrate or not visible. **Leaves** usually deciduous with scape, withering from tip at anthesis, 2, basally sheathing, sheaths not extending much above soil surface; blade solid, flat or very broadly channeled, ± falcate, 15–40 cm × 2–18 mm, margins entire. **Scape** usually forming abcission layer and deciduous with leaves after seeds mature, frequently breaking at this level after pressing, solitary, erect, solid, terete, 6–23 cm × 1–3 mm. **Umbel** persistent, erect, compact, 20–100-flowered, globose to hemispheric, bulbils unknown; spathe bracts persistent, 2–4, 7–9-veined, ovate, ± equal, apex acuminate. **Flowers** campanulate, 7–15 mm; tepals erect, white to pink with darker midveins, linear-oblong, ± equal, becoming membranous in fruit, margins entire, apex acute; stamens ± equaling tepals; anthers yellow or purple; pollen yellow; ovary crested; processes 3, 2-lobed, minute, margins entire; style short-exserted, linear; stigma capitate, scarcely thickened, unlobed; pedicel 7–34 mm. **Seed coat** dull; cells ± smooth. $2n = 14$.

Flowering May–Jun. Moist soil along cracks and margins of large metamorphic outcrops; of conservation concern; 800–2200 m; Calif.

Allium yosemitense is known only from the central Sierra Nevada.

89. Allium hoffmanii Ownbey ex Traub, Pl. Life 28: 63. 1972 [E]

Bulbs usually solitary, not clustered on stout primary rhizomes, ovoid, 1.5–2.5 × 1–1.5 cm; outer coats enclosing single bulb, brown or gray, membranous, lacking cellular reticulation or cells arranged in only 2–3 rows distal to roots, ± quadrate, without fibers; inner coats white or pink, cells not visible. **Leaves** usually deciduous with scape, withering from tip at anthesis, 1, basally sheathing, sheath not extending much above soil surface; blade solid, flat or broadly channeled, ± falcate, 10–22 cm × 4–8 mm, margins entire. **Scape** usually forming abcission layer and deciduous with leaves after seeds mature, frequently breaking at this level after pressing, solitary, erect, solid, terete, 5–10 cm × 0.5–2 mm. **Umbel** persistent, erect, compact, 10–40-flowered, globose to conic, bulbils unknown; spathe bracts persistent, 3–4, 3-veined, ovate, ± equal, apex acuminate. **Flowers** conical to campanulate, 8–10 mm; tepals strictly erect, pink to purplish with prominent greenish midvein, linear-lanceolate, ± equal, becoming membranous, withering in fruit, margins entire, apex acuminate; stamens exserted; filaments papillose-glandular proximally; anthers purple; pollen yellow or gray; ovary crested; processes 6, obscure, central, low, rounded, margins entire; style linear, equaling stamens; stigma capitate, scarcely thickened, obscurely 3-lobed; pedicel 8–15 mm. **Seed coat** dull; cells smooth. $2n = 14$.

Flowering Jun–Jul. Serpentine clay; 1100–1800 m; Calif.

90. Allium burlewii Davidson, Bull. S. Calif. Acad. Sci. 15: 17. 1916 [E] [F]

Bulbs usually solitary, not clustered on stout, primary rhizome, ovoid, 1.5–2.5 × 1–1.5 cm; outer coats enclosing single bulb, brown or grayish, membranous, lacking cellular reticulation or cells arranged in only 2–3 rows distal to roots, ± quadrate, without fibers; inner coats white, cells not visible. **Leaves** usually deciduous with scape, withering from tip at anthesis, 1, basally sheathing, sheath not extending much above soil surface; blade solid, flat or broadly channeled, ± falcate, 8–33 cm × 1–10 mm, margins entire. **Scape** usually forming abcission layer and deciduous with leaves after seeds mature, frequently breaking at this level after pressing, solitary, erect, solid, terete, 2–20 cm × 1–3 mm. **Umbel** persistent, erect, compact, 8–20(–30)-flowered, globose to conic, bulbils unknown; spathe bracts persistent, 3, 9–13-veined, broadly ovate, ± equal, apex abruptly acuminate. **Flowers** conical to campanulate, 7–10 mm; tepals strictly erect, dull purplish with darker midveins, ovate, ± equal, becoming membranous, withering in fruit, margins entire, apex obtuse; stamens equaling tepals or ± exserted; filaments smooth proximally; anthers purple; pollen yellow or gray; ovary crestless or crested; processes 6, central, rounded, minute, margins entire; style linear, slightly longer than stamens; stigma capitate, scarcely thickened, unlobed; pedicel 6–10 mm. **Seed coat** shining; cells smooth. $2n = 14$.

Flowering late Apr–Jul. Granite sands, on dry slopes and ridges; 1800–2800 m; Calif.

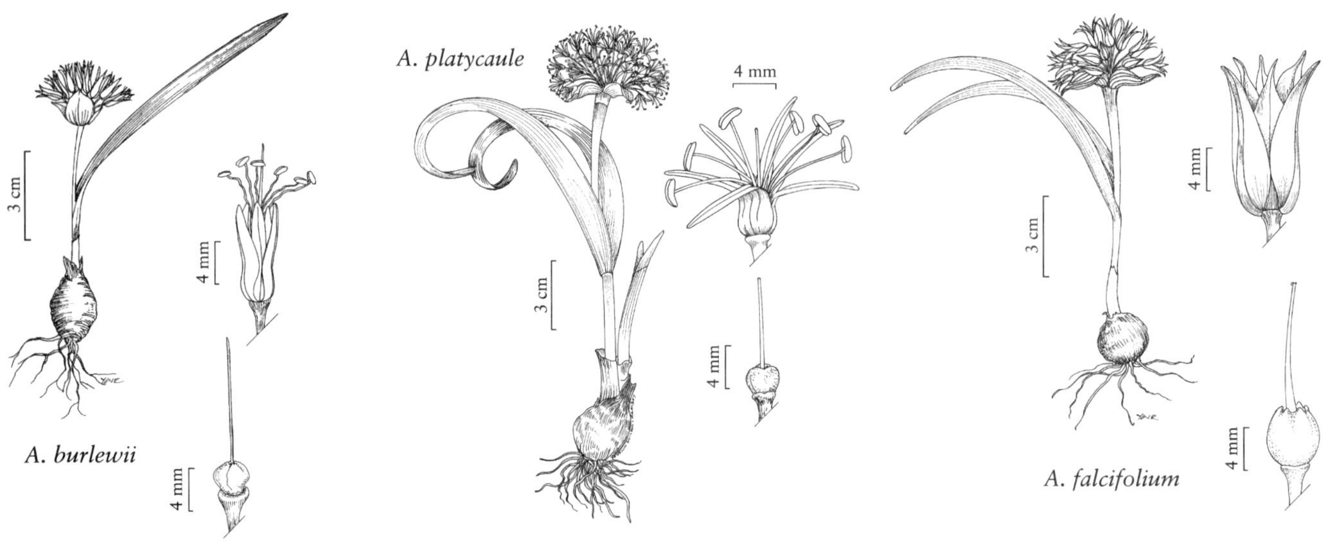

ALLIUM

91. Allium crenulatum Wiegand, Bull. Torrey Bot. Club 26: 135, plate 355, fig. 1. 1899 [E]

Allium cascadense M. Peck; *A. vancouverense* J. Macoun; *A. watsonii* Howell

Bulbs 1–6+, replaced annually with new bulbs borne terminally on rhizomes; rhizomes 1–2, secondary, short; parent bulb disappearing by anthesis except for still-functional roots and shriveled bulb coat, oblique-ovoid, 1–2 × 0.8–1.8 cm; outer coats not or only partially enclosing bulbs, brown to gray-brown, membranous, lacking cellular reticulation or cells arranged in only 2–3 rows distal to roots, ± quadrate, without fibers; inner coats white, cells very obscurely quadrate or not visible. **Leaves** usually deciduous with scape, withering from tip at anthesis, (1–)2, basally sheathing, sheaths not extending much above soil surface; blade solid, flat, falcate, 10–33 cm × 1.5–10 mm, margins sometimes minutely denticulate. **Scape** usually forming abcission layer and deciduous with leaves after seeds mature, frequently breaking at this level after pressing, solitary, ± erect, solid, flattened, winged distally, wings frequently crenulate proximal to umbel, 5–15 cm × 1–5 mm. **Umbel** persistent, erect, compact, 10–25-flowered, conic to hemispheric, bulbils unknown; spathe bracts persistent, 2, 8–10-veined, lanceolate, ± equal, apex acute. **Flowers** campanulate, 6–12 mm; tepals erect, pinkish with deeper pink midveins, lanceolate, ± equal, becoming papery in fruit, margins entire, apex acute; stamens included; anthers yellow or purple; pollen yellow; ovary obscurely crested; processes 3, central, 2-lobed, minute, margins entire; style linear, equaling stamens; stigma capitate, scarcely thickened, unlobed; pedicel 6–16 mm. **Seed coat** shining; cells smooth. $2n = 14$.

Flowering late May–Jul. Talus slopes and clay soils, including serpentine, on bald summits and ridges; 600–2500 m; B.C.; Oreg., Wash.

Allium crenulatum is known only from west of the Cascade Mountains from Vancouver Island to southwestern Oregon, in Jefferson Park, Oregon, and in the Wenatchee Mountains, central Washington.

The disjunct populations of *Allium crenulatum* in western Oregon are markedly different among themselves and from the more typical representatives to the north. It has thus far proven impossible to draw meaningful taxonomic distinctions among these populations, hence we have followed historical precedent and have placed them all in a single, highly variable species.

92. Allium platycaule S. Watson, Proc. Amer. Acad. Arts 14: 234. 1879 [E] [F]

Bulbs 1–5+, not clustered on stout primary rhizomes, renewal bulbs formed within coats of parent bulb, ovoid, 2–3 × 1.4–2.5 cm, increase bulbs ± equaling parent bulbs, rhizomes absent; outer coats enclosing 1 or more bulbs, gray or brown, membranous, lacking cellular reticulation or cells arranged in only 2–3 rows distal to roots, ± quadrate, without fibers; inner coats white, cells obscure, quadrate to rectangular, or not visible. **Leaves** usually deciduous with scape, withering from tip at anthesis, 2, basally sheathing, sheaths not extending much above soil surface; blade solid, flat, falcate, 10–30-cm × 8–18 mm, margins entire. **Scape** usually forming abcission layer and deciduous with leaves after seeds mature, frequently breaking at this level after pressing, solitary, erect, solid, strongly flattened, winged distally, 7–25 cm × 2–7 mm. **Umbel** persistent, erect, loose, 30–90-flowered, globose, bulbils unknown; spathe bracts persistent, 2–5, 10–15-veined, lanceolate to broadly ovate, ± equal, apex acuminate. **Flowers** stellate, 8–15 mm; tepals spreading, bright pink to rose, narrowly lanceolate, ± equal, becoming erect, rigid, constricted distal to ovary in fruit, margins entire, apex long-acuminate; stamens exserted; anthers yellow; pollen gray; ovary crestless; style linear, equaling stamens; stigma capitate, scarcely thickened, unlobed, apex acute; pedicel 10–25 mm. **Seed coat** dull or shining; cells smooth. $2n = 14$.

Flowering May–Jun. Rocky, sandy slopes; 1500–2500 m; Calif., Nev., Oreg.

93. Allium falcifolium Hooker & Arnott, Bot. Beechey Voy., 400. 1841 [E] [F]

Allium breweri S. Watson; *A. falcifolium* var. *breweri* (S. Watson) M. E. Jones

Bulbs 1–5+, not clustered on stout primary rhizomes, rhizomes absent, increase bulbs absent or ± equaling parent bulbs, never appearing as basal cluster, ovoid, 1.5–2.5 × 1–1.8 cm; outer coats enclosing renewal and increase bulbs, brown to reddish brown, membranous, lacking cellular reticulation or cells arranged in only 2–3 rows distal to roots, ± quadrate, without fibers; inner coats white or pink, cells obscure, ± quadrate. **Leaves** usually deciduous with scape, withering from tip at anthesis, 2, basally sheathing, sheaths not extending much above soil surface; blade solid, flat, falcate, 8–21 cm × 2–8 mm, margins entire. **Scape** usually forming abcission layer and deciduous with leaves after seeds mature, frequently breaking at this level after pressing, solitary, erect, solid, strongly flattened, winged distally, 5–25 cm × 1–4 mm. **Umbel** persistent, erect, compact to ± loose, 10–30-flowered, hemispheric, bulbils unknown; spathe bracts persistent, 2, 6–9-veined, lanceolate, ± equal, apex acuminate. **Flowers** conic to campanulate, 9–15 mm; tepals erect, reddish purple or dingy white, lanceolate, ± equal, rigid and membranous in fruit, at least inner margins denticulate with minute glands, apex long-acuminate; stamens included; anthers purple or yellow; pollen yellow or white; ovary crested; processes 3, central, low, rounded, broad, margins entire; style linear, equaling stamens; stigma capitate, unlobed; pedicel 8–15 mm. **Seed coat** dull; cells ± smooth. $2n = 14$.

Flowering Apr–Jun. Heavy, rocky, clay soils, including serpentine; 100–2100 m; Calif., Oreg.

94. Allium siskiyouense Ownbey ex Traub, Pl. Life 28: 63. 1972 [E]

Allium falcifolium Hooker & Arnott var. *demissum* Jepson

Bulbs 1–5, not clustered on stout primary rhizome, increase bulbs absent or ± equaling parent bulbs, never appearing as basal cluster, rhizomes absent, ovoid, 0.8–2 × 0.8–2 cm; outer coats enclosing renewal bulbs, grayish to brown, membranous, lacking cellular reticulation or cells arranged in only 2–3 rows distal to roots, ± quadrate, without fibers; inner coats pink to red, cells obscure, ± quadrate, or not visible. **Leaves** usually deciduous with scape, withering from tip at anthesis, 2, basally sheathing, sheaths not extending much above soil surface; blade solid, flat, falcate, 8–18 cm × 3–5 mm, margins entire. **Scape** usually forming abcission layer and deciduous with leaves after seeds mature, frequently breaking at this level after pressing, solitary, erect, solid, flattened, winged distally, wing margins commonly minutely serrulate proximal to umbel, 3–8 cm × 1–4 mm. **Umbel** persistent, erect, compact, 10–35-flowered, hemispheric, bulbils unknown; spathe bracts persistent, 2, 8–10-veined, ovate, ± equal, apex acuminate. **Flowers** conic to campanulate, 8–11 mm; tepals erect, pink, usually with darker pink midveins, lanceolate to elliptic, ± equal, becoming membranous and rigid in fruit, outer margins entire, inner margins usually minutely denticulate distally, apex obtuse, margins becoming ± involute, appearing acuminate in age; stamens included; anthers yellow to light purple; pollen yellow; ovary crested; processes 3, central, 2-lobed, minute, margins entire; style linear, equaling stamens; stigma capitate, scarcely thickened, unlobed; pedicel 5–16 mm. **Seed coat** shining; cells smooth. $2n = 14$.

Flowering late Apr–Jun. Heavy, rocky, clay soils, including serpentine; 900–2500 m; Calif., Oreg.

Allium siskiyouense is known only from the Siskiyou Mountains.

95. Allium scilloides Douglas ex S. Watson, Proc. Amer. Acad. Arts 14: 229. 1879 [C] [E]

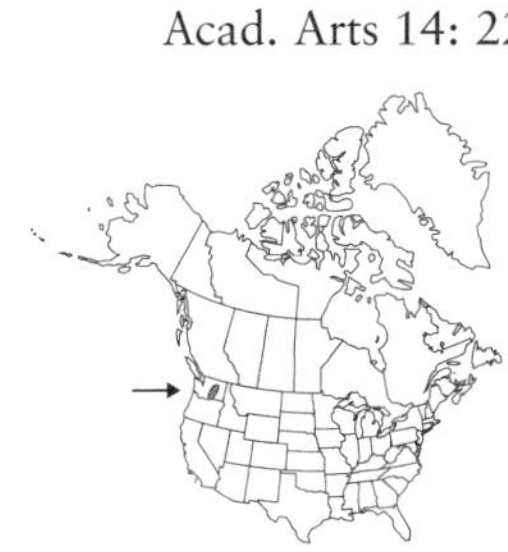

Bulbs 1–5, not clustered on stout, primary rhizome, increase bulbs ± equaling parent bulbs, never appearing as basal cluster, rhizomes absent, globose to ovoid, 1–2 cm × 8–20 mm; outer coats enclosing renewal bulbs, reddish or brownish, membranous, lacking cellular reticulation or cells arranged in only 2–3 rows distal to roots, ± quadrate, without fibers; inner coats white or light brown, cells absent or obscure and ± quadrate. **Leaves** deciduous with scape after seeds mature, 2, basally sheathing, sheaths not extending much above soil surface; blade solid, flat, falcate, 6–15 cm × 2–4 mm, margins entire or obscurely papillose. **Scape** usually forming abcission layer and deciduous with leaves after seeds mature, frequently breaking at this level after pressing, solitary, erect, solid, strongly flattened, 2-edged or winged distally, wings obscurely papillose, 4–8 cm × 1–3 mm. **Umbel** decisuous with scape, erect, compact, 5–12-flowered, hemispheric, bulbils unknown; spathe bracts persistent, 2–3, 3-veined, broadly ovate, ± equal, apex acute. **Flowers** campanulate, 6–8 mm; tepals erect, white or pinkish with green midribs, becoming reddish purple, broadly elliptic-oblong, ± equal, becoming rigid, papery and ± connivent over fruit, margins entire, apex obtuse, not involute at tip; stamens included; anthers purple; pollen white to gray; ovary crestless or obscurely crested; processes 3, central, rounded, margins entire; style linear, equaling stamens; stigma capitate, unlobed or minutely 3-lobed; pedicel 2–10.5(–18.5 in fruit) mm. **Seed coat** dull; cells smooth. $2n = 14$.

Flowering Mar–Apr. Barren, gravelly soils, arid interior slopes and ridges well back from Columbia River; of conservation concern; 300–1300 m; Wash.

Allium scilloides is known only from east of the Cascades.

96. Allium robinsonii L. F. Henderson, Rhodora 32: 22. 1930 [E]

Bulbs 1–3+, not clustered on stout primary rhizome, rhizomes absent, renewal bulbs formed within coats of parent bulb, increase bulbs absent or 1–2, ± equaling parent bulbs, never appearing as basal cluster, ovoid, 1–2 cm × 8–15 mm; outer coats grayish to brownish, membranous, lacking cellular reticulation or cells arranged in only 2–3 rows distal to roots, ± quadrate, without fibers; inner coats white, cells absent or ± quadrate. **Leaves** usually deciduous with scape, 2, basally sheathing, sheaths not extending much above soil surface; blade solid, flat, strongly falcate, 5–18 cm × 1–2 mm, margins entire. **Scape** usually forming abcission layer and deciduous with leaves after seeds mature, frequently breaking at this level after pressing, solitary, erect, solid, flattened, slightly 2-edged, 3–7 cm × 1–3 mm. **Umbel** deciduous with scape, erect, compact, 5–12-flowered, hemispheric, bulbils unknown; spathe bracts persistent, 2, 5–8-veined, ovate, ± equal, apex abruptly acute. **Flowers** campanulate, 7–9 mm; tepals erect, white to pale pink with red midribs, oblong, ± equal, becoming papery and not carinate in fruit, margins entire, apex obtuse; stamens included; anthers purple; pollen gray or yellow; ovary ± distinctly crested; processes 3 or 6, central, low, rounded, margins entire; style included, linear; stigma capitate, unlobed; pedicel 3–12 mm. **Seed coat** dull; cells smooth. $2n = 14$.

Flowering Apr–May. Sand and gravel deposits along bottom and lower benches of Columbia River; 50–200 m; Oreg., Wash.

Allium robinsonii has been found along the Columbia River from Ferry County, northeastern Washington, to about the mouth of the John Day River, north-central Oregon, and is now possibly extirpated from Oregon.

45. NOTHOSCORDUM Kunth, Enum. Pl. 4: 457. 1843 • False onion, false garlic [Greek *nothos*, false, and *scordon*, garlic]

T. D. Jacobsen

Dale W. McNeal Jr.

Herbs, perennial, scapose, from bulbs. **Bulbs** in general appearance very similar to some species of *Allium*, without alliaceous odor; outer coats membranous. **Leaves** usually present at flowering time, basal, imbricate, sheathing basally; blade filiform to linear. **Inflorescences** umbellate, subtended by spathe bracts; bracts 2, membranous. **Flowers** fastigiate, withering-persistent; tepals 6, 2-whorled, connate proximal $^{1}/_{3}$, 1-veined, subequal; stamens 6, adnate to tepal bases, included; filaments distinct, ± dilated basally, subulate and entire apically; anthers dorsifixed, oblong, introse; ovary superior, sessile, 3-locular; ovules several(–12); style filiform; stigma terminal, small. **Fruits** capsular, 3-lobed, membranous, dehiscence loculicidal. **Seeds** black, angled [compressed or almost flat].

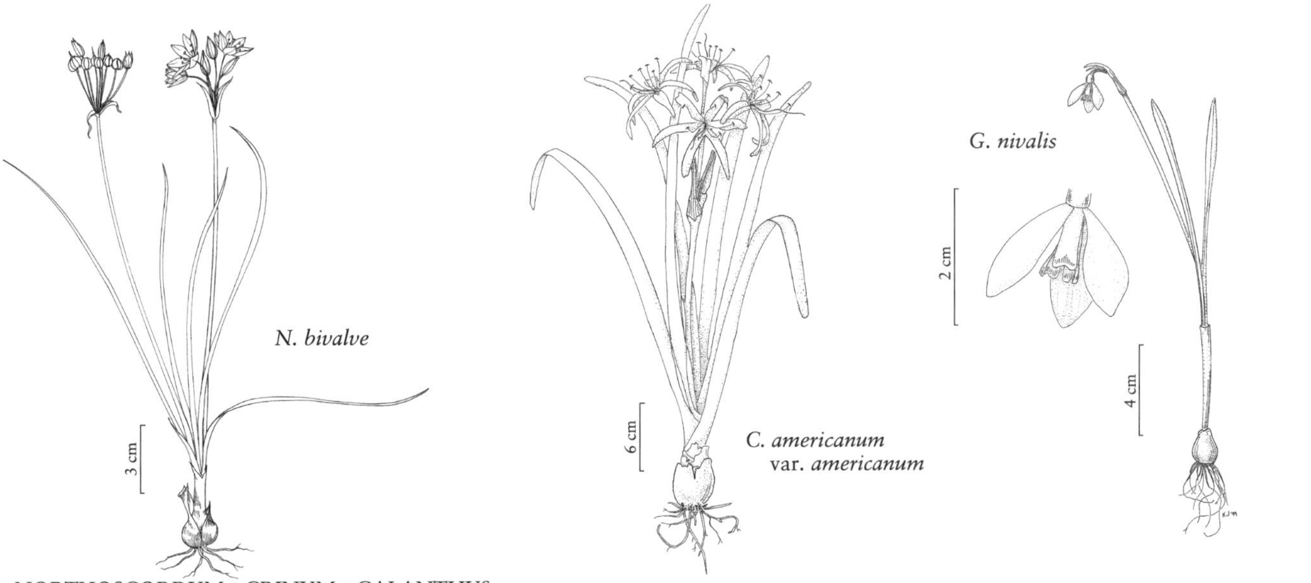

NORTHOSCORDUM ◦ CRINUM ◦ GALANTHUS

Species ca. 19 (2 in the flora): North America, South America.

All species of *Nothoscordum* are native to the Americas. *Nothoscordum gracile* has become naturalized in Europe, Africa, Asia, and Australia; it spreads rapidly by seeds and bulblets.

1. Leaves 4–12 mm wide; flowers fragrant; tepals connate to 1/3 their length . . . . 1. *Nothoscordum gracile*
1. Leaves 1–4(–5) mm wide; flowers not fragrant; tepals distinct or nearly so . . . . 2. *Nothoscordum bivalve*

1. Nothoscordum gracile (Dryander) Stearn, Taxon 35: 335. 1986 [I]

Allium gracile Dryander in W. Aiton, Hort. Kew. 1: 429. 1789; *A. fragrans* Ventenat; *Nothoscordum fragrans* (Ventenat) Kunth

Bulbs ovoid, ca. 1.5 cm diam.; outer coats brown; bulblets 10–20+. **Scapes** 1–2, terete, 30–60 cm × 2–3 mm. **Leaves** 2–9, sheaths enveloping neck of bulb, hyaline, reduced apically; blade linear, 20–40(–50) cm × 4–12 mm, margins entire. **Umbel** 10–20-flowered, often asymmetrical, to 5 cm diam.; bracts persistent, 2, 1.2–2 cm × 5–8 mm, bases fused, imbricate, margins scarious. **Flowers** fragrant; tepals white adaxially with pink midvein and greenish bases, oblanceolate, 9–15 × 4 mm, apex obtuse; filaments simple, adnate to tepals, 7–10 mm; anthers dark brown; ovary crestless; style persistent in fruit, ± equaling stamens; stigma unlobed, equaling perianth; pedicel erect to spreading, 2–6 cm, unequal. **Capsules** obovoid, ca. 6–7 × 6–7 mm. **Seeds** 8–12 per locule.

Flowering Mar–Dec. Roadsides, waste places; 0–100 m; introduced; Ala., Calif., Fla., Ga., La., Miss., S.C.; South America.

There is considerable controversy over the correct epithet. Based on the information available, the authors feel that *Nothoscordum borbonicum* Kunth is a misapplied name and that the correct name is *Nothoscordum gracile*.

In the flora area, *Nothoscordum gracile* is an escape from cultivation.

SELECTED REFERENCES Ravenna, P. 1991. *Nothoscordum gracile* and *N. borbonicum* (Alliaceae). Taxon 40: 485–487. Small, J. K. 1928. *Nothoscordum fragrans*. Addisonia 33: plate 433. Stearn, W. T. 1986. *Nothoscordum gracile*, the correct name of *N. fragrans* and *N. inodorum* of authors (Alliaceae). Taxon 35: 335–338.

2. Nothoscordum bivalve (Linnaeus) Britton in N. L. Britton and A. Brown, Ill. Fl. N. U.S. 1: 415. 1896 [F] [W2]

Ornithogalum bivalve Linnaeus, Sp. Pl. 1: 306. 1753; *Allium bivalve* (Linnaeus) Kuntze; *A. striatum* Jacquin; *Nothoscordum striatum* (Jacquin) Kunth; *N. texanum* M. E. Jones

Bulbs globose to subglobose, to 1–1.5 cm diam.; outer coats brown; bulblets absent. **Scape** solitary (rarely 2), terete, (10–)20–40 cm × 1–3 mm. **Leaves** 1–4, sheaths enveloping neck of bulb, hyaline, subtruncate apically; blade filiform or linear, to 30 cm × 1–4(–5) mm, margins entire. **Umbel** mostly 3–6(–10)-flowered, often asymmetrical, 1–3 cm diam.; bracts persistent, 2, 1–2 cm × 4–8 mm, base coherent, imbricate, margins scarious or hyaline, apex acuminate. **Flowers** not fragrant; tepals whitish to cream, at least outer ones with red or purplish red midvein, elliptic, (8–) 10–12(–15) × 3–4.5 mm, apex acute or acuminate; filaments simple, adnate to tepals, 5–6 mm; anthers yellow; ovary crestless; style persistent, equaling stamens; stigma unlobed, 3/4 perianth; pedicel erect, 1–2 cm, elongating as flowers develop, but often of several lengths even in fruit. **Capsules** subglobose or obovoid, 6–8 × 6–8 mm. **Seeds** 4–7 per locule.

Flowering Mar–Dec. Open woods, prairies, barrens; 0–1700 m; Ala., Ariz., Ark., Fla., Ga., Ill., Ind., Kans., Ky., La., Miss., Mo., N.Mex., N.C., Ohio, Okla., S.C., Tenn., Tex., Va.; s to South America.

46. CRINUM Linnaeus, Sp. Pl. 1: 291. 1753; Gen. Pl. ed. 5, 141. 1754 • String-lily, swamp-lily [Greek *krinon*, a lily]

Walter C. Holmes

Herbs, perennial, scapose, from bulbs with elongate columnar apices. **Leaves** basal, thick; blade lorate or ensiform, not narrowed near base. **Scape** solid. **Inflorescences** umbellate, few- to many-flowered, subtended by 2 large, lanceolate, scarious bracts. **Flowers** sessile or pedicellate; perianth connate proximally, red to white, often striped, streaked, or overlaid with red abaxially, funnelform to salverform to semicampanulate, tube straight to curved, ca. same length as limb lobes; stamens inserted on perianth tube throat; filaments thin, often declinate; ovary inferior, globose; style slender; stigma capitate. **Fruits** capsular, globose or subglobose, usually prominently beaked. **Seeds** fleshy, testa often corky.

Species ca. 100 (4 in the flora): tropical and warm regions worldwide, mostly in Africa.

SELECTED REFERENCES Moldenke, H. N. 1962. Amaryllid genera and species. Pl. Life 18: 48–50. Wunderlin, R. P., B. F. Hansen, and D. W. Hall. 1985. The vascular flora of central Florida: Taxonomic and nomenclatural changes, additional taxa. Sida 11: 232–244.

1. Flowers sessile or nearly so.
   2. Perianth white . . . . 1. *Crinum americanum*
   2. Perianth purplish red, particularly abaxially . . . . 4. *Crinum zeylanicum*
1. Flowers distinctly pedicellate.
   3. Perianth pink to red; leaves 3–5 cm wide; umbels 8–13-flowered . . . . 3. *Crinum bulbispermum*
   3. Perianth white; leaves 7.5–12 cm wide; umbels 20–100-flowered . . . . 2. *Crinum asiaticum*

1. **Crinum americanum** Linnaeus, Sp. Pl. 1: 292. 1753

E F

Bulbs 3.5–6 cm diam., neck 4–6 cm. **Leaves** 10–15 dm × 1–4 (–7) cm; blade lorate, margins usually scabrous. **Scape** equaling leaves. **Umbels** 2–7-flowered. **Flowers** sessile; perianth white, salverform, tube 6–15 cm, limb lobes linear, narrowly elliptic, or oblanceolate, 5–15 × 1–1.5 cm. **Capsules** 2.5–4.5 × 2.5–4 cm, beak 1–8 cm.

Varieties 2 (2 in the flora): se, sc United States.

1. Umbels 2–5-flowered . . . . 1a. *Crinum americanum* var. *americanum*
1. Umbels 6–7-flowered . . . . 1b. *Crinum americanum* var. *traubii*

1a. Crinum americanum Linnaeus var. **americanum** [E] [F]

Crinum herbertianum Schultes f.; *C. strictum* Herbert; *C. texanum* L. S. Hannibal

Leaves slightly shorter than scape; blade pale green. **Umbels** 2–5-flowered.

Flowering late winter–autumn (all year in Fla.). Swamps, marshes, stream banks; 0–100 m; Ala., Fla., Ga., La., Miss., S.C., Tex.

Variety *americanum* is the most widespread and abundant *Crinum* in the United States. It is quite variable, particularly in the size of floral parts and width of leaves.

1b. Crinum americanum Linnaeus var. **traubii** (Moldenke) L. S. Hannibal, Bull. Louisiana Soc. Hort. Res. 13: 308. 1972 [E]

Crinum strictum Herbert var. *traubii* Moldenke, Pl. Life 18: 49. 1962

Leaves longer than scape; blade dark green. **Umbels** 6–7-flowered.

Flowering summer–autumn. Swamps, marshes; 0–100 m; Tex.

Variety *traubii* is known only from specimens grown from bulbs collected in Jefferson and Hardin counties.

2. Crinum asiaticum Linnaeus, Sp. Pl. 1: 292. 1753 • Poison bulb, tree crinum [I]

Bulbs 25 cm or more × 10–12 cm. **Leaves** numerous, 10 dm or more × 7.5–12 cm; blade ensiform, gradually tapering to apex. **Scape** stout, 2-edged. **Umbels** 20–100-flowered. **Flowers:** perianth white, salverform, tube 3–7.5 cm, limb lobes linear, 7.5 × 1–1.5 cm; pedicel 1–2 cm. **Capsules** to 5 cm diam., beak long.

Flowering summer–fall. Wet areas; 0–100 m; introduced; Fla., La.; tropical Asia.

Crinum asiaticum is a robust and highly variable species that is distinguished by its very broad leaves. The bulb is reputed to be poisonous.

3. Crinum bulbispermum (Burman f.) Milne-Redhead & Schweickerdt, J. Linn. Soc., Bot. 52: 161. 1939 [I]

Amaryllis bulbisperma Burman f., Fl. Indica, 9. 1768

Bulbs 8–10 × 6–8 cm. **Leaves** 5–8 dm × 3–5 cm; blade lorate. **Scape** 4–7.5 dm. **Umbels** 8–13-flowered. **Flowers:** perianth pink to red, funnelform, tube narrow, 5–10 cm, limb lobes lanceolate-linear, lanceolate-elliptic, or lanceolate-ovate, 6–11 × 1–1.7 cm; pedicel (2–)4–6 cm. **Capsules** not seen at maturity, beak very short to absent.

Flowering spring–summer. Margins of wet areas and disturbed sites; 0–200 m; introduced; Fla., La., N.C., Tex.; S. Africa.

Crinum bulbispermum is apparently the nonnative species of *Crinum* that is most commonly cultivated outdoors. The name *C. longifolium* (Linnaeus) Thunberg was long misapplied to this species.

4. Crinum zeylanicum (Linnaeus) Linnaeus, Syst. Nat. ed. 12, 2: 236. 1767 [I]

Amaryllis zeylanica Linnaeus, Sp. Pl. 1: 293. 1753; *Crinum latifolium* Linnaeus var. *zeylanicum* (Linnaeus) Hooker f.

Bulbs 12–15 cm diam., neck short. **Leaves** several, 5–10 dm × 3–5 cm; blade lorate. **Scape** to 7.5 dm or more. **Umbels** 8–12-flowered. **Flowers** sessile; perianth white with broad purplish-red keels (most pronounced adaxially), funnelform, tube 7.5–15 cm, limb lobes oblong-elliptic, 7.5–18 × 1–2 cm. **Capsules** not seen at maturity, globose, beak 8–10 cm. **Seeds** warty.

Flowering late spring–summer. Fields, wasteplaces, wet areas; 0–200 m; introduced; Fla., La.; tropical Asia.

Crinum zeylanicum is extensively cultivated outdoors in the lower Gulf coastal plain.

47. GALANTHUS Linnaeus, Sp. Pl. 1: 288. 1753; Gen. Pl. ed. 5, 140. 1754 • Snowdrop, perce-neige [Greek *gala*, milk, and *anthos*, flower, alluding to the color of the flowers] [I]

Gerald B. Straley†

Frederick H. Utech

Herbs, perennial, scapose, from brown, tunicate, ovoid to globose bulbs; offset bulbs often present. **Leaves** 2(–3), basal, opposite, with sheathing blade, vernation flat and parallel, or convolute; nonsheathing blade erect to recurving at maturity, grayish green, linear-oblanceolate, glaucous; sheathing blade white, tubular, membranous, enclosing leaf bases and scape. **Scape** erect in flower, prostrate in fruit, green, solid. **Inflorescences** pendulous, 1-flowered, spathaceous; spathe bracteate, membranous; bracts 2, connate, split on 1 side. **Flowers** nodding, fragrant; perianath 2.5 cm or shorter; tepals 6, distinct, unequal; outer tepals spreading, white, narrowly obovate to almost orbicular, larger than inner; inner tepals overlapping, appearing tubular, green-spotted at apex only or apex and base, straight to semiorbicular, apex notched; stamens 6, inserted at bases of tepals, distinct; anthers basifixed, longer than filaments, bases lobed, apices tapered, dehiscense introrse, via terminal slits; ovary inferior, green, 3-locular, globose, septal nectaries present; style, white, unbranched, filiform; stigma indistinct to minutely capitate; pedicel wiry, short, slender. **Fruits** capsular, green, globose, fleshy, dehiscence loculicidal. **Seeds** 18–36, light brown, 3.5 mm, oblong to obtuse, elaiosomes fleshy. $x = 12$.

Species 14–17 (2 in the flora): introduced; Europe, sw Asia (Asia Minor, Iranian Caucasus, and Caspian Sea regions); introduced and naturalized elsewhere.

A number of *Galanthus* species, especially *G. nivalis*, are commonly cultivated for their late-winter and early-spring flowers, which emerge through the snow. Species of *Galanthus* are sometimes confused with *Leucojum*, snowflakes, a spring-flowering relative. The plants of *Leucojum* are usually taller, bear 2–3 flowers per stem, and their tepals are all equal.

SELECTED REFERENCES Davis, A. P. 1999. The Genus *Galanthus*. Portland. Stern, F. C. 1956. Snowdrops and Snowflakes—A Study of the Genera *Galanthus* and *Leucojum*. London. Yeo, P. F. 1975. The hybrid origin of some cultivated snowdrops (*Galanthus*–Amaryllidaceae). Baileya 19: 157–162.

1. Inner tepals blotched with green at base and apex or with green stripe from base to apex; leaves convolute within basal sheath . . . . 1. *Galanthus elwesii*
1. Inner tepals blotched with green at apex only; leaves flat and parallel within basal sheath . . . 2. *Galanthus nivalis*

1. Galanthus elwesii Hooker f., Bot. Mag. 101: plate 6166. 1875 • Giant snowdrop [I]

Plants to 9–15(–20) cm; bulbs 2–3 × 1.5–2 cm. **Leaves:** vernation convolute, remaining so within basal sheath; blade narrowly oblanceolate, 10–30 × 0.5–2.5(–3.5) cm. **Spathe** 3–5 cm. **Flowers:** outer tepals white, elliptic to broadly obovate or nearly orbiculate, 2–2.5 × 1–2 cm; inner tepals white with green blotch at base and apex, or with green stripe from base to apex, narrowly obovate, 10–12 × (4–)6–7 mm; anthers 4–6 mm; ovary 5–8 × 4–5 mm; style 8–10 mm; pedicel (1.5–)3–4 cm.

Flowering late winter–spring. Roadsides, open forests, abandoned gardens; 0–300 m; introduced; N.Y., Pa.; se Europe; sw Asia (Turkey); expected elsewhere.

2. Galanthus nivalis Linnaeus, Sp. Pl. 1: 288. 1753 [F] [I]

Plants to 7–20(–40) cm; bulbs 1.5–2.5 × 1–1.5(–2) cm. **Leaves:** vernation flat and parallel, remaining so within basal sheath; blade linear to ligulate, 5–15 × 0.3–0.7 cm. **Spathe** 2–3.5 cm. **Flowers:** outer tepals white, oblong to broadly obovate, 1.5–2(–2.5) × 0.6–1.1 cm; inner tepals white with green blotch at apex only, narrowly obovate or oblong, 7–12 × 4–6 mm; anthers 3–5 mm; ovary 5–6 × 3–4 mm; style 6–8 mm; pedicel 1.2–3(–4) cm.

Flowering late winter–early spring. Roadsides, open forests, abandoned gardens; 0–500 m; introduced; N.B., Nfld. and Labr. (Nfld.), Ont.; Md., Mass., Mich., N.J., N.Y., N.C., Ohio, Pa., Utah, Va.; Europe (Spain to w Russia); expected elsewhere.

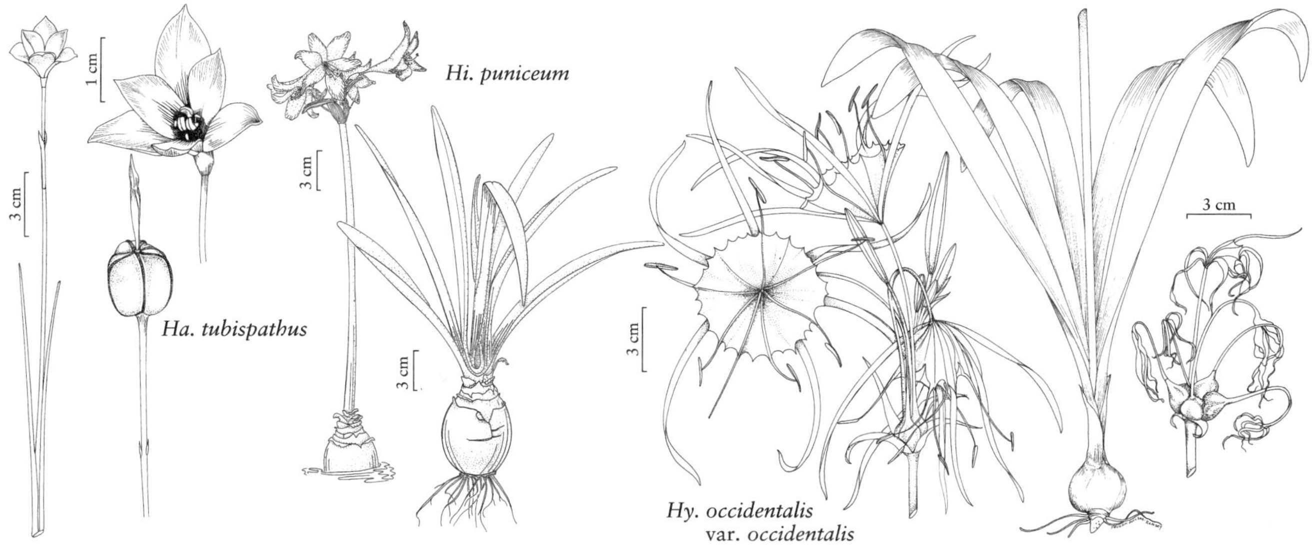

HABRANTHUS ◦ HIPPEASTRUM ◦ HYMENOCALLIS

48. HABRANTHUS Herbert, Bot. Mag. 51: plate 2464. 1824 • Rain-lily [Greek *habros*, delicate or splendid, and *anthos*, flower]

Raymond O. Flagg

Gerald L. Smith

Walter S. Flory†

Herbs, perennial, scapose, from bulbs. **Bulbs** black or brown, tunicate, ovoid or globose, sometimes with long neck. **Leaves** deciduous, sessile, erect or recumbent, bases overlapping, sheathing; blade linear, rarely exceeding 2 cm wide, smooth. **Scape** hollow. **Inflorescences** umbellate, 1[–4]-flowered, spathaceous; spathe proximally tubular; bracteoles sometimes present. **Flowers** declinate [to suberect], somewhat zygomorphic; perianth connate proximally, funnelform, shortly tubular basally, 2–8 cm; tepals subequal; filaments inserted on tepals distal to perianth tube, fasciculate [semifasciculate], declinate and recurving distally, filiform, in sets of 4 different lengths; anthers submedially dorsifixed, usually parallel with floral axis, linear-oblong; ovary inferior; style filiform; stigma 3-fid, lobes linear; pedicel rarely absent, hollow. **Fruits** capsular, 3-locular, thin-walled, subglobose or ± oblate. **Seeds** numerous, dark brown to black, flat, D-shaped, obliquely winged, lustrous. $x = 6$.

Species ca. 30 (2 in the flora): se to sc United States, Mexico, South America.

Habranthus species have at times been included in the genus *Zephyranthes*. The somewhat zygomorphic flower of *Habranthus* is clearly distinguished by filaments of 4 different lengths, which are fasciculate to semifasciculate, and always longer than the perianth tube. The actinomorphic flower of *Zephyranthes* is distinguished by filaments of two lengths (sometimes subequal), which are not fasciculate and are spreading except when shorter than the perianth tube.

All measurements were taken from pressed, dried specimens. Generally, the maximum width of a fresh leaf would be about one and one half times that reported here.

SELECTED REFERENCES Flory, W. S. and R. O. Flagg 1958. A cytological study of the genus *Habranthus*. Nucleus (Calcutta) 1: 267–280. Flory, W. S. and G. L. Smith. 1980. High chromosome numbers in several Zephyrantheae taxa. Pl. Life 36: 63–72. Sealy, J. R. 1937. *Zephyranthes*, *Pyrolirion*, *Habranthus* and *Hippeastrum*. J. Roy. Hort. Soc. 62: 195–209.

1. Perianth bright yellow or orangish yellow, less than 4 cm; stigma borne near top of stamens; leaf blade not exceeding 4 mm wide . . . . 1. *Habranthus tubispathus*
1. Perianth pink, 6+ cm; stigma borne ca. 1 cm beyond stamens; leaf blade more than 5 mm wide . . . . 2. *Habranthus robustus*

1. Habranthus tubispathus (L'Héritier) Traub, Pl. Life 7: 42. 1951 • Copper-lily [F]

Amaryllis tubispatha L'Héritier, Sert. Angl., 9. 1789; *Atamosco texana* (Herbert) Greene; *Habranthus andersonianus* Herbert; *H. texanus* (Herbert) Steudel; *Zephyranthes texana* Herbert

Leaf blade dull green, to 4 mm wide. **Inflorescence** 1-flowered; spathe 1.9–3.1 cm. **Flower:** perianth 2.1–3.1 cm, tube tinged green, ca. 0.2 cm, distinct portions of tepals somewhat spreading, not reflexed, bright yellow or orangish yellow, sometimes tinged burgundy abaxially, color deepening proximally; filaments 0.8–1.8 cm; anthers (2–)4–5 mm; style 1.8–2 cm; stigma equaling or rarely extending more than 1 mm beyond anthers; pedicel (2–)3.3–4.5(–5.6) mm, exceeding spathe. $2n = 24$.

Flowering summer–early fall (Jun–Oct). Grasslands, prairies, disturbed areas; 0–300 m; Ala., Fla., La., Tex.; South America.

In addition to positional differences in the stamens, the trifid, linear stigma of *Habranthus tubispathus* distinguishes it in its North American range from late summer- and fall-blooming, yellow-flowered, sympatric *Zephyranthes* species, which have capitate stigmas.

It is not clear whether *Habranthus tubispathus* is native to North America or was introduced, but it is a disjunct species also found in Argentina, Brazil, central and southern Chile, and southern Uruguay. Another such species is *Cienfuegosia drummondii* (A. Gray) Lewton (Malvaceae), which occurs in northern Argentina, southernmost Brazil, Paraguay, and southern Texas. These taxa are two of a small group of bicentric species showing disjunct distributions between eastern North America and temperate South America (P. A. Fryxell 1967). All are perennial and fully self-compatible, with a high reproductive potential, so the establishment of a single individual could accomplish disjunction. However, *H. tubispathus* lacks what all the others share: an impermeable seed coat that allows for a degree of dormancy and longevity.

In the 1930s, Harold H. Hume (J. R. Sealy 1937) suggested that Spanish missionaries had introduced *Habranthus tubispathus* to Texas around 1700. W. C. Holmes and C. J. Wells (1980) have recently supported this idea. Nevertheless, this means of distribution is questionable since, as W. S. Flory (1938) pointed out, the records indicate that Spaniards out of Mexico founded the Texas missions, and *H. tubispathus* is not found in Mexico. A. R. Diamond et al. (1998) reported historical evidence for the introduction of *H. tubispathus* into Alabama in the late 1800s by seeds being sent from Texas. A similar mechanism could have introduced the species to Texas from South America.

2. Habranthus robustus Herbert ex Sweet, Brit. Fl. Gard., ser. 2, 1: plate 14. 1831 [I]

Leaf blade dull green, to 18 mm wide. **Inflorescence** 1-flowered; spathe 3.8–4.1 cm. **Flower:** perianth 6.8–8 cm, tube light green, ca. 0.3 cm, distinct portions of tepals somewhat spreading, reflexed, pink; filaments 1.3–3.7 cm; anthers 6–13 mm; style 4.5–5.5 cm; stigma exserted ca. 1 cm beyond anthers; pedicel 4–6.5 cm, usually exceeding spathe. $2n = 12$.

Flowering summer–fall (Jun–Nov). Sandy loam, lawns; 0–100 m; introduced; Fla.; South America.

Habranthus robustus persists and is locally adventive in Florida.

49. HIPPEASTRUM Herbert, Appendix, 31. 1821 • Barbados-lily, naked lady, red spider-lily [Greek *hippeus*, rider, and *astron*, star, the allusion obscure] [I]

Walter C. Holmes

Herbs, perennial, scapose, glabrous, from globose bulbs. **Leaves** few, basal; blade liguliform, fleshy, parallel-veined, margins entire, apex tapering. **Scape** hollow. **Inflorescences** umbellate, bracteate; bracts 2, scarious. **Flowers** spreading to slightly drooping, syntepalous; perianth connate proximally, funnelform to campanulate, with minute corona reduced to small crown inserted on throat of tube; tepals 6 in 2 whorls of 3, outer slightly shorter than inner; stamens inserted on perianth tube, declinate, subequal; filaments slender; ovary inferior, ellipsoid; style slender, ca. equaling tepals; stigma capitate or slightly 3-lobed. **Fruits** capsular, dehiscence loculicidal. **Seeds** few to many.

Species ca. 75 (1 in the flora): c America, West Indies, South America, w Africa.

1. **Hippeastrum puniceum** (Lamarck) Voss, Vilm. Blumengärtn. ed. 3, 1: 1033. 1895 [F] [I]

Amaryllis punicea Lamarck in J. Lamarck et al., Encycl. 1: 122. 1783

Bulbs 6–10 cm diam. **Leaves** 6–8, appearing after flowering, 50 × 3–5 cm. **Scape** to 1 m. **Inflorescences** 2–4-flowered; bracts 5 cm. **Flowers** slightly zygomorphic; perianth reddish to salmon, with whitish midstripe on adaxial surface of each outer tepal, tube 3 cm; outer tepals lanceolate to subrhombic, 12 cm or more, apex acuminatenate. **Capsules** ellipsoid to ovoid, 2 cm. **Seeds** black, compressed-globose or -subglobose.

Flowering spring–early summer. Disturbed sites and old gardens, spreading or persisting from cultivation; 0–100 m; introduced; La., Tex.; Mexico; West Indies; Central America; South America.

Hippeastrum puniceum is the "amaryllis" of commerce that is extensively cultivated outdoors in the southern United States and indoors elsewhere.

50. HYMENOCALLIS Salisbury, Trans. Hort. Soc. London 1: 338. 1812 • Spider-lily [Greek *hymên*, membrane, and *kallos*, beauty, in reference to the corona]

Gerald L. Smith

Walter S. Flory†

Herbs, perennial, scapose, from bulbs. **Bulb** 1, ovoid or globose, tunicate, often extending into neck of clasping, distichous leaf bases. **Leaves** 2–16, deciduous or evergreen, sessile, rarely petiolate; blade narrowly to widely liguliform or oblanceolate, rarely ovate to elliptic. **Scape:** bracts 2–3, triangular, ovate, or lanceolate. **Inflorescences** umbellate, bracteate; each flower with subtending, often narrowly lanceolate bract. **Flowers** 1–16, usually sessile, erect or slightly diverging, large and starlike, fragrant; perianth connate basally into short or long tube, surmounted by conspicuous staminal corona; tepals extending from base of corona, free portions reflexed or ascending, often distally recurved, linear; stamens adnate basally into showy funnelform or rotate corona, margins between free portions of filaments often dentate or lacerate, portions of filaments inserted on margin of corona, erect to incurved, filiform; anthers versatile, introrse, pollen yellow, often golden, or orange; ovary inferior, globose, ovoid, oblong, or pyriform, ovules 2–10 per locule; style exserted beyond stamens, deflexed laterally, filiform; stigma capitate. **Fruits** capsular, green, subglobose to elongate, 3-locular, large, leathery. **Seeds** large, green, fleshy. $x = 20, 23$.

Species ca. 50 (15 in the flora): se and sc United States, West Indies, Central America, and South America.

The species of *Hymenocallis* are some of the most difficult to identify from herbarium specimens. Distinguishing field characteristics such as position, surface, texture, and color of the leaves, and three-dimensional shape and margin of the staminal corona, are undeterminable on most herbarium specimens. These characteristics, as observed in the field, are discussed in the treatment of each species, and distinguishing features are emphasized for their usefulness in identification. With the exception of dimensions for fruits, seeds, and some bulbs, all measurements presented are from pressed, dried specimens. Variations in habitat and distribution are also emphasized, and they are often used in the key to support character distinctions.

The first eight *Hymenocallis* species treated here were classified by H. P. Traub (1962) in the Caroliniana Alliance. Species in this group primarily have 1–3 ovules per locule and deciduous, sessile, liguliform or oblanceolate leaves. They are distributed in northern Florida and throughout wetland areas of the southeastern United States. Species studied to date have a base chromosome number of $x = 20$.

Hymenocallis rotata, *H. godfreyi*, *H. puntagordensis*, *H. palmeri*, and *H. henryae* (species 9–14) were all classified in Traub's Henryae Alliance. They have distinctly larger ovaries than the species classified in Traub's Caroliniana Alliance, having 4–8 ovules per locule, as well as coriaceous, suberect to erect, liguliform leaves. All species in the Henryae Alliance treated here occur in Florida. There is considerable chromosome diversity among these taxa, but the base number of many is $x = 23$.

Hymenocallis latifolia is the only spider-lily in the flora that is classified in Traub's Caribaea Alliance. Species in this alliance are characterized by sessile, liguliform, evergreen leaves. Many have 2–3 ovules per locule and a base number of $x = 23$.

D. S. Correll and M. C. Johnston (1970) and D. S. Correll and H. B. Correll (1972) recognized three species of spider-lilies in Texas, and L. H. Shinners (1951) recognized two. We recommend a thorough investigation of *Hymenocallis* in Texas and adjacent states to achieve a clearer understanding of the southwestern spider-lily species.

Like many other amaryllids, *Hymenocallis* species contain various alkaloids. It is not recommended that plant parts be eaten or even touched by allergic individuals (J. A. Bauml 1979).

The precise localities of the bulb collections of *Hymenocallis* collected by Mary G. Henry from which Traub described numerous new species of *Hymenocallis* were determined from the field diaries of Mary G. Henry, courtesy of Josephine de N. Henry, President Emerita of the Henry Foundation for Botanical Research. This treatment would not have been possible without this critical information.

SELECTED REFERENCES Flory, W. S. 1976. Distribution, chromosome numbers and types of various species and taxa of *Hymenocallis*. Nucleus (Calcutta) 19: 204–227. Flory, W. S. 1978. Known distributions of *Hymenocallis* Salisbury in North and Middle America and the West Indies. Pl. Life 34: 47–59. Morton, C. V. 1935. A check list of the bulbous Amaryllidaceae native to the United States. Year Book Amer. Amaryllis Soc. 2: 80–83. Sealy, J. R. 1954. Review of the genus *Hymenocallis*. Kew Bull. 1954: 201–240. Shinners, L. H. 1951. The north Texas species of *Hymenocallis* (Amaryllidaceae). Field & Lab. 19: 102–104. Traub, H. P. 1962. Key to the subgenera, alliances and species of *Hymenocallis*. Pl. Life 18: 55–72.

1. Leaves evergreen; perianth tube (6–)7.5–20 cm; s Florida and West Indies.
    2. Leaves 5–10 dm × 4–9 cm, blade broadly liguliform; margins between free portions of filaments wavy, prominent projections absent; ovary ovoid, 0.9–1.6 cm × 5–10 mm . . . . . 15. *Hymenocallis latifolia*
    2. Leaves 3.5–7.5 dm × 1.5–3 cm, blade narrowly liguliform; margins between free portions of filaments with 1 or 2 prominent lacerations; ovary pyriform, 1.5–2.4 cm × ca. 10 mm . . . . . 12. *Hymenocallis puntagordensis*
1. Leaves deciduous; perianth tube rarely exceeding 13 cm; se United States including n Florida.
    3. Ovules 4–8 per locule; ovary 1.4–3 cm × 6–15 mm.
        4. Corona 6 cm or wider . . . . . 9. *Hymenocallis rotata*
        4. Corona rarely wider than 6 cm.
            5. Tepals yellow-green to pale green.
                6. Flowers 1 per inflorescence; tepals ascending, equal to or shorter than perianth tube . . . . . 13. *Hymenocallis palmeri*
                6. Flowers 2 (rarely 3) per inflorescence; tepals long-spreading, nearly always longer than perianth tube . . . . . 14. *Hymenocallis henryae*
            5. Tepals white.
                7. Longest leaf not exceeding 4 dm at anthesis; free filaments to 3 cm; anthers 1.2–1.5 cm . . . . . 11. *Hymenocallis godfreyi*
                7. Longest leaf exceeding 4 dm at anthesis; free filaments to 4 cm; anthers 1.5–2 cm . . . . . 10. *Hymenocallis tridentata*
    3. Ovules 1–3(–4) per locule; ovary 0.7–1.5 cm × 5–10 mm.

8. Staminal cup 4.5 cm or longer; habitat rocky river shoals in Piedmont . . . . 2. *Hymenocallis coronaria*
8. Staminal cup 4.5 cm or shorter; habitat wetland areas or mesic upland woods.
   9. Leaf blades oblanceolate, distinctly wider beyond middle before tapering.
      10. Leaves noncoriaceous, glaucous; scape bracts 4–7 cm, apex long-acuminate; bulbs nonrhizomatous . . . . 1. *Hymenocallis occidentalis*
      10. Leaves coriaceous, not glaucous; scape bracts 3–4(–6) cm, apex acute but not long-acuminate; bulbs rhizomatous . . . . 4. *Hymenocallis choctawensis*
   9. Leaf blades liguliform to narrowly lanceolate or narrowly oblanceolate.
      11. Flowers (3–)5–12; leaf blade occasionally narrowly lanceolate . . . . 3. *Hymenocallis liriosme*
      11. Flowers 1–3, rarely more; leaf blade typically liguliform, occasionally narrowly oblanceolate.
         12. Staminal cup rotate at full anthesis; leaves chiefly arching low, often appearing prostrate; leaf blade occasionally narrowly oblanceolate . . . . 7. *Hymenocallis duvalensis*
         12. Staminal cup funnelform at full anthesis, gradually spreading in time; leaves suberect to erect; leaf blade typically liguliform.
            13. Tepals 5–7 cm; leaves 1.5–4 dm; bogs and stream banks, outer Coastal Plain, Carolinas . . . . 6. *Hymenocallis pygmaea*
            13. Tepals (6–)7–11.5 cm; leaves (2.4–)2.8–6.8 dm; bogs, stream banks, alluvial margins, coastal Carolinas, Florida.
               14. Scape bracts narrowly lanceolate, 2.5–5 cm × 7–12 mm; chiefly bogs and stream banks, outer Coastal Plain, Carolinas to ne Florida . . . . 5. *Hymenocallis crassifolia*
               14. Scape bracts lanceolate, 3–4.5 cm × 10–15 mm; alluvial margins, lower Ochlockonee River system, Florida panhandle . . . . 8. *Hymenocallis franklinensis*

1. Hymenocallis occidentalis (J. Le Conte) Kunth, Enum. Pl. 5: 856. 1850 • Woodland, hammock, or northern spider-lily [E] [F]

Pancratium occidentale J. Le Conte, Ann. Lyceum Nat. Hist. New York 3: 146. 1836

Bulb nonrhizomatous, globose, 3.5–5.5 × 3–4.5 cm; basal plate 1–3 cm; neck 1.5–4 cm; tunic dark brown. **Leaves** deciduous, 5–12, arching to suberect, 3.5–6 dm × 2–6 cm, noncoriaceous; blade oblanceolate, shallowly channeled, distinctly wider beyond middle, tapering to petiolelike base, glaucous when young, apex subacute to acute. **Scape** (4–)5–7 dm, 2-edged, glaucous; scape bracts 2, enclosing buds, 4–7 × 1–1.5 cm, apex long-acuminate; subtending floral bracts 2.5–4.5 cm × 5–10 mm. **Flowers** 3–9, opening sequentially, with heavy, sweet fragrance; perianth tube green, slender, 7–13.5 cm; tepals slightly ascending, white, green-striped on keel, (7–) 8.5–11.5 cm × 5–10 mm; corona white with small, yellowish green eye, funnelform, shortly tubulose proximally, 2.5–4(–4.5) × 4–5.5 cm, margins between free portions of filaments often irregularly 3-dentate; free portions of filaments inserted on flat sinal base, nearly erect, white, 2.5–4 cm; anthers 1.3–2 cm, pollen golden; ovary ovoid, 0.8–1.5 cm × 4–7 mm, ovules 2–3 per locule; style green in distal 1/3, fading into white proximally, 13–23 cm. **Capsules** broadly triangular to subglobose, ca. 2.5 × 2 cm. **Seeds** subglobose, 1.5–2.1 × 1.2–1.6 cm.

Varieties 2 (2 in the flora): North America.

1. Leaves fresh at anthesis . . . . 1a. *Hymenocallis occidentalis* var. *occidentalis*
1. Leaves withering before anthesis . . . . 1b. *Hymenocallis occidentalis* var. *eulae*

1a. Hymenocallis occidentalis (J. Le Conte) Kunth var. **occidentalis** [E] [F]

Hymenocallis bidentata Small; *H. moldenkiana* Traub

Leaves appearing in late winter, fresh at anthesis, then dying off, arching outward, to 6 cm wide. **Scape bracts** lanceolate, 4.5–7 cm. 2*n* = 52, 54.

Flowering summer–early fall. Floodplain forests, hammocks, meadows, wooded hillsides; 0–1100 m; Ala., Ark., Fla., Ga., Ill., Ind., Ky., La., Miss., Mo., N.C., S.C., Tenn.

Hymenocallis occidentalis var. *occidentalis* has a long history of cultivation in southern gardens. It is unique in that it grows well in upland mesic forests, sometimes a considerable distance from a river or stream. In southern hammocks, its glaucous leaves appear in midwinter, long before flowering. Flowering may occur as soon as early July in northern Florida but is more usual in late July, August, and September. A single scape will often bear 4–9 flowers, producing an impressive display. Distinctive features are the glaucous, oblanceolate leaves; lanceolate bracts with long-acuminate apex; long, slender perianth tube; and funnelform corona with distinctly

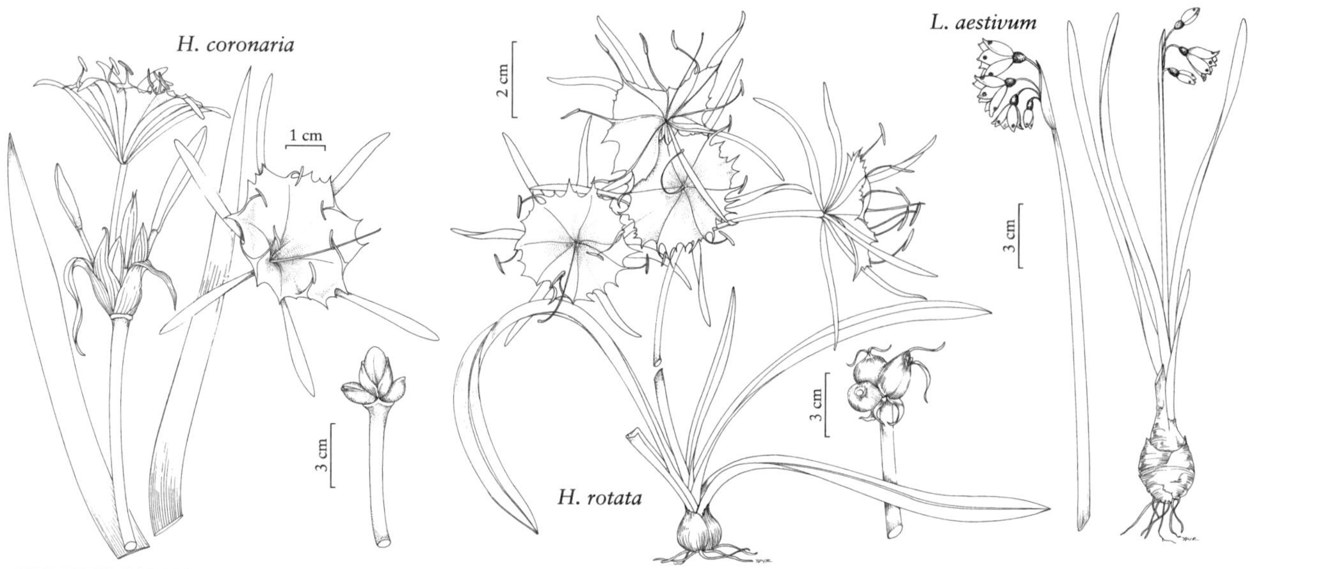

HYMENOCALLIS ◦ LEUCOJUM

dentate margin between the free portions of filaments.

This variety is the most widely distributed spider-lily taxon (G. L. Smith et al. 1993) and is the hardiest member of the genus. For over 40 years, a collection made by the late Mary G. Henry near Evergreen, Alabama, has put on an annual, prolific floral display in late August/September at the Henry Foundation's southern garden in Gladwyne, Pennsylvania.

Plants of *Hymenocallis occidentalis* often have been identified as *H. caroliniana* (Linnaeus) Herbert on herbarium sheets and in floras. However, the latter name is based on *Pancratium carolinianum* Linnaeus, and its type, as drawn by Catesby, is clearly *P. maritimum* Linnaeus.

1b. Hymenocallis occidentalis (J. Le Conte) Kunth var. **eulae** (Shinners) Gerald L. Smith & Flory, Novon 11: 230. 2001 [E]

Hymenocallis eulae Shinners, Field & Lab. 19: 103. 1951

Leaves appearing in late winter, withering in late spring before anthesis, suberect, to 5 cm wide. **Scape bracts** lanceolate, 4–5 cm. $2n = 52$.

Flowering mid summer–early fall. Heavy soils near streams that periodically overflow and seepage slopes; 0–100 m; Okla., Tex.

Hymenocallis occidentalis var. *eulae* is in need of additional study. The type collection is in poor condition, and only a small number of other collections were available for study. The floral data now available does not justify recognizing this entity as a separate species. However, the presence of suberect leaves and their senescence before anthesis do indicate a degree of genetic difference that we consider sufficient to support varietal rank.

2. Hymenocallis coronaria (J. Le Conte) Kunth, Enum. Pl. 5: 855. 1850 • Shoals spider-lily, cahaba-lily [E] [F]

Pancratium coronarium J. Le Conte, Ann. Lyceum Nat. Hist. New York 3: 145. 1836

Bulb nonrhizomatous, globose, 4–8 × 3.5–7.5 cm; basal plate 1–2 cm; neck 2.5–3.5 cm; tunic brown. **Leaves** deciduous, 3–6, erect, (3.8–)5–9.5 dm × 1.5–4.5 cm, coriaceous; blade prominently channeled proximally, becoming expanded and reflexed in distal 1/3, liguliform, apex acute. **Scape** 4–12.5 dm, 2-edged, glaucous; scape bracts 2, enclosing buds, 4–6.5 × 1.5–2 cm; subtending floral bracts 3.5–4 × ca. 1 cm. **Flowers** 3–7(–12), opening sequentially, fragrant; perianth tube green, 4.5–7.5(–9) cm; tepals ascending, white, tinged green at base, 6.5–10 cm × 5.5–9 mm; corona white with prominent yellowish green eye, broadly funnelform to chalicelike, tubulose proximally, 4.5–7 × 5–8.5 cm, margins between free portions of filaments often with 2 prominent dentations; free portions of filaments inserted on flat sinal base, incurved, white, 1–2 cm; anthers 1–1.7 cm, pollen golden; ovary ovoid, 1–1.8 cm × 5–10 mm, ovules 2–4 per locule; style green in distal 1/3, fading into white, 11–18 cm. **Capsules** subglobose, 3 × 3 cm. **Seeds** elongate, 2.5–4.3 × 1.2–2.1 cm. $2n = 44$.

Flowering late spring–early summer. Rocky shoals of

Piedmont river systems above and at the Fall Line; 100–200 m; Ala., Ga., S.C.

Hymenocallis coronaria is under consideration for federal protection. It may be locally abundant in some places, but at other sites where it was once abundant, whole populations have been wiped out by dams (J. K. Small 1933; L. J. Davenport 1990b, 1996). The environments where populations still exist should be protected to ensure their continued existence.

Hymenocallis coronaria is one of the most magnificent plants in nature. In 1783 William Bartram was the first botanist to observe this species, as the "odoriferous *Pancratium fluitans* which almost alone possesses the little rocky islets" (W. Bartram 1791). His sighting was at the cataracts of the Savannah River, Augusta, Georgia. Distinguishing features are channeled leaves that may reach over a meter in length, and staminal coronas that are among the largest of the genus, being chalicelike at full anthesis. These characteristics make it highly desirable for cultivation in streams and brooks of our southern gardens.

There has been some confusion in the literature as to the identity of *Hymenocallis coronaria* (A. E. Radford et al. 1968). However, L. J. Davenport (1990) and G. L. Smith et al. (1990) have confirmed its distinctive characteristics and unique habitat. An interesting discovery by Davenport that *H. coronaria* seeds sink offers a mechanism for its local dispersion and establishment in favorable environments. Support for its inclusion with *H. occidentalis* in Traub's Caroliniana Alliance has been provided by the production of viable F_1 hybrids in a greenhouse cross between *H. occidentalis* (seed parent, $2n$ = 54) and *H. coronaria* (pollen parent, $2n$ = 44). As expected, the hybrids have $2n$ = 49 (D. B. Joye and G. L. Smith 1993), and after several years are still thriving in the High Point University greenhouse.

3. Hymenocallis liriosme (Rafinesque) Shinners, Field & Lab. 19: 102. 1951 • Western marsh spider-lily, Louisiana marsh spider-lily [E]

Pancratium liriosme Rafinesque, Fl. Ludov., 19. 1817; *Choretis galvestonensis* Herbert; *Hymenocallis galvestonensis* (Herbert) Baker; *H. galvestonensis* subsp. *angustifolia* Traub

Bulb nonrhizomatous, subglobose to ovoid, 4.5–6 × 5–6 cm; basal plate 2–3 cm; neck 3–5 cm; tunic grayish brown. **Leaves** deciduous, 5–8, suberect, 3.2–8.5(–10.6) dm × 1.7–3.6 cm, highly coriaceous; blade shiny green, liguliform to narrowly lanceolate, channeled, apex acute. **Scape** 3.6–6.8(–8.8) dm, 2-edged, glaucous; scape bracts 2, enclosing buds, 4.5–8 × 1.5–2 cm; subtending floral bracts 3–4 cm × 5–10 mm. **Flowers** (3–)5–12, opening sequentially, sweetly fragrant; perianth tube green, 4.5–8(–10.5) cm; tepals slightly ascending from base of corona, white, green-striped on keel, 6–9 cm × 4–8 mm; corona white with prominent, yellowish green eye, becoming rotate, shortly tubulose proximally, 2–3.5 × 4–5 cm, margins between free portions of filaments wavy to coarsely dentate; free portions of filaments inserted on flat sinal base, suberect, 1.5–3 cm; anthers 1–2.3 cm, pollen yellow; ovary subglobose, 1.2–1.8 cm × ca. 10 mm, ovules 2–3 per locule; style green in distal 1/2, fading to white proximally, 10–17 cm. **Capsules** subglobose, ca. 2 × 2 cm. **Seeds** ca. 2 × 1.5 cm. $2n$ = 40, 42.

Flowering early–mid spring. Ditches, margins of marshes and ponds, swamps, alluvial woods; 0–200 m; Ala., Ark., La., Miss., Okla., Tex.

Hymenocallis liriosme has become a favorite for gardens west of Mobile Bay, Alabama. Its leaves are numerous, liguliform to narrowly lanceolate, highly coriaceous, and shiny green. The numerous, small, highly fragrant flowers with the center of the staminal corona having a prominent, yellowish green eye are characteristics useful in distinguishing this species from *H. occidentalis* (W. S. Flory 1950) and from *H. choctawensis* (R. Gooch and G. L. Smith 1994; G. L. Smith and N. C. Coile 1999).

Numerous herbarium specimens of *Hymenocallis liriosme* were identified by H. P. Traub as *H. galvestonensis*. However, close examination of Rafinesque's original publication of *Pancratium liriosme* leaves no doubt that this earlier name does apply to this taxon and that therefore its epithet has priority.

4. Hymenocallis choctawensis Traub, Pl. Life 18: 70. 1962 • Florida panhandle spider-lily [E]

Bulb rhizomatous, ovoid to subglobose, 3–6 × 1.5–5 cm; basal plate 1.5–5 cm; neck 2–8 cm; tunic grayish brown. **Leaves** deciduous, 5–9, arching upward to suberect, 3.5–7.5(–8.5) dm × 2.5–6 cm, coriaceous; blade shiny green, oblanceolate, channeled proximally, distinctly wider beyond middle before tapering gradually towards base, not glaucous, apex acute. **Scape** 3.5–6.2 dm, 2-edged, glaucous; scape bracts 2–3, enclosing buds, 3–4(–6) × 1.5–2 cm; subtending floral bracts 2.8–3.7 cm × 10–15 mm. **Flowers** 2–8(–12), opening sequentially, highly fragrant; perianth tube green, 6–11 cm; tepals slightly ascending to spreading, white, tinged green on keel and at base, 8–12 cm × 5–9 mm; corona white with small, yellowish green eye, funnelform, gradually spreading, shortly tubulose proximally, 2.5–4(–4.6) × 4–5.5 cm, margins between free portions of filaments irregularly 2- or 3-dentate, projections ranging from slight to prominent; free portions of filaments inserted on flat sinal base, suberect, white, 2.5–4 cm; anthers 1.5–2.5

cm, pollen golden; ovary ovoid, 1–1.5 cm × 5–10 mm, ovules 1–3 per locule; style green in distal 1/2, fading into white proximally, 13–19 cm. **Capsules** broadly conic to subglobose, 3–4 × 3 cm. **Seeds** obovoid to elongate, 2–3(–4) × 1.4–2.3 cm. 2*n* = 44.

Flowering spring–early summer. Swamp and floodplain forests, stream and river banks; 0 m; Ala., Fla., Ga., La., Miss.

Hymenocallis choctawensis is not well known (G. L. Smith et al. 1991). It is most abundant along the stream banks and swamp forests of the central and western Florida panhandle. It has great horticultural potential due to its large, oblanceolate, shiny green leaves and showy inflorescences that may have up to but rarely more than eight flowers. The long, spreading tepals and the funnelform staminal corona that gradually spreads in time are especially distinctive (G. L. Smith and N. C. Coile 1999; G. L. Smith and M. A. Garland 1996).

Most herbarium specimens of *Hymenocallis choctawensis* encountered in this study have been misidentified as either *H. caroliniana* or *H. occidentalis.* It is readily distinguished from *H. occidentalis* on herbarium sheets by its rhizomatous bulb and triangular bracts that are not long-tapering.

5. Hymenocallis crassifolia Herbert, Appendix, 44. 1821 • Coastal Carolina spider-lily [E]

Hymenocallis palusvirensis Traub; *Pancratium crassifolium* (Herbert) Schultes

Bulb rhizomatous, narrowly ovoid, 2–5 × 1.5–3 cm; basal plate 1–2.5 cm; neck 3–9 cm; tunic light gray. **Leaves** deciduous, 3–5, nearly erect, (2.4–)3.5–6.8 dm × 1–2.5 cm, coriaceous; blade shiny green, liguliform, channeled in proximal 1/2, apex acute. **Scape** 3–5 dm, 2-edged, glaucous; scape bracts 2, enclosing buds, 2.5–5 cm × 7–12 mm; subtending floral bracts 2–3 cm × ca. 5 mm. **Flowers** 2–3, opening sequentially, fragrant; perianth tube green, 5–8(–10) cm; tepals slightly ascending, white, tinged green on keel, especially at base, (6–)7–9(–11.5) cm × 3–6 mm; corona white with small green eye, funnelform at full anthesis, then gradually spreading, shortly tubulose proximally, (2–)2.5–3.5 × 3.5–4.5 cm (unbroken on herbarium sheets), margins between free portions of filaments irregularly dentate, projections not prominent; free portions of filaments inserted on flat sinal base, suberect, white, 1.5–3 cm; anthers 1–2 cm, pollen golden; ovary green, ovoid, 0.8–1.5 cm × ca. 5 mm, ovules 2–3 per locule; style green in distal 1/2, rapidly fading to white proximally, (10–)13–18 cm. **Capsules** subglobose, 2–3 × 2–3 cm. **Seeds** elongate, 1.5–2 × 1–1.5 cm. 2*n* = 40.

Flowering mid–late spring. Ditches, margins of streams and rivers, bogs, brackish marshes, alluvial woods; 0 m; Fla., Ga., N.C., S.C.

Hymenocallis crassifolia was described from a collection made near St. Mary, Georgia, by Herbert. The type specimen appears to have been lost for some time. However, collections that have been made from the nearby Satilla River fit the description well, as do other collections made from wetlands in the outer Coastal Plain of the Carolinas.

This species is distinguished in the field by coriaceous, liguliform leaves that are nearly erect throughout their length. The staminal corona is funnelform but gradually spreads with age, and the margin of the cup has small, irregular dentations.

Hymenocallis crassifolia is difficult to distinguish on herbarium sheets from the closely allied *H. duvalensis.* They are separated by distribution and habitat, and it is sometimes apparent on the herbarium sheets that *H. crassifolia* tends to have a more slender perianth tube, smaller funnelform corona, bracts that are not noticeably long-acuminate, and often a less elongated bulb neck.

6. Hymenocallis pygmaea Traub, Pl. Life 18: 70. 1962 • Dwarf spider-lily [E]

Bulb rhizomatous, narrowly ovoid, 1.3–3 × 1–2 cm; basal plate ca. 1 cm; neck 2–5 cm; tunic gray. **Leaves** deciduous, 2–5, suberect, 1.5–4 dm × 0.5–1.3(–2) cm, coriaceous; blade shiny green, narrowly liguliform, apex subacute. **Scape** 1.2–3(–4.2) dm, 2-edged, glaucous; scape bracts 2, enclosing buds, 2–3 × 0.5–1 cm; subtending floral bracts 1.5–2.5 cm × 3–5 mm. **Flowers** 1–2, opening sequentially, with lemony fragrance; perianth tube green, 4–6 cm; tepals slightly ascending, white, tinged green on keel, 5–7 cm × 3–6 mm; corona white with small green eye, funnelform at full anthesis, then gradually spreading, shortly tubulose proximally, 2–3 × ca. 3 cm, margins between free portions of filaments irregularly dentate, projections small; free portions of filaments inserted on flat sinal base, suberect, white, 1.8–2.8 cm; anthers 1–1.5 cm, pollen golden; ovary ovoid, ca. 1 cm × 5 mm, ovules 2 per locule; style green in distal 1/3, fading to white proximally, 10–13 cm. **Capsules** subglobose, ca. 2 × 2 cm. **Seeds** obovoid, ca. 1.5 × 1 cm. 2*n* = 40, 41.

Flowering mid spring. Bogs and stream banks; 0 m; N.C., S.C.

Hymenocallis pygmaea was described from bulbs collected by Mary G. Henry along a small tributary of the Waccamaw River in South Carolina. Populations that we have studied and vouchered from the Waccamaw

River drainage are characterized by extremely narrow, coriaceous, nearly erect, shiny leaves and small flowers. This species could be considered a cytological and/or ecological variant of *H. crassifolia*, but because of its smaller stature and features it is maintained here as a separate species. Additional studies on the dwarf spider-lilies of the Atlantic Coastal Plain should be undertaken to clarify their taxonomic status.

7. Hymenocallis duvalensis Traub, Pl. Life 23: 66. 1967 • White sands spider-lily, Dixie spider-lily [E]

Bulb rhizomatous, narrowly ovoid, 2–4 × 1–3 cm; basal plate 1–2 cm; neck 3–10 cm; tunic light gray. **Leaves** deciduous, 4–7, arching low, often appearing prostrate, 2.4–5.2 dm × 1–2.5 cm, not highly coriaceous; blade narrowly liguliform or narrowly oblanceolate, apex acute. **Scape** 2.4–4 dm, 2-edged, glaucous; scape bracts 2, enclosing buds, 3–5.5 cm × 5–12 mm; subtending floral bracts 2.7–3 cm × 3–4 mm. **Flowers** 2–3, rarely more, opening sequentially, intensely fragrant; perianth tube green, 5–9(–10.5) cm; tepals slightly ascending, tinged green on keel, 7–10 cm × 4–6 mm; corona white with small green eye, becoming rotate at full anthesis, shortly tubulose proximally, (2.5–)3–4 × 4–5 cm, margins between free portions of filaments irregularly and coarsely dentate; free portions of filaments inserted on flat sinal base, suberect, white, 2.5–3 cm; anthers 1–1.7 cm, pollen golden; ovary green, ovoid, ca. 1 cm × 5 mm, ovules 2–4 per locule; style green in distal $1/3$, fading into white proximally, 13–17 cm. **Capsules** subglobose, ca. 2.5 × 2.5 cm. **Seeds** elongate, 1.3–2.1 × 0.7–1.2 cm. $2n = 42$.

Flowering early–mid spring. Floodplain forests and sandy stream and river banks; 0 m; Fla., Ga.

Hymenocallis duvalensis is another southeastern spider-lily that is not well known (S. Puffenbarger and G. L. Smith 1993; G. L. Smith et al. 1999). It forms dense colonies along the interior floodplains of streams in north Florida, extending to the Ochlockonee River and up into south-central Georgia. Its arching, deep green, liguliform or sometimes narrowly oblanceolate leaves and rotate staminal corona at full anthesis make it easily identifiable in the field. Newly opened flowers are intensely fragrant. This species would make a handsome addition to the gardens of the South.

Herbarium specimens of *Hymenocallis duvalensis* may be hard to distinguish from those of *H. crassifolia*. The less coriaceous, arching leaves and rotate staminal corona are excellent field characteristics, but they are not readily discernible on herbarium specimens. Characteristics that are more reliable for identification on some herbarium specimens are the narrowly oblanceolate leaves, the width of the corona (unbroken), and the width of the perianth tube. The tube is often more robust than in *H. crassifolia*, measuring 1.5–3.5 mm wide. Numerous herbarium sheets have been misidentified as either *H. crassifolia* or *H. rotata*.

8. Hymenocallis franklinensis Gerald L. Smith, L. C. Anderson & Flory, Novon 11: 235, figs. 1, 5, 6. 2001 • Cow Creek spider-lily [E]

Bulb rhizomatous, narrowly ovoid, 2.5–4 × 2–3.5 cm; basal plate 1–2 cm; neck 3.5–7 cm; tunic grayish brown. **Leaves** deciduous, 3–5, nearly erect, 2.8–4.8 dm × 1.8–2.7 cm, coriaceous; blade lustrous green, liguliform, distinctly channeled, apex acute. **Scape** 3–4.5 dm, 2-edged, glaucous; scape bracts 2, enclosing buds, 3–4.5 cm × 10–15 mm; subtending floral bracts 2.5–4.5 cm × 5–12 mm. **Flowers** 2–3, opening sequentially, moderately fragrant; perianth tube green, 6.5–9 cm; tepals slightly ascending, white, tinged green on keel, 8.8–11.5 cm × 5–7 mm; corona white with small, yellowish green proximal eye, funnelform at full anthesis, then spreading with age, shortly tubulose proximally, 3–4 × 4–5.5 cm, margins between free portions of filaments with irregularly dentate projections; free portions of filaments inserted on flat sinal base, suberect, white, 2.5–3.5 cm; anthers 1.5–1.8 cm, pollen golden; ovary ovoid, 1–1.5 cm × 5 mm, ovules 2–3 per locule; style green in distal $1/2$, fading to white proximally, 14–17 cm. **Capsules** subglobose, 2–2.5 × 2–2.5 cm. **Seeds** obovoid, 1.6–2 × 1.2–1.5 cm. $2n = 43, 44$.

Flowering early–mid spring. Alluvial margins of slightly brackish rivers and streams; 0 m; Fla.

Hymenocallis franklinensis is a narrow endemic of the lower Ochlockonee River system of the Florida panhandle, possibly including the Sopchoppy River. Only a few collections have been made (*L. C. Anderson 13,382, 18,299*; *G. L. Smith, M. A. Garland & G. R. Knight 1631*; *G. L. Smith & L. C. Anderson 1717*). The plants bear a strong resemblance to *H. crassifolia*. The leaves of the two species are nearly identical in form in the field and on herbarium sheets, but the flowers of *H. franklinensis* are larger, and the scape bracts are broader at the base and taper in the distal half.

9. Hymenocallis rotata (Ker Gawler) Herbert, Appendix, 44. 1821 • Spring-run spider-lily [C] [E] [F]

Pancratium rotatum Ker Gawler, Bot. Mag. 21: plate 827. 1805; *Hymenocallis floridana* (Rafinesque) C. V. Morton; *H. floridana* subsp. *amplifolia* Traub; *H. laciniata* Small

Bulb rhizomatous, broadly ovoid, 6–8 × 3–6 cm; basal plate 3–6 cm; neck 3–5 cm; tunic grayish brown. **Leaves** deciduous, 3–9, arching upward to suberect, 4–10 dm × 2–5(–7) cm, coriaceous; blade deep green, liguliform, strongly channeled proximally, apex acute, glaucous at base. **Scape** 3.5–7 dm, 2-edged, glaucous; scape bracts 3, enclosing buds, 4–7.5 × 1.3–2 cm; subtending floral bracts 3.5–6 cm × 5–10 mm. **Flowers** 2–4, opening nearly at same time, fragrant; perianth tube green, (7.5–)9–12 cm; tepals extending nearly horizontally from base of corona, white, green-striped along keel and at base and apex, 9–14.5 cm × 7–15 mm; corona white with small, yellowish green proximal eye, rotate at full anthesis, distinctly tubulose proximally, 4–6 × 6–9 cm, margins between free portions of filaments with 2 or 3 prominent lacerate projections; free portions of filaments inserted on flat base, suberect, white except green proximally, 2.5–4 cm; anthers 1.5–2 cm, pollen yellow; ovary oblong to pyramidal, 2–3 cm × ca. 10 mm, ovules 4–6 per locule; style green in distal 1/2, becoming white proximally, 16–26 cm. **Capsules** elongate, 4–6 × 2–3 cm. **Seeds** elongate, 2–3 × 1.5–2 cm. $2n = 48$.

Flowering late spring–early summer. Margins of spring-run streams; of conservation concern; 0 m; Fla.

Hymenocallis rotata is known chiefly from the spring runs south of Tallahassee, and is found also along other springs to the east and down the peninsula. Although it may be locally abundant, it is restricted to the immediate vicinity (including ditches and lakes) of the spring runs.

A. F. Clewell (1985) treated *Hymenocallis rotata* as the sole representative of the genus in the Florida panhandle, concurring with the decision made by R. K. Godfrey and J. W. Wooten (1979). Collections often have been misidentified as either *H. crassifolia* or *H. coronaria*, and the name *H. floridana* has been applied to some specimens.

Hymenocallis rotata is a striking species with its large, liguliform, deep green leaves, and flowers with large, rotate staminal coronas. In width, the corona is similar to that of *H. coronaria*, but it is not chalicelike in shape, and its free tepal tips are widely spreading rather than ascending. In cultivation as a garden ornamental since the seventeenth century, *H. rotata* seems to thrive in most semitropical areas, especially along the banks of streams or wet ditches.

10. Hymenocallis tridentata Small, Man. S.E. Fl., 323. 1933 • Florida spider-lily [E]

Hymenocallis traubii Moldenke

Bulb rhizomatous, ovoid, 4–6 × 2–4 cm; basal plate 2–3 cm; neck 2–5 cm; tunic grayish brown. **Leaves** deciduous, 3–7, arching, reflexed at apex, 2–5.5 dm × 1–1.8(–2.5) cm, longest leaf exceeding 4 dm at anthesis, coriaceous; blade shiny green, narrowly liguliform, channeled proximally, apex acute. **Scape** 3.5–5 dm, 2-edged, glaucous; scape bracts 2, enclosing buds, 4–5.5(–6.5) × 1–1.5 cm; subtending floral bracts 2.5–5 cm × 4–8 mm. **Flowers** (1–)2(–3), one opening slightly before the other, fragrant; perianth tube green, 6–8(–9.5) cm; tepals slightly ascending, white, distal green-striped on keel, 7–11(–12.5) cm × 5–8 mm; corona white with small, yellowish green proximal eye, broadly funnelform to rotate, tubulose proximally, 3–4.5 × 4–6 cm, margins between free portions of filaments with 2–3 prominent lacerate projections; free portions of filaments inserted on flat sinal base, suberect, white, 2–4 cm; anthers 1.5–2 cm, pollen yellow; ovary oblong, 1.5–2 cm × 10 mm, ovules 4–6 per locule; style green in distal 1/2, fading to white proximally, 15–19 cm. **Capsules** subglobose, 3–4 × 2.5–3 cm. **Seeds** obovoid, 2.4–2.7 × 1.5–1.9 cm.

Flowering early–mid spring. Marshes, wet prairies; 0 m; Fla.

Little information about *Hymenocallis tridentata* was available before a study by A. Herndon (1987), who contrasted its traits (under the name *H. floridana*) with those of *H. palmeri*, remarking that as one proceeds northward, the plants bearing two flowers show an increase in floral size. Additional study is needed to further resolve the distribution of *H. tridentata* and to clarify its relationship to *H. rotata*. Our data indicate that *H. tridentata* is distinctly smaller than *H. rotata* in all features.

11. Hymenocallis godfreyi Gerald L. Smith & Darst, Novon 4: 396. 1994 • Godfrey's spider-lily, St. Mark's marsh spider-lily [C] [E]

Bulb basally rhizomatous, ovoid, 3–4.5 × 2.5–3.5 cm; basal plate 1.5–2 cm; neck 2–3 cm; tunic grayish brown. **Leaves** deciduous, 3–6, nearly erect, (1.3–)2–3.8 dm × 1–2.5 cm, coriaceous; blade yellowish green, liguliform, channeled proximally, apex obtuse to subacute. **Scape** 2–3 dm, 2-edged, glaucous; scape bracts 2, enclosing buds, 4–5 × 1–1.5 cm; subtending floral bracts 3.5–5 cm × 5–10 mm. **Flowers** 2, one opening slightly before the other, fragrant; perianth tube green, 4.5–7.5(–8.5) cm; tepals slightly ascending, white, distal green-striped on keel, 7–10.5 cm × 5–8 mm; corona white with small, yellowish green eye, broadly funnelform, shortly tubulose proximally, 3–4.5 × 5 cm, margins between free portions of filaments erect, with 2 prominent lacerate projections; free portions of filaments inserted on flat depression, slightly incurved, white, 1.5–3 cm; anthers 1.2–1.5 cm, pollen golden; ovary oblong to slightly pyriform, 1.5–2 cm × 10 mm, ovules 4–6 per locule; style green in distal $^{1}/_{2}$, fading to white proximally, 13–17 cm. **Capsules** subglobose, ca. 3.5 × 2.5 cm. **Seeds** obovoid, 2–2.5 × 1.3–1.8 cm. $2n = 48$.

Flowering early–mid spring. Brackish marshes; of conservation concern; 0 m; Fla.

Hymenocallis godfreyi is endemic to the marsh area near Ft. San Marcos de Apalache along the St. Mark's River. It is somewhat abundant locally, but because of its highly restricted range it should be recommended for conservation. G. L. Smith and M. Darst (1994) have proposed a relationship to *H. rotata*, from which it likely evolved. It is distinguished from that species by its short scape, short, erect coriaceous, yellowish green leaves, broadly funnelform staminal corona with prominent lacerate projections, and unique basal rhizome.

12. Hymenocallis puntagordensis Traub, Pl. Life 18: 71. 1962 • Small cup spider-lily [C] [E]

Bulb nonrhizomatous, narrowly ovoid, 5–8 × 2–4.5 cm; basal plate 1–2 cm; neck (5–) 8–12 cm; tunic dark brown. **Leaves** evergreen, 5–8, suberect, 3.5–7.5 dm × 1.5–3 cm, highly coriaceous; blade shiny green, narrowly liguliform, channeled proximally, apex acute. **Scape** 5–6 dm, distinctly 2-edged, glaucous; scape bracts 2, enclosing buds, 4–6 × ca. 1.5 cm; subtending floral bracts 3–5 cm × 5–10 mm. **Flowers** 3–5, opening sequentially, fragrant; perianth tube green, (6–)7.5–12 cm; tepals ascending, green on keel, (7–)9.5–11.5 cm × 4–6 mm; corona white with small, yellowish green eye, funnelform, 2–2.8 × 2–3 cm, margins between free portions of filaments with 1 or 2 prominent lacerate projections; free portions of filaments inserted on flat base, slightly curved, green in distal $^{1}/_{2}$, fading to white proximally, (2.5–)3.5–4.5 cm; anthers 1.3–2 cm, pollen yellow; ovary pyriform, 1.5–2.4 cm × ca. 10 mm, ovules 5–9 per locule; style green in distal $^{1}/_{2}$, fading to white proximally, 13–22 cm. **Capsules** unknown. **Seeds** unknown. $2n = 46$.

Flowering late summer. Roadsides and margins of disturbed pine flatwoods; of conservation concern; 0 m; Fla.

Hymenocallis puntagordensis is one of the most poorly known spider-lilies (G. L. Smith et al. 1994; G. L. Smith and M. A. Garland 1996). Apparently it occurs only in the vicinity of Punta Gorda. It was known only from the type collection until collections were made in 1993 (*G. L. Smith & M. A. Garland 1625*) and 1995 (*G. L. Smith & M. A. Garland 1664–1668*).

This species has been designated in the horticultural trade as narrow-leaved *Hymenocallis latifolia*, but it is easily distinguished from *H. latifolia* by its highly coriaceous, suberect, narrowly liguliform leaves and its small staminal corona with prominently lacerate margin.

13. Hymenocallis palmeri S. Watson, Proc. Amer. Acad. Arts 14: 301. 1879 • Alligator-lily [E]

Hymenocallis humilis S. Watson

Bulb nonrhizomatous, ovoid, 2.5–4 × 1.5–3.5(–5) cm; basal plate 1–3.5 cm; neck 3–7 cm; tunic grayish brown. **Leaves** deciduous, 2–6, erect, 1.5–4(–5–6.5) dm × 4–10 mm, coriaceous; blade bluish green, linear, channeled, margins hyaline, apex acute, glaucous. **Scape** 1.8–4.5(–6) dm, suborbiculate, glaucous; scape bracts 2, not enclosing flower bud, 4–6 cm × 4–7 mm; subtending floral bracts 1–2, 3.7–5 cm × 3–5 mm. **Flowers** 1, rarely 2, then opening sequentially, faintly fragrant; perianth tube green, 7.5–12 cm; tepals ascending, yellow-green to pale green, linear, 7.5–12.5 cm × 3–7 mm (unfolded); corona white with faint green eye, funnelform, margins with some gradual spreading in time, shortly tubulose proximally, 2.5–3.5(–5) × 3–5 cm, margins between free portions of filaments irregularly dentate, with some filiform projections; free portions of filaments inserted on flat sinal base, erect, white, 2.5–4 cm; anthers 1.2–2 cm, pollen deep yellow; ovary oblong to slightly pyriform, 1.4–2.2 cm × 6–12 mm, ovules (3–)5–8 per locule; style green in distal $^{1}/_{2}$, fading to white proximally, 13–20 cm. **Capsules** ellipsoid, 3.5–4 × ca. 2.5 cm. **Seeds** obovoid, 1.5–2 × 1–1.4 cm. $2n = 42, 46, 48$.

Flowering late spring–summer. Cypress swamps, wet prairies, open pine flatwoods, wet roadsides; 0 m; Fla.

Hymenocallis palmeri is most closely allied to *H. henryae*, having similar leaf and floral characteristics, but it is distinguished by its linear leaves, single flower, and perianth tube that is often decidedly longer than the tepals (G. L. Smith and W. S. Flory 1990).

14. Hymenocallis henryae Traub, Pl. Life 18: 71. 1962 • Green spider-lily, Henry's spider-lily [C] [E]

Bulb nonrhizomatous, broadly ovoid, 4–7(–9) × 4.5–6.5(–8) cm; basal plate 1–3 cm; neck 4–6 cm; tunic grayish brown. **Leaves** deciduous, 3–9, erect, (2–)3–6.7 dm × (0.8–)1.5–3.2 cm, coriaceous; blade bluish green to deep green, narrowly liguliform, broadly channeled, margins hyaline, apex acute, glaucous to slightly glaucous, or not glaucous. **Scape** (3.5–)4.5–7.5 dm, suborbiculate, glaucous; scape bracts 2, not enclosing flower buds, 4.5–5.5 × 1–1.5 cm; subtending floral bracts 3.8–6.1 cm × 5–10 mm. **Flowers** 2(–3 rarely), opening sequentially, mildly fragrant; perianth tube green, robust, 6–10(–12) cm; tepals slightly ascending to long-spreading, greenish white, 8.8–16 cm × 4.5–9.5(–11) mm; corona white with faint green proximal eye, funnelform, gradually spreading in time, shortly tubulose proximally, 3–4 × 5–6 cm, margins between free portions of filaments irregularly dentate; free portions of filaments inserted on flat sinal base, slightly incurved, white, 2.8–4.5 cm; anthers 1.5–2.5 cm, pollen yellow; ovary oblong to pyriform, 1.5–3 cm × 10 mm, ovules 4–8 per locule; style green in distal 1/2, fading to white proximally, (13–)16–20 cm. **Capsules** subglobose, shortly beaked, 3.5–4.5 × 3–4 cm. **Seeds** obovoid, 1.5–2.2 × 1.2–1.5 cm. 2*n* = 38.

Varieties 2 (2 in the flora): Florida.

Hymenocallis henryae is being considered for federal protection. It has received the highest endangerment ranking by the Florida Natural Areas Inventory (Gary Knight, FNAI Director, pers. comm.) and is known only from scattered localities in Liberty, Gulf, Bay, and Walton counties. It is readily distinguished in the field by consistently having two flowers per plant and long, pale green tepals radiating from the base of a white, funnelform staminal corona. The erect, liguliform leaves vary in size and glaucousness over its range (G. L. Smith and J. N. Henry 1999).

Hymenocallis henryae was designated "*H. viridiflora*" by J. K. Small (1933). This name is listed as number 10 in Small's key, but as a result of a clerical error (C. V. Morton 1935), species number 10 was described by Small under the name *H. rotatum* Le Conte. Even if Small had applied the name *H. viridiflora* to the description, as he evidently had intended, that name still would be invalid, since it would have been a later homonym of *H. viridiflora* (Ruiz & Pavón) R. W. Wallace from Peru.

1. Leaf blades not glaucous or only slightly so; plants growing singly or loosely clumped . . . . 14a. *Hymenocallis henryae* var. *henryae*
1. Leaf blades glaucous; plants often growing in dense clumps . . . . . 14b. *Hymenocallis henryae* var. *glaucifolia*

14a. Hymenocallis henryae Traub var. **henryae** C E

Plants growing singly or in loose, scattered clumps. **Leaves** 3–9, erect, (2–)3–6.7 dm × 0.8–3.2 cm; blade liguliform, not glaucous, or slightly glaucous to deep shiny green. **Tepals** no longer than 13 cm.

Flowering late spring. Cypress drainages and stringers associated with wet pine flatwoods; of conservation concern; 0 m; Fla.

Hymenocallis henryae var. *henryae* is only found in Bay, Gulf, and Walton counties.

14b. Hymenocallis henryae Traub var. **glaucifolia** J. N. Henry & Gerald L. Smith, Herbertia 54: 114. 1999 C E

Plants often growing in dense clumps. **Leaves** 4–7, erect, 3–6 dm × 1.9–2.9 cm; blade blue-green in appearance due to glaucousness. **Tepals** to 16 cm.

Flowering late spring. Cypress depressions at edges of pine flatwoods; of conservation concern; 0 m; Fla.

Hymenocallis henryae var. *glaucifolia* occurs only in Liberty County and represents an apparent divergence of the populations of *H. henryae* east of the Apalachicola River. In these plants, a heavy glaucousness causes the surface of the leaves to appear blue-green. This characteristic, along with the clumping habit and tepal length to 16 cm, supports recognition of these Liberty County populations as a distinct variety.

15. Hymenocallis latifolia (Miller) M. Roemer, Fam. Nat. Syn. Monogr. 4: 168. 1847 • Broad-leaf spider-lily, mangrove spider-lily E

Pancratium latifolium Miller, Gard. Dict. ed. 8, Pancratium no. 8. 1768; *Hymenocallis caymanensis* Herb; *H. collieri* Small; *H. keyensis* Small; *H. kimballiae* Small

Bulb nonrhizomatous, forming numerous bulblets, enlarging with age, 4–8 × 2–5 cm; basal plate 2–4 cm; neck 4–6 cm; tunic grayish brown. **Leaves** evergreen, numerous, arching outwards, spreading distally, 5–10 dm × 4–9 cm, coriaceous; blade broadly liguliform, gradually tapering proximally, apex subacute to acute. **Scape** 4–10 dm, slightly compressed, glaucous; scape bracts 2, enclosing buds, 4–7 × 1.5–2.5 cm; subtending floral bracts 4–5 cm × 10–15 mm. **Flowers** 9–15, opening sequentially, fragrant; perianth tube green, 9–20 cm; tepals extending nearly horizontally, reflexed distally, white, 7.5–13 cm × 4–6 mm; corona white with small green eye, funnelform, 2–3 × 3–4 cm, margins between free portions of filaments wavy, without prominent projections; free portions of filaments inserted on flat base, slightly incurved, green except becoming white near insertion, 4–6 cm; anthers 1.2–2 cm, pollen orange; ovary ovoid, 0.9–1.6 cm × 5–10 mm, ovules 2 per locule; style green in distal 1/2, fading to white proximally, 19–23 cm. **Capsules** subglobose, ca. 3 × 3 cm. **Seeds** obovoid, 1.8–2.4 × 1.4–1.8 cm. 2*n* = 46, 70.

Flowering summer. Coastal dunes, sand ridges, edges of hammocks and flatwoods, mangrove swamps of southern Florida peninsula and keys, rarely disjunct along coastal nw Florida and Gulf coastal states; 0 m; Fla.

The long perianth tube in *Hymenocallis latifolia* contrasts sharply with the short, funnelform staminal corona. This species is a yard favorite throughout Florida.

51. LEUCOJUM Linnaeus, Sp. Pl. 1: 289. 1753; Gen. Pl. ed. 5, 140. 1754 • Snowflake, nivéole [Greek *leukos*, white, and *ion*, violet, alluding to the color and scent of the flowers] [I]

Gerald B. Straley†

Frederick H. Utech

Herbs, perennial, scapose, from brown, globose to ovoid, tunicate bulb. **Leaves** several; blade linear-ligulate, base sheathing. **Scape** stout, hollow [slender, solid]. **Inflorescences** umbellate, 2–5(–7)-flowered, spathaceous; spathe bracts 2, free or entirely adnate on 1 side, appearing monophyllous. **Flowers** nodding; perianth campanulate; tepals 6, distinct, oblanceolate to ovate, equal; stamens 6, distinct; anthers basifixed, conic, longer than filaments, blunt apically, dehiscing by terminal pores; ovary inferior, green, 3-locular, globose, septal nectaries present; style filiform or clavate, exceeding anthers; stigma minutely capitate. **Fruits** capsular, erect, pyriform to subglobose, dehiscence loculicidal; pericarp somewhat fleshy. **Seeds** numerous, black, appendages absent. x = 7, 8, 9, 10, 11, 12.

Species ca. 10 (1 in the flora): introduced; c, sw Europe, nw Africa, and sw Asia (Armenia, Crimea); introduced elsewhere.

Several species of *Leucojum* are cultivated for their flowers. Although other species such as *L. autumnale* Linnaeus (autumn snowflake) and *L. vernum* Linnaeus (spring snowflake) may persist in old gardens, only the commonly cultivated *L. aestivum* (summer snowflake) is known definitely to be naturalized in the flora. There are unconfirmed reports that *L. vernum* is naturalized in the panhandle area of Florida.

Leucojum is sometimes confused with *Galanthus*, a spring-flowering, Eurasian relative, but it differs in that *Leucojum* has perianth segments that are all equal in size, and hollow stems that are usually taller and bear 2–5 flowers. In the southern states, the common name snowdrop, which usually refers to *Galanthus* species, is applied to *Leucojum*.

Leaves and bulbs of *Leucojum* contain the alkaloids lycorine and galanthamine, and are poisonous.

SELECTED REFERENCES Crespo, M. B. M., D. Lledo, M. F. Fay, and M. W. Chase. 1996. Molecular phylogeny of *Leucojum* based on ITS sequences. [Abstract.] Amer. J. Bot. 83(6, suppl.): 149. Stern, F. C. 1956. Snowdrops and Snowflakes—A Study of the Genera *Galanthus* and *Leucojum*. London.

1. Leucojum aestivum Linnaeus, Syst. Nat. ed. 10, 2: 975. 1759 • Summer snowflake [F] [I]

Plants to 30–60 cm; bulb 2–4 × 2–3(–4) cm. **Leaves:** sheaths 3–6 mm, membranous; blade deep green, linear to ligulate, 30–50(–70) × 0.5–2 cm, apex obtuse, glossy, well developed at anthesis. **Scape** stout, hollow, 2-winged, compressed, (20–)30–60 cm. **Spathe bract** solitary, fused, green, lanceolate, 3–5 cm. **Flowers:** tepals white with yellowish green spot near apex, broadly ovate, (10–)13–22 × 10–12 mm; filaments white, slender, 2–5 mm; anthers 5–6 mm; ovary oblong-ovate, 6–10 × 5–8 mm; style clavate, 5–10 mm, exceeding stamens; stigma truncate, inconspicuous; pedicel 2.5–7 cm, longest equaling or exceeding spathe. **Seeds** black, 5–7 mm; coat loose, with air pockets at each end. $2n$ = 22, 24.

Flowering spring–early summer. Roadsides, fields, waste places; 0–500 m; introduced; N.S.; Ala., Ark., Conn., Del., Ga., Ill., Ind., Ky., La., Maine, Md., Mass., Miss., N.Y., N.C., Ohio, Pa., S.C., Tenn., Tex., Va.; c, s Europe; naturalized throughout temperate areas.

52. NARCISSUS Linnaeus, Sp. Pl. 1: 289. 1753; Gen. Pl. ed. 5, 141. 1754 • Narcisse [from Greek *Narkissos*, mythological youth who fell in love with his own reflection and changed into a flower] [I]

Gerald B. Straley†

Frederick H. Utech

Herbs perennial, scapose, from ovoid, tunicate bulbs. **Leaves** (1–)several; blade linear to ligulate, flat to semiterete, fleshy. **Inflorescences** umbellate in clusters of 2–20, or solitary, spathaceous; spathe 1-valved, enclosing buds, membranous or papery. **Flowers** pedicellate or sessile, erect or declinate, often fragrant; tepals 6, connate proximally, distinct and reflexed to ascending distally, yellow and/or white; perianth tube surmounted by a cupular to trumpetlike corona with margins often frilled; stamens 6, epitepalous, often of 2 lengths; filaments separate from corona; anthers basifixed; ovary inferior, 3-locular; style often exserted; stigma minutely 3-lobed. **Fruits** capsular, 3-locular, papery to leathery, dehiscence loculidical. **Seeds** numerous, subglobose, often with elaiosomes; testa black. $x = 7, 11$.

Species ca. 26 (5 in the flora): introduced; Europe, n Africa, Asia; introduced and naturalized elsewhere.

Narcissus species and especially a vast array of their natural hybrids and garden cultivars are among the most popular spring flowers (A. Huxley et al. 1992). Many species are extremely variable due to horticultural selection and naturalization. Besides the following species, many of the cultivars also may persist around old gardens, although they never fully naturalize.

All parts of the plant are poisonous, especially the bulb, due to phenanthridine alkaloids such as narcissine and lycorine (G. E. Burrows and R. J. Tyrl 2001).

SELECTED REFERENCES Blanchard, J. W. 1990. *Narcissus*—A Guide to Wild Daffodils. Woking. Jefferson-Brown, M. J. 1991. *Narcissus*. Portland. Meyer, F. G. 1966. *Narcissus* species and wild hybrids. Amer. Hort. Mag. 45: 47–76.

1. Inflorescences 1-flowered.
   2. Corona tubular, ± equal in length to free portions of tepals; tepals yellow; stamens uniseriate . . . . 1. *Narcissus pseudonarcissus*
   2. Corona cup-shaped, much shorter than free portions of tepals; tepals white; stamens biseriate . . . . 2. *Narcissus poeticus*
1. Inflorescences umbellate, (1–)2–20-flowered.
   3. Leaf blades nearly terete, 2–4 mm wide; flowers uniformly golden yellow . . . . 3. *Narcissus jonquilla*
   3. Leaf blades flat, 6–15(–20) mm wide; flowers white, or yellow and white.
      4. Tepals and corona white . . . . 4. *Narcissus papyraceus*
      4. Tepals white to cream, corona yellow . . . . 5. *Narcissus tazetta*

1. Narcissus pseudonarcissus Linnaeus, Sp. Pl. 1: 289. 1753 (as pseudo narcissus) • Daffodil, trumpet narcissus [F] [I]

Bulbs ovoid, 3–5 × 2–3 cm, tunic pale brown. **Leaves** 3–4; blade flat, 20–45 cm × 5–12(–15) mm, glaucous. **Inflorescences** 1-flowered, 25–50 cm; spathe pale brown, 2–3 cm, papery. **Flowers** fragrant; perianth white, 5–7 cm wide; perianth tube 1.5–2 cm, tapering abruptly to base; distinct portions of tepals erect to spreading, yellow, often twisted, oblanceolate, 2.5–3.5 × 1–1.5 cm, apex acute; corona yellow, tubular, 30–35 × 15–25 mm, apex flared and ruffled; stamens uniseriate, exserted to ca. midlength of corona; style exserted 2–5 mm beyond anthers; pedicel 5–10 mm. $2n = 14$.

Flowering spring. Roadsides, fields, waste places; 0–1000 m; introduced; B.C., Nfld. and Labr. (Nfld.), Ont.; Ala., Ark., Conn., Ga., Ill., Ind., Ky., La., Md., Mass., Mich., Miss., Mo., N.J., N.Y., N.C., Ohio, Oreg., Pa., R.I., S.C., Tenn., Tex., Va., Wash.; w Europe; expected naturalized elsewhere.

Narcissus pseudonarcissus is the most variable species in the genus and includes many elements that some-

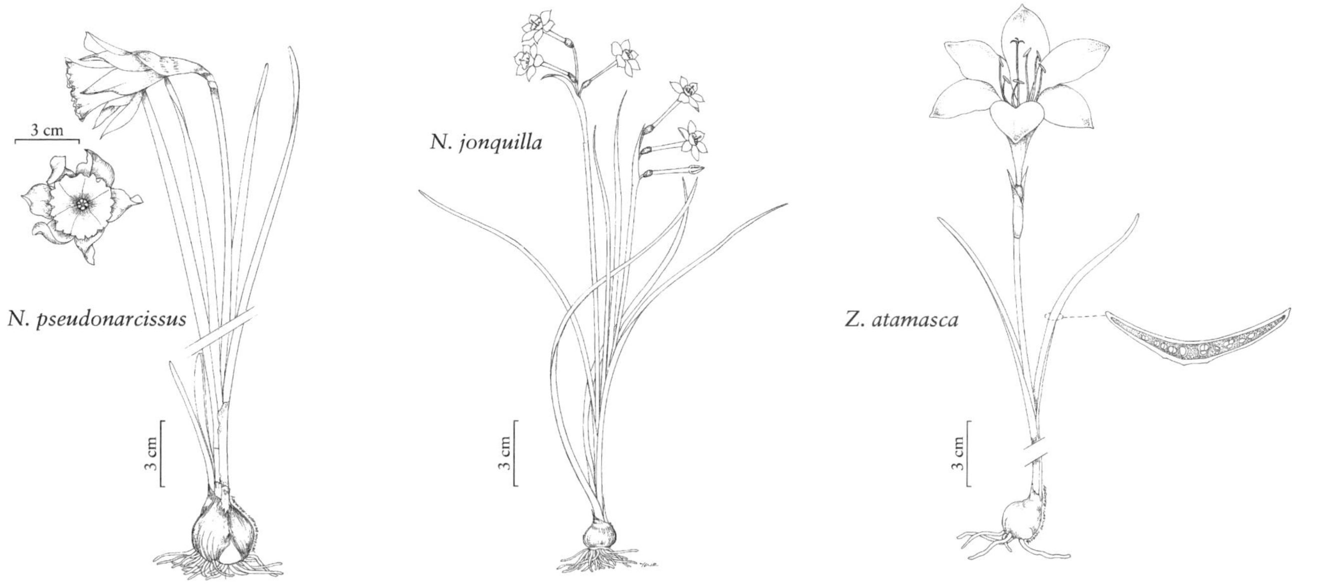

NARCISSUS ◦ ZEPHYRANTHES

times have been recognized as separate species (e.g., H. W. Pugsley 1933). An old cultivated variety, "Telemonius Plenus," with highly doubled flowers, commonly persists, although it does not reseed. Natural hybrids between *N. pseudonarcissus* and *N. poeticus* (*N.* ×*incomparabilis* Miller) have 1-flowered inflorescences and yellow flowers with the corona about half as long as the distinct portions of the tepals. They are known to persist in Kentucky, Louisiana, Mississippi, New Jersey, New York, North Carolina, Oregon, Pennsylvania, Texas, and Virginia. Natural hybrids between *N. pseudonarcissus* and *N. jonquilla* (*N.* ×*odorus* Linnaeus) have 1–4-flowered inflorescences and bright yellow flowers with the corona one-half to three-fourths as long as the distinct portions of the tepals. They are known to persist in Louisiana, South Carolina, Texas, and Virginia.

2. Narcissus poeticus Linnaeus, Sp. Pl. 1: 289. 1753
• Poet's narcissus, pheasant's-eye narcissus [I]

Bulbs ovoid, 3–4 × 1.5–2 cm, tunic brown. **Leaves** 4; blade flat, 20–40 cm × 6–10 mm, green to glaucous. **Inflorescences** 1-flowered, 30–45 cm; spathe pale brown, 4–5 cm, papery. **Flowers** fragrant; perianth 5–7 cm wide; perianth tube 2–3 cm, gradually tapering to base; tepals overlapping, distinct portions spreading to reflexed, white, ovate-orbicular, 1.5–2.5 × 1.5–2 cm, apex mucronate; corona yellow with red, crenulate margin, cup-shaped, 3–5 × 10–15 mm; 3 short stamens included in perianth tube, 3 longer stamens and style exserted into mouth of corona; pedicel 2–3 cm. $2n$ = 14, 21.

Flowering late spring. Roadsides, fields, waste places; 0–500 m; introduced; B.C., N.B., Nfld. and Labr. (Nfld.), Ont., Que.; Ala., Ark., Conn., Ga., Ill., Ky., La., Maine, Md., Mich., Miss., Mo., N.J., N.Y., N.C., Ohio, Oreg., Pa., S.C., Tenn., Utah, Vt., Va., Wash., Wis.; c, s Europe; expected naturalized elsewhere.

Natural hybrids between *Narcissus poeticus* and *N. tazetta* have been given the name *N.* ×*medioluteus* Miller (*N. biflorus* Curtis). These are more or less intermediate between the parents, and the inflorescence is usually 2-flowered. The tepals are white to pale yellow, and the short corona is dark yellow. These hybrids may persist in the flora area, as known from Alabama, Arkansas, Louisiana, Maryland, North Carolina, South Carolina, and Virginia.

3. Narcissus jonquilla Linnaeus, Sp. Pl. 1: 290. 1753
• Jonquil, jonquille [F] [I]

Bulbs ovoid, 2–3 × 2–2.5 cm, tunic dark brown. **Leaves** 2–3(–4); blade dark green, channeled adaxially, nearly terete, 30–40 cm × 2–4 mm. **Inflorescences** umbellate, 1–2(–4)-flowered, 25–35 cm; spathe pale brown, 2–4 cm, papery. **Flowers** strongly fragrant; perianth 1.5–2.5 cm wide; perianth tube 2–2.5 cm, tapering gradually to base; distinct portions of tepals spreading to reflexed, bright golden yellow, ovate to oblanceolate, 0.4–0.6 × 0.8–1 cm, apex acute to mucronate; corona golden yellow, cup-shaped, 2–4 × 5–8 mm, apex ruffled; 3 short stamens included in perianth tube, 3 longer stamens and style exserted into mouth of corona; pedicel of variable length, to 4 cm. $2n$ = 14.

Flowering late winter–early spring. Roadsides, fields, waste places; 0–100 m; introduced; B.C.; Ala., Ark., Ga., Ill., La., Md., Miss., N.C., Ohio, S.C., Tex., Utah; sw Europe (c, s Spain; e, s Portugal); expected naturalized elsewhere.

Natural hybrids between *Narcissus jonquilla* and *N. tazetta* have been given the name *N.* ×*intermedius* Louisel. The leaves of these plants are intermediate in width between those of the two parents. The inflorescence is 3–6-flowered, and the flowers are uniformly bright yellow or with a darker corona. Such hybrids are known to persist in Louisiana and can be expected elsewhere in the flora area.

4. **Narcissus papyraceus** Ker Gawler, Bot. Mag. 24: plate 947. 1806 • Paper-white narcissus [I]

Bulbs ovoid, 4–6 × 3–5 cm, tunic dark brown. **Leaves** 4; blade flat, keeled, 20–35(–45) cm × 6–15(–20) mm, glaucous. **Inflorescences** umbellate, 5–15(–20)-flowered, 25–35 cm; spathe pale brown, 4–6 cm, papery. **Flowers** strongly fragrant; perianth 3–5 cm wide; perianth tube 1.5–2 cm, gradually tapering to base; distinct portions of tepals spreading to reflexed, white, narrowly ovate to oblanceolate, 1–2.5 × 0.8–1 cm, apex acute; corona white, cup-shaped, 2–4 × 5–8 mm, apex crenulate to ruffled; 3 short stamens included in perianth tube, 3 longer stamens and style exserted into mouth of corona; pedicel of variable length, to 6 cm. $2n = 22$.

Flowering winter–spring. Roadsides, waste places; 0–100 m; introduced; Calif., La.; w Europe (se France, sw Spain, Portugal), Mediterranean region; n Africa; expected naturalized elsewhere.

Although *Narcissus papyraceus* and *N. tazetta* are distinct and easily distinguished as fresh specimens, they are very difficult to separate in herbarium material. Many early authors considered the two synonymous, making literature references to *N. tazetta* unreliable.

5. **Narcissus tazetta** Linnaeus, Sp. Pl. 1: 290. 1753 • Polyanthus narcissus, bunch-flowered narcissus [I]

Bulbs ovoid, 4–6 × 3–5 cm, tunic pale to dark brown. **Leaves** 4; blade flat, 25–35 cm × 8–15(–20) mm, glaucous. **Inflorescences** umbellate, 5–15-flowered, 25–35 cm; spathe pale brown, 4–6 cm, papery. **Flowers** strongly fragrant; perianth 2–4 cm wide; perianth tube 1.5–2 cm, gradually tapering to base; distinct portions of tepals spreading to reflexed, white to cream, linear-ovate to oblanceolate, 1–2 × 0.5–1 cm, apex acute; corona yellow, cup-shaped, 3–5 × 5–10 mm, apex crenulate to ruffled; 3 shorter stamens included within perianth tube, 3 longer stamens and style exserted into mouth of corona; pedicel of variable length, to 8 cm. $2n = 22$.

Flowering late winter–spring. Roadsides, waste places; 0–100 m; introduced; Ala., Ark., Calif., Fla., La., Miss., N.C., Oreg., S.C., Tex., Va.; w Europe (s Portugal), Mediterranean region; sw Asia (Iran); naturalized in Kashmir, China, and Japan; expected naturalized elsewhere.

53. ZEPHYRANTHES Herbert, Appendix, 36. 1821, name conserved • Rain-lily, fairy-lily, zephyr-lily [Greek *Zephyros*, west wind, and *anthos*, flower]

Raymond O. Flagg

Gerald L. Smith

Walter S. Flory†

Atamosco Adanson 1763, name rejected; *Cooperia* Herbert

Herbs, perennial, scapose, from bulbs. **Bulbs** black or brown, tunicate, ovoid or globose, sometimes with long neck. **Leaves** sessile, erect or recumbent, with overlapping sheathing bases; blade linear, rarely exceeding 1 cm wide, smooth. **Scape** hollow. **Inflorescence** 1-flowered (rarely 2-flowered in *Z. drummondii*), spathaceous, otherwise ebracteate; spathe proximally tubular. **Flowers** erect to declinate, actinomorphic; perianth subrotate to funnelform to salverform, connate basally into tube, 2–16 cm; tepals subequal; stamens 6, of 2 different lengths, appearing equal or subequal (anthers in 2 overlapping sets of 3) to unequal (anthers of the sets not overlapping in *Z. longifolia*); filaments inserted just above perianth tube, erect, diverging except when shorter than tube, long-filiform to short-subulate, those inserted on distal tepals usually 1+ mm longer than those inserted on proximal ones; anthers submedially dorsifixed, usually parallel with floral axis, linear-oblong; ovary inferior; style filiform; stigma capitate or 3-fid with lobes linear; pedicel sometimes absent, hollow. **Fruits** capsular, thin-walled, 3-locular, subglobose or ± oblate. **Seeds** numerous, black, flat, D- or wedge-shaped, lustrous. $x = 6$.

Species ca. 70 (16 in the flora): se and sc United States, Mexico, West Indies, Central America, South America.

In *Zephyranthes*, the degree of fusion of the perianth into a tube is reflected in flower form, filament length, staminal position, and the opening of the bud. Flowers with much-abbreviated perianth tubes are subrotate; with increasing degrees of perianth fusion, flower form shifts from subrotate to funnelform to salverform. The more the perianth is fused, the shorter the filaments in relative length. The longer the perianth tube, the less the stamens diverge, to the point of fasciculation. The timing of the opening of the bud and the expansion of the flower runs across this morphological spectrum from early morning (shortest tubes) through the day to the evening (longest tubes).

The phylogeny of *Zephyranthes* is not well understood. The species with long perianth tubes and fasciculate stamens are sometimes segregated in the genus *Cooperia*. While there is considerable diversity within *Zephyranthes*, splitting the genus into two genera by the degree of fusion of the perianth tube does not fit well with the continuity of the morphological spectrum and with the apparent hybrid origins of species in Texas and Mexico. By and large, species with linear stigmatic lobes bloom mid winter–spring–summer, while those with capitate stigmas bloom summer–fall. Many species with linear stigmatic lobes have leaves over 5 mm wide, but such wide leaves are not known in species with capitate stigmas. Whether there is particular generic significance in differences in stigmatic lobes (linear or capitate) or in filaments (filiform or subulate, and apically acute or blunt) remains to be discovered.

The perianth tube as a portion of the perianth, the relative lengths of the filaments and perianth tube, the type of stigma, the spatial relationships of the stamens, and the position of the stigma relative to the anthers are critical characteristics in the separation of species of *Zephyranthes*. These characteristics, especially the capitate stigma, can be very difficult to determine in herbarium specimens, but reasonable determinations can usually be achieved with strong backlighting and practice. For consistency, almost all measurements given herein are from herbarium specimens. The width of the leaf shrinks in pressing and drying; generally the maximum width of a fresh leaf could be about one and one half that reported here.

It has been thought that pedicels are consistently either present or absent in species of *Zephyranthes*. This is belied by the occurrence of both pedicellate and sessile flowers within each of the three species (*Z. atamasca*, *Z. treatiae*, and *Z. simpsonii*) native to the southeastern United States.

SELECTED REFERENCES Flagg, R. O. 1961. Investigations in the Tribe Zephyrantheae of the Amaryllidaceae. Ph.D. dissertation. University of Virginia. Flagg, R. O. and W. S. Flory. 1976. Origins of three Texas species of *Zephyranthes*. Pl. Life 32: 67–80. Flory, W. S. and R. O. Flagg. 1981. *Pyrolirion* and *Zephyranthes*: Distinct genera. Bol. Soc. Brot., ser. 2, 53: 1197–1210. Sealy, J. R. 1937. *Zephyranthes, Pyrolirion, Habranthus* and *Hippeastrum*. J. Roy. Hort. Soc. 62: 195–209. Spencer, L. B. 1973. A Monograph of the Genus *Zephyranthes* (Amaryllidaceae) in North and Central America. Ph.D. dissertation. Wake Forest University.

1. Stigma capitate.
   2. Perianth white, sometimes tinged pink.
      3. Perianth tube shorter than spathe, shorter than filaments .... 16. *Zephyranthes candida*
      3. Perianth tube longer than spathe, longer than filaments.
         4. Stigma exserted 2 mm or more beyond anthers .... 14. *Zephyranthes traubii*
         4. Stigma among or very near anthers .... 15. *Zephyranthes chlorosolen*
   2. Perianth yellow.
      5. Perianth tube shorter than filaments.
         6. Stamens in 2 distinctly subequal sets .... 9. *Zephyranthes citrina*
         6. Stamens appearing equal .... 10. *Zephyranthes pulchella*
      5. Perianth tube longer than filaments.
         7. Perianth tube longer than spathe, more than 7 times as long as filaments ...
            .... 13. *Zephyranthes jonesii*

7. Perianth tube usually shorter than spathe, less than 6 times as long as filaments.
   8. Perianth tube ca. 2 times as long as filaments . . . . 11. *Zephyranthes refugiensis*
   8. Perianth tube ca. 3–5 times as long as filaments . . . . 12. *Zephyranthes smallii*

1. Stigma 3-fid.
   9. Perianth yellow . . . . 8. *Zephyranthes longifolia*
   9. Perianth white or pink.
      10. Stigma included within perianth tube . . . . 5. *Zephyranthes drummondii*
      10. Stigma exserted beyond perianth tube.
         11. Anthers 13–22 mm . . . . 4. *Zephyranthes grandiflora*
         11. Anthers 8 mm or shorter.
            12. Perianth tube 0.4 cm or shorter.
               13. Perianth rose pink; distal tepals not erect; stamens in 2 slightly subequal sets . . . . 6. *Zephyranthes rosea*
               13. Perianth white, sometimes flushed pink abaxially; distal tepals erect; stamens in 2 distinctly subequal sets . . . . 7. *Zephyranthes insularum*
            12. Perianth tube 0.8 cm or longer.
               14. Stigma very near, among, or below anthers; perianth tube 1/3 or more as long as perianth . . . . 3. *Zephyranthes simpsonii*
               14. Stigma exserted more than 2 mm beyond anthers; perianth tube 1/3 or less as long as perianth.
                  15. Perianth tube less than 1/4 as long as perianth, ca. 1/2 as long as filaments . . . . 1. *Zephyranthes atamasca*
                  15. Perianth tube 1/4 or more as long as perianth, ± equaling filaments . . . . 2. *Zephyranthes treatiae*

1. Zephyranthes atamasca (Linnaeus) Herbert, Appendix, 36. 1821 (as atamasco) • Atamasco-lily, Carolina-lily, Easter-lily, naked-lady, occidental swamp-lily, Virginia-lily [E] [F]

Amaryllis atamasca Linnaeus, Sp. Pl. 1: 292. 1753; *Atamosco atamasca* (Linnaeus) Greene (as atamasco)

Leaf blade shiny green, to 8 mm wide. **Spathe** (2–)2.2–3.2(–3.6) cm. **Flowers** erect to slightly inclined; perianth mostly white, sometimes tinged or veined pink, more so with age, funnelform, (5.5–)6.6–9(–10) cm; perianth tube green, (0.8–)1–2(–2.1) cm, increasing in diam., less than 1/4 perianth length, ca. 1/2 (1/3–2/3) filament length, ca. 1/2 (1/3–3/4) spathe length; tepals usually reflexed; stamens diverging, appearing equal; filaments filiform, (2.9–)3.2–4.4(–4.7) cm; anthers 3–6(–8) mm; style longer than perianth tube; stigma 3-fid, exserted more than 2 mm beyond anthers; pedicel (0–)0.4–1.6(–3.4) cm, usually shorter than spathe. $2n = 24$.

Flowering mid winter–spring (Jan–May). Rich, mixed forests, moist clearings, meadows, moist to wet pastures, coastal plains, piedmonts; 0–700 m; Ala., Fla., Ga., Md., Miss., N.C., S.C., Va.

2. Zephyranthes treatiae S. Watson, Proc. Amer. Acad. Arts 14: 300. 1879 [E]

Atamosco treatiae (S. Watson) Greene; *Zephyranthes atamasca* (Linnaeus) Herbert var. *treatiae* (S. Watson) Meerow

Leaf blade dull green, to 4 mm wide. **Spathe** (1.8–)2–3.3(–3.6) cm. **Flowers** erect to inclined; perianth mostly white to pink, color deepening with age, funnelform, (6–)6.6–9.5(–11) cm; perianth tube green, (1.7–)2–3(–3.1) cm, increasing in diam., at least 1/4 perianth length, ± equaling (2/3–1 1/4) filament length, ca. 2/3–1 1/3 spathe length; tepals usually reflexed; stamens diverging, appearing equal; filaments filiform, (1.6–)2–4(–4.6) cm; anthers 3–6(–8) mm; style longer than perianth tube; stigma 3-fid, exserted more than 2 mm beyond anthers; pedicel 0–1(–1.2) cm, shorter than spathe. $2n = 24$.

Flowering mid winter–spring (Jan–Apr). Peaty-sandy soil, usually associated with wet inlands or former pineland sites; 0–100 m; Fla., Ga.

It has generally been thought that *Zephyranthes atamasca* and *Z. treatiae* are distinguishable only by differences in leaves and habitat, and that their flowers are not significantly different. However, the greater length of the perianth tube compared with that of the filaments in *Z. treatiae* readily separates that species from *Z. atamasca*.

3. Zephyranthes simpsonii Chapman, Fl. South. U.S. ed. 2 repr. 2, 696. 1892 [C] [E]

Atamosco simpsonii (Chapman) Greene (as simpsoni)

Leaf blade dull green, to 4 mm wide. **Spathe** (1.7–)2–4(–4.2) cm. **Flowers** erect; perianth mostly white proximally, often strongly infused pink or purple distally, more so with age, funnelform, (4–)4.3–8.5(–10) cm; perianth tube green, (1.4–)1.6–3.3(–4) cm, increasing in diam., ca. 1/3 or more perianth length, ± equaling or seldom less than (3/4–1 1/4) filament length, ± equaling (2/3–1 1/4) spathe length; tepals slightly diverging, rarely reflexed; stamens diverging, appearing equal; filaments filiform, (1.5–)1.6–3(–3.2) cm; anthers 3–6(–8) mm; style longer than perianth tube; stigma 3-fid, usually among or below anthers, rarely exserted more than 1 mm beyond anthers; pedicel absent or to 0.6(–1.5) cm and shorter than spathe. $2n = 48$.

Flowering late winter–spring (Feb–May). Peaty-sandy soil, coastal plains, rarely piedmont; of conservation concern; 0–100 m; Ala., Fla., Ga., Miss., N.C., S.C.

Although the perianth of *Zephyranthes simpsonii* was originally described as pale pink, white forms are quite common. Any of the North American white-flowered species of *Zephyranthes* may be variously flushed with pink, increasingly so with age.

4. Zephyranthes grandiflora Lindley, Bot. Reg. 11: plate 902. 1825 [I]

Leaf blade shiny green, to 7 mm wide. **Spathe** 3–4 cm. **Flowers** usually inclined; perianth pink, funnelform, 5.6–9 cm; perianth tube green, (1.2–)1.8–2.3 cm, increasing in diam., ca. 1/4 (1/5–1/3) perianth length, usually less than or equaling filament length, ca. 1/3–1/2 spathe length; tepals reflexed; stamens diverging, appearing equal; filaments filiform, 2–3 cm; anthers 13–22 mm; style longer than perianth tube; stigma 3-fid, exserted more than 2 mm beyond anthers; pedicel 1.3–3.8 cm, usually shorter than spathe. $2n = 48$.

Flowering spring–early summer (Apr–Jul). Open areas in or near moist woods; 0–100 m; introduced; Ala., Fla., La., Miss., Tex.; e Mexico; West Indies; Central America.

Zephyranthes grandiflora frequently displays flowers with extra parts and has been widely distributed as an ornamental. Flowers with extra parts have been seen in *Z. atamasca, Z. candida, Z. rosea,* and *Z. treatiae. Zephyranthes grandiflora* also has been found to have a variable number of chromosomes in cells from single plants (R. O. Flagg 1961).

As a rule, the flowers of *Zephyranthes* species having high chromosome numbers, $2n = 48$ or more, are built for self-fertilization, with the stigma borne very near (within 2 mm), among, or beneath the anthers (R. O. Flagg and W. S. Flory 1962). However, *Zephyranthes grandiflora* is an exception and often exhibits poor fertility.

5. Zephyranthes drummondii D. Don in R. Sweet, Brit. Fl. Gard., ser. 2, 4: plate 328. 1836 • Cebolleta

Cooperia pedunculata Herbert

Leaf blade glaucous-green, to 8 mm wide. **Spathe** (3–)4–5 cm. **Flowers** erect; perianth white, sometimes flushed pink abaxially, more so with age, broadly funnelform, 6–9 cm; perianth tube white to green, 3–4(–4.7) cm, diam. primarily uniform, ca. 1/2 perianth length, more than 15 times filament length, equaling (3/4–1 1/4) spathe length; tepals rarely reflexed; stamens fasciculate, distinctly subequal; filaments subulate, 0.1–0.2 cm, apex acute; anthers 5–8 mm; style notably shorter than perianth tube; stigma 3-fid, included in perianth tube; pedicel (0.2–)0.5–2(–3.3) cm, shorter than spathe. $2n = 48, 72$.

Flowering mid spring–mid summer (Mar–Aug). Sandy/rocky, usually calcareous soils; 0–900 m; Fla., La., Tex.; ne Mexico.

Specimens of *Zephyranthes drummondii* with rare biflowered inflorescences were collected by B. C. Tharp in 1939 and 1946 in Austin, Texas. The species has naturalized in Florida.

6. Zephyranthes rosea Lindley, Bot. Reg. 10: plate 821. 1824 [I]

Leaf blade dull green, to 5 mm wide. **Spathe** 2–2.8 cm. **Flowers** erect to slightly inclined; perianth rose pink, spreading, funnelform, 2.7–2.8 cm; perianth tube green, 0.2–0.3 cm, increasing in diam., ca. 1/10 or less of perianth length, ca. 1/5 filament length, ca. 1/8 spathe length; tepals not reflexed, distalmost not erect; stamens diverging, in 2 slightly subequal sets with anthers overlapping ca. 90 percent; filaments filiform, 1–1.7 cm; anthers 3–6 mm; style 1.8–2.7 cm, longer than perianth tube; stigma 3-fid, exserted ca. 1–5 mm beyond anthers; pedicel 1.5–3.2 cm, usually exceeding spathe. $2n = 24$.

Flowering spring–summer (Mar–Jul). Sandy loam; 0–50 m; introduced; Fla.; Mexico; West Indies; Central America.

Zephyranthes rosea has been widely distributed horticulturally. It persists and is locally adventive in Florida.

7. Zephyranthes insularum H. H. Hume ex Moldenke, Pl. Life 8: 152. 1952 [I]

Leaf blade dull green, to 5 mm wide. **Spathe** 2–2.8 cm. **Flowers** erect to slightly inclined; perianth white, sometimes flushed pink abaxially, more so with age, funnelform, 3.6–4.6 cm; perianth tube green, 0.2–0.4 cm, increasing in diam., 1/10 or less of perianth length, ca. 1/4 filament length, ca. 1/6 spathe length; tepals not reflexed, distal 3 more erect than proximal 3; stamens diverging, in 2 distinctly subequal sets with anthers overlapping ca. 50 percent; filaments filiform, 1–1.3 cm; anthers 4–7 mm; style 2.2–2.5 cm, longer than perianth tube; stigma 3-fid, exserted ca. 10 mm beyond anthers; pedicel 2.3–3.3 cm, usually exceeding spathe. $2n = 28$.

Flowering summer (Jun–Aug). Sandy loam; 0–50 m; introduced; Fla.; Mexico; West Indies (Cuba).

Zephyranthes insularum was based on plants from a garden in Key West, Florida, and from Santiago de las Vegas, Cuba. Its native locale is uncertain. It persists and is locally adventive in Florida.

8. Zephyranthes longifolia Hemsley, Diagn. Pl. Nov. Mexic., 55. 1880 • Cebolleta

Leaf blade dull green, to 1(–2) mm wide. **Spathe** 2.1–3.1 cm. **Flowers** erect to slightly inclined; perianth yellow, funnelform, 2.1–2.8 cm; perianth tube green, 0.2–0.6 cm, increasing in diam., no more than 1/4 perianth length, ca. 1/2 (1/4–3/4) filament length, no more than 1/4 spathe length; tepals rarely reflexed; stamens diverging, in 2 distinctly unequal sets; filaments filiform, 0.7–1.1 cm; anthers in 2 nonoverlapping sets, 3–6 mm; style longer than perianth tube; stigma 3-fid, usually among anthers; pedicel 0.7–1.5 cm, shorter than spathe. $2n = 48$.

Flowering late spring–summer (May–Aug). Sandy, gravelly, calcareous, alkaline soils in highlands; 40–2400 m; Ariz., N.Mex., Tex.; Mexico.

9. Zephyranthes citrina Baker, Bot. Mag. 108: plate 6605. 1882 • Citron rain-lily [I]

Leaf blade dull green, to 4 mm wide. **Spathe** 1.6–2.6 cm. **Flowers** erect; perianth lemon yellow, funnelform, 3.1–5 cm; perianth tube green, 0.7–1 cm, increasing in diam., less than 1/3 perianth length, ca. 1/2 (1/3–3/4) filament length, less than 1/2 spathe length; tepals rarely reflexed; stamens diverging, in 2 distinctly subequal sets; filaments filiform, subequal, 1.2–2 cm; anthers 5–7 mm; style longer than perianth tube; stigma capitate, usually among or below anthers, not exserted more than 1 mm beyond anthers; pedicel 2.3–4.4 cm, usually longer than spathe. $2n = 48$.

Flowering summer (Jul–Sep). Moist, sandy loam; 0–100 m; introduced; Ala., Fla., Miss.; West Indies; Central America; South America.

In the original description of *Zephyranthes citrina*, Baker stated that Messrs. Veitch (Chelsea, England) brought it to him in flower and that they believed it had come to them from Demerara (Guyana). The type specimen is in the Kew Herbarium, where there are also specimens of this species from areas where it clearly is not native: tropical Africa, the Malay Peninsula, and India. It is also found in the West Indies (Cuba) and Central America (Panama). This species has naturalized widely from cultivation, and considerably more work is needed to determine its native distribution.

10. Zephyranthes pulchella J. G. Smith, Rep. (Annual) Missouri Bot. Gard. 6: 114, plate 52. 1895 • Cebolleta [F]

Leaf blade dull green, to 3 mm wide. **Spathe** 1.9–3.8 cm. **Flowers** erect; perianth cadmium yellow, funnelform, (1.9–)2.2–4(–6) cm; perianth tube green, (0.2–) 0.5–0.9(–1) cm, increasing in diam., less than 1/4 perianth length, ca. 1/2 (1/5–2/3) filament length, ca. 1/6 (1/8–1/4) spathe length; tepals rarely reflexed; stamens diverging, appearing equal; filaments filiform, (0.5–)0.7–1.1(–1.5) cm; anthers 3–7 mm; style longer than perianth tube; stigma capitate, usually among anthers; pedicel (0.3–)1.4–3.3(–4.4) cm, sometimes longer than spathe. $2n = 48$.

Flowering late spring–late fall (May–Dec). Primarily low, sandy-loam open fields, swales, ditches, coastal plains; 0–400 m; Tex.; ne Mexico.

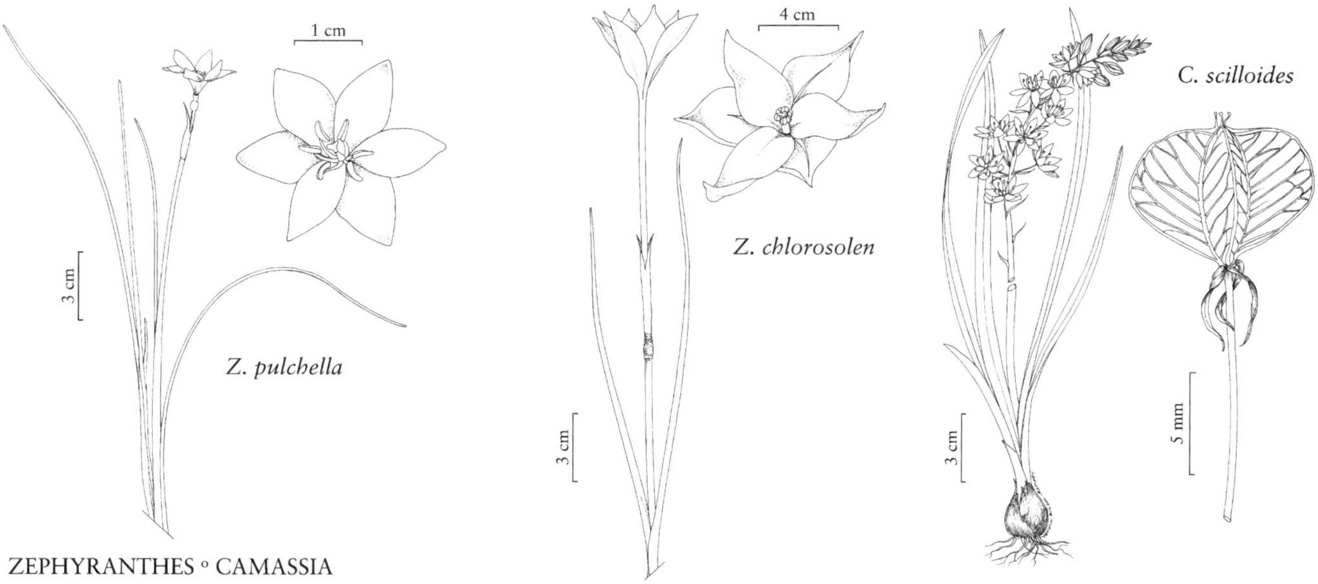

ZEPHYRANTHES ◦ CAMASSIA

11. Zephyranthes refugiensis F. B. Jones, Rhodora 63: 214. 1961 [C] [E]

Leaf blade dull green, to 4 mm wide. **Spathe** 1.9–3.2 cm. **Flowers** erect; perianth dark lemon yellow, funnelform, 2.5–5 cm; perianth tube green, (1–) 1.2–2 (–2.4) cm, increasing in diam., less than 1/2 perianth length, ca. 2 times filament length, ca. 1/2–3/4 spathe length; tepals sometimes reflexed; stamens diverging to semifasciculate, appearing equal; filaments filiform, (0.5–)0.6–0.8 (–1.1) cm; anthers ca. 7 mm; style longer than perianth tube; stigma capitate, usually among anthers; pedicel (0.1–)0.2–1.6(–1.8) cm, shorter than spathe. $2n$ = 46, 48.

Flowering late summer–early fall (Aug–Oct). Primarily low, sandy loam, open fields, swales, ditches; of conservation concern; 10–30 m; Tex.

Morphology, cytology, and distribution suggest that *Zephyranthes refugiensis* arose as a hybrid between *Z. pulchella* and *Z. jonesii*. It is known only from Goliad and Refugio counties.

12. Zephyranthes smallii (Alexander) Traub, Pl. Life 7: 42. 1951 [E]

Cooperia smallii Alexander, Addisonia 21: 7, plate 676. 1939

Leaf blade dull green, to 5 mm wide. **Spathe** 2.2–4.6 cm. **Flowers** erect; perianth lemon yellow, salverform to broadly funnelform, 2.5–9 cm; perianth tube yellowish green, 1.1–4.4 cm, diam. uniform, ca. 1/2 perianth length, ca. 3–5 times filament length, somewhat shorter than spathe; tepals usually reflexed; stamens fasciculate to semifasciculate, appearing equal; filaments subulate, 0.4–0.9 cm, apex acute; anthers 3–7 mm; style longer than perianth tube; stigma capitate, among or exserted more than 2 mm beyond anthers; pedicel (0–)0.1–3.5 cm, shorter than spathe. $2n$ = 53, 54, 58, 70, 72.

Flowering early fall (Sep–Oct). Low, sandy loam, open fields, swales, ditches; 0–10 m; Tex.

Morphology, cytology, and distribution suggest that *Zephyranthes smallii* (predominantly $2n$ = 54) arose from hybridization between *Z. pulchella* ($2n$ = 48) and *Z. chlorosolen* ($2n$ = 60 in Cameron County) in Brownsville.

13. Zephyranthes jonesii (Cory) Traub, Pl. Life 7: 42. 1951 [E]

Cooperia jonesii Cory, Field & Lab. 18: 45. 1950

Leaf blade dull green, to 4.5 mm wide. **Spathe** 2.1–4 cm. **Flowers** erect; perianth pale lemon yellow, salverform to funnelform, 2–12.5 cm; perianth tube yellow with some green, 1.5–7.2 cm, diam. primarily uniform, ca. $^{2}/_{3}$–$^{3}/_{4}$ perianth length, ca. 8–10 times filament length, longer than spathe; tepals often reflexed; stamens fasciculate, appearing equal; filaments subulate, 0.15–0.7 cm, apex acute; anthers 4–7 mm; style longer than perianth tube; stigma capitate, among to exserted beyond anthers; pedicel 0–0.6 cm, shorter than spathe. $2n$ = 48, 72.

Flowering mid summer–early fall (Jul–Oct). Primarily low, sandy loam, open fields, swales, ditches, coastal bends; 0–30 m; Tex.

Morphology, cytology, and distribution suggest that *Zephyranthes jonesii*, like *Z. smallii*, arose from hybridization between *Z. pulchella* ($2n$ = 48) and *Z. chlorosolen* ($2n$ = 48).

14. Zephyranthes traubii (W. Hayward) Moldenke, Pl. Life 7: 79. 1951

Cooperia traubii W. Hayward, Herbertia 3: 63, 65 (fig. s.n.), 66 (figs. 1–3). 1936

Leaf blade dull green, to 1 mm wide. **Spathe** (2–)3–4 cm. **Flowers** erect; perianth white, sometimes tinged and veined pink, salverform, (11–)12–15.4 cm; perianth tube primarily white, 9–12.6 cm, diam. uniform, $^{3}/_{4}$ or more perianth length, at least 15 times filament length, 3–5 times spathe length; tepals often reflexed; stamens fasciculate, appearing equal; filaments subulate, 0.2–0.4 cm, apex blunt; anthers 4–8 mm; style longer than perianth tube; stigma capitate, exserted more than 2 mm beyond anthers; pedicel absent. $2n$ = 24.

Flowering early summer–mid fall (Jul–Nov). Primarily sandy loam, open fields, coastal plains; 0–100 m; Tex.; ne Mexico.

15. Zephyranthes chlorosolen (Herbert) D. Dietrich, Syn. Pl. 2: 1176. 1840 [F]

Cooperia chlorosolen Herbert, Edwards's Bot. Reg. 22: plate 1835. 1836; *C. drummondii* Herbert; *C. kansensis* W. Stevens; *Zephyranthes brazosensis* Traub; *Z. herbertiana* D. Dietrich

Leaf blade dull green, to 5 mm wide. **Spathe** (3–)4–5(–5.7) cm. **Flowers** erect; perianth white, sometimes tinged or veined pink, salverform, (7.3–)9–14(–16) cm; perianth tube primarily white, sometimes pale green proximally, (7–)8–12(–13) cm, diam. uniform, $^{3}/_{4}$ or more perianth length, at least 15 times filament length, ca. 2–4 times spathe length; tepals rarely reflexed; stamens fasciculate, appearing equal; filaments subulate, 0.2–0.5 mm, apex blunt; anthers 4–8(–9) mm; style longer than perianth tube; stigma capitate, among or very near anthers, exserted no more than 2 mm beyond anthers; pedicel absent or rarely 0.1 cm and much shorter than spathe. $2n$ = 48, 60, 68, 72.

Flowering late spring–early fall (May–Oct). Wide range of soils and conditions, usually not very wet; 0–2400 m; Ala., Ark., Kans., La., Miss., Okla., Tex.; e Mexico.

16. Zephyranthes candida (Lindley) Herbert, Bot. Mag. 53: plate 2607. 1826 • Autumn rain-lily, Peruvian swamp-lily [I]

Amaryllis candida Lindley, Bot. Reg. 9: plate 724. 1823; *Argyropsis candida* (Lindley) M. Roemer; *Atamosco candida* (Lindley) Sasaki; *Plectronema candida* (Lindley) Rafinesque

Leaf blade glossy deep green, to 3 mm wide. **Spathe** (1.8–)2–4 cm. **Flowers** erect; perianth white, sometimes pinkish abaxially, subrotate, 3–4.5 cm; perianth tube green, 0.1–0.4 cm, increasing in diam., less than $^{1}/_{4}$ times perianth length, $^{1}/_{5}$–$^{1}/_{3}$ times filament length, ca. $^{1}/_{10}$ times spathe length; tepals not reflexed; stamens diverging, subequal; filaments filiform, 1–1.4 cm; anthers 5–8 mm; style longer than perianth tube; stigma capitate, usually among or exserted less than 2 mm beyond anthers; pedicel (0.4–)1–2.5 cm, usually shorter than spathe. $2n = 38$.

Flowering summer–mid fall (Jun–Nov). Sandy humus soil, coastal plains; 0-200 m; introduced; Ala., Fla., Ga., La., Miss., N.C., S.C., Tex.; South America (Argentina, Peru, Uruguay).

W. Herbert (1837) suggested that *Zephyranthes candida* might belong in a segregate genus, and within ten years Rafinesque and M. Roemer each separated it from *Zephyranthes*. Its leaves are about twice as thick as those of other species in the genus, and they persist through winter frosts and snow, a rare, if not unique, characteristic in *Zephyranthes*. The stigmatic lobes are not globose, as in *Z. chlorosolen*, but are somewhat erect and might be described as very abbreviated linear lobes; a careful study of fresh stigmas of "capitate" species is in order. Also, the chromosome complement of *Z. candida* appears to be anomalous within *Zephyranthes*. The species has been maintained in *Zephyranthes* for more than a century, but were Herbert, Rafinesque, and Roemer correct after all?

54. CAMASSIA Lindley, Edwards's Bot. Reg. 18: plate 1486. 1832, name conserved • Camas, quamash [Shoshone name *camas* or *quamash*] E

Tom A. Ranker

Tim Hogan

Herbs, perennial, from bulbs. **Bulbs** solitary or clustered, tunicate, ovoid to globose; tunic black or brown. **Leaves** basal, appearing whorled; blade linear, keeled. **Inflorescences** appearing terminal, racemose, bracteate; bracts sterile or subtending flowers, narrowly lanceolate. **Flowers** actinomorphic or zygomorphic; tepals 6, persistent, ± equal in 2 whorls of 3, distinct, violet, blue, or white, each 3–9-veined, lanceolate, ± twisted in drying; stamens 6; filaments inserted on receptacles at base of tepals, slender; anthers versatile, dehiscence introrse; ovary 3-locular, septal nectaries present, ovules 6–36; style filiform; stigma 3-lobed; pedicel spreading to incurving-erect in fruit. **Fruits** capsular, ovoid to ellipsoid or subglobose, dehiscence loculicidal. **Seeds** 6–36, lustrous black, obpyriform to ovoid-ellipsoid, 2–4 mm. $x = 15$.

Species 6 (6 in the flora): North America.

Camassia has been associated with other western North American genera of Liliaceae such as *Schoenolirion*, *Hastingsia*, and especially *Chlorogalum* (F. Speta 1998; M. Pfosser and F. Speta 1999), but recent molecular evidence (D. J. Bogler and B. B. Simpson 1996; M. F. Fay and M. W. Chase 1996) suggests that it may be related instead to the Agavaceae. Furthermore, the bimodal, $2n = 30$ karyology of *Camassia* (A. Fernandez and J. R. Davina 1991) is similar to that of Agavaceae (D. Satô 1935) and not that of *Chlorogalum*.

Camassia bulbs have been an important food staple for native Americans, especially in the Pacific Northwest (G. R. Downing and L. S. Furniss 1968; N. J. Turner and H. V. Kuhnlein 1983), where bulbs were dug and traded on large encampment meadows. Similarity to the poisonous bulbs of *Zigadenus* ("death camas") is a concern where ranges of the two genera overlap. Several *Camassia* species are cultivated and represent a major horticultural contribution from the native flora.

Variation and intergradation of *Camassia angusta* and *C. scilloides* have been reviewed by T. A. Ranker and A. F. Schnabel (1986), as well as J. A. Steyermark (1961), R. O. Erickson (1941), and F. W. Gould (1942).

SELECTED REFERENCES Gould, F. W. 1942. A systematic treatment of the genus *Camassia* Lindl. Amer. Midl. Naturalist 28: 712–742. Ranker, T. A. and A. F. Schnabel. 1986. Allozymic and morphological evidence for a progenitor-derivative species pair in *Camassia* (Liliaceae). Syst. Bot. 11: 433–445.

1. Tepals connivent over capsules after anthesis and deciduous as capsules develop, or withering separately.
   2. Fruiting pedicels usually incurving-erect (often with capsules closely appressed to raceme axes); flowers actinomorphic or zygomorphic [5 tepals curving upward, the 6th downward; usually (4–)10–35(–58) blooming simultaneously, except on few-flowered individuals]; tepals long-persistent on fruiting racemes .... 1. *Camassia quamash* (in part)
   2. Fruiting pedicels usually spreading-erect (capsules not appressed to raceme axes); flowers actinomorphic (usually 1–3 blooming simultaneously); tepals connivent over capsules after anthesis, deciduous as capsules develop.
      3. Capsules dull green, ovoid to ellipsoid, 10–25 mm; seeds 6–12 per locule; s British Columbia to c California .... 2. *Camassia leichtlinii*
      3. Capsules shiny green, subglobose, 5–10 mm; seeds 2–5 per locule; sw Oregon... .... 3. *Camassia howellii*
1. Tepals mostly withering separately at base of capsules after anthesis, sometimes connivent over capsules, not deciduous.
   4. Capsules subglobose or ovoid-ellipsoid; e Great Plains to Appalachians and Great Lakes to s United States.
      5. Inflorescences 19–47 cm, with 0–5 sterile bracts; fruiting pedicels mostly spreading-erect; capsules subglobose; flowering earlier than sympatric populations of *Camassia angusta*... 4. *Camassia scilloides*
      5. Inflorescences 27–87 cm, with 3–28 sterile bracts; fruiting pedicels mostly incurving-erect; capsules ovoid-ellipsoid; flowering later than sympatric populations of *Camassia scilloides*... 5. *Camassia angusta*
   4. Capsules ovoid or ellipsoid; s British Columbia, s Alberta, nw United States.
      6. Bulbs usually clustered, ellipsoid, 2–7 cm diam.; leaves rarely fewer than 10, 2–5 cm wide; ne Oregon and adjacent Idaho .... 6. *Camassia cusickii*
      6. Bulbs seldom clustered, globose, 1–5 cm diam.; leaves usually fewer than 10, 4–20 mm wide; widespread in Pacific Northwest..................... 1. *Camassia quamash* (in part)

1. Camassia quamash (Pursh) Greene, Man. Bot. San Francisco, 313. 1894 [E]

Phalangium quamash Pursh, Fl. Amer. Sept. 1: 226. 1814

Bulbs seldom clustered, globose, 1–5 cm diam. **Leaves** usually fewer than 10, 1–6 dm × 4–20 mm. **Inflorescences** 20–80 cm; sterile bracts absent, bracts subtending flowers usually equaling or exceeding pedicel. **Flowers** usually zygomorphic, sometimes actinomorphic; tepals withering separately or connivent over capsules after anthesis, long-persistent on fruiting racemes, blue or bluish violet, each 3–9-veined, 12–35 × 1.5–8 mm; anthers usually yellow, sometimes bluish violet, violet, or brown, 2.5–7 mm; fruiting pedicel mostly incurving-erect, occasionally spreading-erect, 5–70 mm. **Capsules** not deciduous, pale green to pale brown, ovoid, 6–19 mm. **Seeds** 5–10 per locule. $2n = 30$.

Camassia quamash is highly variable morphologically. Although there tend to be distinct geographical variants, here recognized as subspecies following F. W. Gould (1942), there is much overlap among them. The subspecific status of these taxa is retained to highlight the extreme morphological variability and geographical patterns within the species. A detailed biosystematic study of this complex is needed.

Subspecies 8 (8 in the flora): w United States and Canada.

1. Tepals connivent over capsules after anthesis, usually forced apart again by capsule growth.
   2. Leaves 6–15 mm wide, not glaucous; tepals 5-, 7-, or 9-veined, or occasionally 3-veined in outer whorls; n California coast ranges .... 1f. *Camassia quamash* subsp. *linearis*
   2. Leaves 10–20 mm wide, glaucous adaxially; tepals usually 3-veined, or 5-veined, in inner whorls; w and nw United States.
      3. Tepals bright blue to deep bluish violet, 15–20 mm; anthers bright yellow; ec California, n to Washington .... 1g. *Camassia quamash* subsp. *breviflora*
      3. Tepals pale to deep bluish violet, 16–31 mm; anthers dull yellow to violet; s Idaho, n Utah .... 1h. *Camassia quamash* subsp. *utahensis*
1. Tepals withering separately after anthesis, with no tendency to be connivent over capsules.
   4. Fruiting pedicels with capsules held away from raceme axes.
      5. Tepals light bluish violet; (bulbs 20–35 × 15–25 mm, shallowly buried in) light, well-drained prairie soils; sw Washington.... .... 1b. *Camassia quamash* subsp. *azurea*
      5. Tepals bright to deep bluish violet; (bulbs 15–65 × 14–50 mm, deeply buried in) wet

meadows, fields, and rocky coastal bluffs, sw British Columbia to nw Oregon . . . . . 1c. *Camassia quamash* subsp. *maxima* (in part)

4. Fruiting pedicels with capsules usually closely appressed to raceme axes (except subsp. *maxima*).
   6. Pedicels usually 5–10 mm; flowers actinomorphic; tepals 12–20 mm; sw Oregon . . . . 1d. *Camassia quamash* subsp. *walpolei*
   6. Pedicels 10–70 mm; flowers slightly zygomorphic or actinomorphic; tepals 15–35 mm; sw Canada, w Oregon and Washington e to Rocky Mountains.
      7. Leaves not glaucous; anthers bright yellow; Oregon only . . . . 1e. *Camassia quamash* subsp. *intermedia*
      7. Leaves glaucous adaxially; anthers dull yellow, bluish violet, violet, or brown; extreme sw Canada, nw United States. [8. Shifted to left margin.—Ed.]

8. Anthers never yellow, bluish violet to brown only . . . . 1a. *Camassia quamash* subsp. *quamash*
8. Anthers dull yellow, violet, or brown . . . . 1c. *Camassia quamash* subsp. *maxima* (in part)

1a. Camassia quamash (Pursh) Greene subsp. **quamash** [E]

Leaves 6–15 mm wide, glaucous adaxially. **Flowers** slightly zygomorphic or actinomorphic; tepals withering separately after anthesis, pale to deep blue or bluish violet, each 3-veined, sometimes 5-veined in inner whorls, 15–30 × 3–5 mm; anthers dark bluish violet to brown, 2.5–4 mm; fruiting pedicel incurving-erect, with capsules appressed to raceme axes, 25–70 mm. **Capsules** 8–16 mm. **Seeds** 5–10 per locule.

Flowering mid–late spring. Wet meadows; 300–2500 m; Alta., B.C.; Idaho, Mont., Oreg., Wash., Wyo.

1b. Camassia quamash (Pursh) Greene subsp. **azurea** (A. Heller) Gould, Amer. Midl. Naturalist 28: 733. 1942 [E]

Quamasia azurea A. Heller, Bull. Torrey Bot. Club 26: 547. 1899

Leaves 4–13 mm wide, not glaucous. **Flowers** slightly zygomorphic or actinomorphic; tepals withering separately after anthesis, light bluish violet, each 5–9-veined, 3-veined in outer whorls, 12–27 × 1.5–3 mm; anthers yellow, 2.5–4 mm; fruiting pedicel spreading-erect, with capsules held away from raceme axes, 25–60 mm. **Capsules** 7–15 mm. **Seeds** 5–10 per locule.

Flowering mid–late spring. Alluvial plains, meadows, light, well-drained prairie soils; 0–200 m; Wash.

1c. Camassia quamash (Pursh) Greene subsp. **maxima** Gould, Amer. Midl. Naturalist 28: 732, fig. 7. 1942 [E]

Leaves 6–15 mm wide, slightly glaucous adaxially. **Flowers** zygomorphic or actinomorphic; tepals withering separately after anthesis, bright to deep bluish violet, each 5-, 7-, or 9-veined, occasionally 3-veined in outer whorls, 18–28 × 2–5 mm; anthers dull yellow, violet, or brown, 2.5–4 cm; fruiting pedicel spreading- or incurving-erect, with capsules held away from raceme axes, 10–25 mm. **Capsules** 9–17 mm. **Seeds** 5–10 per locule. $2n = 30$.

Flowering mid–late spring. Wet meadows, fields, and rocky, coastal bluffs; 0–600 m; B.C.; Oreg., Wash.

1d. Camassia quamash (Pursh) Greene subsp. **walpolei** (Piper) Gould, Amer. Midl. Naturalist 28: 735. 1942 [E]

Quamasia walpolei Piper, Proc. Biol. Soc. Wash. 29: 81. 1916; *Camassia walpolei* (Piper) J. F. Macbride

Leaves 4–13 mm wide, not glaucous. **Flowers** actinomorphic; tepals withering separately after anthesis, pale blue or blue-violet, each 3-veined, sometimes 5-veined in outer whorls, 12–20 × 3–5 mm; anthers bright yellow, 3–5 mm; fruiting pedicel incurving-erect, with capsules closely appressed to raceme axes, 5–10 mm. **Capsules** 8–15 mm. **Seeds** 3–6 per locule.

Flowering mid–late spring. Wet meadows; 100–1500 m; Oreg.

1e. Camassia quamash (Pursh) Greene subsp. **intermedia** Gould, Amer. Midl. Naturalist 28: 734, figs. 7, 10c. 1942 [E]

Leaves 6–13 mm wide, not glaucous. **Flowers** usually zygomorphic; tepals withering separately after anthesis, pale bluish violet, each usually 5- or 7-veined, occasionally 3-veined in outer whorls, 20–35 × 5 mm; anthers bright yellow, 5–7 mm; fruiting pedicel strongly incurving, with capsules closely appressed to raceme axes, 35–60 mm. **Capsules** 7–19 mm. **Seeds** 6–10 per locule.

Flowering mid–late spring. Fields, pastures; 300–1000 m; Oreg.

1f. Camassia quamash (Pursh) Greene subsp. **linearis** Gould, Amer. Midl. Naturalist 28: 736, figs. 5(5–7), 7, 9. 1942 [E]

Leaves 6–15 mm wide, not glaucous. **Flowers** slightly zygomorphic; tepals connivent over capsules after anthesis, usually forced apart again by capsule growth, deep bluish violet, each 5-, 7-, or 9- veined, occasionally 3-veined in outer whorls, 20–35 × 5–8 mm; anthers dull yellow to violet, 4–7 mm; fruiting pedicel incurving-erect, 10–25 mm. **Capsules** 6–18 mm. **Seeds** 5–10 per locule.

Flowering mid–late spring. Wet meadows; 0–800 m; Calif.

1g. Camassia quamash (Pursh) Greene subsp. **breviflora** Gould, Amer. Midl. Naturalist 28: 737, figs. 7, 10a, b. 1942 [E]

Leaves 10–17 mm wide, glaucous adaxially. **Flowers** actinomorphic or zygomorphic; tepals connivent over capsules after anthesis, usually forced apart again by capsule growth, bright blue to deep bluish violet, each 3- or occasionally 5-veined, 15–20 × 3–5 mm; anthers bright yellow, 4–5 mm; fruiting pedicel incurving-erect, 5–15 mm. **Capsules** 7–16 mm. **Seeds** 5–10 per locule.

Flowering mid–late spring. Wet meadows; 700–2500 m; Calif., Nev., Oreg., Wash.

1h. Camassia quamash (Pursh) Greene subsp. **utahensis** Gould, Amer. Midl. Naturalist 28: 740, fig. 7. 1942 [E]

Leaves 10–20 mm wide, glaucous adaxially. **Flowers** usually zygomorphic; tepals connivent over capsule after anthesis, usually forced apart again by capsule growth, pale to deep bluish violet, each 3- or 5-veined, 16–31 × 3–5 cm; anthers dull yellow to violet, 2.5–4 mm; fruiting pedicel incurving-erect, 10–20 mm. **Capsules** 9–18 mm. **Seeds** 5–10 per locule.

Flowering mid–late spring. Wet meadows; 1000–2400 m; Idaho, Mont., Utah, Wyo.

2. Camassia leichtlinii (Baker) S. Watson, Proc. Amer. Acad. Arts 20: 376. 1885 • Leichtlin camas [E]

Chlorogalum leichtlinii Baker, Gard. Chron., n. s. 1: 689. 1874; *Camassia esculenta* Lindley var. *leichtlinii* (Baker) Baker; *Quamasia leichtlinii* (Baker) Coville

Bulbs usually solitary, ovoid, 1.5–3 cm diam. **Leaves** 3–9, 2–6 dm × 5–25 mm. **Inflorescences** 20–130 cm; sterile bracts 0, bracts subtending flowers usually shorter than pedicel. **Flowers** actinomorphic; tepals strongly connivent over capsules after anthesis, deciduous as capsules develop, bluish violet to bright blue or creamy white, each 5-, 7-, or 9-veined, 20–40 × 4–10 mm; anthers yellow, 4–7 mm; fruiting pedicel spreading-erect, 15–50 mm. **Capsules** often deciduous from pedicel, dull green, ovoid to ellipsoid, 10–25 × 6–12 mm. **Seeds** 6–12 per locule. $2n = 30$.

Subspecies 2 (2 in the flora): w United States and Canada.

1. Perianth always creamy white . . . . 2a. *Camassia leichtlinii* subsp. *leichtlinii*
1. Perianth usually bluish violet to bright blue, occasionally white . . . . 2b. *Camassia leichtlinii* subsp. *suksdorfii*

2a. Camassia leichtlinii (Baker) S. Watson subsp. **leichtlinii** [E]

Perianth creamy white.

Flowering late spring. Wet meadows; 100–200 m; Oreg.

2b. Camassia leichtlinii (Baker) S. Watson subsp. **suksdorfii** (Greenman) Gould, Amer. Midl. Naturalist 28: 723. 1942 [E]

Camassia suksdorfii Greenman, Bot. Gaz. 34: 307. 1902; *Quamasia suksdorfii* (Greenman) Piper

Perianth usually bluish violet to bright blue, occasionally white.

Flowering late spring. Wet meadows; 600–2400 m; B.C.; Calif., Oreg., Wash.

3. Camassia howellii S. Watson, Proc. Amer. Acad. Arts 25: 135. 1889 • Howell camas [C][E]

Quamasia howellii (S. Watson) Coville

Bulbs sometimes clustered, ovoid, 1.5–3 cm diam. **Leaves** 4–7, 2–6 dm × 5–12 mm. **Inflorescences** 15–40 cm; sterile bracts 0, bracts subtending flowers shorter than or equaling pedicel. **Flowers** actinomorphic; tepals connivent over capsules after anthesis, deciduous as capsules develop, deep bluish violet, each 3- or 5-veined, 10–20 × 3–5 mm; anthers yellow, 2.5–3.5 mm; fruiting pedicel spreading, 15–25 mm. **Capsules** deciduous, shiny green, subglobose, 5–10 mm. **Seeds** 2–5 per locule. $2n = 30$.

Flowering late spring. Wet meadows; of conservation concern; 200–400 m; Oreg.

Camassia howellii flowers one to two weeks later than sympatric populations of *C. leichtlinii*.

4. Camassia scilloides (Rafinesque) Cory, Rhodora 38: 405. 1936 (as scillioides) • Eastern camas, wild hyacinth [E] [F]

Cyanotris scilloides Rafinesque, Amer. Monthly Mag. & Crit. Rev. 3: 356. 1818; *Quamasia hyacinthina* (Rafinesque) Britton; *Schoenolirion texanum* (Scheele) A. Gray

Bulbs sometimes clustered, ovoid, 1–3 cm diam. **Leaves** 3–8, 2–6 dm × 5–20 mm. **Inflorescences** 19–47 cm; sterile bracts 0–3(–5), bracts subtending flowers shorter than or equaling pedicel. **Flowers** actinomorphic; tepals usually withering separately after anthesis, not deciduous, light blue, occasionally whitish, each 3- or 5-veined, 7–15 × 2.6–4.2 mm; anthers bright yellow, 1.3–3.2 mm; fruiting pedicel mostly spreading to spreading-erect, 5–30 mm. **Capsules** deciduous, pale green to light brown, subglobose, 6–10 mm. **Seeds** 2–5 per locule. $2n = 30$.

Flowering mid–late spring. Prairies; 100–1000 m; Ont.; Ala., Ark., Ga., Ill., Ind., Iowa, Kans., Ky., La., Md., Mich., Miss., Mo., Ohio, Okla., Pa., S.C., Tenn., Tex., Va., W.Va., Wis.

Camassia scilloides flowers two to three weeks earlier than sympatric populations of *C. angusta*.

The name *Schoenolirion texanum* was long misapplied to a taxon now correctly known as *S. wrightii* Sherman.

5. Camassia angusta (Engelmann & A. Gray) Blankinship, Rep. (Annual) Missouri Bot. Gard. 18: 195. 1907 • Prairie camas [E]

Scilla angusta Engelmann & A. Gray, Boston J. Nat. Hist. 5: 237. 1845; *Camassia fraseri* Torrey var. *angusta* (Engelmann & A. Gray) Torrey; *Quamasia angusta* (Engelmann & A. Gray) Piper

Bulbs sometimes clustered, globose, 1–3 cm diam. **Leaves** 3–11, 2–6 dm × 5–20 mm. **Inflorescences** 27–87 cm; sterile bracts 3–19(–28), bracts subtending flowers shorter than or equaling pedicel. **Flowers** actinomorphic; tepals withering separately, sometimes connivent over capsules, light blue to lavender, each 3- or 5-veined, 6–10 × 2.2–3.6 mm; anthers bright yellow, 1.5–2 mm; fruiting pedicel incurving-erect, 6–12 mm. **Capsules** deciduous, pale green to light brown, ovoid-ellipsoid, 6–10 mm. **Seeds** 2–5 per locule. $2n = 30$.

Flowering mid–late spring. Prairies; 100–500 m; Ark., Ill., Ind., Iowa, Kans., Mo., Okla., Tex.

Camassia angusta flowers two to three weeks later than sympatric populations of *C. scilloides*.

6. Camassia cusickii S. Watson, Proc. Amer. Acad. Arts 22: 479. 1887 • Cusick camas [E]

Quamasia cusickii (S. Watson) Coville

Bulbs clustered, ellipsoid, 2–7 cm diam. **Leaves** rarely fewer than 10, 2.5–5 dm × 2–5 cm. **Inflorescences** 50–80 cm; sterile bracts 0, bracts subtending flowers exceeding pedicel. **Flowers** slightly zygomorphic; tepals withering separately after anthesis, pale blue or bluish violet, each 3(–5)-veined, 25–35 × 3–5 mm; anthers yellow, 3.5–4(–5) mm; fruiting pedicel incurving-erect or slightly spreading, 15–25 mm. **Capsules** not deciduous, light brown, ovoid or ellipsoid, 15–25 mm. **Seeds** 5–10 per locule. $2n = 30$.

Flowering mid–late spring. Hillsides; 1000–2000 m; Idaho, Oreg.

55. CHLOROGALUM Kunth, Enum. Pl. 4: 681. 1843, name conserved • Soap plant, amole [Greek *chloros*, green, and *gala*, milk, alluding to the lather-producing juice of the bulbs]

Judy Jernstedt

Laothoë Rafinesque 1837, name rejected

Herbs, perennial, from tunicate bulbs. **Leaves** basal; blade linear, reduced to scarious bracts in inflorescence. **Inflorescences** paniculate. **Flowers** scattered at intervals along branches, 1–several per node, each subtended by small bract; perianth white, purple, or pinkish; tepals 6, distinct, linear to oblong, with nectaries at base, persisting in fruit and twisting together above capsule; stamens 6, inserted on bases of tepals; anthers versatile; style slightly 3-lobed at apex; pedicel articulate. **Fruits** capsular, 3-valved, dehiscence loculicidal. **Seeds** 1 or 2 per locule.

Species 5 (5 in the flora): w North America, mainly California.

Based on floral characters, chromosome numbers, and karyotypes, *Chlorogalum* probably is closely related to *Camassia*, from which it differs primarily in the number of ovules per locule (1 or 2 in *Chlorogalum*, more than 2 in *Camassia*).

SELECTED REFERENCES Hoover, R. F. 1940. A monograph of the genus *Chlorogalum*. Madroño 5: 137–176. Jernstedt, J. A. 1980. Ultraviolet absorbtion by flowers of *Chlorogalum* (Liliaceae). Bull. Torrey Bot. Club 107: 163–171.

1. Flowers diurnal.
   2. Tepals white or pinkish; flowers or buds 2 or more per axil . . . . 4. *Chlorogalum parviflorum*
   2. Tepals deep blue to purple; flowers and buds occurring singly along inflorescence axis . . . . 5. *Chlorogalum purpureum*
1. Flowers vespertine.
   3. Leaves 2–5 mm wide, margins not strongly undulate. . . . . 3. *Chlorogalum angustifolium*
   3. Leaves 6–25 mm wide, margins strongly undulate.
      4. Pedicel 5–35 mm, ca. equal to or exceeding perianth . . . . 1. *Chlorogalum pomeridianum*
      4. Pedicel 2–5 mm, much shorter than perianth . . . . 2. *Chlorogalum grandiflorum*

1. Chlorogalum pomeridianum (de Candolle) Kunth, Enum. Pl. 4: 682. 1843 E F W1

Scilla pomeridiana de Candolle, Cat. Pl. Hort. Monsp., 143. 1813; *Anthericum pomeridianum* (de Candolle) Ker Gawler; *Laothoë pomeridiana* (de Candolle) Rafinesque; *Phalangium pomeridianum* (de Candolle) Sweet

Bulbs 7–15 cm, tunic thick, with numerous coarse, brown fibers of old bulb scales (except var. *minus*). **Leaves** 2–7 dm × 6–25 mm, margins usually strongly undulate. **Panicles** 3–25 dm, branches many. **Flowers** vespertine, open 1 night only, closed by next morning; tepals spreading and recurved at anthesis, white with green or purple midveins, linear, 1.5–2.5 cm; stamens shorter than tepals; anthers purple or yellow, ca. 3 mm; pollen yellow or whitish; style 10–15 mm, from shorter to longer than tepals; pedicel slender, 5–35 mm, usually ca. equal to or exceeding perianth. **Capsules** 5–7 mm.

Varieties 3 (3 in the flora): California, Oregon.

1. Bulb tunic membranous or with few coarse fibers . . . . 1c. *Chlorogalum pomeridianum* var. *minus*
1. Bulb tunic not membranous, with numerous coarse fibers.
   2. Panicles erect, 5–25 dm . . . . 1a. *Chlorogalum pomeridianum* var. *pomeridianum*
   2. Panicles nearly prostrate or with branches diverging widely from base, 3–4 dm or shorter . . . 1b. *Chlorogalum pomeridianum* var. *divaricatum*

1a. Chlorogalum pomeridianum (de Candolle) Kunth var. **pomeridianum** E F

Bulb tunic not membranous, with numerous coarse fibers. **Panicles** erect, 5–25 dm, with many branches. $2n = 30, 36$.

Flowering May–Aug. Common on open hills and plains, less often in woods and chaparral; 0–1500 m; Calif., Oreg.

Two chromosome numbers have been found in *Chlorogalum pomeridianum* var. *pomeridianum*: $n = 18$ for plants west of the Central Valley of the San Joaquin and Sacramento rivers, and $n = 15$ for populations east of the rivers.

1b. Chlorogalum pomeridianum (de Candolle) Kunth var. **divaricatum** (Lindley) Hoover, Madroño 5: 137. 1940 E

Ornithogalum divaricatum Lindley, Edwards's Bot. Reg. 27: misc. 53. 1841; *Chlorogalum divaricatum* (Lindley) Kunth; *Laothoë divaricata* (Lindley) Greene

Bulb tunic not membranous, with numerous coarse fibers. **Panicles** nearly prostrate or with branches diverging widely from base, 3–4 dm or shorter. $2n = 36$.

Flowering May–Jul. Found along immediate coast, on wind-swept bluffs and hills; 0–100 m; Calif.

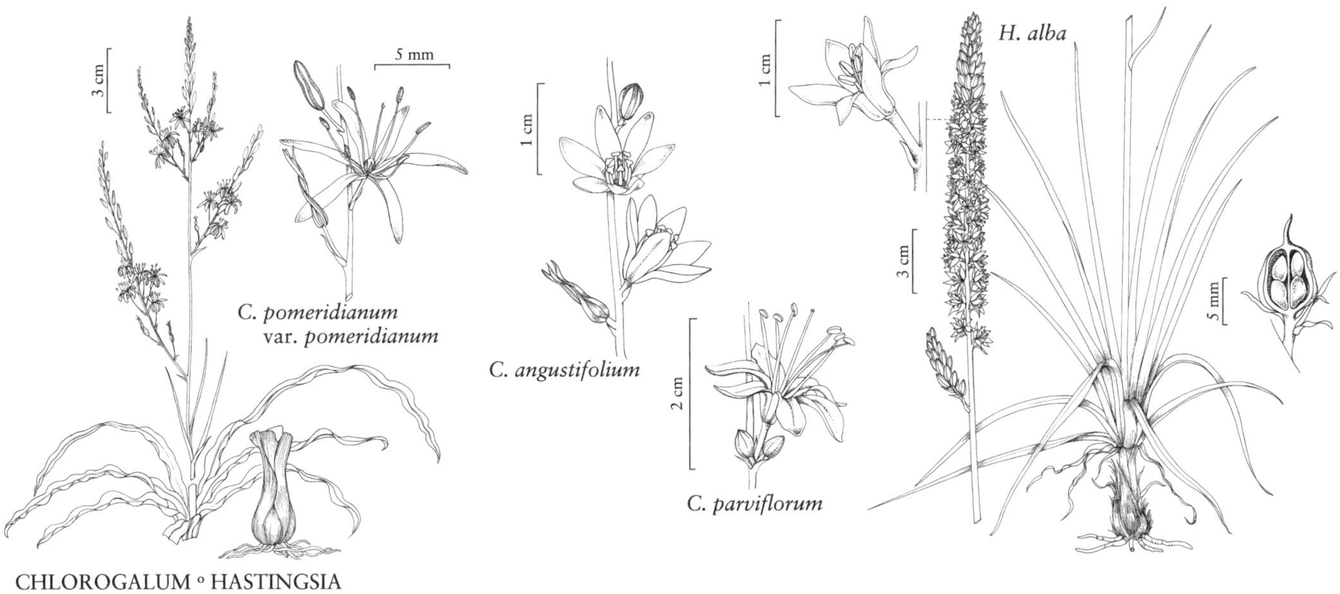

CHLOROGALUM ◦ HASTINGSIA

1c. Chlorogalum pomeridianum (de Candolle) Kunth var. **minus** Hoover, Madroño 5: 144. 1940 • Small soap plant [C] [E]

Bulb tunic membranous, or with few coarse fibers. **Panicles** erect or with ascending branches, 3–4 dm. $2n = 36$.

Flowering May–Jun. Serpertine areas in chaparral; of conservation concern; usually less than 800 m; Calif.

Chlorogalum pomeridianum var. *minus* differs from the other two varieties mainly in the absence of numerous persistent coarse fibers coating the bulb and in its comparatively diminutive stature. It may be confused with *C. grandiflorum*, which also has membranous bulb coats and undulate leaf blades, but var. *minus* can be distinguished by the presence of slender pedicels that are as long or nearly as long as the perianth.

2. Chlorogalum grandiflorum Hoover, Leafl. W. Bot. 2: 128. 1938 • Red hills soap plant [C] [E]

Bulbs ovoid, 5–7 cm, tunic reddish to brown, membranous, outer scales with few delicate fibers. **Leaves** 1–3 dm × 4–12 mm, margins strongly undulate. **Panicles** 3–10 dm, branches ascending. **Flowers** vespertine, closed by following morning; tepals recurved at anthesis, white with purplish midvein, linear, 1.5–3 cm; stamens typically slightly shorter than tepals; anthers yellow, 3 mm; pollen yellow; style often exserted after anthesis, 12–28 mm, equal to or slightly longer than tepals; pedicel stout, 2–5 mm, much shorter than perianth. **Capsules** 5–8 mm.

Flowering May–Jun. Among rocks (usually serpentinite) on open brushy or wooded hills; of conservation concern; 300–500 m; Calif.

Chlorogalum grandiflorum is considered by the California Native Plant Society to be endangered in a portion of its range. It is often confused with *C. pomeridianum* var. *minus* because of bulb characters, but it can be distinguished by its short, stout pedicels and larger flowers.

3. Chlorogalum angustifolium Kellogg, Proc. Calif. Acad. Sci. 2: 104. 1863 [E] [F]

Laothoë angustifolia (Kellogg) Greene, Leafl. Bot. Observ. Crit. 1: 91. 1904

Bulbs 3–5 cm, tunic reddish brown, membranous, with delicate fibers. **Leaves** 1–3 dm × 2–5 mm, margins not strongly undulate. **Panicles** 3–7 dm, branches ascending. **Flowers** vespertine, closing by following morning; tepals spreading but not strongly recurved, white with greenish yellow midveins, oblong, 8–12 mm at anthesis; stamens 8–12 mm, not exceeding tepals; anthers yellow, 1.5–3 mm; style long, 6–8 mm, or short, 2–4 mm; pedicel slender, 2–3 mm. **Capsules** 1.5–3 mm. $2n = 34$.

Flowering Apr–Jul. Heavy, usually rocky soils, valleys and hills; 0–500 m; Calif., Oreg.

4. Chlorogalum parviflorum S. Watson, Proc. Amer. Acad. Arts 14: 243. 1879 [F]

Laothoë parviflora (S. Watson) Greene, Leafl. Bot. Observ. Crit. 1: 91. 1904

Bulbs 4–7 cm, tunic dark brown, membranous. **Leaves** 1–2 dm × 3–9 mm, margins undulate. **Panicles** 3–9 dm, branches usually ascending, erect. **Flowers** diurnal, closing after 1 day, 2 or more per axil; tepals spreading from above base, white or pinkish, with darker midvein, subrotate, 7–8 mm; stamens 3–4 mm, shorter than tepals; anthers yellow, 1.5–2 mm; pollen yellow; style 7–9 mm, longer than tepals; pedicels typically 2 or more in axil of each bract, slender, 2–8 mm, usually shorter than perianth. **Capsules** 3–4 mm. $2n = 60$.

Flowering May–Aug. Dry open areas of coastal sage scrub; 0–800 m; Calif.; Mexico (Baja California).

5. Chlorogalum purpureum Brandegee, Zoë 4: 159. 1893 • Purple amole [C] [E]

Laothoë purpurea (Brandegee) Greene, Leafl. Bot. Observ. Crit. 1: 91. 1904

Bulbs 2.5–3 cm, tunic white to brown, with membranous scales. **Leaves** 1 cm × 2–5 mm, margins undulate. **Panicles** 1–4 dm, branches few. **Flowers** diurnal, occurring singly along inflorescence axis; tepals recurved, deep blue to purple, subrotate, 5–7 mm; stamens ca. equal to tepals; anthers yellow, 1 mm; pollen yellow; style exserted, especially in old flowers, 5–6 mm; pedicel slender, 4–10 mm, typically longer than perianth. **Capsules** 3 mm.

Varieties 2 (2 in the flora): California.

1. Panicles 2.5–4 dm . . . . 5a. *Chlorogalum purpureum* var. *purpureum*
1. Panicles 1–2 dm . . . . 5b. *Chlorogalum purpureum* var. *reductum*

5a. Chlorogalum purpureum Brandegee var. **purpureum** [C] [E]

Panicles 2.5–4 dm. $2n = 60$.

Flowering May–Jun. Open woodlands; of conservation concern; 300 m; Calif.

Chlorogalum purpureum var. *purpureum* is known from only three or four localities on the east side of the Santa Lucia Mountains in the South Coast Ranges. The California Native Plant Society considers it to be endangered throughout its range, and the U. S. Fish and Wildlife Service has listed it as threatened.

5b. Chlorogalum purpureum Brandegee var. **reductum** Hoover, Leafl. W. Bot. 10: 121. 1964 • Camatta canyon amole [C] [E]

Panicles 1–2 dm.

Flowering May–Jun. Serpentine woodlands; of conservation concern; 600 m; Calif.

Chlorogalum purpureum var. *reductum* is known only from two localities in the La Panza Range, South Coast Ranges. The California Native Plant Society considers it to be endangered throughout its range, the California Department of Fish and Game has designated it rare, and the U. S. Fish and Wildlife Service has listed it as threatened.

This variety seems to be identical to *Chlorogalum purpureum* var. *purpureum* except for its compact, dwarf habit of growth. The two are generally distinct in their extreme forms, but may intergrade. In his original description of var. *reductum*, Hoover noted that the two varieties retain their distinctive habits when grown in a common garden. However, fertile seeds are produced when they are crossed artificially.

56. HASTINGSIA S. Watson, Proc. Amer. Acad. Arts 14: 217, 242. 1879 • [For S. Clinton Hastings of San Francisco, supporter of S. Watson et al. (1876–1880) on California botany] [E]

Rudolf W. Becking

Herbs, perennial, scapose, glabrous, from bulbs. **Bulb** solitary, ovoid to ellipsoid, fleshy, with densely packed lunate scales, vein remnants sometimes persisting as blackish, fibrous tunics; no rootstocks or vertical rhizomes. **Leaves** predominantly basal, grasslike; blade often prominently keeled or V-shaped; cauline leaves widely spaced, progressively reduced distally, intergrading with floral bracts. **Scape** borne singly, arising from top of bulb, 25–99 cm, sometimes with 1–3(–5) shorter, ascending branches; scape and branches becoming hollow and brittle. **Inflorescences** racemose, with flowers (10–)24–65(–78) per 10 cm of raceme. **Flowers:** perianth at anthesis closed (not exposing stamens) or partially to fully open (partially or fully exposing stamens); tepals 6, persistent, withering and shriveling to base, distinct, pure white, yellowish or greenish white, or purplish black, narrowly lanceolate to oblong-lanceolate, sometimes each with greenish, yellowish, or purplish central vein, 4–12 mm, often with flattened, triangular apex covered adaxially with minute, glandular hairs; stamens 6; filaments dimorphic, 4–8 mm, 3 longer and with dehisced anthers at anthesis, 3 shorter and with undehisced anthers, all becoming equal after anthesis; ovary superior, ± sessile, 3-lobed, globose, ovules 2 per locule; placentation axile; style 1; stigma distinctly 3-lobed; pedicel 2–3 mm. **Fruits** capsular, broadly 3-lobed, ellipsoid to ellipsoid-ovoid, slightly constricted 1/3 from apex. **Seeds** gray-green to black or yellowish brown, fusiform, usually flattened adaxially; coat rough, reticulate. $x = 26, 27$.

Species 4 (4 in the flora): w North America.

SELECTED REFERENCE Sherman, H. L. and R. W. Becking. 1991. The generic distinctness of *Schoenolirion* and *Hastingsia*. Madroño 38: 130–138.

1. Tepals pure white to slightly yellowish white, or whitish to greenish or yellowish white with greenish or light purplish central vein, narrowly lanceolate, 4–7.5 × 1–2 mm; stamens exposed fully or distally at anthesis.
   2. Bulb 26–56 × 17–31 mm, sometimes with blackish, fibrous tunic; leaves (28–)35–41(–53) × 7–14 mm; scape 40–89 cm; terminal raceme (5–)14–27(–40) cm; flowers (24–)44–51(–78) per 10 cm of raceme; tepals 5–7.5 mm; stamens exposed only distally at anthesis. . . . . 1. *Hastingsia alba*
   2. Bulb 23–41 × 13–21 mm, usually without blackish, fibrous tunic; leaves (19–)21–27(–35) cm × 2–6 mm; scape 29–51 cm; terminal raceme (4–)7–12(–35) cm; flowers (12–)24–35(–48) per 10 cm of raceme; tepals 4–6.5 mm; stamens fully exposed at anthesis . . . . 4. *Hastingsia serpentinicola*
1. Tepals purplish black with pale green central vein, or yellowish white with pale yellowish or purplish central vein, oblong-lanceolate, 8–12 × 2 mm; stamens seldom exposed at anthesis.
   3. Leaves 25–55 cm × 6–12 mm, blade bluish green; terminal raceme (8–)12–35(–49) cm; flowers (20–)30–36(–46) per 10 cm of raceme; tepals purplish black with pale green central vein, 9–12 × 2 mm . . . . 2. *Hastingsia atropurpurea*
   3. Leaves 24–52 cm × 4–8 mm, blade grayish green; terminal raceme (5–)8–16(–32) cm; flowers (10–)25–30(–40) per 10 cm of raceme; tepals yellowish white with pale yellowish or purplish central vein, 8–12 × 2 mm . . . . 3. *Hastingsia bracteosa*

1. Hastingsia alba (Durand) S. Watson, Proc. Amer. Acad. Arts 14: 242. 1879 [E] [F]

Schoenolirion album Durand, J. Acad. Nat. Sci. Philadelphia, ser. 2, 3: 103. 1855

Bulb ellipsoid, 26–56 × 17–31 mm, sometimes with blackish, fibrous tunic. **Leaves** (28–)35–41(–53) cm × 7–14 mm, mature plants with blackish, shriveled foliage persisting at base of scape; blade glaucous-green, aging to light green. **Scape** 40–89 cm, often with 1–3 ascending branches, 3–5 mm thick at base. **Inflorescences:** terminal raceme (5–)14–27(–40) cm; flowers (24–)44–51(–78) per 10 cm of raceme. **Seeds** dark gray-green to black, 4–6 mm. $2n = 52$.

Flowering Jun–Jul, fruiting late Jul–Sep. Serpentine, granite, and diorite sites, open rocky seepage areas with year-round water supply, bogs, wet meadows; 500–2300 m; Calif., Oreg.

At high elevations, *Hastingsia alba* is often stunted and has smaller bulbs, shorter scapes with shorter and narrower leaves, and spreading-rotate tepal tips.

2. Hastingsia atropurpurea Becking, Madroño 33: 175, figs. 1, 2. 1986 [C] [E]

Bulb ellipsoid or infrequently ellipsoid-ovoid, 28–54 × 18–30 mm, usually with blackish, fibrous tunic. **Leaves** 25–55 cm × 6–12 mm, mature plants with blackish, shriveled foliage at base of scape; blade bluish green. **Scape** 71–99 cm, often with 2–5 ascending branches, 4–6 mm thick at base. **Inflorescences:** terminal raceme (8–)12–35(–49) cm; flowers (20–) 30–36(–46) per 10 cm of raceme. **Flowers:** tepals purplish black with pale green central vein, oblong-lanceolate, 9–12 × 2 mm; stamens seldom exposed at anthesis; filaments 5–8 mm; anthers purple. **Capsules** ellipsoid to ellipsoid-ovoid, 7–11 × 5–8 mm. **Seeds** black, 5 mm.

Flowering Jun–Jul, fruiting Jul–Aug. Ultramafic riverbeds that have year-round water in rooting horizon and wet, open, sunny bogs; of conservation concern; 500–700 m; Oreg.

3. Hastingsia bracteosa S. Watson, Proc. Amer. Acad. Arts 20: 377. 1885 [C] [E]

Bulb ellipsoid-ovoid, 26–43 × 15–27 mm, sometimes with blackish tunic. **Leaves** 24–52 cm × 4–8 mm, mature plants without blackish, shriveled foliage at base of scape; blade grayish green. **Scape** 47–87 cm, rarely branched, 2–4 mm thick at base. **Inflorescences:** terminal raceme (5–)8–16(–32) cm; flowers (10–)25–30(–40) per 10 cm of raceme. **Flowers:** tepals yellowish white with pale yellowish or purplish central vein, oblong-lanceolate, 8–12 × 2 mm; stamens seldom exposed at anthesis; filaments 5–7 mm; anthers pale yellowish. **Capsules** ellipsoid to obovoid-ellipsoid, 9–11 × 6–8 mm. **Seeds** dark grayish green, sometimes yellowish brown to black, 4–5 mm.

Flowering Jun–Jul, fruiting Jul–Aug. Bogs and wet seepage areas of serpentine outcroppings; of conservation concern; 400–500 m; Oreg.

SELECTED REFERENCE Becking, R. W. 1986. *Hastingsia atropurpurea* (Liliaceae: Asphodeleae). A new species from southwestern Oregon. Madroño 33: 175–181.

4. Hastingsia serpentinicola Becking, Madroño 36: 208, figs. 1, 3. 1989 [E]

Bulb narrowly ellipsoid, 23–41 × 13–21 mm, usually without blackish, fibrous tunic. **Leaves** (19–)21–27(–35) cm × 2–6 mm, mature plants often without shriveled foliage at base of scape; blade grayish green. **Scape** 29–51 cm, rarely with 1–3 ascending branches, 1–2 mm thick at base. **Inflorescences:** terminal raceme (4–)7–12(–35) cm; flowers (12–)24–35 (–48) per 10 cm of raceme. **Flowers:** tepals whitish to greenish or yellowish white with greenish or light purplish central vein, narrowly lanceolate, 4–6.5 × 1–2 mm; stamens fully exposed at anthesis; filaments 4–7 mm; anthers often light purple. **Capsules** ellipsoid, 5–7 × 3–6 mm. **Seeds** gray-green to black, 3.5–4.7 mm. $2n$ = 52, 54.

Flowering May–Jun, fruiting Jul–Aug. Ultramafic or serpentine rock outcroppings, bogs; 300–2200 m; Calif., Oreg.

Hastingsia serpentinicola grows in open sites that are moist in the early spring and dry out rapidly in the summer. It often occurs along the edges of *Darlingtonia* bogs or on the drier islands within the bog environment. At high elevations, it occurs on exposed serpentine ridges, usually close to wet habitats.

SELECTED REFERENCE Becking, R. W. 1989. Segregation of *Hastingsia serpentinicola* sp. nov. from *Hastingsia alba* (Liliaceae: Asphodeleae). Madroño 36: 208–216.

57. SCHOENOLIRION Torrey ex Durand, J. Acad. Nat. Sci. Philadelphia, ser. 2, 3: 103. 1855, name conserved • Rush-lily [Greek *schoenos*, rush, and *lirion*, white lily] [E]

H. L. Sherman

Amblystima Rafinesque 1837, name rejected; *Oxytria* Rafinesque 1837, name rejected

Herbs, perennial, scapose, glabrous, from rootstocks; rootstock vertical, 1–12 cm, fleshy, sometimes bulbose; principal roots contractile, arising from tops of rootstocks. **Bulbs,** when present, ovoid or elongate, scales lunate. **Leaves** 2–7, basal; blade flat to nearly terete, sometimes keeled. **Inflorescences** racemose or 1–3(–6)-branched, bracteate, flowers initially crowded, internodes elongating after anthesis. **Flowers** soon deciduous if unfertilized; tepals persistent, withering separately to base, distinct, yellow or white, each 3–7-veined, equal, 2–6 mm; stamens 6; filaments equal, shorter than tepals, 1–2 mm, inner 3 nectariferous at base; anthers introrse; ovary superior, globose, 3-lobed, 3-locular, placentation axile, ovules 2 per locule; style 1; stigma unlobed or slightly 3-lobed; pedicel 6–15(–30 mm). **Fruits** capsular, dehiscent, flattened or indented at top, prominently 3-lobed. **Seeds** glossy black, globose, flattened on 1 side, smooth. $x = 12$.

Species 3 (3 in the flora): se United States, including e Texas.

SELECTED REFERENCES Gray, A. 1876. *Schoenolirion* Torr. Amer. Naturalist 10: 426–427. Pollard, C. L. 1897. The genus *Oxytria* of Rafinesque. Bull. Torrey Bot. Club 24: 405–406. Sherman, H. L. 1969. A Systematic Study of the Genus *Schoenolirion* (Liliaceae). Ph.D. thesis. Vanderbilt University. Sherman, H. L. and R. W. Becking. 1991. The generic distinctness of *Schoenolirion* and *Hastingsia*. Madroño 38: 130–138.

1. Leaves without fleshy bases, arising from tops of vertical rootstocks, withering to persistent fibers; inflorescences 1–3(–6)-branched; tepals white or greenish white . . . . 3. *Schoenolirion albiflorum*
1. Leaves with fleshy bases, arising from bulbs at tops of vertical rootstocks, not withering to persistent fibers; inflorescences seldom branched; tepals white or yellow.
    2. Tepals yellow . . . . 1. *Schoenolirion croceum*
    2. Tepals white . . . . 2. *Schoenolirion wrightii*

1. Schoenolirion croceum (Michaux) Alph. Wood, Amer. Bot. Fl., 344. 1870 • Sunnybells, swamp candle [E] [F]

Phalangium croceum Michaux, Fl. Bor.-Amer. 1: 196. 1803; *Oxytria crocea* (Michaux) Rafinesque

Plants with prominent bulbs at tops of vertical rootstocks; bulbs ovoid to elongate, to 17 mm diam. **Leaves** 3–7, 20–32 cm × 5–8 mm, not withering to persistent fibers; blade flattened or somewhat keeled, usually equaling or exceeding inflorescence, base fleshy. **Inflorescences** seldom branched; bracts ovate-obtuse to lanceolate-acuminate. **Flowers:** tepals seldom strongly recurved, yellow, each with green or reddish abaxial stripe, 3–5-veined, ovate to ovate-oblong, 4.5–6 mm, apex obtuse; ovary green or greenish yellow. $2n$ = 24, 30, 32.

Flowering mid Mar–mid Apr in south and west; Apr–mid May in north; dormant by late Jun. Rock outcrops, moist pinelands; 0–400 m; Ala., Fla., Ga., La., N.C., S.C., Tenn., Tex.

Schoenolirion croceum is very localized on limestone outcrops in central Tennessee and northern Alabama, sandstone outcrops of the Alabama plateau region, granite outcrops of the Georgia and Carolina Piedmont, Selma chalk outcrops in western Alabama, and wet pinelands and boggy areas in more southern and western parts of its range. All of these habitats are very wet in spring and often very dry in summer.

2. Schoenolirion wrightii Sherman, SouthW. Naturalist 24: 125. 1979 [E]

Plants with prominent bulbs at tops of vertical rootstocks; bulbs ovoid to elongate, to 17 mm diam. **Leaves** 2–7, 14–34 cm × 2–6 mm, not withering to persistent fibers; blade flattened or somewhat keeled, usually shorter than scape and inflorescence, base fleshy. **Inflorescences** seldom branched; bracts mostly lanceolate-acuminate, sometimes ovate-obtuse. **Flowers:** tepals seldom strongly recurved, white, each with green abaxial stripe, mostly 3(–5)-veined, ovate to ovate-lanceolate, 4.4–6.5 mm, apex obtuse; ovary and style green. $2n$ = 24.

Flowering late Mar–mid Apr in Texas, Arkansas; mid Apr–early May in Alabama; dormant by mid Jun. Sandstone outcrops, moist pinelands, or boggy areas; 0–400 m; Ala., Ark., La., Tex.

Schoenolirion wrightii occurs in habitats similar to those of *S. croceum* in the Alabama plateau region and in the western part of its range.

Treatment of these white-flowered plants as a species distinct from *Schoenolirion croceum* is very tenuous; tepal color seems to be the only consistent difference between the two. At the time of the last systematic study of the genus (H. L. Sherman 1969), there seemed to be a significant difference in chromosome number between the two "color forms" ($2n$ = 24 for *S. wrightii* and $2n$ = 30 or 32 for *S. croceum*). Since that time, one population of *S. croceum* with $2n$ = 24 has been found. In future treatments of the genus, *S. wrightii* may be considered to be a color form or variety of *S. croceum*.

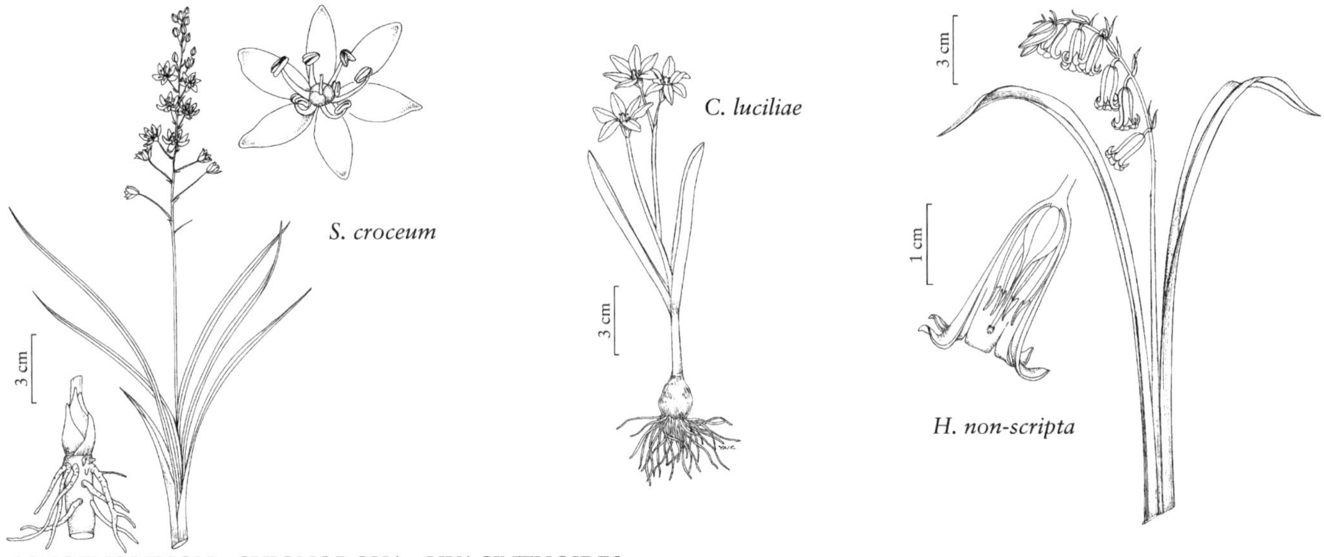

SCHOENOLIRION ◦ CHIONODOXA ◦ HYACINTHOIDES

3. **Schoenolirion albiflorum** (Rafinesque) R. R. Gates, J. Linn. Soc., Bot. 44: 167. 1918 [E]

Amblostima albiflora Rafinesque, Fl. Tellur. 2: 26. 1837; *Oxytria albiflora* (Rafinesque) Pollard

Plants with coarsely fibrous leaves arising directly from top of vertical rootstock; bulbs absent. **Leaves** 4–7, 26–45(–72) cm × 2–4 mm, fibers withering to usually persistent; blade flattened or ± terete with prominent adaxial groove, usually shorter than scape and inflorescence, base not fleshy. **Inflorescences** 1–3(–6)-branched (occasionally simple or with secondary branches), flowers separated up to 6 cm or distally clustered; bracts lanceolate-acuminate to ovate-acuminate, basal usually much longer than distal. **Flowers:** tepals often strongly recurved, white or greenish white, each with greenish abaxial stripe, 5–7-veined, ovate, 3–4.5 mm, apex obtuse; ovary and style green. $2n = 48?, 49$.

Flowering May in south; late May–early Jun in north. Marshy pinelands, cypress bogs, wet savannas; 0–100 m; Fla., Ga.

Excluded species:

Schoenolirion texanum (Scheele) A. Gray and *Oxytria texana* (Scheele) Pollard are names based upon *Ornithogalum texanum* Scheele, which is the same as *Camassia scilloides* (Rafinesque) Cory, and does not belong to *Schoenolirion*.

58. CHIONODOXA Boissier, Diagn. Pl. Orient. 1(5): 61. 1844 • Glory-of-the-snow [Greek *chion*, snow, and *doxa*, glory or repute] [I]

J. McNeill

Herbs, perennial, scapose, from tunicate bulbs. **Bulbs** perennial, ovoid to globose, composed of free scales, progressively renewed annually; tunics brown. **Leaves** usually 2, basal, erect or spreading. **Inflorescences** racemose, open, 1–few-flowered; bracts none or 1 subtending each flower. **Flowers:** perianth blue or blue and white; tepals shortly connate to form urceolate tube, lobes spreading to reflexed; stamens 6; filaments inserted at apex of perianth tube, unequal, ± equaling anthers; anthers dorsifixed, introrse; ovary superior, 3-locular, septal nectaries present; style simple. **Fruits** capsular, subglobose, 3-lobed, dehiscence loculicidal. **Seeds** 3–18, globose to ellipsoid, not winged, elaiosomes present. $x = 9$.

Species 6 (1 in the flora): introduced; e Mediterranean area, 1 or 2 species naturalized in temperate areas.

Chionodoxa, traditionally separated from *Scilla* because of its partially fused tepals, is very similar to species of *Scilla* subg. *Scilla*, especially *S. bifolia* Linnaeus, and hybrids between species of the two groups (×*Chionoscilla allenii* Nicholson) occur frequently. As a result, the species of *Chionodoxa* are sometimes included within *Scilla* subg. *Scilla* (F. Speta 1971, 1976).

SELECTED REFERENCES Meikle, R. D. 1970. *Chionodoxa luciliae*, a taxonomic note. J. Roy. Hort. Soc. 95: 21–24. Rix, E. M. 1986. *Chionodoxa*. In: S. M. Walters et al., eds. 1984+. European Garden Flora.... 6+ vols. Cambridge etc. Vol. 1, pp. 214–215. Speta, F. 1971. Beitrag zur Systematik von *Scilla* L. subgen. *Scilla* (inklusive *Chionodoxa* Boiss.). Oesterr. Bot. Z. 119: 6–18. Speta, F. 1976. Über *Chionodoxa* Boiss., ihre Gliederung und Zugehörigkeit zu *Scilla* L. Naturk. Jahrb. Stadt Linz 21: 9–79.

1. Chionodoxa luciliae Boissier, Diagn. Pl. Orient. 1(5): 61. 1844 [F] [I]

Scilla luciliae (Boissier) Speta

Plants 10–20 cm; bulbs ovoid, 1.5–2 cm. **Leaves** 2–4, often recurved, broadly linear, 7–20 × 1–2 cm. **Scapes** usually solitary. **Inflorescences** 1–2-flowered. **Flowers:** perianth deep blue with white central zone, tube 2.5–4 mm; tepals 12–15(–20) mm; filaments white; pedicel erect, equaling or shorter than perianth. **Capsules** ± globose, 4–6 mm. **Seeds** with white appendages. $2n = 18$.

Flowering early to mid spring. Lawns, open mesic areas; [1600–2000 m in Turkey]; introduced; Mass., Mich., Utah; Europe (w Turkey).

Chinodoxa luciliae, a spring-blooming, garden species, has been reported established on lawns in Houghton and Borrago counties, Michigan (E. G. Voss 1972–1985, vol. 1). Because *C. forbesii* Baker, distinguished by its taller scapes (to 30 cm), with more numerous (4–10), slightly larger, and slightly pendent flowers, is more widely cultivated, often as "*C. luciliae*," both it and *C. luciliae* are expected to be naturalized elsewhere.

59. HYACINTHOIDES Heister ex Fabricius, Enum., 2. 1759 • Bluebell [genus *Hyacinthus* and Greek *oides*, resembling] [I]

J. McNeill

Endymion Dumortier

Herbs, perennial, scapose, from bulbs. **Bulb** annual, ovoid, composed of few tubular coalescent scales, completely renewed annually. **Leaves** few, basal. **Inflorescences** racemose, bracteate; bracts 2 subtending each flower, linear-lanceolate. **Flowers:** tepals shortly connate basally, each 1-veined, 15– 20 mm; stamens 6; filaments inserted on perianth; anthers dorsifixed, versatile, introrse; ovary superior, 3-locular, septal nectaries present, ovules 1–10 per locule; style simple; stigma swollen distally. **Fruits** capsular, subglobose, 3-lobed, dehiscence loculicidal. **Seeds** 3–30, black, globose to ellipsoid, unwinged. $x = 8$.

Species 3–4 (2 in the flora): introduced; w Europe, nw Africa.

Occasional garden escapes of *Hyacinthoides* occur in North America. Although the species are still often referred to as *Scilla* in North American floras, the group has been recognized separately for some 70 years (originally as *Endymion*), distinguished from *Scilla* by a number of developmental features as well as the partially fused perianth.

Hybrids between *Hyacinthoides non-scripta* and *H. hispanica* are widespread in gardens and are widely naturalized in Washington state, and probably elsewhere, occurring in the absence of either parent. These hybrids, which have been called *H.* ×*variabilis* P. D. Sell, show a range of intermediacy between the parents, and, in particular, have slightly one-sided racemes, slightly recurved tepals, stamens that are variable in length, and bluish or whitish anthers (P. D. Sell and G. Murrell 1996).

SELECTED REFERENCES Chouard, P. 1930. Types de Développement de l'Appareil Végétatif Chez les Scillées. D.Sc. thesis. University of Paris. [Preprinted from Ann. Sci. Nat., Bot., sér. 10, 13: 131–322. 1931.] Speta, F. 1987. Die verwandschaftlichen Beziehungen von *Brimeura* Salisb.: Ein Vergleich mit den Gattungen *Oncostema* Rafin., *Hyacinthoides* Medic. und *Camassia* Lindl. (Hyacinthaceae). Phyton (Horn) 26: 247–310. Wilson, J. Y. 1958. Cytogenetics of triploid bluebells, *Endymion non-scripta* (L.) Garcke and *E. hispanica* (Mill.) Chouard. Cytologia 23: 435–446.

1. Raceme 1-sided; flowers all becoming pendent, fragrant; tepals strongly recurved at apex; stamens unequal, outer inserted just above middle of tepals, inner near base; anthers cream . . . . 1. *Hyacinthoides non-scripta*
1. Raceme not 1-sided; at least distal flowers remaining erect, not fragrant; tepals spreading but not recurved; stamens equal, inserted below middle of tepals; anthers blue . . . . 2. *Hyacinthoides hispanica*

1. Hyacinthoides non-scripta (Linnaeus) Chouard, Bull. Soc. Bot. France 81: 625. 1934 • English bluebell F I

Hyacinthus non-scriptus Linnaeus, Sp. Pl. 1: 316. 1753 (as non scriptus); *Endymion non-scriptus* (Linnaeus) Garcke; *Scilla non-scripta* (Linnaeus) Hoffmannsegg & Link; *S. nutans* Smith

Plants 20–30 cm; bulb 1–1.5 mm diam. **Leaves** 3–6; blade linear to linear-lanceolate, 20–50 cm × 7–15(–25) mm. **Racemes** 6–12-flowered, 1-sided, apex drooping. **Flowers** erect in bud, all becoming pendent, fragrant; perianth cylindrical-campanulate; tepals erect proximally, reflexed distally, violet-blue or rarely white or pink, oblong-lanceolate, 1.5–2 cm; stamens unequal, outer inserted near middle of perianth, inner near base; anthers cream; pedicel 4–10 mm, ± equaling perianth. $2n = 16$, 24.

Flowering late spring. Escaped from gardens; 0–1500 m; introduced; B.C.; Ind., Ky., N.Y., Ohio, Pa., Va., Wash.; w Europe.

2. Hyacinthoides hispanica (Miller) Rothmaler, Feddes Repert. Spec. Nov. Regni Veg. 53: 14. 1944 • Spanish bluebell I

Scilla hispanica Miller, Gard. Dict. ed. 8, Scilla no. 8. 1768; *Endymion hispanicus* (Miller) Chouard; *Hyacinthoides non-scripta* (Linnaeus) Chouard subsp. *hispanica* (Miller) Kerguélen

Plants 20–40 cm; bulb 1–2 mm diam. **Leaves** 4–8; blade linear-lanceolate to lanceolate, 20–50 cm × 10–35 mm. **Racemes** 6–8-flowered, not 1-sided, apex not drooping. **Flowers** erect in bud, at least distal ones remaining so, not fragrant; perianth campanulate; tepals spreading, not reflexed distally, blue, oblong-lanceolate, 1.5–2 cm; stamens equal, inserted below middle of perianth; anthers blue; pedicel 4–10 mm, ± equaling perianth. $2n = 16$, 24.

Flowering late spring. Escaped from gardens; 0–1000 m; introduced; Va.; sw Europe, nw North Africa.

60. MUSCARI Miller, Gard. Dict. Abr. ed. 4, vol. 2. 1754 • Grape-hyacinth [Greek *moschos*, musk, alluding to the scent of the flowers] I

Gerald B. Straley†

Frederick H. Utech

Herbs perennial, scapose, from brown, tunicate, ovoid bulbs, with or without offsets (bulblets). **Leaves** (1–)2–7, basal; blade linear, sometimes sulcate, glabrous, rather fleshy. **Scape** terete. **Inflorescences** terminally racemose, many-flowered, dense, bracteate, usually elongating in fruit; distal flowers smaller, sterile, differing in color, forming a tuft (coma); bracts minute. **Flowers** fragrant; perianth tubular to urceolate, usually constricted basally; tepals 6, connate most of their length, distal portions distinct, reflexed, short, toothlike; stamens 6, epitepalous, in 2 rows, included; anthers dark blue, dorsifixed, globose; ovary superior, green, 3-locular, inner sepal nectaries present; style 1; stigma 3-lobed. **Fruits** capsular, obtusely 3-angled, papery, dehiscence loculicidal. **Seeds** 6, black, globose, wrinkled to reticulate. $x = 9$.

Species ca. 30 (3 in the flora): introduced; temperate Europe, n Africa, sw Asia; expected introduced elsewhere.

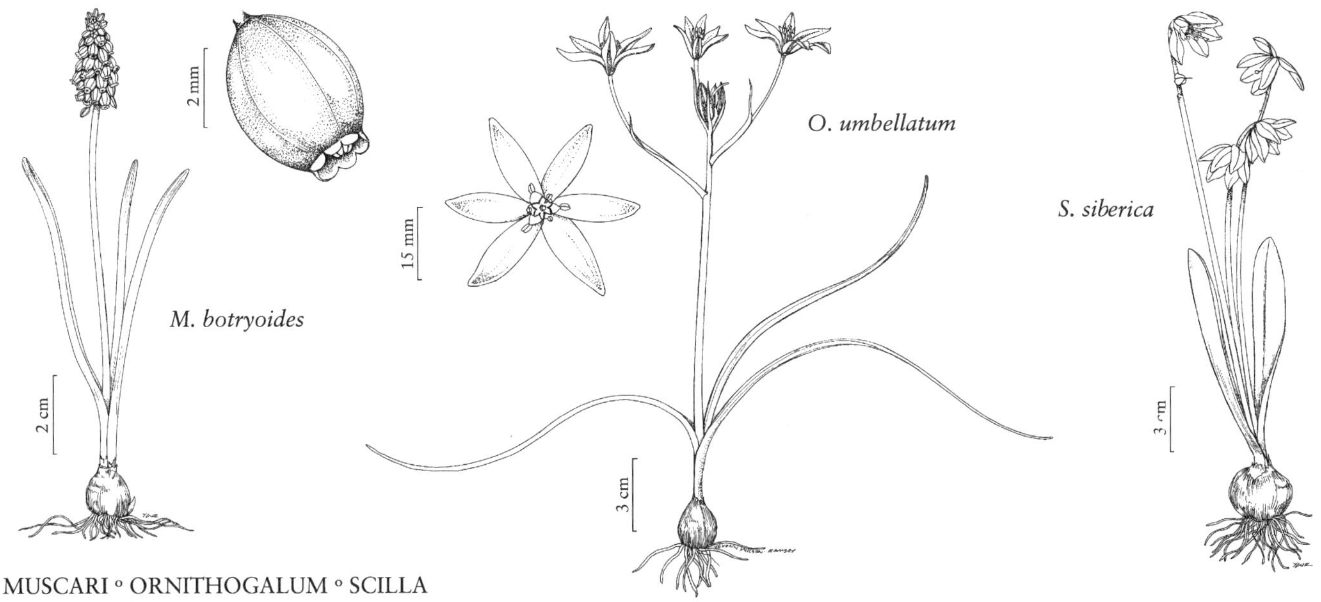

MUSCARI ◦ ORNITHOGALUM ◦ SCILLA

Various species and cultivated forms of *Muscari* are commonly grown for their early spring flowers. They may reseed in the flora area, but they are mostly transported in soil containing the bulblets.

Muscari armeniacum Baker has been attributed to the flora, but no definite records of naturalized plants have been found. Herbarium specimens of that species are difficult to distinguish from those of *M. neglectum*, but live specimens of *M. armeniacum* have much paler blue flowers (A. Huxley et al. 1992).

1. Tepals of fertile flowers pale to olive brown, sterile flowers bright violet, shorter than the 6–25 mm ascending pedicels . . . . 1. *Muscari comosum*
1. Tepals of fertile and sterile flowers blue, longer than the 1–4(–5) mm declined, nodding, or spreading pedicels.
    2. Racemes 12–20-flowered; leaf blades 3–8 mm wide; perianth tubes of fertile flowers globose, sky blue . . . . 2. *Muscari botryoides*
    2. Racemes 20–40-flowered; leaf blades 2–4(–5) mm; perianth tubes of fertile flowers obovoid to oblong-urceolate or cylindric, blackish blue . . . . 3. *Muscari neglectum*

1. Muscari comosum (Linnaeus) Miller, Gard. Dict. ed. 8, Muscari no. 2. 1768 • Tassel grape-hyacinth [I]

Hyacinthus comosus Linnaeus, Sp. Pl. 1: 318. 1753

Plants to 15(–30) cm. **Bulbs** ovoid, 3–3.5 × 2.5–3 cm, offsets absent, tunics pink. **Leaves** 3–5(–7); blade channeled, linear-lanceolate, 20–40 cm × 6–20(–30) mm, apex long-attenuate. **Scape** (15–)30–50(–80) cm, usually longer than leaves. **Racemes** 40–100-flowered. **Flowers:** perianth tube oblong-urceolate, those of fertile flowers pale to olive-brown, 5–8 × 2–4 mm, teeth lighter brown or white; sterile flowers bright violet, 2–6 × 1–2 mm; pedicel of fertile flowers ascending, 4–10(–15) mm, ± equaling perianth; pedicel of sterile flowers ascending, 6–25(–40) mm. **Capsules** 6–12 × 6–8 mm. 2*n* = 18, 27.

Flowering late spring. Roadsides, open woods, abandoned gardens; 0–800 m; introduced; Ga., Ill., Ky., Mo., N.C., Ohio, Oreg., Pa., Tenn., Va.; s Europe; n Africa; sw Asia; expected elsewhere.

2. Muscari botryoides (Linnaeus) Miller, Gard. Dict. ed. 8, Muscari no. 1. 1768 • Common grape-hyacinth, muscari [F] [I] [W2]

Hyacinthus botryoides Linnaeus, Sp. Pl. 1: 318. 1753

Plants to 20(–30) cm. **Bulbs** ovoid, 1.5–2.5 × 1–2 cm, offsets absent, tunics translucent to pale brown. **Leaves** 2–4(–5); blade prominently ribbed, linear-spatulate, 15–35(–40) cm × 3–8(–12) mm, apex abruptly contracted. **Scape** 20–35(–40) cm, usually slightly exceeding leaves.

Racemes 12–20-flowered. **Flowers:** perianth tube sky blue, globose to ovoid, 2–4 × 2–3 mm, teeth white; fertile and sterile flowers ± equal (sterile may be slightly smaller and paler); pedicel spreading, 1–3(–5) mm. **Capsules** 4–6 × 4–6 mm. $2n$ = 18, 36.

Flowering early–mid spring. Roadsides, fields, woods, abandoned gardens; 0–1500 m; introduced; B.C., Nfld. and Labr. (Nfld.), N.S., Ont.; Ala., Ark., Calif., Conn., Del., D.C., Ill., Ind., Kans., Ky., Md., Mass., Mich., Minn., Miss., Mo., N.H., N.J., N.Y., Ohio, Okla., Oreg., Pa., Tenn., Tex., Utah, Vt., Va., Wash., W.Va.; c, se Europe; expected elsewhere.

Muscari botryoides is the commonest and most cold-hardy of the *Muscari* species in the flora.

3. **Muscari neglectum** Gussone ex Tenore, Syll. Pl. Fl. Neapol., App. 5: 13. 1842 [I]

Hyacinthus racemosus Linnaeus 1753, not *Muscari racemosum* Miller 1768; *M. atlanticum* Boissier & Reuter

Plants to 20(–30) cm. **Bulbs** ovoid, 2–3 × 2–2.5 cm, offsets usually present, tunics dark brown. **Leaves** 3–6; blade channeled to subterete, narrowly linear, (10–)15–30(–40) cm × 2–5(–8) mm, apex abruptly contracted. **Scape** 8–25(–30) cm, usually ± equaling leaves. **Racemes** 20–40-flowered. **Flowers:** perianth tube blackish blue, obovoid to oblong-urceolate or cylindric, 4–6 × 2–3 mm, teeth white; fertile and sterile flowers ± equal (sterile may be slightly smaller and paler); pedicel declined or nodding, 3–5 mm. **Capsules** 5–7(–10) × 5–7(–10) mm. $2n$ = 18, 36, 45, 54, 72.

Flowering early–mid spring. Roadsides, fields, woods, abandoned gardens; 0–1500 m; introduced; Ala., Ark., Conn., Del., D.C., Ga., Ill., Ind., La., Md., Mass., Mich., Miss., N.Mex., N.Y., N.C., Ohio, Pa., S.C., Tenn., Tex., Va., W.Va.; s Europe; n Africa; sw Asia; expected elsewhere.

W. T. Stearn (1990) typified *Hyacinthus racemosus* and discussed the taxonomically different applications of that name and *Muscari racemosum*. The latter binomial, misattributed to (Linnaeus) Miller, sometimes has been used for what is correctly called *M. neglectum*.

61. ORNITHOGALUM Linnaeus, Sp. Pl. 1: 306. 1753; Gen. Pl. ed. 5, 145. 1754 • Star-of-Bethlehem, ornithogale, dame-d'onze-heures, étoile de Bethléem [Greek, *ornis*, bird, and *gala*, milk, alluding to the color of the flowers] [I]

Gerald B. Straley†

Frederick H. Utech

Herbs, perennial, scapose, poisonous, from ovoid, tunicate bulbs; bulb tunics white to pale brown, papery. **Leaves** few to several, basal; blade linear to lanceolate, margins smooth or hairy. **Inflorescences** racemose or corymbose, 2–many-flowered, bracteate; bracts white, membranous. **Flowers:** tepals 6, widely spreading, distinct, equal to slightly unequal; stamens 6, distinct, dimorphic; filaments simple or 3-dentate, flattened; anthers dorsifixed, introrse; ovary superior, green, 3-locular, cylindric to globose, 6-angled, septal nectaries present; style 1; stigma margins entire or indistinctly 3-lobed. **Fruits** capsular, angled, papery, dehiscence loculicidal. **Seeds** numerous, globose to ovoid. x = 3, 5, 6, 7, 8, 9, 11.

Species 100–150 (2 in the flora): introduced; Europe, Africa, w Asia.

Ornithogalum is a large genus centered in South Africa and around the Mediterranean, with numerous species of horticultural note. *Ornithogalum pyrenaicum* Linnaeus was apparently naturalized for a number of years at Elkins, West Virginia (E. E. Hutton and R. E. Clarkson 1961), but it is now extirpated from that site. *Ornithogalum arabicum* Linnaeus, Arabian star flower, *O. caudatum* Aiton, sea-onion, and *O. thyrsoides* Jacquin, chincherinchee, are frequently cultivated in the warmer parts of the flora area.

The bulbs of *Ornithogalum* species are poisonous due to the presence of a variety of cardiotoxic cardenolides (G. E. Burrows and R. J. Tyrl 2001), and the bulbs should not be confused with those of wild onions.

SELECTED REFERENCES Moret, J. and N. Galland. 1992. Phenetic, biogeographical, and evolutionary study of *Ornithogalum* subg. *Heliocharmos* (Hyacinthaceae) in the western Mediterranean Basin. Pl. Syst. Evol. 181: 179–202. Müller-Doblies, U. and D. Müller-Doblies. 1996. Revisionula incompleta *Ornithogalorum* Austro-Africanorum. Feddes Repert. 107: 361–548. Zahariadi, C. 1962. Caractères morphologiques, anatomiques et biologiques dans la taxonomie du genre *Ornithogalum*. Rev. Biol. (Bucharest), n. s. 7: 5–41.

1. Leaves 5–10 mm wide; inflorescences racemose, cylindrical, 5–12(–18)-flowered; lower pedicels to 1 cm; flowers nodding, especially after anthesis; filaments 3-dentate .... 1. *Ornithogalum nutans*
1. Leaves 3–5 mm wide; inflorescences corymbose, flat-topped, (4–)8–20-flowered; lower pedicels 2–6 cm; flowers erect; filaments simple .... 2. *Ornithogalum umbellatum*

1. Ornithogalum nutans Linnaeus, Sp. Pl. 1: 308. 1753 • Nodding star-of-Bethlehem [I] [W1] [W2]

Plants (15–)30–50(–60) cm; bulbs progressively renewed over 3–4 years, 2–3 × 2.5–3.5 cm; bulblets numerous. **Leaves** (3–)4–6; blade with white adaxial stripe, 30–60 cm × (3–)5–10(–15) mm. **Scape** 2–5 dm. **Inflorescences** racemose, 5–12(–18)-flowered, cylindrical; bracts 3–4 cm. **Flowers** nodding, especially after anthesis; perianth campanulate; tepals white with wide green abaxial stripe, lanceolate, 20–30 × 9–12 mm, outer 3 slightly gibbous at base, inner 3 slightly wider; stamens: outer 6–8 × 3–4 mm, inner 13–15 × 4–5 mm; filaments 3-dentate, flattened; anthers 3–4 mm; ovary ovoid to subglobose, slightly 6-angled, 5–6 mm; style 5–6 mm; lower pedicels to 1 cm. **Capsules** broadly ovoid. $2n = 45$.

Flowering spring. Fields, waste places, abandoned gardens; 0–1500 m; introduced; Ont.; Conn., Del., D.C., Ill., Md., Mich., Mo., N.J., N.Y., N.C., Ohio, Oreg., Pa., Va., W.Va.; Europe; sw Asia; expected elsewhere.

Though widely cultivated and naturalized, *Ornithogalum nutans* is not as common as *O. umbellatum*.

2. Ornithogalum umbellatum Linnaeus, Sp. Pl. 1: 307. 1753 • Nap-at-noon, sleepy dick [F] [I] [W1] [W2]

Plants (10–)20–30(–40) cm; bulbs renewed each year, 1–2 × 1–2.5(–3.5) cm; bulblets numerous. **Leaves** 4–6(–9); blade with white adaxial stripe, 20–30 cm × 3–5(–8) mm. **Scape** 1–3 dm. **Inflorescences** corymbose, (4–)8–20-flowered, flat-topped; bracts to 4 cm. **Flowers** erect; perianth opening flat or bowl-shaped in sun, closing at night, remaining closed on cloudy days; tepals white with wide green abaxial stripe, lanceolate to oblong-lanceolate, 15–22(–30) × 7–8 mm; stamens: outer 5–7 × 2–3 mm, inner 6–8 × 2–3 mm; filaments simple, flattened; anthers 2–4 mm; ovary ovoid to obovoid, strongly 6-angled, 5–6 mm; style 3–4 mm; lower pedicels 2–6 cm. **Capsules** oblong-ovoid. $2n = 18, 20, 27, 28, 35, 36, 42, 44, 45, 54, 72, 90, 108$.

Flowering spring. Roadsides, open forests, waste places, dumps; 0–1500 m; introduced; B.C., N.B., Nfld. and Labr.(Nfld.), N.S., Ont., Que.; Ala., Ark., Conn., Del., D.C., Ga., Idaho, Ill., Ind., Iowa, Kans., Ky., La., Maine, Md., Mass., Mich., Miss., Mo., Nebr., N.H., N.J., N.Y., N.C., Ohio, Okla., Oreg., Pa., R.I., S.C., S.Dak., Tenn., Tex., Utah, Vt., Va., Wash., W.Va., Wis.; Europe; n Africa; Asia; Middle East; expected elsewhere.

Planted as a garden ornamental, *Ornithogalum umbellatum* produces many offsetting bulblets that are transported in soil and can become rampant weeds. Adding to the vegetative vigor of this species may be its aneuploid-polyploid karyology (T. W. J. Gadella and L. van Raamsdonk 1981; L. van Raamsdonk 1984). The flowers are noteworthy for their regularity in opening just before noon and closing again before sunset.

Two digitalis-like glycosides, convallatoxin and convalloside, poisonous to humans and livestock, are found throughout the plant, but are concentrated in the bulbs and the flowers (W. H. Blackwell 1990; K. F. Lampe and M. A. McCann 1985; D. G. Spoerke Jr. and S. C. Smolinske 1990).

62. SCILLA Linnaeus, Sp. Pl. 1: 308. 1753; Gen. Pl. ed. 5, 146. 1754 • Squill [Greek *skilla*, squill, the officinal or its source, *Urginea maritima*, syn. *Scilla maritima*] [I]

J. McNeill

Herbs, perennial, scapose, from bulbs. **Bulbs** perennial, ovoid to globose, composed of free scales, progressively renewed annually. **Leaves** few, basal. **Inflorescences** racemose or cymose, 1–many-flowered, sometimes bracteate; bracts none or 1, subtending each flower. **Flowers:** perianth usually blue or purple, rarely white; tepals distinct to base, each 1-veined; stamens 6; filaments inserted at base of perianth, distinct; anthers dorsifixed, introrse; pistil 1, 3-carpellate; ovary superior, 3-locular, septal nectaries present, ovules 1–10 per locule; style simple. **Fruits** capsular, 3-lobed, subglobose, dehiscence loculicidal. **Seeds** 3–30, not winged, globose to ellipsoid, elaiosomes present. x = 5, 6, 7, 8, 9, 10, 11, 12.

Species ca. 50 (1 in the flora): introduced; Eurasia, especially Mediterranean area and sw Asia; s Africa.

A number of species of *Scilla* are commonly grown for their early, showy spring flowers, and present the possibility of becoming naturalized. In particular, *S. bifolia* Linnaeus, two-leaved squill, has been reported in Michigan (E. G. Voss 1972–1985, vol. 1) and northwestern Indiana (F. Swink and G. S. Wilhelm 1994). The summer-flowering hyacinth squill, *S. hyacinthoides* Linnaeus [*Nectaroscilla hyacinthoides* (Linnaeus) Parlatore], has been collected along roadsides near Mooringsport, Louisiana, and in Navarro County, Texas; it is readily distinguished from the spring-flowering species by its tall scapes (30–80 cm) with more than 40 flowers, and its more numerous leaves (8–10).

F. Speta (1998, 1998b) drastically split *Scilla*, placing the Eurasian members into 10–12 mostly small genera on the basis of molecular (M. Pfosser and F. Speta 1999) and karyological (J. Greilhuber 1982; J. Greilhuber et al. 1981; F. Speta 1979) studies, as well as morphological data. Some of these segregate genera correspond to subgenera and sections recognized in other Eurasian treatments over the past 75 years (e.g., P. Chouard 1930; J. McNeill 1980; E. V. Mordak 1984), but others represent even finer splitting.

SELECTED REFERENCES Chouard, P. 1930. Types de Développement de l'Appareil Végétatif Chez les Scillées. D.Sc. thesis. University of Paris. [Preprinted from Ann. Sci. Nat., Bot., sér. 10, 13: 131–322. 1931.] McNeill, J. 1980. *Scilla*. In: T. G. Tutin et al., eds. 1964–1980. Flora Europaea. 5 vols. Cambridge. Vol. 5, pp. 41–43. Mordak, E. V. 1984. *Scilla*. In: P. H. Davis, ed. 1965–1988. Flora of Turkey and the East Aegean Islands. 10 vols. Edinburgh. Vol. 8, pp. 214–224. Speta, F. 1998b. Systematische Analyse der Gattung *Scilla* L. (Hyacinthaceae). Phyton (Horn) 38: 1–141.

1. Scilla siberica Haworth, Bot. Repos. 6: plate 365. 1804 • Siberian squill [F] [I]

Othocallis siberica (Haworth) Speta

Plants 10–20(–30) cm; bulbs tunicate, ovoid, 1.5–2 cm; tunics dark purplish brown. **Leaves** 2–4; blade broadly linear, 10–15 × 0.5–2 cm. **Scapes** 1–4. **Inflorescences** racemose, 1–2(–5)-flowered, bracteate; bract 1–2 mm. **Flowers:** perianth deep blue, 12–16 × 4–6 mm; pedicel drooping, 8–12 mm, equaling or shorter than perianth. **Capsules** 4–6 mm. **Seeds** pale brown, with long white appendage, globose, 2–3 mm. $2n$ = 12.

Flowering early to mid spring. Escaping from gardens; 0–1500 m; introduced; B.C., N.B., Ont., Que.; Ill., Ind., Ky., Mass., Mich., Mo., N.J., N.Y., Ohio, Pa., Utah, Wis.; Russia, north to 55°, Caucasia, and possibly n Iran; naturalized in c Europe.

Naturalization of *Scilla siberica* is expected elsewhere.

63. BRODIAEA Smith, Trans. Linn. Soc. London 10: 2. 1811, name conserved • Cluster-lily [for James Brodie, 1744–1824, Scottish cryptogamic botanist]

J. Chris Pires

Hookera Salisbury [1 March] 1808, name rejected, not *Hookeria* Smith [1 June] 1808, name conserved

Herbs, perennial, scapose, from fibrous-coated corms. **Leaves** 1–6, basal; blade linear, crescent-shaped in cross section. **Scape** solitary, cylindrical, usually slender, occasionally stout, rigid. **Inflorescences** umbellate, open, bracteate; bracts scarious, not enclosing flower buds. **Flowers:** perianth 6-tepaled, distinctly connate proximally into tube, shiny, abaxial perianth usually bluish purple, tube narrowly campanulate or funnelform, outer 3 lobes narrower than inner 3; stamens 3, epitepalous, opposite inner perianth lobes, alternating with 3 staminodia (staminodia absent in *B. orcuttii*) opposite outer perianth lobes; filaments adnate to perianth tube, linear, base sometimes dilated to form triangular flap, or sometimes with abaxial wings or appendages; anthers basifixed, appressed to style; pistil 3-carpellate; ovary superior, green (purple in *B. jolonensis*), sessile, 3-locular, ovules several; style erect; stigma 3-lobed, lobes distinctly spreading and recurved; pedicel erect, articulate at base. **Fruits** capsular, ovoid, dehiscence loculicidal. **Seeds** black, rounded to flattened, coat with crust with longitudinal surface striations. x = 6, 8, 12, 16, 18, 20, or 24.

Species 14 (14 in the flora): w North America including Mexico (Baja California).

Two schools of thought have existed regarding generic limits within the complex of species recognized under *Brodiaea*, *Triteleia*, and *Dichelostemma*: a single large genus (*Brodiaea* s.l.) with three subgenera (S. Watson 1879; W. L. Jepson 1923–1925; P. A. Munz 1959), or three separate genera (E. L. Greene 1886; R. F. Hoover 1939; G. Keator 1967, 1989, 1993; T. F. Niehaus 1971, 1980). Recent molecular, anatomical, and developmental evidence supports neither of these views. *Bloomeria* is related to *Triteleia*, and *Brodiaea* is closely related to *Dichelostemma*, with the only hexandrous species, *D. capitatum*, being sister to the rest of the three-staminate *Brodiaea/Dichelostemma* clade (R. Y. Berg 1978, 1996; J. C. Pires 2000). This recent evidence also suggests that the sections presently established within *Brodiaea* are in need of revision; thus a sectional classification is not utilized in this treatment (R. F. Hoover 1939b; T. F. Niehaus 1971; J. C. Pires 2000).

Polyploidy and ecological specialization to serpentine and other unique substrates is common in *Brodiaea*, resulting in several rare and endangered species. Eleven of the fourteen species are restricted to California, where the flowering date is highly dependent on the amount of moisture in the early spring. Several species are exceedingly variable. Corms of some species were eaten by native Americans.

Among the most important diagnostic characters within *Brodiaea* are features of the androecium, particularly the size and shape of the staminodia and apical filament appendages. These characters are easily seen with a hand lens in the field. When collecting flowering specimens, one should make a point of mounting a few dissected flowers in a manner that displays these critical characters.

SELECTED REFERENCE Niehaus, T. F. 1971. A biosystematic study of the genus *Brodiaea* (Amaryllidaceae). Univ. Calif. Publ. Bot. 60.

1. Staminode apex subulate or cuspidate, or staminodia absent.
   2. Staminodia absent . . . . 10. *Brodiaea orcuttii*

2. Staminodia present.
   3. Staminode apex subulate; perianth tube splitting in fruit . . . . 5. *Brodiaea filifolia*
   3. Staminode apex cuspidate; perianth tube not splitting in fruit . . . . 8. *Brodiaea kinkiensis*

1. Staminode apex rounded or notched (2-lobed in *B. insignis*).
   4. Flowers 24–38 mm.
      5. Staminodia narrowly linear, margins 1/4–1/2 involute; perianth tube splitting in fruit.
         6. Filaments with 2 threadlike appendages at base; perianth lobes usually less than twice length of perianth tube; anthers obcordate, 3–6 mm . . . . 1. *Brodiaea appendiculata*
         6. Filaments without appendages at base, or appendages rudimentary; perianth lobes usually more than twice length of perianth tube; anthers linear, 9–12 mm . . . . 2. *Brodiaea californica*
      5. Staminodia broad, margins mostly involute or planate; perianth tube not splitting in fruit.
         7. Perianth tube ovoid to narrowly campanulate; staminodia curving inward toward stamens overall but curving outward at apex, margins 3/4 involute . . . 3. *Brodiaea coronaria*
         7. Perianth tube funnelform; staminodia erect or recurved distally, distant from stamens, margins planate or 1/4 involute . . . . 4. *Brodiaea elegans*
   4. Flowers 14–24(–33) mm.
      8. Staminodia leaning inward, margins 1/4 involute, apex broad, hooded; perianth tube cylindrical or campanulate.
         9. Scape 0.5–20 cm; ovary green . . . . 14. *Brodiaea terrestris*
         9. Scape 6–15 cm; ovary light purple . . . . 7. *Brodiaea jolonensis*
      8. Staminodia erect, margins 1/2–3/4 involute (1/4 involute in *B. stellaris*), apex notched (2-lobed in *B. insignis*); perianth tube usually urceolate, sometimes campanulate or narrowly cylindrical.
         10. Filaments with 2 conspicuous, broad, white, abaxial appendages appearing as wings behind anthers . . . . 13. *Brodiaea stellaris*
         10. Filaments lacking broad white appendages, or appendages narrow and inconspicuous.
             11. Perianth tube not constricted above ovary; staminodia margins 3/4 involute . . . . 6. *Brodiaea insignis*
             11. Perianth tube slightly to strongly constricted above ovary, staminodia margins 1/2–3/4 involute.
                 12. Anthers obcordate . . . . 11. *Brodiaea pallida*
                 12. Anthers linear.
                     13. Scape 2–10 cm; inner perianth lobes 5–7 mm wide; style 5–6 mm . . . . 9. *Brodiaea minor*
                     13. Scape 10–25 cm; inner perianth lobes 4–5 mm wide; style 7–9 mm . . . . 12. *Brodiaea purdyi*

1. Brodiaea appendiculata Hoover, Madroño 4: 130, fig. 1. 1937 [E]

Scape 10–45 cm, stout. **Flowers** 24–38 mm; perianth violet purple, tube cylindrical, 8–12 mm, translucent, splitting in fruit, lobes ascending, recurved distally, 15–22 mm, usually less than twice length of tube; filaments 4–7 mm, base not triangular, with 2 threadlike, forked appendages; anthers obcordate, 3–6 mm, apex hooked; staminodia erect, usually white, narrowly linear, 8–15 mm, margins 1/2 involute, wavy, apex rounded; ovary 5–6 mm; style 9–12 mm; pedicel 4–10 cm. $2n = 12$.

Flowering spring (Apr–May). Grasslands, open woodlands, gravelly clay soils; 0–600 m; Calif.

This uncommon species is found at low altitudes in the Sierra foothills of central and northern California, usually in stony, red clay soils that become baked very hard during the flowering season. It is rare or extirpated from coast-range foothills.

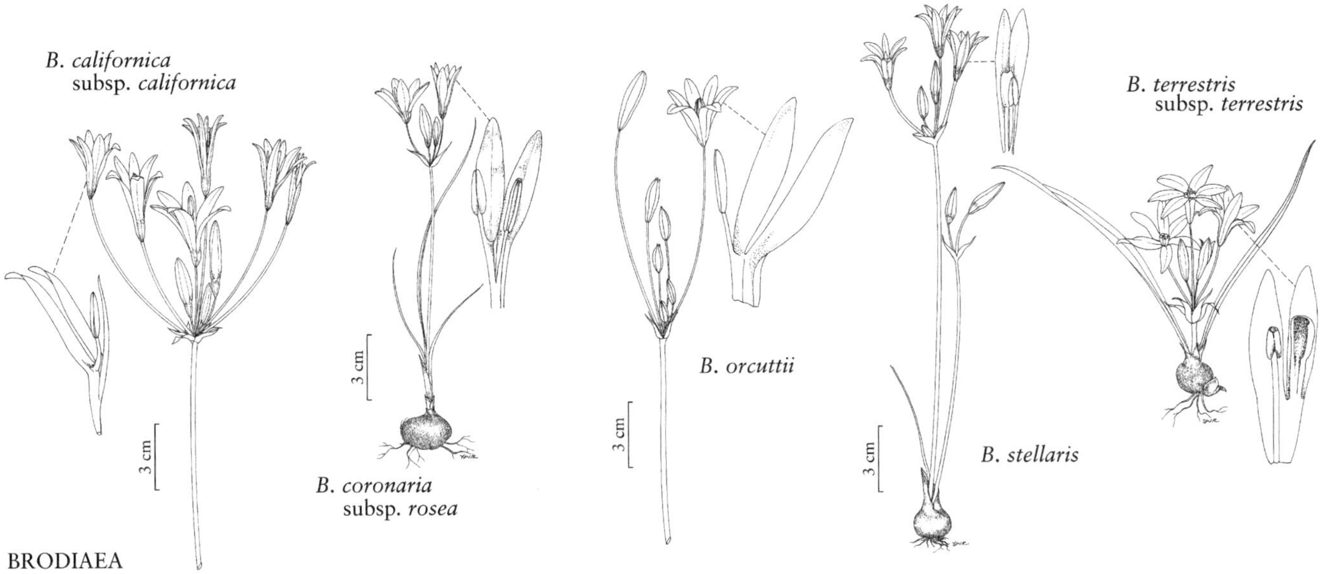

BRODIAEA

2. Brodiaea californica Lindley, J. Hort. Soc. London 4: 84. 1849 [E] [F]

Hookera californica (Lindley) Greene

Scape 20–70 cm, stout. **Flowers** 24–38 mm; perianth pale lilac or sometimes white, tube cylindrical, 9–12 mm, translucent, splitting in fruit, lobes ascending, recurved distally, 20–30 × 4–10 mm, usually more than twice length of tube; filaments 7–10 mm, base not triangular, appendages absent or rudimentary; anthers linear, 9–12 mm, apex rounded; staminodia erect, white to pale lilac, narrowly linear, usually flat, 16–27 mm, margins $^{1}/_{4}$–$^{1}/_{2}$ involute, apex rounded; ovary 5–11 mm; style 15–23 mm; pedicel 2–10 cm.

Subspecies 2 (2 in the flora): California.

1. Perianth lobes 6–10 mm wide; ovary 7–11 mm . . . . 2a. *Brodiaea californica* subsp. *californica*
1. Perianth lobes 4–7 mm wide; ovary 5–7 mm . . . . 2b. *Brodiaea californica* subsp. *leptandra*

2a. Brodiaea californica Lindley subsp. **californica** [E] [F]

Brodiaea grandiflora Smith var. *elatior* Baker

Flowers: perianth lobes 6–10 mm wide; staminodia margins $^{1}/_{4}$ involute; ovary 7–11 mm. $2n = 12$.

Flowering spring (late May–early Jun). Grasslands, open woodlands, gravelly clay soils; 0–800 m; Calif.

Subspecies *californica* is found in the foothills and grasslands of northern California. There are unverified reports of occurrence in southern Oregon.

2b. Brodiaea californica Lindley subsp. **leptandra** (Greene) J. C. Pires, Taxon 50: 201. 2001 [E]

Hookera leptandra Greene, Pittonia 1: 74. 1887; *Brodiaea californica* var. *leptandra* (Greene) Hoover; *B. leptandra* (Greene) Baker; *Hookera synandra* A. Heller

Flowers: perianth lobes 4–7 mm wide; staminodia margins $^{1}/_{2}$ involute; ovary 5–7 mm. $2n = 10, 12, 24$.

Flowering summer (Jun). Open mixed evergreen forests, chaparral, gravelly clay soils, often on serpentines; 0–900 m; Calif.

This subspecies is uncommon, occuring in the Coast Ranges of northern California. T. F. Niehaus (1971, 1980) recognized *Brodiaea leptandra* as a separate species related to *B. californica* and *B. appendiculata*. G. Keator (1993) regarded it as a variety that is geographically disjunct and slightly smaller.

3. Brodiaea coronaria (Salisbury) Engler, Notizbl. Königl. Bot. Gart. Berlin 2: 317. 1899 [E] [F]

Hookera coronaria Salisbury, Parad. Lond. 2: plate 98. 1808

Scape 4–25 cm, slender. **Flowers** 24–38 mm; perianth bluish violet, bluish purple, rosy purple, or rose, tube ovoid to campanulate, 6–13 cm, opaque, not splitting in fruit, lobes ascending, recurved distally, 12–25 mm; filaments 3–4 mm, base dilated to form triangular flap; anthers linear, 5–7 mm, apex hooked or rounded; staminodia curving inward toward stamens overall but curving outward at apex, white or pink, broad, 10–11 mm, margins 3/4 involute, apex rounded; ovary 6–9 mm; style 6–11 mm; pedicel 1–5 cm.

Subspecies 2 (2 in the flora): w North America.

Most of the plants that W. L. Jepson (1923–1925) placed under this name have been transferred to *Brodiaea elegans* (T. F. Niehaus 1971, 1980).

1. Perianth bluish violet or bluish purple; corm coat heavily fibrous; nw United States, British Columbia . . . . 3a. *Brodiaea coronaria* subsp. *coronaria*
1. Perianth rosy purple or rose; corm coat thin; only in California Coast Ranges . . . . 3b. *Brodiaea coronaria* subsp. *rosea*

3a. Brodiaea coronaria (Salisbury) Engler subsp. **coronaria** [E]

Brodiaea grandiflora Smith; *B. howellii* Eastwood 1938, not S. Watson 1879; *B. synandra* (A. Heller) Jepson

Corm coat heavily fibrous. **Scape** 5–25 cm. **Flowers:** perianth bluish violet or bluish purple, tube 6–13 cm, lobes 15–25 mm; anther apex hooked; staminodia white; ovary 8–9 mm; style 9–11 mm. 2*n* = 12, 24, 42.

Flowering spring (Apr–Jun). Grasslands, volcanic mesas; 0–1600 m; B.C.; Calif., Oreg., Wash.

Subspecies *coronaria*, like *Brodiaea elegans* subsp. *elegans*, is very widespread and grows well in a variety of conditions. It is found primarily in the grasslands of California valleys and in the prairies of western Oregon and Washington. The 2*n* = 12 populations are found in the northern (Oregon and Washington) and southern (central California) parts of the range, while the 2*n* = 24 populations are found in the middle part of the range (northern California and southern Oregon).

3b. Brodiaea coronaria (Salisbury) Engler subsp. **rosea** (Greene) T. F. Niehaus, Univ. Calif. Publ. Bot. 60: 41. 1971 • Indian valley brodiaea [E] [F]

Hookera rosea Greene, Bull. Calif. Acad. Sci. 2: 137. 1886; *Brodiaea coronaria* var. *rosea* (Greene) Hoover; *B. rosea* (Greene) Baker

Corm coat thin. **Scape** 4–7 cm. **Flowers:** perianth rosy purple or rose, tube 7–9 cm, lobes 12–18 mm; anther apex rounded; staminodia white to pink; ovary 6–8 mm; style 6–8 mm. 2*n* = 12.

Flowering spring (late May–Jun). Grasslands, often on serpentine; 0–100 m; Calif.

This is a rare taxon known only from three localities in the northern Coast Ranges of California. It is in cultivation.

4. Brodiaea elegans Hoover, Amer. Midl. Naturalist 22: 555. 1939 • Harvest brodiaea, elegant cluster-lily [E]

Scape 10–50 cm, stout. **Flowers** 24–38 mm; perianth bluish purple to violet, tube funnelform, 8–19 mm, opaque, not splitting in fruit, lobes ascending, recurved distally, 15–30 mm; filaments 4–6 mm, base not triangular; anthers linear, 4–10 mm, apex rounded; staminodia erect or slightly recurved distally, distant from stamens, white to pale lilac, broad, equaling stamens, 6–9 mm, margins planate or 1/4 involute, apex rounded; ovary 9–15 mm; style 7–15 mm; pedicel 5–10 cm.

Subspecies 2 (2 in the flora): California, w Oregon.

Documentation is needed to verify that this species ranges as far north as Polk County, Oregon, and as far south as San Diego County, California. Most of the plants included in *Brodiaea coronaria* by W. L. Jepson (1923–1925) have been moved to *B. elegans* (R. F. Hoover 1939b; T. F. Niehaus 1980; P. A. Munz 1959).

1. Staminodia erect distally, margins planate and straight; California and sw Oregon . . . . 4a. *Brodiaea elegans* subsp. *elegans*
1. Staminodia recurved distally, margins 1/4 involute; Oregon . . . . 4b. *Brodiaea elegans* subsp. *hooveri*

4a. Brodiaea elegans Hoover subsp. **elegans** [E]

Brodiaea coronaria (Salisbury) Jepson var. *mundula* Jepson; *B. elegans* var. *mundula* (Jepson) Hoover

Flowers: perianth tube 8–19 mm, lobes 15–30 mm; anthers 4–10 mm; staminodia erect distally, margins planate, straight; ovary 9–15 mm; style 7–15 mm. $2n$ = 12, 16, 24, 32.

Flowering spring–summer (May–Jul). Grasslands, meadows, open woodlands; 0–2200 m; Calif., Oreg.

Subspecies *elegans*, like *Brodiaea coronaria* subsp. *coronaria*, is very widespread and grows well in a variety of conditions. It has $2n$ = 16 in the upper foothills of the Sierra Nevada, but elsewhere $2n$ = 32 or, in a zone of contact with *B. coronaria* near Sonora, California, $2n$ = 12 and 24, apparently resulting from hybridization. This subspecies is in cultivation.

4b. Brodiaea elegans Hoover subsp. **hooveri** T. F. Niehaus, Univ. Calif. Publ. Bot. 60: 48, fig. 12a–d. 1971 [E]

Flowers: perianth tube 13–17 mm, lobes 22–30 mm; anthers 6–8 mm; staminodia slightly recurved distally, margins 1/4 involute; ovary 9–11 mm; style 12–15 mm. $2n$ = 40.

Flowering spring. Grasslands; 0–100 m; Oreg.

Subspecies *hooveri* occurs primarily in the Willamette Valley of Oregon. Documentation of occurrence in northern California is lacking.

5. Brodiaea filifolia S. Watson, Proc. Amer. Acad. Arts 17: 381. 1882 [C] [E]

Hookera filifolia (S. Watson) Greene

Scape 20–30 cm, slender. **Flowers** 14–20 mm; perianth violet-reddish purple, tube narrowly cylindrical, 6–8 mm, transparent, splitting in fruit, lobes widely spreading, 10–14 mm; filaments 0.5–1 mm, base not triangular, with narrow abaxial wings; anthers linear, 3–5 mm, apex widely notched; staminodia inconspicuous, reflexed against perianth, purple, threadlike, 2–4 mm, apex subulate; ovary 4–5 mm; style 6–7 mm; pedicel 1–4 cm. $2n$ = 24.

Flowering spring (Apr–May). Grasslands, vernal pools; of conservation concern; 0–300 m; Calif.

Brodiaea filifolia is endangered. It grows on clay soils at the edges of vernal pools and flood plains in southern California. It has been extirpated from Los Angeles and San Bernadino counties and is seriously threatened by development, vehicles, and agriculture in Riverside and San Diego counties. It is in cultivation.

6. Brodiaea insignis (Jepson) T. F. Niehaus, Univ. Calif. Publ. Bot. 60: 52. 1971 • Kaweah brodiaea [C] [E]

Brodiaea synandra (A. Heller) Jepson var. *insignis* Jepson, Fl. Calif. 1: 288. 1921

Scape 7–24 cm, slender. **Flowers** 14–24 mm; perianth rose to pinkish purple, rotate, tube narrowly cylindrical, not constricted above ovary, 6–9 mm, opaque, not splitting in fruit, lobes widely spreading, 11–15 mm; filaments linear, 1–2 mm, base dilated to form triangular flap, appendages absent or narrow and inconspicuous; anthers linear, 2–4 mm, apex hooked; staminodia erect, held close to stamens, white, broad, 6–8 mm, margins 3/4 involute, apex 2-lobed; ovary 4–5 mm; style 4–5 mm; pedicel 2–9 cm. $2n$ = 32.

Flowering spring (May–Jun). Foothill woodland openings; of conservation concern; 200–500 m; Calif.

Brodiaea insignis is endangered. It is endemic to three localities along the Kaweah and Tule river drainages and is threatened by development, road maintenance, and grazing. It is in cultivation.

7. Brodiaea jolonensis Eastwood, Leafl. W. Bot. 2: 111. 1938 • Mesa brodiaea, chaparral cluster-lily

Scape 5–15 cm, slender. **Flowers** 14–24 mm; perianth bluish violet, tube cylindrical or narrowly campanulate, 7–9 mm, thick, opaque, not splitting in fruit, lobes ascending, recurved distally, 11–18 mm; filaments 1–2 mm, base dilated to form narrow 2-lobed triangular wing; anthers linear, 4–6 mm, apex with V-shaped notch; staminodia leaning inward toward stamens, violet, broad, 5–6 mm, margins 1/4 involute, apex broad, hooded, slightly notched; ovary light purple, 5–6 mm; style 5–7 mm; pedicel 1–4 cm. $2n$ = 12, 36.

Flowering spring (Apr–May). Grasslands, foothill woodlands, coastal prairies, on clay flats; 0–300 m; Calif.; Mexico (Baja California).

Brodiaea jolonensis grows along the southern coast of California as well as on Santa Cruz, Santa Rosa, and Santa Catalina islands. T. F. Niehaus (1971) was impressed by the distinctive flavonoid chemistry of this plant compared to that of other members of the genus.

8. Brodiaea kinkiensis T. F. Niehaus, Madroño 18: 233, fig. 1. 1966 • San Clemente Island brodiaea C E

Scape 2–3 cm, slender. **Flowers** 14–24 mm; perianth purple or violet, tube narrowly campanulate, 9–12 mm, opaque, not splitting in fruit, lobes widely spreading, 12–18 mm, inner round, with widely rounded apex; filaments 1.5–2 mm, base not triangular, with narrow abaxial wings; anthers linear, 4–7 mm, apex widely notched; staminodia erect, held away from stamens, broad, 3–7 mm, margins linear, apex cuspidate; ovary 6–9 mm; style 5–7 mm; pedicel 1–4 cm. $2n = 32$.

Flowering spring (May–Jun). Grasslands on clay flats; of conservation concern; 100–200 m; Calif.

Brodiaea kinkiensis is endemic to the mesa tops of San Clemente Island. It is the only single-leaved *Brodiaea*. It is in cultivation.

9. Brodiaea minor (Bentham) S. Watson, Proc. Amer. Acad. Arts 14: 236. 1879 • Dwarf brodiaea E

Brodiaea grandiflora Smith var. *minor* Bentham, Pl. Hartw., 340. 1857; *B. minor* var. *nana* (Hoover) Hoover; *B. nana* Hoover; *Hookera minor* (Bentham) Britten ex Greene

Scape 2–10 cm, slender. **Flowers** 14–24 mm; perianth pale bluish to lilac, rotate, tube urceolate, strongly constricted above ovary, 6–10 mm, tough, opaque, not splitting in fruit, lobes widely spreading, 11–17 mm, inner 5–7 mm wide; filaments 0.5–4 mm, base not triangular, with narrow abaxial wings, appendages absent; anthers linear, 1–5 mm, apex hooked; staminodia erect, held close to stamens, white, broad, 5–8 mm, margins 1/2 involute at mid-length, apex notched; ovary 3–4 mm; style 5–6 mm; pedicel 1–3 cm. $2n = 12$.

Flowering spring (Apr–early May). Grasslands, gravelly clay soils; 0–500 m; Calif.

Brodiaea minor grows along the low foothills of the Sierra Nevada range. It is very similar to *B. purdyi*, which may be only a subspecies of it.

10. Brodiaea orcuttii (Greene) Baker, Gard. Chron., ser. 3, 20: 214. 1896 • Orcutt's brodiaea E F

Hookera orcuttii Greene, Bull. Calif. Acad. Sci. 2: 138. 1886; *Brodiaea filifolia* S. Watson var. *orcuttii* (Greene) Jepson; *Hookera multipedunculata* Abrams

Scape 8–25 cm, slender. **Flowers** 14–20 mm; perianth violet, tube funnelform, 3–7 mm, transparent, splitting in fruit, lobes widely spreading, 12–19 mm; filaments 4–6 mm; anthers linear, 4–6 mm, apex with V-shaped notch; staminodia absent; ovary 4–6 mm; style 7–11 mm; pedicel 1–5 cm. $2n = 24$.

Flowering spring (May). Grasslands near streams, vernal pools; 0–1600 m; Calif.

Brodiaea orcutti is rare and endemic to southern California. It was once abundant on Kearney Mesa north of San Diego, but most of that area has been developed for housing, and only those portions on military reservations remain undisturbed. This is the only species in the genus with no staminodia.

11. Brodiaea pallida Hoover, Leafl. W. Bot. 2: 129. 1938 • Chinese Camp brodiaea C E

Scape 10–20 cm, slender. **Flowers** 14–24 mm; perianth pale purple or lilac, rotate, tube urceolate, slightly contricted above ovary, 9–11 mm, thin, opaque, not splitting in fruit, lobes ascending to strongly recurved, 9–11 mm; filaments 4–5 mm, base not triangular, with narrow abaxial wings, appendages absent; anthers ± obcordate, 2–3 mm, apex notched into wide V; staminodia erect, held close to stamens, white, broad, 8–11 mm, margins 1/2 involute at mid length, apex deeply notched; ovary 4–5 mm; style 8–11 mm; pedicel 5–30 cm. $2n = 12$.

Flowering spring (late May–early Jun). Foothill woodlands in open areas along intermittent streambeds, serpentine soils; of conservation concern; 300–400 m; Calif.

Brodiaea pallida is endangered. It forms a single population 10–20 feet wide for approximately one-fourth of a mile along a stream near Chinese Camp, Tuolumne County. It putatively hybridizes with *B. elegans* and is threatened by cattle-grazing and development. It is in cultivation.

12. Brodiaea purdyi Eastwood, Proc. Calif. Acad. Sci., ser. 2, 6: 427, plate 58. 1896 • Purdy's brodiaea, Sierran cluster-lily [E]

Hookera purdyi (Eastwood) A. Heller

Scape 10–25 cm, slender. **Flowers** 14–24 mm; perianth bluish violet, rotate, tube urceolate, strongly constricted above ovary, 6–8 mm, opaque, not splitting in fruit, lobes spreading, 9–19 mm, inner 4–5 mm wide; filaments 1–2 mm, base not triangular, with narrow abaxial wings, appendages absent; anthers linear, 3–4 mm, apex notched into V; staminodia erect, held close to stamens, white, broad, 6–8 mm, margins 3/4 involute, apex notched; ovary 4–5 mm; style 7–9 mm; pedicel 1–3 cm. 2*n* = 24, 32, 48.

Flowering spring (Jun). Open foothill woodlands, often on serpentine; 100–600 m; Calif.

Brodiaea purdyi is very similar to *B. minor* and might be just a subspecies of it.

13. Brodiaea stellaris S. Watson, Proc. Amer. Acad. Arts 17: 381. 1882 • Star brodiaea, star-flower cluster-lily [E] [F]

Hookera stellaris (S. Watson) Greene

Scape 2–6 cm, slender. **Flowers** 14–24 mm; perianth bluish purple, tube campanulate, 7–10 mm, transparent, not splitting in fruit, lobes ascending, recurved distally, 7–15 mm; filaments 1–3 mm, base not triangular, apex forked with 2 conspicuous, broad, white, abaxial appendages appearing as wings behind anthers; anthers linear, 4–6 mm, apex notched; staminodia erect, held close to stamens, white, broad, 4–8 mm, wide, margins 1/4 involute, apex widely notched; ovary 6–9 mm; style 4–5 mm; pedicel 1–5 cm. 2*n* = 12.

Flowering spring (May–Jun). Openings in coastal forests, on serpentine; 0–900 m; Calif.

Brodiaea stellaris is a serpentine endemic found in mixed evergreen and redwood forests in the North Coast Ranges of California. The white, glossy, forked appendages on the filaments of this species are very different from those found in *B. appendiculata* and *B. californica* and perhaps are not homologous. Also, the capsule is unique in the genus in that it does not split at maturity and is transparent. All other members of the genus have mature capsules that either split and are transparent, or do not split and are opaque.

14. Brodiaea terrestris Kellogg, Proc. Calif. Acad. Sci. 2: 6. 1859 (as terrestria) • Dwarf brodiaea [E] [F]

Hookera terrestris (Kellogg) Britten ex Greene

Scape 0.5–20 cm, slender. **Flowers** 16–33 mm; perianth pinkish purple or rose, tube narrowly campanulate, 6–13 mm, opaque, not splitting in fruit, lobes ascending, recurved distally, 10–20 mm; filaments 2–3 mm, base dilated to form triangular flap; anthers linear, 3–6 mm, apex separated; staminodia leaning inward slightly but held away from stamens, violet to whitish, broad, 4–5 mm, margins 1/4 involute, apex broad, slightly hooded, 2-notched; ovary 7–9 mm; style 4–9 mm; pedicel 1–15 cm.

Subspecies 2 (2 in the flora): California, s Oregon.

1. Scape 0.5–7 cm; flowers 16–25 mm; pedicel 3–15 cm; leaves lunate in cross section . . . . 14a. *Brodiaea terrestris* subsp. *terrestris*
1. Scape 2–20 cm; flowers 26–33 mm; pedicel 1–7 cm; leaves not very lunate in cross section . . . . 14b. *Brodiaea terrestris* subsp. *kernensis*

14a. Brodiaea terrestris Kellogg subsp. **terrestris** [E] [F]

Brodiaea coronaria var. *macropoda* (Torrey) Hoover; *B. grandiflora* Smith var. *macropoda* Torrey; *B. torreyi* Alph. Wood; *Hookera macropoda* (Torrey) Kuntze

Leaves lunate in cross section. **Scape** 0.5–7 cm. **Flowers** 16–25 mm; perianth tube 6–9 mm, lobes 10–16 mm; anthers 3–4 mm; ovary 7–9 mm; style 4–5 mm; pedicel 3–15 cm. 2*n* = 12, 36.

Flowering spring (Apr–Jun). Coastal prairies, foothill woodlands; 0–400 m; Calif., Oreg.

14b. Brodiaea terrestris Kellogg subsp. **kernensis** (Hoover) T. F. Niehaus, Univ. Calif. Publ. Bot 60: 50. 1971 [E]

Brodiaea coronaria (Salisbury) Engler var. *kernensis* Hoover, Amer. Midl. Naturalist 22: 561. 1939; *B. elegans* Hoover var. *australis* Hoover

Leaves not very lunate in cross section. **Scape** 2–20 cm. **Flowers** 26–33 mm; perianth tube 11–13 mm, lobes 15–20 mm; anthers 5–6 mm; ovary 8–9 mm; style 7–9 mm;

pedicel 1–7 cm. $2n = 48$.

Flowering spring (Apr–Jun). Grasslands, open foothill woodlands; 0–1500 m; Calif.

Subspecies *kernensis* is common and widespread in central and southern California. It appears to be an octoploid version of subsp. *terrestris*.

64. DICHELOSTEMMA Kunth, Enum. Pl. 4: 469. 1843 • [Greek *dichelos*, split hoof, and *stemma*, crown or garland, alluding to the bifid perianth appendages that form a corona]

J. Chris Pires

Brevoortia Alph. Wood; *Brodiaea* Smith subg. *Dichelostemma* (Kunth) Eastwood; *Dipterostemon* Rydberg; *Macroscapa* Kellogg; *Rupalleya* Morière; *Stropholirion* Torrey

Herbs, perennial, scapose, from fibrous-coated corms. **Leaves** 1–5, basal; blade narrowly lanceolate, usually keeled and channeled, margins entire. **Scape** solitary, usually weak, curved to twining, cylindrical, smooth to scabrous. **Inflorescences** umbellate or racemose, usually dense, 2–20-flowered, bracteate; bracts 2–4, ± papery, not enclosing flower buds. **Flowers:** perianth 6-tepaled, distinctly connate proximally into tube, tube cylindrical, ovoid, or campanulate, occasionally globose or urceolate, soft, limb lobes similar; perianth appendages arising from intersection of perianth tube and limb lobes, leaning toward or away from anthers, forming corona; anthers basifixed, held close to style; stamens 3 (6 in *Dichelostemma capitatum*), epitepalous; filaments entirely adnate to perianth tube; staminodia absent (except in *D. volubile*); pistil 3-carpellate; ovary superior, sessile or stipitate, 3-locular, ovules several; style 1; stigma weakly 3-lobed; pedicel erect or flexuous, articulate beneath perianth, usually shorter than flowers. **Fruits** capsular, 3-angled, usually ovoid, firm, dehiscence loculicidal. **Seeds** black, sharply angled, coat with crust. $x = 9$ (8 in *D. ida-maia*).

Species 5 (5 in the flora): w North America including n Mexico (Baja California, Sonora).

For discussion of relationships, see under *Brodiaea*.

Although *Hookera pulchella* Salisbury was the first name published for an entity in the group of species here treated under *Dichelostemma*, its application, and therefore that of *Brodiaea pulchella* (Salisbury) Greene, *D. pulchellum* (Salisbury) A. Heller, and *Dipterostemon pulchella* (Salisbury) Rydberg, is uncertain (G. Keator 1967). Some authors have associated it with *Dichelostemma congesta*, others with *D. capitatum* or with a yet different species.

The corms of some *Dichelostemma* species were eaten by native Americans.

SELECTED REFERENCES Hoover, R. F. 1940b. The genus *Dichelostemma*. Amer. Midl. Naturalist 24: 463–476. Keator, G. 1967. Ecological and Taxonomic Studies on the Genus *Dichelostemma*. Ph.D. thesis. University of California, Berkeley. Keator, G. 1991. Studies in the genus *Dichelostemma*. Four Seasons 9: 24–39. Lenz, L. W. 1976. The nature of the floral appendages in four species of *Dichelostemma*. Aliso 8: 379–381.

1. Stamens 6, smaller 3 on outer tepals alternating with larger 3 on inner tepals . . . . . 1. *Dichelostemma capitatum*
1. Stamens 3, all equal.
    2. Perianth tube 20 mm or longer; flowers nodding at anthesis, perianth tube red, limb lobes yellow-green . . . . 3. *Dichelostemma ida-maia*
    2. Perianth tube 10 mm or shorter; flowers horizontal or erect, entire perianth pink to bluish purple, rarely white.
        3. Scape not self-supporting, twining; perianth pink, rarely white, tube 6-angled . . . . 5. *Dichelostemma volubile*
        3. Scape self-supporting, with occasional bends; perianth pink to bluish purple, tube unangled.
            4. Apex of perianth appendages deeply 2-fid; inflorescences racemose; perianth tube slightly constricted above ovary . . . . 2. *Dichelostemma congestum*
            4. Apex of perianth appendages truncate or rounded; inflorescences umbellate; perianth tube strongly constricted above ovary . . . . 4. *Dichelostemma multiflorum*

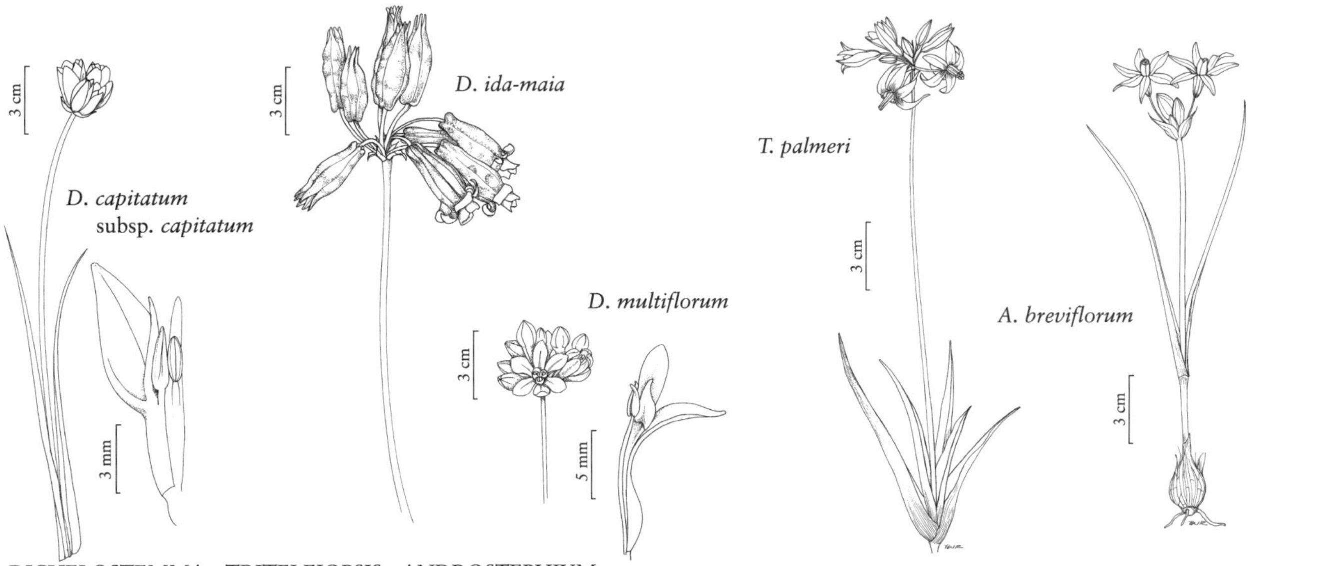

DICHELOSTEMMA ◦ TRITELEIOPSIS ◦ ANDROSTEPHIUM

1. **Dichelostemma capitatum** (Bentham) Alph. Wood, Proc. Acad. Nat. Sci. Philadelphia 20: 173. 1868 • Blue dicks, covenna, grass nuts, Indian hyacinth, fool's onion, crow poison [F]

Brodiaea capitata Bentham, Pl. Hartw., 339. 1857; *Dichelostemma pulchellum* (Salisbury) A. Heller var. *capitatum* (Bentham) Reveal; *Dipterostemon capitatus* (Bentham) Rydberg; *Hookera capitata* (Bentham) Kuntze; *Milla capitata* (Bentham) Baker

Leaves 2–3, 10–70 cm; blade barely keeled. **Scape** shorter than 65 cm, smooth. **Inflorescences** umbellate, dense, 2–15-flowered; bracts whitish to dark purple, ovate to lanceolate, 7–20 mm. **Flowers** horizontal or erect; perianth blue, bluish purple, pinkish purple, or white, tube narrowly cylindrical to short-campanulate, not constricted above ovary, 3–12 mm, lobes usually ascending or widely spreading, 7–12 mm; perianth appendages slightly reflexed distally, leaning toward anthers to form corona, white, lanceolate, 4–6 mm, apex deeply notched; stamens 6, smaller 3 on outer tepals alternating with larger 3 on inner tepals; outer filaments wider than inner at base, 2 mm; outer anthers 2–3 mm, inner 3–4 mm; ovary sessile, ovoid, 4–8 mm; style 4–8 mm; pedicel 1–35 mm.

Subspecies 2 (2 in the flora): w United States; n Mexico.

Dichelostemma capitatum is the most wide-ranging species in the genus, with great variation in habitat, morphology, and chromosome number (G. Keator 1992). Two subspecies are recognized here, after G. Keator (1991, 1993b), but many other subspecies and even separate species have been recognized. For example, *D. insulare* has been recognized because it is endemic to the California Channel Islands and Guadalupe Island and is larger than mainland plants. *Dichelostemma capitatum* ranges from northern Mexico through California, Oregon, Nevada, New Mexico, and possibly north to Idaho and Washington. Despite its great intraspecific variability, it is quite distinct from other species in the genus. It is the only species with six fertile stamens and epigeous germination, and it has unique corm, seed, and ovule characteristics. It also does not form hybrids with any other species, and it has even been placed in its own genus, *Dipterostemon* (P. A. Rydberg 1912). Recent anatomical and molecular data support the idea that this species is not directly related to the other members of *Dichelostemma* and perhaps is best treated as its own genus (R. Y. Berg 1996; J. C. Pires 2000).

1. Bracts usually dark purple; inflorescences 6–15-flowered; pedicel 1–15 mm, usually shorter than bracts; habitat variable . . . . 1a. *Dichelostemma capitatum* subsp. *capitatum*
1. Bracts whitish or streaked purple; inflorescences 2–5-flowered; pedicel 6–35 mm, longer than bracts; desert habitat . . . . 1b. *Dichelostemma capitatum* subsp. *pauciflorum*

1a. Dichelostemma capitatum (Bentham) Alph. Wood subsp. **capitatum** [F]

Brodiaea capitata var. *insularis* (Greene) J. F. Macbride; *B. insularis* Greene; *Dichelostemma insulare* (Greene) Burnham; *D. lacuna-vernalis* L. W. Lenz; *Dipterostemon insularis* (Greene) Rydberg

Inflorescences 6–15-flowered; bracts usually dark purple, sometimes paler and striped dark purple. **Flowers:** perianth lobes usually ascending; pedicel 1–15 mm, usually shorter than bracts. $2n$ = 18, 36, 54, 72, 90.

Flowering mostly spring (Feb–Apr; as early as Dec in the south, as late as Jun–Jul for high-elevation populations). Most abundant in open woodlands and grasslands, but found in many habitat types; 0–2300 m; Calif., Oreg.; n Mexico (Baja California, including islands).

Many populations of subsp. *capitatum* may merit separate subspecific if not specific status. For example, the plants sometimes recognized separately as *Dichelostemma lacuna-vernalis* have a distinct morphology and favor vernal pools (L. W. Lenz 1974). However, they are no more distinct in morphology or habitat than are several other populations, and further study is needed (G. Keator 1992).

1b. Dichelostemma capitatum (Bentham) Alph. Wood subsp. **pauciflorum** (Torrey) Keator, Four Seasons 9(2): 30. 1992

Brodiaea capitata Bentham var. *pauciflora* Torrey in W. H. Emory, Rep. U. S. Mex. Bound. 2(1): 218. 1859; *B. pulchella* (Salisbury) Greene var. *pauciflora* (Torrey) C. V. Morton; *Dichelostemma pauciflorum* (Torrey) Standley; *D. pulchellum* (Salisbury) A. Heller var. *pauciflorum* (Torrey) Hoover; *Dipterostemon pauciflorus* (Torrey) Rydberg; *Hookera pauciflora* (Torrey) Tidestrom; *Milla capitata* (Bentham) Baker var. *pauciflora* (Torrey) Baker

Inflorescences 2–5-flowered; bracts whitish or streaked purple. **Flowers:** perianth lobes widely spreading; pedicel 6–35 mm, longer than bracts. $2n$ = 18, 36, 54.

Flowering time highly variable with elevation. Deserts, xeric grasslands, open scrub; 300–2100 m; Ariz., Calif., Nev., N.Mex., Utah; n Mexico (Sonora).

The characterisics listed for subsp. *pauciflorum* may be found to some degree also in subsp. *capitatum*, intergrading especially along distributional boundaries. However, the adaptation of subsp. *pauciflorum* to a desert-type environment usually distinguishes it from most populations of subsp. *capitatum*, which typically is found in more mesic habitats. Since the two taxa are not clearly distinct, G. Keator (1991) concluded that subspecific status is appropriate for them.

2. Dichelostemma congestum (Smith) Kunth, Enum. Pl. 4: 470. 1843 • Fork-toothed ookow [E]

Brodiaea congesta Smith, Trans. Linn. Soc. London 10: 3, plate 1. 1811; *Hookera congesta* (Smith) Jepson

Leaves 3–4, 4–35 cm; blade strongly keeled, ± glaucous. **Scape** self-supporting, with occasional bends, 30–80(–90) cm, ± scabrous. **Inflorescences** racemose, very dense, 6–15-flowered; bracts pale purple to green, widely lanceolate, 6–12 mm. **Flowers** horizontal or erect; perianth bluish purple, tube unangled, narrowly ovoid, slightly constricted above ovary, 8–10 mm, lobes ascending, 8–10 mm; perianth appendages 1 per stamen, each coalescent to an inner and outer tepal, leaning away from inner anthers to form corona, erect, purplish, narrowly lanceolate, 5–6 mm, apex deeply 2-fid into 2 wings; stamens 3, equal; anthers 4–5 mm; ovary sessile, 4–6 mm; style 5–6 mm; pedicel 1–6 mm. $2n$ = 18, 36.

Flowering spring (late Mar–May). Open woodlands, grasslands near coast; 0–2000 m; Calif., Oreg., Wash.

Dichelostemma congestum can be recognized by its congested racemose inflorescence and deeply bifid perianth appendages that stand away from the anthers to form a corona.

3. Dichelostemma ida-maia (Alph. Wood) Greene, Man. Bot. San Francisco, 318. 1894 • Firecracker flower [E] [F]

Brevoortia ida-maia Alph. Wood, Proc. Acad. Nat. Sci. Philadelphia 19: 82. 1867; *B. coccinea* (A. Gray) S. Watson; *B. venusta* Greene; *Brodiaea coccinea* A. Gray; *B. ida-maia* (Alph. Wood) Greene; *B. venusta* (Greene) Greene; *Dichelostemma venustum* (Greene) Hoover

Leaves 3–5, 30–50 cm; blade keeled, glaucous. **Scape** 30–90 cm, smooth. **Inflorescences** umbellate, open, 6–20-flowered; bracts reddish, lanceolate, 15–25 mm, apex obtuse. **Flowers** nodding at anthesis, erect in fruit; perianth tube red, cylindrical, not constricted above ovary, 20–27 mm, with 6 sac-like angles on proximal 1/3, lobes recurved, erect in fruit, yellow-green, 7–9 mm; perianth appendages each coalescent with an inner and outer tepal, forming conspicuous corona, erect, whitish, 3–4 mm, margins entire, involute, apex rounded; stamens 3, equal; anthers 6–8 mm; ovary stipitate, narrowly ovoid, 7–12 mm, stipe 3–6 mm; style 7–20 mm; pedicel 10–50 mm. $2n$ = 28, 48.

Flowering spring (May–early Jul). Forest edges, grasslands near coast; 0–2000 m; Calif., Oreg.

Dichelostemma ida-maia is so distinctive that it has typified a segregate genus, *Brevoortia*. This species stands apart from all other members of *Dichelostemma* by its long, bright red perianth tube and green perianth lobes, nodding flowers, long anthers, stipitate ovary, and large seeds.

Dichelostemma venustum, the rose firecracker flower, may be a hybrid between *D. ida-maia* and *D. congestum* or *D. multiflorum*. It is intermediate in perianth tube length and color (pink), and gynophore length (L. W. Lenz 1971; G. Keator 1991, 1993b). It is restricted to relatively mesic habitats of northwestern California and southwestern Oregon.

4. **Dichelostemma multiflorum** (Bentham) A. Heller, Muhlenbergia 2: 15. 1905 • Wild hyacinth, round-toothed ookow [E] [F]

Brodiaea multiflora Bentham, Pl. Hartw., 339. 1857; *B. grandiflora* Smith var. *brachypoda* Torrey; *B. parviflora* Torrey & A. Gray; *Dichelostemma parviflorum* (Torrey & A. Gray) Keator; *Hookera multiflora* (Bentham) Britten; *H. parviflora* (Torrey & A. Gray) Kuntze

Leaves 3–4, 30–85 cm; blade unkeeled, glaucous. **Scape** self-supporting, with occasional bends, 20 (rarely –90) cm, scabrous. **Inflorescences** umbellate, dense, 6–15-flowered; bracts streaked purple, ovate, 7–12 mm, apex acuminate. **Flowers** horizontal or erect; perianth pink to bluish purple, tube unangled, cylindrical, strongly constricted above ovary, 8–10 mm, lobes widely spreading, 8–10 mm; perianth appendages each coalescent to inner and outer tepal, arching slightly inward (but outward distally) toward anthers, forming corona, erect, white or pale purple, 5–6 mm, margins entire or denticulate, strongly involute, apex truncate or rounded; stamens 3, equal; anthers 4–5 mm; ovary sessile, globose, 4–6 mm; style 5–8 mm; pedicel 3–15 mm, not hidden by bracts. $2n$ = 18, 36, 45, 54.

Flowering spring (Apr–early Jun). Open woodlands, foothill grasslands, scrubs; 0–2000 m; Calif., Oreg.

Dichelostemma multiflorum is similar to *D. capitatum* and *D. congestum*. *Dichelostemma capitatum* is distinguished by its six fertile stamens, and *D. congestum* by perianth appendages that are deeply bifid and lean away from the inner anthers; racemose inflorescences; and a weakly constricted perianth tube.

5. **Dichelostemma volubile** (Kellogg) A. Heller, Bull. S. Calif. Acad. Sci. 2: 65. 1903 • Twining brodiaea, snakelily [E]

Macroscapa volubilis Kellogg, Pacific (San Francisco) 3(34): 132. 1854; *Brodiaea californica* (Torrey) Jepson 1911, not Lindley 1849; *B. volubilis* (Kellogg) Baker; *Dichelostemma californicum* (Torrey) Alph. Wood; *Hookera volubilis* (Kellogg) Jepson; *Rupalleya volubilis* (Kellogg) Morière; *Stropholirion californicum* Torrey

Leaves 3–4, 30–70 cm; blade strongly keeled. **Scape** weak, not self-supporting except distal 10 cm erect, otherwise twining, 40–150 cm, scabrous. **Inflorescences** umbellate, dense, 6–20-flowered; bracts pinkish, widely ovate, 12–15 mm, apex acute. **Flowers** horizontal or erect; perianth pink or rarely white, tube globose or urceolate, constricted above ovary, 5–7 mm, with 6 saclike angles, lobes widely spreading, ascending in fruit, 5–7 mm; perianth appendages 2 per stamen, folded inward toward anthers, hiding them and forming corona, white, narrowly lanceolate, 3–4 mm, apex 2-fid into 2 wings; stamens 3, equal; anthers 3–4 mm; staminodia 3, opposite outer tepals, creamy whitish, linear-oblong, 2.5–3 mm, margins ciliate-dentate, involute, apex usually shallowly notched; ovary sessile or short-stipitate, subglobose, 4–5 mm; style 3–4 mm; pedicel 10–40 mm. $2n$ = 18, 36.

Flowering spring (Apr–early Jun). Foothill woodlands, chaparral, scrub; 100–1600 m; Calif.

Dichelostemma volubile is distinguishable by its very long, weak, twining scape; small pink flowers; strongly urceolate perianth tube with well-developed saccate angles; and possession of both perianth appendages and true staminodia (sterile stamens) opposite the outer limb segments, as found in most species of *Brodiaea*. It is confined to a narrow belt in the foothill areas adjacent to the Central Valley, and is known to hybridize with *D. multiflorum* in Tuolumne County.

65. TRITELEIOPSIS Hoover, Amer. Midl. Naturalist 25: 98. 1941 • Baja-lily [genus *Triteleia* and Greek *opsis*, like]

J. Chris Pires

Herbs, perennial, scapose, from fibrous-coated corms. **Leaves** alternate, crowded at base; blade planate, keeled, linear, base expanded, margins entire. **Scape** solitary (rarely 2), cylindrical, stout, 4.5–6 dm × 7–15 mm, pithy, glaucous distally. **Inflorescences** umbellate, open, many-flowered, bracteate; involucral bracts 7–10, scarious, apex acuminate. **Flowers:** perianth 6-tepaled, funnelform, distinctly connate proximally into tube; perianth appendages arising at intersection of perianth tube and lobes; stamens 6, epitepalous; filaments inserted in distal part of perianth tube; anthers basifixed, distant from style; pistil 3-carpellate; ovary superior, stipitate, 3-locular, ovules several; style erect, slender; stigma not evidently lobed; pedicel articulate, 1.5–5 cm. **Fruits** capsular, rounded at apex, equaling withered perianth, dehiscence loculicidal. **Seeds** black, elongate, thin, flat, surface minutely and irregularly roughened, coat with crust. $2n = 33$.

Species 1: sw Arizona, n Mexico (Baja California, Sonora).

G. Keator (1989) considered *Triteleiopsis* to be part of a group of Mexican genera (*Milla*, *Bessera*, *Behria* Greene), while H. E. Moore Jr. (1953) and R. F. Hoover (1941) considered it to be allied with *Androstephium*, *Bloomeria*, *Brodiaea*, *Dichelostemma*, *Muilla*, and *Triteleia*. Molecular and morphological evidence indicates that it is related to *Dichelostemma* and *Brodiaea*, but not to *Triteleia* (J. C. Pires 2000).

SELECTED REFERENCES Hoover, R. F. 1941. A systematic study of *Triteleia*. Amer. Midl. Naturalist 25: 73–100. Lenz, L. W. 1975. A biosystematic study of *Triteleia* (Liliaceae): 1. Revision of the species of section *Calliprora*. Aliso 8: 221–258.

1. Triteleiopsis palmeri (S. Watson) Hoover, Amer. Midl. Naturalist 25: 99. 1941 [F]

Brodiaea palmeri S. Watson, Proc. Amer. Acad. Arts 24: 78. 1889; *Triteleia palmeri* (S. Watson) Greene

Corms straw-colored, 1–2 cm diam., young cormlets 3–6 mm, freely dividing within old corm coat and in leaf axils. **Leaves** 3–10, 15–30 cm × 3–20 mm wide at base. **Flowers:** perianth tube 9–10 mm, lobes 7–8 mm, alternating with transverse perianth appendages, having no apparent connection with stamens; filaments erect, adnate to perianth tube and forming internal ridges at tube base, terete, 5–7 mm, papillose, apex slender.

Flowering spring. Deserts, sand dunes; 100–1500 m; Ariz.; Mexico (Baja California, Sonora).

Triteleiopsis palmeri flowers are similar to those of *Triteleia*, but differ in certain details of structure (anthers basifixed, papillate; presence of internal transverse perianth appendages; stigma not lobed) and especially in vegetative characters. The leaves, although keeled, are not channeled, and they are not basal but rather alternate and crowded at the base of the stem. The scape is stout (7–10 mm in diameter) compared to that of *Triteleia* (1–5 mm). The plants grow in loose, sandy soil. They are now in danger of extirpation in Arizona and thus the United States.

66. ANDROSTEPHIUM Torrey in W. H. Emory, Rep. U.S. Mex. Bound. 2(1): 218. 1859 • Funnel-lily [Greek *andros*, stamen, and *stephanos*, crown, alluding to the apical appendages of the united filaments]

J. Chris Pires

James. L. Reveal

Herbs, perennial, scapose, from fibrous-coated corms. **Leaves** several, basal; blade linear, channeled. **Scape** solitary, cylindrical. **Inflorescences** umbellate, terminal, bracteate; bracts 3, lanceolate. **Flowers:** perianth 6-tepaled, distinctly connate proximally into tube, tube funnelform, slightly less to ca. 1/2 overall length of tepals; tepals similar; stamens 6, epitepalous; filaments erect, dilated their entire length, conate into a nectariferous tube with erect, 2-fid apical filament appendages forming a crown between anthers; anthers basifixed, introrse; pistil 3-carpellate; ovary superior, sessile, 3-locular, ovules several; style persistent, long, slender; stigma small, 3-lobed; pedicel not articulate, ± stout. **Fruits** capsular, 3-angled, subglobose, dehiscence loculicidal. **Seeds** black, flat, coat with crust.

Species 3 (2 in the flora): w United States, n Mexico.

The distribution of the species in this genus is poorly documented (S. D. White et al. 1996). Recent molecular and morphological evidence suggests that *Androstephium* is closely related to *Muilla* (J. C. Pires 2000).

1. Perianth white to light violet-purple, lobes longer than tube; scape scabrous basally; w United States . . . . 1. *Androstephium breviflorum*
1. Perianth light blue to violet-purple, lobes shorter than to ca. as long as tube; scape entirely glabrous; c United States . . . . 2. *Androstephium coeruleum*

1. Androstephium breviflorum S. Watson, Amer. Naturalist 7: 303. 1873 • Pink funnel-lily, small-flowered androstephium [E] [F]

Bessera breviflora (S. Watson) Jepson; *Brodiaea paysonii* A. Nelson; *B. breviflora* (S. Watson) J. F. Macbride

Leaves 10–30 × 0.15–0.2 cm. **Scape** 1–3.5 dm, scabrous basally. **Inflorescences** 3–12-flowered. **Flowers:** perianth white to light violet-purple, 1.5–2 cm, lobes longer than tube; anthers 2.5–3 mm; pedicel 1.5–3 cm. **Capsules** 1–1.5 cm. **Seeds** 6–9 mm.

Flowering spring (Mar–Jun). Desert, arid scrub, dry sandy to rocky soil; 100–1600 m; Ariz., Calif., Colo., Nev., N.Mex., Utah, Wyo.

Some populations of this species are threatened by off-road vehicle recreation.

2. Androstephium coeruleum (Scheele) Greene, Pittonia 2: 57. 1890 • Blue funnel-lily [E]

Milla coerulea Scheele, Linnaea 25: 260. 1852

Leaves 10–30(–35) × 0.15–0.3 cm. **Scape** 1–2(–3.5) dm, entirely glabrous. **Inflorescences** 2–6(–9)-flowered. **Flowers:** perianth light bluc to violet-purple, 1.5–2.5(–3) cm, lobes shorter than to ca. as long as tube; anthers 2–4 mm; pedicel 1–2(–2.5) cm. **Capsules** 1–1.5 cm. **Seeds** 7–9 mm.

Flowering spring (Feb–May). Prairies, grassy slopes, frequently in rocky areas; 100–800 m; Kans., Okla., Tex.

67. MUILLA S. Watson ex Bentham in G. Bentham and J. D. Hooker, Gen. Pl. 3: 801. 1883 • [Anagram of *Allium*] [E]

J. Chris Pires

James L. Reveal

Herbs, perennial, scapose, from fibrous-coated corms. **Leaves** 1–10, strictly basal; blade linear, elongate, flat, not keeled or channeled, surrounded by fibrous sheath. **Scapes** 1–2, cylindrical. **Inflorescences** umbellate, terminal, bracteate; bracts several, white or greenish white, scarious. **Flowers:** perianth subrotate, narrowly oblong; tepals 6, spreading, mostly distinct, only shortly connate at base, white to greenish white or bluish, sometimes tinged with lilac, subequal; stamens 6, epitepalous, inserted near base of tepals; filaments filiform, subulate, or conspicuously dilated, sometimes overlapping or connate; anthers versatile, introrse; pistil 3-carpellate; ovary superior, sessile, 3-locular, ovules several; style persistent, clavate; stigma 3-lobed, minute; pedicel slender, not articulate. **Fruits** capsular, slightly lobed, globose, dehiscence loculicidal. **Seeds** black, angled, coat with crust. x = 7, 8, 10.

Species 3 (3 in the flora): w North America.

The distribution of the species in this genus is poorly documented (L. W. Lenz 1975b; J. R. Shevock 1984). R. F. Hoover (1955) transferred *Bloomeria clevelandii* to *Muilla*. However, molecular and morphological evidence indicates that *B. clevelandii*, *B. crocea*, and *B. humilis*, all treated here under *Bloomeria*, are more closely related to *Triteleia*, and that the three species treated here under *Muilla* are more closely related to *Androstephium* (R. Y. Berg and J. R. Maze 1966; J. C. Pires 2000).

SELECTED REFERENCES Ingram, J. W. 1953. A monograph of the genera *Bloomeria* and *Muilla* (Liliaceae). Madroño 12: 19–27. Lenz, L. W. 1975b. The chromosomes of *Bloomeria* and *Muilla* (Liliaceae) and range extensions for *Muilla coronata* and *Muilla transmontana*. Aliso 8: 259–262.

1. Filaments filiform or subulate; anthers green, blue, or purple; tepals greenish white with brownish midveins; coast and valleys of California . . . . 1. *Muilla maritima*
1. Filaments conspicuously dilated; anthers yellow; tepals white often tinged with lilac, or whitish or bluish with green abaxial midvein; mountains and deserts of California and Nevada.
    2. Filaments dilated entire length; scape 3–5 cm; leaves 1(rarely 2); tepals bluish or whitish, with green abaxial midvein; perianth lobes 3–6 mm; deserts of California and Nevada . . . . 2. *Muilla coronata*
    2. Filaments dilated only basally; scape 15–50 cm; leaves 3–5; tepals white, often tinged with lilac; perianth lobes 6–8 mm; mountains of California and Nevada . . . . 3. *Muilla transmontana*

1. Muilla maritima (Torrey) S. Watson, Proc. Amer. Acad. Arts 14: 235. 1879 [E]

Hesperoscordum maritimum Torrey, Pacif. Railr. Rep. 4(5): 148. 1857; *Allium maritimum* (Torrey) Bentham; *Bloomeria maritima* (Torrey) J. F. Macbride; *B. maritima* var. *serotina* (Greene) J. F. Macbride; *Milla maritima* (Torrey) S. Watson; *Muilla serotina* Greene; *M. tenuis* Congdon

Plants 1–5 dm; corms 1.2–2 cm in diam. **Leaves** 3–10, 10–60 × 0.08–0.2 cm. **Scape** 8–38 cm. **Flowers** (4–)10–70; tepals greenish white with brownish midveins, mostly lanceolate; perianth tube 0.3–0.5 mm, lobes 3–6 × 2–3.5 mm; stamens 2–5 mm; filaments erect, filiform or subulate; anthers green, blue, or purple, 0.6–0.9 mm; pedicel 1–5 cm. **Capsules** 5–8 mm. **Seeds** 2–3 mm. $2n$ = 14, 16, 20.

Flowering spring (Mar–Jun). Coastal scrub, grasslands, woodlands, valleys; 0–2300 m; Calif.

Muilla maritima grows in a wide range of habitats and soil types compared to the other species of the genus. In most floras, the larger form found in Los Angeles and San Bernadino counties has been called *M. serotina*.

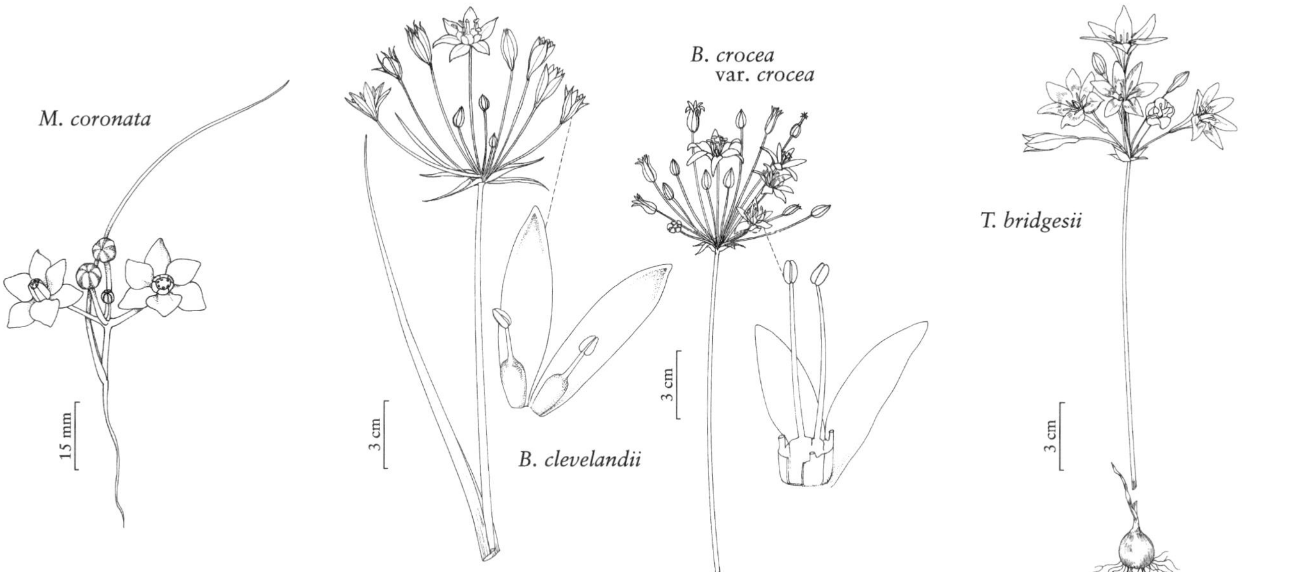

MUILLA ◦ BLOOMERIA ◦ TRITELEIA

2. Muilla coronata Greene, Pittonia 1: 165. 1888 E F

Plants 0.3–1.5 dm; corms 1.2–2 cm in diam. **Leaves** 1 (rarely 2), 7–18 × 0.07–0.1 cm. **Scape** 3–5 cm. **Flowers** 2–5(–10); tepals whitish or rarely bluish, with green abaxial midvein, broadly lanceolate; perianth tube 0.3–0.5 mm, lobes 3–6 × 1.5–2.5 mm; stamens 2–4 mm; filaments erect, conspicuously dilated entire length, distinct but with broad overlapping margins forming cylindrical nectariferous tube with apical crown; anthers yellow, 0.6–0.9 mm; pedicel 0.5–3 cm. **Capsules** 3–7 mm. **Seeds** 2–3 mm. $2n = 14$.

Flowering spring (Mar–Jun). Desert scrub; 1000–1600 m; Calif., Nev.

Muilla coronata is found in deserts of southern California and southwestern Nevada. The scape is very short compared to that of other *Muilla* species. The erect, dilated filaments form a cylindrical crown similar to that found in some species of *Triteleia* and *Androstephium*.

SELECTED REFERENCE Shevock, J. R. 1984. Redescription and distribution of *Muilla coronata* (Liliaceae). Aliso 10: 621–627.

3. Muilla transmontana Greene, Pittonia 1: 73. 1887 E

Bloomeria transmontana (Greene) J. F. Macbride

Plants 1.5–5 dm; corms 1–2.5(–3) cm in diam. **Leaves** 3–5, 5–30 × 0.05–0.4 cm. **Scape** 8–50 cm. **Flowers** 12–30; tepals white, often tinged with lilac, mostly lanceolate; perianth tube 1–1.5 mm, lobes 5–8 × 1.5–2 mm; stamens 4–6 mm; filaments erect, conspicuously dilated basally and connate to form nectariferous cup; anthers yellow, 1–1.5 mm; pedicel 1–2.5 cm. **Capsules** 8–10 mm. **Seeds** 4–5 mm. $2n = 20$.

Flowering spring and early summer (May–Jul). Desert scrub, coniferous woodlands; 1400–2500 m; Calif., Nev.

Muilla transmontana is found in coniferous forests in mountains of northern California and northwestern Nevada, and infrequently on sagebrush flats of lower elevations. The filaments are dilated and connate at the base to form a nectariferous cup similar to that seen in *Bloomeria crocea*.

68. BLOOMERIA Kellogg, Proc. Calif. Acad. Sci. 2: 11. 1863 • Golden star [for H. G. Bloomer, 1821–1874, early San Francisco botanist and one-time botanical curator at the California Academy of Sciences]

J. Chris Pires

Herbs, perennial, scapose, from fibrous-coated corms. **Leaves** 1–8, basal; blade linear-lanceolate, keeled, margins entire. **Scape** slender, cylindrical, rigid. **Inflorescences** umbellate, open, 10–35-flowered, bracteate; bracts 2–4, scarious, membranous, not enclosing flower buds. **Flowers:** tepals 6, persistent, widely spreading, distinct or barely connate at base, golden yellow, striped brownish or green, nearly equal, oblong-linear, subrotate at anthesis; stamens 6, epitepalous, slightly shorter than and inserted at base of tepals; filaments filiform distally, dilated basally, ca. 6 mm, dilated bases sometimes connate into nectariferous cup, cup sometimes having basal filament appendages arising from apex; anthers subbasifixed, versatile; pistil 3-carpellate; ovary superior, sessile, 3-locular, ovules anatropous, several per locule; style persistent, splitting with capsule, filiform or clavate, 5 mm; stigma 3-lobed; pedicel long, erect, ray-like, base and apex articulate. **Fruits** sessile, capsular, 3-angled, subglobose, 5–6 mm, dehiscence loculicidal. **Seeds** black, angular, subovoid, wrinkled, coat with crust. $x = 9$ (except for *B. clevelandii* $x = 14$).

Species 3 (3 in the flora): c and s California, Mexico (n Baja California).

J. W. Ingram (1953) considered *Bloomeria* to be closely related to *Brodiaea*, *Muilla*, and *Allium*. R. F. Hoover (1941, 1955), on the other hand, considered it to be closely related to *Triteleia*, on the basis of corm morphology, keeled leaves, umbel structure, versatile anthers distant from the style, filament appendages, a stigma with three small lobes, similar seeds, and the same base chromosome number. However, *Bloomeria* differs generally from *Triteleia* in that it has distinct tepals and a sessile ovary, although the tepals of some *B. crocea* are slightly connate at the base, and *T. ixioides* has a very short perianth tube. Hoover kept the two genera separate on the basis of the stipitate ovary and geographic distribution. *Triteleia* rarely occurs south of the Tehachapi and the Santa Lucia Mountains in California, where *Bloomeria* occurs.

Among the most important diagnostic characters within *Bloomeria* are features of the androecium, particularly the orientation of the stamens relative to the style, and the appendages at the dilated bases of the filaments. These characters are easily seen in the field with a hand lens. When collecting flowering specimens, one should make a point of preparing a few dissected flowers in a manner that displays these critical characters.

SELECTED REFERENCES Hoover, R. F. 1955. Further observations on *Brodiaea* and some related genera. Pl. Life 11: 13–22. Ingram, J. W. 1953. A monograph of the genera *Bloomeria* and *Muilla* (Liliaceae). Madroño 12: 19–27.

1. Filaments leaning away from style, leaving dilated bases separated and therefore not forming nectariferous cup; style shorter than or equal to ovary; leaves 2–8, 1–3 mm wide . . 3. *Bloomeria clevelandii*
1. Filaments parallel to style, dilated bases connate into nectariferous cup; style longer than ovary; leaves 1–2, 3–15 mm wide.
   2. Scape 15–70 cm; tepals abruptly spreading at base; leaves usually 1 . . . . 1. *Bloomeria crocea*
   2. Scape 5–10 cm; tepals ascending at base, then gradually spreading; leaves usually 1–2 . . . . 2. *Bloomeria humilis*

1. Bloomeria crocea (Torrey) Coville, Contr. U.S. Natl. Herb. 4: 203. 1893 • Common golden star [F]

Allium croceum Torrey in W. H. Emory, Rep. U. S. Mex. Bound. 2(1): 218. 1859

Leaves usually 1, 6–30 cm × 3–15 mm. **Scape** 15–70 cm, minutely scabrous. **Flowers:** tepals abruptly spreading at base, distinct, yellow with brown midvein, 5–13 mm; filaments parallel to style, 3–10 mm, dilated bases connate into nectariferous cup, cup with basal filament appendages arising from apex, appendages shallowly to strongly 2-cuspidate at apex, cusps linear; anthers 1.5–3 mm; ovary 2–3 mm; style 6 mm; pedicel 0.5–6.5 cm.

Varieties 3 (3 in the flora): California, n Mexico.

1. Basal filament appendages not aristate . . . . 1a. *Bloomeria crocea* var. *crocea*
1. Basal filament appendages aristate.
   2. Cusps of basal filament appendages 1–2 mm, less than 1/2 length of filaments, apex acute . . . . 1b. *Bloomeria crocea* var. *aurea*
   2. Cusps of basal filament appendages 3–3.5 mm, ca. 1/2 length of filaments, apex attenuate . . . . 1c. *Bloomeria crocea* var. *montana*

1a. Bloomeria crocea (Torrey) Coville var. **crocea** [F]

Bloomeria gracilis Borzí

Flowers: tepals yellowish orange, nearly always striped with 2 dark parallel lines, 5–12 mm; basal filament appendages shallowly 2-cuspidate, cusps shorter than 1 mm. $2n = 18$.

Flowering spring (Apr–Jun). Dry flats, hillsides, often in heavy soil, chaparral, coastal sage scrub, valley grasslands, oak woodlands; 0–1700 m; Calif.; Mexico (Baja California).

In the flora area, var. *crocea* is restricted to the southern Coast Ranges and the Channel Islands. The plants previously recognized as *Bloomeria gracilis* are simply small versions of var. *crocea* (J. W. Ingram 1953). The variety is in cultivation.

1b. Bloomeria crocea (Torrey) Coville var. **aurea** (Kellogg) J. W. Ingram, Madroño 12: 22. 1953 [E]

Bloomeria aurea Kellogg, Proc. Calif. Acad. Sci. 2: 11. 1863; *Nothoscordum aureum* (Kellogg) Hooker f.

Flowers: tepals yellow, 11–12 mm; basal filament appendages 2-cuspidate, aristate, cusps 1–2 mm, less than 1/2 length of filaments, apex acute.

Flowering spring (Apr–Jun). Foothill woodlands; 0–1700 m; Calif.

Variety *aurea* is restricted to the southern Coast Ranges.

1c. Bloomeria crocea (Torrey) Coville var. **montana** (Greene) J. W. Ingram, Madroño 12: 23. 1953 • Mountain golden star, inland golden star [E]

Bloomeria montana Greene, Bull. Calif. Acad. Sci. 1: 281. 1885

Flowers: tepals yellow, 11–13 mm; basal filament appendages 2-cuspidate, aristate, cusps 3–3.5 mm, ca. 1/2 length of filaments, apex attenuate. $2n = 18$.

Flowering spring (Apr–Jun). Chaparral, yellow pine forests; 300–1000 m; Calif.

Variety *montana* is restricted to the southern Coast Ranges and the Tehachapi Mountains.

2. Bloomeria humilis Hoover, Pl. Life 11: 21, plate 2 (right), fig. 1. 1955 • Dwarf golden star [C] [E]

Leaves usually 1–2, 5–30 cm × 3–15 mm. **Scape** 5–10 cm, minutely scabrous. **Flowers:** tepals ascending at base, then gradually spreading, connate basally for ca. 1 mm, yellow with brownish purplish midvein, 7–11 mm; filaments parallel to style, dilated bases 2.5–4 mm, often smooth, connate into nectariferous cup; anthers 1.5–1.8 mm; ovary 2 mm; style 6 mm; pedicel 1–5 cm.

Flowering spring. Grasslands, chaparral edges, open mesas on ocean bluffs; of conservation concern; 0–100 m; Calif.

Bloomeria humilis is rare, known only from a single location on the California coast.

3. Bloomeria clevelandii S. Watson, Proc. Amer. Acad. Arts 20: 376. 1885 (as clevelandi) • San Diego golden star [C] [F]

Muilla clevelandii (S. Watson) Hoover

Leaves 2–8, 6–15 cm × 1–3 mm. **Scape** 15–70 cm, minutely scabrous. **Flowers:** tepals abruptly spreading from base, distinct, yellow with green midvein, 5–10 mm; filaments leaning away from style, leaving dilated bases separated and therefore not forming nectariferous cup, 3–5 mm; anthers 0.4–1 mm; ovary 2–3 mm; style 2–3 mm, equal to or shorter than ovary; pedicel 0.5–6.5 cm.

Flowering spring (Apr–Jun). Coastal scrub; 0–100 m; of conservation concern; Calif.; n Mexico (Baja California).

Bloomeria clevelandii is restricted to western San Diego County and northern Baja California. Recent evidence suggests that this species is not closely related to *Muilla*, but rather to *Bloomeria* and *Triteleia*, and probably should be placed in a separate, monotypic genus (J. C. Pires 2000). Previous taxonomic keys have focused on the shape of the dilated filament bases, but the orientation of the stamens and the unfused filament bases are more reliably diagnostic.

The species is in cultivation, and the corms were probably eaten by native Americans.

69. TRITELEIA Douglas ex Lindley, Edwards's Bot. Reg. 15: under plate 1293. 1830 • Triteleia, triplet-lily [Greek *tri,* three, and *teleios,* perfect, alluding to all flower parts in threes]

J. Chris Pires

Brodiaea Smith sect. *Calliprora* (Lindley) S. Watson; *B.* subg. *Calliprora* (Lindley) Eastwood; *B.* subg. *Hesperoscordum* (Lindley) Eastwood; *B.* sect. *Seubertia* (Kunth) S. Watson; *B.* subg. *Triteleia* (Douglas ex Lindley) Eastwood; *Calliprora* Lindley; *Hesperoscordum* Lindley; *Seubertia* Kunth

Herbs, perennial, scapose, from fibrous-coated corms. **Leaves** 1–3, basal; blade narrowly lanceolate (linear in *Triteleia ixioides*), keeled, channeled, glabrous, margins entire. **Scape** erect, cylindrical, 1–5 mm diam., rigid. **Inflorescences** umbellate, open, bracteate; bracts green (purplish in *T. lemmoniae*), ± lanceolate, scarious. **Flowers:** perianth 6-tepaled, connate proximally into tube of varying length and shape, usually funnelform, lobes similar, usually ascending to spreading; stamens 6, epitepalous; filaments distinct, adnate to perianth tube in 1 or 2 rows, equal or of 2 unequal lengths, free portions flattened, sometimes dilated at base to form triangle, apical appendages usually absent, when present sometimes forming a crown; anthers versatile, usually curving away from stigma; pistil 3-carpellate; ovary superior, green or colored like perianth (yellow in *T. peduncularis*, white in *T. clementina*), stipitate, 3-locular, ovules anatropous, 2–several per locule; style 2–4 mm; stigma weakly 3-lobed; pedicel ± erect, often articulate, usually longer than perianth (shorter in *T. crocea*). **Fruits** capsular, ovoid, dehiscence loculicidal. **Seeds** black, ridged on 1 side, subglobose, rounded, coarsely and irregularly pitted, minutely granulate or granulate-reticulate, coat with crust. $x = 7, 8$.

Species 15 (14 in the flora): w North America, n Mexico.

For discussion of relationships, see under *Brodiaea*.

Molecular evidence (J. C. Pires 2000) suggests the artificiality of the subgenera and sections that have been recognized within *Triteleia*, consistent with R. F. Hoover (1941), who recognized sections for reasons of convenience only. Thus, those infrageneric taxa are not utilized here.

Several species of *Triteleia* are exceedingly variable, and polyploidy is common: multiples of both $x = 7$ and $x = 8$ occur, suggesting that chromosomal changes have played a significant evolutionary role within the genus (M. P. Burbanck 1941).

Triteleia is widely distributed west of the Rocky Mountains, but its greatest diversity is in the "Klamath area" of northwestern California and southern Oregon. The corms of some species were eaten by native Americans.

Among the most important diagnostic characters within *Triteleia* are features of the androecium, particularly stamen height and insertion relative to the perianth, and the presence of apical filament appendages. These characters are easily seen in the field with a hand lens. When collecting flowering specimens, one should make a point of mounting a few dissected flowers in a manner that displays these critical characters.

The only *Triteleia* species that does not occur in the flora, *T. guadalupensis* L. W. Lenz, is endemic to Guadalupe Island off Baja California.

SELECTED REFERENCES Hoover, R. F. 1941. A systematic study of *Triteleia*. Amer. Midl. Naturalist 25: 73–100. Lenz, L. W. 1975. A biosystematic study of *Triteleia* (Liliaceae): 1. Revision of the species of section *Calliprora*. Aliso 8: 221–258. Lenz, L. W. 1976b. A biosystematic study of *Triteleia* (Liliaceae): 2. Chromosome numbers and karyotypes of the species of section *Calliprora*. Aliso 8: 353–377.

1. Stamens attached alternately at 2 levels on perianth tube, forming 2 rows of 3.
   2. Perianth tube obtuse and rounded at base; ovary twice as long as stipe . . . . 5. *Triteleia grandiflora*
   2. Perianth tube acute or attenuate at base; ovary shorter than or equal to stipe (slightly longer than stipe in *T. crocea*).
      3. Stamens unequal, alternately long and short.
         4. Pedicel 0.7–2 cm; perianth 12–19 mm, bright yellow or pale blue; ovary green . . 3. *Triteleia crocea*
         4. Pedicel 2–10(–18) cm; perianth 15–28 mm, white, often flushed violet or lilac abaxially; ovary bright yellow . . . . 14. *Triteleia peduncularis*
      3. Stamens equal in length or nearly so.
         5. Perianth 16–27 mm, lavender; anthers purple, 1.5 mm; endemic to San Clemente Island . . . . 2. *Triteleia clementina*
         5. Perianth 18–47 mm, pale blue, sometimes deep bluish purple or white; anthers white to bluish, 2–5 mm; widespread in California . . . . 9. *Triteleia laxa*
1. Stamens all attached at same level at throat of perianth tube.
   6. Stamens unequal, alternately long and short.
      7. Longer filaments rounded apically . . . . 12. *Triteleia lugens*
      7. Longer filaments not rounded apically.
         8. Apical filament appendages short, blunt, or absent; perianth tube ca. equal to lobes . . . . 4. *Triteleia dudleyi*
         8. Apical filament appendages pointed, conspicuous; perianth tube much shorter than or equal to lobes . . . . 8. *Triteleia ixioides*
   6. Stamens equal in length or nearly so.
      9. Perianth 27–45 mm, tube strongly attenuate, lobes shorter than tube . . . . 1. *Triteleia bridgesii*
      9. Perianth 7–23 mm, tube turbinate, bowl-shaped, or funnelform and moderately attenuate at base, lobes 2–3 times longer than tube.
         10. Perianth tube funnelform, attenuate at base, lobes ca. twice as long as tube.
            11. Perianth 12–17 mm; filaments more than 1/2 as long as perianth lobes . . . . 13. *Triteleia montana*
            11. Perianth 18–26 mm; filaments less than 1/2 as long as perianth lobes . . . 6. *Triteleia hendersonii*
         10. Perianth tube turbinate or shallowly bowl-shaped, lobes 2–3 times longer than tube.
            12. Perianth bright yellow to deep orange, turbinate; Arizona . . . . 11. *Triteleia lemmoniae*
            12. Perianth white, blue, or lilac, shallowly bowl-shaped; not in Arizona.
               13. Filaments usually triangular-dilated, hyaline vesicles absent from perianth tube; usually moist soils, widespread . . . . 7. *Triteleia hyacinthina*
               13. Filaments always linear, hyaline vesicles present in perianth tube; dry rocky outcrops, volcanic hills and mesas of n California . . . . 10. *Triteleia lilacina*

1. **Triteleia bridgesii** (S. Watson) Greene, Bull. Calif. Acad. Sci. 2: 141. 1886 • Bridges' triteleia E F

Brodiaea bridgesii S. Watson, Proc. Amer. Acad. Arts 14: 237. 1878; *Hookera bridgesii* (S. Watson) Kuntze

Leaves 20–55 cm × 3–10 mm. **Scape** 10–60 cm, smooth except sometimes scabrous near base. **Flowers:** perianth lilac, bluish purple, pink, or reddish purple, 27–45 mm, tube strongly attenuate with slender base, 17–25 mm, hyaline vescicles present in tube, lobes abruptly spreading, 10–20 mm, shorter than tube; stamens attached at 1 level, equal; filaments triangular, widened toward base, 3–4 mm, apical appendages absent; anthers bluish, 3.5–4.5 mm; ovary 1/4–1/3 length of stipe; pedicel 2–9 cm. $2n = 16$.

Flowering spring (Apr–Jun). Foothills, yellow pines, mixed evergreen forests, often at forest edges and on rocks, dry bluffs, hillsides, mainly areas of serpentine; 0–100 m; Calif., Oreg.

Triteleia bridgesii is similar to *T. laxa* and in herbarium specimens can be distinguished only by the stamens. However, in fresh flowers, the perianth lobes of *T. bridgesii* spread abruptly from the throat of the perianth tube, unlike those of *T. laxa*, and the flowers of *T. bridgesii* are erect, while those of *T. laxa* are oriented horizontally. *Triteleia bridgesii* flowers a month earlier than *T. laxa* where they both occur in the Sierra Nevada foothills. The flowers in specimens of *T. bridgesii* from Humboldt County are unusually long, which accounts for the wide ranges of measurements for the perianth.

2. **Triteleia clementina** Hoover, Amer. Midl. Naturalist 25: 82. 1941 • San Clemente Island triteleia C E

Brodiaea clementina (Hoover) Munz

Leaves 30–100 cm × 4–30 mm. **Scape** 30–90 cm, smooth. **Flowers:** perianth lavender, 16–27 mm, tube funnelform-campanulate, acute at base, 7–12 mm, lobes erect, 9–15 mm; stamens attached alternately at 2 levels, equal; filaments triangular, widest at base, 2 mm, apical appendages absent; anthers purple, 1.5 mm; ovary white, equal to stipe; pedicel 3–8 cm. $2n = 16$.

Flowering spring (Mar–Apr). Damp clefts, rocky walls, coastal sage scrub; of conservation concern; 0–200 m; Calif.

This is the only species of *Triteleia* occurring on any of the islands off the southern California coast; it is endemic to San Clemente Island.

3. **Triteleia crocea** (Alph. Wood) Greene, Bull. Calif. Acad. Sci. 2: 141. 1886 • Yellow triteleia E

Seubertia crocea Alph. Wood, Proc. Acad. Nat. Sci. Philadelphia 20: 172. 1868 (as Subertia); *Brodiaea crocea* (Alph. Wood) S. Watson; *B. crocea* var. *modesta* (H. M. Hall) Munz; *B. modesta* H. M. Hall; *Hookera crocea* (Alph. Wood) Kuntze; *Milla crocea* (Alph. Wood) Baker; *Triteleia crocea* var. *modesta* (H. M. Hall) Hoover; *T. modesta* (H. M. Hall) Abrams

Leaves 9–40 cm × 2–10 mm. **Scape** 10–30 cm, smooth except weakly scabrous near base. **Flowers:** perianth bright yellow or pale blue, 12–19 mm, tube attenuate at base, 5–10 mm, lobes widely spreading, striped greenish, 5–11 mm; stamens attached alternately at 2 levels, unequal, those of proximal row very short; filaments linear or barely wider at base, 1 or 3 mm, apical appendages absent; anthers yellow or blue, 1–2 mm; ovary green, equal to or longer than stipe; pedicel 0.7–2 cm, usually shorter than perianth. $2n = 16$.

Flowering spring (May–Jun). Open conifer/yellow pine forests, dry slopes; 1200–2200 m; Calif., Oreg.

Plants of *Triteleia crocea* from the Trinity Mountains of northern California differ from others of the species in having pale blue perianths instead of yellow ones, with lobes slightly fringed toward the apex instead of entire. These plants first were assigned to a separate species, and subsequently have been transferred to this one, where clearly they belong. Formal recognition at subspecific rank may prove desirable in future.

4. **Triteleia dudleyi** Hoover, Amer. Midl. Naturalist 25: 86. 1941 • Dudley's triteleia E

Brodiaea dudleyi (Hoover) Munz

Leaves 10–30 cm × 3–11 mm. **Scape** 10–35 cm, smooth. **Flowers:** perianth pale yellow, drying purplish, 18–24 mm, tube ± cylindrical to narrowly funnelform, 8–12 mm, lobes spreading, lanceolate, 8–12 mm; stamens attached at 1 level, unequal, alternately long and short; filaments dilated entire length or narrowly triangular, 2 or 3.5 mm, sometimes with short, blunt apical appendages; anthers lavender, 1 mm; ovary equal to or longer than stipe; pedicel slender, 1.5–4 cm. $2n = 16$.

Flowering summer (Jul). Subalpine forests, black soils; 3000–3500 m; Calif.

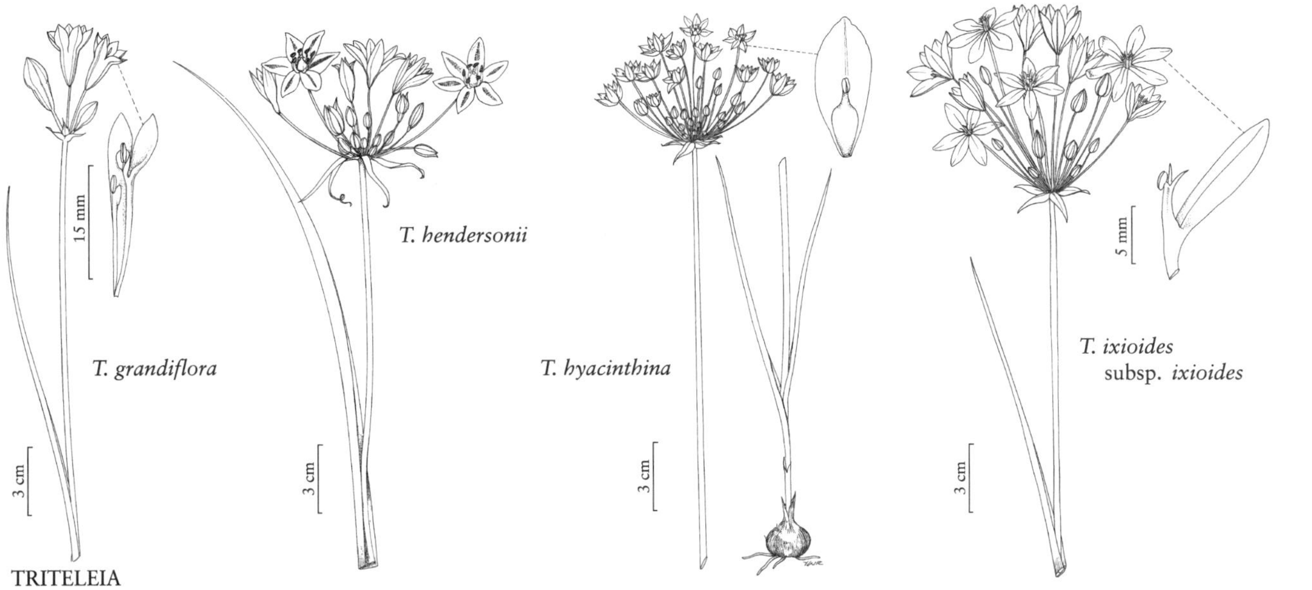

TRITELEIA

5. **Triteleia grandiflora** Lindley, Edwards's Bot. Reg. 15: under plate 1293. 1830 • Large-flowered triteleia, blue umber-lily, blue-lily, wild hyacinth [E] [F]

Brodiaea bicolor Suksdorf; *B. douglasii* S. Watson; *B. douglasii* var. *howellii* (S. Watson) M. Peck; *B. grandiflora* (Lindley) J. F. Macbride 1918, not Smith 1811; *Brodiaea howellii* S. Watson; *Hookera bicolor* (Suksdorf) Piper; *H. douglasii* (S. Watson) Piper; *H. grandiflora* (Lindley) Kuntze; *H. howellii* (S. Watson) Piper; *Milla grandiflora* (Lindley) Baker; *Triteleia bicolor* (Suksdorf) A. Heller; *T. grandiflora* var. *howellii* (S. Watson) Hoover; *T. howellii* (S. Watson) Greene

Leaves 20–70 cm × 4–10 mm. **Scape** 20–75 cm, smooth. **Flowers:** perianth bluish purple to white, 17–35 mm, tube obtuse and rounded at base, 8–20 mm, lobes spreading, 9–13 mm; stamens attached alternately at 2 levels, unequal; filaments slender and somewhat triangular, wider toward base, or broad, 1–4 mm, apical appendages absent or present; anthers yellow or purple, 2–4 mm; ovary twice as long as stipe; pedicel 1–4 cm. $2n$ = 16, 24, 32, 40, 48, 56.

Flowering spring–summer (Apr–Jul). Grasslands, sagebrush, pinyon-juniper woodlands, pine forests and hills; 100–3000 m; B.C; Calif., Idaho, Mont., Oreg., Utah, Wash., Wyo.

Triteleia grandiflora is the type species of the genus and, along with *T. hyacinthina*, is its most widely distributed member. Found throughout the region between the Cascade Range and the northern Rocky Mountains, in sagebrush steppe and adjacent woodlands, it is easily recognized by the shape of the perianth, which is rounded at the base instead of tapered as in other *Triteleia* species.

M. E. Barkworth (1975, 1977) studied variation within *Triteleia grandiflora* in relation to ploidy level. Polyploid plants are larger, flower later, and have more effective vegetative reproduction by cormlets and contractile roots than their diploid progenitors.

Plants of *Triteleia grandiflora* from the area west of the Cascade Range and extending into the Columbia River valley and the Klamath Lake region that have been distinguished as var. *howellii* differ from others of the species only in the shape of the filaments. L. Abrams and R. S. Ferris (1923–1960, vol. 1) used relative perianth length as a key character, but this is not consistent (R. F. Hoover 1941), and several specimens from the part of the Columbia River valley where both filament morphologies occur appear to represent intergrades (R. F. Hoover 1955). Thus it seems inadvisable to recognize infraspecific taxa. Plants assignable to var. *howellii* have not been found in California or southwestern Oregon in recent decades, and may be extirpated there.

Triteleia bicolor is merely a color form having a perianth with a blue tube and white lobe.

6. Triteleia hendersonii Greene, Pittonia 1: 164. 1888 (as hendersoni) • Henderson's triteleia, yellow tiger-lily E F

Brodiaea hendersonii (Greene) S. Watson; *B. leachiae* M. Peck; *Triteleia hendersonii* var. *leachiae* (M. Peck) Hoover

Leaves 15–40 cm × 3–12 mm. **Scape** 10–35 cm, smooth or slightly scabrous near base. **Flowers:** perianth yellow or white, often tinged or fading blue, 18–26 mm, tube slenderly funnelform, moderately attenuate at base, 6–10 mm, lobes widely spreading, with conspicuous dark purple midvein, 12–16 mm, ca. twice as long as tube; stamens attached at 1 level, nearly equal; filaments narrowly subulate, 3–4 mm, apical appendages absent; anthers blue or sometimes white, 1.5–2 mm; ovary 1/2 as long as stipe; pedicel 1.5–4 cm. $2n = 32$.

Flowering spring (May–Jul). Dry slopes; 100–3000 m; Calif., Oreg.

Triteleia hendersonii is common within its restricted range. The plants previously recognized as var. *leachiae*, or separately as *T. leachiae*, differ primarily in having a white perianth and are limited to Curry County, Oregon. Intermediates have been found, and further study of this variation is required to determine whether formal recognition of infraspecific taxa is warranted.

7. Triteleia hyacinthina (Lindley) Greene, Bull. Calif. Acad. Sci. 2: 142. 1886 • White brodiaea, fool's onion, hyacinth brodiaea, wild hyacinth E F

Hesperoscordum hyacinthinum Lindley, Edwards's Bot. Reg. 15: under plate 1293. 1829; *Allium lacteum* (Lindley) Bentham; *A. tilingii* Regel; *Brodiaea dissimulata* M. Peck; *B. hyacinthina* (Lindley) Baker; *B. hyacinthina* var. *lactea* (Lindley) Baker; *B. hyacinthina* var. *lilacina* (S. Watson) Jepson; *B. lactea* (Lindley) S. Watson; *B. lactea* var. *lilacina* S. Watson; *Hesperoscordum lacteum* Lindley; *H. lewisii* Hooker; *Hookera hyacinthina* (Lindley) Kuntze; *H. hyacinthina* var. *lactea* (Lindley) Jepson; *Milla hyacinthina* (Lindley) Baker; *M. hyacinthina* var. *lactea* (Lindley) Baker; *Triteleia lactea* (Lindley) Davidson & Moxley; *Veatchia crystallina* Kellogg

Leaves 10–40 cm × 4–22 mm. **Scape** 30–60 cm, smooth to scabrous. **Flowers:** perianth white, sometimes flushed purple abaxially, 9–16 mm, tube shallowly bowl-shaped, 2–4 mm, lobes ascending to spreading, with green midvein, 7–12 mm, 2–3 times longer than tube; stamens attached at 1 level, equal; filaments usually triangular-dilated, 2–4 mm, apical appendages absent; anthers whitish to yellow, rarely blue, 1–2 mm; ovary twice as long as stipe; pedicel 0.5–5 cm. $2n = 28, 84$.

Flowering spring (Mar–Jun). Grasslands, vernally wet meadows, occasionally on drier slopes; 0–2000 m; B.C; Calif., Idaho, Nev., Oreg., Wash.

Triteleia hyacinthina is one of the most widely distributed members of the genus, common in low, moist places such as meadows, vernal pools, and along streams, and found occasionally in drier valley grasslands, foothill woodlands, and closed-cone pine forests. It is cytologically and morphologically variable, with large-flowered forms found in wet places; smaller, scabrous forms found in dry places; forms with rather long pedicels (sometimes recognized as var. *lactea*); and forms with filaments lacking dilated triangular bases. Most plants have white flowers, but a few interesting forms with restricted distributions have lilac, blue, or even purple flowers. *Triteleia* ×*versicolor* Hoover is a sterile hybrid of *T. ixioides* and *T. hyacinthina* found in Monterey. *Triteleia lilacina* is recognized here as a species distinct from *T. hyacinthina* because of its different base chromosome number ($2n = 16$) and presence of hyaline vesicles (R. F. Hoover 1955).

8. Triteleia ixioides (W. T. Aiton) Greene, Bull. Calif. Acad. Sci. 2: 142. 1886 • Yellow-brodiaea E F

Ornithogalum ixioides W. T. Aiton in W. Aiton and W. T. Aiton, Hortus Kew. 2: 257. 1811; *Brodiaea ixioides* (W. T. Aiton) S. Watson 1879, not Sims 1823; *Calliprora ixioides* (W. T. Aiton) Greene; *Hookera ixioides* (W. T. Aiton) Kuntze; *Milla ixioides* (W. T. Aiton) Baker; *Themis ixioides* (W. T. Aiton) Salisbury

Leaves 1–2, 10–50 cm × 3–15 mm; blade linear. **Scape** 10–80 cm, usually smooth, sometimes scabrous near base. **Flowers:** perianth straw-colored or golden yellow, or sometimes white flushed with purple, usually similarly colored both adaxially and abaxially, 12–27 mm, tube acute at base, 3–10 mm, much shorter than or equal to lobes, lobes ascending to reflexed, with green, brown, or purple midveins, 6–20 mm; stamens attached at 1 level, held close to pistil, unequal, alternately long and short; filaments dilated entire length, noticeably flattened, apical appendages present, forked, those of longer filaments pointed, conspicuous; anthers white, cream, yellow, or blue, 1–2 mm; ovary longer than stipe; pedicel 1–9 cm (–12 cm in subsp. *cookii*). $2n = 10, 14, 16, 24, 32, 40, 42, 50$.

Subspecies 5 (5 in the flora): California, s Oregon.

As defined by L. W. Lenz (1975, 1976b), *Triteleia ixioides* consists of five groups of plants that, although readily recognizable as a rule, often cannot be separated by constant characters. Subspecies *ixioides* and subsp. *cookii* are strictly coastal and widely separated geographically from the three subspecies of the interior. They flower from May to August (a month earlier than subsp. *scabra*) and usually are found in the shade of oaks, cypresses, or pines. Polyploidy is rampant and perhaps indicative of the geographical and ecological differentiation of these morphologically similar subspecies.

1. Leaf 1, relatively short . . . . 8e. *Triteleia ixioides* subsp. *unifolia*
1. Leaves 1–2.
    2. Perianth white or pale straw-colored, sometimes with darker midvein color extending into distal portion of tepal lobe, tube nearly equal to lobes . . . . 8c. *Triteleia ixioides* subsp. *cookii*
    2. Perianth yellow, tube shorter than tepal lobes.
        3. Anthers blue or cream; filaments with short, spreading, straight, or incurved apical appendages . . . . 8b. *Triteleia ixioides* subsp. *anilina*
        3. Anthers cream or yellow, occasionally blue; filaments with long, slender, often recurving apical appendages.
            4. Perianth tube 7–10 mm; coastal California . . . . 8d. *Triteleia ixioides* subsp. *ixiodes*
            4. Perianth tube 3–7 mm; c, n California . . . . 8d. *Triteleia ixioides* subsp. *scabra*

8a. Triteleia ixioides (W. T. Aiton) Greene subsp. **ixioides** • Golden-brodiaea [E] [F]

Brodiaea lutea (Lindley) C. V. Morton; *Calliprora lutea* Lindley

Leaves 1–2, 10–40 cm. **Scape** 20–80 cm, smooth or scabrous. **Flowers:** perianth yellow, tube 7–10 mm, much shorter than lobes, lobes spreading, 10–15 mm; short filaments 3 mm, long filaments 4–5.5 mm, apical appendages long, slender, often recurving; anthers cream or yellow, occasionally blue; pedicel 1–7 cm. $2n = 14$.

Flowering May–Aug. Sandy soil in shade of oaks or pines; 0–1500 m; Calif.

Subspecies *ixioides* occurs in coastal California, sometimes on serpentine (San Mateo to San Luis Obispo counties). In the past it hybridized with *Triteleia hyacinthina* in Monterey to produce *T.* ×*versicolor* Hoover, which probably is extinct due to successional changes.

8b. Triteleia ixioides (W. T. Aiton) Greene subsp. **anilina** (Greene) L. W. Lenz, Aliso 8: 233. 1975 • Pretty face, Sierra-brodiaea [E]

Calliprora scabra Greene var. *anilina* Greene, Erythea 3: 126. 1895; *Brodiaea lutea* (Lindley) C. V. Morton var. *anilina* (Greene) Munz; *B. scabra* (Greene) Baker var. *anilina* (Greene) Baker; *Calliprora anilina* (Greene) A. Heller; *Triteleia anilina* (Greene) Hoover; *T. ixioides* var. *anilina* (Greene) Hoover

Leaves 1–2, 10–25 cm. **Scape** shorter than 30 cm, smooth or scabrous. **Flowers:** perianth yellow, tube 4–7 mm, much shorter than lobes, lobes widely spreading, 7–12 mm; short filaments 1–2 mm or nearly absent, long filaments 3–3.5 mm, apical appendages straight, spreading, or incurved, short; anthers usually blue, sometimes cream; pedicel 1–7 cm. $2n = 10, 14, 40, 42, 50$.

Flowering spring (May–Jul). Coniferous forest edges, often in moist gravel or sand; 600–3000 m; Calif., Oreg.

Subspecies *anilina* might be considered a reduced, "high-altitude" form of subsp. *scabra*, but although it reaches an elevation of 3000 meters, it is sometimes found in the chaparral belt as low as 600 meters, while subsp. *scabra* extends as high as 2200 meters.

8c. Triteleia ixioides (W. T. Aiton) Greene subsp. **cookii** (Hoover) L. W. Lenz, Aliso 8: 237. 1975 • Cook's triteleia [E]

Triteleia ixioides var. *cookii* Hoover, Pl. Life 1: 19, plate 2(left). 1955; *Brodiaea lutea* (Lindley) C. V. Morton var. *cookii* (Hoover) Munz

Leaves 1–2, 10–20 cm. **Scape** 20–25 cm, slightly scabrous. **Flowers:** perianth white to pale straw-colored, sometimes with darker midvein color extending into distal portion of tepal lobe, tube 6–10 mm, nearly equal to lobes, lobes strongly reflexed, 6–10 mm; short filaments 3 mm, long filaments 4 mm, apical appendages straight or slightly recurved; anthers white; pedicel 1–12 cm. $2n = 14, 28$.

Flowering spring. Stream sides, wet ravines on serpentine, often near cypresses; 0–500 m; Calif.

Subspecies *cookii* is restricted to San Luis Obispo County and is recognized by its long pedicels and relatively long perianth tube in relation to the length of the perianth lobes.

8d. Triteleia ixioides (W. T. Aiton) Greene subsp. **scabra** (Greene) L. W. Lenz, Aliso 8: 240. 1975 [E]

Calliprora scabra Greene, Erythea 3: 126. 1895; *Brodiaea ixioides* (W. T. Aiton) S. Watson var. *scabra* (Greene) Smiley; *B. lutea* (Lindley) C. V. Morton var. *scabra* (Greene) Munz; *B. scabra* (Greene) Baker; *Calliprora aurantea* Kellogg; *Triteleia ixioides* var. *scabra* (Greene) Hoover; *T. scabra* (Greene) Hoover

Leaves 1–2, 10–50 cm. **Scape** 20–50 cm, usually scabrous. **Flowers:** perianth usually straw-colored to pale yellow, tube 3–7 mm, much shorter than lobes, lobes ascending to reflexed, 10–20 mm; short filaments 4–5 mm, long filaments 5–7 mm, apical appendages long, slender, straight to recurved; anthers cream or yellow, occasionally blue; pedicel 1–9 cm. $2n$ = 10 (11), 16, 24, 32.

Flowering spring (Mar–May, later at higher altitudes). Scrub edges, mixed or coniferous forests, foothill woodlands, valley grasslands, in clay and granite soils; 200–2200 m; Calif.

8e. Triteleia ixioides (W. T. Aiton) Greene subsp. **unifolia** L. W. Lenz, Aliso 8: 243, figs. 15, 25. 1975 [E]

Leaves usually only 1, relatively short, 10–20 cm. **Scape** 10–30 cm, smooth or slighty scabrous. **Flowers:** perianth dull yellow, sometimes with darker midvein color extending into abaxial portion of perianth, tube 5–7 mm, ca. 1/2 length of lobes, lobes spreading, 9–12 mm; short filaments 2–4 mm, long filaments 3–5 mm, apical appendages short, straight or incurved; anthers cream or yellow; pedicel 1–4 cm. $2n$ = 14, 16.

Flowering spring. Coniferous forests, in clay and granite soils; 600–2000 m; Calif., Oreg.

9. Triteleia laxa Bentham, Trans. Hort. Soc. London, ser. 2, 1: 413, plate 15, fig. 2. 1835 • Ithuriel's spear, grass nut, common triteleia [E]

Brodiaea candida (Greene) Baker; *B. laxa* (Bentham) S. Watson; *B. laxa* var. *candida* (Greene) Jepson; *B. laxa* var. *nimia* Jepson; *B. laxa* var. *tracyi* Jepson; *Hookera laxa* (Bentham) Kuntze; *Milla laxa* (Bentham) Baker; *Seubertia laxa* (Bentham) Kunth; *S. obscura* Borzí; *Triteleia angustiflora* A. Heller; *T. candida* Greene

Leaves 20–40 cm × 4–25 mm. **Scape** 10–70 cm, smooth or scabrous near base. **Flowers:** perianth usually pale blue, sometimes deep bluish purple or white, 18–47 mm, tube attenuate at base, 12–25 mm, lobes gradually spreading, 8–20 mm; stamens attached alternately at 2 levels, horizontal and curved upward at apex, nearly equal; filaments linear, 3–6 mm, apical appendages present near base of adnate portion of filaments, membranous; anthers white to bluish, 2–5 mm, apex blunt to tapered; ovary 1/3–1/2 as long as stipe, either central or along proximal side of flower; pedicel ascending or spreading, often bent at apex, 1–10 cm. $2n$ = 16, 18, 28 (30), 32, 42, 48.

Flowering spring (Apr–Jun). Open forests, mixed conifer or foothill woodlands, grasslands on clay soil; 0–1500 m; Calif.

Triteleia laxa is very common at low altitudes in central California, growing under a variety of ecological conditions, with wide attendant variation, especially in flower size. It could be considered a species complex (like *T. ixioides*) made up of many forms, with consistent differences in chromosome number and in the size and shape of the anthers (M. P. Burbanck 1944). The flowers can be quite large or small, and usually are pale blue, but sometimes are white or deep bluish purple. These morphologies often intergrade, making it difficult to recognize formal varieties.

This species may occur in southwestern Oregon, but documentation of its presence there is lacking.

Triteleia ×*tubergenii* L. W. Lenz is a cultivated amphidiploid hybrid between *T. laxa* and *T. peduncularis* (L. W. Lenz 1970).

10. Triteleia lilacina Greene, Bull. Calif. Acad. Sci. 2: 143. 1886 • Lilac-flowered wild hyacinth [E]

Brodiaea hyacinthina (Lindley) Baker var. *greenei* (Hoover) Munz; *B. lilacina* (Greene) Baker; *Hesperoscordum lilacinum* (Greene) A. Heller ex Abrams; *Triteleia hyacinthina* (Lindley) Greene var. *greenei* Hoover

Leaves 10–40 cm × 4–22 mm. **Scape** 30–60 cm, smooth to scabrous. **Flowers:** perianth white, unstriped, 7–10 mm, tube shallowly bowl-shaped, 2–4 mm, hyaline vescicles present in tube, lobes 7–12 mm, 2–3 times longer than tube; stamens attached at 1 level, equal; filaments linear, 2–4 mm, apical appendages absent; anthers bluish purple or bluish purple-dotted, 1–2 mm; ovary twice as long as stipe; pedicel 0.5–5 cm. $2n = 16$.

Flowering spring (Mar–Jun). Dry rocky outcrops, volcanic hills and mesas; 100–200 m; Calif.

Triteleia lilacina differs from *T. hyacinthina* in base chromosome number and in that the interior of the perianth tube is covered with hyaline vesicles (making a glassy nectary). In addition, *T. lilacina* has linear filaments and bluish purple anthers at anthesis, whereas *T. hyacinthina* tends to have triangular-dilated filaments and whitish to yellow (or rarely blue) anthers at anthesis.

11. Triteleia lemmoniae (S. Watson) Greene, Bull. Calif. Acad. Sci. 2: 141. 1886 (as lemmonae) • Lemmon's triteleia, Oak Creek triteleia [E]

Brodiaea lemmoniae S. Watson, Proc. Amer. Acad. Arts 20: 376. 1885 (as lemmonae)

Leaves 10–40 cm × 2–6 mm. **Scape** 8–30 cm, smooth or scabrous near base; bracts purplish. **Flowers:** perianth bright yellow to deep orange, fading to purple, 9–12 mm, tube turbinate, 2.5–3 mm, lobes ascending to slightly spreading, 7–9 mm, 2–3 times longer than tube; stamens attached at 1 level, equal; filaments linear, 3 mm, apical appendages absent; anthers yellow, 2 mm; ovary 3 times longer than stipe; pedicel 0.7–2.5 cm.

Flowering spring–summer (late May–Aug). Yellow pine belts, mountains; 1000–3000 m; Ariz.

Triteleia lemmoniae is the only representative of the genus in Arizona. Molecular data suggest that it is related to *T. montana*, found in the Sierra Nevada of California (J. C. Pires 2000), and not to the morphologically similar *T. hyacinthina* (R. F. Hoover 1941).

12. Triteleia lugens Greene, Bull. Calif. Acad. Sci. 2: 142. 1886 • Coast Range triteleia [E]

Brodiaea ixioides (W. T. Aiton) S. Watson var. *lugens* (Greene) Jepson; *B. lugens* (Greene) Baker; *B. lutea* (Lindley) C. V. Morton var. *lugens* (Greene) C. V. Morton; *Calliprora ixioides* (W. T. Aiton) Greene var. *lugens* (Greene) Abrams; *C. lugens* (Greene) Greene; *Hookera ixioides* (W. T. Aiton) Kuntze var. *lugens* (Greene) Jepson

Leaves 10–40 cm × 3–10 mm. **Scape** 10–40 cm, smooth. **Flowers:** perianth pale yellow to golden, 12–15 mm, tube funnelform, rather broad, 4–6 mm, lobes spreading, never rotate, with dark midvein, 6–9 mm; stamens attached at 1 level, unequal, alternately long and short; filaments dilated entire length, 1–2 or 2–4 mm, inner filaments equal to lobes, longer filaments rounded apically, apical appendages absent; anthers yellow or blue, 1.5–2 mm; ovary equal to or longer than stipe; pedicel 1–2.5 cm. $2n = 16$.

Flowering spring (Mar–Jun). Edges of chaparral, mixed forests (foothill woodlands), along streams; 100–1000 m; Calif.

Triteleia lugens has a disjunct distribution, being found in Lake, Napa, and Solano counties north of San Francisco Bay (with pale yellow flowers), and 140 miles away in Monterey and San Benito counties south of San Francisco Bay (with bright yellow flowers).

13. Triteleia montana Hoover, Amer. Midl. Naturalist 25: 95. 1941 • Mountain triteleia [E]

Brodiaea gracilis S. Watson, Proc. Amer. Acad. Arts 14: 238. 1879; *Hookera gracilis* (S. Watson) Kuntze; *Triteleia gracilis* (S. Watson) Greene 1886, not Philippi 1873

Leaves 10–30 cm × 2–5 mm. **Scape** 5–20(–25) cm, ± scabrous. **Flowers:** perianth yellow, aging purple, 12–17 mm, tube slender, narrowly funnelform, attenuate at base, 4–7 mm, lobes somewhat spreading, with brown midvein, 8–10 mm, ca. twice as long as tube; stamens attached at 1 level, equal; filaments linear, 5–6 mm, more than 1/2 as long as perianth lobes, apical appendages absent; anthers cream or blue, 1–1.5 mm; ovary equal to stipe; pedicel 0.5–3 cm. $2n = 16$.

Flowering summer (Jun–Jul). Open montane coniferous forest, gravelly plains, granite ridges; 1200–3000 m; Calif.

Locally rather plentiful, *Triteleia montana* appears to have a disjunct distribution in the Sierra Nevada range north and south from Yosemite, though future collections may bridge the gap between the distributions currently indicated. Molecular data suggest that it is related to *T. lemmoniae* of Arizona (J. C. Pires 2000).

14. Triteleia peduncularis Lindley, Edwards's Bot. Reg. 20: under plate 1685. 1835 • Long-rayed brodiaea [E]

Brodiaea peduncularis (Lindley) S. Watson; *Hookera peduncularis* (Lindley) Kuntze; *Milla peduncularis* (Lindley) Baker

Leaves 20–40 cm × 5–15 mm. **Scape** 10–80 cm, smooth. **Flowers:** perianth white, often flushed violet or lilac abaxially, 15–28 mm, tube broadly funnelform-campanulate, acute at base, 7–11 mm, lobes 10–16 mm; stamens attached alternately at 2 levels, unequal, those of proximal row shorter; filaments nearly linear, slightly wider at base, 1–1.5 or 2–3 mm, apical appendages absent; anthers white, 2–4 mm; ovary bright yellow in flower, equal to stipe; pedicel 2–10(–18) cm. $2n$ = 14, 28.

Flowering spring (May–Jul). Low fields, wet grasslands, vernal streams and pools, closed cone pine forests, mixed evergreens, foothill woodlands, often on serpentine; 0–800 m; Calif.

Triteleia peduncularis has a wide distribution in the northern California Coast Ranges, but it is usually not common and is quite rare south of San Francisco. The yellow ovary contrasts notably with the white perianth. The long-ascending pedicels are also distinctive. *Triteleia* ×*tubergenii* L. W. Lenz is a cultivated amphidiploid hybrid between *T. laxa* and *T. peduncularis* (L. W. Lenz 1970).

70. MILLA Cavanilles, Icon. 2: 76, plate 196. 1793 • Mexican star

Dale W. McNeal Jr.

Herbs, perennial, scapose, from corms. **Corm** solitary, coated with membrane. **Leaves** 2–7, basal; blade linear, channeled to terete; veins sometimes scabrous, particularly proximally. **Scapes** 1 or rarely 2, persistent, terete, sometimes scabrous, particularly proximally. **Inflorescences** falsely umbellate, 1–30-flowered, subtended by 4 narrowly triangular bracts that do not enclose flowers in bud. **Flowers** sessile but appearing pedicellate, 4–15 cm; perianth salverform; tepals 6, connate basally into long, slender tube; perianth lobes white with green, pink, or blue abaxial stripes, shorter than tube; stamens 6, inserted on perianth tube, distinct; ovary superior, 3-locular, long-stipitate, stipe adnate to perianth tube on 3 angles; style usually exserted; stigma obscurely 3-lobed; pseudopedicel formed by elongate perianth tube ± enlarged around ovary. **Fruits** capsular, beaked by persistent style base, dehiscence loculicidal. **Seeds** numerous, black, flattened, minutely cellularly reticulate.

Species 6–7 (1 in the flora): sw United States, Mexico, Central America (Guatemala).

SELECTED REFERENCE Moore, H. E. Jr. 1953. The genus *Milla* (Amaryllidaceae–Allieae) and its allies. Gentes Herb. 8: 262–294.

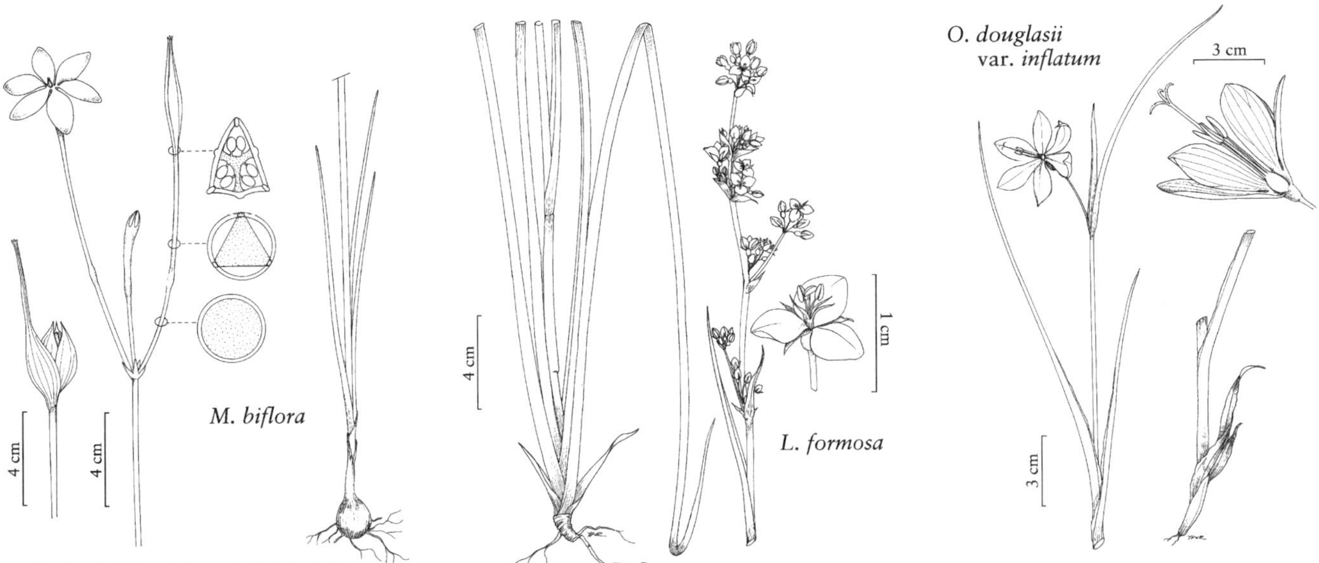

MILLA ∘ LIBERTIA ∘ OLSYNIUM

1. **Milla biflora** Cavanilles, Icon. 2: 76, plate 196. 1793 [F]

Askolame biflora (Cavanilles) Rafinesque

Corm ovoid, 1–2 cm; coat brown or reddish brown, minutely striate, splitting from base into narrow strips that shred and appear fibrous. **Leaves** persistent, 2–7 (–10), 1 mm wide, 1/2 as long as to equaling scape; blade channeled. **Scape** 4–55 cm, scabrous on proximal veins. **Inflorescences** 1–9-flowered, if solitary, one or more undeveloped flower buds often present; bracts 4 in 2 whorls, 5–12 mm, apex acute. **Flowers** 4.5–18 cm (appearing 2.5–4 cm due to pseudopedicel); perianth tube with lacuna between stipe angles to within 0.5–5 cm of base, then tube and stipe completely fused into pseudopedicel; perianth lobes spreading, white with green abaxial stripe, 3–5-veined, elliptic, becoming papery and persisting in fruit, eventually split by developing capsule, 1.5–2.5 cm; outer lobes 5–10 mm wide, apex subacute, inner lobes 8–12 mm wide, apex rounded; filaments triangular, 1 mm; anthers yellow, lanceolate, sagittate at base, 3–5 mm, suture margins minutely denticulate, ± crisped; ovary proximally adnate to perianth tube, ovoid to obovoid, 1 cm, stipe 3–16 cm; style exserted; stigma capitate, minutely 3-lobed. **Capsules** ovoid, 1.5–2 cm; beak 1.5–2.5 mm.

Flowering mid Aug–Sep. Most volcanic soils, dry hillsides, ridges; 1000–2700 m; Ariz., N.Mex.; Mexico; Central America (Guatemala).

Persistent reports of the occurrence of *Milla biflora* in trans-Pecos Texas are due to a collection made by Charles Wright "On the San Pedro, West Texas" in 1851–1852. Wright was not in west Texas during the blooming period of *M. biflora* in either year. This collection most likely came from along the San Pedro River below Benson, Cochise County, Arizona.

223. IRIDACEAE Jussieu
• Iris Family

Peter Goldblatt

Herbs, perennial, rarely annual [or shrubs with woody caudex], evergreen or seasonal, sometimes cespitose; rootstock a rhizome, bulb, or corm. **Flowering stems** aerial (or subterranean in *Romulea*), simple or branched, terete or variously compressed, angled or winged. **Leaves** basal and cauline, distichous; proximal 2–3 sometimes membranous, not reaching much above ground; others with open or closed sheaths, usually unifacial [bifacial or terete], oriented edgewise to the stem; blade parallel-veined, plane or pleated, channeled. **Inflorescences** umbellate, monochasial cymes (rhipidia), spikes, or solitary flowers; rhipidia enclosed in 2, opposed, usually large, leafy to dry bracts (spathes); flowers except for the first subtended by 1 floral bract; spike flowers each subtended by 2, opposed bracts. **Flowers** usually pedicellate [± sessile]; perianth actinomorphic or zygomorphic, petaloid, with 2 equal or unequal whorls of 3 tepals each [1 whorl of 6]; tepals usually large, showy, distinct or connate in tube; stamens 3 [2], inserted at base of outer tepals or in tube, symmetrically arranged or unilateral, arcuate [declinate]; filaments distinct or partly to completely connate, sometimes weak, unable to support anthers; anthers with 2 pollen sacs, extrorse, occasionally latrorse, usually dehiscing longitudinally [rarely apically]; ovary inferior [superior in Tasmanian *Isophysis*], 3-locular [1-locular]; placentation axile [parietal]; ovules 2–few, anatropous; style single, filiform at least proximally, usually 3-branched or 3-lobed, branches either filiform, distally expanded, sometimes each divided in distal 1/2, stigmatic toward apices, or branches thickened, or flattened, petaloid, stigmas then abaxial below apices. **Fruits** capsular, loculicidal, rarely indehiscent, firm to cartilaginous, occasionally woody. **Seeds** globose to angular (prismatic) or discoid, sometimes broadly winged; seed coat usually dry (rarely fleshy); endosperm hard, with reserves of hemicellulose, oil, and protein; embryo small.

Genera ca. 65, species ca. 1810 (16 genera, 92 species; 9 genera, 19 species introduced, and various hybrid complexes in the flora): nearly worldwide but rare in tropical lowlands, best represented in southern Africa.

Iridaceae are currently divided into four subfamilies (P. Goldblatt 1990, 1991). Subfamily Isophysidoideae Takhtajan is monotypic, comprising the Tasmanian *Isophysis* T. Moore with a superior ovary. Only subfamily Iridoideae is native in North America. The remaining

Nivenioideae Schulze ex Goldblatt and Ixioideae Klatt are centered in Africa south of the Sahara. Iridaceae are of considerable economic importance in ornamental horticulture and the cut-flower industry, especially *Iris*, *Gladiolus*, and *Freesia*. Several other genera (e.g., *Crocus*, *Dietes*, *Sparaxis*, *Tritonia*, *Watsonia*) are cultivated in gardens in both tropical and temperate areas. *Moraea* and *Homeria* are poisonous and pose significant problems in cattle- and sheep-raising areas, notably in southern Africa.

In addition to the several genera and species escaped from cultivation and dealt with in detail below, the following are widely grown in areas of mild winter and may persist in and near abandoned gardens, sometimes reproducing successfully: *Dietes* Salisbury [*D. iridioides* (Linnaeus) Salisbury ex Klatt, *D. grandiflora* N. E. Brown]; *Ixia* Linnaeus (*I. maculata* Linnaeus, *I. polystachya* Linnaeus); *Crocus* cultivars and even some wild species.

SELECTED REFERENCES Arber, A. 1921. The leaf structure of the Iridaceae. Ann. Bot. (Oxford) 35: 301–336. Goldblatt, P. 1975. Revision of the bulbous Iridaceae of North America. Brittonia 27: 373–385. Goldblatt, P. 1990. Phylogeny and classification of Iridaceae. Ann. Missouri Bot. Gard. 77: 607–627. Goldblatt, P. 1991. An overview of the systematics, phylogeny and biology of the southern African Iridaceae. Contr. Bolus Herb. 13: 1–74. Reeves, G. et al. 2001. Molecular systematics of Iridaceae: Evidence from four plastid DNA regions. Amer. J. Bot. 88: 2074–2087.

1. Inflorescences spicate, 2–many-flowered, or flowers solitary and sessile, each subtended by pair of opposed bracts; rootstock a corm; flowers actinomorphic or zygomorphic; tepals tubular-connate.
    2. Flowers solitary on aerial axes; leaves all inserted below ground; blade filiform, oval to terete in cross section, with 4 longitudinal grooves . . . . 15. *Romulea*, p. 407
    2. Flowers in 2–many-flowered spikes; leaves some at least inserted above ground level; blade usually expanded, plane (rarely linear, cruciform in cross section).
        3. Flowers actinomorphic, upright; bracts scarious, translucent, with brown streaking; perianth tube shorter than limb . . . . 13. *Sparaxis*, p. 404
        3. Flowers zygomorphic, upright or facing side; bracts not scarious; perianth tube shorter or longer than limb.
            4. Style branches each divided for about 1/2 its length, filiform throughout.
                5. Spikes inflexed, ± horizontal, flowers secund; seeds globose, without wings . . . . 14. *Freesia*, p. 405
                5. Spikes erect, flowers distichous; seeds angular, 1–2-winged . . . . 10. *Watsonia*, p. 401
            4. Style branches each simple or apically notched, filiform throughout or expanded terminally.
                6. Tepals variously colored; style branches expanded terminally, not filiform throughout; seeds usually broadly winged, globose or angular . . . . 16. *Gladiolus*, p. 407
                6. Tepals shades of orange to reddish; style branches apically notched, not expanded above; seeds not winged, globose.
                    7. Perianth tube obliquely funnel-shaped; tepals subequal (dorsal only slightly larger) . . . . 11. *Crocosmia*, p. 402
                    7. Perianth tube narrow and cylindric proximally, abruptly widened and broadly cylindric distally; dorsal tepal ± 2 times others . . . . 12. *Chasmanthe*, p. 403
1. Inflorescences rhipidiate, (1–)2–several-flowered; spathes 2, paired, opposed, often leafy; rootstock a rhizome, bulb, or indistinct; flowers actinomorphic, tepals distinct or tubular-connate.
    8. Leaves plicate, occasionally so narrow that pleats are obscure; rootstock a bulb with brown, dry tunics.
        9. Tepals unequal, inner whorl less than 1/2 outer . . . . 6. *Herbertia*, p. 396
        9. Tepals subequal or inner whorl only slightly smaller than outer.
            10. Style eccentric and recurving; style branches fairly short, undivided, ascending . . . . 9. *Calydorea*, p. 400
            10. Style straight, central; style branches each deeply divided into filiform arms.
                11. Style branches divided for 1/2 their length, rarely more; style arms arching over or between anthers; anthers pandurate . . . . 7. *Alophia*, p. 397
                11. Style branches divided almost to base; style arms extending horizontally between anthers; anthers linear-oblong . . . . 8. *Nemastylis*, p. 398

8. Leaves plane; rootstock a rhizome or indistinct.
    12. Style branches broad, petaloid, terminating in paired crests; anthers appressed to style branches .... 4. *Iris*, p. 371
    12. Style branches not broad or petaloid; anthers not appressed to style branches.
        13. Filaments distinct; style branches flattened, relatively short, not extending between stamens .... 5. *Belamcanda*, p. 395
        13. Filaments partly connate; style branches filiform, relatively short or long, extending between stamens.
            14. Tepals white or yellow, spreading from base; style branches relatively long, extending between stamens below anthers.
                15. Tepals white, inner exceeding outer .... 1. *Libertia*, p. 350
                15. Tepals yellow, inner ± equaling to slightly smaller than outer .... 3. *Sisyrinchium* (in part), p. 351
            14. Tepals blue, purple, pink, or white; style branches relatively short, extending between anthers.
                16. Leaves centric, somewhat inflated; stem simple; seeds brown, angular .... 2. *Olsynium*, p. 350
                16. Leaves plane; stem simple or branched; seeds blackish, globose, with broad concave indentation in chalazal area .... 3. *Sisyrinchium* (in part), p. 351

1. LIBERTIA Sprengel, Syst. Veg. 1: 127. 1824, name conserved • [for Marie A. Libert, 1782–1863, Belgian botanist and student of liverworts] [I]

Elizabeth McClintock

Herbs, perennial, from creeping rhizomes. **Stems** branched or rarely simple. **Leaves** mostly basal; blade flat, ensiform. **Inflorescences** rhipidiate; spathes green. **Flowers** lasting one day, actinomorphic; tepals spreading, distinct, white [blue], unequal, outer whorl smaller than inner; filaments shortly connate basally; anthers not appressed to style branches; style dividing into 3 filiform branches, extending between stamens below anthers. **Capsules** ± ovoid, cartilaginous to woody. **Seeds** many, angular; seed coat light to dark brown. $x = 19$.

Species ca. 11 (1 in the flora): introduced; Australasia, South America (Andes).

1. Libertia formosa Graham, Edinburgh New Philos. J. 15: 383. 1833 [F] [I]

Stems 60–125 cm. **Leaf blades** linear, 15–45 cm × 5–15 mm. **Rhipidia** fewer than 10-flowered. **Tepals** white; outer suberect, flushed greenish or brownish abaxially, lanceolate, less than 10 mm; inner spreading, obovate, 12–18 mm; pedicel 6–12 mm. **Capsules** round to oblong, ca. 1 mm wide. **Seeds** ca. 25.

Flowering spring. Disturbed, urban, coastal areas; 0–50 m; introduced; Calif.; South America (Chile).

Libertia formosa is cultivated as an ornamental and is an uncommon naturalized escape in California.

2. OLSYNIUM Rafinesque, New Fl. 1: 72. 1836 • Grass widow [said by Rafinesque to mean hardly united, alluding to the stamens]

Anita F. Cholewa

Douglass M. Henderson†

Herbs, perennial, strongly cespitose, rootstock indistinct. **Stems** simple, terete. **Leaves** 2–4, cauline, alternate, reduced basally; blade basally clasping or terete. **Inflorescences** rhipidiate, 1–5-flowered; spathes 2, green, leaflike, unequal, apex acute or, rarely, obtuse. **Flowers** not fragrant, actinomorphic; tepals distinct, reddish purple to magenta, occasionally pink or white, not clawed, ± equal; stamens symmetrically arranged; filaments connate to 1/2 their length; anthers parallel, not appressed to style branches; style branches 3, filiform, short, extending between anthers. **Capsules** globose to obovoid or broadly turbinate, ± smooth to distended around seeds, apex broadly rounded, basal 0.2 mm of styles sometimes persistent. **Seeds** many, obconic, angular; seed coat brown, coarsely rugulose. $x = 32$.

Species 12 (1 in the flora): North America, South America.

SELECTED REFERENCES Goldblatt, P., P. J. Rudall, and J. E. Henrich. 1990. The genera of the *Sisyrinchium* alliance (Iridaceae: Iridoideae): Phylogeny and relationships. Syst. Bot. 15: 497–510. Henderson, D. M. 1972. A Biosystematic Study of the Genus *Sisyrinchium* (Iridaceae) in the Pacific Northwest. Ph.D. dissertation. University of Washington.

1. Olsynium douglasii (A. Dietrich) E. P. Bicknell, Bull. Torrey Bot. Club 27: 237. 1900 [E] [F]

Sisyrinchium douglasii A. Dietrich, Sp. Pl. 2: 504. 1833, based on *S. grandiflorum* Douglas ex Lindley, Edwards's Bot. Reg. 16: plate 1364. 1830, not Cavanilles 1790

Stems to 4.1 dm × 1–4.6 mm. **Leaf blades** 1.5–4.5 mm wide distal to midlength. **Spathes** 4.3–7 mm wide, outer 2–6.2 cm; hyaline margins 0.2–0.5(–1.6) mm wide, extending to just proximal to, or occasionally exceeding, green apex. **Tepals** 12.5–33 mm; filaments 6–12.2 mm; anthers 3–10.2 mm; ovary 2–5.6 mm; pedicel 2–3.8 cm. **Capsules** 4.5–11.3 mm. $2n = 64$.

Varieties 2 (2 in the flora): w North America.

1. Filament column abruptly and broadly flared at base . . . . 1a. *Olsynium douglasii* var. *inflatum*
1. Filament column slightly or not flared . . . . 1b. *Olsynium douglasii* var. *douglasii*

1a. Olsynium douglasii (A. Dietrich) E. P. Bicknell var. **inflatum** (Suksdorf) Cholewa & Douglass M. Henderson, Madroño 38: 232. 1991 [E] [F]

Olsynium inflatum Suksdorf, Werdenda 1: 8. 1923

Filament columns abruptly and broadly flared at base.

Flowering early spring–early summer. Open, often rocky areas moist in spring, coniferous forest, sagebrush, margins of lakes; 80–2400 m; B.C.; Idaho, Nev., Oreg., Utah, Wash.

1b. Olsynium douglasii (A. Dietrich) E. P. Bicknell var. **douglasii** [E]

Filament columns tapering evenly to base or slightly flared.

Flowering late winter–early summer. Open, often rocky areas moist in spring, oak woodlands, coniferous forests, sagebrush, lake margins; 0–2200 m; B.C.; Calif., Oreg., Wash.

3. SISYRINCHIUM Linnaeus, Sp. Pl. 2: 954. 1753; Gen. Pl. ed. 5, 409. 1754 • Blue-eyed grass [Greek *sys*, pig, and *rynchos*, snout, alluding to swine grubbing the roots for food]

Anita F. Cholewa

Douglass M. Henderson†

Herbs, annual or perennial, often cespitose, rhizomatous (sometimes only obscurely, especially when cespitose) or not, sometimes with thickened, fleshy roots. **Stems** scapelike or branched, compressed, and 2-winged. **Leaves** 2–6, basal or basal and cauline, alternate, basally equitant; blade plane, ensiform, usually glabrous. **Inflorescences** rhipidiate, usually terminal (basal flowers occasionally produced in some montane populations), 1–11(–15)-flowered; spathes 2, opposed, green or with purplish tinge, equitant, equal or unequal, smooth to scabrous, margins hyaline, apex undifferentiated, acute to obtuse or bifid, margins of outer spathe usually connate basally. **Flowers** not fragrant, actinomorphic; tepals widely spreading to reflexed (*S. minus* and *S. rosulatum* with campanulate bases), ± distinct, bluish violet to light blue, white, lavender to pink, magenta, purple, or yellow, not clawed, subequal; stamens symmetrically arranged; filaments distinct, connate basally or into tube, tapering evenly to apex (basally inflated in *S. rosulatum*); anthers parallel, surrounding but not appressed to style branches; styles 3, erect, connate at least basally, filiform, not broad and petaloid, long, extending between stamens usually beyond anthers. **Fruits** capsular, ± globose, smooth to roughened by underlying seeds, apex usually rounded. **Seeds** many, globose to obconic or hemispheric; seed coat black, granular to rugulose. $x = 8$.

Species ca. 80 (37 in the flora): New World, Hawaii; 1 species in New Zealand and 1 in Great Britain probably naturalized.

Sisyrinchium is a complex polyploid taxon in which the species are not always easily distinguished. When immature, plants of branched species appear to be simple-stemmed (internodes do not elongate until just before anthesis) and those of usually simple-stemmed species occasionally are branched. White flowers may occur in otherwise blue-flowered species, and vivipary occasionally occurs. Furthermore, vegetative characteristics, while distinctive in some species, may overlap greatly in wide-ranging species. Writers of past floras sometimes were unaware of such phenotypic plasticity, or were inconsistent in their use of terminology (spathes, for instance, have variously been called leaves or valves). Some taxonomists have thought differences too subtle and chosen to lump species (e.g., *S. angustifolium* and *S. montanum* have been considered synonymous by several authors).

Flowers and spathes provide the best characters but require fresh material or extreme care during pressing. Many older herbarium collections have been misidentified due to improper collecting/pressing techniques that obscured critical characters. Additionally, variability within populations makes accurate identification based on only one individual nearly impossible.

Chromosome counts can be useful in differentiating species but consistent numbers seem to be obtained only with meiotic counts. Some of the unusual counts reported in the literature were obtained with mitotic material or were from misidentified plants. A definite ploidal series seems to be evident, however, at least among the blue-flowered species. Limited breeding studies have shown that ploidal levels are strong barriers to reproduction (A. F. Cholewa and D. M. Henderson 1984; D. M. Henderson 1976). There is some indication, however, that true hybrids may exist (D. B. Ward 1959; D. S. Correll and M. C. Johnston 1970; G. Ajilvsgi 1984). Much more work, especially with molecular data and cladistic analyses, needs to be done to understand species relationships.

In eastern Texas (especially) and adjacent states, intermediates between *Sisyrinchium ensigerum*, *S. pruinosum*, *S. langloisii*, and perhaps *S. sagittiferum* can be found. These have been referred to *S. texanum* E. P. Bicknell (D. S. Correll and M. C. Johnston 1970), and Bicknell's original description of that entity does mention a high degree of variability. Many of the specimens that we examined that were previously labeled *S. texanum* exhibit character states intermediate between those of *S. langloisii* and *S. pruinosum* or *S. ensigerum*. The types of *S. texanum* that we examined (*Bray s.n.*, NY; *Bush 32*, NY; *Hall 636*, NY) would key to *S. langloisii* or *S. pruinosum* but exhibit several intermediate character states. Further, although K. L. Hornberger (1987, 1991) and D. S. Correll and M. C. Johnston (1970) considered *S. texanum* synonymous with *S. sagittiferum* (which they described as a branched species), E. P. Bicknell (1899, 1901) clearly indicated that the latter is unbranched, and thus it probably does not play a role in *S. texanum*. Much more work is needed to resolve the proper disposition of *S. texanum* and the true nature of this complex of southern species.

Characters critical for distinguishing species of *Sisyrinchium* are often found in floral material, requiring extra care in collecting and pressing. Accurate identification requires examining more than one individual in order to discount uncommon or atypical character states (generally not accounted for in this key). In branched individuals, leaves will be associated with one or more branches or peduncles that are narrower than the primary stem. Peduncles are morphologically indistinguishable from branches and are, hereafter, referred to simply as branches.

SELECTED REFERENCES Cholewa, A. F. and D. M. Henderson. 1984. Biosystematics of *Sisyrinchium* section *Bermudiana* (Iridaceae) of the Rocky Mountains. Brittonia 36: 342–363. Goldblatt, P., P. J. Rudall, and J. E. Henrich. 1990. The genera of the *Sisyrinchium* alliance (Iridaceae: Iridoideae): Phylogeny and relationships. Syst. Bot. 15: 497–510. Henderson, D. M. 1972. A Biosystematic Study of the Genus *Sisyrinchium* (Iridaceae) in the Pacific Northwest. Ph.D. dissertation. University of Washington. Henderson, D. M. 1976. A biosystematic study of Pacific Northwestern blue-eyed grasses (*Sisyrinchium*, Iridaceae). Brittonia 28: 149–176. Hornberger, K. L. 1987. Systematics of the Genus *Sisyrinchium* (Iridaceae) in the Southeastern United States. Ph.D. dissertation. University of Arkansas. Hornberger, K. L. 1991. The blue-eyed-grasses (*Sisyrinchium*: Iridaceae) of Arkansas. Sida 14: 597–604. Rudall, P. J., A. Y. Kenton, and T. J. Lawrence. 1986. An anatomical and chromosomal investigation of *Sisyrinchium* and allied genera. Bot. Gaz. 147: 466–477. Ward, D. B. 1968. The nomenclature of *Sisyrinchium bermudiana* and related North America species. Taxon 17: 270–276.

1. Tepals yellow to orange; seeds usually hemispherical with depression on flattened side, or globose with or without obvious depression; filaments distinct or connate only basally.
    2. Stems 2–8 mm wide; capsules 6–19 mm; tepals usually 12+ mm.
        3. Stems simple; seeds 0.7–1.5 mm, hemispherical with shallow depression . . . . 1. *Sisyrinchium californicum*
        3. Stems branched; seeds 1.5–2.2 mm, globose, lacking depression . . . . 2. *Sisyrinchium arizonicum*
    2. Stems 0.5–2(–2.3) mm wide; capsules 3–9 mm; tepals usually less than 12 mm.
        4. Plants annual; tepals 2.6–5 mm; seeds 0.9–1.1 mm, with deep depression . . . . 5. *Sisyrinchium cernuum*
        4. Plants perennial; tepals 7.5–11 mm; seeds 1–1.7 mm, with shallow depression.
            5. Fruiting pedicels recurved to spreading; capsules globose or turbinate to broadly fusiform . . . . 3. *Sisyrinchium elmeri*
            5. Fruiting pedicels ascending to erect; capsules slightly turbinate to ± globose . . . . 4. *Sisyrinchium longipes*
1. Tepals purple to light blue, white, lavender, pink, or magenta, occasionally yellow; seeds globose to obconic, depression absent or occasionally slight; filaments connate ± full length (only basally in *S. rosulatum*).
    6. Perianth campanulate basally, flaring distally; plants annual or occasionally short-lived perennial; tepals lavender, pink, magenta, yellow, or occasionally white.
        7. Stems with 3–6 nodes; filaments connate ± full length, tapering evenly, stipitate-glandular on proximal 0.8–1.2 mm; capsule turbinate to broadly fusiform, uniformly light brown . . . . 6. *Sisyrinchium minus*
        7. Stems with 1–2(–3) nodes; filaments connate basally or occasionally to 1/2 length, inflated basally, stipitate-glandular on proximal 0.5–0.8 mm; capsule ± globose, bicolored, sutures and sometimes apex purplish, intrasuture surface tan . . . . 7. *Sisyrinchium rosulatum*
    6. Perianth widely flaring or reflexed from base, not campanulate; plants perennial; tepals purple to violet, blue, or white.
        8. Rhipidia usually paired, rarely borne singly, sessile or rarely outer with branch to 7 mm; rhipidia and associated spathes closely subtended by enveloping bractlike leaf (because the 2 rhipidia are nearly enveloped by the leaf, these species can be mistaken as having a single rhipidium).
            9. Stems 1.5–3.4 mm wide, obviously winged, not wiry . . . . 8. *Sisyrinchium albidum*
            9. Stems 0.5–1 mm wide, hardly or not winged, wiry . . . . 9. *Sisyrinchium capillare*
        8. Rhipidia borne singly, terminating stem or branch; rhipidium and associated spathes not subtended by leaf.
            10. Stems branched or populations predominantly with branched individuals (see discussion).
                11. Stems with 2–5 nodes; spathes ± equaling branches in width . . . 10. *Sisyrinchium dichotomum*
                11. Stems with 1–3 nodes; spathes much wider than branches.

[12. Shifted to left margin.—Ed.]

12. Main stems usually more than 2 mm wide.
    13. First internode equaling or shorter than longest leaf, if equaling leaves then apex of hyaline margins of inner spathe acute to acuminate.
        14. Stems and leaves glabrous; hyaline margins of inner spathe broadly acute to obtuse or truncate apically, sometimes projecting as lobes . . . . 11. *Sisyrinchium ensigerum*
        14. Stems and leaves usually scabrous, at least apically on margins; hyaline margins of inner spathe tapering to acute to acuminate apex, never projecting as lobes.
            15. Leaf bases persistent as fibrous tufts; outer spathe 0.5–1.8 mm longer than inner . . . . 12. *Sisyrinchium xerophyllum*
            15. Leaf bases not persistent or fibrous; outer spathe 2.5–5.5 mm longer than inner . . . . 13. *Sisyrinchium pruinosum* (in part)
    13. First internode equaling or longer than longest leaf, if equaling leaves then hyaline margins of inner spathe broadly obtuse or acute apically.
        16. Leaf bases persistent as fibrous tufts . . . . 14. *Sisyrinchium nashii*
        16. Leaf bases not persistent or fibrous.
            17. Hyaline margins of inner spathe 0.1–0.3 mm wide; e of the Rocky Mountains.
                18. Capsules dark brown or black; plants usually drying dark olive green, bronze, or blackish . . . . 15. *Sisyrinchium angustifolium*
                18. Capsules tan to nearly white; plants drying light green to yellowish green . . . . 16. *Sisyrinchium strictum* (in part)
            17. Hyaline margins of inner spathe (0.1–)0.4–1.1 mm wide; w of the Rocky Mountains.
                19. Hyaline margins of inner spathe usually equaling or longer than green apex; foliage strongly glaucous; tepals pale blue . . . . 17. *Sisyrinchium funereum* (in part)
                19. Hyaline margins of inner spathe usually shorter than green apex; foliage various but usually not strongly glaucous; tepals usually bluish violet to violet-purple.
                    20. Outer tepals elliptic to oblanceolate; stem margins white or translucent cartilaginous; filament column usually stipitate-glandular . . . . 18. *Sisyrinchium radicatum* (in part)
                    20. Outer tepals often broadly cuneate; stem margins same color and texture as stem body; filament column usually not stipitate-glandular . . . . 19. *Sisyrinchium bellum* (in part)
12. Main stems usually 2 mm or less wide.
    21. Margins of outer spathe connate basally (1.5–)2–4 mm (–5 mm in *S. langloisii*).
        22. Outer spathe 2.5–5.5 mm longer than inner; spathes usually scabrous . . . . 13. *Sisyrinchium pruinosum* (in part)
        22. Outer spathe shorter or no more than 2.7 mm longer than inner; spathe usually glabrous.
            23. Leaf bases persistent as thick, fibrous tufts . . . . 20. *Sisyrinchium fuscatum*
            23. Leaf bases not persistent as fibrous tufts.
                24. Ovary and capsule black, contrasting with much lighter foliage; hyaline margins of inner spathe broadly rounded or truncate apically . . . . 21. *Sisyrinchium atlanticum* (in part)
                24. Ovary and capsule not contrasting with foliage; hyaline margins of inner spathe usually acuminate to acute or sometimes obtuse apically . . . . 22. *Sisyrinchium langloisii*
    21. Margins of outer spathe connate basally 3.5–9 mm.
        26. Ovaries and capsules black, contrasting with much lighter foliage . . . . 21. *Sisyrinchium atlanticum* (in part)
        26. Ovaries and capsules not contrasting with foliage.
            27. Hyaline margins of inner spathe broadly rounded or truncate apically.
                28. Hyaline margins of inner spathe usually equaling or longer than green apex; foliage strongly glaucous; tepals pale blue . . . . 17. *Sisyrinchium funereum* (in part)

28. Hyaline margins of inner spathe usually shorter than green apex; foliage various, not strongly glaucous; tepals bluish violet to violet-purple.
    29. Outer tepals elliptic to oblanceolate; stem margins white or translucent-cartilaginous; filament column usually stipitate-glandular . . . . 18. *Sisyrinchium radicatum* (in part)
    29. Outer tepals often broadly cuneate; stem margins similar to stem body; filament column usually not stipitate-glandular . . . . . . 19. *Sisyrinchium bellum* (in part)

27. Hyaline margins of inner spathe usually acute or acuminate apically.
    30. First internode equaling or shorter than leaves.
        31. Plants drying brownish olive to bronze; rhizomes obvious . . 23. *Sisyrinchium miamiense*
        31. Plants drying yellowish green to olive, occasionally bronze; rhizomes scarcely discernable . . . . 24. *Sisyrinchium biforme* (in part)
    30. First internode longer than leaves.
        32. Capsules 4–7.5+; sw United States . . . . 25. *Sisyrinchium demissum*
        32. Capsules 3.5–4.9 mm; Great Lakes region . . . . 16. *Sisyrinchium strictum* (in part)

[10. Shifted to left margin.—Ed.]

10. Stems simple or populations predominantly with simple-stemmed individuals.
    33. Margins of outer spathe distinct basally or rarely connate to 1 mm, and then stems greater than 1.5 mm wide, glabrous or scabrous . . . . 26. *Sisyrinchium campestre*
    33. Margins of outer spathe obviously connate basally 1 mm or more (rarely connate less than 1 mm and then spathes tinged strongly purple), usually glabrous or rarely slightly scabrous.
        34. Leaf bases persisting in fibrous tufts . . . . 35. *Sisyrinchium sagittiferum*
        34. Leaf bases not persisting in fibrous tufts.
            35. Rhizomes obvious and elongate; tepals purple to dark reddish purple or very rarely bluish violet, base purple . . . . 30. *Sisyrinchium hitchcockii*
            35. Rhizomes scarcely discernable; tepals bluish violet, occasionally purple or purplish blue, to light blue or white, base yellow.
                36. Spathes subequal, or outer up to 4.5 mm longer than inner.
                    37. Capsules 2.2–4 mm; hyaline margins of inner spathe broadly rounded to truncate apically, often extending as 2 lobes beyond green apex; w United States. . . . . 31. *Sisyrinchium halophilum*
                    37. Capsules 5–7 mm; hyaline margins of inner spathe acute to acuminate or occasionally lobed and slightly exceeding green apex; s United States and n Mexico . . . . 24. *Sisyrinchium biforme* (in part)
                36. Spathes distinctly unequal, outer at least 6 mm longer than inner.
                    38. Tepals pale blue; outer tepals usually rounded apically to slightly emarginate.
                        39. Outer tepals greater than 10 mm; Columbia River Gorge . . . . 33. *Sisyrinchium sarmentosum*
                        39. Outer tepals 10 mm or less; elsewhere.
                            40. Outer spathes usually connate more than 2.6 mm; outer tepals slightly emarginate to rounded apically; Colorado, Wyoming . . . . 28. *Sisyrinchium pallidum*
                            40. Outer spathes connate less than 2.6 mm; outer tepals rounded apically, never emarginate; elsewhere.
                                41. Capsules dark brown to black; Greenland . . . . 36. *Sisyrinchium groenlandicum*
                                41. Capsules beige to light brown; w Canada, Washington . . . . 37. *Sisyrinchium septentrionale*
                    38. Tepals blue-violet to purplish blue, occasionally purple; outer tepals emarginate, truncate or occasionally rounded apically.
                        42. Stems with very narrow or scarcely discernable wings, almost wiry, 0.9–2 mm wide . . . . 29. *Sisyrinchium mucronatum*
                        42. Stems obviously winged, 1–3.7 mm wide.

43. Capsules 6.2–8.1 mm; outer tepals usually rounded to slightly truncate apically . . . . 32. *Sisyrinchium littorale*
43. Capsules 3–6.8 mm; outer tepals usually emarginate or truncate apically (rarely rounded and capsules shorter than 6 mm).
    44. Outer spathes usually at least (12–)16 mm longer than inner; stems usually (1.5–)2–3.7 mm wide; keel of inner spathe ± gibbous basally; seed coat rugulose . . 27. *Sisyrinchium montanum*
    44. Outer spathes no more than 16 mm longer than inner; stems 1–2.5 mm wide; keel of inner spathe not gibbous; seed coat usually granular . . . . 34. *Sisyrinchium idahoense*

1. Sisyrinchium californicum (Ker Gawler) Dryander in W. Aiton and W. T. Aiton, Hortus Kew. 4: 135. 1812 [E] [F]

Marica californica Ker Gawler, Bot. Mag. 25: plate 983. 1807; *Hydastylus borealis* E. P. Bicknell; *H. brachypus* E. P. Bicknell; *H. californicus* (Ker Gawler) Salisbury; *Sisyrinchium boreale* (E. P. Bicknell) J. K. Henry; *S. brachypus* (E. P. Bicknell) J. K. Henry; *S. flavidum* Kellogg; *S. lineatum* Torrey

Herbs, perennial, cespitose, dark brown to black when dry, to 6.2 dm, not glaucous. **Stems** simple, 2–6.8 mm wide, glabrous, margins entire, similar in color and texture to stem body. **Leaf blades** glabrous, bases not persistent in fibrous tufts. **Inflorescences** borne singly; spathes mostly greenish, glabrous, keels entire; outer 13–53 mm, 2.2–9 mm longer than inner, tapering evenly towards apex, margins basally connate 3–8 mm; inner with keel straight to evenly curved, hyaline margins 0.5–1 mm, apex usually rounded, occasionally erose, ending 0.2–1 mm proximal to green apex. **Flowers:** tepals medium to bright yellow with brownish veins; outer tepals 12–18 mm, apex rounded or acute, aristate; filaments ± distinct, glabrous; ovary similar in color to foliage. **Capsules** dark brown to black, broadly fusiform or slightly turbinate, 6–13 mm; pedicel erect to ascending. **Seeds** hemispherical, with shallow depression on flattened side, 0.7–1.5 mm, rugulose. $2n = 32, 34, 36$.

Flowering spring–late summer. Moist areas near coast; 0–700 m; B.C.; Calif., Oreg., Wash.

2. Sisyrinchium arizonicum Rothrock, Bot. Gaz. 2: 125. 1877 [F]

Herbs, perennial, cespitose, dark olive to blackish when dry, to 6 dm, not glaucous; roots conspicuously fleshy-thickened. **Stems** branched, with 2–4 nodes, 3.5–8 mm wide, glabrous to scabrous, margins denticulate at least apically, similar in color and texture to stem body; first internode 14–27 cm, shorter than leaves; distalmost node with 1–2 branches. **Leaf blades** glabrous to scabrous, bases not persistent in fibrous tufts. **Inflorescences** borne singly; spathes green, obviously wider than supporting branch, scabrous or glabrous, keels denticulate; outer 30–55 mm, equaling or to 3.7 mm longer than inner, tapering evenly towards apex, margins basally connate 5–6 mm; inner with keel evenly curved or abruptly gibbous, hyaline margins 0.3–0.8 mm wide, apex acuminate or rounded, ending just proximal to green apex. **Flowers:** tepals yellow to orange, usually with brownish veins; outer tepals 11–23 mm, apex rounded to acute, aristate; filaments connate only basally, base glabrous or occasionally stipitate-glandular; ovary similar in color to foliage. **Capsules** black or blackish brown, obovoid to turbinate, 8–19 mm; pedicel erect to ascending. **Seeds** globose, lacking depression, 1.5–2.2 mm, rugulose. $2n = 34, 36, 72$.

Flowering mid–late summer. Moist meadows, stream banks, open places in coniferous forest; 1600–3000 m; Ariz., N.Mex.; nw Mexico.

3. Sisyrinchium elmeri Greene, Pittonia 2: 106. 1890 [E]

Hydastylus elmeri (Greene) E. P. Bicknell; *H. rivularis* E. P. Bicknell

Herbs, perennial, cespitose, dark green to ashy olive when dry, to 3.2 dm, not glaucous. **Stems** simple, 0.6–1.7 mm wide, glabrous, margins usually entire, similar in color and texture to stem body. **Leaf blades** glabrous, bases not persistent in fibrous tufts. **Inflorescences** borne singly; spathes green, glabrous, keels entire; outer 13–20 mm, 1–2.4 mm longer than inner, tapering evenly towards apex, margins basally connate 3.8–5.5 mm; inner with keel ± straight, hyaline margins 0.2–0.7 mm wide, apex extending as 2 rounded or broadly erose lobes, ending just proximal to or slightly exceeding green apex. **Flowers:** tepals deep yellow to yellowish orange with brownish veins; outer tepals 7.5–8.5 mm, apex obtuse or rounded; filaments ± distinct or connate to 1/2 their length, glabrous; ovary similar in color to foliage. **Capsules** dark brown, globose to turbinate or broadly fusiform, 5–7.9 mm; pedicel recurved to spreading. **Seeds**

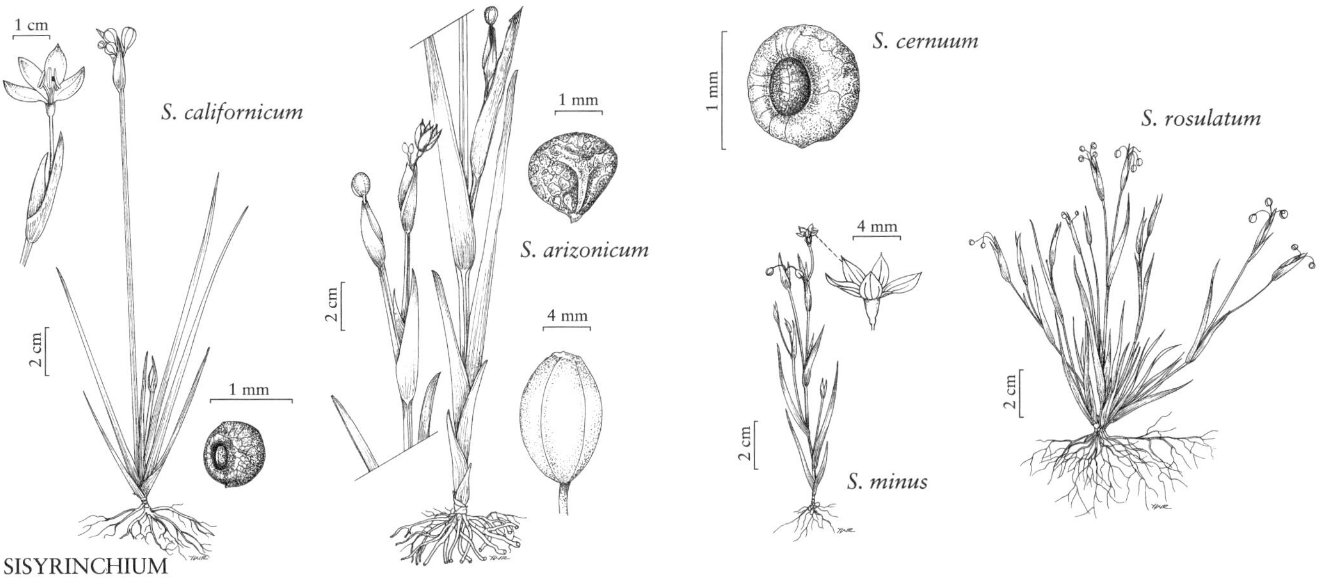

SISYRINCHIUM

hemispherical, with slight depression on one side, 1–1.2 mm, smooth. 2*n* = 34.

Flowering late spring–late summer. Moist montane areas, bogs, edges of marshes, stream banks, meadows; 1100–2700 m; Calif.

4. Sisyrinchium longipes (E. P. Bicknell) Kearney & Peebles, J. Wash. Acad. Sci. 29: 474. 1939 [C] [E]

Hydastylus longipes E. P. Bicknell, Bull. Torrey Bot. Club 27: 382. 1900

Herbs, perennial, cespitose, green to medium olive when dry, to 4.6 dm, not glaucous. **Stems** simple, 0.6–1.7(–2.3) mm wide, glabrous, margins entire, similar in color and texture to stem body. **Leaf blades** glabrous, bases not persistent in fibrous tufts. **Inflorescences** borne singly; spathes green, glabrous, keels entire; outer 12.5–36 mm, 0–16 mm longer than inner, tapering evenly towards apex, margins basally connate 3.5–9 mm; inner with keel evenly curved to straight, hyaline margins 0.4–0.6 mm wide, apex usually extending as 2 rounded, sometimes erose lobes, ending 0–1.5 mm proximal to green apex. **Flowers:** tepals yellow to orange, usually with brownish or occasionally purplish veins; outer tepals 7–11 mm, apex acute to rounded; filaments connate basally, glabrous; ovary similar in color to foliage. **Capsules** dark brown to black, slightly turbinate to ± globose, 4.5–9 mm; pedicel ascending to erect. **Seeds** ± hemispherical, with slight depression on flattened side, 1–1.7 mm, rugulose. 2*n* = 34.

Flowering summer. Wet to moist meadows, stream banks, moist open areas in forests; of conservation concern; 2200–3200 m; Ariz., Calif., N.Mex.

Plants of *Sisyrinchium longipes* that are disjunct in the San Bernadino Mountains, California, were previously identified as *S. elmeri.*

5. Sisyrinchium cernuum (E. P. Bicknell) Kearney, Leafl. W. Bot. 5: 197. 1949 [F]

Hydastylus cernuus E. P. Bicknell, Bull. Torrey Bot. Club 27: 384. 1900

Herbs, annual, cespitose, medium olive when dry, to 2 dm, not glaucous. **Stems** simple, 0.5–2 mm wide, glabrous, margins entire to denticulate apically, similar in color and texture to stem body. **Leaf blades** glabrous, bases not persistent in fibrous tufts. **Inflorescences** borne singly; spathes green, glabrous, keels entire; outer 14–41 mm, 5–12 mm longer than inner, tapering evenly towards apex, margins basally connate 3.3–5 mm; inner with keel evenly curved to straight, hyaline margins 0.1–0.5 mm wide, apex acute to rounded, ending 0–1.1 mm proximal to green apex. **Flowers:** tepals yellow to orange, usually with brownish veins; outer tepals 2.6–5 mm, apex acute to rounded; filaments connate basally, glabrous; ovary similar in color to foliage. **Capsules** light to dark brown, globose, 3–8 mm; pedicel recurved. **Seeds** hemispherical, with deep depression on flattened side, 0.9–1.1 mm, rugulose.

Flowering early spring–late summer. Moist areas, meadows, stream banks, woods; 1000–2400 m; Ariz.; Mexico (Chihuahua, Durango).

6. Sisyrinchium minus Engelmann & A. Gray, Boston J. Nat. Hist. 5: 263. 1845 (as Sysirinchium) F

Sisyrinchium bermudianum Linnaeus var. *minus* (Engelmann & A. Gray) Klatt; *S. filiforme* Rafinesque 1833, not Sprengel 1824; *S. flexuosum* Rafinesque 1840, not (Linnaeus f.) Sprengel 1824; *S. thurowii* J. M. Coulter & Fisher

Herbs, annual, occasionally cespitose, olive green or darker when dry, to 2.6 dm, not glaucous. **Stems** branched, with 3–6 nodes, 0.5–2.1 mm wide, glabrous, margins entire, similar in color and texture to stem body; first internode 1–4 cm, shorter than leaves; distalmost node with 1–2 branches. **Leaf blades** glabrous, bases not persistent in fibrous tufts. **Inflorescences** borne singly; spathes green, wider than supporting branch, glabrous, keels denticulate; outer 10.6–31 mm, 6–11 mm longer than inner, constricted proximal to apex, margins basally connate 2.7–4.7 mm; inner with keel evenly curved to straight, hyaline margins 0.2–0.5 mm wide, apex acute, ending 1.1–5.4 mm proximal to green apex. **Flowers:** perianth flaring distally, campanulate basally; tepals lavender-pink to rosy purple, magenta, or white, very rarely yellow or bluish, bases yellow; outer tepals 5–6.3 mm, apex rounded to acute or slightly emarginate, aristate; filaments connate ± entirely, tapering evenly, basally stipitate-glandular 0.8–1.2 mm; ovary similar in color to foliage. **Capsules** light brown, turbinate to broadly fusiform, 3.8–5.6 mm; pedicel recurved to pendent. **Seeds** ± globose, with slight depression on one side, 0.5–0.8 mm, rugulose. $2n = 10$.

Flowering spring. Moist, sandy areas in prairies and meadows; 0–70 m; La., Miss., Tex.; n Mexico.

Morphologically *Sisyrinchium minus* appears closely related to and may even be conspecific with *S. tinctorium* of Mexico, but further study is needed for resolution.

7. Sisyrinchium rosulatum E. P. Bicknell, Bull. Torrey Bot. Club 26: 228. 1899 E F W2

Sisyrinchium brownei Small; *S. exile* E. P. Bicknell

Herbs, annual or short-lived perennial, cespitose, yellowish green to medium or dark olive green when dry, to 3.6 dm, not glaucous. **Stems** branched or rarely simple, with 1–2(–3) nodes, 0.7–2 mm wide, glabrous, margins usually entire, similar in color and texture to stem body; first internode 2–10 cm, usually shorter than leaves; distalmost node with 1–2 branches. **Leaf blades** glabrous, bases not persistent in fibrous tufts. **Inflorescences** borne singly; spathes green, obviously wider than supporting branch, glabrous, keels usually entire; outer 13–32.2 mm, 2–8.2 mm longer than inner, tapering evenly towards apex or occasionally slightly constricted proximal to apex, margins basally connate 2.3–6.7 mm; inner with keel evenly curved to straight, hyaline margins 0.2–0.3 mm wide, apex acute, occasionally erose, ending 0.3–4.6 mm proximal to green apex. **Flowers:** perianth flaring distally, campanulate basally; tepals maroon or pink to lavender-rose with purple stripes, or yellow with rosy purple bases; outer tepals 5–11 mm, apex acute, rarely aristate; filaments connate basally, occasionally to 1/2 their length, basally inflated and stipitate-glandular 0.5–0.8 mm; ovary similar in color to foliage. **Capsules** tan with purplish sutures and sometimes apex, ± globose, 2.1–4.2 mm; pedicel spreading to recurved. **Seeds** ± globose, sometimes with slight depression, 0.5–1 mm, rugulose. $2n = 32$.

Flowering spring–fall. Roadsides, old fields, and other disturbed areas, stream banks, wet areas bordering woods; 0–80 m; Ala., Ark., Fla., Ga., La., Miss., N.C., S.C., Tex.

Sisyrinchium rosulatum is apparently weedy throughout much of its range.

8. Sisyrinchium albidum Rafinesque, Atlantic J. 1: 17. 1832 E F W2

Sisyrinchium asheianum E. P. Bicknell; *S. bermudianum* Linnaeus var. *albidum* (Rafinesque) A. Gray; *S. floridanum* Rafinesque; *S. hastile* E. P. Bicknell; *S. niveum* Rafinesque; *S. scabrellum* E. P. Bicknell

Herbs, perennial, cespitose, yellowish green to light olive when dry, to 4 dm, not glaucous. **Stems** apparently simple, not wiry, obviously winged, 1.5–3.4 mm wide, glabrous to scabrous, margins entire to denticulate apically, similar in color and texture to stem body. **Leaf blades** usually glabrous, bases not persistent in fibrous tufts. **Inflorescences** paired, closely subtended by long, bractlike leaf that often obscures first (inner) inflorescence, sessile or occasionally second (outer) with branch to 5 mm; spathes green or occasionally with purplish tinge, glabrous or occasionally scabrous, keels usually denticulate; outer spathe of outer rhipidium 14–30 mm, 0.3–3.7 mm longer than inner, outer spathe of inner rhipidium 1–3.5 mm shorter than inner, both tapering evenly towards apex, margins distinct or rarely connate basally to 1 mm; inner with keel straight or somewhat curved, hyaline margins 0.1–0.5 mm wide, apex acute to obtuse or rarely truncate, ending 0.2–2.5 mm proximal to green apex. **Flowers:** tepals white or blue, bases yellow; outer tepals 6–12.5 mm, apex emarginate,

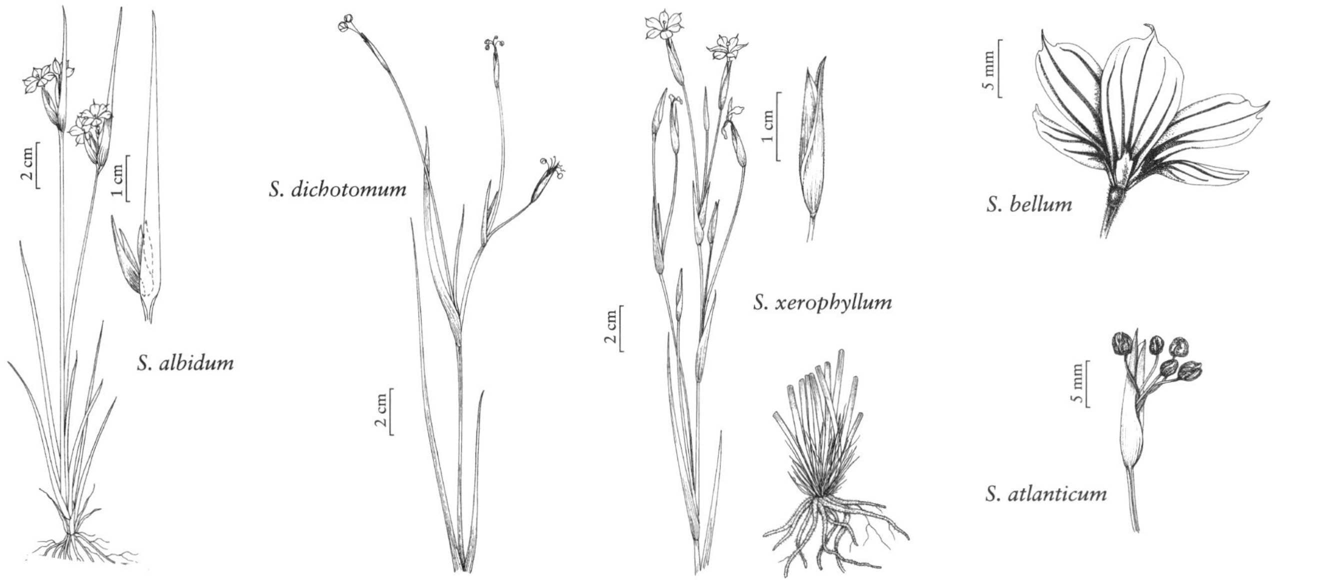

SISYRINCHIUM

aristate; filaments connate ± entirely, slightly stipitate-glandular basally; ovary similar in color to foliage. **Capsules** tan to light brown, ± globose, 2.8–4 mm; pedicel spreading or ascending. **Seeds** globose to obconic, lacking obvious depression, 0.8–1.1 mm, granular to rugulose. $2n = 32$.

Flowering spring–early summer. Roadsides and open slopes, prairies, or rich open woods, often on shallow rocky or sandy soil; 0–600 m; Ont.; Ala., Ark., Fla., Ga., Ill., Ind., Ky., La., Maine, Mich., Miss., Mo., N.Y., N.C., Ohio, Okla., Pa., S.C., Tenn., Tex., Va., Wis.

9. Sisyrinchium capillare E. P. Bicknell, Bull. Torrey Bot. Club 26: 608. 1899 [E]

Herbs, perennial, cespitose, yellowish green to olive when dry, to 4.5 dm, not glaucous. **Stems** apparently simple, wiry, wings present as slight ridges, 0.5–1 mm wide, glabrous, margins minutely denticulate, similar in color and texture to stem body. **Leaf blades** glabrous, bases not persistent in fibrous tufts. **Inflorescences** usually paired, closely subtended by long, bractlike leaf that often obscures first (inner) inflorescence, sessile or second (outer) with branch to 6.6 mm; spathes often purplish, glabrous, keels entire; outer 12.1–14.7 mm, 0.9 mm shorter to 2.7 mm longer than inner, usually tapering evenly towards apex, margins distinct or connate basally to 1.5 mm; inner with keel straight or evenly curved, hyaline margins 0.2–0.4 mm wide, apex acute to occasionally obtuse, ending 0.2–1.4 mm proximal to green apex. **Flowers:** tepals pale blue to bluish violet, bases yellow; outer tepals 7.7–12 mm, apex emarginate, aristate; filaments connate ± entirely, glabrous; ovary similar in color to foliage. **Capsules** tan or light brown, ± globose, 2–3.3 mm, glabrous; pedicel loosely erect or spreading. **Seeds** globose to obconic, lacking obvious depression, 0.8–1.2 mm, rugulose.

Flowering spring. Open woods; 0–200 m; Ala., Ga., N.C., S.C., Va.

Sisyrinchium capillare has, at times, been considered synonymous with *S. albidum* (e.g., D. T. MacRoberts 1984). The very narrow, almost wingless stems and glabrous filaments will help distinguish it from *S. albidum*, which is much more wide-ranging.

10. Sisyrinchium dichotomum E. P. Bicknell, Bull. Torrey Bot. Club 26: 609. 1899 [C] [E] [F]

Herbs, perennial, cespitose, yellowish green to light green or dark olive when dry, to 4 dm, not glaucous. **Stems** branched, with 2–5 nodes, 2.6–3.5 mm wide, glabrous, margins minutely denticulate near nodes, similar in color and texture to stem body; first internode 6–12 cm, shorter than leaves; distalmost node with 2 branches. **Leaf blades** glabrous, bases not persistent in fibrous tufts. **Inflorescences** borne singly; spathes sometimes with purplish tinge, ± equaling supporting branch in width, glabrous, keels slightly denticulate; outer 14–21.5 mm, 3.3–5.3 mm longer than inner, tapering evenly towards apex, margins basally connate 2–3.4 mm; inner with keel straight, hyaline margins to 0.1 mm wide, apex acute to acuminate, ending 0.7–1.7 mm proximal to green apex. **Flowers:** tepals white, bases yellow; outer tepals 3.3–7.5 mm, apex apparently rounded, aristate; filaments connate ± entirely, glabrous; ovary similar in color to foliage. **Capsules** pale to medium brown, ± globose, 2.1–3.4 mm; pedicel erect. **Seeds** globose, lacking obvious depression, 1–3 mm, rugulose. $2n = 32$.

Flowering summer. Dry to moist oak-hickory woods and edges; of conservation concern; 400–1000 m; N.C., S.C.

Sisyrinchium dichotomum is endemic to the Piedmont-Blue Ridge escarpment of the Carolinas, known from only seven populations according to the U.S. Fish and Wildlife Service. The fertility and fecundity of this species are very low. In an average individual, only 63 percent of the pollen is viable, and mature capsules typically have only 1 or 2 seeds (K. L. Hornberger 1987).

11. Sisyrinchium ensigerum E. P. Bicknell, Bull. Torrey Bot. Club 28: 580. 1901

Sisyrinchium amoenum E. P. Bicknell; *S. colubriferum* E. P. Bicknell

Herbs, perennial, cespitose, olive green to yellowish bronze when dry, to 3.2 dm, occasionally glaucous. **Stems** branched, with 1–2 nodes, 1.9–4.8 mm wide, glabrous, margins variously serrulate, denticulate, or entire, similar in color and texture to stem body; first internode 2.6–19 cm, shorter than leaves; distalmost node with 2–3 erect to arcuate branches. **Leaf blades** glabrous, bases not persistent in fibrous tufts. **Inflorescences** borne singly; spathes green, obviously wider than supporting branch, glabrous, keels entire to denticulate; outer 15–24.5 mm, 1.4 mm shorter to 2.2 mm longer than inner, tapering evenly towards apex, margins basally connate 2.5–4.5 mm; inner with keel evenly curved, hyaline margins 0.3–0.6 mm wide, apex broadly acute to obtuse or lobed, 0.8 mm shorter to 0.4 mm longer than green apex. **Flowers:** tepals blue to light bluish violet, bases yellow; outer tepals 8–15 mm, apex rounded to occasionally slightly emarginate, aristate; filaments connate ± entirely, stipitate-glandular basally; ovary similar in color to foliage. **Capsules** light to dark brown, obovoid to ± globose, 3.8–5.3 mm; pedicel erect to spreading. **Seeds** globose to obconic, with slight depression on one side, 1–1.5 mm, rugulose. $2n = 32$.

Flowering spring. Sandy roadsides, meadows, prairies; 0–300 m; Kans., Okla., Tex.; ne Mexico.

12. Sisyrinchium xerophyllum Greene, Pittonia 4: 32. 1899 [E] [F]

Sisyrinchium solstitiale E. P. Bicknell

Herbs, perennial, dark to blackish olive when dry, to 5.7 dm, not glaucous; rhizomes slightly elongate. **Stems** branched, with 2–3 nodes, 2.4–5.7 mm wide, scabrous at least apically, margins denticulate to entire, similar in color and texture to stem body; first internode 15–39 cm, subequal to leaves; distalmost node with 2–3 branches. **Leaf blades** scabrous, bases becoming fibrous, persisting in tufts. **Inflorescences** borne singly; spathes green, obviously wider than supporting branch, slightly scabrous, keels entire; outer 14.8–24 mm, 0.5–1.8 mm longer than inner, tapering evenly towards apex, basally connate 3.8–5.6 mm; inner with keel evenly curved to straight, hyaline margins 0.1–0.3 mm wide, apex acute to acuminate, ending 0.3–7.5 mm proximal to green apex. **Flowers:** tepals pale blue to deep bluish violet, bases yellow; outer tepals 8.1–12.3 mm, apex rounded or emarginate, aristate; filaments connate ± entirely, stipitate-glandular basally; ovary similar in color to foliage. **Capsules** dark brown to black, ± globose, 3.8–8 mm; pedicel ascending to spreading. **Seeds** globose to obconic, lacking obvious depression, 1.3–3 mm, rugulose. $2n = 32$.

Flowering year round. Sandy soil in scrub oak barrens and pinelands; 0–100 m; Fla., Ga., N.C.

13. Sisyrinchium pruinosum E. P. Bicknell, Bull. Torrey Bot. Club 28: 578. 1901 [E]

Sisyrinchium brayi E. P. Bicknell; *S. bushii* E. P. Bicknell; *S. helleri* E. P. Bicknell; *S. varians* E. P. Bicknell

Herbs, perennial, cespitose, ashy to bronze olive when dry, to 2.5 dm, not glaucous. **Stems** branched, with 1–2 nodes, 1–2.8 mm wide, scabrous at least apically, margins entire to denticulate apically, similar in color and texture to stem body; first internode 5.3–12.5 cm, equaling or shorter than leaves; distalmost node with 2–3 branches. **Leaf blades** usually scabrous, bases not persistent in fibrous tufts. **Inflorescences** borne singly; spathes green, obviously wider than supporting branch, scabrous, keels denticulate or entire; outer 18.6–26.5 mm, 2.5–5.5 mm longer than inner, tapering evenly towards apex, margins basally connate 2.5–3.9 mm; inner with keel evenly curved or straight, hyaline margins 0.2–0.4 mm wide, apex acuminate to acute, ending 0–1.5 mm proximal to green apex. **Flowers:** tepals bluish violet to purplish blue, bases yellow; outer tepals 8.3–10.5 mm, apex rounded to truncate or emarginate, aristate; filaments connate ± entirely, slightly stipitate-glandular basally; ovary similar in color to foliage. **Capsules** tan to light brown, ± globose, 3–5 mm; pedicel ascending to spreading. **Seeds** globose to obconic, lacking obvious depression, rugulose. $2n = 32$.

Flowering spring. Prairies, open woods; 20–200 m; Ala., Ark., Kans., Nebr., Okla., Tex.

Sisyrinchium pruinosum can be expected also in Louisiana, Missouri, and Mississippi.

14. Sisyrinchium nashii E. P. Bicknell, Bull. Torrey Bot. Club 26: 223. 1899 [E]

Sisyrinchium bicknellianum Fernald; *S. fibrosum* E. P. Bicknell; *S. floridanum* E. P. Bicknell 1899, not Rafinesque 1840; *S. tortum* E. P. Bicknell

Herbs, perennial, cespitose, yellowish green or light olive green when dry, to 4.9 dm, sometimes glaucous. **Stems** branched, with 1–2 nodes, 2.3–4.4 mm wide, glabrous, margins often denticulate apically, similar in color and texture to stem body; first internode 12–30 cm, longer than leaves; distalmost node with 2–3 branches. **Leaf blades** glabrous, bases becoming fibrous, persisting in tufts. **Inflorescences** borne singly; spathes with purplish tinge along hyaline margins, obviously wider than supporting branch, glabrous, keels usually entire; outer 18–25.5 mm, 1 mm shorter to 55 mm longer than inner, tapering evenly towards apex, margins basally connate 2–3.8 mm; inner with keel straight, hyaline margins 0.3–0.5 mm wide, apex obtuse to truncate, ending at green apex or occasionally extending beyond as 2 obtuse or truncate lobes. **Flowers:** tepals blue to light blue or light purple, bases yellow; outer tepals 7.2–12.5 mm, apex rounded to emarginate, aristate; filaments connate ± entirely, slightly stipitate-glandular basally; ovary similar in color to foliage. **Capsules** tan to beige, ± globose to ± obovoid, 2.5–4.8 mm; pedicel ascending to erect. **Seeds** globose to obconic, lacking obvious depression, 1–1.2 mm, rugulose. $2n = 16$.

Flowering ± year round. Sandy, rocky, or rich soil, upland pines, rich deciduous lowland woods, pine-oak woodlands; 0–300 m; Ala., Fla., Ga., Miss., N.C., S.C., Tenn., Va.

15. Sisyrinchium angustifolium Miller, Gard. Dict. ed. 8, Sisyrinchium no. 2. 1768 • Bermudienne à feuilles étroites [E] [W1]

Sisyrinchium graminoides E. P. Bicknell

Herbs, perennial, cespitose, dark olive green to bronze or blackish when dry, to 4.5 dm, not glaucous. **Stems** branched, with 1–2 nodes, 2.3–5 mm wide, glabrous, margins often minutely denticulate especially basally, similar in color and texture to stem body; first internode 10–30 cm, usually longer than leaves; distalmost node with 1–3 branches. **Leaf blades** glabrous, bases not persistent in fibrous tufts. **Inflorescences** borne singly; spathes usually green, obviously wider than supporting branch, glabrous, keels denticulate to entire; outer 18–38 mm, 2–9.5 mm longer than inner, usually tapering evenly towards apex, margins basally connate 4–6 mm; inner with keel evenly curved or straight, hyaline margins 0.1–0.3 mm wide, apex acuminate to acute, ending 0.2–0.7 mm proximal to green apex. **Flowers:** tepals pale blue to violet, occasionally white, bases yellow; outer tepals 7.7–12.5 mm, apex rounded or emarginate, aristate; filaments connate ± entirely, stipitate-glandular basally; ovary similar in color to foliage. **Capsules** dark brown or black, sometimes with purplish tinge, ± globose, 4–7 mm; pedicel spreading or ascending. **Seeds** globose to obconic, lacking obvious depression, 0.5–1.2 mm, rugulose. $2n = 96$.

Flowering spring–early summer. Moist meadows, stream banks, swamp edges, sandy meadows, moist open woods; 0–800 m; Nfld. and Labr. (Labr.), N.S., Ont., Que.; Ala., Ark., Conn., Del., D.C., Fla., Ga., Ill., Ind., Iowa, Kans., Ky., La., Maine, Md., Mass., Mich., Minn., Miss., Mo., Nebr., N.H., N.J., N.Y., N.C., Ohio, Okla., Pa., R.I., S.C., Tenn., Tex., Vt., Va., W.Va., Wis.

Sisyrinchium membranaceum E. P. Bicknell probably belongs here; Bicknell indicated that its relationship was "with *S. graminoides*" and his description falls within that of *S. angustifolium*, except for slightly shorter spathe bracts.

In previous floras, *Sisyrinchium angustifolium* often has been confused with *S. montanum*, especially when *S. graminoides* was segregated. Branching seems to be the primary point of confusion. The original descriptions of *S. angustifolium* and *S. graminoides* clearly indicated branching while that of *S. montanum* indicates it to be single-stemmed. There is some slight similarity between *S. montanum* var. *crebrum* and *S. angustifolium* with respect to spathe connation and dry color, and chromosome counts indicate that both have $2n = 96$, but there is some indication that breeding barriers may exist (D. B. Ward 1959).

16. Sisyrinchium strictum E. P. Bicknell, Bull. Torrey Bot. Club 26: 299. 1899 [E]

Herbs, perennial, cespitose, generally light green to yellowish green when dry, to 4.8 dm, not glaucous. **Stems** branched, with 1 node, 1.4–2.3 mm wide, glabrous, margins usually minutely denticulate, similar in color and texture to stem body; internode with 1–2 branches, 22–42 cm, usually longer than leaves. **Leaf blades** glabrous, bases not persistent in fibrous tufts. **Inflorescences** borne singly; spathes usually with purplish tinge on margins, obviously wider than supporting branch, glabrous, keels entire to occasionally denticulate; outer 16–22.8 mm, 0.2 mm shorter to 4 mm longer than inner, tapering evenly towards apex, margins basally connate 4.2–6.9 mm; inner with keel usually straight, hyaline margins 0.1–0.3 mm, apex acute, occasionally obtuse, sometimes separated as 2 lobes, ending 0–0.4 mm proximal to green apex. **Flowers:** tepals pale to deep bluish violet, bases yellow; outer tepals 9–19.5 mm, apex rounded or truncate, aristate; filaments connate ± entirely, usually stipitate-glandular basally; ovary similar in color to foliage. **Capsules** tan to nearly white, occasionally light brown, sometimes with purplish apex, ± globose to ± turbinate, 3.5–4.9 mm; pedicel erect or ascending. **Seeds** globose to obconic, lacking obvious depression, 0.9–1.2 mm, rugulose or granular.

Flowering early–mid summer. Moist meadows, stream banks, edges of open woods, mostly on sandy soil; 150–200 m; Mich., Wis.

17. Sisyrinchium funereum E. P. Bicknell, Bull. Torrey Bot. Club 31: 387. 1904 [C] [E]

Herbs, perennial, cespitose, green to ashy olive when dry, to 7.6 dm, strongly glaucous; roots somewhat fleshy-thickened. **Stems** branched, with 1 node, 1–4 mm wide, glabrous, margins white-cartilaginous; internode 16–65 cm, longer than leaves, with 2–3 branches. **Leaf blades** glabrous, bases not persistent in fibrous tufts. **Inflorescences** borne singly; spathes green, obviously wider than supporting branch, glabrous, keels entire; outer 13.5–24 mm, 3 mm shorter to 1 mm longer than inner, tapering evenly towards apex, margins basally connate 5–6.5 mm; inner with keel evenly curved to straight, hyaline margins 0.6–1.1 mm wide, apex extending as 2 broadly rounded and erose lobes, ending at or slightly beyond green apex. **Flowers:** tepals pale blue to light bluish violet, bases yellow; outer tepals 9–15 mm, apex truncate to occasionally rounded, aristate; filaments connate ± entirely, glabrous; ovary similar in color to foliage. **Capsules** beige, globose, 5–6 mm; pedicel erect to ascending. **Seeds** globose to obconic, lacking obvious depression, 1.1–1.5 mm, slightly granular. $2n = 32$.

Flowering early spring. Moist, grassy areas along streams and springs where soil strongly alkaline; of conservation concern; 0–800 m; Calif., Nev.

Sisyrinchium funereum is endemic to the Death Valley-Ash Meadows region.

18. Sisyrinchium radicatum E. P. Bicknell, Bull. Torrey Bot. Club 28: 576. 1901 [C] [E]

Herbs, perennial, cespitose, pale olive green when dry, to 3.4 dm, rarely glaucous. **Stems** branched, with 1–2 nodes, 1.7–3.2 mm wide, glabrous, margins white or translucent-cartilaginous; first internode 11–41 cm, longer than leaves; distalmost node with 2 branches. **Leaf blades** glabrous, bases not persistent in fibrous tufts. **Inflorescences** borne singly; spathes green, obviously wider than supporting branch, glabrous, keels entire to denticulate; outer 14–18 mm, 3.8 mm shorter to 0.5 mm longer than inner, tapering evenly towards apex, basally connate 4–7 mm; inner with keel evenly curved to straight, hyaline margins 0.5–0.8 mm wide, apex abruptly broadened, ending 0.5–2.3 mm proximal to green apex. **Flowers:** tepals bluish violet, bases yellow; outer tepals elliptic to oblanceolate, 9–12 mm, apex rounded to slightly emarginate, aristate; filaments connate ± entirely, stipitate-glandular basally; ovary similar in color to foliage. **Capsules** beige to tan, globose, 4.3–6 mm; pedicel erect to ascending. **Seeds** globose to obconic, lacking obvious depression, 0.8–1.3 mm, rugulose. $2n = 32$.

Flowering late spring–mid summer. Moist, sometimes alkaline meadows, stream banks, borders of springs; of conservation concern; 600–1300 m; Nev., Utah.

Sisyrinchium radicatum has been confused with *S. demissum*: S. L. Welsh and G. Moore (1973) called all branched *Sisyrinchium* plants in Utah *S. radicatum*, while S. Goodrich and E. Neese (1986) called such plants *S. demissum* with *S. radicatum* a synonym. *Sisyrinchium radicatum* differs in having white or cartilaginous margins on the stem and a broad apex to the hyaline margin of the inner spathe; it is apparently restricted to the St. George–Las Vegas region, and is to be expected in the adjacent northwest corner of Arizona.

19. Sisyrinchium bellum S. Watson, Proc. Amer. Acad. Arts 12: 277. 1877 [F]

Sisyrinchium angustifolium Miller var. *bellum* (S. Watson) Baker; *S. eastwoodiae* E. P. Bicknell; *S. greenei* E. P. Bicknell; *S. hesperium* E. P. Bicknell; *S. maritimum* A. Heller

Herbs, perennial, cespitose, green to ashy olive when dry, to 3.8 dm, not glaucous. **Stems** branched, with 1–2 nodes, 1.5–5.3 mm wide, glabrous, margins entire to denticulate apically, similar in color and texture to stem body; first internode 11–33 cm, usually equaling or longer than leaves; distalmost node with 2–3 branches. **Leaf blades** glabrous, bases not persistent in fibrous tufts. **Inflorescences** borne singly; spathes green, obviously wider than supporting branch, glabrous, keels entire to denticulate; outer 14–30 mm, usually 2–7 mm longer than inner, tapering evenly towards apex, margins basally connate 4–7.6 mm; inner with keel evenly curved to straight, hyaline margins 0.1–0.7 mm wide, apex abruptly widening, ending 0.2–1.3 mm proximal to green apex. **Flowers:** tepals light blue to dark bluish violet or rarely purplish or white, bases yellow; outer tepals often broadly cuneate, 10–17 mm, apex rounded, truncate, or occasionally emarginate, aristate; filaments connate ± entirely, usually glabrous; ovary similar in color to foliage. **Capsules** tan to dark brown, globose to obovoid, 3–5 mm; pedicel erect to ascending. **Seeds** globose to obconic, lacking obvious depression, 1.5–2.2 mm, rugulose. $2n = 32$.

Flowering early spring–late summer. Open, mostly moist, grassy areas and woods; 0–2400 m; Calif., Oreg.; Mexico (Baja California).

Sisyrinchium bellum is the most common species of *Sisyrinchium* in California; it is highly variable in habit, flower color, and size. Plants of serpentine areas are intermediate with *S. idahoense* var. *idahoense*; plants of the Transverse Ranges are intermediate with *S. demissum*.

20. Sisyrinchium fuscatum E. P. Bicknell, Bull. Torrey Bot. Club 26: 225. 1899 [E]

Sisyrinchium arenicola E. P. Bicknell; *S. farwellii* E. P. Bicknell; *S. incrustatum* E. P. Bicknell; *S. rufipes* E. P. Bicknell; *S. tenellum* E. P. Bicknell

Herbs, perennial, cespitose, yellowish green to olive green when dry, to 5.3 dm, not glaucous. **Stems** branched, with 1 node, 1–1.7 mm wide, glabrous, margins entire, similar in color and texture to stem body; internode (20–)23–47 cm, longer than leaves, with 1–5 branches. **Leaf blades** glabrous, bases becoming fibrous, persistent in tufts. **Inflorescences** borne singly; spathes green, obviously wider than supporting branch, glabrous, keels entire; outer 12–18 mm, 0.9 mm shorter to 1.7 mm longer than inner, tapering evenly towards apex, margins basally connate 2.1–4 mm; inner with keel straight, hyaline margins 0.2–0.4 mm wide, apex acute to obtuse, ending at or to 0.5 mm proximal to green apex. **Flowers:** tepals blue to bluish violet, bases yellow; outer tepals 7.9–11.6 mm, apex emarginate, aristate; filaments connate ± entirely, stipitate-glandular basally; ovary similar in color to foliage. **Capsules** light to medium brown, ± globose, 2.9–4.3 mm; pedicel erect to ascending. **Seeds** globose to obconic, lacking obvious depression, 1–1.5 mm, rugulose. $2n = 32$.

Flowering spring–early summer. Marshy areas, moist pine barrens; 0–200 m; N.S.; Ala., Conn., Del., Fla., Ga., Md., Mass., Miss., N.J., N.Y., N.C., S.C., Va.

21. Sisyrinchium atlanticum E. P. Bicknell, Bull. Torrey Bot. Club 23: 134. 1896 [E] [F]

Sisyrinchium apiculatum E. P. Bicknell; *S. apiculatum* var. *mesochorum* Nieuwland; *S. corymbosum* E. P. Bicknell; *S. flexile* E. P. Bicknell; *S. mucronatum* Michaux var. *atlanticum* (E. P. Bicknell) H. E. Ahles; *S. scoparium* E. P. Bicknell; *S. tracyi* E. P. Bicknell; *S. violaceum* E. P. Bicknell

Herbs, perennial, cespitose, yellowish to light olive when dry, to 5.7 dm, not glaucous. **Stems** branched, with 1 or 2 nodes, 0.8–1.9 mm wide, usually glabrous, margins entire, similar in color and texture to stem body; first internode 11–36 cm, longer than leaves; distalmost node with 2–3 branches. **Leaf blades** usually glabrous, bases occasionally becoming fibrous, but not persistent in tufts. **Inflorescences** borne singly; spathes green or occasionally with purplish tinge on margins, obviously wider than supporting branch, glabrous or slightly scabrous, keels entire; outer 12–16.1 mm, 1.4 mm shorter to 1.5 mm longer than inner, tapering evenly towards apex, margins basally connate 3–5.2 mm; inner with keel evenly curved or straight, hyaline margins 0.2–0.6 mm wide, apex broadly rounded to truncate, usually erose, ending 0–0.5 mm proximal to green apex or, occasionally, exceeding it by up to 0.5 mm. **Flowers:** tepals light blue or bluish violet to purple or occasionally white, bases yellow; outer tepals 6.3–11 mm, apex emarginate to truncate, aristate; filaments connate ± entirely, stipitate-glandular basally; ovary blackish, in contrast with much lighter foliage. **Capsules** dark brown to black or purplish black, ± globose to obovoid, 2–4.1 mm; pedicel ascending to erect. **Seeds** globose to obconic, lacking obvious depression, 0.5–1.2 mm, rugulose or occasionally granular. 2*n* = 16, 32.

Flowering spring–early summer. Moist meadows and coastal dunes in sandy, peaty, or rich, loamy soil; 0–300 m; N.S.; Ala., Ark., Conn., Del., D.C., Fla., Ga., Ill., Ind., Ky., La., Maine, Md., Mass., Mich., Miss., Mo., N.H., N.J., N.Y., N.C., Ohio, Pa., R.I., S.C., Tenn., Vt., Va., W.Va., Wis.

22. Sisyrinchium langloisii Greene, Pittonia 4: 32. 1899 [E]

Sisyrinchium canbyi E. P. Bicknell; *S. flaccidum* E. P. Bicknell; *S. furcatum* E. P. Bicknell

Herbs, perennial, cespitose, olive to bronze-olive when dry, to 3.2 dm, not glaucous. **Stems** branched, with 1 or 2 nodes, often purplish, 0.5–2(–2.2) mm wide, glabrous, margins entire to denticulate, similar in color and texture to stem body; first internode 3.2–7.2 cm, usually shorter than leaves; distalmost node with 1–3 branches. **Leaf blades** glabrous, bases not persistent in fibrous tufts. **Inflorescences** borne singly; spathes purplish-tinged basally and sometimes along margins, obviously wider than supporting branch, glabrous, keels entire to occasionally denticulate; outer 12.5–25 mm, 0.9–2.7 mm longer than inner, tapering evenly towards apex, margins basally connate 2.2–3.8 (–5) mm; inner with keel evenly curved or straight, hyaline margins 0.2–0.4 mm wide, apex acuminate to acute or occasionally obtuse, ending 0–1.5 mm proximal to green apex. **Flowers:** tepals pale blue to bluish violet or white, bases yellow; outer tepals 6.1–10 mm, apex rounded to truncate, aristate; filaments connate ± entirely, glabrous or sparsely stipitate-glandular basally; ovary similar in color to foliage. **Capsules** medium brown to black, ± globose, 3.1–4.7 mm; pedicel spreading or ascending. **Seeds** globose to obconic, lacking obvious depression, 0.9–1.1 mm, rugulose. 2*n* = 32.

Flowering spring. Moist prairies, roadsides, open woods; 0–200 m; Ala., Ark., Ga., La., Miss., Okla., Tenn., Tex.

Sisyrinchium implicatum E. P. Bicknell probably belongs here; the type (*Hilgard s.n.*, Apr 1858, Calhoun Co., Mississippi, MO) was not seen.

Sisyrinchium langloisii is to be expected in northeastern Mexico.

23. Sisyrinchium miamiense E. P. Bicknell, Bull. Torrey Bot. Club 26: 226. 1899 [E]

Sisyrinchium flagellum E. P. Bicknell; *S. recurvatum* E. P. Bicknell

Herbs, perennial, brownish or bronze-olive when dry, to 4 dm, not glaucous; rhizomes slightly elongated. **Stems** branched, with 1 or 2 nodes, 1.3–2(–2.5) mm wide, glabrous, margins usually entire, similar in color and texture to stem body; first internode 9–30 cm, equaling or shorter than leaves; distalmost node with 1–3 branches. **Leaf blades** glabrous, bases not persistent in fibrous tufts. **Inflorescences** borne singly; spathes green to occasionally purplish, obviously wider than supporting branch, glabrous, keels usually entire; outer 12.5–23 mm, 2.3 mm shorter to 3.7 mm longer than inner, tapering evenly towards apex, margins basally connate 3.9–6.5 mm; inner with keel straight to evenly curved, hyaline margins 0.1–0.3 mm wide, apex acute or obtuse, ending at or to 1 mm proximal to aristate green apex. **Flowers:** tepals pale blue to deep bluish violet, bases yellow; outer tepals 8.3–12.8 mm, apex rounded to emarginate or truncate, aristate; filaments connate ± entirely, stipitate-glandular basally or glabrous; ovary similar in color to foliage. **Capsules** dark brown or nearly black, ± globose to obovoid, 3.7–4.5 mm; pedicel ascending to spreading. **Seeds** globose to obconic, lacking obvious depression, 1–1.2 mm, rugulose. 2*n* = 32.

Flowering late winter–summer. Roadside ditches, hammocks, open woods; 0–50 m; Ala., Fla., Ga., Miss.

24. Sisyrinchium biforme E. P. Bicknell, Bull. Torrey Bot. Club 28: 575. 1901

Sisyrinchium dimorphum R. L. Oliver

Herbs, perennial, cespitose, pale to olive green or occasionally bronze when dry, to 4.7 dm, not glaucous; rhizomes scarcely discernable. **Stems** simple or branched, with 1 node, 1.1–1.9 dm wide, glabrous, margins entire, usually same color and texture as stem body but occasionally drying lighter; internode 17.2–21.3 cm, equaling or shorter than leaves, with 2 branches. **Leaf blades** glabrous, bases not persistent in fibrous tufts. **Inflorescences** borne singly; spathes green, obviously wider than supporting branch, glabrous, keels entire; outer 13–19.6 mm on branched stems, 19.3–25.6 mm on simple stems, 1.2 mm longer to 2 mm shorter than inner on branched stems, 1.5–4.2 mm longer than inner on simple stems, tapering evenly towards apex, margins basally connate 3.9–6.1 mm; inner with keel

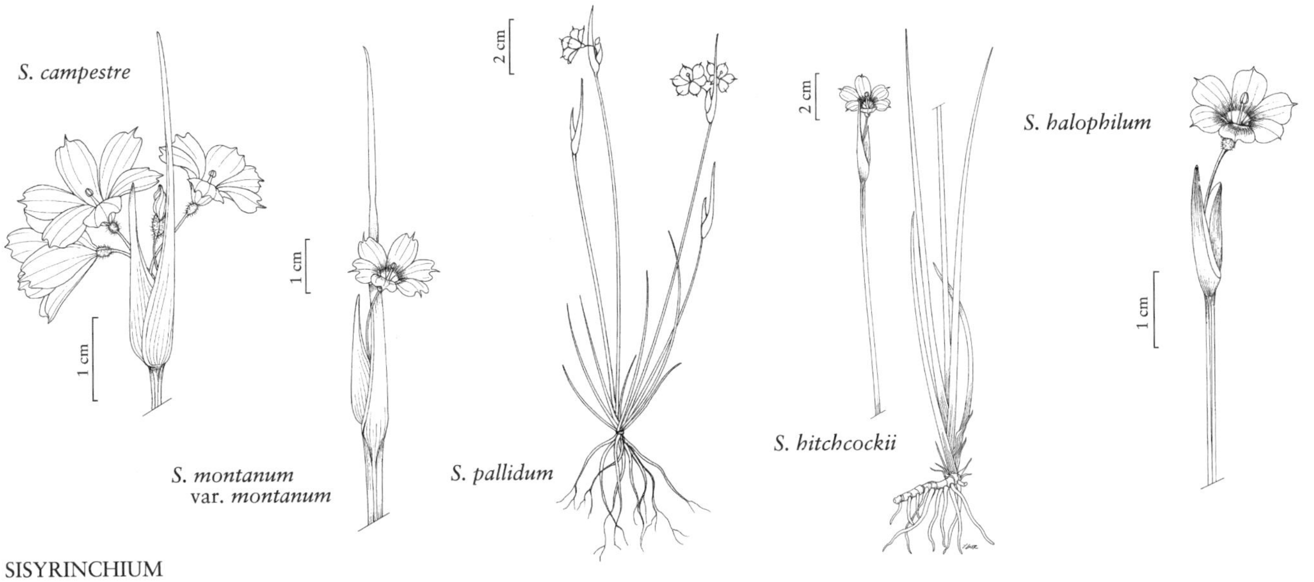

SISYRINCHIUM

evenly curved, hyaline margins 0.2–0.5 mm wide, apex acuminate to acute, ending 0.2–0.8 mm proximal to green apex on branched stems, sometimes lobed and exceeding green apex on simple stems. **Flowers:** tepals light blue, bases yellow; outer tepals 5–13 mm, apex rounded, short-aristate; filaments connate ± entirely, stipitate-glandular basally; ovary similar in color to foliage. **Capsules** light to dark brown, ± globose to ± fusiform, 5–7 mm; pedicel usually erect. **Seeds** globose to obconic, occasionally with slight depression, 1–1.3 mm, rugulose. $2n = 96$.

Flowering spring–early summer. Meadows or swales within stabilized coastal dunes, sandy river banks inland; 0–200 m; La., Tex.; adjacent coastal n Mexico.

25. Sisyrinchium demissum Greene, Pittonia 2: 69. 1890

Sisyrinchium amethystinum E. P. Bicknell; *S. demissum* var. *amethystinum* (E. P. Bicknell) Kearney & Peebles; *S. longipedunculatum* E. P. Bicknell

Herbs, perennial, cespitose, light green to olive when dry, to 5 dm, often somewhat glaucous. **Stems** branched, with 1 or 2 nodes, 1–2(–2.3) mm wide, glabrous, margins entire to denticulate apically, occasionally white-cartilaginous; first internode 8–34 cm, usually longer than leaves; distalmost node with 2–3 branches. **Leaf blades** glabrous, bases not persistent in fibrous tufts. **Inflorescences** borne singly; spathes green, obviously wider than supporting branch, glabrous, keels entire to denticulate; outer 11–26 mm, 2.5 mm shorter to 5 mm longer than inner, tapering evenly towards apex, margins basally connate 3.3–8 mm; inner with keel evenly curved to straight, hyaline margins 0.2–0.4 mm wide, apex acute to acuminate, ending 0.1–1.1 mm proximal to green apex. **Flowers:** tepals dark bluish violet, bases yellow; outer tepals 6–15 mm, apex rounded to acute or occasionally emarginate, aristate; filaments connate ± entirely, stipitate-glandular basally; ovary similar in color to foliage. **Capsules** beige to tan, globose, 4–7.5+ mm; pedicel erect to ascending. **Seeds** globose to obconic, lacking obvious depression, 0.8–2 mm, granular or slightly rugulose. $2n = 32, 64$.

Flowering mid spring–early fall. Moist areas, springs, stream banks, meadows, forest seeps; 500–2900 m; Ariz., Colo., Nev., N.Mex., Tex., Utah; n Mexico.

Previously, we (A. F. Cholewa and D. M. Henderson 1984) misidentified the octoploid northern Utah populations of *Sisyrinchium demissum* as *S. radicatum*. New data indicate that these northern Utah populations rightly belong here as a polyploid race of *S. demissum*. Also see discussion under 18. *S. radicatum*.

26. Sisyrinchium campestre E. P. Bicknell, Bull. Torrey Bot. Club 26: 341. 1899 [E] [F]

Sisyrinchium campestre var. *kansanum* E. P. Bicknell; *S. flaviflorum* E. P. Bicknell; *S. kansanum* (E. P. Bicknell) Alexander

Herbs, perennial, cespitose, light green to olive or ashy olive, sometimes with purplish base when dry, to 4 dm, not glaucous. **Stems** simple, 1.2–2.5 mm wide, glabrous or scabrous, margins usually entire, similar in color and texture to stem body. **Leaf blades** usually glabrous, bases occasionally becoming fibrous but not persistent in tufts. **Inflorescences** borne singly; spathes green to purplish, glabrous or scabrous, keels usually entire; outer 22–50 mm, 10–32 mm longer than inner, tapering evenly towards apex or occasionally slightly constricted proximal to apex, margins distinct or rarely connate basally to 1 mm; inner with keel usually gibbous, hyaline margins 0.1–0.3 mm wide, apex acuminate to acute, ending 0.4–3.5 mm proximal to green apex. **Flowers:** tepals white to pale blue, bases yellow; outer tepals 7–12.7 mm, apex rounded to emarginate, aristate; filaments connate ± entirely, stipitate-glandular basally; ovary similar in color to foliage. **Capsules** light to dark brown, globose, 3–5 mm; pedicel spreading to ascending. **Seeds** globose to obconic, lacking obvious depression, 0.5–1 mm, granular to rugulose. $2n = 32$.

Flowering spring–early summer. Prairies, meadows, roadsides; 20–500 m; Man.; Ark., Ill., Ind., Iowa, Kans., La., Mich., Minn., Mo., Nebr., N.Dak., Ohio, Okla., S.Dak., Tex., Wis.

27. Sisyrinchium montanum Greene, Pittonia 4: 33. 1899 • Bermudienne montagnarde [E] [F] [W1]

Herbs, perennial, cespitose, pale to olive green or dark brown to bronze when dry, to 5 dm, not glaucous; rhizomes scarcely discernable. **Stems** simple, obviously winged, (1.5–)2–3.7 mm wide, glabrous, margins entire (in eastern populations) to denticulate (in western populations) apically, similar in color and texture to stem body. **Leaf blades** glabrous, bases not persistent in fibrous tufts. **Inflorescences** borne singly; spathes usually green or bronze, rarely with purplish margins, glabrous, keels usually denticulate; outer 36–76 mm, 12–46 mm longer than inner, slightly constricted proximal to apex, margins basally connate 2–5.7 mm; inner with keel ± gibbous basally, sinuous proximally, hyaline margins 0.1–0.3 mm wide, apex acuminate to acute, ending 0.9–4.3 mm proximal to recurved green apex. **Flowers:** tepals dark bluish violet, bases yellow; outer tepals 9–14.5 mm, apex emarginate to retuse, aristate; filaments connate ± entirely, stipitate-glandular basally; ovary similar in color to foliage. **Capsules** tan to dark brown, sometimes with purplish tinge apically, ± globose to obovoid, 4–6.8 mm; pedicel erect to spreading. **Seeds** globose to obconic, lacking obvious depression, 0.9–1.5 mm, rugulose.

Varieties 2 (2 in the flora): North America.

Eastern coastal populations of *Sisyrinchium montanum* appear to have some slight affinity to *S. angustifolium* (e.g., long connation of outer spathe) and some previous floras have combined the two taxa. Some taxonomists have questioned the recognition of varieties within *S. montanum*, but we feel that the differences between them are no more subtle than those between varieties generally recognized elsewhere in the genus. Living material was not available to us to investigate breeding barriers. Although *Sisyrinchium montanum* is considered weedy by D. T. Patterson et al. (1989), I have seen many populations all through the western states and Great Lakes areas and would not consider it weedy in any of these portions of the range.

1. Margins of outer spathe connate basally 4–5.7 mm; plants drying dark brown or bronze . . . . 27a. *Sisyrinchium montanum* var. *crebrum*
1. Margins of outer spathe connate basally 1–3.5 mm; plants drying green to olive . . . . 27b. *Sisyrinchium montanum* var. *montanum*

27a. Sisyrinchium montanum Greene var. **crebrum** Fernald, Rhodora 48: 159, plate 1044, fig. 3. 1946 [E]

Sisyrinchium montanum subsp. *crebrum* (Fernald) Böcher

Plants bronze to dark brown when dry. **Spathes:** outer 30–56.7 mm, 12–25 mm longer than inner, connate basally 4–5.7 mm; inner with hyaline margins ending 0.9–1.9 mm proximal to green apex. $2n = 96$.

Flowering late spring–early summer. Moist, rocky crevices, moist meadows especially near coast and along Saint Lawrence Seaway; 0–200 m; St. Pierre and Miquelon; N.B., Nfld. and Labr. (Nfld.), N.S., Ont., P.E.I., Que.; Conn., Maine, Mass., N.H., N.Y., R.I., Vt.

27b. Sisyrinchium montanum Greene var. **montanum** [E] [F]

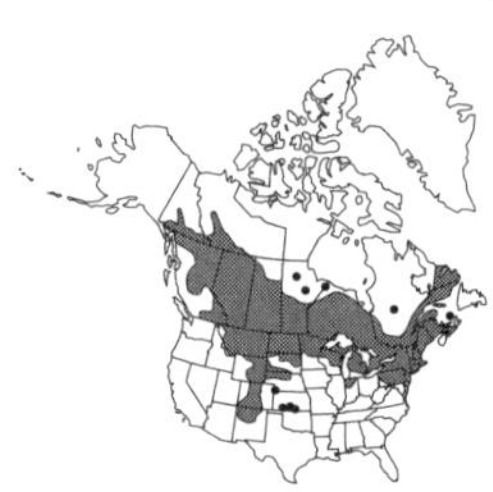

Sisyrinchium alpestre E. P. Bicknell; *S. heterocarpum* E. P. Bicknell

Plants pale green to olive when dry. **Spathe:** outer 36–76 mm, 14–46 mm longer than inner, connate basally 1–3.5 mm; inner with hyaline margins ending 0.9–4.3 mm proximal to green apex. $2n = 96$.

Flowering late spring–early summer. Moist meadows, stream banks, open woods; 0–3100 m; Alta., B.C., Man., N.B., Nfld. and Labr. (Labr.), N.W.T., N.S., Ont., P.E.I., Que., Sask., Yukon; Colo., Conn., Idaho, Ind., Kans., Maine, Mass., Mich., Minn., Mont., Nebr., N.H., N.J., N.Mex., N.Y., N.Dak., Pa., R.I., S.Dak., Vt., Wis., Wyo.

Populations of var. *montanum* in Kansas and Nebraska have an unusually high proportion of branched-stem plants mixed with simple-stemmed ones—even within the same clump (Great Plains Flora Association 1986). Although branched stems are found occasionally elsewhere in the taxon's range, the high incidence in this region is unusual and may be related to the distributional approach of *S. angustifolium*, with which it often has been confused.

Canadian populations of var. *montanum* tend to have narrower bracts and other less pronounced interspecific differences than do populations elsewhere. These populations can easily be confused with *S. mucronatum*, which consistently has some purple in the spathes and usually has a shorter outer spathe, and with *S. septentrionale*, which lacks emarginate outer tepals and has a very long and narrow outer spathe.

28. Sisyrinchium pallidum Cholewa & Douglass M. Henderson, Brittonia 36: 361, fig. 7E–G. 1984 [C] [E] [F]

Herbs, perennial, cespitose, green to olive when dry, to 3 dm, not glaucous; rhizomes scarcely discernable. **Stems** simple, 1–2 mm wide, glabrous, margins entire, similar in color and texture to stem body. **Leaf blades** glabrous, bases not persistent in fibrous tufts. **Inflorescences** borne singly; spathes green, glabrous, keels entire; outer 28–38 mm, 7–26 mm longer than inner, basally connate 2.6–4.3 mm, often very slightly constricted proximal to apex; inner with keels evenly curved, hyaline margins 0.1–0.5 mm wide, apex usually acute, ending 0–1.5 mm proximal to green apex. **Flowers:** tepals pale blue, bases yellow; outer tepals 7.6–10 mm, apex slightly emarginate or rounded, aristate; filaments connate ± entirely, stipitate-glandular basally; ovary similar in color to foliage. **Capsules** beige to dark brown, globose, 3–5 mm; pedicel ascending. **Seeds** globose to obconic, lacking obvious depression, 0.8–1.2 mm, granular or somewhat rugulose. $2n = 64$.

Flowering mid summer. Poorly drained, montane meadows; of conservation concern; 2600–3000 m; Colo., Wyo.

29. Sisyrinchium mucronatum Michaux, Fl. Bor.-Amer. 2: 33. 1803 [E]

Sisyrinchium angustifolium Miller var. *mucronatum* (Michaux) Baker; *S. intermedium* E. P. Bicknell; *S. versicolor* E. P. Bicknell

Herbs, perennial, cespitose, dull green to olive when dry, to 4.2 dm, not glaucous; rhizomes scarcely discernable. **Stems** simple, often purplish, almost wiry, with very narrow or scarcely discernable wings, 0.9–2 mm wide, usually glabrous, margins usually entire to denticulate apically, similar in color and texture to stem body. **Leaf blades** usually glabrous, bases not persistent in fibrous tufts. **Inflorescences** borne singly; spathes mostly purplish, glabrous to slightly scabrous (in southern populations), keels entire; outer 34–52 mm, 14–31 mm longer than inner, usually tapering evenly towards apex, margins basally connate (0.7–)1–2.7 mm; inner with keel evenly curved to occasionally gibbous basally, hyaline margins 0.1–0.3 mm wide, apex acuminate to acute, ending 0.5–3.2 mm proximal to green apex. **Flowers:** tepals dark blue or bluish violet, occasionally white, bases yellow; outer tepals 9–12.5 mm, apex emarginate to rounded, aristate; filaments connate ± entirely, stipitate-glandular basally; ovary similar in color to foliage. **Capsules** tan to light brown with purplish tinge apically, ± globose, 3.2–5.5 mm; pedicel spreading to erect. **Seeds** globose to obconic, lacking obvious depression, 0.8–1.5 mm, usually granular. $2n = 32$.

Flowering spring–early summer. Prairies, roadsides, moist open woods, rocky and sandy open shores; 0–700 m; Man., Ont., Que., Sask.; Conn., Del., D.C., Ga., Maine, Md., Mass., Mich., Minn., N.H., N.J., N.Y., N.C., N.Dak., Ohio, Pa., R.I., S.C., S.Dak., Vt., Va., W.Va., Wis.

Sisyrinchium mucronatum can be a difficult species to identify, especially in the north. Herbarium specimens seem to indicate that morphologically it may approach *S. septentrionale* in central Canada, and narrow specimens of *S. montanum* in the northeast. Fresh material, however, reveals that *S. septentrionale* tends to have much paler flowers with no emarginate apices and much longer outer spathes, and that *S. montanum* usually has much wider stems, leaves, and bracts, and the outer tepals have emarginate to retuse apices. Southeastern populations of *S. mucronatum* tend to be more scabrous while northwestern populations have less purple in the spathes.

30. Sisyrinchium hitchcockii Douglass M. Henderson, Brittonia 28: 170, fig. 4. 1976 C E F

Herbs, perennial, green to dark olive when dry, to 5 dm, not glaucous; rhizomes elongate. **Stems** mostly simple, 2.5–4 mm wide, glabrous, margins usually entire, similar in color and texture to stem body. **Leaf blades** glabrous, bases not persistent in fibrous tufts. **Inflorescences** borne singly; spathes green, glabrous, keels entire or denticulate; outer 23–37 mm, 6–17 mm longer than inner, tapering evenly towards apex, margins basally connate 5–6 mm; inner with keel evenly curved, hyaline margins 0.1–0.2 mm wide, apex obtuse or truncate, ending 0.1–0.3 mm proximal to green apex. **Flowers:** tepals purple to dark reddish purple, rarely bluish violet, bases darker purple; outer tepals 15–19 mm, apex truncate or emarginate, aristate; filaments connate ± entirely, stipitate-glandular basally; ovary similar in color to foliage. **Capsules** beige to light brown, globose, 6–7 mm; pedicel ascending. **Seeds** globose to obconic, lacking obvious depression, 1.9–2.5 mm, rugulose. $2n = 64$.

Flowering spring–early summer. Grassy areas, openings in woods, mostly where somewhat dry later in season; of conservation concern; 200–1000 m; Calif., Oreg.

Sisyrinchium hitchcockii is endemic to the southern Willamette Valley and the Roseburg area of Douglas County, Oregon, and to Cape Ridge, Humboldt County, California. It often occurs in mixed populations with *S. bellum*. The lack of yellow tepal bases and the extensive rhizomes are the most distinctive features.

31. Sisyrinchium halophilum Greene, Pittonia 4: 34. 1899 E F

Sisyrinchium leptocaulon E. P. Bicknell

Herbs, perennial, cespitose, green to ashy olive when dry, to 2.6 dm, often strongly glaucous; rhizomes scarcely discernable. **Stems** simple, 0.5–3 mm wide, glabrous, margins entire, similar in color and texture to stem body. **Leaf blades** glabrous, bases not persistent in fibrous tufts. **Inflorescences** borne singly; spathes usually green, glabrous, keels entire; outer 13–27 mm, equaling or 3 mm longer than inner, tapering evenly towards apex, margins basally connate 2.7–5.5 mm; inner with keel evenly curved to straight, hyaline margins 0.4-0.6 mm wide, apex broadly rounded or truncate, often erose, often extending as 2 lobes beyond green apex. **Flowers:** tepals light to dark bluish violet, bases yellow; outer tepals 9–12 mm, apex truncate, aristate; filaments connate ± entirely, stipitate-glandular basally; ovary similar in color to foliage. **Capsules** beige, ± globose, 2.2–4 mm; pedicel erect. **Seeds** globose to obconic, lacking obvious depression, 1–1.7 mm, rugulose. $2n = 32$.

Flowering spring–summer. Moist, grassy areas, stream banks, springs, meadows, especially in alkaline soil; 1100–2600 m; Calif., Nev., Oreg., Utah.

Sisyrinchium halophilum is the most common species of the genus in Nevada and eastern California.

32. Sisyrinchium littorale Greene, Pittonia 4: 33. 1899 E F

Herbs, perennial, slightly cespitose, dark olive when dry, to 4.2 dm, not glaucous; rhizomes scarcely discernable. **Stems** simple, obviously winged, 2.2–3.5 mm wide, glabrous, margins denticulate, similar in color and texture to stem body. **Leaf blades** glabrous, bases not persistent in fibrous tufts. **Inflorescences** borne singly; spathes green, glabrous, keels entire or denticulate; outer 36–58 mm, 6–39 mm longer than inner, tapering evenly towards apex, margins basally connate 4.5–7 mm; inner with keel evenly curved, hyaline margins 0.2–0.5 mm wide, apex acute, ending 0.2–2 mm proximal to green apex. **Flowers:** tepals light blue to usually deep bluish violet, bases yellow; outer tepals 11–16 mm, apex rounded to slightly truncate, aristate; filaments connate ± entirely, stipitate-glandular basally; ovary similar in color to foliage. **Capsules** beige to dark brown, ± globose, 6.2–8.1 mm; pedicel erect to ascending. **Seeds** globose to obconic, lacking obvious depression, 1.2–1.7 mm, rugulose. $2n = 96$.

Flowering spring–mid summer. Moist, sandy or grassy areas near coast, interdunal depressions, edges of streams and lakes; 0–300 m; B.C.; Alaska, Wash.

Populations of *Sisyrinchium littorale* in coastal Washington and on Vancouver Island, British Columbia, intergrade with *S. idahoense* var. *segetum*. *Sisyrinchium littorale*, however, has larger capsules and wider stems and leaves.

33. Sisyrinchium sarmentosum Suksdorf ex Greene, Erythea 3: 121. 1895 [C] [E]

Herbs, perennial, cespitose, green to olive green when dry, to 3.2 dm, not glaucous; rhizomes scarcely discernable. **Stems** simple, 1.7–2.8 mm wide, glabrous, margins entire, similar in color and texture to stem body. **Leaf blades** glabrous, bases not persistent in fibrous tufts. **Inflorescences** borne singly; spathes green, glabrous, keels entire; outer 27–48 mm, 14–17 mm longer than inner, tapering evenly towards apex, margins basally connate 3–5 mm; inner with keel evenly curved, hyaline margins 0.2–0.5 mm wide, apex obtuse to acute, ending 0.2–2.3 mm proximal to green apex. **Flowers:** tepals pale blue, bases yellow; outer tepals 10–14 mm, apex rounded, aristate; filaments connate ± entirely, slightly stipitate-glandular basally; ovary similar in color to foliage. **Capsules** tan to medium brown, globose, 4–5 mm; pedicel erect to spreading. **Seeds** globose to obconic, lacking obvious depression, 1–1.5 mm, rugulose. $2n = 96$.

Flowering early–mid summer. Moist, grassy areas; of conservation concern; 500–1000 m; Oreg., Wash.

Sisyrinchium sarmentosum is known from only three or four populations in the Columbia River Gorge area of the Cascade Mountains. The pale blue flowers with rounded apices on the outer tepals set this species apart from others in the region.

34. Sisyrinchium idahoense E. P. Bicknell, Bull. Torrey Bot. Club 26: 445. 1899 [E] [F]

Herbs, perennial, cespitose, green to ashy olive when dry, to 4.5 dm, not glaucous; rhizomes scarcely discernable. **Stems** simple, obviously winged, 1–2.5 mm wide, glabrous, margins entire to denticulate apically, similar in color and texture to stem body. **Leaf blades** glabrous, bases not persistent in fibrous tufts. **Inflorescences** borne singly; spathes usually green, glabrous, keels entire to denticulate; outer 14–55(–65) mm, 13–16 mm longer than inner, basally connate 2.6–7 mm, tapering evenly towards apex; inner with keel evenly curved to straight, hyaline margins 0.1–0.5 mm wide, apex acuminate to acute, ending 0.5–2.1 mm proximal to green apex. **Flowers:** tepals light to deep bluish violet or occasionally purple, bases yellow; outer tepals 8–20 mm, apex rounded, truncate, or emarginate, aristate; filaments connate ± entirely, stipitate-glandular basally; ovary similar in color to foliage. **Capsules** beige to light or dark brown, occasionally with purple blotches on apex, ± globose to pyriform, 3–6 mm; pedicel erect to ascending. **Seeds** globose to obconic, lacking obvious depression, 0.8–1.8 mm, usually granular.

Varieties 4 (4 in the flora): w North America.

1. Stems less than 1.5 mm wide.
   2. Outer tepals 18–20 mm; outer spathe at least 2 times inner; margins of stem and keels of spathes entire . . . . 34a. *Sisyrinchium idahoense* var. *macounii*
   2. Outer tepals 10–17 mm; outer spathe less than 2 times inner; margins of stem and keels of spathes denticulate . . . . 34b. *Sisyrinchium idahoense* var. *segetum*
1. Stems 1.5 mm or more wide.
   3. Outer tepals 8–13 mm; inner spathe 12–20 mm; stem 1.5–2 mm wide . . . . 34c. *Sisyrinchium idahoense* var. *occidentale*
   3. Outer tepals 13–17(–20) mm; inner spathe 20–30 mm; stem 2–2.5 mm wide . . . . 34d. *Sisyrinchium idahoense* var. *idahoense*

34a. Sisyrinchium idahoense E. P. Bicknell var. **macounii** (E. P. Bicknell) Douglass M. Henderson, Brittonia 28: 172. 1976 [E]

Sisyrinchium macounii E. P. Bicknell, Bull. Torrey Bot. Club 27: 245. 1900

Stems less than 1.5 mm wide, margins entire. **Spathes:** keels entire, outer at least 2 times inner, basally connate 2.6–7 mm. **Outer tepals** 18–20 mm; filaments purple. $2n = 64$.

Flowering late spring–summer. Moist, open, grassy areas; 0–250 m; B.C.; Wash.

Var. *macounii* is especially common in the San Juan Islands.

34b. Sisyrinchium idahoense E. P. Bicknell var. **segetum** (E. P. Bicknell) Douglass M. Henderson, Brittonia 28: 174. 1976 [E]

Sisyrinchium segetum E. P. Bicknell, Bull. Torrey Bot. Club 26: 449. 1899

Stems less than 1.2 mm wide, margins denticulate. **Spathes:** keels denticulate; outer less than 2 times inner, basally connate 5–6 mm. **Outer tepals** 10–17 mm; filaments pale blue to deep violet or purple. $2n = 64$.

Flowering late spring–early summer. Moist to dry, gravelly prairies, open, grassy areas; 0–800 m; B.C.; Wash.

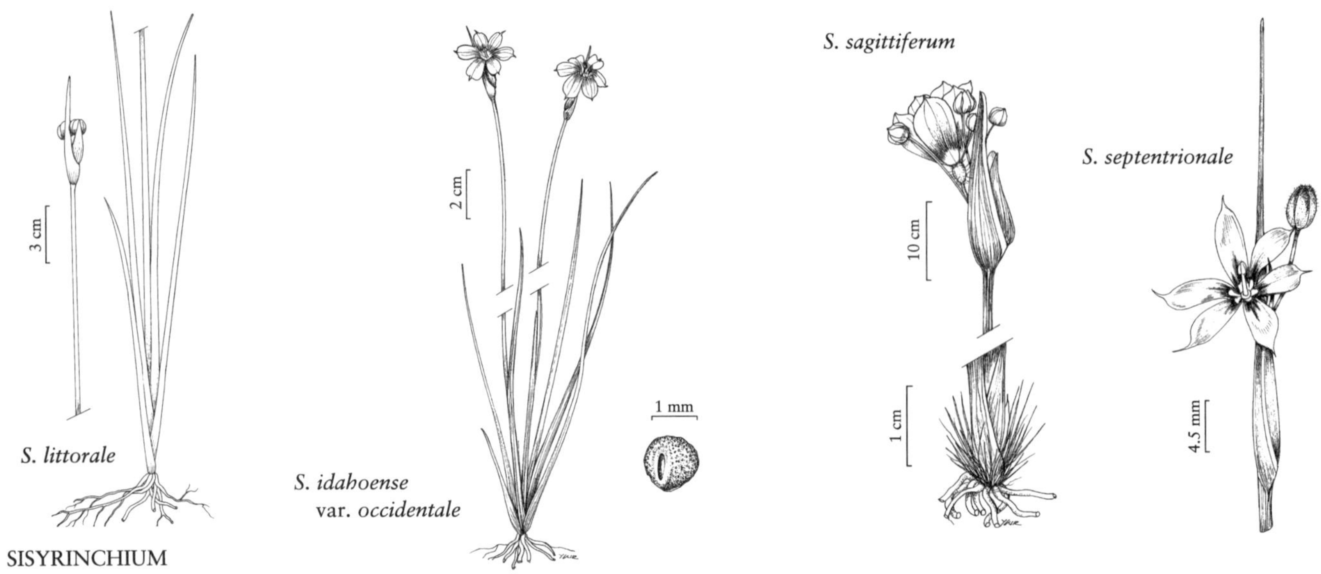

SISYRINCHIUM

34c. Sisyrinchium idahoense E. P. Bicknell var. **occidentale** (E. P. Bicknell) Douglass M. Henderson, Brittonia 28: 174. 1976 [E] [F]

Sisyrinchium occidentale E. P. Bicknell, Bull. Torrey Bot. Club 26: 447. 1899; *S. juncellum* Greene

Stems 1.5–2 mm wide, margins entire. **Spathes:** keels entire, outer less than 2 times inner, basally connate 3–6 mm; inner 12–20 mm. **Outer tepals** 8–13 mm; filaments bluish violet. $2n = 64$.

Flowering summer. Moist meadows, stream banks, edges of springs; 400–3100 m; Ariz., Calif., Colo., Idaho, Mont., Nev., N.Mex., Oreg., Utah, Wash., Wyo.

34d. Sisyrinchium idahoense E. P. Bicknell var. **idahoense** [E]

Sisyrinchium birameum Piper; *S. idahoense* var. *birameum* (Piper) J. K. Henry; *S. oreophilum* E. P. Bicknell

Stems 2–2.5 mm wide, margins denticulate. **Spathes:** keels denticulate; outer less than 2 times inner, basally connate 3.8–7 mm; inner 20–30 mm. **Outer tepals** 13–17(–20) mm; filaments bluish violet. $2n = 64, 96$.

Flowering late spring–summer. Moist meadows, stream banks; 0–2200 m; B.C.; Calif., Idaho, Mont., Oreg., Wash., Wyo.

The few dodecaploid populations of var. *idahoense* in eastern Oregon are morphologically indistinguishable from the far more common octoploid plants.

35. Sisyrinchium sagittiferum E. P. Bicknell, Bull. Torrey Bot. Club 26: 230. 1899 [E] [F]

Herbs, perennial, cespitose, dark brown to reddish brown or brownish olive when dry, to 4.5 dm, not glaucous. **Stems** simple, 0.9–2.1 mm wide, glabrous, margins usually entire, usually not hyaline, similar in color and texture to stem body. **Leaf blades** glabrous, bases becoming fibrous, persisting in tufts. **Inflorescences** borne singly; spathes often purplish brown, glabrous, keels entire; outer 15–33 mm, 6.2–13 mm longer than inner, basally connate 1–2.7 mm, tapering evenly towards apex; inner with keel gibbous basally, hyaline margins 0.5–0.6 mm wide, apex rounded or truncate, somewhat erose, usually extending beyond green apex as 2 lobes. **Flowers:** tepals blue to bluish violet, bases yellow; outer tepals 7.8–13 mm, apex rounded to slightly emarginate, aristate; filaments connate ± entirely, slightly stipitate-glandular basally; ovary similar in color to foliage. **Capsules** dark reddish brown, ± globose, 2.7–4.3 mm; pedicel spreading to ascending. **Seeds** globose to obconic, lacking obvious depression, 0.5–1.1 mm, rugulose. $2n = 32$.

Flowering spring, sometimes again in fall. Low, moist areas, sandy grasslands or pine woods, roadsides; 0–30 m; Ala., La., Tex.

K. L. Hornberger (1987, 1991) and D. S. Correll and M. C. Johnston (1970) considered *Sisyrinchium texanum* to be conspecific with *S. sagittiferum* (Correll and Johnston called it a "hybrid form"). E. P. Bicknell, however, described *S. sagittiferum* as being unbranched, and the single (poor) type specimen (*Riddell s.n.*, NY) agrees. We follow Bicknell's original concept of a single-stemmed *S. sagittiferum*. *Sisyrinchium texanum*, being branched, is not considered a synonym here.

36. **Sisyrinchium groenlandicum** Böcher, Bot. Tidsskr. 61: 286, figs. 3a, e–i, 5–7. 1966 E

Herbs, perennial, cespitose, green to olive or dark olive when dry, to 3.3 dm, not glaucous; rhizomes scarcely discernable. **Stems** simple, often with purplish tinge apically, 1–2 mm wide, glabrous, margins entire, similar in color and texture to stem body. **Leaf blades** glabrous, bases not persistent in fibrous tufts. **Inflorescences** borne singly; spathes tinged purplish, at least on hyaline margins, glabrous, keels entire; outer 25.4–44 mm, 10.9–23.3 mm longer than inner, usually tapering evenly towards apex, margins basally connate 1–2.5 mm; inner with keel evenly curved to straight, hyaline margins 0.3–0.6 mm wide, apex acute to acuminate, ending 0.3–1.6 mm proximal to green apex. **Flowers:** tepals pale blue to whitish, bases yellow; outer tepals 5.7–10.3 mm, apex rounded, aristate; filaments connate ± entirely, glabrous or slightly stipitate-glandular basally; ovary similar in color to foliage. **Capsules** dark brown to black, slightly turbinate to subglobose, 2.8–6.5 mm; pedicel usually ascending. **Seeds** globose to obconic, lacking obvious depression, 0.7–1.2 mm, granular. $2n = 32$.

Flowering summer. Southerly exposures in steppe/heath vegetation; 30–300 m; Greenland.

37. **Sisyrinchium septentrionale** E. P. Bicknell, Bull. Torrey Bot. Club 26: 452. 1899 E F

Herbs, perennial, cespitose, green to pale olive when dry, to 4.3 dm, not glaucous; rhizomes scarcely discernable. **Stems** simple, 0.8–2 mm wide, glabrous, margins usually entire to denticulate apically, similar in color and texture to stem body. **Leaf blades** glabrous, bases not persistent in fibrous tufts. **Inflorescences** borne singly; spathes usually green, glabrous, keels entire to slightly denticulate; outer 20–63 mm, 17–42 mm longer than inner, tapering evenly towards apex, basally connate 1.5–2.5 mm; inner with keel evenly curved, hyaline margins 0.1–0.3 mm wide, apex acute to acuminate, ending 0.4–2.3 mm proximal to green apex. **Flowers:** tepals pale blue to light bluish violet, rarely white, bases yellow; outer 8–9.1 mm, apex usually rounded, aristate; filaments connate ± entirely, stipitate-glandular basally; ovary similar in color to foliage. **Capsules** beige to light brown, ± globose, 3–5 mm; pedicel spreading to erect. **Seeds** globose to obconic, lacking obvious depression, 0.5–1.2 mm, rugulose. $2n = 32$.

Flowering early–mid summer. Mesic to dry meadows, stream banks, often in gravelly soil; 500–1600 m; Alta., B.C., N.W.T., Sask.; Wash.

Sisyrinchium septentrionale is widespread but apparently not common in western Canada. In central Canada it intergrades with *S. mucronatum*, to which it appears closely related (see discussion, p. 367). It is confused also with *S. montanum* but can be distinguished by its very slender, very long outer spathe and nongibbous inner spathe. Fresh material will show lighter blue flowers and outer tepals with rounded apices.

4. IRIS Linnaeus, Sp. Pl. 1: 38. 1753; Gen. Pl. ed. 5, 24. 1754 • [Greek *iris*, rainbow]

Norlan C. Henderson

Herbs, perennial, from rhizomes [bulbs or fascicles of fleshy roots]; rhizomes homogeneous with branches like primary in size and texture, or heterogeneous, branches cordlike with scalelike leaves, enlarging at apex to produce vegetative leaves, additional cordlike branches, and flowering stems. **Flowering stems** simple or branched, erect (declining or semierect and obviously zigzag in *I. brevicaulis* and some of its hybrids), solid or hollow, terete or slightly flattened. **Leaves:** basal 3–10, in fan; blade monofacial (except at base), smooth or ridged, sometimes centrally thickened, veins obscure to prominent; cauline 0–4 on branched stems, usually similar to basal, subtending each branch, decreasing in length distally, sometimes bracteiform and sheathing stem. **Inflorescences** rhipidiate, units 1 or more-flowered, spathaceous; spathes 2, herbaceous with scarious tips, or completely scarious, with or without distinct keel, often persisting and enclosing capsule(s) at maturity. **Flowers** lasting 1–4 days, upward-facing, usually somewhat fragrant (odor extremely unpleasant in *I. foetidissima*), pedicellate or sessile; perianth epigynous, white, yellow, tan, brown, copper-red, maroon, blue, blue-violet, or purple, often with markings of contrasting colors, differentiated into sepals and petals, actinomorphic, 4–18 cm diam.; floral tube distal to ovary, terete or occasionally ridged, solid proximally, hollow distally; sepals 3, spreading or reflexed, expanding either gradually or abruptly from claw into broader limb, with signal of prominent ridge, crest, distinct lines or dots, pubescence, or band of multicellular hairs (beard) along midline of claw and for short distance along midrib of limb; petals 3, erect, spreading or rarely reflexed, sometimes very reduced and mostly hidden by sepal bases; stamens opposite sepals, free but appressed to style branches; style dividing distally into petaloid branches, these arching outward and over stamens and claws of sepals, dividing at apex into 2 rounded or triangular lobes (style crests); stigma a lip of tissue on adaxial surface of style arm at base of crest; ovary terete or roundly 3- or 6-angled or -grooved. **Fruits** capsular, wall papery or becoming dry and hardened, sometimes indehiscent (in *I. giganticaerulea*). **Seeds** 4–20, in 1–2 rows per locule, often flattened in contact with others, rounded on outer surface; seed coat tan to dark brown (red in *I. foetidissima*), thin, membranous, and smooth, or conspicuously roughened to extensively corky (usually in wetland species), with or without obvious aril. x = uncertain.

Species ca. 280 (34 in the flora): Northern Hemisphere.

Because irises have large, handsome flowers, have many different forms, colors, and color patterns, and are easily grown and propagated, they have become popular garden ornamentals. Many non-native irises may be found in the flora area, not only in specialized gardens, but in casual plantings and persisting around old dwellings, in cemeteries, and along roadsides. These include not only the more common bearded irises, but members of the Sibirian, Spuria, Japanese, and Dutch groups. In the Pacific Coast and southeastern areas of the flora, repeated hybridization, both natural and artificial, of some of the local species has produced hundreds of garden forms that have added greatly to the popularity of this genus. Asexual reproduction in many *Iris* species may be more important than sexual reproduction in their persistence, and many hybrid clones may persist for decades in sites no longer cultivated.

The species of *Iris* differ from members of other native genera of the Iridaceae in North America in three ways: 1) the sets of outer and inner perianth members (sepals and petals herein) differ from each other in shape, orientation, and, in some cases, coloration; 2) the parts of the flowers—petals, sepals, stamens, and styles—are united basally into a floral tube that surmounts the ovary; and 3) the styles are petaloid and modified with some specialized structures to insure cross pollination.

T. Holm (1929) restricted the term rhizome to a horizontal, usually subterranean, stem that produces roots from its lower surface and green leaves from its apex, developed directly from the plumule of the embryo. He recognized stolons as axillary, subterranean branches that do not bear green leaves but only membranaceous, scalelike ones. The rhizomes of *Iris* are of two basic kinds: homogeneous, with the branches like the primary one in size and texture, and heterogeneous, with the branches of the primary rhizome like stolons with scalelike leaves. These enlarge at the apex to produce green leaves and more stolonlike branches.

According to B. Mathew (1990), *Iris* includes six subgenera. Species of subgenera *Iris* and *Limniris* are rhizomatous; the others are either bulbous or perennial from a fascicle of fleshy roots. Three species of subg. *Iris* are sometimes found naturalized from cultivation in the flora area. Other species of the subgenus are found here only in cultivation or as waifs. The native

North American species of *Iris* all belong to subg. *Limniris*, usually referred to as the beardless irises. The four other subgenera, which are found in the flora area only in cultivation or as waifs, have all been recognized sometimes as separate genera: subg. *Nepalensis* (Dykes) Lawrence as *Junopsis* Schulze; subg. *Xiphium* (Miller) Spach as *Xiphion* Miller; subg. *Scorpiris* Spach as *Juno* Trattinnick; and subg. *Hermodactyloides* Spach as *Iridodictyum* Rodionenko. F. Köhlein (1987) accepted these four segregate genera; however, most current authors circumscribe *Iris* in the broad sense, as here.

Elevation is unimportant in the taxonomy and ecology of *Iris*, and very rarely is it recorded on specimen labels. Consequently, elevations are not reported in the species treatments that follow.

SELECTED REFERENCES Anderson, E. 1936. The species problem in *Iris*. Ann. Missouri Bot. Gard. 23: 457–509. British Iris Society. 1997. A Guide to Species Irises: Their Identification and Cultivation. Cambridge and New York. Dykes, W. R. 1913. The Genus Iris. Cambridge. Foster, M. 1889. On irises. J. Roy. Hort. Soc. 11: 131–149. Foster, R. C. 1937. A cytotaxonomic survey of the North American species of *Iris*. Contr. Gray Herb. 119: 3–79. Henderson, N. C. 1991. *Iris* ×*pacifica*. Bull. Amer. Iris Soc. 280: 85–87. Henderson, N. C. 1994. The Louisiana irises. Bull. Amer. Iris Soc. 293: 73–82. Köhlein, F. 1987. Iris, transl. M. C. Peters. Portland. Mathew, B. 1990. The Iris, rev. ed. Portland. Wister, J. C. 1927. The Iris: A Treatise on the History, Development and Culture of the Iris for the Amateur Gardener. New York and London.

1. Sepals with multicellular hairs (beard) along midrib of claw and base of blade.
    2. Stems absent or not more than 1 cm; floral tubes 6–10 cm; capsules borne on tips of rhizomes . . . . . 1. *Iris pumila*
    2. Stems 6–12 dm; floral tubes 1–2.5 cm; capsules borne on ends of stems and branches.
        3. Spathes entirely scarious, silvery white . . . . . 2. *Iris pallida*
        3. Spathes herbaceous with narrow scarious margins and tip, green, sometimes purple at base . . . . . 3. *Iris germanica*
1. Sepals without beard, usually with signal of contrasting color, fine pubescence, ridges, or cockscomblike crest.
    4. Rhizomes heterogeneous, innovations appearing at some distance from parent rhizome.
        5. Sepals with 3-ridged, toothed crests.
            6. Stems 2.5–4.5 cm; floral tubes filiform, 4–8 cm; spathes sharply keeled . . . . . 4. *Iris cristata*
            6. Stems 0.8–4 cm; floral tubes funnelform, 1–2 cm; spathes slightly keeled . . . . . 5. *Iris lacustris*
        5. Sepals without crests or crests inconspicuous.
            7. Cordlike portions of rhizomes 4 dm; stems 3–8 dm; pedicels 3–7 cm; floral tubes 2–3 mm . . . . . 34. *Iris prismatica*
            7. Cordlike portions of rhizomes 0.1–2 dm; stems 0–3.5 dm; pedicels 0.4–3.5 cm; floral tubes 0.3–6.5 cm.
                8. Cordlike portions of rhizomes 10–20 cm; stems 1–2-branched; leaf blades 15 mm wide; pedicels 0.4–1 cm . . . . . 6. *Iris tenuis*
                8. Cordlike portions of rhizomes 1–15 cm; stems simple; leaf blades 3–13 mm wide; pedicels 1–3 cm . . . . . 7. *Iris verna*
    4. Rhizomes homogeneous; innovations appearing close to parent rhizome.
        9. Petals 1–2 cm.
            10. Petals without apical bristle; se United States . . . . . 8. *Iris tridentata*
            10. Petals with apical bristle 3–8 mm.
                11. Stems branched; Alaska, British Columbia, and Yukon . . . . . 9. *Iris setosa*
                11. Stems usually simple, with single, clasping leaf at midstem; Maritime Provinces and e Maine . . . . . 10. *Iris hookeri*
        9. Petals 2–9.5 cm.
            12. Stems hollow . . . . . 11. *Iris sibirica*
            12. Stems solid.
                13. Capsules 6-angled or -ridged.
                    14. Perianths copper colored or reddish brown, rarely yellow; petals spreading or declining with sepals . . . . . 29. *Iris fulva*
                    14. Perianths blue-violet or rarely white; petals erect or spreading-erect.

15. Stems declining or semierect, sharply zigzag . . . . 30. *Iris brevicaulis*
15. Stems erect, not obviously zigzag.
16. Capsules with 3 plane faces and 3 faces with 2 rounded ridges; strongly hexagonal in cross section . . . . 28. *Iris hexagona*
16. Capsules approximately circular in cross section, only slightly if at all hexagonal, with 6 narrow, winglike ridges or 6 broadly rounded lobes.
17. Capsules with 6 sharp, winglike ridges, dehiscent . . . . . 31. *Iris savannarum*
17. Capsules with 6 broad, rounded lobes, indehiscent . . . . 32. *Iris giganticaerulea*

[13. Shifted to left margin.—Ed.]

13. Capsules 3-angled or nearly round in cross section.
18. Rhizomes stout, 1–4 cm diam.; roots fleshy.
19. Perianths yellow or white.
20. Perianths entirely yellow . . . . 25. *Iris pseudacorus*
20. Perianths white with large yellow patch at base of sepals . . . . 33. *Iris orientalis*
19. Perianths blue-violet or rarely white.
21. Stigmas unlobed; leaf blades prominently veined; e, c United States and Canada.
22. Sepals with yellow pubescent signal at base of blade . . . . 27. *Iris virginica*
22. Sepals with pubescent, green or greenish yellow patch surrounded by heavily veined, purple-on-white signal at base of blade . . . . 26. *Iris versicolor*
21. Stigmas 2-lobed; leaf blades not prominently veined; w United States.
23. Pedicels slender; spathes scarious, 3.5–7 cm; perianths veined deep blue-violet on lighter background; capsules 4–5 cm; w Great Plains and mountains . . . . 23. *Iris missouriensis*
23. Pedicels stout; spathes herbaceous, 8–15 cm; perianths pale blue-violet or nearly white with prominent yellow ridge on sepal claws; capsules 8–9 cm; coastal nc California . . . . 24. *Iris longipetala*
18. Rhizomes slender, 3–12 mm diam.; roots fibrous.
24. Stems 1–4-branched . . . . 13. *Iris douglasiana*
24. Stems simple.
25. Blades of fan leaves with margins thickened . . . . 17. *Iris purdyi*
25. Blades of fan leaves with margins not thickened.
26. Cauline leaves bractlike, imbricated, blade inflated.
27. Floral tubes 0.8–0.9 cm . . . . 16. *Iris bracteata*
27. Floral tubes 3–12 cm.
28. Floral tubes linear; capsules 2–3 cm; pedicels absent or 0.5–1 cm at anthesis . . . . 18. *Iris chrysophylla*
28. Floral tubes funnelform; capsules 3–4 cm; pedicels 0.8–1.8 cm at anthesis . . . . 19. *Iris tenuissima*
26. Cauline leaves foliaceous, spreading, blade not inflated.
29. Floral tubes greater than 1.5 cm.
30. Floral tubes 1.5–3 cm . . . . 20. *Iris innominata*
30. Floral tubes 3–8 cm.
31. Spathes broadly lanceolate; leaf blades 7–8 mm wide; perianths creamy yellow . . . . 21. *Iris fernaldii*
31. Spathes linear-lanceolate; leaf blades 4–6 mm wide; perianths indigo blue, purple, lavender, white, cream, or yellow . . . . 14. *Iris macrosiphon*
29. Floral tubes 0.5–1.5 cm (occasionally to 2 cm in *I. tenax*).
32. Rhizomes creeping, not producing dense clumps. . . . . 15. *Iris hartwegii*
32. Rhizomes many-branched, producing dense clumps.
33. Stems 1.5–2.7 dm; inflorescence units 1–2-flowered; style crests with margins crenate or incised. . . . . 12. *Iris tenax*
33. Stems 6–7 dm; inflorescence units 3(–4)-flowered; style crests with margins entire or obscurely lobed . . . . 22. *Iris munzii*

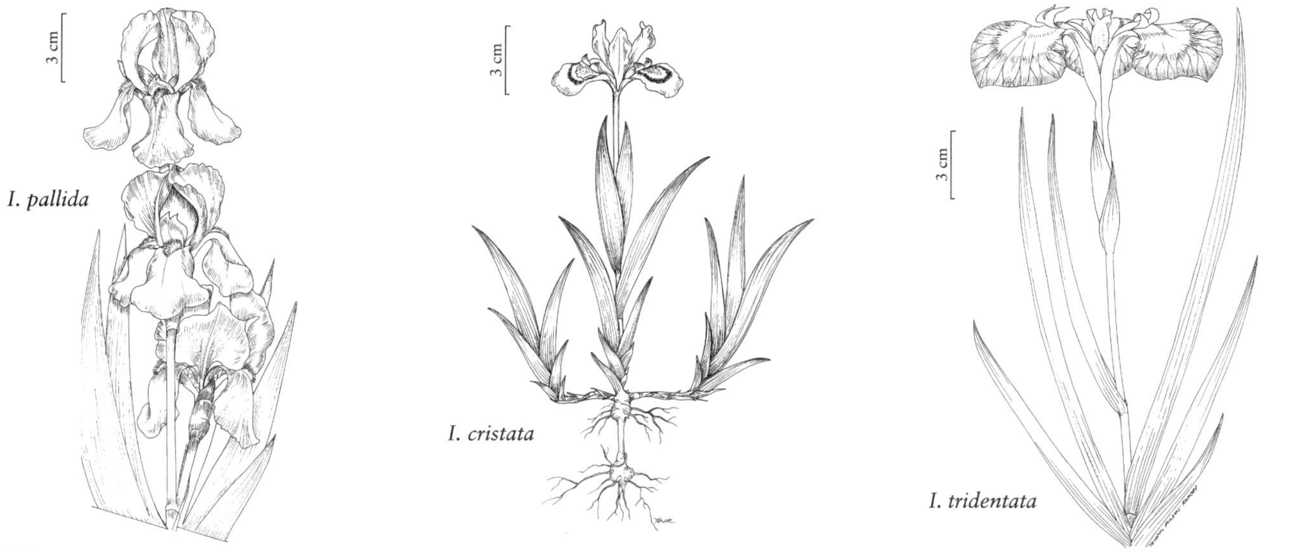

IRIS

4a. Iris Linnaeus subg. Iris • Bearded iris [I]

Sepals with signal of multicellular hairs (beard) [or papillae] along midrib of claw and base of limb. **Seeds** without [with] fleshy aril.

Species ca. 90 (3 in the flora): introduced; Eurasia.

The species of subg. *Iris* that are established in the flora outside cultivation all belong to sect. *Iris*.

1. Iris pumila Linnaeus, Sp. Pl. 1: 38. 1753 • Dwarf iris [I]

Rhizomes freely branching, producing dense clumps, 1–1.5 cm diam. **Stems** absent or not more than 1 cm. **Leaves** ascending-spreading; blade grayish green, ensiform or slightly falcate, 8–10 cm × 6–15 mm at anthesis, increasing to 15–20 cm in length on innovations after flowering, dying back in autumn, new growth only in spring, glaucous. **Inflorescence units** 1-flowered, sessile in spathes; spathes 5–10 cm; outer darker green than inner, inner closely shielding floral tube. **Flowers:** floral tube with 3 purple stripes in line with midrib of petals, slender, 6–10 cm × 2–3 mm, lifting expanded portion of flower out of spathes; sepals yellow, blue, or purple, with yellowish or bluish beard, often with contrasting spot around beard at base of limb, 4–5 × 1.2–2 cm; petals with darker veins near base, rounded-ovate, broader than sepals, 4–5 × 2.2–2.7 cm, narrowing abruptly to brown-veined claw, apex emarginate; ovary rounded-trigonal, ca. 1 cm × 4–5 mm; style crests triangular, margins serrate distally; stigmas rounded on distal margin, margins entire; pedicel absent. **Capsules** borne on tips of rhizomes at ground level or slightly below, rounded-trigonal, with short portion of dried floral tube forming tip, smooth, 2–3.5 × 1.5–2 cm, dehiscing while green. **Seeds** usually in 2 rows per locule, light brown, subspherical, 3–5 mm diam., wrinkled. $2n = 30, 31, 32, 36$.

Flowering late Mar–mid May. Dry, grassy areas; introduced; Ill., Ind., Mich., Ohio; c, e Europe.

Iris pumila has been shown to be a natural amphidiploid hybrid between *I. pseudopumila* Boissier & Heldreich and *I. attica* Tineo, both $2n = 16$. Like many other such hybrids, its range of variability is so extensive that many of its forms have been given specific names. In spite of the variability in color and size, it is now recognized as a distinct species. During the past 50 years, it has been crossed repeatedly with several other species, including *I. germanica* and its mixed progeny, and this has introduced patterns from the tall-bearded iris into the dwarfs, and vice versa. The ranges of flowering season and plant height are now almost continuous from one extreme to the other, and intermediate groups are now recognized.

Iris pumila is widely grown in gardens and often persists for years after cultivation around old dwellings and along roadsides. It is thus extremely difficult to document and map its spontaneous occurrence. Although most, if not all, of the modern dwarf irises are of hybrid parentage, a number of herbarium specimens from the nineteenth century seem to represent wild forms of *I. pumila*.

2. Iris pallida Lamarck in J. Lamarck et al., Encycl. 3: 294. 1789 • Sweet iris [F] [I]

Rhizomes branching, forming extensive clumps, 1.5–2 cm diam., smooth except for narrow leaf scars that completely encircle rhizome. **Stems** 1–3-branched, 6–10 dm. **Leaves** erect to spreading, blade green (some forms with white or yellow longitudinal stripes), ensiform, 3–6 dm × 3–4 cm, very glaucous, smooth. **Inflorescences** with terminal unit 3-flowered, branch units 2–3-flowered; distal branches subtended by scarious bracts, lower one to 15 cm, herbaceous, green; spathes completely silvery white, 2–3.5 cm, scarious. **Flowers** often very fragrant; perianth light blue-violet to mauve-purple, one form with blue pigment limited as stipples or stitches along margin and/or along veins, with ground color white on both petals and sepals; floral tube 1–1.3 cm; sepals slightly reflexed, obovate, 7.8–8.3 × 5–5.3 cm, with wedge-shaped claw marked with brown veins, beard yellow with white tips; petals blue-violet, in some forms lighter than sepals, with some brown veining at base and on claw, obovate, 8 × 5 cm; ovary trigonal, 6-grooved, 1–1.5 cm; style violet, fading paler along margins, keeled along midrib, 2–2.4 × 1 cm, crests rounded, 1.4 cm, apex acute; stigmas rounded, margins entire; pedicel very short, 2–3 mm. **Capsules** borne on ends of stems and branches, oblong, trigonal, 6-grooved, 4–5 × 1.5–2 cm. **Seeds** in 2 rows per locule, dark brown, compressed, cubical, wrinkled. $2n = 24$.

Flowering Apr–mid Jun. Widely cultivated and often persisting around old dwellings; introduced; Calif., Ga., Ind., Mo.; Europe.

Iris pallida is considered to be one of the parents of *I. germanica*, to which it contributed blue pigment (delphinin) as well as the recessive white allele responsible for white-flowered forms. It also contributed the recessive distribution-pattern allele that limits the blue pigment to veins and to stipples or larger dots along the margins of the petals and sepals, the plicata pattern so popular in the garden today.

3. Iris germanica Linnaeus, Sp. Pl. 1: 38. 1753 • German iris, flags [I]

Rhizomes homogeneous, usually many-branched, light brown, 1.2–2 cm diam., smooth, nodes marked by rings around rhizome, branches may arise in the fan or as many as 15–20 nodes produced prior to active leaves. **Stems** green, 2–3-branched, solid, 6–12 dm × 1–1.5 cm, glaucous. **Leaves** equitant, blade sometimes purplish at base, ensiform, to 4.5 dm × 3.5 cm, glaucous. **Inflorescences** with terminal unit 2–3-flowered, branch units 1–2-flowered; spathes green, sometimes with purple base, 2–5 cm, herbaceous with narrow, scarious margins and tip. **Flowers:** perianth shades of blue-violet, yellow, brown, or white with various patterns of pigment distribution; floral tube 1–2.5 cm; sepals spreading, drooping, or somewhat reflexed, blue-violet, yellow, brown, or white with patterned overlay of darker blue-violet, with white or yellow beard along midrib of claw and lower part of limb, obovate limb tapering gradually to claw, 6–7.5 × 4–5.5 cm; petals alternating with sepals, erect, obovate, 5–7 × 4–5.5 cm, with short, 1.5–2 cm, channeled claw; ovary roundly trigonal, 1.5–2.5 cm, slightly wider than floral tube. **Capsules** borne on ends of stems and branches, roundly 3-lobed, 3–5 × 2.5 cm, apex with short remnant of floral tube. **Seeds** in 2 rows per locule, red-brown, oval, 3–4 mm, wrinkled. $2n = 24$.

Flowering mid Apr–mid Jun. Widely grown, may persist after cultivation; introduced; c, s Europe.

Iris germanica is considered to have been a natural hybrid between *I. pallida* and *I. variegata* Linnaeus, both of which also have the chromosome number $2n = 24$. Whenever an interbreeding population of a fertile natural hybrid occurs, the plants can be expected to exhibit characters varying from one parental extreme to the other, and this is certainly true in the case of *I. germanica*. *Iris pallida* has striking silvery spathes subtending the flowers, whereas *I. variegata* has completely green or purplish, herbaceous spathes. Usually *I. germanica* has the spathes green or purplish at the base with scarious margins. Since *I. pallida* is up to 1 m tall and *I. variegata* is only up to 4 dm tall, *I. germanica* would be expected to be intermediate in height (this is true), but the forms have subsequently been selected for greater height and thus have become established as the "tall bearded irises."

The perianths of *Iris pallida* are primarily blue, but one form had the blue pigment limited to stipples or stitches on the margins of both sepals and petals, and this recessive pattern was carried over into *I. germanica*. The perianths of *I. variegata* are yellow, with the veins of the sepals marked with a dark blue-violet that gives, along with the yellow background, a purple or brown effect. In some forms, the pigment of these veins spreads over most of the surface of the sepals, leaving only a small band of background color along the margins. In *I. germanica*, if this pattern introduced from *I. variegata* is added to a light blue-violet background, the petals remain light blue and the sepals become a very dark violet with light blue margins; this form has been called *I. neglecta* Hornemann. If the pattern is added to a white flower, the result has been called *I. amoena* de Candolle. A form in which the dark blue veins were replaced by white has been named *I. leucographa* Kerner. Another form in which this white pigment had spread to produce a large white area on the sepals with a narrow yellow border has been called *I. flavescens* Delile. All of these forms were known early in the breeding of the garden cultivars of *I. germanica* and added greatly to their popularity.

J. C. Wister (1927) said that *Iris pallida* and *I. variegata* were the only species involved in the production of *I. germanica* until about 1889, when Sir Michael Foster of England had several large-flowered irises sent to him from the eastern part of the Mediterranean: *I. cypriana* Foster & Baker, *I. trojana* A. Kerner ex Stapf, and *I. mesopotamica* Dykes. These he crossed with the best of the *I. germanica* forms that he had. He didn't know it at the time, but these Mediterranean species were tetraploids, and with them he began to produce larger plants and more different patterns, which changed the entire direction of iris breeding. Since that time, many other species from Europe and Asia have been brought into the breeding: *I. subbiflora* Brotero, *I. humilis* M. Bieberstein, *I. reichenbachii* Heuffel, *I. imbricata* Lindley, *I. attica* Boissier & Heldreich, *I. aphylla* Linnaeus, *I. albicans* Lange, and many others. The plants resulting from hybridization with those other species can not be considered as *I. germanica*, but must have another name: *I. ×conglomerata* N. C. Henderson has been proposed.

SELECTED REFERENCES Henderson, N. C. 1992. What is *Iris germanica*? Bull. Amer. Iris Soc. 286: 6–11. Henderson, N. C. 1993. *Iris ×conglomerata*. Bull. Amer. Iris Soc. 290: 17–22.

4b. Iris Linnaeus subg. Limniris (Tausch) Spach, Ann. Sci. Nat., Bot., sér. 3, 5: 99. 1846 • Beardless iris

Iris sect. *Limniris* Tausch, Hort. Canal. 1. 1823; *Limniris* (Tausch) Fuss

Sepals without beard, but with signal of crests or ridges, distinct lines or dots, or a patch of contrasting color.

Species ca. 82 (31 in the flora): North America, Eurasia.

4b.1. Iris Linnaeus (subg. Limniris) sect. Lophiris Tausch, Hort. Canal. 1. 1823

Evansia Bentham

Rhizomes heterogeneous, widely spreading. **Sepals** with signal of one or more ridges or crests.

Species 12 (3 in the flora): North America, e Asia.

SELECTED REFERENCE Hannan, G. L. and M. W. Orick. 2000. Isozyme diversity in *Iris cristata* and the threatened glacial endemic *I. lacustris*. Amer. J. Bot. 87: 293–301.

4. Iris cristata Solander ex Aiton, Hort. Kew. 1: 70. 1789 • Crested iris [E] [F]

Neubeckia cristata (Solander ex Aiton) Alefeld

Rhizomes producing fleshy roots, and 2–8 cordlike branches at apex, cordlike portion 2–3 dm × 1–2 mm, gradually enlarging to 8–12 mm diam., nodes with brown, scalelike leaves and rarely roots. **Stems** simple, 2.5–4.5 cm. **Leaves:** basal 6–8, proximal 2–3 sheathing, blade light brown with darker brown line along midrib at base, falcate, scarious, distal 4–5 not sheathing, blade green or yellowish green, with few, subprominent veins, broadly ensiform, slightly falcate, to 1.5 dm × 1–2.5 cm, enlarging to 4 dm after anthesis; cauline 2–3, sheathing, proximal very similar to basal leaves, distal 1 or 2 reduced, herbaceous, blade falcate, not inflated. **Inflorescence units** 1–2-flowered; spathes green, sharply keeled, somewhat inflated, 2–6 cm, unequal, outer shorter than inner. **Flowers:** perianth blue, lilac-purple, or white; floral tube filiform, widening distally, 4–8 cm, lifting expanded portion of flower out of spathes; sepals spreading, with 3 parallel, toothed, crested ridges on white signal bordered with purple, tapering gradually into claw, 3–6 × 1.5–2.5 cm, base gradually attenuate, apex rounded, emarginate; petals spreading, same color as sepals, oblanceolate, 3–4 × 1–2 cm; ovary triangular, with shallow groove along each face, 0.6–1 cm; style 1.5 cm, crests narrowly triangular, 6–9 mm; stigmas oblong, margins entire; pedicel 0.7–1.8 cm. **Capsules** usually enclosed in spathes, oval, sharply triangular, each angle ridged, 1–1.5 cm. **Seeds** yellowish brown,

3.2–3.5 mm, smooth, with narrow, white appendage wrapped around seed, 3.4–4 mm, quickly drying upon exposure to air. $2n = 24, 32$.

Flowering Apr–Jul. Rich woods, ravines, bluffs, usually in calcareous soil; Ala., Ark., D.C., Ga., Ill., Ind., Iowa, Ky., Md., Miss., Mo., N.C., Ohio, Okla., Pa., S.C., Tenn., Va., W.Va.

5. **Iris lacustris** Nuttall, Gen. N. Amer. Pl. 1: 23. 1818 • Dwarf lake iris [E]

Iris cristata Solander ex Aiton subsp. *lacustris* (Nuttall) H. H. Iltis; *I. cristata* var. *lacustris* (Nuttall) Dykes

Rhizomes with proximal cordlike portion 0.8–2 dm × 2 mm, bearing brown, scalelike leaves and rarely roots, distal portion enlarging to 3–5 × 0.8–1.2 cm, producing fibrous roots, 2–3 brown, sheathing, bractlike leaves, then 8–12 green foliage leaves; usually 2 cordlike rhizomes produced from apex as new growth begins in spring. **Stems** simple, 0.8–4 cm. **Leaves:** basal with blade broadly linear at anthesis, slightly falcate, 4–6 cm × 6–8 mm, enlarging to 16 cm × 10 mm; cauline with proximal 1–2 resembling basal leaves, 4–6 cm, distal 2 more sheathing of stem, exceeding spathes. **Inflorescence units** 1(–2)-flowered; spathes green, outer closely sheathing inner and enclosing ovary, slightly keeled at midrib, 4.5 cm, ± equal, margins scarious. **Flowers:** perianth sky blue; floral tube dull yellow, funnelform, 1–2 cm, dilated upward to 4–5 mm diam.; sepals 2–2.3 × 0.8 cm, tapering gradually into claw, apex emarginate, signal white with deep violet margin, with 3 yellow and white, toothed, low ridges; petals spreading, narrowly oblanceolate, 1–1.5 × 0.4–0.5 cm, shorter and narrower than sepals, apex emarginate; ovary sharply trigonal, 0.8–1 cm; style 1–2 cm, crests linear to semiovate, 4 mm, margins crenate; stigmas rounded, margins entire; pedicel 0.6–1.5 cm. **Capsules** roundly triangular, 1.2 × 8 mm, enclosed in spathes. **Seeds** dark brown, with white appendage spiraled around seed, ca. 3 mm, quickly drying when exposed to air. $2n = 32, 42$.

Flowering May. Moist, sandy gravel, limestone shores of lakes; Ont.; Mich., Wis.

Several authors have described *Iris lacustris* simply as being very similar to *I. cristata*, but smaller in every detail. W. R. Dykes (1913) said, "This name may be retained as that of a local form or variety but it hardly deserves specific rank." He treated the taxon as *I. cristata* var. *lacustris*. R. C. Foster (1937) discussed the relationship between *I. cristata* and *I. lacustris* in detail, and recognized the latter at specific rank. He stated, "It is interesting to note that it has been recorded in several localities on or near the Bruce Peninsula, Ontario, that home of so many pre-Wisconsin relics. Professor Fernald informs me that he has seen it growing there in some abundance on the upper levels of the sandy beaches. Quite probably, it is present there, not as a newcomer, but as a relic." However, G. L. Hannan and M. W. Orick (2000) concluded that *I. lacustris* is a product of some post-glacial invasion or introduction of a plant or seed of *I. cristata* that has developed its own very stable and quite unvariable character.

6. **Iris tenuis** S. Watson, Proc. Amer. Acad. Arts 17: 380. 1882 • Clackamas iris [C] [E]

Rhizomes superficial or very shallow, cordlike portion 1–2 dm × 2 mm, expanding to 10–15 mm diam., nodes with brown, scalelike leaves and few to no roots. **Stems** 1–2-branched, 3–3.2 dm. **Leaves:** basal with blade pale green, 3.2 dm × 1.5 cm, margins scarious basally, apex acute, proximal 2 semi-sheathing, blade 5 cm, scarious; cauline 1–2, sheathing branch and stem for ca. 1/2 their length, blade ca. 5 cm, semimembranous. **Inflorescences** 2–3-branched, each unit with single flower, all borne at approximately the same level; spathes 2–3 cm × 5 mm, subequal, scarious except basally and along midrib. **Flowers:** perianth white or blue tinged with deep violet lines; floral tube funnelform, 0.3 cm; sepals oblong-spatulate, 2.8 × 1 cm, apex deeply emarginate, signal an inconspicuous crest with low, yellow, undissected ridge; petals bluish white, oblanceolate-spatulate, base gradually attenuate into claw; ovary elliptical, 0.4–0.7 cm; style 1.8 cm, crests broadly obovate, 0.7 cm, margins erose; stigmas triangular-acuminate, margins entire; pedicel 0.4–1 cm, not lifting flower clear of spathes. **Capsules** depressed-globose, roundly triangular, 0.9–1.5 × 1.2 cm. **Seeds** pale brown, with whitish raphe, D-shaped, pitted. $2n = 28$.

Flowering May. Open, wooded slopes in leafy soil with Douglas fir and dense shrubs; of conservation concern; Oreg.

Iris tenuis was originally placed in ser. *Californicae*, but R. C. Foster (1937) stated, "It is with some hesitation that I leave it in association with them. The broad, pale green leaves are much like those of a giant *I. cristata*." F. H. Smith and Q. D. Clarkson (1956) said, "It clearly does not belong in the subsection with the other members of the *Californicae*," and Clarkson in a later treatment (1958) created a new subsection, the *Oregonae*, for it. L. W. Lenz (1959b) moved this species into subsect. *Evansia* (= sect. *Lophiris*), with which it shows many relationships.

4b.2. IRIS Linnaeus (subg. LIMNIRIS) sect. LIMNIRIS Tausch, Hort. Canal. 1. 1823

Rhizomes usually homogeneous (heterogeneous in ser. *Prismaticae*; heterogeneous or torulose in ser. *Vernae*), spreading or compact. **Sepals** with signal of conspicuous lines or patch of contrasting color.

Species ca. 70 (28 in the flora): North America, Eurasia.

4b.2a. IRIS Linnaeus (subg. LIMNIRIS, sect. LIMNIRIS) ser. VERNAE (Diels) G. H. M. Lawrence, Gentes Herb. 8: 363. 1953 [E]

Iris subsect. *Vernae* Diels in H. G. A. Engler and K. Prantl, Nat. Pflanzenfam. ed. 2, 15a: 502. 1930

Rhizomes heterogeneous or torulose. **Sepals and petals** subequal, signal yellow or orange basal band, margins not fimbriate.

Species 1: se United States.

7. Iris verna Linnaeus, Sp. Pl. 1: 39. 1753 • Dwarf violet iris [E]

Neubeckia verna (Linnaeus) Small

Rhizomes heterogeneous, whitish, cordlike portions 0.1–1.5 dm × 2–4 mm, enlarging to 6–8 mm diam., densely covered with brown, scale-like leaves, roots absent, or torulose with roots borne along entire rhizome. **Stems** simple, 5–15 cm. **Leaves:** basal with blade light green, ensiform, 0.3–1.5 dm × 0.3–1.3 cm, enlarging to 3.5 dm, glaucous; cauline 5–9, sheathing, imbricate, blade light green, obovate, 1.3–5 cm, increasing in length, proximal shortest, membranous, apex acute. **Inflorescence units** 1–2-flowered; spathes divergent, exposing floral tube, green, lanceolate, 2–2.5 cm, apex acuminate. **Flowers:** floral tube pale violet, filiform, 2.5–6.5 cm, expanding somewhat distally to 2.5–4 cm diam.; sepals widely spreading, blue to violet with yellow or orange longitudinal papillose band at base of blade, obovate, 2–6 × 0.8–2 cm, base gradually attenuate into claw, not crested; petals erect, arching inward at tip, spatulate, 2–7 × 1–2 cm, base abruptly attenuate into claw; ovary linear, to 1.3 cm; styles pale violet, 2.5–4 cm, crests linear-acute, narrow, 0.7 cm; stigmas rounded, margins entire; pedicel 1–3 cm, increasing to 25 cm as capsule matures. **Capsules** 3-angled with single ridge at each angle, almost hidden in bases of spathes, 1–3.2 × 0.8–1.5 cm, tapering into beak consisting of dried remnant of floral tube. **Seeds** dark brown, ribbed, 2.8–3.2 mm, lustrous, with small, fleshy aril basally.

Varieties 2 (2 in the flora): se United States.

1. Rhizomes 5–15 cm between offshoots; leaf blades 0.3–0.8 cm wide; capsules 1.2–1.8 cm . . . . 7a. *Iris verna* var. *verna*
1. Rhizomes 1.3 cm between offshoots; leaf blades 5–13 cm wide; capsules 2–3.2 cm . . . . 7b. *Iris verna* var. *smalliana*

7a. Iris verna Linnaeus var. **verna** [E]

Rhizomes 5–15 cm between offshoots. **Leaf blades** 0.3–0.8 cm wide. **Capsules** 1.2–1.8 cm. $2n = 42$.

Flowering late Mar–early May. Dry to moist, acid, sandy, or peaty soil; D.C., Ga., Md., N.C., S.C., Va.

7b. Iris verna Linnaeus var. **smalliana** Fernald, Rhodora 49: 214. 1947 • Upland violet iris [E]

Rhizomes 1.3 cm between offshoots, forming very dense clumps. **Leaf blades** 0.5–1.3 cm wide. **Capsules** 2–3.2 cm. $2n = 42$.

Flowering mid Apr–mid May. Acid soil, semishaded or open woodlands; Ala., Ark., Fla., Ga., Ky., Md., Miss., N.C., Ohio, Pa., S.C., Tenn., Va., W.Va.

J. K. Small (1929b) illustrated this taxon as the "mountain form" of *Iris verna*, without giving it an infraspecific name. In 1933 he did not mention it at all and placed the species in *Neubeckia*.

SELECTED REFERENCE Small, J. K. 1929b. *Iris verna* (mountain form). Addisonia 14: 15–16, plate 456.

4b.2b. IRIS Linnaeus (subg. LIMNIRIS, sect. LIMNIRIS) ser. TRIPETALAE (Diels) G. H. M. Lawrence, Gentes Herb. 8: 363. 1953

Iris subsect. *Tripetalae* Diels in H. G. A. Engler and K. Prantl, Nat. Pflanzenfam. ed. 2, 15a: 502. 1930

Petals much reduced, often ending in bristles.

Species 3 (3 in the flora): North America, e Asia.

8. Iris tridentata Pursh, Fl. Amer. Sept. 1: 30. 1814 • Savannah iris [E] [F]

Iris tripetala Walter 1788, not Linnaeus f. 1782

Rhizomes extensively branching, to 12 × 0.5–0.7 cm, becoming 1.5–2 cm diam., with coarse, strongly ribbed, brown, scalelike leaves. **Stems** simple or rarely 1-branched, 3–7 dm. **Leaves:** basal with blade green, often with red-brown margins, lightly ribbed, linear-ensiform, 3–5 dm × 1.5–2.3 cm, glaucous, apex acute; cauline 2–3, spreading, blade narrowly linear, 0.5–0.7 cm wide, apex acute. **Inflorescence units** 1-flowered, branch unit (if present) 1-flowered; spathes covering pedicel, ovary, floral tube, and base of sepals, unequal, outer 3 cm × 10 mm, inner 6 cm × 10 mm. **Flowers:** perianth blue-violet; floral tube with brown striations, funnelform, 2–2.5 cm; sepals blue-violet with deeper veins, with yellow-white signal, orbicular, 7–8 × 3–4 cm, base abruptly attenuate into whitish claw with yellow-brown veining; petals mostly hidden by bases of sepals, much reduced, 1.5 × 0.3–0.5 cm, bristle absent; ovary trigonal, with shallow groove at each angle, 1.5 cm; style 3.5–4 cm, crests linear to subquadrate, 1–1.8 cm, margins sometimes incised; stigmas semicircular, margins entire; pedicel 2–3.5 cm. **Capsules** globose to oblong, obtusely 3-angled, with rounded angles, 2.5–4 × 2 cm, rounded basally, abruptly contracted into beak apically. **Seeds** in 1 row per locule, dark red-brown, semicircular, flattened, 6–8 mm; seed coat thick. $2n = 40$.

Flowering May–Jun(–Oct). Rich, swampy, shaded places, usually along coastal plain; Ala., Fla., Ga., La., N.C., S.C., Tenn.

Iris tridentata is quite different from the other two species of ser. *Tripetalae* that occur in the flora. The rhizome has been described by W. R. Dykes (1913) as “almost stoloniferous,” by J. K. Small (1933), “the cordlike rootstocks are peculiar,” and by R. K. Godfrey and J. W. Wooton (1979), “clothed with coarse, strongly many-ribbed, brown, overlapping scales.” The brown scalelike leaves are produced very close together and are long enough that they appear as a small fan; as the internodes elongate, the scales are pulled apart but still overlap along the rhizome. Roots are produced from the lower side of the rhizome at the nodes. Branches may appear at any node along the narrow portion of the rhizome rather than just from the broader apex.

9. Iris setosa Pallas ex Link, Jahrb. Gewächsk. 1(3): 71. 1820 • Beach-head iris

Iris arctica Eastwood

Rhizomes many-branched, producing dense clumps, 3–6 × 1.5–2 cm, covered with remnants of old leaves. **Stems** 1–2-branched, 1.5–9 dm, exceeding leaves; branches 1.25–1.5 dm proximal to terminal, but bearing flowers on nearly the same level. **Leaves:** basal with blade green with purple tinge basally, prominently ribbed, ensiform, 0.2–0.45 dm × 1–2 cm, ± glaucous distally; cauline subtending each lateral branch, blade reduced, 0.5–1.5 dm. **Inflorescence units** 2–3-flowered, lateral units usually 2-flowered; spathes green flushed with purple, foliaceous, ovate-lanceolate, 2.5–4.5 cm × 5–7 mm, unequal, inner longer than outer, sometimes scarious. **Flowers:** perianth dark blue-violet to red-purple with darker veins; floral tube ca. 0.7 cm; sepals 4–6 × 3–5 cm, glabrous, base abruptly attenuate to broad claw with undulate margins; petals much reduced, 1–2 × 0.3–0.4 cm, widest basally, mostly hidden by bases of outer sepals, apex acuminate, with 3–8 mm bristle; ovary conspicuously 3-angled, inflated at anthesis, 1.3 cm; style 2.5 cm, crests overlapping, subquadrate, 0.7 cm, margins incised; stigmas semicircular, margins entire; pedicel 2.5–4.5 cm, ± equaling spathes. **Capsules** roundly trigonal, 2.5 × 1.3–1.5 cm. **Seeds** in 2 rows per locule, light brown, with prominent raphe, D-shaped, 2–3 mm, very smooth. $2n = 34, 36, 38$.

Flowering Jul–Aug. Boggy meadows, shores, and dunes; B.C., Yukon; Alaska; e coastal Asia to Japan.

Iris setosa was originally described from Siberia; whether our populations are an extension from or were formerly continuous with the Asian ones is not known. Two varieties from Japan, var. *hondoensis* Honda and var. *nasuensis* Hara, with $2n = 54$ may be triploids.

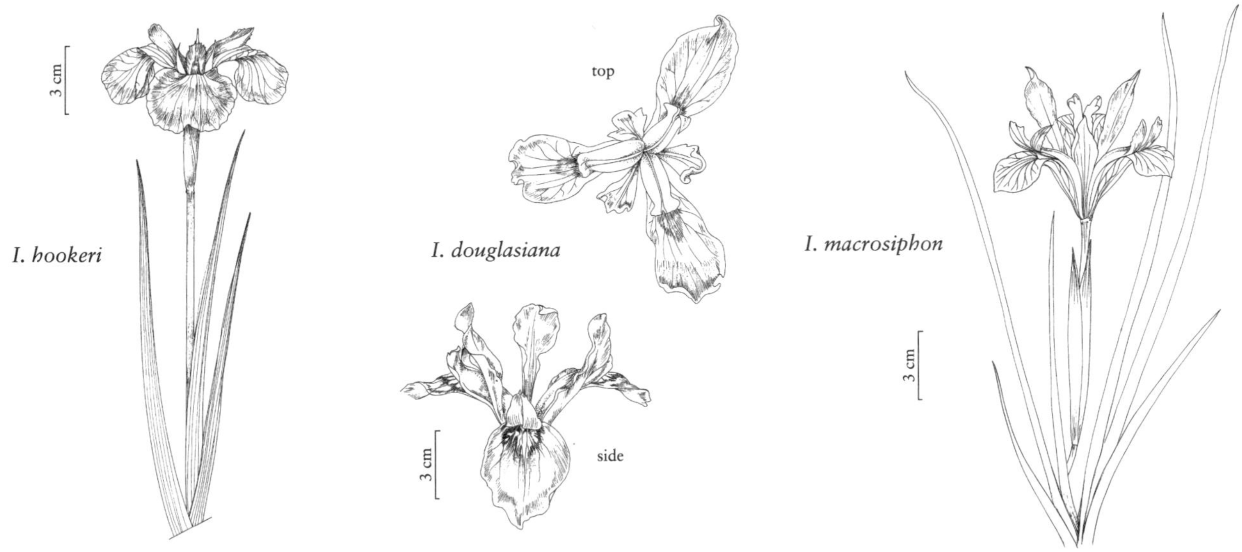

IRIS

10. Iris hookeri Penny ex G. Don in J. C. Loudon, Hort. Brit. ed. 2, 591. 1832 • Beach-head iris, iris de Hooker [E] [F]

Iris canadensis (Foster) Peckham; *I. setosa* Pallas ex Link var. *canadensis* Foster; *I. setosa* subsp. *pygmaea* C. E. Lundström

Rhizomes many-branched, forming dense, cespitose clumps with many crowded fans with persistent old leaf bases, 3–5 × 1.3–1.5 cm. **Stems** several to many from a single clump of fans, simple, 0.5–6 dm. **Leaves:** basal erect or strongly ascending, blade 1–5.2 dm × 0.5–1.4 cm; cauline 2–4, proximal 2–3 similar to basal leaves, distalmost leaf occasionally borne nearly midway on stem, clasping, blade bracteiform, lanceolate, 0.5–1.5 dm. **Inflorescence units** 1–2-flowered; spathes herbaceous, lanceolate to oblong-ovate, 2.5–6 cm, firm, outer spathe apex acute, inner similar or scarious at apex. **Flowers:** perianth deep to pale blue or blue-violet; floral tube 0.5–0.75 cm; sepals broadly rounded, 2.5–4 cm wide, base strongly attenuate to claw, signal diffuse white basal patch; petals much reduced to insignificant, involute or tubular rudiments, 1–2 cm, apex with short bristle, mostly hidden by sepal bases; ovary green or flushed purple, acutely trigonal; style white with purple keel, 2–5 cm, crests overlapping, 2-lobed, subquadrate, margins coarsely serrate; stigmas rounded-triangular, margins entire; pedicel 2.5–4 cm. **Capsules** thin walled, trigonal, with rounded angles and grooved sides, 2–4 cm, apex blunt. **Seeds** in 2 rows per locule, dark brown with prominent white raphe, compressed-pyriform, 4–6 mm, lustrous. $2n = 38$.

Flowering Jun–Jul. Grassy headlands, upper borders of beaches, dunes, and other coast formations, within reach of ocean spray; N.B., Nfld. and Labr., N.S., P.E.I., Que.; Maine.

Some authors still treat this taxon as *Iris setosa* var. *canadensis*, but others believe that it merits recognition as a separate species. I have seen *I. hookeri* flowering in its native habitat in both Nova Scotia and Prince Edward Island, and have examined dozens of herbarium specimens of both it and *I. setosa*, from which it is clearly distinct.

4b.2c. IRIS Linnaeus (subg. LIMNIRIS, sect. LIMNIRIS) ser. SIBIRICA (Diels) G. H. M. Lawrence, Gentes Herb. 8: 358. 1953 [I]

Iris subsect. *Sibiricae* Diels in H. G. A. Engler and K. Prantl, Nat. Pflanzenfam. ed. 2, 15a: 501. 1930

Sepals with 2 flanges at base of blade; stigmas triangular. **Capsules** 3-cornered or almost round in cross section.

Species ca. 8 (1 in the flora): introduced; c, e Europe, Asia.

11. Iris sibirica Linnaeus, Sp. Pl. 1: 39. 1753 • Siberian iris [I]

Rhizomes compact, freely branching, forming dense clumps, 0.9–1.2 cm diam., covered with remnants of old leaves. **Stems** simple or 1–3-branched, hollow, 6–12 dm. **Leaves** dying back in winter, blade dark green, often tinged pink at base, 4–8 dm × 0.4–0.6 cm. **Inflorescence units** 3–5-flowered, lateral units 2–3-flowered; spathes brown, to 4 cm, narrow, papery, apex acute. **Flowers:** perianth light to dark blue-violet to white; floral tube with indistinct ribs, circular, ca. 1 cm; sepals flaring or curving downward apically, widely orbiculate, 5–7 × 2–2.5 cm, base abruptly attenuate into claw with two narrow flanges basally, signal white, semicircular, with dark violet veins basally; petals erect, narrowly elliptic-obovate, 4.5–5.5 × 1.5–1.8 cm; ovary roundly triangular, spindle-shaped, 1.5–2 cm; style pale blue, bluntly keeled, 4–5 cm, crests overlapping, triangular, margins crenate; stigmas tonguelike projections, triangular; pedicel 1–15 cm, unequal, later flowers in each spathe with longer pedicel. **Capsules** roundly triangular with low ridges at angles, 3–4.5 × 1–1.3 cm, smooth, apex with extremely short tip, opening only in upper $^1/_4$–$^1/_3$ of capsule. **Seeds** in 2 rows per locule, dark brown, D-shaped, flattened, 5 × 3 mm, slightly roughened by small, rounded protuberances. $2n = 28$.

Flowering May–Jul. Widely cultivated, found along roadsides; introduced; Ont.; Calif., Conn., Maine, Mass., N.Y., Pa., Vt.; expected elsewhere; Eurasia.

Many forms of *Iris sibirica* have been cultivated widely across North America, where it is quite hardy and persistent.

4b.2d. IRIS Linnaeus (subg. LIMNIRIS, sect. LIMNIRIS) ser. CALIFORNICAE (Diels) G. H. M. Lawrence, Gentes Herb. 8: 360. 1953 [E]

Iris subsect. *Californicae* Diels in H. G. A. Engler and K. Prantl, Nat. Pflanzenfam. ed. 2, 15a: 501. 1930

Stigmas triangular or bilobed. **Capsules** 3-ribbed, 3-cornered, or cylindrical. **Seeds** without aril or raphe.

Species 11 (11 in the flora): w United States.

Series *Californicae* presents some of the most complex taxonomic problems in all of our American irises. All but one species of the series have produced naturally occurring primary hybrids with other species of the group, the one exception being *Iris munzii*. Two have hybridized with only one other species of the group: *I. bracteata* with *I. chrysophylla*, and *I. hartwegii* with *I. macrosiphon*.

Horticulturally, primary hybrids involving every one of the species have been produced and introduced into gardens. *Iris douglasiana* has been hybridized with all ten of the others. Only one of the primary hybrids has been given a nothospecific name: *I.* ×*thompsonii* (*I. douglasiana* × *I. innominata*).

All of these species have $2n = 40$. The primary hybrids that have been studied cytologically have shown no significant abnormalities during meiosis, and most of them have had a high degree of fertility. In turn, primary hybrids have been crossed with others, as well as with yet other species, over and over again. Many of these multispecific hybrids have been given cultivar names, and have been propagated vegetatively and widely dispersed over those areas of the country in which they are hardy.

L. W. Lenz (1959) considered very carefully the role of introgressive hybridization within this group. Since every species in the series has the same chromosome number and a great percentage of the hybrids are fertile, as the ecological barriers are decreased or eliminated by the efforts of man, introgression will increase as more and more are brought into contact with others. In considering *Iris tenax* subsp. *klamathensis*, Lenz remarked, "This subspecies belongs with the *I. tenax–hartwegii–munzii* complex and it is most closely related to *I. tenax*. Indeed in many ways it might be considered as a member of that species. However, there are certain morphological differences, most important of which is the difference in perianth-tube length." Are the differences due to introgression with some of the other species? Should this be separated as a nothospecies? The same questions could be asked regarding the four subspecies of *I. hartwegii*. Lenz concluded that "From present available evidence it would appear that primary speciation in the *Californicae* has taken place principally, though perhaps not exclusively, through ecological isolation of sympatric forms and there has been little or no chromosomal re-patterning within the group. As a result, the species possess the potential of producing vigorous and fertile hybrids when any breakdown occurs in the ecological barriers separating the species. When this does happen, one of the possible results is the merging of separate species into a single interbreeding population providing the isolating barriers are altered sufficiently." This he termed "secondary speciation." He went on to observe that "Opposed to the mass fusion of two species into a single interbreeding population is the mere trickle of genes across an interspecific barrier which is manifest in the production of an occasional hybrid. Between the two extremes there can be every intermediate stage."

SELECTED REFERENCES Lenz, L. W. 1958. A revision of the Pacific Coast irises. Aliso 4: 1–72. Lenz, L. W. 1959. Hybridization and speciation in the Pacific Coast irises. Aliso 4: 237–309.

12. Iris tenax Douglas ex Lindley, Edwards's Bot. Reg. 15: plate 1218. 1829 • Tough-leaf iris, Oregon flag [E]

Iris gormanii Piper; *I. tenax* var. *gormanii* (Piper) R. C. Foster; *I. tenax* subsp. *klamathensis* L. W. Lenz

Rhizomes many-branched, not creeping, forming dense clumps, slender, 0.3–0.8 cm diam.; roots fibrous. **Stems** simple, somewhat angular, solid, 1.5–2.7 dm. **Leaves:** basal somewhat lax, overtopping stem, blade light green, paling to pink or straw color basally, finely ribbed, linear-acute to linear, ensiform, 4.5 dm × 0.5 cm, margins not thickened; cauline 1–3, sheathing for 1/2 length then spreading, foliaceous, blade linear-lanceolate, not inflated, narrow, reduced, to 15 cm. **Inflorescence units** 1–2-flowered; spathes distant basally by 3 cm in some cases, keeled, linear-lanceolate to lanceolate-acuminate, 5–7 cm × 2–4 mm, unequal, outer longer than inner, herbaceous, margins scarious. **Flowers:** perianth color variable, purple, pink, lavender, cream, yellow, or rarely white; floral tube funnelform, 0.6–2 cm; sepals veined with color of limb, obovate, 5.8–6 × 1.6–2 cm, base gradually attenuate into white claw with slight yellow ridge, apex emarginate to bluntly rounded; petals same color as sepals, not prominently veined, lanceolate to oblanceolate, 5 × 1 cm; ovary 1–2 cm, slightly wider distally, base very gradually attenuate; style 2.2–3.2 cm, crests subquadrate, 8–12 mm, margins crenate or incised; stigmas triangular, margins entire; pedicel 1–5 cm, longer in second flower (when present) than in first. **Capsules** oblong, triangular in cross section, prominently ribbed, beaked, 3–5 cm. **Seeds** brown, D-shaped to irregular, wrinkled.

Flowering Jun–Aug. Dry soils in fields and open woods; Calif., Oreg., Wash.

Plants described as subsp. *klamathensis* differ from others of the species in having floral tubes 11–20 mm instead of 6–10 mm, style crests somewhat longer and narrower, and flower color and markings more like those of *Iris bracteata* or *I. innominata*. The latter species has an even longer floral tube, 15–30 mm, which could be evidence of introgression. Hybrids are known to occur in the area common to both species, in Douglas County, Oregon.

Iris tenax hybridizes with *I. bracteata*, *I. chrysophylla*, *I. douglasiana*, *I. hartwegii*, *I. innominata*, *I. macrosiphon*, *I. purdyi*, and *I. tenuissima*.

13. Iris douglasiana Herbert in W. J. Hooker and G. A. W. Arnott, Bot. Beechey Voy., 395. 1840 • Mountain iris [E] [F]

Iris beecheyana Herbert; *I. douglasiana* var. *alpha* Dykes; *I. douglasiana* var. *altissima* Purdy ex Jepson; *I. douglasiana* var. *beecheyana* (Herbert) Baker; *I. douglasiana* var. *bracteata* Herbert; *I. douglasiana* var. *major* Torrey; *I. douglasiana* var. *mendocinensis* Eastwood; *I. douglasiana* var. *nuda* Herbert; *I. douglasiana* var. *oregonensis* R. C. Foster; *I. watsoniana* Purdy

Rhizomes freely branching, forming large colonies, slender, 0.8–0.9 cm diam., covered with remains of old leaves; roots fibrous. **Stems** 1–4-branched, solid, 1.5–7 dm. **Leaves:** basal with blade yellow-green, occasionally darker green, sometimes flushed pink or red basally, prominently ribbed, linear, 4.5–10 dm × 2 cm, apex acute; cauline 1–3, reduced. **Inflorescence units** (2–)3-flowered, branch units 2–3-flowered; spathes opposite or separated, divergent, green, sometimes flushed purple basally, lanceolate-acuminate, 6–12 cm × 7–12 mm. **Flowers:** perianth deep red-purple, lavender, gray-blue, cream, or white, with gold signal and blue or purple veins; floral tube 1.5–2.8 cm, usually widening to bowl shape at base of flower; sepals oblanceolate to obovate, 5–9 × 1.4–3 cm, base gradually attenuate, apex obtusely rounded; petals oblanceolate, 4.5–7 × 0.9–1.8 cm, base attenuate to narrow claw; ovary elliptic-oval, sharply triangular in cross section, 3–4 cm; style 1.7–3.5 cm; crests overlapping, subquadrate, 1–2 cm, margins coarsely toothed; stigmas triangular; pedicel 2–5 cm. **Capsules** sharply triangular in cross section with ridge at each angle, tapering at both ends, remnant of floral tube forming tip at apex, 2.5–5 cm. **Seeds** dark brown, pyriform, wrinkled. $2n = 40$.

Flowering Apr–Jun. Open woods, sunny slopes and fields; Calif., Oreg.

R. C. Foster (1937) named several varieties of *Iris douglasiana*, about which L. W. Lenz (1954) said, "This is a widespread and extremely variable species whose total variability is being increased due to introgressive hybridization between it and other species with which it has come into contact. Well marked and distinct geographic races cannot be detected; however, pronounced variations are to be found within a single population. For these reasons no attempt is made here to segregate taxa within such a polymorphic species."

Iris douglasiana hybridizes with *I. bracteata*, *I. chrysophylla*, *I. fernaldii*, *I. hartwegii*, *I. innominata*, *I. macrosiphon*, *I. munzii*, *I. purdyi*, *I. tenax*, and *I. tenuissima*. The natural hybrid between *I. douglasiana* and *I. innominata* has been designated as *Iris* ×*thompsonii* R. C. Foster and the garden hybrid as *Iris* ×*aureonympha* E. H. English.

14. Iris macrosiphon Torrey, Pacif. Railr. Rep. 4(5): 144. 1857 • Ground iris [E] [F]

Iris elata Piper

Rhizomes many-branched, forming tufts, with fibrous remains of old leaves at nodes, slender, 0.6–0.8 cm diam.; roots few, fibrous. **Stems** simple, solid, almost absent or to 2.5 dm. **Leaves:** basal longer than stem, blade light green, finely veined, narrowly linear, 3–4 dm × 0.4–0.6 cm, glaucous, margins not thickened, apex acute; cauline 1–2, spreading, sheathing for about 1/2 length, foliaceous, blade not inflated, 0.7–1 dm. **Inflorescence units** (1–)2-flowered; spathes nearly opposite, connivent, linear-lanceolate, 4–9.5 cm × 4–6 mm, subequal or outer longer. **Flowers:** perianth indigo, purple, lavender, white, cream, or yellow; floral tube linear, gradually dilating apically, 6 cm; sepals usually with fine, dark veins basally, becoming coarser on claw, oblanceolate or obovate, 3.9–7 × 2 cm, base abruptly attenuate into claw; petals oblanceolate, 3.5–6 × 0.5–1.6 cm, base gradually attenuate; ovary ovoid, ca. 1 cm; style 2–3.5 cm, crests overlapping, reflexed, semiovate, 0.8–1.8 cm, margins denticulate; stigmas triangular, margins entire; pedicel 1.5–2 cm. **Capsules** oblong to ovoid, somewhat 3-angled in cross section, 2.5–3 cm. **Seeds** dark brown, angular, finely wrinkled. $2n = 40$.

Flowering Apr–Jun. Sunny hillsides, meadows, roadsides; Calif.

Iris macrosiphon hybridizes with *I. chrysophylla*, *I. douglasiana*, *I. fernaldii*, *I. hartwegii*, *I. innominata*, *I. munzii*, *I. purdyi*, *I. tenax*, and *I. tenuissima*.

The invalid name "Iris californica" Leichtlin has sometimes been applied to a portion of this species.

15. Iris hartwegii Baker, Gard. Chron., n. s. 6: 323. 1876 • Sierra iris [E]

Iris hartwegii subsp. *australis* (Parish) L. W. Lenz; *I. hartwegii* var. *australis* Parish; *I. hartwegii* subsp. *columbiana* L. W. Lenz; *I. hartwegii* subsp. *pinetorum* (Eastwood) L. W. Lenz; *I. pinetorum* Eastwood

Rhizomes creeping, not producing dense clumps, covered with remains of old leaves, slender to moderately thick, 0.5–0.9 cm diam.; roots fibrous. **Stems** simple, solid, 0.5–3 dm. **Leaves:** basal deciduous, blade pale green, not pink basally, 2–4.5 dm × 0.2–0.6(–1) cm, sometimes glaucous, margins not thickened; cauline usually 1–several, spreading, sheathing for about 1/2 length, foliaceous, blade not inflated. **Inflorescence units** 1–2(–3)-flowered; spathes divergent, separated by 1.5–4 cm, linear to linear-lanceolate, unequal, outer 5–11 cm × 4–7 mm, inner 5–6 cm × 2–3 mm, herbaceous, apex acute. **Flowers:** perianth lavender, cream, or pale to deep yellow; floral tube funnelform, 0.5–1(–1.5) cm; sepals oblanceolate, 4–7 × 1.4–2 cm, base gradually attenuate; petals narrowly oblanceolate, 3.5–6 × 0.5–1.1 cm, widest at about 1/2 their length, base abruptly attenuate, claw 1–2 mm wide; ovary nearly cylindrical, 1–2 cm; style 1.6–3 cm, crests overlapping, obtusely angled, 0.5–1.1 cm; stigmas acutely triangular, margins entire; pedicel 0.5–2.5 cm at anthesis, lengthening to 3.5–7.5 cm at maturity. **Capsules** oblong-oval, 3-angled, tapering abruptly at either end, 2–3 cm. **Seeds** brown, irregularly D-shaped, wrinkled. $2n = 40$.

Flowering Jun. Yellow-pine forests on sunny or partially shaded slopes; Calif.

Plants with pedicels 2.8–4.7 cm, and flowers light purple or bluish violet have been recognized as subsp. *australis*. Plants having a 3-flowered inflorescence unit, leaves about 1 cm wide, and pale yellow flowers with golden yellow veining have been called subsp. *columbiana*. Plants having both flowers open at the same time rather than consecutively, pedicels averaging only 1 cm, long narrow style arms and crests, and floral tubes 1.2–1.5 cm have been named subsp. *pinetorum*.

Iris hartwegii hybridizes with *I. douglasiana*, *I. innominata*, *I. macrosiphon*, *I. munzii*, *I. tenax*, and *I. tenuissima*. It is known from Butte County to Kern County.

16. Iris bracteata S. Watson, Proc. Amer. Acad. Arts 20: 375. 1885 • Siskiyou iris [E]

Rhizomes sheathed with old leaf bases, slender, 0.6–0.9 cm diam.; roots few, fibrous. **Stems** simple, solid, 1.5–3 dm. **Leaves:** basal with abaxial surface of blades deep glossy green on one side of fan, light yellow-green on other side, pink or red-tinged basally, strongly ribbed, 4–6 dm × 0.8–1 cm, rigid, margins not thickened; cauline 3–6, imbricated, closely sheathing stem ca. 2/3 length, spreading distally, bractlike, blade inflated, often shorter than basal leaves. **Inflorescences** simple, units 1–2-flowered; spathes closed tightly around pedicel and ovary, lanceolate, 5.2–9 cm × 6–8 mm, subequal, herbaceous, margins scarious, apex acuminate. **Flowers:** perianth cream to buff-yellow; floral tube 0.8–0.9 cm; sepals with deeper yellow signal, veined with purple or brown, obovate-lanceolate, 6.5 × 2.5 cm, base gradually attenuate into wide claw; petals narrowly oblanceolate, 7–9 × 0.8–2 cm, base gradually attenuate; ovary nearly circular in cross section, 1.5–2.5 cm, base gradually attenuate into pedicel, apex abruptly acuminate into floral tube; style 2.2–3 cm, crests spreading, yellow, not veined, 1.2 × 0.9–1.7 cm, margins toothed; stigmas triangular or tongue-shaped, margins entire; pedicel 3–6.2 cm. **Capsules** nearly circular in cross section, tapering abruptly at each end, 2–2.5 × 1–1.5 cm. **Seeds** dark brown, irregular in shape, wrinkled. $2n = 40$.

Flowering May–Jun. Shady forests, particularly ponderosa pine; Calif., Oreg.

Iris bracteata is limited to one county each in northern California and southern Oregon. It hybridizes with *I. chrysophylla*, *I. douglasiana*, *I. innominata*, *I. munzii*, *I. purdyi*, and *I. tenax*.

17. Iris purdyi Eastwood, Proc. Calif. Acad. Sci., ser. 3, 1: 78, plate 7, fig. 2. 1897 • Purdy's iris [E]

Iris lansdaleana Eastwood; *I. macrosiphon* Torrey var. *purdyi* (Eastwood) Jepson

Rhizomes much branched, forming dense clumps, dark red-brown, very slender, 0.3–0.6 cm diam., covered with remains of old leaves; roots fibrous. **Stems** simple, solid, 1.5–2.5 dm. **Leaves:** basal few, laxly spreading, longer than stem, blade bright dark green adaxially, flushed pink basally, veins subprominent, linear, 2.8–4.8 dm × 0.5–0.8 cm, rather glaucous abaxially, margins thickened, apex acute; cauline imbricated, sheathing, free only at tips, bracteiform, blade green edged with pink, strongly striate, inflated, apex acuminate. **Inflores-**

cence units 1–2-flowered; spathes green with prominent red margins, inflated, broadly lanceolate-ovate, 5.6–7 cm × 8–13 mm, unequal, outer shorter than inner, herbaceous, apex acuminate. **Flowers:** perianth pale creamy yellow flushed with pale lavender, with conspicuous brownish purple lines; floral tube linear, 3–5 cm, somewhat dilated apically; sepals widely spreading, veined and dotted with deeper purple on claw and limb, oblanceolate, 5.5–8.4 × 1.6–2.7 cm; petals spreading, lanceolate, 5–7 × 1–2 cm, margins sinuate; ovary trigonal in cross section with groove along each angle, narrow, 1–1.5 cm; style 2–3 cm, crests narrowly semiovate or nearly linear, laciniate, 1–2 cm; stigmas rounded-truncate to 2-lobed, never triangular, margins minutely denticulate; pedicel 1–2 cm. **Capsules** oblong-ovoid, trigonal, somewhat beaked, 2–3 cm. **Seeds** light brown, D-shaped, oblong-ovoid, thick, finely wrinkled. $2n = 40$.

Flowering Apr–Jul. Open woods of redwood region; Calif.

Iris purdyi hybridizes with *I. bracteata*, *I. chrysophylla*, *I. douglasiana*, *I. innominata*, *I. macrosiphon*, *I. tenax*, and *I. tenuissima*.

18. Iris chrysophylla Howell, Fl. N. W. Amer., 633. 1902 • Yellow-leaf iris [E]

Rhizomes compact, dark brown, slender, 0.5–0.6 cm diam., covered with old leaf bases; roots few, fibrous. **Stems** simple, solid, slender, 0.25–2 dm. **Leaves:** basal with blade light green, pink or red-tinged basally, finely ribbed, linear-acute, 2.5–5 dm × 0.3–0.5 cm, somewhat glaucous, margins not thickened; cauline 1–3, imbricated, blade bractlike, inflated. **Inflorescence units** usually 2-flowered; spathes sometimes flushed pink or purple apically, lanceolate to linear-lanceolate, 5–10 cm × 3–7 mm, unequal, outer much shorter and narrower than inner, herbaceous, margins often scarious, apex acute. **Flowers:** perianth cream or very pale yellow, with deeper yellow or lavender veins; floral tube linear, slender, 4.5–12 cm; sepals with bluish tinge and veins, oblanceolate, 4.6–7 × 1–2 cm, base gradually attenuate; petals lanceolate, 3–5.5 × 0.6–1.2 cm, base gradually attenuate, apex emarginate; ovary 1–1.8 cm, base abruptly attenuate into pedicel (when present), apex gradually acuminate into floral tube; style 2–3 cm, crests 1.5–2 cm, sometimes exceeding arms, apex narrowly pointed; stigmas tongue-shaped, margins entire; pedicel absent or 0.5–1 cm at anthesis. **Capsules** sometimes on elongated pedicel, oblong, round in cross section, sharply beaked, 2–3 cm. **Seeds** buff, slightly compressed. $2n = 40$.

Flowering Apr–Jun. Open, coniferous forests; Calif., Oreg.

Iris chrysophylla hybridizes with *I. bracteata*, *I. douglasiana*, *I. innominata*, *I. macrosiphon*, *I. purdyi*, *I. tenax*, and *I. tenuissima*.

19. Iris tenuissima Dykes, Gard. Chron., ser. 3, 51: 18. 1912 • Long-tube iris [E]

Iris citrina Eastwood; *I. humboldtiana* Eastwood; *I. tenuissima* subsp. *purdyiformis* (R. C. Foster) L. W. Lenz; *I. tenuissima* var. *purdyiformis* R. C. Foster

Rhizomes freely branching, producing dense clumps, slender, 0.5–0.7 cm diam., covered with bases of old leaves; roots fibrous. **Stems** simple, solid, 2–3 dm. **Leaves:** basal with blade sometimes pink to red basally, finely ribbed, linear, 3–3.5 dm × 0.5 cm, subglaucous, margins not thickened, apex acute; cauline 1–3, imbricated, mostly free, bractlike, blade linear-lanceolate, inflated. **Inflorescence units** (1–)2-flowered; spathes often flushed with pink or red, lanceolate, 4–8 cm × 5–10 mm, subequal or unequal and inner somewhat longer than outer, rigid, apex acuminate. **Flowers:** perianth pale cream with distinct veins of lavender, reddish brown, or brown, margins often distinctly crisped; floral tube funnelform, 3–5.8 cm, slender, then expanding abruptly to base of flower; sepals spreading outward from base and remaining in horizontal position for entire length, narrowly lanceolate to oblanceolate, 4–6.5 × 1.1–1.8 cm, base attenuate into long claw; petals somewhat spreading, not strictly erect, 4.5–6.5 × 0.6–1.4 cm, base gradually attenuate; ovary 1–2 cm, gradually tapering to each end; style 2–3 cm, crests linear, undulate near tip, 1–2.5 cm; stigmas triangular; pedicel 0.8–1.8 cm at anthesis. **Capsules** oblong, round in cross section, 3–4 cm, base abruptly attenuate into pedicel, apex gradually acuminate into distinct beak apically. **Seeds** grayish brown, pyriform to oval, 3 mm, wrinkled. $2n = 40$.

Flowering Jun–Jul. Dry, sunny woods; Calif.

Iris tenuissima hybridizes with *I. chrysophylla*, *I. douglasiana*, *I. hartwegii*, *I. macrosiphon*, *I. purdyi*, and *I. tenax*.

20. Iris innominata L. F. Henderson, Rhodora 32: 23. 1930 • Del Norte County iris [E]

Rhizomes freely branching, forming compact tufts, slender, 0.3–0.4 cm diam., covered with remains of old leaves; roots fibrous. **Stems** simple, solid, slender, 2–2.5 dm. **Leaves:** basal evergreen, abundant, blade dark, shining green on one side, lighter green on the other, pink to deep purple basally, 3.5 dm × 0.2–0.4 cm, longer than flowering stem, margins not thickened; cauline 2–4, sheathing stem for about 2/3 length, foliaceous, spreading, blade not inflated, 4–9 cm. **Inflorescence units** 1(–2)-flowered; spathes closely enclosing ovary and floral tube, broadly lanceolate to ovate, 3.3–6 cm × 5–7 mm, herbaceous, margins scarious. **Flowers:** perianth usually dark golden yellow, cream, pink, lavender, or purple, heavily veined purple or brown; floral tube 1.5–3 cm; sepals usually yellow with heavy veining, broadly oblanceolate, 4.5–6 × 1.7–3 cm, base gradually attenuate, margins often wavy; petals same color as sepals with lighter veining, narrowly oblanceolate, 4–5.7 × 0.9–1.6 cm, base gradually attenuate; ovary roundly triangular in cross section, 1.5–1.8 cm; style 2–2.6 cm, crests overlapping, subquadrate to semiovate, 0.9–1.4 cm, margins irregularly toothed; stigmas triangular, margins entire; pedicel 0.4–1.3 cm at anthesis. **Capsules** oblong-oval, circular in cross section, 2.5 × 1.2 cm. **Seeds** in 2 rows per locule, dark brown, oval, sharply angled, 3 mm, wrinkled, pitted. $2n = 40$.

Flowering May–Jun. Dry, sunny woods; Calif., Oreg.

Iris innominata hybridizes with *I. bracteata*, *I. chrysophylla*, *I. douglasiana*, *I. fernaldii*, *I. macrosiphon*, *I. munzii*, *I. purdyi*, and *I. tenax*. It is known only from southwestern Oregon and Del Norte County in northwestern California.

21. Iris fernaldii R. C. Foster, Iridis Sp. Nov., 1. 1938 • Fernald's iris [E]

Rhizomes many-branched, producing dense clumps, compact, slender, ca. 0.6 cm diam., base covered with remains of old leaves; roots fibrous. **Stems** simple, solid, 2–4 dm. **Leaves:** basal with blade gray-green, usually brilliantly colored basally, drying to unusual gray-green, veins fairly prominent, to 4 dm × 0.7–0.8 cm, often quite glaucous, margins not thickened; cauline 2–several, spreading, sheathing stem for about 1/2 length, foliaceous, blade not inflated. **Inflorescence units** 2-flowered; spathes opposite, connivent, often flushed red basally, broadly lanceolate, 5–9 cm × 6–11 mm, subequal. **Flowers:** perianth pale creamy yellow with gold or gray veins; floral tube funnelform, 3–6.2 cm, spreading apically to form wide throat; sepals horizontally spreading, claw with deep yellow median line, oblanceolate to spatulate, 4.5–7 × 1–2 cm, base gradually attenuate into broad claw; petals narrowly oblanceolate, 4.3–6 × 0.6–1.4 cm, base gradually attenuate into narrow claw; ovary elliptical, nearly circular in cross section, 1.5–2.3 cm; style 2.2–3 cm, crests divergent, linear to narrowly oblong, 1–1.7 cm; stigmas triangular, margins entire; pedicel 0.9–2.2 cm at anthesis. **Capsules** oblong, roundly trigonal, distinctly beaked, 2.5–3.5 cm. **Seeds** dark brown, wrinkled. $2n = 40$.

Flowering Apr–Jul. Mixed evergreen forest; Calif.

Iris fernaldii hybridizes with *I. douglasiana*, *I. innominata*, and *I. macrosiphon*.

22. Iris munzii R. C. Foster, Iridis Sp. Nov., 2. 1938 • Munz's iris [E]

Rhizomes freely branching, forming large clumps, slender, 1–1.2 cm diam., covered with remains of old leaves; roots fibrous. **Stems** simple, solid, 6–7 dm. **Leaves:** basal persistent, blade gray-green, green and without pink coloration basally, 6–7 dm × 1.5–2 cm, glaucous, margins not thickened; cauline 1–2, free from stem for ca. 1/2 length, foliaceous, spreading, blade not inflated. **Inflorescence units** 3(–4)-flowered; spathes widely separated, sometimes by 9 cm, divergent, unequal, outer 6.5–15 cm × 8–14 mm, inner much shorter. **Flowers:** perianth pale powder blue to lavender or violet, often frilled and veined in violet or darker blue; floral tube funnelform, 0.7–1 cm; sepals oblong-ovate to broadly oblanceolate, 6.2–9 × 1.8–3.7 cm, base gradually attenuate, apex emarginate; petals oblong to spatulate, 5–9.5 × 1.2–2.1 cm, base attenuate, apex apiculate; ovary rounded in cross section, 1.4–3 cm, base gradually attenuate into pedicel, apex abruptly acuminate into floral tube; style 3 cm, crests reflexed, subquadrate, shallowly lobed, 1.1–2 cm, margins entire or obscurely and shallowly lobed; stigmas triangular, margins entire; pedicels of varying lengths, 0.8–4.8 cm on first flower, those of later flowers longer. **Capsules** oblong, rounded in cross section, tapering abruptly at either end, to 5 cm. **Seeds** brown, D-shaped or irregular, coarsely wrinkled. $2n = 40$.

Flowering Apr–May. Partially shaded areas and along stream banks or on moist slopes; Calif.

No natural hybrids of *Iris munzii* are known, but horticultural hybrids have been made with *I. bracteata*, *I. douglasiana*, *I. hartwegii*, *I. innominata*, and *I. macrosiphon*. It is known only from Tulare County, California.

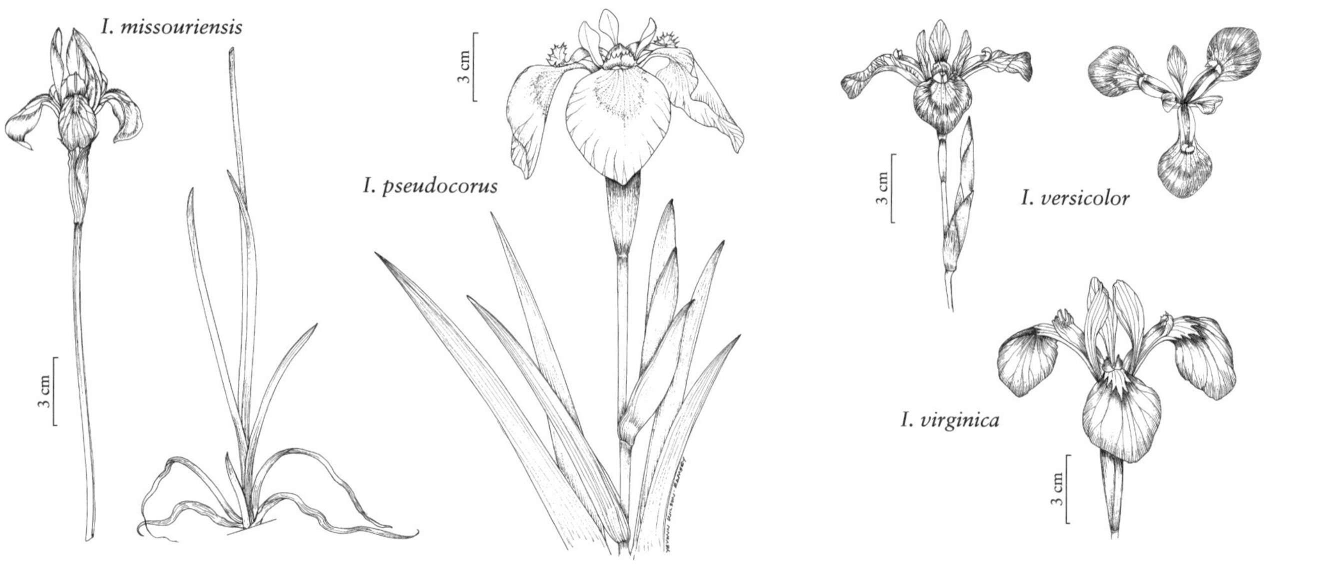

IRIS

4b.2e. Iris Linnaeus (subg. Limniris, sect. Limniris) ser. Longipetalae (Diels) G. H. M. Lawrence, Gentes Herb. 8: 362. 1953

Iris subsect. *Longipetalae* Diels in H. G. A. Engler and K. Prantl, Nat. Pflanzenfam. ed. 2, 15a: 502. 1930

Stigmas 2-toothed. **Capsules** 6-ribbed, tapering at both ends.

Species 2 (2 in the flora): w North America (incl. n Mexico).

23. Iris missouriensis Nuttall, J. Acad. Nat. Sci. Philadelphia 7: 58. 1834 • Rocky Mountain iris F W2

Iris arizonica Dykes; *I. longipetala* Herbert var. *montana* Baker; *I. montana* Nuttall ex Dykes; *I. pariensis* S. L. Welsh; *I. pelogonus* Goodding; *I. tolmieana* Herbert

Rhizomes freely branching, producing large colonies, stout, 2–3 cm diam., clothed with dark remnants of old leaves; roots fleshy. **Stems** simple or 1–2-branched, solid, 2.5–6 dm. **Leaves:** basal with blade light green, often white basally, not prominently veined, linear, 4.5–6 dm × 0.3–1.2 cm, glaucous, apex acute; cauline 2–3, often shed with spent stem, similar to basal leaves, distal may subtend branch. **Inflorescences** with terminal and lateral units (when present) (1–)2–3-flowered; spathes subopposite to distant, to as much as 3–7.5 cm apart, lanceolate or ovate, keeled, outer usually shorter, 3.5–4 cm, inner 5.5–7 cm, scarious with herbaceous areas basally and along keel, apex acuminate. **Flowers:** perianth blue to lavender to white, veined deeper violet; floral tube green with purple veins in line with midrib of petals, funnelform, constricted above ovary, 0.5–1.2 cm; sepals deeply veined lilac-purple, with yellow-white signal at base of limb, claw yellowish white, veined and dotted with purple, obovate, 3.7–7 × 1.2–3.2 cm; petals slightly divergent, oblanceolate to spatulate, 3.6–7 × 0.5–1.2 cm, base gradually attenuate, apex acuminate or rarely emarginate; ovary trigonal with 6 shallow ridges, 1–2 cm; style 2.4–4 cm, crests quadrate, 0.7–1.4 cm, margins irregularly toothed; stigmas 2-lobed, margins entire; pedicel slender, 1–6 cm at anthesis, increasing to 7.5–8 cm at maturity. **Capsules** almost circular in cross section, with 6 obvious, equidistant ridges, 4–5 × 1.5 cm. **Seeds** light brown, nearly globular to pyriform, 4–4.5 mm, wrinkled. $2n = 38$.

Flowering May–Jul. Wet meadows, roadside ditches, margins of streams; 10–3000 m; Alta., B.C.; Ariz., Calif., Colo., Idaho, Mont., Nebr., Nev., N.Mex., Oreg., S.Dak., Utah, Wash., Wyo.; n Mexico.

The ecological range of *Iris missouriensis* is probably more varied than that of any other North American species of the genus, extending from almost sea level in southern California to 3000 m in Montana and Wyoming.

There is correspondingly wide variation in a number of characters, which has caused much confusion as to taxonomic circumscription. Homer Metcalf (pers. comm.) made a detailed study of this species. The basic requirement for its success seems to be an extremely wet area before flowering and then almost desertlike conditions for the rest of the summer.

In large populations, sometimes covering hundreds of acres, *Iris missouriensis* may be found with either simple or branched stems, leaves from 4 mm to more than 1 cm wide, shorter than the stem or longer, only one flower to as many as three on a stem, and colors from deep blue to almost pure white. A single plant found on the Pariah Plateau in Kane County, Utah, with leaves only 3–4 mm wide and a single flower stem only 4 cm long, which meant that the flower was at almost ground level, was named *Iris pariensis*. No other such specimen has been located, and this entity must be considered as just an aberrant form that was due to the desertlike conditions in which it was growing.

Iris missouriensis is known to hybridize with *I. longipetala*.

24. Iris longipetala Herbert in W. J. Hooker and G. A. W. Arnott, Bot. Beechey Voy., 395. 1840 • Long-petaled iris [E]

Rhizomes with short, creeping branches, stout, 2–2.5 cm diam., covered with bases of old leaves; roots numerous, fleshy. **Stems** simple or 1–2-branched, solid, 3–6 dm. **Leaves:** basal persistent, blade dark green, turning gray or yellow-brown when drying, veins not prominent, linear, 4–7 dm × 0.6–1 cm, usually longer than stem, glaucous or subglaucous, apex acute; cauline 2–3, reduced. **Inflorescence units** 3–6-flowered; spathes distant, inner sometimes 10 cm distal to outer, green, narrowly linear-lanceolate, 8–15 cm × 20–40 mm, herbaceous, margins and apex scarious. **Flowers:** perianth pale blue-violet or nearly white, with lilac-purple veins; floral tube funnelform, 0.5–1.3 cm; sepals spreading, then drooping at apex, white, flecked and veined with violet, obovate, 8–10 × 3–5 cm, base attenuate into narrow claw with prominent yellow ridge, signal a yellowish basal patch; petals semierect, oblong, 8–9 × 1.5–2 cm, base gradually attenuate, apex emarginate, bluntly rounded; ovary trigonal with central ridge in each face, 2–5 cm; style narrow basally, widening distally, 3.5–4 × 0.6–1 cm, crests overlapping basally, subquadrate, 1–1.5 cm, margins incised; stigmas obscurely to prominently 2-lobed, margins crenate; pedicels stout, 3–9 cm, unequal at anthesis, becoming ± equal at maturity. **Capsules** 6-ribbed, oblong-ovoid, tapering to either end, almost round in cross section, 8–9 cm. **Seeds** dark brown, globular, wrinkled. $2n = 86, 88$.

Flowering Mar–Apr. Wet, heavy soil along coast; Calif.

Iris longipetala hybridizes with *I. missouriensis* and is found from Mendocino County south to Monterey.

4b.2f. Iris Linnaeus (subg. Limniris, sect. Limniris) ser. Laevigatae (Diels) G. H. M. Lawrence, Gentes Herb. 8: 361. 1953

Iris subsect. *Laevigatae* Diels in H. G. A. Engler and K. Prantl, Nat. Pflanzenfam. ed. 2, 15a: 502. 1930

Stigmas bilobed. **Capsules** 3-ribbed, splitting irregularly. **Seeds** shiny.

Species 5 (3 in the flora): North America, Eurasia.

25. Iris pseudacorus Linnaeus, Sp. Pl. 1: 38. 1753 • Yellow iris, pale-yellow iris, yellow flag, yellow water iris, fleur-de-lis, iris jaune [F] [I] [W1] [W2]

Rhizomes pink, freely branching, producing extensive clumps, 2–3 cm diam., with fibrous remains of old leaves; roots fleshy. **Stems** usually 1-branched, solid, 7–15 dm. **Leaves:** basal deciduous, at first erect, then recurved, blade dark green, with prominent median thickening, 4–10 dm × 2–3 cm, slightly glaucous basally; cauline equaling inflorescence unit. **Inflorescence units** 4–12-flowered; spathes green with brown margins, outer spathe strongly keeled, inner without keel, 6–9 cm, subequal, margins not scarious. **Flowers:** perianth bright yellow; floral tube 0.6–0.8 cm, with no constriction into ovary; sepals bright yellow or cream colored, lanceolate to ovate or suborbiculate, 5–7.5 × 3–4 cm, base abruptly attenuate, claw ca. 1/2 length of limb, signal a darker yellow basal patch limited by short, brown lines; petals without veining, lanceolate to spatulate, 2–3 cm; ovary triangular in cross section with concave sides and narrow groove at each angle, 1.5 cm; style keeled, 3–4 cm, crests spreading, 1–1.2 cm, laciniate at apex; stigmas rounded with prominent tongue; pedicel 2.5–7 cm. **Capsules** prismatic to oblong-ovoid, obscurely 3-angled with obvious groove at each angle, 3.5–6 cm, beak 5 mm. **Seeds** D-shaped, flattened, 6–7 mm, corky, lustrous. $2n = 34$.

Flowering Apr–Jun. Swamps, wet shores of rivers and lakes; introduced; B.C., Man., N.B., Nfld. and Labr. (Nfld.), N.S., Ont., P.E.I., Que.; Ala., Ark., Calif., Del., D.C., Fla., Ga., Ill., Ind., Iowa, Kans., Ky., La., Maine, Mass., Miss., N.H., N.Y., N.C., Ohio, Pa., R.I., S.C., Tenn., Va., Wash., W.Va.; Eurasia, n Africa.

26. Iris versicolor Linnaeus, Sp. Pl. 1: 39. 1753 • Blue flag, iris versicolore [E] [F] [W1] [W2]

Rhizomes pale pinkish white, freely branching, forming large clumps, 1–2.5 cm diam., clothed with remnants of old leaves; roots fleshy. **Stems** 1–2-branched, solid, 2–6 dm. **Leaves:** basal with blade green to grayish green, often purplish basally, centrally thickened in mature leaves, prominently veined, narrowly ensiform, 1–8 dm × 1–3 cm; cauline 1–2, blade linear-lanceolate, seldom equaling stem. **Inflorescences** compact, units 2–4-flowered; spathes never foliaceous, 3–6 cm, unequal, outer shorter than inner, thickly chartaceous to scarious, margins shiny, darker in color. **Flowers:** perianth violet-blue to rarely white; floral tube funnelform, constricted above ovary, 1–1.2 cm; sepals ovate to reniform, 4–7.2 × 1.8–4 cm, base abruptly attenuate, signal a pubescent, greenish or greenish yellow patch surrounded by heavily veined purple on white at base of blade; petals lanceolate to oblong-lanceolate, 2–5 × 0.5–2 cm, much shorter than sepals, firm, apex rarely emarginate; ovary rounded-triangular in cross section, somewhat inflated, 0.8–2 cm; style 3–3.5 cm, base not auriculate, margins entire or toothed, crests reflexed, 0.7–1.5 cm; stigmas unlobed, triangular or rounded-triangular, margins entire; pedicel 2–8 cm, frequently exceeding spathe. **Capsules** often persistent over winter, ovoid to oblong-ellipsoid, conspicuously beaked, obtusely triangular in cross section, 1.5–6 cm, tardily dehiscent. **Seeds** dark brown, D-shaped, 5–8 mm, shiny, thin, hard, regularly pebbled, not corky. $2n = 108$.

Flowering May–Jul. Marshy places, along roadsides, shores, and along mountains; Man., N.B., Nfld. and Labr., N.S., Ont., P.E.I., Que.; Conn., Maine, Md., Mass., Mich., Minn., N.H., N.J., N.Y., Ohio, Pa., R.I., Vt., Va., Wis.

E. Anderson (1936) showed rather conclusively that *Iris versicolor* arose as an amphidiploid between *I. virginica* ($n = 35$) and *I. hookeri* (*I. setosa* var. *canadensis*) ($n = 19$). Back-cross hybrids have been produced both ways: *I. virginica* × *I. versicolor* producing *Iris* ×*robusta* E. S. Anderson, and *I. versicolor* × *I. hookeri* producing *I.* ×*sancti-cyri* J. Rousseau.

Iris versicolor is becoming a weed in New Brunswick, Prince Edward Island, and Nova Scotia. Livestock will not eat iris foliage, but feed voraciously on the competition, thus giving the irises plenty of room to expand.

27. Iris virginica Linnaeus, Sp. Pl. 1: 39. 1753 • Southern blue flag, iris de Virginie [E] [F]

Iris caroliniana S. Watson; *I. georgiana* Britton; *I. shrevei* Small; *I. virginica* var. *shrevei* (Small) E. S. Anderson

Rhizomes many-branched, forming dense clumps, 2–4 cm diam., usually covered with remnants of old leaves; roots fleshy. **Stems** rather weak, often falling over after flowering, solid, usually 1-branched, 5–10 dm. **Leaves:** basal erect or often flexible, blade gray-green to bright green, buff to purplish basally, with several prominent ribs in mature leaves, linear-ensiform, 6–8 dm × 2.5–3 cm, apex acute. **Inflorescence units** 2–3-flowered, branch units 1–2-flowered; spathes compact, often with brown striations, ridged, unequal, outer 3–8 cm, inner 8–14 cm, firm, herbaceous. **Flowers:** perianth lavender to violet, rarely white; floral tube constricted above ovary, 1–2 cm; sepals spreading and arched, pale blue to purple with darker blue or purple lines, obovate to oval, 4–8.4 × 1.6–4 cm, base abruptly attenuate, claw green in median, bordered by yellow ground with blue or purple lines, yellow extending onto base of limb as finely pubescent signal patch; petals oblong-lanceolate to oblong-spatulate, 3–7 × 1–3 cm, claw greenish yellow with blue or purplish lines, apex often emarginate; ovary trigonal, 1.3–3.8 cm; style inwardly auriculate at convergences, 3–4.5 cm, crests reflexed, 0.7–2 cm; stigmas unlobed, with prominent triangular tongues, margins entire; pedicel 2.5–8 cm. **Capsules** ovoid, ellipsoid, or long-cylindric, trigonal or polygonal in cross section, 3–6 × 1–2 cm. **Seeds** in 2 rows per locule, pale brown, usually D-shaped, 5–8 mm, pitted, corky. $2n$ = 70, 72.

Flowering May–Jun. Wetlands, margins of lakes and streams; Ont., Que.; Ala., Ark., D.C., Fla., Ga., Ill., Ind., Iowa, Kans., Ky., La., Md., Mich., Minn., Miss., Mo., Nebr., N.Y., N.C., Ohio, Okla., Pa., S.C., Tenn., Tex., Va., W.Va., Wis.

Plants of *Iris virginica* from the southeastern and south-central states having stems 2–3-branched and seldom falling to the ground after flowering, and with capsules long-cylindric have been recognized as var. *shrevei*.

4b.2g. Iris Linnaeus (subg. Limniris, sect. Limniris) ser. Hexagonae (Diels) G. H. M. Lawrence, Gentes Herb. 8: 362. 1953 [E]

Iris subsect. *Hexagonae* Diels in H. G. A. Engler and K. Prantl, Nat. Pflanzenfam. ed. 2, 15a: 502. 1930

Inflorescences with leaflike spathes. **Capsules** 6-ribbed or -lobed. **Seeds** large, corky.

Species 5 (5 in the flora): se, sc United States.

Series *Hexagonae* has had a long and interesting taxonomic history, yet is still one of the most confusing and controversial groups of North American irises. *Iris hexagona*, *I. fulva*, and *I. brevicaulis* were the only ones of the series described prior to 1924, when J. K. Small began to study the group along the eastern and Gulf coasts. He described a number of new species, including *I. giganticaerulea* from Louisiana, which is one of the most interesting of the series. J. K. Small and E. J. Alexander (1931) then proposed an additional 76 species from the Gulf States, all within this series. This created considerable interest among botanists and resulted in a large number of papers with different interpretations, many of which are still in question. P. Viosca (1935) stated, "A whole new series of hybrids has sprung up. Thus, theoretically, the number of possibilities is infinite, and one could just as well describe 500 or a thousand as a hundred species."

It is quite possible that many of the 76 "new species" that Small and Alexander listed may actually be a part of the population of one of the primary hybrids. If that can be demonstrated, the pertinent names should be moved into synonymy under the particular hybrid. Tentatively, they are here placed in *Iris* ×*louisianica* N. C. Henderson, including: (published by Small) *I. alticristata, I. atroenantha, I. auralata, I. bifurcata, I. brevipes, I. chlorolopha, I. chrysolopha, I. citricristata, I. citriviola, I. cyanochrysea, I. dewinkeleri, I. fluviatilis, I. fumifulva, I. fuscaurea, I. fuscirosea, I. fuscivenosa, I. iochroma, I. iocyanea, I. ludoviciana, I. moricolor, I. oenantha, I. parvirosea, I. phoenicis, I. pseudocristata, I. purpurissata, I. regalis, I. rhodochrysea, I. rosiflora, I. rubricunda, I. salmonicolor, I. subfulva, I. thomasii, I. tyriana, I. violivenosa, I. wherryana;* (published by Alexander) *I. albilinea, I. amnicola, I. aurilinea, I. callirhodea, I. cerasioides, I. crocinubia, I. cyanantha, I. ecristata, I. fumiflora, I. fuscisanguinea, I. gentilliana, I. ianthina, I. iodantha, I. ioleuca, I. iophaea, I. lancipetala, I. lilacinaurea, I. marplei, I. oölopha, I. pallidirosea, I. paludicola, I. parvicaerulea, I. pyrrholopha, I. regifulva, I. rhodantha, I. rosilutea, I. rosipurpurea, I. rubea, I. rubrolilacina, I. schizolopha, I. venulosa, I. violilutea, I. viridis.*

SELECTED REFERENCES Small, J. K. and E. J. Alexander. 1931. Botanical interpretations of the iridaceous plants of the Gulf States. Contr. New York Bot. Gard. 327: 325–358. Viosca, P. 1935. The irises of southeastern Louisiana, a taxonomic and ecological interpretation. Bull. Amer. Iris Soc. 57: 3–56.

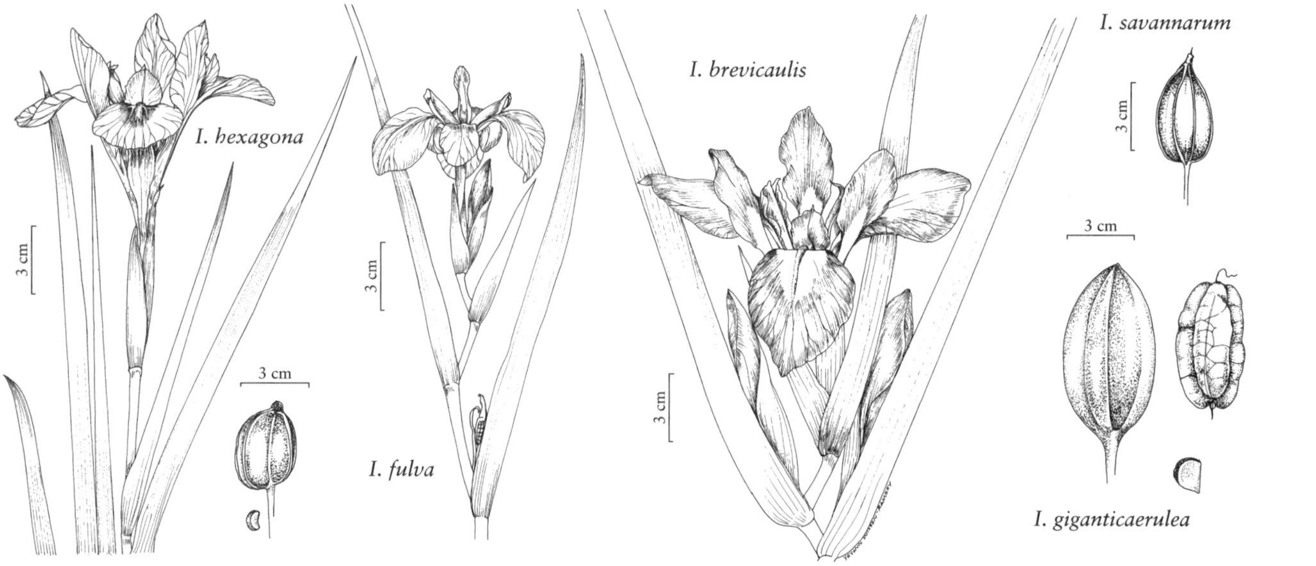

IRIS

28. Iris hexagona Walter, Fl. Carol., 66. 1788 • Carolina iris [E] [F]

Rhizomes greenish with brown leaf scars, branching from older buds quite some distance proximal to apex, forming widely scattered colonies, 2–2.5 cm diam. **Stems** 1–2-branched, solid, 3–9 dm. **Leaves:** basal stiffly erect, blade yellow-green, lightly ribbed, 8–9 dm × 2–3 cm; cauline 1–2, foliaceous, blade 1.2–2 dm, exceeding subtended flower. **Inflorescence units** 1–2-flowered, branch units 1-flowered; spathes lanceolate, subequal or unequal, apex acute; outer foliaceous, 15–20 cm, usually exceeding flower; inner 8–10 cm, herbaceous, with scarious margins. **Flowers:** perianth blue to violet, rarely white; floral tube funnelform, grooved in line with grooves of ovary, 2–3 cm; sepals spreading horizontally, obovate to oval, 8–8.5 × 4 cm, claw greenish with prominent yellow midrib which is pubescent on each side at base, apex rounded; petals erect or spreading-erect, oblanceolate to spatulate, 7–9 × 2 cm, claw veined with green; ovary roundly triangular in cross section, with wide grooves at angles and concave sides, 2–3 cm; style convex laterally from central reddish ridge, 3.5–5 cm, narrower than claw of sepals, crests erect, overlapping, semiovate to triangular, 1–5 cm, margins coarsely toothed; stigmas 2-lobed, lobes triangular or rounded-deltoid, margins entire; pedicel 2.5–3.5 cm. **Capsules** hexagonal in cross section, with 3 alternate sides plane, others with 2 rounded ridges with shallow groove between them, 2.5–3.5 × 2–2.5 cm. **Seeds** in 2 rows per locule, light brown, D-shaped or irregularly rounded, 4–6 mm, corky. $2n = 44$.

Flowering Apr–May. Margins of wetlands, low roadsides; Fla., S.C.

Iris hexagona is one of the rarest of all our native irises. We had seen herbarium specimens from seven localities in five counties of South Carolina. In searching for living plants at each of these sites we found that six of the seven are now under the water of Lake Moultrie. After extensive searching, one small population has been located at the other site, in a state park at Charleston, which is being watched carefully by the rangers of the park. A good-sized population has since been found in Dixie County, Florida, north of the small town of Shamrock, and another in neighboring Taylor County. These two populations are in the drainage system of Georgia, which in turn is fed by some of the streams from South Carolina. So far, though, *I. hexagona* has not been found in Georgia.

29. Iris fulva Ker Gawler, Bot. Mag. 36: plate 1496. 1812 • Copper iris [E] [F]

Iris fulvaurea Small

Rhizomes compact, greenish brown or sometimes red-tinged, many-branched, with ringlike scars of old leaves, 1.5–2 cm diam. **Stems** simple or sometimes 1–2-branched, solid, 3–9 dm. **Leaves:** basal arching distally, blade bright green, lightly ribbed, linear-ensiform, 6–10 dm × 1.5–2.5 cm; cauline subtending branches, blade 4.5–6 dm. **Inflorescence units**

1–2-flowered; spathes unequal, outer green, 10–12 cm, herbaceous, inner 6–8 cm, partly scarious. **Flowers:** perianth copper colored or reddish brown (yellow in forma *fulvaurea*); floral tube hollow to ovary, 2–2.5 cm; sepals widely spreading or arching downward, obovate, with 1–3 prominent veins, 4.5–5.5 cm, glabrous, often with lighter yellow basal signal; petals spreading or declining with sepals, 4–5 × 1.5–2 cm, base gradually attenuate, apex deeply emarginate; ovary green, hexagonal, 1.5–1.7 cm; style convex, not keeled, 1.8–2 cm, crests reflexed, rounded-triangular, margins shallowly toothed; stigmas 2-lobed, lobes pointed, margins entire; pedicel 2–4 cm. **Capsules** remaining green even after seeds mature, oblong-elliptic, hexagonal in cross section, with 6 equally spaced ribs, short beak, 4.5–8 × 2.5 cm. **Seeds** in 2 rows per locule, irregular, flattened, 10–15 mm, corky. 2*n* = 42.

Flowering Apr–Jun. Shallow water or low wet areas; Ark., Ill., Ky., La., Miss., Mo., Tenn.

Iris fulva hybridizes with *I. brevicaulis* to produce *I.* ×*fulvala* Dykes, which has reddish purple sepals; with *I. giganticaerulea* to produce *I.* ×*vinicolor* Small; and with *I. savannarum* to produce *I.* ×*cacique* (J. Berry) N. C. Henderson.

30. Iris brevicaulis Rafinesque, Fl. Ludov., 20. 1817 • Short-stem iris [E] [F]

Iris alabamensis Small; *I. foliosa* Mackenzie & Bush; *I. mississippiensis* Alexander

Rhizomes freely branching, producing extensive clumps, 1–2.5 cm diam. **Stems** declining or semi-erect, sharply zigzag, with short branch at each of 2–3 bends, leaf at every bend, flower bud at every bend, solid, 2.5–3 dm. **Leaves:** basal erect, spreading, or prostrate, blade deep green, ensiform, 3.5–7 dm × 1.5–3.5 cm, somewhat glaucous; cauline at each of 2–3 nodes, subtending and exceeding short branch and flower, blade 2–6 dm. **Inflorescences** with terminal unit 2-flowered, each short branch unit 1-flowered; spathes foliaceous, lanceolate, 3–5 cm, subequal, apex acute; outer green, exceeding inner; inner partly scarious. **Flowers:** perianth rich blue or blue-violet; floral tube funnelform, 1–2 cm; sepals ovate, 7.5–9 × 2.5–3 cm, base abruptly attenuate, claw light greenish yellow with darker veins, with prominent yellow midrib and large, yellow-white signal at union of limb and claw; petals erect or spreading-erect, lighter blue than sepals, oblanceolate, 5–6.5 × 1.5–2 cm; ovary prominently 6-angled, 1–1.3 cm; style greenish, 3.5–4 cm, crests reflexed, semiovoid to subquadrate, 1.2–1.5 cm, margins irregularly toothed or entire; stigmas 2-lobed, lobes rounded-deltoid; pedicel 1–1.3 cm. **Capsules** hexagonal to almost circular in cross section with 6 equally placed ribs, with short beak at apex, 3–5 × 2–2.5 cm. **Seeds** in 2 rows per locule, light brown, irregularly circular, flattened, 3–5 mm, very corky. 2*n* = 42, 44.

Flowering Apr–Jun. Open, rather moist woodlands; Ont.; Ala., Ark., Fla., Ill., Ind., Kans., Ky., La., Miss., Mo., Okla., Tenn.

Iris brevicaulis hybridizes with *I. fulva* to produce *I.* ×*fulvala* Dykes, and with *I. giganticaerulea* to produce *I.* ×*flexicaulis* Small.

31. Iris savannarum Small, Addisonia 9: 57, plate 317. 1925 • Prairie iris [E] [F]

Iris albispiritus Small; *I. hexagona* Walter var. *savannarum* (Small) R. C. Foster; *I. kimballiae* Small; *I. rivularis* Small

Rhizomes greenish, freely branching, forming large, open clumps, 2–2.5 cm diam., fleshy, with fibrous bases of old leaves not covering rhizome but persisting on ringlike scars. **Stems** 1-branched, solid, 3–10 dm. **Leaves:** basal usually dying back shortly after anthesis, blade bright green, 6–10 dm × 0.8–2 cm, sometimes glaucescent; cauline with blade elongate, linear-attenuate, 15–20 cm, exceeding flowers. **Inflorescence units** 1–3-flowered, branch units 1–2-flowered; spathes lanceolate, subequal or unequal, apex acute; outer 8–10 cm, herbaceous, margins scarious; inner 6–9 cm, completely scarious. **Flowers:** perianth blue to blue-violet (white in forma *albispiritis*); floral tube funnelform, 1.2–1.5 cm; sepals blue or pale violet with flecks of white and lines of deeper blue on either side of ridge, spatulate, elliptic to ovate, 7–8 cm, apex acuminate, ridge yellow, linear, extending to about middle of limb, finely pubescent, claw light green, striated, ribbed, 1–1.2 cm; petals erect or spreading-erect, deep blue, linear to narrowly linear-spatulate, 6–7.5 cm, base abruptly attenuate into claw, claw channeled, green with darker lines, margins pale; ovary bluntly 3-angled, ridged at each angle and on faces between angles; style linear-acute, 5–6 cm; crests ovate, 1–2 cm, margins sharply and irregularly toothed; stigmas broadly 2-lobed, margins crenate; pedicel 1.6–2 cm. **Capsules** ellipsoid, round in cross section, roundly 6-lobed, with 6 sharp, winglike, equally spaced ridges converging to form stout beak at apex, 6–10 cm, dehiscence loculicidal. **Seeds** in 1 row per locule, brown, circular or uneven, flattened, 8–12 mm diam, very corky. 2*n* = 44.

Flowering late Mar–early Apr. Wet ditches, margins of lakes, streams, and swamps; Ala., Fla., Ga.

Iris savannarum has often been included in *I. hexagona*, but differs in several ways. The capsules of *I. hexagona* are more obviously hexagonal, with smooth faces alternating with those that have two rounded lobes, whereas the capsules of *I. savannarum* are more rounded but with six obvious ridges, almost of winglike proportions. *Iris hexagona* has yellow-green leaves instead of the brighter green of *I. savannarum*. The leaves of *I. savannarum* die back after anthesis, while those of *I. hexagona* remain green. The flowering period of *I. savannarum* is almost one month earlier than that of *I. hexagona* in the area where both are known to grow. The sepals of *I. savannarum* are more acute at the apex than those of *I. hexagona*.

Iris savannarum has been hybridized with *Iris fulva* to produce *I.* ×*cacique* (J. Berry) N. C. Henderson.

32. Iris giganticaerulea Small, Addisonia 14: 5, plate 451. 1929 • Giant blue iris [E] [F]

Iris elephantina Small; *I. hexagona* Walter var. *giganticaerulea* (Small) R. C. Foster; *I. miraculosa* Small

Rhizomes green, extensively branched, forming large clumps, 2–4 cm diam., with pale leaf scars. **Stems** 2–3-branched, solid, 10–15(–20+) dm. **Leaves:** basal 4–6, erect, blade bright green, ensiform, 10–13 dm × 2–3 cm, apex slightly curved; cauline with blade foliaceous, appearing as elongated spathe, 5–6.5 dm × 2–3 cm, not exceeding flowers. **Inflorescence units** 2–3-flowered, branch units 1–2-flowered; spathes tightly enclosing ovary and pedicel, 15–20 cm, subequal; outer narrowly lanceolate, keeled, margins pale, narrow, inner margins translucent, broad, scarious. **Flowers:** perianth blue or blue-purple; floral tube tubular-prismatic, 9-ribbed and -grooved, 4–5 cm; sepals blue or blue-purple with white streaks surrounding signal, oval to orbicular-oval, 9.5–11 × 4–5 cm, base abruptly attenuate into claw, apex obtusely angled, signal with raised, yellow, pubescent central ridge, claw green, striate, 4–4.5 × 1–1.2 cm; petals erect or spreading-erect, spatulate, 7–8.5 × 2–3 cm, base cuneate, apex emarginate; ovary bluntly 6-angled, 4–4.5 cm; style blue-violet, keeled, 3.5–4 cm, raised above sepals, crests recurved, 1.5–2 cm, margins sharply and irregularly serrate; stigmas 2-lobed with 2 triangular teeth; pedicel columnar, 3.5–5 cm. **Capsules** drooping, bright green, ellipsoid, hexagonal in cross section, with 6 broad, rounded lobes, 8–10 × 3 cm, indehiscent; after anthesis, flower stem elongates, arching downward to push capsule into water, where walls are broken down and as funiculus is broken, seed rises to surface. **Seeds** in 2 rows per locule, brown, D-shaped, 8–10 mm wide, very corky. $2n = 44$.

Flowering Mar–Apr. Shallow water or very wet roadside ditches; Ala., La., Miss.

4b.2h. IRIS Linnaeus (subg. LIMNIRIS, sect. LIMNIRIS) ser. SPURIAE (Diels) G. H. M. Lawrence, Gentes Herb. 8: 361. 1953 [I]

Iris subsect. *Spuriae* Diels in H. G. A. Engler and K. Prantl, Nat. Pflanzenfam. ed. 2, 15a: 502. 1930

Stigmas 2-toothed. **Capsules** with 2 ribs at each of the 3 corners. **Seeds** with loose, shiny coat.

Species ca. 10 (1 in the flora): introduced; Eurasia.

33. Iris orientalis Miller, Gard. Dict. ed. 8, Iris no. 9. 1768 [I]

Rhizomes sparingly short-branched, forming dense clumps, 1–1.5 cm diam., hard, with old leaf bases at nodes; roots fleshy. **Stems** slightly flattened, with 1–2 short branches, solid, 4–12 dm. **Leaves:** basal erect, blade with slight spiral twist and central ridge, 3.5–8 dm × 1–2 cm, stiff, harsh, fibrous, glaucous; cauline 2–3, 1–2 subtending floral clusters, blade reduced. **Inflorescence units** clustered, 2–4-flowered; spathes white, 3–5 cm, subequal, papery. **Flowers:** perianth white; floral tube funnelform, 1–2.5 cm; sepals spreading and arching downward, with large yellow basal area, broadly orbicular, 8–10 × 3–6 cm, apex rounded, deeply emarginate; petals white, spatulate, 4–6 × 1–1.5 cm, base gradually attenuate, apex emarginate; ovary triangular in cross section with 2 ribs at each angle, 2–2.5 cm; style white, with parallel sides, 4–5 cm, crests erect, triangular, 1–2 cm; stigmas 2-lobed; pedicel 2.5–7.5 cm. **Capsules** ovoid to oblong-elliptic, triangular in cross section, each angle 2-ribbed, 4–5 × 2–2.5 cm. **Seeds** in 2 rows per locule, white, flattened or wedge-shaped, 4–5 mm, papery, wrinkled. $2n = 40$.

Flowering May–Jul. Persisting after cultivation or discarded along roadsides; introduced; Calif., Conn., Mo.; expected elsewhere; Greece, Turkey.

4b.2i. IRIS Linnaeus (subg. LIMNIRIS, sect. LIMNIRIS) ser. PRISIMATICAE (Small) G. H. M. Lawrence, Gentes Herb. 8: 362. 1953 [E]

Iris [unranked] *Prismaticae* Small, Man. S.E. Fl., 336. 1933

Rhizomes heterogeneous, thin, creeping. **Capsules** sharply 3-cornered. **Seeds** smooth.

Species 1: e North America.

34. Iris prismatica Pursh, Fl. Amer. Sept. 1: 30. 1814 • Slender blue iris [E]

Rhizomes superficial or only slightly buried in peaty soil, cordlike portions usually 1-branched, 40 cm × 2–5 mm, bearing scalelike leaves at nodes; nodal roots absent. **Stems** simple or 1–2-branched, 3–8 dm. **Leaves:** basal with blade slightly ribbed, 3–6 dm × 0.2–0.5 cm; cauline 1–4, erect. **Inflorescence units** 1–3-flowered, branch units 1–2-flowered; spathes pale brown, narrowly lanceolate, 2–4 cm, scarious or partially membranaceous. **Flowers:** perianth pale blue or blue-violet; floral tube 0.2–0.3 cm; sepals pale violet, veined deep violet basally, ovate, 4–5 × 1.3–2 cm, base abruptly attenuate; petals lavender, oblanceolate, 3.5–4.5 × 0.7–1.5 cm; ovary trigonal, sharply angled; style arched, narrow, 2–3 cm, crests divergent, quadrate, 0.5–0.7 cm, margins serrate; stigmas sharply triangular; pedicel somewhat flattened, 3–7 cm, exserted from spathe. **Capsules** sharply 3-angled, almost winged, 3–4 × 1.2–1.4 cm, concave faces 6–14 mm wide. **Seeds** in 1 row per locule, buff to dark brown, pyriform, with convex sides, 3–4 mm, smooth. $2n = 42$.

Flowering Apr–Jul. Swampy, peaty soil; Ala., Conn., Del., Ga., Ky., Maine, Md., Mass., N.H., N.J., N.Y., N.C., R.I., S.C., Tenn., Va.

5. BELAMCANDA Adanson, Fam. Pl. 2: 60. 1763 (as Belamkanda), name conserved • Blackberry-lily [apparently based on a vernacular name in western India] [I]

Peter Goldblatt

Gemmingia Kuntze; *Pardanthus* Ker Gawler

Herbs, perennial, from creeping rhizomes. **Stems** few- to several-branched. **Leaves** several; blade plane, ensiform. **Inflorescences** rhipidiate, several-flowered; spathes green, membranous distally. **Flowers** fleeting, erect, unscented, actinomorphic; tepals erect, basally connate into vestigial tube, light orange to reddish (rarely yellow), with scattered spots of darker pigment, obscurely clawed, ± equal, outer whorl slightly larger than inner; stamens diverging; filaments distinct; anthers diverging, not appressed to style branches; style slender, short, not extending between stamens, branching distally into lobes; lobes 3, flattened, stigmatic distally, stigmatic surfaces each subtended adaxially by paired flaps of tissue. **Capsules** ovoid-truncate, ± woody, apex obtuse. **Seeds** several, globose; seed coat blackish. $x = 16$.

Species 1 or 2 (1 in the flora): introduced; Asia including Japan; introduced in tropical Asia, the Pacific Islands, and parts of South America.

SELECTED REFERENCE Mathew, B. 1990. The Iris, rev. ed. Portland.

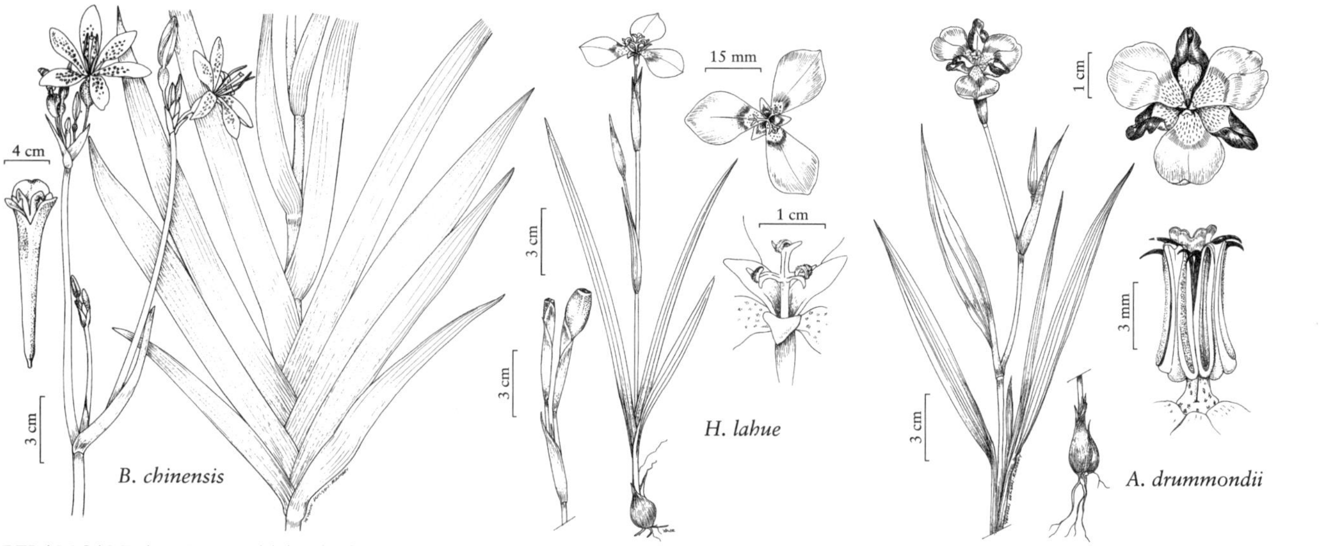

BELAMCANDA ◦ HERBERTIA ◦ ALOPHIA

1. **Belamcanda chinensis** (Linnaeus) de Candolle in P. J. Redouté, Liliac. 3(21): plate 121. 1805 [F] [I]

Ixia chinensis Linnaeus, Sp. Pl. 1: 36. 1753; *Gemmingia chinensis* (Linnaeus) Kuntze

Plants 0.6–1 m. **Rhizomes** just below ground level, horizontal, flesh usually orange. **Stems** usually 3–5-branched. **Leaves** 8–14 per stem; blade narrowly lanceolate, 1/2 to ± equaling stem, 10–20 mm wide, glaucous. **Rhipidia** usually 3–6-flowered; spathes ± equal, 10–20 mm. **Tepals** ascending proximally, ± spreading distally, light orange to reddish (rarely yellow), with large, irregularly scattered spots of darker pigment, lanceolate, 16–35 mm; filaments (6–)10–12 mm; anthers 6–8 mm; ovary narrowly ovoid-3-angled, (4–)6–8 mm; style 3-angled proximally, thickened in distal 1/2, dividing opposite anther apices, branches ca. 2 mm. **Capsules** ovoid, 18–26 mm, apex truncate. **Seeds** persisting on placentas after capsule dehiscence, black, ca. 4 mm diam., shiny.

Flowering Jun–Aug. Grassland, pastures, woodland clearings, disturbed limestone glades, rocky outcrops; introduced; Ala., Ark., Conn., Del., D.C., Fla., Ga., Ill., Ind., Iowa, Kans., Ky., La., Md., Mich., Miss., Mo., Nebr., N.J., N.Y., N.C., Ohio, Okla., Pa., S.C., S.Dak., Tenn., Va., W.Va.; w Asia.

6. HERBERTIA Sweet, Brit. Fl. Gard. 3: plate 222. 1827 • Pleat-leaf iris [for William Herbert, 1778–1847, prominent British botanist and specialist in bulbous plants]

Peter Goldblatt

Trifurcia Herbert

Herbs, perennial, from tunicate, ovoid bulbs; tunic brown, dry, brittle, papery. **Stems** simple or branched. **Leaves** few, basal larger; blade pleated, linear-lanceolate. **Inflorescences** rhipidiate, few-flowered; spathes green, unequal, inner spathe exceeding outer, apex brown, acute, usually dry. **Flowers** short-lived, erect, unscented, actinomorphic; tepals spreading, distinct, blue to mauve with white markings, unequal, outer whorl more than 2 times inner; filaments connate; anthers diverging, appressed to style branches; style slender, branching at apex of filament column; branches diverging from base, flattened, divided apically into 2 slender lobes, apically stigmatic. **Capsules** ovoid, apex truncate. **Seeds** many, prismatic; seed coat brown. $x = 7$.

Species ca. 5 (1 in the flora): North America, temperate South America.

The name *Alophia* has been misapplied to this genus because of an erroneous interpretation of the identity of the type species of *Alophia* (P. Goldblatt 1975).

1. **Herbertia lahue** (Molina) Goldblatt, Ann. Missouri Bot. Gard. 64: 379. 1978 [F]

Ferraria lahue Molina, Sag. Stor. Nat. Chili ed. 2, 110. 1810; *Alophia lahue* (Molina) Espinosa; *A. lahue* subsp. *caerulea* (Herbert) Ravenna; *Cypella drummondii* Graham; *Herbertia caerulea* (Herbert) Herbert; *H. drummondiana* Herbert; *H. lahue* subsp. *caerulea* (Herbert) Goldblatt; *H. watsonii* Baker; *Iris brachystigma* E. Scheele; *Trifurcia caerulea* Herbert; *T. lahue* (Molina) Goldblatt; *T. lahue* subsp. *caerulea* (Herbert) Goldblatt

Plants (6–)8–15(–24) cm. **Bulbs** (10–)15–20 mm diam. **Stems** simple or 1–3 branched, branching usually from base. **Leaves** 4–6, mostly basal, ± reaching base of inflorescence, cauline leaves often entirely sheathing; blade ± linear, 4–6 mm wide. **Rhipidia** (1–)2–5-flowered; outer spathe ca. $^2/_3$ to ± equaling inner, inner (35–)40–50 mm, apex becoming dry. **Tepals** bluish purple to violet; outer spreading, lanceolate, 23–28 × 15–18 mm, flaccid, claws white at base, speckled with violet, 6–7 mm; inner violet, darkest on claws, ca. 8–12 × 3 mm; filament column ca. 5 mm; anthers recurving soon after dehiscence, 7–10 mm; ovary oblong, 5–7 mm; style branches 5–6 mm, forked apically for ca. 2.5 mm. **Capsules** ovoid–oblong-truncate, 15–25 mm. **Seeds** ca. 2.5 mm.

Flowering mid Mar–early May. Woodlands and prairies, most common near coast; Fla., La., Tex.; South America (Argentina, Brazil, Chile, Uruguay).

Herbertia lahue is probably introduced in Florida.

The taxonomy of *Herbertia lahue* has been much confused. Plants from North America were originally treated as *H. caerulea*, separate from the South American members of the genus. P. Ravenna (1968) regarded differences between this and two South American species, *H. lahue* and *H. amoena,* as trivial and not sufficient to allow anything more than infraspecific separation. He considered plants from northern Argentina to be indistinguishable from those from North America and united them under subsp. *caerulea*. Without ennumerating their differences, Ravenna treated plants from coastal southern Brazil, Uruguay, and Argentina with slightly smaller flowers as subsp. *amoena* (Grisebach) Goldblatt, and those from Chile as subsp. *lahue*. The variation in North American populations makes it impossible to maintain even subspecies in *H. lahue* on the basis of available information.

7. ALOPHIA Herbert, Bot. Mag. 66: under plate 3779. 1840 • Pinewoods-lily, purple pleat-leaf [Greek *a*, not, and *lophos*, crest, referring to the absence of style crests]

Peter Goldblatt

Eustylis Engelmann & A. Gray

Herbs, perennial, from tunicate, ovoid bulbs; tunic brown, dry, brittle, papery. **Stems** simple or branched. **Leaves** few, basal larger; blade pleated, linear-lanceolate. **Inflorescences** rhipidiate, few-flowered; spathes green, unequal, outer shorter than inner, apex acute. **Flowers** short-lived, ± nodding [sometimes drooping], actinomorphic; tepals spreading, distinct, blue to purple, clawed, unequal, outer whorl larger than [± equaling] inner, claw surface sometimes folded or pinched basally forming lobes that partly obscure nectary zone, often with band of nectar-bearing hairs; stamens closely appressed to style; filaments distinct [occasionally connate], usually weak, unable to support anthers; anthers fused lightly to style, fiddle-shaped; pollen sacs marginal, connective visible; style slender proximally, gradually widening and branched distally; branches fairly deeply divided into 2 recurved arms, arching over or between anthers, filiform, apically stigmatic. **Capsules** ovoid to oblong, cartilaginous, apex truncate. **Seeds** many, irregularly globose; seed coat brown. $x = 7$.

Species ca. 5 (1 in the flora): North America (incl. Mexico), Central America, tropical South America.

SELECTED REFERENCE Goldblatt, P. and T. M. Howard. 1992. Notes on *Alophia* (Iridaceae) and a new species, *A. veracruzana*, from Vera Cruz, Mexico. Ann. Missouri Bot. Gard. 79: 901–905.

1. Alophia drummondii (Graham) R. C. Foster, Contr. Gray Herb. 155: 34. 1945 • Propeller flower [F]

Cypella drummondii Graham, Edinburgh New Philos. J. 20: 190. 1836; *Calydorea punctata* (Herbert) Baker; *Eustylis punctata* (Herbert) Ravenna; *E. purpurea* (Herbert) Engelmann & A. Gray; *Gelasine punctata* Herbert; *Herbertia drummondii* (Graham) Small; *Nemastylis punctata* (Herbert) Hemsley; *N. purpurea* Herbert; *Tigridia buccifera* S. Watson; *T. purpurea* (Herbert) Shinners

Plants 15–45 cm. **Bulbs** (5–)8–15 mm diam.; tunic dark brown. **Stems** simple or 2-branched. **Leaves** 3–6, basal leaves 2–3; blade narrowly lanceolate, ± equaling or slightly exceeding flowers; distalmost cauline blade reduced, often ± bractlike. **Rhipidia** 2–4-flowered; outer spathe 24–30 mm, inner spathe (30–)35–42 mm. **Tepals** dark purple; outer broadly ovate, 20–25 × 17–19 mm; inner marked with yellow shortly distal to claw, lanceolate, 16–20 × 10–12 mm; claw spreading, forming shallow cup, dull cream basally, speckled with reddish brown to black, broad; filaments ca. 2 mm wide basally; anthers 6–8 mm; ovary ovoid, 4–5 mm; style branching opposite middle of anthers; pedicel curved at base of ovary. **Capsules** 15–30 mm. **Seeds** ovoid, ca. 2.5 × 2 mm. $2n = 28$.

Flowering mostly May–Jul(–Sep). Open woodlands, grassy pastures, often sandy soils; Ark., La., Okla., Tex.; Mexico; South America (Guyana).

The name *Alophia drummondii* has been misapplied to *Herbertia lahue* by several authors. The two genera have frequently been confused with one another nomenclaturally (P. Goldblatt 1975). They are quite different in both vegetative and floral details.

8. NEMASTYLIS Nuttall, Trans. Amer. Philos. Soc., n. s., 5: 157. 1835 • Celestial-lily, shell-flower [Greek *nema*, thread, and *stylos*, pillar or rod, alluding to the style with thread-like arms]

Peter Goldblatt

Chlamydostylus Baker

Herbs, perennial, from tunicate, ovoid bulbs; tunic brown, dry, brittle, papery. **Stems** simple or branched. **Leaves** few; blade pleated, lanceolate to ± linear. **Inflorescences** rhipidiate, few-flowered; spathes green, unequal, outer shorter than inner, apex usually brown, acute, dry. **Flowers** short-lived, actinomorphic; tepals spreading from base, distinct, blue [to mauve, or white], ± equal [outer whorl slightly larger than inner]; filaments distinct or partly to completely connate; anthers initially erect, collapsing in spiral fashion after dehiscing; style branching ± at or shortly below base of anthers; branches divided ± to base, 2-armed; arms filiform, extending horizontally on either side of subtending anther, stigmatic apically. **Capsules** ovoid to oblong, cartilaginous, apex truncate to turbinate. **Seeds** many, ± prismatic, or compressed, winged; seed coat brown. $x = 7$.

Species 5 (4 in the flora): North America (incl. Mexico), Central America.

SELECTED REFERENCE Foster, R. C. 1945. Studies in the Iridaceae, III. Contr. Gray Herb. 155: 3–54.

1. Cauline leaf blades well developed, 4–20 mm wide; filaments distinct or connate basally; anthers 11–15 mm . . . . 1. *Nemastylis geminiflora*
1. Cauline leaf blades narrow, less than 4 mm wide, or bractlike; filaments connate entirely or only in proximal 1/2, rarely connate only basally; anthers 1.5–10 mm.
    2. Plants 40–150 cm; stems usually 3–6-branched; style branches ca. 5 mm; blooming mainly Aug–Oct . . . . 3. *Nemastylis floridana*
    2. Plants 10–40 cm; stems simple or occasionally 1-branched; style branches 2–4(–5) mm; blooming mainly Mar–Jul.
        3. Filaments connate, 1–2 mm; style branching above base of anthers; Arkansas, Missouri, Oklahoma . . . . 2. *Nemastylis nuttallii*
        3. Filaments connate or distinct in part, 1.5–3.5 mm; style branching below base of anthers; Arizona, Texas, Mexico . . . . 4. *Nemastylis tenuis*

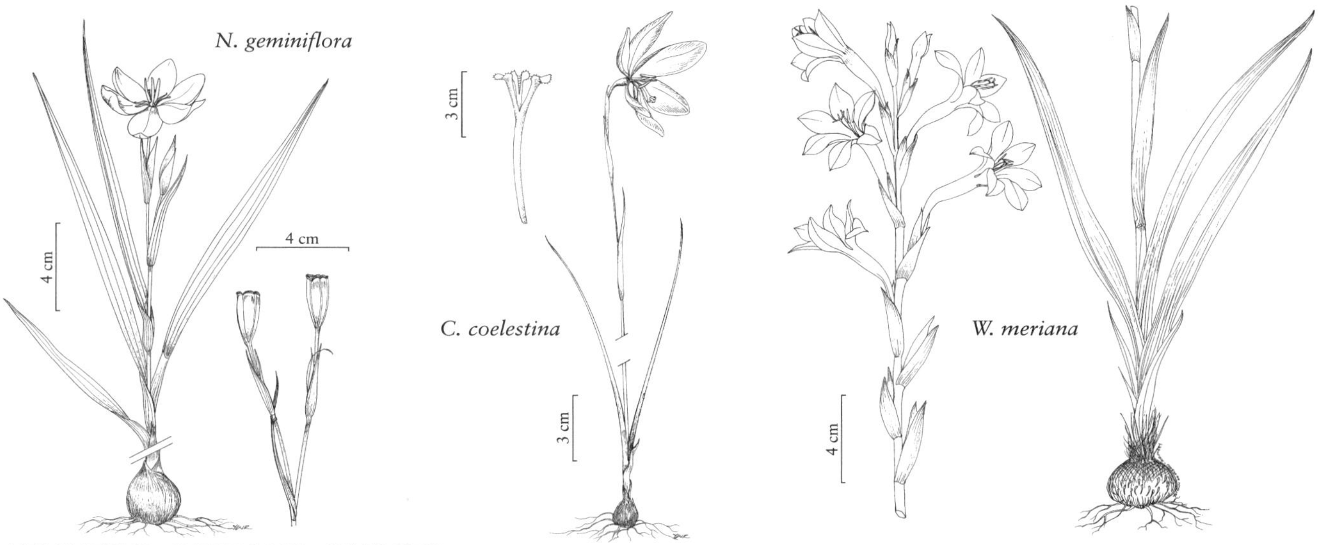

NEMASTYLIS ◦ CALYDOREA ◦ WATSONIA

1. Nemastylis geminiflora Nuttall, Trans. Amer. Philos. Soc., n. s., 5: 157. 1837 • Prairie celestial [E] [F]

Calydorea texana (Herbert) Baker; *Ixia acuta* 1822, not Lichtenstein ex Roemer & Schultes 1817; *Nemastylis acuta* Herbert; *N. texana* Whitehouse

Plants slender, 12–40 cm. **Bulbs** 20–30 mm diam. **Stems** sturdy, usually branched, mainly from base. **Leaves** usually 3; blade linear to ensiform, usually exceeding rhipidia, 4–20 mm wide. **Rhipidia** usually 2-flowered; outer spathe 20–35 mm, inner 30–55 mm, apex dry. **Tepals** blue, white in center, white zone sometimes edged with dark blue, ovate-lanceolate, 2.5–3 cm; filaments distinct or connate basally, ca. 4 mm; anthers 11–15 mm; ovary conic, 4–5 mm; style branching above base of anthers; branches 5–6 mm. **Capsules** ovoid, truncate, (10–)14–20 mm. **Seeds** angular. $2n = 56$.

Flowering Apr–Jun. Woodlands, prairies, and pastures; Ala., Ark., Kans., La., Miss., Mo., Okla., Tex.

The flowers of *Nemastylis geminiflora* open midmorning and fade before sundown.

There seems to be no merit in recognizing *Nemastylis texana* (E. Whitehouse 1936) for plants from southern Texas, said to have steel-blue flowers, more frequently found in oak woodlands than in prairies. Reports in the literature of *N. geminiflora* from Tennessee appear to be incorrect, possibly based on an early record from "E. Tennessee" (*Dr. Eight s.n.*, NY).

2. Nemastylis nuttallii Pickering ex R. C. Foster, Contr. Gray Herb. 155: 36. 1945 [E]

Plants slender, 18–35 cm. **Bulbs** ca. 10 mm diam. **Stems** simple, rarely 1-branched. **Leaves** usually 3–4, basal leaves (0–)1–2; blade linear, ± reaching base of rhipidia, proximalmost cauline leaf blade linear, others reduced, bractlike. **Rhipidia** 1–2-flowered; outer spathe 18–22 mm, inner 25–30 mm, apex dry. **Tepals** pale blue, lanceolate, ca. 20 mm; filaments connate into tube, 1–2 mm; anthers (4–)6–8 mm; ovary oblong, 6–8 mm; style branching ca. 2 mm above apex of filament tube; branches ca. 2.5 mm. **Capsules** oblong, truncate, 10–15(–20) mm. **Seeds** compressed with winged angles, ca. 2.5 × 1.5 mm.

Flowering Apr–Jul. Limestone barrens, bluffs, wet prairies; Ark., Mo., Okla.

The flowers of *Nemastylis nuttallii* open ca. 5 P.M., and close ca. 7 P.M. The name *N. coelestina* (W. Bartram) Nuttall has long been used for this species, but in fact that name properly applies to a different entity, *Ixia coelestina* W. Bartram.

3. Nemastylis floridana Small, J. New York Bot. Gard. 32: 266, fig. 2. 1931 • Fall-flowering ixia, Florida celestial [C] [E]

Plants slender, 40–150 cm. **Bulbs** 7–9(–15) mm diam. **Stems** slender, usually 3–6-branched, rarely simple. **Leaves** (3–)5–7, basal leaves 2–3, longest; blade ± linear, 3–10 mm wide; cauline leaves progressively reduced distally, becoming bractlike. **Rhipidia** usually 2-flowered; outer spathe ± $1/2$ or slightly less than inner, inner 25–30 mm, apex dry. **Tepals** dark blue, lanceolate, 20 × 7.5 mm; filaments connate in proximal $1/2$ or entirely, ca. 2 mm; anthers ca. 8 mm; ovary ovoid, ca. 2.5 mm; style branching at apex of filament tube; branches ca. 5 mm. **Capsules** ovoid, truncate, 6–9 mm. **Seeds** angular-prismatic, ca. 2 mm. $2n = 56$.

Flowering (Jul–)Aug–Oct. Wet, calcareous, pine flatwoods, wet hummocks; of conservation concern; Fla.

The flowers of *Nemastylis floridana* open ca. 4 P.M. and close ca. 6 P.M.

4. Nemastylis tenuis (Herbert) S. Watson, Proc. Amer. Acad. Arts 18: 160. 1883

Nemastylis coelestina (W. Bartram) Nuttall var. *tenuis* Herbert, Bot. Mag. 66: under plate 3779. 1840

Subspecies 4 (1 in the flora): North America (incl. Mexico), Central America.

4a. Nemastylis tenuis (Herbert) Baker subsp. **pringlei** (S. Watson) Goldblatt, Brittonia 27: 379. 1975

Nemastylis pringlei S. Watson, Proc. Amer. Acad. Arts 24: 85. 1888; *N. tenuis* var. *pringlei* (S. Watson) R. C. Foster

Plants slender, 10–30 cm. **Bulbs** 15–20 mm diam. **Stems** usually simple. **Leaves** 3–4, cauline leaves smaller than basal; blade linear, to 25 cm. **Rhipidia** 1-flowered; outer spathe 20–25 mm, inner 18–40 mm, apex often dry. **Tepals** pale blue, lanceolate, 20–30 × 6–9 mm; filaments connate ca. $2/3$ length, 2.5–3.5 mm; anthers ca. 6 mm; ovary ovoid, 4–6 mm; style branching distal to filament column; branches 3.5–5 mm. **Capsules** oblong, truncate to turbinate, 15–20 mm. **Seeds** angular, ca. 2 mm. $2n = 14$.

Flowering Jul–Aug. Rocky, open ground in mountains; Ariz., Tex.; Mexico; Central America.

In the flora area, subsp. *pringlei* is restricted to the mountains of western Texas and southeastern Arizona.

9. CALYDOREA Herbert, Edwards's Bot. Reg. 29(misc.): 85. 1843 • [Greek *caly*, sheathed or covered, and *dory*, spear, most likely alluding to the spear-shaped buds enclosed until anthesis within the rhipidial spathes]

Peter Goldblatt

Cardiostigma Baker; *Catila* Ravenna; *Salpingostylis* Small; *Sphenostigma* Baker

Herbs, perennial, from tunicate, ovoid bulbs; tunic brown, dry, brittle, papery. **Stems** simple or branched. **Leaves** few, basal larger; blade pleated, linear-lanceolate. **Inflorescences** rhipidiate, few-flowered; spathes green, unequal, outer shorter than inner, apex usually brown, acute, dry. **Flowers** short-lived, erect, actinomorphic; tepals spreading from base, distinct, blue to mauve, ± equal [outer whorl considerably larger than inner]; stamens distinct or filaments variously connate; anthers sometimes connate basally; style eccentric [central] when flower fully open, recurving, slender, branching between middle of filaments and anther apices [branching distal to anthers or 3-lobed apically]; branches [lobes] ascending, undivided, short, stigmatic apically. **Capsules** ovoid to oblong, cartilaginous, apex truncate. **Seeds** many, prismatic; seed coat brown. $x = 7$.

Species ca. 12 (1 in the flora): North America (incl. Mexico), South America.

The generic position of the single North American species of *Calydorea* has long been unsettled. Its segregation in the monotypic *Salpingostylis* (J. K. Small 1931b) is unsatisfactory. This Florida species differs little from those of Mexico, which in turn show strong similarities with others in South America. Some of these species were included by J. G. Baker (1892) in *Sphenostigma*, but the type of that genus was shown to be a species of *Gelasine* by P. Ravenna (1977), who then placed the Floridian and Mexican species that had been referred to *Sphenostigma* in *Cardiostigma*. P. Goldblatt and J. E. Henrich (1991) preferred to include all these species in a slightly expanded *Calydorea*, other species of which have the style dividing near the anther bases and the style branches somewhat longer than in the Mexican and Floridian species.

SELECTED REFERENCES Goldblatt, P. and J. E. Henrich. 1991. *Calydorea* Herbert (Iridoideae–Tigridieae): Notes on this New World genus and reduction to synonymy of *Salpingostylis*, *Cardiostigma*, *Itysa*, and *Catila*. Ann. Missouri Bot. Gard. 78: 504–511. Small, J. K. 1931b. Bartram's *Ixia coelestina* rediscovered. J. New York Bot. Gard. 32: 155–161.

1. **Calydorea coelestina** (W. Bartram) Goldblatt & Henrich, Ann. Missouri Bot. Gard. 78: 510. 1991 • Celestial-lily, Bartram's ixia [C] [E] [F]

Ixia coelestina W. Bartram, Travels Carolina, 155, unnumbered plate. 1791; *Beatonia coelestina* (W. Bartram) Klatt; *Marica coelestina* (W. Bartram) Ker Gawler; *Salpingostylis coelestina* (W. Bartram) Small; *Sphenostigma coelestina* (W. Bartram) R. C. Foster; *Trichonema coelestina* (W. Bartram) Sweet

Plants slender, 15–30 cm. **Bulbs** dark brown, 12 mm diam. **Stems** simple, rarely 1-branched. **Leaves** 3–4; basal leaves usually 2, blade narrowly lanceolate to linear, ca. 1/2 stem, 1–3 mm wide; cauline leaves 1(–2), distalmost often entirely sheathing, shorter than basal. **Rhipidia** usually 1–2-flowered; outer spathe 24–36 mm, inner (33–)40–55 mm. **Tepals** blue, lanceolate-ovate, ± equal, outer whorl slightly larger than inner, 24–32 × 14–17 mm; filaments distinct, ca. 3.5 mm; anthers 6–7 mm; ovary conic, often included in spathes, ca. 4 mm; style 10–12 mm; branches 3, flat, less than 2 mm. **Capsules** ovoid, 7–12 mm. **Seeds** ca. 2 mm. $2n = 28$.

Flowering mostly May–Jul. Pine flatwoods; of conservation concern; Fla.

Calydorea coelestina blooms well only after a fire. The flowers open at sunrise and last about three hours. The species is rare, but not currently threatened. A single collection from Camden County, southern Georgia (*Schallert s.n.*, UC) requires confirmation.

10. WATSONIA Miller, Fig. Pl. Gard. Dict., 184, plate 276. 1758, name conserved • [For William Watson, 1715–1787, British botanist] [I]

Peter Goldblatt

Herbs, perennial, from corms. **Stems** simple or branched. **Leaves** several; blade plane, lanceolate to linear, usually coarse, fibrotic. **Inflorescences** spicate, erect, many-flowered; bracts green, often flushed with red, unequal, usually outer exceeding inner, apex acute, inner forked apically, firm to leathery. **Flowers** short-lived, odorless [rarely fragrant], zygomorphic [actinomorphic], distichous; tepals horizontal or suberect, connate into tube, orange, red, or purple [pink, rarely white], ± equal [equal]; perianth tube funnel-shaped or elongate, expanded distally into wide, horizontal upper part; stamens unilateral [symmetrical], arcuate [declinate], extended horizontally below dorsal tepal; anthers parallel [diverging]; style arching below or above filaments [central], dividing opposite to [beyond] anthers into 3 filiform branches each divided for ca. 1/2 their length, apically stigmatic. **Capsules** [globose to] oblong, wood-textured, rounded [acute or attenuate]. **Seeds** several to many, angular, 1- or 2-winged [prismatic]; seed coat light brown. $x = 9$.

Species 52 (1 in the flora): introduced; s Africa.

Several species of *Watsonia* are cultivated in the flora area where the winters are mild, especially in California; only *W. meriana* is truly naturalized. The following have been recorded as persisting for some years around abandoned dwellings, in cemeteries and garbage dumps, and along roads and highways: *W. borbonica* (Pourret) Goldblatt (both pink- and white-flowered forms), *W. fourcadei* J. W. Mathews & L. Bolus, and *W. marginata* (Linnaeus f.) Ker Gawler.

SELECTED REFERENCE Goldblatt, P. 1989. The genus *Watsonia*. A systematic monograph. Ann. Kirstenbosch Bot. Gard. 17.

1. Watsonia meriana (Linnaeus) Miller, Gard. Dict. ed. 8, Watsonia no. 1. 1768 F I

Gladiolus merianus Linnaeus, Syst. Nat. ed. 10, 2: 863. 1759; *Watsonia bulbillifera* J. W. Mathews & L. Bolus

Plants sometimes growing in dense clumps, 50–200 cm. **Corms** tunicate, depressed-globose, 30–45 mm diam.; tunic fibrous, fibers coarse, reticulate. **Stems** simple or 1–2-branched, often with large cormlets in proximal axils, occasionally also in distal axils of stem, spike. **Leaves** 4–6, basal leaves 3–4, blade lanceolate, 12–35 mm wide; cauline leaves smaller than basal, blade with margins and midribs hyaline, heavily thickened. **Spikes** 8–12(–25)-flowered; spathes often flushed with red, outer 18–30 mm, dry in distal 3–5 mm, inner slightly longer to slightly shorter than outer. **Tepals** orange, red, or purple, lanceolate, outer slightly narrower than inner, 22–27 × 8–15 mm; perianth tube proximally slender, 18–25 mm wide, abruptly expanded distally, 20–25 mm wide; filaments 35–45 mm; anthers 8–12 mm; ovary oblong, 3–4 mm; style branching opposite middle of anthers; branches 6–7 mm. **Capsules** 22–30 mm. **Seeds** prominently winged distally, weakly or not winged proximally, 10–18 mm.

Flowering May–Jul. Roadsides and grassy banks; introduced; Calif.; South Africa.

Most *Watsonia* plants recorded from North America correspond to the triploid, sexually sterile form of *W. meriana*, also known as *W. bulbillifera*, which reproduces by means of cormlets produced at the aerial nodes. Most of the flowers do not produce capsules, but a few, containing viable seeds, are occasionally formed. Chromosome number in the Californian plants has not been determined.

11. CROCOSMIA Planchon, Fl. Serres Jard. Eur. 7: 161. 1851 • [Greek *krokos*, crocus, and *osme*, scent, because the dried flowers boiled in water smell like the spice saffron obtained from that plant] I

Peter Goldblatt

Curtonus N. E. Brown

Herbs, perennial, from corms. **Stems** usually branched, branching often strongly divaricate. **Leaves** several, forming fan; blade plane [or plicate], ensiform. **Inflorescences** spicate, axes inclined to ± horizontal, 4–8-flowered[–many-flowered]; bracts orange to reddish [green], unequal, outer bract usually exceeding inner, inner bifid, firm textured. **Flowers** odorless, zygomorphic [actinomorphic]; tepals connate into tube, orange to red, subequal [unequal], outer tepals slightly larger than inner with dorsal largest [dorsal tepal much larger than others]; stamens unilateral; perianth tube funnel-shaped [nearly cylindric]; anthers parallel [radially disposed]; style arching over filaments [central], dividing into 3 filiform branches, usually notched apically [undivided]. **Capsules** globose, 3-lobed, cartilaginous. **Seeds** 2–4 per locule, globose [compressed]; seed coats hard, shiny or matte. $x = 11$.

Species 9 (1 in the flora): introduced; sub-Saharan Africa, Madagascar.

In addition to *Crocosmia* ×*crocosmiiflora*, now naturalized, *C. paniculata* (Klatt) Goldblatt and *C. masonorum* (L. Bolus) N. E. Brown are grown in gardens. They may readily be distinguished by their broad, plicate leaves.

CROCOSMIA ◦ CHASMANTHE ◦ SPARAXIS

An artificial hybrid between two southern African species, *Crocosmia aurea* (Pappe ex Hooker f.) Planchon and *C. pottsii* (Macnab ex Baker) N. E. Brown, is now widely naturalized in the Neotropics, Madagascar, and locally in the Pacific Islands.

SELECTED REFERENCE De Vos, M. P. 1984. The African genus *Crocosmia* Planchon. J. S. African Bot. 50: 463–502.

1. **Crocosmia ×crocosmiiflora** (Lemoine) N. E. Brown, Trans. Roy. Soc. South Africa 20: 264. 1932 (as crocosmiflora) • Montbretia [F] [I]

Montbretia ×crocosmiiflora Lemoine, Garden (London 1871–1927) 18: 188. 1881; *Tritonia ×crocosmiiflora* (Lemoine) G. Nicholson

Plants 50–100 cm; corms globose, 15–25 mm diam. **Stems** usually 2–4-branched, often curving distally. **Leaves** 5–8, mostly basal, basal much larger than cauline; blade lanceolate, 8–20 mm wide. **Spikes:** spathes 6–10 mm, apex brown, dry, outer apex acute, inner apex 2-keeled, bifurcate; flowers distichous. **Tepals** spreading, dorsal weakly arcuate, paler in center, sometimes with darker markings at base, subequal, outer whorl slightly larger than inner, 16–25 × 6–9 mm; perianth tube 12–15 mm; filaments 15–22 mm; anthers 6–8 mm; ovary oblong, ca. 3 mm; style arching over stamens, branching at or shortly beyond anther apices; branches ca. 4 mm, apically bifid. **Capsules** ca. 8 mm. **Seeds** blackish, ca. 2.5 mm diam. (or aborted and plants partly or completely sterile). $2n = 22$.

Flowering mainly Jul–Aug. Roadsides, thickets, forest margins; 50 m; introduced; B.C.; Calif., Fla., N.C., Oreg., S.C., Tex., Wash.; South Africa.

Crocosmia ×crocosmiiflora is an escape from cultivation.

12. CHASMANTHE N. E. Brown, Trans. Roy. Soc. South Africa 20: 272. 1932 • [Greek *chasme*, gap, and *anthos*, flower, alluding to the shape of the flower] [I]

Peter Goldblatt

Herbs, perennial, from corms. **Stems** simple or branched. **Leaves** several; blade plane, ensiform. **Inflorescences** spicate, axes usually lightly flexed, many-flowered; bracts green or purplish, ± equal, coriaceous. **Flowers** unscented, zygomorphic; tepals connate into tube, orange to scarlet (rarely yellow), unequal, dorsal tepal horizontal, more than 2 times others, other tepals spreading to reflexed; perianth tube slender, cylindric proximally, abruptly expanded, longer, wider, horizontal distally; stamens unilateral, extended horizontally below dorsal tepal; anthers parallel; style arching over filaments, branching opposite to or beyond anthers; branches 3, filiform, shallowly notched apically. **Capsules** irregularly globose, cartilaginous. **Seeds** 2–4 per locule, globose; seed coat bright orange, sometimes somewhat fleshy. $x = 10$.

Species 3 (1 in the flora): introduced; South Africa.

SELECTED REFERENCE De Vos, M. P. 1985. Revision of the South African genus *Chasmanthe* (Iridaceae). S. African J. Bot. 51: 253–261.

1. Chasmanthe floribunda (Salisbury) N. E. Brown, Trans. Roy. Soc. South Africa 20: 274. 1932 [F] [I]

Antholyza floribunda Salisbury, Trans. Hort. Soc. London 1: 324. 1812; *A. prealta* Redouté

Plants 60–120 cm; corm depressed-globose, 40–70 mm diam. **Stems** 1–2-branched. **Leaves** 8–10, mostly basal, basal longer than cauline; blade lanceolate, ± reaching base of spike, 18–35 mm wide. **Spikes** 25–40-flowered; outer spathe 13–15 mm, apex obtuse to truncate; inner ± equaling outer, apex bifurcate; flowers distichous. **Tepals:** perianth tube 35–45 mm, 9–12 mm proximally, often twisted, ca. 40 mm distally, base pouched; dorsal tepal horizontal, 28–33 × 7–9 mm, much exceeding others; lateral tepals patent or recurved, 12–15 × 4–7 mm; lower median tepal slightly smaller than laterals; filaments 50–55 mm; anthers 7–8 mm; ovary 6–9 mm, style branching shortly below and opposite to anthers (or exceeding them); branches 7–10 mm. **Capsules** 10–15 mm. **Seeds** 5–7 mm diam.

Flowering mostly Feb–May. Road verges, grassy slopes; 0–100 m; introduced; Calif.; South Africa.

Chasmanthe floribunda has escaped from gardens and has become naturalized locally. It is native to the winter-rainfall region of South Africa. The species has been confused in North America with *C. aethiopica* (Linnaeus) N. E. Brown, which has similar flowers but is a smaller plant with unbranched stems, normally flowering in the early winter months.

13. SPARAXIS Ker Gawler, Ann. Bot. (König & Sims) 1: 225. 1804 • [Greek *sparasso*, to tear, alluding to the sometimes lacerated bracts] [I]

Peter Goldblatt

Streptanthera Sweet; *Synnotia* Sweet

Herbs, perennial, from corms. **Stems** simple or branched. **Leaves** several, sometimes prostrate; blade plane, ensiform to oblong. **Inflorescences** spicate, spiral or secund, 2–5-flowered[–many-flowered]; bracts pale to light brown, irregularly streaked with dark brown, unequal, outer usually exceeding inner, scarious, acute or 3-cuspidate, inner apex 2-furcate to 2-cuspidate. **Flowers** odorless [fragrant], actinomorphic [zygomorphic]; perianth tube funnel-shaped, sometimes obliquely so; tepals spreading, connate into tube, variously colored, sometimes with strongly contrasting markings, subequal, outer whorl slightly larger than inner [unequal, dorsal tepal largest]; stamens symmetrically arranged [asymmetrical], unilateral; anthers straight [coiled]; style [unilateral, often arching over filaments] dividing into 3 short branches, apices spatulate [long, filiform]. **Capsules** globose to oblong, cartilaginous. **Seeds** 8–15 per locule, globose; seed coat light to dark brown, hard, shiny. $x = 10$.

Species 15 (1 in the flora): introduced; s Africa.

SELECTED REFERENCES Goldblatt, P. 1969. The genus *Sparaxis*. J. S. African Bot. 35: 219–252. Goldblatt, P. 1992. Phylogenetic analysis of the South African genus *Sparaxis* (including *Synnotia*) (Iridaceae–Ixioideae), with two new species and a review of the genus. Ann. Missouri Bot. Gard. 79: 143–159. Goldblatt, P. 1999. *Sparaxis*. In: Department of Agricultural Technical Services [South Africa]. 1963+. Flora of Southern Africa. 19+ vols. (some partial). Pretoria. Vol. 7, part 2, fasc. 1, pp. 151–169.

1. **Sparaxis tricolor** (Schneevoogt) Ker Gawler, Ann. Bot. (König & Sims) 1: 225. 1804 • Harlequin flower [F] [I]

Ixia tricolor Schneevoogt, Icon. Pl. Rar., plate 39. 1794

Plants 10–40 cm. **Corms** tunicate; tunic with fine, whitish fibers. **Leaves** several, ± reaching base of spike. **Spikes:** spathes pale brown with darker brown streaks, subequal, dry, crinkled, outer with entire or lightly lacerate margins, inner forked apically. **Tepals** bright orange, yellow in center, with heart-shaped yellow mark in proximal 1/3, outlined distally with dark reddish brown band, lanceolate, subequal, 25–33 mm; perianth tube funnel-shaped, ca. 8 mm; stamens symmetrical; filaments contiguous, 6–7 mm; anthers yellow, linear, 8–9 mm; style slender, branching opposite anthers; branches 1–2 mm, expanded apically.

Flowering Mar–Apr. Near gardens, dump sites, abandoned dwellings; 0–100 m; introduced; Calif.; s Africa (winter-rainfall region).

14. FREESIA Ecklon ex Klatt, Linnaea 34: 672. 1866, name conserved • Freesia [for F. H. T. Freese, d. 1876, student of C. F. Ecklon, 1795–1868, who first used the name (as Freesea), although in a different sense] [I]

Peter Goldblatt

Anomatheca Ker Gawler

Herbs, perennial, from corms. **Corms** tunicate, conic, ca. 10 mm diam.; tunics fibrous; fibers fine, pale, reticulate. **Stems** simple or branched. **Leaves** several, sometimes prostrate; blade plane, lanceolate [ensiform or oblong]. **Inflorescences** spicate, secund, usually strongly flexed basally, ± horizontal, several–many-flowered; bract green [pale straw colored], coriaceous or membranous. **Flowers** often strongly fragrant, zygomorphic; tepals connate into tube, mostly white or yellow, sometimes pink or red, subequal, outer whorl slightly larger than inner; perianth tube cylindric or funnel-shaped, short or long proximally, abruptly expanded into short to long tube or flared distally; stamens asymmetrically disposed, unilateral; anthers usually parallel; style arching over filaments, dividing into 3 filiform branches. **Capsules** irregularly globose, usually rugose. **Seeds** several per locule, globose, without wings; seed coat light to dark brown, hard, shiny. $x = 11$.

Species 15 (2 in the flora): introduced; South Africa.

Critical phylogenetic study has shown *Freesia* to be nested in *Anomatheca*, and the two genera have been united.

SELECTED REFERENCES Goldblatt, P. 1982. Systematics of *Freesia* Klatt (Iridaceae). J. S. African Bot. 48: 39–91. Goldblatt, P. and J. C. Manning. 1995. Phylogeny of the African genera *Anomatheca* and *Freesia* (Iridaceae–Ixioideae), and a new genus *Xenoscapa*. Syst. Bot. 20: 161–178.

1. Perianth hypocrateriform; tepals pink to red with dark red markings on lower 3 tepals; tube cylindric throughout . . . . 1. *Freesia laxa*
1. Perianth funnel-shaped; tepals predominantly white, sometimes with yellow markings on lower tepals, tepals and tube lightly flushed with purple abaxially; distal part of tube longer than basal part, very gradually flared . . . . 2. *Freesia alba*

1. **Freesia laxa** (Thunberg), Goldblatt & J. C. Manning, Syst. Bot. 20: 171. 1995 [I]

Gladiolus laxus Thunberg, Fl. Cap. ed. 2, 15. 1823; *Anomatheca cruenta* Lindley; *A. laxa* (Thunberg) Goldblatt; *Freesia cruenta* (Lindley) Klatt; *Lapeirousia laxa* (Thunberg) N. E. Brown

Plants 20–35 cm. **Stems** erect or inclined, flexed outward below base of spike, usually branched, smooth. **Leaves** several, erect; blades usually shortly exceeding spike. **Spike** nearly horizontal, 2–6-flowered; spathes 6–8(–13) mm, outer slightly larger than inner, becoming membranous above, apex often dark brown, bifid.

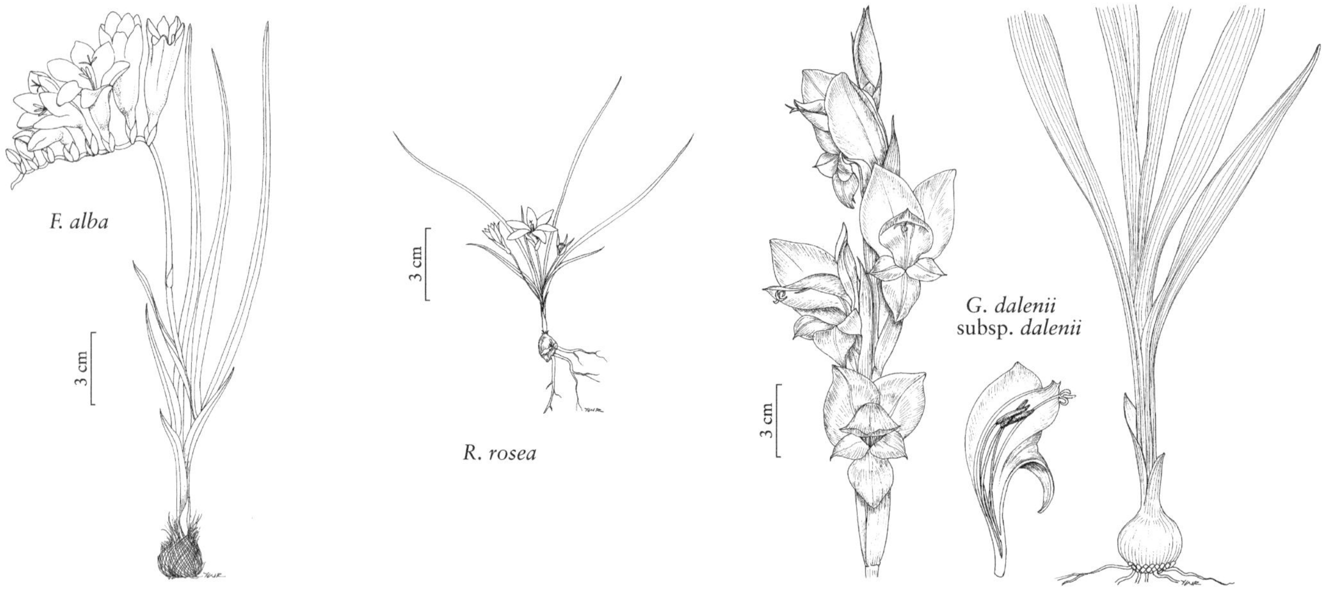

FREESIA◦ROMULEA◦GLADIOLUS

Flowers unscented, hypocrateriform; perianth tube cylindric, widening slightly at apex, 18–33 mm; tepals spreading at right angle to tube, pink to red [pale blue to white] with dark red [blue-violet] marks at base of lower 3 tepals, ovate to oblong, 9–13 mm; filaments exserted 1.5–2 mm from tube; anthers 3–4 mm; style branching between base and middle of anthers; branches ca. 2.5 mm, often tangled among anthers. **Capsules** 9–12 × 9–10 mm, sparsely papillose. **Seeds** 2–3 mm diam.

Flowering mainly Apr. Disturbed sites, abandoned gardens; introduced; Fla.; s, e Africa.

Freesia laxa has been in cultivation for nearly 200 years, although it has never been particularly popular. For years it was known by the later synonyms *Lapeirousia cruenta* or *L. laxa*. Two subspecies are recognized: subsp. *azurea* (Goldblatt & Hutchings) Goldblatt & J. C. Manning from Mozambique and Natal, South Africa, having white tepals with blue to violet markings; and subsp. *laxa*, having pink to red tepals with red markings. Only the latter is found in North America.

2. Freesia alba (G. L. Meyer) Gumbleton, Gard. Chron., ser. 3, 19: 392. 1896 F I

Freesia refracta (Jacquin) Klatt var. *alba* G. L. Meyer, Z. Schweiz. Gartenbauver. 1: 37, plate 3. 1881

Plants (5–)12–40 cm. **Stems** erect, simple or branched, minutely pubescent. **Leaves** (4–)6–8, erect or inclined; blades slightly shorter than to ± equaling spike. **Spikes** horizontal, (2–)4–6(–8)-flowered; spathes 5–8 mm, outer usually longer than inner, firm-textured. **Flower** strongly scented, funnel-shaped; perianth tube very gradually flared, distal portion (15–)20–40 mm, basal portion slender, 6–12 mm; tepals spreading, white, often with yellow markings on lower tepals, tepals and tube flushed with purple abaxially, oblong, 15–18; filaments included, 15–25 mm; anthers 6–9 mm, unilateral, sometimes parallel; ovary 2–3 mm; style branching opposite anther apices. **Capsules** ca. 10 mm, lightly papillose. **Seeds** ca. 2 mm diam.

Flowering mainly Mar. Roadsides, abandoned gardens, grassy slopes; 50 m; introduced; Calif.; South Africa.

Freesia alba often has been confused with *F. refracta* in older literature, and consequently the latter has been erroneously listed in regional floras in North America. *Freesia refracta* is rarely cultivated; *F. alba* and cultivars derived from crosses between it and other species are widely grown, as both ornamentals and cut flowers.

15. ROMULEA Maratti, Pl. Romul. Saturn., 13. 1772, name conserved • [For Romulus, one of the mythical founders of Rome, the type species of the genus being common around that city] [I]

Peter Goldblatt

Herbs, perennial, from corms. **Stems:** true stem aerial [subterranean and only peduncles of individual flowers aerial], simple or branched. **Leaves** 1–several; blade linear, oval to terete in cross section, 4-grooved (2 on each surface, on either side of thickened midrib) [2-grooved]. **Inflorescences** solitary flowers, interpreted as sessile, thus without pedicels; bracts inserted at ovary base, green or flushed with purple, unequal, outer exceeding inner, firm, inner bract margins membranous to scarious (rarely entirely scarious). **Flowers** odorless [fragrant], actinomorphic; tepals forming wide cup, connate into tube, subequal, outer whorl often slightly larger than inner; perianth tube funnel-shaped [or cylindric]; stamens symmetrical; filaments distinct [connate]; anthers usually erect, contiguous; style branching opposite or beyond anthers into 3 slender branches divided for ± 1/2 length. **Capsules** ovoid-oblong, walls firm, cartilaginous. **Seeds** many, globose; seed coat light to dark brown, often smooth, shiny. x = 13 or 14.

Species ca. 90 (1 in the flora): introduced; Africa, s Europe, Middle East.

SELECTED REFERENCES De Vos, M. P. 1972. The genus *Romulea* in South Africa. J. S. African Bot., suppl. 9. De Vos, M. P. 1983. *Romulea*. In: Department of Agricultural Technical Services [South Africa]. 1963+. Flora of Southern Africa.... 19+ vols. (some partial). Pretoria. Vol. 7, part 2, fasc. 2, pp. 10–73. Manning, J. C. and P. Goldblatt. 2001. A synoptic review of *Romulea* (Iridaceae: Crocoideae) in sub-Saharan Africa, the Arabian Peninsula and Socotra including new species, biological notes, and a new infrageneric classification. Adansonia, sér. 3, 23: 59–108.

1. Romulea rosea (Linnaeus) Ecklon, Topogr. Verz. Pflanzensamml. Ecklon, 19. 1827 [F] [I]

Ixia rosea Linnaeus, Syst. Nat. ed. 12, 2: 75. 1767

Plants 3–12(–30) cm. **Corm** tunicate, 5–15 mm diam.; tunic hard, woody, splitting above and below into acuminate segments, those below bent backward. **Stems** usually branched below ground level; branches (peduncles) aerial, suberect, becoming falcate, ultimately erect. **Leaves** (2–)6–9, much exceeding stem; blade 10–30 cm × ca. 1.5 mm. **Spathes** green or flushed with purple; outer 12–25 mm, margin narrow, membranous; inner margin brown-streaked, broad, membranous. **Tepals** pink to purple, usually pale yellow in cup, outer pale abaxially, main veins dark green to purple, lanceolate, 13–22 × 3–4 mm; perianth tube 2–3.5 mm; filaments ca. 5 mm; anthers ca. 4 mm; style branching opposite upper 1/3 anthers; branches ca. 2 mm, shorter than anther apices. **Capsules** 10–15 mm. **Seeds** ca. 2 mm diam.

Flowering mainly Mar–Apr. Abandoned dwellings, along paths, in meadows; 0–50 m; introduced; Calif.; South Africa.

M. P. de Vos (1972) recognized five varieties of *Romulea rosea*; the plants naturalized in North America correspond to var. *australis* (Ewart) M. P. de Vos, which has become a common weed of lawns, pathways, and roadsides in Australia, where it is a pest. To date, weediness does not seem to be the case in the parts of California where the species occurs.

16. GLADIOLUS Linnaeus, Sp. Pl. 1: 36. 1753; Gen. Pl. ed. 5, 23. 1754 • Corn-flag [Latin *gladiolus*, little sword, alluding to the leaf shape] [I]

Peter Goldblatt

Herbs, perennial, from corms. **Stems** simple or branched. **Leaves** 1–9; blade lanceolate to linear, plane or margins and/or midribs variously raised and thickened (then H- or X-shaped in cross section), or evidently terete, midribs and margins much thickened, grooved; grooves 4, narrow, longitudinal. **Inflorescences** spicate, partly to fully secund or with flowers weakly distichous; bracts green, sometimes flushed grayish purple, unequal, outer usually exceeding inner, acute or inner forked or notched apically. **Flowers** somewhat fragrant, zygomorphic [actinomorphic]; tepals basally connate into tube, variously colored, usually with contrasting markings comprising nectar guide on outer tepals, usually unequal, dorsal tepal largest, arched to hooded over stamens, outer 3 tepals narrower; perianth tube obliquely funnel-shaped to cylindric; stamens usually unilateral; anthers usually parallel; style usually arching over stamens, dividing into 3 filiform branches, these distally expanded. **Capsules** usually slightly inflated, oblong to ellipsoid or globose [rarely nearly cylindric], softly cartilaginous. **Seeds** usually many, broadly winged; rarely few, wingless, globose or angular; seed coat light to dark brown. x = 15.

Species ca. 260 (4 in the flora): introduced; Africa, Madagascar, Eurasia.

Hybrid strains of *Gladiolus* are widely cultivated, and some wild species are grown outdoors in areas with mild winters. The common, large-flowered hybrids are the product of crossing between four or five species, followed by selection. Named variously *G. gandavensis* L. van Houtte, *G. lemoinei* Baker, and *G. hortulanus* L. H. Bailey, these plants are occasionally found in and near gardens and cemeteries and are easily recognized by their large, brightly colored flowers with the inner tepals 60–70 mm long and the outer tepals about 50 mm long. They reproduce mainly vegetatively and persist for some years where winters are mild; they show no evidence of spreading into native vegetation.

SELECTED REFERENCES Goldblatt, P. 1996. *Gladiolus* in Tropical Africa. Portland. Goldblatt, P. and J. C. Manning. 1998. *Gladiolus* in Southern Africa. Cape Town.

1. Tepals pink to reddish or light purple with white markings on outer 3, 30–65 mm (perianth tube plus dorsal tepal).
    2. Anthers 10–13 mm, shorter than to ± equaling filaments; capsules oblong, 18–24 mm; seeds winged . . . . 1. *Gladiolus communis*
    2. Anthers ca. 15 mm, exceeding filaments; capsules globose, 10–12 mm; seeds not winged . . . . 2. *Gladiolus italicus*
1. Tepals white to cream or orange to red, 60–95 mm (perianth tube plus dorsal tepal).
    3. Leaf blade cruciform in cross section; tepals white to cream with darker greenish yellow or purple on midlines of each tepal . . . . 3. *Gladiolus tristis*
    3. Leaf blade plane; tepals red to orange with yellow markings on outer tepals . . . . 4. *Gladiolus dalenii*

1. Gladiolus communis Linneaus, Sp. Pl. 1: 36. 1753 • False corn-flag, Turkish corn-flag [I]

Gladiolus byzantinus Miller; *G. communis* subsp. *byzantinus* (Miller) A. P. Hamilton

Plants 50–100 cm. **Corms** tunicate, ca. 20 mm diam.; tunic coriaceous, fragmenting into irregular pieces, rarely ultimately becoming fibrous. **Stems** simple. **Leaves** ± reaching base of spike; blade plane, lanceolate, sometimes narrowly so, 5–22 mm wide. **Spikes** 10–20-flowered; spathes unequal, outer 25–35(–50) mm, inner ± 2/3 outer. **Flowers** unscented, weakly distichous; perianth tube obliquely funnel-shaped, 10–12 mm; tepals reddish purple with narrow median white streak on outer 3 tepals, unequal, dorsal tepal 30–40 × 14–19 mm, inner lateral tepals joined to outer tepals for ca. 3 mm, 28–35 mm, outer 3 tepals connate for ca. 2 mm, outer lateral tepals 18–25 mm, outer median tepal 24–28 mm; filaments 12–15 mm; anthers 10–13 mm; style branching opposite distal 1/3 of anthers; branches ca. 2 mm. **Capsules** oblong, 18–24 mm. **Seeds** broadly winged, 4–6 mm diam.

Flowering May–Jul. Roadsides, abandoned gardens, disturbed sites; introduced; Ala., Ark., Ga., Ky., La., Miss., N.C., S.C., Tenn., Tex.; sw Europe, n Africa.

Gladiolus communis is a garden escape. In regional floras it is sometimes confused with the southern African *G. papilio* Hooker; the resemblance is entirely superficial. Plants of *G. communis* found in North America have traditionally been treated as *G. byzantinus*, which differs little from *G. communis* except in degree of robustness. Distinction at even subspecific rank does not seem warranted.

2. Gladiolus italicus Miller, Gard. Dict. ed. 8, Gladiolus no. 2. 1768 [I]

Gladiolus segetum Ker Gawler

Plants 50–100 cm. **Corms** tunicate, ca. 20 mm diam.; tunic fibrous. **Stems** usually simple. **Leaves** 3–5, ± reaching base of spike; blade plane, lanceolate, sometimes narrowly so, 8–22 mm wide. **Spikes** 6–16-flowered; spathes unequal, outer 30–40(–50) mm, inner 1/2–2/3 outer.

Flowers unscented, weakly distichous; perianth tube obliquely funnel-shaped, 10–12 mm; tepals pink to light purple with narrow median white streak on outer 3 tepals, unequal, dorsal tepal 45–50 × ca. 16 mm, inner lateral tepals ca. 30 × 8 mm, outer 3 tepals connate for ca. 5 mm, outer lateral tepals ca. 25 mm, outer median tepal ca. 20 mm; filaments ca. 12 mm; anthers ca. 15 mm; style branching opposite level of anther apices; branches ca. 2.5 mm. **Capsules** globose, 10-12 mm. **Seeds** globose or lightly angled, 2–3 mm diam.

Flowering mostly Apr–May. Roadsides, crop fields; 500 m; introduced; Calif.; probably native to Near East.

Gladiolus italicus is a weed of cultivation, not known certainly as a truly wild plant. Occasionally, small-flowered plants with aborted anthers occur in some populations, but such gynodioecious individuals have not been recorded in North America.

3. **Gladiolus tristis** Linneaus, Sp. Pl. ed. 2, 1: 53. 1762 [I]

Plants 50–90(–150) cm. **Corms** tunicate, 10–20 mm diam.; tunics ± woody, broken into segments proximally and distally. **Stems** simple. **Leaves** 3, basal longest, usually ± reaching middle of stem, sheathing in lower 1/2, distalmost 2 leaves largely to entirely sheathing; blade cruciform in cross section, 2–3 mm diam. **Spikes** 2–4 (–8)-flowered, secund; sheaths sometimes flushed grayish purple, subequal, 30–40 mm, inner slightly shorter than to ± equaling outer. **Flowers** strongly scented of carnations and cloves in evening; perianth tube obliquely funnel-shaped, 45–50 mm, narrow cylindrical section ca. 25 mm; tepals white to cream, with darker greenish yellow or purple on midlines, margins transparent in lower 10 mm, especially between dorsal and inner lateral tepals, lanceolate, unequal, dorsal inclined to ± horizontal, 22–25 × 16 mm, outer tepals fused with inner lateral tepals for ca. 2 mm, ± straight and directed forward, or gently curving outward in distal 1/2, ca. 22 × 10 mm; filaments ca. 20 mm, included in tube; anthers ca. 12 mm, partly included in tube; style branching opposite distal 1/3 of anthers; branches ca. 6 mm, reaching beyond anthers. **Capsules** oblong-ellipsoid, 34–36 mm. **Seeds** broadly winged, 5 × 4 mm.

Flowering mostly Mar–Apr. Roadsides, grassy banks; 800 m; introduced; Calif.; s Africa.

4. **Gladiolus dalenii** Van Geel, Sert. Bot. 1: unnumbered plate. 1828 (as daleni) [F] [I]

Gladiolus psittacinus Hooker

Subspecies 3 (1 in the flora): introduced; sub-Saharan Africa, Madagascar, Arabia.

4a. **Gladiolus dalenii** Van Geel subsp. **dalenii** [F] [I]

Plants (50–)70–120(–150) cm. **Corms** tunicate, (15–)20–30 mm diam.; tunic layered, brown, brittle, cartilaginous, outer layer becoming irregularly broken to fibrous, usually bearing numerous tiny cormlets basally. **Stems** simple. **Leaves** contemporary with flowering stem [foliage leaves borne later on separate shoots], 4–6(–7), at least proximalmost 2 ± basal, distalmost 1–2 cauline, ± sheathing, ± 1/2 length of spike; blade narrowly lanceolate to ± linear, (5–)10–20(–30) mm wide, midribs and margins notably thickened. **Spikes** (2–)3–7(–14)-flowered, secund; sheaths subequal, (2.5–)4–7 cm, inner slightly shorter than outer, apex sometimes dry, pale. **Flowers** unscented; perianth tube ± cylindrical, curving outward in distal 1/2, (25–)35–45 mm; tepals red to orange with yellow mark on outer 3 tepals (or yellow to greenish and often with red to brown streaks on inner tepals), unequal, inner 3 tepals broadly elliptic, dorsal tepal horizontal to downcurved, 35–50 mm, to 22–30 mm wide, inner lateral tepals usually curving outward distally, ± equaling dorsal, outer 3 tepals curving downward, 20–25(–30) mm, outermost tepal somewhat longer and narrower than outer lateral tepals; filaments ca. 25 mm; anthers 12–16 mm; style branching near apex of anthers; branches (4–)5–6 mm. **Capsules** ellipsoid to ovoid, (18–)25–35 × 12–14 mm. **Seeds** broadly winged, 8–12 × 5–9 mm; seed coats light brown, glossy.

Flowering mainly May–Jul. Introduced; Ala., La.; sub-Saharan Africa, Madagascar, Arabia.

224. ALOACEAE Batsch
• Aloe Family

Walter C. Holmes
Heather L. White

Trees, shrubs, and **succulents**, perennial, simple to sparsely branched, rhizomatous, some tuberous-thickened. **Leaves** simple, alternate, usually crowded at bases of stems or ends of branches, sessile; blade fleshy, margins often prickly, venation parallel. **Inflorescences** terminal, axillary, or lateral, spicate, racemose, or paniculate. **Flowers** 3-merous, short- to long-pedicellate, rarely sessile; perianth red, brown, yellow, orange, or whitish; tepals petaloid, connivent or connate basally to almost entirely into tube, sometimes fleshy; stamens sometimes 3, usually 6, exserted or included; anthers dorsifixed, dehiscence antrorse; pollen grains monosulcate; ovary 3-carpellate, placentation axile, usually with septal nectaries; style terminal; stigmas punctate, discoid, or 3-lobed. **Fruits** capsular, rarely baccate, dehiscence loculicidal, apical. **Seeds** usually winged or flattened.

Genera 5, species ca. 700 (1 genus, 2 species in the flora): all introduced; Africa, Madagascar, Arabia, and Atlantic islands.

Aloaceae are closely related to and included by some authors in Liliaceae.

The juice of some *Aloe* species is used to make a purgative called bitter aloe; active ingredients include aloin and other anthraquinones. Additionally, the thick, mucilaginous gel of some species is widely used to treat minor thermal burns, itching, and sunburn.

1. ALOE Linnaeus, Sp. Pl. 1: 319. 1753; Gen. Pl. ed. 5, 150. 1754 • Aloe [Arabic *alloeh*, a name for these or similar plants] [I]

Plants succulent, shrubby or arborescent, scapose. **Stems** erect, clambering or ascending, branched or not. **Leaves** succulent, crowded, often rosulate or distichous; blade margins spiny-toothed or entire. **Inflorescences** axillary or terminal, paniculate to more often racemose, dense, bracteate. **Flowers** usually nodding; perianth red to yellow; tepals connate basally to almost

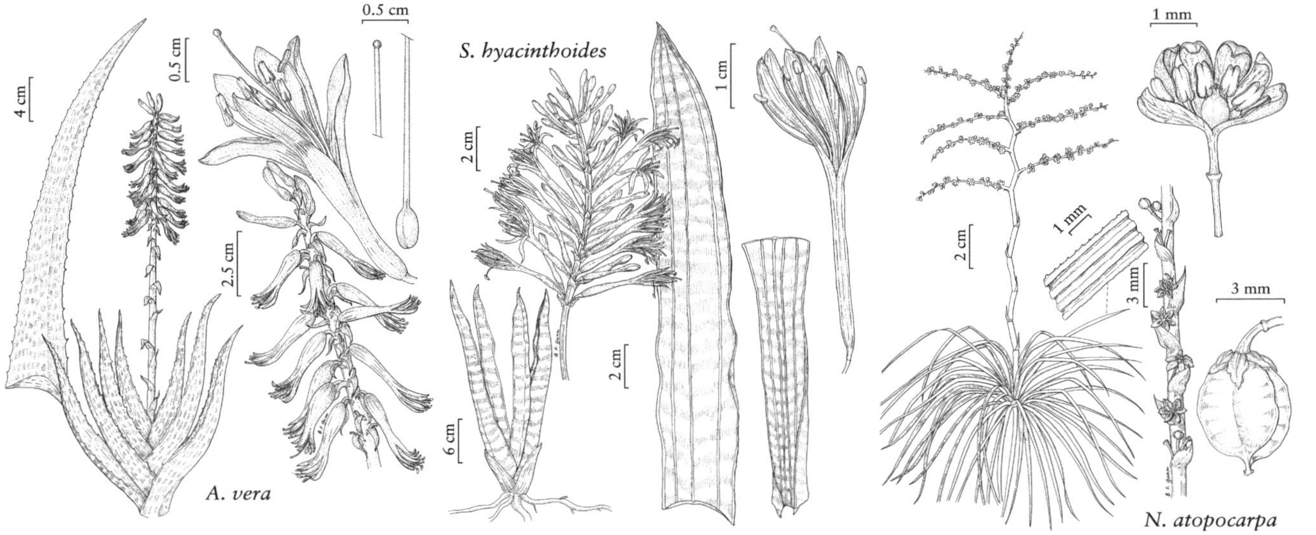

ALOE ◦ SANSEVIERIA ◦ NOLINA

entirely into tube; stamens 3 or 6; style slender; pedicel not articulate. **Capsules** papery to woody. $x = 7$.

Species 300 or more (2 in the flora): introduced; primarily s and tropical Africa; also Madagascar, Arabian peninsula, and Atlantic islands (Madeira, Canary, and Cape Verde); naturalized in the Mediterranean region, India, and China.

Aloe saponaria (Aiton) Haworth, distinguished by its yellow sap and glaucous red flowers with yellow throats, is cultivated in the southwestern United States and has been observed to escape. Apparently it persists only when supplementary water is available.

SELECTED REFERENCES Moran, R. 1992. *Aloe* wild in California. Cact. Succ. J. (Los Angeles) 64: 55–56. Reynolds, G. W. 1982. The Aloes of South Africa, ed. 4. Rotterdam.

1. Perianth yellow; inflorescences unbranched or rarely branched; leaf blade margins green . . . . 1. *Aloe vera*
1. Perianth red; inflorescences divided distally into 5–10 arching branches; leaf blade margins narrowly whitish . . . . 2. *Aloe* ×*schonlandii*

1. Aloe vera (Linnaeus) Burman f., Fl. Indica, 83. 1768 • Barbados aloe, Curaçao aloe, medicinal aloe, unguentine cactus, burn plant [F] [I]

Aloe perfoliata Linnaeus var. *vera* Linnaeus, Sp. Pl. 1: 320. 1753; *A. barbadensis* Miller

Plants short-stemmed, woody-based, stoloniferous. **Stems** to 50 cm; scarious leaf sheaths persistent. **Leaves** alternate, rosulate to distichous, 10–50 × 10–70 cm; blade glaucous-green to variegated with small white or glaucous dots, irregular bands, or blotches, often reddish near apex or margins, lanceolate to ensate, tapering from base to apex, glabrous, margins green, spiny-toothed, teeth 1–1.5 cm apart. **Inflorescences** terminal, usually unbranched, racemose, 10–15 dm, usually covered with scalelike bracts; racemes cylindrical, dense, 0.5 m; bracts glabrous or puberulent, with 3 prominent purple veins that are confluent at tips. **Flowers:** perianth yellow; tepals prominently 3-veined, connate basally for 1/2 their length, lobes broadly linear to oblong-lanceolate, apex rounded; stamens 6, included to slightly exserted, slightly unequal; filaments 2–2.5 cm; anthers 2.5–4 mm; style usually exserted; stigmas not expanded; pedicel 2.2–3.3 cm. **Capsules** somewhat elongate. $2n = 14$.

Flowering spring–winter, occasionally at other times. Hammocks, sandy areas, roadsides, and similar places in full sun; 0 and 1300 m; introduced; Ariz., Fla., Tex.;

Mediterranean region and Atlantic islands (Canary, Madeira, and Cape Verde).

This is the aloe of commerce and source of bitter aloe. Most of the world's supply is grown in southern Texas and adjacent northwestern Mexico and the West Indies. The species is thought to be native to the Atlantic islands and is widely used as an indoor ornamental. It is often cultivated outdoors in the southwestern United States, where it occasionally escapes.

2. **Aloe ×schonlandii** Baker, Gard. Chron., ser. 3, 32: 430. 1902, as species [I]

Plants erect (to decumbent in age). **Stems** branching, to 40 cm. **Leaves** rosulate, 20–25 × 9–12 cm; blade light green to slightly glaucous, white-spotted, triangular-ovate to lanceolate, glabrous, margins whitish, spiny-toothed, teeth ca. 5 mm apart. **Inflorescences** racemose, maroon to slightly glaucous, divided distally into 5–10 arching branches, 4–9 dm; each branch with racemes 5–8 cm; bracts scarious, persistent. **Flowers:** perianth red, 2.5–3.2 cm; tepals almost entirely connate, limb lobes ovate; stamens 6, included, slightly unequal, as long as perianth tube; stigmas not expanded; pedicel 1.5–3 cm. **Capsules** brown, oblong, 2–3 cm. **Seeds** 3-winged, ca. 5 mm, sterile.

Flowering spring. Open sage scrub; 0–100 m; introduced; Calif.; South Africa.

Aloe ×schonlandii is a common natural hybrid, found growing wild with its parents, *A. saponaria* (Aiton) Haworth and *A. striata* Haworth, both of South Africa. It has long been in cultivation and is reported to have first been planted in the flora area in La Jolla, California, in the earlier part of the twentieth century. Since then it has spread and become firmly established in that area. The plants are apparently sterile and propagate only vegetatively.

225. AGAVACEAE Dumortier
• Agave or Century Plant Family

Susan Verhoek
William J. Hess

Plants usually perennial, occasionally epiphytic, sometimes monocarpic or polycarpic, monoecious, dioecious, or polygamodioecious, small to gigantic, sometimes arborescent, usually scapose. **Stems** subterranean or aboveground, sometimes branched. **Leaves** simple, annual or long-lived, in terminal rosettes or occasionally cauline, sessile or occasionally pseudo-petiolate; blade linear, lanceolate, oblanceolate, ovate, or elliptic, fibrous, thin and flexible, thick and rigid or succulent, or fibrous, often glaucous, margins entire, serrulate, dentate, denticulate, corneous, or filiferous, apex rigid or flexible, sometimes pungent, often with short or long spine. **Inflorescences** terminal or axillary spikes, racemose or paniculate, sometimes umbellate, bracteate, often huge; bracts ascending or erect, occasionally reflexed, leaflike proximally, scalelike distally. **Flowers** 6-merous, bisexual or functionally unisexual; perianth of 2 similar petallike whorls, semisucculent; tepals distinct or connate into tube, apex glandular or glandular-pubescent; stamens included or exserted; filaments often broadened and succulent, glabrous, pubescent, or papillose; anthers versatile, dehiscence longitudinal; ovary superior or inferior, 3-locular or occasionally 1-locular, 3-angled, ovoid, or cylindrical, with axillary or rarely parietal placentation; style included or exserted; stigmas 1 or 3, 3-lobed or capitate; pedicel usually distinct, articulate or not, rarely absent. **Fruits** occasionally baccate, usually capsular and sometimes winged or lobed, or indehiscent and dry or fleshy. **Seeds** 1–3(–many) per locule, flattened, 3-angled, hemispheric, ovoid, obovoid, or globose.

Genera 17 or 18, species ca. 550 (9 genera, 84 species in the flora; 2 genera, 6 species introduced): worldwide, primarily arid, semitropical, subtropical, and tropical regions.

There is little agreement on the treatment of Agavaceae. The group containing *Agave*, *Yucca*, *Furcraea*, *Hesperaloe*, and *Manfreda* generally has been accepted as the core of Agavaceae, or as subfamilies Agavoideae and Yuccoideae, but treatment of *Dracaena*, *Sansevieria*, *Cordyline*, *Nolina*, and *Dasylirion* has been varied. A. L. Takhtajan (1987) and R. F. Thorne (1992b) placed these genera in Dracaenaceae but treated them at different levels. Takhtajan located them in the subfamily Dracaenoideae within sections Nolineae (*Nolina* and *Dasylirion*), Sansevierieae (*Sansevieria*), and Dracaeneae (*Cordyline* and *Dracaena*). Thorne, on the other hand, placed these same groupings at the subfamily level. R. M. T. Dahlgren et al. (1985) recognized them as separate families, Nolinaceae (*Nolina* and *Dasylirion*), Dracaenaceae (*Sansevieria* and *Dracaena*), and Asteliaceae (*Cordyline*), in addition to the Agavaceae (*Yucca*, *Hesperaloe*, *Agave*, *Manfreda*, and *Furcraea*).

A. Cronquist (1981) based his broadly circumscribed Agavaceae on a common xerophytic habit. However, the karyotype of 5 long and 25 short chromosomes for the Agavoideae and Yuccoideae is distinct from the karyotypes of the other subfamilies that Cronquist included in the Agavaceae. Current research on the phylogenetics of moncotyledons, using DNA sequences of *rbc*L, support the separation of *Dracaena*, *Nolina*, and *Dasylirion* from Agavaceae (M. R. Duvall et al. 1993b). We believe that a broad interpretation of the Agavaceae unites groups that should be recognized as separate.

Many genera in Agavaceae are economically important. All genera in the Agavoideae and Yuccoideae contain steroidal sapogenins; some have been used in folk medicine, and locally and commercially as soap (G. Blunden et al. 1978; S. E. Verhoek 1978; M. Wall et al. 1957). They provide fibers for cordage, baskets, and hats, as well as food and drink for many indigenous peoples of the southwestern United States (H. S. Gentry 1982). They are also used as commercial fiber and beverage crops in Latin America and the Old World (H. Brucher 1989). In the southern United States, some species in each genus are cultivated and represented in the flora, and at least one species of *Yucca* is now grown as far north as Canada. Collectors should record the uses of these plants in their notes along with the critical information on plant habit and morphology. Photographs are often important tools for the identification of these plants, and, with the advent of digital cameras, are now much easier to obtain and process.

SELECTED REFERENCES Bogler, D. J. and B. B. Simpson. 1995. A chloroplast DNA study of the Agavaceae. Syst. Bot. 20: 191–205. Bogler, D. J. and B. B. Simpson. 1996. Phylogeny of Agavaceae based on ITS rDNA sequence variation. Amer. J. Bot. 83: 1225–1235.

1. Ovary inferior.
    2. Stems subterranean; leaf apex flexible . . . . . 9. *Manfreda*, p. 462
    2. Stems aboveground; leaf apex rigid.
        3. Leaf apex abruptly narrowed, with long spine; flowers erect or recurved; style subulate, filaments filiform . . . . . 7. *Agave*, p. 442
        3. Leaf apex blunt; flowers drooping; style and filaments abruptly swollen in proximal half . . . . . 8. *Furcraea*, p. 461
1. Ovary superior.
    4. Flowers functionally unisexual; fruits capsular; tepals no longer than 5 mm.
        5. Capsules 3-locular, dehiscent; seeds globose; leaf blade not fibrous, margins entire or serrulate . . . . . 2. *Nolina*, p. 415
        5. Capsules 1-locular, indehiscent; seeds 3-angled; leaf blade fibrous, margins with sharp, curved prickles . . . . . 3. *Dasylirion*, p. 422
    4. Flowers bisexual; fruits capsular or baccate; tepals longer than 5 mm.
        6. Fruits baccate; seeds 1–3 . . . . . 1. *Sansevieria*, p. 415
        6. Fruits capsular, rarely baccate; seeds many.
            7. Perianth narrowly tubular to broadly campanulate; filaments glabrous; capsules ovoid . . . . . 6. *Hesperaloe*, p. 441
            7. Perianth campanulate or globose; filaments smooth or papillose; capsules almost never ovoid.
                8. Fruits erect or pendent, usually indehiscent, occasionally dehiscent and septicidal; scape usually less than 2.5 cm diam.; bracts usually ascending . . . . . 4. *Yucca*, p. 423
                8. Fruits erect, dehiscent and loculidical; scape usually more than 2.5 cm diam.; bracts usually reflexed . . . . . 5. *Hesperoyucca*, p. 439

1. SANSEVIERIA Thunberg, Prod. Pl. Cap. 1: [xii], 65. 1794, name conserved • Bowstring hemp, devil's-tongue [for Raimond de Sangro, Prince of Sanseviero] [I] [W2]

William J. Hess

Herbs, perennial, erect, from thick rhizomes. **Leaves** basal, erect; blade often variegated and mottled, elongate, thick, flat or cylindrical, rigid. **Inflorescences** paniculate, racemose, or spicate. **Flowers** bisexual; tepals basally connate into tube, white, yellowish white, pale pink, or greenish white, tips narrow; ovary superior. **Fruits** baccate. **Seeds** 1–3. x = 9, 19, 20, 21, 33.

Species 60 (1 in the flora): introduced; Africa, Sri Lanka.

Many species of *Sansevieria* are used as house and greenhouse plants. Some besides *S. hyacinthoides* have reportedly escaped in southern Florida but I have seen no specimens.

1. Sansevieria hyacinthoides (Linnaeus) Druce, Rep. Bot. Exch. Club Soc. Brit. Isles 3: 423. 1914 • African bowstring hemp [F] [I] [W2]

Aloe hyacinthoides Linnaeus, Sp. Pl. 1: 321. 1753; *A. guineensis* (Linnaeus) Jacquin; *Cordyline guineensis* (Linnaeus) Britton ex Small; *Sansevieria guineensis* (Linnaeus) Willdenow; *S. thyrsiflora* Thunberg

Leaves 4–10 × 0.5–0.9 dm; blade with transverse white- or yellow-mottled bands, flat, glabrous, margins often yellow, entire, tufted. **Inflorescences** paniculate, surpassing leaves, 6 dm. **Flowers:** tepals greenish white to pale pink, 25–40 mm, limb lobes recurved, linear, 14–18 mm; filaments surpassing lobes; ovary 5 mm; style filiform, to 4 mm; stigma capitate. **Berries** red, globose, 8 mm diam.

Flowering spring. Near old dwelling sites; 0–100 m; introduced; Fla.; Africa.

The taxonomic disposition of plants here treated as *Sansevieria hyacinthoides* has varied. R. W. Long and O. Lakela (1971) considered them *S. thyrsiflora*, while J. K. Small (1933) treated them as *Cordyline guineensis*, and L. H. Bailey et al. (1976) as *S. hyacinthoides*.

2. NOLINA Michaux, Fl. Bor.-Amer. 1: 207. 1803 • Beargrass [for Abbé C. P. Nolin, eighteenth-century French arboriculturist and director of the royal nurseries]

William J. Hess

Plants perennial, cespitose or arborescent, acaulescent to short-caulescent, scapose, from branched, woody caudices or bulblike structures; usually forming colonies with few to many rosettes. **Stems** to 25 dm. **Leaves** forming rosettes; blade linear, not rigid or fibrous, bases broadly expanding, margins serrulate or entire. **Scape** 0.5–25 dm. **Inflorescences** paniculate, rarely racemose, 3–18 dm; bracts caducous or occasionally persistent. **Flowers** 2–5 per node, functionally unisexual, pistillate flowers with staminodes, staminate flowers with reduced pistils; tepals white to cream or tan, 1.3–5 mm, apex glandular; ovary superior; pedicel jointed near middle. **Fruits** capsular, 3-locular, 3-lobed, thin-walled or sometimes firm-walled, often inflated, mostly notched at base and apex or rounded distally; dehiscent, often splitting irregularly. **Seeds** closely or loosely invested in capsules, globose, turgid. x = 19.

Species ca. 30 (14 in the flora): North America (se and sw United States), n and nc Mexico.

In areas of west Texas, New Mexico, and Arizona, the species of *Nolina* are not always well defined. There is variation in the presence or absence of marginal teeth on the leaves and placement of the inflorescences within or beyond the leaves. Collectors should take good notes about the plant habit and morphology and include these with their collections. Further study is needed on *Nolina* throughout its range. Some species of *Nolina* are extremely infrequent.

Some are on federal and/or state rare and endangered species lists, and possibly some of those listed with a state should be listed federally.

J. C. Dice (1988) studied section *Arborescens* of *Nolina* in the United States, and presented extensive descriptions and discussion of *N. bigelovii, N. parryi, N. cismontana,* and *N. interrata.*

SELECTED REFERENCES Dice, J. C. 1988. Systematic Studies in the *Nolina bigelovii–N. parryi* (Nolinaceae) Complex. M.S. thesis. San Diego State University. Trelease, W. 1911. The desert group Nolineae. Proc. Amer. Philos. Soc. 50: 404–443.

1. Plants caulescent.
   2. Leaf margins entire, filiferous . . . . . 14. *Nolina bigelovii*
   2. Leaf margins serrulate, not filiferous.
      3. Inflorescence bracts caducous; Texas . . . . . 5. *Nolina erumpens* (in part)
      3. Inflorescence bracts persistent; California.
         4. Stems to 4 dm, occasionally to 15 dm; leaf blades 12–30 mm wide; scape 14–35 mm diam. at base . . . . . 12. *Nolina cismontana*
         4. Stems 5–21 dm; leaf blades 20–40 mm wide; scape 26–90 mm diam. at base . . . . 13. *Nolina parryi*
1. Plants acaulescent.
   5. Subterranean caudices horizontal; leaf blades glaucous, 1.2–3.5 cm wide; California . . . . . 11. *Nolina interrata*
   5. Subterranean caudices vertical; leaf blades not glaucous, or if glaucous then no more than 8 mm wide; Arizona, Colorado, Florida, Georgia, New Mexico, Oklahoma, South Carolina, Texas.
      6. Leaf margins entire or only slightly roughened, with widely separated, noncartilaginous teeth.
         7. Inflorescences usually conspicuously tinged purple, main rachis and divisions slender, flexible; fruiting pedicels slender distal to joint, typically to 2.5 mm, articulate near middle, not noticeably dilated; tepals 1.9–3.2 mm; se New Mexico, w Texas . . . . . 8. *Nolina micrantha*
         7. Inflorescences not purple or rarely so, main rachis and divisions thick, rigid; fruiting pedicels thick distal to joint, (1–)4–7 mm, articulate near base, noticeably dilated into perianths; tepals 2–3.5 mm; s New Mexico, Texas.
            8. Leaf blades slightly concavo-convex toward base, triangular, 2–4(–7) mm wide; inflorescences 25–70 cm, lateral branches ascending; seeds 2.6–3.4 mm diam.; rocky habitats of limestone; s New Mexico, Texas . . . . . 9. *Nolina texana*
            8. Leaf blades flat or concavo-convex, 5–10 mm wide; inflorescences 40–70 cm, lateral branches spreading; seeds 4.3 mm diam.; sandy soils or dunes; w Texas . . . . . 10. *Nolina arenicola*
      6. Leaf margins serrulate, with close-set, cartilaginous teeth, or occasionally entire.
         9. Leaf blades flattened.
            10. Capsules 6–11 mm diam. . . . . . 2. *Nolina lindheimeriana*
            10. Capsules 3.5–7 mm diam.
                11. Capsules asymmetrical, up to 5.5 mm diam.; seeds closely invested in capsules; Florida . . . . . 1. *Nolina atopocarpa*
                11. Capsules symmetrical, 5–7 mm diam.; seeds loosely invested in capsules; Georgia, South Carolina . . . . . 3. *Nolina georgiana*
         9. Leaf blades concavo-convex.
            12. Capsules 8–12 mm diam. . . . . . 4. *Nolina brittoniana*
            12. Capsules 3.5–6.4 mm diam.
                13. Bracts persistent; capsules distinctly notched distally; se Colorado, e New Mexico, w Oklahoma . . . . . 7. *Nolina greenei*
                13. Bracts caducous; capsules indistinctly notched apically; Arizona, New Mexico, Texas.
                    14. Fruiting pedicels distal to joint to 2.5 mm; inflorescences not surpassing leaves; seeds 1.5–2.5 mm; w Texas . . . . . 5. *Nolina erumpens* (in part)
                    14. Fruiting pedicels distal to joint 3–6 mm; inflorescences surpassing leaves; seeds 2.2–3.4 mm; Arizona, New Mexico . . . . . 6. *Nolina microcarpa*

1. Nolina atopocarpa Bartlett, Rhodora 11: 81. 1909 • Florida beargrass [E] [F]

Plants acaulescent, cespitose; rosettes from bulblike bases, with vertical, subterranean caudices. **Leaf blades** wiry, lax or stiff, grasslike, flattened, 45–85 cm × 1.5–4.5 mm, not glaucous; margins serrulate, with close-set, cartilaginous teeth, rarely entire; inflorescence leaf blades 1.5–6.5 cm. **Scape** 2.5–6 dm. **Inflorescences** racemose, sometimes branched, 6.5–9 dm × 0.7–2.8(–20) cm; bracts caducous, 1.5–3.5 mm, apex acute, fragile. **Flowers:** tepals 1.3–2.5 mm; fertile stamens: anthers up to 1 mm; pistil not ridged; pedicel recurved in age, not dilated, proximal to joint 1–1.5 mm, distal to joint 1.2–2 mm. **Capsules** asymmetrical, rounded, inflated, 4–4.5 × (3.5–)4–5.5 mm, tapering at base. **Seeds** closely invested in capsule, rounded, 3–4.1 × 2.4–3.2 mm.

Flowering summer. Sandy loam, often with peat in pine flatwoods; 0–50 m; Fla.

Nolina atopocarpa is fire-tolerant and possibly fire-dependent. It is found primarily on the east side of the Apalachicola River in Liberty and Franklin counties and scattered counties in eastern Florida. The plants are extremely rare and are listed as endangered by the state of Florida.

2. Nolina lindheimeriana (Scheele) S. Watson, Proc. Amer. Acad. Arts 14: 247. 1879 • Lindheimer's beargrass [E]

Dasylirion lindheimerianum Scheele, Linnaea 25: 262. 1852; *Beaucarnea lindheimeriana* (Scheele) Baker

Plants acaulescent; caudices short, vertical, subterranean, with many rosettes. **Leaf blades** not wiry, lax, grasslike, flattened, 30–100 cm × (3.5–)4–12 mm, not glaucous; margins serrulate, with close-set, cartilaginous, corneous teeth; inflorescence leaf blades 25–60 cm. **Scape** 3–7 dm. **Inflorescences** paniculate, 3–10 × 1–3.5 dm; bracts caducous, 1–15 cm; bractlets to 2 mm, margins hyaline. **Flowers:** tepals yellow-green, 1.8–2.5 mm, margins hyaline; fertile stamens: filaments 1.2–1.5 mm, anthers 0.7–1 mm; pedicel pendulous or recurved, proximal to joint 1–2 mm, distal to joint 2–5(–6) mm. **Capsules** winged, slightly inflated, 6–10 × (6–)7–11 mm, notched basally and apically. **Seeds** loosely invested in capsules, rounded, 3.1–4.2 × 2.4–3.3 mm.

Flowering spring. Limestone hills of open woodlands and scrubland; 400–600 m; Tex.

Nolina lindheimeriana is quite infrequent and becoming more so as its habitat is destroyed through development or overgrazing.

3. Nolina georgiana Michaux, Fl. Bor.-Amer. 1: 208. 1803 • Georgia beargrass [E] [F]

Plants acaulescent, cespitose; rosettes from bulblike bases. **Leaf blades** grasslike, flexible, flattened, 50–130 cm × 3–8 mm, glaucous; margins serrulate, with close-set, cartilaginous teeth; inflorescence leaf blades 2–25 cm. **Scape** 6–14 dm, glaucous. **Inflorescences** paniculate, 9–13 dm × 10–25 cm; bracts caducous, 4–10 mm, glaucous; bractlets white to tan, to 2.5 mm. **Flowers:** tepals yellow-green to tan, 1.9–2.4 mm; fertile stamens: filaments 1.4–1.9 mm, anthers 0.7–1 mm; pedicel recurved in age, proximal to joint 1–3.5 mm, distal to joint 3–11 mm. **Capsules** symmetrical, winged, obovate, 6.5–8 × 5–7 mm, rounded basally, notched apically. **Seeds** loosely invested in capsules, rounded, 3.2–4.7 × 2–3 mm.

Flowering spring. Sandy soil in pinelands, savanna, turkey-oak woods; 0–100 m; Ga., S.C.

Nolina georgiana is fire-tolerant and is probably fire-dependent. Nowhere is this species common and it should be investigated for listing on state rare and endangered species lists.

4. Nolina brittoniana Nash, Bull. Torrey Bot. Club 22: 158. 1895 • Britton's beargrass [C] [E]

Plants acaulescent, cespitose; rosettes from bulblike bases, with vertical, subterranean caudices. **Leaf blades** lax, grasslike, flattened, 70–100 cm × 5–9(–11) mm, not glaucous; margins serrulate, with close-set, cartilaginous teeth; inflorescence leaf blades reduced, 4–35 cm. **Scape** 7–12 dm. **Inflorescences** paniculate, 5–10.5 × 0.7–1.5 dm; bracts caducous, 1.2–4 cm; bractlets white to tan, 2–4 mm. **Flowers:** tepals 1.7–2.5 mm; fertile stamens: anthers 0.7–1.2 mm; pedicel recurved in age, proximal to joint 1.5–3.5 mm, distal to joint 6–10 mm. **Capsules** symmetrical, broadly winged, 7.2–9 × 8–10(–12) mm, distinctly notched basally and apically. **Seeds** loosely invested in capsules, rounded, 3.4–4.8 × 2.6–3.3 mm.

Flowering early spring. Sandy soil in oak and pine scrublands; of conservation concern; 0–30 m; Fla.

Nolina brittoniana is fire-tolerant and is probably fire-dependant. It is restricted to central Florida in Marion, Lake, Orange, Hernando, Polk, Osceola, and Highlands

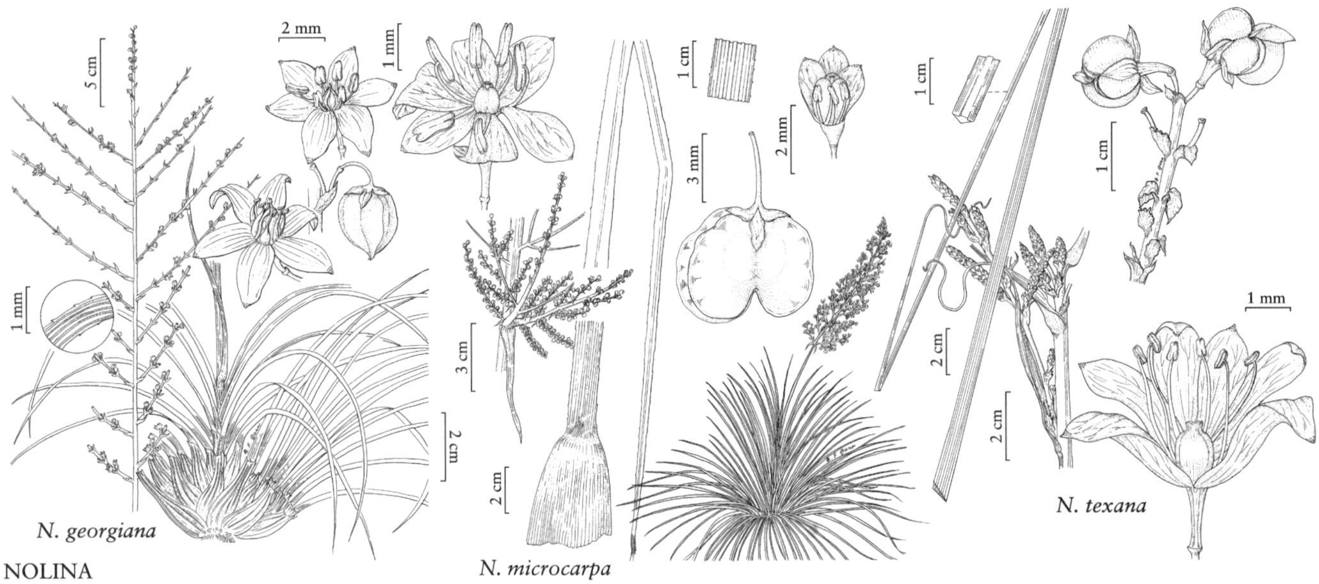

NOLINA

counties. The plants are rare and listed as endangered by Florida and the U.S. Fish and Wildlife Service under the Federal Endangered Species Act.

5. **Nolina erumpens** (Torrey) S. Watson, Proc. Amer. Acad. Arts 14: 248. 1879 • Mesa sacahuista

Dasylirion erumpens Torrey in W. H. Emory, Rep. U.S. Mex. Bound. 2(1): 216. 1859; *Beaucarnea erumpens* (Torrey) Baker

Plants acaulescent, occasionally arborescent, cespitose; rosettes from vertical, subterranean, branched caudices. **Stems** 0–2.5 m. **Leaf blades** stiff, somewhat carinate, concavo-convex, 80–170 cm × 9–18 mm, not glaucous; margins serrulate, with close-set cartilaginous teeth; apex lacerate. **Scape** 1.5–3 dm, to 1.2 cm diam. **Inflorescences** paniculate, 4–9 dm × 7–18 cm, contained within rosettes, not surpassing leaves; bracts caducous; bractlets to 2 mm, margins hyaline. **Flowers:** tepals 1.6–2.2 mm; fruiting pedicel ascending, proximal to joint 1–1.5 mm, distal to joint 1.5–2.5 mm. **Capsules** thin-walled, inflated, 2.6–4.4 × 3.5–5.7 mm, indistinctly notched at apex. **Seeds** closely invested in capsules, rounded, bursting ovary walls, 1.5–2.5 × 1–1.5 mm.

Flowering late spring. Rocky hillsides of limestone or igneous slopes in shrub-grasslands; 900–2300 m; Tex.; n Mexico.

The leaves of *Nolina erumpens* have teeth that are at right angles to the margin. Compared with *N. texana* in common range, *N. erumpens* flowers much later in the spring.

6. **Nolina microcarpa** S. Watson, Proc. Amer. Acad. Arts. 14: 247. 1879 • Sacahuista [F]

Nolina caudata Trelease

Plants acaulescent, cespitose; rosettes from vertical, subterranean, branched caudices. **Leaf blades** wiry, lax, concavo-convex, 80–130 cm × 5–12 mm, not glaucous; margins serrulate, with close-set, cartilaginous teeth; apex lacerate; inflorescence leaf blades curling distally, 10–50 cm. **Scape** 3–15 dm, 1.2–2.5 cm diam. **Inflorescences** paniculate, 4–12 dm × 10–30 cm, surpassing leaves; bracts caducous, rarely persistent; bractlets 2–5 mm, slightly erose. **Flowers:** tepals white, 1.5–3.3 mm; fertile stamens: filaments 1.6–1.9 mm, anthers 1.2–1.4 mm; infertile stamens: filaments 0.9–1.2 mm, anthers 0.3–0.5 mm; fruiting pedicel erect, proximal to joint 1–2 mm, distal to joint 3–6 mm. **Capsules** hyaline, thin-walled, inflated, 4.2–6 × 5.4–6.4 mm, indistinctly notched at apex. **Seeds** loosely invested in capsules, compressed, 2.2–3.4 × 1.5–3 mm. $2n = 38$.

Flowering mid–late spring. Rocky hillsides, desert grasslands, oak and pinyon pine-juniper woodlands; 900–1900 m; Ariz., N.Mex.; n Mexico.

D. S. Correll and M. C. Johnston (1970) included *Nolina microcarpa* in their flora of Texas; however, they reported that they had seen no specimen from that state, nor have I. This species is found primarily from western New Mexico through central Arizona. It forms large clumps up to 2 m in diameter and inflorescences that generally are exserted from the basal leaf rosettes. Considerable variation occurs, some of it geographically restricted to southeastern Arizona and southwestern New Mexico. Most such plants from the latter areas have been referred to *N. texana* or *N. caudata*, but are here included in *N. microcarpa*. In the Grand Canyon area, there are variants that have been referred to *N. parryi* because in width the leaves approach those of *N. parryi* and they are serrulate. These plants, however, are acaulescent and also are here included in *N. microcarpa*. B. J. Albee et al. (1988) reported *N. microcarpa* on rocky slopes in canyons in Washington County, Utah, but the more recent online version of that work excludes it from Utah.

7. Nolina greenei S. Watson ex Trelease, Proc. Amer. Philos. Soc. 50: 418. 1911 [E]

Plants acaulescent, cespitose; rosettes from vertical, subterranean, branched caudices. **Leaf blades** wiry, stiff or slightly lax, concavo-convex, 45–90(–110) cm × (4–)5–8 mm, not glaucous; margins remotely serrulate with close-set, cartilaginous teeth, or sometimes entire; apex lacerate; scape and inflorescence leaf blades curling, 10–40 cm. **Scape** curling, 0.5–2 dm. **Inflorescences** paniculate, (3–)3.5–6(–6.5) dm × 8–14(–20) cm; bracts persistent, lower bracts to 40 cm; bractlets 2–5 mm, apex lacerate. **Flowers:** tepals white, sometimes with purple midveins, 2.2–3.3 mm, margins hyaline; pedicel erect, proximal to joint 1–1.5 mm, distal to joint 2.5–4.5 mm. **Capsules** hyaline, thin-walled, inflated, 2.1–3.8 × 3.8–5.2 mm, distinctly notched distally. **Seeds** becoming bronze with red tinge, rounded, bursting ovary walls, 3–3.9 mm diam.

Flowering mid–late spring. Rocky, limestone hillsides, volcanic flows, open juniper-pinyon pine-oak woodlands and adjacent grasslands; 1200–1900 m; Colo., N.Mex., Okla.

Nolina greenei is resurrected for the plants of *Nolina* that occur from central New Mexico north to the Black Mesa region of Oklahoma and just into southeastern Colorado. They are similar to *N. texana* with respect to the inflorescence contained within the leaves, persistent elongated bracts, and seeds that burst the ovary wall and remain attached. They differ primarily in their broader, slightly serrulate leaves (although some leaves may be entire), copper-colored seeds, and an open woodland-grassland habitat.

8. Nolina micrantha I. M. Johnston, J. Arnold Arbor. 24: 91. 1943 [E]

Plants acaulescent, cespitose; rosettes from vertical, subterranean, branched caudices. **Leaf blades** stiff, wiry, concavo-convex, 80–130 cm × 4–6 mm, not glaucous; margins entire or remotely serrulate, with widely separated, noncartilaginous teeth; apex lacerate. **Scape** 0.5–2 dm. **Inflorescences** paniculate, usually conspicuously tinged purple, 3.5–7.5 dm × 10–20 cm, held partially within rosettes; main rachis and divisions slender, flexible; bracts mostly persistent, to 30 cm, apex curled; bractlets erose, 1.5–3 mm, margins hyaline. **Flowers:** tepals 1.9–3.2 mm; fertile stamens: filaments 0.9–1.3 mm, anthers 0.7–0.9 mm; infertile stamens: filaments 0.5–0.7 mm, anthers 0.4–0.5 mm; fruiting pedicel erect, slender, articulate near middle, not noticeably dilated, proximal to joint to 1.5 mm, distal to joint to 2.5 mm. **Capsules** firm-walled, inflated, 3–4 × 4.3–5.5 mm. **Seeds** closely invested in capsules, rounded, bursting ovary walls, 3–4 mm diam.

Flowering late spring–early summer. Rocky limestone slopes or sandy soils, in grasslands; 1100–1400 m; N.Mex., Tex.

Nolina micrantha is similar to *N. texana* except for the purple pigment throughout its inflorescences, later flowering dates, and less robust habit.

9. Nolina texana S. Watson, Proc. Amer. Acad. Arts 14: 248. 1879 • Texas sacahuiste, bunchgrass [F]

Nolina erumpens (Torrey) S. Watson var. *compacta* Trelease; *N. texana* var. *compacta* (Trelease) I. M. Johnston

Plants acaulescent, cespitose; rosettes from vertical, subterranean branched caudices. **Leaf blades** wiry, stiff, triangular, slightly concavo-convex toward base, 40–90 cm × 2–4(–7) mm, not glaucous; margins entire, rarely remotely serrate with widely separated, noncartilaginous teeth, tips dying; inflorescence leaf blades curling distally, 10–50 cm. **Scape** curling distally, 0.5–2 dm. **Inflorescences** paniculate, rarely purple, 2.5–7 dm × 5–17 cm, held completely or partially within rosettes; main rachis and divisions thick, rigid, lateral branches ascending; bracts persistent, to 10–40 cm; bractlets laciniate, 1–2 mm. **Flowers:** tepals yellow-green, 2.5–3.5 mm; fertile stamens: filaments 0.9–1.4 mm, anthers 0.8–1 mm; infertile stamens: filaments 0.6–0.8 mm, anthers 0.4–0.5 mm; fruiting pedicel ascending, thick, articulate near base, noticeably dilated into perianth, proximal to joint 2.5–6 mm, distal to joint

1.5–2 mm. **Capsules** thin-walled, inflated, 3–4 × 4.5–5.5(–8) mm. **Seeds** closely invested in capsules, rounded, bursting ovary walls, 2.6–3.4 mm diam. 2*n* = 38.

Flowering late winter–early spring. Rocky hillsides, limestone, granite, grasslands, shrublands; 200–2000 m; N.Mex., Tex.; n Mexico.

Nolina texana is centered primarily on the Edwards Plateau west through the trans-Pecos region of Texas. *Nolina texana* var. *compacta* is not recognized as distinct from var. *texana*. They have the same distributions. Variety *compacta* has compact inflorescences with functionally pistillate flowers, which is typical of the functionally pistillate plants of var. *texana*. Those plants with elongate secondary branches have functionally staminate flowers and are characteristic of both varieties. The dark, round seeds are quite distinct in the open fruit chambers on inflorescences hidden within the basal leaves. Some plants from the high plains of eastern New Mexico, northwestern Oklahoma, and southern Colorado were considered to be *N. texana* (see 7. *N. greenei*). Further study of these populations may help determine if they truly are a part of this complex.

10. Nolina arenicola Correll, Madroño 19: 187. 1968 • Sand sacahuista [C] [E]

Plants acaulescent, cespitose; rosettes from vertical, subterranean, branched caudices. **Leaf blades** wiry, carinate, flat or concavo-convex, 100–200 cm × 5–10 mm, not glaucous; margins remotely serrulate, with widely separated, noncartilaginous teeth, to entire; apex lacerate. **Scape** 1–3 dm. **Inflorescences** paniculate, rarely purple, 4–7 dm × 12–18 cm; main rachis and divisions thick, rigid, lateral branches spreading; bracts persistent, to 45 cm, apices curling; bractlets slightly erose, 1.5–5 mm, apex aristate. **Flowers:** tepals 2–2.8 mm; infertile stamens: filaments to 1 mm, anthers to 0.4 mm; fruiting pedicel erect, thick, articulate near base, noticeably dilated into perianth, proximal to joint 2–3.5 mm, distal to joint to 1.5 mm. **Capsules** firm-walled, inflated, 2.5–3.3 × 5–6.2 mm, broadly rounded distally. **Seeds** closely invested in capsules, bursting ovary walls, 3 × 4.3 mm diam.

Flowering late spring. Open sandy soils or dunes in brushlands; of conservation concern; 1200 m; Tex.

Nolina arenicola is endemic to the trans-Pecos region of Texas and is listed as threatened and endangered by the U.S. Fish and Wildlife Service. *Nolina arenicola* is similar to *N. texana* except that in *N. arenicola* the clumps are much more robust, with wider leaves, serrulate near the point of attachment, and are restricted to sandy areas.

11. Nolina interrata Gentry, Madroño 8: 181, fig. 1, plate 19. 1946 • Dehesa beargrass [C]

Plants acaulescent; rosettes from branching, horizontal, subterranean caudices from lignotubers. **Leaf blades** 30–150 cm × 12–35 mm, glaucous; bases 15–70 mm wide; margins serrulate. **Scape** 0.7–9 dm, 5–18 mm diam. **Inflorescences** compound paniculate, branched distally, 5–11 dm × 10–50 cm; bracts persistent, 2–15 cm; bractlets laciniate. **Flowers:** tepals 2–3.5 mm; fertile stamens: anthers 1.1–1.7 mm; pedicel erect, proximal to joint 1–2(–4.5) mm, distal to joint 1–2 mm. **Capsules** thin-walled, 7–12 × 9–15 mm, notched basally and apically. **Seeds** bursting ovary walls, 4–6 × 3–4 mm.

Flowering early spring–early summer. Rocky hillsides of chaparral; of conservation concern; 200–700 m; Calif.; Mexico (Baja California).

Nolina interrata is very rare and is listed by the U.S. Fish and Wildlife Service as rare and endangered. Its habitat is threatened by development.

12. Nolina cismontana Dice, Novon 5: 162. 1995 • Chaparral nolina, chaparral beargrass [C] [E]

Plants caulescent; rosettes from woody, branched caudices and forming small colonies. **Stems** to 3 dm, occasionally to 15 dm. **Leaves** 30–90 per rosette; blade wiry, upright, or lax, 50–140 cm × 12–30 mm, occasionally glaucous; bases broad, spoon-shaped, 30–85 mm; margins serrulate, not filiferous. **Scape** 4–15 dm, 14–35 mm diam. at base. **Inflorescences** compound paniculate, 9–18 dm × 10–40 cm; bracts persistent, conspicuous; bractlets laciniate. **Flowers:** tepals 2–5 mm; fertile stamens: filaments 2–4 mm, anthers to 1.2 mm; infertile stamens: filaments 1–1.2 mm, anthers 0.4–0.6 mm; pedicel erect, proximal to joint to 2 mm, distal to joint 1–3 mm. **Capsules** thin-walled, 8.2–11.2 × 9–12 mm, notched basally and apically. **Seeds** reddish brown, ovoid, bursting ovary wall, 4–5 × 3–4 mm.

Flowering early–mid spring. Rocky hillsides in dry chaparral of coastal mountains; of conservation concern; 200–1300 m; Calif.

P. A. Munz and J. C. Roos (1950) mistakenly believed that this coastal entity was the basis of *Nolina parryi* S. Watson (J. C. Dice 1988). Dice determined that it is not "typical" *N. parryi* and is sufficiently distinct to recognize separately. Few populations of *N. cismontana* exist, and they are of conservation concern.

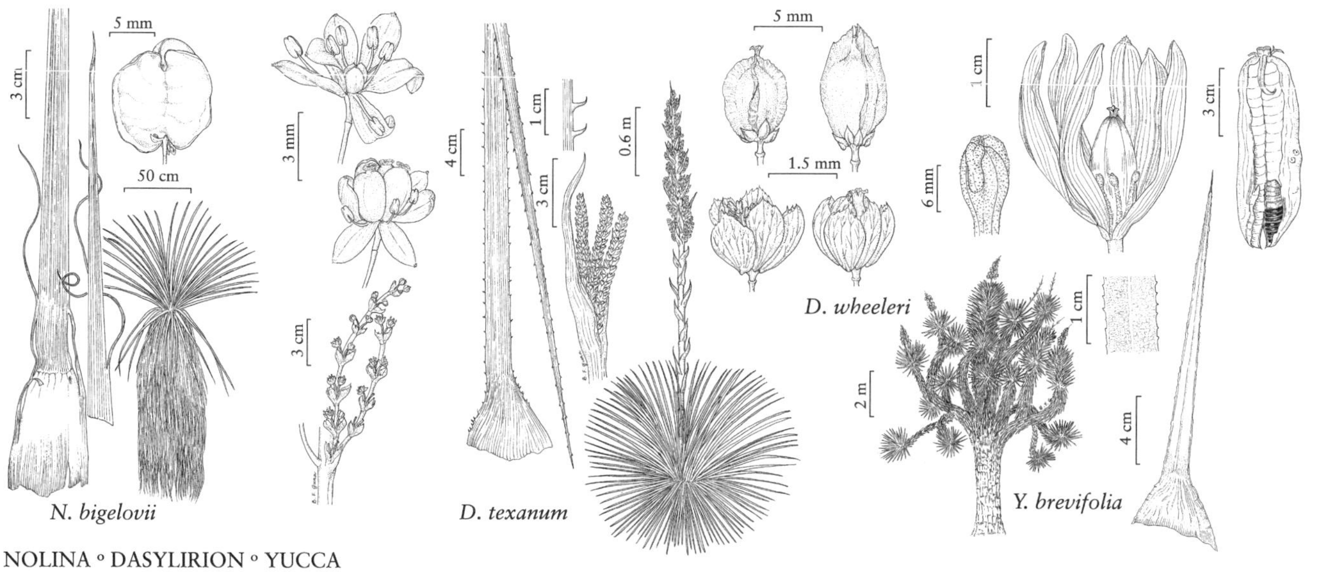

NOLINA ◦ DASYLIRION ◦ YUCCA

13. Nolina parryi S. Watson, Proc. Amer. Acad. Arts 14: 247. 1879 • Parry nolina, giant nolina [E]

Nolina bigelovii subsp. *parryi* (S. Watson) E. Murray; *N. bigelovii* (Torrey) S. Watson var. *parryi* (S. Watson) L. D. Benson; *N. bigelovii* subsp. *wolfii* (Munz) E. Murray; *N. bigelovii* var. *wolfii* (Munz) L. D. Benson; *N. parryi* subsp. *wolfii* Munz; *N. wolfii* (Munz) Munz

Plants caulescent; rosettes from woody, branched caudices and forming small colonies. **Stems** 5–21 dm. **Leaves** 65–220 per rosette; blade wiry, upright, or lax, 50–140 cm × 20–40 mm, rarely glaucous; bases broad, spoon-shaped, 5–16.5 cm wide; margins serrulate, not filiferous. **Scape** 4–17 dm, 26–90 mm diam. at base. **Inflorescences** compound paniculate, 9–18 × 20–130 cm; bracts persistent, 15–36 cm, papery; bractlets laciniate, to 10 mm, fragile. **Flowers:** tepals 2.5–5 mm; fertile stamens: filaments 3–4 mm, anthers 1.3–1.8 mm; infertile stamens: filaments 1–2.5 mm, anthers 0.4–0.5 mm; pedicel erect, proximal to joint 0.5–2 mm, distal to joint 1–2 mm. **Capsules** thin-walled, 9–13 × 8–11 mm, notched basally and apically. **Seeds** reddish brown, ovoid to oblong, bursting ovary wall, 3–4 × 2–3 mm. $2n = 38$.

Flowering spring. Rocky slopes of desert and pinyon-juniper woodlands; 900–2100 m; Calif.

When P. A. Munz and J. C. Roos (1950) named *Nolina parryi* subsp. *wolfii*, a mountain woodland plant, the related coastal entity became known as *N. parryi* subsp. *parryi*. J. C. Dice (1988) determined that the mountain entity, not the coastal one, was probably the basis for Watson's *N. parryi*. Consequently, Dice reduced subsp. *wolfii* to synonymy under subsp. *parryi*, which left the coastal plants without a distinguishing name. Dice has named the plants as a separate species, *N. cismontana*.

14. Nolina bigelovii (Torrey) S. Watson, Proc. Amer. Acad. Arts 14: 247. 1879 • Bigelow nolina, Bigelow beargrass [F]

Dasylirion bigelovii Torrey in W. H. Emory, Rep. U.S. Mex. Bound. 2(1): 151. 1857; *Beaucarnea bigelovii* (Torrey) Baker

Plants caulescent; rosettes from woody, branched caudices and forming small colonies. **Stems** 10–25 dm. **Leaves** 34–160 per rosette; blade stiff, linear-lanceolate, firm, 50–150 cm × 12–48 mm, glaucous, scabrous abaxially; bases spoon-shaped, 35–110 mm wide; margins entire, filiferous. **Scape** 6–24 dm, 15–45 cm diam. **Inflorescences** compound paniculate, 7–13 dm × 13–70 (–110) cm; bracts caducous, 3.5–16 cm; bractlets short-laciniate. **Flowers:** tepals cream to white, 2–4 mm; fertile stamens: 2–3 mm, anthers to 1 mm; infertile stamens: filaments 0.9–1.4 mm, anthers 0.4–0.6 mm; pedicel erect, proximal to joint 0.5–1 mm, distal to joint 0.5–2 mm. **Capsules** thin-walled, 8–12 mm × 8–12 mm, notched basally and apically. **Seeds** grayish, ovoid to oblong, 2.5–3.5 mm.

Flowering mid spring. Rocky hillsides and flats of the southern Mojave and Sonoran deserts; 300–1500 m; Ariz., Calif., Nev.; Mexico (Baja California, Sonora).

3. DASYLIRION Zuccarini, Allg. Gartenzeitung 6: 258. 1838 • Sotol [Greek *dasy*-, thick- or dense-, and *lirion*, white lily, alluding to the compact arrangement of flowers in the inflorescence]

David J. Bogler

Plants perennial, polycarpic; roots spreading, 2–8 mm diam. **Stems** short or elongate, often with erect or reclining trunk. **Leaves** persistent, numerous, forming rosettes, spreading; blade long-linear, fibrous, glabrous, sometimes waxy-glaucous, bases expanded, overlapping, spoon-shaped, margins with sharp, curved prickles, apex fibrous. **Inflorescences** paniculate, racemose, bracteate; panicles narrow, borne on long, woody stalks; bracts leaflike proximally, stramineous distally, lanceolate; fascicles of dense racemes in axils of bracts. **Flowers** small, functionally unisexual, some plants with only male flowers, others with only female flowers; floral bracts laciniate, membranous; tepals 6, distinct, whitish, greenish, or purple, obovate, margins denticulate; stamens 6, rudimentary in pistillate flowers; ovary superior, 3-angled, abortive in staminate flowers; style short, ± strap-shaped, slightly expanded at tip; stigma 3-lobed, lobes weakly connate into tube; pedicel of pistillate flowers articulate. **Fruits** capsular, 1-locular, dry, indehiscent, 3-winged. **Seed** 1, golden brown, 3-angled, turbinate. $x = 19$.

Species 17 (3 in the flora): arid, mountainous regions of sw United States and Mexico.

Dasylirion was once a major food source in the desert southwest. The bulbs were roasted in rock-lined pits and pounded into cakes. A potent alcoholic beverage, known as "sotol," is distilled from the baked and fermented bulbs. Leaves of *Dasylirion* are widely used to weave baskets.

SELECTED REFERENCES Bogler, D. J. 1994. Taxonomy and Phylogeny of *Dasylirion* (Nolinaceae). Ph.D. dissertation. University of Texas. Trelease, W. 1911. The desert group Nolineae. Proc. Amer. Philos. Soc. 50: 404–443.

1. Leaves whitish or bluish green, waxy-glaucous, papillose, dull; marginal prickles all antrorse; fascicles of flowers elongate, branches lateral, pendent, 10–20 cm from base to tip; capsules obovoid, 5–8 × 4–5(–7) mm, distal wing lobes 2–2.5 mm, often indented; Arizona, New Mexico, extreme w Texas . . . . 1. *Dasylirion wheeleri*
1. Leaves bright green, glabrous, not waxy, smooth, shiny; marginal prickles antrorse, retrorse, or mix of both; fascicles of flowers short, condensed, branches mostly basal, not pendent, 5–12 cm from base to tip; capsules ellipsoid or obovoid, 5–7 × 4–5 mm, distal wing lobes less than 2 mm, rounded, not indented; New Mexico, Texas.
    2. Majority of prickles antrorse; leaves 1–2 cm wide; capsules ellipsoid, 5.5–6 × 4–5 mm, distal wing lobes 1–1.5 mm; s edge of Edwards Plateau, Texas . . . . 2. *Dasylirion texanum*
    2. Majority of leaf prickles retrorse; leaves 1.5–3 cm wide; capsules ellipsoid to obovoid, 6–7 × 4–4.5 mm, distal wing lobes 1.5–2 mm; sw Texas, s New Mexico . . . . . . 3. *Dasylirion leiophyllum*

1. Dasylirion wheeleri S. Watson ex Rothrock, Rep. U.S. Geogr. Surv., Wheeler, 272. 1878 • Desert spoon [F]

Plants robust, with large crowns, to 40 cm diam.; trunks to 1.5 m, usually reclining. **Leaves** stout, rigid; blade whitish or bluish green, 35–100 × 2–3 cm wide above broadened base, densely waxy-glaucous, papillose, dull; prickles all antrorse. **Inflorescences** often massive, to 5 m; stalk 3–6 diam. at base; branches lateral, pendent in fruit, 3–10 cm; bracts wedge-shaped, attenuate; fascicles of flowers spreading, 10–20 cm from base to tip; primary axes 4–14 cm. **Flowers** with receptacles 0.2–0.5 mm; tepals sometimes tinged purple, 2.4 × 1–1.5 mm; style 0.2–0.3 mm, becoming swollen and golden brown in fruit; stigma lobes 0.4 mm; pedicel 3–3.5 mm in fruit. **Capsules** broadly obovoid or rounded in cross section, not indented, 5–8 × 4–5(–7) mm; distal wing lobes 2–2.5 mm, often indented on side. $2n = 38$.

Flowering mostly late May–Jun. Open, rocky slopes; 1200–1900 m; Ariz., N.Mex., Tex.; Mexico (Chihuahua, Sonora).

Morphologically, *Dasylirion wheeleri* is fairly uniform within its range in the United States, with some minor variation in fruit size and receptacle length.

2. Dasylirion texanum Scheele, Linnaea 23: 140. 1850 [F]

Plants typically small, with small crowns; trunks usually less than 0.5 m. **Leaves** narrow, arching gently downward, often twisting outward toward tip; blade usually shiny, bright green, 70–130 × 1–2 cm wide above broadened base, glabrous, not waxy, smooth; prickles mostly antrorse. **Inflorescences** 2–4 m; stalk 1.5–3 cm diam. at base; branches mostly basal, erect, 2.5–6 cm; bracts lanceolate; fascicles of flowers condensed, 5–8 cm from base to tip; primary axes 0.5–1 cm. **Flowers** with receptacles ca. 0.2 mm; tepals whitish or greenish, 1.5 × 1 mm; style ca. 0.2 mm; stigma lobes 0.2–0.3 mm, loosely united; pedicel 2.2–3 mm in fruit. **Capsules** ellipsoid, 5.5–6 × 4–5 mm, distal wing lobes 1–1.5 mm, evenly rounded in cross section, not indented. $2n = 38$.

Flowering May–Jul. Open, rocky limestone slopes, arroyos and canyons, dry limestone hills; 600–1800 m; Tex.; Mexico (Coahuila).

Dasylirion texanum is often small, but can become quite large in moist situations. It is relatively homogeneous in the northeastern portion of its range in east-central Texas. In west Texas, plants are larger and have both antrorse and retrorse prickles. These have been included in *D. heteracanthum* I. M. Johnston, but they are probably hybrids between *D. texanum* and *D. leiophyllum*, in which the prickles are retrorse. There is little mixing of these two species except in a narrow zone extending northward through the Chisos, Davis, and Guadalupe mountains in west Texas.

3. Dasylirion leiophyllum Engelmann ex Trelease, Proc. Amer. Philos. Soc. 50: 433, fig. 1911

Dasylirion heteracanthum I. M. Johnston

Plants typically large and robust, with small crowns; trunks to 1 m, erect or reclining. **Leaves** elongate, rigid; blade bright green, 90–110 × (1.5–)2–3 cm wide above broadened base, usually smooth and shiny, glabrous, not waxy; prickles mostly retrorse. **Inflorescences** 2.5–5 m; stalk 3–5 cm diam. at base; branches mostly basal, erect, 4–10 cm; bracts lanceolate; fascicles of flowers condensed, 8–12 cm from base to tip; primary axes 2–5 cm. **Flowers** with receptacles 0.2–0.4 mm; tepals whitish or greenish, 1.8–2 × 1 mm; style 0.2–0.3 mm; stigma lobes 0.3–0.4 mm; pedicel 2–2.4 mm in fruit. **Capsules** ellipsoid to obovoid, 6–7 × 4–4.5 mm; distal wing lobes 1.5–2 mm, rounded in cross section, not indented.

Flowering mainly Jun–Jul. Gravelly limestone and igneous slopes, bajadas, canyons, and arroyos of mountains; 1200–1800 m; N.Mex., Tex.; Mexico (Chihuahua, Coahuila).

Dasylirion leiophyllum hybridizes with *D. texanum* in the area of overlap. Specimens with mixed prickle curvature have been treated as *D. heteracanthum*, but on the whole these resemble *D. leiophyllum*, and so *D. heteracanthum* is reduced to synonomy here.

4. YUCCA Linnaeus, Sp. Pl. 1: 319. 1753; Gen. Pl. ed. 5, 150. 1754 • Spanish-bayonet [Carib Indian name for *Manihot*, erroneously applied]

William J. Hess

R. Laurie Robbins

Clistoyucca (Engelmann) Trelease; *Samuela* Trelease

Plants perennial, acaulescent or caulescent, sometimes subscapose, sometimes arborescent, usually branching extensively, from woody, subterranean or aboveground caudices, or single stems. **Leaves** sessile, in rosettes on caudices or at branch ends; blade linear-lanceolate, expanded basally, usually rigid, occasionally fleshy, margins entire or denticulate, often filiferous and separating into elongated fibers, corneous, apex mostly sharp-pointed. **Scape**, when present, usually less than 2.5 cm diam. **Inflorescences** erect or rarely pendent, paniculate or racemose, sometimes paniculate proximally and racemose distally, bracteate, occasionally pubescent; bracts ascending, erect, or rarely reflexed; peduncle sometimes scapelike, sometimes extending beyond leaves, sometimes pubescent. **Flowers** bisexual; perianth campanulate or globose; tepals 6, similar, fleshy, distinct to or connate at base, whitish to cream or tinged slightly with green or purple, occasionally pubescent; stamens 6; filaments flattened, as wide as anthers, smooth, papillose, or granular, fleshy; pistil obovoid or oblong-cylindrical; ovary superior, usually green, 3-locular or 6-locular with false septa, 6-lobed; style white to dark green, often thick; stigmas usually 3, sometimes 1 and subcapitate, white to pale green, 1–2 mm. **Fruits** erect or pendent, capsular or baccate. **Seeds** many per locule, usually black, occasionally gray, flattened, round, rarely obovate or ovate. $x = 25, 30$.

Species ca. 35–40 (28 in the flora): e coastal plain, se, sc, w United States, n, c, w Mexico, n Central America.

W. Trelease (1902) recognized *Hesperoyucca*, *Clistoyucca*, and *Samuela* as segregates from, but closely related to, *Yucca*. S. D. McKelvey (1938–1947) returned them to *Yucca*, recognizing sections *Hesperoyucca* and *Clistoyucca*, and placing *Samuela* in section *Sarcocarpa*. However, McKelvey indicated that there is ample justification for recognizing *Hesperoyucca* at genus level, "since a number of flower and fruit characters differ from those in all other sections." Recent DNA evidence provides strong support for separate recognition of *Hesperoyucca* (M. A. Hanson 1993; D. J. Bogler 1994; D. J. Bogler and B. B. Simpson 1995, 1996; K. H. Clary 1997). Clary's data indicate that it is more closely related to *Hesperaloe* than to *Yucca*, and that *Clistoyucca* and *Samuela* should be retained in *Yucca*.

Yuccas are often cultivated in many places in the United States, particularly in the south. They are not easily identified, particularly in herbarium specimens, and collectors should be certain to record information about the habits and sizes of plants, and the colors of ovaries, styles, and stigmas. If possible, field photographs of the pistil and stamens should accompany the specimens.

SELECTED REFERENCES Bogler, D. J., J. L. Neff, and B. B. Simpson. 1995. Multiple origins of the *Yucca*–yucca moth association. Proc. Natl. Acad. Sci. U.S.A. 92: 6864–6867. Clary, K. H. 1997. Phylogeny, Character Evolution, and Biogeography of *Yucca* L. (Agavaceae) As Inferred from Plant Morphology and Sequences of the Internal Transcribed Spacer (ITS) Region of the Nuclear Ribosomal DNA. Ph.D. dissertation. University of Texas. Hanson, M. A. 1993. Dispersed Unidirectional Introgression from *Yucca schidigera* into *Y. baccata* (Agavaceae). Ph.D. dissertation. Claremont Graduate School. Lenz, L. W. and M. A. Hanson. 2000. Yuccas (Agavaceae) of the International Four Corners: Southwestern USA and northwestern Mexico. Aliso 19: 165–179. McKelvey, S. D. 1938–1947. Yuccas of the Southwestern United States. 2 vols. Jamaica Plain. Robbins, R. L. 1983. A Systematic Study of the Indehiscent-fruited Yuccas in the Chihuahuan Desert. Ph.D. dissertation. Texas Tech University. Trelease, W. 1907. Additions to the genus *Yucca*. Rep. (Annual) Missouri Bot. Gard. 18: 225–230. Trelease, W. 1902. The Yucceae. Rep. (Annual) Missouri Bot. Gard. 13: 27–133. Webber, J. M. 1953. Yuccas of the Southwest. Washington. [U.S.D.A., Agric. Monogr. 17.]

1. Fruits indehiscent, pendent or occasionally erect.
    2. Fruits dry and spongy or leathery.
        3. Leaf blade 15–35 × 0.7–1.5 cm, rigid; fruits 6–8.5 cm; sw United States . . . . 1. *Yucca brevifolia*
        3. Leaf blade 40–100 × 3.5–6 cm, rigid or flexible; fruits 2.5–8 cm; se United States . . . 8. *Yucca gloriosa*
    2. Fruits fleshy and succulent.
        4. Tepals connate basally for 1 mm or more; pistil 2.8–8 cm.
            5. Mature plants taller than 2.5 m; tepals 3.9–10.8 cm. 2. *Yucca faxoniana*
            5. Mature plants shorter than 2.5 m; tepals 4.5–13 cm . . . . 3. *Yucca baccata*
        4. Tepals distinct, or connate basally for less than 1 mm; pistil 1.5–4 cm.
            6. Leaf blade thin, flexible, margins entire or occasionally filiferous with slender fibers . . . . 4. *Yucca madrensis*
            6. Leaf blade thick, rigid, margins entire, denticulate, or filiferous with slender or coarse fibers.

7. Leaf margins sharply denticulate or entire, rarely filiferous with slender fibers . . . . . 7. *Yucca aloifolia*
7. Leaf margins entire, filiferous with coarse fibers.
8. Style 2–8 mm . . . . . 5. *Yucca treculeana*
8. Style 1–2 mm . . . . . 6. *Yucca schidigera*

1. Fruits dehiscent, erect.
9. Leaf margins entire or often denticulate, sometimes wavy, lemon yellow to orangish red or reddish brown; pistil 2.5–4.5 cm.
10. Mature plants caulescent; rosettes each with more than 100 leaves.
11. Mature plants 2.5–3.6 m, not including inflorescence; leaf blade 25–60 × 1.2–1.7 cm, widest considerably above middle, smooth . . . . . 9. *Yucca rostrata*
11. Mature plants 0.7–2.5 m, not including inflorescence; leaf blade 20–30(–45) × 0.7–1.2 cm, widest at or above middle, ± scabrous . . . . . 10. *Yucca thompsoniana*
10. Mature plants acaulescent, usually forming colonies; rosettes each with fewer than 100 leaves.
12. Plants in colonies of 10–30 rosettes each; mature leaf blade straight, flat except becoming concave near apex, glaucous, margins wavy, yellow; inflorescences with wide-spreading branchlets . . . . . 11. *Yucca pallida*
12. Plants in open colonies of 2–15 rosettes each, or sometimes solitary; leaf blade flattened or twisted and concave, glaucous or not, margins yellow, dark orange, or reddish brown; inflorescences narrowly ovoid.
13. Mature leaf blade twisted, strongly concave, 1.7–4 cm at greatest width, flexible, margins orange or reddish brown . . . . . 12. *Yucca rupicola*
13. Mature leaf blade straight, slightly concave, 1–2(–2.5) cm at greatest width, rigid, margins yellow or reddish brown . . . . . 13. *Yucca reverchonii*
9. Leaf margins entire, filiferous or becoming so, not wavy, whitish to gray, light green, or light brown; pistil 1.5–3.8 cm.
14. Mature plants caulescent, arborescent, with thick stems 0.6–2.5 m, rarely acaulescent; rosettes usually large.
15. Mature plants distinctly arborescent, solitary or forming small colonies; stems 1–7 per colony, erect, 1–2.5 m . . . . . 14. *Yucca elata*
15. Mature plants not distinctly arborescent, forming large colonies; stems usually 7 or more per colony, often procumbent, 0.6–1.3 m . . . . . 15. *Yucca utahensis*
14. Mature plants acaulescent or occasionally caulescent with stems to 1 (rarely –2) m, sometimes indistinctly arborescent; rosettes usually small.
16. Leaf blade usually soft and limp, rarely erect and rigid, 1–5 cm wide, proximal leaves often becoming reflexed; inflorescences paniculate.
17. Tepals 5–7 cm; inflorescences glabrous . . . . . 16. *Yucca filamentosa*
17. Tepals 3–5 cm; inflorescences pubescent or glabrous.
18. Inflorescences pubescent . . . . . 17. *Yucca flaccida*
18. Inflorescences glabrous . . . . . 25. *Yucca necopina*
16. Leaf blade rigid or sometimes flexible, including proximal leaves, usually less than 2 cm wide, or if wider, then inflorescences racemose and partially held within rosettes; inflorescences racemose or paniculate.
19. Inflorescences paniculate (and racemose distally in *Y. constricta*); leaf blade plano-convex or sometimes plano-keeled.
20. Distance from leaf tips to proximal inflorescence branches less than twice leaf length when fully expanded; peduncle 0.5–1 m; leaf blade 0.3–0.7(–1.5) cm wide . . . . . 28. *Yucca campestris*
20. Distance from leaf tips to proximal inflorescence branches more than twice leaf length when fully expanded; peduncle 1–2 m; leaf blade (0.3–)0.7–2 cm wide.
21. Capsules not constricted; leaf blade mostly recurving, 1–2 cm wide . . . . . 26. *Yucca tenuistyla*
21. Capsules deeply constricted near middle; leaf blade mostly linear, (0.3–)0.7–1.5 cm wide . . . . . 27. *Yucca constricta*

19. Inflorescences primarily racemose, occasionally paniculate proximally; leaf blade concave or concavo-convex, except for those less than 0.9 cm wide, then plano-convex or plano-keeled.
    22. Peduncle (0.1–)0.3–2.5 m; inflorescences arising well beyond rosettes.
        23. Leaf blade usually linear, 20–80 (–150) cm, not glaucous; inflorescences (4–)8–20 dm . . . . 18. *Yucca angustissima*
        23. Leaf blade spatulate-lanceolate, 15–46 cm, rather glaucous; inflorescences 4–7 dm . . . . 21. *Yucca neomexicana* (in part)
    22. Peduncle 0.1–0.6 m; inflorescences arising within or just beyond rosettes.
        24. Capsules 3–5.5(–7) × 2–3 cm.
            25. Peduncle shorter than 0.2 m . . . . 20. *Yucca harrimaniae* (in part)
            25. Peduncle (0.1–)0.2 m or longer.
                26. Style pale green or rarely white; Colorado, New Mexico, Oklahoma panhandle . . . . 21. *Yucca neomexicana* (in part)
                26. Style dark green; Arkansas, Kansas, Missouri, Oklahoma, Texas . . . . 22. *Yucca arkansana* (in part)
        24. Capsules (3–)5–7(–9) cm × 2–4(–5) cm.
            27. Leaf blade 0.7–4.3 cm wide.
                28. Style pale to bright green; leaf blade 1.8–4.3 cm wide; Arizona, Colorado, Nevada, New Mexico, Utah . . . . 20. *Yucca harrimaniae* (in part)
                28. Style dark green; leaf blade 0.7–2.5 cm wide; Arkansas, Kansas, Missouri, Oklahoma, Texas . . . . 22. *Yucca arkansana* (in part)
            27. Leaf blade 0.5–2(–2.5) cm wide.
                29. Style dark green.
                    30. Inflorescences 5–10 dm; leaf blade 0.8–1.2 cm wide, rigid . . . . 24. *Yucca glauca*
                    30. Inflorescences 3–6(–8) dm; leaf blade 0.7–2(–2.5) cm wide, flexible . . . . 22. *Yucca arkansana* (in part)
                29. Style white or pale yellowish green.
                    31. Inflorescences 2.5–4.5(–8.5) dm; peduncle shorter than 0.2 m . . . . 19. *Yucca bailey*
                    31. Inflorescences (5–)7–13 dm; peduncle 0.1–0.5 m . . . . 23. *Yucca intermedia*

1. Yucca brevifolia Engelmann in S. Watson, Botany (Fortieth Parallel), 496. 1871 • Joshua tree [E] [F]

Yucca draconis Linnaeus var. *arborescens* Torrey, Pacif. Railr. Rep. 5(4): 147. 1857; *Cleistoyucca arborescens* (Torrey) Trelease; *C. brevifolia* (Engelmann) Rydberg; *Yucca arborescens* (Torrey) Trelease; *Y. brevifolia* var. *herbertii* (J. M. Webber) Munz; *Y. brevifolia* var. *jaegeriana* McKelvey

Plants usually solitary, erect, arborescent, to 15 m. **Stems** usually 1, occasionally several–many, simple or usually branched distally; branches 1–3 m. **Leaf blade** green, 15–35 × 0.7–1.5 cm, rigid, smooth, glabrous, margins nearly entire, minutely denticulate, yellow. **Inflorescences** erect, paniculate, densely crowded, ca. 1/2 enclosed in rosettes, ellipsoid-ovoid, 3–5.5 × 3–3.8 dm, glabrous. **Flowers** erect, 4–7 cm; perianth oblong-ellipsoid to globose; tepals barely connate at base, greenish white to cream, lanceolate to oblong, 2.5–7 × 1.1–2.2 cm; filaments 1–1.2 cm; anthers 3–3.2 mm, short-pubescent to papillate; pistil 2–3.5 cm; ovary ovoid, 2.5–3 cm, longer than broad; style ca. 1.6 mm; stigmas distinct. **Fruits** pendent, capsular, indehiscent, deeply furrowed, ellipsoid, 6–8.5 cm, dry, spongy. **Seeds** dull black, thin, 8–12 mm diam., smooth.

Flowering late winter–mid spring. Deserts; 400–1800 m; Ariz., Calif., Nev., Utah.

There is disagreement regarding the status of entities within *Yucca brevifolia*. S. D. McKelvey (1935) described var. *jaegeriana*, the miniature Joshua tree, and L. D. Benson and R. A. Darrow (1981) recognized the variety as well; however, J. L. Reveal (1977c) declined to recognize it, a decision with which we agree.

2. Yucca faxoniana Sargent, Man. Trees, 121. 1905 • Faxon yucca, Spanish dagger

Samuela faxoniana Trelease; *Yucca australis* Trelease 1893, not (Engelmann) Trelease 1902; *Y. macrocarpa* Coville 1893, not Engelmann 1881

Plants solitary, erect, arborescent, 2.5–6.9 m, including inflorescence. **Stems** 1, simple or with 2–4 branches, to 5.1 m, average diam. 32 cm. **Leaf blade** erect, yellowish green, 43–115 × 3.1–8.4 cm, rigid, smooth, glabrous, margins conspicuous, curling, filiferous, brown. **Inflorescences** erect, paniculate, often with proximal branches arising beyond rosettes, broadly ovoid, 5.5–25.5 dm, glabrous; peduncle 0.3–0.6 m. **Flowers** pendent, 4.4–12.4 cm; perianth campanulate; tepals connate basally into floral cup 1–32 mm, white to greenish white, ovate, 3.9–10.8 cm; filaments averaging 2.2 cm from base of tepals, glabrous; anthers 1–6 mm; pistil 2.8–8 × 0.7 cm; ovary ca. 4.5–5 times longer than wide; style 4.5 mm; stigmas distinct. **Fruits** pendent, baccate, indehiscent, elongate, 3.6–13.6 × 1.8 –3.6 cm, fleshy, succulent. **Seeds** black, 7.7 mm diam., 2.9 mm thick, smooth.

Flowering late winter–spring. Rocky slopes, flat plains; 800–2100 m; Tex.; Mexico (Chihuahua, Coahuila).

Trelease described the genus *Samuela* based on two species, *Samuela faxoniana* and *S. carnerosana*. K. H. Clary's DNA study (1997) shows them to be closely related but genetically distinct.

Yucca faxoniana is often used for landscaping in arid and semiarid regions of Texas and New Mexico.

3. Yucca baccata Torrey in W. H. Emory, Rep. U.S. Mex. Bound. 2(1): 221. 1859 [F]

Plants often forming open colonies of rosettes, acaulescent or short-caulescent, shorter than 2.5 m. **Stems,** if present, decumbent, 1–24, aerial or subterranean, simple or sometimes branched, to 2 m. **Leaf blade** erect, bluish green, concavo-convex, 30–100 × 2–6 cm, rigid, scabrous or glaucous, margins brown. **Inflorescences** erect, paniculate, dense, arising completely within to mostly extending beyond rosettes, ovoid, 6–8.2 dm, glabrous, rarely slightly pubescent; peduncle scapelike, to 0.8 m. **Flowers** pendent, 5–13 cm, nearly as wide; perianth campanulate; tepals connate basally to form shallow floral cup 7–12 mm, usually cream-colored, occasionally tinged with purple, 4.5–13 cm; filaments connate proximally into collarlike structure, 3.2–12 mm, fleshy, pubescent; anthers 5–7 mm; pistil 4.5–8 × 0.7–1.2 cm (usually 4–5(–7) times longer than broad); ovary 0.7–1.2 cm; style 5–7 mm; stigmas distinct. **Fruits** pendent, baccate, indehiscent, elongate, 5–23 × 4–7.5 cm, fleshy, succulent. **Seeds** dull black, obovate, 7–11 mm diam., 3 mm thick, rugose.

Varieties 2 (2 in the flora): North America, Mexico (Chihuahua, Sonora).

1. Plants acaulescent or caulescent; stems 1–6, aerial or subterranean, shorter than 0.3 m; leaf margins coarse, curling . . . . 3a. *Yucca baccata* var. *baccata*
1. Plants caulescent; stems 1–24, aerial, often branched, some reaching 2 m; leaf margins filiferous . . . . 3b. *Yucca baccata* var. *brevifolia*

3a. Yucca baccata Torrey var. **baccata** • Banana yucca, blue yucca, datil [F]

Yucca baccata var. *vespertina* McKelvey; *Y. vespertina* (McKelvey) S. L. Welsh

Plants acaulescent or caulescent. **Stems,** if present, 1–6, aerial or subterranean, shorter than 0.3 m. **Leaf margins** coarse, curling. **Inflorescences** arising completely within or extending 1/4 beyond rosettes; peduncle 0.6–0.8 m.

Flowering spring–early summer. Rocky slopes, pinyon, oak, and juniper woodlands, grasslands; 400–2500 m; Ariz., Calif., Colo., Nev., N.Mex., Tex., Utah; Mexico (Chihuahua).

3b. Yucca baccata Torrey var. **brevifolia** L. D. Benson & R. A. Darrow, Amer. J. Bot. 30: 234. 1943 • Thornber yucca

Yucca brevifolia Schott ex Trelease, Rep. (Annual) Missouri Bot. Gard. 13: 100, plates 57–59. 1902, not Engelmann 1871; *Y. arizonica* McKelvey; *Y. confinis* McKelvey; *Y. thornberi* McKelvey; *Y. treleasei* J. F. Macbride 1918, not Sprenger 1906

Plants caulescent. **Stems** 1–24, aerial, often branched, some reaching 2 m. **Leaf margins** filiferous. **Inflorescences** arising almost completely within to mostly extending beyond rosettes; peduncle to 0.3 m.

Flowering spring–early summer. Hills, mesas, flats in Sonoran Desert, desert grasslands, oak woodlands, 500–2000 m; Ariz., N.Mex.; Mexico (Sonora).

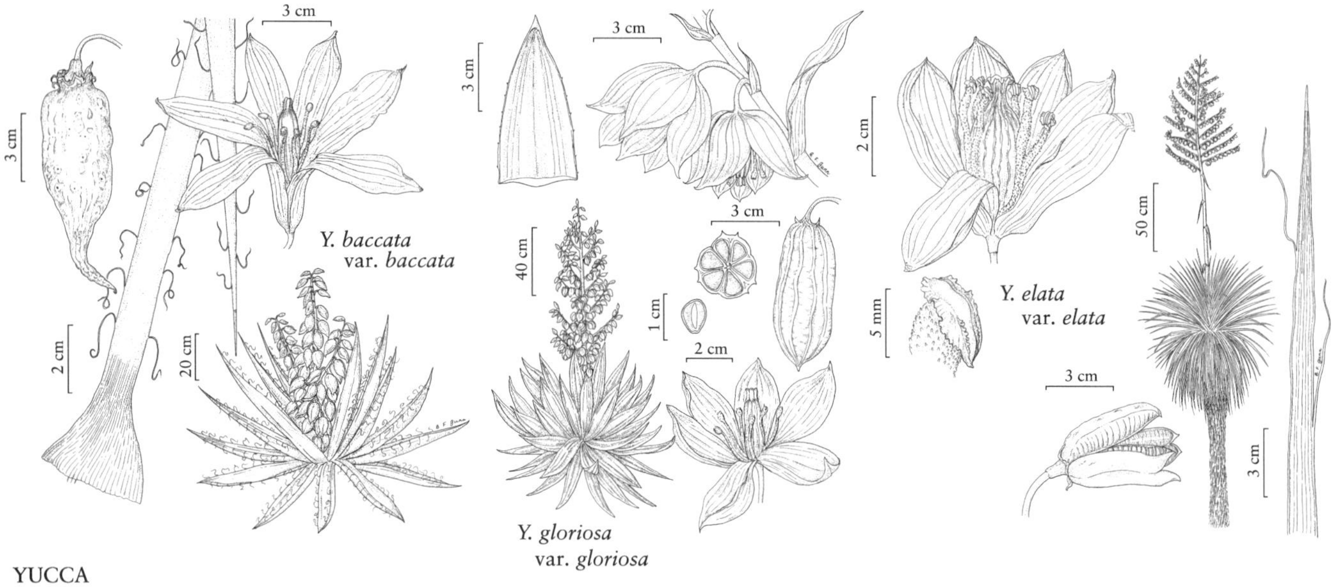

YUCCA

4. Yucca madrensis Gentry, Agave Fam. Sonora, 159, fig. 62. 1972 • Sierra Madre yucca

Plants solitary, caulescent, shorter than 3 m. **Stems** 1, simple, unbranched, shorter than 2 m. **Leaf blade** erect to reflexing with age, bluish-glaucous or green to yellow-green, thin, flat to conduplicate, flexible, margins entire or occasionally filiferous with slender fibers, brown to grayish, smooth. **Inflorescences** erect, paniculate, somewhat open, arising 1/4–1/2 within rosettes, elongate-ovoid, to 8 dm, sparsely pubescent. **Flowers** pendent, 3.5 cm; perianth ovoid; tepals distinct, or barely connate basally, white, ovate-lanceolate to lanceolate, ca. 3 × 1.4 cm; filaments ca. 1.5 cm, glabrous; pistil 1.5–4 cm; ovary ca. 2.5 cm. **Fruits** becoming pendent, baccate, indehiscent, 6–12.5 × 2.5–3.8 cm, fleshy, succulent. **Seeds** gray, 7–10 mm diam., 5–8 mm thick.

Flowering in summer with rains. Madrean pine-oak forest; 1400–1600 m; Ariz.; Mexico (Chihuahua, Sonora).

Yucca madrensis is closely related to *Y. jaliscensis* to the south and the *Y. schottii* complex to the north. We follow L. W. Lenz and M. A. Hanson (2000b), who have determined that plants with long, wide, flexible, blue leaves that are now usually identified as *Y. schottii* Engelmann (not congruent with Engelmann's original description) are more correctly referred to *Y. madrensis*. Lenz and Hanson further suggested that what is now called *Y. schottii* represents a hybrid complex among *Y. baccata*, *Y. elata*, and *Y. madrensis*. This putative hybridization is conjectural, and without firm evidence we are uncertain as to the appropriate disposition of the name *Y. schottii*. We are somewhat skeptical of the potential for crosses between baccate species (*Y. baccata* and *Y. madrensis*) and capsular species (*Y. elata*).

SELECTED REFERENCE Lenz, L. W. and M. A. Hanson. 2000b. Typification and change in status of *Yucca schottii* (Agavaceae). Aliso 19: 93–98.

5. Yucca treculeana Carrière, Rev. Hort., 580. 1858 • Spanish dagger, palma pita

Yucca argospatha Verlot; *Y. aspera* Engelmann; *Y. baccata* Torrey var. *australis* Engelmann; *Y. baccata* var. *macrocarpa* Torrey; *Y. canaliculata* Hooker; *Y. canaliculata* var. *pendula* K. Koch; *Y. crassifila* Engelmann; *Y. longifolia* Buckley 1863, not Karwinsky ex Schultes f. 1830; *Y. torreyi* Shafer; *Y. treculeana* var. *succulenta* McKelvey; *Y. undulata* K. Koch

Plants often forming colonies of rosettes, arborescent, to 7 m; rosettes with leaves hanging at various angles, giving an overall ragged appearance. **Stems** 1–8, occasionally 2–5-branched, 14–15 cm diam. **Leaf blade** erect, yellowish to bluish green, usually U- or V-shaped in cross section, thick, 36–128 × 1.6–7 cm, rigid, scabrous, margins entire, filiferous with straight, coarse fibers, light brown. **Inflorescences** erect, paniculate, arising mostly within rosettes, variable in shape, usually ovoid, 18 dm, glabrous, rarely slightly pubescent; peduncle scapelike, 0.3 m or longer. **Flowers** pendent; perianth globose; tepals distinct, cream-colored, occasionally tinged with purple, ovate, 2.7–8.1 × 1–3.4 cm, apex rounded or acute; filaments 1–2.7 cm, short-pubescent proximally;

anthers 1–6 mm; pistil 1.5–3.5 × 0.5–1 cm; ovary 1.3–3.3 cm; style 2–8 mm; stigmas distinct. **Fruits** pendent, baccate, indehiscent, 4.4–18.7 × 1.8–4.6 cm, fleshy, succulent. **Seeds** black, 5–14 mm diam., 1–5 mm thick.

Flowering mid winter–spring. Grassy or rocky slopes or mesas, brushland, chaparral; 0–1600 m; N.Mex., Tex.; n Mexico (Chihuahua, Coahuila, Tamaulipas, Zacatecas).

6. **Yucca schidigera** Roezl ex Ortgies, Gartenflora 20: 110. 1871 • Mohave yucca

Yucca californica Nuttall ex Baker; *Y. mohavensis* Sargent

Plants forming colonies of rosettes, shrubby, caulescent, to 5 m. **Stems** 1–8, simple or sparingly branched. **Leaf blade** erect, yellowish green, sword-shaped, broadest at middle, markedly concave, thick, 33–130 × 3–5 cm, rigid, glabrous, margins entire, filiferous with coarse fibers. **Inflorescences** erect, paniculate, dense, arising entirely within or scarcely beyond rosettes, obovoid or ellipsoid and flat distally, 3–6 dm; peduncle scapelike, to 4 dm. **Flowers** pendent; perianth globose; tepals spreading, distinct, white or cream, lanceolate or broadly lanceolate, 3–5(–7.5) × 1–1.8 cm; filaments 2–2.5 cm, papillose at base; anthers 3.2 mm; pistil 1.8 × 2.5 cm; ovary 1.7–2.7 × 0.7–1.2 cm, apex abruptly tapered; style 1–2 mm; stigmas distinct. **Fruits** pendent, baccate, indehiscent, elongate, cylindrical, 5–11.5 × 3–4 cm, fleshy, succulent. **Seeds** gray, 8–11 mm diam., 6–9 mm thick, rugose.

Flowering late winter–early spring. Deserts and chaparral, rocky or gravelly mountain slopes; 300–1200 m; Ariz., Calif., Nev.; Mexico (Baja California).

Yucca schidigera is usually used for landscaping in the southwest and by Native Americans for food and fiber. Capsules made from inflorescence stalks are sold as natural health products.

7. **Yucca aloifolia** Linnaeus, Sp. Pl. 1: 319. 1753 • Dagger plant [W2]

Yucca serrulata Haworth

Plants forming colonies of rosettes, arborescent, to 7 m. **Stems** 1–3, erect or somewhat declining, simple or sparingly branched. **Leaf blade** erect, dark green, flattened or slightly concave, thick, 12–40 × 2.5–6 cm, rigid, margins sharply denticulate or entire, rarely filiferous with straight fibers. **Inflorescences** pendent, paniculate, arising 1/4–1/2 within rosettes, somewhat conical, 4.5–6.1 dm, glabrous or slightly pubescent; peduncle scapelike, to 3 dm. **Flowers** pendent, to 7 cm; perianth globose; tepals spreading, connate basally for less than 1 mm, creamy white tinged with green or purple near base, lanceolate, 3–4 × 1.2–2.2 cm; filaments ca. 2 cm; anthers 2–3 mm; pistil light green, 3–4 × 0.8–1 cm; ovary stipitate, 2–5 cm; style 5 mm; stigmas distinct. **Fruits** pendent, baccate, without core, indehiscent, 3.5–5 × 2–2.6 cm, with fleshy, succulent, purple pulp. **Seeds** dull black, round-ovate, 5–7 mm diam., 2.5 mm thick.

Flowering fall. Sand dunes or shell mounds near coasts; to 1800 m; Ala., Fla., Ga., La., Miss., N.C., S.C., Tex., Va.; Mexico; West Indies; Central America.

Yucca aloifolia has been widely cultivated, and horticultural forms (or varieties, depending upon the source) differ in the striping of yellow and white on the leaves. Results of DNA studies by K. H. Clary (1997) show a close relationship between this species and *Y. gloriosa*.

8. **Yucca gloriosa** Linnaeus, Sp. Pl. 1: 319. 1753 [E] [F]

Plants forming colonies of rosettes, caulescent, arborescent, simple or more often branching. **Stems** erect, to 5 m. **Leaf blade** erect or recurving, green or blue-green, lanceolate or sword-shaped, flattened, concave distally, thin, 40–100 × 3.5–6 cm, rigid or flexible, glaucous at least when young, margins entire or roughly and minutely denticulate, often becoming filiferous with straight fibers, yellow or brown, opaque. **Inflorescences** paniculate, arising partly within to well beyond rosettes, ovoid to ellipsoid, 5–12 × 4.5 dm, glabrous or pubescent; peduncle scapelike, 0.9–1.5 m. **Flowers** pendent; perianth globose to campanulate; tepals distinct, white to creamy white or greenish white, sometimes tinged with purple, elliptic to narrowly ovate, 4–5 × 2–2.5 cm; filaments ca. 2.6 cm, hispid or slightly papillose; anthers ca. 3.5 mm; pistil light green, ca. 3.6 cm; ovary sessile, ca. 2.8 cm; style ca. 9 mm; stigmas separate; pedicel to 2 cm, often arching. **Fruits** erect or pendent, baccate, with core, indehiscent, 6-winged or 6-ribbed, elongate, 2.5–8 cm, leathery. **Seeds** black, lustrous, ovate, thin, 5–8 mm diam. $2n = 50$.

Varieties 2 (2 in the flora): se United States.

Yucca gloriosa has a growth habit similar to that of *Y. aloifolia*, except that the former appears more moundlike due to the terminal branching mode, whereas the latter appears more open because the branching is more median on the trunk.

1. Leaf blade mostly erect, rigid; inflorescences extending well beyond rosettes; berries pendent, 5.5–8 cm . . . . 8a. *Yucca gloriosa* var. *gloriosa*
1. Leaf blade mostly recurving, flexible; inflorescences barely extending beyond rosettes; berries mostly erect, 2.5–4.5 cm . . . 8b. *Yucca gloriosa* var. *recurvifolia*

8a. Yucca gloriosa Linnaeus var. **gloriosa** • Mound-lily yucca [E] [F]

Plants caulescent, branching near summits, to 5 m. **Leaf blade** mostly erect, blue-green, lanceolate, 40–70 × 4–6 cm, rigid, becoming more flexible with age, glaucous, margins brown. **Inflorescences** extending 4–5 dm beyond rosettes, ovoid to ellipsoid, 7.5–12 dm; peduncle 0.9–1.5 m. **Flowers:** tepals white to creamy white, 4–5 × 2–2.5 cm; filaments granular or powdery, pubescent. **Berries** pendent, 6-ribbed, 5.5–8 cm.

Flowering spring. Coastal dunes; 0 m; Ga., N.C., S.C.

Yucca gloriosa var. *gloriosa* is usually cultivated in the southern United States and persists at old homesites. Named forms have been based on habit and leaf morphology.

8b. Yucca gloriosa Linnaeus var. **recurvifolia** (Salisbury) Engelmann, Trans. Acad. Sci. St. Louis 3: 39. 1873 • Curve-leaf yucca [E]

Yucca recurvifolia Salisbury, Parad. Lond., plate 31. 1806; *Y. pendula* Groenland; *Y. recurva* Haworth

Plants arborescent, simple or branching, to 2 m. **Leaf blade** mostly recurving, green, sword-shaped, 50–100 × 3.5–5 cm, flexible, margins narrowly yellow or brown. **Inflorescences** barely extending beyond rosettes, narrowly ellipsoid, 5–10 dm; peduncle 0.9–1.1 m. **Flowers:** tepals white or slightly greenish white, 4–5 × 2–2.5 cm. **Berries** mostly erect, oblong, 6-winged or 6-ribbed, 2.5–4.5 cm.

Flowering spring. Sandy soils of Gulf coast plains; 0–100 m; Ala., Fla., Ga., La., Miss.

Yucca gloriosa var. *recurvifolia* is poorly known and may not be distinct from var. *gloriosa*. It has been treated also at species and forma levels. Additional studies are needed to clarify its status.

9. Yucca rostrata Engelmann ex Trelease, Rep. (Annual) Missouri Bot. Gard. 13: 68, plates 40–42, plate 84, fig. 3, plate 93, fig. 2. 1902 • Beaked yucca

Yucca rostrata var. *linearis* Trelease

Plants solitary or forming colonies of rosettes, caulescent, arborescent, 2.5–3.6 m, not including inflorescence, 1.8–3.2 dm diam; rosettes each with more than 100 leaves. **Stems** 1–several, erect, mostly simple, occasionally 1–3-branched. **Leaf blade** linear, often twisted, flat to concavo-convex, widest considerably beyond middle, 25–60 × 1.2–1.7 cm, glaucous, smooth, margins minutely denticulate, lemon yellow, hyaline, apex spinose, spine tipped. **Inflorescences** paniculate, arising just within or beyond rosettes, ovoid, 3–10 dm; branches up to 3.8 dm; bracts erect; peduncle sometimes scapelike, 0.3–1 m, less than 2.5 cm diam., glabrous or glabrescent. **Flowers** pendent; perianth globose to campanulate; tepals distinct, white, narrowly ovate, 4.2–5.2 × 1.1–2 cm, apex sharply acuminate; filaments 1.7–2 cm; pistil 2.5–3.5 cm; style white, 6–14 mm; stigmas lobed. **Fruits** erect, capsular, dehiscent, ovoid to ellipsoid, rarely constricted, 4–7 × 1.8–2.5 cm, dehiscence septicidal.

Flowering spring. Rocky mountain slopes, canyon bottoms; 700 m; Tex.; n Mexico (Coahuila).

In the United States, *Yucca rostrata* is restricted to Brewster County, Texas. It is closely related to *Y. thompsoniana* (K. H. Clary 1997), which is perhaps just a northern variant of this species.

10. Yucca thompsoniana Trelease, Rep. (Annual) Missouri Bot. Gard. 22: 101, plates 104–107. 1911 • Thompson yucca, bayoneta

Plants solitary or forming colonies of rosettes, caulescent, arborescent, mostly asymmetrical, 0.7–2.5 m, not including inflorescence, to 30 cm diam.; rosettes each with more than 100 leaves. **Stems** 1, erect, 1–3-branched. **Leaf blade** linear, flat or concavo-convex or slightly keeled, widest at or above middle, 20–30(–45) × 0.7–1.2 cm, flexible, ± scabrous adaxially and abaxially, margins denticulate, yellow or orangish red, corneous, apex sharp-pointed. **Inflorescences** paniculate, racemose at apex, arising beyond rosettes, 5–8 dm; branches to 2 dm; bracts erect, 10–13(–17) cm; peduncle sometimes scapelike, (0.3–)0.4–0.7 m, 1.3–2 cm diam., glabrous or glabrescent. **Flowers**

pendent; perianth globose to campanulate; tepals distinct, white, narrowly elliptic, 3.5–6.5 × 1.2–3.5 cm, glossy, apex sharply acuminate; filaments 1.7–3.3 cm; pistil 2.5–3.8 × 0.4–0.8 cm; style white, 6–18 mm; stigmas lobed. **Fruits** erect, capsular, dehiscent, ovoid, rarely constricted, 3.5–7 × 2–2.5 cm, dehiscence septicidal. **Seeds** dull black, thin, 6–8 mm diam.

Flowering spring. Rocky slopes and hills; 200–1400 m; Tex.; n Mexico (Chihuahua, Coahuila).

Yucca thompsoniana occurs in the trans-Pecos area of Texas. J. M. Webber (1953) characterized it as a dwarflike form of *Y. rostrata,* and the differences are perhaps not sufficient to merit separate recognition. Webber also suggested that it may hybridize with *Y. reverchonii.*

11. Yucca pallida McKelvey, Yuccas Southw. U.S. 2: 57, plates 13, 14. 1947 • Pale yucca [E]

Yucca pallida var. *edentata* (Trelease) Cory; *Y. rupicola* Scheele var. *edentata* Trelease

Plants forming loose colonies of rosettes, acaulescent, with branching subterranean caudices; rosettes 10–30 per colony, each rosette with fewer than 100 leaves. **Leaf blade** lanceolate, straight, flat except becoming concave near apex, widest above middle, 20–50 × 1–4.5 cm, flexible, glaucous, margins denticulate or sometimes wavy, yellow. **Inflorescences** paniculate, often distally racemose, arising beyond rosettes, 7–12 dm, with wide-spreading branchlets 1.5–2.3 dm; bracts erect; peduncle scapelike, 0.6–1.3 m, less than 2.5 cm diam., glaucous. **Flowers** pendent; perianth campanulate; tepals distinct, greenish white, elliptic to ovate, 5–6.5 × 2–3.2 cm; filaments 1.8–3.2 cm; pistil 3.2–4 cm; style white, 13–20 mm; stigmas lobed. **Fruits** erect, capsular, dehiscent, oblong-cylindric, 4.5–5.5 × 1.3–2 cm, dehiscence septicidal. **Seeds** dull black, thin, 5–7 mm diam.

Flowering spring. Rocky prairies and uplands, rarely limestone hills; 100–400 m; Tex.

Yucca pallida is endemic to the blackland prairies of Texas and was originally included in *Y. rupicola*. J. M. Webber (1953) indicated that he found plants of *Y. rupicola* with key features similar to those of *Y. pallida*. Occasionally, plants with entire margins are found; these have been called *Y. pallida* var. *edentata*, which S. D. McKelvey (1938–1947) suggested might be a hybrid between *Y. pallida* and *Y. arkansana*. K. H. Clary (1997) indicated that DNA evidence confirms the close relationship of *Y. pallida* and *Y. rupicola*.

12. Yucca rupicola Scheele, Linnaea 23: 143. 1850 • Twisted-leaf yucca [E]

Yucca rupicola var. *tortifolia* Engelmann

Plants usually forming open colonies of rosettes, acaulescent, with branching subterranean caudices; rosettes 2–15 per colony, each rosette with fewer than 100 leaves. **Leaf blade** lanceolate, twisted, strongly concave, widest at middle, distinctly narrowing toward base, broadened at stem attachments, 35–60 × 1.7–4 cm, somewhat fleshy, flexible, margins denticulate or wavy, dark orange or reddish brown, apex pointed. **Inflorescences** paniculate, arising beyond rosettes, narrowly ovoid, 2.5–10 dm; branches 0.7–2 dm; bracts erect; peduncle scapelike, 0.3–1.5 m, 1.5–2 cm diam. **Flowers** pendent; perianth campanulate; tepals distinct, white or greenish white, ovate, 4–7 × 1.5–3 cm, apex sharply acuminate; filaments 1.8–3.2 cm; pistil 2.5–4.5 cm; style white or greenish, 12–20 mm; stigmas lobed. **Fruits** erect, capsular, dehiscent, ellipsoid to cylindric, 4–5.5 × 2–3 cm, dehiscence septicidal. **Seeds** dull black, thin, 6–8 mm diam.

Flowering late spring. Rocky hillsides of limestone ledges, open plains, woodlands; 400–900 m; Tex.

Yucca rupicola is endemic to southeastern Edwards Plateau. It is the only species in the flora with flaccid, twisted leaves.

13. Yucca reverchonii Trelease, Rep. (Annual) Missouri Bot. Gard. 22: 102, plate 108. 1911 (as reverchoni) San Angelo yucca [E]

Plants solitary or forming dense, open colonies of rosettes, acaulescent, with branching subterranean caudices; rosettes 2–10 per colony, each rosette with fewer than 100 leaves. **Leaf blade** lanceolate, straight, slightly concave, nearly flattened, widest at middle, 25–60 × 1–2(–2.5) cm, rigid, slightly glaucous, margins denticulate, yellow or reddish brown, hyaline. **Inflorescences** narrowly paniculate, arising beyond rosettes, narrowly ovoid, 3.5–10 dm, densely pubescent; bracts erect; peduncle sometimes scapelike, 0.5–1.1 m, less than 2.5 cm diam. **Flowers** pendent; perianth campanulate; tepals distinct, white or greenish white, ovate, 4–6 × 1.5–3 cm, apex sharply acuminate; filaments 1.8–3.2 cm; pistil 3–4.5 cm × 4–6 mm; style white or greenish, 10–20 mm; stigmas lobed. **Fruits** erect, capsular, dehiscent, ellipsoid, 4–6 × 1.8–3.2 cm, dehiscence septicidal. **Seeds** dull black, thin, 5–7 mm diam.

Flowering late spring. Rocky hillsides, limestone ledges, gravelly plains with dense brush; 400–900 m; Tex.

Yucca reverchonii is endemic to the Edwards Plateau. J. M. Webber (1953) reported apparent hybrids of this species with *Y. rupicola* and *Y. thompsoniana*. S. D. McKelvey (1938–1947) indicated that *Y. reverchonii* occupies an area of the Plateau between those of *Y. rupicola* and *Y. thompsoniana*. K. H. Clary (1997) presented DNA evidence that confirms a close relationship among these three species as distinct entities.

14. Yucca elata Engelmann, Bot. Gaz. 7: 17. 1882 • Palmella, soap-tree yucca, soap-weed yucca F W1

Plants solitary or forming small colonies of rosettes, caulescent or rarely acaulescent, distinctly arborescent, mostly few-branched, 1.2–4.5 m; rosettes usually large, symmetrical or asymmetrical. **Stems** 1–7 per colony, erect, thick, 1–1.5(–2.5) m. **Leaf blade** pale green, linear, widest near middle, 25–95 × 0.2–1.3 cm, flexible, margins entire, curled, filiferous, whitish, apex tapering to short spine. **Inflorescences** mostly paniculate, sometimes distally racemose, arising beyond rosettes, mostly narrowly ovoid to ovoid, 7–15 × 2.5–6.5 dm; branches 0.7–3.5 dm; bracts erect; peduncle sometimes scapelike, 1–2 m, 2.5–5.5 cm diam. **Flowers** pendent; perianth campanulate or globose; tepals distinct, creamy white, often tinged green or pink, narrow to broadly elliptic or ovate, 3.2–5.7 × 1.3–3.2 cm; filaments 1.5–2.5(–3.2) cm, pubescent; anthers (2–)2.5–4.8 mm; pistil 2–3.2 × 0.6–1 cm; style white or pale green, 6–11 mm; stigmas lobed. **Fruits** erect, capsular, dehiscent, oblong-cylindric, symmetrical or rarely constricted, 4–8.2 × 2–4 cm, dehiscence septicidal. **Seeds** dull black, thin, 7–11(–14) mm diam.

Varieties 2 (2 in the flora): North America, Mexico.

J. L. Reveal (1977c) reduced S. D. McKelvey's (1938–1947) *Yucca utahensis* and *Y. verdiensis* to varieties of *Y. elata* based primarily u!pon growth forms. J. M. Webber (1953) considered that these taxa are populations of hybrids between members of the *Y. glauca* alliance of the Great Plains and *Y. elata* of the American Southwest. Reveal believed that Webber did not provide adequate justification for his hybrid hypothesis. K. H. Clary (1997, pers. comm.) believes that *Y. utahensis* is genetically distinct from *Y. elata* and *Y. verdiensis*, based on DNA evidence and the morphological characters of style, stigma, fruit, and leaf. Her DNA evidence shows that *Y. elata* and *Y. verdiensis* are sister taxa, while *Y. utahensis* is not. Although there is a great range of variation within *Y. elata* as circumscribed here, the two varieties recognized are difficult to distinguish.

1. Capsules 5–8.2 cm; pistil 2.8–3.2 cm; leaf blade 30–95 cm . . . . 14a. *Yucca elata* var. *elata*
1. Capsules 4–4.5 cm; pistil 2–2.5 cm; leaf blade 25–45 cm . . . . 14b. *Yucca elata* var. *verdiensis*

14a. Yucca elata Engelmann var. **elata** F W1

Yucca angustifolia Engelmann var. *elata* Engelmann; *Y. angustifolia* var. *radiosa* Engelmann; *Y. radiosa* (Engelmann) Trelease

Plants caulescent, arborescent, 1.2–4.5 m. **Stems** 1–2.5 m. **Leaf blade** 30–95 × 0.4–1.3 cm. **Inflorescences** paniculate, 7–15 dm; branches 0.7–2.5 dm; peduncle 1–2 m. **Flowers:** tepals 4.5–5.7 × 1.3–3.2 cm; filaments 2–2.5(–3.2) cm; anthers 3.2–4.8 mm; pistil 2.8–3.2 cm. **Capsules** 5–8.2 × 2.5–4 cm. $2n = 30$.

Flowering spring–early summer. Desert hillsides, grasslands; 400–1900 m; Ariz., N.Mex., Tex.; n Mexico (Chihuahua, Coahuila, Sonora).

Yucca elata var. *elata* is quite variable according to both S. D. McKelvey (1938–1947) and J. M. Webber (1953). The plants are arborescent with peduncles and inflorescences shorter than to as long as the stems, and are found from west Texas through southern New Mexico and southeastern and central Arizona. Hybrids between *Y. elata* and *Y. glauca*, *Y. constricta*, and *Y. baileyi* are thought to occur in areas of sympatry (J. M. Webber 1953).

14b. Yucca elata Engelmann var. **verdiensis** (McKelvey) Reveal in A. Cronquist et al., Intermount. Fl. 6: 533. 1977 E W1

Yucca verdiensis McKelvey, Yuccas Southw. U.S. 2: 98, plate 35. 1947

Plants primarily caulescent, rarely acaulescent, 1.5–3.5 m. **Stems** 1–1.5 m. **Leaf blade** 25–45 × 0.2–0.6(–1.3) cm. **Inflorescences** paniculate, distally racemose, 7–13 dm; branches 1–3.5 dm; peduncle 1–1.5 m. **Flowers:** tepals 3.2–4.5 × 1.3–2 cm; filaments 1.5–2.5 cm; anthers (2–)2.5–4.8 mm; pistil 2–2.5 cm. **Capsules** 4–4.5 × 2–4 cm.

Flowering late spring. Rocky hillsides, canyons; 300–1900 m; Ariz.

S. D. McKelvey (1938–1947) wondered if *Yucca elata* var. *verdiensis* might be a mountain variety of *Y. angustissima*. On the other hand, she pointed out the differences between the two taxa and stated that the features of the former bear "no resemblance to those of *Y. angustissima*." J. M. Webber (1953) suggested that var. *verdiensis* is a hybrid between var. *elata* and *Y. angustissima*.

15. Yucca utahensis McKelvey, Yuccas Southw. U.S. 2: 94, plates 32–34. 1947 [E] [W1]

Yucca elata Engelmann var. *utahensis* (McKelvey) Reveal

Plants forming large colonies, caulescent or appearing acaulescent, not distinctly arborescent, 1.2–2.8 m; rosettes usually large. **Stems** often procumbent, usually 7 or more per colony, thick, 0.6–1.3 m. **Leaf blade** yellow-green, linear-lanceolate, plano-convex or plano-keeled, widest near middle, 20–70 × 0.7–2.2 cm, flexible, margins entire, curled, filiferous, whitish, apex tapering to short spine. **Inflorescences** paniculate, sometimes distally racemose, arising beyond rosettes, narrowly ellipsoid, 10–15 dm; branches 1–2 dm; bracts erect; peduncle scapelike, 1–1.6 m, 2.5–4.5 cm diam. **Flowers** pendent; perianth campanulate; tepals distinct, creamy white, broad-elliptic or ovate, 4–5 × 2–2.5 cm; filaments 2–2.5 cm; anthers 2–3 mm; pistil 2.5–3.5 cm; style white, 5–10 mm; stigmas lobed. **Fruits** erect, capsular, dehiscent, oblong-cylindric, 5–6 × 2–2.5 cm, dehiscence septicidal. **Seeds** dull black, thin.

Flowering spring. Desert and desert hillsides and canyons, often sandy places; 700–2000 m; Ariz., Nev., Utah.

S. D. McKelvey (1938–1947) suggested that *Yucca utahensis* is most closely related to *Y. elata*, from which it differs primarily in growth habit, capsule size, and leaf width, although with overlaps. S. L. Welsh et al. (1993) considered it equally or more closely related to *Y. angustissima* var. *kanabensis*, but DNA evidence (K. H. Clary 1997) suggests a closer relationship with *Y. elata*. J. M. Webber (1953) believed that *Y. utahensis* is a hybrid between *Y. elata* and *Y. angustissima*. The plants that are caulescent or subcaulescent were recognized by J. L. Reveal (1977c) as *Y. elata* var. *utahensis*; those that are acaulescent, or infrequently caulescent and with racemose inflorescences, he included in *Y. angustissima*.

16. Yucca filamentosa Linnaeus, Sp. Pl. 1: 319. 1753 • Adam's needle, beargrass, needle palm, silkgrass [E]

Plants acaulescent or rarely caulescent, indistinctly arborescent, 1–4.6 m; rosettes usually small. **Stems** simple, 0–0.5 m. **Leaf blade** occasionally erect, proximal leaves often becoming reflexed near middle, lanceolate, flattened, abruptly narrowed and furrowed to apex, thin, widest near middle, 50–75 × 2–4 cm, usually soft and limp, scabrous, margins entire, long and curling, filiferous. **Inflorescences** paniculate, arising beyond rosettes, ovoid, 7.5–15 dm, glabrous; bracts erect; peduncle scapelike, 1–3 m, less than 2.5 cm diam. **Flowers** pendent; perianth globose; tepals distinct, nearly white, ovate, 5–7 × 2–3 cm, glabrous, apex short-acuminate; filaments shorter than pistil; pistil 1.5–3.8 cm; stigmas lobed. **Fruits** erect, capsular, dehiscent, oblong, 3.8–5 × 2 cm, dehiscence septicidal. **Seeds** dull black, thin, 6 mm diam.

Flowering mid spring–early summer. Sandy soil; Ala., Fla., Ga., La., Md., Miss., N.C., S.C., Tenn., Va., W.Va.

Yucca filamentosa is often cultivated and has become naturalized in some areas beyond its native range. Varieties that have been described are rarely recognized in recent literature. *Yucca filamentosa* and *Y. flaccida* are very closely related and perhaps are not distinct species.

17. Yucca flaccida Haworth, Suppl. Pl. Succ., 34. 1819 • Adam's needle, beargrass [E]

Yucca concava Haworth; *Y. filamentosa* Linnaeus var. *concava* (Haworth) Baker; *Y. filamentosa* var. *flaccida* (Haworth) Engelmann; *Y. filamentosa* var. *glaucescens* (Haworth) Baker; *Y. filamentosa* var. *puberula* (Haworth) Baker; *Y. filamentosa* var. *smalliana* (Fernald) H. E. Ahles; *Y. flaccida* var. *glaucescens* (Haworth) Trelease; *Y. flaccida* var. *major* (Baker) Rehder; *Y. freemanii* Shinners; *Y. glaucescens* Haworth; *Y. louisianensis* Trelease; *Y. orchioides* Carrière var. *major* Baker; *Y. puberula* Haworth; *Y. smalliana* Fernald

Plants cespitose, forming small colonies, erect, acaulescent or rarely short-caulescent, 1–3 m; rosettes usually small, slowly dying after flowering. **Stems** simple, to 0.3 m. **Leaf blade** erect, proximal leaves becoming reflexed at middle in age, lanceolate, gradually tapering to apex, thin, widest near middle, 40–80 × 1–4(–5) cm, rigid or soft and limp, glabrous, margins entire, filiferous, apex spinose. **Inflorescences** paniculate, arising beyond rosettes, narrowly ovoid, 4–15 dm, mostly pubescent; bracts erect, proximal to 25 cm, distal 2–3 cm; peduncle scapelike, 0.5–2.8 m, 0.5–4 cm diam. **Flowers** pendent; tepals distinct, white, creamy white, or light greenish white, lanceolate to elliptic, 3–5 × 1–3 cm, mostly pubescent, apex obtuse; filaments 1.2–2.2 cm; pistil 1.5–3.8 cm; ovary pale green, 1.5 cm; style greenish white to white, papillate; stigmas lobed; pedicel 1.5–3 cm. **Fruits** erect, capsular, dehiscent, oblong, obpyriform, or conical, 3.5–4 × 1.5–2 cm, dehiscence septicidal. **Seeds** dull black, thin, 6–8 × 5–6 mm.

Flowering spring. Sand pine-scrub oak, mixed pine-hardwood woodlands, old fields, coastal sands, open or semiopen sites and pine plantations; Ont.; Ala., Ark., Fla., Ga., Ill., Ind., Kans., La., Md., Mich., Miss., Mo., N.C., Ohio, Okla., Pa., S.C., Tenn., Tex., Wis.

Perhaps *Yucca flaccida* should be considered a variety of *Y. filamentosa*. The morphological differences are minor. The former has thinner, narrower leaves, and smaller, narrower flowers 4–5 cm long, whereas *Y. filamentosa* has thick, rigid leaves and flowers 5–7 cm long. *Yucca freemanii* was described from plants growing in northeastern Texas and southwestern Arkansas. These plants were reported to be glabrous, but some plants that are otherwise attributable to *Y. freemanii* are mostly pubescent through the infloresecence, and overall they fall within the variation range of *Y. flaccida*. *Yucca flaccida* is cultivated in some of the northern states and Canada, where it sometimes has naturalized.

The results of DNA studies by K. H. Clary (1997) are incongruent with our treatment of *Yucca flaccida*, which warrants further research. It is the only dry-fruited species that groups with the fleshy-fruited ones in Clary's consensus tree, where *Y. flaccida* and *Y. filamentosa* are quite distant. DNA variation indicates that *Y. louisianensis* is genetically distinct and more closely related to *Y. filamentosa* than to *Y. flaccida*.

18. Yucca angustissima Engelmann ex Trelease, Rep. (Annual) Missouri Bot. Gard. 13: 58, plate 23, fig. 1, plate 24, fig.1, plate 83, fig. 6, plate 93, fig. 1. 1902 [E]

Plants solitary or forming small to large colonies of rosettes, acaulescent or rarely caulescent, to 3 m diam.; rosettes usually small. **Stems** procumbent, 0.1–0.4 m, or erect, 1–2 m. **Leaf blade** rigidly spreading, including distal leaves, linear, lanceolate, concavo-convex, or plano-keeled, widest near middle, 20–80(–150) × 0.4–2 cm, rigid or flexible, not glaucous, margins entire, becoming filiferous, white, becoming brownish, gray, or green, apex long-acuminate, spinose, spine acicular, short, 3–7 mm. **Inflorescences** racemose, occasionally paniculate proximally, arising well beyond rosettes, (4–)8–20 dm, glabrous or finely pubescent; branches, when present, 0.5–1 dm; bracts erect, linear, proximal 10–20 × 1–2 cm, distal 3–8 × 1–2 cm; peduncle scapelike, 0.3–2.5 m, 1–2 cm diam. **Flowers** pendent, 3–5.5(–6.5) cm; perianth campanulate to globose; tepals distinct, white to cream or greenish white, often tinged pink or brown, broad to narrowly elliptic to lanceolate-elliptic or orbiculate, 3–6.5 × 1.3–2.5 cm; filaments 0.7–2.8 cm; anthers 3.2–6 mm; pistil 1.5–3.5 cm; style white to pale green, 3–10(–13) mm; stigmas lobed. **Fruits** erect, capsular, dehiscent, moderately to deeply constricted, 3.5–7.5 × 2–3 cm, dehiscence septicidal. **Seeds** dull black, thin, 7–9 × 5–7 mm.

Varieties 4 (4 in the flora): North America.

Our treatment of *Yucca angustissima* reflects the concepts of J. L. Reveal (1977c). Each variety is well isolated geographically, but they overlap with one another morphologically. S. L. Welsh et al. (1993) treated the taxa in this complex at species level, with the exception of *Y. angustissima* var. *aria*, considered a high-altitude extreme of the typical variety. K. H. Clary (1997) presented DNA evidence that supports Welsh et al.'s treatment of this complex, in that *Y. angustissima*, *Y. kanabensis*, and *Y. toftiae* sort out distinctly from one another in her consensus tree. However, Welsh et al. indicated significant intergradation among these taxa, which makes their recognition at varietal rank seem most appropriate.

1. Capsules deeply constricted, 3.5–5.5(–5.7) cm; inflorescence and peduncle 0.8–1.5(–1.8) m.
   2. Flowers 4.5–5.5 cm; leaf blade 25–45 cm; style 10–13 mm . . . . 18a. *Yucca angustissima* var. *angustissima*
   2. Flowers 3.5–4.5(–5.2) cm; leaf blade 40–60 cm; style 7–10 mm . . . . 18b. *Yucca angustisima* var. *avia*
1. Capsules moderately constricted, 4.5–7.5 cm; inflorescence and peduncle 1–2.5(–4.5) m.
   3. Flowers 3–5.2 cm; leaf blade 20–75 cm; pistil 1.5–3.2 cm; style 3–10 mm; stem plus peduncle and inflorescence 2.3–4.5 m . . . . 18c. *Yucca angustissima* var. *toftiae*
   3. Flowers 5.5–6.5 cm; leaf blade 45–80(–150) cm; pistil 3–3.5 cm; style 5–8 mm; stem plus peduncle and inflorescence 2–3.5 m . . . . 18d. *Yucca angustissima* var. *kanabensis*

18a. Yucca angustissima Engelmann ex Trelease var. **angustissima** [E]

Plants acaulescent or caulescent. **Stems** procumbent, 0.1–0.4 m. **Leaf blade** narrow, long-tapering and primarily flat or concave distally, 25–45 × 0.5–0.8 cm, mostly flexible, smooth adaxially and abaxially. **Inflorescences** racemose, 8–10 dm, glabrous; peduncle 0.3–0.5 m. **Flowers** 4.5–5.5 cm; tepals pale green, often tinged rose or rosy purple; filaments 0.7–2.5 cm; anthers 3.2 mm; pistil 2.5–2.8 cm; style 10–13 mm. **Capsules** often deeply constricted near center, 3.5–5(–5.7) cm.

Flowering spring–early summer. Sandy places, sandstone outcrops, rocky hillsides of deserts; 900–2200 m; Ariz., N.Mex., Utah.

18b. Yucca angustissima Engelmann ex Trelease var. **avia** Reveal in A. Cronquist et al., Intermount. Fl. 6: 534. 1977 [E]

Plants acaulescent to short-caulescent. **Stems** erect, to 2 m. **Leaf blade** lanceolate, 40–60 cm, apex long-tapering. **Inflorescences** racemose, 8–18 dm; peduncle 0.3–0.5 m. **Flowers** 3.5–4.5(–5.2) cm; style 7–10 mm. **Capsules** deeply constricted, 3.5–5.5 cm.

Flowering late spring–early summer. Loamy, rocky soils; 2500 m; Utah.

J. L. Reveal (1977c) separated var. *avia* from var. *kanabensis*, and believed it to be more closely related to var. *angustissima*. S. L. Welsh et al. (1993) considered var. *avia* to be a northern, high elevation variant at the end of a cline within the species. It occurs in loamy, rocky soils instead of the sandy soils that the other varieties inhabit. Little is known about this variety.

18c. Yucca angustissima Engelmann ex Trelease var. **toftiae** (S. L. Welsh) Reveal in A. Cronquist et al., Intermount. Fl. 6: 534. 1977 [E]

Yucca toftiae S. L. Welsh, Great Basin Naturalist 34: 308, fig. 3. 1975

Plants acaulescent or short-caulescent. **Stems** erect, to 1 m, 2.3–4.5 m with peduncle and inflorescence. **Leaf blade** linear, 20–75 × 0.4–1.7 cm, rigid. **Inflorescences** mostly racemose, sometimes paniculate proximally, 4–20 dm; peduncle 1.2–2.5 m. **Flowers** 3–5.2 cm; tepals creamy white, elliptic to lanceolate-ovate; filaments 0.7–2.4 cm; pistil 1.5–3.2 cm; style 3–10 mm. **Capsules** moderately constricted, 4.5–6 cm.

Flowering spring–early summer. Sandy hillsides of desert mesas; 2000 m; Utah.

18d. Yucca angustissima Engelmann ex Trelease var. **kanabensis** (McKelvey) Reveal in A. Cronquist et al., Intermount. Fl. 6: 534. 1977 [E]

Yucca kanabensis McKelvey, Yuccas Southw. U.S. 2: 122, plates 48, 49. 1947

Plants forming small to extremely large colonies, caulescent or short-procumbent. **Stems** short-procumbent, 0.1–0.4 m, 2–3.5 m with peduncle and inflorescence. **Leaf blade** plano-convex or plano-keeled, 45–80(–150) × 1.2–2 cm, stiff but somewhat flexible, smooth abaxially, slightly roughened adaxially. **Inflorescences** racemose, 10–20 dm; peduncle 1–1.5 m. **Flowers** 5.5–6.5 × 2.3–4 cm; tepals often tinged pink or brown, elliptic to orbiculate; filaments 2.5–2.8 cm; anthers 5–6 mm; pistil oblong-cylindric, 3–3.5 × 0.7–0.9 cm; style 5–8 mm. **Capsules** moderately constricted, 6.5–7.5 cm.

Flowering spring–early summer. Sandy places of juniper - pinyon pine - oak woodlands, desert canyon hillsides; 1400–1900 m; Ariz., Utah.

19. Yucca baileyi Wooton & Standley, Contr. U.S. Natl. Herb. 6: 114. 1913 • Alpine yucca [E]

Yucca baileyi var. *navajoa* (J. M. Webber) J. M. Webber; *Y. navajoa* J. M. Webber; *Y. standleyi* McKelvey

Plants solitary or cespitose, forming colonies 1.3–2 m diam., acaulescent, erect or semierect, up to 0.2 m; rosettes usually small, symmetrical. **Leaf blade** yellowish green, plano-convex or plano-keeled, occasionally falcate, widest near middle, 25–45(–50) × 0.6–0.9 cm, rigid, smooth adaxially and abaxially, margins entire, recurved, filiferous, whitish, apex spinose, spine acicular, to 3.2 mm. **Inflorescences** racemose, arising within or just beyond rosettes, 2.5–4.5(–8.5) dm; bracts erect, purplish; peduncle scapelike, 0.1–0.2 m, less than 2.5 cm diam. **Flowers** pendent; perianth campanulate; tepals distinct, ovate to obovate or elliptic, 5–6.5 × 1.5–3.2 cm; filaments to 2 cm, finely pubescent; pistil green, 2.5–3.2 × 0.8 cm; style white, 7 mm; stigmas lobed. **Fruits** erect, capsular, dehiscent, oblong-cylindric, not usually constricted, 5 × 2.5 cm, dehiscence septicidal. **Seeds** dull black, thin, 6–10 mm.

Flowering spring. Mountains, adjacent woodlands and grasslands; 1300–2500 m; Ariz., Colo., N.Mex., Utah.

Some populations of *Yucca baileyi* comprise compact colonies of semierect, branched plants instead of scattered, cespitose individuals. These populations were recognized as var. *navajoa* by J. M. Webber (1953), who noted that *Y. baileyi* possibly hybridizes with *Y. glauca* and *Y. angustissima*.

20. Yucca harrimaniae Trelease, Rep. (Annual) Missouri Bot. Gard. 13: 59, plates 28, 29, plate 83, fig. 10, plate 93, fig. 1. 1902 [E]

Yucca gilbertiana (Trelease) Rydberg; *Y. harrimaniae* var. *gilbertiana* Trelease; *Y. harrimaniae* var. *sterilis* Neese & S. L. Welsh

Plants cespitose, forming dense to open colonies, acaulescent or short-caulescent; rosettes usually small, asymmetrical or symmetrical. **Stems** to 0.3 m. **Leaf blade** pale green, linear- or spatulate-lanceolate, concavo-convex, widest near middle, 30–50 × 1.8–4.3 cm, rigid, margins entire, filiferous, white or brown, apex pungent. **Inflorescences** racemose, rarely paniculate proximally, arising within or just beyond rosettes, 3.5–7 dm; branches, when present, few, short; bracts erect, proximal to 20 cm, distal 5–8 cm; peduncle scapelike, 0.1–0.2 m, less than 2.5 cm diam. **Flowers** pendent; perianth broadly campanulate; tepals distinct, yellow or greenish yellow, usually tinged purple, broadly lanceolate, 4–5(–5.3) × 1.6–3.4 cm; filaments 1.6–2.8 cm; pistil pale green, 2.7–3.8 × 0.4–0.8 cm; style pale to bright green, 9–13 mm; stigmas lobed. **Fruits** erect, capsular, dehiscent, cylindrical, usually deeply constricted toward center, 3–5(–5.5) × 2–3 cm, dehiscence septicidal. **Seeds** dull black, thin, 5–8 × (3.5–)4–8 mm.

Flowering spring–early summer. Desert slopes, foothills, and plateaus in limestone and volcanic outcrops; 1000–2500 m; Ariz., Colo., Nev., N.Mex., Utah.

S. D. McKelvey (1938–1947) included *Yucca neomexicana* in *Y. harrimaniae*, as did J. L. Reveal (1977c), who gave it varietal status. K. H. Clary's (1997) DNA studies contributed to a consensus tree in which *Y. harrimaniae* and *Y. neomexicana* are discrete and widely separated. These taxa are allopatric, separated by the San Juan Mountains of Colorado, and this along with Clary's evidence supports their recognition as separate species.

21. Yucca neomexicana Wooton & Standley, Contr. U.S. Natl. Herb. 16: 115. 1913 [E]

Yucca harrimaniae Trelease var. *neomexicana* (Wooton & Standley) Reveal

Plants cespitose, forming single or open colonies, each separated by 35–60 cm, acaulescent or rarely caulescent; rosettes usually small, mostly asymmetrical. **Stems,** when present, primarily rhizomatous, shorter than 1 m. **Leaf blade** spreading, including distal leaves, spatulate-lanceolate, concavo-convex, thin, widest near middle, 15–46 × 0.7–2 cm, flexible, rather glaucous, margins entire, filiferous, brown or straw-colored. **Inflorescences** racemose, rarely paniculate proximally, arising within or more often 0–20 cm beyond rosettes, 4–7 dm; branches, when present, few, short; bracts reflexed when mature, purplish, narrowly triangular, proximal to 10 cm, distal 2–3 cm, tapering to pungent apex; peduncle sometimes scapelike, 0.1–0.4 m. **Flowers** pendent; perianth campanulate; tepals distinct, white to somewhat greenish white, usually tinged pink or purple, broadly lanceolate, 3–4.7 × 1.5–3 cm; filaments 1.3–1.7 cm; pistil 2.4–3.2 cm × 0.5–0.8 cm; style pale green or rarely white, 9–13 mm; stigmas lobed. **Fruits** erect, capsular, dehiscent, cylindrical, deeply constricted near middle, 3–4.2 × 2–2.7 cm, dehiscence septicidal. **Seeds** dull black, thin, 6–8 × 5–8 mm.

Flowering spring–early summer. Exposed rocky ledges in woodlands and grasslands; 1500–2700 m; Colo., N.Mex., Okla.

Although J. L. Reveal (1977c) reduced *Yucca neomexicana* to a variety of *Y. harrimaniae*, the two taxa are morphologically and geographically distinct, and K. H. Clary's (1997) DNA evidence supports their recognition as separate species.

22. Yucca arkansana Trelease, Rep. (Annual) Missouri Bot. Gard. 13: 63. 1902 • Arkansas yucca [E]

Yucca angustifolia Pursh var. *mollis* Engelmann; *Y. arkansana* var. *paniculata* McKelvey; *Y. glauca* Nuttall var. *mollis* Engelmann ex Branner & Coville

Plants forming small colonies, acaulescent or caulescent; rosettes usually small. **Stems** decumbent, short, to 0.2 m. **Leaf blade** mostly yellowish green, flattened, grasslike, concavo-convex, widest near middle, 20–60(–70) × 0.7–2(–2.5) cm, flexible, margins entire, curled, filiferous, apex long, tapering to short spines 1.6–3.2 mm. **Inflorescences** racemose, occasionally paniculate proximally, arising within rosettes or at rosette level, 3–6(–8) dm, glabrous; bracts erect; peduncle scapelike, 0.2–0.5(–0.6) m, 0.3–0.7(–1.3) cm diam. **Flowers** pendent; perianth globose; tepals distinct, greenish white, elliptic to orbicular or oblong, 3.2–6.5 × 2–5 cm; filaments 1.3–2.5 cm; anthers 3.2 mm; pistil 2.5–2.8(–3.2) cm; style dark green, 7–13 mm, tumid; stigmas lobed. **Fruits** erect, capsular, dehiscent, oblong-cylindric to obovoid, constricted near middle, stout, 4–6.5(–7) × 2–3 cm, dehiscence septicidal. **Seeds** dull black, thin, ca. 1 cm diam.

Flowering spring. Gravelly soil, limestone outcrops, rocky hillsides, prairies; 100–400 m; Ark., Kans., Mo., Okla., Tex.

Yucca arkansana shows considerable variation, particularly in the eastern part of its range. S. D. McKelvey (1938–1947) described var. *paniculata* and suggested that it is an eastern extension of the species with a taller, paniculate, and pubescent inflorescence. *Yucca arkansana* approaches *Y. louisianensis*, which we have reduced to synonymy under *Y. flaccida*. K. H. Clary's (1997) DNA consensus tree places *Y. arkansana* and *Y. louisianensis* adjacent to one another.

23. Yucca intermedia McKelvey, Yuccas Southw. U.S. 2: 116, plates 44–47. 1947 [E]

Yucca baileyi Wooton & Standley var. *intermedia* (McKelvey) Reveal; *Y. intermedia* var. *ramosa* McKelvey

Plants cespitose, forming widely separated colonies, acaulescent or occasionally caulescent; rosettes 1–5 per colony, usually small. **Stems** erect, simple, shorter than 1 m. **Leaf blade** linear, plano-convex or plano-keeled, widest near middle, 33–65 × 0.5–0.8 cm, slightly flexible, margins entire, white to grayish. **Inflorescences** mostly green, racemose, sometimes paniculate proximally, occasionally paniculate and long-racemose distally, arising within rosettes, (5–)7–13 dm, glabrous; bracts erect; peduncle scapelike, 0.1–0.5 m, to 2.5 cm diam. **Flowers** pendent; perianth campanulate or rarely globose; tepals distinct, cream or greenish, often tinged rose or rosy brown, narrowly elliptic to broadly elliptic, 5.5–7 × 2–3.2 cm; filaments to 2.5 cm; anthers 3.2–4.8 mm; pistil 1.5–3.2 cm; style white or pale yellowish green, 7 mm; stigmas lobed. **Fruits** erect, capsular, dehiscent, oblong-cylindric, occasionally constricted at middle, 5–5.7 × 2–2.5 cm, dehiscence septicidal. **Seeds** glossy or dull black, thin, 6–10 mm.

Flowering spring. Juniper-pinyon pine woodlands to adjacent grasslands; 1100–2100 m; N.Mex.

J. M. Webber (1953) believed *Yucca intermedia* to be a hybrid between *Y. glauca* and *Y. angustissima* or *Y. baileyi*. S. D. McKelvey (1938–1947) recognized var. *ramosa* based primarily on paniculate inflorescences. Plants of this type are found in the southeastern area of the species range, where populations with only paniculate inflorescences occur, as well as populations with racemose and paniculate inflorescences.

24. Yucca glauca Nuttall, Cat. Pl. Upper Louisiana, no. 89. 1813 • Beargrass, Great Plains yucca, soapweed yucca [E] [W1]

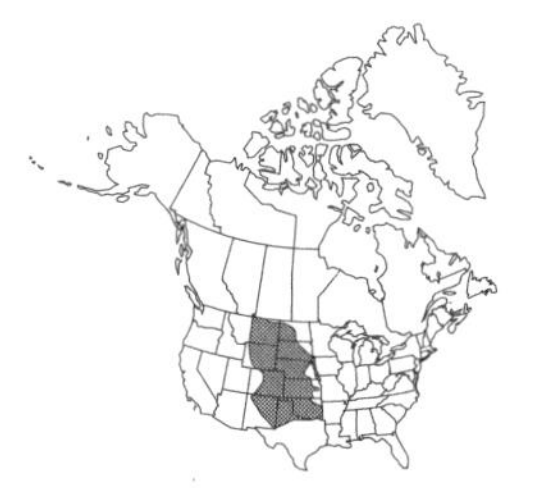

Yucca angustifolia Pursh; *Y. glauca* var. *gurneyi* McKelvey; *Y. glauca* var. *stricta* (J. Sims) Trelease; *Y. stricta* J. Sims

Plants forming small to moderate colonies, acaulescent or caulescent and arborescent, occasionally branched; rosettes 1–15 per colony, usually small. **Stems** erect, to 0.4 m. **Leaf blade** linear to linear-lanceolate, concave to concavo-convex, widest near middle, 40–60 × 0.8–1.2 cm, rigid, margins entire, filiferous, white, apex blunt to acicular. **Inflorescences** racemose, occasionally paniculate proximally, arising within or just beyond rosettes, 5–10 dm; bracts erect, leaflike, 2–5 cm, reduced toward apex; peduncle scapelike, 0.2–0.5 m, less than 2.5 cm diam. **Flowers** pendent; tepals distinct, greenish white to white, elliptic, 5–5.3 × 2.6–3.5 cm, apex acute; filaments white, 1.7–1.9 cm; anthers yellow, 4 mm; pistil green, obovoid, 3–3.7 × 1.7 cm; style dark green, 10 mm; stigmas lobed. **Fruits** erect, capsular, dehiscent, cylindric to obovoid, rarely constricted, 5–8(–9) × 3–4.5(–5) cm, dehiscence septicidal. **Seeds** black, slightly glossy, thin, 9–12 × 8–9 mm.

Flowering spring. Prairies and waste areas in sandy or limestone soils; 500–2600 m; Alta.; Colo., Iowa, Kans., Mo., Mont., Neb., N.Mex., N.Dak., Okla., S.Dak., Tex., Wyo.

Yucca glauca has the most extensive distribution of any North American *Yucca*. Its inflorescences are primarily racemose, but some plants exhibit branched inflorescences and varietal names have been given them. *Yucca glauca* and *Y. arkansana* are very similar. The leaves of *Y. glauca* are uniform in size, rigid, linear or linear-lanceolate, and up to 1.2 cm wide. In *Y. arkansana*, the young leaves immediately surrounding the peduncle are distinctly shorter than the outer leaves of the rosette, and mature leaves are generally somewhat flexible, lanceolate, and up to 2.5 cm wide. K. H. Clary's (1997) DNA evidence does not indicate as close a relationship as the morphological characters suggest. J. M. Webber (1953) believed that *Y. glauca* forms hybrids with *Y. baileyi*, *Y. elata*, *Y. constricta*, and *Y. angustissima*.

25. Yucca necopina Shinners, Spring Fl. Dallas-Fort Worth, 91, 408. 1958 [E]

Plants cespitose, forming small colonies of rosettes, acaulescent or caulescent; rosettes usually small, each with ca. 50–85 leaves. **Stems** erect, to 0.4 m. **Leaf blade** erect, proximal becoming reflexed, slightly twisted, plano-convex, widest near middle, 50–80 × (1.5–)2–4 cm, rigid, margins entire, filiferous, white, apex acicular. **Inflorescences** paniculate, racemose distally or entirely, beginning mostly beyond rosettes, ovoid, 50–120 cm, glabrous; proximal branches to 15 cm; bracts erect, distal reduced; peduncle 0.8–1.6 m, less than 2.5 cm diam. **Flowers** pendent; perianth globose; tepals greenish white, 4–4.5 × 1.5–3 cm; filaments 12–15 mm, shorter than pistil; pistil 1.5–3.8 cm; stigmas lobed. **Fruits** erect, capsular, dehiscent, not conspicuously constricted, dehiscence septicidal.

Flowering spring. River terraces, deep sand; 200–300 m; Tex.

Yucca necopina was originally described based only on plants from a single locality in Somervell County. Shinners' sketchy description compared it with *Y. arkansana*, and he suggested that it might be a hybrid between *Y. pallida* and *Y. arkansana*. More recently, G. M. Diggs et al. (1999) reported new locations of *Y. necopina* from Hood, Parker, and Tarrant counties, and noted that molecular evidence provided to them by K. H. Clary supports its separate recognition.

26. Yucca tenuistyla Trelease, Rep. (Annual) Missouri Bot. Gard. 13: 53, plate 17, fig. 2, plates 18, 19, plate 83, fig. 3, plate 92, fig. 1. 1902 [E]

Plants forming open colonies, acaulescent or short-caulescent; rosettes usually small. **Stems** erect, to 0.5 m. **Leaf blade** mostly recurving, lanceolate, plano-convex, widest near middle, 40–70 × 1–2 cm, rigid, margins entire, filiferous, whitish, apex scarcely pungent. **Inflorescences** paniculate, arising beyond rosettes, ovoid, to 10 dm, distance from leaf tips to proximal inflorescence branches more than twice leaf length when fully expanded, glabrous or slightly pubescent; bracts erect; peduncle scapelike, 1–1.7 m, less than 2.5 cm diam. **Flowers** pendent; tepals narrow, apex acute; filaments shorter than pistil; pistil 1.5–3.8 cm; ovary white; style white or green, oblong, often deeply lobed; stigmas lobed. **Fruits** erect, capsular, dehiscent, cylindrical, symmetrical, not constricted, stout, 5–6.5 × 2.5–3 cm, dehiscence septicidal. **Seeds** glossy black, thin, 8–10 × 7–8 mm.

Flowering spring. Brushlands to coast; 0–200 m; Tex.

Very little is known about *Yucca tenuistyla*, and further study is needed. Along with *Y. louisianensis*, it was treated by S. D. McKelvey (1938–1947) as a part of *Y. constricta*. The characteristics separating these three taxa are tenuous, and it is possible that *Y. tenuistyla* is just a variant of *Y. flaccida*, which includes *Y. louisianensis*.

27. Yucca constricta Buckley, Proc. Acad. Nat. Sci. Philadelphia 14: 8. 1863 [F]

Plants forming small to large, open colonies, acaulescent or occasionally caulescent and arborescent; rosettes usually small, asymmetrical. **Stems** procumbent, to 0.4 m. **Leaf blade** divergent, spreading, linear or somewhat wider near middle, mostly straight, flattened or plano-convex, 25–65 × (0.3–)0.7–1.5 cm, soft, rigid, margins entire, becoming filiferous, whitish or light green. **Inflorescences** paniculate, racemose distally, arising 2–4.6 dm beyond rosettes, ovoid, 4.5–13 dm, distance from leaf tips to proximal inflorescence branches more than twice leaf length when fully expanded, glabrous; raceme 20–25 cm; proximal branches 15–25 cm; bracts erect; peduncle scapelike, 1–2 m, less than 2.5 cm diam. **Flowers** pendent; perianth hemispheric; tepals distinct, pale greenish white, thin, 2.5–4.8 × 1.1–2.6 cm, apex acute; filaments 1.7–2.2 cm, pubescent; anthers 3.2 mm; pistil slender, cylindric, 2.5–3.8 × 0.5–0.6 cm; style whitish or pale green, 8–11 mm; stigmas lobed. **Fruits** erect, capsular, dehiscent, oblong-cylindric, deeply constricted near middle, (3.5–)4.6–6.3 × (1.5–)2.5–4.3 cm, dehiscence septicidal. **Seeds** glossy black, thin, 10–15 mm diam.

Flowering spring. Brushwood, grassland areas; 300–700 m; Tex.; ne Mexico.

S. D. McKelvey (1938–1947) treated *Yucca tenuistyla* and *Y. louisianensis* as synonyms of *Y. constricta*. The major differences among these entities involve vestiture in the inflorescences.

28. Yucca campestris McKelvey, Yuccas Southw. U.S. 2: 173, plates 62, 63. 1947 • Plains yucca [E]

Plants forming small or large, open colonies, acaulescent or occasionally caulescent and arborescent, rhizomatous; rosettes usually small. **Stems** 0.6–1 m. **Leaf blade** linear, plano-convex or plano-keeled, widest near middle, 40–65 × 0.3–0.7(–1.5) cm, rigid, margins entire, filiferous, white,

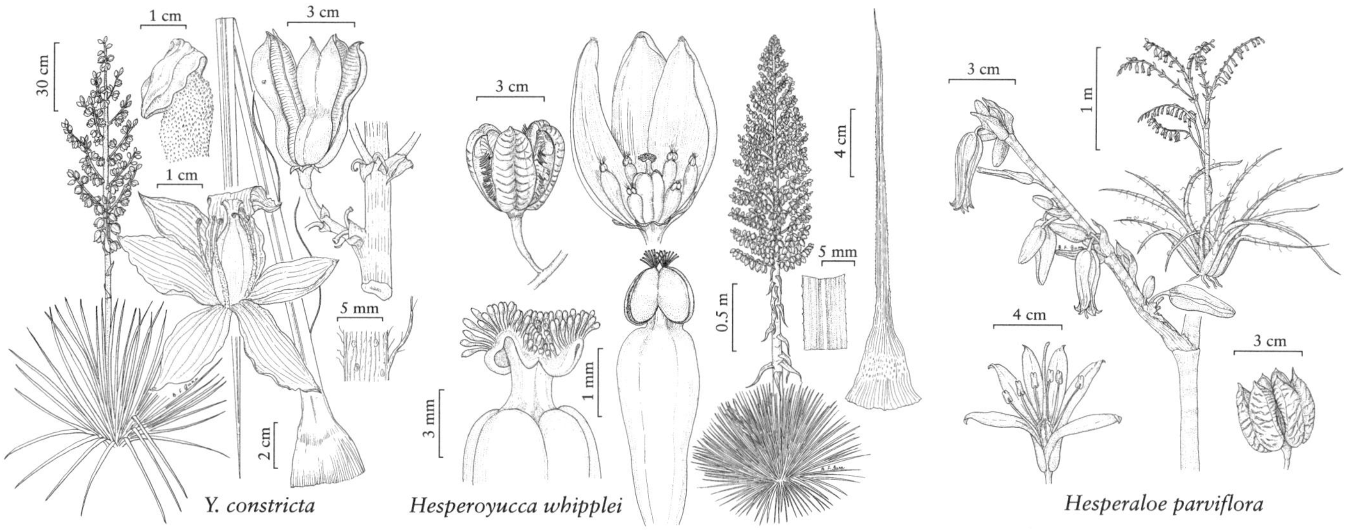

YUCCA ◦ HESPEROYUCCA ◦ HESPERALOE

apex spinose, spine acicular, 7 mm. **Inflorescences** paniculate, arising within or occasionally beyond rosettes, narrowly ellipsoid, 6–10 dm, distance from leaf tips to proximal inflorescence branches less than twice leaf length when fully expanded, glabrous; branches to 13 cm; bracts erect; peduncle scapelike, 0.5–1 m, less than 2.5 cm diam. **Flowers** pendent; perianth globose; tepals connate, dull green, sometimes tinged pink, 4.1–6.5 × 1.5–2.5 cm; filaments shorter than pistil, flaccid; anthers 3.2 mm; pistil ovoid to obovoid, 2.5–3 × 0.5–0.9 cm; style bright green; stigmas lobed. **Fruits** erect, capsular, dehiscent, symmetrical or rarely constricted, 4.5–5.5 (–6.3) × 3–5 cm, dehiscence septicidal. **Seeds** glossy black, thin, 11–14 × 8–11 mm.

Flowering spring. Deep sands; 800–900 m; Tex.

Yucca campestris is endemic to the plains region in the southern panhandle counties of western Texas. S. D. McKelvey (1938–1947) discussed its variation in relation to its distribution, and K. H. Clary's (1997) DNA studies support its recognition as a distinct species. J. M. Webber (1953) considered *Y. campestris* to be a hybrid between *Y. constricta* and *Y. elata*, and reported a distribution from west Texas into southern New Mexico, and possibly into northwestern New Mexico. Additional study within these regions may help resolve the relationships and origin of this species.

5. HESPEROYUCCA (Engelmann) Baker, Bull. Misc. Inform. Kew 1892: 8. 1892 • Our Lord's candle, Quixote plant

Karen H. Clary

Yucca Linnaeus [unranked] *Hesperoyucca* Engelmann in S. Watson, Botany (Fortieth Parallel), 497. 1871

Plants simple or cespitose, erect, acaulescent, scapose, rosulate, monocarpic or polycarpic; rosettes solitary or in colonies. **Leaf blades** linear, rarely narrowly lanceolate, widest at base, tapering to apex, glaucous, somewhat flexible when young, rigid at maturity, margins pale yellow, mostly denticulate, corneous, apex distinctly spinose. **Scape** well developed, extending beyond rosettes, usually more than 2.5 cm in diam., glabrous. **Inflorescences** paniculate, cylindrical, bracteate, glabrous; rachis and peduncle reddish purple; bracts reflexed, deltate, abruptly narrowing to linear, sharp-pointed apex. **Flowers** bisexual; perianth campanulate or globose; tepals distinct, white or creamy white to greenish or purple-tinged, broadly lanceolate, 3.2–4.5(–6) cm; filaments ca. 1.3 cm, equal to or longer than pistil, papillose; anthers reniform; pistil 1–1.3 × 0.6–1 cm; ovary superior; style white, 2–3 mm; stigmas green, capitate. **Fruits** erect, capsular, obovoid, 3–5 × 1.5–4 cm, dehiscence loculicidal. **Seeds** many per locule, dull black, thin, flattened, 6–8 mm diam. $x = 30$.

Species 3 (2 in the flora): w United States, nw Mexico.

DNA evidence (M. A. Hanson 1993; D. J. Bogler 1994; K. H. Clary 1997) supports W. Trelease's (1893) recognition of *Hesperoyucca* at the genus level and reinforces S. D. McKelvey's (1938–1947, vol. 2) statement that several characteristics of *Hesperoyucca whipplei* are not typical of *Yucca* and could justify its removal from *Yucca*.

J. M. Webber (1953) believed that the differences in leaf, inflorescence, and capsule that McKelvey used to separate *Hesperoyucca newberryi* from *H. whipplei* are weak and within the normal ranges of variation in the latter species. F. Hochstätter (2000) treated *H. newberryi* as a subspecies of *H. whipplei*. DNA evidence (K. H. Clary 1997) supports specific status for *H. newberryi*, *H. whipplei*, and a third species from Mexico, *H. peninsularis* (McKelvey) Clary. Conclusive DNA analyses have not been performed on the proposed varieties and subspecies of *H. whipplei* (K. H. Clary 2001), nor has there been any recent systematic morphological work that would support those earlier proposals. For these reasons, only the species of *Hesperoyucca* are described. This treatment follows S. D. McKelvey's and K. H. Clary's interpretations.

SELECTED REFERENCES Clary, K. H. 1997. Phylogeny, Character Evolution, and Biogeography of *Yucca* L. (Agavaceae) As Inferred from Plant Morphology and Sequences of the Internal Transcribed Spacer (ITS) Region of the Nuclear Ribosomal DNA. Ph.D. dissertation. University of Texas. McKelvey, S. D. 1938–1947. Yuccas of the Southwestern United States. Jamaica Plain. Trelease, W. 1893. Further studies of *Yucca* and their pollination. Rep. (Annual) Missouri Bot. Gard. 4: 181–225. Webber, J. M. 1953. Yuccas of the Southwest. Washington. [U.S.D.A. Agric. Monogr. 17.]

1. Mature capsules with conspicuous placental wings; plants monocarpic and rosettes single, or plants polycarpic and rosettes cespitose; sw California .... 1. *Hesperoyucca whipplei*
1. Mature capsules with slight or inconspicuous placental wings; plants monocarpic, rosettes single; Arizona .... 2. *Hesperoyucca newberryi*

1. Hesperoyucca whipplei (Torrey) Trelease, Rep. (Annual) Missouri Bot. Gard 4: 208. 1893 [F]

Yucca whipplei Torrey in J. C. Ives, Rep. Colorado R. 4: 29. 1861; *Hesperoyucca whipplei* var. *graminifolia* (Alph. Wood) Trelease; *Yucca californica* Groenland; *Y. graminifolia* Alph. Wood; *Y. nitida* C. Wright ex S. Watson; *Y. ortgensiana* Roezl ex Ortgies; *Y. whipplei* subsp. *caespitosa* (M. E. Jones) A. L. Haines; *Y. whipplei* var. *caespitosa* M. E. Jones; *Y. whipplei* subsp. *intermedia* A. L. Haines; *Y. whipplei* var. *intermedia* (A. L. Haines) J. M. Webber; *Y. whipplei* subsp. *parishii* (M. E. Jones) A. L. Haines; *Y. whipplei* var. *parishii* M. E. Jones; *Y. whipplei* subsp. *percursa* A. L. Haines; *Y. whipplei* var. *percursa* (A. L. Haines) J. M. Webber

Plants monocarpic or polycarpic, sometimes rhizomatous, some with secondary rosettes at base or stems branching to form new rosettes after flowering; rosettes single or cespitose, in small to very large, compact or open communities, or occasionally solitary. **Leaves** 20–90(–125) × 0.7–2 cm wide at base in narrow forms, 2–2.5 cm in broader forms. **Scape** 0.9–3 m × 2.5–15 cm diam. near base. **Inflorescences** 1 when rosette solitary, 1–many per rosette when rosettes in colonies, 2.5–28 × 1.5–7 dm at widest point. **Capsules** with conspicuous placental wings at maturity.

Flowering spring. Desert, chaparral, desert woodland coastal sage; 300–2500 m; Calif.; Mexico (Baja California).

Within *Hesperoyucca whipplei*, W. Trelease (1893) recognized two varieties, A. L. Haines (1941) and P. A. Munz (1959) each recognized five subspecies, and J. M. Webber (1953) recognized four varieties based on growth form. Others (S. D. McKelvey 1938–1947; K. K. McKinney and J. C. Hickman 1993) have argued that growth form in this species is highly variable, and have questioned the desirability of recognizing any infraspecific taxa at all in *H. whipplei*.

SELECTED REFERENCE Haines, A. L. 1941. Variation in *Yucca whipplei*. Madroño 6: 33–45.

2. **Hesperoyucca newberryi** (McKelvey) Clary, Sida 19: 845. 2001 [E]

Yucca newberryi McKelvey, Yuccas Southw. U.S. 2: 49. 1947; *Y. whipplei* Torrey subsp. *newberryi* (McKelvey) Hochstätter

Plants monocarpic; rosettes single. **Leaves** 50–60 × 0.7–2 cm wide at base in narrow forms, 2–2.5 cm in broader forms. **Scape** 1.4–1.6 m × 10–14 cm diam. **Inflorescences** 14–16 dm, 4–5.5 dm at widest point. **Capsules** with slight or inconspicuous placental wings at maturity.

Flowering spring. Desert, rocky granite, western slopes of Colorado River Canyon; 400–1300 m; Ariz.

Hesperoyucca newberryi is found in Mohave County, in the vicinity of the Colorado River.

6. HESPERALOE Engelmann in S. Watson, Botany (Fortieth Parallel), 497. 1871 • [Greek *hesperos*, western, and *aloe*, a kind of plant]

R. Laurie Robbins

Plants perennial, cespitose, acaulescent, semisucculent, primarily short- to long-rhizomatous. **Leaves** in basal rosettes; blade linear, thick and striate-ridged abaxially, margins with thread-like, detaching filaments, apex frayed or with hard spine. **Inflorescences** paniculate [racemose], loose, 3–8-branched. **Flowers** bisexual; perianth narrowly tubular to broadly campanulate; tepals 6; stamens 6, inserted on receptacle or at tepal bases, included; filaments glabrous; anthers sagittate, with septal nectaries; receptacle fleshy; ovary superior; style slender, barely exceeding tepals. **Fruits** capsular, ovoid, dehiscence septicidal. **Seeds** many, black, flattened. x = 30.

Species 5 (2 in the flora): arid regions of Texas and Mexico (Coahuila, Nuevo León, San Luis Potosí, Sonora).

Hesperaloe parviflora, and to a much lesser extent *H. funifera*, are cultivated in the semiarid and arid Southwest. The original distribution of *H. parviflora* is likely much obscured because of a long history of roadside and ornamental plantings throughout the area.

SELECTED REFERENCES Engard, R. G. 1980. *Hesperaloe*: Potential for arid lands. In: Centro de Investigación en Química Aplicada. 1970. *Yucca*. Saltillo. Pp. 115–123. Starr. G. 1995. *Hesperaloe*: Aloes of the West. Desert Pl. 11(4): 3–8. Starr, G. 1997. A revision of the genus *Hesperaloe* (Agavaceae). Madroño 44: 282–296.

1. Flowers green, tinged with purple, broadly campanulate; leaf blade yellowish green, 1–2 m × 2–5 cm . . . . 1. *Hesperaloe funifera*
1. Flowers rosy red to salmon, narrowly tubular; leaf blade dark green, 1–1.25 m × 2.5 cm . . . . 2. *Hesperaloe parviflora*

1. **Hesperaloe funifera** (K. Koch) Trelease, Rep. (Annual) Missouri Bot. Gard. 13: 36. 1902

Yucca funifera K. Koch, Belgique Hort. 12: 132. 1862; *Agave funifera* Lemaire; *Hesperaloe davyi* Baker f.

Leaf blades yellowish green, slightly constricted near base, widest near middle, 10–20 × 0.2–0.5 dm, margins with coarse threads. **Panicles** 2–2.5 m. **Flowers:** perianth broadly campanulate; tepals green, tinged with purple, ca. 2.5 cm. **Capsules** 8–9 × 5–7 mm. **Seeds** 8–9 × 5–7 mm.

Flowering spring. Rocky slopes and open plains in arid and semiarid regions; 500–1000(–2100) m; Tex.; Mexico (Coahuila, Nuevo León, San Luis Potosí).

Hesperaloe funifera is occasionally cultivated, although far less frequently than *H. parviflora*.

2. Hesperaloe parviflora (Torrey) J. M. Coulter, Contr. U.S. Natl. Herb. 2: 436. 1894 • Red yucca, red hesperaloe [F]

Yucca parviflora J. Torrey in W. H. Emory, Rep. U.S. Mex. Bound. 2(1): 221. 1859; *Aloe yuccaefolia* A. Gray; *Hesperaloe engelmannii* Krauskopf; *H. parviflora* var. *engelmannii* (Krauskopf) Trelease; *H. yuccaefolia* (A. Gray) Engelmann

Leaf blades dark green, barely constricted along length, 5–12.5 × 0.1–0.25 dm, margins with slender threads. **Panicles** 1–2.5 m. **Flowers:** perianth narrowly tubular; tepals rosy red to salmon, 2.5–3.5 cm. **Capsules** 2.5–3 cm. **Seeds** 9–10 × 6–7 mm.

Flowering late winter–early summer. Prairies, rocky slopes, and mesquite thickets; to 600 m; Tex.; Mexico (Coahuila).

Hesperaloe parviflora is widely cultivated in arid and semiarid regions, both as roadside plantings and as landscaping elements.

7. AGAVE Linnaeus, Sp. Pl. 1: 323. 1753; Gen. Pl. ed. 5, 150. 1754 • Century plant [Greek *agave*, noble or admirable]

James L. Reveal

Wendy C. Hodgson

Plants short-stemmed pachycauls, perennial, often flowering after 8–20+ years, monocarpic or polycarpic, acaulescent or caulescent, scapose, forming succulent rosettes on thick, fibrous-rooted crowns, often rhizomatous. **Stems** aboveground, unbranched or, less often, branched. **Leaves** evergreen in rosette; blade light green to green and occasionally with lighter patterns of white ("cross-zoned") or imprinted with white ("bud-prints"), linear-lanceolate to ovate, firm to rigid, often thick and fleshy, margins entire, filiferous, or armed with marginal teeth and short to long, sharp-pointed apical spine. **Scapes,** with inflorescences, much exceeding foliage. **Inflorescences** terminal atop a semiwoody stalk, spicate, racemose, or paniculate, open to dense, bracteate, occasionally bulbiferous, with flowers borne singly, in pairs, or in umbellike clusters of 2–40+ on peduncles or the lateral branches borne by the peduncle. **Flowers** protandrous, erect or recurved, showy; perianth mostly yellow, infrequently whitish or reddish, funnelform to tubular; tepals 6, connate basally into tube atop a typically constricted neck; limb lobes erect or curved, equal to unequal in length and/or width, linear to oblong or deltate, often papillate at recurved or hooded apex; stamens 6, exserted, attached atop or within perianth tube; filaments mostly filiform; anthers versatile, linear; ovary inferior, greenish at anthesis, 3-locular, succulent, thick-walled, ovules numerous; style subulate; stigma 3-lobed, glandular, capitate, papillate. **Fruits** capsular, oblong to ovoid, mostly thick-walled and fleshy, dehiscence loculicidal. **Seeds** black, flattened, obovoid, becoming globose distally. $x = 30$ (5 large, 25 small).

Species ca. 200 (27 in the flora): warmer regions in the Americas from the sw United States and Florida s to the Caribbean Islands and into Mexico, Central America, and tropical South America; widely cultivated.

Specimens of *Agave* are planted occasionally for their horticultural value in warmer regions. Native Americans have used most of our native species for food and fiber. Mescal and tequila are distilled liquors derived from the meristem and leaf base of agave. Tequila comes mainly from *A. tequilana* F. A. C. Weber, whereas *A. angustifolia* Haworth is the usual source for "commerical" mescal. Some variants of *A. angustifolia* are also important sources of fiber. Commercially, most agave fiber comes from *A. sisalana* in what is now a worldwide industry. *Agave fourcroydes* Lemaire is also used but more locally in Latin America including Mexico. Both species are sterile hybrids of unknown origin.

Native Americans who used agave plants for food and fiber developed numerous cultivars, including many that may be the results of hybridization or mutation events resulting in fully or partially sterile clones. Some have been sufficiently studied cytologically and morphologically so that the origin of the hybrid species can be ascertained. Naturally occurring hybrids also are known, and some sterile recombinants are so distinctive that they have been formally named (e.g., *Agave ×ajoensis*, *A. ×arizonica*, *A. ×glomeruliflora*, and probably *A. schottii* var. *treleasei*). There remain some species that may be of hybrid origin, or at least now are sterile or partially sterile, but little is known about their origin. Essentially, all the species are known today from pre-Columbian agricultural sites, and most appear to have had a Mexican origin (e.g., the many expressions of *A. americana* subsp. *americana*, *A. decipiens*, *A. desmettiana*, *A. neglecta*, *A. delamateri, A. phillipsiana*, *A. sisalana,* and *A. weberi*). Some rare species associated with pre-Columbian sites are fertile (e.g., *A. murpheyi*); whether they were locally selected or brought from elsewhere is unknown. Some hybrid expressions, such as *A. gracilipes*, seem to be well established. It is a matter of taxonomic judgment whether the species status for *A. gracilipes* should be altered to hybrid status; we have not proposed that now because the cytological work to confirm the morphological hypothesis of a hybrid origin has not been done.

Indigenous people long have used agaves. Those containing high amounts of sapogenins are bitter and generally not used for food. Agave plants used for food are harvested when they show signs of flowering. Rosettes are collected and the leaves cut away, leaving the "heart" or "head," which is pit-baked for many hours. It has been known for a long time that agaves have been cultivated in the warmer regions of North America, but it was only recently discovered that cultivation occurred north of the Mexican border. Such agaves are typically found near agricultural features and possess morphological characteristics that reflect desirable agronomic attributes. People no doubt traded agave products and live plants, thereby expanding the distribution of many expressions of agave. Agaves whose distributions directly reflect their movement by man include *Agave murpheyi, A. delamateri, A. phillipsiana, A. applanata, A. karatto* Miller, and *A. decipiens.* It is also suggested that *A. americana* subsp. *americana* var. *expansa, A. desmettiana, A. sisalana, A. neglecta,* and *A. weberi* may have been ancient cultivars originating in Latin America or the Caribbean, although at present there is little information to support this hypothesis.

Identification of *Agave* species is often difficult, especially with nonflowering material. See H. S. Gentry (1982, p. 47) for information on how to prepare specimens properly.

SELECTED REFERENCES Berger, A. 1915. Die Agaven. Jena. Breitung, A. J. 1968. The Agaves.... Edited by Charles Glass and Robert A. Foster. Reseda. Gentry, H. S. 1972. The Agave Family in Sonora. Washington. [Agric. Handb. 399]. Gentry, H. S. 1978. The agaves of Baja California. Occas. Pap. Calif. Acad. Sci. 130. Gentry, H. S. 1982. Agaves of Continental North America. Tucson. Hodgson, W. C. 1999. Vascular plants of Arizona: Agavaceae. J. Arizona-Nevada Acad. Sci. 32: 1–21. Hodgson, W. C. 2001. Food Plants of the Sonoran Desert. Tucson.

1. Inflorescences spicate, subspicate, or sometimes narrowly racemose-paniculate, rarely racemose; flowers solitary, in pairs, or in clusters, borne on peduncles or lateral branches 10 cm or shorter (subg. *Littaea* (Tagliabue) Baker).
    2. Leaf margins entire or filiferous, sometimes with inconspicuous teeth near leaf base; limb lobes no more than twice length of perianth tube.
        3. Rosettes shorter than 1.5 dm; leaves 6–10 cm, marginal fibers tightly curled, apex abruptly acuminate; flowers 1.4–1.7 cm; limb lobes 2–3.5 mm; sc Arizona . . . . . 1. *Agave parviflora*

3. Rosettes 1–6 dm high (rarely shorter in *A. toumeyana*); leaves (3–)8.5–50 cm, marginal fibers curled, straight, or absent, apex acute to long-acuminate; flowers 1.6–5.3 cm; limb lobes 6–20 mm; Arizona, New Mexico.
4. Flowers 1.6–2.8 cm; perianth tube 3–4.5 mm, limb lobes 6.5–10 mm; marginal leaf fibers conspicuous, curled, infrequently sparse or absent; c Arizona . . . . 2. *Agave toumeyana*
4. Flowers (2.9–)3–5.3 cm; perianth tube (4–)6–21 mm, limb lobes (7–)10–20 mm; marginal leaf fibers erect, spreading, or absent.
5. Leaves 0.7–2.5 cm wide, widest near base, blade yellowish green or deep green, usually with conspicuous bud-prints, linear, margins filiferous or not; se Arizona, sw New Mexico . . . . 3. *Agave schottii*
5. Leaves 1.5–3 cm wide, widest near middle, blade light green, without bud-prints, linear-lanceolate, margins nonfiliferous; sw Arizona . . . . 4. *Agave* ×*ajoensis*
2. Leaf margins conspicuously armed, nonfiliferous; limb lobes more than twice length of perianth tube.
6. Limb lobes 5–6.5 times longer than perianth tube.
7. Leaf margins undulate to crenate, not easily detached, armed, teeth mostly 1–2 cm apart, apical spine 1–2 cm; perianth greenish white to yellowish green; s Texas . . . . 5. *Agave univittata*
7. Leaf margins straight, easily detached, armed, teeth mostly (1–)2–4 cm apart, apical spine 1.5–4.5 cm; perianth yellow, frequently tinged with red or purple; sc New Mexico, w Texas . . . . 6. *Agave lechuguilla*
6. Limb lobes 2–4 times longer than perianth tube.
8. Leaves (5–)7–10(–12.5) cm wide; flowers 4–4.7 cm; w Texas . . . . 7. *Agave* ×*glomeruliflora*
8. Leaves 1–5.5 cm wide; flowers shorter than 4.3 cm.
9. Leaf margins horny, horns extending from apical spine to below mid-blade; limb lobes erect to ascending, not connivent; flowers cylindric-urceolate, in clusters of 10–20; filaments light yellow, inserted mid perianth tube; ovary 0.8–1.5 cm; c Arizona . . . . 8. *Agave* ×*arizonica*
9. Leaf margins horny, horns extending from apical spine to distalmost marginal teeth; limb lobes erect, connivent; flowers urceolate, in clusters of 2–12; filaments white, inserted near base of perianth tube; ovary 1.2–2.9 cm; se California, s Nevada, nw Arizona, sw Utah . . . . 9. *Agave utahensis*
1. Inflorescences paniculate; flowers in clusters, borne on peduncles terminating in lateral branches longer than 10 cm (subg. *Agave*).
10. Leaf blade linear-lanceolate, 10 times longer than wide; limb lobes drying reflexed on perianth tubes; capsules ellipsoid to oblong; s Florida . . . . 10. *Agave decipiens*
10. Leaf blade linear-lanceolate, oblanceolate, lanceolate to ovate, or spatulate, less than 10 times longer than wide; limb lobes not drying reflexed on perianth tubes; capsules variable; Florida, Texas, sw United States.
11. Rosettes 10 dm or longer (sometimes only 7 dm in *A. asperrima*); leaves (80–)120–250 cm (exc. 60–110 cm in *A. asperrima*).
12. Leaves mostly reflexed, 60–110 cm, scabrous; marginal teeth 8–15 mm or more, usually deflexed along lower leaf edge; flowers 6–8 cm, peduncles 15–18; s Texas . . . . 11. *Agave asperrima*
12. Leaves occasionally reflexed, (80–)100–200 cm, smooth, marginal teeth 5–10 mm, nearly straight or undulate to crenate; flowers 7–10 cm, peduncles 15–35; s Texas, sc Arizona . . . . 12. *Agave americana*
11. Rosettes usually shorter than 16 dm (up to 20 dm in *A. sisalana* and *A. shawii*); leaves typically shorter than 160 cm.
13. Leaf margins unarmed or with teeth reduced to prickles 1–2 mm.
14. Scape 2.5–3 m; leaves 50–90 cm; panicles compact; flowers 4–6 cm; s Florida . . . . 13. *Agave desmettiana*
14. Scape 5–10(–13) m; leaves 80–160 cm; panicles open; flowers 5.5–8 cm; wc Florida, s Texas.

15. Mature leaves 9–12 cm wide, margins typically unarmed, apical spine dark brown, 2–2.5 cm; flowers 5.5–6.5 cm; s Florida . . . . 14. *Agave sisalana*
15. Mature leaves 12–25 cm wide, margins typically armed with minute teeth, apical spine dark brown or brown to grayish, 2.5–4.5 cm; flowers 5.5–8 cm; wc Florida, s Texas.
    16. Scape 8–10(–13) m; flowers 5.5–6 cm; apical leaf spine 2.5 cm; wc Florida . . . . 15. *Agave neglecta*
    16. Scape 6.5–8 m; flowers 7–8 cm; apical leaf spine 3–4.5 cm; s Texas . . . . 16. *Agave weberi*

13. Leaf margins armed with well-defined teeth longer than 2 mm.
    17. Limb lobes 17–40 mm; new plants budding from leaf axils, forming fragmented supine clones or sometimes single rosettes; mature rosettes 0.8–20 dm; scape 2–4 m; sw California . . . . 17. *Agave shawii*
    17. Limb lobes (6–)9–27 mm; plants forming single rosettes, or new plants budding from rhizomes and forming clones; mature rosettes up to 15 dm; scape (1.7–)2–7(–7.2) m; sw United States.
        18. Leaf blade spatulate, light glaucous-green to yellow green, apical spine 2 cm or shorter; inflorescences prolifically bulbiferous; c Arizona . . . . 18. *Agave murpheyi*
        18. Leaf blade linear-lanceolate, lanceolate, oblanceolate, or obovate, variously colored, apical spine usually longer than 2 cm; inflorescences not bulbiferous, or producing bulbils only when damaged; Arizona, California, New Mexico, Texas.
            19. Leaves of rosettes somewhat open to dense; leaf blade linear to lanceolate to broadly ovate; inflorescences dense, usually with 20–40 lateral branches (fewer than 20 in *A. havardiana* and sometimes *A. parryi*).
                20. Perianth tube 14–22 mm, ca. equal to or slightly longer than limb lobes (18–24 mm); filaments inserted ca. mid perianth tube; leaves 30–60(–70) × 15–27 cm, marginal teeth (5–)7–10 mm, apical spine 3–5(–10) mm; flowers 6.8–9 cm; w Texas . . . . 21. *Agave havardiana*
                20. Perianth tube 4–18 mm, 1/2 as long or less than limb lobes (13–27 mm); filaments inserted above mid perianth tube to just below rim; leaves 7–65 × 4.5–20 cm, marginal teeth 2–8 mm, apical spine 1.5–5 cm; flowers 4–8.1 cm; Arizona, New Mexico, Texas.
                    21. Perianth tube ca. 1/2 (6–18 mm) as long or less than limb lobes (13–27 mm); filaments inserted above mid-perianth tube to just below rim; leaves 7–65 × 4.5–20 cm, apical spine 1.5–4 cm; inflorescences broadly paniculate; flowers 4.3–8.1 cm; plants blooming spring to summer; c and se Arizona, sw New Mexico . . . . 19. *Agave parryi*
                    21. Perianth tube much shorter (4–7 mm) than limb lobes (14–18 mm); filaments inserted just below rim of perianth tube; leaves 18–30 × 4.5–7 cm, apical spine 2.5–5 cm; inflorescences narrowly to broadly paniculate; flowers 4–5.5 (–5.8) cm; plants blooming summer to early fall; se New Mexico, w Texas . . . . 20. *Agave gracilipes*
            19. Leaves of rosettes usually open; leaf blade linear-lanceolate to lanceolate or oblanceolate; inflorescences open, usually with 6–26 (–32) lateral branches.
                22. Perianth tube shallow, 2–10 mm, much shorter than limb lobes (12–20 mm); filaments inserted at base of perianth tube or near rim; limb lobes ascending or spreading, occasionally erect, equal or subequal, wilting soon after anthesis; flowers 3–6 cm.

23. Leaves (20–)25–70 × 4.5–10 cm, widest at base, marginal teeth weakly attached; limb lobes 13–20 mm; s California, sw Arizona . . . . 22. *Agave deserti*
23. Leaves 17.5–40 × 2.8–5 cm, broadest at middle, marginal teeth firmly attached; limb lobes 11–13 mm; wc Arizona . . . . 23. *Agave mckelveyana*

22. Perianth tube not shallow, 6–20 mm, frequently equaling or exceeding limb lobes (6–22 mm); filaments inserted near perianth tube base to ca. mid tube (inserted at 2 levels in *A. palmeri*, mid-tube and near rim, inserted near rim in *A. phillipsiana*); limb lobes erect to erect-ascending, unequal, persistent and often leathery during and after anthesis; flowers 3.5–8.6 cm.
24. Perianth uniformly golden yellow; leaf margins armed with teeth 4–10 mm, (0.8–)1–3(–4) cm apart, interstitial teeth (0–)1–3(–5) on distal $^2/_3$ of margins; c, se Arizona . . . . 24. *Agave chrysantha*
24. Perianth greenish cream to cream to pale yellow or light green, apex of limb lobes and filaments flushed with maroon; leaf margins armed with teeth 3–7 mm, 0.2–2.5 cm apart, interstitial teeth (2–)3–12 on distal $^2/_3$ of margins; Arizona, sw New Mexico.
25. Leaves erect or erect-ascending, blade glaucous-gray to -bluish, margins armed, interstitial teeth (3–)6–12 on distal $^2/_3$, apex conspicuously incurved; filaments inserted at 1 level ca. mid perianth tube; c Arizona . . . . 25. *Agave delamateri*
25. Leaves ascending to spreading, blade variously colored, margins armed, interstitial teeth (2–)3–7 on distal $^2/_3$ of margins, apex not conspicuously incurved; filaments inserted at 2 levels or subequally on perianth tube; Arizona, New Mexico.
26. Flowers 4.5–7.5 cm; limb lobes (6–)9–18 mm; filaments inserted at 2 levels on perianth tube; scape (1.7–)4–6.5(–7.2) m; leaf margins armed with firmly attached teeth (0.5–)1–2 cm apart; rosettes solitary; plants rarely suckering; se Arizona, sw New Mexico . . . . 26. *Agave palmeri*
26. Flowers 7.4–8.6 cm; limb lobes 15–22 mm; filaments inserted subequally below rim of tube; scape 2.7–5.5 m; leaf margins armed with brittle teeth, 1–2.5 cm apart; rosettes solitary or cespitose; plants freely suckering; nc Arizona . . . . 27. *Agave phillipsiana*

1. Agave parviflora Torrey in W. H. Emory, Rep. U.S. Mex. Bound. 2(1): 214. 1859

Subspecies 2 (1 in the flora): sw United States, nw Mexico.

1a. Agave parviflora Torrey subsp. **parviflora** • Small-flowered agave

Plants acaulescent, freely suckering; rosettes solitary or cespitose, 1–1.5 × 1.5–2 dm. **Leaves** ascending to erect, 6–10 × 0.8–1 cm; blade dark green with white bud-prints on both surfaces, oblong-linear, firm, adaxially plane toward apex, abaxially convex toward base; margins straight, denticulate near base, filiferous, fibers white, tightly curled; apex abruptly acuminate, spine brown to grayish white, subulate, 0.5–0.8 cm. **Scape** 1–1.8 m. **Inflorescences** spicate, laxly flowered on distal 1/2; bracts caducous, narrowly triangular, 1–3 cm; peduncle shorter than 2 mm. **Flowers** 2–4 per cluster, strongly recurved, 1.4–1.7 cm; perianth cream to pale yellow, tube urceolate, 5–7 × 4.5–5.5 mm, limb lobes erect to incurved, subequal, 2–3.5 mm; stamens slightly exserted; filaments inserted 1 mm above base of perianth tube, erect, cream, 0.9–1.4 cm; anthers yellow, 5–8 mm; ovary 0.6–0.9 cm, neck slightly constricted, 2 mm. **Capsules** sessile to short-pedicellate, globose, 0.6–1 cm, apex short-beaked. **Seeds** 2.5–3 mm. 2*n* = 60.

Flowering late spring–late summer. Open rocky slopes, mostly in desert grasslands and oak woodlands; 900–1500 m; Ariz.; nw Mexico.

2. Agave toumeyana Trelease, Contr. U.S. Natl. Herb. 23: 140. 1920 [E] [F]

Plants acaulescent, freely suckering; rosettes densely cespitose, (0.3–)1–5 × 1.5–6 dm. **Leaves** ascending to erect, (3–)9–46 × 0.5–2 cm; blade light green to yellowish green or dark green with white bud-prints on both surfaces, linear to linear-lanceolate, firm, adaxially plane toward apex, abaxially convex toward base; margins straight, infrequently denticulate, filiferous, fibers conspicuous, white, curled, infrequently sparse or absent, teeth sparse or absent, in proximal 1/2 of blade or near base; apex acute to long-acuminate, spine brown to grayish, subulate to filiform, 0.6–2 cm. **Scape** 1.5–3 m. **Inflorescences** spicate, laxly to rather densely flowered on distal 1/3; bracts caducous, linear-triangular, 1–3.5 cm; peduncle 5 mm or shortcr. **Flowcrs** 1 or 2, slightly rccurvcd, 1.6–2.8 cm; perianth greenish white, tube campanulate, 3–4.5 × 6–10 mm, limb lobes erect to incurved, subequal, 6.5–10 mm; stamens slightly exserted; filaments inserted 1–3.5 mm above base of perianth tube, erect, white, 1.1–1.7 cm; anthers pale yellow, 7–16 mm; ovary 0.8–1.6 cm, neck slightly constricted, 3–7 mm. **Capsules** sessile, oblong, 0.9–1.5 cm, apex short-beaked. **Seeds** 2.5–3 mm.

Varieties 2 (2 in the flora): Arizona.

Both varieties hybridize with *Agave chrysantha*.

1. Leaves 40–70 per mature rosette, 19–46 × 1.2–2 cm, margins dentate, or denticulate only near base . . . . 2a. *Agave toumeyana* var. *toumeyana*
1. Leaves 100–200 per mature rosette, (3–)9–20(–27) × 0.5–1.2 cm, margins denticulate below mid leaf . . . . 2b. *Agave toumeyana* var. *bella*

2a. Agavc toumcyana Trelease var. **toumcyana**
Toumey agave [E]

Rosettes each with 40–70 leaves at maturity, light green to yellowish, round-topped. **Leaves** 19–46 × 1.2–2 cm; margins edentate, or denticulate only near and at base. **Flowers** 2.1–2.8 cm; limb lobes 8–10 mm; filaments 1.2–1.7 cm; ovary 1–1.6 cm. 2*n* = 120.

Flowering late spring–summer. Open gravelly to rocky limestone or basalt slopes, mostly with desert scrub, chaparral, and pinyon-juniper woodlands; 600–1500 m; Ariz.

2b. Agave toumeyana Trelease var. **bella** Breitung, Cact. Succ. J. (Los Angeles) 32: 81, fig. 41. 1960 • Fairy-ring agave [E] [F]

Agave toumeyana subsp. *bella* (Breitung) Gentry

Rosettes each with 100–200 leaves at maturity, light green, flat-topped. **Leaves** (3–)9–20(–27) × 0.5–1.2 cm; margins denticulate below mid leaf. **Flowers** 1.6–2.1 cm; limb lobes 6.5–7 mm; filaments 1.1–1.3 cm; ovary 0.8–1.2 cm. 2*n* = 60.

Flowering late spring. Open gravelly to rocky limestone or basalt slopes, mostly with desert scrub, chaparral, and pinyon-juniper woodlands; 800–1700 m; Ariz.

3. Agave schottii Engelmann, Trans. Acad. Sci. St. Louis 3: 305. 1875 (as schotti)

Plants acaulescent, freely suckering; rosettes solitary or densely cespitose, 3–6 × 6–12 dm. **Leaves** mostly erect, widest near base, 20–40(–50) × 0.7–2.5 cm; blade yellowish green or deep green, with or without conspicuous white bud-prints on both surfaces, linear, firm, adaxially plane or somewhat concave toward apex, abaxially convex to deeply convex toward base; margins straight, filiferous or not, unarmed, fibers erect to spreading, sparse, white, filiform, brittle; apex acuminate to long-acuminate, spine brown or grayish, acicular, 0.8–1.9 cm. **Scape** 1.6–4 m. **Inflorescences** spicate or subspicate to narrowly racemose-paniculate, laxly flowered on distal 1/4–1/2; bracts caducous, narrowly triangular, 2–4 cm; peduncle 4 cm or shorter.

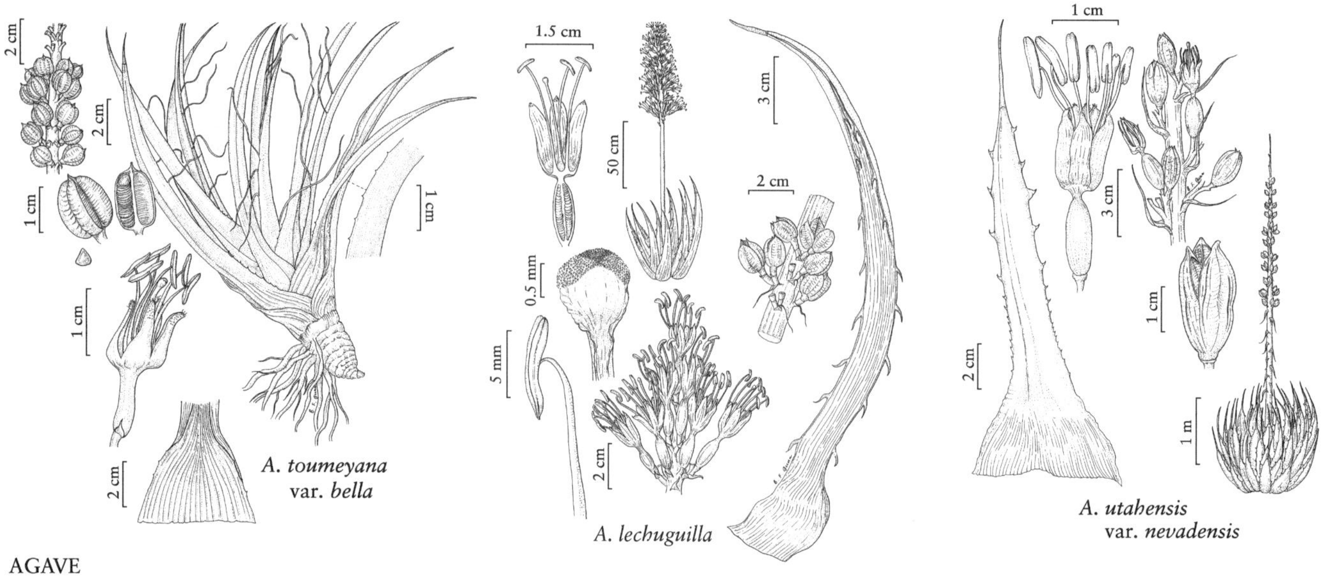

AGAVE

Flowers 1 or 2–3(–6) per cluster, erect, (2.9–)3–5 cm; perianth yellow, tube funnelform, 8–11(–14) × 5–8(–13) mm, limb lobes erect to incurved, subequal, (7–)11–16 mm; stamens slightly exserted; filaments inserted high in perianth tube, 6–13 mm above tube base at 1 or 2 levels, erect, yellow, 1.5–2.2 cm; anthers light to deep yellow, (7.5–)10–17 mm; ovary 0.8–1.6 cm, neck constricted, 3–9 mm. **Capsules** short-pedicellate, obovoid, 1–2.2 cm, apex rounded to beaked. **Seeds** 3–3.5 mm.

Varieties 2 (2 in the flora): sw United States, nw Mexico.

1. Leaves 0.7–1.2 cm wide, blade yellowish green, with conspicuous bud-prints, margins filiferous, apical spine 0.8–1.2 cm; flowers 2.9–4.2 cm; inflorescences spicate ..... 3a. *Agave schottii* var. *schottii*
1. Leaves 1.2–2.5 cm wide, blade deep green, without bud-prints, margins nonfiliferous, apical spine 1.2–1.9 cm; flowers 3.5–5 cm; inflorescences subspicate to narrowly racemose-paniculate ... . 3b. *Agave schottii* var. *treleasei*

3a. Agave schottii Engelmann var. **schottii** • Schott agave

Leaves 0.7–1.2 cm wide; blade yellowish green with conspicous bud-prints on both surfaces, margins filiferous, apical spine grayish, weak and brittle, 0.8–1.2 cm. **Scape** 1.6–2.5 m. **Inflorescences** spicate; peduncle shorter than 2 mm. **Flowers** 1 or 2–3 per cluster, 2.9–4.2 cm; perianth tube 8–11(–14) × 5–8 mm, limb lobes light yellow; filaments inserted 6–13 mm above perianth base; ovary 0.8–1.5 cm. 2*n* = 60, 120.

Flowering late spring–late summer. Gravelly to rocky places, mostly in desert scrub, grasslands, juniper and oak woodlands; 900–2000 m; Ariz., N.Mex.; Mexico (Sonora).

Agave schottii var. *schottii* hybridizes with *A. deserti* var. *simplex*, *A. chrysantha*, and possibly *A. palmeri* or *A. parryi* var. *parryi*.

3b. Agave schottii Engelmann var. **treleasei** (Toumey) Kearney & Peebles, J. Wash. Acad. Sci. 29: 474. 1939 • Trelease agave [E]

Agave treleasei Toumey, Rep. (Annual) Missouri Bot. Gard. 12: 75, plates 32, 33. 1901 (as treleasii)

Leaves 1.2–2.5 cm wide; blade deep green, without bud-prints, margins nonfiliferous, apical spine brown, stout and firm, 1.2–1.9 cm. **Scape** 2–4 m. **Inflorescences** subspicate to narrowly racemose-paniculate; peduncle 0.4–4 cm. **Flowers** 2(–6) per cluster, 3.5–5 cm; perianth tube 11 × 13 mm, limb lobes deep yellow; filaments inserted 8 mm above perianth base; ovary 1.6 cm.

Flowering late spring–summer. Gravelly to rocky places, mostly in desert scrub, grasslands, juniper and oak woodlands; 1100–2000 m; Ariz.

Agave schottii var. *treleasei* may be of a hybrid origin involving *A. chrysantha* or *A. palmeri* and *A. schottii* var. *schottii*.

4. **Agave** ×**ajoensis** W. C. Hodgson, Novon 11: 414, fig. 2. 2001 • Ajo agave [E]

Plants acaulescent, freely suckering; rosettes solitary or more commonly cespitose, 1.9–4 × 2.5–4.5 dm. **Leaves** spreading, widest near middle, 16–34 × 1.5–3 cm; blade light green, without "bud-prints" but sometimes with a lighter green midstripe, linear-lanceolate, rigid, adaxially shallowly concave toward apex, abaxially thickly convex toward base; margins entire, straight, nonfiliferous, unarmed except for small interstitial teeth near leaf base; apex acute to long-acuminate, spine reddish brown, aging gray, subulate, 1.5–2.5 cm. **Scape** 2.5–3.5 m. **Inflorescences** subspicate to narrowly racemose-paniculate on distal 1/3; bracts caducous, narrowly triangular, 2–3 cm; lateral branches 30–40, ascending, 4 cm or shorter. **Flowers** (3–)4–6 per cluster, erect, straight, (3.2–)3.5–5.3 cm; perianth yellow, tube funnelform, (4–)6–21 × 8–13 mm, limb lobes erect to ascending, slightly unequal, 10–20 mm; stamens slightly exserted; filaments inserted equally or subequally 6–16 mm above base of perianth tube, erect, light yellow, (1.7–)2–3 cm; anthers yellow, 9–14 mm; ovary 1–1.7 cm, neck constricted, 2–6 mm. **Capsules** short-pedicellate, ellipsoid, 1.2–2.1 cm, apex short-beaked. **Seeds** unknown. $2n = 90$.

Flowering late spring–summer. Rocky slopes in desert scrub and grasslands; 900–1100 m; Ariz.

Agave ×*ajoensis* appears to be of hybrid origin involving *A. deserti* var. *simplex* and *A. schottii* var. *schottii*. It is a triploid, having a chromosome number of $2n = 90$; in the same area, var. *simplex* is a diploid ($2n = 60$) and var. *schottii* is a tetraploid ($2n = 120$).

5. **Agave univittata** Haworth, Philos. Mag. Ann. Chem. 10: 415. 1831 • Thorn-crested agave

Agave heteracantha Zuccarini; *A. lophantha* Schiede ex Kunth

Plants acaulescent, usually suckering; rosettes densely cespitose, 3–6 × 5–10 dm. **Leaves** mostly erect, 30–40(–70) × 3–4.5(–5) cm; blade light green to yellowish green, without bud-prints, linear-lanceolate, stiff, adaxially plane to concave toward apex, abaxially convex toward base; margins undulate to crenate, nonfiliferous, not easily detached, conspicuously armed, teeth single or occasionally double, 4–6(–8) mm, 1–2 cm apart; apical spine reddish to grayish, subulate, 1–1.5(–2) cm. **Scape** 3–4.5(–5) m. **Inflorescences** spicate, densely flowered on distal 1/2; bracts caducous, acicular, 3–5 cm; peduncle 4 cm or shorter. **Flowers** 1 on short pedicel, or 2–7 per cluster, erect, 3.5–4(–4.5) cm; perianth greenish white to yellowish green, tube campanulate, 2–4 × 8–10 mm, limb lobes erect to slightly spreading, subequal, 14–20 mm; stamens long-exserted; filaments inserted on rim of perianth tube, spreading, greenish or lavender, 3–4.5 cm; anthers pale yellow, 15–20 mm; ovary 1.8–2.2 cm, neck constricted 5–7 mm. **Capsules** sessile or short-pedicellate, oblong and 1.8–2.5 cm or globose and 1.5–2 cm, apex beaked. **Seeds** 5–6 mm. $2n = 60$.

Flowering spring. Sandy flats; 0–100 m; Tex.; Mexico.

The name *Agave lophantha* was mentioned by C. J. W. Schiede (Linnaea 4: 582. 1829) without a description. C. S. Kunth (Enum. Pl. 5: 838. 1850) validated the name, but not before *A. univittata* was proposed. Neither name is typified nor is any authentic material known. H. S. Gentry (1982) incorrectly retained *A. lophantha* and it is now widely used. A. J. Breitung (1959) was the first to adopt *A. univittata*, and that name is taken up here.

6. **Agave lechuguilla** Torrey in W. H. Emory, Rep. U.S. Mex. Bound. 2(1): 213. 1859 • Lechuguilla [F]

Plants acaulescent, frequently suckering; rosettes openly cespitose, 3–4 × 5–6 dm. **Leaves** mostly ascending to erect, (25–)30–50 × 2–4(–5.2) cm; blade light green to yellowish green, sometimes checkmarked but without bud-prints, linear-lanceolate, stiff, adaxially concave toward apex, abaxially convex toward base; margins straight, easily detached, nonfiliferous, conspicuously armed, teeth single 2–6 mm, mostly (1–)2–4 cm apart, rarely absent; apical spine grayish, conical to subulate, 1.5–4.5 cm. **Scape** (2–)2.5–3.5 m. **Inflorescences** spicate, densely flowered on distal 1/2; bracts caducous, linear, 1–3 cm; peduncle 2–5 mm, rarely 20–150 mm. **Flowers** 2–3 per cluster, erect to slightly recurved, (2.4–)3–4.5 cm; perianth yellow, frequently tinged with red or purple, tube campanulate, 1.5–4 × 6–12 mm, limb lobes ascending, subequal, 11–20 mm; stamens long-exserted; filaments inserted on rim of perianth tube, spreading, yellow to reddish, 2.5–4.2 cm; anthers pale yellow, (11–)15–20 mm; ovary (0.8–)1.5–2.2 cm, neck constricted (2–)4–8.5 mm. **Capsules** sessile or short-pedicellate, oblong, 1.8–2.5(–3) cm, apex beaked. **Seeds** 4.5–6 mm. $2n = 110–120$.

Flowering mid spring–late summer. Gravelly to rocky calcareous places in desert scrub; 500–1400 m; N.Mex., Tex.; n, e Mexico.

Agave lechuguilla is the principal source of "istle" or "ixtle," a hard fiber used for rope and known by the trade name "Tampico fibre." The plant is poisonous to cattle, goats, and sheep. This species is the dominant agave on the Chihuahuan Desert. It hybridizes with *A. havardiana*, *A. neomexicana*, *A. gracilipes*, and *A.* ×*glomeruliflora*.

7. Agave ×glomeruliflora (Engelmann) A. Berger, Agaven, 95. 1915, as species • Chisos agave [C]

Agave heteracantha Zuccarini forma *glomeruliflora* Engelmann, Gard. Chron., n. s. 19: 48, fig. 6. 1883; *A. chisosensis* C. H. Muller

Plants acaulescent, frequently suckering; rosettes cespitose, 4–6 × 5–9 dm. **Leaves** mostly spreading; blade light to dark green, without bud-prints, linear and 40–55 × 5–6 cm or lanceolate to deltate and 35–50 × (5–)7–10(–12.5) cm, rigid, adaxially concave toward apex, abaxially convex toward base; margins straight, nonfiliferous, conspicuously armed, teeth single, 3–10 (–15) mm, 1–3 cm apart; apical spine grayish, conical to subulate, 2.5–4.5 cm. **Scape** 4–6 m. **Inflorescences** subspicate to narrowly racemose-paniculate, sparsely flowered; bracts persistent, triangular to narrowly triangular, 1–4 cm; lateral branches 70–90+, comprising distal 1/3–1/2 of inflorescence, 10 cm or shorter. **Flowers** (6–)10–12 per cluster, erect, 4–4.7 cm; perianth yellow to yellowish lime green, tube funnelform, (3.2–)6–8 × 10–13 mm, limb lobes ascending, subequal, 16–23 mm; stamens long-exserted; filaments inserted on rim of perianth tube, spreading, yellow to lime green or maroon, 4–5.8 cm; anthers yellow, 20–21 mm; ovary 1.3–2(–2.2) cm, neck constricted, 3.5–6 mm. **Capsules** sessile or short-pedicellate, oblong, 2.5–4 cm, apex beaked. **Seeds** 5–6 mm.

Flowering mid spring–early fall. Gravelly calcareous slopes in grasslands and oak-juniper woodlands; of conservation concern; 600–1600 m; Tex.; n Mexico.

H. S. Gentry (1982) applied *Agave* ×*glomeruliflora* to a series of hybrids involving *A. lechuguilla* (subg. *Littaea*) and *A. gracilipes*, *A. havardiana*, and *A. parryi* var. *neomexicana* (subg. *Agave*), representing “all those cases of the racemose inflorescence seconded by the loose corneous margins of leaves intermediate in width between the wide leaves of the above cited species and *A. lechuguilla*.” However, *A.* ×*glomeruliflora* should probably be more appropriately applied only to crosses and back-crosses between *A. lechuguilla* and *A. havardiana*.

Unlike species of subg. *Littaea*, this hybrid (and others that result from intersubgeneric crosses) exhibits the more open inflorescence types of subg. *Agave* in that the scape is dominated by clusters of flowers on more or less well-defined lateral branches.

8. Agave ×arizonica Gentry & J. H. Weber, Cact. Succ. J. (Los Angeles) 42: 223, figs. 1, 2. 1970, as species • Arizona agave [C] [E]

Plants acaulescent, sparingly suckering; rosettes solitary or sparingly cespitose, 2–3.5 × 3–4 dm. **Leaves** spreading, 13–31 × 2–3.2 cm; blade dark green, without bud-prints, linear-lanceolate, rigid, adaxially slightly concave toward apex, abaxially convex toward base; margins straight, nonfiliferous, conspicuously armed, horny from below mid-blade to apical spine, teeth single, (1–)2–5 mm, 1–3(–5) cm apart; apical spine dark brown to gray, subulate, 1–2.5 cm. **Scape** 2.5–4 m. **Inflorescences** subspicate to racemose-paniculate, openly flowered; bracts caducous, triangular, 8–12 cm; lateral branches 30–40, comprising distal 1/4–1/3 of inflorescence, 4 cm or shorter. **Flowers** 10–20 per cluster, erect, cylindric-urceolate, 2.5–3.5 cm; perianth pale yellow, sometimes fused with red, tube campanulate, 3.5–7.5 × 8–9 mm, limb lobes not connivent, erect to ascending, subequal, 8–11 mm; stamens long-exserted; filaments inserted 4 mm above base of perianth tube, erect, light yellow, 1.6–3 cm; anthers pale yellow to yellow, 7–11 mm; ovary 0.8–1.5 cm, neck constricted, 4–6 mm. **Capsules** short-pedicellate, ellipsoid to obovoid, 1.5–2.2 cm, apex beaked. **Seeds** 3–4.5 mm. $2n = 60$.

Flowering mid spring–early summer. Rocky slopes in chaparral and juniper-dominated grasslands; of conservation concern; 900–1800 m; Ariz.

Agave ×*arizonica* is a hybrid population involving *A. chrysantha* and *A. toumeyana* var. *bella*. Only about 60 clones are known.

9. Agave utahensis Engelmann in S. Watson, Botany (Fortieth Parallel), 497. 1871 (as utahense) [E] [F]

Plants acaulescent, suckering or not; rosettes solitary or sparingly cespitose, 1.5–6 × 1.5–10 dm. **Leaves** ascending, 12–50 × 1–5.5 cm; blade grayish green to yellowish green or green, without bud-prints, linear-lanceolate or lanceolate, stiff, adaxially plane or concave toward apex, abaxially convex toward base; margins straight or undulate, nonfiliferous, conspicuously armed, horny from distalmost teeth to apical spine, teeth single, 2–12(–15) mm, 1–4 cm apart, rarely absent; apical spine light gray to ivory white, acicular or subulate, 2–20 cm. **Scape** 1.7–5 m. **Inflorescences** spicate, racemose or narrowly paniculate, open to densely flowered; bracts caducous, narrowly triangular, 4–8 cm; lateral branches 50+, horizontal to slightly ascending, comprising distal $^1/_2$–$^2/_3$ of inflorescence, shorter than 10 cm. **Flowers** 2–12 per cluster, erect, urceolate, 2.3–4.3 cm; perianth yellow, tube campanulate, 1–4 × 6.5–11.5 mm, limb lobes connivent, erect, subequal, 7–12 mm; stamens slightly exserted; filaments inserted at or 1–3 mm above base of perianth tube, erect, white, 1.3–2.4 cm; anthers pale yellow, 5–12 mm; ovary 1.2–2.9 cm, neck constricted, 3.5–8.5(–10) mm. **Capsules** pedicellate, oblong to ovoid, 1–2.5 cm, apex beaked. **Seeds** 2–4 mm.

Subspecies 2 (2 in the flora): sw United States.

1. Leaves 3–5.5 cm wide; rosettes 4–10 dm wide; flowers (2.6–)2.9–4.3 cm . . . . 9a. *Agave utahensis* subsp. *kaibabensis*
1. Leaves 1–3 cm wide; rosettes 1.5–4 dm wide; flowers 2.3–3.2 cm . . . . 9b. *Agave utahensis* subsp. *utahensis*

9a. Agave utahensis Engelmann subsp. **kaibabensis** (McKelvey) Gentry, Agaves Continental N. Amer., 259. 1982 • Kaibab agave [E]

Agave kaibabensis McKelvey, J. Arnold Arbor. 30: 227, plate 1. 1949; *A. utahensis* var. *kaibabensis* (McKelvey) Breitung

Rosettes (2–)3–6 × 4–10 dm. **Leaves** (25–)27–50 × 3–5.5 cm; blade lanceolate; marginal teeth (2–)3–5 mm, 2–4 cm apart; apical spine stout, subulate, 2–4 cm. **Scape** 3.5–5 m. **Inflorescences** narrowly paniculate; longer lateral branches (1–)4–10 cm. **Flowers** 4–12 per cluster, (2.6–)2.9–4.3 cm; ovary 1.5–2.9 cm. $2n = 60$.

Flowering mid spring–early summer. Calcareous or sandstone outcrops in desert scrub, pinyon-juniper, or adjacent conifer woodlands; 700–2500 m; Ariz.

9b. Agave utahensis Engelmann subsp. **utahensis** [E] [F]

Rosettes 1.5–3 × 1.5–4 dm. **Leaves** 12–30 × 1–3 cm; blade linear-lanceolate; marginal teeth 2–12(–15) mm, 1–2.5 cm apart; apical spine slender or stout, acicular, 2–20 cm. **Scape** 1.7–4 m. **Inflorescences** spicate, racemose or narrowly paniculate; longer lateral branches 0.5–5 cm. **Flowers** 2–8 per cluster, 2.3–3.2 cm; ovary 1.2–2.2 cm.

Varieties 3 (3 in the flora): sw United States.

1. Apical leaf spine grayish, 2–4 cm, marginal teeth 2.5–4 mm . . . . 9b.1. *Agave utahensis* var. *utahensis*
1. Apical leaf spine brown to whitish or ivory white, 4–20 cm, marginal teeth 4–12(–15) mm.
   2. Apical spine brown to whitish, 4–8 cm . . . . 9b.2. *Agave utahensis* var. *nevadensis*
   2. Apical spine ivory white, 10–20 cm. . . . . 9b.3. *Agave utahensis* var. *eborispina*

9b.1. Agave utahensis Engelmann var. **utahensis** • Utah agave [E]

Agave newberryi Engelmann; *A. scaphoidea* Greenman & Roush

Rosettes 1.8–3 dm. **Leaves** 12–30 × 1–2.5 cm; marginal teeth 2.5–4 mm; apical spine grayish, slender, 2–4 cm. **Scape** 1.7–4 m.

Flowering late winter–early summer. Calcareous outcrops with desert scrub; 600–1400 m; Ariz., Nev., Utah.

9b.2. Agave utahensis Engelmann var. **nevadensis** Engelmann ex Greenman & Roush, Ann. Missouri Bot. Gard. 16: 390, plate 32. 1929 • Clark Mountain agave [E] [F]

Agave nevadensis (Engelmann ex Greenman & Roush) Hester

Rosettes 1.5–2.5 dm. **Leaves** 15–25 × 1.5–3 cm; marginal teeth 4–6 mm; apical spine brown to whitish, slender, 4–8 cm. **Scape** 2–4 m. $2n = 60$.

Flowering late winter–early summer. Calcareous outcrops with desert scrub; 1200–1900 m; Calif., Nev.

9b.3. Agave utahensis Engelmann var. **eborispina** (Hester) Breitung, Cact. Succ. J. (Los Angeles) 32: 22. 1960 • Ivory-spined agave [E]

Agave eborispina Hester, Cact. Succ. J. (Los Angeles) 15: 131, fig. 59. 1943

Rosettes 1.5–3 dm. **Leaves** 15–30 × 1.5–3 cm; marginal teeth 6–12 (–15) mm; apical spine ivory white, stout, 10–20 cm. **Scape** 2–4 m.

Flowering late winter–spring. Calcareous outcrops with desert scrub; 1100–1900 m; Calif., Nev.

10. Agave decipiens Baker, Bull. Misc. Inform. Kew 1892: 184. 1892 • Florida agave [E]

Plants arborescent, trunks 1–3 (–4) m, frequently suckering; rosettes not cespitose, 10–15 × 15–20 dm. **Leaves** spreading to recurved, 70–100(–200) × 7–10 cm; blade green, without bud-prints, linear-lanceolate, rigid, adaxially concave toward apex, abaxially convex toward base; margins slightly wavy, armed, teeth single, 2–3 mm, 1–2 cm apart, often with some interstitial smaller teeth; apical spine dark brown, conical, 1–2 cm. **Scape** 3–5 m. **Inflorescences** paniculate, often bulbiferous; bracts caducous; lateral branches 10–18, ascending, comprising distal 1/2 of inflorescence, longer than 4 cm. **Flowers** 15 or more per cluster, erect, 6–8 cm; perianth greenish yellow, tube funnelform, 11–13 × 4–6 mm, limb lobes incurved, drying reflexed on tube, subequal, 18–22 mm; stamens long-exserted; filaments inserted on 2 levels at or slightly above mid perianth tube, erect, yellow, 4–5 cm; anthers yellow, 22–25 mm; ovary 4–5 cm, neck absent. **Capsules** short-pedicellate, ellipsoid to oblong, 3.5–5 cm, apex beaked. **Seeds** not seen. $2n$ = 120, 150, 160, 180, 190.

Flowering mid fall–early winter. Sandy soil in hammocks; 0 m; Fla.

Agave decipiens might have been introduced from Latin America by Native Americans; it is not otherwise known from the wild. The proliferation of chromosome numbers suggests prolonged human propagation and a probable hybrid origin. Until demonstrated otherwise, the species is considered a rare and localized native species known only from a few sites in southern Florida. H. S. Gentry (1982) discounted the suggestion that *A. decipiens* was found on the Yucatán peninsula of Mexico. He considered such plants to be *A. angustifolia.*

11. Agave asperrima Jacobi, Hamburger Garten-Blumenzeitung 20: 561. 1864

Varieties 4 (1 in the flora): sc United States, n Mexico.

11a. Agave asperrima Jacobi var. **asperrima** • Rough agave, maguey cenizo

Agave caeciliana A. Berger

Plants acaulescent, frequently suckering; rosettes not cespitose, 7–10 × 15–20 dm. **Leaves** spreading to recurved and mostly reflexed, 60–110 × 12–18 cm; blade light green to grayish, without bud-prints, lanceolate, scabrous, rigid, adaxially plane and deeply guttered from mid blade to base, abaxially convex; margins straight, toothed, teeth single, usually deflexed along lower leaf edge, 8–15 mm, 1 cm apart; apical spine dark brown to blackish, subulate to acicular, 3–6 cm. **Scape** 4–7 m. **Inflorescences** paniculate, not bulbiferous; bracts persistent, triangular, 3–10 cm; lateral branches 8–14, slightly ascending, comprising distal 1/3 of inflorescence, longer than 10 cm. **Flowers** 14–18 per cluster, erect, 6–8 cm; perianth yellow, tube cylindric, 13–18 × 14–16 mm, limb lobes erect, subequal, 18–26 mm; stamens long-exserted; filaments inserted just above mid perianth tube, erect, yellow, 5–6.5 cm; anthers yellow, 24–30 mm; ovary 3–4 cm, neck constricted, 4–6 mm. **Capsules** short-pedicellate, oblong, 4–5 cm, apex beaked. **Seeds** 6–7 mm. $2n$ = 174.

Flowering late spring–summer. Sandy to gravelly, often calcareous places in desert scrub; 500–1500(–1900) m; Tex.; n Mexico.

M. S. Cave (1964) reported chromosome counts of n = 74–93; H. S. Gentry (1982) stated that only one of Cave's counts applied to var. *asperrima* and he gave it as n = 87.

12. Agave americana Linnaeus, Sp. Pl. 1: 323. 1753

Plants acaulescent or short-stemmed, commonly suckering, trunks less than 2 m; rosettes not cespitose, 10–20 × 20–37 dm. **Leaves** erect, spreading to ascending, occasionally reflexed, 80–200 × 15–25 cm; blade light green to green or glaucous-gray, sometimes variegated or cross-zoned, narrowly to broadly lanceolate, smooth, rigid; margins nearly straight or undulate to crenate, armed, teeth single, 5–10 mm, 1–4 cm apart; apical spine dark brown to grayish, conical or subulate, 2–6 cm. **Scape** 5–9 m. **Inflorescences** paniculate, not bulbiferous; bracts persistent, triangular, 5–15 cm; lateral branches 15–35, horizontal to slightly ascending, comprising distal 1/3–1/2 of inflorescence, longer than 10 cm. **Flowers** erect, 7–10.5 cm; perianth yellow, tube funnelform to cylindric, 8–20 × 12–20 mm, limb lobes erect, subequal, 20–35 mm; stamens long-exserted; filaments inserted above mid perianth tube, erect, yellow, 6–9 cm; anthers yellow, 25–35 mm; ovary 3–4.5 cm, neck constricted, 3–6(–8) mm. **Capsules** short-pedicellate, oblong, 3.5–8 cm, apex beaked. **Seeds** 6–8 mm.

Subspecies 2 (2 in the flora): sw United States, Mexico.

Various chromosome numbers have been reported for *Agave americana* under a variety of names, typically without regard to the plant's origin or its precise taxonomic disposition. Nonetheless, the species is most certainly a polyploid complex based on x = 30, with reports of n = 30 and $2n$ = 60, 120, and 180 documented by S. D. McKelvey and K. Sax (1933), H. Matsuura and T. Sutô (1935), E. B. Granick (1944), A. K. Sharma and U. C.

Bhattacharyya (1962), M. S. Cave (1964), S. Banerjee and A. K. Sharma (1987), Huang S. F. et al. (1989) and B. Vijayavalli and P. M. Mathew (1990). Various dysploids have also been reported (A. F. Dyer et al. 1970; J. L. Strother and G. L. Nesom 1997). See H. S. Gentry (1982) for details.

1. Leaves 80–135 cm, 4–6 times longer than wide; capsules 3.5–4 cm . . . . 12a. *Agave americana* subsp. *protamericana*
1. Leaves 100–200 cm, 6–10 times longer than wide; capsules 4–8 cm . . . . 12b. *Agave americana* subsp. *americana*

12a. Agave americana Linnaeus subsp. **protamericana** Gentry, Agaves Continental N. Amer., 287, figs. 12.1, 12.3–12.5, 12.8, 12.12, 12.14, 12.15, plate 12.1. 1982 • Wild century plant

Plants acaulescent or short-stemmed, trunks less than 1 m; rosettes open. **Leaves** frequently reflexed, 80–135 × 17–22 cm; blade light green to glaucous-gray, sometimes cross-zoned, broadly lanceolate, adaxially plane or guttered, abaxially convex; margins crenate, teeth, 5–10 mm; apical spine subulate, 3–6 cm. **Scape** 6–8 m. **Inflorescences:** lateral branches 15–20. **Flowers** 7.5–9 cm; perianth tube 15–20 mm; ovary 4–4.5 cm. **Capsules** 3.5–4 cm. **Seeds** 7–8 mm.

Flowering early spring–early summer. Sandy places in desert scrub; 200 m; Tex.; ne Mexico.

The Starr County, Texas, plants of subsp. *protamericana* are smaller in stature than those seen in Mexican populations but appear to belong to this wild taxon.

12b. Agave americana Linnaeus subsp. **americana**

Plants acaulescent or short-stemmed, trunks less than 2 m; rosettes open or dense. **Leaves** erect, ascending, or frequently reflexed, 100–200 × 15–25 cm; blade light green to green or glaucous-gray, sometimes variegated yellow or whitish, especially in cultivated forms, sometimes cross-zoned, lanceolate, adaxially plane and somewhat guttered near base, abaxially convex near base; margins straight or undulate to crenate, teeth 5–10 mm; apical spine conical or subulate, 2–5 cm. **Scape** 5–9 m. **Inflorescences:** lateral branches 15–35. **Flowers** 7–10 cm; perianth tube 8–20 × 12–20 cm; ovary 3–4.5 cm. **Capsules** 4–8 cm. **Seeds** unknown.

Varieties 3 (2 in the flora): sw United States, Mexico.

1. Apical leaf spine 3–5 cm, leaves frequently reflexed; Texas . . . . 12b.1. *Agave americana* var. *americana*
1. Apical leaf spine 2–3 cm, leaves erect; Arizona . . . . 12b.2. *Agave americana* var. *expansa*

12b.1. Agave americana Linnaeus var. **americana** • Century plant, maguey

Agave gracilispina Engelmann ex Trelease

Plants acaulescent or short-stemmed; rosettes rather dense. **Leaves** frequently reflexed, 100–200 × 15–25 cm; blade light green to glaucous-gray, sometimes variegated yellow or whitish; margins undulate to crenate, teeth 5–10 mm; apical spine subulate, 3–5 cm. **Scape** 5–9 m. **Inflorescences:** lateral branches 15–35. **Flowers** 7–10 cm; perianth tube 8–20 mm; ovary 3–4.5 cm. **Capsules** 4–5 cm. **Seeds** unknown. $2n = 120$.

Flowering mid spring–early summer. Sandy places in desert scrub; 200 m; Tex.; e Mexico.

The cultivar of *Agave americana* var. *americana*, technically a variant of *A. americana* subsp. *americana* derived from subsp. *protamericana*, is planted widely in the warmer regions of the United States and elsewhere in the world; only in the south Texas region of the flora is the plant native. At least five additional variants have been named based on leaf variegation patterns, but none is worthy of formal taxonomic recognition.

12b.2. Agave americana Linnaeus var. **expansa** (Jacobi) Gentry, Agave Fam. Sonora, 80. 1972 • Spreading century plant

Agave expansa Jacobi, Abh. Schles. Ges. Vaterl. Cult., Abth. Naturwiss. (Naturwiss.-Med. Abth.) 1868: 151. 1868; *A. abrupta* Trelease

Plants short-stemmed; rosettes dense. **Leaves** erect, 120–150 × 18–24 cm; blade glaucous-gray to green, typically cross-zoned; margins nearly straight to crenate, teeth 5–8 mm; apical spine conical, 2–3 cm. **Scape** 7–9 m. **Inflorescences:** lateral branches 20–30. **Flowers** 7–8.5 cm; perianth tube 13–14 mm; ovary 3.5–4 cm. **Capsules** not seen. **Seeds** unknown. $2n = 120$.

Flowering late spring–early summer. Sandy places in desert scrub and grasslands; 500–1300 m; Ariz.; w Mexico.

The wild expression of *Agave americana* var. *expansa* is technically a variant of subsp. *americana*. A cultivar derived from var. *expansa* was introduced into western Europe, where it was first named. That cultivar is planted occasionally in the warmer regions of the United States; it is not naturalized. The var. *expansa* has never been found in fruit, but it is likely that it itself is an early cultivar developed by pre-Columbian people who established it far to the west of the species range in eastern Mexico. Reproduction in var. *expansa* is strictly vegetative.

13. Agave desmettiana Jacobi, Hamburger Garten-Blumenzeitung 22: 217. 1866 • Regel agave [I]

Agave regeliana Jacobi

Plants short-stemmed, commonly suckering, trunk less than 0.5 m; rosettes not cespitose, 7–9 × 7–9 dm. **Leaves** spreading and arching, 50–90 × 7–12 cm; blade green, not cross-zoned, linear-lanceolate, firm, adaxially guttered, abaxially slightly convex; margins straight, finely and sparsely fibrous, typically unarmed, teeth single prickles when present, 1–2 mm, 1–2+ cm apart; apical spine dark brown, subulate, 2–3 cm. **Scape** 2.5–3 m. **Inflorescences** paniculate, compact, often bulbiferous; bracts persistent, triangular, 0.5–2 cm; lateral branches 20–36, ascending, comprising distal 1/2–2/3 of inflorescence, longer than 10 cm. **Flowers** 15–21 per cluster, erect, 4–6 cm; perianth pale yellow, tube salverform, 10–12 × 14–17 mm, limb lobes erect, equal, 13–15 mm; stamens long-exserted; filaments inserted near top of perianth tube, erect, yellow, 3–4 cm; anthers yellow, 13–26 mm; ovary 1.5–2.5 cm, neck not constricted, 1–3 mm. **Capsules** not seen. **Seeds** unknown. $2n = 60$.

Flowering late spring–early summer. Sandy places along roadsides and in hammocks; 0 m; introduced; Fla.; Mexico.

Agave desmettiana is an attractive and frequently cultivated plant, probably native to Sinaloa, Mexico. The species may be an ancient cultivar, possibly derived from *A. sisalana* or *A. kewensis* Jacobi. The plant reproduces only by vegetative means; capsules and seeds have never been seen. H. S. Gentry (1982) indicated that the plant produces a fine but weak fiber. In Mexico it is grown only as an ornamental.

14. Agave sisalana Perrine, Trop. Pl., 87. 1838 • Sisal [I]

Plants short-stemmed, commonly suckering with elongate rhizomes, trunks 0.4–1 m; rosettes not cespitose, 15–20 × 15–25 dm. **Leaves** spreading, 90–130 × 9–12 cm; blade green, slightly cross-zoned when immature, lanceolate, firm, adaxially plane, abaxially slightly convex; margins straight, finely fibrous, typically unarmed, teeth single prickles when present, 1–2 mm, 2–5+ cm apart; apical spine dark brown, subulate, 2–2.5 cm. **Scape** 5–6 m. **Inflorescences** paniculate, open, often bulbiferous; bracts persistent, triangular, 0.5–2 cm; lateral branches 10–15(–25), ascending, comprising distal 1/2 of inflorescence, longer than 10 cm. **Flowers** erect, 5.5–6.5 cm; perianth greenish yellow, tube urceolate, 15–18 × 10–12 mm, limb lobes erect, equal, 17–18 mm; stamens long-exserted; filaments inserted ca. mid perianth tube, erect, yellow, 5–6 cm; anthers yellow, 20–25 mm; ovary 2–2.5 cm, neck slightly constricted, 2–4 mm. **Capsules** not seen. **Seeds** unknown. $2n = 138, 147, 149, 150$.

Flowering winter–early spring. Sandy places along roadsides and in hammocks; 0 m; introduced; Fla.; s Mexico.

Agave sisalana is frequently cultivated for its fiber and ornamental value. The plant is not known from the wild. As with *A. desmettiana*, capsules and seeds of this species are unknown. Capsules are known from *A. kewensis* (native to Chiapas), *A. neglecta*, and *A. weberi*, although no mature seeds have been observed. It may be that all are anciently selected cultivars that now persist only by vegetative means. None save *A. kewensis* is thought to occur in the wild. Plants similar to those found in Florida are known from Chiapas. *Agave sisalana* is an important source of fiber and probably was widely distributed by pre-Columbian people.

15. Agave neglecta Small, Fl. S.E. U.S., 289. 1903 • Small agave [C] [E]

Plants short-stemmed, commonly suckering, without rhizomes, trunks 0.3–0.4 m; rosettes not cespitose, 13–15 × 15–20 dm. **Leaves** ascending to arching or reflexed, 100–150 × 15–25 cm; blade light green, probably not cross-zoned, broadly lanceolate, firm, adaxially plane, abaxially slightly convex; margins nearly straight, not fibrous, minutely armed near base, teeth single prickles, 1–2 mm, 2–3+ cm apart; apical spine dark brown, acicular, 2.5 cm. **Scape** 8–10(–13) m. **Inflorescences** paniculate, open, often bulbiferous; bracts persistent, triangular, 0.5–2 cm; lateral branches 12–20, slightly ascending, comprising distal 1/3–1/2 of inflorescence, longer than 10 cm. **Flowers** 12–22 per cluster, erect, 5.5–6 cm; perianth greenish yellow, tube urceolate, 8–10 × 4–7 mm, limb lobes erect, equal, 20–30 mm; stamens long-exserted; filaments inserted ca. mid perianth tube or above, erect, yellow, 4–5 cm; anthers yellow, 20–25 mm; ovary 2–3 cm, neck slightly constricted. **Capsules** oblong to ovoid, 3 cm. **Seeds** 5–6 mm.

Flowering early summer. Sandy places near beaches; of conservation concern; 0 m; Fla.

Agave neglecta continues to live up to its name, being

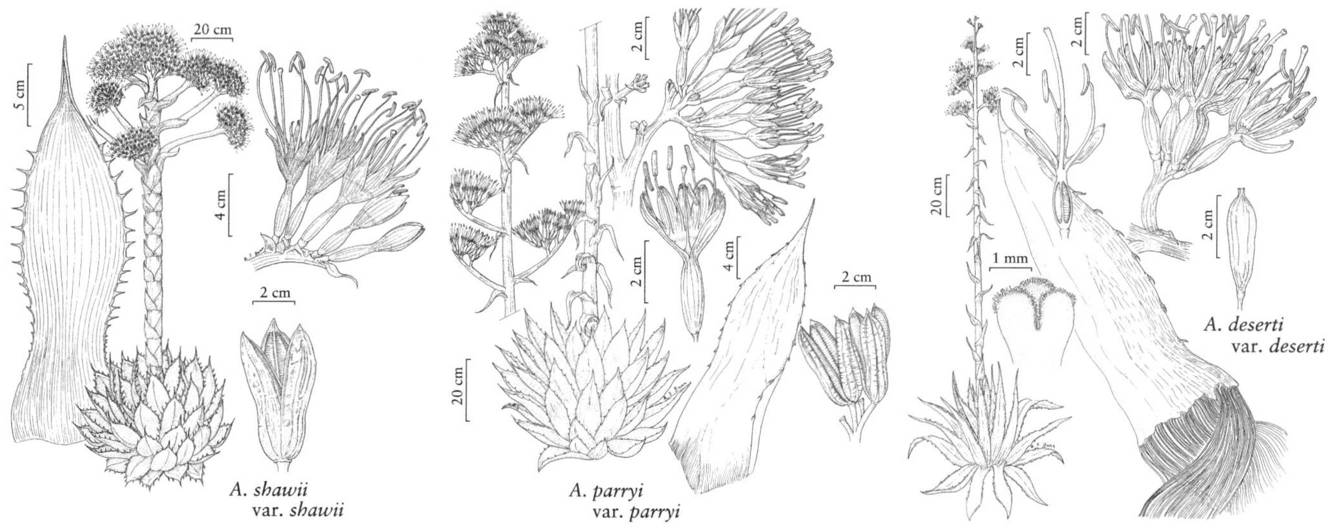

AGAVE

known only from a series of photographs and incomplete, sterile herbarium specimens. Therefore, the above measurements are only an approximation, and all are subject to modification if and when fresh, reproductive material can be studied. Attempts to see this rare species in the field have been thwarted by development and hurricanes. H. S. Gentry (1982) suggested that it might be related to *A. weberi*, a freely suckering species that produces fiber of an excellent quality. Nothing is known about the uses of *A. neglecta*. The plant may well be a cultivar of *A. sisalana* or *A. kewensis* and represent an ancient introduction from Mexico.

16. Agave weberi F. Cels ex J. Poisson, Bull. Mus. Natl. Hist. Nat. 7: 231. 1901 • Weber agave, maguey liso [I]

Plants short-stemmed, commonly suckering, without rhizomes, trunks 0.4–1 m; rosettes not cespitose, 12–14 × 20–30 dm. **Leaves** erect or recurving, 110–160 × 12–18 cm; blade green or grayish green or yellowish green, not cross-zoned, lanceolate, pliable, adaxially concave or guttered at least toward apex, abaxially slightly convex; margins straight, finely fibrous, minutely armed only near base or teeth absent, teeth single prickles, 1–2 mm, less than 1 cm apart; apical spine brown to grayish, subulate, 3–4.5 cm. **Scape** 6.5–8 m. **Inflorescences** paniculate, open, sometimes bulbiferous; bracts persistent, triangular, 0.5–2 cm; lateral branches 15–25, horizontal to slightly ascending, comprising distal 1/3–1/2 of inflorescence, longer than 10 cm. **Flowers** 15–27 per cluster, erect, 7–8 cm; perianth yellow, tube urceolate, 15–20 × 14–19 mm, limb lobes erect, subequal, 20–24 mm; stamens long-exserted; filaments inserted ca. mid perianth tube, erect, yellow to yellow-green, 5.5–6 cm; anthers yellow, 30–31 mm; ovary 3.3–4 cm, neck unconstricted, 2–4 mm. **Capsules** short-pedicellate, oblong, 5.5 cm, apex beaked. **Seeds** unknown.

Flowering late spring–early summer. Sandy places with grasses and low shrubs; 100–300 m; introduced; Tex.; ne and c Mexico.

Agave weberi is locally naturalized in southern Texas. It is frequently cultivated elsewhere for its ornamental value, pulque, and fiber.

17. Agave shawii Engelmann, Trans. Acad. Sci. St. Louis 3: 314. 1875 [F]

Varieties 2 (1 in the flora): California, nw Mexico.

17a. Agave shawii Engelmann var. **shawii** • Shaw agave [F]

Agave orcuttiana Trelease; *A. pachyacantha* Trelease

Plants short-stemmed, budding from leaf axils, forming fragmented supine clones or sometimes single rosettes, trunks erect to decumbent, shorter than 2 m; rosettes solitary or cespitose, 0.8–20 × 10–25 dm. **Leaves** ascending, 20–50 × 8–20 cm; blade glossy light to dark green, not cross-zoned, narrowly ovate, rigid, adaxially plane to slightly hollowed at least toward apex, abaxially convex; margins straight, armed, teeth single, well defined, mostly 5–6 mm, 1–2+ cm apart; apical spine dark reddish brown to gray, acicular, 2–4 cm. **Scape** 2–4 m. **Inflorescences** paniculate, not bulbiferous; bracts persistent, lanceolate to triangular, 10–25 cm; lateral branches 8–14, horizontal to ascending, comprising distal 1/4–1/2 of inflorescence, longer than 10 cm. **Flowers** 35–75 per cluster, erect, 6–10 cm; perianth yellow to reddish, especially upon withering, tube broadly funnelform, 12–19 × 15–22 mm, limb lobes erect, unequal, 17–40 mm; stamens long-exserted; filaments inserted near mid perianth tube, erect, yellow, 4.3–7 cm; anthers yellow, 20–35 mm; ovary 3–5 cm, neck slightly constricted, 6–15 mm. **Capsules** short-pedicellate, obovoid to oblong, 5.5–7 cm, apex short- to long-beaked. **Seeds** 4–7 mm. $2n = 60$.

Flowering winter–mid spring. Coastal bluffs with coastal sage scrub and chaparral; 0–100 m; Calif.; Mexico (Baja California).

Agave shawii was abundant in southern California at the beginning of the twentieth century. Since then, unrestricted collecting and habitat destruction reduced the populations significantly. By 1969 the species was believed to occur in California only within the U.S. Navy area at Point Loma (*Moran 16707*, DES, SD). At present, native populations are known only from Border Fields State Park north of Tijuana, although plants have been introduced at other locales.

18. Agave murpheyi F. Gibson, Contr. Boyce Thompson Inst. Pl. Res. 7: 83, fig. 1. 1935 • Murphey agave, Hohokam agave [C]

Plants acaulescent, freely suckering; rosettes cespitose, 6–12 × 8–14 dm, open. **Leaves** ascending, 50–80 × 6–10 cm; blade light glaucous green to yellow green, frequently lightly cross-zoned, spatulate, firm, adaxially concave toward apex, abaxially convex toward base; margins undulate, armed, teeth single, well defined, mostly 3–4 mm, ca. 1–2 cm apart; apical spine dark brown to grayish, conical, 1.2–2 cm. **Scape** 3–4 m. **Inflorescences** narrowly paniculate, prolifically bulbiferous; bracts persistent, triangular, 10–15 cm; lateral branches 10–16, slightly ascending, comprising distal 1/4 of inflorescence, longer than 10 cm. **Flowers** 12–21 per cluster, erect, (5.1–)6–7.5 cm; perianth cream, apex purplish or brownish, tube urceolate, 14–20 × (11–)14–19 mm, limb lobes erect, unequal, (14–)15–20 mm; stamens long-exserted; filaments inserted unequally at or slightly above mid perianth tube, erect, yellow, (3.3–)4.5–5 cm; anthers yellow, (16–)22–25 mm; ovary (1.8–)2.2–4 cm, neck slightly constricted, (0.5–)4–6 mm. **Capsules** short-pedicellate, obovate to oblong or ovate, 5–7 cm, apex short-beaked. **Seeds** (7–)9–11 mm. $2n = 60$.

Flowering late winter–spring. Near drainage systems in desert scrub; of conservation concern; 400–1000 m; Ariz.; nw Mexico.

Agave murpheyi, used by the Hohokam for food, is known only from around pre-Columbian agricultural and settlement areas. The plants (including bulbils) are easily transported and transplanted. *Agave murpheyi* hybridizes with *A. chrysantha* in Arizona.

19. Agave parryi Engelmann, Trans. Acad. Sci. St. Louis 3: 311. 1875 [F]

Plants acaulescent, freely suckering; rosettes solitary or cespitose, 3.5–7.5 × 4–8.5 dm, somewhat open to dense. **Leaves** ascending to erect, 7–65 × 4–20 cm; blade glaucous-gray to light green, not cross-zoned, linear-lanceolate to lanceolate or broadly ovate, rigid, adaxially nearly plane to concave toward apex, abaxially convex; margins straight or undulate, rarely unarmed, teeth single, well defined, 3–7(–8) mm, 1–3 cm apart; apical spine dark brown to gray, subulate or acicular, 1.5–4 cm. **Scape** (2–)3–6 m. **Inflorescences** broadly paniculate, somewhat open to dense, not bulbiferous; bracts persistent, triangular, 3–10 cm; lateral branches (7–)10–36, horizontal to slightly ascending, comprising distal 1/2 of inflorescence, longer than 10 cm. **Flowers** 27–48 per cluster, erect, 4.3–8.1 cm; perianth pink to red or red to orange in bud, yellow to yellowish green at anthesis, tube campanulate, 6–18 × 11–21 mm, limb lobes erect to ascending, unequal, 13–27 mm; stamens long-exserted; filaments inserted (sometimes irregularly) above mid perianth tube to just below rim, erect, yellow or green, 3.5–5.8 cm; anthers yellow, 12–34 mm; ovary (2–)3–4.5(–4.8) cm, neck slightly constricted, 4–9 mm. **Capsules** short-pedicellate, ovoid or oblong to obovoid, 2.5–5 cm, apex beaked. **Seeds** 7–8 mm.

Subspecies 2 (2 in the flora): sw United States, n Mexico.

1. Rosettes flat-topped; scape (2–)3–4.5 m; perianth tube 12–18 mm, limb lobes 3–4 mm wide; capsules 2.5–3.5 cm . . . . 19a. *Agave parryi* subsp. *neomexicana*
1. Rosettes globose; scape 4–6 m; perianth tube 6–12 mm, limb lobes 4–6(–7) mm wide; capsules 3.5–5 cm . . . . 19b. *Agave parryi* subsp. *parryi*

19a. Agave parryi Engelmann subsp. **neomexicana** (Wooton & Standley) B. Ullrich, Sida 15: 259. 1992 • New Mexico agave [E]

Agave neomexicana Wooton & Standley, Contr. U.S. Natl. Herb. 16: 115, plate 48. 1913

Rosettes flat-topped, 3.5–4.5 × 4–6 dm. **Leaves** ascending, 17–45 × 4–12 cm; blade lanceolate; margins straight, teeth 5–7 mm, 1–3 cm apart; apical spine dark brown, subulate or acicular, 2.5–3 cm. **Scape** (2–)3–4.5 m. **Inflorescences:** lateral branches (7–)10–17, slightly ascending. **Flowers** 5.5–6.7 cm; perianth red to orange in bud, yellow to yellowish green at anthesis, tube 12–18 mm, limb lobes erect to ascending, 15–20 × 3–4 mm; filaments inserted (sometimes irregularly) above mid perianth tube, green, 3.5–4.5 cm; ovary neck 4–7 mm. **Capsules** oblong, 2.5–3.5 cm. $2n = 120$.

Flowering mid spring–early summer. Gravelly to rocky places in grasslands and desert scrub; 1600–2100 m; N.Mex., Tex.

The Mescalero Apaches were so named because of their use of *Agave parryi* subsp. *neomexicana*. The rosettes were baked to produce a sweet, sticky food source. Overuse probably accounts in part for the limited distribution and infrequent occurrence of the species. H. S. Gentry (1982) suggested "the extensive stands of agaves along the eastern bajada of the Sierra Guadalupe" in New Mexico and Texas today may have been "a tribal no-man's-land" between the warring Comanche and Apache peoples. E. B. Granick (1944) reported the plant as "tetraploid" without actually providing a chromosome number.

19b. Agave parryi Engelmann subsp. **parryi** [F]

Rosettes globose, 3.5–7.5 × 4–8.5 dm. **Leaves** ascending to erect, (18–)25–65 × (4.5–)6.5–20 cm; blade linear to lanceolate to broadly ovate; margins straight or undulate, teeth 3–7(–8) mm, 1–2 cm apart; apical spine dark brown to gray, acicular, 1.5–3 cm. **Scape** 4–6 m. **Inflorescences:** lateral branches 20–36, horizontal to slightly ascending. **Flowers** 4.3–8.1 cm; perianth pink to red in bud, yellow at anthesis, tube 6–12 mm, limb lobes erect, 13–27 × 4–6(–7) mm; filaments inserted near rim of perianth tube, 3.3–5.8 cm; ovary neck 6–9 mm. **Capsules** ovoid or oblong to obovoid, 3.5–5 cm.

Varieties 4 (3 in the flora): sw United States, nw Mexico.

The following variants are weakly differentiated but appear to have some geographic distinction. B. Ullrich (1992) reduced all to synonymy under *Agave parryi* subsp. *parryi*.

1. Flowers 4.3–5.5(–6) cm; filaments 3.3–4.4 cm . . . . 19b.3. *Agave parryi* var. *couesii*
1. Flowers 6–8.1 cm; filaments 4–5.8 cm.
    2. Perianth tube 9–12 mm; leaves (4.5–)8–12 cm wide . . . . 19b.1. *A. parryi* var. *parryi*
    2. Perianth tube 8–9 mm; leaves 10–20 cm wide . . . . 19b.2. *A. parryi* var. *huachucensis*

19b.1. Agave parryi Engelmann var. **parryi** • Parry agave, mescal [F]

Agave scabra Salm-Dyck

Rosettes (3.5–)4–6 × 6–7.5 dm. **Leaves** (18–)25–50 × (4.5–)8–12 cm; blade linear to ovate. **Flowers** 6–7.7 cm; perianth tube 9–12 mm, limb lobes (17–)19–24 mm; filaments 4–5.4 cm; ovary (2.7–)3–4.7 cm.

Flowering mid spring–mid summer. Gravelly to rocky places in grasslands, desert scrub, chaparral, pinyon-juniper, and oak woodlands; 1200–2800 m; Ariz., N.Mex.; nw Mexico.

B. Ullrich (1992) showed that the earlier name *Agave scabra* Salm-Dyck was based on a collection attributable to *A. parryi*, and that the correct name for what H. S. Gentry (1982) called *A. scabra* is *A. asperrima*. Fortunately, the Salm-Dyck name is illegitimate, being a later homonym, and so *A. parryi* remains the correct one. Variety *parryi* hybridizes with *A. chrysantha* and *A. schottii* var. *schottii* in Arizona.

19b.2. Agave parryi Engelmann var. **huachucensis** (Baker) Little, Amer. J. Bot. 30: 235. 1943 • Huachuca agave

A. huachucensis Baker, Handb. Amaryll., 172. 1888

Rosettes 4.5–7.5 × 7.5–8.5 dm. **Leaves** 32–65 × 10–20 cm; blade ovate. **Flowers** 6.2–8.1 cm; perianth tube 8–9 mm, limb lobes 20–27 mm; filaments 4.6–5.8 cm; ovary 3.4–4.7 cm.

Flowering mid spring–early summer. Gravelly places in oak and pine woodlands; 1500–2200 m; Ariz.; nw Mexico.

Agave parryi var. *huachensis* is widely cultivated.

19b.3. Agave parryi Engelmann var. **couesii** (Engelmann ex Trelease) Kearney & Peebles, J. Wash. Acad. Sci. 29: 474. 1939 • Coues agave E

A. couesii Engelmann ex Trelease, Rep. (Annual) Missouri Bot. Gard. 22: 94, plates 94–97. 1911

Rosettes 3.5–5.5 × 4–6.5 dm. **Leaves** 25–42(–47) cm × 6.5–11 cm; blade ovate. **Flowers** 4.3–5.5(–6) cm; perianth tube 6–9 mm, limb lobes 13–21 mm; filaments 3.3–4.4 mm; ovary 2–3.4 cm. $2n$ = 120.

Flowering late spring–early summer. Gravelly places in grasslands, chaparral, pinyon-juniper, and oak woodlands; 1100–2100 m; Ariz.

Variety *couesii* hybridizes with *A. chrysantha* in Arizona.

20. Agave gracilipes Trelease, Rep. (Annual) Missouri Bot. Gard. 22: 95, plates 98, 99. 1912 • Slim-footed agave

Plants acaulescent, sparsely suckering; rosettes usually solitary, 3–4 × 7–8 dm, dense. **Leaves** ascending, 18–30 × 4.5–7 cm; blade glaucous and yellow- or gray-green, not cross-zoned, lanceolate to broadly lanceolate, rigid, adaxially concave toward apex, abaxially convex; margins straight, armed, teeth single, well defined, 2–8 mm, 1.5–2.5 cm apart; apical spine reddish brown to gray, acicular, 2.5–5 cm. **Scape** (1.8–)4–5 m. **Inflorescences** narrowly to somewhat broadly paniculate, dense, not bulbiferous; bracts persistent, lanceolate, 1–3(–5) cm; lateral branches 30–40, slightly ascending, comprising distal 1/3–1/2 of inflorescence, longer than 10 cm, or short-scaped individuals (1.8–3.5 m) with only 7–20 lateral branches on distal 1/4. **Flowers** 6–45 per cluster, erect, 4–5.5(–5.8) cm; perianth red in bud, yellow to yellow-green at anthesis, tube broadly campanulate, 4–7 × 9–15 mm, limb lobes spreading to ascending, slightly unequal, 14–18 mm; stamens long-exserted; filaments inserted just below rim of perianth tube, erect, yellow, (2.8–)3–4.5 cm; anthers yellow, 16–20 mm; ovary 2.5–3.2 cm, neck constricted, 3–5 mm. **Capsules** sessile or short-pedicellate, oblong, 2.7–4.4 cm, apex beaked. **Seeds** 5–6.5 mm. $2n$ = 60.

Flowering summer–early fall. Gravelly to rocky, often calcareous places in grasslands, desert scrub, and pinyon-juniper woodlands; 1200–1900 m; N.Mex., Tex.; Mexico (Chihuahua).

Agave gracilipes probably represents an ancient hybrid involving *A. lechuguilla* and *A. parryi* subsp. *neomexicana*. In fact, hybrids between *A. gracilipes* and *A. lechuguilla* appear to occur at present. However, until populations of *A. gracilipes* are thoroughly studied cytologically, we are reluctant to designate this taxon as a hybrid species (e.g., see 8. *A.* ×*arizonica*).

21. Agave havardiana Trelease, Rep. (Annual) Missouri Bot. Gard. 22: 91, plates 84–86. 1912 • Havard agave

Plants acaulescent, sparsely suckering; rosettes usually solitary, (4–)5–8 × (5–)10–15 dm, rather open. **Leaves** ascending, 30–60(–70) × 15–27 cm; blade glaucous-gray to gray-green, not cross-zoned, lanceolate to broadly lanceolate, rigid, adaxially concave, abaxially convex; margins straight, armed, teeth single, well defined, (5–)7–10 mm, 1.5–2 cm apart; apical spine dark brown to gray, subulate, 3–5(–10) cm. **Scape** 2–7 m. **Inflorescences** paniculate, not bulbiferous, dense; bracts persistent, lanceolate, (3–)5–10 cm; lateral branches 12–20, slightly ascending, comprising distal 1/2–2/3 of inflorescence, longer than 10 cm. **Flowers** 21–48 per cluster, erect, 6.8–9 cm; perianth yellow to yellow-green, tube funnelform, 14–22 × 15–22 mm, limb lobes erect, slightly unequal, 18–24 mm; stamens long-exserted; filaments inserted irregularly ca. mid perianth tube, erect, yellow, 5–6.5 cm; anthers yellow, 25–30 mm; ovary 3–4 cm, neck constricted, 2–8 mm. **Capsules** short-pedicellate, oblong to obovoid, 4–5.7 cm, apex beaked. **Seeds** 6–7 mm.

Flowering summer–early fall. Gravelly to rocky, often calcareous places in grasslands, desert scrub, pinyon-juniper, and oak woodlands; 1200–2000 m; Tex.; nw Mexico.

Agave havardiana hybridizes with *A. lechuguilla* to form *A.* ×*glomeruliflora*, with which it may also hybridize.

22. Agave deserti Engelmann, Trans. Acad. Sci. St. Louis 3: 310. 1875 F

Plants acaulescent or short-stemmed, sparsely or prolifically suckering, trunks 0.2–0.4 m; rosettes solitary or numerous, 3–7 × 4–8 dm, compact to rather open. **Leaves** ascending, widest at base, (20–)25–70 × 4.5–10 cm; blade grayish, gray- or blue-glaucous, yellowish green, or greenish, occasionally cross-zoned, linear-lanceolate to lanceolate, rigid, adaxially concave, abaxially convex; margins straight or undulate to crenate, armed, teeth single, weakly attached, well defined, of 2 types, 2–3 mm or (5–)6–8 mm, (1–)1.5–3 cm apart; apical spine light or dark brown to gray, subulate to acicular, 2–4 cm. **Scape** 2–6

m. **Inflorescences** paniculate, not bulbiferous, open; bracts persistent, triangular, 8–15 cm; lateral branches 6–15, slightly ascending, comprising distal 1/5–1/4 of inflorescence, longer than 10 cm. **Flowers** 12–48 per cluster, erect, 3–6 cm; perianth yellow, tube shallow, campanulate or funnelform, 3–10 × 9–15 mm, limb lobes wilting soon after anthesis, spreading, equal, 13–20 mm; stamens long-exserted; filaments inserted unequally above middle or at rim of perianth tube, erect, yellow, 2.5–3.5 (–4.2) cm; anthers yellow, 13–21 mm; ovary 1.6–4 cm, neck slightly constricted, 4–6 mm. **Capsules** short-pedicellate, ovoid to oblong or obovoid, 3.5–5.5 cm, apex beaked. **Seeds** 5–6 mm.

Varieties 3 (2 in the flora): sw United States, nw Mexico.

1. Perianth tube campanulate, 3–5 mm; filaments inserted at top of tube; plants sparsely to prolifically suckering; rosettes numerous . . . . 22a. *Agave deserti* var. *deserti*
1. Perianth tube funnelform, (3–)5–10 mm; filaments inserted above middle but well within tube; plants rarely with 1–3 suckers; rosettes usually solitary . . . . 22b. *Agave deserti* var. *simplex*

22a. Agave deserti Engelmann var. **deserti** • Desert agave [F]

Agave consociata Trelease

Plants sparse to prolifically suckering; rosettes numerous, rather open. **Leaves** 25–40 × 6–8 cm; blade grayish, linear-lanceolate to lanceolate; margins straight; apical spine 2–4 cm. **Scape** 2–4 m. **Perianth tube** campanulate, 3–5 mm; filaments inserted at top of tube. $2n$ = 60, 118.

Flowering spring. Sandy to gravelly or rocky places in desert scrub and pinyon-juniper woodlands; 500–1500 m; Calif.; nw Mexico.

22b. Agave deserti Engelmann var. **simplex** (Gentry) W. C. Hodgson & Reveal, Novon 11: 413. 2001 • Simple desert agave

Agave deserti subsp. *simplex* Gentry, Occas. Pap. Calif. Acad. Sci. 130: 22. 1978

Plants rarely with 1–3 suckers; rosettes usually solitary, compact. **Leaves** (20–)25–60 × 4.5–10 cm; blade greenish, lanceolate; margins undulate to crenate; apical spine 3–4 cm. **Scape** 4–6 m. **Perianth tube** funnelform, (3–)5–10 mm; filaments inserted above middle but well within tube. $2n$ = 60.

Flowering spring. Sandy to gravelly or rocky places in desert scrub and pinyon-juniper woodlands; 300–1500 m; Ariz., Calif.; nw Mexico.

Variety *simplex* hybridizes with *A. schottii* var. *schottii* and possibly with *A. mckelveyana*.

23. Agave mckelveyana Gentry, Cact. Succ. J. (Los Angeles) 42: 225, figs. 4–6. 1970 • McKelvey agave [E]

Plants acaulescent, solitary or sparsely suckering; rosettes 2–4.5 × 3–6 dm, rather open. **Leaves** spreading, broadest at middle, 17.5–40 × 2.8–5 cm; blade light glaucous green or yellowish green to dark green, usually cross-zoned, linear-lanceolate to lanceolate, rigid, adaxially concave, abaxially convex; margins straight or undulate to crenate, armed, teeth single, firmly attached, well defined, 4–8 mm, 1–3 cm apart; apical spine reddish brown to gray, subulate, 1.5–4 cm. **Scape** 2–5 m. **Inflorescences** paniculate, not bulbiferous, open; bracts persistent, triangular, 0.5–2 cm; lateral branches 10–19, ascending, comprising distal (1/4–)1/3–1/2 of inflorescence, longer than 10 cm. **Flowers** 6–12 per cluster, erect, 3–4 cm; perianth light yellow, tube shallow, broadly campanulate, 2–4.5 × 8–10.5(–18) mm, limb lobes wilting soon after anthesis, spreading, equal or subequal, 11–13 mm; stamens long-exserted; filaments inserted near or at top of perianth tube, erect, yellow, 2.5–3.3 cm; anthers yellow, (7–)9–16 mm; ovary 1.6–2.2 cm, neck slightly constricted, 1–2.5 mm. **Capsules** short-pedicellate, narrowly oblong to oblong, 3–4.5 cm, apex beaked. **Seeds** 5–6.5 mm.

Flowering mid spring–mid summer. Sandy to gravelly or rocky places with desert scrub, chaparral and pinyon-juniper woodlands; 800–2200 m; Ariz.

Agave mckelveyana may hybridize with *A. deserti* var. *simplex* and *A. utahensis* var. *utahensis* in Arizona.

24. Agave chrysantha Peebles, Proc. Biol. Soc. Wash. 48: 139. 1935 • Golden-flowered agave [E]

Agave palmeri Engelmann var. *chrysantha* (Peebles) Little

Plants acaulescent, occasionally suckering; rosettes solitary or infrequently cespitose, 0.5–1.2 × 0.8–1 dm, rather open to somewhat dense. **Leaves** spreading to ascending, 40–75(–82) × (4–)8–10(–11) cm; blade pale to glaucous or yellowish green to green, not cross-zoned, linear-lanceolate to lanceolate, rigid, adaxially deeply guttered, abaxially convex; margins undulate to wavy, armed, teeth single, well defined, 4–10 mm, (0.8–)1–3(–4) cm apart, interstitial teeth (0–)1–3(–5), mostly along distal $^2/_3$ of margins; apical spine reddish brown or brown to gray, acicular, 2.5–4.5 cm. **Scape** 4–7 m. **Inflorescences** narrowly to broadly paniculate, open, not bulbiferous; bracts persistent, triangular, 1–5 cm; lateral branches (6–)8–18, slightly ascending, comprising distal $^1/_4$–$^1/_3$ of inflorescence, longer than 10 cm. **Flowers** 12–21 per cluster, erect, 3.5–6.5(–6.7) cm; perianth golden yellow, tube not shallow, campanulate, 6–18 × (7–)10–21(–26) mm, limb lobes erect persistent and often leathery during and after anthesis, strongly unequal, (6–)9–15(–18) mm; stamens long-exserted; filaments inserted at 2 levels ca. mid perianth tube, erect, pale yellow, (3.2–)3.5–5 cm; anthers yellow, 7–20 mm; ovary (1.4–)2.2–3(–3.3) cm, neck slightly constricted, (0.5–)4–6(–8) mm. **Capsules** sessile or short-pedicellate, narrowly oblong to obovoid, 3.5–5 cm, apex short-beaked. **Seeds** 6–7 mm. $2n$ = 60.

Flowering late spring–summer. Sandy to gravelly places on granitic or volcanic soils in desert scrub, grasslands, pinyon-juniper, and oak woodlands; 700–2100 m; Ariz.

Agave chrysantha hybridizes with *A. murpheyi, A. palmeri, A. parryi* var. *couesii,* and *A. delamateri.*

25. Agave delamateri W. C. Hodgson & Slauson, Haseltonia 3: 133, figs. 1, 3–5. 1995 • Tonto Basin agave [C] [E]

Plants acaulescent, freely suckering; rosettes solitary to cespitose, 9–10 × 9–10 dm, open. **Leaves** erect or erect-ascending, 50–63(–73) × 7.5–9 cm; blade glaucous-gray to -bluish, cross-zoned, lanceolate or oblanceolate, rigid, adaxially concave, abaxially convex; margins straight or repand, armed, teeth single, well defined, 3.5–5 mm, 1–3 cm apart, interstitial teeth (3–)6–12, mostly along distal $^2/_3$ of margins; apex conspicuously incurved, spine brownish gray, slender, 2.8–3.5(–4.9) cm. **Scape** 4.5–6 m. **Inflorescences** broadly paniculate, not bulbiferous, open; bracts persistent, triangular, 1–5.5 cm; lateral branches 12–17, horizontal, comprising distal $^3/_5$–$^5/_8$ of inflorescence, longer than 10 cm. **Flowers** 14–20 per cluster, erect, 4.7–6.7(–7) cm; perianth cream, tinged with light green, tube not shallow, campanulate, 11–16 × 11–16 mm, limb lobes persistent and often leathery during and after anthesis, spreading, unequal, 9–18 mm, apex flushed with maroon; stamens long-exserted; filaments inserted at 1 level ca. mid perianth tube, erect, yellow, 3–5.3 cm, apex flushed with maroon; anthers yellow, 11–20 mm; ovary 2.1–2.9 cm, neck slightly constricted, 1–3.5 mm. **Capsules** not seen. **Seeds** unknown.

Flowering early summer. Gravelly places with desert scrub, rarely in chaparral or pinyon-juniper woodlands; of conservation concern; 700–1600 m; Ariz.

Agave delamateri is known only from about 90 individual clones and always in association with Mogollon or Salado settlement features. The plant is probably a cultivar derived from *A. palmeri* or a closely related taxon by pre-Columbian people. *Agave delamateri* hybridizes with *A. chrysantha*, acting as a pollen donor.

26. Agave palmeri Engelmann, Trans. Acad. Sci. St. Louis 3: 319. 1875 • Palmer agave

Plants acaulescent, rarely suckering; rosettes usually solitary, not cespitose, 4–13 × 7–13 dm, open. **Leaves** ascending to spreading, 35–92 × 3.5–19 cm; blade pale to glaucous-green or green, sometimes tinged reddish, not cross-zoned, linear-lanceolate to lanceolate, rigid, adaxially concave toward apex, abaxially convex toward base; margins straight or somewhat undulate, armed, teeth single, well defined, firmly attached, 3–6 mm, (0.5–)1–2 cm apart, interstitial teeth 4–6, mostly along distal $^2/_3$ of margins; apex not conspicuously incurved, spine reddish brown to brown, acicular, 3–6 cm. **Scape** (1.7–)4–6.5(–7.2) m. **Inflorescences** broadly paniculate, not bulbiferous, open; bracts persistent, triangular, 1–5 cm; lateral branches 8–26(–32), slightly ascending, comprising distal $^1/_3$ of inflorescence, longer than 10 cm. **Flowers** 8–16 per cluster, erect, 4.5–7.5 cm; perianth cream to pale yellow or light green, tube not shallow, urceolate, 10–18 × 10–16 mm, limb lobes persistent and often leathery during and after anthesis, erect, strongly unequal, (6–)9–18 mm, apex often flushed with maroon; stamens long-exserted; filaments inserted at 2 levels, above mid perianth tube and near rim, erect, pale yellow or reddish, 4–5.5(–5.8) cm, apex often flushed with maroon; anthers yellow, 11–25 mm; ovary 1.8–3.6 cm, neck slightly constricted, (0.5–)4–6 mm. **Capsules** short-pedicellate, narrowly oblong to oblong, 3.5–6 cm, apex short- to long-beaked. **Seeds** 5–7 mm. $2n$ = 60.

Flowering late spring–summer. Sandy to gravelly places on limestone in oak woodlands and grassy plains; 900–2000 m; Ariz., N.Mex.; nw Mexico.

Agave palmeri hybridizes with *A. chrysantha* in Arizona.

27. **Agave phillipsiana** W. C. Hodgson, Novon 11: 410, fig. 1. 2001 • Phillips agave E

Plants acaulescent, freely suckering; rosettes solitary to cespitose, 7.5–10 × 7.5–10 dm, open. **Leaves** erect, 76–78 × 10–11 cm; blade glaucous-green to dark green, lightly cross-zoned, lanceolate, rigid, adaxially concave toward apex, abaxially convex at base; margins straight or undulate, armed, teeth single, well defined, brittle, 4–7 mm, 1–2.5 cm apart, interstitial teeth (2–)3–7, mostly along distal 2/3 of margins; apex not conspicuously incurved, spine brownish gray, slender, 2.5–4 cm. **Scape** 2.7–5.5 m. **Inflorescences** narrowly paniculate, not bulbiferous, open; bracts persistent, triangular, 1–2+ cm; lateral branches 9–16, ascending to nearly perpendicular, comprising distal 1/3–1/2 of inflorescence, longer than 10 cm. **Flowers** 32–45 per cluster, erect, 7.4–8.6 cm; perianth greenish cream, tube campanulate, 15.5–20 × 15–23 mm, limb lobes persistent and often leathery during and after anthesis, spreading, unequal, 15–22 mm, apex often flushed with maroon; stamens long-exserted; filaments inserted subequally below rim of perianth tube, erect, yellow, 4.8–6.4 cm, apex flushed with maroon; anthers yellow, 17–25 mm; ovary 3.3–4.6 cm, neck slightly constricted, 4–8 mm. **Capsules** not seen. **Seeds** unknown. $2n$ = ca. 120.

Flowering early summer. Sandy to gravelly places with desert scrub; 700–1100 m; Ariz.

Agave phillipsiana is known from only four sites within Grand Canyon National Park. All are found on terraces along permanent waterways. Three of the sites are near pre-Columbian agricultural features or habitation sites. Two sites occur near a Kayenta-Anasazi Pueblo cliff dwelling (A.D. 1000–1150), on a terrace presumably once farmed by the early inhabitants. *Agave phillipsiana* possesses traits that would be advantageous to those harvesting it for food, since it is a plant that freely offsets and has open rosettes with large, long leaves that are easily cut. The species is possibly an ancient cultivar selected by pre-Columbian people from populations related to *A. palmeri*, *A. colorata* Gentry, or other closely related taxa.

8. FURCRAEA Ventenat, Bull. Sci. Soc. Philom. Paris 1: 65. 1793 • Mauritius hemp, pita, cabuya, cahum [for Antoine François de Fourcroy, 1755–1809, French chemist who helped establish the system of chemical nomenclature] I

Susan Verhoek

Plants massive multiannuals, scapose; trunks usually absent, to 6 m when present. **Stems** aboveground. **Leaves** up to 50, broad or narrowly lanceolate, to 3.3 m, rigid or flexible, very fibrous, margins with small or large corneous teeth, mature apex a firm, blunt point. **Inflorescences** paniculate, to 13 m, frequently producing bulblets. **Flowers** in clusters of 2–5, drooping; tepals spreading, distinct except at base, greenish or white, ovate to oblong, longer than stamens and styles; filaments abruptly expanded below middle; ovary inferior; style dilated and 3-lobed proximal to middle; stigma 3-lobed. **Fruits** capsular, globose to cylindrical, to 8 cm, infrequently produced. **Seeds** many, black, flat, 2 rows per locule. x = 30.

Species ca. 22 (2 in the flora): introduced; Mexico, Central America, West Indies, South America.

A number of *Furcraea* species are cultivated in warm climates worldwide as ornamentals and for cordage from leaf fibers. Reproduction by bulblets from the inflorescence is common.

Furcraea was divided by J. R. Drummond (1907) into three variable groups based on leaf margins. Plants of two of these groups are reported to persist or reproduce locally in southern Florida. Specimens are scarce. Within *Furcraea* there is a large synonymy, including names that are difficult to apply with certainty because they are based on incomplete specimens or European cultivated material.

R. W. Long and O. Lakela (1971) mentioned *Furcraea macrophylla* Baker as an ornamental that possibly persists in old sites, but I have seen no herbarium specimens that document its presence in Florida. That species is marked by larger leaves with larger, horny, hooked teeth.

SELECTED REFERENCES Drummond, J. R. 1907. The literature of *Furcraea* with a synopsis of the known species. Rep. (Annual) Missouri Bot. Gard. 18: 25–75. Trelease, W. 1910. Observations on *Furcraea*. Ann. Jard. Bot. Buitenzorg, Suppl. 3: 905–915.

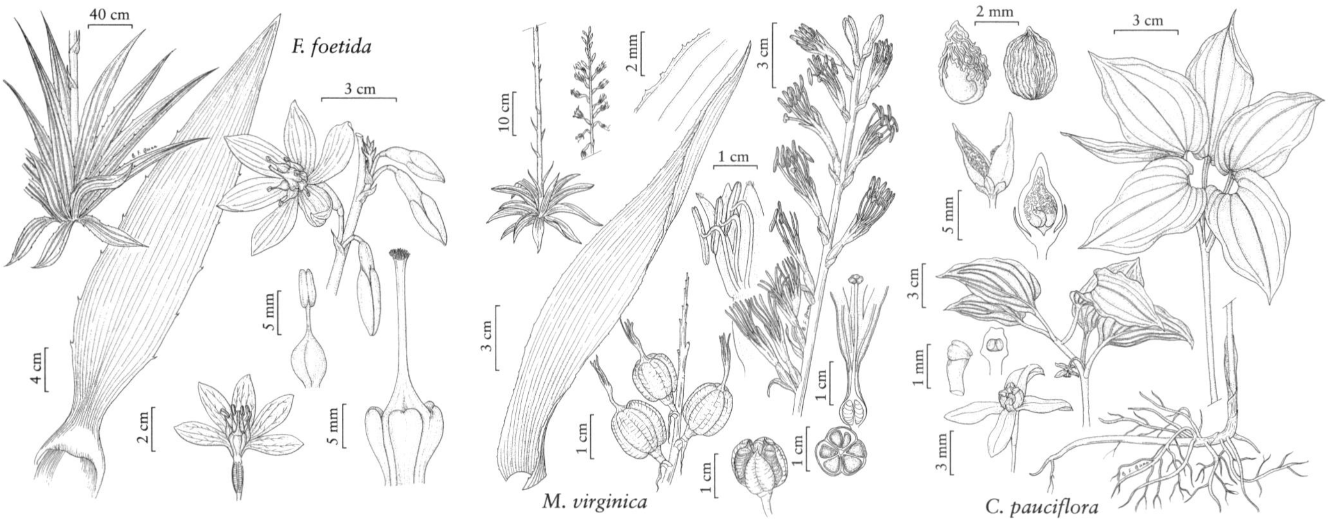

FURCRAEA ◦ MANFREDA ◦ CROOMIA

1. Leaves smooth or striate, margins usually entire at least in distal half . . . . 1. *Furcraea foetida*
1. Leaves roughened by small points, margins with hard, curved teeth . . . . 2. *Furcraea selloa*

1. Furcraea foetida (Linnaeus) Haworth, Syn. Pl. Succ., 73. 1812 [F] [I]

Agave foetida Linnaeus, Sp. Pl. 1: 323. 1753; *Furcraea gigantea* Ventenat

Leaves lanceolate, to 2.4 m × 20 cm, thick, firm, smooth or striate; margins hard and smooth, usually entire at least in distal half. **Flowers** 4–5 cm; tepals greenish white, 2.3–3 cm; ovary 1.2–1.5 cm. $2n = 60$.

Flowering fall–mid winter. Old homesteads in disturbed areas; 0 m; introduced; Fla.; Central America; West Indies; widely introduced elsewhere.

Specimens from Florida referred to in the literature as *Furcraea cabuya* Trelease with entire leaf margins (var. *integra* Trelease) probably belong here.

2. Furcraea selloa K. Koch, Wochenschr. Vereines Beförd. Gartenbaues Königl. Preuss. Staaten 3: 22. 1860 [I]

Leaves lanceolate, to 1.2 m × 10–15 cm, firm, roughened abaxially by small points, margins white with hard, curved teeth ca. 6 mm. **Flowers** ca. 4 cm; tepals white, 2.5–2.7 cm; ovary 1.4–1.8 cm. $2n = 60$.

Flowering mid fall–mid winter. Old homesteads in disturbed areas; 0 m; introduced; Fla.; Mexico; Central America; South America.

9. MANFREDA Salisbury, Gen. Pl., 78. 1866 • [For Manfredus de Monte Imperiale, fourteenth-century Italian writer on medical simples]

Agave Linnaeus sect. *Manfreda* (Salisbury) Bentham & Hooker f.

Susan Verhoek

Plants perennial, scapose, rhizomatous; rhizomes upright, globose, or cylindrical. **Roots** fleshy or fibrous. **Stems** subterranean. **Leaves** in rosettes, flexible, thin or succulent; blade linear-lanceolate to oblanceolate, often glaucous, sometimes mottled, base membranous [fibrous in species of s Mexico], margins entire, papillose, or dentate, with cartilaginous teeth, apex a soft point. **Inflorescences** racemose or spikate, bracteate; proximal flowers occasionally pedicellate. **Flowers** usually solitary at node, nocturnal; tepals connate basally into tube, green, sometimes tinged with brown, or pale yellow, or white to pink; perianth tube narrowly to broadly funnelform; limb lobes erect or recurved, oblong; stamens 6; filaments inserted on tube, in bud straight or bent, usually exserted at anthesis; ovary inferior; style usually exserted at maturity; stigma capitate, 3-angled and furrowed, or 3-lobed with lobes reflexed. **Fruits** capsular, globose to cylindrical, dehiscence loculicidal. **Seeds** flat, triangular, 0.3–0.6 cm on radial margins. $x = 30$.

Species 26 (5 in the flora): se and sc United States, nw Mexico to Honduras and El Salvador.

Most *Manfreda* species occur in Mexico. All species grow in rocky or sandy ground, frequently on slopes. At anther maturity, the style is shorter than the filaments, or bent downward away from the anthers. By stigma maturity on the third day, the style has elongated (and straightened). *Manfreda* has been included in both *Agave* and *Polianthes* in other floras.

SELECTED REFERENCES Verhoek, S. E. 1975. A Study of the Tribe Polianthеae (Including *Manfreda*) and Revisions of *Manfreda* and *Prochnyanthes* (Agavaceae). Ph.D. thesis. Cornell University. Pina L., I. 1985–1986. Consideraciones sobre el género *Manfreda*. Cact. Suc. Mex. 30: 27–32, 56–64, 84–90; 31: 12–18.

1. Limb lobes erect; leaves semisucculent; se and c United States . . . . 1. *Manfreda virginica*
1. Limb lobes spreading or recurved; leaves succulent; s Texas.
   2. Perianth tube funnelform-campanulate; filaments and styles exserted by 3 cm or more; stigmas clavate, lobes not reflexed; tepals green or brownish.
      3. Leaves erect-arching, blade linear-lanceolate; inflorescences lax; flowers spreading; filaments wide-spreading . . . . 2. *Manfreda variegata*
      3. Leaves spreading, blade broadly lanceolate; inflorescences crowded; flowers nearly erect; filaments nearly erect . . . . 3. *Manfreda sileri*
   2. Perianth tube funnelform; filaments and styles included or shortly exserted to less than 2.2 cm; stigma not clavate, deeply 3-lobed, lobes recurved; tepals white or pale yellow to pink.
      4. Filaments exserted; ovary 6–19 mm; perianth tube not narrow, 0.6–2.6 cm; leaf margins minutely denticulate . . . . 4. *Manfreda maculosa*
      4. Filaments included; ovary 4–6 mm; perianth tube narrow, 2.3–3.6 cm; leaf margins coarsely denticulate . . . . 5. *Manfreda longiflora*

1. Manfreda virginica (Linnaeus) Salisbury ex Rose, Contr. U.S. Natl. Herb. 8: 19. 1903 • Virginian agave, rattlesnake master, false aloe [F]

Agave virginica Linnaeus, Sp. Pl. 1: 323. 1753; *A. lata* Shinners; *A. tigrina* (Engelmann) Cory; *A. virginica* var. *tigrina* Engelmann; *Manfreda tigrina* (Engelmann) Small; *M. virginica* subsp. *lata* (Shinners) O'Kennon, Diggs & Lipscomb; *M. virginica* var. *tigrina* (Engelmann) Rose; *Polianthes lata* (Shinners) Shinners; *P. virginica* (Linnaeus) Shinners

Rhizomes cylindrical. **Leaves** spreading, semisucculent, 8–40(–47) × 0.5–6.5(–9.3) cm; blade usually spotted or speckled with maroon, shallowly channeled, oblanceolate to linear-lanceolate, margins entire or with cartilaginous prickles. **Scape** 4.5 × 13.8 dm. **Inflorescences** 14–68 cm, bearing 10–61 closely spaced flowers. **Flowers** sessile or pedicellate, nearly erect, slender, with sweet, fruity odor; tepals green; perianth tube 0.9–2.3 × 0.3–0.6 cm; limb lobes erect, 0.4–0.8 cm; filaments inserted near base of tube, bent in bud, exceeding tube by 1.2–3.1 cm; ovary 4–10 mm; style shorter than stamens, exceeding tube by 0.6–2.3 cm; stigma white, 3-lobed, lobes reflexed. **Capsules** globose, 1–1.7 cm diam. $2n = 60$.

Flowering summer–late summer, rarely in spring; fruiting late summer–early fall. Glades and open woods, on rocky and sandy soils, often on slopes; 0–600 m; Ala., Ark., Fla., Ga., Ill., Ind., Ky., La., Miss., Mo., N.C., Ohio, Okla., S.C., Tenn., Tex., Va., W.Va.; Mexico (Nuevo León, Tamaulipas).

Leaf shape and size in *Manfreda virginica* vary with soil type, amount of shade, length of cold period, and position of leaf in the rosette. Speckles and spots occur frequently on some leaves in most populations, and some authors have used the informal designation "forma tigrina" for such variants. Pollination is primarily by sphinx moths (S. E. Verhoek 1978).

2. Manfreda variegata (Jacobi) Rose, Contr. U.S. Natl. Herb. 8: 20. 1903 • Huaco, amole

Agave variegata Jacobi, Hamburger Garten- Blumenzeitung 21: 459. 1865; *Manfreda tamazunchalensis* Matuda; *M. xilitlensis* Matuda; *Polianthes variegata* (Jacobi) Shinners

Rhizomes globose. **Leaves** erect-arching, succulent, 14–45 × 1–2.7(–3.9 in cultivated specimens) cm; blade sometimes spotted with elongate brown or green markings, sometimes glaucous, deeply channeled, linear-lanceolate, margins usually with small, distantly spaced teeth. **Scape** 4.7–10.8 dm (–13.3 in cultivation). **Inflorescences** lax, (0.75–)1.4–2.9 dm (–4.8 dm in cultivation), bearing 7–41 distantly spaced flowers. **Flowers** sessile, spreading, with cooked onion odor; tepals green or mahogany brown; perianth tube broadly funnelform-campanulate, 0.6–1.6 cm; limb lobes recurved, yellowish green or brownish, (0.6–)0.9–1.6 cm; filaments wide-spreading at maturity, bent near middle in bud, exceeding tube by 3–6.5(–8.2) cm; ovary 9–19 mm; style ± equal to stamens; stigma pale green or brownish, clavate-triangular, 3-angled, furrowed apically, lobes not reflexed. **Capsules** globose to cylindrical, 1.6–1.8(–2.5) × 1.3–1.6 cm.

Flowering Feb–Jun; flowering period of inflorescences extending up to 40 days; fruiting early summer–mid fall, ca. 75 days after flowering begins. Dry chaparral or moist situations, on rocky slopes, open or dense oak woods in Mexico; 0–2000 m; Tex.; Mexico (Nuevo León, San Luis Potosí, Tamaulipas, and s along e Sierra Madre Oriental to Veracruz, Yucatán).

Manfreda variegata is occasionally cultivated as an ornamental (S. E. Verhoek 1978).

3. Manfreda sileri Verhoek, Brittonia 30: 168, figs. 4–6. 1978

Rhizomes globose. **Leaves** spreading, succulent, (14–)25–39 × 2.2–4.8 cm; blade light green, glaucous, spotted with green or brown round to oval markings up to 1 × 0.5 cm, channeled and undulate, or flat, broadly lanceolate, margins with teeth of several sizes, irregularly spaced, often retrorse. **Scape** 21.4–21.95 dm. **Inflorescences** crowded, 2.8–3.95 dm, bearing 27–46(–81) flowers. **Flowers** nearly erect, with cooked onion odor; tepals golden green adaxially, glaucous and green abaxially; perianth tube broadly funnelform-campanulate, 0.7–1.5(–2.2) cm; limb lobes recurved, usually longer than tube, 0.7–2.1 cm; filaments nearly erect at maturity, exserted by 4–6.6(–9.7) cm, bent near middle in bud; ovary 10–20 mm; style slightly longer than stamens; stigma pale green, clavate, 3-angled, furrowed apically, lobes not reflexed. **Capsules** cylindrical, 2.3–3.1 × 1.6–1.9 cm.

Flowering spring–mid summer. Growing separately or in colonies, in open areas, on clay soil; 0–100 m; Tex.; Mexico (Tamaulipas).

4. Manfreda maculosa (Hooker) Rose, Contr. U.S. Natl. Herb. 8: 17. 1903 • Huaco, wild-tuberose, amole plant, spice-lily, soap plant

Agave maculosa Hooker, Bot. Mag. 85: plate 5122. 1859; *A. maculosa* var. *brevituba* Engelmann; *A. maculosa* var. *minor* Jacobi; *Polianthes maculosa* (Hooker) Shinners

Rhizomes cylindrical. **Leaves** sprawling, succulent, to 36 × 0.6–2.5 cm; blade spotted with green or brown, glaucous, channeled, linear-lanceolate, brittle, margins with small, sparse, cartilaginous teeth. **Scape** 4.2–13.1 dm. **Inflorescences** open, to 4 dm, 7–33-flowered. **Flowers** erect, usually sessile, with strong, sweet odor; tepals pale yellow or white, becoming pink to rose; perianth tube funnelform, equal to or longer than ovary, 0.6–2.6 cm; limb lobes recurved, 0.9–1.9 × 0.3–0.5 cm; filaments bent near tip in bud, inserted near top of tube, exceeding tube by 0.8–2.2 cm; ovary 6–19 mm; style exceeding tube by 0.5–2.2 cm, usually shorter than stamens; stigma white, not clavate, deeply 3-lobed, lobes reflexed. **Capsules** ellipsoid, 1.1–2.4 × 1–1.5 cm.

Flowering spring–early fall. Sandy or clay soils, in mesquite scrubland, in dry or moist sites; 0–300 m; Tex.; Mexico (Coahuila, Nuevo León, Tamaulipas).

Chopped rhizomes of *Manfreda maculosa* have been used as a source of soap and shampoo in Texas, and plants are occasionally sold as ornamentals (S. E. Verhoek 1978b). Leaf size varies with growing conditions; flower size and ratio of floral parts are variable. The epithet *maculata*, which properly applies to a different Mexican species, has been used erroneously for this species.

5. **Manfreda longiflora** (Rose) Verhoek, Baileya 19: 163. 1975 • Huaco, Runyon's huaco [C]

Runyonia longiflora Rose, Addisonia 7: 39, plate 244. 1922; *Agave longiflora* (Rose) G. D. Rowley; *Polianthes runyonii* Shinners

Rhizomes cylindrical. **Leaves** spreading, succulent, to 26.5 × 1.4 cm; blade green with darker green or brown narrowly elongate spots, channeled, lanceolate, margins with coarse, distantly spaced, cartilaginous teeth. **Scape** to 3 dm (–6 dm in cultivation). **Inflorescences** dense or lax, 0.8–2(–3.5) dm, 10–21-flowered. **Flowers** semierect, sessile; tepals white, maturing to rose; perianth tube narrowly funnelform, 2.3–3.6 cm; limb lobes recurved, 0.8–1.9 × 0.2–0.35 cm; filaments included, attached at top of tube; anthers nearly sessile; ovary 4–6 mm; style included, 1.4–2.8 cm; stigma white, deeply 3-lobed, lobes reflexed. **Capsules** depressed-globose, 1 × 1–1.3 cm. $2n = 60$.

Flowering late summer–early fall. Clay slopes, dry gravelly hills or sandy prairies; of conservation concern; 0–100 m; Tex.; Mexico (Tamaulipas).

226. STEMONACEAE Engler
• Stemona Family

R. David Whetstone

Herbs [herbaceous vines, or low shrubs], caulescent [stemless], from rhizomes [tuberous roots]. **Leaves** simple, stipules absent. **Basal leaves** sheathing. **Cauline leaves** present [absent], alternate [whorled], petiolate, venation ± palmate [arcuate-pinnate with strongly developed midvein], transverse parallel veinlets connecting primary veins. **Inflorescences** axillary, cymose [racemose or unifloral]. **Flowers** bisexual, radially [bilaterally] symmetric; perianth hypogynous [to epigynous]; tepals 4[–5], distinct [connate]; stamens 4[–5], opposite tepals; ovary 1-locular; placentation apical [basal]; style 1 [absent]; stigmas [1]–3. **Fruits** capsular, 2[–3]-valved. **Seeds** globose [to ellipsoid].

Genera 4, species 30–35 (1 genus, 1 species in the flora): North America, e and se Asia, n Australia.

SELECTED REFERENCES Rogers, G. K. 1982. The Stemonaceae in the southeastern United States. J. Arnold Arbor. 63: 327–336. Whetstone, R. D. 1984. Notes on *Croomia pauciflora* (Stemonaceae). Rhodora 86: 131–137.

1. CROOMIA Torrey in J. Torrey and A. Gray, Fl. N. Amer. 1: 663. 1840 • [For Hardy B. Croom, 1797–1837, the discoverer]

Herbs perennial. **Stems** erect. **Basal leaves** scalelike, venation parallel. **Cauline leaves** 3–6 near stem tips; blade cordate to broadly elliptic, venation palmate. **Cymes** 2–3-flowered. **Flowers:** ovary globose [to broadly conic]; style short, thick; stigmas 2–3, capitate; pedicel jointed, bracteolate. **Seeds** 1–4.

Species 3 (1 in the flora): North America, se Asia (China, Japan).

1. Croomia pauciflora (Nuttall) Torrey in J. Torrey and A. Gray, Fl. N. Amer. 1: 663. 1840 E F

Cissampelos pauciflora Nuttall, J. Acad. Nat. Sci. Philadelphia 7: 114. 1834

Rhizomes extensive, internodes 8–10 cm. **Stems** to 28 cm. **Basal leaves** 1–2, from rhizomes and proximal 5 cm of aerial stems, striate (from parallel veins). **Cauline leaves** distichous; petiole channeled, 3–4.5 cm; blade ovate, 7–8 cm, base cordate, margin scabrous, apex cuspidate-obtuse. **Flowers:** perianth ca. 1 cm wide; tepals caducous, greenish, often purple-tinged, ligulate to ovate; filaments thick; ovary ca. 1 × 1.5 mm; pedicel 8–11 mm, joint ca. $^{2}/_{3}$ distance from base. **Capsules** globose, to 1 cm.

Flowering spring. Rich, sandy or rocky loam of mesic, wooded slopes and bottoms, usually in nearly neutral soils; 20–300 m; Ala., Fla., Ga., La.

The attribution to Louisiana (based on a single nineteenth-century collection) is doubtful (R. D. Thomas, pers. comm.). Apparently rare, *Croomia pauciflora* sometimes may be overlooked because of its superficial resemblance to juvenile forms of *Smilax ecirrhata* S. Watson and *S. herbacea* Linnaeus.

227. SMILACACEAE Ventenat • Catbrier Family

Walter C. Holmes

Shrubs, herbs, or vines, perennial, rhizomatous. **Stems** erect or climbing, usually prickly, sometimes unarmed. **Leaves** opposite or alternate, prominently 3-veined, reticulate between veins, usually bearing tendrils, usually leathery. **Inflorescences** umbellate [or racemose or spicate]. **Flowers** unisexual, staminate and pistillate on different plants; tepals 6, distinct, rarely united into perianth tube; stamens 2–3-whorled, anthers 1-locular; pistillate flowers bearing staminodes, pistil 3-carpellate; ovary 2-locular, 1–2 ovules per locule. **Fruits** baccate. **Seeds** 1–3.

Genera 4(–12), species ca. 375 (1 genus, 20 species in the flora): worldwide, mainly tropical to subtropical, a few temperate.

The leaves of Smilacaceae are atypical of monocotyledons in being reticulate between major veins. The family is closely related to and sometimes included in Liliaceae. It differs mainly in leaf characteristics and in being dioecious.

SELECTED REFERENCES Arber, A. 1920. Tendrils of *Smilax*. Bot. Gaz. 69: 438–442. Coker, W. C. 1944. The woody smilaxes of the United States. J. Elisha Mitchell Sci. Soc. 60: 27–69, plates 9–39. Duncan, W. H. 1975. Woody Vines of the Southeastern United States. Athens, Ga. Fernald, M. L. 1944. Overlooked species, transfers, and novelties in the flora of eastern North America. Rhodora 46: 1–28, 32–60. Judd, W. S. 1998. The Smilacaceae in the southeastern United States. Harvard Pap. Bot. 3: 147–169. Mangaly, J. K. 1968. A cytotaxonomic study of the herbaceous species of *Smilax*: Section *Coprosmanthus*. Rhodora 70: 55–82, 247–273. Morong, T. 1894. The Smilaceae [sic] of North and Central America. Bull. Torrey Bot. Club 21: 419–448. Pennell, F. W. 1916b. Notes on plants of the southern United States—II. Bull. Torrey Bot. Club 43: 409–421.

1. SMILAX Linnaeus, Sp. Pl. 2: 1028. 1753; Gen. Pl. ed. 5, 455. 1754 • Catbrier, greenbrier, sarsaparilla [ancient Greek name of an evergreen oak]

Coprosmanthus (Torrey) Kunth; *Nemexia* Rafinesque

Shrubs, vines, or herbs; rhizomes tuberous or stoloniferous, woody; roots filiform. **Stems** erect, sprawling or, more often, climbing, simple or branching, unarmed or armed with prickles; woody or herbaceous. **Leaves** deciduous or evergreen, alternate; stipules present; tendrils often present (few or rudimentary in *S. hugeri* and *S. ecirrhata*, absent in *S. biltmoreana*), paired, originating from petioles; blade linear, oblong, ovate, or, sometimes, reduced to scales in herbaceous species, base sometimes lobed. **Inflorescences** umbellate, axillary to leaves or bracts, loose to dense, pedunculate. **Flowers** unisexual; tepals 6, greenish, yellow, or bronze, ovate to elliptic; staminate flowers sometimes with pistillode, stamens 6, anthers basifixed, dehiscence introrse; pistillate flowers with 6 staminodes, style short or absent, stigmas 3, recurved, ligulate. **Berries** black, blue, purple, red, or orange. x = 13–16.

Species ca. 350 (20 in the flora): worldwide in the tropics and subtropics, with extensions into temperate areas.

The North American herbaceous species of *Smilax* (numbers 2, 5, 8, 9, 10, 11, 12, 14, and 15 in this treatment) traditionally have been placed in sect. *Nemexia* (Rafinesque) A. de Candolle. J. K. Mangaly (1968) concluded that the correct name for this group at that rank is sect. *Coprosmanthus* (Torrey) Bentham. The remaining North American species, all more or less woody, belong to sect. *Smilax*. The relatively small number of species (20) present in the flora does not warrant the elaboration of an updated subgeneric classification, which should take into account all species of the genus on a worldwide basis.

The leaves of *Smilax* are very unusual. A. Arber (1918, 1920) believed that the "blade of *Smilax* is not equivalent to the lamina of a dicotyledon but is merely a 'pseudolamina' representing an expansion of the upper region of the petiole." In this view, tendrils are also proliferations of the petiole and are not homologous to tendrils of dicotyledons. However, D. R. Kaplan (1973) remarked that unifacial monocotyledonous leaves never exhibit a lamina rudiment at the apex, and therefore there is no convincing argument that their apices are simply petiolar. He suggested that the terete leaf axis of monocotyledons is not merely an expanded petiole but is positionally equivalent to the lamina region of a dicotyledonous leaf. *Smilax* leaves lack an abscission layer, but the distal portion of the petiole undergoes a soft disintegration and the "blade" falls, leaving a rough end on the stub (W. C. Coker 1944).

Smilax has numerous uses. Sarsaparilla, a beverage and medicinal used against rheumatism, is obtained from the rhizomes of various species, mainly from Mexico and Central America. A jelly can be made from the rhizomes. The fleshy rhizomes of several vining species, most notably *S. smallii,* which have a texture of firm, crisp apples, were used by Native Americans and early settlers in the same manner as were potatoes, or else in making bread or mush. The young, succulent stems of several species are cooked and used as asparagus or the tender stems may be used in salads. Seeds were sometimes used as beads ("Indian coral") and a brown dye can be made from the roots of various species. Woody rhizomes were reportedly used by Native Americans and settlers in making pipes. Some species have been used in Native American (D. E. Moerman 1986) and folk medicine. All species of *Smilax* are excellent wildlife food and are also browsed, or the rhizomes dug and eaten, by domestic stock.

1. Stems annual, herbaceous, without prickles; ovules (1–)2 in each locule.
    2. Plants usually less than 1 m, erect; tendrils few and short or absent; umbels few, mostly axillary to bract.
        3. Leaf blades glabrous, glaucous abaxially . . . . 2. *Smilax biltmoreana*
        3. Leaf blades pubescent, not glaucous abaxially.
            4. Leaf blades broadly ovate, base cordate, apex acuminate . . . . 5. *Smilax ecirrhata*
            4. Leaf blades narrowly ovate or elliptic, base mostly rounded, apex broadly rounded, obtuse, or abruptly acute.
                5. Leaves disposed distally, ± equal; blade oblong, oblong-ovate, or oval; tendrils mostly absent; petiole shorter than blade . . . . 9. *Smilax hugeri*
                5. Leaves evenly distributed, progressively smaller proximally; blade narrowly ovate; tendrils usually present, but few; petiole usually equaling or longer than blade . . . . 10. *Smilax illinoensis*

2. Plants greater than 1 m, climbing; tendrils numerous, long, functional; umbels numerous, axillary to leaf.
   6. Leaf blades pubescent abaxially, with transparent trichomes.
      7. Leaf blades abaxially dark green, lustrous; petiole ± equaling blade . . . . . 15. *Smilax pulverulenta*
      7. Leaf blades abaxially pale green, not lustrous; petiole shorter than blade . . . . 12. *Smilax lasioneura*
   6. Leaf blades glabrous abaxially.
      8. Leaf blades never hastate; tepals 3.5–4.5 mm; anthers much shorter than filaments . . . . 8. *Smilax herbacea*
      8. Leaf blades sometimes hastate; tepals 1.5–2.5 mm; anthers ± equaling filaments.
         9. Plants less than 2 m; leaf blades coriaceous; peduncles (often 3 per axil) shorter than to ca. 1.5 as long as subtending leaf; berry diameter equaling length of pedicel; e United States . . . . 14. *Smilax pseudochina*
         9. Plants 2–3 m; leaf blades membranous; solitary pistillate umbel with peduncle longer than subtending leaf, staminate with peduncle equaling subtending leaf; berry diameter much less than length of pedicel; nw California . . . . 11. *Smilax jamesii*

1. Stems perennial, woody, usually prickly (especially at base); ovule 1 per locule.
   10. Stems densely pubescent; leaf blades densely pubescent abaxially; fruit red, apex pointed . . . . 16. *Smilax pumila*
   10. Stems glabrous (except for prickles); leaf blades glabrous or minutely pubescent abaxially; fruit red, orange, black, or purple, globose, ovoid, or flattened.
       11. Leaf blades abaxially silvery to grayish, glaucous . . . . 6. *Smilax glauca*
       11. Leaf blades abaxially green, not glaucous.
           12. Peduncle 1.5 or more times as long as petiole of subtending leaf.
               13. Leaf blade margins with thickened, cartilaginous band, often revolute and appearing as prominent vein parallel to margins, entire to remotely spinose-ciliate, frequently lobed at base; prickles rigid, broad-based; leaves drying to uniform tan . . . . 3. *Smilax bona-nox*
               13. Leaf blade margins thin, flat, without cartilaginous band, uniformly minutely serrulate in lower half or entire, never lobed; prickles flexible, bristlelike; leaves drying to dull ashy green.
                   14. Leaf blade margins minutely serrulate in basal half; e United States . . . . 19. *Smilax tamnoides*
                   14. Leaf blade margins entire; California, Oregon . . . . 4. *Smilax californica*
           12. Peduncle less than 1.5 times as long as petiole of subtending leaf.
               15. Leaf blade margins spinose-dentate, teeth firm, spiny, or margins sometimes entire with thick edges; s Florida including the Keys . . . . 7. *Smilax havanensis*
               15. Leaf blade margins entire to lobed basally, without teeth or thick edges; widely distributed.
                   16. Leaves deciduous or partially evergreen; blade ovate, ovate-oblong, narrowly ovate, suborbicular, or reniform, base rounded to cordate.
                       17. Fruit bright red to orange; perianth brownish yellow; leaf blades ovate-lanceolate to ovate-oblong; in mud or shallow water . . . . . 20. *Smilax walteri*
                       17. Fruit black to blue; perianth yellowish green to bronze; leaf blades usually ovate to broadly ovate; dry sites . . . . 17. *Smilax rotundifolia*
                   16. Leaves evergreen; blade oblong, oblong-lanceolate, oblong-linear, lance-ovate, or narrowly ovate, base cuneate to attenuate (sometimes rounded in *S. laurifolia*).
                       18. Distal leaf blades usually less than 5 cm, base auriculate, pandurate, or rounded . . . . 1. *Smilax auriculata*
                       18. Distal leaf blades (5–)7+ cm, base unlobed.
                           19. Leaf blades lanceolate-ovate to narrowly ovate, prominently reticulate, often variegated, 5-veined from base, apex gradually narrowed, acute, or acuminate . . . . 18. *Smilax smallii*

19. Leaf blades linear, oblong, lance-oblong, or narrowly elliptic, sometimes broadly ovate, not prominently reticulate, 3-veined from base, apex abruptly narrowed to acute point or rounded. . . . . 13. *Smilax laurifolia*

1. Smilax auriculata Walter, Fl. Carol., 245. 1788 • Wild-bamboo [W1]

Smilax beyrichii Kunth; *S. lata* Small

Vines; rhizomes linear or dense masses of potatolike tubers. **Stems** perennial, climbing, branching zigzag, terete, 5–9 m × 5–8 mm, woody, glabrous; prickles numerous, sparse or absent distally, flattened, rigid, stout, to 4 mm. **Leaves** evergreen; petiole 0.5–1.2 cm; blade green abaxially, drying to brownish green, narrowly ovate to ovate-elliptic, with 3 prominent veins, secondary veins obscure to prominent, 4.5–6(–8.5) × 2–3.5 cm, not glaucous, glabrous or minutely pubescent abaxially, base auriculate, pandurate, or rounded, cuneate at insertion of petiole, margins entire, apex acute to abruptly deflexed point. **Umbels** 3–8, terminal, axillary to leaves, 5–8 (–25)-flowered, loose; peduncle 0.2–1.5 cm. **Flowers:** perianth green; tepals: staminate 6–8 mm, pistillate 3–4 mm; ovule 1 per locule; pedicel 0.2–1 cm. **Berries** purplish maroon, purple, or black, ovoid to flattened, 5–7 mm, glaucous.

Flowering Apr–Jul. Dunes and sandy flatwoods, full sun; 0–100 m; Ala., Fla., Ga., La., Miss., N.C., S.C.; West Indies (Bahamas).

2. Smilax biltmoreana (Small) J. B. Norton ex Pennell, Bull. Torrey Bot. Club 43: 413. 1916 [E]

Nemexia biltmoreana Small, Fl. S.E. U.S., 281, 1328. 1903; *Smilax ecirrhata* (Engelmann ex Kunth) S. Watson var. *biltmoreana* (Small) H. E. Ahles

Herbs. Stems annual, erect, 0.2–0.6 m, herbaceous, glabrous, glaucous abaxially; prickles absent. **Leaves** few, approximate distally; tendrils absent; blade ovate, 6–15 × 1.5–8 cm; glabrous and glaucous abaxially, base cordate, margins entire, apex acute to acuminate. **Umbels** 1–few, proximalmost axillary to leaves or bracts, few-flowered, open; peduncle 5–10 cm. **Flowers:** perianth greenish; tepals 2.5–3.5 mm; anthers longer than filaments; ovules (1–)2 per locule; pedicel slender, ca. 1 cm. **Berries** black to bluish black, globose, ca. 8 mm diam., not glaucous. $2n = 26$.

Flowering May–Jun. Rich, open woods in ravines, along streams, and at bases of bluffs; 0–1500 m; Ala., Fla., Ga., Ky., N.C., S.C., Va.

Smilax biltmoreana is distinguished by its ovate leaves with glabrous and glaucous abaxial surfaces.

3. Smilax bona-nox Linnaeus, Sp. Pl. 2: 1030. 1753 (as bona nox) • Zarzaparrilla [F] [W1]

Smilax bona-nox var. *exauriculata* Fernald; *S. bona-nox* var. *hastata* (Willdenow) A. de Candolle; *S. bona-nox* var. *hederifolia* (Beyrich ex Kunth) Fernald; *S. bona-nox* var. *littoralis* Coker; *S. hastata* Willdenow; *S. hederifolia* Beyrich ex Kunth; *S. renifolia* Small; *S. variegata* Walter

Vines; rhizomes tuberous, woody, or stoloniferous. **Stems** perennial, climbing, often zigzag distally, branched, terete to 4-angled, stout, to 5+ m × 5 mm, woody, glabrous or infrequently with stellate trichomes; prickles often absent distally, tips black, flattened, broad-based, stout, 4–9 mm, rigid. **Leaves** evergreen, ± evenly dispersed; petiole 0.7–1.5 cm; blade pale green, often with white blotches, drying to uniform tan, thickish, broadly ovate to lanceolate-ovate or hastate to pandurate, with 3(–5) ± prominent veins, 3–10 × 2.5–9 cm, not glaucous, glabrous or minutely pubescent abaxially, base cordate to truncate, frequently lobed; margins entire to remotely spinose-ciliate, thickened by ribbed, cartilaginous band, often revolute and appearing as prominent vein parallel to margins, apex rounded to short-apiculate. **Umbels** few to numerous, axillary to leaves, 10–15+-flowered, moderately dense; peduncle 1.5–6+ cm. **Flowers:** perianth pale green; tepals 3–4.5 mm; anthers shorter than to ± equaling filaments; ovule 1 per locule; pedicel 0.8–1.2 cm. **Berries** black, ovoid to spherical, 6–8 mm, shiny to dull, sometimes glaucous. $2n = 32$.

Flowering Apr–May. Well-drained to wet areas in woods, fields, thickets, hedgerows, floodplain forests, etc., full to partial sun; 0–1000 m; Ala., Ark., Del., D.C., Fla., Ga., Ill., Ind., Kans., Ky., La., Md., Miss., Mo., N.C., Ohio, Okla., S.C., Tenn., Tex., Va.; Mexico; West Indies.

Numerous varieties, based mainly on differences in leaf shape, have been proposed for *Smilax bona-nox*. Variation is so great even in individual plants that recognition of these varieties is untenable. J. A. Steyermark (1963) suggested that leaf variation may be correlated with stages of plant maturity. The species often may be considered weedy, occurring in very dense, tangled masses.

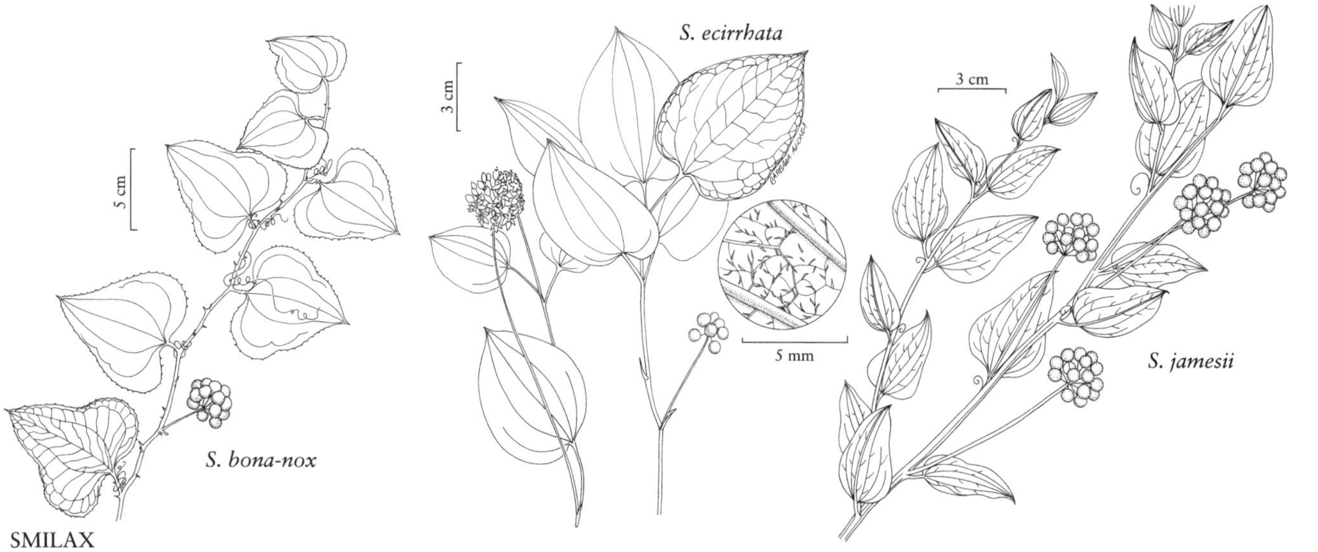

SMILAX

4. **Smilax californica** (A. de Candolle) A. Gray in S. Watson, Bot. California 2: 186. 1880 [E]

Smilax rotundifolia Linnaeus var. *californica* A. de Candolle in A. L. P. P. de Candolle and C. de Candolle, Monogr. Phan. 1: 75. 1878

Shrubs or vines; rhizomes short, knotty. **Stems** perennial, climbing or not, to 12 m, woody, glabrous; prickles sometimes absent distally, bristlelike, 3–11 mm, flexible. **Leaves** evergreen, ± evenly dispersed; petiole 0.5–2 cm; blade dull green, drying to dull, ashy green, ovate to broadly ovate, conspicuously veined, 4–11 × 3–8 cm, not glaucous, glabrous, base cordate to subcordate; margins entire, thin, flat, not banded, never lobed; apex acute, often apiculate. **Umbels** axillary to distal leaves, (2–)8–13(–19)-flowered; peduncle 2–5 cm, longer than petiole of subtending leaf. **Flowers:** perianth green; tepals 3–6 mm; ovule 1 per locule; pedicel thin, 1–1.5 cm. **Berries** black, ovoid, 7–9 mm.

Flowering May–Jun. Yellow pine and mixed evergreen forests, often in thickets along rivers, streams, and springs; partial–full sun; 250–1200 m; Calif., Oreg.

Smilax californica is apparently closely related to the more eastern *S. tamnoides*. It lacks the minute serrulations characteristic of the latter's leaves.

5. **Smilax ecirrhata** S. Watson in A. Gray et al., Manual ed. 6, 520. 1890 [E] [F]

Coprosmanthus ecirrhatus (S. Watson) Chapman; *C. herbaceus* (Linnaeus) Kunth var. *ecirratus* Engelmann ex Kunth; *Nemexia ecirrata* (Engelmann ex Kunth) Small (as ecirrhata); *Smilax herbacea* Linnaeus var. *ecirrata* (Engelmann ex Kunth) A. de Candolle

Herbs. **Stems** annual, erect to leaning, with bladeless bracts proximally, leaves distally, 0.3–0.8 m, herbaceous; prickles absent. **Leaves** mostly distal (upper $^1/_2$–$^2/_3$ of plant); petiole thin, shorter than blade; tendrils few and short or absent; blade thin, broadly elliptic-ovate to subrotund, 9–12 × 5–9 cm, pubescent and not glaucous abaxially, base cordate to truncate, margins entire, convex, apex round to acuminate. **Umbels** 1–3, proximalmost axillary to bracts below leaves, to 25-flowered, hemispherical to globose. **Flowers:** perianth green; tepals 3.5–4 mm; anthers ± equaling filaments; ovules (1–)2 per locule; pedicel 0.5–1.3 cm. **Berries** purplish black, globose, 9–11 mm, not glaucous. $2n = 52$.

Flowering May–Jun. Open woods and thickets; 100–800 m; Ont.; Ark., Ill., Ind., Iowa, Kans., Ky., Mich., Minn., Mo., Nebr., Ohio, Okla., S.Dak., Tenn., Wis.

Smilax ecirrhata is similar to *S. huberi* and *S. biltmoreana*, both with fewer leaves of different shapes.

6. **Smilax glauca** Walter, Fl. Carol., 245. 1788 • Wild sarsaparilla, sawbrier [E] [W1]

Smilax glauca var. *leurophylla* S. F. Blake; *S. spinulosa* Smith

Vines; rhizomes tuberous, or spinose, linear. **Stems** perennial, climbing, branching, green, often mottled, terete, to 5+ m, woody, glaucous, glabrous; prickles, when present, thin, 1–5 mm. **Leaves** deciduous to semi evergreen, ± evenly disposed; petiole 0.5–1.5 cm; blade green to glaucous-green, often mottled adaxially, silvery grayish abaxially, drying to brownish tan adaxially, broadly ovate, elliptic to reniform, with 3 (or 5) conspicuous veins, 4.5–11 × 2.5–6.6 cm, glabrous and glaucous abaxially, base truncate, subcordate, or attenuate, margins entire, apex rounded, tapering, or short-acuminate. **Umbels** few to many, axillary to leaves, 5–12+-flowered, open, umbellate to hemispherical; peduncle 2–5 cm. **Flowers:** perianth yellow to bronze; tepals 3–7 mm; anthers longer than filaments; ovule 1 per locule; pedicel 0.5–1 cm. **Berries** blue to black, subglobose, 8–10 mm, shiny black at maturity, glaucous. $2n = 28, 32$.

Flowering May–Jul. Dry to wet woods, thickets, hedgerows, roadsides; 0–800 m; Ala., Ark., Conn., Del., D.C., Fla., Ga., Ill., Ind., Ky., La., Md., Miss., Mo., N.J., N.Y., N.C., Ohio, Okla., Pa., S.C., Tenn., Tex., Va., W.Va.

Smilax glauca is easily recognized by its glaucous to whitened abaxial leaf surfaces, which, however, may be altered by heat in drying. It is reportedly the most weedy species of the genus. The plants tend to be evergreen in the more southern part of the distribution.

7. **Smilax havanensis** Jacquin, Enum. Syst. Pl., 33. 1760

Smilax coriacea Sprengel; *S. coriacea* var. *ilicifolia* O. E. Schulz; *S. guianensis* Vitman var. *subarmata* O. E. Schulz; *S. havanensis* var. *portoricensis* A. de Candolle; *S. ilicifolia* Desvaux ex Hamilton; *S. subarmata* (O. E. Schulz) O. E. Schulz

Vines; rhizomes short. **Stems** perennial, climbing, branching, ± woody, glabrous, ± unarmed, to stunted, rigid (shrublike), densely armed; prickles pale green with dark tips, recurved, flattened, stout, 1–2 mm; stems (especially in nodal areas) with small, waxlike nodules often connected by arachnoid filaments. **Leaves** evergreen, ± evenly disposed; petiole 0.5–2 cm; blade variable, dull green, thick, narrowly oblong-elliptic, elliptic-ovate, or oval, conspicuously veined, 2–19 × 0.6–13.5 cm, not glaucous or pubescent, base pointed to rounded to narrowly cordate, margins spinose, or sometimes entire with thick edges, apex abruptly mucronate, spiny, or emarginate. **Umbels** few to numerous, axillary to leaves, few to numerous-flowered, ± open; peduncle to 1.6 cm, ± equaling or shorter than subtending petiole. **Flowers:** perianth green; tepals ca. 1.5 mm; ovule 1 per locule. **Berries** purplish black, globose, sometimes pointed, ca. 6 mm, not glaucous.

Flowering year-round. Pinelands and hammocks; 0–100 m; Fla.; West Indies.

Smilax havanensis is unusually variable in habit, leaf shape, amount of spines (particularly on the leaf margins), and shape of fruit.

8. **Smilax herbacea** Linnaeus, Sp. Pl. 2: 1030. 1753 • Carrion-flower, Jacob's-ladder, smilax herbacé [E]

Coprosmanthus herbaceus (Linnaeus) Kunth; *C. peduncularis* (Muhlenberg ex Willdenow) Kunth; *Nemexia cerulea* Rafinesque; *N. herbacea* (Linnaeus) Small; *N. nigra* Rafinesque; *Smilax herbacea* subsp. *crispifolia* Pennell; *S. herbacea* var. *peduncularis* (Muhlenberg ex Willdenow) A. de Candolle; *S. herbacea* var. *simsii* A. de Candolle; *S. peduncularis* Muhlenberg ex Willdenow

Vines. **Stems** annual, climbing, branching, to 2.5 m, herbaceous, glabrous; prickles absent. **Leaves:** petiole thin, 1–6 cm; tendrils numerous, long, functional; blade oblong-ovate, ovate, or round, 4.5–12 × 3–9 cm, glabrous abaxially, base cordate to truncate, margins entire, apex obtuse to acute; proximal cauline leaves narrower and smaller. **Umbels** many, axillary to leaves, 20–100+-flowered, globose; peduncle to 30 cm, progressively shorter distally. **Flowers:** perianth greenish, carrion-scented; tepals 3.5–4.5 mm; anthers much shorter than filaments; ovules (1–)2 per locule; pedicel 0.5–2 cm. **Berries** blue, subglobose, ca. 10 mm diam., glaucous. $2n = 26$.

Flowering May–Jun. Higher elevations in rich woods, alluvial thickets, and meadows, often in calcareous soils; 100–800 m; N.B., Ont., Que; Ala., Ga., Ky., Maine, Md., Mass., N.H., N.Y., N.C., Ohio, Pa., S.C., Tenn., Vt., Va., W.Va.

The leaves and habit of *Smilax herbacea* are quite variable.

9. Smilax hugeri (Small) J. B. Norton ex Pennell, Bull. Torrey Bot. Club 43: 420. 1916 [E]

Nemexia hugeri Small, Fl. S.E. U.S., 280, 1328. 1903; *Smilax ecirrhata* S. Watson var. *hugeri* (Small) H. E. Ahles

Herbs; rhizomes knotty, slender. **Stems** annual, erect, 2–5 m, herbaceous, glabrous; prickles absent. **Leaves** few, disposed distally or evenly distributed, subequal; petiole shorter than blade; tendrils absent or rudimentary; blade oblong, oblong-ovate, or sometimes oval, 2.5–12 × 1.5–7.5 cm, not glaucous, puberulent abaxially, base rounded to slightly cordate, margins entire, apex broadly rounded to obtuse or abruptly pointed. **Umbels** 1–3, proximalmost axillary to bracts, 5–12-flowered. **Flowers:** perianth greenish; tepals 3–3.5 mm; anthers shorter than filaments; ovules (1–)2 per locule. **Berries** globose, 8–10 mm diam., glaucous. $2n = 26$.

Flowering Apr–May. Moist, mixed woods and wooded slopes; Ala., Fla., Ga., N.C., S.C., Tenn.

Smilax hugeri is distinguished from *S. ecirrhata* and *S. biltmoreana* mainly by leaf morphology. Specimens from Tennessee are not sharply distinct from *S. ecirrhata*. J. K. Mangaly (1968) cited this as possible evidence of divergence from that species.

10. Smilax illinoensis Mangaly, Rhodora 70: 263, fig. 25A. 1968 [E]

Herbs. Stems annual, erect, unbranched, 0.5–1 m, herbaceous; prickles absent. **Leaves** numerous, evenly distributed, larger distally; petiole thin, equaling or longer than blade; tendrils distal, few, short; blade narrowly ovate, pubescent and not glaucous abaxially, base rounded to truncate, margins convex, apex acute to acuminate. **Umbels** 3–10, axillary to leaves and bracts, 10–50-flowered, semiglobose; proximalmost peduncle longer than distalmost. **Flowers:** tepals 3.5–4.5 mm; anthers usually shorter than filaments; ovules (1–)2 per locule. **Berries** blue to black, globose. $2n = 26$.

Flowering May–Jun. Roadside thickets, woods; 150–600 m; Ont.; Ill., Ind., Iowa, Mich., Minn., Mo., Ohio, Wis.

Smilax illinoensis is intermediate between *S. ecirrhata* and *S. lasioneura*.

11. Smilax jamesii G. A. Wallace, Brittonia 31: 416, fig. 1. 1979 [C] [E] [F]

Herbs; rhizomes brown or black, zigzag. **Stems** annual, climbing, branched, 2–3 m, herbaceous, glabrous; prickles absent. **Leaves** evenly distributed, proximalmost reduced to bracts; petiole shorter than blade; tendrils numerous, long, functional; blade dark green, triangular, ovate, or slightly hastate, 2.5–8.5 × 4–7.5 cm, membranous, glabrous and glaucous abaxially, base truncate, cordate, or hastate, lobes rounded; margins entire; apex cuspidate. **Umbels** axillary to leaves, staminate to 20-flowered, pistillate few–40-flowered, moderately dense, subspherical; peduncle of staminate umbel ± equaling subtending leaf, pistillate to 13.5 cm. **Flowers:** perianth light green; tepals 1.5–2 mm; anthers slightly shorter than to equaling filaments; ovules (1–)2 per locule; pedicel thin, ca. 1.5 cm. **Berries** blue, ovoid, 6–8 mm.

Flowering May–Jul. Alder thickets, lake and stream sides, bracken fern slopes; of conservation concern; 1200–2500 m; Calif.

Smilax jamesii is the only herbaceous species of *Smilax* known to occur in western North America. It is found in the Klamath Mountains.

SELECTED REFERENCE Wallace, G. A. 1979. An overlooked new species of *Smilax* (Smilacaceae) from northern California. Brittonia 31: 416–421.

12. Smilax lasioneura Hooker, Fl. Bor.-Amer. 2: 173, plate 187A. 1840 (as lasioneuron) [E] [F]

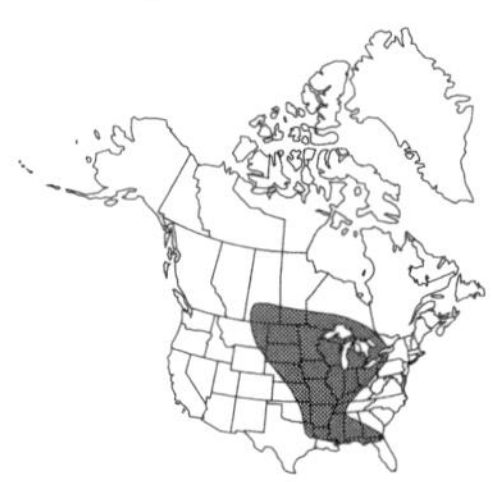

Coprosmanthus lasioneuron (Hooker) Kunth; *Nemexia herbacea* (Linnaeus) Small subsp. *melica* A. Nelson; *N. lasioneura* (Hooker) Rydberg; *N. tenuis* (Small) Small; *Smilax diversifolia* Small; *S. herbacea* Linnaeus var. *inodora* M. E. Jones; *S. herbacea* subsp. *lasioneura* (Hooker) Á. Löve & D. Löve; *S. herbacea* var. *lasioneura* (Hooker) A. de Candolle; *S. tenuis* Small

Herbs. Stems annual, erect to ascending, branched, 2–2.5 m, herbaceous, glabrous; prickles absent. **Leaves** evenly distributed, proximalmost smaller, narrower; petiole 1.5–9 cm, shorter than blade; tendrils numerous, long, functional; blade not lustrous, pale green abaxially, ovate to round, 4–8 × 3–6 cm, not glaucous, pubescent abaxially, with transparent trichomes, base cordate, margins entire, convex, apex acuminate to rounded and cuspidate. **Umbels** many, axillary to leaves, to 35-flowered, dense, globose; peduncle to 12+ cm, short. **Flow-**

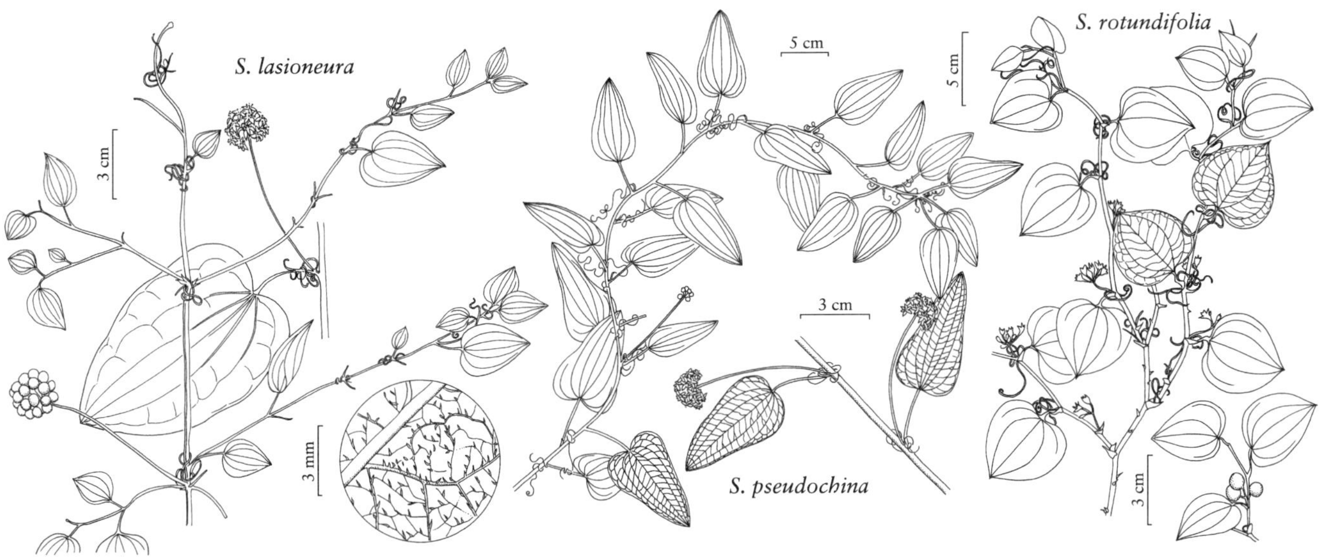

SMILAX

ers: perianth greenish; tepals 35–45 mm; anthers equaling or shorter than filaments; ovules (1–)2 per locule; pedicel 0.5–1.5 cm. **Berries** bluish black to black, subglobose, 8–10 mm, glaucous. 2*n* = 26.

Flowering May–Jun. Rich, alluvial woods, thickets, borders; 300–700 m; Man., Ont., Sask.; Ala., Ark., Fla., Ga., Ill., Ind., Iowa, Kans., Ky., La., Mich., Minn., Miss., Mo., Mont., Nebr., N.Y., N.Dak., Ohio, Okla., Pa., S.Dak., Tenn., Tex., Wis., Wyo.

Smilax lasioneura is primarily distributed in the central plains and lower elevations of the Appalachian and Rocky mountains. Plants distributed in the southern humid areas tend to be larger than those of more northern areas.

13. Smilax laurifolia Linnaeus, Sp. Pl. 2: 1030. 1753 • Bamboo vine, blaspheme vine, laurel greenbrier [W1]

Smilax alba Pursh; *S. lanceolata* Linnaeus

Vines forming extensive colonies; rhizomes irregularly branched, tuberous, woody. **Stems** perennial, climbing, branching, terete, 5+ m × 15 mm , woody, glaucous, glabrous; prickles dark, flat, to 12 mm, rigid. **Leaves** evergreen, ± evenly arranged; petiole 0.5–1.5 cm; blade abaxially green, drying to pale brown to brownish green, oblong-elliptic, lance-elliptic, or, sometimes, linear or broadly ovate, not prominently reticulate, 3-veined from base, 7–13 × 1.3–4(–6) cm, coriaceous, not glaucous, glabrous or minutely pubescent abaxially, base attenuate to rounded; margins entire, often revolute, teeth absent; apex acute to rounded, often mucronate. **Umbels** numerous, axillary to leaves, generally on short branches, 5–12(–25)-flowered, ± open, spherical; peduncle 0.5–1.5 cm, shorter than subtending leaf. **Flowers:** perianth yellow, cream, or white; tepals 4–5 mm; anthers exceeding filaments; ovule 1 per locule; pedicel 0.5-1 cm. **Berries** black, ovoid, 5–8 mm, shining, glaucous.

Flowering Aug–Oct. Bays, bogs, pocosins, swamp margins, marshy banks; Ala., Ark., Del., Fla., Ga., La., Md., Miss., N.C., Okla., S.C., Tenn., Tex., Va.; West Indies (Bahamas, Cuba).

The stems of *Smilax laurifolia* are viciously armed.

14. Smilax pseudochina Linnaeus, Sp. Pl. 2: 1031. 1753 (as pseudo china) [E] [F]

Coprosmanthus tamnifolius (Michaux) Kunth; *Nemexia tamnifolia* (Michaux) Small; *Smilax inermis* Walter; *S. leptanthera* Pennell; *S. tamnifolia* Michaux

Herbs. Stems annual, immature erect, older climbing, 1–2 m, occasionally branching, herbaceous, glabrous; prickles absent. **Leaves** numerous, evenly distributed; tendrils numerous, long, functional; blade deltate-ovate, pandurate, or cordate-ovate, the larger hastate, thin, 5–12 × 2–5 cm, coriaceous, glabrous, glaucous abaxially, base truncate to subcordate; apex round to cuspidate. **Umbels** solitary or in sets of 3, axillary to leaves, 10–35-flowered, loose, subglobose; peduncle shorter than to 1.5 as long as subtending leaf. **Flowers:** perianth greenish; tepals 1.5–2.5 mm; anthers equaling or slightly larger than filaments; ovules (1–)2 per locule; pedicel 0.3–1.5 cm. **Berries** blue to black, subglobose, 4–6 mm, glaucous. $2n = 30$.

Flowering May–Jul. Bogs, wet woods, marshes, pocosins; Del., D.C., Ga., Md., N.J., N.Y., N.C., Pa., S.C., Va.

15. Smilax pulverulenta Michaux, Fl. Bor.-Amer. 2: 238. 1803 [E]

Smilax herbacea Linnaeus var. *pulverulenta* (Michaux) A. Gray

Herbs. Stems annual, climbing, branching, to 2.5 m, herbaceous, glabrous; prickles absent. **Leaves** evenly distributed; petiole ± equaling blade; tendrils numerous, long, functional; blade lustrous, dark green abaxially, narrowly to broadly ovate, 8–16 × 3.5–9 cm, often with minute, whitish pubescence abaxially mostly on veins, with transparent trichomes, base cordate, margins entire, convex, apex long-acuminate; leaves on branches with oblong, smaller blade. **Umbels** numerous, axillary to leaves, 10–35-flowered, globose; peduncle equaling or shorter than subtending leaf. **Flowers:** perianth greenish; tepals 3.5–5 mm; anthers shorter than filaments, ovules (1–)2 per locule; pedicel 0.5–2.2 cm. **Berries** black, subglobose, 8–10 mm, not glaucous. $2n = 26$.

Flowering Apr–Jun. Rich woods, thickets, usually in calcareous soils; 0–900 m; Ark., Del., D.C., Ill., Ind., Ky., Md., Minn., Mo., Nebr., N.J., N.Y., N.C., Pa., Tenn., Va., W.Va.

The distribution of *Smilax pulverulenta* is disjunct. The eastern distribution is centered in the mid-Atlantic United States, while the western distribution is centered in the Ozark Plateau of Missouri.

16. Smilax pumila Walter, Fl. Carol., 244. 1788 • Sarsaparilla vine [E] [W1]

Smilax humilis Miller, name proposed for rejection; *S. pubera* Michaux; *S. puberula* Kunth

Subshrubs or vines; rhizomes black, knotted, 5–6 × 2 cm, often with white to pinkish stolons. **Stems** perennial, prostrate to clambering, branching, slender, to 1 m, ± woody, densely woolly-pubescent, usually prickly (especially at base). **Leaves** mostly evergreen, ± evenly disposed; petiole 0.05–0.25 cm, often longer on sterile shoots; blade gray-green, drying to ashy gray-green, obovate to ovate-lanceolate, with 3 prominent veins, 6–10.5 × 5–8 cm, glabrous adaxially, densely puberulent abaxially, base cordate to deeply notched, margins entire, apex bluntly pointed. **Umbels** 1–7, axillary to leaves, 5–16-flowered, loose, spherical; peduncle 0.2–0.8 cm, shorter than to 1.5 as long as petiole of subtending leaf. **Flowers:** perianth yellowish; tepals 3–4 mm; anthers much shorter than filaments; ovule 1 per locule; pedicel thin, 0.1–0.4 cm. **Berries** red, ovoid, 5–8 mm, with acute beaks, not glaucous.

Flowering Oct–Nov. Woods, along streams, sandy soil; Ala., Ark., Fla., Ga., La., Miss., S.C., Tex.

The red, pointed fruits and densely pubescent herbage of *Smilax pumila* are distinctive. In Louisiana, the dried leaves are used to prepare a tea for upset stomach.

The name *Smilax humilis* Miller, which predates *S. pumila* by 20 years and recently has been determined to apply also to this species, has been proposed for rejection (J. L. Reveal 2000). If that proposal is not adopted, the correct name will be *S. humilis*.

17. Smilax rotundifolia Linnaeus, Sp. Pl. 2: 1030. 1753 • Common greenbrier or catbrier, bullbrier, horsebrier [E] [F] [W1] [W2]

Smilax caduca Linnaeus; *S. quadrangularis* Muhlenberg ex Willdenow; *S. rotundifolia* var. *crenulata* Small & A. Heller; *S. rotundifolia* var. *quadrangularis* (Muhlenberg ex Willdenow) Alph. Wood

Vines; rhizomes linear. **Stems** perennial, climbing, branching, terete to quadrangular, 5–6+ m × 6 mm, woody, glabrous; prickles green with dark tips, stout, to 12 mm. **Leaves** deciduous to evergreen, ± evenly disposed; petiole 0.5–1.5 cm; tendrils numerous; blade variable, bright green, drying to pale to brownish green, usually ovate to broadly ovate, with 3 (or 5) ± prominent veins, 4–17 × 4–16 cm, lustrous, not glaucous, glabrous abaxially, base cordate to rounded with acute insertion at petiole, margins entire, apex abruptly pointed. **Umbels** numerous, axillary to leaves, 5–12(–20)-flowered, open to dense, hemispherical to spherical; peduncle to 1.5 cm, longer or shorter than petiole of subtending leaf. **Flowers:** perianth pale yellowish green to bronze; tepals 3–4 mm; anthers shorter than to ± equaling filaments; ovule 1 per locule; pedicel 0.2–1.5 cm. **Berries** blue-black to black, globose, 5–8 mm, glaucous. $2n = 32$.

Flowering Apr–Jun. Dry to moist, sometimes riparian woods, borders, hedgerows, thickets; 0–200 m; N.S.,

Ont.; Ala., Ark., Conn., Del., D.C., Fla., Ga., Ill., Ind., Kans., Ky., La., Maine, Md., Mass., Mich., Miss., Mo., N.H., N.J., N.Y., N.C., Ohio, Okla., Pa., R.I., S.C., Tenn., Tex., Va., W.Va.

Normally, the peduncle of *Smilax rotundifolia* is about the same length as the petiole of the subtending leaf. In exceptional cases, the peduncle may be considerably longer, thereby making this widely distributed species difficult to distinguish from *S. bona-nox* and *S. tamnoides*. It lacks the marginal cartilaginous band found on the leaves of the former species and the hispid prickles of the stem of the latter. Specimens of *S. tamnoides* lacking prickles may be distinguished by their more strongly ridged stems.

18. Smilax smallii Morong, Bull. Torrey Bot. Club 21: 430. 1894 • Cantaque [E] [W1]

Smilax cinnamomiifolia Small; *S. ovata* Pursh 1814, not Duhamel 1803

Vines; rhizomes irregularly tuberous, branched, thick, to 1.5+ m. **Stems** perennial, climbing, greenish or reddish brown, terete, 10+ m × 27 mm, woody, glabrous, glaucous when young; prickles few, scattered, recurved, flattened, 3–4 mm. **Leaves** evergreen, distally disposed (± evenly disposed on immature plants); petiole 0.3–1.3 cm; blade deep green, often variegated, drying to pale green, lanceolate-elliptic to narrowly ovate, prominently reticulate, 5-veined from base, 5–6.6 × 1.7–3 cm, abaxially glabrous, not glaucous, base rounded to cuneate, margins entire to undulate, apex acuminate. **Umbels** many, axillary to leaves, 7–15-flowered, loose to dense, hemispherical to spherical; peduncle 0.5–2.3 cm. **Flowers:** perianth green; tepals 3–6 mm; anthers ± 1/5 as long as filaments; ovule 1 per locule; pedicel (0.3–)0.5–0.7(–1) cm. **Berries** black, globose to compressed, 5.5–8 mm, glaucous.

Flowering Jun–Jul. Rich woods, pinewoods, streambanks, edges of fields, swamp margins, ditches, well-drained but not dry soils; 0–600 m; Ala., Ark., Del., D.C., Fla., Ga., La., Md., Miss., N.C., Okla., S.C., Tex., Va.

Smilax smallii is the highest-climbing species of *Smilax* within the flora and it is particularly conspicuous during the winter. It has been used as an ornamental and as a winter decoration, primarily in the Christmas trade. The starchy rhizomes and succulent, immature stems were used for food by Native Americans and early settlers.

19. Smilax tamnoides Linnaeus, Sp. Pl. 2: 1030. 1753 • China root, hellfetter, bristly greenbrier [E] [F]

Smilax grandifolia Buckley; *S. hispida* Muhlenberg ex Torrey; *S. hispida* var. *australis* Small; *S. hispida* var. *montana* Coker; *S. tamnoides* var. *hispida* (Muhlenberg ex Torrey) Fernald

Vines; rhizomes knotty, short. **Stems** perennial, climbing, branching, terete, to 7+ m × 9 mm, woody, glabrous, with prickles proximally, prickles usually absent distally; prickles blackish, unequal, 3–10+ mm, bristly, flexible. **Leaves** ± persistent in southern part of range, deciduous in north; petiole 1–2 cm; blade green, drying to light olive-gray, ovate-lanceolate, ovate, or sometimes pandurate, prominently reticulate, thin, 5–13 × 3–10 cm, not glaucous, glabrous (to prickly on major veins abaxially), base rounded to cordate, margins entire apically, minutely serrulate basally, thin, flat, not banded, not lobed, apex pointed. **Umbels** many, axillary to leaves, to 25-flowered, open to dense, spherical; peduncle often drooping, 1.5–6.5 cm. **Flowers:** perianth green to bronze; tepals 4–5 mm; anthers ± equaling to shorter than filaments; ovule 1 per locule; pedicel 0.4–1.2 cm. **Berries** black, globose, 6–10 mm, not glaucous.

Flowering May–Jun. Wet to dry woods, thickets, bottomlands; 0–400 m; Ont.; Ala., Ark., Conn., Del., D.C., Fla., Ga., Ill., Ind., Iowa, Kans., Ky., La., Md., Mass., Mich., Minn., Miss., Mo., Nebr., N.J., N.Y., N.C., Ohio, Okla., Pa., S.C., S.Dak., Tenn., Tex., Vt., Va., W.Va., Wis.

Smilax tamnoides is marked by its dark, flexible, and unequal bristles and leaves with minutely serrulate margins basally. The far-western *S. californica* appears to be closely related but lacks the serrulate margins of the leaf bases.

20. Smilax walteri Pursh, Fl. Amer. Sept. 1: 249. 1814 • Red-berried greenbrier, red-berried bamboo [E] [W1]

Vines; rhizomes slender, running. **Stems** perennial, climbing to clambering, branching, slender, 2–6 m, woody, glabrous; prickles scattered, subulate. **Leaves** deciduous to semievergreen; petiole 0.5–1 cm; blade abaxially green, drying to lightly orange-tinted brown, ovate-oblong, to ovate-lanceolate,

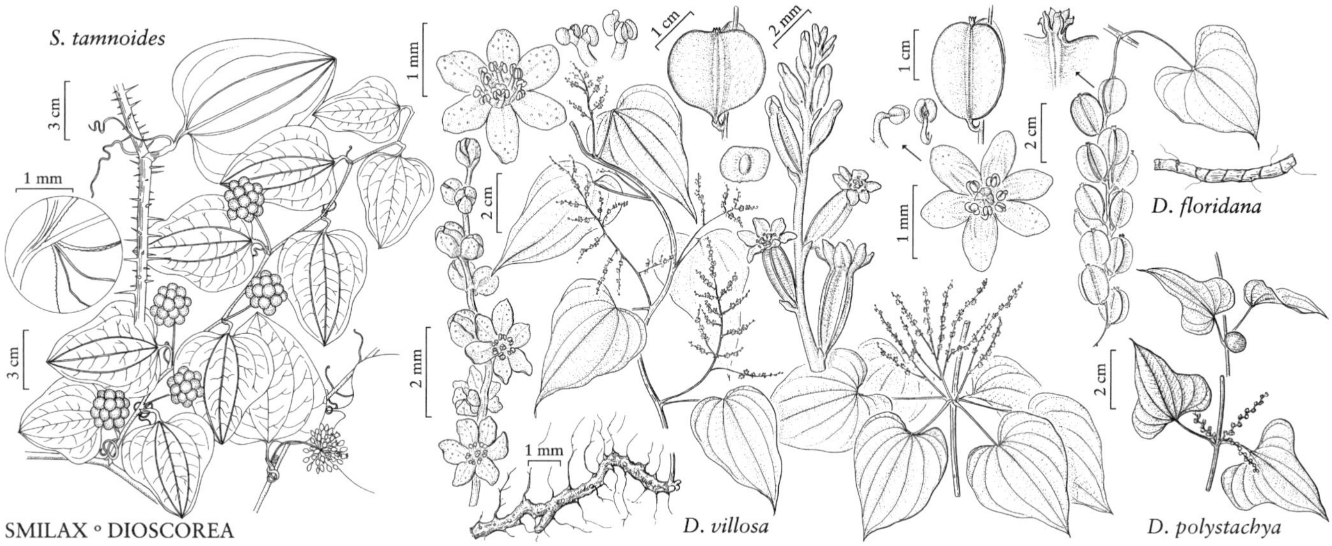

SMILAX ◦ DIOSCOREA

with 3 prominent veins, thin, 6–10 × 3–7 cm, glabrous or minutely pubescent abaxially, not glaucous, base rounded, subcordate, or, sometimes, cuneate at petiole insertion, margins entire, apex acute, mucronate. **Umbels** few to many, axillary to leaves, few-flowered, loose, hemispherical; peduncle 0.5–2 cm, generally shorter than petiole of subtending leaf. **Flowers:** perianth brownish yellow; tepals 3–6 mm; ovule 1 per locule; pedicel 0.5–1 cm. **Berries** bright red to orange, globose, 7–9 mm, shining.

Flowering Apr–Jun. Wet thickets, low pinelands, swamps, boggy areas; 0–350 m; Ala., Ark., Del., Fla., Ga., La., Md., Miss., N.C., S.C., Tex., Va.

The brightly colored fruits make *Smilax walteri* conspicuous in the winter.

228. DIOSCOREACEAE R. Brown
• Yam Family

Lauren Raz

Vines [herbs], geophytes perennating from fleshy rhizomes or tubers. **Stems** renewed annually or occasionally persisting for more than a single growing season, twining-climbing, procumbent, or seldom erect; vascular bundles commonly arranged in 2 concentric circles. **Leaves** alternate, opposite, or whorled at basal nodes; blades simple or occasionally digitately compound, typically cordate, reticulate-veined; principal veins arcuate, ascending from base to apex; margins entire or sometimes palmately lobed; stipules absent, rarely stipulate, never with tendrils; petioles typically with distinct pulvinus at each end. **Inflorescences** axillary, solitary or in fascicles of 2–7, branched and paniculate, or unbranched and spicate or racemose, ultimate inflorescence unit cymose, sessile or pedunculate, often reduced to 1 flower. **Flowers** unisexual, staminate and pistillate flowers on different plants, rarely staminate and pistillate flowers on same plant or bisexual; perianth epigynous, rotate, campanulate, or funnelform; tepals 6, petaloid, in 2 similar whorls of 3, connate at least basally; stamens 6, in 2 whorls of 3, inner whorl sometimes sterile or absent; filaments distinct [connate basally or completely into tube], inserted at base of perianth; anthers 2-locular, tetrasporangiate, dehiscing by longitudinal slits, thecae distinct or sometimes connate, rarely divided by a long-appendiculate connective; pistils 3; ovary (1–)3-locular, ovules anatropous, 2(–many) per locule, placentation axile or parietal; styles 3, ± distinct. **Fruits** capsular, rarely baccate or samarate, dehiscence loculicidal. **Seeds** flattened or globose, winged or unwinged, embryo small, endosperm copious, starchless.

Genera (3–)7(–more than 20), species ca. 650 (1 genus, 6 species in the flora; 4 species introduced): pantropical, of both humid and arid regions, with few temperate outliers in Europe, Asia, and North America.

Circumscription of the Dioscoreaceae remains controversial. A strict interpretation of the family, adhering to the original taxonomic concept of R. Brown, would include only *Dioscorea* Linnaeus (ca. 600 species), *Tamus* Linnaeus (5 species, Mediterranean), and *Rajania* Linnaeus (25 species, West Indies); recent systematic studies by L. R. Caddick et al. (2000) portray *Dioscorea* as subsuming the latter two genera. In its broad sense (I. H. Burkill 1960; L. R. Caddick et al. 2000; R. Knuth 1924), the family includes the bisexual genera *Avetra* H. Perrier

(1 species, Madagascar), *Trichopus* Gaertner (1 species, Indo-Malaysia), and *Stenomeris* Planchon (2 species, Malaysia), each of which has also been recognized at the family level (R. M. T. Dahlgren et al. 1985; J. Hutchinson 1973; A. L. Takhtajan 1997). *Tacca* J. R. Forster & G. Forster (10 species, Malaysia; Taccaceae) is closely allied with *Avetra* and *Trichopus*, (L. R. Caddick et al. 2000) and the broadest interpretation of the Dioscoreaceae (adopted here) includes *Tacca* as well. Only *Dioscorea* is represented in the flora.

SELECTED REFERENCES Al-Shehbaz, I. A. and B. G. Schubert. 1989. The Dioscoreaceae in the southeastern United States. J. Arnold Arbor. 70: 57–95. Ayensu, E. S. 1972. Dioscoreales. In: C. R. Metcalfe, ed. 1960+. Anatomy of the Monocotyledons. 8+ vols. Oxford. Vol. 6. Bouman, F. 1995. Seed structure and systematics in Dioscoreales. In: P. J. Rudall et al., eds. 1995. Monocotyledons: Systematics and Evolution. 2 vols. Kew. Vol. 1, pp. 139–156. Burkill, I. H. 1960. The organography and evolution of the Dioscoreaceae, the family of yams. J. Linn. Soc., Bot. 56: 319–407. Caddick, L. R. et al. 1998. Microsporogenesis and pollen morphology in Dioscoreales and allied taxa. Grana 37: 321–336. Caddick, L. R. et al. 2000. Yams and their allies: Systematics of Dioscoreales. In: K. L. Wilson and D. A. Morrison, eds. 2000. Monocots: Systematics and Evolution. Melbourne. Pp. 475–487. Coursey, D. G. 1967. Yams: An Account of the Nature, Origins, Cultivation and Utilisation of the Useful Members of the Dioscoreaceae. London. Huber, H. 1998. Dioscoreaceae. In: K. Kubitzky et al., eds. 1990+. The Families and Genera of Vascular Plants. 4+ vols. Berlin etc. Vol. 3, pp. 216–235. Knuth, R. 1924. Dioscoreaceae. In: H. G. A. Engler, ed. 1900–1953. Das Pflanzenreich.... 107 vols. Berlin. Vol. 87 [IV,43], pp. 1–387. Queva, C. 1894. Recherches sur l'Anatomie de l'Appareil Vegetatif des Taccacées & Dioscorées. Lille.

1. DIOSCOREA Linnaeus, Sp. Pl. 2: 1032. 1753; Gen. Pl. ed. 5, 456. 1754 • Yam, ñame [for Dioscorides, ca. 40–90, Greek physician, author of *De Materia Medica*]

Vines, typically herbaceous; rhizomatous or tuberous. **Stems** twining clockwise or counterclockwise, branched or not, smooth or winged, polygonal or terete, glabrous or sometimes bearing prickles; some species producing bulbils (branches modified into small aerial tubers) in leaf axils. **Leaves** initiated in spiral pattern, modified in some species and appearing opposite or whorled. **Inflorescences:** staminate nodes at ultimate flowering spicate or paniculate, individual flowers grouped in bracteolate cymes of (1–)3(–8) sessile or pedicellate flowers; pistillate solitary or rarely fasciculate, 1 flower per node, sessile or subsessile, unbranched, subtended by pair of minute bracts. **Flowers** unisexual, staminate and pistillate flowers on different plants or rarely staminate and pistillate flowers on same plant; tepals glandular or not, 1–3(–30) mm; staminate flowers with filaments distinct (connate basally only in *D. polystachya*), gynoecium rudimentary; pistillate flowers with staminodes present or sometimes absent, ovary inferior, usually 2–3 times length of perianth, style 3-branched, branches 2-fid; pedicel, when present, not articulate. **Fruits** capsular, 3-winged, dehiscence loculicidal, perianth and styles generally persistent. **Seeds** (1–)2 per locule, compressed, usually winged unilaterally, bilaterally, or with wings expanded circumferentially; tannins, saponins present. $x = 9, 10$.

Species ca. 600 (6 in the flora): worldwide.

The most comprehensive treatment to date (R. Knuth 1924) divides *Dioscorea* into 60+ sections. The native North American species are assigned to sect. *Macropoda* Uline: stems twining counter-clockwise, staminate flowers aggregated into more or less sessile cymes and with six fertile stamens, tepals united only at the base, and capsules as broad as or broader than long, reflexed at maturity, with seeds winged all around. This section also includes disjunct species from the Balkan Peninsula, the Himalayas, temperate East Asia, and the Caucasus Mountains.

The taxonomy of *Dioscorea* is notoriously problematic. Many of the species are poorly known, and in the absence of comparative studies there has been an unchecked proliferation of names in the genus. That *Dioscorea* exhibits considerable diversity across its expansive geographic range is not contested, but a great many of the names in current use are very narrowly applied and lack any corroboration from field, laboratory, or herbarium studies. At present there is also no phylogenetic framework from which to interpret the variation that has been documented. Segregate genera have been erected, only to be subsumed again. A robust classification will ultimately emerge from rigorous systematic investigation, now in progress at research institutions around the world (L. R. Caddick et al. 2000; H. Huber 1998; P. Wilkin 1999; C. C. Xifreda 2000).

Dioscorea species are cultivated circumtropically, especially in West Africa and the West Indies, for their starchy tubers (yams), the cultivars having been derived from about ten species, including three of the four taxa naturalized in the flora. The rhizomes/tubers of many noncomestible species accumulate varying concentrations of steroidal saponins (F. W. Martin 1969), and *Dioscorea* species of Mexican, South African, and Asian origin have been utilized extensively in the industrial synthesis of cortisone and human sex hormones. Much lower saponin yields are obtained from the native North American species than from the species harvested commercially elsewhere. Saponin content varies as a function of plant age and time of harvest, as well as phylogenetic position (L. Degras 1993). The rhizomes of *D. villosa* are included in some Native American pharmacopeias and are used to ease the pain of childbirth (H. H. Smith 1928). An alcohol extract of the "root" was widely administered in nineteenth-century eclectic medicine as a remedy for colic (H. H. Bartlett 1910).

SELECTED REFERENCES Bartlett, H. H. 1910. The source of the drug Dioscorea, with a consideration of the *Dioscoreae* found in the United States. U.S.D.A. Bur. Pl. Industr. Bull. 189: 1–29. Degras, L. 1993. The Yam: A Tropical Root Crop. London, New York, and Wageningen. Martin, F. W. 1969. The species of *Dioscorea* containing sapogenin. Econ. Bot. 23: 373–379. Wilkin, P. 1999. A revision of the compound-leaved yams of Africa. Kew Bull. 54: 19–39. Xifreda, C. C. 2000. Evaluation of pollen and vegetative characters in the systematics of South American species of *Dioscorea* (Dioscoreaceae). In: K. L. Wilson and D. A. Morrison, eds. 2000. Monocots: Systematics and Evolution. Melbourne. Pp. 488–496.

1. Plants rhizomatous; bulbils never produced in leaf axils; leaf margins entire or repand; petiole base never clasping.
    2. Rhizomes brownish, nodes not articulate; staminate inflorescences solitary in leaf axils; perianth of both staminate and pistillate flowers greenish white; stamens erect; thecae distinct, widely spreading . . . . . 1. *Dioscorea villosa*
    2. Rhizomes yellow, nodes articulate, each bearing 1 dark, contrasting, ± deltate scale leaf or the dark vertical scar remaining once leaf falls; staminate inflorescences (1–)2–5 in leaf axils; perianth of both staminate and pistillate flowers orangish yellow; stamens inwardly curved; thecae connate . . . . . 2. *Dioscorea floridana*
1. Plants tuberous; bulbils produced in leaf axils; leaf margins lobed or entire; petiole base sometimes clasping.
    3. Leaf margins 3–5-lobed.
        4. Leaf blade hastate-cordate, margins 3-lobed, apex acute or mucronate; petiole base not clasping; staminate plants flowering annually . . . . . 3. *Dioscorea polystachya*
        4. Leaf blade reniform, margins irregularly 3–5-lobed, apex conspicuously caudate; petiole base clasping; plants rarely flowering . . . . . 6. *Dioscorea sansibarensis*
    3. Leaf margins entire.
        5. Stems broadly winged, 4-angular; distal leaves decussate, blade base sagittate; seeds winged all around . . . . . 4. *Dioscorea alata*
        5. Stems terete, usually unwinged; leaves alternate throughout, blade base orbicular; seeds unilaterally winged . . . . . 5. *Dioscorea bulbifera*

1. Dioscorea villosa Linnaeus, Sp. Pl. 2: 1032. 1753 • Wild yam, yam root, colic root [E] [F]

Dioscorea cliffortiana Lamarck; *D. glauca* Muhlenberg ex Bartlett; *D. hexaphylla* Rafinesque; *D. hirticaulis* Bartlett; *D. longifolia* Rafinesque; *D. lloydiana* E. H. L. Krause; *D. megaptera* Rafinesque; *D. paniculata* Michaux; *D. paniculata* var. *glabrifolia* Bartlett; *D. pruinosa* Kunth; *D. quaternata* J. F. Gmelin; *D. quaternata* var. *glauca* (Muhlenberg ex Bartlett) Fernald; *D. quinata* J. F. Gmelin; *D. repanda* Rafinesque; *D. villosa* var. *glabra* Lloyd; *D. villosa* subsp. *glabrifolia* (Bartlett) W. Stone; *D. villosa* var. *glabrifolia* (Bartlett) S. F. Blake; *D. villosa* subsp. *glauca* (Muhlenberg ex Bartlett) R. Knuth; *D. villosa* subsp. *hirticaulis* (Bartlett) R. Knuth; *D. villosa* var. *hirticaulis* (Bartlett) H. E. Ahles; *D. villosa* subsp. *paniculata* (Michaux) R. Knuth; *D. villosa* subsp. *quaternata* (J. F. Gmelin) R. Knuth; *D. villosa* var. *vera* Prain & Burkill; *D. waltheri* Desfontaines

Plants rhizomatous; rhizomes brownish, unbranched or highly branched, linear to irregularly contorted, 0.5–1.5 cm or more diam., nodes not articulate. **Stems** longitudinally grooved or sometimes narrowly winged, usually terete in cross section, 1–7 m, ± rigid proximally, or flexible, glabrous or rarely with sparse pubescence, wings when present less than 1 mm and stems polygonal. **Leaves** alternate, subopposite, subverticillate, or in verticels of 3–7 proximally, due to suppression of proximal internodes (esp. in woodland understory), always alternate distally, 3–13 × 2–13 cm, ca. as long as wide; petiole ridged or narrowly winged, 3–14 cm, glabrous or puberulent at pulvinus, base not clasping; blade green to ± glaucous, (7–)9–11-veined, ovate-cordate, abaxial surface sometimes ± glandular, or sparsely or sometimes densely pubescent to glabrous, base with sinus rounded, acute, or ± truncate basally, margins entire or repand, apex acute to acuminate, occasionally mucronate. **Staminate inflorescences** solitary in leaf axils, rarely terminal, spicate or branched; cymes sessile, bearing 1–3 sessile flowers, braceolate, internodes between cymes 1–8 mm, bracteoles ca. 1 mm; rachis 2–30 cm, secondary axes to 15 cm, robust plants occasionally branched to third order, axes subtended by linear-lanceolate bracts 1–3 mm. **Pistillate inflorescences** solitary, 4–18-flowered, 4–20 cm, internodes 6–12 mm. **Staminate flowers:** perianth greenish white, appearing darker in some specimens due to presence of irregularly distributed tannin crystals, rotate-campanulate to funnelform, 1–2(–3) mm diam.; tepals ± glandular, ovate-elliptic, margins hyaline, apex rounded or acute; stamens in 2 subequal whorls, erect; anthers ca. ½ length of filaments, thecae distinct, widely spreading. **Pistillate flowers:** perianth greenish white, rotate-campanulate, 2–4 mm wide; tepals as in staminate flowers; staminodes 6, differentiated into anthers and filaments, less than 1/2 length of fertile stamens. **Capsules** greenish gold, ovoid to obovoid to obreniform, 1–3 × 1–3.5 cm, varying continously in size, occasionally ± glaucous. **Seeds** generally 2 per locule, rarely 1, 5–18 mm. $2n$ = 20, 36, 54, 60.

Flowering mid spring–summer; fruiting late summer. Borders of bogs, swamps, marshes, river and lake margins, creek bottoms, sandy or rocky soils, moist or dry woods, hammocks, thickets, limestone or talus slopes, roadsides; 0–1500 m; Ont.; Ala., Ark., Conn., Del., D.C., Fla., Ga., Ill., Ind., Iowa, Kans., Ky., La., Md., Mass., Mich., Minn., Miss., Mo., Nebr., N.J., N.Y., N.C., Ohio, Okla., Pa., R.I., S.C., Tenn., Tex., Vt., Va., W.Va., Wis.

Dioscorea villosa is a highly polymorphic species, exhibiting complex patterns of variation across its geographic range. Characters that have been used previously to delineate taxonomic boundaries within this complex—pubescence, glaucousness, rhizome thickness and shape, length of internodes within the inflorescence, arrangement of proximal leaves, geometry of the stem, and fruit and seed size/shape—fail when individuals from all parts of the range and specimens representing both apical and basal portions of single stems can be examined. At its morphological extremes, *D. villosa* comprises 1) small vines with tightly congested inflorescences, winged stems, and variously pubescent leaves, occurring in bogs and branch swamps; and 2) robust plants, rigid at the base, the proximal leaves verticillate with large, glaucous blades, from the axils of which arise lax spikes or panicles, inhabiting rocky, upland woods and steep talus slopes. As one ascends from the Atlantic Coastal Plain through the Appalachians, continuing westward to the Great Lakes region, south to the Ozarks, and east to the branch swamps of Georgia, particular morphologies are associated with particular ecological conditions, independent of geography. As well as the morphological extremes, every intermediate condition of leaf, stem, and inflorescence architecture can be found, in all combinations, and variation may be encountered even within individual plants. What sort of genetic structure underlies these patterns of morphological diversity remains an open question. That there is a significant degree of genetic variability within the complex is evident from the chromosome counts thus far reported. Further research is needed to shed light on patterns of gene flow in the complex, and garden studies would be instructive as to the limits of individual plasticity. At present, I can find no natural gaps in the variation between the plants that have been called (albeit ambiguously; see H. H. Bartlett 1910) *D. villosa* and those called *D. quaternata*, and therefore I am treating the complex as a single species.

2. Dioscorea floridana Bartlett, U.S.D.A. Bur. Pl. Industr. Bull. 189: 18. 1910 [E] [F]

Dioscorea villosa Linnaeus subsp. *floridana* (Bartlett) R. Knuth; *D. villosa* var. *floridana* (Bartlett) H. E. Ahles

Plants rhizomatous; rhizomes yellow, unbranched, horizontal, 0.5–1 cm or more diam., nodes articulate, each segment bearing 1 dark, contrasting, ± deltate scale leaf, scales brown and papery, caducous, each leaving dark, linear scar perpendicular to growing axis. **Stems** twining counterclockwise, ± polygonal to terete, 1–4 m, surface longitudinally grooved. **Leaves** alternate proximally (variation expected) and distally, 3–11 × 3–9 cm; petiole 3–8 cm, base not clasping; blade sometimes glandular, 7–9-veined, ovate-cordate to deltate, glabrous, margins entire or repand, apex acuminate. **Staminate inflorescences** (1–)2–5 in leaf axils, terminal, spicate to highly branched; cymes sessile, bearing 1–3 sessile flowers, bracteolate, internodes between cymes 2–5 mm, bracteoles linear-lanceolate, ca. 1 mm; rachis to 30 cm, secondary axes to 15 cm, subtended by linear-lanceolate bracts ca. 1–2 mm. **Pistillate inflorescences** borne singly, 4–20 or more-flowered, 4–25 cm, internodes 7–12 mm. **Staminate flowers:** perianth yellowish orange, rotate-campanulate to funnelform, 1–2 mm; tepals ± glandular, oblong, margins hyaline, apex (acute to) rounded or truncate; stamens in 2 equal whorls, inwardly curved; anthers ca. 1/3 length of filaments, thecae connate. **Pistillate flowers:** perianth as in staminate flowers; staminodes 6, differentiated into anthers and filaments, less than 1/2 size of fertile stamens. **Capsules** greenish gold, ovoid, obreniform, or oblong, 1.5–2.8 × 1.5–2.5 cm. **Seeds** 9–13 mm.

Flowering late spring–mid summer; fruiting late summer. Edges of swamps, in hammocks, moist and dry (pine) forests and forest margins, sandy soils, lowlands; 0–100 m; Fla., Ga., S.C.

Dioscorea floridana is undoubtedly a close relative of *D. villosa*, but because it is identifiable using characters that vary discretely, with states not manifest in plants occupying similar habitats outside of its range, I have chosen to retain it at the rank of species.

3. Dioscorea polystachya Turczaninow, Bull. Soc. Imp. Naturalistes Moscou 10(7): 158. 1837 • Cinnamon vine, Chinese yam [F] [I]

Dioscorea batatas Decaisne

Plants tuberous; tubers 1–many, on long stalks and deeply buried, spindle-shaped. **Stems** twining clockwise, unwinged, terete, climbing to 5 m, producing small bulbils (less than 2 cm diam.) in leaf axils. **Leaves** alternate proximally, usually opposite distally, occasionally ternate, 3–9 × 3–11 cm; petiole as long as blade, base not clasping; blade 7(–9)-veined, hastate-cordate, glabrous, margins 3-lobed, basal lobes spreading, rounded to ± rhombic, median lobe deltate to ovate, apex acuminate or mucronate; primary veins 7(–9); basal leaves often less distinctly lobed or entire and ovate-cordate. **Staminate inflorescences** axillary, in fascicle of (1–)3–5 spikes, each spike shorter than 5 cm, or primary inflorescence axis elongate and bearing multiple fascicles; spikes bearing flowers singly, flowers bracteate, distributed along rachis in markedly zigzag pattern, internodes less than 2 mm. **Pistillate inflorescences** 1 per leaf axil, few-flowered, shorter than 5 cm, internodes between mature flowers greater than 1 cm. **Staminate flowers:** perianth drying dark, tightly cup-shaped; tepals ovate, ca. 1 mm, margins hyaline, apex acute; fertile stamens 6; filaments connate basally; anthers as long as or longer than filaments, thecae distinct, weakly spreading. **Pistillate flowers:** perianth as in staminate flowers; staminodes 6 in 2 subequal whorls, anthers and filaments more strongly differentiated in inner whorl, filaments less developed in outer whorl, all smaller than fertile stamens. **Capsules** not reflexed at maturity, broadly ovate, (1.2–)1.7–2 cm wide. **Seeds** winged all around. $2n = 140$.

Staminate plants flowering summer. Thickets, ravines, stream banks, creek bottoms, limesinks, granite outcrops, alluvial woods, roadsides, drainage canals, waste places, fence rows; 0–500 m; introduced; Ala., Ark., Conn., D.C., Fla., Ga., Ky., Md., Mass., Miss., Mo., N.J., N.C., Ohio, Pa., S.C., Tenn., Va., W.Va.; e Asia.

Dioscorea polystachya is native to eastern Asia, where it is cultivated for its edible tubers. In North America, it is planted as a garden ornamental and has become naturalized throughout much of the eastern United States.

Only one pistillate specimen was encountered from the flora area, the plants apparently propagating vegetatively.

Dioscorea polystachya is the earliest published name for this species. The name *D. oppositifolia* Linnaeus, which has been misapplied to this species, actually refers to an ovate-leaved species that is native to the Indian Subcontinent and is not found in the flora area. For a more detailed nomenclatural history of *D. polystachya*, see Ting C. T. and M. G. Gilbert (2000).

4. Dioscorea alata Linnaeus, Sp. Pl. 2: 1033. 1753 • White yam, greater yam, water yam I

Plants tuberous; tubers 1–many, stalked and ± deeply buried, elongate (highly polymorphic), often massive, weighing up to several kg, flesh starchy, white or variously colored. **Stems** twining clockwise, climbing up to 20 m, broadly winged, 4-angular, producing bulbils up to 4 cm in diam. in leaf axils, wings often purplish. **Leaves** alternate proximally, opposite and ultimately decussate distally, 6–16 × 4–13 cm; petiole ca. as long as blade, winged, base clasping, basal lobes stipulate, growing as extensions of wings, less than 1 mm wide; blade 5–7-veined, ovate, glabrous, base typically sagittate, margins entire, apex acute to acuminate. **Staminate inflorescences** axillary, 1–2 per axil, paniculate, fasciculate; panicles bearing flowers singly, bracteolate, in a zigzag pattern along rachis, internodes less than 2 mm; rachis to 25 cm, secondary axes 1–3(–6), fasciculate, less than 3 cm, each subtended by deltate-ovate bracteole shorter than 1 mm. **Pistillate inflorescences** solitary, 4–8(–20)-flowered, 6–35 cm, internodes ca. 1 cm. **Staminate flowers:** perianth whitish, cup-shaped; tepals ca. 1 mm, connate at base in 2 subequal whorls, outer widely ovate, inner narrower, apex obtuse to rounded in both; fertile stamens 6 in 2 equal whorls; anthers longer than filaments, thecae distinct, not spreading. **Pistillate flowers:** perianth lightly colored; tepals broadly ovate, ca. 1 mm, fleshy; staminodes 6, smaller than fertile stamens. **Capsules** not reflexed at maturity, broadly ovate, ca. 2 × 3.5 cm. **Seeds** winged all around. $2n$ = 30, 40, 50, 60, 70, 80.

Flowering late summer; fruiting into late fall. Near ponds, marshes, drainage canals, waste areas; 0 m; introduced; Fla., Ga.; Southeast Asia; cultivated worldwide.

Dioscorea alata is the most extensively cultivated of the edible yams. There are hundreds of cultivars worldwide (L. Degras 1993; D. G. Coursey 1967), varying in the color and consistency of the flesh. The bulbils may also be eaten, but the large tubers are favored. Typically the plants are propagated vegetatively. The species is not known to exist in the wild, and its putative origin is from Southeast Asia. It has escaped from cultivation in the southeastern United States, recorded here from Florida and Georgia, although it is likely to be found elsewhere in the Gulf states. The plants seldom flower in the flora area; I observed no staminate flowers at all, and very few pistillate ones. Although the fruits do set seed, the seeds are frequently unviable.

5. Dioscorea bulbifera Linnaeus, Sp. Pl. 2: 1033. 1753 • Aerial yam, potato yam I W1

Plants tuberous or not; tubers, when present, buried just below ground surface, not stalked, globose, weighing less than 1 kg. **Stems** twining counter-clockwise, climbing to more than 20 m, often flecked with purple, unwinged or rarely narrowly winged, terete, producing axillary bulbils frequently greater than 5 cm in diam. in leaf axils. **Leaves** alternate throughout, 5–25 × 5–26 cm; petiole usually somewhat shorter than blade, base clasping, basal lobes stipulelike, 1–4 mm wide; blade 5–11-veined, broadly ovate-cordate, glabrous, base orbicular, margins entire, apex long-acuminate. **Staminate inflorescences** axillary, borne singly, spicate or paniculate, cymose, to 70 cm; cymes reduced to 1 sessile bracteolate flower, internodes ca. 2 mm; secondary axes to 6 per node, fasciculate, subtended by deltate bracteoles or sometimes leafy bracts, 3–20 cm. **Pistillate inflorescences** borne singly or fasciculate, to 6 per axil, spicate; spikes bearing to 50 flowers, 6–40 cm, subopposite to 8 mm apart. **Staminate flowers** fragrant; perianth white, becoming purple; tepals in 2 similar whorls, spreading at anthesis, lanceolate, (1–)2–5 mm; fertile stamens 6 in 2 equal whorls; anthers as long as or longer than filaments; thecae distinct, not spreading. **Pistillate flowers:** perianth greenish white, not changing color; tepals as in staminate flowers; staminodes 6, smaller than fertile stamens. **Capsules** not reflexed at maturity, longer than wide, 1.8–2.8 × 1–1.5 cm. **Seeds** unilaterally winged, 12–20 mm. $2n$ = 80, 100.

Flowering late summer–early fall. Near ponds, marshes, drainage canals, disturbed woods and thickets, waste areas; 0 m; introduced; Fla.; tropical Asia; tropical Africa.

Dioscorea bulbifera is the most widely distributed of all *Dioscorea* species. It is found throughout the tropics and is thought to be native to both Africa and Asia. The tubers are either small and bitter or do not form at all. The bulbils are eaten, although this species is not widely consumed outside of Oceania. It is weedy in Florida and probably elsewhere in the Gulf States, but I have not seen any specimens to document its range outside Florida. Plants in the flora area rarely flower, and those that do are usually pistillate; I observed no staminate flowers.

6. **Dioscorea sansibarensis** Pax, Bot. Jahrb. Syst. 15: 146. 1892 • Zanzibar yam [I]

Dioscorea macabiha Jumelle & H. Perrier; *D. macroura* Harms; *D. welwitschii* Rendle

Plants tuberous; tubers buried just below soil surface, irregularly lobed, globose. **Stems** twining counterclockwise, climbing to more than 7 m, terete, grooved, or weakly angled with variable number of longitudinal ridges raised less than 1 mm adaxially, producing axillary bulbils frequently greater than 5 cm diam. in leaf axils. **Leaves** alternate at basal nodes, opposite distally, 6–27 × 7–42 cm; petiole 6–20 cm, as long as or slightly longer than blade, base clasping, basal lobes stipulate, 1–8 mm wide; blade 7–11-veined, reniform to somewhat deltate, glabrous, margins irregularly 3–5-lobed, apex conspicuously caudate. **Staminate inflorescences** 1–2(–4) per axil, spicate or terminally paniculate, cymose; cymes subsessile, bearing (1–) 2–4 flowers subtended by ovate bracteoles, internodes ca. 1 cm; rachis 20–40 cm, subtended by ovate bracts ca. 1 mm. **Pistillate inflorescences** 1–3 per axil, to 80 cm, internodes 2–2.5 cm. **Staminate flowers:** perianth yellowish; tepals in 2 similar whorls, narrowly spreading at anthesis, lanceolate, 3–6 mm; fertile stamens 6 in 2 subequal whorls; anthers less than ½ length of filaments, thecae distinct, not spreading. **Pistillate flowers:** perianth white with purple veins; tepals as in staminate flowers; staminodes 6, smaller than fertile stamens; style branches irregularly 2-fid. **Capsules** not reflexed at maturity, longer than wide. **Seeds** bilaterally winged.

Not known to flower in the flora area. Roadsides, hammocks, swamps; 0 m; introduced; Fla.; tropical Africa.

Dioscorea sansibarensis, an ornamental species native to tropical Africa, is widely cultivated for its large, unusually shaped leaves. It has recently been reported growing amongst native vegetation in Miami-Dade and Collier counties in southern Florida. Axillary bulbils shed from mature plants were observed to be growing vigorously. Both the tubers and axillary bulbils of this species are toxic.

229. BURMANNIACEAE Blume
• Burmannia Family

Deborah Q. Lewis

Herbs, annual or perennial, mycotrophic and white or purplish, or autotrophic and green. **Roots** subterranean, mycorrhizal, filiform or stoutly terete. **Stems** monopodial, unbranched or with few branches, erect, slender; rhizomes present or absent. **Leaves** present or absent, cauline or basal, alternate, simple, sessile, often scalelike, margins entire; stipules absent. **Inflorescences** terminal cymes, sometimes appearing racemose or capitate, or flowers solitary; each flower subtended by scalelike floral bract. **Flowers:** tepals 6, partially or wholly connate, tube persistent or caducous in fruit, limb lobes 3 in 1 whorl or 6 in 2 whorls, greenish, purple, or white, sometimes blue-tinged, often ribbed or winged, throat with or without annulus; stamens 3 or 6, sessile or subsessile; pollen sacs separated on appendaged connective or connective broadened and connate, forming ring proximal to annulus; pistil 1, ovary inferior, 1- or 3-locular; placentation parietal or axile; ovules numerous; style 1, 3-branched apically; stigmas 3. **Fruits** capsules, ovoid, obconic, or cup-shaped; dehiscence transverse, longitudinal by 3 valves, or irregular. **Seeds** numerous, minute.

Genera ca. 15, species ca. 125 (3 genera, 5 species in the flora): mostly tropical and subtropical, worldwide.

Although Burmanniaceae have traditionally been placed in the Orchidales, it has been suggested that the family may be more closely related to the Melanthiales (= Liliales in the broad sense, as traditionally circumscribed) (R. M. T. Dahlgren et al. 1985). More recently, molecular and morphological studies by L. R. Caddick et al. (2000, 2000b) support placement of the family in the Dioscoreales.

Burmanniaceae traditionally have been characterized as saprophytic, but the life form is more accurately described as mycotrophic or myco-heterotrophic (J. R. Leake 1994). Even the photosynthetic species have been shown to be endomycorrhizal (P. J. M. Maas et al. 1986b; C. E. Wood Jr. 1983) and have been described as hemisaprophytic. Saprophytism in the Burmanniaceae and other achlorophyllous families has been fully discussed by J. R. Leake (1994).

SELECTED REFERENCES Jonker, F. P. 1938. A monograph of the Burmanniaceae. Meded. Bot. Mus. Herb. Rijks Univ. Utrecht 51: 1–279. Leake, J. R. 1994. Tansley review no. 69. The biology of myco-heterotrophic ('saprophytic') plants. New Phytol. 127: 171–216. Maas, P. J. M. et al. 1986. Burmanniaceae. In: Organization for Flora Neotropica. 1968+. Flora Neotropica. 75+ nos.

New York. No. 42. Wood, C. E. Jr. 1983. The genera of Burmanniaceae in the southeastern United States. J. Arnold Arbor. 64: 293–307.

1. Leaves and stems green; ovary 3-locular; placentation axile . . . . 1. *Burmannia*, p. 487
1. Leaves and stems white or purplish; ovary 1-locular; placentation parietal.
    2. Flowers purplish or white with purplish longitudinal stripes and apices of perianth lobes; inflorescences usually racemose cymes; stem much longer than 1 cm; perianth lobes all erect, not connate at apex . . . . 2. *Apteria*, p. 488
    2. Flowers white with faint blue-green markings on lobes; flowers solitary; stem 1 cm or less; perianth lobes 3 recurved, 3 inflexed and convergent to connate at apex . . . . 3. *Thismia*, p. 489

1. BURMANNIA Linnaeus, Sp. Pl. 1: 287. 1753; Gen. Pl. ed. 5, 139. 1754 • [For Johannes Burman, 1707–1779, Dutch botanist]

Plants annual, autotrophic, at least partly green. **Rhizomes** absent. **Roots** filiform. **Stems** usually unbranched, green. **Leaves** cauline, sometimes basal, green, purplish, or pale, scalelike to linear. **Inflorescences** 2–25-flowered cymes or flowers solitary; floral bracts sometimes appearing imbricate if internodes of cyme rachis are very short; pedicels 0–1 mm. **Flowers** erect, 3-ribbed to 3-winged; perianth persistent in fruit; annulus absent; stamens 3, proximal to inner perianth lobes, sessile; ovary 3-locular; placentation axile. **Capsules** obovoid to ellipsoid, transversely dehiscent.

Species ca. 60 (3 in the flora): mostly tropical and subtropical, Western Hemisphere, Africa, Asia (China).

1. Flowers 3-ribbed or slightly 3-winged; inflorescences usually appearing capitate; inner perianth lobes usually less than 2/3 length of outer lobes . . . . 1. *Burmannia capitata*
1. Flowers distinctly 3-winged; inflorescences loosely flowered cymes or flowers solitary; inner perianth lobes 2/3 to nearly equaling length of outer lobes.
    2. Flowers bluish with perianth lobes paler, often cream-colored; basal leaves absent . . . . 2. *Burmannia biflora*
    2. Flowers yellow or greenish; basal leaves present . . . . 3. *Burmannia flava*

1. Burmannia capitata (Walter ex J. F. Gmelin) Martius, Nov. Gen. Sp. Pl. 1: 12. 1823 F

Vogelia capitata Walter ex J. F. Gmelin, Syst. Nat. 1: 107. 1791

Stems 3–24(–33) cm. **Leaves:** basal leaves 0–3, ± same size as cauline leaves; cauline leaves subulate-lanceolate, 2–8 × 0.3–1.2 mm. **Inflorescences** 2–25-flowered cymes, usually appearing capitate (rarely less contracted), or occasionally flower solitary; cyme rachis usually absent, rarely to 6 mm; floral bracts subulate to lanceolate, 1.5–4 mm. **Flowers** 3-ribbed or slightly 3-winged, 2–6 mm; perianth cream-colored to white, sometimes tinged with blue or green; perianth tube (3- or) 6-lobed, to 0.8 mm wide; outer lobes usually incurved, triangular, 0.3–0.5 mm, margins involute; inner lobes absent or linear to elliptic, to 0.3 mm; pedicel 0–1 mm. **Capsules** 1.5–3 mm.

Flowering mostly late summer–fall. Low woods, pond margins, savannas, bogs, swamps, ditches; 0–100 m; Ala., Fla., Ga., La., Miss., N.C., Okla., S.C., Tex.; Mexico; West Indies; Central America; South America (to Argentina, Paraguay, Bolivia, Brazil).

Chromosome counts of this species are not known from our region. A count of $2n$ = ca. 136 was reported for collections from Argentina.

2. Burmannia biflora Linnaeus, Sp. Pl. 1: 287. 1753 F

Stems 3–18 cm. **Leaves:** basal leaves absent; cauline leaves subulate to lanceolate, 1–4 × 0.2–1 mm. **Inflorescences** loosely 2–12–flowered cymes, 2-cleft, or flowers solitary; cyme rachis usually distinct, to 2.5 cm; floral bracts subulate to lanceolate, 1.5–3.2 mm. **Flowers** 3-winged, 3–6 mm, wings blue, to

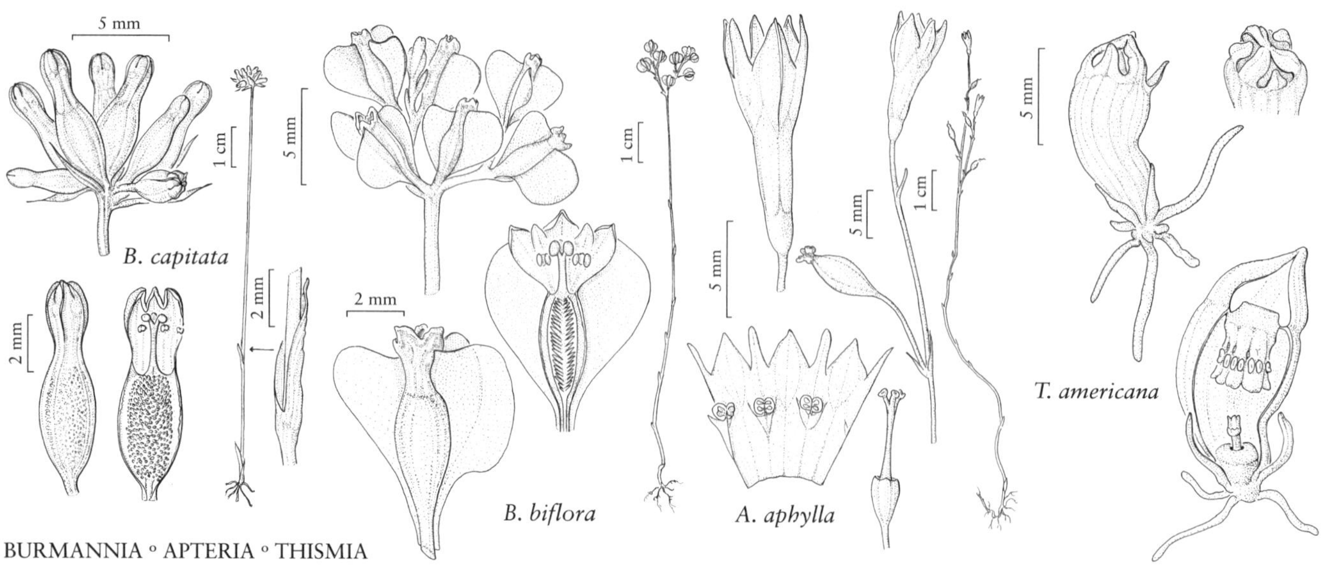

BURMANNIA ∘ APTERIA ∘ THISMIA

2.5 mm wide; perianth blue to violet, often with cream-colored lobes; perianth tube 6-lobed; outer lobes triangular, 0.5–1 mm; inner lobes erect, elliptic, 0.4–0.8 mm; pedicel 0–1 mm. **Capsules** 2–3.5 mm.

Flowering summer–fall. Low woods, pond margins, savannas, bogs, swamps, ditches, pine barrens; 0–100 m; Ala., Fla., Ga., La., Miss., N.C., S.C., Tex., Va.; West Indies (Cuba).

3. **Burmannia flava** Martius, Nov. Gen. Sp. Pl. 1: 11, plate 5, fig. 3. 1823

Stems 3.5–15 cm. **Leaves:** basal leaves 2–7, linear, 3–11 × 0.8–1 mm; cauline leaves subulate to linear, 2–12 × 0.3–1 mm. **Inflorescences** 1–2-flowered; floral bracts lanceolate to ovate, 2–5 mm. **Flowers** 3-winged, 5–10(–14) mm, wings yellow, to 1.8 mm wide; perianth yellow or greenish; perianth tube 6-lobed; outer lobes erect, triangular, 1.3–1.8 mm; inner lobes linear or elliptic, 1–1.5 mm. **Capsules** 3–6 mm.

Flowering fall (Oct). Coastal plain; 0–50 m; Fla.; Mexico; West Indies (Cuba); Central America; South America (to Argentina, Bolivia, Paraguay, Brazil).

2. APTERIA Nuttall, J. Acad. Nat. Sci. Philadelphia 7: 64. 1834 • [Greek *a*, without, and *pteron*, wing]

Plants possibly perennial, mycotrophic, not green. **Rhizomes** present. **Roots** filiform. **Stems** sometimes branched. **Leaves** cauline, white or purplish, scalelike to lanceolate. **Inflorescences** 2–6-flowered racemose cymes or flowers solitary; floral bracts not imbricate. **Flowers** erect to nodding, neither ribbed nor winged; perianth persistent and rolled in fruit; annulus absent; stamens 3, proximal to inner perianth lobes, sessile; ovary 1-locular; placentation parietal; pedicel

4–12(–18) mm. **Capsules** broadly ellipsoid; dehiscence longitudinal by valves, leaving 3 placental ribs.

Species 1: tropical North America, Mexico, West Indies, Central America, South America.

1. **Apteria aphylla** (Nuttall) Barnhart ex Small, Fl. S.E. U.S., 309. 1903 • Nodding-nixie [F]

Lobelia aphylla Nuttall, Amer. J. Sci. Arts 5: 297. 1822; *Apteria setacea* Nuttall

Stems often purplish, 3–27 cm. **Leaves** often purplish, 1.5–4 × 0.5–1.5 mm. **Inflorescences:** floral bracts 2–4 mm. **Flowers** 8–14 mm; perianth purple or white with purplish longitudinal stripes inside throat and purplish perianth lobes; lobes erect, incurved and converging but not connate; outer lobes ovate to triangular, 1.5–4 mm; inner lobes lanceolate, ± as long as outer. **Capsules** 3–6 mm.

Flowering late summer–fall. Bogs, alluvial and upland rich woods; 0–100 m; Ala., Fla., Ga., La., Miss., Tex.; Mexico; West Indies; Central America; South America (to Peru, Paraguay, Bolivia, Brazil).

3. THISMIA Griffith, Proc. Linn. Soc. London 1: 221. 1845 • [Anagram, for Thomas Smith, English microscopist, died ca. 1825]

Plants perennial, mycotrophic, not green. **Rhizomes** absent. **Roots** horizontal, stoutly terete. **Stems** unbranched, very short. **Leaves** cauline, bractlike, white. **Inflorescences** solitary flowers; floral bracts imbricate. **Flowers** erect, neither ribbed nor winged; perianth caducous in fruit; annulus present in perianth throat immediately proximal to lobes; stamens 6, pendent, forming ring proximal to annulus; ovary 1-locular; placentation parietal; pedicel absent. **Capsules** cup-shaped; dehiscence irregular by breaking down of wall.

Species ca. 32 (1 in the flora): non-North American species mostly pantropical.

1. **Thismia americana** N. Pfeiffer, Bot. Gaz. 57: 123, plates 7–11. 1914 [C][E][F]

Stems white, 0.3–1 cm. **Leaves** appressed, white, subulate to ovate, 2–4 × 1.5–3 mm. **Flowers** white with faint blue-green lobes, 8–15 mm; perianth tube slightly urceolate, 6-lobed; outer lobes recurved, ovate, 2.5–4 mm; inner lobes erect, convergent to connate at apex, linear; annulus 0.5–1.5 mm wide; staminal connectives dilated, connate, forming tube proximal to annulus. **Capsules** 2–3 mm.

Flowering summer–early fall. Prairie; of conservation concern; 200 m; Ill.

Thismia americana was observed and collected between 1912–1916 from a single site, in a prairie near Chicago, Illinois. It is now possibly extinct. Numerous attempts to relocate the species have been unsuccessful (L. A. Masters 1995); because only a small part of the minute plant is above the level of soil and moss, it could be easily overlooked. This species is believed to be most closely related to *T. rodwayi* F. von Müller of Australia and New Zealand, presenting a puzzling pattern of distribution (P. J. M. Maas et al. 1986).

Some authors (e.g., R. M. T. Dahlgren et al. 1985; Angiosperm Phylogeny Group 1998) have proposed Thismiaceae to include *Thismia* and allied African and South American genera.

SELECTED REFERENCE Pfeiffer, N. E. 1914. Morphology of *Thismia americana*. Bot. Gaz. 57: 122–135.

230. ORCHIDACEAE Jussieu
• Orchid Family

Gustavo A. Romero-González
Germán Carnevali Fernández-Concha
Robert L. Dressler
Lawrence K. Magrath
George W. Argus

Herbs or rarely vines, perennial, rarely annual, strongly mycotrophic, epiphytic, terrestrial, lithophytic, or rarely aquatic or subterranean, usually green and photosynthetic, some without chlorophyll and saprophytic. **Roots** subterranean or aerial, tuberoid or stolonoid, usually with spongy, multilayered velamen. **Stems** erect or pendent or modified into creeping rhizomes, simple or sympodially or monopodially branched, delicate to stout, or thickened as corms or pseudobulbs, or greatly reduced, sometimes proliferous (especially diverse in sympodial orchids). **Leaves** solitary, several, or reduced to scales, basal or cauline, alternate, distichous, or sometimes opposite or whorled, either convolute or duplicate, simple, sessile or petiolate; stipules absent; blade articulate or not, plicate or conduplicate, cylindric, triangular, or laterally flattened, margins entire. **Inflorescences** terminal or lateral, racemes, spikes, panicles, or rarely cymose, erect or variously pendent, 1–many-flowered, lax or dense, flowering successively or simultaneously. **Flowers** bisexual [rarely unisexual], epigynous, resupinate or not, pedicellate or sessile, 3-merous, usually bilaterally symmetric [rarely nearly radially symmetric], with abscission layer between pedicel and peduncle, rarely between ovary and perianth or ovary and pedicel; perianth of 6 tepals in 2 whorls, all petaloid or sepals sometimes greener and more foliaceous in texture; sepals alike or not, lateral sepals often connate (forming synsepal), or all 3 sepals variously connate and/or adnate or distinct and/or free; petals 3, median petal modified as lip, commonly larger or differing in form and color, lateral petals commonly but not always similar to sepals; nectaries of various sorts; extrafloral nectaries sometimes present on pedicels, bracts, or leaf sheaths; stamens usually 1–2(–3, if 3 the 3d modified into sterile staminode), all on side opposite lip, fully or partially adnate to style, forming column; pollen grains in monads or tetrads, usually in 2–8 pollinia, sometimes subdivided into small packets, rarely granular, sometimes pollinia with caudicles and/or stipes; gynoecium 3-carpellate, connate, forming compound, inferior, 1- or 3-locular ovary; style variously adnate to filaments; stigmas usually 3-lobed, concave to convex, part of median stigma lobe modified into rostellum, often separating anther from fertile portions of stigma, commonly preventing or in some cases facilitating self-pollination; ovules numerous, anatropous, minute. **Fruits** capsules, opening (dehiscing) by longitudinal slits, rarely fleshy and indehiscent berries. **Seeds** numerous (millions in some species), minute; endosperm absent.

Genera ca. 800, species 22,000–35,000 (70[1] genera, 208 species in the flora; 1 genus, 6 species introduced): worldwide except Antarctica, most diverse in tropical forests.

The overall count for orchid genera in the flora includes *Spathoglottis plicata* Blume, which was recently reported from Palm Beach County, Florida. The plants, known locally since 1982, are apparently widely naturalized in old shellpits. The number of species in the flora includes one newly recognized species in *Habenaria* that is morphologically described, but not fully treated here.

Orchidaceae are by far the largest and most diverse monocot family and rank among the largest families of flowering plants. An accurate account of the number of genera and species has eluded orchid scientists, and species counts published in the last 20 years range from 15,000 to 35,000. New species are continually being described. In addition, numerous natural and artificial hybrids exist.

Although orchids are important in horticulture, most of the plants traded in the national and international market belong to a small number of species and their hybrids in only a few genera; the majority of orchids are not commonly cultivated. Few orchids are economically important outside the horticultural trade: the fruits of several species of *Vanilla* are the source of the spice vanilla, and the dry roots of some species of *Dactylorhiza*, *Eulophia*, and *Orchis* are made into salep, a flour consumed in northern Africa, the Middle East (especially Turkey), and Asia. Some species are locally used for medicinal purposes; the mucilage from pseudobulbs of several species is sometimes used as glue; and in the Far East the stems of some species of *Dendrobium* are split into strips used to weave handicrafts. A few orchids have been found to cause contact dermatitis (e.g., *Cypripedium reginae*).

Orchids range vegetatively from Lilliputian plants a few millimeters long (*Bulbophyllum* Thouars and *Platystele* Schlechter) to gigantic clusters weighing several hundred kilograms (*Grammatophyllum* Blume) to some as much as 13.4 meters in height (*Sobralia altissima* D. E. Bennett & Christenson, a recently described species from Peru). Likewise, flowers vary in size from less than 1 mm and barely visible to the naked eye (*Platystele* Garay), to 15–20 cm diameter (some *Paphiopedilum* Pfitzer, *Phragmipedium* Rolfe, and *Cattleya* Lindley spp.), and ultimately to 76 cm [*Phragmipedium caudatum* (Lindley) Rolfe]. Weight can vary from a fraction of a gram (many *Pleurothallus* R. Brown spp.) to nearly 100 grams (*Coryanthes* Hooker spp.). Their fragrances vary from delightful (*Cattleya* Lindley) to repulsive and unbearable (in some species of *Bulbophyllum* Thouars). The plants colonize habitats ranging from some of the driest and hottest places on earth to the wettest and coolest, literally occurring from polar regions to the equator.

Within the monocots, the most important diagnostic features of *Orchidaceae* are reduction of adaxial stamens, fusion of the remaining stamens to the gynoecium forming the column, aggregation of pollen into compact pollinia (present elsewhere only in the dicots, in

[1]As we were going to press, we received word of the presence of *Spathoglottis plicata* in Palm Beach County, Florida. Apparently the plants are widely naturalized in old shellpits and have been locally known since 1982. The genus and species are included in the overall count for orchid genera in the flora.—Ed.

Asclepiadaceae), differentiation of the median petal into the lip, a sometimes complex organ, and the exceedingly small size of the seed, which lacks endosperm. Among other distinguishing characteristics: pollen in the pollinia is usually not available as a nutrient-source (*Cleistes* Richard ex Lindley being a notable exception), and the often complex interaction with pollinators culminates in the phenomenon of pseudocopulation in several genera (e.g., *Ophrys* Linnaeus, *Caladenia* R. Brown sect. *Calonema, Drakaea* Lindley). In the latter process, the flower mimics the appearance, the smell, and often the movements of a female wasp, attracting a male of a suitable species that tries to copulate with the flower. It usually only succeeds in becoming attached to a pollinium, which will then be transferred if the male tries to copulate with another flower.

Roots of orchids may be covered with *velamen*, spongy layers derived from the epidermis; fleshy thickenings of roots are *tuberoids* (tubers being restricted to stems). Stems may be swollen or thickened, underground *corms* or aerial *pseudobulbs*. Flowers are often resupinate: the lip (modified median petal) is "lowermost," usually as a result of the pedicel being twisted or bent in its development by 180°. *Pedicellate ovary*, usually used in reference to length, refers to the combined pedicel and ovary. Flowers are not always borne on pedicels; when they are, it is sometimes difficult to distinguish between a slender ovary and the pedicel. Consequently, because of their slender ovaries, flowers of a "racemose spike" appear to be pedicellate even though they are sessile, while a "spicate raceme" has pedicels so short that they appear to be absent. Orchid flowers often have a modified median sepal, the *dorsal sepal*. Sepals coalescing at their tips form a *synsepal*. The middle portion of the upper (adaxial) face of the lip is the *disc*: it may be a thickened *callus* and may bear hairs, papillae, or other ornamentation. In orchids the style, stigmas, filaments, and one or more anthers are united to form a *column*; appendages projecting laterally from the stigma are column wings; the lip may be attached to the protrusion at the base of the column to form a *column foot*; lateral sepals that are also attached to the foot form a *mentum* (chin). In most orchids the column bears a single anther at its apex; the clinandrium is the cavity within which the anther is borne or embedded. Pollen is borne in discrete masses (*pollinia*). Genera with mealy (*sectile*) pollinia may have pollinia within the anther tapering into a *caudicle* (stalk), which is attached to a sticky *viscidium*. Those with waxy pollinia have pollinia attached to one or two stipes (of stigmatic origin and formed outside the anther), which in turn are attached to a viscidium. The various aggregations of pollinia, caudicles, stipes, and viscidium form a *pollinarium*, the pollination unit carried by pollinators. The median stigma lobe may have a slender extension or little beak (*rostellum*), which aids in gluing the pollinarium to the pollinator.

Taxa excluded from the flora as neither native nor definitely naturalized are discussed below.

Bletia florida (Salisbury) R. Brown, grows in gardens near Homestead, Florida. This species has escaped and may persist but does not spread to the surrounding area (P. M. Brown 2002).

Bletia patula Hooker is an apparent garden escape (P. M. Brown 2000). A single vegetative specimen, which later flowered, was collected in 1947. No specimens that can be referred to this species have been collected or seen since then (P. M. Brown 2002).

Bletilla striata (Thunberg) Reichenbach f. is a garden escape (P. M. Brown 2000, 2002) that has been reported growing along roadsides and in lawns in Escambia County, Florida (G. S. Wilhelm 1984). A specimen is located in FLAS but its naturalized status cannot be established (R. Wunderlin, pers. comm.).

Brassavola nodosa (Linnaeus) Lindley is a dubious record (C. McCartney 1997; P. M. Brown 2000, 2002).

Encyclia rufa (Lindley) Britton & Millspaugh has been excluded because only one specimen has been recorded. R. P. Sauleda and R. M. Adams (1983) considered it to have been introduced.

Epidendrum blancheanum Urban is endemic to Hispaniola. It is not confirmed for Florida.

Epidendrum radicans Pavón ex Lindley has been reported to be a persistent escape in Lee County, Florida (P. M. Brown 2000). Two specimens previously identified as this species from Captiva Island, Florida, are probably hybrids and not the pure species, which is very easily identified (E. Hágsater, pers. comm.).

Gymnadenia conopsea (Linnaeus) R. Brown was reported for Litchfield, Connecticut, by R. C. Bean et al. (1951), who noted that a specimen was in the herbarium of the Torrey Botanical Club but that the station has not been relocated. Apparently the species did not persist (P. M. Brown 1997), and it is not a likely naturalization.

Laelia rubescens Lindley was discovered in 1999, growing on a large branch of a live oak in Hammock County Park, Miami, Miami-Dade County, Florida. It was believed to have originated from seed rather than an intentional planting because the plant could be traced from a pea-sized pseudobulb for at least 8 growth periods, and other seedlings were found in the area (P. M. Brown 2000, 2002). The species may be a possible naturalization.

Leochilus labiatus (Swartz) Kuntze was reported by C. A. Luer (1972) to be naturalized in Florida based on second-hand evidence (C. McCartney 1997). The purported occurrence in the Fakahatchee Swamp may have been a deliberate deception at Luer's expense (R. L. Hammer, pers. comm.). All of Luer's photographs were of cultivated plants and there is no herbarium voucher. Its reported naturalization in Florida is dubious (P. M. Brown 2000).

Lophiaris carthagenensis (Jacquin) Braem was reported in Florida [as *Oncidium carthagenense* (Jacquin) Swartz] based upon a single specimen collected by J. K. Small in 1916, at Coot Bay (near Flamingo in Everglades National Park), Monroe County, Florida (D. S. Correll 1950; C. A. Luer 1972). The specimen, in New York, includes a partial inflorescence and lacks foliage. A number of attempts were made to relocate the species at Coot Bay but all were unsuccessful (R. P. Sauleda and R. M. Adams 1989; R. L. Hammer, pers. comm.). Sauleda and Adams suggested that the specimen collected by Small had come from a garden in southern Florida where he kept plants until they flowered. They stated that Small probably collected the specimen under the mistaken belief that the plant was collected in Florida although it had actually come from Central America. Others, including P. M. Brown (2000), simply consider it to be a possible introduction but not naturalized.

Maxillaria sanguinea Rolfe has reportedly become naturalized, but naturalization of the species in the flora area is dubious (C. McCartney 1997; P. M. Brown 2000).

Restrepiella ophiocephala (Lindley) Garay & Dunsterville was reported to occur in North America. Apparently, it has not become naturalized in the flora area (P. M. Catling, pers. comm.). It is considered a dubious record (C. McCartney 1997; P. M. Brown 2000).

Tetramicra canaliculata (Aublet) Urban was treated as naturalized in the flora of Florida (C. A. Luer 1972). Recent evidence (C. A. Luer, pers. comm.) indicates that the species is not native and has not become naturalized. Moreover, it has not been reported for Jamaica or Cuba and reports for the Bahamas are erroneous (D. S. Correll and H. B. Correll 1982). Other citations (L. A. Garay and H. R. Sweet 1974; J. G. Williams and A. E. Williams 1983) also are erroneous, probably based on C. A. Luer (1972). The scenarios that could have led to this and other cases of dubious reports of naturalized orchids are discussed in C. McCartney (1997).

SELECTED REFERENCES Bentley, S. L. Native Orchids of the Southern Appalachian Mountains. Chapel Hill. Brown, P. M. 1997. Wild Orchids of the Northeastern United States: A Field Guide. Ithaca, N.Y. Brown, P. M. 2000. The Florida

Native Orchid Project. Palmetto 20: 6–10. Brown, P. M. 2002. Wild Orchids of Florida. Gainesville. Burns-Balogh, P. and V. A. Funk. 1986. A phylogenetic analysis of the Orchidaceae. Smithsonian Contr. Bot. 61. Case, F. W. 1987. Orchids of the western Great Lakes region, rev. ed. Bull. Cranbrook Inst. Sci. 48. Coleman, R. A. 1995. The Wild Orchids of California. Ithaca, N.Y. Correll, D. S. 1950. Native Orchids of North America North of Mexico.... Waltham, Mass. Dressler, R. L. 1981. The Orchids: Natural History and Classification. Cambridge, Mass. Dressler, R. L. 1993. Phylogeny and Classification of the Orchid Family. Portland. Homoya, M. A. 1993. Orchids of Indiana. Bloomington. Liggio, J. and A. Liggio. 1999. Wild Orchids of Texas. Austin. Luer, C. A. 1972. The Native Orchids of Florida. Bronx. Luer, C. A. 1975. The Native Orchids of the United States and Canada, Excluding Florida. Bronx. Magrath, L. K. 1973. The Native Orchids of the Prairies and Plains Region of North America. Ph.D. dissertation. University of Kansas. Morris, F. and E. A. Eames. 1929. Our Wild Orchids: Trails and Portraits. New York. Pridgeon, A. M., P. J. Cribb, and F. N. Rasmussen. 1999+. Genera Orchidacearum. 1+ vol. Oxford. Rasmussen, F. N. 1985. Orchids. In: R. M. T. Dahlgren et al. 1995. The Families of the Monocotyledons: Structure, Evolution, and Taxonomy. Berlin etc. Pp. 249–274. Sheehan, T. J. and M. Sheehan. 1994. An Illustrated Survey of Orchid Genera. Portland. Smith, W. R. 1993. Orchids of Minnesota. Minneapolis. Szlachetko, D. L. 1995. Systema orchidalium. Fragm. Florist. Geobot., suppl. 3. Whiting, R. E. and P. M. Catling. 1986. Orchids of Ontario: An Illustrated Guide. Ottawa. Williams, J. G. and A. E. Williams. 1983. Field Guide to Orchids of North America. New York.

Key to the Genera of Orchidaceae

Lawrence K. Magrath
Robert L. Dressler
John T. Atwood
John Beckner

1. Lip inflated, slipper (calceolate) or sac-shaped (saccate).
  2. Lateral anthers 2, fertile, dorsal anther a large, subapical staminode, partially closing entrance to lip . . . . 1. *Cypripedium*, p. 499
  2. Anthers 1, fertile, staminode absent.
    3. Apex of lip 2-horned; flowers solitary; leaf solitary . . . . 51. *Calypso*, p. 622
    3. Apex of lip blunt or acute; flowers 5–72, in spikes; leaves more than 1, in basal rosettes . . . . 6. *Goodyera* (in part), p. 514
1. Lip not inflated.
  4. Plants without leaves at anthesis, or without green blade either diverging from axis or reduced to bract.
    5. Plants epiphytes or vines.
      6. Plants long, thick, succulent vines; lip without spur . . . . 2. *Vanilla* (in part), p. 507
      6. Plants epiphytes with conspicuous grayish green roots, stems much reduced; lip with spur.
        7. Lip middle lobe terminating in 2 arching, twisting, caudate lobes, 45–70 mm . . . . 49. *Dendrophylax*, p. 620
        7. Lip middle lobe not as above, less than 10 mm.
          8. Flowers more than 12, in 2 rows, in spike . . . . 48. *Campylocentrum*, p. 619
          8. Flowers fewer than 12, not in 2 rows, in lax racemes or panicles . . . 50. *Harrisella*, p. 621
    5. Plants terrestrial, never vines.
      9. Plants pure white except for lip with yellow markings . . . . 31. *Cephalanthera*, p. 583
      9. Plants green or not, never entirely white with yellow-marked lip.
        10. Anther acute with terminal viscidium; petals often adherent to dorsal sepal and difficult to separate . . . . 17. *Sacoila* (in part), p. 528
        10. Anther not acute, terminal viscidium absent; petals not adherent to dorsal sepal.
          11. Flowers asymmetric; spur longer than lip . . . . 52. *Tipularia* (in part), p. 624
          11. Flowers symmetric; spur absent or shorter than lip.
            12. Spur broad-based and conic . . . . 62. *Galeandra*, p. 643
            12. Spur narrow-based, or a mentum, or absent.
              13. Inflorescences corymbose racemes; perianth closed to slightly gaping, not resupinate; pollinia 2 . . . . 34. *Triphora* (in part), p. 592
              13. Inflorescences spikes, racemes, or panicles; perianth spreading, resupinate; pollinia 4 or 8 (2 half anthers only in *Piperia*).

14. Plants achlorophyllous.
   15. Rhizomes annulate; lip 3-lobed with 3–5 lamellae; lateral sepals distinct at base; pollinia 8 . . . . 39. *Hexalectris*, p. 603
   15. Rhizomes corraloid; lip 1–3-lobed with 2 lamellae; sepals connate at base forming mentum; pollinia 4 . . . . 56. *Corallorhiza*, p. 633
14. Plants chlorophyllous.
   16. Lips with clavate to filiform spurs; leaves not plicate . . . . 27. *Piperia* (in part), p. 571
   16. Lips without spurs; leaves plicate.
      17. Leaves basal, solitary, elliptic, withering by anthesis; column 7 mm; pollinia 4; mature capsules pendent . . . . 55. *Aplectrum* (in part), p. 632
      17. Leaves basal and cauline, 3–5, linear, narrowly elliptic, or lanceolate, present at anthesis; column 8–25 mm; pollina 8; mature capsules erect to suberect or pendulous . . . . 38. *Bletia* (in part), p. 602

4. Plants with leaves at anthesis, or at least with green blade diverging from axis.
   18. Leaves articulate, deciduous (persistent in *Oeceoclades*).
      19. Leaves plicate, thin, never conduplicate.
         20. Pseudobulbs fusiform, 15–50 cm . . . . 61. *Cyrtopodium*, p. 642
         20. Pseudobulbs ovoid to subglobose, not fusiform, less than 10 cm.
            21. Flowers mostly pink, rosy purple, or deep purple, rarely white; lip with elongate crests; leaf bases almost wiry . . . . 38. *Bletia* (in part), p. 602
            21. Flowers mostly white or tinged with purple; lip without crests; leaf sheaths multiple, concentric, polygonal, keeled . . . . 57. *Govenia*, p. 638
      19. Leaves not plicate, not thin, distinctly conduplicate or not.
         22. Inflorescences distinctly lateral, not continuous with stem axis.
            23. Flowers solitary from leaf axils . . . . 63. *Maxillaria*, p. 644
            23. Flowers in racemes or spikes or panicles.
               24. Plants terrestrial; spurs 4–5 mm . . . . 59. *Oeceoclades*, p. 640
               24. Plants epiphytic; spurs absent.
                  25. Inflorescences spikes; flowers sessile . . . . 47. *Bulbophyllum*, p. 618
                  25. Inflorescences racemes or panicles; flowers pedicellate.
                     26. Lips unlobed or 2-lobed.
                        27. Lips 2-lobed, 6–17 mm; sepals 3–7 mm; column 3.5 mm; flowers pink to rose, pink-lavender, or white with pink flush . . . . 65. *Ionopsis*, p. 646
                        27. Lips unlobed, 14–27 mm; sepals 30–200 mm; column 2 mm; flowers yellow-green to yellow-orange with red-brown markings . . . . 64. *Brassia*, p. 645
                     26. Lips 3-lobed.
                        28. Racemes pendent; column 8 mm; middle lobe of lip nearly linear, strongly reflexed; callus of 3 shallow keels . . . . 66. *Macradenia*, p. 647
                        28. Racemes or panicles erect to arching or pendent; column 5–6 mm; middle lobe of lip expanded, not reflexed; callus tuberculate.
                           29. Callus prominent tuberculate-lobulate; pseudobulbs reduced to absent . . . . 69. *Tolumnia*, p. 651
                           29. Callus prominent warty-tuberculate-fringed; pseudobulbs conspicuous . . . . 67. *Oncidium* (in part), p. 648
         22. Inflorescences distinctly terminal, continuous with stem axis.
            30. Pedicel (± 1 mm) and ovary articulate; capsule deciduous.
               31. Stem sheaths flared and hispid apically . . . . 44. *Lepanthopsis*, p. 615
               31. Stem sheaths neither flared nor hispid apically . . . . 45. *Pleurothallis*, p. 616

30. Pedicel and ovary not articulate; capsule not deciduous.
32. Lip adnate to column throughout, forming nectary tube . . . 42. *Epidendrum*, p. 609
32. Lip free from column or adnate for less than 1/2 column length.
33. Flowers not resupinate; column foot present or absent.
34. Flowers rosy pink to white; plants terrestrial . . . . 37. *Calopogon*, p. 597
34. Flowers various but not rosy pink to white; plants epiphytic or lithophytic or if terrestrial, then with yellow-green flowers.
35. Column foot present, conspicuous . . . . 46. *Polystachya*, p. 617
35. Column foot absent . . . . 43. *Prosthechea* (in part), p. 613
33. Flowers resupinate; column foot absent.
36. Plants terrestrial; flowers incompletely opening; leaves 1–2, basal . . . . 40. *Basiphyllaea*, p. 607
36. Plants epiphytic or lithophytic; flowers completely opening; leaves 1–3, apical from pseudobulbs.
37. Pseudobulbs ovoid-pyriform, not flattened . . . . 41. *Encyclia*, p. 608
37. Pseudobulbs fusiform or suborbicular to ovoid or ellipsoid, flattened . . . . 43. *Prosthechea* (in part), p. 613
18. Leaves not articulate, persistent (except in *Trichocentrum*).
38. Leaves distinctly plicate.
39. Leaves cauline.
40. Inflorescences racemes . . . . 32. *Epipactis*, p. 584
40. Inflorescences panicles . . . . 35. *Tropidia*, p. 595
39. Leaves basal.
41. Plants 6–60 cm; petals filiform, linear-oblanceolate, narrowly spatulate, or linear, less than 1/2 width of sepals . . . . 53. *Liparis* (in part), p. 624
41. Plants 30–170 cm; petals elliptic, obovate, or oblong, equal to or more than 1/2 width of sepals.
42. Column 3–4 mm; lip without lamellae; mature capsules erect . . . . 60. *Pteroglossaspis*, p. 641
42. Column 7–10 mm; lip with lamellae; mature capsules pendent . . . . 58. *Eulophia*, p. 639
38. Leaves convolute, flat, or conduplicate, seldom plicate.
43. Plants vines . . . . 2. *Vanilla* (in part), p. 507
43. Plants various but not vines.
44. Leaves opposite, subopposite, or whorled at terminal node proximal to inflorescence.
45. Leaves 2(–3); flowers, 2–100, in racemes . . . . 33. *Listera*, p. 586
45. Leaves whorled, usually 5; flowers 1(–2, rarely) . . . . 4. *Isotria*, p. 511
44. Leaves various but not opposite, subopposite (except for *Malaxis monophyllos* var. *monophyllos*), or whorled.
[46. Shifted to left margin.—Ed.]
46. Flowers with protruding spur at base of lip, spur free from ovary.
47. Flowers asymmetric; corms present; tuberoids absent . . . . 52. *Tipularia* (in part), p. 624
47. Flowers bilaterally symmetric; corms absent; tuberoids sometimes present.
48. Leaves basal or radical.
49. Leaves 1(–2).
50. Petiole conspicuous, erect; lip margin fringed . . . . 13. *Eltroplectris*, p. 524
50. Petiole inconspicuous, sheathing or clasping stem; lip margin entire to crenate to erose.
51. Lip deeply 3-lobed . . . . 25. *Amerorchis*, p. 550
51. Lip unlobed . . . . 26. *Platanthera* (in part), p. 551
49. Leaves 2 or more.
52. Leaves more than 2 . . . . 27. *Piperia* (in part), p. 571
52. Leaves 2.
53. Stems conspicuously angled . . . . 24. *Galearis*, p. 550

53. Stems terete.
54. Tuberoids ellipsoid-ovoid; lateral sepals adnate to base of lip; lip with midrib thickened . . . . 27. *Piperia* (in part), p. 571
54. Tuberoids elongate; lateral sepals free; lip flat . . . . . 26. *Platanthera* (in part), p. 551
48. Leaves basal and/or cauline.
55. Plants from creeping rhizomes; leaves petiolate, with inflated tubular sheaths . . . . 7. *Platythelys*, p. 517
55. Plants from fusiform, fleshy or tuberous roots; leaves sessile, without inflated tubular sheaths.
56. Flowers purple, lavender, yellow, or orange.
57. Anther cells parallel or nearly so, viscidia lying close together . . . 28. *Dactylorhiza*, p. 577
57. Anther cells divergent, viscidia separated . . . . 26. *Platanthera* (in part), p. 551
56. Flowers white to green, never purple, lavender, yellow, or orange.
58. Pollinia parallel or nearly so, viscidia lying close together.
59. Leaves basal; lateral sepals adnate to base of lip; lip with midrib thickened; column adnate to base of lip . . . . 27. *Piperia* (in part), p. 571
59. Leaves cauline; lateral sepals free; lip flat; column diverging from lip . . . . 26. *Platanthera* (in part), p. 551
58. Pollinia divergent, viscidia separated.
60. Viscidia covered by membranes; mouth of spur covered by membrane with small aperture . . . . 29. *Coeloglossum*, p. 580
60. Viscidia exposed; mouth of spur widely open.
61. Stigmatic processes fleshy, well developed; petals often 2-partite; lip 3-partite into linear lobes or small basal teeth; tuberoids usually spheroid . . . . 30. *Habenaria*, p. 581
61. Stigmatic processes absent or poorly developed; petals not 2-partite; lip not 3-partite into linear divisions; tuberoids elongate, tapering.
62. Sepals less than 5 mm; lip apex with 3 subequal, lanceolate or oblong lobes . . . . 23. *Pseudorchis*, p. 549
62. Sepals more than 5 mm; lip various but not as above . . . . 26. *Platanthera* (in part), p. 551
46. Flowers without spur at base of lip or spur adnate to ovary.
63. Plants arising from corms, pseudobulbs, or tuberoids, not from fascicles of fleshy roots.
64. Plants epiphytic; some leaves at least 40 cm; inflorescences panicles.
65. Leaves linear-lanceolate; pseudobulbs prominent . . . . 67. *Oncidium* (in part), p. 648
65. Leaves broadly elliptic to ovate-lanceolate; pseudobulbs highly reduced . . . . 68. *Trichocentrum*, p. 649
64. Plants terrestrial; all leaves much less than 40 cm (except in *Spiranthes odorata*); inflorescences solitary flowers, spikes, or racemes, never panicles.
66. Sepals more than 20 mm . . . . 36. *Arethusa*, p. 596
66. Sepals less than 20 mm.
67. Leaves basal, plicate.
68. Leaves 1, withering or absent at flowering; column straight, compressed, 7 mm; mature capsules pendent; lip 3-lobed . . . . 55. *Aplectrum* (in part), p. 632
68. Leaves 2–7, present at flowering; column curved, apically winged, 3–5 mm; mature capsules erect; lip unlobed . . . . 53. *Liparis* (in part), p. 624
67. Leaves basal and/or cauline, various, but not plicate.
69. Spurs present . . . . 26. *Platanthera* (in part), p. 551
69. Spurs absent.
70. Lip 3-lobed.
71. Column 5–10 mm . . . . 34. *Triphora* (in part), p. 592
71. Column 0.2–1 mm . . . . 54. *Malaxis* (in part), p. 627
70. Lip unlobed or 3-dentate.

72. Flowers in racemes.
73. Column 0.2–1 mm . . . . 54. *Malaxis* (in part), p. 627
73. Column 1.5–5 mm . . . . 53. *Liparis* (in part), p. 624
72. Flowers in spirally twisted or secund spikes.
74. Lip white, 3.5–5 mm . . . . 19. *Spiranthes* (in part), p. 530
74. Lip white with central cinnabar-red disc and 3 green stripes near apex, 5.5–8 mm . . . . 18. *Schiedeella*, p. 530
63. Plants arising from slender fibrous roots or fascicles of thickened fleshy roots or creeping rhizomes, not from corms, pseudobulbs, or tuberoids (*Cleistes* occasionally has nodular tuberoids).
75. Inflorescences of 1(–3, rarely) flowers.
76. Lip margin laciniate; petals and sepals pink or rarely white or bluish; sepals oblong, elliptic, or narrowly lanceolate, 14–23 mm . . . . 5. *Pogonia*, p. 513
76. Lip margin crenulate; petals rose-pink to white; sepals olive-green, brown, or maroon, linear-lanceolate, 24–65 mm . . . . 3. *Cleistes*, p. 510
75. Inflorescences of 4 or more flowers.
77. Plants arising from creeping rhizomes . . . . 6. *Goodyera* (in part), p. 514
77. Plants not arising from creeping rhizomes.
78. Flowers white with yellow lip.
79. Lip pandurate, apex slightly broadened to winglike, margins entire; leaves cauline . . . . 8. *Zeuxine*, p. 518
79. Lip ovate, apex rounded, recurved, margin crisped, crenulate; leaves basal . . . . 19. *Spiranthes* (in part), p. 530
78. Flowers various, not white with yellow lip.
80. Flowers not resupinate.
81. Column 0.3 mm; lip 1–2 mm; dorsal sepal 1–2.2 mm, reflexed . . . . 9. *Prescottia*, p. 519
81. Column 1.5–5 mm; lip 2–7 mm; dorsal sepal 3–7 mm, not reflexed.
82. Petals obliquely triangular; lip obovate to suborbiculate, deeply concave, white with 1–2 dark green spots; column slightly winged, 2.5–5 mm . . . . 22. *Ponthieva*, p. 547
82. Petals linear; lip ovate-subquadrangular, shallowly concave, white with green or pale yellow spots; column prominently winged, 1.5–2 mm . . . . 21. *Cranichis*, p. 546
80. Flowers resupinate.
83. Flowers yellow-orange to scarlet.
84. Petals and sepals strongly recurved; lip 15–20 × 3–5 mm; lateral sepals free, not forming mentum . . . . 14. *Dichromanthus*, p. 524
84. Petals and sepals not recurved; lip 20–31 × 3–8 mm; lateral sepals forming mentum with base of lip . . . . 17. *Sacoila* (in part), p. 528
83. Flowers various colors, not orange-yellow to scarlet.
85. Flower colors green, greenish brown, or yellowish brown.
86. Petioles of leaves 12–16.5 cm, reddish purple . . . . 16. *Pelexia*, p. 526
86. Petioles of leaves 10 cm or less, green.
87. Column 11 mm; lip 20–24 mm . . . . 17. *Sacoila* (in part), p. 528
87. Column 2–4 mm; lip 5–7 mm.
88. Flowers greenish or yellowish green; apices of petals and sepals not recurved; lip oblong, apex slightly flared or pandurate to oblong-ovate with apex usually fan-shaped, 5–7 mm; column 3.5–4.5 mm . . . . 10. *Cyclopogon*, p. 520
88. Flowers brownish green to coppery; apices of petals and sepals spreading to recurved; lip elliptic to ovate-lanceolate, apex obtuse-rounded, 3.5–6 mm; column 2–2.5 mm . . . . 15. *Mesadenus*, p. 525

85. Flowers of various colors, not green, greenish brown, or yellowish green, green markings present in some.
    89. Petals and sepals without green lines or veins . . . . 19. *Spiranthes* (in part), p. 530
    89. Petals and sepals with green lines or veins.
        90. Flowers pink or dull brown to purplish; column 8 mm . . . . 12. *Deiregyne*, p. 523
        90. Flowers various, not pink or dull brown to purplish; column 2–10 mm.
            91. Column 8–10 mm; dorsal sepal 15–18 mm; lip 8–16 mm . . . . 20. *Stenorrhynchos*, p. 545
            91. Column 2–4 mm; dorsal sepal 4–8 mm; lip 4.5–7 mm.
                92. Lip white with central green stripe, 4.5 mm, sagittate, 3-lobed, lateral lobes rounded, middle lobe elongate-rounded, margins finely denticulate . . . . 11. *Beloglottis*, p. 522
                92. Lip white, centrally yellow to yellowish or greenish white, 5–7 mm, ovate to ovate-oblong, unlobed, truncate, obtuse to deeply laciniate, margins erose-crisped . . . . . 19. *Spiranthes* (in part), p. 530

230a. ORCHIDACEAE Jussieu subfam. CYPRIPEDIOIDEAE Lindley, Coll. Bot., App. 1821

1. CYPRIPEDIUM Linnaeus, Sp. Pl. 2: 951. 1753; Gen. Pl. ed. 5, 408. 1754 • Lady's-slipper, moccasin-flower, cypripède, sabot de la Vierge [Greek *Kypris*, Aphrodite, and Latin *pes*, foot, perhaps an orthographic error for Greek *pedilon*, slipper]

Charles J. Sheviak

Herbs, perennial. **Roots** closely to widely spaced along rhizome, slender, fleshy; rhizomes short to elongate. **Stems** leafy or scapose. **Leaves** alternate, in single radical pair, or subopposite near midpoint of stem, ascending to spreading, plicate, bases sheathing stem. **Inflorescences** terminal, solitary; flowers solitary or 2–several in lax racemose spike; bracts large, foliaceous. **Flowers** resupinate, showy; sepals distinct or lateral sepals usually connate proximal to lip forming synsepal; petals entire; lip inflated, slipper- or sac-shaped, with adaxial orifice; pollinaria absent; loose granular pollen in 2 lateral anthers, dorsal anther a large subapical staminode; stigma free, 2–3-lobed. **Fruits** capsules, ellipsoid to oblong-ellipsoid.

Species ca. 45 (12 in the flora): primarily temperate Eurasia and North America, ranging from arctic to subtropics in East Asia and Cordilleran North America and Central America.

SELECTED REFERENCES Cribb, P. J. 1997. The Genus *Cypripedium*. Portland. Sheviak, C. J. 1992. Natural hybridization between *Cypripedium montanum* and its yellow-lipped relatives. Amer. Orchid Soc. Bull. 61: 546–559.

1. Lip orifice a longitudinal fissure, length of lip; leaves 2, radical, stem a leafless scape . . . . 5. *Cypripedium acaule*
1. Lip orifice basal, 4–37 mm; leaves 2–several, cauline.
    2. Lip appearing angular, adaxial surface swollen near middle, apex abruptly deflected downward; lateral sepals distinct . . . . 1. *Cypripedium arietinum*

2. Lip not angular, slender slipper-shaped to globose sac, apex not deflected; lateral sepals connate or nearly distinct.
   3. Leaves 2, subopposite to alternate, near midpoint of stem (very rarely, leaves 3, closely alternate near midpoint of stem); lip prominently marked reddish or brownish between veins.
      4. Flowers (1–)2–4 in short nodding spike, dull yellow to reddish brown or dark purple, lip mostly 8–14(–25) mm . . . . 2. *Cypripedium fasciculatum*
      4. Flowers solitary, erect, white to yellowish or greenish, mottled with pink or brownish; lip 15–30 mm.
         5. Petals slightly shorter than to equaling globose to obovate lip, constricted near apex, lanceolate-subpandurate in general outline; markings pink to magenta . . . . 3. *Cypripedium guttatum*
         5. Petals much shorter than oblance-fusiform to oblance-ovate lip, constricted near middle, acuminate-subpandurate to pandurate in general outline; markings brownish or tan . . . . 4. *Cypripedium yatabeanum*
   3. Leaves 3 or more (very rarely 2), alternate on basal to middle portion of stem or along length; lip yellow, ivory, or white, without extensive markings between veins.
      6. Petals flat, oblong to ovate, unmarked, white or green, yellow-green, or brownish.
         7. Flowers large; lip 25–53 mm . . . . 6. *Cypripedium reginae*
         7. Flowers small; lip 11–20 mm.
            8. Plants usually less than 35 cm; flowers 1(–2); lateral sepals nearly distinct to connate almost to apex . . . . 7. *Cypripedium passerinum*
            8. Plants usually much more than 35 cm; flowers 3–18(–22); lateral sepals connate almost to apex . . . . 8. *Cypripedium californicum*
      6. Petals spirally twisted or undulate (rarely flat), linear-lanceolate, or rarely linear-oblong, spotted and striped (sometimes minutely so) or suffused with darker reddish brown or madder on yellowish or greenish ground color (rarely without markings).
         9. Lip orifice 27–37 mm; flowers very large; lip (41–)53–65 mm, ivory to pale yellow . . . . 12. *Cypripedium kentuckiense*
         9. Lip orifice usually less than 23 mm (rarely to 27 mm, then lip deep yellow); flowers smaller; lip mostly 15–45 mm (rarely to 54 mm, then lip deep yellow), white or yellow.
            10. Lip white, rarely suffused with magenta; staminode lanceoloid or oblong-lanceoloid to ellipsoid.
                11. Sepals and petals green or yellowish, usually spotted and striped with reddish brown or madder; dorsal sepal 15–35 mm; leaves lanceolate or elliptic to oblanceolate, ascending from insertions toward base or middle of stem . . . . 9. *Cypripedium candidum*
                11. Sepals and petals suffused with reddish brown or madder; dorsal sepal 33–60 mm; leaves suborbiculate to ovate or elliptic-lanceolate, ascending to spreading, inserted toward base or scattered along stem . . . . 10. *Cypripedium montanum* (in part)
            10. Lip white or yellow; staminode cordiform-ovoid, deltoid to lance-ovoid, or ovoid-oblong (rarely staminode narrower or ellipsoid, then lip yellow).
                12. Lip white, 19–33 mm; sepals and petals suffused with reddish brown or madder, often intensely so (very rarely clear green), long and slender; dorsal sepal 33–60 mm, ratio of length to width 2.5–6.5; cordilleran . . . . 10. *Cypripedium montanum* (in part)
                12. Lip yellow (very rarely white), 15–54 mm; sepals and petals suffused with reddish brown or madder or more commonly spotted and striped (often minutely so), rarely unmarked, short and broad to long and slender; dorsal sepal 19–80 mm, ratio of length to width 1.3–3.7; widespread . . . . 11. *Cypripedium parviflorum*

1. Cypripedium arietinum R. Brown in W. Aiton and W. T. Aiton, Hortus Kew. 5: 222. 1813 • Ram's-head [E]

Plants erect, 10–35 cm. **Leaves** 3–4, along middle portion of stem, spiraled to alternate, ascending to spreading; blade narrowly elliptic to ovate-lanceolate or oblong, 5–11 × 1.3–3.5 cm. **Flowers** usually solitary (rarely 2 in forma *biflorum* P. M. Brown); sepals green with reddish brown markings; dorsal sepal broadly elliptic to ovate-lanceolate, 15–25 × 5–10 mm; lateral sepals distinct, 12–20 × 1.5–4 mm; petals somewhat spreading, same color as sepals, somewhat spirally twisted, linear- to linear-lanceolate, 11–24 × 1–2 mm; lip white with green apex, usually with extensive reddish reticulations often merging toward apex, adaxially swollen near middle, 10–16 mm, apex abruptly deflected downward, orifice basal, 7–12 mm; staminode suborbicular. 2*n* = 20.

Flowering May–Jun. Dry to moist open coniferous and mixed forests, coniferous-forested fens, beach thickets; mostly 0–400 m; Man., N.S., Ont., Que., Sask.; Conn., Maine, Mass., Mich., Minn., N.H., N.Y., Vt., Wis.

2. Cypripedium fasciculatum Kellogg ex S. Watson, Proc. Amer. Acad. Arts 17: 380. 1882 • Clustered lady's-slipper, brownie lady's-slipper [E]

Cypripedium knightiae A. Nelson

Plants erect, 6–35 cm in flower, taller in fruit; inflorescences nodding in flower, erect in fruit. **Leaves** 2, initially near middle of stem, subopposite, wide-spreading, distal portion elongating greatly in fruit; blade elliptic to ovate-orbiculate, 4–12 × 2.5–7.5 cm. **Flowers** (1–)2–4; sepals dull yellow, marked and suffused often intensely with reddish brown or dark purple; dorsal sepal lanceolate, 13–25 × 3–8 mm; lateral sepals connate, synsepal 11–23 × 4–9 mm; petals porrect-spreading, same color as sepals, flat, ovate-lanceolate to lance-acuminate, 10–23 × 6–17 mm; lip color of sepals, usually less intensely marked, mostly obovoid, 8–14(–25) mm, orifice basal, 4–5 mm; staminode ellipsoid to oblong. 2*n* = 20.

Flowering Apr–Aug. Moist to dry coniferous forests and thickets; 0–3200 m; Calif., Colo., Idaho, Mont., Oreg., Utah, Wash., Wyo.

SELECTED REFERENCE Brownell, V. R. and P. M. Catling. 1987. Notes on the distribution and taxonomy of *Cypripedium fasciculatum* Kellogg ex Watson (Orchidaceae). Lindleyana 2: 53–57.

3. Cypripedium guttatum Swartz, Kongl. Vetensk. Acad. Nya Handl. 21: 251. 1800 • Spotted lady's-slipper [F]

Plants erect, 12–35 cm. **Leaves** 2 (–3,very rarely), on middle half of stem, alternate to subopposite, wide-spreading; blade lance-ovate to ovate-suborbiculate, 5–15 × 1.5–8 cm. **Flowers** solitary, erect; sepals white with pink to reddish or magenta markings; dorsal sepal ovate- to suborbiculate-elliptic, 12–28 × 6–19 mm; lateral sepals connate, synsepal 12–21 × 3–8 mm; petals spreading, same color as sepals, lanceolate-subpandurate (constricted near apex), flat, 10–20 × 4–9 mm, slightly shorter than to equaling lip, margins undulate-revolute to slightly spiraled; lip similarly colored, subglobose to obovoid, 15–30 mm; orifice basal, 10–24 mm; staminode oblong-quadrangular to broadly ellipsoid or ovoid. 2*n* = 20.

Flowering Jun–Jul. Moist to dry open deciduous and spruce forest, tundra, meadows, scree; 0–800 m; N.W.T., Yukon; Alaska; Asia.

Cypripedium guttatum has been reported from British Columbia (D. S. Correll 1950), apparently based on a single collection of *C. parviflorum* (*M. G. Henry 119* GH, PH). See the discussion under 4. *C. yatabeanum*.

4. Cypripedium yatabeanum Makino, Bot. Mag. (Tokyo) 13: 91. 1899

Cypripedium guttatum Swartz subsp. *yatabeanum* (Makino) Hultén; *C. guttatum* var. *yatabeanum* (Makino) Pfitzer

Plants erect, 17–32 cm. **Leaves** 2 (–3, very rarely), on middle half of stem, alternate to subopposite, wide-spreading; blade elliptic-lanceolate to lance-ovate or ovate-suborbiculate, 6–12 × 3–7 cm. **Flowers** solitary; sepals white to yellowish or greenish with brownish or tan markings; dorsal sepal obovate to ovate, 15–23 × 9–18 mm; petals spreading, same color as sepals, acuminate-subpandurate to acuminate-pandurate (constricted near middle), flat, 10–14 × 5–8 mm, much shorter than lip, margins undulate-revolute; lip similarly colored, oblance-fusiform to oblance-ovoid, 17–32 mm; orifice basal, 10–22 mm; staminode oblong-quadrangular to broadly ellipsoid or ovoid. 2*n* = 20.

Flowering Jun–Aug. Mesic tundra, swales, marsh borders, beach dune lags; mostly 0–200 m; Alaska; Asia (n Japan, Kamchatka).

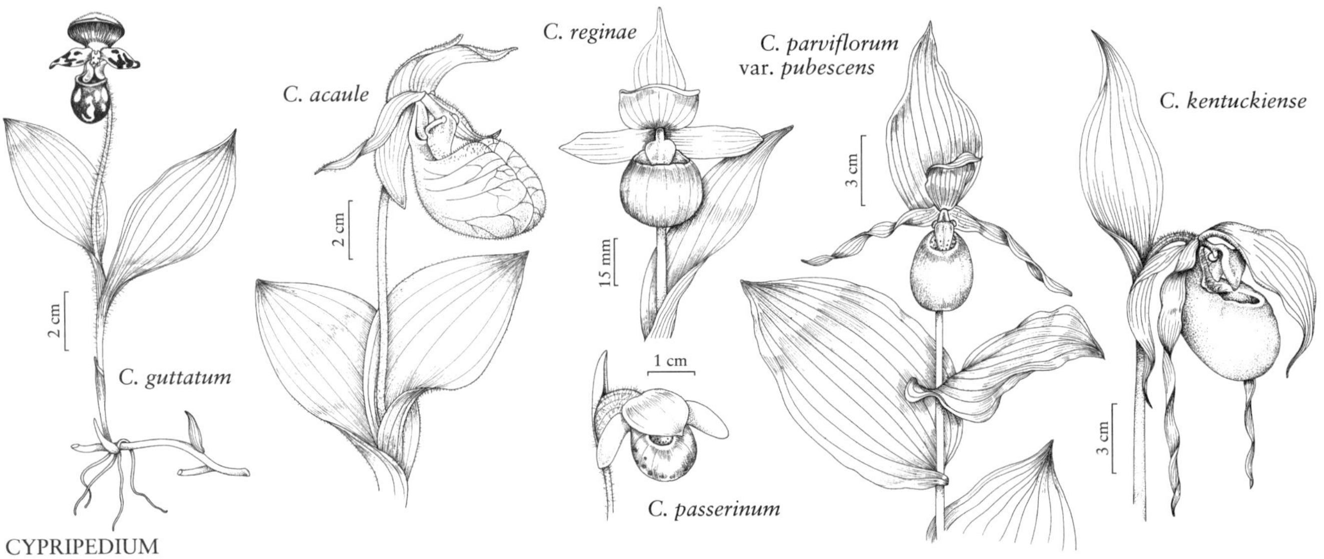

CYPRIPEDIUM

Cypripedium yatabeanum is known from only a few collections in North America and hence the description may be too restrictive. From the much more common and widespread *C. guttatum*, it is distinguished by flowers with pale brown markings and a markedly narrow aspect, with comparatively narrow lip, short, often pandurate petals constricted toward the middle, and narrower dorsal sepals commonly broadest toward the apex. Asiatic plants are sometimes considerably larger than those of *C. guttatum*, and robust Japanese plants sometimes bear three leaves. Plants typical of Asiatic *C. yatabeanum* occur on some of the Aleutian Islands, while on other islands, plants with brownish markings appear morphologically intermediate between these taxa, or more typical of *C. guttatum*. Elsewhere in southern Alaska, populations of typical *C. guttatum* support some plants with pale brownish pink markings. This situation has suggested to some that these taxa were only infraspecifically distinct. The distribution of the intermediate plants, however, suggests that geographically limited introgression during an earlier, more extensive presence of *C. yatabeanum* is an appropriate interpretation. Evident introgressants have been named *C.* ×*alaskanum* P. M. Brown, and most reports and illustrations of *C. yatabeanum* in Alaska are in fact based on them.

5. **Cypripedium acaule** Aiton, Hort. Kew. 3: 303. 1789 • Small pink lady's-slipper, pink moccasin-flower, stemless lady's-slipper, cypripède acaule [E] [F]

Fissipes acaulis (Aiton) Small

Plants erect, 15–61 cm; stems leafless scapes. **Leaves** 2, radical, arising directly from rhizome, ascending to spreading; blade broadly elliptic to oblong-, ovate-, or obovate-elliptic, 9–30 × 2.5–15 cm. **Flowers** solitary; sepals reddish brown to green; dorsal sepal lanceolate to lance-ovate or elliptic, 19–52 × 5–22 mm; lateral sepals connate, synsepal 17–49 × 6–25 mm; petals deflexed to somewhat spreading, somewhat spirally twisted, same color as sepals, linear- to ovate-lanceolate, 24–60 × 4–17 mm; lip magenta to white, obovoid to oblance-ovoid, 30–67 mm, orifice a longitudinal fissure, length of lip; staminode quadrangular to suborbicular-rhomboid or ovoid-deltoid. $2n = 20$.

Flowering Apr–Jul. Dry to wet forests, bogs, brushy barrens, heath, and roadsides on highly acidic soil; mostly 0–1200 m; Alta., Man., N.B., Nfld. and Labr. (Nfld.), N.W.T., N.S., Ont., P.E.I., Que., Sask.; Ala., Conn., Del., D.C., Ga., Ill., Ind., Ky., Maine, Md., Mass., Mich., Minn., N.H., N.J., N.Y., N.C., Ohio, Pa., R.I., S.C., Tenn., Vt., Va., W.Va., Wis.

6. Cypripedium reginae Walter, Fl. Carol., 222. 1788 • Showy lady's-slipper, queen lady's-slipper, cypripède royal [E] [F]

Cypripedium spectabile Salisbury

Plants erect, 21–90 cm. **Leaves** 3–9, along length of stem, alternate, ascending to spreading; blade broadly elliptic to ovate or elliptic-lanceolate, 10–27 × 5–16 cm. **Flowers** 1–3(–4); sepals white; dorsal sepal suborbiculate to ovate-orbiculate, 25–45 × 18–42 mm; lateral sepals connate, synsepal 24–42 × 15–37 mm; petals spreading, same color as sepals, oblong to elliptic or ovate-lanceolate, flat, 25–47 × 6–17 mm; lip white, usually suffused around orifice and variably beyond with pale pink to crimson magenta, subglobose-obovoid, 25–53 mm; orifice basal, 18–28 mm; staminode cordiform-ovoid to ovoid-lanceoloid. $2n = 20$.

Flowering May–Aug. Coniferous and hardwood fen forests, openings, fen meadows, hillside seeps, meadows, prairies, dune lags, seeping cliffs; mostly 0–600 m; Man., N.B., Nfld. and Labr. (Nfld.), N.S., Ont., P.E.I., Que., Sask.; Ark., Conn., Ill., Ind., Iowa, Ky., Maine, Md., Mass., Mich., Minn., Mo., N.H., N.J., N.Y., N.C., N.Dak., Ohio, Pa., Tenn., Vt., Va., W.Va., Wis.

7. Cypripedium passerinum Richardson in J. Franklin et al., Narr. Journey Polar Sea, 762. 1823 • Sparrow's-egg lady's-slipper, Franklin's lady's-slipper [E] [F]

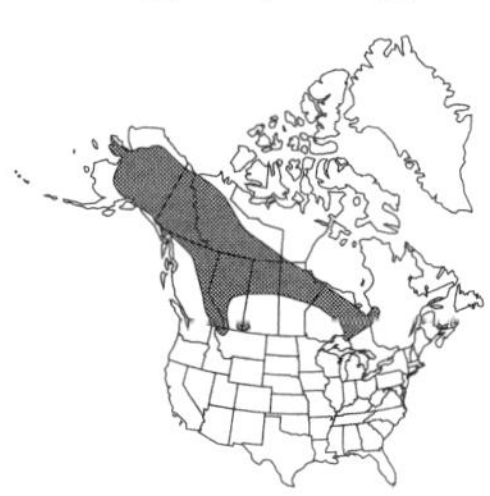

Cypripedium passerinum var. *minganense* Victorin

Plants erect, 12–38[–50] cm. **Leaves** 3–7, along length of stem, alternate, ascending to spreading; blade elliptic to elliptic-lanceolate or ovate-lanceolate, 5–19 × 1.5–6 cm. **Flowers** 1(–2); sepals white or green; dorsal sepal suborbiculate to obovate-orbiculate, rarely elliptic, 11–20 × (7–)9–15 mm; lateral sepals connate to nearly distinct, 6–12 mm, segments 3–6 mm wide; synsepal 6–15 mm wide; petals spreading, white, oblong to linear-elliptic, flat, 12–20 × 3–6 mm; lip white, rarely pinkish, obovoid, 11–20 mm; orifice basal but often proportionally very large, 9–13 mm; staminode cordiform-ovoid to ovoid-oblong. $2n = 20$.

Flowering Jun–Jul. Moist coniferous forests, thickets, openings, tundra, especially in coarse soils of terraces and stream banks; 0–2200 m; Alta., B.C., Man., N.W.T., Nunavut, Ont., Que., Sask., Yukon; Alaska, Mont.

8. Cypripedium californicum A. Gray, Proc. Amer. Acad. Arts 7: 389. 1868 • California lady's-slipper [E]

Plants erect, 25–120 cm. **Leaves** 5–10, along length of stem, alternate, ascending to spreading; blade elliptic-lanceolate to broadly elliptic, 5–16 × 1.5–6.5 cm. **Flowers** 3–18(–22); sepals yellow-green to pale brownish yellow; dorsal sepal elliptic, 14–20 × 7–13 mm; lateral sepals connate almost to apex; synsepal 12–20 × 10–12 mm; petals spreading, same color as sepals or more yellowish, linear-oblong to linear-lanceolate, flat, 14–16 × 3–5 mm; lip white, sometimes pinkish, obovoid, 15–20 mm; orifice basal, 11–14 mm; staminode suborbicular-subauriform.

Flowering May–Jun. Forest openings, especially on steep slopes, in seeps, springy marshes; 0–1600 m; Calif., Oreg.

9. Cypripedium candidum Muhlenberg ex Willdenow, Sp. Pl. 4(1): 142. 1805 • White lady's-slipper [E]

Plants erect, 11–40 cm. **Leaves** 3–4(–5), on proximal or middle portion of stem, alternate, erect-ascending; blade lanceolate or elliptic to oblanceolate, 7–20 × 0.9–5.3 cm. **Flowers** 1(–2); sepals green to pale brownish yellow, usually spotted and striped with reddish brown or madder; dorsal sepal ovate to ovate-lance-acuminate, 15–35 × 7–13 mm; lateral sepals connate; synsepal 13–35 × 7–15 mm; petals spreading to somewhat deflexed, same color as sepals, spirally twisted or spiral-undulate, lanceolate to linear-lanceolate, 23–46 × 3–5 mm; lip white, obovoid or oblance-ovoid to oblance-fusiform, 17–27 mm; orifice basal, 10–15 mm; staminode lanceoloid or oblong-lanceoloid to ellipsoid. $2n = 20$.

Flowering Apr–Jul. Mesic to wet prairies and fen meadows, very rarely open wooded slopes; 70–700 m; Man., Ont., Sask.; Ala., Ill., Ind., Iowa., Ky., Md., Mich., Minn., Mo., Nebr., N.J., N.Y., N.Dak., Ohio, Pa., S.Dak., Wis.

Cypripedium candidum begins blooming while the shoots are still emerging from the soil; at that time they are small, with the leaves clustered near the base. The plants enlarge greatly during and after anthesis, and the leaves may assume a more median position. The orifice of the lip in *C. candidum* is distinctive, the apical margin forming an acute angle. This feature is shared with C.

montanum and differs from the commonly obtuse margin in *C. parviflorum*; hence, it can aid determination of discolored herbarium specimens. Hybrids of *C. candidum* and *C. parviflorum* have been designated *C.* ×*andrewsii* A. M. Fuller. See 11. *C. parviflorum* for a general discussion of hybridization and variation within and between related species.

10. Cypripedium montanum Douglas ex Lindley, Gen. Sp. Orchid. Pl., 528. 1840 • Mountain lady's slipper [E]

Plants erect, 25–71 cm. **Leaves** 4–6, inserted toward base or scattered along stem, alternate, erect, ascending, or spreading; blade suborbiculate or broadly ovate to elliptic-lanceolate, 3.3–17 × 2.5–9.5 cm. **Flowers** 1–3; sepals greenish, suffused, often heavily, with reddish brown or madder, or rarely clear green; dorsal sepal lance-acuminate to elliptic-lance-acuminate, 33–60 × 8–16 mm; lateral sepals connate; synsepal 30–60 × 6–18 mm; petals spreading-deflexed, same color as sepals, spirally twisted, linear to linear-lanceolate, 36–77 × 3–5 mm; lip white, rarely suffused with magenta, obovoid or oblance-ovoid to oblance-fusiform, 19–33 mm; orifice basal, 13–22 mm; staminode lanceoloid to broadly ovoid or ellipsoid-ovoid.

Flowering Feb–Sep. Mesic to dry (rarely wet) coniferous, deciduous, and broadleaf evergreen forests, openings, and thickets, around shrubs on open slopes; 0–2400 m; Alta., B.C.; Alaska, Calif., Idaho, Mont., Oreg., Wash., Wyo.

Plants of *Cypripedium montanum* grown in exposed, relatively sunny situations have the ascending leaves inserted along the basal portion of the stem and the flowers displayed well above the leaves. In shadier, especially sheltered sites, the spreading leaves may be more evenly scattered along the stem. In this species the apical margin of the orifice of the lip is usually acute, in common with *C. candidum*, and in contrast to the usually obtuse margin in *C. parviflorum*; this difference can aid determination of discolored herbarium specimens. Hybrids of *C. montanum* and *C. parviflorum* have been designated *C.* ×*columbianum* Sheviak. See 11. *C. parviflorum* for a general discussion of hybridization and variation within and between related species.

11. Cypripedium parviflorum Salisbury, Trans. Linn. Soc. London 1: 77, plate 2, fig. 2. 1791 [E] [F]

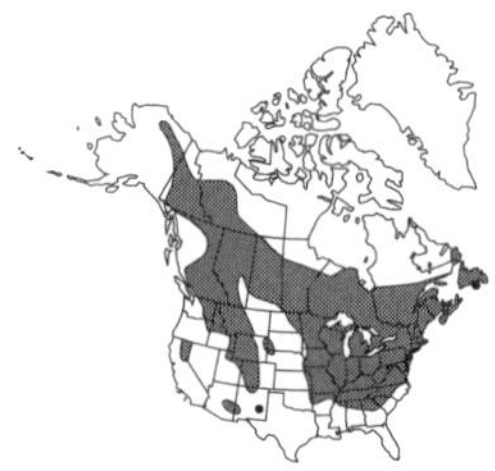

Plants erect, 70–700 cm. **Flowers:** sepals greenish or yellowish (often obscured by darker markings); dorsal sepal suborbiculate or ovate to ovate-lance-acuminate, 19–80 × 7–40 mm; lateral sepals connate; synsepal 11–80 × 5–34 mm; petals horizontal to strongly descending, same color as sepals, commonly spirally twisted or undulate, sometimes flat, linear-lanceolate to lance-ovate or oblong, 24–97 × 3–12 mm; lip rather pale to deep yellow, very rarely white, rarely with reddish spots or suffusion on adaxial external surface, 15–54 mm; orifice basal; staminode cordiform-ovoid, deltoid, lance-ovoid, or ovoid-oblong.

Varieties 3 (3 in the flora): North America.

Cypripedium parviflorum is extremely variable, reflecting individual phenotypic plasticity, infraspecific differentiation, and hybridization with related species. In particular, var. *pubescens* is difficult to delimit. In exposed situations, especially in calcareous soils, plants are low-growing with ascending, often narrow leaves and rather small flowers; in exposed boreal and arctic sites, plants and flowers may be very small with scarcely spiraled to flat petals. Such plants from Newfoundland were originally described as *C. parviflorum* var. *planipetalum*. When transplanted to less severe conditions, those plants often develop into larger plants with larger flowers of more common form; indeed, the holotype sheet of var. *planipetalum* includes a range in habit and floral morphology and includes a plant rather typical of boreal var. *pubescens*. The lips of smaller flowers often are markedly compressed laterally, with parallel sides, but when larger flowers are produced by the same plant, the proportions of the lip often change, the lip then being very broad below and tapering toward the adaxial surface (trapezoidal in cross section) or sometimes dorsiventrally compressed and broader than high.

The southeastern var. *parviflorum* differs from var. *pubescens* primarily in flower size and color, and the two might be merely forms. Most works dealing with *Cypripedium parviflorum* have treated the primarily boreal var. *makasin* as var. *parviflorum*, either including all small-lipped plants within var. *parviflorum*, or in some cases restricting the name to the northern variety and excluding the southeastern plants described by Salisbury as *C. parviflorum*. Fernald's original publication on *C. calceolus* var. *parviflorum* actually treated var. *makasin*, citing a description of that variety and clearly discussing the northern plant. Additionally, although geographically accommodating Salisbury's plant, Fernald excluded most of the range of the southeastern var. *parviflorum*, thereby referring most plants of var. *parviflorum* to var. *pubescens*, and further restricted var. *pubescens* to the east, thereby assigning most plants of that variety to his northern var. *parviflorum*, i.e., var. *makasin*. Consequently, most published illustrations of var. *parviflorum* are in fact var. *makasin*. Variety *parviflorum* has been dealt with primarily in publications on the southeastern flora. In the east, var. *makasin* is quite distinct, but in the west it becomes difficult to separate from very small plants of var. *pubescens* that are common there; in that area, fragrance is often the least equivocal character. In the northwest it seems to merge with *C.* ×*columbianum*, and in fact the northwestern elements of the species are only artificially accommodated within the variety. Very rarely plants apparently referable to var. *makasin* or var. *parviflorum* bear white lips. In some cases that may reflect past gene flow, but in others the plants appear to be color forms.

Hybrids of *Cypripedium parviflorum* with *C. candidum* are *C.* ×*andrewsii* A. M. Fuller, and different varietal parentages are recognized as nothovar. *andrewsii* [*C. candidum* × *C. parviflorum* var. *makasin*], nothovar. *favillianum* (J. T. Curtis) B. Boivin [*C. candidum* × *C. parviflorum* var. *pubescens*], and nothovar. *landonii* (Garay) B. Boivin [*C. parviflorum* var. *parviflorum* × *C.* ×*andrewsii* nothovar. *favillianum*]. Hybrids of *C. parviflorum* with *C. montanum* are *C.* ×*columbianum* Sheviak; the type was evidently derived from a cross with var. *pubescens*. Hybrids of var. *pubescens* commonly exhibit small lips and thus obscure varietal limits: in the northwest, *C.* ×*columbianum* merges with var. *makasin* and var. *pubescens*; in the midwest, *C.* ×*andrewsii* nothovar. *favillianum* may simulate var. *parviflorum*, and the delimitation of those two entities is unclear. Additionally, many plants of putative *C.* ×*andrewsii* nothovar. *favillianum* suggest *C.* ×*columbianum*. In general, those hybrids exhibit vegetative and floral morphology and color intermediate between those of their parents, or combinations of their parental characteristics. In particular, lips are commonly creamy, ivory, or yellow; often lips are yellow when the flower first opens and fade to white over the period of bloom. Consequently, different flowers on the same plant frequently exhibit a range of lip colors. Lip color furthermore sometimes varies from year to year in individual plants. Additionally, *C. montanum* and *C. parviflorum* var. *makasin* commonly contribute the dark coloration of their sepals and petals to hybrids with *C. parviflorum* and *C. candidum*, respectively. The apical margin of the orifice in *C. candidum* and *C. montanum* is typically acute, forming a sharp angle directed toward the apex of the lip. In *C. parviflorum*, this is a variable feature, but typically the margin is obtuse. Variation in this feature in yellow-lipped plants may in some circumstances aid the recognition of hybrids.

SELECTED REFERENCE Wallace, L. E. and M. A. Case. 2000. Contrasting allozyme diversity between northern and southern populations of *Cypripedium parviflorum* (Orchidaceae): Implications for Pleistocene refugia and taxonomic boundaries. Syst. Bot. 25: 281–296.

1. Abaxial surface of distalmost sheathing bract (and often next) sparsely and inconspicuously pubescent to glabrous when young; flowers small; lip 15–29 mm; sepals and petals usually suffused with dark reddish brown or madder, or in the west often spotted and blotched; scent intense, sweet .... 11c. *Cypripedium parviflorum* var. *makasin*
1. Abaxial surface of distalmost sheathing bract (and often next) densely and conspicuously silvery-pubescent when young (later sometimes glabrescent); flowers large to small, lip 20–54 mm; sepals and petals variably spotted, striped, blotched, and reticulately marked with reddish brown or madder (rarely unmarked); scent moderate to faint, rose or musty.
   2. Flowers small, lip 22–34 mm; sepals and petals densely and minutely spotted with dark reddish brown or madder and appearing uniformly dark (rarely coarsely spotted and blotched); s New England to Kansas and southward .... 11a. *Cypripedium parviflorum* var. *parviflorum*
   2. Flowers commonly large, lip to 54 mm, but very small in some boreal and northern cordilleran specimens (as small as 20 mm); sepals unmarked to commonly spotted, striped, and reticulately marked with reddish brown or madder, rarely extensively blotched .... 11b. *Cypripedium parviflorum* var. *pubescens*

11a. Cypripedium parviflorum Salisbury var. **parviflorum** • Small yellow lady's-slipper [E]

Bracts: abaxial surface of distalmost sheathing bract (and often the next) densely and conspicuously silvery-pubescent when young (later sometimes glabrescent). **Leaves** 4–5, rather evenly spaced along stem, alternate, spreading; blade orbiculate to lance-elliptic to ovate or obovate, 9–19 × 2.5–9 cm. **Flowers** 1–2, small, scent moderate to faint, rose or musty; sepals and petals usually minutely but densely spotted with reddish brown or madder and appearing uniformly dark, rarely only coarsely spotted and blotched; lip oblance-ovoid to calceolate (slipper-shaped), 22–34 mm; orifice 12–19 mm.

Flowering Apr–Jun. Mesic to dry deciduous and deciduous-hemlock forests, usually on slopes; mostly 0–1200 m; Ala., Ark., Conn., Del., Ga., Ill., Kans., Md., Mass., Mo., Nebr., N.J., N.Y., N.C., Ohio, Okla., R.I., Tenn., Va.

11b. Cypripedium parviflorum Salisbury var. **pubescens** (Willdenow) O. W. Knight, Rhodora 8: 93. 1906 • Cypripède pubescent [E] [F]

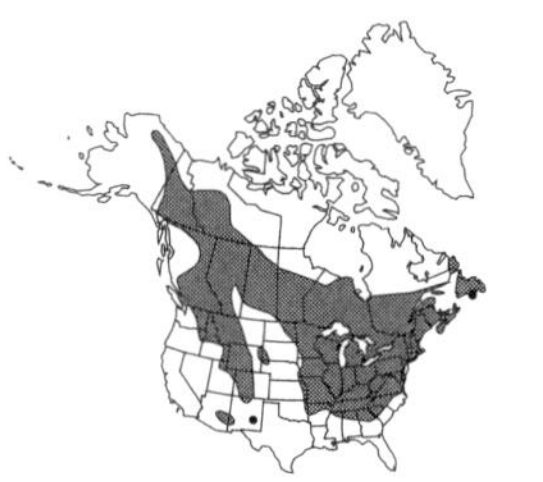

Cypripedium pubescens Willdenow, Hort. Berol. 1: plate 13. 1804; *C. calceolus* Linnaeus var. *planipetalum* (Fernald) Victorin & J. Rousseau; *C. calceolus* var. *pubescens* (Willdenow) Correll; *C. flavescens* de Candolle; *C. parviflorum* var. *planipetalum* Fernald; *C. veganum* Cockerell, P. Barker & M. Barker

Bracts: abaxial surface of distalmost sheathing bract (and often the next) densely and conspicuously silvery-pubescent when young (later sometimes glabrescent). **Leaves** 3–5, on proximal portion of or more evenly spaced along stem, alternate, erect to spreading; blade orbiculate or broadly ovate to elliptic-lanceolate or oblanceolate, 7.9–20.9 × 1.5–12 cm. **Flowers** 1–2, large to rather small (very small in some boreal and northern cordilleran specimens), scent moderate to faint, rose or musty; sepals commonly spotted, striped, and reticulately marked with reddish brown or madder, rarely extensively blotched or wholly unmarked; lip oblance-ovoid to calceolate or subglobose, 20–54 mm; orifice 10–23(–27) mm. 2*n* = 20.

Flowering Apr–Aug. Mesic deciduous and coniferous forest, openings, thickets, prairies, meadows, fens; 0–2900 m; Alta., B.C., Man., N.B., Nfld. and Labr., N.W.T., N.S., Ont., Que., Sask.; Ariz., Colo., Conn., D.C., Ga., Ill., Ind., Iowa, Kans., Ky., Maine, Md., Mass., Mich., Minn., Miss., Mo., Mont., N.H., N.J., N.Mex., N.Y., N.C., Ohio, Oreg., Pa., R.I., Tenn., Utah, Vt., Va., Wash., W.Va., Wis.

11c. Cypripedium parviflorum Salisbury var. **makasin** (Farwell) Sheviak, Amer. Orchid Soc. Bull. 62: 403. 1993 • Cypripède mocassin [E]

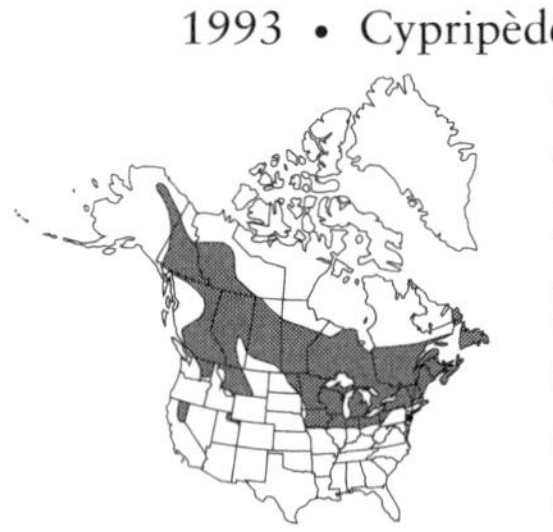

Cypripedium pubescens Willdenow var. *makasin* Farwell, Rep. (Annual) Michigan Acad. Sci. 20: 198. 1918

Bracts: abaxial surface of distalmost sheathing bract (and often the next) sparsely and inconspicuously pubescent or glabrous when young. **Leaves** (2–)3–5, on proximal portion of or more evenly spaced along stem, alternate, erect to spreading; blade orbiculate or broadly ovate to elliptic-lanceolate, 5.2–18.5 × 1.6–14.3 cm. **Flowers** 1–2(–3), small, scent intense, sweet; sepals and petals usually suffused with dark reddish brown or madder, or in west often spotted and blotched; lip oblance-ovoid to calceolate, 15–29 mm; orifice 10–17 mm. 2*n* = 20.

Flowering May–Aug. Mesic to wet fens, prairies, meadows, thickets, open coniferous and mixed forest; mostly 0–1500 m; Alta., B.C., Man., N.B., Nfld. and Labr., N.W.T., Ont., P.E.I., Que., Sask., Yukon; Alaska, Calif., Conn., Ill., Ind., Iowa, Maine, Mass., Mich., Minn., Mont., N.J., N.Y., N.Dak., Ohio, Pa., Utah, Vt., Wash., Wis.

12. Cypripedium kentuckiense C. F. Reed, Phytologia 48: 426, fig. 1981 • Ivory lady's-slipper, purloined slipper [E] [F]

Plants erect, 35–97 cm. **Leaves** 3–6, rather evenly spaced along stem, alternate, spreading; blade broadly ovate to ovate-lanceolate or ovate-elliptic, 13–24 × 4.3–15 cm. **Flowers** 1–2; sepals greenish or yellowish, heavily spotted, striped, reticulately marked with dark reddish brown or madder; dorsal sepal broadly ovate to ovate and elliptic, 61–126 × 24–65 mm; lateral sepals connate; synsepal 55–103 × 12–40 mm; petals spreading-deflexed, same color as sepals, spirally twisted, linear, 84–156 × 7–15 mm; lip ivory to pale yellow, obovoid, (41–)53–65 mm; orifice basal, 27–37 mm; staminode broadly ovoid, ovoid-cordiform, or ovoid-deltoid.

Flowering Apr–Jun. Rich, mesic to dry, deciduous forests on well-drained alluvium and bases of slopes, mucky seeps; mostly 0–400 m; Ala., Ark., Ga., Ky., La., Miss., Okla., Tenn., Tex., Va.

The brief nomenclatural history of *Cypripedium kentuckiense* is remarkably confused. The plant was originally described, without Latin diagnosis, as *C. daultonii* Soukup; this *nomen nudum* came into general usage prior to Reed's publication. Subsequently, an earlier name, *C. furcatum* Rafinesque, was proposed as conspecific, but Rafinesque's description is not adequate for a positive determination.

Cypripedium kentuckiense is very distinctive. In addition to the very large flowers and pale coloring of the lip, the form of the orifice is unique. In related species the orifice is a restricted opening *in* the adaxial surface of the lip; in *C. kentuckiense* the orifice *replaces* the basal portion of the adaxial surface, the sides of the lip terminating abruptly at the orifice without curving toward the horizontal. In herbarium specimens this detail is obscured, but the cavernous nature of the orifice is emphasized by the adaxial surface descending from the apical margin of the orifice toward the apex of the lip; the obovoid lip appears to hang from the margin of the orifice, and the lip is not particularly slipper-shaped. In contrast, in related species, the adaxial surface of the lip surrounds the orifice and extends forward toward the apex, forming a more convincing slipper. These distinctions hold virtually throughout the known populations of *C. kentuckiense*; only in two Arkansas populations is the lip form suggestive of related species. The Arkansas populations may reflect very limited introgression from *C. parviflorum* var. *pubescens*.

SELECTED REFERENCE Weldy, T. W. et al. 1996. The current status of *Cypripedium kentuckiense* (Orchidaceae) including a morphological analysis of a newly discovered population in eastern Virginia. Sida 17: 423–435.

230b. ORCHIDACEAE Jussieu subfam. VANILLOIDEAE (Lindley) Szlachetko, Fragm. Florist. Geobot., suppl. 3: 48. 1995

230b.1. ORCHIDACEAE Jussieu (subfam. VANILLOIDEAE) tribe VANILLEAE Blume, Rumphia 1: 196. 1837

230b.1a. ORCHIDACEAE Jussieu (tribe VANILLEAE) subtribe VANILLINAE Lindley, Gen. Sp. Orchid. Pl., 429. 1840

2. VANILLA Miller, Gard. Dict. Abr. ed. 4, vol. 3. 1754 • [Spanish *vainilla*, little pod or capsule, referring to long, podlike fruits]

James D. Ackerman

Vines, terrestrial or hemiepiphytic, monopodial. **Roots** at each stem node, gray-green, slender, and glabrous when free, thick and villous on contact with substrate. **Stems** scandent, branching, naked, terete, thick, succulent, glabrous. **Leaves** persistent when large or deciduous when scalelike, distichous, articulate, sheathless, fleshy or leathery. **Inflorescences** on short lateral branches or peduncles, racemes, bracteate, densely flowered. **Flowers** ephemeral, resupinate, large, showy, opening sequentially; sepals spreading, distinct and free; petals distinct and/or free, keeled; lip adnate to base of column, simple or lobed, margins basally involute; disc variously ornamented; column elongate, semiterete, footless, often pubescent proximally; anther terminal, versatile, incumbent; pollinia 4, soft, mealy, composed of monads, without accessory structures, appearing triangular when removed as a unit; ovaries articulate proximally and distally; stigma lobes confluent; rostellum undeveloped. **Fruits** berries, elongate, leathery, indehiscent. **Seeds** small; seed coat hard.

Species 100 (5 in the flora): tropical and subtropical regions, North America, Mexico, West Indies, Central America, South America, Africa, Southeast Asia, w Pacific Islands.

Vanilla pompona Schiede has escaped cultivation in Miami-Dade County, Florida (P. M. Brown 2002). It may be distinguished from the other *Vanilla* species by the following combination of characteristics: persistent leaves that are much longer than the internodes; flat or straight sepal and petal margins; a simple, yellow-green to yellow-orange lip with a tuft of retrorse hairs; and floral bracts that are at least 12 mm. In addition to the characteristics above, *V. pompona* generally has thicker leaves and stems than the two species for which it may be confused, *V. planifolia* and *V. phaeantha*.

SELECTED REFERENCE Nielsen, L. R. and H. R. Siegismund. 1999. Interspecific differentiation and hybridization in *Vanilla* species (Orchidaceae). Heredity 83: 560–567.

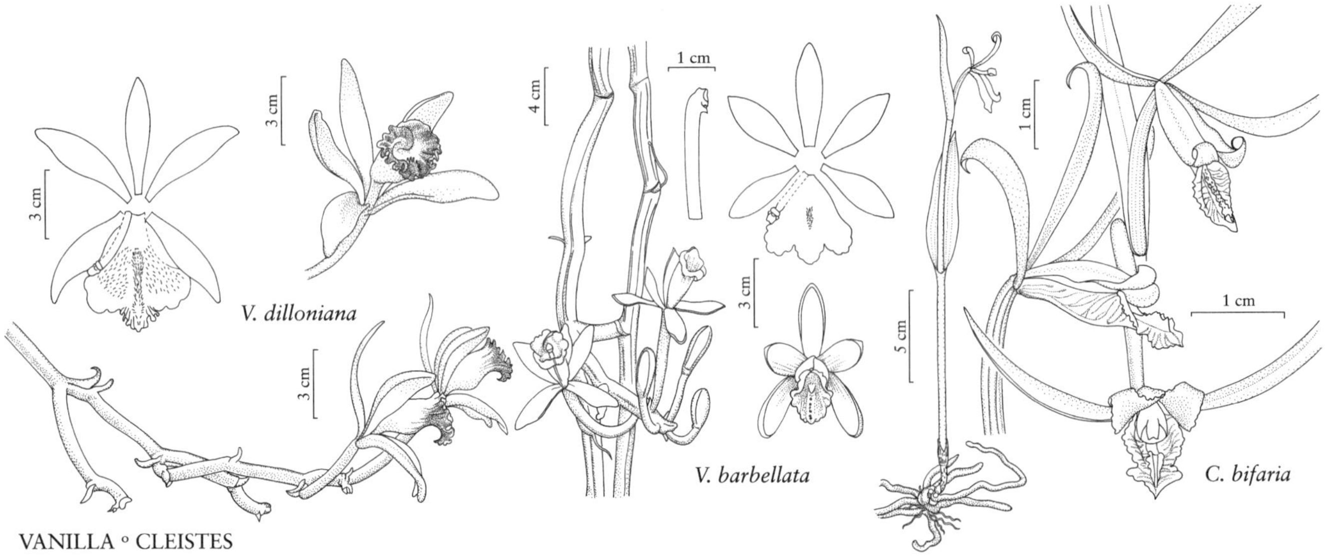

VANILLA ◦ CLEISTES

1. Leaves much shorter than internodes, early deciduous or persistent.
   2. Lip largely reddish purple, no yellow present (in flora); most leaves early deciduous, a few persistent, margins revolute, apex hooked . . . . 1. *Vanilla dilloniana*
   2. Lip deep red adaxially, shading to white margins, with broad yellow midrib; all leaves early deciduous, margins not revolute, apex not hooked . . . . 2. *Vanilla barbellata*
1. Leaves nearly as long as or longer than internodes, persistent.
   3. Sepal and petal margins undulate, apex reflexed; lip disc 3-keeled toward apex, tuft of hairs absent; stems less than 5 mm diam . . . . 3. *Vanilla mexicana*
   3. Sepal and petal margins straight, apex spreading; lip disc without keels, tuft of scalelike hairs present; stems more than 6 mm diam.
      4. Sepals 7–9 cm; fruits less than 11 cm. . . . . 4. *Vanilla phaeantha*
      4. Sepals 3.5–5.5 cm; fruits 15–25 cm . . . . 5. *Vanilla planifolia*

1. Vanilla dilloniana Correll, Amer. Orchid Soc. Bull. 15: 331, figs. 1–7. 1946 • Leafless vanilla, Mrs. Lott's vanilla [F]

Roots usually 1 per node, gray, 1–1.5 mm diam., villous and thicker when in contact with substrate, glabrous when aerial. **Stems** 4–8 mm diam., thick, smooth, internodes to 20 cm. **Leaves** with at least some persistent, fleshy; blade to 8 cm, margins somewhat involute, apex hooked. **Inflorescences** axillary, 15-flowered racemes, short-pedunculate, to 9 cm excluding peduncle; floral bracts broadly ovate, 2–13 mm, fleshy. **Flowers:** sepals and petals distinct and free, light green; dorsal sepal narrowly oblong-elliptic, slightly concave, 4.8–5.5 × 0.8–1.2 cm, apex obtuse to acute; lateral sepals elliptic-oblanceolate, slightly falcate, 4.5–5 × 0.9–1.3 cm; petals narrowly oblong-oblanceolate, abaxially keeled, 4.1–5 × 1.1–1.4 cm, apex acute; lip arching over column, reddish purple, becoming very pale at margin edge and within throat, broadly obovate, obscurely 3-lobed, 4–5 × 4–5 cm when spread, margins basally involute; claw and basal margins adnate to column for more than 1/2 its length, free portion crenulate; disc with broad thickened midrib extending length of lip, distal 1.5 cm of midrib with erect, scattered, thick, sometimes forked hairs, 1–2.5 mm, with hinged, stiff beard ± 1 cm at middle of disc midrib, midrib glabrous proximal to beard; disc flanking midrib glabrous but with numerous short fleshy hairs extending to lateral margins of throat, within 5–10 mm of apical margin; column arched, slender, 3.2 cm; pollinia yellow; pedicellate ovary 4.7–5 cm. **Berries** pendent, green, darkening at maturity, fusiform-clavate, 16 × 1.2 cm.

Flowering May–Jun. Hammocks; 0–20 m; Fla.; West Indies.

Vanilla dilloniana is known in the flora from Paradise Key, Everglades National Park, Miami-Dade County, Florida (P. M. Brown 2002).

2. Vanilla barbellata Reichenbach f., Flora 48: 274. 1865 • Worm vine F

Vanilla articulata Northrop

Roots usually 1–2 per node, gray, 1–3 mm diam. when aerial, thicker and villous when in contact with substrate, glabrous. **Stems** occasionally branched, 3–9 mm diam., smooth. **Leaves** early deciduous; blade broad basally, otherwise narrowly lanceolate, relatively thin, to 4 × 0.8 cm. **Inflorescences** axillary on short lateral branches, several- to many-flowered racemes, 1.5–6 cm excluding peduncle; floral bracts broadly ovate, 4–12 mm, fleshy. **Flowers:** sepals and petals green, somewhat spreading, distinct and free; sepals oblong-oblanceolate, 3–4 × 0.9–1.2 cm; petals oblong-oblanceolate, slightly falcate, dorsally keeled, 3–4 × 1–1.3 cm, apex acute to obtuse; lip greenish abaxially, deep red adaxially, shading to white margin, with broad, yellow midrib, overall triangular-obovate, medially thickened, apex 3-lobed, lateral lobes arching over column, orbiculate, margins involute, sinuses 4–5 mm deep, middle lobe reflexed, fleshy; disc with tuft of rigid, retrorse bristles; claw and basal margins adnate to proximal 1/2 of column; column straight, semiterete, 2.3–3.3 cm; pedicellate ovary 3–4.5 cm. **Berries** pendent, fusiform-cylindric, slightly curved, 7–9 cm × 9–13 mm.

Flowering Apr–Jun. Mangroves, coastal hammocks, bay tree islands, rocky pinelands; 0–20 m; Fla.; West Indies.

Vanilla barbellata is known in the flora area from several keys in Everglades National Park, Miami, Miami-Dade and Monroe counties, Florida (P. M. Brown 2002).

3. Vanilla mexicana Miller, Gard. Dict. ed. 8, Vanilla no. 1. 1768

Epidendrum vanilla Linnaeus; *Vanilla anaromatica* Grisebach; *V. aromatica* Swartz; *V. inodora* Schiede; *V. vanilla* (Linnaeus) Britton

Roots usually 1 per node, aerial portions 1–2 mm diam. **Stems** occasionally branched, leafy, very slender, 2–3 mm diam., smooth. **Leaves** persistent; blade flat, elliptic to broadly ovate, longer than internodes, to 25 × 12 cm, leathery, apex acuminate; distal leaves smaller, proportionately narrower. **Inflorescences** axillary, several-flowered racemes, short-pedunculate, to 17 cm excluding peduncle; floral bracts foliaceous, ovate-elliptic, to 2.5 × 1.5 cm. **Flowers:** sepals and petals spreading, green, thick, rigid; sepals oblong-lanceolate to lanceolate-elliptic, to 6.5 × 2 cm, margins undulate, apex reflexed, acute; petals similar, abaxially keeled, slightly smaller; lip basally adnate to column for 5 mm, white with yellow crest, cuneate, to 5 × 3 cm when spread, distally 3-lobed, lateral lobes arching over column, rounded, margins involute, middle lobe broadly 3-angular, margins undulate, apex obtuse to rounded; disc with thick longitudinal crest, becoming 3-keeled toward apex, 5 mm wide, glabrous; column white, slender, 20–25 mm; pollinia yellow; pedicellate ovary to 5 cm. **Berries** cylindric, slender, to 27 cm.

Flowering May–Jun. Cypress swamps, hammocks; 0–20 m; Fla.; Mexico; West Indies; Central America (Guatemala, Nicaragua); South America.

Considerable confusion has swirled around the identity of *Vanilla mexicana*, including the notion that the epithet refers to the commercial vanilla, *V. planifolia*. Lectotypification of the Linnaean basionym, *Epidendrum vanilla*, has settled the matter (S. Cafferty and C. E. Jarvis 1999).

4. Vanilla phaeantha Reichenbach f., Flora 48: 274. 1865 • Leafy vanilla, oblong-leaved vanilla

Roots usually 1 per node, aerial portions 1–2 mm diam. **Stems** occasionally branched, leafy, thick, 8–12 mm diam., smooth. **Leaves** persistent; blade oblong-elliptic, flat, equal to or shorter than internodes, to 15 × 2.5–5 cm, fleshy-leathery, tapering to acute apex. **Inflorescences** axillary, 12-flowered racemes, short-pedunculate, ± to 8 cm excluding peduncle; floral bracts foliaceous, ovate, cucullate, to 2.5 × 1.5 cm, fleshy. **Flowers:** sepals and petals spreading, green, fleshy, rigid; sepals narrowly oblanceolate, 7–9 × 1.3–2 cm, margins straight, apex obtuse to acute, spreading; petals obliquely narrowly oblanceolate, abaxially keeled, thinner than sepals, 6.5–8 ×1–1.5 cm, apex acute to subacuminate; lip adnate to column for 3 cm, white becoming yellow-cream toward apex, lamina gulletlike, cuneate, obovate, clawed, obscurely 3-lobed distally, to 7.5–9 × 3.5–4 cm when spread, lateral lobes arching over column, margins involute, apex rounded, middle lobe broadly subquadrangular, margins undulate, apex retuse; disc with multiple longitudinal ribs, central tuft of retrorse scales, thickened broad rib extending from tuft to apex; column curved distal to middle, white, slender, 6.5 cm, adaxially bearded; pollinia yellow; pedicellate ovary 4–8 cm. **Berries** cylindric, 8–10 × 1 cm.

Flowering Apr–Jul. Cypress swamps, hammocks; 0–20 m; Fla.; West Indies (Trinidad).

Vanilla phaeantha is known in the flora area only from Collier County, Florida (P. M. Brown 2002).

5. **Vanilla planifolia** Jackson, Bot. Repos. 8: plate 538. 1808 • Vanilla, commercial vanilla, vanilla vine

Myrobroma fragrans Salisbury; *Vanilla fragrans* (Salisbury) Ames

Roots usually 1 per node, aerial portions 2–3 mm diam. **Stems** occasionally branched, leafy, thick, 5–10 mm diam., smooth. **Leaves** persistent; blade flat, oblong-elliptic to ovate, longer than internodes, 15–25 × 5–8 cm, fleshy-leathery, apex acute to acuminate. **Inflorescences** axillary, 15-flowered racemes, short-pedunculate, to 5 cm excluding peduncle; floral bracts broadly triangular-ovate, 7–10 × 7–10 cm, leathery. **Flowers:** sepals and petals erect-spreading, yellow-green, fleshy, rigid; sepals oblanceolate, 3.5–5.5 × 1.1–1.3 cm, margins straight, apex acute to obtuse; petals elliptic-oblanceolate, abaxially keeled, thinner than sepals, 3.5–5.5 × 1.1–1.3 cm, apex acute to obtuse; lip adnate to column for 1.5–2 cm, yellow-green, becoming dark yellow toward apex, lamina gulletlike, cuneate, rhomboid, 4–5 × ± 3 cm, with apical retuse lobule; disc with central tuft of retrorse scales, several lines of short, fleshy hairs extending to apex; column white, slender, 3–3.5 cm, margins slightly sinuate, adaxially bearded; pollinia yellow; pedicellate ovary 3–5 cm. **Berries** cylindric, 15–25 × 0.8–1 cm.

Flowering Apr. Cypress swamps, hammocks; 0–20 m; Fla.; Mexico; West Indies (Trinidad); Central America; n South America.

The long, slender black fruits of *Vanilla planifolia* are the vanilla "beans" of commerce.

The natural distribution of *Vanilla planifolia* is most likely tropical evergreen forests of eastern Mexico and the Caribbean watersheds of Guatemala, Belize, and Honduras (M. Soto Arenas, pers. comm.). It has been cultivated and escaped or persisted in many areas of the tropics, including south Florida. It is known in the flora area from Long Pine Key, Everglades National Park, Miami, Miami-Dade County, Florida (P. M. Brown 2002).

Pollinators are euglossine bees (J. D. Ackerman 1983), which do not occur in Florida. Natural pollination has been recorded in Florida, although very rarely (C. A. Luer 1972).

230b.1b. ORCHIDACEAE Jussieu (tribe VANILLEAE) subtribe POGONIINAE Pfitzer, Entwurf Anord. Orch., 97. 1887

3. CLEISTES Richard ex Lindley, Gen. Sp. Orchid. Pl., 409. 1840 • Spreading pogonia [Greek *kleistos*, closed, alluding to lip and petals that diverge only near apex, forming tube for most of their length, the flower thus appearing closed]

Katharine B. Gregg

Paul M. Catling

Herbs, terrestrial. **Roots** 1–2 mm wide, sometimes with wider nodular tuberoids, frequently rather brittle. **Leaves** 1–3 on vegetative plants, 1(–2) centrally on flowering stems, sessile; blade oblong-lanceolate or elliptic-oblong. **Inflorescences** terminal,1–3-flowered; floral bracts leaflike. **Flowers** resupinate, erect, sessile; sepals olive-green, brown, or maroon, linear-lanceolate, apex acuminate; petals rose-pink to white, spatulate or oblanceolate, usually forming tube with lip; lip rose-pink to white, with rose or purple veins, oblanceolate, 3-lobed, middle lobe medially winged or tuberculate, central keel grooved from base to 3/4 its length; column free, white, apex lacerate, truncate; anther terminal, hinged; pollinia 2, without viscidium, soft, mealy; pollen grains in tetrads. **Fruits** capsules, erect, cylindric.

Species ca. 25 (2 in the flora): tropical and subtropical regions, mostly South America, 2 in Central America.

Authors of some neotropical floras have questioned the separation of *Cleistes* from *Pogonia*; most recent North American authors have followed O. Ames (1905–1922) in recognizing *Cleistes*, and we prefer to do the same until a detailed study suggests otherwise. The two North American species can reproduce asexually by root shoots.

SELECTED REFERENCES Ames, O. 1905–1922. Orchidaceae: Illustrations and Studies of the Family Orchidaceae Issuing from the Ames Botanical Laboratory.... 7 vols. Boston and New York. Vol. 7. Catling, P. M. and K. B. Gregg. 1992. Systematics of the genus *Cleistes* in North America. Lindleyana 7: 57–73. Fernald, M. L. 1946b. Some orchids of the Manual range. Rhodora 48: 161–162, 184–197, plates 1045–1048.

1. Column 13–19 mm; lip 21–33(–38) mm . . . . 1. *Cleistes bifaria*
1. Column 21–25(–29) mm; lip (26–)34–55 mm . . . . 2. *Cleistes divaricata*

1. Cleistes bifaria (Fernald) Catling & Gregg, Lindleyana 7: 65. 1992 • Smaller spreading pogonia [E] [F]

Cleistes divaricata (Linnaeus) Ames var. *bifaria* Fernald, Rhodora 48: 186, plate 1048. 1946

Plants (12–)15–46(–64) cm (in flower). **Leaves:** blades on flowering stems, (2.5–)45–145 × (4–)6–25 mm. **Flowers:** sepals 24–55 × (2–)3–5 mm; petals 21–36 × 6–10(–12) mm; lip 21–33(–38) × 13–16 mm, central keel relatively thick, 2.4–2.6 mm wide, very slightly grooved, groove 0.3 mm deep and verrucose or braided with 5–7 discontinuous ridges; column 13–19 mm. **Capsules** (9–)27–41(–81) × (3.2–)4–8(–8.5) mm. $2n = 18$.

Flowering Apr–May (coastal plain) –Jul (mountains). Savannas, meadows, openings in oak or pine woodlands, mountain habitat often xeric, in acidic soil; 0–1000 m; Ala., Fla., Ga., Ky., La., Miss., N.C., S.C., Tenn., Va., W.Va.

A station in Barbour County, West Virginia, is 240 km north of other stations and may be a result of introduction (K. B. Gregg 1989).

Cleistes bifaria has been treated as a variety of the following species; it is readily separated by its relatively short column, which is correlated with the relatively small size of other floral parts, and differences in lip keel. It is, to a large extent, isolated from the following taxon by geographic distribution and flowering time where their distributions overlap (P. M. Catling and K. B. Gregg 1992).

SELECTED REFERENCE Gregg, K. B. 1989. Reproductive biology of the orchid *Cleistes divaricata* (L.) Ames var. *bifaria* Fernald growing in a West Virginia meadow. Castanea 54: 57–78.

2. Cleistes divaricata (Linnaeus) Ames, Orchidaceae 7: 21, plate 108. 1922 [E]

Arethusa divaricata Linnaeus, Sp. Pl. 2: 951. 1753; *Pogonia divaricata* (Linnaeus) R. Brown

Plants (15–)31–76 cm (in flower). **Leaves:** blades on flowering stems, 33–150 × 5–23 mm. **Flowers:** sepals 31–65 × 3–5.5 mm; petals 27–53 × 5–19 mm; lip (26–)34–55 × 18–24 mm, central keel relatively thin or evidently thin-winged, 3–4 mm wide, deeply and broadly grooved, groove 1.5 mm deep, smooth with 1–3 low, continuous ridges; column 21–25(–29) mm. **Capsules** (35–)40–65(–80) × 5–10 mm. $2n = 18$.

Flowering May (Fla.)–Jul (Del.). Coastal Plain savannas, in acidic soil; 0–200 m; Del., Fla., Ga., Md., N.J., N.C., S.C., Va.

4. ISOTRIA Rafinesque, Med. Repos., hexade 2, 5: 357. 1808 • Whorled pogonia orchid [Greek *iso*, equal, and *tri*, 3; probably referring to 3 sepals of equal size and shape] [E]

Loyal A. Mehrhoff

Michael A. Homoya

Herbs, perennial, terrestrial, winter dormant. **Roots** long, slender, hairy with mycorrhizae. **Stems** simple, hollow, fleshy, smooth, glaucous. **Leaves** (2–)5(–6), appearing whorled. **Inflorescences** terminal; bracts absent. **Flowers** 1–2, resupinate, erect, yellowish green to white, yellow, and purple, subsessile to pedicellate; sepals of same color, linear-oblanceolate to lanceolate, of equal size; petals oblanceolate, elliptic-obovate, or elliptic-lanceolate, enclosing column; lip apically 3-lobed, lateral lobes triangular, margins involute; callus forming fleshy medial crest; column white, apex denticulate; anther operculate; pollinia 2, hinged, soft, mealy; rostellum obscure; pollen in tetrads, granular; ovary green, slender. **Fruits** capsules, commonly persistent through following growing season, erect, ellipsoid-cylindric, dehiscent in fall. **Seeds** 1.2 × 0.2 mm. $x = 9$.

Species 2 (2 in the flora): e North America.

SELECTED REFERENCES Ames, O. 1905–1922. Orchidaceae: Illustrations and Studies of the Family Orchidaceae Issuing from the Ames Botanical Laboratory.... 7 vols. Boston and New York. Vol. 7. Homoya, M. A. 1977. The Distribution and Ecology of the Genus *Isotria* in Illinois. Master's thesis. Southern Illinois University. Mehrhoff, L. A. 1983. Pollination in the genus *Isotria* (Orchidaceae). Amer. J. Bot. 70: 1444–1453.

1. Sepals green to light green, less than 30 mm; lip light yellowish green to pale greenish white, streaked with green; pedicel of mature capsule 5–17(–20) mm; stem and leaves pale grayish green . . . . 1. *Isotria medeoloides*
1. Sepals purplish brown, greater than 30 mm; lip yellowish green to white, streaked with purple; pedicel of mature capsule 20–55 mm; stem purplish to brownish green; leaves green to dark green . . . . 2. *Isotria verticillata*

1. Isotria medeoloides (Pursh) Rafinesque, Fl. Tellur. 4: 47. 1838 • Small whorled pogonia orchid [C] [E]

Arethusa medeoloides Pursh, Fl. Amer. Sept. 2: 591. 1814; *Pogonia affinis* Austin ex A. Gray

Herbs, singly or rarely in small clumps, short-rhizomatous, 4–25 cm. **Stems** pale grayish green, glaucous; scales 2–4, near soil line, white to light green. **Leaves** pale grayish green; blade elliptic-obovate, 17–85 × 8–40 mm, apex acuminate, glaucous when young. **Flowers** subsessile, without nectar and fragrance; sepals arching, green to light green, linear-oblanceolate, 12–25 × 2–3 mm; petals light green, oblanceolate, 12–18 × 2–4 mm; lip light yellowish green to pale greenish white, streaked with green, obovate, 11–16 × 4–5 mm, lateral lobes narrowly triangular, margins involute, middle lobe rounded, margins slightly revolute, undulate; callus yellowish green, longitudinal, becoming elongate, with fleshy, wartlike papillae toward apex; column 8–10 mm; ovary 10–15 mm; rostellar flap reduced. **Capsules** 12–28 × 2–10 mm; pedicel of mature capsule elongating to 5–17(–20, rarely) mm. $2n = 18$.

Flowering May–Jun; capsules mature fall. Acidic soils, in dry to mesic 2d-growth, deciduous or deciduous-coniferous forests; typically light to moderate leaf litter, open herb layer (occasionally in dense ferns), moderate to light shrub layer, relatively open canopy; frequently on flats or slope bases near canopy breaks; of conservation concern; 30–1000 m; Ont.; Conn., Del., Ga., Ill., Maine, Md., Mass., Mich., Mo., N.H., N.J., N.Y., N.C., Ohio, Pa., R.I., S.C., Tenn., Vt., Va.

Isotria medeoloides should be sought in Alabama, Arkansas, Indiana, Kentucky, and West Virginia.

Isotria medeoloides is self-pollinating (L. A. Mehrhoff 1983). Nonflowering plants commonly have a white, arrested floral bud (1–2 mm).

SELECTED REFERENCES Brumback, W. E. and C. W. Fyler. 1988. Monitoring of *Isotria medeoloides* in New Hampshire. Wild Fl. Notes 3: 32–40. Dyer, R. W. 1986. Small whorled pogonia. Audubon Wildlife Rep. 1986: 780–789. Mehrhoff, L. A. 1989. Reproductive vigor and environmental factors in populations of an endangered North American orchid, *Isotria medeoloides* (Pursh) Rafinesque. Biol. Conservation 47: 281–296. Mehrhoff, L. A. 1989b. The dynamics of declining populations of an endangered orchid, *Isotria medeoloides*. Ecology 70: 783–786. Oettingen, S. L. 1992. Small Whorled Pogonia (*Isotria medeoloides*) Recovery Plan: First Revision. Washington. Paulos, P. G. 1985. Small Whorled Pogonia Recovery Plan. [Newton Corner, Mass.]

2. Isotria verticillata (Muhlenberg ex Willdenow) Rafinesque, Med. Repos., hexade 2, 5: 357. 1808 • Large whorled pogonia orchid [E] [F]

Arethusa verticillata Muhlenberg ex Willdenow, Sp. Pl. 4(1): 81. 1805; *Pogonia verticillata* (Muhlenberg ex Willdenow) Nuttall

Herbs, clonal, rhizomatous, 4–40 cm. **Stems** purplish to brownish green, glaucous; scales 2–4, typically below soil line, purplish. **Leaves** green adaxially; blade oblong-lanceolate, obovate, or elliptic, 25–100 × 10–53 mm, apex obtuse to acuminate, sometimes glaucous abaxially. **Flowers** pedunculate, showy, sweetly fragrant, without nectar; sepals spreading, purplish brown, narrowly lanceolate, 34–67 × 2–4 mm; petals yellowish green, elliptic-obovate to elliptic-lanceolate, 15–25 × 3–7 mm; lip yellowish green to white, streaked with purple, oblong, 15–25 × 8–9 mm, lateral lobes streaked with purple,

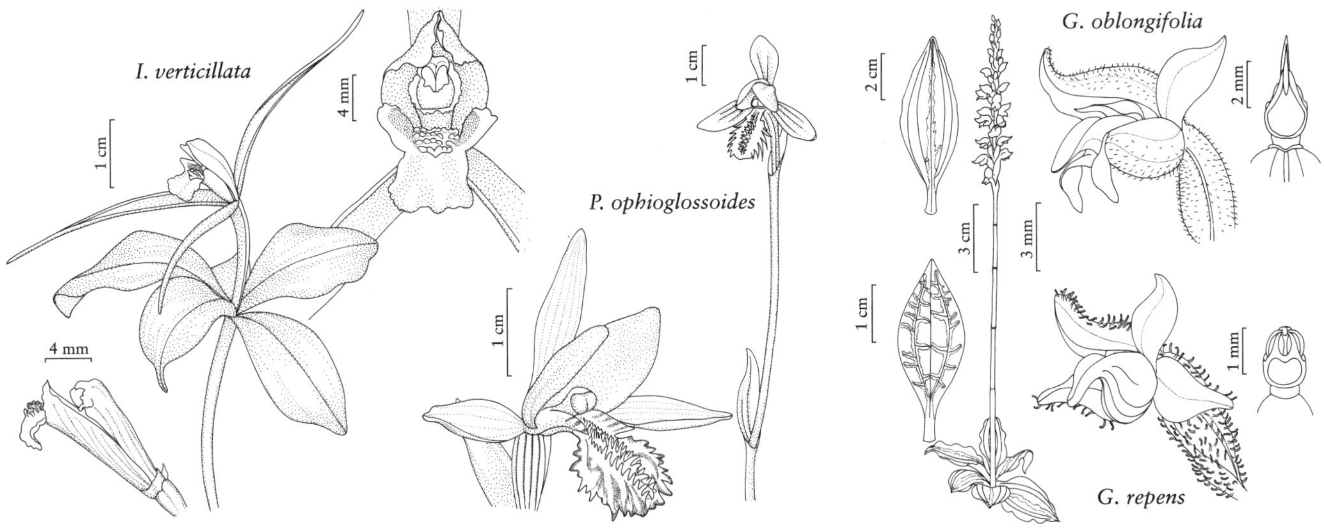

ISOTRIA ◦ POGONIA ◦ GOODYERA

margins involute, middle lobe rounded, margins revolute, undulate; callus green, longitudinal, fleshy; column 8–12 mm; ovary 20–30 mm; rostellar flap prominent. **Capsules** 20–42 × 5–10 mm; pedicel of mature capsule elongating to 20–55 mm. $2n = 18$.

Flowering Apr–Jun. Acidic soils, in dry to mesic forests, seeps, sphagnum bogs; 10–2000 m; Ont.; Ala., Ark., Conn., Del., Fla., Ga., Ill., Ind., Ky., La., Maine, Md., Mass., Mich., Miss., Mo., N.H., N.J., N.Y., N.C., Ohio, Okla., Pa., R.I., S.C., Tenn., Tex., Vt., Va., W.Va.

Isotria verticillata frequently forms extensive clones with hundreds of stems. It is pollinated by solitary bees of the Andrenidae, Anthophoridae, and Halictidae; plants are apparently self-compatible (L. A. Mehrhoff 1983). Nonflowering plants only rarely have a white, arrested floral bud (1–2 mm).

5. POGONIA Jussieu, Gen. Pl., 65. 1789 • Beard-flower [Greek *pogonias*, bearded, alluding to bearded lip]

Charles J. Sheviak

Paul M. Catling

Herbs, perennial, terrestrial, rather succulent. **Roots** few, scattered along slender rootstock producing shoots at wide intervals, slender, fibrous. **Stems** slender, solitary leaves inserted toward middle. **Leaves** ascending; blade fleshy-leathery. **Inflorescences** terminal, solitary flower or 2-flowered racemes; floral bracts foliaceous, prominent. **Flowers** resupinate, showy; lip bearing prominent adaxial fleshy crests divided into distinct cylindric processes, margins laciniate; pollinia 2, viscidia absent; stigma flat. **Fruits** capsules.

Species 3 (1 in the flora): temperate regions, North America, e Asia.

SELECTED REFERENCE Ames, O. 1905–1922. Orchidaceae: Illustrations and Studies of the Family Orchidaceae Issuing from the Ames Botanical Laboratory.... 7 vols. Boston and New York. Vol. 7.

1. **Pogonia ophioglossoides** (Linnaeus) Ker Gawler, Bot. Reg. 2: plate 148. 1816 • Rose pogonia, adder's-mouth, pogonie langue-de-serpent [E] [F]

Arethusa ophioglossoides Linnaeus, Sp. Pl. 2: 951. 1753; *Pogonia ophioglossoides* var. *brachypogon* Fernald

Plants 4–70 cm. **Leaves:** blade elliptic, narrowly lanceolate, lance-ovate, or oblong, 14–120 × 4–32 mm. **Inflorescences:** floral bracts conspicuous, foliaceous, 7–37 × 2–12 mm. **Flowers** solitary or 2-flowered raceme, pink or rarely white or bluish, gaping, showy; sepals oblong, oblong-elliptic, or narrowly lanceolate, 14–23 × 3–7 mm; petals elliptic to lance-obovate, 13–25 × 3–11 mm; lip spatulate, 12–25 × 4–10 mm, margins involute, lacerate; crest often white, yellow, or green toward base. **Capsules** erect, ellipsoid, 14–30 × 4–8 mm. 2*n* = 18.

Flowering spring–summer. Sphagnum bogs, poor fens, moist acidic sandy meadows and prairies, open wet woods, wet pine flatwoods, pine savannas, cypress swamps, sandy-peaty stream banks, seepage slopes, ditches, roadcuts, rarely calcareous fens; 0–1100 m; St. Pierre and Miquelon; Man., N.B., Nfld. and Labr. (Nfld.), N.S., Ont., P.E.I., Que.; Ala., Ark., Conn., Del., D.C., Fla., Ga., Ill., Ind., La., Maine, Md., Mass., Mich., Minn., Miss., Mo., N.H., N.J., N.Y., N.C., N. Dak., Ohio, Okla., Pa., R.I., S.C., Tenn, Tex., Vt., Va., W.Va., Wis.

230c. ORCHIDACEAE Jussieu subfam. ORCHIDOIDEAE

230c.1. ORCHIDACEAE Jussieu (subfam. ORCHIDOIDEAE) tribe CRANICHIDEAE Endlicher, Gen. Pl., suppl. 2: 18. 1842

230c.1a. ORCHIDACEAE Jussieu (tribe CRANICHIDEAE) subtribe GOODYERINAE Klotzsch in F. G. Hayne et al., Getreue Darstell. Gew. 14: plate 24. 1846

6. GOODYERA R. Brown in W. Aiton and W. T. Aiton, Hortus Kew. 5: 197. 1813 • Rattlesnake-plantain, lattice-leaf, goodyérie [for John Goodyer, 1592–1664, British botanist]

Jacquelyn A. Kallunki

Herbs, terrestrial, rhizomatous, scapose, glabrous except for rather sticky, multicellular hairs on peduncles, bracts, sepals, and ovaries. **Roots** arising from nodes of rhizome, fibrous. **Stems** erect, with rosette of leaves, not succulent. **Leaves** evergreen, more than 1, in basal rosette, petiolate; blade commonly marked with white to pale green. **Inflorescences** terminal, 5–72-flowered spikes, erect; peduncles with sheathing bracts. **Flowers** resupinate, white, sometimes tinged green, ivory, or brown, sessile; sepals distinct, nearly equal; dorsal sepal and petals forming hood; lip free from column, fleshy, base concave to saccate, apex ligulate or pointed; anther 1, erect or inflexed; pollinia 2, sectile; rostellum notched or 2-pronged. **Fruits** capsules, erect, dehiscing along 3 ribs.

Species 40–100 (4 in the flora): nearly worldwide, primarily Southeast Asia; ca. 16 species in Western Hemisphere.

The four species of *Goodyera* in the flora are sometimes difficult to distinguish, especially without flowers. This difficulty is compounded, even with flowers, by the intermediate nature of *Goodyera tesselata*, which is likely an allotetraploid derived from *G. repens* with white-reticulate leaves and *G. oblongifolia*, and by the presence of triploid hybrids in some mixed populations of the three species.

SELECTED REFERENCES Kallunki, J. A. 1976. Population studies in *Goodyera* (Orchidaceae) with emphasis on the hybrid origin of *G. tesselata*. Brittonia 28: 53–75. Kallunki, J. A. 1981. Reproductive biology of mixed-species populations of *Goodyera* (Orchidaceae) in northern Michigan. Brittonia 33: 137–155.

1. Lip scrotiform, apex reflexed, no fleshy callosities on inner surface; rostellum notched; inflorescences cylindric (equally dense on all sides); leaves with midrib bordered by broad white bands and lateral veins traced with narrower white bands, both sharply demarcated from adjacent green tissue . . . . 1. *Goodyera pubescens*
1. Lip deeply concave or saccate, apex spreading or recurved, fleshy callosities on inner surface; rostellum with 2-pronged beak; inflorescences loosely spiraled or secund (infrequently cylindric); leaves uniformly green or reticulate with white or pale green on midrib and/or lateral veins.
    2. Leaf blades usually with only midrib whitened (infrequently with lateral veins, especially those near midrib, lightly penciled in white); sepals 5.7–7.8 mm; lip 4.9–7.9 mm; rostellar beak 2.3–3.6 mm; lip apex short, spreading or slightly arching with upright or involute margins . . . . 2. *Goodyera oblongifolia*
    2. Leaf blades uniformly green or reticulate with white or pale green on lateral veins and sometimes midrib (very infrequently only midrib whitened in *G. tesselata*); sepals 3–6 mm; lip 1.8–5.5 mm; rostellar beak 1.7 mm or less; lip apex spreading or recurved with spreading margins.
        3. Lip narrowly saccate with elongate, recurved apex; rostellar beak 0.2–0.6 mm, shorter than body of stigma . . . . 3. *Goodyera repens*
        3. Lip deeply concave with short spreading or recurved apex; rostellar beak 0.6–1.7 mm, equal to or longer than body of stigma . . . . 4. *Goodyera tesselata*

1. Goodyera pubescens (Willdenow) R. Brown in W. Aiton and W. T. Aiton, Hortus Kew. 5: 198. 1813 • Downy rattlesnake-plantain, goodyérie pubescente [E]

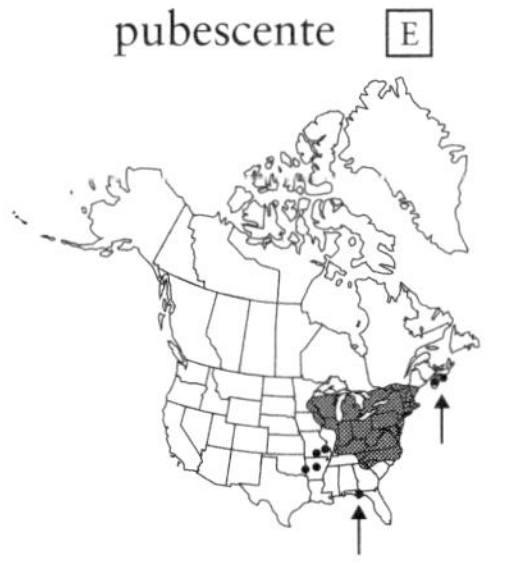

Neottia pubescens Willdenow, Sp. Pl. 4(1): 76. 1805; *Peramium pubescens* (Willdenow) MacMillan

Leaves: blade distinctly lined with white or greenish white on veins throughout, broadly elliptic to broadly ovate, 2.1–6.2 × 1.3–3 cm, apex acute or obtuse. **Inflorescences** 10–57-flowered, cylindric (equally dense on all sides); peduncle 11–35 cm. **Flowers:** lateral sepals 3.1–5.3 mm; petals distinct; hood 3.6–5.7 mm; lip scrotiform, 2.5–4.2 × 2.2–3.5 mm, apex reflexed, outer surface slightly tuberculate, inner surface with somewhat thickened veins, without glandular processes; anther inflexed, immersed in cup-shaped clinandrium, apex blunt; pollinia blunt; rostellum with marginal notch 0.1–0.4 mm deep; viscidium orbiculate. 2*n* = 26.

Flowering mid Jul–early Sep. Almost any wooded habitat with acid surface conditions; most frequent on moist humus soils in shady, upland woods of hemlock, pine, oak, or maple, less frequent in lowland woods, bogs, swamps; 0–1600 m; N.S., Ont., Que.; Ala., Ark., Conn., Del., D.C., Fla., Ga., Ill., Ind., Iowa, Ky., Maine, Md., Mass., Mich., Minn., Mo., N.H., N.J., N.Y., N.C., Ohio, Okla., Pa., R.I., S.C., Tenn., Vt., Va., W.Va., Wis.

Goodyera pubescens is very different in its lip and column morphology, phenolic constituents, and chromosome number (2*n* = 26) from other North American *Goodyera* species, indicating that it is not closely related to them.

2. Goodyera oblongifolia Rafinesque, Herb. Raf., 76. 1833 • Menzies's giant rattlesnake-plantain, giant rattlesnake-plantain, goodyérie à feuilles oblongues [F]

Goodyera decipiens (Hooker) F. T. Hubbard; *G. oblongifolia* var. *reticulata* B. Boivin; *Peramium decipiens* (Hooker) Piper

Leaves: blade usually streaked with white only along midrib, sometimes with fine white lateral veins, especially near midrib, narrowly elliptic to ovate, 2.5–10.2 × 1.3–3.5 cm, apex acute. **Inflorescences** densely to loosely spiraled or secund, 10–48-flowered; peduncle 7–38 cm. **Flowers:** lateral sepals 5.7–7.8 mm; petals connivent, hood 5–10 mm; lip deeply concave, boat-shaped, 4.9–

7.9 × 1.3–3.2 mm, margins upright or involute, apex spreading or slightly recurved, blunt, inner surface with 4 unequal rows of glandular papillae; anther erect, base 1/3–1/2 immersed in cup-shaped clinandrium, apex acuminate; pollinia acuminate; rostellar beak 2-pronged, 2.3–3.6 mm, longer than body of stigma; viscidium elongate. $2n = 30$.

Flowering mid Jul–mid Sep. Moist or dry coniferous or mixed woods, in East infrequent in cedar swamps, in s Rocky Mountains confined to high elevation spruce-fir forests; 0–3400 m; Alta., B.C., N.B., Nfld. and Labr. (Nfld.), N.S., Ont., Que., Sask.; Alaska, Ariz., Calif., Colo., Idaho, Maine, Mich., Mont., Nebr., N.Mex., Oreg., S.Dak., Utah, Vt., Wash., Wis., Wyo.; Mexico.

In eastern North America, *Goodyera oblongifolia* is restricted to formerly glaciated areas.

Plants with leaves white-reticulate on the lateral veins have been described as *Goodyera oblongifolia* var. *reticulata*. This segregate, essentially coastal in distribution, occurs from northern California to southeastern Alaska and is less frequent inland from British Columbia to New Mexico and in Michigan and Wisconsin. Because garden transplant experiments (J. A. Calder and R. L. Taylor 1968, vol. 1) have shown that both reticulate and non-reticulate leaves are found within the same clone, varieties are not recognized.

3. Goodyera repens (Linnaeus) R. Brown in W. Aiton and W. T. Aiton, Hortus Kew. 5: 198. 1813 • Dwarf rattlesnake-plantain, lesser rattlesnake-plantain, goodyérie rampante [F]

Satyrium repens Linnaeus, Sp. Pl. 2: 945. 1753; *Goodyera repens* var. *ophioides* Fernald; *Peramium ophioides* (Fernald) Rydberg

Leaves: blade uniformly green or with green lateral veins bordered by white or greenish white tissue, narrowly to broadly ovate, 1.1–3.2 × 0.5–1.8 cm, apex acute or obtuse. **Inflorescences** secund, infrequently loosely spiraled, 7–36-flowered; peduncle 3–18 cm. **Flowers:** lateral sepals 3–5.2 mm; petals distinct; hood 3–5.5 mm; lip narrowly saccate, lanceolate, 1.8–4.8 × 1.4–3.2 mm, apex acute, recurved or reflexed, inner surface with 2 or 4 glandular ridges; anther inflexed, not immersed in shallowly concave clinandrium, apex apiculate; pollinia blunt; rostellar beak 2-pronged, 0.2–0.6 mm, shorter than body of stigma; viscidium orbiculate. $2n = 30$.

Flowering early Jul–early Sep. Shady, moist, coniferous or mixed woods, on mossy or humus-covered ground, sometimes in bogs or cedar swamps; 0–2900 m; St. Pierre and Miquelon; Alta., B.C., Man., N.B., Nfld. and Labr., N.W.T., N.S., Ont., P.E.I., Que., Sask., Yukon; Alaska, Ariz., Colo., Conn., Maine, Md., Mass., Mich., Minn., Mont., N.H., N.Mex., N.Y., N.C., Pa., S.Dak., Tenn., Vt., Va., W.Va., Wis.; Eurasia.

Two varieties of *Goodyera repens* have been recognized: individuals with plain green leaves as var. *repens* and those with leaf veins bordered by broad white bands as var. *ophioides* Fernald. Many specimens from western North America are intermediate in this character: the bands bordering the veins are extremely narrow and almost indistinguishable by color from the leaf blade proper. Most specimens from eastern North America have obviously white-reticulate leaves, and most specimens from western Canada have plain green leaves (some with darker green veins). At several localities in western North America, plants with plain green leaves and plants with faintly reticulate leaves are found together (W. J. Cody 1961). Individuals with faintly reticulate leaves and those with plain green leaves are often found on the same herbarium sheet and, infrequently, both kinds of leaves occur on the same plant. Because of this variation in the degree of white reticulation, the varieties are not recognized here. Individuals with plain green leaves, obviously white-reticulate leaves, and intermediate leaves occur in Alaska.

4. Goodyera tesselata Loddiges, Bot. Cab. 10: plate 952. 1824 • Checkered rattlesnake-plantain, goodyérie panachée [E]

Peramium tesselatum (Loddiges) A. Heller

Leaves: blade with lateral veins and sometimes midrib prominently to faintly bordered with bands of white, gray, or paler green tissue frequently merging, giving leaf rather blotched appearance, especially near margins and apex where veins are closer together, infrequently uniformly green or with only midrib whitened, narrowly elliptic to ovate, 1.4–5.5 × 0.9–2.6 cm, apex usually acute. **Inflorescences** densely to loosely spiraled or secund, 5–72-flowered; peduncle 6–23 cm. **Flowers:** lateral sepals 3.8–6 mm; petals connivent; hood 3.9–7.1 mm; lip deeply concave to saccate, 3–5.5 × 1.2–3.1 mm, apex spreading or recurved, blunt or acute, inner surface with 2 or 4 unequal rows of glandular papillae; anther erect, base 0–1/3 immersed in shallowly concave to cup-shaped clinandrium, apex apiculate to acuminate; pollinia acute or short-acuminate; rostellar beak 2-pronged, 0.6–1.7 mm, equal to or longer than body of stigma; viscidium elliptic. $2n = 45, 60$.

Flowering mid Jul–early Sep. More common in dry or moist, upland, coniferous or mixed woods, less frequent in white-cedar swamps, margins of spruce-tamarack bogs; to 300 m; Man., N.B., Nfld. and Labr. (Nfld.), N.S., Ont.,

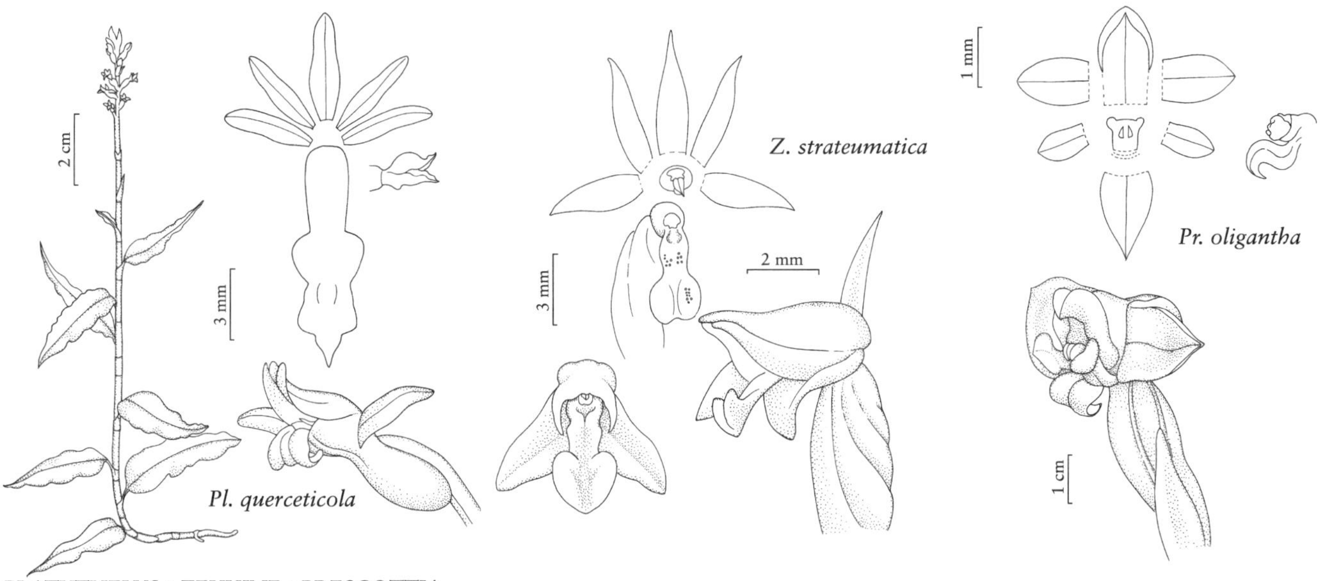

PLATYTHELYS ◦ ZEUXINE ◦ PRESCOTTIA

P.E.I., Que.; Conn., Maine, Md., Mass., Mich., Minn., N.H., N.J., N.Y., Ohio, Pa., R.I., Vt., Wis.

Goodyera tesselata is almost completely restricted to once-glaciated areas in the Great Lakes region, the northeastern United States, and adjacent Canada.

Goodyera tesselata tends to be taller and to have larger leaves, more cauline bracts, more flowers, and longer perianths than *Goodyera repens*. Although overlap exists in any count or measurement, the two are, for the most part, easily distinguishable. Some of the difficulty in identifying ambiguous specimens is explained by the hypothesis (J. A. Kallunki 1976) that *G. tesselata* itself is intermediate between and a probable allotetraploid of *G. repens* with white-reticulate leaves and *G. oblongifolia*. Because the tetraploids and the triploid hybrids ($2n$ = ca. 45) cannot be distinguished with certainty by their morphology, the description here includes individuals of both tetraploid and triploid individuals. In the extreme, individuals of *G. tesselata* (in the broad sense) approach *G. repens* more often than *G. oblongifolia*, and it is sometimes impossible to distinguish some triploids from *G. repens* with white-reticulate leaves.

7. PLATYTHELYS Garay, Bradea 2: 196. 1977 • [Greek *platy*, broad, and *thely(s)*, woman, alluding to the broad, flat rostellum]

James D. Ackerman

Herbs, terrestrial. **Roots** few, from nodes, villous. **Stems** elongate, terminal shoots ascending. **Leaves** cauline, petiolate, not articulate. **Inflorescences** terminal, racemes, pedunculate. **Flowers** resupinate, erect to somewhat spreading, small; sepals subsimilar, distinct and free, margins entire; petals connivent with sepals; lip 2-parted, fleshy, base spurred, ventricose, hemispheric, concave, margins notched, apex ± spreading to arcuate, rough; column subsessile, cuneate; stigmas 2, approximate, horizontal; rostellum flat, elliptic to suborbiculate, broad, apex notched to 2-denticulate, generally bursting; viscidium small; anther erect; pollinia sectile, attached to short caudicles; ovary short-pedicellate. **Fruits** capsules, ellipsoid to nearly cylindric.

Species 9 (1 in the flora): subtropics and tropics, North America, Mexico, West Indies, Central America, South America.

SELECTED REFERENCE Garay, L. A. 1977. Systematics of the Physurinae (Orchidaceae) in the New World. Bradea 2: 191–204.

1. Platythelys querceticola (Lindley) Garay, Bradea 2: 197. 1977 • Jug orchid F

Physurus querceticola Lindley, Gen. Sp. Orchid. Pl., 505. 1840; *Erythrodes jamaicensis* (Fawcett & Rendle) Fawcett & Rendle; *E. querceticola* (Lindley) Ames; *E. sagreana* (A. Richard) Ames; *Physurus commelynifolius* Reichenbach f.; *P. jamaicensis* Fawcett & Rendle; *P. sagreanus* A. Richard; *Platythelys sagreana* (A. Richard) Garay

Plants creeping. **Roots** 1–2 at each rhizome node, 1–2.5 mm diam. **Stems** rarely branched, succulent, glabrous; rhizome internodes 5–22 × 2–5 mm; shoots ascending, 12–32 cm, excluding inflorescences. **Leaves** 3–6, spirally arranged, green, sometimes whitish veined, glabrous; sheath inflated, enclosing stem, 5–12 mm, scarious; petiole distinct, 3–8 mm; blade lanceolate to ovate, thin, 14–65 × 8–21 mm, apex acute to acuminate, distalmost often bractlike. **Inflorescences:** scapes erect, 1–3 bracteate, 3–8 cm; racemes 4–22-flowered, 1.5–6 cm; floral bracts broadly ovate, 3.5–6 mm, scarious, apex acuminate. **Flowers** white to greenish, somewhat tubular, glabrous; sepals 1-veined, 3–4.5 × 1–2.5 mm, margins entire, apex obtuse; dorsal sepal concave, narrowly oblong-ovate to linear-oblong; lateral sepals ovate-elliptic, lanceolate to oblong, slightly falcate; petals slightly oblique, linear, linear-lanceolate, or subspatulate, 3–4.5 × 1–1.3 mm, margins entire, apex acute to obtuse; lip basally extended into descending saccate nectar spur, 2.5–3.5 mm, lamina basally concave, constricted distally, apically expanded and 3-lobed, lateral lobes spreading to projecting forward, ovate to suborbiculate, middle lobe reflexed, ovate, triangular, ligulate, or apiculate, 3–4 × 2–2.5 mm across lateral lobes; column stout, 2 × 1.5 mm; pollinia yellow; pedicellate ovary arcuate-spreading, cylindric, 5–6.5 mm, glabrous. **Capsules** 6–9 × 3–4 mm, glabrous.

Flowering Sep–early Mar. In humus of swamps, hardwood forests, hammocks; 0–200 m; Fla., La.; Mexico; West Indies; Central America; South America.

Platythelys querceticola was reported from Texas based on a sight observation (O. Ames 1924b), but its occurrence in that state has not been substantiated, and it has been excluded from *The Vascular Plants of Texas...* (M. C. Johnston 1988).

The general distribution of *Platythelys querceticola* may be much more restricted than what is given here. It has been reported for Mexico, but in at least one region, the plants identified as *Platythelys querceticola* are now thought to be another species (R. McVaugh and W. R. Anderson 1974+, vol. 16), and the record for Belize may also be a misdetermination (P. M. Catling and V. R. Catling 1988). Because R. L. Dressler (1993; L. O. Williams et al. 1980) excluded the species from Costa Rica and Panama, the South American records are suspect. Furthermore, J. T. Atwood (pers. comm.) suggested that it may be altogether absent from Mexico and Central America.

8. ZEUXINE Lindley, Orchid. Scelet., 9. 1826, orthography conserved • [Greek *zeuxis*, a yoking or joining, referring to partial union of lip and column, or possibly to fusion of pollinia] I

James D. Ackerman

Herbs, terrestrial, sympodial. **Roots** fleshy, villous. **Stems** succulent, glabrous. **Leaves** cauline, sessile [petiolate], not articulate; blade convolute, conduplicate, often membranous. **Inflorescences** terminal, dense spikes [racemes]; flowers spirally arranged; floral bracts mostly longer than flowers, membranaceous, scarious. **Flowers** few–many, resupinate, white with yellow lip, small, scarcely opening; sepals subequal; dorsal sepal erect, connivent with petals, forming hood over column; lateral sepals distinct and free, enclosing base of lip; lip pandurate, broadly rounded, margins entire, apex slightly broadened to winglike; column short; anther abaxial, 2-chambered, membranous; pollinia 2, sometimes 2-lobed, pyriform, sectile, caudicles present or absent; stigmas 2, on either side of column, convex; rostellum large, deeply divided. **Fruits** capsules, erect, ovoid [subglobose].

Species 30 (1 in the flora): introduced; tropical and subtropical regions, Africa, Asia, Australia, Pacific Islands; 1 species has escaped in parts of the Neotropics.

1. **Zeuxine strateumatica** (Linnaeus) Schlechter, Bot. Jahrb. Syst. 45: 394. 1911 • Lawn orchid, soldier orchid F I

Orchis strateumatica Linnaeus, Sp. Pl. 2: 943. 1753

Plants 4–25 cm. **Roots** clustered at nodes of subterranean part of stem and rhizome, short. **Stems** from decumbent base, erect or ascending, greenish, tinged with purple or brown, slender. **Leaves** 5–12, spirally arranged, sessile, sheathing; blade erect, dark green, linear to narrowly lanceolate, keeled, 1–9 × 0.3–0.8 cm, apex long-acuminate. **Inflorescences** 8–50-flowered; rachis 1–8 cm; floral bracts green, ovate-lanceolate to lanceolate, 15 × 4 mm. **Flowers** white with prominent yellow lip; dorsal sepal concave, ovate-oblong to oblong-elliptic, 4–7 × 2–3 mm, apex obtuse; lateral sepals oblique, ovate-oblong to oblong-elliptic, 4–6 × 2–3 mm, apex obtuse to acute; petals oblong-lanceolate, falcate, 4–6 × 1.5–2 mm, converging with dorsal sepal; lip adnate to base of column, sessile, base concave, subsaccate or cymbiform, containing 2 glands, lamina ± contracted in middle, abruptly dilated at apex, not extending beyond lateral sepals, 4 × 3 mm, fleshy; column 1.5 mm; anther between 2 large stigmatic processes, scarious; pollinia yellow; ovary sessile, stout, 7 mm. **Capsules** suberect, ovoid to ellipsoid, 7 mm. $2n$ = 22–64, 30–56, 50, 100.

Flowering fall–winter (Oct–Jan), occasionally spring. Moist areas in a variety of habitats, including lawns, roadsides, nurseries, farm fields, occasionally hammocks and pinelands; 0–30 m; introduced; Ala.(?), Fla., Ga., La., Miss., Tex.; West Indies; native, Asia; Pacific Islands.

Given the weedy nature of these plants, the apparently high fruit set (C. A. Luer 1972, plate 36), and the spontaneous reproduction in greenhouses (O. Ames 1938), *Zeuxine strateumatica* is very likely autogamous or apomictic. Five cytotypes whose floral morphology overlapped considerably have been identified (S. P. Vij and N. Vohra 1974). In all but the $2n$ = 20 cytotype, meiotic behavior was abnormal.

230c.1b. ORCHIDACEAE Jussieu (tribe CRANICHIDEAE) subtribe PRESCOTTIINAE Dressler, Lindleyana 5: 115. 1990

9. PRESCOTTIA Lindley in W. J. Hooker, Exotic Fl. 2: plate 115. 1824 (as Prescotia) • [For John Prescott (d. 1837), a British botanist resident in Russia who traveled widely in northern Asia]

James D. Ackerman

Herbs, terrestrial. **Roots** fasciculate, fibrous or thick and fleshy, villous. **Stems:** rhizomes sometimes branched. **Leaves** basal, petiolate or sessile, not articulate, membranous. **Inflorescences** terminal, many-flowered spikes; peduncles partially covered by sheathing bracts. **Flowers** not resupinate, sessile; sepals spreading or reflexed, rarely distinct, basally connate forming short cup, thin; petals adnate to column and sepal cup, narrow, thin; lip distalmost in flower, attached to column foot, clawed, often basally auriculate distal to claw, deeply concave, often enclosing column, very fleshy; column and foot adnate to sepal cup when present, minute, column blunt; anther abaxial, erect; pollinia 4, slightly flattened, soft, mealy; caudicles absent; stigmas entire. **Fruits** capsules, ovoid to ellipsoid.

Species 24 (1 in the flora): tropical and subtropical regions, North America, Mexico, West Indies, Central America, South America.

Originally the name was spelled as "*Prescotia.*" Since then and until recently, most authors, with the exception of W. Voth (1976) and J. D. Ackerman (1989), spelled it with two *t*s. "*Prescotia*" should be treated as an orthographic error (J. D. Ackerman 1995), and the spelling with two *t*s should prevail

1. Prescottia oligantha (Swartz) Lindley, Gen. Sp. Orchid. Pl., 454. 1840 [F]

Cranichis oligantha Swartz, Prodr., 120. 1788; *C. micrantha* Sprengel; *Prescottia gracilis* Schlechter; *P. myosurus* Reichenbach ex Grisebach; *P. panamensis* Schlechter

Plants to 40 cm, glabrous (except roots). **Roots** fasciculate, fingerlike, 3–8 mm diam., fleshy. **Leaves** 2–4; petiole 1–3 cm; blade elliptic to suborbiculate, 1.5–8 × 0.8–3 cm, base cuneate, margins entire, apex acute to rounded. **Inflorescences** slender; scape 12–31 cm, partially covered by sheathing bracts; spike to 17 cm, densely many-flowered; floral bracts ovate to lanceolate, clasping base of ovary, 1.5–4 mm, apex acuminate. **Flowers** minute, glabrous; sepals reflexed-spreading, pinkish, 1-veined, 1–2.2 × 1 mm; dorsal sepal reflexed, ovate, curled; lateral sepals adnate with lip forming short mentum, triangular to deltate, slightly larger than dorsal sepal; petals reflexed, white or pink, narrowly obovate to oblong, 1–1.5 × 0.5 mm; lip erect, white, basally auriculate, hoodlike, enclosing column, shallowly saccate, 1–2 mm, fleshy, apiculate; column adaxially adnate to sepal tube, minute (0.3 mm), apex winged; ovary sessile, 1.5–3.5 mm. **Capsules** erect, ellipsoid, 4 mm.

Flowering Dec–early Mar. Hammocks; 0–20 m; Fla.; Mexico; West Indies; Central America; South America (Galapagos Islands).

230c.1c. ORCHIDACEAE Jussieu (tribe CRANICHIDEAE) subtribe SPIRANTHINAE Lindley, Gen. Sp. Orchid. Pl., 441, 462. 1840

SELECTED REFERENCES Burns-Balogh, P. 1982. Generic redefinition in subtribe Spiranthinae (Orchidaceae). Amer. J. Bot. 69: 1119–1132. Burns-Balogh, P. and H. Robinson. 1983. Evolution and phylogeny of the *Pelexia* alliance (Orchidaceae: Spiranthoideae: Spiranthinae). Syst. Bot. 8: 263–268. Garay, L. A. 1980[1982]. A generic revision of the Spiranthinae. Bot. Mus. Leafl. 28: 277–425. Greenwood, E. W. 1982. Tipos de viscidio en Spiranthinae. Orquidea (Mexico City) 8: 283–310.

10. CYCLOPOGON C. Presl, Reliq. Haenk. 1: 93, plate 13. 1827 • [Greek *cyclo*, circular, and *pogon*, beard, perhaps in reference to pubescent bases of sepals of the type species]

James D. Ackerman

Beadlea Small

Herbs, terrestrial, sympodial. **Roots** fasciculate, fleshy, villous. **Stems** simple, rhizomatous. **Leaves** few to many, basal, petiolate; blade not articulate, convolute, mostly ovate to elliptic, soft. **Inflorescences** terminal, many-flowered spikes or racemes, erect; scapes bracteate. **Flowers** resupinate, horizontal, greenish or yellowish green, small; sepals subparallel, distinct or connate at base, forming obscure mentum with base of column or sepaline nectar tube; petals connivent with dorsal sepal; lip clawed, sagittate to cordate, constricted proximal to apex; lateral margins appressed to sides of column; column erect; pollinia 2, clavate-oblong, mealy; stylar canal entrance central; stigma lobes 2, distinct or approximate; rostellum longer than wide; viscidium relatively large, disc-shaped; ovary sessile or subsessile. **Fruits** capsules.

Species 70 (2 in the flora): tropical and subtropical regions, s North America, Mexico, West Indies, Central America, South America.

All species of *Cyclopogon* except the holotype were placed in the genus *Beadlea* (L. A. Garay 1980[1982]) because their perianth base was not united into a tube. That characteristic was rejected as a valid generic distinction but the taxon was recognized as *Cyclopogon* sect. *Beadlea* (P. Burns-Balogh 1982; P. Burns-Balogh and H. Robinson 1983). The perianth tube character is a matter of degree and not one of presence or absence (D. L. Szlachetko 1993).

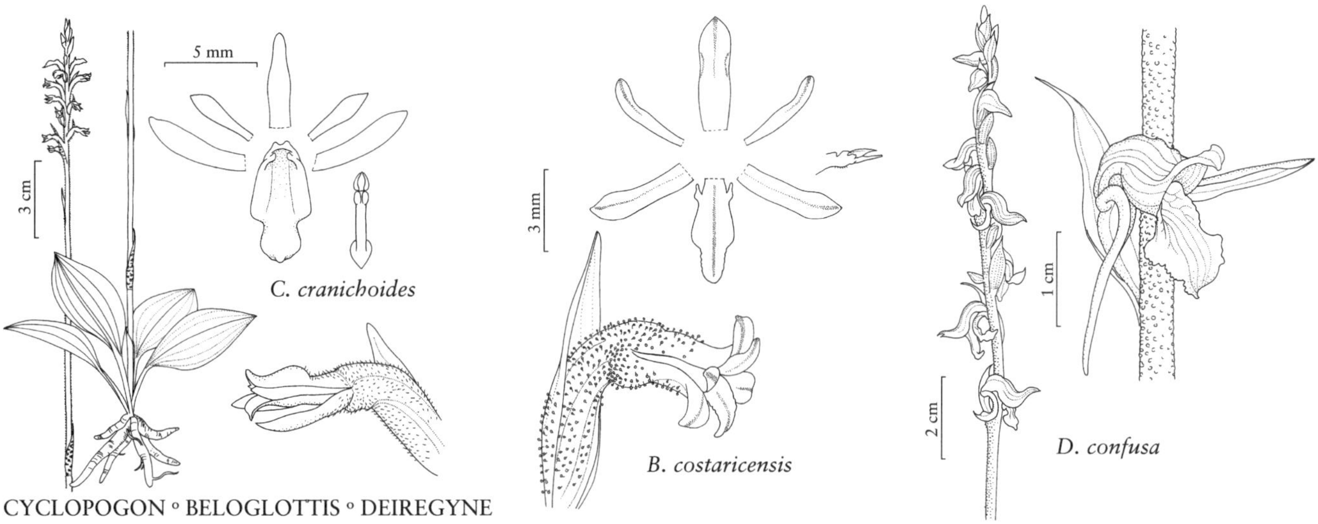

CYCLOPOGON ◦ BELOGLOTTIS ◦ DEIREGYNE

1. Midvein and margins of petals darkened distally; lip broadest proximal to apical constriction; leaf blades purplish abaxially, dark purplish green adaxially; petioles spreading to suberect, 1.1–3.5(–5.5) cm . . . . 1. *Cyclopogon cranichoides*
1. Midvein and margins of petals unmarked distally; lip broadest distal to apical constriction or distal and proximal parts of lip equal in width; leaf blades green; petioles erect, 3–10 cm . . . . 2. *Cyclopogon elatus*

1. Cyclopogon cranichoides (Grisebach) Schlechter, Beih. Bot. Centralbl. 37: 387. 1920 [F]

Pelexia cranichoides Grisebach, Cat. Pl. Cub., 269. 1866; *Beadlea cranichoides* (Grisebach) Small; *B. storeri* (Chapman) Small; *Sauroglossum cranichoides* (Grisebach) Ames; *Spiranthes cranichoides* (Grisebach) Cogniaux; *S. storeri* Chapman

Herbs small, to 40 cm. **Roots** fusiform, to 4.5 cm × 4–8 mm, fleshy. **Stems** subterranean, short. **Leaves** 3–8; petiole spreading to suberect, 1.1–3.5(–5.5) cm; blade abaxially purplish, adaxially dark purplish green, sometimes with whitish markings, broadly elliptic to ovate, 2.5–8 × 1.3–3.8 cm, apex obtuse to acute. **Inflorescences** mottled, purplish; scape pubescent; raceme continuously elongating as fruits mature, to 15 cm, pubescent; floral bracts mottled, lanceolate, longer than ovary. **Flowers** tubular-funnelform; sepals greenish brown, pubescent; dorsal sepal midvein and margins dark near apex, narrowly elliptic to narrowly pandurate, 3.5–5 × 1.2–2 mm, apex acute; lateral sepals often reflexed, narrowly elliptic to narrowly pandurate, 4–6 × 1–1.5 mm, apex acute; petals greenish-brownish proximally, greenish white distally with midvein and apical margins dark brown, narrowly spatulate-oblanceolate, 4.5 × 0.5–1 mm, apex acute, oblique; lip white, canaliculate, oblong, constricted distal to middle, broadest proximal to constriction, to 5 × 2.5 mm, base gibbose, with pair of tubercles, margins entire, apex slightly flared; column slender, 3.5 mm; pollinia yellow; pedicellate ovary 4–7 mm. **Capsules** erect, ellipsoid, 6–8 × 3.5–4.5 mm.

Flowering Feb–Mar. In humus of shady moist hammocks; 0–100 m; Fla.; West Indies; Central America; n South America.

Fruit production in *Cyclopogon cranichoides* is pollen limited (R. N. Calvo 1990). Natural fruit set was 26–62% over a 4-year period, whereas hand pollinations gave better than 96% fruit set. Contrary to most other studies of orchids, Calvo did not detect a cost of reproduction associated with resource constraints.

SELECTED REFERENCE Calvo, R. N. 1990. Four-year growth and reproduction of *Cyclopogon cranichoides* (Orchidaceae) in south Florida. Amer. J. Bot. 77: 736–741.

2. Cyclopogon elatus (Swartz) Schlechter, Repert. Spec. Nov. Regni Veg. Beih. 8: 53. 1919

Satyrium elatum Swartz, Prodr., 119. 1788; *Beadlea elata* (Swartz) Small ex Britton; *Sauroglossum richardii* Ames; *Spiranthes elata* (Swartz) Richard

Herbs to 75 cm. **Roots** fusiform-cylindric, 13 cm × 3–10 mm. **Leaves** 2–6; petiole erect, 3–10 cm; blade green abaxially and adaxially, narrowly to broadly elliptic, or elliptic-lanceolate, 3.7–13 × 1.6–6.4 cm, apex obtuse to acute-acuminate. **Inflorescences** immaculate, green to greenish brown; scape to 57 cm × 1–5 mm, glabrous or sparsely pubescent; racemes 7–50-flowered, 7–23 cm, pubescent; floral bracts erect, immaculate or minutely speckled, lanceolate, 8–19 mm. **Flowers** green to coppery brown, sometimes purplish, tubular-funnelform; sepals basally connate for 0.1 mm, abaxially pubescent; dorsal sepal elliptic-lanceolate, 4.5–7 × 1.5–2 mm, apex acuminate; lateral sepals slightly sinuate-falcate, narrowly oblong, 5–7.5 × 1–2 mm; petals linear to narrowly spatulate, 4–6.5 × 0.5–0.75 mm, glabrous; lip white, canaliculate, pandurate to oblong-ovate, then constricted distal to middle, distal portion flabellate to reniform, usually broadest, or distal and proximal portions equal in width, 5–7 × 2–4 mm, base laterally thickened, with or without pair of tubercles, margins entire or crenulate; column slender, 3.5–4.5 mm; pollinia yellow. **Capsules** erect-spreading, ellipsoid, 6–12 × 4–6 mm. $2n = 28, 30, 45$.

Flowering Feb–Mar. In humus of moist, rocky hammocks; 0–20 m; Fla.; Mexico; West Indies; Central America; n South America.

Morphologic variation is substantial in *Cyclopogon elatus*. Variants may be sympatric and locally distinct (J. D. Ackerman 1995), or they may form a continuous series within or among regions. I suspect that inbreeding (via self-pollination) is prevalent and has generated and maintained local variation.

11. BELOGLOTTIS Schlechter, Beih. Bot. Centralbl. 37: 364. 1920 • [Greek (Latinized) *belos*, arrow, and *glotta*, tongue, possibly alluding to sagittate lip]

James D. Ackerman

Herbs, terrestrial, sympodial. **Roots** fasciculate, fleshy, villous. **Stems** solitary, erect, rhizomatous, short. **Leaves** several, basal, petiolate; blade not articulate, involute when immature, margins entire. **Inflorescences** terminal, racemes, erect, scapose, laxly to densely flowered, slender. **Flowers** resupinate; perianth tubular; sepals subparallel, distinct or connate at base, apex spreading; lateral sepals with oblique, subdecurrent base; petals parallel and adnate to dorsal sepal; lip clawed, blade white with central green stripe, canaliculate, sagittate, 3-lobed, lateral lobes rounded, midlobe elongate-rounded, margins finely denticulate, lateral margins adnate to sides of column; column short, clinandrium inflated; stigmas 2, adjacent; rostellum erect, 2-lobed, 2-cleft, or 2-dentate, short; anther ovate, concave, base slightly cordate, apex acute; pollinia 4, clavate, soft, mealy; viscidium narrowly elliptic. **Fruits** capsules, erect, ellipsoid.

Species 10 (1 in the flora): tropical and subtropical regions, North America, Mexico, West Indies, Central America, South America.

1. Beloglottis costaricensis (Reichenbach f.) Schlechter, Beih. Bot. Centralbl. 37: 365. 1920 [F]

Spiranthes costaricensis Reichenbach f., Bonplandia (Hanover) 3: 214. 1855; *Gyrostachys costaricensis* (Reichenbach f.) Kuntze

Plants to 40 cm. **Roots** 1–4, 1–6 cm × 3–8 mm. **Stems** subterranean or nearly so, short. **Leaves** fugacious, present or absent at time of flowering, 2–5, delicate, glabrous; petiole erect, slender, 2–4 cm; blade spreading, elliptic, 2–4.5 × 1.3–2 cm, apex acute. **Inflorescences** 6–35-flowered racemes, pubescent distally; scape 10–12.5 cm × 0.6–1 mm; bracts 5–6; floral bracts ovate, 5–12 mm, apex acuminate. **Flowers** white with green midveins, subsalverform; sepals connate at base, 1-veined, lanceolate, margins entire, apex acute to acuminate, abaxially pubescent; dorsal sepal 4–6 × 1.5 mm; lateral sepals oblique, 4.2–6 × 1.2 mm; petals 1-veined, linear, slightly falcate, 4 × 0.6 mm, apex acute, glabrous; lip adnate to base of column foot, 3-veined, 4.5 × 1.5–2 mm, glabrous; claw 1.5 mm, auricles subulate, 0.7–0.8 mm, projecting basally, flanking column and column foot; column slender, 2.5 mm, foot extending obliquely along ovary 1.5 mm without forming mentum; pedicellate ovary 3–4 mm, sparsely pubescent. **Capsules** 4–6 mm.

Flowering Mar–Apr. Hammocks, in humus; 0–10 m; Fla.; Mexico; West Indies; Central America.

Beloglottis costaricensis has been questionably recorded in South America.

12. DEIREGYNE Schlechter, Beih. Bot. Centralbl. 37: 426. 1920 • [Greek *deire*, neck, and *gyne*, pistil or woman, referring to sepals that sit on top of ovary and form a necklike extension]

Paul Martin Brown

Herbs, perennial, terrestrial. **Roots** tuberous, fascicled, fleshy. **Stems** with foliaceous sheaths. **Leaves** usually absent at anthesis, basal or cauline, base cuneate. **Inflorescences** terminally flowered spikes. **Flowers** resupinate; perianth parts distinct and/or free; sepals similar, basally connivent, forming inflated nectary; lip free, oblong-ovate, arcuate, base conduplicate, apex acute; column partially adnate to dorsal sepal, not inflated adaxially at base; anther ovate-cucullate, margins entire; viscidium rounded; rostellum longer than wide; stigma 2-lobed at apex. **Fruits** capsules.

Species 14 (1 in the flora): tropical regions, United States, Mexico, Central America (Guatemala).

1. Deiregyne confusa Garay, Bot. Mus. Leafl. 28: 283, plate 18C. 1982 F

Schiedeella confusa (Garay) Espejo & López-Ferrari; *Spiranthes confusa* (Garay) Kartesz & Gandhi

Plants 2.5–5 dm. **Roots** 7–12, tuberous, fasciculate, large, 12 × 1 cm, fleshy. **Stems** with 6–9 bladeless cauline sheaths. **Leaves** absent at anthesis, 2–3, mostly basal; petiole not distinct; blade linear-lanceolate, 15–30 × 2–3.5 cm. **Inflorescences** to 15 deflexed flowers, secund or loosely spiraled, 6–18 cm, glandular-pubescent; floral bracts ovate-lanceolate, striate-veined, 1–2.5 cm, apex acuminate. **Flowers** pink or dull brown to purplish with green veins; sepals recurved, 3-veined, glandular-pubescent; dorsal sepal elliptic-obovate, abruptly tapering to apex, concave, slightly constricted to form sac 12–19 × 4.2–5.2 mm; lateral sepals linear-lanceolate, somewhat sigmoid, 11–17 × 2–3.5 mm, apex acute, glandular-pubescent on outer surface; petals linear-oblanceolate, sigmoid, 3–5-veined; lip whitish pink with green veins, 13–20 mm, strongly pandurate, basal portion orbiculate, 7–10 mm wide, pubescent, apical portion suborbiculate, 5–6 mm wide, margins crenulate or erose; column flattened, 8 mm; rostellum cuspidate; anther abaxial; pedicellate ovary green, stout, 10 × 3 mm.

Flowering May–Jul. Open grassy slopes in rim rock; 2000 m; Tex.; Mexico.

Deiregyne confusa is known in the United States only from the original collections made in 1931 in the Chisos Mountains of Brewster County, Texas. Searches since then have failed to find any more plants (C. A. Luer 1975; J. Liggio and A. Liggio 1999). The Texas plants were originally identified as *Spiranthes durangensis* Ames & C. Schweinfurth (D. S. Correll 1950). They were subsequently placed in *Deiregyne* and described as *D. confusa* Garay. L. A. Garay (1980[1982]) stated that *D. durangensis* does not occur in the United States.

13. ELTROPLECTRIS Rafinesque, Fl. Tellur. 2: 51. 1837 • [Greek *eleutheros*, free, and *plectron*, spur, referring to free spur of sepal

James D. Ackerman

Centrogenium Schlechter; *Stenorrhynchos* Richard ex Sprengel sect. *Eltroplectris* (Rafinesque) Burns-Balogh

Herbs, terrestrial, sympodial. **Roots** fasciculate, fleshy, densely villous. **Stems** erect, rhizomatous. **Leaves** 1–2, basal; blade not articulate, long-petiolate. **Inflorescences** terminal, few- to many-flowered racemes, erect; peduncle slender, bracteate. **Flowers** resupinate; sepals spreading, distinct; lateral sepals longer than dorsal, attached along column foot forming spur; petals appressed to dorsal sepal, basally decurrent on column; lip arcuately recurved, clawed, shorter than sepals, membranous; column erect, short, slender; column foot partially adnate to ovary, elongate; anther persistent; pollinia 2, clavate, mealy; stigmas distinct; rostellum subulate to linear, pointed, rigid; viscidium oblong. **Fruits** capsules, ellipsoid-cyclindric.

Species 12 (1 in the flora): subtropical and tropical regions, North America, Mexico, West Indies.

Eltroplectris has been treated as *Stenorrhynchos* Richard ex Sprengel sect. *Eltroplectris* (P. Burns-Balogh 1982); others have treated *Eltroplectris* as distinct (e.g., L. A. Garay 1980[1982]; D. L. Szlachetko 1995).

SELECTED REFERENCES Burns-Balogh, P. 1982. Generic redefinition in subtribe Spiranthinae (Orchidaceae). Amer. J. Bot. 69: 1119–1132. Szlachetko, D. L. 1995. Systema orchidalium. Fragm. Florist. Geobot., suppl. 3: 1–152.

1. Eltroplectris calcarata (Swartz) Garay & H. R. Sweet, J. Arnold Arbor. 53: 390. 1972 • Longclaw orchid, spurred neottia [F]

Neottia calcarata Swartz, Fl. Ind. Occid. 3: 1413, plate 28. 1806; *Centrogenium setaceum* (Lindley) Schlechter; *Pelexia calcarata* (Swartz) Cogniaux; *P. setacea* Lindley; *P. setacea* var. *glabra* Cogniaux; *Spiranthes calcarata* (Swartz) Jimenez; *Stenorrhynchus calcaratus* (Swartz) Richard

Plants to 70 cm. **Roots** numerous, 18 cm × 2–6 mm. **Stems** subterranean, obscured by roots. **Leaves** glabrous; petiole erect, reddish, slender, 9–20 cm, usually longer than blade; blade green, lanceolate to broadly elliptic, 9–15 × 3.5–7 cm, base cuneate, margins entire to crenulate, apex acute to acuminate. **Inflorescences:** peduncle reddish, satinlike, glabrous proximally, pubescent distally, with a few distant, sheathing bracts, 28–65 cm; racemes 3–11-flowered. **Flowers** spindly; sepals green, narrowly lanceolate, 21–30 × 6–7 mm, apex acuminate; dorsal sepal projecting over column; lateral sepals spreading; petals green, linear, 19–21 × 3 mm, apex acuminate; lip recurved, attached to column foot, white, ovate, lamina notched distal to middle, margins fringed, 2 cm, apex acuminate; spur adnate to ovary for 8–9 mm, cylindric, 15–17 mm, free portion curved, 7–9 mm; column semiterete, 8–10 mm, apex pointed; pedicellate ovary 1.5–2.3 cm, glabrous or sparsely pubescent. **Capsules** 18–25 mm; spur persistent.

Flowering Jan–Mar. In humus of moist hardwood hammocks; 0–30 m; Fla.; West Indies; n South America.

14. DICHROMANTHUS Garay, Bot. Mus. Leafl. 28: 313. 1982 • [Greek *di*, two, *chroma*, color, and *anthos*, flower, indicating 2-colored nature of flowers]

Paul Martin Brown

Herbs, perennial, terrestrial. **Roots** tuberous, fascicled, fleshy. **Stems** with foliaceous sheaths. **Leaves** present at anthesis, basal or cauline. **Inflorescences** terminal, spikes, glabrous proximally, glabrous or pubescent distally. **Flowers** resupinate; perianth parts distinct and/or free,

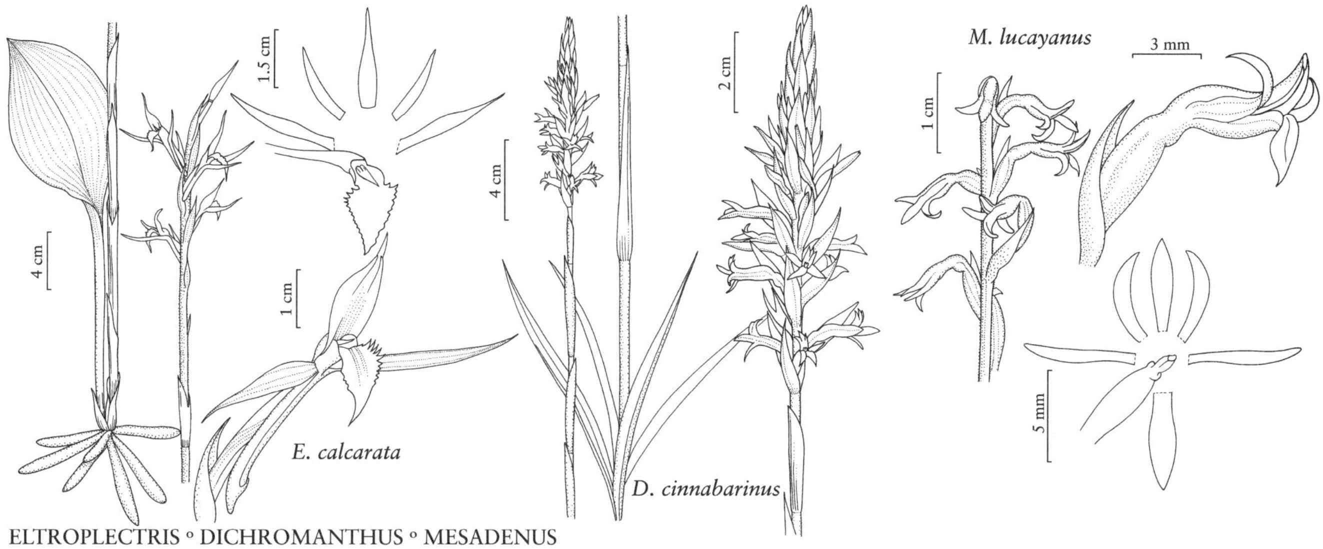

ELTROPLECTRIS ∘ DICHROMANTHUS ∘ MESADENUS

strongly recurved, yellow-orange to scarlet; sepals similar, basally connivent, with flared apex; lip sessile, base conduplicate; column partially adnate to dorsal sepal, with decurrent foot, short, pubescent; anther ovate-cucullate, margins entire; viscidium oblong-linear; rostellum oblong; stigma 2-lobed at apex. **Fruits** capsules.

Species 1: tropical regions, Western Hemisphere.

1. **Dichromanthus cinnabarinus** (La Llave & Lexarza) Garay, Bot. Mus. Leafl. 28: 314. 1982 [F]

Neottia cinnabarina La Llave & Lexarza, Nov. Veg. Descr. 2: 3. 1825; *Spiranthes cinnabarina* (La Llave & Lexarza) Hemsley; *Stenorrhynchus cinnabarinus* (La Llave & Lexarza) Lindley

Plants 50–90 cm. **Roots** 7–12, tuberous, fasciculate, large, 12 × 1 cm, fleshy. **Stems** with 6–9 bladeless cauline sheaths, green. **Leaves** present at anthesis, 3–4, ascending stem, dark green, linear-oblanceolate, 10–20 × 1.5–3 cm, passing into orange bracts with green stripes. **Inflorescences** congested, 30–40-flowered spikes, 4–17 cm, minutely papillose; floral bracts red, lanceolate, 3.5–8 mm. **Flowers** yellow-orange to scarlet, tubular; perianth apices strongly recurved; ovaries red, stout, 10 × 4 mm, pubescent; sepals sparingly pubescent; dorsal sepal recurved, vermilion, lanceolate, 15–20 × 3–4 mm; lateral sepals linear-lanceolate, 15–22 × 2–3.5 mm, apex acute; petals yellow with vermilion apex, linear, falcate, 15–20 × 2–2.5 mm; lip yellow with vermilion apex, lanceolate, concave proximal to middle, 15–20 × 3–5 mm; column 9 mm, thick, papillose; rostellum flat, slender, 2–3 mm; anther abaxial.

Flowering Jul–Oct. Rocky mountain slopes, grassy hills and canyon meadows, often in limestone; 1800 m; Tex.; Mexico; Central America (Guatemala).

15. MESADENUS Schlechter, Beih. Bot. Centralbl. 37: 367. 1920 • [Greek (Latinized) *meso*, middle, and *aden*, gland]

James D. Ackerman

Herbs, terrestrial, sympodial. **Roots** fasciculate, fleshy, villous. **Stems** erect, rhizomatous, short. **Leaves** several, basal, petiolate; blade not articulate, involute when immature, margins entire. **Inflorescences** terminal, laxly to densely flowered racemes, erect, scapose, slender. **Flowers** resupinate; perianth parts partially connate, apices spreading to recurved; sepals subsimilar, subparallel; dorsal sepal adherent to base of column; lateral sepals attached to column foot; petals adnate to dorsal sepal, falcate; lip auriculate, canaliculate; column arcuate, column foot present, clinandrium inflated; stigmas 2, confluent; rostellum narrow, notched; anther stalked, cordate, curved backward during anthesis, surpassing rostellum; pollinia 4, clavate, soft, mealy; viscidium ovate to subelliptic. **Fruits** capsules, erect, obliquely fusiform to ovoid.

Species 7 (1 in the flora): tropical and subtropical regions, North America, Mexico, West Indies, Central America, South America.

1. **Mesadenus lucayanus** (Britton) Schlechter, Beih. Bot. Centralbl. 37: 368. 1920 [F]

Ibidium lucayanum Britton, Bull. New York Bot. Gard. 5: 312. 1907 [preprint]; *Spiranthes lucayana* (Britton) Cogniaux

Plants to 60 cm. **Roots** several, 1.5–5 cm × 5–15 mm. **Stems** subterranean. **Leaves** spreading, pale green to gray-green, 3–5, glabrous; petiole 0.5–5 cm; blade lanceolate, elliptic to oblanceolate, 2–10.9 × 0.8–3.3 cm, apex obtuse, acute, or acuminate. **Inflorescences** glabrous; scapes 6.5–30 cm; bracts numerous, overlapping proximally, distant distally; racemes many-flowered, 6.5–29 cm; floral bracts erect, lanceolate, 3–4.5 mm, apex acuminate. **Flowers** reflexed, brownish green to coppery, slightly swollen basally, subsalverform, glabrous; sepals linear-lanceolate; dorsal sepal 4–6 × 0.7–1.2 mm; lateral sepals basally connate 1–1.5 mm, falcate, 4–6.5 × 0.7–1.2 mm; petals connate and adnate to dorsal sepal, falcate, linear-lanceolate, 4–5 × 0.5–1.2 mm; lip sometimes auriculate, short-unguiculate, elliptic-lanceolate to ovate-lanceolate, 3.5–6 × 1–2.5 mm, apex obtuse-rounded; column clavate, 2–2.5 mm; pedicellate ovary erect, subsessile, 2.5–4 mm, apically geniculate. **Capsules** ovoid, 4–6 mm.

Flowering winter–early spring (late Dec–Mar). Oak forests, in rocky soil; 0–30 m; Fla.; Mexico; West Indies.

Mesadenus lucayanus is known in the flora area only from a few small populations in Citrus, Duval, and Miami-Dade counties (R. P. Wunderlin et al. 1996). Recent collections from Sarasota County are available, but the population has been extirpated (P. M. Brown 2002).

Mesadenus lucayanus has often been regarded as a synonym of *M. polyanthus*; however, I have decided that the two taxa are distinct. *Mesadenus polyanthus* is a robust Mexican species from dry, volcanic habitats at high elevations (1500–2250 m). Plants may reach 95 cm and the lower stem bracts are spreading, to 7.5 cm. The inflorescence is very congested, 10 mm thick. The perianth is attenuate (lateral sepals to 10 mm) with a narrowly lanceolate lip giving the flowers a very spidery appearance. Morphologic variation is subdued. On the other hand, *M. lucayanus* occurs at low elevations over calcareous soils. The plants are smaller, more slender, and the stem bracts are erect. The inflorescence is laxly flowered, and the perianth is generally smaller than that of *M. polyanthus*. The lip is elliptic-lanceolate to narrowly ovate. Although *M. lucayanus* is highly variable in many features, including flower color, the two taxa appear to differ substantially in both morphologic features as well as habitat requirements.

16. PELEXIA Poiteau ex Lindley, Bot. Reg. 12: plate 985. 1826, name conserved • [Greek *pelex*, helmet, in reference to dorsal sepal, which is united with petals to form narrow hood]

James D. Ackerman

Herbs, terrestrial, rarely subaquatic, sympodial. **Roots** fasciculate, thick, fleshy, villous. **Leaves** 1–several, basal; petiole not articulate, reddish purple, long. **Inflorescences** terminal, many-flowered racemes or spikes, erect, pedunculate, bracteate. **Flowers** resupinate, green; dorsal sepal connivent with petals forming hood over column; lateral sepals decurrent on column foot; lip sagittate to auriculate, fleshy, basal margins adnate to sides of column; column cylindric, elongate, puberulent to pilose; column foot elongate; anther abaxial, ovate-cordate; pol-

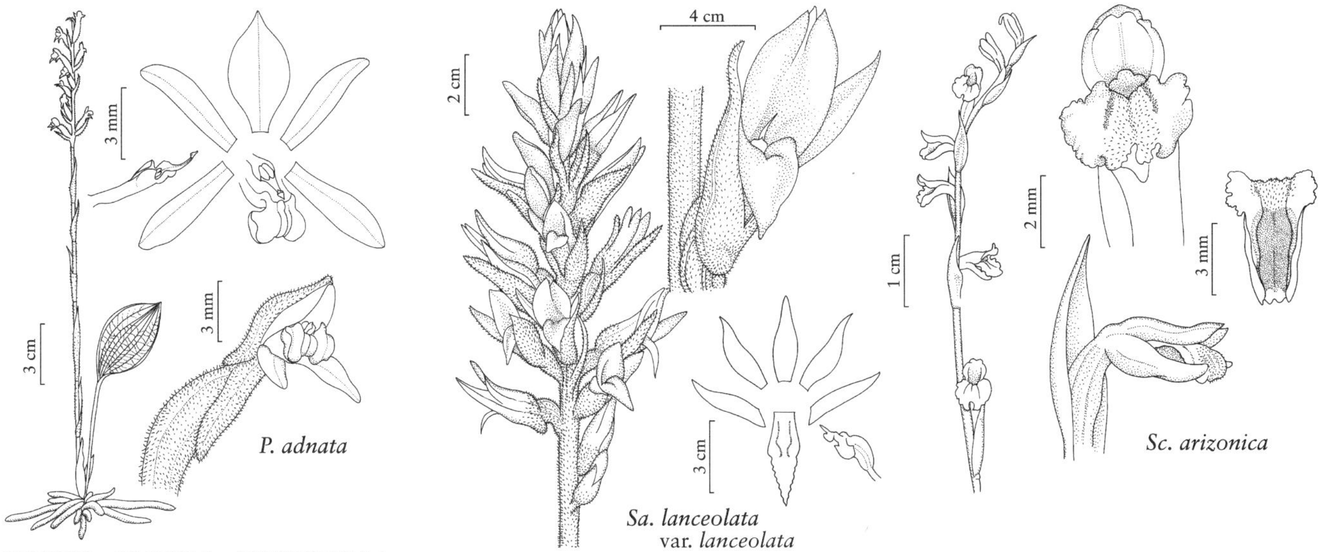

PELEXIA ◦ SACOILA ◦ SCHIEDEELLA

linia 2, oblong-clavate, mealy; stigmas 2, adjacent; rostellum elongate; viscidium ovate to suborbiculate; ovary sessile to subsessile. **Fruits** capsules, erect.

Species 67 (1 in the flora): subtropical and tropical regions, North America, Mexico, West Indies, Central America, South America.

1. Pelexia adnata (Swartz) Sprengel, Syst. Veg. 3: 704. 1826 • Hachuela F

Satyrium adnatum Swartz, Prodr., 118. 1788; *Neottia adnata* (Swartz) Swartz; *Spiranthes adnata* (Swartz) Bentham ex Fawcett

Plants cespitose, to 70 cm. **Roots** 3–8 mm diam. **Leaves** 2–7, erect; petiole 12–16.5 cm; blade elliptic to lanceolate, 7.5–14 × 3–5.5 cm, margins entire, apex acuminate. **Inflorescences** scapose racemes, pubescent; floral bracts linear-lanceolate, as long as flower or longer, apex acuminate. **Flowers:** dorsal sepal green, broadly ovate to elliptic, 5–6 × 3–4 mm, apex obtuse, pubescent; sepals gently recurved, narrowly elliptic, 5–6 × 1.5 mm, apex obtuse, pubescent; lateral sepals adnate to column foot forming spurlike mentum, mentum adnate to ovary, only partially free near tip, 6–7 mm; petals green, linear to linear-oblanceolate, 5 mm, apex obtuse, glabrous; lip white with yellowish throat, constricted at or distal to middle; apical half abruptly geniculate, fleshy, 5–6 mm, tip recurved, glabrous; column to 6 mm, column foot produced basally, decurrent along ovary, to 7 mm; viscidium ovate; pedicellate ovary erect-spreading, 10–11 mm, pubescent. **Capules** ellipsoid, 12–17 mm.

Flowering Mar–May. Understory of moist, hardwood hammocks; 0–20 m; Fla.; Mexico; West Indies; Central America; n South America.

Phenology is based on observations in the Greater Antilles.

Plants of *Pelexia adnata* formerly found in a tropical hardwood hammock in Florida have not been seen recently and are probably extirpated (Edwin Bridges, pers. comm.).

17. SACOILA Rafinesque, Fl. Tellur. 4: 86. 1838 • [Latin *saccus*, sac, and *labium*, lip, referring to sac-shaped mentum]

Paul Martin Brown

Paul M. Catling

Herbs, terrestrial. **Roots** fascicled, fleshy. **Stems** with abbreviated rhizome. **Leaves** several, in basal rosette; blade oblong-lanceolate to oblong-elliptic. **Inflorescences** densely flowered racemes. **Flowers** resupinate, with prominent, free-tipped mentum formed by backwardly extended column foot, lip, and lateral sepals; anther abaxial; pollinarium with linear viscidium; pollinia 2 pairs, pale yellow, soft, mealy; stigma nearly perpendicular to viscidium and rostellum; rostellum linear-needlelike, sharp-pointed, rigid. **Fruits** capsules, erect, ovoid.

Species ca. 10 (2 in the flora): tropical and subtropical regions, mostly South America.

1. Rachis pubescent with bubble-shaped, glandular-capitate, and transitional hairs . . . . . . 1. *Sacoila lanceolata*
1. Rachis scurfy, with white dots . . . . 2. *Sacoila squamulosa*

1. Sacoila lanceolata (Aublet) Garay, Bot. Mus. Leafl. 28: 352. 1982 [F]

Limodorum lanceolatum Aublet, Hist. Pl. Guiane 2: 821. 1775; *Satyrium lanceolata* (Aublet) León; *Spiranthes lanceolata* (Aublet) León

Plants 23–60 cm. **Stems:** sheaths 5–9, cauline, bladeless. **Leaves** 2–7; petiole not distinct; blade elliptic, 5–25(–35) × 2–8 cm. **Inflorescences** 3–18 cm; rachis pubescent with bubble-shaped, glandular-capitate, and transitional hairs; floral bracts lanceolate. **Flowers** usually orange-red, occasionally pale green or golden bronze; dorsal sepal free, lanceolate, ascending at apex; lateral sepals basally connate, lanceolate, decurrent along lateral ribs of ovary; petals lanceolate, falcate, slightly decurrent basally; lip lanceolate, base sessile, grooved, apex acute to acuminate, with linear, nearly marginal, pubescent calli near base; anther cap 6–7 mm. **Capsules** ascending, ovoid, expanded portion 9–16 × 6–10 mm. **Seeds** 0.5–1.4 mm.

Varieties 2 (2 in the flora): Fla.; Mexico; West Indies; Central America; n South America.

In Florida the two varieties differ in their breeding systems, *Sacoila lanceolata* var. *lanceolata* producing seed by adventitious embryony and var. *paludicola* by autopollination (P. M. Catling 1987). Plants of this species from Florida, referred to *Spiranthes lanceolata* var. *luteoalba* (Reichenbach) Luer (C. A. Luer 1972), represent only a very restricted clone with more ascending flowers whose characters intergrade with those of plants from other parts of the range. Because all green-flowered plants addressed by Luer do not consistently possess the distinctive features he noted and do not conform to the description of var. *luteoalba* (Reichenbach f.) Luer, they are best recognized at the rank of forma as forma *albidaviridis* Catling & Sheviak (P. M. Catling and C. J. Sheviak 1993).

Sacoila lanceolata is widespread in tropical and subtropical America. Synonyms for *Sacoila lanceolata* in the restricted sense are given by L. A. Garay (1980[1982]) and in the very broad sense by C. A. Luer (1972). Two color forms associated with var. *lanceolata* have been named: forma *albidaviridis* Catling & Sheviak, with green and white flowers, and forma *folsomii* P. M. Brown with golden bronze flowers.

SELECTED REFERENCES Brown, P. M. 1999. A striking new color form of *Sacoila lanceolata*. N. Amer. Native Orchid J. 5: 169–173. Catling, P. M. 1987. Notes on the breeding systems of *Sacoila lanceolata* (Aublet) Garay (Orchidacae). Ann. Missouri Bot. Gard. 74: 58–68. Luer, C. A. 1971. A variety of *Spiranthes lanceolata* in Florida. Florida Orchidist 14: 19.

1. Dorsal sepal 19–23 mm; leaves usually absent at flowering . . . . 1a. *Sacoila lanceolata* var. *lanceolata*
1. Dorsal sepal 9–11 mm; leaves present at flowering . . . . 1b. *Sacoila lanceolata* var. *paludicola*

1a. Sacoila lanceolata (Aublet) Garay var. **lanceolata** • Leafless beaked orchid [F]

Neottia aphylla Hooker; *Satyrium orchioides* Swartz; *Spiranthes jaliscana* S. Watson; *S. orchioides* (Swartz) A. Richard

Leaves usually absent at flowering time. **Inflorescences** 4–18 cm; floral bracts green to reddish brown, lanceolate, 15–20(–25) mm. **Flowers** 9–50, coral to orange-red; dorsal sepal 19–23 × 7–9 mm; lateral sepals 23–31 × 3–8 mm, connate for 5–7 mm over basal portion of mentum, free portion of mentum 4–6 mm; petals 21–26 × 6–8 mm; lip pale red or pinkish white to white, 23–31 × 8–11 mm; column 9–10 mm from attachment of dorsal sepal to tip, foot extending back 11–12 mm; pollinaria 8–9 mm; viscidia 4–5 mm; pedicellate ovary 10–20 mm; rostellum 4–5 mm. **Capsules** 14 × 8 mm. **Seeds** largely polyembryonic. $2n = 46$.

Flowering mid Mar–early Jul. Open, dry to moist habitats including poorly drained pinelands, roadside ditches, pastures, open woods in sandy, ± acid substrates; 0–30 m; Fla.; Mexico; West Indies; Central America; n South America.

1b. Sacoila lanceolata (Aublet) Garay var. **paludicola** (Luer) Sauleda, Wunderlin & B. F. Hansen, Phytologia 56: 308. 1984 • Swamp beaked orchid

Spiranthes lanceolata (Aublet) León var. *paludicola* Luer, Florida Orchidist 14: 19. 1971

Leaves present at flowering time. **Inflorescences** 3–15 cm; floral bracts red, lanceolate, 12–15 mm. **Flowers** 9–29, deep red; dorsal sepal 9–11 × 3.5–4.5 mm; lateral sepals 12–14 × 33–42 mm, connate 1–2 mm over basal portion of mentum, distinct portion of mentum 0.8–1.2 mm; petals 9–10.5 × 2.2–2.8 mm; lip red to pale red, 11–13 × 3.8–4.5 mm; column 6–7.5 mm from attachment of dorsal sepal to tip, foot extending back 3–4 mm; pollinaria 3–3.5 mm; viscidium 2.9–3.1 mm; pedicellate ovary 12–15 mm; rostellum 2.3–2.8 mm. **Capsules** 12 × 6 mm. **Seeds** monoembryonic.

Flowering late Jan–early Apr. Moist to wet hardwood forests on sandy or organic substrates; 0–10 m; Fla.; West Indies.

Sacoila lanceolata var. *paludicola* is very uncommon. It is local in Collier County and introduced in Miami-Dade and Broward counties, Florida.

2. Sacoila squamulosa (Kunth) Garay, Bot. Mus. Leafl. 28: 352. 1982 • Hoary beaked orchid

Neottia squamulosa Kunth in A. von Humboldt et al., Nov. Gen. Sp. 1: 266, plate 71. 1815; *Sacoila lanceolata* var. *squamulosa* (Kunth) Szlachetko; *Spiranthes squamulosa* (Kunth) León; *Stenorrhynchos squamulosum* (Kunth) Sprengel

Herbs, terrestrial or lithophytic, 30–60(–75) cm. **Leaves** absent at flowering time, 2–8, short-petiolate; blade elliptic to broadly lanceolate, 30 × 5 cm, apex acute to obtuse. **Inflorescences** 8–18 cm; rachis scurfy with white dots; floral bracts red, lanceolate, 15–25 mm, exceeding ovary, scurfy. **Flowers** brick red, 15–50; dorsal sepal 15–18 × 5 mm; lateral sepals 24 × 3.5 mm, connate for 5–7 mm over basal portion of mentum, distinct portion of mentum 15–18 mm, about 2/3 length of ovary; petals 18–24 × 3–6 mm; lip brick red to pale red, 20–24 × 6 mm; column 11 mm from attachment of dorsal sepal to tip, foot extending back 11–12 mm; pollinaria 4–5 mm, viscidia 3–3.5 mm; pedicellate ovary 10–13 mm; rostellum 4–5 mm. **Capsules** 10–13 × 7 cm, scurfy. **Seeds** largely monoembryonic.

Flowering Apr–May. Open or wooded, dry–moist including poorly drained pinelands on calcareous substrates; 0–30 m; Fla.; Mexico; West Indies; Central America; n South America.

Sacoila squamulosa is widespread in tropical and subtropical America. Although often overlooked in Florida, it was included in the range of *Stenorrhynchos squamulosum* (M. Nir 2000), and details were given for its inclusion in the Florida flora (P. M. Brown 2000b). That resulted in determining that a large stand of nonconforming *Sacoila* in Marion County was indeed this species. It is a tall and handsome plant that, upon close examination, has the unmistakable scurfy or hoary dots instead of the short pubescence of *Sacoila lanceolata*. This is one of the most recent additions to the orchid flora of Florida and the United States. Plants have been either overlooked or not well distinguished from the more common *S. lanceolata* var. *lanceolata*. Knowledge of the range in Florida is incomplete, pending further study.

SELECTED REFERENCE Brown, P. M. 2000b. Recent taxonomic and distributional notes from Florida 8: *Sacoila squamulosa*, verified for the United States. N. Amer. Native Orchid J. 5: 333–334.

18. SCHIEDEELLA Schlechter, Beih. Bot. Centralbl. 37: 379. 1920 • [For Christian Julius Wilhelm Schiede, 1798–1836, a German naturalist and plant collector in Mexico]

Paul Martin Brown

Herbs, perennial, terrestrial. **Roots** tuberous, fleshy. **Stems** erect, with slender sheaths. **Leaves** senescing before or at anthesis, petiolate; blade broadly elliptic. **Inflorescences** spikes, few–many-flowered, in spiral or sloping downward from center in all directions. **Flowers** resupinate; sepals similar, connivent; dorsal sepal adnate to back of column, concave; lateral sepals adnate to column foot; petals linear; lip with distinct claw, central cinnabar-red disc, and 3 green stripes near apex; column slender, widened toward apex, with decurrent foot at base; stigmas 2, confluent; rostellum linear-triangular, apex acuminate; anther ovate-cordate; pollinia clavate, viscidium oblong, short; ovary cyclindric. **Fruits** capsules.

Species 9 (1 in the flora): montane tropical regions, Western Hemisphere.

1. Schiedeella arizonica P. M. Brown, N. Amer. Native Orchid J. 6: 3, figs. 1, 2. 2000 E F

Herbs slender, to 20 cm. **Roots** thickened tuber, to 3 cm. **Leaves** not present at anthesis, 3–5, in basal rosette, petoliate; blade deep green, oval, to 3 × 6 cm. **Infloresecenses:** stems greenish yellow to rosy, slender; stem bracts several, prominent, 10–15 mm. **Flowers** to 12, in slender, loose, 1-ranked spiral, whitish; floral bracts translucent white, lanceolate-acuminate, 6–10 mm; dorsal sepal oblong-obtuse, 5–7 × 1.5–2 mm; lateral sepals oblong-oblique, 5–7 × 2 mm, apex obtuse; petals spatulate, tapering to narrow base, 4.5–5 × 1.5 mm; lip oblong, constricted above middle, 5.5–8 × 3.5 mm, base with poorly defined tuberosities clasping green column; margins crenulate-lacerate (outer portion strongly recurved in living material), apex with 3 green stripes, expanded, finely pubescent; disc in center of lip, rich cinnabar-red, especially if viewed from beneath, elliptic, thickened; ovary green, 3–5 mm, pubescent.

Flowering May–early Jul. Dry coniferous forest, hillsides, creek canyons; 1500–4600 m; Ariz., N.Mex., Tex.

Plants of *Schiedeella arizonica* from the southwestern United States were originally identified as *Spiranthes parasitica* A. Richard & Galeotti by C. A. Luer (1975), and as *Schiedeella parasitica* (A. Richard & Galeotti) Schlechter by J. Liggio and A. Liggio (1999). Meanwhile P. M. Brown (1996) treated them as *S. fauci-sanguinea* (Dod) Burns-Balogh. Neither of those species actually occurs in the United States, and Brown later described the taxon as *Schiedeella arizonica*.

19. SPIRANTHES Richard, De Orchid. Eur., 20, 28, 36. 1817, name conserved • [Greek *speira*, coil, and *anthos*, flower, in allusion to the spirally arranged inflorescence]

Charles J. Sheviak

Paul Martin Brown

Herbs, perennial, terrestrial. **Roots** fleshy, fasciculate, slender to tuberous. **Stems** with foliaceous sheaths. **Leaves** basal or occasionally cauline. **Inflorescences** terminal spikes, flowers inserted in lax to dense, ± evident spiral, pubescent to glabrous, often glandular. **Flowers** resupinate, often somewhat apically recurved, white, cream, or yellow (pink in *S. sinensis*); perianth parts distinct or sepals basally connate, connivent; lip lanceolate to ovate or pandurate, rarely linear, fleshy to somewhat membranaceous, base with pair of calli, apex usually ± crisped; mentum or spur absent; column short, cylindric; anther cordate, apex acute or obtuse; pollinia clavate with slender viscidium; ovary sessile, cylindric.

Species 45 (23 in the flora): North America, South America, Eurasia, Australia.

Two extra-limital species of *Spiranthes* have been reported from the flora. The Central American *S. graminea* Lindley was reported from southern Arizona (C. A. Luer 1975), based on plants subsequently described as *S. delitescens* Sheviak. The Asiatic *S. sinensis* Persoon was reported from Florida (J. Beckner 1996) based on plants of *S. vernalis* with unusual orange markings on the lips.

SELECTED REFERENCES Catling, P. M. 1980b. Systematics of *Spiranthes* L. C. Richard in Northeastern North America. Ph.D. thesis. University of Toronto. Catling, P. M. 1981. Taxonomy of autumn-flowering *Spiranthes* species of southern Nova Scotia. Canad. J. Bot. 59: 1253–1270. Catling, P. M. 1982. Breeding systems of northeastern North American *Spiranthes* (Orchidaceae). Canad. J. Bot. 60: 3017–3039. Catling, P. M. and J. R. Brown. 1983. Morphometrics and ecological isolation in sympatric *Spiranthes* (Orchidaceae) in southwestern Ontario. Canad. J. Bot. 61: 2747–2759.

1. Lip pandurate, subapical constriction deep, apex broadly dilated, glabrous adaxially; veins typically 3, lateral veins with branches wide-spreading toward margins; rachis glabrous or very sparsely pubescent . . . . 19. *Spiranthes romanzoffiana*
1. Lip linear to lanceolate, ovate, oblong, or elliptic, at most only slightly constricted or scarcely to not at all differentiated from petals, glabrous to pubescent or papillate adaxially; veins various; rachis glabrous or densely pubescent.
   2. Viscidium oval (2–2.5 times as long as wide); lip bright yellow or orange-yellow with greenish veins centrally; sepals and petals white; sepals basally connate . . . . 18. *Spiranthes lucida*
   2. Viscidium lanceolate to linear (3–12 times as long as wide), or absent; lip white or if yellowish or orange, sepals and petals creamy or ivory to yellowish, not pure white, or lip without greenish veins, or perianth green, or lip scarcely to not at all differentiated from petals; sepals basally connate to distinct throughout.
      3. Rachis densely pubescent with all or many trichomes tapered, either pointed and not glandular or obscurely glandular; terminal glands equaling or narrower than stipes.
         4. Rachis with obscurely glandular trichomes, glands equaling or narrower than stipes; veins of lip divergent toward margin; southernmost Arizona . . . 22. *Spiranthes delitescens*
         4. Rachis with all or many trichomes pointed, wholly not glandular; veins of lip straight, branches parallel; widespread, c, e flora . . . . 1. *Spiranthes vernalis*
      3. Rachis glabrous or densely pubescent with conspicuously glandular trichomes; terminal glands markedly broader than stipes, non-glandular trichomes of uniform thickness, blunt or truncate, not tapered or pointed.
         5. Lip with conspicuous, terminally widened, green (rarely cream-colored) diverging veins extending to near tip; lip glabrous on adaxial distal surface; leaves linear to linear-lanceolate . . . . 17. *Spiranthes praecox*
         5. Lip without conspicuously widened veins, veins sometimes greenish, either straight or divergent, not also widened; lip either pubescent or glabrous on adaxial distal surface; leaves linear to ovate.
            6. Root solitary (rarely divided near tip), vertical, turbinate (4–12 mm wide, ratio of length/width about 5); rachis entirely glabrous; flowers gaping from near middle, tubular portion less than 3 mm; lip pure white . . . . 9. *Spiranthes tuberosa*
            6. Roots few to numerous, horizontal to strongly descending, slender to tuberously thickened, elongate (mostly 2–8 mm wide, rarely to 15 mm, ratio of length/width more than 5); rachis conspicuously or minutely and sparsely pubescent or glabrous; flowers gaping only from 2/3–3/4 their lengths from base, tubular portion more than 3 mm; lip white, creamy, yellow, or green in center, or often wholly green if undifferentiated from petals (i.e., flowers peloric).
               7. Petals spatulate to obovate, suborbiculate, or ovate-elliptic, conspicuously shorter than sepals.
                  8. Lip with margins fleshy, ± inflexed at apex, apex acute; petals creamy white to greenish white; Nova Scotia . . . . 15. *Spiranthes casei* (in part)
                  8. Lip truncate to broadly rounded at apex, apical margin thin, delicately crisped; petals whitish to yellow-green with longitudinal cen-

tral green stripe; Texas . . . . 16. *Spiranthes parksii*

7. Petals slender, linear, linear-lanceolate, linear-oblong, slenderly elliptic, or linear-oblanceolate, slightly shorter to longer than sepals.
   9. Sepals basally connate for more than 0.4 mm; rachis glabrous or sparsely (rarely densely) pubescent.
      10. Lip broadest near middle, narrowed to base from between center and basal 1/3; perianth parts with basal 1/4 (or more) thick, fleshy, bright green, the remainder yellowish; lip variably orange centrally; sepals 4–6 mm; rachis glabrous, often farinose . . . . 21. *Spiranthes infernalis*
      10. Lip broadest near base, narrowed from below basal 1/4; perianth parts uniformly white to yellowish, membranaceous (except sometimes lip centrally fleshy, more deeply colored); sepals 4.6–15 mm; rachis glabrous or sparsely to densely pubescent, not farinose.
         11. Rachis glabrous or sparsely pubescent, longest trichomes 0.18 mm, glands often sessile; lip lanceolate to lance-ovate, distal surface adaxially with dense cushion of peg-shaped trichomes; flowers creamy to markedly yellowish, slenderly tubular, apex of lip recurved less than to scarcely more than reflexed apices of sepals and petals . . . . 20. *Spiranthes porrifolia*
         11. Rachis sparsely (rarely densely) pubescent, longest trichomes 0.2–0.4 mm, glands obviously stalked; lip ovate to ovate-oblong, distal surface adaxially without dense cushion of peg-shaped trichomes; flowers white to ivory, strongly gaping with lip prominently diverging from sepals and petals . . . . 23. *Spiranthes diluvialis* (in part)
   9. Sepals not basally connate or connate for less than 0.3 mm; rachis sparsely to densely pubescent.
      [12. Shifted to left margin.—Ed.]

12. Inflorescences with 5 or more flowers per cycle of spiral; spiral usually open and obvious.
    13. Lateral sepals widely diverging from base, 8–10 mm; lip dilated at base, oblong toward apex, yellow centrally; flowers gaping from near middle; leaves fugaceous or persisting through anthesis; inflorescence secund to twisted usually no more than 180°; fall–winter blooming . . . . 3. *Spiranthes longilabris*
    13. Lateral sepals somewhat spreading to appressed, not widely diverging, 3.8–10 mm; lip ovate to oblong-quadrangular, basal dilation not distinct from oblong apex, white or creamy centrally; flowers gaping from near middle to merely apically; leaves persistent to fugaceous; spiral of inflorescence usually with few–several cycles of 360°, sometimes secund; spring–fall blooming.
        14. Lip with distal margin laciniate-dentate; leaves usually linear with length/width ratio greater than 30 (rarely broader on stout, leafy plants nearly to more than 1 m), present at flowering . . . . 2. *Spiranthes laciniata*
        14. Lip with distal margin undulate to crisped, sometimes ragged, not laciniate-dentate; leaves lanceolate to ovate to obovate (linear in *S. torta*), with length/width ratio less than 30, often fugaceous and absent at flowering.
            15. Flowers comparatively large, stout, perianth 5–12 mm, white to yellowish, lip often darker centrally, not green nor greenish yellow, ovary often equaling perianth in freshly opened flowers; leaves ascending, relatively slender, not ovate nor obovate, present at flowering; rachis conspicuously and densely glandularly capitate-pubescent.
                16. Lip distally truncate, 5–7.5 mm; flowers yellowish, cream, or ivory, arcuate-nodding for most or all length of perianth . . . . 15. *Spiranthes casei* (in part)
                16. Lip distally acute to rounded, not truncate, usually more than 7.5 mm (in shorter flowers ovary often much shorter than perianth); flowers white to yellowish, arcuate-nodding only at base of perianth . . . . . . 14. *Spiranthes cernua* (in part)

15. Flowers comparatively small, slender, perianth usually less than 5.5 mm (if longer, lip green centrally and leaves ascending, slender, often fugaceous), ovary less than 1/2 length of perianth in freshly opened flowers; leaves strictly basal, spreading and ovate to obovate or oblanceolate, or ascending and slender, and often on base of stem in *S. torta*, often absent at flowering; rachis glabrous or sparsely to densely, minutely pubescent.
    17. Rachis densely pubescent; flowers cream to pale yellow, lip yellow centrally . . . . 4. *Spiranthes brevilabris*
    17. Rachis glabrous or sparsely and minutely pubescent; flowers yellowish to white, lip yellowish or green centrally.
        18. Flowers creamy yellow to pale greenish yellow, lip yellow centrally . . . . 5. *Spiranthes floridana*
        18. Flowers white, lip green centrally.
            19. Leaves ascending, linear, often absent at flowering; lateral sepals arching downward; s Florida . . . . 8. *Spiranthes torta*
            19. Leaves spreading, ovate to obovate, oblanceolate, or elliptic, present or absent at flowering; lateral sepals slightly spreading to appressed (not arching downward); various locales.
                20. Leaves oblanceolate, withering at anthesis; sepals spatulate, green at base; spring-blooming . . . . 7. *Spiranthes eatonii*
                20. Leaves obovate, spreading, present or absent at anthesis; lateral sepals acuminate, white throughout; summer/fall-blooming . . . . 6. *Spiranthes lacera*

12. Inflorescences with 3 or 4 flowers per cycle of spiral, spiral usually tight and obscure, then with 3 or 4 secondary vertical ranks evident.
    21. Petals 6 mm; proximal portion of stem with recurved-spreading leaves . . . . 10. *Spiranthes ovalis*
    21. Petals 7.5 mm or longer; leaves wholly basal, or proximal portion of stem with ascending to recurved-spreading blades, or leaves absent at flowering.
        22. Lip only slightly or not at all differentiated from petals (flowers peloric); often buds failing to open, setting seed agamospermically; column normal or abnormally formed or aborted; leaves usually absent at flowering . . . . 14. *Spiranthes cernua* (in part)
        22. Lip clearly differentiated from petals; buds opening into normal flowers; column normal; leaves present or absent at flowering.
            23. Lip with variably evident median constriction, apical 1/3 of lip usually about 1/2 as wide as basal 1/3; veins of lip often with laterally diverging branches; flowers ascending . . . . 23. *Spiranthes diluvialis* (in part)
            23. Lip without conspicuous median constriction, apical 1/3 of lip 2/3 or more as wide as basal 1/3; veins of lip with branches parallel.
                24. Basal calli relatively short and conic, as wide as high, mostly to 1 mm; lateral sepals distinct, spreading, often over top of flower; leaves absent at flowering (rarely persisting in Northern and High Plains, and New Mexico) . . . . 11. *Spiranthes magnicamporum*
                24. Basal calli 1–2 mm; lateral sepals ± appressed, very rarely spreading over top of flower; leaves present at flowering (commonly absent in prairie populations).
                    25. Distal margin of lateral sepals obviously separated from adjacent margin of dorsal sepal, separation abrupt from base, commonly by about 1 mm; lip strongly curving from claw (resulting angle 20–60°), base cuneate; perianth creamy, yellowish, or greenish white . . . . 12. *Spiranthes ochroleuca*
                    25. Distal margin of lateral sepals touching, or nearly touching, adjacent margin of dorsal sepal, or only gradually separated with distance from base; lip not strongly curving from claw (angle less than 30°), base cordate to truncate; perianth white to creamy or ivory, center of lip ivory to pale yellowish or rarely greenish.

26. Plants to 50 cm; leaves comparatively slender, ascending to spreading (flaccidly so because of membranaceous blades with thickened midrib); petioles of basal leaves less than 6 mm wide; leaves wholly basal or lower sheaths with ascending-spreading blades; perianth usually 8–11 mm; lip membranaceous to fleshy, less than 7 mm wide . . . . 14. *Spiranthes cernua* (in part)
26. Plants to 100 cm or more; leaves broad, ascending to spreading (rigidly so because of aerenchymatous thickening of blade); petioles of basal leaves 7 mm or more wide; blades spreading-recurved on proximal cauline sheaths, frequently also on distal, leaves then extending to inflorescence; perianth (4–)10–18 mm (shorter in young or depauperate plants); lip fleshy, (4–)7–9.5 mm . . . . 13. *Spiranthes odorata*

1. Spiranthes vernalis Engelmann & A. Gray, Boston J. Nat. Hist. 5: 236. 1845 [E] [F]

Ibidium vernale (Engelmann & A. Gray) House

Plants 20–65 cm. **Roots** numerous, spreading, mostly to 1 cm diam., stout. **Leaves** persisting through anthesis, to 4–5, basal, reduced to sheathing bracts upward on stem, spreading, linear-lanceolate, keeled, 5–25 × 1 cm. **Inflorescences:** spikes loosely to tightly spiraled, 3–7 or more flowers per cycle of spiral, sometimes nearly secund; rachis densely pubescent, trichomes articulate, pointed, capitate glands absent. **Flowers** nodding to somewhat ascending, white to cream, mostly gaping; sepals distinct to base, lanceolate, 6–10 × 2–3 mm; lateral sepals spreading; petals oblong, 6–9 × 2 mm, apex obtuse; lip creamy yellow centrally or some individuals with 2 brown-orange spots, ovate, 5–8 × 4 mm; veins of lip straight, branches parallel; basal calli conic, to 1 mm; viscidium linear-lanceolate; ovary mostly 8 mm. **Seeds** monoembryonic. $2n = 30$.

Flowering Jan (Fla.)–Oct (north). Dry to moist meadows, dune hollows, prairies, old fields, roadsides, cemeteries, lawns; 0–1000 m; Ala., Ark., Conn., Del., D.C., Fla., Ga., Ill., Ind., Iowa, Kans., Ky., La., Md., Mass., Miss., Mo., Nebr., N.H., N.J., N.Y., Ohio, Okla., Pa., R.I., S.C., Tenn., Tex., Va., W.Va.

The habit of *Spiranthes vernalis* is variable: inflorescences range from secund to loosely and even densely coiled, and flower shapes vary as well. The most consistent diagnostic character is the presence in inflorescences of copious articulate, pointed trichomes that readily distinguish *S. vernalis* from similar species.

2. Spiranthes laciniata (Small) Ames, Orchidaceae 1: 120. 1905 [E]

Gyrostachys laciniata Small, Fl. S.E. U.S., 318, 1329. 1903; *Ibidium laciniatum* (Small) House

Plants 20–95(–100+) cm. **Roots** numerous, spreading, mostly to 1 cm diam., slender. **Leaves** persisting through anthesis, to 5, basal, reduced to sheathing bracts upward on stem, spreading, linear-lanceolate, keeled, 5–40 × 1.7 cm, rigid. **Inflorescences:** spikes loosely to tightly spiraled, 5–9 flowers per cycle of spiral, sometimes nearly secund; rachis moderately to densely pubescent, some trichomes capitate, glands obviously stalked. **Flowers** mostly ascending, white to cream, slenderly tubular, gaping toward apex; sepals distinct to base, lanceolate, tapering in distal 1/3, 6–10 × 2–3 mm; lateral sepals spreading; petals linear, falcate, 7–9 × 2 mm, apex apiculate or sometimes crenulate; lip yellow centrally, ovate, 6–9 × 4 mm, apical margin laciniate-dentate, crisped; basal calli incurved, long-tapered, to 1 mm; viscidium linear-lanceolate; ovary mostly 5 mm. **Seeds** monoembryonic.

Flowering May (Fla.)–Sep (north). Primarily on coastal plain in swamps, marshes, meadows, dry to damp roadsides, ditches, fields, cemeteries, lawns; occasionally in standing water; 0–50 m; Ala., Del., Fla., Ga., La., Md., Miss., N.J., N.C., S.C., Tex., Va.

Spiranthes laciniata is easily distinguished from *S. vernalis*, which it superficially resembles, by its capitate trichomes. It typically flowers later than *S. vernalis* where the two are sympatric.

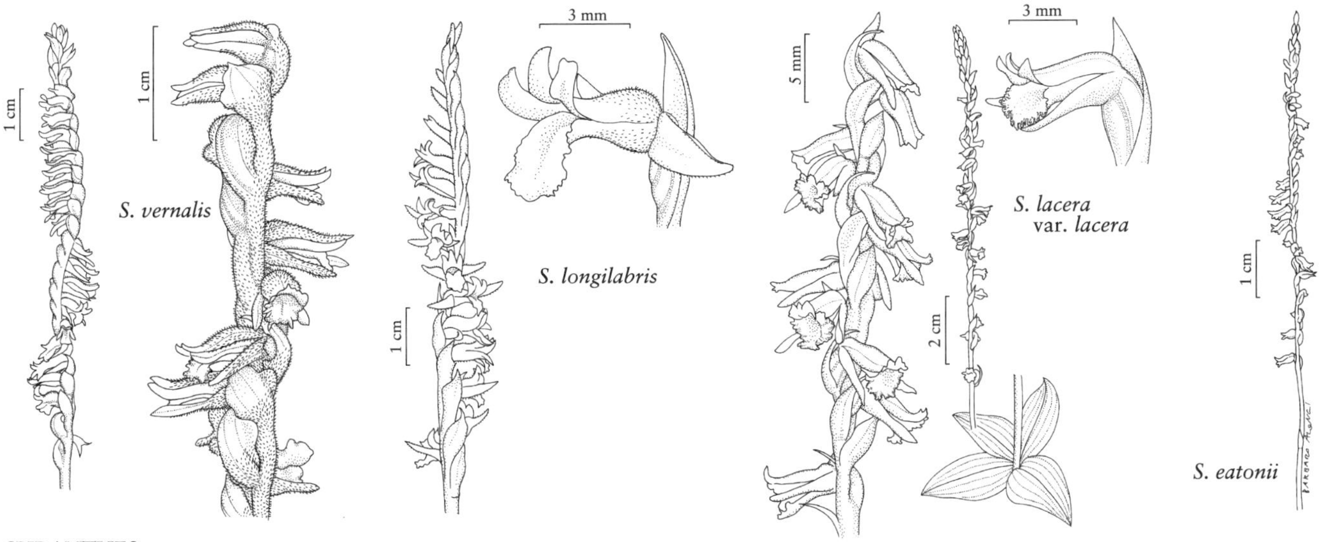

SPIRANTHES

3. **Spiranthes longilabris** Lindley, Gen. Sp. Orchid. Pl., 467. 1840 [E] [F]

Ibidium longilabre (Lindley) House

Plants 15–50 cm. **Roots** numerous, mostly to 1 cm diam., stout. **Leaves** fugacious or persisting through anthesis, to 3–5, cauline, reduced to sheathing bracts upward on stem, spreading, linear-lanceolate, keeled, 5–15 × 0.5 cm, rigid. **Inflorescences:** spikes loosely to tightly spiraled, usually no more than 180°, 5–9 flowers per cycle of spiral, sometimes secund; rachis sparsely pubescent, some trichomes capitate, glands obviously stalked. **Flowers** mostly horizontal, strongly gaping, white to cream; sepals distinct to base, lanceolate, tapering in distal 1/3, (8–)9–10 × 2–3 mm; lateral sepals widely spreading; petals linear, slightly falcate, 7–9 × 2–2.5 mm; lip yellow centrally, ovate in general outline, base rounded-dilated, apical portion descending, oblong, 8–11 × 4 mm, apex recurved with margin crisped; basal calli erect, tapered, to 1 mm; viscidium linear-lanceolate; ovary mostly 7 mm. **Seeds** monoembryonic.

Flowering Oct–Dec. Primarily on coastal plain in dry to moist meadows, pine flatwoods and savannas, fields, roadsides; 0–50 m; Ala., Fla., Ga., La., Miss., N.C., S.C., Tex., Va.

Hybrids of *Spiranthes longilabris* and *S. odorata* are known as *Spiranthes* ×*folsomii* P. M. Brown.

4. **Spiranthes brevilabris** Lindley, Gen. Sp. Orchid. Pl., 471. 1840 [E]

Spiranthes gracilis (Bigelow) Beck var. *brevilabris* (Lindley) Correll

Plants 7–40 cm. **Roots** few, spreading to descending, mostly to 0.5 cm diam., slender. **Leaves** usually persisting through anthesis, 3–5, basal, spreading, oval-oblanceolate, 2–6 × 1–2 cm. **Inflorescences:** spikes secund to loosely spiraled, 8–10 flowers per cycle of spiral; rachis densely pubescent, some trichomes capitate to clavate, glands obviously stalked. **Flowers** cream to pale yellow; sepals distinct to base, 5 × 1 mm; lateral sepals appressed; petals linear to lance-oblong, 5 × 1 mm, apex acute to obtuse; lip yellow centrally, ovate to oblong, 5 × 2.5 mm, apex dilated, yellow, pubescent, apical margin crisped, finely lacerate; veins several, branches very short; basal calli long-pointed, mostly to 1 mm; viscidium linear-lanceolate; ovary mostly 3 mm. **Seeds** monoembryonic.

Flowering Feb–Apr. Dry to moist roadsides and fields; 0–100 m; Ala., Fla., Ga., La., Miss., Tex.

Spiranthes brevilabris and *S. floridana* are often and easily confused, although the degree of pubescence is an excellent diagnostic tool in the field.

This species has dramatically declined, with only a single extant population known in 1998–2000.

5. Spiranthes floridana (Wherry) Cory, Rhodora 38: 405. 1936 [E]

Ibidium floridanum Wherry, J. Wash. Acad. Sci. 21: 49, fig. 1. 1931; *Spiranthes brevilabris* Lindley var. *floridana* (Wherry) Luer; *S. gracilis* (Bigelow) Beck var. *floridana* (Wherry) Correll

Plants 7–40 cm. **Roots** few–numerous, spreading to descending, mostly to 0.5 cm diam., stout. **Leaves** usually persisting through anthesis, 3–5, basal, spreading, oval-oblanceolate, 2–6 × 1–2 cm. **Spikes** secund to loosely spiraled, 8–10 flowers per cycle of spiral; rachis glabrous to sparsely pubescent, trichomes commonly minute, abortive; any glands stalked, trichomes then capitate. **Flowers** creamy yellow to pale greenish yellow; sepals 5 × 1 mm, distinct to base, appressed; petals elliptic, slightly oblique, linear to lance-oblong, 5 × 1 mm, apex acute to obtuse; lip commonly yellow centrally, ovate to oblong, 5 × 2.5 mm, apex dilated, yellow and papillate-pubescent, its margin crisped, finely lacerate; veins several, branches very short; basal calli long-pointed, mostly to 1 mm; viscidium linear-lanceolate; ovary mostly 3 mm. **Seeds** monoembryonic.

Flowering Feb–Apr. Dry to moist roadsides, fields; 0–100 m; Ala., Fla., Ga., La., Miss., N.C., S.C., Tex.

Spiranthes floridana and *S. brevilabris* are often and easily confused, although the degree of pubescence is an excellent diagnostic tool in the field.

This species has become very uncommon, with only a single extant population known in 1998–2000.

6. Spiranthes lacera (Rafinesque) Rafinesque, Herb. Raf., 44. 1833 [E] [F] [W1]

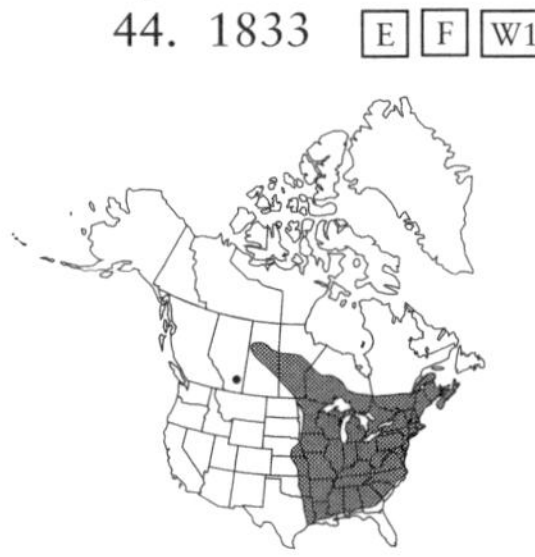

Neottia lacera Rafinesque, Amer. Monthly Mag. & Crit. Rev. 2: 206. 1818

Plants 15–65 cm. **Roots** few–several, spreading to descending, mostly to 0.75 cm diam., slender to stout. **Leaves,** basal, spreading, obovate, 2–5 × 1–2 cm. **Spikes** secund to tightly spiraled, 8–10 flowers per cycle of spiral; rachis glabrous to sparsely pubescent, trichomes capitate to clavate, glands obviously stalked. **Flowers** white; sepals somewhat spreading, distinct to base, elliptic, 5 × 1 mm; lateral sepals white throughout, apex acuminate; petals lanceolate, slightly oblique, 5 × 1 mm; lip white with green or yellowish green center, ovate to oblong, 5 × 2.5 mm, apex dilated, its margin crisped-crenate, pubescent; veins obscure; basal calli erect, short-pointed, mostly to 1 mm; viscidium linear-lanceolate; ovary mostly 3 mm. **Seeds** monoembryonic.

Varieties 2 (2 in the flora): e North America.

1. Flowers laxly arranged on spike (ratio of spike length/flower number equal to or greater than 2.3); inflorescence capitate-pubescent; leaves usually persisting through anthesis . . . . 6a. *Spiranthes lacera* var. *lacera*
1. Flowers densely arranged on spike (ratio of spike length/flower number less than 2.3); entire plant essentially glabrous; leaves usually absent at anthesis . . . . 6b. *Spiranthes lacera* var. *gracilis*

6a. Spiranthes lacera (Rafinesque) Rafinesque var. **lacera** [E] [F]

Leaves usually persisting through anthesis, obovate. **Inflorescences** secund or loosely spiraled, lax, with distinct space between proximal flowers (ratio of spike length/flower number equal to or greater than 2.3), sparsely pubescent, some trichomes capitate, glands obviously stalked. **Flowers:** lip 3–6 mm, broadest at apex; central portion green to yellowish green, with narrow, crisp white apron clearly defined to obscure. $2n = 30$.

Flowering Jun–Sep. Dry to moist meadows, open woods, barrens, old fields, roadsides, 0–700 m; Alta., Man., N.B., N.S., Ont., P.E.I., Que, Sask.; Ark., Conn., Ill., Ind., Iowa, Kans., Ky., Maine, Md., Mass., Mich., Minn., Mo., N.H., N.J., N.Y., N.C., Ohio, Pa., R.I., Tenn., Vt., Va., W.Va., Wis.

Hybrids of *Spiranthes lacera* var. *lacera* with *S. romanzoffiana* are known as *Spiranthes* ×*simpsonii* Catling & Sheviak.

6b. Spiranthes lacera (Rafinesque) Rafinesque var. **gracilis** (Bigelow) Luer, Native Orchids U.S. & Canada, 112. 1975 [E]

Neottia gracilis Bigelow, Fl. Boston ed. 2, 322. 1824; *Ibidium beckii* (Lindley) House; *I. gracile* (Bigelow) House; *Spiranthes beckii* Lindley; *S. gracilis* (Bigelow) Beck

Leaves usually absent at anthesis. **Inflorescences** loosely to very tightly spiraled, sometimes secund, flowers approximate, mostly without distinct space between proximal flowers (ratio of spike length/flower number less than 2.3), glabrous. **Flowers:** lip 4–7.5 mm, rounded at apex; central portion green with clearly defined, broad crisp white apron. 2*n* = 30.

Flowering Jul–Oct(–Nov). Dry to moist meadows, prairies, barrens, open woods, dunes, old fields, roadsides, cemeteries, lawns; 0–1100 m; N.S., Ont.; Ala., Ark., Conn., Del., D.C., Ga., Ill., Ind., Iowa, Kans., Ky., La., Maine, Md., Mass., Mich., Minn., Mo., Ohio, Okla., N.H., N.J., N.Y., N.C., Pa., R.I., S.C., Tenn., Tex., Vt., Va., W.Va., Wis.

Hybrids of *Spiranthes lacera* var. *gracilis* with *S. vernalis* are known as *Spiranthes* ×*intermedia* Ames.

7. Spiranthes eatonii Ames ex P. M. Brown, N. Amer. Native Orchid J. 5: 9, figs. on pp 14, 15. 1999 E F

Plants 15–55 cm. **Roots** mostly to 1 cm diam., few, stout. **Leaves** withering at anthesis, 3–7, basal, spreading, oblanceolate, 5.5 × 0.75–1 cm. **Spikes** secund to tightly spiraled, 8–10 flowers per cycle of spiral; rachis pubescent, some trichomes capitate, glands obviously stalked. **Flowers** white; sepals green at base, spatulate, 3–4.5 × 1 mm; petals green at base, linear to lance-oblong, 3–4.5 × 1 mm, apex acute to obtuse; lip with distinct green central portion, ovate to oblong, 3–5 × 2–3 mm, narrowed to rounded apex; veins several, divergent; basal calli pointed outward, thickened, mostly to 1 mm; viscidium linear-lanceolate; ovary mostly 3 mm. **Seeds** monoembryonic.

Flowering Feb–May. Se coastal plain and Gulf Coast in dry to moist fields, pine flatwoods, wood roads, cemeteries; 0–100 m; Ala., Fla., Ga., La., Miss., N.C., S.C., Tex., Va.

Spiranthes eatonii is easily confused with *S. lacera* var. *lacera* (and in herbarium specimens with *S. floridana*, *S. brevilabris*, *S. tuberosa,* and *S. torta*), except that it flowers in the winter and spring. It is the only white-flowered, basal-leaved *Spiranthes* within its range to bloom at that time of year. The narrow, oblanceolate leaves are distinctive within this group.

8. Spiranthes torta (Thunberg) Garay & H. R. Sweet in R. A. Howard, Fl. Less. Antill. 1: 77. 1974

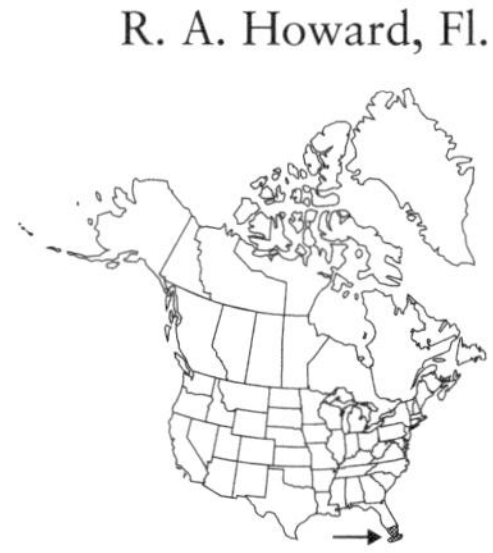

Ophrys torta Thunberg, Mus. Nat. Acad. Upsal. 9: 136. 1791; *Ibidium tortile* (Swartz) House; *Spiranthes tortilis* (Swartz) Richard

Plants 7–50 cm. **Roots** few, spreading to descending, slender to somewhat tuberous, mostly to 1 cm diam. **Leaves** fugacious, 2 or 3, basal, reduced to bracts on stem, ascending, narrowly oblanceolate, 5–20 × 0.5 cm. **Spikes** secund to loosely spiraled, 8–10 flowers per cycle of spiral; rachis sparsely pubescent, some trichomes capitate, glands obviously stalked. **Flowers** nodding, white; sepals elliptic, 5 × 1 mm; lateral sepals distinct to base, curving slightly downward, oblique; petals green at base, oblique, elliptic, 5 × 1.5 mm; lip green with white apex, ovate, 4 × 2 mm; veins parallel; basal calli very short, stout, to 0.3 mm; viscidium linear-lanceolate; ovary mostly 3 mm. **Seeds** monoembryonic.

Flowering May–Jul. Dry pinelands; 0–10 m; Fla.; West Indies (Bahamas); Bermuda; Central America.

Spiranthes torta is restricted to the rocky pinelands in southern Florida and is easily recognized by its nodding flowers and downward-pointing lateral sepals. It could only be confused with *S. tuberosa,* typically a much shorter plant with pure white flowers.

Spiranthes amesiana Schlechter is often included as a synonym of *S. torta*; recent studies indicate that it may be a distinct species. It is said to differ from *S. torta* in its denser inflorescence, spade-shaped lip with a crenulate-ciliate margin, and densely pubescent basal calli. Within the area covered by the flora, it is restricted to extreme southern Florida; it also has been collected in the Bahamas and Nicaragua.

9. Spiranthes tuberosa Rafinesque, Herb. Raf., 45. 1833 E

Spiranthes grayi Ames; *S. simplex* A. Gray; *S. tuberosa* var. *grayi* (Ames) Fernald

Plants 5–30 cm. **Roots** solitary, vertical, tuberous, turbinate, mostly to 1 cm diam. **Leaves** fugacious, 3–5, basal, spreading, oval-oblanceolate, 2–6 × 1–2 cm. **Spikes** loosely spiraled, 4–7 flowers per cycle of spiral; rachis glabrous. **Flowers** pure white, gaping from near middle, tubular portion less than 3 mm; sepals distinct to base, 5 × 1 mm; lateral sepals slightly spreading; petals linear to lance-oblong, 5 × 1 mm, apex acute to obtuse; lip 5 × 2.5 mm, ovate to oblong, apex dilated with broad crisped, finely lacerate margin; veins several, branches very short; basal calli long-pointed, mostly to 1 mm; viscidium linear-lanceolate; ovary mostly 3 mm. **Seeds** monoembryonic.

Flowering Jun–Sep. Dry to open woods, outcrops, old fields, roadsides, cemeteries; 0–400 m; Ala., Ark., Conn., Del., D.C., Fla., Ga., Ill., Ind., Kans., Ky., La., Md., Mass., Mich., Miss., Mo., N.J., N.Y., N.C., Ohio, Okla., Pa., R.I., S.C., Tenn., Tex., Va., W.Va.

The nomenclatural history of *Spiranthes tuberosa* is rather complex, and among the names applied to it are *Spiranthes beckii* Lindley and *Ibidium beckii* (Lindley) House. See D. S. Correll (1950) for a discussion.

This species is easily recognized by its pure white flowers, broad crisped lip, and fugacious leaves.

10. Spiranthes ovalis Lindley, Gen. Sp. Orchid. Pl., 466. 1840 [E]

Ibidium ovale (Lindley) House

Plants 5–40 cm. **Roots** few, slender, spreading to descending, mostly to 0.5 cm diam. **Leaves** persisting through anthesis, few–several, basal, usually on proximal 1/2 stem, recurved-spreading, oblanceolate, 3–15 × 0.5–1.5 cm. **Spikes** tightly spiraled, usually 3 flowers per cycle of spiral, rarely loosely spiraled in depauperate plants; rachis pubescent, some trichomes clavate to capitate, glands obviously stalked. **Flowers** pure white; sepals distinct to base, slightly spreading, 3.5–6.1 × 1–1.5 mm; petals lanceolate, 6 × 1 mm; lip ovate, 5 × 3 mm, apex with finely crisped margin; veins few, branches parallel; basal calli spreading, slender, mostly to 0.75 mm; ovary mostly 3–8 mm. **Seeds** monoembryonic.

Varieties 2 (2 in the flora): e North America.

1. Rostellum and viscidium present . . . . 10a. *Spiranthes ovalis* var. *ovalis*
1. Rostellum and viscidium absent . . . . 10b. *Spiranthes ovalis* var. *erostellata*

10a. Spiranthes ovalis Lindley var. **ovalis** [E]

Gyrostachys parviflora (Chapman) Small; *Spiranthes cernua* var. (Linnaeus) Richard *parviflora* Chapman; *Spiranthes parviflora* (Chapman) Ames

Plants allogamous. **Flowers:** sepals 4–6.1 mm; rostellum well developed; viscidium linear-lanceolate.

Flowering Oct–Dec. Moist, rich woodlands, thickets, wooded hillsides; 0–300 m; Ala., Ark., Fla., Ga., La., Miss., Tex.

Hybrids with *Spiranthes odorata* are known as *S. ×itchetuckneensis* P. M. Brown.

10b. Spiranthes ovalis Lindley var. **erostellata** Catling, Brittonia 35: 123, figs. 1 [right], 2, 3. 1983 [E]

Plants allogamous. **Flowers:** sepals 3.5–5 mm; rostellum and viscidium absent. $2n = 30$.

Flowering late Aug–Oct. Moist, rich woodlands, thickets, old fields, 2d-growth woodlands, wooded hillsides; 0–900 m; Ont.; Ala., Ark., Fla., Ga., Ill., Ind., Iowa, Kans., Ky., La., Md., Mich., Miss., Mo., Ohio, Okla., N.C., Pa., S.C., Tenn., Tex., Va., W.Va.

In *Spiranthes ovalis* var. *ovalis*, all plants have fully open flowers, and the ovaries swell progressively. In most plants of *S. ovalis* var. *erostellata,* the flowers are never quite fully open, and the ovaries are simultaneously swollen on each flower. Because the varieties are largely allopatric, identification can be aided by locality.

11. Spiranthes magnicamporum Sheviak, Bot. Mus. Leafl. 23: 287, plate 22. 1973 [E]

Plants 7–60 cm. **Roots** few, descending, tuberous, mostly to 0.8 cm diam. **Leaves** fugaceous or rarely persisting to anthesis, basal, ascending, linear-lanceolate to oblanceolate, to 16 × 1.5 cm. **Spikes** usually very tightly spiraled, 3–4 flowers per cycle of spiral; rachis moderately pubescent, some trichomes capitate, glands obviously stalked (longest trichomes 0.2–0.52 mm). **Flowers** abruptly nodding from base, white to ivory, gaping, lip not strongly curving from claw, not urceolate; sepals distinct to base, 5–14 mm; lateral sepals wide-spreading, commonly ascending above flower; petals linear to lance-oblong, 4.9–13 mm, apex acute to obtuse; lip commonly yellow centrally, ovate to oblong, 4.9–12 × 3.3–7 mm, margins crenulate, glabrous; veins several, branches parallel; basal calli short-conic, mostly to 1 mm; viscidia linear-lanceolate; ovary 4–10 mm. **Seeds** monoembryonic. $2n = 30$.

Flowering Aug–Nov. Dry to wet prairies and fens; 0–1900 m; Man., Ont.; Ala., Ark., Ill., Ind., Iowa, Kans., Ky., La., Mich., Minn., Miss., Mo., Nebr., N.Mex., N.Dak., Ohio, Okla., Pa., S.Dak., Tex., Va., Wis.

Leaves typically senesce some weeks before anthesis, usually before the inflorescence appears. Occasionally at the northern and western range limits of the species, however, especially in wetter habitats, they may persist into anthesis. See notes on gene flow and apomixis under 14. *Spiranthes cernua.*

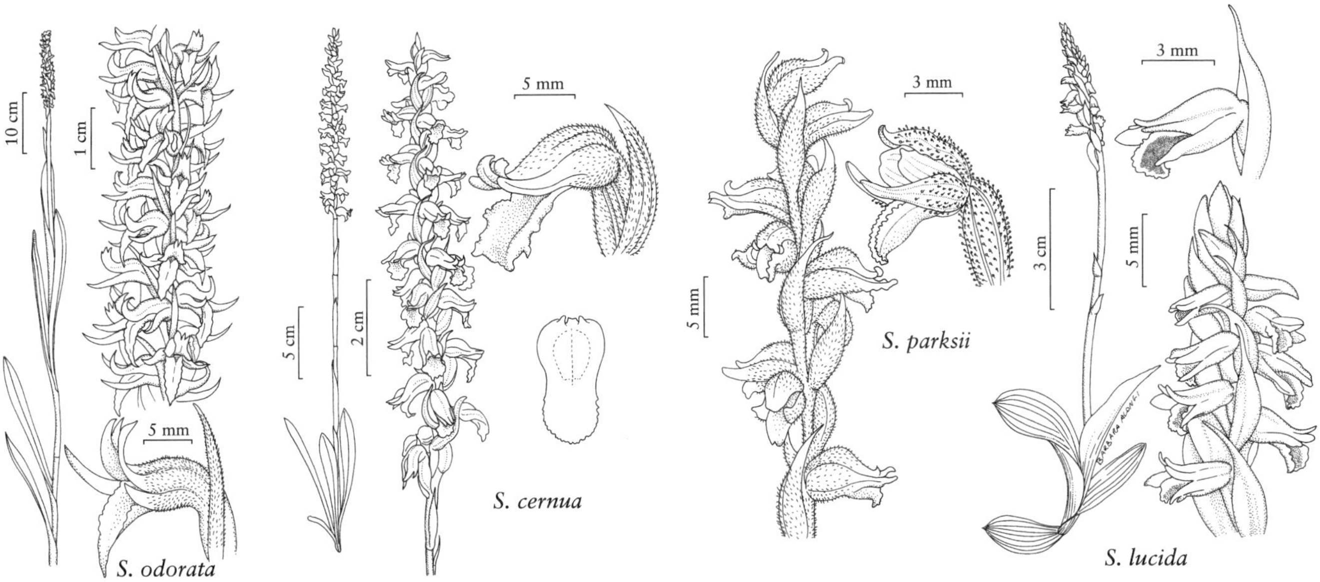

SPIRANTHES

12. Spiranthes ochroleuca (Rydberg) Rydberg, Fl. Plains N. Amer., 241. 1932 [E]

Gyrostachys ochroleuca Rydberg in N. L. Britton, Man. Fl. N. States, 300. 1901; *Spiranthes cernua* (Linnaeus) Richard var. *ochroleuca* (Rydberg) Ames; *Spiranthes* ×*steigeri* Correll

Plants 10–55 cm. **Roots** few, horizontally spreading, slender, mostly to 0.4 cm diam. **Leaves** persisting through anthesis, basal, sometimes on basal portion of stem, spreading, linear-oblanceolate to narrowly elliptic, to 21 × 2 cm. **Spikes** rather loosely to very tightly spiraled, 3–4 flowers per cycle of spiral, occasionally 5 or more in very loose spiral; rachis moderately pubescent, some trichomes capitate, glands obviously stalked. **Flowers** ochroleucous, ivory, creamy, yellowish, or greenish white, ascending, urceolate to somewhat gaping, claw of lip at pronounced (20–60°) angle to lamina, base cuneate, appearing strongly inflated; sepals distinct to base, 6–14 mm; lateral sepals appressed to petals and lip, straight, separated from dorsal sepal by mostly 0.7–1 mm; petals linear to linear-oblong, 6–13 mm, apex acute to obtuse; lip often more yellow centrally, oblong to ovate, 7–12 × 4–7 mm, margins crenulate, glabrous; veins several, branches parallel; basal calli incurved, prominent, very rarely conic, 1–2 mm; viscidia linear; ovary 3–7 mm. **Seeds** monoembryonic. $2n = 30$.

Flowering Aug–Nov. Dry to mesic open woodland, thickets, meadows, barrens, ledges, outcrops, banks and roadsides, old fields; 0–1700 m; N.B., N.S., Ont., P.E.I.; Conn., Ind., Ky., Maine, Md., Mass., Mich., N.H., N.J., N.Y., N.C., Ohio, Pa., R.I., S.C., Tenn., Vt., Va., W.Va.

While leaves of *Spiranthes ochroleuca* are usually present at anthesis, in prairie populations they are commonly absent.

Spiranthes ×*steigeri* was described as a hybrid of *S. cernua* and *S. romanzoffiana*, but additional study (P. M. Catling 1984) has shown the plants to be referable to *S. ochroleuca*.

See notes on gene flow and apomixis under 14. *Spiranthes cernua*.

Hybrids of *Spiranthes ochroleuca* with *S. casei* are known as *S.* ×*borealis* P. M. Brown.

SELECTED REFERENCES Catling, P. M. 1984. A re-evaluation of *Spiranthes* ×*steigeri*. Rhodora 86: 469–473. Sheviak, C. J. and P. M. Catling. 1980. The identity and status of *Spiranthes ochroleuca* (Rydberg) Rydberg. Rhodora 82: 525–562.

13. Spiranthes odorata (Nuttall) Lindley, Gen. Sp. Orchid. Pl., 467. 1840 [E] [F]

Neottia odorata Nuttall, J. Acad. Nat. Sci. Philadelphia 7: 98. 1834; *Spiranthes cernua* (Linnaeus) Richard var. *odorata* (Nuttall) Correll

Plants 10–100+ cm. **Roots** several, horizontally spreading, slender, mostly to 0.3 cm diam. **Leaves** persisting through anthesis, basal, gradually or occasionally abruptly reduced upward on stem, base of blade sheathing stem or narrowed to broad petiole over 7 mm wide, ascending to spreading (rigidly so because of aerenchymatous thickening of blade); blades on proximal cauline sheaths spreading-recurved, frequently also on distal (then extending to inflorescence), elliptic to oblong- or linear-oblanceolate, to 52 × 4 cm. **Spikes** tightly spiraled, 3–4 flowers per cycle of spiral; rachis moderately pubescent, some trichomes capitate, glands obviously stalked. **Flowers** nodding, white, urceolate to gaping, claw of lip at low (less than 30°) angle to lamina, base appearing only slightly inflated; sepals distinct to base, 4–18 mm; lateral sepals loosely associated with dorsal and petals or spreading, sometimes arching above flower; petals lanceolate to elliptic-lanceolate, 4–18 mm, apex acute to obtuse or emarginate; lip often yellow or green centrally, rhombic-ovate to ovate or lance-ovate, 4–16 × (4–)7–9.5 mm, fleshy, margins lacerate-crenulate, glabrous; veins several, branches parallel; basal calli incurved, prominent, sometimes conic in small flowers, 0.2–2.5 mm; viscidia linear to linear-lanceolate; ovary 4–8 mm. **Seeds** monoembryonic. $2n = 30$.

Flowering Sep–Dec. Cypress and hardwood swamps, marshes, prairies, riverbanks, ditches; 0–300 m; Ala., Ark., Fla., Ga., Ky., La., Md., Miss., N.C., Okla., S.C., Tenn., Tex., Va.

Spiranthes odorata typically occurs in seasonally inundated sites and may bloom while emerging from shallow water. The rather thick, broad leaves are aerenchymatous and give the plant a distinctive vegetative habit. The very long, wide-spreading roots produce vegetative offshoots often 30 cm from the parent shoot, giving rise to extensive clonal colonies. At its extreme southern range limit in Florida, individual plants may bloom sporadically throughout much of the year.

The northern range limit of the species along the Atlantic Coast is uncertain; see notes on gene flow and apomixis under 14. *Spiranthes cernua*.

14. Spiranthes cernua (Linnaeus) Richard, De Orchid. Eur., 37. 1817 E F

Ophrys cernua Linnaeus, Sp. Pl. 2: 946. 1753; *Ibidium cernuum* (Linnaeus) House

Plants 10–50 cm. **Roots** few, slender to tuberously thickened, horizontally spreading to descending, mostly to 1 cm diam. **Leaves** persisting through anthesis or fugaceous, basal, sometimes on proximal portion of stem, ascending to spreading (flaccidly so because of membranaceous blades with thickened midribs); petiole usually slender, less than 6 mm wide; blade linear-lanceolate to linear-oblanceolate, oblanceolate, obovate, or elliptic, to 26 × 2 cm. **Inflorescences:** spikes very tightly to rather loosely spiraled, 3–4 flowers per cycle of spiral, rarely in loose spiral with 5 or more flowers per cycle; rachis moderately to densely pubescent, some trichomes capitate, glands obviously stalked. **Flowers** white to ochroleucous, creamy, or ivory, occasionally green or rarely yellowish, nodding only at base of perianth or occasionally ascending, urceolate to gaping, base usually appearing only slightly inflated, claw of lip at low (less than 30°) angle to lamina, or occasionally not opening; sepals distinct to base, 6–12 mm; lateral sepals appressed to spreading, apices incurved or sometimes straight, sometimes arching above flower; petals linear-lanceolate, 6–12 mm, apex acute to obtuse, occasionally broad and liplike; lip often yellow centrally, ovate to oblong in general outline, usually constricted at middle, 6–10.5 × 2–7.5 mm, base ± prominently dilated, suborbiculate, orbiculate, or linear, margins crenulate, lacerate or sometimes entire, apex acute to rounded distally, glabrous; veins several, branches parallel; basal calli usually incurved, prominent to conic, small, mostly 0.5–1.2 mm, or sometimes lip undifferentiated from petals; viscidia linear to linear-lanceolate; ovary 3–14 mm. **Seeds** polyembryonic, often also monoembryonic, rarely wholly monoembryonic; typically showing irregularities such as embryos developing through or outside testa or naked embryos mixed with whole seed. $2n = 45, 60$.

Flowering late Aug–Nov. Wet to dry open sites in fens, marshes, meadows, swales, dunes, prairies, open woodlands, riverbanks, shores, ditches, roadsides, old fields, cemeteries, lawns; 0–1800 m; N.B., N.S., Ont., P.E.I., Que.; Ala., Ark., Conn., Ga., Ill., Ind., Iowa, Kans., Ky., La., Maine, Md., Mass., Mich., Minn., Miss., Mo., Nebr., N.H., N.J., N.Y., N.C., Ohio, Okla., Pa., R.I., S.C., Tenn., Tex., Vt., Va., W.Va., Wis.

Spiranthes cernua is a facultatively agamospermic polyploid compilospecies in which unidirectional gene flow from related diploids generates a wide range of novel forms and races. Chromosomal segregation has given rise to some forms that very nearly approximate the diploid donor, and meiotic irregularities produce peloria and other abnormal conditions. Apomixis then permits reproduction in pure lines, some of which are so distinctive as to appear to be species. Determination of this and related species is consequently often challenging.

Adventitious embryony in this species commonly gives rise to polyembryonic seeds and seeds exhibiting irregularities such as embryos erupting through or developing outside the testa and naked embryos lacking testae. Often a capsule contains a mixture of polyembryonic and monoembryonic seeds. Rarely all or most seed is monoembryonic, yet irregularities can still be seen.

Polyembryony is also characteristic of *Spiranthes casei* and *S. parksii*, species that are probably derivatives of *S. cernua*. For the most part, at least, related diploids and all other species are strictly sexual and characterized by wholly monoembryonic seed. Some indication of apomixis and polyembryony in *S. magnicamporum*, *S.*

ochroleuca, and *S. odorata* in the northeastern portion of the range of the complex has been reported (P. M. Catling 1981, 1982), but the difficulty in distinguishing between diploid donors and tetraploid chromosomal segregates makes interpretation difficult (C. J. Sheviak 1982, 1991). In particular, the range of *S. odorata* northward along the Atlantic Coast is uncertain. Plants that appear typical of *S. odorata* in all respects but with polyembryonic seed occur from Virginia northward and quickly replace monoembryonic plants toward the north. Such plants from as far south as Virginia, however, have proven to be tetraploids and furthermore to assume morphology typical of *S. cernua* when grown away from the Coast.

Gene flow from *Spiranthes magnicamporum* generates cytogenetically based peloria, a common element in portions of the prairie region. Most commonly differentiation of the lip is suppressed, but in some forms the petals are elaborated and approach the lip in form. Additionally, buds may fail to open and a functional column may be lacking, yet seed is produced apomictically. See the discussion under 16. *S. parksii*.

SELECTED REFERENCES Sheviak, C. J. 1982. Biosystematic study of the *Spiranthes cernua* complex. Bull. New York State Mus. Sci. Serv. 448. Sheviak, C. J. 1991. Morphological variation in the compilospecies *Spiranthes cernua* (L.) L. C. Rich.: Ecologically-limited effects of gene flow. Lindleyana 6: 228–234.

15. Spiranthes casei Catling & Cruise, Rhodora 76: 527, figs. 1–4. 1974 [C] [E]

Plants 7–44 cm. **Roots** few–several, descending, slender to somewhat tuberously thickened, mostly to 0.5 cm diam. **Leaves** persisting through anthesis, basal and on proximal portion of stem, ascending to spreading, ovate-lanceolate to linear-lanceolate to oblanceolate, to 20 × 2 cm. **Spikes** loosely spiraled, usually 5 or more flowers per cycle of spiral; rachis moderately to densely pubescent, some trichomes capitate, glands obviously stalked. **Flowers** ivory to yellowish white or greenish cream, strongly nodding, curved downward throughout lengths, scarcely gaping, urceolate-tubular; sepals distinct to base, 5.2–8 mm; lateral sepals slightly spreading, apex straight; petals creamy white to greenish white, ovate-elliptic to obovate, 3.9–7.6 mm, apex obtuse; lip often darker centrally, 4.1–8 × 3.2–5.1 mm, glabrous; veins several, branches parallel; basal calli incurved, prominent, 0.7–1.1 mm; viscidia linear to linear-lanceolate; ovary 4–7 mm. **Seeds** partly or wholly polyembryonic.

Varieties 2 (2 in the flora): e, c North America along Canadian and Unites States border.

1. Lip ovate to ovate-oblong, apex truncate, margin thin, delicately crisped and fringed . . . . 15a. *Spiranthes casei* var. *casei*
1. Lip ovate, apex acute, margin thick and inflexed . . . . 15b. *Spiranthes casei* var. *novaescotiae*

15a. Spiranthes casei Catling & Cruise var. **casei** [C] [E]

Lip ovate to ovate-oblong, 5–7.5 mm, margins thin, delicately crisped, fringed, apex truncate.

Flowering Aug–Sep. Mesic to dry open sites in meadows, barrens, open woodlands, outcrops, roadsides, sand pits, old fields; of conservation concern; 0–550 m; N.S., Ont., Que.; Maine, Mich., N.H., N.Y., Pa., Vt., Wis.

15b. Spiranthes casei Catling & Cruise var. **novaescotiae** Catling, Canad. J. Bot. 59: 1261. 1981 [C] [E]

Lip ovate, 4.1–5(–8) mm, margins inflexed, thick, apex acute.

Flowering Aug–Sep. Open barrens and roadsides; of conservation concern; 0 m; N.S.

The plants from southern Nova Scotia segregated as *Spiranthes casei* var. *novaescotiae* tend to be smaller, with smaller flowers and broader petals than those of the more widespread var. *casei*.

Putative hybrids of *Spiranthes casei* var. *novaescotiae* with *S. ochroleuca* have been designated *S.* ×*borealis* P. M. Brown.

16. Spiranthes parksii Correll, Amer. Orchid Soc. Bull. 16: 400, figs. 1–6. 1947 [C] [E] [F]

Plants 15–33 cm. **Roots** few, spreading to descending, slender to somewhat tuberously thickened, mostly to 0.8 cm diam. **Leaves** fugaceous, basal, ascending to spreading, linear-lanceolate to somewhat broader. **Spikes** loosely spiraled, usually 5 flowers per cycle of spiral; rachis moderately to densely pubescent, some trichomes capitate, glands obviously stalked. **Flowers** horizontal to ascending, white to pale yellowish green, scarcely gaping, urceolate-tubular; sepals distinct to base, 5–8 mm; lateral sepals directed nearly forward, apex straight or apical 1–2 mm curved upward; petals whitish to yellow-green with longitudinal central green stripe, ovate, obovate, or suborbiculate, 4.5–6.7 mm, apex obtuse to emarginate-erose; lip centrally yellow to yellowish or greenish white, ovate to ovate-oblong, 5–7 × 3.8–4.9 mm, apex truncate, obtuse to deeply emarginate, erose-crisped, glabrous; veins several, branches parallel; basal calli incurved to elongate, prominent; viscidia linear-lanceolate; ovary 4–6 mm. **Seeds** partly or wholly polyembryonic. $2n = 60$.

Flowering Oct–Nov. Dry open to lightly shaded sites in post-oak savanna; of conservation concern; 100 m; Tex.

The tetraploid chromosome number and apomictic development of polyembryonic seeds indicate that *Spiranthes parksii* is a member of the *S. cernua* complex. The broad petals with central green stripe, several veins (instead of the three typical of the group), and erose-emarginate apical margin furthermore evidently represent partial peloria. Peloria is common in *S. cernua*, especially in the prairies, although in most cases it involves the suppression of the lip rather than the elaboration of the petals to a condition approximating the lip, as is the case in *S. parksii*. With a very limited distribution in east-central Texas, *S. parksii* might therefore represent merely a local, minor form of *S. cernua*. Other characteristics, however, including the small flower size and often upturned lateral sepal apices, lie outside the normal range of variation in *S. cernua* and suggest that the plants represent a distinct species.

17. Spiranthes praecox (Walter) S. Watson in A. Gray et al., Manual ed. 6, 503. 1890 E

Limodorum praecox Walter, Fl. Carol., 221. 1788; *Ibidium praecox* (Walter) House

Plants 20–75 cm. **Roots** numerous, slender, mostly to 0.8 cm diam. **Leaves** persisting through anthesis, to 7, primarily basal reduced to sheathing bracts upward on stem, spreading, linear to linear-lanceolate, 10–25 × 1–5 cm. **Spikes** loosely to tightly spiraled, 4–7 flowers per cycle of spiral or nearly secund; rachis sparsely pubescent, some trichomes capitate, glands obviously stalked. **Flowers** white, rarely green; sepals distinct to base, 5.5–10 mm × 2–3 mm; lateral sepals clasping lip, appearing tubular; petals adhering to dorsal sepal, linear, 5–10 × 1–2 mm, apex subacute; lip oblong, 5.5–11 × 2–6 mm, apex dilated, margins often crenulate or toothed, glabrous on adaxial distal surface; veins raised, green, rarely cream colored, prominent, branches divergent, extending nearly to apex; basal calli straight, long-pointed, slender, mostly to 1 mm; viscidium linear-lanceolate; ovary mostly 5 mm. **Seeds** monoembryonic.

Flowering Feb–Jun (Sep, north). Primarily on coastal plain and Gulf Coast in dry to moist roadsides, fields, pine flatwoods, pine savannas, cemeteries; 0–400 m; Ala., Ark., Del., Fla., Ga., La., Md., Miss., N.J., N.C., Okla., S.C., Tex., Va.

Typical plants are easily recognized by the stark white flowers with green venation of the lip. In forma *albolabia* P. M. Brown & C. McCartney, the lip appears pure white but the raised veins are actually a pale yellow. Plants with creamy green to green flowers, usually in tight spirals, and incurved lateral sepals are local to frequent in north-central Florida and scattered westward in the Gulf counties to eastern Texas.

Hybrids of *Spiranthes praecox* with *S. vernalis* are known as *Spiranthes* ×*meridionalis* P. M. Brown.

18. Spiranthes lucida (H. H. Eaton) Ames, Orchidaceae 2: 258. 1908 E F

Neottia lucida H. H. Eaton, Transylvania J. Med. Assoc. Sci. 5: 107. 1832; *Ibidium plantagineum* (Rafinesque) House; *Spiranthes plantaginea* Rafinesque

Plants 4–37 cm. **Roots** spreading to descending, slender, mostly to 0.6 cm diam. **Leaves** persisting through and long after anthesis, 3–4, basal, ascending-spreading, elliptic-lanceolate, 3–12 × 0.5–1.5 cm. **Spikes** rather openly spiraled, 3–4 flowers per cycle of spiral; rachis sparsely to moderately pubescent, some trichomes capitate, glands obviously stalked. **Flowers** nearly horizontal to abruptly nodding from base, white, tubular; sepals linear-oblong; dorsal sepal connate at base, 4.5–5.5 mm; lateral sepals appressed to petals and lip, straight, 5–6 mm × 1–2 mm; petals linear-oblong, 5–6 × 2–4 mm, apex obtuse; lip brilliant yellow or yellow-orange, centrally greenish yellow, oblong, 5–6 × 2–4 mm, margins crenulate, veins several, greenish, branches few, very short, mostly widely diverging, sometimes forming reticulum with primary veins; basal calli spreading, 1 mm; viscidium oval; ovary 3–8 mm. **Seeds** monoembryonic. $2n = 44$.

Flowering May–Aug. Rocky and sandy riverbanks, calcareous seeps, fens, 0–900 m; N.B., N.S., Ont., Que.; Conn., Del., Ill., Ind., Kans., Ky., Maine, Mass., Mich., Mo., Nebr., N.H., N.J., N.Y., Ohio, Pa., R.I., Vt., Va., W.Va., Wis.

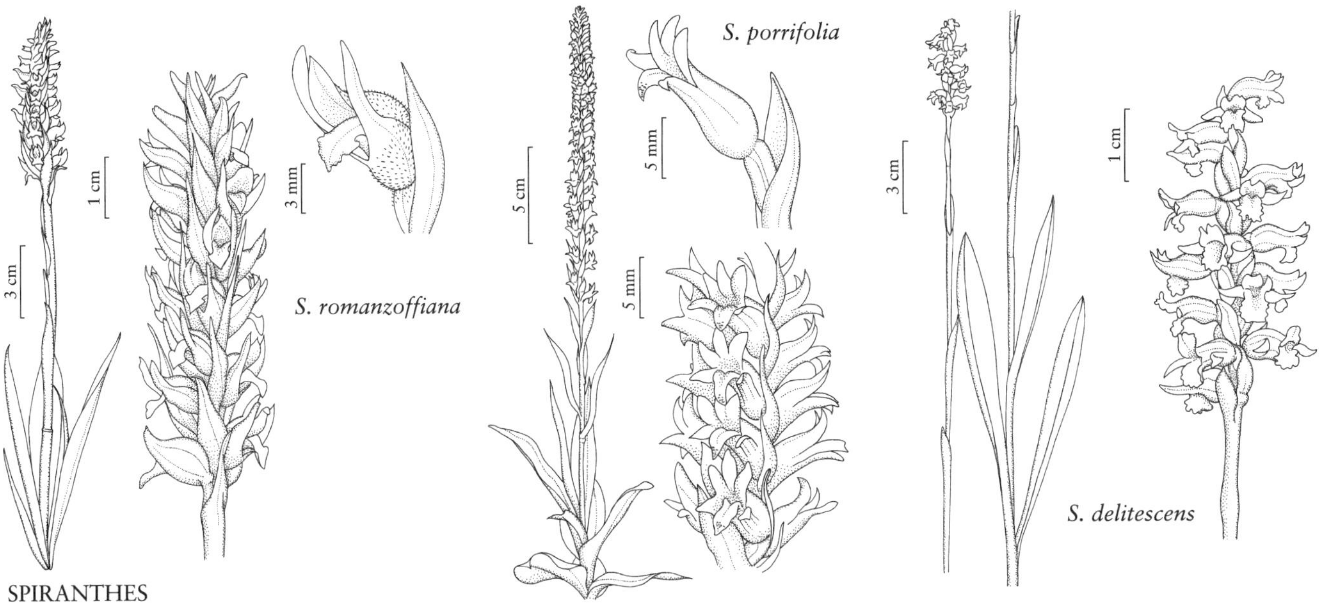

SPIRANTHES

19. Spiranthes romanzoffiana Chamisso, Linnaea 3: 32. 1828 [F]

Gyrostachys stricta Rydberg; *Ibidium strictum* (Rydberg) House; *Spiranthes stricta* (Rydberg) A. Nelson

Plants 8–55 cm. **Roots** few–several, spreading to descending, tuberous, mostly to 1 cm diam. **Leaves** persisting through anthesis or fugaceous, basal, often on proximal portion of stem, ascending to spreading, linear to linear-lanceolate, elliptic, or oblanceolate, to 26 × 3 cm. **Spikes** usually very tightly spiraled, 3 flowers per cycle of spiral, rarely loosely spiraled or with more than 5 flowers per cycle; rachis glabrous to very sparsely pubescent, capitate glands short-stalked or often sessile (longest trichomes 0.03–0.18 mm). **Flowers** ascending, white to ivory, rarely yellowish white, tubular; sepals and petals connivent, forming hood above abruptly reflexed lip; sepals connate at base, 5.3–12.5 mm; petals linear to ovate, apex obtuse to subacute; lip pandurate, 4.8–10.2 × 1.6–6.8 mm, base broadly expanded, apex broadly dilated, separated by narrow isthmus, apical dilation variable, rarely lip nearly ovate, glabrous or with a few scattered trichomes on adaxial surface; veins typically 3, branches wide-spreading; viscidia linear-lanceolate; ovary 2–7 mm. **Seeds** monoembryonic. $2n = 44, 66, 88$.

Flowering Jun–Sep(–Oct). Moist to wet meadows, tundra, marshes, fens, prairies, stream banks, seeps, coastal bluffs, dunes; 0–3400 m; Alta., B.C., Man., N.B., Nfld. and Labr. (Nfld.), N.W.T., N.S., Nunavut, Ont., P.E.I., Que., Sask., Yukon; Alaska, Ariz., Calif., Colo., Conn., Idaho, Ill., Ind., Iowa, Maine, Mass., Mich., Minn., Mont., Nebr., Nev., N.H., N.Mex., N.Y., N.Dak., Oreg., Pa., S.Dak., Utah, Vt., Wash., Wis., Wyo.; Europe (Great Britain, n Ireland).

Plants of *Spiranthes romanzoffiana* vary considerably in habit but are usually quite consistent in floral morphology. The strongly hooded, ascending flowers with abruptly reflexed lips provide a distinctive geometric precision. The pandurate lip with typically three veins, the lateral with abruptly wide-spreading branches, is a key feature. In some areas, however, especially at the edges of the range of the species, some variation is apparent and is sometimes coincident with variability in ploidy level. In particular, in California and adjacent southwestern Oregon variability reaches its peak, with plants variously exhibiting yellowish flowers, loosely spiraled inflorescences, and spreading lateral sepals. Some of this variation may result from gene flow from *S. porrifolia*, but with various ploidy levels common in this area, even within populations, the situation is apparently more complex than simple hybridization.

Hybrids of *Spiranthes romanzoffiana* with *S. lacera* var. *lacera* are known as *S.* ×*simpsonii* Catling & Sheviak.

20. Spiranthes porrifolia Lindley, Gen. Sp. Orchid. Pl., 467. 1840 [E] [F]

Spiranthes romanzoffiana Chamisso var. *porrifolia* (Lindley) Ames & Correll

Plants 15–60 cm. **Roots** few–several, spreading to descending, tuberous, mostly to 1 cm diam. **Leaves** persisting through anthesis or fugaceous, basal, often on proximal portion of stem, ascending, linear to linear-lanceolate, linear-oblanceolate, oblanceolate or elliptic, to 34 × 3.5 cm. **Spikes** usually very tightly spiraled, rarely loosely spiraled, usually 3 flowers per cycle of spiral; rachis glabrous to sparsely pubescent, trichomes less (usually much less) than 0.18 mm, capitate glands often sessile. **Flowers** ascending, creamy to markedly yellowish, slenderly tubular; sepals connate at base, 4.6–11 mm, apex reflexed-spreading; petals linear-lanceolate to linear-oblanceolate, apex reflexed-spreading, obtuse to subacute or subemarginate; lip lance-ovate to lanceolate, 4–9.5 × 1.9–4.3 mm, apex recurved less than to scarcely more than apices of sepals and petals, only slightly or not at all dilated, with subapical dense cushion of peg-shaped trichomes on adaxial surface; veins few to several, branches wide-spreading or often parallel; viscidia linear to linear-elliptic; ovary mostly 2–7.5 mm. **Seeds** monoembryonic. $2n = 44$ [66].

Flowering May–Aug. Moist to wet riparian meadows, stream banks, marshes, fens, seeping banks; 100–2600 m; Calif., Idaho, Nev., Oreg., Wash.

21. Spiranthes infernalis Sheviak, Rhodora 91: 226, figs. 1, 2A, 3a, b. 1989 [C] [E]

Plants 18–40 cm. **Roots** few, strongly descending, tuberous, to 1.5 cm diam. **Leaves** persisting into anthesis, basal, often on proximal 1/2 of stem, ascending, linear to linear-lanceolate or oblanceolate, to 15 × 0.9 cm. **Spike** usually very tightly spiraled, rarely loosely spiraled, 3 flowers per cycle of spiral; rachis glabrous, commonly minutely and sparsely farinose. **Flowers** ascending, ochroleucous, yellowish white, and ochraceous (tinged yellowish brown), with basal 1/4 of sepals and petals and to 1/2 of lip green and fleshy, tubular; sepals connate at base, 4–6 mm; lateral sepals with apices spreading; petals lanceolate, apex spreading, obtuse; lip variably orange centrally, broadly or occasionally narrowly elliptic, 5–6 × 3.8–6.7 mm, often divided by ± evident constrictions into abruptly narrowed apical and basal parts narrower than middle, the apical part minutely but conspicuously puberulent adaxially; veins few to several, branches parallel to wide-spreading; viscidia linear to linear-elliptic; ovary mostly 2–4 mm. **Seeds** monoembryonic. $2n = 44$.

Flowering Jun–Jul. Seasonally wet riparian meadows and spring runs; of conservation concern; 700 m; Nev.

Spiranthes infernalis is known only from Ash Meadows, Nye County, Nevada, and it is probably endemic. The description is based on correspondingly few specimens and may prove to be too restrictive.

SELECTED REFERENCE Sheviak, C. J. 1989. A new *Spiranthes* (Orchidaceae) from Ash Meadows, Nevada. Rhodora 91: 225–234.

22. Spiranthes delitescens Sheviak, Rhodora 92: 215, figs. 1, 2. 1990 [C] [E] [F]

Plants 20–48 cm. **Roots** few–several, descending, slenderly tuberous, to 5 mm diam. **Leaves** persisting through anthesis, basal, on proximal 1/2 of stem, ascending to spreading, linear-lanceolate to oblanceolate, to 18 × 1.5 cm (rarely wider). **Spikes** loosely to tightly spiraled, 3 flowers per cycle of spiral; rachis rather densely pubescent, trichomes without glands pointed, glandular obviously stalked, glands equaling or narrower than stipes. **Flowers** curving through horizontal, white, somewhat gaping and open with wide-spreading lateral sepals; sepals usually connate at base, 6.5–8(–9.9) mm; lateral sepals curving outward throughout lengths, usually directed somewhat downward; petals linear to linear-oblanceolate, apex obtuse; lip creamy or very pale yellow centrally, oblong, with variably evident median constriction, 5.8–7.9 × 3.2–4.9 mm, apex broadly expanded with extensive transverse pleats, distal surface mostly glabrous adaxially; basal calli prominent, stout; veins few, branches mostly wide-spreading; viscidia linear to linear-lanceolate; ovary 3–6 mm. **Seeds** monoembryonic. $2n = 74$.

Flowering Jul–Aug. Cienegas, wet riparian meadows, stream banks; of conservation concern; 1400–1500 m; Ariz.

Spiranthes delitescens is an amphiploid product of hybridization of species in the $n = 15$ and $n = 22$ lineages. Although its origin is unknown, on morphologic grounds C. J. Sheviak (1990) proposed *S. vernalis* and *S. porrifolia* as possible parents.

SELECTED REFERENCE Sheviak, C. J. 1990. A new *Spiranthes* (Orchidaceae) from the cienegas of southernmost Arizona. Rhodora 92: 213–231.

23. Spiranthes diluvialis Sheviak, Brittonia 36: 11, figs. 1C, D, 2A–C, F. 1984 [C] [E] [F]

Spiranthes romanzoffiana Chamisso var. *diluvialis* (Sheviak) S. L. Welsh

Plants 20–62 cm. **Roots** few–several, horizontal to descending, slenderly tuberous, to 1 cm diam. **Leaves** persisting through anthesis, usually restricted to base of stem, ascending, linear-lanceolate, to 28 × 1.5 cm. **Spikes** usually tightly spiraled, 3 flowers per cycle of spiral, rarely loosely spiraled with more than 4 flowers per cycle; rachis sparsely (rarely densely) pubescent, some trichomes capitate, glands obviously stalked (longest trichomes 0.2–0.4 mm).

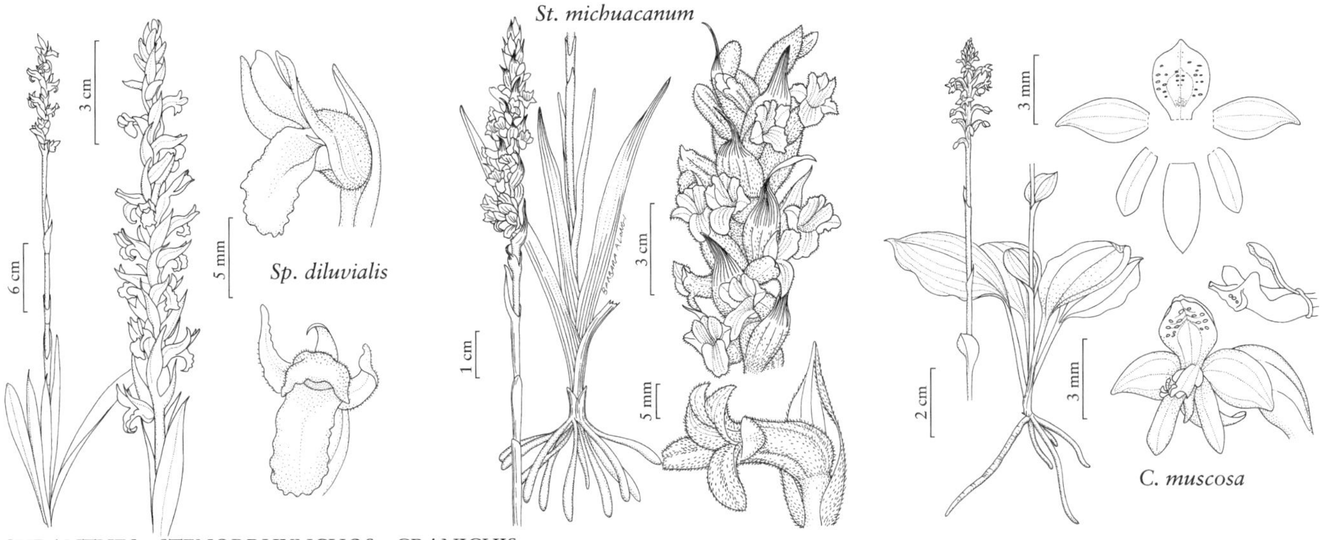

SPIRANTHES ◦ STENORRHYNCHOS ◦ CRANICHIS

Flowers white or ivory, ascending, strongly gaping from near base (lip prominently diverging from sepals and petals); sepals distinct or connate at base, 7.5–15 mm; lateral sepals broadly spreading to loosely incurved, often raised above rest of flower or often appressed; petals linear, apex acuminate; lip often yellow centrally, ovate, lanceolate, or oblong, with variably evident median constriction, occasionally subpandurate, 7–12 × 2.5–6.8 mm, margins crisped, entire, or apically dentate, distal surface mostly glabrous adaxially; basal calli prominent; veins few to several, branches parallel to wide-spreading; viscidia linear to linear-lanceolate; ovary mostly 4–10 mm. **Seeds** monoembryonic. $2n = 74$.

Flowering Jul–Aug(–Sep). Mesic to wet riparian meadows, marshes, stream banks; of conservation concern; 1300–1800 m; Colo., Idaho, Mont., Nebr., Nev., Utah, Wash., Wyo.

Spiranthes diluvialis is an amphiploid product of hybridization of *S. romanzoffiana* and *S. magnicamporum*. It is correspondingly somewhat variable in individual characters, but specimens can be determined through consideration of a combination of features. The position of lateral sepals commonly varies within an inflorescence, and on the same flower one is often tightly appressed while the other is widespreading and ascending. Lip venation in particular can be helpful, as most branches are parallel, but often a few are wide-spreading.

SELECTED REFERENCE Sheviak, C. J. 1984. *Spiranthes diluvialis* (Orchidaceae), a new species from the western United States. Brittonia 36: 8–14.

20. STENORRHYNCHOS Richard ex Sprengel, Syst. Veg. 3: 709. 1826 [Greek *stenos*, narrow, and *rhynchos*, snout, alluding to the narrow rostellum on the column]

Paul M. Catling

Paul Martin Brown

Herbs, perennial, terrestrial. **Roots** fascicled, fleshy. **Stems** with foliaceous sheaths. **Leaves** flat at maturity. **Inflorescences** densely flowered racemes. **Flowers** resupinate, without mentum or spur; perianth parts distinct and free, basally connivent, forming tube, apically recurved; lip elliptic-lanceolate, base truncate or slightly cordate, apex free, acute; calli barely raised on either side of basal claw extending onto base of lamina of lip where hairy; column not inflated adaxially at base, extended obliquely on top of ovary only very slightly; anther entire; viscidium narrowly elliptic, sheathing rostellum; stigma nearly terminal; rostellum well developed, needlelike. **Fruits** capsules.

Species 9 (1 in the flora): tropical regions, Western Hemisphere.

SELECTED REFERENCE Ames, O. and C. Schweinfurth. 1935. Nomenclatural studies in *Malaxis* and *Spiranthes*. Bot. Mus. Leafl. 3: 113–133.

1. **Stenorrhynchos michuacanum** (La Llave & Lexarza) Lindley, Gen. Sp. Orchid. Pl., 478. 1840 [F]

Neottia michuacana La Llave & Lexarza, Nov. Veg. Descr. 2: Orchid. Opusc., 3. 1825; *Spiranthes michuacana* (La Llave & Lexarza) Hemsley

Plants 13–80 cm. **Roots** 4–10 × 3–6 mm. **Stems:** sheaths 3–6, cauline, bladeless, striate, scarious, glabrous. **Leaves** 0–5, mostly basal; distinct petiole absent; blade lanceolate, 10–50 × 1–3 cm. **Inflorescences** 5–19 cm; floral bracts green basally, becoming reddish brown apically, elliptic, 21–32 × 15 mm, apex acute or acuminate; basal portions of bracts, rachis, and outer surfaces of sepals densely hairy with many-septate, tapered, tortuous hairs. **Flowers** cream or white; perianth parts with 1 or more longitudinal green stripes or markings; sepals lanceolate, 9–20 × 3.6–7 mm, apex acute to mucronate; dorsal sepal 15–18 mm; petals lanceolate, 9–20 × 4–5 mm, base arcuate, apex acute; lip 8–16 × 5.2–7.9 mm, with short basal claw ca. 1 mm; column 8–10 mm; anther cap 7–8 mm; pollinaria 8 mm; rostellum needlelike, 4 mm; pedicellate ovary 7–10 mm. **Capsules** ascending, ovoid.

Flowering Sep–Feb. Open grassy slopes, sand pine-oak woodlands, especially in zones of periodic seepage; 1900–2200 m; Ariz., Tex.; Mexico.

In the flora *Stenorrhynchos michuacanum* grows in the Huachuca and Santa Catalina mountains of Arizona and the Chisos and Chinati mountains of Texas. This rather robust species has stems 2–10 mm thick. The flowers are densely hairy externally with many-septate, tapered, tortuous hairs. Recent field work (R. A. Coleman 1996) has shown that the species is more frequent than previously thought.

SELECTED REFERENCE Coleman, R. A. 1996. *Stenorrhynchos michuacanum*. Orchids 65: 1284–1286.

230c.1d. ORCHIDACEAE Jussieu (tribe CRANICHIDEAE) subtribe CRANICHIDINAE Lindley, Gen. Sp. Orchid. Pl., 441, 444. 1840

21. CRANICHIS Swartz, Prodr., 8, 120. 1788 • Helmet orchid [Greek *kranos*, helmet, for helmetlike appearance of lip]

James D. Ackerman

Herbs terrestrial, sympodial. **Roots** several, fasciculate, thick, villous. **Stems** simple, rhizomatous. **Leaves** 1–many, often forming basal rosette; blade petiolate, not articulate, thin. **Inflorescences** terminal, several- to many-flowered racemes, erect; scape with a few sheathing bracts. **Flowers** not resupinate; sepals and petals spreading, distinct, thin; lip concave, fleshy; column cylindric, short; anther erect, abaxial; pollinia 4, clavate, mealy, breaking apart in chunks, not powdery, attached to caudicle; stigma entire; rostellum apical, viscidium terminal; ovary pedicellate. **Fruits** capsules, erect.

Species 60 (1 in the flora): subtropical and tropical regions, North America, Mexico, West Indies, Central America, South America.

1. Cranichis muscosa Swartz, Prodr., 120. 1788 [F]

Plants to 50 cm. **Roots** 2–3 mm diam. **Leaves** 4–8, basal; petiole erect; blade spreading, broadly elliptic with several prominent arching veins, 2–9 × 1–4 cm. **Inflorescences** glabrous; scape bracts leafy, short; raceme 1–15 cm; floral bracts lanceolate, shorter than ovary. **Flowers** mostly white, glabrous; dorsal sepal 3-veined, elliptic-lanceolate, 3–4 × 1.5–2 mm; lateral sepals 3-veined, obliquely ovate-lanceolate to ovate, 4 × 2 mm; petals 1-veined, linear, 2–3.5 × 1 mm, margins facing lip minutely ciliate, apex blunt; lip white, spotted green or pale yellow, ovate-subquadrangular, to 4 × 2.5 mm, distal margins undulate; column prominently winged, 1.5–2 mm; pollinia yellow, elongate; pedicellate ovary 3–6 mm. **Capsules** erect-spreading, ellipsoid-cylindric, 8–10 × 3–4 mm.

Flowering Jan–Mar. Hardwood hammocks; 0–20 m; Fla.; Mexico; West Indies; Central America; South America.

When plants are grown isolated from potential pollinators, nearly all flowers form fruits. After the anther dehisces, the pollinarium bends over and contacts the stigma (J. D. Ackerman 1995). An attempt to relocate historical localities of *Cranichis muscosa* in Miami-Dade and Lee counties was unsuccessful (C. A. Luer 1972), and more recent searches have also been unsuccessful.

22. PONTHIEVA R. Brown in W. Aiton and W. T. Aiton, Hortus Kew. 5: 199. 1813 • [For Henri de Ponthieu, a West Indian merchant who sent plant collections to Sir Joseph Banks in 1778]

James D. Ackerman

Herbs, terrestrial, rarely epiphytic, sympodial. **Roots** fasciculate, fleshy, villous. **Stems** simple, rhizomatous. **Leaves** mostly basal, subsessile to petiolate; blade not articulate, thin, delicate. **Inflorescences** terminal, few- to many-flowered racemes; peduncles bracteate, glandular-pubescent. **Flowers** not resupinate, erect or suberect, dishlike; perianth spreading; dorsal sepal lightly adherent to petals at apex; lateral sepals distinct or connate; petals free or adnate to basal flanks of column, obliquely triangular, clawed, conspicuous, often forming "pseudolip"; lip adnate to base of column, distinctly to obscurely clawed, deeply concave, fleshy; column semiterete, dilated and slightly winged distally, short, apex pointed; anther adaxial, erect behind rostellum; pollinia 4, bright yellow, clavate, joined in pairs, firm but somewhat mealy, attached to caudicles; stigma entire; viscidium terminal; pedicellate ovary pubescent. **Fruits** capsules, suberect, ellipsoid [obovoid].

Species 25 (2 in the flora): tropical and subtropical, North America, Mexico, West Indies, Central America, South America.

1. Perianth more than 5 mm; petals conspicuously striped green, 3.5–5 mm wide; lip distinctly clawed, suborbiculate; leaves present at anthesis; shady broadleaf forests . . . . 1. *Ponthieva racemosa*
1. Perianth less than 5 mm; petals not conspicuously striped green, 1.5 mm wide; lip indistinctly clawed, obovate; leaves withering or absent at anthesis; open pinelands . . . . . . 2. *Ponthieva brittoniae*

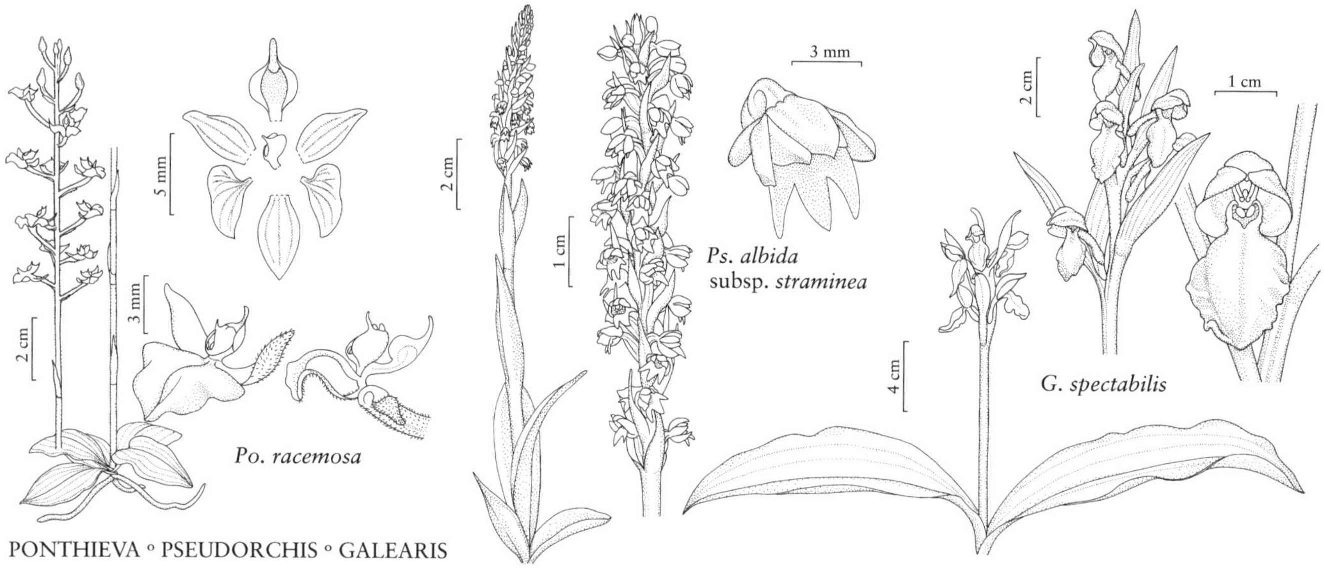

PONTHIEVA ∘ PSEUDORCHIS ∘ GALEARIS

1. Ponthieva racemosa (Walter) C. Mohr, Contr. U.S. Natl. Herb. 6: 460. 1901 • Shadow witch [F]

Arethusa racemosa Walter, Fl. Carol., 222. 1788; *Neottia glandulosa* Sims; *Ponthieva glandulosa* (Sims) R. Brown

Plants 30–60 cm. **Roots** 2–3 mm diam. **Stems** subterranean, short. **Leaves** green at anthesis, 2–6, subpetiolate; blade elliptic to oblanceolate, 5–17 × 2–5 cm, apex rounded to acute. **Inflorescences:** peduncle to 25 cm, partially enclosed by tubular sheaths, proximalmost sometimes leafy; rachis laxly 20–35-flowered, 5–25 cm; floral bracts narrowly lanceolate, clasping base of ovary, to 10 mm, apex acuminate, pubescent. **Flowers:** sepals greenish white, adaxially pubescent; dorsal sepal distinct, ovate-elliptic to elliptic-lanceolate, 4–7 × 2–3 mm, apex obtuse; lateral sepals broadly, obliquely ovate, 4.3–6.5 × 2.5–3.5 mm, apex acute to obtuse; petals white, green-veined, obliquely triangular, 4–6 × 3.5–5 mm, margins minutely ciliate or entire, glabrous; lip white with green, deeply concave center, distinctly clawed, suborbiculate, 5–7 × 5–7 mm, apex short-caudate; column white, 4–5 mm; pedicellate ovary 10–20 mm. **Capsules** 8–13 mm.

Flowering fall–winter (Sep–Feb). Moist, shady hammocks, swamps, ravines, wet savannas, pine forests; 0–50 m; Ala., Fla., Ga., La., Miss., N.C., S.C., Tex., Va.; Mexico; West Indies; Central America; South America.

In Florida, *Ponthieva racemosa* is self-compatible but not autogamous. Natural fruit-set in one population in northern Florida was 35% (J. D. Ackerman 1995). In Florida, small halictid bees were observed visiting the flowers (C. A. Luer 1972).

2. Ponthieva brittoniae Ames, Torreya 10: 90. 1910 (as brittonae) • Mrs. Britton's shadow witch

Ponthieva racemosa (Walter) C. Mohr var. *brittoniae* (Ames) Luer

Plants 30–50 cm. **Roots** unknown. **Stems** subterranean, short. **Leaves** withering at or during anthesis, 3–12; petiole 1–2.5 cm; blade elliptic to ovate, 4–10 × 1.5–3.5 cm, apex obtuse to acute. **Inflorescences:** peduncle 26–37 cm, partially enclosed by tubular sheaths, proximalmost larger but not leafy; rachis laxly 20–35-flowered, 10–18 cm; floral bracts lanceolate, clasping base of ovary, 5–10 mm, apex acuminate, pubescent. **Flowers:** sepals green, adaxially pubescent; dorsal sepal distinct, oblong-lanceolate, 4–5 × 2–3 mm, apex acute to obtuse; lateral sepals obliquely ovate-lanceolate, 4–5 × 2.5 mm, apex acute to obtuse; petals recurved, white, indistinctly clawed, obliquely triangular-ovate, 4–4.5 × 1.5 mm, margins entire, apex acute, pubescent; lip subsessile, white with 2 dark green blotches, obovate, 4.5 × 3.5 mm, apex short-caudate; column white, 2.5 mm; pedicellate ovary 7–9 mm. **Capsules** 8–13 mm.

Flowering Dec–early Mar. Open rocky pinelands; 0–20 m; Fla.; West Indies (Bahamas).

Most of the original habitat of *Ponthieva brittoniae* has succumbed to urbanization. Only two locations are known to exist. One consisted of only a single specimen, and plants of the other have not reappeared since 1986 and may have been destroyed by roadwork (C. McCartney 1997).

SELECTED REFERENCES McCartney, C. 1997. Mrs. Britton's shadow witch. Palmetto 17: 10–11. Sauleda, R. P. and R. M. Adams. 1980. The taxonomic status of *Ponthieva brittoniae* Ames (Orchidaceae). Bull. Torrey Bot. Club 107: 544–545.

230c.2. ORCHIDACEAE Jussieu (subfam. ORCHIDOIDEAE) tribe ORCHIDEAE

230c.2a. ORCHIDACEAE Jussieu (tribe ORCHIDEAE) subtribe ORCHIDINAE

23. PSEUDORCHIS Séguier, Pl. Veron. 3: 254. 1754 • Small white orchid [Greek *pseudo*, false, and the generic name *Orchis*]

Charles J. Sheviak

Herbs, perennial, rather succulent. **Roots** fasciculate, fleshy, both slender and digitately divided tuberoids. **Stems** leafy, leaves gradually reduced apically, generally to bracts proximal to inflorescence. **Leaves** few to several, ascending-spreading, conduplicate, bases sheathing. **Inflorescences** terminal, solitary, rather dense spikes; floral bracts foliaceous, prominently exserted to reduced and inconspicuous. **Flowers** several to many, resupinate; lip deeply 3-lobed, lobes entire, nearly equal, base spurred; anther cells parallel; pollinia 2, pollinaria 2, separated by thick, fleshy, laminar projection of midlobe of rostellum, free viscidia sheltered within pockets formed by auriculate lateral rostellum lobes; stigma concave. **Fruits** capsules.

Species 2 or 3 (1 in the flora): ne North America, Europe, n Atlantic Islands.

1. Pseudorchis albida (Linnaeus) Á. Löve & D. Löve, Taxon 18: 312. 1969 [F]

Satyrium albidum Linnaeus, Sp. Pl. 2: 944. 1753

Subspecies 2 (1 in the flora): North America, Iceland, Europe.

1a. Pseudorchis albida (Linnaeus) Á. Löve & D. Löve subsp. **straminea** (Fernald) Á. Löve & D. Löve, Taxon 18: 312. 1969 [F]

Habenaria straminea Fernald, Rhodora 28: 174. 1926; *H. albida* (Linnaeus) R. Brown var. *straminea* (Fernald) Morris & Eames; *Leucorchis albida* (Linnaeus) E. Meyer ex Schur var. *subalpina* (Neuman) Hylander; *Platanthera albida* (Linnaeus) Lindley var. *straminea* (Fernald) Luer

Plants erect, 5–35 cm. **Leaves:** blades oblong-obovate to elliptic or lanceolate, 20–90 × 6–40 mm, distal blades narrower. **Inflorescences:** bracts lanceolate, to 15 × 3 mm, proximal bracts often spreading, conspicuously exceeding flowers. **Flowers** dull greenish yellow or creamy; perianth nodding, ovaries ascending; dorsal sepal concave, forming hood over column, 2.3–4 × 1–2 mm; lateral sepals wide-spreading, ovate-lanceolate, 3–4 × 1.2–1.8 mm; petals narrowly ovate, 3 × 1.5 mm, associated with dorsal sepal; lip 3-lobed to 1/2 its length, 2–5 × 2–3 mm, lateral lobes lanceolate, slightly falcate, middle lobe oblong, slightly longer than laterals, spur cylindric to clavate, 2–3 mm. **Capsules** ascending, ellipsoid-obovoid, 6 × 4 mm. $2n$ = 40, 42.

Flowering summer. Wet to rather dry tundra and limestone barrens; 0–700 m; Greenland; Nfld. and Labr. (Nfld.), Que.; n Europe; Atlantic Islands (Faeroe Islands, Iceland).

The proper status of the American plants is not clear. Originally described as *Habenaria straminea* Fernald, the plants of North America, Iceland, and the Faeroe Islands were distinguished by a more yellowish and slightly larger perianth than the common European form. These seemingly minor differences are inconsistent, however, and the plants have generally been treated as only infraspecifically distinct. Because of the occurrence of plants in northern and western Scandinavia, E. Hultén (1958) treated them as *Leucorchis albida* var. *subalpina*, and D. M. Moore (1980) treated them as *Pseudorchis albida* subsp. *straminea*. Recently, however, L.-G. Reinhammer (1995) presented evidence supporting the recognition of two species. His sample, limited to western Scandinavia and a single Icelandic site, may not be representative of American plants, and our populations should be studied accordingly.

24. GALEARIS Rafinesque, Herb. Raf., 71. 1833 • Showy orchis [Latin *galea*, helmet]

Charles J. Sheviak

Paul M. Catling

Herbs, perennial, scapose, rather succulent. **Roots** not tuberously thickened, fascicled, fleshy. **Stems** bractless proximal to inflorescence, conspicuously angled. **Leaves** [1–]2, basal, spreading, conduplicate, gradually narrowed to sheathing petioles; blade oblance-ovate to obovate, elliptic, or suborbiculate, apex rounded-obtuse. **Inflorescences** terminal, solitary, racemose spikes; bracts large, foliaceous. **Flowers** 2–15, resupinate, showy; lip protruding downward from hood formed by sepals and enclosing petals, base produced into spur; pollinaria 2; pollinia 2; viscidia within single 2-lobed bursicle; stigma concave, hidden behind bursicle. **Fruits** capsules, erect.

Species 2 (1 in the flora): e North America, 1 in e Asia.

1. Galearis spectabilis (Linnaeus) Rafinesque, Herb. Raf., 71. 1833 E F

Orchis spectabilis Linnaeus, Sp. Pl. 2: 943. 1753; *Galeorchis spectabilis* (Linnaeus) Rydberg

Plants 5–20 cm. **Leaves:** blades 9–20 × 2–10 cm. **Inflorescences:** floral bracts lanceolate to elliptic, 15–80 × 5–20 mm. **Flowers** showy; sepals pink-magenta (rarely white), ovate to elliptic, concave, 10–20 × 4–6 mm; lip white (rarely pink-magenta), rhombic- to lance-ovate, 10–18 × 6–15 mm, margins entire or irregularly cuneate, spur conspicuous, clavulate, 9–20 mm, apex obtuse. **Capsules** erect, ellipsoid, somewhat triangular in cross section. 2*n* = 42.

Flowering spring–early summer. Moist, calcareous woodlands, thickets, and old fields; 0–1300 m; N.B., Ont., Que.; Ala., Ark., Conn., Del., D.C., Ga., Ill., Ind., Iowa, Kans., Ky., Maine, Md., Mass., Mich., Minn., Miss., Mo., Nebr., N.H., N.J., N.Y., N.C., Ohio, Okla., Pa., R.I., S.C., Tenn., Vt., Va., W.Va., Wis.

Galearis spectabilis forma *gordinierii* (House) Whiting & Catling has white flowers; in forma *willeyi* (Seymour) P. M. Brown, the flowers are entirely pink.

25. AMERORCHIS Hultén, Ark. Bot., n. s. 7: 34. 1968 • Round-leaved orchis, amérorchis, Amérorchis à feuille ronde [America plus *orchis*, from the American distribution of this close relative of Eurasian *Orchis*] E

Charles J. Sheviak

Paul M. Catling

Herbs, perennial, rather succulent. **Roots** scattered along rhizome, slender, fleshy. **Stems** scapose. **Leaves** solitary, basal, spreading, conduplicate; petioles inconspicuous, sheathing or clasping stem; blade orbiculate to broadly elliptic, ovate, obovate, or lance- or oblance-elliptic. **In-**

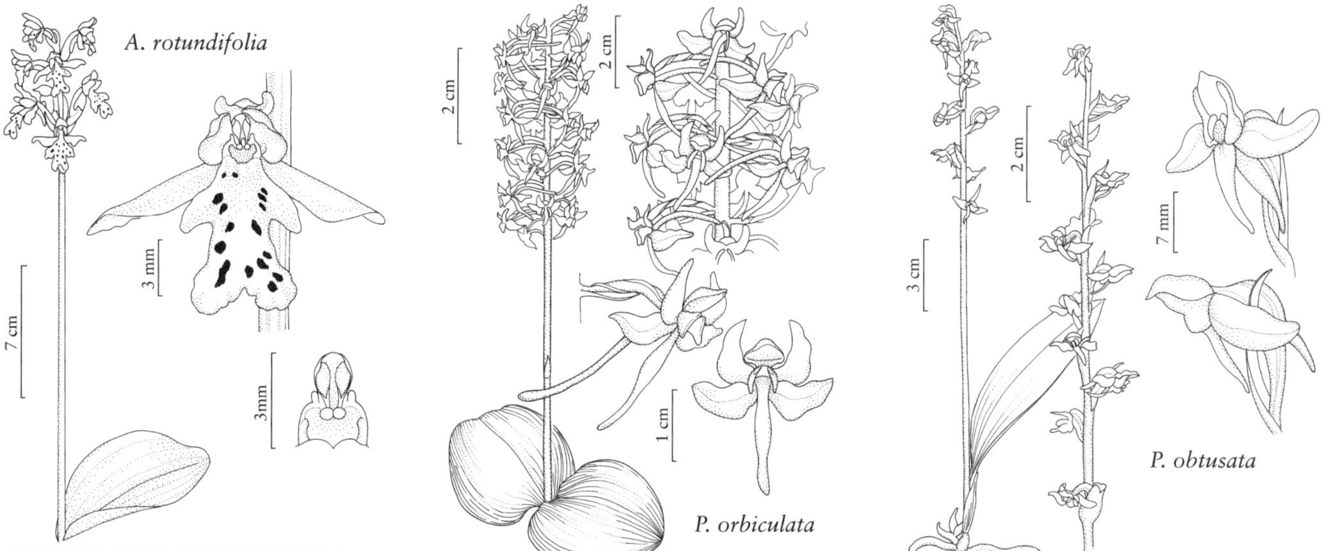

AMERORCHIS ∘ PLATANTHERA

florescences terminal, solitary spikes, lax; floral bracts foliaceous. **Flowers** few to several, resupinate, showy; lip spurred at base, deeply 3-lobed, middle lobe notched, margins entire to crenate or erose; pollinaria 2; pollinia 2; viscidia within single 2-lobed bursicle; stigma flat, narrowly reniform to obcordate, ends flanking bursicle. **Fruits** capsules.

Species 1: North America.

1. **Amerorchis rotundifolia** (Banks ex Pursh) Hultén, Ark. Bot., n. s. 7: 34. 1968 [E] [F]

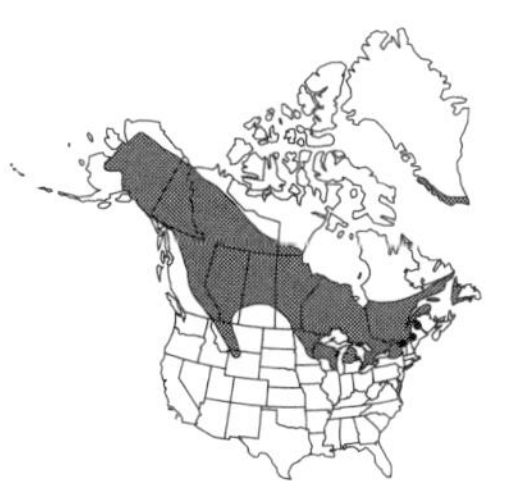

Orchis rotundifolia Banks ex Pursh, Fl. Amer. Sept. 2: 588. 1814; *Ponerorchis rotundifolia* (Banks ex Pursh) Soó

Plants 7–33 cm. **Leaves:** blades 2.7–11 × 1.2–8 cm. **Inflorescences:** floral bracts lanceolate to linear, 5–15 × 1–5 mm, apex acute to acuminate. **Flowers** showy; sepals white to pale magenta, ovate to elliptic-oblong, 6–10 × 3–4 mm; petals white to pale magenta, ovate to lance-oblong, 5–6 mm, less than 1/2 as wide; lip white, prominently magenta-spotted, rarely with broad, longitudinal, reddish bars, ovate in outline, 6–9 × 4–8 mm, lateral lobes spreading, spur conspicuous, 2/3 as long as to equaling lip, slender. **Capsules** erect, ellipsoid. $2n = 42$.

Flowering Jun–Jul. Moist, often calcareous coniferous forests, thickets, fens, tundra; 0–1200 m; Greenland; Alta., B.C., Man., N.B., Nfld. and Labr., N.W.T., Nunavut, Ont., Que., Sask., Yukon; Alaska, Maine, Mich., Minn., Mont., N.Y., Vt., Wis., Wyo.

26. PLATANTHERA Richard, De Orchid. Eur., 20, 26, 35. 1817, name conserved • Bog orchid, fringed orchid, rein orchid [Greek *platys*, broad, and *anthera*, anther]

Charles J. Sheviak

Herbs, perennial, erect to somewhat decumbent, rather succulent. **Roots** fasciculate, both slender and tuberous, fleshy; if tuberous, then lance-fusiform. **Stems** leafy or leafless, terete; **Leaves** 1–several, strictly basal or gradually reduced to bracts toward inflorescence, conduplicate, ascending to spreading, bases sheathing stem. **Inflorescences** solitary, terminal, lax to dense spikes. **Flowers** few to many, usually resupinate (not resupinate in *P. nivea*), sometimes showy; petals entire to fringed or emarginate; lip lobed, 3-partite, spurred at base, margins entire to fringed; pollinaria 2; pollinia 2; viscidia free; stigma entire. **Fruits** capsules, ellipsoid to cylindric.

Species ca. 200 (32 species in the flora): primarily north temperate (a few tropical).

SELECTED REFERENCES Ames, O. 1905–1922. Orchidaceae: Illustrations and Studies of the Family Orchidaceae Issuing from the Ames Botanical Laboratory.... 7 vols. Boston and New York. Vol. 4. Folsom, J. P. 1984. Una reinterpretación del estatus y relaciones de las taxa del complejo de *Platanthera ciliaris*. (A reinterpretation of the status and relationships of taxa of the yellow-fringed orchid complex.) Orquidea (Mexico City) 9: 321–345. Reddoch, A. H. and J. M. Reddoch. 1993. The species pair *Platanthera orbiculata* and *P. macrophylla* (Orchidaceae): Taxonomy, morphology, distributions and habitats. Lindleyana 8: 171–187. Stoutamire, W. P. 1974. Relationships of purple-fringed orchids *Platanthera psycodes* and *P. grandiflora*. Brittonia 26: 42–58.

1. Lip lobed, fringed, eroded, lacerate, or emarginate.
  2. Lip deeply and subequally 3-lobed (lateral lobes each in area equaling approximately 1/2 middle lobe, often much more).
    3. Petals linear-oblong to oblong-spatulate, usually entire; lip greenish white . . . . 24. *Platanthera lacera*
    3. Petals oblong-obovate to flabellate, lacerate toward apex; lip pure white or pink-magenta to lavender.
      4. Lobes of lip dentate to dentate-lacerate, not fringed; rostellum lobes spreading . . . . 19. *Platanthera peramoena*
      4. Lobes of lip fringed; rostellum lobes spreading or parallel.
        5. Flowers pink-magenta to lavender, rarely white; pollinia erect; pollinaria nearly straight; column appearing truncate in lateral view.
          6. Rostellum lobes spreading, angular in lateral view . . . . 20. *Platanthera grandiflora*
          6. Rostellum lobes parallel, rounded in lateral view . . . . 21. *Platanthera psycodes*
        5. Flowers white; pollinia directed forward; pollinaria geniculate; column appearing hooded in lateral view.
          7. Rostellum lobes spreading, angular in lateral view . . . . 22. *Platanthera praeclara*
          7. Rostellum lobes parallel, rounded in lateral view . . . . 23. *Platanthera leucophaea*
  2. Lip not deeply, subequally 3-lobed, merely auriculate, fringed, eroded, lacerate, or emarginate, or apically shallowly 3-lobed.
    8. Lip basally auriculate, or apically emarginate or shallowly lobed.
      9. Lip with basal pair of auricules flanking prominent tubercle on adaxial surface, apex obtuse to emarginate . . . . 18. *Platanthera flava*
      9. Lip without basal pair of auricles or tubercle, apex shallowly 3-lobed . . . 31. *Platanthera clavellata*
    8. Lip fringed or eroded.
      10. Lip not conspicuously fringed, merely eroded to lacerate.
        11. Flowers white; lip 10–15 mm . . . . 26. *Platanthera integrilabia*
        11. Flowers yellow to orange; lip less than 6 mm . . . . 32. *Platanthera integra*
      10. Lip conspicuously fringed.
        12. Rostellum lobes triangular, apices directed forward (column appearing acute in lateral view).
          13. Flowers white . . . . 25. *Platanthera blephariglottis*
          13. Flowers orange . . . . 27. *Platanthera ciliaris*
        12. Rostellum lobes slender, directed downward toward apices (column appearing truncate in lateral view), straight to distally retrorse.
          14. Spur 8–17 mm, its mouth nearly circular; rostellum lobes strongly retrorsely curved toward apices; viscidia presented downward and approximately parallel to lip . . . . 29. *Platanthera chapmanii*
          14. Spur 4–10 mm, its mouth triangular or keyhole-shaped; rostellum lobes scarcely to strongly curved; viscidia presented forward, nearly perpendicular to approximately parallel to lip . . . . 28. *Platanthera cristata*
1. Lip entire, at most abruptly dilated toward base (in some robust Alaskan material of *P. dilatata* var. *albiflora* basal dilation of lip is elaborated forward, taking form of lateral lobules).

15. Leaves 1–2; scape naked or with 1(–2, very rarely) bracts.
  16. Spur saccate to stoutly cylindric; lip broadly elliptic-suborbiculate to obovate, concave . . . . . 17. *Platanthera chorisiana*
  16. Spur slender; lip linear to rhombic-lanceolate, ± flat.
    17. Leaf 1, or if 2 alternate and widely spaced on stem, erect-spreading.
      18. Lip linear to linear-rhombic-lanceolate (lanceolate in subsp. *oligantha*); spur tapering from broad base, ± as long as lip; widespread . . . . . 4. *Platanthera obtusata*
      18. Lip elliptic to elliptic-oblong or elliptic-lanceolate; spur cylindric to clavate, ± 2 times as long as lip; w Aleutian Islands . . . . . 5. *Platanthera tipuloides*
    17. Leaves 2 in basal pair, wide-spreading, commonly lying on ground.
      19. Scape naked; lip directed forward, lanceolate; spur tapering to apex from broad base . . . . . 3. *Platanthera hookeri*
      19. Scape with 1 or more bracts; lip pendent, linear to linear-oblong; spur clavate.
        20. Spur less than 28 mm; pollinaria less than 4.6 mm . . . . . 1. *Platanthera orbiculata*
        20. Spur equal to or greater than 28 mm; pollinaria equal to or greater than 4.6 mm . . . . . 2. *Platanthera macrophylla*
15. Leaves (1–)3–several; scape with gradually reduced bracts distially.
  21. Lip uppermost (flowers not resupinate); flowers white . . . . . 30. *Platanthera nivea*
  21. Lip lowermost (flowers resupinate); flowers white, green, or yellowish.
    22. Viscidia linear to linear-oblong; flowers white; lip usually with suborbiculate to orbiculate basal dilation . . . . . 6. *Platanthera dilatata* (in part)
    22. Viscidia oblong to orbiculate; flowers greenish, yellowish, or white; lip linear to linear-rhombic, elliptic, or lanceolate, base undilated or with obscure to orbiculate basal dilation.
      23. Column comparatively large, occupying ca. 2/3 hood formed by dorsal sepal and petals; rostellum lobes widely spaced, diverging, and together with stigma and connective forming hemispheric chamber; lip linear to linear-oblong or linear-lanceolate, or narrowly elliptic, base undilated.
        24. Leaves reduced; blades rigidly erect, loosely sheathing stem, scarcely spreading beyond width of inflorescence, sometimes absent; dry to mesic forest . . . . . 14. *Platanthera brevifolia*
        24. Leaves well developed; blades prominent, spreading well beyond sheathing bases and width of inflorescence; wet-mesic to wet sites.
          25. Spur 6–14 mm, 0.7–1.6 times length of lip; lip with prominent basal median ridge; leaves ascending or wide-spreading, scattered along stem . . . . . 13. *Platanthera sparsiflora* (in part)
          25. Spur 12–17 mm, 1.3–2.6 times length of lip; lip without prominent median ridge, but often with series of obscure parallel ridges; leaves wide-spreading, usually limited to proximal portion of stem . . . . . 15. *Platanthera zothecina*
      23. Column comparatively small, occupying less than 1/2 hood formed by dorsal sepal and petals; rostellum lobes parallel, diverging, or converging, but scarcely elevated, at most separated by narrow slit, not forming hemispheric chamber; lip linear to linear-lanceolate, rhombic-linear, rhombic-lanceolate, or elliptic, base undilated or with obscure to orbiculate basal dilation.
        26. Anther low, appearing to lie atop stigma; anther sacs widely diverging from apices scarcely separated by obscure connective; flowers autogamous; pollinia rotated forward, commonly free of anther sacs and/or fragmenting into loose masses that trail downward onto stigma; spur clavate, mostly slightly shorter than lip; lip rhombic-lanceolate, yellowish to yellowish green . . . . . 10. *Platanthera aquilonis*

[26. Shifted to left margin.—Ed.]

26. Anther high, rising above stigma; anther sacs more nearly parallel, converging, or diverging from apices separated by evident connective; flowers not autogamous (except evidently autogamous in *P. hyperborea*; see also note under *P. huronensis*); pollinia remaining within anther sacs; spur scrotiform or saccate to slenderly clavate or filiform, much shorter to much longer than lip; lip linear, elliptic, or broadly lanceolate, or with abrupt basal dilation, white to green or yellowish.
   27. Spur 8–25 mm, much longer than lip, filiform; lip usually with small median basal thickening, linear-oblong to linear-elliptic, rarely rhombic-ovate . . . . 16. *Platanthera limosa*
   27. Spur 2–12 mm, much shorter to slightly longer than lip, saccate or scrotiform to capitate, clavate, or slenderly cylindric; lip linear-rhombic, linear-lanceolate, or rhombic-lanceolate, or with pronounced orbiculate to suborbiculate basal dilation.
      28. Pollinaria and rostellum lobes parallel to converging toward orbiculate viscidia; spur scrotiform or saccate to spatuloid-capitate or strongly clavate, much shorter than lip . . . . 12. *Platanthera stricta*
      28. Pollinaria and rostellum lobes diverging toward oblong or sometimes (in *P. purpurascens*) orbiculate viscidia; spur slenderly cylindric to strongly clavate, saccate, or scrotiform, slightly longer to much shorter than lip.
         29. Lip oblong or rarely lanceolate, dull yellowish; rostellum lobes usually projecting forward, separated by narrow slit; leaves clustered toward base of stem . . . . 13. *Platanthera sparsiflora* (in part)
         29. Lip broadest at base, lanceolate, ovate, or linear, or with pronounced basal dilation, white, greenish or yellowish; rostellum lobes low, not projecting forward; column without evident slit; leaves scattered along stem.
            30. Lip pure white, usually with pronounced suborbiculate to orbiculate basal dilation or rarely shallowly 3-lobed; spur clavate to slightly capitate, somewhat to very much shorter than lip . . . . 6. *Platanthera dilatata* (in part)
            30. Lip greenish or yellowish, mostly lanceolate, ovate, or linear, sometimes obscurely or prominently dilated at base; spur slenderly cylindric to strongly clavate, saccate, or scrotiform, shorter than to slightly longer than lip.
               31. Spur saccate or scrotiform to clavate-inflated (widest at apex), ca. 1/2 length of lip; lip yellowish green to dull yellowish or intensely bluish green, sometimes marked with red, linear-lanceolate to prominently rounded-dilated at base; musty scented . . . . 11. *Platanthera purpurascens*
               31. Spur slenderly cylindric to slenderly clavate or clavate, not inflated, merely obtuse to subacute, shorter to slighter longer than lip; lip whitish green to perhaps yellowish green in *P. hyperborea*; sweet scented.
                  32. Lip linear-rhombic, with straight sides, not at all rounded-dilated at base; spur ca. 2/3 to slightly shorter than lip; Aleutian Islands, adjacent Alaska . . . . 9. *Platanthera convallariifolia*
                  32. Lip lanceolate to ovate, usually obscurely to prominently rounded-dilated at base; spur slightly shorter to slightly longer than lip.
                     33. Lip lanceolate, usually obscurely rounded-dilated at base, whitish green; spur slenderly clavate to slenderly cylindric; flowers not autogamous; pollinia retained within anther sacs (but see note); viscidia oblong; widespread . . . . 7. *Platanthera huronensis*
                     33. Lip ovate, rarely lanceolate, abruptly and broadly dilated at base, apparently yellowish green; spur markedly clavate; flowers autogamous; pollinia rotating forward and downward from anther sacs or fragmenting and pollen masses trailing onto stigma; viscidia linear to linear-oblong; Greenland . . . 8. *Platanthera hyperborea*

1. Platanthera orbiculata (Pursh) Lindley, Gen. Sp. Orchid. Pl., 286. 1835 [E] [F]

Orchis orbiculata Pursh, Fl. Amer. Sept. 2: 588. 1814; *Habenaria orbiculata* (Pursh) Torrey; *H. orbiculata* var. *lehorsii* Fernald; *H. orbiculata* var. *menziesii* (Lindley) Fernald; *Platanthera orbiculata* var. *lehorsii* (Fernald) Catling

Plants 17–62 cm. **Leaves** 2, in subequal basal pair lying on ground; bracts (very rarely, 0–)1–6, scattered along stem; blade broadly elliptic to orbiculate or oblate, 5–21 × 3–22 cm. **Spikes** lax to dense. **Flowers** resupinate, rather showy; calyx mostly greenish white; corolla mostly white; lateral sepals reflexed to somewhat spreading; petals lanceolate-falcate, margins entire; lip descending to somewhat reflexed, linear-oblong to linear-lanceolate, without basal thickening, 7–17 × 1–2.5 mm, margins entire; spur slenderly clavate, 14–27 mm; rostellum lobes directed strongly forward, wide-spreading, angular; pollinaria nearly straight, 3–4.5(–4.8, very rarely) mm; pollinia remaining enclosed in anther sacs; viscidia orbiculate; ovary slender to rather stout, mostly 10–26 mm. $2n = 42$.

Flowering Jun–Aug(–Sep). Mesic to wet coniferous and deciduous forest, fen forest; 0–1500 m; Alta., B.C., Man., N.B., Nfld. and Labr. (Nfld.), N.W.T., N.S., Ont., P.E.I., Que., Sask.; Alaska, Idaho, Ill., Ind., Maine, Md., Mass., Mich., Minn., Mont., Ohio, Oreg., Pa., N.H., N.J., N.Y., N.C., S.Dak., Tenn., Vt., Va., Wash., W.Va., Wis., Wyo.

The key will permit determination of most specimens of *Platanthera orbiculata* and *P. macrophylla*. More precise discrimination of ambiguous specimens is possible by use of the formula from A. H. Reddoch and J. M. Reddoch (1993): spur length + (2 × pollinarium length), which for *P. orbiculata* is less than 38 and for *P. macrophylla* is equal to or greater than 38.

Considerable variation in size and shape of leaves occurs, and although to some extent regional in nature, intergradation is complete; recognition of infraspecific taxa is unwarranted. A few collections from isolated areas on the Pacific Coast of Canada are noteworthy, however. Those are small, few-flowered plants with rather narrow leaves borne alternately or suboppositely toward the base of the stem, as in some Asiatic species. They are in some respects very similar to *Platanthera freynii* Kränzlin, an Asiatic species distinguished primarily by its abruptly narrowed petals, in contrast to the generally broader, but variable, petals in North American plants. These western plants warrant further study to establish their identity and to elucidate relationships between North American and Asiatic species.

2. Platanthera macrophylla (Goldie) P. M. Brown, Wild Fl. Notes 3: 23. 1988 [E]

Habenaria macrophylla Goldie, Edinburgh Philos. J. 6: 331. 1822; *Platanthera orbiculata* (Pursh) Lindley var. *macrophylla* (Goldie) Luer

Plants 23–63 cm. **Leaves** 2, in subequal basal pair lying on ground; bracts (very rarely, 0–)1–6, scattered along stem; blade broadly elliptic, orbiculate, or oblate, 7–24 × 5–19 cm. **Spikes** lax. **Flowers** resupinate, rather showy; calyx mostly greenish white; corolla mostly white; lateral sepals reflexed to somewhat spreading; petals lanceolate-falcate, margins entire; lip descending to somewhat reflexed, linear-oblong to linear-lanceolate, without basal thickening, 10–23 × 1–2.5 mm, margins entire; spur slenderly clavate, 28–46 mm; rostellum lobes directed strongly forward, wide-spreading, angular; pollinaria nearly straight, (4.6–)4.7–6.8 mm; pollinia remaining enclosed in anther sacs; viscidia orbiculate; ovary slender to rather stout, mostly 16–29 mm.

Flowering Jun–Aug. Mesic to wet coniferous and deciduous forest; mostly 0–1000 m; N.B., Nfld. and Labr., N.S., Ont., P.E.I., Que.; Conn., Maine, Mass., Mich., Minn., N.H., N.Y., Pa., Vt.

3. Platanthera hookeri (Torrey ex A. Gray) Lindley, Gen. Sp. Orchid. Pl., 286. 1835 [E]

Habenaria hookeri Torrey ex A. Gray, Ann. Lyceum Nat. Hist. New York 3(first printing): 228. 1835; *H. hookeri* var. *abbreviata* Fernald; *Platanthera hookeri* var. *abbreviata* (Fernald) W. J. Schrenk

Plants 18–45 cm. **Leaves** 2, in subequal basal pair, spreading-ascending or lying on ground; bracts 0(–1, very rarely); blade orbiculate, broadly elliptic, or obovate, 5–17 × 4–13 cm. **Spikes** lax. **Flowers** resupinate, not showy; calyx green; corolla commonly yellowish green; lateral sepals reflexed; petals narrowly lanceolate-falcate, margins entire; lip projecting to porrect, lanceolate to narrowly lance-rhombic, without basal thickening, 8–23 × 1–6 mm, margins entire; spur slenderly conic, 11–27 mm; rostellum lobes very wide-spreading, angular; pollinaria nearly straight; pollinia remaining enclosed in anther sacs; viscidia suborbiculate; ovary slender to rather stout, mostly 8–20 mm. $2n = 42$.

Flowering May–Aug. Dry to mesic coniferous and deciduous forest; mostly 0–500 m; Man., N.B., Nfld. and Labr. (Nfld.), N.S., Ont., P.E.I., Que.; Conn., Ill., Ind., Iowa, Maine, Mass., Mich., Minn., Ohio, Pa., N.H., N.J., N.Y., R.I., Vt., Wis.

4. Platanthera obtusata (Banks ex Pursh) Lindley, Gen. Sp. Orchid. Pl., 284. 1835 [F]

Orchis obtusata Banks ex Pursh, Fl. Amer. Sept. 2: 588. 1814; *Habenaria obtusata* (Banks ex Pursh) Richardson; *H. obtusata* var. *collectanea* Fernald

Plants 5.5–35 cm. **Leaves** 1(–2, rarely), spreading-ascending on base of stem; bracts 0(–1, rarely); blade linear-oblanceolate, elliptic, or broadly obovate, 3.5–15 × 0.8–5 cm. **Spikes** lax. **Flowers** resupinate, not showy, greenish white to yellowish green; corolla commonly whiter than calyx; lateral sepals reflexed; petals rhombic, lanceolate-falcate, margins entire; lip descending, linear, narrowly lance-rhombic, or linear-hastate, with median basal thickening, unlobed, 2.5–8(–10) × less than 1–2 mm, margins entire; spur slenderly conic, 3–8(–10) mm; rostellum lobes directed forward, wide-spreading, rounded-angular; pollinaria straight; pollinia remaining enclosed in anther sacs; viscidia orbiculate; ovary rather slender to stout, mostly 3–10 mm. $2n$ = 42, 63 (American, as *Habenaria obtusata*), 126 (Eurasian, as *Habenaria obtusata*).

Flowering Jun–Aug. Mesic to wet coniferous forest, forested fens, sphagnum bogs, stream banks, tundra, moist roadsides; 0–3500 m; Alta., B.C., Man., N.B., Nfld. and Labr., N.W.T., N.S., Nunavut, Ont., P.E.I., Que., Sask., Yukon; Alaska, Colo., Idaho, Maine, Mass., Mich., Minn., Mont., N.H., N.Y., Oreg., Utah, Vt., Wash., Wis., Wyo.; Eurasia [*Platanthera obtusata* subsp. *oligantha*].

The rare Eurasian *Platanthera obtusata* subsp. *oligantha* (Turczaninow) Hultén differs from the North American subsp. *obtusata* in its smaller dimensions and rhombic-lanceolate lip. It is also said to be densely few-flowered, although some Siberian material is comparable to American plants. Supposedly intermediate plants are reported from Alaska, and much material from that area is reduced in stature and with smaller flowers than typical of American plants. In most cases, however, lips are relatively slender, and the plants seem merely to be stunted by their environment. One or two collections from the Alaskan Peninsula and Aleutians, however, seem entirely referable to subsp. *oligantha* with dense, few-flowered inflorescences of very small flowers with rhombic-lanceolate lips and shorter curved spurs. Eurasian plants are reported to be hexaploid or perhaps sometimes triploid, and if the apparent ploidy differences delimit the taxa, then it should be possible to unequivocally identify Alaskan plants.

5. Platanthera tipuloides (Linnaeus f.) Lindley, Gen. Sp. Orchid. Pl., 285. 1835

Orchis tipuloides Linnaeus f., Suppl. Pl., 401. 1782

Varieties 2 (1 in the flora): Alaska; Aleutian Islands; e Asia.

5a. Platanthera tipuloides (Linnaeus f.) Lindley var. **behringiana** (Rydberg) Hultén, Ark. Bot., n. s. 7: 35. 1968

Limnorchis behringiana Rydberg, Bull. Torrey Bot. Club 28: 620. 1901; *Habenaria behringiana* (Rydberg) Ames

Plants 7.5–18 cm. **Leaves** 1–2, ascending on base or middle of stem, rather abruptly reduced to 1–2 bracts distally; blade lanceolate to ovate-lanceolate or elliptic-oblong, 2.5–8 × 0.8–2 cm. **Spikes** rather dense. **Flowers** resupinate, rather conspicuous, yellowish; corolla commonly richer yellow than more greenish calyx; lateral sepals spreading, slightly reflexed; petals ovate- to broadly lance-falcate, margins entire; lip descending, elliptic, elliptic-oblong, or elliptic-lanceolate, without basal thickening, 4–8 × 1–3 mm, margins entire; spur slenderly cylindric to clavate, 9–13 mm; rostellum lobes directed downward, closely spaced, parallel, very swollen adjacent to orifice of spur, small, obscure, rounded; pollinaria appearing straight; pollinia apparently remaining enclosed in anther sacs; viscidia oblong, perhaps varying to orbiculate; ovary stout, mostly 6–7 mm.

Flowering Jul–Aug. Exposed, wet tundra; 0–200 m; Alaska (w Aleutian Islands); Asia (Komandorskiye Islands).

Only a few specimens of *Platanthera tipuloides* var. *behringiana* have been seen, and the description is consequently somewhat imprecise.

6. Platanthera dilatata (Pursh) Lindley ex L. C. Beck, Bot. North. Middle States, 347. 1833 [E]

Orchis dilatata Pursh, Fl. Amer. Sept. 2: 588. 1814; *Habenaria dilatata* (Pursh) Hooker; *Limnorchis dilatata* (Pursh) Rydberg

Plants 11–130 cm or more. **Leaves** few–several, ascending to recurved-spreading, scattered along stem, gradually reduced to bracts distally; blade linear, linear-lanceolate, oblong, or oblanceolate, 3.5–32 × 0.3–7 cm. **Spikes** very lax to very dense. **Flowers** resupinate, rather showy, conspicuous, white; lateral sepals spreading to reflexed; petals ovate- to linear-lance-falcate, margins entire; lip descending or apex adhering to dorsal sepal and petal apices, linear-lanceolate to lanceolate, without basal thickening, 4–11 × 2–5 mm, base slightly rounded-dilated to orbiculate, rarely rhombic, margins entire; rostellum lobes parallel to slightly divergent, directed downward, very small, obscure, rounded; pollinaria straight; pollinia remaining enclosed in anther sacs; ovary rather slender to stout, 5–15 mm. $2n = 42$.

An intense clove scent distinguishes *Platanthera dilatata* from related species across most of its range, but in the far northwest a more complex blend of spicy fragrances predominates.

Varieties 3 (3 in the flora): North America.

1. Spur shorter than lip, clavate to slightly capitate . . . . 6b. *Platanthera dilatata* var. *albiflora*
1. Spur nearly equaling to exceeding lip, slender to slightly clavate.
    2. Spur about equal to lip (mostly within 2 mm) . . . . 6a. *Platanthera dilatata* var. *dilatata*
    2. Spur markedly exceeding lip (commonly 1½ or more times length of lip) . . . . 6c. *Platanthera dilatata* var. *leucostachys*

6a. Platanthera dilatata (Pursh) Lindley ex L. C. Beck var. **dilatata** E

Lip usually with basal dilation moderately to strongly developed, 5–10 mm; spur slenderly cylindric to slightly clavate, 4–12 mm (± as long as lip, mostly within 2 mm), apex obtuse to tapered; viscidia linear to linear-oblong. $2n = 42$.

Flowering May–Aug. Wet meadows, tundra, marshes, fens, stream banks, shores, ditches, seeping slopes, roadsides; 0–3100 m; Alta., B.C., Man., N.B., Nfld. and Labr., N.S., Ont., P.E.I., Que., Yukon; Alaska, Colo., Idaho, Ill., Ind., Maine, Mass., Mich., Minn., N.H., N.Y., Utah, Vt., Wash.

6b. Platanthera dilatata (Pursh) Lindley ex L. C. Beck var. **albiflora** (Chamisso) Ledebour, Fl. Ross. 4: 71. 1853 E

Habenaria borealis Chamisso var. *albiflora* Chamisso, Linnaea 3: 28. 1828; *H. dilatata* (Pursh) Hooker var. *albiflora* (Chamisso) Correll

Lip commonly with pronounced basal dilation, sometimes lateral margins elaborated along axes, diverging forward and forming lobules, 6–10 mm; spur clavate to slightly capitate, 2–7 mm (markedly shorter than lip), apex obtuse; viscidia linear-oblong to oblong. $2n = 42$.

Flowering Jun–Aug. Wet meadows, tundra, marshes, fens, stream banks, shores, ditches, seeping slopes, roadsides; 0–3700 m; Alta., B.C., Yukon; Alaska, Colo., Idaho, Mont., Nev., Oreg., Utah, Wash., Wyo.

6c. Platanthera dilatata (Pursh) Lindley ex L. C. Beck var. **leucostachys** (Lindley) Luer, Native Orchids U.S. & Canada, 225. 1975 E

Platanthera leucostachys Lindley, Gen. Sp. Orchid. Pl., 288. 1835; *Habenaria dilatata* (Pursh) Hooker var. *leucostachys* (Lindley) Ames; *H. leucostachys* (Lindley) S. Watson; *Limnorchis leucostachys* (Lindley) Rydberg

Lip often slender with dilation absent or only moderately developed, 4–11 mm; spur slenderly cylindric to filiform, 8–20 mm (markedly longer than lip), apex usually slenderly tapered; viscidia linear to linear-oblong. $2n = 42$.

Flowering May–Aug. Wet meadows, marshes, fens, stream banks, shores, ditches, seeping slopes, roadsides; 0–3400 m; B.C.; Alaska, Calif., Mont., Nev., Oreg., Utah, Wash.

Platanthera dilatata traditionally has been divided on the basis of spur length into three varieties, one of which, var. *leucostachys*, is sometimes treated as a distinct species. These infraspecific taxa seem to reflect differing pollination pressures. The moderate spur length and diurnal fragrance of var. *dilatata* suggests adaptation to diurnal Lepidoptera; the long spurs and primarily nocturnal fragrance of var. *leucostachys* indicates specialization for moth pollination, and the short spurs and often broader viscidia of var. *albiflora* suggest a broader range of pollinators or, in extreme cases, specialization for bee or fly pollination. Alone, these characteristics might support recognition at the specific level, but intermediates and populations with variable spur lengths are abundant. In some plants in western Canada, in particular, spurs that are very short when the flower is young grow to equal the lip as the flower ages, and in some they may eventually greatly exceed the lip, thereby simulating all three varieties. Plants with short spurs, either thick or variably slender, occur occasionally across the range of the species. In the southern Rocky Mountains spur reduction reaches an extreme, yet populations with moderate-length spurs occur there as well. In the broad sense, then, *P. dilatata* forms a cohesive unit in which spur length varies greatly, apparently in response to differing pollination pressures. The northwest is the center of variability of the species, and as it ranges eastward through the boreal forest, and southward down the Rockies and the more western ranges, it appears to have specialized for different pollinators. The recognized varieties of *P. dilatata* are evidentally merely endpoints in a very complex variation pattern. They have some utility for dis-

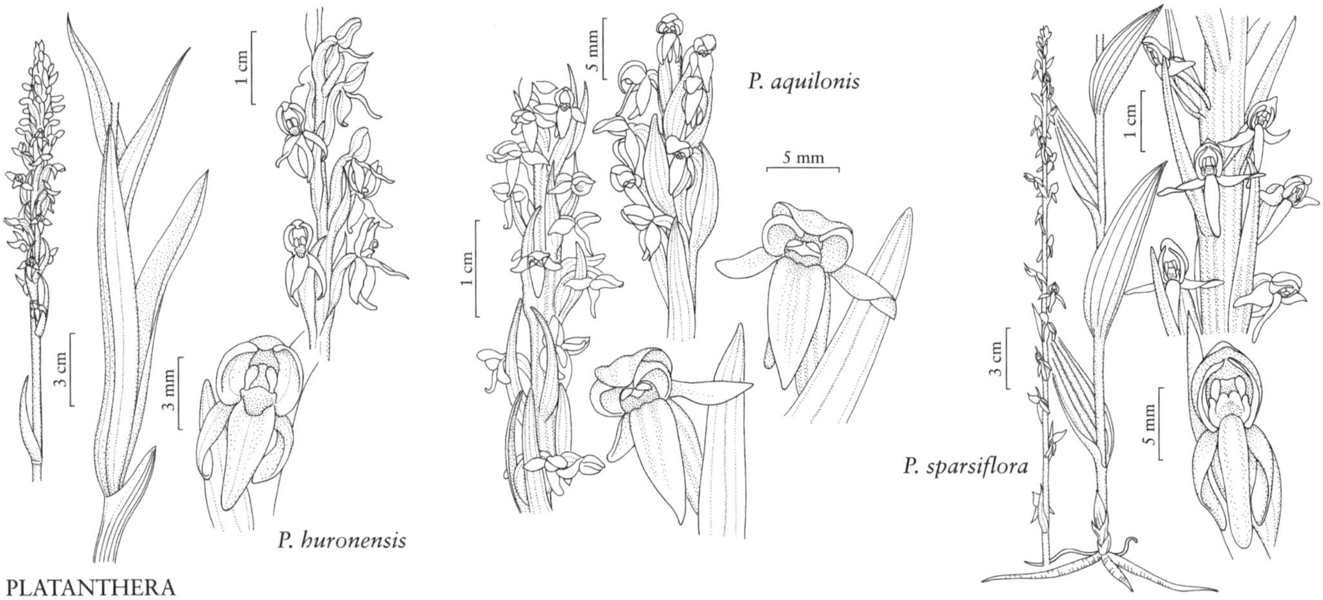

PLATANTHERA

cussion purposes, but they are very simplistic representations of the underlying situation. Variety *leucostachys* and, to a lesser extent, var. *dilatata* appear to be real entities that have emerged from a background of variability that continues to produce similar plants. This variability is here treated within var. *albiflora*.

Platanthera dilatata hybridizes with *P. huronensis* and *P. purpurascens*, and perhaps also does so with other related species; see the note under 10. *P. aquilonis*.

7. **Platanthera huronensis** (Nuttall) Lindley, Gen. Sp. Orchid. Pl., 288. 1835 E F

Orchis huronensis Nuttall, Gen. N. Amer. Pl. 2: 189. 1818; *Habenaria hyperborea* (Linnaeus) R. Brown var. *huronensis* (Nuttall) Farwell; *H.* ×*media* (Rydberg) Niles; *Limnorchis media* Rydberg; *Platanthera hyperborea* (Linnaeus) Lindley var. *huronensis* (Nuttall) Luer; *P.* ×*media* (Rydberg) Luer

Plants 10–100 cm or more. **Leaves** few–several, ascending, scattered along stem, gradually reduced to bracts distally; blade oblong to linear-lanceolate, 5–30 × 0.6–7 cm. **Spikes** very lax to very dense. **Flowers** resupinate, not showy but sometimes conspicuous, whitish green; corolla often whiter than calyx; lateral sepals spreading to slightly reflexed; petals ovate- to lance-falcate, margins entire; lip descending or apex adhering to dorsal sepal and petal apices, lanceolate to nearly linear, without basal thickening, 5–12 × 2–4 mm, base slightly to rather markedly rounded-dilated, margins entire; spur slenderly cylindric to clavate, 4–12 mm, apex usually slenderly tapered; rostellum lobes divergent, directed downward, very small, obscure, rounded; pollinaria straight; pollinia remaining enclosed in anther sacs; viscidia oblong; ovary rather slender to stout, mostly 6–15 mm. $2n = 84$.

Flowering Jun–Aug. Wet meadows, tundra, marshes, fens, stream banks, shores, ditches, seeping slopes, roadsides; 0–3300 m; Alta., B.C., Man., N.B., Nfld. and Labr., N.S., Ont., P.E.I., Que.,Yukon; Alaska, Colo., Conn., Idaho, Maine, Mass., Mich., Minn., Mont., N.H., N.J., N.Mex., N.Y., Oreg., R.I., S.Dak., Utah, Vt., Wash., Wis., Wyo.

Northwestern plants commonly treated as *Platanthera hyperborea* var. *viridiflora* (Chamisso) Kitamura (note Kitamura's priority over Luer) are *P. huronensis*; Chamisso's name furthermore is synonymous with *P. stricta*. Aleutian and coastal Alaskan plants are often short, stout, and broad-leaved, and they have incorrectly been referred to 9. *P. convallariifolia*.

Platanthera huronensis as here delimited does not autopollinate in the manner of *P. aquilonis*. Occasional plants and populations that may be referable to *P. huronensis*, however, exhibit the movement of pollinia typical of *P. aquilonis*. These plants might reflect infraspecific variation within an allotetraploid species, result from hybridization, or constitute a distinct taxon. The relationship of some of these plants to *P. hyperborea* needs study.

Platanthera huronensis is typically intensely fragrant with the sweet, pungent scent of some related species.

Platanthera huronensis is known to hybridize with *P. dilatata*; it may hybridize with other species as well. Although hybrids of *P. dilatata* and *P. aquilonis* may occur, the name traditionally used for them, *P.* ×*media* (Rydberg) Luer is a synonym of *P. huronensis*. See notes under 10. *P. aquilonis* and 8. *P. hyperborea*.

8. Platanthera hyperborea (Linnaeus) Lindley, Gen. Sp. Orchid. Pl., 287. 1835

Orchis hyperborea Linnaeus, Mant. Pl., 121. 1767; *Habenaria hyperborea* (Linnaeus) R. Brown; *Limnorchis hyperborea* (Linnaeus) Rydberg

Plants 7–35 cm. **Leaves** few–several, ascending to arcuate-spreading, scattered along stem or clustered at base, sometimes in basal rosette, gradually reduced to bracts distally; blade elliptic-oblong to ovate-oblong or linear-lanceolate, 3–14 × 0.4–4 cm. **Spikes** lax to very dense. **Flowers** resupinate, not showy but sometimes conspicuous, yellowish or perhaps whitish green; lateral sepals spreading to slightly reflexed; petals ovate-falcate, margins entire; lip descending, usually ovate with abruptly dilated base or occasionally lanceolate with less prominent dilation, without basal thickening, 4–5.5(–6) × (1.5–)2–2.5 mm; spur clavate, somewhat tapering toward obtuse apex, 4–6 mm; rostellum lobes divergent, directed downward, very small, obscure, rounded; pollinaria straight; pollinia often protruding from anther sacs, commonly fragmenting and pollen masses trailing onto stigma; viscidia linear to linear-oblong; ovary rather slender to stout, mostly 6–10 mm. $2n = 84$.

Flowering Jul–Aug. Wet tundra, stream banks; 0 m; Greenland; Iceland.

The relationship between *Platanthera hyperborea* and *P. huronensis* is unclear. Typical Icelandic material differs from *P. huronensis* in its shorter, broader, more extremely dilated, and evidently often yellowish lip; shorter, more strongly clavate spur; and narrower viscidia. The plants furthermore appear to be autogamous. Most Greenland material appears to be entirely comparable to these Icelandic plants. In both areas, however, considerable variation occurs, and some plants suggest *P. huronensis*. Whether this variation reflects the occurrence in these areas of two taxa or is within *P. hyperborea* is unknown.

Platanthera hyperborea refers to a bewildering complex that has defied satisfactory treatment. The number and delimitation of species, indeed, the nature of species, in the group is unclear. The system proposed here utilizes more characters than past attempts and it is to be hoped it is an improvement over its predecessors, but it is far from definitive. Hybridization is frequently invoked as an explanation for much of the variation encountered in the group, and some of the possible crossings have been given names. Until species can be properly delimited, however, it is premature to recognize putative hybrids with formal epithets.

9. Platanthera convallariifolia Fischer ex Lindley, Gen. Sp. Orchid. Pl., 287. 1835 (as convallariaefolia)

Limnorchis convallariifolius (Fischer ex Lindley) Rydberg

Plants 15–27 cm. **Leaves** few–several, ascending to rather wide spreading, scattered along stem or commonly loosely clustered on distal portion, gradually reduced to bracts distally; blade elliptic-lanceolate to elliptic-oblong, 4.5–12 × 0.7–2.5 cm. **Spikes** moderately dense. **Flowers** resupinate, not showy, sometimes conspicuous, whitish green, perhaps more yellowish on lip; lateral sepals spreading; petals ovate-falcate, margins entire; lip descending, linear to lance-linear, without basal thickening, 4–9 × 1–2 mm, margins entire; spur slenderly to markedly clavate, rarely more slender, 4–8 mm, apex obtuse; rostellum lobes widely divergent, directed downward, slightly angular, very small, obscure; pollinaria straight; pollinia remaining enclosed in anther sacs but apparently often fragmenting; viscidia oblong to oblong-spatulate or elliptic-suborbiculate; ovary slender to stout, mostly 6–10 mm.

Flowering late Jun–early Aug. Wet meadows, tundra, marshes, fens, stream banks, shores, seeping slopes; 0–400 m; Alaska; Asia (Kamchatka to Hokkaido).

Platanthera convallariifolia is commonly reported from the Aleutians and adjacent coastal Alaska based on short, very broad-leaved plants with rather large flowers in dense, thick inflorescences. Those plants, however, apparently represent robust, maritime phenotypes of *P. huronensis* or in some cases *P. stricta* or *P. dilatata* comparable to those seen in *Piperia elegans* and *Spiranthes romanzoffiana*. In fact, typical Asiatic *Platanthera convallariifolia* is a slender-leaved plant entirely comparable in habit to *P. huronensis*. The two species are very similar, and they share a tetraploid chromosome number. Typical Asiatic *P. convallariifolia* is characterized by a linear lip and clavate spur about 2/3 to nearly the length of the lip. The lip occasionally varies to slightly lanceolate and the spur to somewhat saccate. The column appears somewhat broader than in *P. huronensis*, and viscidia also are apparently rather broad. A few specimens from the Aleutians and Alaskan peninsula exhibit these floral characteristics. Vegetatively, however, some Alaskan material is distinctive in the position and orientation of the leaves. In floral and vegetative characters, Alaskan plants furthermore intergrade with *P. huronensis*. Whether Alaskan plants are truly referable to *P. convallariifolia* is uncertain, and the status of the Asiatic species also needs study. Both *P. convallariifolia* and the Alaskan plants here referred to it appear to combine features of *P. huronensis* and *P. stricta*. *Platanthera convallariifolia* may have arisen through hybridization of these two species, or of *P. stricta* and *P. dilatata*, thus accounting for the difficulty in identification. The description presented here is based on few Alaskan plants and is apt to be too restrictive; therefore the poor understanding of the plant makes the present delimitation and distinctions from *P. huronensis* rather arbitrary.

10. Platanthera aquilonis Sheviak, Lindleyana 14: 193, figs. 1–5. 1999 [E] [F]

Plants 5–60 cm. **Leaves** few–several, ascending to arcuate-spreading, scattered along stem, gradually reduced to bracts distally; blade oblong to linear-lanceolate, 2.7–23 × 0.4–4 cm. **Spikes** lax to very dense. **Flowers** resupinate, not showy, yellowish green with dull yellowish lip, or more whitish green under cool conditions; lateral sepals spreading to reflexed; petals rhombic-ovate- to lance-falcate, margins entire; lip descending, projecting, or apex adhering to dorsal sepal and petal apices, rhombic-lanceolate to lanceolate, without basal thickening, 2.5–6 × 1–1.5 mm, base not rounded-dilated, margins entire; spur clavate or sometimes rather cylindric, 2–5 mm, apex usually broadly obtuse; rostellum lobes divergent, directed downward, very small, rounded, obscure; pollinaria straight; pollinia rotating forward and/or fragmenting, loose pollen masses free of anther sacs and trailing downward onto stigma; viscidia orbiculate; ovary rather slender to stout, mostly 5–13 mm. $2n = 42$.

Flowering May–Aug. Wet meadows, tundra, marshes, fens, stream banks, shores, ditches, seeping slopes, roadsides, borrow pits, mesic deciduous forest slopes; 0–2500 m; Alta., B.C., Man., N.B., Nfld. and Labr., N.W.T., N.S., Ont., P.E.I., Que., Sask., Yukon; Alaska, Colo., Conn., Idaho, Ill., Ind., Iowa, Maine, Mass., Mich., Minn., Mont., Nebr., N.H., N.J., N.Mex., N.Y., N.Dak., Ohio, Oreg., Pa., R.I., S.Dak., Utah, Vt., Wash., Wis., Wyo.

Flowers of *Platanthera aquilonis* are usually scentless, but in the far northwest they have a sweet, pungent scent, like that of some related species. The flowers are commonly self-pollinating: the pollinia rotate forward and downward, contacting the stigma, and/or the pollen masses dissociate and are deposited on the stigma as if they had sifted downward.

Platanthera aquilonis is a North American diploid species long confused with the tetraploid Icelandic *P. hyperborea* (Linnaeus) Lindley. Flowers of both species autopollinate, although the details of the mechanisms may differ. The two species differ in column structure and lip and viscidium shape. True *P. hyperborea* is similar to *P. huronensis*, and the relationship of these two species needs further study.

11. Platanthera purpurascens (Rydberg) Sheviak & W. F. Jennings, N. Amer. Native Orchid J. 3: 445. 1997 [E]

Limnorchis purpurascens Rydberg, Bull. Torrey Bot. Club 28: 269. 1901; *Platanthera hyperborea* (Linnaeus) Lindley var. *purpurascens* (Rydberg) Luer

Plants 24–80 cm. **Leaves** few–several, abruptly diverging or sometimes ascending, scattered along stem; blade oblong to ovate or lanceolate, 5.5–16(–28) × 1–4.5 cm. **Spikes** very lax to dense, flowers sometimes irregularly grouped in fascicles. **Flowers** resupinate, not showy, green to yellowish green, often suffused with blue or lip sometimes with red; lateral sepals spreading to reflexed; petals ovate- to lance-falcate, margins entire; lip descending to somewhat reflexed or projecting, lanceolate to ovate, sometimes with prominent suborbiculate basal dilation, without basal thickening, 4–8 × 1.5–3 mm; spur strongly clavate to scrotiform, 2–3 mm, apex obtuse; rostellum lobes strongly diverging, very small, rounded, obscure; pollinaria straight; pollinia remaining enclosed in anther sacs; viscidia orbiculate to oblong; ovary rather slender to stout, mostly 5–9 mm. $2n = 42$ [63].

Flowering Jul–Aug. Stream banks, shores, ditches, fens, seeping slopes, roadsides, commonly in thickets, open woods; 2500–3300 m; Ariz., Calif., Colo., N.Mex., Wyo.

Flowers of *Platanthera purpurascens* are strongly semen scented, and their musty scent is markedly different than the sweet and spicy scents of other fragrant species.

In some areas this species hybridizes with *Platanthera dilatata* var. *albiflora*, producing sweetly fragant plants with lanceolate green lips. At the southern edge of its range, however, it often occurs with similar plants in the absence of another putative parent. The identity of these plants is uncertain. Their rather generalized morphology might mask polyphyletic origins.

12. Platanthera stricta Lindley, Gen. Sp. Orchid. Pl., 288. 1835 [E]

Habenaria borealis Chamisso var. *viridiflora* Chamisso; *H. saccata* Greene; *Limnorchis stricta* (Lindley) Rydberg; *Platanthera gracilis* Lindley; *P. hyperborea* (Linnaeus) Lindley var. *viridiflora* (Chamisso) Kitamura; *P. hyperborea* var. *viridiflora* (Chamisso) Luer; *P. saccata* (Greene) Hultén

Plants 18–100 cm. **Leaves** few–several, abruptly diverging or sometimes ascending, scattered along stem; blade oblong to ovate or rarely linear-lanceolate, 3–32 × (0.6–)1–4.5 cm. **Spikes** very lax to dense. **Flowers** resupinate, not showy, green to yellowish green; lateral sepals spreading to reflexed; petals ovate- to lance-falcate, margins entire; lip descending to somewhat reflexed or projecting, linear to lanceolate, without basal thickening, 3–9 × less than 1–3 mm; spur strongly clavate, capitate-inflated, or scrotiform, 2–6 mm, apex obtuse; rostellum lobes approximately parallel to convergent, directed downward, very small, rounded, obscure; pollinaria straight; pollinia remaining enclosed in anther sacs; viscidia orbiculate; ovary rather slender to stout, mostly 5–10 mm. $2n = 42$.

Flowering Jun–Aug. Wet meadows, tundra, marshes, fens, stream banks, shores, ditches, seeping slopes, roadsides; 0–2300 m; Alta., B.C., Yukon; Alaska, Calif., Idaho, Mont., Oreg., Wash, Wyo.

Flowers of *Platanthera stricta* are evidently mostly scentless or only faintly spicy.

The plants here treated as *Platanthera stricta* have in common more or less saccate spurs, orbiculate viscidia, and leaves that abruptly diverge from the stem, often at angles approaching 90° (this feature is sometimes obscured in sheltered, deeply shaded habitats). The plants described as *P. gracilis* Lindley are florally typical of the slender-spurred extreme of *P. stricta*; they differ only in peculiarly reduced, slenderly oblong but nonetheless abruptly wide-spreading leaves. The plants figured by C. A. Luer (1975) as *P. hyperborea* var. *gracilis* (Lindley) Luer are not referable to *P. stricta* but rather are apparently hybrids of *P. stricta* and *P. dilatata*. Critical study of the description of *Habenaria borealis* var. *viridiflora* Chamisso and an evident isotype show this plant to be referable to *P. stricta*, although the name has been applied to *P. huronensis* in the Northwest and to *P. convallariifolia* in Japan. See also the discussion under 9. *P. convallariifolia*.

13. Platanthera sparsiflora (S. Watson) Schlechter, Bull. Herb. Boissier 7: 538. 1899 [F]

Habenaria sparsiflora S. Watson, Proc. Amer. Acad. Arts 12: 276. 1877; *H. sparsiflora* var. *laxiflora* (Rydberg) Correll; *Limnorchis ensifolia* Rydberg; *L. laxiflora* Rydberg; *L. sparsiflora* (S. Watson) Rydberg; *Platanthera sparsiflora* var. *ensifolia* (Rydberg) Luer

Plants 20–125 cm or more. **Leaves** few–several, ascending to recurved-spreading, scattered along stem or sometimes clustered near base, gradually to abruptly reduced to bracts distally; blade ovate-, oblong-, or oblanceolate-elliptic to linear, 6.5–30 × 0.8–5 cm. **Spikes** very lax to dense. **Flowers** resupinate, not showy, green to yellowish green; lateral sepals reflexed or sometimes spreading; petals ovate- to lance-falcate, margins entire; lip descending or reflexed, oblong or linear to linear-oblong or linear-lanceolate, a thickened median basal ridge present or absent, 4.5–11 × 0.6–3 mm, margins entire; spur slightly clavate, cylindric, or tapering toward usually rather acute apex; rostellum lobes somewhat divergent, directed forward, rounded-subangular, rather prominent, to strongly divergent, very small, rounded, obscure; pollinaria straight; pollinia remaining enclosed in anther sacs; viscidia orbiculate to oblong; ovary rather slender to stout, mostly 5–14 mm. $2n = 42$.

Flowering Apr–Sep. Wet meadows, marshes, fens, stream banks, shores, seeping slopes; 0–3500 m; Ariz., Calif., Colo., Nev., N.Mex., Oreg., Utah; Mexico (Baja California).

The columns of *Platanthera sparsiflora*, *P. brevifolia*, and *P. zothecina* differ from those of superficially similar species in their comparatively large, diverging, and forward-directed rostellum lobes; they are typical of columns of *Platanthera* species that deposit pollinaria on insects' compound eyes. The resulting broad column fills a large portion (generally about 2/3) of the width of the hood formed by the dorsal sepal and petals, and in live and well-prepared herbarium specimens this feature separates these species from their relatives with proportionately smaller, narrower columns. Selection for more generalized pollination mechanisms, as is known elsewhere in the genus, however, has apparently produced derivatives of *P. sparsiflora* with reduced columns that simulate members of the *P. hyperborea* complex. The use of column structure, as has been traditionally employed to delimit *P. sparsiflora*, therefore actually recognizes a pollination syndrome and may not be adequate to represent the evolutionary and systematic status of this group.

Platanthera sparsiflora is highly variable in both vegetative habit and flower structure. Plants may be rather short, stout, with numerous broad leaves scattered along the stem and a rather dense inflorescence of comparatively large flowers. Others are tall and very slender, with few small, ascending leaves widely spaced on the proximal portions of the stem and very lax inflorescences of small flowers (*Limnorchis laxiflora* Rydberg). Others, with long, usually dense inflorescences of small flowers distal to basal clusters of leaves, have recently been treated as *Limnorchis ensifolia* by W. A. Weber (1990). These last, also discussed at length by D. S. Correll (1950) under *Habenaria sparsiflora* var. *laxiflora*, are characterized by smaller columns in which the reduced rostellum lobes appear to be folded forward, replacing the typical hemispheric chamber with a narrower fissure. Flower size in some plants furthermore may vary greatly in different years. Spurs vary from uniformly tapering to clavate, extending stiffly along the ovary to curving forward, and in length range from slightly shorter than the lip to about 1.6 times its length. In both vegetative and floral characters various degrees of intermediacy and combinations of character extremes occur. The complex including the three species recognized here is currently under critical study.

14. Platanthera brevifolia (Greene) Kraenzlin, Orchid. Gen. Sp. 1: 639. 1899

Habenaria brevifolia Greene, Bot. Gaz. 6: 218. 1881; *H. sparsiflora* var. *brevifolia* (Greene) Correll; *Limnorchis brevifolia* (Greene) Rydberg; *Platanthera sparsiflora* var. *brevifolia* (Greene) Luer

Plants 16–60 cm. **Leaves** few, ascending-sheathing, little more than bracts, scarcely reduced distally; bracts often lacking; blade ovate to elliptic-ovate, 2.5–6 × 0.7–2 cm. **Spikes** lax to rather dense. **Flowers** resupinate, not showy, green to yellowish green; lateral sepals reflexed or sometimes spreading; petals ovate- to lance-falcate, margins entire; lip descending or reflexed, linear to linear-oblong or linear-lanceolate, 6–10(–14) mm, margins entire, often with series of low parallel ridges on adaxial surface; spur cylindric or tapering toward apex, 9–20 mm, apex usually rather acute; rostellum lobes somewhat divergent, directed forward, rounded-subangular, rather prominent; pollinaria straight; pollinia remaining enclosed in anther sacs; viscidia orbiculate to broadly elliptic-oblong; ovary rather slender to stout, mostly 6–12 mm.

Flowering Jul–Sep. Dry to moist open, usually coniferous, forest, sometimes seasonally moist banks of intermittent streams; 2100–2750 m; N.Mex.; Mexico.

The greatly reduced herbage of *Platanthera brevifolia* is distinctive. Robust plants may lack developed leaves altogether, with merely a few reduced bracts at the base of the stem and an inflorescence comprising nearly the entire height of the plant. A strong mycotrophic mode is suggested by the occurrence in some colonies of truly albino plants without pigmentation.

Platanthera brevifolia largely replaces *P. sparsiflora* in Mexico and is very distinctive at its northern limit in the United States. Although *P. sparsiflora* is exceedingly variable in floral dimensions, in habit, and in the number, size, and placement of leaves, the pattern seen in *P. brevifolia* seems to lie outside this range of variation. With a generally longer spur and very different ecologic requirements, it is here treated at the specific level. Nonetheless, *P. brevifolia* is reported to intergrade with *P. sparsiflora* in Mexico (R. McVaugh and W. R. Anderson. 1974+, vol. 16, pp. 254–256), yet the latter species probably does not range there beyond a possible occurrence in Baja California (M. A. Soto 1988).

15. Platanthera zothecina (L. C. Higgins & S. L. Welsh) Kartesz & Gandhi, Phytologia 69: 134. 1990 [C] [E]

Habenaria zothecina L. C. Higgins & S. L. Welsh, Great Basin Naturalist 46: 259. 1986

Plants 24–38 cm. **Leaves** few, scattered along stem, sometimes restricted to proximal portion, wide-spreading, gradually to abruptly reduced distally; bracts 1 or absent; blade oblanceolate, narrowly elliptic, oblong, or ovate-elliptic, 7–17 × 1.8–3.5 cm. **Spikes** rather lax. **Flowers** resupinate, not showy, green to yellowish green; lateral sepals reflexed to somewhat spreading; petals ovate-falcate, margins entire; lip descending or reflexed, linear-lanceolate to linear-lance-elliptic, 5–12 × 1–3 mm, margins entire, midline scarcely (perhaps variably) thickened toward base; spur slenderly cylindric, 12–17 mm; rostellum lobes divergent, directed forward, rounded-subangular, rather prominent; pollinaria straight; pollinia remaining enclosed in anther sacs; viscidia orbiculate to suborbiculate; ovary rather slender to stout, 9–16 mm.

Flowering May–Aug. "Hanging garden" communities on moist to wet, dripping sandstone cliffs and ledges, occasionally springy sites, riparian meadows; of conservation concern; 1200–1700 m; Ariz., Colo., Utah.

Only recently described, the original publication separated *Platanthera zothecina* from *P. sparsiflora* solely on the basis of the former's spur/lip length ratio of 1.5 and few-flowered habit. In fact, the range in ratios overlaps, with a range of at least 1.3–2.6 in *P. zothecina* and 0.7–1.6 in *P. sparsiflora*. Actual spur lengths are more useful for determination, and perhaps more significant, because they may indicate specialization for different pollinators; the very limited overlap in spur lengths supports recognition of *P. zothecina*. Other features support the separation. The thickened median ridge toward the base of the lip characteristic of *P. sparsiflora* seems to be lacking in *P. zothecina*, which instead seems to display a parallel series of low ridges, but this distinction unfortunately cannot be established with certainty from the limited sample available. The columns should be investigated in detail, for a limited sample suggests differences in orientation of anther sacs and rostellum lobes. *Platanthera sparsiflora* is often fewer flowered than *P. zothecina*, but the typically few-flowered inflorescences of the latter distal to the broad, spreading, and commonly succulent leaves of an often peculiar whitish green color, contribute a markedly distinctive appearance to the plant.

Because of the restricted distribution of *Platanthera zothecina* and its recent description, the species is poorly known; the description here is based on few specimens and hence apt to be too restrictive.

16. Platanthera limosa Lindley, Ann. Nat. Hist. 4: 381. 1840

Habenaria limosa (Lindley) Hemsley; *H. thurberi* A. Gray; *Limnorchis arizonica* Rydberg

Plants 30–165 cm. **Leaves** few to several, spreading-ascending on base of stem, gradually reduced to bracts upwards; blade lanceolate, 9–28 × 1.2–3.5 cm. **Spikes** dense to lax. **Flowers** resupinate, inconspicuous, green; lateral sepals reflexed to spreading; petals lanceolate-, ovate-oblong-, or ovate-falcate, margins entire; lip descending to reflexed, linear-oblong, linear-elliptic, or rhombic-ovate, usually with inconspicuous to scarcely discernable (possibly sometimes absent) median basal thickening, 2.5–8.5 × less than 1–3.5 mm, margins entire; spur filiform, tapering toward apex, very rarely filiform-conic from slightly stout base, 8–25 mm; rostellum lobes mostly parallel, closely spaced, directed downward, very small, rounded, obscure; pollinaria straight; pollinia remaining enclosed in anther sacs; viscidia evidently elliptic-oblong; ovary rather slender to stout, mostly 4.5–11 mm.

Flowering Jun–Aug. Open to lightly forested springy marshes, seeps, stream banks; 1800–2500 m (- 4000 m, Central America); Ariz., N.Mex.; Mexico; Central America (to Guatemala).

Platanthera limosa is traditionally distinguished from other slender-spurred plants and from *P. sparsiflora* by the presence on the lip of a median thickening that culminates in a small basal process. Although diagnostic when present, this feature is often obscure, especially in live material, and seems to be lacking in some plants. *Platanthera limosa* is most consistently separated from other green-flowered species by the combination of very small, narrow column and slender spur much longer than the lip. These features are in contrast to the very large, broad columns of *P. sparsiflora* and *P. zothecina* and much shorter spurs of the *P. hyperborea* complex. The lips of most plants are elliptic to elliptic-oblong, unlike any other species except the Aleutian *P. tipuloides*, and such plants are readily determined by lip shape and spur characters. Within populations plants may vary greatly in flower size, spur shape and length, and the ratio of spur to lip length. Individual plants may vary markedly in size of the flowers and density of the inflorescence in different years. That situation warrants additional study, as it may obscure the presence of other taxa.

17. Platanthera chorisiana (Chamisso) Reichenbach f. in H. G. L. Reichenbach et al., Icon. Fl. Germ. Helv. 13–14: 128. 1851

Habenaria chorisiana Chamisso, Linnaea 3: 31. 1828; *Limnorchis chorisiana* (Chamisso) J. P. Anderson; *Pseudodiphryllum chorisianum* (Chamisso) Nevski

Plants 4–20(–30) cm. **Leaves** (1–)2(–4), spreading, subopposite toward base of stem; bracts (0–)1(–2) distally; blade elliptic, lanceolate, to suborbiculate, 2–9 × 0.9–4 cm. **Spikes** rather lax to dense. **Flowers** incompletely resupinate, inconspicuous, greenish; lateral sepals porrect; petals broadly ovate, margins entire; lip projecting forward, concave, orbiculate, ovate, or elliptic, without basal thickening, 1.5–2.5 × 1.2–2 mm, margins entire; spur stoutly cylindric to saccate, obtuse, 0.7–1.25 mm; rostellum lobes spreading, directed downward, minute, obscure; pollinaria straight; pollinia evidently remaining enclosed in anther sacs; viscidia oblong to suborbiculate-quadrangular; ovary rather stout, mostly 2–4 mm. $2n = 42$.

Flowering Jul–Aug. Wet tundra, meadows, bogs; 0–400 m; B.C.; Alaska, Wash.; Asia (Kamchatka to Japan).

18. Platanthera flava (Linnaeus) Lindley, Gen. Sp. Orchid. Pl., 293. 1835 E F

Orchis flava Linnaeus, Sp. Pl. 2: 942. 1753; *Habenaria flava* (Linnaeus) R. Brown; *Perularia scutellata* (Nuttall) Small

Plants 13–65 cm. **Leaves** (1–)2–3(–4), spreading to ascending, gradually to rather abruptly reduced to bracts distally; blade lanceolate to elliptic-oblong, 5–35 × 1–7 cm. **Spikes** lax to dense. **Flowers** resupinate, inconspicuous, yellowish green; lateral sepals porrect to reflexed; petals oblong-orbiculate to ovate, apex somewhat crenulate; lip descending to commonly reflexed, 2–6 × 1.8–5 mm, base auriculate, with prominent central tubercle, distal margin entire to slightly crenulate; spur cylindric to strongly clavate, 4–8(–11) mm, apex rarely 2-lobed; rostellum lobes mostly parallel to slightly diverging, directed downward, minute, obscure; pollinia remaining enclosed in anther sacs; viscidia mostly quad-

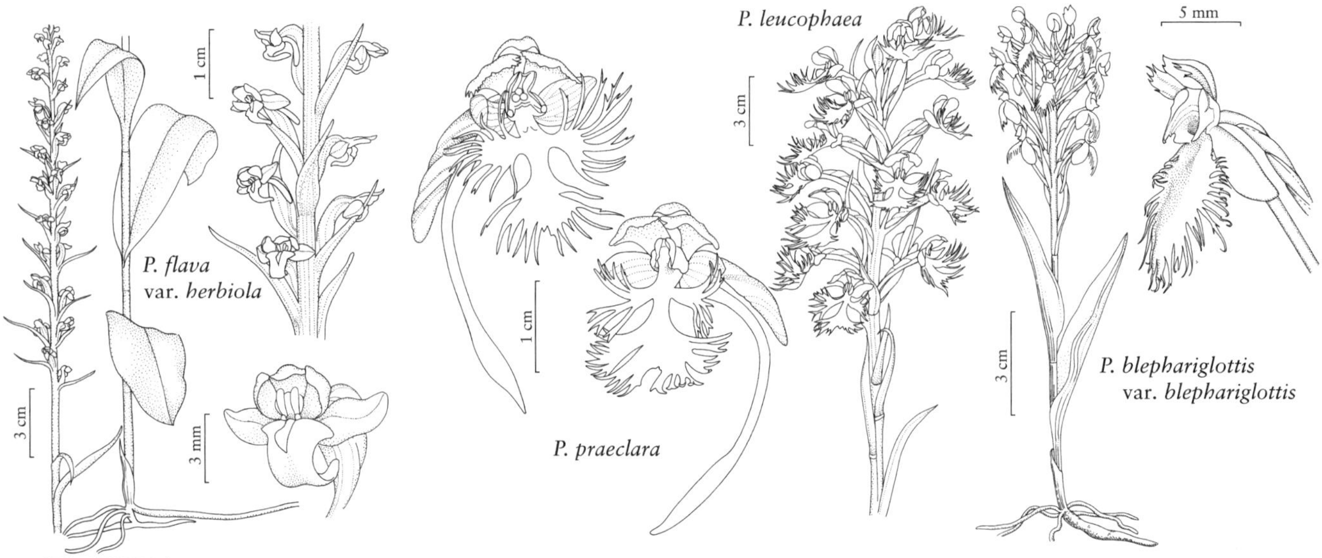

PLATANTHERA

rangular, oblong-quadrangular, orbiculate, or broadly asymmetric, commonly canaliculate to nearly cylindric and difficult to assess; ovary rather slender to stout, mostly 8–10 × 2 mm.

Varieties 2 (2 in the flora): e North America.

1. Lip (excluding auricles) subquadrangular, suborbiculate, or broadly ovate, often emarginate; inflorescences usually rather lax, slender (this enhanced by floral bracts mostly shorter than flowers) . . . . 18a. *Platanthera flava* var. *flava*
1. Lip (excluding auricles) oblong, usually obtuse, (rarely triangular, acute); inflorescences usually rather dense, stout (this emphasized by floral bracts usually long, often greatly exceeding flowers) . . . . 18b. *Platanthera flava* var. *herbiola*

18a. Platanthera flava (Linnaeus) Lindley var. **flava** [E]

Inflorescences lax, slender (in part from floral bracts usually shorter than flowers); lip (excluding auricles) subquadrangular, suborbiculate, or broadly ovate, obtuse or often emarginate.

Flowering Mar–Oct. Floodplain forests, hardwood, white-cedar, and cypress swamps, riparian thickets, wet meadows; 0–300 m; N.S., Ont.; Ala., Ark., D.C., Fla., Ga., Ill., Ind., Ky., La., Miss., Mo., N.C., Okla., S.C., Tenn., Tex., Va.

18b. Platanthera flava (Linnaeus) Lindley var. **herbiola** (R. Brown) Luer, Native Orchids U.S. & Canada, 214. 1975 [E] [F]

Habenaria herbiola R. Brown in W. Aiton and W. T. Aiton, Hortus Kew. 5: 193. 1813; *H. flava* (Linnaeus) R. Brown var. *herbiola* (R. Brown) Ames & Correll

Inflorescences rather dense, stout (in part from floral bracts usually long, often greatly exceeding flowers); lip (excluding auricles) oblong, obtuse or rarely emarginate, rarely triangular and acute. 2*n* = 42.

Flowering May–Aug. Alluvial forests, riparian thickets, wet meadows, wet prairies, seeps, salt marshes; 0–900 m; N.B., N.S., Ont., Que.; Ark., Conn., Del., Ill., Ind., Iowa, Ky., Maine, Md., Mass., Mich., Minn., Mo., N.H., N.J., N.Y., N.C., Ohio, Pa., R.I., Tex., Vt., Va., W.Va., Wis.

Tuberoids and the over-wintering bud typically form at a marked distance from the parent shoot, resulting in clonal colonies dominated by sterile plants.

The southern *Platanthera flava* var. *flava* and the northern var. *herbiola* are very distinct in their extreme forms, but they seem to intergrade, at least in the herbarium. They are usually as readily distinguished by plant habit as by lip shape. The slender, laxly flowered habit of the former is compounded by usually short floral bracts, whereas in the latter bracts are typically longer than the flowers and often exceed them, adding to the congested and broader aspect of the inflorescence. Further complicating determination are a few specimens from the region from the southern tip of Lake Michigan south and southwestward. Those have markedly triangular lips with prominent triangular auricles or basal lobes. These stout but rather laxly flowered plants seem to occupy mesic wooded slopes and may constitute a distinct taxon; comparisons should be made with similar Asiatic species.

19. Platanthera peramoena (A. Gray) A. Gray, Manual, 473. 1848 [E]

Habenaria peramoena A. Gray, Amer. J. Sci. Arts 38: 310. 1840

Plants 35–105 cm. **Leaves** 2–5, spreading to ascending, scattered along stem, gradually reduced to bracts distally; blade lanceolate, elliptic, oblong-elliptic, or oblanceoblong, 7–27 × 1.2–5 cm. **Spikes** lax to dense. **Flowers** resupinate, showy, rose-purple; lateral sepals somewhat reflexed; petals oblong-linear to spatulate, distally crenate to entire; lip descending to somewhat porrect, deeply 3-lobed, without basal thickening, 11–20 × 12–23 mm, distal margins of lobes dentate-lacerate, rarely nearly entire, lateral lobes narrowly cuneate, middle lobe broadly cuneate-flabellate, emarginate to 2-fid; spur clavate, 20–30 mm; rostellum lobes directed forward, spreading, angular; pollinaria straight to geniculate; pollinia directed forward; viscidia orbiculate; ovary slender, 13–23 mm.

Flowering (Jun–)Jul–Aug(–Oct). Alluvial forests, wet wooded flats, stream banks, seeping slopes, marshes, moist prairies, old fields and pastures, ditches, thickets; 0–800 m; Ala., Ark., Del., Ga., Ill., Ind., Ky., Md., Miss., Mo., N.J., N.C., Ohio, Pa., S.C., Tenn., Va., W.Va.

20. Platanthera grandiflora (Bigelow) Lindley, Gen. Sp. Orchid. Pl., 294. 1835 [E]

Orchis grandiflora Bigelow, Fl. Boston ed. 2, 321. 1824; *Habenaria fimbriata* (Dryander) R. Brown; *H. psycodes* (Linnaeus) Sprengel var. *grandiflora* (Bigelow) A. Gray

Plants 27–120 cm. **Leaves** 2–6, wide-spreading and recurved to somewhat ascending, scattered along stem, gradually reduced to bracts distally; blade lanceolate, elliptic, oblong-elliptic, or oblong-obovate, 13–24 × 2.5–9 cm. **Spikes** lax to rather dense. **Flowers** resupinate, showy, lavender- to rose-purple, rarely white; lateral sepals spreading to somewhat reflexed; petals ovate, oblong, elliptic-oblong, or oblong-obovate, margins crenate to dentate-lacerate, sometimes nearly entire; lip descending to somewhat porrect, deeply 3-lobed, without basal thickening, 10–25 × 14–26 mm, distal margins of lobes fringed, usually deeply, lateral lobes sometimes perpendicular to middle lobe, lateral lobes cuneate, often broadly so, middle lobe broadly cuneate-flabellate, often emarginate to 2-fid; spur clavate, 15–35 mm; rostellum lobes directed forward, spreading, angular; pollinaria nearly straight; pollinia remaining enclosed in anther sacs; viscidia orbiculate; ovary slender to stout, 12–25 mm. $2n = 42$.

Flowering Jun–Aug. Alluvial and swamp forests, stream banks, seeping slopes, marshes, sphagnum bog forests, mesic wooded slopes, old fields; 0–1800 m; N.B., Nfld. and Labr. (Nfld.), N.S., Ont., P.E.I., Que.; Conn., Maine, Md., Mass., N.H., N.J., N.Y., N.C., Ohio, Pa., R.I., Tenn., Vt., Va., W.Va.

Platanthera grandiflora is known to hybridize with *P. lacera* in the northeasternmost portion of its range, where the species bloom simultaneously. The name *P.* ×*keenanii* P. M. Brown has recently been proposed for these hybrids, but the complex synonymy of this group dictates that types be checked for possible priority. See the discussion under *P. lacera*. A few specimens also suggest very rare hybridization with *P. psycodes* (W. P. Stoutamire 1974).

Rare white-flowered plants may be separated from 23. *Platanthera leucophaea* by the orientation of the pollinia.

21. Platanthera psycodes (Linnaeus) Lindley, Gen. Sp. Orchid. Pl., 294. 1835 [E]

Orchis psycodes Linnaeus, Sp. Pl. 2: 943. 1753; *Habenaria psycodes* (Linnaeus) Sprengel

Plants 14–101 cm. **Leaves** 2–5, wide-spreading, recurved to somewhat ascending, scattered along stem, gradually reduced to bracts distally; blade lanceolate, oblong-elliptic, or oblong-obovate, 5–22 × 1.5–7 cm. **Spikes** lax to very dense. **Flowers** resupinate, showy, lavender- to rose-purple, rarely white; lateral sepals reflexed to somewhat spreading; petals spatulate to broadly obovate or cuneate-oblong, margins dentate-lacerate to sparsely fringed; lip descending to somewhat porrect, deeply 3-lobed, without basal thickening, 5–13 × 5–17 mm, distal margins of lobes fringed, sometimes deeply, lateral lobes often reflexed or slightly elevated above middle lobe, broadly to narrowly cuneate, middle lobe sometimes reflexed, broadly cuneate-flabellate, often emarginate to 2-fid; spur slenderly cylindric to slightly clavate, 12–22 mm; rostellum lobes nearly parallel, directed downward, short, rounded; pollinaria nearly straight; pollinia remaining enclosed within anther sacs; viscidia suborbiculate to broadly elliptic; ovary slender to stout, 9–16 mm. $2n = 42$.

Flowering Jun–Aug(–Sep). Alluvial and swamp forests, stream banks, riparian meadows, moist and seeping slopes, marshes, roadside banks, ditches, old fields; 0–2000 m; Man., N.B., Nfld. and Labr. (Nfld.), N.S., Ont., P.E.I., Que.; Conn., Ga., Ill., Ind., Iowa, Ky., Maine, Mass., Mich., Minn., N.H., N.J., N.Y., N.C., Ohio, Pa., R.I., Tenn., Vt., Va., W.Va., Wis.

Platanthera psycodes is known to hybridize with *P. lacera* in the northeasternmost portion of its range, where the species bloom simultaneously. Such hybrids are known as *P. ×andrewsii* (M. White) Luer; see the discussion under *P. lacera*. A few specimens also suggest very rare hybridization with *P. grandiflora*.

22. Platanthera praeclara Sheviak & M. L. Bowles, Rhodora 88: 278, figs. 1a–d, 2–4. 1986 C E F

Habenaria leucophaea (Nuttall) A. Gray var. *praeclara* (Sheviak & Bowles) Cronquist

Plants 38–85 cm. **Leaves** several to many, ascending, scattered along stem, imperceptibly reduced to bracts distally; blade lanceolate to ovate-lanceolate, to 26 × 5 cm. **Spikes** lax. **Flowers** resupinate, showy, creamy white; lateral sepals somewhat porrect; petals flabellate, truncate, sometimes emarginate, apically lacerate; lip descending, deeply 3-lobed, without basal thickening, 17–32 × 20–39 mm, distal margins of lobes deeply incised and fringed, lateral lobes flabellate, sometimes broadly and overlapping middle lobe, middle lobe flabellate, sometimes very broadly, emarginate; spur slenderly clavate, 36–55 mm; rostellum lobes directed strongly forward, wide-spreading, angular; pollinaria geniculate; pollinia directed forward (column appearing hooded), remaining enclosed in anther sacs; viscidia orbiculate; ovary slender, mostly 20–30 mm. $2n = 42$.

Flowering (May–)Jun–Aug. Mesic to wet prairies; of conservation concern; 200–800(–1600 m); Man.; Iowa, Kans., Minn., Mo., Nebr., N.Dak., Okla., S.Dak., Wyo.

In *Platanthera praeclara* the wide-spreading rostellum lobes separate the viscidia by 6–7.5 mm, whereas in *P. leucophaea* the short rostellum lobes are more nearly parallel, and the viscidia are separated only by about 1–3.5 mm. These differences are immediately obvious on living material, but the measurements are nearly always impossible in herbarium specimens.

In Wyoming *Platanthera praeclara* is known from a single, much higher station outside the usual range.

23. Platanthera leucophaea (Nuttall) Lindley, Gen. Sp. Orchid. Pl., 294. 1835 C E F

Orchis leucophaea Nuttall, Trans. Amer. Philos. Soc., n. s. 5: 161. 1834; *Habenaria leucophaea* (Nuttall) A. Gray

Plants 32–112 cm. **Leaves** several to many, ascending, scattered along stem, imperceptibly reduced to bracts distally; blade lanceolate to ovate-lanceolate, usually to 20 × 4 cm. **Spikes** lax to moderately dense. **Flowers** resupinate, showy, corolla white, calyx green to whitish green; lateral sepals porrect; petals obovate to rarely flabellate, apically lacerate; lip descending to horizontally projecting, deeply 3-lobed, without basal thickening, 14–22 × 15–29 mm, distal margins of lobes deeply incised, fringed, lateral lobes flabellate, usually broad, overlapping middle lobe, middle lobe flabellate, sometimes very broadly, emarginate; spur slenderly clavate, 28–47 mm; rostellum lobes nearly parallel, directed downward, short, rounded; pollinaria geniculate; pollinia directed forward (column appearing hooded), remaining enclosed in anther sacs; viscidia orbiculate; ovary slender, mostly 15–30 mm. $2n = 42$.

Flowering Jun–Aug. Mesic to wet prairies, marshes, fens, lake shores, old fields; of conservation concern; 80–300 m; Ont.; Ill., Ind., Iowa, La., Maine, Mich., Mo., Nebr., N.J., N.Y., Ohio, Okla., Va., Wis.

A very rare hybrid with *Platanthera psycodes*, known only from Ontario, is *P. ×reznicekii* Catling, Brownell & G. Allen.

See the discussion under 22. *Plantanthera praeclara*.

24. Platanthera lacera (Michaux) G. Don in R. Sweet, Hort. Brit. ed. 3, 650. 189 E

Orchis lacera Michaux, Fl. Bor.-Amer. 2: 156. 1803; *Habenaria lacera* (Michaux) R. Brown

Plants 14–80 cm. **Leaves** 1–4, wide-spreading to ascending, scattered along stem, gradually to abruptly reduced to bracts distally; blade linear-lanceolate, linear-oblong, or oblong-obovate, 7–23 × 0.7–5 cm. **Spikes** lax to dense. **Flowers** resupinate, not showy to inconspicuous, whitish green or yellowish green, lip often paler; lateral sepals reflexed; petals linear to narrowly oblong-spatulate, margins entire or apex rarely slightly lacerate; lip descending, without basal thickening, deeply 3-lobed, 10–17 × 13–17 mm, lobes extensively and deeply divided, commonly into filiform divisions, margins rarely merely fringed or middle lobe merely lacerate-fringed, lateral lobes narrowly to broadly cuneate, middle lobe cuneate-flabellate, usually narrow; spur slenderly to markedly clavate, 11–23 mm; rostellum lobes nearly parallel, directed downward, short, rounded; pollinaria nearly straight; pollinia remaining enclosed in anther sacs; viscidia linear; ovary slender to stout, mostly 8–20 mm. $2n = 42$.

Flowering May–Aug. Sphagnum bogs, alluvial and swamp forests, stream banks, riparian meadows, sand flats, moist and seeping slopes, prairies, roadside banks, ditches, old fields, borrow pits; 0–1200 m; Man., N.B., Nfld. and Labr. (Nfld.), N.S., Ont., P.E.I., Que.; Ala., Ark., Conn., Del., D.C., Ill., Ind., Kans., Ky., Maine, Md., Mass., Mich., Minn., Miss., Mo., N.H., N.J., N.Y., N.C., Ohio, Okla., Pa., R.I., S.C., Tex., Vt., Va., W.Va., Wis.

Hybrids with *Platanthera grandiflora* (*P.* ×*keenanii* P. M. Brown; see note under 20. *P. grandiflora*) and *P. psycodes* [*P.* ×*andrewsii* (M. White) Luer] are variably intermediate between the parents, with various purple-green shadings and lobes of the lip generally narrower and more dissected than in the purple-flowered parents. The linear viscidia of *P. lacera* and the large anther sacs of *P. grandiflora* provide means for determination. The ratio of viscidium length/width of 1–3.2 distinguishes hybrids and purple-flowered parents from *P. lacera*, which is characterized by a ratio of 2.7–4.6. The parentage of hybrids can be determined by different sized anther sacs: *P.* ×*andrewsii*, 1.4–2.3 mm; *P.* ×*keenanii*, 2.5–4.2 mm. Those hybrids can be distinguished from their purple-flowered parents by less intense coloring and the ratio of viscidium length/width: *P. psycodes* 1–1.5; *P.* ×*andrewsii*, 1.3–3; *P. grandiflora*, 1–1.5; *P.* ×*keenanii*, 1.5–3.2. The plants known as *Platanthera lacera* var. *terrae-novae* (Fernald) Luer (*Habenaria lacera* var. *terrae-novae* Fernald) apparently represent swarms of *P.* ×*andrewsii* and introgressed *P. lacera* (P. M. Catling and V. R. Catling 1994).

SELECTED REFERENCE Catling, P. M. and V. R. Catling. 1994. Identification of *Platanthera lacera* hybrids (Orchidaceae) from New Brunswick and Nova Scotia. Lindleyana 9: 19–32.

25. Platanthera blephariglottis (Willdenow) Lindley, Gen. Sp. Orchid. Pl., 291. 1835 [E] [F]

Orchis blephariglottis Willdenow, Sp. Pl. 4(1): 9. 1805; *Habenaria blephariglottis* (Willdenow) Hooker

Plants 8–110 cm. **Leaves** 2–several, spreading to ascending, scattered along stem, rather abruptly to gradually reduced distally; blade linear-lanceolate, ovate-lanceolate, elliptic-lanceolate, 5–35 × 1–5 cm. **Spikes** dense to rather lax. **Flowers** resupinate, showy, white; lateral sepals reflexed; petals linear to oblanceolate, apex fringed to nearly entire; lip reflexed to porrect, without basal thickening, ovate-oblong to oblong-elliptic, rarely quadrangular-suborbiculate, margin fringed; spur essentially cylindric, slender; rostellum lobes curving forward, slightly spreading, angular-elongate; pollinaria with stalks curved forward; pollinia remaining enclosed in anther sacs; viscidia elliptic-suborbiculate to orbiculate; ovary slender, 10–28 mm.

Varieties 2 (2 in the flora): e North America.

1. Spur less than 26 mm, to 1.25 times length of ovary .... 25a. *Platanthera blephariglottis* var. *blephariglottis*
1. Spur at least 30 mm, commonly about 2 times length of ovary .... 25b. *Platanthera blephariglottis* var. *conspicua*

25a. Platanthera blephariglottis (Willdenow) Lindley var. **blephariglottis** [E] [F]

Plants 8–80 cm. **Flowers:** lip 6–13 × 2–9 mm; spur slenderly cylindric, 15–25 mm; ovary 10(–13)–24 mm. $2n = 42$.

Flowering Jun–Aug. Open sphagnum bog mats and moist sandy and peaty meadows, pine savannas, openings; 0–600 m; N.B., Nfld. and Labr. (Nfld.), N.S., Ont., P.E.I., Que.; Conn., Del., Ga., Ill., Maine, Md., Mass., Mich., N.H., N.J., N.Y., N.C., Ohio, Pa., R.I., S.C., Vt., Va.

25b. Platanthera blephariglottis (Willdenow) Lindley var. **conspicua** (Nash) Luer, Native Orchids Florida, 140. 1972 [E]

Habenaria conspicua Nash, Bull. Torrey Bot. Club 23: 100. 1896

Plants 35–110 cm. **Flowers:** lip 9–26 × 4–9 mm; spur 30(–48)–50 mm, slenderly cylindric to commonly somewhat clavate; ovary 18–31 mm.

Flowering Aug–Sep. Moist to wet sandy and peaty meadows, marshes, pine savannas, open woods; 0–400 m; Ala., Fla., Ga., Miss., N.C., S.C., Tex.

The widespread *Platanthera blephariglottis* var. *blephariglottis* tends to have a dense inflorescence of smaller flowers with the descending to reflexed lips rather simply fringed (in extreme cases scarcely more than deeply lacerate). In contrast, the typically open, lax inflorescence of var. *conspicua* bears larger flowers often with an exquisitely elaborate filiform fringe on the porrect lips. Although typical plants of the two varieties are quite distinctive, most characters vary considerably within and between the varieties and can be misleading. In contrast, spur length seems to be unambiguous, and insofar as it reflects specialization for different pollinators, upon further study it may be shown to delimit distinct species. Additionally, in the limited area in which the two varieties occur together, var. *conspicua* may bloom somewhat later than var. *blephariglottis*.

Hybrids of *Platanthera blephariglottis* with *P. ciliaris* are *P.* ×*bicolor* (Rafinesque) Luer and with *P. cristata* are *P.* ×*canbyi* (Ames) Luer. The unusual and rare hybrid *P. blephariglottis* × *P. clavellata* is *P.* ×*vossii* Case.

26. Platanthera integrilabia (Correll) Luer, Native Orchids U.S. & Canada, 186. 1975 [C] [E]

Habenaria blephariglottis (Willdenow) Hooker var. *integrilabia* Correll, Bot. Mus. Leafl. 9: 153. 1941; *H. correlliana* Cronquist

Plants 50–80 cm. **Leaves** 2(–3), spreading to ascending, on basal portion of stem, abruptly reduced to bracts distally; blade lanceolate to elliptic, 12–23 × 1–3.2 cm. **Spikes** rather lax. **Flowers** resupinate, showy, white; lateral sepals reflexed; petals oblong, margins entire; lip porrect, spatulate to lance-spatulate, without basal thickening, 10–15 × 2.5–4 mm, margins slightly lacerate to subentire; spur slenderly cylindric, scarcely clavate, 35–60 mm; rostellum lobes directed forward, spreading, angular; rostellum lobes slenderly curving forward, slightly spreading, angular-elongate, long; pollinaria with stalks curved forward; pollinia remaining enclosed in anther sacs; viscidia orbiculate; ovary slender, ca.16–22 mm.

Flowering (Jun–)Jul–Aug. Wet wooded flats, seeping slopes, marshes, sphagnum bogs; of conservation concern; mostly 100–700 m; Ala., Ga., Ky., Miss., N.C., S.C., Tenn.

The description of *Platanthera integrilabia* is based on a limited sample and is likely too restrictive.

Flowers of this species are nocturnally sweet-scented. The vegetative habit is distinctive, with leaves restricted to the base of the stem and a relatively few-flowered inflorescence borne atop a tall bracted scape. In addition to the normal tuberoid, roots often enlarge distally to form bud-bearing tuberoids at a marked distance from the parent stem, giving rise to diffuse clonal colonies.

27. Platanthera ciliaris (Linnaeus) Lindley, Gen. Sp. Orchid. Pl., 292. 1835 [E]

Orchis ciliaris Linnaeus, Sp. Pl. 2: 939. 1753; *Habenaria ciliaris* (Linnaeus) R. Brown

Plants 24–100 cm. **Leaves** 2–4, spreading to ascending, gradually to somewhat abruptly reduced to bracts distally; blade lanceolate to lance-elliptic, 5–40 × 0.6–6 cm. **Spikes** rather dense. **Flowers** resupinate, showy, orange; dorsal sepal entire or rarely emarginate; lateral sepals reflexed; petals linear to linear-cuneate, margins apically fringed; lip porrect to descending, oblong to broadly ovate-spatulate, without basal thickening, 8–19 × 4–18 mm, margins deeply filiform-fringed; spur slenderly cylindric, rarely slightly clavate, 20–35 mm; rostellum lobes scarcely to markedly curving forward, slightly spreading, angular-elongate; pollinaria with stalks nearly straight to curved forward; pollinia remaining enclosed in anther sacs; viscidia orbiculate; ovary slender, 12–27 mm.

Flowering (Jun–)Jul–Sep. Moist sandy and peaty meadows, marshes, prairies, pine savannas, open woods, wet wooded flats, seeping slopes, roadsides, dry wooded slopes, sphagnum bogs; 0–1700 m; Ont.; Ala., Ark., Conn., Del., D.C., Fla., Ga., Ill., Ind., Ky., La., Md., Mass., Mich., Miss., Mo., N.J., N.Y., N.C., Ohio, Okla., Pa., R.I., S.C., Tenn., Tex., Va., W.Va.

Selection for longer spurs in response to local pollination pressures on the southeastern Coastal Plain has been documented in *Platanthera ciliaris*, but the extent of variation is minimal, and the situation does not appear to be comparable to that in *P. blephariglottis*.

Hybrids of *Platanthera ciliaris* with *P. blephariglottis* are *P.* ×*bicolor* (Rafinesque) Luer and with *P. cristata* are *P.* ×*channellii* Folsom; until recently the latter was confused with *P. chapmanii* (J. P. Folsom 1984).

28. Platanthera cristata (Michaux) Lindley, Gen. Sp. Orchid. Pl., 291. 1835 [E]

Orchis cristata Michaux, Fl. Bor.-Amer. 2: 156. 1803; *Habenaria cristata* (Michaux) R. Brown; *Platanthera pallida* P. M. Brown

Plants 18–90 cm. **Leaves** 2–4, spreading to ascending, gradually reduced to bracts distally; blade linear-lanceolate to lance-oblong, 5–21 × 1–3 cm. **Spikes** dense.

Flowers resupinate, showy, orange; dorsal sepal entire or sometimes emarginate; lateral sepals spreading; petals obovate to oblong-elliptic, margins fringed throughout to entire; lip porrect, ovate to ovate-oblong, without basal thickening, 4–8 × 4–8 mm, margins deeply filiform-fringed; spur slenderly cylindric (mouth triangular or keyhole-shaped), 4–10 mm; rostellum lobes scarcely to strongly curved downward, rarely distally retrorse, digitate; pollinaria with stalks curved downward; pollinia remaining enclosed in anther sacs; viscidia presented forward, nearly perpendicular to ± parallel to lip, orbiculate; ovary slender, 7–13 mm. $2n = 42$.

Flowering (Jun–)Jul–Aug(–Sep). Moist sandy and peaty meadows, marshes, prairies, pine savannas, wet wooded flats, seeping slopes, sphagnum bogs; 0–800 m; Ala., Ark., Del., Fla., Ga., La., Md., Mass., Miss., N.J., N.Y., N.C., Pa., S.C., Tenn., Tex., Va.

The rostellum lobes in *Platanthera cristata* normally are only slightly curved, presenting the viscidia in a generally forward position. Considerable variation is apparent, however, and sometimes a marked curvature similar to that in *P. chapmanii* is seen, apparently presenting the viscidia downward. These orientations are difficult to assess in distorted herbarium material, and the species is in need of further study. The triangular mouth of the spur of *P. cristata* is distinctive, however, and can aid in determination of confusing material (see also below and the discussion under 29. *P. chapmanii*).

Platanthera pallida P. M. Brown was recently proposed to accommodate two populations of pale-flowered plants on Long Island, New York. Their distinctive nature has long been apparent. In the field, pollination normally is rapid and the flowers senesce while small and yellowish, but in cultivation in the absence of pollinators, flowers remain open for an extended period during which the lips elongate markedly and the flowers fade to a dull white. Hence they suggest *P.* ×*canbyi* (Ames) Luer [*P. blephariglottis* × *P. cristata*], but they differ in their much shorter spurs.

Plants of *Platanthera pallida* were reported to differ from *P. cristata* in shorter spurs, entire dorsal sepals, and lateral sepal and lip orientation. The range of spur lengths in *P. cristata*, however, is much greater than specified and in fact includes the range of *P. pallida*. Dorsal sepals with entire margins are commonly seen also in *P. cristata*, and those of *P. pallida* may be deeply emarginate. Lateral sepal and lip orientation tend to differ in the two taxa, but the conditions typical of each taxon can be found in the other. The keyhole shaped orifice of the spur in *P. pallida* mentioned by Brown is a unique feature, but it is present in only a small percentage of individuals and completely intergrades to the triangular orifice of *P. cristata*. Furthermore, superficially similar pale-flowered plants from farther west on Long Island are more similar to typical *P. cristata* and hence possibly indicative of a cline. The plants described as *P. pallida* therefore appear to represent neither a distinct species nor are they merely hybrids; they seem to be partially stabilized introgression products and potentially useful subjects for evolutionary study.

Hybrids of *Platanthera cristata* with *P. blephariglottis* are *P.* ×*canbyi* (Ames) Luer and with *P. ciliaris* are *P.* ×*channellii* Folsom; until recently the latter was confused with *P. chapmanii.*

29. **Platanthera chapmanii** (Small) Luer, Native Orchids Florida, 151. 1972, as hybrid E

Blephariglottis chapmanii Small, Fl. S.E. U.S., 314, 1329. 1903; *Habenaria* ×*chapmanii* (Small) Ames

Plants 30–77 cm. **Leaves** 1–3, spreading to ascending, abruptly or gradually reduced to bracts distally; blade linear-lanceolate or lanceolate to lance-oblong, 15–26 × 1–3 cm. **Spikes** rather dense. **Flowers** resupinate, showy, orange to pale dull yellow; dorsal sepal entire or emarginate; lateral sepals reflexed; petals linear to obovate, apically fringed to merely lacerate-dentate; lip porrect, ovate to ovate-orbiculate, without basal thickening, 5–10 × 4–9 mm, margins deeply filiform-fringed; spur slenderly cylindric, 8–17 mm; rostellum lobes directed downward, distally retrorse, digitate; pollinaria with stalks recurved; pollinia remaining enclosed in anther sacs; viscidia orbiculate to suborbiculate; ovary slender, 10–19 mm.

Flowering Jul–Aug. Wet to dry flatwoods, pine barrens, savannas; 0–375 m; Fla., Ga., Tex.

The retrorse rostellum lobes of *Platanthera chapmanii* usually separate small-flowered plants of this species from *P. cristata*, which normally presents the viscidia forward on only slightly curved lobes. Additionally, the nearly circular mouth of the spur of *P. chapmanii* is in contrast to the triangular opening of *P. cristata*. Until recently *P. chapmanii* was confused with the hybrid of *P. ciliaris* and *P. cristata*, *P.* ×*channellii* Folsom, which is readily recognized by its angular rostellum lobes that are directed forward and outward, presenting the viscidia forward, as in its parents (J. P. Folsom 1984).

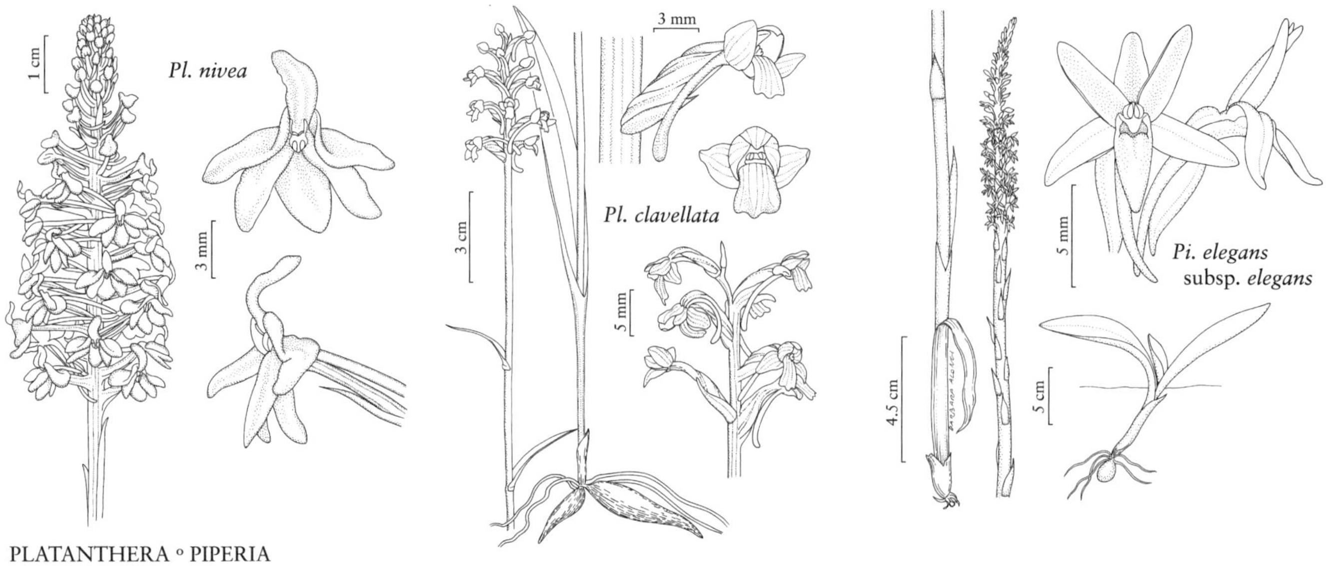

PLATANTHERA ◦ PIPERIA

30. Platanthera nivea (Nuttall) Luer, Native Orchids Florida, 146. 1972 [E] [F]

Orchis nivea Nuttall, Gen. N. Amer. Pl. 2: 188. 1818; *Habenaria nivea* (Nuttall) Sprengel

Plants 17–90 cm. **Leaves** 1–3, ascending to spreading, abruptly or gradually reduced to bracts distally, often fugaceous and withered at anthesis; blade linear to linear-lanceolate, 3–31 × 0.3–2 cm. **Spikes** dense. **Flowers** not resupinate, showy, white; lateral sepals spreading; petals linear-oblong to oblong or elliptic, falcate, basally somewhat dilated, margins entire; lip reflexed ± at middle, linear-elliptic to linear-oblong, without basal thickening, 3–8 × 1–3 mm, margins entire; spur slenderly cylindric, 10–18 mm; rostellum lobes directed downward, very short, obscure, rounded; pollinaria straight, pollinia remaining enclosed in anther sacs; viscidia oblong to linear-oblong; ovary slender, 4–12 mm.

Flowering summer (May–Sep). Moist pine barrens, pine savannas, meadows, wet sandy woods, bogs, and cypress swamps; 0–500 m; Ala., Ark., Del., Fla., Ga., La., Miss., N.J., N.C., S.C., Tex.

The column of *Platanthera nivea* is similar to that of *P. clavellata* and *P. integra* and unlike that of other *Platanthera* species. The columns of these three species bear two pairs of appendages. In *P. nivea*, on either side of the column and flanking the anther sacs, a cushionlike flap simulates a pollinium; below these structures, elongated slender processes parallel the basal margins of the lip, similar to the situation in *Habenaria*. Together with its non-resupinate flower and perhaps elongate-spheroidal tuberoids, these characteristics suggest that this species should not be included in *Platanthera*.

31. Platanthera clavellata (Michaux) Luer, Native Orchids Florida, 148. 1972 [E] [F]

Orchis clavellata Michaux, Fl. Bor.-Amer. 2: 155. 1803; *Habenaria clavellata* (Michaux) Sprengel; *H. clavellata* var. *ophioglossoides* Fernald

Plants 8–47 cm. **Leaves** 1(–2), on basal half of stem, ascending to spreading, usually abruptly reduced to bracts distally; blade oblanceolate to oblong, linear-oblong, or elliptic, 3–19 × 0.8–3.5 cm. **Spikes** moderately dense. **Flowers** resupinate, often incompletely so and held at angle, not showy, rather inconspicuous pale green to dull yellowish green; lateral sepals porrect; petals ovate to obovate; lip oblong, obscurely 3-lobed, without basal thickening, 3–7 × 3–4 mm, margins sometimes dentate-lacerate, apex truncate; spur clavate, 7–13 mm; rostellum lobes directed downward, very short, truncate; pollinaria straight or slightly curved laterally; pollinia fragmenting, pollen masses trailing down onto stigma; viscidia linear to linear lanceolate; ovary rather stout, 6–11 mm. $2n = 42$.

Flowering summer (Jun–Aug). Sphagnum bogs, sphagnous seeps and meadows, wet sandy and peaty meadows, marshes, low woods, wet prairies, and roadsides; 0–2000 m; N.B., Nfld. and Labr. (Nfld.), N.S., Ont.,

P.E.I., Que.; Ala., Ark., Conn., Del., D.C., Fla., Ga., Ill., Ind., Iowa, Ky., La., Maine, Md., Mass., Mich., Minn., Miss., N.H., N.J., N.Y., N.C., Ohio, Okla., Pa., R.I., S.C., Tenn., Tex., Vt., Va., W.Va., Wis.

In common with *Platanthera nivea* and *P. integra*, the column of *P. clavellata* bears a series of lateral projections that are directed forward; the upper pair is elaborately adorned and may be glandular. It appears that this species is inappropriately placed in *Platanthera*. See note under 30. *P. nivea*.

The broader leaves cited as distinguishing *Platanthera clavellata* var. *ophioglossoides* are more prevalent in the North, but occur throughout the range of the species. In some areas populations commonly display a complete range of leaf shape; this feature alone is of no taxonomic significance.

The unusual and infrequent hybrid *Platanthera blephariglottis* × *P. clavellata* is *P.* ×*vossii* Case.

32. Platanthera integra (Nuttall) A. Gray ex L. C. Beck, Bot. North. Middle States ed. 2, 348. 1848 [E]

Orchis integra Nuttall, Gen. N. Amer. Pl. 2: 188. 1818; *Habenaria integra* (Nuttall) Sprengel

Plants 20–75 cm. **Leaves** 1–3, ascending to spreading, rather abruptly or gradually reduced to bracts distally; blade lance-oblong to linear-lanceolate, 4.5–32 × 1–3 cm. **Spikes** dense. **Flowers** resupinate, showy, pale yellow-orange to pale orange; dorsal sepal entire or rarely apically dentate; lateral sepals spreading; petals elliptic to linear-oblong, margins entire; lip descending, ovate-elliptic to obovate or oblong, 3–5 × 1.5–4 mm, margins eroded to lacerate or rarely entire, with basal pair of fleshy ridges on adaxial surface; spur tapering from broad base to slender tube, 5–10 mm; rostellum lobes curved downward, short, rounded; pollinaria essentially straight; large pollinia protruding forward; viscidia orbiculate to suborbiculate; ovary slender to somewhat stout, 5–11 mm.

Flowering Jul–Sep. Wet pine barrens, peaty depressions in pine savannas, wet sandy woods; 0–900 m; Ala., Del., Fla., Ga., La., Miss., N.J., N.C., S.C., Tenn., Tex.

As in *Platanthera clavellata* and *P. nivea*, the column of *P. integra* bears two pairs of lateral processes. In *P. integra*, the distal structures are essentially sessile and cushionlike, as in *P. nivea*, but the proximal are short, stout, clublike, and bear several stout horns. The tuberoids of *P. integra* are abruptly swollen into oblong-cylinders, somewhat like those of *P. nivea*. These three species evidently form a group apart from *Platanthera*. See note under 30. *P. nivea*.

27. PIPERIA Rydberg, Bull. Torrey Bot. Club 28: 269. 1901 • [For C. V. Piper, American botanist of the Pacific Northwest]

James D. Ackerman

Randall Morgan

Herbs, perennial, terrestrial, glabrous. **Roots** few, fleshy; tuberoids ellipsoid-ovoid. **Stems** simple. **Leaves** ephemeral, 2–6, basal; blade sessile, lanceolate to oblanceolate; cauline bracts ovate to linear-lanceolate. **Inflorescences** many-flowered racemes or spicate racemes. **Flowers** resupinate, white, yellow-green to green, nearly sessile; sepals 1–3-veined; lateral sepals similar, adnate to lip and slightly longer than free dorsal sepal; lip base adnate to column, blade simple, midrib thickened, broad; spur clavate to filiform; column short, auricles absent; anther 1, erect, connective narrow, cells close and parallel; pollinaria 2; pollinia 2, sectile; caudicles inconspicuous, attached to round to oblong viscidia; stigmas confluent, concave; rostellum 3-lobed. **Fruits** capsules, erect, nearly sessile, ellipsoid to sausage-shaped. $x = 21$.

Species 10 (10 in the flora): temperate regions, North America (including Mexico).

Piperia is perhaps the most taxonomically complex orchid genus in North America. Considerable variation occurs within most species, and distinctions among taxa are often subtle. In pressed, dried specimens important characteristics that otherwise may be obvious in the field are frequently obscured. Our views on the taxonomy of the genus evolved as more extensive field and garden studies have helped us interpret geographic patterns of variation and recognize additional unifying or distinguishing characteristics.

Leaf, bract, and floral measurements in the following descriptions represent the largest dimensions on a given plant.

SELECTED REFERENCE Ackerman, J. D. 1977. Biosystematics of the genus *Piperia* Rydb. (Orchidaceae). Bot. J. Linn. Soc. 75: 245–270.

1. Spur 7–18 mm; viscidia ovate to oblong.
  2. Viscidia oblong, at least 2 times as long as wide; flowers white with green or yellow-green midvein; stem base usually thickened toward tuberoid.
    3. Spurs of most flowers straight, horizontal; stem 0.7–4.5 mm diam.; floral fragrance clovelike . . . . 1. *Piperia transversa*
    3. Spurs of most flowers curved, deflexed; stem (2–3)–12 mm diam.; floral fragrance musky or pleasant, not clovelike . . . . 2a. *Piperia elegans* subsp. *elegans*
  2. Viscidia ovate to broadly elliptic or broadly oblong, usually less than 2 times as long as wide; flowers green; stem base uniform in diameter or attenuate toward tuberoid.
    4. Seeds usually blackish brown; lip triangular-ovate; petals erect-spreading; stem attenuate toward tuberoid . . . . 8. *Piperia michaelii*
    4. Seeds pale brown to cinnamon brown; lip lanceolate to broadly triangular; petals usually strongly erect-recurved; stem basally attenuate or not.
      5. Petals falcate, ovate-attenuate, about 2 mm wide at base and 2–3 times as long; lip less than 2 times as long as wide; spur (7–)9–15(–18) mm; widespread . . . . 9. *Piperia elongata*
      5. Petals straight, linear, about 1 mm wide at base and 4–5 times as long; lip about 2 times as long as wide; spur 4–9 mm; not common, California . . . 10. *Piperia leptopetala* (in part)
1. Spur 1–6 mm; viscidia broadly elliptic-ovate to rectangular-ovate.
  6. Lip strongly recurved toward spur with age; flowers white or white and green; raceme rachis usually shorter than peduncle.
    7. Stem bracts usually fewer than 6; raceme ± secund; lip triangular-ovate; petals straight, white or with faint green midvein; c California to se Alaska . . . . 4. *Piperia candida*
    7. Stem bracts usually more than 6; raceme cylindric; lip triangular-lanceolate; petals erect, curved, green with white border; California.
      8. Petals falcate, outer white border much broader than inner; stem base attenuate toward tuberoid; flowers weakly fragrant, never cinnamonlike, diurnal; Monterey Bay area, California . . . . 3. *Piperia yadonii*
      8. Petals straight-sided, inner and outer white border subequal; stem base swollen; flowers strongly cinnamon-scented, nocturnal; Point Reyes, California . . . . 2b. *Piperia elegans* subsp. *decurtata*
  6. Lip straight or upwardly curved; flowers green; raceme rachis usually at least as long as peduncle.
    9. Spur shorter than lip, 1–2.5 mm; flowers scentless; leaves grasslike, erect, less than 1 cm wide . . . . 5. *Piperia colemanii*
    9. Spur at least as long as lip, 2–6 mm; flowers scented, at least at night; leaves not grasslike, prostrate to erect-spreading, usually over 1 cm wide.
      10. Petals linear with slightly expanded base, about 4 times longer than wide, strongly upwardly recurved; lip narrowly lanceolate; spur attenuate, 4–9 mm; flowers with lemony scent or otherwise pleasant, not musky or honeylike . . . . 10. *Piperia leptopetala* (in part)
      10. Petals lanceolate to triangular-ovate, less than 3.5 times longer than wide, usually projecting to ± erect; lip triangular-ovate to lanceolate-elliptic; spur narrowly cylindric, blunt, 2–6 mm; flowers musky- or honey-scented.

11. Mature seeds blackish brown; flowers honey-scented; raceme rachis ± length of peduncle; stem attenuate toward tuberoid; Santa Monica Mountains to n Baja California, below 1500 m . . . . 7. *Piperia cooperi*
11. Mature seeds tan to cinnamon brown; flowers ± musk-scented; raceme rachis usually longer than peduncle; stem diameter uniform or increasing toward tuberoid; widespread in North America but in s California occurring only above 1500 m . . . . 6. *Piperia unalascensis*

1. Piperia transversa Suksdorf, Allg. Bot. Z. Syst. 12: 43. 1906 [E]

Plants 12–57 cm. **Stems** basally swollen above tuberoid, 0.7–4.5 mm diam. distal to leaves; bracts 3–15. **Leaves** prostrate; blade 6–19 × 1–4.5 cm. **Inflorescences** ± densely flowered, 3.5–32 cm; rachis ± length of peduncle; bracts 3–12 mm. **Flowers** ± white with green or yellow-green midveins, fragrance nocturnal, strong, clovelike; sepals 2.5–4.5 × 1–2 mm; dorsal sepal projecting forward between petals, oblong to lanceolate; lateral sepals reflexed-spreading, obliquely oblong-lanceolate; petals reflexed-spreading, oblique to falcate, oblong, 2–5 × 1–2 mm; lip straight to ± deflexed, oblong to ovate-elliptic, 2.2–5.3 × 1–2.5 mm; spur ± horizontal, straight, 7–12 mm; viscidia oblong, 0.5–0.7 × 0.2–0.3 mm; rostellum blunt. **Capsules** 4–13 mm. **Seeds** cinnamon brown. $2n = 42$.

Flowering late May–Aug. Conifer and mixed evergreen forest, oak woodlands, chaparral; 0–2600 m; B.C.; Calif., Oreg., Wash.

Piperia transversa is pollinated by geometrid moths (J. D. Ackerman 1977).

2. Piperia elegans (Lindley) Rydberg, Bull. Torrey Bot. Club 28: 270 1901 [E] [F]

Platanthera elegans Lindley, Gen. Sp. Orchid. Pl., 285. 1835

Plants 12–73(–100) cm. **Stems** swollen at base (rarely attenuate), 2–12 mm diam. distal to leaves; bracts (4–)12–37. **Leaves** prostrate; blade 6–20(–38) × 1–9 cm. **Inflorescences** usually densely flowered, cylindric, 2.5–40 cm; rachis shorter than peduncle; bracts 6–24 mm. **Flowers** mostly white, fragrance strong at night; sepals white with green midvein, 3–7 × 1.6–3 mm; dorsal sepal ovate to oblong; lateral sepals usually widespreading, ovate-elliptic to oblong-lanceolate; petals spreading, recurved, straight, ovate to oblong-lanceolate, (3–)4–6 × 1.5–2.5 mm; lip recurved, pale greenish to white, triangular-lanceolate to lanceolate, 2.5–7 × 1.5–3.5 mm; spur usually deflexed, curved, (3–)7–14 mm; viscidia 0.4–1 × 0.15–0.5 mm; rostellum blunt. **Capsules** 5–12 mm. **Seeds** cinnamon brown. $2n = 42$.

Subspecies 2 (2 in the flora): w North America.

[Key to subspecies contained in key to species.—Ed.]

2a. Piperia elegans (Lindley) Rydberg subsp. **elegans** [E] [F]

Habenaria elegans (Lindley) Bolander var. *maritima* (Greene) Ames; *H. elegans* var. *multiflora* (Rydberg) Peck; *H. greenei* Jepson; *H. maritima* Greene 1892, not Rafinesque 1833; *H. unalascensis* (Sprengel) S. Watson subsp. *maritima* (Greene) Calder & Roy L. Taylor; *H. unalascensis* var. *maritima* (Greene) Correll; *Piperia maritima* Rydberg; *P. multiflora* Rydberg

Plants 12–73(–100) cm. **Flowers** pleasantly scented or musky; petals and lip pale green to white; spur 7–14 mm; viscidia oblong, 0.6–1 mm, usually more than 2 times as long as wide.

Flowering Jun–Sep/Oct. Conifer and mixed evergreen forests, coastal scrub, exposed coastal bluff; 0–1700 m; B.C.; Calif., Idaho, Mont., Oreg., Wash.

Plants on exposed coastal bluffs are shorter and stouter than plants in sheltered or inland localities. Plants in warmer interior locations flower earlier than those on the coast and have fewer stem bracts. The flowers are pollinated by noctuid moths (J. D. Ackerman 1977).

A population in San Francisco, California, is also unusual by having basally attenuate stems and by flowering very late (Sep–Oct).

2b. Piperia elegans (Lindley) Rydberg subsp. **decurtata** Rand. Morgan & Glicenstein, Lindleyana 8: 93, figs. 1, 3. 1993 [E]

Plants 13–35 cm. **Flowers** strongly cinnamon-scented; petals greenish with broad white margin, outer and inner white border subequal; lip white; spur (3–)5–6 mm; viscidia rectangular-ovate, 0.15–0.75 mm, nearly 2 times as long as wide.

Flowering Aug–Sep. Coastal

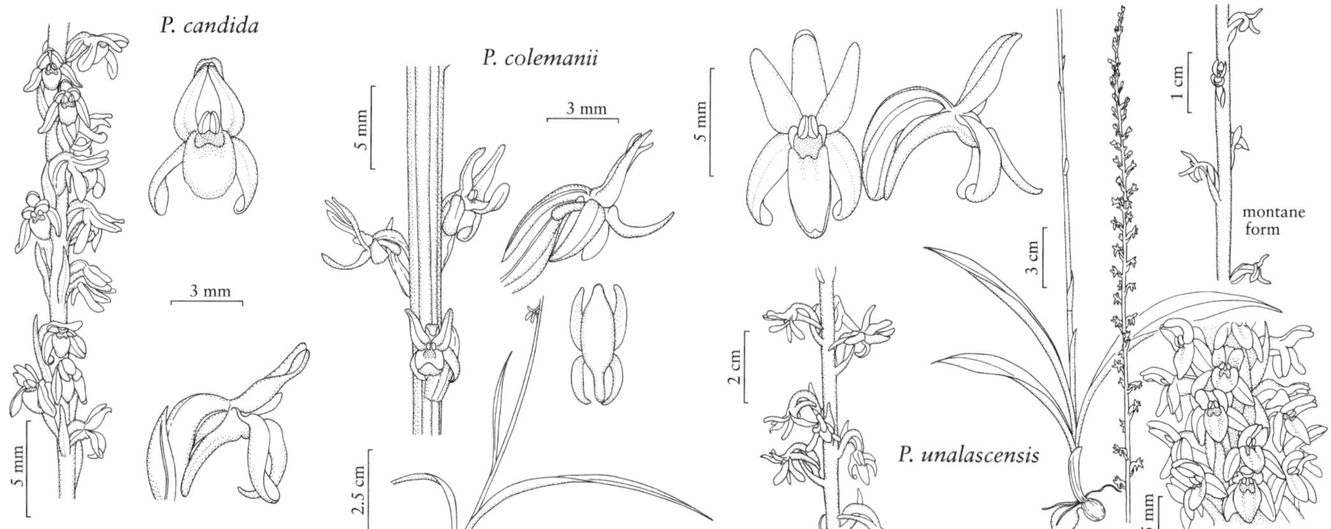

PIPERIA

scrub and prairie, exposed coastal bluffs; 100 m; Calif.

Piperia elegans subsp. *decurtata* is an uncommon subspecies known only from two small populations at the tip of the Pt. Reyes peninsula, California, and separated from *P. elegans* subsp. *elegans* by only 14 km.

3. Piperia yadonii Rand. Morgan & Ackerman, Lindleyana 5: 209, figs. 1A–G, 2. 1990 [C] [E]

Plants 10–50(–80) cm. **Stems** attenuate toward tuberoid, 0.5–3.5 mm diam.; bracts (4–)7–20(–26). **Leaves** prostrate; blade 11–17 × 2.2–3.9 cm. **Inflorescences** densely flowered, cylindric, (2–)5–15(–30) cm; rachis shorter than peduncle; bracts (3–)5–8(–11) mm. **Flowers** green and white, fragrance faint, harsh to honeylike, diurnal; sepals 3–5.5 × 1–2.5 mm; dorsal sepal green with white margins, elliptic-lanceolate; lateral sepals spreading to recurved, white, lanceolate; petals erect-recurved, green with broad outer, narrow inner white borders, lanceolate, falcate, 3–5 × 1.5 mm, inner margins often forming U, apices often connivent; lip recurved, triangular-lanceolate, 2.5–5 × 1.2–2.5 mm; spur deflexed, 2–5 mm; viscidia ovate to broadly elliptic, 0.4–0.5 × 0.3–0.4 mm; rostellum blunt. **Capsules** 5–9 mm. **Seeds** cinnamon brown.

Flowering late May–early Aug. Monterey pine forest, maritime chaparral; of conservation concern; 0–150 m; Calif.

Piperia yadonii, a narrow endemic (Monterey Bay area), bears a superficial resemblance to *P. elegans*. Much of the *P. yadonii* habitat has been preempted for residential and golf course development.

4. Piperia candida Rand. Morgan & Ackerman, Lindleyana 5: 207, figs. 1H–N, 2. 1990 [E] [F]

Plants 10–60 cm. **Stems** uniform toward base, 0.7–3.5 mm diam. distal to leaves; bracts 2–5(–10). **Leaves** prostrate to erect-spreading; blade 5–18 × 1.1–3.5 cm. **Inflorescences** sparsely flowered, ± secund, (2–)10–30 cm; rachis shorter than or equal to peduncle; bracts 3–10 mm. **Flowers** white, fragrance faint, harsh to honeylike; sepals oblong-elliptic, 1.8–3.5(–4.5) × 1–1.9 mm; dorsal sepal projecting to ± recurved; lateral sepals deflexed, ± twisted; petals usually projecting forward and ± connivent, white, midvein faintly green, asymmetrically lanceolate, 2–4 × 1 mm; lip recurved, triangular-ovate, (1.5–)2–3(–4 mm) × 1–2.8 mm; spur curved, tapered, 1.5–4 mm; viscidia broadly elliptic to ovate, 0.3–0.6 × 0.2–0.4 mm; rostellum ± elongate. **Capsules** 5–9 mm. **Seeds** cinnamon brown. $2n = 42$.

Flowering late spring–summer (May–Aug). Conifer and mixed evergreen forests, chaparral, sometimes on serpentine soils, not more than 150 km from coast; 0–1500 m; B.C.; Alaska, Calif., Oreg., Wash.

The flowers in *Piperia candida* are more completely white and more ephemeral than in any other member of the genus.

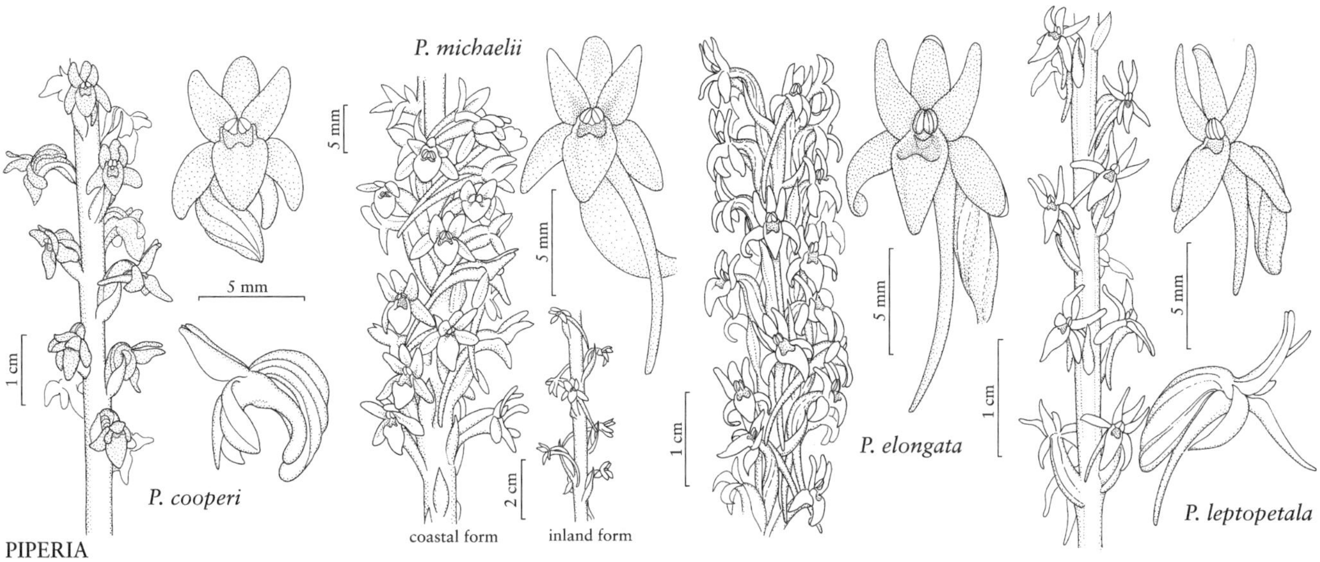

PIPERIA

5. **Piperia colemanii** Rand. Morgan & Glicenstein, Lindleyana 8: 89, figs. 1, 2. 1993 [E] [F]

Plants 16–53 cm. **Stems** attenuate toward tuberoid, 0.9–3(–3.7) mm diam. distal to leaves; bracts 0–2(–4). **Leaves** erect, folded, narrow, rather grasslike; blade 6–16 × 0.3–1(–1.9) cm. **Inflorescences** usually sparsely flowered, (6–)9–30(–40) cm; rachis usually longer than peduncle; bracts 3–16(–25) mm. **Flowers** translucent green, unscented; sepals 2–4 × 0.8–2 mm; dorsal sepal projecting between petals; lateral sepals strongly recurved; petals erect-recurved to sigmoid, ± falcate, linear-lanceolate, 2.2–4.5 × 0.7–1.7 mm; lip projecting, strongly upcurved, subhastate, triangular-lanceolate, 2–3(–5) × 1.2(–3) mm, apex sometimes touching dorsal sepal; spur ± club-shaped, 1–2.5 mm, always shorter than lip; viscidia broadly ovate, 0.2–0.25 × 0.15–0.2 mm; rostellum blunt. **Capsules** 3–9 mm. **Seeds** cinnamon brown.

Flowering Jun–Aug. Open coniferous forests and chaparral; 1300–2000 m; Calif.

Piperia colemanii is very uncommon.

In addition to the characteristics indicated in the key, *Piperia colemanii* may be distinguished from its more common relative, *P. unalascensis*, by having generally fewer stem bracts (0–2 instead of 2–5), by a deep-seated tuberoid usually 6–11 cm underground (less than 6 cm in *P. unalascensis*), and by a more hastate, triangular-lanceolate lip than *P. unalascensis*. The two species are partially sympatric and often occur in proximity, but no intermediates are known.

6. **Piperia unalascensis** (Sprengel) Rydberg, Bull. Torrey Bot. Club 28: 270. 1901 [E] [F]

Spiranthes unalascensis Sprengel, Syst. Veg. 3: 708. 1826; *Habenaria schischmareffiana* Chamisso; *H. unalaschensis* (Sprengel) S. Watson; *Platanthera foetida* Geyer ex Hooker; *P. unalaschcensis* (Sprengel) Kurtz

Plants 9–70 cm. **Stems** swollen to uniform diameter toward base, 0.7–6.1 mm diam. distal to leaves; bracts 1–8. **Leaves** prostrate; blade 5–16(–20) × 1.3–3.5 cm. **Inflorescences** usually sparsely flowered, 3–44 cm; rachis usually longer than peduncle; bracts 3–23 mm. **Flowers** ± translucent green, fragrance nocturnal but lingering during day, musky or soapy, sometimes honeylike; sepals ± translucent, 2–4.2 × 1–2.6 mm; dorsal sepal ovate to oblong; lateral sepals spreading to strongly recurved; petals usually projecting to erect, ± straight-sided, asymmetrically ovate to linear-lanceolate, 2–5.5 × 0.6–2 mm; lip ± deflexed, broadly ovate to lanceolate-elliptic, 2–5 × 1–3 mm, apex usually somewhat upcurved; spur horizontal to decurved, 2–5.5 mm, ± equal to lip; viscidia broadly elliptic-ovate, 0.15–0.5 × 0.1–0.3 mm; rostellum blunt. **Capsules** 3.5–10.5 mm. **Seeds** tan to cinnamon brown.

Flowering Jun–Aug; Apr–May, lowlands of n Calif. Coniferous and mixed evergreen forests, generally dry sites, rarely coastal bluffs; 0–3000 m; Alta., B.C., Nfld. and Labr. (Nfld.), Ont., Que.; Alaska, Calif., Colo., Idaho, Mich., Mont., Nev., N.Mex., Oreg., S.Dak., Utah, Wash., Wyo.

The racemes in *Piperia unalascensis* are usually slender and sparsely flowered; racemes of uncommon coastal populations (including the type) are short, stout, and densely flowered. Plants of the coast ranges and the Pacific Northwest are stouter and have broader sepals and petals than do interior and montane forms. Two sympatric forms appear to be in the Sierra Nevada, differing in lip morphology and scent.

7. Piperia cooperi (S. Watson) Rydberg, Bull. Torrey Bot. Club 28: 636. 1901 [F]

Habenaria cooperi S. Watson, Proc. Amer. Acad. Arts 12: 276. 1877; *Piperia lancifolia* Rydberg

Plants 14–90 cm. **Stems** attenuate toward tuberoid, fistulous, 1–8 mm diam. distal to leaves; bracts 3–12. **Leaves** erect-spreading; blade 9–20(–27) × 1.5–3.1 cm. **Inflorescences** sparsely flowered, (3–)10–40(–56) cm; rachis at least as long as peduncle; bracts 3–16 mm. **Flowers** green, fragrance nocturnal, honeylike; sepals 2.3–4 × 1.2–2.5 mm; dorsal sepal spreading to recurved, oblong to ovate; lateral sepals lanceolate to oblong; petals projecting to ± erect-spreading, asymmetrically triangular-ovate, 2.6–3.8 × 1.2–2 mm; lip deflexed, triangular-ovate, 1.6–4 × 1.6–3.5 mm; spur horizontal to deflexed, narrowly cylindric, 3–6(–9) mm; viscidia ovate to oblong, 0.35–0.6 × 0.2–0.4 mm; rostellum blunt. **Capsules** 4–12 mm. **Seeds** blackish brown.

Flowering Mar–May. Chaparral and coastal scrub; 0–1500 m; Calif.; Mexico (Baja California).

Piperia cooperi and *P. michaelii* have very similar flowers, and their blackish brown seeds are unique in the genus. A few specimens from the Santa Monica Mountains, California, described as *P. lancifolia* Rydberg, have flowers with spurs of intermediate length (6–9 mm).

8. Piperia michaelii (Greene) Rydberg, Bull. Torrey Bot. Club 28: 640. 1901 (as michaeli) [E] [F]

Habenaria michaelii Greene, Bull. Calif. Acad. Sci. 1: 282. 1885; *Piperia elongata* Rydberg subsp. *michaelii* (Greene) Ackerman

Plants 9–70 cm. **Stems** attenuate toward tuberoid, fistulous, 1.5–12(–17) mm diam. distal to leaves; bracts 3–50. **Leaves** prostrate to erect-spreading; blade 7–22(–30) × 1.5–4(–6.2) cm. **Inflorescences** ± densely flowered, 3–33 cm; rachis usually shorter than peduncle; bracts 4–17(–22) mm. **Flowers** green, fragrance nocturnal, strong, pleasant; sepals 2–4.8 × 1.3–3.7 mm; dorsal sepal ovate-elliptic to ovate; lateral sepals spreading to recurved, oblong-lanceolate to oblong-ovate; petals erect-spreading, asymmetrically triangular-ovate, 2.2–4.5 × 1–2.2 mm; lip straight to ± bent, ovate-triangular, 1.7–5 × 1.3–5 mm; spurs mostly deflexed, sometimes mixed with horizontal ones, 9–16 mm; viscidia broadly oblong, 0.4–0.75 × 0.15–0.4 mm; rostellum blunt. **Capsules** 5–13 mm. **Seeds** blackish brown, rarely brown.

Flowering late Apr–mid Aug. Coastal bluffs and scrub, chaparral, oak woodland; 0–700 m; Calif.

Coastal plants of *Piperia michaelii* are stouter, have more stem bracts and denser inflorescences, and are later blooming than interior populations. Occasionally the seeds of plants in interior locations are brown. Where *P. michaelii* and *P. elongata* are sympatric, the former blooms earlier than the latter.

9. Piperia elongata Rydberg, Bull. Torrey Bot. Club 28: 270. 1901 [E] [F]

Habenaria elegans (Lindley) Bolander var. *elata* Jepson; *H. unalascensis* (Sprengel) S. Watson subsp. *elata* (Jepson) Calder & Roy L. Taylor; *H. unalascensis* var. *elata* (Jepson) Correll; *Piperia elegans* (Lindley) Rydberg var. *elata* (Jepson) Luer

Plants (9–)14–130 cm. **Stems** ± uniform toward tuberoid, 1–12 mm diam. distal to leaves; bracts 4–22. **Leaves** prostrate; blade 7.5–32 × 1–8.4 cm. **Inflorescences** sparsely to densely flowered, (3.5–)6–70 cm; rachis usually longer than peduncle; bracts 5–20 mm. **Flowers** green; fragrance when present nocturnal, faint, harsh to honeylike; sepals 2.4–6 × 1.1–3 mm; dorsal sepal ovate; lateral sepals spreading to recurved, ± obliquely lanceolate to oblong; petals erect-spreading, ± falcate, ovate-attenuate, 3–6 × 1.5–2.5 mm, inner margins forming U; lip becoming strongly deflexed, triangular-lanceolate, 2.1–5 × 1.2–3.4 mm; spur curved, mostly deflexed, (7–)9–15(–18) mm; viscidia broadly elliptic-ovate, 0.4–0.6 × (0.2–)0.3–0.4 mm; rostellum blunt. **Capsules** 4–11 mm. **Seeds** cinnamon brown. $2n = 42$.

Flowering May–Sep. Forest chaparral, dry, open sites, usually away from the immediate coast; 0–1600 m; B.C.; Calif., Idaho, Mont., Oreg., Wash.

Plants of the California Channel Islands are smaller, have narrower viscidia than typical plants, and flower earlier (May–June).

10. Piperia leptopetala Rydberg, Bull. Torrey Bot. Club 28: 637, fig. 29. 1901 [E] [F]

Plants 13–70 cm. **Stems** uniform toward base, 0.7–3.4 mm diam.; bracts 2–6(–12). **Leaves** ± prostrate; blade 6.5–15 × 1.5–3 cm. **Inflorescences** densely flowered, 4–40 cm; rachis usually longer than peduncle; bracts 5–11 mm. **Flowers** pale green, translucent; fragrance nocturnal, delicate, ± lemony or otherwise pleasant; sepals oblong, 2.5–5 × 0.7–1.6 mm; lateral sepals recurved; petals erect, strongly upwardly recurved, straight, linear, spreading to form a V, 3–6 × 0.6–1.3 mm, apex long-acuminate; lip projecting to ± deflexed, narrowly lanceolate, 2.5–5 × 0.8–2.5 mm; spur deflexed, curved, tapered, 4–9 mm; viscidia broadly elliptic-ovate, 0.25–0.45 × 0.2–0.3 mm; rostellum slightly elongate. **Capsules** 4–9 mm. **Seeds** cinnamon brown.

Flowering May–Jul. Woodlands and chaparral, sometimes on serpentine soils; 100–2200 m; Calif.

Piperia leptopetala is uncommon but widely distributed in California. Its flowers have the narrowest sepals and petals of the genus; this gives the inflorescence a lacy appearance. *Piperia leptopetala* blooms earlier than sympatric *P. transversa* and *P. elongata*, two species with which it is sometimes confused.

28. DACTYLORHIZA Necker ex Nevski, Trudy Bot. Inst. S.S.S.R., Ser. 1, Fl. Sist. Vyssh. Rast. 4: 332. 1937, conservation proposed • Orchis [Greek *dactylos*, finger, and *rhiza*, root, in reference to the fingerlike tuberoids of the more primitive species]

Charles J. Sheviak

Paul M. Catling

Susan J. Meades

Richard M. Bateman

Herbs, perennial, terrestrial, rather succulent, glabrous. **Roots** from base of stem fascicled tuberoids, usually palmately divided with 2–5 lobes, fleshy. **Stems** leafy. **Leaves** several, ascending to recurved, not enfolded around spike, with or without purplish spots; base sheathing in proximal leaves, distal leaves bractlike, not sheathing. **Inflorescences** terminal, spikes; floral bracts foliaceous, prominent. **Flowers** few to many, resupinate; dorsal sepal, sometimes lateral sepals, and petals connivent, forming hood distal to lip; petals ± obliquely dilated basally; lip 3-lobed, base spurred, margins occasionally entire, nectarless; pollinaria 2, each with 1 pollen mass; viscidia within single 2-lobed bursicle; stigma reniform or obcordate, concave with median ridge, hidden behind bursicle. **Fruits** capsules, ascending, ellipsoid.

Species ca. 75 (2 in the flora): Alaska, Canada, mostly Eurasian.

Dactylorhiza is a taxonomically complex genus in which closely related species have been combined into species aggregates (P. Vermeulen 1947; R. M. Bateman et al. 1997). Recognition of the aggregate taxa alone reduces the number of species by more than half. Recent research synthesizing morphometric and allozyme data to circumscribe species (R. M. Bateman and I. Denholm 1983, 1985, 1989; M. Hédren 1996), and DNA sequences and chromosome studies to determine the relationships of those species (R. M. Bateman et al. 1997; A. M. Pridgeon et al. 1997), is shedding much light on the evolution of the genus. The diploid lineage ($2n$ = 40) appears to have evolved in Asia, migrating and speciating to both the west and northeast. Several alloploidy events (hybridization followed by chromosome doubling) occurred recently in Europe, apparently between the distinct diploids *D. fuchsii* and *D. incarnata*. That generated a highly complex suite of poorly distinguishable "prospecies" of $2n$ = 80, treated as a single species by some authorities and as many species by others (L. V. Averyanov 1990). Of the two North American species, *D. aristata* is a native diploid originating during the northeasterly migration, and *D. majalis* is an allotetraploid that originated in Europe and presumably is naturalized in North America (H. J. Clase and S. J. Meades 1996).

Despite an extensive literature, much taxonomic work still remains to be done.

SELECTED REFERENCES Averyanov, L. V. 1990. A review of the genus *Dactylorhiza*. In: J. Arditti, ed. 1977+. Orchid Biology: Reviews and Perspectives. 7+ vols. Ithaca, N.Y. and New York. Vol. 5, pp. 159–206. Bateman, R. M. and I. Denholm. 1983. A reappraisal of the British and Irish dactylorchids: 1. The tetraploid marsh orchids. Watsonia 14: 347–376. Bateman, R. M. and I. Denholm. 1985. A reappraisal of the British and Irish dactylorchids: 2. The diploid marsh orchids. Watsonia 15: 321–355. Bateman, R. M. and I. Denholm. 1989. A reappraisal of the British and Irish dactylorchids: 3. The spotted orchids. Watsonia 17: 319–349. Bateman, R. M., A. M. Pridgeon, and M. W. Chase. 1997. Phylogenetics of subtribe Orchidinae (Orchidoideae, Orchidaceae) based on nuclear ITS sequences: 2. Infrageneric relationships and taxonomic revision to achieve monophyly of *Orchis* sensu stricto. Lindleyana 12: 113–141. Hédren, M. 1996. Genetic differentiation, polyploidization and hybridization in northern European *Dactylorhiza* (Orchidaceae): Evidence from allozyme markers. Pl. Syst. Evol. 201: 31–55. Pridgeon, A. M. et al. 1997. Phylogenetics of subtribe Orchidinae (Orchidoideae, Orchidaceae) based on nuclear ITS sequences: 1. Intergeneric relationships and polyphyly of *Orchis* sensu lato. Lindleyana 12: 89–109. Stace, C. A. 1991. New Flora of the British Isles. Cambridge and New York. Pp. 1166–1173. Tyteca, D. and J.-L. Gathoye. 1993. On the morphological variability of *Dactylorhiza praetermissa* (Druce) Soó (Orchidaceae). Belg. J. Bot. 126: 81–99. Vermeulen, P. 1947. Studies on Dactylorchids. Utrecht.

1. Sepals and petals awned; lip often magenta- to red-spotted .... 1. *Dactylorhiza aristata*
1. Sepals and petals obtuse, hoodlike, or acute, not awned; lip with red-violet loops (continuous or discontinuous), dashes, and spots .... 2. *Dactylorhiza majalis*

1. Dactylorhiza aristata (Fischer ex Lindley) Soó, Nom. Nov. Gen. Dactylorhiza, 5. 1962 • Fischer's orchid

Orchis aristata Fischer ex Lindley, Gen. Sp. Orchid. Pl., 262. 1835; *Dactylorhiza aristata* var. *kodiakensis* Luer & G. M. Luer

Plants 5–40 cm. **Leaves** 2–5; blade often purple-spotted or suffused with purple, proximal blade 3–18 × 1–6 cm. **Inflorescences:** floral bracts lanceolate, 15–50 × 5–15 mm, the larger often suffused with purple, extending well beyond flowers. **Flowers** magenta, pink, or white; sepals prominently oblique and erect at base, upwardly curving forward or downward, ovate-lanceolate, 9–11 (–13) × 3–6 mm, awned; petals ovate-lanceolate, 6–9 × 3–4 mm, awned; lip often magenta- to red-spotted, suborbiculate to obcordate, 8–12 × 8–12 mm, shallowly 3-lobed or entire, tip of middle lobe apiculate; spur 5.5–7.5 mm, $\frac{1}{2}$–$\frac{2}{3}$ as long as ovary. $2n = 40, 42$.

Flowering late spring–mid summer. Moist open tundra, meadows, bogs; 0–800 m; Alaska; e Asia (Russian Far East).

Dactylorhiza aristata var. *kodiakensis* has been separated primarily on the basis of its being uniform instead of variable like the typical variety (C. A. Luer 1975); supplementary characters all partially overlap. While this variety is not recognized here, variability in the Alaskan Fischer's orchids merits detailed taxonomic study.

2. Dactylorhiza majalis (Reichenbach) P. F. Hunt & Summerhayes, Watsonia 6: 130. 1965 [F] [I]

Orchis majalis Reichenbach, Iconogr. Bot. Pl. Crit. 6(2): 7. 1828

Varieties ca. 8 (1 in the flora): introduced; ne Canada; native Eurasia.

2a. Dactylorhiza majalis (Reichenbach) P. F. Hunt & Summerhayes var. **junialis** (Vermeulen) Senghas, Jahresber. Naturwiss. Vereins Wuppertal 21–22: 126. 1968 • Leopard marsh orchid [F] [I]

Orchis latifolia Linnaeus var. *junialis* Vermeulen, Ned. Kruidk. Arch. 43: 404. 1933; *Dactylorhiza praetermissa* (Druce) Soó subsp. *junialis* (Vermeulen) Soó; *D. praetermissa* var. *junialis* (Vermeulen) Vermeulen; *Orchis pardalina* Pugsley

Plants 20–55(–70) cm. **Stems** 4–10.5 mm diam. **Leaves:** proximal leaves 4–5, sheathing; distal leaves 2–3, not sheathing, all dark green, heavily marked with purplish spots (some annular) on adaxial surface only, keeled abaxially, margins entire, proximal leaves well spaced along stem, lance-elliptic to oblong, slightly hooded at apex, largest 12–20+ × 1.5–5(–6) cm, distal leaves sessile, narrowly lanceolate, apex acute to acuminate. **Inflorescences:** floral bracts green to suffused with purple, often spotted, narrowly lanceolate, 6.5–15

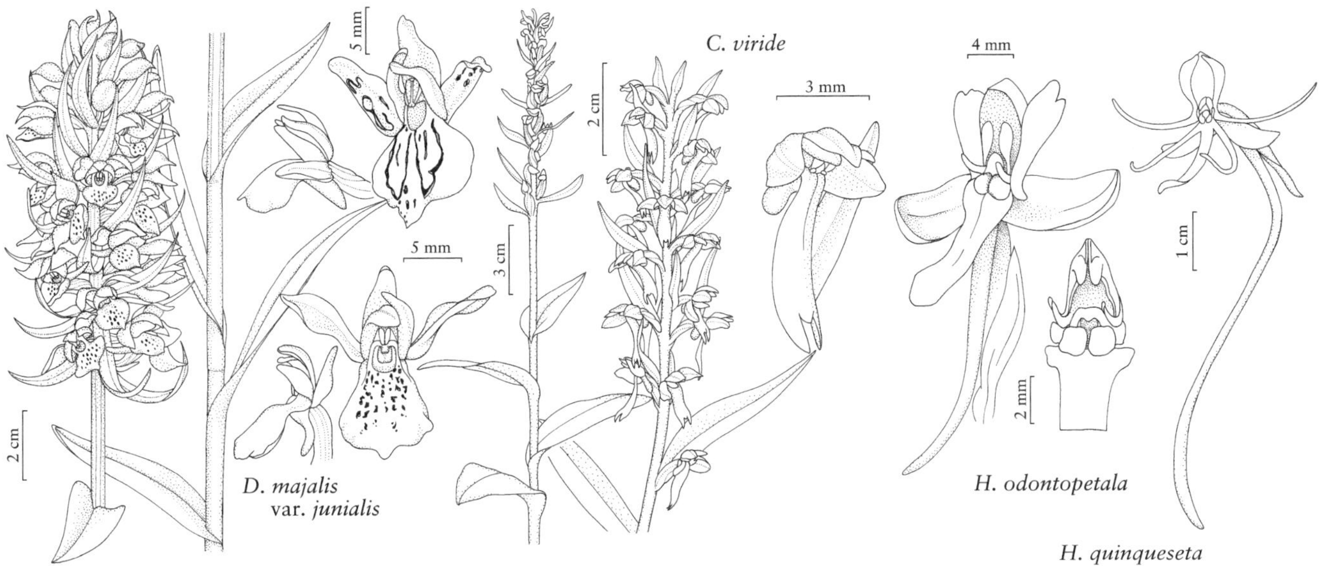

DACTYLORHIZA ◦ COELOGLOSSUM ◦ HABENARIA

× 2–5 mm, extending slightly beyond flowers. **Flowers** mauve to lavender; sepals and petals lanceolate, 6–10 × 3–4.5 mm, apex obtuse to acute; sepals suberect, prominently oblique at base, dark spotted; distal petals hooded; lip marked with red-violet loops (continuous or discontinuous), dashes, and spots, obcordate, shallowly 3-lobed, 7–11 × 9.5–14 mm, widest at middle; lateral lobes slightly to less often strongly reflexed, with shallow sinus on either side of small toothlike middle lobe 1–3.5 mm, apex obtuse to acute, usually exceeding lateral lobes by ca. 1 mm; spur straight to slightly down-curved, 4.5–7 × 2–4 mm, $\frac{1}{2}$–$\frac{2}{3}$ as long as ovary. $2n = 80$.

Flowering late Jun (Ont.)–late Jul (Nfld.). Shores, seepage slopes, bogs; to 200 m; introduced; Nfld. and Labr. (Nfld.), Ont.; Europe (Belgium, s England, n France, Luxembourg, Netherlands).

Dactylorhiza majalis var. *junialis* occurs on seepage slopes in Ontario and in bogs in Newfoundland.

First discovered in 1959 in the town of Timmins, Ontario, and reported by F. Cowell to H. Andrews (1961), *Dactylorhiza majalis* var. *junialis* became known as the Timmins orchid. Originally identified as *Orchis purpurella* T. Stephenson & T. A. Stephenson, it was later assigned to *Dactylorhiza maculata* (Linnaeus) Soó (C. A. Luer 1975; P. M. Catling and V. R. Catling 1991), *D. fuchsii* (P. M. Catling and C. J. Sheviak 1993), and most recently *D. majalis* subsp. *praetermissa* var. *junialis* by R. M. Bateman (in S. J. Meades 1999), who suggested that the Timmins orchid and a population of *Dactylorhiza* in St. John's, Newfoundland, may be the same taxon but from a different source (H. J. Clase and S. J. Meades 1996). The St. John's plants differ from those in Timmins by having larger spurs and less defined lip markings. Although the original lakeshore habitat in Timmins was destroyed, plants transplanted in the local area have become naturalized along the Mattagami River.

Another controversial naturalized *Dactylorhiza* population occurs at Tilt Cove (Baie Verte Peninsula, Newfoundland), near the site of an abandoned nickel and copper mine. It differs from the previous *Dactylorhiza* populations by its unspotted leaves; smaller, less marked flowers with short, conic spurs; and wetter, more basic (ultramafic) habitat. According to local accounts, its presence dates to the early 1900s, when mining equipment, packed in hay, was imported from England (S. J. Meades 1999). It remained unknown to the botanical community until found by S. G. Hay, A. Bouchard, and L. Brouillet in 1988. First identified as *D. incarnata* (S. G. Hay et al. 1990), it is known locally as the Tilt Cove orchid. Recently, R. M. Bateman tentatively identified the taxon as *D. majalis* subsp. *praetermissa* (southern marsh-orchid). He noted, however, that while the Tilt Cove plants most closely resemble subsp. *praetermissa*, they do not match exactly any European *Dactylorhiza* species. Additional studies are underway to obtain a more confident identification.

29. COELOGLOSSUM Hartman, Handb. Skand. Fl., 329. 1820 • Frog-orchis, long-bracted orchid [Greek *koilos*, hollow, and *glossa*, tongue]

Charles J. Sheviak

Paul M. Catling

Herbs, perennial, rather succulent. **Roots** both slender and palmately divided tuberoids, fascicled, fleshy. **Stems** leafy. **Leaves** several, alternate, ascending to recurved, not petiolate; blade obovate to oblanceolate to lanceolate, base sheathing. **Inflorescences** terminal, spicate racemes, lax to rather dense; floral bracts prominently exserted, foliaceous, often large. **Flowers** few to many, resupinate; lip notched to 3-lobed; lateral lobes prominent, middle lobe reduced, base elongated into short, saccate to scrotiform spur; pollinaria 2; pollinia 2; viscidia enclosed within membrane; stigmas 2-labiate, concave. **Fruits** capsules, ascending, ellipsoid.

Species 1: circumpolar.

1. Coeloglossum viride (Linnaeus) Hartman, Handb. Skand. Fl., 329. 1820 F

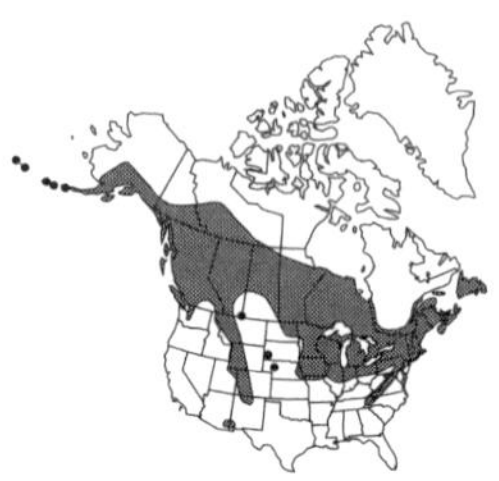

Satyrium viride Linnaeus, Sp. Pl. 2: 944. 1753; *Coeloglossum viride* subsp. *bracteatum* (Muhlenberg) Hultén; *C. viride* var. *islandicum* (Lindley) M. Schulze; *C. viride* var. *virescens* (Muhlenberg) Luer; *Dactylorhiza viridis* (Linnaeus) R. M. Bateman, Pridgeon & M. W. Chase; *Habenaria bracteata* (Muhlenberg) R. Brown; *H. viridis* (Linnaeus) R. Brown; *H. viridis* subsp. *bracteata* (Muhlenberg) Clausen; *H. viridis* var. *bracteata* (Muhlenberg) A. Gray; *H. viridis* var. *interjecta* Fernald

Plants 6–80 cm. **Leaves** 2–several; proximal blades obovate to oblanceolate, the distal oblong-lanceolate, becoming reduced to bracts, 2–18 × 1–7 cm. **Inflorescences:** floral bracts usually widely spreading, sometimes ascending (especially northward), conspicuous. **Flowers** greenish, often suffused with dull red or brown; sepals ovate, 3–8 × 2–5 mm, forming hood over column; petals linear, 3–5 × 0.5 mm; lip descending beneath hood, 3–10 × 0.5–4 mm; spur inconspicuous. **Capsules** 7–14 × 4–5 mm. $2n = 40$.

Flowering late spring–summer. Moist to wet coniferous and hardwood forests, tundra, prairies, meadows, thickets, coastal heaths, and bogs; 0–2800 m; Alta., B.C., Man., N.B., Nfld. and Labr. (Nfld.), N.W.T., N.S., Ont., P.E.I., Que., Sask., Yukon; Alaska, Ariz., Colo., Conn., Ill., Ind., Iowa, Maine, Md., Mass, Mich., Minn., Mo., Mont., Nebr., N.H., N.J., N.Mex., N.Y., N.C., N.Dak., Ohio, Pa., R.I., S.Dak., Tenn., Utah, Vt., Va., Wash., W.Va., Wis., Wyo.; Eurasia.

Recent molecular studies, in which nuclear ribosomal DNA internal transcribed spacer (ITS) sequences were analyzed, showed that within the palmate-tuber clade *Coeloglossum* is embedded within a strongly supported, monophyletic *Dactylorhiza* (A. M. Pridgeon et al. 1997); subsequently the monotypic *Coeloglossum* was formally transferred to *Dactylorhiza* (R. M. Bateman et al. 1997). The latter authors note that although *Coeloglossum* differs from *Dactylorhiza* in many morphologic characters, relatively little molecular divergence has occurred. In this treatment we continue to recognize them as separate but closely related genera.

Coeloglossum viride is quite variable across its range, and most American plants are referable to *Coeloglossum viride* (Linnaeus) Hartman var. *virescens* (Muhlenberg) Luer. These plants have been distinguished from the largely Eurasian subsp. *viride* by very long, spreading floral bracts greatly exceeding the flowers. The often large, stout plants with prominently bracted inflorescences and obscure flowers with mostly greenish lips are indeed distinctive. Similar plants also occur in Europe and eastern Asia, however. Small plants referable to subsp. *viride*, with reduced bracts and few-flowered inflorescences, often with wholly reddish lips, occur in Alaska, where nonetheless most plants are typical of the widespread American plant. Supposedly intermediate plants from the northern part of the American range of the species were described as *Habenaria viridis* (Linnaeus) R. Brown var. *interjecta* Fernald. At least in part these differ from long-bracted American plants merely in the long bracts being more nearly appressed to the rachis rather than widespeading. Very likely the variation in this species should be recognized at some level, but the focus on a single character and its variable expression on different continents does not provide an adequate basis for taxonomy.

230c.2b. ORCHIDACEAE Jussieu (tribe ORCHIDEAE) subtribe HABENARIINAE Bentham, J. Linn. Soc., Bot. 18: 288. 1881

30. HABENARIA Willdenow, Sp. Pl. 4(1): 5, 44. 1805 • Rein orchid [Latin *habena*, rein or strap]

Charles J. Sheviak

Herbs perennial, terrestrial or semiaquatic, often rather succulent. **Roots** scattered along basal portion of stem, both slender and tuberous, fleshy; tuberoids usually spheroid. **Stems:** leaves basal, abruptly reduced to bracts or scattered, gradually reduced toward inflorescence. **Leaves** several, alternate, ascending to spreading, conduplicate, bases sheathing stem. **Inflorescences** terminal, lax to rather dense spicate racemes; bracts large and foliaceous to reduced. **Flowers** few to many, resupinate, pedicellate, sometimes showy; petals 2-lobed, lateral lobe on abaxial margin; lip 3-lobed, spurred at base; pollinaria 2; pollinia 2; viscidia free; stigma with 2 fleshy processes flanking or encircling mouth of spur. **Fruits** capsules, ellipsoid.

Species ca. 600 (4 in the flora): pantropical.

A recently identified species, *Habenaria macroceratitis,* is described but not fully treated here.

SELECTED REFERENCE Ames, O. 1905–1922. Orchidaceae: Illustrations and Studies of the Family Orchidaceae Issuing from the Ames Botanical Laboratory.... 7 vols. Boston and New York. Vol. 4.

1. Lateral lobes of lip and petals greatly reduced; lip hastate-auriculate; petals auriculate on abaxial margin . . . . 1. *Habenaria odontopetala*
1. Lateral lobes of lip and petals prominently filiform, nearly equal to or exceeding other lobes; lip conspicuously 3-lobed; petals conspicuously 2-lobed.
    2. Spur more than 4 cm (often much more) . . . . 2. *Habenaria quinqueseta*
    2. Spur at most 2 cm (often much shorter).
        3. Leaves basal, spreading, broadly elliptic to ovate or obovate; spur slenderly clavate . . . . 3. *Habenaria distans*
        3. Leaves scattered along length of stem, ascending, linear-lanceolate to lanceolate, narrowly elliptic, or oblanceolate; spur slenderly cylindric to scarcely clavate . . . 4. *Habenaria repens*

1. Habenaria odontopetala Reichenbach f., Linnaea 18: 407. 1844 [F]

Habenaria strictissima Reichenbach f. var. *odontopetala* (Reichenbach f.) L. O. Williams; *Habenella odontopetala* (Reichenbach f.) Small

Plants terrestrial, erect, 13–100 cm. **Leaves** scattered along stem and gradually reduced distally, or rarely grouped toward base and abruptly reduced to bracts; blade wide-spreading to ascending, elliptic to lanceolate, 7–20 × 2–5 cm. **Inflorescences:** floral bracts spreading to appressed, lanceolate, 8–50 × 2–12 mm. **Flowers** widely diverging, not showy; sepals green; dorsal sepal deeply concave, forming hood over column, 4–7 × 4–5 mm; lateral sepals wide-spreading, 5–9 × 4–5 mm; petals yellowish, oblong-quadrangular, 3–5 × 1–2 mm, lateral lobe greatly reduced, petals appearing auriculate; lip yellowish green, linear to linear-oblong, 6–12 × 3–4 mm, lateral lobes greatly reduced, middle lobe appearing hastate-auriculate; spur cylindric to prominently clavate, slender, 1.3–2.8 mm; ovaries 2 cm. **Capsules** on prominent pedicel, ascending, 8–19 × 4–10 mm.

Flowering Sep–Feb. Dry to wet forests, forested swamps, and thickets; 0–80 m; Fla.; Mexico; West Indies; Central America.

2. Habenaria quinqueseta (Michaux) Eaton, Man. Bot. ed. 5: 253. 1829 • Long-horned habenaria, Michaux's orchid [F]

Orchis quinqueseta Michaux, Fl. Bor.-Amer. 2: 155. 1803; *Habenaria habenaria* (Linnaeus) Small; *H. macroceratitis* Willdenow; *H. quinqueseta* var. *macroceratitis* (Willdenow) Luer

Plants terrestrial, erect, 2–9 dm. **Leaves** scattered along stem and gradually reduced distally, or occasionally grouped toward base and abruptly reduced to bracts; blade wide-spreading to ascending, elliptic-oblong to oblanceolate, 5–25 × 2–6 cm. **Inflorescences:** floral bracts ascending to appressed, foliaceous, ovate to lance-acuminate, 15–28 × 7–15 mm. **Flowers** ascending, showy; sepals green; dorsal sepal shallowly concave, 6–13 × 5–10 mm; lateral sepals descending-spreading, 8–16 × 4–7 mm; petals white, lamina ascending, falcate, 6–15 × 2 mm, lateral lobe spreading-recurved, filiform, nearly equal to more than 2 times length of petal; lip white, middle lobe descending, linear, 8–20 × 2–3 mm, lateral lobes spreading, descending, filiform, nearly equal to more than 2 times middle lobe, apices recurved; spur slender, often markedly clavate, 4–18 cm; ovaries 2–3 cm. **Capsules** on short pedicel, nearly erect, 15–30 × 5–10 mm.

Flowering Aug. Dry to wet pine savannas and mixed flatwoods, hammocks, swamps, meadows, and roadsides; 0–100 m; Ala., Fla., Ga., La., Miss., S.C., Tex.; Mexico; West Indies; Central America; South America.

Spur length is extraordinarily variable in *Habenaria quinqueseta* as here delimited. Recognition of *H. quinqueseta* var. *macroceratitis*, a long-spurred variety with a more tropical distribution, has been proposed (C. A. Luer 1972), restricting var. *quinqueseta* to the United States. Spur length varies continuously, and recognition of segregates based on this feature alone is not warranted (D. S. Correll 1950). Recent study, however, has shown *Habenaria macroceratitis* Wildenow to be distinct (P. M. Brown 2000c). Both *H. quinqueseta* and *H. macroceratitis* vary in spur lengths, and this situation has obscured other differences. This information did not appear in time to be accommodated fully in the present treatment.

The two species may be distinguished as follows (P. M. Brown 2000c):

1. Distal division of lateral petals less than 2 times (10–18 mm) length of proximal division (6–9 mm); spur in living material typically less than 10 cm; plants of open pinelands, hedgerows, and fields .... *Habenaria quinqueseta*
1. Distal division of lateral petals more than two times (20–24 mm) length of proximal division (8–11 mm); spur in living material often greater than 10 cm; plants of rich hardwood hammocks .... *Habenaria macroceratitis*

SELECTED REFERENCE Brown, P. M. 2000c. Recent taxonomic and distributional notes from Florida 7. The genus *Habenaria*. N. Amer. Native Orchid J. 6: 148–158.

3. Habenaria distans Grisebach, Cat. Pl. Cub., 270. 1866

Plants terrestrial, erect, 14–33 cm. **Leaves** in basal rosette, abruptly reduced to bracts along stem; blade wide-spreading, often lying on soil surface, broadly elliptic to ovate or obovate, 6–15 × 2–5 cm. **Inflorescences:** floral bracts ascending to appressed, lance-acuminate, 10–25 × 3–7 mm. **Flowers** ascending, not showy; sepals greenish; dorsal sepal descending-reflexed, concave, 5–9 × 4–6 mm; lateral sepals descending-reflexed, 5–10 × 2–3 mm; petals greenish, lamina ascending, falcate, 6–9 × 1 mm, lateral lobe spreading, filiform, ± equal to petal; lip greenish, middle lobe descending, linear, 7–11 × 1 mm, lateral lobes ascending-spreading, filiform, 8–15 mm; spur slenderly club-shaped, 10–17 mm; ovaries 10–15 mm. **Capsules** on short pedicels, nearly erect, 1–1.5 cm × 5 mm.

Flowering Aug–Sep. Dense, moist hardwood forests; 0–10 m; Fla.; Mexico; West Indies; Central America; South America (Venezuela).

4. Habenaria repens Nuttall, Gen. N. Amer. Pl. 2: 190. 1818 • Water-spider orchid, Floating orchid, water orchid [F]

Habenaria nuttallii Small

Plants terrestrial or semiaquatic, erect to decumbent, 10–90 cm. **Leaves** scattered along stem and gradually reduced to sheathing bracts; blade ascending, linear-lanceolate to lanceolate, narrowly elliptic, or oblanceolate, 3–25 × 1–4.5 cm. **Inflorescences:** floral bracts spreading to ascending, lanceolate, 15–90 × 5–12 mm. **Flowers** ascending, not showy; sepals 3–7 × 3–4 mm; dorsal sepal shallowly concave; lateral sepals reflexed-spreading; petals greenish, lamina ascending, falcate, 3–7 × 1 mm, lateral lobe arcuately spreading-ascending, filiform, slightly to markedly exceeding petal; lip greenish, middle lobe descending, linear, 4–7 × 1 mm, lateral lobes ascending-spreading, filiform, 5–11 mm; spur slenderly cylindric to scarcely club-shaped, 0.8–1.4 cm; ovaries 9–15 mm. **Capsules** on short pedicel, nearly erect, 8–15 × 3–7 mm,

Flowering primarily summer–fall, sporadically (Apr–Dec). Marshes, wet meadows, bogs, margins of streams, ditches, and ponds; commonly an emergent aquatic in shallow water and in floating mats alone or with other vegetation; 0–100 m; Ala., Ark., Fla., Ga., La., Miss., N.C., Okla., S.C., Tex.; Mexico; West Indies; Central America; South America.

Habenaria repens is remarkable in sometimes being truly aquatic. Often forming floating mats, the plants then are commonly decumbent, at least basally, and new shoots and slender roots arise abundantly from much of the length of the stem. A few spheroid tuberoids are sometimes produced from roots arising at wide intervals. Other roots bear new shoots some decimeters from the parent stem; the distal portion of the root then commonly enlarges into a slenderly lance-fusiform tuberoid.

230d. ORCHIDACEAE Jussieu subfam. EPIDENDROIDEAE Lindley, Coll. Bot., App. 1826

230d.1. ORCHIDACEAE Jussieu (subfam. EPIDENDROIDEAE) tribe NEOTTIEAE Lindley, Coll. Bot., App. 1826

230d.1a. ORCHIDACEAE Jussieu (tribe NEOTTIEAE) subtribe LIMODORINAE Bentham, J. Linn. Soc., Bot. 18: 288. 1881

31. CEPHALANTHERA Richard, De Orchid. Eur., 29. 1817 • Helleborine [Greek *kephale*, head, and *anthera*, anther]

Charles J. Sheviak

Paul M. Catling

Herbs, perennial, achlorophyllous. **Roots** fleshy, slender, scattered along slender rhizome. **Stems** leafy. **Leaves** several, alternate; blade spreading, plicate, or reduced to appressed sheathing bracts. **Inflorescences** terminal, solitary spikes, lax; floral bracts often foliaceous, large to minute, scarious distally. **Flowers** few–many, resupinate, sessile, showy; lip white with yellow markings, divided by median constriction, saccate, distal end with fleshy parallel adaxial lamellae; pollinaria absent; pollinia 2; viscidia absent. **Fruits** capsules.

Species ca. 15 (1 in the flora): widespread, w North America, Eurasia.

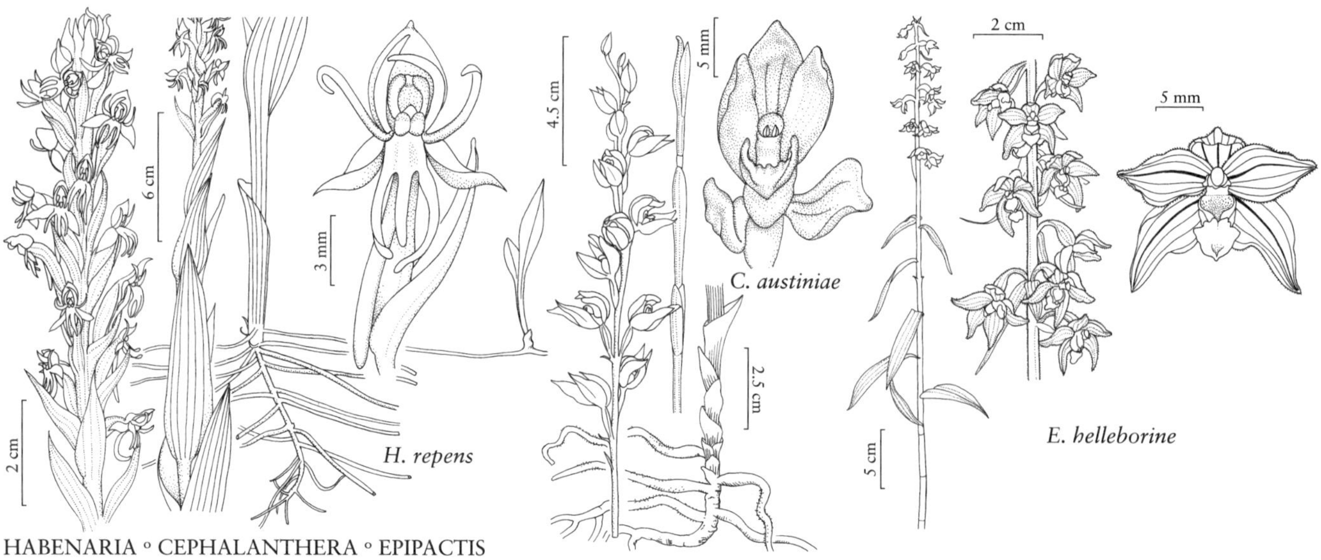

HABENARIA ◦ CEPHALANTHERA ◦ EPIPACTIS

1. **Cephalanthera austiniae** (A. Gray) A. Heller, Cat. N. Amer. Pl. ed. 2, 4. 1900 (as austinae) • Phantom orchid, snow orchid [E] [F]

Chloraea austiniae A. Gray, Proc. Amer. Acad. Arts 12: 83. 1877 (as austinae); *Eburophyton austiniae* (A. Gray) A. Heller

Plants pure white, 19–65 cm. **Leaves** proximally reduced to scarious sheathing bracts, to 10 cm, distally reduced to bracts, pure white, becoming brown. **Inflorescences:** floral bracts scarious, the proximal often foliaceous, distal usually reduced, lanceolate, apex acuminate. **Flowers** ascending, loosely open orbs, white with yellow markings; sepals lance-elliptic, 12–20 × 4–7 mm, base spread, curving forward to meet but not touch, apex obtuse to subacute; petals oblanceolate, arcuate, 10–17 × 3–6 mm, associated with dorsal sepal; lip 8–12 × 9–14 mm, lateral lobes flanking column, distal end deflexed, yellow centrally. **Capsules** erect, ellipsoid-oblanceoloid, 15 × 10 mm.

Flowering summer. Mineral soil in dry to moist coniferous forests; 0–2200 m; B.C.; Calif., Idaho, Oreg., Wash.

32. EPIPACTIS Zinn, Cat. Pl. Hort. Gott., 85. 1757, name conserved • [Ancient Greek name used by Theophrastus for plant used to curdle milk]

Paul Martin Brown

George W. Argus

Herbs perennial, terrestrial, glabrous or puberulent to glabrate. **Roots** from short rhizome, fibrous. **Stems** leafy. **Leaves** several, alternate, sheathing, plicate, grading into floral bracts. **Inflorescences** terminal racemes; floral bracts leaflike, often exceeding flowers. **Flowers** few to many, resupinate, pedicellate; perianth free, spreading or loosely connivent, subequal; sepals ovate to ovate-lanceolate, concave; petals often shorter than sepals; lip sessile at base of column, fleshy, divided into 2 parts by central constriction, proximal part (hypochile) deeply concave to saccate, distal part (epichile) triangular to oblanceolate, with pair of calli near base; column curved over lip, broadened distally, short; anther terminal, sessile; pollinia 4, mealy. **Fruits** capsules, pendent to spreading, obovoid to ellipsoid.

Species ca. 25 (3 in the flora): North America, Mexico, Eurasia, North Africa.

1. Lip 3-lobed, distal part linear-oblanceolate to narrowly spatulate-oblanceolate; lateral sepals 16–24 mm .... 2. *Epipactis gigantea*
1. Lip not 3-lobed, distal part triangular-ovate to broadly triangular; lateral sepals 6–13 mm.
    2. Lip 9–12 mm, distal part triangular-ovate, calli at base not rugose; ovary glabrous; flowers green to pink to purple to yellowish .... 3. *Epipactis helleborine*
    2. Lip 5.5–6.5 mm, distal part broadly triangular, calli at base rugose; ovary tomentose; flowers dark wine to cranberry red .... 1. *Epipactis atrorubens*

1. Epipactis atrorubens (Hoffmann ex Bernhardi) Besser, Prim. Fl. Galiciae Austriac. 2: 220. 1809 • Dark red helleborine [I]

Serapias atrorubens Hoffmann ex Bernhardi in G. F. Hoffmann, Deutchl. Fl. 4: 182. 1804; *Epipactis helleborine* (Linnaeus) Crantz subsp. *atrorubens* (Hoffmann ex Bernhardi) Syme; *E. rubiginosa* (Crantz) W. D. J. Koch

Plants 20–60(–106) cm, sparsely hairy. **Leaves** 5–10; blade orbiculate, elliptic, or narrowly lanceolate, 4–10 × 1.5–4.5 cm. **Inflorescences** lax racemes, usually secund; floral bracts lanceolate, 10–35 mm, often exceeding flowers. **Flowers** 8–18, dark wine to cranberry red, small; sepals dark wine red, rarely dull rose or greenish abaxially; lateral sepals 6–7 mm; petals similar to sepals in color, ovate, 4–7 × 2–4.5 mm; lip indistinctly veined, 5.5–6.5 mm, constricted at middle into 2 parts, proximal part greenish with red edges, deeply concave, adaxial surface spotted with violet-red or brown, distal part brighter and darker, broadly triangular or transversely elliptic, minutely toothed with small reflexed tip; calli 2, brownish, brighter, darker in color than lip, rugose; column very short, 2–3 mm; anther yellow; pollinia 2 pairs, yellow, mealy; ovary tomentose. **Capsules** ovoid, 7–9 mm, moderately to densely tomentose. 2*n* = 40.

Flowering Aug. Partial shade in abandoned serpentine/asbestos quarry; 700 m; introduced; Vt.; Europe; Asia (Iran).

Epipactis atrorubens is apparently naturalized in Vermont where a small population persists (P. M. Brown 1997).

2. Epipactis gigantea Douglas ex Hooker, Fl. Bor.-Amer. 2: 202, plate 202. 1839 • Chatterbox, stream orchid, giant helleborine

Amesia gigantea (Douglas ex Hooker) A. Nelson & J. F. Macbride; *Helleborine gigantea* (Douglas ex Hooker) Druce

Plants to 1.4 m, essentially glabrous. **Leaves** 4–14, ovate, ovate-elliptic to narrowly lanceolate, 5–20 × 2–7 cm. **Inflorescences** lax racemes; floral bracts lanceolate to oblong, 7–127 mm. **Flowers** 2–32, rather showy; sepals greenish to rose-colored with rose-colored to purple veins; lateral sepals 16–24 × 8–9 mm, apex very oblique; petals pale pink to rose-colored to orange with red or purple veins, broadly ovate to ovate-lanceolate, 13–17 × 6–8 mm; lip marked with red or purple, strongly veined, distinctly 3-lobed, constricted at middle into 2 parts, proximal part papillose, calli red, minute, warty, lateral lobes prominent subtriangular wings, distal part linear-oblanceolate to narrowly spatulate-oblanceolate, grooved to tip, 14–20 mm, calli near base, erect, orange or yellow, winglike; column erect, short, stout, with pair of lateral processes, 5–10 × 3 mm; anther green; pollinia 2 pairs, yellow, soft. **Capsules** ellipsoid, glabrate or sparsely pubescent, 20–25 mm. 2*n* = 40.

Flowering throughout summer (Mar–Aug). Wet gravelly and sandy stream shores and bars, may occur in riparian willow, box elder, and river birch woodlands or in chaparral, seepages, marshes, wet cliffs, hot springs; 0–3000 m; B.C.; Ariz., Calif., Colo., Idaho, Mont., Nev., N.Mex., Okla., Oreg., S.Dak., Tex., Utah, Wash., Wyo.; Mexico; Asia (China, India, s Japan, Tibet).

Two very distinct color forms of this species have been published: *Epipactis gigantea* forma *rubrifolia* P. M. Brown, with deep red stems and leaves, and *E. gigantea* forma *citrina* P. M. Brown, with lemon-yellow flowers. Both are known from California.

The occurrence of this wide-ranging species in India and Tibet is based on *Epipactis royaleana* Lindley ex Royle being given as a synonym in a study of the Monocotyledoneae of Karakorum (W. B. Dickoré 1995) and an embryologic study (S. P. Vij et al. 1999).

3. **Epipactis helleborine** (Linnaeus) Crantz, Stirp. Austr. Fasc. ed. 2, 2: 467. 1769 • Broad-leaved helleborine [F] [I]

Serapias helleborine Linnaeus, Sp. Pl. 2: 949. 1753; *Epipactis latifolia* (Linnaeus) Allioni

Plants 25–80(–100) cm, sparsely to densely pubescent. **Leaves** 3–10, orbiculate, elliptic to narrowly lanceolate, 4–18 × 1.5–8.5 cm. **Inflorescences** racemes, loose to moderately dense, commonly 1-sided; floral bracts spreading, linear to narrowly lanceolate, 10–40(–70) mm, often exceeding flowers. **Flowers** 15–50, small; sepals greenish, often suffused with purple; lateral sepals 10–13 × 5–6 mm, apex oblique; petals ovate, pale green, pink, purple, or yellowish, 9–11 × 4–6 mm; lip indistinctly veined, constricted at middle into 2 parts, proximal part purplish to brownish, deeply concave, not papillose, 9–12 × 8 mm, distal part recurved, pink, broadly triangular-ovate, ± flat to tip, 5 × 5 mm; calli 2, near base, brownish, not rugose; column 3–6 mm; ovary glabrous. **Capsules** obovoid, 9–14 mm, glabrate to densely pubescent. $2n$ = 36, 38, 40, 44.

Flowering Jun–Oct. Moist to dry, rocky, shaded, deciduous to mixed woods; cedar swamps and forested stream margins; often in disturbed places such as lawns, and cracks in concrete sidewalks; 0–1300 m; introduced; B.C., N.B., Nfld. and Labr. (Nfld.), Ont., Que.; Ark., Calif., Conn., Ill., Ind., Ky., Maine, Md., Mass., Mich., Minn., Mo., Mont., N.H., N.J., N.Mex., N.Y., N.C., Ohio, Oreg., Pa., R.I., Tenn., Vt., Va., Wash., W.Va., Wis.; Eurasia; North Africa.

Epipactis helleborine is known to have been thoroughly naturalized in North America since 1879 (C. A. Luer 1975).

Extensive and striking color variations in *Epipactis helleborine* have led to the description of numerous varieties and forms. Most of the varieties originated in Europe, but the following forms originated in North America. With the exception of forma *luteola*, which has been noted only from Vermont, the other forms occur randomly throughout the range of the species: forma *alba* (Webster) B. Boivin, white-flowered form; forma *luteola* P. M. Brown, yellow-flowered; forma *monotropoides* (Mousley) Scoggan, albino; forma *variegata* (Webster) B. Boivin, variegated, the leaves with white and cream markings; forma *viridens* A. Gray, green-flowered.

33. LISTERA R. Brown in W. Aiton and W. T. Aiton, Hortus Kew. 5: 201. 1813 • [For Martin Lister (1638–1711), noted English physician and naturalist]

Lawrence K. Magrath

Ronald A. Coleman

Herbs, perennial, terrestrial. **Roots** slender, fibrous. **Stems** slender to stout, glabrous proximal to leaves; cauline bracts 2(–3), enclosing base of stem. **Leaves** 2(–3, rarely), at apex of stem, opposite or subopposite, sessile, glabrous. **Inflorescences** terminal, 2–100-flowered racemes; peduncle and rachis densely glandular-pubescent or glabrate; floral bracts inconspicuous, lanceolate, elliptic, suborbiculate-ovate, rhombic-ovate, or oblong. **Flowers** resupinate, maroon-purple, yellowish green to dark green, blue-green, or pinkish tan; dorsal sepal ovate-elliptic, elliptic-obovate, linear-elliptic, or lanceolate; lateral sepals semiorbiculate-elliptic, ovate, linear-elliptic, or oblong-lanceolate, often falcate-recurved; petals reflexed, spreading, or connivent, linear, linear-oblong to lanceolate, or elliptic, falcate; lip prominently deflexed or not, sessile or clawed, linear-oblong to obovate, suborbiculate-ovate, or ovate-reniform, base of lip with or without prominent auricles or lobes, apex deeply 2-lobed, dilated to rounded, apiculate; calli various, papillose, pair of horns, or with 1–2 lamellae; column arcuate, thick, short, apex expanded or not; anther on adaxial side of column near apex; pollinia 2, yellow, soft; ovaries pedicellate. **Fruits** capsules, horizontal to semierect, ellipsoid, ovoid, or subglobose, glabrous or glabrate to glandular-pubescent.

Species 25 (8 in the flora): cool temperate regions, Northern and Southern hemispheres.

1. Lip cleft 1/2–2/3 its length into 2 pointed, linear-lanceolate apical lobes; column 0.5–1 mm.
    2. Lip with basal auricles rounded and projecting distally, partially surrounding column; flowers smaller, dorsal and lateral sepals 1.5 mm . . . . 1. *Listera australis*
    2. Lip with basal auricles or lobes pointed and projecting outward like horns, not partially surrounding column; flowers larger, dorsal and lateral sepals 2–3 mm . . . . 2. *Listera cordata*
1. Lip expanded at apex or if cleft, then less than 1/2 its length and apical lobes rounded; column 1.5–4 mm.
    3. Lip ± as broad at apex as at base, with basal auricles or lobes extending beyond column.
        4. Auricles curving around and clasping base of column; pedicels glabrous, rarely glabrate; dorsal sepal 3–3.5 mm, lateral sepals 3–4 mm . . . . 3. *Listera auriculata*
        4. Auricles divergent from column; pedicels glandular-pubescent; dorsal sepal 4–6 mm, lateral sepals 4.5–7 mm . . . . 4. *Listera borealis*
    3. Lip broader at apex than at base, without basal auricles or lobes extending beyond column.
        5. Lip with short claw, not sessile . . . . 5. *Listera convallarioides*
        5. Lip sessile, without claw.
            6. Lip deflexed, base of lip without teeth or lobules; pedicels stout, 3–4 mm; capsules 10 mm . . . . 6. *Listera ovata*
            6. Lip not deflexed, base of lip with teeth or lobules; pedicels slender, 4–12 mm; capsules 5 mm.
                7. Lip rounded or retuse at apex, base with 2 small, pointed auricles; pedicels glandular-puberulent . . . . 7. *Listera caurina*
                7. Lip deeply notched at apex into 2 broadly rounded lobes, base with rounded lobules on each side; pedicels glabrous . . . . 8. *Listera smallii*

1. Listera australis Lindley, Gen. Sp. Orchid. Pl., 456. 1840 • Southern twayblade [E]

Bifolium australe (Lindley) Nieuwland; *Diphryllum australe* Kuntze; *Neottia australis* (Lindley) Szlachetko; *Ophrys australis* (Lindley) House

Plants 8–29 cm. **Stems** purplish green, succulent, glabrous. **Leaves:** blade dark green, ovate, ovate-oblong, or elliptic, 1.3–4 × 0.5–2.1 cm, base often subcordate, apex obtuse to apiculate. **Inflorescences** 5–25-flowered, lax, 45–110 mm; floral bracts suborbiculate-ovate, 2 × 1 mm, apex obtuse; peduncle and rachis slightly glandular-puberulent; bracts, pedicel, and ovary glabrous. **Flowers** reddish purple; pedicel filiform, 4–6 mm; dorsal sepal ovate-elliptic, concave, 1.5 × 1 mm, apex obtuse; lateral sepals ovate, falcate, 1.5 × 1 mm, apex obtuse; petals recurved, oblong, 1.5 × 0.5 mm, apex obtuse; lip linear, 6–12 × 1.5–2 mm, base extended into 2 rounded, linear lobes, projecting distally, partially surrounding column, distal half split into 2 linear-lanceolate lobes with minute tooth in sinus; column 0.5–1 × 0.5–1 mm. **Capsules** horizontal, ovoid, 8 × 5 mm, with relatively few large seeds.

Flowering Jan (Fla.)–Jul. Rich humus of low moist woods, marshes, sphagnum bogs, usually in association with rhizomes of cinnamon fern (*Osmunda cinnamomea*) and royal fern (*O. regalis*); 0–1000 m; N.B., N.S., Ont., Que.; Ala., Ark., Fla., Ga., La., Miss., N.J., N.Y., N.C., Okla., Pa., S.C., Tenn., Tex., Vt.

Perianth parts of *Listera australis* persist as capsules mature. This is such a small orchid that it can easily be overlooked; it may have a wider distribution than recorded here.

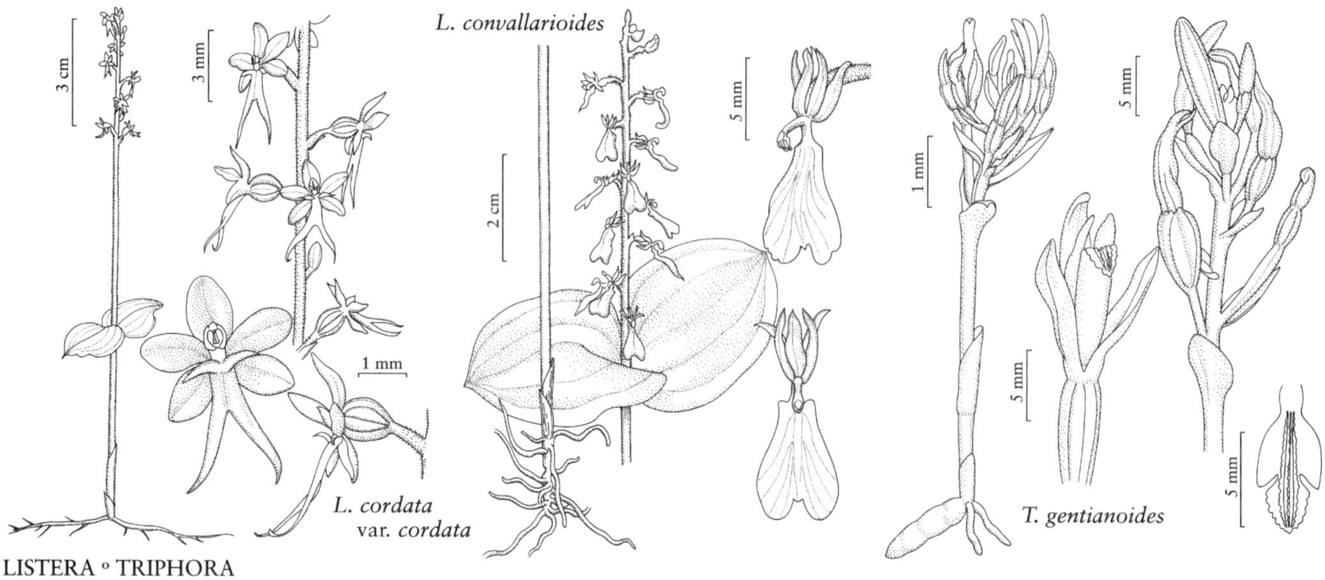

LISTERA ◦ TRIPHORA

2. Listera cordata (Linnaeus) R. Brown in W. Aiton and W. T. Aiton, Hortus Kew. 5: 201. 1813 • Heart-leaved twayblade [F]

Ophrys cordata Linnaeus, Sp. Pl. 2: 946. 1753; *Bifolium cordatum* (Linnaeus) Nieuwland; *Diphryllum cordatum* (Linnaeus) Kuntze; *Distomaea cordata* (Linnaeus) Spenner; *Neottia cordata* (Linnaeus) Richard; *Pollinirhiza cordata* (Linnaeus) Dulac

Plants 5–33 cm. **Stems** green to reddish purple, succulent, glabrous. **Leaves:** blade broadly to narrowly ovate-cordate or deltate, 0.9–2(–4) × 0.7–2(–3.8) cm, apex mucronate. **Inflorescences** 5–25-flowered, lax to dense, 20–100 mm, slender; floral bracts ovate, 1–1.5 × 1 mm; peduncle and rachis slightly glandular-puberulent or glabrate; bracts, pedicel, and ovary glabrous. **Flowers** yellow-green, green, or reddish purple; pedicel slender, 2–3 mm; dorsal sepal ovate-oblong to oblong-elliptic, 2–3 × 1 mm, apex obtuse; lateral sepals ovate-oblong to oblong-elliptic, slightly falcate, 2–3 × 0.5–1.5 mm, apex obtuse; petals elliptic to oblong-linear, 1.5–2.5 × 0.5–1 mm, apex obtuse; lip linear-oblong, cleft 1/2 –2/3 its length into 2 linear-lanceolate lobes, 3–4 × 1–1.5 mm; disc with pair of spreading, linear lobes, apices acute; column 0.5 × 0.5 mm. **Capsules** semierect, subglobose, 5 × 4 mm.

Listera cordata is the most widespread species in the genus.

Varieties 2 (2 in the flora): n North America, Europe, Asia (Japan), Iceland.

SELECTED REFERENCE Ackerman, J. D. and M. R. Mesler. 1979. Pollination biology of *Listera cordata* (Orchidaceae). Amer. J. Bot. 66: 820–824.

1. Leaf blades 0.7–2 cm wide; lip 3–4 mm; flowers yellow-green, green, or reddish purple . . . . 2a. *Listera cordata* var. *cordata*
1. Leaf blades 1.8–3.8 cm wide; lip 5–6 mm; flowers green to yellow-green . . . . 2b. *Listera cordata* var. *nephrophylla*

2a. Listera cordata (Linnaeus) R. Brown var. **cordata** [F]

Listera cordata var. *chlorantha* Beauverd

Leaves: blade 0.9–2 × 0.7–2 cm. **Flowers** yellow-green, green, or reddish purple; pedicels 2 m; sepals 2–3 × 0.5–1.5 mm; petals 1.5–2.5 × 0.5–1 mm; lip 3–4 × 1–1.5 mm. $2n$ = 36, 38, 40, 42.

Flowering late May–Aug. Mountainous regions in mossy, damp, coniferous or mixed coniferous-hardwood forests, sphagnum bogs and evergreen swamps, and subacid humus in moist woodlands; 0–1200 m; Greenland; St. Pierre and Miquelon; Alta., B.C., Man., N.B., Nfld. and Labr., N.W.T., N.S., Ont., P.E.I., Que., Sask., Yukon; Conn., D.C., Maine, Md., Mass., Mich., Minn., N.H., N.J., N.Y., N.C., Ohio, Pa., R.I., Vt., Va., W.Va., Wis.; Europe; Asia (Japan); Iceland.

2b. Listera cordata (Linnaeus) R. Brown var. **nephrophylla** (Rydberg) Hultén, Fl. Aleut. Isl., 145. 1937 • Western Heart-leaved twayblade [E]

Listera nephrophylla Rydberg, Mem. New York Bot. Gard. 1: 108. 1900; *Neottia nephrophylla* (Rydberg) Szlachetko; *Ophrys nephrophylla* (Rydberg) Rydberg

Leaves: blade 2–4 × 1.8–3.8 cm. **Flowers** green to yellow-green; pedicels slender, 3 mm; sepals 4 × 1.5 mm; petals 1.5–2.5 × 0.5–1 mm; lip 5–6 × 1–1.5 mm. $2n$ = 36, 38

Flowering Jun–Jul. Shaded humus of rich, well-drained, coniferous or coniferous-hardwood forest; 1500–3500 m; Alta., B.C., N.W.T., Yukon; Alaska, Calif., Colo., Idaho, Mont., N.Mex., Oreg., Wash., Wyo.

Taxonomists have differed about whether these two varieties deserve to be recognized. The arguments for maintaining the varieties as distinct (C. A. Luer 1975) and against maintaining them as distinct (D. S. Correll 1950; C. L. Hitchcock et al. 1955–1969, vol. 5; J. A. Calder and R. L. Taylor 1968; E. Hultén 1941–1950, vol. 3) have been summarized (R. A. Coleman 1995). Although leaf size and shape overlap, the leaves of the western plants are broader than long, those of the eastern plants longer than broad. Nonetheless, the distinction is not sufficient to maintain separate varieties (J. A. Calder and R. L. Taylor 1968). In the present treatment we have taken a conservative approach and include the two varieties. We recognize, however, that in North America, *Listera cordata* is a highly variable species: although some plants easily fit one or the other of the described varieties, a host of intermediates do not fit either. Further study of this species "complex" is needed.

3. Listera auriculata Wiegand, Bull. Torrey Bot. Club 26: 166, plate 356, fig. 2. 1899 • Auricled twayblade [E]

Bifolium auriculatum (Wiegand) Nieuwland; *Neottia auriculata* (Wiegand) Szlachetko; *Ophrys auriculata* (Wiegand) House

Plants 5–25 cm. **Stems** pale green, succulent, glabrous. **Leaves:** blade pale green, suborbiculate to suborbiculate-ovate or elliptic-ovate, 2.5–6 × 1.5–4.2 cm, apex subobtuse. **Inflorescences** 5–20-flowered, lax, 20–100 mm; floral bracts broadly elliptic to oblong-lanceolate, 2–7 × 1–2 mm, apex obtuse; peduncle and rachis densely glandular-puberulent; bracts, pedicel, and ovary glabrous to rarely glabrate. **Flowers** pale green to blue-green, fading whitish with age; pedicels stout, 2.5–5 mm, glabrous, rarely glabrate; sepals and petals reflexed away from column and lip; dorsal sepal elliptic-obovate, 3–3.5 × 1.5–2 mm, apex subobtuse; lateral sepals elliptic, oblong, or ovate-lanceolate, strongly falcate, 3–4 × 1–1.5 mm, apex subobtuse to acute; petals linear-oblong to linear, falcate, 3–3.7 × 0.8 mm, apex obtuse; lip obovate to oblong, base of lip with auricles curving around and clasping base of column, apex slightly expanded, cleft approximately 1/5–1/3 its length, forming 2 broadly rounded lobes, margins ciliate; disc 3-veined with lateral veins branched, base with ridge in center; column arcuate, 2.5–3.3 × 1 mm, dilated at distal end. **Capsules** horizontal to semi-erect, ellipsoid, 8 × 4 mm.

Flowering late Jun–Aug. Alluvial soils in low, moist, hardwood forests or mixed hardwood-coniferous forests, shrub swamps, sphagnum bogs; requires soil that remains cool throughout summer, with high degree of acidity; 0–1000 m; Man., N.B., Nfld. and Labr., N.S., Ont., P.E.I., Que.; Maine, Mich., Minn., N.H., N.Y., Vt., Wis.

Leaves of *Listera auriculata* forma *trifolia* (Lepage) Lepage are three in a whorl.

Listera auriculata crosses with *L. convallarioides* to produce a natural hybrid *L.* ×*veltmanii* Case, originally discovered by H. S. Veltman and F. W. Case in 1962. The hybrid is almost exactly intermediate between its putative parents, and it is usually associated with one or the other parent (P. M. Catling 1976). Occasionally it appears to backcross with the parent species. Plants of the hybrid, although intermediate between the two putative parents, frequently differ from the parents in being much taller (to 2.5 dm). Also, the glandular pubescence of the pedicels and ovaries of the hybrid is shorter and sparser than that of the rachis; in *L. convallarioides* the glandular pubescence is as dense and long on pedicels as on the rachis, continuing on the ribs of the ovary. Pedicels and ovaries of *L. auriculata* are glabrous. The hybrids may be distinguished by using the following key.

1. Lip not clawed, sessile or essentially so; auricles or lobes at base of lip curving around and clasping base of column . . . . *Listera auriculata*
1. Lip clawed, not sessile; auricles or lobes absent or projecting laterally away from column.
    2. Auricles or lobes at base of lip absent . . . . *Listera convallarioides*
    2. Auricles or lobes present at base of lip, projecting laterally away from column . . . . *Listera ×veltmanii*

Listera ×veltmanii flowers in July and August. Its habitat is intermediate between those typical of the parents and apparently is often disturbed; 500–800 m. It is known from New Brunswick, Newfoundland and Labrador, Quebec, Ontario, Michigan, and Wisconsin.

4. Listera borealis Morong, Bull. Torrey Bot. Club 20: 31. 1893 • Northern twayblade [E]

Neottia borealis (Morong) Szlachetko; *Ophrys borealis* (Morong) Rydberg

Plants 4–26 cm. **Stems** green, slender to stout, slightly 4-angled, succulent, glabrous. **Leaves:** blade green to dark bluish green, elliptic, ovate-elliptic, or lanceolate, 1.3–6 × 0.7–3 cm, apex obtuse to rounded. **Inflorescences** 5–20-flowered, lax, 20–90 mm; floral bracts lanceolate, ovate, or oblong, 1–3 × 0.5–1 mm, apex obtuse; peduncle and rachis glandular-pubescent. **Flowers** pale green, yellowish green, or bluish green with veins darker green; pedicel filiform, 3.5–7 mm, glandular-pubescent; sepals and petals strongly reflexed away from lip and column; dorsal sepal elliptic-lanceolate to linear-elliptic, 4–6 × 1.5–2.2 mm, apex obtuse to rounded; lateral sepals linear-elliptic to oblong-lanceolate, falcate, 4.5–7 × 1.4–2.3 mm, apex obtuse to rounded; petals linear to linear-oblong, falcate, 4.5–5 × 0.7–1.5 mm, apex obtuse; lip oblong, slightly narrowed in center, base with broadly rounded and bluntly angled divergent auricles, apex slightly dilated, cleft into 2 oblong or semiorbiculate lobes, with apicule in sinus; disc 3-veined, lateral veins branched and purplish, base darker green, with ridge in center, thickened along center, 7–12 × 4.2–6.5 mm, margins ciliate; column arcuate, stout, 3–4 × 1 mm. **Capsules** ellipsoid, 8 × 5 mm. 2*n* = 56.

Flowering Jun–Aug. In moist, rich humus of mossy coniferous or mixed hardwood forests, swamps, often along cold streams fed by melting snow, prefers high acidic soils; 1500–3000 m; Alta., B.C., Man., Nfld. and Labr., N.W.T., Nunavut, Ont., Que., Sask., Yukon; Alaska, Colo., Idaho, Mont., Oreg., Utah, Wash., Wyo.

In Japan *Listera borealis* is replaced by *L. yatabei* Makino, which is nearly identical except for short basal auricles. *Listera borealis* and *L. auriculata* are very similar in overall appearance; the ovaries and pedicels in *L. borealis* are glandular-pubescent, and in *L. auriculata* they are glabrous.

Leaves occur three in a whorl in *Listera borealis* forma *trifolia* Lepage.

5. Listera convallarioides (Swartz) Elliott, Sketch Bot. S. Carolina 2: 494. 1823 • Broad-leaved twayblade [E] [F]

Epipactis convallarioides Swartz, Kongl. Vetensk. Acad. Nya Handl. 21: 232. 1800; *Bifolium convallarioides* (Swartz) Nieuwland; *Diphryllum convallarioides* (Swartz) Kuntze; *Listera eschscholziana* Chamisso; *Ophrys convallarioides* (Swartz) W. Wight ex House

Plants 5–37 cm, somewhat stoloniferous. **Stems** green, succulent, glabrous. **Leaves:** blade green, broadly ovate to elliptic, oval or suborbiculate, 2–7 × 1.5–5.8 cm, apex obtuse to rarely acute or apiculate. **Inflorescences** 5–20-flowered, lax, 20–120 mm; floral bracts semitranslucent, rhombic-ovate to lanceolate, 3–5 × 1–2.5 mm, apex acute to acuminate; peduncle and rachis densely whitish glandular-pubescent. **Flowers** yellowish green, sometimes faintly tinged with purple; pedicels slender, 2.5–7 mm, glandular-pubescent; ovary glandular-pubescent, becoming glabrate as it matures; sepals and petals reflexed away from column and lip; dorsal sepal ovate-lanceolate to elliptic, 4.5–5 × 1.5–2 mm, apex obtuse to acute; lateral sepals lanceolate, falcate, 4.5–5.5 × 1.5–1.8 mm, apex acute to obtuse; petals linear, falcate, 4–5 × 0.8–1 mm; lip cuneate, apex dilated and shallowly notched, lateral lobes rounded, with small tooth in sinus, tapering to short, slender claw, with indistinct triangular tooth on either side of base above claw, 8–13 × 5–7 mm, margins minutely bristly-ciliate; column arcuate, slender, 2.5–3.5(–4) × 1 mm, apex slightly expanded. **Capsules** ellipsoid, 8 × 5 mm, essentially glabrous. 2*n* = 36.

Flowering Jun–Sep. Rich humus in open woods to boggy meadows, soil circumneutral or at most only mildly acid, prefers cool soil; widespread across northern part of continent, extending southward only at high elevations; 100–2600 m; St. Pierre and Miquelon; Alta., B.C., N.B., Nfld. and Labr. (Nfld.), N.S., Ont., P.E.I., Que.; Alaska (including Aleutian Islands), Ariz., Calif., Colo., Idaho, Maine, Mich., Minn., Mont., Nev., N.H., N.Y., Oreg., S.Dak., Utah, Vt., Wash., Wis., Wyo.

Listera convallarioides has a relatively large flower compared to most of the other listeras. It crosses with *L. auriculata* to produce a natural hybrid *L.* ×*veltmanii* Case, which is discussed under 3. *L. auriculata*.

6. Listera ovata (Linnaeus) R. Brown in W. Aiton and W. T. Aiton, Hortus Kew. 5: 201. 1813 • Common twayblade [I]

Ophrys ovata Linnaeus, Sp. Pl. 2: 946. 1753; *Diphryllum ovatum* (Linnaeus) Kuntze; *Distomaea ovata* (Linnaeus) Spenner; *Epipactis ovata* (Linnaeus) Crantz; *Helleborine ovata* (Linnaeus) F. W. Schmidt; *Neottia latifolia* Richard; *N. ovata* (Linnaeus) Bluff & Fingerhut

Plants 20–60 cm. **Stem** green, stout, succulent, glabrous. **Leaves:** blade yellow-green to green, ovate-elliptic, 10–17 × 10–12 cm, apex acute. **Inflorescences** 10–100-flowered, lax to dense, 10–50 cm; floral bracts lanceolate, 3 × 1 mm; peduncle and rachis glandular-pubescent. **Flowers** yellowish green; pedicels stout, 3–4 mm, slightly glandular-pubescent; sepals and petals projecting forward, connivent, forming hood over column; dorsal sepal ovate, concave, 5–6 × 2–3 mm, apex obtuse; lateral sepals ovate, concave, falcate, 4 × 2–3 mm, apex obtuse; petals linear, concave, 4 × 1 mm, apex obtuse; lip acutely deflexed near base, sessile, linear, apical 1/2 expanded, cleft into bluntly rounded lobes separated by tooth in sinus; disc with longitudinal thickened ridge leading to deflection, 8–10 × 4 mm; column short, 2 × 1.5 mm. **Capsules** semierect, ellipsoid, 10 × 6 mm. $2n$ = 32, 34, 35, 36, 37, 38, 40, 42.

Flowering late Jun–Jul. Moist rich humus, or in disturbed areas; 700–800 m; introduced; Ont.; Eurasia.

Listera ovata is a large, robust, aggressive, and exceedingly common orchid weed found in many kinds of habitat throughout Europe into Siberia and India. It may have the potential to become a weedy orchid in North America just as *Epipactis helleborine*. *Listera ovata* was used by Charles Darwin in his investigation and description of the method of cross-fertilization in the genus *Listera*.

7. Listera caurina Piper, Erythea 6: 32. 1898 • Northwest twayblade [E]

Listera retusa Suksdorf; *Neottia caurina* (Piper) Szlachetko; *Ophrys caurina* (Piper) Rydberg

Plants 5–30 cm. **Stems** green, succulent, glabrous. **Leaves:** blade green, ovate to elliptic, 2.5–7 × 1.8–4.5 cm, apex rounded to obtuse or subacute. **Inflorescences** 5–30-flowered, lax, 50–130 mm; floral bracts rhombic-ovate to lanceolate, apex acute to acuminate, glabrous to slightly glandular; peduncle and rachis glandular-pubescent. **Flowers** yellowish green, small; pedicel filiform, 4–12 mm, glandular-puberulent; sepals and petals distinct and free, spreading, only slightly reflexed; dorsal sepal linear-lanceolate to lanceolate, 2.8–4 × 0.5–0.8 mm, apex acute to acuminate; lateral sepals linear-lanceolate to lanceolate, falcate, 2.8–4 × 0.5–0.8 mm, apex acute to acuminate; petals linear to linear-lanceolate, slightly falcate, 2.8–3.5 × 0.5–0.8 mm, apex acute to acuminate; lip sessile, cuneate to obovate, 4.5–6 × 2–4.5 mm, apex rounded to retuse, apiculate, apical margin slightly erose or crenulate; disc with pair of dark green stripes, each swollen at base and with small horn 1 mm from top of each swelling; column short, 1.5–2 × 1 mm. **Capsules** semierect, ovoid to subglobose, 5 × 4 mm, glabrous. $2n$ = 34.

Flowering late Apr–Sep. Dry litter of mixed and coniferous forests, dense moist, coniferous forests, and on boggy wooded slopes, also damp to wet conditions on moss-covered rocks or in mossy soil; 100–2200 m; Alta., B.C.; Alaska, Calif., Idaho, Mont., Oreg., Wash.

8. Listera smallii Wiegand, Bull. Torrey Bot. Club 26: 169. 1899 • Appalachian twayblade, kidney-leaf twayblade [E]

Listera reniformis Small, Bull. Torrey Bot. Club 24: 334. 1897, not D. Don 1825; *Bifolium smallii* (Wiegand) Nieuwland; *Neottia smallii* (Wiegand) Szlachetko; *Ophrys smallii* (Wiegand) House

Plants 5–35 cm. **Stems** slender, succulent, glabrous. **Leaves:** blade dark green, ovate-reniform, 2–4 × 1.5–3.5 cm, apex acute, mucronate, or apiculate to short-acuminate. **Inflorescences** 5–15-flowered, lax, 40–100 mm; floral bracts ovate, 3–4 × 1.5 mm, apex acute; peduncle and rachis glandular-pubescent. **Flowers** green, purple-brown, to pinkish tan; pedicel slender, 6–7 mm, glabrous; sepals strongly reflexed; dorsal sepal lanceolate to linear-lanceolate, 3–4 × 1 mm, apex acute; lateral sepals lanceolate to linear-lanceolate, 3–4 × 1 mm, apex acute; petals slightly reflexed, linear to linear-lanceolate, margins slightly revolute, apex acute; lip sessile, broadly obovate to cuneate, 6–10 × 5–7 mm, base with rounded lobule on each side, apex dilated and deeply cleft into pair of ± divergent broadly rounded lobes, shallowly toothed in sinus; column slightly arcuate, short, 1.5–3 × 1–2 mm. **Capsules** semierect, ellipsoid, 5 × 3 mm. $2n = 38$.

Flowering Jun–Jul. Damp humus in shady forests of Appalachian Mountains, often beneath *Rhododendron* in acidic soil, also in sphagnous thickets and bogs; 600–1300 m; Ga., Ky., Md., N.J., N.C., Pa., S.C., Tenn., Va., W.Va.

Listera nipponica Makino, a species similar to *L. smallii*, occurs in the mountains of Japan.

230d.2. ORCHIDACEAE Jussieu (subfam. EPIDENDROIDEAE) tribe TRIPHOREAE Dressler, Selbyana 5: 204. 1979

34. TRIPHORA Nuttall, Gen. N. Amer. Pl. 2: 192. 1818 • Three-birds orchid [Greek *tri*, 3-fold, and *phoros*, bearing, possibly in reference to the few-flowered inflorescence or the 3 crests on the lip of the type species]

Max E. Medley

Herbs, perennial, terrestrial, apparently epiphytic in one species, photosynthetic to mycotrophic. **Roots** fleshy, with a few filiform roots from stem base, some terminating in tuberoids. **Stems** erect, usually simple, terete, succulent, glabrous. **Leaves** 1–10, alternate and distichous, or solitary at midstem, without stipules, glabrous, sessile, not articulate, clasping stem; blade lance-ovate to ovate, or reduced to sheathing bracts. **Inflorescences** terminal, solitary or 1–10-flowered racemes, flowers axillary or terminal, arrangement spiral; floral bracts consisting of smaller proximalmost leaves, ovate to cordate, foliose. **Flowers** resupinate or not, short-pedicellate; sepals and petals distinct and free; dorsal sepal lanceolate to oblanceolate; lateral sepals lanceolate, slightly falcate; petals linear-lanceolate to lanceolate; lip clawed, 3-lobed; disc with 3 lines or crests; column slender, 5–10 mm, footless; anther erect, white or pale green (with magenta margins in *Triphora trianthophora*), short-stalked, not articulate, rigid; pollinia 2; pollen in tetrads, soft, mealy; ovary fusiform, slender; stigma proximal to anther, entire or 2-lobed; rostellum simple or absent. **Fruits** capsules, erect or pendent, ellipsoid-ovoid or obovoid, with 6 low, narrow keels.

Species ca. 25 (5 in the flora): e North America, West Indies, Central America, South America.

Most species of *Triphora* exist in small, scattered populations and exhibit synchronous, ephemeral flowering; flowers last only one day. Synchrony enhances the chances for fertilization. Isolated flowers are rarely fertilized.

SELECTED REFERENCES Ames, O. 1905–1922. Orchidaceae: Illustrations and Studies of the Family Orchidaceae Issuing from the Ames Botanical Laboratory.... 7 vols. Boston and New York. Vol. 7. Medley, M. E. 1996. Notes on collections of *Triphora* (Orchidaceae) from the American tropics. Selbyana 17: 83–87.

1. Flowers not resupinate, lip facing stem at anthesis, always erect; column yellowish.
   2. Leaves bractlike, ascending, usually partially sheathing stem; inflorescences corymbose racemes . . . . 1. *Triphora gentianoides*
   2. Leaves not bractlike, diverging at ± right angles from stem; inflorescences racemes, not corymbose . . . . 2. *Triphora yucatanensis*
1. Flowers resupinate, lip facing away from stem at anthesis, erect or nodding; column white or pale green.

3. Leaves deep purple abaxially, often faded on dried specimens, ovate to cordate, not reniform; flowers 1–3; lip white with irregular purple spots, disc with 3 parallel white longitudinal crests, not papillose; pollinia yellow . . . . 5. *Triphora craigheadii*
3. Leaves green to yellow-green on both surfaces, sometimes veined with purple abaxially, broadly ovate to nearly reniform; flowers 1–6; lip white, disc with 3 green longitudinal crests, papillose; pollinia yellow or magenta.
    4. Flowers (1–)3–6(–20), lip crests bright green; leaves ovate; pollinia magenta . . . . 3a. *Triphora trianthophora* subsp. *trianthophora*
    4. Flowers 1, lip crests pale green; leaves broadly ovate to reniform; pollinia yellow . . . . 4. *Triphora amazonica*

1. Triphora gentianoides (Swartz) Ames & Schlechter in O. Ames, Orchidaceae 7: 5. 1922 [F]

Limodorum gentianoides Swartz, Prodr., 119. 1788; *Triphora cubensis* (Reichenbach f.) Ames

Plants 8–26 cm. **Roots** from cylindric tuberoids, 30–60 × 8–10 mm. **Stems** green, often tinged with brown or purplish, unbranched or rarely 2–3-branched from base. **Leaves** 3–10, ascending, usually partially sheathing stem; blade green, often tinged with brown or purple, bractlike, ovate, 1–1.8 × 0.5–1 cm, margins entire. **Inflorescences** 3–10(–15)-flowered corymbose racemes. **Flowers** not resupinate, erect, closed to slightly gaping, pale green, sometimes tinged with brown or reddish brown; flowering successive, anthesis overlapping; dorsal sepal linear-lanceolate, 6–11 × 2–2.5 mm; lateral sepals linear-lanceolate, falcate, 6–11 × 2–2.5 mm; petals linear-lanceolate, falcate, 8–11 × 1–2 mm; lip facing stem, white to pale green, ovate to obovate, clawed, 3-lobed, 8–10 × 3–4 mm, middle lobe ovate, lateral lobes lance-ovate, margins erose to nearly lacerate; disc with 3 green, raised, denticulate crests; column yellowish white, clavate, slender, 7 mm; pollinia 2, yellow. **Capsules** erect, ellipsoid, 10–20 × 5–10 mm.

Flowering summer. Sandy pinelands in shade or full sun, persisting in sandy lawns, or rarely in shade of live-oak hammocks; 0–10 m; Fla.; se Mexico; West Indies (Cuba, Dominican Republic, Haiti); Central America (Belize, Guatemala, Honduras, Panama); South America (Ecuador, Venezuela).

The flowers are presumed to be self-fertilizing. The anther color has been reported as purple for Venezuelan material (G. C. K. Dunsterville and L. A. Garay [1959]–1976, vol. 3).

2. Triphora yucatanensis Ames, Orchidaceae 7: 39, plate 109. 1922

Triphora rickettii Luer

Plants 5–20 cm. **Roots** from ovoid-cylindric tuberoids, 30–50 × 6–10 mm. **Stems** green, sometimes suffused with purple, unbranched. **Leaves** 3–6, diverging at ± right angles from stem; blade dark green on both surfaces, not bractlike, broadly ovate to cordate, 1–2 × 1–2 cm, margins undulate. **Inflorescences** 1–4(–8)-flowered racemes. **Flowers** not resupinate, erect, closed to somewhat gaping, yellowish green abaxially, white or pale green adaxially, flowering apparently successive, anthesis overlapping; dorsal sepal oblanceolate, 8 × 1.5 mm; lateral sepals linear-lanceolate, falcate, 8 × 1.5 mm; petals linear-lanceolate, falcate, 7 × 1.5 mm; lip facing stem, white, linear-ovate to ovate, 5–6 × 2.5–3 mm, middle lobe nearly rectangular, lateral lobes broadly triangular, margins sinuate to nearly lacerate; disc with 3 yellow, raised, denticulate crests; column yellowish white, clavate, slender, 5 mm; pollinia yellow. **Capsules** erect, ellipsoid, 15 × 8 mm.

Flowering summer. Upland hardwood hammocks, in rich humus and rotting wood; 0–20 m; Fla.; Mexico.

3. Triphora trianthophora (Swartz) Rydberg in N. L. Britton, Man. Fl. N. States, 298. 1901 • Nodding pogonia [F]

Arethusa trianthophoros Swartz, Kongl. Vetensk. Acad. Nya Handl. 21: 230. 1800

Subspecies 2 (1 in the flora): e North America, Mexico, Central America.

SELECTED REFERENCE Medley, M. E. 1979. Some Aspects of the Life History of *Triphora trianthophora* (Sw.) Rydb. (Three Birds Orchid) with Special Reference to Its Pollination. M.A. thesis. Andrews University.

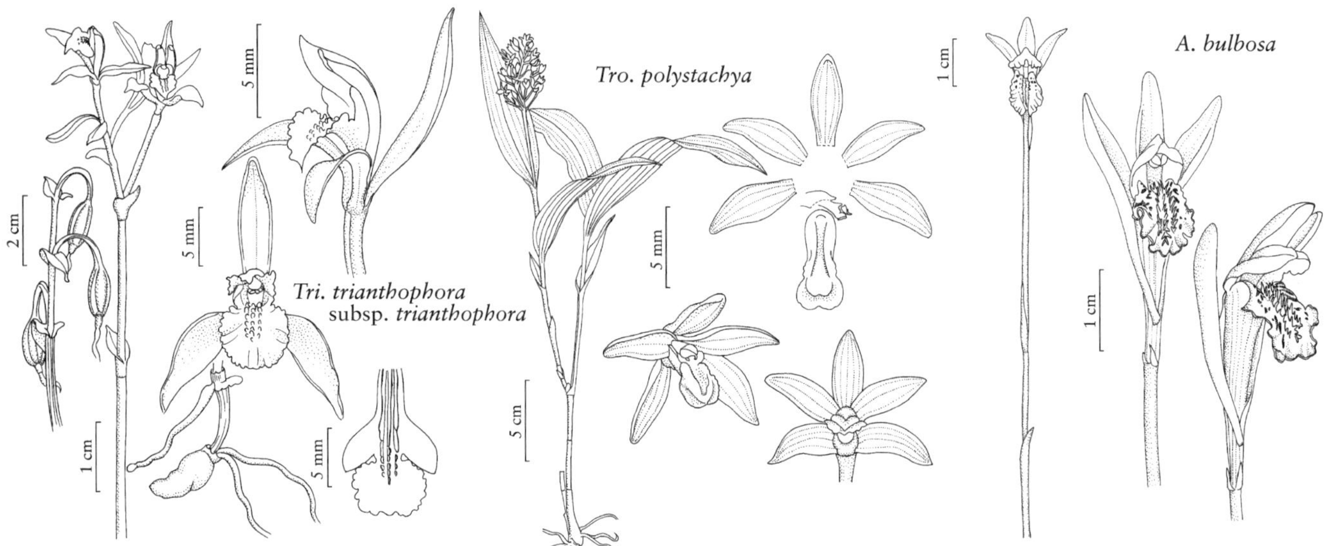

TRIPHORA ◦ TROPIDIA ◦ ARETHUSA

3a. Triphora trianthophora (Swartz) Rydberg subsp. **trianthophora** [F]

Triphora trianthophora var. *schaffneri* Camp

Plants 7–20(–30) cm. **Roots** from spheroid to ovoid or nearly cylindric tuberoids, 10–30 × 5–15 mm. **Stems** purple or sometimes green apically, unbranched or rarely branched from base. **Leaves** 2–6; blade abaxially green or suffused or veined with purple, adaxially green, not bractlike, broadly ovate, margins entire. **Inflorescences** (1–)3–6(–20)-flowered racemes. **Flowers** resupinate, nodding to nearly erect, gaping, white, often tinged with pale pink or fading to pale pink with age; flowering synchronous and ephemeral, occasionally flowers may persist for 2–4 days; dorsal sepal oblanceolate, 11–15 × 3–4 mm; lateral sepals oblanceolate, falcate, 11–15 × 3–4 mm; petals oblanceolate, falcate, 11–14 × 3–4 mm; lip facing away from stem, white, obovate, clawed, 3-lobed, 8–20 × 6–10 mm, middle lobe ovate to nearly circular, lateral lobes ovate to ovate-triangular, margins sinuate-denticulate; disc with 3 bright green, papillose crests; column white, clavate, slender, 10 × 2 mm; pollinia magenta. **Capsules** pendent or erect, ellipsoid, 10–15 × 5–10 mm. $2n = 44$.

Flowering summer–late fall. Dry-mesic to mesophytic forests over sandstone or limestone, sandy oak-mixed hardwood forests, seasonally wet, sandy flatwoods, Great Lakes dune forests, coniferous forests, tamarack swamps, rhododendron thickets, floodplain forests, wet muck in glacial lake bed forest, and in seasonally flooded sinkhole swamps; 0–2000 m; Ont.; Ala., Ark., Conn., Del., D.C., Fla., Ga., Ill., Ind., Iowa, Kans., Ky., La., Maine, Md., Mass., Mich., Miss., Mo., Nebr., N.H., N.J., N.Y., N.C., Ohio, Okla., Pa., S.C., Tenn., Tex., Vt., Va., W.Va., Wis.; nc Mexico.

Pollination is by small (mostly halictid) bees (M. E. Medley 1979). Pollination by small bees, mostly *Augochlora pura*, has also been reported (P. M. Catling and V. R. Catling 1991). Flowers of *Triphora trianthophora* subsp. *trianthophora* usually last only one day, sometimes 2–4 days near its western limit.

Triphora trianthophora subsp. *mexicana* (S. Watson) Medley occurs in Mexico and in Central America (El Salvador, Guatemala, Honduras, Panama).

4. Triphora amazonica Schlechter, Beih. Bot. Centralbl. 42: 75. 1925

Triphora latifolia G. M. Luer

Plants delicate, 4–9 cm, glabrous. **Roots** from fleshy ovoid tuberoids, 10–20 × 5–8 mm. **Stems** erect, unbranched, green, slender, leafy. **Leaves** 2–5, widely spaced along stem; blade green to yellow-green, broadly ovate to nearly reniform, 4–8 × 7–13 mm, margins entire, apex acute to rounded; distal leaves narrower than proximal, 4–8 × 4–13 mm, margins entire. **Inflorescences** terminal, solitary; floral bracts leaflike, ovate-lanceolate, to 8 mm. **Flowers** resupinate, nodding to nearly erect, gaping to somewhat so, white or pink; dorsal sepal oblong-lanceolate, 10–10.5 × 2.5–2.8 mm; lateral sepals oblong-lanceolate, falcate, 10–10.5 × 2.5–2.8 mm; petals oblong-lanceolate, slightly falcate, 10 × 2–2.3 mm; lip facing away from stem, ovate, clawed, 3-lobed, 8.5 × 6 mm, middle lobe finely erose, undulate, ovate to suborbiculate, lateral lobes triangular, falcate; disc with 3 pale green, denticulate to papillose, longitudinal crests; column greenish, slender, clavate, 8 mm, anther stalked; pollinia yellow. **Capsules** erect or pendent, ellipsoid, 8–11 mm.

Flowering summer. Upland hardwood hammocks, in rich, well-drained, moist humus; 0–20 m; Fla.; West Indies (Dominican Republic, Puerto Rico, Trinidad); South America (Brazil, French Guiana, Guyana).

Originally J. D. Ackerman (1995) determined vegetative material from Puerto Rico as the Florida endemic, *Triphora latifolia*. After a comparison with it, and some others from Trinidad, with Schlechter's illustration, the only difference that could be detected between *T. latifolia* and *T. amazonica* was that the middle lobe of the lip may be rounded in *T. latifolia* and acute in *T. amazonica* (J. D. Ackerman 2000). The distribution of *T. latifolia* as known in 1995 (Florida, Puerto Rico, and Brazil) was unusual, but recent information showing that *T. amazonica* ranges from Brazil to Guyana, French Guiana, Surinam, Trinidad, and Dominican Republic makes the link with Puerto Rico and Florida more reasonable (J. D. Ackerman 2000) and supports treating *T. amazonica* and *T. latifolia* as conspecific. Ackerman observed that such geographic patterns suggest that rare species may have more extensive distributions than presently known, or that additional specimens may show that the disjunct populations are not conspecific.

5. **Triphora craigheadii** Luer, Brittonia 18: 243, figs. 5–8. 1966 [E]

Plants 4–8 cm. **Roots** from ovoid tuberoids, 15 × 5 mm. **Stems** purplish, unbranched. **Leaves** 1–4; blade dark green adaxially, purple abaxially, broadly ovate-cordate, not reniform, 9–10 × 11–12 mm, margins undulate. **Inflorescences** 1–3-flowered racemes. **Flowers** resupinate, nodding to nearly erect, somewhat gaping, flowering synchronous and ephemeral; sepals light green; dorsal sepal obovate, 6 × 2 mm; lateral sepals ovate, slightly falcate, 6 × 2 mm; petals white or suffused with pink, lanceolate-ovate, falcate, 6 × 2 mm; lip facing away from stem, white with irregular purple spots, broadly triangular, 3-lobed, 5.5 × 4 mm, middle lobe broadly truncate, lateral lobes ± triangular, margins sinuate; disc with 3 white, parallel crests, not papillose; column pale green, clavate, slender, 5 mm; pollinia 2, yellow. **Capsules** pendent or erect, ellipsoid, 10 × 5 mm.

Flowering summer. Low, poorly drained hardwoods in rich humus, often on or near decaying logs; 0–20 m; Fla.

230d.3. ORCHIDACEAE Jussieu (subfam. EPIDENDROIDEAE) tribe TROPIDIEAE Dressler, Telopea 2: 422. 1983

35. TROPIDIA Lindley, Edwards's Bot. Reg. 19: text for plate 1618. 1833 • [Greek *tropideion*, keel, referring to the boat-shaped floral lip of many species]

James D. Ackerman

Tomotris Rafinesque

Herbs, terrestrial, cespitose, erect. **Roots** branched. **Rhizomes** short, rigid. **Stems** erect, branched or simple, reedlike, slender, hard. **Leaves** few, cauline; sheaths not articulate; blade convolute, plicate, thin. **Inflorescences** terminal or occasionally from axils of distal leaves, few- to many-flowered panicles [racemes, rarely glomerate], pedunculate [sessile]. **Flowers** resupinate, small; sepals subequal; lateral sepals distinct [connate at base], forming small mentum; petals similar to sepals, slightly falcate, shorter than sepals; lip sessile, base saccate; column short, fleshy, with short foot; anther abaxial, erect, subequal to rostellum, more than 1/2 length of column; pollinia 2, sectile; stigma entire; viscidium terminal. **Fruits** capsules, oblong, cylindric.

Species 20 (1 in the flora): tropical and subtropical regions, North America, Mexico, West Indies, Central America, South America, Asia, w Pacific.

SELECTED REFERENCE Rasmussen, F. N. 1977. The genus *Corymborkis* Thou. (Orchidaceae): A taxonomic revision. Bot. Tidsskr. 71: 161–192.

1. **Tropidia polystachya** (Swartz) Ames, Orchidaceae 2: 262. 1908 • Young palm orchid [F]

Serapias polystachya Swartz, Prodr., 119. 1788; *Chloidia polystachya* (Swartz) Reichenbach f.; *C. vernalis* Lindley; *Corymbis polystachya* (Swartz) Bentham ex Fawcett; *Corymborchis polystachya* (Swartz) Kuntze; *Macrostylis vernalis* (Lindley) Reichenbach f.; *Neottia polystachya* (Swartz) Swartz; *Polystachya membranacea* A. Richard; *Stenorrhynchos polystachyon* (Swartz) Sprengel; *Tomotris polystachya* (Swartz) Rafinesque; *Tropidia eatonii* Ames

Plants to 70 cm. **Roots** several from rhizome, brown, 0.5–3 mm diam., villous. **Rhizomes** 2.5–4 mm diam. **Stems** partially concealed by bladeless tubular sheaths proximally when young, naked when older, covered by leaf sheaths distally, usually branched, terete, 31–55 cm × 2–4 mm. **Leaves** several; sheaths clasping stem, 4–4.5 cm; blade oblong-lanceolate, 6–27 × 2.4–6.7 cm, apex acuminate. **Inflorescences:** peduncles 3.5–14 cm; panicles 3–48-flowered, dense, subglobular, 2.5–9 × 3–6.5 cm; floral bracts linear to lanceolate, 4–10 mm, apex acuminate to subulate. **Flowers** 8–12 mm; sepals greenish white, 5-veined; dorsal sepal concave, oblong to elliptic, 6.5–7 × 2–2.5 mm, apex acute to obtuse; lateral sepals somewhat gibbous, lanceolate-oblong, slightly falcate, 5.5–6.5 × 1.7–2 mm, apex acute; petals white, 3-veined, somewhat falcate, lanceolate to elliptic-oblanceolate, 5–6 × 1.5–2 mm, apex acute; lip embracing column, white with yellow spot on disc, oblong, 4.5–6 × 2 mm, constricted at middle, apex retuse, sometimes obscurely 4-lobed; disc 2-lamellate and flabellate distally; column 4–4.5 mm. **Capsules** ribbed, 10–14 × 3–6 mm.

Flowering Sep–Mar. Moist hardwood hammocks; 0–20 m; Fla.; Mexico; West Indies; Central America; South America (Galapagos Islands).

230d.4. ORCHIDACEAE Jussieu (subfam. EPIDENDROIDEAE) tribe ARETHUSEAE Lindley, Coll. Bot., App. 1821

SELECTED REFERENCE Goldman, D. H. et al. 2001. Phylogenetics of Arethuseae (Orchidaceae) based on plastid *mat*K and *rbc*L sequences. Syst. Bot. 26: 670–695.

230d.4a. ORCHIDACEAE Jussieu (tribe ARETHUSEAE) subtribe ARETHUSINAE Lindley, Gen. Sp. Orchid. Pl., 385. 1840

36. ARETHUSA Linnaeus, Sp. Pl. 2: 950. 1753; Gen. Pl. ed. 5, 407. 1754 • Arethusa, dragon's-mouth, aréthuse [Greek *Arethusa*, mythical river nymph] [E]

Charles J. Sheviak

Paul M. Catling

Herbs, perennial. **Roots** few, slender, fleshy. **Stems** scapose, at base a bulbous corm. **Leaves** solitary, basal; blade subtended by 2–3 tubular sheaths, obscurely plicate, linear-lanceolate, developing during and after anthesis. **Inflorescences** terminal; floral bracts minute. **Flowers** solitary (rarely 2), resupinate, erect, showy; anther with pair of soft yellow-green pollinia; stigma emergent, with free lateral lobes. **Fruits** capsules, erect.

Species 1: North America.

1. **Arethusa bulbosa** Linnaeus, Sp. Pl. 2: 950. 1753 E F

Plants 2–40 cm. **Leaves:** blade 5–23 cm × 3–12 mm. **Inflorescences:** floral bracts 2, inconspicuous. **Flowers** pink to magenta, rarely white or bluish, gaping, large; sepals erect, narrowly oblanceolate to narrowly oblong, 20–55 × 3–9 mm; petals arcuate, 23–49 × 4–10 mm; lip obovate to oblong, arcuate to reflexed near middle, often obscurely 3-lobed, 19–35 ×10–19 mm; disc with crest of yellow lamellae and fleshy processes. **Capsules** erect, ovoid to ellipsoid, 15–25 mm. $2n = 40$.

Flowering Jun–Jul. Sphagnum bogs, coniferous swamps, calcareous treed to open fens, and moist, acid, sandy meadows; 0–1100 m; Man., N.B., Nfld. and Labr. (Nfld.), N.S., Ont., P.E.I., Que., Sask.; Conn., Del., Ind., Maine, Md., Mass., Mich., Minn., N.H., N.J., N.Y., N.C., Ohio, Pa., R.I., S.C., Tenn., Vt., W.Va., Wis.

The occurrence of *Arethusa bulbosa* in Cook County, Illinois (F. Swink and G. S. Wilhelm 1994), and in Maryland are unconfirmed.

230d.4b. ORCHIDACEAE Jussieu (tribe ARETHUSEAE) subtribe BLETIINAE Bentham, J. Linn. Soc., Bot. 18: 287. 1881

37. CALOPOGON R. Brown in W. Aiton and W. T. Aiton, Hortus Kew. 5: 204. 1813 • Grass pink, calopogon [Greek *kalos*, beautiful, and *pogon*, beard, alluding to hairlike protuberances on lamellae]

Douglas H. Goldman

Lawrence K. Magrath

Paul M. Catling

Herbs, perennial, terrestrial, scapose, sympodial. **Roots** from globose, ellipsoid, oblong, or forked corms, 0.5–2 mm wide. **Stems:** glabrous; cauline bracts 2–3, sheathing. **Leaves** 1(–3), sessile; blade linear, lanceolate, or elliptic-lanceolate, conduplicate, frequently prominently veined. **Inflorescences** terminal, 1(–3) racemose spikes, 1–25-flowered (ovaries not pedicellate despite appearance), scapose; flowers opening nearly simultaneously to sequentially; floral bracts subulate, lanceolate, or ovate, typically small. **Flowers** not resupinate, magenta, deep rose pink, pale pink, or white, sometimes fragrant, sessile; dorsal sepal lanceolate; lateral sepals ovate to nearly lanceolate, falcate, apex apiculate to acuminate; petals linear, ovate, elliptic, lanceolate, or oblanceolate, sometimes pandurate, clawed; lip basally hinged, obscurely 3-lobed, middle lobe distally dilated into triangular or broadly rounded lamina, large, lateral lobes 2, small; disc with 2–3 longitudinal lamellae grading distally into brush of hairlike protuberances; protuberances white, yellow, or orange to magenta, elongate, filiform to clavellate; column arcuate, broadly dilated at distal end; anther terminal, green to crimson; pollinia 4 in 2 pairs, soft and mealy; viscidium absent; pollen grains in tetrads. **Fruits** capsules, erect, cylindric, ellipsoid, or obconic; column persistent in mature capsule.

Species 5 (5 in the flora): e North America, West Indies (Bahamas, Cuba).

Based on morphology and allozymes, D. W. Trapnell (1995) determined that all taxa she sampled, which did not include *Calopogon oklahomensis* and *C. tuberosus* var. *latifolius*, were properly circumscribed, although results were not strongly conclusive about relationships among the species. Based on DNA data, D. H. Goldman (2000) determined that *C. multiflorus* and *C. pallidus* are most closely related, although the relationships of the other taxa were not as certain. DNA evidence did support the circumscription of all taxa except *C. tuberosus* var. *latifolius*, which was in agreement with the work of P. M. Catling and Z. Lucas (1987). Plastid DNA sequences suggested that *Calopogon* is most closely related to the eastern North American genus *Arethusa*, and the Asian genera *Anthogonium*, *Arundina*, and *Eleorchis* (D. H. Goldman 2000; D. H. Goldman et al. 2001).

Calopogon appears to exhibit deceit pollination (L. van der Pijl and C. H. Dodson 1966; L. B. Thien and B. G. Marcks 1972; D. H. Firmage and F. R. Cole 1988; J. T. Boland and P. J. Scott 1991). This occurs when pollinating bees land on the lip and try to gather pollen off the large hairlike protuberances that resemble anthers. The weight of the pollinator causes the lip to fall, dropping the pollinator backwards onto the column; pollinia adhere to the back of the bee and subsequently are carried to another flower. Soft pollinia in *Calopogon* had been noted to be an evolutionary reversal to facilitate pollination by hairy bees (W. P. Stoutamire 1971).

SELECTED REFERENCES Goldman, D. H. 2000. Systematics of *Calopogon* and the Tribe Arethuseae (Orchidaceae). Ph.D. dissertation. University of Texas. Thien, L. B. 1973. Isolating mechanisms in the genus *Calopogon*. Amer. Orchid Soc. Bull. 42: 794–797. Trapnell, D. W. 1995. Systematic Review of *Calopogon* (Orchidaceae). Master's thesis. University of Georgia.

1. Petals wider at distal half than at proximal half (distinctly pandurate); lip usually as wide as or wider than long; inflorescences dark purple; flowers pungent . . . . 1. *Calopogon multiflorus*
1. Petals equal to or narrower at distal half than at proximal half; lip usually narrower than long; inflorescences green or slightly purple at base, rarely entirely purple; flowers scentless to mildly fragrant.
   2. Leaf appressed to inflorescence at flowering; flowers less than 1 cm apart, not fragrant, opening nearly simultaneously . . . . 3. *Calopogon barbatus*
   2. Leaf not appressed to inflorescence at flowering; flowers more than 1 cm apart, variously fragrant, opening nearly simultaneously to sequentially.
      3. Petals lanceolate to weakly pandurate, falcate; floral bracts 1.5–4.5 mm; lateral sepals 10–15 mm, falcate, widely spreading . . . . 2. *Calopogon pallidus*
      3. Petals obpandurate, slightly falcate to straight; floral bracts 3–12(–30) mm; lateral sepals 15–28 mm, weakly falcate to straight, occasionally spreading.
         4. Flowers opening sequentially; dilated distal portion of middle lip lobe usually much wider than long, typically anvil-shaped; stigma at angle to column surface; corms globose to elongate, not forked . . . . 4. *Calopogon tuberosus*
         4. Flowers opening nearly simultaneously; dilated distal portion of middle lip lobe usually much narrower than long, triangular to broadly rounded; stigma typically flat against column surface; corms elongate, forked . . . . 5. *Calopogon oklahomensis*

1. Calopogon multiflorus Lindley, Gen. Sp. Orchid. Pl., 425. 1840 • Many-flowered grass-pink [E]

Calopogon barbatus (Walter) Ames var. *multiflorus* (Lindley) Correll; *Helleborine multiflora* (Lindley) Kuntze; *Limodorum multiflorum* (Lindley) C. Mohr; *L. pinetorum* Small

Plants 17–33(–50) cm. **Corms** elongate, forked, 18–40 mm. **Leaves:** blade linear, 3–15 cm × 3–9 mm at flowering, later elongating to 6–41 cm. **Inflorescences** dark purple, becoming green after flowering, 17–33(–50) cm; floral bracts ovate, ovate-lanceolate, or subulate, 3–8 mm. **Flowers** 2–15, opening nearly simultaneously, crimson, magenta, to rarely light pink, strongly fragrant to pungent; dorsal sepal obovate to oblanceolate, 9–17 × 5–8 mm, apex acuminate; lateral sepals reflexed only at tip or straight, ovate to lanceolate, slightly falcate to straight, 10–13.5 × 6–9 mm; petals pandurate, rarely obovate, 8–14 × 4–8 mm, apex obtuse; lip 7–12 mm, middle lobe with dilated apex triangular, rounded to truncate, 7–13 mm wide; column 5.5–7.5 × 1–2 mm, distal end 5–7 mm wide; rostellum present. **Capsules** ellipsoid to obconic, 13–22 × 4–7 mm. $2n = 40, 42$.

Flowering Mar–Apr, rarely Nov–Aug (s Fla. only), usually several weeks following fires. Sandy, relatively dry pine savannas and grasslands; 0–60 m; Ala., Fla., Ga., La., Miss., N.C., S.C.

Calopogon multiflorus is nearly extirpated outside of central Florida.

Distinctive features of *Calopogon multiflorus* are the purple rachis, forked corm, strong floral fragrance, pandurate petals, and a lip usually as wide or wider than long. Although previously considered to be a variety of *Calopogon barbatus* (D. S. Correll 1940), *C. multiflorus* is readily distinguished from that species and all other *Calopogon* species by its features. The biology of *C. multiflorus* has been discussed in detail by D. H. Goldman and S. L. Orzell (2000).

2. Calopogon pallidus Chapman, Fl. South. U.S., 457. 1860 • Pale grass-pink [E]

Helleborine pallida (Chapman) Kuntze; *Limodorum pallidum* (Chapman) C. Mohr

Plants 15–50(–67) cm. **Corms** globose to elongate, rarely forked, 11–26 mm. **Leaves** not appressed to inflorescence at flowering; blade linear, 7–23 cm × 2–6 mm. **Inflorescences** green or slightly purple at base, becoming entirely green after flowering, 20–62 cm; floral bracts ovate to ovate-lanceolate, 1.5–4.5 mm. **Flowers** 3–20, more than 1 cm apart, opening sequentially, magenta, pale pink, or white, faintly fragrant; dorsal sepal oblanceolate, 12–20 × 4.5–6 mm; lateral sepals widely spreading (reflexed), ovate to lanceolate, falcate, 10–15 × 5–9 mm, apex acuminate; petals lanceolate to weakly pandurate, falcate, 9–18 × 3–5.5 mm, apex acuminate; lip 9–13 mm, middle lobe with dilated apex triangular, 5–11 mm wide, apex truncate, retuse to apiculate; column 7–11 × 1–1.5 mm, distal end 6–9 mm wide; rostellum present. **Capsules** ellipsoid to obconic, 11–23 × 4–8 mm. $2n = 40, 42$.

Flowering Mar–Jun, rarely Jul–Aug. Moist to wet acid pine savannas and grasslands; flowering abundantly several weeks after late winter to early spring fires; 0–100 m; Ala., Fla., Ga., La., Miss., N.C., S.C., Va.

Specimens of plants from western Louisiana reported as *Calopogon pallidus* (A. M. Pridgeon and L. E. Urbatsch 1977) are actually *C. oklahomensis.*

Natural hybrids of *Calopogon multiflorus* and *C. pallidus* have been observed in central Florida, and collections of other such hybrids are known from peninsular Florida (D. H. Goldman 2000; D. H. Goldman and L. S. Orzell 2000).

Calopogon pallidus is distinguished by its relatively widely spaced flowers in a long spike, one to three flowers opening simulataneously, lateral sepals typically falcate and reflexed, and petals entire to barely obpandurate and falcate. The blooming period is longer, and *C. pallidus* flowers later than all species of *Calopogon*, with the exception of *C. tuberosus*.

3. Calopogon barbatus (Walter) Ames, Orchidaceae 2: 272. 1908 • Bearded grass-pink [E]

Ophrys barbata Walter, Fl. Carol., 221. 1788, name conserved; *Calopogon parviflorus* Lindley; *C. parviflorus* Rafinesque; *C. pulchellus* (Salisbury) R. Brown var. *graminifolius* Elliott; *Helleborine graminifolia* (Elliott) Kuntze; *Limodorum graminifolium* (Elliott) Small; *L. parviflorum* (Lindley) Nash

Plants 11–34 cm at flowering, 13–57 cm at senescence. **Corms** globose to elongate, rarely forked, 13–26 mm. **Leaves** closely appressed to inflorescence at flowering; blade linear, 7–22 cm × 2–5.5 mm, later elongating to 12–54 cm. **Inflorescences** green or slightly purple at base, rarely entirely purple, becoming entirely green after flowering, 11–34 cm, later elongating to 13–57 cm; floral bracts 2–4.5 mm. **Flowers** 1–12, less than 1 cm apart, opening nearly simultaneously, magenta, pink, or rarely white, without fragrance; dorsal sepal obovate to oblanceolate, 12–20 × 4–9 mm, apex acuminate; lateral sepals distally reflexed to straight, ovate to moderately lanceolate, weakly falcate to straight, 11–16 × 5–9 mm, apex apiculate to acuminate; petals obpandurate, 11–16 × 4–6 mm, apex obtuse; lip 7–13.5 mm, dilated apex of middle lobe triangular to quadrangular, broadly rounded, 7–11 mm wide, apex retuse or apiculate; column 6–9 × 1.5–2 mm, distal end 5–8 mm wide; rostellum present. **Capsules** ovoid to ellipsoid, 12–17 × 4.5–6 mm. $2n = 40, 42$.

Flowering spring, seldom winter (s Fla.), late spring in north (Jan–Jun). Moist, acidic, sandy pine savannas and grasslands; 0–100 m; Ala., Fla., Ga., La., Miss., N.C., S.C.

Calopogon barbatus can be distinguished from other *Calopogon* species by its narrow leaf closely appressed to the inflorescence at flowering, relatively closely spaced flowers that open nearly simultaneously, general lack of floral fragrance, and the significant elongation of leaves and aboveground stems over the growing season. *Calopogon barbatus* also flowers earlier than other species within its range except *C. oklahomensis*. Specimens of plants from Texas and most from western Louisiana that have been identified as *C. barbatus* are actually *C. oklahomensis* (D. H. Goldman 1995). *Calopogon barbatus*, however, is found in western Louisiana, although local and most infrequent (W. C. Holmes and P. S. Mathies 1980).

The type specimen for the basionym of *Calopogon barbatus*, *Ophrys barbata* Walter, is actually *C. multiflorus* (D. H. Goldman 1998). The name *Ophrys barbata*, however, has been proposed for conservation (R. K. Brummitt 2000), in which case no nomenclatural change will be necessary.

4. Calopogon tuberosus (Linnaeus) Britton, Sterns & Poggenberg, Prelim. Cat., 52. 1888 [F]

Limodorum tuberosum Linnaeus, Sp. Pl. 2: 950. 1753

Plants 4–110(–135) cm. **Corms** globose to elongate, not forked, 8–31 mm. **Leaves** not appressed to inflorescences at flowering; blade linear, lanceolate, or seldom elliptic-lanceolate, 2–50 cm × 2–35 (–50) mm, margins slightly to strongly curled transversely. **Inflorescences** green or slightly purple at base, becoming entirely green after flowering, 4–135 cm; floral bracts ovate to ovate-lanceolate, subulate, 3–30 mm. **Flowers** 1–25, opening sequentially, magenta, pink, to white, faintly fragrant; dorsal sepal oblanceolate, 15–31 × 5–18 mm, apex acuminate; lateral sepals occasionally reflexed distally, ovate to lanceolate, slightly falcate, 13–26 × 5–16 mm, apex apiculate to acuminate; petals obpandurate to seldom elliptic, slightly falcate, 15–28 × 4–14 mm, apex obtuse; lip 11–23 mm, middle lobe with dilated end typically anvil-shaped, seldom triangular, 5.5–21 mm wide; column 12–25 × 1–2 mm, distal end 4.5–10 mm wide; rostellum present or absent; stigma at angle to column surface. **Capsules** ovoid to ellipsoid, 13–30 × 5–10 mm. $2n = 26$, 40, 42.

Varieties 2 (2 in the flora): e North America, West Indies (Bahamas, Cuba).

Calopogon tuberosus is the most variable and widespread species in the the genus. It occurs from Manitoba and Newfoundland south to Texas, Florida, the Bahamas, and Cuba.

Calopogon tuberosus can be distinguished from other species in the genus where sympatric by its relatively robust size and elongated lip with wide, anvil-shaped dilated distal end. Plants with a triangular-shaped middle lip lobe are uncommon in the south and occur toward the northern part of the range, and those with long but narrowed lips occur in *C. tuberosus* var. *simpsonii*, the latter plants generally being much larger and occurring in a different habitat than all other species of *Calopogon*.

1. Leaf blade slightly curled transversely; dilated distal portion of middle lip lobe broadly expanded and not whitened except in entirely white flowers; acidic wetlands . . . . 4a. *Calopogon tuberosus* var. *tuberosus*
1. Leaf blade strongly curled transversely; dilated distal portion of middle lip lobe narrowly to widely expanded, whitened; wet, alkaline savannas . . . . 4b. *Calopogon tuberosus* var. *simpsonii*

4a. Calopogon tuberosus (Linnaeus) Britton, Sterns & Poggenberg var. **tuberosus** • Calopogon tubéreux [E] [F]

Calopogon pulchellus (Salisbury) R. Brown; *C. pulchellus* var. *latifolius* (H. St. John) Fernald; *C. tuberosus* var. *latifolius* (H. St. John) B. Boivin; *Cathea pulchella* Salisbury; *C. tuberosa* (Linnaeus) Morong; *Cymbidium pulchellum* (Salisbury) Swartz; *Helleborine tuberosa* (Linnaeus) Kuntze; *Limodorum barbatum* Lamarck; *L. pulchellum* Salisbury; *L. tuberosum* Linnaeus var. *nanum* Nieuwland

Plants 4–110(–135) cm. **Corms** globose to elongate, 8–31 mm. **Leaves:** blade slightly curled transversely, linear, lanceolate, or seldom elliptic-lanceolate, 3–50 cm × 2–35(–50) mm. **Inflorescences** green or slightly purple at base, becoming entirely green after flowering, 4–110 (–135) cm; floral bracts ovate to ovate-lanceolate, subulate, 3–8(–30) mm. **Flowers** 1–15(–25); dorsal sepal 15–31 × 5–18 mm; lateral sepals reflexed distally, 13–26 × 5–16 mm, apex apiculate to acuminate; petals 15–28 × 4–14 mm; lip 11–23 mm, dilated end of middle lobe same color as rest of flower (except for brushlike lamellae patch), typically anvil-shaped, seldom triangular or reniform, 9–21 mm wide; column 12–25 × 1–2 mm, distal end 6–10 mm wide; rostellum usually present. **Capsules** 13–30 × 5–10 mm. $2n = 26$, 40, 42.

Flowering Nov (s Fla.)–Aug (north). Acidic soils in fens, bogs, pine and oak savannas, grasslands, interdune swales; 0–1200 m; St. Pierre and Miquelon; Man., N.B., Nfld. and Labr. (Nfld.), N.S., Ont., P.E.I., Que.; Ala., Ark., Conn., Del., D.C., Fla., Ga., Ill., Ind., Iowa, Ky., La., Maine, Md., Mass., Mich., Minn., Miss., Mo., N.H., N.J., N.Y., N.C., Ohio, Okla., Pa., R.I., S.C., Tenn., Tex., Vt., Va., W.Va., Wis.

At the northeastern edge of the range, tiny plants 4–20 cm have 1–5 deep magenta flowers, lips 13–18 mm with pale trichomes, and leaves 30–110 × 5–30 mm; they have been considered to be *Calopogon tuberosus* var. *latifolius* (H. St. John) B. Boivin. The variety has been reported in coastal bogs from Nova Scotia, the Magdalen Islands in the Gulf of St. Lawrence, Sable Island off Nova Scotia, Newfoundland, and Great Wass Island in the Gulf of Maine (C. A. Luer 1975; P. M. Catling and Z. Lucas 1987; L. M. Eastman 1988). Catling and Lucas found that plants corresponding to *C. tuberosus* var. *latifolius*

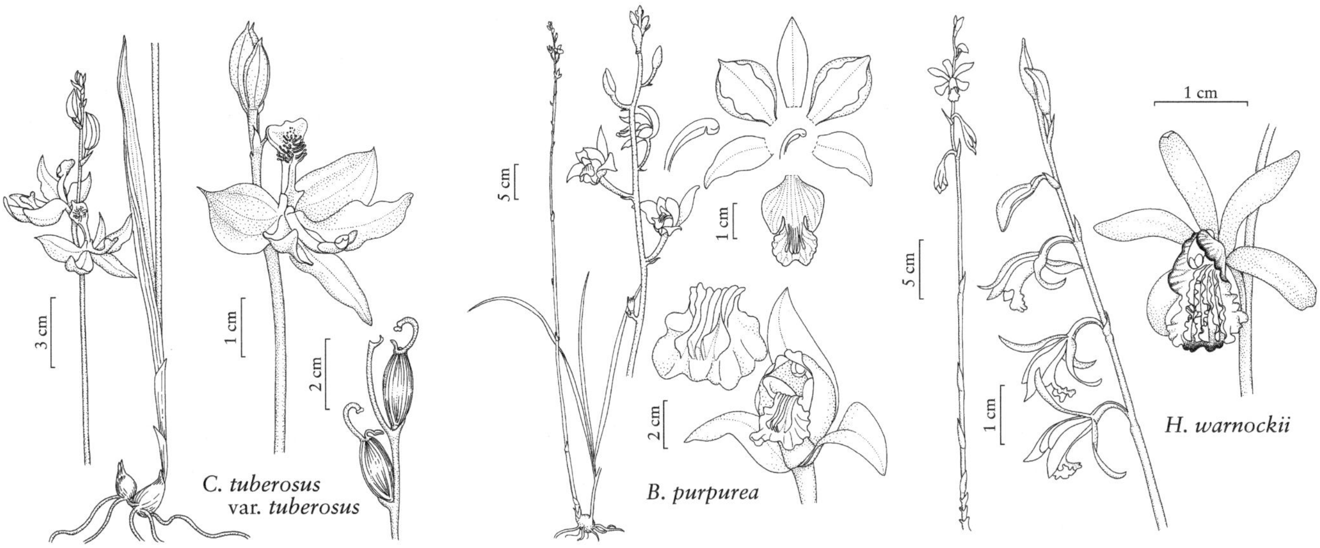

CALOPOGON ◦ BLETIA ◦ HEXALECTRIS

represented the end of a continuum of variation and recommended against the recognition of the variety; those results were corroborated by DNA data (D. H. Goldman 2000). In contrast to the plants in the northeastern range, plants in the southern range of this variety can exceed 1 m in height and produce flowers close to 5 cm diameter.

4b. Calopogon tuberosus (Linnaeus) Britton, Sterns & Poggenberg var. **simpsonii** (Small) Magrath, Sida 13: 371. 1989

Limodorum simpsonii Small, Fl. S.E. U.S., 322, 1329. 1903; *Calopogon pulchellus* (Salisbury) R. Brown var. *simpsonii* (Small) Ames

Plants 20–100 cm or more. **Corms** globose, 8–22 mm. **Leaves:** blade strongly curled transversely, linear, 25–50 cm × 3.5–15 mm. **Inflorescences** green or seldom slightly purple at base, becoming entirely green after flowering, 20–100 cm or more; floral bracts ovate to ovate-lanceolate, subulate, 3–12(–30) mm. **Flowers** 1–20(–25); dorsal sepal 18–25 × 5–10 mm; lateral sepals sometimes reflexed distally, often narrowed, 15–20 × 7.5–13 mm, apex acuminate; petals 18–25 × 5–10 mm; lip 12–23 mm, dilated end of middle lobe paler color than rest of flower, narrowly anvil-shaped to triangular, 5.5–16 mm wide; column 15–25 × 1.5–2 mm, distal end 4.5–8 mm wide; rostellum absent or present. **Capsules** ovoid to ellipsoid, 15–25 × 5–8 mm. 2*n* = 40.

Flowering Nov–Jul, most commonly mid–late spring. Seasonally wet, alkaline tropical and subtropical conifer savannas, meadows, and grasslands; 0–10 m; Fla.; West Indies (Bahamas, Cuba).

Calopogon tuberosus var. *simpsonii* can be distinguished from the typical variety of the species based on its strongly transversely curled leaves, the frequently narrowed and pale apex of the middle lip lobe, and its habitat in wet, marly soils. Based on morphology and allozymes (D. W. Trapnell 1995) and on nuclear and chloroplast DNA (D. H. Goldman 2000), *C. tuberosus* var. *simpsonii* was determined to be a distinct taxon.

5. Calopogon oklahomensis D. H. Goldman, Lindleyana 10: 37, figs. 1, 2. 1995 [E]

Plants 11–35 cm. **Corms** elongate, forked, 18–29 mm. **Leaves** not appressed to inflorescences at flowering; blade linear-lanceolate, 7–35 cm × 5–15 mm. **Inflorescences** green, 11.8–34 cm; floral bracts ovate-lanceolate to subulate, 3–8 mm. **Flowers** 2–7(–11), more than 1 cm apart, opening nearly simultaneously, magenta, pink, to white, fragrant; dorsal sepal strongly spreading to straight, obovate to oblanceolate, 14–25 × 5–8 mm, apex acuminate or acute; lateral sepals strongly spreading to straight, ovate to broadly lanceolate, slightly falcate, 15–22 × 8–10 mm, apex acuminate to apiculate; petals slightly falcate to straight, oblong to obpandurate, 11–20 × 5–8 mm, apex obtuse; lip 10–17 mm, middle lobe with dilated apex triangular, broadly rounded, 5–12 mm wide, apex acute or retuse; column 8.5–13 × 1.5–2 mm, distal end 5–8.5 mm wide; rostellum present; stigma typically flat against column surface. $2n = 120$.

Flowering Mar–Jul. Mesic, acidic, sandy to loamy prairies, pine savannas, oak woodlands, edges of bogs, and frequently mowed meadows; 0–300 m; Ala., Ark., Ga., Ill., Ind., Iowa, Kans., La., Minn., Miss., Mo., Okla., S.C., Tenn., Tex., Wis.

Calopogon oklahomensis originally was thought to be restricted to the south-central states; it is much more widespread. It is, however, nearly extirpated outside the south-central states. DNA evidence, as well as its hexaploid chromosome number, suggest that this species may be an alloploid, possibly derived from ancient hybridization between *C. barbatus* and *C. tuberosus* (D. H. Goldman 2000).

SELECTED REFERENCE Goldman, D. H. 1995. A new species of *Calopogon* from the midwestern United States. Lindleyana 10: 37–42.

38. BLETIA Ruiz & Pavón, Fl. Peruv. Prodr., 119, plate 26. 1794 • [For Luis Blet, a Catalonian apothecary of the eighteenth century who accompanied Ruiz and Pavón on their New World explorations]

James D. Ackerman

Plants herbs, terrestrial, cespitose, sympodial. **Roots** fleshy, velamentous. **Stems** short, stout rhizomes and pseudobulbs. **Leaves** deciduous, 3–5, basal and cauline, petiolate; blade articulate, pleated, thin. **Inflorescences** lateral, racemes or sparsely branched panicles, erect, scapose, 3–80-flowered; floral bracts distant, tubular. **Flowers** resupinate; sepals and petals subsimilar, distinct; lip attached to base of column, margins entire to 3-lobed; disc smooth, lamellate or with 5–7 denticulate crests; column elongate, arcuate, sometimes winged distally; anther terminal, incumbent; pollinia 8 in 2 sets of 4, yellow, compressed, hard, waxy or rarely mealy; stigma entire; ovary pedicellate. **Fruits** capsules, cylindric.

Species 31 (2 in the flora): s Florida, Mexico, West Indies, Central America, South America.

SELECTED REFERENCES Dodson, C. H. and P. M. Dodson. 1980. *Bletia purpurea*. Icon. Pl. Trop. 1: plate 7. Dressler, R. L. 1968. Notes on *Bletia* (Orchidaceae). Brittonia 20: 182–190.

1. Sepals (12–)15–26 mm; pedicellate ovary 9–18 mm; column foot present . . . . 1. *Bletia purpurea*
1. Sepals 30–40 mm; pedicellate ovary 20–35 mm; column foot absent . . . . 2. *Bletia patula*

1. Bletia purpurea (Lamarck) de Candolle, Mém. Soc. Phys. Genève 9: 97, 100. 1841 • Pine-pink, purple bletia, sharp-petaled bletia [F]

Limodorum purpureum Lamarck in J. Lamarck et al., Encycl. 3: 515. 1791; *Bletia acutipetala* Hooker; *B. havanensis* Lindley; *B. verecunda* (Salisbury) R. Brown; *Cymbidium verecundum* (Salisbury) Swartz; *Gyas verecunda* (Salisbury) Salisbury; *Limodorum trifidum* Michaux; *L. verecundum* Salisbury; *Thiebautia nervosa* Colla

Plants to 180 cm. **Roots** numerous, 1–1.5 mm diam. **Stems:** pseudobulbs ovoid, 2–4 cm diam. **Leaves:** blade linear to narrowly elliptic-lanceolate, 20–100 × 1–5 cm, apex acuminate. **Inflorescences** 25–170 cm; floral bracts ovate-triangular to ovate-lanceolate, 2–9 mm, apex acute to acuminate. **Flowers** 3–80, mostly pink, rosy purple, or deep purple, rarely almost white; sepals spreading; dorsal sepal oblong-elliptic to ovate-lanceolate, 15–26 × 5–9 mm, apex subobtuse to acute; lateral sepals obliquely ovate-oblong to elliptic-oblong, 12–20 × 5–8 mm, apex acute to acuminate; petals erect, forming hood over column, oblong-ovate to elliptic or oblong-lanceolate, 12–21 × 7–11 mm, apex obtuse to acute; lip adnate to column foot, purple-veined with dark margin especially distally, cordate to ovate, 3-lobed, 10–18 × 8–14 mm, middle lobe recurved, suborbiculate, truncate to deeply emarginate, margins undulate-crenulate, lateral lobes incurved, base rounded, apex triangular-obtuse; disc lamellae yellow, 5–7; column white or greenish, clavate, 8–12 mm; column foot present; pedicellate ovary slender, 9–18 mm. **Capsules** erect to suberect, 2–4.5 cm × 8–10 mm. $2n = 60$ (as *B. verecunda* from Mexico).

Flowering Dec–May, sporadically all year. Pinelands and hammocks, in humus over limestone, or in swamps on logs, stumps, or base of cypress trees above high water level; 0–20 m; Fla.; Mexico; West Indies; Central America; South America.

2. **Bletia patula** Graham, Edinburgh New Philos. J. 21: 155. 1836 • Haitian pine-pink

Plants to 100 cm. **Roots** numerous, 1.5–3 mm diam. **Stems** pseudobulbs, ovoid to subglobose, 2–5 cm diam. **Leaves:** blade narrowly elliptic to narrowly oblanceolate, 25–75 × 1.5–4 cm, apex acuminate. **Inflorescences** 39–118 cm; floral bracts ovate-triangular to lanceolate, 1–3 mm, apex acute to acuminate. **Flowers** 5–22, pink to rosy purple, rarely entirely white; dorsal sepal elliptic-oblanceolate, 30–40 × 6–8 mm, apex acute to acuminate; lateral sepals similar but falcate; petals erect, oblong-elliptic, 30–35 × 5–6 mm, thinner than sepals, apex obtuse; lip adnate to base of column, clawed, 3-lobed, 30–35 × 27–31 mm, middle lobe subquadrangular, flabellate, emarginate, margins crispate-undulate, lateral lobes erect, flanking column, rounded to subquadrangular, broadest distally; disc lamellae 5, middle 3 white; column clavate, 23–25 mm, column foot absent; pedicellate ovary slender, 20–35 mm. **Capsules** pendulous, 3–4.5 cm × 8–10 mm.

Flowering Mar–May. Pinelands; 0–20 m; Fla.; Mexico; West Indies; Central America; South America.

Bletia patula was originally collected from Miami-Dade County, Florida, in 1947. The locality has since become part of the greater Miami area (C. A. Luer 1972), and the persistence of the populations is doubtful.

39. HEXALECTRIS Rafinesque, Neogenyton, 4. 1825 • [Greek *hex*, six, and *alectryon*, rooster, alluding to six longitudinal fleshy crests on the floral lip]

Douglas H. Goldman

Ronald A. Coleman

Lawrence K. Magrath

Paul M. Catling

Herbs, perennial, terrestrial, leafless, rootless, mycotrophic, solitary or colonial. **Rhizomes** branching, articulate, annular, stout. **Stems** erect, yellowish brown, pink, pinkish red, or reddish purple, glabrous; sheathing bracts 2–5, reduced. **Inflorescences** racemes, consisting of entire aboveground portion of plant. **Flowers** 2–25, resupinate, tan, brown, magenta, purple, or maroon; dorsal sepal lanceolate; lateral sepals falcate to lanceolate, base distinct and free; petals linear, lanceolate, or elliptic; lip 3-lobed; lamellae 3–5(–7), prominent; column arcuate, apex narrowly winged; anthers terminal, white, yellow, or maroon; pollinia 8, firm, in 4 pairs, 2 pairs per anther cell; stigma subterminal. **Fruits** capsules, pendent, ellipsoid; seeds 0.5–0.7 mm.

Species 7 (5 in the flora): neotemperate and neotropical, mostly Mexican.

Rafinesque's choice of the name *Hexalectris* was probably inappropriate because plants in the genus tend to have five or seven crests on the lip (C. A. Luer 1975; J. Liggio and A. Liggio 1999).

Examination of material, both living and preserved, suggests that a species complex exists that involves *Hexalectris nitida*, *H. revoluta*, *H. spicata* in the broad sense, and the Mexican and Guatemalan *H. parviflora*. This is a complicated group, some plants of each species resembling plants of another species within the complex. *Hexalectris spicata* var. *arizonica* possibly is a hybrid-derived taxon from *H. nitida* and *H. spicata* var. *spicata*, or the latter species and *H. revoluta* (P. M. Catling and V. S. Engel 1993), although plants with cleistogamous flowers of *H. nitida* are often similar to *H. spicata* var. *arizonica*. A number of recent significant range extensions have been noted for *H. nitida* (V. S. Engel 1987), *H. revoluta* (R. A. Coleman), and *H. warnockii* (V. S. Engel 1987; G. A. Salazar-Chavez 1991), suggesting that members of this genus lurk unsuspected over a very large geographic area; they may be more common than currently assumed. Because relatively few herbarium collections have been made for this genus, particularly in the western United States and Mexico, it is clear that more has yet to be understood about the biology and taxonomy of the members of this complicated genus. Based on plastid DNA sequences, it has been estimated that *Hexalectris* is most closely related to the neotropical genera *Basiphyllaea* and *Bletia* (D. H. Goldman 2000; D. H. Goldman et al. 2001).

1. Lateral sepals 7–12 mm; lip 7–11 mm; column 6–8 mm; anthers yellow . . . . 3. *Hexalectris nitida*
1. Lateral sepals longer than 12 mm; lip 10–20 mm; column 9–18 mm; anthers whitish.
    2. Petals and sepals dark purple to maroon, not recurved; lip with 5 yellow-margined, undulate lamellae; anthers reddish to maroon . . . . 1. *Hexalectris warnockii*
    2. Petals yellowish brown, pinkish brown, or pink to rose pink, weakly to strongly recurved; lip with 3–7 white or purple, smooth, ± straight lamellae; anthers white to yellow.
        3. Lip with 5–7 typically purple to white lamellae; lip shallowly 3-lobed, fissure less than 2 mm deep . . . . 2. *Hexalectris spicata*
        3. Lip with 3–7 white to yellow lamellae; lip deeply 3-lobed, fissure more than 3 mm deep.
            4. Petals and sepals pink to crimson, slightly recurved; lip with 5 white lamellae . . . . 4. *Hexalectris grandiflora*
            4. Petals and sepals pinkish brown, strongly recurved; lip with 3–7 obscure white to yellow lamellae . . . . 5. *Hexalectris revoluta*

1. Hexalectris warnockii Ames & Correll, Bot. Mus. Leafl. 11: 8, plate 2. 1943 • Texas purple-spike [C] [F]

Stems dark red-purple, 15–40 cm; sheathing bracts 2–4. **Inflorescences:** floral bracts lanceolate to ovate, 3–9 × 2–4 mm, apex acute. **Flowers** 3–10, pedicellate, chasmogamous; sepals and petals spreading, not recurved, dark purple to maroon; dorsal sepal linear-oblong, 12–19 × 2–4 mm; lateral sepals linear-lanceolate, falcate, 12–18 × 2.5–4.5 mm; petals oblanceolate to narrowly obovate, falcate, 12–20 × 1.5–3.8 mm, apex obtuse to acute; lip ovate to ovate-cuneate, deeply 3-lobed, 12–18 × 9–16 mm, margins maroon, centrally white with purple or maroon veins, middle lobe white, rectangular to obcordate, 4–6 × 6–11 mm, margins undulate, lateral lobes incurved, obtuse to broadly rounded, at least 1/2 length of middle lobe, 5 × 5 mm, margins entire or nearly so; lamellae 5, undulate, yellow-margined, prominent; column whitish, 9–12 mm; anthers red to maroon. **Capsules** 15 × 5 mm.

Flowering Jun–Sep. Shaded slopes and dry, rocky creek beds in canyons, in leaf mold in oak-juniper-pinyon pine woodlands; of conservation concern; 0–400 m (c Tex.), 600–2000 m (west); Ariz., Tex.; Mexico (Baja California Sur).

Hexalectris warnockii grows in the Chiricahua Mountains of Arizona and in the Davis and Chisos mountains, the Edwards Plateau, and the Dallas area of Texas.

2. Hexalectris spicata (Walter) Barnhart, Torreya 4: 121. 1904 • Crested coral-root

Arethusa spicata Walter, Fl. Carol., 222. 1788; *Corallorhiza spicata* (Walter) Tidestrom

Stems: sheathing bracts 3–5. **Inflorescences:** floral bracts lanceolate to triangular-ovate, 5–10 × 2–6 mm. **Flowers** 5–25; pedicels 8–20 mm; lateral sepals oblong-ovate to oblong-lanceolate, falcate, apex acute; petals weakly to strongly recurved, elliptic, oblong-obovate, oblanceolate, or obovate, falcate; lip ovate to obovate, clawed, shallowly 3-lobed, fissure between lobes less than 2 mm deep, middle lobe ovate to suborbiculate, margins undulate, apically rounded to truncate, lateral lobes incurved, broadly rounded to obtuse, margins entire, apically triangular to rounded; lamellae 5(–7), central, purple to white; column white to yellowish white, apically winged, 11–18 mm; anthers whitish to yellow.

Varieties 2 (2 in the flora): c, e North America, Mexico.

1. Flowers chasmogamous; rostellum present; lamellae 0.7–1 mm; petals greater than 14 × 5 mm . . . . 2a. *Hexalectris spicata* var. *spicata*
1. Flowers cleistogamous or nearly so, occasionally chasmogamous; rostellum reduced or absent; lamellae 0.2–0.7 mm; petals less than 16 × 5 mm . . . . 2b. *Hexalectris spicata* var. *arizonica*

2a. Hexalectris spicata (Walter) Barnhart var. **spicata** • Crested coral-root

Bletia aphylla Nuttall; *Hexalectris aphylla* (Nuttall) Rafinesque; *H. squamosa* Rafinesque

Stems yellow-brown, pink-brown, to purple, 25–80 cm. **Flowers** chasmogamous; sepals and petals often apically recurved, yellow-tan to purple-brown with prominent purple or brown veins; dorsal sepal oblong-elliptic, obtuse, 15–24 × 4–8 mm; lateral sepals 14–20 × 5.5–9 mm; petals 14–23 × 5–9 mm; lip tan to purple-white, rarely entirely white, 13–20 × 8–16 mm; lamellae central, purple, 0.7–1 mm; column 13–18 mm; rostellum present; anthers yellow. **Capsules** 16–30 × 8–20 mm.

Flowering Apr–Aug. Humus, typically in mesic to dry soil over limestone or sandstone, in vicinity of *Juniperus*, *Pinus*, or *Quercus*; 0–600 m (c Texas eastward and northward), 1600–2000 m (west); Ala., Ariz., Ark., Fla., Ga., Ill., Ind., Kans., Ky., La., Md., Miss., Mo., N.Mex., N.C., Ohio, Okla., S.C., Tenn., Tex., Va., W.Va.; Mexico.

Hexalectris spicata var. *spicata* occurs in Missouri in calcareous soil in dry forests and limestone glades, often in association with *Juniperus* in the latter habitat (G. Yatskievych 1999). Detailed descriptions of habitat and associated species for *H. spicata* var. *spicata* in Illinois are given by C. J. Sheviak (1974) and for Indiana by M. A. Homoya (1993).

In Oklahoma, populations of *Hexalectris spicata* var. *spicata* growing in decaying *Juniperus* needle litter over sandstone are far more robust.

2b. Hexalectris spicata (Walter) Barnhart var. **arizonica** (S. Watson) Catling & V. S. Engel, Lindleyana 8: 122. 1993 • Arizona crested coral-root

Corallorhiza arizonica S. Watson, Proc. Amer. Acad. Arts 17: 379. 1882

Stems pinkish red, pale pink-yellow, or pink-white, 15–70 cm; sheathing bracts 3–5. **Flowers** cleistogamous or chasmogamous, apparently autogamous; sepals and petals usually not apically recurved on chasmogamous flowers, yellow-tan, pinkish, red-brown, or purple-brown with dull magenta, faint pink-brown to purple-brown veins, margins pale; dorsal sepal lanceolate to oblong, 15–18 × 2.7–4.5 mm; lateral sepals 13–15 × 4–7 mm; petals 14–16 × 4–5 mm; lip 12–14.5 × 10–12 mm, white or creamy white with darkly colored veins and paler margin; lamellae central, pale, 0.2–0.7 mm; column 11–14 mm; rostellum absent; anthers whitish. **Capsules** 16–25 × 8–15 mm.

Flowering May–Aug. In organic mesic to dry soil over limestone or sandstone, in juniper, pine, and oak woodlands; 100–600 m (c Tex.), 1600–2000 m (west); Ariz., N.Mex., Tex.; Mexico.

This autogamous and florally reduced segregate of *Hexalectris spicata* has been thought to be of hybrid origin, derived either from *H. spicata* var. *spicata* and *H. nitida*, or less likely from the former taxon and *H. revoluta* (P. M. Catling and V. S. Engel 1993). In Texas this plant tends to begin flowering later than the typical variety, from June and July (P. M. Catling and V. S. Engel 1993), as it does in Arizona and New Mexico, where it flowers in July and August (R. A. Coleman).

SELECTED REFERENCE Catling, P. M. and V. S. Engel. 1993. Systematics and distribution of *Hexalectris spicata* var. *arizonica* (Orchidaceae). Lindleyana 8: 119–125.

3. Hexalectris nitida L. O. Williams, J. Arnold Arbor. 25: 81. 1944 • Glass Mountain coral-root, shining cock's-comb

Stems pinkish, pale red, or brown-purple, 10–32 cm; sheathing bracts 3–5. **Inflorescences:** floral bracts ovate-oblong to lanceolate, 3–7 × 3 mm. **Flowers** 6–24, pedicellate; sepals and petals spreading and recurved, cleistogamous to chasmogamous; dorsal sepal narrowly oblong-elliptic, 8–13 × 3–4.5 mm, apex obtuse; lateral sepals obliquely elliptic to oblanceolate, rounded, slightly falcate, 7–12 × 3–4.5 mm; petals obovate to oblanceolate, falcate, 8–11 × 2–3.5 mm, apically rounded; lip shallowly 3-lobed, 7–11 × 4–8 mm, middle lobe suborbiculate, pink to purple, lateral lobes 1/3 length of middle lobe; minimal lamellae or slightly raised veins 5 or 7, near base of middle lobe, mostly obscured apically; column white, 6–8 mm; anther yellow. **Capsules** 15 × 5 mm.

Flowering Jun–Aug. Moist canyons in oak-juniper-pinyon pine woodlands growing in humus, often in decaying juniper needle litter; 200–1500 m; N.Mex., Tex.; Mexico.

The Texas plants occur in the Chisos and Glass mountains, on the eastern and southern Edwards Plateau, and in the Dallas area. Most of the plants on the Edwards Plateau appear to have cleistogamous flowers, and only occasionally display open flowers (J. Liggio and A. Liggio 1999). Plants of *Hexalectris nitida* examined from the

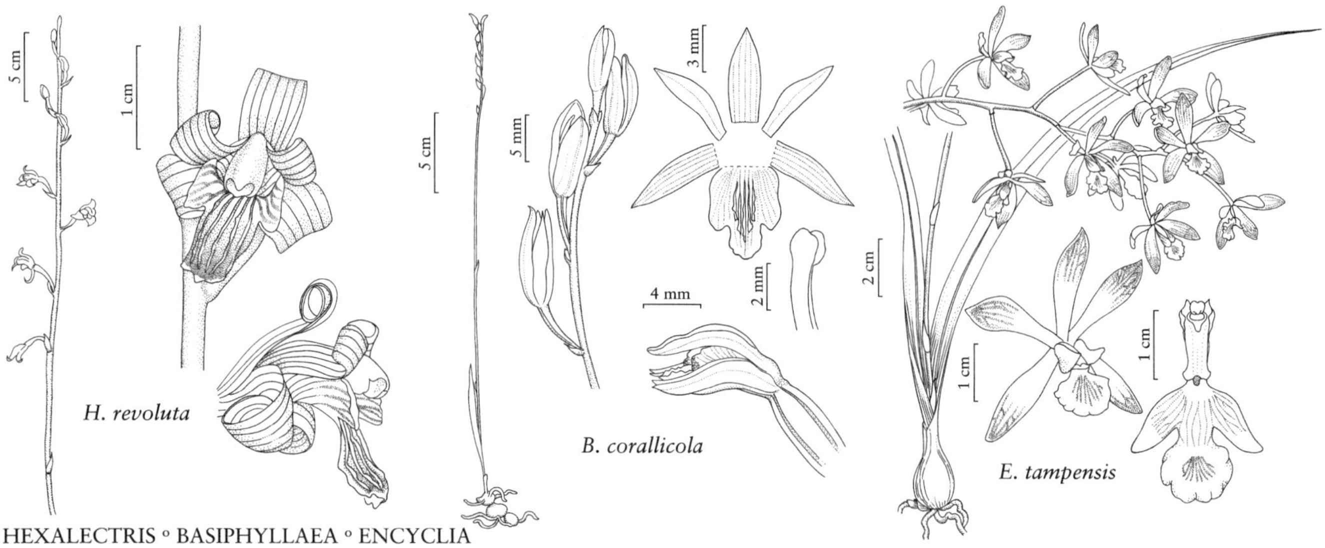

HEXALECTRIS ◦ BASIPHYLLAEA ◦ ENCYCLIA

United States have been self-pollinating because of a reduced rostellum. In central Texas, in particular, flowers of this species are cleistogamous. They also are found under *Juniperus* in that area (V. S. Engel 1987) and flower well after ample late-spring rain.

4. **Hexalectris grandiflora** (A. Richard & Galeotti) L. O. Williams, J. Arnold Arbor. 25: 81. 1944 • Greenman's hexalectris, giant coral-root, Greenman's cock's-comb

Corallorhiza grandiflora A. Richard & Galeotti, Ann. Sci. Nat. Bot., sér. 3, 3: 19. 1845; *Hexalectris mexicana* Greenman; *Neottia grandiflora* (A. Richard & Galeotti) Kuntze

Stems pink, pink-red, or red-purple, 10–60 cm; sheathing bracts 3–5. **Inflorescences:** floral bracts ovate to lanceolate, 5–12 × 3–5 mm. **Flowers** 5–20, pedicellate, chasmogamous; sepals and petals slightly recurved, pink, magenta to crimson; dorsal sepals linear-oblong to oblong-elliptic, 14–27 × 4–7 mm, apex obtuse to acute; lateral sepals oblong-elliptic, slightly falcate, 12–23 × 4–7 mm, apex obtuse to subacute; petals oblanceolate to elliptic, falcate, 13–23 × 5–7 mm; lip ovate to obovate, clawed, deeply 3-lobed, 10–16 × 5–12 mm, fissure between lobes more than 3 mm deep, middle lobe cuneate, suborbiculate, to flabellate, rounded, apex obtuse to apiculate, lateral lobes ovate to semiorbiculate, 2/3 length of middle lobe, apex obtuse to subtruncate; lamellae 5, white; column pink with white apex, 12–14 mm; anther white to pale yellow. **Capsules** 25 × 15 mm.

Flowering Jun–Aug. Moist canyons in pine-oak-juniper woodlands in leaf litter and humus; 700–2500 m; Tex.; Mexico.

Hexalectris grandiflora grows in the Davis and Chisos mountains in Texas.

5. **Hexalectris revoluta** Correll, Bot. Mus. Leafl. 10: 19, fig. 2. 1941 • Curly coral-root, Correll's cock's-comb [C] [F]

Stems tan, pale pink, to pale brown-purple, 30–50 cm; sheathing bracts 3–5. **Inflorescences:** floral bracts lanceolate to ovate, 5–15 × 3–4 mm. **Flowers** 5–15(–20), pedicellate, chasmogamous; sepals and petals strongly recurved, tan to pinkish brown; dorsal sepal oblong-elliptic to lanceolate, 15–25 × 3–8 mm; lateral sepals obliquely elliptic to elliptic-lanceolate, 15–22 × 3–8 mm; petals narrowly elliptic to obovate or oblanceolate, slightly falcate, 15–19 × 4.5–7.5 mm; lip broadly elliptic, deeply 3-lobed, 13–18 × 9–13 mm, fissure between lobes more than 3 mm deep, middle lobe white with purple, pale yellow near base, obovate-cuneate, apex truncate to retuse, lateral lobes incurved, oblong, obtuse, 1/2–2/3 length of middle lobe; lamellae 3, 5, or 7, white to yellow, obscure; column purple abaxially, white-purple adaxially, 9–15 mm; anther yellow. **Capsules** 20 × 5 mm.

Flowering May–Aug. Oak-juniper-pinyon pine woodlands in leaf litter and humus, occasionally in rocky, open terrain; of conservation concern; 1000–1600 m; Ariz., Tex.; ne Mexico.

In Arizona *Hexalectris revoluta* grows in Cochise and Pima counties; in Texas, in the Guadalupe and Chisos mountains.

Hexalectris revoluta has a less densely flowered raceme than *H. nitida*, and more revolute sepals and petals, and more deeply lobed labellum than *H. spicata*, both species with which it is occasionally confused. The plants are occasionally associated with *Acacia*, *Juglans*, and *Prosopis* in Arizona (R. A. Coleman).

230d.5. ORCHIDACEAE Jussieu (subfam. EPIDENDROIDEAE) tribe EPIDENDREAE Kunth in A. von Humboldt et al., Nov. Gen. Sp. 1: 269. 1815

230d.5a. ORCHIDACEAE Jussieu (tribe EPIDENDREAE) subtribe LAELIINAE Bentham, J. Linn. Soc., Bot. 18: 287. 1881

40. BASIPHYLLAEA Schlechter, Repert. Spec. Nov. Regni Veg. 17: 77. 1921 · [Greek *basis*, base, and *phyllon*, leaf, referring to the single basal leaf]

James D. Ackerman

Herbs, terrestrial, sympodial, glabrous. **Roots** few, velamentous, fleshy, villous. **Stems** subterranean tubers, rhizomes, and erect, delicate, aerial shoots. **Leaves** deciduous, 1–2, basal; blade articulate, conduplicate, margins entire. **Inflorescences** terminal, few-flowered racemes or sometimes branched, erect, scapose. **Flowers** resupinate; perianth not fully spreading, distinct and free; sepals lanceolate to linear-lanceolate; petals linear-elliptic, oblong-elliptic, or oblanceolate; lip 3-lobed, middle lobe small, obovate to subquadrangular, apex notched to retuse, lateral lobes clasping column, broad, rounded; disc lamellate; column elongate, arcuate, slender; anther terminal, operculate; pollinia 8, hard, waxy; ovary pedicellate; stigma entire. **Fruits** capsules, erect, ellipsoid.

Species 6 (1 in the flora): s Florida, West Indies.

SELECTED REFERENCE McCartney, C. 1992. Stalking the elusive *Basiphyllaea corallicola*. Part 2. Florida Orchidist 34: 195–211.

1. Basiphyllaea corallicola (Small) Ames, Schedul. Orchid. 7: 1. 1924 • Carter's orchid [C] [F]

Carteria corallicola Small, Torreya 10: 188. 1910; *Basiphyllaea angustifolia* Schlechter

Plants to 40 cm. **Roots** 1–1.5 mm diam. **Stems** tubers, horizonal, ellipsoid-conic, 10–20 × 5–10 mm. **Leaves:** blade linear to linear-elliptic, 9–24 cm × 1–6.5 mm, slightly fleshy. **Inflorescences** scapose, 14–36 cm × 0.5–1 mm; bracts 2–3; racemes to 10 cm; floral bracts lanceolate, 2–3 mm. **Flowers** 1–10, nodding; sepals and petals yellow-green; sepals flat, 5-veined, linear-lanceolate, 8–13 × 2–3 mm, margins entire, apex acute; lateral sepals slightly falcate, otherwise similar; petals 4-veined, linear-elliptic, slightly thinner than sepals, 8–12.5 × 2–2.5 mm; lip white, sometimes with pink markings along margin, 7–10 × 5–6 mm when spread, middle lobe obovate, 3 × 3 mm, apex retuse, broadly apiculate, lateral lobes rounded, 1–1.5 mm short of lip apex, sinuses narrow, 1–1.5 mm deep; disc lamellae 5, middle 3 extending to middle lobe; column semiterete, 4–5 mm, basal wings flanking shallow, lyre-shaped cavity; pedicellate ovary 5–7 mm. **Capsules** 10–15 mm.

Flowering Sep–Nov. Open pineland savannas and hammocks, in humus over limestone; of conservation concern; 0–20 m; Fla.; West Indies.

41. ENCYCLIA Hooker, Bot. Mag. 55: plate 2831. 1828 • [Greek *enkyklos*, to encircle, referring to the lateral lobes of the lip, which encircle the column]

Eric Hágsater

Herbs, epiphytic. **Roots** fleshy, glabrous. **Stems** aggregate, erect, forming ovoid-pyriform pseudobulbs. **Leaves** apical, sessile, articulate; blade linear-elliptic, leathery. **Inflorescences** apical, racemes to panicles, flowering only once; spathe absent. **Flowers** resupinate, pedicellate, simultaneous; sepals subequal; petals similar; lip appressed to column, adnate only at base, callus boat-shaped; pollinia 4, obovoid, laterally compressed, subequal; caudicles 4 in 2 pairs; column winged; rostellum semiorbiculate, entire, abaxially covered with viscous substance. **Fruits** capsules, ellipsoid, 3-ribbed, 1-locular.

Species over 750 (1 in the flora): neotropical.

Stems of the encyclias are variable outside the flora area: rarely they are canelike and do not form pseudobulbs; leaves are sometimes semiterete. The genus has been misrepresented as epidendrums having pseudobulbs (O. Ames et al. 1936), as most of them do; some epidendrums also do.

The Florida species were segregated by J. K. Small (1933) into four genera: *Encyclia*, *Hormidium*, *Anacheilium*, and *Epicladium*. Most later authors recognized only *Epidendrum* or *Epidendrum* and *Encyclia*. Recent attempts for the reestablishment of *Anacheilium* and *Hormidium* (G. F. Pabst et al. 1981; R. P. Sauleda et al. 1984) were based mainly on Brazilian species and did not take Mexican species into consideration. *Prosthechea* has been recently recognized (W. E. Higgins 1997[1998]), thus leaving only one species of *Encyclia* in the flora area.

Encyclia rufa (Lindley) Britton & Millspaugh has been excluded from the flora because only one specimen has been recorded, and it was probably introduced (R. P. Sauleda 1983). The species is endemic to the Bahamas and Cuba.

SELECTED REFERENCES Ames, O., F. T. Hubbard, and C. Schweinfurth. 1936. The Genus *Epidendrum* in the United States and Middle America. Cambridge, Mass. Dressler, R. L. and G. E. Pollard. 1976. The Genus *Encyclia* in Mexico, ed. 2. Mexico City. Pabst, G. F., J. L. Moutinho, and A. V. Pinto. 1981. An attempt to establish the correct statement for genus *Anacheilium* Hoffmgg. and revision of the genus *Hormidium* Lindl. ex Heynh. Bradea 3: 173–186. Sauleda, R. P. and R. M. Adams. 1983. The genus *Encyclia* Hook. (Orchidaceae) in the Bahama Archipelago. Rhodora 85: 127–174.

1. Encyclia tampensis (Lindley) Small, Fl. Miami, 56. 1913 (as tampense) • Florida butterfly orchid [F]

Epidendrum tampense Lindley, Edwards's Bot. Reg. 33: plate 35. 1847; *Encyclia tampensis* (Lindley) Schlechter; *Epidendrum tampense* var. *albolabium* A. D. Hawkes

Plants to 50 cm. **Pseudobulbs** aggregate, dark green, ovoid-pyriform, 1–7 × 1–2.5 cm. **Leaves** 1–3, linear-lanceolate, 8–40 × 0.5–2 cm, leathery. **Inflorescences** racemes to panicles, lax, 10–80 cm. **Flowers** 3–45, green, yellow, or brown, often suffused with purple; sepals and petals similar, extended, oblanceolate-spatulate, 12–22 × 4–6.5 mm, apex obtuse; lip white, deeply 3-lobed, 12–18 × 12–18 mm when spread, middle lobe usually with large purple spot, suborbiculate, 6–10 mm wide, margins undulate, lateral lobes purple-veined, ovate-triangular; callus on isthmus; anthers 1, yellow; column straight, prominent wings on sides of stigmatic cavity, 1 cm. **Capsules:** pedicel 15–17 mm, body 17–27 × 15 mm, beak 3 mm.

Flowering May–Sep; fruiting throughout year. Epiphytic on many different trees and palms in forests and hammocks; 0–100 m; Fla.; West Indies (Bahamas).

Flowers of *Encyclia tampensis* are fragrant, with a sweet or honey odor, starting about noon and peaking in early afternoon when small bees in genera *Auglochlora* and *Halictus* pollinate the flowers.

42. EPIDENDRUM Linnaeus, Syst. Nat. ed. 10, 2: 1246. 1759 • [Greek *epi-*, on, and *dendron*, tree, alluding to the epiphytic habit]

Eric Hágsater

Amphiglottis Salisbury; *Larnandra* Rafinesque; *Nyctosma* Rafinesque; *Spathiger* Small; *Tritelandra* Rafinesque

Herbs, epiphytic, rarely lithophytic, cespitose. **Roots** fleshy, glabrous. **Stems** erect, repent, or hanging, canelike, simple or branching. **Leaves** alternate; petiole tubular, sheathing, articulate; blade ovate to elliptic or lanceolate, fleshy-leathery. **Inflorescences** terminal [lateral or basal], racemose to nearly corymbose or distichous, producing flowers only once, or during several years from same or new racemes from old axis; flowers pedicellate or not. **Flowers** resupinate or not, if not then distichous or spirally arranged, simultaneous or successive; sepals nearly equal; lip adnate to column throughout, forming nectary tube, tube penetrating ovary, occasionally producing nectar; column straight or slightly arched; clinandrium hood partly covering anther; anther 4-celled; pollinia 4, obovoid, laterally compressed, nearly equal; rostellum slit, producing semiliquid viscidium attached to caudicles of pollinarium. **Fruits** capsules, ellipsoid to subglobose, sometimes with pedicel or beak, 3-ribbed, 1-locular.

Species over 1000 (7 in the flora): tropical regions, Western Hemisphere.

Outside the flora area *Epidendrum* is highly varied (E. Hágsater 1984), sometimes producing pseudobulbs or thickened stems, semiterete to membranaceous leaf blades, and apical, lateral, or basal, racemose to nearly corymbose to paniculate inflorescences. The lip of the flower is adnate to the column, forming a nectary tube, but it rarely produces nectar. The lip is exceptionally distinct and free. The rostellum is always slit, producing a transparent or whitish semiliquid viscidium. The pollinarium usually has 4 pollinia, sometimes with 2 very reduced, rarely with only 2.

The Florida species were segregated into 2 genera, *Amphiglottis* and *Spathiger* (J. K. Small 1933). Later authors have recognized only *Epidendrum*.

SELECTED REFERENCES Ames, O., F. T. Hubbard, and C. Schweinfurth. 1936. The Genus *Epidendrum* in the United States and Middle America. Cambridge, Mass. Hágsater, E. 1984. Towards an understanding of the genus *Epidendrum*. In: K. W. Tan, ed. 1984. Proceedings of the Eleventh World Orchid Conference: March 1984, Miami, Florida, U.S.A. Miami. Pp. 195–201.

1. Plants creeping or hanging, ± branching; inflorescences distichous.
    2. Plants hanging, main stem greater than 40 cm; inflorescences distichous-imbricate; sepals greater than 12 mm . . . . . 1. *Epidendrum acunae*
    2. Plants creeping, main stem less than 20 cm; inflorescences distichous-elongate or distichous-imbricate; sepals less than 8 mm.
        3. Inflorescences distichous-elongate, equal to or greater than 5 cm; flowers mostly more than 5, sepals 5 mm . . . . . 6. *Epidendrum rigidum*
        3. Inflorescences distichous-imbricate, less than 3 cm; flowers usually 3, sepals 3 mm . . . . . 7. *Epidendrum strobiliferum*
1. Plants cespitose, aerial stems erect, unbranched; inflorescences nearly racemose or nearly corymbose.
    4. Inflorescences 3–50 cm (excluding flowers), racemose or compound racemose; sepals 5–11 mm.
        5. Inflorescences racemose; flowers spread out along apical 1/2 of inflorescence; sepals 6–11 mm . . . . . 4. *Epidendrum magnoliae*
        5. Inflorescences racemose, compound racemose with time; flowers in compact short racemes at end or from nodes of long peduncle; sepals 5.5–7.5 mm . . . 2. *Epidendrum amphistomum*

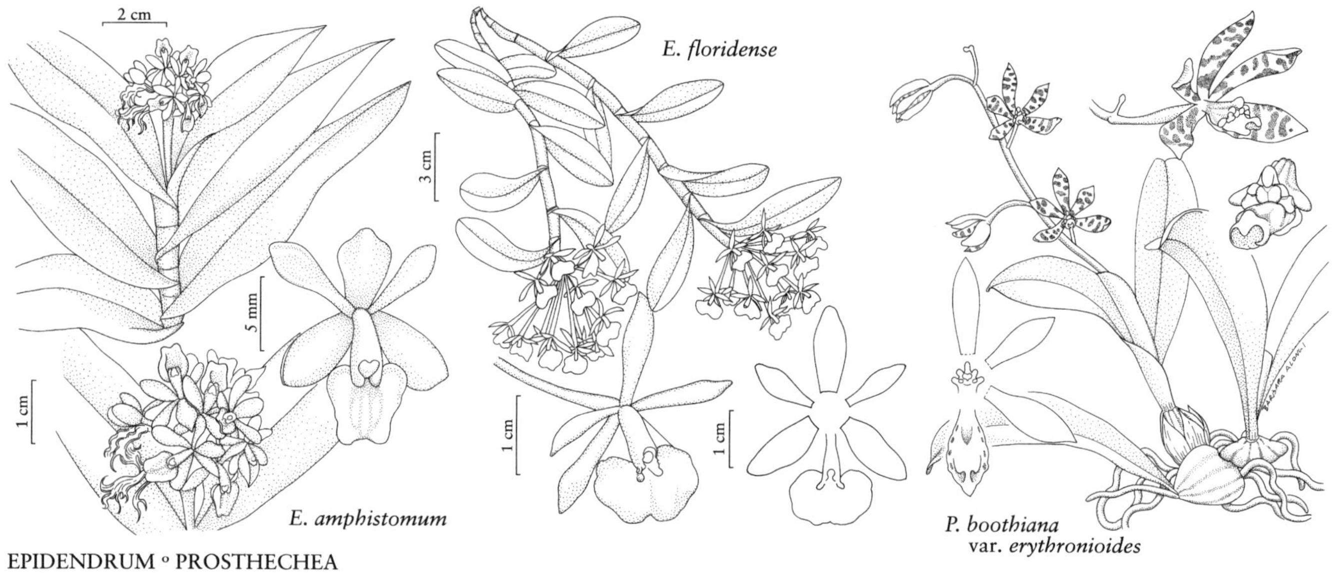

EPIDENDRUM ∘ PROSTHECHEA

4. Inflorescences less than 3 cm (excluding flowers), nearly corymbose; sepals equal to or greater than 12 mm.
   6. Sepals 60 mm; flowers yellowish and white, flowering successively, 1–2 at a time . . . . 5. *Epidendrum nocturnum*
   6. Sepals 12–15 mm; flowers green, flowering simultaneously, 4–14 . . . . 3. *Epidendrum floridense*

1. Epidendrum acunae Dressler, Amer. Orchid Soc. Bull. 28: 358, figs. 1–4. 1959 (as acuñae) • Galé's epidendrum

Plants hanging, 50–160 cm. **Roots** basal or from proximal branches, 2–3 mm diam. **Stems** branched, sinuous, terete, main stem to 160 cm, leaves 100–120 mm × 12–22 mm; vegetative branches to 60 cm, leaves 58–110 × 10–20 mm; flowering branchlets 3–8 cm, leaves 20–60 × 9–13 mm, flowering branchlets produced from both main stem and vegetative branches. **Leaves** many, distributed throughout stems; petiole to 4 cm; blade narrowly lanceolate, 2-lobed, fleshy-leathery. **Inflorescences** distichous-imbricate; peduncle 1 cm; bracts 2–3, conduplicate, 8–18 mm. **Flowers** 1–3, not resupinate, simultaneous, brownish green; sepals spreading, lanceolate, abaxially keeled, many veined, 12–21 × 3–6 mm, apex acute; petals partly open, elliptic, many veined, 12–18 × 2.5–5 mm, apex obtuse to rounded; lip cordate, callus V-shaped, prolonged into central keel, 9–14 × 5–9 mm; column suffused with purple, 5 mm; clinandrium erose, short; anther obovate, 2-cleft; ovary 15–20 mm. **Capsules** ovoid, body 18–24 × 10–12 mm, tapered beak 5–6 mm.

Flowering May–Jun. Epiphytic on main trunks of larger trees in Fahkahatchee Swamp; 0–30 m; Fla.; Mexico; West Indies (Cuba); Central America.

Epidendrum acunae, described from Cuba, was discovered in Florida in 1962, as a single population, and it has apparently become extirpated because of overcollecting (R. Sauleda, pers. comm.). Only one specimen from Florida has been found (*Sauleda 942*, FTG), of which a photograph of the live flower is at AMO, apart from the record in C. A. Luer (1972). The description has been prepared from Luer and from Cuban and Mexican material.

2. Epidendrum amphistomum A. Richard in R. de la Sagra, Hist. Fis. Cuba 11: 240. 1853 [F]

Plants cespitose, to 110 cm. **Roots** basal, 1.5–4 mm diam. **Stems** unbranched, straight, compressed, to 60 cm. **Leaves** 5–13; petiole 20–36 mm; blade elliptic, obtuse, middle ones somewhat larger than basal or apical, 4–18 × 1–4 cm, leathery. **Inflorescences** racemes, subcapitate, densely flowered, 10–

50 cm; 1st raceme terminal, producing additional racemes from nodes of long peduncle over years; peduncle elongate, laterally compressed, 9–48 cm. **Flowers** to 25 per raceme, not resupinate, spirally oriented, yellow-green to red-brown; sepals obovate-elliptic, 5.5–7.5 × 3–4 mm, fleshy, apex obtuse; petals narrowly oblanceolate, thin, 5–7 × 0.5–1 mm, apex obtuse; lip cordate, apex 2-lobed, 4–6 × 4.5–7 mm, fleshy, with prominent longitudinal ridge terminating in fleshy mucro; column 5 mm; anther ovoid, apex saddle-shaped; ovary 7–14 mm. **Capsules** subglobose; pedicel 6.5 mm; body 7.5 × 6 mm; beak obsolete.

Flowering Dec–Jul. Widespread and common in forests and hammocks of the Big Cypress Swamp and Everglades, on deciduous trees and palms; 0–30 m; Fla.; West Indies (Cuba).

Epidendrum amphistomum grows in Martin, Lee, Collier, Miami-Dade, and Monroe counties, Florida.

The species is pollinated by male night moths: *Eucereon carolina*, *Lymire edwardsii*, *Cisseps fulvicollis*, and *Oxydia vesulia*. Its reproduction is obligately allogamous. Flowers produce a fragrance reminiscent of overly ripe vegetables, heaviest between late afternoon and dawn (R. M. Adams and G. J. Goss 1976).

This name has been generally regarded as synonymous with *Epidendrum anceps* Jacquin, which L. A. Garay and H. R. Sweet (1974) considered synonymous with *Epidendrum secundum* Jacquin (E. Hágsater 1993).

Outside the flora area are several closely related species, all of them having been considered by recent authors under the name *Epidendrum anceps*.

SELECTED REFERENCES Adams, R. M. and G. J. Goss. 1976. The reproductive biology of the epiphytic orchids of Florida 3. *Epidendrum anceps* Jacquin. Amer. Orchid Soc. Bull. 43: 488–492. Hágsater, E. 1993. *Epidendrum anceps* or *Epidendrum secundum*? Orquidea (Mexico City) 13: 153–158.

3. Epidendrum floridense Hágsater, Icon. Orchid. 2(1): plate 133. 1993 [E] [F]

Amphiglottis difformis (Jacquin) Britton

Plants cespitose, 6–26 cm. **Roots** basal, 1–2 mm diam. **Stems** unbranched, sinuous, terete, 5–23 cm. **Leaves** throughout length of stem; petiole to 25 mm; blade narrowly elliptic, 2-lobed, 2–6.7 × 0.7–2.3 cm, leathery-succulent. **Inflorescences** nearly corymbose; peduncle less than 1 cm; floral bracts triangular-ovate, to 12 mm. **Flowers** 4–14, resupinate, simultaneous, green; sepals wide open, narrowly obovate-elliptic, 5-veined, 12–15 × 3–5 mm, nearly leathery, apex acute to rounded; petals linear-oblanceolate, 3-veined, 11–12 × 2–2.5 mm, membranaceous, apex acute; lip widely cordate, 5–9 × 9–18.5 mm, usually with pair of shallow sinuses in front, fleshy-thickened toward center, apex somewhat notched, 2-callose; column 7–8.5 mm; clinandrium entire, erose, covering anther; anther transversely reniform, with prominent beak in front; ovary 22–28 mm. **Capsules** nearly globose; pedicel 20 mm; body 16–20 × 10–13 mm.

Flowering Aug–Oct; fruiting Dec–Apr. On trees in forests and swamps in and around Everglades and Lake Okeechobee; 0–30 m; Fla.

Epidendrum floridense is pollinated by male night moths: *Anticarsia gemmatilis*, *Phyprosopus callitrichoides*, and *Lymire edwardsii* (G. J. Goss 1977). The flowers produce copious nectar within the nectar tube; in some cases nectar overflows, and a drop is found in front of the callus. The species is threatened because of increasing intensities of winter freezes.

Previous authors have cited this species as *Epidendrum difforme* Jacquin and considered it to be highly variable and widespread throughout the Neotropics. Field and herbarium studies have shown that it is not a single highly variable species but rather a complex of about 50 species (L. M. Sánchez and E. Hágsater 1997). Most of these can be easily recognized by a series of characters including: terete or compressed stems, simultaneous or successive flowering, few or many flowers, shape of the lip and petals, veining, fimbriation of the clinandrium, shape of the nectarium, presence or absence of nectar, distribution, and ecological and elevational preferences. Thus many names previously relegated to synonymy are being recognized, and many new species described. The Florida plants are clearly distinct from the true *Epidendrum difforme*, which, among other characteristics, has stems that are strongly compressed laterally and is endemic to the Windward Antilles.

SELECTED REFERENCES Goss, G. J. 1977. The reproductive biology of the epiphytic orchids of Florida 5. *Epidendrum difforme* Jacquin. Amer. Orchid Soc. Bull. 46: 630–636. Sánchez, L. M. and E. Hágsater. 1997. Taxonomic study of the *Epidendrum difforme* group (Orchidaceae). In: C. E. de B. Pereira, ed. 1997. Proceedings of the 15th World Orchid Conference: Rio de Janeiro 1996. Turriers. Pp. 235–244.

4. Epidendrum magnoliae Muhlenberg, Cat. Pl. Amer. Sept., 81. 1813 • Green-fly orchid

Amphiglottis conopsea (R. Brown) Small; *Epidendrum conopseum* R. Brown; *Larnandra conopsea* (R. Brown) Rafinesque; *L. magnoliae* (Muhlenberg) Rafinesque

Plants cespitose, 4.5–38 cm. **Roots** basal, 3–5 mm diam. **Stems** unbranched, straight, terete, 2–5 cm. **Leaves** 2–3, near apex of stem;

petiole to 1.5 cm; blade narrowly elliptic, 1–9.5 × 0.4–1.3 cm, nearly leathery, apex acute. **Inflorescences** erect, racemose, 3–26 cm; peduncle 2–10 cm; bracts triangular, to 1/2 length of ovary, 12–18 mm, apex acuminate. **Flowers** 6–14, spread along apical 1/2 of inflorescence, resupinate, simultaneous, pale green to bronze-colored; sepals wide open, narrowly obovate, 5-veined, 6–11 × 2–3 mm, margins revolute, apex obtuse; petals narrowly oblanceolate, 1-veined, 6–11 × 1–2 mm, apex rounded; lip base cordate, 3-lobed, middle lobe triangular, apex rounded to slightly notched, 2-callose, with low midrib, 4–6 × 5–7 mm, lateral lobes semiorbiculate; column 11–12 mm; clinandrium hood erose, covering anther; anther ovate, with low keel along front; ovary 12–18 mm. **Capsules** ellipsoid; pedicel 7–11 mm; body 14–23; beak 5 mm.

Flowering Jun–Jan, sporadically in fall; fruiting Oct–Jan. Evergreen and deciduous woodlands; 0–100 m; Ala., Fla., Ga., La., Miss., N.C., S.C.; Mexico (Nuevo León, San Luis Potosí, Tamaulipas).

The flowers of *Epidendrum magnoliae* produce a sweet-oily fragrance, especially at night.

5. Epidendrum nocturnum Jacquin, Enum. Syst. Pl., 29. 1760 • Night-smelling epidendrum

Amphiglottis nocturna (Jacquin) Britton; *Auliza nocturna* (Jacquin) Small; *Epidendrum carolinianum* Lamarck; *Nyctosma nocturna* (Jacquin) Rafinesque

Plant cespitose, to 100 cm. **Roots** basal, 1–2.5 mm diam. **Stems** unbranched, straight, terete proximally, compressed distally, to 90 cm. **Leaves** 4–10, evenly distributed on distal 1/2 of stem; petiole to 46 mm; blade elliptic, 2-lobed, 6–15 × 0.5–3 cm, leathery. **Inflorescences** nearly corymbose; peduncle zigzag, branching with time, greater than 5 mm, producing flowers over several years. **Flowers** resupinate, produced in succession, 1–2 at a time; sepals yellowish, narrowly linear-lanceolate, to 60 × 6 mm, apex acute; petals yellowish, narrowly linear-lanceolate, to 58 × 2 mm; lip white, deeply 3-lobed, middle lobe narrowly linear-acuminate, overall to 24 × 40 mm, lateral lobes obliquely semiovate; callus yellow; column 18 mm; clinandrium hood surpassing anther, erose; anther ovoid; ovary 45–50 mm. **Capsules** ellipsoid; pedicel 7–14 mm; body 25–32 × 10–15 mm; beak 16 mm.

Flowering Sep–Mar; fruiting Jan–Apr, sporadically throughout year. Epiphytic on trees and palms in forests, hammocks, and swamps; 0–30 m; Fla.; Mexico; West Indies; Central America; South America.

Epidendrum nocturnum is widespread and common throughout the tropical regions of the Western Hemisphere. Outside the flora area, the size of stems, leaves, and flowers is somewhat variable, and plants or whole populations often have cleistogamous flowers. A number of good species occur in the Neotropics. The length of the ovary and the pedicel of the capsule often are good specific characters.

Flowers are very fragrant at night, with an almost pungent scent.

6. Epidendrum rigidum Jacquin, Enum. Syst. Pl., 29. 1760 • Rigid epidendrum

Spathiger rigidus (Jacquin) Small

Plants creeping, to 40 cm. **Roots** from nodes along creeping rhizome, 1–2 mm diam. **Stems** 2 cm apart, ascending, compressed, to 20 cm. **Leaves** 2–7, near apex of stem; petiole 15–26 mm, compressed; blade elliptic, 3–8 × 0.8–1.5 cm, leathery, apex obtuse. **Inflorescences** distichous-elongate, usually erect, rarely arching, sinuous, 5–15 cm; peduncle 1–3 cm. **Flowers** 3–10, with underside of lip always facing rachis, not resupinate, somewhat successive, green; sepals ovate, 5 × 2.5 mm, apex acute to obtuse; petals linear, 5 × 1 mm; lip ovate to nearly quadrangular, 3 × 3 mm, apex obtuse to rounded, 2-callose; column 3 mm; clinandrium hood denticulate; anther ovoid, with low keel in front; ovary 10–11 mm, covered by prominent, conduplicate bracts often somewhat longer than ovary. **Capsules** ovoid; pedicel 1–2 mm; body 10–12 × 7–9 mm; beak 3–4 mm.

Flowering Sep–Dec; fruiting spring. Growing on trees in swamps [common in lowlands]; 0–30 m; Fla.; West Indies; Bermuda; South America.

Epidendrum rigidum is the most widespread and common epidendrum in the tropical regions of the Western Hemisphere. In Mexico and Central America it is replaced by a closely related species, *E. cardiophorum* Schlechter.

7. Epidendrum strobiliferum Reichenbach f., Ned. Kruidk. Arch. 4: 333. 1859 • Pine-cone epidendrum

Spathiger strobiliferus (Reichenbach f.) Small

Plants creeping to pendent, to 30 cm. **Roots** basal or from proximal branches, 0.5–1.5 mm. **Stems** 1, compressed distally, branched, to 23 cm. **Leaves** 2–8, near apex of stem; petiole 6–8 mm; blade lanceolate-elliptic, 2-lobed, 1–2.8 × 0.3–0.7 cm, leathery. **Inflorescences** from branches, distichous, imbricate, usually pendent, 10–17 mm; peduncle greater than 5 mm. **Flowers** 1–4, usually 3, with underside of lip always facing rachis, not resupinate, simultaneous, pale yellow, occasionally marked with reddish; dorsal sepals ovate, apex acute; lateral sepals obdolabriform, somewhat wider at base, 3–5.4 × 1.2–2.1 mm; petals linear-elliptic, somewhat falcate, 3–5.2 × 0.5–0.8 mm; lip triangular-ovate, 3–4 × 2–2.4 mm, base with pair of lamellae, apex acute to acuminate; column 2 mm, with prominent tooth on each side, 2 smaller ones distally; anther ovoid, 2-fid; ovary with small swollen vesicle underneath, behind lateral sepals, 4 mm, covered by conduplicate, subimbricating bracts, 6–7.5 mm. **Capsules** ovoid-ellipsoid, 9–10 × 5–6 mm; pedicel and beak absent.

Flowering Oct–Nov, Mar. On smooth barked trees in and around cypress swamps and heads; 0–30 m; Fla.; Central America; South America.

Epidendrum strobiliferum grows throughout the Neotropics to Peru and Brazil.

43. PROSTHECHEA Knowles & Westcott, Fl. Cab. 2: 111. 1838 • [Greek *prostheke*, appendix, in reference to appendage on back of column]

Eric Hágsater

Anacheilium Hoffmannsegg; *Epicladium* Small; *Epithecia* Knowles & Westcott; *Hormidium* Lindley ex Heynhold

Herbs, epiphytic. **Roots** fleshy, glabrous. **Stems** aggregate or spaced, erect, forming flattened or fusiform pseudobulbs. **Leaves** 1–3, apical, sessile, articulate; blade ovate to lanceolate, thinly leathery. **Inflorescences** apical, paniculate racemes, flowers simultaneous or successive, produced only once, pedunculate or sessile. **Flowers** resupinate or not; sepals subequal; petals sometimes much narrower; lip appressed to column, adnate to proximal $1/2$, with callus; pollinia 4, obovoid, laterally compressed, subequal; caudicles 4 in 2 pairs; column not winged, apically 3- or 5-toothed; rostellum semiorbiculate, entire, covered abaxially with viscous substance. **Fruits** capsules, ellipsoid to ovoid, prominently 3-winged, 1-locular.

Species ca. 100 (3 in the flora): neotropical regions.

Prosthechea has been variously treated as *Epidendrum*, *Encyclia*, *Hormidium*, and *Anacheilium*, and variously circumscribed to include the cockle-shell orchids and others. It has been recently reestablished by W. E. Higgins (1997[1998]) to include not only the cockle-shell species but also their less showy Mexican relatives. Recent molecular data (C. van den Berg et al. 2000) support this status.

The Florida species of *Prosthechea* were segregated (J. K. Small 1933) into four genera: *Encyclia*, *Hormidium*, *Anacheilium*, and *Epicladium*. Most later authors recognize only *Epidendrum* or *Epidendrum* and *Encyclia*. The proposed reestablishment of *Anacheilium* and *Hormidium* (G. F. Pabst et al. 1981; R. P. Sauleda et al. 1984), was based mainly on Brazilian species and did not take Mexican species into consideration. The Mexican species have been dealt with as *Encyclia* subg. *Osmophytum* sections *Osmophytum* and *Hormidium* (R. L. Dressler and G. E. Pollard 1976).

Prosthechea is recognized by the presence of a spathe at the base of the inflorescence, lip adnate to proximal half of column, 3-winged capsule, and presence of glycoside crystals. *Encyclia* has no spathe, the column is usually adnate at base of column, the capsule is fusiform, not winged, and no glycoside crystals are present. In *Epidendrum* the lip is adnate to the column throughout, and the rostellum is slit; stems are usually canelike.

SELECTED REFERENCES Higgins, W. E. 1997[1998]. A reconsideration of the genus *Prosthechea* (Orchidaceae). Phytologia 82: 370–383. Pabst, G. F., J. L. Moutinho, and A. V. Pinto. 1981. An attempt to establish the correct statement for genus *Anacheilium* Hoffmgg. and revision of the genus *Hormidium* Lindl. ex Heynh. Bradea 3: 173–186.

1. Pseudobulbs not elongate, suborbicular, strongly flattened . . . 1a. *Prosthechea boothiana* var. *erythronioides*
1. Pseudobulbs mostly elongate, rarely ovoid, somewhat flattened.
   2. Inflorescences elongate, to 45 cm, pedunculate; sepals more than 20 mm . . . . 2a. *Prosthechea cochleata* var. *triandra*
   2. Inflorescences short, 0.5 cm, sessile; sepals less than 8 mm . . . . 3. *Prosthechea pygmaea*

1. Prosthechea boothiana (Lindley) W. E. Higgins, Phytologia 82: 376. 1998 F

Epidendrum boothianum Lindley, Edwards's Bot. Reg. 24: Misc. 5, no. 7. 1838

Varieties 2 (1 in the flora): s Florida, Mexico, West Indies (Bahamas, Cuba), Central America (Belize).

1a. Prosthechea boothiana (Lindley) W. E. Higgins var. **erythronioides** (Small) W. E. Higgins, N. Amer. Native Orchid J. 5: 18. 1999 • Florida dollar-orchid, dollar orchid [F]

Epidendrum erythronioides Small, Fl. S.E. U.S., 328, 1329. 1903; *Encyclia bidentata* (Lindley) Hágsater & Soto Arenas subsp. *erythronioides* (Small) Hágsater; *E. boothiana* (Lindley) Dressler var. *erythronioides* (Small) Luer; *Epicladium boothianum* (Lindley) Small var. *erythronioides* (Small) Acuña

Plants to 30 dm. **Stems:** pseudobulbs aggregate, suborbicular, strongly flattened, 2.5–3.8 × 1–3 cm, smooth, glossy. **Leaves** 1–3, oblanceolate, 6–18 × 1–3.5 cm, apex obtuse to acute. **Inflorescences** racemes, to 25 cm; sheath conduplicate, elongate. **Flowers** 12, resupinate, simultaneous, greenish tan with reddish purple blotches; sepals oblanceolate, 1–1.4 × 3–3.5 mm, apex acute; petals similar, somewhat smaller, 1–1.3 × 1–2 mm; lip white to pale green, occasionally marked with magenta, rhombic to obscurely 3-lobed, 10 × 6–7 mm, lateral margins revolute; callus 3-dentate, middle tooth extended into thickened termination near apex of lip; anthers 3, middle anther with 4 pollinia, laterals with 2 each, yellow; column green at base, white distally, apex 3-toothed with smaller tooth in each sinus, 6–7 mm. **Capsules** 2.5–3 × 1.5–2 cm.

Flowering Jul–Sep; fruiting Nov–Dec or throughout year. In hammocks and thickets, on small trees; 0–30 m; Fla.; West Indies (Bahamas).

The autogamous, 3-anthered *Prosthechea boothiana* var. *erythronioides* is the only variety found in Florida, probably because of the absence of a pollinator. Throughout the rest of the distribution of the species, the 3-anthered plants seldom occur. As most other autogamous tropical orchids in Florida, it has been able to survive there only because of the self-pollinating feature. The 1-anthered variety, var. *boothiana*, occurs in Cuba, Bahamas, Mexico, and Belize.

2. Prosthechea cochleata (Linnaeus) W. E. Higgins, Phytologia 82: 377. 1998

Epidendrum cochleatum Linnaeus, Sp. Pl. ed. 2, 2: 1351. 1763

Varieties 2 (1 in the flora): s Florida, Mexico, West Indies, Central America, n South America.

2a. Prosthechea cochleata (Linnaeus) W. E. Higgins var. **triandra** (Ames) W. E. Higgins, N. Amer. Native Orchid J. 5: 18. 1999 • Florida clamshell orchid, clam-shell orchid [E]

Epidendrum cochleatum Linnaeus var. *triandrum* Ames, Contr. Orch. Fl. S. Florida, 16, plate 8. 1904; *Anacheilium cochleatum* (Linnaeus) Hoffmannsegg var. *triandrum* (Ames) Sauleda, Wunderlin & B. F. Hansen; *Encyclia cochleata* (Linnaeus) Dressler var. *triandra* (Ames) Dressler; *E. cochleata* (Linnaeus) Lemée subsp. *triandra* (Ames) Hágsater; *Epidendrum triandrum* (Ames) House

Plants to 60 cm. **Stems:** pseudobulbs slightly spaced, ovoid to ellipsoid, usually elongate, flattened, 3–15 × 1–3.5 cm. **Leaves** 1–3, oblong-lanceolate to linear-lanceolate, 5–40 × 1–6 cm. **Inflorescences** racemes, rarely paniculate, pedunculate, elongate, loosely few-flowered, to 45 cm, base with 1 or 2 spathes. **Flowers** not resupinate, successive, usually 3–4 open at a time, sometimes with capsules present, green to yellowish green with a few purplish spots near base; sepals and petals similar, hanging, twisted, linear-lanceolate, 25–35 × 3–6 mm; lip deep purple, orbiculate-cordate, concave, 1.2–2.3 × 1.5–3 mm, base whitish marked with conspicuous purple veins, calli 2, yellowish, cushionlike, apex sharply mucronate; anthers 3, middle anther with 4 pollinia, laterals with 2 each, orange; column short, stout, 6–10 mm, apex 5-toothed. **Capsules** 2–4 × 0.8–1.5 cm.

Flowering Oct–Mar; fruiting Nov–May. In hammocks and cypress swamps, on trees and rocks; 0–30 m; Fla.

The autogamous, 3-anthered *Prosthechea cochleata* var. *triandra* is the only variety found in Florida (Broward, Collier, Miami-Dade, Lee, Martin, Monroe counties, and Florida Keys), probably because of the absence of a pollinator. Throughout the rest of the distribution of the species, 3-anthered, abnormal individuals seldom occur, and no populations have been reported. The 1-anthered var. *cochleata* occurs in Mexico, West Indies, Central America, and northern South America.

3. **Prosthechea pygmaea** (Hooker) W. E. Higgins, Phytologia 82: 380. 1998 • Dwarf butterfly orchid, dwarf encyclia, dwarf epidendrum

Epidendrum pygmaeum Hooker, Bot. Mag. 60: plate 3233. 1822; *Aulizeum pygmaeum* (Hooker) Bentham & Hooker f.; *A. pygmaeum* (Hooker) Lindley ex Stein; *Encyclia pygmaea* (Hooker) Dressler; *Hormidium pygmaeum* (Hooker) Bentham & Hooker f. ex Hemsley

Plants creeping, to 12 cm. **Stems:** pseudobulbs spaced 3–4 cm, fusiform, somewhat flattened, slender, 20–40 × 3–8 mm. **Leaves** 2(–3, rarely), elliptic, 25–55 × 9–13 mm. **Inflorescences** racemes, sessile, 0.5 cm. **Flowers** 1–3, resupinate, successive, flowers and fruits may be present with capsules, light yellowish green to green or brownish green; sepals oblanceolate, apex acute; lateral sepals oblique, 5–6 × 2 mm; petals linear, 4 × less than 1 mm, apex acute; lip white with purple blotch on middle lobe, obreniform, middle lobe small, triangular, apex acute, lateral lobes suborbiculate, embracing column, 4 × 2 mm; anthers 1, orange; column white, 2 mm, apex 3-toothed. **Capsules** 1 cm.

Flowering Oct–Feb; fruiting Feb–May. In swamps, infrequent but individual plants form large patches; 0–30 m; Fla.; Mexico; West Indies; Central America; n South America.

In Florida, flowers of some plants of *Prosthechea pygmaea* have been reported to be cleistogamous (C. A. Luer 1972).

230d.5b. ORCHIDACEAE Jussieu (tribe EPIDENDREAE) subtribe PLEUROTHALLIDINAE Lindley, Gen. Sp. Orchid. Pl., 3. 1830

SELECTED REFERENCE Luer, C. A. 1986. Icones Pleurothallidinarum I. Systematics of the Pleurothallidinae. Monogr. Syst. Bot. Missouri Bot. Gard. 15.

44. LEPANTHOPSIS (Cogniaux) Ames, Bot. Mus. Leafl. 1(9): 3. 1933 • [Greek *lepis*, scale, and *anthos*, flower, referring to small, scalelike flowers]

C. A. Luer

Pleurothallis R. Brown sect. *Lepanthopsis* Cogniaux in C. F. P. von Martius et al., Fl. Bras. 3(4): 591. 1896

Herbs, epiphytic. **Roots** fibrous. **Stems** erect, slender; sheaths enclosing stems, ± imbricate, ribbed, tubular, dilated at mouth, ± ciliate or scabrous on ribs and margin of mouth. **Leaves** 1, erect, subsessile; blade elliptic, leathery. **Inflorescences** terminal, 1–several-clustered racemes, distichous. **Flowers** opening simultaneously; sepals unequal; lateral sepals variously connate, membranous, glabrous [pubescent]; lip usually simple [3-lobed], not callous, base inflexibly attached to obsolete column foot; column short, broad, wingless, column foot rudimentary, apex hooded; anther apical; pollinia 2, free or adnate to common viscidium, obovoid; ovary articulate with minute, curved pedicel; stigma apical, 2-lobed; rostellum apical. **Fruits** capsules, deciduous at pedicel (ca. 1 mm), ovoid, 3-valved.

Species ca. 35 (1 in the flora): tropics, Mexico, West Indies (Greater Antilles), Central America, South America.

SELECTED REFERENCE Luer, C. A. 1991. Systematics of the genus *Lepanthopsis* (Orchidaceae). Monogr. Syst. Bot. Missouri Bot. Gard. 39: 1–78.

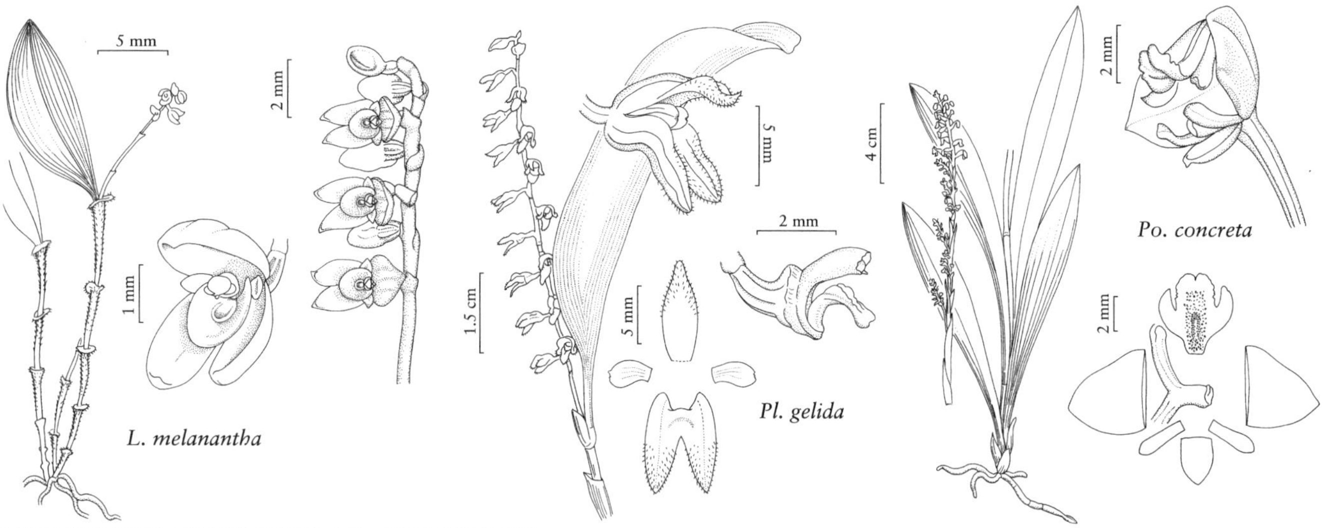

LEPANTHOPSIS ∘ PLEUROTHALLIS ∘ POLYSTACHYA

1. Lepanthopsis melanantha (Reichenbach f.) Ames, Bot. Mus. Leafl. 1: 19. 1933 [F]

Pleurothallis melanantha Reichenbach f., Flora 48: 275. 1865; *Lepanthes brevipetala* Fawcett & Rendle; *L. harrisii* Fawcett & Rendle

Plants cespitose, 4–7 cm. **Roots** slender. **Stems** 1.5–5 cm; sheaths 4–8, closely appressed, ciliate. **Leaves:** petiole 2–4 mm; blade 12–25 × 6–13 mm, base cuneate, apex subacute to obtuse. **Inflorescences** from near apex of stem, few- to several-flowered racemes, ± congested, to 10 mm; spathe absent; peduncle filiform, 15–30 mm; bracts 0.75–1 mm; pedicels 0.5–1 mm. **Flowers** dark red-purple, glabrous; sepals slightly keeled; dorsal sepal free, broadly ovate, concave, 1.25–1.5 × 1.25–1.5 mm, apex obtuse; lateral sepals connate proximal to middle, oblong, 1.5–1.8 × 0.8–1 mm, apex obtuse; petals oblong-obovate, 0.5–0.6 × 0.2–0.4 mm, apex obtuse; lip 1 × 1 mm, base broadly cordate, apex obtuse to rounded, lobes broadly rounded, embracing column; disc slightly concave with distinct, smooth, shallow depression immediately distal to base; column 0.4 × 0.5 mm; ovary 0.5 mm.

Flowering sporadically throughout year. Moist forests; 0–60 m; Fla.; West Indies (Greater Antilles).

Lepanthopsis melanantha is recognized by its small stature and shoots with ciliate sheaths with dilated mouths; a relatively short, distichous raceme of few or several, simultaneous, dark red-purple flowers, usually surpassing the leaf; and a broadly ovate, concave, dorsal sepal, with oblong lateral sepals connate proximal to the middle, minute petals, and a broadly cordate, obtuse lip with a smooth, shallow depression at the base. It is known in the flora area only from the Fahkahatchee Swamp in Collier County, Florida.

45. PLEUROTHALLIS R. Brown in W. Aiton and W. T. Aiton, Hortus Kew. 5: 211. 1813 • [Greek *pleuron*, rib, and thallos, branch, referring to cespitose, slender, aerial shoots]

C. A. Luer

Herbs epiphytic. **Roots** fibrous. **Stems** erect, stout; sheaths tubular, variously enclosing stems. **Leaves** 1, erect, sessile or petiolate; blade leathery. **Inflorescences** terminal [axillary], 1–several lax or clustered racemes [solitary flower]. **Flowers** sometimes distichous, resupinate, opening simultaneously; sepals distinct [variously connate], subequal, membranous [thickly fleshy], pubescent [smooth or verrucose]; petals glabrous [verrucose, papillose, pubescent, ciliate, or fringed]; lip simple [2–5-lobed], membranous [thickly fleshy], callous, base loosely articulate to base of column or apex of column foot [sometimes inflexibly adnate]; column semiterete, stout [slender], winged [wingless], with or without column foot, apex with anther exposed or hooded; anther apical or adaxial; ovary articulate between ovary and minute curved pedicel; pollinia 2, pyriform with viscidium [spheric without viscidium] and caudicles [naked, with or without viscidium]; stigma adaxial [apical], 1–2-lobed; rostellum apical or adaxial. **Fruits** capsules, deciduous at pedicel (ca. 1 mm), ellipsoid, 3-valved.

Species ca. 2500 (1 in the flora): tropics, s Florida, s Mexico, West Indies, Central America, South America.

Pleurothallis is the largest genus in the subtribe Pleurothallidinae, and perhaps in the Orchidaceae.

1. Pleurothallis gelida Lindley, Edwards's Bot. Reg. 27: Misc. 91. 1841 [F]

Plants cespitose, 10–45 cm. **Roots** coarse. **Stems** 10–25 cm; sheaths 2–3, loose, inflated. **Leaves:** petiole 1–4 cm; blade narrowly elliptic to oblanceolate, 10–20 × 2.5–10 cm, base cuneate, apex obtuse. **Inflorescences** apical, many-flowered racemes, simultaneous, 10–25 cm; spathe 1–2 cm; bracts 2–3 mm; pedicels 2–4 mm. **Flowers:** sepals distinct, light yellow, white-pubescent, apex obtuse; dorsal sepal oblong to obovate, concave, 5–8 × 2.5–3 mm; lateral sepals connate near base to form cup, oblong, ovate, oblique, 5–7.5 × 1.5–2.5 mm, apex obtuse; petals translucent white, 3-veined, oblong to obovate, 3–3.5 × 1.75–2 mm, apex obtuse, erose or notched; lip greenish white, 3-veined, oblong to nearly pandurate, curved, 2–2.5 × 0.75–2.25 mm, apex rounded to truncate; disc with pair of lamellae near middle; column 2.5 mm, hooded; ovary 2 mm.

Flowering winter and spring. Moist forests; 0–60 m; Fla.; West Indies (Greater Antilles); Central America; South America.

Pleurothallis gelida is recognized by a loosely sheathed, stout stem with a solitary, obtuse leaf; several long racemes of small, yellowish flowers that are pubescent within; and fused lateral sepals. It is uncommon in the Fahkahatchee Swamp in Florida.

230d.5c. ORCHIDACEAE Jussieu (tribe EPIDENDREAE) subtribe POLYSTACHYINAE Schlechter, Orchideen, 292. 1914

46. POLYSTACHYA Hooker, Exot. Fl. 2: plate 103. 1824 • [Greek *poly*, many, and *stachys*, spike, alluding to many branchlets of the inflorescence in some species]

James D. Ackerman

Herbs epiphytic, lithophytic, or terrestrial. **Roots** velamentous. **Stems** slender or pseudobulbous, clustered or distant along rhizome. **Leaves** 1–several, conduplicate, thin [fleshy], leathery. **Inflorescences** terminal, simple or paniculate racemes with ± secund branches; peduncle enclosed by scarious sheaths. **Flowers** few to many, mostly not resupinate, generally small; dorsal sepal free; lateral sepals adnate to column foot forming mentum; petals smaller than sepals, free; lip 3-lobed [simple], adnate to and articulate with column foot, forming part of mentum; disc usually callose, pubescent; column semiterete, short; foot distinct; stigmas confluent, transverse under rostellum, viscidium ovate to elliptic; anther terminal, incumbent, operculate; pollinia 4, globose or ellipsoid, hard, waxy, attached to single short stipe. **Fruits** capsules.

Species 150 (1 in the flora): tropical and subtropical regions, especially rich in Africa; s Florida, Mexico, West Indies, Central America, South America, Asia, Australia, Pacific Islands.

1. Polystachya concreta (Jacquin) Garay & H. R. Sweet in R. A. Howard, Fl. Less. Antill. 1: 178. 1974 • Yellow spike orchid [F]

Epidendrum concretum Jacquin, Enum. Syst. Pl., 30. 1760; *E. minutum* Aublet; *Polystachya extinctoria* Reichenbach f.; *P. minuta* (Aublet) Britton

Plants cespitose. **Roots** numerous, 2–3 mm diam. **Stems** thickened at base, hidden by leaf sheaths, composed of several internodes, to 4 cm. **Leaves** 2–4; blade thin, narrowly oblong to elliptic, 3.5–20 × 1.1–3 cm, apex mostly acute. **Inflorescences** usually paniculate racemes, simple racemes in small plants; peduncle compressed; floral bracts spreading to reflexed, 1–2 mm. **Flowers** yellow or pale green with white lip; dorsal sepal ovate-elliptic, 2.5–4 × 1.5–2.5 mm; lateral sepals triangular, 2.5–5 mm; petals oblanceolate, rounded, 2–3 × 1 mm; lip 3-lobed, 3–5 mm, middle lobe quadrangular to oblong, lateral lobes falcate; disc with linear, densely farinose callus (pseudopollen); column cylindric, short; column foot 2–3 mm; pedicellate ovary to 7 mm. **Capsules** brown at maturity, ellipsoid, 6–12 mm.

Flowering Aug–early Dec. Cypress swamps, hammocks, mangroves; 0–20 m; Fla.; pantropical (see discussion below).

Most plants of *Polystachya concreta* flower from August to December; some flower at other times of the year. In Florida, *Polystachya concreta* [*P. flavescens* (Blume) J. J. Smith] is pollinated by the halictid bee, *Dialictus* aff. *creberrimus* (G. J. Goss 1977b). The bee collects the detachable, starchy lip hairs (pseudopollen).

An extensive list of synonomy is available (G. C. K. Dunsterville and L. A. Garay [1959]–1976, vol. 3), but after discussions with Phillip Cribb (pers. comm.), we suspect that the African and Asian synonyms represent different species (see also P. J. Cribb 1984). For this reason, we have excluded them from the list, even though some of them are often used for neotropical plants. Furthermore, *Polystachya concreta* is not as widespread in the Neotropics as formerly believed; it is absent from Mexico and Central America (J. T. Atwood, G. Carnevali F.-C., R. L. Dressler, pers. comm.).

Cranichis luteola Swartz and *Polystachya luteola* (Swartz) Hooker are not legitimate names.

230d.6. ORCHIDACEAE Jussieu (subfam. EPIDENDROIDEAE) tribe DENDROBIEAE Endlicher, Gen. Pl. 3: 190. 1837

47. BULBOPHYLLUM Thouars, Hist. Orchid., table 3. 1822 • [Greek *bolbos*, bulb, and *phyllon*, leaf, referring to its leafy pseudobulb]

James D. Ackerman

Herbs, epiphytic or lithophytic, sympodial. **Roots** velamentous, glabrous. **Stems** rhizomatous, covered by scarious sheaths; pseudobulbs often angled, ovoid, stout. **Leaves** conduplicate, articulate, thin to leathery. **Inflorescences** lateral from base of pseudobulb, spikes [racemes], erect to pendent, pedunculate, rachis swollen, fleshy. **Flowers** resupinate, sessile; dorsal sepal distinct and free; lateral sepals adnate to column foot forming mentum; petals distinct and free, smaller than sepals; lip fleshy; column erect with aristate terminal teeth or wings; anther terminal, incumbent, operculate; pollinia 2 or 4, hard, waxy; stigma entire; ovary sessile [pedicellate]. **Fruits** capsules, beakless. $x = 19, 20$.

Species 1000–1200 (1 in the flora): pantropical, mostly Asia and Africa.

SELECTED REFERENCES Vermeulen, J. J. 1987. A taxonomic revision of the continental African Bulbophyllinae. Orchid Monogr. 2. Vermeulen, J. J. 1993. A taxonomic revision of *Bulbophyllum*, sections *Adelopetalum*, *Lepanthanthe*, *Macrouris*, *Pelma*, *Peltopus*, and *Uncifera* (Orchidaceae). Orchid Monogr. 7.

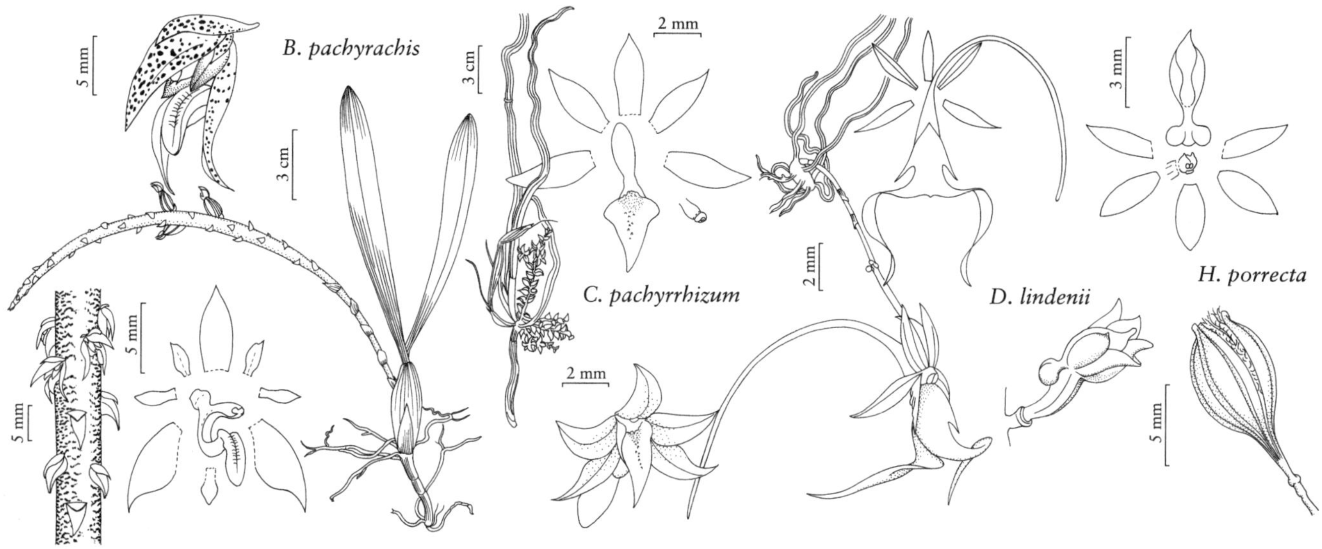

BULBOPHYLLUM ◦ CAMPYLOCENTRUM ◦ DENDROPHYLAX ◦ HARRISELLA

1. **Bulbophyllum pachyrachis** (A. Richard) Grisebach, Fl. Brit. W. I., 613. 1864 • Rat-tail orchid [F]

Pleurothallis pachyrachis A. Richard in R. de la Sagra, Hist. Fis. Cuba 11: 234. 1850 (as pachyrhachis)

Plants creeping. **Roots** numerous, yellowish, often branched, 0.5–1 mm diam. **Stems** stout; rhizomes 8–55 × 2.5–7 mm; pseudobulbs 3–4-angled, 1–5 × 0.7–2.5 cm. **Leaves** 1–3; blade oblong to oblong-oblanceolate, 6–20 × 7–28 mm, leathery, apex acute to obtuse. **Inflorescences:** peduncle partially concealed by ensheathing, scarious bracts, 4–24 cm × 1.5–2.5 mm; spikes 12–28-flowered, 4–25 cm × 2–7 mm; floral bracts triangular-ovate, 2–5.5 mm. **Flowers** yellow-green, spotted red, sessile, glabrous; sepals 3-veined, concave, triangular-ovate or ovate-lanceolate, 4–7 × 1.5–3 mm, fleshy, apex acuminate-caudate; petals oblong-elliptic, 1–2.5 × 0.5–1.5 mm, apex obtuse to rounded; lip adnate to and articulate with column foot, linguiform, 2.5–3 mm, very thick, fleshy, margins entire, verrucose; column stout, 2–2.5 mm, apex 3-dentate, middle tooth acute, 0.5 mm, lateral teeth winglike, oblong, subulate, 1 mm; column foot slender, 1–1.5 mm; pollinia 2; ovary sessile, partially embedded in pits on rachis. **Capsules** cylindric, ribbed, 8–11 mm, glabrous.

Flowering Nov–Dec. On hardwoods in swamps; 5–20 m; Fla.; Mexico; West Indies; Central America; n South America.

The Florida population of *Bulbophyllum pachyrachis* reported by C. A. Luer (1972) no longer exists (E. Christenson, pers. comm.).

230d.7. ORCHIDACEAE Jussieu (subfam. EPIDENDROIDEAE) tribe VANDEAE Lindley, Coll. Bot., App. 1826

48. CAMPYLOCENTRUM Bentham, J. Linn. Soc., Bot. 18: 337. 1881 • [Greek *kampylos*, crooked, and *kentros*, spur, alluding to the floral lip with a long, slender, sharply curved spur]

James D. Ackerman

Todaroa A. Richard & Galeotti, Ann. Sci. Nat., Bot., sér. 3, 3: 28. 1845, not Parlatore 1843

Herbs, epiphytic, monopodial. **Roots** velamentous, glabrous. **Stems** rhizomatous, very reduced [elongate]. **Leaves** early deciduous (“leafless species”), articulate, minute [persistent, distichous, leathery]. **Inflorescences** axillary, spikes [racemes], often secund. **Flowers** more than 12, distichous, resupinate [not resupinate], small to minute; sepals and petals connivent with spreading apices, distinct; lip distinct and free, basally spurred, margins entire or lobed; column very short; clinandrium prominent, truncate; anther terminal, operculate; pollinia 2, globose, waxy, hard, attached to 1 or 2 filiform stipes; stigma entire; rostellum deeply notched; ovary sessile [pedicellate]. **Fruits** capsules.

Species 40–45 (1 in the flora): North America, Mexico, West Indies, Central America, tropical South America.

SELECTED REFERENCE Todzia, C. A. 1980. A revision of the Costa Rican species of *Campylocentrum* (Orchidaceae). Brenesia 18: 117–136.

1. Campylocentrum pachyrrhizum (Reichenbach f.) Rolfe, Orchid Rev. 11: 246. 1903 • Leafless bentspur orchid [F]

Aeranthus pachyrrhizus Reichenbach f., Flora 48: 279. 1865; *A. spathaceus* Grisebach

Plants appressed to substrate, inconspicuous, leafless at anthesis. **Roots** gray-green, fascicled, 2.5–5 mm wide when flattened against substrate. **Stems** hidden by roots, to 5 mm. **Leaves** early deciduous, minute, scarious. **Inflorescences** densely flowered spikes, erect to pendent, stout, to 3.5 cm; floral bracts broadly cordate, 1–3 mm, scarious, margins erose-denticulate. **Flowers** resupinate, funnelform; sepals brownish orange, elliptic to elliptic-lanceolate, 4 × 1.5 mm, apex acute; petals creamy white with pink flush, otherwise similar to sepals, 3.5 × 1.5 mm; lip same color as petals, 3-lobed, 2–3 × 2 mm, middle lobe triangular-acuminate, lateral lobes triangular-rounded; spur saccate, 2.5 mm; column less than 0.5 mm; pollinia greenish; ovary sessile, 1 mm. **Capsules** brownish orange, ellipsoid, 6–9 × 3–4 mm, pubescent.

Flowering Sep–Oct. Cypress swamps and wet hammocks; 0–20 m; Fla.; West Indies; n South America.

49. DENDROPHYLAX Reichenbach f. in W. G. Walpers and K. Müller berol., Ann. Bot. Syst. 6: 903. 1864 • [Greek *dendro*, tree, and *phylax*, epiphyte or guardian, in reference to the epiphytic habit]

James D. Ackerman

Polyradicion Garay; *Polyrrhiza* Pfitzer

Herbs epiphytic, monopodial. **Roots** velamentous, fleshy, glabrous. **Stems** very short. **Leaves** early deciduous, conduplicate, articulate, sessile, minute. **Inflorescences** axillary, 1–10-flowered racemes; peduncle slender. **Flowers** resupinate, large; sepals and petals similar, spreading, distinct and free, lanceolate; lip free, 3-lobed, small to very large, lateral lobes relatively short, middle lobe entire or deeply lobed, basal spur elongate; column very short, without foot; clinandrium prominent, truncate; anthers terminal, incumbent, operculate; pollinia 2, globose, waxy, hard, attached to single or separate stipes; stigma entire; rostellum deeply notched; ovary pedicellate. **Fruits** capsules, narrowly cylindric, ribbed.

Species 10 (1 in the flora): s Florida, West Indies.

The Florida representative of *Dendrophylax* has been variously placed in *Polyrrhiza* Pfitzer (D. S. Correll 1950; C. A. Luer 1972) or *Polyradicion* Garay (L. A. Garay 1969). The type of *Polyrrhiza* is the Jamaican *P. funalis* (Swartz) Pfitzer, which is regarded as a *Dendrophylax* (W. Fawcett and A. E. Rendle 1910–1936, vol. 1; C. D. Adams 1972). Garay proposed *Polyradicion* to accommodate *P. lindenii* and another species. Recent research by M. Whitten (pers. comm.) on the molecular systematics of neotropical angraecoid taxa has shown that *Polyradicion* is phylogenetically embedded in *Dendrophylax*. Consequently, both *Polyrrhiza* and *Polyradicion* are best treated as synonyms of *Dendrophylax*.

1. **Dendrophylax lindenii** (Lindley) Bentham ex Rolfe, Gard. Chron., ser. 3, 4: 533. 1888 • Ghost orchid, frog orchid, palm-polly, Linden's angurek, white butterfly orchid [F]

Angraecum lindenii Lindley, Gard. Chron. 1846: 135. 1846; *Aeranthus lindenii* (Lindley) Reichenbach f.; *Polyradicion lindenii* (Lindley) Garay; *Polyrrhiza lindenii* (Lindley) Cogniaux

Plants appressed to substrate, seemingly leafless. **Roots** gray-green, fascicled, to 50 cm × 3–5 mm. **Stems** inconspicuous, obscured by roots. **Leaves** bractlike, inconspicuous, often absent. **Inflorescences:** peduncles arcuate-ascending, 6–25 cm; bracts tubular, sheathing, 5–7 mm; racemes 1–10-flowered; floral bracts lanceolate, 4 × 3 mm, scarious. **Flowers** opening in succession, usually 1–2 at a time, nocturnally fragrant; sepals and petals spreading, white to creamy green or pale yellow-brown; dorsal sepal lanceolate to elliptic-lanceolate, 20–21 × 5–6 mm, apex acute; lateral sepals slightly oblique, lanceolate to narrowly lanceolate, 25–30 × 4–5.5 mm, apex obtuse to acute; petals falcate-recurved, narrowly lanceolate, 22–27 × 4–4.5 mm, apex acute to acuminate; lip white, basally concave, 3-lobed, middle lobe with subquadrangular isthmus, terminating in 2 arching, twisting, caudate lobes 45–70 mm, apiculate, lateral lobes spreading, 18–25 mm wide when spread, apex obtuse to acute; callus with narrow, denticulate keel; spur arcuate, slender, 11–17 cm; column stout, with 2 triangular wings projecting toward lip, 2–3 × 5 mm diam.; pollinia orange; pedicellate ovary slender, 30 × 1.5 mm. **Capsules** 6–10 × 0.5 cm.

Flowering May–Aug, occasionally as early as Feb. Epiphytic in cypress swamps, wet hammocks; 0–20 m; Fla.; West Indies.

50. HARRISELLA Fawcett & Rendle, J. Bot. 47: 265. 1909 • [For William H. Harris, 1860–1920, F.L.S., British botanist and prolific collector of Jamaican plants]

James D. Ackerman

Herbs epiphytic, monopodial, miniature, glabrous. **Roots** velamentous, slender. **Stems** very short, inconspicuous. **Leaves** absent or scalelike. **Inflorescences** axillary, often sparsely branched. **Flowers** fewer than 12, not resupinate, minute, globose; sepals and petals similar, distinct and free; lip simple, with basal, saccate spur; column short; anthers terminal, incumbent, operculate; pollinaria 2, each with 1 hard, waxy pollinium attached to stipe; viscidia 2, elongate; ovary subsessile. **Fruits** capsules, dehiscent, 3 broad valves alternating with 3 linear ribs, all 6 segments separating from distal end and reflexing.

Species 1: subtropical and tropical regions, North America, Mexico, West Indies, Central America.

Harrisella merits careful systematic study. Although L. O. Williams (1965) asserted that the genus has no characteristics by which it can be distinguished from *Campylocentrum*, among the Caribbean species *Harrisella* is readily distinguished by several key characteristics: capsules that dehisce from the distal end and separate into 6 parts; relatively long-pedicellate ovaries; and slender, sometimes branched, laxly flowered inflorescences. In addition, *Harrisella* may be distinguished from *Campylocentrum* by having two pollinaria instead of one, each possessing only one pollinium. It seems likely that the genus may be more related to the Caribbean *Dendrophylax*.

1. Harrisella porrecta (Reichenbach f.) Fawcett & Rendle, J. Bot. 47: 266. 1909 [F]

Aeranthus porrectus Reichenbach f., Flora 48: 279. 1865; *Campylocentrum porrectum* (Reichenbach f.) Rolfe; *Harrisella amesiana* Cogniaux; *H. uniflora* H. Dietrich

Plants leafless, inconspicuous. **Roots** numerous, silvery gray-green, 0.25–1 mm diam., covering stems. **Stems** erect with a few reddish brown, scarious scales, to 15 mm. **Inflorescences** few to many, lax racemes or panicles arising through mass of roots, 2–5 cm; floral bracts less than 1 mm. **Flowers** pale yellow; dorsal sepal elliptic, obovate, 2–2.5 × 1.3 mm, apex obtuse to rounded; lateral sepals obliquely elliptic-ovate, 2–2.5 × 1–1.2 mm, apex obtuse to rounded; petals ovate, obtuse, clawed, slightly smaller than sepals; lip broadly ovate, concave, 2.7–4 mm, base with short, apically expanded, scrotiform spur 1 mm wide, margins involute, apex obtuse to rounded; disc with erect tubercle; column stout, 0.25 mm; ovary subsessile, 2 mm. **Capsules** reddish brown when mature, ovoid, 4.5–6 mm, sparsely verruculose.

Flowering Aug–Sep. On twigs of shrubs and small trees, swamps, mesic hardwood hammocks; 0–20 m; Fla.; Mexico; West Indies; Central America (El Salvador).

Four species have been recognized in *Harrisella*: *H. porrecta*, *H. filiformis* (Swartz) Cogniaux, *H. monteverdi* (Reichenbach f.) Cogniaux, and *H. uniflora* Dietrich (H. Dietrich 1982). *Harrisella porrecta* is sometimes confused with *H. filiformis* (e.g., N. L. Britton et al. [1923–] 1930, vol. [1], part 2); a comparison of the types clearly shows that they are distinct. *Harrisella filiformis* and *H. monteverdi* are synonymous and belong to the related genus *Campylocentrum* (J. D. Ackerman 1995). The Cuban *H. uniflora* appears to be a depauperate *H. porrecta*. Nonetheless, the genus may not be monotypic. Mexican populations of *H. porrecta* do have some distinctive features and may represent at least one additional species (N. M. Warford 1997).

230d.8. ORCHIDACEAE Jussieu (subfam. EPIDENDROIDEAE) tribe CALYPSOEAE Dressler, Selbyana 5: 204. 1979

51. CALYPSO Salisbury, Parad. Lond., plate 89. 1807, name conserved • Calypso, fairy-slipper [Greek *kalypso*, a figure in Homer's *Odyssey*]

Charles J. Sheviak

Paul M. Catling

Herbs, perennial, rather succulent. **Roots** few, slender, fleshy. **Stems** scapose; corm slender to stout, fleshy; sheathing bracts usually 2, partially cloaking stem. **Leaves** produced in autumn, withering spring, solitary, arising from corm, plicate, leathery. **Inflorescences** terminal, each with 1 flower, arising from corm; floral bract with color of stem or sepals, reduced to prominent. **Flowers** solitary, resupinate, horizontal to slightly nodding, large, showy; sepals and petals ascending to erect; lip slipper-shaped, with basal orifice and 2 horns near apex, margin of lip dilated, forming bearded, apronlike lamina; pollinarium solitary; pollinia 4, flattened, superposed in 2 pairs flanking axis of pollinarium; viscidia large, quadrangular; stigma concave. **Fruits** capsules.

Species 1: circumboreal.

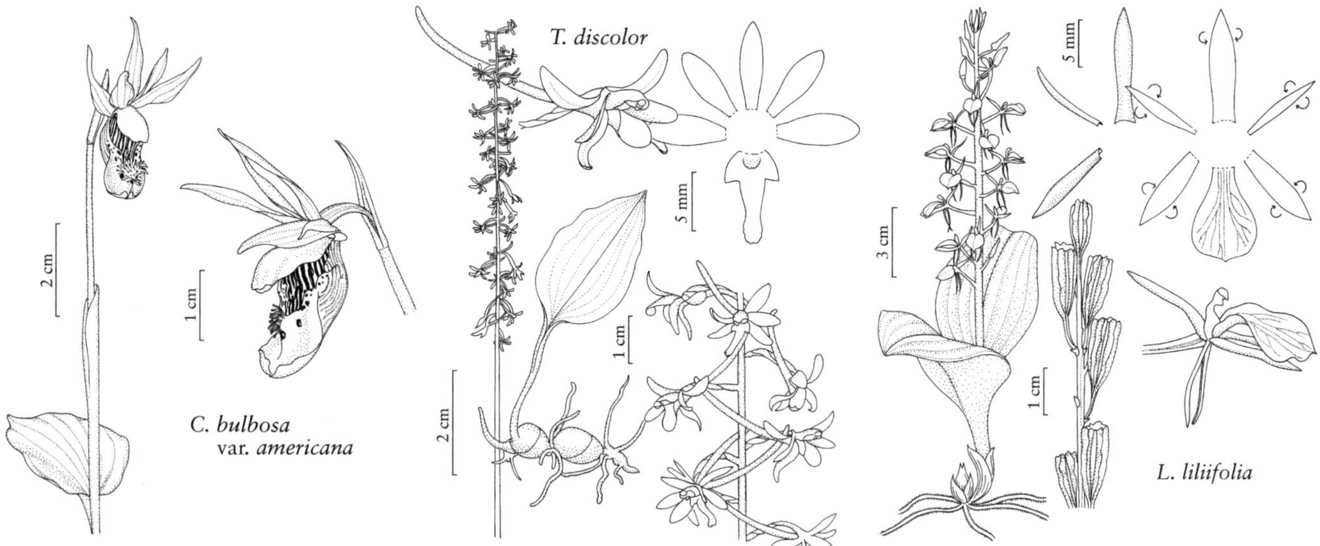

CALYPSO ◦ TIPULARIA ◦ LIPARIS

1. Calypso bulbosa (Linnaeus) Oakes in Z. Thompson, Nat. Hist. Vermont 1: 200. 1842 [F]

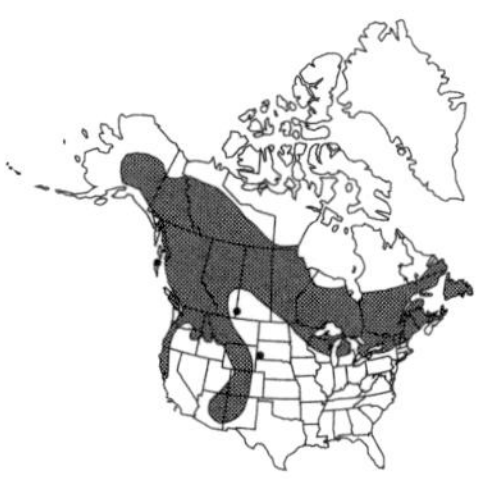

Cypripedium bulbosum Linnaeus, Sp. Pl. 2: 951. 1753; *Cytherea bulbosa* (Linnaeus) House

Plants 4.5–22 cm. **Leaves:** blade elliptic to suborbiculate or ovate, often cordate, 10–65 × 12–52 mm. **Inflorescences:** floral bracts lance-acuminate, 5–28 mm. **Flowers** pink, magenta, or rarely white; sepals and petals erect-spreading, linear-lanceolate to linear-oblong, 10–24 × 1.5–5 mm; lip 13–23 × 4–13 mm, lamina shorter than to longer than apical horns, basal bristles sparse to extensive, brightly and contrastingly colored to dull and obscure. **Capsules** erect, ellipsoid to lanceoloid, 2–3 × 1–1.5 cm. $2n = 28$ [var. *bulbosa*].

Varieties 3 (2 in the flora): circumpolar, North America, Europe, Asia (China, Japan, Russia).

1. Bristles of lip yellow, extensive; lamina white or suffused with pink, usually equaling or longer than horns . . . . 1a. *Calypso bulbosa* var. *americana*
1. Bristles of lip whitish or color of lamina, sparse; lamina commonly with reddish markings on dull white, usually shorter than horns . . . . 1b. *Calypso bulbosa* var. *occidentalis*

1a. Calypso bulbosa (Linnaeus) Oakes var. **americana** (R. Brown) Luer, Native Orchids U.S. & Canada, 336. 1975 [E] [F]

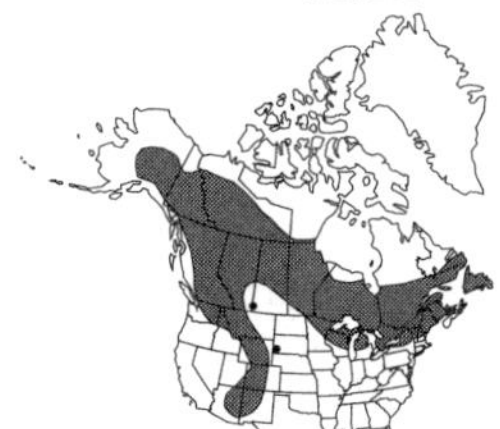

Calypso americana R. Brown in W. Aiton and W. T. Aiton, Hortus Kew. 5: 208. 1813

Flowers with prominent tuft of contrasting yellow bristles at base of lamina; lamina white or suffused with pink, equaling or longer than conspicuous horns. $2n = 32$.

Flowering spring–early summer. Mesic to wet coniferous forests, mixed forests, and bogs; 0–3100 m; Alta., B.C., Man., N.B., Nfld. and Labr., N.W.T., N.S., Nunavut, Ont., P.E.I., Que., Sask., Yukon; Alaska, Ariz., Colo., Idaho, Maine, Mich., Minn., Mont., N.H., N.Mex., N.Y., S.Dak., Utah, Vt., Wash., Wis., Wyo.

1b. Calypso bulbosa (Linnaeus) Oakes var. **occidentalis** (Holzinger) B. Boivin, Naturaliste Canad. 94: 522. 1967 [E]

Calypso bulbosa forma *occidentalis* Holzinger, Contr. U.S. Natl. Herb. 3: 251. 1895

Flowers with inconspicuous, sparse bristles at base of lamina, these whitish or with color of lamina; lamina with dull white and reddish markings, usually shorter than horns.

Flowering spring–early summer. Moist to dry, shady coniferous forests; 0–1600 m; B.C.; Alaska, Calif., Idaho, Mont., Oreg., Wash.

52. TIPULARIA Nuttall, Gen. N. Amer. Pl. 2: 195. 1818 • Crane-fly orchid [Latin *tippula*, crane-fly, alluding to the resemblance of the flowers to crane-flies]

Paul M. Catling

Charles J. Sheviak

Herbs, terrestrial. **Roots** 3–10, from bases of newer corms, spongy. **Stems** produced in summer, solitary, subapical, opposite leaf, arising from ovate, angular corms. **Leaves** produced in autumn, persisting over winter, withering in spring, solitary; blade broad, ovate, pleated. **Inflorescences** lax racemes. **Flowers** resupinate, bilaterally asymmetric; lip with long, free basal spur; column slightly arcuate; anther terminal, deciduous but retained on pollinaria for short time; pollinia 4, in 2 pairs, yellow, attached to elastic caudicle 1 mm with basal viscidium; stigma entire; rostellum a 2-fid flap enclosing viscidium. **Fruits** capsules, pendent, ovoid. **Seeds** 0.5–0.6 mm.

Species 3 (1 in the flora): temperate regions; 1 in e North America, 1 in the Himalayas, 1 in Japan.

SELECTED REFERENCES Catling, P. M. and V. R. Catling. 1990. Anther cap retention in *Tipularia discolor*. Lindleyana 6: 113–116. Snow, A. A. and D. F. Whigham. 1989. Costs of flower and fruit production in *Tipularia discolor* (Orchidaceae). Ecology 70: 1286–1293. Stoutamire, W. P. 1978. Pollination of *Tipularia discolor*, an orchid with modified symmetry. Amer. Orchid Soc. Bull. 47: 413–415. Whigham, D. F. and M. McWethy. 1980. Studies on the pollination ecology of *Tipularia discolor* (Orchidaceae). Amer. J. Bot. 67: 550–555.

1. Tipularia discolor (Pursh) Nuttall, Gen. N. Amer. Pl. 2: 195. 1818 [E] [F]

Orchis discolor Pursh, Fl. Amer. Sept. 2: 586. 1814; *Tipularia unifolia* (Muhlenberg) Britton, Sterns & Poggenberg

Plants 10–65 cm. **Corms** 7–30 (–50) mm diam. **Stems** scapose, glabrous, basally 1–2-sheathed. **Leaves** persisting over winter; petioles 3–15 cm; blade purple abaxially, green or greenish purple adaxially, ovate, 5–10.5 × 2.5–7 cm. **Inflorescences** 8–28 cm; floral bracts 0.2 mm. **Flowers** green, pale greenish yellow, or greenish purple, (5–)10–55; sepals distinct and free, oblong-elliptic to oblanceolate, 5–8 × 1.5–2.8 mm; petals linear-oblong to oblong-elliptic to linear-oblanceolate, 4–7 × 1–1.8 mm; lip 5–8 × 2.5–3 mm, with 2 prominent basal lobes, spur 10–23 mm; column 2.5–4 mm. **Capsules** 9–12 × 4–5 mm.

Flowering Jun (north)–Sep (south). In humus-rich soil of deciduous woodlands; frequently in sandy, acid oak-pine woods of Southeast, often in depressions under sweet gum; 0–1000 m; Ala., Ark., Del., Fla., Ga., Ill., Ind., Ky., La., Md., Mass., Mich., Miss., Mo., N.J., N.Y., N.C., Ohio, Okla., Pa., S.C., Tenn., Tex., Va., W.Va.

Tipularia discolor is pollinated by noctuid moths, the pollinaria attaching to either the left or right compound eye depending on whether the column of a particular flower is slightly twisted to the left or to the right (W. P. Stoutamire 1978).

230d.9. ORCHIDACEAE Jussieu (subfam. EPIDENDROIDEAE) tribe MALAXIDEAE Lindley, Coll. Bot., App. 1826

53. LIPARIS Richard, De Orchid. Eur., 21, 30, 38. 1817, name conserved • [Greek *liparos*, fat, greasy, or shining, referring to the almost oily feel and luster of the leaves typical of plants in this genus]

Lawrence K. Magrath

Herbs, perennial, terrestrial [lithophytic or epiphytic]. **Roots** slender, fibrous; pseudobulbs green, conic, ovoid, or ellipsoid, slightly compressed, soft (2 pseudobulbs typically present, parent pseudobulb enveloped in old leaf-bases, connected laterally by short rhizome to daughter pseudobulb, which gives rise to current year's stem, parent pseudobulb often retaining dried stem of previous year). **Stems** somewhat angled, slender, glabrous; cauline bracts 2(–3), enclosing base of stem. **Leaves** 2 or 3–7, basal, sheathing pseudobulb and proximal part of stem, conduplicate or plicate, glabrous. **Inflorescences** terminal, 3–40-flowered racemes, lax; peduncle and rachis glabrous; floral bracts subulate to triangular-lanceolate. **Flowers** resupinate, maroon-purple, yellowish green, or green, pedicellate; dorsal sepal oblong-elliptic, linear-oblong, oblong-lanceolate, or lanceolate, margins revolute, apex obtuse to acute; lateral sepals oblong-lanceolate, lanceolate, ovate-oblong, or elliptic, falcate or not, margins revolute or involute, apex obtuse to acute; petals filiform, linear, linear-oblanceolate, or narrowly spatulate, tubular or strongly revolute; lip strongly arcuate-recurved, obcordate, oblong-flabellate, or suborbiculate, base broadly cuneate, slightly auriculate, apex emarginate or subtruncate, mucronate or not, apical margins erose-serrulate to crenulate-wavy; column arcuate, slender to stout, weakly to strongly winged at apex; anther apical, anther cap yellow or green; pollinia 2 pairs, yellow, waxy. **Fruits** capsules, erect, ellipsoid to ovoid, obscurely to prominently winged, glabrous.

Species 250 (3 in the flora): nearly worldwide.

1. Leaves 3–7, membranaceous, plicate; anthers bright green; pedicels stout . . . . 3. *Liparis nervosa*
1. Leaves 2, succulent, conduplicate; anthers yellow; pedicels slender.
    2. Lip 8–12 mm, mauve purplish, rarely green; column 3–4 mm; pedicels of fruiting capsules 11–18 mm . . . . 1. *Liparis liliifolia*
    2. Lip 4–5.5 mm, green or yellowish green to yellowish white; column 2–3 mm; pedicels of fruiting capsules 3–7 mm . . . . 2. *Liparis loeselii*

1. Liparis liliifolia (Linnaeus) Richard ex Lindley, Bot. Reg. 11: plate 882. 1825 • Lily-leaved twayblade, large twayblade, mauve sleekwort [E] [F]

Ophrys liliifolia Linnaeus, Sp. Pl. 2: 946. 1753 (as lilifolia); *Leptorchis liliifolia* (Linnaeus) Kuntze; *Malaxis liliifolia* (Linnaeus) Swartz; *Ophrys trifolia* Walter

Plants 9–25 cm. **Pseudobulbs** ovoid, 20 ×10 mm, sheathed by bracts and persistent leaf bases, previous year's pseudobulb usually present, connected by short rhizome. **Stems** bright green, with or without purplish brown tinge, angled, sometimes obscurely winged or fluted distally. **Leaves** 2; blade conduplicate, dark green, glossy, ovate-elliptic, keeled abaxially, 4–18 × 2–8.5 cm, succulent, apex obtuse to acute. **Inflorescences** 4–15 cm; floral bracts subulate, 2 × 1 mm, apex acute; pedicels slender, 5–7 mm. **Flowers** 5–31, mauve-purple and green; dorsal sepal oblong-lanceolate to narrowly lanceolate, 8–11.5 × 1.2–2 mm, apex obtuse to acute; lateral sepals oblong-lanceolate to narrowly lanceolate, 8–11.5 × 1.2–2 mm, apex obtuse to acute; petals pendent, curved, tubular, narrowly linear to filiform, 8.5–12 × 0.2–0.3 mm, margins strongly revolute; lip translucent, mauve to pale purple or rarely green with prominent purplish veining, cuneate-obovate to suborbiculate, 8–12 × 6–10 mm, base slightly auriculate, apical margin erose-serrulate, apex subtruncate, mucronate; column winged apically, 3–4 × 1–1.5 mm, with 2 blunt tubercles on inner surface near base; anthers yellow; pollinia yellow. **Capsules:** pedicel 11–18 mm; body ellipsoid, 15 × 5 mm, veins often slightly winged.

Flowering May–Jul (north). Mature mesic to moist deciduous forests, pine woods, rich moist humus, often colonizing previously open, disturbed habitats during early and middle stages of reforestation; 10–1500 m; Ont.; Ala., Ark., Conn., Del., Ga., Ill., Ind., Iowa, Ky., Md., Mass., Minn., Miss., Mo., N.H., N.J., N.Y., N.C., Ohio, Okla., Pa., R.I., S.C., Tenn., Vt., Va., W.Va., Wis.

Liparis liliifolia has what appear to be extremely similar relatives in Japan [*L. japonica* (Miquel) Maximowicz and *L. makinoana* Schlechter] and China (*L. pauliana* Handel-Mazzetti and *L. cathcartii* Hooker f.). Further work is needed to clarify the relationships among these five species. It could be a case of either divergent evolution from a common ancestor or convergent evolution from two or more ancestors.

2. Liparis loeselii (Linnaeus) Richard, De Orchid. Eur., 38. 1817 • Loesel's twayblade, fen orchid

Ophrys loeselii Linnaeus, Sp. Pl. 2: 947. 1753; *Leptorchis loeselii* (Linnaeus) MacMillan; *Liparis correana* (Barton) Sprengel; *Malaxis correana* W. Barton; *M. longifolia* W. Barton

Plants 6–26 cm. **Pseudobulbs** ovoid, 10 × 5 mm, sheathed by bracts and persistent leaf bases, previous year's pseudobulb usually present, connected by short rhizome. **Stems** pale green or yellowish green, angled, sometimes obscurely winged or fluted distally. **Leaves** 2; blade conduplicate, green, glossy, oblong-elliptic to elliptic-lanceolate, keeled abaxially, 3.7–18 × 1–4 cm, succulent, apex obtuse to subacute. **Inflorescences** 2–10 cm; floral bracts subulate, minute, 2 × 1 mm, apex acute; pedicels slender, 3–5 mm. **Flowers** 2–15, green or yellowish green to yellowish white or greenish white; dorsal sepal oblong-lanceolate to narrowly lanceolate, 4.5–6 × 1–2 mm, apex obtuse to acute; lateral sepals oblong-lanceolate to narrowly lanceolate, 4.5–5.5 × 1–2.2 mm, apex obtuse to acute; petals pendent, slightly curved, tubular, filiform, 4.5–5.5 × 0.5–1 mm, margins strongly revolute; lip translucent to opaque, green or yellowish green to yellowish white, obovate to oblong to suborbiculate, 4–5.5 × 2.2–3.5 mm, base narrowly cuneate, slightly auriculate, margins crenulate-wavy, apex subtruncate to obtuse, mucronate; disc with central elongate thickening; column short, stout, 2–3 × 0.5–1 mm, winged apically; anthers yellow; pollinia yellow. **Capsules:** pedicel 3–7 mm; body obovate to ellipsoid, 9–13 × 3–6 mm, veins often slightly winged. $2n = 32, 26$.

Flowering May–Aug. Cool, moist ravines, bogs, or fens, wet peaty or sandy meadows, and exposed sand along edges of lakes, often colonizing previously open and disturbed habitats during early and middle stages of reforestation; 100–1100 m; B.C., Man., N.B., N.W.T., N.S., Ont., P.E.I., Que., Sask.; Ala., Ark., Conn., Del., Ill., Ind., Iowa, Kans., Ky., Maine, Md., Mass., Mich., Minn., Miss., Mo., Mont., Nebr., N.H., N.J., N.Y., N.C., Ohio, Pa., R.I., Tenn., Vt., Va., Wash., W.Va., Wis.; w, c Europe.

Liparis loeselii is such a small orchid that it can be easily overlooked; it may have a wider distribution than has been reported. Rain-assisted self-pollination in this species was described by P. M. Catling (1980).

3. Liparis nervosa (Thunberg ex Murray) Lindley, Gen. Sp. Orchid. Pl., 26. 1830 • Tall liparis

Ophrys nervosa Thunberg ex Murray, Syst. Veg. ed. 14, 814. 1784; *Leptorchis elata* (Lindley) Kuntze; *Liparis bituberculata* Reichenbach f.; *L. elata* Lindley; *L. elata* var. *latifolia* Ridley; *Sturmia bituberculata* Reichenbach f.

Plants 12–60 cm. **Pseudobulbs** conic, slightly compressed, 5–7 × 2–3 cm, sheathed by bracts and persistent leaf bases, previous year's pseudobulb usually present, connected by short rhizome, producing numerous fibrous roots. **Stems** pale green or green, sometimes suffused with purple-maroon, angled, prominently winged distally. **Leaves** 3–7, sheathing stem and pseudobulb, reduced distally to minute bracts; blade green, glossy, plicate, ovate or elliptic to oblong-lanceolate, 6–30 × 5.5–12 cm, membranaceous, apex broadly rounded to abruptly acuminate. **Inflorescences** 3–26 cm; floral bracts subulate to triangular-lanceolate, 6–12 × 3 mm, apex acuminate; pedicels stout, 3–5 mm. **Flowers** 10–40, greenish purple; dorsal sepal oblong-elliptic to linear-oblong, 5–8 × 1.8–4 mm, margins strongly revolute, apex broadly rounded to obtuse; lateral sepals falcate, ovate-oblong to elliptic, 4–7 × 2–3.5 mm, apex obtuse to subacute; petals falcate, linear-oblanceolate to narrowly spatulate, nearly tubular, 5–7.5 × 1–2 mm, margins strongly revolute; lip strongly arcuate-recurved, obcordate to broadly cuneate or oblong-flabellate, base obsoletely auriculate, apex emarginate, sometimes mucronate; callus with 2 fleshy tubercles on basal portion, maroon-purple, 4–5.5 ×3.5–5.3 mm; column stout, 3.5–5 × 1–1.5 mm, winged apically; anthers bright green; pollinia yellow. **Capsules** obovoid to ellipsoid, 12–15 × 5–6 mm.

Flowering Jul–Sep. Cypress swamps on decaying logs and stumps, in rich humus of hammock or in wet muck, also dense hardwood forests; 0–1500 m; Fla.; Mexico; West Indies; Central America; South America; Asia (Formosa, India, Japan, Thailand); tropical Africa; Pacific Islands (Philippines).

Liparis nervosa, a pantropical orchid, is probably the most widespread orchid in the world. *Liparis nervosa* has over fifty different published synonyms from tropical areas.

54. MALAXIS Solander ex Swartz, Prodr., 8, 119. 1788 • Adder's-mouth [Greek *malaxis*, softening, in reference to soft and tender texture of leaves]

Paul M. Catling

Lawrence K. Magrath

Herbs, terrestrial to semiepiphytic, glabrous. **Roots** few, fibrous, 0.3–1 mm wide. **Stems** swollen at base into pseudobulb, ± globose, glabrous. **Leaves** 1–3(–5), with sheathing base; blade elliptic, ovate, or lanceolate. **Inflorescences** terminal, racemes, spicate racemes, corymbose racemes, or subumbellate racemes; floral bracts inconspicuous, lanceolate, subulate, or triangular-acuminate. **Flowers** 2–160, resupinate or not, erect or spreading, sessile or minutely to strongly pedicellate; sepals spreading, distinct or lateral sepals basally connate, ovate, elliptic, or lanceolate, 1–6 mm, margins revolute or not; petals spreading or recurved, filiform to linear, lanceolate, or triangular, usually much narrower than sepals; lip ovate or lanceolate, cordate, unlobed to 3-lobed, concave or saccate, widest proximal to middle, base auriculate or truncate; column free; anther terminal; pollinaria 4, waxy; 1 pollinarium or 2 separate hemipollinaria; viscidia yellow or orange. **Fruits** capsules; previous year's fruiting stem and capsules frequently present during current year's anthesis. x = 14, 15, 18, ca. 20, ca. 21, 22.

Species ca. 250 (10 in the flora): widespread, mostly in Asia and East Indies.

SELECTED REFERENCES Ames, O. and C. Schweinfurth. 1935. Nomenclatural studies in *Malaxis* and *Spiranthes*. Bot. Mus. Leafl. 3: 113–133. Catling, P. M. 1991. Systematics of *Malaxis bayardii* and *M. unifolia*. Lindleyana 6: 3–23.

1. Leaves 2–3(–5).
    2. Lip 2.5–2.9(–4) mm, base prominently cordate-auriculate . . . . 1. *Malaxis spicata*
    2. Lip 1.2–2.5 mm, base truncate or cuneate.
        3. Leaves basal; base of lip cuneate; flowers not resupinate . . . . 2. *Malaxis paludosa*
        3. Leaves cauline; base of lip truncate, 3-lobed, lateral lobes auriculate; flowers resupinate . . . . 3b. *Malaxis monophyllos* var. *brachypoda*
1. Leaves 1(–2, rarely).
    4. Lip 3-dentate at apex, middle tooth smallest, sometimes so inconspicuous that apex appears 2-dentate.
        5. Flowers apparently sessile, pedicels 1.3–1.7 mm; flowers not resupinate . . . . 4. *Malaxis soulei*
        5. Flowers clearly pedicellate, pedicels 3.4–10(–13) mm; flowers resupinate.
            6. Lip with auricles 0.6 or more times as long as distance from base of lip to apex of middle lobe . . . . 5. *Malaxis bayardii*
            6. Lip with auricles less than 0.6 times as long as distance from base of lip to apex of middle lobe . . . . 6. *Malaxis unifolia*
    4. Lip unlobed (with or without single point) at apex.
        7. Lip base truncate with prominent, forward-directed auricles; flowers not resupinate . . . . 3a. *Malaxis monophyllos* var. *monophyllos*
        7. Lip base cordate or hastate-auriculate; flowers resupinate.
            8. Inflorescences 1.5–3 cm; pedicels crowded . . . . 7. *Malaxis corymbosa*
            8. Inflorescence 2.6–25 cm; pedicels not crowded.
                9. Sepals 4–6 mm . . . . 9. *Malaxis abieticola*
                9. Sepals 1.8–2.4 mm.
                    10. Sepals glabrous, not papillose; apex of lip broadly acuminate, auricles at base narrow and parallel . . . . 10. *Malaxis porphyrea*
                    10. Sepals papillose, not glabrous; apex of lip narrowly acuminate, auricles at base broad and diverging . . . . 8. *Malaxis wendtii*

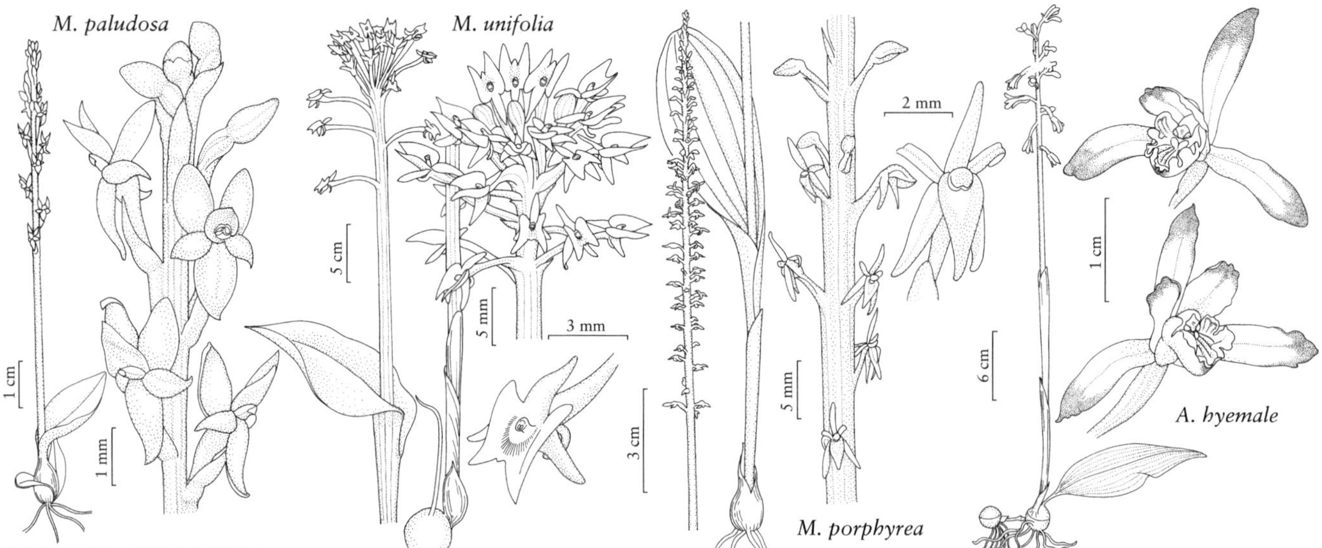

MALAXIS ◦ APLECTRUM

1. **Malaxis spicata** Swartz, Prodr., 119. 1788 • Florida adder's-mouth

Achroanthes floridana (Chapman) Greene; *Malaxis floridana* (Chapman) Kuntze; *Microstylis spicata* (Swartz) Lindley

Plants (7–)10–29(–45) cm. **Pseudobulbs** 5–20 mm diam. **Leaves** 2–3, within proximal 1/3 of stem; blade ovate, suborbiculate, or elliptic, 2.2–9(–10) × 1.4–5 (–5.5) cm. **Inflorescences** racemes, 1.6–13(–24) cm; floral bracts triangular, 1–3 mm; pedicels 5–10 mm. **Flowers** 6–150, not resupinate, green with lip also green or pale yellow, brown, orange, or vermillion; dorsal sepal reflexed, ovate, 3–3.5 × 1.5–2 mm, apex acuminate; lateral sepals reflexed, elliptic or lanceolate, 2.5–3 × 1–1.5 mm, apex acuminate; petals strongly reflexed, linear, crossing behind flower, 1.5–2.5 × 0.2–0.3 mm, apex rounded; lip ovate, 2.5–2.9(–4) × 2–3.3 mm, base prominently cordate-auriculate, nearly surrounding column, apex unlobed, acute; column 0.6–1 × 0.6–1 mm; pollinia yellow. **Capsules** erect, ovoid, 8 × 4 mm.

Flowering winter–early fall. Hammocks and swamps, sometimes semiepiphytic on tree bases; 0–100 m; Fla., Ga., N.C., S.C., Va.; n West Indies.

To some extent the lip color is dependent upon flower age. Plants bloom in Florida from August to February; in Virginia, in June, July, and August; and in Pennsylvania in September. The extent to which the different flowering times are associated with groups that differ in other respects has yet to be determined, and the nature of variation in this species deserves more study.

2. **Malaxis paludosa** (Linnaeus) Swartz, Kongl. Vetensk. Acad. Nya Handl. 21: 235. 1800 • Bog adder's-mouth [F]

Ophrys paludosa Linnaeus, Sp. Pl. 2: 947. 1753; *Epipactis paludosa* (Linnaeus) F. W. Schmidt; *Hammarbya paludosa* (Linnaeus) Kuntz; *Sturmia paludosa* (Linnaeus) Reichenbach

Plants 3–23 cm. **Roots** fine. **Pseudobulbs** 4–8 mm diam. **Leaves** 2–3, basal; blade elliptic or narrowly elliptic, 0.3–3.5 × 0.1–1.5 cm. **Inflorescences** spicate racemes, 0.5–9 cm; floral bracts lanceolate, 1.4–3.3 mm; pedicels 2–3 mm. **Flowers** 2–55, not resupinate, green or yellowish green; dorsal sepal ovate-lanceolate, (1.6–)2–2.5 × 1–1.6 mm, apex acuminate; lateral sepals reflexed, elliptic, falcate, (1.6–)2.5 × 1–1.6 mm, apex acuminate; petals strongly recurved, ovate-lanceolate, 1.4–1.9 × 0.5–1 mm, apex acuminate; lip ovate with dark green veins, 1.2–1.8 × 0.7– 1 mm, base cuneate, apex unlobed, acuminate; column (0.3–)0.5–0.7 × (0.3–)0.5–0.7 mm; pollinia yellow. **Capsules** ascending, ellipsoid, 4 × 2 mm. $2n = 28$.

Flowering summer. Open sphagnum bogs, swampy woods; 10–300 m; Alta., B.C., Man., N.W.T., Ont., Sask. Yukon; Alaska, Minn.; Eurasia.

Malaxis paludosa sometimes reproduces asexually by production of gemmae on the leaf tips (R. L. Taylor 1967).

SELECTED REFERENCES Baldwin, W. K. W. 1961. *Malaxis paludosa* (L.) Sw. in the Hudson Bay lowlands. Canad. Field-Naturalist 75: 74–77. Taylor, R. L. 1967. The foliar embryos of *Malaxias paludosa*. Canad. J. Bot. 45: 1553–1556.

3. Malaxis monophyllos (Linnaeus) Swartz, Kongl. Vetensk. Acad. Nya Handl. 21: 234. 1800

Ophrys monophyllos Linnaeus, Sp. Pl. 2: 947. 1753

Plants 3–25(–30) cm. **Pseudobulbs** 5–10 mm diam. **Leaves** within proximal 1/3 of stem; blade narrowly to broadly elliptic. **Inflorescences** spicate racemes, 1–12 cm. **Flowers** 5–80, resupinate or not; sepals lanceolate to ovate-lanceolate to oblong-lanceolate, margins revolute or not, apex acuminate; petals reflexed, filiform to linear or linear-lanceolate, apex rounded; lip ovate to broadly triangular, concave, base truncate, with forward-directed lobes, apex unlobed, acuminate; column 0.4–0.8 × 0.4–0.8 mm; pollinia yellow. **Capsules** ascending, ellipsoid, 5 × 3 mm.

Varieties 2 (2 in the flora): n North America, Eurasia.

1. Flowers not resupinate . . . . 3a. *Malaxis monophyllos* var. *monophyllos*
1. Flowers resupinate . . . . 3b. *Malaxis monophyllos* var. *brachypoda*

3a. Malaxis monophyllos (Linnaeus) Swartz var. **monophyllos** • Eurasian white adder's-mouth

Malaxis diphyllos Chamisso; *M. monophyllos* (Linnaeus) Swartz var. *diphyllos* (Chamisso) Luer

Plants 3–25 cm. **Pseudobulbs** 4–10 mm diam. **Leaves** 1–2(–3), subopposite, yellow-green, oblong-elliptic, base sheathing stem, apex acute; blade 2–10 × 0.5–4.1 cm. **Inflorescences:** floral bracts lanceolate, 1–4 mm; pedicels 2.5–4.5 mm. **Flowers** not resupinate, green, yellowish green, or yellow; dorsal sepal slightly reflexed, ovate-lanceolate, 2.5–2.8 × 0.9–1.4 mm, margins revolute, apex acuminate; lateral sepals slightly reflexed, lanceolate, 1.5–2.5 × 0.9–1.4 mm, apex acuminate; petals strongly reflexed, filiform to narrowly linear, 1.8–2.2 × 0.3–0.4 mm, apex rounded; lip ovate, concave, 3-lobed, middle lobe ovate, 1.5–2.5 × 1–2 mm, apex acuminate, lateral lobes basal, curved up on either side of middle lobe, narrow, short; column greenish yellow to yellow, 0.5–0.8 × 0.5–0.8 mm. $2n = 30$ (Eurasia).

Flowering summer. Moist or dry slopes or banks, shaded or open, bogs and conifer woods; 0–400 m; B.C. (Queen Charlotte Islands); Alaska; Eurasia.

SELECTED REFERENCE Beese, W. J. 1983. White malaxis, *Malaxis monophyllos* var. *diphyllos*, an addition to the orchids of Canada from the Queen Charlotte Islands, British Columbia. Canad. Field-Naturalist 97: 215–216.

3b. Malaxis monophyllos (Linnaeus) Swartz var. **brachypoda** (A. Gray) F. Morris & E. A. Eames, Our Wild Orchids, 358. 1929 • North American white adder's-mouth, malaxis à pédicelles courts [E]

Microstylis brachypoda A. Gray, Ann. Lyceum Nat. Hist. New York 3: 228. 1835; *Malaxis brachypoda* (A. Gray) Fernald

Plants 3–30 cm. **Pseudobulbs** 4–8 mm diam. **Leaves** 1(–2, rarely), petiolate base sheathing stem; blade light green, ovate-elliptic, keeled abaxially, 1.5–9.5 × 1–5 cm, apex acute. **Inflorescences:** floral bracts lanceolate, 1.5–2 mm; pedicels 2–4.5 mm. **Flowers** resupinate, green or greenish white; dorsal sepal ovate-lanceolate, 1.5–2.5 × 1–1.4 mm, margins revolute, apex acuminate; lateral sepals oblong-lanceolate, slightly falcate, 1.5–2.5 × 0.5–1.2 mm, apex acuminate; petals strongly reflexed, filiform to narrowly linear-lanceolate, 1.4–2.5 × 0.3–0.4 (–0.5) mm, apex rounded; lip broadly triangular, concave, 3-lobed, middle lobe ovate, apex acuminate, lateral lobes auriculate, thickened, curved upward; disc with 2 low, thickened, elongate calli, 1.3–2.2 × 1.2–1.8(–2) mm; column 0.4–0.6 × 0.4–0.6 mm. $2n = 28$.

Flowering summer. Swamps, bogs; 0–400 m; Alta., B.C., Man., N.B., Nfld. and Labr., N.S., Ont., P.E.I., Que., Sask., Yukon; Alaska, Calif., Colo., Ill., Ind., Maine, Mass., Minn., N.H., N.Y., Pa., Vt., Wis.

4. Malaxis soulei L. O. Williams, Ann. Missouri Bot. Gard. 21: 343. 1934 • Mountain malaxis

Microstylis montana Rothrock, Rep. U.S. Geogr. Surv., Wheeler, 264. 1879; *Achroanthes montana* (Rothrock) Greene; *Malaxis montana* (Rothrock) Kuntze 1891, not Blume 1826

Plants 11–60 cm. **Pseudobulbs** 5–15 mm diam. **Leaves** 1, at middle or proximal 1/3 of stem, sheathing leaf bases persistent, shredding; blade ovate-elliptic or oblanceolate, keeled abaxially, 2.4–15 × 1–5.2(–6) cm, apex acute. **Inflorescences** spicate racemes, rachis thick, slightly ridged or fluted, 6–22 cm; floral bracts triangular to lanceolate, 0.5–1 mm; pedicels 1.3–1.7 mm. **Flowers** 40–160, not resupinate, green to yellowish green; dorsal sepal ovate to elliptic, 1.5–2.8 × 0.8–1.3 mm, apex acuminate; lateral sepals ovate to elliptic, falcate, 1.5–2.8 × 0.8–1.3 mm, apex acuminate; petals strongly reflexed, filiform to linear, falcate, 1.3–2 × 0.2–0.3 mm, apex acuminate; lip suborbiculate-ovate to subquadrangular-ovate, 1.5–2.8 × 1.3–2.2(–2.5) mm, base hastate-auriculate, apex obliquely 3-dentate or retuse with apiculum in sinus; disc deeply concave, 5-veined; column 0.2–0.5 × 0.2–0.5 mm; pollinia yellow. **Capsules** ascending, ellipsoid, 7 × 3 mm.

Flowering Jul–Sep. Moist, wooded canyons and ravines, rocky open slopes, pine savannas; 2000–3000 m; Ariz., N.Mex., Tex.; Mexico; Central America.

Malaxis soulei grows in the Santa Catalina, Santa Rita, and Chiricahua mountains in Arizona, and in the Davis Mountains in Texas.

The flowers of *Malaxis soulei* are retained on the axis until the capsules are fully developed, unlike *M. corymbosa* and some other species where the flowers fall quickly if not fertilized. The apparently sessile flowers are a very distinctive characteristic of this species. The central apiculum of the lip may be obsolete, and the apex thus retuse.

The name *Malaxis macrostachya* has been applied to *M. soulei* (C. A. Luer 1975). It is a *nomen confusum*, however, and it is not clearly applicable to this species (L. O. Williams 1965).

5. **Malaxis bayardii** Fernald, Rhodora 38: 402, fig. 1; plate 446, figs. 1, 2. 1936 (as bayardi) • Bayard's malaxis [C] [E]

Plants 9.3–25.6 cm. **Pseudobulbs** 5–20 mm diam. **Leaves** 1(–2, rarely), near middle of stem; blade bright green, glossy, narrowly to broadly ovate, keeled abaxially, 1.7–3.8(–7) × 0.7–3.5(–4.3) cm, apex acute. **Inflorescences** racemes, 1–12 cm; floral bracts triangular, 0.9–1.6 mm; pedicels (3.8–)5–10(–13) mm. **Flowers** 10–70, resupinate, green or yellowish green; dorsal sepal oblong-elliptic, 1.4–2 × 0.6–1 mm, apex acuminate; lateral sepals oblong-elliptic, 1.4–2 × 0.6–1 mm, apex acuminate; petals linear, slightly falcate, 1–1.8 × 0.1–0.2 mm; lip rhombic-deltate to cordate-ovate or oblong-elliptic, 1.8–2.5 × 1.9–2.5 mm, base with auricles 0.6 or more times as long as distance from base of lip to apex of middle lobe, apex 3-dentate; column 0.5–0.7 × 0.5–0.7 mm; pollinia yellow. **Capsules** spreading to slightly drooping, subglobose to ovoid, 6 × 3 mm.

Flowering summer. Dry, open woods, shale barrens and sandy pine barrens; of conservation concern; 10–600 m; N.S.; Conn., Mass., N.J., N.Y., N.C., Ohio, Pa., R.I., S.C., Vt., Va., W.Va.

6. **Malaxis unifolia** Michaux, Fl. Bor.-Amer. 2: 157. 1803 • Green adder's-mouth, malaxis unifolié [F]

Malaxis ophioglossoides Muhlenberg ex Willdenow; *M. thlaspiformis* A. Richard & Galeotti; *Microstylis grisebachiana* Fawcett & Rendle

Plants 3–50 cm. **Pseudobulbs** 5–20 mm diam. **Leaves** 1(–2, rarely), near middle of stem; blade bright green, glossy, narrowly to broadly ovate, keeled abaxially, 1.6–10 × 0.5–5 cm, apex acute. **Inflorescences** racemes, appearing umbellate before elongation of rachis, 1–13 cm; floral bracts triangular, 0.16–1.4 mm; pedicels (3.8–)5–10(–13) mm. **Flowers** 10–160, resupinate, green; dorsal sepal oblong-elliptic, 1.1–1.9(–2.2) × 0.5–1 mm, apex acuminate; lateral sepals oblong-elliptic, 1.1–1.9(–2.2) × 0.5–1 mm, apex acuminate; petals strongly recurved, linear to filiform, slightly falcate, 0.8–1.7(–3) × 0.1–0.2(–0.3) mm; lip rhombic-deltate to cordate-ovate or oblong-elliptic, 1.1–2.3 × 1–2.2 mm, base cordate or auriculate, auricles less than 0.6 times as long as distance from base of lip to apex of middle lobe, apex 3-dentate; column 0.4–0.6 × 0.4–0.6 mm; pollinia yellow. **Capsules** spreading to slightly drooping, subglobose to ellipsoid, 8 × 4 mm.

Flowering spring–fall (south–north). Swamps, bogs, sand barrens, heathlands, and dry woods; 0–600 m; Man., N.B., Nfld. and Labr. (Nfld.), N.S., Ont., P.E.I., Que.; Ala., Ark., Conn., Del., D.C., Fla., Ga., Ill., Ind., Iowa, Kans., Ky., La., Maine, Md., Mass., Mich., Minn., Miss., Mo., N.H., N.J., N.Y., N.C., Ohio, Okla., Pa., R.I., S.C., Tenn., Tex., Vt., Va., W.Va., Wis.; Mexico; West Indies; n Central America.

7. **Malaxis corymbosa** (S. Watson) Kuntze, Revis. Gen. Pl. 2: 673. 1891

Microstylis corymbosa S. Watson, Proc. Amer. Acad. Arts 18: 195. 1883; *Achroanthes corymbosa* (S. Watson) Greene

Plants 6–40 cm. **Pseudobulbs** 5–15 mm diam. **Leaves** 1(–2, rarely), near middle of stem; blade green, glossy, oval or broadly elliptic, cordate, 3–9 × (1.2–)3–6.4 cm, apex acute. **Inflorescences** corymbose racemes, 1.5–3 cm; floral bracts triangular, 0.8–1.2(–2.5) mm, apex acute; pedicels crowded. **Flowers** 10–33, resupinate, green; sepals recurved at apex, ovate-elliptic or lanceolate, (2.8–)3.6–4.2 × 1.4–1(–2) mm, apex subobtuse to acute; petals strongly reflexed, crossing behind flower, linear to filiform, (2–)3.5–4 × 0.2–0.6 mm, apex obtuse; lip triangular, 3–4 × 1.5–2.3 mm, base upcurved, cordate or hastate-auriculate, apex entire, narrowly obtuse to somewhat abruptly acuminate; column 0.4–0.6(–1) × 0.4–0.6(–1) mm; pollinia yellow. **Capsules** horizontal to erect, ellipsoid, 12 × 4 mm.

Flowering summer. Shaded mountain canyons; 2000 m; Ariz.; Mexico; n Central America.

Malaxis corymbosa grows in the Santa Rita, Huachuca, and Chiricahua mountains in Arizona. It differs from some of its close relatives in Mexico by having a more flattened and less fleshy lip that is less deeply concave proximal to anthers. Plants have 10–33 flowers open at one time and may produce a much larger number.

The hemipollinaria are apically widely divergent.

8. Malaxis wendtii Salazar, Orquidea (Mexico City) 13: 281, fig. 1. 1993

Plants 16–45 cm. **Pseudobulbs** 5–10 mm diam. **Leaves** 1, within proximal 1/3 of stem; blade orbiculate-ovate or ovate-lanceolate, (3–)4.1–8.5(–10) × 0.2–0.45 (–0.65) cm, apex obtuse to acute. **Inflorescences** racemes, 6–25 cm; rachis slightly angled; floral bracts triangular to lanceolate, 1–2 × 0.5 mm, apex acute; pedicels not crowded, 2.5–5 mm. **Flowers** 16–133, resupinate, deep maroon or greenish maroon; sepals papillose, not glabrous; dorsal sepal lanceolate, 1.8–2.4 × 0.8–1.2 mm, margins revolute, apex acute; lateral sepals lanceolate-elliptic, falcate, 1.8–2.4 × 0.8–1.2 mm, apex acute; petals strongly recurved, linear to filiform, slightly falcate, 1.8–2.2 × 0.2–0.4 mm, apex obtuse; lip linear to linear-lanceolate, 1.8–2.3 × 1.3–1.8 mm, base with auricles broad and diverging, apex narrowly acuminate; column 0.5–0.8 × 0.5–0.8 mm; pollinia yellow. **Capsules** ascending, ellipsoid, 5 × 3 mm.

Flowering summer (Jul–Sep). Open coniferous and mixed forests on dry slopes; 2000 m; Tex.; Mexico.

The confusion among *Malaxis porphyrea*, *M. ehrenbergii*, and *M. wendtii* is discussed under 10. *M. porphyrea*.

9. Malaxis abieticola Salazar & Soto Arenas, Lindleyana 16: 149. 2001 • Slender-flowered malaxis

Malaxis tenuis (S. Watson) Ames 1922, not Reichenbach f. 1861; *Microstylis tenuis* S. Watson

Plants 11–25 cm. **Pseudobulbs** 5–10 mm diam. **Leaves** 1, usually near middle of stem; blade bright green, glossy, elliptic to broadly ovate, keeled abaxially, 3–8 × 1.4–4.4 cm, base cordate to truncate, apex rounded to obtuse. **Inflorescences** racemes, 2.6–8 cm; floral bracts triangular, 0.8–1.2 mm, apex acute; pedicels not crowded, 7–11 mm. **Flowers** 13–63, resupinate, green to greenish yellow; dorsal sepal light geen, lanceolate, 4–6 × 0.9–2 mm, apex acuminate; lateral sepals distinct or sometimes connate to middle, light green edged in dark green, lanceolate, falcate, 4–6 × 0.9–2 mm, apex acuminate; petals filiform to filiform-linear, falcate, 3.5–4.5 × 0.2–0.3 mm, apex rounded; lip triangular-lanceolate, concave abaxially, 3–4.2 × 1.5–2 mm, base slightly cordate, apex long-acuminate to attenuate; disc with 2 broad green, low, longitudinal ridges; column 0.6–0.9 × 0.6–0.9 mm; pollinia orange. **Capsules** subhorizontal, ellipsoid, 12 × 3 mm.

Flowering summer–early fall. In humus of pine and fir forests on dry slopes; 2000 m; Ariz., N.Mex.; Mexico.

10. Malaxis porphyrea (Ridley) Kuntze, Revis. Gen. Pl. 2: 673. 1891 [F]

Microstylis porphyrea Ridley, J. Linn. Soc., Bot. 24: 320. 1888; *M. purpurea* S. Watson

Plants 16–45 cm. **Pseudobulbs** 5–10 mm diam. **Leaves** 1, within proximal 1/3 of stem; blade orbiculate-ovate or ovate-lanceolate, (3–)4.1–8.5(–10) × 0.2–0.45 (–0.65) cm, apex obtuse to acute. **Inflorescences** racemes, 6–25 cm; rachis slightly angled; floral bracts triangular to lanceolate, 1–2 × 0.5 mm, apex acute; pedicels not crowded, 2.5–5 mm. **Flowers** 16–133, resupinate, deep maroon or greenish maroon; sepals glabrous, not papillose; dorsal sepal lanceolate, 1.8–2.4 × 0.8–1.2 mm, margins revolute, apex acute; lateral sepals lanceolate-elliptic, falcate, 1.8–2.4 × 0.8–1.2 mm, apex acute; petals strongly recurved, linear to filiform, slightly falcate, 1.8–2.2 × 0.2–0.4 mm, apex obtuse; lip triangular to triangular-lanceolate, 1.8–2.3 × 1.3–1.8 mm, base with auricles narrow and nearly parallel, apex broadly acuminate; column 0.5–0.8 × 0.5–0.8 mm; pollinia yellow. **Capsules** ascending, ellipsoid, 5 × 3 mm.

Flowering Jul–Sep. Open coniferous and mixed forests on dry slopes; 2000 m; Ariz., N.Mex.; Mexico.

Both *Malaxis porphyrea* and *M. wendtii* have been included incorrectly in *M. ehrenbergii* (D. S. Correll 1950; C. A. Luer 1975). The confusion among the taxa is discussed by T. K. Todsen (1997). *Malaxis ehrenbergii* is a Mexican species with its northernmost limit in southern Hidalgo, about 1200 km to the south of the ranges of the other two taxa, the southernmost ranges of which are in northern Chihuahua and Sonora. The following criteria are used to separate *M. wendtii* from *M. porphyrea*: the sepals of *M. wendtii* are papillose while the sepals of *M. porphyrea* are essentially glabrous, although they have prominent cell structures that may appear to be very small papillae; the lip of *M. wendtii* is linear to linear-lanceolate with a narrowly acuminate apex, and the auricles at the base of the lip are broad and diverging, while the lip of *M. porphyrea* is nearly triangular to triangular-lanceolate with a broadly acuminate apex, and the auricles at the base of the lip are narrow and nearly parallel. Also, *M. wendtii* has a denser inflorescence and the flowers are not quite as intensely colored as in *M. porphyrea* (R. A. Coleman, W. Jennings, pers. comm.).

230d.10. ORCHIDACEAE Jussieu (subfam. EPIDENDROIDEAE) tribe CYMBIDIEAE Pfitzer, Entwurf Anord. Orch., 105. 1887

230d.10a. ORCHIDACEAE Jussieu (tribe CYMBIDIEAE) subtribe CORALLORHIZINAE E. G. Camus, Bergon & A. Camus, Monogr. Orchid., 378. 1908

55. APLECTRUM Nuttall, Gen. N. Amer. Pl. 2: 197. 1818 • Putty-root, Adam-and-Eve, aplectrum d'hiver [Greek *a*, without, and *plectron*, spur]

Charles J. Sheviak

Paul M. Catling

Herbs, perennial. **Roots** in fascicles at base of pseudobulb, slender, fibrous. **Corms** connected by slender rhizomes. **Stems** scapes, cormose. **Leaves** produced in late summer–fall, withering in spring, solitary, borne at distal end of corm, plicate. **Inflorescences** solitary, lax racemes from lateral nodes on corm; floral bracts inconspicuous; cauline bracts sheathing, large. **Flowers** several, resupinate; sepals to 15 mm; lip 3-lobed, with 3 parallel lamellae; column compressed; pollinarium 1; pollinea 4, attached to viscidium by short stipe; stigma deeply concave. **Fruits** capsules, pendent, straight, 7 mm.

Species 2 (1 in the flora): e North America, e Asia (Japan).

1. Aplectrum hyemale (Muhlenburg ex Willdenow) Nuttall, Gen. N. Amer. Pl. 2: 198. 1818 [E] [F]

Cymbidium hyemale Muhlenburg ex Willdenow, Sp. Pl. 4(1): 107. 1805

Plants 18–50 cm. **Leaves** persistent, petiolate; blade narrowly to broadly elliptic, prominently and irregularly plicate, 10–20 × 3–8 cm, abaxially usually maroon or greenish purple, adaxially dark green with whitish veins. **Inflorescences:** floral bracts inconspicuous, 3–7 × 1.5 mm, apex subulate to acuminate. **Flowers** prominent but inconspicuously colored; sepals spreading, greenish or yellowish suffused with magenta or purple-brown, oblong-oblanceolate to narrowly elliptic, 10–15 × 1.8–4 mm; petals parallel to column, greenish or yellowish, suffused with magenta or purple-brown, oblong-oblanceolate, 9–13 × 1.8–3.5 mm; lip white with magenta spots, obovate in outline, 9–12 × 7–9 mm, apex 3-lobed, lateral lobes ascending and flanking column, middle lobe broadly expanded, crenate; column 7 × 2 mm. **Capsules** ellipsoid, strongly ribbed, 15–30 × 7–12 mm.

Flowering late spring. Moist, deciduous, upland to swampy forests; 0–1200 m; Ont., Que.; Ala., Ark., Conn., Del., D.C., Ga., Ill., Ind., Iowa, Kans., Ky., Md., Mass., Mich., Minn., Mo., N.J., N.Y., N.C., Ohio, Okla., Pa., Tenn., Vt., Va., W.Va., Wis.

56. CORALLORHIZA Gagnebin, Acta Helv. Phys.-Math. 2: 61. 1755, orthography conserved • Coral-root, corallorhize [Greek *korallion*, coral, and *rhiza*, root, referring to coral-like appearance of branching, underground rhizome]

Lawrence K. Magrath

John V. Freudenstein

Cladorhiza Rafinesque; *Corallorhiza* Châtelain; *Rhizocorallon* Gagnebin

Herbs, perennial, solitary or colonial, mycotrophic, rootless, leafless. **Stems** erect, yellowish green or brown to purple, glabrous; sheathing bracts few, reduced. **Rhizomes** branching, coralloid; scales minute. **Inflorescences** laxly to densely flowered racemes, subtended by 3–4 tubular bracts reduced to closed sheaths, glabrous. **Flowers** 2–41, inconspicuous to showy; perianth spreading to closed; sepals and petals lanceolate to oblanceolate, apices acute to nearly obtuse; lip broadly obovate, with 2 basal lamellae, often with 2 lateral lobes, narrowed to claw; column curved toward lip, with or without basal auricles; anther incumbent, operculate; pollinia 4, superimposed, yellow, unequal; rostellum triangular; stipe narrowly triangular; ovary narrowly ellipsoid-obovoid, often with small mentum adaxially at distal end. **Fruits** capsules, pendent, ellipsoid to slightly obovoid, 3-ribbed. **Seeds** cylindric-fusiform. $x = 20, 21$.

Species 11 (7 in the flora): mostly temperate regions, North America, Central America, 1 circumboreal.

Sporadically occurring color forms are known in all *Corallorhiza* species in the flora, and they range from plants with unspotted petals to completely yellow plants, apparently devoid of anthocyanin. Various genera of fungi are known to be associated with *Corallorhiza.*

SELECTED REFERENCE Freudenstein, J. V. 1997. A monograph of *Corallorhiza* (Orchidaceae). Harvard Pap. Bot. 1(10): 5–51.

1. Lip with upturned margins giving boat-shaped appearance, especially toward apex; mentum absent; lamellae connate for at least part of length.
    2. Callus at base of lip less than 1/3 length of lip; flowers chasmogamous . . . . 1. *Corallorhiza striata*
    2. Callus at base of lip greater than 1/3 length of lip; flowers cleistogamous . . . . 2. *Corallorhiza bentleyi*
1. Lip without upturned margins, not giving boat-shaped appearance toward apex; mentum present (although sometimes inconspicuous); lamellae distinct.
    3. Lip unlobed; lateral sepals upcurved.
        4. Dorsal sepal less than 4.5 mm, 1-veined; flowering Aug–Oct . . . . 3. *Corallorhiza odontorhiza*
        4. Dorsal sepal greater than 4.5 mm, 3-veined; flowering Jan–Jul . . . . 4. *Corallorhiza wisteriana*
    3. Lip with 2 small lateral lobes or teeth; lateral sepals spreading or downcurved.
        5. Sepals 1-veined . . . . 5. *Corallorhiza trifida*
        5. Sepals 3-veined.
            6. Tip with rounded lateral lobes; mentum obscure, adnate to edge of ovary; dorsal sepal 4.7–15 mm; column 3.3–7.8 mm . . . . 6. *Corallorhiza maculata*
            6. Tip with acute lateral teeth; mentum prominent, protruding backward along but free from ovary; dorsal sepal 6–12 mm; column 5–8.2 mm . . . . 7. *Corallorhiza mertensiana*

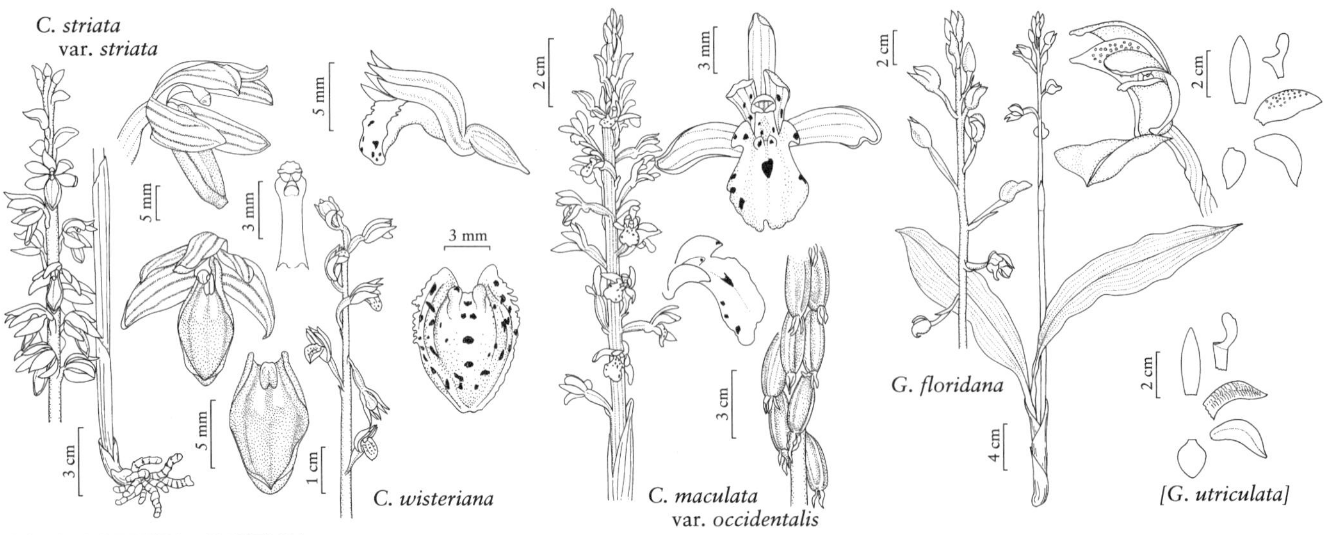

CORALLORHIZA ◦ GOVENIA

1. Corallorhiza striata Lindley, Gen. Sp. Orchid. Pl., 534. 1840 • Striped coral-root, corallorhize striée [F]

Stem ± strongly thickened, base not bulbous. **Inflorescences:** racemes lax to dense, 10–67 × 1.2–4.5 cm. **Flowers** 2–35, chasmogamous, conspicuous to showy; perianth open and spreading to connivent; sepals and petals salmon reddish purple with darker veins or pure yellow, lanceolate to oblanceolate, 3–5-veined; lip obovate, unlobed, 3.2–16.5 × 2–6.2 mm, thickened, margins deep red with white or yellow at base, or pure yellow, upturned, giving boat-shaped appearance especially toward apex; callus formed by partial connation of 2 basal lamellae, less than 1/3 length of lip; column pure yellow or yellow spotted with red, curved toward lip, 3–6 mm, base without auricles or ridges; ovary 3–8.5 mm; mentum absent. **Capsules** ellipsoid, 11–30 × 5–10 mm.

Varieties 3 (2 in the flora): North America, Mexico.

1. Lip 7–16.5 mm; column 4–6.5 mm; sepals and petals open, spreading .... 1a. *Corallorhiza striata* var. *striata*
1. Lip 2–7 mm; column 1.5–4 mm; sepals and petals connivent around column .... 1b. *Corallorhiza striata* var. *vreelandii*

1a. Corallorhiza striata Lindley var. **striata** [E] [F]

Corallorhiza macraei A. Gray

Racemes 10–67 cm. **Flowers** 5–35, showy; sepals and petals open, spreading, usually not nodding; sepals 8–18 mm; petals 7–18 mm; lip 7–16.5 × 4.4–8.5 mm; column 4–6.5 mm. **Capsules** 12–30 mm. $2n = 42$.

Flowering spring–early summer. Coniferous, deciduous, and mixed woods, lakeshores, swamps; 0–2600 m; Alta., B.C., Man., Nfld. and Labr. (Nfld.), Ont., Que., Sask.; Calif., Colo., Idaho, Mich., Minn., Mont., Nev., N.Mex., N.Y., N.Dak., Oreg., S.Dak., Tex., Utah, Wash., Wis., Wyo.

A parasitic wasp (*Pimplinae*, *Ichneumonidae*) has been found visiting flowers and removing pollinia of *Corallorhiza striata* var. *striata*, our most striking and largest flowered coral-root (J. V. Freudenstein 1997).

1b. Corallorhiza striata Lindley var. **vreelandii** (Rydberg) L. O. Williams, Ann. Missouri Bot. Gard. 21: 343. 1934 (as Corallorrhiza)

Corallorhiza vreelandii Rydberg, Bull. Torrey Bot. Club 28: 271. 1901; *C. bigelovii* S. Watson; *C. ochroleuca* Rydberg; *C. striata* var. *flavida* Todsen & T. A. Todsen

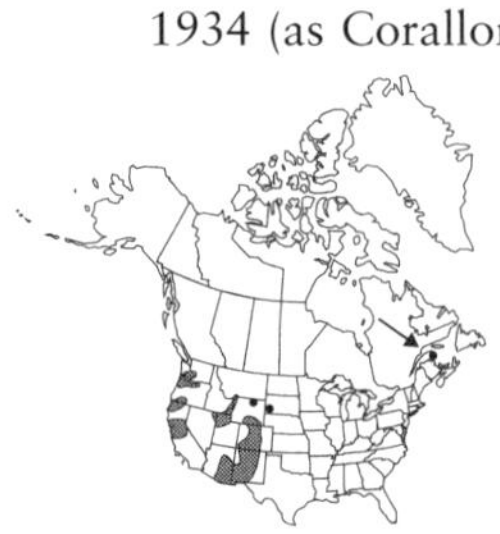

Racemes 10–44 cm. **Flowers** 2–28, conspicuous; sepals and petals often connivent around column, nodding; sepals 5–11 mm; lateral

sepals often broadened at base, nearly auriculate; petals 5–11 mm; lip 2–7 × 1.5–4.5 mm; column 1.5–4 mm. **Capsules** 11–23 mm.

Flowering spring–summer. Coniferous woods; 400–2600 m; Que.; Ariz., Calif., Colo., Idaho, Nev., N.Mex., Oreg., S.Dak., Utah, Wash., Wyo.; Mexico.

The two varieties of *Corallorhiza striata* intergrade in California and Oregon, although they are generally distinct. Small specimens from Quebec will key to this variety; they are not thought to be closely related to populations from the southwest.

2. Corallorhiza bentleyi Freudenstein, Novon 9: 511, fig. 1. 1999 • Bentley's coral-root [E]

Stems reddish to yellowish, to 20 cm. **Inflorescences:** racemes lax. **Flowers** 2–20, cleistogamous, inconspicuous; sepals and petals reddish to yellow, 3-veined; dorsal sepal lanceolate, 5.9–6.1 × 2.1–2.5 mm; lateral sepals lanceolate, somewhat falcate, strongly boat-shaped, 5.5–6.1 × 2.1–2.5 mm; petals narrowly ovate, weakly boat-shaped, 5–5.8 × 2.2–3 mm, apex acute; lip unlobed, ovate, 4.9–5.7 mm, margins involute, thickened, prominent, apex acute; callus formed from 2 connate lamellae at base of lip, ± 1.2 as long as lip, 2.5–2.8 × 1.5–1.8 mm; column curved forward, 3–3.8 × 0.8–1 mm, with small depression adaxially at base; pedicellate ovary narrowly obovoid, 6–9 × 2.5–3 mm; mentum absent. **Capsules** obovoid-ellipsoid.

Flowering mid–late July. Appalachian deciduous forest, often at edges of forest in somewhat disturbed sites; 800 m; Va., W.Va.

Corallorhiza bentleyi appears to be most similar to the small Mexican variants of *C. striata* that have been called *C. involuta* Greenman but were most recently recognized as C. *striata* var. *involuta* (Greenman) Freudenstein (J. V. Freudenstein 1999).

SELECTED REFERENCE Freudenstein, J. V. 1999. A new species of *Corallorhiza* (Orchidaceae) from West Virginia, U.S.A. Novon 9: 511–513.

3. Corallorhiza odontorhiza (Willdenow) Poiret in F. Cuvier, Dict. Sci. Nat. ed. 2, 10: 375. 1818 • Fall coral-root, small-flowered coral-root, corallorhize d'automne

Cymbidium odontorhizon Willdenow, Sp. Pl. 4(1): 110. 1805; *Corallorhiza micrantha* Chapman; C. *pringlei* Greenman

Stems yellow, greenish, or purplish brown, base strongly thickened, bulbous. **Inflorescences:** racemes lax, 11–46 × 1–2.5 cm. **Flowers** 5–26, inconspicuous; perianth open or closed; sepals and petals reddish purple to brown, often suffused with green, broadly lanceolate, 1–2-veined; lateral sepals strongly curved upward and directed forward, 3–4.5 mm; petals 2.5–3.8 mm, connivent with dorsal sepal to form hood over column, or sepals and petals connivent to produce closed flower; lip white, often spotted with purple, ovate to orbiculate-quadrangular, unlobed, 2.7–4.6 ×1.8–3.7 mm, thin, with 2 distinct basal lamellae, margins entire to erose-fringed; column white basally, often purple apically, 1.9–2.5 mm, straight or curved toward lip, with 2 distinct auricles in open-flowered plants; ovary 2.5–5 mm; mentum inconspicuous. **Capsules** broadly ellipsoid, 5.5–8 × 3.5–5 mm.

Varieties 2 (2 in the flora): North America, Mexico, Central America.

1. Perianth tending to be closed, column without or with only poorly developed auricles at base on adaxial surface; stigmatic surface 0.3–0.5 mm wide; lip 1.7–2.2(–3) mm wide . . . . 2a. *Corallorhiza odontorhiza* var. *odontorhiza*
1. Perianth open, column with 2 prominent auricles at base on adaxial surface; stigmatic surface 0.7–1 mm wide; lip 2–3.7 mm wide . . . . 2b. *Corallorhiza odontorhiza* var. *pringlei*

3a. Corallorhiza odontorhiza (Willdenow) Poiret var. **odontorhiza** [E]

Corallorhiza micrantha Chapman

Flowers: perianth tending to be closed; lip distinctly different from lateral sepals, 2.6–3.8 × 1.7–2.2 (–3) mm; column 1.9–2.5 × 0.8–1 mm or with only poorly developed auricles at base on adaxial surface; stigmatic surfaces 0.2–0.4 × 0.3–0.5 mm.

Flowering mid Aug–Oct. Rich deciduous woods, mixed woods, and conifer plantations; 0–2800 m; Ont., Que.; Ala., Ark., Conn., Del., D.C., Fla., Ga., Ill., Ind., Iowa, Kans., Ky., La., Maine, Md., Mass., Mich., Minn., Miss., Mo., Nebr., N.H., N.J., N.Mex., N.Y., N.C., Ohio, Okla., Pa., R.I., S.C., S.Dak., Tenn., Tex., Vt., Va., W.Va., Wis.

3b. Corallorhiza odontorhiza (Willdenow) Poiret var. **pringlei** (Greenman) Freudenstein, Harvard Pap. Bot. 1(10): 24. 1997

Corallorhiza pringlei Greenman, Proc. Amer. Acad. Arts 33: 475. 1898

Flowers: perianth open; lip similar to petals (narrowly obovate-spatulate), 2.7–4.6 × 2–3.7 mm; column 1.8–2.4 × 0.8–1.7 mm, base with 2 prominent auricles on adaxial surface; stigmatic surface 0.3–0.5 × 0.7–1 mm.

Flowering late summer–fall. Rich deciduous woods, mixed woods, and conifer plantations; 0–2800 m; Ont.; Conn., Ga., Ind., Iowa, Mich., N.J., N.Y., N.C., Pa., Tenn., Wis.; Mexico; Central America (El Salvador, Guatemala, Honduras, Nicaragua).

In *Corallorhiza odontorhiza* both chasmogamous and cleistogamous flowers exist; they occur on separate plants, although sometimes in the same populations. The cleistogamous form is by far the more frequent.

4. Corallorhiza wisteriana Conrad, J. Acad. Nat. Sci. Philadelphia 6: 145. 1829 • Wister's coral-root [F]

Corallorhiza hortensis Suksdorf

Stems red-purple to yellowish brown to yellow, base not strongly thickened nor bulbous. **Inflorescences:** racemes lax to dense, 10–55 × 1.3–3.5 cm. **Flowers** 2–25, inconspicuous; perianth open; sepals and petals purplish brown, often greenish distally, or pure yellow, 3-veined; sepals lanceolate, 4.5–10 mm; lateral sepals upcurved, connivent with dorsal sepal and petals, forming hood over column, somewhat falcate; petals often spotted with purple, broadly lanceolate, 4–7.2 mm; lip white, often spotted with purple, ovate to suborbiculate, unlobed, 4–7.5 × 2.5–7.5 mm, with 2 distinct basal lamellae, margins erose-denticulate to nearly entire; column curved toward lip, yellow to white, often with purplish apex, 2.4–5 mm, with basal auricles; ovary 3–8 mm; mentum present. **Capsules** ellipsoid-obovoid, 7–14 × 3.5–6.2 mm.

Flowering Jan–Jul, according to latitude. Deciduous and coniferous woods, often preferring richer habitats than other coral-roots; 0–3100 m; Ala., Ariz., Ark., Colo., Del., D.C., Fla., Ga., Idaho, Ill., Ind., Kans., Ky., La., Md., Miss., Mo., Mont., Nebr., N.J., N.Mex., N.C., Ohio, Okla., Pa., S.C., S.Dak., Tenn., Tex., Utah, Va., Wash., W.Va., Wyo.; Mexico.

Western collections of *Corallorhiza wisteriana* have frequently been misidentified as *C. maculata*. Anthocyanin-free forms from the central states have been misidentified as *C. trifida*.

5. Corallorhiza trifida Châtelain, Specim. Inaug. Corallorhiza, 8. 1760 • Early coral-root, northern coral-root, corallorhize trifide

Corallorhiza corallorhiza (Linnaeus) MacMillan; *C. corallorhiza* var. *coloradensis* Cockerell; *C. innata* R. Brown; *C. innata* var. *virescens* Farr; *C. trifida* var. *verna* (Nuttall) Fernald; *C. trifida* var. *virescens* (Farr) Fernald; *C. verna* Nuttall; *C. wyomingensis* Hellmayr & K. Hellmayr

Stems ± strongly thickened, base not bulbous. **Inflorescences:** racemes lax to dense, 8–35 × 1–3 cm. **Flowers** 3–18, inconspicuous; perianth open; sepals spreading, yellowish to green, lanceolate, 1-veined, 3.5–7 mm; lateral sepals often strongly recurved; petals arching over and clasping column, light yellow-green, often spotted with purple, lanceolate, 1-veined, 3–5.5 mm; lip white, often purple spotted, 2.5–4 × 1.5–3 mm, thin, with 2 small lateral lobes or teeth, middle lobe oblong, often somewhat dilated near apex, with 2 distinct basal lamellae; column curved toward lip, yellowish green, sometimes spotted purple basally, with shallow adaxial channel, 2–3.3 mm; ovary 2.1–6.5 mm; mentum inconspicuous. **Capsules** ellipsoid, 4.5–15 × 4.3–6 mm. $2n = 42$.

Flowering spring–summer. Coniferous, deciduous, and mixed woods, swamps, mountain woods, upland habitats northward, restricted to cool, moist habitats southward; 0–3100 m; Greenland; St. Pierre and Miquelon; Alta., B.C., Man., N.B., Nfld. and Labr., N.W.T., N.S., Nunavut, Ont., P.E.I., Que., Sask., Yukon; Alaska, Calif., Colo., Conn., Del., D.C., Idaho, Ill., Ind., Maine, Md., Mass., Mich., Minn., Mont., N.H., N.J., N.Mex., N.Y., N.Dak., Ohio, Oreg., Pa., R.I., S.Dak., Utah, Vt., Wash., W.Va., Wis., Wyo.; Eurasia.

Corallorhiza trifida is largely autogamous, although a syrphid fly (*Syrphus cinctellus*) was reported as a pollinator by F. Silen (1906). Various floral morphs exist, some with weak geographic correlation; they do not appear to warrant taxonomic recognition. Variants of *C. odontorhiza*, *C. wisteriana*, and *C. maculata* without red and purple pigments in sepals and petals are occasionally misidentified as *C. trifida*.

6. Corallorhiza maculata (Rafinesque) Rafinesque, Amer. Monthly Mag. & Crit. Rev. 2: 119. 1817 • Spotted coral-root, corallorhize maculée [F]

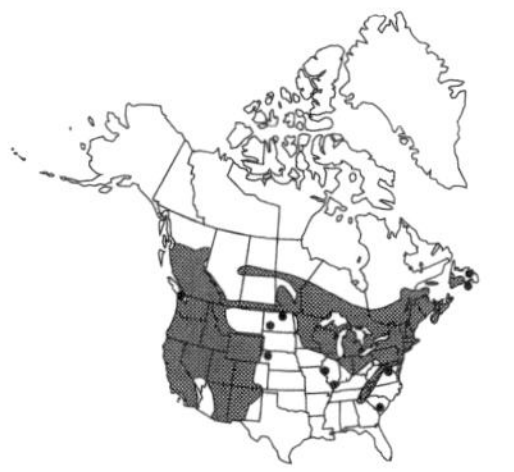

Cladorhiza maculata Rafinesque, Amer. Monthly Mag. & Crit. Rev. 1: 429. 1817

Stems ± strongly thickened, base not bulbous. **Inflorescences:** racemes dense to lax, 10–65 × 1.5–6.5 cm. **Flowers** 6–41, conspicuous; perianth open; sepals directed forward to somewhat spreading, brown, tan, red, or yellow, often darker distally, lanceolate, 3-veined, 4.7–15 mm; petals frequently curved toward column, nearly adhering to it, tan to yellowish, often red distally, usually spotted with purple, lanceolate to oblanceolate, 4.5–11.5 mm; lip white, usually spotted with purple, obovate to elliptic, thin, lateral lobes 2, rounded, middle lobe oblong to broadly dilated toward apex, with 2 distinct basal lamellae, 4–9 × 1.5–6 mm; column whitish yellow, spotted with purple, usually curved, 3.3–7.8 mm, basally with ridges or poorly developed auricles; ovary 5–14 mm; mentum obscure, adnate to edge of ovary. **Capsules** pendulous, ellipsoid, 9–24 × 5–9 mm. $2n = 42$.

Several color forms exist in *Corallorhiza maculata*, often occurring in proximity and sometimes forming large clumps. White-flowered southern plants have occasionally been misidentified as *Corallorhiza trifida*. Dance-flies (*Empis*) have been reported as pollinators (J. L. Kipping 1971).

Varieties 3 (2 in the flora): North America, Mexico, Central America (Guatemala).

SELECTED REFERENCE Freudenstein, J. V. 1987. A preliminary study of *Corallorhiza maculata* (Orchidaceae) in eastern North America. Contr. Univ. Michigan Herb. 16: 145–153.

1. Middle lobe of lip expanded slightly or not at all distally, ratio of widths of dilated part to base of middle lobe less than 1.5; floral bracts averaging 0.5–1 mm . . . . . 6a. *Corallorhiza maculata* var. *maculata*
1. Middle lobe of lip distinctly expanded, ratio of widths of dilated part to base of middle lobe more than 1.5; floral bracts averaging 1–2.8 mm . . . . 6b. *Corallorhiza maculata* var. *occidentalis*

6a. Corallorhiza maculata (Rafinesque) Rafinesque var. **maculata** [E]

Corallorhiza multiflora Nuttall; *C. multiflora* var. *sulphurea* Suksdorf; *C. vancouveriana* Finet

Inflorescences: floral bracts 0.5–1 × 0.5–9 mm, usually entire. **Flowers:** middle lobe of lip expanded slightly or not at all distally, ratio of width of dilated part to base of middle lobe less than 1.5; lip 3.3–7 × 1.5–4.5 mm; column curved, 3.3–5.5 × 0.7–1 mm at stigma; stigmatic surface elliptic-oblong to nearly square, 0.7 × 1 mm.

Flowering May–Aug. Moist to dry coniferous and deciduous woods, and conifer plantations, often in florests with little other herbaceous cover; 0–3700 m; St. Pierre and Miquelon (variety unconfirmed); Alta., B.C., Man., N.B., Nfld. and Labr. (Nfld.), N.S., Ont., P.E.I., Que., Sask.; Ariz., Calif., Colo., Conn., Del., D.C., Ga., Idaho, Ill., Ind., Iowa, Maine, Md., Mass., Mich., Minn., Mont., Nev., N.H., N.J., N.Mex., N.Y., N.C., Ohio, Oreg., Pa., R.I., S.C., Tenn., Utah, Vt., Va., Wash., W.Va., Wis., Wyo.

6b. Corallorhiza maculata (Rafinesque) Rafinesque var. **occidentalis** (Lindley) Ames, Enum. Orchids U.S. & Canada, 22. 1924 [E] [F]

Corallorhiza multiflora Nuttall var. *occidentalis* Lindley, Gen. Sp. Orchid. Pl., 534. 1840; *C. grabhamii* Cockerell; *C. leimbachiana* Suksdorf; *C. maculata* subsp. *occidentalis* (Lindley) Cockerell; *C. maculata* var. *fusca* H. H. Bartlett; *C. maculata* var. *immaculata* M. Peck; *C. maculata* var. *intermedia* Farwell; *C. maculata* var. *punicea* Bartlett; *C. multiflora* var. *flavida* Peck

Inflorescences: floral bracts averaging 1–2.8 mm, often forked. **Flowers:** middle lobe of lip distinctly expanded, ratio of width of dilated part to base of middle lobe greater than than 1.5; column curved, 3.5–7.8 × 1–1.3 mm at stigma; stigmatic surface oblong-pentagonal, 0.8 × 1.3 mm.

Flowering Apr–Jul. Coniferous and mixed woods, and conifer plantations; 0–3700 m; St. Pierre and Miquelon (variety unconfirmed); Alta., B.C., Man., N.B., Nfld. and Labr. (Nfld.), N.S., Ont., P.E.I., Que., Sask.; Ariz., Calif., Colo., Conn., Idaho, Ill., Maine, Mass., Mich., Minn., Mont., Nebr, Nev., N.H., N.Mex., N.Y., Oreg., Pa., S.Dak., Utah, Vt., Va., Wash., Wis., Wyo.

Where they occur together, *Corallorhiza maculata* var. *occidentalis* typically flowers 2–4 weeks earlier than *C. maculata* var. *maculata*.

7. **Corallorhiza mertensiana** Bongard, Mém. Acad. Imp. Sci. St. Pétersbourg, Sér. 6, Sci. Math. 2: 165. 1832 • Merten's coral-root, western coral-root [E]

Corallorhiza maculata (Rafinesque) Rafinesque subsp. *mertensiana* (Bongard) Calder & Roy L. Taylor; *C. purpurea* L. O. Williams

Stems ± strongly thickened, base not bulbous. **Inflorescences:** racemes dense, 35–65 × 1.5–4 cm. **Flowers** 8–35, showy; perianth open; sepals reddish purple, sometimes yellowish near base, or completely yellow, lanceolate, 3-veined, 6–12 mm; dorsal sepal arching over column, nearly adhering to it; lateral sepals strongly spreading; petals arching over column, connivent with dorsal sepal, often yellowish basally, and streaked with purple, or completely yellow suffused with purple toward apex; lip red-purple, white, or white with purple streaks or spots, narrowly obovate, 4.8–9.5 × 2.5–5 mm, thin, usually with small (0.7 mm) tooth on each side, margins undulate-denticulate; column curved somewhat toward lip, yellow, often flushed with purple or white basally, and streaked or spotted with purple, 5–8.2 mm; ovary 5.9–10 mm; mentum prominent, protruding backward along ovary but free from it. **Capsules** ellipsoid, 10–25 × 6–9 mm. $2n = 40$.

Flowering late spring–summer. Moist to dry coniferous and mixed woods; 0–2300 m; Alta., B.C.; Alaska, Calif., Idaho, Mont., Oreg., Wash., Wyo.

In the Pacific Northwest *Corallorhiza mertensiana* is largely sympatric with *C. maculata* and occasionally intergrades with it. It frequently forms large clumps.

230d.10b. ORCHIDACEAE Jussieu (tribe CYMBIDIEAE) subtribe GOVENIINAE Dressler, Lindleyana 5: 124. 1990

57. GOVENIA Lindley, Gen. Sp. Orchid. Pl., 153. 1832 • [For J. R. Gowen, English collector in Assam]

Edward W. Greenwood†

Herbs, terrestrial. **Roots** spreading, slender, fleshy. **Stems** reduced to short, nearly ovoid pseudobulbs. **Leaves** petiolate, articulate, from multiple, concentric, basal sheaths; blade pleated. **Inflorescences** lateral from within sheaths, racemes, erect, loose. **Flowers** resupinate; columns fleshy, arcuate 90°, prominently winged, foot prominent [absent]; anthers terminal; pollinaria with 2 pairs of pollinia, stipe, and viscidium.

Species 30 or more (1 in the flora): tropical and subtropical regions, North America, Central America, South America.

SELECTED REFERENCES Brown, P. M. 2000d. *Govenia floridana* (Orchidaceae), a new species endemic to southern Florida, U.S.A. N. Amer. Native Orchid J. 6: 231–238. Correll, D. S. 1947. Revision of the genus *Govenia*. Lloydia 10: 218–278. Fawcett, W. and A. B. Rendle. 1910–1936. Flora of Jamaica.... 5 vols. London. Vol. 1, p. 113. Greenwood, E. W. 1982b. *Govenia* in Mexico, an introductory note. Orquidea (Mexico City) 8: 114–120. Greenwood, E. W. 1991. The Florida *Govenia*. Amer. Orchid Soc. Bull. 69: 867–869.

1. **Govenia floridana** P. M. Brown, N. Amer. Native Orchid J. 6: 233, figs. 1, 2, 3 [left], 4. 2000 [E] [F]

Plants to 50 cm. **Roots** 6–10 × 5–10 cm. **Pseudobulbs** 3–8 cm. **Leaves** 2; petioles concentrically arising from terminal nodes of pseudobulb, inside 3 concentric, polygonal, keeled sheaths, 3–30 cm; blade elliptic, 20–35 × 8–11 cm, apex acute. **Inflorescences** to 50 cm. **Flowers** opening partially, white; sepals arcuate; dorsal sepal obovate, concave, 19 × 6 mm; lateral sepals falcate, 15 × 5 mm; petals internally purple-flecked, arcuate-concave, oblique, 13 × 7 mm; lip with 3–5 purplish spots on apical margins, nearly elliptic, arcuate, clawed, 16 × 11 mm. **Capsules** pendent, ellipsoid, 3 × 1 cm.

Flowering Nov–Dec. In leaf litter in shade of mixed tropical forest; 0–100 m; Fla.

Govenia was discovered in Everglades National Park in 1957 by Frank C. Craighead (C. A. Luer 1972). Intensive searches in 1988 and 1989 failed to find any plants;

in 1990 R.P. Sauleda reported it from nearby Osteen Hammock, and in 2000, Paul M. Brown found four small, immature plants at the original site (P. M. Brown 2000d).

The genus *Govenia* is unusual in that morphology, as seen preserved in pressed specimens, is not always a reliable character for distinguishing species. With this in mind, the Florida plant was compared with known similar species. It was first identified as *G. utriculata* (Swartz) Lindley (C. A. Luer 1972) but *G. floridana* differs from that species in that the sheaths are not inflated, it is nearly polygonal in cross section, the column is prominently winged, and the petals are spotted (E. W. Greenwood 1991). The racemes and flowers of the Florida *Govenia* are similar to those of the eastern Mexican *G. alba* A. Richard & Galeotti; they differ in the nature of the column foot. The foot of *G. floridana* is concave and winged and emerges from the column at an oblique angle, whereas the foot of *G. alba* shows no indication of wings and emerges from the column at a right angle (cf. P. M. Brown 2000d, fig. 2). Brown has taken those differences to indicate that the Florida *Govenia* is a new species endemic to Florida. For the purposes of this flora we are accepting that decision but recognize that proof of its uniqueness will only come when fresh flowers can be studied and compared with those of other species. The immature plants found in Everglades National Park may one day make this possible.

230b.10c. ORCHIDACEAE Jussieu (tribe CYMBIDIEAE) subtribe EULOPHIINAE Bentham, J. Linn. Soc., Bot. 18: 287. 1881

58. EULOPHIA R. Brown ex Lindley, Bot. Reg. 8: plate 686. 1823 • [Greek *eu*, well, and *lophos*, plume, crest, alluding to the crest on the lip]

Gustavo A. Romero-González

Platypus Small & Nash

Herbs, terrestrial [rarely epiphytic], cespitose. **Stems** modified into corms [pseudobulbs], concealed by leaf sheaths. **Leaves** usually caducous, 1–several [sometimes much reduced, scale-like], basal, distichous, imbricate near base, clustered distally, nearly [distinctly] petiolate; blade plicate, not articulate with leaf sheaths, leathery. **Inflorescences** lateral, arising from base of corm [pseudobulb], racemes [rarely paniculate], erect. **Flowers** resupinate, showy; sepals and petals distinct [lateral sepals sometimes connate at base] and free, sepals erect to spreading; lip adnate or articulate to column foot, concave, saccate [sometimes spurred], 3-lobed; disc with erect lamellae; column erect, slightly arcuate, semiterete, long, more than 10 mm [short], basally produced into conspicuous foot, with shallow lateral wings distally; anther terminal, incumbent, 1-locular, operculate, 2-lobed at apex [entire]; pollinarium with 2 pollinia, yellow, nearly triangular to nearly circular, sulcate, cartilaginous; stipe short, inconspicuous [conspicuous, ligulate], viscidium semicircular [semilunar]. **Fruits** capsules, pendent when mature.

Species 215 (1 in the flora): pantropical.

SELECTED REFERENCES Cribb, P. J. 1987. The genus *Eulophia* in Africa. In: K. Kondo and K. Hashimoto, eds. 1987. World Orchid Hiroshima Symposium.... [Sekai Ran Kaigi Hiroshima Taikai.] Hiroshima. Pp. 97–103. Cribb, P. J. 1989. *Eulophia*. In: W. B. Turrill et al., eds. 1952+. Flora of Tropical East Africa. 152+ vols. London and Rotterdam. Orchidaceae, part 3, pp. 422–480.

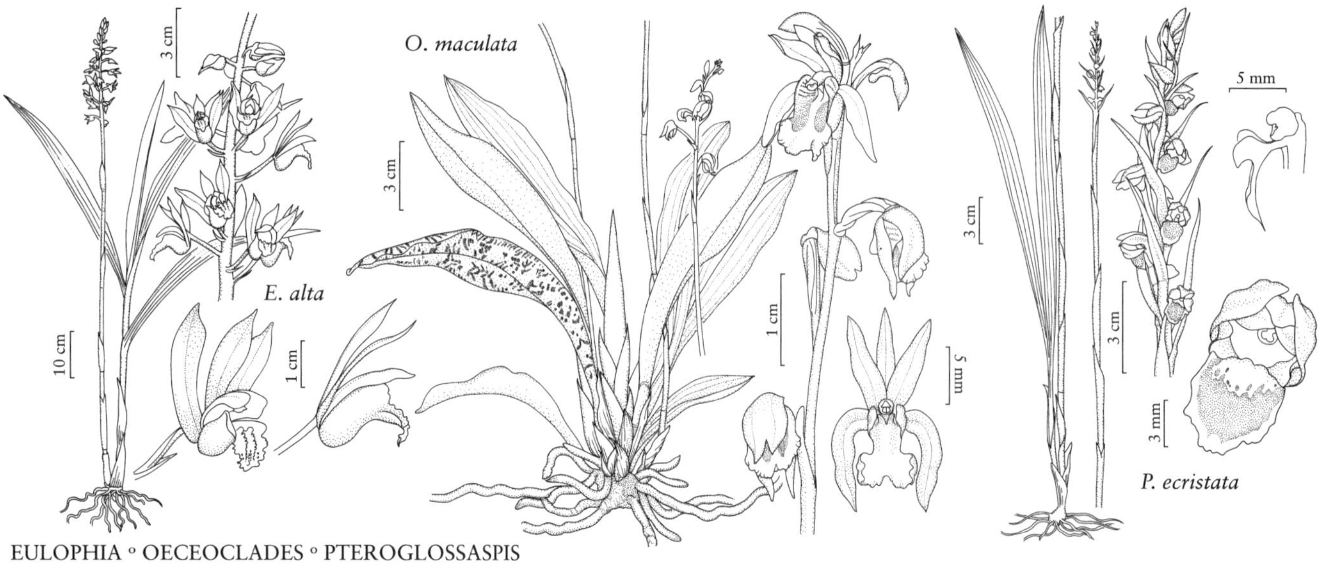

EULOPHIA ◦ OECEOCLADES ◦ PTEROGLOSSASPIS

1. **Eulophia alta** (Linnaeus) Fawcett & Rendle, Fl. Jamaica 1: 112. 1910 • Wild coco [F]

Limodorum altum Linnaeus, Syst. Nat. ed. 12, 594. 1767; *Cyrtopera longifolia* (Kunth) Reichenbach f.; *C. woodfordii* (Sims) Lindley; *Cyrtopodium woodfordii* Sims; *Dendrobium longifolium* Kunth; *Eulophia woodfordii* (Sims) Rolfe; *Platypus altus* (Linnaeus) Small; *P. papilliferus* Small

Plants 70–170 cm. **Leaves** linear-elliptic to narrowly ovate, 20–120 × 3–11 cm, apex acuminate. **Inflorescences** erect, to 1 m; bracts narrowly ovate to filiform, 1–3 cm. **Flowers** many, 1.5–2.5 cm wide; sepals light green, very narrowly elliptic to narrowly obovate, apex acuminate to apiculate; lateral sepals slightly oblique, adnate to column foot; petals spreading, green or green with pink overflush, 13 × 7 mm, broader than sepals; lip red-maroon, concave, saccate, 11–17 × 18–28 mm, slightly to conspicuously 3-lobed, middle lobe apically obtuse, margins undulate, lateral lobes partially enclosing column; disc with erect lamellae; column 7–10 mm. **Capsules** ellipsoid, to 6 cm.

Flowering summer–fall. Moist sand and rich soils, in shaded to sunny habitats, swamps, pastures, and roadside ditches; 10–200[–2000] m; Fla., Ga.; Mexico; West Indies; Central America; South America; tropical Africa.

59. OECEOCLADES Lindley, Edwards's Bot. Reg. 18: plate 1522. 1832 · [Greek *oikeios*, of a household (Latin *oeceos*), and Latin *clades*, destruction, possibly alluding to breaking up of existing classification]

Roger L. Hammer

Herbs, terrestrial (rarely on rocks or epiphytic), erect. **Stems** modified into pseudobulbs, often densely congested, concealed by overlapping fibrous sheaths, 1[–3]-leaved. **Leaves** persistent; blade not imbricate or plicate, usually leathery, often patterned. **Inflorescences** lateral, arising from base of pseudobulb, racemes or panicles, erect. **Flowers** resupinate, inconspicuous; sepals and petals distinct and free, spreading; lip deeply 3-lobed, base spurred; column erect, curved, produced into conspicuous foot, oblique at base; anther terminal, operculate; pollinia 2, with short or aborted caudicles attached directly to viscidium. **Fruits** capsules.

Species 31 (1 in the flora): tropical regions, North America, West Indies, Central America, South America, Africa (including Madagascar, the Mascarene Islands, and the Seychelles).

SELECTED REFERENCE Garay, L. A. and P. Taylor. 1976. The genus *Oeceoclades* Lindl. Bot. Mus. Leafl. 24: 249–274.

1. Oeceoclades maculata (Lindley) Lindley, Gen. Sp. Orchid. Pl., 237. 1833 [F]

Angraecum maculatum Lindley, Coll. Bot., plate 15 and text on facing page. 1821; *Eulophia maculata* (Lindley) Reichenbach f.; *Eulophidium maculatum* (Lindley) Pfitzer

Plants 20–43 cm. **Roots** fibrous, white, stout. **Pseudobulbs** dark green, ovoid to orbicular, 2–5 × 1–3 cm, often concealed by imbricate fibrous sheaths or 1 closely appressed sheath, 1-leaved at apex. **Leaves:** blade nearly erect, olive green with dark green mottling, oblong-elliptic, abaxially keeled, 12–32 × 3.5–5.5 cm, succulent leathery, apex acute. **Inflorescences** lax, 10–42 cm; peduncles erect, slender, with few remote sheaths. **Flowers** 5–15, 9–15 mm wide; sepals and petals light brown to pinkish green; sepals 8–12 × 2–3 mm; dorsal sepal linear-oblong, apex acute; lateral sepals slightly falcate; petals oblong-elliptic, 12 × 4 mm, apex acute; lip white with 2 parallel pink blotches, 9–14 × 8–12 mm, lateral lobes erect, white with thin, radiating purple lines, orbiculate, spur curved, 5 × 2 mm; column curved, white, 5–7 mm. **Capsules** pendent, elliptic-oblong, to 3.5 cm.

Flowering Aug–Nov. In humus under shady forests, in orchards and residential areas; 0–10 m; Fla.; Mexico; West Indies; Central America; South America; Africa.

Oeceoclades maculata was first discovered in the United States in Miami-Dade County, Florida in 1974, and is spreading rapidly in central and southern Florida. It is not known whether it escaped from cultivation or arrived in Florida via windblown seeds from the Greater Antilles or the nearby Bahama Archipelago.

60. PTEROGLOSSASPIS Reichenbach f., Otia Bot. Hamburg., 67. 1878 • [Greek *pteron*, wing, *glossa*, tongue, and *aspis*, shield]

Gustavo A. Romero-González

Triorchos Small & Nash

Herbs, terrestrial, cespitose. **Stems** cormlike pseudobulbs. **Leaves** wilting at end of growing season, basal, nearly petiolate; blade plicate, not articulate with leaf sheaths. **Inflorescences** lateral, from base of pseudobulb, **racemes**, erect. **Flowers** resupinate, inconspicuous; sepals and petals converging [usually spreading], distinct and free, nearly equal; lip spreading, sessile at base of column, 3-lobed; disc smooth, without spurs; column erect, wingless, stout, 3–4 mm, base 2-auriculate, apex blunt; anther terminal, incumbent, operculate, 1-locular; pollinia 2, yellow, spheric, sulcate, cartilaginous; stipe semilunar; viscidium present. **Fruits** capsules, erect when mature.

Species 7–10 (1 in the flora): neotropical regions; Africa.

SELECTED REFERENCES Rolfe, R. A. 1898. *Pteroglossaspis*. In: D. Oliver et al., eds. 1868–1937. Flora of Tropical Africa.... 10 vols. London. Vol. 7, pp. 99–100. Romero-G., G. A. 1993. Notes on *Pteroglossaspis* (Orchidaceae), a new genus for the flora of Colombia. Orquidea (Mexico City) 13: 275–280. Wood, J. J. 1989. *Pteroglossaspis*. In: W. B. Turrill et al., eds. 1952+. Flora of Tropical East Africa. 152+ vols. London and Rotterdam. Orchidaceae, part 3, pp. 480–482.

1. Pteroglossaspis ecristata (Fernald) Rolfe, Orchid Rev. 12: 136. 1904 [C] [F]

Cyrtopodium ecristatum Fernald, Bot. Gaz. 24: 433. 1897; *Bletia verecunda* Chapman; *Cyrtopodium strictum* Grisebach; *Eulophia ecristata* (Fernald) Ames; *Triorchos ecristatus* (Fernald) Small; *T. strictus* (Grisebach) Acuña

Plants 30–130 cm. **Pseudobulbs** underground. **Leaves:** blade linear-lanceolate, 15–70 × 1–3.5 cm, apex long-acuminate. **Inflorescences** erect, densely flowered toward apex, to 170 cm. **Flowers** yellowish green with purplish brown markings, 0.75–1 cm wide; sepals and petals converging over column and disc of lip in natural position, yellowish green; sepals oblong, 10–15 × 4–5 mm, apex acuminate; lateral sepals slightly oblique; petals oblong-elliptic, 8–15 × 4–5 mm; lip dark purplish brown with green margins, 3-lobed, middle lobe strongly reflexed in natural position, nearly circular, 5–10 × 5–10 mm including lateral lobes, apex obtuse, lateral lobes incurved in natural position, truncate to obtuse; disc green; column green, 3–4 mm, apex blunt. **Capsules** ovoid, to 2 cm.

Flowering summer–fall. Old fields, pine savanna, and scrub oak lands; of conservation concern; 0–200[–500] m; Ala., Fla., La., N.C., S.C.; West Indies (Cuba); South America (Colombia).

Cyrtopodium stricta, described by Grisebach in 1866, is referable to this species (G. A. Romero-G. 1993). Despite the nomenclatural priority of the epithet *stricta* over *ecristata*, it cannot now be taken up in *Pteroglossaspis* because of the existence of *P. stricta* Schlechter, an African species described in 1915. Moreover, when Rolfe transferred *Cyrtopodium ecristatum* to *Pteroglossaspis* in 1904, he did not include *Cyrtopodium strictum* Grisebach within his concept of the species, and so his name *Pteroglossaspis ecristata* is legitimate. *Pteroglossaspis ecristata* has a spotty distribution throughout its range, perhaps because of its rather inconspicuous habit (D. S. Correll 1950).

230d.10d. ORCHIDACEAE Jussieu (tribe CYMBIDIEAE) subtribe CYRTOPODIINAE Bentham, J. Linn. Soc., Bot. 18: 288. 1881

61. CYRTOPODIUM R. Brown in W. Aiton and W. T. Aiton, Hortus Kew. 5: 216. 1813 • [Greek *kyrtos*, curved swelling, and *podium*, foot, probably alluding to conspicuous column foot]

Gustavo A. Romero-González

Herbs, terrestrial or epiphytic, cespitose. **Stems** pseudobulbs, ovate-fusiform. **Leaves** deciduous, distichous; blade articulate, plicate, leathery, apex acuminate. **Inflorescences** from base of pseudobulbs, racemes to panicles, erect; floral bracts large, showy, margins undulate. **Flowers** resupinate, showy; sepals and petals membranaceous, margins often conspicuously undulate; lip adnate to apex of column foot, subsessile to conspicuously clawed, conspicuously 3-lobed, disc with tuberculate, cristate, or verrucose callus; column slightly arcuate, sometimes nearly club-shaped, semiterete, forming conspicuous foot; anther terminal, incumbent, operculate, imperfectly 2-locular; pollinia 2, yellow, nearly triangular, sulcate, cartilaginous, attached to short, trowel-shaped viscidium. **Fruits** capsules, pendent, pear-shaped.

Species 15–30 (2 in the flora): tropical and subtropical regions, North America, West Indies, Central America, South America.

1. Lip middle lobe margin not conspicuously verruculose-tuberculate; lateral lobes bright yellow, shorter than middle lobe; plants terrestrial . . . . 1. *Cyrtopodium polyphyllum*
1. Lip middle lobe margin conspicuously verruculose-tuberculate, lateral lobes spotted or solid dark orange or maroon, longer than middle lobe; epiphytic . . . . 2. *Cyrtopodium punctatum*

1. Cyrtopodium polyphyllum (Vellozo) Pabst ex F. Barrios, Acta Bot. Brasil. 8: 12. 1994 F I

Epidendrum polyphyllum Vellozo, Fl. Flumin., plate vol. 9, 17. 1831; *Cyrtopodium paranaense* Schlechter

Plants terrestrial, 50–150 cm. **Pseudobulbs** elongate, fusiform. **Inflorescences** erect, racemes to panicles, 100–200 cm. **Flowers** bright yellow, 3–4 cm wide; lip with middle lobe margins not conspicuously verruculose-tuberculate, lateral lobes bright yellow, shorter than middle lobe. **Capsules** to 10 cm.

Flowering winter. Pinelands; 10–100[–1500] m; introduced; Fla.; South America (Brazil).

Cyrtopodium polyphyllum was reported for Florida (C. A. Luer 1972) as *C. andersonii* (Lambert) R. Brown. The only available herbarium specimen [*McCormack sub G. N. Avery 1155* (FTG)] is referable to *C. polyphyllum*. *Cyrtopodium andersonii* differs from *C. polyphyllum* in having narrowly ovate rather than broadly elliptic sepals (G. A. Romero-G. and G. Carnevali F.-C. 1999).

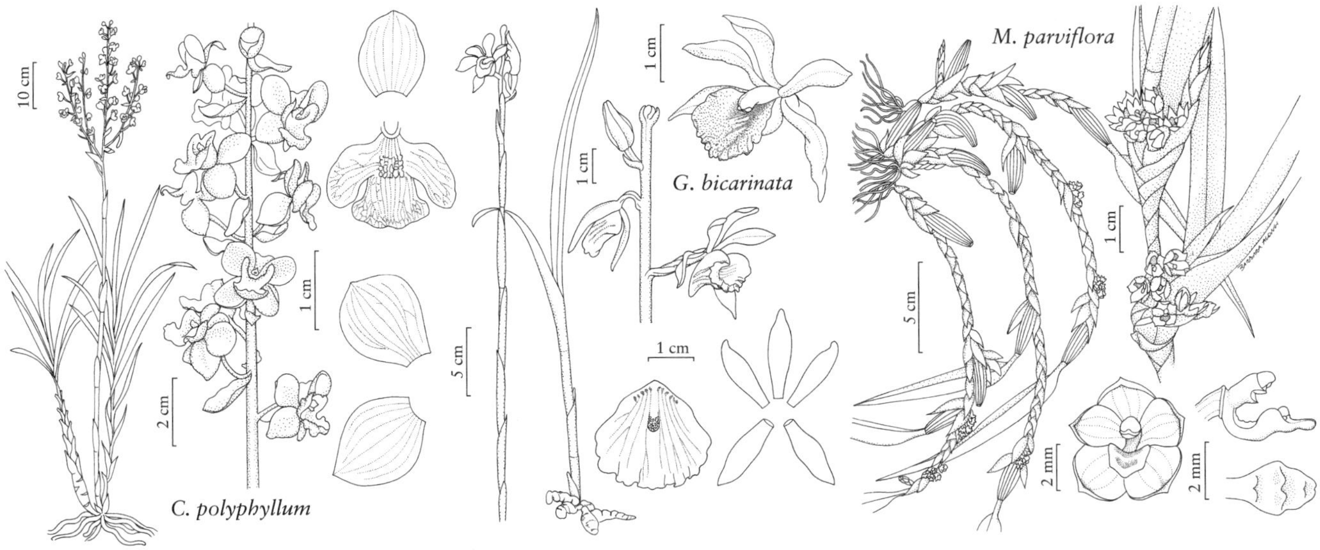

CYRTOPODIUM ◦ GALEANDRA ◦ MAXILLARIA

2. **Cyrtopodium punctatum** (Linnaeus) Lindley, Gen. Sp. Orchid. Pl., 188. 1833 • Cowhorn orchid, cigar orchid, bee-swarm orchid

Epidendrum punctatum Linnaeus, Syst. Nat. ed. 10, 2: 1246. 1759

Plants epiphytic, 50–150 cm. **Pseudobulbs** elongate, fusiform, to 50 cm. **Inflorescences** erect, paniculate, to 90 cm. **Flowers** bright yellow with maroon and green markings, 4–5 cm wide; lip with middle lobe margins conspicuously verruculose-tuberculate, lateral lobes spotted or solid dark orange or maroon, longer than middle lobe. **Capsules** to 10 cm. $2n = 46$.

Flowering winter–early spring. Coastal mangrove swamps and cypress swamps; 0–50[–500] m; Fla.; West Indies; South America (Peru, Venezuela).

The pseudobulbs of *Cyrtopodium punctatum* were used in the past as a source of glue for book binding.

SELECTED REFERENCES Ames, O. 1905–1922. Orchidaceae: Illustrations and Studies of the Family Orchidaceae Issuing from the Ames Botanical Laboratory.... 7 vols. Boston and New York, Vol., 1, pp. 55–58.. Hawkes, A. D. 1952. Habitat notes on *Cyrtopodium punctatum*. Orchid J. 1: 296–297.

62. GALEANDRA Lindley in F. A. Bauer and J. Lindley, Ill. Orch. Pl. 2: plate 8. 1832 • [Latin *galea*, helmet, and Greek *-andrus*, male, referring to shape of anther]

Gustavo A. Romero-González

Herbs, terrestrial [epiphytic], cespitose. **Stems** reedlike or more often pseudobulbs. **Leaves** 1–2 [several], not articulate [articulate], plicate, absent [present] at flowering. **Inflorescences** terminal, racemes, erect [lightly arching to pendent]; floral bracts lanceolate, scarious. **Flowers** resupinate, showy; sepals and petals membranaceous; lip sessile, simple [obscurely 2–3-lobate], base with conspicuous spur adnate to column foot, margins tomentose [glabrous]; spur shortly conic [sometimes apically recurved, funnel-shaped]; disc basally 2[–4]-keeled; column lightly arcuate, shallowly winged, semiterete, produced into conspicuous foot; anther terminal, incumbent, operculate, imperfectly 2-locular; pollinia 2, yellow, nearly spheric, cartilaginous. **Fruits** capsules.

Species 20–24 (1 in the flora): tropical and semitropical regions, North America, Mexico, West Indies, Central America, South America.

SELECTED REFERENCE Rolfe, R. A. 1892. The genus *Galeandra*. Gard. Chron., ser. 3, 12: 430–431.

1. **Galeandra bicarinata** G. A. Romero & P. M. Brown, N. Amer. Native Orchid J. 6: 78, fig. 1. 2000 [F]

Plants apparently saprophytic, leafless at anthesis, to 60 cm. **Roots** numerous, white, thick, fleshy. **Stems** abbreviated rhizomes; pseudobulbs subterranean, cormlike, ovoid, covered with scales. **Leaves:** petiole to 15 cm; blade 3-ribbed, linear-lanceolate, 25 × 3 cm. **Inflorescences** borne after leafy shoot, sometimes branched, to 20-flowered, 100 cm, base with several small, imbricating bracts; peduncle purple, with several leaflike bracts to 7 cm; rachis green; floral bracts to 2 cm. **Flowers** resupinate, light green; sepals spreading, oblanceolate, keeled, to 2.4 × 0.7 cm; petals spreading, oblanceolate, slightly oblique, 2 × 0.7 cm; lip subquadrangular, slightly 3-lobate at apex when spread, 2 × 2.5 cm, base produced into nearly cylindric, slightly hooked, short, obtuse spur, margins crenate, covered with short trichomes, apex reflexed; disc with a narrow callus, covered with short trichomes, bicarinate, keels merging toward base, diverging toward apex; column erect, slightly arcuate, 3-angled, to 1 × 0.3 cm, with prominent foot; anther green, cucullate, acute, the margins and apex papillose, 2 × 0.3 cm; pollinaria with no discernible viscidium or stipe; pollinia 2, yellow, spheric, 1 × 1 mm; pedicellate ovary to 2.5 cm. **Capsules** pendent, ellipsoid, 3 cm.

Flowering fall. Hammocks; 0–100[–1500] m; Fla.; West Indies (Cuba).

Galeandra bicarinata was previously identified as *G. beyrichii* Reichenbach f., a similar species with four pronounced keels on the lip.

230d.11. ORCHIDACEAE Jussieu (subfam. EPIDENDROIDEAE) tribe MAXILLARIEAE Pfitzer, Entwurf Anordn. Orch., 105. 1887

230d.11a. ORCHIDACEAE Jussieu (tribe MAXILLARIEAE) subtribe MAXILLARIINAE Lindley, Edwards's Bot. Reg. 29: Misc. 12. 1843

63. MAXILLARIA Ruiz & Pavón, Fl. Peruv. Prodr., 116, plate 25. 1794 • [Latin *maxilla*, jawbone; apparently an allusion to the open-mouth appearance of the flower when viewed laterally]

John T. Atwood

Camaridium Lindley; *Ornithidium* Salisbury; *Pseudomaxillaria* Hoehne; *Sepalosaccus* Schlechter

Herbs, epiphytic, rarely terrestrial, cespitose to climbing. **Stems** reduced [elongate], usually terminated with pseudobulb. **Leaves** 1–many; blade conduplicate, linear, lanceolate, or elliptic, margins entire. **Inflorescences** solitary flowers borne laterally within bract or leaf axil of rhizome of mature, immature, or apparently abortive shoots; floral bracts triangular, usually acute. **Flowers:** sepals and petals nearly same; dorsal sepal erect, concave; lateral sepals adnate to column foot; lip 3-lobed or simple; column not winged, semiterete, with foot at base; pollinia 4, waxy; stigmatic cavity round; rostellum not prominent. **Fruits** capsules, ellipsoid to obovoid.

Species ca. 650 (2 in the flora): tropical regions, Western Hemisphere.

SELECTED REFERENCES Atwood, J. T. 1993. A revision of the *Maxillaria neglecta* complex (Orchidaceae) in Mesoamerica. Lindleyana 8: 25–31. Brieger, F. G. 1977. On the Maxillariinae (Orchidaceae) with sepaline spur. Bot. Jahrb. Syst. 97: 548–574. Hammer, R. L. 1981. Finding new orchids: A contribution to the Orchidaceae of Florida. Fairchild Trop. Gard. Bull. 36(3): 16–18.

1. Plants conspicuously rhizomatous; pseudobulbs conspicuous; sepals less than 1 cm . . . 1. *Maxillaria parviflora*
1. Plants cespitose; pseudobulbs absent or hidden from view; sepals more than 1 cm . . . 2. *Maxillaria crassifolia*

1. **Maxillaria parviflora** (Poeppig & Endlicher) Garay, Bot. Mus. Leafl. 21: 258. 1967 [F]

Scaphyglottis parviflora Poeppig & Endlicher, Nov. Gen. Sp. Pl. 1: 58, plate 97. 1836; *Maxillaria conferta* (Grisebach) C. Schweinfurth ex León; *Ornithidium confertum* Grisebach

Plants epiphytic, climbing. **Stems** with pseudobulbs distributed 2–8 cm along conspicuous rhizome; pseudobulbs nearly cylindric to ovoid, 1.5–3 cm. **Leaves** 1 per shoot; blade lanceolate, 9–20 × 0.9–2 cm, almost leathery, apex acute. **Inflorescences** several, at base of mature shoots, less than 1 cm. **Flowers** white to light yellow [purplish in Jamaican populations] with yellow-orange lip; sepals ovate, 6–7 × 3–3.5 mm, leathery, apex acute; lateral sepals 1/3 connate; petals ovate-elliptic, 4.5–5 × 2–2.5 mm, apex keeled, acute; lip continuous with column foot, deltate to 3-lobate, 2.5–4 × 3 mm when spread, lateral lobes somewhat erect; callus a simple transverse ridge connecting lateral lobes; column cylindric, stout, less than 2.5 mm, foot less than 1 mm. **Capsules** 7–8 mm.

Flowering fall. Epiphytic on *Fraxinus caroliniana* (pop-ash); 0–10[–1200] m; Fla.; Mexico; West Indies; Central America (to Honduras); South America.

Maxillaria parviflora flowers in September in Florida and in November in Central America.

The Florida occurrence is based on a collection by Roger Hammer from a single host tree; other unconfirmed sightings have been made. The several large plants reported by Hammer, the occurrence in Cuba, and the presence of other widespread tropical species locally leaves little doubt of the identity and presence of this outlier in the flora.

The types of *Scaphyglottis parviflora* and *Ornithidium confertum* agree well with the Florida plant often recognized as *Maxillaria conferta*.

2. **Maxillaria crassifolia** (Lindley) Reichenbach f., Bonplandia (Hanover) 2: 16. 1854

Heterotaxis crassifolia Lindley, Bot. Reg. 12: plate 1028. 1826; *Dicrypta baueri* Lindley; *D. crassifolia* (Lindley) Lindley ex Loudon; *Epidendrum sessile* Swartz; *Maxillaria sessilis* (Swartz) Fawcett & Rendle

Plants epiphytic, cespitose, to 20 cm. **Stems** without pseudobulbs, or apparently so. **Leaves** 4–5 per shoot; blade linear-elliptic, 7–27 × 1.3–3 cm, leathery, apex unequally 2-lobed. **Inflorescences** several per leaf axil, successively borne and barely emerging, apparently supported on short, lateral rhizomes. **Flowers** yellow; sepals elliptic to ovate, 13–15 × 4–6 mm, apex acute, thickened; petals oblanceolate, 11–14 × 3–4 mm, apex acute; lip articulate with column foot, rhombic, 12–15 × 5–6 mm, lateral lobes inflexed toward column, constricted distal to middle; callus sticky, linear; column arching, cylindric, 9–10 mm, foot 1–2 mm. **Capsules** 2.5–3 cm.

Flowering throughout year (fall–winter, Fla.). Epiphytic; 0–10[–1300] m; Fla.; Mexico; West Indies; Central America; South America.

Maxillaria crassifolia is self-pollinating in Florida. It occurs much more commonly in the rest of its range.

230d.11b. ORCHIDACEAE Jussieu (tribe MAXILLARIEAE) subtribe ONCIDIINAE Bentham, J. Linn. Soc., Bot. 18: 288. 1881

64. BRASSIA R. Brown in W. Aiton and W. T. Aiton, Hortus Kew. 5: 215. 1813 • Spider orchid [for William Brass, an eighteenth-century British botanical illustrator and collector]

Mark W. Chase

Herbs, epiphytic, ± cespitose. **Stems** either reduced and covered with sheathing bracts or elongate with prominent, single-noded pseudobulb. **Leaves** 1(–2) per sympodium, from apex of pseudobulb, often with 1–3 short foliaceous bracts subtending single-noded pseudobulb; blade conduplicate, oblong-lanceolate, dorsiventrally flattened, margins entire. **Inflorescences** lateral, racemes; bracts narrowly triangular. **Flowers:** sepals and petals distinct and free, long, attenu-

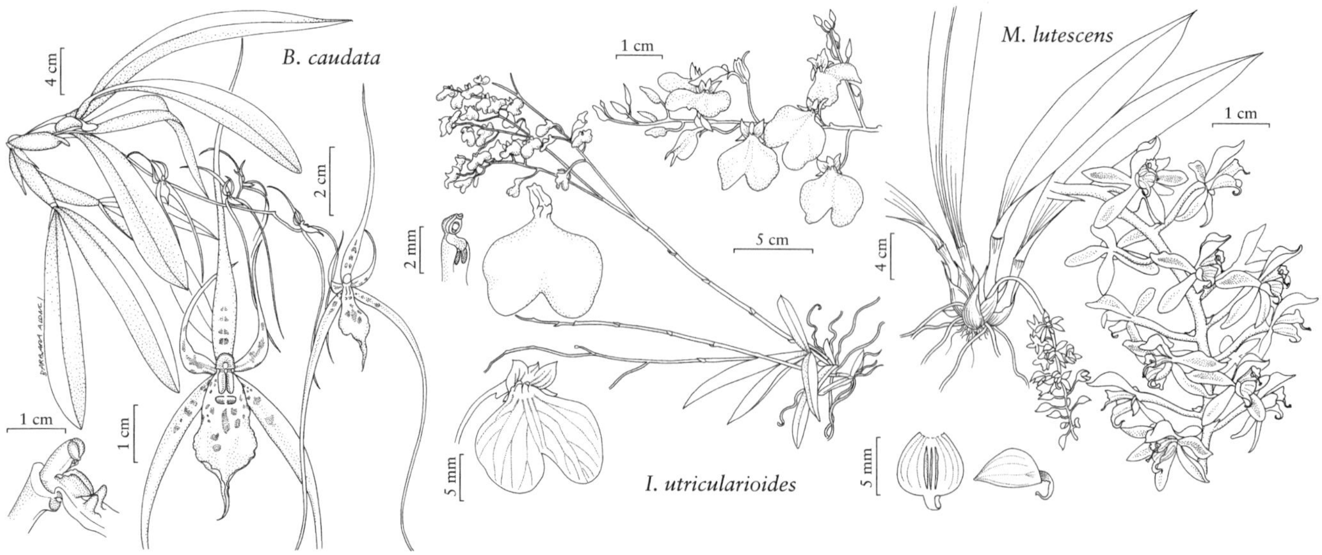

BRASSIA ◦ IONOPSIS ◦ MACRADENIA

ate-caudate; lip unlobed; callus prominent, 2-lamellate; column without appendages, footless; stigmatic cavity ± round; rostellum not prominent. **Fruits** capsules.

Species 35 (1 in the flora): subtropics and tropics, Western Hemisphere.

SELECTED REFERENCES Kooser, R. G. and G. C. Kennedy. 1979. The genus *Brassia* R. Brown section *Eubrassia* Lindley. Orchid Digest 43: 164–172. Williams, N. H. 1972. A reconsideration of *Ada* and the glumaceous brassias. Brittonia 24: 93–110.

1. **Brassia caudata** (Linnaeus) Lindley, Bot. Reg. 10: plate 832. 1824 [F]

Epidendrum caudatum Linnaeus, Syst. Nat. ed. 10, 2: 1246. 1759

Plants epiphytic, with prominent pseudobulbs, base covered by sheathing bracts, to 50 cm (exclusive of inflorescence). **Leaves** 1–2 with 1–3 foliaceous bracts; blade 12–40 × 2–8 cm, leathery. **Inflorescences** to 50 cm; bracts 0.5–1.3 cm. **Flowers** yellow to yellow-orange or yellow-green with red-brown spots on all perianth parts; sepals 30–200 × 3–6 mm, apex acuminate-setaceous; petals shorter than sepals, 14–37 × 3–5 mm, apex acuminate; lip yellow to yellow-green with several large red-brown spots near lip callus, ovate-lanceolate, 14–27 × 8–14 mm, apex abruptly acuminate; callus of 2 pubescent lamellae, each with distal tooth; column yellow, 2 mm. **Capsules** 1–15 cm.

Flowering winter–summer (Jan–Aug). Epiphytic in hammocks; 0[–1000] m; Fla.; Mexico; West Indies; Central America; n South America.

This widespread species is extremely variable in the dimensions of the sepals and petals. Plants in collections from Florida are generally smaller than those from other parts of the range.

65. IONOPSIS Kunth in A. von Humboldt et al., Nov. Gen. Sp. 1: 279, plate 83. 1816 • [Greek *ion*, violet, and *opsis*, having the appearance of]

Mark W. Chase

Herbs, epiphytic, cespitose [sometimes with elongate rhizome]. **Stems** reduced, covered with leafy bracts; pseudobulbs reduced, inconspicuous. **Leaves** 2–6 basal per sympodium, 0 apical, not recurved [1 apical, 2–4 terete, basal to pseudobulb]; blade conduplicate, linear-lanceolate, dorsiventrally flattened margins, entire. **Inflorescences** panicles [racemes]; bracts tubular. **Flowers:** petals and sepals shortly oblong-elliptic, apiculate; lateral sepals connate; lip 2-lobed; callus lobulate-lamellate; nectar cavity basal; column short (3.5 mm), without appendages; stigmatic cavity ± round; rostellum not prominent. **Fruits** capsules.

Species 2 (1 in the flora): tropical and subtropical regions, Western Hemisphere.

1. Ionopsis utricularioides (Swartz) Lindley, Coll. Bot., plate 39, fig. A. 1825 [F]

Epidendrum utricularioides Swartz, Prodr., 122. 1788

Plants to 15 cm (excluding inflorescence). **Leaves** often suffused with purple-brown; blade 3–13 cm × 6–18 mm. **Inflorescences** 1–3, erect-arching, to 20 cm; bracts 2–3 mm. **Flowers** pale pink to rose, pink-lavender, or white with pink flush, with darker rose-colored veins; sepals 3–7 × 1–2 mm; lateral sepals forming asymmetrically pouched cavity; petals 5–7 × 3–4 mm; lip pink-lavender with yellow patch near callus, 6–17 × 6–18 mm; callus of 2 lamellae and 2 lateral teeth, with short, basal, nectar horn, enclosed by nectar spur formed by lateral sepals.

Flowering Mar–Apr. Epiphytic on small branches and twigs in hammocks; 0–10[–1200] m; Fla.; Mexico; West Indies; Central America; South America.

Ionopsis utricularioides is one of the most widely ranging species of neotropical orchids. It occurs in several different variants, all of which intergrade. The various extreme forms, some with relatively enormous flowers, have been given form names, but little other than size separates them.

66. MACRADENIA R. Brown, Bot. Reg. 8: plate 612. 1822 • [Greek *makros*, large, and *aden*, gland, probably referring to the prominent viscidium, which is often referred to as a "gland"]

Mark W. Chase

Herbs, epiphytic, cespitose. **Stems** with prominent, elongate pseudobulbs partially covered by bracts. **Leaves** per sympodium 1, rarely 2, apical, leathery; blade conduplicate, oblong-lanceolate, dorsiventrally flattened, margins entire. **Inflorescences** racemes, arising from base of pseudobulb, pendent; bracts narrowly triangular. **Flowers:** petals and sepals oblong-elliptic, distinct and free; lip 3-lobed, without nectar cavity, callus shallow; column with prominent dorsal hood projecting over anther cap, otherwise without wings or arms; stigmatic cavity longitudinal; rostellum prominent, beaklike, projecting outward. **Fruits** capsules, ellipsoid, ± triangular in cross section.

Species 10–12 (1 in the flora): tropical and subtropical regions of Western Hemisphere.

1. Macradenia lutescens R. Brown, Bot. Reg. 8: plate 612. 1822 [F]

Plants to 18 cm (excluding inflorescence). **Leaves:** blade 8–15 × 1–2.5 cm. **Inflorescences** 1–2, to 12 cm; bracts 3–10 mm. **Flowers** cream to pale orange with red-brown to red-purple spots; sepals 6–120 × 3–6 mm, apex acute; petals 6–120 × 3–4 mm, apex acute; lip white to pale yellow with yellow patch near callus, 7–10 × 6–8 mm, lateral lobes nearly circular, middle lobe nearly linear, apex acute and strongly reflexed; callus of 3 shallow keels; column white, often with 2 red-brown streaks, 8 mm. **Capsules** 15–20 mm.

Flowering Oct–Dec. Epiphytic in shaded sites on small branches and twigs; 0–10[–500] m; Fla.; West Indies; n South America.

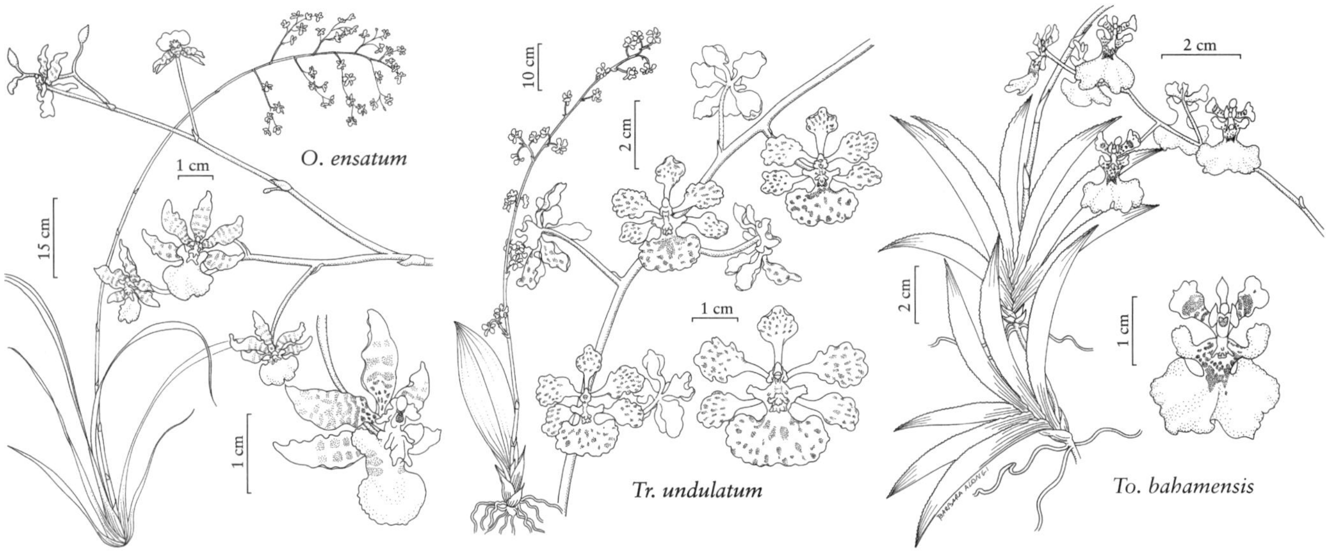

ONCIDIUM ◦ TRICHOCENTRUM ◦ TOLUMNIA

67. ONCIDIUM Swartz, Kongl. Vetensk. Acad. Nya Handl. 21: 239. 1800 • [Greek *oncos*, swelling, and *-idium*, diminutive, in reference to prominent lip callus]

Mark W. Chase

Herbs, epiphytic or terrestrial, ± cespitose. **Stems** pseudobulbs, green, ovoid. **Leaves** persistent, 1–3, not recurved; blade conduplicate, linear-lanceolate, dorsiventrally flattened, margins entire. **Inflorescences** paniculate racemes, lateral branches to 10[–42] cm; bracts narrowly triangular. **Flowers** resupinate; sepals and petals distinct and free, oblong-elliptic, obovate to nearly quadrangular or nearly circular, apex retuse to obtuse, sometimes apiculate; lip pandurate, 3-lobed; callus prominent warty-tuberculate-fringed; column with wings, arms, or fingerlike projections; stigmatic cavity ± round; rostellum not prominent. **Fruits** capsules.

Species 800 (1 in the flora): tropical and subtropical, Western Hemisphere.

Oncidium is a large polymorphic, and probably paraphyletic, assemblage.

SELECTED REFERENCES Chase, M. W. 1987. Systematic implications of pollinarium morphology in *Oncidium* Sw., *Odontoglossum* Kunth, and allied genera (Orchidaceae). Lindleyana 2: 8–28. Garay, L. A. and J. E. Stacy. 1974. A revision of *Oncidium*. Bradea 1: 393–428. Sauleda, R. P. and R. M. Adams. 1989. The orchid genera *Oncidium* Sw. and *Tolumnia* Raf. in Florida. Rhodora 91: 188–200.

1. Oncidium ensatum Lindley, Edwards's Bot. Reg. 28: Misc. 17. 1842 F

Oncidium floridanum Ames

Plants mostly epiphytic, sometimes terrestrial, to 100 cm (excluding inflorescence). **Stems** modified into prominent, compressed pseudobulbs, subtended by 1–3 leafy bracts. **Leaves** 3, apical [2, lateral] on pseudobulb, 2 lateral; blade linear-lanceolate, 60–100 × 3–5 cm, not leathery. **Inflorescences** erect-arching to pendent, 100–250 cm; bracts 5–8 mm. **Flowers** yellow with numerous small brown spots; sepals reflexed, oblong-elliptic, clawed, 16 × 3.5–5.5 mm, margins somewhat undulate, apex acute; petals reflexed at least at apex, oblong-elliptic, clawed,10–14 × 5–7 mm, apex acute; lip mostly yellow with brown patches near lip callus, 11–15 × 8.5–11 mm, apiculate, middle lobe about as wide as combined lateral lobes, isthmus 5–6 mm wide; callus of central ridge with 3–5 paired lateral ridges; column yellow, with pair of erect, axe-shaped, apical wings; pedicellate ovary [15–]24 mm. **Capsules** pendent, ellipsoid, 18–20 × 8–20 mm.

Flowering throughout year, especially summer–fall (May–Aug; Nov–Dec). In rich humus of dry hammocks and epiphytic on bases and knees of cypress trees in pockets of decaying debris; 0[–1800] m; Fla.; Mexico; West Indies (Bahamas, Cuba); Central America.

In describing *Oncidium floridanum*, O. Ames (1924) stated that the Florida plants were distinct from *O. sphacelatum* Lindley but failed to recognize the affinities of the Florida plants to the Mexican and Central American *O. ensatum*. Several differences between the plants from Florida and a cultivated specimen of *O. ensatum* from Guatemala were cited by R. P. Sauleda and R. M. Adams (1989). My study of *Oncidium ensatum* throughout its range reveals no features consistent enough to warrant recognizing the Florida populations as a separate taxon.

68. TRICHOCENTRUM Poeppig & Endlicher, Nov. Gen. Sp. Pl. 2: 11, plate 115. 1836 • Mule-ear oncidium [Greek *trichos*, hair, and *kentron*, spur, referring to the slender nectarless spur found in some species of the genus]

James D. Ackerman

Mark W. Chase

Herbs, epiphytic, lithophytic, or terrestrial, cespitose or creeping, sympodial. **Roots** numerous, white, velamentous. **Stems** highly reduced, rhizomatous, pseudobulbs, short-cylindric, laterally compressed, covered by scalelike sheathing bracts. **Leaves** deciduous, solitary, conduplicate, articulate; blade elliptic to broadly oblanceolate, large, fleshy, leathery, margins entire. **Inflorescences** axillary, from base of pseudobulb, racemes or panicles, loosely or densely few- to many-flowered; floral bracts without sheathing base, ovate to linear-lanceolate, small, apex acute. **Flowers** resupinate, generally large, fleshy, showy; sepals and petals spreading, distinct and free, often spotted; lip spotted or immaculate, unlobed to pandurate and 2–3-lobed; basal crest sometimes present, and if present variable in shape, often large, some species with slender spur formed by base of lip; column erect, winged; stigma simple, concave; anther operculate; pollinarium composed of small triangular to reniform viscidium, short convex stipe, and 2 obovoid pollinia. **Fruits** capsules, pendent, ellipsoid.

Species 54 (1 in the flora): subtropical and tropical regions, North America, Mexico, West Indies, Central America, South America.

Most of the species included in this expanded concept of *Trichocentrum*, including the single species in the flora, languished in *Oncidium* for a long time even though the vegetative morphology and chromosome counts indicated that it might be a good genus, or that it should be segregated from *Oncidium* (M. W. Chase 1986; R. L. Dressler 1993). Need for segregation of the group was reaffirmed by plastid DNA work (M. W. Chase and J. D. Palmer 1992). Those data, however, did not clearly indicate how those elements of the "Lophiaris clade" should be classified. It was suggested that the mule-ear and rat-tailed oncidiums should be grouped together along with the genus *Trichocentrum* (R. L. Dressler 1993). Additional molecular work (N. H. Williams et al. 2001) reinforced that opinion, making the concept of *Trichocentrum* broader than that of either *Lophiaris* sensu G. J. Braem (1993) or *Trichocentrum* sensu F. Pupulin (1995).

SELECTED REFERENCES Ackerman, J. D. 2000. Notes on the Caribbean orchid flora. II. Lindleyana 15: 89–95. Braem, G. J. 1993. Studies in the Oncidiinae: Discussion of some taxonomic problems with description of *Gudrunia* Braem, gen. nov., and reinstatement of the genus *Lophiaris* Rafinesque. Schlechteriana 1–2: 8–29. Chase, M. W. 1986. A reappraisal of the oncidioid orchids. Syst. Bot. 11: 477–491. Chase, M. W. and J. D. Palmer. 1992. Floral morphology and chromosome number in subtribe Oncidiinae (Orchidaceae): Evolutionary insights from a phylogenetic analysis of chloroplast DNA restriction sites variation. In: D. E. Soltis et al., eds. 1992. Molecular Systematics of Plants. New York. Pp. 324–339. Pupulin, F. 1995. A revision of the genus

Trichocentrum (Orchidaceae: Oncidiinae). Lindleyana 10: 183–210. Williams, N. H., M. W. Chase, T. Fulcher, and M. W. Whitten. 2001. Molecular systematics of the Oncidiinae based on evidence from four DNA sequence regions: Expanded circumscriptions of *Cyrtochilum, Erycina, Otogolossum,* and *Trichocentrum* and a new genus (Orchidaceae). Lindleyana 16: 113–139.

1. Trichocentrum undulatum (Swartz) Ackerman & M. W. Chase, Lindleyana 16: 225. 2001 [F]

Epidendrum undulatum Swartz, Prodr., 122. 1788; *Cymbidium undulatum* (Swartz) Swartz; *Lophiaris lurida* (Lindley) Braem; *L. maculata* (Aublet) Ackerman; *Oncidium guttatum* (Lindley not Linnaeus) Reichenbach f. var. *intermedium* (Lindley) Reichenbach f.; *O. intermedium* Knowles & Westcott; *O. lindenii* Loddiges ex Lindley; *O. luridum* Lindley; *O. luridum* var. *intermedium* Lindley; *O. maculatum* (Aublet) Urban; *O. undulatum* (Swartz) Salisbury

Plants epiphytic, lithophytic, or sometimes terrestrial, cespitose, to 3 m. **Stems:** pseudobulbs laterally compressed, 1–2 cm. **Leaves** from apex of pseudobulb, green, sometimes spotted or suffused with purple, elliptic to ovate-lanceolate, 10–85 × 5–12 cm, margins entire. **Inflorescences** arcuate-pendent, 0.8–3 m (including peduncle); peduncle terete, 3.5–8 mm diam.; bracts distant, tubular sheaths, 6–15 mm, scarious; panicles open, branches few- to several-flowered, 1–42 cm; floral bracts linear-lanceolate, 3–5(–10) mm, scarious. **Flowers** yellow (infrequently brown to green) with reddish brown blotchy spots or not, fragrant or odorless; sepals clawed, margins undulate; dorsal sepal suborbiculate, rounded, 180–190 × 9–11 mm; lateral sepals broadly elliptic, 16–17 × 6–7 mm; petals clawed, broadly elliptic, 16–20 × 8–11 mm, margins undulate, apex obtuse; lip 3-lobed, pandurate, 16–19 mm, middle lobe ± reniform, 18–22 mm wide, margins undulate, apex retuse, lateral lobes basal, spreading, appearing auriculate, small, margins recurved, apex retuse; disc red with prominent, yellow, multi-horned callus composed of 2 pairs of tuberculate protuberances separated by lobed median ridge; column erect, white, wings prominent, 5–6 × 7 mm; ovary 24–30 mm. **Capsules** pendent, green, ellipsoid, 30–60 × 15–20 mm. $2n = 32$.

Flowering (Mar–)Apr–Oct. In understory of hilly broadleaf forests on tree trunks, rocks, or in leaf mold on limestone rocks; 10–950 m; Fla.; Mexico; West Indies (Bahamas, Greater Antilles, Lesser Antilles); Central America; South America (to Colombia, n Ecuador and Peru, Venezuela, French Guiana, Surinam, Guyana, n Brazil).

The nomenclature of this species has a complex and confusing history. Some have argued that Jacquin based his *Epidendrum altissimum* on an illustration published by Sloane, which is represented by the current concept of *L. maculata*. Consequently, this species was regarded as *Oncidium altissimum* (Jacquin) Swartz (L. A. Garay and H. R. Sweet 1972, 1974). Others, however, argued that Jacquin's name was based on Jacquin's collections and that *Epidendrum altissimum* does not refer to this species (R. L. Dressler and N. H. Williams 1975, 1982). The latter view has prevailed (R. K. Brummitt 1985).

Two names have been commonly associated with this concept. *Oncidium luridum* Lindley appears in floras and popular books (e.g., H. León 1946; E. Foldats 1969–1970, part 5; C. D. Adams 1972; C. A. Luer 1972; R. L. Dressler 1993; H. Bechtel et al. 1981; A. Gloudon and C. Tobisch 1995) even though several earlier names were well known. The second name, *Oncidium guttatum* (Linnaeus) Fawcett & Rendle, has been frequently placed in synonymy with *Oncidium luridum* but the Linnaean name applies to what has been known as *Oncidium tetrapetalum* (Jacquin) Willdenow or *Tolumnia tetrapetala* (Jacquin) Braem (M. Nir 1994).

Some authors have treated *Oncidium undulatum* as a synonym of *Trichocentrum luridum,* but R. P. Sauleda and R. M. Adams (1989) argued that they represent two distinct species. Although not explicitly stated, *O. undulatum* supposedly occurs in Florida, Bahamas, Cuba, and Jamaica whereas *O. luridum* represents plants of Mexico, Central America, and South America. The authors gave no indication as to the identity of those plants in the Lesser Antilles, and presumably Trinidad would be included with the South American plants. *Oncidium luridum* has a reniform lip, short-clawed and obovate to broadly elliptic sepals and petals, and simple column wings, whereas *O. undulatum* has a narrowly reniform lip, long-clawed, spatulate to obovate sepals and petals, and 2-lobed column wings. *Oncidium undulatum* has longer and wider leaves and inflorescences twice as long as those of *O. luridum.* Furthermore, the sepals, petals, and lip of *Oncidium undulatum* are yellow with reddish brown spots, and the sepals and petals of *O. luridum* are yellowish brown with brown blotches that usually overlap, while the lip is brownish orange without spots. None of the numerous characteristics that Sauleda and Adams use to differentiate the two taxa hold up to scrutiny except perhaps flower color.

Differences may yet be identified between the two, but the patterns of variation certainly need to be clarified. Some taxa are simply widespread and variable (cf. J. D. Ackerman and M. Galarza-Pérez 1991), and *Trichocentrum undulatum* could be one of those (F. R. Stermitz 1982, for Mexican populations). The oldest name available for the broader concept of this species is *T. undulatum,* which we use here, but our use of this epithet does not imply that we accept that the material in the Caribbean is distinct from the Central and South American material that is typically referred to *T. luridum.*

69. TOLUMNIA Rafinesque, Fl. Tellur. 2: 101. 1837 • [Probably for Tolumnius, a Rutulian mentioned by Virgil]

Mark W. Chase

James D. Ackerman

Xaritonia Rafinesque

Herbs, epiphytic or rarely terrestrial, ± climbing, fan-shaped. **Rhizomes** often elongate, covered with bracts; pseudobulbs reduced to absent. **Leaves:** basal 3–7, apical not more than 1; blade recurved, lanceolate, laterally flattened–3-gonal, margins often serrulate. **Inflorescences** racemes or weakly paniculate; bracts narrowly triangular. **Flowers:** sepals and petals spatulate-clawed; lateral sepals sometimes partially connate; lip 3-lobed, pandurate, without nectar cavity; callus prominent tuberculate-lobulate; column with pair of outstretched apical wings, without hood; stigmatic cavity round; rostellum not prominent. **Fruits** capsules, ellipsoid.

Species 30 (1 in the flora): restricted to s Florida, West Indies.

Tolumnia does not belong to the main body of *Oncidium* (its traditional placement) even though many of the species have floral morphologies nearly identical to that of *Oncidium*. It shares, with other twig epiphytes, a fan-shaped habit and seeds with prominent extensions of the testa.

SELECTED REFERENCES Braem, G. J. 1986. *Tolumnia*—Der neue, aber doch alte, Name für die "Variegaten Oncidien." Orchidee (Hamburg) 37: 55–59. Chase, M. W. and J. S. Pippen. 1988. Seed morphology in the Oncidiinae and related subtribes (Orchidaceae). Syst. Bot. 13: 313–323. Williams, N. H., M. W. Chase, T. Fulcher, and W. M. Whitten. 2001. Molecular systematics of the Oncidiinae based on evidence from four DNA sequence regions: Expanded circumscriptions of *Cyrtochilum*, *Erycina*, *Otoglossum*, and *Trichocentrum* and a new genus (Orchidaceae). Lindleyana 16: 113–139.

1. Tolumnia bahamensis (Nash ex Britton & Millspaugh) Braem, Orchidee (Hamburg) 37: 59. 1986 [F]

Oncidium bahamense Nash ex Britton & Millspaugh, Bahama Fl., 97. 1920; *O. variegatum* (Swartz) Swartz; *O. variegatum* subsp. *bahamense* (Nash ex Britton & Millspaugh) Withner

Plants to 12 cm (excluding inflorescence). **Leaves** semiterete 3–10 × 0.4–1 cm. **Inflorescences** 1–4, erect, to 50 cm, side branches few or none, to 4.5 cm; bracts 2–6 mm. **Flowers** white with red-brown spots at bases of perianth parts; sepals spatulate, 4–6 × 1–2 mm, apex apiculate; lateral sepals connate to $^{3}/_{4}$ length; petals spatulate to pandurate, 5 × 2 mm, apex rounded to retuse, often apiculate; lip with large red-brown spots near yellow callus, 8–12 × 8–13 mm, middle lobe about $^{1}/_{3}$ wider than combined lateral lobes, apex retuse-apiculate, isthmus 0.2–0.4 cm wide with serrate margins; callus with 2 sets of tuberculate processes, 3 proximal, broader, 2 distal, smaller; column whitish to pale lavender with yellow base and white to rose-purple column wings 5–6 mm. **Capsules** 1.5 cm.

Flowering Apr–Jun. Terrestrial or epiphytic on shrubs and scrub trees in dry thickets; 0–10 m; Fla.; West Indies (Bahamas).

Tolumnia bahamensis has been regarded by some authors as a synonym or variety of *T. variegata* (D. S. Correll 1950; C. L. Withner 1980). Flower morphology helps little in the *T. variegata* complex (J. D. Ackerman and M. Galarza-Pérez 1991); *T. bahamensis* differs, however, by being a tetraploid (84 chromosomes) and having semiterete leaves. These features also exist in *T. sylvestris* (G. J. Braem 1986b), a species of eastern Cuba generally found in high elevation pine forests, a habitat somewhat different from that of either Florida or the Bahamas. Whether or not *T. sylvestris* and *T. bahamensis* are conspecific awaits further studies.

Literature Cited

Robert W. Kiger, Editor

This is a consolidated list of all works cited in volume 26, whether as selected references, in text, or in nomenclatural contexts. In citations of articles, both here and in the taxonomic treatments, and also in nomenclatural citations, the titles of serials are rendered in the abbreviated forms recommended in G. D. R. Bridson and E. R. Smith (1991). Cross references to the corresponding full serial titles are interpolated here alphabetically by abbreviated form. In nomenclatural citations (only), book titles are rendered in the abbreviated forms recommended in F. A. Stafleu and R. S. Cowan (1976–1988) and F. A. Stafleu and E. A. Mennega (1992+). Here, those abbreviated forms are indicated parenthetically following the full citations of the corresponding works, and cross references to the full citations are interpolated in the list alphabetically by abbreviated form. Two or more works published in the same year by the same author or group of coauthors will be distinguished uniquely and consistently throughout all volumes of *Flora of North America* by lower-case letters (b, c, d, ...) suffixed to the date for the second and subsequent works in the set. The suffixes are assigned in order of editorial encounter and do not reflect chronological sequence of publication. The first work by any particular author or group from any given year carries the implicit date suffix "a"; thus, the sequence of explicit suffixes begins with "b". Some citations in this list have dates suffixed "b" but are not preceded by citations of "[a]" works for the same year, or have dates suffixed "c," "d," or "e" but are not preceded by citations of "[a]," "b," "c," and/or "d" works for that year. In these cases, the missing "[a]," "b," "c," and "d" works are ones cited (and encountered first from) elsewhere in the *Flora* that are not pertinent here.

A. S. B. Bull. = A S B Bulletin.

Abh. Königl. Akad. Wiss. Berlin = Abhandlungen der Königlichen Akademie der Wissenschaften in Berlin.

Abh. Königl. Ges. Wiss. Göttingen = Abhandlungen der Königlichen Gesellschaft der Wissenschaften zu Göttingen.

Abh. Naturwiss. Vereine Bremen = Abhandlungen herausgegeben vom Naturwissenschaftlichen Vereine zu Bremen.

Abh. Schles. Ges. Vaterl. Cult., Abth. Naturwiss. (Naturwiss.-Med. Abth.) = Abhandlungen der Schlesischen Gesellschaft für Vaterländische Cultur. Abtheilung für Naturwissenschaften un Medicin (later Naturwissenschaftlich-medicinische Abtheilung).

Abramova, L. I. 1975. On the taxonomical structure of the genus *Polygonatum* Mill. Bot. Zhurn. (Moscow & Leningrad) 60: 490–497.

Abrams, L. and R. S. Ferris. 1923–1960. Illustrated Flora of the Pacific States: Washington, Oregon, and California. 4 vols. Stanford. (Ill. Fl. Pacific States)

Ackerman, J. D. 1977. Biosystematics of the genus *Piperia* Rydb. (Orchidaceae). Bot. J. Linn. Soc. 75: 245–270.

Ackerman, J. D. 1983. Specificity and mutual dependency of the orchid-euglossine bee interaction. Biol. J. Linn. Soc. 20: 301–314.

Ackerman, J. D. 1989. *Prescotia* and *Cranichis* of Puerto Rico and the Virgin Islands. Lindleyana 4: 42–47.

Ackerman, J. D. 1995. An orchid flora of Puerto Rico and the Virgin Islands. Mem. New York Bot. Gard. 73.

Ackerman, J. D. 2000. Notes on the Caribbean orchid flora. II. Lindleyana 15: 89–95.

Ackerman, J. D. and M. Galarza-Pérez. 1991. Patterns and maintenance of extraordinary variation in the Caribbean orchid, *Tolumnia (Oncidium) variegata*. Syst. Bot. 16: 182–194.

Ackerman, J. D. and M. R. Mesler. 1979. Pollination biology of *Listera cordata* (Orchidaceae). Amer. J. Bot. 66: 820–824.

Acta Bot. Austro Sin. = Acta Botanica Austro Sinica; Contributions from the South China Institute of Botany, Academia Sinica. [Zhongguo Kexueyuan Huanan Zhiwu Yanjiusuo Jikan.]

Acta Bot. Brasil. = Acta Botanica Brasilica.

Acta Bot. Mex. = Acta Botanica Mexicana.

Acta Helv. Phys.-Math. = Acta Helvetica, Physico-mathematico-anatomico-botanico-medica.

Acta Horti Berg. = Acta Horti Bergiani.

Acta Phytotax. Geobot. = Acta Phytotaxonomica et Geobotanica. [Shokubutsu Bunrui Chiri.]

Acta Phytotax. Sin. = Acta Phytotaxonomica Sinica. [Chih Wu Fen Lei Hsüeh Pao.]

Adams, C. D. 1972. Flowering Plants of Jamaica. Mona.

Adams, R. M. and W. J. Dress. 1982. Nodding *Lilium* species of eastern North America (Liliaceae). Baileya 21: 165–188.

Adams, R. M. and G. J. Goss. 1976. The reproductive biology of the epiphytic orchids of Florida III. *Epidendrum anceps* Jacquin. Amer. Orchid Soc. Bull. 43: 488–492.

Adams, S. P. et al. 2001. Loss and recovery of *Arabidopsis*-type telomere repeat sequences 5'-(TTTAGGG)(n)-3' in the evolution of a major radiation of flowering plants. Proc. Roy. Soc. Biol. Sci. Ser. B 268: 1541–1546.

Adanson, M. 1763[–1764]. Familles des Plantes. 2 vols. Paris. [Vol. 1, 1764; vol. 2, 1763.] (Fam. Pl.)

Adansonia = Adansonia; Recueil Périodique d'Observations Botaniques.

Addisonia = Addisonia; Colored Illustrations and Popular Descriptions of Plants.

Aden, P. 1988. The *Hosta* Book. Portland.

Agave Fam. Sonora—See: H. S. Gentry 1972

Agaven—See: A. Berger 1915

Agaves Continental N. Amer.—See: H. S. Gentry 1982

Aiton, W. 1789. Hortus Kewensis; or, a Catalogue of the Plants Cultivated in the Royal Botanic Garden at Kew. 3 vols. London. (Hort. Kew.)

Aiton, W. and W. T. Aiton. 1810–1813. Hortus Kewensis; or a Catalogue of the Plants Cultivated in the Royal Botanic Garden at Kew. 5 vols. London. (Hortus Kew.)

Ajilvsgi, G. 1984. Wildflowers of Texas. Bryan, Tex.

Aker, S. and W. Healy. 1990. The phytogeography of the genus *Alstroemeria*. Herbertia 46: 76–87.

Alabama Herit. = Alabama Heritage.

Albee, B. J., L. M. Shultz, and S. Goodrich. 1988. Atlas of the Vascular Plants of Utah. Salt Lake City.

Alexander, E. J. 1937. Pontederiaceae. In: N. L. Britton et al., eds. 1905+. North American Flora.... 47+ vols. New York. Vol. 19, pp. 51–60.

Allg. Bot. Z. Syst. = Allgemeine Botanische Zeitschrift für Systematik, Floristik, Pflanzengeographie etc.

Allg. Gartenzeitung = Allgemeine Gartenzeitung.

Al-Shehbaz, I. A. and B. G. Schubert. 1989. The Dioscoreaceae in the southeastern United States. J. Arnold Arbor. 70: 57–95.

Alvarez, A. and E. Köhler. 1987. Morfología del polen de las Agavaceae y algunos géneros afines. Grana 26: 25–46.

Ambrose, J. D. 1975. Comparative Anatomy and Morphology of the Melanthioideae (Liliaceae). Ph.D. dissertation. Cornell University.

Ambrose, J. D. 1980. A re-evaluation of the Melanthioideae (Liliaceae) using numerical analyses. In: C. D. Brickell et al., eds. 1980. Petaloid Monocotyledons: Horticultural and Botanical Research. London and New York. Pp. 65–81.

Ambrose, J. D. 1985. *Lophiola*, familial affinity with the Liliaceae. Taxon 34: 140–150.

Amer. Bot. (Binghamton) = The American Botanist; A Monthly Journal for the Plant Lover.

Amer. Bot. Fl.—See: A. Wood 1870

Amer. Hort. Mag. = American Horticultural Magazine.

Amer. J. Bot. = American Journal of Botany.

Amer. J. Pharm. = The American Journal of Pharmacy.

Amer. J. Sci. Arts = American Journal of Science, and Arts.

Amer. Midl. Naturalist = American Midland Naturalist; Devoted to Natural History, Primarily That of the Prairie States.

Amer. Monthly Mag. & Crit. Rev. = American Monthly Magazine and Critical Review.

Amer. Naturalist = American Naturalist....

Amer. Orchid Soc. Bull. = American Orchid Society Bulletin.

Ames, O. 1904. A Contribution to Our Knowledge of the Orchid Flora of Southern Florida.... Cambridge, Mass. (Contr. Orch. Fl. S. Florida)

Ames, O. 1905–1922. Orchidaceae: Illustrations and Studies of the Family Orchidaceae Issuing from the Ames Botanical Laboratory.... 7 vols. Boston and New York. (Orchidaceae)

Ames, O. 1924. Additions to the orchid flora of tropical America. Schedul. Orchid. 7.

Ames, O. 1924b. An Enumeration of the Orchids of the United States and Canada.... Boston. (Enum. Orchids U.S. & Canada)

Ames, O. 1938. *Zeuxine strateumatica* in Florida. Bot. Mus. Leafl. 6: 37–45.

Ames, O., F. T. Hubbard, and C. Schweinfurth. 1936. The Genus *Epidendrum* in the United States and Middle America. Cambridge, Mass.

Ames, O. and C. Schweinfurth. 1935. Nomenclatural studies in *Malaxis* and *Spiranthes*. Bot. Mus. Leafl. 3: 113–133.

Amoroso, J. L., ed. 1997. Natural Heritage Program List of the Rare Plant Species of North Carolina. Raleigh.

Anales Inst. Biol. Univ. Nac. Auton. Mexico, Bot. = Anales del Instituto de Biológia de la Universidad Nacional Autónoma de México. Série Botánica.

Anderson, E. 1936. The species problem in *Iris*. Ann. Missouri Bot. Gard. 23: 457–509.

Anderson, E. and T. W. Whitaker. 1934. Speciation in *Uvularia*. J. Arnold Arbor. 15: 28–42.

Anderson, L. C. and G. L. Smith. 1994. Carpels and chromosomes of the endangered Harper's beauty (*Harperocallis flava*). [Abstract.] A. S. B. Bull. 41: 103.

Andersson, L. and M. W. Chase. 2001. Phylogeny and classification of Marantaceae. Bot. J. Linn. Soc. 135: 275–287.

Andrews, H. 1961. Two new orchid records for Ontario. Rhodora 63: 175–176.

Angelo, R. and D. E. Boufford. 2000. Atlas of the flora of New England: Monocots except Poaceae and Cyperaceae. Rhodora 102: 1–119.

Angiosperm Phylogeny Group. 1998. An ordinal classification for the families of flowering plants. Ann. Missouri Bot. Gard. 85: 531–553.

Ann. Bot. (König & Sims) = Annals of Botany. [Edited by König & Sims.]

Ann. Bot. (Oxford) = Annals of Botany. (Oxford.)

Ann. Bot. Soc. Zool.-Bot. Fenn. "Vanamo" = Annales Botanici Societatis Zoologicae-Botanicae Fennicae "Vanamo."

Ann. Bot. Syst.—See: W. G. Walpers and K. Müller berol. 1848–1871

Ann. Carnegie Mus. = Annals of the Carnegie Museum.

Ann. Kirstenbosch Bot. Gard. = Annals of Kirstenbosch Botanic Gardens.

Ann. Lyceum Nat. Hist. New York = Annals of the Lyceum of Natural History of New York.

Ann. Missouri Bot. Gard. = Annals of the Missouri Botanical Garden.

Ann. Mus. Goulandris = Annales Musei Goulandris; Contributiones ad Historiam Naturalem Graeciae et Regionis Mediterraneae.

Ann. Mus. Natl. Hist. Nat. = Annales du Muséum National d'Histoire Naturelle. ["National" dropped with vol. 5.]

Ann. Nat. Hist. = Annals of Natural History; or, Magazine of Zoology, Botany and Geology.

Ann. Sci. Nat., Bot. = Annales des Sciences Naturelles. Botanique.

Annual Rev. Ecol. Syst. = Annual Review of Ecology and Systematics.

Appendix—See: W. Herbert 1821

Applegate, E. I. 1935. The genus *Erythronium*: A taxonomic and distributional study of the western North American species. Madroño 3: 58–113.

Aquatic Bot. = Aquatic Botany; International Scientific Journal Dealing with Applied and Fundamental Research on Submerged, Floating and Emergent Plants in Marine and Freshwater Ecosystems.

Arber, A. 1918. The phyllode theory of the monocotyledonous leaf, with special reference to anatomical evidence. Ann. Bot. (Oxford) 32: 465–501.

Arber, A. 1920. Tendrils of *Smilax*. Bot. Gaz. 69: 438–442.

Arber, A. 1920b. Water Plants: A Study of Aquatic Angiosperms. Cambridge.

Arber, A. 1921. The leaf structure of the Iridaceae. Ann. Bot. (Oxford) 35: 301–336.

Arber, A. 1925. Monocotyledons. A Morphological Study. Cambridge.

Arch. Bot. (Leipzig) = Archiv für die Botanik. (Leipzig.)

Arch. Gewächsk.—See: L. Trattinnick [1811–]1812–1818

Arch. Hydrobiol. = Archiv für Hydrobiologie.

Arditti, J., ed. 1977+. Orchid Biology: Reviews and Perspectives. 7+ vols. Ithaca, N.Y. and New York.

Ark. Bot. = Arkiv för Botanik Utgivet av K. Svenska Vetenskapsakademien.

Artz, L. 1951. *Xerophyllum asphodeloides* (L.) Nutt. in the Massanutten Mountains in Virginia. Castanea 16: 124–125.

Ascherson, P. F. A. 1889. Potamogetonaceae. In: H. G. A. Engler and K. Prantl, eds. 1887–1915. Die natürlichen Pflanzenfamilien.... 254 fasc. Leipzig. Fasc. 26[II,1], pp. 194–214.

Asmussen, C. B., W. J. Baker, and J. Dransfield. 2000. Phylogeny of the palm family (Arecaceae) based on *rps*16 intron and *trn*L-*trn*F plastid DNA sequences. In: K. L. Wilson and D. A. Morrison, eds. 2000. Monocots: Systematics and Evolution. Melbourne. Pp. 525–535.

Asmussen, C. B. and M. W. Chase. 2001. Coding and noncoding plastid DNA in palm systematics. Amer. J. Bot. 88: 1103–1117.

Aston, H. 1977. Aquatic Plants of Australia. Melbourne.

Atlantic J. = Atlantic Journal, and Friend of Knowledge.

Atwood, J. T. 1993. A revision of the *Maxillaria neglecta* complex (Orchidaceae) in Mesoamerica. Lindleyana 8: 25–31.

Aublet, J. B. 1775. Histoire des Plantes de la Guiane Françoise.... 4 vols. Paris. [Vols. 1 and 2: text, paged consecutively; vols. 3 and 4: plates.] (Hist. Pl. Guiane)

Audubon Wildlife Rep. = Audubon Wildlife Report.

Austral. J. Bot. = Australian Journal of Botany.

Austral. Syst. Bot. = Australian Systematic Botany.

Autik. Bot.—See: C. S. Rafinesque 1840

Averyanov, L. V. 1990. A review of the genus *Dactylorhiza*. In: J. Arditti, ed. 1977+. Orchid Biology: Reviews and Perspectives. 7+ vols. Ithaca, N.Y. and New York. Vol. 5, pp. 159–206.

Ayensu, E. S. 1972. Dioscoreales. In: C. R. Metcalfe, ed. 1960+. Anatomy of the Monocotyledons. 8+ vols. Oxford. Vol. 6.

Ayensu, E. S. 1973. Phytogeography and evolution of the Velloziaceae. In: B. J. Meggers et al., eds. 1973. Tropical Forest Ecosystems in Africa and South America: A Comparative Review. Washington. Pp. 105–119.

Ayensu, E. S. 1974. Leaf anatomy and systematics of New World Velloziaceae. Smithsonian Contr. Bot. 15.

Badawi, A. and Z. Elwan. 1986. A taxonomic study of Liliaceae sensu lato: I. Numerical analysis. Phytologia 60: 201–213.

Badawi, A. and Z. Elwan. 1986b. A taxonomic study of Liliaceae sensu lato: II. Evaluation of Engler's subfamilies. Phytologia 60: 214–221.

Baer, R. G. and P. C. Quimby Jr. 1981. Laboratory rearing

and life history of *Arzama densa,* a potential native biological control agent against waterhyacinth. J. Aquatic Pl. Managem. 19: 22–26.

Bahama Fl.—See: N. L. Britton and C. F. Millspaugh 1920

Bailey, L. H. 1930. *Hosta:* The plantain-lilies. Gentes Herb. 2: 119–142.

Bailey, L. H., E. Z. Bailey, and Bailey Hortorium Staff. 1976. Hortus Third. A Concise Dictionary of Plants Cultivated in the United States and Canada. New York.

Baileya = Baileya; a Quarterly Journal of Horticultural Taxonomy.

Baker, C. A. 1951. Pontederiaceae. In: C. G. G. J. van Steenis, ed. 1950+. Flora Malesiana.... Series I. Spermatophyta. 11+ vols. in parts. Djakarta and Leiden. Vol. 4, part 3, pp. 255–261.

Baker, J. G. 1879. A synopsis of Colchicaceae and the aberrant tribes of Liliaceae. J. Linn. Soc., Bot. 17: 405–510.

Baker, J. G. 1888. Handbook of the Amaryllideae, Including the Alstroemerieae and Agaveae. London. (Handb. Amaryll.)

Baker, J. G. 1892. Handbook of the Irideae. London and New York.

Baker, W. J. et al. 1999. A phylogenetic study of the palm family (Palmae) based on chloroplast DNA sequences from the *trn*L-*trn*F region. Pl. Syst. Evol. 219: 111–126.

Baker, W. J. and J. Dransfield. 2000. Towards a biogeographic explanation of the calamoid palms. In: K. L. Wilson and D. A. Morrison, eds. 2000. Monocots: Systematics and Evolution. Melbourne. Pp. 545–553.

Baker, W. J., J. Dransfield, and T. A. Hedderson. 2000. Phylogeny, character evolution, and a new classification of the calamoid palms. Syst. Bot. 25: 297–322.

Baldwin, W. K. W. 1961. *Malaxis paludosa* (L.) Sw. in the Hudson Bay lowlands. Canad. Field-Naturalist 75: 74–77.

Ballantyne, O. H. 1983. Lilies of the Pacific coast. Lily Yearb. N. Amer. Lily Soc. 36: 11–26.

Balslev, H. 1998. Juncaceae. In: K. Kubitzki et al., eds. 1990+. The Families and Genera of Vascular Plants. 4+ vols. Berlin etc. Vol. 4, pp. 252–260.

Banerjee, S. and A. K. Sharma. 1987. Cytophotometric estimation of nuclear DNA in different species and varieties of *Agave.* Cytologia 52: 85–90.

Banks, J. A. 1980. The reproductive biology of *Erythronium propullans* Gray and sympatric populations of *E. albidum* Nutt. (Liliaceae). Bull. Torrey Bot. Club 107: 181–188.

Barker, N. P., H. P. Linder, and E. F. Harley. 1995. Polyphyly in the Arundinoideae (Poaceae): Evidence from *rbc*L. Syst. Bot. 20: 423–435.

Barkman, T. J. et al. 2000. Independent and combined analyses of sequences from all three genomic compartments converge on the root of flowering plant phylogeny. Proc. Natl. Acad. Sci. U.S.A. 97: 13166–13171.

Barksdale, L. 1938. The pedicellate species of *Trillium* found in the southern Appalachians. J. Elisha Mitchell Sci. Soc. 54: 271–296.

Barkworth, M. E. 1975. Intraspecific Variation in *Brodiaea douglasii* Watson, Liliaceae. Ph.D. thesis. Washington State University.

Barkworth, M. E. 1977. Intraspecific variation in *Brodiaea douglasii* Watson (Liliaceae). NorthW. Sci. 51: 79–90.

Barnard, C. 1960. Floral histogenesis in the monocotyledons. IV. The Liliaceae. Austral. J. Bot. 8: 213–225.

Barrett, S. C. H. 1977. Tristyly in *Eichhornia crassipes* (Mart.) Solms (water hyacinth). Biotropica 9: 230–238.

Barrett, S. C. H. 1978. The floral biology of *Eichhornia azurea* (Swartz) Kunth (Pontederiaceae). Aquatic Bot. 5: 217–228.

Barrett, S. C. H. 1979. The evolutionary breakdown of tristyly in *Eichhornia crassipes* (Mart.) Solms (water hyacinth). Evolution 33: 499–510.

Barrett, S. C. H. 1980. Sexual reproduction in <Eichhornia crassipes> (water hyacinth) I. Fertility of clones from diverse regions. J. Appl. Ecol. 17: 101–112.

Barrett, S. C. H. 1980b. Sexual reproduction in *Eichhornia crassipes* (water hyacinth) II. Seed production in natural populations. J. Appl. Ecol. 17: 113–124.

Barrett, S. C. H. 1988. Evolution of breeding systems in *Eichhornia* (Pontederiaceae): A review. Ann. Missouri Bot. Gard. 75: 741–760.

Barrett, S. C. H. 1989. Waterweed invasions. Sci. Amer. 261(4): 90–97.

Barrett, S. C. H. and I. W. Forno. 1982. Style morph distribution in New World populations of *Eichhornia crassipes* (Mart.) Solms-Laubach (water hyacinth). Aquatic Bot. 13: 299–306.

Barrett, S. C. H. and D. E. Seaman. 1980. The weed flora of California rice fields. Aquatic Bot. 9: 351–376.

Barrows, E. M. 1979. Flower biology and arthropod associates of *Lilium philadelphicum.* Michigan Bot. 18: 109–115.

Bartlett, H. H. 1910. The source of the drug Dioscorea, with a consideration of the *Dioscoreae* found in the United States. U.S.D.A. Bur. Pl. Industr. Bull. 189: 1–29.

Bartonia = Bartonia; a Botanical Annual.

Bartram, W. 1791. Travels through North and South Carolina, Georgia, East and West Florida, the Cherokee Country, the Extensive Territories of the Muscogulges, or Creek Confederacy, and the Country of the Chactaws.... Philadelphia. (Travels Carolina)

Bateman, R. M. and I. Denholm. 1983. A reappraisal of the British and Irish dactylorchids: 1. The tetraploid marsh orchids. Watsonia 14: 347–376.

Bateman, R. M. and I. Denholm. 1985. A reappraisal of the British and Irish dactylorchids: 2. The diploid marsh orchids. Watsonia 15: 321–355.

Bateman, R. M. and I. Denholm. 1989. A reappraisal of the British and Irish dactylorchids: 3. The spotted orchids. Watsonia 17: 319–349.

Bateman, R. M., A. M. Pridgeon, and M. W. Chase. 1997. Phylogenetics of subtribe Orchidinae (Orchidoideae, Orchidaceae) based on nuclear ITS sequences: 2. Infrageneric relationships and taxonomic revision to achieve monophyly of *Orchis* sensu stricto. Lindleyana 12: 113–141.

Bauer, F. A. and J. Lindley. 1830–1838. Illustrations of Orchidaceous Plants.... 4 parts. London. (Ill. Orch. Pl.)

Bauml, J. A. 1979. A Study of the Genus *Hymenocallis* (Amaryllidaceae in Mexico. M.S. thesis. Cornell University.

Bayer, E. 1998. Alstroemeriaceae. In: K. Kubitzki et al., eds.

1990+. The Families and Genera of Vascular Plants. 4+ vols. Berlin etc. Vol. 3, pp. 79–83.

Bean, R. C., C. H. Knowlton, and A. F. Hill. 1951. Tenth report of the Committee on Plant Distribution. Rhodora 53: 79–89.

Bechtel, H., P. Cribb, and E. Launert. 1981. The Manual of Cultivated Orchid Species. Cambridge, Mass.

Beck, C. H. 1951. Fritillaries: A Gardener's Introduction to the Genus *Fritillaria*. London.

Beck, L. C. 1833. Botany of the Northern and Middle States; or, a Description of the Plants found in the United States, North of Virginia.... Albany. (Bot. North. Middle States)

Beck, L. C. 1848. Botany of the United States North of Virginia..., ed. 2. New York. (Bot. North. Middle States ed. 2)

Becking, R. W. 1986. *Hastingsia atropurpurea* (Liliaceae: Asphodeleae). A new species from southwestern Oregon. Madroño 33: 175–181.

Becking, R. W. 1989. Segregation of *Hastingsia serpentinicola* sp. nov. from *Hastingsia alba*. Madroño 36: 208–216.

Beckner, J. 1996. *Spiranthes sinensis* (Persoon) Ames found in Florida. N. Amer. Native Orchid J. 2: 154–157.

Bedell, H. G. and J. L. Reveal. 1982. Amended outlines and indices for six recently published systems of angiosperm classification. Phytologia 51: 65–156.

Beese, W. J. 1983. White malaxis, *Malaxis monophyllos* var. *diphyllos*, an addition to the orchids of Canada from the Queen Charlotte Islands, British Columbia. Canad. Field-Naturalist 97: 215–216.

Beetle, D. F. 1944. A monograph of the North American species of *Fritillaria*. Madroño 7: 133–159.

Behnke, H.-D. et al. 2000. Systematics and evolution of Velloziaceae, with special reference to sieve-element plastids and *rbc*L sequence data. Bot. J. Linn. Soc. 134: 93–129.

Beih. Bot. Centralbl. = Beihefte zum Botanischen Centralblatt. Original Arbeiten.

Beitr. Biol. Pflanzen = Beiträge zur Biologie der Pflanzen.

Belg. J. Bot. = Belgian Journal of Botany: Bulletin de la Société Royale de Botanique de Belgique.

Belgique Hort. = Belgique Horticole; Journal des Jardins, des Serres et des Vergers.

Bell, A. D. 1974. Rhizome organization in relation to vegetative spread in *Medeola virginiana*. J. Arnold Arbor. 55: 458–468.

Bentham, G. 1839[–1857]. Plantas Hartwegianas Imprimis Mexicanas.... London. [Issued by gatherings with consecutive signatures and pagination.] (Pl. Hartw.)

Bentham, G. and J. D. Hooker. 1862–1883. Genera Plantarum ad Exemplaria Imprimis in Herbariis Kewensibus Servata Definita. 3 vols. London. (Gen. Pl.)

Bentley, S. L. 2000. Native Orchids of the Southern Appalachian Mountains. Chapel Hill.

Ber. Deutsch. Bot. Ges. = Berichte der Deutschen botanischen Gesellschaft.

Berg, R. Y. 1958. Seed dispersal, morphology, and phylogeny of *Trillium*. Skr. Norske Vidensk.-Akad. Oslo, Mat.-Naturvidensk. Kl. 1958(1): 1–36.

Berg, R. Y. 1959. Seed dispersal, morphology and taxonomic position of *Scoliopus*, Liliaceae. Skr. Norske Vidensk.-Akad. Oslo, Mat.-Naturvidensk. Kl. 1959(4): 1–56.

Berg, R. Y. 1962. Morphology and taxonomic position of *Medeola*, Liliaceae. Skr. Norske Vidensk.-Akad. Oslo, Mat.-Naturvidensk. Kl., n. s. 3: 1–56.

Berg, R. Y. 1962b. Contribution to the comparative embryology of the Liliaceae: *Scoliopus, Trillium, Paris* and *Medeola*. Skr. Norske Vidensk.-Akad. Oslo, Mat.-Naturvidensk. Kl., n. s. 4: 1–64.

Berg, R. Y. 1978. Development of ovule, embryo sac, and endosperm in *Brodiaea* (Liliales). Norweg. J. Bot. 25: 1–7.

Berg, R. Y. 1996. Development of ovule, embryo sac, and endosperm in *Dipterostemon* and *Dichelostemma* (Alliaceae) relative to taxonomy. Amer. J. Bot. 83: 790–801.

Berg, R. Y. and J. R. Maze. 1966. Contribution to the embryology of *Muilla*, with a remark on the taxonomic position of the genus. Madroño 18: 143–151.

Berger, A. 1915. Die Agaven. Jena. (Agaven)

Bernhardt, P. 1977. The pollination ecology of a population of *Erythronium americanum* Ker (Liliaceae). Rhodora 79: 278–282.

Besser, W. S. J. G. von. 1809. Primitiae Florae Galiciae Austriacae Utriusque. 2 vols. Vienna. (Prim. Fl. Galiciae Austriac.)

Best, C. A. 1962. Studies in the Eastern North American Native *Lilium* Species. Ph.D. dissertation. University of Toronto.

Bicknell, E. P. 1899. Studies in *Sisyrinchium*—I: Sixteen new species from the southern states. Bull. Torrey Bot. Club 26: 217–231.

Bicknell, E. P. 1901. Studies in *Sisyrinchium*—IX: The species of Texas and the southwest. Bull. Torrey Bot. Club 28: 570–592.

Bigelow, J. 1824. Florula Bostoniensis. A Collection of Plants of Boston and Its Vicinity..., ed. 2. Boston. (Fl. Boston. ed. 2)

Biltmore Bot. Stud. = Biltmore Botanical Studies; a Journal of Botany Embracing Papers by the Director and Associates of the Biltmore Herbarium.

Biochem. Syst. & Ecol. = Biochemical Systematics and Ecology.

Biol. Conservation = Biological Conservation; International Quarterly Journal Devoted to Scientific Protection of Plant and Animal Wildlife and All Nature throughout the World....

Biol. J. Linn. Soc. = Biological Journal of the Linnean Society.

Blackwell, W. H. 1990. Poisonous and Medicinal Plants. Englewood Cliffs.

Blain, M. T. 1997. The Taxonomic Status of *Clintonia alleghaniensis* Harned (Liliaceae): A Thesis in Applied Ecology and Conservation Biology. M.S. thesis. Frostburg State University.

Blanchard, J. W. 1990. *Narcissus*—A Guide to Wild Daffodils. Woking.

Blume, C. L. 1835[1836]–1848[1849]. Rumphia, sive Commentationes Botanicae Imprimis de Plantis Indiae Orientalis.... 4 vols. in 40 parts. Leiden etc. [Parts paged consecutively within volumes.] (Rumphia)

Blumea = Blumea; Tidjschrift voor die Systematiek en die Geografie der Planten (A Journal of Plant Taxonomy and Plant Geography).

Blunden, G., Yi Y., and K. Jewers. 1978. Steroidal sapogenins from leaves of Agaveae species. Phytochemistry 17: 1923–1925.

Blunt, W. 1950. Tulipomania. Harmondsworth.

Bodkin, N. L. 1978. A Revision of North American *Melanthium* L. (Liliaceae). Ph.D. dissertation. University of Maryland.

Bodkin, N. L. 1979. *Melanthium hybridim* Walt.—a misapplied name. Virginia J. Sci. 30: 54.

Bogler, D. J. 1994. Taxonomy and Phylogeny of *Dasylirion* (Nolinaceae). Ph.D. dissertation. University of Texas.

Bogler, D. J., J. L. Neff, and B. B. Simpson. 1995. Multiple origins of the *Yucca*-yucca moth association. Proc. Natl. Acad. Sci. U.S.A. 92: 6864–6867.

Bogler, D. J. and B. B. Simpson. 1995. A chloroplast DNA study of the Agavaceae. Syst. Bot. 20: 191–205.

Bogler, D. J. and B. B. Simpson. 1996. Phylogeny of Agavaceae based on ITS rDNA sequence variation. Amer. J. Bot. 83: 1225–1235.

Bogner, J. and S. J. Mayo. 1998. Acoraceae. In: K. Kubitzki et al., eds. 1990+. The Families and Genera of Vascular Plants. 4+ vols. Berlin etc. Vol. 4, pp. 7–10.

Bogner, J. and D. H. Nicolson. 1991. A revised classification of Araceae with dichotomous keys. Willdenowia 21: 35–50.

Boissier, P. E. 1842–1859. Diagnoses Plantarum Orientalium Novarum. 3 vols. in 19 parts. Geneva etc. [Parts paged independently.] (Diagn. Pl. Orient.)

Boivin, B. 1967c. Énumération des plantes du Canada. V. Naturaliste Canad. 94: 131–157.

Boivin, B. and W. J. Cody. 1956. The variations of *Lilium canadense* Linnaeus. Rhodora 58: 14–20.

Bol. Soc. Brot. = Boletim da Sociedade Broteriana.

Boland, J. T. and P. J. Scott. 1991. Ecological aspects of *Arethusa bulbosa, Calopogon tuberosus,* and *Pogonia ophioglossoides* (Orchidaceae) in eastern Newfoundland: I. Flowering and fruiting patterns. Rhodora 93: 248–255.

Bonplandia (Corrientes) = Bonplandia. Corrientes.

Bonplandia (Hanover) = Bonplandia; Zeitschrift für die gesammte Botanik.

Boston J. Nat. Hist. = Boston Journal of Natural History.

Bot. Beechey Voy.—See: W. J. Hooker and G. A. W. Arnott [1830–]1841

Bot. Cab. = Botanical Cabinet; Consisting of Coloured Delineations of Plants from All Countries.... [Edited by G. Loddiges.]

Bot. California—See: S. Watson 1876–1880

Bot. Gaz. = Botanical Gazette; Paper of Botanical Notes.

Bot. J. Linn. Soc. = Botanical Journal of the Linnean Society.

Bot. Jahrb. Syst. = Botanische Jahrbücher für Systematik, Pflanzengeschichte und Pflanzengeographie.

Bot. Mag. = Botanical Magazine; or, Flower-garden Displayed.... [Edited by Wm. Curtis.] [With vol. 15, 1801, title became Curtis's Botanical Magazine; or....]

Bot. Mag. (Tokyo) = Botanical Magazine. [Shokubutsu-gaku Zasshi.] (Tokyo.)

Bot. Mus. Leafl. = Botanical Museum Leaflets. [Harvard University.]

Bot. North. Middle States—See: L. C. Beck 1833

Bot. North. Middle States ed. 2—See: L. C. Beck 1848

Bot. Reg. = Botanical Register....

Bot. Repos. = Botanists' Repository, for New, and Rare Plants.

Bot. Rev. (Lancaster) = Botanical Review, Interpreting Botanical Progress.

Bot. Tidsskr. = Botanisk Tidsskrift.

Bot. Zeitung (Berlin) = Botanische Zeitung. (Berlin.)

Bot. Zhurn. (Moscow & Leningrad) = Botanicheskii Zhurnal. (Moscow and Leningrad.)

Botanical Society of America. 1978. Program and Abstracts of Papers to be Presented at the Meetings of the Botanical Society of America and Certain Affiliated Groups at Virginia Polytechnic Institute and State University, Blacksburg, Virginia, 25–29 June 1978. Blacksburg. [Bot. Soc. Amer. Misc. Ser. 156.]

Botanist = The Botanist; Containing Accurately Coloured Figures, of Tender and Hardy Ornamental Plants, with Descriptions....

Botany (Fortieth Parallel)—See: S. Watson 1871

Bothalia = Bothalia; a Record of Contributions from the National Herbarium, Union of South Africa.

Botschantzeva, Z. P. 1982. Tulips: Taxonomy, Morphology, Cytology, Phytogeography and Physiology, transl. and ed. H. Q. Varekamp. Rotterdam.

Bouman, F. 1995. Seed structure and systematics in Dioscoreales. In: P. J. Rudall et al., eds. 1995. Monocotyledons: Systematics and Evolution. 2 vols. Kew. Vol. 1, pp. 139–156.

Bouman, F. and N. Devente. 1992. A comparison of the structure of ovules and seeds in *Stemona* (Stemonaceae) and *Pentastemona* (Pentastemonaceae). Blumea 36: 501–514.

Brackett, A. E. 1923. A revision of the American species of *Hypoxis.* Rhodora 25: 120–147, 151–155.

Bradea = Bradea; Boletim do Herbarium Bradeanum.

Braem, G. J. 1986. *Tolumnia*—Der neue, aber doch alte, Name für die "Variegaten Oncidien." Orchidee (Hamburg) 37: 55–59.

Braem, G. J. 1986b. The Taxonomy of Plants Formerly Referred to *Oncidium* Section *Oncidium* (Orchidaceae) in the Caribbean Islands. Dissertation. University of Newcastle upon Tyne.

Braem, G. J. 1993. Studies in the Oncidiinae: Discussion of some taxonomic problems with description of *Gudrunia* Braem, gen. nov., and reinstatement of the genus *Lophiaris* Rafinesque. Schlechteriana 1–2: 8–29.

Braun, E. L. 1967. The Monocotyledonae: Cat-tails to Orchids. Columbus. [The Vascular Flora of Ohio, vol. 1.]

Breitung, A. J. 1959. Cultivated and native agaves in the southwestern United States. Cact. Succ. J. (Los Angeles) 31: 114–117.

Breitung, A. J. 1968. The Agaves.... Edited by Charles Glass and Robert A. Foster. Reseda.

Brickell, C. D., D. F. Cutler, and M. Gregory, eds. 1980. Petaloid Monocotyledons: Horticultural and Botanical Research. London and New York.

Bridson, G. D. R. and E. R. Smith. 1991. B-P-H/S. Botanico-Periodicum-Huntianum/Supplementum. Pittsburgh.

Brieger, F. G. 1977. On the Maxillariinae (Orchidaceae) with sepaline spur. Bot. Jahrb. Syst. 97: 548–574.

Briggs, B. G. 2000. What is significant—the Wollemi pine or the southern rushes? Ann. Missouri Bot. Gard. 87: 72–80.

Briggs, B. G. and L. A. S. Johnson. 2000. Hopkinsiaceae and Lyginiaceae, two new families of Poales in western Australia, with revisions of *Hopkinsia* and *Lyginia*. Telopea 8: 477–502.

Briggs, B. G., A. D. Marchant, S. Gilmore, and C. L. Porter. 2000. A molecular phylogeny of Restionaceae and allies. In: K. L. Wilson and D. A. Morrison, eds. 2000. Monocots: Systematics and Evolution. Melbourne. Pp. 661–671.

Brinker, R. 1942. Monograph of *Schoenocaulon*. Ann. Missouri Bot. Gard. 29: 287–309.

Brit. Fl. Gard.—See: R. Sweet 1823–1837

British Iris Society. 1997. A Guide to Species Irises: Their Identification and Cultivation. Cambridge and New York.

Britt, R. F. 1967. A Revision of the Genus *Hypoxis* in the United States and Canada. Ph.D. dissertation. University of North Carolina.

Britton, N. L. 1901. Manual of the Flora of the Northern States and Canada. New York. (Man. Fl. N. States)

Britton, N. L. et al., eds. 1905+. North American Flora.... 47+ vols. New York. [Vols. 1–34, 1905–1957; ser. 2, vols. 1–13, 1954+.]

Britton, N. L. and A. Brown. 1896–1898. An Illustrated Flora of the Northern United States, Canada and the British Possessions from Newfoundland to the Parallel of the Southern Boundary of Virginia, and from the Atlantic Ocean Westward to the 102d Meridian.... 3 vols. New York. (Ill. Fl. N. U.S.)

Britton, N. L. and C. F. Millspaugh. 1920. The Bahama Flora. New York. (Bahama Fl.)

Britton, N. L., E. E. Sterns, J. F. Poggenburg, et al. 1888. Preliminary Catalogue of Anthophyta and Pteridophyta Reported As Growing Spontaneously within One Hundred Miles of New York City. New York. [Authorship often attributed as B.S.P. in nomenclatural contexts.] (Prelim. Cat.)

Britton, N. L., P. Wilson, and W. R. Maxon. [1923–]1930. Botany of Porto Rico and the Virgin Islands. 2 vols. New York. [Constitutes vols. 5 and 6 of New York Academy of Sciences. Scientific Survey of Porto Rico and the Virgin Islands.]

Brittonia = Brittonia; a Journal of Systematic Botany....

Brown, P. M. 1993. Wild Orchids of the Northeastern United States: A Field Guide. Ithaca, N.Y.

Brown, P. M. 1995. Checklist of the Orchids of North America North of Mexico. Brighton, Mass. [N. Amer. Native Orchid J. Special Publ.]

Brown, P. M. 1996. *Schiedeella fauci-sanguinea:* A new name for an old plant. N. Amer. Native Orchid J. 2: 66–68.

Brown, P. M. 1997. Wild Orchids of the Northeastern United States.... Ithaca, N.Y.

Brown, P. M. 1999. A striking new color form of *Sacoila lanceolata*. N. Amer. Native Orchid J. 5: 169–173.

Brown, P. M. 2000. The Florida Native Orchid Project. Palmetto 20: 6–10.

Brown, P. M. 2000b. Recent taxonomic and distributional notes from Florida 8: *Sacoila squamulosa,* verified for the United States. N. Amer. Native Orchid J. 5: 333–334.

Brown, P. M. 2000c. Recent taxonomic and distributional notes from Florida 7. The genus *Habenaria*. N. Amer. Native Orchid J. 6: 148–158.

Brown, P. M. 2000d. *Govenia floridana* (Orchidaceae), a new species endemic to southern Florida, U.S.A. N. Amer. Native Orchid J. 6: 231–238.

Brown, P. M. 2002. Wild Orchids of Florida. Gainesville.

Browne, E. T. Jr. and R. Athey. 1992. Vascular Plants of Kentucky: An Annotated Checklist. Lexington, Ky.

Brownell, V. R. and P. M. Catling. 1987. Notes on the distribution and taxonomy of *Cypripedium fasciculatum* Kellogg ex Watson (Orchidaceae). Lindleyana 2: 53–57.

Brucher, H. 1989. Useful Plants of Neotropical Origin and Their Wild Relatives. New York.

Brumback, W. E. and C. W. Fyler. 1988. Monitoring of *Isotria medeoloides* in New Hampshire. Wild Fl. Notes 3: 32–40.

Brummitt, R. K. 1985. Report of the Committee for Spermatophyta: 29. Taxon 34: 659–708.

Brummitt, R. K. 2000. Report of the Committee for Spermatophyta: 49. Taxon 49: 261–278.

Brummitt, R. K. and C. E. Powell, eds. 1992. Authors of Plant Names. A List of Authors of Scientific Names of Plants, with Recommended Standard Forms of Their Names, Including Abbreviations. Kew.

Brummitt, R. K. and M. S. M. Sosef. 1998. Paraphyletic taxa are inherent in Linnaean classification—a reply to Freudenstein. Taxon 47: 411–412.

Buchneau, F. 1872. Ueber einege von Liebmann in Mexiko gesammelte Pflanzen. Abh. Naturwiss. Vereine Bremen 3: 339–350.

Bull. Amer. Iris Soc. = Bulletin of the American Iris Society.

Bull. Amer. Rock Gard. Soc. = Bulletin of the American Rock Garden Society.

Bull. Calif. Acad. Sci. = Bulletin of the California Academy of Sciences.

Bull. Cranbrook Inst. Sci. = Bulletin of the Cranbrook Institute of Science.

Bull. Herb. Boissier = Bulletin de l'Herbier Boissier.

Bull. Jard. Bot. État Bruxelles = Bulletin du Jardin Botanique de l'État à Bruxelles.

Bull. Jard. Bot. Natl. Belg. = Bulletin du Jardin Botanique National de Belgique.

Bull. Kanagawa Pref. Mus., Nat. Sci. = Bulletin of the Kanagawa Prefectural Museum. Natural science. [Kanagawa Kenritsu Hakubutsukan Kenkyu Hokoku, Shizen Kagaku.]

Bull. Lab. Nat. Hist. Iowa State Univ. = Bulletin of the Laboratories of Natural History, Iowa State University.

Bull. Louisiana Soc. Hort. Res. = Bulletin of the Louisiana Society for Horticultural Research.

Bull. Misc. Inform. Kew = Bulletin of Miscellaneous Information, Royal Gardens, Kew.

Bull. Mus. Life Sci. Louisiana State Univ. = Bulletin of the Museum of the Life Sciences, Louisiana State University.

Bull. Mus. Natl. Hist. Nat. = Bulletin du Muséum National d'Histoire Naturelle.
Bull. Natl. Mus. Canada = Bulletin of the National Museum of Canada.
Bull. New York Bot. Gard. = Bulletin of the New York Botanical Garden.
Bull. New York State Mus. Sci. Serv. = Bulletin of the New York State Museum and Science Service.
Bull. Public Mus. Milwaukee = Bulletin of the Public Museum of Milwaukee.
Bull. S. Calif. Acad. Sci. = Bulletin of the Southern California Academy of Sciences.
Bull. Sci. Soc. Philom. Paris = Bulletin des Sciences, par la Société Philomatique (de Paris).
Bull. Soc. Bot. France = Bulletin de la Société Botanique de France.
Bull. Soc. Imp. Naturalistes Moscou = Bulletin de la Société Impériale des Naturalistes de Moscou.
Bull. Torrey Bot. Club = Bulletin of the Torrey Botanical Club.
Bull. Univ. Mus. Univ. Tokyo = Bulletin, University Museum, University of Tokyo. [Sogo Kenkyu Shiryokan.]
Burbanck, M. P. 1941. Cytological and taxonomic studies in the genus *Brodiaea.* I. Bot. Gaz. 103: 247–265.
Burbanck, M. P. 1944. Cytological and taxonomic studies in the genus *Brodiaea.* II. Bot. Gaz. 105: 339–345.
Burger, W. C. 1977. The Piperales and the monocots: Alternate hypotheses for the origin of monocotyledonous flowers. Bot. Rev. (Lancaster) 43: 345–393.
Burkill, I. H. 1960. The organography and evolution of the Dioscoreaceae, the family of yams. J. Linn. Soc., Bot. 56: 319–407.
Burman, N. L. 1768. Flora Indica ... Nec Non Prodromus Florae Capensis. Leiden and Amsterdam. (Fl. Indica)
Burns-Balogh, P. 1982. Generic redefinition in subtribe Spiranthinae (Orchidaceae). Amer. J. Bot. 69: 1119–1132.
Burns-Balogh, P. and V. A. Funk. 1986. A phylogenetic analysis of the Orchidaceae. Smithsonian Contr. Bot. 61.
Burns-Balogh, P. and H. Robinson. 1983. Evolution and phylogeny of the *Pelexia* alliance (Orchidaceae: Spiranthoideae: Spiranthinae). Syst. Bot. 8: 263–268.
Burrows, G. E. and R. J. Tyrl. 2001. Toxic Plants of North America. Ames.
Bush, B. F. 1927. The species of *Polygonatum.* Amer. Midl. Naturalist 10: 385–400.
Buxbaum, F. 1925. Vergleichende Anatomie der Melanthioideae. Repert. Spec. Nov. Regni Veg. Beih. 29: 1–80.
Buxbaum, F. 1927. Nachtrage zur vergleichenden Anatomie der Melanthioideae I. Beih. Bot. Centralbl. 44: 255–263.
Buxbaum, F. 1958. Der morphologische Typus und die systematische Stellung der Gattung *Calochortus.* Beitr. Biol. Pflanzen 34: 405–452.
Buzgo, M. 2001. Flower structure and development of Araceae compared with alismatids and Acoraceae. Bot. J. Linn. Soc. 136: 393–425.
Cabe, P. R. 1995. The *Trillium pusillum* Michaux (Liliaceae) complex in Virginia. I. Morphological investigations. Castanea 60: 1–14.
Cabe, P. R. and C. R. Werth. 1995. The *Trillium pusillum* Michaux (Liliaceae) complex in Virginia. II. Isozyme evidence. Castanea 60: 15–29.
Cact. Suc. Mex. = Cactaceas y Suculentes Mexicanas.
Cact. Succ. J. (Los Angeles) = Cactus and Succulent Journal.
Caddick, L. R. et al. 2000. Yams and their allies: Systematics of Dioscoreales. In: K. L. Wilson and D. A. Morrison, eds. 2000. Monocots: Systematics and Evolution. Melbourne. Pp. 475–487.
Caddick, L. R. et al. 2000b. Floral morphology and development in Dioscoreales. Feddes Repert. 111: 189–230.
Caddick, L. R. et al. 2002. Phylogenetics of Dioscoreales based on combined analyses of morphological and molecular data. Bot. J. Linn. Soc. 138: 123–144.
Caddick, L. R. et al. 2002b. Yams reclassified: A recircumscription of Dioscoreaceae and Dioscoreales. Taxon 51: 103–114.
Caddick, L. R., C. A. Furness, K. L. Stobart, and P. J. Rudall. 1998. Microsporogenesis and pollen morphology in Dioscoreales and allied taxa. Grana 37: 321–336.
Cafferty, S. and C. E. Jarvis. 1999. Typification of Linnaean specific and varietal names in the Orchidaceae. Taxon 48: 45–50.
Calder, J. A. and R. L. Taylor. 1965. New taxa and nomenclatural changes with respect to the flora of the Queen Charlotte Islands, British Columbia. Canad. J. Bot. 43: 1392–1393.
Calder, J. A. and R. L. Taylor. 1968. Flora of the Queen Charlotte Islands. 2 vols. Ottawa.
Calif. Fl.—See: P. A. Munz 1959
Callihan, R. H., S. L. Carson, and R. T. Dobbins. 1995. NAWEEDS, Computer-aided Weed Identification for North America. Illustrated User's Guide plus Computer Floppy Disk. Moscow, Idaho.
Calvo, R. N. 1990. Four-year growth and reproduction of *Cyclopogon cranichoides* (Orchidaceae) in south Florida. Amer. J. Bot. 77: 736–741.
Cameron, K. M. et al. 1999. A phylogenetic analysis of the Orchidaceae: Evidence from *rbc*L nucleotide sequences. Amer. J. Bot. 86: 208–224.
Camus, E. G., P. Bergon, and A. A. Camus. [1908.] Monographie des Orchidées.... Paris. (Monogr. Orchid.)
Canad. Field-Naturalist = Canadian Field-Naturalist.
Canad. J. Bot. = Canadian Journal of Botany.
Canad. J. Pl. Sci. = Canadian Journal of Plant Science.
Candolle, A. L. P. P. de and C. de Candolle, eds. 1878–1896. Monographiae Phanerogamarum.... 9 vols. Paris. (Monogr. Phan.)
Candolle, A. P. de. 1813b. Catalogus Plantarum Horti Botanici Monspeliensis.... Monpellier, Paris, and Strasbourg. (Cat. Pl. Hort. Monsp.)
Candolle, A. P. de. [1817]1818–1821. Regni Vegetabilis Systema Naturale.... 2 vols. Paris, Strasbourg, and London.
Candollea = Candollea; Organe du Conservatoire et du Jardin Botaniques de la Ville de Genève.
Carlquist, S. and E. L. Schneider. 1998. Origins and nature of vessels in monocotyledons. 3. Lowiaceae, with comments on rhizome anatomy. Blumea 43: 219–224.
Caro, J. A. 1982. Sinopsis taxonomica de las gramineas Argentinas. Dominguezia 4: 1–51.
Carr, B. L. 1986. Autotriploidy in *Erythronium rostratum* (Liliaceae): Reproduction, distribution, and origin. Cas-

tanea 51: 73–87.
Caryologia = Caryologia; Giornale di Citologia, Citosistematica e Citogenetica.
Case, F. W. 1987. Orchids of the western Great Lakes region, rev. ed. Bull. Cranbrook Inst. Sci. 48.
Case, F. W. and G. L. Burrows. 1962. The genus *Trillium* in Michigan: Some problems of distribution and taxonomy. Pap. Michigan Acad. Sci. 47: 180–200.
Case, F. W. and R. B. Case. 1993. *Trillium erectum* and its hybrids. Bull. Amer. Rock Gard. Soc. 51: 162–168.
Case, F. W. and R. B. Case. 1997. Trilliums. Portland.
Castanea = Castanea; Journal of the Southern Appalachian Botanical Club.
Cat. N. Amer. Pl. ed. 2—See: A. A. Heller 1900
Cat. Pl. Amer. Sept.—See: G. H. E. Muhlenberg 1813
Cat. Pl. Cub.—See: A. H. R. Grisebach 1866
Cat. Pl. Hort. Gott.—See: J. G. Zinn 1757
Cat. Pl. Hort. Monsp.—See: A. P. de Candolle 1813b
Cat. Pl. Upper Louisiana—See: T. Nuttall 1813
Catalan, P., E. A. Kellogg, and R. G. Olmstead. 1997. Phylogeny of Poaceae subfamily Pooideae based on chloroplast *ndh*F gene sequences. Molec. Phylogen. Evol. 8: 150–166.
Catling, P. M. 1976. On the geographical distribution, ecology and distinctive features of *Listera* ×*veltmanii* Case. Rhodora 78: 261–269.
Catling, P. M. 1980. Rain-assisted autogamy in *Liparis loeselii* (L.) L. C. Rich. (Orchidaceae). Bull. Torrey Bot. Club 107: 525–529.
Catling, P. M. 1980b. Systematics of *Spiranthes* L. C. Richard in Northeastern North America. Ph.D. thesis. University of Toronto.
Catling, P. M. 1981. Taxonomy of autumn-flowering *Spiranthes* species of souithern Nova Scotia. Canad. J. Bot. 59: 1253–1270.
Catling, P. M. 1982. Breeding systems of northeastern North American *Spiranthes* (Orchidaceae). Canad. J. Bot. 60: 3017–3039.
Catling, P. M. 1984. A re-evaluation of *Spiranthes* ×*steigeri*. Rhodora 86: 469–473.
Catling, P. M. 1987. Notes on the breeding systems of *Sacoila lanceolata* (Aublet) Garay (Orchidaceae). Ann. Missouri Bot. Gard. 74: 58–68.
Catling, P. M. 1991. Systematics of *Malaxis bayardii* and *M. unifolia*. Lindleyana 6: 3–23.
Catling, P. M. and J. R. Brown. 1983. Morphometrics and ecological isolation in sympatric *Spiranthes* (Orchidaceae) in southwestern Ontario. Canad. J. Bot. 61: 2747–2759.
Catling, P. M. and V. R. Catling. 1988. Lista comentada de las orquídeas de Belice. Orquidea (Mexico City) 11: 63–102.
Catling, P. M. and V. R. Catling. 1990. Anther cap retention in *Tipularia discolor*. Lindleyana 6: 113–116.
Catling, P. M. and V. R. Catling. 1991. A synopsis of breeding systems and pollination in North American orchids. Lindleyana 6: 187–210.
Catling, P. M. and V. R. Catling. 1994. Identification of *Platanthera lacera* hybrids (Orchidaceae) from New Brunswick and Nova Scotia. Lindleyana 9: 19–32.
Catling, P. M. and V. S. Engel. 1993. Systematics and distribution of *Hexalectris spicata* var. *arizonica* (Orchidaceae). Lindleyana 8: 119–125.
Catling, P. M. and K. B. Gregg. 1992. Systematics of the genus *Cleistes* in North America. Lindleyana 7: 57–73.
Catling, P. M. and Z. Lucas. 1987. The status of *Calopogon tuberosus* var. *latifolius* with comments on the application of varietal rank. Rhodora 89: 401–413.
Catling, P. M. and C. J. Sheviak. 1993. Taxonomic notes on some North American orchids. Lindleyana 8: 77–81.
Cavanilles, A. J. 1791–1801. Icones et Descriptiones Plantarum, Quae aut Sponte in Hispania Crescunt, aut in Hortis Hospitantur. 6 vols. Madrid. (Icon.)
Cave, M. S. 1948. Sporogenesis and embryo-sac development of *Hesperocallis* and *Leucocrinum* in relation to their systematic position. Amer. J. Bot. 35: 343–349.
Cave, M. S. 1964. Cytological observations on some genera of the Agavaceae. Madroño 17: 163–170.
Cave, M. S. 1966. The chromosomes of *Scoliopus* (Liliaceae). Madroño 18: 211–213.
Cave, M. S. 1970. Chromosomes of the California Liliaceae. Univ. Calif. Publ. Bot. 57: 1–58.
Cellule = La Cellule; Recueil [later Revue] de Cytologie et d'Histologie Générale.
Center, T. D. and N. R. Spencer. 1981. The phenology and growth of water hyacinth (*Eichhornia crassipes* (Mart.) Solms) in a eutrophic north-central Florida lake. Aquatic Bot. 10: 1–32.
Center, T. D. and T. K. Van, T. K. 1989. Alteration of water hyacinth (*Eichhornia crassipes* (Mart.) Solms) leaf dynamics and phytochemistry by insect damage and plant density. Aquatic Bot. 35: 181–195.
Centro de Investigación en Química Aplicada. 1970. *Yucca*. Saltillo.
Chapman, A. W. 1860. Flora of the Southern United States.... New York. (Fl. South. U.S.)
Chapman, A. W. [1892.] Flora of the Southern United States..., ed. 2 reprinted. New York, Cincinnati, and Chicago. [Second Supplement added.] (Fl. South. U.S. ed. 2 repr. 1892)
Chapman, A. W. 1892. Flora of the Southern United States..., ed. 2 reprint 2. New York, Cincinnati, and Chicago. (Fl. South. U.S. ed. 2 reprint 2)
Chase, M. W. 1986. A reappraisal of the oncidioid orchids. Syst. Bot. 11: 477–491.
Chase, M. W. 1987. Systematic implications of pollinarium morphology in *Oncidium* Sw., *Odontoglossum* Kunth, and allied genera (Orchidaceae). Lindleyana 2: 8–28.
Chase, M. W. et al. 1993. Phylogenetics of seed plants: An analysis of nucleotide sequences from the plastid gene *rbc*L. Ann. Missouri Bot. Gard. 80: 528–580.
Chase, M. W. et al. 1995. Molecular phylogenetics of Lilianae. In: P. J. Rudall et al., eds. 1995. Monocotyledons: Systematics and Evolution. 2 vols. Kew. Vol. 1, pp. 109–137.
Chase, M. W. et al. 1995b. Monocot systematics: A combined analysis. In: P. J. Rudall et al., eds. 1995. Monocotyledons: Systematics and Evolution. 2 vols. Kew. Vol. 2, pp. 685–730.
Chase, M. W. et al. 2000. Higher-level systematics of the monocotyledons: An assessment of current knowledge and a new classification. In: K. L. Wilson and D. A. Morrison,

eds. 2000. Monocots: Systematics and Evolution. Melbourne. Pp. 3–16.

Chase, M. W. et al. 2000b. Phylogenetics of the Asphodelaceae (Asparagales): An analysis of plastid *rbc*L and *trn*L-F DNA sequences. Ann. Bot. (Oxford) 86: 935–951.

Chase, M. W. and J. D. Palmer. 1992. Floral morphology and chromosome number in subtribe Oncidiinae (Orchidaceae): Evolutionary insights from a phylogenetic analysis of chloroplast DNA restriction sites variation. In: D. E. Soltis et al., eds. 1992. Molecular Systematics of Plants. New York. Pp. 325–339.

Chase, M. W. and J. S. Pippen. 1988. Seed morphology in the Oncidiinae and related subtribes (Orchidaceae). Syst. Bot. 13: 313–323.

Chase, M. W., P. J. Rudall, and J. G. Conran. 1996. New circumscriptions and a new family of asparagoid lilies: Genera formerly included in Anthericaceae. Kew Bull. 51: 667–680.

Chase, M. W., P. J. Rudall, M. F. Fay, and K. L. Stobart. 2001. Xeronemataceae, a new family of asparagoid lilies from New Caledonia and New Zealand. Kew Bull. 55: 865–870.

Châtelain, J. J. 1760. Specimen Inaugurale de *Corallorhiza*.... [Basel.] (Specim. Inaug. Corallorhiza)

Cheadle, V. I. and H. Kosakai. 1971. Vessels in Liliaceae. Phytomorphology 21: 320–333.

Chen, S. C. and H. Takahashi. 2000. *Veratrum*. In: Wu Z. and P. H. Raven, eds. 1994+. Flora of China. 7+ vols. Beijing and St. Louis. Vol. 24, pp. 82–85.

Chiariello, N., C. Field, H. A. Mooney, and J. Seemann. 1980. Architecture and thermal relations of *Veratrum californicum* (Liliaceae), a snowbank emergent. Madroño 27: 113–121.

Chigbo, F., Ralph W. Smith, and F. Shore. 1982. Uptake of arsenic, cadmium, lead and mercury from polluted waters by the water hyacinth *Eichhornia crassipes*. Environm. Pollut., A 27: 31–36.

Cholewa, A. F. and D. M. Henderson. 1984. Biosystematics of *Sisyrinchium* section *Bermudiana* (Iridaceae) of the Rocky Mountains. Brittonia 36: 342–363.

Chouard, P. 1930. Types de Dévelopement de l'Appareil Végétatif Chez les Scillées. D.Sc. thesis. University of Paris. [Preprinted from Ann. Sci. Nat., Bot., sér. 10, 13: 131–322. 1931.]

Chung, M. G. 1990. A Biosystematic Study of the Genus *Hosta* Tratt. (Liliaceae / Funkiaceae) in Korea. Ph.D. dissertation. University of Georgia.

Chung, M. G. and S. B. Jones. 1989. Pollen morphology of *Hosta* Tratt. and related genera. Bull. Torrey Bot. Club 116: 31–44.

Chung, M. G. and J. W. Kim. 1991. The genus *Hosta* Tratt. (Liliaceae) in Korea. Sida 14: 411–420.

Cirillo, D. M. L. 1788–1792. Plantarum Rariorum Regni Neapolitani. 2 fasc. Naples. (Pl. Rar. Neapol.)

Cladistics = Cladistics; the International Journal of the Willi Hennig Society.

Clark, L. G. et al. 2000. The Puelioideae, a new subfamily of Poaceae. Syst. Bot. 25: 181–187.

Clark, L. G., Zhang W. P., and J. F. Wendel. 1995. A phylogeny of the grass family (Poaceae) based on *ndh*F sequence data. Syst. Bot. 20: 436–460.

Clarkson, Q. D. 1958. *Iris* section *Apogon* subsection *Oregonae*. Subsect. nov. Madroño 14: 246–247.

Clary, K. H. 1997. Phylogeny, Character Evolution, and Biogeography of *Yucca* L. (Agavaceae) As Inferred from Plant Morphology and Sequences of the Internal Transcribed Spacer (ITS) Region of the Nuclear Ribosomal DNA. Ph.D. dissertation. University of Texas.

Clary, K. H. 2001. The genus *Hesperoyucca* (Agavaceae) in the western United States and Mexico: New nomenclatural combinations. Sida 19: 839–847.

Clase, H. J. and S. J. Meades. 1996. Dactylorchids in Newfoundland. N. Amer. Native Orchid J. 2: 208–216.

Class-book Bot. ed. 2—See: A. Wood 1847

Clayton, W. D. and S. A. Renvoize. 1986. Genera Graminum: Grasses of the World. London. [Kew Bull. Addit. Ser. 13.]

Clewell, A. F. 1985. Guide to the Vascular Plants of the Florida Panhandle. Tallahassee.

Clifford, H. T., R. J. F. Henderson, and J. G. Conran. 1998. Hemerocallidaceae. In: K. Kubitzki et al., eds. 1990+. The Families and Genera of Vascular Plants. 4+ vols. Berlin etc. Vol. 3, pp. 245–253.

Cody, W. J. 1961. New records from the upper Mackenzie River Valley, Mackenzie District, Northwest Territories. Canad. Field-Naturalist 75: 55–69.

Coker, W. C. 1944. The woody smilaxes of the United States. J. Elisha Mitchell Sci. Soc. 60: 27–69, plates 9–39.

Coleman, R. A. 1995. The Wild Orchids of California. Ithaca, N.Y.

Coleman, R. A. 1996. *Stenorrhynchos michuacana*. Orchids 65: 1284–1286.

Coll. Bot.—See: J. Lindley 1821[–1826]

Conran, J. G. 1994. The Geitonoplesiaceae Dahlgren ex Conran (Liliiflorae, Asparagales), a new family of monocotyledons. Telopea 6: 39–40.

Conran, J. G. 1998. Anthericaceae. In: K. Kubitzki et al., eds. 1990+. The Families and Genera of Vascular Plants. 4+ vols. Berlin etc. Vol. 3, pp. 114–121.

Conran, J. G., M. W. Chase, and P. J. Rudall. 1997. Two new monocotyledon families, Anemarrhenaceae and Behniaceae (Lilianae, Asparagales). Kew Bull. 52: 995–999.

Conran, J. G. and M. N. Tamura. 1998. Convallariaceae. In: K. Kubitzki et al., eds. 1990+. The Families and Genera of Vascular Plants. 4+ vols. Berlin etc. Vol. 3, pp. 186–198.

Conservation Biol. = Conservation Biology; Journal of the Society for Conservation Biology.

Contr. Bolus Herb. = Contributions from the Bolus Herbarium.

Contr. Boyce Thompson Inst. Pl. Res. = Contributions of the Boyce Thompson Institute for Plant Research.

Contr. Dudley Herb. = Contributions from the Dudley Herbarium of Stanford University.

Contr. Gray Herb. = Contributions from the Gray Herbarium of Harvard University. [Some numbers reprinted from (or in?) other periodicals, e.g. Rhodora.]

Contr. New York Bot. Gard. = Contributions of the New York Botanical Garden.

Contr. Ocas. Mus. Hist. Nat. Colegio "De La Salle" = Contribuciones Ocasionales del Museo de Historia Natu-

ral del Colegio "De La Salle".
Contr. Orch. Fl. S. Florida—See: O. Ames 1904
Contr. U.S. Natl. Herb. = Contributions from the United States National Herbarium.
Contr. Univ. Michigan Herb. = Contributions from the University of Michigan Herbarium.
Contr. W. Bot. = Contributions to Western Botany.
Cook, C. D. K. 1989. Taxonomic revision of *Monochoria* (Pontederiaceae). In: K. Tan, ed. 1989. The Davis & Hedge Festschrift...Plant Taxonomy, Phytogeography, and Related Subjects. Edinburgh. Pp. 149–184.
Cook, C. D. K. 1998. Hydrocharitaceae. In: K. Kubitzki et al., eds. 1990+. The Families and Genera of Vascular Plants. 4+ vols. Berlin etc. Vol. 4, pp. 234–248.
Cooke, D. A. 1987. Hydatellaceae. In: R. Robertson et al. eds. 1981+. Flora of Australia. 23+ vols. Canberra. Vol. 45, pp. 1–5.
Cooney-Sovetts, C. and R. Sattler. 1986. Phylloclade development in the Asparagaceae: An example of homeosis. Bot. J. Linn. Soc. 94: 327–371.
Corbet, A. S., H. Chapman, and N. Saville. 1988. Vibratory pollen collection and flower from: bumblebees on *Actinidia, Symphytum, Borago* and *Polygonatum*. Funct. Ecol. 2: 146–156.
Correll, D. S. 1940. A contribution to our knowledge of the orchids of the southeastern United States. Bot. Mus. Leafl. 8: 69–92.
Correll, D. S. 1947. Revision of the genus *Govenia*. Lloydia 10: 218–278.
Correll, D. S. 1950. Native Orchids of North America North of Mexico.... Waltham, Mass.
Correll, D. S. and H. B. Correll. 1972. Aquatic and Wetland Plants of Southwestern United States. Washington. [Reissued 1975 in 2 vols., Stanford.]
Correll, D. S. and H. B. Correll. 1982. Flora of the Bahama Archipelago. Vaduz.
Correll, D. S. and M. C. Johnston. 1970. Manual of the Vascular Plants of Texas. Renner, Tex.
Corrigan, M. J. G., E. G. Gordon, D. F. Blaxell, and W. T. Upton, eds. 1971. Proceedings of the 6th World Orchid Conference. Sydney.
Cotton, A. D. 1936. On the names of *Lilium bolanderi* and *Lilium kelloggii*. Lily Year-Book 5: 6–8.
Coursey, D. G. 1967. Yams: An Account of the Nature, Origins, Cultivation and Utilisation of the Useful Members of the Dioscoreaceae London.
Cox, P. A. and C. J. Humphries. 1993. Hydrophilous pollination and breeding system evolution in seagrasses: A phylogenetic approach to the evolutionary ecology of the Cymodoceaceae. Bot. J. Linn. Soc. 113: 217–226.
Crantz, H. J. N. von. 1769. Stirpium Austriarum Fasciculus I [–VI]. 2 vols. in 6 fasc. Vienna. [Volumes paged consecutively.] (Stirp. Austr. Fasc. ed. 2)
Crawford, D. J. 2000. Plant macromolecular systematics in the past 50 years: One view. Taxon 49: 479–501.
Crayn, D. M., R. G. Terry, J. A. C. Smith, and K. Winter. 2000. Molecular systematic investigations in Pitcairnioideae (Bromeliaceae) as a basis for understanding the evolution of Crassulacean Acid Metabolism (CAM). In: K. L. Wilson and D. A. Morrison, eds. 2000. Monocots: Systematics and Evolution. Melbourne. Pp. 569–579.
Crespo, M. B. M., D. Lledo, M. F. Fay, and M. W. Chase. 1996. Molecular phylogeny of *Leucojum* based on ITS sequences. [Abstract.] Amer. J. Bot. 83(6, suppl.): 149.
Cribb, P. J. 1984. *Polystachya*. In: W. B. Turrill et al., eds. 1952+. Flora of Tropical East Africa. 152+ vols. London and Rotterdam. Orchidaceae, part 2, pp. 333–400.
Cribb, P. J. 1987. The genus *Eulophia* in Africa. In: K. Kondo and K. Hashimoto, eds. 1987. World Orchid Hiroshima Symposium.... [Sekai Ran Kaigi Hiroshima Taikai.] Hiroshima. Pp. 97–103.
Cribb, P. J. 1989. *Eulophia*. In: W. B. Turrill et al., eds. 1952+. Flora of Tropical East Africa. 152+ vols. London and Rotterdam. Orchidaceae, part 3, pp. 422–480.
Cribb, P. J. 1997. The Genus *Cypripedium*. Portland.
Cronquist, A. 1957. Outline of a new system of families and orders of dicotyledons. Bull. Jard. Bot. État Bruxelles 27: 13–40.
Cronquist, A. 1978b. The Zingiberidae, a new subclass of Liliopsida (monocotyledons). Brittonia 30: 505.
Cronquist, A. 1981. An Integrated System of Classification of Flowering Plants. New York.
Cronquist, A. 1983. Some realignments in the dicotyledons. Nordic J. Bot. 3: 75–83.
Cronquist, A. 1988. The Evolution and Classification of Flowering Plants, ed. 2. Bronx.
Cronquist, A. 1993. A commentary on the general system of classification of flowering plants. In: Flora of North America Editorial Committee, eds. 1993+. Flora of North America North of Mexico. 5+ vols. New York and Oxford. Vol. 1, pp. 272–293.
Cronquist, A. et al. 1972+. Intermountain Flora. Vascular Plants of the Intermountain West, U.S.A. 5+ vols. New York and London. [Vol. 1, 1972; vol. 3, part B, 1989; vol. 4, 1984; vol. 5, 1994; vol. 6, 1977.] (Intermount. Fl.)
Cronquist, A. and M. Ownbey. 1977. *Allium*. In: A. Cronquist et al. 1972+. Intermountain Flora. Vascular Plants of the Intermountain West, U.S.A. 5+ vols. New York and London. Vol. 6, pp. 508–522.
Cruden, R. W. 1991. A revision of *Isidrogalvia* (Liliaceae): Recognition for Ruíz and Pavón's genus. Syst. Bot. 16: 270–282.
Cruden, R. W. 1994. *Echeandia*. In: G. Davidse et al., eds. 1994+. Flora Mesoamericana. 2+ vols. Mexico City. Vol. 6, pp. 27–30.
Cruden, R. W. 1999. A new subgenus and fifteen new species of *Echeandia* (Anthericaceae) from Mexico and the United States. Novon 9: 325–338.
Cruden, R. W. and R. McVaugh. 1989. *Echeandia*. In: R. McVaugh and W. R. Anderson, eds. 1974+. Flora Novo-Galiciana: A Descriptive Account of the Vascular Plants of Western Mexico. 8+ vols. Ann Arbor. Vol. 15, pp. 178–197.
Curr. Genet. = Current Genetics; Eukaryotes with Emphasis on Yeasts, Fungi, Protists, Cell Organelles.
Currie, H. E. 1992. A Biosystematic Study of the Genus *Hosta* Tratt. (Liliaceae / Hostaceae) in Eastern Asia. Ph.D. dissertation. University of Georgia.
Cuvier, F., ed. 1816–1845. Dictionnaire des Sciences Naturelles..., ed. 2. 61 vols. Paris. (Dict. Sci. Nat. ed. 2)

Cycl.—See: A. Rees [1802–]1819–1820

Cytologia = Cytologia; International Journal of Cytology.

Dahlgren, G. 1989. An updated angiosperm classification. Bot. J. Linn. Soc. 100: 197–203.

Dahlgren, G. 1989b. The last Dahlgrenogram: System of classification of the dicotyledons. In: K. Tan, ed. 1989. The Davis & Hedge Festschrift...Plant Taxonomy, Phytogeography, and Related Subjects. Edinburgh. Pp. 249–260.

Dahlgren, R. M. T. 1980. A revised system of classification of the angiosperms. Bot. J. Linn. Soc. 80: 91–124.

Dahlgren, R. M. T. 1983. General aspects of angiosperm evolution and macrosystematics. Nordic J. Bot. 3: 119–149.

Dahlgren, R. M. T. and K. Bremer. 1985. Major clades of the angiosperms. Cladistics 1: 349–368.

Dahlgren, R. M. T. and H. T. Clifford. 1982. The Monocotyledons: A Comparative Study. New York.

Dahlgren, R. M. T., H. T. Clifford, and P. F. Yeo. 1985. The Families of the Monocotyledons: Structure, Evolution, and Taxonomy. Berlin etc.

Dahlgren, R. M. T. and F. N. Rasmussen. 1983. Monocotyledon evolution: Characters and phylogenetic estimation. Evol. Biol. 16: 255–395.

Dahlgren, R. M. T., S. Rosendal-Jensen, and B. J. Nielsen. 1981. A revised classification of the angiosperms with comments on correlation between chemical and other characters. In: D. A. Young and D. S. Seigler, eds. 1981. Phytochemistry and Angiosperm Phylogeny. New York. Pp. 149–204.

Dahlgren, R. M. T. and A. E. van Wyk. 1988. Structures and relationships of families endemic to or centered in southern Africa. Monogr. Syst. Bot. Missouri Bot. Gard. 25.

Dash, M. 1999. Tulipomania: The Story of the World's Most Coveted Flower and the Extraordinary Passions It Aroused. New York.

Davenport, L. J. 1990. Distribution, habitat requirements and distinct nature of *Hymenocallis coronaria*. [Abstract.] A. S. B. Bull. 37: 56.

Davenport, L. J. 1990b. The cahaba lily. Alabama Herit. 26: 24–29.

Davenport, L. J. 1996. The cahaba lily: Its distribution and status in Alabama. J. Alabama Acad. Sci. 67: 222–233.

Davidse, G., M. Sousa S., A. O. Chater, et al., eds. 1994+. Flora Mesoamericana. 2+ vols. Mexico City.

Davies, D. 1992. Alliums, the Ornamental Onions. Portland.

Davis, A. P. 1999. The Genus *Galanthus*. Portland.

Davis, J. S. 1956. Natural Pollination of California Lilies. M.S. thesis. Claremont Graduate School.

Davis, J. S. 1958. Reproductive Biology of California Lilies. Ph.D. dissertation. Claremont Graduate School.

Davis, P. H., ed. 1965–1988. Flora of Turkey and the East Aegean Islands. 10 vols. Edinburgh.

De Orchid. Eur.—See: L. C. M. Richard 1817

de Vos, M. P. 1972. The genus *Romulea* in South Africa. J. S. African Bot., suppl. 9.

de Vos, M. P. 1983. *Romulea*. In: Department of Agricultural Technical Services [South Africa]. 1963+. Flora of Southern Africa.... 19+ vols. (some partial). Pretoria. Vol. 7, part 2, fasc. 2, pp. 10–73.

de Vos, M. P. 1984. The African genus *Crocosmia* Planchon. J. S. African Bot. 50: 463–502.

de Vos, M. P. 1985. Revision of the South African genus *Chasmanthe* (Iridaceae). S. African J. Bot. 51: 253–261.

Deam, C. C., R. C. Friesner, R. Kriebel, and T. G. Yuncker. 1941. Indiana plant distribution records, 1940. Proc. Indiana Acad. Sci. 50: 72–78.

Degras, L. 1993. The Yam: A Tropical Root Crop. London, New York, and Wageningen.

Denison, S. S. and D. W. McNeal Jr. 1989. A reevaluation of the *Allium sanbornii* (Alliaceae) complex. Madroño 36: 122–130.

Department of Agricultural Technical Services [South Africa]. 1963+. Flora of Southern Africa.... 19+ vols. (some partial). Pretoria.

Desert Pl. = Desert Plants.

Deutschl. Fl.—See: G. F. Hoffmann [1791–1804]

Diagn. Pl. Nov. Mexic.—See: W. B. Hemsley 1878–1880

Diagn. Pl. Orient.—See: P. E. Boissier 1842–1859

Diamond, A. R., C. P. Chapman, and J. Brummett. 1998. *Habranthus tubispathus* (Liliaceae) new to the flora of Alabama. Sida 18: 353–355.

Dice, J. C. 1988. Systematic Studies in the *Nolina bigelovii–N. parryi* (Nolinaceae) Complex. M.S. thesis. San Diego State University.

Dickoré, W. B. 1995. Systematische Revision und chorologische Analyse der Monocotyledoneae des Karakorum (Zentralasien, West-Tibet). Flora Karakorumensis 1, Angiospermae, Monocotyledoneae. Stapfia 39: 1–298.

Dict. Sci. Nat. ed. 2—See: F. Cuvier 1816–1845

Dietrich, A. G. 1831–1833. Caroli a Linné Species Plantarum.... Olim Curante Carolo Ludovico Willdenow. Editio Sexta. 2 vols. Berlin. (Sp. Pl.)

Dietrich, D. N. F. 1839–1852. Synopsis Plantarum.... 5 vols. Weimar. (Syn. Pl.)

Dietrich, H. 1982. *Harrisella uniflora*—eine neue blattlose Orchidee für Cuba. Orchidee (Hamburg) 33: 18–19.

Dietz, R. A. 1952. Variation in the perfoliate uvularias. Ann. Missouri Bot. Gard. 39: 219–247.

Diggs, G. M., B. L. Lipscomb, and R. J. O'Kennon. 1999. Shinners' and Mahler's Illustrated Flora of North Central Texas. Fort Worth. [Sida Bot. Misc. 16.]

Dingle, H. J. and J. Hegmann, eds. 1982. Evolution and Genetics of Life Histories. New York.

Dodson, C. and P. M. Dodson. 1980. *Bletia purpurea*. Icon. Pl. Trop. 1: plate 7.

Dorrat Haaksma, E. and H. P. Linder. 2000. Restios of the Fynbos. Cape Town.

Douzery, E. J. P. et al. 1999. Molecular phylogenetics of Diseae (Orchidaceae): A contribution from nuclear ribosomal ITS sequences. Amer. J. Bot. 86: 887–899.

Doweld, A. B. 2001. Tentamen Systematis Plantarum Vascularium (Tracheophytorum). [Opyt Systemy Sosudistykh Rastenij (Trakheofitov).] Moscow.

Downing, G. R. and L. S. Furniss. 1968. Some observations on camas digging and baking among present-day Nez Perce. Tebiwa 11: 48–59.

Dransfield, J. and N. W. Uhl. 1998. Palmae. In: K. Kubitzki et al., eds. 1990+. The Families and Genera of Vascular Plants. 4+ vols. Berlin etc. Vol. 4, pp. 306–389.

Dress, W. J. 1955. *Hemerocallis lilio-asphodelus:* The yellow day-lily. Baileya 3: 107–108.

Dressler, R. L. 1961. The Orchids: Natural History and Classification. Cambridge, Mass.

Dressler, R. L. 1968. Notes on *Bletia* (Orchidaceae). Brittonia 20: 182–190.

Dressler, R. L. 1993. Phylogeny and Classification of the Orchid Family. Portland.

Dressler, R. L. 1993b. Field Guide to the Orchids of Costa Rica and Panama. Ithaca, N.Y.

Dressler, R. L. and G. E. Pollard. 1976. The Genus *Encyclia* in Mexico, ed. 2. Mexico City.

Dressler, R. L. and N. H. Williams. 1975. Jacquin names—again. Taxon 24: 647–650.

Dressler, R. L. and N. H. Williams. 1982. Proposal for the conservation of the generic name 1779 *Oncidium* Swartz (Orchidaceae) with a conserved type species, *Oncidium altissimum* Sw. Taxon 31: 752–754.

Drummond, J. R. 1907. The literature of *Furcraea* with a synopsis of the known species. Rep. (Annual) Missouri Bot. Gard. 18: 25–75.

Duke, J. A. and E. S. Ayensu. 1985. Medicinal Plants of China. 2 vols. Algonac, Mich.

Duncan, W. H. 1975. Woody Vines of the Southeastern United States. Athens, Ga.

Dunsterville, G. C. K. and L. A. Garay. [1959]–1976. Venezuelan Orchids Illustrated. 6 vols. London.

Duvall, M. R. et al. 1993b. Phylogenetic hypotheses for the monocotyledons constructed from *rbc*L data. Ann. Missouri Bot. Gard. 80: 607–619.

Duvall, M. R. and B. R. Morton. 1996. Molecular phylogenetics of Poaceae: An expanded analysis of *rbc*L sequence data. Molec. Phylogen. Evol. 5: 352–358.

Duvall, M. R., J. D. Noll, and A. H. Minn. 2001. Phylogenetics of Paniceae (Poaceae). Amer. J. Bot. 88: 1988–1992.

Duyfjes, B. E. E. 1991. Stemonaceae and Pentastemonaceae; with miscellaneous notes on members of both families Blumea 36: 239–252.

Duyfjes, B. E. E. 1992. Formal description of the family Pentastemonaceae with some additional notes on Pentastemonaceae and Stemonaceae. Blumea 36: 551–552.

Dyer, A. F., K. Jong, and J. A. Ratter. 1970. Aneuploidy: A redefinition. Notes Roy. Bot. Gard. Edinburgh 30: 177–182.

Dyer, R. W. 1986. Small whorled pogonia. Audubon Wildlife Rep. 1986: 780–789.

Dykes, W. R. 1913. The Genus Iris. Cambridge.

Eastman, L. M. 1988. *Calopogon tuberosus* var. *latifolius* (St. John) Boivin (Orchidaceae) new to the United States. Rhodora 90: 101.

Eaton, A. 1829. A Manual of Botany for the Northern States. Albany. (Man. Bot. ed. 5)

Ebinger, J. E. 1993. False hellebore (*Veratrum woodii,* Liliaceae) populations in Illinois. Trans. Illinois State Acad. Sci. 86: 85–91.

Eckenwalder, J. E. and S. C. H. Barrett. 1986. Phylogenetic systematics of Pontederiaceae. Syst. Bot. 11: 373–391.

Ecklon, C. F. 1827. Topographisches Verzeichniss der Pflanzensammlung.... Esslingen. (Topogr. Verz. Pflanzensamml. Ecklon)

Ecol. Food Nutr. = Ecology of Food and Nutrition.

Ecol. Monogr. = Ecological Monographs.

Ecology = Ecology, a Quarterly Journal Devoted to All Phases of Ecological Biology.

Econ. Bot. = Economic Botany; Devoted to Applied Botany and Plant Utilization.

Edinburgh New Philos. J. = Edinburgh New Philosophical Journal.

Edinburgh Philos. J. = Edinburgh Philosophical Journal.

Edwards's Bot. Reg. = Edwards's Botanical Register....

Eichhornia—See: C. S. Kunth 1842

Eldenas, P. K. and H. P. Linder. 2000. Congruence and complementarity of morphological and *trn*L-*trn*F sequence data and the phylogeny of the African Restionaceae. Syst. Bot. 25: 692–707.

Elliott, S. [1816–]1821–1824. A Sketch of the Botany of South-Carolina and Georgia. 2 vols. in 13 parts. Charleston. (Sketch Bot. S. Carolina)

Elwes, H. J. 1880. A Monograph of the Genus *Lilium.* London.

Emory, W. H. 1857–1859. Report on the United States and Mexican Boundary Survey, Made under the Direction of the Secretary of the Interior. 2 vols. in parts. Washington. (Rep. U.S. Mex. Bound.)

Encycl.—See: J. Lamarck et al. 1783–1817

Endlicher, S. L. 1836–1840[–1850]. Genera Plantarum Secundum Ordines Naturales Disposita. 18 parts + 5 suppl. in 6 parts. Vienna. [Paged consecutively through suppl. 1(2); suppls. 2–5 paged independently.] (Gen. Pl.)

Endress, P. K., P. Baas, and M. Gregory. 2000. Systematic plant morphology and anatomy—50 years of progress. Taxon 49: 401–434.

Engard, R. G. 1970. *Hesperaloe:* Potential for arid lands. In: Centro de Investigación en Química Aplicada. 1970. *Yucca.* Saltillo. Pp. 115–123.

Engel, V. S. 1987. Saprophytic orchids of Dallas. Amer. Orchid Soc. Bull. 56: 831–835.

Engler, H. G. A. 1897. Übersicht über die Unterrabteilungen, Klassen, Reihen, Unterreihen und Familien der Embryophyta Siphonogama. In: H. G. A. Engler and K. Prantl, eds. 1887–1915. Die natürlichen Pflanzenfamilien.... 254 fasc. Leipzig. Fasc. 165[Nachtr. 1], pp. 341–357.

Engler, H. G. A., ed. 1900–1953. Das Pflanzenreich.... 107 vols. Berlin. [Sequence of volume (Heft) numbers (order of publication) is independent of the sequence of series and family (Roman and Arabic) numbers (taxonomic order).]

Engler, H. G. A. et al., eds. 1924+. Die natürlichen Pflanzenfamilien..., ed. 2. 26+ vols. Leipzig and Berlin. (Nat. Pflanzenfam. ed. 2)

Engler, H. G. A. and K. Prantl, eds. 1887–1915. Die natürlichen Pflanzenfamilien.... 254 fasc. Leipzig. [Sequence of fascicle (Lieferung) numbers (order of publication) is independent of the sequence of division (Teil) and subdivision (Abteilung) numbers (taxonomic order).] (Nat. Pflanzenfam.)

Entwurf Anordn. Orch.—See: E. H. H. Pfitzer 1887

Enum.—See: P. C. Fabricius 1759

Enum. Hort. Berol. Alt.—See: J. H. F. Link 1821–1822

Enum. Orchids U.S. & Canada—See: O. Ames 1924b

Enum. Pl.—See: C. S. Kunth 1833–1850

Enum. Syst. Pl.—See: N. J. Jacquin 1760
Environm. Exp. Bot. = Environmental and Experimental botany; an International Journal.
Environm. Pollut., A = Environmental Pollution. Series A, Ecological and Biological.
Erhardt, W. 1992. *Hemerocallis:* Daylilies. Portland.
Erickson, R. O. 1941. *Camassia scilloides.* Ann. Missouri Bot. Gard. 28: 293–298.
Erigenia = Erigenia; Journal of the Southern Illinois Native Plant Society.
Eriksson, R. 1994. Phylogeny of the Cyclanthaceae. Pl. Syst. Evol. 190: 31–47.
Erythea = Erythea; a Journal of Botany, West American and General.
Evans, T. M. et al. 2000. Phylogenetic relationships in the Commelinaceae: I. A cladistic analysis of morphological data. Syst. Bot. 25: 668–691.
Evans, T. M. et al. 2000b. Homoplasy in the Commelinaceae: A comparison of different classes of morphological characters. In: K. L. Wilson and D. A. Morrison, eds. 2000. Monocots: Systematics and Evolution. Melbourne. Pp. 557–566.
Evol. Biol. = Evolutionary Biology.
Evolution = Evolution, International Journal of Organic Evolution.
Exot. Bot.—See: J. E. Smith 1804–1805[–1808]
Exot. Fl.—See: W. J. Hooker [1822–]1823–1827
Exped. Great Salt Lake—See: H. Stansbury 1852
Fabricius, P. C. 1759. Enumeratio Methodica Plantarum Horti Medici Helmstadiensis.... Helmstedt. (Enum.)
Fairchild Trop. Gard. Bull. = Fairchild Tropical Garden Bulletin.
Fam. Nat. Syn. Monogr.—See: M. J. Roemer 1846–1847
Fam. Pl.—See: M. Adanson 1763[–1764]
Farrens, B. M. 1947. A Taxonomic Study of the North American Species of *Fritillaria.* M.A. thesis. Stanford University.
Fassett, N. C. 1935. Notes from the herbarium of the University of Wisconsin—XII. A study of *Streptopus.* Rhodora 37: 88–113.
Fawcett, W. and A. B. Rendle. 1910–1936. Flora of Jamaica.... 5 vols. London. (Fl. Jamaica)
Fay, M. F. et al. 2000. Phylogenetic studies of Asparagales based on four plastid DNA regions. In: K. L. Wilson and D. A. Morrison, eds. 2000. Monocots: Systematics and Evolution. Melbourne. Pp. 360–371.
Fay, M. F. and M. W. Chase. 1996. Resurrection of Themidaceae for the *Brodiaea* alliance, and recircumscription of Alliaceae, Amaryllidaceae and Agapanthoideae. Taxon 45: 441–451.
Fay, M. F. and M. W. Chase. 2000. Modern concepts of Liliaceae with a focus on the relationships of *Fritillaria.* Bot. Mag., [n. s.] 17: 146–149.
Feddes Repert. = Feddes Repertorium.
Feddes Repert. Spec. Nov. Regni Veg. = Feddes Repertorium Specierum Novarum Regni Vegetabilis.
Fernald, M. L. 1921. The Gray Herbarium expedition to Nova Scotia, 1920. Rhodora 23: 153–171, 223–245.
Fernald, M. L. 1925. *Pontederia* versus *Unisema.* Rhodora 27: 76–81.
Fernald, M. L. 1928c. Contributions from the Gray Herbarium of Harvard University, LXXXII. Rhodora 30: 183–184.
Fernald, M. L. 1943e. Virginia botanizing under restrictions. Part II. Range extensions, technical notes and revisions. Rhodora 45: 86.
Fernald, M. L. 1944. Overlooked species, transfers, and novelties in the flora of eastern North America. Rhodora 46: 1–28, 32–60.
Fernald, M. L. 1944b. Specific distinctions between *Polygonatum biflorum* and *P. canaliculatum.* Rhodora 46: 9–12.
Fernald, M. L. 1944c. The indigenous Alleghenian *Convallaria.* Rhodora 46: 12–14.
Fernald, M. L. 1946b. Some orchids of the Manual range. Rhodora 48: 161–162, 184–197, plates 1045–1048.
Fernald, M. L. 1946c. *Stenanthium* in the eastern United States. Rhodora 48: 148–152.
Fernald, M. L. 1950. Gray's Manual of Botany, ed. 8. New York.
Fernandez, A. and J. R. Davina. 1991. Heterochromatin and genome size in *Fortunatia* and *Camassia* (Hyacinthaceae). Kew Bull. 46: 307–316.
Field & Lab. = Field & Laboratory.
Fig. Pl. Gard. Dict.—See: P. Miller [1755–]1760
Firmage, D. H. and F. R. Cole. 1988. Reproductive success and inflorescence size of *Calopogon tuberosus* (Orchidaceae). Amer. J. Bot. 75: 1371–1377.
Fish and Wildlife Service [U.S.D.I.]. 1998. Recovery Plan for the Endangered Western Lily (*Lilium occidentale*). Portland.
Fitzsimons, R. E. and R. H. Vallejos. 1986. Growth of water hyacinth (*Eichhornia crassipes* (Mart.) Solms) in the middle Paraná River (Argentina). Hydrobiologia 131: 257–260.
Fl. & Sylva = Flora and Sylva; a Monthly Review for Lovers of Garden, Woodland, Tree or Flower; New and Rare Plants, Trees, Shrubs, and Fruits; the Garden Beautiful, Home Woods, and Home Landscape.
Fl. Aleut. Isl.—See: E. Hultén 1937
Fl. Amer. Sept.—See: F. Pursh 1814
Fl. Angl.—See: W. Hudson 1762
Fl. Angl. ed. 2—See: W. Hudson 1778
Fl. Austriac.—See: N. J. Jacquin 1773–1778
Fl. Bor.-Amer.—See: W. J. Hooker [1829–]1833–1840; A. Michaux 1803
Fl. Boston. ed. 2—See: J. Bigelow 1824
Fl. Bras.—See: C. F. P. von Martius et al. 1840–1906
Fl. Brit. W. I.—See: A. H. R. Grisebach [1859–]1864
Fl. Cab. = The Floral Cabinet, and Magazine of Exotic Botany.
Fl. Calif.—See: W. L. Jepson 1909–1943
Fl. Cap. ed. 2—See: C. P. Thunberg 1823
Fl. Carol.—See: T. Walter 1788
Fl. Flumin.—See: J. M. Vellozo 1825–1827[1829–1831]
Fl. Franç. ed. 3—See: J. Lamarck and A. P. de Candolle 1805[–1815]
Fl. Germ. Excurs.—See: H. G. L. Reichenbach 1830[–1832]
Fl. Ind. Occid.—See: O. P. Swartz 1797–1806
Fl. Indica—See: N. L. Burman 1768
Fl. Jamaica—See: W. Fawcett and A. B. Rendle 1910–1936
Fl. Less. Antill.—See: R. A. Howard 1974–1989

Fl. Ludov.—See: C. S. Rafinesque 1817
Fl. N. Amer.—See: J. Torrey and A. Gray 1838–1843
Fl. N. W. Amer.—See: T. J. Howell 1897–1903
Fl. Peruv.—See: H. Ruiz López and J. A. Pavón 1798–1802
Fl. Peruv. Prodr.—See: H. Ruiz López and J. A. Pavón 1794
Fl. Plains N. Amer.—See: P. A. Rydberg 1932
Fl. Ross.—See: C. F. von Ledebour [1841]1842–1853
Fl. S.E. U.S.—See: J. K. Small 1903
Fl. S.E. Washington—See: C. V. Piper and R. K. Beattie 1914
Fl. Saghalin—See: K. Miyabe and T. Miyake 1915
Fl. Serres Jard. Eur. = Flore des Serres et des Jardins de l'Europe.
Fl. South. U.S.—See: A. W. Chapman 1860
Fl. South. U.S. ed. 2 repr. 1892—See: A. W. Chapman [1892]
Fl. South. U.S. ed. 2 reprint 2—See: A. W. Chapman 1892
Fl. Tellur.—See: C. S. Rafinesque 1836[1837–1838]

Flagg, R. O. 1961. Investigations in the Tribe Zephyrantheae of the Amaryllidaceae. Ph.D. dissertation. University of Virginia.

Flagg, R. O. and W. S. Flory. 1962. Correlation between chromosome numbers and breeding habits in northern hemisphere *Zephyranthes.* [Abstract.] Amer. J. Bot. 49: 659.

Flagg, R. O. and W. S. Flory. 1976. Origins of three Texas species of *Zephyranthes.* Pl. Life 32: 67–80.

Flexner, S. B. and L. C. Hauck, eds. 1987. The Random House Dictionary of the English Language, ed. 2 unabridged. New York.

Flora = Flora; oder (allgemeine) botanische Zeitung. [Vols. 1–16, 1818–33, include "Beilage" and "Ergänzungsblätter"; vols. 17–25, 1834–42, include "Beiblatt" and "Intelligenzblatt."]

Flora of North America Editorial Committee, eds. 1993+. Flora of North America North of Mexico. 5+ vols. New York and Oxford.

Flory, W. S. 1938. Cytotaxonomic notes on the genus *Habranthus.* Herbertia 5: 151–153.

Flory, W. S. 1950. Taxonomical and cytological distinctions between *Hymenocallis occidentalis* and *H. galvestonensis.* [Abstract.] J. Tennessee Acad. Sci. 25: 221.

Flory, W. S. 1976. Distribution, chromosome numbers and types of various species and taxa of *Hymenocallis.* Nucleus (Calcutta) 19: 204–227.

Flory, W. S. 1978. Known distributions of *Hymenocallis* Salisbury in North and Middle America and the West Indies. Pl. Life 34: 47–59.

Flory, W. S. and R. O. Flagg. 1958. A cytological study of the genus *Habranthus.* Nucleus (Calcutta) 1: 267–280.

Flory, W. S. and R. O. Flagg. 1981. *Pyrolirion* and *Zephyranthes:* Distinct genera. Bol. Soc. Brot., ser. 2, 53: 1197–1210.

Flory, W. S. and G. L. Smith. 1980. High chromosome numbers in several Zephyrantheae taxa. Pl. Life 36: 63–72.

Foldats, E. 1969–1970. Orchidaceae. 6 parts. Caracas. [Flora de Venezuela, vol. 15.]

Folsom, J. P. 1984. Una reinterpretación del estatus y relaciones de las taxa del complejo de *Platanthera ciliaris.* (A reinterpretation of the status and relationships of taxa of the yellow-fringed orchid complex.) Orquidea (Mexico City) 9: 321–345.

Forest, H. S., C. N. Horn, and B. A. Marcus. 1986. New observations of galls on *Zosterella dubia* (Jacq.) Small (*Heteranthera dubia* (Jacq.) MacM.) infected by *Membranosorus heterantherae* Ostenfeld and Peterson. Environm. Exp. Bot. 26: 377–384.

Foster, M. 1889. On irises. J. Roy. Hort. Soc. 11: 131–149.

Foster, R. C. 1937. A cytotaxonomic survey of the North American species of *Iris.* Contr. Gray Herb. 119: 3–79.

Foster, R. C. 1938. Iridis Species Novae. Cambridge, Mass. (Iridis Sp. Nov.)

Foster, R. C. 1945. Studies in the Iridaceae, III. Contr. Gray Herb. 155: 3–54.

Fragm. Florist. Geobot. = Fragmenta Floristica et Geobotanica.

Frame, D. 1989. *Schoenocaulon.* In: R. McVaugh and W. R. Anderson, eds. 1974+. Flora Novo-Galiciana: A Descriptive Account of the Vascular Plants of Western Mexico. 8+ vols. Ann Arbor. Vol. 15, pp. 263–275.

Frame, D. 2001. Chromosome studies in *Schoenocaulon* (Liliaceae: Melanthieae) a relict genus. Anales Inst. Biol. Univ. Nac. Auton. Mexico, Bot. 72: 123–129.

Frame, D., A. Espejo, and A. R. López-F. 1999. A conspectus of Mexican Melanthiaceae including a description of new taxa of *Schoenocaulon* and *Zigadenus.* Acta Bot. Mex. 48: 27–50.

Franklin, J. et al. 1823. Narrative of a Journey to the Shores of the Polar Sea, in the Years 1819, 20, 21 and 22. London. [Richardson: Appendix VII. Botanical appendix, pp. [729]–778, incl. bryophytes by Schwägrichen, algae and lichens by Hooker. Brown: Addenda [to Appendix VII], pp. 779–784.] (Narr. Journey Polar Sea)

Freeman, J. D. 1975. Revision of *Trillium* subgenus *Phyllantherum* (Liliaceae). Brittonia 27: 1–62.

Fremontia = Fremontia: Journal of the California Native Plant Society.

French, J. C., Chung M. G., and Hur Y. K. 1995. Chloroplast DNA phylogeny of the Ariflorae. In: P. J. Rudall et al., eds. 1995. Monocotyledons: Systematics and Evolution. 2 vols. Kew. Vol. 1, pp. 255–275.

Freudenstein, J. V. 1987. A preliminary study of *Corallorhiza maculata* (Orchidaceae) in eastern North America. Contr. Univ. Michigan Herb. 16: 145–153.

Freudenstein, J. V. 1997. A monograph of *Corallorhiza* (Orchidaceae). Harvard Pap. Bot. 1(10): 5–51.

Freudenstein, J. V. 1998. Paraphyly, ancestors, and classification—a response to Sosef and Brummitt. Taxon 47: 95–104.

Freudenstein, J. V. 1999. A new species of *Corallorhiza* (Orchidaceae) from West Virginia, U.S.A. Novon 9: 511–513.

Freudenstein, J. V. and M. W. Chase. 2001. Analysis of mitochondrial *nad1*b-c intron sequences in Orchidaceae: Utility and coding of length-change characters. Syst. Bot. 26: 643–657.

Freudenstein, J. V. and F. N. Rasmussen. 1999. What does morphology tell us about orchid relationships? A cladistic analysis. Amer. J. Bot. 86: 225–248.

Friedman, W. E. and S. K. Floyd. 2001. Perspective: The origin of flowering plants and their reproductive biology—a tale of two phylogenies. Evolution 55: 217–231.

Frye, T. C. and G. B. Rigg. 1912. Northwest Flora. Seattle. (N.W. Fl.)

Fryxell, P. A. 1967. The interpretation of disjunct distributions. Taxon 16: 316–324.

Fujita, N. 1976. The genus *Hosta* in Japan. Acta Phytotax. Geobot. 27: 66–96.

Fukuhara, T. and Z. K. Shinwari. 1994. Seed-coat anatomy in Uvulariaceae (Liliales) of the Northern Hemisphere: Systematic implications. Acta Phytotax. Geobot. 45: 1–14.

Funct. Ecol. = Functional Ecology.

Fuse, S. and M. N. Tamura, M. N. 2000. A phylogenetic analysis of the plastid *mat*K gene with emphasis on Melanthiaceae sensu lato. Pl. Biol. (Stuttgart) 2: 415–427.

Gadella, T. W. J. and L. van Raamsdonk. 1981. Biosystematic studies on *Ornithogalum umbellatum* L. s.l. Bol. Soc. Brot., ser. 2, 53: 745–791.

Garay, L. A. 1969. Notes on West Indian orchids. J. Arnold Arbor. 50: 462–468.

Garay, L. A. 1977. Systematics of the Physurinae (Orchidaceae) in the New World. Bradea 2: 191–204.

Garay, L. A. 1980[1982]. A generic revision of the Spiranthinae. Bot. Mus. Leafl. 28: 277–425.

Garay, L. A. and J. E. Stacy. 1974. A revision of *Oncidium*. Bradea 1: 393–428.

Garay, L. A. and H. R. Sweet. 1972. Notes on West Indian orchids, III. J. Arnold Arbor. 53: 515–530.

Garay, L. A. and H. R. Sweet. 1974. Orchidaceae. In: R. A. Howard. 1974–1989. Flora of the Lesser Antilles: Leeward and Windward Islands. 6 vols. Jamaica Plain. Vol. 1.

Garay, L. A. and P. Taylor. 1976. The genus *Oeceoclades* Lindl. Bot. Mus. Leafl. 24: 249–274.

Gard. Chron. = Gardener's Chronicle.

Gard. Dict. Abr. ed. 4—See: P. Miller 1754

Gard. Dict. ed. 8—See: P. Miller 1768

Gard. J. New York Bot. Gard. = Garden Journal of the New York Botanical Garden.

Garden (London 1871–1927) = The Garden; an Illustrated Weekly Journal of Gardening in All Its Branches.

Gartenflora = Gartenflora; Monatsschrift für deutsche und schweizerische Garten- und Blumenkunde.

Gates, R. R. 1917. A revision of the genus *Polygonatum* in North America. Bull. Torrey Bot. Club 44: 117–126.

Gates, R. R. 1917b. A systematic study of the North American genus *Trillium*, its variability and its relation to *Paris* and *Medeola*. Ann. Missouri Bot. Gard. 4: 43–92.

Gates, R. R. 1918. A systematic study of the North American Melanthaceae from a genetic standpoint. J. Linn. Soc., Bot. 44: 131–172.

Gaut, B. S.. A. S. Peek, B. R. Morton, and M. T. Clegg. 1999. Patterns of genetic diversification within the Adh gene family in the grasses (Poaceae). Molec. Biol. Evol. 16: 1086–1097.

Geerinck, D. 1969. Genera des Haemodoraceae et des Hypoxidaceae. Bull. Jard. Bot. Natl. Belg. 39: 47–82.

Gen. N. Amer. Pl.—See: T. Nuttall 1818

Gen. Pl.—See: G. Bentham and J. D. Hooker 1862–1883; S. L. Endlicher 1836–1840[–1850]; A. L. de Jussieu 1789; R. A. Salisbury 1866

Gen. Pl. ed. 5—See: C. Linnaeus 1754

Gen. Sp. Orchid. Pl.—See: J. Lindley 1830–1840

Gentes Herb. = Gentes Herbarum; Occasional Papers on the Kinds of Plants.

Gentry, H. S. 1972. The Agave Family in Sonora. Washington. [Agric. Handb. 399.] (Agave Fam. Sonora)

Gentry, H. S. 1978. The agaves of Baja California. Occas. Pap. Calif. Acad. Sci. 130.

Gentry, H. S. 1982. Agaves of Continental North America. Tucson. (Agaves Continental N. Amer.)

Gervais, C. 1979. Liste annotée de nombres chromosomiques de la flore vasculaire du nord-est de l'Amérique. I. Naturaliste Canad. 106: 451–461.

Ges. Naturf. Freunde Berlin Mag. Neuesten Entdeck. Gesammten Naturk. = Der Gesellschaft naturforschender Freunde zu Berlin Magazin für die neuesten Entdeckungen in der gesammten Naturkunde.

Ges. Naturf. Freunde Berlin Neue Schriften = Der Gesellschaft naturforschender Freunde zu Berlin, neue Schriften.

Getreue Darstell. Gew.—See: F. G. Hayne et al. 1805–1846

Gilbertson, T. J. 1993. Zygacine and zygadenine: The major alkaloids from *Zygadenus gramineus*. Phytochemistry 12: 2079–2080.

Gilmartin, A. J. and G. K. Brown. 1987. Bromeliales, related monocots and resolution of relationships among Bromeliaceae subfamilies. Syst. Bot. 12: 493–500.

Giuletti, A. M. et al. 2000. Multidisciplinary studies of neotropical Eriocaulaceae. In: K. L. Wilson and D. A. Morrison, eds. 2000. Monocots: Systematics and Evolution. Melbourne. Pp. 580–589.

Givnish, T. J. et al. 2000. Molecular evolution, adaptive radiation, and geographic diversification in the amphiatlantic family Rapateaceae: Evidence from *ndh*F sequences and morphology. Evolution 54: 1915–1937.

Givnish, T. J., T. M. Evans, J. C. Pires, and K. J. Sytsma. 1999. Polyphyly and convergent evolution in Commelinales and Commelinidae: Evidence from *rbc*L sequence data. Molec. Phylogen. Evol. 12: 360–385.

Gleason, H. A. and A. Cronquist. 1991. Manual of Vascular Plants of Northeastern United States and Adjacent Canada, ed. 2. Bronx.

Gloudon, A. and C. Tobisch. 1995. Orchids of Jamaica. Kingston.

Gmelin, J. F. 1791[–1792]. Caroli à Linné...Systema Naturae per Regna Tria Naturae.... Tomus II. Editio Decima Tertia, Aucta, Reformata. 2 parts. Leipzig. (Syst. Nat.)

Godfrey, R. K. [1979.] Endangered: Harper's beauty. In: D. B. Ward, ed. [1979.] Rare and Endangered Biota of Florida. Vol. 5. Plants. Gainesville. Pp. 32–33.

Godfrey, R. K. and J. W. Wooten. 1979. Aquatic and Wetland Plants of Southeastern United States: Monocotyledons. Athens, Ga.

Godt, M. J. W., J. Walker, and J. L. Hamrick. 1997. Genetic diversity in the endangered lily *Harperocallis flava* and a close relative *Tofieldia racemosa*. Conservation Biol. 11: 361–366.

Goetghebeur, P. 1998. Cyperaceae. In: K. Kubitzki et al., eds. 1990+. The Families and Genera of Vascular Plants. 4+ vols. Berlin etc. Vol. 4, pp. 141–190.

Goldblatt, P. 1969. The genus *Sparaxis*. J. S. African Bot. 35: 219–252.

Goldblatt, P. 1975. Revision of the bulbous Iridaceae of North America. Brittonia 27: 373–385.

Goldblatt, P. 1982. Systematics of *Freesia* Klatt (Iridaceae). J. S. African Bot. 48: 39–91.
Goldblatt, P. 1989. The genus *Watsonia*. A systematic monograph. Ann. Kirstenbosch Bot. Gard. 17.
Goldblatt, P. 1990. Phylogeny and classification of Iridaceae. Ann. Missouri Bot. Gard. 77: 607–627.
Goldblatt, P. 1991. An overview of the systematics, phylogeny and biology of the southern African Iridaceae. Contr. Bolus Herb. 13: 1–74.
Goldblatt, P. 1992. Phylogenetic analysis of the South African genus *Sparaxis* (including *Synottia*) (Iridaceae-Ixioideae), with two new species and a review of the genus. Ann. Missouri Bot. Gard. 79: 143–159.
Goldblatt, P. 1995. The status of R. Dahlgren's orders Liliales and Melanthiales. In: P. J. Rudall et al., eds. 1995. Monocotyledons: Systematics and Evolution. 2 vols. Kew. Vol. 1, pp. 181–200.
Goldblatt, P. 1996. *Gladiolus* in Tropical Africa. Portland.
Goldblatt, P. 1999. *Sparaxis*. In: Department of Agricultural Technical Services [South Africa]. 1963+. Flora of Southern Africa.... 19+ vols. (some partial). Pretoria. Vol. 7, part 2, fasc. 1, pp. 151–169.
Goldblatt, P. and J. E. Henrich. 1991. *Calydorea* Herbert (Iridoideae–Tigridieae): Notes on this New World genus and reduction to synonymy of *Salpingostylis*, *Cardiostigma*, *Itysa*, and *Catila*. Ann. Missouri Bot. Gard. 78: 504–511.
Goldblatt, P. and T. M. Howard. 1992. Notes on *Alophia* (Iridaceae) and a new species, *A. veracruzana*, from Vera Cruz, Mexico. Ann. Missouri Bot. Gard. 79: 901–905.
Goldblatt, P. and J. C. Manning. 1995. Phylogeny of the African genera *Anomatheca* and *Freesia* (Iridaceae–Ixioideae), and a new genus *Xenoscapa*. Syst. Bot. 20: 161–178.
Goldblatt, P. and J. C. Manning. 1998. *Gladiolus* in Southern Africa. Cape Town.
Goldblatt, P., P. J. Rudall, and J. E. Henrich. 1990. The genera of the *Sisyrinchium* alliance (Iridaceae: Iridoideae): Phylogeny and relationships. Syst. Bot. 15: 497–510.
Goldman, D. H. 1995. A new species of *Calopogon* from the midwestern United States. Lindleyana 10: 37–42.
Goldman, D. H. 1998. Proposal to conserve the name *Ophrys barbata* (Orchidaceae) with a conserved type. Taxon 47: 161–162.
Goldman, D. H. 2000. Systematics of *Calopogon* and the Tribe Arethuseae (Orchidaceae). Ph.D. dissertation. University of Texas.
Goldman, D. H. et al. 2001. Phylogenetics of Arethuseae (Orchidaceae) based on plastid *mat*K and *rbc*L sequences. Syst. Bot. 26: 670–695.
Goldman, D. H. and S. L. Orzell. 2000. Morphological, geographical and ecological re-evaluation of *Calopogon multiflorus* (Orchidaceae). Lindleyana 15: 237–251.
Gooch, R. and G. L. Smith. 1994. Distinctions between *Hymenocallis liriosme* (Raf.) Shinners and *Hymenocallis choctawensis* Traub. [Abstract.] J. Elisha Mitchell Sci. Soc. 110: 152–153.
Goodrich, S. and E. Neese. 1986. Uinta Basin Flora. Ogden.
Gorham, A. L. 1953. The question of fertilization in *Smilacina racemosa* (L.) Desf. Phytomorphology 3: 44–50.
Goss, G. J. 1973. Pollination Biology in the Orchidaceae: *Polystachya flavescens, Epidendrum difforme* and *Eulophia alta* from South Florida; *Encyclia gracilis, Encyclia altissima* and *Encyclia rufa* from Great Inagua, Bahamas. Master's thesis. Florida Atlantic University.
Goss, G. J. 1977. The reproductive biology of the epiphytic orchids of Florida 5. *Epidendrum difforme* Jacquin. Amer. Orchid Soc. Bull. 46: 630–636.
Goss, G. J. 1977b. The reproductive biology of the epiphytic orchids of Florida 6. *Polystachya flavescens* (Lindley) J. J. Smith. Amer. Orchid Soc. Bull. 46: 990–994.
Gould, F. W. 1942. A systematic treatment of the genus *Camassia* Lindl. Amer. Midl. Naturalist 28: 712–742.
Graham, S. W. et al. 1998. Phylogenetic congruence and discordance among one morphological and three molecular data sets from Pontederiaceae. Syst. Biol. 47: 545–567.
Graham, S. W. and R. G. Olmstead. 2000. Evolutionary significance of an unusual chloroplast DNA inversion found in two basal angiosperm lineages. Curr. Genet. 37: 183–188.
Grana = Grana; an International Journal of Palynology Including World Pollen and Spore Flora.
Granick, E. B. 1944. A karyosystematic study of the genus *Agave*. Amer. J. Bot. 31: 283–298.
Grant, J. R. and G. Zijlstra. 1998. An annotated catalogue of the generic names of the Bromeliaceae. Selbyana 19: 91–121.
Gravendeel, B. et al. 2001. Molecular phylogeny of *Coelogyne* (Epidendroideae: Orchidaceae) based on plastid RFLPS, *mat*K, and nuclear ribosomal ITS sequences: Evidence for polyphyly. Amer. J. Bot. 88: 1915–1927.
Gray, A. 1848. A Manual of the Botany of the Northern United States.... Boston, Cambridge, and London. (Manual)
Gray, A. 1876. *Schoenolirion* Torr. Amer. Naturalist 10: 426–427.
Gray, A., S. Watson, and J. M. Coulter. 1890. A Manual of the Botany of the Northern United States..., ed. 6. New York and Chicago. (Manual ed. 6)
Grayum, M. H. 1990. Evolution and phylogeny of the Araceae. Ann. Missouri Bot. Gard. 77: 628–697.
Great Plains Flora Association. 1986. Flora of the Great Plains. Lawrence, Kans.
Green, P. S. 1986. The correct name for *Asparagus sprengeri*. Plantsman 7: 249–250.
Greene, E. L. 1886. Studies in the botany of California and parts adjacent. I. Some genera which have been confused with the name *Brodiaea*. Bull. Calif. Acad. Sci. 2: 125–144.
Greene, E. L. 1894. Manual of the Botany of the Region of San Francisco Bay.... San Francisco. (Man. Bot. San Francisco)
Greenwood, E. W. 1982. Tipos de viscidio en Sprianthinae. Orquidea (Mexico City) 8: 283–310.
Greenwood, E. W. 1982b. *Govenia* in Mexico, an introductory note. Orquidea (Mexico City) 8: 114–120.
Greenwood, E. W. 1991. The Florida *Govenia*. Amer. Orchid Soc. Bull. 69: 867–869.
Gregg, K. B. 1989. Reproductive biology of the orchid *Cleistes divaricata* (L.) Ames var. *bifaria* Fernald growing in a West Virginia meadow. Castanea 54: 57–78.

Gregory, M. et al. 1998. Nomenclator *Alliorum: Allium* Names and Synonyms, a World Guide. Kew.

Greilhuber, J. 1982. Trends in der Chromosomenevolution von *Scilla* (Liliaceae). Stapfia 10: 11–51.

Greilhuber, J., B. Deumling, and F. Speta. 1981. Evolutionary aspects of chromosome banding, heterochromatin, satellite DNA, and genome size in *Scilla* (Liliaceae). Ber. Deutsch. Bot. Ges. 94: 249–266.

Grenfell, D. 1996. The Gardener's Guide to Growing Hostas. Portland.

Grenfell, D. 1998. *Hosta:* The Flowering Foliage Plant. London.

Greuter, W. et al., eds. 2000. International Code of Botanical Nomenclature (St. Louis Code): Adopted by the Sixteenth International Botanical Congress, St. Louis, Missouri, July–August 1999. Königstein. [Regnum Veg. 138.]

Grisebach, A. H. R. [1859–]1864. Flora of the British West Indian Islands. 7 parts. London. [Parts paged consecutively.] (Fl. Brit. W. I.)

Grisebach, A. H. R. 1866. Catalogus Plantarum Cubensium Exhibens Collectionem Wrightianam Aliasque Minores ex Insula Cuba Missas. Leipzig. (Cat. Pl. Cub.)

Grosvenor, G. 1999. Daylilies for the Garden. Portland.

Gunther, P. P. 1972. The Phenodynamics of *Clintonia borealis* (Ait.) Raf. and *C. umbellulata* (Michx.) Morong (Liliaceae) As Related to Their General Distribution. Ph.D. dissertation. Pennsylvania State University.

Hágsater, E. 1984. Towards an understanding of the genus *Epidendrum.* In: K. W. Tan, ed. 1984. Proceedings of the Eleventh World Orchid Conference: March 1984, Miami, Florida, U.S.A. Miami. Pp. 195–201.

Hágsater, E. 1993. *Epidendrum anceps* or *Epidendrum secundum*? Orquidea (Mexico City) 13: 153–158.

Hahn, W. J. 2002. A molecular phylogenetic study of the Palmae (Arecaceae) based on *atp*B, *rbc*L, and 18S nrDNA. Syst. Biol. 51: 92–112.

Hahn, W. J., W. J. Kress, and E. A. Zimmer. 1994. Molecular sequence phylogenetics of the Palmae. [Abstract.] Amer. J. Bot. 81(6, suppl.): 158–159.

Haines, A. L. 1941. Variation in *Yucca whipplei.* Madroño 6: 33–45.

Hall, A. D. 1940. The Genus *Tulipa.* London.

Hallier, H. 1905. Phylogenetic studies of flowering plants. New Phytol. 5: 151–162.

Hamann, U. 1976. Hydatellaceae: A new family of Monocotyledoneae. New Zealand J. Bot. 14: 193–196.

Hamann, U. 1998. Hydatellaceae. In: K. Kubitzki et al., eds. 1990+. The Families and Genera of Vascular Plants. 4+ vols. Berlin etc. Vol. 4, pp. 231–234.

Hamann, U., K. Kaplan, and T. Rübsamen. 1979. Über die Samenschalenstruktur der Hydatellaceae (Monocotyledoneae) und die systematische Stellung von *Hydatella filamentosa.* Bot. Jahrb. Syst. 100: 555–563.

Hamburger Garten- Blumenzeitung = Hamburger Garten- und Blumenzeitung.

Hammer, R. L. 1981. Finding new orchids: A contribution to the Orchidaceae of Florida. Fairchild Trop. Gard. Bull. 36(3): 16–18.

Handb. Amaryll.—See: J. G. Baker 1888

Handb. Skand. Fl.—See: C. J. Hartman 1820

Hanelt, P. 1992. Infrageneric grouping of *Allium*—The Gatersleben approach. In: P. Hanelt et al., eds. 1992. The Genus *Allium:* Taxonomic Problems and Genetic Resources.... Gatersleben. Pp. 107–123.

Hanelt, P., K. Hammer, and H. Knupffer, eds. 1992. The Genus *Allium:* Taxonomic Problems and Genetic Resources.... Gatersleben.

Hannan, G. L. and M. W. Orick. 2000. Isozyme diversity in *Iris cristata* and the threatened glacial endemic *I. lacustris.* Amer. J. Bot. 87: 293–301.

Hannibal, L. S. 1976. *Leucocrinum montanum* Nuttall. Pl. Life 3: 25–27.

Hanson, M. A. 1993. Dispersed Unidirectional Introgression from *Yucca schidigera* into *Y. baccata* (Agavaceae). Ph.D. dissertation. Claremont Graduate School.

Hara, H. 1968. A revision of the genus *Chionographis* (Liliaceae). J. Jap. Bot. 43: 257–267.

Hara, H. 1988. A revision of the Asiatic species of the genus *Disporum* (Liliaceae). Bull. Univ. Mus. Univ. Tokyo 31: 163–209, plates 7–47.

Hara, H. and S. Kurosawa. 1962. Morphological and chromosomal variations in *Chionographis japonica* Maxim. Acta Phytotax. Geobot. 20: 3–38.

Harborne, J. B. and H. Baxter. 1993. Phytochemical Dictionary. London.

Harborne, J. B., C. A. Williams, B. G. Briggs, and L. A. S. Johnson. 2000. Flavonoid patterns and phylogeny of the Restionaceae. In: K. L. Wilson and D. A. Morrison, eds. 2000. Monocots: Systematics and Evolution. Melbourne. Pp. 672–675.

Harley, M. M. and W. J. Baker. 2001. Pollen aperture morphology in Arecaceae: Application within phylogenetic analyses, and a summary of the fossil record of palm-like pollen. Grana 40: 45–77.

Harling, G., B. Sparre, and L. Andersson, eds. 1973+. Flora of Ecuador. 60+ nos. Göteborg. [Nos. 1—4 published as Opera Bot., B, 1—4.]

Harned, J. E. 1931. Wild Flowers of the Alleghenies. Oakland, Md.

Hartman, C. J. 1820. Handbok i Skandinaviens Flora, Innefettande Sveriges och Norriges Vexter, till och med Mossorna. Stockholm. (Handb. Skand. Fl.)

Harvard Pap. Bot. = Harvard Papers in Botany.

Haseltonia = Haseltonia; Yearbook of the Cactus and Succulent Society of America.

Haw, S. G. 1986. The Lilies of China. Portland.

Hawkes, A. D. 1952. Habitat notes on *Cyrtopodium punctatum.* Orchid J. 1: 296–297.

Haworth, A. H. 1812. Synopsis Plantarum Succulentarum.... London. (Syn. Pl. Succ.)

Haworth, A. H. 1819. Supplementum Plantarum Succlentarum.... London. (Suppl. Pl. Succ.)

Hay, S. G., A. Bouchard, and L. Brouillet. 1990. Additions to the flora of the island of Newfoundland. Rhodora 92: 277–293.

Hayashi, K. et al. 1998. Molecular systematics of the genus *Uvularia* and selected Liliales based upon *mat*K and *rbc*L gene sequence data. Pl. Spec. Biol. 13: 129–146.

Hayashi, K. and S. Kawano. 2000. Molecular systematics of *Lilium* and allied genera (Liliaceae): Phylogenetic relationships among *Lilium* and related genera based on *rbc*L and *mat*K gene sequence data. Pl. Spec. Biol. 15: 73–93.

Hayashi, K., S. Yoshida, F. H. Utech, and S. Kawano. 2001. Molecular systematics in the genus *Clintonia* and related taxa based on *rbc*L and *mat*K gene sequence data. Pl. Spec. Biol. 16: 119–137.

Hayne, F. G., J. F. Brandt, J. T. C. Ratzeburg, and J. F. Klotzsch. 1805–1846. Getreue Darstellung und Beschreibung der in der Arzneykunde gebräuchlichen Gewächse.... 14 vols. Berlin. (Getreue Darstell. Gew.)

Hédren, M. 1996. Genetic differentiation, polyploidization and hybridization in northern European *Dactylorhiza* (Orchidaceae): Evidence from allozyme markers. Pl. Syst. Evol. 201: 31–55.

Hegnauer, R. 1962+. Chemotaxonomie der Pflanzen.... 11+ vols. Basel and Boston.

Heller, A. A. 1900. Catalogue of North American Plants North of Mexico, Exclusive of the Lower Cryptogams, ed. 2. [Lancaster, Pa.] (Cat. N. Amer. Pl. ed. 2)

Heller, A. A. 1905. The western veratrums. Muhlenbergia 1: 119–125.

Hemsley, W. B. 1878–1880. Diagnoses Plantarum Novarum...Mexicanarum et Centrali-americanarum. 3 parts. London. [Parts paged consecutively.] (Diagn. Pl. Nov. Mexic.)

Henderson, D. M. 1972. A Biosystematic Study of the Genus *Sisyrinchium* (Iridaceae) in the Pacific Northwest. Ph.D. dissertation. University of Washington.

Henderson, D. M. 1976. A biosystematic study of Pacific Northwestern blue-eyed grasses (*Sisyrinchium*, Iridaceae). Brittonia 28: 149–176.

Henderson, N. C. 1991. *Iris* ×*pacifica*. Bull. Amer. Iris Soc. 280: 85–87.

Henderson, N. C. 1992. What is *Iris germanica*? Bull. Amer. Iris Soc. 286: 6–11.

Henderson, N. C. 1993. *Iris* ×*conglomerata*. Bull. Amer. Iris Soc. 290: 17–22.

Henderson, N. C. 1994. The Louisiana irises. Bull. Amer. Iris Soc. 293: 73–82.

Herb. Raf.—See: C. S. Rafinesque 1833

Herbert, W. 1821. An Appendix [to the Botanical Magazine and the Botanical Register]. London. (Appendix)

Herbert, W. 1837. Amaryllidaceae.... London.

Heredity = Heredity; an International Journal of Genetics.

Herndon, A. 1987. A morphometric comparison of *Hymenocallis palmeri* and *Hymenocallis floridana* (Amaryllidaceae) in southern Florida. Sida 12: 295–305.

Hess, W. J. and R. C. Sivinski. 1995. A new species of *Zigadenus* (Liliaceae) from New Mexico, with additional comments on the section *Anticlea*. Sida 16: 389–400.

Hickman, J. C., ed. 1993. The Jepson Manual. Higher Plants of California. Berkeley, Los Angeles, and London.

Higgins, W. E. 1997[1998]. A reconsideration of the genus *Prosthechea* (Orchidaceae). Phytologia 82: 370–383.

Hilu, K. W. and L. A. Alice. 2001. A phylogeny of Chloridoideae (Poaceae) based on *mat*K sequences. Syst. Bot. 26: 386–405.

Hilu, K. W., L. A. Alice, and H. P. Liang. 1999. Phylogeny of Poaceae inferred from *mat*K sequences. Ann. Missouri Bot. Gard. 86: 835–851.

Hist. Fis. Cuba—See: R. de la Sagra 1840–1855

Hist. Orchid.—See: L. M. A. du P. Thouars 1822

Hist. Pl. Guiane—See: J. B. Aublet 1775

Hitchcock, C. L. 1944. The *Tofieldia glutinosa* complex of western North America. Amer. Midl. Naturalist 31: 487–498.

Hitchcock, C. L., A. Cronquist, M. Ownbey, and J. W. Thompson. 1955–1969. Vascular Plants of the Pacific Northwest. 5 vols. Seattle. (Vasc. Pl. Pacif. N.W.)

Hochstätter, F. 2000. Het geslacht *Yucca* (Agavaceae). Deel 5. Sectie *Hesperoyucca, Yucca whipplei.* Succulenta (Netherlands) 79: 32–44.

Hodgson, W. C. 1999. Vascular plants of Arizona: Agavaceae. J. Arizona-Nevada Acad. Sci. 32: 1–21.

Hodgson, W. C. 2001. Food Plants of the Sonoran Desert. Tucson.

Hoffmann, G. F. [1791–1804.] Deutschland Flora oder botanisches Taschenbuch.... 4 vols. Erlangen. (Deutschl. Fl.)

Hollingsworth, E. B. 1966. Waterstargrass as an aquatic weed. Univ. Arkansas Coll. Agric. Arkansas Agric. Exp. Sta. Bull. 705.

Holm, T. 1929. The application of the term "rhizome." Rhodora 31: 6–17.

Holmes, W. C. and P. S. Mathies. 1980. Notes on the distribution of two Louisiana orchids. Castanea 45: 72.

Holmes, W. C. and C. J. Wells. 1980. The distribution of *Habranthus tubispathus* (L'Her.) Traub in South America and North America—Texas and Louisiana. Sida 8: 328–333.

Holt, P. C., ed. 1971. The Distributional History of the Biota of the Southern Appalachians. Part 2. Flora. Blacksburg. [Virginia Polytechnic Inst. and State Univ., Res. Div. Monogr. 2.]

Homoya, M. A. 1977. The Distribution and Ecology of the Genus *Isotria* in Illinois. Master's thesis. Southern Illinois University.

Homoya, M. A. 1993. Orchids of Indiana. Bloomington.

Hooker, W. J. [1822–]1823–1827. Exotic Flora, Containing Figures and Descriptions of New, Rare, or Otherwise Interesting Exotic Plants.... 3 vols. in 38 fasc. Edinburgh. (Exot. Fl.)

Hooker, W. J. [1829–]1833–1840. Flora Boreali-Americana; or, the Botany of the Northern Parts of British America.... 2 vols. in 12 parts. London, Paris, and Strasbourg. (Fl. Bor.-Amer.)

Hooker, W. J. and G. A. W. Arnott. [1830–]1841. The Botany of Captain Beechey's Voyage; Comprising an Account of the Plants Collected by Messrs Lay and Collie, and Other Officers of the Expedition, during the Voyage to the Pacific and Bering's Strait, Performed in His Majesty's Ship Blossom, under the Command of Captain F. W. Beechey...in the Years 1825, 26, 27, and 28. 10 parts. London. [Parts paged and plates numbered consecutively.] (Bot. Beechey Voy.)

Hooker's Icon. Pl. = Hooker's Icones Plantarum; or Figures, with Brief Descriptive Characters and Remarks of New or Rare Plants.

Hooper, G. R., F. W. Case, and R. Myers. 1971. Mycoplasma-like bodies associated with a flower greening disorder of a wild flower, *Trillium grandiflorum.* Pl. Dis. Reporter 55: 824–828.

Hoover, R. F. 1939. A definition of the genus *Brodiaea.* Bull. Torrey Bot. Club 66: 161–166.

Hoover, R. F. 1939b. A revision of the genus *Brodiaea.* Amer. Midl. Naturalist 22: 551–574.

Hoover, R. F. 1940. A monograph of the genus *Chlorogalum.* Madroño 5: 137–176.

Hoover, R. F. 1940b. The genus *Dichelostemma.* Amer. Midl. Naturalist 24: 463–476.

Hoover, R. F. 1941. A systematic study of *Triteleia.* Amer. Midl. Naturalist 25: 73–100.

Hoover, R. F. 1955. Further observations on *Brodiaea* and some related genera. Pl. Life 11: 13–22.

Hopper, S. D., M. F. Fay, M. Rossetto, and M. W. Chase. 1999. A molecular phylogenetic analysis of the bloodroot and kangaroo paw family, Haemodoraceae: Taxonomic, biogeographic and conservation implications. Bot. J. Linn. Soc. 131: 285–299.

Horn, C. N. 1981. Life History of *Heteranthera dubia* (Jacq.) MacM. (Water Star-grass) (Pontederiaceae) with Respect to Seasonal and Environmental Effects on Morphology. Columbus, Ohio. [Center Lake Erie Area Res. Techn. Rep. 214.]

Horn, C. N. 1983. The annual growth cycle of *Heteranthera dubia* in Ohio. Michigan Bot. 23: 29–34.

Horn, C. N. 1985. A Systematic Revision of the Genus *Heteranthera* (Sensu Lato; Pontederiaceae). Ph.D. dissertation. University of Alabama.

Horn, C. N. 1987. Pontederiaceae. In: G. Harling et al., eds. 1973+. Flora of Ecuador. 60+ nos. Göteborg. Vol. 29, pp. 1–20.

Horn, C. N. 1988. Developmental heterophylly in the genus *Heteranthera* (Pontederiaceae). Aquatic Bot. 31: 197–209.

Horn, C. N. and R. R. Haynes. 1987. Typification of *Pontederia* and its Linnaean species names (Pontederiaceae). Taxon 36: 621–625.

Hornberger, K. L. 1987. Systematics of the Genus *Sisyrinchium* (Iridaceae) in the Southeastern United States. Ph.D. dissertation. University of Arkansas.

Hornberger, K. L. 1991. The blue-eyed grasses (*Sisyrinchium:* Iridaceae) of Arkansas. Sida 14: 597–604.

Horres, R., G. Zizka, G. Kahl, and K. Weising. 2000. Molecular phylogenetics of Bromeliaceae: Evidence from *trn*L (UAA) intron sequences of the chloroplast genome. Pl. Biol. (Stuttgart) 2: 306–315.

Hort. Berol.—See: C. L. Willdenow 1803–1816

Hort. Brit. ed. 2—See: J. C. Loudon 1832

Hort. Brit. ed. 3—See: R. Sweet 1839

Hort. Canal.—See: I. F. Tausch 1823–1825

Hort. Kew.—See: W. Aiton 1789

Hortus Kew.—See: W. Aiton and W. T. Aiton 1810–1813

Howard, R. A. 1974–1989. Flora of the Lesser Antilles: Leeward and Windward Islands. 6 vols. Jamaica Plain. (Fl. Less. Antill.)

Howell, T. J. 1897–1903. A Flora of Northwest America. 1 vol. in 8 fasc. Portland. [Fasc. 1–7 (text) paged consecutively, fasc. 8 (index) independently.] (Fl. N. W. Amer.)

Hsiao, C., S. W. L. Jacobs, N. P. Barker, and N. J. Chatterton. 1998. A molecular phylogeny of the subfamily Arundinoideae (Poaceae) based on sequences of rDNA. Austral. Syst. Bot. 11: 41–52.

Hsiao, C., S. W. L. Jacobs, N. J. Chatterton, and K. H. Asay. 1999. A molecular phylogeny of the grass family (Poaceae) based on the sequences of nuclear ribosomal DNA (ITS). Austral. Syst. Bot. 11: 667–688.

Hu, S. Y. 1968. The species of *Hemerocallis.* Amer. Hort. Mag. 47: 86–111.

Huang, S. F. et al. 1989. Chromosome counts on one species and infraspecific taxa. Acta Bot. Austro Sin. 5: 161–176.

Huber, H. 1969. Die Samenmerkmale und Verwandtschaftsverhältnisse der Liliifloren. Mitt. Bot. Staatssamml. München 8: 219–538.

Huber, H. 1998. Dioscoreaceae. In: K. Kubitzki et al., eds. 1990+. The Families and Genera of Vascular Plants. 4+ vols. Berlin etc. Vol. 3, pp. 216–235.

Hudson, W. 1762. Flora Anglica.... London. (Fl. Angl.)

Hudson, W. 1778. Flora Anglica..., ed. 2. 2 vols. London. (Fl. Angl. ed. 2)

Hultén, E. 1937. Flora of the Aleutian Islands and Westernmost Alaska Peninsula with Notes on the Flora of Commander Islands.... Stockholm. (Fl. Aleut. Isl.)

Hultén, E. 1941–1950. Flora of Alaska and Yukon. 10 vols. Lund and Leipzig. [Vols. paged consecutively and designated as simultaneous numbers of Lunds Univ. Årsskr. (= Acta Univ. Lund.) and Kungl. Fysiogr. Sällsk. Handl.]

Hultén, E. 1958. The Amphi-Atlantic Plants and Their Phytogeographical Connections. Stockholm.

Humboldt, A. von, A. J. Bonpland, and C. S. Kunth. 1815[1816]–1825. Nova Genera et Species Plantarum Quas in Peregrinatione Orbis Novi Collegerunt, Descripserunt.... 7 vols. in 36 parts. Paris. (Nov. Gen. Sp.)

Hutchinson, J. 1926–1934. The Families of Flowering Plants. 2 vols. London.

Hutchinson, J. 1959. The Families of Flowering Plants, ed. 2. 2 vols. Oxford.

Hutchinson, J. 1969. Evolution and Phylogeny of Flowering Plants. London and New York.

Hutchinson, J. 1973. The Families of Flowering Plants, ed. 3. Oxford.

Hutton, E. E. and R. B. Clarkson. 1961. Two plants new for North America and some new or otherwise interesting plants in West Virginia. Castanea 26: 84–85.

Huxley, A., M. Griffiths, and M. Levy. 1992. The New Royal Horticultural Society Dictionary of Gardening. 4 vols. London, etc.

Hydrobiologia = Hydrobiologia. Acta Hydrobiologica, Limnologica et Protistologica.

Hylander, N. 1954. The genus *Hosta* in Swedish gardens. With contributions to the taxonomy, nomenclature and botanical history of the genus. Acta Horti Berg. 16: 331–420.

Hylander, N. 1960. The genus *Hosta.* J. Roy. Hort. Soc. 85: 356–369.

I. A. W. A. J. = I A W A Journal; International Association of Wood Anatomists.

Icon.—See: A. J. Cavanilles 1791–1801

Icon. Fl. Germ. Helv.—See: H. G. L. Reichenbach et al. 1835–

1914
Icon. Orchid. = Icones Orchidacearum.
Icon. Pl. Rar.—See: G. V. Schneevoogt [1792–]1793[–1795]
Icon. Pl. Trop. = Icones Plantarum Tropicarum.
Iconogr. Bot. Pl. Crit.—See: H. G. L. Reichenbach 1823–1832
Igersheim, A., M. Buzgo, and P. K. Endress. 2001. Gynoecium diversity and systematics in basal monocots. Bot. J. Linn. Soc. 136: 1–65.
Ill. Fl. N. U.S.—See: N. L. Britton and A. Brown 1896–1898
Ill. Fl. Pacific States—See: L. Abrams and R. S. Ferris 1923–1960
Ill. Orch. Pl.—See: F. A. Bauer and J. Lindley 1830–1838
Ingram, J. W. 1953. A monograph of the genera *Bloomeria* and *Muilla* (Liliaceae). Madroño 12: 19–27.
Int. J. Pl. Sci. = International Journal of Plant Sciences.
Intermount. Fl.—See: A. Cronquist et al. 1972+
International Botanical Congress. 1993. Abstracts. XV International Botanical Congress. Congress Center of Pacifico Yokohama, Japan, August 28 to September 3, 1993. [Tokyo.]
Iridis Sp. Nov.—See: R. C. Foster 1938
Irmisch, T. 1863. Beiträge zur vergleichend Morphologie der Pflanzen. *Lloydia serotina.* Bot. Zeitung (Berlin) 21: 161–164, 169–173.
Ito, M. et al. 1999. Phylogenetic relationships of Amaryllidaceae based on *mat*K sequence data. J. Pl. Res. 112: 207–216.
Ives, J. C. 1861. Report Upon the Colorado River of the West, Explored in 1857 and 1858 by Lieutenant Joseph C. Ives.... 5 parts, appendices. Washington. (Rep. Colorado R.)
Iwatsuki, K. and P. H. Raven, eds. 1997. Evolution and Diversification of Land Plants. Tokyo and New York.
J. Acad. Nat. Sci. Philadelphia = Journal of the Academy of Natural Sciences of Philadelphia.
J. Alabama Acad. Sci. = Journal of the Alabama Academy of Science.
J. Amer. Chem. Soc. = Journal of the American Chemical Society.
J. Amer. Pharm. Assoc. = Journal of the American Pharmaceutical Association.
J. Appl. Ecol. = Journal of Applied Ecology.
J. Aquatic Pl. Managem. = Journal of Aquatic Plant Management.
J. Arizona-Nevada Acad. Sci. = Journal of the Arizona-Nevada Academy of Science.
J. Arnold Arbor. = Journal of the Arnold Arboretum.
J. Bot. = Journal of Botany, British and Foreign.
J. Ecol. = Journal of Ecology.
J. Elisha Mitchell Sci. Soc. = Journal of the Elisha Mitchell Scientific Society.
J. Fac. Sci. Hokkaido Imp. Univ., Ser. 5, Bot. = Journal of the Faculty of Science of the Hokkaido Imperial University. Series 5, Botany. [Rigaku-bu Kiyo.]
J. Fac. Sci. Univ. Tokyo, Sect. 3, Bot. = Journal of the Faculty of Science, University of Tokyo. Section III. Botany.
J. Hort. Soc. London = Journal of the Horticultural Society of London.
J. Indian Bot. Soc. = Journal of the Indian Botanical Society.
J. Jap. Bot. = Journal of Japanese Botany.
J. Linn. Soc., Bot. = Journal of the Linnean Society. Botany.
J. Molec. Evol. = Journal of Molecular Evolution.
J. New York Bot. Gard. = Journal of the New York Botanical Garden.
J. Palynol. = Journal of Palynology.
J. Phys. Chim. Hist. Nat. Arts = Journal de Physique, de Chimie, d'Histoire Naturelle et des Arts.
J. Pl. Res. = Journal of Plant Research. [Shokubutsu-gaku zasshi.]
J. Roy. Hort. Soc. = Journal of the Royal Horticultural Society.
J. S. African Bot. = Journal of South African Botany.
J. Soc. Centr. Hort. France = Journal de la Société Centrale d'Horticulture de France.
J. Tennessee Acad. Sci. = Journal of the Tennessee Academy of Science.
J. Wash. Acad. Sci. = Journal of the Washington Academy of Sciences.
Jacobsen, T. D. 1978. A Comparative Study of Three Alliances of the Genus *Allium.* Ph.D. thesis. Washington State University.
Jacquin, N. J. 1760. Enumeratio Systematica Plantarum, Quas in Insulis Caribaeis Vicinaque Americes Continente Detexit Novas.... Leiden. (Enum. Syst. Pl.)
Jacquin, N. J. 1764–1771. Observationum Botanicarum.... 4 parts. Vienna. (Observ. Bot.)
Jacquin, N. J. 1773–1778. Florae Austriaceae.... 5 vols. Vienna. (Fl. Austriac.)
Jahrb. Gewächsk. = Jahrbücher der Gewächskunde.
Jahresber. Naturwiss. Vereins Wuppertal = Jahresbericht des Naturwissenschaftlichen Vereins in Wuppertal.
Jap. J. Bot. = Japanese Journal of Botany.
Jap. J. Genet. = Japanese Journal of Genetics.
Jefferson-Brown, M. J. 1991. *Narcissus.* Portland.
Jepson, W. L. 1909–1943. A Flora of California.... 3 vols. in 12 parts. San Francisco etc. [Pagination consecutive within each vol.; vol. 1 page sequence independent of part number sequence (chronological); part 8 of vol. 1 (pp. 1–32, 579–index) never published.] (Fl. Calif.)
Jepson, W. L. [1923–1925.] A Manual of the Flowering Plants of California.... Berkeley. (Man. Fl. Pl. Calif.)
Jernstedt, J. A. 1980. Ultraviolet absorbtion by flowers of *Chlorogalum* (Liliaceae). Bull. Torrey Bot. Club 107: 163–171.
Jessop, J. P. 1966. The genus *Asparagus* in southern Africa. Bothalia 9: 31–96.
Johnson, R. G. 1968. Notes on the distribution of *Disporum maculatum* (Buckl.) Britton. Castanea 33: 262–266.
Johnson, R. G. 1969. A Taxonomic and Floristic Study of the Liliaceae and Allied Families in the Southeastern United States. Ph.D. dissertation. West Virginia University.
Johnston, I. M. 1923. Diagnoses and notes relating to the spermatophytes chiefly of North America. Contr. Gray Herb. 68: 81.
Johnston, M. C. 1990. Vascular Plants of Texas: A List, Updating the Manual of the Vascular Plants of Texas, ed. 2. Austin. (Vasc. Pl. Texas ed. 2)
Jones, A. G. 1979. A study of wild leek, and the recognition of *Allium burdickii* (Liliaceae). Syst. Bot. 4: 29–43.
Jones, B. and C. Gliddon. 1999. Reproductive biology and genetic structure in *Lloydia serotina.* Pl. Ecol. 141: 151–

161.

Jones, Q. 1951. A cytotaxonomic study of the genus *Disporum* in North America. Contr. Gray Herb. 173: 1–39.

Jonker, F. P. 1938. A monograph of the Burmanniaceae. Meded. Bot. Mus. Herb. Rijks Univ. Utrecht 51: 1–279.

Joyce, J. C. and W. T. Haller. 1985. Effect of gibberellic acid and 2,4-D on water hyacinth. Aquatic Bot. 23: 119–125.

Judd, W. S. 1997. The Asphodelaceae in the southeastern United States. Harvard Pap. Bot. 2: 109–123.

Judd, W. S. 1998. The Smilacaceae in the southeastern United States. Harvard Pap. Bot. 3: 147–169.

Judd, W. S. 2001. The Asparagaceae in the southeastern United States. Harvard Pap. Bot. 6: 223–244.

Judd, W. S., C. S. Campbell, E. A. Kellogg, and P. F. Stevens. 1999. Plant Systematics. A Phylogenetic Approach. Sunderland, Mass.

Judziewicz, E. J., F. H. Utech, and W. Mackinnon. 1997. *Prosartes* (*Disporum*) *trachycarpa* (Liliaceae) in Isle Royale National Park: New to Michigan and the eastern United States. Michigan Bot. 36: 63–72.

Jussieu, A. L. de. 1789. Genera Plantarum.... Paris. (Gen. Pl.)

Kallunki, J. A. 1976. Population studies in *Goodyera* (Orchidaceae) with emphasis on the hybrid origin of *G. tesselata*. Brittonia 28: 53–75.

Kallunki, J. A. 1981. Reproductive biology of mixed-species populations of *Goodyera* (Orchidaceae) in northern Michigan. Brittonia 33: 137–155.

Kaplan, D. R. 1973. The problem of leaf morphology and evolution in the monocotyledons. Quart. Rev. Biol. 48: 437–457.

Kapoor, B. M. 1973. Cytology and supernumerary chromosome fragments in *Clintonia borealis* (Ait.) Raf. Caryologia 26: 457–467.

Karol, K. G., R. M. McCourt, M. T. Cimino, and C. F. Delwiche. 2001. The closest living relatives of land plants. Science, ser. 2, 294: 2351–2353.

Kartesz, J. T. and C. A. Meacham. 1999. Synthesis of the North American Flora, ver. 1.0. Chapel Hill. [CD-ROM.]

Kato, H. et al. 1995. Evolutionary biology of *Trillium* and related genera (Trilliaceae). 1. Restriction site mapping and variation of chloroplast DNA and its systematic implications. Pl. Spec. Biol. 10: 7–30.

Kato, H. et al. 1995b. Molecular systematics of the Trilliaceae sensu lato as inferred from *rbc*L sequence data. Molec. Phylogen. Evol. 4: 184–193.

Kato, H. et al. 1996. Chloroplast DNA variation in *Veratrum* L. (Liliaceae) based on restriction site analysis. Acta Phytotax. Geobot. 47: 203–211.

Kaul, M. K. 1986. Weed Flora of Kashmir Valley. Jodhpur.

Kaul, R. B. and C. N. Horn. 1986. Pontederiaceae. In: Great Plains Flora Association. 1986. Flora of the Great Plains. Lawrence, Kans. Pp. 1239–1241.

Kawano, S., M. Ihara, and M. Suzuki. 1968. Biosystematic studies on *Maianthemum* (Liliaceae-Polygonateae [sic]). II. Geography and ecological life history. Jap. J. Bot. 20: 35–65.

Kawano, S., M. Ihara, and M. Suzuki. 1968b. Biosystematic studies on *Maianthemum* (Liliaceae-Polygonateae [sic]). IV. Variation in gross morphology of *M. kamtschaticum*. Bot. Mag. (Tokyo) 81: 473–490.

Kawano, S., M. Ihara, M. Suzuki, and H. H. Iltis. 1967. Biosystematic studies on *Maianthemum* (Liliaceae). I. Somatic chromosome number and morphology. Bot. Mag. (Tokyo) 80: 345–352.

Kawano, S. and H. H. Iltis. 1963. Cytotaxonomy of the genus *Polygonatum* (Liliaceae) I. Karyotype analysis of some eastern North American species. Cytologia 28: 321–330.

Kawano, S. and H. H. Iltis. 1964. Cytotaxonomic and geographic notes on *Uvularia* (Liliaceae). Bull. Torrey Bot. Club 91: 13–23.

Kawano, S. and H. H. Iltis. 1966. Cytotaxonomy of the genus *Smilacina*. II. Chromosome morphology and evolutionary consideration of the New World species. Cytologia 31: 12–28.

Kawano, S. and J. Masuda. 1980. The reproductive biology of flowering plants. VII. Resource allocation and reproductive capacity in wild populations of *Heloniopsis orientalis* (Thunb.) C. Tanaka (Liliaceae). Oecologia 45: 307–317.

Kawano, S. and M. Suzuki. 1971. Biosystematic studies on *Maianthemum* (Liliaceae-Polygonateae [sic]). VI. Variation in gross morphology of *M. bifolium* and *M. canadense* with special reference to their taxonomic status. Bot. Mag. (Tokyo) 84: 349–361.

Kawano, S., M. Suzuki, and S. Koyima. 1971. Biosystematic studies on *Maianthemum* (Liliaceae-Polygonateae [sic]). V. Variation in gross morphology, karyology and ecology of North American populations of *M. dilatatum* sensu lato. Bot. Mag. (Tokyo) 84: 299–318.

Kazempour Osaloo, S. and S. Kawano. 1999. Molecular systematics of Trilliaceae II. Phylogenetic analyses of *Trillium* and its allies using sequences of *rbc*L and *mat*K genes of cpDNA and internal transcribed spacers (ITS) of 18S-26S nrDNA. Pl. Spec. Biol. 14: 75–94.

Kazempour Osaloo, S., F. H. Utech, M. Ohara, and S. Kawano. 1999. Molecular systematics of Trilliaceae I. Phylogenetic analyses of *Trillium* using *mat*K gene sequences. J. Pl. Res. 112: 35–49.

Kearney, T. H. and R. H. Peebles. 1960. Arizona Flora, ed. 2. Berkeley.

Keator, G. 1967. Ecological and Taxonomic Studies on the Genus *Dichelostemma*. Ph.D. thesis. University of California, Berkeley.

Keator, G. 1989. The brodiaeas. Four Seasons 8: 4–11.

Keator, G. 1991. Studies in the genus *Dichelostemma*. Four Seasons 9: 24–39.

Keator, G. 1992. Blue dicks brodiaea (*Dichelostemma capitatum*): A common but problematic species. Four Seasons 9: 24–39.

Keator, G. 1993. *Brodiaea*. In: J. C. Hickman, ed. 1993. The Jepson Manual. Higher Plants of California. Berkeley, Los Angeles, and London. Pp. 1180–1183.

Keator, G. 1993b. *Dichelostemma*. In: J. C. Hickman, ed. 1993. The Jepson Manual. Higher Plants of California. Berkeley, Los Angeles, and London. Pp. 1190–1192.

Kelchner, S. A. and L. G. Clark. 1997. Molecular evolution and phylogenetic utility of the chloroplast *rpl*16 intron in *Chusquea* and the Bambusoideae (Poaceae). Molec. Phylogen. Evol. 8: 385–397.

Kellogg, A. M. 1872. On *Lilium bloomerianum.* Proc. Calif. Acad. Sci. 4: 160.
Kellogg, E. A. 1998. Relationships of cereal crops and othere grasses. Proc. Natl. Acad. Sci. U.S.A. 125: 1198–2010.
Kellogg, E. A. 2000. The grasses: A case study in macroevolution. Annual Rev. Ecol. Syst. 31: 217–238.
Kellogg, E. A. 2001. Evolutionary history of the grasses. Pl. Physiol. (Lancaster) 125: 1198–1205.
Kellogg, E. A. and H. P. Linder. 1995. Phylogeny of the Poales. In: P. J. Rudall et al., eds. 1995. Monocotyledons: Systematics and Evolution. 2 vols. Kew. Vol. 2, pp. 511–542.
Kew Bull. = Kew Bulletin.
Kiger, R. W. and D. M. Porter. 2001. Categorical Glossary for the Flora of North America Project. Pittsburgh.
Kingsbury, J. M. 1964. Poisonous Plants of the United States and Canada. Englewood Cliffs.
Kipping, J. L. 1971. Pollination Studies of Native Orchids. M.A. thesis. San Francisco State College.
Kirkpatrick, N. S. 1993. A Chloroplast DNA Restriction Mapping Study of the Genus *Hosta* (Liliaceae). Ph.D. dissertation. Miami University.
Kitchingman, R. M. 1985. Some species and cultivars of *Hemerocallis.* Plantsman 7: 68–89.
Knuth, R. 1924. Dioscoreaceae. In: H. G. A. Engler, ed. 1900–1953. Das Pflanzenreich.... 107 vols. Berlin. Vol. 87[IV,43], pp. 1–387.
Kocyan, A. and P. K. Endress. 2001. Floral structure and development and systematic aspects of some 'lower' Asparagales. Pl. Syst. Evol. 229: 187–216.
Kocyan, A. and P. K. Endress. 2001b. Floral structure and development of *Apostasia* and *Neuwiedia* (Apostasioideae) and their relationships to other Orchidaceae. Int. J. Pl. Sci. 162: 847–867.
Köhlein, F. 1987. Iris, transl. M. C. Peters. Portland.
Komar, G. A. 1983. Morphology of Liliaceae ovules. Bot. Zhurn. (Moscow & Leningrad) 68: 417–427.
Kondo, K. and K. Hashimoto, eds. 1987. World Orchid Hiroshima Symposium.... [Sekai Ran Kaigi Hiroshima Taikai.] Hiroshima.
Kongl. Vetensk. Acad. Nya Handl. = Kongl[liga]. Vetenskaps Academiens Nya Handlingar.
Kooser, R. G. and G. C. Kennedy. 1979. The genus *Brassia* R. Brown...section *Eubrassia* Lindley. Orchid Digest 43: 164–172.
Korean J. Pl. Taxon. = Korean Journal of Plant Taxonomy.
Kores, P. J. et al. 2001. A phylogenetic analysis of Diurideae (Orchidaceae) based on plastid DNA sequence data. Amer. J. Bot. 88: 1903–1914.
Kosenko, V. N. 1987. Pollen morphology in Tofieldieae, Narthecieae, Melanthieae, Xerophylleae (Melanthiaceae). Bot. Zhurn. (Moscow & Leningrad) 72: 1318–1330.
Kosenko, V. N. 1988. Pollen morphology in Chionographideae, Uvularieae, Tricyrtideae, Scolopeae, Anguillarieae, Iphigenieae, Glorioseae, Colchiceae (Melanthiaceae). Bot. Zhurn. (Moscow & Leningrad) 73: 172–185.
Kosenko, V. N. 1991. Palynomorphology of the family Liliaceae s. str. Bot. Zhurn. (Moscow & Leningrad) 76: 1696–1710.
Kraenzlin, F. W. L. [1897–]1901–1904. Orchidacearum Genera et Species. 2 vols. Berlin. (Orchid. Gen. Sp.)
Kral, R. 1981. Some distributional reports of weedy or naturalized foreign species of vascular plants for the southern states, particularly Alabama and middle Tennessee. Castanea 46: 334–339.
Kral, R., ed. 1983. A Report on Some Rare, Threatened, or Endangered Forest-related Vascular Plants of the South. 2 vols. Washington. [U.S.D.A. Forest Serv., Techn. Publ. R8-TP 2.]
Kress, W. J. 1990. The phylogeny and classification of the Zingiberales. Ann. Missouri Bot. Gard. 77: 698–721.
Kress, W. J. 1995. Phylogeny of the Zingiberanae: Morphology and molecules. In: P. J. Rudall et al., eds. 1995. Monocotyledons: Systematics and Evolution. 2 vols. Kew. Vol. 2, pp. 443–460.
Kress, W. J., L. M. Prince, W. J. Hahn, and E. A. Zimmer. 2001. Unraveling the evolutionary radiation of the families of Zingiberales using morphological and molecular evidence. Syst. Biol. 50: 926–944.
Kromosomo = La Kromosomo. [Senshokutai.]
Kubitzki, K. 1998. *Hesperocallis.* In: K. Kubitzki et al., eds. 1990+. The Families and Genera of Vascular Plants. 4+ vols. Berlin etc. Vol. 3, p. 260.
Kubitzki, K. 1998b. Hostaceae. In: K. Kubitzki et al., eds. 1990+. The Families and Genera of Vascular Plants. 4+ vols. Berlin etc. Vol. 3, pp. 256–260.
Kubitzki, K. et al., eds. 1990+. The Families and Genera of Vascular Plants. 4+ vols. Berlin etc.
Kubitzki, K. and P. J. Rudall. 1998. Asparagaceae. In: K. Kubitzki et al., eds. 1990+. The Families and Genera of Vascular Plants. 4+ vols. Berlin etc. Vol. 3, pp. 125–129.
Kudoh, H. et al. 1999. Genetic structure and determinants of clonal structure in a temperate deciduous woodland herb, *Uvularia perfoliata.* J. Ecol. 87: 244–257.
Kunkel, G. 1984. Plants for Human Consumption. Königstein.
Kunth, C. S. 1833–1850. Enumeratio Plantarum.... 5 vols. in 6. Stuttgart and Tübingen. (Enum. Pl.)
Kunth, C. S. 1842. *Eichhornia,* Genus Novum e Familia Pontederiacearum. Berlin. (Eichhornia)
Kuntze, O. 1891–1898. Revisio Generum Plantarum Vascularium Omnium atque Cellularium Multarum.... 3 vols. Leipzig etc. (Revis. Gen. Pl.)
Kuo, J. and A. J. McComb. 1998. Cymodoceaceae. In: K. Kubitzki et al., eds. 1990+. The Families and Genera of Vascular Plants. 4+ vols. Berlin etc. Vol. 4, pp. 133–140.
Kuo, J. and A. J. McComb. 1998b. Posidoniaceae. In: K. Kubitzki et al., eds. 1990+. The Families and Genera of Vascular Plants. 4+ vols. Berlin etc. Vol. 4, pp. 404–407.
Kupchan, S. M., J. H. Zimmerman, and A. Afonso. 1961. The alkaloids and taxonomy of *Veratrum* and related genera. Lloydia 24: 1–26.
La Llave, P. de, and J. M. de Lexarza. 1824–1825. Novorum Vegetabilium Descriptiones. 2 fasc. Mexico City. [Fasc. 2 includes Orchidianum Opusculum, paged separately.] (Nov. Veg. Descr.)
LaBerge, W. E. 1987. A revision of the bees of the genus *Andrena* of the western hemisphere. Part XII. Subgenera *Leucandrena, Ptilandrena, Scoliandrena,* and

Melandrena. Trans. Amer. Entomol. Soc. 112: 191–248.

LaFrankie, J. V. 1984. Anatomy of stem abcission in the genus *Smilacina* (Liliaceae). J. Arnold Arbor. 75: 563–570.

LaFrankie, J. V. 1986. Transfer of the species of *Smilacina* Desf. to *Maianthemum* Wiggers (Liliaceae). Taxon 35: 584–589.

LaFrankie, J. V. 1986b. Morphology and taxonomy of the New World species of *Maianthemum* (Liliaceae). J. Arnold Arbor. 67: 371–439.

Lamarck, J. et al. 1783–1817. Encyclopédie Méthodique. Botanique.... 13 vols. Paris and Liège. [Vols. 1–8, suppls. 1–5.] (Encycl.)

Lamarck, J. and A. P. de Candolle. 1805[–1815]. Flore Française, ou Descriptions Succinctes de Toutes les Plantes Qui Croissent Naturellement en France..., ed. 3. 5 tomes in 6 vols. Paris. [Tomes 1–4(2), 1805; tome 5, 1815.] (Fl. Franç. ed. 3)

Lamarck, J. and J. Poiret. 1791–1823. Tableau Encyclopédique et méthodique des Trois Règnes de la Nature. Botanique.... 6 vols. Paris. [Vols. 1–2 = tome 1; vols. 3–5 = tome 2; vol. [6] = tome 3. Vols. paged consecutively within tomes.] (Tabl. Encycl.)

Lampe, K. F. and M. A. McCann. 1985. AMA Handbook of Poisonous and Injurious Plants. Chicago.

Landolt, E. 1986 The family of Lemnaceae—A monographic study, vol. 1. Veröff. Geobot. Inst. E. T. H. Stiftung Rübel Zürich 71.

Landolt, E. 1998. Lemnaceae. In: K. Kubitzki et al., eds. 1990+. The Families and Genera of Vascular Plants. 4+ vols. Berlin etc. Vol. 4, pp. 264–270.

Lazarte, J. E. and B. F. Palser. 1979. Morphology, vascular anatomy and embryology of pistillate and staminate flowers of *Asparagus officinalis*. Amer. J. Bot. 66: 753–764.

Leafl. Bot. Observ. Crit. = Leaflets of Botanical Observation and Criticism.

Leafl. W. Bot. = Leaflets of Western Botany.

Leake, J. R. 1994. Tansley review no. 69. The biology of myco-heterotrophic ('saprophytic') plants. New Phytol. 127: 171–216.

Ledebour, C. F. von. [1841]1842–1853. Flora Rossica sive Enumeratio Plantarum in Totius Imperii Rossici Provinciis Europaeis, Asiaticis, et Americanis Hucusque Observatarum.... 4 vols. Stuttgart. (Fl. Ross.)

Lee, N. S. 1985. A taxonomic study of Korean *Veratrum*. Korean J. Pl. Taxon. 15: 49–65.

Lee, N. S. 1985b. A cytotaxonomic study of the Korean *Veratrum* species. Korean J. Pl. Taxon. 15: 155–161.

Leinfellner, W. 1961. Zur Kenntnis des Monokotyledonen—Perigons. III. Die Perigonblatter einiger weiterer Melanthioideen (*Melanthium, Zygadenus, Anticlea, Toxicoscordion, Veratrum* und *Kreysigia*). Oesterr. Bot. Z. 108: 108–210.

Lenz, L. W. 1958. A revision of the Pacific Coast irises. Aliso 4: 1–72.

Lenz, L. W. 1959. Hybridization and speciation in the Pacific Coast irises. Aliso 4: 237–309.

Lenz, L. W. 1959b. *Iris tenuis* S. Wats., a new transfer to the subsection *Evansia*. Aliso 4: 311–319.

Lenz, L. W. 1970. *Triteleia tubergenii*, an amphidiploid of garden origin. Aliso 7: 157–160.

Lenz, L. W. 1971. Experimental evidence for hybrid origin of *Dichelostemma venustum* (Liliaceae). Aliso 7: 309–312.

Lenz, L. W. 1974. A new species of *Dichelostemma* (Liliaceae) from California. Aliso 8: 129–131.

Lenz, L. W. 1975. A biosystematic study of *Triteleia* (Liliaceae): 1. Revision of the species of section *Calliprora*. Aliso 8: 221–258.

Lenz, L. W. 1975b. The chromosomes of *Bloomeria* and *Muilla* (Liliaceae) and range extensions for *Muilla coronata* and *Muilla transmontana*. Aliso 8: 259–262.

Lenz, L. W. 1976. The nature of the floral appendages in four species of *Dichelostemma*. Aliso 8: 379–381.

Lenz, L. W. 1976b. A biosystematic study of *Triteleia* (Liliaceae): 2. Chromosome numbers and karyotypes of the species of section *Calliprora*. Aliso 8: 353–377.

Lenz, L. W. and M. A. Hanson. 2000. Yuccas (Agavaceae) of the International Four Corners: Southwestern USA and northwestern Mexico. Aliso 19: 165–179.

Lenz, L. W. and M. A. Hanson. 2000b. Typification and change in status of *Yucca schottii* (Agavaceae). Aliso 19: 93–98.

León (Hermano). 1946. Flora de Cuba. Vol. 1. Gymnospermas - Monocotiledoneas. Contr. Ocas. Mus. Hist. Nat. Colegio "De La Salle" 8.

Leonard, S. W. and W. W. Baker. 1983. Additional populations of *Harperocallis flava* McDaniel (Liliaceae). Castanea 48: 151–152.

Les, D. H. et al. 1997. Phylogenetic studies in Alismatidae, II. Evolution of marine angiosperms (seagrasses) and hydrophily. Syst. Bot. 22: 443–463.

Les, D. H. et al. 1997b. Systematics of the Lemnaceae (duckweeds): Inferences from micromolecular and morphological data. Pl. Syst. Evol. 204: 161–177.

Les, D. H. and R. R. Haynes. 1995. Systematics of subclass Alismatidae: A synthesis of approaches. In: P. J. Rudall et al., eds. 1995. Monocotyledons: Systematics and Evolution. 2 vols. Kew. Vol. 2, pp. 353–377.

Les, D. H. and E. L. Schneider. 1995. The Nympheales, Alismatidae, and the theory of an aquatic monocotyledon origin. In: P. J. Rudall et al., eds. 1995. Monocotyledons: Systematics and Evolution. 2 vols. Kew. Vol. 1, pp. 23–42.

Lewis, C. E. and J. J. Doyle. 2001. Phylogenetic utility of the nuclear gene malate synthase in the palm family (Arecaceae). Molec. Phylogen. Evol. 19: 409–420.

Lewis, W. H. and M. P. F. Elvin-Lewis. 1977. Medical Botany: Plants Affecting Man's Health. New York.

L'Héritier de Brutelle, C.-L. 1788[1789–1792]. Sertum Anglicum.... 4 fasc. Paris. [All text in fasc. 1; plates numbered consecutively.] (Sert. Angl.)

Li, S. F. and Chang Z. Y. 1996. A cytogeographical study on *Clintonia udensis* (Liliaceae). Acta Phytotax. Sin. 34: 29–38.

Li, S. F., Chang Z. Y., and Yuan Y. M. 1996. The origin and dispersal of the genus *Clintonia* Raf. (Liliaceae): Evidence from its cytogeography and morphology. Caryologia 49: 125–135.

Liang, H. P. and Hilu K. W. 1996. Application of the *mat*K gene sequences to grass systematics. Canad. J. Bot. 74: 125–134.

Liang, S. Y. and Dai L. K. 1992. Pollen morphology and generic phylogenetic relationships in Ophiopogonoideae (Liliaceae). Acta Phytotax. Sin. 28: 228–236.

Liggio, J. and A. Liggio. 1999. Wild Orchids of Texas. Austin.

Liliac.—See: P. J. Redouté 1802[–1815]

Lily Yearb. N. Amer. Lily Soc. = Lily Yearbook of the North American Lily Society.

Lily Year-Book = Lily Year-Book.

Linder, H. P. 2000. Vicariance, climate change, anatomy and phylogeny of Restionaceae. Bot. J. Linn. Soc. 134: 159–177.

Linder, H. P., B. G. Briggs, and L. A. S. Johnson. 2000. Restionaceae—a morphological phylogeny. In: K. L. Wilson and D. A. Morrison, eds. 2000. Monocots: Systematics and Evolution. Melbourne. Pp. 653–660.

Linder, H. P. and E. A. Kellogg. 1995. Phylogenetic patterns in the commelinid clade. In: P. J. Rudall et al., eds. 1995. Monocotyledons: Systematics and Evolution. 2 vols. Kew. Vol. 2, pp. 473–496.

Lindley, J. 1821[–1826]. Collectanea Botanica; or, Figures and Botanical Illustrations of Rare and Curious Exotic Plants.... 8 parts. London. (Coll. Bot.)

Lindley, J. 1826. Orchidearum Sceletos. London. (Orchid. Scelet.)

Lindley, J. 1830–1840. The Genera and Species of Orchidaceous Plants.... 7 parts. London. [Parts paged consecutively.] (Gen. Sp. Orchid. Pl.)

Lindleyana = Lindleyana; the Scientific Journal of the American orchid Society.

Link, J. H. F. 1821–1822. Enumeratio Plantarum Horti Regii Berolinensis Altera.... 2 parts. Berlin. (Enum. Hort. Berol. Alt.)

Linnaea = Linnaea. Ein Journal für die Botanik in ihrem ganzen Umfange.

Linnaeus, C. 1753. Species Plantarum.... 2 vols. Stockholm. (Sp. Pl.)

Linnaeus, C. 1754. Genera Plantarum, ed. 5. Stockholm. (Gen. Pl. ed. 5)

Linnaeus, C. 1758[–1759]. Systema Naturae per Regna Tria Naturae..., ed. 10. 2 vols. Stockholm. (Syst. Nat. ed. 10)

Linnaeus, C. 1762. Planta *Alströmeria*.... Uppsala. (Pl. Alströmeria)

Linnaeus, C. 1762–1763. Species Plantarum..., ed. 2. 2 vols. Stockholm. (Sp. Pl. ed. 2)

Linnaeus, C. 1766–1768. Systema Naturae per Regna Tria Naturae..., ed. 12. 3 vols. Stockholm. (Syst. Nat. ed. 12)

Linnaeus, C. 1767[–1771]. Mantissa Plantarum. 2 parts. Stockholm. [Mantissa [1] and Mantissa [2] Altera paged consecutively.] (Mant. Pl.)

Linnaeus, C. f. 1781[1782]. Supplementum Plantarum Systematis Vegetabilium Editionis Decimae Tertiae, Generum Plantarum Editionis Sextae, et Specierum Plantarum Editionis Secundae. Braunschweig. (Suppl. Pl.)

Lloyd, J. U. 1897. *Veratrum viride*. W. Druggist 19: 447–449.

Lloydia = Lloydia; a Quarterly Journal of Biological Science.

Loconte, H. and D. W. Stevenson. 1990. Cladistics of the Spermatophyta. Brittonia 42: 197–211.

Loesener, O. 1926. Studien über die Gattung *Veratrum* und ihre Verbreitung. Verh. Bot. Vereins Prov. Brandenburg 68: 105–166.

Loesener, O. 1927. Übersicht über die Arten der Gattung *Veratrum*. Teil I. Repert. Spec. Nov. Regni Veg. 24: 61–72.

Loesener, O. 1928. Übersicht über die Arten der Gattung *Veratrum*. Schluss. Repert. Spec. Nov. Regni Veg. 25: 1–10.

Long, R. W. and O. Lakela. 1971. A Flora of Tropical Florida: A Manual of the Seed Plants and Ferns of Southern Peninsular Florida. Coral Gables. [Reprinted 1976, Miami.]

Loptien, H. 1979. Identification of the sex chromosome pair in asparagus (*Asparagus officinalis* L.). Z. Pflanzenzücht. 82: 162–173.

Loudon, J. C. 1832. Hortus Britannicus. A Catalogue of All the Plants Indigenous, Cultivated in, or Introduced to Britain, ed. 2. London. (Hort. Brit. ed. 2)

Löve, D. and H. Harries. 1963. *Streptopus oreopolus* Fern., a hybrid taxon. Rhodora 65: 310–317.

Lowden, R. M. 1973. Revision of the genus *Pontederia* L. Rhodora 75: 426–487.

Luer, C. A. 1971. A variety of *Spiranthes lanceolata* in Florida. Florida Orchidist 14: 19.

Luer, C. A. 1972. The Native Orchids of Florida. Bronx. (Native Orchids Florida)

Luer, C. A. 1975. The Native Orchids of the United States and Canada, Excluding Florida. Bronx. (Native Orchids U.S. & Canada)

Luer, C. A. 1986. Icones Pleurothallidinarum I. Systematics of the Pleurothallidinae. Monogr. Syst. Bot. Missouri Bot. Gard. 15.

Luer, C. A. 1991. Systematics of the genus *Lepanthopsis* (Orchidaceae). Monogr. Syst. Bot. Missouri Bot. Gard. 39: 1–78.

Maas, P. J. M. et al. 1986. Burmanniaceae. In: Organization for Flora Neotropica. 1968+. Flora Neotropica. 75+ nos. New York. No. 42.

Maas, P. J. M. et al. 1986b. General introduction [to Saprophytes pro parte]. In: Organization for Flora Neotropica. 1968+. Flora Neotropica. 75+ nos. New York. Preceding nos. 40–42, pp. 1–5.

Maas, P. J. M. and H. Maas. 1993. Haemodoraceae. In: Organization for Flora Neotropica. 1968+. Flora Neotropica. 75+ nos. New York. No. 61.

Mabberly, D. J. 1997. The Plant-book: A Portable Dictionary of the Vascular Plants..., ed. 2. Cambridge.

MacMillan, C. 1892. The Metaspermae of the Minnesota Valley. List of the Higher Seed-producing Plants Indigenous to the Drainage-basin of the Minnesota River.... Minneapolis. (Metasp. Minnesota Valley)

MacRoberts, D. T. 1984. The vascular flora of Louisiana. Bull. Mus. Life Sci. Louisiana State Univ. 6.

Madroño = Madroño; Journal of the California Botanical Society [from vol. 3: a West American Journal of Botany].

Maekawa, F. 1940. The genus *Hosta*. J. Fac. Sci. Univ. Tokyo, Sect. 3, Bot. 5: 317–425.

Maekawa, F. and K. Kaneko. 1968. Evolution of karyotype in *Hosta* (Liliaceae). J. Jap. Bot. 43: 132–140.

Magee, D. W. and H. E. Ahles. 1999. Flora of the Northeast: A Manual of the Vascular Flora of New England and Adjacent New York. Amherst.

Magrath, L. K. 1973. The Native Orchids of the Prairies and Plains Region of North America. Ph.D. dissertation. University of Kansas.

Maki, M. 1992. Allozyme variation in Japanese species of *Chionographis* (Liliaceae). Bot. Mag. (Tokyo) 105: 149–160.

Maki, M. 1992b. Fixation indices and genetic diversity in hermaphroditic and gynodioecious populations of Japanese *Chionographis* (Liliaceae). Heredity 68: 329–336.

Maki, M. 1993. Outcrossing and fecundity advantage of females in gynodioecious *Chionographis japonica* var. *kurohimensis* (Liliaceae). Amer. J. Bot. 80: 629–634.

Maki, M. and M. Masusa. 1993. Spatial autocorrelation of genotypes in a gynodioecious population of *Chionographis japonica* var. *kirohimensis* (Liliaceae). Int. J. Pl. Sci. 154: 467–472.

Maki, M. and M. Masusa. 1994. Spatial structure within two populations of a self-incompatible perennial, *Chionographis japonica* var. *japonica* (Liliaceae). J. Pl. Res. 107: 283–387.

Makinoa = Makinoa; Sive, Plantarum Notae Miscellaneae. [Makino Shokubutsu Konkonroku.]

Malcomber, S. T. and Sebsebe D. 1993. The status of *Protasparagus* and *Myrsiphyllum* in the Asparagaceae. Kew Bull. 48: 63–78.

Man. Bot. ed. 5—See: A. Eaton 1829

Man. Bot. San Francisco—See: E. L. Greene 1894

Man. Fl. N. States—See: N. L. Britton 1901

Man. Fl. Pl. Calif.—See: W. L. Jepson [1923–1925]

Man. S.E. Fl.—See: J. K. Small 1933

Man. Trees—See: C. S. Sargent 1905

Mangaly, J. K. 1968. A cytotaxonomic study of the herbaceous species of *Smilax*:Section *Coprosmanthus*. Rhodora 70: 55–82, 247–273.

Manning, J. C. and P. Goldblatt. 2001. A synoptic review of *Romulea* (Iridaceae: Crocoideae) in sub-Saharan Africa, the Arabian Peninsula and Socotra including new species, biological notes, and a new infrageneric classification. Adansonia, sér. 3, 23: 59–108.

Mant. Pl.—See: C. Linnaeus 1767[–1771]

Manual—See: A. Gray 1848

Manual ed. 6—See: A. Gray et al. 1890

Maratti, G. F. 1772. Plantarum Romuleae, et Saturniae.... Rome. (Pl. Romul. Saturn.)

Marchant, C. J. 1981. *Fritillaria* in British Columbia. Davidsonia 12: 19–25.

Marsh, C. D. and A. B. Clawson. 1922. The death camas species, *Zygadenus paniculatus* and *Z. elegans,* as poisonous plants. U.S.D.A. Bull. (1915-23) 1012.

Marsh, C. D., A. B. Clawson, and H. Marsh. 1918. Staggergrass (*Chrosperma muscaetoxicum*) as a poisonous plant. U.S.D.A. Bull. (1915-23) 710: 1–14.

Martin, F. W. 1969. The species of *Dioscorea* containing sapogenin. Econ. Bot. 23: 373–379.

Martius, C. F. P. von. [1823–]1824–1832. Nova Genera et Species Plantarum Quas in Itinere per Brasiliam Annis MDCCCXVII–MDCCCXX.... 3 vols. in parts. Munich. (Nov. Gen. Sp. Pl.)

Martius, C. F. P. von, A. W. Eichler, and I. Urban, eds. 1840–1906. Flora Brasiliensis. 15 vols. in 40 parts, 130 fasc. Munich, Vienna, and Leipzig. [Volumes and parts numbered in systematic sequence, fascicles numbered independently in chronological sequence.] (Fl. Bras.)

Mason-Gamer, R. J. and E. A. Kellogg. 1996. Testing for phylogenetic conflict among molecular data sets in the tribe Triticeae (Gramineae). Syst. Biol. 45: 524–545.

Masters, L. A. 1995. *Thismia americana* N. E. Pfeiffer, a history. Erigenia 14: 8–14.

Mathew, B. 1988. Hostaceae, a new name for the invalid Funkiaceae. Kew Bull. 43: 302.

Mathew, B. 1989. A review of *Veratrum.* Plantsman 11: 35–61.

Mathew, B. 1990. The Iris, rev. ed. Portland.

Mathew, B. 1992. A taxonomic and horticultural review of *Erythronium* L. (Liliaceae). Bot. J. Linn. Soc. 109: 453–471.

Mathew, B. 1996. A Review of *Allium* Sect. *Allium.* Kew.

Mathews, S. and R. A. Sharrock. 1996. The phytochrome gene family in grasses (Poaceae): A phylogeny and evidence that grasses have a subset of the loci found in dicot angiosperms. Molec. Biol. Evol. 13: 1141–1150.

Mathews, S., Tsai R. C., and E. A. Kellogg. 2000. Phylogenetic structure in the grass family (Poaceae): Evidence from the nuclear gene phytochrome B. Amer. J. Bot. 87: 96–107.

Matsura, H. and H. Toyokuni. 1963. A karyological and taxonomical study of *Fritillaria camschatcensis.* Sci. Rep. Tohoku Imp. Univ., Ser. 4, Biol. 29: 239–245.

Matsuura, H. and T. Sutô. 1935. Contributions to the idiogram study in phanerogamous plants. I. J. Fac. Sci. Hokkaido Imp. Univ., Ser. 5, Bot. 5: 33–75.

Matthews, V. A. 1986. *Leucocrinum.* In: S. M. Walters et al., eds. 1984+. The European Garden Flora.... 6+ vols. Cambridge etc. Vol. 1, p. 144.

Maule, S. M. 1959. *Xerophyllum tenax,* squawgrass, its geographic distribution and its behavior on Mount Rainier, Washington. Madroño 15: 39–48.

Mayo, S. J., J. Bogner, and P. C. Boyce. 1997. The Genera of Araceae. 1 vol. + laser disc. [London.]

Mayo, S. J., J. Bogner, and P. C. Boyce. 1998. Araceae. In: K. Kubitzki et al., eds. 1990+. The Families and Genera of Vascular Plants. 4+ vols. Berlin etc. Vol. 4, pp. 26–74.

McCartney, C. 1992. Stalking the elusive *Basiphyllaea corallicola.* Part 2. Florida Orchidist 34: 195–211.

McCartney, C. 1997. Mrs. Britton's shadow witch. Palmetto 17: 10–11.

McCartney, C. 1997b. The *Tetramicra* mystery. Palmetto 17: 12, 14–15.

McDaniel, S. 1968. *Harperocallis,* a new genus of the Liliaceae from Florida. J. Arnold Arbor. 49: 35–40.

McDearman, W. W. 1984. Systematics of *Zigadenus densus* and *Z. leimanthoides* and phylogenetic implications of breeding systems in the Veratreae. [Abstract.] A. S. B. Bull. 31: 71.

McFarlane, R. M. 1978. On the taxonomic status of *Fritillaria phaeanthera* Eastw. (Liliaceae). Madroño 25: 93–100.

McKelvey, S. D. 1935. Notes on *Yucca.* J. Arnold Arbor. 16: 269.

McKelvey, S. D. 1938–1947. Yuccas of the Southwestern United States. 2 vols. Jamaica Plain. (Yuccas Southw. U.S.)

McKelvey, S. D. and K. Sax. 1933. Taxonomic and cytological relationships of *Yucca* and *Agave.* J. Arnold Arbor. 14: 76–81.

McKinney, K. K. and J. C. Hickman. 1993. *Yucca.* In: J. C. Hickman, ed. 1993. The Jepson Manual. Higher Plants of California. Berkeley, Los Angeles, and London. P. 1210.

McNeal, D. W. Jr. 1970. Comparative Studies of the *Allium acuminatum* Alliance. Ph.D. thesis. Washington State University.

McNeal, D. W. Jr. 1982. Taxonomy of the *Allium lacunosum* complex (Liliaceae). Madroño 29: 79–86.

McNeal, D. W. Jr. 1992. A revision of the *Allium fimbriatum* (Alliaceae) complex. Aliso 13: 411–426.

McNeal, D. W. Jr. 1992b. A reappraisal of *Allium cristatum* (Alliaceae) and its allies. Madroño 39: 83–89.

McNeal, D. W. Jr. 1993. *Hesperocallis.* In: J. C. Hickman, ed. 1993. The Jepson Manual. Higher Plants of California. Berkeley, Los Angeles, and London. P. 1198.

McNeill, J. 1980. *Scilla.* In: T. G. Tutin et al., eds. 1964–1980. Flora Europaea. 5 vols. Cambridge. Vol. 5, pp. 41–43.

McVaugh, R. and W. R. Anderson, eds. 1974+. Flora Novo-Galiciana: A Descriptive Account of the Vascular Plants of Western Mexico. 8+ vols. Ann Arbor.

Meades, S. J. 1999. Fred and Vi Cowell, champions of the Timmins *Dactylorhiza.* Wake Robin 46(3): 9–10.

Meagher, T. R. 1978. The Evolutionary Consequences of Dioecy in *Chamaelirium luteum,* a Perennial Plant Species. Ph.D. dissertation. Duke University.

Meagher, T. R. 1980. Population biology of *Chamaelirium luteum,* a dioecious lily. I. Spatial distributions of males and females. Evolution 34: 1127–1137.

Meagher, T. R. 1981. Population biology of *Chamaelirium luteum,* a dioecious lily. II. Mechanisms governing sex ratios. Evolution 35: 557–567.

Meagher, T. R. 1982. The population biology of *Chamaelirium luteum,* a dioecious member of the lily family: Two-sex population projections and stable population structure. Ecology 63: 1701–1711.

Meagher, T. R. 1984. Sexual dimorphism and ecological differentiation of male and female plants. Ann. Missouri Bot. Gard. 71: 254–264.

Meagher, T. R. and J. Antonovics. 1982. Life history variation in dioecious plant populations: A case study of *Chamaelirium luteum.* In: H. J. Dingle and J. Hegmann, eds. 1982. Evolution and Genetics of Life Histories. New York. Pp. 137–154.

Meagher, T. R. and J. Antonovics. 1982b. The population biology of *Chamaelirium luteum,* a dioecious member of the lily family: Life history studies. Ecology 63: 1690–1700.

Meagher, T. R. and E. Thompson. 1987. Analysis of parentage for naturally established seedlings of *Chamaelirium luteum* (Liliaceae). Ecology 68: 803–812.

Med. Fl.—See: C. S. Rafinesque 1828[–1830]

Med. Repos. = Medical Repository.

Meddeland. Soc. Fauna Fl. Fenn. = Meddelanden af Societas pro Fauna et Flora Fennica.

Meded. Bot. Mus. Herb. Rijks Univ. Utrecht = Mededeelingen van het Botanisch Museum en Herbarium van de Rijks Universiteit te Utrecht.

Medley, M. E. 1979. Some Aspects of the Life History of *Triphora trianthophora* (Sw.) Rydb. (Three Birds Orchid) with Special Reference to Its Pollination. M.A. thesis. Andrews University.

Medley, M. E. 1996. Notes on collections of *Triphora* (Orchidaceae) from the American tropics. Selbyana 17: 83–87.

Meerow, A. W. et al. 1999. Systematics of Amaryllidaceae based on cladistic analysis of plastid *rbc*L and *trn*L-F sequence data. Amer. J. Bot. 86: 1325–1345.

Meerow, A. W., C. L. Guy, Li Q. B., and Yang S. L. 2000. Phylogeny of the American Amaryllidaceae based on nrDNA ITS sequences. Syst. Bot. 25: 708–726.

Meerow, A. W. and D. A. Snijman. 1998. Amaryllidaceae. In: K. Kubitzki et al., eds. 1990+. The Families and Genera of Vascular Plants. 4+ vols. Berlin etc. Vol. 3, pp. 83–110.

Meerow, A. W. and D. A. Snijman. 2001. Phylogeny of Amaryllidaceae tribe Amaryllideae based on nrDNA ITS sequences and morphology. Amer. J. Bot. 88: 2321–2330.

Meggers, B. J., E. S. Ayensu, and W. D. Duckworth, eds. 1973. Tropical Forest Ecosystems in Africa and South America: A Comparative Review. Washington.

Mehrhoff, L. A. 1983. Pollination in the genus *Isotria* (Orchidaceae). Amer. J. Bot. 70: 1444–1453.

Mehrhoff, L. A. 1989. Reproductive vigor and environmental factors in populations of an endangered North American orchid, *Isotria medeoloides* (Pursh) Rafinesque. Biol. Conservation 47: 281–296.

Mehrhoff, L. A. 1989b. The dynamics of declining populations of an endangered orchid, *Isotria medeoloides.* Ecology 70: 783–786.

Meikle, R. D. 1970. *Chionodoxa luciliae,* a taxonomic note. J. Roy. Hort. Soc. 95: 21–24.

Melchior, H., ed. 1964. A. Engler's Syllabus der Pflanzenfamilien..., ed. 12. 2 vols. Berlin.

Mém. Acad. Imp. Sci. St.-Pétersbourg, Sér. 6, Sci. Math. = Mémoires de l'Académie Impériale des Sciences de St.-Pétersbourg. Sixième Série. Sciences Mathématiques, Physiques et Naturelles.

Mem. Fac. Sci. Kyoto Univ., Ser. Biol. = Memoirs of the Faculty of Science, Kyoto University. Series of Biology.

Mem. New York Bot. Gard. = Memoirs of the New York Botanical Garden.

Mém. Soc. Phys. Genève = Mémoires de la Société de Physique et d'Histoire Naturelle de Genève.

Mem. Torrey Bot. Club = Memoirs of the Torrey Botanical Club.

Mem. Wern. Nat. Hist. Soc. = Memoirs of the Wernerian Natual History Society.

Meney, K. A. 2000. Reproductive patterns in relation to generic classification: Habitat and growth form in Australian Restionaceae. In: K. L. Wilson and D. A. Morrison, eds. 2000. Monocots: Systematics and Evolution. Melbourne. Pp. 676–684.

Meney, K. A. and J. S. Pate, eds. 1999. Australian Rushes: Biology, Identification and Conservation of Restionaceae and Allied Families. Nedlands.

Merriam-Webster. 1988. Webster's New Geographical Dictionary. Springfield, Mass.

Metasp. Minnesota Valley—See: C. MacMillan 1892

Metcalfe, C. R., ed. 1960+. Anatomy of the Monocotyledons. 8+ vols. Oxford.

Metcalfe, C. R. and L. Chalk. 1979+. Anatomy of the Dicotyledons, ed. 2. 4+ vols. Oxford and New York.

Meyer, F. G. 1966. *Narcissus* species and wild hybrids. Amer. Hort. Mag. 45: 47–76.

Michaux, A. 1803. Flora Boreali-Americana.... 2 vols. Paris and Strasbourg. (Fl. Bor.-Amer.)

Michigan Bot. = Michigan Botanist.

Midland Coll. Monthly = Midland College Monthly.

Miller, P. 1754. The Gardeners Dictionary.... Abridged..., ed. 4. 3 vols. London. (Gard. Dict. Abr. ed. 4)

Miller, P. [1755–]1760. Figures of...Plants Described in the Gardeners Dictionary.... 2 vols. London. [Volumes paged consecutively, plates numbered consecutively.] (Fig. Pl. Gard. Dict.)

Miller, P. 1768. The Gardeners Dictionary..., ed. 8. London. (Gard. Dict. ed. 8)

Mingrone, L. V. 1968. A Comparative Study of the *Allium falcifolium* Alliance. Ph.D. thesis. Washington State University.

Mitt. Bot. Mus. Univ. Zürich = Mitteilungen aus dem Botanischen Museum der Universität Zürich.

Mitt. Bot. Staatssamml. München = Mitteilungen (aus) der Botanischen Staatssammlung München.

Miyabe, K. and T. Miyake. 1915. Flora of Saghalin. N.p. (Fl. Saghalin)

Moerman, D. E. 1986. Medicinal Plants of Native America. 2 vols. Ann Arbor. [Univ. Michigan, Mus. Anthropol., Techn. Rep. 19.]

Moldenke, H. N. 1962. Amaryllid genera and species. Pl. Life 18: 48–50.

Molec. Biol. Evol. = Molecular Biology and Evolution.

Molec. Phylogen. Evol. = Molecular Phylogenetics and Evolution.

Molina, G. I. 1810. Saggio sulla Storia Naturale del Chili..., ed. 2. Bologna. (Sag. Stor. Nat. Chili ed. 2)

Monogr. Orchid.—See: E. G. Camus et al. [1908]

Monogr. Phan.—See: A. L. P. P. de Candolle and C. de Candolle 1878–1896

Monogr. Syst. Bot. Missouri Bot. Gard. = Monographs in Systematic Botany from the Missouri Botanical Garden.

Moore, D. M. 1980. *Pseudorchis*. In: T. G. Tutin et al., eds. 1964–1980. Flora Europaea. 5 vols. Cambridge. Vol. 5, p. 332.

Moore, H. E. Jr. 1953. The genus *Milla* (Amaryllidaceae–Allieae) and its allies. Gentes Herb. 8: 262–294.

Moran, R. 1992. *Aloe* wild in California. Cact. Succ. J. (Los Angeles) 64: 55–56.

Mordak, E. V. 1984. *Scilla*. In: P. H. Davis, ed. 1965–1988. Flora of Turkey and the East Aegean Islands. 10 vols. Edinburgh. Vol. 8, pp. 214–224.

Moret, J. and N. Galland. 1992. Phenetic, biogeographical, and evolutionary study of *Ornithogalum* subg. *Heliocharmos* (Hyacinthaceae) in the western Mediterranean Basin. Pl. Syst. Evol. 181: 179–202.

Morley, T. 1988. Observations on colonies and on seedling growth of apparent hybrids between *Erythronium albidum* and *E. propullans*. Phytologia 40: 381–389.

Morong, T. 1894. The Smilaceae [sic] of North and Central America. Bull. Torrey Bot. Club 21: 419–448.

Morris, F. 1929. Our Wild Orchids: Trails and Portraits. New York. (Wild Orchids)

Mortola, W. R. and D. W. McNeal Jr. 1985. Taxonomy of the *Allium tribracteatum* (Alliaceae) complex. Aliso 11: 27–35.

Morton, B. R., B. S. Gaut, and M. T. Clegg. 1996. Evolution of alcohol-dehydrogenase genes in the palm and grass families. Proc. Natl. Acad. Sci. U.S.A. 93: 11735–11739.

Morton, C. V. 1935. A check list of the bulbous Amaryllidaceae native to the United States. Year Book Amer. Amaryllis Soc. 2: 80–83.

Moyle, J. B. 1945. Some chemical factors influencing the distribution of aquatic plants in Minnesota. Amer. Midl. Naturalist 34: 402–420.

Muasya, A. M. et al. 2000. Suprageneric phylogeny of Cyperaceae: A combined phylogeny. In: K. L. Wilson and D. A. Morrison, eds. 2000. Monocots: Systematics and Evolution. Melbourne. Pp. 593–601.

Muasya, A. M. et al. 2000b. Phylogenetic relationships within the heterogeneous *Scirpus* s. lat. (Cyperaceae) inferred from *rbc*L and *trn*L-F sequence data. In: K. L. Wilson and D. A. Morrison, eds. 2000. Monocots: Systematics and Evolution. Melbourne. Pp. 610–614.

Muasya, A. M., D. A. Simpson, M. W. Chase, and A. Culham. 1998. An assessment of suprageneric phylogeny in Cyperaceae using *rbc*L DNA sequences. Pl. Syst. Evol. 211: 257–271.

Muasya, A. M., D. A. Simpson, M. W. Chase, and A. Culham. 2001. A phylogeny of *Isolepis* (Cyperaceae) inferred using plastid *rbc*L and *trn*L-F sequence data. Syst. Bot. 26: 342–353.

Muhlenberg, G. H. E. 1813. Catalogus Plantarum Americae Septentrionalis.... Lancaster, Pa. (Cat. Pl. Amer. Sept.)

Muhlenbergia = Muhlenbergia; a Journal of Botany.

Müller-Doblies, U. and D. Müller-Doblies. 1996. Revisionula incompleta *Ornithogalorum* Austro-Africanorum. Feddes Repert. 107: 361–548.

Mulligan, G. A. and D. B. Munro. 1987. The biology of Canadian weeds. 77. *Veratrum viride* Ait. Canad. J. Pl. Sci. 67: 777–786.

Munro, S. L. and H. P. Linder. 1997. The embryology and systematic relationships of *Prionium serratum* (Juncaceae, Juncales). Amer. J. Bot. 84: 850–860.

Munro, S. L. and H. P. Linder. 1998. The phylogenetic position of *Prionium* (Juncaceae) within the order Juncales based on morphological and *rbc*L sequence data. Syst. Bot. 23: 43–55.

Munson, R. W. Jr. 1989. *Hemerocallis:* The Daylily. Portland.

Munz, P. A. 1959. A California Flora. Berkeley and Los Angeles. (Calif. Fl.)

Munz, P. A. 1968. Supplement to A California Flora. Berkeley and Los Angeles. (Suppl. Calif. Fl.)

Munz, P. A. and J. C. Roose. 1950. California miscellany II. Aliso 2: 217–238.

Murray, J. A. 1784. Caroli à Linné Equitis Systema Vegetabilium.... Editio Decima Quarta.... Göttingen. (Syst. Veg. ed. 14)

Mus. Nat. Acad. Upsal.—See: C. P. Thunberg 1787–1827

Mycologia = Mycologia; in Continuation of the Journal of Mycology.

N. Amer. Native Orchid J. = North American Native Orchid Journal.

N.W. Fl.—See: T. C. Frye and G. B. Rigg 1912

Nair, P. K. K. and M. Sharma. 1965. Pollen morphology of Liliaceae. J. Palynol. 1: 38–61.

Nakai, T. 1937. Japanese species of *Veratrum* (I). J. Jap. Bot. 13: 631–645.

Nakai, T. 1937b. Japanese species of *Veratrum* (II). J. Jap. Bot. 13: 701–713.

Nakai, T. 1943. Chosakuronbun Mokuroku. [Ordines, Familiae, Tribi, Genera, Sectiones, Species, Varietates, Formae et Combinationes Novae.] Tokyo.

Nandi, O. I., M. W. Chase, and P. K. Endress. 1998. A combined cladistic analysis of angiosperms using *rbc*L and non-molecular data sets. Ann. Missouri Bot. Gard. 85: 137–212.

Narr. Journey Polar Sea—See: J. Franklin et al. 1823

Nat. Hist. Vermont—See: Z. Thompson 1842

Nat. Pflanzenfam.—See: H. G. A. Engler and K. Prantl 1887–1915

Nat. Pflanzenfam. ed. 2—See: H. G. A. Engler et al. 1924+

Native Orchids Florida—See: C. A. Luer 1972

Native Orchids U.S. & Canada—See: C. A. Luer 1975

Naturaliste Canad. = Naturaliste Canadien. Bulletin de Recherches, Observations et Découvertes se Rapportant à l'Histoire Naturelle du Canada.

Nature = Nature; a Weekly Illustrated Journal of Science.

Naturk. Jahrb. Stadt Linz = Naturkundliches Jahrbuch der Stadt Linz.

Ned. Kruidk. Arch. = Nederlandsch Kruidkundig Archief; Verslagen en Mededelingen der Nederlandsche Botanische Vereeniging.

Neogenyton—See: C. S. Rafinesque 1825

Ness, B. D. 1989. Seed morphology and taxonomic relationships in *Calochortus* (Liliaceae). Syst. Bot. 14: 495–505.

Ness, B. D. 1993. *Trillium*. In: J. C. Hickman, ed. 1993. The Jepson Manual. Higher Plants of California. Berkeley, Los Angeles, and London. P. 1206.

Ness, B. D., D. E. Soltis, and P. S. Soltis. 1990. An examination of polyploidy and putative introgression in *Calochortus* subsection *Nudi* (Liliaceae). Amer. J. Bot. 77: 1519–1531.

Neuss, N. 1953. A new alkaloid from *Amianthium muscaetoxicum* Gray. J. Amer. Chem. Soc. 75: 2772–2773.

New Fl.—See: C. S. Rafinesque 1836[–1838]

New Phytol. = New Phytologist; a British Botanical Journal.

New Zealand J. Bot. = New Zealand Journal of Botany.

Neyland, R. 2002. A phylogeny inferred from large-subunit (26S) ribosomal DNA sequences suggests that Burmanniales are polyphyletic. Austral. Syst. Bot. 15: 19–28.

Neyland, R., L. E. Urbatsch, and A. M. Pridgeon. 1995. A phylogenetic analysis of subtribe Pleurothallidinae (Orchidaceae). Bot. J. Linn. Soc. 117: 13–28.

Nickrent, D. L. and D. E. Soltis. 1995. A comparison of angiosperm phylogenies from nuclear 18S rDNA and *rbc*L sequences. Ann. Missouri Bot. Gard. 82: 208–234.

Niehaus, T. F. 1971. A biosystematic study of the genus *Brodiaea* (Amaryllidaceae). Univ. Calif. Publ. Bot. 60.

Niehaus, T. F. 1980. The *Brodiaea* complex. Four Seasons 6: 11–21.

Nielsen, L. R. and H. R. Siegismund. 1999. Interspecific differentiation and hybridization in *Vanilla* species (Orchidaceae). Heredity 83: 560–567.

Nir, M. 1994. Taxonomic changes in Caribbean orchids. Lindleyana 9: 147–151.

Nir, M. 2000. Orchidaceae Antillanae. New York.

Nishikawa, T. et al. 1999. A molecular phylogeny of *Lilium* in the internal transcribed spacer region of nuclear ribosomal DNA. J. Molec. Evol. 49: 238–249.

Nom. Nov. Gen. Dactylorhiza—See: R. Soó 1962

Nordenstam, B. 1998. Colchicaceae. In: K. Kubitzki et al., eds. 1990+. The Families and Genera of Vascular Plants. 4+ vols. Berlin etc. Vol. 3, pp. 175–185.

Nordenstam, B., G. El-Ghazaly, and M. Kassas, eds. 2000. Plant Systematics for the 21st Century: Proceedings from a Symposium.... London.

Nordic J. Bot. = Nordic Journal of Botany.

NorthW. Sci. = Northwest Science.

Norweg. J. Bot. = Norwegian Journal of Botany.

Not. Mens. Mus. Nac. Hist. Nat. = Noticiário Mensual, Museo Nacional de História Natural.

Notes Roy. Bot. Gard. Edinburgh = Notes from the Royal Botanic Garden, Edinburgh.

Notizbl. Königl. Bot. Gart. Berlin = Notizblatt des Königlichen botanischen Gartens und Museums zu Berlin.

Nov. Gen. Sp.—See: A. von Humboldt et al. 1815[1816]–1825

Nov. Gen. Sp. Pl.—See: C. F. P. von Martius [1823–]1824–1832; E. F. Poeppig and S. L. Endlicher 1835–1845

Nov. Pl. Descr. Dec.—See: C. G. Ortega 1797–1800

Nov. Veg. Descr.—See: P. de La Llave and J. M. de Lexarza 1824–1825

Novon = Novon; a Journal for Botanical Nomenclature.

Nucleus (Calcutta) = Nucleus; International Journal of Cytology and Allied Topics.

Nuttall, T. 1813. A Catalogue of New and Interesting Plants Collected in Upper Louisiana.... London. (Cat. Pl. Upper Louisiana)

Nuttall, T. 1818. The Genera of North American Plants, and Catalogue of the Species, to the Year 1817.... 2 vols. Philadelphia. (Gen. N. Amer. Pl.)

Observ. Bot.—See: N. J. Jacquin 1764–1771

Occas. Pap. Calif. Acad. Sci. = Occasional Papers of the California Academy of Sciences.

Oesterr. Bot. Z. = Oesterreichische botanische Zeitschrift. Gemeinützige Organ für Botanik.

Oettingen, S. L. 1992. Small Whorled Pogonia (*Isotria medeoloides*) Recovery Plan: First Revision. Washington.

Oganezowa, G. G. 1982. On the anatomical structure of seed and fruit of some Liliaceae in connection with the systematics of the family. II. Scilloideae. Bot. Zhurn. (Moscow & Leningrad) 67: 729–742.

Oganezowa, G. G. 1984. Morphological and anatomical specific features of seed and fruit in some representatives of the subfamily Melanthioideae (Liliaceae) in relation with their systematics and phylogeny. Bot. Zhurn. (Moscow & Leningrad) 69: 772–781.

Oganezowa, G. G. 1986. Morphological and anatomical characters of seed and fruit in some members of the subfamily Allioideae (Liliaceae) in relation to their systematics and phylogeny. Bot. Zhurn. (Moscow & Leningrad) 71: 300–310.

Oliver, D., W. T. Thiselton-Dyer, D. Prain, and A. W. Hill, eds. 1868–1937. Flora of Tropical Africa.... 10 vols. London.

Orchid J. = Orchid Journal.

Orchid Monogr. = Orchid Monographs.

Orchid Rev. = The Orchid Review; an Illustrated Monthly Journal Devoted to Orchidology in All Its Branches.

Orchid. Gen. Sp.—See: F. W. L. Kraenzlin [1897–]1901–1904

Orchid. Scelet.—See: J. Lindley 1826

Orchidaceae—See: O. Ames 1905–1922

Orchidee (Hamburg) = Die Orchidee; Zeitschrift der Deutschen Orchideen-Gesellschaft.

Orchideen—See: F. R. R. Schlechter 1914–1915

Orchids = Orchids; the Magazine of the American Orchid Society.

Organization for Flora Neotropica. 1968+. Flora Neotropica. 75+ nos. New York.

Ornduff, R. 1966. The breeding system of *Pontederia cordata* L. Bull. Torrey Bot. Club 93: 407–416.

Ornduff, R. and M. S. Cave. 1975. Geography of pollen and chromosomal heteromorphism in *Leucocrinum montanum* (Liliaceae). Madroño 23: 65–67.

Orquidea (Mexico City) = Orquidea; Organo Oficial de la Sociedad Mexicana “Amigos de las Orquídeas.”

Ortega, C. G. 1797–1800. Novarum, aut Rariorum Plantarum Horti Reg. Botan. Matrit. Descriptionum Decades.... 10 decades in 4 parts. Madrid. (Nov. Pl. Descr. Dec.)

Osol, A., R. Pratt, and M. D. Altschule. 1960. The United States Dispensatory and Physicians’ Pharmacology, ed. 26. Philadelphia.

Otia Bot. Hamburg.—See: H. G. Reichenbach 1878–1881

Ownbey, M. 1940. A monograph of the genus *Calochortus*. Ann. Missouri Bot. Gard. 27: 371–560.

Ownbey, M. 1947. The genus *Allium* in Arizona. Res. Stud. State Coll. Wash. 15: 211–232.

Ownbey, M. 1950. The genus *Allium* in Idaho. Res. Stud. State Coll. Wash. 18: 3–39.

Ownbey, M. 1950b. The genus *Allium* in Texas. Res. Stud. State Coll. Wash. 18: 181–222.

Ownbey, M. 1969. *Calochortus*. Univ. Wash. Publ. Bot. 17: 765–779.

Ownbey, M. 1969b. *Allium*. In: C. L. Hitchcock et al. 1955–1969. Vascular Plants of the Pacific Northwest. 5 vols. Seattle. Vol. 1, pp. 739–760.

Ownbey, M. and H. C. Aase. 1955. Cytotaxonomic studies in *Allium*. I. The *Allium canadense* alliance. Res. Stud. State Coll. Wash., suppl. 1: 1–106.

Ownbey, R. P. 1944. The liliaceous genus *Polygonatum* in North America. Ann. Missouri Bot. Gard. 31: 373–413.

Pabst, G. F., J. L. Moutinho, and A. V. Pinto. 1981. An attempt to establish the correct statement for genus *Anacheilium* Hoffmgg. and revision of the genus *Hormidium* Lindl. ex Heynh. Bradea 3: 173–186.

Pacif. Railr. Rep. 4(5)—See: J. Torrey 1857

Pacific (San Francisco) = The Pacific.

Packer, J. G. 1993. Two new combinations in *Triantha* (Liliaceae). Novon 3: 278–279.

Pahuja, A. N. and V. Kumar. 1971. Embryo sac development, cytology and systematic position of *Clintonia*. Phytomorphology 20: 97–102.

Palmer, M. E., J. Travis, and J. Antonovics. 1989. Temporal mechanisms influencing gender expression and pollen flow within a self-incompatible perennial, *Amianthium muscaetoxicum* (Liliaceae). Oecologia 78: 231–236.

Pap. Michigan Acad. Sci. = Papers of the Michigan Academy of Sciences, Arts and Letters.

Parad. Lond.—See: R. A. Salisbury 1805[–1808]

Parks, C. R. and J. W. Hardin. 1963. Yellow erythroniums of the eastern United States. Brittonia 15: 245–259.

Patrick, T. S. 1973. Observations on the Life History of *Trillium grandiflorum* (Michaux) Salisbury. M.A. thesis. Cornell University.

Patrick, T. S. 1984. *Trillium sulcatum* (Liliaceae), a new species of the southern Appalachians. Brittonia 36: 26–36.

Patterson, D. T. et al. 1989. Composite List of Weeds. Champaign.

Patterson, T. B. 1998. Phylogeny, Biogeography, and Evolutionary Trends in the Core Liliales and *Calochortus* (Calochortaceae): Insights from DNA Sequenced Data. Ph.D. dissertation. University of Wisconsin.

Patterson, T. B. and T. J. Givnish. 1998. Phylogeny and evolutionary trends in Liliaceae s.s., Calochortaceae, and Uvulariaceae: Insights from *ndh*F sequence data. [Abstract.] Amer. J. Bot. 85(6, suppl.): 149–150.

Patterson, T. B. and T. J. Givnish. 2002. Phylogeny, concerted convergence, and phylogenetic niche conservatism in the core Liliales: Insights from *rbc*L and *ndh*F sequence data. Evolution 56: 233–252.

Paulos, P. G. 1985. Small Whorled Pogonia Recovery Plan. [Newton Corner, Mass.]

Pavord, A. 1999. The Tulip. New York, London, and Bloomsbury.

Penfound, W. T. and T. T. Earle. 1948. The biology of the water hyacinth. Ecol. Monogr. 18: 448–472.

Pennell, F. W. 1916b. Notes on plants of the southern United States—II. Bull. Torrey Bot. Club 43: 407–421.

Pereira, C. E. de B., ed. 1997. Proceedings of the 15th World Orchid Conference: Rio de Janeiro 1996. Turriers.

Perrine, H. 1838. 25th Congress, 2d Session [Rep. no. 564]...Tropical Plants...in the House of Representatives, March 8, 1838.... Washington. (Trop. Pl.)

Persoon, C. H. 1805–1807. Synopsis Plantarum.... 2 vols. Paris and Tubingen. (Syn. Pl.)

Petersen, G. and O. Seberg. 1997. Phylogenetic analysis of the Triticeae (Poaceae) based on *rpo*A sequence data. Molec. Phylogen. Evol. 7: 217–230.

Peterson, L. 1978. A Field Guide to Edible Wild Plants. Boston.

Peterson, P. M., C. R. Annable, and L. H. Rieseberg. 1988. Systematic relationships and nomenclatural changes in the *Allium douglasii* complex (Alliaceae). Syst. Bot. 13: 207–214.

Pfeiffer, N. E. 1914. Morphology of *Thismia americana*. Bot. Gaz. 57: 122–135.

Pfitzer, E. H. H. 1887. Entwurf einer natürlichen Anordnung der Orchideen.... Heidelberg. (Entwurf Anordn. Orch.)

Pfosser, M. and F. Speta. 1999. Phylogenetics of Hyacinthaceae based on plastid DNA sequences. Ann. Missouri Bot. Gard. 86: 852–875.

Philos. Mag. Ann. Chem. = The Philosophical Magazine, or Annals of Chemistry, Mathematics, Astronomy, Natural History, and General Science; New and United Series of the Philosophical Magazine and Annals of Philosophy.

Phytologia = Phytologia; Designed to Expedite Botanical Publication.

Phytomorphology = Phytomorphology; an International Journal of Plant Morphology.

Phyton (Horn) = Phyton; Annales Rei Botanica.

Pijl, L. van der and C. Dodson. 1966. Orchid Flowers: Their Pollination and Evolution. Coral Gables.

Pina L., I. 1985–1986. Consideraciones sobre el género *Manfreda*. Cact. Suc. Mex. 30: 27–32, 56–64, 84–90; 31: 12–18.

Piper, C. V. 1914. Flora of Southeastern Washington and Adjacent Idaho.... Lancaster, Pa. (Fl. S.E. Washington)

Pires, J. C. 2000. Biosystematics and Molecular Phylogenetics of *Brodiaea* (Themidaceae) and Related Lilioid Monocots. Ph.D. dissertation. University of Wisconsin.

Pires, J. C. et al. 2001. Molecular and morphological phylogenetic analyses of Themidaceae (Asparagales). Kew Bull. 56: 601–626.

Pl. Alströmeria—See: C. Linnaeus 1762

Pl. Biol. (Stuttgart) = Plant Biology. [Stuttgart.]

Pl. Dis. Reporter = Plant Disease Reporter.

Pl. Ecol. = Plant Ecology.

Pl. Hartw.—See: G. Bentham 1839[–1857]

Pl. Life = Plant Life.

Pl. Physiol. (Lancaster) = Plant Physiology.

Pl. Rar. Neapol.—See: D. M. L. Cirillo 1788–1792

Pl. Romul. Saturn.—See: G. F. Maratti 1772

Pl. Spec. Biol. = Plant species Biology; an International Journal.

Pl. Syst. Evol. = Plant Systematics and Evolution.

Pl. Veron.—See: J. F. Séguier 1745–1754

Plunkett, G. M., D. E. Soltis, P. S. Soltis, and R. E. Brooks. 1995. Phylogenetic relationships between Juncaceae and Cyperaceae: Insights from *rbc*L sequence data. Amer. J. Bot. 82: 520–525.

Poeppig, E. F. and S. L. Endlicher. 1835–1845. Nova Genera ac Species Plantarum Quas in Regno Chilensi Peruviano et in Terra Amazonica.... 3 vols. Leipzig. (Nov. Gen. Sp. Pl.)

Pojar, J. and A. MacKinnon, eds. 1994. Plants of the Pacific Northwest Coast: Washington, Oregon, British Columbia, & Alaska. Redmond.

Pollard, C. L. 1897. The genus *Oxytria* of Rafinesque. Bull. Torrey Bot. Club 24: 405–406.

Porsild, A. E. 1951. Botany of southeastern Yukon adjacent to the Canol Road. Bull. Natl. Mus. Canada 121.

Powell, J. A. and R. A. Mackie. 1966. Biologic interrelationships of moths and *Yucca whipplei*. Univ. Calif. Publ. Entomol. 42: 1–59.

Preece, S. J. 1956. A Cytotaxonomic Study of the Genus *Zigadenus*. Ph.D. dissertation. State College of Washington [Washington State University].

Prelim. Cat.—See: N. L. Britton et al. 1888

Presl, C. B. 1825–1835. Reliquiae Haenkeanae seu Descriptiones et Icones Plantarum, Quas in America Meridionali et Boreali, in Insulis Philippinis et Marianis Collegit Thaddeus Haenke.... 2 vols. in 7 parts. Prague. (Reliq. Haenk.)

Price, S. D. and S. C. H. Barrett. 1982. Tristyly in *Pontederia cordata* (Pontederiaceae). Canad. J. Bot. 60: 897–905.

Price, S. D. and S. C. H. Barrett. 1984. The function and adaptive significance of tristyly in *Pontederia cordata* L. (Pontederiaceae). Biol. J. Linn. Soc. 21: 315–329.

Pridgeon, A. M. et al. 1997. Phylogenetics of subtribe Orchidinae (Orchidoideae, Orchidaceae) based on nuclear ITS sequences: 1. Intergeneric relationships and polyphyly of *Orchis* sensu lato. Lindleyana 12: 89–109.

Pridgeon, A. M., P. J. Cribb, and F. N. Rasmussen. 1999+. Genera Orchidacearum. 1+ vol. Oxford.

Pridgeon, A. M., R. Solano, and M. W. Chase. 2001. Phylogenetic relationships in Pleurothallidinae (Orchidaceae): Combined evidence from nuclear and plastid DNA sequences. Amer. J. Bot. 88: 2286–2308.

Pridgeon, A. M. and L. E. Urbatsch. 1977. Contributions to the flora of Louisiana. 2. Distribution and identification of Orchidaceae. Castanea 42: 293–304.

Prim. Fl. Galiciae Austriac.—See: W. S. J. G. von Besser 1809

Prim. Fl. Holsat.—See: F. H. Wiggers [1780]

Proc. Acad. Nat. Sci. Philadelphia = Proceedings of the Academy of Natural Sciences of Philadelphia.

Proc. Amer. Acad. Arts = Proceedings of the American Academy of Arts and Sciences.

Proc. Amer. Philos. Soc. = Proceedings of the American Philosophical Society.

Proc. Biol. Soc. Wash. = Proceedings of the Biological Society of Washington.

Proc. Calif. Acad. Sci. = Proceedings of the California Academy of Sciences.

Proc. Davenport Acad. Nat. Sci. = Proceedings of the Davenport Academy of Natural Sciences.

Proc. Indiana Acad. Sci. = Proceedings of the Indiana Academy of Science.

Proc. Linn. Soc. London = Proceedings of the Linnean Society of London.

Proc. Natl. Acad. Sci. U.S.A. = Proceedings of the National Academy of Sciences of the United States of America.

Proc. Roy. Soc. Biol. Sci. Ser. B = Proceedings of the Royal Society. Biological Sciences Series B.

Procter, W. 1839. Observations on *Xerophyllum setifolium*. Amer. J. Pharm. 11: 183–186.

Proctor, M. C. F. and P. F. Yeo. 1972. The Pollination of Flowers. New York.

Proctor, M. C. F., P. F. Yeo, and A. J. Lack. 1996. The Natural History of Pollination. Portland.

Prodr.—See: O. P. Swartz 1788

Prodr. Pl. Cap.—See: C. P. Thunberg 1794–1800

Progr. Bot. = Progress in Botany; Morphology, Physiology, Genetics, Taxonomy, Geobotany.

Prychid, C. J. and P. J. Rudall. 1999. Calcium oxalate crystals in the monocotyledons: A review of their structure and systematics. Ann. Bot. (Oxford) 84: 725–739.

Puffenbarger, S. and G. L. Smith. 1993. Biosystematic investigations of *Hymenocallis crassifolia* Herbert and *Hymenocallis duvalensis* Traub in the southeastern United States. [Abstract.] J. Elisha Mitchell Sci. Soc. 109: 227.

Pugsley, H. W. 1933. A monograph of *Narcissus*, subgenus *Ajax*. J. Roy. Hort. Soc. 58: 17–93.

Pupulin, F. 1995. A revision of the genus *Trichocentrum* (Orchidaceae: Oncidiinae). Lindleyana 10: 183–210.

Purdy, C. 1937. Western American lilies. Lily Year-Book 6: 20–24.

Pursh, F. 1814. Flora Americae Septentrionalis; or, a Systematic Arrangement and Description of the Plants of North America. 2 vols. London. (Fl. Amer. Sept.)

Qiu, Y. L. et al. 1999. The earliest angiosperms: Evidence from mitochondrial, plastid and nuclear genomes. Nature 402: 404–407.

Qiu, Y. L. et al. 2000. Phylogeny of basal angiosperms: Analyses of five genes from three genomes. Int. J. Pl. Sci. 161: S3–S27.

Quart. Rev. Biol. = Quarterly Review of Biology.

Queva, C. 1894. Recherches sur l'Anatomie de l'Appareil Vegetatif des Taccacées & Dioscorées. Lille.

Radford, A. E., H. E. Ahles, and C. R. Bell. 1968. Manual of the Vascular Flora of the Carolinas. Chapel Hill.

Raffauf, R. F. 1970. A Handbook of Alkaloids and Alkaloid-containing Plants. New York.

Rafinesque, C. S. 1817. Florula Ludoviciana; or, a Flora of the State of Louisiana. Translated, Revised, and Improved, from the French of C. C. Robin.... New York. (Fl. Ludov.)

Rafinesque, C. S. 1825. Neogenyton, or Indication of Sixty-six New Genera of Plants of North America. [Lexington, Ky.] (Neogenyton)

Rafinesque, C. S. 1828[–1830]. Medical Flora; or, Manual of the Medical Botany of the United States of North America. 2 vols. Philadelphia. (Med. Fl.)

Rafinesque, C. S. 1833. Herbarium Rafinesquianum. .... Herbals; or Botanical Collections of C. S. Rafinesque...the Labor of a Whole Life! 3 parts. Philadelphia. [Parts designated as extras of Atlantic J., paged consecutively.] (Herb. Raf.)

Rafinesque, C. S. 1836[–1838]. New Flora and Botany of North America.... 4 parts. Philadelphia. (New Fl.)

Rafinesque, C. S. 1836[1837–1838]. Flora Telluriana.... 4 vols. Philadelphia. (Fl. Tellur.)

Rafinesque, C. S. 1840. Autikon Botanikon. 3 parts. Philadelphia. [Parts paged consecutively.] (Autik. Bot.)

Rahn, K. 1998. Alliaceae. In: K. Kubitzki et al., eds. 1990+. The Families and Genera of Vascular Plants. 4+ vols. Berlin etc. Vol. 3, pp. 70–78.

Rahn, K. 1998b. Themidaceae. In: K. Kubitzki et al., eds. 1990+. The Families and Genera of Vascular Plants. 4+ vols. Berlin etc. Vol. 3, pp. 436–441.

Ranker, T. A. and A. F. Schnabel. 1986. Allozymic and morphological evidence for a progenitor-derivative species pair in *Camassia* (Liliaceae). Syst. Bot. 11: 433–445.

Ranker, T. A., D. E. Soltis, and A. J. Gilmartin. 1990. Subfamilial phylogenetic relationships of the Bromeliaceae: Evidence from chloroplast DNA restriction site variation. Syst. Bot. 15: 425–434.

Rasmussen, F. N. 1977. The genus *Corymborkis* Thou. (Orchidaceae): A taxonomic revision. Bot. Tidsskr. 71: 161–192.

Rasmussen, F. N. 1985. Orchids. In: R. M. T. Dahlgren et al. 1985. The Families of the Monocotyledons: Structure, Evolution, and Taxonomy. Berlin etc. Pp. 249–274.

Raup, H. M. 1947. The botany of southwestern Mackenzie. Sargentia 6: 1–275.

Ravenna, P. 1968. Notas sobre Iridaceae. III. Bonplandia (Corrientes) 2: 273–291.

Ravenna, P. 1977. Notas sobre Iridaceae V. Not. Mens. Mus. Nac. Hist. Nat. 21(249): 7–9.

Ravenna, P. 1991. *Nothoscordum gracile* and *N. borbonicum* (Alliaceae). Taxon 40: 485–487.

Rayner, D. A. et al. 1979. Native Vascular Plants Endangered, Threatened or Otherwise in Jeopardy in South Carolina. Columbia.

Reddoch, A. H. and J. M. Reddoch. 1993. The species pair *Platanthera orbiculata* and *P. macrophylla* (Orchidaceae): Taxonomy, morphology, distributions and habitats. Lindleyana 8: 171–187.

Redmon, A. M., L. E. Robbins, and J. Travis. 1989. The effects of pollination distance on seed production in three populations of *Amianthium muscaetoxicum* (Liliaceae). Oecologia 79: 260–264.

Redouté, P. J. 1802[–1815]. Les Liliacées.... 8 vols. in 80 parts. Paris. [Parts and plates numbered consecutively.] (Liliac.)

Rees, A. [1802–]1819–1820. The Cyclopaedia; or, Universal Dictionary of Arts, Sciences, and Literature.... 39 vols. in 79 parts. London. [Pages unnumbered.] (Cycl.)

Reeves, G. et al. 2001. Molecular systematics of Iridaceae: Evidence from four plastid DNA regions. Amer. J. Bot. 88: 2074–2087.

Reichenbach, H. G. 1878–1881. Otia Botanica Hamburgensia.... 2 parts. Hamburg. [Parts paged consecutively.] (Otia Bot. Hamburg.)

Reichenbach, H. G. L. 1823–1832. Iconographia Botanica seu Plantae Criticae. 10 vols. Leipzig. [Vols. 6 and 7 each published in two half-centuries paged independently.] (Iconogr. Bot. Pl. Crit.)

Reichenbach, H. G. L. 1830[–1832]. Flora Germanica Excursoria.... 2 parts. Leipzig. [Parts paged consecutively.] (Fl. Germ. Excurs.)

Reichenbach, H. G. L. et al. 1835–1914. Icones Florae Germanicae et Helveticae.... 25 vols. Leipzig and Gera. [Vol. 1 originally published under the title: Agrostographia Germanica....] (Icon. Fl. Germ. Helv.)

Reinhammer, L.-G. 1995. Evidence for two distinctive species of *Pseudorchis* (Orchidaceae) in Scandinavia. Nordic J. Bot. 15: 469–481.

Reliq. Haenk.—See: C. B. Presl 1825–1835

Rendle, A. B. 1930–1938. The Classification of Flowering

Plants, ed. 2. 2 vols. Cambridge.

Rep. (Annual) Michigan Acad. Sci. = Report (Annual) of the Michigan Academy of Science, (Arts, and Letters).

Rep. (Annual) Missouri Bot. Gard. = Report (Annual) of the Missouri Botanical Garden.

Rep. Bot. Exch. Club Soc. Brit. Isles = (Report,) Botanical Exchange Club and Society of the British Isles.

Rep. Colorado R.—See: J. C. Ives 1861

Rep. U.S. Geogr. Surv., Wheeler—See: J. T. Rothrock 1878[1879]

Rep. U.S. Mex. Bound.—See: W. H. Emory 1857–1859

Repert. Spec. Nov. Regni Veg. = Repertorium Specierum Novarum Regni Vegetabilis.

Repert. Spec. Nov. Regni Veg. Beih. = Repertorium specierum novarum regni vegetabilis. Beihefte.

Res. Stud. State Coll. Wash. = Research Studies of the State College of Washington.

Rev. Biol. (Bucharest) = Revue de biologie.

Rev. Hort. = Revue Horticole; Journal d'Horticulture Pratique.

Reveal, J. L. 1977c. *Yucca.* In: A. Cronquist et al. 1972+. Intermountain Flora. Vascular Plants of the Intermountain West, U.S.A. 5+ vols. New York and London. Vol. 6, pp. 527–536.

Reveal, J. L. 1993d. *Streptopus lanceolatus* (Aiton) Reveal, a new name for *Streptopus roseus* Michx. (Convallariaceae). Phytologia 74: 185–189.

Reveal, J. L. 1993e. Flowering plant families: An overview. In: Flora of North America Editorial Committee, eds. 1993+. Flora of North America North of Mexico. 5+ vols. New York and Oxford. Vol. 1, pp. 294–330.

Reveal, J. L. 2000. Proposal to reject the name *Smilax humilis* (Smilacaceae). Taxon 49: 297.

Reveal, J. L. and W. B. Zomlefer. 1998. Two new orders of monocotyledonous plants. Novon 8: 176–177.

Revis. Gen. Pl.—See: O. Kuntze 1891–1898

Reynolds, G. W. 1982. The Aloes of South Africa, ed. 4. Rotterdam.

Rhoads, A. F. and W. M. Klein. 1993. The Vascular Flora of Pennsylvania: Annotated Checklist and Atlas. Philadelphia.

Rhodora = Rhodora; Journal of the New England Botanical Club.

Richard, L. C. M. 1817. De Orchideis Europaeis Annotationes.... Paris. (De Orchid. Eur.)

Richardson, M. W. and D. Ugent. 1974. The genus *Polygonatum* (Solomon's seal) in southern Illinois. Trans. Illinois State Acad. Sci. 67: 285–302.

Rickabaugh, H. 1975. Two lilies of Modoc County. Fremontia 3: 15–17.

Rickett, H. W. [1966–1973.] Wild Flowers of the United States. 6 vols. in 14 parts. New York.

Riddell, J. L. 1835. A Synopsis of the Flora of the Western States. Cincinnati. (Syn. Fl. West. States)

Rix, E. M. 1986. *Chionodoxa.* In: S. M. Walters et al., eds. 1984+. The European Garden Flora.... 6+ vols. Cambridge etc. Vol. 1, pp. 214–215.

Rix, E. M. and D. Rast. 1975. Nectar sugars and subgeneric classification in *Fritillaria.* Biochem. Syst. & Ecol. 2: 207–209.

Roalson, E. H., J. T. Columbus, and E. A. Friar. 2001. Phylogenetic relationships in Cariceae (Cyperaceae) based on ITS (nrDNA) and *trn*T-L-F (cpDNA) region sequences: Assessment of subgeneric and sectional relationships in *Carex* with emphasis on section *Acrocystis.* Syst. Bot. 26: 318–341.

Roalson, E. H. and E. A. Friar. 2000. Infrageneric classification of *Eleocharis* (Cyperaceae) revisited: Evidence from the internal transcribed spacer (ITS) region of nuclear ribosomal DNA. Syst. Bot. 25: 323–336.

Robbins, R. L. 1983. A Systematic Study of the Indehiscent-fruited Yuccas in the Chihuahuan Desert. Ph.D. dissertation. Texas Tech University.

Robertson, K. R. 1966. The genus *Erythronium* (Liliaceae) in Kansas. Ann. Missouri Bot. Gard. 53: 197–204.

Robertson, K. R. 1976. The genera of Haemodoraceae in the southeastern United States. J. Arnold Arbor. 57: 205–216.

Robertson, R. et al., eds. 1981+. Flora of Australia. 23+ vols. Canberra.

Robinson, B. L. and M. L. Fernald. 1908. Gray's New Manual of Botany: A Handbook of the Flowering Plants and Ferns of the Central and Northeastern United States and Adjacent Canada, ed. 7. New York, Cincinnati, and Chicago.

Robinson, H. and D. C. Taylor. 1999. The status of the pitcairnioid genera of the Bromeliaceae. Harvard Pap. Bot. 4: 195–202.

Roemer, J. J., J. A. Schultes, and J. H. Schultes. 1817[–1830]. Caroli a Linné...Systema Vegetabilium...editione XV.... 7 vols. Stuttgart. (Syst. Veg.)

Roemer, M. J. 1846–1847. Familiarum Naturalium Regni Vegetabilis Synopses Monographicae.... 4 fasc. Weimar. (Fam. Nat. Syn. Monogr.)

Rogers, G. K. 1982. The Stemonaceae in the southeastern United States. J. Arnold Arbor. 63: 327–336.

Rolfe, R. A. 1892. The genus *Galeandra.* Gard. Chron., ser. 3, 12: 430–431.

Rolfe, R. A. 1898. *Pteroglossaspis.* In: D. Oliver et al., eds. 1868–1937. Flora of Tropical Africa.... 10 vols. London. Vol. 7, pp. 99–100.

Romero-G., G. A. 1993. Notes on *Pteroglossaspis,* a new genus for the flora of Columbia. Orquidea (Mexico City) 13: 275–280.

Romero-G., G. A. and G. Carnevali F.-C. 1999. Notes on the species of *Cyrtopodium* (Cyrtopodiinae, Orchidaceae) from Florida, the Greater Antilles, Mexico, Central and northern South America. Harvard Pap. Bot. 4: 327–341.

Rosatti, T. J. 1987. The genera of Pontederiaceae in the southeastern United States. J. Arnold Arbor. 68: 35–71.

Rosendahl, C. O. 1919. Variations in the flowers of *Erythronium propullans* Gray. Torreya 19: 43–47.

Rostlinnictví—See: D. Sloboda 1852

Rothrock, J. T. 1878[1879]. Report upon United States Geographical Surveys West of the One Hundredth Meridian, in Charge of First Lieut. Geo. M. Wheeler.... Vol. 6—Botany. Washington. (Rep. U.S. Geogr. Surv., Wheeler)

Rudall, P. J. 1994. Anatomy and systematics of Iridaceae. Bot. J. Linn. Soc. 114: 1–21.

Rudall, P. J. 2002. Homologies of inferior ovaries and septal nectaries in monocotyledons. Int. J. Pl. Sci. 163: 261–

276.

Rudall, P. J. et al., eds. 1995. Monocotyledons: Systematics and Evolution. 2 vols. Kew.

Rudall, P. J. et al. 1998. Embryology, cytology and systematics of *Hemiphylacus, Asparagus* and *Anemarrhena* (Asparagales). Pl. Syst. Evol. 211: 181–199.

Rudall, P. J. et al. 1998b. Anatomical and molecular systematics of Asteliaceae and Hypoxidaceae. Bot. J. Linn. Soc. 127: 1–42.

Rudall, P. J. et al. 2000. Systematics of Ruscaceae/Convallariaceae: A combined morphological and molecular investigation. Bot. J. Linn. Soc. 134: 73–92.

Rudall, P. J. et al. 2000b. Consider the lilies—systematics of Liliales. In: K. L. Wilson and D. A. Morrison, eds. 2000. Monocots: Systematics and Evolution. Melbourne. Pp. 347–359.

Rudall, P. J. and M. W. Chase. 1996. Systematics of Xanthorrhoeaceae sensu lato: Evidence of polyphyly. Telopea 6: 629–647.

Rudall, P. J. and D. F. Cutler. 1995. Asparagales: A reappraisal. In: P. J. Rudall et al., eds. 1995. Monocotyledons: Systematics and Evolution. 2 vols. Kew. Vol. 1, pp. 157–168.

Rudall, P. J. and A. Eastman. 2002. The questionable affinities of *Corsia* (Corsiaceae): Evidence from floral anatomy and pollen morphology. Bot. J. Linn. Soc. 138: 315–324.

Rudall, P. J., C. A. Furness, M. W. Chase, and M. F. Fay. 1997. Microsporogenesis and pollen sulcus type in Asparagales (Lilianae). Canad. J. Bot. 75: 408–430.

Rudall, P. J., A. Y. Kenton, and T. J. Lawrence. 1986. An anatomical and chromosomal investigation of *Sisyrinchium* and allied genera. Bot. Gaz. 147: 466–477.

Rudall, P. J., D. W. Stevenson, and H. P. Linder. 1999. Structure and systematics of *Hanguana,* a monocotyledon of uncertain affinity. Austral. Syst. Bot. 12: 311–330.

Ruiz López, H. and J. A. Pavón. 1794. Flora Peruvianae, et Chilensis Prodromus.... Madrid. (Fl. Peruv. Prodr.)

Ruiz López, H. and J. A. Pavón. 1798–1802. Flora Peruviana, et Chilensis, sive Descripciones, et Icones Plantarum Peruvianarum, et Chilensium.... 3 vols. [Madrid.] (Fl. Peruv.)

Rumphia—See: C. L. Blume 1835[1836]–1848[1849]

Rydberg, P. A. 1900. Studies on the Rocky Mountain flora II. The Rocky Mountain species of Melanthaceae. Bull. Torrey Bot. Club 27: 528–538.

Rydberg, P. A. 1912. Studies on the Rocky Mountain flora. Bull. Torrey Bot. Club 39: 110.

Rydberg, P. A. 1932. Flora of the Prairies and Plains of Central North America.... New York. (Fl. Plains N. Amer.)

S. African J. Bot. = South African Journal of Botany.

Sag. Stor. Nat. Chili ed. 2—See: G. I. Molina 1810

Sagra, R. de la. 1840–1855. Historia Fisica Politica y Natural de la Isla de Cuba.... 12 vols. Paris and Madrid. (Hist. Fis. Cuba)

Salazar-Chavez, G. A. 1991. *Hexalectris warnockii* (Orchidaceae): Primer registro para México. Acta Bot. Mex. 16: 1–5.

Salisbury, R. A. 1805[–1808]. The Paradisus Londinensis.... 2 vols. London. [Plates numbered consecutively.] (Parad. Lond.)

Salisbury, R. A. 1866. The Genera of Plants.... London. [Published posthumously.] (Gen. Pl.)

Samejima, K. and J. Samejima. 1987. *Trillium* Genus Illustrated. Sapporo.

Sánchez, L. M. and E. Hágsater. 1997. Taxonomic study of the *Epidendrum difforme* group (Orchidaceae). In: C. E. de B. Pereira, ed. 1997. Proceedings of the 15th World Orchid Conference: Rio de Janeiro 1996. Turriers. Pp. 235–244.

Santana, D. O. 1984. Morphological and Anatomical Observations on the Bulbs and Tepals of *Fritillaria* (Liliaceae) Section *Liliorhiza* (Kellogg) Watson, Their Taxonomic Implications with a Synopsis and the Reproductive Biology of the Section. Ph.D. dissertation. University of California, Davis.

Sargent, C. S. 1905. Manual of the Trees of North America (Exclusive of Mexico).... Boston and New York. (Man. Trees)

Sargent, E. 1903. A theory of the origin of monocotyledons, founded on the structure of their seedlings. Ann. Bot. (Oxford) 17: 1–92.

Sargent, E. 1904. The evolution of monocotyledons. Bot. Gaz. 37: 325–345.

Sargentia = Sargentia; Continuation of the Contributions from the Arnold Arboretum of Harvard University.

Satô, D. 1935. Analysis of the karyotypes in *Yucca, Agave,* and related genera with special reference to the phylogenetic significance. Jap. J. Genet. 11: 272–278.

Sauleda, R. P. and R. M. Adams. 1980. The taxonomic status of *Ponthieva brittoniae* Ames (Orchidaceae). Bull. Torrey Bot. Club 107: 544–545.

Sauleda, R. P. and R. M. Adams. 1983. The genus *Encyclia* Hook. (Orchidaceae) in the Bahama Archipelago. Rhodora 85: 127–174.

Sauleda, R. P. and R. M. Adams. 1989. The orchid genera *Oncidium* Sw. and *Tolumnia* Raf. in Florida. Rhodora 91: 188–200.

Sauleda, R. P., R. P. Wunderlin, and B. F. Hansen. 1984. Transfers in Florida orchids. Phytologia 56: 308.

Sauza-Chies, T. T. et al. 1997. Phylogenetic analysis of Iridaceae with parsimony and distance methods using the plastid gene *rps*4. Pl. Syst. Evol. 204: 109–123.

Savile, D. B. O. 1961. Some fungal parasites of Liliaceae. Mycologia 53: 31–52.

Savolainen, V. et al. 2000. Phylogenetics of flowering plants based on combined analysis of plastid *atp*B and *rbc*L gene sequences. Syst. Biol. 49: 306–362.

Schedul. Orchid. = Schedulae Orchidianae.

Scheepen, J. van. 1996. Classified List and International Register of Tulip Names. Hellegon.

Schlechter, F. R. R. 1914–1915. Die Orchideen.... 10 parts. Berlin. [Parts paged consecutively.] (Orchideen)

Schlechteriana = Schlechteriana; Zeitschrift für Orchideenkunde.

Schlittler, J. 1940. Monographie der Liliaceengattung *Dianella* Lam. Mitt. Bot. Mus. Univ. Zürich 163: 1–283.

Schlittler, J. 1948. Unsere gegenwaertige Kenntnis uber die Liliaceengattung *Dianella* in Malesien. Blumea 6: 200–228.

Schlittler, J. 1957. Die Verbreitung der Liliaceengattung *Dianella* Lam. in Zusammenhang mit der Organdifferenzierung und der Areabildung. Mitt. Bot. Mus. Univ. Zürich 207: 1–38.

Schmid, W. G. 1991. The Genus *Hosta*. Portland.

Schmitz, D. C. 1990. The invasion of exotic aquatic and wetland plants in Florida: History and efforts to prevent new introductions. Aquatics 12(2): 6–13, 24.

Schmutz, E. M. and L. B. Hamilton. 1979. Plants That Poison: An Illustrated Guide for the American Southwest. Flagstaff.

Schneevoogt, G. V. [1792–]1793[–1795]. Icones Plantarum Rariorum. 16 parts. Haarlem. [Plates numbered consecutively.] (Icon. Pl. Rar.)

Schulze, W. 1975. Beiträge zur Taxonomie der Liliifloren: I. Asphodelaceae. Wiss. Z. Friedrich-Schiller-Univ. Jena, Math.-Naturwiss. Reihe 24: 403–415.

Schulze, W. 1975b. Beiträge zur Taxonomie der Liliifloren: II. Colchicaceae. Wiss. Z. Friedrich-Schiller-Univ. Jena, Math.-Naturwiss. Reihe 24: 417–428.

Schulze, W. 1978. Beiträge zur Taxonomie der Liliifloren: III. Alstroemeriaceae. Wiss. Z. Friedrich-Schiller-Univ. Jena, Math.-Naturwiss. Reihe 27: 79–85.

Schulze, W. 1978b. Beiträge zur Taxonomie der Liliifloren: IV. Melanthiaceae. Wiss. Z. Friedrich-Schiller-Univ. Jena, Math.-Naturwiss. Reihe 27: 87–95.

Schulze, W. 1980. Beiträge zur Taxonomie der Liliifloren: V. Alliaceae. Wiss. Z. Friedrich-Schiller-Univ. Jena, Math.-Naturwiss. Reihe 29: 595–606.

Schulze, W. 1980b. Beiträge zur Taxonomie der Liliifloren: VI. Der Umfang der Liliaceae. Wiss. Z. Friedrich-Schiller-Univ. Jena, Math.-Naturwiss. Reihe 29: 607–636.

Schulze, W. 1982. Beiträge zur Taxonomie der Liliifloren: X. Asparagaceae. Wiss. Z. Friedrich-Schiller-Univ. Jena, Math.-Naturwiss. Reihe 31: 309–330.

Schwartz, F. C. 1994. Molecular Systematics of *Zigadenus* Section *Chitonia* (Liliaceae). Ph.D. dissertation. University of Washington.

Sci. Amer. = Scientific American.

Sci. Rep. Tohoku Imp. Univ., Ser. 4, Biol. = Science Reports of the Tohoku Imperial University. Ser. 4, Biology. [Tohoku Teikoku-daigaku Rikwa Hokoku.]

Science = Science; an Illustrated Journal [later: a Weekly Journal Devoted to the Advancement of Science]. [American Association for the Advancement of Science.]

Scoggan, H. J. 1978–1979. The Flora of Canada. 4 parts. Ottawa. [Natl. Mus. Nat. Sci. Publ. Bot. 7.]

Sealy, J. R. 1937. *Zephyranthes, Pyrolirion, Habranthus* and *Hippeastrum*. J. Roy. Hort. Soc. 62: 195–209.

Sealy, J. R. 1954. Review of the genus *Hymenocallis*. Kew Bull. 1954: 201–240.

Séguier, J. F. 1745–1754. Plantae Veronenses, seu Stirpium Quae in Agro Veronensi Reperiuntur Methodica Synopsis.... 3 vols. Verona. (Pl. Veron.)

Selbyana = Selbyana; Journal of the Marie Selby Botanical Gardens.

Sell, P. D. and G. Murrell. 1996+. Flora of Great Britain and Ireland. 1+ vol. Cambridge and New York.

Sen, S. 1975. Cytotaxonomy of Liliales. Feddes Repert. 86: 255–305.

Sert. Angl.—See: C.-L. L'Héritier de Brutelle 1788[1789–1792]

Sert. Bot.—See: P. C. Van Geel [1827–]1828–1832[–1836]

Seymour, F. C. 1993. The Flora of New England: A Manual for the Identification of All Vascular Plants...Growing without Cultivation..., ed. 2. [N. p.]

Sharma, A. K. and U. C. Bhattacharyya. 1962. A cytological study of the factors influencing evolution in *Agave*. Cellule 62: 257–279.

Sheehan, T. J. and M. Sheehan. 1994. An Illustrated Survey of Orchid Genera. Portland.

Sherman, H. L. 1969. A Systematic Study of the Genus *Schoenolirion* (Liliaceae). Ph.D. thesis. Vanderbilt University.

Sherman, H. L. and R. W. Becking. 1991. The generic distinctness of *Schoenolirion* and *Hastingsia*. Madroño 38: 130–138.

Sheviak, C. J. 1974. An Introduction to the Ecology of the Illinois Orchidaceae. Springfield, Ill.

Sheviak, C. J. 1982. Biosystematic study of the *Spiranthes cernua* complex. Bull. New York State Mus. Sci. Serv. 448.

Sheviak, C. J. 1984. *Spiranthes diluvialis* (Orchidaceae), a new species from the western United States. Brittonia 36: 8–14.

Sheviak, C. J. 1990. A new *Spiranthes* (Orchidaceae) from the cienegas of southernmost Arizona. Rhodora 92: 213–231.

Sheviak, C. J. 1991. Morphological variation in the compilospecies *Spiranthes cernua* (L.) L. C. Rich.: Ecologically-limited effects of gene flow. Lindleyana 6: 228–234.

Sheviak, C. J. 1992. Natural hybridization between *Cypripedium montanum* and its yellow-lipped relatives. Amer. Orchid Soc. Bull. 61: 546–559.

Sheviak, C. J. and P. M. Catling. 1980. The identity and status of *Spiranthes ochroleuca* (Rydberg) Rydberg. Rhodora 82: 525–562.

Shevock, J. R. 1984. Rediscription and distribution of *Muilla coronata* (Liliaceae). Aliso 10: 621–627.

Shevock, J. R., J. A. Bartel, and G. A. Allen. 1990. Distribution, ecology and taxonomy of *Erythronium* (Liliaceae) in the Sierra Nevada of California. Madroño 37: 261–273.

Shinners, L. H. 1951. The North Texas species of *Hymenocallis* (Amaryllidaceae). Field & Lab. 19: 102–104.

Shinners, L. H. 1958. Spring Flora of the Dallas-Fort Worth Area, Texas. Dallas. (Spring Fl. Dallas-Fort Worth)

Shinners, L. H. 1962b. *Eichhornia azurea* (Pontederiaceae) in the Texas coastal bend: New to the United States. Sida 1: 99.

Shinwari, Z. K. et al. 1994. Recognition of the New World *Disporum* section *Prosartes* as *Prosartes* (Liliaceae) based on the sequence of the *rbc*L gene. Taxon 43: 353–366.

Shinwari, Z. K. et al. 1994b. Phylogenetic relationships among genera in the Liliaceae-Asparagoides-Polygonatae s.l. inferred from *rbc*L gene sequence data. Pl. Syst. Evol. 192: 263–277.

Sida = Sida; Contributions to Botany.

Sierra Club Bull. = Sierra Club Bulletin.

Sierra, S. E. C., B. Gravendeel, and E. F. de Vogel. 2000. Revision of *Coelogyne* section *Verrucosae* (Orchidaceae): A new sectional delimitation based on morphological and molecular evidence. Blumea 45: 275–318.

Silen, F. 1906. Blombiologisk iakttagelser i Kallila Lappmark. Meddeland. Soc. Fauna Fl. Fenn. 31: 30–99.

Silliman, F. E. 1957. *Chamaelirium luteum* (L.) Gray: A Biological Study. Ph.D. dissertation. University of North Carolina.

Simpson, D. 1995. Relationships within Cyperales. In: P. J. Rudall et al., eds. 1995. Monocotyledons: Systematics and Evolution. 2 vols. Kew. Vol. 2, pp. 459–509.

Simpson, M. G. 1990. Phylogeny and classification of the Haemodoraceae. Ann. Missouri Bot. Gard. 77: 722–784.

Simpson, M. G. 1993. Septal nectary anatomy and phylogeny of the Haemodoraceae. Syst. Bot. 18: 593–613.

Sketch Bot. S. Carolina—See: S. Elliott [1816–]1821–1824

Skinner, M. W. 1988. Comparative Pollination Ecology and Floral Evolution in Pacific Coast *Lilium*. Ph.D. dissertation. Harvard University.

Skr. Norske Vidensk.-Akad. Oslo, Mat.-Naturvidensk. Kl. = Skrifter Utgitt av det Norske Videnskaps-Akademi i Oslo. Matematisk-naturvidenskapelig Klasse.

Sloboda, D. 1852. Rostlinnictví Čili Návod k Snadnéma Určení a Pojmenování Rostlin v Cechách, Moravě.... Prague. (Rostlinnictví)

Small, J. K. 1903. Flora of the Southeastern United States.... New York. (Fl. S.E. U.S.)

Small, J. K. 1913. Flora of the Southeastern United States..., ed. 2. New York.

Small, J. K. 1913b. Flora of Miami.... New York.

Small, J. K. 1928. *Nothoscordum fragrans*. Addisonia 33: plate 433.

Small, J. K. 1929b. *Iris verna* (mountain form). Addisonia 14: 15–16, plate 456.

Small, J. K. 1931b. Bartram's *Ixia coelestina* rediscovered. J. New York Bot. Gard. 32: 155–161.

Small, J. K. 1933. Manual of the Southeastern Flora, Being Descriptions of the Seed Plants Growing Naturally in Florida, Alabama, Mississippi, Eastern Louisiana, Tennessee, North Carolina, South Carolina and Georgia. New York. (Man. S.E. Fl.)

Small, J. K. and E. J. Alexander. 1931. Botanical interpretations of the iridaceous plants of the Gulf States. Contr. New York Bot. Gard. 327: 325–358.

Small, J. K. and J. J. Carter. 1913. Flora of Lancaster County. New York.

Smith, C. F. 1976. A Flora of the Santa Barbara Region, California. Santa Barbara.

Smith, F. H. 1943. Megagametophyte of *Clintonia*. Bot. Gaz. 105: 263–267.

Smith, F. H. and Q. D. Clarkson. 1956. Cytological studies of interspecific hybridization in *Iris*, subsection *Californicae*. Amer. J. Bot. 43: 582–588.

Smith, G. F. and B. E. van Wyk. 1998. Asphodelaceae. In: K. Kubitzki et al., eds. 1990+. The Families and Genera of Vascular Plants. 4+ vols. Berlin etc. Vol. 3, pp. 130–140.

Smith, G. L. 1993. Studies on *Hymenocallis occidentalis* (Le Conte) Kunth. [Abstract.] A. S. B. Bull. 40: 247.

Smith, G. L. and N. C. Coile. 1999. Distinctions between *Hymenocallis choctawensis* and *Hymenocallis liriosme*. [Abstract.] A. S. B. Bull. 46: 242.

Smith, G. L. and M. Darst. 1994. A new species of *Hymenocallis* (Amaryllidaceae) in the Florida panhandle. Novon 4: 396–399.

Smith, G. L. and W. S. Flory. 1990. Studies on *Hymenocallis henryae* (Amaryllidaceae). Brittonia 42: 210–220.

Smith, G. L., W. S. Flory, and M. A. Garland. 1991. Studies on *Hymenocallis choctawensis* Traub in the Florida panhandle. [Abstract.] A. S. B. Bull. 38: 122.

Smith, G. L., W. S. Flory, and J. B. Nelson. 1990. Cytotaxonomic studies on *Hymenocallis coronaria* (Le Conte) Kunth in South Carolina piedmont river systems. [Abstract.] A. S. B. Bull. 37: 141.

Smith, G. L. and M. A. Garland. 1996. Taxonomic status of *Hymenocallis choctawensis* and *Hymenocallis puntagordensis* (Amaryllidaceae). Sida 17: 305–319.

Smith, G. L., M. A. Garland, and W. S. Flory. 1994. Rediscovery of *Hymenocallis puntagordensis* Traub. [Abstract.] A. S. B. Bull. 41: 239.

Smith, G. L. and J. N. Henry. 1999. *Hymenocallis henryae*, a rare endemic of the Florida panhandle. Herbertia 54: 110–117.

Smith, G. L., C. C. Moretz, and F. T. Yeats. 1999. Spider-lilies of the Atlantic coastal plain. [Abstract.] A. S. B. Bull. 46: 240.

Smith, Hu. H. 1928. Ethnobotany of the Meskawi Indians. Bull. Public Mus. Milwaukee 4: 175–326.

Smith, J. E. 1791–1792. Spicilegium Botanicum. 2 fasc. London. [Fascicles paged consecutively.] (Spic. Bot.)

Smith, J. E. 1804–1805[–1808]. Exotic Botany.... 2 vols. London. (Exot. Bot.)

Smith, J. F., W. J. Kress, and E. A. Zimmer. 1993. Phylogenetic analysis of the Zingiberales based on *rbc*L sequences. Ann. Missouri Bot. Gard. 80: 620–630.

Smith, L. B. and E. S. Ayensu. 1974. Classification of Old World Velloziaceae. Kew Bull. 29: 181–205.

Smith, L. B. and E. S. Ayensu. 1976. A revision of American Velloziaceae. Smithsonian Contr. Bot. 30.

Smith, L. B. and R. J. Downs. 1974. Pitcairnioideae. In: Organization for Flora Neotropica. 1968+. Flora Neotropica. 75+ nos. New York. No. 14(1).

Smith, L. B. and W. Till. 1998. Bromeliaceae. In: K. Kubitzki et al., eds. 1990+. The Families and Genera of Vascular Plants. 4+ vols. Berlin etc. Vol. 4, pp. 74–99.

Smith, R. W. 1911. The tetranucleate embryo sac of *Clintonia*. Bot. Gaz. 52: 209–217.

Smith, W. R. 1993. Orchids of Minnesota. Minneapolis.

Snijman, D. A. and H. P. Linder. 1996. Phylogenetic relationships, seed characters, and dispersal system evolution in Amaryllideae (Amaryllidaceae). Ann. Missouri Bot. Gard. 83: 362–386.

Snow, A. A. and D. F. Whigham. 1989. Costs of flower and fruit production in *Tipularia discolor* (Orchidaceae). Ecology 70: 1286–1293.

Snyder, D. B. and V. E. Vivian. 1981. Rare and Endangered Vascular Plant Species in New Jersey. Washington.

Solms-Laubach, H. 1883. Pontederiaceae. In: A. L. P. P. de Candolle and C. de Candolle, eds. 1878–1896. Monographiae Phanerogamarum.... 9 vols. Paris. Vol. 4, pp. 501–535.

Soltis, D. E. et al. 1997. Angiosperm phylogeny inferred from 18S ribosomal DNA sequences. Ann. Missouri Bot. Gard. 84: 1–49.

Soltis, D. E. et al. 1997b. Molecular phylogenetic relationships among angiosperms: An overview based on *rbc*L and 18S rDNA sequences. In: K. Iwatsuki and P. H. Raven, eds. 1997. Evolution and Diversification of Land Plants. Tokyo and New York. Pp. 157–178.

Soltis, D. E. et al. 2000. Angiosperm phylogeny inferred from 18S rDNA, *rbc*L, and *atp*B sequences. Bot. J. Linn. Soc. 133: 381–461.

Soltis, D. E., P. S. Soltis, and J. J. Doyle, eds. 1992. Molecular Systematics of Plants. New York.

Soltis, P. S. and D. E. Soltis. 2001. Molecular systematics: Assembling and using the Tree of Life. Taxon 50: 663–677.

Soó, R. 1962. Nomina Nova Generis Dactlorhiza. [Budapest.] (Nom. Nov. Gen. Dactylorhiza)

Soper, J. H. 1952. Phytogeographic studies in Ontario. I. The genus *Uvularia* in southern Ontario. Rhodora 54: 58–67.

Soper, J. H. 1962. Some genera of restricted range in the Carolinian flora of Canada. Trans. Roy. Canad. Inst. 34: 3–56.

Sosef, M. S. M. 1997. Hierarchical models, reticulate evolution and the inevitability of paraphyletic supraspecific taxa. Taxon 46: 75–85.

Soto, M. A. 1988. Listado actualizado de las orquídeas de México. Orquidea (Mexico City) 11: 233–277.

Southon, I. W. and J. Buckingham, eds. 1989. Dictionary of Alkaloids. 2 vols. New York.

SouthW. Naturalist = Southwestern Naturalist.

Sp. Pl.—See: A. G. Dietrich 1831–1833; C. Linnaeus 1753; C. L. Willdenow et al. 1797–1830

Sp. Pl. ed. 2—See: C. Linnaeus 1762–1763

Spangler, R., B. Zaitchik, E. Russo, and E. A. Kellogg. 1999. Andropogoneae evolution and generic limits in *Sorghum* (Poaceae) using *ndh*F sequences. Syst. Bot. 24: 267–281.

Specht, C. D., W. J. Kress, D. W. Stevenson, and R. DeSalle. 2001. A molecular phylogeny of Costaceae (Zingiberales). Molec. Phylogen. Evol. 21: 333-345.

Specim. Inaug. Corallorhiza—See: J. J. Châtelain 1760

Spellenberg, R., R. Worthington, P. Knight, and R. Fletcher. 1986. Additions to the flora of New Mexico. Sida 11: 455–470.

Spencer, L. B. 1973. A Monograph of the Genus *Zephyranthes* (Amaryllidaceae) in North and Central America. Ph.D. dissertation. Wake Forest University.

Speta, F. 1971. Beitrag zur Systematik von *Scilla* L. subgen. *Scilla* (inclusive *Chionodoxa* Boiss.). Oesterr. Bot. Z. 119: 6–18.

Speta, F. 1976. Über *Chionodoxa* Boiss., ihre Gliederung und Zugehörigkeit zu *Scilla* L. Naturk. Jahrb. Stadt Linz 21: 9–79.

Speta, F. 1979. Karyological investigation in *Scilla* in regard to their importance for taxonomy. Webbia 34: 419–431.

Speta, F. 1987. Die verwandschaftlichen Beziehungen von *Brimeura* Salisb.: Ein Vergleich mit den Gattungen *Oncostema* Rafin., *Hyacinthoides* Medic. und *Camassia* Lindl. (Hyacinthaceae). Phyton (Horn) 26: 247–310.

Speta, F. 1998. Hyacinthaceae. In: K. Kubitzki et al., eds. 1990+. The Families and Genera of Vascular Plants. 4+ vols. Berlin etc. Vol. 3, pp. 261–285.

Speta, F. 1998b. Systematische Analyse der Gattung *Scilla* L. (Hyacinthaceae). Phyton (Horn) 38: 1–141.

Spic. Bot.—See: J. E. Smith 1791–1792

Spoerke, D. G. Jr. and S. C. Smolinske. 1990. Toxicity of House-plants. Boca Raton.

Sprengel, K. [1824–]1825–1828. Caroli Linnaei...Systema Vegetabilium. Editio Decima Sexta.... 5 vols. Göttingen. [Vol. 4 in 2 parts, each paged independently; vol. 5 by A. Sprengel.] (Syst. Veg.)

Spring Fl. Dallas-Fort Worth—See: L. H. Shinners 1958

Stabursvik, A. 1959. A Phytochemical Study of *Narthecium ossifragum* (L.) Huds., with Additional Chapters on the Botany and the Veterinary History of the Plant. Trondheim.

Stace, C. A. 1991. New Flora of the British Isles. Cambridge and New York.

Stafleu, F. A. 1963. The tulip and the Netherlands. Gard. J. New York Bot. Gard. 13: 79–83, 108.

Stafleu, F. A. and R. S. Cowan. 1976–1988. Taxonomic Literature: A Selective Guide to Botanical Publications and Collections with Dates, Commentaries and Types, ed. 2. 7 vols. Utrecht, Antwerp, The Hague, and Boston.

Stafleu, F. A. and E. A. Mennega. 1992+. Taxonomic Literature. A Selective Guide to Botanical Publications and Collections with Dates, Commentaries and Types. Supplement. 6+ vols. Königstein.

Stansbury, H. 1852. An Expedition to the Valley of the Great Salt Lake of Utah.... Philadelphia. [Botanical appendix by J. Torrey, pp. 381–397, plates 1–9.] (Exped. Great Salt Lake)

Stapf, O. 1934. *Lilium, Notholirion,* and *Fritillaria.* Bull. Misc. Inform. Kew 1934: 94–96.

Stapfia = Stapfia; Publikation der Botanischen Arbeitsgemeinschaft am O. Ö. Landesmuseum, Linz.

Starr, G. 1995. *Hesperaloe:* Aloes of the West. Desert Pl. 11(4): 3–8.

Starr, G. 1997. A revision of the genus *Hesperaloe* (Agavaceae). Madroño 44: 282–296.

Starr, J. R., R. J. Bayer, and B. A. Ford. 1999. The phylogenetic position of *Carex* section *Phyllostachys* and its implications for phylogeny and subgeneric circumscription in *Carex* (Cyperaceae). Amer. J. Bot. 86: 563–577.

Stearn, F. C. 1956. Snowdrops and Snowflakes—A Study of the Genera *Galanthus* and *Leucojum.* London.

Stearn, W. T. 1931. The hostas and funkias: A revision of the plantain lilies. Gard. Chron., ser. 3, 90: 27, 47–49, 88–89, 110.

Stearn, W. T. 1947. The nomenclature and synonymy of *Tofieldia calyculata* and *T. pusilla.* J. Linn. Soc., Bot. 53: 194–204.

Stearn, W. T. 1986. *Nothoscordum gracile,* the correct name of *N. fragrans* and *N. inodorum* of authors (Alliaceae). Taxon 35: 335–338.

Stearn, W. T. 1990. The Linnaean species of *Hyacinthus* (Liliaceae : Hyacinthaceae). Ann. Mus. Goulandris 8: 181–222.

Stebbins, G. L. 1974. Flowering Plants: Evolution above the Species Level. Cambridge, Mass.

Steenis, C. G. G. J. van, ed. 1950+. Flora Malesiana.... Series I. Spermatophyta. 11+ vols. in parts. Djakarta and Leiden.

Sterling, C. 1977. Comparative morphology of the carpel in the Liliaceae: Uvularieae. Bot. J. Linn. Soc. 74: 345–354.

Sterling, C. 1979. Comparative morphology of the carpel in the Liliaceae: Tofieldieae. Bot. J. Linn. Soc. 79: 321–332.

Sterling, C. 1980. Comparative morphology of the carpel in the Liliaceae: Helonieae. Bot. J. Linn. Soc. 80: 341–356.

Sterling, C. 1982. Comparative morphology of the carpel in the Liliaceae: Veratreae. Bot. J. Linn. Soc. 84: 57–77.

Stermitz, F. R. 1982. *Oncidium andreanum* (Cogn.) Garay—A newly described mule-ear *Oncidium*. Amer. Orchid Soc. Bull. 51: 606–607.

Stern, W. L. and W. S. Judd. 2000. Comparative anatomy and systematics of the orchid tribe Vanilleae excluding *Vanilla*. Bot. J. Linn. Soc. 134: 179–202.

Stern, W. L. and W. S. Judd. 2001. Comparative anatomy and systematics of Catasetinae (Orchidaceae). Bot. J. Linn. Soc. 136: 153–178.

Stevenson, D. W. 1998. Mayacaceae. In: K. Kubitzki et al., eds. 1990+. The Families and Genera of Vascular Plants. 4+ vols. Berlin etc. Vol. 4, pp. 294–296.

Stevenson, D. W. et al. 2000. A phylogenetic analysis of the monocotyledons based on morphological and molecular character sets, with comments on the placement of *Acorus* and Hydatellaceae. In: K. L. Wilson and D. A. Morrison, eds. 2000. Monocots: Systematics and Evolution. Melbourne. Pp. 17–24.

Stevenson, D. W. and H. Loconte. 1995. Cladistic analysis of monocot families. In: P. J. Rudall et al., eds. 1995. Monocotyledons: Systematics and Evolution. 2 vols. Kew. Vol. 2, pp. 543–578.

Stewart, R. N. 1947. The morphology of somatic chromosomes in *Lilium*. Amer. J. Bot. 34: 9–27.

Stewart, R. N. and R. Bamford. 1942. The chromosomes and nuclei of *Medeola virginiana*. Amer. J. Bot. 29: 301–303.

Steyermark, J. A. 1961. A neglected *Camassia*. Brittonia 13: 206–211.

Steyermark, J. A. 1963. Flora of Missouri. Ames.

Stirp. Austr. Fasc. ed. 2—See: H. J. N. von Crantz 1769

Stockey, R. A., G. L. Hoffmann, and G. W. Rothwell. 1997. The fossil monocot *Limnobiophyllum scutatum*: Resolving the phylogeny of Lemnaceae. Amer. J. Bot. 84: 355–368.

Stout, A. B. 1934. Daylilies. New York. [Reprinted 1986, London.]

Stoutamire, W. P. 1971. Pollination in temperate American orchids. In: M. J. G. Corrigan et al., eds. 1971. Proceedings of the 6th World Orchid Conference. Sydney. Pp. 233–243.

Stoutamire, W. P. 1974. Relationships of purple-fringed orchids *Platanthera psycodes* and *P. grandiflora*. Brittonia 26: 42–58.

Stoutamire, W. P. 1978. Pollination of *Tipularia discolor*, an orchid with modified symmetry. Amer. Orchid Soc. Bull. 47: 413–415.

Streveler, B. E. 1966. A Taxonomic Study of the Genus *Convallaria* (Liliaceae). M.S. thesis. University of Wisconsin.

Strother, J. L. and G. L. Nesom. 1997. Conventions for reporting plant chromosome numbers. Sida 17: 829–831.

Succulenta (Netherlands) = Succulenta; Orgaan van de Nederl. Vereeniging van Vetplantenverzamelaars. Maanblad van de Nederlands-Belgische Vereniging van Liefhebbers van Cactussen en Andere Vetplanten.

Sullivan, V. I. 1973. Biosystematics of *Aletris lutea* Small, *Aletris obovata* Nash and natural hybrids. Brittonia 25: 294–303.

Suppl. Calif. Fl.—See: P. A. Munz 1968

Suppl. Pl.—See: C. Linnaeus f. 1781[1782]

Suppl. Pl. Succ.—See: A. H. Haworth 1819

Sutter, R. D. 1984. The status of *Helonias bullata* L. (Liliaceae) in the southern Appalachians. Castanea 49: 9–16.

Swartz, O. P. 1788. Nova Genera & Species Plantarum seu Prodromus.... Stockholm, Uppsala, and Åbo. (Prodr.)

Swartz, O. P. 1797–1806. Flora Indiae Occidentalis.... 3 vols. Erlangen and London. (Fl. Ind. Occid.)

Sweet, R. 1823–1837. The British Flower Garden.... 7 vols. London. [Vols. 4–7 also issued as vols. 1–4 of series 2.] (Brit. Fl. Gard.)

Sweet, R. 1839. Hortus Britannicus..., ed. 3, edited by G. Don. London. (Hort. Brit. ed. 3)

Swink, F. and G. S. Wilhelm. 1994. Plants of the Chicago Region: An Annotated Checklist of the Vascular Flora..., ed. 4. Indianapolis.

Syll. Pl. Fl. Neapol.—See: M. Tenore 1831–1842

Syn. Fl. West. States—See: J. L. Riddell 1835

Syn. Pl.—See: D. N. F. Dietrich 1839–1852; C. H. Persoon 1805–1807

Syn. Pl. Succ.—See: A. H. Haworth 1812

Synge, P. M. 1980. Lilies: A Revision of Elwes' Monograph of the Genus *Lilium* and Its Supplements. New York.

Syst. Biol. = Systematic Biology.

Syst. Bot. = Systematic Botany; Quarterly Journal of the American Society of Plant Taxonomists.

Syst. Nat.—See: J. F. Gmelin 1791[–1792]

Syst. Nat. ed. 10—See: C. Linnaeus 1758[–1759]

Syst. Nat. ed. 12—See: C. Linnaeus 1766[–1768]

Syst. Veg.—See: J. J. Roemer et al. 1817[–1830]; K. Sprengel [1824–]1825–1828

Syst. Veg. ed. 14—See: J. A. Murray 1784

Sytsma, K. J. and W. J. Hahn. 1996. Molecular systematics: 1994–1995. Progr. Bot. 58: 470–499.

Sytsma, K. J. and W. J. Hahn. 2000. Molecular systematics: 1997–1999. Progr. Bot. 62: 307–339.

Szlachetko, D. L. 1993. Two new species of the genus *Cyclopogon* (Orchidaceae) from South America. Candollea 48: 431–435.

Szlachetko, D. L. 1995. Systema orchidalium. Fragm. Florist. Geobot., suppl. 3.

Tabl. Encycl.—See: J. Lamarck and J. Poiret 1791–1823

Takahashi, H. 1976. Comparison of *Streptopus streptopoides* subsp. *brevipes* with its relatives, and chromosome number of *S. streptopoides* subsp. *brevipes*. Bull. Kanagawa Pref. Mus., Nat. Sci. 9: 1–5.

Takahashi, M. 1984. Pollen morphology in *Paris* and its related genera. Bot. Mag. (Tokyo) 97: 233–245.

Takahashi, M. 1988. The pollination biology of *Heloniopsis orientalis* (Thunb.) C. Tanaka (Liliaceae). Pl. Spec. Biol. 3: 117–123.

Takahashi, M. and S. Kawano. 1989. Pollen morphology of the Melanthiaceae and its systematic implications. Ann. Missouri Bot. Gard. 76: 863–876.

Takahashi, M. and K. Sohma. 1982. Pollen morphology of the genus *Clintonia* (Liliaceae). Sci. Rep. Tohoku Imp. Univ., Ser. 4, Biol. 38: 157–164.

Takahashi, M. and K. Sohma. 1983. Pollen morphology of the genus *Smilacina* (Liliaceae). Sci. Rep. Tohoku Imp. Univ., Ser. 4, Biol. 38: 191–218.

Takhtajan, A. L. 1959. Die Evolution der Angiospermen. Jena.

Takhtajan, A. L. 1966. A System and Phylogeny of the Flowering Plants. [Sistema i Filogeniia Tsvetkovykh Rastenii.] Moscow.

Takhtajan, A. L. 1969. Flowering Plants: Origin and Dispersal. Edinburgh.

Takhtajan, A. L. 1970. Origin and Dispersal of the Flowering Plants. Leningrad.

Takhtajan, A. L. 1980. Outline of the classification of flowering plants (Magnoliophyta). Bot. Rev. (Lancaster) 46: 225–359.

Takhtajan, A. L. 1983. The systematic arrangement of dicotyledonous families. In: C. R. Metcalfe and L. Chalk. 1979+. Anatomy of the Dicotyledons, ed. 2. 4+ vols. Oxford and New York. Vol. 2, pp. 180–201.

Takhtajan, A. L. 1986. Floristic Regions of the World. Berkeley and Los Angeles.

Takhtajan, A. L. 1987. Systema Magnoliophytorum. [Sistema Magnoliofitov.] Leningrad.

Takhtajan, A. L. 1994. Six new families of flowering plants. Bot. Zhurn. (Moscow & Leningrad) 79(1): 96–97.

Takhtajan, A. L. 1994b. New families of the monocotyledons. Bot. Zhurn. (Moscow & Leningrad) 79(12): 65–66.

Takhtajan, A. L. 1997. Diversity and Classification of Flowering Plants. New York.

Takhtajan, A. L. and A. G. Vouk, eds. 1985+. Sravnitelnaia Anatomiia Semian. [Anatomia Seminum Comparativa.] 4+ vols. Leningrad.

Tamura, M. N. 1993. Biosystematic studies on the genus *Polygonatum* (Liliaceae) III. Morphology of staminal filaments and karyology of eleven Eurasian species. Bot. Jahrb. Syst. 115: 1–26.

Tamura, M. N. 1995. A karyological review of the orders Asparagales and Liliales (Monocotyledonae). Feddes Repert. 106: 83–111.

Tamura, M. N. 1998. Melanthiaceae. In: K. Kubitzki et al., eds. 1990+. The Families and Genera of Vascular Plants. 4+ vols. Berlin etc. Vol. 3, pp. 369–380.

Tamura, M. N. 1998b. Nartheciaceae. In: K. Kubitzki et al., eds. 1990+. The Families and Genera of Vascular Plants. 4+ vols. Berlin etc. Vol. 3, pp. 381–392.

Tamura, M. N. 1998c. Liliaceae. In: K. Kubitzki et al., eds. 1990+. The Families and Genera of Vascular Plants. 4+ vols. Berlin etc. Vol. 3, pp. 343–353.

Tamura, M. N. 1998d. Calochortaceae. In: K. Kubitzki et al., eds. 1990+. The Families and Genera of Vascular Plants. 4+ vols. Berlin etc. Vol. 3, pp. 165–172.

Tamura, M. N. 1998e. Trilliaceae. In: K. Kubitzki et al., eds. 1990+. The Families and Genera of Vascular Plants. 4+ vols. Berlin etc. Vol. 3, pp. 444–452.

Tamura, M. N., A. E. Schwarzbach, S. Kruse, and R. Reski. 1997. Biosystematics studies on the genus *Polygonatum* (Convallariaceae) IV. Molecular phylogenetic analysis based on restriction site mapping of the chloroplast gene *trn*K. Feddes Repert. 108: 159–168.

Tamura, M. N., F. H. Utech, and S. Kawano. 1992. Biosystematic studies in *Disporum* (Liliaceae-Polygonateae). Iv. Karyotype analysis of some Asiatic and North American taxa with special reference to their systematic status. Pl. Spec. Biol. 7: 103–120.

Tan, K., ed. 1989. The Davis & Hedge Festschrift...Plant Taxonomy, Phytogeography, and Related Subjects. Edinburgh.

Tan, K. W., ed. 1984. Proceedings of the Eleventh World Orchid Conference: March 1984, Miami, Florida, U.S.A. Miami.

Tanaka, N. 1997. Taxonomic significance of tome floral characters in *Helonias* and *Ypsilandra* (Liliaceae). J. Jap. Bot. 72: 110–116.

Tanaka, N. 1997b. Evolutionary significance of the variation of the floral structure of *Heloniopsis*. J. Jap. Bot. 72: 131–138.

Tanaka, N. 1997c. Phylogenetic and taxonomic studies on *Helonias, Ypsilandra* and *Heloniopsis*. I. Comparison of character states (1). J. Jap. Bot. 72: 221–228.

Tanaka, N. 1997d. Phylogenetic and taxonomic studies on *Helonias, Ypsilandra* and *Heloniopsis*. I. Comparison of character states (2). J. Jap. Bot. 72: 286–292.

Tanaka, N. 1997e. Phylogenetic and taxonomic studies on *Helonias, Ypsilandra* and *Heloniopsis*. II. Evolution and geographical distribution. J. Jap. Bot. 72: 329–336.

Tanaka, N. 1998. Phylogenetic and taxonomic studies on *Helonias, Ypsilandra* and *Heloniopsis*. III. Taxonomic revision. J. Jap. Bot. 73: 102–115.

Tanaka, N. 2001. Taxonomic revision of the family Cannaceae in the New World and Asia. Makinoa, n. s. 1: 1–74.

Tanaka, N., H. Setoguchi, and J. Murata. 1997. Phylogeny of the family Hydrocharitaceae inferred from *rbc*L and *mat*K gene sequence data. J. Pl. Res. 110: 329–337.

Tanaka, N. Y. and N. Tanaka. 1985. Fluorescent banding in the chromosomes of *Chionographis* and *Chamaelirium*. Kromosomo, ser. 2, 37: 1161.

Tardiff, B. and P. Morisset. 1990. Clinal morphological variation of *Allium schoenoprasum* in eastern North America. Taxon 39: 417–429.

Tausch, I. F. 1823–1825. Hortus Canalius.... 2 parts. Prague. (Hort. Canal.)

Taxon = Taxon; Journal of the International Association for Plant Taxonomy.

Taylor, C. A. 1956. Culture of false hellebore. Econ. Bot. 10: 155–165.

Taylor, C. A. 1956b. Alkaloid yield of *Veratrum fimbriatum* as influenced by site, season and other factors. Econ. Bot. 10: 166–173.

Taylor, R. L. 1967. The foliar embryos of *Malaxis paludosa*. Canad. J. Bot. 45: 1553–1556.

Taylor, T. M. C. and A. F. Szczawinski. 1974. *Trillium ovatum* Pursh forma *hibbersonii* Taylor et Szczawinski f. nov. Syesis 7: 250.

Tebiwa = Tebiwa; Miscellaneous Papers of the Idaho State University Museum of Natural History.

Tenore, M. 1831–1842. Sylloge Plantarum Vascularium Florae Neapolitanae.... 1 vol. + 2 vols. app. Naples. [App. nos. 1–3 included in main vol., nos. 4 and 5 issued separately.] (Syll. Pl. Fl. Neapol.)

Tepidino, V. J. 1982. Effects of defoliation on reproduction of a toxic range plant, *Zigadenus paniculatus*. Great Basin Naturalist 42: 524–528.

Terry, R. G., G. K. Brown, and R. G. Olmstead. 1997. Phylogenetic relationships in subfamily Tillandsioideae (Bromeliaceae) using *ndh*F sequences. Syst. Bot. 22: 333–345.

Terry, R. G., G. K. Brown, and R. G. Olmstead. 1997b. Examination of subfamilial phylogeny in Bromeliaceae using comparative sequencing of the plastid locus *ndh*F. Amer. J. Bot. 84: 664–670.

Therman, E. 1950. Chromosome numbers in American *Polygonatum* species. Amer. J. Bot. 37: 407–413.

Therman, E. 1953. Chromosomal evolution in the genus *Polygonatum*. Hereditas (Lund) 39: 277–288.

Therman, E. 1953b. On the cytology of the genus *Polygonatum* groups *Verticillata* and *Oppositifolia*. Ann. Bot. Soc. Zool.-Bot. Fenn. "Vanamo" 25: 1–26.

Thien, L. B. 1973. Isolating mechanisms in the genus *Calopogon*. Amer. Orchid Soc. Bull. 42: 794–797.

Thien, L. B. and B. G. Marcks. 1972. The floral biology of *Arethusa bulbosa*, *Calopogon tuberosus*, and *Pogonia ophioglossoides* (Orchidaceae). Canad. J. Bot. 50: 2319–2325.

Thieret, J. W. 1972b. Observations on some aquatic plants in northwestern Minnesota. Michigan Bot. 10: 117–124.

Thompson, Z. 1842. Natural History of Vermont. Burlington. [Botany by W. Oakes, vol. 1, pp. 173–221, reprinted separately as Catalogue of Vermont Plants.... with independent pagination.] (Nat. Hist. Vermont)

Thomson, J. D. and D. A. Stratton. 1985. Floral morphology and cross-pollination of *Erythronium grandiflorum* (Liliaceae). Amer. J. Bot. 72: 433–437.

Thomson, J. D. and B. A. Thomson. 1989. Dispersal of *Erythronium grandiflorum* pollen by bumblebees: Implications for gene flow and reproductive success. Evolution 43: 657–661.

Thorne, R. F. 1968. Synopsis of a putatively phylogenetic classification of the flowering plants. Aliso 6: 57–66.

Thorne, R. F. 1973. The "Amentiferae" or Hamamelidae as an artificial group: A summary statement. Brittonia 25: 395–405.

Thorne, R. F. 1974. A phylogenetic classification of the Annoniflorae. Aliso 8: 147–209.

Thorne, R. F. 1976. A phylogenetic classification of the Angiospermae. Evol. Biol. 9: 35–106.

Thorne, R. F. 1977. Some realignments in the Angiospermae. Nordic J. Bot. 3: 85–117.

Thorne, R. F. 1981. Phytochemistry and angiosperm phylogeny: A summary statement. In: D. A. Young and D. S. Seigler, eds. 1981. Phytochemistry and Angiosperm Phylogeny. New York. Pp. 233–295.

Thorne, R. F. 1983. Proposed new realignments in the angiosperms. Nordic J. Bot. 3: 85–117.

Thorne, R. F. 1992. An updated phylogenetic classification of the flowering plants. Aliso 13: 365–389.

Thorne, R. F. 1992b. Classification and geography of the flowering plants. Bot. Rev. (Lancaster) 58: 225–348.

Thorne, R. F. 2000. The classification and geography of the monocotyledon subclasses Alismatidae, Liliidae and Commelinidae. In: B. Nordenstam et al., eds. 2000. Plant Systematics for the 21st Century: Proceedings from a Symposium.... London. Pp. 75–124.

Thorsch, J. A. and V. I. Cheadle. 1996. Vessels in Eriocaulaceae. I. A. W. A. J. 17: 183–204.

Thouars, L. M. A. du P. 1822. Histoire Particulière des Plantes Orchidées Recueillies sur les Trois Îles Australes d'Afrique, de France, de Bourbon et de Madagascar.... Paris. (Hist. Orchid.)

Thunberg, C. P. 1787–1827. Museum Naturalium Academiae Upsaliensis. 33 parts, 26 app., 2 add. Uppsala. (Mus. Nat. Acad. Upsal.)

Thunberg, C. P. 1794–1800. Prodromus Plantarum Capensium, Quas in Promontorio Bonae Spei Africes, Annis 1772–1775, Collegit.... 2 parts. Upsala. (Prodr. Pl. Cap.)

Thunberg, C. P. 1823. Flora Capensis.... Edidit et Praefatus est J. A. Schultes.... 2 vols. Stuttgart etc. [Volumes paged consecutively.] (Fl. Cap. ed. 2)

Ting, C. T. and M. G. Gilbert. 2000. Dioscoreaceae. In: Wu Z. and P. H. Raven, eds. 1994+. Flora of China. 7+ vols. Beijing and St. Louis. Vol. 24, pp. 276–296.

Todsen, T. K. 1997. Naming a southwestern *Malaxis* (Orchidaceae). Sida 14: 636–637.

Todzia, C. A. 1980. A revision of the Costa Rican species of *Campylocentrum* (Orchidaceae). Brenesia 18: 117–136.

Topogr. Verz. Pflanzensamml. Ecklon—See: C. F. Ecklon 1827

Torrey, J. 1857. Explorations and Surveys for a Railroad Route from the Mississippi River to the Pacific Ocean. War Department. Route Near the Thirty-fifth Parallel, Explored by Lieutenant A. W. Whipple, Topographical Engineers, in 1853 and 1854. Report on the Botany of the Expedition. Washington. [Pacif. Railr. Rep. 4(5)]

Torrey, J. and A. Gray. 1838–1843. A Flora of North America.... 2 vols. in 7 parts. New York, London, and Paris. (Fl. N. Amer.)

Torreya = Torreya; a Monthly Journal of Botanical Notes and News.

Trans. Acad. Sci. St. Louis = Transactions of the Academy of Science of St. Louis.

Trans. Amer. Entomol. Soc. = Transactions of the American Entomological Society.

Trans. Amer. Philos. Soc. = Transactions of the American Philosophical Society Held at Philadelphia for Promoting Useful Knowledge.

Trans. Hort. Soc. London = Transactions, of the Horticultural Society of London.

Trans. Illinois State Acad. Sci. = Transactions of the Illinois State Academy of Science.

Trans. Kansas Acad. Sci. = Transactions of the Kansas Academy of Science.

Trans. Linn. Soc. London = Transactions of the Linnean Society of London.

Trans. Roy. Canad. Inst. = Transactions of the Royal Canadian Institute.

Trans. Roy. Soc. South Africa = Transactions of the Royal Society of South Africa.

Transylvania J. Med. Assoc. Sci. = The Transylvania Journal of Medicine and the Associate Sciences.

Trapnell, D. W. 1995. Systematic Review of *Calopogon* (Orchidaceae). Master's thesis. University of Georgia.

Trattinnick, L. [1811–]1812–1818. Archiv der Gewächskunde.... 5 parts. Vienna. [Parts paged erratically, plates numbered consecutively throughout.] (Arch. Gewächsk.)

Traub, H. P. 1962. Key to the subgenera, alliances and species of *Hymenocallis*. Pl. Life 18: 55–72.

Traub, H. P. 1972. The order Alliales. Pl. Life 28: 129–138.

Traub, H. P. 1972b. Genus *Allium*—subgenera, sections, and subsections. Pl. Life 28: 132–137.

Travels Carolina—See: W. Bartram 1791

Travis, J. 1984. Breeding system, pollination, and pollinator limitation in a perennial herb, *Amianthium muscaetoxicum* (Liliaceae). Amer. J. Bot. 71: 941–947.

Trelease, W. 1893. Further studies of *Yucca* and their pollination. Rep. (Annual) Missouri Bot. Gard. 4: 181–225.

Trelease, W. 1902. The Yucceae. Rep. (Annual) Missouri Bot. Gard. 13: 27–133.

Trelease, W. 1907. Additions to the genus *Yucca*. Rep. (Annual) Missouri Bot. Gard. 18: 225–230.

Trelease, W. 1911. The desert group Nolineae. Proc. Amer. Philos. Soc. 50: 404–443.

Trop. Pl.—See: H. Perrine 1838

Trudy Bot. Inst. Akad. Nauk S.S.S.R, Ser. 1, Fl. Sist. Vyssh. Rast. = Trudy Botanicheskogo Instituta Akademii Nauk S S S R. Ser. 1, Flora i Sistematika Vysshikh Rastenii.

Trudy Imp. S.-Peterburgsk. Bot. Sada = Trudy Imperatorskago S.-Peterburgskago Botanicheskago Sada.

Tucker, A. O. et al. 1979. Rare and Endangered Vascular Plant Species in Delaware. Newton Corner, Mass.

Turner, N. J. and H. V. Kuhnlein. 1983. Camas (*Camassia* spp.) and riceroot (*Fritillaria* spp.): Two liliaceous "root" foods of the Northwest Coast Indians. Ecol. Food Nutr. 13: 199–219.

Turrill, W. B. 1950. Character combinations and distribution in the genus *Fritillaria* and allied genera. Evolution 4: 1–6.

Turrill, W. B., E. Milne-Redhead, C. E. Hubbard, and R. M. Polhill, eds. 1952+. Flora of Tropical East Africa. 152+ vols. London and Rotterdam. [Volumes by family, unnumbered, some in parts.]

Turrill, W. B. and J. R. Sealy. 1980. Studies in the genus *Fritillaria* (Liliaceae). Hooker's Icon. Pl. 39.

Tutin, T. G. et al., eds. 1964–1980. Flora Europaea. 5 vols. Cambridge.

Tyteca, D. and J.-L. Gathoye. 1993. On the morphological variability of *Dactylorhiza praetermissa* (Druce) Soó (Orchidaceae). Belg. J. Bot. 126: 81–99.

Tzvelev, N. N. 1989. The system of grasses (Poaceae) and their evolution. Bot. Rev. (Lancaster) 55: 141–204.

U.S.D.A. Bull. (1915-23) = U S Department of Agriculture Bulletin.

U.S.D.A. Bur. Pl. Industr. Bull. = U S Department of Agriculture. Bureau of Plant Industry. Bulletin.

U.S.D.A. Div. Bot. Bull. = U S Department of Agriculture. Division of Botany. Bulletin.

Uhl, N. W. et al. 1995. Phylogenetic relationships among palms: Cladistic analysis of morphological and chloroplast DNA restriction site variation. In: P. J. Rudall et al., eds. 1995. Monocotyledons: Systematics and Evolution. 2 vols. Kew. Vol. 2, pp. 623–662.

Uhl, N. W. and J. Dransfield. 1987. Genera Palmarum. Lawrence, Kans.

Ullrich, B. 1992. On the history of *Agave asperrima* and *A. scabra* (Agavaceae) as well as some taxa of the Parryanae. Sida 15: 241–261.

Ultsch, G. R. 1973. The effects of water hyacinths (*Eichhornia crassipes*) on the microenvironment of aquatic environments. Arch. Hydrobiol. 72: 460–473.

Univ. Arkansas Coll. Agric. Arkansas Agric. Exp. Sta. Bull. = University of Arkansas. The College of Agriculture. Arkansas Agricultural Experiment Station. Bulletin.

Univ. Calif. Publ. Bot. = University of California Publications in Botany.

Univ. Calif. Publ. Entomol. = University of California Publications in Entomology.

Univ. Wash. Publ. Bot. = University of Washington Publications in Botany.

University of Chicago Press. 1993. The Chicago Manual of Style, ed. 14. Chicago.

Uphoff, J. C. T. 1952. A review of the genus *Alstroemeria*. Pl. Life 8: 37–53.

Utech, F. H. 1973. A Biosystematic Study of the Genus *Clintonia* Raf. (Liliaceae: Polygonatae). Ph.D. thesis. Washington University.

Utech, F. H. 1975. Biosystematic studies in *Clintonia* (Liliaceae-Polygonatae) III. Cytogeography, chromosome numbers and chromosomal morphology of the North American species of *Clintonia*. Cytologia 40: 765–786.

Utech, F. H. 1978. Floral vascular anatomy of *Pleea tenuifolia* Michx. (Liliaceae-Tofieldieae) and its reassignment to *Tofieldia*. Ann. Carnegie Mus. 47: 423–454.

Utech, F. H. 1978b. Vascular floral anatomy of *Helonias bullata* (Liliaceae–Helonieae), with a comparison to the Asian *Heloniopsis orientalis*. Ann. Carnegie Mus. 47: 169–191.

Utech, F. H. 1978c. Comparison of the vascular floral anatomy of *Xerophyllum asphodeloides* (L.) Nutt. and *X. tenax* (Pursh) Nutt. (Liliaceae–Melanthioideae). Ann. Carnegie Mus. 47: 147–167.

Utech, F. H. 1978d. Somatic karyotype analysis of *Uvularia floridana* Chapman (Liliaceae). Cytologia 43: 671–678.

Utech, F. H. 1978e. Floral vascular anatomy of *Medeola virginiana* L. (Liliaceae-Parideae = Trilliaceae) and tribal note. Ann. Carnegie Mus. 47: 13–28.

Utech, F. H. 1979. Karyotype analysis, palynology, and external seed morphology of *Tofieldia tenuifolia* (Michx.) Utech (Liliaceae-Tofieldieae). Ann. Carnegie Mus. 48: 171–184.

Utech, F. H. 1979b. Floral vascular anatomy of *Scoliopus bigelovii* Torrey (Liliaceae-Parideae = Trilliaceae) and tribal note. Ann. Carnegie Mus. 48: 43–71.

Utech, F. H. 1980. Somatic karyotype analysis of *Helonias bullata* L. (Liliaceae), with a comparison to the Asian *Heloniopsis orientalis* (Thunb.) C. Tanaka. Ann. Carnegie Mus. 49: 153–160.

Utech, F. H. 1986. Floral morphology and vascular anatomy of *Amianthium muscaetoxicum* (Walter) A. Gray (Liliaceae–Veratreae) with notes on distribution and taxonomy. Ann. Carnegie Mus. 55: 481–504.

Utech, F. H. 1987. Biosystematic studies in *Stenanthium* (Liliaceae: Veratreae). I. Floral morphology, floral vascular anatomy, geography and taxonomy of *S. occidentale* A. Gray. Ann. Carnegie Mus. 56: 113–135.

Utech, F. H. 1987b. Biosystematic studies in *Stenanthium* (Liliaceae: Veratreae). II. Floral morphology, floral vascular anatomy, geography and taxonomy of the Mexican *S. frigidum* (Schlecht. & Cham.) Kunth. Ann. Carnegie Mus. 56: 197–212.

Utech, F. H. 1992. Biology of *Scoliopus* (Liliaceae). I. Systematics and phytogeography. Ann. Missouri Bot. Gard. 79: 126–142.

Utech, F. H. 1993. Floral vascular anatomy of *Harperocallis flava* McDaniel and its systematic relationship within the Tofieldieae (Liliaceae). [Abstract.] In: International Botanical Congress. 1993. Abstracts. XV International Botanical Congress. Congress Center of Pacifico Yokohama, Japan, August 28 to September 3, 1993. [Tokyo.] P. 207.

Utech, F. H. and S. Kawano. 1975. Biosystematic studies in *Streptopus* (Liliaceae-Polygonateae) I. Morphological variation of the Asian *S. streptopoides* (Ledeb.) Frye & Rigg. Acta Phytotax. Geobot. 27: 1–20.

Utech, F. H. and S. Kawano. 1976. Biosystematic studies on *Maianthemum* (Liliaceae). VIII. Floral anatomy of *M. dilatatum, M. bifolium, M. canadense*. Bot. Mag. (Tokyo) 89: 145–157.

Utech, F. H. and S. Kawano. 1976b. Biosystematic studies in *Streptopus* (Liliaceae-Polygonateae) II. Taxonomy of the Asian *S. streptopoides* (Ledeb.) Frye & Rigg. Acta Phytotax. Geobot. 17: 97–105.

Utech, F. H. and S. Kawano. 1976c. Floral vascular anatomy of *Convallaria majalis* L. and *C. keiskei* Miq. (Liliaceae-Convallarieae). Bot. Mag. (Tokyo) 89: 173–182.

Utech, F. H. and S. Kawano. 1981. Vascular floral anatomy of the east Asian *Heloniopsis orientalis* (Thunb.) C. Tanaka (Liliaceae–Helonieae). Bot. Mag. (Tokyo) 94: 295–311.

Utech, F. H., Z. K. Shinwari, and S. Kawano. 1995. Biosystematic studies in *Disporum* (Liliaceae-Asparagoideae-Polygonateae). VI. Recognition of the North American section *Prosartes* as an autonomous genus. Mem. Fac. Sci. Kyoto Univ., Ser. Biol. 16: 1–41.

Utech, F. H. and L. B. Thien. 1973. Chromosome numbers in *Clintonia borealis* (Liliaceae). Michigan Bot. 12: 122–128.

Valentine, D. H. and H. M. Hassan. 1971. Cytotaxonomy of the genus *Maianthemum*. J. Indian Bot. Soc. 50: 437–446.

van den Berg, C. et al. 2000. A phylogenetic analysis of Laeliinae (Orchidaceae) based on sequence data from internal transcribed spacers (ITS) of nuclear ribosomal DNA. Lindleyana 15: 96–114.

Van Geel, P. C. [1827–]1828–1832[–1836]. Sertum Botanicum. 4 vols. Brussels. [Pages and plates unnumbered.] (Sert. Bot.)

van Raamsdonk, L. 1984. Biosystematic Studies on the *umbellatum/angustifolium* Complex in the Genus *Ornithogalum* L. Ph.D. dissertation. Rijksuniversiteit Utrecht.

Varadarajan, G. S. and A. J. Gilmartin. 1988. Phylogenetic relationships of groups of genera within the subfamily Pitcairnioideae (Bromeliaceae). Syst. Bot. 13: 283–293.

Vasc. Pl. Pacif. N.W.—See: C. L. Hitchcock et al. 1955–1969

Vasc. Pl. Texas ed. 2—See: M. C. Johnston 1990

Vellozo, J. M. 1825–1827[1829–1831]. Florae Fluminensis, seu Descriptionum Plantarum Praefectura Fluminensi Sponte Nascentium.... 1 vol. text + 11 vols. plates + indexes. Rio de Janeiro. (Fl. Flumin.)

Verh. Bot. Vereins Prov. Brandenburg = Verhandlungen des Botanischen Vereins der Provinz Brandenburg.

Verhoek, S. E. 1975. A Study of the Tribe Poliantheae (Including *Manfreda*) and Revisions of *Manfreda* and *Polianthes* (Agavaceae). Ph.D. thesis. Cornell University.

Verhoek, S. E. 1978. Floral biology of *Manfreda virginica* (L.) Rose (Agavaceae). [Abstract.] In: Botanical Society of America. 1978. Program and Abstracts of Papers to be Presented at the Meetings of the Botanical Society of America and Certain Affiliated Groups at Virginia Polytechnic Institute and State University, Blacksburg, Virginia, 25–29 June 1978. Blacksburg. P. 82.

Verhoek, S. E. 1978b. Huaco and amole: A survey of the uses of *Manfreda* and *Prochnyanthes*. Econ. Bot. 32: 124–130.

Vermeulen, J. J. 1987. A taxonomic revision of the continental African Bulbophyllinae. Orchid Monogr. 2.

Vermeulen, J. J. 1993. A taxonomic revision of *Bulbophyllum*, sections *Adelopetalum, Lepanthanthe, Macrouris, Pelma, Peltopus*, and *Uncifera* (Orchidaceae). Orchid Monogr. 7.

Vermeulen, P. 1947. Studies on Dactylorchids. Utrecht.

Veröff. Geobot. Inst. E. T. H. Stiftung Rübel Zürich = Veröffentlichungen des Geobotanischen Institutes der Eidg. Techn. Hochschule, Stiftung Rübel, in Zürich.

Vij, S. P., P. Kaur, and R. K. Bhamura. 1999. Embryological studies in *Epipactis gigantea* (Orchidaceae). Lindleyana 14: 160–167.

Vij, S. P. and N. Vohra. 1974. Cytomorphological studies in *Zeuxine strateumatica* (L.) Schltr. (= *Z. sulcata* L.). Cytologia 39: 411–417.

Vijayavalli, B. and P. M. Mathew. 1990. Cytotaxonomy of the Liliaceae and Allied Families. Kerala.

Vilm. Blumengärtn. ed. 3—See: A. Voss [1894–]1896

Vinnersten, A. and K. Bremer. 2001. Age and biogeography of major clades in Liliales. Amer. J. Bot. 88: 1695–1703.

Viosca, P. 1935. The irises of southeastern Louisiana, a taxonomic and ecological interpretation. Bull. Amer. Iris Soc. 57: 3–56.

Virginia J. Sci. = Virginia Journal of Science.

Voss, A. [1894–]1896. Vilmorin's Blumengärtnerei.... 2 vols. Berlin. (Vilm. Blumengärtn. ed. 3)

Voss, E. G. 1972–1996. Michigan Flora.... 3 vols. Bloomfield Hills and Ann Arbor.

Voth, W. 1976. *Prescotia plantaginifolia* Lindl. ex Hook. Orchidee (Hamburg) 27: 148–153.

Vyshenskaya, T. D. 1985. Hydatellaceae. In: A. L. Takhtajan and A. G. Vouk, eds. 1985+. Sravnitelnaia Anatomiia Semian. [Anatomia Seminum Comparativa.] 4+ vols. Leningrad. Vol. 1, pp. 190–191.

W. Druggist = Western Druggist.

Wake Robin = Wake Robin; Newsletter of the Sault Naturalists of Ontario and Michigan.

Walker, R. I. 1944. Chromosome number, megasporogenesis, and development of embryo-sac of *Clintonia.* Bull. Torrey Bot. Club 71: 528–535.

Wall, M. et al. 1957. Steroidal sapogenins XLIII. J. Amer. Pharm. Assoc. 46: 653–684.

Wallace, G. A. 1979. An overlooked new species of *Smilax* (Smilacaceae) from northern California. Brittonia 31: 416–421.

Wallace, L. E. and M. A. Case. 2000. Contrasting allozyme diversity between northern and southern populations of *Cypripedium parviflorum* (Orchidaceae): Implications for Pleistocene refugia and taxonomic boundaries. Syst. Bot. 25: 281–296.

Walpers, W. G. and K. Müller berol. 1848–1871. Annales Botanices Systematicae.... 7 vols. Leipzig. (Ann. Bot. Syst.)

Walsh, O. S. 1940. A Systematic Study of the Genus *Zigadenus* Michx. Ph.D. dissertation. University of California, Berkeley.

Walter, T. 1788. Flora Caroliniana, Secundum Systema Vegetabilium Perillustris Linnaei Digesta.... London. (Fl. Carol.)

Walters, S. M. et al., eds. 1984+. The European Garden Flora.... 6+ vols. Cambridge etc.

Ward, D. B. 1959. Relationships Among Certain Species of *Sisyrinchium* in Northeastern North America. Ph.D. dissertation. Cornell University.

Ward, D. B. 1968. The nomenclature of *Sisyrinchium bermudiana* and related North American species. Taxon 17: 270–276.

Ward, D. B., ed. [1979.] Rare and Endangered Biota of Florida. Vol. 5. Plants. Gainesville.

Warford, N. M. 1997. *Harrisella* cf. *porrecta:* Leafless and lethal near Puerto Vallarta. Orchid Digest 61: 148–154.

Watson, L. and M. J. Dallwitz. 1992. The Grass Genera of the World. Wallingford.

Watson, S. 1871. United States Geological Expolration [sic] of the Fortieth Parallel. Clarence King, Geologist-in-charge. [Vol. 5] Botany. By Sereno Watson.... Washington. [Botanical portion of larger work by C. King.] [Botany (Fortieth Parallel)]

Watson, S. 1879. Contributions to American botany. I. Revision of the North American Liliaceae. Proc. Amer. Acad. Arts 14: 213–288.

Watson, S., W. H. Brewer, and A. Gray. 1876–1880. Geological Survey of California.... Botany.... 2 vols. Cambridge, Mass. (Bot. California)

Watsonia = Watsonia; Journal of the Botanical Society of the British Isles.

Webber, H. J. 1897. The water hyacinth and its relations to navigation in Florida. U.S.D.A. Div. Bot. Bull. 18: 1–20.

Webber, J. M. 1953. Yuccas of the Southwest. Washington. [U.S.D.A., Agric. Monogr. 17.]

Webbia = Webbia; Raccolta di Scritti Botanici.

Weber, W. A. 1990. Colorado Flora: Eastern Slope. Niwot, Colo.

Weldy, T. W., H. T. Mlodozeniec, L. E. Wallace, and M. A. Case. 1996. The current status of *Cypripedium kentuckiense* (Orchidaceae) including a morphological analysis of a newly discovered population in eastern Virginia. Sida 17: 423–435.

Welsh, S. L., N. D. Atwood, S. Goodrich, and L. C. Higgins, eds. 1993. A Utah Flora, ed. 2. Provo.

Welsh, S. L. and G. Moore. 1973. Utah Plants: Tracheophyta, ed. 3. Provo.

Werdenda = Werdenda. Beiträge zur Pflanzenkunde.

Westbrooks, R. G. 1990. Interstate sale of aquatic federal noxious weeds as ornamentals in the United States. Aquatics 12(2): 16–18, 24.

Wherry, E. T. 1946. A key to the eastern North American lilies. Bartonia 24: 5–8.

Wherry, E. T., J. M. Fogg Jr., and H. A. Wahl. 1979. Atlas of the Flora of Pennsylvania. Philadelphia.

Whetstone, R. D. 1984. Notes on *Croomia pauciflora* (Stemonaceae). Rhodora 86: 131–137.

Whigham, D. F. 1974. An ecological life history study of *Uvularia perfoliata* L. Amer. Midl. Naturalist 91: 343–359.

Whigham, D. F. and M. McWethy. 1980. Studies on the pollination ecology of *Tipularia discolor* (Orchidaceae). Amer. J. Bot. 67: 550–555.

Whigham, D. F. and R. L. Simpson, R. L. 1982. Germination and dormancy studies of *Pontederia cordata* L. Bull. Torrey Bot. Club 109: 524–528.

White, D. J., R. V. Maher, and G. W. Argus. 1982. Atlas of the Rare Vascular Plants of Ontario. Ottawa.

White, S. D., A. C. Sanders, and M. D. Wilcox. 1996. Noteworthy collections. Madroño 43: 334–338.

Whitehouse, E. 1936. Texas Flowers in Natural Colors.... Austin.

Whiting, R. E. and P. M. Catling. 1986. Orchids of Ontario: An Illustrated Guide. Ottawa.

Whittaker, T. W. 1934. Chromosome constitution in certain monocotyledons. J. Arnold Arbor. 15: 135–143.

Whitten, W. M., N. H. Williams, and M. W. Chase. 2000. Subtribal and generic relationships of Maxillarieae (Orchidaceae) with emphasis on Stanhopeinae: Combined with molecular evidence. Amer. J. Bot. 87: 1842–1856.

Wiggers, F. H. [1780.] Primatiae Florae Holsaticae. Kiel. (Prim. Fl. Holsat.)

Wiggins, I. L. 1980. Flora of Baja California. Stanford.

Wijesinghe, D. K. and D. F. Whigham. 1997. Costs of producing clonal offspring and the effects of plant size on population dynamics of the woodland herb *Uvularia perfoliata* (Liliaceae). J. Ecol. 85: 907–919.

Wijesinghe, D. K. and D. F. Whigham. 2001. Nutrient foraging in woodland herbs: A comparison of three species of *Uvularia* (Liliaceae) with contrasting belowground morphologies. Amer. J. Bot. 88: 1071–1079.

Wilbur, R. L. 1963. A revision of the North American genus *Uvularia* (Liliaceae). Rhodora 65: 158–188.

Wild Fl. Notes = Wild Flower Notes.

Wild Orchids—See: F. Morris and E. A. Eames 1929

Wilhelm, G. S. 1984. Vascular Plants of the Pensacola Region. Ph.D. dissertation. Southern Illinois University.

Wilkin, P. 1999. A revision of the compound-leaved yams of Africa. Kew Bull. 54: 19–39.

Willdenow, C. L. 1803–1816. Hortus Berolinensis.... 2 vols. in 10 fasc. Berlin. [Fascicles and plates numbered consecutively.] (Hort. Berol.)

Willdenow, C. L., C. F. Schwägrichen, and J. H. F. Link. 1797–1830. Caroli a Linné Species Plantarum.... Editio Quarta.... 6 vols. Berlin. [Vols. 1–5(1), 1797–1810, by Willdenow; vol. 5(2), 1830, by Schwägrichen; vol. 6, 1824–1825, by Link.] (Sp. Pl.)

Williams, C. A. 1975. Biosystematics of the Monocotyledoneae—Flavonoid patterns in leaves of the Liliaceae. Biochem. Syst. & Ecol. 3: 229–244.

Williams, J. G. and A. E. Williams. 1983. Field Guide to Orchids of North America. New York.

Williams, L. O. 1965. The Orchidaceae of Mexico. Edited by Abdul Bari Awan. Tegucigalpa. [Reprinted from Ceiba 2(1–4), 1951.]

Williams, L. O., P. H. Allen, and R. L. Dressler. 1980 Orchids of Panama: A facsimile reprint of the Orchidaceae, Flora of Panama...with a checklist of the orchids of Panama as known today.... Monogr. Syst. Bot. Missouri Bot.Gard. 4.

Williams, N. H. 1972. A reconsideration of *Ada* and the glumaceous brassias. Brittonia 24: 93–110.

Williams, N. H., M. W. Chase, T. Fulcher, and W. M. Whitten. 2001. Molecular systematics of the Oncidiinae based on evidence from four DNA sequence regions: Expanded circumscriptions of *Cyrtochilum, Erycina, Otoglossum,* and *Trichocentrum* and a new genus (Orchidaceae). Lindleyana 16: 113–139.

Wilson, J. Y. 1958. Cytogenetics of triploid bluebells, *Endymion non-scripta* (L.) Garcke and *E. hispanica* (Mill.) Chouard. Cytologia 23: 435–446.

Wilson, K. L. and D. A. Morrison, eds. 2000. Monocots: Systematics and Evolution. Melbourne.

Wiss. Z. Friedrich-Schiller-Univ. Jena, Math.-Naturwiss. Reihe = Wissenschaftliche Zeitschrift der Friedrich-Schiller-Universität Jena/Thüringen. Mathematisch-naturwissenschaftliche Reihe.

Wister, J. C. 1927. The Iris: A Treatise on the History, Development and Culture of the Iris for the Amateur Gardener. New York and London.

Withner, C. L. 1980. The equitant or variegatum oncidiums, their varieties, forms, and natural hybrids. Orchid Digest 44: 84–94.

Witzman, J. A. 1991. The Biology of *Fritillaria pluriflora* (Liliaceae): A Rare Endemic of the California Flora. M.S. thesis. California State University, Chico.

Wochenschr. Vereines Beförd. Gartenbaues Königl. Preuss. Staaten = Wochenschrift des Vereines zur Beförderung des Gartenbaues in den Königlich Preussischen Staaten für Gärtnerei und Pflanzenkunde.

Wood, A. 1847. A Class-book of Botany..., ed 2. Boston and Claremont, N.H. (Class-book Bot. ed. 2)

Wood, A. 1870. The American Botanist and Florist; Including Lessons in the Structure, Life, and Growth of Plants; together with a Simple Analytical Flora, Descriptive of the Native and Cultivated Plants Growing in the Atlantic Division of the American Union.... New York and Chicago. (Amer. Bot. Fl.)

Wood, C. E. Jr. 1971. Some floristic relationships between the southern Appalachians and western North America. In: P. C. Holt, ed. 1971. The Distributional History of the Biota of the Southern Appalachians. Part 2. Flora. Blacksburg. Pp. 331–404.

Wood, C. E. Jr. 1983. The genera of Burmanniaceae in the southeastern United States. J. Arnold Arbor. 64: 293–307.

Wood, D. A. 1996. Florida's Endangered Species, Threatened Species and Species of Special Concern. Official Lists. Tallahassee.

Wood, J. J. 1989. *Pteroglossaspis.* In: W. B. Turrill et al., eds. 1952+. Flora of Tropical East Africa. 152+ vols. London and Rotterdam. Orchidaceae, part 3, pp. 480–482.

Woodcock, H. D. and W. T. Stearn. 1950. Lilies of the World: Their Cultivation and Classification. London.

Woodhead, N. 1951. Biological flora of the British Isles. *Lloydia serotina* (L.) Rehb. J. Ecol. 39: 198–203.

Woodland, D. W. 2000. Contemporary Plant Systematics, ed. 3. Berrien Springs, Mich.

Wu, Z. and P. H. Raven, eds. 1994+. Flora of China. 7+ vols. Beijing and St. Louis.

Wunderlin, R. P., B. F. Hansen, and E. L. Bridges. 1996. Atlas of Florida Vascular Plants. Tampa. [CD-ROM.]

Wunderlin, R. P., B. F. Hansen, and D. W. Hall. 1985. The vascular flora of central Florida: Taxonomic and nomenclatural changes, additional taxa. Sida 11: 232–244.

Wylie, R. B. 1917. Cleistogamy in *Heteranthera dubia.* Bull. Lab. Nat. Hist. Iowa State Univ. 7: 48–58.

Xifreda, C. C. 2000. Evaluation of pollen and vegetative characters in the systematics of South American species of *Dioscorea* (Dioscoreaceae). In: K. L. Wilson and D. A. Morrison, eds. 2000. Monocots: Systematics and Evolution. Melbourne. Pp. 488–496.

Yamashita, J. and M. N. Tamura. 2000. Molecular phylogeny of the Convallariaceae (Asparagales). In: K. L. Wilson and D. A. Morrison, eds. 2000. Monocots: Systematics and Evolution. Melbourne. Pp. 387–400.

Yatskievych, G. 1999+. Steyermark's Flora of Missouri. 1+ vol. Jefferson City.

Year Book Amer. Amaryllis Soc. = Year Book of the American Amaryllis Society.

Yen, A. C. and R. G. Olmstead. 2000. Molecular systematics of Cyperaceae tribe Cariceae based on two chloroplast DNA regions: *ndh*F and *trn*L intron-intergenic spacer. Syst. Bot. 25: 479–494.

Yen, A. C. and R. G. Olmstead. 2000b. Phylogenetic analysis of *Carex* (Cyperaceae): Generic and subgeneric relationships based on chloroplast DNA. In: K. L. Wilson and D. A. Morrison, eds. 2000. Monocots: Systematics and Evolution. Melbourne. Pp. 602–609.

Yeo, P. F. 1975. The hybrid origin of some cultivated snowdrops (*Galanthus*–Amaryllidaceae). Baileya 19: 157–162.

Young, D. A. and D. S. Seigler, eds. 1981. Phytochemistry and Angiosperm Phylogeny. New York.

Yuccas Southw. U.S.—See: S. D. McKelvey 1938–1947

Z. Pflanzenzücht. = Zeitschrift für Pflanzenzüchtung; zugleich Organ der Gesellschaft zur Förderung Deutscher Pflanzenzucht und der Österreichischen Gesellschaft für Pflanzenzuchtung.

Zahariadi, C. 1962. Caractères morphologiques, anatomiques et biologiques dans la taxonomie du genre *Ornithogalum.* Rev. Biol. (Bucharest), n. s. 7: 5–41.

Zavada, M., Xu X. L., and J. M. Edwards. 1983. On the taxonomic status of *Lophiola aurea* Ker-Gawler. Rhodora 85: 73–81.

Zhang, W. P. 2000. Phylogeny of the grass family (Poaceae) from *rpl*16 intron sequence data. Molec. Phylogen. Evol. 15: 135–146.

Zimmerman, J. H. 1958. A Monograph of *Veratrum.* Ph.D. dissertation. University of Wisconsin.

Zinn, J. G. 1757. Catalogus Plantarum Horti Academici et Agri Gottingensis.... [Göttingen.] (Cat. Pl. Hort. Gott.)

Zoë = Zoë; a Biological Journal.

Zomlefer, W. B. 1994. Guide to Flowering Plant Families. Chapel Hill.

Zomlefer, W. B. 1996. The Trilliaceae in the southeastern United States. Harvard Pap. Bot. 1(9): 91–120.

Zomlefer, W. B. 1997. The genera of Nartheciaceae in the southeastern United States. Harvard Pap. Bot. 2: 195–211.

Zomlefer, W. B. 1997b. The genera of Melanthiaceae in the southeastern United States. Harvard Pap. Bot. 2: 133–177.

Zomlefer, W. B. 1997c. The genera of Tofieldiaceae in the southeastern United States. Harvard Pap. Bot. 2: 179–194.

Zomlefer, W. B. 1998. The genera of Hemerocallidaceae in the southeastern United States. Harvard Pap. Bot. 3: 113–145.

Zomlefer, W. B. 1999. Advances in angiosperm systematics: Examples from the Liliales and Asparagales. Bull. Torrey Bot. Club 126: 58–62.

Zomlefer, W. B. and W. S. Judd. 2002. Resurrection of segregates of the polyphyletic genus *Zigadenus* s.l. (Liliales: Melanthiaceae) and resulting new combinations. Novon 12: 299–308.

Zomlefer, W. B., N. H. Williams, W. M. Whitten, and W. S. Judd. 2001. Generic circumscription and relationships in the tribe Melanthieae (Liliales, Melanthiaceae), with emphasis on *Zigadenus:* Evidence from ITS and *trn*L-F sequence data. Amer. J. Bot. 88: 1657–1669.

Index

Names in *italics* are synonyms, casually mentioned hybrids, or plants not established in the *Flora*. Page numbers in **boldface** indicate the primary entry for a taxon. Page numbers in *italics* indicate an illustration. Roman type is used for all other entries, including author names, vernacular names, and accepted scientific names for plants treated as established members of the flora.

Political Map of North America North of Mexico

Canadian Provinces

Alta.	*Alberta*
B.C.	*British Columbia*
Man.	*Manitoba*
N.B.	*New Brunswick*
Nfld. and Labr.	*Newfoundland and Labrador*
N.W.T.	*Northwest Territories*
N.S.	*Nova Scotia*
Nunavut	
Ont.	*Ontario*
P.E.I.	*Prince Edward Island*
Que.	*Quebec*
Sask.	*Saskatchewan*
Yukon	

United States

Ala.	*Alabama*
Alaska	
Ariz.	*Arizona*
Ark.	*Arkansas*
Calif.	*California*
Colo.	*Colorado*
Conn.	*Connecticut*
Del.	*Delaware*
D.C.	*District of Columbia*
Fla.	*Florida*
Ga.	*Georgia*
Idaho	
Ill.	*Illinois*
Ind.	*Indiana*
Iowa	
Kans.	*Kansas*
Ky.	*Kentucky*
La.	*Louisiana*
Maine	
Md.	*Maryland*
Mass.	*Massachusetts*
Mich.	*Michigan*
Minn.	*Minnesota*
Miss.	*Mississippi*
Mo.	*Missouri*
Mont.	*Montana*
Nebr.	*Nebraska*
Nev.	*Nevada*
N.H.	*New Hampshire*
N.J.	*New Jersey*
N.Mex.	*New Mexico*
N.Y.	*New York*
N.C.	*North Carolina*
N.Dak.	*North Dakota*
Ohio	
Okla.	*Oklahoma*
Oreg.	*Oregon*
Pa.	*Pennsylvania*
R.I.	*Rhode Island*
S.C.	*South Carolina*
S.Dak.	*South Dakota*
Tenn.	*Tennessee*
Tex.	*Texas*
Utah	
Vt.	*Vermont*
Va.	*Virginia*
Wash.	*Washington*
W.Va.	*West Virginia*
Wis.	*Wisconsin*
Wyo.	*Wyoming*

Four factors have been recognized as useful in determining if a non-custodial encounter has escalated into a custodial encounter:

- Whether probable cause to arrest exists
- Subjective intent of the police to hold the suspect
- Subjective belief of a defendant as to the status of his freedom
- Whether or not the focus of the investigation has finally centered on the defendant

State v. Stevenson, 958 S.W.2d 824 (Tex. Crim. App. 1997).

These factors are relevant only to the extent they are manifested to the suspect through the words and actions of law enforcement officials; the custody determination is based entirely upon objective circumstances. Stevenson.

A noncustodial situation is not converted to one in which Miranda applies simply because even in the absence of any formal arrest or restraint on freedom of movement, the questioning took place in a coercive environment. Estrada v. State, 313 S.W.3d 274, 294 (Tex. Crim. App. 2010).

A suspect's age or experience is not relevant to the Miranda custody analysis. Yarborough v. Alvarado, 541 U.S. 652, 124 S.Ct. 2140, 158 L.Ed.3d 938 (2004).

However, where it was known to the interviewing officer at the time of police questioning that the interviewee was a juvenile, or would have been objectively apparent to a reasonable officer, the inclusion in the Miranda custody analysis of that fact is consistent with the objective nature of the custody test. While a child's age will not be a determinative, or even a significant, factor in every case, in some cases, it will prove an important factor in deciding if the child was in custody for Miranda purposes. J.D.B. v. North Carolina, 564 U.S. ___, 131 S.Ct. 2394, 2406, 180 L.Ed.2d 310 (2011).

CCP Art. 15.22 states that "[a] person is arrested when he has been actually placed under restraint or taken into custody by an officer or person executing a warrant of arrest, or by an officer or person arresting without a warrant." *See* Dancy v. State, 728 S.W.2d 772 (Tex. Crim. App. 1987).

The handcuffing of a suspect does not always mean that he is under arrest. Rhodes v. State, 945 S.W.2d 115 (Tex. Crim. App. 1997) (officer testified that, during a car chase incident, he handcuffed the defendant for his safety and the safety of other officers and, during this time, the defendant dropped to the ground the drugs which formed the basis of the prosecution). *See also*, Mays v. State, 726 S.W.2d 937 (Tex. Crim. App. 1986).

Incarceration does not always constitute "custody" for Miranda purposes when an inmate is questioned by law enforcement officials regarding an offense separate and distinct from the offense for which he was incarcerated. Herrera v. State, 241 S.W.3d 520 (Tex. Crim. App. 2007) (where the court found that the defendant who was incarcerated on an unrelated warrant was not in custody for purposes of interrogation on a separate offense).

In evaluating whether an incarcerated person is in "custody" for Miranda purposes, the following circumstances surrounding the interrogation should be considered: (1) the language used to summon the inmate; (2) the physical surroundings of the interrogation; (3) the extent to which the inmate is confronted with evidence of his or her guilt; (4) the additional pressure exerted to detain the inmate or the change in the surroundings of the inmate which results in an added imposition on the inmate's freedom of movement; and (5) the inmate's freedom to leave the scene and the purpose, place, and length of the questioning. Herrera.

When a prisoner is questioned, the determination of custody should focus on all of the features of the interrogation. Where the facts of the case are consistent with an environment in which a reasonable person would have felt free to terminate the interview and leave, a person serving time in a prison is not in custody for Miranda purposes. Howes v. Fields, 565 U.S. ___, 132 S.Ct. 1181, 182 L.Ed.2d 17 (2012).

§6:21.4 Subjective Intent of Police

The subjective intent of the police and the subjective belief of the defendant become a part of the analysis if the subjective intent of the police has been manifested in their words or actions. Dowthitt v. State, 931 S.W.2d 244 (Tex. Crim. App. 1996), citing Stansbury v. California, 511 U.S. 318, 114 S.Ct. 1526, 128 L.Ed.2d 293 (1994).

A police officer's unarticulated plan to take a motorist into custody following a traffic stop has no bearing on whether the motorist was in custody at a particular time. The relevant inquiry is how a reasonable man in the motorist's place would have understood the situation. Berkemer v. McCarty, 468 U.S. 420, 104 S.Ct. 3138, 82 L.Ed.2d 317 (1984).

An officer's unarticulated plan and subjective beliefs about what may have occurred have no bearing on whether or not appellant was in custody at a particular time. Ard v. State, 418 S.W.3d 256 (Tex. App.—Houston [14th Dist.] 2013, *no pet.*).

§6:21.5 Restriction on Movement

A person is in custody only if, under the circumstances, a reasonable person would believe that his or her freedom of movement was restrained to the degree associated with a formal arrest. Dowthitt, *supra*, citing Stansbury v. California, 511 U.S. 318, 114 S.Ct. 1526, 128 L.Ed.2d 293 (1994). The "reasonable person" standard presupposes an innocent person. Dowthitt, citing Florida v. Bostick, 501 U.S. 429, 111 S.Ct. 2382, 115 L.Ed.2d 389 (1991).

The restriction on the suspect's freedom of movement must amount to the degree associated with an arrest, as opposed to an investigative detention. Dowthitt, citing Stansbury.

The defendant was in custody for Miranda purposes where police officers blocked her car into a car wash stall and pointed a gun at her during questioning. Taylor v. State, 874 S.W.2d 362 (Tex. App.—Fort Worth 1994, *no pet.*).

§6:21.6 Probable Cause

Where the defendant was questioned for approximately 12 hours, during which time he admitted being present during the murders in question (which gave the police probable cause to arrest him), and during which time the police exercised control over him (accompanying him to the restroom and not allowing him to see his wife), the defendant was found to be in custody. Dowthitt v. State, 931 S.W.2d 244 (Tex. Crim. App. 1996).

If the defendant is trying to show that he was in custody because officers had probable cause to arrest him, he must show that the officers manifested that probable cause to him. Such manifestation can occur if the officers relate information substantiating probable cause to the suspect or the suspect relates such information to the officers. Furthermore, custody is established only if the manifestation of probable cause is combined with other circumstances which would lead a reasonable person to believe that he is under restraint to the degree associated with an arrest. Dowthitt, citing Stansbury v. California, 511 U.S. 318, 114 S.Ct. 1526, 128 L.Ed.2d 293 (1994).

§6:21.7 Location of Questioning

The fact that the questioning of the defendant took place at the police station does not in and of itself establish that the defendant was in custody. *See* Oregon v. Mathiason, 429 U.S. 492, 97 S.Ct. 711, 50 L.Ed.2d 714 (1977) (suspect came in voluntarily, was told he was not under arrest, participated in a 30-minute interview, and then left); Meek v. State, 790 S.W.2d 618 (Tex. Crim. App. 1990) (suspect came in voluntarily, was allowed to go to his car unhindered during the interview, and then left after the statements were taken); and Dancy v. State, 728 S.W.2d 772 (Tex. Crim. App. 1987) (suspect voluntarily accompanied officers to the station, answered questions, gave hair samples, and allowed officers to test his shoes, and arrest was not made until a later point).

Cases decided under Miranda have made clear that a "compelling atmosphere"—the phenomenon that marks a "custodial interrogation" and against which warnings are required—may develop beyond the confines of the station house. Orozco v. Texas, 394 U.S. 324, 89 S.Ct. 1095, 22 L.Ed.2d 311 (1969).

§6:21.8 Focus of Investigation

The fact that an investigation has focused on the defendant does not trigger the need for Miranda warnings in a non-custodial setting. Minnesota v. Murphy, 465 U.S. 420, 104 S.Ct. 1136, 79 L.Ed.2d 409 (1984).

The mere fact that a suspect has become the "focus" of the investigation, without more, does not convert a roadside stop into an arrest. State

v. Stevenson, 958 S.W.2d 824 (Tex. Crim. App. 1997). In order to amount to an arrest, enforcement officials must manifest to the suspect that he is the focus of the investigation. Stevenson.

Whether the defendant is the "focus of the investigation" for Miranda purposes involves questioning by law enforcement officers after the defendant has been taken into custody or otherwise deprived of his freedom of action in any significant way. Beckwith v. United States, 425 U.S. 341, 96 S.Ct. 1612, 48 L.Ed.2d 1 (1976).

§6:21.9 Traffic Stops

Generally, a routine traffic stop does not place a person in custody for Miranda purposes. But a traffic stop may escalate from a non-custodial detention into a custodial detention when formal arrest ensues or a detainee's freedom of movement is restrained to the degree associated with a formal arrest. Appellate courts evaluate whether a person has been detained to the degree associated with arrest on a case-by-case basis. In making the custody determination, the primary question is whether a reasonable person would perceive the detention to be a restraint on his movement comparable to formal arrest, given all the objective circumstances. State v. Ortiz, 382 S.W.3d 367, 372 (Tex. Crim. App. 2012).

In evaluating whether a reasonable person would believe his freedom has been restrained to the degree of formal arrest, the appellate courts look only to the objective factors surrounding the detention. The subjective beliefs of the detaining officer are not included in the calculation of whether a suspect is in custody. But if the officer manifests his belief to the detainee that he is a suspect, then that officer's subjective belief becomes relevant to the determination of whether a reasonable person in the detainee's position would believe he is in custody. Conversely, any undisclosed subjective belief of the suspect that he is guilty of an offense should not be taken into consideration—the reasonable person standard presupposes an innocent person. State v. Ortiz, 382 S.W.3d at 372-3.

§6:22 Admissible Custodial Statements

§6:22.1 Statements to State Caseworkers

Where a defendant is in custody of law enforcement but makes incriminating statements to civil child abuse caseworkers who are not acting as agents of law enforcement or pursuant to law enforcement practices, the statements are not the product of custodial interrogation and are admissible. Cantu v. State, 817 S.W.2d 74 (Tex. Crim. App. 1991); Paez v. State, 681 S.W.2d 34 (Tex. Crim. App. 1984); Berry v. State, 233 S.W.3d 847 (Tex. Crim. App. 2007).

However, where the investigator is acting for the express purpose of investigating allegations of child abuse and then turning the product over to the proper authority for prosecution, a custodial interview of the defendant will be found to be custodial interrogation. Cates v. State, 776 S.W.2d 170 (Tex. Crim. App. 1989).

Where a suspect is not in custody at the time he makes a statement to a caseworker, the resulting statement is not a product of custodial interrogation and is admissible even in the absence of Miranda compliance. Wicker v. State, 740 S.W.2d 779 (Tex. Crim. App. 1987).

§6:22.2 Questioning by Civilians

Custodial statements made to news reporters who are not acting as state agents are admissible at the defendant's subsequent trial. Escamilla v. State, 143 S.W.3d 814 (Tex. Crim. App. 2004).

§6:22.3 Spontaneous Oral Statements

Where a defendant in custody makes a spontaneous admission during the reading of Miranda rights to effect that he killed a jailer, the statement is admissible at trial. Hernandez v. State, 819 S.W.2d 806 (Tex. Crim. App. 1991).

Where a defendant in custody following the giving of Miranda warnings, observes his brother also in custody and states, "He didn't do it; I did it," the statement is admissible because it is not the product of interrogation. Richards v. State, 524 S.W.2d 67 (Tex. Crim. App. 1975).

Where a suspect made the unsolicited remark, "[y]ou have the right one; I am Pete," to arresting officers, the remark was admissible because it was not the product of custodial interrogation. Sosa v. State, 769 S.W.2d 909 (Tex. Crim. App. 1989).

Statements beyond the scope of an officer's custodial interrogation are admissible where the circumstances demonstrate that the officer's question was not leading or suggestive, that the suspect was excited or emotionally stimulated, and that the suspect had been informed of his Miranda rights. Etheridge v. State, 903 S.W.2d 1 (Tex. Crim. App. 1994) (where the suspect proceeded to admit his guilt after the officer asked if he knew why he was being arrested).

Where a defendant (who was in custody) made an admission in an exchange with the judge initiated by himself in the noncoercive atmosphere of open court with his lawyers present, it is not considered the result of "interrogation" in any sense of how that term is described in Miranda. Williams v. State, 270 S.W.3d 112 (Tex. Crim. App. 2008).

§6:22.4 General and Routine Questions

An inquiry of a driving-while-intoxicated suspect being transported by police to a police station as to whether or not he would take a breath test is not interrogation for Miranda purposes. Shepherd v. State, 915 S.W.2d 177 (Tex.App.—Fort Worth 1996, *pet. ref'd*).

General on-the-scene questioning as to the facts surrounding a crime or other general questioning of citizens in a fact-finding process is not affected by Miranda rules. Cannon v. State, 691 S.W.2d 664 (Tex. Crim. App. 1985).

A police officer's questions asked for the purpose of obtaining routine information such as name, address or social security number does not constitute custodial interrogation. Garner v. State, 779 S.W.2d 498 (Tex. App.—Fort Worth 1989, *pet. ref'd*).

Requiring a suspect to reveal his name during a valid Terry stop does not violate his Fifth Amendment rights where the disclosure does not implicate any articulable real and appreciable fear on the suspect's part that his name would be used to incriminate him or that it would furnish a link in the chain of evidence needed to prosecute him. Hiibel v. Sixth Judicial Dist. Court of Nev., Humboldt Cty., 542 U.S. 177, 124 S.Ct. 2451, 159 L.Ed.2d 292 (2004).

§6:22.5 Booking Questions

There is a routine booking question exception to Miranda that exempts from Miranda's coverage questions to secure the biographical data necessary to complete booking or pretrial services. Alford v. State, 358 S.W.3d 647 (Tex. Crim. App. 2012), *citing* Pennsylvania v. Muniz, 496 U.S. 582, 600-02, 110 S.Ct. 2638, 110 L.Ed.2d 528 (1990) (Brennan, J., *plurality op.*).

Questions that are asked for record-keeping purposes only and which are reasonably related to the police's administrative concerns fall outside the protections of Miranda. Alford; Pennsylvania v. Muniz.

Although termed a "booking" exception, this exception appears to be somewhat broader than booking and encompasses routine administrative processing associated with arrest or custody. The point of the exception is that questions that fall within the exception serve a legitimate administrative need. State v. Cruz, 461 S.W.3d 531, 537 (Tex. Crim. App. 2015).

To determine whether a particular question falls within the booking exception, courts ask whether the question reasonably relates to a legitimate administrative concern, applying an objective standard. State v. Cruz, 461 S.W.3d at 537.

If a question is a legitimate administrative question, then it does not matter whether the officer should know that the question is reasonably likely to elicit an incriminating response. That is, a legitimate administrative question is exempt from the "should know" test and is deemed not to be interrogation. State v. Cruz, 461 S.W.3d at 537.

Whether a question reasonably relates to an administrative concern must be ascertained by both the content of the question and the circumstances in which the question is asked. State v. Cruz, 461 S.W.3d at 540.

In deciding whether the questions are legitimate booking questions or a pretext for eliciting incriminating information the following factors can be considered:

- the nature of the questions, such as whether they seek merely identifying data necessary for booking;
- the context of the interrogation, such as whether the questions were asked during a non-investigative clerical booking process and pursuant to a standard booking form or questionnaire;
- the knowledge and intent of the government agent asking the questions;
- the relationship between the question asked and the crime the defendant was suspected of committing;
- the administrative need for the information sought;
- and any other indications that the questions were designed, at least in part, to elicit incriminating evidence and merely asked under the guise or pretext of seeking routine biographical information. State v. Cruz, 461 S.W.3d at 538.

A question that seeks to elicit biographical data may be deemed not interrogation under either of two theories. First, such a question may be deemed not interrogation because it does not meet the general test for interrogation, i.e., it does not meet the "should know" test. That is, a court could decide that a particular question about biographical data was so innocuous that its tendency to produce an incriminating response was not something the police officer should have known. Second a biographical question may be deemed not interrogation under the booking (or administrative) exception. That is, a court could decide that a particular question about biographical data was a routine administrative inquiry. State v. Cruz, 461 S.W.3d at 537-8.

The biographical nature of a question does not alone exempt it from being interrogation because such questions may constitute interrogation under the "should know" test. State v. Cruz, 461 S.W.3d at 538-9.

Questioning that seeks to elicit biographical data does not become interrogation simply because the identifying information may later aid law enforcement in conducting a criminal investigation. State v. Cruz, 461 S.W.3d at 539.

However, the elicitation of biographical data that has incriminating value in and of itself and which does not simply lead to other incriminating evidence can constitute interrogation. State v. Cruz, 461 S.W.3d at 539 (in this case, the suspect's name could be considered incriminating because he had previously given a false name; his cell phone number could be considered incriminating because it ultimately put him near the scene of the offense).

Routine booking questions are, by definition, questions normally attendant to arrest and custody and do not, by their very nature, involve the psychological intimidation that Miranda is designed to prevent. At the same time, they serve a legitimate administrative need. Alford, citing United States v. Doe, 878 F.2d 1546, 1551 (1st Cir. 1989).

Questions directed to the name, address, height, weight, eye color, date of birth, and current age of the arrestee are Miranda-exempt because these questions are reasonably related to the police's administrative concerns. Alford; Pennsylvania v. Muniz.

Questions regarding a suspect's name, address, weight, height, place of employment, and physical disabilities are normally attendant to arrest and custody. Townsend v. State, 813 S.W.2d 181, 186 (Tex.App.—Houston [14th Dist.] 1991, *pet. ref'd*)

Conversely, questions that do not reasonably relate to a legitimate administrative concern are not booking questions within the exception. Alford, citing Branch v. State, 932 S.W.2d 577, 581 (Tex. App.—Tyler 1995, *no pet.*) (asking defendant "where he had been going at the time he was pulled over, when and what he had last eaten, and whether he had drunk an alcoholic beverage... went beyond the scope of that normally attending arrest and custody and amounted to custodial interrogation").

In deciding the admissibility of a statement under the exception, a trial court must examine whether, under the totality of the circumstances, a question is reasonably related to a legitimate administrative concern. In deciding this, the trial court applies an objective standard. An appellate court generally reviews de novo the objective reasonableness of a question's stated administrative purpose, but defers to the trial court's resolution of disputed facts. Alford.

§6:22.6 Overheard Statements

Because society is not prepared to recognize a legitimate expectation of privacy in conversations between arrestees who are in custody in a county law enforcement building, statements made by the arrestees that are tape-recorded without their knowledge by law enforcement officers are admissible in the absence of any deliberate misrepresentations by law enforcement. State v. Scheineman, 77 S.W.3d 810 (Tex. Crim. App. 2002).

The fact that a police officer was stationed in an interview room when a recently detained murder defendant confessed the murder to his wife did not elevate the situation to the functional equivalent of interrogation where the purpose for the presence of the officer was to protect the wife, even though the officers were aware of the possibility that the defendant would confess. Arizona v. Mauro, 481 U.S. 520, 107 S.Ct. 1931, 95 L.Ed.2d 458 (1987).

The contents of a telephone conversation between a defendant in custody and his mother are admissible because they are not the product of custodial interrogation. Autry v. State, 626 S.W.2d 758 (Tex. Crim. App. 1982).

§6:22.7 Public Safety Exception

The Miranda warnings are not necessary even if the suspect is in custody where the interrogation is necessary to ensure the safety of the public or the officer. New York v. Quarles, 467 U.S. 649, 104 S.Ct. 2626, 81 L.Ed.2d 550 (1984), Bryant v. State, 816 S.W.2d 554 (Tex.App.—Fort Worth 1991, *no pet.*).

§6:22.8 Custodial Statements for Impeachment Purposes

CCP Art. 38.22 §5 allows the admission of "a *voluntary* oral statement, whether or not the result of custodial interrogation that has a bearing upon the credibility of the accused as a witness." *See* Roberts v. State, 672 S.W.2d 570 (Tex.App.—Fort Worth 1984); Girndt v. State, 623 S.W.2d 930 (Tex. Crim. App. 1981); Appling v. State, 904 S.W.2d 912 (Tex.App.—Corpus Christi 1995, *pet. ref'd*).

The use of oral statements for impeachment under CCP Art. 38.22 §5 does not require the statement to have been recorded in accordance with §3. Jones v. State, 859 S.W.2d 537 (Tex.App.—Houston [1st Dist.] 1993, *pet. ref'd*).

TIP TO THE PROSECUTION: Always interview police officers who had contact with the defendant to learn of any statements made by the defendant that were not recorded in writing or on audiotape. These statements can be admitted to impeach the defendant's trial testimony.

TIP TO THE DEFENSE: If the state seeks to offer a custodial statement of the defendant as impeachment evidence, request a hearing outside the presence of the jury to determine its voluntariness.

Prior inconsistent custodial statements can be used to impeach the testimony of an accused even though the statements were inadmissible in the state's case-in-chief because of a failure to comply with Miranda as long as the evidence satisfies legal standards for trustworthiness. Harris v. New York, 401 U.S. 222, 91 S.Ct. 643, 28 L.Ed.2d 1 (1971); Lykins v. State, 784 S.W.2d 32 (Tex. Crim. App. 1989); Baker v. State, 956 S.W.2d 19 (Tex. Crim. App. 1997).

Likewise, the Sixth Amendment does not forbid the use of statements taken in violation where they are offered for impeachment purposes only. This is because the constitutional violation occurs at the time the statement is elicited, not when it is offered at trial for impeachment purposes. Kansas v. Ventris, 556 U.S. ___, 129 S.Ct. 1841, 173 L.Ed.2d 801 (2009).

Miranda and the Fifth Amendment do not prohibit impeachment of a testifying defendant with statements he made in custody after the administration of the warnings. Anderson v. Charles, 447 U.S. 404, 100 S.Ct. 2180, 65 L.Ed.2d 222 (1980).

The failure to administer the Miranda warnings creates a presumption of compulsion and consequently, unwarned statements that are otherwise voluntary within the meaning of the Fifth Amendment must be excluded from evidence under Miranda; but the presumption, though not rebuttable for purposes of the state's case-in-chief, does not require that the statements and their fruits be discarded as inherently tainted. Oregon v. Elstad, 470 U.S. 298, 105 S.Ct. 1285, 84 L.Ed.2d 222 (1985).

However, even though U.S. constitutional grounds may not prohibit the use of an illegally obtained confession, its use may be prohibited under CCP Art. 38.23 which prohibits the use of evidence obtained in violation of the laws of the State of Texas. Robbins v. State, 696 S.W.2d 689 (Tex.App.—Dallas 1985).

Statements made by a defendant during plea bargain negotiations are not later admissible against him for impeachment purposes. Wayne v. State, 756 S.W.2d 724 (Tex. Crim. App. 1988).

Even if the confession is used for impeachment, failure to make a determination of voluntariness outside the presence of the jury, when requested, is reversible error. Madden v. State, 691 S.W.2d 688 (Tex. Crim. App. 1985).

If the prosecution violates a defendant's Fifth Amendment rights by introducing an inadmissible confession at trial compelling the defendant to testify in rebuttal, the use of the compelled testimony is prohibited at a retrial. Harrison v. U.S., 392 U.S. 219, 88 S.Ct. 2008, 20 L.Ed.2d 1047 (1968).

§6:23 Interrogation in Non-Custodial Setting

Where a defendant is not under arrest or in custody, his voluntary statement made without threat, promise, or persuasion, is admissible in evidence against him. Miranda v. Arizona, 384 U.S. 436, 86 S.Ct. 1602, 16 L.Ed.2d 694 (1966); CCP Art. 38.22 §5.

The Texas statutory warning is not a prerequisite to the admission of a voluntary statement that does not stem from custodial interrogation. CCP Art. 38.22 §5.

The U.S. Supreme Court has held that the Miranda warning is not a pre-requisite to admissibility of a confession or statement made while not under custodial interrogation. Beckwith v. U.S., 425 U.S. 341, 96 S.Ct. 1612, 48 L.Ed.2d 1 (1976).

Even though a non-custodial statement will not be excluded for a failure to comply with the prophylactic warning requirements of Miranda and CCP Art. 38.22, the procedures surrounding the giving of the statement are still subject to due process and due course of law scrutiny. Wolfe v. State, 917 S.W.2d 270 (Tex. Crim. App. 1996).

Miranda is not implicated where a suspect who is not under arrest voluntarily appears at a police station and is allowed to leave unhindered after a brief interview. California v. Beheler, 463 U.S. 1121, 103 S.Ct. 3517, 77 L.Ed.2d 1275 (1983).

When a defendant is not in custody, law enforcement officers are not required to scrupulously honor his invocation of rights, even where the defendant has been read his rights. Estrada v. State, 313 S.W.3d 274 (Tex. Crim. App. 2010).

Where the initial detention of a defendant is lawful and not unreasonably prolonged, no additional Fourth Amendment justification is necessary for law enforcement officers to ask questions of the defendant. Muehler v. Mena, 544 U.S. ___, 125 S.Ct. 1465, 161 L.Ed.2d 299 (2005).

§6:24 Admissible Non-Custodial Statements

§6:24.1 Statements to Probation Officers

The Fifth and Fourteenth Amendments do not prohibit the introduction of evidence of a defendant's admissions to a probation officer made during a non-custodial monthly report. A probationer's responsibility to provide truthful answers to the probation officer's questions does not convert voluntary statements into compelled ones. Minnesota v. Murphy, 465 U.S. 420, 104 S.Ct. 1136, 79 L.Ed.2d 409 (1984).

The fact that a probationer is ordered as a condition of supervision to submit to a polygraph examination does not elevate the polygraph interview to a custodial situation. Marcum v. State, 983 S.W.2d 762 (Tex.App.—Houston [14th Dist.] 1998, *pet. ref'd*).

Where the defendant has no outstanding warrants and no probable cause to arrest him exists, admissions made to his probation officer during monthly reports are admissible. Payne v. State,

579 S.W.2d 932 (Tex. Crim. App. 1979); Holmes v. State, 752 S.W.2d 700 (Tex.App.—Waco 1988, *no pet.*).

However, where the defendant is being detained by the probation officer prior to his arrest on a probation revocation warrant, any statement made to the probation officer which is the product of interrogation should be suppressed. Creeks v. State, 542 S.W.2d 849 (Tex. Crim. App. 1976).

§6:24.2 Statements to State Caseworkers

An interview conducted by a state social caseworker investigating whether or not to instigate proceedings to terminate the parent-child relationship is not custodial interrogation. Davis v. State, 687 S.W.2d 78 (Tex. App.—Dallas 1985, *pet. ref'd*); Wicker v. State, 740 S.W.2d 779 (Tex. Crim. App. 1987).

However, where the caseworker is acting as an agent of law enforcement, the Miranda warnings are a necessary predicate to any interview of the defendant. Wilkerson v. State, 173 S.W.3d 521 (Tex. Crim. App. 2005).

§6:24.3 Traffic Stops

Where a person detained on a routine traffic stop is not subjected to restraints comparable to those associated with a formal arrest, he is not in custody for Miranda purposes. Therefore, questions asked by police during the traffic stop are not custodial interrogation. Berkemer v. McCarty, 468 U.S. 420, 104 S.Ct. 3138, 82 L.Ed.2d 317 (1984). *See also,* State v. Stevenson, 958 S.W.2d 824 (Tex. Crim. App. 1997).

The request of a motorist to perform a roadside balancing test at a location visible to passing motorists, without more, does not elevate the traffic stop to a custodial interrogation. Pennsylvania v. Bruder, 488 U.S. 9, 109 S.Ct. 205, 102 L.Ed.2d 172 (1988).

§6:25 Agency Relationship

§6:25.1 Statements to Inmates

Where law enforcement agents use a civilian to deliberately elicit incriminating statements from a defendant, the defendant's Sixth Amendment right to counsel has been violated. Massiah v. U.S., 377 U.S. 201, 84 S.Ct. 1199, 12 L.Ed.2d 246 (1964).

Three factors are examined to determine if the defendant's statement was deliberately elicited: (1) whether the inmate was acting under instructions as a paid informant for law enforcement; (2) whether the inmate was ostensibly no more than a fellow inmate of the defendant; and (3) whether the defendant was in custody and under indictment at the time he was engaged in conversation by the fellow inmate. U.S. v. Henry, 447 U.S. 264, 100 S.Ct. 2183, 65 L.Ed.2d 115 (1980).

Deliberate elicitation simply involves some conduct designed to obtain incriminating state-ments. Such conduct has not occurred if the informant is merely a passive listening post who relates any statements the defendant makes of his own volition. Rubalcado v. State, 424 S.W.3d 560, 576 (Tex. Crim. App. 2014).

The person claiming that there is an agency relationship between the interrogator and police has the burden of proving it. Wilkerson v. State, 173 S.W.3d 521, 529 (Tex. Crim. App. 2005).

The law does not presume an agency relationship between foreign and U.S. police. Alvarado v. State, 853 S.W.2d 17 (Tex. Crim. App. 1991). The agency relationship may be express or implied from the actions of the parties. Alvarado.

Where there is no evidence that U.S. police had any involvement with the apprehension or interrogation of the suspect other than informing foreign authorities that the suspect had crossed the border, the agency relationship was not shown to have existed. Alvarado.

A prisoner's confession to a fellow inmate who was acting as a government agent and where the confession is made as a result of a promise by the agent to protect the prisoner/defendant from fellow inmates is deemed to be coerced, and therefore inadmissible under the Fourteenth Amendment. Arizona v. Fulminante, 499 U.S. 279, 111 S.Ct. 1246, 113 L.Ed.2d 302 (1991).

However, any information obtained by the inmate before he enters into an agreement with the State is free of any Sixth Amendment concerns. Wesbrook v. State, 29 S.W.3d 103, (Tex. Crim. App. 2000) (*plurality op.*); Thompson v. State, 93 S.W.3d 16 (Tex. Crim. App. 2001).

Where an inmate who was a government agent did not take any action that would have constituted the functional equivalent of interrogation, there is no Sixth Amendment violation. Kuhlmann v. Wilson, 477 U.S. 436, 106 S.Ct. 2616, 91 L.Ed.2d 364 (1986).

Likewise, there is no Sixth Amendment violation where the inmate was not found to be a government agent. Manns v. State, 122 S.W.3d 171 (Tex. Crim. App. 2003).

A letter from a defendant in jail to someone outside of jail is not custodial interrogation; therefore, a letter admitting a crime is admissible. Hightower v. State, 629 S.W.2d 920 (Tex. Crim. App. 1982).

Evidence obtained through the use of a government agent can be found to be harmless error at trial. Wesbrook.

§6:25.2 Statements to Private Citizens

The person claiming that there is an agency relationship between a private citizen and law enforcement has the burden of proving it. Wilkerson v. State, 173 S.W.3d 521, 529 (Tex. Crim. App. 2005).

To determine whether an agency relationship exists between a private citizen and law enforcement, courts will look to the following factors: (1) the relationship between the private citizen and law enforcement; (2) the purpose of the citizen's interview with the defendant; and (3) whether a reasonable person in the defendant's position could have believed that the interviewer was a law enforcement agent. Elizondo v. State, 382 S.W.3d 389 (Tex. Crim. App. 2012) (where the court found that a loss-prevention officer was not an agent of law enforcement).

§6:30 Magistration and Sufficiency of Miranda Warnings

§6:31 Statutory Law

CCPArt. 2.09. Who Are Magistrates

> *Each of the following officers is a magistrate within the meaning of this Code: The justices of the Supreme Court, the judges of the Court of Criminal Appeals, the justices of the Courts of Appeals, the judges of the District Court, the magistrates appointed by the judges of the district courts of Bexar County, Dallas County, or Tarrant County that give preference to criminal cases, the criminal law hearing officers for Harris County appointed under Subchapter L, Chapter 54, Government Code, the criminal law hearing officers for Cameron County appointed under Subchapter BB, Chapter 54, Government Code, the magistrates or associate judges appointed by the judges of the district courts of Lubbock County, Nolan County, or Webb County, the magistrates appointed by the judges of the criminal district courts of Dallas County or Tarrant County, the associate judges appointed by the judges of the district courts and the county courts at law that give preference to criminal cases in Jefferson County, the associate judges appointed by the judges of the district courts and the statutory county courts of Brazos County, Nueces County, or Williamson County, the magistrates appointed by the judges of the district courts and statutory county courts that give preference to criminal cases in Travis County, the criminal magistrates appointed by the Brazoria County Commissioners Court, the criminal magistrates appointed by the Burnet County Commissioners Court, the criminal magistrates appointed by the Burnet County Commissioners Court, the county judges, the judges of the county courts at law, judges of the county criminal courts, the judges of statutory probate courts, the associate judges appointed by the judges of the statutory probate courts under Chapter 54A , Government Code, the associate judges appointed by the judge of a district court under Chapter 54A, Government Code, the magistrates appointed under Subchapter JJ, Chapter 54, Government Code, as added by H.B. No. 2132, Acts of the 82nd Legislature, Regular Session, 2011, the justices of the peace, and the mayors and recorders and the judges of the municipal courts of incorporated cities or towns.*

CCP Art. 14.06. Must Take Offender Before Magistrate

(a) Except as otherwise provided by this article, in each case enumerated in this Code, the person making the arrest or the person having custody of the person arrested shall take the person arrested or have him taken without unnecessary delay, but not later than 48 hours after the person is arrested, before the magistrate who may have ordered the arrest, before some magistrate of the county where the arrest was made without an order, or, to provide more expeditiously to the person arrested the warnings described by Article 15.17 of this Code, before a magistrate in any other county of this state. The magistrate shall immediately perform the duties described in Article 15.17 of this Code.

(b) A peace officer who is charging a person, including a child, with committing an offense that is a Class C misdemeanor, other than an offense under Section 49.02, Penal Code, may, instead of taking the person before a magistrate, issue a citation to the person that contains written notice of the time and place the person must appear before a magistrate, the name and address of the person charged, the offense charged, and the following admonishment, in boldfaced or underlined type or in capital letters:

"If you are convicted of a misdemeanor offense involving violence where you are or were a spouse, intimate partner, parent, or guardian of the victim or are or were involved in another, similar relationship with the victim, it may be unlawful for you to possess or purchase a firearm, including a handgun or long gun, or ammunition, pursuant to federal law under 18 U.S.C. Section 922(g)(9) or Section 46.04(b), Texas Penal Code. If you have any questions whether these laws make it illegal for you to possess or purchase a firearm, you should consult an attorney."

(c) If the person resides in the county where the offense occurred, a peace officer who is charging a person with committing an offense that is a Class A or B misdemeanor may, instead of taking the person before a magistrate, issue a citation to the person that contains written notice of the time and place the person must appear before a magistrate of this state as described by Subsection (a), the name and address of the person charged, and the offense charged.

(d) Subsection (c) applies only to a person charged with committing an offense under:

(1) Section 481.121, Health and Safety Code, if the offense is punishable under Subsection (b)(1) or (2) of that section;

(1-a) Section 481.1161, Health and Safety Code, if the offense is punishable under Subsection (b)(1) or (2) of that section;

(2) Section 28.03, Penal Code, if the offense is punishable under Subsection (b)(2) of that section;

(3) Section 28.08, Penal Code, if the offense is punishable under Subsection (b)(1) of that section;

(4) Section 31.03, Penal Code, if the offense is punishable under Subsection (e)(2)(A) of that section;

(5) Section 31.04, Penal Code, if the offense is punishable under Subsection (e)(2) of that section;

(6) Section 38.114, Penal Code, if the offense is punishable as a Class B misdemeanor; or

(7) Section 521.457, Transportation Code.

CCP Art. 15.17. Duties of Arresting Officer and Magistrate

(a) In each case enumerated in this Code, the person making the arrest or the person having custody of the person arrested shall without unnecessary delay, but not later than 48 hours after the person is arrested, take the person arrested or have him taken before some magistrate of the county where the accused was arrested or, to provide more expeditiously to the person arrested the warnings described by this article, before a magistrate in any other county of this state. The arrested person may be taken before the magistrate in person or the image of the arrested person may be presented to the magistrate by means of an electronic broadcast system. The magistrate shall inform in clear language the person arrested, either in person or through the electronic broadcast system, of the accusation against him and of any affidavit filed therewith, of his right to retain counsel, of his right to remain silent, of his right to have an attorney present during any interview with peace officers or attorneys representing the state, of his right to terminate the interview at any time, and of his right to have an examining trial. The magistrate shall also inform the person arrested of the person's right to request the appointment of counsel if the person cannot afford counsel. The magistrate shall inform the person arrested of the procedures for requesting appointment of counsel. If the person does not speak and understand the English language or is deaf, the magistrate shall inform the person in a manner consistent with Articles 38.30 and 38.31, as appropriate. The magistrate shall ensure that reasonable assistance in completing the necessary forms for requesting appointment of counsel is provided to the person at the same time. If the person arrested is indigent and requests appointment of counsel and if the magistrate is authorized under Article 26.04 to appoint counsel for indigent defendants in the county, the magistrate shall appoint counsel in accordance with Article 1.051. If the magistrate is not authorized to appoint counsel, the magistrate shall without unnecessary delay, but not later than 24 hours after the person arrested requests appointment of counsel, transmit, or cause to be transmitted to the court or to the court's designee authorized under Article 26.04 to appoint counsel in the county, the forms requesting the appointment of counsel. The magistrate shall also inform the person arrested that he is not required to make a statement and that any statement made by him may be used against him. The magistrate shall allow the person arrested reasonable time and opportunity to consult counsel and shall, after determining whether the person is currently on bail for a separate criminal offense, admit the person arrested to bail if allowed by law. A recording of the communication between the arrested person and the magistrate shall be made. The recording shall be preserved until the earlier of the following dates: (1) the date on which the pretrial hearing ends; or (2) the 91st day after the date on which the recording is made if the person is charged with a misdemeanor or the 120th day after the date on which the recording is made if the person is charged with a felony. The counsel for the defendant may obtain a copy of the recording on payment of a reasonable amount to cover costs of reproduction. For purposes of this subsection, "electronic broadcast system" means a two-way electronic communication of image and sound between the arrested person and the magistrate and includes secure Internet videoconferencing.

(b) After an accused charged with a misdemeanor punishable by fine only is taken before a magistrate under Subsection (a) and the magistrate has identified the accused with certainty, the magistrate may release the accused without bond and order the accused to appear at a later date for arraignment in the applicable justice court or municipal court. The order must state in writing the time, date, and place of

the arraignment, and the magistrate must sign the order. The accused shall receive a copy of the order on release. If an accused fails to appear as required by the order, the judge of the court in which the accused is required to appear shall issue a warrant for the arrest of the accused. If the accused is arrested and brought before the judge, the judge may admit the accused to bail, and in admitting the accused to bail, the judge should set as the amount of bail an amount double that generally set for the offense for which the accused was arrested. This subsection does not apply to an accused who has previously been convicted of a felony or a misdemeanor other than a misdemeanor punishable by fine only.

(c) When a deaf accused is taken before a magistrate under this article or Article 14.06 of this Code, an interpreter appointed by the magistrate qualified and sworn as provided in Article 38.31 of this Code shall interpret the warning required by those articles in a language that the accused can understand, including but not limited to sign language.

(d) If a magistrate determines that a person brought before the magistrate after an arrest authorized by Article 14.051 of this code was arrested unlawfully, the magistrate shall release the person from custody. If the magistrate determines that the arrest was lawful, the person arrested is considered a fugitive from justice for the purposes of Article 51.13 of this Code, and the disposition of the person is controlled by that article.

(e) In each case in which a person arrested is taken before a magistrate as required by Subsection (a), a record shall be made of:

(1) the magistrate informing the person of the person's right to request appointment of counsel;

(2) the magistrate asking the person whether the person wants to request appointment of counsel; and

(3) whether the person requested appointment of counsel.

(f) A record required under Subsection (e) may consist of written forms, electronic recordings, or other documentation as authorized by procedures adopted in the county under Article 26.04(a).

(g) If a person charged with an offense punishable as a misdemeanor appears before a magistrate in compliance with a citation issued under Article 14.06(b) or (c), the magistrate shall perform the duties imposed by this article in the same manner as if the person had been arrested and brought before the magistrate by a peace officer. After the magistrate performs the duties imposed by this article, the magistrate except for good cause shown may release the person on personal bond. If a person who was issued a citation under Article 14.06(c) fails to appear as required by that citation, the magistrate before whom the person is required to appear shall issue a warrant for the arrest of the accused.

CCP Art. 15.16. How Warrant Is Executed

(a) The officer or person executing a warrant of arrest shall without unnecessary delay take the person or have him taken before the magistrate who issued the warrant or before the magistrate named in the warrant, if the magistrate is in the same county where the person is arrested. If the issuing or named magistrate is in another county, the person arrested shall without unnecessary delay be taken before some magistrate in the county in which he was arrested.

(b) Notwithstanding Subsection (a), to provide more expeditiously to the person arrested the warnings described by Article 15.17, the officer or person

executing the arrest warrant may as permitted by that article take the person arrested before a magistrate in a county other than the county of arrest.

§6:32 Delay in Taking Before Magistrate

CCP Art. 15.17(a) requires that the person making an arrest shall without unnecessary delay take the arrestee before a magistrate. Where a defendant is taken before a magistrate within 16 hours of his arrest the "without unnecessary delay" requirement is satisfied, even where there was a magistrate available at the time of the arrest. Jenkins v. State, 912 S.W.2d 793 (Tex. Crim. App. 1995).

Absent a showing of a causal connection between the failure to take an accused before a magistrate and the accused's confession, the validity of the confession is not affected. Boyd v. State, 811 S.W.2d 105 (Tex. Crim. App. 1991).

Where the record shows that the defendant voluntarily confessed within two hours of his arrest, there is no evidence of a causal connection between any CCP Art. 15.17 violation and the defendant's voluntary decision to confess. Jenkins.

It is well-settled that an unreasonable delay in bringing an arrestee before a magistrate will not invalidate an otherwise voluntary confession if the arrestee was properly advised of his Miranda rights prior to making the statement. Garcia v. State, 191 S.W.3d 870, 876 (Tex.App.—Houston [14th Dist.] 2006, *no pet.*).

Delay in taking a defendant before a magistrate will not invalidate a confession where the delay was due to reasonable investigative and processing procedures necessitated by the particular facts of the case. Cantu v. State, 842 S.W.2d 667, 680 (Tex. Crim. App. 1992).

§6:33 Sufficiency of Warnings

§6:33.1 In General

A defendant in custody has a right to certain procedural safeguards designed to effectively secure his privilege against self-incrimination. Miranda v. Arizona, 384 U.S. 436, 86 S.Ct. 1602, 16 L.Ed.2d 694 (1966).

In Texas, these procedural safeguards are codified as a set of warnings in CCP Art. 38.22 §§2 and 3. *See* Herrera v. State, 241 S.W.3d 520 (Tex. Crim. App. 2007).

The Miranda warnings and a valid waiver thereof are a constitutional predicate to the admission of a custodial statement in state and federal courts. Dickerson v. U.S., 530 U.S. 428, 120 S.Ct. 2326, 147 L.Ed.2d 405 (2000).

The Miranda warnings are not constitutionally mandated, but are procedural safeguards necessary to prevent governmental coercion. Alvarado v. State, 853 S.W.2d 17 (Tex. Crim. App. 1991).

In order for a subsequent, custodial confession to be admissible, it must be shown that the defendant was advised of his rights and given an opportunity to exercise them before any questioning by law enforcement officers. Miranda, and CCP Art. 38.22. In other words, the prophylactic requirements of Miranda and CCP Art. 38.22 are prerequisite to the admission of a custodial confession. Baker v. State, 956 S.W.2d 19 (Tex. Crim. App. 1997).

An accused held in custody must be given the required warnings prior to questioning. A failure to do so results in forfeiture of the use of any statement obtained during that interrogation by the prosecution during its case-in-chief. Jones v. State, 119 S.W.3d 766 (Tex. Crim. App. 2003) (where the resulting statement was admitted in the punishment phase of a capital murder trial and the constitutional error was found to be harmless beyond a reasonable doubt).

Two interrogation sessions separated by a gap in time can be considered part of a single interview for the purpose of Article 38.22 and Miranda; thus sufficient warnings given at the beginning of the first session will carry over and apply to the second session. Bible v. State, 162 S.W.3d 234 (Tex. Crim. App. 2005).

However, a break in the interrogation can sometimes require the giving of new warnings. Flemming v. State, 949 S.W.2d 876 (Tex.App.—Houston [14th Dist.] 1997, *no pet.*).

The failure to Mirandize the defendant before his initial interrogation will render later Miranda warnings ineffective and make a subsequent statement inadmissible. Martinez v. State, 272 S.W.3d 615 (Tex. Crim. App. 2009).

The prosecution bears the burden of proving admissibility when a Miranda violation is found. Martinez.

§6:33.2 Lack of Miranda Warnings

The failure to administer the Miranda warnings creates a presumption of compulsion and consequently, unwarned statements that are otherwise voluntary within the meaning of the Fifth Amendment must be excluded from evidence under Miranda; but the presumption, though not rebuttable for purposes of the state's case-in-chief, does not require that the statements and their fruits be discarded as inherently tainted. Oregon v. Elstad, 470 U.S. 298, 105 S.Ct. 1285, 84 L.Ed.2d 222 (1985); Baker v. State, 956 S.W.2d 19 (Tex. Crim. App. 1997).

The absence of Miranda warnings does not require the suppression of non-testimonial fruits of a defendant's statement where the statement was made voluntarily and was uncoerced. U.S. v. Patane, 542 U.S. 630, 124 S.Ct. 2620, 159 L.Ed.2d 667 (2004) (*plurality opinion*) (where the defendant was prosecuted for possessing a gun that was found as a result of his statement).

When a defendant is not in custody, law enforcement officers are not required to scrupulously honor his invocation of rights, even where the defendant has been read his rights. Estrada v. State, 313 S.W.3d 274 (Tex. Crim. App. 2010).

As a general rule, the Miranda warnings are not a prerequisite to the admission of a statement obtained in a foreign country by foreign officials. Alvarado. Despite the inapplicability of Miranda to confessions obtained by foreign officials, the confession will still be excluded if the circumstances surrounding the taking of the confession shock the conscience of the U.S. court. Alvarado.

§6:33.3 Incomplete or Inaccurate Warnings

The warnings are sufficient where they reasonably convey to a suspect his rights under Miranda. Florida v. Powell, 559 U.S. ___, 130 S.Ct. 1195, 175 L.Ed.2d 1009 (2010).

CCP Art. 38.22 warnings do not have to be given verbatim to the suspect in order for a resulting confession to be admissible. Where the warnings substantially comply with CCP Art. 38.22, the trial court does not err in admitting the confession. Cockrell v. State, 933 S.W.2d 73 (Tex. Crim. App. 1996) (suspect was warned that the statement could be used against him in "court" and that it could be used against him at "trail" instead of "trial"); Sosa v. State, 769 S.W.2d 909 (Tex. Crim. App. 1989).

Warning a suspect that he had a right to an attorney "prior to *or* during any questioning," sufficiently warned the suspect that he had a right to an attorney "prior to *and* during any questioning." Penry v. State, 691 S.W.2d 636 (Tex. Crim. App. 1985), citing California v. Prysock, 453 U.S. 355, 101 S.Ct. 2806, 69 L.Ed.2d 696 (1981) (emphasis added).

A warning to a suspect that, "anything I say may be used against me," is sufficient even though no reference is made to using the statement in court or at a trial. Williams v. State, 883 S.W.2d 317 (Tex. App.—Dallas 1994, *pet. ref'd*).

§6:33.4 Warnings for Written Statements

The warnings required to be given to a suspect in custody are contained in CCP Art. 38.22 §2 for written statements and in §3 for oral statements. The §2 warnings and a waiver thereof must appear on the face of any resulting written confession. Westley v. State, 754 S.W.2d 224 (Tex. Crim. App. 1988); Heiselbetz v. State, 906 S.W.2d 500 (Tex. Crim. App. 1995).

The warnings provided in CCP Art. 38.22 §2(a) are virtually identical to the Miranda warnings, with the exception that the warning that an accused "has the right to terminate the interview at any time" as set out in §2(a)(5) is not required by Miranda. Herrera v. State, 241 S.W.3d 520, 526 (Tex. Crim. App. 2007).

TIP TO THE PROSECUTION: Compose a form for the taking of a suspect's statement, which contains on its face the CCP Art. 38.22 warnings and a valid waiver thereof, and distribute it to all local police agencies for their use.

Where the statutory warnings are considered as being incorporated into the confession rather than appearing on its face, the facial warning requirement of CCP Art. 38.22 will be deemed satisfied. White v. State, 779 S.W.2d 809 (Tex. Crim. App. 1989).

The CCP Art. 38.22 §2(a) requirement that the warnings must be given by a magistrate or the person to whom the statement is made is satisfied so long as the officer taking the confession was present when the warnings were given. Dowthitt v. State, 931 S.W.2d 244 (Tex. Crim. App. 1996).

Because a written statement is not "obtained" until it is signed, giving the required warnings before the accused signs the statement meets the statutory requirements. Dowthitt, 931 S.W.2d at 258-9 (where the defendant was not in custody when he began giving the statement).

§6:33.5 Warnings for Oral Statements

Article 38.22, §3 of the Code of Criminal Procedure provides that no oral statement of an accused made as a result of custodial interrogation shall be admissible against the accused in a criminal proceeding unless (1) the statement was recorded and (2) prior to the statement but during the recording, the accused was warned of his rights and knowingly, intelligently, and voluntarily waived those rights. Joseph v. State, 309 S.W.3d 20, 23-4 (Tex. Crim. App. 2010).

The statute contains two distinct elements pertaining to a statement's admissibility: the defendant's receipt of the prescribed warning and his waiver of the rights set out in the warning. Joseph, 309 S.W.3d at 24.

The CCP Art. 38.22 §2(a) requirement that the warnings must be given by a magistrate or the person to whom the statement is made does not apply to oral confessions that are governed by §3. Dowthitt v. State, 931 S.W.2d 244, 258-9 (Tex. Crim. App. 1996).

§6:34 "Two-Step" Confessions

The failure to give Miranda warnings before an oral confession does not automatically render a subsequent written confession involuntary. Corwin v. State, 870 S.W.2d 23 (Tex. Crim. App. 1993); Oregon v. Elstad, 470 U.S. 298, 105 S.Ct. 1285, 84 L.Ed.2d 222 (1985).

In a custodial situation, the failure to give Miranda warnings before a first statement renders a second statement that repeats the substance of the first statement inadmissible; even where Miranda warnings are given before the second statement. Missouri v. Seibert, 542 U.S. ___, 124 S.Ct. 2601, 159 L.Ed.2d 643 (2004) (*plurality op.*).

The taint from a first confession can be sufficiently attenuated to allow admission of a subsequent confession where the defendant is brought before a magistrate for Miranda warnings and the further interrogation done by a second law enforcement officer. Sterling v. State, 800 S.W.2d 513 (Tex. Crim. App. 1990); Elstad. *See also*, Rodriguez v. State, 968 S.W.2d 554 (Tex.App—Houston [14th Dist.] 1998 *no pet.*).

Where questioning of the defendant that occurred pre-Miranda was inadvertent (rather than deliberate) and the defendant's post-Miranda statements were voluntary, the trial court does not err in allowing the admission of the post-Miranda statements by finding that there was not an illegal two-step interrogation. Carter v. State, 309 S.W.3d 31 (Tex. Crim. App. 2010).

The taint of an earlier inducement can be sufficiently attenuated even in the absence of a specific repudiation of the earlier inducement. Sterling. It is also not necessary to inform the suspect that his earlier, oral statement cannot be used against him in order for the subsequent written statement to be admissible. Elstad.

The question of whether the rationale of Elstad applies when a suspect makes incriminating statements after a knowing and voluntary waiver of his right to counsel notwithstanding earlier police questioning in violation of the Sixth Amendment has not yet been decided. Fellers v. U.S., 540 U.S. 519, 124 S.Ct. 1019, 157 L.Ed.2d 1016 (2004).

The appropriate remedy when a two-step interrogation technique has been impermissibly employed is the suppression of both the unwarned and the warned statements. Vasquez v. State, 483 S.W.3d 550, 555 (Tex. Crim. App. 2016).

A defendant's objection that he was subject to an impermissible two-step interrogation technique must be sufficiently specific for the trial court to understand the objection. Vasquez v. State, 483 S.W.3d at 556.

§6:35 Out-of-State Confessions

Under basic conflict-of-law principles, the law of the forum in which the judicial proceeding is held governs the admissibility of evidence. Therefore, CCP Art. 38.22 (being a procedural evidentiary rule), governs the admissibility of out-of-state confessions offered in Texas trials. Nonn v. State, 41 S.W.3d 677 (Tex. Crim. App. 2001).

Out-of-state confessions must pass a substantial compliance test rather than a strict compliance test in order to be admissible under CCP Art. 38.22 §2. Nonn v. State, 41 S.W.3d 677 (Tex. Crim. App. 2001).

However, oral statements offered under §5 of Art. 38.22 must strictly comply in order to be admissible because of a legislative mandate in §5. Nonn v. State, 41 S.W.3d 677 (Tex. Crim. App. 2001).

Out-of-state confessions made after September 1, 2001 will be governed by CCP Art. 38.22 §8 which reads as follows:

> ***§8.*** *Notwithstanding any other provision of this article, a written, oral, or sign language statement of an accused made as a result of a custodial interrogation is admissible against the accused in a criminal proceeding in this state if:*
>
> *(1) the statement was obtained in another state and was obtained in compliance with the laws of that state or this state; or*
>
> *(2) the statement was obtained by a federal law enforcement officer in this state or another state and was obtained in compliance with the laws of the United States.*

§6:40 Invocation and Waiver of Rights

§6:41 Invocation of Rights

The Supreme Court has held that the Miranda warnings and a valid waiver thereof are a constitutional predicate to the admission of a custodial statement in state and federal courts. Dickerson v. U.S., 530 U.S. 428, 120 S.Ct. 2326, 147 L.Ed.2d 405 (2000).

Although the Fifth Amendment right to interrogation counsel is triggered by the Miranda warnings that police must give before beginning any custodial questioning and the Sixth Amendment right to trial counsel is triggered by judicial arraignment or Article 15.17 magistration each right to counsel applies to post-magistration custodial interrogation, and each is invoked and waived in exactly the same manner—under the Fifth Amendment prophylactic Miranda rules. Pecina v. State, 361 S.W.3d 68, 71 (Tex. Crim. App. 2012).

The accused must unambiguously request counsel during a custodial interrogation; thus, he must articulate his desire to have counsel present sufficiently clearly that a reasonable police officer in the circumstances would understand the statement to be a request for an attorney. Pecina, 361 S.W.3d at 79 (citing Davis v. United States, 512 U.S. 452, 458-59, 114 S.Ct. 2350, 129 L.Ed.2d 362 (1994)).

The Supreme Court has held that, in deciding whether an accused actually invoked his right to counsel, reviewing courts must use an objective standard "[t]o avoid difficulties of proof and to provide guidance to officers conducting interrogations[.]" Pecina, 361 S.W.3d at 79 (citing Davis v. United States).

To protect the Fifth Amendment privilege against self-incrimination, the police may not continue or re-initiate custodial interrogation of a suspect who has previously requested assistance of counsel after the police informed him of his right to counsel at the beginning of a custodial interrogation. Pecina, 361 S.W.3d at 80.

When a defendant is not in custody, law enforcement officers are not required to scrupulously honor his invocation of rights, even where the defendant has been read his rights. Estrada v. State, 313 S.W.3d 274 (Tex. Crim. App. 2010).

In reviewing claims concerning Miranda violations and the admission of statements made as the result of custodial interrogation, appellate courts afford almost total deference to the trial court's rulings on questions of historical fact and on application of law to fact questions that turn upon credibility and demeanor while reviewing *de novo* the trial court's rulings on application of law to fact questions that do not turn upon credibility and demeanor. Pecina v. State, 361 S.W.3d at 79.

§6:42 Waiver of Rights

§6:42.1 Case Law on Waiver

§6:42.1.1 In General

Miranda and its prophylactic safeguards only apply to custodial statements. Wolfe v. State, 917 S.W.2d 270 (Tex. Crim. App. 1996).

Texas Code of Criminal Procedure Art. 38.22 §§ 2 and 3 provide that a defendant's custodial-interrogation statement is not admissible unless, prior to making the statement, he received the warnings provided in Art. 15.17 or Art. 38.22, §2(a) or Sec. 3(a) (which incorporate the requirements of Miranda), and he knowingly, intelligently, and voluntarily waived those rights. Oursbourn v. State, 259 S.W.3d 159, 171-2 (Tex. Crim. App. 2008).

A defendant may waive his Miranda and CCP Art. 38.22 rights provided the waiver is made voluntarily, knowingly, and intelligently. Jacobs v. State, 787 S.W.2d 397 (Tex. Crim. App. 1990), citing Moran v. Burbine, 475 U.S. 412, 106 S.Ct. 1135, 89 L.Ed.2d 410 (1986).

A waiver is an intentional and knowing relinquishment of a known right. Miranda v. Arizona, 384 U.S. 436, 86 S.Ct. 1602, 16 L.Ed.2d 694 (1966).

"Knowing" in the context of waiver is an intellectual, conscious understanding, by thought process and reason, of the releasing of, or absolute surrender of, a right, or of a constitutionally granted guarantee; and, the understanding of the effects of waiving such a right of a constitutionally granted guarantee. Combs v. State, 631 S.W.2d 534 (Tex. App.—Houston [1st Dist.] 1982), *rev'd on other grounds* 643 S.W.2d 709 (Tex. Crim. App. 1982).

It is the State's burden to establish a valid waiver of Miranda rights by a preponderance of the evidence. Leza v. State, 351 S.W.3d 344, 349 (Tex. Crim. App. 2011); Colorado v. Connelly, 479 U.S. 157, 168, 107 S.Ct. 515, 93 L.Ed.2d 473 (1986).

There are two facets to any inquiry with respect to the adequacy of a purported waiver of Miranda rights: First, the waiver must be voluntary in the sense that it was the product of a free and deliberate choice rather than intimidation, coercion, or deception. Second, the waiver must be made with a full awareness both of the nature of the right being abandoned and the consequences of the decision to abandon it. Leza, 351 S.W.3d at 349; Moran v. Burbine, 475 U.S. 412, 421, 106 S.Ct. 1135.

Only if the totality of the circumstances surrounding the interrogation reveals both an uncoerced choice and the requisite level of comprehension may a court properly conclude that the Miranda rights have been waived. Joseph v. State, 309 S.W.3d 20, 25 (Tex. Crim. App. 2010); Moran v. Burbine, 475 U.S. 412, 421, 106 S.Ct. 1135.

The totality-of-the-circumstances approach requires the consideration of "all the circumstances surrounding the interrogation," including the defendant's experience, background, and conduct. Joseph, 309 S.W.3d at 25.

It will suffice to render a waiver knowing and intelligent that the accused has been made aware, and fully comprehends, that he has the right to remain silent in the face of police interrogation and to discontinue the dialogue at any time, and that the consequence of his waiver is that his words may be used against him later in a court of law. Leza, 351 S.W.3d at 350.

A suspect's awareness of all the possible subjects of questioning in advance of interrogation is not relevant to determining whether the suspect voluntarily, knowingly, and intelligently waived his Fifth Amendment privilege under Miranda. Leza, 351 S.W.3d at 350; *citing* Colorado v. Spring, 479 U.S. 564, 574, 107 S.Ct. 851, 857, 93 L.Ed.2d 954 (1987).

Similarly, the failure to immediately inform a defendant of the subject of the interrogation impacts neither the voluntariness nor the knowing and intelligent nature of any waiver of his statutory rights—as a matter of law. Leza, 351 S.W.3d at 350.

The U.S. Constitution does not require that a criminal suspect knows and understands every possible consequence of a waiver of his Fifth Amendment privilege. Colorado v. Spring, 479 U.S. at 577, 107 S.Ct. at 859.

Only if the totality of the circumstances surrounding the interrogation reveal both an uncoerced choice and the requisite level of comprehension may a court properly conclude that the Miranda and CCP Art. 38.22 rights have been waived. Jacobs.

The mere failure of the suspect to invoke his rights is not a valid waiver thereof. Miranda v. Arizona.

Waivers will not be presumed from silence. Miranda v. Arizona, 384 U.S. at 475; Joseph v. State, 309 S.W.3d 20, 24 (Tex. Crim. App. 2010). The fact that an incriminating statement was eventually made will not presume a valid waiver. Miranda v. Arizona, 384 U.S. at 475; Joseph, 309 S.W.3d at 24.

§6:42.1.2 Implied Waivers

A suspect who has received and understood his Miranda warnings, and has not invoked his Miranda rights, implicitly waives the right to remain silent by making an uncoerced statement to the police. Berghuis v. Thompkins, 560 U.S. ___, 130 S.Ct. 2250, 176 L.Ed.2d 1098 (2010).

An implicit waiver of the right to remain silent is sufficient to admit a defendant's statement into evidence. An express waiver is not necessary. Berguis v. Thompkins.

Police are not required to obtain a waiver of the suspect's right to remain silent before interrogating the suspect. Berguis v. Thompkins.

Neither a written nor an oral express waiver is required. Joseph v. State, 309 S.W.3d 20, 24 (Tex. Crim. App. 2010).

In some cases, a waiver can be inferred from the actions and words of the person interrogated. North Carolina v. Butler, 441 U.S. 369, 373, 99 S.Ct. 1755, 60 L.Ed.2d 286 (1979); Joseph, 309 S.W.3d at 24.

Whether a waiver is shown must depend in each case upon the particular facts and circumstances surrounding the case, including the background, experience, and conduct of the accused. Muniz v. State, 851 S.W.2d 238 (Tex. Crim. App. 1993), *citing* Johnson v. Zerbst, 304 U.S. 458, 58 S.Ct. 1019, 82 L.Ed.2d 1461 (1938).

§6:42.1.3 Waivers in Written Statements

For a written statement to unambiguously meet the waiver requirements of CCP Art. 38.22 §2(b), the preferable practice is for the statement to contain the following language, near or adjacent to the signature of the individual giving the statement: "I knowingly, voluntarily and intelligently waived the rights described above before and during the making of this statement." Garcia v. State, 919 S.W.2d 370, 387 (Tex. Crim. App. 1994).

The phrase, "I do hereby voluntarily waive these rights," on each page of a written confession has been found to constitute a valid waiver. Cockrell v. State, 933 S.W.2d 73 (Tex. Crim. App. 1996).

Where a written confession of defendant does not contain any statement whatever to the effect that defendant, prior to making the statement, voluntarily, knowingly, and intelligently waived his rights, the confession should not be admitted in evidence. Thibedeaux v. State, 729 S.W.2d 137, 139 (Tex.App.—Beaumont 1987, *pet. ref'd*).

However, the circumstances of the taking of the statement along with other writings and recitations can add up to the conclusion that the defendant executed a valid waiver of his rights. Garcia, 919 S.W.2d at 387.

§6:42.2 Voluntariness of Waivers

§6:42.2.1 Voluntariness of Waivers Under Miranda

For purposes of the Fifth Amendment, waiver of the privilege against compelled self-incrimination during custodial questioning can be deemed involuntary only if it is a product of official coercion, intimidation, or deception. Leza v. State, 351 S.W.3d 344, 352 (Tex. Crim. App. 2011).

Where a defendant's drug intoxication is not due to any causative action on the part of the police, his assertion that his intoxication rendered his Miranda waiver involuntary is foreclosed as a matter of federal constitutional law because the Fifth Amendment privilege against self-incrimination is not concerned with moral and psychological pressures to confess emanating from sources other than official coercion. Leza, 351 S.W.3d at 350.

Drug intoxication can, however, be a factor that is relevant to determining whether a Miranda waiver was knowing and intelligent. Leza, 351 S.W.3d at 351.

On direct appeal, the courts measure the propriety of the trial court's ruling with respect to alleged Miranda violations under the totality of the circumstances, almost wholly deferring to the trial court on questions of historical fact and credibility, but reviewing *de novo* all questions of law and mixed questions of law and fact that do not turn on credibility determinations. Leza, 351 S.W.3d at 349.

§6:42.2.2 Voluntariness of Waivers Under Art. 38.22

A claim that a purported waiver of the state statutory rights enumerated in Article 38.22 is involuntary need not be predicated on police overreaching. Leza v. State, 351 S.W.3d 344, 349 (Tex. Crim. App. 2011) (citing Oursbourn v. State, 259 S.W.3d 159, 172 (Tex. Crim. App. 2008)).

Circumstances not attributable to the police that nevertheless adversely impact an accused's ability to resist reasonable police entreaties to waive his statutory rights, such as intoxication, are factors in the voluntariness inquiry, though they are usually not enough, by themselves, to render a statement inadmissible under Article 38.22. Leza, 351 S.W.3d at 352 (citing Oursbourn, 259 S.W.3d at 172).

Drug intoxication can be a factor that is relevant to determining whether a waiver of statutory rights under Art. 38.22 was made knowingly and intelligently. Leza, 351 S.W.3d at 351.

Claims of involuntariness under Art. 38.22 can be, but need not be, predicated on police overreaching. Claims of involuntariness that are not themselves relevant to due process claims could involve sweeping inquiries into the state of mind of a criminal defendant who has confessed. Oursbourn, 259 S.W.3d at 172.

§6:42.2.3 Jury Charge on the Voluntariness Issue

§6:42.2.3.1 Jury Charges and Miranda Issues

Due process and Miranda claims may warrant both general and specific voluntariness jury instructions; while Texas statutory claims warrant only a general voluntariness jury instruction. It is the defendant's responsibility to delineate which type of "involuntariness" he is claiming—a general (perhaps subjective) lack of voluntariness or a specific police-coerced lack of voluntariness—because the jury instruction is very different depending upon the type of claim. Oursbourn v. State, 259 S.W.3d 159, 174 (Tex. Crim. App. 2008).

A statement that is involuntary as a matter of constitutional law is also involuntary under Article 38.22, but the converse need not be true. Oursbourn, 259 S.W.3d at 169.

Because the failure of police to honor a defendant's rights under Miranda is not itself a constitutional violation, a defendant is not entitled to an Art. 38.23 charge when evidence at trial shows that the police did not comply with Miranda. Under these circumstances, the defendant is entitled to a charge under Art. 38.22. Contreras v. State, 312 S.W.3d 566 (Tex. Crim. App. 2010).

§6:42.2.3.2 Charging Art. 38.22 Issues

The obvious purpose of Art. 38.22, §7 is to authorize and require jury instructions regarding the warnings and safeguards for written and oral statements outlined in Article 38.22, §§2 and 3 (warnings on the right to remain silent, right to counsel, etc). Oursbourn v. State, 259 S.W.3d 159, 175 (Tex. Crim. App. 2008).

As with §6, the trial judge's §7 jury instructions are general ones that set out the pertinent law and legal requirements of §§2 and 3 (or, in an appropriate case, those of Article 15.17). Oursbourn, 259 S.W.3d at 176.

The §7 instruction sets out the requirements of 38.22, §2 or §3 and asks the jury to decide whether all of those requirements were met. Oursbourn, 259 S.W.3d at 173.

The issue to be decided is whether there was compliance with the statutory warnings set out in both Articles 15.17 (Duties of Arresting Officer and Magistrate) and 38.22, §§2 and 3, and the voluntariness of the defendant's waiver of these rights. For the issue to be raised by the evidence there must be a genuine factual dispute, just as is true under Article 38.23 issues. Oursbourn, 259 S.W.3d at 176.

The same procedures—including a hearing outside the presence of the jury and the entry of written findings—that apply to a general voluntariness challenge under §6, also apply to a challenge that is made to the sufficiency of warnings and voluntary waiver of the rights communicated by those warnings. Oursbourn, 259 S.W.3d at 176.

§6:43 Invocation and Waiver of Specific Rights

§6:43.1 Invocation and Waiver of Right to Remain Silent

§6:43.1.1 General Rules on Invocation of Right to Remain Silent

Persons subject to questioning by law enforcement officers have a right to remain silent. Miranda and CCP Art. 38.22. Before these persons are questioned they have a right to be warned that they have such right. Miranda and CCP Art. 38.22.

The failure of law enforcement to cut off questioning when a suspect invokes his right to remain silent renders any subsequent statement inadmissible. Dowthitt v. State, 931 S.W.2d 244 (Tex. Crim. App. 1996) (citing Michigan v. Mosley, 423 U.S. 96, 96 S.Ct. 321, 46 L.Ed.2d 313 (1975)).

The suspect's right to remain silent must be scrupulously honored. The failure to scrupulously honor this right will result in the suppression of a resulting confession. Michigan v. Mosley.

A law enforcement agent is not required to cease his questioning of a suspect unless the suspect's invocation of his right to remain silent is unambiguous. Dowthitt; citing Davis v. United States, 512 U.S. 452, 114 S.Ct. 2350, 129 L.Ed.2d 362 (1994); Berghuis v. Thompkins, 560 U.S. ___, 130 S.Ct. 2250, 176 L.Ed.2d 1098 (2010).

A law enforcement agent is not required to clarify ambiguous remarks, but may continue to question in order to do so. Dowthitt.

§6:43.1.2 Sufficient Invocation of Right to Remain Silent

Where a suspect tells an interrogating officer that he doesn't want to talk to him and doesn't want to talk about it anymore, he has sufficiently invoked his right to remain silent. Ramos v. State, 245 S.W.3d 410 (Tex. Crim. App. 2007).

A suspect's statement that he "didn't see how he could help himself by talking to [officers]," has been found to be a sufficient invocation of the right to remain silent. Murphy v. State, 766 S.W.2d 246 (Tex. Crim. App. 1989).

A suspect's statement to police when asked if he wished to discuss an offense that he "wanted a little time" to think about the matter, has been found to be a sufficient invocation of the right to remain silent. Phillips v. State, 701 S.W.2d 875 (Tex. Crim. App. 1985).

No magic word or phrase is necessary to invoke the right to remain silent. Prolonged silence and lack of cooperation with authorities will serve to invoke it. Watson v. State, 762 S.W.2d 591 (Tex. Crim. App. 1988).

Where the defendant concluded his handwritten statement with the phrase, "I have nothing else to say," he validly invoked his right to remain silent. Autry v. State, 626 S.W.2d 758 (Tex. Crim. App. 1982).

§6:43.1.3 Insufficient Invocation of Right to Remain Silent

A law enforcement agent is not required to cease questioning of a suspect unless the suspect's invocation of his right to remain silent is unambiguous. Dowthitt v. State, 931 S.W.2d 244 (Tex. Crim. App. 1996), citing Davis v. United States, 512 U.S. 452, 114 S.Ct. 2350, 129 L.Ed.2d 362 (1994).

A suspect's statement that "I can't say any more than that. I need to rest," is not an unambiguous invocation of the right to remain silent. Dowthitt.

A suspect's statement that "[m]aybe I should talk to someone," is not an unambiguous invocation of the right to remain silent. Dinkins v. State, 894 S.W.2d 330 (Tex. Crim. App. 1995).

§6:43.1.4 Valid Waivers of Right to Remain Silent

A voluntary, knowing, and intelligent waiver of the right to remain silent must be proven by a preponderance of the evidence. Robinson v. State, 851 S.W.2d 216 (Tex. Crim. App. 1991) (citing Colorado v. Connelly, 479 U.S. 157, 107 S.Ct. 515, 93 L.Ed.2d 473 (1986)).

§6:43.2 Invocation and Waiver of Right to Attorney

§6:43.2.1 Fifth Amendment Right to Attorney

The Fifth Amendment right to counsel is the right to interrogation counsel. Pecina v. State, 361 S.W.3d 68 (Tex. Crim. App. 2012) (citing Montejo v. Louisiana, 556 U.S. 778, 129 S.Ct. 2079, 173 L.Ed.2d 955 (2009)).

The Fifth Amendment right to interrogation counsel is triggered by the Miranda warnings that police must give before beginning any custodial questioning. Pecina; Montejo.

The Fifth Amendment right to counsel applies to post-magistration custodial interrogation. Pecina.

A suspect undergoing custodial interrogation has the right to the assistance of counsel in his dealings with law enforcement agents. This right includes both the right to consult with a lawyer prior to questioning, as well as the right to have a lawyer present during questioning. Miranda v. Arizona, 384 U.S. 436, 86 S.Ct. 1602, 16 L.Ed.2d 694 (1966); CCP Art. 38.22 §2.

A request for counsel that is ambiguous in scope will be afforded a broad rather than a narrow interpretation. Smith v. State, 779 S.W.2d 417 (Tex. Crim. App. 1989); citing Connecticut v. Barrett, 479 U.S. 523, 107 S.Ct. 828, 93 L.Ed.2d 920 (1987).

Further questioning of the suspect may be permitted for the limited purpose of clarifying an equivocal invocation of the right to counsel. Smith v. State; Jamail v. State, 787 S.W.2d 372 (Tex. Crim. App. 1990).

However, where the invocation of the right to counsel is clear, a suspect's post-request responses to further interrogation may not be used to cast retrospective doubt on the clarity of the initial request itself. Smith v. State, citing Smith v. Illinois, 469 U.S. 91, 105 S.Ct. 490, 83 L.Ed.2d 488 (1984).

A suspect who has expressed his desire to deal with the police only through counsel is not subject to further interrogation by the authorities until counsel has been made available to him, unless the accused himself initiates further communication, exchanges, or conversations with the police. Edwards v. Arizona, 451 U.S. 477, 484–485, 101 S.Ct. 1880, 1884–1885, 68 L.Ed.2d 378 (1981).

The Edwards rule holds true whether a contemplated reinterrogation concerns the same or a different offense, or whether the same or different law enforcement authorities are involved in the second investigation; under any scenario, the suspect's invocation of his Fifth Amendment rights bars any further interrogation. Arizona v. Roberson, 486 U.S. 675, 688, 108 S.Ct. 2093, 2101, 100 L.Ed.2d 704 (1988).

Once an accused has invoked his right to counsel, he may not be subjected to further interrogation until counsel is actually present, unless the accused himself initiates contact with the authorities. Murphy v. State, 801 S.W.2d 917 (Tex. Crim. App. 1991), citing Minnick v. Mississippi, 498 U.S. 46, 111 S.Ct. 486, 112 L.Ed.2d 489 (1990). This is true even where the accused has consulted with counsel in the interim. Murphy; Minnick.

Where a suspect has invoked his right to counsel, his rights are violated when police continue interrogation. Whether police reinitiation of contact with the suspect amounts to the functional equivalent of interrogation is judged on the basis of the definition of "interrogation" set out in Rhode Island v. Innis (446 U.S. 291, 100 S.Ct. 1682, 64 L.Ed.2d 297 (1980)). Moran v. State, 213 S.W.3d 917 (Tex. Crim. App. 2007).

This definition of "interrogation" focuses primarily upon the perceptions of the suspect, rather than the intent of the police. Moran (citing Innis).

Where the defendant omits mention of the actions of the police and gives another reason for his continued cooperation with police in his suppression motion testimony, these factors will support a finding that he did not subjectively feel that the police were further interrogating him. Moran.

Under certain narrow circumstances, interrogation may continue where an accused has invoked a limited right to counsel accompanied by an affirmative statement expressing his willingness to otherwise go forward with the interview. Hicks v. State, 860 S.W.2d 419 (Tex. Crim. App. 1993) (citing Connecticut v. Barrett).

A circumstance under which officers may reapproach a suspect who has invoked his right to counsel is that in which officers reapproach to question about a different offense, provided that the suspect's original invocation must have been only as to the specific offense under discussion at the time of the invocation. Arizona v. Roberson, 486 U.S. 675, 108 S.Ct. 2093, 100 L.Ed.2d 704 (1988).

Events occurring outside the presence of the suspect and entirely unknown to him can have no bearing on his capacity to comprehend and knowingly relinquish a constitutional right. Fuller v. State, 253 S.W.3d 220 (Tex. Crim. App. 2008).

Police conduct such as not informing the suspect that his family has retained an attorney for him is irrelevant to the inquiry into the effectiveness of the suspect's waiver of counsel, absent a showing that such conduct prevented the suspect from attaining awareness and comprehension of the information conveyed in the Miranda warnings and the consequences of abandoning those rights. Terrell v. State, 891 S.W.2d 307, 311 (Tex.App.—El Paso 1994, *pet. ref'd*) (citing Moran v Burbine).

One officer's knowledge that a defendant has requested counsel is imputed to every other state agent. Herron v. State, 86 S.W.3d 621, 628 (Tex. Crim. App. 2002) (citing Arizona v. Roberson, 486 U.S. 675, 687-88, 108 S.Ct. 2093, 100 L.Ed.2d 704 (1988)).

Whether a resumption of questioning after a refusal to talk is consistent with scrupulously honoring the right to remain silent depends on the facts of each case. Five factors are considered when examining whether rights were scrupulously honored:

- Whether the suspect was informed of the right to remain silent before the first questioning;
- Whether the suspect was informed of the right to remain silent before the subsequent questioning;
- The length of time between the first and subsequent questioning;
- If the subsequent questioning focused on a different crime; and
- Whether police scrupulously honored the suspect's initial invocation of the right to silence (resolution of this issue does not turn on an evaluation of credibility and demeanor). Michigan v. Mosley, 423 U.S. 96, 96 S.Ct. 321, 46 L.Ed.2d 313 (1975); Maestas; Maestas v. State, 987 S.W.2d 59 (Tex. Crim. App.1999); Herron v. State, 86 S.W.3d 621 (Tex. Crim. App. 2002).

§6:43.2.2 Sixth Amendment Right to Attorney

The Sixth Amendment right to counsel is the right to trial counsel. Pecina v. State, 361 S.W.3d 68 (Tex. Crim. App. 2012) (citing Montejo v. Louisiana, 556 U.S. 778, 129 S.Ct. 2079, 173 L.Ed.2d 955 (2009)).

The Sixth Amendment right to trial counsel is triggered by judicial arraignment or magistration conducted under Code of Criminal Procedure Art. 15.17; and it guarantees a defendant the right to have counsel present at all critical stages of the criminal proceedings. Pecina; Montejo.

Generally, an Article 15.17 initial appearance and magistration marks the initiation of adversarial judicial proceedings in Texas and plainly signals the attachment of a defendant's Sixth Amendment right to counsel. Pecina.

Thus, a Sixth Amendment request for an attorney at an arraignment, initial appearance, or Article 15.17 hearing is a request for the guiding hand of counsel for all judicial criminal proceedings. Pecina.

However, a defendant's invocation of his right to counsel at his arraignment or initial appearance says nothing about his possible invocation of his right to counsel during later police-initiated custodial interrogation. The magistration hearing is not an interrogation event. A defendant who has been arraigned may invoke his Sixth Amendment right to counsel for purposes of custodial interrogation when the police or other law enforcement agents approach him and give him his Miranda warnings. Pecina.

This is because the Sixth Amendment right to counsel applies to post-magistration custodial interrogation. Pecina.

After formal adversary proceedings have begun, the Sixth Amendment right to counsel applies in exactly the same way as the Fifth Amendment right applies to custodial interrogation. Thus, Miranda warnings serve the arrestee's interests in both the Fifth and Sixth Amendment rights to counsel during custodial interrogations conducted after a person has been formally charged. Pecina.

A defendant also has a right to counsel under the Sixth Amendment at any critical stage. *See* Montejo v. Louisiana, 556 U.S. ___, 129 S.Ct. 2079, 173 L.Ed.2d 955 (2009) (overruling Michigan v. Jackson, 475 U.S. 625, 106 S.Ct. 1404, 89 L.Ed.2d 631 (1986)).

The Sixth Amendment does not bar police-initiated interrogation of an accused who has previously asserted his right to counsel during a magistrate's warning. Montejo v. Louisiana; Hughen v. State, 297 S.W.3d 330 (Tex. Crim. App. 2009).

§6:43.2.3 Sixth Amendment Violations

Where a defendant has counsel representing him on a charge, before law enforcement can question him on a factually interwoven charge they are under an obligation to contact his attorney and gain his permission. State v. Frye, 897 S.W.2d 324 (Tex. Crim. App. 1995); Upton v. State, 853 S.W.2d 548 (Tex. Crim. App. 1993). Note: Both Frye and Upton were decided based on Michigan v. Jackson, 475 U.S. 625, 106 S.Ct. 1404, 89 L.Ed.2d 631 (1986) now overruled by Montejo v. Louisiana, 556 U.S. ___, 129 S.Ct. 2079, 173 L.Ed.2d 955 (2009).

If police initiate interrogation after a defendant's invocation of his right to counsel at any critical stage of the proceeding, any alleged waiver of his right to counsel is invalid under the Sixth Amendment. Montejo v. Louisiana.

The Sixth Amendment does not bar police-initiated interrogation of an accused who has previously asserted his right to counsel during a magistrate's warning where litigation has not reached a critical stage. Montejo v. Louisiana; Hughen v. State, 297 S.W.3d 330 (Tex. Crim. App. 2009).

The defendant can unilaterally waive his Sixth Amendment right to counsel where he initiates the communication with law enforcement. *See* Michigan v. Harvey, 494 U.S. 344, 110 S.Ct. 1176, 108 L.Ed.2d 293 (1990); Ripkowski v. State, 61 S.W.3d 378 (Tex. Crim. App. 2001); State v. Maldonado, 259 S.W.3d 184 (Tex. Crim. App. 2008). Note: Cases were decided based on Michigan v. Jackson, 475 U.S. 625, 106 S.Ct. 1404, 89 L.Ed.2d 631 (1986) (now overruled by Montejo v. Louisiana).

Where the police officer merely introduced himself to a defendant, and the defendant then presented the officer with a letter and said that he had been waiting to talk to somebody, the defendant was found to have initiated the communication with police. Maldonado.

An accused is denied the basic protections of the Sixth Amendment when there is used against him at his trial evidence of his own incriminating words deliberately elicited from him by law enforcement agents after he had been indicted and in the absence of counsel. Fellers v. U.S., 540 U.S. 519, 124 S.Ct. 1019, 157 L.Ed.2d 1016 (2004).

The deliberate-elicitation test is consistently applied in Sixth Amendment cases; as opposed to the custodial interrogation test applied in Fifth Amendment cases. Fellers.

The Sixth Amendment does not forbid the use of statements taken in violation where they are offered for impeachment purposes only. This is because the constitutional violation occurs at the time the statement is elicited, not when it is offered at trial for impeachment purposes. Kansas v. Ventris, 556 U.S. 586, 129 S.Ct. 1841, 173 L.Ed.2d 801 (2009).

Where the state recorded the defendant's telephone calls from jail to his attorney, a Sixth Amendment violation requiring dismissal of the indictment was not established absent evidence that the defendant was prejudiced by the State obtaining information which led to admissible evidence against the defendant or obtaining any information that would give the State a tactical advantage. Woodruff v. State, 330 S.W.3d 709 (Tex.App.—Texarkana, 2010, *pet. ref'd*).

§6:43.2.4 Sufficient Invocation of Right To Attorney

Whether or not an accused has sufficiently invoked his right to counsel is measured on a totality of the circumstances basis. Smith v. State, 779 S.W.2d 417 (Tex. Crim. App. 1989).

The phrase "I don't want to give up any right, though, if I don't got no lawyer," is an unequivocal invocation of right to counsel. State v. Gobert, 275 S.W.3d 888 (Tex. Crim. App. 2009).

Where a suspect simply asks for "his attorney," he has sufficiently invoked his right to counsel. Smith v. State, 779 S.W.2d 417 (Tex. Crim. App. 1989).

The statement "I don't want to answer those questions without my lawyer," accompanied by statements such as "No answer" and "I don't want to answer that," is sufficient to invoke the right to counsel. Hicks v. State, 860 S.W.2d 419 (Tex. Crim. App. 1993).

The phrase "I think I want a lawyer" is sufficient to invoke the right to counsel. Jones v. State, 742 S.W.2d 398 (Tex. Crim. App. 1982).

A formal or absolute demand for an attorney is not required to sufficiently invoke the right. All that is necessary is that the suspect indicates in some manner to authorities that he wishes to exercise his right. Ochoa v. State, 573 S.W.2d 796 (Tex. Crim. App. 1978).

Where the suspect surrenders to authorities in the presence of his attorney and advises them through his attorney that he did not wish to be questioned in the absence of counsel, he sufficiently invoked his right to counsel. Stone v. State, 612 S.W.2d 542 (Tex. Crim. App. 1981).

§6:43.2.5 Reinitiation of Contact by Accused

If an accused expresses his desire to deal with the police only through counsel, he is not subject to further interrogation by authorities until counsel has been made available to him; unless the accused himself initiates further communications, exchanges, or conversations with the police. Herron v. State, 86 S.W.3d 621, 628 (Tex. Crim. App. 2002); citing Edwards v. Arizona, 451 U.S. 477, 101 S.Ct. 1880, 68 L.Ed.2d 378 (1981).

Where the suspect has invoked his right to remain silent, he may only be reapproached in a manner that scrupulously honors his continuing right to cut off questioning. Murphy v. State, 766 S.W.2d 246 (Tex. Crim. App. 1989); citing Michigan v. Mosley, 423 U.S. 96, 96 S.Ct. 321, 46 L.Ed.2d 313 (1975).

A passage of time and the repeating of the suspect's Miranda rights are two factors in determining if the right to remain silent has been scrupulously honored. Murphy.

Where the police continue to question the suspect after the invocation of the right to remain silent in an effort to persuade him to change his mind, the right has not been scrupulously honored. Hearne v. State, 534 S.W.2d 703 (Tex. Crim. App. 1976). *See also*, Faulder v. State, 611 S.W.2d 630 (Tex. Crim. App. 1979); Dowthitt v. State, 931 S.W.2d 244 (Tex. Crim. App. 1996).

Even if the defendant validly invokes his right to an attorney, he can later waive the right by resuming conversations with officers of his own volition. Etheridge v. State, 903 S.W.2d 1 (Tex. Crim. App. 1994).

A defendant who has invoked his Fifth Amendment right to counsel can later waive that right by reinitiating contact with law enforcement authorities. Hicks v. State, 860 S.W.2d 419 (Tex. Crim. App. 1993); citing Edwards v. Arizona, 451 U.S. 477, 101 S.Ct. 1880, 68 L.Ed.2d 378 (1981). *See also*, Baldree v. State, 784 S.W.2d 676 (Tex. Crim. App. 1989).

When a suspect has invoked his right to counsel, but then voluntarily reinitiates conversation with the police and expressly waives his right to counsel, the Edwards rule has been satisfied and police are free to directly deal with the suspect until a subsequent invocation. Cross v. State, 144 S.W.3d 521 (Tex. Crim. App. 2004).

A valid waiver of the Fifth Amendment right will not be found by showing only that he responded to further police-initiated custodial interrogation. Hicks, citing Edwards v. Arizona.

The circumstances must show that the accused initiated the conversation by remarks which fairly can be said to represent a desire to open up a more generalized discussion relating directly or indirectly to the investigation. Oregon v. Bradshaw, 462 U.S. 1039, 103 S.Ct. 2830, 77 L.Ed.2d 405 (1983) (*plurality op.*). Where after invoking his right to remain silent, a suspect asked a police officer, "[w]ell, what is going to happen to me now?" he was found to have reinitiated contact such that officers could begin interrogation. Bradshaw.

A suspect's statement that, "I'm ready. I told you I would tell you about it if you would let me talk to [the attorney]," constituted a valid reinitiation of contact where the suspect had, in fact, talked to the attorney. Mays v. State, 726 S.W.2d 937 (Tex. Crim. App. 1986).

After reinitiating conversation with the police, the suspect must also validly waive the right to counsel. Oregon v. Bradshaw.

A suspect is deemed to have waived his previously-invoked right to counsel only when (1) the suspect himself initiates further communication with police, and (2) after he re-initiates communication with police, the suspect validly waives the right to counsel. Oregon v. Bradshaw.

Once communication is reinitiated, police are free to reinitiate any future communications and obtain any further statements as long as each statement is voluntarily made after the waiver of Miranda rights. Cross.

Federal constitutional law does not prohibit the police from clarifying whether a suspect wishes to waive his rights in a situation where the suspect has reinitiated contact with police after earlier invoking his rights. Marshall v. State, 210 S.W.3d 618 (Tex. Crim. App. 2006).

Where the police officer merely introduced himself to a defendant, and the defendant then presented the officer with a letter and said that he had been waiting to talk to somebody, the defendant was found to have initiated the communication with police. State v. Maldonado, 259 S.W.3d 184 (Tex. Crim. App. 2008).

A suspect's post-invocation request for a polygraph examination is a sufficient reinitiation of contact to allow further interrogation. Wyrick v. Fields, 459 U.S. 42, 103 S.Ct. 394, 74 L.Ed.2d 214 (1982).

The question, "[h]ow is everyone in Georgia doing?" asked by the defendant to police officers investigating his case, constituted initiation of further communication with police following invocation of the Fifth Amendment by the defendant. Janecka v. State, 739 S.W.2d 813 (Tex. Crim. App. 1987).

Where the defendant asked for counsel at his arraignment then was immediately asked if he wanted to speak to detectives, he did not unilaterally reinitiate communications with police as required by the Edwards rule. Pecina v. State, 268 S.W.3d 564 (Tex. Crim. App. 2008).

The Edwards rule is inapplicable where the defendant has experienced a break in Miranda custody for a period of at least 14 days before he is re-approached by law enforcement officers. Maryland v. Shatzer, 559 U.S. ___, 130 S.Ct. 1213, 175 L.Ed.2d 1045 (2010) (where the defendant's release into the general prison population constituted a "break in custody").

A circumstance under which officers may reapproach a suspect who has invoked his right to counsel is that in which officers reapproach to question about a different offense, provided that the suspect's original invocation must have been as to the specific offense under discussion at the time of the invocation. Arizona v. Roberson, 486 U.S. 675, 108 S.Ct. 2093, 100 L.Ed.2d 704 (1988).

§6:43.2.6 Insufficient Invocation of Right to Attorney

A suspect's question, "[d]o I need to talk to a lawyer before I sign?" is not a sufficient invocation of the right to remain silent. Robinson v. State, 851 S.W.2d 216 (Tex. Crim. App. 1991).

A suspect's statement in a letter to a sheriff that, "I'm here in by myself and still can't talk to a lawyer on this," is not a sufficient invocation of the right to counsel. Lucas v. State, 791 S.W.2d 35 (Tex. Crim. App. 1989).

A suspect's statement that, "[m]aybe I should talk to someone," is not a sufficient invocation of the right to counsel. Dinkins v. State, 894 S.W.2d 330 (Tex. Crim. App. 1995).

The defendant's statement, "I should have an attorney," was not an unambiguous request for an attorney under the totality of the circumstances. Davis v. State, 313 S.W.3d 317 (Tex. Crim. App. 2010).

A suspect's inquiry as to whether he would have a lawyer appointed to him upon reaching another city, when the inquiry dealt with charges pending in that other city is an insufficient invocation of the right to counsel. Collins v. State, 727 S.W.2d 565 (Tex. Crim. App. 1987).

Where the suspect at first invokes his right to counsel then voluntarily withdraws the invocation in the absence of interrogation, there is a valid waiver. Muniz v. State, 851 S.W.2d 238 (Tex. Crim. App. 1993) (where defendant stated, "No, I will talk to him later" when the officer began to dial the attorney's telephone number).

The defendant's refusal to sign the bottom of a magistrate warning form does not constitute the invocation of the right to counsel. Jacobs v. State, 787 S.W.2d 397 (Tex. Crim. App. 1990).

Where the suspect merely told his brother and grandmother to contact his attorney, even when done in the presence of a law enforcement officer, he did not sufficiently personally invoke his right to an attorney. Casias v. State, 678 S.W.2d 140 (Tex. App.—Waco 1984, *no pet.*).

Likewise, a defendant's request of a friend to call the defendant's attorney is not a sufficient invocation of the right to counsel. Nichols v. State, 754 S.W.2d 185 (Tex. Crim. App. 1988).

§6:43.2.7 Invocation of Right by Suspect's Attorney

The circumstances under which an attorney may invoke a client's Fifth Amendment rights are:

- Attorney-client relationship exists
- Attorney invokes the right after conferring with the client or in the client's presence
- Accused does nothing to contradict the attorney, and
- Officers agree not to question the accused in the attorney's absence

Janecka v. State, 739 S.W.2d 813 (Tex. Crim. App. 1987).

Efforts by an attorney, occurring outside the presence of the defendant and entirely unknown to him, do not constitute an invocation of his right to counsel. Fuller v. State, 253 S.W.3d 220 (Tex. Crim. App. 2008) (citing Moran v. Burbine, 475 U.S. 412, 106 S.Ct. 1135, 89 L.Ed.2d 410 (1986)).

§6:43.2.8 Valid Waivers of Right to Attorney

A voluntary, knowing, and intelligent waiver of the right to remain silent must be proven by a preponderance of the evidence. Robinson v. State, 851 S.W.2d 216 (Tex. Crim. App. 1991) (citing Colorado v. Connelly, 479 U.S. 157, 107 S.Ct. 515, 93 L.Ed.2d 473 (1986)).

If the prosecution shows that a suspect's decision not to rely on his Miranda rights was not coerced, that he at all times knew he could stand mute and consult a lawyer, and that he was aware the State could use his statement against him, the State's burden of proof is met and waiver is shown as a matter of law. Robinson (citing Moran v. Burbine, 475 U.S. 412, 106 S.Ct. 1135, 89 L.Ed.2d 410 (1986)).

Under certain narrow circumstances, interrogation may continue where an accused has invoked a limited right to counsel accompanied by an affirmative statement expressing his willingness to otherwise go forward with the interview. Hicks v. State, 860 S.W.2d 419 (Tex. Crim. App. 1993) (citing Connecticut v. Barrett, 479 U.S. 523, 107 S.Ct. 828, 93 L.Ed.2d 920 (1987)).

Where a suspect invokes his right to counsel only as to a "formal" statement, interrogation may continue informally, and the suspect may later fully waive the right to counsel by informing law enforcement officers that he wishes to make a "formal" statement without benefit of counsel. Corwin v. State, 870 S.W.2d 23 (Tex. Crim. App. 1993).

A suspect's statement to officer that, "I'm ready. I told you I would tell you about it if you would let me talk to [the attorney]," was a sufficient waiver of his rights (and also constituted a reinitiation of contact with officers). Mays v. State, 726 S.W.2d 937 (Tex. Crim. App. 1986).

§6:43.2.9 Invalid Waivers of Right to Attorney

The mere failure of the suspect to invoke his rights is not a valid waiver thereof. Miranda v. Arizona, 384 U.S. 436, 86 S.Ct. 1602, 16 L.Ed.2d 694 (1966).

Waivers will not be presumed from silence. Miranda.

Where a suspect did not expressly waive his rights and did not indicate that he had ceased to rely on an agreement with police to not question him in his attorney's absence, a resulting confession was involuntarily made. Stone v. State, 612 S.W.2d 542 (Tex. Crim. App. 1981).

Where the written confession of defendant does not contain any statement whatever to the effect that defendant, prior to making the statement, voluntarily, knowingly, and intelligently waived his rights, the confession should not have been admitted in evidence. Thibedeaux v. State, 729 S.W.2d 137 (Tex.App.—Beaumont 1987, *pet. ref'd*).

§6:43.3 Right to Terminate Interrogation

§6:43.3.1 In General

A defendant has the right to terminate a custodial interrogation with law enforcement at any time. Miranda v. Arizona, 384 U.S. 436, 86 S.Ct. 1602, 16 L.Ed.2d 694 (1966); CCP Art. 38.22.

A suspect's invocation of the right to terminate the interrogation must be scrupulously honored. The failure to scrupulously honor this right will result in the suppression of a resulting confession. Michigan v. Mosley, 423 U.S. 96, 96 S.Ct. 321, 46 L.Ed.2d 313 (1975); Faulder v. State, 611 S.W.2d 630 (Tex. Crim. App. 1979).

Where a suspect has invoked his right to remain silent, he may only be re-approached in a manner that scrupulously honors his continuing right to cut off questioning. Murphy v. State, 766 S.W.2d 246 (Tex. Crim. App. 1989), citing Michigan v. Mosley.

Whether a resumption of questioning after a refusal to talk is consistent with scrupulously honoring the right to remain silent depends on the facts of each case. Five factors are considered when examining whether rights were scrupulously honored:

- Whether the suspect was informed of the right to remain silent before the first questioning;
- Whether the suspect was informed of the right to remain silent before the subsequent questioning;
- The length of time between the first and subsequent questioning;
- If the subsequent questioning focused on a different crime; and
- Whether police scrupulously honored the suspect's initial invocation of the right to silence (resolution of this issue does not turn on an evaluation of credibility and demeanor). Michigan v. Mosley, 423 U.S. 96, 96 S.Ct. 321, 46 L.Ed.2d 313 (1975); Maestas; Maestas v. State, 987 S.W.2d 59 (Tex. Crim. App.1999); Herron v. State, 86 S.W.3d 621 (Tex. Crim. App. 2002).

§6:43.4 Right to Contact Consulate

The Vienna Convention on Consular Relations grants a foreign national who has been arrested, imprisoned or taken into custody the right to contact his consulate and requires the arresting government authorities to inform the individual of this right "without delay." Vienna Convention, Art. 36(1)(b), 21 U.S.T. at 100-101; 595 U.N.T.S. at 292; Maldonado v. State, 998 S.W.2d 239 (Tex. Crim. App. 1999).

However, if the Vienna treaty is not followed in the case of a foreign national, CCP Art. 38.23 does not require the exclusion of a resulting confession because a treaty violation is not a "law" violation under CCP Art. 38.23. Rocha v. State, 16 S.W.3d 1 (Tex. Crim. App. 2000).

Several courts have held that the Vienna Convention does not create individually enforceable rights. Sorto v. State, 173 S.W.3d 469 (Tex. Crim. App. 2005).

In order to prevail on a Vienna Convention treaty violation claim, the defendant must show a causal connection between the violation and his resulting confession. Sorto.

Where the evidence shows that law enforcement officers faxed notice to the defendant's consulate within 48 hours of his arrest, the defendant has failed to show a deliberate delay under the treaty. Sorto.

Even if the Vienna Convention treaty creates a judicially enforceable right (which the Court assumed without deciding), suppression of a confession is not an appropriate remedy. Sanchez-Llamas v. Oregon, 548 U.S. 331, 126 S.Ct. 2669, 165 L.Ed.2d 557 (U.S. 2006); Sierra v. State, 218 S.W.3d 85 (Tex. Crim. App. 2007).

PRACTICE TIP: Sanchez-Llamas suggests using the failure of law enforcement to notify the defendant's national consulate as part of a broader attack on the voluntariness of the confession.

The International Court of Justice Avena decision and a subsequent presidential memorandum urging state enforcement of the Avena decision do not constitute binding federal law preempting Art. 11.071, §5 of the Texas Code of Criminal Procedure under the Supremacy Clause of the United States Constitution and neither qualify as a previously unavailable factual or legal basis under §5(a) (1). Ex parte Medellin, 223 S.W.3d 315 (Tex. Crim. App. 2006).

§6:50 Voluntariness of Confessions

Under Texas Code of Criminal Procedure Article 38.21, "A statement of an accused may be used in evidence against him if it appears that the same was freely and voluntarily made without compulsion or persuasion." Oursbourn v. State, 259 S.W.3d 159, 169 (Tex. Crim. App. 2008).

§6:51 In General

Article 38.22 of the Code of Criminal Procedure sets out the rules governing admissibility of a defendant's written and/or oral statement that is the product of custodial interrogation. Oursbourn v. State, 259 S.W.3d 159, 171 (Tex. Crim. App. 2008).

The prosecution is under no burden to prove a confession is voluntary until the defendant presents some evidence raising a question of voluntariness. Terrazas v. State, 4 S.W.3d 720 (Tex. Crim. App. 1999); Romero v. State, 800 S.W.2d 539 (Tex. Crim. App. 1990).

The prosecution must then prove voluntariness by a preponderance of the evidence, when the issue is litigated before the judge as to the admissibility of the statement. Terrazas; Romero.

The state need not rebut a defendant's assertions but only must only controvert them. Muniz v. State, 851 S.W.2d 238 (Tex. Crim. App. 1993).

Voluntariness is determined under a totality of the circumstances test, weighing the particular facts and circumstances of each specific case. Schneckloth v. Bustamonte, 412 U.S. 218, 93 S.Ct. 2041, 36 L.Ed.2d 854 (1973); Mason v. State, 116 S.W.3d 248 (Tex.App.—Houston [14th Dist.] 2003, *pet. ref'd*).

A confession is involuntary or coerced if the totality of the circumstances demonstrate that the confessor did not make the decision to confess of his own free will. Arizona v. Fulminante, 499 U.S. 279, 111 S.Ct. 1246, 113 L.Ed.2d 302 (1991). A finding of coercion need not be based on actual violence by a government agent, a credible threat is sufficient. Fulminante.

Claims of involuntariness under Article 38.22 can be, but need not be, predicated on police overreaching, and they could involve the sweeping inquiries into the state of mind of a criminal defendant who has confessed that are not of themselves relevant to due process claims. Oursbourn, 259 S.W.3d at 172.

The voluntariness determination must be made by the trial court without regard to the truth or falsity of the alleged confession. Rogers v. Richmond, 365 U.S. 524, 81 S.Ct. 735, 5 L.Ed.2d 760 (1961).

In order for a confession to be found voluntary, it does not have to be in the exact language of the defendant. In fact, it can be handwritten by the interrogating officer. Henson v. State, 794 S.W.2d 385 (Tex.App.—Dallas 1990, *pet. ref'd*).

A police officer telling a defendant in a non-custodial setting "what had to be" in her statement is not the type of practice that has been held to be so inherently coercive as to make a statement involuntary. Terrazas.

Circumstances relevant to determining if a confession was voluntarily made include:

- Length of the detention and interrogation
- Whether the defendant was permitted access to family or attorney
- Presence or absence of physical brutality

Lane v. State, 933 S.W.2d 504 (Tex. Crim. App. 1996).

The circumstances support the voluntariness of the confession where:

- Defendant initiated a chain of events leading to his interviews with police
- Defendant initiated numerous attempts to mislead police with false alibis
- Police action was not coercive because the interviews were low-keyed and non-confrontational
- Defendant was repeatedly "Mirandized"

Lane.

Under Articles 38.21 and 38.22, fact scenarios that can raise a state-law claim of involuntariness (even though they do not raise a federal constitutional claim) include the following: (1) the suspect was ill and on medication and that fact may have rendered his confession involuntary; (2) the suspect was mentally retarded and may not have "knowingly, intelligently, and voluntarily" waived his rights; (3) the suspect "lacked the mental capacity to understand his rights"; (4) the suspect was intoxicated, and he "did not know what he was signing and thought it was an accident report"; (5) the suspect was confronted by the brother-in-law of his murder victim and beaten; (6) the suspect was returned to the store he broke into for questioning by several persons armed with six-shooters. Oursbourn, 259 S.W.3d at 172-3.

Although the courts have held that youth, intoxication, mental retardation, and other disabilities are usually not enough, by themselves, to render a statement inadmissible under Article 38.22, they are factors that a jury, armed with a proper instruction, is entitled to consider. Oursbourn, 259 S.W.3d at 173.

A defendant may claim that his statement was not freely and voluntarily made and, thus, may not be used as evidence against him under several different theories: (1) Article 38.22, §6, General Voluntariness; (2) Miranda v. Arizona as expanded in Article 38.22, §§2 and 3 (the Texas Confession Statute); or (3) the Due Process Clause. It may be involuntary under one, two, or all three theories. The theory of involuntariness determines whether and what type of an instruction may be appropriate. Oursbourn, 259 S.W.3d at 169.

Where the voluntariness issue is submitted to the jury, the state's burden is to prove voluntariness beyond a reasonable doubt. Oursbourn, 259 S.W.3d at 175.

Under Texas statutory law, there are three types of jury instructions that relate to the taking of confessions: (1) a "general" Article 38.22, §6 voluntariness instruction; (2) a "general" Article 38.22, §7 warnings instruction (involving warnings given under §§2 and 3); and (3) a "specific" Article 38.23(a) exclusionary-rule instruction. Oursbourn, 259 S.W.3d at 173.

Obviously, the evidence must raise a voluntariness issue, and the defendant should request a jury instruction that relates to his theory of involuntariness. But if the defendant never presents a proposed jury instruction (or fails to object to the lack of one), any potential error in the charge is reviewed only for egregious harm under Almanza; Oursbourn, 259 S.W.3d at 174.

§6:52 Art. 38.22 §6 General Voluntariness Issues

Under Texas Code of Criminal Procedure Art. 38.22 §6, the voluntariness of any statement of an accused can be challenged whether the statement was made in or out of custody. Terrazas v. State, 4 S.W.3d 720 (Tex. Crim. App. 1999); Oursbourn v. State, 259 S.W.3d 159, 171 (Tex. Crim. App. 2008).

When evidence raises an issue of the voluntariness of a defendant's statement under Art. 38.22, the trial judge must give a general voluntariness instruction in the jury charge. Oursbourn, 259 S.W.3d at 175.

This is the sequence of events that seems to be contemplated by §6: (1) a party notifies the trial judge that there is an issue about the voluntariness of the confession (or the trial judge raises the issue on his own); (2) the trial judge holds a hearing outside the presence of the jury; (3) the trial judge decides whether the confession was voluntary; (4) if the trial judge decides that the confession was voluntary, it will be admitted, and a party may offer evidence before the jury suggesting that the confession was not in fact voluntary; (5) if such evidence is offered before the jury, the trial judge shall give the jury a voluntariness instruction. Oursbourn, 259 S.W.3d at 175.

Section 6 expressly dictates the content of that instruction to be as follows: "unless the jury believes beyond a reasonable doubt that the statement was voluntarily made, the jury shall not consider such statement for any purpose nor any evidence obtained as a result thereof." Because §6 contains its own jury-instruction provision, it is not governed by the jury-instruction provision found in §7. Oursbourn, 259 S.W.3d at 175.

An interpretation of §6 that requires some sort of litigation before it becomes law applicable to the case accords not only with the statutory language but also with common sense. The broad range of voluntariness issues covered by §6 could easily be implicated by evidence that would also be relevant for other purposes, and §6 does not even require the existence of a factual dispute that might at least obliquely alert the trial judge to the need for an instruction. The §6 requirement that voluntariness be litigated in some manner before a jury instruction becomes necessary ensures that the trial judge is on notice that the instruction is required. Oursbourn, 259 S.W.3d at 176.

Article 38.22, §6 is "the law applicable" to any case in which a question is raised and litigated as to the general voluntariness of a statement of an accused. Under that statute, the trial judge must then (1) make an independent determination that the statement was made under voluntary conditions; and then (2) instruct the jurors that they shall not consider the statement for any purpose unless they believe, beyond a reasonable doubt, that the statement was made voluntarily. Oursbourn, 259 S.W.3d at 180-1.

Article 38.22 is aimed at protecting suspects from police overreaching. Section 6 of that article also protects people from themselves because the sole focus of the section is upon whether the defendant voluntarily made the statement freely and voluntarily, without compulsion or persuasion. Or, in the case of a custodial-interrogation statement, whether the suspect "knowingly, intelligently, and voluntarily" waived the rights set out in Article 38.22 §2(a) or §(3)(a). These inquiries do not turn solely on police overreaching. The behavior of the police may or may not be a factor. A confession given under the duress of hallucinations, illness, medications, or even a private threat, for example, could be involuntary under Article 38.21 and the Texas confession statute. Oursbourn, 259 S.W.3d at 172.

Thus, a claim that a purported waiver of the statutory rights enumerated in Article 38.22 is involuntary need not be predicated on police overreaching. Leza v. State, 351 S.W.3d 344, 352 (Tex. Crim. App. 2011) (citing Oursbourn, 259 S.W.3d at 172).

Circumstances unattributable to the police that nevertheless adversely impact an accused's ability to resist reasonable police entreaties to waive his statutory rights, such as intoxication, are factors in the voluntariness inquiry, though they are usually not enough, by themselves, to render a statement inadmissible under Article 38.22. Leza, 351 S.W.3d at 352 (citing Oursbourn, 259 S.W.3d at 173).

The exclusion of expert witness testimony that is so general that it has little, if any, relevance to a jury's voluntariness determination is not an abuse of trial court discretion. Lucio v. State, 351 S.W.3d 878, 902 (Tex. Crim. App. 2011).

§6:53 Miranda and Art. 38.22 §§2 and 3 Voluntariness Issues

Sections 2 and 3 of Art. 38.22 apply to a defendant's custodial-interrogation statements and provide that a defendant's custodial-interrogation statement is not admissible unless, prior to making the statement, he received the warnings provided in Art. 15.17 or Art. 38.22, §2(a) or §3(a) (which incorporate the requirements of Miranda), and he knowingly, intelligently, and voluntarily waived those rights. Oursbourn v. State, 259 S.W.3d 159, 171-2 (Tex. Crim. App. 2008).

It is the State's burden to establish a valid waiver of Miranda rights by a preponderance of the evidence. Leza v. State, 351 S.W.3d 344 (Tex. Crim. App. 2011).

There are two facets to any inquiry with respect to the adequacy of a purported waiver of Miranda rights: First, the waiver must be voluntary in the sense that it was the product of a free and deliberate choice rather than intimidation, coercion, or deception. Second the waiver must be made with a full awareness both of the nature of the right being abandoned and the consequences of the decision to abandon it. Leza.

Before it may be said that a waiver of a Miranda right is involuntary, however, there must be some element of official intimidation, coercion, or deception. And, with respect to the requirement that the waiver also be knowing and intelligent, once it is determined that a suspect at all times knew he could stand mute, and that he was aware of the State's intention to use his statements to secure a conviction, the analysis is complete and the waiver is valid as a matter of law. Leza.

It will suffice to render a waiver knowing and intelligent, in other words, that the accused has been made aware, and fully comprehends, that he has the right to remain silent in the face of police interrogation and to discontinue the dialogue at any time, and that the consequence of his waiver is that his words may be used against him later in a court of law. Leza.

A suspect's awareness of all the possible subjects of questioning in advance of interrogation is not relevant to determining whether the suspect voluntarily, knowingly, and intelligently waived his Fifth Amendment privilege under Miranda. Leza (citing Colorado v. Spring, 479 U.S. 564, 577, 107 S.Ct. 851, 93 L.Ed.2d 954 (1987)).

Similarly, the failure to immediately inform a defendant of the subject of the interrogation impacts neither the voluntariness nor the knowing and intelligent nature of any waiver of his statutory rights—as a matter of law. Leza.

Where a defendant's drug intoxication is not due to any causative action on the part of the police, his assertion that his intoxication rendered his Miranda waiver involuntary is foreclosed as a matter of federal constitutional law because the Fifth Amendment privilege against self-incrimination is not concerned with moral and psychological pressures to confess emanating from sources other than official coercion. Leza.

Drug intoxication can, however, be a factor that is relevant to determining whether a Miranda waiver was knowing and intelligent. Leza.

Similarly, drug intoxication can be a factor that is relevant to determining whether a waiver of statutory rights under Art. 38.22 was knowing and intelligent. Leza.

The obvious purpose of Art. 38.22, §7 is to authorize and require jury instructions regarding the warnings and safeguards for written and oral statements outlined in Article 38.22, §2 & §3 (warnings on the right to remain silent, right to counsel, etc). Oursbourn, 259 S.W.3d at 175.

As with §6, the trial judge's §7 jury instructions are general ones that set out the pertinent law and legal requirements of §§2 and 3 (or, in an appropriate case, those of Article 15.17). Oursbourn, 259 S.W.3d at 176.

Due process and Miranda claims may warrant both general and specific voluntariness jury instructions; Texas statutory claims warrant only a general voluntariness jury instruction. It is the defendant's responsibility to delineate which type of "involuntariness" he is claiming—a general (perhaps subjective) lack of voluntariness or a specific police-coerced lack of voluntariness—because the jury instruction is very different depending upon the type of claim. Oursbourn, 259 S.W.3d at 174.

The §7 instruction sets out the requirements of 38.22, §2 or §3 and asks the jury to decide whether all of those requirements were met. Oursbourn, 259 S.W.3d at 173.

The issue to be decided is whether there was compliance with the statutory warnings set out in both Articles 15.17 (Duties of Arresting Officer and Magistrate) and 38.22, §§2 and 3, and the voluntariness of the defendant's waiver of the rights. For the issue to be raised by the evidence there must be a genuine factual dispute, just as is true under Article 38.23 issues. Oursbourn, 259 S.W.3d at 176.

The same procedures—including a hearing outside the presence of the jury and the entry of written findings—that apply to a general voluntariness challenge under §6, also apply to a challenge that is made to the sufficiency of warnings and voluntary waiver of the rights communicated by those warnings. Oursbourn, 259 S.W.3d at 176.

Claims of involuntariness under Art. 38.22 can be, but need not be, predicated on police overreaching. Claims of involuntariness that are not themselves relevant to due process claims could involve sweeping inquiries into the state of mind of a criminal defendant who has confessed. Oursbourn, 259 S.W.3d at 172.

A statement that is involuntary as a matter of constitutional law is also involuntary under Article 38.22, but the converse need not be true. Oursbourn, 259 S.W.3d at 169.

Miranda and its prophylactic safeguards only apply to custodial statements. Wolfe v. State, 917 S.W.2d 270 (Tex. Crim. App. 1996).

There is no violation of a defendant's Miranda rights that apply to custodial-interrogation statements unless there is police misconduct causally related to a waiver of rights. Oursbourn, 259 S.W.3d at 170.

Because the failure of police to honor a defendant's rights under Miranda is not itself a constitutional violation, a defendant is not entitled to an Art. 38.23 charge when evidence at trial shows that the police did not comply with Miranda. Under these circumstances, the defendant is entitled to a charge under Art. 38.22. Contreras v. State, 312 S.W.3d 566 (Tex. Crim. App. 2010).

The failure to administer the Miranda warnings creates a presumption of compulsion and consequently, unwarned statements that are otherwise voluntary within the meaning of the Fifth Amendment must be excluded from evidence under Miranda, but the presumption, though not rebuttable for purposes of the state's case-in-chief, does not require that the statements and their fruits be discarded as inherently tainted. Oregon v. Elstad, 470 U.S. 298, 105 S.Ct. 1285, 84 L.Ed.2d 222 (1985); Baker v. State, 956 S.W.2d 19 (Tex. Crim. App. 1997).

The absence of Miranda warnings does not require the suppression of non-testimonial fruits of a defendant's statement where the statement was made voluntarily and was uncoerced. U.S. v. Patane, 542 U.S. 630, 124 S.Ct. 2620, 159 L.Ed.2d 667 (2004) (*plurality op.*) (where the defendant was prosecuted for possessing a gun that was found as a result of his statement).

Despite the inapplicability of Miranda to confessions obtained by foreign officials, the confession will still be excluded if the circumstances surrounding the taking of the confession shock the conscience of the U.S. court. Alvarado v. State, 853 S.W.2d 17 (Tex. Crim. App. 1991).

On direct appeal, the courts measure the propriety of the trial court's ruling with respect to alleged Miranda violations under the totality of the circumstances, almost wholly deferring to the trial court on questions of historical fact and credibility, but reviewing *de novo* all questions of law and mixed questions of law and fact that do not turn on credibility determinations. Leza.

§6:54 Due Process Issues

Due process may warrant both general and specific voluntariness jury instructions. It is the defendant's responsibility to delineate which type of involuntariness he is claiming—a general (perhaps subjective) lack of voluntariness or a specific police-coerced lack of voluntariness—because the jury instruction is very different depending upon the type of claim. Oursbourn v. State, 259 S.W.3d 159, 174 (Tex. Crim. App. 2008).

A confession may be involuntary under the Due Process Clause only when there is police overreaching. Absent police misconduct causally related to the confession, there is simply no basis for concluding that any state actor has deprived a criminal defendant of due process of law. Oursbourn, 259 S.W.3d at 169-170.

Involuntary confessions offend due process only when they flow from improper conduct of law enforcement officials. Lane v. State, 933 S.W.2d 504 (Tex. Crim. App. 1996).

A statement is involuntary if there was official, coercive conduct of such a nature that any statement obtained thereby was unlikely to have been the product of an essentially free and unconstrained choice by its maker. Alvarado v. State, 912 S.W.2d 199 (Tex. Crim. App. 1995).

The test is whether the defendant's will was overborne by police coercion. Guardiola v. State, 20 S.W.3d 216 (Tex.App.—Houston [14th Dist.] 2000, *pet. ref'd*); Mason v. State, 116 S.W.3d 248 (Tex.App.—Houston [14th Dist.] 2003, *pet. ref'd*).

Coercive police activity is a necessary predicate to the finding that a confession is not voluntary within the meaning of the Due Process Clause of the Fourteenth Amendment. Colorado v. Connelly, 479 U.S. 157, 107 S.Ct. 515, 93 L.Ed.2d 473 (1986).

In determining whether police conduct is improper, the court should take into account police knowledge of the suspect's special weaknesses, including youth and low intelligence. Lane; Gallegos v. Colorado, 370 U.S. 49, 82 S.Ct. 1209, 8 L.Ed.2d 325 (1962).

Certain interrogation techniques, either in isolation or as applied to the unique characteristics of a particular suspect, will render a confession involuntary and thus, inadmissible. Miller v.

Fenton, 474 U.S. 104, 106 S.Ct. 445, 88 L.Ed.2d 410 (1985).

However, factors bearing on the characteristics of the accused are irrelevant in the absence of any state action that would support a finding of police misconduct or overreaching. Terrazas v. State, 4 S.W.3d 720 (Tex. Crim. App. 1999); Colorado v. Connelly.

Both physical and mental compulsion remove the voluntariness element from a defendant's decision to incriminate himself. Thomas v. State, 723 S.W.2d 696 (Tex. Crim. App. 1986).

Physical compulsion includes such obvious force as physical torture or extended deprivation of food and water. Thomas (citing U.S. v. Carignan, 342 U.S. 36, 72 S.Ct. 97, 96 L.Ed.2d 48 (1951)).

Mental compulsion includes the more subtle force associated with offering a defendant two choices, one of which results in a penalty, punishment or detriment from which the defendant is entitled to be free. Thomas (citing South Dakota v. Neville, 459 U.S. 553, 103 S.Ct. 916, 74 L.Ed.2d 748 (1983)).

Statements that have been found to be involuntary under Miranda or the Due Process Clause involve the crucial element of police overreaching and involve fact scenarios such as the following: (1) the suspect was subjected to a four-hour interrogation while incapacitated and sedated in an intensive-care unit; (2) the suspect, while on medication, was interrogated for over eighteen hours without food, medication, or sleep; (3) the police officers held a gun to the head of the wounded suspect to extract a confession; (4) the police interrogated the suspect intermittently for 16 days using coercive tactics while he was held incommunicado in a closed cell without windows and was given limited food; (5) the suspect was held for four days with inadequate food and medical attention until he confessed; (6) the suspect was subjected to five days of repeated questioning during which police employed coercive tactics; (7) the suspect was held incommunicado for three days with little food, and the confession was obtained when officers informed him that their chief was preparing to admit a lynch mob into the jail; (8) the suspect was questioned by relays of officers for 36 hours without an opportunity for sleep. Oursbourn, 259 S.W.3d at 170-1 (citing Colorado v. Connelly).

Also, a confession obtained under threat of removal from office of the defendants as police officers if they refused to answer (even though they were given a warning that anything they said would be used against them and that they could refuse to answer if it would tend to incriminate them) was coerced and invalid in a subsequent prosecution for conspiracy to obstruct administration of traffic laws. Garrity v. New Jersey, 385 U.S. 493, 87 S.Ct. 616, 17 L.Ed.2d 562 (1967).

As is evident from these fact scenarios, due process and Miranda claims of involuntariness generally do not require sweeping inquiries into the state of mind of a criminal defendant who has confessed. They involve an objective assessment of police behavior. The Constitution leaves voluntariness claims based on the defendant's state of mind to be resolved by state laws governing the admission of evidence. In Texas, that state law is Article 38.22, the Texas Confession Statute. Oursbourn, 259 S.W.3d at 171.

§6:55 Art. 38.23 Issues

Article 38.23 provides that evidence obtained in violation of the Constitution or laws of the United States or Texas may not be admitted in a criminal case. The statute also provides that a jury instruction should be submitted if a fact issue is raised about whether such a violation has occurred. The jury instruction requires the jury to disregard evidence that it finds was obtained in violation of the United States or Texas Constitution or laws. Hamal v. State, 390 S.W.3d 302 (Tex. Crim. App. 2012); Robinson v. State, 377 S.W.3d 712 (Tex. Crim. App. 2012).

To be entitled to an Article 38.23 jury instruction, three predicates must be met: (1) the evidence heard by the jury must raise an issue of fact; (2) the evidence on that fact must be affirmatively contested; and (3) the contested factual issue must be material to the lawfulness of the challenged conduct. Hamal, 390 S.W.3d at 306; Robinson, 377 S.W.3d at 719.

When a disputed, material issue of fact is successfully raised, the terms of the statute are mandatory, and the jury must be instructed accordingly. Evidence to justify an Article 38.23(a) instruction can derive from any source, no matter whether strong, weak, contradicted, unimpeached, or unbelievable. But it must, in any event, raise a *factual* dispute about how the evidence was obtained. Where the issue raised by the evidence at trial does *not* involve controverted historical facts, but only the proper application of the law to undisputed facts, that issue is properly left to the determination of the trial court. Robinson, 377 S.W.3d at 719.

If there is a disputed fact issue about whether an inherently coercive practice was employed—by either an officer or a private citizen—to wring a confession out of a suspect against his will, a specific exclusionary-rule instruction under Article 38.23 is appropriate. Oursbourn, 259 S.W.3d at 177.

The Article 38.23(a) "specific" instruction is fact-based: For example, "Do you believe that Officer Obie held a gun to the defendant's head to extract his statement? If so, do not consider the defendant's confession." Oursbourn, 259 S.W.3d at 173-4.

§6:56 Involuntary Confessions

If a confession is involuntarily obtained, it is error for either the confession or evidence obtained from it to be admitted at trial. Sossamon v. State, 816 S.W.2d 340 (Tex. Crim. App. 1991); citing Jackson v. Denno, 378 U.S. 368, 84 S.Ct. 1774, 12 L.Ed.2d 908 (1964); Baker v. State, 956 S.W.2d 19 (Tex. Crim. App. 1997).

There is no statutory exception in Art. 38.22 (as there is under Art. 38.23) allowing the admission of statements that constitute a crime committed after law enforcement officials violate Art. 38.22 while attempting to obtain evidence of a previously committed crime. Nguyen v. State, 292 S.W.3d 671 (Tex. Crim. App. 2009).

The express language of Art. 38.22 does not include an exception after a 38.22 violation for statements that are not confessional in nature, do not implicate the accused for the offense prosecuted, or constitute an offense. Nguyen.

§6:56.1 Factors Rendering a Confession Involuntary

§6:56.1.1 Promises

A promise will render a confession "involuntary" if it is:

- of some benefit to the accused
- positive
- made or sanctioned by a person in authority; and
- of such a character as would likely influence the accused to speak untruthfully

Smith v. State, 779 S.W.2d 417 (Tex. Crim. App. 1989); Long v. State, 823 S.W.2d 259 (Tex. Crim. App. 1991).

To determine if the promise of a benefit was likely to influence a suspect to speak untruthfully, courts must look to whether the circumstances surrounding the promise made the suspect inclined to admit a crime he did not commit. Sossamon v. State, 816 S.W.2d 340 (Tex. Crim. App. 1991).

A promise will not render a resulting confession inadmissible unless someone in authority makes the promise. Dixon v. State, 624 S.W.2d 373 (Tex. App.—Dallas 1981, *no pet.*) (promise by victim's mother to drop charges if the defendant confessed did not render the confession involuntary); Bordman v. State, 56 S.W.3d 63 (Tex.App.—Houston [14th Dist.] 2001, *pet. ref'd*) (where the defendant confessed to a church elder).

Promises made to the defendant before he became a suspect do not affect the voluntariness of a subsequent confession. Arnold v. State, 873 S.W.2d 27 (Tex. Crim. App. 1993).

A defendant does not have to show that his confession was false in order to prevail on the issue of whether a promise by law enforcement induced him to confess untruthfully. Martinez v. State, 127 S.W.3d 792 (Tex. Crim. App. 2004).

In situations where there was testimony that the defendant could have taken a police officer's statement to be a promise, the admission of the confession will be upheld where the appellate courts find there was no positive promise made to the defendant. Martinez.

A prediction about future events (such as a detective saying to the defendant that things will go better for you if you cooperate) is not the same as a promise because the prediction about future events depends on the actions of the defendant rather than the detective. Mason v. State, 116 S.W.3d 248 (Tex. App.—Houston [14th Dist.] 2003, *pet. ref'd*).

§6:56.1.1.1 Promises Not Affecting Voluntariness

A promise to allow a suspect to take a polygraph is not the type of promise that will induce a person to falsely inculpate himself. Smith v. State, 779 S.W.2d 417 (Tex. Crim. App. 1989).

Where a detective told the defendant that he would do what he could to help the defendant if he gave further information on another suspect's involvement in other crimes, the statements did not render the subsequent confession involuntary. Garcia v. State, 919 S.W.2d 370 (Tex. Crim. App. 1994).

Where a police officer made the general statement he would help the defendant if the defendant cooperated with him, the resulting confession was voluntary. Dykes v. State, 657 S.W.2d 796 (Tex. Crim. App. 1983).

Where the detective told the defendant that he would inform the district attorney as to his cooperation or lack of same, the resulting confession was voluntary. Sorola v. State, 674 S.W.2d 809 (Tex.App.—San Antonio 1984), affirmed at 693 S.W.2d 417 (Tex. Crim. App. 1985).

Promises to seek the appropriate charge based upon the facts and circumstances of the case do not render a subsequent confession involuntary. Jacobs v. State, 787 S.W.2d 397 (Tex. Crim. App. 1990); Freeman v. State, 723 S.W.2d 727 (Tex. Crim. App. 1986).

Where the promise made was for a benefit that the defendant should know would occur anyway as a matter of course, the promise will not render a subsequent confession involuntary. Jacobs (where the defendant was allowed a visit from his girlfriend).

A promise to contact charitable authorities on behalf of the defendant's family where the contact could have been made by anyone, including defendant's family, does not render a subsequent confession involuntary. Muniz v. State, 851 S.W.2d 238 (Tex. Crim. App. 1993).

A statement by a prosecutor to effect that the case seemed like a good death penalty case and that the state would do its best to get the death penalty is not a promise that would likely induce an accused to speak untruthfully. Long v. State, 823 S.W.2d 259 (Tex. Crim. App. 1991).

§6:56.1.1.2 Promises and Other Factors Affecting Voluntariness

Where a defendant was promised a reduction of time to be served in another case it was found to be likely to induce him to speak untruthfully and rendered the resulting confession involuntary. Sossamon v. State, 816 S.W.2d 340 (Tex. Crim. App. 1991).

If a suspect is told that things might go easier on him if he confesses, this will be an important factor in deciding against voluntariness. Sterling v. State, 800 S.W.2d 513 (Tex. Crim. App. 1990).

A confession made after a representation that the state would not seek the death penalty in exchange, is inadmissible. McMahon v. State, 582 S.W.2d 786 (Tex. Crim. App. 1978).

A promise to drop charges against the defendant's mother in exchange for his confession rendered the resulting confession involuntary. Walker v. State, 626 S.W.2d 777 (Tex. Crim. App. 1982).

§6:56.1.2 "For or Against"

A confession is not admissible if the uncontradicted evidence shows the person who obtained the confession informed the accused prior to the interrogation that his confession might be used "for or against" him. Dinkins v. State, 894 S.W.2d 330 (Tex. Crim. App. 1995).

A confession resulting from a law enforcement agent's pre-interrogation statement that it can be used "for or against" the defendant is inadmissible as a matter of law because it constitutes an improper inducement and because it does not comply with the statutory warnings in CCP Art. 38.22. Dunn v. State, 721 S.W.2d 325 (Tex. Crim. App. 1986) (*plurality opinion*).

Where the proper statutory warning is given and a suspect is told at a different time that a statement can be used "for or against" him, there is no statutory violation.

The voluntariness of the statement may be affected, but it is not rendered automatically inadmissible. Nenno v. State, 970 S.W2d 549 (Tex. Crim. App. 1998).

The *per se* rule of inadmissibility as stated in Dinkins and Dunn does not apply to an officer's remarks which were made during the subsequent interrogation. Creager v. State, 952 S.W.2d 852 (Tex. Crim. App. 1997).

Where the representation that the confession can be used "for or against" the suspect occurs during the interrogation, the issue is one of voluntariness of the confession. Creager.

However, where there is a factual discrepancy as to whether such a representation was made, the trial judge is responsible for determining whether the confession is admissible. Dinkins.

Evidence that a defendant received an improper warning need not be wholly rebutted, but merely contradicted in order that a confession be admissible. Muniz v. State, 851 S.W.2d 238 (Tex. Crim. App. 1993).

§6:56.1.3 Intoxication

Intoxication is but one relevant factor in determining whether a suspect understood his rights. Ripkowski v. State, 61 S.W.3d 378 (Tex. Crim. App. 2001).

The fact that the person making the statement was intoxicated at the time he confessed is not alone sufficient to render the confession involuntary. Garcia v. State, 919 S.W.2d 370 (Tex. Crim. App. 1994).

In order for a resulting confession to be involuntary there must also be some evidence of police coercion. Ripkowski v. State, 61 S.W.3d 378 (Tex. Crim. App. 2001).

If the suspect is intoxicated to the point that he is incapable of making an independent, informed choice of free will, then his confession is involuntary. Nichols v. State, 754 S.W.2d 185 (Tex. Crim. App. 1988).

Where the record supported the conclusion that the defendant was not so intoxicated on cocaine as to render his confession involuntary the trial court's finding will be upheld. Jenkins v. State, 912 S.W.2d 793 (Tex. Crim. App. 1995).

§6:56.1.4 Pain and/or Torture

Where the record did not demonstrate that the defendant was in pain to the point he could not make an informed decision, his subsequent confession will be found voluntary. Jenkins v. State, 912 S.W.2d 793 (Tex. Crim. App. 1995).

A confession resulting from brutal torture was obtained in violation of the Due Process Clause. Brown v. Mississippi, 297 U.S. 278, 56 S.Ct. 461, 80 L.Ed. 682 (1936).

Where police officers held a gun to the head of the defendant to obtain a confession, the resulting confession was involuntarily obtained. Beecher v. Alabama, 389 U.S. 35, 88 S.Ct. 189, 19 L.Ed.2d 35 (1967).

§6:56.1.5 Deception or Misrepresentation

In determining the voluntariness of a defendant's written statement, police falsehoods are relevant. Mason v. State, 116 S.W.3d 248 (Tex.App.—Houston [14th Dist.] 2003, *pet. ref'd*); Frazier v. Cupp, 394 U.S. 731, 89 S.Ct. 1420, 22 L.Ed.2d 684 (1969).

Trickery or deception does not make a statement involuntary unless the method is calculated to produce an untruthful confession or is offensive to due process. Creager v. State, 952 S.W.2d 852 (Tex. Crim. App. 1997); Mason.

The effect of a lie must be analyzed in the context of all the circumstances of the interrogation. Mason.

Some types of police deception employed during custodial interrogation designed to elicit a confession from an accused are permissible. The focus is whether the behavior of the police was such as to overbear the will of the accused. Rogers v. Richmond, 365 U.S. 534, 81 S.Ct. 735, 5 L.Ed.2d 760 (1961).

Misrepresentations made by police to a suspect during an interrogation are a relevant factor in assessing whether the suspect's confession was voluntary, but they are insufficient to render an otherwise voluntary confession inadmissible. Green v. State, 934 S.W.2d 92 (Tex. Crim. App. 1996) (citing Frazier v. Cupp, 394 U.S. 731, 89 S.Ct. 1420, 22 L.Ed.2d 684 (1969)).

The misrepresentation must be viewed in the context of the totality of the circumstances. Green (citing Frazier v. Cupp).

Of the numerous types of police deception, a misrepresentation relating to an accused's connection to the crime is the least likely to render a confession involuntary. Green.

A police misrepresentation that they had an eyewitness to the murder for which the suspect was being interrogated did not rise to the level of coercion contemplated in Lynumn v. Illinois, 372 U.S. 528, 83 S.Ct. 917, 9 L.Ed.2d 922 (1963). Green.

It is not impermissible for the police to falsely tell the suspect that his accomplice had given a complete confession. Frazier v. Cupp.

It is not impermissible for police to falsely tell the suspect that the victim, on his deathbed, had identified the suspect as the perpetrator. Rodriquez v. State, 934 S.W.2d 881 (Tex.App.—Waco 1996, *no pet.*).

Article 38.23 of the Code of Criminal Procedure bars the admissibility of a confession where the interrogating officer fabricates documentary evidence in violation of Texas Penal Code §37.09 and uses it to persuade a suspect to confess. Wilson v. State, 311 S.W.3d 452 (Tex. Crim. App. 2010) (where the officer fabricated a forensic fingerprint report).

Other types of deception which are likely to lead to involuntary confessions are based on extrinsic considerations. Green.

For example, a threat to take away a suspect's welfare benefits and custody of her children was found to be inherently coercive because it did more than affect the suspect's belief concerning her actual guilt or innocence; it impaired her free choice, thereby casting doubt upon the reliability of the resulting confession. Lynumn v. Illinois.

It is impermissible for the police to falsely suggest that a close police officer friend of the suspect would lose his job if the suspect did not cooperate and confess. Spano v. New York, 360 U.S. 315, 79 S.Ct. 1202, 3 L.Ed.2d 1265 (1959).

Mere silence by law enforcement officials as to the subject matter of the interrogation is not "trickery" sufficient to invalidate a suspect's waiver of Miranda rights. Colorado v. Spring, 479 U.S. 564, 107 S.Ct. 851, 93 L.Ed.2d 954 (1987).

§6:56.1.6 Prolonged Interrogation; Lack of Food or Sleep

In order to make a voluntariness claim at trial where alleging a lack of sleep, a defendant must show that there is a connection between his lack of sleep and coercive government conduct. Either the police must have caused the defendant to go without sleep or they must have somehow exploited the lack of sleep to overbear the defendant's will. Contreras v. State, 312 S.W.3d 566, 575 (Tex. Crim. App. 2010).

Where the defendant successfully raises this issue, he will be entitled to an Art. 38.23 instruction in the jury charge. Contreras, 312 S.W.3d at 575.

Holding a defendant for a prolonged period without allowing him to call relatives or an attorney can also result in an involuntary confession. Contreras, 312 S.W.3d at 578.

A two-hour interrogation is not a sufficiently lengthy period to render a resulting confession involuntary. Contreras, 312 S.W.3d 579.

Lack of sleep for 16 hours will not alone render a suspect's confession involuntary. Barney v. State, 698 S.W.2d 114 (Tex. Crim. App. 1985).

Where the suspect's inability to eat or sleep was brought about by his conscience and not by the police, there is no official coercion and, therefore, a resulting confession is not involuntary. Chambers v. State, 866 S.W.2d 9 (Tex. Crim. App. 1993).

Questioning by relays of officers over a 36-hour period without adequate opportunity for sleep rendered a resulting confession involuntary. Ashcraft v. Tennessee, 322 U.S. 143, 64 S.Ct. 921, 88 L.Ed.2d 1192 (1944).

Where a defendant was held for four days without adequate food and medical attention, the resulting confession was involuntary. Reck v. Pate, 367 U.S. 433, 81 S.Ct. 1541, 6 L.Ed.2d 948 (1961).

Where a defendant was held for five days of repeated questioning during which police employed coercive tactics, a resulting confession was found to be involuntary. Culombe v. Connecticut, 367 U.S. 568, 81 S.Ct. 1860, 6 L.Ed.2d 1037 (1961).

Where a defendant was held incommunicado for three days with little food, and was informed by officers that the chief of police was ready to admit a lynch mob, the resulting confession was found to be involuntary. Payne v. Arkansas, 356 U.S. 560, 78 S.Ct. 844, 2 L.Ed.2d 975 (1958).

Where a defendant on medication was interrogated for over 18 hours without food or sleep, a resulting confession was found involuntary. Greenwald v. Wisconsin, 390 U.S. 519, 88 S.Ct. 1152, 20 L.Ed.2d 77 (1968).

Where a defendant was incapacitated and sedated in an intensive-care unit, a four-hour interrogation was found to render a subsequent confession involuntary. Mincey v. Arizona, 437 U.S. 385, 98 S.Ct. 2408, 57 L.Ed.2d 290 (1978).

§6:56.1.7 Force or Violence

A threat of violence made by a police officer in order to obtain a confession may render a resulting confession inadmissible; however, where no express or implied threat is made by police, a suspect's belief that a confession will relieve the danger will not render a subsequent confession inadmissible. Roberts v. State, 545 S.W.2d 157 (Tex. Crim. App. 1977).

A confession resulting from brutal torture has been obtained in violation of the Due Process Clause. Brown v. Mississippi, 297 U.S. 278, 56 S.Ct. 461, 80 L.Ed. 682 (1936); Connor v. State, 640 S.W.2d 374 (Tex.App.—1982), affirmed at 773 S.W.2d 13 (Tex. Crim. App. 1989).

Where police officers held a gun to the head of the defendant to obtain a confession, the resulting confession was involuntarily obtained. Beecher v. Alabama, 389 U.S. 35, 88 S.Ct. 189, 19 L.Ed.2d 35 (1967).

§6:56.1.8 Psychological Tactics

There is no *per se* rule against the use of psychological tactics in interrogations. Mason v. State, 116 S.W.3d 248 (Tex.App.—Houston [14th Dist.] 2003, *pet. ref'd*).

Interrogation tactics that include praying with and praying for the accused should be considered as a factor in the totality of the circumstances test of the voluntariness of a confession. Davis v. North Carolina, 384 U.S. 737, 86 S.Ct. 1761, 16 L.Ed.2d 895 (1966).

§6:56.1.9 Lack of Intellect or Mental Impairment

The totality of the circumstances standard for assessing the voluntariness of a confession given by a person of normal mentality is also the appropriate standard to apply when a confession is made by someone suffering from mental retardation and mental illness. Delao v. State, 235 S.W.3d 235 (Tex. Crim. App. 2007).

There is no requirement that a person be literate before his confession can be admissible. Westley v. State, 754 S.W.2d 224 (Tex. Crim. App. 1988).

Where the evidence shows that a suspect's lack of literacy did not prevent him from understanding the warnings and his rights, a resulting confession will be admissible. Westley.

In order for a defendant's mental condition to render a confession involuntary it must be shown in connection with official coercion. Walker v. State, 842 S.W.2d 301 (Tex.App.—Tyler 1992, *no pet.*).

A suspect's limited intelligence is a factor to be considered in a voluntariness analysis; but does not mandate a finding of involuntariness as a matter of law. King v. State, 831 S.W.2d 891 (Tex.App.—Houston [14th Dist.] 1992, *no pet.*).

A mental defect may be significant enough to render a confession inadmissible. Page v. State, 614 S.W.2d 819 (Tex. Crim. App. 1981).

Where the bulk of the evidence shows that the defendant had the basic reasoning skills necessary to understand his warnings, a resulting confession will be found voluntary. Penry v. State, 903 S.W.2d 715 (Tex. Crim. App. 1995).

Insanity will negate the admissibility of a confession where this weakness is exploited by coercive tactics. Blackburn v. Alabama, 361 U.S. 199, 80 S.Ct. 274, 4 L.Ed.2d 242 (1960).

Where a police physician administered a "truth-serum" drug to the defendant, the resulting confession was rendered involuntary. Townsend v. Sain, 372 U.S. 293, 83 S.Ct. 745, 9 L.Ed.2d 770 (1963).

Emotional confusion and debilitation of a suspect may call into question the voluntariness of a confession. Ballestero v. State, 640 S.W.2d 423 (Tex.App.—San Antonio 1982, *no pet.*).

§6:56.1.10 Language Barriers

An ability to understand the police officers, understand the oral and written warnings and to read the confession will defeat a claim of inability to sufficiently understand the English language. Farmah v. State, 789 S.W.2d 665 (Tex.App.—Houston [1st Dist.] 1990), overruled on other grounds 883 S.W.2d 674 (Tex. Crim. App. 1994).

Where the evidence shows that the law enforcement agents went to great lengths to translate everything several times to the defendant in order to ensure that he understood what he was doing, a trial court does not abuse its discretion in ruling the resulting confession voluntary. Montoya v. State, 810 S.W.2d 160 (Tex. Crim. App. 1989).

A statement dictated in Spanish but recorded in English is not rendered an involuntary oral confession where the Code of Criminal Procedure is otherwise followed. Mestiza v. State, 923 S.W.2d 720 (Tex.App.—Corpus Christi 1996, *no pet.*).

§6:56.1.11 Use of Grand Jury Subpoena

Where a defendant has invoked any of his Miranda or CCP Art. 38.22 rights, the use of a grand jury subpoena to bring him in for further questioning renders a subsequent confession involuntary. Guardiola v. State, 20 S.W.3d 216 (Tex.App.—Houston [14th Dist.] 2000 *pet. ref'd*).

§6:56.1.12 Threats Against Family Members

Threats against family members can result in an involuntary confession. Contreras v. State, 312 S.W.3d 566, 578 (Tex. Crim. App. 2010).

Threats to arrest a family member can result in an involuntary confession where police do not have probable cause to arrest the family member. Where they have sufficient probable cause, the threats will not vitiate the voluntariness of the confession. Contreras, 312 S.W.3d at 578.

§6:57 The Voluntariness Hearing

§6:57.1 In General

If, at any point during trial, an issue arises as to the voluntariness of a custodial statement made by the defendant, the trial court must hold a hearing on the issue. Jackson v. Denno, 378 U.S. 368, 84 S.Ct. 1774, 12 L.Ed.2d 908 (1964). The failure to hold a voluntariness hearing where the record indicates a voluntariness issue is reversible error. Hugley v. State, 505 S.W.2d 914 (Tex. Crim. App. 1974).

A voluntariness hearing must also be held where the defendant challenges a non-custodial statement made to a law enforcement officer. Terrazas v. State, 4 S.W.3d 720 (Tex. Crim. App. 1999).

Even if the confession is used for impeachment, failure to make a determination of voluntariness outside the presence of the jury, when requested, is reversible error. Madden v. State, 691 S.W.2d 688 (Tex. Crim. App. 1985). The voluntariness hearing must be held outside the presence of the jury. Rule 104(c) Texas Rules of Evidence. The trial court is not required to hold the voluntariness hearing pretrial but may do so at any time before the statement is presented to the jury. Ross v. State, 678 S.W.2d 491 (Tex. Crim. App. 1984).

As the movant in a motion to suppress a confession, the defendant has the burden to produce evidence that defeats the presumption of proper police conduct; thereby shifting the burden of proof to the state. Chavarria v. State, 876 S.W.2d 388 (Tex.App.—El Paso 1994). At a hearing to determine the voluntariness of a confession, the state need not rebut the defendant's assertions, but only controvert them. Muniz v. State, 851 S.W.2d 238 (Tex. Crim. App. 1993).

By taking the stand during the voluntariness hearing, the accused does not subject himself or herself to cross-examination as to other issues in the case. Rule 104(d) Texas Rules of Evidence.

The trial court may ask questions at the hearing in order to determine specific issues, as long as there is no indication that the court is not impartial. Richardson v. State, 632 S.W.2d 700 (Tex.App.—Fort Worth 1982).

TIP TO THE PROSECUTION: If, during the voluntariness hearing, the defendant presents testimony on a particular issue or develops the issue on cross-examination, present testimony controverting the defendant's assertions. The failure to do so will result in a finding at trial or on appeal that the confession should not be admitted.

The burden is on the defendant, as the moving party in a motion to suppress evidence obtained in violation of the law under Art. 38.23, to produce evidence demonstrating a causal connection between the alleged statutory violation and the resulting evidence. The burden then shifts to the state to either disprove the evidence the defendant has produced, or bring an attenuation-of-taint argument to demonstrate that the causal chain asserted by the defendant was in fact broken. Pham v. State, 175 S.W.3d 767 (Tex. Crim. App. 2005).

A defendant who moves for suppression under Art. 38.23 due to the violation of a statute has the burden of producing evidence of a statutory violation. Only when this burden is met does the State bear a burden to prove compliance. State v. Robinson, 334 S.W.3d 776, 779 (Tex. Crim. App. 2011).

The Rules of Evidence do not apply at suppression hearings. This is because former Rule 1101(d)(4) that previously dictated to the contrary no longer exists. Granados v. State, 85 S.W.3d 217 (Tex. Crim. App. 2002).

Therefore, Rule 702 is not applicable at a suppression hearing. Hall v. State, 297 S.W.3d 294 (Tex. Crim. App. 2009).

In a pre-trial motion to suppress hearing, a trial court may rely upon any relevant, reliable, and credible information, even though it may be unsworn hearsay (in this case a police report found to be sufficiently reliable). The language of CCP Art. 28.01 is permissive, not restrictive, and allows for reports to be used. Ford v. State, 305 S.W.3d 530 (Tex. Crim. App. 2009).

It is within the trial court's discretion to grant the State's request to reopen a motion-to-suppress hearing to supplement the record with additional testimony, outside the jury's presence, even over the objection of the defendant. Black v. State, 362 S.W.3d 626 (Tex. Crim. App. 2012).

Appellate courts review a trial court's ruling on a motion to suppress for an abuse of discretion and overturn that ruling only if it is outside the zone of reasonable disagreement. Martinez v. State, 348 S.W.3d 919, 922 (Tex. Crim. App. 2011). They conduct their review with two standards, giving almost total deference to a trial court's determination of historic facts and mixed questions of law and fact that rely on the credibility of a witness, but applying a *de novo* standard of review to pure questions of law and mixed questions that do not depend on credibility determinations. Martinez, 348 S.W.3d. at 922–23.

§6:57.2 Findings of Fact and Conclusions of Law

If the trial court finds beyond a reasonable doubt that the confession is voluntarily made and admissible as a matter of law and fact, the judge prepares and signs an order stating these findings. The order is filed among the papers of the cause. CCP Art. 38.22 §6.

TIP TO THE BENCH: Findings of fact and conclusions of law must be filed in the record before the record is filed in the appellate court (i.e., within 75 days of the verdict). Green v. State, 906 S.W.2d 937 (Tex. Crim. App.1995).

The requirement that the trial court file written findings of fact and conclusions of law applies to bench trials as well as jury trials. Wicker v. State, 740 S.W.2d 779 (Tex. Crim. App. 1987). The findings must be filed whether or not the defendant requests them, and if they are not filed the appeal will be abated for their preparation. Wicker; Urias v. State, 155 S.W.3d 141 (Tex. Crim. App. 2004).

Art. 38.22, sec. 6 clearly requires that the trial court make findings of fact and conclusions of law in all cases concerning voluntariness, whether or not requested by a party. The statute has no exceptions. Vasquez v. State, 411 S.W.3d 918, 920 (Tex. Crim. App. 2013).

A trial court satisfies the requirements of Art. 38.22 when it dictates its findings and conclusions to the court reporter, and they are transcribed and made a part of the statement of facts, filed with the district clerk, and made a part of the appellate record. Murphy v. State, 112 S.W.3d 592 (Tex. Crim. App. 2003).

As a constitutional matter, the trial court's conclusion as to the voluntariness of a confession must appear in the record with unmistakable clarity. Bonham v. State, 644 S.W.2d 5 (Tex. Crim. App. 1983).

TIP TO THE BENCH: When making written findings of fact and conclusions of law, it is the better practice to make an express finding on the issue of waiver of rights. Stanley v. State, 866 S.W.2d 306 (Tex. App.—Houston [14th Dist.] 1993).

The findings of fact and conclusions must be made by the judge who heard the testimony at the hearing and not by a succeeding judge. Garcia v. State, 15 S.W.3d 533 (Tex. Crim. App. 2000).

TIP TO THE BENCH: When writing findings of fact and conclusions of law regarding the voluntariness of a confession, the trial court is not obligated to outline testimony which does not support its conclusions. Guidry v. State, 12 S.W.3d 15 (Tex. Crim. App. 1999). However, the findings must be sufficiently detailed to enable the appellate court to determine the basis for the trial court's ruling and to assist the appellate court in determining the sufficiency of the evidence to support whatever unstated findings of fact were made by the fact finder. Guidry.

Where the trial court finds that the statement was not voluntary, CCP Art. 38.22 §6 does not require the entry of written findings of fact and conclusions of law. Terrazas v. State, 4 S.W.3d 720 (Tex. Crim. App. 1999).

PRACTICE TIP: Where a trial court has ruled a statement voluntary, be sure to request that the trial court enter written findings of fact and conclusions of law. The right to findings and conclusions is a statutory right which is forfeited by a party's failure to insist upon its implementation. Terrazas v. State, 4 S.W.3d 720 (Tex. Crim. App. 1999).

TIP TO THE BENCH: The findings of fact and conclusions of law must be made by the judge that heard the testimony. If that judge is unavailable for any reason to prepare them, the case must be abated for another voluntariness hearing. Garcia v. State, 15 S.W.3d 533 (Tex. Crim. App. 2000).

Where the trial court's written findings and conclusions of law are ambiguous and lack a credibility determination, an appellate court can remand the case for supplemental findings. State v. Mendoza, 365 S.W.3d 666. 673 (Tex. Crim. App. 2012).

§6:57.3 Voluntariness Issue to Jury

§6:57.3.1 Statutory Provisions

CCP Art. 38.22 §6

In all cases where a question is raised as to the voluntariness of a statement of an accused, the court must make an independent finding in the absence of the jury as to whether the statement was made under voluntary conditions. If the statement has been found to have been voluntarily made and held admissible as a matter of law and fact by the court in a hearing in the absence of the jury, the court must enter an order stating its conclusion as to whether or not the statement was voluntarily made, along with the specific finding of facts upon which the conclusion was based, which order shall be filed among the papers of the cause. Such order shall not be exhibited to the jury nor the finding thereof made known to the jury in any manner. Upon the finding by the judge as a matter of law and fact that the statement was voluntarily made, evidence pertaining to such matter may be submitted to the jury and it shall be instructed that unless the jury believes beyond a reasonable doubt that the statement was voluntarily made, the jury shall not consider such statement for any purpose nor any evidence obtained as a result thereof. In any case where a motion to suppress the statement has been filed and evidence has been submitted to the court on this issue, the court within its discretion may reconsider such evidence in its finding that the statement was voluntarily made and the same evidence submitted to the court at the hearing on the motion to suppress shall be made a part of the record the same as if it were being presented at the time of trial. However, the state or the defendant shall be entitled to present any new evidence on the issue of the voluntariness of the statement prior to the court's final ruling and order stating its findings.

CCP Art. 38.23. Evidence not to be used

(a) No evidence obtained by an officer or other person in violation of any provision of the Constitution or laws of the State of Texas, or of the Constitution or laws of the United States of America, shall be admitted in evidence against the accused on the trial of any criminal case. In any case where the legal evidence raises an issue hereunder, the jury shall be instructed that if it believes, or has a reasonable doubt, that the evidence was obtained in violation of the provisions of this Article, then and in such event, the jury shall disregard any such evidence so obtained.

(b) It is an exception to the provisions of Subsection (a) of this Article that the evidence was obtained by a law enforcement officer acting in objective good faith reliance upon a warrant issued by a neutral magistrate based on probable cause.

Art. 38.23 requires the exclusion of evidence where there is a causal connection between the legal violation and the obtaining of the evidence. Johnson v. State, 871 S.W.2d 744 (Tex. Crim. App. 1994)

§6:57.3.2 Application of Law

The issue of the voluntariness of a confession can be submitted to the jury under two different theories depending on the proof adduced at trial. If there is a factual dispute as to how the evidence was obtained, the jury should be instructed under art. 38.23. If the defense introduces evidence at trial from which a reasonable jury could conclude that the statement was not voluntary, a voluntariness charge under Art. 38.22 §6 is required. Vasquez v. State, 225 S.W.3d 541 (Tex. Crim. App. 2007).

When the evidence raises an issue of the voluntariness of a defendant's statement under Article 38.22, the trial judge must give a general voluntariness instruction under §§6 and 7 of that article because the law is applicable to the case. The law is applicable to the case because the statute requires an instruction in this situation. Oursbourn v. State, 259 S.W.3d 159, 165 (Tex. Crim. App. 2008).

However, when the defendant does not request this statutorily mandated instruction, the trial court's failure to include it is reviewed only for egregious harm under Almanza; Oursbourn.

The first step in deciding upon an appropriate jury instruction is identifying the theory of involuntariness because this will determine what type of instruction will be appropriate. Oursbourn, at 169.

Even though the trial court has a *sua sponte* duty to include an Art. 38.23 voluntariness instruction in the charge where raised by the evidence, it is incumbent on the defendant to explain to the trial court what he desires in the way of an instruction. Contreras v. State, 312 S.W.3d 566 (Tex. Crim. App. 2010).

Under Texas statutory law, there are three types of instructions that relate to the taking of confessions: (1) a general Article 38.22, §6, Voluntariness Instruction; (2) a general Article 38.22, §7, Warnings Instruction (involving warnings given under §§2 and 3; and (3) a specific Article 38.23(a), Exclusionary-Rule Instruction. Oursbourn, at 173.

In essence, the §6 general instruction asks the jury: "Do you believe, beyond a reasonable doubt, that the defendant's statement was voluntarily made? If it was not, do not consider the defendant's confession." The §7 instruction sets out the requirements of Art. 38.22, §2 or §3, and asks the jury to decide whether all of those requirements were met. The Article 38.23(a) specific instruction is fact-based: For example, "Do you believe that Officer Obie held a gun to the defendant's head to extract his statement? If so, do not consider the defendant's confession." Oursbourn, at 173.

Due process and Miranda claims may warrant both "general" and "specific" voluntariness instructions; Texas statutory claims warrant only a "general" voluntariness instruction. It is the defendant's responsibility to delineate which type of "involuntariness" he is claiming—a general (perhaps subjective) lack of voluntariness or a specific police-coerced lack of voluntariness—because the jury instruction is very different depending upon the type of claim. Oursbourn, at 174.

If the defendant never presents a proposed jury instruction (or fails to object to the lack of one), any potential error in the charge is reviewed only for "egregious harm" under Almanza. Oursbourn, at 174.

Article 38.22, §6 expressly dictates the content of that instruction to be as follows: "Unless the jury believes beyond a reasonable doubt that the statement was voluntarily made, the jury shall not consider such statement for any purpose nor any evidence obtained as a result thereof." Thus, Art. 38.22, §6 is "the law applicable" to any case in which a question is raised and litigated as to the general voluntariness of a statement of a defendant. Oursbourn, at 175.

Because §6 contains its own jury-instruction provision, it is not governed by the jury-instruction provision found in §7. The obvious purpose of §7 is to authorize and require jury instructions regarding the warnings and safeguards for written and oral statements outlined in Art. 38.22, §§2 and 3 (warnings on the right to remain silent, right to counsel, etc.). Oursbourn, at 175.

A §6 instruction becomes "law applicable to the case" only if the parties actually litigate a §6 voluntariness issue before the trial judge. If such litigation occurs (on the admissibility of evidence, for example), a jury instruction need not be specifically requested, although a request would still be necessary to obtain the most beneficial harm analysis under Almanza. Oursbourn, at 175.

If the defendant made his statement as the result of custodial interrogation, he is also entitled, when the issue is raised by the evidence, to have the jury decide whether he was adequately warned of his rights and knowingly and intelligently waived these rights. The phrase "the issue" in §7 refers to compliance with the statutory warnings set out in both Articles 15.17 (Duties of Arresting Officer and Magistrate) and 38.22, §§2 and 3, and the voluntariness of the defendant's waiver of the rights. For it to be "raised by the evidence" there must be a genuine factual dispute, just as is true under Article 38.23 issues. As with §6, the trial judge's §7 jury instructions are "general" ones that set out the pertinent law and legal requirements of §§2 and 3 (or, in an appropriate case, those of Article 15.17). Oursbourn, at 176.

Under Art. 38.23, a jury instruction is required only if there is a genuine dispute about a material fact. A defendant must establish three foundation requirements to trigger an Article 38.23 instruction: (1) the evidence heard by the jury must raise an issue of fact; (2) the evidence on that fact must be affirmatively contested; and (3) the contested factual issue must be material to the lawfulness of the challenged conduct in obtaining the statement claimed to be involuntary. The defendant must offer evidence that, if credited, would create a reasonable doubt as to a specific factual matter essential to the voluntariness of the statement. This factual dispute can be raised only by affirmative evidence, not by mere cross-examination questions or argument. Oursbourn, at 177.

If raised, Art. 38.23 is "the law applicable" to the case. Because such an instruction is statutorily mandated, the trial judge must include it in the jury instructions when the voluntariness of a defendant's statement is at issue. Oursbourn, at 181.

Because the failure of police to honor a defendant's rights under Miranda is not itself a constitutional violation, a defendant is not entitled to an Art. 38.23 charge when evidence at trial shows that police did not comply with Miranda. Contreras v. State, 312 S.W.3d 566 (Tex. Crim. App. 2010). Under these circumstances, the defendant is entitled to a charge under Art. 38.22.

At trial, the defendant should be permitted to show the circumstances under which the confession was allegedly made so that the jury may evaluate its reliability and credibility. Crane v. Kentucky, 476 U.S. 683, 106 S.Ct. 2142, 90 L.Ed.2d 636 (1986). The defendant may offer witnesses before the jury to establish an issue of non-voluntariness and it is error to refuse him this right. Ross v. State, 504 S.W.2d 862 (Tex. Crim. App. 1974).

However, where the defendant is claiming that his confession was involuntary because he requested a lawyer but does not claim that his confession was false or otherwise unreliable, the trial court does not err in refusing to allow him to put on testimony alleging that the police were not credible. Guidry v. State, 9 S.W.3d 133 (Tex. Crim. App. 1999).

Before the trial court is required to instruct the jury on the law pertaining to the voluntariness of confessions, some evidence must be presented to the jury that raises the issue. Butler v. State, 872 S.W.2d 227 (Tex. Crim. App. 1994).

Absent some positive evidence that the statement of the defendant was coerced by law enforcement, the trial court does not abuse its discretion in refusing an Art. 38.23 instruction. It is not enough that the jury may disbelieve certain testimony. Janecka v. State, 937 S.W.2d 456 (Tex. Crim. App. 1996); Perry v. State, 158 S.W.3d 438 (Tex. Crim. App. 2004).

Where there is no factual dispute regarding the necessary probable cause for the arrest of the defendant, he is not entitled to an Art. 38.23 instruction. Jackson v. State, 33 S.W.3d 828 (Tex. Crim. App. 2000); Wesbrook v. State, 29 S.W.3d 103 (Tex. Crim. App. 2000).

Even where there is evidence raising the issue of voluntariness, the defendant is not entitled to a specific instruction as to whether the statement met all technical and mechanical requirements, but rather is entitled to submission of an instruction surrounding the general issue of whether he was adequately warned. White v. State, 779 S.W.2d 809 (Tex. Crim. App. 1989); Bell v. State, 582 S.W.2d 800 (Tex. Crim. App. 1979); Dinkins v. State, 894 S.W.2d 330 (Tex. Crim. App. 1995).

An instruction which focuses on a particular factor that may render a statement involuntary is an impermissible comment on the weight of the evidence. Rocha v. State, 16 S.W.3d 1 (Tex. Crim. App. 2000); Gallo v. State, 239 S.W.3d 757 (Tex. Crim. App. 2007) (defendant claimed that mental retardation made his confession involuntary).

The trial court is allowed as necessary to give instruction to the jury clarifying the law surrounding the issue. Cockrell v. State, 933 S.W.2d 73 (Tex. Crim. App. 1996).

Where the defense raises the voluntariness issue in front of the jury, the state has the burden to prove voluntariness beyond a reasonable doubt. CCP Art. 38.22 §6.

TIP TO THE BENCH: For an example of the proper jury charge to be given on the voluntariness of a confession, see Mendoza v. State, 88 S.W.3d 236 (Tex. Crim. App. 2002).

§6:60 Illegal Arrests and Resulting Confessions

§6:61 In General

The primary taint of an illegal arrest bears so strongly on the will of a suspect that evidence (including confessions) obtained as a result of the illegality may not be used. Wong Sun v. U.S., 371 U.S. 471, 83 S.Ct. 407, 9 L.Ed.2d 441 (1963). This rationale applies under Texas law as CCP Art. 38.23 prohibits use of any evidence "obtained" in violation of the laws. Bell v. State, 724 S.W.2d 780 (Tex. Crim. App. 1986).

Where a defendant is illegally arrested and there is a causal connection between the arrest and a subsequent confession, the confession is inadmissible. De Bolt v. State, 604 S.W.2d 164 (Tex. Crim. App. 1980).

Where a suspect is illegally arrested and then confesses, the resulting confession is inadmissible unless the state can show that the taint of the illegal arrest was sufficiently attenuated to the point that the confession is sufficiently an act of free will. Gregg v. State, 667 S.W.2d 125 (Tex. Crim. App. 1984).

The "fruit of the poisonous tree" doctrine is not applicable where knowledge or possession of the evidence in question is obtained from a source independent of the state's wrongful act. Wicker v. State, 667 S.W.2d 137 (Tex. Crim. App. 1984), citing Costello v. U.S., 365 U.S. 265, 81 S.Ct. 534, 5 L.Ed.2d 551 (1961).

A confession should not be suppressed where the law violation committed by the police does not bear any relation to the purposes the law is to serve. New York v. Harris, 495 U.S. 14 (1990) (where the law violation had to do with an illegal warrantless in-home arrest of the defendant and the confession was later taken at the police station).

§6:62 Attenuation Analysis

A four-prong analysis is applied to determine if the taint of the illegal arrest is attenuated to the point that the confession should be admitted:

- Giving of Miranda warnings
- Temporal proximity of the arrest and the confession
- Presence of intervening circumstances
- Purpose and flagrancy of the official misconduct

Brown v. Illinois, 422 U.S. 590, 95 S.Ct. 2254, 45 L.Ed.2d 416 (1975); Monge v. State, 315 S.W.3d 35 (Tex. Crim. App. 2010).

Texas has adopted the Brown test for the purposes of analyzing the admissibility of confessions made following illegal arrests under state statutory law. Hankins v. State, 132 S.W.3d 380 (Tex. Crim. App. 2004).

The burden is on the defendant, as the moving party in a motion to suppress evidence obtained in violation of the law under Art. 38.23, to produce evidence demonstrating a causal connection between the alleged statutory violation and the resulting evidence. The burden then shifts to the state to either disprove the evidence the defendant has produced, or bring an attenuation-of-taint argument to demonstrate that the causal chain asserted by the defendant was in fact broken. Pham v. State, 175 S.W.3d 767 (Tex. Crim. App. 2005).

§6:62.1 Giving Miranda Warnings

The giving of Miranda warnings alone, even repeated Miranda warnings, are not enough to purge the taint of an otherwise illegal arrest. Bell (citing Taylor v. Alabama, 457 U.S. 687, 102 S.Ct. 2664, 73 L.Ed.2d 314 (1982)).

However, the giving of Miranda warnings can be an important factor when combined with other factors. Crutsinger v. State, 206 S.W.3d 607 (Tex. Crim. App. 2006).

The less flagrant the police misconduct surrounding the illegal arrest, the more important the giving of the Miranda warnings become as tending to attenuate the taint of the illegal arrest. Self v. State, 709 S.W.2d 662 (Tex. Crim. App. 1986); Little v. State, 758 S.W.2d 551 (Tex. Crim. App. 1988).

However, the giving of Miranda warnings before the confession is a Brown factor that weighs in favor of the state. Monge v. State, 315 S.W.3d 35 (Tex. Crim. App. 2010).

§6:62.2 Temporal Proximity

The temporal proximity of the arrest and the confession is generally not a strong determining factor *per se*. Bell v. State, 724 S.W.2d 780 (Tex. Crim. App. 1986). For examples of cases in which the time span between illegal arrest and the confession has been found too short to attenuate the taint, *see*:

- Ussery v. State, 651 S.W.2d 767 (Tex. Crim. App. 1983) (one and a half hours)
- Green v. State, 615 S.W.2d 700 (Tex. Crim. App. 1980) (about two hours)
- Beasley v. State, 674 S.W.2d 762 (Tex. Crim. App. 1982) (time period of two to three days apparently, from poorly developed record)

For examples of cases in which the time span between illegal arrest and the confession has been found sufficiently long to attenuate the taint, *see*:

- Dowdy v. State, 534 S.W.2d 336 (Tex. Crim. App. 1976) (three hours)
- Coleman v. State, 643 S.W.2d 947 (Tex. Crim. App. 1982) (five hours)
- Townsley v. State, 652 S.W.2d 791 (Tex. Crim. App. 1983) (24 hours)

- Gant v. State, 649 S.W.2d 30 (Tex. Crim. App. 1983) (29 hours)
- Alonzo v. State, 591 S.W.2d 842 (Tex. Crim. App. 1979) (two days)
- Crutsinger v. State, 206 S.W.3d 607 (Tex. Crim. App. 2006) (approximately one hour where there were significant intervening circumstances)

The examples referenced in the preceding lists demonstrate that the temporal proximity of the confession to the illegal arrest is analyzed on a case-by-case basis and is, at best, an ambiguous factor. Self v. State, 709 S.W.2d 662 (Tex. Crim. App. 1986).

However, a short period of time between the illegal arrest and the confession is a Brown factor that weighs in favor of the defendant. Monge v. State, 315 S.W.3d 35 (Tex. Crim. App. 2010).

§6:62.3 Intervening Circumstances

The importance of intervening circumstances to the analysis depends on the origin and nature of the intervening circumstances. Intervening circumstances that are themselves products of the illegal arrest militate strongly against the attenuation of the taint. Taylor v. Alabama, 457 U.S. 687, 102 S.Ct. 2664, 73 L.Ed.2d 314 (1982) (police obtained an arrest warrant during the interrogation but it was itself based on information obtained from the tainted interrogation).

Where the intervening circumstances are the product of police investigation independent of the illegal arrest, they militate in favor of attenuation of the taint. Townsley v. State, 652 S.W.2d 791 (Tex. Crim. App. 1983) (probable cause was obtained by examining the body of the deceased and by interviewing neighbors). The intervening circumstances do not have to actually occur after the illegal arrest, but rather must manifest themselves so as to significantly intervene and thus attenuate the taint of the illegal arrest. Foster v. State, 677 S.W.2d 507 (Tex. Crim. App. 1984). The state's burden to establish intervening circumstances that attenuate the taint is heavy. Foster. The defendant's status as a lawyer is an insufficient intervening circumstance. Foster.

Allowing the suspect to read the confession of a co-defendant during the interrogation is, alone, an insufficient intervening circumstance. Gregg v. State, 667 S.W.2d 125 (Tex. Crim. App. 1984). Merely allowing the suspect to drink coffee and talk with his alcohol program counselor during the interrogation are insufficient intervening circumstances to attenuate the taint of an illegal arrest. Ussery v. State, 651 S.W.2d 767 (Tex. Crim. App. 1983).

Confronting a suspect with evidence that is derived independently of the illegal arrest is a factor that weighs in favor of the state. Monge v. State, 315 S.W.3d 35 (Tex. Crim. App. 2010).

Sufficient intervening circumstances are present when five days elapse between the illegal arrest and the confession, and the suspect is allowed to consult with attorneys and family. Wicker v. State, 667 S.W.2d 137 (Tex. Crim. App. 1984).

An extremely important intervening circumstance is the release of the defendant from custody between the illegal arrest and the confession. Maixner v. State, 753 S.W.2d 151 (Tex. Crim. App. 1988).

Taking a defendant before a neutral magistrate is an important intervening circumstance. Hankins v. State, 132 S.W.3d 380 (Tex. Crim. App. 2004).

The fact that the defendant became emotional, stated that he had "messed up," and asked to speak to the detective is a product of the defendant's free will and an intervening circumstance which weighs heavily in favor of the state. Crutsinger v. State, 206 S.W.3d 607 (Tex. Crim. App. 2006).

§6:62.4 Police Misconduct

The fourth factor enunciated in Brown, that of police conduct, is one of the most important factors to be considered. Bell v. State, 724 S.W.2d 780 (Tex. Crim. App. 1986); and is perhaps the most important of the four factors. Self v. State, 709 S.W.2d 662 (Tex. Crim. App. 1986). *See* §6:82.

The proof of attenuation required should be the clearest where the police conduct is the most flagrant. Examples of flagrant police conduct include:

- Reliance on factors in making an arrest which were so lacking in indicia of probable cause as to render belief in its existence entirely unreasonable
- Arrest effectuated as a pretext for collateral objectives

- Arrest which is unnecessarily intrusive on personal privacy

Bell.

An arrest made solely on the basis of an unsubstantiated tip that identified the defendant only as an acquaintance of the murder victim is an example of flagrant police misconduct. Brown. An arrest made with probable cause but without the formality of an arrest warrant is an example of non-flagrant police conduct lacking the quality of purposefulness of an expedition for evidence indicative of flagrant misconduct. Maixner; Self.

Where the police had probable cause to obtain a warrant to arrest the defendant for an offense, although they arrested the defendant without a warrant for another offense, their conduct was found to be not purposeful or flagrant. Crutsinger v. State, 206 S.W.3d 607 (Tex. Crim. App. 2006); Monge v. State, 315 S.W.3d 35 (Tex. Crim. App. 2010).

§6:70 Oral Confessions

§6:71 Types

Oral confessions are admissible under several circumstances:

- Made in a non-custodial situation (CCP Art. 38.22 §5)
- Statement is electronically recorded and the proper predicate is laid under CCP Art. 38.22 §3
- Statement is made by the accused in open court at his trial, before a grand jury, or at an examining trial (CCP Art. 38.22 §5)
- Statement is res gestae of the arrest or of the offense (CCP Art. 38.22 §5)
- Statement has a bearing upon the credibility of the accused as a witness (CCP Art. 38.22 §5)
- Statement is admissible under law (CCP Art. 38.22 §5)
- Statement contains assertions of facts or circumstances that are found to be true and which conduce to establish the guilt of the accused, such as the finding of secreted or stolen property or the instrument with which he states the offense was committed (CCP Art. 38.22 §3(c))

§6:72 In General

A non-custodial oral statement is admissible as substantive evidence as well as for its impeachment value. Chambliss v. State, 647 S.W.2d 257 (Tex. Crim. App. 1983). The law surrounding the voluntariness of confessions applies to oral confessions to the same extent as written confessions. Boles v. Stevenson, 379 U.S. 43, 85 S.Ct. 174, 13 L.Ed.2d 109 (1964); Terrazas v. State, 4 S.W.3d 720 (Tex. Crim. App. 1999).

Where the voluntariness of an oral confession is questioned, a voluntariness hearing must be held. It is reversible error to not hold such a hearing. Moore v. State, 505 S.W.2d 887 (Tex. Crim. App. 1974).

A defendant can invoke the right to an attorney as to a written statement while waiving the right as to oral statements. Connecticut v. Barrett, 479 U.S. 523, 107 S.Ct. 828, 93 L.Ed.2d 920 (1987).

Illegally obtained oral confessions are not admissible in the punishment phase of capital murder trials. Rumbaugh v. State, 589 S.W.2d 414 (Tex. Crim. App. 1979).

When the State introduces a part of an oral confession that inculpates defendant, the court should allow the defendant to elicit from the state witness that part of the confession that exculpates defendant. Texas Rules of Evidence 107.

An oral confession is sufficient evidence to convict if the evidence otherwise proves the corpus delicti. Gordon v. State, 667 S.W.2d 544 (Tex. App.—Beaumont 1983, *pet. ref'd*).

TIP TO THE PROSECUTION: When presenting an oral confession to the jury, prepare a written transcript. This will allow the jury to better follow the wording of the confession and is permissible when accompanied by the proper instruction. Moore v. State, 882 S.W.2d 844 (Tex. Crim. App. 1994).

§6:72.1 Form: Motion to Reduce Oral Statements to Writing

See our companion book *Texas Criminal Forms* (James Publishing) Form 6-1.

§6:73 Admissibility

The requirements for the admissibility of oral statements are found in CCP Art. 38.22 §§3 and 5. Generally, the requirements for the admissibility for admission of oral statements are the same as that for written statements. However, there are some exceptions.

The oral confession statute does not require that a recorded statement contain an express verbal statement from an accused that he waives his rights prior to giving a statement. Barefield v. State, 784 S.W.2d 38 (Tex. Crim. App. 1989); Etheridge v. State, 903 S.W.2d 1 (Tex. Crim. App. 1994).

The CCP Art. 38.22 §2(a) requirement that the warnings must be given by a magistrate or the person to whom the statement is made does not apply to oral confessions, which are governed by §3. Dowthitt v. State, 931 S.W.2d 244 (Tex. Crim. App. 1996).

The plain language of Art. 38.22, §3(e) requires strict compliance with all portions of §3(a), which sets out the specific conditions that must be satisfied before such statements may be used as evidence against a defendant. Art. 38.22, §3(a)(1) makes no distinction between in-state and out-of-state oral statements, made as a result of custodial interrogation, which are not in compliance with its dictates; and therefore, Texas law will always govern admissibility. Davidson v. State, 25 S.W.3d 183 (Tex. Crim. App. 2000).

> *CCP Art. 38.22. When Statements May Be Used*
>
> *§8. Notwithstanding any other provision of this article, a written, oral, or sign language statement of an accused made as a result of a custodial interrogation is admissible against the accused in a criminal proceeding in this state if:*
>
> *(1) the statement was obtained in another state and was obtained in compliance with the laws of that state or this state*
>
> *(2) the statement was obtained by a federal law enforcement officer in this state or another state and was obtained in compliance with the laws of the United States.*

Where a defendant is a juvenile at the time of the commission of an offense, the admissibility of the defendant's statement is decided under Title 3 of the Texas Family Code, not Article 38.22; and therefore Davidson is inapplicable. Vega v. State, 84 S.W.3d 613 (Tex. Crim. App. 2002).

§6:73.1 Assertions of Truth and Discovery of Evidence

CCP Art. 38.22 §3(c) allows admission of oral statements that contain assertions of fact or circumstances conducive to establishing the guilt of the accused and places no limitation on the manner in which the facts asserted are found to be true. Port v. State, 791 S.W.2d 103 (Tex. Crim. App. 1990). The examples set out in CCP Art. 38.22 §3(c) are illustrative only and are not a limitation on the admission of oral statements under this Section. Gunter v. State, 858 S.W.2d 430 (Tex. Crim. App. 1993).

The statement is admissible if some of the facts therein are found to be true after the statement is made. Port. Even if only one of the assertions contained within an oral statement is found to be true, the entire confession is admissible. Port; Gunter. There is no requirement that the defendant clearly admit commission of the crime in order for the entire confession to be admissible. Anderson v. State, 701 S.W.2d 868 (Tex. Crim. App. 1985).

In Port, the defendant's statement was found to be admissible because a subsequent autopsy corroborated his statement that he used a particular handgun to cause the death of the deceased; and the fact that the handgun was already in the custody of the police did not render the statement inadmissible.

In a case where the defendant orally confessed that he used a particular knife to commit a murder but it was never established that the particular knife could have been the murder weapon, the statement was found to be inadmissible. Romero v. State, 800 S.W.2d 539 (Tex. Crim. App. 1990).

Where the defendant's statement led to the discovery of the victim's body, the entire statement is admissible. Anderson.

The phrase "[f]ound to be true," as contained in Art. 38.22 §3(c), means facts about which the police were unaware at the time of the confession that are later, after the confession, found to be true. Dansby v. State, 931 S.W.2d 297 (Tex. Crim. App. 1996). The statute places no limitation on the manner in which the facts are found to be true. Gunter.

Statements that are nothing more than an assertion of guilt are not admissible under this section. Almanza v. State, 839 S.W.2d 817 (Tex. Crim. App. 1992) (defendant merely pointed to a piece of paper later found to contain heroin and stated that it was his "personal stuff").

Even where the most incriminating part of the defendant's long, continuous statement was given after the earlier statement that led to evidence which conduced to establish his guilt, the entire statement was nevertheless admissible. Marini v. State, 593 S.W.2d 709 (Tex. Crim. App. 1980).

The giving of the Miranda warnings is not a prerequisite to the admission of a statement admitted under this theory; even where the statement was made as the result of custodial interrogation. Perillo v. State, 758 S.W.2d 567 (Tex. Crim. App. 1988); Ex parte Stansbery, 702 S.W.2d 643 (Tex. Crim. App. 1986). However, at least one other court has held that a valid Miranda waiver is necessary to the admis-sion of a statement under this theory, where the statement was made while the suspect was in custody. Hutchison v. State, 424 S.W.3d 164, 182 (Tex. App.—Texarkana 2014, *no pet.*).

§6:73.2 Electronically Recorded Statements

Under Article 38.22, §3(a)(2) of the Code of Criminal Procedure, before an oral recorded statement may be admitted into evidence, the State must show, *inter alia,* that prior to the statement but during the recording the accused is given the warning in Subsection (a) of §2 above and the accused knowingly, intelligently, and voluntarily waives any rights set out in the warning. Leza v. State, 351 S.W.3d 344 (Tex. Crim. App. 2011).

Art. 38.22 §3(a)(5) prohibits the admission of a recorded oral statement if a copy of the statement is not provided to defense counsel at least twenty (20) days before the proceeding. For purposes of this section, "proceeding" means all steps between official accusation and final judgment. Thus, a copy must be served upon defense counsel at least twenty (20) days before voir dire in the case begins. Tigner v. State, 928 S.W.2d 540 (Tex. Crim. App. 1996). Transcripts of the electronic recording do not constitute copies of the recordings for purposes of the statute; the copy itself must be an electronic recording. Tigner.

The proceeding to which the 20-day requirement applies is the same proceeding at which the evidence is rendered inadmissible, when there is no compliance. Thus, failure to provide the recording twenty days before a pretrial hearing renders the oral confession inadmissible at the pretrial hearing. Sells v. State, 121 S.W.3d 748 (Tex. Crim. App. 2003).

The defendant must object at a pretrial hearing to preserve error in the state's failure to deliver to him copies of any electronically recorded confession at least twenty (20) days prior to a "proceeding." Lane v. State, 933 S.W.2d 504 (Tex. Crim. App. 1996) (plurality opinion). The statute requires only that the defendant have access to the tapes at least twenty (20) days before the proceeding, not that copies actually be delivered to him. Lane.

The practical remedy for the failure to comply with §3(a)(5) with regard to the pretrial hearing is a 20-day continuance of the hearing. Lane; Sells.

Since the legislature deleted the applicable provision in 1989, the oral confession statute does not require that the suspect be informed that his confession is being tape-recorded. Moore v. State, 882 S.W.2d 844 (Tex. Crim. App. 1994). A law enforcement officer's subjective reasoning for intentionally not informing a suspect of the fact that the confession is being recorded is no longer independently relevant to the admissibility of the oral confession. Moore.

A confession recorded without the knowledge of the suspect will not be found involuntary even though the detective ignored the suspect's specific request that the confession not be recorded, even where the suspect would not have made the confession had he known it was being recorded. Flemming v. State, 949 S.W.2d 876 (Tex.App.—Houston [14th Dist.] 1997).

Accidental alterations in the audiotape do not render it inadmissible if the alteration is sufficiently explained such that the reliability and trustworthiness of the recording are intact. Quinones v. State, 592 S.W.2d 933 (Tex. Crim. App. 1980).

Where brief interruptions in the tape recording occur accidentally, obscuring nothing of value in the dialogue, and there is adequate evidence to support the conclusion that the breaks were merely inadvertent and did not affect the overall reliability of the tape, the tape will be admissible. Maldonado v. State, 998 S.W.2d 239 (Tex. Crim. App. 1999).

Circumstantial evidence can be sufficient to establish the authenticity and voluntariness of a sound recording. Wallace v. State, 782 S.W.2d 854 (Tex. Crim. App. 1989).

An electronically recorded oral statement is not required to contain an express waiver of rights by the defendant before it can be admissible. Rocha v. State, 16 S.W.3d 1 (Tex. Crim. App. 2000).

A claim that certain portions of an audio recording of a defendant's statement to the police are inaudible does not mean that these portions of the court reporter's record are lost or destroyed for purposes of TEX.R.App.P. 34.6(f). Lucio v. State, 351 S.W.3d 878 (Tex. Crim. App. 2011).

For an example of a case where it was held that audio recordings that were incomplete and inaccurately represented a conversation between a peace officer and a defendant were inadmissible pursuant to article 38.22, section 3(a)(3) of the Code of Criminal Procedure see Flores v. State, (No. PD-1189-15, 2018 WL 2327162, May 23, 2018, unpublished).

§6:73.3 Statements Res Gestae of Arrest

Where a police officer was looking for someone who had just shot the deceased, had the defendant at gunpoint, and was told by the defendant, "I shot her," the statement was admissible as *res gestae* of the arrest under CCP Art. 38.22 §5. Bryant v. State, 816 S.W.2d 554 (Tex.App.—Fort Worth 1991, *no pet.*).

Under the above circumstances, the officer is not required to warn the defendant of his rights under Miranda before the statement can be admitted. Bryant (citing New York v. Quarles, 467 U.S. 649, 104 S.Ct. 2626, 81 L.Ed.2d 550 (1984)).

§6:74 Prior Testimony of the Defendant

Reading a portion of the defendant's grand jury testimony, which contained a confession to a crime, is permitted under CCP Art. 38.22 §5. Fuller v. State, 827 S.W.2d 919 (Tex. Crim. App. 1992).

§6:80 Corroboration of Extrajudicial Confessions

Under the common law, the corpus delicti rule concerns the usability in a criminal case of a confession made by the defendant outside of court. The old corpus-delicti usability rule has, however, been superceded by the due-process sufficiency of the evidence model set out in Jackson v. Virginia. Carrizales v. State, 414 S.W.3d 737, 742 (Tex. Crim. App. 2015); Nisbett v. State, 552 S.W.3d 244, 263 (Tex. Crim. App. 2018).

No criminal conviction can be based upon a defendant's extrajudicial confession unless the confession is corroborated by independent evidence tending to establish the *corpus delicti*. Fisher v. State, 851 S.W.2d 298 (Tex. Crim. App. 1993).

The *corpus delicti* rule requires that there be some independent evidence tending to show the essential nature of the charged crime. That is, there must be evidence that an offense was committed and that it was the result of criminal conduct rather than accident. Salazar v. State, 86 S.W.3d 640 (Tex. Crim. App. 2002). The rule does not require corroboration of every element and descriptive allegation in the indictment. Salazar.

The corpus delicti rule requires corroboration of two elements of a crime—"an injury or loss and a criminal agent"—but there need not be any independent evidence that the defendant was the criminal culprit. Bradford v. State, 515 S.W.3d 433, 437 (Tex. App.—Houston [14th Dist.] 2017, pet.ref'd).

The corpus delicti rule applies to admissions of incriminating facts that do not amount to a full confession of guilt. Bradford v. State, 515 S.W.3d at 437.

The corpus delicti rule will apply to crimes of omission as well as crimes of commission. Bradford v. State, 515 S.W.3d at 442 (where appellant was charged with the offense of failing to report child abuse).

In a capital murder case, the evidence must also establish the *corpus delicti* of the underlying offense making the murder a capital crime. Gribble v. State, 808 S.W.2d 65 (Tex. Crim. App. 1990) (*plurality op.*). The corroborating evidence need not, however, be sufficient by itself to establish the underlying offense. Gribble.

Texas criminal law recognizes a closely related crimes exception to the corpus delicti rule in cases with an extrajudicial confession, and it is implicated when the temporal connection between the offenses confessed to is sufficiently close that introduction of the confession does not violate the purposes of the corpus delicti rule. Miller v. State, 457 S.W.3d 919, 927 (Tex. Crim. App. 2015).

The closely related crime exception comes into play where an accused is charged with more than one crime, and the accused makes a statement related to all the crimes charged, but the prosecution is only able to establish the corpus delicti of one of the crimes charged. Under those circumstances where the relationship between the crimes is sufficiently close so that the introduction of the statement will not violate the purpose underlying the corpus delicti rule, the statement of the accused will be admissible as to all the crimes charged. Miller, 919 S.W.3d at 926.

The *corpus delicti* doctrine does not apply to extraneous offenses offered at the punishment phase of a capital murder trial. Bible v. State, 162 S.W.3d 234 (Tex. Crim. App. 2005).

However, an uncorroborated confession is sufficient evidence to revoke probation. Bush v. State, 506 S.W.2d 603 (Tex. Crim. App. 1974).

A judicial confession does not require corroboration. Dinnery v. State, 592 S.W.2d 343 (Tex. Crim. App. 1979).

§6:90 Juvenile Confessions

Family Code §51.09. Waiver of Rights

> *Unless a contrary intent clearly appears elsewhere in this title, any right granted to a child by this title or by the constitution or laws of this state or the United States may be waived in proceedings under this title if:*
>
> *(1) the waiver is made by the child and the attorney for the child;*
>
> *(2) the child and the attorney waiving the right are informed of and understand the right and the possible consequences of waiving it;*
>
> *(3) the waiver is voluntary; and*
>
> *(4) the waiver is made in writing or in court proceedings that are recorded.*

Family Code §51.095. Admissibility of a Statement of a Child

> *(a) Notwithstanding Section 51.09, the statement of a child is admissible in evidence in any future proceeding concerning the matter about which the statement was given if:*
>
> *(1) the statement is made in writing under a circumstance described by Subsection (d) and:*
>
> *(A) the statement shows that the child has at some time before the making of the statement received from a magistrate a warning that:*
>
> *(i) the child may remain silent and not make any statement at all and that any statement that the child makes may be used in evidence against the child;*
>
> *(ii) the child has the right to have an attorney present to advise the*

child either prior to any questioning or during the questioning;

(iii) if the child is unable to employ an attorney, the child has the right to have an attorney appointed to counsel with the child before or during any interviews with peace officers or attorneys representing the state; and

(iv) the child has the right to terminate the interview at any time;

(B) and:

(i) the statement must be signed in the presence of a magistrate by the child with no law enforcement officer or prosecuting attorney present, except that a magistrate may require a bailiff or a law enforcement officer if a bailiff is not available to be present if the magistrate determines that the presence of the bailiff or law enforcement officer is necessary for the personal safety of the magistrate or other court personnel, provided that the bailiff or law enforcement officer may not carry a weapon in the presence of the child; and

(ii) the magistrate must be fully convinced that the child understands the nature and contents of the statement and that the child is signing the same voluntarily, and if a statement is taken, the magistrate must sign a written statement verifying the foregoing requisites have been met;

(C) the child knowingly, intelligently, and voluntarily waives these rights before and during the making of the statement and signs the statement in the presence of a magistrate; and

(D) the magistrate certifies that the magistrate has examined the child independent of any law enforcement officer or prosecuting attorney, except as required to ensure the personal safety of the magistrate or other court personnel, and has determined that the child understands the nature and contents of the statement and has knowingly, intelligently, and voluntarily waived these rights;

(2) the statement is made orally and the child makes a statement of facts or circumstances that are found to be true and tend to establish the child's guilt, such as the finding of secreted or stolen property, or the instrument with which the child states the offense was committed;

(3) the statement was res gestae of the delinquent conduct or the conduct indicating a need for supervision or of the arrest;

(4) the statement is made:

(A) in open court at the child's adjudication hearing;

(B) before a grand jury considering a petition, under Section 53.045, that the child engaged in delinquent conduct; or

(C) at a preliminary hearing concerning the child held in compliance with this code,

other than at a detention hearing under Section 54.01; or

(5) subject to Subsection (f), the statement is made orally under a circumstance described by Subsection (d) and the statement is recorded by an electronic recording device, including a device that records images, and:

(A) before making the statement, the child is given the warning described by Subdivision (1)(A) by a magistrate, the warning is a part of the recording, and the child knowingly, intelligently, and voluntarily waives each right stated in the warning;

(B) the recording device is capable of making an accurate recording, the operator of the device is competent to use the device, the recording is accurate, and the recording has not been altered;

(C) each voice on the recording is identified; and

(D) not later than the 20th day before the date of the proceeding, the attorney representing the child is given a complete and accurate copy of each recording of the child made under this subdivision.

(b) This section and Section 51.09 do not preclude the admission of a statement made by the child if:

(1) the statement does not stem from interrogation of the child under a circumstance described by Subsection (d); or

(2) without regard to whether the statement stems from interrogation of the child under a circumstance described by Subsection (d), the statement is:

(A) voluntary and has a bearing on the credibility of the child as a witness; or

(B) recorded by an electronic recording device, including a device that records images, and is obtained:

(i) in another state in compliance with the laws of that state or this state; or

(ii) by a federal law enforcement officer in this state or another state in compliance with the laws of the United States.

(c) An electronic recording of a child's statement made under Subsection (a)(5) or (b)(2)(B) shall be preserved until all juvenile or criminal matters relating to any conduct referred to in the statement are final, including the exhaustion of all appeals, or barred from prosecution.

(d) Subsections (a)(1) and (a)(5) apply to the statement of a child made:

(1) while the child is in a detention facility or other place of confinement;

(2) while the child is in the custody of an officer; or

(3) during or after the interrogation of the child by an officer if the child is in the possession of the Department of Family and Protective Services and is suspected to have engaged in conduct that violates a penal law of this state.

(e) A juvenile law referee or master may perform the duties imposed on a magistrate under this section without the approval of the juvenile court if the

juvenile board of the county in which the statement of the child is made has authorized a referee or master to perform the duties of a magistrate under this section.

(f) A magistrate who provides the warnings required by Subsection (a)(5) for a recorded statement may at the time the warnings are provided request by speaking on the recording that the officer return the child and the recording to the magistrate at the conclusion of the process of questioning. The magistrate may then view the recording with the child or have the child view the recording to enable the magistrate to determine whether the child's statements were given voluntarily. The magistrate's determination of voluntariness shall be reduced to writing and signed and dated by the magistrate. If a magistrate uses the procedure described by this subsection, a child's statement is not admissible unless the magistrate determines that the statement was given voluntarily.

Family Code §51.17. Procedure and Evidence

(a) Except for the burden of proof to be borne by the state in adjudicating a child to be delinquent or in need of supervision under Section 54.03(f) or otherwise when in conflict with a provision of this title, the Texas Rules of Civil Procedure govern proceedings under this title.

(b) Discovery in a proceeding under this title is governed by the Code of Criminal Procedure and by case decisions in criminal cases.

(c) Except as otherwise provided by this title, the Texas Rules of Evidence apply to criminal cases and Articles 33.03 and 37.07 and Chapter 38, Code of Criminal Procedure, apply in a judicial proceeding under this title.

(d) When on the motion for appointment of an interpreter by a party or on the motion of the juvenile court, in any proceeding under this title, the court determines that the child, the child's parent or guardian, or a witness does not understand and speak English, an interpreter must be sworn to interpret for the person as provided by Article 38.30, Code of Criminal Procedure.

(e) In any proceeding under this title, if a party notifies the court that the child, the child's parent or guardian, or a witness is deaf, the court shall appoint a qualified interpreter to interpret the proceedings in any language, including sign language, that the deaf person can understand, as provided by Article 38.31, Code of Criminal Procedure.

(f) Any requirement under this title that a document contain a person's signature, including the signature of a judge or a clerk of the court, is satisfied if the document contains the signature of the person as captured on an electronic device or as a digital signature. Article 2.26, Code of Criminal Procedure, applies in a proceeding held under this title.

(g) Articles 21.07, 26.07, 26.08, 26.09, and 26.10, Code of Criminal Procedure, relating to the name of an adult defendant in a criminal case, apply to a child in a proceeding held under this title.

(h) Articles 57.01 and 57.02, Code of Criminal Procedure, relating to the use of a pseudonym by a victim in a criminal case, apply in a proceeding held under this title.

(i) Except as provided by Section 56.03(f), the state is not required to pay any cost or fee otherwise imposed for court proceedings in either the trial or appellate courts.

Family Code §52.02. Release or Delivery to Court

(a) Except as provided by Subsection (c), a person taking a child into custody, without unnecessary delay and without first taking the child to any place other than a juvenile processing office designated under Section 52.025, shall do one of the following:

(1) release the child to a parent, guardian, custodian of the child, or other responsible adult upon that person's promise to bring the child before the juvenile court as requested by the court;

(2) bring the child before the office or official designated by the juvenile board if there is probable cause to believe that the child engaged in delinquent conduct, conduct indicating a need for supervision, or conduct that violates a condition of probation imposed by the juvenile court;

(3) bring the child to a detention facility designated by the juvenile board;

(4) bring the child to a secure detention facility as provided by Section 51.12(j);

(5) bring the child to a medical facility if the child is believed to suffer from a serious physical condition or illness that requires prompt treatment;

(6) dispose of the case under Section 52.03; or

(7) if school is in session and the child is a student, bring the child to the school campus to which the child is assigned if the principal, the principal's designee, or a peace officer assigned to the campus agrees to assume responsibility for the child for the remainder of the school day.

(b) A person taking a child into custody shall promptly give notice of the person's action and a statement of the reason for taking the child into custody, to:

(1) the child's parent, guardian, or custodian; and

(2) the office or official designated by the juvenile board.

(c) A person who takes a child into custody and who has reasonable grounds to believe that the child has been operating a motor vehicle in a public place while having any detectable amount of alcohol in the child's system may, before complying with Subsection (a):

(1) take the child to a place to obtain a specimen of the child's breath or blood as provided by Chapter 724, Transportation Code; and

(2) perform intoxilyzer processing and videotaping of the child in an adult processing office of a law enforcement agency.

(d) Notwithstanding Section 51.09(a), a child taken into custody as provided by Subsection (c) may submit to the taking of a breath specimen or refuse to submit to the taking of a breath specimen without the concurrence of an attorney, but only if the request made of the child to give the specimen and the child's response to that request is videotaped. A videotape made under this subsection must be maintained until the disposition of any proceeding against the child relating to the arrest is final and be made available to an attorney representing the child during that period.

§6:91 In General

Although juvenile delinquency proceedings are considered civil proceedings, they are quasi-criminal in nature. In Re J.R., 907 S.W.2d 107 (Tex.App.—Austin 1995, *no writ*); C.E.J. v. State, 788 S.W.2d 849 (Tex.App.—Dallas 1990, *writ denied*); Smith v. Rankin, 661 S.W.2d 152 (Tex. App.—Houston [1st Dist.] 1983, *orig. proceeding*).

Numerous aspects of juvenile proceedings are governed by criminal procedural rules; including:

- State's burden of proof
- Presumption of innocence
- Right to trial by jury
- Privilege against self-incrimination
- Right to confrontation of witnesses
- Right to representation by counsel

Family Code §54.03; In Re J.R.; In Re M.A.G., 541 S.W.2d 899 (Tex. Civ. App. Corpus Christi 1976, *writ ref'd n.r.e.*); In Re R.C.S., 940 S.W.2d 750 (Tex.App.—El Paso 1997) (*op. denying rehearing*).

Where the defendant was a juvenile at the time of his arrest, the provisions of the Texas Family Code control issues regarding the admissibility of a confession (even where the defendant is later certified to stand trial as an adult). Roquemore v. State, 60 S.W.3d 862 (Tex. Crim. App. 2001). The appellate courts have established a policy of strict compliance with the Family Code. Roquemore.

Other than these matters, civil procedural rules apply. Family Code §51.17(a) & (b); In Re R.C.S.

Family Code §51.095 is the juvenile equivalent of CCP Art. 38.22.

Where a defendant is a juvenile at the time of the commission of an offense, the admissibility of the defendant's statement is decided under Title 3 of the Family Code, not Art. 38.22. Vega v. State, 84 S.W.3d 613 (Tex. Crim. App. 2002).

Where it was known to the interviewing officer at the time of police questioning that the interviewee was a juvenile, or would have been objectively apparent to a reasonable officer, the inclusion in the Miranda custody analysis of that fact is consistent with the objective nature of the custody test. While a child's age will not be a determinative, or even a significant, factor in every case, in some cases, it will prove an important factor in deciding if the child was in custody for Miranda purposes. J.D.B. v. North Carolina, 564 U.S. ___, 131 S.Ct. 2394, 2406, 180 L.Ed.2d 310 (2011).

Before the enactment of Family Code §51.17 the Family Code applied to all proceedings until transfer to adult court, such that the Family Code governed procedure surrounding the taking of confessions. Travis v. State, 921 S.W.2d 559 (Tex. App.—Beaumont 1996); Griffin v. State, 765 S.W.2d 422 (Tex. Crim. App. 1989).

Certain statements by juveniles are exempted from use by the state as confessions. Family Code §53.03(c) (statements made by a juvenile during an intake conference); Family Code §54.01 (statements made by a juvenile during a detention hearing).

Where the police had no reason to disbelieve the defendant's assertions that he was 17 years of age at the time he gave a confession, and it is later learned that the defendant was actually 15, the defendant has waived any error resulting from the police not complying with the Family Code rules for taking juvenile confessions. Childs v. State, 21 S.W.3d 631 (Tex.App.—Houston [14th Dist.] 2000, *pet. ref'd*).

§6:92 Custodial Statements

Before a police officer can take the custodial confession of a juvenile, a decision to further detain the juvenile must be made by the proper official designated by the juvenile court. Comer v. State, 776 S.W.2d 191 (Tex. Crim. App. 1989), citing Family Code §51.09(b)(1) (as it existed before the 1997 amendments).

Failure to take the juvenile before the proper authority before subjecting him to further custody will render a subsequent confession inadmissible. Comer.

Under the plain language of Family Code §52.02(b)(1), an officer must inform a child's parent of "the reason" for taking the child into custody. That is, police must provide a parent with the legal justification for the officer's action. However, the section does not require police to re-notify parents concerning their suspicions of criminal conduct or any other further suspicions. Hampton v. State, 86 S.W.3d 603 (Tex. Crim. App. 2002).

Family Code §52.02(a) has three requirements once a juvenile is taken into custody:

- the officer must do one of the six acts enumerated in the statute;
- without unnecessary delay; and

- without first taking the child to any place other than a juvenile processing office.

Roquemore v. State, 60 S.W.3d 862 (Tex. Crim. App. 2001) (where officers violated the statute by taking the juvenile to recover stolen property first).

PRACTICE TIP: Effective September 1, 2007, there are seven requirements an officer must fulfill under §52.02(a) upon taking a juvenile into custody. See §6:160, above.

Family Code §52.02(a) requires that after being warned by a magistrate that a juvenile suspect must be interrogated only in an area designated for such by a juvenile court. Le v. State, 993 S.W.2d 650 (Tex. Crim. App. 1999).

Under Family Code §51.09(d)(2), if a statement does not stem from a custodial interrogation, the statement is admissible. Roquemore.

Once a defendant produces evidence that §52.02(a) was violated, the burden then shifts to the state to show compliance with the statute. Roquemore.

Evidence obtained in violation of §52.02(a) should be excluded under CCP Art. 38.23 where there is a causal connection between the illegality and the evidence. Roquemore.

The burden is on the defendant, as the moving party in a motion to suppress evidence obtained in violation of the law under Art. 38.23, to produce evidence demonstrating a causal connection between the alleged statutory violation and the resulting evidence. The burden then shifts to the state to either disprove the evidence the defendant has produced, or bring an attenuation-of-taint argument to demonstrate that the causal chain asserted by the defendant was in fact broken. Pham v. State, 175 S.W.3d 767 (Tex. Crim. App. 2005).

Before a juvenile's statement will be excluded from evidence under CCP Art. 38.22 for a violation of the Family Code §52.02(b) that requires a detaining authority to notify the juvenile's parent, there must be a causal connection between the failure to notify the parent and the giving of the statement. Gonzales v. State, 67 S.W.3d 910 (Tex. Crim. App. 2002).

Where a juvenile's statement is taken in another state, a conflict-of-laws analysis is conducted to determine which state's laws govern the admissibility of the statement. Vega v. State, 84 S.W.3d 613 (Tex. Crim. App. 2002).

§6:93 Voluntariness

The state carries the burden of proving that a confession was given voluntarily under the due process clause; at least once accused has alleged it was not. Griffin v. State, 765 S.W.2d 422 (Tex. Crim. App. 1989). The voluntariness of a waiver of the right against self-incrimination and due process rights must be established by a preponderance of the evidence. Griffin.

The proper test in determining whether a juvenile has validly waived his privilege against self-incrimination is the totality of the circumstances test. Spears v. State, 801 S.W.2d 571 (Tex.App.—Fort Worth 1990). If the circumstances indicate that the juvenile defendant was threatened, coerced, promised something in exchange for his confession, or incapable of understanding his rights and warnings, the trial court must conclude the confession is involuntary. In re R.M., 880 S.W.2d 297 (Tex.App.—Fort Worth 1994).

A confession was found to be voluntary where on the day following the arrest the 16-year-old suspect was removed from the juvenile detention center to the police department, and after being warned of his rights by a magistrate was taken upstairs to a designated juvenile facility; but, on finding that facility unavailable, was removed to homicide division and was questioned while isolated from all adult offenders and was then taken back downstairs and warned of his rights a second time and then signed the confession in the magistrate's presence. Matthews v. State, 677 S.W.2d 282 (Tex.App.—Fort Worth 1984).

The fact that a juvenile's oral statement to police was not obtained in compliance with the Family Code did not render a subsequent written confession inadmissible in a criminal trial of the juvenile as an adult, where the subsequent written statement was given after the juvenile waived her Miranda rights; and where a second magistrate's warning was a significant intervening circumstance. Griffin. *See also*, In Re J.T.H., 779 S.W.2d 954 (Tex.App.—Austin 1989).

§6:94 Sufficiency of Warnings

Family Code §51.09 requires that a magistrate certify that the juvenile "knowingly, intelligently and voluntarily" waived his rights prior to accepting the juvenile's confession. In re R.M., 880 S.W.2d 297 (Tex.App.—Fort Worth 1994).

A warning which was given to a child following his detention in connection with a burglary, which expressly warned him of his right to counsel prior to or during interrogation and which contained the substantial equivalent of the Miranda language, was adequate to meet the Miranda requirements despite the fact that the child was not warned that an attorney would be appointed to represent him if he were too poor to employ an attorney himself. In re B.J.N., 667 S.W.2d 347 (Tex.App.—Amarillo 1984).

The juvenile warning statute does not require that the child's parent be informed of the juvenile's right to court-appointed counsel before a confession is given, nor does the statute require that a child's parent be present when the juvenile waives right to counsel. In re B.J.N.

A juvenile's statement that was written before the juvenile was given the statutory warnings by a magistrate, but not signed until after a magistrate gave the juvenile the warnings, where the magistrate informed the juvenile that he did not have to give a statement at all, the magistrate was convinced that the juvenile understood the nature and contents of the statement, and the juvenile signed it voluntarily, was admissible. Petree v. State, 778 S.W.2d 507 (Tex.App.—Dallas 1989).

The reading of the magistrate's warnings by someone other than the magistrate will not render a confession inadmissible where the magistrate was present during the reading of the warnings, made inquiry of the juvenile as to whether he understood them, and then magistrate then signed the required certification. In re S.D.W., 811 S.W.2d 739 (Tex. App.—Houston [1st Dist.] 1991).

Texas Family Code subsection 51.095(a)(1)(A) does not explicitly prohibit the presence of law-enforcement officers when a magistrate reads the required statutory rights to a juvenile. Herring v. State, 395 S.W.3d 161, 165 (Tex. Crim. App. 2013).

TIP TO THE PROSECUTION: Compose a form for the taking of a suspect's statement which contains on its face the Family Code §51.095 warnings and a valid waiver thereof, and distribute it to all local police agencies for their use.

Further revisions to chapter 6

6:22.8; 6:34; 6:35; 6:57

§6:95 Oral Statements

Where the oral confession of a juvenile led to the recovery of stolen property, the statement was admissible. Salazar v. State, 648 S.W.2d 421 (Tex. App.—3 1983). *See also*, Littlefield v. State, 720 S.W.2d 254 (Tex.App.—Beaumont 1986); Spears v. State, 801 S.W.2d 571 (Tex.App.—Fort Worth 1990).

If even one assertion in an oral statement is found to be true, the entire statement is admissible. S.D.J. v. State, 879 S.W.2d 370 (Tex.App.—Eastland 1994).

§6:100 Objections to Admissibility

§6:101 Requirement of Objection

Lodging an objection is required if the defendant wishes to complain against the admission of a confession. Ross v. State, 678 S.W.2d 491 (Tex. Crim. App. 1984); Harris v. State, 827 S.W.2d 949 (Tex. Crim. App. 1992); Wainwright v. Sykes, 433 U.S. 72, 97 S.Ct. 2497, 53 L.Ed.2d 594 (1977).

A defendant who objected to the admission of his confession under state law was not required to specifically object to its admission on a particular statutory ground in order to invoke the state exclusionary rule. He was only required to object to its admissibility on the ground that it was obtained through illegal search or seizure. Polk v. State, 738 S.W.2d 274 (Tex. Crim. App. 1987).

However, an objection to a technical aspect of the taking of the confession must be specific enough to apprise the trial court of the nature of the complaint. For example, an objection that a confession is not voluntary is not specific enough to preserve appeal on the issue of whether the confession is invalid because the suspect was informed that the confession could be used "for or against" him. Ramirez v. State, 815 S.W.2d 636 (Tex. Crim. App. 1991).

An objection as to an "improper predicate" is insufficient to preserve the issue. Walsh v. State, 658 S.W.2d 285 (Tex.App.—Fort Worth 1983).

The failure to object to a statement first used to impeach the defendant waives any error in its subsequent admission during the state's case in rebuttal. Ross.

Appellate courts review a trial court's ruling on a motion to suppress for an abuse of discretion and overturn that ruling only if it is outside the zone of reasonable disagreement. Martinez v. State, 348 S.W.3d 919, 922 (Tex.Crim.App.2011). They conduct their review with two standards, giving almost total deference to a trial court's determination of historic facts and mixed questions of law and fact that rely on the credibility of a witness, but applying a *de novo* standard of review to pure questions of law and mixed questions that do not depend on credibility determinations. Martinez, 348 S.W.3d. at 922–23.

§6:102 Nature of Objection

There are many different general objections which can be lodged to the admissibility of a confession, each with many separate subdivisions. Some of the general areas and their subdivisions are as follows.

§6:102.1 Lack of Voluntariness

- State has not proven voluntariness, generally (has not sustained its burden of proof). Terrazas v. State, 4 S.W.3d 720 (Tex. Crim. App. 1999).
- Confession was not taken in compliance with Miranda or CCP Art. 38.22. Wolfe v. State, 917 S.W.2d 270 (Tex. Crim. App. 1996).
- Due process or due course of law violation. Wolfe.
- Decision to confess was not made of the defendant's own free will. Lane v. State, 933 S.W.2d 504 (Tex. Crim. App. 1996).
- Mental or physical compulsion was employed. Colorado v. Connelly, 479 U.S. 157, 107 S.Ct. 515, 93 L.Ed.2d 473 (1986).
- Coercive police activity was employed. Connelly.
- Confession was involuntarily obtained. Ochoa v. State, 573 S.W.2d 796 (Tex. Crim. App. 1978).
- Confession is not the language of the defendant. Henson v. State, 794 S.W.2d 385 (Tex.App.—Dallas 1990, *pet. ref'd*).
- Promise made to the defendant by law enforcement officials rendered the confession involuntary. Sossamon v. State, 816 S.W.2d 340 (Tex. Crim. App. 1991).
- Defendant was told that the statement could be used "for or against" him. Dinkins v. State, 894 S.W.2d 330 (Tex. Crim. App. 1995).
- Defendant was intoxicated to the point that he was incapable of making an independent, informed choice of his free will. Nichols v. State, 754 S.W.2d 185 (Tex. Crim. App. 1988).
- Law enforcement officials inflicted pain on the defendant or tortured him such that his resulting confession was involuntary. Jenkins v. State, 912 S.W.2d 793 (Tex. Crim. App. 1995).
- Deception and/or misrepresentations by police rendered the confession involuntary. Green v. State, 934 S.W.2d 92 (Tex. Crim. App. 1996).
- Prolonged interrogation that rendered the confession involuntary. Barney v. State, 698 S.W.2d 114 (Tex. Crim. App. 1985).
- Defendant was deprived of food or sleep to the point that his resulting confession was rendered involuntary. Greenwald v. Wisconsin, 390 U.S. 519, 88 S.Ct. 1152, 20 L.Ed.2d 77 (1968).
- Confession was the product of force or violence such that it was involuntary. Roberts v. State, 545 S.W.2d 157 (Tex. Crim. App. 1977).

- Police employed prayer tactics which rendered the confession involuntary. Davis v. North Carolina, 384 U.S. 737, 86 S.Ct. 1761, 16 L.Ed.2d 895 (1966).
- Defendant's lack of intellect or his mental impairment was used through official coercion to render the confession involuntary. Walker v. State, 842 S.W.2d 301 (Tex.App.—Tyler 1992).
- Defendant was unable to understand English to the point that the resulting confession was rendered involuntary. Farmah v. State, 789 S.W.2d 665 (Tex. App.—Houston [1st Dist.] 1990).

§6:102.2 Inadequate Warnings

- Warnings did not comply with CCP Art. 38.22 or Miranda. Cockrell v. State, 933 S.W.2d 73 (Tex. Crim. App. 1996).
- Warnings do not appear on the face of the resulting written confession. Heiselbetz v. State, 906 S.W.2d 500 (Tex. Crim. App. 1995).
- Language of the warnings did not substantially comply with the wording of Miranda or CCP Art. 38.22. Cockrell.
- Warnings were not given by a magistrate. Dowthitt v. State, 931 S.W.2d 244 (Tex. Crim. App. 1996).
- Warnings were not given by the person to whom the statement was given. Dowthitt.
- Hiatus took place in interrogation and the warnings were not properly re-administered. Dowthitt.
- Foreign national defendant not warned of his right to contact his country consulate under the Vienna Convention on Consular Relations. Maldonado v. State, 998 S.W.2d 239 (Tex. Crim. App. 1999).

§6:102.3 Invalid Waiver

- Waiver was not voluntary, knowing or intelligent. Robinson v. State, 851 S.W.2d 216 (Tex. Crim. App. 1991).
- Waiver was not the product of a free and deliberate choice. Moran v. Burbine, 475 U.S. 412, 106 S.Ct. 1135, 89 L.Ed.2d 410 (1986).
- Waiver was the product of coercion. Robinson.
- Background of the accused was such that the waiver was involuntary. Colorado v. Spring, 479 U.S. 564, 107 S.Ct. 851, 93 L.Ed.2d 954 (1987).
- Written statement contains no waiver of rights. Garcia v. State, 919 S.W.2d 370 (Tex. Crim. App. 1994).

§6:102.4 Lack of Compliance With CCP Art. 38.22

- Warnings were insufficient under CCP Art. 38.22. Cockrell.
- Defendant invoked his right to counsel. Smith v. State, 779 S.W.2d 417 (Tex. Crim. App. 1989).
- Defendant invoked his right to remain silent. Dowthitt.
- Defendant invoked his right to terminate the interrogation. Michigan v. Mosley, 423 U.S. 96, 96 S.Ct. 321, 46 L.Ed.2d 313 (1975).
- Defendant invoked his right to a court-appointed attorney. Smith.
- Defendant was told that his statement could be used "for or against" him. Dinkins v. State, 894 S.W.2d 330 (Tex. Crim. App. 1995).

§6:103 Suppression Under Art. 38.23

A defendant who moves for suppression under Art. 38.23 due to the violation of a statute has the burden of producing evidence of a statutory violation. Only when this burden is met does the State bear a burden to prove compliance. State v. Robinson, 334 S.W.3d 776, 779 (Tex. Crim. App. 2011).

Searches conducted in objectively reasonable reliance on binding appellate precedent are not subject to the Fourth Amendment exclusionary rule. Davis v. U.S., 564 U.S. ___, 131 S.Ct. 2419, 2429, 180 L.Ed.2d 285 (2011).

PRACTICE TIP: Although Texas has the statutory exclusionary rule contained in Art. 38.23, the State can use the Davis case to argue that the evidence was not "obtained" in violation of the law where legal precedent at the time of police conduct allowed the police action.

The Fourth Amendment does not apply to the actions of private individuals, and reading Article 38.23 to apply the Fourth Amendment to private individuals would lead to an absurdity. Ruiz v. State, ___ S.W.3d ___ at *2 (PD-1348-17, 2019 WL 2814867, Tex. Crim. App. July 3, 2019).

It is within the trial court's discretion to grant the State's request to reopen a motion-to-suppress hearing to supplement the record with additional testimony, outside the jury's presence, even over the objection of the defendant. Black v. State, 362 S.W.3d 626 (Tex. Crim. App. 2012).

(This page intentionally left blank.)

Chapter 7

RULES OF STATUTORY AND LEGAL INTERPRETATION

Table of Contents

§7:10 Hierarchical Governance

§7:11 Case Law

A statute cannot be superseded by a rule. Igo v. State, 210 S.W.3d 645 (Tex. Crim. App. 2006) (appellate rule relating to new trials superseded by statute setting out harm analysis standard for jury charge error); Rent v. State, 982 S.W.2d 382 (Tex. Crim. App. 1998) (appellate rule relating to new trials superseded by statute relating to remands for new trials).

§7:20 Rules of Statutory and Rules Construction

§7:21 Basic Rules

§7:21.1 Statutory Construction

When discerning the meaning of a statute, the reviewing court should start its analysis by looking at the plain language of the statute in question. Ex parte Whiteside, 12 S.W.3d 819 (Tex. Crim. App. 2000).

Appellate courts are required to construe a statute in accordance with the plain meaning of its literal text unless the language of the statute is ambiguous or the plain meaning leads to absurd results. Boykin v. State, 818 S.W.2d 782 (Tex. Crim. App. 1991); Hampton v. State, 86 S.W.3d 603 (Tex. Crim. App. 2002).

Consequently, the courts focus on the text of the statute and interpret it in a literal manner, attempting to discern the fair, objective meaning of the text. Boykin.

It is the court's duty when interpreting the statute to give the ordinary and plain meaning to the language of the Legislature. Boykin.

Statutory construction is a question of law, and appellate courts review the record de novo. Ramos v. State, 303 S.W.3d 302, 306 (Tex.Crim.App.2009).

When the Legislature meets, after a particular statute has been judicially construed, without changing that statute, the courts will presume the Legislature intended the same construction should continue to be applied to that statute. Marin v. State, 891 S.W.2d 267 (Tex. Crim. App. 1994); Dyar v. State, 125 S.W.3d 460 (Tex. Crim. App. 2003).

In construing whether a statute is vague and/ or overbroad, there is an elementary principle of statutory construction: a statute is interpreted in accordance with the plain meaning of its language unless the language is ambiguous or the plain meaning leads to absurd results. Boykin.

A statutory construction that would create a needless conflict between separate articles of the Code of Criminal Procedure is to be avoided, if at all possible. Joslin v. State, 722 S.W.2d 725 (Tex. App.—Dallas 1986), *vacated on other grounds*, 761 S.W.2d 376 (Tex. Crim. App. 1988).

In interpreting a prior law, the courts generally accord little weight to subsequent legislative enactments. Volosen v. State, 227 S.W.3d 77 (Tex. Crim. App. 2007) *citing* Ex parte Schroeter, 958 S.W.2d 811 (Tex. Crim. App. 1997).

The rule of *ejusdem generis* provides that where general words follow specific words in a statutory enumeration, the general words are construed to embrace only objects similar in nature to those objects enumerated by the preceding specific words. It is one of those canons of statutory construction that may prove useful in determining the plain meaning of statutory language: that plain meaning constituting the best indicator of legislative intent to which we are bound to give effect. Shipp v. State, 331 S.W.3d 433 (Tex. Crim. App. 2011).

§7:21.2 Rules Construction

The strictures that apply to statutory construction do not apply to an appellate court's construction of the Rules of Criminal Evidence. Ludwig v. State, 931 S.W.2d 239 (Tex. Crim. App. 1996).

Thus, appellate courts may consider extratextual sources to interpret a rule even if the plain language is unambiguous. Henderson v. State, 962 S.W.2d 544 (Tex. Crim. App. 1997).

§7:22 Statutory Intentions

§7:22.1 Statutory Law

Tex.Gov't.Code §311.021. Intention in Enactment of Statutes.

> *In enacting a statute, it is presumed that:*
>
> *(1) compliance with the constitutions of this state and the United States is intended;*
>
> *(2) the entire statute is intended to be effective;*
>
> *(3) a just and reasonable result is intended;*
>
> *(4) a result feasible of execution is intended; and*

(5) public interest is favored over any private interest.

§7:22.2 Case Law

When interpreting statutes, the courts seek to effectuate the collective intent or purpose of the legislators who enacted the legislation. Boykin v. State, 818 S.W.2d 782 (Tex. Crim. App. 1991); Camacho v. State, 765 S.W.2d 431 (Tex. Crim. App. 1989).

Statutes are to be construed, if at all possible, so as to give effect to all of their parts, and so that no part is to be construed as void or redundant. Badgett v. State, 42 S.W.3d 136 (Tex. Crim. App. 2001).

It is presumed in the enactment of a statute that the entire statute and all words in the statute are intended to be effective, and the language therein will create a just and reasonable result. Tex Gov't Code §311.021; State v. Mancuso, 919 S.W.2d 86 (Tex. Crim. App. 1996).

A reviewing court must consider the statute's context and presume that the legislature not only intended that a particular sentence would work a just and reasonable result, but that the intended result would be feasible of execution. State ex rel. White v. Bradley, 956 S.W.2d 725 (Tex.App.—Fort Worth 1997), *rev'd on other grounds*, 990 S.W.2d 245 (Tex. 1999).

§7:23 Statute Construction Aids

§7:23.1 Statutory Law

Tex Gov't Code §311.023. Statute Construction Aids.

In construing a statute, whether or not the statute is considered ambiguous on its face, a court may consider among other matters the:

(1) object sought to be attained;

(2) circumstances under which the statute was enacted;

(3) legislative history;

(4) common law or former statutory provisions, including laws on the same or similar subjects;

(5) consequences of a particular construction;

(6) administrative construction of the statute; and

(7) title (caption), preamble, and emergency provision.

§7:23.2 Case Law

In conducting an extratextual inquiry, courts may consider, among other matters: 1) the object sought to be attained; 2) the circumstances under which the statute was enacted; 3) legislative history; 4) common law or former statutory provisions, including laws on the same or similar subjects; 5) the consequences of a particular construction; 6) administrative construction of the statute; and 7) the title (caption), preamble, and emergency provision. Tex Gov't Code §311.023; Jordan v. State, 36 S.W.3d 871 (Tex. Crim. App. 2001); Brown v. State, 943 S.W.2d 35 (Tex. Crim. App. 1997).

The courts can also consider previous executive interpretations of the statute in question. Hines v. State, 75 S.W.3d 444 (Tex. Crim. App. 2002).

The courts may also look to other offense provisions contained in the same statutory section. Long v. Texas, 931 S.W.2d 285 (Tex. Crim. App. 1996) (comparing stalking provision to other provisions contained in Penal Code §42.07).

In construing a statute, whether or not the statute is considered ambiguous on its face, a court may consider among other matters the consequences of a particular construction. Griego v. State, 853 S.W.2d 664 (Tex.App.—Houston [14th Dist.] 1993, *no pet.*)

§7:24 Irreconcilable Statutes and Amendments

§7:24.1 Statutory Law

Tex Gov't Code §311.025. Irreconcilable Statutes and Amendments.

(a) Except as provided by §311.031(d), if statutes enacted at the same or different sessions of the legislature are irreconcilable, the statute latest in date of enactment prevails.

(b) Except as provided by §311.031(d), if amendments to the same statute are enacted at the same session of the legislature, one amendment without reference to another, the amendments shall be harmonized, if possible, so that effect may be given to each. If the amendments are irreconcilable, the latest in date of enactment prevails.

(c) In determining whether amendments are irreconcilable, text that is reenacted because of the requirement of Article III, §36, of the Texas Constitution is not considered to be irreconcilable with additions or omissions in the same text made by another amendment. Unless clearly indicated to the contrary, an amendment that reenacts text in compliance with that constitutional requirement does not indicate legislative intent that the reenacted text prevail over changes in the same text made by another amendment, regardless of the relative dates of enactment.

(d) In this section, the date of enactment is the date on which the last legislative vote is taken on the bill enacting the statute.

(e) If the journals or other legislative records fail to disclose which of two or more bills in conflict is latest in date of enactment, the date of enactment of the respective bills is considered to be, in order of priority:

(1) the date on which the last presiding officer signed the bill;

(2) the date on which the governor signed the bill; or

(3) the date on which the bill became law by operation of law.

§7:24.2 Case Law

Statutes that have the same purpose or relate to the same subject are "in pari material," and must be construed together if possible. Petta v. State, 840 S.W.2d 721 (Tex.App.—Corpus Christi 1992, *pet. ref'd*).

Statutory provisions must first be irreconcilable before one of them will take precedence over the other. H & C Communications, Inc. v. Reed's Food International, Inc., 887 S.W.2d 475 (Tex.App.—San Antonio 1994, *no pet.*).

When a harmonious construction of statutes dealing with the same or similar subject matter is not possible, the more recent legislative enactment should prevail. State v. McKinney, 803 S.W.2d 374 (Tex.App.—Houston [14th Dist.] 1990, *no pet.*).

§7:30 Definitions of Words, Phrases and Terms

§7:31 Words and Phrases

§7:31.1 Statutory Law

CCP Art. 3.01. Words and Phrases.

All words, phrases and terms used in this Code are to be taken and understood in their usual acceptation in common language, except where specially defined.

§7:31.2 Case Law

In determining plain meaning, "[w]ords and phrases shall be read in context and construed according to the rules of grammar and usage." Tex Gov't Code §311.011(a); Dowthitt v. State, 931 S.W.2d 244 (Tex. Crim. App.1996).

Words and phrases are to be read in context and given their common, everyday meaning unless they have acquired technical or particular meaning. R.R.E. v. Glenn, 884 S.W.2d 189 (Tex.App.—Fort Worth 1994, *writ denied*).

The meaning of words and phrases used in a statute is a question of pure law; and the application of the scope of a statute to specific, undisputed historical facts is a mixed question of law and fact. Both of these matters are reviewed de novo at each appellate level. Hampton v. State, 86 S.W.3d 603 (Tex. Crim. App. 2002).

Words that are not statutorily defined are to be given their usual meanings and no specific jury instructions are required. Garcia v. State, 887 S.W.2d 846 (Tex. Crim. App.1994), *cert. denied*, 514 U.S. 1005, 115 S.Ct.1317, 131 L.Ed.2d 198 (1995).

Jurors are presumed to know and apply the common and ordinary meanings. Wheeler v. State, 35 S.W.3d 126 (Tex.App.—Texarkana 2000, *pet. ref'd*).

§7:32 Common and Technical Terms

§7:32.1 Statutory Law

Tex Gov't Code §311.011. Common and Technical Usage of Words.

> *(a) Words and phrases shall be read in context and construed according to the rules of grammar and common usage.*
>
> *(b) Words and phrases that have acquired a technical or particular meaning, whether by legislative definition or otherwise, shall be construed accordingly.*

§7:32.2 Case Law

Words and phrases are to be read in context and given their common, everyday meaning unless they have acquired technical or particular meaning. R.R.E. v. Glenn, 884 S.W.2d 189 (Tex.App.—Fort Worth 1994, *writ denied*).

The canons of construction dictate that words and phrases possessing a technical meaning are generally to be considered as having been used in their technical sense. This applies to those terms that have a known and established legal meaning, or which have acquired a peculiar and appropriate meaning in the law, as where the words used have a well-known common law meaning. However, in some cases justice is best served, and more consistently applied, if jurors are provided a precise, uniform definition to guide their determination of the particular circumstances at issue. Medford v. State, 13 S.W.3d 769 (Tex. Crim. App. 2000).

The legislature has obligated the courts to construe statutory words or phrases having technical or particular meaning, such as terms of art or science, in accordance with their technical meaning. State v. Howard, 908 S.W.2d 602 (Tex. App.—Amarillo 1995, *no pet.*).

§7:33 Interpretations of Certain Words

Tex Gov't Code §311.016. "May," "Shall," "Must," etc.

> *The following constructions apply unless the context in which the word or phrase appears necessarily requires a different construction or unless a different construction is expressly provided by statute:*
>
> *(1) "May" creates discretionary authority or grants permission or a power.*
>
> *(2) "Shall" imposes a duty.*
>
> *(3) "Must" creates or recognizes a condition precedent.*
>
> *(4) "Is entitled to" creates or recognizes a right.*
>
> *(5) "May not" imposes a prohibition and is synonymous with "shall not."*
>
> *(6) "Is not entitled to" negates a right.*
>
> *(7) "Is not required to" negates a duty or condition precedent.*

§7:40 Headings

Tex Gov't Code §311.024. Headings.

> *The heading of a title, subtitle, chapter, subchapter, or section does not limit or expand the meaning of a statute.*

§7:50 Prospective Operation of Statutes

Tex Gov't Code §311.022. Prospective Operation of Statutes.

> *A statute is presumed to be prospective in its operation unless expressly made retrospective.*

§7:60 Special or Local Provision Prevails Over General

§7:61 Statutory Law

Tex Gov't Code §311.026. Special or Local Provision Prevails Over General.

(a) If a general provision conflicts with a special or local provision, the provisions shall be construed, if possible, so that effect is given to both.

(b) If the conflict between the general provision and the special or local provision is irreconcilable, the special or local provision prevails as an exception to the general provision, unless the general provision is the later enactment and the manifest intent is that the general provision prevail.

§7:62 Case Law

If a general provision conflicts with a specific provision, the provisions shall be construed, if possible, so that effect is given to both. Tex Gov't Code §311.026(a); Dillehey v. State, 815 S.W.2d 623 (Tex. Crim. App. 1991).

If the statutes are unable to be reconciled, the specific statute will prevail as an exception to the general statute, unless the general statute is the later enactment and the manifest intent is that the general provision prevail. Tex Gov't Code §311.026(b); State v. Mancuso, 919 S.W.2d 86 (Tex. Crim. App. 1996).

If the definition of an offense does not prescribe a culpable mental state, a culpable mental state is nevertheless required unless the definition plainly dispenses with any mental element. Penal Code §6.02(b); Harvey v. State, 78 S.W.3d 368 (Tex. Crim. App. 2002).

§7:63 Doctrine of In Pari Materia

The rule of *in pari materia* is a principle of statutory interpretation, a means of divining and giving full effect to legislative intent. Two statutes that are *in pari materia* are to be construed together, each enactment in reference to the other, as though they were part of one and the same law. Any conflict between their provisions will be harmonized, if possible, and effect will be given to all the provisions of each act if they can be made to stand together and have concurrent efficacy. Mills v. State, 722 S.W.2d 411 (Tex. Crim. App. 1986).

Statutes may be said to be *in pari materia* when they relate to the same person or thing, to the same class of persons or things, or have the same purpose or object. Characterization of the object or purpose is more important than characterization of subject matter in determining whether different statutes are closely enough related to justify interpreting one in light of the other. Mills. *See also* Strickland v. State, 193 S.W.3d 662 (Tex.App.—Fort Worth 2006, *pet. ref'd*) (the doctrine of *in pari materia* applies if two statutes share a common purpose or object).

A defendant has a due process right to be prosecuted under a "special" statute that is *in pari materia* with a broader statute when these statutes irreconcilably conflict. Ex parte Smith, 185 S.W.3d 887 (Tex. Crim. App. 2006).

To determine whether two statutes share a common purpose, we consider whether the two statutes (1) are contained in the same legislative act, (2) require the same elements of proof, (3) involve different penalties, and (4) were clearly written to achieve the same objective. Strickland.

Because the doctrine of *in pari material* is applicable only to a comparison of statutes (passed by the same governmental body), the doctrine does not apply to a comparison between a statute and a court-made rule. State v. Vasilas, 253 S.W.3d 268 (Tex. Crim. App. 2008).

A defendant's *in pari materia* claim is generally not cognizable as a motion to quash because generally the indictment will be sufficient on its face. The claim is also not cognizable on a pre-trial writ of habeas corpus. A decision on an *in pari materia* claim would be premature before the state has had an opportunity to develop a complete factual record during a trial. Ex parte Smith.

Convictions in a single trial for the offenses of fraudulent use or possession of identifying information and failure to identify do not violate the doctrine of *in pari materia*. Jones v. State, 396 S.W.3d 558, 563 (Tex. Crim. App. 2013).

PRACTICE TIP: The in pari materia claim is cognizable at the jury charge stage of the trial, in the nature of a request for a lesser included offense in the charge. See Ex parte Smith.

§7:70 Construction of Penal Offenses

Penal Code §1.05. Construction of Code

> *(a) The rule that a penal statute is to be strictly construed does not apply to this code. The provisions of this code shall be construed according to the fair import of their terms, to promote justice and effect the objectives of the code.*
>
> *(b) Unless a different construction is required by the context, §§311.011, 311.012, 311.014, 311.015, and 311.021 through 311.032 of Chapter 311, Government Code (Code Construction Act), apply to the construction of this code.*
>
> *(c) In this code:*
>
> *(1) a reference to a title, chapter, or section without further identification is a reference to a title, chapter, or section of this code; and*
>
> *(2) a reference to a subchapter, subsection, subdivision, paragraph, or other numbered or lettered unit without further identification is a reference to a unit of the next-larger unit of this code in which the reference appears.*

§7:71 Applicability of Penal Code to Offenses Under Other Codes

Penal Code §1.03(b) controls the application of Penal Code provisions to criminal offenses defined outside the Penal Code. It directs the export of the provisions contained only in Titles 1, 2, and 3 of the Penal Code to criminal offenses defined outside the Penal Code and contemporaneously bars the import of extra-Penal Code offenses to offenses defined in Titles 4 through 11 of the Penal Code. The offenses defined in Title 4 of the Penal Code apply to offenses defined outside the Penal Code only where the Legislature has designated that Title 4 is applicable via legislative action like the amendments made to the Controlled Substances Act, the Simulated Controlled Substances Act, the Dangerous Drugs Act, the Abusable and Volatile Chemicals Act, and the Election Code. State v. Colyandro, 233 S.W.3d 870 (Tex. Crim. App. 2007).

§7:72 Construction of Offenses Outside Penal Code

Tex. Gov't Code §311.035. CONSTRUCTION OF STATUTE OR RULE INVOLVING CRIMINAL OFFENSE OR PENALTY.

> *(a) In this section, "actor" and "element of offense" have the meanings assigned by Section 1.07, Penal Code.*
>
> *(b) Except as provided by Subsection (c), a statute or rule that creates or defines a criminal offense or penalty shall be construed in favor of the actor if any part of the statute or rule is ambiguous on its face or as applied to the case, including:*
>
> *(1) an element of offense; or*
>
> *(2) the penalty to be imposed.*
>
> *(c) Subsection (b) does not apply to a criminal offense or penalty under the Penal Code or under the Texas Controlled Substances Act.*
>
> *(d) The ambiguity of a part of a statute or rule to which this section applies is a matter of law to be resolved by the judge.*

§7:80 Constitutionality of Statutes

§7:81 In General

In construing a statute, the courts begin with the presumption that it is constitutional. White v. State, 50 S.W.3d 31 (Tex.App.—Waco 2001, *pet. ref'd*).

The courts must uphold the statute if it can be afforded a reasonable construction that renders it constitutional and effectuates legislative intent. White.

In construing whether a law is vague and/or overbroad, the courts keep in mind the elementary principle of statutory construction: they interpret a statute in accordance with the plain meaning of its language unless the language is ambiguous or the plain meaning leads to absurd results. Sanchez v. State, 995 S.W.2d 677 (Tex. Crim. App. 1999).

There is no basis under Texas law for a trial court to conduct a pretrial evidentiary hearing to determine the "as applied" constitutionality of a state penal or criminal procedural statute or to make a pretrial declaratory judgment on same. St. ex rel. Lykos v. Fine, 330 S.W.3d 904 (Tex. Crim. App. 2011) (where the trial court sought to convene a hearing to determine the constitutionality of the Texas death penalty scheme).

§7:82 Basic Constitutional Principles

It is well-established that criminal laws must be sufficiently clear in at least three respects:

- A person of ordinary intelligence must be given a reasonable opportunity to know what is prohibited.
- The law must establish determinate guidelines for law enforcement.
- Where First Amendment freedoms are implicated, the law must be sufficiently definite to avoid chilling protected expression.

Grayned v. City of Rockford, 408 U.S. 104, 92 S.Ct.2294, 33 L.Ed.2d 222 (1972); Long v. Texas, 931 S.W.2d 285 (Tex. Crim. App. 1996).

Absent First Amendment considerations, the Due Process Clause requires that a law be specific enough to: (1) give a person of ordinary intelligence a reasonable opportunity to know what is prohibited, and (2) establish determinate guidelines for law enforcement. Long.

A statute will survive attack as being unconstitutionally vague provided it gives fair warning to the defendant as to what conduct it proscribes. New York v. Ferber, 458 U.S. 747, 102 S.Ct.3348, 73 L.Ed.2d 1113 (1982).

In order to mount a successful attack on the basis that a statute is facially unconstitutional, a defendant must show that there is no conceivable set of circumstances in which the statute is capable of a constitutional application. Briggs v. State, 789 S.W.2d 918 (Tex. Crim. App. 1990).

Constitutional errors may be waived by failure to object at trial. Briggs.

§7:83 First Amendment Considerations

When a statute is capable of reaching First Amendment freedoms, the doctrine of vagueness demands a greater degree of specificity than in other contexts. Long v. Texas, 931 S.W.2d 285 (Tex. Crim. App. 1996).

Greater specificity is required to preserve adequately the right of free expression because uncertain meanings inevitably lead citizens to steer far wider of the unlawful zone than if the boundaries of the forbidden areas were clearly marked. Grayned v. City of Rockford, 408 U.S. 104, 92 S.Ct.2294, 33 L.Ed.2d 222 (1972); Long.[SEP]Moreover, when a vagueness challenge involves First Amendment considerations, a criminal law may be held facially invalid even though it may not be unconstitutional as applied to the defendant's conduct. Gooding v. Wilson, 405 U.S. 518, 92 S.Ct.1103, 31 L.Ed.2d 408 (1972); Long.

Where First Amendment freedoms are implicated, the law must be sufficiently definite to avoid chilling protected expression. Long.

Under the First Amendment, a statute may be subject to a facial challenge even though it may have some legitimate application. Long.

A defendant can challenge a statute on a First Amendment claim where the statute's deterrent effect on legitimate expression is both real and substantial, and the statute is not readily subject to a narrowing construction by the state courts. Young v. American Mini Theatres, Inc., 427 U.S. 50, 96 S.Ct.2440, 49 L.Ed.2d 310 (1976); White v. State, 50 S.W.3d 31 (Tex.App.—Waco 2001, *pet. ref'd*).

A statute places a real and substantial burden on protected speech when there is a realistic danger that the statute itself will significantly compromise recognized First Amendment protections of parties not before the court. Members of City Council

v. Taxpayers for Vincent, 466 U.S. 789, 104 S.Ct.2118, 80 L.Ed.2d 772 (1984); White.

Even if the statute substantially burdens these constitutional guarantees, the courts will not conclude that it is unconstitutionally vague if they can narrowly construe it in such a manner as to remove the seeming threat or deterrence to constitutionally protected expression. Broadrick v. Oklahoma, 413 U.S. 601, 93 S.Ct.2908, 37 L.Ed.2d 830 (1973); White.

A statute impermissibly compels a person's speech in violation of the First Amendment when the statute requires him "to be an instrument for fostering public adherence to an ideological point of view he finds unacceptable." Wooley v. Maynard, 430 U.S. 705, 97 S.Ct.1428, 51 L.Ed.2d 752 (1977); White.

§7:84 Due Process Claims

Absent First Amendment considerations, the Due Process Clause requires that a law be specific enough to: (1) give a person of ordinary intelligence a reasonable opportunity to know what is prohibited, and (2) establish determinate guidelines for law enforcement. Long v. Texas, 931 S.W.2d 285 (Tex. Crim. App. 1996).

It is a basic principle of due process that a statute is void for vagueness if its prohibitions are not clearly defined. The void-for-vagueness doctrine requires that a penal statute define the criminal offense with sufficient definiteness that ordinary people can understand what conduct is prohibited and in a manner that does not permit arbitrary and discriminatory enforcement. Although a statute is not impermissibly vague because it fails to define words or phrases, it is invalid if it fails to give a person of ordinary intelligence a reasonable opportunity to know what conduct is prohibited. Moreover, where a statute does not substantially implicate constitutionally protected conduct or speech, it is valid unless it is impermissibly vague in all applications. State v. Holcombe, 187 S.W.3d 496 (Tex.Crim.App.2006); Watson v. State, 369 S.W.3d 865, 870 (Tex. Crim. App. 2012).

Due process requires that a criminal statute give fair notice of that conduct which is criminally prohibited. Bynum v. State, 767 S.W.2d 769 (Tex. Crim. App. 1989).

Where a due process vagueness claim does not involve a First Amendment issue the courts will not address the merits of the defendant's claims unless he first makes a showing that the statute is unconstitutional as applied to him. Vuong v. State, 830 S.W.2d 929 (Tex. Crim. App. 1992).

Where no First Amendment rights are involved, the reviewing court need only examine the statute to determine whether it is impermissibly vague as applied to the defendant's specific conduct. Bynum.

The natural corollary to this principle is that a defendant to whom a statute clearly applies may not successfully challenge it for vagueness. Vuong.

Accordingly, it is the defendant's burden to establish that the statute is unconstitutional as applied to him; that it might be unconstitutional to others is not sufficient. Rodriguez v. State, 47 S.W.3d 86 (Tex.App.—Houston [14th Dist.] 2001, *no pet.*).

Absent a First Amendment claim, a facial challenge can be mounted successfully only if the statute is vague in all of its applications. Long v. Texas, 931 S.W.2d 285 (Tex. Crim. App. 1996).

§7:85 Specific Statutory Provisions

The absence of a "reasonable person" requirement is not necessarily fatal to a statute's constitutionality. Long v. State, 931 S.W.2d 285 (Tex. Crim. App. 1996).

In some situations an affirmative defense for constitutionally protected activity can save a statute from a constitutionality challenge. *See* Long.

A savings clause may give courts more latitude to place a narrowing construction upon a statute that would save it from a constitutionality challenge. Long.

§7:86 The Single Subject Rule

§7:86.1 Constitutional Provision

TEX. CONST. art. III, § 35

> *(a) No bill (except general appropriation bills, which may embrace the various subjects and accounts, for and on account of which moneys are appropriated) shall contain more than one subject.*

(b) The rules of procedure of each house shall require that the subject of each bill be expressed in its title in a manner that gives the legislature and the public reasonable notice of that subject. The legislature is solely responsible for determining compliance with the rule.

(c) A law, including a law enacted before the effective date of this subsection, may not be held void on the basis of an insufficient title.

§7:86.2 Case Law

Under TEX. CONST. art. III, § 35(a) a statute is facially unconstitutional where the bill in which it was enacted or amended contained more than one subject. Ex parte Jones, 440 S.W.3d 628, 629 (Tex. Crim. App. 2014) (where the court found that a 2011 amendment to *TEX. PENAL CODE* §38.04(b)(2)(A), the evading arrest statute, did not violate the single subject rule).

§7:90 Conflict of Laws Principles

Traditional conflict-of-law principles prescribe that issues that are strictly procedural in nature are governed by the laws of the forum state, whereas issues that are substantive in nature require an analysis of which state has the most significant relationship with the communication in question. Gonzalez v. State, 45 S.W.3d 101 (Tex. Crim. App. 2001) *citing* Restatement (Second) of Conflict of Laws §139 (1971).

A substantive right has been defined by the Texas Court of Criminal Appeals as a right to the equal enjoyment of fundamental rights, privileges, and immunities or a right that can be protected or enforced by law. Gonzalez.

A procedural right is a right that helps in the protection or enforcement of a substantive right. Gonzalez; Vega v. State, 84 S.W.3d 613 (Tex. Crim. App. 2002).

There are, under Texas conflict-of-law principles, several factors to consider in determining which jurisdiction has the most significant relationship to the case, including: (1) where the injury or unlawful conduct occurred; (2) the place where the relationship between the parties is the strongest; (3) the number and nature of contacts that the non-forum state has with the parties and with the transaction involved; (4) the relative materiality of the evidence that is sought to be excluded; and (5) the fairness to the parties. Gonzalez; Vega.

§7:100 Ex Post Facto Laws

There are four categories of ex post facto laws: (1) Every law that makes an action done before the passing of the law, and which was innocent when done, criminal; and punishes such action; (2) Every law that aggravates a crime, or makes it greater than it was, when committed; (3) Every law that changes the punishment, and inflicts a greater punishment, than the law annexed to the crime, when committed; (4) Every law that alters the legal rules of evidence, and receives less, or different, testimony, than the law required at the time of the commission of the offence, in order to convict the offender. Phillips v. State, 362 S.W.3d 606, 610 (Tex. Crim. App. 2011)(overruled on other grounds by Ex parte Heilman, 456 S.W.3d 159 (Tex. Crim. App. 2015)).

The Ex Post Facto Clause applies not only to laws that are facially retroactive, but also to laws that are applied retroactively. Phillips (overruled on other grounds by Ex parte Heilman, 456 S.W.3d 159 (Tex. Crim. App. 2015)).

The right to be free from ex post facto laws or the ex post facto application of a law is an absolute right—a first category Marin right and nnot be waived or forfeited. Phillips (overruled on other grounds by Ex parte Heilman, 456 S.W.3d 159 (Tex. Crim. App. 2015)).

A defendant can raise a claim for the first time on appeal that his convictions were barred by the Ex Post Facto Clause. Ieppert v. State, 908 S.W.2d 217 (Tex. Crim. App. 1995).

A statute of limitations may be changed or repealed without violating constitutional prohibitions against ex post facto laws in any case where a right to acquittal has not been absolutely acquired by the completion of the period of limitations. Phillips (overruled on other grounds by Ex parte Heilman, 456 S.W.3d 159 (Tex. Crim. App. 2015)).

However, a facially retroactive law that revives a previously time-barred offense violates the Ex Post Facto Clause. Phillips; (overruled on other grounds by Ex parte Heilman, 456 S.W.3d 159

(Tex. Crim. App. 2015)); Stogner v. California, 539 U.S. 607, 123 S.Ct. 2446 156 L.Ed.2d 544 (2003).

The Texas constitutional prohibition against a retroactive law has never been made applicable to statutes merely affecting matters of procedure that do not disturb vested substantive rights. Grimes v. State, 807 S.W.2d 582 (Tex. Crim. App. 1991).

Where a defendant asserts that a particular statute is an ex post facto law, the courts must determine: (1) whether the statute is being applied retroactively; and (2) whether its provisions constitute punishment for constitutional purposes. Rodriguez v. State, 93 S.W.3d 60 (Tex. Crim. App. 2002).

The question whether a statute constitutes punishment for constitutional purposes is answered by application of what is known as the "intent-effects" test. Rodriguez.

Under the "intent-effects" test, a reviewing court must first ask whether the legislature intended the statute to be a criminal punishment. Rodriguez.

Whether a particular punishment is criminal or civil is, at least initially, a matter of statutory construction. A court must first ask whether the legislature, in establishing the penalizing mechanism, indicated either expressly or impliedly a preference for one label or the other. Hudson v. United States, 522 U.S. 93, 118 S.Ct.488, L.Ed.2d 450 (1997); Rodriguez.

A reviewing court must afford a high level of deference to the legislature's stated aims in passing the statute. Hudson; Rodriguez.

Where the legislature manifests an expressly punitive intent, the inquiry is at an end and the statute is a violation of the ex post facto clause. Kennedy v. Mendoza-Martinez, 372 U.S. 144, 83 S.Ct.554, 9 L.Ed.2d 644 (1963); Rodriguez.

If the legislature intends to establish a civil remedy, a reviewing court then examines "whether the statutory scheme [is] so punitive either in purpose or effect as to transform what was clearly intended as a civil remedy into a criminal penalty." Hudson; Rodriguez.

The legislature's manifest intent will be rejected only where the party challenging the statute provides the clearest proof that the statute is actually criminally punitive in operation. Kansas v. Hendricks, 521 U.S. 346, 117 S.Ct.2072, 138 L.Ed.2d 501 (1997); Rodriguez.

To evaluate whether the effects of a statute are criminally punitive, courts generally consider:

- Whether the sanction involves an affirmative disability or restraint
- Whether it has traditionally been regarded as a punishment
- Whether it comes into play only on a finding of scienter
- Whether its operation will promote the traditional aims of punishment—retribution and deterrence
- Whether the behavior to which it applies is already a crime
- Whether an alternative purpose to which it may rationally be connected is assignable to it
- Whether it appears excessive in relation to the alternative purpose assigned

Kennedy v. Mendoza-Martinez, 372 U.S. 144, 83 S.Ct.554, 9 L.Ed.2d 644 (1963); Rodriguez.

A clear example of an ex post facto law is a law enacted after expiration of a previously applicable limitations period when it is applied to revive a previously time-barred prosecution. Stogner v. California, 539 U.S. 607, 123 S.Ct.2446, 156 L.Ed.2d 544 (2003).

The use of a successfully completed deferred adjudication to enhance the punishment for a later offense when the statute permitting such an enhancement did not exist at the time adjudication was deferred is an ex post facto violation. Scott v. State, 55 S.W.3d 593 (Tex. Crim. App. 2001).

Authorizing conviction of certain sexual offenses on the victim's testimony alone where corroborating evidence had previously been required is an ex post facto violation. Carmell v. Texas. 529 U.S. 513, 120 S.Ct. 1620, 146 L.Ed.2d 577 (2000).

§7:110 Separation of Powers

The separation of powers clause of Tex.Const., Art. II, §1, provides for the separation of powers of the three departments of Texas government: Judicial, Legislative, and Executive. Ex parte Davis, 947 S.W.2d 216 (Tex. Crim. App. 1996).

The separation of powers provision may be violated in either of two ways: 1) when one branch of government assumes, or is delegated, to whatever degree, a power that is more properly attached to another branch; or 2) when one branch unduly interferes with another branch so that the other branch cannot effectively exercise its constitutionally assigned powers. Ex parte Perry, 483 S.W.3d 884, 894-5 (Tex. Crim. App. 2016). A separation of powers claim is cognizable on pretrial habeas. Ex parte Perry, 483 S.W.3d at 895.

§7:120 Computation of Time

Tex.Gov't.Code §311.014. Computation of Time.

> *(a) In computing a period of days, the first day is excluded and the last day is included.*
>
> *(b) If the last day of any period is a Saturday, Sunday, or legal holiday, the period is extended to include the next day that is not a Saturday, Sunday, or legal holiday.*
>
> *(c) If a number of months is to be computed by counting the months from a particular day, the period ends on the same numerical day in the concluding month as the day of the month from which the computation is begun, unless there are not that many days in the concluding month, in which case the period ends on the last day of that month.*

§7:130 Doctrine of Stare Decisis

Under the doctrine of stare decisis, the courts hold that it is often better to be consistent than right. This doctrine promotes judicial efficiency and consistency, fosters reliance on judicial decisions, and contributes to the actual and perceived integrity of the judicial process. Bawcom v. State, 78 S.W.3d 360 (Tex. Crim. App. 2002).

However, overruling precedent is acceptable under certain circumstances. Some factors that support overruling precedent are: 1) when the original rule is flawed from the outset, 2) when older precedent conflicts with a newer decision that is found to be more soundly reasoned, and 3) when the rule consistently creates unjust results or places unnecessary burdens upon the system. Bawcom.

The achievement of harmony between criminal and civil precedent may also be considered in determining what rule to fashion. Awadelkariem v. State, 974 S.W.2d 721 (Tex. Crim. App. 1998).

In determining whether to retain or reject a particular rule of law, the courts may also consider whether justifications proffered for the rule have been undercut with the passage of time. Awadelkariem.

§7:140 Unpublished and Plurality Appellate Court Opinions

§7:141 Court of Appeals Unpublished Opinions

Tex.R.App.Pro. Rule 47.7. Unpublished Opinions

> *(a) Criminal Cases. Opinions and memorandum opinions not designated for publication by the court of appeals under these or prior rules have no precedential value but may be cited with the notation, "(not designated for publication)."*
>
> *(b) Civil Cases. Opinions and memorandum opinions designated "do not publish" under these rules by the courts of appeals prior to January 1, 2003 have no precedential value but may be cited with the notation, "(not designated for publication)." If an opinion or memorandum opinion issued on or after that date is erroneously designated "do not publish," the erroneous designation will not affect the precedential value of the decision.*

§7:142 Court of Criminal Appeals Unpublished Opinions

Tex.R.App.Pro. Rule 77.3. Unpublished Opinions

> *Unpublished opinions have no precedential value and must not be cited as authority by counsel or by a court.*

§7:143 Plurality Opinions

Plurality opinions are not binding precedent. State v. Hardy, 963 S.W.2d 516 (Tex. Crim. App. 1997).

§7:150 Estoppel

A party may be estopped from asserting a claim that is inconsistent with that party's prior conduct. Arroyo v. State, 117 S.W.3d 795 (Tex. Crim. App. 2003) (the state turned over a rap sheet on a witness and later tried to assert that the defendant had not shown the witness to be the same person named in the rap sheet).

A defendant who requests and receives a jury charge on a lesser-included offense is estopped from complaining that his conviction of that lesser-included offense was barred by limitations. State v. Yount, 853 S.W.2d 6 (Tex. Crim. App. 1993).

A defendant, who affirmatively requests that the trial court not submit to the jury one of the special issues statutorily required for capital sentencing, is estopped from arguing on appeal that the trial court had erred in not submitting the special issue. Prystash v. State, 3 S.W.2d 522 (Tex. Crim. App. 1999).

A defendant who agrees to stipulate to a fact is estopped from complaining on appeal that the evidence is insufficient even where the stipulation is not formally introduced into evidence. Bryant v. State, 187 S.W.3d 397 (Tex. Crim. App. 2005).

Where the defendant's objection prevented the state from fulfilling its burden of proof at a suppression hearing, the defendant is prevented under the invited-error doctrine from asserting this ground on appeal. Vennus v. State, 282 S.W.3d 70 (Tex. Crim. App. 2009).

Where a defendant requests an improper lesser included charge or has some responsibility for its inclusion, he is estopped on appeal from complaining of its inclusion in the charge. Woodard v. State, 322 S.W.3d 648 (Tex. Crim. App. 2010).

A defendant who declines to have a witness attached and brought to court is estopped from arguing that her confrontation rights were violated on the grounds that the witness was absent from court. Woodall v. State, 336 S.W.3d 634 (Tex. Crim. App. 2011).

§7:160 Statutory Presumptions

§7:161 Statutory Law

Texas Penal Code § 2.05. Presumption

> *(a) Except as provided by Subsection (b), when this code or another penal law establishes a presumption with respect to any fact, it has the following consequences:*
>
> *(1) if there is sufficient evidence of the facts that give rise to the presumption, the issue of the existence of the presumed fact must be submitted to the jury, unless the court is satisfied that the evidence as a whole clearly precludes a finding beyond a reasonable doubt of the presumed fact; and*
>
> *(2) if the existence of the presumed fact is submitted to the jury, the court shall charge the jury, in terms of the presumption and the specific element to which it applies, as follows:*
>
> *(A) that the facts giving rise to the presumption must be proven beyond a reasonable doubt;*
>
> *(B) that if such facts are proven beyond a reasonable doubt the jury may find that the element of the offense sought to be presumed exists, but it is not bound to so find;*
>
> *(C) that even though the jury may find the existence of such element, the state must prove beyond a reasonable doubt each of the other elements of the offense charged; and*
>
> *(D) if the jury has a reasonable doubt as to the existence of a fact or facts giving rise to the presumption, the presumption fails and the jury shall not consider the presumption for any purpose.*
>
> *(b) When this code or another penal law establishes a presumption in favor of the*

defendant with respect to any fact, it has the following consequences:

(1) if there is sufficient evidence of the facts that give rise to the presumption, the issue of the existence of the presumed fact must be submitted to the jury unless the court is satisfied that the evidence as a whole clearly precludes a finding beyond a reasonable doubt of the presumed fact; and

(2) if the existence of the presumed fact is submitted to the jury, the court shall charge the jury, in terms of the presumption, that:

(A) the presumption applies unless the state proves beyond a reasonable doubt that the facts giving rise to the presumption do not exist;

(B) if the state fails to prove beyond a reasonable doubt that the facts giving rise to the presumption do not exist, the jury must find that the presumed fact exists;

(C) even though the jury may find that the presumed fact does not exist, the state must prove beyond a reasonable doubt each of the elements of the offense charged; and

(D) if the jury has a reasonable doubt as to whether the presumed fact exists, the presumption applies and the jury must consider the presumed fact to exist.

§7:162 Case Law

When charging the jury on a statutory presumption, the court must charge jurors that they may use the presumption to find the defendant guilty only if the State proves the facts giving rise to the presumption beyond a reasonable doubt. Hollander v. State, 414 S.W.3d 746, 753 (Tex. Crim. App. 2013).

(This page intentionally left blank.)

Chapter 8

DOUBLE JEOPARDY

DOUBLE JEOPARDY

Table of Contents

§8:10 Basic Double Jeopardy Principles

§8:11 Constitutional and Statutory Provisions

U.S. Constitution, Fifth Amendment: " ... nor shall any person be subject for the same offense to be twice put in jeopardy of life or limb."

Texas Constitution, Art. 1 §14: "No person, for the same offense, shall be twice put in jeopardy of life or liberty, nor shall a person be again put upon trial for the same offense, after a verdict of not guilty in a court of competent jurisdiction."

CCP Art. 1.10. JEOPARDY

"No person for the same offense shall be twice put in jeopardy of life or liberty; nor shall a person be again put upon trial for the same offense, after a verdict of not guilty in a court of competent jurisdiction."

The double jeopardy clause is a guarantee against being twice put to trial for the same offense. Abney v. U.S., 431 U.S. 651, 97 S.Ct. 2034, 52 L.Ed.2d 651 (1977); Phillips v. State, 787 S.W.2d 391 (Tex. Crim. App. 1990).

The Fifth Amendment guarantee against being placed in double jeopardy is applicable to the states through the Fourteenth Amendment. Benton v. Maryland, 395 U.S. 784, 89 S.Ct. 2056, 23 L.Ed.2d 707 (1969).

The Texas and United States Constitutions' double jeopardy provisions provide substantially identical protections. Ex parte Mitchell, 977 S.W.2d 575 (Tex. Crim. App. 1997), *cert. denied*, 525 U.S. 873 (1998).

The only exception is in cases where the prosecutor causes a mistrial. State v. Webb, 980 S.W.2d 924 (Tex.App.—Fort Worth 1998), *affirmed at* 12 S.W.3d 808 (Tex. Crim. App. 2000).

§8:12 Basic Double Jeopardy Protections

Both the United States and Texas Constitutional provisions speak of double jeopardy in terms of the same offense rather than the same transaction. Phillips v. State, 787 S.W.2d 391 (Tex. Crim. App. 1990).

Protection against double jeopardy consists of three separate constitutional protections:

- Against a second prosecution for the same offense after acquittal
- Against a second prosecution for the same offense after conviction
- Against multiple punishments for the same offense.

U.S. v. Halper, 490 U.S. 435, 109 S.Ct. 1892, 104 L.Ed.2d 487 (1989); Ex parte Stover, 946 S.W.2d 343 (Tex. Crim. App. 1997).

§8:13 Blockburger Test

Texas employs the Blockburger test for deciding when two offenses are the same for double jeopardy purposes. The applicable rule is that where the same act or transaction constitutes a violation of two distinct statutory provisions, the test to be applied to determine whether there are two offenses or only one, is whether each provision requires proof of a fact which the other does not. Blockburger v. U.S., 284 U.S. 299, 52 S.Ct. 180, 76 L.Ed.2d 306 (1932); U.S. v. Dixon, 509 U.S. 688, 113 S.Ct. 2849, 125 L.Ed.2d 556 (1993); Ervin v. State, 991 S.W.2d 804 (Tex. Crim. App. 1999).

If application of the Blockburger test reveals that the offenses have identical statutory elements or that one is a lesser included offense of the other, then the two offenses are the same for double jeopardy purposes. Granger v. State, 850 S.W.2d 513 (Tex. Crim. App. 1993).

In the multiple punishments context, the Blockburger test is simply a rule of statutory construction, which is useful in attempting to ascertain legislative intent. Hence, the Supreme Court has concluded that the Blockburger test cannot negate a clearly expressed legislative intent to impose multiple punishments. Missouri v. Hunter, 459 U.S. 359, 366, 103 S.Ct. 673, 74 L.Ed.2d 535 (1983); Ervin v. State, *supra*.

The Blockburger test's status as a rule of statutory construction raises an inverse conclusion as well: the Blockburger test cannot authorize two punishments where the legislature clearly intended only one. This raises the possibility that there exist other rules of statutory construction that may be employed to help ascertain whether a legislature intended multiple punishments. Ex parte Ervin, *supra*.

For double jeopardy purposes, the same offense means the identical criminal act, not the same offense by name. Ex parte Goodbread, 967 S.W.2d 859 (Tex. Crim. App. 1998).

As an illustration, successive prosecutions for possession of a pornographic photograph and also a pornographic videotape resulting from single incident of possession did not place a defendant in double jeopardy because each offense called for proof of an element which the other did not. Vineyard v. State, 958 S.W.2d 834 (Tex. Crim. App. 1998).

The offenses of assault on a public servant and resisting arrest are not the same offense for double jeopardy purposes because they are not the same offense under the Blockburger test. Ortega v. State, 171 S.W.3d 895 (Tex. Crim. App. 2005).

Since double jeopardy analysis is concerned with the actual offense, the state may not prosecute a defendant under a second indictment that merely substitutes the name of a victim for the phrase "unknown to the grand jury" which had been alleged in the first indictment. Ex parte Coleman, 940 S.W.2d 946 (Tex. Crim. App. 1996).

The Court of Criminal Appeals' opinion in Hall v. State, 225 S.W.3d 524 (Tex. Crim. App. 2007) dealing with lesser included offense analysis reiterated that the Blockburger test remains the applicable test for double jeopardy analysis. Ex parte Watson. 306 S.W.3d 259 (Tex. Crim. App. No. 294-08, May 6, 2009).

§8:14 Legislative Intent Test

When multiple offenses are prosecuted at a single trial, the Double Jeopardy Clause prevents the sentencing court from prescribing greater punishment than the legislature intended. Missouri v. Hunter, 459 U.S. 359, 366, 103 S.Ct. 673, 74 L.Ed.2d 535 (1983); Ervin v. State, 991 S.W.2d 804 (Tex. Crim. App. 1999).

A defendant suffers multiple punishments in violation of the Double Jeopardy Clause when he is convicted of more offenses than the legislature intended. Ball v. United States, 470 U.S. 856, 105 S.Ct. 1668, 84 L.Ed.2d 740 (1985); Ervin v. State, *supra*.

The Blockburger test is a useful tool for ascertaining legislative intent, but it is not the only tool. Other (nonexclusive) considerations relevant to determining whether the Legislature intended multiple punishments are: whether the offenses' provisions are contained within the same statutory section, whether the offenses are phrased in the alternative, whether the offenses are named similarly, whether the offenses have common punishment ranges, whether the offenses have a common focus (*i.e.*, whether the "gravamen" of the offense is the same) and whether that common focus tends to indicate a single instance of conduct, whether the elements that differ between the offenses can be considered the "same" under an imputed theory of liability which would result in the offenses being considered the same under Blockburger analysis (*i.e.*, a liberalized Blockburger standard utilizing imputed elements), and whether there is legislative history containing an articulation of an intent to treat the offenses as the same or different for double jeopardy purposes. Ervin v. State, *supra*.

For example, because the Legislature intended to make possession of each individual controlled substance within the same penalty group a separate and distinct offense, convictions for both heroin and cocaine possession are permitted under double jeopardy analysis even where they arise from a single incident of possession. Watson v. State, 900 S.W. 2d 60 (Tex. Crim. App. 1995).

§8:15 Allowable Units of Prosecution Analysis

The Double Jeopardy Clause is offended if a defendant is successively prosecuted for the same offense. The legislature defines whether offenses are the same. It does so by prescribing the allowable unit of prosecution, which is a distinguishable discrete act that is a separate violation of the statute. The discovery of the allowable unit of prosecution is a task of statutory construction. Sanabria v. United States, 437 U.S. 54, 98 S.Ct. 2170, 57 L.Ed.2d 43 (1978); Ex parte Hawkins, 6 S.W.3d 554 (Tex. Crim. App. 1999).

The "allowable units of prosecution" analysis

is undertaken in cases in which the convictions at issue arose from the same statutory provision. Ex parte Milner, 394 S.W.3d 502 (Tex. Crim. App. 2013) (where the defendant was successively prosecuted under the capital murder statute).

Absent an express statement defining the allowable unit of prosecution, the gravamen of an offense best describes the allowable unit of prosecution. Loving v. State, 401 S.W.3d 642, 647 (Tex. Crim. App. 2013).

The gravamen of an offense can be: (1) the result of the conduct; (2) the nature of the conduct; or (3) the circumstances surrounding the conduct. To determine the gravamen of a statutory provision, the appellate courts have used various tools, including a focus on grammar, and when examining the statute, focusing on sentence syntax and whether the statute refers to an item in the singular or plural (usually the direct object). Loving.

The unit of prosecution can be defined by the element of the offense requiring a completed act and that, if each statutory provision protects a victim from a different type of harm, there is evidence that the Legislature intended for each commission of a prohibited act to be punished separately. Loving.

Thefts from two separate people during the same transaction can both be prosecuted because they involve two separate violations of the theft statute; *i.e.*, two allowable units of prosecution. Inglehart v. State, 837 S.W.2d 122 (Tex. Crim. App. 1992).

The allowable unit of prosecution in a burglary is the unlawful entry; therefore, it offends the double jeopardy clause for a defendant to be convicted of a burglary against two separate complainants for a single entry. Ex parte Cavazos, 203 S.W.3d 333 (Tex. Crim. App. 2006).

Because the allowable unit of prosecution in an assaultive offense is each victim, each person threatened in a robbery case can be the subject of an individual prosecution. Ex parte Hawkins, 6 S.W.3d 554 (Tex. Crim. App. 1999) (*overruling* Ex parte Crosby, 703 S.W.2d 683 (Tex. Crim. App. 1986); Simmons v. State, 745 S.W.2d 348 (Tex. Crim. App. 1987); and Cook v. State, 840 S.W.2d 384 (Tex. Crim. App. 1992)).

A defendant can be prosecuted for each separate piece of child pornography in his possession because each is an allowable unit of prosecution. Vineyard v. State, 958 S.W.2d 834 (Tex. Crim. App. 1998).

Under this theory a defendant who commits unlawful acts while operating a motor vehicle may be convicted and punished for each person injured and/or murdered in violation of a single statute. Phillips v. State, 787 S.W.2d 391 (Tex. Crim. App. 1990).

The same is true for acts prosecuted under the official oppression statute. Haight v. State, 137 S.W.3d 48 (Tex. Crim. App. 2004).

However, in a capital murder prosecution where the aggravating element is the murder of multiple victims (Penal Code §19.03(a)(7)(A)), the allowable unit of prosecution is at minimum two victims so that a defendant is subjected to double jeopardy where he is subjected to more than one prosecution arising out of a single set of facts. Saenz v. State, 166 S.W.3d 270 (Tex. Crim. App. 2005); Ex parte Milner; 394 S.W.3d 508.

For the offense of making false statement to obtain property or credit contained in Penal Code sec. 32.32, the appropriate unit of prosecution for double jeopardy analysis is the "materially false or misleading statement," not the loan application. Each materially false or misleading statement constitutes a separate offense. Jones v. State, 323 S.W.3d 885 (Tex. Crim. App. 2010).

The clear language of §21.11(a)(2)(A) (indecency by exposure to a child) indicates that the exposure, not the number of children present, constitutes the unit of prosecution. Therefore, no matter how many children are present, a defendant can only be prosecuted for one indecent exposure to a child per exposure. Harris v. State, 359 S.W.3d 625 (Tex. Crim. App. 2011).

The text of the criminal-attempt statute does not define an allowable unit of prosecution, nor does it change the allowable unit of prosecution of the offense attempted. Given that a criminal attempt to commit an offense is simply an act amounting to more than mere preparation of the intended offense, criminal-attempt offenses acquire their allowable unit of prosecution from the offense attempted. Ex parte Milner, 394 S.W.3d 508-09.

Indecency with a child by exposure and indecency with a child by contact are separate and distinct offenses. Therefore, a conviction for each offense in a single trial is permissible under double jeopardy principles because the Legislature intended to allow separate punishments under these circumstances. Loving.

Even though aggravated assault on a public servant and intoxication assault are not the "same offense," they have the same allowable unit of prosecution (the injury of a single individual) and therefore a conviction in a single trial for each offense is prohibited under double jeopardy principles. Shelby v. State, 448 S.W.3d 431, 439 (Tex. Crim. App. 2014).

§8:16 Dual-Sovereignty

Under the dual-sovereignty doctrine, an individual may be tried and sentenced in both state and federal court for the same conduct because a crime under one sovereign's laws is not the same offense as a crime under the laws of another sovereign. Gamble v. United States, 587 U.S. ___, ___ S.Ct. ___, ___ L.Ed.2d ___ (No. 17-646, 2019 WL 2493923, June 17, 2019).

§8:20 Previous Jeopardy and Subsequent Prosecutions

Although the double jeopardy clause speaks of being twice placed in jeopardy of life or limb, it is also applicable to imprisonment and monetary penalties. Dept. of Revenue v. Kurth Ranch, 511 U.S. 767, 114 S.Ct. 1937, 128 L.Ed.2d 767 (1994); Ex parte Chappell, 959 S.W.2d 627 (Tex. Crim. App. 1998).

§8:21 When Jeopardy Attaches

Before the double jeopardy clause is invoked, an individual must have been placed in jeopardy for a particular offense. Ex parte George, 913 S.W.2d 523 (Tex. Crim. App. 1995).

In a jury trial, jeopardy attaches when the jury is sworn; Pierson v. State, 426 S.W.3d 763 (Tex. Crim. App. 2014); Martinez v. Illinois, 572 U.S. ___, 134 S.Ct. 2070, 188 L.Ed.2d 1112 (2014).

Jeopardy attaches in a trial to the court when the defendant pleads to the indictment. State v. Torres, 805 S.W.2d 418 (Tex. Crim. App. 1991).

Jeopardy attaches in a negotiated plea bargain proceeding when the trial judge accepts the plea bargain. Ortiz v. State, 933 S.W.2d 523 (Tex. Crim. App. 1996).

Where the state abandons or waives a count or paragraph of a charging instrument after jeopardy has attached, a retrial is barred on that count or paragraph. State v. Florio, 845 S.W.2d 849 (Tex. Crim. App. 1992). *See also* Brown v. State, 900 S.W.2d 805 (Tex.App.—San Antonio 1995, *pet. ref'd*).

If the State dismisses, waives, or abandons a charge after a jeopardy has attached (after a jury is impaneled and sworn in a jury trial), it is tantamount to an acquittal, as the State is barred from later litigating those allegations. If a charge is abandoned in this fashion, it should not be included in a judgment as a conviction. Duran v. State, 492 S.W.3d 741, 745 (Tex. Crim. App. 2016).

The prosecution must take affirmative steps on the record to validly waive, dismiss or abandon that portion of a charging instrument they wish to reserve for a possible later prosecution. The mere failure to read the applicable portions of the charging instrument to the trier of fact, without more, results in the attachment of jeopardy to those portions. Ex parte Preston, 833 S.W.2d 515 (Tex. Crim. App. 1992).

§8:22 When Jeopardy Does Not Attach

Jeopardy does not attach and bar retrial where the first conviction is void. Hoang v. State, 872 S.W.2d 694 (Tex. Crim. App. 1993) (where the first convictions were void because the defendant was a juvenile at the time of their commission and had not been certified by a juvenile court).

Jeopardy does not attach where a defendant is convicted under an unconstitutional statute. Lowry v. State, 671 S.W.2d 601 (Tex.App.—Dallas 1984), *affirmed in part, reversed in part* 692 S.W.2d 86 (Tex. Crim. App. 1985).

A fundamentally defective indictment does not confer jurisdiction on the trial court and therefore, jeopardy does not attach. Cook v. State, 902 S.W.2d 471 (Tex. Crim. App. 1995).

Jeopardy does not attach to counts or paragraphs that are abandoned before jeopardy attaches. Proctor v. State, 841 S.W.2d 1 (Tex. Crim. App. 1992).

Jeopardy does not attach to a trial if an appellate mandate has not yet issued in the case from a prior appeal. Ex parte Macias, 541 S.W.3d 782, 785-6 (Tex. Crim. App. 2017).

PRACTICE TIP: A writ of habeas corpus is the proper means to attack a trial court judgment that is jurisdictionally void. Ex parte Sanchez, 918 S.W.2d 526 (Tex. Crim. App. 1996).

§8:23 Previous Convictions and Subsequent Prosecutions

CCP Art. 28.13. Former Acquittal or Conviction

> *A former judgment of acquittal or conviction in a court of competent jurisdiction shall be a bar to any further prosecution for the same offense, but shall not bar a prosecution for any higher grade of offense over which said court had not jurisdiction, unless such judgment was had upon indictment or information, in which case the prosecution shall be barred for all grades of the offense.*

The double jeopardy clause protects against multiple prosecutions for the same offense. Jeffers v. U.S., 432 U.S. 137, 97 S.Ct. 2207, 53 L.Ed.2d 168 (1977); Ex parte Kopecky, 821 S.W.2d 957 (Tex. Crim. App. 1992).

For double jeopardy purposes, the same offense means the identical criminal act, not the same offense by name. Ex parte Goodbread, 967 S.W.2d 859 (Tex. Crim. App. 1998).

When it cannot be determined from the state's pleadings whether the offenses prosecuted are the same, the court must look to the proof offered at trial. Goodbread, *supra*.

When an indictment permits the state to obtain only one conviction, a defendant is not placed in jeopardy for more than one criminal act unless the State offers multiple instances of conduct in support of the indictment. Goodbread, *supra*.

If evidence of more than one offense is admitted and a conviction for either could be had under the indictment, and neither the state nor the court elects, a plea of former conviction is good upon a prosecution based upon one of said offenses, it being uncertain for which one the conviction was had. Goodbread, *supra*.

Consequently, where the state dismisses an indictment after jeopardy attaches without offering proof of any instance of conduct, only the act alleged in the indictment is jeopardy barred. Ex parte Goodman, 152 S.W.3d 67 (Tex. Crim. App. 2004) (this is true even though there were multiple instances of conduct that the state could have offered in support of the indictment).

In order to ensure that the state does not offer jeopardy barred proof at a subsequent trial, the trial court may require the state to prove, by a preponderance of the evidence, that it is not relying upon such proof. If the state is unable to so prove, then the trial court must select, from among the constituent acts that the state does prove, and strike from the jury's consideration, the one constituent act that most closely resembles the act alleged in the first indictment. Ex parte Goodman.

A double jeopardy violation occurs if the defendant is tried twice for the same offense in a single trial even if the sentences run concurrently. Ervin v. State, 991 S.W.2d 804 (Tex. Crim. App. 1999) (where the offenses were manslaughter and intoxication manslaughter).

§8:24 Greater Inclusive and Lesser Included Offenses

CCP Art. 37.14. Acquittal of Higher Offense as Jeopardy

> *"If a defendant, prosecuted for an offense which includes within it lesser offenses, be convicted of an offense lower than that for which he is indicted, and a new trial be granted him, or the judgment be arrested for any cause other than the want of jurisdiction, the verdict upon the first trial shall be considered an acquittal of the higher offense; but he may, upon a second trial, be convicted of the same offense of which he was before convicted, or any other inferior thereto."*

See also CCP Art. 28.13, above.

Greater inclusive and lesser included offenses are the same offense for jeopardy purposes. Brown v. Ohio, 432 U.S. 161, 97 S.Ct. 2221, 53 L.Ed.2d 187 (1977); Parrish v. State, 869 S.W.2d 352 (Tex. Crim. App. 1994).

Double jeopardy rules prohibit the State from obtaining a conviction for an offense when the defendant has previously been convicted of a lesser-included offense of that offense. Ex parte Amador 326 S.W.3d 202 (Tex. Crim. App. 2010).

Generally, if a defendant is acquitted of a greater offense, he cannot be tried for a lesser included offense. State v. Engelking, 817 S.W.2d 64 (Tex. Crim. App. 1991) (where there was insufficient evidence to prove the aggravating element of the greater offense).

However, where the state requests that the jury be instructed on the lesser included offense, or the court instructs on the lesser included offense without objection from the state, the defendant may be retried on the lesser charge where he has been acquitted on appeal of the greater charge by reason of insufficient evidence. Granger v. State, 850 S.W.2d 513 (Tex. Crim. App. 1993); Hampton v. State, 165 S.W.3d 691 (Tex. Crim. App. 2005).

An exception to the general rule that conviction of a lesser offense bars prosecution for a greater offense is where all the events necessary to the greater crime have not taken place at the time the prosecution for the lesser has begun. Brown v. Ohio, *supra*.

The Double Jeopardy Clause does not prohibit a retrial for a lesser included offense when the first trial was before the court and there was sufficient evidence on which the trial court could have found the defendant guilty of the lesser offense. Shute v. State, 877 S.W.2d 314 (Tex. Crim. App. 1994).

The double jeopardy clause also prohibits a subsequent prosecution where a defendant has been convicted or acquitted of an offense that "subsumes" another offense. Ex parte Pruitt, 233 S.W.3d 338 (Tex. Crim. App. 2007) (defendant's acquittal of aggravated sexual assault of a child by penetration barred a subsequent prosecution for aggravated sexual assault of a child by genital-to-genital contact).

§8:25 Previous Acquittals and Subsequent Prosecutions

CCP Art. 1.11 Acquittal a Bar

> *An acquittal of the defendant exempts him from a second trial or a second prosecution for the same offense, however irregular the proceedings may have been; but if the defendant shall have been acquitted upon trial in a court having no jurisdiction of the offense, he may be prosecuted again in a court having jurisdiction.*

See also CCP Art. 28.13, above.

An acquittal of charges bars retrial of the same charges on double jeopardy principles. Smalis v. Pennsylvania, 476 U.S. 140, 106 S.Ct. 1745, 90 L.Ed.2d 116 (1986); Ex parte Kopecky, 821 S.W.2d 957 (Tex. Crim. App. 1992).

A jury verdict that necessarily decides a critical issue of ultimate fact in a defendant's favor protects him from prosecution for any charge for which that is an essential element. Yeager v. U.S., 557 U.S. ___, 129 S.Ct. 2360, 174 L.Ed.2d 78 (2009).

An instructed verdict is the functional equivalent of an acquittal for double jeopardy purposes. Smalis v. Pennsylvania, *supra*; Ex parte Coleman, 940 S.W.2d 946 (Tex. Crim. App. 1996).

This is true even where the state has refused to present evidence during the trial. Martinez v. Illinois, 572 U.S. ___, 134 S.Ct. 2070, ___ L.Ed.2d ___ (2014).

Whether the acquittal is based on a jury verdict of not guilty or on a ruling by the court that the evidence is insufficient to convict, any further prosecution, including an appeal by the prosecution that would lead to a second trial, is prohibited by the double jeopardy clause. State v. Blackshere, 344 S.W.3d 400 (Tex. Crim. App. 2011).

Where a court acquits a defendant of a charge during trial and trial proceeds on other charges, the court cannot reinstate the acquitted charge in the absence of statutory authority to do so. Smith v. Massachusetts, 543 U.S. 462, 125 S.Ct. 1129, 160 L.Ed.2d 914 (2005).

Where the trial court grants a motion for new trial based on insufficient evidence the state is barred from retrying the defendant on the same charge. Moore v. State, 749 S.W.2d 54 (Tex. Crim. App. 1988).

An acquittal bars retrial even where the charging instrument was defective. *See* Studer v. State, 799 S.W.2d 263 (Tex. Crim. App. 1990) (a defective charging instrument confers jurisdiction on the trial court).

Even an erroneous acquittal (such as where the trial court misconstrues a statute or erroneously adds an extra element to the proof required) will serve as a jeopardy bar to a subsequent trial. Evans v. Michigan, 568 U.S. ___, 133 S.Ct. 1069, 1073, 185 L.Ed.2d 124 (2013).

A defendant who has been acquitted under an indictment alleging an offense against a particular victim can be retried for the same course of conduct if the subsequent indictment alleges a different victim. Bailey v. State, 87 S.W.3d 122 (Tex. Crim. App. 2002).

Where the trial court dismisses an indictment on the State's motion after jeopardy attaches, then the Double Jeopardy Clause prohibits the State from reprosecuting the defendant for the offense for which he had been placed in jeopardy of conviction. Proctor v. State, 841 S.W.2d 1 (Tex. Crim. App. 1992).

Where the state has not elected a primary act for conviction, any acts that were offered as proof, and would allow a conviction under the allegations of the indictment at that trial, are jeopardy-barred at a subsequent trial. Ex parte Pruitt, 233 S.W.3d 338 (Tex. Crim. App. 2007).

Where a defendant has been acquitted of capital murder in the course of burglary, a subsequent prosecution for the underlying burglary is jeopardy-barred, but a prosecution for aggravated assault on the same facts is not. Ex parte Castillo, 469 S.W.3d 165, 172 (Tex. Crim. App. 2015).

§8:30 Multiple Punishments

A multiple punishments double jeopardy claim can arise in two contexts: (1) the lesser-included offense context, in which the same conduct is punished twice -- once for the basic conduct, and a second time for that same conduct plus more (for example, attempted assault of Y and assault of Y; assault of X and aggravated assault of X); and (2) punishing the same criminal act twice under two distinct statutes when the legislature intended the conduct to be punished only once (for example, causing a single death by committing both intoxication manslaughter and involuntary manslaughter). Langs v. State, 183 S.W.3d 680 (Tex. Crim. App. 2006).

§8:31 Multiple Punishments in Single Trial

When multiple offenses are prosecuted at a single trial, the Double Jeopardy Clause prevents the sentencing court from prescribing greater punishment than the legislature intended. Missouri v. Hunter, 459 U.S. 359, 103 S.Ct. 673, 74 L.Ed.2d 535 (1983); Ervin v. State, 991 S.W.2d 804 (Tex. Crim. App. 1999).

A defendant suffers multiple punishments in violation of the Double Jeopardy Clause when he is convicted of more offenses than the legislature intended. Ball v. United States, 470 U.S. 856, 105 S.Ct. 1668, 84 L.Ed.2d 740 (1985); Ervin v. State, *supra*.

The Blockburger case addresses two multiple-punishment issues: the "continuous action vs. separate and distinct acts" issue, and the "one act violates separate distinct statutes" issue. Aekins v. State, 447 S.W.3d 270, 277 (Tex. Crim. App. 2014).

The first part of the Blockburger test asks whether each criminal act is a separate and distinct one, separated by time. If the offense is a single continuous act, with a single impulse, in which several different statutory provisions are necessarily violated along that continuum, the offenses merge together. This is variously called "the merger doctrine," "the single impulse doctrine," or, here in Texas, "the doctrine of subsumed acts." If more than one statutory offense is necessarily committed by that single criminal act and impulse, then the offenses merge and the defendant may be punished only once. Aekins, 447 S.W.3d at 274-5.

Where two acts are the products of separate impulses, even though they may occur closely together in time, the defendant can be convicted of each act in a single trial. Aekins, 447 S.W.3d at 283.

Cumulative punishment, consistent with the double jeopardy clause, may be imposed where separate offenses occur in the same transaction, as long as each conviction requires proof of an additional element that the other does not. Blockburger v. U.S., 284 U.S. 299, 52 S.Ct. 180, 76 L.Ed.2d 306 (1932); Phillips v. State, 787 S.W.2d 391 (Tex. Crim. App. 1990).

In certain limited circumstances, the Blockburger "same elements" test is not the sole test for analyzing a multiple-punishment claim. Because legislative intent, rather than the presence or absence of the "same elements" is the ultimate inquiry, the "same elements" test, although the usual method for discerning legislative intent, cannot authorize two punishments where the legislature clearly intended only one, nor, conversely, negate a clearly expressed legislative intent to impose multiple punishments. Ervin.

Under Ervin, additional non-exclusive factors should be examined to determine if two offenses are the same in the context of multiple punishment: whether offenses are in the same statutory section; whether the offenses are phrased in the alternative; whether the offenses are named similarly; whether the offenses have common punishment ranges; whether the offenses have a common focus; whether the common focus tends to indicate a single instance of conduct; whether the elements that differ between the two offenses can be considered the same under an imputed theory of liability that would result in the offenses being considered the same under Blockburger; and whether there is legislative history containing an articulation of an intent to treat the offenses as the same or different for double-jeopardy purposes. Ervin; Bigon v. State, 252 S.W.3d 360 (Tex. Crim. App. 2008).

A multiple-punishments double-jeopardy violtion may arise either in the context of lesser-included offenses (when the same conduct is punished under both a greater and a lesser-included statutory offense) or when the same criminal act is punished under two distinct statutory provisions, but the legislature intended only one punishment. Aekins v. State, 447 S.W.3d at 274.

A multiple punishments claim can arise in a lesser-included offense context, in which the same conduct is punished twice under two distinct statutes when the legislature intended the conduct to be punished only once. Langs v. State, 183 S.W.3d 680 (Tex. Crim. App. 2006).

The double jeopardy clause does not bar cumulative punishments imposed in a single proceeding. Ohio v. Johnson, 467 U.S. 493, 104 S.Ct. 2536, 81 L.Ed.2d 425 (1984); Phillips, *supra.*

Manslaughter and intoxication manslaughter are the same offense for double jeopardy purposes when they involve the same victim, and imposing convictions for both in this situation violates the Double Jeopardy Clause. A double jeopardy violation occurs even when, as in this case, the sentences are concurrent. Ball v. United States, *supra*; Ervin v. State, *supra.*

Convictions for both felony murder and intoxication manslaughter in the same trial, which arise from a single offense, can violate the double jeopardy clause. Bigon.

An offer to sell drugs and the possession of drugs with intent to complete that particular sale are one offense for double jeopardy purposes because they are both part of the same criminal "impulse." Lopez v. State, 108 S.W.3d 293 (Tex. Crim. App. 2003).

Multiple punishments for possession with intent to deliver and manufacturing a controlled substance for a single quantity of controlled substance violate double jeopardy because the legislature has established that both offenses are alternative means of punishing a single offense in the continuum of drug distribution. Weinn v. State, 326 S.W.3d 189 (Tex. Crim. App. 2010).

Attempted capital murder (attempting to cause the death of a peace officer) and aggravated robbery offenses arising out of the same transaction against the same victim can each be prosecuted in a single trial because the legislature has determined that each be a separate offense. Cervantes v. State, 815 S.W.2d 569 (Tex. Crim. App. 1991); *see also* Garrett v. United States, 471 U.S. 773, 105 S.Ct. 2407, 85 L.Ed.2d 764 (1985).

However, a defendant cannot be prosecuted for both felony murder with aggravated robbery as the underlying offense and aggravated robbery because the aggravated robbery is a lesser included offense of the felony murder as charged and the legislature has not expressed an intention that multiple punishments should be allowed in this context. Littrell v. State, 271 S.W.3d 273 (Tex. Crim. App. 2008).

A defendant can be prosecuted for injury to a child and any other penal code offense in the same trial because the legislature has authorized multiple punishments within the injury to a child statute, Texas Penal Code §22.04(h). Desormeaux v. State, 362 S.W.3d 233 (Tex.App.—Beaumont 2012, *no pet.*).

It is permissible for a jury to convict a defendant for both the offense of capital murder and the separate offense of engaging in organized criminal activity by committing capital murder as a member of a criminal street gang, and punish the defendant for both where they are prosecuted in a single trial, even where these two are the same offense under the Blockburger test. Garza v. State, 213 S.W.3d 338 (Tex. Crim. App. 2007).

Under the pleadings in this particular case, the offense of criminal solicitation was subsumed within the proof necessary to establish the intent to commit capital murder under the attempted capital murder indictment. Consequently, under the pleadings in this case, attempted capital murder and solicitation of capital murder are the same offense under the Blockburger test. Bien v. State, 550 S.W.3d 180, 187 (Tex. Crim. App. 2018).

Because of the "fundamental nature of the double jeopardy protections, a double jeopardy claim may be raised for the first time on appeal or even for the first time on collateral attack" if two conditions are met: (1) "the undisputed facts show the double jeopardy violation is clearly apparent on the face of the record"; and (2) "when enforcement of the usual rules of procedural default serves no legitimate state interest. Gonzalez v. State, 8 S.W.3d 640 (Tex. Crim. App. 2000).

When offenses, one of which could give rise to a multiple-punishment double-jeopardy violation, are listed disjunctively in a jury charge, the burden is upon the defendant to preserve, in some fashion a double jeopardy objection at or before the time the charge is submitted to the jury, or else the record is insufficient to meet the burden under Gonzalez. Langs v. State, 183 S.W.3d 680 (Tex. Crim. App. 2006).

Serious bodily injury committed against the same victim at the same time should be considered the same offense for purposes of the double-jeopardy prohibition against multiple punishments regardless of whether that injury to that victim resulted from the actor's act, his omission, or by a combination of his act and omission. Villaneuva v. State, 227 S.W.3d 744 (Tex. Crim. App. 2007) (the defendant had been prosecuted for both injury to a child by act and injury to a child by omission).

There is no double jeopardy violation when a defendant is convicted of both aggravated sexual assault by penetration of the vagina and aggravated sexual assault by penetration of the anus occurring in the same transaction because the legislature intended for these to be separate offenses. Gonzales v. State, 304 S.W.3d 838 (Tex. Crim. App. 2010).

The principles of double jeopardy prohibit the State from obtaining a conviction for indecency with a child and aggravated sexual assault of a child based on a single act of molestation because indecency with a child is a lesser-included offense of aggravated sexual assault of a child when both offenses are predicated on the same act. Evans v. State, 299 S.W.3d 138 (Tex. Crim. App. 2009).

Convictions in a single trial for aggravated robbery and aggravated assault are double jeopardy violations where the aggravated assault as plead is a lesser-included offense of aggravated robbery. Ex parte Denton, 399 S.W.3d 540, 547 (Tex. Crim. App. 2013).

However, depending on how the offenses are charged a defendant can be convicted of both aggravated robbery and aggravated assault in a single trial because a "units of prosecution" analysis can show that the legislature intended to allow multiple punishments. Garfias v. State, 424 S.W.3d 54, 61 (Tex. Crim. App. 2014).

Convictions in a single trial for aggravated robbery by threats and aggravated robbery by causing bodily injury against the same victim violates the Double Jeopardy clause. Cooper v. State, 430 S.W.3d 426, 434 (Tex. Crim. App. 2014).

The legislature's intent was to permit one punishment where continuous sexual abuse is alleged against a single victim within a specified time frame. This intent extends to the continuous sexual abuse statute's enumerated predicate offenses and to criminal attempts to commit those predicate offenses. Price v. State, 434 S.W.3d 601, 610 (Tex. Crim. App. 2014).

Because felony DWI and intoxication assault do

not have the same type of focus, the same actual focus or gravamen, and they do not have the same unit of prosecution convictions for each in the same trial does not violate double jeopardy principles. Ex parte Benson, 459 S.W.3d 67, 89 (Tex. Crim. App. 2015).

The Texas Legislature's intent in sexual assault cases is to punish each discrete assault. Separate criminal acts committed during a single sexual encounter may be punished separately, but a criminal act (such as exposure or contact) that is committed as part of a continuing sexual assault that results in one complete, ultimate act of penetration may not be punished along with that complete, ultimate act. Steps along the way to one sexual assault merge into the completed act. Aekins v. State, 447 S.W.3d 270, 277-8 (Tex. Crim. App. 2014).

A person who commits more than one sexual act against the same person may be convicted and punished for each separate and discrete act, even if those acts were committed in close temporal proximity. The key is that one act ends before another act begins and is thus, motivated by a separate impulse. This is true for acts violating not only different statutes, but different subsections of a single statute, and even different discretely prohibited acts within the same subsections. Aekins, 447 S.W.3d at 278.

When evaluating multiple convictions for sexual offenses in the double jeopardy context the following rules apply: (1) when a single exposure or contact offense is incident to and subsumed by a penetration offense, the offenses are the same for double-jeopardy purposes, and (2) the Texas Legislature has not manifested its intent to allow multiple punishments for those same offenses, so (3) multiple convictions for those same offenses violate double-jeopardy principles. Aekins, 447 S.W.3d at 279.

In other words, where two crimes are such that the one cannot be committed without necessarily committing the other, then they stand in the relationship of greater and lesser offenses, and the defendant cannot be convicted or punished for both. Aekins, 447 S.W.3d at 280.

The Legislature intended only one conviction for one completed sexual assault. This means that multiple convictions for exposure, contact and penetration for one complete, ultimate sexual assault violates the Double Jeopardy Clause. Patterson v. State, 152 S.W.3d 88, 91-2 (Tex. Crim. App. 2004).

Separate convictions for sexual assault based on defendant's contact of victim's sexual organ with his mouth and penetration of victim's sexual organ with his mouth and/or tongue arising out of same criminal act violate the prohibition against double jeopardy. Aekins, 447 S.W.3d at 281.

A person can be convicted of indecency by exposure and indecency by contact in a single trial because even though these acts are criminalized in the same statute, they require two distinct acts. Loving v. State, 401 S.W.3d 642, 646-9 (Tex. Crim. App. 2013).

The indecency with a child statute criminalizes three separate contact acts—touching of the (1) anus, (2) breast, or (3) genitals. A person who touches a child's anus, breast, and genitals may be convicted of three offenses because each touching is a separate act. Pizzo v. State, 235 S.W.3d 711, 719 (Tex. Crim. App. 2007).

A single count alleging sexual contact is not subsumed by a count alleging penetration where there is evidence of multiple incidents of both contact and penetration which could have formed the basis for each count. Maldonado v. State, 461 S.W.3d 144, 150 (Tex. Crim. App. 2015).

The legislature intended that both theories of indecency with a child (exposure and fondling) may be pled and punished even when the exposure precedes the contact and even when both acts occur within the same transaction when the evidence presented actually supports conviction and punishment under each theory of the offense. Speights v. State, 464 S.W.3d 719, 724 (Tex. Crim. App. 2015).

A sexually violent predator civil-commitment order violation is a circumstances-surrounding-the-conduct crime. A violation arises only by the circumstance that the person has been adjudicated a sexually violent predator who has been civilly committed. It is the circumstance coupled with the conduct that results in a criminal offense. Therefore, a person can only be convicted of a single violation of the order in a single trial. Stevenson v. State, 499 S.W.3d 842, 851 (Tex. Crim. App. 2016).

See also sec. 17:72

§8:32 Multiple Punishments in Successive Trials

A defendant waives his double jeopardy claim as to multiple prosecutions when his motion to sever is granted. Jeffers v. United States, 432 U.S. 137, 97 S.Ct. 2207, 53 L.Ed.2d 168 (1977); Gongora v. State, 916 S.W.2d 570 (Tex.App.—Houston [1st Dist.] 1996, *pet. ref'd*); Currier v. Virginia, 585 U.S. ___, 138 S.Ct. 2144, 201 L.Ed.2d 650 (2018).

A defendant ordinarily may not be punished or tried twice for a greater-inclusive and a lesser-included offense without violating double jeopardy because lesser-included offenses typically do not pass muster under the *Blockburger* test because the elements of the lesser offense are wholly subsumed by the elements of the greater offense. Ex parte Chaddock, 369 S.W.3d 880, 882-3 (Tex. Crim. App. 2012).

An exception to this rule is where the legislature has specifically authorized cumulative punishment under two statutes in a single prosecution. Ex parte Chaddock, at page 2, citing Missouri v. Hunter, 459 U.S. 359, 103 S.Ct. 673, 74 L.Ed.2d 535 (1983).

The Hunter rule, while applicable in a multiple punishment single trial context, is not applicable in the successive prosecution context; and therefore, any statute that purports to authorize a second conviction for a lesser included offense in a successive trial operates unconstitutionally. Ex parte Chaddock, at page 4.

Under the dual-sovereignty doctrine, an individual may be tried and sentenced in both state and federal court for the same conduct because a crime under the laws of one sovereign is not the same offense as a crime under the laws of another sovereign. Gamble v. United States, 587 U.S. ___, ___ S.Ct. ___, ___ L.Ed.2d ___ (No. 17-646, 2019 WL 2493923, June 17, 2019).

§8:40 Retrials and Double Jeopardy Clause

§8:41 Retrials After Mistrials

A mistrial is a device used to halt trial proceedings when the error is so prejudicial that expenditure of further time and expense would be wasteful or futile. A trial court may properly exercise its discretion to declare a mistrial if an impartial verdict cannot be reached, or if a verdict of conviction could be reached but would have to be reversed on appeal due to an obvious procedural error. Ladd v. State, 3 S.W.3d 547 (Tex. Crim. App. 1999).

Despite the general prohibition against jeopardy-barred trials, there are two exceptions when a criminal defendant may be tried a second time without violating double-jeopardy principles if the prosecution ends prematurely as the result of a mistrial: (1) if the criminal defendant consents to retrial or (2) there was a manifest necessity to grant a mistrial. Pierson v. State, 426 S.W.3d 763 (Tex. Crim. App. 2014).

The consent of the defendant to the mistrial must appear in the record as a deliberate relinquishment on the defendant's part—an exercise of her primary control—if it is to be relied upon as a reason to allow the State to re-prosecute her. Consent need not be expressed, but may be implied from the totality of circumstances attendant to a declaration of mistrial. Ex parte Garrels, 559 S.W.3d 517, 523 (Tex. Crim. App. 2018).

The initial burden is on the defendant to go forth with evidence in support of the allegation of former jeopardy. In order to satisfy this prima facie burden, the defendant need only establish that she was tried for the same offense after a mistrial. Once the defendant makes this initial showing, the burden shifts to the State to prove that appellant consented to the order terminating her first trial. Ex parte Garrels, 559 S.W.3d at 524.

Where a jury is unable to reach a decision in a case (and the foreman does not sign a verdict form), the Double Jeopardy clause does not bar a second trial. Blueford v. Arkansas, 566 U.S. ___, 132 S.Ct. 2044, 182 L.Ed.2d 937 (2012).

A jury note indicating that the jury has acquitted of a particular charge is not an informal verdict of acquittal under Texas law where it lacks the "finality necessary to amount to an acquittal" as required under Blueford v. Arkansas. Traylor v. State, 567 S.W.3d 741, 748 (Tex. Crim. App. 2018).

The asking of an improper question will seldom call for a mistrial because in most cases an instruction to disregard will cure any harm. A mistrial is only required where the improper question is clearly prejudicial to the defendant and is of such character that an instruction to disregard will not cure the impression made on the minds of the jurors.

Ladd, *supra.*

The determination of whether a given error necessitates a mistrial must be made on the facts of the particular case on trial. Ladd, *supra.*

§8:42 Manifest Necessity

As a general rule, the prohibition against double jeopardy will prevent a retrial if the defendant is placed in jeopardy and the jury is discharged without the defendant's consent. Green v. United States, 355 U.S. 184, 78 S.Ct. 221, 2 L.Ed.2d 199 (1957); Alvarez v. State, 864 S.W.2d 64 (Tex. Crim. App. 1993).

However, a retrial is not jeopardy barred if the mistrial was based on manifest necessity. Oregon v. Kennedy, 456 U.S. 667, 102 S.Ct. 2083, 72 L.Ed.2d 416 (1982); Alvarez, *supra.*

Manifest necessity exists when the circumstances render it impossible to arrive at a fair verdict, when it is impossible to continue with trial, or when the verdict would be automatically reversed on appeal because of trial error. Hill v. State, 90 S.W.3d 308 (Tex. Crim. App. 2002).

The initial burden of proving a double jeopardy violation is on the defendant. He meets that burden by proving that he was tried for the same offense after a mistrial. The burden then shifts to the state to prove manifest necessity for the mistrial. Hill; Pierson v. State, 426 S.W.3d 763 (Tex. Crim. App. 2014).

Before granting a mistrial on grounds of manifest necessity, a trial court must determine whether alternative courses of action are available and, if so, choose one that is less drastic than a mistrial. Ex parte Brown v. State, 907 S.W.2d 835 (Tex. Crim. App.1995). The trial court's objective in those circumstances is to choose the alternative that best preserves the defendant's right to have the trial completed before the jury that has already been seated. Brown, *supra.*

Where a trial judge grants a mistrial despite the available option of less drastic alternatives, there is no manifest necessity prompting the decision and a retrial is barred by double jeopardy. Ex parte Little, 887 S.W.2d 62 (Tex. Crim. App. 1994).

The Fifth Amendment is not offended by a successive prosecution for the same offense when the earlier proceeding was terminated at the defendant's request, unless the attorney representing the government deliberately set out to provoke the defendant's motion for mistrial. Oregon v. Kennedy, 456 U.S. 667, 102 S.Ct. 2083, 72 L.Ed.2d 416 (1982); Bauder v. State, 921 S.W.2d 696 (Tex. Crim. App. 1996).

Manifest necessity issues decided under the Texas Constitution will also be decided using the Oregon v. Kennedy standard. Ex parte Lewis, 219 S.W.3d 335 (Tex. Crim. App. 2007) (overruling Bauder II, (974 S.W.2d 729 (Tex. Crim. App. 1998) and its progeny, including Ex parte Peterson, 117 S.W.3d 804 (Tex. Crim. App. 2003) and State v. Lee, 15 S.W.3d 921 (Tex. Crim. App. 2000)).

In determining whether the prosecutorial misconduct warranted a mistrial, the court should balance three factors: (1) the severity of the misconduct (the magnitude of the prejudicial effect of the prosecutor's remarks), (2) the measures adopted to cure the misconduct (the efficacy of any cautionary instruction by the judge), and (3) the certainty of conviction absent the misconduct (the strength of the evidence supporting the conviction). Mosley v. State, 983 S.W.2d 249 (Tex. Crim. App. 1998), *cert. denied*, 526 U.S. 1070, 143 L. Ed. 2d 550, 119 S.Ct. 1466 (1999); Ramon v. State, 159 S.W.3d 927 (Tex. Crim. App. 2004).

TIP TO THE BENCH: Although it is not legally required, a trial judge is well-advised to set out his factual findings on the record in support of his ruling on a double jeopardy motion. Ex parte Lewis.

Manifest necessity exists where leaving a juror on the jury makes it impossible for the jury to arrive at a fair verdict Parrish v. State, 38 S.W.3d 831 (Tex.App.—Houston [14th Dist.] 2001, *pet. ref'd*).

Where a mistrial granted at the defendant's request was not mandated under the facts of the case and an instruction to disregard would have been sufficient, a future prosecution for the same offense is not jeopardy barred because the request for mistrial could not have been prompted by the state's deliberate or reckless misconduct. Ex parte Primrose, 950 S.W.2d 775 (Tex.App.—Fort Worth 1997, *pet. ref'd*) (i.e., the first prong of the Kennedy analysis is a threshold determination).

Where a defendant is harmed by the state's failure to reveal exculpatory evidence and a mistrial has been declared, a retrial may be barred by double jeopardy. Ex parte Masonheimer, 220 S.W.3d 494 (Tex. Crim. App. 2007).

There is manifest necessity for a mistrial where the jury is unable to reach a verdict and the trial court has employed less drastic alternatives such as issuing an Allen charge, allowing the jury more time to deliberate and verifying the jury's deliberation status. Ex parte Perusquia, 336 S.W.3d 270 (Tex.App.—San Antonio 2010, *pet. ref'd*).

Manifest necessity for a retrial can exist where the original mistrial was caused by an improper question asked by defense counsel and an instruction to disregard was not a viable alternative to the granting of the mistrial. Pierson v. State, 426 S.W.3d 763 (Tex. Crim. App. 2014).

The trial court's ruling on an issue of manifest necessity is reviewed on an abuse of discretion standard with the appellate court assessing the objective facts in the light most favorable to the trial judge's ruling. Ex parte Wheeler, 203 S.W.3d 317 (Tex. Crim. App. 2006).

§8:43 Retrials Following Appeals

Underlying the application of double jeopardy principles to instances where convictions have been reversed on appeal is the concept of continuing jeopardy. Continuing jeopardy means, in effect, an accused may be retried after his conviction has been reversed on appeal because the proceedings against him never stopped. Montana v. Hall, 481 U.S. 400, 107 S.Ct. 1825, 95 L.Ed.2d 354 (1987); Ex parte Mitchell, 977 S.W.2d 575 (Tex. Crim. App. 1997).

The double jeopardy clause does not bar retrial where the defendant has managed through appeal or some other procedure to have his conviction set aside because of reversible error. United States v. Dinitz, 424 U.S. 600, 96 S.Ct. 1075, 47 L.Ed.2d 267 (1976); Granger v. State, 850 S.W.2d 513 (Tex. Crim. App. 1993).

A retrial is barred on double jeopardy grounds after appeal if there is legally insufficient evidence to support the conviction. Burks v. United States, 437 U.S. 1, 98 S.Ct. 2141, 57 L.Ed.2d 1 (1978); Granger, *supra*.

However, where the conviction is set aside due to factually insufficient evidence the double jeopardy clause does not prohibit retrial of the defendant on the same charges. Tibbs v. Florida, 457 U.S. 31, 102 S.Ct. 2211, 72 L.Ed.2d 652 (1982).

The double jeopardy clause of the Texas Constitution does not bar retrial of an accused whose conviction for the same offense has been reversed on appeal, or as the result of a post-conviction writ of habeas corpus proceeding, due to prosecutorial misconduct. Ex parte Davis, 957 S.W.2d 9 (Tex. Crim. App. 1997); Ex parte Mitchell, *supra*.

A capital murder case that was reversed on the defendant's appeal can be retried where the jury originally deadlocked on the punishment issues and a life sentence was imposed by the trial judge as mandated by law in that situation. Sattazahn v. Pennsylvania, 537 U.S. 101, 123 S.Ct. 732, 154 L.Ed.2d 588 (2003).

The issue-preclusion component of the Double Jeopardy Clause does not bar the Government from retrying defendants after a jury has returned irreconcilably inconsistent verdicts of conviction and acquittal, and the convictions are later vacated for legal error unrelated to the inconsistency. Bravo-Fernandez v. U.S., 580 U.S. ___, 137 S.Ct. 352, 362-3, 196 L.Ed.2d 242 (2016).

§8:50 Raising Double Jeopardy Claims

§8:51 Habeas Corpus Writs

A pretrial writ of habeas corpus is usually the procedural vehicle by which a defendant should raise a successive prosecution for the same offense double jeopardy claim. Ex parte Robinson, 641 S.W.2d 552 (Tex. Crim. App. 1982); Abney v. U.S., 431 U.S. 651, 97 S.Ct. 2034, 52 L.Ed.2d 651 (1977); Apolinar v. State, 820 S.W.2d 792 (Tex. Crim. App.1991).

§8:52 Special Pleas

CCP Art. 27.05. Defendant's Special Plea

> *"A defendant's only special plea is that he has already been prosecuted for the same or a different offense arising out*

DOUBLE JEOPARDY

of the same criminal episode that was or should have been consolidated into one trial, and that the former prosecution:

1. resulted in acquittal;

2. resulted in conviction;

3. was improperly terminated; or

4. was terminated by a final order or judgment for the defendant that has not been reversed, set aside, or vacated and that necessarily required a determination inconsistent with a fact that must be established to secure conviction in the subsequent prosecution."

The statutorily prescribed procedure utilized on a special plea mandates that the defendant's claim in the special plea shall not be determined before the trial on the merits. *See* CCP Art. 27.07.

Trial courts are required to submit the special plea to the trier of fact unless, assuming all the facts alleged in the plea to be true, the special plea does not present a legally sufficient former jeopardy claim. Arredondo v. State, 582 S.W.2d 457 (Tex. Crim. App. 1979).

§8:53 On Appeal

In order to raise the issue of a double jeopardy violation on appeal, the record must explicitly show the violation. Gonzalez v. State, 8 S.W.3d 640 (Tex. Crim. App. 2000) (where the record merely reflected a general verdict of guilty from an indictment alleging several offenses).

§8:60 Remedy for Double Jeopardy Violations

Where a defendant files a pretrial habeas writ claiming a double jeopardy violation and the appeals courts find such a violation, the case will be remanded to the trial court with instructions to dismiss the pending prosecution. State v. Engelking, 817 S.W.2d 64 (Tex. Crim. App. 1991).

If a defendant has been found guilty in one trial of two offenses that are determined to be the same for double jeopardy purposes, the most serious punishment test is used to determine which offense to dismiss and which to retain. That test requires retaining the offense with the most serious punishment and vacating any remaining offenses that are the same for double jeopardy purposes. Landers v. State, 957 S.W.2d 558 (Tex. Crim. App.1997).

§8:70 Double Jeopardy and Sentencing Proceedings

Double jeopardy analysis is inapplicable to non-capital sentencing proceedings. Monge v. California, 524 U.S. 721, 118 S.Ct. 2246, 141 L.Ed.2d 615 (1998); Bell v. State, 994 S.W.2d 173 (Tex. Crim. App. 1999).

Use of unadjudicated offenses at the punishment phase of a trial does not bar their later prosecution. Witte v. U.S., 515 U.S. 389, 115 S.Ct. 2199, 132 L.Ed.2d 351 (1995); Ex parte Broxton, 888 S.W.2d 23 (Tex. Crim. App. 1994).

Where the trial court pronounces a statutorily authorized sentence, any attempt to alter the sentence after the day of the original sentencing violates the defendant's double jeopardy rights. Harris v. State, 153 S.W.3d 394 (Tex. Crim. App. 2005).

The Double Jeopardy Clause does not preclude retrial on a prior conviction allegation in the noncapital sentencing context where in the original proceeding the trial judge pronounced an illegal sentence that was based on a misinterpretation of the law. Wooldridge v. State, 237 S.W.3d 714 (Tex. Crim. App. 2007).

§8:80 Administrative Hearings and Double Jeopardy

§8:81 Administrative License Revocation Proceedings

The Texas Department of Public Safety and a district attorney are not same parties for collateral estoppel purposes, and therefore, a district attorney is not prohibited from relitigating an adverse finding in a prior administrative proceeding for revocation of a driver's license in a subsequent criminal prosecution for DWI. Reynolds v. State, 4 S.W.3d 15 (Tex. Crim. App. 1999); State v. Brabson, 976 S.W.2d 859 (Tex. Crim. App. 1998).

Furthermore, collateral estoppel principles do not prohibit our Legislature from expressly providing that the determination of an administrative judge in a civil administrative proceeding to revoke a person's driver's license does not preclude litigation of the same or similar facts in a criminal prosecution. Reynolds, *supra*.

An administrative driver's license suspension does not constitute punishment for purposes of the Double Jeopardy Clause of the Fifth Amendment, and therefore, a subsequent prosecution for driving while intoxicated does not implicate the protection under the Double Jeopardy Clause against multiple punishments for the same offense. Tharp v. State, 935 S.W.2d 157 (Tex. Crim. App.1996).

This is true even though under Blockburger analysis, the Court of Criminal Appeals has found that a driver's license suspension under Tex. Rev. Civ. Stat. Art. 6687b-1, §2(a) and the offense of Driving While Intoxicated under P.C. §49.01 are the same offense for double jeopardy purposes. Voisinet v. State, 935 S.W.2d 424 (Tex. Crim. App. 1996).

§8:82 Probation Revocation Proceedings

The result of a hearing on a motion to revoke probation is neither a conviction nor acquittal for purposes of double jeopardy analysis. Salinas v. State, 1 S.W.3d 700 (Tex.App.—Amarillo 1999, *pet. ref'd*). This is because they are administrative proceedings rather than criminal trials. Cobb v. State, 851 S.W.2d 871 (Tex. Crim. App.1993).

Because jeopardy does not attach in revocation proceedings for any criminal offense that is alleged as a violation of the terms of community supervision, double jeopardy principles cannot properly apply to bar a subsequent prosecution under these circumstances. State v. Waters, 560 S.W.3d 651, 658 (Tex. Crim. App. 2018).

The double jeopardy clauses of the U.S. and Texas constitutions are not offended by the use in a subsequent trial of a new offense allegation that was used successfully or unsuccessfully to revoke a probation. Chambers v. State, 700 S.W.2d 597 (Tex. Crim. App. 1985).

Even though no amendment is allowed once evidence has been taken in a revocation hearing, the state is not prevented by statute or double jeopardy principles from dismissing the petition and filing a subsequent motion. Johnson v. State, 633 S.W.2d 687 (Tex.App.—Amarillo 1982, *pet. ref'd*).

A dismissal of a new offense by the state does not preclude the state from seeking revocation by proof of the conduct that was the basis of the new offense. Ex parte Moffett, 542 S.W.2d 184 (Tex. Crim. App. 1976).

§8:83 Parole Revocation Proceedings

A determination by a hearing examiner of insufficient evidence to revoke parole does not operate to collaterally estop prosecution in state district court. Ex parte Daniel, 781 S.W.2d 412 (Tex.App.— Houston [1st Dist.] 1989 *pet. ref'd*); Salinas v. State, 1 S.W.3d 700 (Tex.App.—Amarillo 1999, *pet. ref'd*).

§8:84 Jail Disciplinary Proceedings

Administrative sanctions such as jail discipline are not criminal punishment for double jeopardy purposes when they are not grossly unrelated to the remedial interests of the sanctioning body. Ex parte Hernandez, 953 S.W.2d 275 (Tex. Crim. App. 1997).

§8:90 Prosecutions by Separate Sovereigns

A defendant may be prosecuted in both federal and state court for the same offense. Bartkus v. Illinois, 351 U.S. 121, 79 S.Ct. 676, 3 L.Ed.2d 684 (1956); Ex parte Gary, 895 S.W.2d 465 (Tex. App.—Amarillo 1995, *pet. ref'd*).

Two separate states can prosecute a defendant for the same offense. Heath v. Alabama, 474 U.S. 82, 106 S.Ct. 433, 88 L.Ed.2d 387 (1985).

§8:100 Civil Penalties and Double Jeopardy

The mere assessment of a drug tax against a defendant is not a punishment for the purposes of the double jeopardy clause. Ex parte Chappell, 959 S.W.2d 627 (Tex. Crim. App. 1998).

In fact, absent full payment of the drug tax or

an arrangement with the collecting authority for payment of the amount due, there is no punishment for double jeopardy purposes. Ex parte Ward, 964 S.W.2d 617 (Tex. Crim. App. 1998).

Furthermore, the attachment of a tax lien against a defendant's property without an attempt to enforce the lien is not punishment for double jeopardy purposes. Ex parte Ward, *supra*.

However, a criminal conviction for possession of drugs is punishment for double jeopardy purposes when an assessment is later made and collection sought for a drug tax for the same drugs. Dept. of Revenue v. Kurth Ranch, 511 U.S. 767, 114 S.Ct. 1937, 128 L.Ed.2d 767 (1994).

§8:110 Double Jeopardy and Contempt Proceedings

It is well established that a contempt conviction is criminal if it punishes for past violations, and civil if it attempts to coerce future action. Ex parte Williams, 799 S.W.2d 304 (Tex. Crim. App. 1990).

Double jeopardy principles generally do not apply to an order assessing solely coercive confinement when a contemnor can obtain release by purging contempt. Yates v. United States, 355 U.S. 66, 78 S.Ct. 128, 2 L.Ed.2d 95 (1957); Ex parte Jones, 36 S.W.3d 139 (Tex.App.—Houston [1st Dist.] 2000, *pet. ref'd*).

Where a defendant has already been prosecuted for criminal contempt of court for an act, a subsequent criminal prosecution is barred by the Double Jeopardy Clause of the U.S. Constitution. U.S. v. Dixon, 509 U.S. 688, 113 S.Ct. 2849, 125 L.Ed.2d 556 (1993); Ex parte Rhodes, 974 S.W.2d 735 (Tex. Crim. App. 1998).

§8:115 Double Jeopardy and Protective Orders

Code of Criminal Procedure Art. 17.292 is not criminally punitive and therefore jeopardy does not attach following the issuance of a protective order under that section. Ex parte Necessary, 333 S.W.3d 782 (Tex.App.—Houston [1st Dist.], 2010, *no pet.*).

§8:120 Basic Collateral Estoppel Principles

The collateral estoppel doctrine provides that when an issue of ultimate fact has been determined by a valid and final judgment, that issue cannot again be litigated between the same parties in any future lawsuit. Ashe v. Swenson, 397 U.S. 436, 90 S.Ct. 1189, 25 L.Ed.2d 469 (1970); Headrick v. State, 988 S.W.2d 226 (Tex. Crim. App. 1999).

Collateral estoppel bars any retrial of specific and discrete facts that have been fully and fairly adjudicated. Ex parte Watkins, 73 S.W.3d 264 (Tex. Crim. App. Nos. 1460-01 and 1461-01, April 3, 2002).

The mere possibility that a fact may have been determined in a former trial is insufficient to bar relitigation of that same fact in a second trial. In applying the doctrine of collateral estoppel, courts must first determine whether the jury determined a specific fact, and if so, how broad in terms of time, space and content was the scope of its finding. Watkins.

In deciding whether a party is collaterally estopped from litigating an issue, a reviewing court must examine the record of a prior proceeding, taking into account the pleadings, evidence, charge, and other relevant matters, and conclude whether a rational jury could have grounded its verdict upon an issue other than that which the defendant seeks to foreclose from consideration. Ashe v. Swenson, *supra*; Ex parte Mathes, 830 S.W.2d 596 (Tex. Crim. App. 1992); State v. Sauceda, 980 S.W.2d 642 (Tex. Crim. App. 1998).

The doctrine of collateral estoppel does not depend on slight differences in statutory language or an alternate "manner and means." Rather, it depends on the scope of a specific factual finding in a particular case. Ex parte Taylor, 101 S.W.3d 434 (Tex. Crim. App. 2002) (the state was estopped from retrying the defendant for intoxication manslaughter by reason of marijuana intoxication, where a jury had previously acquitted him of intoxication manslaughter by reason of alcohol intoxication arising out of the same incident).

Collateral estoppel deals only with specific factual determinations, not legal claims or legal conclusions. State v. Stevens, 235 S.W.3d 736 (Tex. Crim. App. 2007) (citing Guajardo v. State, 109 S.W.3d 456 (Tex. Crim. App. 2003)).

Courts must determine exactly what facts were necessarily decided in the first proceeding and whether those necessarily decided facts constitute essential elements of the offense in the second trial. Ex parte Taylor.

Mere issues of the admissibility of evidence are not elements essential to proving the commission of an offense; therefore, where a court has granted a defendant's motion to suppress, that issue can be relitigated in another case arising from the same transaction. Murphy v. State, 239 S.W.3d 791 (Tex. Crim. App. 2007).

That is, the State is not barred by the Double Jeopardy Clause from relitigating a suppression issue that was not an ultimate fact in the first prosecution and was not an ultimate fact in the second prosecution. York v. State, 342 S.W.3d 528 (Tex. Crim. App. 2011).

Collateral estoppel operates only if the very fact or point in issue in the second case was determined in the prior proceeding. It must be precisely the same issue in both cases. Thus, issue preclusion is limited to cases where the legal and factual situations are identical. Collateral estoppel may apply to a phase, an issue of fact, or congeries of fact, depending on the particular circumstances. Ex parte Taylor.

On the other hand, issue preclusion cannot be defeated simply by advancing new or different evidence to support the same issue already litigated. Ex parte Taylor.

An exception to the application of traditional double jeopardy analysis may exist where the additional facts necessary to sustain a charge have not occurred or have not been discovered despite the exercise of due diligence. *See* Ashe v. Swenson, *supra* (Brennan, J., concurring); Ex parte Mathes, *supra* (fn.4).

A conviction that is reversed on appeal cannot be used to bar a subsequent prosecution on collateral estoppel principles. Garcia v. State, 768 S.W.2d 726 (Tex. Crim. App. 1987).

§8:130 Collateral Estoppel in Punishment Proceedings

The state is barred from relitigating the issue of "sudden passion" in a subsequent trial where a jury has found against the state on this issue in a previous trial of an offense arising out of the same criminal transaction. Ex parte Watkins, 73 S.W.3d 264 (Tex. Crim. App. 2002).

Where a jury returns a verdict finding against the state on a special issue in a death penalty case, the state is collaterally estopped from seeking the death penalty in the trial of a case involving a second victim of the same offense. Ex parte Mathes, 830 S.W.2d 596 (Tex. Crim. App. 1992).

However, where a jury merely is unable to answer a special issue, the state is not foreclosed from litigating it in a second proceeding because the issue has not been finally determined against the state. Padgett v. State, 717 S.W.2d 55 (Tex. Crim. App. 1986).

§8:140 Collateral Estoppel and Probation Revocation Proceedings

Collateral estoppel does not apply to bar a subsequent criminal prosecution following a trial judge's finding of "not true" at a revocation hearing. State v. Waters, 560 S.W.3d 651, 663 (Tex. Crim. App. 2018) (overruling Ex parte Tarver, 725 S.W.2d 195 (Tex. Crim. App. 1986) and Wafer v. State, 58 S.W.3d 158 (Tex. App.—Amarillo 2001)).

An acquittal of a new offense will not bar subsequent prosecution of the allegation in a motion to revoke probation. Ex parte Tarver, 725 S.W.2d 195, 200 (Tex. Crim. App. 1986).

§8:150 Raising Collateral Estoppel Claims

It is appropriate for a defendant to apply for a pretrial writ of habeas corpus where the claims concern collateral estoppel and constitutional double jeopardy violations. Ex parte Robinson, 641 S.W.2d 552 (Tex. Crim. App. 1982); Ex parte Watkins, 73 S.W.3d 264 (Tex. Crim. App. 2002).

Collateral estoppel claims not alleging constitutional double jeopardy violations are not cognizable on application for pretrial writ of habeas corpus. This is because there is an adequate remedy at law by way of direct appeal. Headrick v. State, 988 S.W.2d 226 (Tex. Crim. App. 1999) (where the defendant sought to assert collateral estoppel based on an administrative judge's decision).

In order to obtain appellate review of a collateral estoppel claim, a defendant must introduce a record of the first proceeding in the second proceeding and include that record on appeal. Guajardo v. State, 109 S.W.3d 456 (Tex. Crim. App. 2003).

Chapter 9

BAIL AND BOND ISSUES

Table of Contents

§9:10 Right to Bail

§9:11 Statutory and Constitutional Law

CCP Art. 1.07. Right to bail

All prisoners shall be bailable unless for capital offenses when the proof is evident. This provision shall not be so construed as to prevent bail after indictment found upon examination of the evidence, in such manner as may be prescribed by law.

CCP Art. 1.09. Cruelty forbidden

Excessive bail shall not be required, nor excessive fines imposed, nor cruel or unusual punishment inflicted.

CCP Art. 17.38. Rules applicable to all cases of bail

The rules in this Chapter respecting bail are applicable to all such undertakings when entered into in the course of a criminal action, whether before or after an indictment, in every case where authority is given to any court, judge, magistrate, or other officer, to require bail of a person accused of an offense, or of a witness in a criminal action.

Texas Constitution Art.1 §11. Bail

§11. All prisoners shall be bailable by sufficient sureties, unless for capital offenses, when the proof is evident; but this provision shall not be so construed as to prevent bail after indictment found upon examination of the evidence, in such manner as may be prescribed by law.

Texas Constitution Art. 1, §13. Excessive bail or fines; cruel and unusual punishment; remedy by due course of law

§13. Excessive bail shall not be required, nor excessive fines imposed, nor cruel or unusual punishment inflicted. All courts shall be open, and every person for an injury done him, in his lands, goods, person or reputation, shall have remedy by due course of law.

§9:12 In General

With narrow exceptions, Article I, §11 of the Texas Constitution provides all prisoners a right to bail pending trial. Smith v. State, 829 S.W.2d 885 (Tex.App.—Houston [1st Dist.] 1992, *pet. ref'd*).

A defendant has a liberty interest in bail, which requires the protection of the due process clause of the Fourteenth Amendment. Smith v. State, 993 S.W.2d 408 (Tex.App.—Houston [14th Dist.] 1999, *pet. ref'd*); Robinson v. State, 700 S.W.2d 710 (Tex. App.—Houston [14th Dist.] 1985, *no pet.*).

The primary purpose of bond is to ensure the defendant's presence in court at all proceedings. Estrada v. State, 594 S.W.2d 445 (Tex. Crim. App. [Panel Op.] 1980). The decision regarding a proper bail amount lies within the sound discretion of the trial court. Ex parte Brown, 959 S.W.2d 369 (Tex.App.—Fort Worth 1998, *no pet.*); Ex parte Green, 940 S.W.2d 799 (Tex.App.—El Paso 1997, *no pet.*). There is no precise standard for reviewing bail bond settings. Ex parte Watson, 940 S.W.2d 733 (Tex.App.—Texarkana 1997, *no pet.*); Ex parte Pemberton, 577 S.W.2d 266 (Tex. Crim. App. [Panel Op.] 1979).

The trial court does not have inherent authority to impose conditions on a defendant's pre-trial bond that are not authorized by statute; and further, Art. 17.15 does not implicitly authorize other conditions not expressly stated. Ex parte Tucker, 977 S.W.2d 713 (Tex.App.—Fort Worth 1998, *pet. dism'd*).

Absent legislative intent to the contrary, a trial court has inherent power in misdemeanor and felony appeals to impose conditions on bail that directly or indirectly relate to the purpose of assuring the defendant's continued appearance. Dallas v. State, 983 S.W.2d 276 (Tex. Crim. App. 1998). It is not necessary that the bond condition in question relate directly to securing the defendant's presence in court. Estrada, *supra*. It is sufficient if the condition indirectly increases the likelihood that the defendant will appear. Rodriguez v. State, 744 S.W.2d 361 (Tex.App.—Corpus Christi 1988, *no pet.*); Estrada.

Determining the "reasonableness" of a bail condition must be done in light of its purpose, to assure the presence of the defendant. Valenciano v.

State, 720 S.W.2d 523 (Tex. Crim. App. 1986). In each individual case where conditions are imposed, those conditions must strike a balance between society's interest in assuring that the defendant will appear if and when his conviction becomes final and the defendant's interest in remaining free pending appeal. Valenciano.

The trial court's decision on pretrial bond conditions is reviewable under an abuse of discretion standard. Miller v. State, 855 S.W.2d 92 (Tex.App.—Houston [14th Dist.] 1993, *pet. ref'd*).

A trial court does not have the discretion to set a differential bail bond amount depending upon whether a cash bond or a surety bond is used. Professional Bondsmen of Texas v. Carey, 762 S.W.2d 691 (Tex.App.—Amarillo 1988, *no pet.*). The trial court does not have the authority to require a cash bond or surety bond to the exclusion of the other. Carey; Ex parte Deaton, 582 S.W.2d 151 (Tex. Crim. App.1979).

The practice of a grand jury recommending the amount of bail is not supported in the statutes and is not a commendable practice. Ex parte Dueitt, 529 S.W.2d 531 (Tex. Crim. App. 1975).

A defendant whose adjudication of guilt has been deferred is entitled to bail pending an adjudication hearing. Gutierrez v. State, 927 S.W.2d 783 (Tex.App.—Houston [14th Dist.] 1996, *no pet.*); Ex parte Laday, 594 S.W.2d 102 (Tex. Crim. App. 1980).

If a trial court makes a finding favorable to a convicted person under CCP Art. 64.04, the trial court may release the person on bail after a hearing. V.A.C.C.P Art. 17.48.

§9:13 Form: Sample Application for Writ of Habeas Corpus

See our companion book *Texas Criminal Forms* (James Publishing) Form 9-2.

§9:20 Defendant's Burden to Object to Improper Bond Conditions

The party complaining of the bond condition has the burden of demonstrating that the trial court abused its discretion by imposing an excessive condition of bond. Ex parte Rubac, 611 S.W.2d 848 (Tex. Crim. App. [Panel Op.] 1981).

A defendant must object to a bond condition when it is imposed or he fails to preserve error for appeal. Smith v. State, 993 S.W.2d 408 (Tex. App.—Houston [14th Dist.] 1999, *pet. ref'd*); Ex parte Sotelo, 878 S.W.2d 179 (Tex.App.—Fort Worth 1993, *pet. ref'd*). Additionally, in the *habeas corpus* context, a defendant's failure to object to the conditions at the time that they were imposed also prevents him from challenging the constitutionality of the conditions via a post-conviction collateral attack. Smith.

While the failure to object to the imposition of conditions of a bond waives the right to complain of constitutional error, a *habeas* applicant may assert a complaint concerning the reasonableness of the conditions arising from an order revoking bond. Smith.

A defendant may at any time litigate the reasonableness of a bond condition by way of a *habeas corpus* action. A bail condition that places an improper infringement on a defendant's freedom of action is within the scope of *habeas corpus* relief because it is a restraint. Valenciano v. State, 720 S.W.2d 523 (Tex. Crim. App. 1986).

Where the record reflects that the defendant did not object to the trial court's action, did not develop a record, and did not obtain an adverse ruling, he has waived appeal of the trial court's decision to not set an appeal bond. Hill v. State, 902 S.W.2d 57 (Tex.App.—Houston [1st Dist.] 1995).

§9:30 Fixing Amount of Bail

CCP Art. 17.15 Rules for Fixing Amount of Bail

> *The amount of bail to be required in any case is to be regulated by the court, judge, magistrate or officer taking the bail; they are to be governed in the exercise of this discretion by the Constitution and by the following rules:*
>
> *1. The bail shall be sufficiently high to give reasonable assurance that the undertaking will be complied with.*
>
> *2. The power to require bail is not to be so used as to make it an instrument of oppression.*

BAIL AND BOND ISSUES

3. The nature of the offense and the circumstances under which it was committed are to be considered.

4. The ability to make bail is to be regarded, and proof may be taken upon this point.

5. The future safety of a victim of the alleged offense and the community shall be considered.

CCP Art. 56.02. Crime victims' rights

(a) A victim, guardian of a victim, or close relative of a deceased victim is entitled to the following rights within the criminal justice system:

... (2) the right to have the magistrate take the safety of the victim or his family into consideration as an element in fixing the amount of bail for the accused;

The decision regarding a proper bail amount lies within the sound discretion of the trial court. Ex parte Brown, 959 S.W.2d 369 (Tex.App.—Fort Worth 1998, *no pet.*); Ex parte Green, 940 S.W.2d 799 (Tex.App.—El Paso 1997, *no pet.*).

The amount of bail must be high enough to give reasonable assurance that the accused will appear as required, but the amount should not be oppressively high. Ex parte McDonald, 852 S.W.2d 730 (Tex.App.—San Antonio 1993, *no pet.*).

In addition to the considerations set forth in Arts. 17.15 and 56.02, other factors may be considered by the trial court in setting bail, such as:

- Accused's work record, family and community ties, length of residency, prior criminal record, if any, and conformity with the conditions of any previous bond
- Existence of outstanding bonds, if any
- Alleged aggravated circumstances involved in the offense

Smith v. State, 829 S.W.2d 885 (Tex.App.—Houston [1st Dist.] 1992, *pet. ref'd*).

TIP TO THE STATE: When the state presents no evidence on a particular enumerated factor in Art. 17.15 or when it presents evidence that is indecipherable, the courts will assume they consider these factors unworthy of attention. Ex parte Prelow, 929 S.W.2d 54 (Tex.App.—San Antonio 1996, no pet.).

A court of appeals has no jurisdiction to hear interlocutory appeals regarding excessive bail or the denial of bail because there is no constitutional or statutory authority granting this authority. Ragston v. State, 424 S.W.3d 49, 52 (Tex. Crim. App. 2014).

§9:31 Existence of Outstanding Bonds

The existence of other bonds may be taken into consideration when setting bond. Ex parte Watson, 940 S.W.2d 733 (Tex.App.—Texarkana 1997, *no pet.*); Nguyen v. State, 881 S.W.2d 141 (Tex. App.—Houston [1st Dist.] 1994, *no pet.*).

CCP Art. 15.17 requires a magistrate setting bail to determine whether the person is currently on bail for a separate criminal offense; and implies that if the person is on bail for another offense, bail should not be set on the current case.

§9:32 Conformity With Previous Bonds

The trial court may also consider the fact that the accused has previously failed to appear in court while on bail. Ex parte Poindexter, 511 S.W.2d 529 (Tex. Crim. App. 1974).

§9:33 Defendant's Criminal Record

Prior felony convictions and the potential punishment in the case for which bond is being set are two considerations. Ex parte Runo, 535 S.W.2d 188 (Tex. Crim. App.1976).

§9:34 Circumstances of Offense

The amount of bail also must be based on the nature of the offense and the circumstances under which it was committed. Ex parte Brown, 959 S.W.2d 369 (Tex.App.—Fort Worth 1998, *no pet.*); Ex parte

Davila, 623 S.W.2d 408 (Tex. Crim. App. [Panel Op.] 1981).

§9:35 Ability to Make Existing Bond Amount

In seeking to reduce the amount of bail ordered by the trial court, the defendant has the burden to show that the amount is excessive. Ex parte Watson, 940 S.W.2d 733 (Tex.App.—Texarkana 1997, *no pet.*); Ex parte Rubac, 611 S.W.2d 848 (Tex. Crim. App. [Panel Op.] 1981). The defendant must show that he has attempted and failed to make bail in the amount fixed. Ex parte Welch, 729 S.W.2d 306 (Tex.App.—Dallas 1987, *no pet.*); Ex parte Sellers, 516 S.W.2d 665 (Tex. Crim. App. 1974).

The defendant's ability to make bail is a relevant factor, but is not dispositive. Esquivel v. State, 922 S.W.2d 601 (Tex.App.—San Antonio 1996, *no pet.*). A defendant's mere inability meet the bond set by the trial court does not automatically render the bail excessive. Ex parte Brown, 959 S.W.2d 369 (Tex.App.—Fort Worth 1998, *no pet.*).

Like indigence, the ability of a bail bondsman to post bond should not be a controlling factor for the fixing of bail by the trial court. Wright v. State, 976 S.W.2d 815 (Tex.App.—Houston [1st Dist.] 1998, *no pet.*).

§9:36 Future Safety of Victim

The court also should consider the future safety of the victim of the alleged offense and the community at large. Ex parte Wood, 952 S.W.2d 41 (Tex.App.—San Antonio 1997, *no pet.*). The Court of Criminal Appeals has stated that under Art. 17.15(5) the "victim of the alleged offense" includes only the complainant in the charged offense. Ludwig v. State, 812 S.W.2d 323 (Tex. Crim. App. 1991).

§9:37 Threats and Actions of Defendant

Where the accused has made threats that he "will not be tried" or is otherwise not disposed to regard the obligation of ordinary bonds, the fixing of a higher bond is justifiable. Wright v. State, 976 S.W.2d 815 (Tex.App.—Houston [1st Dist.] 1998, *no pet.*).

Higher bonds are justified where defendant:

- Attempted to obtain fake identification to flee the country
- Committed the offense as part of a plan to escape punishment for another, more serious charge
- Has funds to post a lower bond and is likely to obtain considerably more assets that could be used to accomplish the goal of fleeing the jurisdiction

Wright.

§9:38 Cases on Pretrial Bond Amounts

The Court of Criminal Appeals has yet to condone a bail amount even approaching seven figures, even in a capital case. Ludwig v. State, 812 S.W.2d 323 (Tex. Crim. App. 1991) ($1,000,000 bond was found to be excessive and bond was reduced to $50,000). Where a defendant was alleged to have committed a premeditated retaliation-murder, a $150,000 bond was not unreasonable despite his youthful age and ties to the community. Ex parte Prelow, 929 S.W.2d 54 (Tex.App.—San Antonio 1996, *no pet.*).

A bond of $500,000 was reduced to $100,000 in cases involving murder and injury to a child where the facts surrounding the offenses were not developed at the *habeas* hearing. Smithwick v. State, 880 S.W.2d 510 (Tex.App.—San Antonio 1994, *no pet.*). A bond was reduced in a capital murder case from $1,000,000 to $75,000 where the facts surrounding the offense were not developed at the *habeas* hearing. Ex parte McDonald, 852 S.W.2d 730 (Tex.App.—San Antonio 1993, *no pet.*).

Where a defendant had no criminal record, had ties to the community even though he was not a Texas resident and was returning to California when he was arrested, $60,000 in bonds for three serious felonies was found to be proper. Nguyen v. State, 881 S.W.2d 141 (Tex.App.—Houston [1st Dist.] 1994, *no pet.*). Where a defendant did not show an inability to post a higher bond, exhibited continued illegal activity and had threatened to flee, a $200,000 bond was not excessive in four charges of failure to remit gasoline taxes. Wisenbaker v. State, 782 S.W.2d 534 (Tex.App.—Houston [14th Dist.] 1989, *no pet.*).

$100,000 was not an excessive amount of bail where the defendant was charged in an assault and murder case involving a child and had no ties to the community. O'Neill v. State, 635 S.W.2d 166 (Tex. App.—Houston [1st Dist.] 1982, *no pet.*). Bail of $350,000 in capital murder case and $100,000 in an unrelated aggravated murder case was reduced to $50,000 and $20,000 respectively where the defendant presented several mitigating factors at the bond hearing. Ex parte Wood, 952 S.W.2d 41 (Tex.App.—San Antonio 1997, *no pet.*).

A $250,000 bond for an unemployed armed robbery defendant was reduced to $50,000. Ex parte Ivey, 594 S.W.2d 98 (Tex. Crim. App. 1980). A $100,000 bond for an indigent capital murder defendant who admitted driving the get-away car was reduced to $20,000. Ex parte Vasquez, 558 S.W.2d 477 (Tex. Crim. App. 1977).

§9:40 Specific Bond Conditions

Art. 17.40 gives trial judges the authority to impose conditions on a defendant's bond that are not specifically authorized by statute.

CCP Art. 17.40. Conditions related to victim or community safety

> *(a) To secure a defendant's attendance at trial, a magistrate may impose any reasonable condition of bond related to the safety of a victim of the alleged offense or to the safety of the community.*
>
> *(b) At a hearing limited to determining whether the defendant violated a condition of bond imposed under Subsection (a), the magistrate may revoke the defendant's bond only if the magistrate finds by a preponderance of the evidence that the violation occurred.*

§9:41 Statutorily Authorized Bond Conditions

§9:41.1 Ignition Interlock Devices

CCP Art. 17.441. Conditions requiring motor vehicle ignition interlock

> *(a) Except as provided by Subsection (b), a magistrate shall require on release that a defendant charged with a subsequent offense under §§49.04-49.06, Penal Code, or an offense under §49.07 or 49.08 of that code:*
>
> > *(1) have installed on the motor vehicle owned by the defendant or on the vehicle most regularly driven by the defendant, a device that uses a deep-lung breath analysis mechanism to make impractical the operation of a motor vehicle if ethyl alcohol is detected in the breath of the operator; and*
> >
> > *(2) not operate any motor vehicle unless the vehicle is equipped with that device.*
>
> *(b) The magistrate may not require the installation of the device if the magistrate finds that to require the device would not be in the best interest of justice.*
>
> *(c) If the defendant is required to have the device installed, the magistrate shall require that the defendant have the device installed on the appropriate motor vehicle, at the defendant's expense, before the 30th day after the date the defendant is released on bond.*
>
> *(d) The magistrate may designate an appropriate agency to verify the installation of the device and to monitor the device. If the magistrate designates an agency under this subsection, in each month during which the agency verifies the installation of the device or provides a monitoring service the defendant shall pay a fee to the designated agency in the amount set by the magistrate. The defendant shall pay the initial fee at the time the agency verifies the installation of the device. In each subsequent month during which the defendant is required to pay a fee the defendant shall pay the fee on the first occasion in that month that the agency provides a monitoring service. The magistrate shall set the fee in an amount not to exceed $10 as determined by the county auditor, or by the commissioners court of the county if the county does not have a county auditor, to be sufficient to cover*

the cost incurred by the designated agency in conducting the verification or providing the monitoring service, as applicable in that county.

By conducting a hearing and then determining that the installation of the interlock device was appropriate, the trial court complied with Art. 17.441. Ex parte Elliott, 950 S.W.2d 714 (Tex. App.—Fort Worth 1997, *pet. ref'd*). After the court has determined that the defendant has been previously convicted of DWI and that installation of the device was in the best interest of justice, Art. 17.441 requires installation of the device. Elliot.

Because subsection (b) of Art. 17.441 affords the trial court the discretion to find that the interest of justice mandates against the installation of an interlock device, it does not violate the separation of powers provision and is not unconstitutional. Elliot. The requirement of an interlock device, which is a less severe infringement on the privilege of driving an automobile, does not constitute punishment and is not oppressive. Elliot.

§9:41.2 Conditions Related to Victim or Community Safety

CCP Art. 17.40. Conditions Related to Victim or Community Safety

(a) To secure a defendant's attendance at trial, a magistrate may impose any reasonable condition of bond related to the safety of a victim of the alleged offense or to the safety of the community.

(b) At a hearing limited to determining whether the defendant violated a condition of bond imposed under Subsection (a), the magistrate may revoke the defendant's bond only if the magistrate finds by a preponderance of the evidence that the violation occurred. If the magistrate finds that the violation occurred, the magistrate shall revoke the defendant's bond and order that the defendant be immediately returned to custody. Once the defendant is placed in custody, the revocation of the defendant's bond discharges the sureties on the bond, if any, from any future liability on the bond. A discharge under this subsection from any future liability on the bond does not discharge any surety from liability for previous forfeitures on the bond.

In a case where the defendant was convicted of aggravated assault, a bond condition that the defendant stay away from and not communicate with the victim was found to be reasonable because it was rationally related to the purpose of assuring that the defendant commit no further acts of violence, which in turn increased the likelihood that he would appear in court when ordered. Rodriguez v. State, 744 S.W.2d 361 (Tex.App.—Corpus Christi 1988, *no pet.*).

§9:41.3 Conditions Where Child Is Alleged Victim

CCP Art. 17.41. Condition Where Child Alleged Victim

(a) This article applies to a defendant charged with an offense under any of the following provisions of the Penal Code, if committed against a child younger than 14 years of age:

(1) Chapter 21 (Sexual Offenses) or 22 (Assaultive Offenses);

(2) Section 25.02 (Prohibited Sexual Conduct); or

(3) Section 43.25 (Sexual Performance by a Child).

(b) Subject to Subsections (c) and (d), a magistrate shall require as a condition of bond for a defendant charged with an offense described by Subsection (a) that the defendant not:

(1) directly communicate with the alleged victim of the offense; or

(2) go near a residence, school, or other location, as specifically described in the bond, frequented by the alleged victim.

(c) A magistrate who imposes a condition of bond under this article may

grant the defendant supervised access to the alleged victim.

(d) To the extent that a condition imposed under this article conflicts with an existing court order granting possession of or access to a child, the condition imposed under this article prevails for a period specified by the magistrate, not to exceed 90 days.

§9:41.4 Home Curfew and Electronic Monitoring

CCP Art. 17.43. Home curfew and electronic monitoring as condition

(a) A magistrate may require as a condition of release on personal bond that the defendant submit to home curfew and electronic monitoring under the supervision of an agency designated by the magistrate.

(b) Cost of monitoring may be assessed as court costs or ordered paid directly by the defendant as a condition of bond.

CCP Art. 17.44. Home confinement, electronic monitoring, and drug testing as condition

(a) A magistrate may require as a condition of release on bond that the defendant submit to:

(1) home confinement and electronic monitoring under the supervision of an agency designated by the magistrate; or

(2) testing on a weekly basis for the presence of a controlled substance in the defendant's body.

(b) In this article, "controlled substance" has the meaning assigned by §481.002, Health and Safety Code.

(c) The magistrate may revoke the bond and order the defendant arrested if the defendant:

(1) violates a condition of home confinement and electronic monitoring;

(2) refuses to submit to a test for controlled substances or submits to a test for controlled substances and the test indicates the presence of a controlled substance in the defendant's body; or

(3) fails to pay the costs of monitoring or testing for controlled substances, if payment is ordered under Subsection (e) as a condition of bond and the magistrate determines that the defendant is not indigent and is financially able to make the payments as ordered.

(d) The community justice assistance division of the Texas Department of Criminal Justice may provide grants to counties to implement electronic monitoring programs authorized by this article.

(e) The cost of electronic monitoring or testing for controlled substances under this article may be assessed as court costs or ordered paid directly by the defendant as a condition of bond.

Electronic monitoring is allowed as a condition of bond by CCP Art. 17.44. Grady v. State, 962 S.W.2d 128 (Tex.App.—Houston [1st Dist.] 1997, *pet. ref'd*). Requiring a defendant to wear an electronic monitor and to observe curfew conditions are not unreasonable appeal bond conditions because they increase the likelihood of defendant's appearance in court. Smith v. State, 993 S.W.2d 408 (Tex.App.—Houston [14th Dist.] 1999, *pet. ref'd*); Ly v. State, 908 S.W.2d 598 (Tex. App.—Houston [1st Dist.] 1995, *no pet.*).

Requiring the defendant to submit to electronic monitoring and to pay associated fees is a valid condition of appeal bond. Speth v. State, 939 S.W.2d 969 (Tex.App.—Houston [14th Dist.] 1997, *no pet.*)

§9:41.5 Drug Testing

CCP Art. 17.44. Home confinement, electronic monitoring, and drug testing as condition

(a) A magistrate may require as a condition of release on bond that the defendant submit to:

(1) home confinement and electronic monitoring under the supervision of an agency designated by the magistrate; or

(2) testing on a weekly basis for the presence of a controlled substance in the defendant's body.

(b) In this article, "controlled substance" has the meaning assigned by §481.002, Health and Safety Code.

(c) If a defendant violates a condition of home confinement and electronic monitoring, refuses to submit to a test for controlled substances, or submits to a test for controlled substances and the test indicates the presence of a controlled substance in the defendant's body, the magistrate may revoke the bond and order the defendant arrested.

(d) The community justice assistance division of the Texas Department of Criminal Justice may provide grants to counties to implement electronic monitoring programs authorized by this article.

(e) The cost of electronic monitoring or testing for controlled substances under this article may be assessed as court costs or ordered paid directly by the defendant as a condition of bond.

In a case where the defendant was convicted of possession of heroin, a condition of his appeal bond that required him to submit to urine testing and attend drug counseling was found to be reasonable because the defendant's continued abstinence from heroin was rationally related to his appearance in court when necessary. Estrada v. State, 594 S.W.2d 445 (Tex. Crim. App. [Panel Op.] 1980).

Random urinalysis is allowed as a condition of bond by CCP Art. 17.44. Grady v. State, 962 S.W.2d 128 (Tex.App.—Houston [1st Dist.] 1997, *pet. ref'd*).

§9:41.6 Conditions Requiring AIDS and HIV Instruction

CCP Art. 17.45. Conditions requiring AIDS and HIV instruction

A magistrate may require as a condition of bond that a defendant charged with an offense under §43.02, Penal Code, receive counseling or education, or both, relating to acquired immune deficiency syndrome or human immunodeficiency virus.

§9:41.7 Conditions for Defendant Charged With Stalking

CCP Art. 17.46. Conditions for a defendant charged with stalking

(a) A magistrate may require as a condition of release on bond that a defendant charged with an offense under §42.072, Penal Code, may not:

(1) communicate directly or indirectly with the victim; or

(2) go to or near the residence, place of employment, or business of the victim or to or near a school, day-care facility, or similar facility where a dependent child of the victim is in attendance.

(b) If the magistrate requires the prohibition contained in Subsection (a)(2) of this article as a condition of release on bond, the magistrate shall specifically describe the prohibited locations and the minimum distances, if any, that the defendant must maintain from the locations.

§9:41.8 Submission of DNA Specimen

A magistrate shall require, as a condition of release of a defendant described by Gov. Code §411.1471(a) (basically all sex offenders), that the defendant provide to a law enforcement agency one or more specimens for the purpose of creating a DNA record. V.A.C.C.P Art. 17.47.

§9:41.9 Conditions in Family Violence Cases

Tex. Code Crim. Pro. Art. 17.49. Conditions for Defendant Charged With Offense Involving Family Violence

(a) In this article:

(1) "Family violence" has the meaning assigned by Section 71.004, Family Code.

(2) "Global positioning monitoring system" means a system that electronically determines and reports the location of an individual through the use of a transmitter or similar device carried or worn by the individual that transmits latitude and longitude data to a monitoring entity through global positioning satellite technology. The term does not include a system that contains or operates global positioning system technology, radio frequency identification technology, or any other similar technology that is implanted in or otherwise invades or violates the individual's body.

(b) A magistrate may require as a condition of release on bond that a defendant charged with an offense involving family violence:

(1) refrain from going to or near a residence, school, place of employment, or other location, as specifically described in the bond, frequented by an alleged victim of the offense;

(2) carry or wear a global positioning monitoring system device and, except as provided by Subsection (h), pay the costs associated with operating that system in relation to the defendant; or

(3) except as provided by Subsection (h), if the alleged victim of the offense consents after receiving the information described by Subsection (d), pay the costs associated with providing the victim with an electronic receptor device that:

(A) is capable of receiving the global positioning monitoring system information from the device carried or worn by the defendant; and

(B) notifies the victim if the defendant is at or near a location that the defendant has been ordered to refrain from going to or near under Subdivision (1).

(c) Before imposing a condition described by Subsection (b)(1), a magistrate must afford an alleged victim an opportunity to provide the magistrate with a list of areas from which the victim would like the defendant excluded and shall consider the victim's request, if any, in determining the locations the defendant will be ordered to refrain from going to or near. If the magistrate imposes a condition described by Subsection (b)(1), the magistrate shall specifically describe the locations that the defendant has been ordered to refrain from going to or near and the minimum distances, if any, that the defendant must maintain from those locations.

(d) Before imposing a condition described by Subsection (b)(3), a magistrate must provide to an alleged victim information regarding:

(1) the victim's right to participate in a global positioning monitoring system or to refuse to participate in that system and the procedure for requesting that the magistrate terminate the victim's participation;

(2) the manner in which the global positioning monitoring system technology functions and the risks and limitations of that technology, and the extent to which the system

will track and record the victim's location and movements;

(3) any locations that the defendant is ordered to refrain from going to or near and the minimum distances, if any, that the defendant must maintain from those locations;

(4) any sanctions that the court may impose on the defendant for violating a condition of bond imposed under this article;

(5) the procedure that the victim is to follow, and support services available to assist the victim, if the defendant violates a condition of bond or if the global positioning monitoring system equipment fails;

(6) community services available to assist the victim in obtaining shelter, counseling, education, child care, legal representation, and other assistance available to address the consequences of family violence; and

(7) the fact that the victim's communications with the court concerning the global positioning monitoring system and any restrictions to be imposed on the defendant's movements are not confidential.

(e) In addition to the information described by Subsection (d), a magistrate shall provide to an alleged victim who participates in a global positioning monitoring system under this article the name and telephone number of an appropriate person employed by a local law enforcement agency whom the victim may call to request immediate assistance if the defendant violates a condition of bond imposed under this article.

(f) In determining whether to order a defendant's participation in a global positioning monitoring system under this article, the magistrate shall consider the likelihood that the defendant's participation will deter the defendant from seeking to kill, physically injure, stalk, or otherwise threaten the alleged victim before trial.

(g) An alleged victim may request that the magistrate terminate the victim's participation in a global positioning monitoring system at any time. The magistrate may not impose sanctions on the victim for requesting termination of the victim's participation in or refusing to participate in a global positioning monitoring system under this article.

(h) If the magistrate determines that a defendant is indigent, the magistrate may, based on a sliding scale established by local rule, require the defendant to pay costs under Subsection (b)(2) or (3) in an amount that is less than the full amount of the costs associated with operating the global positioning monitoring system in relation to the defendant or providing the victim with an electronic receptor device.

(i) If an indigent defendant pays to an entity that operates a global positioning monitoring system the partial amount ordered by a magistrate under Subsection (h), the entity shall accept the partial amount as payment in full. The county in which the magistrate who enters an order under Subsection (h) is located is not responsible for payment of any costs associated with operating the global positioning monitoring system in relation to an indigent defendant.

(j) A magistrate that imposes a condition described by Subsection (b)(1) or (2) shall order the entity that operates the global positioning monitoring system to notify the court and the appropriate local law enforcement agency if a defendant violates a condition of bond imposed under this article.

(k) A magistrate that imposes a condition described by Subsection (b) may only allow or require the defendant to execute or be released under a type of bond that is authorized by this chapter.

(l) This article does not limit the authority of a magistrate to impose any other reasonable conditions of bond or enter any orders of protection under other applicable statutes.

§9:42 Pretrial Bond Conditions (Upheld)

Where the defendant was on bond for DWI and was found to have previously been convicted of DWI a condition that he install an interlock device on his vehicle was found to be reasonable because by assuring that the defendant not violate the law, the condition went to assuring his presence in court when required. Ex parte Elliott, 950 S.W.2d 714 (Tex.App.—Fort Worth 1997, *pet. ref'd*).

§9:43 Pretrial Bond Conditions (Not Upheld)

Requiring a defendant to satisfy a civil judgment in a sister court as a condition of bond is improper because it bears no rational relationship to the objective of securing the defendant's appearance before the trial court. Smith v. State, 829 S.W.2d 885 (Tex.App.—Houston [1st Dist.] 1992, *pet. ref'd*).

The trial court lacks the authority to require the defendant to appear with an attorney as a term and condition of pretrial bond. Meador v. State, 780 S.W.2d 836 (Tex.App.—Houston [14th Dist.] 1989, *no pet.*).

§9:44 Forms and Motions

See our companion book *Texas Criminal Forms* (James Publishing):

- Form 9-7.1 Written Objection To Bond Conditions
- Form 9-7.2 Defendant's Motion to Modify Bond Conditions
- Form 9-8 Habeas Corpus Petition for Relief From Unreasonable Bond Condition

§9:50 Increasing Bond Amounts

It is within the trial court's discretion to increase the bail set. Meador v. State, 780 S.W.2d 836 (Tex.App.—Houston [14th Dist.] 1989, *no pet.*). A reviewing court will not disturb the trial court's decision to increase bail unless it finds that the trial court has abused its discretion. Miller v. State, 855 S.W.2d 92 (Tex.App.—Houston [14th Dist.] 1993, *pet. ref'd*). No precise standard exists for determining what constitutes "good and sufficient cause" under Art. 17.09; and therefore, each case must be reviewed on an individual basis. Miller.

Where the accused has made threats that he "will not be tried" or is not otherwise disposed to regard the obligation of ordinary bonds, a higher bond is justified. Wright v. State, 976 S.W.2d 815 (Tex.App.—Houston [1st Dist.] 1998).

Higher bonds are justified where the defendant:

- Attempted to obtain fake identification to flee the country
- Committed the offense as part of a plan to escape punishment for another, more serious charge
- Has funds to post a lower bond and is likely to obtain considerably more assets that could be used to accomplish the goal of fleeing the jurisdiction

Wright.

§9:60 Denial of Bond

CCP Art. 17.152. Denial of Bail for Violation of Certain Court Orders or Conditions of Bond in a Family Violence Case

(a) In this article, "family violence" has the meaning assigned by Section 71.004, Family Code.

(b) Except as otherwise provided by Subsection (d), a person who commits an offense under Section 25.07, Penal Code, related to a violation of a condition of bond set in a family violence case and whose bail in the case under Section 25.07, Penal Code, or in the family violence case is revoked or forfeited for a violation of a condition of bond, may be taken into custody and, pending trial or other court proceedings, denied release on bail if, following a hearing, a judge or magistrate determines by a preponderance of the evidence that the person violated a condition of bond related to:

(1) the safety of the victim of the offense under Section 25.07, Penal Code, or the family violence case, as applicable; or

(2) the safety of the community.

(c) Except as otherwise provided by Subsection (d), a person who commits an offense under Section 25.07, Penal Code, other than an offense related to a violation of a condition of bond set in a family violence case, may be taken into custody and, pending trial or other court proceedings, denied release on bail if, following a hearing, a judge or magistrate determines by a preponderance of the evidence that the person committed the offense.

(d) A person who commits an offense under Section 25.07(a) (3), Penal Code, may be held without bail under Subsection (b) or (c), as applicable, only if, following a hearing, the judge or magistrate determines by a preponderance of the evidence that the person went to or near the place described in the order or condition of bond with the intent to commit or threaten to commit:

(1) family violence; or

(2) an act in furtherance of an offense under Section 42.072, Penal Code.

(e) In determining whether to deny release on bail under this article, the judge or magistrate may consider:

(1) the order or condition of bond;

(2) the nature and circumstances of the alleged offense;

(3) the relationship between the accused and the victim, including the history of that relationship;

(4) any criminal history of the accused; and

(5) any other facts or circumstances relevant to a determination of whether the accused poses an imminent threat of future family violence.

(f) A person arrested for committing an offense under Section 25.07, Penal Code, shall without unnecessary delay and after reasonable notice is given to the attorney representing the state, but not later than 48 hours after the person is arrested, be taken before a magistrate in accordance with Article 15.17. At that time, the magistrate shall conduct the hearing and make the determination required by this article.

Tex. Code Crim Pro. Art. 17.153. Denial of Bail for Violation of Condition of Bond Where Child Alleged Victim

(a) This article applies to a defendant charged with a felony offense under any of the following provisions of the Penal Code, if committed against a child younger than 14 years of age:

(1) Chapter 21 (Sexual Offenses);

(2) Section 25.02 (Prohibited Sexual Conduct); or

(3) Section 43.25 (Sexual Performance by a Child).

(b) A defendant described by Subsection (a) who violates a condition of bond set under Article 17.41 and whose bail in the case is revoked for the violation may be taken into custody and denied release on bail pending trial if, following a hearing, a judge or magistrate determines by a preponderance of the evidence that the defendant violated a condition of bond related to the safety of the victim of the offense or the safety of the community. If the magistrate finds that the violation occurred, the magistrate may revoke the defendant's bond and order that the defendant be immediately returned to custody. Once the defendant is placed in custody, the revocation of the defendant's bond discharges the

sureties on the bond, if any, from any future liability on the bond. A discharge under this subsection from any future liability on the bond does not discharge any surety from liability for previous forfeitures on the bond.

Due process requires the trial court to provide a defendant with reasonable notice that it intends to deny bail pending appeal and to allow a defendant a meaningful opportunity to be heard. Smith v. State, 993 S.W.2d 408 (Tex.App.—Houston [14th Dist.] 1999, *pet. ref'd*).

A defendant who is serving a sentence in another state and has been returned to Texas through the Interstate Agreement on Detainers is not eligible for bail. Mills v. State, 626 S.W.2d 583 (Tex. App.—Amarillo 1981, *pet. ref'd*).

CCP Art. 15.17 requires a magistrate setting bail to determine whether the person is currently on bail for a separate criminal offense; and implies that if the person is on bail for another offense, bail should not be set on the current case.

§9:70 Revocation of Bond

A trial court is permitted to revoke bail on a finding by a preponderance of the evidence of a violation of a valid bond condition. CCP Art. 44.04(c); Smith v. State, 993 S.W.2d 408 (Tex.App.—Houston [14th Dist.] 1999, *pet. ref'd*). Due process protections of notice and a reasonable opportunity to be heard attach to an appeal bond revocation proceeding. Smith. The statute does require presentation of evidence and that the revocation be based on that evidence. Smith.

Bail pending appeal of a conviction may be revoked for the commission of a subsequent offense, even in the absence of a specific condition that no additional offense be committed while on such bail. Oldham v. State, 743 S.W.2d 734 (Tex. App.—Fort Worth 1987, *pet. ref'd*).

A defendant's pretrial bond cannot be revoked if his failure to appear for trial is involuntary due to confinement in a jail or penitentiary. Fernandez v. State, 516 S.W.2d 677 (Tex. Crim. App. 1974).

The trial court may forfeit the bond of a defendant if he does not appear within a reasonable time after his name is called; but the defendant's slight tardiness is not enough to permit forfeiture of his bond. Meador v. State, 780 S.W.2d 836 (Tex. App.—Houston [14th Dist.] 1989, *no pet.*).

§9:80 Personal Bonds

CCP Art. 17.03. Personal bond

(a) Except as provided by Subsection (b) of this article, a magistrate may, in the magistrate's discretion, release the defendant on his personal bond without sureties or other security.

(b) Only the court before whom the case is pending may release on personal bond a defendant who:

(1) is charged with an offense under the following sections of the Penal Code:

(A) Section 19.03 (Capital Murder);

(B) Section 20.04 (Aggravated Kidnapping);

(C) Section 22.021 (Aggravated Sexual Assault);

(D) Section 22.03 (Deadly Assault on Law Enforcement or Corrections Officer, Member or Employee of Board of Pardons and Paroles, or Court Participant);

(E) Section 22.04 (Injury to a Child, Elderly Individual, or Disabled Individual);

(F) Section 29.03 (Aggravated Robbery);

(G) Section 30.02 (Burglary);

(H) Section 71.02 (Engaging in Organized Criminal Activity);

(I) Section 21.02 (Continuous Sexual Abuse of Young Child or Children); or

(J) Section 20A.03 (Continuous Trafficking of Persons);

(2) is charged with a felony under Chapter 481, Health and Safety Code, or Section 485.033, Health and Safety Code, punishable by imprisonment for a minimum term or by a maximum fine that is more than a minimum term or maximum fine for a first degree felony; or

(3) does not submit to testing for the presence of a controlled substance in the defendant's body as requested by the court or magistrate under Subsection (c) of this article or submits to testing and the test shows evidence of the presence of a controlled substance in the defendant's body.

(c) When setting a personal bond under this chapter, on reasonable belief by the investigating or arresting law enforcement agent or magistrate of the presence of a controlled substance in the defendant's body or on the finding of drug or alcohol abuse related to the offense for which the defendant is charged, the court or a magistrate shall require as a condition of personal bond that the defendant submit to testing for alcohol or a controlled substance in the defendant's body and participate in an alcohol or drug abuse treatment or education program if such a condition will serve to reasonably assure the appearance of the defendant for trial.

(d) The state may not use the results of any test conducted under this chapter in any criminal proceeding arising out of the offense for which the defendant is charged.

(e) Costs of testing may be assessed as court costs or ordered paid directly by the defendant as a condition of bond.

(f) In this article, "controlled substance" has the meaning assigned by Section 481.002, Health and Safety Code.

(g) The court may order that a personal bond fee assessed under Section 17.42 be:

(1) paid before the defendant is released;

(2) paid as a condition of bond;

(3) paid as court costs;

(4) reduced as otherwise provided for by statute; or

(5) waived.

CCP Art. 17.031. Release on personal bond

(a) Any magistrate in this state may release a defendant eligible for release on personal bond under Article 17.03 of this code on his personal bond where the complaint and warrant for arrest does not originate in the county wherein the accused is arrested if the magistrate would have had jurisdiction over the matter had the complaint arisen within the county wherein the magistrate presides. The personal bond may not be revoked by the judge of the court issuing the warrant for arrest except for good cause shown.

(b) If there is a personal bond office in the county from which the warrant for arrest was issued, the court releasing a defendant on his personal bond will forward a copy of the personal bond to the personal bond office in that county.

The granting of a personal bond is a matter within the discretion of the court before whom the case is pending. Ex parte Trillo, 540 S.W.2d 728 (Tex. Crim. App. 1976), *overruled on other grounds*; Aguilar v. State, 621 S.W.2d 781 (Tex. Crim. App. 1981). There will be some situations where a personal bond will not be appropriately within the discretion of a trial court, and a personal bond should not be granted. Trillo.

A person who is arrested without a warrant must be released on a personal bond after 24 hours if arrested for a misdemeanor, and after 48 hours if arrested for a felony, if he cannot obtain a surety bond or post bond himself. CCP Art. 17.033.

§9:90 Release of Certain Mentally Ill Defendants

CCP Art. 17.032. Release on Personal Bond of Certain Mentally Ill Defendants

(a) In this article, "violent offense" means an offense under the following sections of the Penal Code:

(1) Section 19.02 (murder);

(2) Section 19.03 (capital murder);

(3) Section 20.03 (kidnapping);

(4) Section 20.04 (aggravated kidnapping);

(5) Section 21.11 (indecency with a child);

(6) Section 22.01(a)(1) (assault), if the offense involved family violence as defined by Section 71.004, Family Code;

(7) Section 22.011 (sexual assault);

(8) Section 22.02 (aggravated assault);

(9) Section 22.021 (aggravated sexual assault);

(10) Section 22.04 (injury to a child, elderly individual, or disabled individual);

(11) Section 29.03 (aggravated robbery);

(12) Section 21.02 (continuous sexual abuse of young child or children); or

(13) Section 20A.03 (continuous trafficking of persons).

(b) Notwithstanding Article 17.03(b), or a bond schedule adopted or a standing order entered by a judge, a magistrate shall release a defendant on personal bond unless good cause is shown otherwise if:

(1) the defendant is not charged with and has not been previously convicted of a violent offense;

(2) the defendant is examined by the local mental health authority, local intellectual and developmental disability authority, or another qualified mental health or intellectual disability expert under Article 16.22;

(3) the applicable expert, in a written assessment submitted to the magistrate under Article 16.22:

(A) concludes that the defendant has a mental illness or is a person with an intellectual disability and is nonetheless competent to stand trial; and

(B) recommends mental health treatment or intellectual disability services for the defendant, as applicable;

(4) the magistrate determines, in consultation with the local mental health authority or local intellectual and developmental disability authority, that appropriate community-based mental health or intellectual disability services for the defendant are available in accordance with Section 534.053 or 534.103, Health and Safety Code, or through another mental health or intellectual disability services provider; and

(5) the magistrate finds, after considering all the circumstances, a pretrial risk assessment, if applicable, and any other credible information provided by the attorney representing the state or the defendant, that release on personal bond would reasonably ensure the defendant's appearance in court as required and the safety of the

community and the victim of the alleged offense.

(c) The magistrate, unless good cause is shown for not requiring treatment, shall require as a condition of release on personal bond under this article that the defendant submit to outpatient or inpatient mental health treatment or intellectual disability services as recommended by the local mental health authority, local intellectual and developmental disability authority, or another qualified mental health or intellectual disability expert if the defendant's:

(1) mental illness or intellectual disability is chronic in nature; or

(2) ability to function independently will continue to deteriorate if the defendant is not treated.

(d) In addition to a condition of release imposed under Subsection (c), the magistrate may require the defendant to comply with other conditions that are reasonably necessary to ensure the defendant's appearance in court as required and the safety of the community and the victim of the alleged offense.

(e) In this article, a person is considered to have been convicted of an offense if:

(1) a sentence is imposed;

(2) the person is placed on community supervision or receives deferred adjudication; or

(3) the court defers final disposition of the case.

§9:100 Regulation of Bail Bond Sureties

For questions regarding the rules which bail bond sureties are required to follow and the statutes regulating county bail bond board activity, *see* Texas Occupations Code Chapter 1704.

§9:110 Release of Defendant Because of Delay

CCP Art. 17.151. Release because of delay

§1. A defendant who is detained in jail pending trial of an accusation against him must be released either on personal bond or by reducing the amount of bail required, if the state is not ready for trial of the criminal action for which he is being detained within:

(1) 90 days from the commencement of his detention if he is accused of a felony;

(2) 30 days from the commencement of his detention if he is accused of a misdemeanor punishable by a sentence of imprisonment in jail for more than 180 days;

(3) 15 days from the commencement of his detention if he is accused of a misdemeanor punishable by a sentence of imprisonment for 180 days or less; or

(4) five days from the commencement of his detention if he is accused of a misdemeanor punishable by a fine only.

§2. The provisions of this article do not apply to a defendant who is:

(1) serving a sentence of imprisonment for another offense while the defendant is serving that sentence;

(2) being detained pending trial of another accusation against the defendant as to which the applicable period has not yet elapsed;

(3) incompetent to stand trial, during the period of the defendant's incompetence; or

> *(4) being detained for a violation of the conditions of a previous release related to the safety of a victim of the alleged offense or to the safety of the community under this article.*

§9:111 In General

Art. 17.151 provides, in relevant part, that a defendant accused of a felony who is detained in jail pending trial must be released, either on personal bond or by reducing the amount of bail required, if the State is not ready for trial within ninety days from the commencement of detention if he is accused of a felony. Ex parte Watson, 940 S.W.2d 733 (Tex.App.—Texarkana 1997, *no pet.*). The article further provides that the provisions do not apply to a defendant who is being detained pending trial of another accusation against him as to which the applicable period has not yet elapsed. Watson.

Article 17.151 does not violate separation of powers, and is not an impermissible encroachment upon prosecutorial discretion. Ex parte Gill, 413 S.W.3d 425, 432 (Tex. Crim. App. 2013); Jones v. State, 803 S.W.2d 712 (Tex. Crim. App. 1991).

The constitutional rule that a defendant being held on a charge of capital murder is not entitled to bail when the proof is evident is subject to the provisions of Art. 17.151. Beckcom v. State, 938 S.W.2d 780 (Tex.App.—Corpus Christi 1997, *no pet.*); Jones.

§9:112 Application of Time Limit

The time limit does not begin when the defendant is charged, but rather when he is actually detained. Balawajder v. State, 759 S.W.2d 504 (Tex.App.—Fort Worth 1988, *pet. ref'd*). The time period under the article does not begin to run until a defendant is certified as an adult and transferred to a criminal court. Ex parte Brosky, 863 S.W.2d 775 (Tex. App.—Fort Worth 1993, *no pet.*).

The state cannot defeat the operation of Art. 17.151 by filing another charge arising out of the same incident against the defendant to keep him in jail after the 90th day of his detention. Beckcom v. State, 938 S.W.2d 780 (Tex.App.—Corpus Christi 1997, *no pet.*). Where a subsequent indictment is for the same offense arising out of the same transaction for which the defendant was originally arrested, announcements of ready by the state carry forward to the new indictment. Behrend v. State, 729 S.W.2d 717 (Tex. Crim. App. 1987).

Article 17.151 contains no provisions excluding certain periods from the statutory time limit to accommodate exceptional circumstances. Rowe v. State, 853 S.W.2d 58 (Tex. Crim. App. 1993); Jones v. State, 803 S.W.2d 712 (Tex. Crim. App. 1991). A defendant's request to testify before the grand jury does not serve to remove him from the protections of Art. 17.151. Rowe.

§9:113 Readiness of State

When a defendant complains that the state was not ready within the statutory time period, the state has the burden to make a *prima facie* showing that it was. Jones v. State, 803 S.W.2d 712 (Tex. Crim. App. 1991). The State may satisfy its burden either by announcing ready within the allotted time or by announcing retroactively that it had been ready within the allotted time. Jones. The state's announcement, "the State is ready now, as it has been at all times," is sufficient to make a *prima facie* case of readiness within the statutory time limit. Applewhite v. State, 872 S.W.2d 32 (Tex. App.—Houston [1st Dist.] 1994, *no pet.*).

A prosecutor's statement at the writ hearing that, "the State has been ready for trial," was not sufficiently definite to show that the state was ready within the designated time. Jones.

The question of the state's "readiness" within the statutory limits refers to the preparedness of the prosecution for trial. Behrend v. State, 729 S.W.2d 717 (Tex. Crim. App. 1987). The concept of "ready" does not demand that the state have a perfect indictment. Behrend. In fact, where there is no showing of bad faith, the state may be prepared for trial even though the indictment that forms the basis for the prosecution of the offense is so defective as to be void. Behrend.

However, the state cannot be ready for trial under Art. 17.151 when no indictment has been returned. Kernahan v. State, 657 S.W.2d 433 (Tex. Crim. App. 1983).

Where the state enumerates before the court sufficient evidence to go to trial it has satisfied its obligation to show that it was ready for trial within the required time frame. Ex parte Watson, 940 S.W.2d 733 (Tex.App.—Texarkana 1997, *no pet.*).

Once the state makes a *prima facie* showing of readiness, it is up to the defendant to rebut it. Barfield v. State, 586 S.W.2d 538 (Tex. Crim. App. 1979). In the absence of a sufficient rebuttal by the defendant, the trial court has the discretion to find the State was timely ready for trial. Ex parte Brosky, 863 S.W.2d 775 (Tex.App.—Fort Worth 1993, *no pet.*).

Evidence to rebut the *prima facie* showing can be from any source, including cross-examination of those responsible for preparing the state's case; and may consist of, among other things, a demonstration that the state did not have a key witness or piece of evidence available by the last day of the applicable time limit. Applewhite, *supra*; Jones, *supra*.

PRACTICE TIP: An Art. 17.151 hearing can and should afford an opportunity to question and cross-examine the prosecutor concerning witness interviews and trial preparation.

To show the state's unreadiness, more than the mere fact that a witness had not been subpoenaed during the 90-day period must be shown. Applewhite, *supra*; Jones, *supra*.

A request by the state for a continuance does not constitute lack of readiness for purposes of Art. 17.151. Ex parte Stearnes, 761 S.W.2d 388 (Tex. App.—Amarillo 1988, *pet. ref'd*).

§9:114 Amount of Bond Set

Art. 17.151 states the accused must be released either on personal bond or by reducing the amount of bail required bond. Where the record reveals that the defendant cannot make any bond, the trial court is required to release him on a personal bond. Rowe v. State, 853 S.W.2d 581 (Tex. Crim. App. 1993).

Article 17.151 precludes a trial judge from considering factors other than the state's readiness for trial, such as the safety of the victim or the community, in determining whether an accused is entitled to release under that section. Ex parte Gill, 413 S.W.3d 425, 430 (Tex. Crim. App. 2013).

§9:115 Burden of Party Seeking Relief

The burden is on the party seeking review to see that a sufficient record is presented to show error requiring reversal. Tex.R.App.P. 50(d); Applewhite v. State, 872 S.W.2d 32 (Tex.App.—Houston [1st Dist.] 1994, *no pet.*).

A defendant's vague references to his inability to make bond are insufficient to require personal bond under Art. 17.151. Balawajder v. State, 759 S.W.2d 504 (Tex.App.—Fort Worth 1988, *pet. ref'd*).

§9:120 Bond in Family Violence Cases

CCP Art. 17.291. Further detention of certain persons

(a) In this article:

(1) "family violence has the meaning assigned to that phrase by §71.004, Family Code; and

(2) "magistrate" has the meaning assigned to it by Article 2.09 of this code.

(b) Article 17.29 does not apply when a person has been arrested or held without a warrant in the prevention of family violence if there is probable cause to believe the violence will continue if the person is immediately released. The head of the agency arresting or holding such a person may hold the person for a period of not more than four hours after bond has been posted. This detention period may be extended for an additional period not to exceed 48 hours, but only if authorized in a writing directed to the person having custody of the detained person by a magistrate who concludes that:

(1) the violence would continue if the person is release; and

(2) if the additional period exceeds 24 hours, probable cause exists to believe that the person committed

the instant offense and that, during the 10-year period preceding the date of the instant offense, the person has been arrested:

(a) on more than one occasion for an offense involving family violence; or

(b) for any other offense, if a deadly weapon, as defined by §1.07, Penal Code, was used or exhibited during commission of the offense or during immediate flight after commission of the offense.

CCP Art. 17.292. Magistrate's Order for Emergency Protection

(a) At a defendant's appearance before a magistrate after arrest for an offense involving family violence or an offense under Section 22.011, 22.021, or 42.072, Penal Code, the magistrate may issue an order for emergency protection on the magistrate's own motion or on the request of:

(1) the victim of the offense;

(2) the guardian of the victim;

(3) a peace officer; or

(4) the attorney representing the state.

(b) At a defendant's appearance before a magistrate after arrest for an offense involving family violence, the magistrate shall issue an order for emergency protection if the arrest is for an offense that also involves:

(1) serious bodily injury to the victim; or

(2) the use or exhibition of a deadly weapon during the commission of an assault.

(c) The magistrate in the order for emergency protection may prohibit the arrested party from:

(c-1) In addition to the conditions described by Subsection (c), the magistrate in the order for emergency protection may impose a condition described by Article 17.49(b) in the manner provided by that article, including ordering a defendant's participation in a global positioning monitoring system or allowing participation in the system by an alleged victim or other person protected under the order.

(1) committing:

(A) family violence or an assault on the person protected under the order; or

(B) an act in furtherance of an offense under Section 42.072, Penal Code;

(2) communicating:

(A) directly with a member of the family or household or with the person protected under the order in a threatening or harassing manner; or

(B) a threat through any person to a member of the family or household or to the person protected under the order;

(3) going to or near:

(A) the residence, place of employment, or business of a member of the family or household or of the person protected under the order; or

(B) the residence, child care facility, or school where a child protected under the order resides or attends; or

(4) possessing a firearm, unless

the person is a peace officer, as defined by Section 1.07, Penal Code, actively engaged in employment as a sworn, full-time paid employee of a state agency or political subdivision.

(d) The victim of the offense need not be present when the order for emergency protection is issued.

(e) In the order for emergency protection, the magistrate shall specifically describe the prohibited locations and the minimum distances, if any, that the party must maintain, unless the magistrate determines for the safety of the person or persons protected by the order that specific descriptions of the locations should be omitted.

(f) To the extent that a condition imposed by an order for emergency protection issued under this article conflicts with an existing court order granting possession of or access to a child, the condition imposed under this article prevails for the duration of the order for emergency protection.

(f-1) To the extent that a condition imposed by an order issued under this article conflicts with a condition imposed by an order subsequently issued under Chapter 85, Subtitle B, Title 4, Family Code, or under Title 1 or Title 5, Family Code, the condition imposed by the order issued under the Family Code prevails.

(f-2) To the extent that a condition imposed by an order issued under this article conflicts with a condition imposed by an order subsequently issued under Chapter 83, Subtitle B, Title 4, Family Code, the condition imposed by the order issued under this article prevails unless the court issuing the order under Chapter 83, Family Code:

(1) is informed of the existence of the order issued under this article; and

(2) makes a finding in the order issued under Chapter 83, Family Code, that the court is superseding the order issued under this article.

(g) An order for emergency protection issued under this article must contain the following statements printed in bold-face type or in capital letters:

"A VIOLATION OF THIS ORDER BY COMMISSION OF AN ACT PROHIBITED BY THE ORDER MAY BE PUNISHABLE BY A FINE OF AS MUCH AS $4,000 OR BY CONFINEMENT IN JAIL FOR AS LONG AS ONE YEAR OR BY BOTH. AN ACT THAT RESULTS IN FAMILY VIOLENCE OR A STALKING OFFENSE MAY BE PROSECUTED AS A SEPARATE MISDEMEANOR OR FELONY OFFENSE. IF THE ACT IS PROSECUTED AS A SEPARATE FELONY OFFENSE, IT IS PUNISHABLE BY CONFINEMENT IN PRISON FOR AT LEAST TWO YEARS. THE POSSESSION OF A FIREARM BY A PERSON, OTHER THAN A PEACE OFFICER, AS DEFINED BY SECTION 1.07, PENAL CODE, ACTIVELY ENGAGED IN EMPLOYMENT AS A SWORN, FULL-TIME PAID EMPLOYEE OF A STATE AGENCY OR POLITICAL SUBDIVISION, WHO IS SUBJECT TO THIS ORDER MAY BE PROSECUTED AS A SEPARATE OFFENSE PUNISHABLE BY CONFINEMENT OR IMPRISONMENT.

"NO PERSON, INCLUDING A PERSON WHO IS PROTECTED BY THIS ORDER, MAY GIVE PERMISSION TO ANYONE TO IGNORE OR VIOLATE ANY PROVISION OF THIS ORDER. DURING THE TIME IN WHICH THIS ORDER IS VALID, EVERY PROVISION OF THIS ORDER IS IN FULL FORCE AND EFFECT UNLESS A COURT CHANGES THE ORDER."

(h) The magistrate issuing an order for emergency protection under this article shall send a copy of the order to the chief of police in the municipality where the member of the family or household or individual protected by the order resides, if the person resides in a municipality, or to the sheriff of the county where the person resides, if the person does not reside in a municipality. If the victim of the offense is not present when the order is issued, the magistrate issuing the order shall order an appropriate peace officer to make a good faith effort to notify, within 24 hours, the victim that the order has been issued by calling the victim's residence and place of employment. The clerk of the court shall send a copy of the order to the victim.

(i) If an order for emergency protection issued under this article prohibits a person from going to or near a child care facility or school, the magistrate shall send a copy of the order to the child care facility or school.

(j) An order for emergency protection issued under this article is effective on issuance, and the defendant shall be served a copy of the order by the magistrate or the magistrate's designee in person or electronically. The magistrate shall make a separate record of the service in written or electronic format. An order for emergency protection issued under Subsection (a) or (b)(1) of this article remains in effect up to the 61st day but not less than 31 days after the date of issuance. An order for emergency protection issued under Subsection (b)(2) of this article remains in effect up to the 91st day but not less than 61 days after the date of issuance. After notice to each affected party and a hearing, the issuing court may modify all or part of an order issued under this article if the court finds that:

(1) the order as originally issued is unworkable;

(2) the modification will not place the victim of the offense at greater risk than did the original order; and

(3) the modification will not in any way endanger a person protected under the order.

(k) To ensure that an officer responding to a call is aware of the existence and terms of an order for emergency protection issued under this article, each municipal police department and sheriff shall establish a procedure within the department or office to provide adequate information or access to information for peace officers of the names of persons protected by an order for emergency protection issued under this article and of persons to whom the order is directed. The police department or sheriff may enter an order for emergency protection issued under this article in the department's or office's record of outstanding warrants as notice that the order has been issued and is in effect.

(l) In the order for emergency protection, the magistrate shall suspend a license to carry a concealed handgun issued under Subchapter H, Chapter 411, Government Code, that is held by the defendant.

(m) In this article:

(1) "Family," "family violence," and "household" have the meanings assigned by Chapter 71, Family Code.

(2) "Firearm" has the meaning assigned by Chapter 46, Penal Code.

(n) On motion, notice, and hearing, or on agreement of the parties, an order for emergency protection issued under this article may be transferred to the court assuming jurisdiction over the criminal act giving rise to the issuance of the emergency order for protection. On transfer, the criminal court may modify all or part of an order issued under this subsection in the same manner and under the same standards as the issuing court under Subsection (j).

§9:130 Appeal Bonds

§9:131 Applicable Statutes

CCP Art. 44.04. Bond pending appeal

(a) Pending the determination of any motion for new trial or the appeal from any misdemeanor conviction, the defendant is entitled to be released on reasonable bail.

(b) The defendant may not be released on bail pending the appeal from any felony conviction where the punishment equals or exceeds 10 years confinement or where the defendant has been convicted of an offense listed under Section 3g(a)(1), Article 42.12, but shall immediately be placed in custody and the bail discharged.

(c) Pending the appeal from any felony conviction other than a conviction described in Subsection (b) of this section, the trial court may deny bail and commit the defendant to custody if there then exists good cause to believe that the defendant would not appear when his conviction became final or is likely to commit another offense while on bail, permit the defendant to remain at large on the existing bail, or, if not then on bail, admit him to reasonable bail until his conviction becomes final. The court may impose reasonable conditions on bail pending the finality of his conviction. On a finding by the court on a preponderance of the evidence of a violation of a condition, the court may revoke the bail.

(d) After conviction, either pending determination of any motion for a new trial or pending final determination of the appeal, the court in which trial was had may increase or decrease the amount of bail, as it deems proper, either upon its own motion or the motion of the State or of the defendant.

(e) Any bail entered into after conviction and the sureties on the bail must be approved by the court where trial was had. Bail is sufficient if it substantially meets the requirements of this code and may be entered into and given at any term of court.

(f) In no event shall the defendant and the sureties on his bond be released from their liability on such bond or bonds until the defendant is placed in the custody of the sheriff.

(g) The right of appeal to the Court of Appeals of this state is expressly accorded the defendant for a review of any judgment or order made hereunder, and said appeal shall be given preference by the appellate court.

(h) If a conviction is reversed by a decision of a Court of Appeals, the defendant, if in custody, is entitled to release on reasonable bail, regardless of the length of term of imprisonment, pending final determination of an appeal by the state or the defendant on a motion for discretionary review. If the defendant requests bail before a petition for discretionary review has been filed, the Court of Appeals shall determine the amount of bail. If the defendant requests bail after a petition for discretionary review has been filed, the Court of Criminal Appeals shall determine the amount of bail. The sureties on the bail must be approved by the court where the trial was had. The defendant's right to release under this subsection attaches immediately on the issuance of the Court of Appeals' final ruling as defined by Tex. Crim. App.R. 209(c).

§9:132 Granting of Bail During Appeal

PRACTICE TIP: Art. 44.04 was changed considerably effective September 1, 1999. The prior law provided that a defendant was eligible for appeal bond unless his sentence was over 15 years. The law in effect for cases where a bond motion is made after September 1, 1999 is that a defendant is not eligible for appeal bond if his sentence is 10 years or more or if he receives any sentence for an article 42.12 §3(g) offense. Note that the language of subsection (b) is very broad and may require confinement while on appeal even if the defendant receives a probated sentence for any 3(g) offense.

The September 1, 1999 amendment to Art. 44.04 applies to any bond request made after that effective date. Cortez v. State, 36 S.W.3d 216 (Tex. App.—Houston [14th Dist.] 2001, *pet. filed*); Ex parte Cole, 43 S.W.3d 713 (Tex.App.—Fort Worth 2001, *no pet.*). The amendment does not violate the *ex post facto* or due process clauses of the Texas Constitution. Cortez.

A person who is sentenced to ten years' imprisonment, but who is placed under community supervision, is entitled to bond pending appeal under Art. 44.04(b). Lebo v. State, 90 S.W.3d 324 (Tex. Crim. App. 2002).

If a defendant is eligible for bail while on appeal, the trial court may continue him on his existing bond or admit him to a reasonable bail until his conviction becomes final. Art. 44.04(c). Ex parte Zigmond, 933 S.W.2d 666 (Tex.App.—San Antonio 1996, *no pet.*). The trial court may increase or decrease the amount of the bail, either on its own motion or on motion of the state. Art. 44.04(d).

Once the trial court determines that a person is eligible for appeal bond, the bail to be set must be reasonable. Perez v. State, 897 S.W.2d 893 (Tex. App.—San Antonio 1995, *no pet.*).

The factors listed in Art. 17.15 should be considered in setting the amount of the appeal bond:

- Bail shall be sufficiently high to give reasonable assurance that the undertaking will be complied with
- Power to require bail is not to be so used as to make it an instrument of oppression
- Nature of the offense and the circumstances under which it was committed are to be considered
- Ability to make bail is to be regarded, and proof may be taken upon this point
- Future safety of a victim of the alleged offense and the community shall be considered

Perez.

The burden is on the person seeking a reduction to show that the bail set is excessive. Ex parte Rubac, 611 S.W.2d 848 (Tex. Crim. App. 1981). While an accused's ability to make bail may be considered, it is not dispositive. Perez; Ex parte Rodriguez, 595 S.W.2d 549 (Tex. Crim. App. 1980).

The following factors should be considered in reviewing the reasonableness of the amount of bail on appeal:

- Punishment assessed
- Nature of the offense
- Appellant's work record, family ties, and length of residency
- Ability to make bond
- Prior criminal record of appellant
- Existence of other outstanding bonds
- Appellant's conformity with previous bond conditions

Jackson v. State, 810 S.W.2d 3 (Tex.App.—Houston [14th Dist.] 1991, *no pet.*); Ex parte Davila, 623 S.W.2d 408 (Tex. Crim. App. 1981).

§9:133 Conditions of Appeal Bond

Art. 44.04(c) authorizes a trial court to set reasonable conditions of bail pending appeal, including pending disposition of a petition for discretionary review filed by the state after an incarcerated defendant's conviction is reversed by a court of appeals and after the Court of Criminal Appeals has set the amount of bail pursuant to art. 44.04(h). Leonard v. State, 376 S.W.3d 886, 890 (Tex. App.—Fort Worth 2012, *pet.ref'd*).

Determining the reasonableness of a bail condition must be done in light of its primary purpose, to assure the defendant's presence in court. In each individual case, when conditions are imposed, those conditions must strike a balance between society's interest in assuring the defendant's appearance, if and when his conviction becomes final, and the defendant's interest in remaining free pending appeal. Cuellar.

Against the interest of a defendant to remain free pending appeal must be balanced the interest of society in enforcing the penal laws. Chief among these, in the cases of those who have been convicted of felonies, is to insure the public safety through:

- the deterrent influence of the penalties provided;
- the rehabilitation of those convicted of violations; and
- such punishment as may be necessary to prevent likely recurrence of criminal behavior. It is reasonable for a condition on bail pending appeal to have the purpose of protecting the public safety.

Ex parte Anderer, 61 S.W.3d 398 (Tex. Crim. App. 2001).

Even though Art. 44.04 does not address misdemeanor cases, a trial court has inherent power in misdemeanor appeals to impose conditions on bail that directly or indirectly relate to the purpose of assuring the defendant's continued appearance. Dallas v. State, 983 S.W.2d 276 (Tex. Crim. App. 1998).

TIP TO THE BENCH: While Art. 44.04(c) may not require the conditions to be in writing, the better practice would be to reduce any conditions imposed on bail to writing and cause a copy of the same to be served on the defendant and a copy attached to the bond. Rodriguez v. State, 744 S.W.2d 361 (Tex.App.—Corpus Christi 1988).

§9:134 Appeal Bond Conditions (Upheld)

An appeal bond condition that a defendant on bond for a criminally negligent homicide conviction who was also involved in another injury accident before trial not operate a motor vehicle while free on bail is reasonable because it has the purpose of protecting the public safety. Ex parte Anderer, 61 S.W.3d 398 (Tex. Crim. App. 2001).

In a case where the defendant was convicted of possession of heroin, a condition of his appeal bond that required him to submit to urine testing and attend drug counseling was found to be reasonable because the defendant's continued abstinence from heroin was rationally related to his appearance in court when necessary. Estrada v. State, 594 S.W.2d 445 (Tex. Crim. App. [Panel Op.] 1980).

In a case where the defendant was convicted of aggravated assault, a bond condition that he stay away from and not communicate with the victim was found to be reasonable because it was rationally related to the purpose of assuring that the defendant commit no further acts of violence, which in turn increased the likelihood that he would appear in court when ordered. Rodriguez v. State, 744 S.W.2d 361 (Tex.App.—Corpus Christi 1988, *no pet.*).

Electronic monitoring and random urinalysis are allowed as conditions of bond by CCP Art. 17.44. Grady v. State, 962 S.W.2d 128 (Tex.App.—Houston [1st Dist.] 1997, *pet. ref'd*). Requiring a defendant to wear an electronic monitor and to observe curfew conditions is not unreasonable because they increase the likelihood of his appearance in court. Smith v. State, 993 S.W.2d 408 (Tex.App.—Houston [14th Dist.] 1999, *pet. ref'd*); Ly v. State, 908 S.W.2d 598 (Tex.App.—Houston [1st Dist.] 1995, *no pet.*).

A condition prohibiting a defendant from associating with persons with whom he was associated in the commission of crime is not an unreasonable appeal bond condition because it increases the likelihood of his appearance in court. Smith; Rodriguez v. State, 744 S.W.2d 361 (Tex. App.—Corpus Christi 1988, *no pet.*) Appeal bond conditions that closely resemble conditions of probation will be upheld where they increase the likelihood of the defendant's appearance in court. Easton v. Rains, 866 S.W.2d 656 (Tex.App.—Houston [1st Dist.] 1993).

§9:135 Appeal Bond Conditions (Not Upheld)

A condition that a defendant not operate a motor vehicle while on appeal bond has been found unreasonable in a criminally negligent homicide case because it fails to advance the purpose of assuring appellant's presence in court when and if his conviction becomes final. Anderer v. State, 7 S.W.3d 245 (Tex.App.—Houston [14th Dist.] 1999, *pet. filed*).

A condition requiring the defendant to abandon his family residence as a condition of an appeal bond where he had been convicted of indecency with a child is unreasonable because it impinges on his freedom without forwarding society's interest in assuring his presence in any way. Valenciano v. State, 720 S.W.2d 523 (Tex. Crim. App. 1986).

In a misdemeanor case, the trial court has no authority to require attendance at Alcoholics Anonymous meetings as a condition of appeal bond. Grady v. State, 962 S.W.2d 128 (Tex.App.—Houston [1st Dist.] 1997, *pet. ref'd*).

In a misdemeanor cruelty to animals case, the trial court abused its discretion in prohibiting the defendant from engaging in the business of training and kenneling dogs as a condition of appeal bond. This condition is not relevant to securing appellant's appearance in court. Dallas v. State, 983 S.W.2d 276 (Tex. Crim. App. 1998).

Requiring a defendant to pay $44.50 in court costs as a condition of appeal bond was an abuse of discretion because it was not reasonably related to ensuring the defendant's appearance in court. Easton v. Rains, 866 S.W.2d 656 (Tex.App.—Houston [1st Dist.] 1993). A trial court abuses its discretion in ordering a payment into the registry of the court where the payment is earmarked for purposes other than securing the defendant's presence when his conviction is affirmed. Cuellar v. State, 985 S.W.2d 656 (Tex.App.—Houston [1st Dist.] 1999, *no pet.*).

A condition of bond that the defendant post the fact of his conviction on his web page was found to not serve the purpose of the defendant avoiding future criminal activity because there was no evidence presented that he used the Internet to commit the offense which was being appealed. Cuellar.

A condition prohibiting a defendant from working as a chiropractor for the duration of his appeal from a conviction for aggravated assault on a police officer is an invalid condition of appeal bond. Speth v. State, 939 S.W.2d 769 (Tex.App.—Houston [14th Dist.] 1997, *no pet.*).

A condition that a defendant submit to twice-weekly urinalysis at his expense as a condition of his appeal bond is unreasonable where the offense is not drug related and there is no showing at trial that the defendant has been involved with controlled substances in any way. Ex parte Sotelo, 878 S.W.2d 179 (Tex.App.—Fort Worth 1993, *pet. ref'd*). However, the same condition became reasonable when it was shown that the defendant tested positive for controlled substances while on the bond. Sotelo.

§9:136 Denial of Bail on Appeal

If the trial court has good cause to believe that the defendant would not appear when his conviction becomes final or that he is likely to commit another offense while on bail, bail may be denied. CCP Art. 44.04(c).

The evidence is sufficient to show that the defendant would be likely to commit another offense where defendant violated the terms of a previous probation, violated the terms of a work release program while in jail and vandalized automobiles after being indicted for the offense on appeal. Short v. State, 923 S.W.2d 168 (Tex.App.—Fort Worth 1996, *no pet.*).

The trial court may consider the nature of the underlying offense in reaching its decision to deny appeal bond. Ex parte Mendiola, 961 S.W.2d 625 (Tex.App.—San Antonio 1998, *no pet.*).

Where the testimony shows that the defendant had molested several children, the trial court did not abuse its discretion in refusing appeal bond on the basis that the defendant was likely to commit future offenses. Mendiola.

Where the state showed that the defendant had on a great number of occasions engaged in the same type of assaultive behavior as that for which he was found guilty by the jury, the evidence was sufficient to support the trial court's implied finding that there was good cause to believe that the defendant was likely to commit another offense while on appeal. Ex parte Castaneda, 739 S.W.2d 456 (Tex.App.—Corpus Christi 1987, *no pet.*).

Where the record reflects that the defendant did not object to the trial court's action, did not develop a record, and did not obtain an adverse ruling, he has waived appeal of the trial court's decision to not set an appeal bond. Hill v. State, 902 S.W.2d 57 (Tex.App.—Houston [1st Dist.] 1995, *pet. ref'd*).

§9:137 Revocation of Appeal Bond

If the trial court finds by a preponderance of the evidence that the defendant has violated a condition of the appeal bond it may revoke the bond. CCP Art. 44.04(c).

A computer-generated printout showing that the defendant violated the terms of his electronic monitoring agreement was sufficient to prove by a preponderance of the evidence that the defendant had violated the condition imposed upon him by the appeal bond. Ly v. State, 908 S.W.2d 598 (Tex. App.—Houston [1st Dist.] 1995, *no pet.*).

If a court finds that the defendant is likely to commit another offense while on bond, the court may revoke the bond even though no specific conditions were placed upon the defendant at the time of making bond. Oldham v. State, 743 S.W.2d 734 (Tex.App.—Fort Worth 1987, *pet. ref'd*); Ex Parte Le Blanc, 615 S.W.2d 724 (Tex. Crim. App. 1981).

Proof that a defendant committed an offense while on bail has been held sufficient to sustain an appeal bond revocation. Oldham; Putnam v. State, 582 S.W.2d 146 (Tex. Crim. App. 1979).

§9:138 Bail When Case Reversed

For the applicable statute *see* CCP Art. 44.04(h) in §9:131 above.

Where a conviction is reversed and the court of appeals is called upon to set bail pending a petition for discretionary review, it shall consider the factors set out in CCP Art. 17.15, but shall give primary consideration to the following three factors: 1) the fact that the conviction has been overturned; 2) the state's ability, if any, to retry the defendant; and 3) the likelihood that the decision of the court of appeals will be overturned on discretionary review. The length of the sentence is no longer a factor. Aviles v. State, 26 S.W.3d 696 (Tex.App.—Houston [14th Dist.] 2000, *pet. ref'd*).

§9:140 Constitutional Provisions

Texas Constitution Art.1 §11. Bail

§11. All prisoners shall be bailable by sufficient sureties, unless for capital offenses, when the proof is evident; but this provision shall not be so construed as to prevent bail after indictment found upon examination of the evidence, in such manner as may be prescribed by law.

Texas Constitution. Art. I, §11a.

(a) Any person (1) accused of a felony less than capital in this State, who has been theretofore twice convicted of a felony, the second conviction being subsequent to the first, both in point of time of commission of the offense and conviction therefor, (2) accused of a felony less than capital in this State, committed while on bail for a prior felony for which he has been indicted, (3) accused of a felony less than capital in this State involving the use of a deadly weapon after being convicted of a prior felony, or (4) accused of a violent or sexual offense committed while under the supervision of a criminal justice agency of the State or a political subdivision of the State for a prior felony, after a hearing, and upon evidence substantially showing the guilt of the accused of the offense in (1) or (3) above, of the offense committed while on bail in (2) above, or of the offense in (4) above committed while under the supervision of a criminal justice agency of the State or a political subdivision of the State for a prior felony, may be denied bail pending trial, by a district judge in this State, if said order denying bail pending trial is issued within seven calendar days subsequent to the time of incarceration of the accused; provided, however, that if the accused is not accorded a trial upon the accusation under (1) or (3) above, the accusation and indictment used under (2) above, or the accusation or indictment used under (4) above within sixty (60) days from the time of his incarceration upon the accusation, the order denying bail shall be automatically set aside, unless a continuance is obtained upon the motion or request of the accused; provided, further, that the right of appeal to the Court of Criminal Appeals of this State

is expressly accorded the accused for a review of any judgment or order made hereunder, and said appeal shall be given preference by the Court of Criminal Appeals.

(b) In this section:

(1) "Violent offense" means:

(A) murder;

(B) aggravated assault, if the accused used or exhibited a deadly weapon during the commission of the assault;

(C) aggravated kidnapping; or

(D) aggravated robbery.

(2) "Sexual offense" means:

(A) aggravated sexual assault;

(B) sexual assault; or

(C) indecency with a child.

§9:141 Denial of Bail in Capital Cases

The judge may deny bail in capital cases where the proof is evident. Angleton v. State, 955 S.W.2d 655 (Tex.App.—Houston [14th Dist.] 1997), *rev'd on other grounds,* 971 S.W.2d 65 (Tex. Crim. App. 1999). *See also,* CCP Art. 16.15.

The term "proof is evident" means the evidence is clear and strong, leading a well-guarded and dispassionate judgment to the conclusions that:

- Offense of capital murder has been committed
- Accused is the guilty party
- Jury will both convict the accused and will return findings requiring a death sentence

Angleton; Ex parte Graves, 853 S.W.2d 701 (Tex.App.—Houston [1st Dist.] 1993, *pet. ref'd*); Beck v. State, 648 S.W.2d 7 (Tex. Crim. App. 1983). The burden of proof is on the state to show that the proof is evident. Angleton; Beck.

The "substantial showing" burden of the accused's guilt at the bail hearing is far less than the trial burden of beyond a reasonable doubt. Angleton; Lee v. State, 683 S.W.2d 8 (Tex. Crim. App. 1985).

The state must also establish that the jury will assess the death penalty. Angleton; Ex parte Collum, 841 S.W.2d 960 (Tex.App.—Fort Worth 1992, *no pet.*).

§9:142 Denial of Bail in Other Cases

Texas Constitution Art. 1, §11a allows a defendant to be held without bond for a limited period where he commits a second criminal offense after being indicted and making bond for a first offense. Pharris v. State, 165 S.W.3d 681 (Tex. Crim. App. 2005).

To hold a person without bond, when that person is accused of a non-capital felony committed while on bail for a prior felony for which he has been indicted, the following sequence of events must occur: (1) A defendant must be indicted for a felony and be out of jail on bond; (2) A defendant must be then accused of committing, while on that bond, a felony less than capital in this state; (3) The defendant must be arrested on the accusation in 2; and (4) A district judge must hold a hearing wherein the state must show substantial evidence of the defendant's guilt for the felony in 2 and the district judge must enter an order denying bond, all within seven days of the defendant's arrest in 2. Pharris.

The state bears the burden of affirmatively showing compliance with the limits and safeguards of §11a. Pharris.

The appellate courts interpret §11a as favoring a defendant's access to bail, and strictly construe all of its provisions in favor of bail. Pharris.

Where denying bail under Art. I, §11(a) of the state constitution, the trial court must enter its order within seven (7) calendar days of the defendant's incarceration. Criner v. State, 878 S.W.2d 162 (Tex. Crim. App. 1994).

If the trial judge enters an order denying bail under Art. I, §11a, the defendant must be accorded a trial within sixty (60) days from his incarceration. Criner; Holloway v. State, 781 S.W.2d 605 (Tex. Crim. App.1989). If the accused is not brought to trial within sixty days, the trial judge's order denying bail is automatically set aside. Criner.

The rule that a defendant being held on a charge of capital murder is not entitled to bail when the proof is evident is subject to the provisions of CCP Art. 17.151. Beckcom v. State, 938 S.W.2d 780 (Tex.App.—Corpus Christi 1997, *no pet.*); Ex parte Jones v. State, 803 S.W.2d 712 (Tex. Crim. App. 1991).

§9:143 Form: Sample Motion to Deny Bail

See our companion book *Texas Criminal Forms* (James Publishing) for the following sample forms:

- Form 9-14.

§9:150 Bond for Post-Trial Habeas Corpus Applicants

§9:151 Statutory Law

CCP Art. 11.65. Bond for certain applicants

(a) This article applies to an applicant for a writ of habeas corpus seeking relief from the judgment in a criminal case, other than an applicant seeking relief from a judgment imposing a penalty of death.

(b) On making proposed findings of fact and conclusions of law jointly stipulated to by the applicant and the state, or on approving proposed findings of fact and conclusions of law made by an attorney or magistrate appointed by the court to perform that duty and jointly stipulated to by the applicant and the state, the convicting court may order the release of the applicant on bond, subject to conditions imposed by the convicting court, until the applicant is denied relief, remanded to custody, or ordered released.

(c) For the purposes of this chapter, an applicant released on bond under this article remains restrained in his liberty.

(d) Article 44.04(b) does not apply to the release of an applicant on bond under this article.

Chapter 10

[RESERVED]

(This page intentionally left blank.)

[RESERVED]

Chapter 11

EXAMINING TRIALS AND GRAND JURY HEARINGS

Table of Contents

I. EXAMINING TRIALS

§11:10 In General

An examining trial is a hearing before a magistrate for the purpose of determining: (1) whether the state has probable cause to hold the defendant under the felony accusation, and the amount of bail which the defendant shall be required to post; or (2) whether the defendant shall be discharged. CCP Arts. 16.01 and 16.17; Taylor v. State, 671 S.W.2d 535 (Tex.App.—Houston [1st Dist.] 1983).

The return of an indictment terminates the accused's right to an examining trial. White v. State, 576 S.W.2d 843 (Tex. Crim. App. 1979); De Leon v. State, 758 S.W.2d 621 (Tex.App.—Houston [14th Dist.] 1988).

Failure to enter an order finding probable cause to hold the defendant within 48 hours of the hearing operates as a discharge of the accused. CCP Art. 16.17.

In cases where a juvenile has been certified to stand trial as an adult, the juvenile may have an absolute right to an examining trial depending on the date of the offense. Rogers, *infra.*

Where the offense for which the juvenile was certified was committed before September 1, 1987, the accused juvenile has the right to an examining trial. The failure to conduct one renders an indictment void. Family Code §54.02(h); Ex parte Rogers, 578 S.W.2d 752 (Tex. Crim. App. 1979).

Where the offense was committed after September 1, 1987 but before January 1, 1996, the accused juvenile has the right to have the trial court consider whether good cause exists to conduct an examining trial. Family Code §54.02; CCP §16.01.

For offenses committed as a juvenile after January 1, 1996, there is no right to an examining trial prior to indictment. Family Code §54.02(h).

A certified juvenile can validly waive his right to an examining trial. Family Code §51.09(a); Ex parte Rogers, *supra.*

The magistrate conducting the examining trial generally has the duties to ensure that the accused has counsel, is properly warned, and that he knows that he has a right to make a voluntary statement. CCP Arts. 16.01 *et seq.*

The Code of Criminal procedure states that the same rules of evidence that govern final trials shall govern examining trials. CCP Art. 16.07. However, the Rules of Evidence state that the Rules are not applicable to hearings on justification for pretrial detention not involving bail. Texas Rules of Evidence 101(d)(1)(F).

PRACTICE TIP: Make all objections to the method of conducting the examining trial (such as lack of confrontation, lack of adequate cross-examination) at the examining trial itself or they are waived. Wilkins v. State, 818 S.W.2d 844 (Tex. App.—Houston [1st Dist.] 1991, pet. ref'd).

§11:20 Use of Witness Testimony at Trial

Before examining trial testimony can be used at a subsequent trial, it must be reduced to writing by or under the direction of the magistrate; and examined (and any corrections made) and signed by the witness. Then, it must be certified by the magistrate. CCP Art. 16.09. In lieu of this procedure, a statement of facts can be authenticated by the state and defense counsel, and approved by the presiding magistrate. *Id.*

However, defense counsel cannot unilaterally exclude examining trial testimony from a subsequent trial merely by failing to authenticate the statement of facts. Wilkins v. State, 818 S.W.2d 844 (Tex.App.—Houston [1st Dist.] 1991, *pet. ref'd*).

Where the above predicate is met, the use of the testimony is further governed by Texas Rules Of Evidence 804 relating to former testimony. In order to admit former testimony under this Rule, the proponent must show that the witness is unavailable. *See* §16:72.18 for discussion of Rule 804.

In order to introduce examining trial testimony in a subsequent trial, the State has the burden of establishing clearly and satisfactorily that:

- Testimony of its witness was given under oath;

EXAMINING TRIALS & GRAND JURY HEARINGS

- Testimony is competent;
- Accused on trial is the same accused who was present, as an accused, at the examining trial;
- Accused had adequate opportunity through counsel to cross-examine the witness.

Russell v. State, 604 S.W.2d 914 (Tex. Crim. App. 1980).

Whether the accused had an adequate opportunity to cross-examine the witness depends on the following:

- Factors which are material to the ultimate finding concerning probable cause;
- Limitations generally imposed by the magistrate as to the scope and depth of permitted cross-examination;
- Constraints deterring defense counsel from extending cross-examination into explorations of details;
- Reflections of a properly authenticated record of cross-examination as actually conducted;
- Intimations of ineffective assistance of counsel;
- Others that may appear from time to time to be significantly weighty.

Russell, *supra*.

§11:30 Use of Defendant Testimony at Trial

In order to use the examining trial testimony of the defendant at a subsequent trial, the state must show that:

- Defendant received the proper examining trial warnings under CCP Art. 16.03
- Defendant was given an opportunity to procure counsel or had counsel appointed under CCP Art. 16.01
- Testimony of the defendant is in the form of a voluntary statement, is reduced to writing, is signed by the defendant, and is properly certified and signed by the magistrate under CCP Art. 16.04
- The examining trial testimony of a defendant is admissible under Texas Rules of Evidence 801(e)(2) as an admission of party-opponent. *See* §16:72.4 for discussion of Rule 801 (e)(2).

§11:40 [Reserved]

II. GRAND JURY HEARINGS

§11:50 In General

The Rules of Evidence, except with respect to privileges, do not apply to grand jury proceedings. Texas Rules of Evidence 101 (d)(1)(B).

§11:60 Use of Defendant Testimony at Trial

The grand jury testimony of a defendant is admissible under Texas Rules of Evidence 801(e)(2) as an admission by a party-opponent where the proper additional predicate is met.

The predicate must show that an accused testifying before a grand jury has been informed of:

- Offense of which he is accused
- County where the offense is said to have been committed
- As nearly as may be the time of the commission of the offense

CCP Art. 20.17.

The accused then is required to be warned orally and in writing as follows:

- Your testimony before this grand jury is under oath
- Any material question that is answered falsely before this grand jury subjects you to being prosecuted for aggravated perjury
- You have the right to refuse to make answers to any question, the answer to which would incriminate you in any manner
- You have the right to have a lawyer present outside this chamber to advise you before making answers to questions you feel might incriminate you
- Any testimony you give may be used against you at any subsequent proceeding

- If you are unable to employ a lawyer, you have the right to have a lawyer appointed to advise you before making an answer to a question, the answer to which you feel might incriminate you

CCP Art. 20.17.

Substantial and not strict compliance with the introductory warnings of Art. 20.17 is all that is required for admissibility of an accused's grand jury testimony. The piecemeal development of the nature of the offense against the accused and the time and place of its commission, through the accused's responses to the questions posed by the prosecutor, is sufficient compliance with the statute. Andino v. State, 645 S.W.2d 615 (Tex. App.—Austin 1983).

The same is true of the formal warnings set out in CCP Art. 20.17(c). *Id.*

The accused must be given a reasonable opportunity to retain counsel or apply to the court for an appointed attorney and to consult with the attorney prior to appearing before the grand jury. CCP Art. 20.17.

Where an accused testifies before a grand jury without first being administered the statutory warning, his testimony may not be used as substantive evidence. Dillard v. State, 640 S.W.2d 85 (Tex.App.—Fort Worth 1982).

§11:70 Appearances Before and Secrecy of Proceedings

Since September 1, 1989, counsel for a person accused or suspected in connection with a grand jury proceeding may address the grand jury about the matter being reviewed. CCP Art. 20.04.

However, counsel may not appear in front of the grand jury to assist a witness or person suspected while that person testifies before the grand jury. CCP Arts. 20.011 and 20.04.

Pursuant to CCP Art. 20.02, the proceedings of the grand jury are deemed to be secret. Article 20.02 (b) provides that any grand juror, bailiff, interpreter, stenographer who discloses anything transpiring before the grand jury shall be subject to contempt. The punishment range for the contempt citation is up to a $500 fine, up to 30 days in jail or both fine and confinement.

Under CCP Art.20.02 (b), counsel for a defendant may petition a court to order disclosure of information made secret pursuant to grand jury proceedings (*i.e.*, a transcript of witness testimony). This can be done by disclosure of a recording or typewritten transcript. The petition must be filed in the district court in which the case is pending. In the event the attorney for the defendant discloses the transcript or information contained therein, the attorney is subject to a contempt citation as noted above.

CCP Art. 20.02 (c) requires that the attorney representing the state shall warn any person who receives information in the form of transcript from a grand jury of the duty to maintain secrecy of information contained therein. Violation of this article is punishable by a contempt citation as noted above.

A subpoena to appear before a grand jury is not a seizure under the Fourth Amendment. U.S. v. Dionisio, 410 U.S. 1, 93 S.Ct. 764, 35 L.Ed.2d 67 (1973).

§11:71 Unauthorized Questioning

Under CCP Art. 20.04 "[n]o person other than the attorney representing the State or a grand juror may question a witness before the grand jury." Mason v. State, 322 S.W.3d 251 (Tex. Crim. App. 2010).

Where an unauthorized person has questioned a witness before the grand jury, the proper subject of a harm analysis is the product of those proceedings: the charging decision. Appellate courts will consider whether the appellant's substantial rights were affected by the violation and whether the unauthorized questioning had a substantial and injurious effect or influence on the grand jury's decision to indict under TRAP 44.2(b). If the record as a whole does not show that the violation influenced the grand jury, or if there was just a "slight effect," then the trial court was correct to deny the motion to quash. Mason.

§11:80 Use of Witness Testimony at Trial

The subsequent admission of testimony of grand jury witnesses other than the defendant is governed by Texas Rule of Evidence 804 concerning former testimony. The witness must be unavailable under

Rule 804 (a) in order for the testimony to be admitted under this provision. Jones v. State, 843 S.W.2d 487 (Tex. Crim. App. 1992). *See* §16:72.18 for discussion of Rule 804.

§11:90 Impact of Selection Procedures

Courts are required to follow the means and methods provided by the legislature in selecting grand juries; and an arbitrary disregard of those statutes in the selection and organization of the grand jury vitiates and renders such grand jury without authority. Ex parte Becker, 459 S.W.2d 442 (Tex. Crim. App. 1970).

However, the Court of Criminal Appeals has held that, absent an allegation of fraud or injury resulting from the selection of the grand jurors, minor deviations from the statutory procedure do not rise to a level requiring reversal. Garcia v. State, 919 S.W.2d 370 (Tex. Crim. App. 1994); Gentry v. State, 770 S.W.2d 780 (Tex. Crim. App. 1988) (where there were allegations that the paperwork was not properly certified and that an oath to a clerk improperly given); Hicks v. State, 630 S.W.2d 829 (Tex.App.—Houston [1st Dist.] 1982 *pet. ref'd*) (where the district court used "loose" procedures); Cantu v. State, 842 S.W.2d 667 (Tex. Crim. App. 1992) (where there was a technical violation of the state nepotism statute).

A defendant must raise his challenge to the composition of the grand jury at his earliest opportunity. Ratcliff v. Estelle, 597 F.2d 474 (C.A.Tex. 1979). Ordinarily the earliest opportunity to contest the composition is when the grand jury is impaneled. Francis v. State, 636 S.W.2d 591 (Tex.App.—San Antonio 1982). If a defendant has an opportunity to challenge the array when it is impaneled and does not do so, he may not challenge it at a later date. Bird v. State, 692 S.W.2d 65 (Tex. Crim. App. 1985).

Where a defendant was indicted after the grand jury had been impaneled and had not had an opportunity to challenge the array under CCP Art. 19.27, he may do so by way of a pretrial motion to quash the indictment. Rodriguez v. State, 597 S.W.2d 917 (Tex. Crim. App. 1980); Romo v. State, 641 S.W.2d 390 (Tex.App.—Tyler 1982).

A challenge to the composition of the grand jury raised for the first time on appeal is untimely. Francis, *supra*.

A defendant's contention that the grand jury was improperly constituted because it included a convicted felon amounts to a contention that the indictment was not returned by a lawfully chosen or impaneled grand jury. Such a contention constitutes an objection to the form of the indictment. Caraway v. State, 911 S.W.2d 400 (Tex.App.—Texarkana 1995).

If the defendant does not object to a defect of form or substance in an indictment before trial on the merits commences, he waives the defect and may not complain of it on appeal or in any other post-conviction proceeding. State v. Yount, 853 S.W.2d 6 (Tex. Crim. App. 1993).

§11:100 Challenges to Indictment on Due Process/Equal Protection Grounds

It is a violation of the U.S. Constitution's equal protection clause for a defendant to be tried under an indictment issued by a grand jury from which persons of his race or color have been purposely excluded by the State. Hernandez v. Texas, 347 U.S. 475, 74 S.Ct. 667, 98 L.Ed.2d 866 (1954); Garcia v. State, 919 S.W.2d 370 (Tex. Crim. App. 1994); Cantu v. State, 842 S.W.2d 667 (Tex. Crim. App. 1992).

A criminal defendant has standing to challenge the system used to select his grand jury on the ground that it arbitrarily excludes from service the members of any race and therefore denies him due process of law, it not being necessary that the defendant show actual harm or bias. Peters v. Kiff, 407 U.S. 493, 92 S.Ct. 2163, 32 L.Ed.2d 83 (1972); Bird v. State, 692 S.W.2d 65 (Tex. Crim. App. 1985).

The gravamen of due process is that a fair cross-section of the community must be represented on grand juries. Taylor v. Louisiana, 419 U.S. 522, 95 S.Ct. 692, 42 L.Ed.2d 690 (1974).

The United States Supreme Court has set forth the elements of a prima facie case of discrimination in the selection of grand jurors. Castaneda v. Partida, 430 U.S. 482, 97 S.Ct. 1272, 51 L.Ed.2d 498 (1977). This standard has been followed in the Texas courts. Ovalle v. State, 13 S.W.3d 774 (Tex. Crim. App. 2000); Gutierrez v. State, 954 S.W.2d 86 (Tex.App.—San Antonio 1997).

Under Castaneda, the defendant must show that the selection system employed resulted in substantial under-representation of his race by establishing:

- Group is a recognizable, distinct class that has been singled out for different treatment under the laws
- Degree to which the group was under-represented by comparing the proportion of the group in the total population to the proportion called to serve as grand jurors over a significant period of time
- Selection process is either susceptible to abuse or is not racially neutral

Id.

Where the defendant is successful in making out a case of purposeful discrimination in the selection of his grand jury, the remedy is dismissal of the indictment. Flores v. State, 783 S.W.2d 793 (Tex. App.—El Paso 1990).

Once appellant establishes the under-representation of a class over a significant period of time by use of gross population statistics, he demonstrates a prima facie case of discriminatory purpose, and the State has the burden to rebut his case. Cerda v. State, 644 S.W.2d 875 (Tex.App.—Amarillo 1982).

Where the defendant fails to show the composition of the grand jury that indicted him, evidence showing the composition of the grand juries for preceding years is irrelevant. If the class to which a defendant belongs is fully represented on the indicting grand jury, the defendant suffers no injury and exclusion of members of the class from earlier grand juries is irrelevant to his case. Only if the defendant's class is substantially under-represented on the indicting grand jury does the make-up of prior grand juries become relevant to explain whether this under-representation on the indicting grand jury is a statistical accident or the result of purposeful discrimination. Espinoza v. State, 604 S.W.2d 908 (Tex. Crim. App. 1980).

§11:101 Recognizable Classes

Spanish-surnamed Americans of Mexican descent are recognized as a clearly identifiable class with a history of being subjected to discriminatory treatment. Hernandez v. Texas, 347 U.S. 475, 74 S.Ct. 667, 98 L.Ed. 866 (1954).

The 18-21 age group is not a discernible class for purposes of proportional representation analysis in the selection of grand juries. Curtis v. State, 573 S.W.2d 219 (Tex. Crim. App. 1978).

Persons over age sixty five are not a distinctive group for purposes of showing fair cross section violations. Weaver v. State, 823 S.W.2d 371 (Tex. App.—Dallas 1992, *pet. ref'd*).

Where the defendant fails to show the proportional representation of a discernible class within the total population of the county in question, he has failed to establish the first prong under Castaneda v. Partida. Ovalle v. State, 13 S.W.3d 774 (Tex. Crim. App. 2000). Burks v. State, 583 S.W.2d 389 (Tex. Crim. App. 1979).

Where the overwhelming preponderance of the evidence did not show that the defendant was a member of a discernible political class, he was unable to show that his grand jury was comprised in an unconstitutional manner. Carrillo v. State, 566 S.W.2d 902 (Tex. Crim. App. 1978).

§11:102 Degree of Under-Representation

The United States Supreme Court never has announced mathematical standards for the demonstration of a denial of constitutional due process and equal protection by the "systematic" exclusion of an identifiable class from jury service. Alexander v. Louisiana, 405 U.S. 625, 92 S.Ct. 1221, 31 L.Ed.2d 536 (1972).

The Court has, however, declared that under a racially non-neutral selection procedure a sufficiently large disparity over a significant period of time supports the presumption of unconstitutional purposeful discrimination raised by the statistical showing. Ovalle v. State, 13 S.W.3d 774 (Tex. Crim. App. 2000); Castaneda v. Partida, 430 U.S. 482, 97 S.Ct. 1272, 51 L.Ed.2d 498 (1977).

The court has held that disparities less than a mean disparity of 16.97% spanning eleven years, is significant enough to establish a prima facie case of discrimination against Mexican-Americans in grand jury selection. Cerda v. State, 644 S.W.2d 875 (Tex.App.—Amarillo 1982).

A disparity of only 1.35% between the percentage of grand jurors summoned and the percentage of Spanish-surnamed individuals in the available county population does not establish an equal protection violation in the discriminatory selection of grand jurors. Gutierrez v. State, 954 S.W.2d 86 (Tex.App.—San Antonio 1997).

Furthermore, a disparity of only 6.35% between the available population and those who actually served on the grand jury in the relevant year does not amount to an equal protection violation. Gutierrez.

Where the record below revealed that Hispanics constituted between 9.9% and 10.9% of the population of the relevant county at the time of indictment; and that during the three years prior to the indictment, Hispanics made up approximately 10.1% of the grand jurors impaneled, there was no systematic under-representation of Hispanics on the county's grand juries and the defendant has failed to satisfy the second prong of the Castaneda test. Cantu v. State, 842 S.W.2d 667 (Tex. Crim. App. 1992).

Where the record revealed that the discrepancy between the expected number and the actual number of Hispanics serving on the grand jury was less than three standard deviations, the defendant has failed to carry his burden. Ovalle, *supra*.

§11:103 Systems Susceptible to Abuse

The key-man system (*i.e.*, the grand jury commissioner system codified in CCP Art. 19.01) still employed in many counties for impaneling grand juries is susceptible to abuse as applied and its proof will satisfy the third prong of the Castaneda test. Carter v. Jury Comm'n., 396 U.S. 320, 90 S.Ct. 518, 24 L.Ed.2d 549 (1970).

§11:104 Second and Subsequent Grand Juries

In the absence of a constitutional or legislative prohibition, a district court acts within its authority in impaneling a second grand jury. State v. Broaddus, 3 S.W.3d 919 (Tex. Crim. App. 1999).

(This page intentionally left blank.)

Chapter 12

PRETRIAL MOTIONS

Table of Contents

§12:10 Filing and Litigating Pretrial Motions

Pretrial preparation and motion practice are the cornerstones to criminal litigation. Every pretrial motion that can advantage the defendant should be filed. However, numbers of motions alone do not equate into benefit for one's client. Evidence supporting the relief requested in the motion is always necessary to fully develop the statutory and constitutional relief sought. Relief should always be requested separately under the Texas and Federal Constitutions. Heitman v. State, 815 S.W.2d 681 (Tex. Crim. App. 1991); Muniz v. State, 852 S.W.2d 520 (Tex. Crim. App. 1993).

PRACTICE TIP: The Code of Criminal Procedure now requires clerks to accept electronic and digital multimedia evidence for filing from defendants where they allow filing of such matters by the prosecutor. See CCP Art. 2.21.

§12:11 Statutes

CCP Art. 27.11. Ten Days Allowed for Filed Pleadings

In all cases the defendant shall be allowed ten entire days, exclusive of all fractions of a day after his arrest, and during the term of the court, to file written pleadings.

CCP Art. 27.12. Time After Service

In cases where the defendant is entitled to be served with a copy of the indictment, he shall be allowed the ten days time mentioned in the preceding article to file written pleadings after such service.

CCP Art. 28.01. Pretrial

Section 1. The court may set any criminal case for a pretrial hearing before it is set for trial upon its merits, and direct the defendant and his attorney, if any of record, and the State's attorney, to appear before the court at the time and place stated in the court's order for a conference and hearing. The defendant must be present at the arraignment, and his presence is required during any pretrial proceeding. The pretrial hearing shall be to determine any of the following matters:

(1) Arraignment of the defendant, if such be necessary; and appointment of counsel to represent the defendant, if such be necessary;

(2) Pleadings of the defendant;

(3) Special pleas, if any;

(4) Exceptions to the form or substance of the indictment or information;

(5) Motions for continuance either by the State or defendant; provided that grounds for continuance not existing or not known at the time may be presented and considered at any time before the defendant announces ready for trial;

(6) Motions to suppress evidence; when a hearing on the motion to suppress evidence is granted, the court may determine the merits of said motion on the motions themselves, or upon opposing affidavits, or upon oral testimony, subject to the discretion of the court;

(7) Motions for change of venue by the State or the defendant; provided, however, that such motions for change of venue, if overruled at the pretrial hearing, may be renewed by the State or the defendant during the voir dire examination of the jury;

(8) Discovery;

(9) Entrapment; and

(10) Motion for appointment of interpreter.

Section 2. When a criminal case is set for such pretrial hearing, any such preliminary matters not raised or filed seven days before the hearing will not thereafter be allowed to be raised or filed, except by permission of the court for good cause shown; provided that the defendant shall have sufficient notice of such hearing to allow him not less than 10 days in which to raise or file such preliminary matters. The record made at such pretrial hearing, the rulings of the court and the exceptions and objections thereto shall become a part of the trial record of the case upon its merits.

Section 3. The notice mentioned in Section 2 above shall be sufficient if given in any one of the following ways:

(1) By announcement made by the court in open court in the presence of the defendant or his attorney of record;

(2) By personal service upon the defendant or his attorney of record;

(3) By mail to either the defendant or his attorney of record deposited by the clerk in the mail at least six days prior to the date set for hearing. If the defendant has no attorney of record, such notice shall be addressed to defendant at the address shown on his bond, if the bond shows such an address; and if not, it may be addressed to one of the sureties on his bond. If the envelope containing the notice is properly addressed, stamped and mailed, the state will not be required to show that it was received.

§12:12 Time Limits for Filing Pretrial Motions

Art. 28.01 §2 requires that all motions be filed seven days prior to the date set for a pretrial hearing so long as the defendant has been given ten days notice of the hearing date. Untimely motions will not thereafter be allowed to be raised or filed, except by permission of the court for good cause shown. Zone v. State, 84 S.W.3d 733 (Tex.App.—Houston [1st Dist.] 2002), affirmed at 118 S.W.3d 776 (Tex. Crim. App. 2003).

The seven-day time limitation under CCP Art. 28.01 Sec. 2 refers to the filing of pretrial motions, not affidavits filed in support of the motion. Zone.

Preliminary matters under Art. 28.01 include the appointment of counsel, pleadings of the defendant, any special pleas, exceptions to the form or substance of the indictment, motions for continuance, motions to suppress evidence, motions to change venue, discovery motions, entrapment issues, and motions for appointment of an interpreter. Zone.

pretrial motions of constitutional dimension may be filed any time prior to trial. A defendant's failure to file a motion of constitutional dimension under the time restrictions of Art. 28.01 does not constitute a waiver of the issue, and the trial court must grant the defendant a hearing. Foster v. State, 779 S.W.2d 845 (Tex. Crim. App. 1989) (defendant sought to pursue a motion for change of venue).

§12:12.1 Misdemeanors

Where no indictment is required, the defendant is allowed ten entire days after his arrest to file written pleadings. CCP Art. 27.11.

When the provisions of CCP Art. 27.11 are either formally or informally invoked, the statutory 10 days must be afforded the defendant. Refusal to grant such time will result in reversible error without the defendant being required to show harm. Ashcraft v. State, 900 S.W.2d 817 (Tex.App.—Corpus Christ 1983, *pet. ref'd*).

§12:12.2 Felonies

CCP Art. 27.12 provides that in felony cases, or anytime the prosecution must be by indictment, the defendant has 10 days from service of the indictment upon him to file written pleadings. Roberts v. State, 93 S.W.3d 528 (Tex.App.—Houston [14th Dist.] 2002, *pet. ref'd*).

Failure to request 10 days to file written pleadings waives any error. Young v. State, 752 S.W.2d. 235 (Tex.App.—Fort Worth 1988) *aff'd at* 796 S.W.2d 195 (Tex. Crim. App. 1990).

§12:13 Pretrial Hearings

The terms of CCP Art. 28.01 are not mandatory, and the question of whether to hold a hearing on a pretrial motion to suppress evidence rests within the sound discretion of the trial judge. Calloway v. State, 743 S.W.2d 645 (Tex. Crim. App. 1988); Morrison

v. State, 71 S.W.3d 821 (Tex.App.—Corpus Christi 2002, *no pet.*).

The trial court is not required by law to hold a pretrial hearing, and has the discretion to determine the merits of such motions as a motion *in limine* or a motion to suppress during the trial itself. Moore v. State, 700 S.W.2d 193 (Tex. Crim. App. 1985).

Article 28.01 requires formal notice of a pre-trial hearing only when the court designates a separate, pre-trial setting for the hearing—not when the judge elects for the hearing to take place on the trial setting itself. State v. Velasquez, 539 S.W.3d 289, 295 (Tex. Crim. App. 2018).

PRACTICE TIP: A technical decision should be made by trial counsel on whether to seek a hearing on a motion to suppress evidence or statement of the defendant, where counsel feels that there is a strong likelihood of a successful suppression. If the suppression issue is heard at a pretrial hearing and relief is granted, the state has a statutory right to appeal this finding if jeopardy has not attached if the counsel for the state can certify that the appeal is not taken for purposes of delay and that the evidence is of substantial importance. Counsel may want to urge the court to carry the motion and object during trial. Thus, the issue could be heard after the jury is seated and sworn and at which time jeopardy is attached. This procedure would effectively prevent a state's appeal of the court's order suppressing evidence.

§12:14 Form: Motion for Additional Pretrial Motions

See our companion book *Texas Criminal Forms* (James Publishing), Form 12-1, Motion to Allow Additional pretrial Motions.

§12:20 Expert Witness Assistance for Indigents

§12:21 Statutes

CCP Art. 26.05. Compensation of Counsel Appointed to Defend

(d) A counsel in a noncapital case, other than an attorney with a public defender's office, appointed to represent a defendant under this code shall be reimbursed for reasonable and necessary expenses, including expenses for investigation and for mental health and other experts. Expenses incurred with prior court approval shall be reimbursed in the same manner provided for capital cases by Articles 26.052(f) and (g), and expenses incurred without prior court approval shall be reimbursed in the manner provided for capital cases by Article 26.052(h).

CCP Art. 26.052. Appointment of Counsel in Death Penalty Case, Reimbursement of Investigative Expenses

(f) Appointed counsel may file with the trial court a pretrial *Ex parte confidential request for advance payment of expenses to investigate potential defenses. The request for expenses must state:*

(1) the type of investigation to be conducted;

(2) specific facts that suggest the investigation will result in admissible evidence; and

(3) an itemized list of anticipated expenses for each investigation.

(g) The court shall grant the request for advance payment of expenses in whole or in part if the request is reasonable. If the court denies in whole or in part the request for expenses, the court shall:

(1) state the reasons for the denial in writing;

(2) attach the denial to the confidential request; and

(3) submit the request and denial as a sealed exhibit to the record.

(h) Counsel may incur expenses without prior approval of the court. On presentation of a claim for reimbursement, the court shall order reimbursement of counsel for the expenses, if the expenses are reasonably necessary and reasonably incurred.

§12:22 Defendant Has a Right to an Expert

An indigent defendant has a right to a court-appointed expert under certain circumstances. Ake v. Oklahoma, 470 U.S. 68, 105 S.Ct. 1087, 84 L.Ed.2d 53 (1985); Robertson v. State, 21 S.W.3d 554 (Tex.App.—Waco 2000, *pet. ref'd*).

A defendant is not guaranteed the right to choose a particular expert witness; rather the defendant has the right to access a competent expert. Ake.

A defendant is also entitled to independent expert assistance, not an expert who is required to report to the court or prosecution. Taylor v. State, 939 S.W.2d 148 (Tex. Crim. App. 1996).

A defendant is entitled to an expert that can serve one or two objectives: (1) play a partisan role by providing the defendant with the means to challenge the state's case; or (2) testify in support of the defense theory, if the expert's opinion supports the theory. Taylor.

Ake clearly establishes that when its threshold criteria are met, a State must provide a mental health professional capable of performing a certain role: conducting an appropriate examination and assisting in evaluation, preparation, and presentation of the defense. Unless a defendant is assured the assistance of someone who can effectively perform these functions, he has not received the minimum to which Ake entitles him. McWilliams v. Dunn, ___ U.S. ___, 137 S.Ct. 1790, 1792, 198 L.Ed.2d 341 (2017).

Access to expert assistance means more than an examination by a neutral expert. It also means the appointment of an expert to provide technical assistance to the defendant, to help evaluate the strength of his defense, to offer expert diagnosis if it is favorable to the defense, and to identify the weaknesses in the state's case by testifying or preparing counsel to cross-examine the opposing expert. Rey.

In capital cases, the defendant does not have the right to receive funds for this purpose; rather counsel has the right to be reimbursed for reasonable expenses incurred with prior court approval. Cantu v. State, 939 S.W.2d 627 (Tex. Crim. App. 1997).

Ake applies to non-psychiatric experts. Rey v. State, 897 S.W.2d 333 (Tex. Crim. App. 1995).

A defendant is entitled to assistance from the following experts where he makes the proper showing:

- A chemist, where analysis of a purported controlled substance is central in the case. McBride v. State, 838 S.W.2d 248 (Tex. Crim. App. 1992).
- A mental health professional, where there is a viable insanity claim. De Freece v. State, 848 S.W.2d 150 (Tex. Crim. App. 1993).
- A pathologist, when manner of death would be a contested issue at trial. Rey v. State, 897 S.W.2d 333 (Tex. Crim. App. 1995).

However, a defendant is not entitled to expert assistance in the following situation:

- A jury selection consultant. Busby v. State, 990 S.W.2d 263 (Tex. Crim. App. 1999).
- A polygraph examiner in a post-trial investigation. Jackson v. State, 992 S.W.2d 469 (Tex. Crim. App. 1999).

Failure to provide an expert after the defendant has made the threshold showing of need constitutes structured error and is not subject to a harm analysis. Rey.

For additional cases pertaining to using and presenting expert witness assistance, *see* §16:60.

§12:23 Defendant Must Show Compelling Need

A defendant must show a compelling need for the assistance of an expert. Matchett v. State, 941 S.W.2d 922 (Tex. Crim. App.1996); Robertson v. State, 21 S.W.2d 554 (Tex.App.—Waco 2000, *pet. ref'd*).

It is crucial that a defendant seeking appointment

of an expert under Ake make a preliminary showing that the expert assistance is necessary to address a significant issue at trial. Moore v. State, 935 S.W.2d 124 (Tex. Crim. App. 1996); Robertson.

The defendant must show a particularized need for the expert assistance. Cantu v. State, 939 S.W.2d 627 (Tex. Crim. App. 1997).

To that extent, a defendant must offer more than undeveloped assertions that the requested assistance would be beneficial. Moore; Robertson.

A trial court must appoint an expert to assist the defendant if a three-part federal due process of law test is met:

- There exists a private interest that will be affected by the action of the state;
- Governmental interest will be affected if the safeguard is provided;
- Probable value of the additional safeguards and the risk of erroneous deprivation of the affected interest if those safeguards are not provided. Rey v. State, 897 S.W.2d 333 (Tex. Crim. App. 1995).

An indigent defendant upon proper request is entitled to make his threshold showing required to obtain expert assistance *ex parte*. Williams v. State, 958 S.W.2d 186 (Tex. Crim. App. 1997).

§12:24 Form: Motions

See our companion book *Texas Criminal Forms* (James Publishing):

- Form 13-31, Motion for Approval of Expert Witness Expenses.
- Form 13-32, Motion for Independent Forensic Testing.
- Form 13-33.1, Motion for Defendant's Access to Physical Evidence.
- Form 13-33.2, Motion for Forensic Identity Testing.

§12:30 Investigative Assistance for Indigents

§12:31 Statutes

CCP Art. 26.05. Compensation of Counsel Appointed to Defend

(d) A counsel in a noncapital case, other than an attorney with a public defender's office, appointed to represent a defendant under this code shall be reimbursed for reasonable and necessary expenses, including expenses for investigation and for mental health and other experts. Expenses incurred with prior court approval shall be reimbursed in the same manner provided for capital cases by Articles 26.052(f) and (g), and expenses incurred without prior court approval shall be reimbursed in the manner provided for capital cases by Article 26.052(h).

CCP Art. 26.052. Appointment of Counsel in Death Penalty Case, Reimbursement of Investigative Expenses

(f) Appointed counsel may file with the trial court a pretrial *Ex parte confidential request for advance payment of expenses to investigate potential defenses. The request for expenses must state:*

(1) the type of investigation to be conducted;

(2) specific facts that suggest the investigation will result in admissible evidence; and

(3) an itemized list of anticipated expenses for each investigation.

(g) The court shall grant the request for advance payment of expenses in whole or in part if the request is reasonable. If the court denies in whole or in part the request for expenses, the court shall:

(1) state the reasons for the denial in writing;

(2) attach the denial to the confidential request; and

(3) submit the request and denial as a sealed exhibit to the record.

(h) Counsel may incur expenses without prior approval of the court. On presentation of a claim for reimbursement, the court shall order reimbursement of counsel for the expenses, if the expenses are reasonably necessary and reasonably incurred.

§12:32 Investigative Expenses Are Within the Discretion of the Court

The allowance of investigative expenses is a matter within the sound discretion of the trial court, and an abuse of that discretion will not be found absent a showing of harm. Castillo v. State, 739 S.W.2d 280 (Tex. Crim. App. 1987).

§12:33 Record Must Show Specific Need

The record must reflect some specific need for the funds, such as the need for a particular expert or witness, and in what manner the defendant will be harmed if the funds are not provided. Phillips v. State, 701 S.W.2d 875 (Tex. Crim. App. 1985).

§12:34 Counsel's Duty to Investigate Mitigation

Defense counsel's failure to investigate the basis of his client's mitigation defense can amount to ineffective assistance of counsel. In determining whether counsel conducted a reasonable investigation, an appellate court's initial inquiry is whether a reasonable investigation should have uncovered the mitigating evidence. Ex parte Gonzales, 204 S.W.3d 391 (Tex. Crim. App. 2006).

This is also true in a non-capital case. *See* Wiggins v. Smith, 539 U.S. 510, 123 S.Ct. 2527, 156 L.Ed.2d 471 (2003).

In a non-death case, counsel can be found ineffective for failure to present character evidence where the record shows that the witnesses would have been able to offer mitigating testimony. Milburn v. State, 15 S.W.3d 267 (Tex.App.—Houston [14th Dist.] 2000, pet. ref'd); Shanklin v. State, 190 S.W.3d 154 (Tex.App.—Houston [1st Dist.] 2005), *pet.dism'd* 211 S.W.3d 315 (Tex. Crim. App. 2007).

A defense attorney's decision not to present particular witnesses at the punishment stage may be a strategically sound decision if the attorney bases it on a determination that the testimony of the witnesses may be harmful, rather than helpful, to the defendant. However, a failure to uncover and present mitigating evidence cannot be justified as a tactical decision when defense counsel has not conducted a thorough investigation of the defendant's background. Shanklin.

If counsel makes a decision to not present potentially mitigating evidence after a full investigation because it could lead to the presentation of aggravating evidence against the defendant, the decision will most likely be upheld as objectively reasonable. Ex parte Woods, 176 S.W.3d 224 (Tex. Crim. App. 2005).

§12:35 Form: Motions

See our companion book *Texas Criminal Forms* (James Publishing):

- Form 13-34.1, Motion for Approval of Funds for Court-Appointed Investigator.
- Form 13-34.2, Motion for Approval of Mitigation Specialist Expenses.

§12:40 Recusal and Disqualification of Trial Judge

A Texas judge may be removed from presiding over a case for one of three reasons: he is constitutionally disqualified; he is subject to a statutory strike (as in civil cases); or he is subject to statutory disqualification or recusal under Texas Supreme Court rules. Gaal v. State, 332 S.W.3d 448 (Tex. Crim. App. 2011).

§12:41 Constitution and Statutes

Tex. Const. Art. 5, §11. Disqualification of Judges; Exchange of Districts; Holding Court for Other Judges

No judge shall sit in any case wherein the judge may be interested, or where either of the parties may be connected with the judge, either by affinity or consanguinity, within such a degree as may be prescribed by law, or when the judge shall have been counsel in the case. When the Supreme Court, the Court of Criminal Appeals, the Court of Appeals, or any member of any of those courts shall be thus disqualified to hear and determine any case or cases in said court, the same shall be certified to the Governor of the State, who shall immediately commission the requisite number of persons learned in the law for the trial and determination of such cause or causes. When a judge of the District Court is disqualified by any of the causes above stated, the parties may, by consent, appoint a proper person to try said case; or upon their failing to do so, a competent person may be appointed to try the same in the county where it is pending, in such manner as may be prescribed by law.

And the District Judges may exchange districts, or hold courts for each other when they may deem it expedient, and shall do so when required by law. This disqualification of judges of inferior tribunals shall be remedied and vacancies in their offices filled as may be prescribed by law.

CCP Art. 30.01. Causes Which Disqualify

No judge or justice of the peace shall sit in any case where he may be the party injured, or where he has been of counsel for the State or the accused, or where the accused or the party injured may be connected with him by consanguinity or affinity within the third degree, as determined under Chapter 573, Government Code.

CCP Art. 30.02. District Judge Disqualified

Whenever any case is pending in which the district judge or criminal district judge is disqualified from trying the case, no change of venue shall be made necessary thereby; but the judge presiding shall certify that fact to the presiding judge of the administrative judicial district in which the case is pending and the presiding judge of such administrative judicial district shall assign a judge to try such case in accordance with the provisions of Article 200a, V.A.C.S.

Tex. Gov't Code §21.005. Disqualification

A judge or a justice of the peace may not sit in a case if either of the parties is related to him by affinity or consanguinity within the third degree, as determined under Chapter 573.

Tex. R. Civ. Pro. 18a. Recusal or Disqualification of Judges

(a) At least 10 days before the date set for trial or other hearing in any court other than the Supreme Court, the Court of Criminal Appeals or the Court of Appeals, any party may file with the clerk of the court a motion stating grounds why the judge before whom the case is pending should not sit in the case. The grounds may include any disability of the judge to sit in the case. The motion shall be verified and must state with particularity the grounds why the judge before whom the case is pending should not sit. The motion shall be made on personal knowledge and shall set forth such facts as would be admissible in evidence provided that facts may be stated upon information and belief if the grounds of such belief are specifically stated.

(b) On the day the motion is filed, copies shall be served on all other parties or their counsel of record, together with a notice that movant expects the motion to be presented to the judge three days after the filing of such motion unless otherwise ordered by the judge. Any other party may file with the clerk an opposing or concurring statement at any time before the motion is heard.

(c) Prior to any further proceedings in the case, the judge shall either recuse himself or request the presiding judge of the administrative judicial district to assign a judge to hear such motion. If the judge recuses himself, he shall enter

an order of recusal and request the presiding judge of the administrative judicial district to assign another judge to sit, and shall make no further orders and shall take no further action in the case except for good cause stated in the order in which such action is taken.

(d) If the judge declines to recuse himself, he shall forward to the presiding judge of the administrative judicial district, in either original form or certified copy, an order of referral, the motion, and all opposing and concurring statements. Except for good cause stated in the order in which further action is taken, the judge shall make no further orders and shall take no further action in the case after filing of the motion and prior to a hearing on the motion. The presiding judge of the administrative judicial district shall immediately set a hearing before himself or some other judge designated by him, shall cause notice of such hearing to be given to all parties or their counsel, and shall make such other orders including orders on interim or ancillary relief in the pending cause as justice may require.

(e) If within 10 days of the date set for trial or other hearing a judge is assigned to a case, the motion shall be filed at the earliest practicable time prior to the commencement of the trial or other hearing.

(f) If the motion is denied, it may be reviewed for abuse of discretion on appeal from the final judgment. If the motion is granted, the order shall not be reviewable, and the presiding judge shall assign another judge to sit in the case.

(g) The Chief Justice of the Supreme Court may also appoint and assign judges in conformity with this rule and pursuant to statute.

(h) If a party files a motion to recuse under this rule and it is determined by the presiding judge or the judge designated by him at the hearing and on motion of the opposite party, that the motion to recuse is brought solely for the purpose of delay and without sufficient cause, the judge hearing the motion may, in the interest of justice, impose any sanction authorized by Rule 215(2)(b).

Tex. R. Civ. Pro. 18b. Grounds for Disqualification and Recusal of Judges

(1) Disqualification

Judges shall disqualify themselves in all proceedings in which:

(a) they have served as a lawyer in the matter in controversy, or a lawyer with whom they previously practiced law served during such association as a lawyer concerning the matter; or

(b) they know that, individually or as a fiduciary, they have an interest in the subject matter in controversy; or

(c) either of the parties may be related to them by affinity or consanguinity within the third degree.

(2) Recusal

A judge shall recuse himself in any proceeding in which:

(a) his impartiality might reasonably be questioned;

(b) he has a personal bias or prejudice concerning the subject matter or a party, or personal knowledge of disputed evidentiary facts concerning the proceeding;

(c) he or a lawyer with whom he previously practiced law has been a material witness concerning it;

(d) he participated as counsel, adviser or material witness in the matter in controversy, or expressed

an opinion concerning the merits of it, while acting as an attorney in government service;

(e) he knows that he, individually or as a fiduciary, or his spouse or minor child residing in his household, has a financial interest in the subject matter in controversy or in a party to the proceeding, or any other interest that could be substantially affected by the outcome of the proceeding;

(f) he or his spouse, or a person within the third degree of relationship to either of them, or the spouse of such a person:

(i) is a party to the proceeding, or an officer, director, or trustee of a party;

(ii) is known by the judge to have an interest that could be substantially affected by the outcome of the proceeding;

(iii) is to the judge's knowledge likely to be a material witness in the proceeding.

(g) he or his spouse, or a person within the first degree of relationship to either of them, or the spouse of such a person, is acting as a lawyer in the proceeding.

(3) A judge should inform himself about his personal and fiduciary financial interests, and make a reasonable effort to inform himself about the personal financial interests of his spouse and minor children residing in his household.

(4) In this rule:

(a) "proceeding" includes pretrial, *trial, or other stages of litigation;*

(b) the degree of relationship is calculated according to the civil law system;

(c) "fiduciary" includes such relationships as executor, administrator, trustee, and guardian;

(d) "financial interest" means ownership of a legal or equitable interest, however small, or a relationship as director, advisor, or other active participant in the affairs of a party, except that:

(i) ownership in a mutual or common investment fund that holds securities is not a "financial interest" in such securities unless the judge participates in the management of the fund;

(ii) an office in an educational, religious, charitable, fraternal, or civic organization is not a "financial interest" in securities held by the organization;

(iii) the proprietary interest of a policyholder in a mutual insurance company, of a depositor in a mutual savings association, or a similar proprietary interest, is a "financial interest" in the organization only if the outcome of the proceeding could substantially affect the value of the interest;

(iv) ownership of government securities is a "financial interest" in the issuer only if the outcome of the proceeding could substantially affect the value of the securities;

(v) an interest as a taxpayer or utility ratepayer, or any similar interest, is not a "financial interest" unless the outcome of the proceeding could substantially affect the liability of the judge or a person related to him within the third degree more than other judges.

(5) The parties to a proceeding may waive any ground for recusal after it is fully disclosed on the record.

(6) If a judge does not discover that he is recused under subparagraphs (2)(e) or (2)(f)(iii) until after he has devoted substantial time to the matter, he is not required to recuse himself if he or the person related to him divests himself of the interest that would otherwise require recusal.

§12:42 Difference Between Recusal and Disqualification

Generally, a motion to recuse seeks to prevent a judge from hearing a case because of a nonconstitutional reason, while grounds for disqualification are limited to those identified in the constitution. Gaal v. State, 332 S.W.3d 448 (Tex. Crim. App. 2011).

Judges may be removed from a particular case either because they are disqualified or because they are recused. The grounds and procedures for each type of removal are fundamentally different. *See* In re Union Pacific Resources Co., 969 S.W.2d 427 (Tex. 1998).

§12:43 Recusal of Trial Judge

§12:43.1 Application of Civil Rule

Rule 18b(2) of the Texas Rules of Civil Procedure sets out the law concerning recusal and includes instances in which a judge must step down from hearing a case for reasons other than the disqualifying grounds listed in the constitution. Rule 18b(2) states, in relevant part, that a "judge shall recuse himself in any proceeding in which: (a) his impartiality might reasonably be questioned; [or] (b) he has a personal bias or prejudice concerning the subject matter or a party, or personal knowledge of disputed evidentiary facts concerning the proceeding[.]" Gaal v. State, 332 S.W.3d 448 (Tex. Crim. App. 2011).

Tex.R.Civ.P. 18a applies to criminal cases absent any explicit or implicit legislative intent indicating otherwise. Arnold v. State, 853 S.W.2d 543 (Tex. Crim. App. 1993); Sanchez v. State, 926 S.W.2d 391 (Tex.App.—El Paso 1996, *pet. ref'd*).

Rule 18a also applies in habeas proceedings that occur before the trial court. Ex parte Sinegar, 324 S.W.3d 578 (Tex. Crim. App. 2010).

The Due Process Clause may sometimes demand recusal even when a judge has no actual bias. Recusal is required when, objectively speaking, the probability of actual bias on the part of the judge or decision maker is too high to be constitutionally tolerable. Rippo v. Baker, 580 U.S. ___, 137 S.Ct. 905, 907, 197 L.Ed.2d 167 (2017).

A judge's remarks during trial that are critical, disapproving, or hostile to counsel, the parties, or their cases, usually will not support a bias or partiality challenge, although they may do so if they reveal an opinion based on extrajudicial information, and they will require recusal if they reveal such a high degree of favoritism or antagonism as to make fair judgment impossible. On the other hand, expressions of impatience, dissatisfaction, annoyance, and even anger, that are within the bounds of what imperfect men and women may display, do not establish bias or partiality. Gaal.

Judicial bias or partiality derived from opinions formed by the judge on the basis of facts introduced or events occurring in the course of the proceedings may be grounds for a recusal motion where they evidence a deep-seated favoritism or antagonism that would make fair judgment impossible. Gaal.

Merely designating someone as a "friend" on Facebook does not show the degree or intensity of a judge's relationship with a person such that the judge's impartiality might reasonably be questioned, or that the judge has a personal bias or prejudice concerning the subject matter or a party to the extent requiring the judge to recuse himself. Youkers v. State, 400 S.W.3d 200, 206 (Tex. App. —Dallas 2013, *pet.ref'd*).

§12:43.2 Motion to Recuse

The procedural requisites for recusal in Rule 18a(a) are mandatory, and a party who fails to conform waives his right to complain of a judge's failure to recuse himself. Arnold v. State, 853 S.W.2d 543 (Tex. Crim. App. 1993)

Rule 18a has a 10-day notice provision. Unless the motion to recuse is filed at least 10 days before trial, it is not sufficient to trigger the provisions of Rule 18a. Arnold.

However, where the defendant does not know of the possible grounds for recusal until well into trial, the 10-day rule of Rule 18a does not apply. Martin v. State, 876 S.W.2d 396 (Tex.App.—Fort Worth 1994, *no pet.*).

A motion to recuse is the proper vehicle to challenge any alleged bias of the trial judge in the case. Sanchez v. State, 926 S.W.2d 391 (Tex. App.—El Paso 1996, *pet. ref'd*).

In order to recuse a judge, any party may file a motion stating grounds why the judge before whom the case is pending should not sit in the case. Tex.R.Civ.P. 18a(a).

The grounds must be stated with particularity. Vargas v. State, 883 S.W.2d 256 (Tex.App.—Corpus Christi 1994, *pet. ref'd*).

A motion to recuse must be verified in order to comply with Rule 18a. Vargas; Moorhead v. State, 972 S.W.2d 93 (Tex.App.—Texarkana 1998, *no pet.*).

Rule 18a(b) requires that, on the day the motion is filed, copies be served upon all other parties or their counsel of record and that the motion be accompanied by "a notice that the movant expects the motion to be presented to the judge three days after the filing of such motion unless otherwise ordered by the judge." Ex parte Sinegar, 324 S.W.3d 578 (Tex. Crim. App. 2010).

§12:43.3 Duty of the Trial Judge

When a recusal motion is filed, the trial judge against whom the motion is directed may properly make an initial decision of whether the motion is sufficient to invoke Rule 18a. *See* Arnold (no error in not referring motion to recuse not timely filed); McDuffie v. State, 854 S.W.2d 195 (Tex.App.—Beaumont 1993, *pet. ref'd*).

Once a sufficient motion to recuse has been filed, before proceeding further in the case, the judge must either recuse himself or request the presiding judge of the administrative judicial district to assign a judge to hear the motion under Tex. Gov't Code §74.059(c)(3). Sanchez.

If a recusal motion is timely filed and facially sufficient, a hearing must be held. Sanchez.

Where it is found that a party was erroneously denied the right to a hearing on their motion to recuse, the remedy is to abate the case back to the trial court so that such hearing can be held. Sanchez.

A judge should recuse himself where he has a financial interest in the outcome of a case, even though the interest is less than what would have been considered personal or direct at common law. Caperton et.al. v. A.T. Massey Coal Co. et.al., 556 U.S. ___, 129 S.Ct. 2252, 173 L.Ed.2d 1208 (2009) citing Tumey v. Ohio, 273 U. S. 510 (1927).

A judge who participated as counsel in an earlier proceeding of the same case should also recuse himself. Caperton et.al. v. A.T. Massey Coal Co. et.al. (citing In re Murchison, 349 U. S. 133, 75 S.Ct. 623, 99 L.Ed. 942 (1955)).

In addition to the above two reasons for recusal, there are objective standards that require recusal under the due process clause when the probability of actual bias on the part of the judge or decisionmaker is too high to be constitutionally tolerable. Caperton et.al. v. A.T. Massey Coal Co. et.al. (citing Withrow v. Larkin, 421 U. S. 35 (1975)).

A judge does not have authority to take any further action after signing a voluntary recusal order. Ex parte Thuesen, 546 S.W.3d 145, 147 (Tex. Crim. App. 2017).

Interim or ancillary orders made by a regional presiding judge regarding the recusal of a trial judge—or the reinstatement or modification of judicial authority following a recusal—must be made in writing, signed by the presiding judge, and entered of record in the case. Ex parte Thuesen, 546 S.W.3d at 156.

§12:43.4 Appeals

If the motion to recuse is denied, the standard for review is abuse of discretion, and the denial may be reviewed on appeal from the final judgment. Tex.R.Civ.P. 18a(f); Thomas v. Walker, 860 S.W.2d 579 (Tex.App.—Waco 1993, *no writ*).

§12:44 Form: Motions and Order

See our companion book *Texas Criminal Forms* (James Publishing):

- Form 15-7, Motion for Recusal of Trial Judge.

- Form 15-8, Motion Concurring in Motion to Recuse Trial Judge.
- Form 15-9, Order of Referral to Presiding Administrative Judge.

§12:45 Disqualification of Trial Judge

§12:45.1 Basis for Motion

The grounds for disqualification of a judge stated in the Texas Constitution and the Texas Code of Criminal Procedure are exclusive. McDuffie v. State, 854 S.W.2d 195 (Tex.App.—Beaumont 1993, *pet. ref'd*).

Under Tex. Const. Art. V, §11 where a party shows that the trial judge has a bias in the case that is of such a nature and to such an extent as to deny the party due process of law, the judge will be disqualified. McClenan v. State, 661 S.W.2d 108 (Tex. Crim. App. 1983); McDuffie.

§12:45.2 Movant's Burden

The movant on a motion to disqualify, alleging that the trial judge is biased, has the burden to provide facts sufficient to establish that a reasonable man, knowing all the circumstances involved, would harbor doubts as to the impartiality of the trial judge. Kemp v. State, 846 S.W.2d 289 (Tex. Crim. App. 1992).

A motion to disqualify the judge must be *prima facie* adequate and state proper grounds upon which disqualification can be based in order for the trial judge to proceed to the next step of disqualifying himself or referring the motion to an administrative judge for hearing. McClenan v. State, 661 S.W.2d 108 (Tex. Crim. App. 1983).

Furthermore, before alleged bias becomes sufficient to warrant the disqualification of a judge, the movant must show that the bias stems from an extrajudicial source and results in an opinion on the merits on some basis other than what the judge learned from his participation in the case. United States v. Grinnell Corp., 384 U.S. 563, 86 S.Ct. 1698, 16 L.Ed.2d 778 (1966); Kemp.

In a situation where the presiding judge is ineligible to preside over the case, the defendant must move to have the judge disqualified as soon as he knows or should have known that an error has occurred. Lackey v. State, 364 S.W.3d 837 (Tex. Crim. App. 2012) (where the defendant preserved error by filing a subsequent pretrial motion to set aside the ineligible judge's rulings).

§12:45.3 Grounds

The opening clause of CCP Art. 30.01 means that a trial court judge, in any particular criminal prosecution, may be the party injured, and is therefore disqualified from presiding, if the evidence shows that he was among the defendant's victims in the criminal transaction or episode at issue, such that a reasonable person would harbor doubts as to the judge's impartiality. Whitehead v. State, 273 S.W.3d 285 (Tex. Crim. App. 2008).

A trial judge's arbitrary refusal to consider the full range of punishment in a case is an instance where a party's due process rights would be abridged and the judge would be disqualified. McClenan v. State, 661 S.W.2d 108 (Tex. Crim. App. 1983).

The mere fact that a trial judge issued a defendant's search or arrest warrant does not establish bias against that defendant in a subsequent criminal proceeding. Kemp v. State, 846 S.W.2d 289 (Tex. Crim. App. 1992).

If a trial judge participated in any manner in the preparation and investigation of the case, he would be a counsel for the state and hence disqualified. Gamez v. State, 737 S.W.2d 315 (Tex. Crim. App. 1987).

Where a judge personally participated as a prosecutor in a hearing on the case and consented to the defendant's waiver of a right during the hearing, she is disqualified as having actively participated in the case. Metts v. State, 510 S.W.3d 1 (Tex. Crim. App. 2016).

A motion to disqualify should be handled under the same procedural requirements as is a motion to recuse. Stafford v. State, 948 S.W.2d 921, (Tex. App.—Texarkana 1997, *pet. ref'd*).

The proceeding imposing probation and the proceeding revoking probation are the same case for the purpose of the disqualification provisions, and a judge is statutorily and constitutionally disqualified from presiding over a revocation hearing if he had prosecuted the original action. Ex parte Miller, 696 S.W.2d 908 (Tex. Crim. App. 1985).

A trial judge is disqualified from acting where

he sua sponte obtained personal knowledge about contested facts and used that personal knowledge in ruling on a motion to suppress. Haworth v. State, 533 S.W.3d 484, 487 (Tex. App.—Corpus Christi 2017, *no pet.*).

§12:45.4 Appeal

The refusal of a defendant's motion to disqualify is reviewable only for abuse of discretion. Tex.R.Civ.P. 18a(f).

The failure to disqualify the trial judge is an issue that can be raised for the first time on appeal. Whitehead v. State, 273 S.W.3d 285 (Tex. Crim. App. 2008).

Thus, an appellate court should not reverse a trial judge whose ruling on the motion was within the zone of reasonable disagreement. Kemp v. State, 846 S.W.2d 289 (Tex. Crim. App. 1992).

Disqualification may be raised for the first time in a collateral attack on the judgment. *See* Lee v. State, 555 S.W.2d 121 (Tex. Crim. App. 1977); In re K.E.M., 89 S.W.3d 814 (Tex.App.—Corpus Christi 2002, *no pet.*).

However, disqualification cannot be raised on post-trial habeas where the defendant was aware of the grounds and could have made the challenge at the trial court or on appeal. Ex parte Richardson, 201 S.W.3d 712 (Tex. Crim. App. 2006).

§12:46 Objecting to Judicial Assignments

A defendant must object to a judicial assignment prior to trial. Wilson v. State, 977 S.W.2d 379 (Tex. Crim. App. 1998).

§12:47 Form: Motions and Order

See our companion book *Texas Criminal Forms* (James Publishing):

- Form 15-10, Motion for Disqualification of Trial Judge.
- Form 15-11, Motion Concurring in Motion for Disqualification of Trial Judge.
- Form 15-12, Order of Referral to Presiding Administrative Judge.

§12:50 Change of Venue

§12:51 Statutes

CCP Art. 27.15. Change of Venue to Plead Guilty

When in any county which is located in a judicial district composed of more than one county, a party is charged with a felony and the maximum punishment therefore shall not exceed 15 years, and the district court of said county is not in session, such party may, if he desires to plead guilty, or enter a plea of nolo contendere, make application to the district judge of such district for a change of venue to the county in which said court is in session, and said district judge may enter an order changing the venue of said cause to the county in which the court is then in session, and the defendant may plead guilty or enter a plea of nolo contendere to said charge in said court to which the venue has been changed.

CCP Art. 31.01. On Court's Own Motion

Whenever in any case of felony or misdemeanor punishable by confinement, the judge presiding shall be satisfied that a trial, alike fair and impartial to the accused and to the State, cannot, from any cause, be had in the county in which the case is pending, he may, upon his own motion, after due notice to accused and the State, and after hearing evidence thereon, order a change of venue to any county in the judicial district in which such county is located or in an adjoining district, stating in his order the grounds for such change of venue. The judge, upon his own motion, after 10 days notice to the parties or their counsel, may order a change of venue to any county beyond an adjoining district; provided, however, an order changing venue to a county beyond an adjoining district shall be grounds for reversal if, upon timely contest by the defendant, the record of the contest affirmatively shows that any county in his own and the adjoining district is not subject to the same conditions which required the transfer.

CCP Art. 31.02. State May Have

Whenever the district or county attorney shall represent in writing to the court before which any felony or misdemeanor case punishable by confinement, is pending, that, by reason of existing combinations or influences in favor of the accused, or on account of the lawless condition of affairs in the county, a fair and impartial trial as between the accused and the State cannot be safely and speedily had; or whenever he shall represent that the life of the prisoner, or of any witness, would be jeopardized by a trial in the county in which the case is pending, the judge shall hear proof in relation thereto, and if satisfied that such representation is well-founded and that the ends of public justice will be subserved thereby, he shall order a change of venue to any county in the judicial district in which such county is located or in an adjoining district.

CCP Art. 31.03. Granted on Motion of Defendant

(a) A change of venue may be granted in any felony or misdemeanor case punishable by confinement on the written motion of the defendant, supported by his own affidavit and the affidavit of at least two credible persons, residents of the county where the prosecution is instituted, for either of the following causes, the truth and sufficiency of which the court shall determine:

(1) That there exists in the county where the prosecution is commenced so great a prejudice against him that he cannot obtain a fair and impartial trial; and

(2) That there is a dangerous combination against him instigated by influential persons, by reason of which he cannot expect a fair trial.

An order changing venue to a county beyond an adjoining district shall be grounds for reversal, if upon timely contest by defendant, the record of the contest affirmatively shows that any county in his own and the adjoining district is not subject to the same conditions which required the transfer.

(b) For the convenience of parties and witnesses, and in the interest of justice, the court upon motion of the defendant and with the consent of the attorney for the state may transfer the proceeding as to him to another district.

(c) The court upon motion of the defendant and with the consent of the attorney for the state may transfer the proceedings to another district in those cases wherein the defendant stipulates that a plea of guilty will be entered.

CCP Art. 31.04. Motion May Be Controverted

The credibility of the persons making affidavit for change of venue, or their means of knowledge, may be attacked by the affidavit of a credible person. The issue thus formed shall be tried by the judge, and the motion granted or refused, as the law and facts shall warrant.

§12:52 Motion Filing Deadlines Do Not Apply

Because a change of venue is a question of constitutional dimension, the pretrial filing deadlines of Art. 28.01 do not apply to change of venue motions. Revia v. State, 649 S.W.2d 625 (Tex. Crim. App. 1983); Gutierrez v. State, 979 S.W.2d 659 (Tex. Crim. App. 1998).

A defendant may file his change of venue motion after a jury has been selected in his case. Bell v. State, 938 S.W.2d 35 (Tex. Crim. App. 1996).

§12:53 Court's Motion for Change of Venue

Under CCP Art. 31.01, whenever a judge is

satisfied that a trial, fair and impartial to the defendant and state, cannot, from any cause, be had in the county in which the case is pending, he may on his own motion, change venue. Brimage v. State, 918 S.W.2d 466 (Tex. Crim. App. 1996) (*op. on rehearing*).

Where the state has filed a motion alternatively asking the court to *sua sponte* change venue, the court's own motion is not required. Aranda v. State, 736 S.W.2d 702 (Tex. Crim. App. 1987).

It will be presumed on appeal that the trial court had a sufficient factual basis for changing venue on its own motion unless the defendant brings proof, formally or informally, to refute the court's finding that a fair trial cannot be had. Brimage.

§12:53.1 Procedure

Under CCP Art. 31.01, the trial court may transfer venue on its own motion where: (1) due notice is given to the state and defendant; (2) the court hears evidence on its motion; (3) venue is changed to any county in the judicial district in which the original county is located or to an adjoining district; (4) if due notice is given to both the state and defendant, venue may be changed to a county beyond the adjoining district; and (5) the court must state in its order the grounds why a fair and impartial trial cannot be had in the county. *See* Brimage v. State, 918 S.W.2d 466 (Tex. Crim. App. 1996) (*op. on rehearing*).

Change of venue beyond an adjoining district shall be grounds for reversal if the defendant, on timely contest, affirmatively shows on the record that any county in his own or the adjoining district is not subject to the same conditions which required the transfer. CCP Art. 31.01.

The purpose of CCP Art. 31.01 is to require a court, once satisfied that a fair trial cannot be had, to give notice to both parties of its intention to change venue and to hold a hearing allowing either party to offer evidence either in support of or against the court's proposed change of venue. The statute does not require the court to offer evidence in support of its own motion, but rather merely offer the parties a chance to be heard on the matter. The court is only required to state in its order the grounds for its decision to change venue. *See* Brimage.

CCP Arts. 31.01 to 31.03 only require that a requested change of venue hearing be held before the trial is completed. A trial court does not commit error by holding in abeyance a ruling in a motion to change venue until after jury selection. Smith v. State, 850 S.W.2d 275 (Tex.App.—Fort Worth 1993, pet. ref'd).

§12:54 State's Motion for Change of Venue

Whenever the prosecuting attorney shall represent in writing that, by reason of existing combinations or influences in favor of the defendant, or on account of the lawless condition of affairs in the county, a fair and impartial trial as between the defendant and the State cannot be fairly and speedily had, or whenever he shall represent that the life of the prisoner or any witness would be jeopardized by a trial in the original county, a change of venue may be had. CCP Art. 31.02.

§12:54.1 Procedure

The procedure for the state obtaining a change of venue is:

(1) Upon the filing of a written motion, the court shall hear proof in relation to the prosecutor's representations.

(2) The court must be satisfied that the prosecutor's representations are well-founded and that the ends of justice will be served by a change of venue.

(3) The court may change venue to any county in the judicial district in which the original county is located or to an adjoining district.

CCP Art. 31.02.

The trial court must conduct an evidentiary hearing on the state's motion before changing venue thereon. Garza v. State, 974 S.W.2d 251 (Tex.App.—San Antonio, *pet. ref'd*).

While the defendant is required to attach supporting affidavits to his motion for change of venue under Art. 31.03, the state is not required to attach such affidavits to its motion under Art. 31.02. Garza.

§12:54.2 Contesting State's Motion

The defendant may contest the state's written representations and should do so by controverting affidavit. CCP Art. 31.04.

§12:55 Defendant's Motion for Change of Venue

Under CCP Art. 31.03, a change of venue may be had on the written motion of the defendant supported by his affidavit and the affidavit of at least two credible persons who are residents of the county where the prosecution is instituted. *See* Willingham v. State, 897 S.W.2d 351 (Tex. Crim. App. 1995).

§12:55.1 Procedure

A trial court may grant a change of venue on the written motion of the defendant in any felony or misdemeanor case punishable by confinement. The motion must be supported by the defendant's own affidavit and the affidavits of two credible persons, residents of the county of prosecution, stating that there exists in the county of prosecution so great a prejudice against him that he cannot obtain a fair and impartial trial. Taylor v. State, 93 S.W.3d 487 (Tex.App.—Texarkana 2002, *pet. ref'd*).

The defendant's motion must be in writing, sworn to by the defendant, supported by affidavits, and timely filed with the court. Cover v. State, 913 S.W.2d 611 (Tex.App.—Tyler 1995, *pet. ref'd*).

§12:55.2 Defendant's Affidavits

The motion must be supported by the affidavit of two credible citizens of the county, otherwise it is defective. Cover v. State, 913 S.W.2d 611 (Tex. App.—Tyler 1995, *pet. ref'd*).

Failure to file the defendant's affidavit and at least two others is a fatal defect. Cover.

The defendant's attorney cannot serve as one of the defendant's affiants. DeLeon v. State, 758 S.W.2d 621 (Tex.App.—Houston [14th Dist.] 1988, *no pet.*).

There must be an assertion in the affidavits that the affiants are residents of the county. Norwood v. State, 486 S.W.2d 776 (Tex. Crim. App. 1972).

A trial court does not abuse its discretion where it denies a change of venue motion that does not contain the required two affidavits. Horner v. State, 129 S.W.3d 210 (Tex.App.—Corpus Christi 2004, pet. ref'd).

A defendant's affidavits must contain more than conclusory allegations that he has been prejudiced or that a dangerous combination is working to deny him a fair trial. Buntion v. State, 482 S.W.3d 58, 74 (Tex. Crim. App. 2016).

§12:55.3 Form: Motion and Affidavit

See our companion book *Texas Criminal Forms* (James Publishing):

- Form 12-7, Change of Venue Motion.
- Form 12-8, Change of Venue Affidavit.

§12:55.4 State's Controverting Affidavit

A controverting affidavit, pursuant to CCP Art. 31.04, is a notarized statement of the opinion of the compurgator that the opposing affiant is not credible, and/or that opposing affiant's means of knowledge are not sufficient. A controverting affidavit is not a witness' sworn statement of fact as to a matter at trial, as contemplated by Rule 602. Burks v. State, 876 S.W.2d 877 (Tex. Crim. App. 1994).

The credibility of the persons making the affidavits, or their means of knowledge, may be attacked by the affidavit of a credible person, which forms a factual issue to be determined by the trial court. Taylor v. State, 93 S.W.3d 487 (Tex. App.—Texarkana 2002, *pet. ref'd*).

Controverting affidavits must assert that the defendant's affiants are not credible and/or that their knowledge is insufficient. Janecka v. State, 937 S.W.2d 456 (Tex. Crim. App. 1996).

The purpose of the controverting affidavits required by Art. 31.04 is to provide a form of pleading which establishes that there is a factual dispute in need of resolution. Busby v. State, 990 S.W.2d 263 (Tex. Crim. App. 1999).

Personal knowledge is not required, and Tex.R.Evid. 602 does not apply to controverting affidavits. Busby.

If the state does not file a controverting affidavit or if the controverting affidavit fails on its face to meet these two requirements, the defendant is entitled to a change of venue as a matter of law. Janecka; Lundstrom v. State, 742 S.W.2d 279 (Tex. Crim. App. 1986); Taylor v. State, 93 S.W.3d 487 (Tex.App.—Texarkana 2002, pet. ref'd).

§12:55.5 Form: State's Controverting Motion and Affidavit

See our companion book *Texas Criminal Forms* (James Publishing):

- Form 12-9, State's Controverting Motion.
- Form 12-10, State's Controverting Affidavit.

§12:55.6 Grounds for Defendant's Motion

A defendant is entitled to a change of venue if he shows that there were community influences which could affect the answers on voir dire or the testimony of witnesses at trial, or that for any other reason a fair and impartial trial could not be had in the county where the offense was committed. Bell v. State, 938 S.W.2d 35 (Tex. Crim. App. 1996).

§12:55.6.1 Prejudice

Under CCP Art. 31.03(a)(1), a change of venue may be granted if the defendant establishes that there exists in the county where the prosecution is commenced so great a prejudice against him that he cannot obtain a fair and impartial trial. *See* Willingham v. State, 897 S.W.2d 351 (Tex. Crim. App. 1995).

To justify a change of venue based upon media attention, a defendant must show that the publicity was pervasive, prejudicial, and inflammatory. Widespread publicity by itself is not considered inherently prejudicial. Indeed, even extensive knowledge of the case or defendant in the community as a result of pretrial publicity is not sufficient if there is not also some showing of prejudicial or inflammatory coverage. Gonzalez v. State, 222 S.W.3d 446 (Tex. Crim. App. 2007); Buntion v. State, 482 S.W.3d 58, 71 (Tex. Crim. App. 2016).

In examining whether the pretrial publicity is prejudicial and inflammatory, a trial court may take three matters into consideration: (1) the nature of the publicity; (2) any evidence presented at a change of venue hearing; and (3) testimony received from venire members at voir dire. Gonzalez.

The mere existence of media attention or publicity is not enough, by itself, to merit a change of venue. Gonzalez.

A defendant must demonstrate an actual, identifiable prejudice attributable to pretrial publicity on the part of the community from which members of the jury will come. Willingham.

The mere fact of media attention and publicity do not automatically establish prejudice or require a change of venue; jurors do not have to be totally ignorant of the facts and issues of a particular case. Teague v. State, 864 S.W.2d 505 (Tex. Crim. App. 1993).

The trial court is within its discretion to believe prospective jurors' statements that they are not unduly influenced and could deliver a fair verdict. Freeman v. State, 340 S.W.3d 717 (Tex. Crim. App. 2011). News stories that are accurate and objective are generally not considered to be prejudicial or inflammatory. Buntion v. State, 482 S.W.3d 58, 74 (Tex. Crim. App. 2016).

Even a large number of venire members who have heard of the case, or who could not set aside their opinions about the case, does not establish that pretrial publicity has permeated the community to such an extent that it is impossible to seat a fair and impartial jury. Freeman.

Racial prejudice may be grounds for changing venue. McWilliams v. State, 496 S.W.2d 630 (Tex. Crim. App. 1973).

The movant has the burden of proof. Willingham.

The defendant must establish that whatever publicity existed permeated the community to such an extent that the veniremembers' initial opinion, if any, could not be set aside. Etheridge v. State, 903 S.W.2d 1 (Tex. Crim. App. 1994).

In deciding whether the trial court abused its discretion in denying a motion to change venue, the reviewing court determines whether there existed such a prejudice in the community that it is doubtful that the defendant received a fair trial by an impartial jury. Moore v. State, 935 S.W.2d 124 (Tex. Crim. App. 1996).

§12:55.6.2 Dangerous Combination

A change of venue is required if there exists a "dangerous combination" against the accused "instigated by influential person, by reason of which" the accused cannot expect a fair trial. CCP Art. 31.03(a)(2).

§12:55.7 Trial Court Must Hold a Hearing

When the defendant files a change of venue motion and affidavits, and the state files a controverting affidavit under Art. 31.04, the trial court must conduct a hearing to resolve the factual disputes raised by the affidavits. Garza v. State, 974 S.W.2d 251 (Tex.App.—San Antonio, *pet. ref'd*); Burks v. State, 876 S.W.2d 877 (Tex. Crim. App. 1994).

The trial court must conduct an evidentiary hearing on the state's motion before changing venue thereon. Garza v. State, 974 S.W.2d 251 (Tex.App.—San Antonio, *pet. ref'd*).

Arts. 31.01-.03 provide only for a requested change of venue hearing to be held before the trial is completed, and a trial court does not commit error by holding in abeyance a ruling on a motion to change venue until after jury selection. Smith v. State, 850 S.W.2d 275 (Tex.App.—Fort Worth 1993, *pet. ref'd).*

The trial court may withhold ruling on a change of venue motion until after the completion of voir dire, but prior to the defendant's plea to the indictment, to see if an impartial jury can be seated in the case. Foster v. State, 770 S.W.2d 845 (Tex. Crim. App. 1989).

The burden is on the movant to require the trial court to hear and rule on the motion for change of venue. Gutierrez v. State, 979 S.W.2d 659 (Tex. Crim. App. 1998).

A trial court may use the jury selection process to help gauge the community climate. Colone v. State, 573 S.W.3d 249, 257 (Tex. Crim. App. 2019).

A trial court can rescind an order changing venue where no steps have been taken to carry out or to comply with the order changing venue, and the record reflects that the order was improperly entered. Colone v. State, 573 S.W.3d at 256.

§12:55.8 Preserving Error

In order to preserve error where the court has refused to grant a requested change of venue, the defendant should exhaust all peremptory challenges, request additional challenges and show that he was forced to take an unacceptable juror. Stiehl v. State, 585 S.W.2d 716 (Tex. Crim. App. 1979).

§12:55.9 Appellate Review

A change of venue under this provision is within the discretion of the court, subject to review only upon a showing that such discretion was abused. Gonzalez v. State, 222 S.W.3d 446 (Tex. Crim. App. 2007); Colone v. State, 573 S.W.3d 249, 257 (Tex. Crim. App. 2019).

If the trial court's decision falls within the zone of reasonable disagreement, it will be upheld. Freeman v. State, 340 S.W.3d 717 (Tex. Crim. App. 2011).

§12:55.10 Changes of Venue by Consent

For the convenience of the parties and witnesses, and in the interest of justice, the court, upon motion of the defendant and with the consent of the attorney for the State, may transfer the proceeding as to him to another district. CCP Art. 31.03(b).

Similarly, by consent, a defendant may have the case transferred to another district where he stipulates that a plea of guilty will be entered. CCP Art. 31.03(c).

§12:60 Speedy Trial

§12:61 Constitution and Statutes

United States Constitution Amend. VI.

> *In all criminal prosecutions, the accused shall enjoy the right to a speedy and public trial by an impartial jury of the State and district wherein the crime shall have been committed ....*

Texas Constitution Art. 1, §10.

> *In all criminal prosecutions the accused shall have a speedy public trial by an impartial jury.*

CCP Art. 1.05. Rights of Accused

In all criminal prosecutions the accused shall have a speedy public trial by an impartial jury.

CCP Art. 28.061. Discharge for Delay

If a motion to set aside an indictment, information, or complaint for failure to provide a speedy trial is sustained, the court shall discharge the defendant. A discharge under this article is a bar to any further prosecution for the offense discharged and for any other offense arising out of the same transaction, other than an offense of a higher grade that the attorney representing the state and prosecuting the offense that was discharged does not have the primary duty to prosecute.

§12:62 Texas Speedy Trial Act

§12:62.1 Statutory Law

Texas Code Crim. Pro. Art. 32A.01. Trial Priorities

(a) Insofar as is practicable, the trial of a criminal action shall be given preference over trials of civil cases, and the trial of a criminal action against a defendant who is detained in jail pending trial of the action shall be given preference over trials of other criminal actions not described by Subsection (b) or (c).

(b) Unless extraordinary circumstances require otherwise, the trial of a criminal action in which the alleged victim is younger than 14 years of age shall be given preference over other matters before the court, whether civil or criminal.

(c) Except as provided by Subsection (b), the trial of a criminal action against a defendant who has been determined to be restored to competency under Article 46B.084 shall be given preference over other matters before the court, whether civil or criminal.

The Texas Speedy Trial Act as found in CCP Art. 32A.02 was declared unconstitutional as a violation of the separation of powers doctrine. Meshell v. State, 739 S.W.2d 246 (Tex. Crim. App. 1987).

§12:62.2 Case Law

§12:63 Constitutional Speedy Trial Motions

Speedy trial guarantees are designed to minimize the possibility of lengthy incarceration before trial, to reduce the lesser impairment of liberty imposed on an accused while released on bail and to shorten the disruption of life caused by arrest and the presence of unresolved criminal charges. United States v. Loud Hawk, 474 U.S. 302, 106 S.Ct. 648, 654, 88 L.Ed.2d 640 (1986).

Constitutional speedy trial motions focus on post-indictment delay. State v. Munoz, 991 S.W.2d 818 (Tex. Crim. App. 1999).

The Sixth Amendment's speedy trial guarantee protects the accused from arrest or indictment through trial, but does not apply once a defendant has been found guilty at trial or has pleaded guilty to criminal charges. Betterman v. Montana, 578 U.S. ___, 136 S.Ct. 1609, 1612, 194 L.Ed.2d 723 (2016).

§12:63.1 Factors for Examination

There are four factors to be examined in determining whether a defendant has been deprived of his constitutional right to a speedy trial:

- The length of the delay, measured from the time the defendant was accused;
- The government's reason for the delay;
- The efforts made by the defendant to obtain a speedy trial; and
- The prejudice to the defendant as a result of the delay.

Barker v. Wingo, 407 U.S. 514, 92 S.Ct. 2182, 33 L.Ed.2d 101 (1972); Cantu v. State, 253 S.W.3d 273, 280 (Tex. Crim. App. 2008)

This balancing test requires weighing case-by-case the conduct of both the prosecution and the defendant. No single factor is a necessary or sufficient condition to the finding of a speedy trial violation. The related factors must be considered together with such other circumstances as may be relevant. Barker; State v. Munoz, 991 S.W.2d 818 (Tex. Crim. App. 1999).

When reviewing an application of the *Barker* test, a reviewing court uses the same burden of proof allocation as in the context of a motion to suppress, giving almost total deference to historical findings of fact of the trial court that the record supports and drawing reasonable inferences from those facts necessary to support the trial court's findings. Appellate courts review *de novo* whether there was sufficient presumptive prejudice to proceed to a *Barker* analysis and the weighing of the Barker factors, which are legal questions. Gonzales v. State, 435 S.W.3d 801, 808-9 (Tex. Crim. App. 2014).

§12:63.2 Length of Delay

The length of delay is a triggering mechanism for analysis of the other factors. Until there is some delay which is presumptively prejudicial, there is no necessity for inquiry into the other factors that go into the balance. Presumptive prejudice does not necessarily indicate a statistical probability of prejudice; it simply marks the point at which courts deem the delay unreasonable enough to trigger the Barker inquiry. Doggett v. United States, 505 U.S. 647, 112 S.Ct. 2686, 120 L.Ed.2d 520 (1992) (f.n.1); State v. Munoz, 991 S.W.2d 818 (Tex. Crim. App. 1999).

To trigger a speedy trial analysis, an accused must allege that the interval between accusation and trial has crossed the threshold dividing ordinary prejudicial delay from presumptively prejudicial delay. If the accused makes this showing, the court must then consider, as one factor among several, the extent to which the delay stretches beyond the bare minimum needed to trigger judicial examination of the claim. Doggett, Munoz.

The length of the delay is measured from the time the accused is arrested or formally accused. Gonzales v. State, 435 S.W.3d 801, 809 (Tex. Crim. App. 2014) (finding that a 6-year delay created a presumptive prejudice and undertaking a full *Barker* evaluation).

When the length of delay stretches well beyond the bare minimum needed to trigger a full *Barker* analysis, the length of a delay weighs against the State, and the longer the delay, the more the defendant's prejudice is compounded. This is due to the presumption that pretrial delay has prejudiced the accused intensifies over time. Gonzales, 435 S.W.3d at 809.

In general, courts deem delay approaching one year to be unreasonable enough to trigger the Barker enquiry. Balderas v. State, 517 S.W.3d 756, 768 (Tex. Crim. App. 2016).

A delay of as little as fourteen months in bringing a defendant to trial is too long a delay where it was found that the defense witnesses' memories were dimming and the state did nothing to refute the prejudice. State v. Perkins, 911 S.W.2d 548 (Tex.App.—Fort Worth 1995, *no pet.*).

However, a delay of eight and one-half years may not result in prejudice to a defendant under another fact situation. Emery v. State, 881 S.W.2d 702 (Tex. Crim. App. 1994).

§12:63.3 Reason for Delay

The third Barker factor—the defendant's assertion of his right to a speedy trial—is entitled to strong evidentiary weight in determining whether the defendant has been deprived of that right. Balderas v. State, 517 S.W.3d 756, 768 (Tex. Crim. App. 2016).

The length of delay can be subdivided into justifiable and unjustifiable reasons for delay depending on the circumstances of the case. Unjustifiable reasons for delay count towards the length of delay, while justifiable reasons for delay do not. Gonzales v. State, 435 S.W.3d 801, 810 (Tex. Crim. App. 2014).

Different weights should be assigned to different reasons for the delay. A deliberate attempt to delay the trial should be weighed heavily against the government. A more neutral reason such as negligence or overcrowded courts should be weighed against the government less heavily. A valid reason for the delay should not be weighed against the government at all. A delay which is attributable in whole or in part to the defendant may even constitute a waiver of a speedy trial claim. Barker v. Wingo, 407 U.S. 514, 92 S.Ct. 2182, 33 L.Ed.2d 101 (1972); State v. Munoz, 991 S.W.2d 818 (Tex. Crim. App. 1999).

Deliberate delay intended to hamper the defense weighs heavily against the State, while more neutral reasons, such as negligence or overcrowded courts, weigh less heavily. Additionally, the courts consider whether the government or the criminal defendant is more to blame for the delay. Balderas v. State, 517 S.W.3d at 771.

Delay caused by good faith plea negotiations is not the result of negligence or a deliberate attempt to delay the trial. Barker, Munoz

A crowded court docket is not a valid reason for delay and must be counted against the State, although not heavily. Shaw v. State, 117 S.W.3d 883 (Tex. Crim. App. 2003).

§12:63.4 The Defendant's Efforts

A defendant must raise a speedy-trial claim in the trial court in order to preserve that issue for appellate review. Henson v. State, 407 S.W.3d 764, 768 (Tex. Crim. App. 2013).

Initially filing a motion to dismiss instead of a motion for a speedy trial weakens a defendant's speedy trial claim because it shows a desire to have no trial instead of a speedy trial. Parkerson v. State, 942 S.W.2d 789 (Tex.App.—Fort Worth 1997, *no pet.*).

Court-appointed counsel, just as retained counsel, act on behalf of their clients, and delays sought by counsel are ordinarily attributable to the defendant when conducting a Barker v. Wingo analysis. Vermont v. Brillon, 556 U.S. ___, 129 S.Ct. 1283, 173 L.Ed.2d 231 (2009).

§12:63.5 Prejudice

A defendant has the burden to make some showing of prejudice although a showing of actual prejudice is not required. When a defendant makes a *prima facie* showing of prejudice, the State carries the obligation of proving that the accused suffered no serious prejudice beyond that which ensued from the ordinary and inevitable delay. State v. Munoz, 991 S.W.2d 818 (Tex. Crim. App. 1999).

There is no speedy trial violation where the defendant fails in his burden to show that he was prejudiced because of the delay. Knox v. State, 934 S.W.2d 678 (Tex. Crim. App. 1996).

Prejudice to the defendant is assessed in the light of the interests that the speedy trial right is designed to protect. These interests are:

- Preventing oppressive pretrial incarceration;
- Minimizing anxiety and concern of the accused; and
- Limiting the possibility that the defense will be impaired.

Of these factors, the most serious is the last, because the inability of a defendant adequately to prepare his case skews the fairness of the entire system. Barker v. Wingo, 407 U.S. 514, 92 S.Ct. 2182, 33 L.Ed.2d 101 (1972); Munoz.

In certain instances, the length of delay may be so excessive that it presumptively compromises the reliability of a trial in ways that neither party can prove or identify. In such instances, the defendant is absolved from the requirement to demonstrate prejudice. Gonzales v. State, 435 S.W.3d 801, 812-3 (Tex. Crim. App. 2014).

Where prejudice has been presumed, the state has the burden to persuasively rebut it. Gonzales, 435 S.W.3d at 815.

§12:63.6 Necessary Proof

An unsworn motion to dismiss for denial of speedy trial rights is not self-proving and cannot serve as proof of any fact. Newman v. State, 331 S.W.3d 447 (Tex. Crim. App. 2011).

The defendant invoking his right to a speedy trial must make some showing of prejudice resulting from the delay. Chapman v. Evans, 744 S.W.2d 133 (Tex. Crim. App. 1988).

A defendant's assertion of his speedy trial right is entitled to strong evidentiary weight in determining whether he has been deprived of his right to a speedy trial. Clarke v. State, 928 S.W.2d 709 (Tex.App.—Fort Worth 1996, *pet. ref'd*).

The trial court as finder of fact on a speedy trial motion is empowered, on the basis of credibility and demeanor evaluations, to completely disregard a witness' testimony, even if that testimony is uncontroverted. Kelly v. State, 163 S.W.3d 722 (Tex. Crim. App. 2005) (citing State v. Ross, 32 S.W.3d 853 (Tex. Crim. App. 2000)).

The burden is on the defendant on appeal to present a sufficient record to show that his speedy trial motion should have been granted. Newman.

§12:63.7 Remedy Is Dismissal

CCP Art. 28.061. Discharge for Delay

> *If a motion to set aside an indictment, information, or complaint for failure to provide a speedy trial is sustained, the court shall discharge the defendant. A discharge under this article is a bar to any further prosecution for the offense discharged and for any other offense arising out of the same transaction, other than an offense of a higher grade that the attorney representing the state and prosecuting the offense that was discharged does not have the primary duty to prosecute.*

If a violation of the right to a speedy trial is established, the proper remedy is dismissal of the prosecution with prejudice. Strunk v. United States, 412 U.S. 434, 93 S.Ct. 2260, 37 L.Ed.2d 56 (1973); Shaw v. State, 117 S.W.3d 883 (Tex. Crim. App. 2003).

§12:63.8 Form: Motions

See our companion book *Texas Criminal Forms* (James Publishing):

- Form 12-11, Motion for Speedy Trial.
- Form 12-12, Motion to Dismiss for Denial of Constitutional Right to Speedy Trial.

12:63.9 Interstate Agreement on Detainers (IAD)

The Interstate Agreement on Detainers (IAD) is a compact between states (and some other jurisdictions) that enables a party state to obtain custody of an out-of-state prisoner for prosecution and imposes duties to ensure a prisoner's quick return to the sending state. Texas has codified the IAD in Code of Criminal Procedure art. 51.14. Hopper v. State, 520 S.W.3d 915, 924 (Tex. Crim. App. 2017). When analyzing speedy trial claims that arise in the context of the IAD, the courts use the Barker factors to analyze the claim. Hopper v. State, 520 S.W.3d at 924

§12:64 Defendant's Right to Speedy Indictment

§12:64.1 Statutes

CCP Art. 32.01. Defendant in Custody and No Indictment Presented

> *When a defendant has been detained in custody or held to bail for his appearance to answer any criminal accusation before the district court, the prosecution, unless otherwise ordered by the court, for good cause shown, supported by affidavit, shall be dismissed and the bail discharged, if indictment or information be not presented against such defendant on or before the last day of the next term of the court which is held after his commitment or admission to bail or on or before the 180th day after the date of commitment or admission to bail, whichever date is later.*

§12:64.2 Application of the Statute

Art. 32.01 prevents citizens from being left in jail or on bail for long periods of time without being indicted. The State must indict within the period set by Art. 32.01, show good cause for the delay, or suffer the dismissal of the charges. Ex parte Martin, 6 S.W.3d 52 (Tex. Crim. App. 1999).

Art. 32.01 is satisfied for all subsequent charges once the state obtains a timely indictment against the defendant for any offense arising out of the same criminal transaction or occurrence. Cameron v. State, 988 S.W.2d 835 (Tex.App.—San Antonio 1999, *pet. ref'd*).

The Barker balancing test has been adopted to determine when pretrial delay denies an accused of his right to a speedy trial under CCP Art. 32.01:

- The length of the delay.
- The reason for the delay.
- The prejudice to the accused.
- The accused's assertion of the right.

Martin.

A motion to dismiss for failure to provide speedy indictment must be filed and heard before indictment as it has no effect if it is presented after indictment. Brooks v. State, 990 S.W.2d 278, 285 (Tex. Crim. App. 1999).

A writ of habeas corpus alleging an Art.

32.01 violation is moot where the state indicts the defendant before the writ is heard. Ex parte Countryman, 226 S.W.3d 435 (Tex. Crim. App. 2007).

§12:64.3 The State's Burden

A showing of good cause to continue the prosecution is unnecessary unless and until the defendant seeks a dismissal pursuant to the statute. Thus, there is no reason for the state to file a good-cause affidavit before the defendant files his motion to dismiss. It is sufficient that the state files its responsive affidavit before the district court rules. Ex parte Mallares, 953 S.W.2d 759 (Tex.App.—Austin 1997, *no pet.*).

§12:64.4 Trial Court May Dismiss Case

Although a trial court generally does not have the power to dismiss a case unless the prosecutor requests it, a trial court does have this power under Art. 32.01. Ex parte Mann, 89 S.W.3d 296 (Tex. App.—Fort Worth 2002, *no pet.*).

§12:64.5 Defendant May Still Be Prosecuted

CCP Art. 28.061, barring further prosecution for a discharged offense, has been amended, and no longer applies to a discharge under Art. 32.01. Therefore, even if a defendant is entitled to discharge from custody under Art. 32.01, he is not free from subsequent prosecution. Ex parte Seidel, 39 S.W.3d 221 (Tex. Crim. App. 2001).

§12:64.6 Form: Motion and Waiver

See our companion book *Texas Criminal Forms* (James Publishing):

- Form 12-13, Motion to Dismiss for Denial of Speedy Indictment.
- Form 12-14, Waiver of Speedy Indictment.

§12:65 Mandamus to Enforce Speedy Trial Rights

§12:65.1 Defendant May Petition for Writ of Mandamus

A defendant who seeks a speedy trial may do so by way of a petition for writ of mandamus. Chapman v. Evans, 744 S.W.2d 133 (Tex. Crim. App. 1988).

To obtain relief through a writ of mandamus, the relator must establish that (1) the act he seeks to compel is ministerial, rather than discretionary, in nature and (2) no other adequate remedy at law is available. Engle v. Coker, 820 S.W.2d 247 (Tex. App.—Beaumont 1991, *no pet.*).

An act is ministerial if it constitutes a duty clearly fixed and required by law, without the exercise of discretion or judgment. Engle v. Coker.

§12:65.2 Mandamus Not Appropriate for Speedy Trial Dismissals

Mandamus is not the proper remedy for a defendant seeking dismissal of his charges based upon an alleged speedy trial violation because he has an adequate remedy at law, that being a motion to dismiss the indictment under CCP Art. 27.03. Smith v. Gohmert, 962 S.W.2d 590 (Tex. Crim. App. 1998).

§12:65.3 Form: Petition for Writ

See our companion book *Texas Criminal Forms* (James Publishing):

- Form 12-15, Petition for Writ of Mandamus.

§12:70 Joinder and Severance

§12:71 Joinder and Severance of Multiple Defendants

§12:71.1 Statutes

CCP Art. 36.09. Severance on Separate Indictments

Two or more defendants who are jointly or separately indicted or complained against for the same offense or any offense growing out of the same transaction may be, in the discretion of the court, tried jointly or separately as to one or more defendants; provided that in any event either defendant may testify for the other or on behalf of the state; and provided further, that in cases in which, upon timely motion to sever, and evidence introduced thereon, it is made known to the court that there is a previous admissible conviction against one defendant or that a joint trial would be prejudicial to any defendant, the court shall order a severance as to the defendant whose joint trial would prejudice the other defendant or defendants.

CCP Art. 36.10. Order of Trial

If a severance is granted, the defendants may agree upon the order in which they are to be tried; but if they fail to agree, the court shall direct the order of the trial.

§12:71.2 Joinder

Under Art. 36.09, the trial court has discretion to join two or more defendants for trial unless statutory grounds for severance exist. Aguilar v. State, 26 S.W.3d 901 (Tex. Crim. App. 2000).

There is a specific statute controlling the joinder of drug offenses into a single trial. It explicitly requires the state to file written notice of joinder. Tex. Health and Safety Code §481.132.

§12:71.3 Severance

Art. 36.09 imposes two basic requirements for showing entitlement to a severance: (1) that the motion for severance be timely, and (2) that at least one of two possible grounds for severance be alleged, with supporting evidence. Qualley v. State, 206 S.W.3d 624 (Tex. Crim. App. 2006) *citing* Zafiro v. United States, 506 U.S. 534, 113 S.Ct. 933, 122 L.Ed.2d 317 (1993).

§12:71.4 Grounds for Severance

The two grounds for severance are: (1) the co-defendant has a previous admissible conviction, and (2) a joint trial would prejudice the (moving) defendant. Qualley v. State, 206 S.W.3d 624 (Tex. Crim. App. 2006) (*citing* Zafiro v. United States, 506 U.S. 534, 113 S.Ct. 933, 122 L.Ed.2d 317 (1993)).

Absent evidence of prejudice to one defendant in a joint trial, or evidence that one of the defendants has a prior admissible conviction, a motion for severance is left to the trial court's discretion. Marsh v. State, 115 S.W.3d 709 (Tex.App.—Austin 2003, pet. ref'd).

In order to establish prejudice entitling him to a severance, a defendant must show a serious risk that a specific trial right would be compromised by a joint trial, or that a joint trial would prevent the jury from making a reliable judgment about guilt or innocence, and that the problem could not be adequately addressed by lesser curative measures, such as a limiting instruction. Prejudice must be analyzed with regard to each defendant. Prejudice may be shown for one defendant in a trial but not another. Qualley.

To establish prejudice on the basis of the right to have a codefendant testify as a witness, the moving party must demonstrate: (1) a bona fide need for the testimony; (2) the substance of the desired testimony; (3) its exculpatory nature and effect; and (4) evidence that the designated co-defendant will in fact testify at a separate trial. Lacy v. State, 901 S.W.2d 518 (Tex.App.—Tyler 1995, *no pet.*).

The "previous admissible conviction" portion of the statute acts only to shield a defendant from the introduction of the co-defendant's conviction. The idea is that a defendant without a conviction may suffer "guilt by association" from the jury's exposure to the co-defendant's previous crime. It is for this reason that severance on this basis is mandatory only when the moving defendant himself has no prior conviction that would be admissible at either the guilt or punishment stage of trial. Qualley.

Failure to grant a severance on this basis is harmless if the co-defendant's previous conviction was not in fact admitted at trial. Qualley.

Claims made under the "prejudice" clause of the statute are not subject to harmless error analysis. Qualley.

When an accused is not entitled to a severance as a matter of right, the denial of a severance motion by the trial court constitutes an abuse of discretion only when the movant satisfies the "heavy burden" of showing "clear prejudice." Mendoza v. State, 61 S.W.3d 418 (Tex.App.—San Antonio 2001) *aff'd at* 88 S.W.3d 236 (Tex. Crim. App. 2002).

An example of a curative measure that can be employed by the trial court in a joint trial is an instruction to the jury to use a confession only against the confessing codefendant and then redacting the confession to eliminate the other defendant's name and any reference to his existence to cure the prejudice caused by admission of a codefendant's statement. Chavez v. State, 6 S.W.3d 56 (Tex.App.—San Antonio 1999, pet. ref'd) *citing* Richardson v. Marsh, 481 U.S. 200, 211, 107 S.Ct. 1702, 95 L.Ed.2d 176 (1987).

§12:71.5 Timeliness of Motion

A motion to sever on the grounds of unfair prejudice under Art. 36.09 is timely if made at the first opportunity or as soon as the grounds for prejudice become apparent or should have become apparent, thus providing the trial court an opportunity to rule on the potentially prejudicial evidence at the time it is introduced. Aguilar v. State, 26 S.W.3d 901 (Tex. Crim. App. 2000).

To be timely, a motion for severance must be made before trial if the factual basis for seeking a severance was known prior to trial. Qualley v. State, 206 S.W.3d 624 (Tex. Crim. App. 2006).

A motion to sever on the grounds of unfair prejudice is untimely if counsel waits until the close of evidence to make the motion. Aguilar.

The mere allegation that prejudice will result is not evidence of, or a sufficient showing of, prejudice under Art. 36.09, particularly when the severance is discretionary with the trial judge. Mendoza v. State, 61 S.W.3d 418 (Tex.App.—San Antonio 2001) *aff'd at* 88 S.W.3d 236 (Tex. Crim. App. 2002).

If the motion to sever is not supported by evidence, its denial is not an abuse of discretion. Marsh v. State, 115 S.W.3d 709 (Tex.App.—Austin 2003, pet. ref'd); Davila v. State, 4 S.W.3d 844 (Tex.App.—Eastland 1999, *no pet.*).

§12:71.6 Form: Motions

See our companion book *Texas Criminal Forms* (James Publishing):

- Form 12-20, Motion for Severance of Defendants' Cases.
- Form 12-21, State's Motion to Consolidate.

§12:72 Joinder and Severance of Multiple Counts

§12:72.1 Statutes

Tex. Penal Code §3.01. Definition

In this chapter, "criminal episode" means the commission of two or more offenses, regardless of whether the harm is directed toward or inflicted upon more than one person or item of property, under the following circumstances:

(1) the offenses are committed pursuant to the same transaction or pursuant to two or more transactions that are connected or constitute a common scheme or plan; or

(2) the offenses are the repeated commission of the same or similar offenses.

Tex. Penal Code §3.02. Consolidation and Joinder of Prosecutions

(a) A defendant may be prosecuted in a single criminal action for all offenses arising out of the same criminal episode.

(b) When a single criminal action is based on more than one charging instrument within the jurisdiction of the trial court, the state shall file written notice of the action not less than 30 days prior to the trial.

(c) If a judgment of guilt is reversed, set aside, or vacated, and a new trial ordered, the state may not prosecute in a single criminal action in the new trial any offense not joined in the former prosecution unless evidence to establish probable guilt for that offense was not known to the appropriate prosecuting

official at the time the first prosecution commenced.

Tex. Penal Code §3.03. Sentences for Offenses Arising Out of Same Criminal Episode

(a) When the accused is found guilty of more than one offense arising out of the same criminal episode prosecuted in a single criminal action, a sentence for each offense for which he has been found guilty shall be pronounced. Except as provided by Subsection (b), the sentences shall run concurrently.

(b) If the accused is found guilty of more than one offense arising out of the same criminal episode, the sentences may run concurrently or consecutively if each sentence is for a conviction of:

(1) an offense:

(A) under Section 49.07 or 49.08, regardless of whether the accused is convicted of violations of the same section more than once or is convicted of violations of both sections; or

(B) for which a plea agreement was reached in a case in which the accused was charged with more than one offense listed in Paragraph (A), regardless of whether the accused is charged with violations of the same section more than once or is charged with violations of both sections;

(2) an offense:

(A) under Section 33.021 or an offense under Section 21.02, 21.11, 22.011, 22.021, 25.02, or 43.25 committed against a victim younger than 17 years of age at the time of the commission of the offense regardless of whether the accused is convicted of violations of the same section more than once or is convicted of violations of more than one section; or

(B) for which a plea agreement was reached in a case in which the accused was charged with more than one offense listed in Paragraph (A) committed against a victim younger than 17 years of age at the time of the commission of the offense regardless of whether the accused is charged with violations of the same section more than once or is charged with violations of more than one section;

(3) an offense:

(A) under Section 21.15 or 43.26, regardless of whether the accused is convicted of violations of the same section more than once or is convicted of violations of both sections; or

(B) for which a plea agreement was reached in a case in which the accused was charged with more than one offense listed in Paragraph (A), regardless of whether the accused is charged with violations of the same section more than once or is charged with violations of both sections;

(4) an offense for which the judgment in the case contains an affirmative finding under Article 42.0197, Code of Criminal Procedure; or

(5) an offense:

(A) under Section 20A.02 or 43.05, regardless of whether the accused is convicted of

violations of the same section more than once or is convicted of violations of both sections; or

(B) for which a plea agreement was reached in a case in which the accused was charged with more than one offense listed in Paragraph (A), regardless of whether the accused is charged with violations of the same section more than once or is charged with violations of both sections.

(b-1) Subsection (b)(4) does not apply to a defendant whose case was transferred to the court under Section 54.02, Family Code.

(6) an offense:

(A) under Section 22.04(a)(1) or (2) or Section 22.04(a-1)(1) or (2) that is punishable as a felony of the first degree, regardless of whether the accused is convicted of violations of the same section more than once or is convicted of violations of more than one section; or

(B) for which a plea agreement was reached in a case in which the accused was charged with more than one offense listed in Paragraph (A) and punishable as described by that paragraph, regardless of whether the accused is charged with violations of the same section more than once or is charged with violations of more than one section.

Tex. Penal Code §3.04. Severance

(a) Whenever two or more offenses have been consolidated or joined for trial under Section 3.02, the defendant shall have a right to a severance of the offenses.

(b) In the event of severance under this section, the provisions of Section 3.03 do not apply, and the court in its discretion may order the sentences to run either concurrently or consecutively.

(c) The right to severance under this section does not apply to a prosecution for offenses described by Section 3.03(b) unless the court determines that the defendant or the state would be unfairly prejudiced by a joinder of offenses, in which event the judge may order the offenses to be tried separately or may order other relief as justice requires.

CCP Art. 21.24. Joinder of Certain Offenses

(a) Two or more offenses may be joined in a single indictment, information, or complaint, with each offense stated in a separate count, if the offenses arise out of the same criminal episode, as defined in Chapter 3 of the Penal Code.

(b) A count may contain as many separate paragraphs charging the same offense as necessary, but no paragraph may charge more than one offense.

(c) A count is sufficient if any one of its paragraphs is sufficient. An indictment, information, or complaint is sufficient if any one of its counts is sufficient.

§12:72.2 Joinder of Counts

An indictment may contain as many separate paragraphs charging the same offense as is necessary to meet the contingencies of the evidence. Where an indictment charges different theories under which a defendant may have committed a single statutory offense, the defendant is not entitled to sever the paragraphs. Hathorn v. State, 848 S.W.2d 101 (Tex. Crim. App. 1992).

Reading the current versions of Art. 21.24(a) of the code of criminal procedure and section 3.01 of the penal code together, the State is authorized to join two or more offenses in a single indictment if the offenses are part of the same transaction or if the offenses are part of a common scheme or plan. Willeford v. State, 72 S.W.3d 820 (Tex.App.—Fort Worth 2002, pet. ref'd).

It is currently permissible for the State to allege multiple nonproperty offenses in the same indictment and obtain convictions on each offense. Willeford.

A defendant does not have the right to consolidate offenses committed in the same criminal episode under Texas Penal Code §3.02; consolidation is permissive, not mandatory. Nelson v. State, 864 S.W.2d 496 (Tex. Crim. App. 1993).

§12:72.3 Severance of Counts

A motion to sever offenses against a single defendant is a pleading of the defendant under CCP Arts. 27.02(8) & 28.01, and must be raised and heard prior to trial. Thornton v. State, 986 S.W.2d 615 (Tex. Crim. App. 1999).

A motion to sever is not self-proving, and must be accompanied by evidence or its denial is not an abuse of discretion. Mulder v. State, 707 S.W.2d 908 (Tex. Crim. App. 1986).

Where the indictment alleges two or more separate offenses, the defendant is entitled to a severance under Texas Penal Code §3.04(a). Graham v. State, 19 S.W.3d 851 (Tex. Crim. App. 2000).

A trial judge's failure to grant a mandatory severance under Penal C §3.04 is subject to a harm analysis, and the error is harmless if it did not adversely affect the defendant's substantial rights Scott v. State, 235 S.W.3d 255 (Tex. Crim. App. 2007).

Any error in failing to grant a severance is harmless where the evidence of each of the crimes would have been admissible in the trial of the others. Tovar v. State, 165 S.W.3d 785 (Tex.App.—San Antonio 2005, *no pet.*).

For further information pertaining to joinder and severance of counts, as well as requiring the state to elect a primary count for conviction purposes, *see* §17:10.

§12:72.4 Form: Motions and Supporting Brief

See our companion book *Texas Criminal Forms* (James Publishing):

- Form 12-18.1, Motion for Severance of Counts.
- Form 12-18.2, Motion for Severance of Defendant's Cases.
- Form 12-19, Brief in Support of Motion for Severance of Counts.

§12:80 Testifying Free From Impeachment by Prior Convictions

§12:81 Statutes and Rules

Tex. R. Evid. 609. Impeachment by Evidence of a Criminal Conviction

> ***(a) In General.*** *Evidence of a criminal conviction offered to attack a witness's character for truthfulness must be admitted if:*
>
> *(1) the crime was a felony or involved moral turpitude, regardless of punishment;*
>
> *(2) the probative value of the evidence outweighs its prejudicial effect to a party; and*
>
> *(3) it is elicited from the witness or established by public record.*
>
> ***(b) Limit on Using the Evidence After 10 Years.*** *This subdivision (b) applies if more than 10 years have passed since the witness's conviction or release from confinement for it, whichever is later. Evidence of the conviction is admissible only if its probative value, supported by specific facts and circumstances, substantially outweighs its prejudicial effect.*
>
> ***(c) Effect of a Pardon, Annulment, or Certificate of Rehabilitation.*** *Evidence of a conviction is not admissible if:*

(1) the conviction has been the subject of a pardon, annulment, certificate of rehabilitation, or other equivalent procedure based on a finding that the person has been rehabilitated, and the person has not been convicted of a later crime that was classified as a felony or involved moral turpitude, regardless of punishment;

(2) probation has been satisfactorily completed for the conviction, and the person has not been convicted of a later crime that was classified as a felony or involved moral turpitude, regardless of punishment; or

(3) the conviction has been the subject of a pardon, annulment, or other equivalent procedure based on a finding of innocence.

(d) Juvenile Adjudications. *Evidence of a juvenile adjudication is admissible under this rule only if:*

(1) the witness is a party in a proceeding conducted under title 3 of the Texas Family Code; or

(2) the United States or Texas Constitution requires that it be admitted.

(e) Pendency of an Appeal. *A conviction for which an appeal is pending is not admissible under this rule.*

(f) Notice. *Evidence of a witness's conviction is not admissible under this rule if, after receiving from the adverse party a timely written request specifying the witness, the proponent of the conviction fails to provide sufficient written notice of intent to use the conviction. Notice is sufficient if it provides a fair opportunity to contest the use of such evidence.*

§12:82 General Points

Once a criminal defendant takes the witness stand, he is subject to the same rules as any other witness and may be impeached, contradicted, made to give evidence against himself, cross-examined on new matters, and treated in every respect as any other witness, except when there are overriding constitutional or statutory provisions. Juhasz v. State, 827 S.W.2d 327 (Tex.App.—Corpus Christi 1992, *pet. ref'd*).

The purpose of Rule 609 is to give an adverse party a fair opportunity to contest the use of a witness' convictions and to prevent ambush. Johnson v. State, 885 S.W.2d 578 (Tex.App.—Dallas 1994, *no pet.*).

When attacking the credibility of witnesses, evidence of prior criminal convictions shall be admitted only if the crime was a felony or involved moral turpitude, regardless of punishment, and the court determines that the probative value outweighs its prejudicial effects to a party. Rule 609(a); Delk v. State, 855 S.W.2d 700 (Tex. Crim. App. 1993).

PRACTICE TIP: If you want to keep the jury from hearing about a defendant's prior conviction, don't ask the defendant about convictions on direct examination. By testifying on direct examination regarding his prior convictions, a defendant will generally waive any error in the admission of his prior convictions. Perea v. State, 870 S.W.2d 314 (Tex. App.—Tyler 1994, no pet.).

§12:83 Test for Admissibility of Priors

In unambiguously plain language, Rule 609(a) provides that witness credibility may be attacked by admitting evidence that the witness previously has been convicted of a felony or crime of moral turpitude if the trial court determines that the probative value of admitting the evidence simply outweighs its prejudicial effect. Meadows v. State, 455 S.W.3d 166, 170 (Tex. Crim. App. 2015).

The trial court should conduct a balancing test to determine if the probative value of the prior convictions is outweighed by their prejudicial effect. Theus v. State, 845 S.W.2d 874 (Tex. Crim. App. 1992).

Although not exclusive, the following factors should be considered: (1) the prior conviction's impeachment value; (2) its temporal proximity to the crime on trial; (3) the similarity between the prior offenses and the present offense; (4) the importance of the defendant's testimony; and (5) the importance of the credibility issue. Theus.

If the prior offenses involved deceptive conduct instead of violence, and if they are recent, the first two factors will favor their admission because they affect credibility and they indicate a propensity for running afoul of the law. Theus.

If the prior offenses are similar to the one on trial, the third factor will militate against their admission because of the danger that the jury might convict on the basis of past conduct rather than on the facts of the case on trial. Edwards v. State, 883 S.W.2d 692 (Tex.App.—Texarkana 1994, *pet. ref'd*).

The effect of the fourth and fifth factors depends on the type of defense presented. Edwards.

The proponent of the evidence has the burden to demonstrate the admissibility of the evidence against the defendant. Dale v. State, 90 S.W.3d 826 (Tex.App.—San Antonio 2002, *pet. ref'd*).

The trial court has wide discretion in deciding the question of admissibility, and its decision will not be reversed unless it is shown that there was a clear abuse of that discretion. Theus.

PRACTICE TIP: Before pursuing a motion to testify free of impeachment by prior convictions, counsel should deliver a request to the state that they give notice under Rule 609 of any prior conviction they intend to use to impeach the defendant at trial. See Chapter 13, Discovery.

§12:84 Convictions More Than 10 Years Old

In unambiguous plain language, Rule 609(b) limits Rule 609(a) by providing that evidence of a prior conviction is inadmissible if more than 10 years has elapsed since the later date of conviction or release of the witness from the confinement imposed for that conviction unless the court determines, in the interest of justice that the probative value of the conviction supported by specific facts and circumstances substantially outweighs its prejudicial effect. Meadows v. State, 455 S.W.3d 166, 170 (Tex. Crim. App. 2015).

In deciding whether, in the interests of justice, the probative value of a remote conviction substantially outweighs its prejudicial effect, a court may consider all the relevant specific facts and circumstances, including whether intervening convictions dilute the prejudice of the remote conviction. Meadows, 455 S.W.3d at 170 (abrogating the "tacking doctrine").

§12:85 Forms: Motions and Supporting Brief

See our companion book *Texas Criminal Forms* (James Publishing):

- Form 12-23.1, Motion to Testify Free of Impeachment.
- Form 12-23.2, Motion to Have Prior Convictions Not Mentioned.
- Form 12-24, Brief in Support of Motion to Testify Free of Impeachment by Prior Convictions.

§12:90 Indictments and Informations

§12:91 Statutes

CCP Art. 21.01. "Indictment"

An "indictment" is the written statement of a grand jury accusing a person therein named of some act or omission which, by law, is declared to be an offense.

CCP Art. 21.02. Requisites of an Indictment

(a) An indictment, information, complaint, or other charging instrument or a related document in a criminal case may be filed in electronic form with a judge or clerk of the court authorized to receive the document.

(b) A judge or clerk of the court is authorized to receive for filing purposes an information, indictment, complaint,

or other charging instrument or a related document in electronic form in accordance with Subchapter I, Chapter 51, Government Code, if:

(1) the document complies with the requirements that would apply if the document were filed in hard-copy form;

(2) the clerk of the court has the means to electronically store the document for the statutory period of record retention;

(3) the judge or clerk of the court is able to reproduce the document in hard-copy form on demand; and

(4) the clerk of the court is able to display or otherwise make the document available in electronic form to the public at no charge.

(c) The person filing the document and the person receiving the document must complete the electronic filing as provided by Section 51.804, Government Code.

(d) Notwithstanding Section 51.806, Government Code, an indictment, information, complaint, or other charging instrument or a related document transmitted in electronic form is exempt from a requirement under this code that the pleading be endorsed by a natural person. The requirement of an oath under this code is satisfied if:

(1) all or part of the document was sworn to; and

(2) the electronic form states which parts of the document were sworn to and the name of the officer administering the oath.

(e) An electronically filed document described by this section may be amended or modified in compliance with Chapter 28 or other applicable law. The amended or modified document must reflect that the original document has been superseded.

(f) This section does not affect the application of Section 51.318, Government Code, Section 118.052(3), Local Government Code, or any other law permitting the collection of fees for the provision of services related to court documents.

CCP Art. 21.03. What Should Be Stated

Everything should be stated in an indictment which is necessary to be proved.

CCP Art. 21.04. The Certainty Required

The certainty required in an indictment is such as will enable the accused to plead the judgment that may be given upon it in bar of any prosecution for the same offense.

§12:92 Indictments in General

The Texas Constitution guarantees to defendants the right to indictment by a grand jury for all felony offenses. Riney v. State, 28 S.W.3d 561 (Tex. Crim. App. 2000); Cook v. State, 902 S.W.2d 471 (Tex. Crim. App. 1995).

An indictment is a written instrument presented to a court by a grand jury charging a person with the commission of an offense. Tex. Const. Art. V, §12(b); Riney.

Indictment by grand jury protects citizens against arbitrary accusations by the government. Riney.

An indictment is essential to vest the trial court with jurisdiction. Riney; Cook.

An indictment provides a defendant notice of the offense charged so that he may prepare, in advance of trial, an informed and effective defense. Riney; Garcia v. State, 981 S.W.2d 683, 685 (Tex. Crim. App.1998).

Alternate pleading of different methods of committing an offense may be charged in one indictment. Kitchens v. State, 823 S.W.2d 256 (Tex. Crim. App. 1991), *cert. denied*, 504 U.S. 958, 112 S.Ct. 2309, 119 L.Ed.2d 230 (1992).

In fact, the indictment may and should be

comprised of as many counts as necessary to meet all of the contingencies in the evidence. Hathorn v. State, 848 S.W.2d 101 (Tex. Crim. App. 1992).

The term "count" is used to charge the offense itself, and a "paragraph" is a portion of a count, which charges a method of committing the offense. The substance of the allegation determines its character as a "count" or a "paragraph," not the terminology used. Dalton v. State, 898 S.W.2d 424 (Tex.App.—Fort Worth 1995, *pet. ref'd*).

An indictment may contain as many separate paragraphs charging the same offense as necessary. CCP Art. 21.24(b).

TIP TO THE PROSECUTION: If the indictment contains an allegation of the commission of an offense and pleads several different methods of the commission of this offense, ask the trial court to submit the allegations to the jury combined into one application paragraph in order to alleviate proving each separate allegation to each member of the jury.

§12:92.1 Allegations Required

The indictment must plead the offense in plain and intelligible words. Everything should be stated in an indictment which is necessary to be proved. The certainty required in an indictment is such that will enable the accused to plead the judgment that may be given upon it in bar of any prosecution for the same offense. State v. Zuniga, 512 S.W.3d 902, 906 (Tex. Crim. App. 2017).

An indictment is sufficient if it charges the commission of the offense in ordinary and concise language in such a manner as to enable a person of common understanding to know what is meant, and with that degree of certainty that will give the defendant notice of the particular offense with which he is charged, and enable the court, on conviction, to pronounce the proper judgment. State v. Zuniga, 512 S.W.3d at 906.

The language of an indictment is sufficient where it tracks the language of the penal statute in question. State v. Zuniga, 512 S.W.3d at 907.

The face of the indictment must charge all the elements of the offense in plain and intelligible words, such that a person of common understanding would know what is meant. State v. Mays, 967 S.W.2d 404 (Tex. Crim. App. 1998).

When a statutory term or element is defined by a statute, the charging instrument does not need to allege the definition of the term or element. Typically the definition of terms and elements are regarded as evidentiary matters. State v. Zuniga, 512 S.W.3d at 907.

However, in some cases, a charging instrument that tracks the statutory language may be insufficient to provide a defendant with adequate notice. This is so when the statutory language fails to be completely descriptive. For example, a statute which uses an undefined term of indeterminate or variable meaning requires more specific pleading in order to notify the defendant of the nature of the charges against him. Likewise, when a statute defines the manner or means of commission in several alternative ways, an indictment will fail for lack of specificity if it neglects to identify which of the statutory means it addresses. In such cases, more particularity is required to provide adequate notice. State v. Zuniga, 512 S.W.3d at 907.

Appellate courts to engage in a two-step analysis when analyzing whether a charging instrument provides adequate notice. First, the reviewing court must identify the elements of the offense. Next, it must consider whether the statutory language is sufficiently descriptive of the charged offense. State v. Jarreau, 512 S.W.3d 352, 354-5 (Tex. Crim. App. 2017).

Tex. Penal Code §1.07(a)(22) defines the elements of the offense to be the forbidden conduct, the required culpability, any required result, and the negation of any exception to the offense. Gray v. State, 152 S.W.3d 125 (Tex. Crim. App. 2004).

An indictment is constitutionally valid when it alleges the essential elements of the offense and charges a defendant with that offense. Duron v. State, 956 S.W.2d 547 (Tex. Crim. App. 1997).

An indictment must allege on its face the facts necessary (1) to show that the offense was committed; (2) to bar a subsequent prosecution for the same offense; and (3) to give the defendant notice of precisely what he is charged with. Weatherby v. State, 61 S.W.3d 733 (Tex.App.—

Fort Worth 2001, *pet. ref'd*).

The indictment must plead the culpable mental state for the offense. Dinkins v. State, 894 S.W.2d 330 (Tex. Crim. App. 1995).

When recklessness is an element of the offense, the charging instrument must allege the circumstances of the act which indicate that the defendant acted in a reckless manner. State v. Rodriguez, 339 S.W.3d 680 (Tex. Crim. App. 2011).

§12:92.2 Allegations Not Required

The elements of constituent offenses need not be pleaded. Robinson v. State, 596 S.W.2d 130 (Tex. Crim. App. 1980). Therefore, an indictment for attempt is not fundamentally defective for the failure to allege the constituent elements of the offense attempted. Epps v. State, 811 S.W.2d 237 (Tex.App.—Dallas 1991, *no pet.*); *see also* In re J.B.M., 157 S.W.3d 823 (Tex.App.—Fort Worth 2005, *no writ*).

Likewise, the state is not required to allege the constituent elements of an offense constituting the aggravating element of a capital murder indictment. Dinkins.

An indictment charging a consummated offense must properly charge all of the elements of that offense, but an indictment charging an attempted offense is not fundamentally defective for failure to allege the constituent elements of the offense attempted. Wood v. State, 560 S.W.3d 162, 168 (Tex. Crim. App. 2018).

The law of parties (under either a Penal Code §7.02(a) or (b) theory) need not be pled in the indictment. Vodochodsky v. State, 158 S.W.3d 502 (Tex. Crim. App. 2005).

In an aggregated theft case, under Texas Penal Code §31.09, the indictment need only allege aggregation, and each separate theft need not be specifically alleged. Kellar v. State, 108 S.W.3d 311 (Tex. Crim. App. 2003).

Allegations that are not essential to constitute the offense and that might be entirely omitted without affecting the charge against the defendant and without detriment to the indictment are treated as mere surplusage, and they may be entirely disregarded. Hernandez v. State, 903 S.W.2d (Tex. App.—Fort Worth 1995, *pet. ref'd*).

Unnecessary words or allegations in an indictment that are descriptive of statutorily required language must be proved by state as alleged. Hernandez.

The definitions of "intoxicated" in Texas Penal Code §49.01(2) are evidentiary and therefore do not need to be alleged in a charging instrument. Therefore, a trial court should not quash a DWI information charging a defendant with DWI due to the state's failure to allege the definition of "intoxicated" that it intends to prove at trial. State v. Barbernell, 257 S.W.3d 248 (Tex. Crim. App. No. 0867-07, July 2, 2008).

Generally, the State need not plead with additional specificity those matters which are statutorily defined in the alternative where the matter is not an act or omission of the defendant. State v. Jarreau, 512 S.W.3d 352, 354 (Tex. Crim. App. 2017).

An indictment charging delivery of a dangerous drug does not have to allege the nature of the dangerous drug as a device or a drug because that is not a manner or means of delivery and does not describe, concern, involve or go to the act of delivery. State v. Jarreau, 512 S.W.3d 352, 356 (Tex. Crim. App. 2017).

In an indictment alleging tampering with evidence, the specific identity of the tampered-with evidence is not an element of the offense. However, more specificity may be required in the indictment in order to provide the defendant with adequate notice of the charged conduct. State v. Zuniga, 512 S.W.3d 902, 909 (Tex. Crim. App. 2017).

§12:92.3 Sufficiency

The sufficiency of an indictment is a question of law reviewed on a *de novo* basis on appeal. State v. Moff, 154 S.W.3d 599 (Tex. Crim. App. 2004).

An indictment or information is sufficient under the Texas Constitution if it accuses the defendant of a crime with enough clarity and specificity to identify the penal statute under which the state intends to prosecute. Duron v. State, 956 S.W.2d 547 (Tex. Crim. App. 1997) (even though the indictment charging the defendant with indecency with a child alleged innocent conduct).

Tex. Code Crim. Pro. article 21.02 sets out what is required to make an indictment "sufficient." If an indictment does not comply with article 21.02, it is considered to be defective. Jenkins v. State, ___ S.W.3d ___ at *5 2018 WL 6332219, overruling Cook v. State, 902 S.W.2d 471 (Tex. Crim. App. 1995).

However, it has been well-settled since 1985 that a defect in an indictment may not prevent it from qualifying as an indictment under article V, section 12(b) of the Texas Constitution. An indictment can be defective, but still be an indictment that vests the court with jurisdiction. In fact, in Duron v. State, the court of criminal appeals held that in order for a defect to render the instrument a non-indictment, the defect must make it impossible for the defendant to know the offense with which he had been charged. Jenkins, ___ S.W.3d at *5.

Thus, with regard to the requirements under the constitutional definition of indictment—that it must (1) charge a person, and (2) charge the commission of an offense—the proper test is whether the face of the charging instrument is clear enough to give an appellant adequate notice of the charge against him. If this is not the case, the indictment is void. Jenkins, ___ S.W.3d at *5.

The face of the indictment can sufficiently put the defendant on notice that he is the defendant referred to in the indictment. Jenkins, ___ S.W.3d at *5.

Texas law requires the defendant to object to any error in the indictment before the day of trial and certainly before the jury is empaneled or any error of sufficiency is waived. Jenkins, ___ S.W.3d at *6.

An indictment is generally sufficient as long as it tracks the language of a penal statute that itself satisfies the constitutional requirement of notice. Lawrence v. State, 240 S.W.3d 912 (Tex. Crim. App. 2007).

Before a felony trial court has subject matter jurisdiction over an indictment, it must put the defendant on notice that the charging of a felony offense was intended. Kirkpatrick v. State, 279 S.W.3d 324 (Tex. Crim. App. 2008).

Where an indictment fails to provide sufficient notice to inform the accused of the specific acts for which he was charged, the trial court does not err in quashing the indictment. State v. Moff (where any of a large number of transactions could have been relied on by the state for conviction).

A statute is void for vagueness if it fails to define the criminal offense with sufficient definiteness so that ordinary people can understand what conduct is prohibited and in a manner that does not permit arbitrary and discriminatory enforcement. Lawrence.

If a statute does not substantially implicate constitutionally protected conduct or speech, it is valid unless it is impermissibly vague in all applications or as applied to the defendant. Lawrence.

When reviewing a trial court's decision to deny a motion to quash an indictment, an appellate court applies a *de novo* standard of review. Lawrence.

§12:92.4 Variance

A variance occurs when there is a discrepancy between the allegations in the charging instrument and the proof at trial. In a variance situation, the state has proven the defendant guilty of a crime, but has proven its commission in a manner that varies from the allegations in the charging instrument. Gollihar v. State, 46 S.W.3d 243 (Tex. Crim. App. 2001).

The widely-accepted rule, regardless of whether viewing variance as a sufficiency of the evidence problem or as a notice-related problem, is that a variance that is not prejudicial to a defendant's substantial rights is immaterial. In determining whether a defendant's substantial rights have been prejudiced in this context, two questions are generally asked: 1) whether the indictment, as written, informed the defendant of the charge against him sufficiently to allow him to prepare an adequate defense at trial; and 2) whether prosecution under the deficiently drafted indictment would subject the defendant to the risk of being prosecuted later for the same crime. Gollihar.

The Texas "immaterial variance" law as set out in Gollihar does not apply to the specific statutory elements alleged in the indictment. Cada v. State, 334 S.W.3d 766 (Tex. Crim. App. 2011) (a retaliation case where there was a variance between the pleading of the element of "a witness" and proof of the element of "an informant"); Byrd v. State, 336 S.W.3d 242 (Tex. Crim. App. 2011)(a theft case where there was a variance between the pleading of the injured party as "Mike Morales," and the proof which showed the injured party to be "Wal-Mart").

Allegations in an indictment that are not a statutory element or an integral part of an essential element of the offense charged fall under the immaterial variance rule of Gollihar. Fuller 73 S.W.3d 250, 254 (Tex. Crim. App. 2002) (a slight variance between the indictment and proof of the complainant's name is not a material variance).

Variances can be classified into three categories, depending upon the type of allegation that the State has pled in its charging instrument but failed to prove at trial. First, a variance involving statutory language that defines the offense always renders the evidence legally insufficient to support the conviction (i.e., such variances are always material). Second, a variance involving a non-statutory allegation that describes an "allowable unit of prosecution" element of the offense may or may not render the evidence legally insufficient, depending upon whether the variance is material (i.e., such variances are sometimes material). Finally, other types of variances involving immaterial non-statutory allegations do not render the evidence legally insufficient. Johnson v. State, 364 S.W.3d 292 (Tex. Crim. App. 2012) (where the precise act or nature of conduct was inconsequential because aggravated assault by causing serious bodily injury is a result-oriented offense). *See also* Ramos v. State, 407 S.W.3d 265, 269 (Tex. Crim. App. 2013).

§12:92.5 Number of Convictions Authorized

An indictment cannot authorize more convictions than there are counts. When the court's charge authorizes convictions on more than one paragraph per count, the trial judge has the responsibility of structuring the judgment to show only one conviction per count. Martinez v. State, 225 S.W.3d 550 (Tex. Crim. App. 2007).

§12:92.6 Election of Primary Act Relied on for Conviction

See §17:11 for a complete analysis of the case law on this subject.

§12:92.7 Duplicity

"Duplicity" is the technical fault of uniting two or more distinct and separate offenses in the same count of an indictment. Lebo v. State, 100 S.W.3d 417 (Tex.App.—San Antonio 2002, *pet. ref'd*).

The rule against duplicitous indictments is based on the proposition that a defendant must receive fair notice of the charge against which he must defend. Lebo.

An indictment that conjunctively alleges more than one culpable mental state in the commission of an offense is not duplicitous. Lebo.

§12:93 Challenges to Charging Instruments

§12:93.1 Statutes

CCP Art. 1.14. Waiver of Rights

> *(a) The defendant in a criminal prosecution for any offense may waive any rights secured him by law except that a defendant in a capital felony case may waive the right of trial by jury only in the manner permitted by Article 1.13(b) of this code.*
>
> *(b) If the defendant does not object to a defect, error, or irregularity of form or substance in an indictment or information before the date on which the trial on the merits commences, he waives and forfeits the right to object to the defect, error, or irregularity and he may not raise the objection on appeal or in any other post-conviction proceeding. Nothing in this article prohibits a trial court from requiring that an objection to an indictment or information be made at an earlier time in compliance with Article 28.01 of this code.*

CCP Art. 21.23.

> *The rules with respect to allegations in an indictment and the certainty required apply also to information.*

CCP Art. 27.03. Motion to Set Aside Indictment

> *In addition to any other grounds authorized by law, a motion to set aside an indictment or information may be based on the following:*

(1) That it appears by the records of the court that the indictment was not found by at least nine grand jurors, or that the information was not based upon a valid complaint;

(2) That some person not authorized by law was present when the grand jury was deliberating upon the accusation against the defendant, or was voting upon the same; and

(3) That the grand jury was illegally impaneled; provided, however, in order to raise such question on motion to set aside the indictment, the defendant must show that he did not have an opportunity to challenge the array at the time the grand jury was impaneled.

CCP Art. 27.04. Motion Tried by Judge

An issue of fact arising upon a motion to set aside an indictment or information shall be tried by the judge without a jury.

CCP Art. 27.08. Exception to Substance of Indictment

There is no exception to the substance of an indictment or information except:

(1) That it does not appear therefrom that an offense against the law was committed by the defendant;

(2) That it appears from the face thereof that a prosecution for the offense is barred by a lapse of time, or that the offense was committed after the finding of the indictment;

(3) That it contains matter which is a legal defense or bar to the prosecution; and

(4) That it shows upon its face that the court trying the case has no jurisdiction thereof.

CCP Art. 27.09. Exception to Form of Indictment

Exceptions to the form of an indictment or information may be taken for the following causes only:

(1) That it does not appear to have been presented in the proper court as required by law;

(2) The want of any requisite prescribed by Articles 21.02 and 21.21.

(3) That it was not returned by a lawfully chosen or empaneled grand jury.

CCP Art. 27.10. Written Pleadings

All motions to set aside an indictment or information and all special pleas and exceptions shall be in writing.

CCP Art. 28.04. Quashing Charge in Misdemeanor

If the motion to set aside or the exception to an indictment or information is sustained, the defendant in a misdemeanor case shall be discharged, but may be again prosecuted within the time allowed by law.

CCP Art. 28.05. Quashing Indictment in Felony

If the motion to set aside or the exception to the indictment in cases of felony be sustained, the defendant shall not therefore be discharged, but may immediately be recommitted by order of the court, upon motion of the State's attorney or without motion; and proceedings may afterward be had against him as if no prosecution had ever been commenced.

§12:93.2 Motion Must Be Filed and Heard Before Trial Date

Motions raising defects in charging instruments must be filed and heard before the scheduled trial date or the motion is waived. The burden is on the defendant to see that this is done. State v. Turner, 898 S.W.2d 303 (Tex. Crim. App. 1995).

Similarly, the defendant must object to a misdemeanor complaint before trial or the issue is waived. Ramirez v. State, 105 S.W.3d 628 (Tex. Crim. App. 2003).

A defendant must raise a claim that the indictment is defective on the grounds that it alleges the manner and means of committing the offense in the disjunctive prior to trial or the claim is forfeited. Swain v. State, 181 S.W.3d 359 (Tex. Crim. App. 2005).

Since the 1985 amendments to Texas Constitution Art. 5, §12(b), any variance between complaint and information, or any defect in the complaint must now be raised pretrial or the claim is waived. The matter cannot be validly raised for the first time on appeal since the mere filing of the complaint vests the trial court with jurisdiction. This amendment did not change the case law interpreting sufficiency of charging instruments, but only mandates that challenges be raised pretrial under Art. 28.10. Aguilar v. State, 846 S.W.2d 318 (Tex. Crim. App. 1993).

Substance exceptions to charging instruments must be raised before trial or the accused forfeits his right to raise objections on appeal or by collateral attack. While omitting an element of an indictment is a defect of substance in an indictment, the instrument does not cease to be an indictment despite the omission of that element. Studer v. State, 799 S.W.2d 263 (Tex. Crim. App. 1990).

A defendant may not raise for the first time on appeal a facial challenge to the constitutionality of a statute. This must be done in the trial court. Karenev v. State, 281 S.W.3d 428 (Tex. Crim. App. 2009).

§12:93.3 Grounds for Challenge

§12:93.3.1 General Challenges

The Texas and United States Constitutions grant a criminal defendant the right to fair notice of the specific charged offense. To provide this fair notice, the charging instrument must convey sufficient information to allow the accused to prepare a defense. In most cases, a charging instrument that tracks the relevant statutory text will provide adequate notice to the accused. State v. Ross, 573 S.W.3d 817, 820 (Tex. Crim. App. 2019).

However, tracking the language of the statute may be insufficient if the statutory language is not completely descriptive of an offense. State v. Ross, 573 S.W.3d at 820.

For example, if the prohibited conduct is statutorily defined to include more than one manner or means of commission, then the State must, upon timely request, allege the particular manner or means it seeks to establish. State v. Ross, 573 S.W.3d at 820.

An indictment can be challenged on the ground that the underlying statute is unconstitutional by not giving fair notice to the defendant of what conduct is criminally prohibited; in other words, that the statute is vague. Bynum v. State, 767 S.W.2d 769 (Tex. Crim. App. 1989).

Indictments based on statutes that regulate speech can be challenged on the ground that the underlying statute is overbroad. Bynum.

Under the First Amendment, a statute may be subject to a facial challenge even though it may have some legitimate application, but absent the First Amendment, a facial challenge could be mounted successfully only if the statute were vague in all of its applications. Sanchez v. State, 995 S.W.2d 677 (Tex. Crim. App. 1999).

For example, a North Carolina statute making it a felony for a registered sex offender to gain access to a number of websites, including commonplace social media websites like Facebook and Twitter, was found to be an impermissible restriction of the First Amendment's Free Speech Clause, applicable to the States under the Due Process Clause of the Fourteenth Amendment. Packingham v. North Carolina, 582 U.S. ___, 137 S.Ct. 1730, 1737, 198 L.Ed.2d 273 (2017).

In situations where a law reaches a substantial amount of constitutionally protected conduct, a facial challenge can be made to the statute even when it conceivably could have some valid applications. State v. Markovich, 77 S.W.3d 274 (Tex. Crim. App. 2002).

To be vague in all of its applications, a statute must necessarily be vague as to the litigant. If the statute is not vague as to the litigant, a Due Process challenge must necessarily fail. A person who engages in some conduct that is clearly proscribed cannot complain of the vagueness of the law as applied to the conduct of others. Sanchez.

Whenever recklessness enters into or is a part

or element of any offense, or it is charged that the defendant acted recklessly in the commission of an offense, the charging instrument must allege, with reasonable certainty, the act or acts relied upon to constitute recklessness or it is subject to a pretrial motion to quash. Smith v. State, 309 S.W.3d 10 (Tex. Crim. App. 2010).

The explication of the act or acts constituting recklessness may be the same as those that constitute the prohibited conduct, but they may also require additional pleading. Smith (explaining that an information charging indecent exposure would require a further allegation that the actor's exposition was done in a reckless manner).

An indictment that fails to allege all of the elements of a statutory offense can be challenged by way of a motion to quash. Grant v. State, 970 S.W.2d 22 (Tex. Crim. App. 1998).

Where the state adds a paragraph to the indictment seeking to toll the statute of limitations, such language must be challenged pretrial via a motion to quash the indictment or a motion to dismiss. This challenge cannot be made in a pretrial writ of habeas corpus where the deficiencies in such allegations are reparable. Ex parte Smith, 178 S.W.3d 797 (Tex. Crim. App. 2005).

When an indictment shows on its face that prosecution is absolutely barred by the statute of limitations and that pleading is not reparable, a defendant may seek relief from further prosecution via a pretrial writ of habeas corpus. Ex parte Smith.

For offenses such as murder with no statute of limitations, a defendant may establish a Fifth Amendment due-process violation where a prosecution has been significantly delayed by proving both prejudice and that an intentional delay was designed to give the state a tactical advantage. State v. Krizan-Wilson, 354 S.W.3d 808 (Tex. Crim. App. 2011).

A defendant's "in pari materia" claim is generally not cognizable as a motion to quash because generally the indictment will be sufficient on its face. The claim is also not cognizable on a pretrial writ of habeas corpus. Ex parte Smith, 185 S.W.3d 887 (Tex. Crim. App. 2006).

An indictment will not be found insufficient and therefore be quashed pretrial on the basis of insufficient evidence to support it. State v. Rosenbaum, 910 S.W.2d 934 (Tex. Crim. App. 1994).

A decision to quash an indictment for lack of certainty in the indictment's allegations is within the sound discretion of the trial court. State v. Draper, 940 S.W.2d 824 (Tex.App.—Austin 1997, *no pet.*).

The sufficiency of a charging instrument presents a question of law. An appellate court therefore reviews a trial judge's ruling on a motion to quash de novo. State v. Ross, 573 S.W.3d at 820.

The trial court did not abuse its discretion in holding a pretrial evidentiary hearing on Hill's motion to dismiss alleging a prosecutorial misconduct and vindictive prosecution. Under the plain language of Article 28.01, the trial court judge had the discretion to hold an evidentiary hearing on Hill's motion to quash and dismiss. State of Texas v. Hill, 499 S.W.3d 853 (Tex. Crim. App. 2016).

§12:93.3.2 Constitutional Challenges

§12:93.3.2.1 Overbreadth Challenges

The First Amendment provides in relevant part that "Congress shall make no law ... abridging the freedom of speech." U.S. Const. amend. I. As a general matter, the First Amendment means that government has no power to restrict expression because of its message, its ideas, its subject matter, or its content. Ashcroft v. Am. Civil Liberties Union, 535 U.S. 564, 573, 122 S.Ct. 1700, 152 L.Ed.2d 771 (2002); Wagner v. State, 539 S.W.3d 298, 310 (Tex. Crim. App. 2017).

Ordinarily, a facial challenge to the constitutionality of a statute can succeed only when it is shown that the statute is unconstitutional in all of its applications. State v. Johnson, 475 S.W.3d 860, 864 (Tex. Crim. App. 2015); Wagner v. State, 539 S.W.3d at 310.

The First Amendment overbreadth doctrine provides an exception to this rule, whereby a litigant may succeed in challenging a law that regulates speech if a substantial number of its applications are unconstitutional, judged in relation to the statute's plainly legitimate sweep. Wash. State Grange v. Wash. State Republican Party, 552 U.S. 442, 449 n.6, 128 S.Ct. 1184, 170 L.Ed.2d 151 (2008); Wagner v. State, 539 S.W.3d at 310.

Thus, the overbreadth doctrine prohibits the government from banning unprotected speech if a substantial amount of protected speech is prohibited or chilled in the process. Ashcroft v. Free Speech Coalition, 535 U.S. 234, 255, 122 S.Ct. 1389, 152 L.Ed.2d 403 (2002); Wagner v. State, 539 S.W.3d at 310.

The overbreadth doctrine is "strong medicine" to be employed with hesitation and only as a last resort. Ex parte Thompson, 442 S.W.3d 325, 349 (Tex. Crim. App. 2014); Wagner v. State, 539 S.W.3d at 310.

Penal Code Section 25.07(a)(2)(A) does not reach a substantial amount of constitutionally protected conduct, judged in relation to its plainly legitimate sweep. Wagner v. State, 539 S.W.3d at 310.

§12:93.3.2.2 Vagueness Challenges

To pass constitutional muster, a law that imposes criminal liability must be sufficiently clear (1) to give a person of ordinary intelligence a reasonable opportunity to know what is prohibited and (2) to establish determinate guidelines for law enforcement. State v. Doyal, ___ S.W.3d ___ at *5 (PD-0254-18, 2019 WL 944022, February 27, 2019) (where the court concluded that § 551.143 of the Texas open meetings act is unconstitutionally vague on its face).

What renders a statute vague is the indeterminacy of precisely what the prohibited conduct is. State v. Doyal, ___ S.W.3d at *5.

It is a basic principle of due process that an enactment is void for vagueness if its prohibitions are not clearly defined. Grayned v. City of Rockford, 408 U.S. 104, 108–09, 92 S.Ct. 2294, 33 L.Ed.2d 222 (1972); Wagner v. State, 539 S.W.3d 298, 313 (Tex. Crim. App. 2017).

A statute which uses an undefined term of indeterminate or variable meaning requires more specific pleading in order to notify the defendant of the nature of the charges against him. State v. Ross, 573 S.W.3d 817, 820 (Tex. Crim. App. 2019).

A conviction fails to comport with due process if the statute under which it is obtained fails to provide a person of ordinary intelligence fair notice of what is prohibited, or is so standardless that it authorizes or encourages seriously discriminatory enforcement. United States v. Williams, 553 U.S. 285, 304, 128 S.Ct. 1830, 170 L.Ed.2d 650 (2008); Wagner v. State, 539 S.W.3d at 313.

Where a statute abuts upon sensitive areas of First Amendment freedoms, the doctrine of vagueness demands a greater degree of specificity than in other contexts. Long v. State, 931 S.W.2d 285, 287–88 (Tex. Crim. App. 1996); Wagner v. State, 539 S.W.3d at 313.

Greater specificity is required to preserve adequately the right of free expression because uncertain meanings inevitably lead citizens to steer far wider of the unlawful zone than if the boundaries of the forbidden areas were clearly marked. Long v. State, 931 S.W.2d at 288; Wagner v. State, 539 S.W.3d at 313-4.

But perfect clarity and precise guidance have never been required, even of regulations that restrict expressive activity. Holder v. Humanitarian Law Project, 561 U.S. 1, 19, 130 S.Ct. 2705, 177 L.Ed.2d 355 (2010); Wagner v. State, 539 S.W.3d at 314.

A statute is not unconstitutionally vague merely because the words or terms used are not specifically defined. Bynum v. State, 767 S.W.2d 769, 774 (Tex. Crim. App. 1989); see also State v. Holcombe, 187 S.W.3d 496, 499 (Tex. Crim. App. 2006); Wagner v. State, 539 S.W.3d at 314.

Rather, the words or phrase must be read in the context in which they are used, and ordinarily the statute must then be construed according to the rules of grammar and common usage. Bynum, 767 S.W.2d at 774; Wagner v. State, 539 S.W.3d at 314.

A statute satisfies vagueness requirements if the statutory language conveys sufficiently definite warning as to the proscribed conduct when measured by common understanding and practices. Jordan v. De George, 341 U.S. 223, 231–32, 71 S.Ct. 703, 95 L.Ed. 886 (1951); Wagner v. State, 539 S.W.3d at 314.

In general, in addressing a vagueness challenge, we will consider whether the statute is vague as applied to a defendant's conduct before considering whether the statute may be vague as applied to the conduct of others. Vill. of Hoffman Estates v. Flipside, Hoffman Estates, Inc., 455 U.S. 489, 495, 102 S.Ct. 1186, 71 L.Ed.2d 362 (1982); Wagner v. State, 539 S.W.3d at 314.

A plaintiff who engages in some conduct that is clearly proscribed cannot complain of the vagueness of the law as applied to the conduct of others. A court should therefore examine the complainant's conduct before analyzing other hypothetical applications of the law. Vill. of Hoffman Estates, 455 U.S. at 495, 102 S.Ct. 1186; Wagner v. State, 539 S.W.3d at 314.

In the context of a challenge to a statute that does not regulate protected speech, a court should uphold a vagueness challenge only if the statute is impermissibly vague in all of its applications. Vill. of Hoffman Estates, 455 U.S. at 494-5, 102 S.Ct. 1186; Wagner v. State, 539 S.W.3d at 314.

Applying the plain language of Section 25.07(a)(2)(A) to the facts of this case, we conclude that the statutory language is sufficiently clear to provide a person of ordinary intelligence a reasonable opportunity to know that appellant's course of conduct would be prohibited. Wagner v. State, 539 S.W.3d at 314.

More clarity is required of a criminal law when that law implicates First Amendment freedoms. State v. Doyal, ___ S.W.3d at *3.

The First Amendment is implicated when the government seeks to impose criminal sanctions on an elected official for communications made in his official capacity. State v. Doyal, ___ S.W.3d at *3.

A facial vagueness challenge to a statute that implicates First Amendment freedoms does not require a showing that there are no possible instances of conduct clearly falling within the statute's prohibitions. State v. Doyal, ___ S.W.3d at *4.

§12:93.4 Defendant's Burden

The defendant generally has the burden of proof on a motion to quash an indictment. Wheat v. State, 537 S.W.2d 20 (Tex. Crim. App. 1976); Farrar v. State, 95 S.W.3d 648 (Tex.App.—Eastland 2002, *no pet.*).

Where the defendant is challenging the constitutionality of a statute his burden is to show beyond a reasonable doubt that the statute is unconstitutional. Luquis v. State, 72 S.W.3d 355 (Tex. Crim. App. 2002).

In most instances, a defendant must make a pretrial objection to a substance defect in the information or indictment (usually by a motion to quash) or forfeit the right to complain about it on appeal. Smith v. State, 309 S.W.3d 10 (Tex. Crim. App. 2010).

The sufficiency of a charging instrument presents a question of law, the trial court's ruling on which is reviewed *de novo* by an appeals court. Smith.

§12:93.5 Form: Motions

See our companion book *Texas Criminal Forms* (James Publishing):

- Form 12-3, Motion to Quash Indictment.
- Form 12-4, Motion to Quash Indictment (Lack of Specificity of Deadly Weapon Allegation).

§12:94 Amendments to the Charging Instrument

§12:94.1 Statutes

CCP Art. 28.10. Amendment of Indictment or Information

(a) After notice to the defendant, a matter of form or substance in an indictment or information may be amended at any time before the date the trial on the merits commences. On the request of the defendant, the court shall allow the defendant not less than 10 days, or a shorter period if requested by the defendant, to respond to the amended indictment or information.

(b) A matter of form or substance in an indictment or information may also be amended after the trial on the merits commences if the defendant does not object.

(c) An indictment or information may not be amended over the defendant's objection as to form or substance if the amended indictment or information charges the defendant with an additional or different offense or if the substantial rights of the defendant are prejudiced.

CCP Art. 28.11. How Amended

All amendments of an indictment or information shall be made with the leave of the court and under its direction.

§12:94.2 What Is an Amendment

An amendment is an alteration to the face of the charging instrument which affects the substance of the charging instrument. Eastep v. State, 941 S.W.2d 130 (Tex. Crim. App. 1997) *overruled on other grounds in part by* Gollihar v. State, 46 S.W.3d 243 (Tex. Crim. App. 2001), *and* Riney v. State, 28 S.W.3d 561 (Tex. Crim. App. 2000).

The alteration of a cause number in an enhancement paragraph is an amendment. Sodipo v. State, 815 S.W.2d 551 (Tex. Crim. App. 1990); Eastep.

The alteration of a weapon in an aggravated assault indictment is an amendment. Hillin v. State, 808 S.W.2d 486 (Tex. Crim. App.1991); Eastep.

The addition of the manner and means of committing an offense is an amendment. Beebe v. State, 811 S.W.2d 604 (Tex. Crim. App.1991); Eastep.

The alteration of the alleged date is an amendment. Garcia v. State, 928 S.W.2d 666 (Tex. App.— Corpus Christi 1996, *no pet.*); Eastep.

The addition of a complainant is an amendment. McCoy v. State, 889 S.W.2d 354 (Tex.App.— Houston [14th Dist.] 1994, pet. ref'd); Eastep.

The alteration of the indictment from "two facsimile machines" to "one facsimile machine" is an amendment. Hilton v. State, 879 S.W.2d 74 (Tex.App.—Houston [14th Dist.] 1994, pet. ref'd); Eastep.

Deleting "delivery" and replacing it with "possession" is an amendment. Hinojosa v. State, 875 S.W.2d 339 (Tex.App.—Corpus Christi 1994, *no pet.*); Eastep.

A change of a name in an indictment pursuant to Art. 26.08 is not an amendment to that indictment under Art. 28.10. Kelley v. State, 823 S.W.2d 300 (Tex. Crim. App. 1992); Wynn v. State, 864 S.W.2d 539 (Tex. Crim. App. 1993).

However, the complete failure to allege the name of the defendant in the indictment renders that instrument not a valid charging instrument. Cook v. State, 902 S.W.2d 471 (Tex. Crim. App. 1995).

Alterations made solely to a separate written confession document are not sufficient to constitute an amendment to the indictment. Puente v. State, 320 S.W.3d 352 (Tex. Crim. App. 2010).

Because punishment enhancement allegations are not required to be plead in the indictment, the addition of an enhancement allegation by written motion does not constitute an amendment to the indictment. Brooks v. State, 957 S.W.2d 30 (Tex. Crim. App. 1997).

There are three situations where an alteration to the face of a charging instrument does not affect its substance, and, is therefore, an abandonment: (1) when it abandons one or more of the alternative means of committing an offense; (2) when it has the effect of reducing the prosecution to a lesser included offense; and (3) when it omits surplusage. Eastep.

Surplusage is unnecessary language not legally essential to constitute the offense alleged in the indictment. Eastep.

§12:94.3 Motion Must Be Filed Before Trial Date

A motion to amend a charging instrument must be made before the date that the trial commences. Sodipo v. State, 815 S.W.2d 551 (Tex. Crim. App. 1990).

Where an indictment is amended over the defendant's objection, the defendant is entitled to a 10-day continuance to prepare for trial. Rent v. State, 838 S.W.2d 548 (Tex. Crim. App. 1992).

Violations of Art. 28.10 are not subject to harm analysis. Brown v. State, 828 S.W.2d 762 (Tex. Crim. App. 1991); Eastep v. State, 941 S.W.2d 130 (Tex. Crim. App. 1997) *overruled on other grounds in part by* Gollihar v. State, 46 S.W.3d 243 (Tex. Crim. App. 2001), *and* Riney v. State, 28 S.W.3d 561 (Tex. Crim. App. 2000).

§12:94.4 Procedure to Amend

The mere granting of a motion to amend a charging instrument is not legally sufficient to constitute a valid amendment. Ward v. State, 829 S.W.2d 787 (Tex. Crim. App. 1992).

One way of validly amending a charging instrument is for the trial court to physically alter the indictment to reflect the amendment. Ward.

Another way to accomplish the amendment is for the prosecution to submit a photocopy of the original charging instrument that contains the amendment for the record. Riney v. State, 28 S.W.3d 561 (Tex. Crim. App. 2000); Perez v. State, 429 S.W.3d 639, 643 (Tex. Crim. App. 2014).

§12:94.5 Form: Motion to Amend

See our companion book *Texas Criminal Forms* (James Publishing):

- Form 12-6, Motion to Amend Indictment.

§12:95 Dismissal of Charging Instrument With Prejudice

A trial court has the power to dismiss a case without the state's consent when a defendant has been denied a right to a speedy trial, when there is a defect in the charging instrument, or pursuant to Art. 32.01, where a defendant is detained and no charging instrument is properly presented. State v. Johnson, 821 S.W.2d 609 (Tex. Crim. App. 1991).

Additionally, a charging instrument may be dismissed to remedy a violation of the Sixth Amendment right to counsel. State v. Frye, 897 S.W.2d 324 (Tex. Crim. App. 1995).

While a trial court may dismiss a charging instrument to remedy a constitutional violation, the dismissal of an indictment is a drastic measure only to be used in the most extraordinary circumstances. Frye.

Although a particular constitutional violation has not yet been recognized as a basis for a trial court to dismiss a charging instrument, this does not preclude a trial court from having authority to dismiss on that ground. State v. Terrazas, 962 S.W.2d 38 (Tex. Crim. App. 1998).

Where there is no constitutional violation, or where the defendant's rights were violated but dismissal of the indictment was not necessary to neutralize the taint of the unconstitutional action, the trial court abuses its discretion in dismissing the charging instrument without the consent of the state. Terrazas.

A trial court abuses its discretion in dismissing a charging instrument without the consent of the state to protect a defendant from retaliation in prison where no constitutional violation has occurred. State v. Mungia, 119 S.W.3d 814 (Tex. Crim. App. 2003).

§12:100 Statute of Limitations

§12:101 Statutes

Tex. Code Crim. Pro. Art. 12.01. Felonies

Except as provided in Article 12.03, felony indictments may be presented within these limits, and not afterward:

(1) no limitation:

(A) murder and manslaughter;

(B) sexual assault under Section 22.011(a)(2), Penal Code, or aggravated sexual assault under Section 22.021(a)(1)(B), Penal Code;

(C) sexual assault, if (i) during the investigation of the offense biological matter is collected and subjected to forensic DNA testing and the testing results show that the matter does not match the victim or any other person whose identity is readily ascertained; or (ii) probable cause exists to believe that the defendant has committed the same or similar sexual offense against five or more victims;

(D) continuous sexual abuse of young child or children under Section 21.02, Penal Code;

(E) indecency with a child under Section 21.11, Penal Code; or

(F) an offense involving leaving the scene of an accident under Section 550.021, Transportation Code, if the accident resulted in the death of a person;

(2) ten years from the date of the commission of the offense:

(A) theft of any estate, real, personal or mixed, by an executor, administrator, guardian or trustee, with intent to defraud any creditor, heir, legatee, ward, distributee, beneficiary or settlor of a trust interested in such estate;

(B) theft by a public servant of government property over which he exercises control in his official capacity;

(C) forgery or the uttering, using or passing of forged instruments;

(D) injury to an elderly or disabled individual punishable as a felony of the first degree under Section 22.04, Penal Code;

(E) sexual assault, except as provided by Subdivision (1); or

(F) arson;

(3) seven years from the date of the commission of the offense:

(A) misapplication of fiduciary property or property of a financial institution;

(B) securing execution of document by deception;

(C) a felony violation under Chapter 162, Tax Code;

(D) false statement to obtain property or credit under Section 32.32, Penal Code;

(E) money laundering;

(F) credit card or debit card abuse under Section 32.31, Penal Code; or

(G) fraudulent use or possession of identifying information under Section 32.51, Penal Code;

(4) five years from the date of the commission of the offense:

(A) theft or robbery;

(B) except as provided by Subdivision (5), kidnapping or burglary;

(C) injury to an elderly or disabled individual that is not punishable as a felony of the first degree under Section 22.04, Penal Code;

(D) abandoning or endangering a child; or

(E) insurance fraud;

(5) if the investigation of the offense shows that the victim is younger than 17 years of age at the time the offense is committed, 20 years from the 18th birthday of the victim of one of the following offenses:

(A) sexual performance by a child under Section 43.25, Penal Code;

(B) aggravated kidnapping under Section 20.04(a)(4), Penal Code, if the defendant committed the offense with the intent to violate or abuse the victim sexually; or

(C) burglary under Section 30.02, Penal Code, if the offense is punishable under Subsection (d) of that section and the defendant committed the offense with the intent to commit an offense described by Subdivision (1)(B) or (D) of this article or Paragraph (B) of this subdivision;

(6) ten years from the 18th birthday of the victim of the offense:

injury to a child under Section 22.04, Penal Code; or

(7) three years from the date of the commission of the offense: all other felonies.

CCP Art. 12.02. Misdemeanors

An indictment or information for any misdemeanor may be presented within two years from the date of the commission of the offense, and not afterward.

CCP Art. 12.03. Aggravated Offenses, Attempt, Conspiracy, Solicitation, Organized Criminal Activity

(a) The limitation period for criminal attempt is the same as that of the offense attempted.

(b) The limitation period for criminal conspiracy or organized criminal activity is the same as that of the most serious offense that is the object of the conspiracy or the organized criminal activity.

(c) The limitation period for criminal solicitation is the same as that of the felony solicited.

(d) Except as otherwise provided by this chapter, any offense that bears the title "aggravated" shall carry the same limitation period as the primary crime.

CCP Art. 12.04. Computation

The day on which the offense was committed and the day on which the indictment or information is presented shall be excluded from the computation of time.

CCP Art. 12.05. Absence From State and Time of Pendency of Indictment, etc., Not Computed

(a) The time during which the accused is absent from the state shall not be computed in the period of limitation;

(b) The time during the pendency of an indictment, information, or complaint shall not be computed in the period of limitation;

(c) The term "during the pendency," as used herein, means that period of time beginning with the day the indictment, information, or complaint is filed in a court of competent jurisdiction, and ending with the day such accusation is, by an order of a trial court having jurisdiction thereof, determined to be invalid for any reason.

CCP Art. 27.08. Exception to Substance of Indictment

There is no exception to the substance of an indictment or information except:

(1) That it does not appear therefrom that an offense against the law was committed by the defendant;

(2) That it appears from the face thereof that a prosecution for the offense is barred by a lapse of time, or that the offense was committed after the finding of the indictment;

(3) That it contains matter which is a legal defense or bar to the prosecution; and(4) That it shows upon its face that the court trying the case has no jurisdiction thereof.

§12:102 Purpose of the Statute of Limitations

The statute of limitations is an act of grace for the benefit of potential defendants, a voluntary surrendering by the people of their right to prosecute. Proctor v. State, 967 S.W.2d 840 (Tex. Crim. App. 1998).

This act of grace serves several objectives:

(1) it protects defendants from having to defend themselves against charges when the basic facts may—or may not—have become obscured by time; (2) it prevents prosecution of those who have been law-abiding for some years; and (3) it lessens the possibility of blackmail. Proctor.

§12:103 Computing the Limitations Period

The period of limitation for presentment of charging instruments begins with the day of commission of the offense and runs to the limits in terms of years respectively prescribed in Arts. 12.01, 12.02 and 12.03. Ex parte Matthews, 933 S.W.2d 134 (Tex. Crim. App. 1996).

For a single offense of aggregate theft, which is the combination of multiple offenses of theft arising from one scheme or continuing course of conduct, the statute of limitations does not begin to run until the date of the commission of the final incident of theft. King v. State, 17 S.W.3d 7 (Tex. App.—Houston [14th Dist.] 2000, *pet. ref'd*).

The statute of limitations for aggravated assault is governed by Tex. Code Crim. Pro. art. 12.03(d), not Article 12.01(7); thus the limitation period is the two year period of the primary crime of misdemeanor assault. The lesser-included offense with the greater limitation period does not control when the lesser-included offenses of the aggravated assault include both misdemeanor assault and a felony. State v. Schunior, 506 S.W.3d 29, 37 (Tex. Crim. App. 2016).

§12:104 Asserting the Statute of Limitations

Statute-of-limitations defenses that are based on facts (challenging a pleading that includes a tolling paragraph, explanatory averments, or even innuendo allegations, that suffice to show that the charged offense is not, at least on the face of the indictment, barred by limitations), are waived if not presented at or before the guilt phase of the trial, and cannot be presented in a pretrial writ. Phillips v. State, 362 S.W.3d 606 (Tex. Crim. App. 2011) (overruled on other grounds by Ex parte Heilman, 456 S.W.3d 159 (Tex. Crim. App. 2015)), *citing* Proctor v. State, 967 S.W.2d 840 (Tex. Crim. App. 1998).

Appeals courts will no longer unquestioningly distinguish between factual and pure-law limitations defenses. Instead, in circumstances lacking any legislative ex post facto violation—and especially when that occurs in the context of a good-faith, arm's length plea agreement—both are Marin category-three forfeitable rights, as under Proctor v. State. Ex parte Heilman, 456 S.W.3d 159 (Tex. Crim. App. 2015).

For offenses such as murder with no statute of limitations, a defendant may establish a Fifth Amendment due-process violation where a prosecution has been significantly delayed by proving both prejudice and that an intentional delay was designed to give the state a tactical advantage. State v. Krizan-Wilson, 354 S.W.3d 808 (Tex. Crim. App. 2011).

§12:104.1 Pretrial Procedure

Before trial, a defendant may assert the statute of limitations defense by filing a motion to dismiss under CCP Art. 27.08(2). Proctor v. State, 967 S.W.2d 840 (Tex. Crim. App. 1998).

The denial of a pretrial motion to dismiss on limitations grounds is not a final judgment from which an interlocutory appeal may be taken. Ahmad v. State, 158 S.W.3d 525 (Tex.App.—Fort Worth 2005, pet. ref'd).

If the state's pleading, on its face, shows that the offense charged is barred by limitations, then the defendant may challenge it by filing for habeas corpus relief. Ex parte Tamez, 38 S.W.3d 159 (Tex. Crim. App. 2001).

Denial of a pretrial writ of habeas corpus making the same ground would be appealable on an interlocutory basis. Ex parte Tamez.

§12:104.2 Procedure During Trial

At trial, the defendant may assert the defense by requesting a jury instruction on limitations if there is some evidence before the jury, from any source, that the prosecution is limitations-barred. If there is some such evidence and the defendant requests a jury instruction on the limitations defense, then the state must prove beyond a reasonable doubt that the prosecution is not limitations-barred. Proctor v. State, 967 S.W.2d 840 (Tex. Crim. App. 1998).

§12:105 Tolling Provisions

There are exceptions to the statute of limitations. One is that the time during the pendency of an indictment, information, or complaint shall not be computed in the period of limitation where the defendant was under indictment for the same conduct, same act, or same transaction. Hernandez v. State, 127 S.W.3d 768 (Tex. Crim. App. 2004); Ex parte Smith, 178 S.W.3d 797 (Tex. Crim. App. 2005).

The tolling provisions of Art. 12.05 operate as a constraint on the limitations defense. White v. State, 50 S.W.3d 31 (Tex.App.—Waco 2001, *pet. ref'd).*

A first indictment tolls the statute of limitations where both the first and subsequent indictments allege the same conduct, act or transaction, even where the offenses alleged are not contained in the same statute. Hernandez v. State, 127 S.W.3d 768 (Tex. Crim. App. 2004); Marks v. State, 560 S.W.3d 169, 170 (Tex. Crim. App. 2018).

However, a subsequent indictment is barred by the statute of limitations if it broadens or substantially amends the charges in the original indictment. Hernandez.

An information is sufficient to toll limitations in a felony case. Ex parte Ulloa, 514 S.W.3d 756, 760 (Tex. Crim. App. 2017).

If a previous indictment or other facts would toll the statute of limitations, the state must plead those facts in the indictment (or subsequent indictment) or the indictment is subject to a defense motion to dismiss. Tita v. State, 267 S.W.3d 33 (Tex. Crim. App. 2008).

A tolling paragraph need not be alleged with the same degree of particularity as we would expect of an allegation of the charged offense. Ex parte Smith; Freda v. State, 704 S.W.2d 41 (Tex. Crim. App. 1986).

Any purported defects of form and substance in the tolling paragraph relate to notice and must be brought to the trial court's attention before trial or they are waived under CCP Art. 1.14(b). They may and must be raised in a motion to quash or motion to dismiss the pleading. Ex parte Smith.

Purported defects in a tolling paragraph are forfeited unless raised in a pretrial motion to quash or dismiss, and such defects may not be raised in a pretrial writ because they do not deprive the trial court of its jurisdiction. Only when an indictment shows on its face that prosecution is absolutely barred by the statute of limitations and that pleading is not reparable, may a defendant seek relief from further prosecution via a pretrial writ of habeas corpus. Ex parte Smith.

A harm analysis under TRAP 44.2(b) is necessary when a trial court fails to grant a motion to quash an indictment that does not toll the statute of limitations. Mercier v. State, 322 S.W.3d 258 (Tex. Crim. App. 2010).

Before the court can charge the jury on the tolling provisions of CCP Art. 12.05, the record must contain sufficient evidence that a former indictment, information or complaint was pending in a court of competent jurisdiction for a sufficient length of time as to make the present charging instrument timely. White.

Presentment of an information charging official oppression vests a district court with jurisdiction over the offense and the defendant, making it a court of competent jurisdiction. State v. Drummond, 501 S.W.3d 78, 84 (Tex. Crim. App. 2016).

§12:106 Form: Motions Based on Statute of Limitations

See our companion book *Texas Criminal Forms* (James Publishing):

- Form 15-13.1, Motion to Dismiss Under Statute of Limitations.
- Form 15-13.2, Petition for Writ of Habeas Corpus Under Statute of Limitations.
- Form 15-14, Motion to Quash and Dismiss Indictment (Insufficient Tolling Allegation).
- Form 12-5, Brief in Support of Motion to Quash Indictment Based on Statute of Limitations.

§12:110 Identification Evidence

§12:111 Statutory Law

Tex. Code Crim. Pro. Art. 2.1386. Eyewitness identification protocols

(a) In this article, "law enforcement agency" means an agency of the state or an agency of a political subdivision of the state authorized by law to employ peace officers.

(b) The Texas Commission on Law Enforcement shall establish a comprehensive education and training program on eyewitness identification, including material regarding variables that affect a witness's vision and memory, practices for minimizing contamination, and effective eyewitness identification protocols.

(c) Each law enforcement agency shall require each peace officer who is employed by the agency and who performs eyewitness identification procedures to complete the education and training described by Subsection (b).

(Note: This requirement is satisfied by completion of Texas Commission on Law Enforcement (TCOLE) course number 3286).

Tex. Code Crim. Pro. Art. 38.20. PHOTOGRAPH AND LIVE LINEUP IDENTIFICATION PROCEDURES

Sec. 1. In this article, "institute" means the Bill Blackwood Law Enforcement Management Institute of Texas located at Sam Houston State University.

Sec. 2. This article applies only to a law enforcement agency of this state or of a county, municipality, or other political subdivision of this state that employs peace officers who conduct photograph or live lineup identification procedures in the routine performance of the officers' official duties.

Sec. 3. (a) Each law enforcement agency shall adopt, implement, and as necessary amend a detailed written policy regarding the administration of photograph and live lineup identification procedures in accordance with this article. A law enforcement agency may adopt:

(1) the model policy adopted under Subsection (b); or

(2) the agency's own policy that, at a minimum, conforms to the requirements of Subsection (c).

(b) The institute, in consultation with large, medium, and small law enforcement agencies and with law enforcement associations, scientific experts in eyewitness memory research, and appropriate organizations engaged in the development of law enforcement policy, shall develop, adopt, and disseminate to all law enforcement agencies in this state a model policy and associated training materials regarding the administration of photograph and live lineup identification procedures. The institute shall provide for a period of public comment before adopting the policy and materials.

(c) The model policy or any other policy adopted by a law enforcement agency under Subsection (a) must:

(1) be based on:

(A) credible field, academic, or laboratory research on eyewitness memory;

(B) relevant policies, guidelines, and best practices designed to reduce erroneous eyewitness identifications and to enhance the reliability and objectivity of eyewitness identifications; and

(C) other relevant information as appropriate; and

(2) address the following topics:

(A) the selection of photograph and live lineup filler photographs or participants;

(B) instructions given to a witness before conducting a photograph or live lineup identification procedure;

(C) the documentation and preservation of results of a photograph or live lineup identification procedure, including the documentation of witness statements, regardless of the outcome of the procedure;

(D) procedures for administering a photograph or live lineup identification procedure to an illiterate person or a person with limited English language proficiency;

(E) for a live lineup identification procedure, if practicable, procedures for assigning an administrator who is unaware of which member of the live lineup is the suspect in the case or alternative procedures designed to prevent opportunities to influence the witness;

(F) for a photograph identification procedure, procedures for assigning an administrator who is capable of administering a photograph array in a blind manner or in a manner consistent with other proven or supported best practices designed to prevent opportunities to influence the witness; and

(G) any other procedures or best practices supported by credible research or commonly accepted as a means to reduce erroneous eyewitness identifications and to enhance the objectivity and reliability of eyewitness identifications.

Sec. 4. (a) Not later than December 31 of each odd-numbered year, the institute shall review the model policy and training materials adopted under this article and shall modify the policy and materials as appropriate.

(b) Not later than September 1 of each even-numbered year, each law enforcement agency shall review its policy adopted under this article and shall modify that policy as appropriate.

Sec. 5. (a) Any evidence or expert testimony presented by the state or the defendant on the subject of eyewitness identification is admissible only subject to compliance with the Texas Rules of Evidence. Evidence of compliance with the model policy or any other policy adopted under this article or with the minimum requirements of this article is not a condition precedent to the admissibility of an out-of-court eyewitness identification.

(b) Notwithstanding Article 38.23 as that article relates to a violation of a state statute, a failure to conduct a photograph or live lineup identification procedure in substantial compliance with the model policy or any other policy adopted under this article or with the minimum requirements of this article does not bar the admission of eyewitness identification testimony in the courts of this state.

§12:112 General Points

An in-court identification is inadmissible when it has been tainted by an impermissibly suggestive pretrial identification procedure. Ibarra v. State, 11 S.W.3d 189 (Tex. Crim. App. 1999).

The test is whether, considering the totality of the circumstances, the identification procedure was so impermissibly suggestive as to give rise to a very substantial likelihood of irreparable misidentification. Simmons v. United States, 390 U.S. 377, 88 S.Ct. 967, 19 L.Ed.2d 1247 (1968); Loserth v. State, 963 S.W.2d 770 (Tex. Crim. App. 1998).

A pretrial identification procedure may be so suggestive and conducive to mistaken identification that subsequent use of that identification at trial would deny the accused due process of law. Brown v. State, 64 S.W.3d 94 (Tex.App.—Austin 2001, *no pet.*); Webb v. State, 760 S.W.2d 263 (Tex. Crim. App. 1988).

There is a Sixth Amendment right to counsel at a post-indictment identification procedure. Kirby v. Illinois, 406 U.S. 682, 92 S.Ct. 1877, 32 L.Ed.2d 411 (1972).

However, the right to counsel does not attach where the lineup was prior to indictment or other initiation of adversary judicial proceedings. *See* U.S. v. Gouveia, 467 U.S. 180, 81 L.Ed.2d 146, fn 5; Winn v. State, 503 S.W.2d 816 (Tex. Crim. App. 1974).

Aside from any question of right to counsel, it is recognized that some identification procedures conducted prior to trial may be so unnecessarily suggestive and conducive to irreparable mistaken identification as to deny due process of law. Stovall v. Denno, 388 U.S. 293, 18 L.Ed.2d 1199, 87 S.Ct. 1967 (1967).

Under Rule 801(e)(1)(c), bolstering is no longer a valid objection to testimony about identification made by another witness who testifies and is subject to cross-examination regarding her pretrial identifications. Smith v. State, 830 S.W.2d 328 (Tex.App.—Houston [14th Dist.] 1992, *no pet.*).

The Due Process Clause does not require a preliminary judicial inquiry into the reliability of an eyewitness identification when the identification was not procured under unnecessarily suggestive circumstances arranged by law enforcement. Perry v. New Hampshire, 565 U.S. ___, 132 S.Ct. 716, 181 L.Ed.2d 694 (2012).

§12:113 Reliability of Pretrial Identification Procedures

If the totality of the circumstances reveals no substantial likelihood of misidentification despite a suggestive pretrial procedure, subsequent identification testimony will be deemed reliable, reliability being the linchpin in determining the admissibility of identification testimony. Ibarra v. State, 11 S.W.3d 189 (Tex. Crim. App. 1999); Webb v. State, 760 S.W.2d 263 (Tex. Crim. App. 1988).

Suggestiveness may be created by the manner in which the pretrial identification procedure is conducted, for example, by police pointing out the suspect or suggesting that a suspect is included in the line-up or photo array. Ibarra; Barley v. State, 906 S.W.2d 27 (Tex. Crim. App. 1995).

Testimony is reliable if the totality of the circumstances reveals no substantial likelihood of misidentification despite a suggestive pretrial procedure. Brown v. State, 64 S.W.3d 94 (Tex. App.—Austin 2001, *no pet.*); Simmons v. United States, 390 U.S. 377, 384, 88 S.Ct. 967, 19 L.Ed.2d 1247 (1968).

The in-court identification testimony is admissible if the witness' ability to identify has an origin independent from the pretrial identification procedure. McFarland v. State, 928 S.W.2d 482 (Tex. Crim. App. 1996).

There are five non-exclusive factors used to assess the reliability of an identification procedure: (1) the opportunity of the witness to view the criminal at the time of the crime; (2) the witness' degree of attention; (3) the accuracy of the witness' prior description of the criminal; (4) the level of certainty demonstrated by the witness at the confrontation; and (5) the length of time between the crime and the confrontation. Loserth v. State, 963 S.W.2d 770 (Tex. Crim. App. 1998); Neil v. Biggers, 409 U.S. 188, 93 S.Ct. 375, 34 L.Ed.2d 401 (1972).

§12:114 Show-Up Identifications

The admission of evidence of a single suspect show-up identification without more does not violate due process. Garza v. State, 633 S.W.2d 508 (Tex. Crim. App. 1982); Neil v. Biggers, 409 U.S. 188, 93 S.Ct. 375, 34 L.Ed.2d 401 (1972).

Some suggestiveness is always present in a show-up identification and is balanced against: 1) the desirability of allowing the witness to view the suspect immediately after the commission of the offense while his memory is fresh and accurate; 2) the fact that the quick confirmation or denial of identification leads to the release of innocent suspects; and 3) the fact that the release of innocent suspects frees the police to continue the search for the perpetrator of the offense. Louis v. State, 825 S.W.2d 752 (Tex.App.—Houston [14th Dist.] 1992, *pet. ref'd*).

The pretrial procedure must be unnecessarily

suggestive and conducive to irreparable mistaken identification before it is objectionable. Garza.

Show-up identification procedures are analyzed under the totality of the circumstances. Garza; Louis.

The totality of the circumstances is analyzed using the following factors: (1) the opportunity to view; (2) the degree of attention; (3) the accuracy of the description; (4) the witness' level of certainty; and (5) the time between the crime and the confrontation. Garza; Louis.

§12:115 Photospread Identification Procedures

An in-court identification is inadmissible when it has been tainted by an impermissibly-suggestive pretrial photographic identification. Ibarra v. State, 11 S.W.3d 189 (Tex. Crim. App. 1999); Simmons v. United States, 390 U.S. 377, 88 S.Ct. 967, 19 L.Ed.2d 1247 (1968).

The test is whether, considering the totality of the circumstances, the photographic identification procedure was so impermissibly suggestive as to give rise to a very substantial likelihood of irreparable misidentification. Ibarra; Simmons.

Under the totality of the circumstances test, any suggestive conduct by law enforcement is weighed against evidence of the reliability of the photospread identification procedure. Luna v. State, 268 S.W.3d 594 (Tex. Crim. App. 2008).

Suggestiveness may be created by the manner in which the pretrial identification procedure is conducted, for example by police pointing out the suspect or suggesting that a suspect is included in the line-up or photo array. Barley v. State, 906 S.W.2d 27 (Tex. Crim. App. 1995); Rogers v. State, 774 S.W.2d 247 (Tex. Crim. App. 1989).

Suggestiveness may also arise where the witness is shown additional photographs or video of the defendant after an initial identification. Santos v. State, 116 S.W.3d 447 (Tex.App.—Houston [14th Dist.] 2003, pet. ref'd.).

Suggestiveness may also be created by the content of the line-up or photo array itself if the suspect is the only individual closely resembling the pre-procedure description. *See* Williams v. State, 675 S.W.2d 754 (Tex. Crim. App. 1984).

Furthermore, an individual procedure may be suggestive, or the cumulative effect of the procedures may be suggestive. Barley.

Every photographic lineup must contain individuals who fit the rough description of the suspect; however, it is not essential that all individuals be identical. Brown v. State, 64 S.W.3d 94 (Tex.App.—Austin 2001, *no pet.*)

There are numerous examples of what does and does not constitute impermissible suggestiveness. For example, a lineup that consists of five persons, two of whom have beards and are not physically close to the defendant in size and hair color, is not impermissibly suggestive. Turner v. State, 600 S.W.2d 927 (Tex. Crim. App. 1980); Brown.

The same holds true even if a defendant appears older than the other persons in the lineup. Williams v. State, 675 S.W.2d 754 (Tex. Crim. App. 1984); Brown.

A lineup is not impermissibly suggestive merely because the defendant is several inches taller than other persons in the lineup. Garcia v. State, 563 S.W.2d 925 (Tex. Crim. App. 1978).

On the other hand, a lineup will be impermissibly suggestive if the defendant was the only white male in the spread and the other pictures were of Mexican American males. Tapley v. State, 673 S.W.2d 284 (Tex.App.—San Antonio 1984, *pet. ref'd*); Brown.

Finally, the use of a lone photograph, without any of the traditional safeguards of a lineup, is inherently suspect. Loserth v. State, 985 S.W.2d 536 (Tex.App.—San Antonio 1998, *pet. ref'd*) (*op. on remand*); Brown.

§12:116 In-Court Identification Procedures

An in-court identification is inadmissible if it has been tainted by an impermissibly suggestive pretrial procedure. Ibarra v. State, 11 S.W.3d 189 (Tex. Crim. App. 1999).

The test is whether, considering the totality of the circumstances, the identification procedure was so impermissibly suggestive as to give rise to a very substantial likelihood of irreparable misidentification. Simmons v. United States, 390 U.S. 377, 88 S.Ct. 967, 19 L.Ed.2d 1247 (1968); Loserth v. State, 963 S.W.2d 770 (Tex. Crim. App. 1998); Barley v. State, 906 S.W.2d 27 (Tex. Crim. App. 1995).

A two-step analysis is used to determine the admissibility of an in-court identification: (1) whether the identification procedure was impermissibly suggestive; and (2) if the identification was the result of a suggestive procedure, whether or not it gives rise to a very substantial likelihood of irreparable misidentification, examining the totality of the circumstances. Delk v. State, 855 S.W.2d 700 (Tex. Crim. App. 1993).

The defendant has the burden to show the identification is unreliable by clear and convincing evidence. Delk; Bond v. State, 29 S.W.3d 169 (Tex. App.—Houston [14th Dist.], *pet. ref'd*).

If the defendant meets this burden, then the in-court identification is inadmissible unless the state can prove by clear and convincing evidence that the identification was of independent origin. Brown v. State, 64 S.W.3d 94 (Tex.App.—Austin 2001, *no pet.*); United States v. Wade, 388 U.S. 218, 240, 87 S.Ct. 1926, 18 L.Ed.2d 1149 (1967).

The following non-exhaustive factors are analyzed in determining the independent reliability of the identification weighing them against the corrupting effect of the suggestive identification itself: (1) the witness' opportunity to view the accused at the time of the crime; (2) the witness' degree of attention; (3) the accuracy of the witness' prior description of the criminal; (4) the level of certainty demonstrated at the pretrial confrontation; and (5) the time between the crime and the confrontation. Delk.

Additional factors that may be considered are: 1) any identification prior to the lineup of another person; 2) the identification by photograph of the defendant prior to the lineup; and 3) any failure to identify the defendant on a prior occasion. Santos v. State, 116 S.W.3d 447 (Tex.App.—Houston [14th Dist.] 2003, pet. ref'd).

The identification testimony will be admissible if the indicia of reliability outweigh the apparent corrupting effect of the unnecessarily suggestive pretrial occurrence. Delk.

The standard of review for a ruling on a motion to suppress an in-court identification procedure is for the appellate courts to afford almost total deference to a trial court's determination of the facts, especially when the trial court's findings are based on an evaluation of credibility and demeanor. They give the same amount of deference to mixed questions of law and fact if the resolution of those ultimate questions turns on an examination of credibility and demeanor. If mixed questions of law and fact do not relate to credibility and demeanor, then they review such determinations *de novo*. Guzman v. State, 955 S.W.2d 85 (Tex. Crim. App. 1997).

On appeal, in reviewing the trial judge's assessment of reliability, appellate courts consider the above seven factors, which are issues of historical fact, deferentially in a light favorable to the trial court's ruling. The courts then weigh them de novo against any corrupting effect of the suggestive pretrial identification procedure. The courts review the evidence adduced at the admissibility hearing as well as the evidence adduced at trial. Balderas v. State, 517 S.W.3d 756, 792 (Tex. Crim. App. 2016).

§12:117 Voice Identification Procedures

Voice is a competent means of identification if the witness had any previous acquaintance with the person identified. Wolf v. State, 674 S.W.2d 831 (Tex.App.—Corpus Christi 1984, *pet. ref'd*).

It is sufficient that the witness had heard such person's voice but once previous to the time in question. Norton v. State, 564 S.W.2d 614 (Tex. Crim. App. 1978); Williams v. State, 747 S.W.2d 812 (Tex.App.—Dallas 1986, *no pet.*).

Requiring the defendant to speak during a lineup or in court for purposes of identification does not violate his Fifth Amendment right against self-incrimination. United States v. Wade, 388 U.S. 218, 87 S.Ct. 1926, 18 L.Ed.2d 1149 (1967); Williams.

The same rule is applicable to challenges made under the Texas Constitution. *See* Olson v. State, 484 S.W.2d 756 (Tex. Crim. App.1969) (article I, section 10 of the Texas Constitution is no broader than the Fifth Amendment of the United States Constitution); Williams.

§12:118 Form: Motion to Suppress

See our companion book *Texas Criminal Forms* (James Publishing):

- Form 12-22, Motion to Suppress In-Court Identification.

§12:120 Invoking the Witness Rule

§12:121 Statutes

CCP Art. 36.03. Invocation of Rule

(a) Notwithstanding Rule 614, Texas Rules of Evidence, a court at the request of a party may order the exclusion of a witness who for the purposes of the prosecution is a victim, close relative of a deceased victim, or guardian of a victim only if the witness is to testify and the court determines that the testimony of the witness would be materially affected if the witness hears other testimony at the trial.

(b) On the objection of the opposing party, the court may require the party requesting exclusion of a witness under Subsection (a) to make an offer of proof to justify the exclusion.

(c) Subsection (a) does not limit the authority of the court on its own motion to exclude a witness or other person to maintain decorum in the courtroom.

(d) In this article:

(1) "Close relative of a deceased victim" and "guardian of a victim" have the meanings assigned by Article 56.01.

(2) "Victim" means a victim of any criminal offense.

(e) At the commencement of a trial, the court shall admonish each witness who is to testify as to those persons whom the court determines the witness may talk to about the case before the trial ends and those persons whom the witness may not talk to about the case. The court may punish as contempt a witness who violates the admonishment provided by the court.

CCP Art. 36.05. Not to Hear Testimony

Witnesses under rule shall be attended by an officer, and all their reasonable wants provided for, unless the court, in its discretion, directs that they be allowed to go at large; but in no case where the witnesses are under rule shall they be allowed to hear any testimony in the case.

CCP Art. 36.06. Instructed by the Court

Witnesses, when placed under rule, shall be instructed by the court that they are not to converse with each other or with any other person about the case, except by permission of the court, and that they are not to read any report of or comment upon the testimony in the case while under rule. The officer who attends the witnesses shall report to the court at once any violation of its instructions, and the party violating the same shall be punished for contempt of court.

Tex. R. Evid. 614. Excluding Witnesses

At a party's request, the court must order witnesses excluded so that they cannot hear other witnesses' testimony. Or the court may do so on its own. But this rule does not authorize excluding:

(a) a party who is a natural person and, in civil cases, that person's spouse;

(b) after being designated as the party's representative by its attorney:

(1) in a civil case, an officer or employee of a party that is not a natural person; or

(2) in a criminal case, a defendant that is not a natural person;

(c) a person whose presence a party shows to be essential to presenting the party's claim or defense; or

(d) the victim in a criminal case, unless the court determines that the victim's testimony would be materially affected by hearing other testimony at the trial.

§12:122 Purpose of the Rule

The Rule serves two purposes. First, it prevents witnesses from tailoring their testimony to fit that of other witnesses. Secondly, it enhances the jury's ability to detect falsehood by exposing inconsistencies in testimony. It prevents corroboration, contradiction, and the influencing of witnesses. White v. State, 958 S.W.2d 460 (Tex. App.—Waco 1997, *no pet.*).

The purpose of the Rule is to prevent the testimony from one witness to influence the testimony of other witnesses. Ladd v. State, 3 S.W.3d 547 (Tex. Crim. App. 1999).

§12:123 Witness Cannot Hear Testimony

The sequestration rules are not discretionary with the trial court. The court shall order the witnesses excluded at the request of a party or on its own motion. White v. State, 958 S.W.2d 460 (Tex.App.—Waco 1997, *no pet.*); Moore v. State, 882 S.W.2d 844 (Tex. Crim. App. 1994).

Once the Rule is invoked, unless a witness is excused from its requirements, he should not be allowed to hear any testimony in the case or talk to any other person about the case without the court's permission. Beasley v. State, 810 S.W.2d 838 (Tex. App.—Fort Worth 1991, *pet. ref'd*).

CCP Arts. 36.05 and 36.06 provide that witnesses placed under the Rule shall not "be allowed to hear any testimony" and are not permitted "to converse with each other or with any other person about the case." Russell v. State, 155 S.W.3d 176 (Tex. Crim. App. 2005).

CCP Art. 36.06 further provides that a witness cannot "read any report of or comment upon the testimony in the case." Russell.

Tex.R.Evid. 614 provides for prospective witnesses to be sworn and removed from the courtroom. Routier v. State, 112 S.W.3d 554 (Tex. Crim. App. 2003).

Under Art. 36.03 the trial court may exclude victims, their guardians, and close relatives of deceased victims if their testimony would be materially affected by hearing other testimony in the trial. CCP Art. 36.03.

§12:124 Exceptions to the Rule

Rule 614 does not authorize the court to exclude certain witnesses from the courtroom. In criminal cases, those witnesses are: (1) a defendant who is a natural person, the representative of a defendant that is not a natural person; (2) a person whose presence a party shows to be essential to the presentation of the party's case; and (3) a victim if the court does not determine that the victim's testimony would be materially affected by hearing other testimony. Russell v. State, 155 S.W.3d 176 (Tex. Crim. App. 2005).

The party seeking to exempt a witness under Rule 614 has the burden of showing that the claimed exception applies. White v. State, 958 S.W.2d 460 (Tex.App.—Waco 1997, *no pet.*); Moore v. State, 882 S.W.2d 844 (Tex. Crim. App. 1994).

There must be evidence in the record to support a finding exempting a witness from the Rule going beyond a mere comment from the court or party that the witness' presence is necessary. the presentation of the party's case. There must be evidence supporting a finding that the witness' presence in the courtroom has a critical importance to the party's case. White.

It is not enough that the reason for lifting the Rule was expediency. The party seeking an exception under Rule 614 has the burden of showing that one of the enumerated sections is met. Moore; Russell (where the state merely designated the witness as a case agent without further proof).

§12:124.1 Presence of the Victim

A sexual assault victim should have been excluded, as requested by defendant, from pretrial hearing on defense motion to suppress victim's in-court identification of defendant notwithstanding her right to "presence" conferred by Tex. Code Crim. Pro. Art. 56.02. Jimenez v. State, 787 S.W.2d 516 (Tex. Crim. App. 1990).

§12:124.2 Essential Witness

There is an exception to the "Rule" for an expert witness where it is demonstrated that such expert is a person whose presence could be essential to the case presentation, such as the situation where an expert's assistance to counsel is essential during testimony by the opponent's expert. Moore v. State, 882 S.W.2d 844 (Tex. Crim. App. 1994).

§12:124.3 Communication With Attorneys

Attorneys are permitted to discuss the testimony with witnesses who have been placed under the Rule. Archer v. State, 703 S.W.2d 664 (Tex. Crim. App. 1986).

§12:125 Violation of the Rule

There are basically two situations that arise under Rule 614 violations. One is when a party complains of the admission of evidence in violation of the Rule. In those cases, courts look at whether the complaining party objected and was harmed. The other situation that can arise under Rule 614 is when a witness is excluded from testifying. Routier v. State, 112 S.W.3d 554 (Tex. Crim. App. 2003).

Rule 614 contains no provision for sanctions for a violation. Routier.

§12:125.1 Disqualification of Witness for Rule Violation

When deciding whether to disqualify a witness from testifying under the Rule, the trial court must balance the interests of the state and the defendant, consider alternative sanctions, and consider the benefit and detriment arising from a disqualification in light of the nature and weight of the testimony to be offered. Webb v. State, 766 S.W.2d 236 (Tex. Crim. App. 1989); Routier.

A decision to disqualify a defense witness for violating the Rule must be balanced against the defendant's constitutional right to call witnesses on his behalf. Davis v. State, 872 S.W.2d 743 (Tex. Crim. App. 1994).

A court should look at two factors in deciding whether to allow the testimony of a witness who has violated the rule. First, the court should ascertain what kind of witness was involved. Second, the court should determine whether the witness actually conferred with or heard the testimony of another witness without court permission, and whether the witness' testimony will contradict the testimony of a witness he actually heard from the opposing side or corroborate the testimony of another witness he actually heard from the same side on an issue of fact bearing upon the issue of guilt or innocence. Minor v. State, 91 S.W.3d 824 (Tex.App.—Fort Worth 2002, *pet. ref'd*).

With respect to the first prong of this test, violations of the Rule involve two main categories of witnesses: 1) witnesses who have been sworn or listed as witnesses in the case and either hear testimony or discuss another's testimony; and 2) persons who were not intended to be witnesses and are not connected with the case-in-chief but who have, due to events during trial, become necessary witnesses. Minor.

If the witness was one who had no connection with either the state's case-in-chief or the defendant's case-in-chief and who, because of a lack of personal knowledge regarding the offense, was not likely to be called as a witness, then the court should allow the witness to testify. Minor.

A third factor in assessing harm is the nature or importance of the witness' testimony. The more important the witness' testimony, the more this militates for admission of the testimony. White v. State, 958 S.W.2d 460 (Tex.App.—Waco 1997, *no pet.*).

When deciding whether a witness for the defense should be prohibited from testifying because the witness was present in the courtroom during the trial, the court should determine: (1) if the rule was violated were there particular circumstances, other than the mere fact of the violation, which would tend to show the defendant or his counsel consented, procured or otherwise had knowledge of the witness' presence in the courtroom, together with knowledge of the content of that witness' testimony; and (2) if no particular circumstances exist to justify disqualification, is the excluded testimony crucial to the defense. Webb v. State, 766 S.W.2d 236 (Tex. Crim. App. 1989); Routier.

The testimony of an excluded witness can be crucial where it corroborates other evidence favorable to the defense that the jury would be more inclined to believe had the excluded testimony been admitted. Davis.

A witness' testimony can also be crucial where the witness' testimony is probative of an accomplice witness' credibility and the defense's theory of the case; where the witness also provides evidence of other witnesses' motives for testifying as they did; and where the excluded witness is the only witness who could corroborate the defendant's claim that another person had been involved in the offense. Webb.

A defense witness' testimony is probably not crucial where it merely contradicts neutral testimony provided by a state's witness. Routier.

§12:126 Appellate Review

If a party fails to make an adequate showing that the witness should be exempted under Rule 614, but the court nevertheless grants the exemption, the ruling is reviewed under an abuse of discretion standard. Martinez v. State, 867 S.W.2d 30 (Tex. Crim. App. 1993).

Violations of the Rule only constitute reversible error when they affect the substantial rights of the defendant under Tex.R.App.Proc. 44.2(b). Ladd v. State, 3 S.W.3d 547 (Tex. Crim. App. 1999).

§12:127 Form: Motions and Supporting Brief

See our companion book *Texas Criminal Forms* (James Publishing):

- Form 15-20, Motion Invoking the Rule.
- Form 15-21, Motion to Exclude Witness From the Rule.
- Form 15-22, Motion to Exclude Testimony of Witness Based on Violation of the Rule.
- Form 15-23, Defendant's Brief in Support of Motion to Exclude Testimony of Witness Based on Violation of the Rule.

§12:130 Entrapment

§12:131 Statutes

Tex. Penal Code §8.06. Entrapment

> *(a) It is a defense to prosecution that the actor engaged in the conduct charged because he was induced to do so by a law enforcement agent using persuasion or other means likely to cause persons to commit the offense. Conduct merely affording a person an opportunity to commit an offense does not constitute entrapment.*
>
> *(b) In this section "law enforcement agent" includes personnel of the state and local law enforcement agencies as well as of the United States and any person acting in accordance with instructions from such agents.*

§12:132 General Points

Entrapment is a defense to prosecution under Texas Penal Code §§2.03 and 8.06(a). State v. Taylor, 886 S.W.2d 262 (Tex. Crim. App. 1994).

To raise the issue of entrapment, the defendant has the burden to produce evidence of an inducement that was likely to cause persons not ready and willing to commit the crime to engage in the conduct charged. State v. Perez, 824 S.W.2d 223 (Tex.App.—Waco 1992, *pet. ref'd*).

In order for a defendant to raise a claim of entrapment, he must initially establish a *prima facie* showing of such a defense. Reese v. State, 877 S.W.2d 328 (Tex. Crim. App. 1994).

When a defendant raises the defense of entrapment at trial, he has the burden of producing evidence to establish every element of that defense. He must present a *prima facie* case that:

- he engaged in the conduct charged
- because he was induced to do so by a law enforcement agent
- who used persuasion or other means, and
- those means were likely to cause persons to commit the offense.

Once the defense makes a *prima facie* showing of each element, the state then has the burden of persuasion to disprove entrapment beyond a reasonable doubt. Hernandez v. State, 161 S.W.3d 491 (Tex. Crim. App. 2005).

§12:133 Pretrial Hearing

The entrapment defense is unique in that the legislature deliberately provided it may be tested and determined at a pretrial hearing pursuant to CCP Art. 28.01. State v. Taylor, 886 S.W.2d 262 (Tex. Crim. App. 1994).

A pretrial motion asserting entrapment as a matter of law is analogous to a motion for summary judgment in civil case in which, if there is no disputed fact issue, the movant is entitled to judgment as a matter of law. A defendant is entitled to dismissal of the charges under §8.06 in the pretrial hearing context only when he can establish entrapment as a matter of law with conflict-free, uncontradicted, uncontested or undisputed evidence. If the facts concerning entrapment are in dispute, there cannot be entrapment as a matter of law determined at the pretrial stage. It is the defendant, not the state, who must establish beyond a reasonable doubt that he was entrapped at the pretrial stage. At that stage, the state has no burden of proof except to raise a fact issue that a jury would be required to resolve. Hernandez v. State, 161 S.W.3d 491 (Tex. Crim. App. 2005).

A defendant's testimony which is not directly contradicted at a pretrial hearing may be disbelieved by the trial judge because as the sole trier of fact, the judge is not required to believe that testimony. Hernandez.

Because a pretrial determination of the defense of entrapment favorable to a defendant does not impact the charging instrument, but is, instead, a finding in the nature of an acquittal, the appropriate court order is one dismissing the prosecution with prejudice. The state cannot appeal this order under CCP Art. 44.01(a)(1). Taylor.

§12:134 Two-Prong Test for Entrapment

The test for entrapment under Texas Penal Code §8.06 is a two-prong test, involving subjective and objective elements. The first prong is a subjective test in which the defendant must show that he was in fact induced by law enforcement to engage in the illegal conduct. This subjective prong requires the defendant to show that he was induced to act because of police persuasion. The second prong is an objective test in which the defendant must show that the conduct that induced him to act would have induced an ordinary person. McGann v. State, 30 S.W.3d 540 (Tex.App.—Fort Worth 2000, *pet. ref'd*).

Under an objective test, once the defendant demonstrates that he was induced, the focus shifts to the police conduct itself. The question becomes whether the persuasion used by law enforcement was such as to cause an ordinarily law-abiding person of average resistance to commit the offense, not whether it was sufficient to cause the defendant to commit the offense. England v. State, 887 S.W.2d 902 (Tex. Crim. App. 1994).

Police conduct merely affording a person an opportunity to commit an offense does not constitute entrapment. Torres v. State, 980 S.W.2d 873 (Tex.App.—San Antonio 1998, *no pet.*).

In Texas, a defendant may not deny commission of an offense and assert entrapment as a defense. Barnes v. State, 70 S.W.3d 294 (Tex.App.—Fort Worth 2002, pet. ref'd).

The defendant need not plead guilty, but he cannot introduce positive evidence that he did not commit the act alleged. Barnes.

Normally, the factual issue of entrapment is a question for the jury unless as a matter the defendant has established beyond a reasonable doubt that he was entrapped. Where the evidence on the issue of entrapment is in conflict, the issue should be submitted to the jury. Melton v. State, 713 S.W.2d 107 (Tex. Crim. App. 1986).

When a definition or instruction on entrapment is given in the abstract portion of the charge, the application paragraph must list the specific conditions under which a jury authorized to acquit. Vega v. State, 394 S.W.3d 514, 520 (Tex. Crim. App. 2013).

§12:135 Form: Motion to Dismiss

See our companion book *Texas Criminal Forms* (James Publishing):

- Form 12-17, Motion to Dismiss Based on Entrapment.

§12:140 Motions in Limine

§12:141 General Points

A motion in limine is a method of raising objection to an area of inquiry prior to the matter reaching the ears of the jury through a posed question, jury argument, or other means. Norman v. State, 523 S.W.2d 669 (Tex. Crim. App. 1975).

As such, it is wider in scope and more effective than the sustaining of an objection made after the objectionable matter has been expressed. Norman.

A true motion in limine does not usually seek an immediate or definite ruling from the trial court. Rather, it seeks a ruling from the court to order the holding of a hearing outside the jury's presence before any mention is made of the evidence. Rawlings v. State, 874 S.W.2d 740 (Tex.App.—Fort Worth 1994, *no pet.*).

It seeks a ruling from the court for a hearing to be held outside the jury presence before any mention of the subject matter is made. Nunfio v. State, 808 S.W.3d 484 (Tex. Crim. App. 1991).

A motion in limine characteristically includes: (1) an objection to a general category of evidence; and (2) a request for an instruction that the proponent of that evidence approach the bench for a hearing on its admissibility before offering it. Rawlings.

A ruling on a motion in limine is merely preliminary because, in most cases, a trial judge is not in a position to rule on the admissibility of evidence before trial begins. Thus, the court's decision on the matter presented in the motion in limine is subject to reconsideration throughout the course of trial. Rawlings.

§12:142 Preservation of Error

The granting or denying of a motion in limine, without more, preserves nothing for appellate review. Fuller v. State, 253 S.W.3d 220 (Tex. Crim. App. 2008).

Instead, a defendant must object on the proper grounds at the time the evidence is offered at trial and must secure a ruling from the court. Snellen v. State, 923 S.W.2d 238 (Tex.App.—Texarkana 1996, pet. ref'd).

Furthermore, a trial court's ruling on a motion in limine is not within the scope of Tex.R.App.Proc. 52(b) (now Tex.R.App.Proc. 33.1), providing that when a court hears objections to offered evidence out of the presence of the jury and rules that the evidence shall be admissible, such objection shall be deemed to apply to such evidence as admitted before the jury without the necessity of repeating those objections. This is because Rule 52(b) requires a specific objection and a motion in limine is merely a general complaint seeking a hearing. Rawlings v. State, 874 S.W.2d 740 (Tex. App.—Fort Worth 1994, *no pet.*).

PRACTICE TIP: As a practical matter, counsel for a defendant should consider raising as many evidentiary complaints in the form of a motion in limine prior to trial as reasonably possible. If a hearing is secured on a motion in limine, the record should be adequately developed as to exactly what the character of the evidence is anticipated to be. Counsel should then formally object to this evidence and secure a ruling on the objection. This procedure will avoid the necessity of objecting in the presence of the jury and/or the jury being excused as the evidence is sought to be introduced at trial.

§12:143 Form: Motions in Limine

See our companion book *Texas Criminal Forms* (James Publishing):

- Form 15-17, Defendant's Motion in Limine.
- Form 15-18, State's Motion in Limine.

§12:150 Pretrial Habeas Corpus Writs

A claim is cognizable in a pretrial writ of habeas corpus if, resolved in the defendant's favor, it would deprive the trial court of the power to proceed and, therefore, result in the defendant's immediate release. State v. Weise, 55 S.W.3d 617 (Tex. Crim. App. 2001); Ex parte Smith, 185 S.W.3d 887 (Tex. Crim. App. 2006).

A defendant may raise certain issues in a pretrial writ of habeas corpus, which, if meritorious, would bar prosecution or conviction. Ex parte Smith, 178 S.W.3d 797 (Tex. Crim. App. 2005); Ex parte Smith, 185 S.W.3d 887 (Tex. Crim. App. 2006).

Overbreadth, a First Amendment doctrine that allows a facial challenge to a statute even though the statute might have some legitimate applications, is cognizable on pretrial habeas. Ex parte Ingram, 533 S.W.3d 887, 894 (Tex. Crim. App. 2017).

An *in pari materia* claim is not cognizable in a pretrial writ of habeas corpus. Ex parte Smith, 185 S.W.3d 887 (Tex. Crim. App. 2006).

A pretrial habeas writ generally may not be used to challenge the sufficiency of the indictment or to construe the meaning and application of the criminal statute defining the charged offense. A pretrial habeas may, however, be used to raise a claim that the statute under which an applicant is being prosecuted is unconstitutional on its face. Ex parte Zavala, 421 S.W.3d 227, 230-1 (Tex. App.—San Antonio 2013, *no pet.*).

Claims that revolve around anti-defensive issues—the mens rea, right to present a defense, and vagueness claims—are not cognizable on pretrial habeas. Ex parte Ingram, 533 S.W.3d at 894.

In order to pursue the remedy of pretrial habeas, a petitioner has to be restrained in his liberty. Being on bond and charged with an offense is sufficient restraint to make habeas application. Ex parte Zavala, 421 S.W.3d at 230.

§12:160 Competency of Defendant to Stand Trial

§12:161 Statutes

The law governing competency issues is at CCP Chapter 46B.

§12:162 Incompetent Person Is Not Subject to Trial

In both Texas and the federal system, it has long been accepted that a person whose mental condition is such that he lacks the capacity to understand the nature and object of the proceedings against him, to consult with counsel, and to assist in preparing his defense may not be subjected to a trial. Drope v. Missouri, 420 U.S. 162, 95 S.Ct. 896, 43 L.Ed.2d 103 (1975); McDaniel v. State, 98 S.W.3d 704 (Tex. Crim. App. 2003).

The conviction of an accused person while he is legally incompetent violates due process. Pate v. Robinson, 383 U.S. 375, 86 S.Ct. 836, 15 L.Ed.2d 815 (1966); McDaniel.

§12:163 Trial Court Must Inquire Into Competence

Thus, to protect a criminal defendant's constitutional rights, a trial court must inquire into the defendant's mental competence once the issue is sufficiently raised. Pate v. Robinson, 383 U.S. 375, 86 S.Ct. 836, 15 L.Ed.2d 815 (1966); McDaniel v. State, 98 S.W.3d 704 (Tex. Crim. App. 2003).

With respect to the evidentiary standard that must be met at the informal inquiry stage, a court must focus on three matters. First, the standard at the informal inquiry stage is whether there is "some evidence" of incompetency to stand trial. Art. 46B.004(c). The statute reads in full: "On suggestion that the defendant may be incompetent to stand trial, the court shall determine by informal inquiry whether there is some evidence from any source that would support a finding that the defendant may be incompetent to stand trial." Boyett v. State, 545 S.W.3d 556, 563-4 (Tex. Crim. App. 2018).

The statutory "some evidence" standard requires "more than none or a scintilla" of evidence that "rationally may lead to a conclusion of incompetency." Boyett v. State, 545 S.W.3d at 564.

Second, a trial court must consider only evidence of incompetency, and it must not weigh evidence of competency against the evidence of incompetency. In making this determination at the informal-inquiry stage, "a trial court must consider only that evidence tending to show incompetency" and "put aside all competing indications of competency." Boyett v. State, 545 S.W.3d at 564.

Third, some evidence must be presented at the informal inquiry stage to show that a defendant's mental illness is the source of his inability to participate in his own defense. There must be "some evidence from which it may rationally be inferred not only 1) that the defendant suffers some degree of debilitating mental illness, and that 2) he obstinately refuses to cooperate with counsel to his own apparent detriment, but also that 3) his mental illness is what fuels his obstinacy." Thus, it is not enough to present evidence of either a defendant's mental illness alone or his refusal to cooperate with counsel—rather, there must be some evidence indicating that the defendant's refusal to rationally engage with counsel is caused by his mental illness. Boyett v. State, 545 S.W.3d at 564.

Under CCP Art. 46B.003(a), "[a] person is incompetent to stand trial if the person does not have: (1) sufficient present ability to consult with the person's lawyer with a reasonable degree of rational understanding; or (2) a rational as well as factual understanding of the proceedings against the person." *See* McDaniel; Greene v. State, 225 S.W.3d 324 (Tex.App.—San Antonio 2007, *no pet.*).

Under Texas cases, the trial court must conduct a non-jury hearing, a Section 2 inquiry, on whether to hold a jury trial on the defendant's competency if evidence of the defendant's incompetency is brought to the attention of the court from any source. A Section 2 or competency inquiry is required only if the evidence brought to the judge's attention raises a bona fide doubt in the judge's mind about the defendant's competency to stand trial. Evidence is usually sufficient to create a bona fide doubt if it shows recent severe mental illness, at least moderate retardation, or truly bizarre acts by the defendant. McDaniel.

Bona fide doubt is still the proper standard under art. 46B.003 for determining whether a trial court should conduct a competency inquiry. If a trial judge has a *bona fide* doubt about the competency of the defendant, he or she shall conduct an informal inquiry to determine if there is evidence that would support a finding of incompetence. A *bona fide* doubt may exist if the defendant exhibits truly bizarre behavior or has a recent history of severe mental illness or at least moderate mental retardation. Montoya v. State, 291 S.W.3d 420 (Tex. Crim. App. 2009).

If sufficient evidence is presented to the court, the informal inquiry must be performed *sua sponte* if not requested by either party. Greene.

In making a determination whether some evidence exists that would support a finding that a defendant may be incompetent to stand trial, a trial court must consider only that evidence tending to show incompetency, putting aside all competing indications of competency, to find whether there is some evidence, a quantity more than none or a scintilla, that rationally may lead to a conclusion of incompetency. If so, then evidence exists to support a finding of incompetency, and the statutory scheme requires the trial court to conduct a formal competency trial. Turner v. State, 422 S.W.3d 676, 692-3 (Tex. Crim. App. 2013).

When such a bona fide doubt does exist, the court must conduct the Section 2 competency inquiry to determine whether there is some evidence to support a finding of incompetency, and, if so, to commence a Section 4 competency hearing. McDaniel.

The mere assertion in a motion that there is an issue as to the defendant's competency to stand trial, or a defendant's bare assertion that "I am incompetent," is not sufficient to raise a bona fide doubt as to the defendant's competency. McDaniel.

Prior hospitalization and treatment for depression do not *per se* warrant the trial court impaneling a separate jury to conduct a competency hearing. To raise the issue of competency by means of the defendant's past mental health history, there generally must be evidence of recent severe mental illness or bizarre acts by the defendant or of moderate retardation. Montoya.

§12:164 Competency Trial

The burden of proof at a competency hearing is on the defendant to prove he is incompetent by a preponderance of the evidence. Morris v. State, 214 S.W.3d 159 (Tex.App.—Beaumont 2007, *pet. filed*).

To prove incompetence, a defendant must prove he does not have a sufficient present ability to consult with his lawyer with a reasonable degree of rational understanding, or that he does not have a rational, as well as a factual, understanding of the proceedings against him. CCP Art. 46B.003(a); Morris.

Where there is a prior unvacated judgment of incompetency in evidence, the burden is on the state to prove the present competency of the defendant beyond a reasonable doubt. Bradford v. State, 172 S.W.3d 1 (Tex.App.—Fort Worth 2005, *no pet.*).

It is error to admit a defendant's statement from the incompetency examination, in violation of former CCP Art. 46.02, §3(g), for any purpose at the guilt/innocence phase of the trial. Mitten v. State, 145 S.W.3d 225 (Tex. Crim. App. 2004).

A jury trial on the issue of competency to stand trial is civil in nature. Williams v. State, 191 S.W.3d 242 (Tex.App.—Austin 2006, *no pet.*).

Procedural issues arising during competency trials are governed by the Rules of Civil Procedure. White v. State, 591 S.W.2d 851 (Tex. Crim. App. 1979).

Under the Rules of Civil Procedure, each side has six peremptory challenges in a competency trial. White.

The rules of evidence apply to hearings to determine a defendant's competency to stand trial. CCP Art. 46B.008.

A claim of amnesia generally will not render a defendant *per se* incompetent to stand trial. Morris v. State, 301 S.W.3d 281 (Tex. Crim. App. 2009).

This is because (1) amnesia is akin to missing evidence; (2) a contrary rule would unduly hamper the State's interest in the prosecution of violators of its criminal laws and jeopardize the safety and security of other citizens; and (3) amnesia can be easily feigned. Gonzales v. State, 313 S.W.3d 840 (Tex. Crim. App. 2010) *citing* Morris.

Also, amnesia does not necessarily affect a defendant's present ability to consult with his lawyer which is the test for competency. Owens v. State, 473 S.W.3d 812, 817 (Tex. Crim. App. 2015).

A trial judge is not required to conduct a competency hearing on his own initiative after hearing evidence that, due to alcohol or drug use, a defendant suffered amnesia with respect to events giving rise to the charged offense. Gonzales.

Should the formal competency trial result in a finding of competency, the trial court is not obliged to revisit the issue later absent a material change of circumstances suggesting that the defendant's mental status has deteriorated. However, especially when there has been a suggestion of incompetency but no formal adjudication of the issue, due process requires the trial court to remain ever vigilant for changes in circumstances that would make a formal adjudication appropriate. Turner v. State, 422 S.W.3d 676, 693 (Tex. Crim. App. 2013).

An appellate court measures the propriety of the competency verdict based on the evidence before the jury at the time of the verdict under the relevant legal standard set out in Art. 46B.003(a). Morris.

PRACTICE TIP: The incompetency statute no longer requires that a jury decide the competency issue if both sides are in agreement. CCP Art. 46B.005(c).

PRACTICE TIP: A defense attorney should be careful, if testifying at an incompetency hearing, not to reveal privileged attorney-client communications. The privilege can only be waived by the client.

§12:165 Appeal

When appealing a conviction on the merits, a defendant may appeal errors made in the incompetency-to-stand-trial hearing. Morris v. State, 214 S.W.3d 159 (Tex.App.—Beaumont 2007, *pet.filed*).

The trial court's order of competency to stand trial is not a final, appealable judgment. Lowe v. State, 999 S.W.2d 537 (Tex.App.—Houston [14th Dist.] 1999, *no pet.*).

Whether incompetency exists at the time of trial is left to the discretion of the trial judge; and this determination is reviewed on appeal under an abuse of discretion standard. Rice v. State, 991 S.W.2d 953 (Tex.App.—Fort Worth 1999, *pet. ref'd*).

The appropriate remedy when there is an error in competency proceedings is to abate the appeal and remand the cause to the trial court to determine the feasibility of a retrospective competency proceeding. Owens v. State, 473 S.W.3d 812, 816 (Tex. Crim. App. 2015). *See also* Turner v. State, 570 S.W.3d 250, 262 (Tex. Crim. App. 2018).

§12:166 Form: Motions and Agreed Judgments Regarding Competency

See our companion book *Texas Criminal Forms* (James Publishing):

- Form 15-3.1, Motion for Competency Examination.
- Form 15-3.2, Defendant's Motion to Vacate Judgment of Competency.
- Form 15-3.3, (State or Defendant's) Motion for Jury on Issue of Defendant's Competency.
- Form 15-3.4, Agreed Judgment of Incompetency to Stand Trial (Non-Violent).
- Form 15-3.5, Agreed Judgment of

Incompetency to Stand Trial (Violent).

- Form 15-3.6, Agreed Judgment of Indefinite Commitment.
- Form 15-3.7, Agreed Judgment of Mental Retardation.

§12:170 Guilty Pleas

§12:171 Statutes

CCP Art. 1.05. Rights of Accused

In all criminal prosecutions the accused shall have a speedy public trial by an impartial jury. He shall have the right to demand the nature and cause of the accusation against him, and to have a copy thereof. He shall not be compelled to give evidence against himself. He shall have the right of being heard by himself, or counsel, or both; shall be confronted with the witnesses against him, and shall have compulsory process for obtaining witnesses in his favor. No person shall be held to answer for a felony unless on indictment of a grand jury.

CCP Art. 1.13. Waiver of Trial by Jury

(a) The defendant in a criminal prosecution for any offense other than a capital felony case in which the state notifies the court and the defendant that it will seek the death penalty shall have the right, upon entering a plea, to waive the right of trial by jury, conditioned, however, that, except as provided by Article 27.19, the waiver must be made in person by the defendant in writing in open court with the consent and approval of the court, and the attorney representing the state. The consent and approval by the court shall be entered of record on the minutes of the court, and the consent and approval of the attorney representing the state shall be in writing, signed by that attorney, and filed in the papers of the cause before the defendant enters the defendant's plea.

(b) In a capital felony case in which the attorney representing the State notifies the court and the defendant that it will not seek the death penalty, the defendant may waive the right to trial by jury but only if the attorney representing the State, in writing and in open court, consents to the waiver.

(c) A defendant may agree to waive a jury trial regardless of whether the defendant is represented by an attorney at the time of making the waiver, but before a defendant charged with a felony who has no attorney can agree to waive the jury, the court must appoint an attorney to represent him.

CCP Art. 1.14. Waiver of Rights

(a) The defendant in a criminal prosecution for any offense may waive any rights secured him by law except that a defendant in a capital felony case may waive the right of trial by jury only in the manner permitted by Article 1.13(b) of this code.

(b) If the defendant does not object to a defect, error, or irregularity of form or substance in an indictment or information before the date on which the trial on the merits commences, he waives and forfeits the right to object to the defect, error, or irregularity and he may not raise the objection on appeal or in any other post-conviction proceeding. Nothing in this article prohibits a trial court from requiring that an objection to an indictment or information be made at an earlier time in compliance with Article 28.01 of this code.

CCP Art. 1.141. Waiver of Indictment for Noncapital Felony

A person represented by legal counsel may in open court or by written instrument voluntarily waive the right to be accused by indictment of any offense other than a capital felony. On waiver as provided in this article, the accused shall be charged by information.

CCP Art. 1.15. Jury in Felony

No person can be convicted of a felony except upon the verdict of a jury duly rendered and recorded, unless the defendant, upon entering a plea, has in open court in person waived his right of trial by jury in writing in accordance with Articles 1.13 and 1.14; provided, however, that it shall be necessary for the state to introduce evidence into the record showing the guilt of the defendant and said evidence shall be accepted by the court as the basis for its judgment and in no event shall a person charged be convicted upon his plea without sufficient evidence to support the same. The evidence may be stipulated if the defendant in such case consents in writing, in open court, to waive the appearance, confrontation, and cross-examination of witnesses, and further consents either to an oral stipulation of the evidence and testimony or to the introduction of testimony by affidavits, written statements of witnesses, and any other documentary evidence in support of the judgment of the court. Such waiver and consent must be approved by the court in writing, and be filed in the file of the papers of the cause.

CCP Art. 26.13. Plea of guilty

(a) Prior to accepting a plea of guilty or a plea of nolo contendere, the court shall admonish the defendant of:

(1) the range of the punishment attached to the offense;

(2) the fact that the recommendation of the prosecuting attorney as to punishment is not binding on the court. Provided that the court shall inquire as to the existence of a plea bargain agreement between the state and the defendant and, if an agreement exists, the court shall inform the defendant whether it will follow or reject the agreement in open court and before any finding on the plea. Should the court reject the agreement, the defendant shall be permitted to withdraw the defendant's plea of guilty or nolo contendere;

(3) the fact that if the punishment assessed does not exceed the punishment recommended by the prosecutor and agreed to by the defendant and the defendant's attorney, the trial court must give its permission to the defendant before the defendant may prosecute an appeal on any matter in the case except for those matters raised by written motions filed prior to trial;

(4) the fact that if the defendant is not a citizen of the United States of America, a plea of guilty or nolo contendere for the offense charged may result in deportation, the exclusion from admission to this country, or the denial of naturalization under federal law;

(5) the fact that the defendant will be required to meet the registration requirements of Chapter 62, if the defendant is convicted of or placed on deferred adjudication for an offense for which a person is subject to registration under that chapter; and

(6) the fact that if the defendant is placed on community supervision, after satisfactorily fulfilling the conditions of community supervision and on expiration of the period of community supervision, the court is authorized to release the defendant from the penalties and disabilities resulting from the offense as provided by Article 42A.701(f).

(b) No plea of guilty or plea of nolo contendere shall be accepted by the court unless it appears that the defendant is mentally competent and the plea is free and voluntary.

(c) In admonishing the defendant as herein provided, substantial compliance by the court is sufficient, unless the defendant affirmatively shows that he was not aware of the consequences of his plea and that he was misled or harmed by the admonishment of the court.

(d) The court may make the admonitions required by this article either orally or in writing. If the court makes the admonitions in writing, it must receive a statement signed by the defendant and the defendant's attorney that he understands the admonitions and is aware of the consequences of his plea. If the defendant is unable or refuses to sign the statement, the court shall make the admonitions orally.

(e) Before accepting a plea of guilty or a plea of nolo contendere, the court shall, as applicable in the case:

(1) inquire as to whether a victim impact statement has been returned to the attorney representing the state and ask for a copy of the statement if one has been returned; and

(2) inquire as to whether the attorney representing the state has given notice of the existence and terms of any plea bargain agreement to the victim, guardian of a victim, or close relative of a deceased victim, as those terms are defined by Article 56.01.

(f) The court must substantially comply with Subsection (e) of this article. The failure of the court to comply with Subsection (e) of this article is not grounds for the defendant to set aside the conviction, sentence, or plea.

(g) Before accepting a plea of guilty or a plea of nolo contendere and on the request of a victim of the offense, the court may assist the victim and the defendant in participating in a victim-offender mediation program.

(h) The court must substantially comply with Subsection (a)(5). The failure of the court to comply with Subsection (a)(5) is not a ground for the defendant to set aside the conviction, sentence, or plea.

(i) Notwithstanding this article, a court shall not order the state or any of its prosecuting attorneys to participate in mediation, dispute resolution, arbitration, or other similar procedures in relation to a criminal prosecution unless upon written consent of the state.

(h-1) The court must substantially comply with Subsection (a)(6). The failure of the court to comply with Subsection (a)(6) is not a ground for the defendant to set aside the conviction, sentence, or plea.

CCP Art. 26.14. Jury on Plea of Guilty

Where a defendant in a case of felony persists in pleading guilty or in entering a plea of nolo contendere, if the punishment is not absolutely fixed by law, a jury shall be impaneled to assess the punishment and evidence may be heard to enable them to decide thereupon, unless the defendant in accordance with Articles 1.13 or 37.07 shall have waived his right to trial by jury.

CCP Art. 27.02. Defendant's Pleadings

The pleadings and motions of the defendant shall be:

(1) A motion to set aside or an exception to an indictment or information for some matter of form or substance;

(2) A special plea as provided in Article 27.05 of this code;

(3) A plea of guilty;

(4) A plea of not guilty;

(5) A plea of nolo contendere, the legal effect of which shall be the same as that of a plea of guilty, except that such plea may not be used against the defendant as an admission in any civil suit based upon or growing out of the act upon which the criminal prosecution is based;

(6) An application for probation, if any;

(7) An election, if any, to have the jury assess the punishment if he is found guilty; and

(8) Any other motions or pleadings that are by law permitted to be filed.

CCP Art. 27.13. Plea of Guilty or Nolo Contendere in Felony

A plea of "guilty" or a plea of "nolo contendere" in a felony case must be made in open court by the defendant in person; and the proceedings shall be as provided in Articles 26.13, 26.14 and 27.02. If the plea is before the judge alone, same may be made in the same manner as is provided for by Articles 1.13 and 1.15.

CCP Art. 27.14. Plea of Guilty or Nolo Contendere in Misdemeanor

(a) A plea of "guilty" or a plea of "nolo contendere" in a misdemeanor case may be made either by the defendant or his counsel in open court; in such case, the defendant or his counsel may waive a jury, and the punishment may be assessed by the court either upon or without evidence, at the discretion of the court.

(b) A defendant charged with a misdemeanor for which the maximum possible punishment is by fine only may, in lieu of the method provided in Subsection (a) of this article, mail or deliver in person to the court a plea of "guilty" or a plea of "nolo contendere" and a waiver of jury trial. The defendant may also request in writing that the court notify the defendant, at the address stated in the request, of the amount of an appeal bond that the court will approve. If the court receives a plea and waiver before the time the defendant is scheduled to appear in court, the court shall dispose of the case without requiring a court appearance by the defendant. If the court receives a plea and waiver after the time the defendant is scheduled to appear in court but at least five business days before a scheduled trial date, the court shall dispose of the case without requiring a court appearance by the defendant. The court shall notify the defendant either in person or by certified mail, return receipt requested, of the amount of any fine assessed in the case and, if requested by the defendant, the amount of an appeal bond that the court will approve. The defendant shall pay any fine assessed or give an appeal bond in the amount stated in the notice before the 31st day after receiving the notice.

(c) In a misdemeanor case for which the maximum possible punishment is by fine only, payment of a fine or an amount accepted by the court constitutes a finding of guilty in open court as though a plea of nolo contendere had been entered by the defendant and constitutes a waiver of a jury trial in writing.

(d) If written notice of an offense for which maximum possible punishment is by fine only or of a violation relating to the manner, time, and place of parking has been prepared, delivered, and filed with the court and a legible duplicate copy has been given to the defendant, the written notice serves as a complaint to which the defendant may plead "guilty," "not guilty," or "nolo contendere." If the defendant pleads "not guilty" to the offense or fails to appear based on the written notice, a complaint shall be filed that conforms to the requirements of Chapter 45 of this code, and that complaint serves as an original complaint. A defendant may waive the filing of a sworn complaint and elect that the prosecution proceed on the written notice of the charged offense if the defendant agrees in writing with the prosecution, signs the agreement, and files it with the court.

(e)(1) Before accepting a plea of guilty or a plea of nolo contendere by a defendant charged with a misdemeanor involving family violence, as defined by Section 71.004, Family Code, the court

shall admonish the defendant by using the following statement:

"If you are convicted of a misdemeanor offense involving violence where you are or were a spouse, intimate partner, parent, or guardian of the victim or are or were involved in another, similar relationship with the victim, it may be unlawful for you to possess or purchase a firearm, including a handgun or long gun, or ammunition, pursuant to federal law under 18 U.S.C. Section 922(g)(9) or Section 46.04(b), Texas Penal Code. If you have any questions whether these laws make it illegal for you to possess or purchase a firearm, you should consult an attorney."

(2) The court may provide the admonishment under Subdivision (1) orally or in writing, except that if the defendant is charged with a misdemeanor punishable by fine only, the statement printed on a citation issued under Article 14.06(b) may serve as the court admonishment required by this subsection.

CCP Art. 27.15. Change of Venue to Plead Guilty

When in any county which is located in a judicial district composed of more than one county, a party is charged with a felony and the maximum punishment therefor shall not exceed fifteen years, and the district court of said county is not in session, such party may, if he desires to plead guilty, or enter a plea of nolo contendere, make application to the district judge of such district for a change of venue to the county in which said court is in session, and said district judge may enter an order changing the venue of said cause to the county in which the court is then in session, and the defendant may plead guilty or enter a plea of nolo contendere to said charge in said court to which the venue has been changed.

Tex. R. Evid. 410. Pleas, Plea Discussions, and Related Statements

(a) Prohibited Uses in Civil Cases. *In a civil case, evidence of the following is not admissible against the defendant who made the plea or was a participant in the plea discussions:*

(1) a guilty plea that was later withdrawn;

(2) a nolo contendere plea;

(3) a statement made during a proceeding on either of those pleas under Federal Rule of Criminal Procedure 11 or a comparable state procedure; or

(4) a statement made during plea discussions with an attorney for the prosecuting authority if the discussions did not result in a guilty plea or they resulted in a later-withdrawn guilty plea.

(b) Prohibited Uses in Criminal Cases. *In a criminal case, evidence of the following is not admissible against the defendant who made the plea or was a participant in the plea discussions:*

(1) a guilty plea that was later withdrawn;

(2) a nolo contendere plea that was later withdrawn;

(3) a statement made during a proceeding on either of those pleas under Federal Rule of Criminal Procedure 11 or a comparable state procedure; or

(4) a statement made during plea discussions with an attorney for the prosecuting authority if the discussions did not result in a guilty or nolo contendere plea or they resulted in a later-withdrawn guilty or nolo contendere plea.

(c) Exception. In a civil case, the court may admit a statement described in paragraph (a)(3) or (4) and in a criminal case, the court may admit a

statement described in paragraph (b) (3) or (4), when another statement made during the same plea or plea discussions has been introduced and in fairness the statements ought to be considered together.

§12:172 Waivers of Rights

A guilty plea is a waiver of several federal constitutional rights, including the protections against self-incrimination, the right to a speedy and public trial by jury, and the right to confrontation. Ex parte Barnaby, 475 S.W.3d 316, 322 (Tex. Crim. App. 2015).

These waivers must be not only voluntary but also a knowing, intelligent act done with sufficient awareness of the relevant circumstances and likely consequences. Ex parte Barnaby, 475 S.W.3d at 322.

A defendant must have sufficient awareness of the relevant circumstances, and must possess an understanding of the law in relation to the facts. To determine whether a defendant's awareness was sufficient at the time of his plea, a reviewing court looks to whether the plea was a voluntary and intelligent choice among the alternative courses of action open to the defendant. Ex parte Palmberg, 491 S.W.3d 804, 807 (Tex. Crim. App. 2015).

But a defendant need not have a comprehensive awareness of the specific impact that relinquishing his constitutional rights may have; sufficient awareness does not require complete knowledge of the prosecution's case. Ex parte Palmberg, b 491 S.W.3d 807 (where the court found that the defendant's plea was voluntary even where he did not know that the alleged controlled substance was consumed during presumptive testing).

The correct question for due process purposes is not whether the defendant knew every fact relevant to the prosecution of his case. Rather, the correct question is whether he was aware of sufficient facts—including an awareness that there are or may be facts that he does not *yet* know—to make an informed and voluntary plea. Ex parte Palmberg, 491 S.W.3d at 809; Ex parte Broussard, 517 S.W.3d 814, 817 (Tex. Crim. App. 2017).

§12:172.1 Right to Waive Trial by Jury

CCP Art. 1.13(a) provides that a defendant has the right, upon entering a plea, to waive the right of trial by jury. This can be done on the condition that such waiver must be made in person by the defendant in writing in open court, with the consent and approval of the court and the attorney representing the state. Johnson v. State, 72 S.W.3d 346 (Tex. Crim. App. 2002).

This is the rule in felony cases. Shaffer v. State, 769 S.W.2d 943 (Tex. Crim. App. 1989).

It is also the rule in misdemeanor cases. Josey v. State, 857 S.W.2d 815 (Tex.App.—Houston [14th Dist.] 1993, *no pet.*).

A defendant may withdraw his jury trial waiver if the withdrawal will not: 1) interfere with the orderly administration of the business of the court; 2) result in unnecessary delay or inconvenience to the witnesses; or 3) prejudice the state. The burden is on the defendant to establish these matters and the state can rebut the defendant's assertions. Green v. State, 36 S.W.3d 211 (Tex.App.—Houston [14th Dist.] 2001, *pet. ref'd*).

The defendant can withdraw his jury trial waiver only where he can show that the withdrawal will not cause adverse consequences to the state. Hobbs v. State, 298 S.W.3d 193 (Tex. Crim. App. 2009).

A defendant does not give up waivable rights unless he does so plainly, freely, and intelligently. If a defendant wants to relinquish a waivable right, he must expressly do so; i.e., he must intentionally relinquish or abandon a known right or privilege. Sometimes the waiver must be in writing; always, it must be on the record. The judge has an independent duty to implement such rights, absent an effective waiver. Marin v. State, 851 S.W.2d 275 (Tex. Crim. App. 1993).

When a defendant pleads guilty to a jury, the trial becomes unitary and the jury will always be called upon to assess punishment. A defendant cannot plead guilty to a jury and then be sentenced by a judge. In re The State of Texas ex rel. Jennifer Tharp, 393 S.W.3d 751, 755 (Tex. Crim. App. 2012).

§12:172.2 Plea Agreements

Because plea bargains are contractual agreements between the state and defendant, the appellate courts apply general contract-law principles. Ex parte Cox, 482 S.W.3d 112, 116 (Tex. Crim. App. 2016).

Although contractual concepts apply to plea bargains, such should not be strictly enforced to the detriment of due process. The terms of plea agreements, being contractual in nature, are left to the parties to determine and agree upon, and the courts will rarely disturb the terms of such agreements. Ex parte Cox, 482 S.W.3d at 116.

It is well established that, after the judge has accepted a plea bargain in open court, a defendant has a right to enforce the state's part of the plea bargain. Ex parte Cox, 482 S.W.3d at 116.

Where a plea bargain fails in whole or in part, the proper remedy is generally to return both parties to their original pre-plea positions. Ex parte Cox, 482 S.W.3d at 118.

This is also true in a case that has been charge-bargained. Thomas v. State, 516 S.W.3d 498, 504 (Tex. Crim. App. 2017).

A defendant may knowingly and intelligently waive his entire appeal as a part of a plea, even when sentencing is not agreed upon, where consideration is given by the State for that waiver. Ex parte Broadway, 301 S.W.3d 694 (Tex. Crim. App. 2009) (state gave up its right to a jury trial in consideration for the defendant waiving his appeal).

The State's relinquishment of its right to a jury trial is sufficient consideration to render a defendant's waiver of his right to appeal knowing and intelligent. Carson v. State, 559 S.W.3d 489, 496 (Tex. Crim. App. 2018).

In a charge-bargain case, T.R.A.P. Rule 25.2(a)(2) applies and the defendant can only appeal matters that were those matters that were raised by written motion filed and ruled on before trial or after getting the trial court's permission to appeal. Kennedy v. State, 297 S.W.3d 338 (Tex. Crim. App. 2009) (state agreed to forego prosecution of one charge in exchange for the defendant's plea to another charge).

The charge-bargain rules also apply where the defendant pleads guilty to one charge and the state agrees to plea in bar a second charge. Shankle v. State, 119 S.W.3d 808 (Tex. Crim. App. 2003).

§12:172.3 Form: Waivers

See our companion book *Texas Criminal Forms* (James Publishing):

- Form 15-1, Written Waiver of Jury Trial.
- Form 15-2, Waiver of Indictment.

§12:173 Guilty Plea Admonishments

§12:173.1 Admonishments Required

Compliance with the admonishments in CCP Art. 26.13 is mandatory. *See* Hughes v. State, 833 S.W.2d 137 (Tex. Crim. App. 1992); Elliott v. State, 874 S.W.2d 238 (Tex.App.—El Paso 1994, *no pet.*).

CCP Art. 26.13 requires the trial court to admonish the defendant of the direct consequences of his guilty plea before accepting the plea. The admonishments are:

- The punishment range;
- The fact that the state's recommendation is not binding on the court;
- The possibility of deportation; and
- The limited right to appeal.

Bessey v. State, 239 S.W.3d 809 (Tex. Crim. App. 2007).

The admonishments can be accomplished orally or in writing. Alvarez v. State, 63 S.W.3d 578 (Tex. App.—Fort Worth 2001, *no pet.*).

After September 1, 2007 courts are required to admonish defendants of the fact that it is unlawful for the defendant to possess or transfer a firearm or ammunition if the defendant is convicted of a misdemeanor involving family violence, as defined by Family Code Art. 26.13(a)(6) §71.004.

TIP TO THE BENCH: Where the defendant is a U.S. citizen, be sure that this fact is reflected in the record. A trial court that fails to admonish a defendant about the immigration consequences of his guilty plea, a silent record on citizenship, or a record that is insufficient to determine citizenship, establishes harm by the standard of Rule 44.2(b). Vannortrick v. State, 227 S.W.3d 706 (Tex. Crim. App. 2007).

§12:173.2 Admonishments Not Required

The trial court is not required to admonish a defendant of all consequences of a guilty plea; only those direct consequences that are punitive in nature or specifically enunciated by law. Mitschke v. State, 129 S.W.3d 130 (Tex. Crim. App. 2004).

The trial court is not required to admonish the defendant on any collateral consequences of the guilty plea. Collateral consequences of a guilty plea are those which are not definite consequences. Ex parte Morrow, 952 S.W.2d 530 (Tex. Crim. App. 1997).

The Code of Criminal Procedure does not require the trial court to admonish the defendant of the consequences of deferred adjudication before he enters an open plea of guilty or nolo contendere. Ray v. State, 919 S.W.2d 125 (Tex. Crim. App. 1996).

§12:173.3 Failure to Properly Admonish

For cases tried before CCP Art. 26.13 specifically required admonishment on the sex offender registration requirement, the failure to admonish the defendant is not reversible because the registration requirement, although a direct consequence, is not a punitive one. Mitschke v. State, 129 S.W.3d 130 (Tex. Crim. App. 2004).

The lack of CCP Art. 26.13 admonishments regarding citizenship is not grounds for relief unless the defendant shows harm. Ex parte Tovar, 901 S.W.2d 484 (Tex. Crim. App. 1995).

Similarly, where the trial court has failed to admonish a defendant who pleads guilty of the registration requirement for sex offenders, the defendant must show harm to prevail on appeal. Anderson v. State, 182 S.W.3d 914 (Tex. Crim. App. 2006); Bessey v. State, 239 S.W.3d 809 (Tex. Crim. App. 2007) (where the court reasoned that the defendant could not show harm because he had to serve 100 years before eligible for parole).

An attorney's performance was not found to be deficient where he failed to object to the lack of CCP Art. 26.13(a)(4) (citizenship) admonishment where the defendant failed to establish he was prejudiced by the omission. Ex parte Akhtab, 901 S.W.2d 488 (Tex. Crim. App. 1995) (where the court had not yet ruled on the effect of the trial court's failure to give a proper citizenship admonishment).

The trial court's failure to admonish a defendant concerning the deportation consequences of a plea of guilty is harmless where the record reflects that the defendant is a citizen of the U.S. Cain v. State, 947 S.W.2d 262 (Tex. Crim. App. 1997).

The admonishments under CCP Art. 26.13(a) are not constitutionally required because their purpose and function is to assist the trial court in making the determination that a guilty plea is knowingly and voluntarily entered. Aguirre-Mata v. State, 992 S.W.2d 495 (Tex. Crim. App. 1999).

§12:173.4 Appealing Improper Admonishments

Because the CCP Art. 26.13 admonishments are a waivable-only right, a defendant can raise the issue for the first time on appeal without having made an objection or raising the issue at trial. However, such an error is subject to a harm analysis. Bessey v. State, 239 S.W.3d 809 (Tex. Crim. App. 2007); Davison v. State, 405 S.W.3d 682, 687-8 (Tex. Crim. App. 2013).

Thus, a trial court commits nonconstitutional error when it fails to admonish a defendant on one of the statutorily required admonishments. Carranza v. State, 980 S.W.2d 653 (Tex. Crim. App. 1998).

The defendant has the burden of proof to show he was unaware of the consequences of his plea and that he was misled or harmed by the admonishment of the trial court. Carranza.

Generally, if a defendant is fully advised of the direct consequences of his plea, his ignorance of a collateral consequence does not render the plea involuntary. State v. Jimenez, 987 S.W.2d 886 (Tex. Crim. App. 1999).

§12:173.5 Form: Admonishments

See our companion book *Texas Criminal Forms* (James Publishing):

- Form 15-4, Written Plea Admonishments.

- Form 20-30, Written Admonishments of Revocation of Probation.

§12:174 Voluntariness of Plea

§12:174.1 Requirements for Voluntary Plea

A guilty plea constitutes a waiver of three constitutional rights: the right to a jury trial, the right to confront one's accusers, and the right not to incriminate oneself. Boykin v. Alabama, 395 U.S. 238, 89 S.Ct. 1709, 23 L.Ed.2d 274 (1969); Kniatt v. State, 206 S.W.3d 657 (Tex. Crim. App. 2006).

Accordingly, a guilty plea, to be consistent with due process of law, must be entered knowingly, intelligently, and voluntarily. Boykin; Kniatt.

To be voluntary, a guilty plea must be the expression of the defendant's own free will and must not be induced by threats, misrepresentations, or improper promises. Brady v. United States, 397 U.S. 742, 90 S.Ct. 1463, 25 L.Ed.2d 747 (1970); Kniatt.

A defendant may legitimately waive his appeal as a part of a plea bargain agreement. Blanco v. State, 18 S.W.3d 218 (Tex. Crim. App. 2000).

This is true even where the sentence is not agreed upon, but where consideration is given by the state for the waiver. Broadway v. State, 301 S.W.3d 694 (Tex. Crim. App. 2009).

A defendant's voluntary plea of no contest or guilty, which is intelligently made in light of the then-applicable law, cannot be invalidated because later judicial decisions indicate that the plea may have rested on a faulty premise. A knowing and intelligent guilty or no-contest plea does not require that all advice offered by the defendant's trial lawyer withstand retrospective examination in a post-conviction hearing. Briggs v. State, 560 S.W.3d 176, 190 (Tex. Crim. App. 2018).

Where the state and defendant agree that the state will not be bound by its recommendation if the defendant commits an offense between the date of his guilty plea and the sentencing hearing, and this agreement is approved by the trial court, it is an enforceable plea bargain agreement. When the defendant breaches this clause of the agreement, he is placed in the position of an open plea at his sentencing hearing. Moore v. State, 240 S.W.3d 248 (Tex. Crim. App. 2007).

A defendant must contemporaneously object to preserve error where the trial court intrudes into the plea-bargaining process. Moore v. State, 295 S.W.3d 329 (Tex. Crim. App. 2009).

PRACTICE TIP: An essential element of an enforceable plea bargain is that it be approved by the trial court. It is important to have the trial court approve the agreement on the record or by writing included in the record for a party to later enforce the terms of the agreement. Moore.

Because prosecutors have no federal constitutional duty to reveal impeaching or affirmative defense information prior to a defendant's guilty plea, a plea is not rendered involuntary where the information is not revealed. U.S. v. Ruiz, 536 U.S. 622, 122 S.Ct. 2450, 153 L.Ed.2d 586 (2002).

§12:174.2 Challenging Voluntariness of Plea

A defendant's sworn representation that his guilty plea is voluntary constitutes a formidable barrier in any subsequent collateral proceedings. Blackledge v. Allison, 431 U.S. 63, 97 S.Ct. 1621, 52 L.Ed.2d 136 (1977); Kniatt.

The voluntariness of a plea cannot be raised on direct appeal, but must be raised on a motion for new trial or on a *habeas corpus* petition. Cooper v. State, 45 S.W.3d 77 (Tex. Crim. App. 2001).

Where the record reflects that the trial court properly admonished the defendant, there is a *prima facie* showing that the plea was knowingly and voluntarily entered. Smith v. State, 857 S.W.2d 71 (Tex.App.—Dallas 1993, *pet. ref'd*).

The burden then shifts to the defendant to establish that he did not understand the consequences of his plea. Miller v. State, 879 S.W.2d 336 (Tex. App.—Houston [14th Dist.] 1994, *pet. ref'd*).

An applicant seeking habeas corpus relief on the basis of an involuntary guilty plea must prove his claim by a preponderance of the evidence. Ex parte Morrow, 952 S.W.2d 530 (Tex. Crim. App. 1997); Kniatt.

An applicant's delay in seeking habeas corpus relief may prejudice the credibility of his claim. Ex parte Young, 479 S.W.2d 45 (Tex. Crim. App. 1972); Kniatt.

The voluntary nature of a defendant's guilty plea can be shown in the record by the overwhelming evidence that the guilty plea was part of a strategy. Gardner v. State, 164 S.W.3d 393 (Tex. Crim. App. 2005).

A defendant's guilty plea may be found involuntary where defense counsel did not advise him of the possibility of removal from the U.S. because removal is nearly an automatic result for a broad class of noncitizen offenders. Padilla v. Kentucky, 559 U.S. 356, 130 S.Ct. 1473, 176 L.Ed.2d 284 (2010).

The collateral consequence of being committed as a sexually violent predator is a civil rather than punitive measure and not so automatic as to render a plea involuntary. Thomas v. State, 365 S.W.3d 537 (Tex.App.—Beaumont 2012, *no pet.*).

§12:174.3 Involuntary Pleas

Misrepresentations that may cause a plea to be involuntary can come from a variety of sources. While many claims allege erroneous advice or misinformation by defense counsel, misrepresentations can also come from the trial court, or the state. Ex parte Barnaby, 475 S.W.3d 316, 322 (Tex. Crim. App. 2015).

A complete failure to admonish a defendant as to the range of punishment he faces renders a plea involuntary without the defendant having to show harm. White v. State, 932 S.W.2d 593 (Tex.App.—Tyler 1995, pet. ref'd).

Improper admonishments do not render a plea involuntary unless the defendant carries his burden to plead and prove that he was harmed. Singleton v. State, 986 S.W.2d 645 (Tex.App.—El Paso 1998, *pet. ref'd*).

Where the statute under which the defendant was charged is unconstitutional and an improper admonishment resulted, the plea of guilty is not voluntarily or knowingly entered where the defendant shows harm. Ex parte Ward, 716 S.W.2d 529 (Tex. Crim. App. 1986).

Where counsel relates incorrect advice to the defendant regarding eligibility for parole, the defendant's resulting plea will be involuntary because parole eligibility is easily determined based on the offense charged and the offense date. Ex parte Moussazadeh, 361 S.W.3d 684 (Tex. Crim. App. 2012) (where the court disavowed the prior test requiring that the defendant show that advice regarding parole eligibility formed an affirmative part or essential element of the plea agreement).

Where the defendant pleads guilty, having been advised that he faces a certain range of punishment, the plea will be involuntary if, based on the indictment, he actually faced a lesser range of punishment. *See* Ex parte McDonald, 568 S.W.3d 688, 689 (Tex. Crim. App. 2018).

Where a defendant pleads guilty to a crime without having been informed of the elements of the crime, his plea is involuntary. Bradshaw v. Stumpf, 545 U.S. 175, 125 S.Ct. 2398, 162 L.Ed.2d 143 (2005).

False evidence can also cause a defendant to be misinformed. The key factor is whether a defendant has sufficient awareness of the relevant circumstances and likely consequences such that his plea is a knowing, intelligent act. If a defendant's knowledge of the relevant circumstances is lacking because of false evidence (instead of, for example, erroneous advice by counsel), his awareness is not sufficient for his plea to be voluntary. But a plea is not involuntary simply because a defendant does not correctly assess every relevant factor entering into his or her decision. Ex parte Barnaby, 475 S.W.3d at 322-3.

To prove a false-evidence habeas corpus claim, a claimant must first show that the evidence in his or her case was false and then that the false evidence was material. Ex parte Barnaby, 475 S.W.3d at 323.

The materiality of false evidence is measured by what impact that false evidence had on the defendant's decision to plead guilty. Ex parte Barnaby, 475 S.W.3d at 325 (where the court found that the value to the defendant of accepting his plea bargain on a drug case outweighed the value of knowing before his plea whether his laboratory report was false).

§12:175 Sufficiency of Pleas of Guilty

§12:175.1 Misdemeanor Cases

In misdemeanor cases before the court or jury, and in felony cases before a jury, the plea is sufficient to support the judgment. Brown v. State,

507 S.W.2d 235 (Tex. Crim. App. 1974); Lenox v. State, 56 S.W.3d 660 (Tex.App.—Texarkana 2001, *pet. ref'd*).

§12:175.2 Felony Cases

Article 1.15 requires substantiation of a guilty plea. By its plain terms it requires evidence in addition to, and independent of, the plea itself to establish the defendant's guilt. Menefee v. State, 287 S.W.3d 9 (Tex. Crim. App. No. 1530-08, July 1, 2009).

A judicial confession is generally sufficient to support a defendant's guilty plea. Dinnery v. State, 592 S.W.2d 343 (Tex. Crim. App. 1979); Honeycutt v. State, 82 S.W.3d 545 (Tex.App.—San Antonio 2002, pet. ref'd).

Under Art. 1.15, the evidence is sufficient to support a plea of guilty if it embraces every essential element of the offense charged. Stone v. State, 919 S.W.2d 424 (Tex. Crim. App. 1996).

Because a plea of guilty in Texas has to be supported by evidence, a valid plea of guilty or *nolo contendre* waives or forfeits the right to appeal a claim of error only when the judgment was rendered independent of, and not supported by, the error. Young v. State, 8 S.W.3d 656 (Tex. Crim. App. 2000) (overruling Helms v. State, 484 S.W.2d 924 (Tex. Crim. App. 1972)).

A defendant's assertion that the allegations in the indictment are true and correct is a judicial confession. Dinnery v. State, 592 S.W.2d 343 (Tex. Crim. App. 1979); Solis v. State, 945 S.W.2d 300 (Tex.App.—Houston [1st Dist.] 1997, *pet. ref'd*).

However, a defendant's sworn affirmation reaffirming his guilty plea to the charges in the indictment does not constitute a judicial confession and does not otherwise supply evidence, in whole or in part, sufficient to support the plea under Article 1.15. Menefee.

Where a guilty plea is entered in open court, it does not have to be orally made by the defendant in order to be voluntary and valid. Costilla v. State, 146 S.W.3d 213 (Tex. Crim. App. 2004).

In Ditto/Ervin v. State, 988 S.W.2d 236 (Tex. Crim. App. 1999), appellants pled guilty under a plea bargain that the state would recommend that the punishment not exceed 10 years confinement. Appellants filed applications for probation. In Ditto's case, the state specifically made no recommendation as to probation. Appellants received 10 years deferred from the judge, who admonished appellant that he was liable for up to 20 years confinement upon violation of probation. Both appellants were later adjudicated and sentenced to 20 years. The plea bargain with a cap of 10 years confinement was fulfilled when appellants were placed on deferred. The words of the statute are "[a]fter an adjudication of guilt, all proceedings ...continue as if the adjudication had not been deferred." After an adjudication of guilt, appellant was not in a position to withdraw his plea. The trial court properly sentenced appellant within the statutory limits for the charged offenses. Ditto/Ervin.

§12:176 Conditional Guilty Pleas

§12:176.1 Appeal Following Plea

Under Tex.R.App.Proc. 40(b)(1) (now Tex.R.App.Proc. 25.2), a defendant may plead guilty pursuant to a plea bargain and reserve for appeal the merits of the denial of written motions or issues on which the trial court had granted permission to appeal. Davis v. State, 870 S.W.2d 43 (Tex. Crim. App. 1994).

A defendant who had entered a guilty plea pursuant to a plea bargain may appeal all written pretrial motions on the grounds stated in the motions, regardless of the manner of proof at the guilty plea hearing. Ellis v. State, 705 S.W.2d 261 (Tex.App.—San Antonio 1986, *no pet.*). This applies to both felony and misdemeanor convictions. McGlynn v. State, 704 S.W.2d 18 (Tex. Crim. App. 1982).

A defendant may appeal the denial of a suppression motion despite the fact that he plead guilty and received a deferred adjudication. Dillehey v. State, 815 S.W.2d 623 (Tex. Crim. App. 1991).

§12:176.2 Deferred Sentencing

Where it is a condition of a plea bargain that if the defendant does not appear for a deferred sentencing date she will face the full range of punishment, the defendant has forfeited any complaint about the trial judge's participation in the plea negotiations and sentence may be pronounced anywhere within the punishment range. Hallmark v. State, 541 S.W.3d 167, 170-1 (Tex. Crim. App. 2017).

A proper objection must be lodged at the time of the plea in order to complain on appeal that the judge exceeded the plea bargain. Hallmark v. State, 541 S.W.3d at 170.

§12:177 Open Pleas of Guilty

Upon entering a plea of guilty in a non-capital felony case, a defendant is not entitled to a bifurcated trial—once the guilty plea is entered, the procedure becomes a "unitary trial" to determine the remaining issue of punishment. There does not exist a separate punishment phase. Appellant's written waiver of her Fifth Amendment right not to self-incriminate applied to the entire criminal adjudication proceeding including the assessment of punishment. Carroll v. State, 42 S.W.3d 129 (Tex. Crim. App. 2001).

When a defendant pleads guilty to a jury, the trial becomes unitary and the jury will always be called upon to assess punishment. A defendant cannot plead guilty to a jury and then be sentenced by a judge. In re The State of Texas ex rel. Jennifer Tharp, 393 S.W.3d 751, 755 (Tex. Crim. App. 2012).

When a defendant enters an open plea of guilty to an offense, he waives the right to appeal any non-jurisdictional defects, other than the voluntariness of his plea, which occurred before entry of the plea. Flowers v. State, 935 S.W.2d 131 (Tex. Crim. App. 1996).

However, a defendant who enters a plea of guilty to a jury without the benefit of a plea bargain waives or forfeits the right to appeal a claim of error only when the judgment was rendered independent of, and not supported by, the error. Young v. State, 8 S.W.3d 656 (Tex. Crim. App. 2000) (overruling Helms v. State, 484 S.W.2d 924 (Tex. Crim. App. 1972)).

Where a defendant pleads guilty without a punishment recommendation but the state agrees to plea in bar another offense, the sentence is considered to be a punishment recommended by the state and the defendant has no right to appeal. Shankle v. State, 119 S.W.3d 808 (Tex. Crim. App. 2003).

However, in a charge-bargain situation, a defendant may appeal matters that were raised by written motion filed and ruled on before trial. Kennedy v. State, 297 S.W.3d 338 (Tex. Crim. App. 2009).

Defendants who plead guilty without the benefit of a plea bargain (open plea) may raise claims of error occurring at or after the entry of the guilty plea. Jack v. State, 871 S.W.2d 741 (Tex. Crim. App. 1994).

Waivers of rights executed pursuant to CCP Art. 1.15 apply only to the guilt stage, and do not constitute a waiver of the right to confrontation and cross-examination at the punishment phase. Stringer v. State, 241 S.W.3d 52 (Tex. Crim. App. 2007).

§12:178 Pleas of Nolo Contendere

A plea of *nolo contendere* (no contest) has the same legal effect as a plea of guilty in a criminal case. Snyder v. State, 629 S.W.2d 930 (Tex. Crim. App. 1982).

However, a *nolo contendere* plea may not be used against the defendant as an admission in any civil suit based upon or growing out of the act upon which the criminal prosecution is based. Art. 27.02(5).

§12:178.1 Form: Waiver and Consent to Stipulate Evidence

See our companion book *Texas Criminal Forms* (James Publishing), Form 15-24.

§12:179 Withdrawal of Plea

CCP Art. 26.13(a)(2) provides that should the court reject any plea bargain agreement, the defendant shall be permitted to withdraw his plea of guilty or *nolo contendere*.

In a situation where the trial court does not accept a plea bargain, the burden is on the defendant to contemporaneously withdraw his guilty plea. Where the defendant does not move to withdraw his plea, the proceeding can be treated as an open plea by the trial court. Moore v. State, 295 S.W.3d 329 (Tex. Crim. App. 2009).

Where evidence inconsistent with guilt is

introduced, the defendant has a right upon timely request to change his guilty plea to a not guilty plea. However, the trial court has no duty to do it on its own motion. Mendez v. State, 138 S.W.3d 645 (Tex. Crim. App. 2004).

A defendant's punishment testimony given at a proceeding where he would be permitted to withdraw his plea of guilty is protected by Rule 410(3). Bowie v. State, 135 S.W.3d 55 (Tex. Crim. App. 2004).

The trial judge is not required to *sua sponte* withdraw the defendant's plea of guilty if he makes contrary statements in a presentence investigation; rather, the trial judge can find the defendant guilty, not guilty, or guilty of a lesser included offense. Solis v. State, 945 S.W.2d 300 (Tex.App.—Houston [1st Dist.] 1997, *pet. ref'd*).

A defendant may withdraw his jury trial waiver if the withdrawal will not: 1) interfere with the orderly administration of the business of the court; 2) result in unnecessary delay or inconvenience to the witnesses; or 3) prejudice the state. The burden is on the defendant to establish these matters and the state can rebut the defendant's assertions. Green v. State, 36 S.W.3d 211 (Tex.App.—Houston [14th Dist.] 2001, *no pet.*).

Tex.R.Evid. 410(1) makes withdrawn guilty and *nolo contendere* pleas inadmissible in a subsequent trial.

§12:180 Guilty Pleas in Absentia

Tex. Code Crim. Pro. Art. 27.19. Plea by Certain Defendants

(a) Notwithstanding any other provision of this code, a court shall accept a plea of guilty or nolo contendere from a defendant who is confined in a penal institution if the plea is made:

(1) in accordance with the procedure established by Article 27.18; or

(2) in writing, including a writing delivered by United States mail or secure electronic or facsimile transmission, before the appropriate court having jurisdiction in the county in which the penal institution is located, provided that:

(A) the defendant is notified by the court of original jurisdiction of the right to counsel and the procedures for requesting appointment of counsel, and is provided a reasonable opportunity to request a court-appointed lawyer;(B) if the defendant elects to proceed without counsel, the defendant must waive the right to counsel in accordance with Article 1.051;

(C) the defendant must waive the right to be present at the taking of the plea or to have counsel present, if the defendant has counsel; and

(D) if the defendant is charged with a felony, judgment and sentence are rendered in accordance with the conditions and the procedure established by Article 42.14(b).

(b) In this article, "penal institution" has the meaning assigned by Section 1.07, Penal Code.

(c) Before accepting a plea submitted under Subsection (a)(2), the court shall verify that the person submitting the plea is:

(1) the defendant named in the information or indictment; or

(2) a person with legal authority to act for the defendant named in the information or indictment.

Tex. Code Crim. Pro. Art. 42.14. In Absence of Defendant

(a) In a misdemeanor case, the judgment and sentence may be rendered in the absence of the defendant.

(b) In a felony case, the judgment and sentence may be rendered in the absence of the defendant only if:

(1) the defendant is confined in a penal institution;

(2) the defendant is not charged with a felony offense:

(A) that is listed in Section 3g(a)(1), Article 42.12; or

(B) for which it is alleged that:

(i) a deadly weapon was used or exhibited during the commission of the offense or during immediate flight from the commission of the offense; and

(ii) the defendant used or exhibited the deadly weapon or was a party to the offense and knew that a deadly weapon would be used or exhibited;

(3) the defendant in writing before the appropriate court having jurisdiction in the county in which the penal institution is located:

(A) waives the right to be present at the rendering of the judgment and sentence or to have counsel present;

(B) affirms that the defendant does not have anything to say as to why the sentence should not be pronounced and that there is no reason to prevent the sentence under Article 42.07;

(C) states that the defendant has entered into a written plea agreement with the attorney representing the state in the prosecution of the case; and

(D) requests the court to pronounce sentence in the case in accordance with the plea agreement;

(4) the defendant and the attorney representing the state in the prosecution of the case have entered into a written plea agreement that is made a part of the record in the case; and

(5) sentence is pronounced in accordance with the plea agreement.

(c) A judgment and sentence may be rendered under this article in the absence of the defendant only after the defendant is notified by the court of original jurisdiction of the right to counsel and the defendant requests counsel or waives the right to counsel in accordance with Article 1.051.

(d) In this article, "deadly weapon" and "penal institution" have the meanings assigned by Section 1.07, Penal Code.

(e) If a defendant enters a plea of guilty or nolo contendere under Article 27.19, the attorney representing the state may request at the time the plea is entered that the defendant submit a fingerprint of the defendant suitable for attachment to the judgment. On request for a fingerprint under this subsection, the county in which the defendant is confined shall obtain a fingerprint of the defendant and use first-class mail or other means acceptable to the attorney representing the state and the county to forward the fingerprint to the court accepting the plea.

§12:190 Pretrial DNA Testing In Capital Cases

§12:191 Statutory Law

Tex. Code Crim. Pro. Art. 38.43. Evidence Containing Biological Material

(a) In this article, "biological

evidence" means:

(1) the contents of a sexual assault examination kit; or

(2) any item that contains blood, semen, hair, saliva, skin tissue, fingernail scrapings, bone, bodily fluids, or any other identifiable biological material that was collected as part of an investigation of an alleged felony offense or conduct constituting a felony offense that might reasonably be used to:

(A) establish the identity of the person committing the offense or engaging in the conduct constituting the offense; or

(B) exclude a person from the group of persons who could have committed the offense or engaged in the conduct constituting the offense.

(b) This article applies to a governmental or public entity or an individual, including a law enforcement agency, prosecutor's office, court, public hospital, or crime laboratory, that is charged with the collection, storage, preservation, analysis, or retrieval of biological evidence.

(c) An entity or individual described by Subsection (b) shall ensure that biological evidence collected pursuant to an investigation or prosecution of a felony offense or conduct constituting a felony offense is retained and preserved:

(1) for not less than 40 years, or until the applicable statute of limitations has expired, if there is an unapprehended actor associated with the offense; or

(2) in a case in which a defendant has been convicted, placed on deferred adjudication community supervision, or adjudicated as having engaged in delinquent conduct and there are no additional unapprehended actors associated with the offense:

(A) until the inmate is executed, dies, or is released on parole, if the defendant is convicted of a capital felony;

(B) until the defendant dies, completes the defendant's sentence, or is released on parole or mandatory supervision, if the defendant is sentenced to a term of confinement or imprisonment in the Texas Department of Criminal Justice;

(C) until the defendant completes the defendant's term of community supervision, including deferred adjudication community supervision, if the defendant is placed on community supervision;

(D) until the defendant dies, completes the defendant's sentence, or is released on parole, mandatory supervision, or juvenile probation, if the defendant is committed to the Texas Juvenile Justice Department; or

(E) until the defendant completes the defendant's term of juvenile probation, including a term of community supervision upon transfer of supervision to a criminal court, if the defendant is placed on juvenile probation.

(d) The attorney representing the state, clerk, or other officer in possession of biological evidence described by Subsection (a) may destroy the evidence, but only if the attorney, clerk, or officer by mail notifies the defendant, the last attorney of record for the defendant, and the convicting court of the decision to destroy the evidence and a written objection is not

received by the attorney, clerk, or officer from the defendant, attorney of record, or court before the 91st day after the later of the following dates:

(1) the date on which the attorney representing the state, clerk, or other officer receives proof that the defendant received notice of the planned destruction of evidence; or

(2) the date on which notice of the planned destruction of evidence is mailed to the last attorney of record for the defendant.

(e) To the extent of any conflict, this article controls over Article 2.21.

(f) The Department of Public Safety shall adopt standards and rules authorizing a county with a population less than 100,000 to ensure the preservation of biological evidence by promptly delivering the evidence to the Department of Public Safety for storage in accordance with Section 411.053, Government Code, and department rules.

(g) The Department of Public Safety shall adopt standards and rules, consistent with best practices, relating to a person described by Subsection (b), that specify the manner of collection, storage, preservation, and retrieval of biological evidence.

(h) A person described by Subsection (b) may solicit and accept gifts, grants, donations, and contributions to support the collection, storage, preservation, retrieval, and destruction of biological evidence.

(i) Before a defendant is tried for a capital offense in which the state is seeking the death penalty, subject to Subsection (j), the state shall require either the Department of Public Safety through one of its laboratories or a laboratory accredited under Article 38.01 to perform DNA testing, in accordance with the laboratory's capabilities at the time the testing is performed, on any biological evidence that was collected as part of an investigation of the offense and is in the possession of the state. The laboratory that performs the DNA testing shall pay for all DNA testing performed in accordance with this subsection.

(j) As soon as practicable after the defendant is charged with a capital offense, or on a motion by the state or the defendant in a capital case, unless the state has affirmatively waived the death penalty in writing, the court shall order the state and the defendant to meet and confer about which biological materials collected as part of an investigation of the offense qualify as biological evidence that is required to be tested under Subsection (i). If the state and the defendant agree on which biological materials constitute biological evidence, the biological evidence shall be tested in accordance with Subsection (i). If the state and the defendant do not agree on which biological materials qualify as biological evidence, the state or the defendant may request the court to hold a hearing to determine the issue. On receipt of a request for a hearing under this subsection, the court shall set a date for the hearing and provide written notice of the hearing date to the state and the defendant. At the hearing, there is a rebuttable presumption that the biological material that the defendant requests to be tested constitutes biological evidence that is required to be tested under Subsection (i). This subsection does not in any way prohibit the state from testing biological evidence in the state's possession.

(k) If an item of biological evidence is destroyed or lost as a result of DNA testing performed under Subsection (i), the laboratory that tested the evidence must provide to the defendant any bench notes prepared by the laboratory that are related to the testing of the evidence and the results of that testing.

(l) The defendant's exclusive remedy for testing that was not performed as

required under Subsection (i) or (j) is to seek a writ of mandamus from the court of criminal appeals at any time on or before the date an application for a writ of habeas corpus is due to be filed in the defendant's case under Section 4(a), Article 11.071. An application for a writ of mandamus under this subsection does not toll any period of limitations applicable to a habeas petition under state or federal law. The defendant is entitled to only one application for a writ of mandamus under this subsection. At any time after the date an application for a writ of habeas corpus is filed in the defendant's case under Section 4(a), Article 11.071, the defendant may file one additional motion for forensic testing under Chapter 64.

(m) A defendant may have another laboratory accredited under Article 38.01perform additional testing of any biological evidence required to be tested under Subsection (i). On an ex parte showing of good cause to the court, a defendant may have a laboratory accredited under Article 38.01 perform testing of any biological material that is not required to be tested under Subsection (i). The defendant is responsible for the cost of any testing performed under this subsection.

§12:192 Case Law

Under 38.43, the trial court has, in the absence of agreement by the state and defendant, the discretion to determine when all necessary biological evidence has been tested. In re Luis Solis-Gonzalez, 489 S.W.3d 459, 462 (Tex. Crim. App. 2016).

(This page intentionally left blank.)

Chapter 13

DISCOVERY

Table of Contents

§13:10 The Basics of Discovery

§13:11 No General Right to Discovery

Criminal defendants do not have a general right to discover evidence in the state's possession, but they have been granted limited discovery by CCP Art. 39.14. Washington v. State, 856 S.W.2d 184 (Tex. Crim. App. 1993); Michaelwicz v. State, 186 S.W.3d 601 (Tex.App.—Austin 2006, pet. ref'd).

The prosecutor is not required to deliver his entire file to defense counsel, but only to disclose evidence favorable to the accused that, if suppressed, would deprive the defendant of a fair trial. Harm v. State, 183 S.W.3d 403 (Tex. Crim. App. 2006), *citing* United States v. Bagley, 473 U.S. 667, 105 S.Ct. 3375, 87 L.Ed.2d 481 (1985).

The state is under no duty to disclose neutral or inculpatory information to the defendant or to share all useful information with the defendant. The prosecution cannot be required to anticipate all possible defenses and provide the defendant with information to bolster those defenses. Michaelwicz.

PRACTICE TIP: The Code of Criminal Procedure now requires clerks to accept electronic and digital multimedia evidence for filing from defendants where they allow filing of such matters by the prosecutor. See CCP Art. 2.21.

§13:12 Scope of Statutory Discovery

§13:12.1 Statutory Law

CCP Art. 39.14. Discovery

(a) Subject to the restrictions provided by Section 264.408, Family Code, and Article 39.15 of this code, as soon as practicable after receiving a timely request from the defendant the state shall produce and permit the inspection and the electronic duplication, copying, and photographing, by or on behalf of the defendant, of any offense reports, any designated documents, papers, written or recorded statements of the defendant or a witness, including witness statements of law enforcement officers but not including the work product of counsel for the state in the case and their investigators and their notes or report, or any designated books, accounts, letters, photographs, or objects or other tangible things not otherwise privileged that constitute or contain evidence material to any matter involved in the action and that are in the possession, custody, or control of the state or any person under contract with the state. The state may provide to the defendant electronic duplicates of any documents or other information described by this article. The rights granted to the defendant under this article do not extend to written communications between the state and an agent, representative, or employee of the state. This article does not authorize the removal of the documents, items, or information from the possession of the state, and any inspection shall be in the presence of a representative of the state.

(b) On a party's request made not later than the 30th day before the date that jury selection in the trial is scheduled to begin or, in a trial without a jury, the presentation of evidence is scheduled to begin, the party receiving the request shall disclose to the requesting party the name and address of each person the disclosing party may use at trial to present evidence under Rules 702, 703, and 705, Texas Rules of Evidence. Except as otherwise provided by this subsection, the disclosure must be made in writing in hard copy form or by electronic means not later than the 20th day before the date that jury selection in the trial is scheduled to begin or, in a trial without a jury, the presentation of evidence is scheduled to begin. On motion of a party and on notice to the other parties, the court may order an earlier time at which one or more of the other parties must make the disclosure to the requesting party.

(c) If only a portion of the applicable document, item, or information is subject to discovery under this article,

the state is not required to produce or permit the inspection of the remaining portion that is not subject to discovery and may withhold or redact that portion. The state shall inform the defendant that a portion of the document, item, or information has been withheld or redacted. On request of the defendant, the court shall conduct a hearing to determine whether withholding or redaction is justified under this article or other law.

(d) In the case of a pro se defendant, if the court orders the state to produce and permit the inspection of a document, item, or information under this subsection, the state shall permit the pro se defendant to inspect and review the document, item, or information but is not required to allow electronic duplication as described by Subsection (a).

(e) Except as provided by Subsection (f), the defendant, the attorney representing the defendant, or an investigator, expert, consulting legal counsel, or other agent of the attorney representing the defendant may not disclose to a third party any documents, evidence, materials, or witness statements received from the state under this article unless:

(1) a court orders the disclosure upon a showing of good cause after notice and hearing after considering the security and privacy interests of any victim or witness; or

(2) the documents, evidence, materials, or witness statements have already been publicly disclosed.

(f) The attorney representing the defendant, or an investigator, expert, consulting legal counsel, or agent for the attorney representing the defendant, may allow a defendant, witness, or prospective witness to view the information provided under this article, but may not allow that person to have copies of the information provided, other than a copy of the witness's own statement. Before allowing that person to view a document or the witness statement of another under this subsection, the person possessing the information shall redact the address, telephone number, driver's license number, social security number, date of birth, and any bank account or other identifying numbers contained in the document or witness statement. For purposes of this section, the defendant may not be the agent for the attorney representing the defendant.

(g) Nothing in this section shall be interpreted to limit an attorney's ability to communicate regarding his or her case within the Texas Disciplinary Rules of Professional Conduct, except for the communication of information identifying any victim or witness, including name, except as provided in Subsections (e) and (f), address, telephone number, driver's license number, social security number, date of birth, and bank account information or any information that by reference would make it possible to identify a victim or a witness. Nothing in this subsection shall prohibit the disclosure of identifying information to an administrative, law enforcement, regulatory, or licensing agency for the purposes of making a good faith complaint.

(h) Notwithstanding any other provision of this article, the state shall disclose to the defendant any exculpatory, impeachment, or mitigating document, item, or information in the possession, custody, or control of the state that tends to negate the guilt of the defendant or would tend to reduce the punishment for the offense charged.

(i) The state shall electronically record or otherwise document any document, item, or other information provided to the defendant under this article.

(j) Before accepting a plea of guilty or nolo contendere, or before trial, each

party shall acknowledge in writing or on the record in open court the disclosure, receipt, and list of all documents, items, and information provided to the defendant under this article.

(k) If at any time before, during, or after trial the state discovers any additional document, item, or information required to be disclosed under Subsection (h), the state shall promptly disclose the existence of the document, item, or information to the defendant or the court.

(l) A court may order the defendant to pay costs related to discovery under this article, provided that costs may not exceed the charges prescribed by Subchapter F, Chapter 552, Government Code.

(m) To the extent of any conflict, this article prevails over Chapter 552, Government Code.

(n) This article does not prohibit the parties from agreeing to discovery and documentation requirements equal to or greater than those required under this article.

13:12.2 Case Law

The Texas Legislature intended Art. 39.14 to constitute a comprehensive pretrial discovery statute, and that criminal discovery orders must fall within the confines of that article's limited authorization. State ex rel. Wade v. Stephens, 724 S.W.2d 141 (Tex.App.—Dallas 1987, *no pet.*).

A corollary of this rule is that trial courts lack inherent authority to order pretrial discovery any greater than that authorized by Art. 39.14. Wade (where the court of appeals declared void a discovery order that was outside the trial court's statutory authority).

When Texas Code Criminal Procedure art. 39.14(f) says that a defendant may view but not obtain copies of discovery materials provided by the state that is exactly what it means. Also, the trial court does not have authority to override the statute and order that the defendant have copies. In re Matthew Powell, Lubbock County District Attorney v. Honorable Mark Hocker, County Court at Law Number One of Lubbock County, 516 S.W.3d 488, 495 (Tex. Crim. App. 2017).

The decision on what is discoverable is left to the discretion of the trial judge. McBride v. State, 838 S.W.2d 248 (Tex. Crim. App. 1992), *citing* Quinones v. State, 592 S.W.2d 933 (Tex. Crim. App. 1980).

Appellate courts will not disturb a trial judge's decision under Art. 39.14 absent an abuse of discretion. McBride.

However, the trial judge is required to permit discovery if the evidence sought is material to the defense of the accused. McBride.

The state has a continuing burden of disclosure once a motion for discovery has been granted. Crane v. State, 786 S.W.2d 338 (Tex. Crim. App. 1990).

It is the opinion of the Texas attorney general that the knowledge of an assistant criminal district attorney working in one division of an office is imputed to prosecutors in other sections of the office for purposes of art. 39.14. Tex. Atty. Gen. Op., No. KP-0213 (2018).

To the extent information provided to an assistant criminal district attorney acting in a civil capacity constitutes an item described by article 39.14(a) but is protected by the attorney-client privilege, the plain language of article (a) would exempt its disclosure to the defendant. However, a court would likely conclude that any exculpatory information meeting the requirements of article 39.14(h) obtained by such an attorney must be disclosed to the defendant, notwithstanding any attorney-client or other evidentiary privilege. Tex. Atty. Gen. Op., No. KP-0213 (2018).

To the extent that information obtained by an assistant criminal district attorney acting in a civil capacity is confidential under section 261.201 of the Family Code, any duty of disclosure in article 39.14(a) of the Code of Criminal Procedure would not be triggered except pursuant to court order obtained under subsection 261.201(b) or (c). A court would likely conclude that any exculpatory information obtained by an assistant criminal district attorney that meets the requirements of article 39.14(h) but that is made confidential by section 261.201 shall be disclosed only pursuant to court order obtained under subsection 261.201 (b) or (c). Tex. Atty. Gen. Op., No. KP-0213 (2018).

§13:13 Defendant's Burden

§13:13.1 General Points

Before a defendant is entitled to discovery, he must show: (1) good cause; (2) materiality; and (3) possession by the State of the information. Medina v. State, 986 S.W.2d 733 (Tex.App.—Amarillo 1999, *pet. ref'd*).

PRACTICE TIP: It is important to have the trial court rule on the motion for discovery and make specific orders because without the rulings, no error can be preserved. See Garcia v. State, 871 S.W.2d 279 (Tex.App.—El Paso 1994, no pet.).

§13:13.2 Defendant Must Show Good Cause

It is the defendant's burden to show that he has good cause under Art. 39.14 before the trial judge is required to permit discovery if the evidence sought is material to the defense of the accused. McBride v. State, 838 S.W.2d 248 (Tex. Crim. App. 1992); Oprean v. State, 201 S.W.3d 724 (Tex. Crim. App. 2006) (Cochran, J., concurring).

When a defendant has shown that the evidence he seeks is material, he has shown good cause. Quinones v. State, 592 S.W.2d 933 (Tex. Crim. App. 1980); McBride.

When a defendant has shown that the evidence he seeks is indispensable to the state's case, he has shown good cause. Bates v. State, 587 S.W.2d 121 (Tex. Crim. App. 1979); McBride.

§13:13.3 Defendant Must Show Materiality

The defendant must demonstrate that the information sought is material. Crane v. State, 786 S.W.2d 338 (Tex. Crim. App. 1990); Lagrone v. State, 942 S.W.2d 602 (Tex. Crim. App. 1997).

Evidence is material if there is a reasonable probability that, had the evidence been disclosed to the defense, the result of the proceeding would have been different. Harm v. State, 183 S.W.3d 403 (Tex. Crim. App. 2006), *citing* Ex parte Kimes, 872 S.W.2d 700 (Tex. Crim. App. 1993).

A reasonable probability is one that is sufficient to undermine confidence in the outcome of the trial. Harm; Ex parte Kimes.

Materiality is determined by examining the alleged error in the context of the entire record and overall strength of the state's case. Harm; Thomas v. State, 841 S.W.2d 399 (Tex. Crim. App. 1992).

The mere possibility that an item of undisclosed information might have helped the defendant or might have affected the outcome of the trial does not establish "materiality" in the constitutional sense. Crane.

§13:13.4 Defendant Must Show Possession by State

In order to show error on the trial court's refusal to order the state to turn over evidence, the defendant must show that the state possessed the evidence. Valdez v. State, 116 S.W.3d 94 (Tex. App.—Houston [14th Dist.] 2002, pet. ref'd).

Evidence held by agents of the state is considered in the possession of the state for discovery purposes. Johnston v. State, 917 S.W.2d 135 (Tex.App.—Fort Worth 1996, *pet. ref'd*).

§13:13.5 Defendant Must Be Specific

A motion for discovery must be specific; and if overly broad, will not support a later constitutional claim that evidence was withheld. State ex rel. Simmons v. Peca, 799 S.W.2d 426 (Tex.App.—El Paso 1990, *no pet.*); Thomas v. State, 511 S.W.2d 302 (Tex. Crim. App. 1974).

A general request for "all exculpatory evidence which the prosecution may have" is equivalent to no request at all. Villarreal v. State, 576 S.W.2d 51 (Tex. Crim. App. 1978).

Requests for "any favorable evidence," "statements of the prosecution witness which are favorable to the defendant," "all evidence favorable to the defendant material either to guilt or to punishment," are equivalent to no requests at all. Showery v. State, 690 S.W.2d 689 (Tex.App.—El Paso 1985, *pet. ref'd*).

Where a general request is made error only exists if there is a showing on appeal that undisclosed evidence exists which creates a reasonable doubt that did not otherwise exist. Green v. State, 587 S.W.2d 167 (Tex. Crim. App. 1979).

§13:13.6 Motion Must Be Timely Filed

Discovery matters are among those that must be litigated at a pre-trial hearing, if such is scheduled by the court. CCP Art. 28.01. *See* Scott v. State, 825 S.W.2d 521 (Tex.App.—Dallas 1992, *pet. ref'd*).

§13:14 Place and Time of Discovery

If discovery is allowed, the Defendant will have the right to inspect, copy or photograph the materials as authorized by the court, and the court's order shall specify the time, place and manner of such inspecting, copying or photocopying. No evidence, however, may be removed from the possession of the state and any inspection shall be in the presence of a representative of the state. Terrell v. State, 521 S.W.2d 618 (Tex. Crim. App. 1975).

In cases where it would be impractical to require the defense to conduct their examination of evidence at a state facility, the evidence or a copy thereof should be delivered to the defense for inspection. The state has a right to have their representative present during the examination of original material. Taylor v. State, 93 S.W.3d 487 (Tex.App.—Texarkana 2002, *pet. ref'd*) (where the evidence in question was a computer hard drive).

Because Art. 39.14 specifies that the trial court shall specify the manner in which the defense may be allowed to copy state's evidence, the trial court has the authority to order the state to make a copy of a DVD of the interview of the complainant at a child advocacy center in its possession. In re Dist. Attorney's Office of the 25th Judicial Dist., 358 S.W.3d 244, 245-6 (Tex. Crim. App. 2011).

Where no time is stated in the court's order for the delivery of discovery materials to the defendant, the state's disclosure is not untimely where it is provided immediately prior to the taking of testimony. Kirksey v. State, 132 S.W.3d 49 (Tex. App.—Beaumont 2004, *no pet.*).

§13:15 Open File Policy

An open-file policy generally satisfies the prosecution's duty to disclose exculpatory evidence. Harm v. State, 183 S.W.3d 403 (Tex. Crim. App. 2006).

However, an open file policy is not sufficient to comply with the notice requirements of Tex.R.Evid. 404(b) and other evidentiary rules that require specific notice of intent to introduce evidence. Buchanan v. State, 911 S.W.2d 11 (Tex. Crim. App. 1995).

§13:16 Form: General Motion for Discovery

See our companion book *Texas Criminal Forms* (James Publishing):

- Form 13-1, General Motion for Discovery.
- Form 13-1.2 Receipt for Discovery.
- Form 13-2.1 Motion for Discovery and Preservation of Specific Evidence.
- Form 13-2.2 Motion to Disclose Documents Signed or Made by Defendant.

- Form 13-2.3 Motion to Discover Search and Arrest Warrants.
- Form 13-2.4 Motion to Discover Evidence of Insanity, Incompetence or Diminished Mental Capacity of Defendant.
- Form 13-3.1 Motion Requesting the Prosecution to File a List of Physical Evidence.
- Form 13-3.2 Defendant's Motion to Designate Location of Physical Evidence.
- Form 13-4.1 Defendant's Motion for Duplicate Photographs.
- Form 13-4.2 Defendant's Motion for Duplicate Audio and Video Recordings.
- Form 13-4.3 Defendant's Motion to Produce Artistic and/or Photographic Likeness Created by Law Enforcement.
- Form 13-5 Defendant's Motion for Discovery and Preservation of Specific Evidence Relating to Canine Drug Search.

§13:20 Particular Types of Evidence

§13:21 Prosecution Witnesses

§13:21.1 Names of Prosecution Witnesses

The names of witnesses should be disclosed, upon proper motion by the defendant, if they will be used by the state at any stage in the trial. Richardson v. State, 744 S.W.2d 65 (Tex. Crim. App. 1987); Martinez v. State, 867 S.W.2d 30 (Tex. Crim. App. 1993).

The state is only required to reveal the names and addresses of the witnesses that it will actually call at trial. Beets v. State, 767 S.W.2d 711 (Tex. Crim. App. 1987).

If the trial judge allows a witness to testify who does not appear on the state's witness list, the courts will consider whether the prosecutor's actions constitute bad faith and whether the defendant could have reasonably anticipated the witness' testimony. Stoker v. State, 788 S.W.2d 1 (Tex. Crim. App. 1989); Martinez.

Reviewing courts have focused on three areas of inquiry in deciding whether the state acted in bad faith: (1) whether the defense shows that the state intended to deceive; (2) whether the state's notice left the defense adequate time to prepare; and (3) whether the state freely provided the defense with information (e.g., by maintaining an open files policy, by providing updated witness lists, or by promptly notifying the defense of new witnesses). Hardin v. State, 20 S.W.3d 84 (Tex. App.—Texarkana 2000, *pet. ref'd*); Martinez v. State, 131 S.W.3d 22 (Tex.App.—San Antonio 2003, *no pet.*).

In determining whether the defense could have anticipated the state's witness, reviewing courts also have examined three areas of inquiry: (1) the degree of surprise to the defendant; (2) the degree of disadvantage inherent in that surprise (e.g., the defendant was aware of what the witness would say, or the witness testified about cumulative or uncontested issues); (3) the degree to which the trial court was able to remedy that surprise (e.g., by granting the defense a recess, postponement, or continuance, or by ordering the State to provide the witness' criminal history). Martinez.

§13:21.1.1 Form: Defense Motion and State's Answer

See our companion book *Texas Criminal Forms* (James Publishing):

- Form 13-18, Motion to List State's Witnesses (All Persons Contacted).
- Form 13-19, State's Answer to Defendant's Motion to List Witnesses.

§13:21.2 Records of Prosecution Witnesses

§13:21.2.1 General Points

The prosecution must disclose criminal records of its witnesses if the defendant establishes:

- Good cause.
- Materiality.
- Possession by the state of the information.

Medina v. State, 986 S.W.2d 733 (Tex.App.—Amarillo 1999, *pet. ref'd*).

This rule also applies to juvenile records. Medina.

Where the defendant meets all the requirements of Art. 39.14 and shows that the requested criminal records of the state's witnesses exist and are within control of the state, it is error for the trial court to deny the request. Aguero v. State, 818 S.W.2d 128 (Tex.App.—San Antonio 1991, *pet. ref'd*).

Under the Fifth Amendment Due Process Clause (*see* Article I, §19, Texas Constitution), the arrest and conviction report of the State's witnesses constitutes impeachment evidence under Brady. U.S. v. Agurs, 427 U.S. 97, 96 S.Ct. 22392, 49 L.Ed.2d 342 (1976).

However, since such evidence serves only to impeach, a "specific" request must be made in order to invoke Brady. U.S. v. Agurs.

§13:21.2.2 Form: Motion

See our companion book *Texas Criminal Forms* (James Publishing):

- Form 13-20, Motion to Discover Criminal History of State's Witnesses.

§13:21.3 Prosecution Witness Statements

§13:21.3.1 Statute

Tex. R. Evid. 615. Producing a Witness's Statement in Criminal Cases

(a) Motion to Produce. *After a witness other than the defendant testifies on direct examination, the court, on motion of a party who did not call the witness, must order an attorney for the state or the defendant and the defendant's attorney to produce, for the examination and use of the moving party, any statement of the witness that is in their possession and that relates to the subject matter of the witness's testimony.*

(b) Producing the Entire Statement. *If the entire statement relates to the subject matter of the witness's testimony, the court must order that the statement be delivered to the moving party.*

(c) Producing a Redacted Statement. *If the party who called the witness claims that the statement contains information that does not relate to the subject matter of the witness's testimony, the court must inspect the statement in camera. After excising any unrelated portions, the court must order delivery of the redacted statement to the moving party. If a party objects to an excision, the court must preserve the entire statement with the excised portion indicated, under seal, as part of the record.*

(d) Recess to Examine a Statement. *Recess to Examine a Statement. On the moving party's request, the court must recess the proceedings to allow time for a party to examine the statement and prepare for its use.*

(e) Sanction for Failure to Produce or Deliver a Statement. *If the party who called the witness disobeys an order to produce or deliver a statement, the court must strike the witness's testimony from the record. If an attorney for the state disobeys the order, the court must declare a mistrial if justice so requires.*

(f) "Statement" Defined. *As used in this rule, a witness's "statement" means:*

(1) a written statement that the witness makes and signs, or otherwise adopts or approves;

(2) a substantially verbatim, contemporaneously recorded recital of the witness's oral statement that is contained in any recording or any transcription of a recording; or

(3) the witness's statement to a grand jury, however taken or recorded, or a transcription of such a statement.

§13:21.3.2 Prosecutor Must Produce Witness Statements

The plain language of Rule 615 requires a prosecutor to produce witness statements that are in the prosecutor's possession, after the witness testifies. Jenkins v. State, 912 S.W.2d 793 (Tex. Crim. App. 1995).

The party contesting production of a statement has the burden to show why the statement cannot be produced. Jenkins.

When a party seeks to invoke production of a prior statement of a witness under Rule 615, the following analysis should take place. First, the trial court should determine whether the writing or recording constitutes a "statement" as that term is defined under Rule 615(f). That determination falls within the sound discretion of the trial court. If the trial court concludes that the writing or recording is not a statement, then Rule 615 cannot mandate production. If, on the other hand, the trial court determines that the writing or recording is a "statement," then production is mandatory under Rule 615(a) if: (1) the statement is in the opposing party's possession, either actual or constructive; and (2) the statement relates to the subject matter concerning which the witness has testified. Proper application of this last requirement would prevent disclosure of mental impressions, personal beliefs, trial strategy, legal conclusions, or anything else that could not fairly be said to be the witness' own statement. In other words, if the trial court determines *in camera* that all or a portion of the statement would otherwise constitute "work product," then it would not "relate to the subject matter concerning which the witness has testified." Finally, if the trial court determines that portions of the statement do not relate to the subject matter of the witness' testimony, then Rule 615(c) mandates that those portions be excised from the writing or recording and the remainder produced for the movant. Jordan v. State, 897 S.W.2d 909 (Tex.App.—Fort Worth 1995, *no pet.)*.

An audio-taped interview, where it meets the definition of a statement in Rule 615, qualifies as a statement. Jordan.

The work product rule and attorney-client privilege cannot be asserted to prevent discovery under Rule 615(a) Tex.R.Evid. Mayfield v. State, 758 S.W.2d 371 (Tex.App.—Amarillo 1988, *no pet.*).

A report by an arresting officer detailing items found after an inventory of the defendant's car can be a statement under Rule 615. Cross v. State, 877 S.W.2d 25 (Tex.App.—Houston [1st Dist.] 1994, *pet. ref'd*).

§13:21.3.3 Other Documents Used by Witness

Tex. R. Evid. 612. Writing Used to Refresh a Witness's Memory

(a) Scope. *This rule gives an adverse party certain options when a witness uses a writing to refresh memory:*

(1) while testifying;

(2) before testifying, in civil cases, if the court decides that justice requires the party to have those options; or

(3) before testifying, in criminal cases.

(b) Adverse Party's Options; Deleting Unrelated Matter. *An adverse party is entitled to have the writing produced at the hearing, to inspect it, to cross-examine the witness about it, and to introduce in evidence any portion that relates to the witness's testimony. If the producing party claims that the writing includes unrelated matter, the court must examine the writing in camera, delete any unrelated portion, and order that the rest be delivered to the adverse party. Any portion deleted over objection must be preserved for the record.*

(c) Failure to Produce or Deliver the Writing. *If a writing is not produced or is not delivered as ordered, the court may issue any appropriate order. But if the prosecution does not comply in a criminal case, the court must strike the witness's testimony or—if justice so requires—declare a mistrial.*

§13:21.3.4 Form: Motion

See our companion book *Texas Criminal Forms* (James Publishing):

- Form 13-21, Motion to Produce Witnesses' Statements.

§13:21.4 Agreements Made With a Witness

§13:21.4.1 State Must Disclose Promises

A failure to disclose this information is a denial of the defendant's due process rights, in that it permits the jury to consider potentially biased information against the defendant without knowing of the bias or of the reason for the bias. Farris v. State, 676 S.W.2d 674 (Tex.App.—Houston [1st Dir.] 1984); *see* Giglio v. United States, 405 U.S. 150, 92 S.Ct. 763, 31 L.Ed.2d 104 (1972).

A plea-bargain agreement that calls for a witness to testify truthfully in another defendant's trial does not violate Rule 3.04 of the Texas State Bar Rules of Professional Conduct, or 18 U.S.C. §§201(b)(1)(A), 201(b)(3) (the federal anti-bribery statute). Castillo v. State, 221 S.W.3d 689 (Tex. Crim. App. 2007).

§13:21.4.2 Form: Motion

See our companion book *Texas Criminal Forms* (James Publishing):

- Form 13-27, Motion to Reveal Agreement with Witness.

§13:22 Photographs and Videorecordings

§13:22.1 Statutory Law

Tex. Code Crim. Pro. Art. 2.139. Video Recordings of Arrests For Intoxication Offenses. A person stopped or arrested on suspicion of an offense under Section 49.04, 49.045, 49.07, or 49.08, Penal Code, is entitled to receive from a law enforcement agency employing the peace officer who made the stop or arrest a copy of any video made by or at the direction of the officer that contains footage of:

(1) the stop;

(2) the arrest;

(3) the conduct of the person stopped during any interaction with the officer, including during the administration of a field sobriety test; or

(4) a procedure in which a specimen of the person's breath or blood is taken.

§13:22.2 Case Law

Photographs are only required to be made available to the defendant where they contain exculpatory evidence. Mize v. State, 754 S.W.2d 732 (Tex.App.—Corpus Christi 1988, *pet. ref'd*).

§13:23 Previous Testimony

§13:23.1 Grand Jury Testimony

§13:23.1.1 Defendant Must Show Special Reason or Particularized Need

CCP Art. 20.02 generally provides that grand jury proceedings are secret. In order to circumvent the traditional policy of grand jury secrecy, the defendant must show that either a special reason or particularized need exists before an inspection of the grand jury proceedings will be allowed. Bynum v. State, 767 S.W.2d 769 (Tex. Crim. App. 1989).

The totality of the circumstances must be examined when determining whether a particularized need exists. Bynum.

A particularized need includes impeachment of a witness, refreshing his recollection and testing his credibility. Dennis v. United States, 384 U.S. 885, 86 S.Ct. 1840, 16 L.Ed.2d 973 (1966).

A particularized need is not shown simply because the requested testimony pertains to a key prosecution witness, or because there is a "need" to locate inconsistencies in the witness' testimony. Bynum; Legate v. State, 52 S.W.3d 797 (Tex. App.—San Antonio 2001, *pet. ref'd*).

§13:23.1.2 Motion for Discovery of Grand Jury Testimony

See our companion book *Texas Criminal Forms* (James Publishing) Form 13-24.

§13:23.2 Previous Trial Testimony

§13:23.2.1 Defendant's Previous Trial Testimony

It is well-settled that an indigent defendant is entitled upon timely request to be furnished without cost, for use at a subsequent trial, a transcription of his earlier mistrial, if it is needed for an effective defense. Britt v. North Carolina, 404 U.S. 226, 92 S.Ct. 431, 30 L.Ed.2d 400, 403 (1971); Brooks v. State, 901 S.W.2d 742 (Tex.App.—Fort Worth 1995, *pet. ref'd*).

In such cases, the accused is presumed to have a need for the transcription of the court reporter's notes from his first trial. Armour v. State, 606 S.W.2d 891 (Tex. Crim. App. 1980).

Furthermore, the defendant does not bear the burden of showing a particularized need for the transcript, nor of showing that the alternatives to acquire the same evidence are inadequate. Brooks.

Where the state does not show that the defendant has no need for the witness testimony from his previous trial, harm to the defendant is presumed where the testimony is not provided. Brooks.

§13:23.2.2 Third-Party Trials

In the absence of a showing of some particularized need, the presumption of need should not be applied to transcriptions of third-party trials. McKibbon v. State, 749 S.W.2d 83 (Tex. Crim. App. 1988).

Rather, as to third-party trial, the defendant shall be required to demonstrate a specific need for the transcript. McKibbon.

Moreover, a simple request for the transcript, without more, is insufficient to show a particularized, specific need. McKibbon.

§13:23.2.3 Form: Motions

See our companion book *Texas Criminal Forms* (James Publishing):

- Form 13-25, Motion for Transcript of Defendant's Previous Trial.
- Form 13-26, Motion for Transcript of Co-Defendant's Trial.

§13:24 Discovering Informant Evidence

§13:24.1 Statute

§13:24.2 Defeating the State's Privilege to Withhold Informer's Identity

The State or the public entity involved in a prosecution has the privilege under Tex. R. Evid. 508(a) to withhold the name of any informer involved in a criminal prosecution. Virtually any public entity involved may claim the informer's privilege. Bodin v. State, 807 S.W.2d 313 (Tex. Crim. App. 1991).

The State's privilege, however, is not absolute. Three exceptions exist where:

1. The state has in some way voluntarily disclosed the identity of the informer. Tex. R. Evid. 508(c)(1).
2. The evidence shows that the informer may be able to give testimony essential to a fair determination of the guilt or innocence of the accused. Tex. R. Evid. 508(c)(2).
3. The court is not satisfied that the informer, who was relied upon to establish the legality of means by which evidence was obtained, is reliable or credible. Tex. R. Evid. 508(c)(3). Bodin.

If a public entity invokes the privilege and refuses to reveal the identity of the informant, it is incumbent on the defendant to file a motion asking that the trial court hold a hearing to determine if the prosecution should be allowed to persist in this refusal. Anderson v. State, 817 S.W.2d 69 (Tex. Crim. App. 1991).

PRACTICE TIP: File a motion to reveal the identity of an informer in every case where the state relies on information gathered from an informant. Even if the state is not ultimately required to reveal the identity, the hearing on the motion can be an excellent discovery device. It can also have the effect of inducing the state to dismiss the case or offer a favorable plea bargain.

A motion to reveal the identity of an informer should be filed in any case in which an informer is involved; its application is not limited to drug cases. Williams v. State, 62 S.W.3d 800 (Tex. App.—San Antonio 2001, *no pet.*) (defendant was charged with murder); Heard v. State, 995 S.W.2d 317 (Tex.App.—Corpus Christi 1999, *pet. ref'd*) (defendant was charged with murder).

Where the public entity validly invokes the privilege, the trial court shall give it an opportunity to show *in camera* facts relevant to determining whether the informer can give the testimony or whether the informer is reliable or credible. Tex. R. Evid. 508(c)(2) & (3).

It is left to the discretion of the trial court to determine when disclosure is warranted under this Rule. In order to show that the trial court erred in not ordering the disclosure, the defendant must show that the trial court abused its discretion. Stokes v. State, 978 S.W.2d 674 (Tex.App.—Eastland 1998, *pet. ref'd*).

Although the preferred method of conducting an *in camera* hearing under this Rule is to review affidavits, the trial court can require testimony if the affidavits will not satisfactorily resolve the matter. Tex. R. Evid. 508(c)(2).

TIP TO THE PROSECUTION: The burden is on the prosecution to request that the trial court hold an in camera hearing to determine the validity of the privilege. Hall v. State, 778 S.W.2d 473 (Tex. App.—Houston [14th Dist.] 1988, pet. ref'd).

TIP TO THE DEFENSE: If the trial court agrees to conduct an in camera hearing, be sure that the record contains the defense's request that the hearing be done on the record and preserved for appellate purposes.

§13:24.3 Defendant's Burden

Only where the defendant demonstrates to the trial court that there is, in fact, an informer is the trial court obligated to hold a hearing on the motion. Smith v. State, 781 S.W.2d 418 (Tex. App.—Houston [1st Dist.] 1989, *no pet.*).

The defendant must explain how the informer's testimony is relevant to his guilt or innocence. Tex. R. Evid. 508 does not contemplate this procedure for the punishment phase of a trial. Mills v. State, 847 S.W.2d 453 (Tex.App.—Eastland 1993, *pet. ref'd*).

The defendant has the burden to make a threshold showing that the informer's identity must be disclosed, and must do so by evidence stronger than mere conjecture or speculation. The mere filing of a motion to disclose the informer's identity is insufficient to make this threshold showing. Bodin v. State, 807 S.W.2d 313 (Tex. Crim. App. 1991).

The defendant's motion must go beyond mere cursory, conclusory allegations. Menefee v. State, 928 S.W.2d 274 (Tex.App.—Tyler 1996, *no pet.*).

Under Rule 508(c)(2), the defendant's burden is to show only that there is a reasonable probability that the informer can give testimony necessary to a fair determination of the issue of guilt or innocence. Anderson v. State, 817 S.W.2d 69 (Tex. Crim. App. 1991).

However, because the defendant may not actually know the nature of the informer's testimony, he or she is only required to make a plausible showing of how the informer's information may be important to merit an *in camera* hearing. A defendant is only required to make a plausible showing that the informer could give testimony necessary to a fair determination of guilt before the privilege is defeated and the identity must be revealed. Anderson; Bodin.

Upon review of such materials, the trial court should order disclosure only where the informer's potential testimony will significantly aid the defendant; mere conjecture or supposition about possible relevancy is insufficient. Bodin.

Where the evidence at the hearing on the motion to reveal the identity of the informer only shows that the police officers involved do not know if the informer can give testimony regarding the guilt or innocence of the defendant, the defendant has failed to carry his burden. Williams v. State,

62 S.W.3d 800 (Tex.App.—San Antonio 2001, *no pet.*).

Where the defendant has made the requisite "plausible showing" and the trial court fails to conduct an *in camera* hearing to determine whether the informer could, in fact, supply testimony necessary to a fair determination of guilt or innocence, the trial court has erred. Loving v. State, 882 S.W.2d 42 (Tex.App.—Houston [1st Dist.] 1994, *no pet.*).

If the trial court finds that there is a reasonable probability that the informer can give the testimony alleged, then the prosecution must reveal the identity of the informer or risk dismissal of the charges by the trial court. Edmond v. State, 911 S.W.2d 487 (Tex.App.—Texarkana 1995, *no pet.*).

Even if the trial court grants the motion and requires the state to reveal the name of the informant, the state has no duty to bring the informant to court. The defendant must issue a subpoena to secure the testimony of the informer. Ashorn v. State, 802 S.W.2d 888 (Tex.App.—Fort Worth 1991, *no pet.*).

§13:24.4 Challenging Reliability or Credibility of Informer

The Rule further provides that where the trial court is not satisfied that an informer, who is being relied on by the state to establish that evidence was legally obtained, is reliable and credible, the trial court may require that the identity of the informer be revealed by the state. Tex. R. Evid. 508(c)(3).

The trial court is not required to hold a hearing on this provision of the Rule where it is satisfied that the informer was credible or reliable. Hall v. State, 778 S.W.2d 473 (Tex.App.—Houston [14th Dist.] 1988, *pet. ref'd*).

TIP TO THE PROSECUTION: The burden is on the prosecution to request that the trial court hold an in camera hearing to determine the validity of the privilege. Hall.

§13:24.5 Mechanics of the Rule

Tex.R.Evid. 508(c)(2) creates a four-step process for resolving prosecution claims of privilege for informers:

1. The evidence must show that the informer may be able to give necessary evidence.
2. The prosecution must invoke the privilege.
3. The trial court must permit the prosecution to show *in camera* whether the witness can give the testimony.
4. If the court determines the informer can give the testimony, and the prosecution does not disclose the informer's identity, the charge must be dismissed.

Lary v. State, 15 S.W.3d 581 (Tex.App.—Amarillo 2000, *pet. ref'd*).

An informant's identity must be disclosed if he or she participated in the offense, was present at the time of the offense or arrest, or is otherwise shown to be a material witness to the transaction. Anderson v. State, 817 S.W.2d 69 (Tex. Crim. App. 1991).

If the informant is an eyewitness to an alleged offense, certainly the informant can give testimony necessary to a fair determination of the issues of guilt or innocence. Anderson.

Where the informer made a "covered buy" which served as the basis for a search warrant of the defendant's apartment, the informer's identity should have been revealed. Bodin v. State, 807 S.W.2d 313 (Tex. Crim. App. 1991).

§13:24.6 Form: Motions to Reveal Identity

See our companion book *Texas Criminal Forms* (James Publishing):

- Form 2-8, Motion to Reveal Identity of Informer (Testimony Relevant to Guilt or Innocence).
- Form 2-9, Motion to Reveal Identity of Informer (Challenge to Legality of Obtaining Evidence).

§13:25 Crime Stoppers Information

§13:25.1 Statutes

Tex. Gov't. Code §414.007. Confidentiality of Council Records

Council records relating to reports of criminal acts are confidential.

Tex. Gov't. Code §414.008. Privileged Information

(a) Except as otherwise provided by this section, evidence of a communication between a person submitting a report of a criminal act to the council or a Crime Stoppers organization and the person who accepted the report on behalf of the council or organization is not admissible in a court or an administrative proceeding.

(b) Records of the council or a Crime Stoppers organization concerning a report of criminal activity may not be compelled to be produced before a court or other tribunal except on a motion:

(1) filed in a criminal trial court by a defendant who alleges that the records or report contains evidence that is exculpatory to the defendant in the trial of that offense; or

(2) filed in a civil case by a plaintiff who alleges that denial of access to the records concerning the report of criminal activity abrogates any part of a cognizable common law cause of action, if the plaintiff is alleging abrogation:

(A) was charged with or convicted of a criminal offense based at least partially on the report and the charges were dismissed, the plaintiff was acquitted, or the conviction was overturned, as applicable; and

(B) in the motion establishes a prima facie case that the plaintiff's abrogated claim is based on injuries from the criminal charge or conviction caused by the wrongful acts of another performed in connection with the report.

(c) On motion of a movant under Subsection (b), the court may subpoena the records or report. The court shall conduct an in camera inspection of materials produced under subpoena to determine whether the materials contain:

(1) evidence that is exculpatory to the defendant; or

(2) information necessary to a plaintiff as described by Subsection (b)(2).

(d) If the court determines that the materials produced contain evidence that is exculpatory to the defendant or information necessary to a plaintiff as described by Subsection (b)(2), the court shall present the evidence to the movant in a form that does not disclose the identity of the person who was the source of the evidence, unless the state or federal constitution requires the disclosure of that person's identity. The court shall execute an affidavit accompanying the disclosed materials swearing that, in the opinion of the court, the materials disclosed represent the evidence the movant is entitled to receive under this section.

(e) The court shall return to the council or Crime Stoppers organization the materials that are produced under this section but not disclosed to the movant. The council or Crime Stoppers organization shall store the materials at least until the first anniversary of the following appropriate date:

(1) the date of expiration of the time for all direct appeals in a criminal case; or

(2) the date a plaintiff's right to appeal in a civil case is exhausted.

§13:25.2 Defendant Has a Right to Crime Stoppers Information

Although the above-quoted Crime Stoppers statutes state that Crime Stoppers information is confidential, a defendant has a constitutional right to production of crime stoppers information in possession of the local Crime Stoppers program, the Crime Stoppers Advisory Council or the District Attorney's office. Thomas v. State, 837 S.W.2d 106 (Tex. Crim. App. 1992).

Because exculpatory information contained within a crime stoppers report is Brady material, there is no burden on the defendant under the Fourteenth Amendment to specifically request the material. Crawford v. State, 892 S.W.2d 1 (Tex. Crim. App. 1994).

§13:25.3 Procedure; In-Camera Review

Both the State's and the defendant's interests can be served by requiring that Crime Stoppers information be inspected by the trial court *in camera*. Neither the attorneys for the State nor the defendant should be present. It will be the responsibility of the court to determine if the produced information contains *Brady* evidence. The court must, in its sound discretion, take steps to ensure that, to the extent possible, the information remains confidential. If information is deemed material at the time it is inspected or at any future stage of the trial, it must be released to the defendant pursuant to well-settled precedent. At the conclusion of the trial, the information shall be sealed and made part of the record. Thomas v. State, 837 S.W.2d 106 (Tex. Crim. App. 1992).

Under the procedure now set forth in the statute, a criminal defendant may file a motion asserting that the Crime Stoppers privileged report or information contains exculpatory material. Tex. Gov't. Code §414.008(b). The trial court may then subpoena the relevant records, review them *in camera*, determine whether they contain evidence exculpatory to the defendant, and present to the defendant any exculpatory information in a form not requiring the disclosure of the identity of the tipster unless required by the state or federal constitution. In Re Hinterlonge, 109 S.W.3d 611 (Tex.App.—Fort Worth 2003, *rehearing of motion for mandamus overruled*).

§13:25.4 Form: Motion and Supporting Brief

See our companion book *Texas Criminal Forms* (James Publishing):

- Form 13-7, Motion to Discover Crime Stoppers Information.
- Form 13-8, Defendant's Brief in Support of Motion to Discover Crime Stoppers Information.

§13:26 Victim and Defendant Statements

§13:26.1 Victim Impact Statements

§13:26.1.1 Statute

CCP Art. 56.03. Victim Impact Statement

> *(g) A victim impact statement is subject to discovery under Article 39.14 of this code before the testimony of the victim is taken only if the court determines that the statement contains exculpatory material.*

§13:26.1.2 Defendant's Right to the Statement

The defendant is entitled to discover exculpatory information contained in a victim impact statement given under Tex. Code of Crim. Pro. Art. 56.03. Jones v. State, 850 S.W.2d 223 (Tex.App.—Fort Worth 1993, *pet. ref'd*).

The victim impact statement is no different from any other statement given by a witness, and the defendant is entitled to inspect the victim impact statement after the testimony of the person giving it, for purposes of cross-examination. Enos v. State, 889 S.W.2d 303 (Tex. Crim. App. 1994).

§13:26.1.3 Defendant Must Request the Statement

The defendant has the burden to request the victim impact statement and make it part of the record for appeal. Harper v. State, 753 S.W.2d 516 (Tex.App.—Houston [1st Dist.] 1988, *pet. ref'd*).

§13:26.1.4 Form: Motion

See our companion book *Texas Criminal Forms* (James Publishing):

- Form 13-9, Motion to Discover Victim Impact Evidence.

§13:26.2 Defendant's Statement

§13:26.2.1 General Points

Under CCP Art. 38.22, admissions by the defendant to police should be disclosed to the defendant where he requests disclosure in a pretrial motion. Cruz v. State, 645 S.W.2d 498 (Tex. App.—San Antonio 1982), *affirmed at* 770 S.W.2d 778 (Tex. Crim. App. 1984).

Written statements of the defendant are discoverable. CCP Art. 39.14.

There is no harm in failing to disclose an inculpatory statement where the defendant filed no pretrial motion requesting such statements. Russeau v. State, 794 S.W.2d 816 (Tex.App.—Tyler 1990, *no pet.*).

A recorded oral statement of the defendant is not admissible unless the prosecution has provided defense counsel with a true, complete, and accurate copy of all recordings of the defendant. CCP Art. 38.22 §3(a)(5); Tigner v. State, 928 S.W.2d 540 (Tex. Crim. App. 1996).

PRACTICE TIP: The discovery of non-recorded oral statements is not provided for by statute. Defense counsel should try to discover such statements by way of a motion to reduce oral statements to writing.

§13:26.2.2 Form: Motion and Notice

See our companion book *Texas Criminal Forms* (James Publishing):

- Form 6-1, Motion to Reduce Oral Statements to Writing.
- Form 6-7, State's Notice of Intent to Offer Defendant's Electronically Recorded Oral Statement.

§13:26.3 Depositions of Elderly or Disabled Persons

Tex. Code Crim. Pro. Art. 39.025. Depositions of Elderly or Disabled Persons

(a) In this article:

(1) "Disabled person" means a person with a disability as defined by Section 3, Americans with Disabilities Act (42 U.S.C. 12102).

(2) "Elderly person" means a person 65 years of age or older.

(b) The court shall order the attorney representing the state to take the deposition of an elderly or disabled person who is the alleged victim of or witness to an offense not later than the 60th day after the date on which the state files an application to take the deposition under Article 39.02.

(c) The attorney representing the state and the defendant or the defendant's attorney may, by written agreement filed with the court, extend the deadline for the taking of the deposition.

(d) The court shall grant any request by the attorney representing the state to extend the deadline for the taking of the deposition if a reason for the request is the unavailability, health, or well-being of the victim or witness.

(e) The Texas Rules of Civil Procedure govern the taking of the deposition, except to the extent of any conflict with this code or applicable court rules adopted for criminal proceedings, in which event this code and the rules for criminal proceedings govern. The attorney representing the state and the defendant or defendant's attorney may agree to modify the rules applicable to the deposition by written agreement filed with the court before the taking of the deposition.

(f) If a defendant is unavailable to attend a deposition because the defendant is confined in a correctional

facility, the court shall issue any orders or warrants necessary to secure the defendant's presence at the deposition. The sheriff of the county in which a deposition under this subsection is to be taken shall provide a secure location for the taking of the deposition and sufficient law enforcement personnel to ensure the deposition is taken safely. The state's application to take a deposition or notice of deposition is not required to include the identity of any law enforcement agents the sheriff assigns to the deposition and may not serve as a basis for the defendant to object to the taking of the deposition.

(g) If a defendant is unavailable to attend a deposition for any reason other than confinement in a correctional facility, the defendant or defendant's attorney shall request a continuance from the court. The court may grant the continuance if the defendant or defendant's attorney demonstrates good cause for the continuance and that the request is not brought for the purpose of delay or avoidance. A defendant's failure to attend a deposition or request a continuance in accordance with this subsection constitutes a waiver of the defendant's right to be present at the deposition.

§13:27 Expert Witness Information

§13:27.1 General Points

§13:27.1.1 Statute

CCP Art. 39.14. Discovery

(b) On a party's request made not later than the 30th day before the date that jury selection in the trial is scheduled to begin or, in a trial without a jury, the presentation of evidence is scheduled to begin, the party receiving the request shall disclose to the requesting party the name and address of each person the disclosing party may use at trial to present evidence under Rules 702, 703, and 705, Texas Rules of Evidence. Except as otherwise provided by this subsection, the disclosure must be made in writing in hard copy form or by electronic means not later than the 20th day before the date that jury selection in the trial is scheduled to begin or, in a trial without a jury, the presentation of evidence is scheduled to begin. On motion of a party and on notice to the other parties, the court may order an earlier time at which one or more of the other parties must make the disclosure to the requesting party.

§13:27.1.2 Reciprocal Discovery

Texas has a reciprocal discovery rule for the exchange of expert witness information.

Art. 39.14(b) expressly authorizes the trial court to order any party to disclose the names of its expert witnesses to the other party upon motion or request, not just the state. Osbourn v. State, 59 S.W.3d 809 (Tex.App.—Austin 2001), *affirmed at* 92 S.W.3d 531 (Tex. Crim. App. 2002).

Apparently no showing of good cause is required, and thus, whether to issue an order is discretionary with the court. Osbourn.

§13:27.1.3 Procedure for Disclosure

The court shall specify the order, the time and manner in which the other party must make the disclosure to the moving party, but in specifying the time in which the other party shall make disclosure, the court shall require the other party to make the disclosure not later than the 20th day before the date the trial begins. Osbourn.

The matter of sanctions is not mentioned, including that of exclusion of the witness' testimony. Osbourn.

The statute is limited and applies only to evidence under Tex.R.Evid. 702, 703, and 705. Osbourn.

§13:27.1.4 Form: Motion and Witness Designation

See our companion book *Texas Criminal Forms* (James Publishing):

- Form 13-28, (State or Defendant's) Motion Requesting Disclosure of Expert Witnesses.
- Form 13-29, Designation of Expert Witness.

§13:27.2 Voir Dire Examination of Expert Witness

§13:27.2.1 Statute

Tex. R. Evid. 705(b) Voir Dire Examination of an Expert About the Underlying Facts or Data. Before an expert states an opinion or discloses the underlying facts or data, an adverse party in a civil case may—or in a criminal case must—be permitted to examine the expert about the underlying facts or data. This examination must take place outside the jury's hearing.

§13:27.2.2 Party Is Entitled to Voir Dire Expert Witness

Under Rule 705(b), a party is entitled, upon timely request, to conduct a voir dire examination directed to the underlying facts or data upon which the opinion of an opposing party's expert is based. Alba v. State, 905 S.W.2d 581 (Tex. Crim. App. 1995).

The trial court must allow this examination to be conducted prior to the expert giving his opinion and out of the hearing of the jury. Alba.

Affording a party the chance to voir dire, the expert witness gives the party the opportunity to determine the foundation of the expert's opinion without fear of eliciting damaging hearsay or other inadmissible evidence in the jury's presence. Goss v. State, 826 S.W.2d 162 (Tex. Crim. App. 1992).

A Rule 705(b) hearing may also supply counsel with sufficient ammunition to make a timely objection to the expert's testimony on the ground that it lacks a sufficient basis for admissibility. Goss.

§13:27.2.3 Timeliness of Request

The request to voir dire outside the jury's presence is timely if lodged at any point before the expert witness testifies as to the underlying facts and data. Harris v. State, 133 S.W.3d 760 (Tex. App.—Texarkana 2004, *pet. ref'd*).

§13:27.2.4 Preservation of Error

A request to voir dire an expert witness on the subject of his qualifications does not preserve error where the court did not allow voir dire on the underlying facts and data. Jenkins v. State, 912 S.W.2d 793 (Tex. Crim. App. 1993).

§13:27.2.5 Form: Motion

See our companion book *Texas Criminal Forms* (James Publishing):

- Form 13-30, Motion for Voir Dire of Expert Witness and for Pretrial Ruling on Admissibility of Expert Testimony.
- Form 13-30.2 Motion for Production of Statements, Reports, Writings and Underlying Data of Expert Witnesses
- Form 13-30.3 Motion for Production of Material Pertaining to Expert Witness

§13:28 Child Abuse Files

§13:28.1 Statutes

Tex.Fam.Code §261.201. Confidentiality and Disclosure of Information

40 Tex. Admin. Code §700.202. Definitions

40 Tex. Admin. Code §700.203. Access to Confidential Information Maintained by the Texas Department of Protective and Regulatory Services (TDPRS)

40 Tex. Admin. Code §700.204. Redaction of Records Prior to Release

40 Tex. Admin. Code §700.205. Procedures for Requesting Access to Confidential Information

40 Tex. Admin. Code §700.206. Videotapes, Audiotapes, and Photographs

40 Tex. Admin. Code §700.207. Charges for Copies of Records

§13:28.2 Discovery Procedures

A defendant is not entitled to obtain copies of Texas Department of Family and Protective Services (DFPS) records under Tex.Fam.Code §261.201 where the trial court has reviewed the records *in camera* and determined that the records would not be useful to the defendant in preparing or conducting his defense or his cross-examination of the witnesses, and where they do not show any prior allegation that was determined to be false. James v. State, 47 S.W.3d 710

(Tex.App.—Texarkana 2001, *no pet.*); *see also* Pennsylvania v. Ritchie, 480 U.S. 39, 107 S.Ct. 989, 94 L.Ed.2d 40 (1987).

PRACTICE TIP: Where the trial court has reviewed copies of DFPS records in camera to determine if they should be released to the defendant, always have the trial court place the records in the record of the case for later review by an appellate court in the event of an appeal. See James.

In the absence of a showing of necessity, the confidential nature of child abuse case reports should be maintained. Coachman v. State, 692 S.W.2d 940 (Tex.App.—Houston [1st Dist.] 1985, *pet. ref'd*).

A regulatory agency such as the Texas Department of Family and Protective Services (DFPS) waives its privilege against disclosure of materials made privileged by the Family Code where it discloses those materials to another state agency or individual. Patmore v. State, 831 S.W.2d 97 (Tex.App.—Eastland 1992, *no pet.*).

If the matter for which a privilege is sought has been disclosed to a third party, thus raising the question of waiver of the privilege, the party asserting the privilege has the burden of proving that no waiver has occurred. Patmore.

Because Art. 39.14 specifies that the trial court shall specify the manner in which the defense may be allowed to copy state's evidence, the trial court has the authority to order the state to make a copy of a DVD of the interview of the complainant at a child advocacy center in its possession. In re Dist. Attorney's Office of the 25th Judicial Dist., 358 S.W.3d 244, 245-6 (Tex. Crim. App. 2011).

§13:28.3 Form: Motion and Records Request

See our companion book *Texas Criminal Forms* (James Publishing):

- Form 17-28, Motion for Discovery of DFPS Records.
- Form 17-29, DFPS Records Request Form.

§13:29 Child Pornography

§13:29.1 Statutes

Art. 39.15. DISCOVERY OF EVIDENCE THAT CONSTITUTES CHILD PORNOGRAPHY.

(a) In the manner provided by this article, a court shall allow discovery under Article 39.14 of property or material:

(1) that constitutes child pornography, as described by Section 43.26(a)(1), Penal Code;

(2) the promotion or possession of which is prohibited under Section 43.261, Penal Code; or

(3) that is described by Section 2 or 5, Article 38.071, of this code.

(b) Property or material described by Subsection (a) must remain in the care, custody, or control of the court or the state as provided by Article 38.45.

(c) A court shall deny any request by a defendant to copy, photograph, duplicate, or otherwise reproduce any property or material described by Subsection (a), provided that the state makes the property or material reasonably available to the defendant.

(d) For purposes of Subsection (c), property or material is considered to be reasonably available to the defendant if, at a facility under the control of the state, the state provides ample opportunity for the inspection, viewing, and examination of the property or material by the defendant, the defendant's attorney, and any individual the defendant seeks to qualify to provide expert testimony at trial.

Art. 38.45. EVIDENCE THAT CONSTITUTES CHILD PORNOGRAPHY.

(a) During the course of a criminal hearing or proceeding, the court may

not make available or allow to be made available for copying or dissemination to the public property or material:

(1) that constitutes child pornography, as described by Section 43.26(a)(1), Penal Code;

(2) the promotion or possession of which is prohibited under Section 43.261, Penal Code; or

(3) that is described by Section 2 or 5, Article 38.071, of this code.

(b) The court shall place property or material described by Subsection (a) under seal of the court on conclusion of the criminal hearing or proceeding.

(c) The attorney representing the state shall be provided access to property or material described by Subsection (a). In the manner provided by Article 39.15, the defendant, the defendant's attorney, and any individual the defendant seeks to qualify to provide expert testimony at trial shall be provided access to property or material described by Subsection (a).

(d) A court that places property or material described by Subsection (a) under seal may issue an order lifting the seal on a finding that the order is in the best interest of the public.

§13:30 Right to Examination of Evidence by Independent Expert

§13:31 General Points

The burden is on the defendant to show good cause before a court must issue an order for production and inspection. Massey v. State, 933 S.W.2d 141, 153 (Tex.Crim.App.1996); Ehrke v. State, 459 S.W.3d 606, 611 (Tex. Crim. App. 2015).

There are two separate grounds under which a criminal defendant might qualify to inspect the state's evidence without having to meet his general discovery burden to show good cause: 1) that the evidence is material to the defense of the accused; and 2) that the evidence is indispensable to the state. Evidence is material to the accused if its omission would create a reasonable doubt that did not otherwise exist. Evidence is indispensable to the state when its exclusion from evidence would affect the essential proof that the defendant committed the crime. Ehrke, 459 S.W.3d at 611.

The right of inspection may be a qualified right when the evidence has been destroyed in the process of analysis. Norton v. State, 930 S.W.2d 101 (Tex. App.—Amarillo 1996, pet. ref 'd) McBride v. State, 838 S.W.2d 248 (Tex. Crim. App. 1992).

In the absence of materiality or indispensability, a defendant may have the additional burden of showing a significant issue of fact, which has been defined as a showing of a material issue sufficient to require the trial court to appoint an independent expert for an indigent defendant. Ehrke, 459 S.W.3d at 615.

In order to meet his burden to show good cause, a defendant should present affidavits or other evidence in support of his defensive theory; an explanation as to what his defensive theory is and why inspection of the evidence would be helpful in establishing that theory; or make a showing that there was a reason to question the state's evidence. The nature of the evidence and the importance and complexity of it may also have a direct bearing on whether the defendant is entitled to inspection. Ehrke, 459 S.W.3d at 615 *7.

However, while all defendants have an absolute right to inspect material and indispensable evidence against them, a defendant does not have an absolute right to a state-appointed expert. The indigent defendant has the additional burden of showing a significant issue of fact, which the Court defined as a showing of a material issue sufficient to require the trial court to appoint an independent expert. Ehrke, 459 S.W.3d at 615.

In order to be entitled to expert assistance for the inspection and analysis of evidence, the indigent must show: a reason to doubt the analysis by the state's expert; a reason to doubt that the evidence in question was other than claimed by the state; how an expert was needed to prepare his defense;information with regard to how important the scientific issue is in his case; or an explanation of how much help a defense expert could provide.

An indigent defendant should present affidavits or other evidence in support of his defensive theory;an explanation as to what his defensive theory is and why expert assistance would be helpful in establishing that theory; or he should make a showing that there was a reason to question the state's expert and proof. The nature of an expert's field and the importance and complexity of the issue may have a direct bearing on whether the appointment of an expert would be helpful. Ehrke, 459 S.W.3d at 615-6.

§13:32 Controlled Substances

A defendant has an absolute right to an independent chemical analysis of alleged contraband in possession cases. Pena v. State, 166 S.W.3d 274 (Tex.App.—Waco 2005, rev'd on other grounds at 191 S.W.3d 133, Tex. Crim. App. 2006) *citing* Terrell v. State, 521 S.W.2d 618 (Tex. Crim. App. 1975).

The indigent defendant has the additional burden of showing a significant issue of fact, which the Court defined as a showing of a material issue sufficient to require the trial court to appoint an independent expert. Ehrke, 459 S.W.3d at 615

§13:33 Tape Recordings

Original tape recordings should be produced and made available to counsel where they are indispensable to the state's case. Bates v. State, 587 S.W.2d 121 (Tex. Crim. App. 1979) (where the failure to allow inspection was not reversible error because the tapes were only one strand of incriminating evidence).

§13:34 Fingerprints

Latent fingerprints of the defendant should be produced. Hollowell v. State, 571 S.W.2d 179 (Tex. Crim. App. 1978) (where the failure to produce the defendant's fingerprints was harmless error under facts of the case). *See also* Helton v. State, 738 S.W.2d 734 (Tex.App.—Beaumont 1987, *pet. ref'd*).

§13:35 Ballistic Tests

The defendant should be allowed to conduct independent ballistic examinations on weapons where evidence is material to the prosecution. Quinones v. State, 592 S.W.2d 933 (Tex. Crim. App. 1980).

§13:36 Blood, Urine, Breath Test

Tex. Trans. Code § 724.018. Furnishing Information Concerning Test Results.

> *On the request of a person who has given a specimen at the request of a peace officer, full information concerning the analysis of the specimen shall be made available to the person or the person's attorney.*

Transportation Code §724.018 provides for discovery of these test results in cases arising under violations of the Motor Vehicle Act.

§13:37 Computer Hard Drive

Where a computer hard drive is indispensable to the state's case, the hard drive or an exact copy of it should be provided to the defense. Taylor v. State, 93 S.W.3d 487 (Tex.App.—Texarkana 2002, *pet. ref'd*).

§13:38 DNA Test Results

The results of DNA testing can be subject to a discovery order issued by the trial judge. Jackson v. State, 17 S.W.3d 664 (Tex. Crim. App. 2000).

§13:40 Failure to Provide Discovery

§13:41 Remedy

When a trial court grants a motion for discovery and the prosecution willfully fails to disclose the evidence ordered disclosed, the evidence will be suppressed. State v. LaRue, 152 S.W.2d 95 (Tex. Crim. App. 2004) (where the conduct was found to be not willful).

Evidence is willfully withheld if the act is done voluntarily and intentionally, with the specific intent to disobey the law. Dunklin v. State, 194 S.W.3d 14 (Tex.App.—Tyler 2006, *no pet.*) *citing* State v. LaRue, 108 S.W.3d 431, 434 (Tex.App.—Beaumont 2003), *aff'd,* 152 S.W.3d 95 (Tex. Crim. App. 2004).

The prosecutor's conduct will be found to be willful where it was a calculated effort to frustrate the defense. Oprean v. State, 201 S.W.3d 724 (Tex. Crim. App. 2006) (the prosecutor failed to turn over a videotape and then sandbagged her intent to introduce it at trial).

In order to establish a due process violation for the failure to comply with a discovery order, the defendant must: 1) show that he was prejudiced pre-trial; 2) show how the failure to comply substantially impaired his actual defensive posture at trial; and 3) explain how his defensive posture could have materially improved had he been alerted to the existence or significance of the evidence earlier. Francis v. State, 428 S.W.3d 850, 859 (Tex. Crim. App. 2014).

An appellate court will review a trial court's ruling to exclude evidence for willful prosecutorial defiance of a discovery order under an abuse-of-discretion standard, deferring to any trial-level ruling that falls within the zone of reasonable disagreement. An appellate court must pay almost absolute deference to the trial court's findings of fact based on determinations of credibility or demeanor. In the absence of any specific findings of fact from the trial court (either oral or written) spread on the record, a reviewing court must assume that the trial court resolved all fact issues in a way that is consistent with its ultimate ruling, so long as those presumed findings of fact are supported by the record. Francis, 428 S.W.3d at 855.

§13:42 Mandamus to Compel Discovery

Where the discovery in question is mandatory, rather than discretionary with the trial court, the courts of appeals have authority to issue a writ of mandamus to compel discovery in a criminal case. The defendant has the burden to show that there is no other adequate remedy and that he is seeking to compel a ministerial act. Dickens v. Ct. of Appeals, 727 S.W.2d 542 (Tex. Crim. App. 1987).

§13:50 Exculpatory and Mitigation Evidence

§13:51 Statutes

CCP Art. 2.01. Duties of District Attorneys

Each district attorney shall represent the State in all criminal cases in the district courts of his district and in appeals therefrom, except in cases where he has been, before his election, employed adversely. When any criminal proceeding is had before an examining court in his district or before a judge upon habeas corpus, and he is notified of the same, and is at the time within his district, he shall represent the State therein, unless prevented by other official duties. It shall be the primary duty of all prosecuting attorneys, including any special prosecutors, not to convict, but to see that justice is done. They shall not suppress facts or secrete witnesses capable of establishing the innocence of the accused.

Tex. Disciplinary R. Prof'l. Conduct Rule 3.09. Special Responsibilities of a Prosecutor

The prosecutor in a criminal case shall:

(a) refrain from prosecuting or threatening to prosecute a charge that the prosecutor knows is not supported by probable cause;

(b) refrain from conducting or assisting in a custodial interrogation of an accused unless the prosecutor has made reasonable efforts to be assured that the accused has been advised of any right to, and the procedure for obtaining, counsel and has been given reasonable opportunity to obtain counsel;

(c) not initiate or encourage efforts to obtain from an unrepresented accused a waiver of important pre-trial, trial or post-trial rights;

(d) make timely disclosure to the defense of all evidence or information known to the prosecutor that tends to negate the guilt of the accused or mitigates the offense, and, in connection with sentencing, disclose to the defense and to the tribunal all unprivileged mitigating information known to the prosecutor, except when the prosecutor is relieved of this responsibility by a protective order of the tribunal; and

(e) exercise reasonable care to prevent persons employed or controlled by the prosecutor in a criminal case

from making an extrajudicial statement that the prosecutor would be prohibited from making under Rule 3.07.

§13:52 Duty of the State

It is axiomatic that the State has an affirmative duty to make available to an accused, in a timely manner, exculpatory evidence which is in its possession. Brady v. Maryland, 373 U.S. 83, 83 S.Ct. 1194, 10 L.Ed.2d 215 (1963); Ex parte Mitchell, 977 S.W.2d 575 (Tex. Crim. App. 1997).

The duty to disclose encompasses both impeachment and exculpatory evidence. Harm v. State, 183 S.W.3d 403 (Tex. Crim. App. 2006).

Brady has been extended to include the required revelation to an accused of material exculpatory evidence in the possession of police agencies and other parts of the prosecutorial team. Kyles v. Whitley, 514 U.S. 419, 115 S.Ct. 1555, 131 L.Ed.2d 490 (1995); Mitchell.

Even if the prosecutor was not personally aware of the evidence, the State is not relieved of its duty to disclose because "the State" includes, in addition to the prosecutor, other lawyers and employees in his office and members of law enforcement connected to the investigation and prosecution of the case. Ex parte Miles, 359 S.W.3d 647 (Tex. Crim. App. 2012), citing Kyles v. Whitley.

The duty to reveal material exculpatory evidence as dictated by Brady overrides the work-product privilege contained in Code of Criminal Procedure 39.14. Ex parte Miles.

The state's duty to reveal Brady material to the defense attaches when the information comes into the state's possession, whether or not the defense requested the information. Harm (citing Thomas v. State, 841 S.W.2d 399 (Tex. Crim. App. 1992)).

An open-file policy generally satisfies the prosecution's duty to disclose exculpatory evidence. Harm.

PRACTICE TIP: If a prosecutor is unsure whether certain evidence must be disclosed, the prosecutor should submit the evidence to the trial court for its consideration. See Thomas v. State, 841 S.W.2d 399 (Tex. Crim. App. 1992).

When police or prosecutors conceal significant exculpatory or impeaching material in the state's possession, it is ordinarily incumbent on the state to set the record straight when contrary evidence is presented at trial. Banks v. Dretke, 540 U.S. 668, 124 S.Ct. 1256, 157 L.Ed.2d 1166 (2004).

§13:53 Exculpatory, Mitigating or Impeaching

To succeed on a Brady claim, a defendant must show that the evidence withheld by the State was "favorable" to his case. Favorable evidence is that which, if disclosed and used effectively, may make the difference between conviction and acquittal. Pena v. State, 353 S.W.3d 797 (Tex. Crim. App. 2011).

Favorable evidence is any evidence that, if disclosed and used effectively, may make a difference between conviction and acquittal and includes both exculpatory and impeachment evidence. Harm v. State, 183 S.W.3d 403 (Tex. Crim. App. 2006) *citing* Thomas v. State, 841 S.W.2d 399 (Tex. Crim. App. 1992).

Exculpatory evidence may justify, excuse, or clear the defendant from fault, while impeachment evidence is that which disputes or contradicts other evidence. Harm; Thomas.

Under Brady, prosecutors must turn over impeachment evidence. Ex parte Richardson, 70 S.W.3d 865 (Tex. Crim. App. 2002).

Under Brady, nondisclosure of favorable evidence violates due process only if it is material to guilt or punishment. The mere possibility that an item of undisclosed information might have helped the defense, or might have affected the outcome of the trial, does not establish materiality in the constitutional sense. Pena.

Evidence is material if there is a reasonable probability that, had the evidence been disclosed to the defense, the result of the proceeding would have been different. Harm (*citing* Ex parte Kimes, 872 S.W.2d 700 (Tex. Crim. App. 1993)).

A reasonable probability is one that is sufficient to undermine confidence in the outcome of the trial. Harm; Ex parte Kimes.

Materiality is determined by examining the alleged error in the context of the entire record and overall strength of the state's case. Harm; Thomas.

In determining whether evidence not disclosed by prosecutors was material, in violation of Brady, the cumulative effect of all suppressed evidence favorable to the defendant should be considered, rather than considering each item of evidence individually. Kyles v. Whitley, 514 U.S. 419, 115 S.Ct. 1555, 131 L.Ed.2d 490 (1995).

Whether the right to Brady information extends to the pretrial stage of prosecution, however, is still an open question. Ex parte Palmberg, 491 S.W.3d 804, 814 (Tex. Crim. App. 2015 fn. 18).

§13:54 Specific Examples of Exculpatory Evidence

The prosecution has a duty to reveal impeaching information that shows that a state's witness was a paid police informant and to disclose information showing that a witness was intensively coached prior to testifying. Banks v. Dretke, 540 U.S. 668, 124 S.Ct. 1256, 157 L.Ed.2d 1166 (2004).

Evidence of a defendant's insanity or incompetency constitutes Brady material requiring disclosure. Ex Parte Lewis, 587 S.W.2d 170 (Tex. Crim. App. 1979); Nickerson v. State, 69 S.W.3d 661 (Tex.App.—Waco 2002, *pet. ref'd*) (evidence of insanity on videotape).

A log book kept by a former police officer showing that the state's star witness was not a truthful person has been found to be favorable evidence. Ex parte Richardson, 70 S.W.3d 865 (Tex. Crim. App. 2002).

An oral statement made by a witness who had previously given a written statement contained in the state's file should have been communicated to the defense where it contained favorable evidence. Flores v. State, 940 S.W.2d 189 (Tex.App.—San Antonio 1996, *no pet.*).

A witness' failure to identify the defendant in a pretrial photo lineup can be material exculpatory or mitigating evidence. Hall v. State, 283 S.W.3d 137 (Tex. Crim. App. 2009).

The state has a duty to reveal material admissible, such as audio recordings of conversations between the defendant and law enforcement. Pena v. State, 353 S.W.3d 797 (Tex. Crim. App. 2011) (where the recording would have impeached the officer's version of his conversation with the defendant).

Undisclosed police reports which identify other potential suspects in the offense can constitute other evidence that is favorable and material where the information can be used for impeachment purposes. Ex parte Miles, 359 S.W.3d 647 (Tex. Crim. App. 2012).

Evidence impeaching an eyewitness can be material where the State's other evidence is not strong enough to sustain confidence in the verdict. Smith v. Cain, 565 U.S. ___, 132 S.Ct. 627, 181 L.Ed.2d 571 (2012) (where the eyewitness was the only evidence linking the defendant to the crime).

Evidence of statements that case doubt on a witnesses' credibility (here statements of motive to "get" the defendant), evidence of a witnesses' attempts to procure a deal for testifying against a defendant and evidence that contradicts a witness' testimony are all examples of Brady material. Wearry v. Cain, 577 U.S. ___, 136 S.Ct. 1002, 1004, 194 L.Ed.2d 78 (2016).

§13:55 Inapplicability of Brady Rule

The Brady rule does not apply when the defendant was already aware of the information. Havard v. State, 800 S.W.2d 195 (Tex. Crim. App. 1989) (overruling an alleged Brady error involving the defendant's prior statement to police); Hayes v. State, 85 S.W.3d 809 (Tex. Crim. App. 2002) (there is no Brady violation where the defendant wrote the letter in question).

The Brady rule does not apply when the evidence in question is inadmissible. Ex parte Kimes, 872 S.W.2d 700 (Tex. Crim. App. 1993).

The state is not required to seek out exculpatory evidence independently on appellant's behalf, or furnish appellant with exculpatory or mitigating evidence that is fully accessible to appellant from other sources. Harm v. State, 183 S.W.3d 403 (Tex. Crim. App. 2006), *citing* Jackson v. State, 552 S.W.2d 798 (Tex. Crim. App. 1976).

Further, the prosecutor is not required to furnish the defendant with exculpatory and mitigating evidence which is equally and fully accessible to the defense. Pena v. State, 353 S.W.3d 797 (Tex. Crim. App. 2011).

Brady and its progeny do not require prosecuting authorities to disclose exculpatory information to defendants that the State does not have in its possession and that is not known to exist. Harm, *citing* Hafdahl v. State, 805 S.W.2d 396, 399 (Tex. Crim. App. 1990).

Where the evidence shows that the state did not come into the possession of the allegedly exculpatory information until after the defendant's trial, the Brady rule is not invoked. Ex parte Reed, 271 S.W.3d 698 (Tex. Crim. App. 2008).

§13:56 Establishing a Brady Violation

A defendant must satisfy three requirements to establish a Brady violation: (1) the state suppressed evidence; (2) the suppressed evidence is favorable to defendant; and (3) the suppressed evidence is material. Harm v. State, 183 S.W.3d 403 (Tex. Crim. App. 2006), *citing* Little v. State, 991 S.W.2d 864 (Tex. Crim. App.1999).

Additionally, we require that the evidence central to the Brady claim be admissible in court. Pena v. State, 353 S.W.3d 797 (Tex. Crim. App. 2011) (citing Ex parte Kimes, 872 S.W.2d 700, 703 (Tex. Crim. App.1993)); Ex parte Miles, 359 S.W.3d 647 (Tex. Crim. App. 2012).

Incorporated into the third prong, materiality, is a requirement that defendant must be prejudiced by the state's failure to disclose the favorable evidence. Harm, *citing* Banks v. Dretke, 540 U.S. 668, 124 S.Ct. 1256, 157 L.Ed.2d 1166 (2004).

Under Brady, the defendant bears the burden of showing that, in light of all the evidence, it is reasonably probable that the outcome of the trial would have been different had the prosecutor made a timely disclosure. Hampton v. State, 86 S.W.3d 603 (Tex. Crim. App. 2002).

The mere possibility that an item of undisclosed information might have helped the defense, or might have affected the outcome of the trial, does not establish "materiality" in the constitutional sense. Hampton.

It is irrelevant whether suppression of the favorable evidence was done willfully or inadvertently. Harm, *citing* Strickler v. Greene, 527 U.S. 263, 119 S.Ct. 1936, 144 L.Ed.2d 286 (1999).

If the defense has the opportunity to cross-examine concerning the allegedly exculpatory material and there is no showing the defense would have pursued a different trial strategy if he or she had known this information sooner, no Brady violation is shown. Taylor v. State, 93 S.W.3d 487 (Tex.App.—Texarkana 2002, *pet. ref'd*), *citing* Williams v. State, 995 S.W.2d 754 (Tex.App.—San Antonio 1999, *no pet.*).

At trial, evidence willfully withheld from disclosure under a discovery order should be excluded from evidence. Oprean v. State, 201 S.W.3d 724 (Tex. Crim. App. 2006).

When reviewing a trial judge's decision to admit evidence that the defense claims was withheld from discovery, an appellate court must determine whether the judge's decision was an abuse of discretion. Unless the trial judge's decision was outside the "zone of reasonable disagreement," an appellate court should uphold the ruling. When a trial judge makes findings of fact based on an evaluation of credibility and demeanor, an appellate court "should show almost total deference" to those findings. When the trial judge fails to enter written or oral findings of fact, an appellate court will "view the evidence in the light most favorable to the trial court's ruling and assume that the trial court made implicit findings of fact that support its ruling as long as those findings are supported by the record." Oprean.

On appeal where it is found that the state failed to divulge potentially exculpatory evidence (whether directly exculpatory or exculpatory as having impeachment value) and where the failure undermined confidence in the trial's outcome resulting in a violation of a defendant's due process right to a fair trial, the remedy prescribed is reversal of the conviction and remand to the trial court for further proceedings. Ex parte Mitchell, 977 S.W.2d 575 (Tex. Crim. App. 1997).

The failure to request a postponement or seek a continuance waives any error urged in an appeal on the basis of surprise or on the timeliness of notice. Martin v. State, 176 S.W.3d 887 (Tex.App.—Fort Worth 2005, *no pet.), citing* Lindley v. State, 635 S.W.2d 541, 544 (Tex. Crim. App. 1982).

§13:57 Form: Motion

See our companion book *Texas Criminal Forms* (James Publishing):

- Form 13-6, Motion to Produce Exculpatory and Mitigating Evidence.

§13:60 Other Discovery Tools

§13:61 Depositions of Witnesses

§13:61.1 Statutes

CCP Art. 39.02. Witness Depositions

> *Depositions of witnesses may be taken by either the state or the defendant. When a party desires to take the deposition of a witness, the party shall file with the clerk of the court in which the case is pending an affidavit stating the facts necessary to constitute a good reason for taking the* witness' *deposition and an application to take the deposition. On filing of the affidavit and application, and after notice to the opposing party, the court shall hear the application and determine if good reason exists for taking the deposition. The court shall base its determination and shall grant or deny the application on the facts made known at the hearing. This provision is limited to the purposes stated in Article 39.01.*

CCP Art. 39.03. Officers Who May Take the Deposition

> *Upon the filing of such an affidavit and application, the court shall appoint, order or designate one of the following persons before whom such deposition shall be taken:*
>
> 1. *A district judge.*
> 2. *A county judge.*
> 3. *A notary public.*
> 4. *A district clerk.*
> 5. *A county clerk.*
>
> *Such order shall specifically name such person and the time when and place where such deposition shall be taken. Failure of a witness to respond thereto, shall be punishable by contempt by the court. Such deposition shall be oral or written, as the court shall direct.*

§13:61.2 Defendant Must Show Good Reason

A defendant may take the deposition of a witness only upon a showing of good reason. May v. State, 738 S.W.2d 261 (Tex. Crim. App. 1987), *cert. den.*, 484 U.S. 872, 108 S.Ct. 206, 98 L.Ed.2d 158 (1987).

The defendant must show that the witness sought to be deposed possesses information critical to a significant factor at trial, or that such witness has information exclusively within that witness' knowledge. Janecka v. State, 937 S.W.2d 456 (Tex. Crim. App. 1996).

Before a trial court will be found to have abused its discretion in declining to permit depositions to be taken, there must be a showing the defendant was harmed by such action. James v. State, 546 S.W.2d 306 (Tex. Crim. App. 1977).

There is not an abuse of discretion in the denying of a motion to take a deposition where the witness testifies at trial, is subject to cross-examination by the defense, and the defendant does not otherwise show he or she was harmed by the trial court's action. Cooks v. State, 844 S.W.2d 697 (Tex. Crim. App. 1992).

CCP Art. 39.02 requires an affidavit, not a verified application. State ex. rel. Simmons v. Moore, 774 S.W.2d 711 (Tex.App.—El Paso 1989, *no pet.*).

§13:61.3 Form: Application to Take Deposition

See our companion book *Texas Criminal Forms* (James Publishing):

- Form 13-23.1, State or Defendant's Application to Take Deposition of Witness.

§13:62 Written Interrogatories

§13:62.1 Statute

CCP Art. 39.06. Written Interrogatories

> *When any such deposition is to be taken by written interrogatories, such written interrogatories shall be filed with the clerk of the court, and a copy of the same served on all other parties or their counsel for the length of time and in the manner required for service of interrogatories in civil action, and the same procedure shall also be followed with reference to cross-interrogatories as that prescribed in civil action.*

§13:62.2 Form: Application to Take Deposition by Written Interrogatories

See our companion book *Texas Criminal Forms* (James Publishing):

- Form 13-23.2, State or Defendant's Application to Take Deposition of Witness By Written Interrogatories.

§13:70 Exceptions to Discovery Rights

§13:71 Police Reports

Police and prosecution reports are not discoverable prior to trial. Michaelwicz v. State, 186 S.W.3d 601 (Tex.App.—Austin 2006, *pet. ref'd*).

Prosecutors are not required to disclose their investigative work. Michaelwicz.

A defendant has a right to have the court inspect a police report in camera to determine whether it contains any Brady material. Crawford v. State, 934 S.W.2d 744 (Tex.App.—Houston [1st Dist.] 1996, *pet. ref'd*).

§13:72 Medical Examiner's Report

Medical examiner's reports are not discoverable under Art. 39.14, since they are public records. Beeson v. State, 651 S.W.2d 944 (Tex.App.—Fort Worth 1983, *no pet.*).

Medical examiner records are subject to required public disclosure in accordance with Chapter 552, Government Code, except that a photograph or x-ray of a body taken during an autopsy is excepted from required public disclosure in accordance with Chapter 552, Government Code, but is subject to disclosure: (1) under a subpoena or authority of other law; or (2) if the photograph or x-ray is of the body of a person who died while in the custody of law enforcement. CCP Art. 49.25.

§13:73 Prior Jury Service

The state has no obligation under the discovery statutes to furnish its prior jury service information to the defense. Martin v. State, 577 S.W.2d 490 (Tex. Crim. App. 1979).

Prosecutors are not obligated to turn over juror information that is available to the defense on voir dire. Armstrong v. State, 897 S.W.2d 361 (Tex. Crim. App. 1995).

The defendant preserves no claim for appellate review where he does not follow the procedures required in the public records act. Etheridge v. State, 903 S.W.2d 1 (Tex. Crim. App. 1994).

PRACTICE TIP: The simplest method to obtaining prior jury service information is to simply ask the veniremen during voir dire. See Espinosa v. State, 653 S.W.2d 446 (Tex.App.—San Antonio 1982), aff'd, 669 S.W.2d 736 (Tex.Crim.App. 1984) (mandamus inappropriate remedy).

§13:80 Requests for Notice of Intention to Introduce Evidence

§13:81 General Points

The purpose of a notice requirement is to prevent surprise to the defendant and apprise him of the evidence the state plans to introduce at trial. Hernandez v. State, 176 S.W.3d 821 (Tex. Crim. App. 2005).

§13:81.1 Defendant Must Request Notice

Under Tex.R.Evid. 404(b) (and all other notice provisions since they reference Rule 404(b)), the defendant must file a timely "request for notice" in order to trigger the state's duty to provide notice under the particular provisions. Mitchell v. State, 982 S.W.2d 425 (Tex. Crim. App. 1998).

Merely filing the request in a discovery motion is not effective until ruled upon by the court. Espinosa v. State, 853 S.W.2d 36 (Tex. Crim. App. 1993).

§13:81.2 Request Must Be Timely

The defendant's request for notice must be timely; that is, notice must be requested in enough time to allow the state to comply in advance of trial. Chimney v. State, 6 S.W.3d 681 (Tex.App.—Waco 1999, *no pet.*).

For more information about timeliness of notice under this provision, *see* §15:113.

§13:81.3 Notice Must Be Timely

The rule requires "reasonable" notice. Hayden v. State, 66 S.W.3d 269 (Tex. Crim. App. 2001).

Generally, notice given at least 10 days in advance of trial is timely. Splawn v. State, 160 S.W.3d 103 (Tex.App.—Texarkana 2005, *pet. ref'd*).

However, the closer to trial the request for notice is made by the defense, the less notice is required by the state for admissibility. Self v. State, 860 S.W.2d 261 (Tex.App.—Fort Worth 1993, *pet. ref'd*) (five days notice reasonable where request made 19 days before trial).

Notice can be insufficient where defense counsel has insufficient time to investigate the subject of the notice. Webb v. State, 36 S.W.3d 164 (Tex. App.—Houston [14th Dist.] 2000, *pet. ref'd*).

Where the state's "notice" does not strictly state an intent to introduce the evidence (such as where the state merely delivers witness statements to the defense), the closer in time the delivery of the evidence is to the time of the request for notice, the more likely the "notice" will be found reasonable. Hayden.

An "open file" policy, by itself, is not sufficient to comply with the rule's notice requirement. Buchanan v. State, 911 S.W.2d 11 (Tex. Crim. App. 1995).

The Rule 404(b) notice provision is a rule of admissibility because Rule 404(b) literally conditions the admissibility of other crimes evidence on the state's compliance with the notice provision. Hernandez v. State, 176 S.W.3d 821 (Tex. Crim. App. 2005).

If an extraneous offense is admitted by the trial court in violation of the Rule 404(b) notice requirement, on appeal harm is analyzed on the basis of whether the defendant was surprised by the admission. Hernandez.

When erroneously admitted Rule 404(b) evidence is substantively admissible, the error is harmless unless (1) it substantially influenced the jury's verdict and (2) the defendant was surprised by the evidence. Hernandez; Allen v. State, 202 S.W.3d 364 (Tex.App.—Fort Worth 2006, *pet. ref'd*).

§13:82 Extraneous Offenses

§13:82.1 Statute

Tex. R. Evid. 404(b) Crimes, Wrongs, or Other Acts.

> *(1) Prohibited Uses. Evidence of a crime, wrong, or other act is not admissible to prove a person's character in order to show that on a particular occasion the person acted in accordance with the character.*
>
> *(2) Permitted Uses; Notice in Criminal Case. This evidence may be admissible for another purpose, such as proving motive, opportunity, intent, preparation, plan, knowledge, identity, absence of mistake, or lack of accident. On timely request by a defendant in a criminal case, the prosecutor must provide reasonable notice before trial that the prosecution intends to introduce such evidence—other than that arising in the same transaction—in its case-in-chief.*

§13:82.2 Evidence Rule Applies Only to Guilt Phase

Since Tex.R.Evid. 404(b) applies to the guilt phase of the trial, a request for 404(b) notice is only requesting notice for evidence sought to be admitted at that phase of the trial. Herring v. State, 752 S.W.2d 169 (Tex.App.—Houston [1st Dist.] 1988), *remanded on other grounds*, 758 S.W.2d 283 (Tex. Crim. App. 1988).

If an extraneous offense is admitted by the trial court in violation of the Rule 404(b) notice requirement, harm is analyzed on appeal on the basis of whether the defendant was surprised by its admission. Hernandez v. State, 176 S.W.3d 821 (Tex. Crim. App. 2005).

PRACTICE TIP: If the trial court allows admission of an extraneous offense in violation of the Rule 404(b) notice requirement, make a record of the surprise suffered by the defendant and his inability to contest, refute or mitigate the extraneous offense because of the lack of timely notice.

For more information about notice of extraneous offenses, *see* §15:113.

§13:82.3 Form: Request for Notice, Motion for Hearing, and Objection to Admissibility

See our companion book *Texas Criminal Forms* (James Publishing):

- Form 13-11.1, Request for Notice of State's Intention to Offer Extraneous Offenses Pursuant to Rule 404(b).
- Form 13-11.2, Motion for Pretrial Evidentiary Hearing on Admissibility of Extraneous Offenses.
- Form 13-11.3, Defendant's Written Objection to Admissibility of Extraneous Offenses, and Request for Findings of Fact and Conclusions of Law.
- Form 13-11.4 Defendant's Motion for Deadline on Notice of Extraneous Offenses

§13:83 Child Abuse Cases

§13:83.1 Statute

CCP Art. 38.37. Evidence of Extraneous Offenses or Acts

Section 3. On timely request by the defendant, the state shall give the defendant notice of the state's intent to introduce the case-in-chief evidence described by Section 2 in the same manner as the state is required to give notice under Rule 404(b), Texas Rules of Evidence.

Sec. 1. This article applies to a proceeding in the prosecution of a defendant for an offense, or an attempt or conspiracy to commit an offense, under the following provisions of the Penal Code:

(1) if committed against a child under 17 years of age:

(A) Chapter 21 (Sexual Offenses);

(B) Chapter 22 (Assaultive Offenses); or

(C) Section 25.02 (Prohibited Sexual Conduct); or

(2) if committed against a person younger than 18 years of age:

(A) Section 43.25 (Sexual Performance by a Child);

(B) Section 20A.02(a)(7) or (8); or

(C) Section 43.05(a)(2) (Compelling Prostitution) .

Sec. 2. Notwithstanding Rules 404 and 405, Texas Rules of Evidence, evidence of other crimes, wrongs, or acts committed by the defendant against the child who is the victim of the alleged offense shall be admitted for its bearing on relevant matters, including:

(1) the state of mind of the defendant and the child; and

(2) the previous and subsequent relationship between the defendant and the child.

Sec. 3. On timely request by the defendant, the state shall give the defendant notice of the state's intent to introduce in the case in chief evidence described by Section 2 in the same manner as the state is required to give notice under Rule 404(b), Texas Rules of Evidence.

Sec. 4. This article does not limit the admissibility of evidence of extraneous crimes, wrongs, or acts under any other applicable law.

§13:83.2 Defendant Must Request Notice

CCP Art. 38.37 provides that upon timely request, the state must give notice of its intent to introduce child abuse extraneous offense or bad act evidence. In order to be entitled to such notice, the defendant must make the proper request under the statute. McCoy v. State, 10 S.W.3d 50 (Tex.App.—Amarillo 1999, *no pet.*).

PRACTICE TIP: A request for notice should be filed in every case where evidence could possibly be admissible under Art. 38.37.

§13:83.3 Request Must Cite Art. 38.37

A request for notice of extraneous offenses under Rule 404(b) will not entitle the defendant to notice of extraneous offenses to be offered under Art. 38.37 and will not preserve error under Art. 38.37. Hitt v. State, 53 S.W.3d 697 (Tex.App.—Austin 2001, *pet. ref'd*).

For further information pertaining to the notice provision of CCP Art. 38.37, *see* §17:22.2.

§13:83.4 Form: Request for Notice

See our companion book *Texas Criminal Forms* (James Publishing):

- Form 13-12, Request for Notice of Extraneous Offenses in Child Abuse Case.

§13:84 Punishment Evidence

§13:84.1 Statute

CCP Art. 37.07. Verdict Must Be General; Separate Hearing on Proper Punishment

Section 3. Evidence of prior criminal record in all criminal cases after a finding of guilty.

(g) On timely request of the defendant, notice of intent to introduce evidence under this article shall be given in the same manner required by Rule 404(b), Texas Rules of Criminal Evidence. If the attorney representing the state intends to introduce an extraneous crime or bad act that has not resulted in a final conviction in a court of record or a probated or suspended sentence, notice of that intent is reasonable only if the notice includes the date on which and the county in which the alleged crime or bad act occurred and the name of the alleged victim of the crime or bad act. The requirement under this subsection that the attorney representing the state give notice applies only if the defendant makes a timely request to the attorney representing the state for the notice.

§13:84.2 Defendant Must Request Notice

CCP Art. 37.07 places the burden on the defendant to make the request for notice of the state. The state has no duty to provide notice until properly requested by the defendant to do so. Espinosa v. State, 853 S.W.2d 36 (Tex. Crim. App. 1993).

For cases on reasonableness and sufficiency of punishment phase notice, *see* §§20:28.3 and 20:28.4.

§13:84.3 Form: Request for Notice and Motion

See our companion book *Texas Criminal Forms* (James Publishing):

- Forms 13-10.1, Request for Notice of State's Intention to Introduce Punishment Evidence Pursuant to Art. 37.07.
- Form 13-10.2, Motion to Have Extraneous Offenses Not Mentioned.

§13:85 Impeachment Evidence

§13:85.1 Statute

Tex. R. Evid. 609(f) Notice. Evidence of a witness's conviction is not admissible under this rule if, after receiving from the adverse party a timely written request specifying the witness, the proponent of the conviction fails to provide sufficient written notice *of intent to use the conviction. Notice is sufficient if it provides a fair opportunity to contest the use of such evidence.*

§13:85.2 Party Must Request Notice

Tex.R.Evid. 609(f) precludes the use of prior convictions against a witness if the proponent of such evidence fails to give advance written notice of intent to use such evidence upon a timely written request. Brown v. State, 880 S.W.2d 249 (Tex. App.—El Paso 1994, *no pet.*).

Where the state does not give timely notice of its intent to use a prior conviction, it is error for the court to admit evidence of it. Bryant v. State, 997 S.W.2d 673 (Tex.App.—Texarkana 1999, *no pet.*).

PRACTICE TIP: Rule 609(f) is broadly written, applying to evidence of conviction of any witness, and to both the state and defense. Therefore, either party can take advantage of its provisions. If timely notice is not provided, the proponent cannot use the prior conviction information.

§13:85.3 Form: Request for Notice

See our companion book *Texas Criminal Forms* (James Publishing):

- Form 13-13, Request for Notice of State's Intention to Introduce Evidence of Impeachment.

§13:90 Public Information Act Requests

§13:91 Statutes

Tex. Gov't. Code §552.002. Definition of Public Information; Media Containing Public Information

(a) In this chapter, "public information" means information that is collected, assembled, or maintained under a law or ordinance or in connection with the transaction of official business:

(1) by a governmental body; or

(2) for a governmental body and the governmental body owns the information or has a right of access to it.

(b) The media on which public information is recorded include:

(1) paper;

(2) film;

(3) a magnetic, optical, or solid state device that can store an electronic signal;

(4) tape;

(5) Mylar;

(6) linen;

(7) silk; and

(8) vellum.

(c) The general forms in which the media containing public information exist include a book, paper, letter, document, printout, photograph, film, tape, microfiche, microfilm, photostat, sound recording, map, and drawing and a voice, data, or video representation held in computer memory.

Tex. Gov't Code §552.0055. Subpoena Duces Tecum or Discovery Request

A subpoena duces tecum or a request for discovery that is issued in compliance with a statute or a rule of civil or criminal procedure is not considered to be a request for information under this chapter.

Tex. Gov't Code §552.022. Categories of Public Information; Examples

(a) Without limiting the amount or kind of information that is public information under this chapter, the following categories of information are public information and not excepted from required disclosure under this chapter unless they are expressly confidential under other law:

(1) a completed report, audit, evaluation, or investigation made of, for, or by a governmental body, except as provided by Section 552.108;

(2) the name, sex, ethnicity, salary, title, and dates of employment of each employee and officer of a governmental body;

(3) information in an account, voucher, or contract relating to the receipt or expenditure of public or other funds by a governmental body;

(4) the name of each official and the final record of voting on all proceedings in a governmental body;

(5) all working papers, research material, and information used to estimate the need for or expenditure of public funds or taxes by a governmental body, on completion of the estimate;

(6) the name, place of business, and the name of the municipality to which local sales and use taxes are credited, if any, for the named person, of a person reporting or paying sales and use taxes under Chapter 151, Tax Code;

(7) a description of an agency's central and field organizations, including:

(A) the established places at which the public may obtain information, submit information or requests, or obtain decisions;

(B) the employees from whom the public may obtain information, submit information or requests, or obtain decisions;

(C) in the case of a uniformed service, the members from whom the public may obtain information, submit information or requests, or obtain decisions; and

(D) the methods by which the public may obtain information, submit information or requests, or obtain decisions;

(8) a statement of the general course and method by which an agency's functions are channeled and determined, including the nature and requirements of all formal and informal policies and procedures;

(9) a rule of procedure, a description of forms available or the places at which forms may be obtained, and instructions relating to the scope and content of all papers, reports, or examinations;

(10) a substantive rule of general applicability adopted or issued by an agency as authorized by law, and a statement of general policy or interpretation of general applicability formulated and adopted by an agency;

(11) each amendment, revision, or repeal of information described by Subdivisions (7)-(10);

(12) final opinions, including concurring and dissenting opinions, and orders issued in the adjudication of cases;

(13) a policy statement or interpretation that has been adopted or issued by an agency;

(14) administrative staff manuals and instructions to staff that affect a member of the public;

(15) information regarded as open to the public under an agency's policies;

(16) information that is in a bill for attorneys' fees and that is not privileged under the attorney-client privilege;

(17) information that is also contained in a public court record; and

(18) a settlement agreement to which a governmental body is a party.

(b) A court in this state may not order a governmental body or an officer for public information to withhold from public inspection any category of public information described by Subsection (a) or to not produce the category of public information for inspection or duplication, unless the category of information is expressly made confidential under other law.

§13:92 Public Is Entitled to Information

The Texas Public Information Act's (TPIA) core provision provides that the public is entitled to information "collected, assembled, or maintained by a governmental body." City of San Antonio v. San Antonio Exp. News, 47 S.W.3d 556 (Tex. App.—San Antonio 2000, *review denied*).

However, there are statutory exceptions to this general principle that governmental information is public. These exceptions include:

- Information confidential by law (statute or judicial decision).
- Information within the informers' privilege.
- Certain personnel information (SSN's, home addresses, home phone numbers, and information relating to family of public employees).
- Information relating to litigation.
- Information relating to bidding.
- Information relating to location or price of property.
- Information within the attorney-client privilege.
- Certain law enforcement records which would interfere with the detection, investigation or prosecution of crime.
- Internal records of a law enforcement agency.
- Criminal history information.
- Trade secrets.
- Commercial or financial information privileged or confidential by law.
- Certain information relating to financial institutions or securities.
- Geological or geophysical information.
- Photographs of peace officers.
- Personal information relating to participants in neighborhood crime watch organizations.
- Crime victim compensation information.
- Family violence shelter center and sexual assault program information.
- Confidentiality of credit card, debit card, charge card and access device numbers.

- Government information related to security issues for computers.

Tex Gov't Code §§552.101-552.137.

§13:93 Statute Contains Specific Exceptions

The TPIA contains specific exceptions from compelled disclosure of information expressly made confidential by law, certain information relating to personnel files, and certain law-enforcement and prosecutors' records. City of San Antonio v. San Antonio Exp. News, 47 S.W.3d 556 (Tex.App.—San Antonio 2000, *review denied*).

The TPIA shall be liberally construed in favor of granting a request for information. City of San Antonio.

§13:94 Form: Request

See our companion book *Texas Criminal Forms* (James Publishing):

- Form 13-35, Public Information Act Letter Request

(This page intentionally left blank.)

CHAPTER 14

JURY SELECTION AND VOIR DIRE

Table of Contents

JURY SELECTION AND VOIR DIRE

§14:10 Composition of Venire

§14:11 Statutory Law

CCP Art. 33.09. Jury drawn

Jury panels, including special venires, for the trial of criminal cases shall be selected and summoned (with return on summons) in the same manner as the selection of panels for the trial of civil cases except as otherwise provided in this Code.

CCP Art. 35.01. Jurors called

When a case is called for trial and the parties have announced ready for trial, the names of those summoned as jurors in the case shall be called. Those not present may be fined not exceeding fifty dollars. An attachment may issue on request of either party for any absent summoned juror, to have him brought forthwith before the court. A person who is summoned but not present, may upon an appearance, before the jury is qualified, be tried as to his qualifications and impaneled as a juror unless challenged, but no cause shall be unreasonably delayed on account of his absence.

CCP Art. 35.11. Preparation of list

The trial judge, on the demand of the defendant or his attorney, or of the State's counsel, shall cause a sufficient number of jurors from which a jury may be selected to try the case to be randomly selected from the members of the general panel drawn or assigned as jurors in the case. The clerk shall randomly select the jurors by a computer or other process of random selection and shall write or print the names, in the order selected, on the jury list from which the jury is to be selected to try the case. The clerk shall deliver a copy of the list to the State's counsel and to the defendant or his attorney.

CCP Art. 35.20. Names called in order

In selecting the jury from the persons summoned, the names of such persons shall be called in the order in which they appear upon the list furnished the defendant. Each juror shall be tried and passed upon separately. A person who has been summoned, but who is not present, may, upon his appearance before the jury is completed, be tried as to his qualifications and impaneled as a juror, unless challenged, but no cause shall be unreasonably delayed on account of such absence.

CCP Art. 35.26. Lists returned to clerk

(a) When the parties have made or declined to make their peremptory challenges, they shall deliver their lists to the clerk. Except as provided in Subsection (b) of this section, the clerk shall, if the case be in the district court, call off the first twelve names on the lists that have not been stricken. If the case be in the county court, he shall call off the first six names on the lists that have not been stricken. Those whose names are called shall be the jury.

(b) In a capital case in which the state seeks the death penalty, the court may direct that two alternate jurors be selected and that the first fourteen names not stricken be called off by the clerk. The last two names to be called are the alternate jurors.

For the technical aspects of the formation of the venire *see* Government Code Chapter 62. Within Chapter 62 are several methods counties are allowed to employ for the formation of jury venires.

§14:12 In General

The Sixth Amendment to the U.S. Constitution requires that a jury venire be drawn from a source fairly representative of the community. Taylor v. Louisiana, 419 U.S. 522, 95 S.Ct. 692, 42 L. Ed. 2d 690 (1975).

Article I, §10 of the Texas Constitution provides the same protection. Lacy v. State, 899 S.W.2d 284 (Tex.App.—Tyler 1995, *no pet.*).

However, there is no rule that requires that the petit jury actually chosen mirror the community and reflect the various distinctive groups in the population. Taylor v. Louisiana, *supra*.

A criminal defendant must make three showings to establish a *prima facie* violation of the Sixth Amendment's fair cross-section requirement. He or she must show: (1) that the group alleged to be excluded is a distinctive group in the community; (2) that the representation of this group in venires from which juries are selected is not fair and reasonable in relation to the number of such persons in the community; and (3) that this underrepresentation is due to systematic exclusion of the group in the jury-selection process. Berghuis v. Smith, 559 U.S. ___, 130 S.Ct. 1382, 176 L.Ed.2d 249 (2010), *citing* Duren v. Missouri, 439 U.S. 357, 99 S.Ct. 664, 58 L.Ed.2d 579 (1979).

A defendant is not entitled to a jury that includes minorities or is comprised of a proportional representation of races so long as the jury selection process was not subject to discrimination. Rousseau v. State, 855 S.W.2d 666 (Tex. Crim. App. 1993), *cert. denied,* 510 U.S. 919, 114 S.Ct. 313, 126 L. Ed. 2d 260 (1993); May v. State, 738 S.W.2d 261 (Tex. Crim. App. 1987), *cert. denied*, 484 U.S. 872, 108 S.Ct. 206, 98 L. Ed. 2d 158 (1987).

A defendant has no right to any particular individual serving on the jury. The defendant's only substantial right is that the jurors who do serve be qualified. Colone v. State, 573 S.W.3d 249, 261 (Tex. Crim. App. 2019).

The Sixth Amendment entitles every defendant to object to a venire that is not designed to represent a fair cross section of the community, whether or not the systematically excluded groups are groups to which the defendant belongs. Holland v. Illinois, 493 U.S. 474, 110 S.Ct. 803, 107 L. Ed. 2d 905 (1990).

A challenge to the array is the proper method of testing the composition of the jury venire. Callaway v. State, 818 S.W.2d 816 (Tex.App.—Amarillo 1991, *pet. ref'd*). *See* §14:13.

The burden of proof is on the movant to show that the juror selection process intentionally, systematically and arbitrarily excludes protected classes from jury service. Garcia v. State, 919 S.W.2d 370 (Tex. Crim. App. 1994).

PRACTICE TIP: In cases where the composition of the venire may be an issue, review the Commissioners Court minutes of the county in question to ascertain which composition plan the county is required to follow and to verify that it is properly enforced.

PRACTICE TIP: If it is anticipated that the composition of the venire may be an issue in a particular case, a party can by written application be present and observe the drawing of the names of prospective jurors and their placement on the jury lists. Government Code §62.005.

Where the trial court errs in some aspect of selecting and empaneling a venire, a defendant can only obtain a reversal by showing some type of harm. Cooks v. State, 844 S.W.2d 697 (Tex. Crim. App. 1992), *cert. denied*, 509 U.S. 927, 113 S.Ct. 3048, 125 L. Ed. 2d 732 (1993).

The Rules of Evidence apply to jury voir dire; thus any evidence offered in support of or opposing challenges for cause must be admissible under the Rules. Watson v. State, 917 S.W.2d 65 (Tex.App.—Fort Worth 1996, *pet. ref'd*).

§14:13 Challenges to Array

In Texas, challenges to the array are conducted under the authority of CCP Arts. 35.06 and 35.07.

CCP Art. 35.06. Challenge to array first heard

> *The court shall hear and determine a challenge to the array before interrogating those summoned as to their qualifications.*

CCP Art. 35.07. Challenge to the array

> *Each party may challenge the array only on the ground that the officer summoning the jury has willfully summoned jurors with a view to securing a conviction or an acquittal. All such challenges must be in writing setting forth distinctly the grounds of*

such challenge. When made by the defendant, it must be supported by his affidavit or the affidavit of any credible person. When such challenge is made, the judge shall hear evidence and decide without delay whether or not the challenge shall be sustained.

CCP Art. 35.08. When challenge is sustained

The array of jurors summoned shall be discharged if the challenge be sustained, and the court shall order other jurors to be summoned in their stead, and direct that the officer who summoned those so discharged, and on account of whose misconduct the challenge has been sustained shall not summon any other jurors in the case.

CCP Art. 35.09. List of new venire

When a challenge to the array has been sustained, the defendant shall be entitled, as in the first instance, to service of a copy of the list of names of those summoned by order of the court.

A challenge to the array is sometimes referred to as a motion to quash the venire. Callaway v. State, 818 S.W.2d 816 (Tex.App.—Amarillo 1991, *pet. ref'd*).

Art. 35.07 provides that a party may challenge an array only on the ground that the officer summoning the jury has willfully summoned jurors with a view to securing a conviction or acquittal. Cooks v. State, 844 S.W.2d 697 (Tex. Crim. App. 1992), *cert. denied*, 509 U.S. 927, 113 S.Ct. 3048, 125 L. Ed. 2d 732 (1993).

A trial court properly overrules the challenge to the array where the challenge alleges any ground other than that specifically authorized by Art. 35.07. Broussard v. State, 910 S.W.2d 952 (Tex. Crim. App. 1995).

A challenge to the array must be: (1) in writing, and (2) supported by affidavit. Garcia v. State, 919 S.W.2d 370 (Tex. Crim. App. 1994).

An instrument must be properly sworn to in order to properly be supported by affidavit. Esquivel v. State, 595 S.W.2d 516 (Tex. Crim. App. 1980), *cert. denied,* 449 U.S. 986, 101 S.Ct. 408, 66 L. Ed. 2d 251 (1980).

CCP Art. 35.06 unambiguously requires a challenge to the array to be made before the panel is qualified. Jackson v. State, 745 S.W.2d 4 (Tex. Crim. App. 1988), *cert. denied*, 487 U.S. 1241, 108 S.Ct. 2916, 101 L. Ed. 2d 947 (1988); Callaway, *supra*.

In order to prevail on a challenge to the array, a party must establish harm, *i.e.*, he must show that the method employed compromised the fairness of the trial. Lewis v. State, 815 S.W.2d 560 (Tex. Crim. App. 1991).

The burden of proof is on the party challenging the array to show that the array should be quashed. Garcia, *supra*.

In order to prevail, the defendant must produce evidence showing that the method employed systematically and arbitrarily excluded a cognizable protected class from selection. Garcia. This can generally be done through the use of statistical comparison between the population of the county at large and those summoned for jury service. Garcia.

A showing that the defendant was forced to accept an objectionable juror or was otherwise denied an impartial jury is required before reversal is merited. Callaway.

See §§11:100 through 11:103 of this book for further information on challenging the make-up of the venire.

§14:13.1 Form: Sample Motion/Challenge to Array

See our companion book *Texas Criminal Forms* (James Publishing) Form 14-1.

§14:14 Writs of Attachment

See CCP Art. 35.01 in §14:11.

In order to preserve error in the case of veniremembers who were summoned for jury duty but did not appear, the defendant must request writs of attachment for those veniremembers. Coleman v. State, 881 S.W.2d 344 (Tex. Crim. App. 1994), *cert.*

denied, 513 U.S. 1096, 115 S.Ct. 763, 130 L. Ed. 2d 660 (1995).

The statute for attaching veniremembers, CCP Art. 35.01 (*see* §14:11), is directory, not mandatory. Dowthitt v. State, 931 S.W.2d 244 (Tex. Crim. App. 1996). In the absence of governmental misconduct in summoning the venire, the failure to grant attachments is not reversible error unless the defendant shows injury. Dowthitt.

The defendant must show that because of the trial court's failure to issue the writ of attachment he was forced to accept an objectionable juror. Starks v. State, 776 S.W.2d 808 (Tex.App.—Fort Worth 1989, *pet. ref'd*).

In determining whether a defendant has been harmed by the trial court's failure to issue attachments for absent jurors, the term "objectionable juror" means "one against whom such cause for challenge exists as would likely affect his competency or his impartiality in the trial." Starks.

TIP TO THE BENCH: Persons who have been summoned for jury duty and who fail to appear are subject to a contempt action punishable by a fine of not less than $100 or more than $1,000. Government Code §62.041.

§14:14.1 Form: Motion for Attachment of Absent Jurors

See our companion book *Texas Criminal Forms* (James Publishing) Form 14-2 and 14-3.

§14:15 Juror Exemptions

CCP Art. 35.04. Claiming exemption

> *Any person summoned as a juror who is exempt by law from jury service may establish his exemption without appearing in person by filing a signed statement of the ground of his exemption with the clerk of the court at any time before the date upon which he is summoned to appear.*

Government Code §62.106. Exemption From Jury Service

> *(a) A person qualified to serve as a petit juror may establish an exemption from jury service if the person:*
>
> *(1) is over 70 years of age;*
>
> *(2) has legal custody of a child younger than 12 years of age and the person's service on the jury requires leaving the child without adequate supervision;*
>
> *(3) is a student of a public or private secondary school;*
>
> *(4) is a person enrolled and in actual attendance at an institution of higher education;*
>
> *(5) is an officer or an employee of the senate, the house of representatives, or any department, commission, board, office, or other agency in the legislative branch of state government;*
>
> *(6) is summoned for service in a county with a population of at least 200,000, unless that county uses a jury plan under Section 62.011 and the period authorized under Section 62.011(b)(5) exceeds two years, and the person has served as a petit juror in the county during the 24-month period preceding the date the person is to appear for jury service;*
>
> *(7) is the primary caretaker of a person who is an invalid unable to care for himself;*
>
> *(8) except as provided by Subsection (b), is summoned for service in a county with a population of at least 250,000 and the person has served as a petit juror in the county during the three-year period preceding the date the person is to appear for jury service; or*

(9) is a member of the United States military forces serving on active duty and deployed to a location away from the person's home station and out of the person's county of residence.

(b) Subsection (a)(8) does not apply if the jury wheel in the county has been reconstituted after the date the person served as a petit juror.

Government Code §62.106(1) which allows juror exemptions for persons over 70 years of age does not violate a defendant's rights to equal protection of the laws and to a fair cross section of the community under either the U.S. or Texas constitutions. Burks v. State, 876 S.W.2d 877 (Tex. Crim. App. 1997), *cert. denied,* 513 U.S. 1114, 115 S.Ct. 909, 130 L. Ed. 2d 791 (1994).

PRACTICE TIP: *In order to preserve error in the case of veniremembers who were summoned for jury duty but did not appear, the defendant must request writs of attachment for those veniremembers. Coleman v. State, 881 S.W.2d 344 (Tex. Crim. App. 1994), cert. denied, 513 U.S. 1096, 115 S.Ct. 763, 130 L. Ed. 2d 660 (1995).*

During jury selection, the trial court has the power to remove a prospective juror on the basis of an excuse, even if it is after the prospective juror has been questioned and accepted. The trial court retains this power in order to rectify problems created by such circumstances as, e.g., a venireperson's sudden realization that an excuse applies to her. Colone v. State, 573 S.W.3d 249, 261 (Tex. Crim. App. 2019).

§14:16 Additional Veniremembers

CCP Art. 34.02. Additional names drawn

In any criminal case in which the court deems that the veniremembers theretofore drawn will be insufficient for the trial of the case, or in any criminal case in which the venire has been exhausted by challenge or otherwise, the court shall order additional veniremembers in such numbers as the court may deem advisable, to be summoned as follows:

(a) In a jury wheel county, the names of those to be summoned shall be drawn from the jury wheel.

(b) In counties not using the jury wheel, the veniremembers shall be summoned by the sheriff.

Although there is not a specific statutory remedy where the court has an insufficient number of veniremembers to complete the selection of a jury, the court may call additional veniremembers from the central jury panel and proceed with the jury selection process. Broussard v. State, 910 S.W.2d 952 (Tex. Crim. App. 1995); Williams v. State, 631 S.W.2d 955 (Tex. Crim. App. 1982).

Where a juror becomes disabled before the swearing of the jury and there is an alternate juror available, the court also has the option of replacing the disabled juror with the alternate. Broussard.

§14:17 Number of Jurors

CCP Art. 33.01. Jury Size

(a) Except as provided by Subsection (b), in the district court, the jury shall consist of twelve qualified jurors. In the county court and inferior courts, the jury shall consist of six qualified jurors.

(b) In a trial involving a misdemeanor offense, a district court jury shall consist of six qualified jurors.

In a district court, a defendant can validly waive his right to be tried by twelve jurors and consent to a trial by a lesser number. Hatch v. State, 958 S.W.2d 813 (Tex. Crim. App. 1997) (where it was discovered that one of the jurors was not a U.S. citizen); Government Code §62.201.

In a county court, a defendant who has executed a valid written jury waiver may be tried by a jury of less than six jurors. Hanley v. State, 909 S.W.2d 117 (Tex.App.—Houston [14th Dist.] 1995, *no pet.*).

Texas statutory law provides two instances in which a trial can proceed with 11 jurors: (1) when the parties consent or (2) if a juror dies or the trial court finds that a juror has become disabled. Tex. Gov't Code Ann. § 62.201 (West 2013); Tex. Code Crim. Proc. Ann. art. 36.29(a) (West 2006 & Supp.2014); Stillwell v. State, 466 S.W.3d 908, 912 (Tex. App.—Fort Worth 2015, no pet.).

In the second instance, when a trial court finds that a juror has become disabled, consent is not necessary to proceed with the 11 jurors remaining on the panel. Stillwell v. State, 466 S.W.3d at 912.

§14:18 Alternate Jurors

CCP Art. 33.011. Alternate jurors

> *(a) In district courts, the judge may direct that not more than four jurors in addition to the regular jury be called and impaneled to sit as alternate jurors. In county courts, the judge may direct that not more than two jurors in addition to the regular jury be called and impaneled to sit as alternate jurors.*
>
> *(b) Alternate jurors in the order in which they are called shall replace jurors who, prior to the time the jury renders a verdict on the guilt or innocence of the defendant and, if applicable, the amount of punishment, become or are found to be unable or disqualified to perform their duties or are found by the court on agreement of the parties to have good cause for not performing their duties. Alternate jurors shall be drawn and selected in the same manner, shall have the same qualifications, shall be subject to the same examination and challenges, shall take the same oath, and shall have the same functions, powers, facilities, security, and privileges as regular jurors. An alternate juror who does not replace a regular juror shall be discharged after the jury has rendered a verdict on the guilt or innocence of the defendant and, if applicable, the amount of punishment.*

CCP Art. 35.26 Lists returned to clerk

> *(b) In a capital case in which the state seeks the death penalty, the court may direct that two alternate jurors be selected and that the first fourteen names not stricken be called off by the clerk. The last two names to be called are the alternate jurors.*

According to CCP Art. 35.15, the state and the defendant shall each be entitled to one peremptory challenge in addition to those otherwise allowed by law if one or two alternate jurors are to be impaneled. Each side is entitled to two peremptory challenges if three or four alternate jurors are to be impaneled. The additional peremptory challenges are to be used against an alternate juror only, and the other peremptory challenges allowed by law may not be used against an alternate juror.

In cases where a juror becomes ill before being sworn, the trial court may replace him with the next qualified member of the venire. Juarez v. State, 890 S.W.2d 184 (Tex.App.—Corpus Christi 1994, *no pet.*).

The same procedural rules that apply to the selection of regular jurors apply to the selection of alternate jurors. Cooks v. State, 844 S.W.2d 697 (Tex. Crim. App. 1992), *cert. denied*, 509 U.S. 927, 113 S.Ct. 3048, 125 L. Ed. 2d 732 (1993).

The trial court does not violate Texas Constitution, Art. V, Sec. 13, by allowing an alternate juror to remain in the jury room during deliberations, as long as the alternate juror does not vote with the 12 members of the jury. Trinidad v. State, 312 S.W.3d 23 (Tex. Crim. App. 2010).

§14:19 Disabled Jurors

CCP Art. 36.29. If a Juror Dies or Becomes Disabled

> *(a) Not less than twelve jurors can render and return a verdict in a felony case. It must be concurred in by each juror and signed by the foreman. Except as provided in Subsection (b), however, after the trial of any felony case begins and a juror dies or, as determined by the judge, becomes disabled from sitting at any time before the charge of the court is read to the jury, the remainder of the jury shall have the power to render the*

verdict; but when the verdict shall be rendered by less than the whole number, it shall be signed by every member of the jury concurring in it.

(b) If alternate jurors have been selected in a capital case in which the state seeks the death penalty and a juror dies or becomes disabled from sitting at any time before the charge of the court is read to the jury, the alternate juror whose name was called first under Article 35.26 of this code shall replace the dead or disabled juror. Likewise, if another juror dies or becomes disabled from sitting before the charge of the court is read to the jury, the other alternate juror shall replace the second juror to die or become disabled.

(c) After the charge of the court is read to the jury, if a juror becomes so sick as to prevent the continuance of the juror's duty and an alternate juror is not available, or if any accident of circumstance occurs to prevent the jury from being kept together under circumstances under which the law or the instructions of the court requires that the jury be kept together, the jury shall be discharged, except that on agreement on the record by the defendant, the defendant's counsel, and the attorney representing the state 11 members of a jury may render a verdict and, if punishment is to be assessed by the jury, assess punishment. If a verdict is rendered by less than the whole number of the jury, each member of the jury shall sign the verdict.

(d) After the jury has rendered a verdict on the guilt or innocence of the defendant and, if applicable, the amount of punishment, the court shall discharge an alternate juror who has not replaced a juror.

Where a juror becomes disabled, CCP Art. 36.29 thus allows a jury of less than 12 to render a verdict in a non-capital case but requires that an alternate replace a disabled juror in a death penalty case.

The language of CCP Art. 36.29 and the cases applying it make clear that the legislative intent was to limit its application to those cases where the juror was physically or mentally impaired in some way that hindered his ability to perform his duty as a juror. Brooks v. State, 990 S.W.2d 278 (Tex. Crim. App. 1999).

The determination as to whether a juror is disabled is within the discretion of the trial court. Absent an abuse of that discretion, no reversible error will be found. Brooks.

An CCP Art. 36.29 disability is not limited to physical disease; it can include a physical illness, mental condition or emotional state. Reyes v. State, 30 S.W.3d 409 (Tex. Crim. App. 2000).

The mere fact that the juror knows the defendant is not a disability except where that knowledge inhibits the juror from fully and fairly performing the functions of a juror. Reyes, (where the juror felt a fear of retaliation from the defendant).

A juror who simply disagrees with other jurors and who refuses to further deliberate for that reason is not disabled under sec. 36.29. Scales v. State, 380 S.W.3d 780, 786 (Tex. Crim. App. 2012).

Although there is not a specific statutory remedy where the court has an insufficient number of veniremembers to complete the selection and swearing of a jury, the court may call additional veniremembers from the central jury panel and proceed with the jury selection process. Broussard v. State, 910 S.W.2d 952 (Tex. Crim. App. 1995); Williams v. State, 631 S.W.2d 955 (Tex. Crim. App.1982).

Where a juror becomes disabled before the swearing of the jury and there is an alternate juror available, the court also has the option of replacing the disabled juror with the alternate. Broussard.

Where a juror has become disabled, the trial judge is required to consider and rule out less drastic alternatives, such as proceeding with a lesser number of jurors, before granting a mistrial. The judge must review the alternatives and choose the one which best preserves the defendant's right to have his trial completed before a particular tribunal. Hill v. State, 90 S.W.3d 308 (Tex. Crim. App. 2002).

Where the juror is not disabled, but merely disqualified, the defendant must waive his right to be tried by a jury of twelve before the trial may proceed to a verdict. Hatch v. State, 958 S.W.2d 813 (Tex. Crim. App. 1997); Government Code §62.201.

In a county court, a defendant who has executed a valid written jury waiver may be tried by a jury of less than six jurors. Hanley v. State, 909 S.W.2d 117 (Tex.App.—Houston [14th Dist.] 1995, *no pet.*).

There is no constitutional or statutory impediment to proceeding with less than six jurors in county court, so long as the defendant waives his right to trial by a complete jury under Article 1.14, and the State and the trial court are willing to consent to do so under Article 1.13(a). Ex parte Garza, 337 S.W.3d 903 (Tex. Crim. App. 2011).

TIP TO THE BENCH: When excusing a juror who has become disabled, the court must ensure that the record reflects the precise reason that the juror is being excused because the appellate courts cannot presume from a silent record that the juror was properly excused. See Scales v. State, 380 S.W.3d 780, 784 (Tex.Crim. App. 2012). Therefore, in order to support its judgment, the trial court must make a finding, sufficiently supported by the record, that the juror was disabled or unable to perform the duties of a juror.

While in a hearing to decide whether to excuse a juror the trial court is the sole fact-finder and judge of the credibility of the testifying jurors, the trial court's decision is subject to a review for abuse of discretion. Absent such an abuse, no reversible error will be found. Scales.

The improper dismissal of a juror under sec. 36.29 can constitute a violation of a defendant's right to a unanimous jury and subject the error to analysis under Texas Rule of Appellate Procedure 44.2(a). Scales v. State, 380 S.W.3d at 786.

§14:19.1 Sufficient Disability to Be Excused

- Intoxication has been found to disable a juror. Griffin v. State, 486 S.W.2d 948 (Tex. Crim. App. 1972).
- Flu has been found to constitute a sufficient disabling factor. Allen v. State, 536 S.W.2d 364 (Tex. Crim. App. 1976).
- Injuries from auto accidents can sufficiently disable a juror. Johnson v. State, 525 S.W.2d 170 (Tex. Crim. App. 1975).
- High blood pressure requiring hospitalization has been found to sufficiently disable a juror. Maciel v. State, 517 S.W.2d 789 (Tex. Crim. App. 1975).
- Death in the juror's family can constitute a sufficient disability. Wells v. State, 762 S.W.2d 673 (Tex.App.—Texarkana 1988, *pet. ref'd*).
- Finding that the juror was so concerned about his ability to support his family while on jury duty that he was unable to concentrate on the trial has been found to be a sufficient disability. Freeman v. State, 838 S.W.2d 772 (Tex.App.—Corpus Christi 1992, *pet. ref'd*).
- Juror who stated that he would be unable to concentrate on the trial due to time pressures associated with moving and starting a new job was found to be properly discharged on grounds that he was disabled. Ramos v. State, 934 S.W.2d 358 (Tex. Crim. App. 1996), *cert. denied*, 520 U.S. 1198, 117 S.Ct. 1556, 137 L. Ed. 2d 704 (1997).
- Inability to find childcare can constitute a sufficient basis for excusal. Edwards v. State, 981 S.W.2d 359 (Tex.App.—Texarkana 1998, *no pet.*).
- Juror's Alzheimer's disease which resulted in her inability to remember any details from her first day of jury service and her resulting emotional

distress have been found to be a sufficient basis for excusal. Timmons v. State, 952 S.W.2d 891 (Tex.App.—Dallas 1997, *no pet.*).

§14:19.2 Insufficient Disability to Be Excused

- Juror's arrest for carrying a handgun into the courthouse does not constitute a sufficient disablement under this article. Brooks v. State, 990 S.W.2d 278 (Tex. Crim. App. 1999).

- Alleged bias and prejudice are not a disability within meaning of this article. Bass v. State, 622 S.W.2d 101 (Tex. Crim. App. 1981), *cert. denied*, 456 U.S. 965, 102 S.Ct. 2046, 72 L. Ed. 2d 491 (1982).

- Juror is not disabled within the meaning of this article where it is discovered that he cannot consider the upper end of the range of punishment. Landrum v. State, 788 S.W.2d 577 (Tex. Crim. App. 1990).

- Juror is not disabled within the meaning of this article where it is discovered that he is not a U.S. citizen. Hatch v. State, 958 S.W.2d 813 (Tex. Crim. App. 1997).

Where the record only reveals that a juror has refused to further deliberate, there is insufficient evidence of disability to excuse the juror. Scales v. State, 380 S.W.3d 780, 786 (Tex. Crim. App. 2012).

§14:20 Juror Qualifications

Gov't Code §62.101. Jury Service

All individuals are competent petit jurors unless disqualified under this subchapter and are liable for jury service except as otherwise provided by this subchapter.

Gov't Code §62.102. General Qualifications For Jury Service

A person is disqualified to serve as a petit juror unless the person:

(1) is at least 18 years of age;

(2) is a citizen of this state and of the county in which the person is to serve as a juror;

(3) is qualified under the constitution and laws to vote in the county in which the person is to serve as a juror;

(4) is of sound mind and good moral character;

(5) is able to read and write;

(6) has not served as a petit juror for six days during the preceding three months in the county court or during the preceding six months in the district court;

(7) has not been convicted of misdemeanor theft or a felony; and

(8) is not under indictment or other legal accusation of misdemeanor theft or a felony.

Prospective jurors who cannot read and write are properly excused *sua sponte* by the trial court. Liveoak v. State, 717 S.W.2d 691 (Tex.App.—San Antonio 1986, *pet. ref'd*).

CCP Art. 33.02. Failure To Register

Failure to register to vote shall not disqualify any person from jury service.

Texas Election Code §11.001. Eligibility to Vote

(a) Except as otherwise provided by law, to be eligible to vote in an election in this state, a person must:

(1) be a qualified voter as defined by Section 11.002 on the day the person offers to vote;

(2) be a resident of the territory covered by the election for the office or measure on which the person desires to vote; and

(3) satisfy all other requirements for voting prescribed by law for the particular election.

(b) For a person who resides on property located in more than one territory described by Subsection (a)(2), the person shall choose in which territory the residence of the person is located.

Sec. 11.002. Qualified Voter

In this code, "qualified voter" means a person who:

(1) is 18 years of age or older;

(2) is a United States citizen;

(3) has not been determined by a final judgment of a court exercising probate jurisdiction to be:

(A) totally mentally incapacitated; or

(B) partially mentally incapacitated without the right to vote;

(4) has not been finally convicted of a felony or, if so convicted, has:

(A) fully discharged the person's sentence, including any term of incarceration, parole, or supervision, or completed a period of probation ordered by any court; or

(B) been pardoned or otherwise released from the resulting disability to vote;

(5) is a resident of this state; and

(6) is a registered voter.

§14:21 Preliminary Testing of Qualifications

At the commencement of the voir dire procedure, and after any challenge to the array has been asserted, it is the responsibility of the trial judge to place the venire under oath according to CCP Art. 35.02, which sets out the statutory oath. CCP Art. 35.10.

After the venire is placed under oath, the trial judge must ask three questions of them to determine if they are qualified under the law. Those three questions are contained in CCP Art. 35.12.

CCP Art. 35.12. Mode of Testing

(a) In testing the qualification of a prospective juror after the juror has been sworn, the juror shall be asked by the court, or under its direction:

(1) Except for failure to register, are you a qualified voter in this county and state under the Constitution and laws of this state?

(2) Have you ever been convicted of theft or any felony?

(3) Are you under indictment or legal accusation for theft or any felony?

(b) In testing the qualifications of a prospective juror, with respect to whether the juror has been the subject of an order of nondisclosure or has a criminal history that includes information subject to that order, the juror may state only that the matter in question has been sealed.

TIP TO THE BENCH: It is advisable to phrase the last two questions to include theft accusations stemming from insufficient funds as a possible disqualifier because many veniremembers do not necessarily equate such an accusation with a theft offense. Therefore, many who legally fit the disqualification may not respond accurately to the question.

Depending on the answers to these questions veniremembers may be qualified, unqualified or absolutely disqualified. Green v. State, 764 S.W.2d 246 (Tex. Crim. App. 1989).

§14:30 Juror Disqualifications

Gov't Code §62.104. Disqualification for Legal Blindness

(a) A person who is legally blind is not disqualified to serve as a juror in a civil case solely because of his legal blindness except as provided by this section.

(b) A legally blind person is disqualified to serve as a juror in a civil case if, in the opinion of the court, his blindness renders him unfit to serve as a juror in that particular case.

(c) In this section, "legally blind" means having:

(1) no more than 20/200 of visual acuity in the better eye with correcting lenses; or

(2) visual acuity greater than 20/200, but with a limitation in the field of vision such that the widest diameter of the visual field subtends an angle no greater than 20 degrees.

Gov't Code §62.1041. Deaf or Hard of Hearing Juror

(a) A deaf or hard of hearing person is not disqualified to serve as a juror solely because of hearing loss except as provided by this section.

(b) A deaf or hard of hearing person is disqualified to serve as a juror if, in the opinion of the court, his hearing loss renders him unfit to serve as a juror in that particular case.

(c) A deaf or hard of hearing person serving as a juror shall be reasonably accommodated in accordance with the Americans with Disabilities Act (42 U.S.C. §12101 et seq.). An interpreter who is assisting a deaf or hard of hearing person serving as a juror may accompany the juror during all proceedings and deliberations in the case.

(d) If an interpreter is provided to a deaf or hard of hearing person serving as a juror in a district, county, or justice court, the county shall pay the cost of obtaining those services.

(e) A deaf or hard of hearing juror may request an auxiliary aid or service for a municipal court proceeding. The city shall honor the request unless the city can demonstrate that another effective means of communication exists. The city shall pay the cost unless the auxiliary aid or service will result in a fundamental alteration of the municipal court proceeding or in undue financial or administrative burdens.

(f) In this section, "deaf or hard of hearing" means having a hearing impairment, regardless of the existence of a speech impairment, that inhibits comprehension of an examination or proceeding or communication with others.

Gov't Code §62.105. Disqualification for Particular Jury

A person is disqualified to serve as a petit juror in a particular case if he:

(1) is a witness in the case;

(2) is interested, directly or indirectly, in the subject matter of the case;

(3) is related by consanguinity or affinity within the third degree, as determined under Chapter 573, to a party in the case;

(4) has a bias or prejudice in favor of or against a party in the case; or

(5) has served as a petit juror in a former trial of the same case or in another case involving the same questions of fact.

Where a venireman is *disqualified* from jury service, rather than absolutely disqualified, a party is required to challenge the venireman for cause to remove him from the jury. Mayo v. State, 4 S.W.3d 9 (Tex. Crim. App. 1999). This includes all reasons for challenge under Art. 35.16 that are not absolute disqualifications.

§14:31 What Constitutes Absolute Disqualification

CCP Art. 35.19. Absolute disqualification

No juror shall be impaneled when it appears that he is subject to the second, third or fourth cause of challenge in Article 35.16, though both parties may consent.

A veniremember is absolutely disqualified if the veniremember:

- Has been convicted of theft or any felony
- Is under indictment or legal accusation for theft or any felony
- Is insane

CCP Art. 35.19 and 35.16(a)(2)(3)(4); Alvarado v. State, 822 S.W.2d 236 (Tex.App.—Houston [14th Dist.] 1991, *pet. ref'd*).

On the other hand, a disqualified veniremember is not absolutely disqualified under CCP Art. 35.19, but is subject to a challenge for cause under CCP Art. 35.16.

A trial judge has no authority on his own motion to excuse a prospective juror for cause under Art. 35.16. Warren v. State, 768 S.W.2d 300 (Tex. Crim. App. 1989); Green v. State, 764 S.W.2d 242 (Tex. Crim. App. 1989). However, the trial court may on its own motion excuse a prospective juror who is absolutely disqualified pursuant to CCP Art. 35.19. Warren.

Whether a venireman is absolutely disqualified is a question of fact for the trial court to determine. Hammond v. State, 799 S.W.2d 741 (Tex. Crim. App. 1990).

The language of CCP Art. 35.19 does not require indisputable certainty on the part of the trial court in making its determination on absolute disqualifications. Rather, Art. 35.19 provides that if it "appears" that the veniremember is absolutely disqualified, then he "shall not be empaneled[.]" Chambers v. State, 903 S.W.2d 21 (Tex. Crim. App. 1995). Thus, where it "appeared" that a veniremember was convicted of a federal felony which turned out to actually be a misdemeanor, the trial court's ruling that the he was absolutely disqualified was not overturned on appeal. Chambers.

Appellate courts apply an abuse of discretion standard when reviewing a trial court's ruling on an absolute disqualification. Chambers; Hammond v. State, 799 S.W.2d 741 (Tex. Crim. App. 1990), *cert. denied*, 501 U.S. 1259, 111 S.Ct. 2912, 115 L. Ed. 2d 1076 (1991).

The parties cannot waive Art. 35.19 disqualifications. Ex parte Smith, 817 S.W.2d 797 (Tex.App.—Amarillo 1991, *no pet.*).

However, failure to object to the impaneling of an absolutely disqualified venireman waives any error on appeal under art. 44.46 (see §14:33). Nelson v. State, 129 S.W.3d 108 (Tex. Crim. App. 2004).

There is some tension between the language of art. 35.19 stating that no absolutely disqualified venireman should be impaneled, and art. 44.46, which says that it is not error to impanel an absolutely disqualified venireman where the defendant does not object. It should be noted that under art. 44.46 the defendant can wait to make the objection at any time before the verdict is entered. Nelson.

TIP TO THE BENCH: If it is shown during voir dire that a venireman is absolutely disqualified, the court should not impanel that venireman unless the defendant affirmatively states that he has no objection. See Nelson.

The juror disqualifications under CCP Art. 35.16 which are not absolute disqualifications under Art. 35.19 are waivable under Art. 35.16. Mayo v. State, 4 S.W.3d 9 (Tex. Crim. App. 1999).

In the absence of other evidence, where the veniremember denies having a disqualifying conviction the trial court abuses its discretion in excusing him. Loughry v. State, 926 S.W.2d 382 (Tex.App.—Fort Worth 1996, *pet. ref'd*).

§14:31.1 Theft Convictions

A conviction for misdemeanor theft is an absolute disqualification. Holland v. State, 761 S.W.2d 307 (Tex. Crim. App. 1988), *cert. denied,* 489 U.S. 1091, 109 S.Ct. 1560, 103 L. Ed. 2d 863 (1989). A conviction for a Class C theft is sufficient for absolute disqualification. Pfeffer v. State, 683 S.W.2d 64 (Tex.App.—Amarillo 1984, *pet. ref'd*).

Even a theft conviction which occurred about twenty years before the jury service in question is sufficient for an absolute disqualification. Frame v. State, 615 S.W.2d 766 (Tex. Crim. App. 1981).

The statutory phrase "other legal accusation" corresponds with the legal definition of "accusation," specifically a formal criminal charge against an individual. Therefore, a civil pleading alleging conversion does not constitute a cause of challenge under Art. 35.19. Staley v. State, 887 S.W.2d 885 (Tex. Crim. App. 1994), *cert. denied,* 514 U.S. 1020, 115 S.Ct. 1366, 131 L. Ed. 2d 222 (1995).

Even where the veniremember cannot remember all of the salient facts of the alleged conviction, the trial court does not err in excusing him because where the record is silent, procedural rules are deemed to have been complied with. Enos v. State, 859 S.W.2d 594 (Tex.App.—Fort Worth 1993), *reversed on other grounds at* 889 S.W.2d 303 (Tex. Crim. App. 1994).

A juror who has previously been given a probated sentence is eligible for juror service after the probation has expired where he has been released from the disabilities of the probation. Payton v. State, 572 S.W.2d 677 (Tex. Crim. App. 1978).

In order to be qualified, a veniremember who has received a probated sentence for any disqualifying offense must have had the case physically dismissed and have been released from all legal disabilities by the court. Hoffman v. State, 922 S.W.2d 663 (Tex. App.—Waco 1996, *pet. ref'd*); Wolfe v. State, 917 S.W.2d 270 (Tex. Crim. App. 1996); CCP Art. 42.12; Smith v. State, 859 S.W.2d 463 (Tex.App.—Fort Worth 1993, *pet. ref'd*).

However, *see* R.R.E. v. Glenn, 884 S.W.2d 189 (Tex.App.—Fort Worth 1994) (a civil action) which holds that the full rights of convicted felons cannot be restored except by executive pardon because of the language of Art. XVI, §2 of the Texas Constitution.

A commutation does not remove the disqualification from serving on a jury. DeBlanc v. State, 799 S.W.2d 701 (Tex. Crim. App. 1990).

§14:31.2 Pending Theft Cases

A deferred adjudication probation is a pending case against a defendant. Brown v. State, 716 S.W.2d 939 (Tex. Crim. App. 1986).

Therefore, a deferred adjudication probation for theft is an absolute disqualification. Thomas v. State, 796 S.W.2d 196 (Tex. Crim. App. 1990). A pending theft charge requires disqualification whether or not the information charging the offense is sufficient under law. Thomas.

§14:31.3 Felony Convictions

A trial court does not abuse its discretion in disqualifying a juror where the record is unclear whether his federal tax conviction is a misdemeanor or a felony. Chambers v. State, 903 S.W.2d 21 (Tex. Crim. App. 1995).

§14:31.4 Pending Felony Cases

A deferred adjudication is deemed a pending case against a defendant, and therefore a veniremember on a deferred adjudication for felony cocaine possession results in the absolute disqualification of that veniremember. State v. Holloway, 886 S.W.2d 482 (Tex.App.—Houston [1st Dist.] 1994, *pet. ref'd*).

§14:31.5 Insanity

A veniremember will be properly excused for cause where he is disqualified because of insanity. Bass v. State, 622 S.W.2d 101 (Tex. Crim. App. 1981), *cert. denied,* 456 U.S. 965, 102 S.Ct. 2046,

72 L. Ed. 2d 491 (1982); Von Byrd v. State, 569 S.W.2d 883 (Tex. Crim. App. 1978).

§14:32 Disqualified Jurors

Jurors are disqualified if they fall into any of the following categories: 1) they are disqualified under Government Code §62.104 (*see* §14:30); 2) they are disqualified under Government Code §62.105 (*see* §14:30); or 3) they are disqualified under CCP Art. 35.16.

A deaf or hard of hearing person is disqualified under Government Code §62.1041 if in the opinion of the court the hearing loss renders him unfit to serve as a juror in that particular trial. Black v. State, 26 S.W.3d 895 (Tex. Crim. App. 2000). The disqualified venireman can be excused outside the presence of the state and defendant. Black.

Government Code, §573.023 specifies the manner of computing the degree of consanguinity between two persons and explicitly names the relationships that fall within the third degree for purposes of deciding if a juror is disqualified under Government Code §62.105(3). Ex parte Fierro, 79 S.W.3d 54 (Tex. Crim. App. 2002).

§14:33 Reversals Due to Disqualified Jurors

CCP Art. 44.46. Reversal of Conviction on the Basis of Service on Jury by a Disqualified Juror

> *A conviction in a criminal case may be reversed on appeal on the ground that a juror in the case was absolutely disqualified from service under Article 35.19 of this code only if:*
>
> > *(1) the defendant raises the disqualification before the verdict is entered; or*
> >
> > *(2) the disqualification was not discovered or brought to the attention of the trial court until after the verdict was entered and the defendant makes a showing of significant harm by the service of the disqualified juror.*

Where a statutorily disqualified juror has served on a jury in contravention of CCP Art. 35.16, Art. 44.46 is not in conflict with Texas Constitution Art. XVI, §2 which provides that, "Laws shall be made to exclude from office, serving on juries, and from the right of suffrage, those who may have been or shall hereafter be convicted of bribery, perjury, forgery, or other high crimes. This is because the phrase "high crimes" was found to be limited to criminal conduct that demonstrates the same type of moral corruption and dishonesty inherent in the offenses specified in the first part of the clause. Perez v. State, 11 S.W.3d 218 (Tex. Crim. App. 2000).

Art. 44.46(1) imposes an affirmative duty on the defendant to raise the issue of an absolutely disqualified juror before the verdict is entered. Where this is done, the defendant is entitled to a reversal of his conviction without a showing of harm. Nelson v. State, 129 S.W.3d 108 (Tex. Crim. App. 2004).

Where the disqualification was not discovered, and thus not brought to the attention of the trial court, until after the verdict was entered, a defendant may be granted a new trial only if the juror's service caused him "significant harm" under Art. 44.46(2). Ristoff v. State, 985 S.W.2d 623 (Tex.App.—Houston [1st Dist.] 1999, *no pet.*).

The defendant cannot call the juror in question during a motion for new trial hearing to establish harm. The testimony of a juror under Rule 606(b) is limited to harm resulting from an outside influence. White v. State, 225 S.W.3d 571 (Tex. Crim. App. 2007)

The defendant is entitled to a new trial only if he demonstrates that the disqualified juror's service caused him significant harm beyond the mere fact of conviction. Hernandez v. State, 952 S.W.2d 59 (Tex.App.—Austin 1997), *vacated on other grounds*, 957 S.W.2d 851 (Tex. Crim. App. 1998).

Where at the motion for new trial hearing the defendant made no attempt to show significant harm, he has failed to meet his burden. State v. Read, 965 S.W.2d 74 (Tex.App.—Austin 1998, *no pet*).

A defendant must object at trial that a juror was not a resident of the county of the trial or the objection is waived for appeal. Mayo v. State, 4 S.W.3d 9 (Tex. Crim. App. 1999). This is because a person who is not a resident of the county of trial is not absolutely disqualified from jury service under Art. 35.19.

Voir dire error in a non-capital case cannot be error affecting the punishment phase only under CCP Art. 44.29(b); therefore, the defendant has the right to begin a retrial in the guilt-innocence phase. Carson v. State, 6 S.W.3d 536 (Tex. Crim. App. 1999).

In capital cases, voir dire error that occurs in the context of a veniremember's inability to follow some facet of the law applying only at the punishment phase of trial constitutes "error affecting punishment only," and does not alone call for reversal of the conviction, but only a remand for a new punishment proceeding under CCP Art. 44.29(c). Clark v. State, 929 S.W.2d 5 (Tex. Crim. App. 1996); Ransom v. State, 920 S.W.2d 288 (Tex. Crim. App. 1996) (Opinion on State's motion for rehearing).

§14:40 Jury Shuffles

CCP Art. 35.11. Preparation of List

> *The trial judge, on the demand of the defendant or his attorney, or of the State's counsel, shall cause a sufficient number of jurors from which a jury may be selected to try the case to be randomly selected from the members of the general panel drawn or assigned as jurors in the case. The clerk shall randomly select the jurors by a computer or other process of random selection and shall write or print the names, in the order selected, on the jury list from which the jury is to be selected to try the case. The clerk shall deliver a copy of the list to the State's counsel and to the defendant or his attorney.*

§14:41 In General

The Texas Court of Criminal Appeals has mandated that consistent with the purpose of the statute—an assurance of the compilation of a random list of jurors—the jury shuffle is an "absolute right" available to either party. Jones v. State, 833 S.W.2d 118 (Tex. Crim. App. 1992), *cert. denied,* 507 U.S. 921, 113 S.Ct. 1285, 122 L. Ed. 2d 678 (1993).

Article 35.11 provides that a defendant is guaranteed that the jury panel will be shuffled once, at either his request or the State's. It does not mandate, however, that a defendant be allowed to reshuffle the panel after the State has caused the panel to be shuffled, absent some misconduct in the State's shuffle. Chappell v. State, 850 S.W.2d 508 (Tex. Crim. App. 1993).

A trial court's *sua sponte* shuffle of the jury list does not foreclose the defendant from requesting a second shuffle under Art. 35.11. Scott v. State, 805 S.W.2d 612 (Tex.App.—Austin 1991, *no pet.*); Wilkerson v. State, 681 S.W.2d 29 (Tex. Crim. App. 1984).

A defendant has a right to a "reshuffle" if the original shuffle was caused by someone other than the State, such as the trial judge or court personnel. Johnson v. State, 977 S.W.2d 137 (Tex. Crim. App. 1998). The defendant has a right to a "reshuffle" where the first shuffle was procedurally flawed. Johnson.

In a trial involving multiple defendants, the defense side of the table is not allowed more than one shake of the names of the members of the assigned jury panel. Latham v. State, 656 S.W.2d 478 (Tex. Crim. App. 1983).

An accused is not entitled to have a shuffle of all the names of the persons summoned for jury service during the week of his trial but is only entitled to have a shuffle of the names of those prospective jurors who make up the assigned jury panel from which will come the jury that will judge the facts of the accusation. Latham.

Where a clerk combines the jury questionnaires in a box, draws some slips out one at a time, but occasionally draws out a handful, and sometimes reshuffles the handful she had drawn, and where the top questionnaire is then removed until the list is complete, the statute has been complied with. McGee v. State, 909 S.W.2d 516 (Tex.App.—Tyler 1995, *pet. ref'd*).

§14:42 Timing of Motion

A motion to shuffle must be made before voir dire commences. Davis v. State, 782 S.W.2d 211

(Tex. Crim. App. 1989), *cert. denied,* 110.S.Ct. 2193, 495 U.S. 940, 109 L. Ed. 2d 520 (1990).

Voir dire in a non-capital case commences when the state begins its examination of the prospective jurors. Williams v. State, 719 S.W.2d 573 (Tex. Crim. App. 1986). In a capital case, the voir dire commences when the trial judge begins his examination of the panel. Yanez v. State, 677 S.W.2d 62 (Tex. Crim. App. 1984).

§14:43 Opportunity to View Venire

Compliance with CCP Art. 35.11 is had when counsel for either the state or the defendant is allowed the opportunity to view the venire seated in the courtroom in proper sequence and is thereafter allowed an opportunity to exercise his or her option to have the names shuffled. Williams v. State, 719 S.W.2d 573 (Tex. Crim. App. 1986).

There is no absolute right to shuffle the jury panel in circumstances where the defendant had the opportunity to view the original panel, declined to exercise his right to shuffle, and the shuffle at the State's request was done in the courtroom. Jones v. State, 833 S.W.2d 118 (Tex. Crim. App. 1992), *cert. denied,* 507 U.S. 921, 113 S.Ct. 1285, 122 L. Ed. 2d 678 (1993).

§14:44 Opportunity to View Jury Questionnaires

The trial court is not required to allow counsel to view jury questionnaires before they make their decision to shuffle. Davis v. State, 782 S.W.2d 211 (Tex. Crim. App. 1989), *cert. denied,* 110.S.Ct. 2193, 495 U.S. 940, 109 L. Ed. 2d 520 (1990).

However, the trial court is permitted to allow counsel to view the questionnaires before the decision to shuffle is made. Garza v. State, 7 S.W.3d 164 (Tex. Crim. App. 1999).

§14:45 Appellate Review

The trial court's denial of the defendant's legitimate request for a shuffle is subject to harmless error analysis on appeal. In order to merit reversal the defendant must probably show that someone unfit for jury service served on the jury or that the jury was constituted in a non-random manner. Ford v. State, 73 S.W.3d 923 (Tex. Crim. App. 2002) (where the error was found to be harmless).

Complaints about jury shuffle error must be preserved. Sanders v. State, 942 S.W.2d 3 (Tex. Crim. App. 1997); Powell v. State, 897 S.W.2d 307 (Tex. Crim. App. 1994). To prevail on appeal, a defendant must raise a specific objection about the jury shuffle at trial. James v. State, 772 S.W.2d 84 (Tex. Crim. App. 1989).

If CCP Art. 35.11 has not been followed, and a defendant requested the trial procedure about which he complains on appeal, appellate review of the impropriety is precluded. Valdez v. State, 472 S.W.2d 754 (Tex. Crim. App. 1971).

§14:46 Form: Sample Motion to Shuffle Venire

See our companion book *Texas Criminal Forms* (James Publishing) Form 14-6.

§14:50 Voir Dire Subject Matter

§14:51 Statutory Law

CCP Art. 35.17. Voir dire examination

> *1. When the court in its discretion so directs, except as provided in Section 2, the state and defendant shall conduct the voir dire examination of prospective jurors in the presence of the entire panel.*
>
> *2. In a capital felony case in which the State seeks the death penalty, the court shall propound to the entire panel of prospective jurors questions concerning the principles, as applicable to the case on trial, of reasonable doubt, burden of proof, return of indictment by grand jury, presumption of innocence, and opinion. Then, on demand of the State or defendant, either is entitled to examine each juror on voir dire individually and apart from the entire panel, and may further question the juror on the principles propounded by the court.*

§14:52 In General

The conduct of voir dire examination rests within the sound discretion of the trial court. Dowden v. State, 758 S.W.2d 264 (Tex. Crim. App. 1988).

The parties have broad latitude to question members of the venire. Battie v. State, 551 S.W.2d 401 (Tex. Crim. App. 1977), *cert. denied,* 434 U.S. 1041, 98 S.Ct. 782, 54 L. Ed. 2d 790 (1978).

Art. 1, §10 of the Texas Constitution guarantees a defendant the right to be heard by counsel. Such right includes the right of counsel to question members of the venire panel in order to intelligently exercise peremptory challenges. Ex parte McKay, 819 S.W.2d 478 (Tex. Crim. App. 1990); Jones v. State, 223 S.W.3d 379 (Tex. Crim. App. 2007)

Where the trial court disallows the defendant to ask a proper question on voir dire, it is a constitutional error. Jones.

Not all rulings limiting defense counsel's voir dire presentation are constitutional error as a *per se* violation of the right to counsel. Most rulings limiting voir dire will be subject to Rule 44.2(b) harm analysis. However, there may be instances when a judge's limitation on voir dire is so substantial as to warrant labeling the error as constitutional error subject to a Rule 44.2(a) harm analysis. Easley v. State, 424 S.W.3d 535, 538-9 (Tex. Crim. App. 2014).

Within reasonable limits, the trial court must allow the state and defense to explore attitudes that may render the veniremember challengeable for cause. Draughon v. State, 831 S.W.2d 331 (Tex. Crim. App. 1992), *cert. denied,* 509 U.S. 926, 113 S.Ct. 3045, 125 L. Ed. 2d 730 (1993).

It is proper for a party to ask questions designed to assess the relative desirability of the veniremember. Davis v. State, 894 S.W.2d 471 (Tex.App.—Fort Worth 1995, *no pet.*).

§14:52.1 Form: Sample Juror Questionnaire

See our companion book *Texas Criminal Forms* (James Publishing) Form 14-7.

§14:52.2 Form: Motion to Submit Jury Questionnaire

See our companion book *Texas Criminal Forms* (James Publishing) Form 14-9.1 and 14-9.2.

§14:53 Particular Inquiries Allowed

§14:53.1 Explanations of Law

During voir dire, a party may explain the law to the venire. Atkins v. State, 951 S.W.2d 787 (Tex. Crim. App. 1997) (remanded on rehearing for harm analysis).

The parties are permitted to discuss general principles of law applicable to the case during voir dire. Nevarez v. State, 832 S.W.2d 82 (Tex.App.—Waco 1992, *pet. ref'd*).

It is permissible for the parties to define terms as they are defined by law. Adanandus v. State, 866 S.W.2d 210 (Tex. Crim. App. 1993), *cert. denied,* 510 U.S. 1215, 114 S.Ct. 1338, 127 L. Ed. 2d 686 (1994).

The parties may question the venire as to any applicable lesser included offenses. Alfano v. State, 780 S.W.2d 494 (Tex.App.—Corpus Christi 1989, *no pet.*). The parties may explain to the jury that if a witness testifies he can be impeached with, among other things, prior convictions. Harris v. State, 656 S.W.2d 481 (Tex. Crim. App. 1983).

The trial court's refusal to allow a defendant to question a veniremember regarding his understanding of the term "reasonable doubt" constitutes reversible error in a non-capital case. Woolridge v. State, 827 S.W.2d 900 (Tex. Crim. App. 1992). The defendant has the right to examine a veniremember on a subject area even though the prosecutor has committed him to certain answers. White v. State, 629 S.W.2d 701 (Tex. Crim. App. 1981).

§14:53.1.1 Misstatements of Law

The right to ask proper questions does not include the right to misstate the law. Powell v. State, 897 S.W.2d 307 (Tex. Crim. App. 1994).

Defense counsel must accurately object when the state misinforms the venire during voir dire because if the point of error is different on appeal nothing will be presented for review. Sterling v. State, 800

S.W.2d 513 (Tex. Crim. App. 1990), *cert. denied,* 501 U.S. 1213, 111 S.Ct. 2816, 115 L. Ed. 2d 988 (1991).

The defendant must further show that he was somehow harmed by the remark before the case will be reversed on appeal. Jackson v. State, 822 S.W.2d 18 (Tex. Crim. App. 1990).

TIP TO THE PROSECUTION: If defense counsel misstates the law during voir dire, object on this ground. If the objection is sustained, ask the trial court to instruct the jury to disregard the misstatement of law. Etheridge v. State, 903 S.W.2d 1 (Tex. Crim. App. 1994).

A party should be allowed to ask the venire questions that seek to determine whether they understand that the standard of proof beyond a reasonable doubt constitutes a level of confidence under the law that is higher than both the preponderance of the evidence and the clear and convincing evidence standards. Fuller v. State, 363 S.W.3d 583 (Tex. Crim. App. 2012).

§14:53.2 Punishment Inquiries

The parties may inform the venire of the range of punishment applicable in the case. Estes v. State, 873 S.W.2d 771 (Tex.App.—Fort Worth 1994, *pet. ref'd*). This includes the range of punishment available for an enhanced offense. Estes.

A prosecutor may inform the venire of the range of punishment applicable if the state were to prove a prior conviction for enhancement purposes. Frausto v. State, 642 S.W.2d 506 (Tex. Crim. App. 1982). It is improper however, for the prosecutor to inform the venire of the particular allegations contained in the indictment of the defendant on trial. Frausto.

When the jury will set punishment in the case, it is error for the trial court to prohibit defendant's counsel from asking veniremembers questions relevant to punishment, testing their philosophy and feelings about punishment. Powell v. State, 631 S.W.2d 169 (Tex. Crim. App. 1982).

§14:53.3 Inquiries on Status of Victim or Witness

The defendant is allowed to discover a veniremember's views on the status of the victim in order that he may intelligently exercise his peremptory challenges. Maddux v. State, 862 S.W.2d 590 (Tex. Crim. App. 1993). Restrictions on a defendant's voir dire questioning designed to discover the veniremember's bias as to the victim's status as a child have been found improper. Cena v. State, 960 S.W.2d 804 (Tex.App.—El Paso 1997); Maddux.

The question "generally speaking, do you feel like you might have a problem convicting a defendant if the only eyewitness is a child" has been found to be a proper type of inquiry. Davis v. State, 894 S.W.2d 471 (Tex.App.—Fort Worth 1995, *no pet.*).

However, a veniremember is not challengeable for cause for an inability to consider the full range of punishment in a case involving a victim of a particular status where he can consider the full punishment range for the offense as set out by law. Sadler v. State, 977 S.W.2d 140 (Tex. Crim. App. 1998).

A defendant is allowed to explore the venire's ability to judge the credibility of police officer witnesses. Hernandez v. State, 508 S.W.2d 853 (Tex. Crim. App. 1974).

The trial court can disallow an inquiry of venire members as to whether they could be fair and impartial in a case in which the victim was nine years old because the inquiry is too broad and global. Barajas v. State, 93 S.W.3d 36 (Tex. Crim. App. 2002).

§14:53.4 Inquiries Into Status and Attitudes of Veniremembers

Counsel should be permitted to question veniremembers about their prior grand jury service. Gonzales v. State, 638 S.W.2d 132 (Tex.App.—Corpus Christi 1982, *no pet.*).

It is not error for the trial court to allow parties to question veniremembers about what verdict they reached during prior jury service. Whiting v. State, 943 S.W.2d 102 (Tex.App.—Houston [1st Dist.] 1997, *pet. ref'd*).

However, it also is not an abuse of discretion for a trial court to refuse to allow defense counsel to inquire of prospective jurors with prior jury service in criminal cases what verdict they reached in those cases. Bolden v. State, 634 S.W.2d 710 (Tex. Crim. App. 1982).

The defendant should be allowed to inquire as to the venire's feeling about moral issues where relevant to the case. Shipley v. State, 790 S.W.2d 604 (Tex. Crim. App. 1990) (where the issue was extra-marital affairs).

Although it is proper to question juror's views and sentiments on social and moral subjects generally, the courts do not permit a hypothetical case to be submitted, nor do they allow questions designed to bring out the juror's views on the case to be tried. Brown v. State, 674 S.W.2d 443 (Tex. App.—Dallas 1984), *affirmed at* 741 S.W.2d 453 (Tex. Crim. App. 1987).

The defense should be allowed to question the venire as to their attitudes about defendants who invoke the insanity defense where this is an issue in the case. Robinson v. State, 720 S.W.2d 808 (Tex. Crim. App. 1986).

The defendant should be allowed to explore any area of questioning which would reveal a possible bias against the defendant. Harrell v. State, 882 S.W.2d 65 (Tex.App.—Houston [14th Dist.] 1994, *pet. ref'd*).

§14:53.5 Inquiries Based on Situational Status of Witness

An inquiry whether the veniremember would be able to convict on the testimony of one eyewitness has been found to be a proper question. Orosco v. State, 827 S.W.2d 575 (Tex.App.—Fort Worth 1992, *pet. ref'd*), *cert. denied,* 506 U.S. 960, 113 S.Ct. 425, 121 L. Ed. 2d 347 (1992). An inquiry whether a veniremember would be able to convict a defendant without hearing directly from a confidential informant has been found to be a proper question. Payton v. State, 830 S.W.2d 722 (Tex.App.—Houston [14th Dist.] 1992, *no pet.*).

Also, whether a veniremember would be able to convict based upon a child-victim's videotaped testimony has been found to a proper inquiry. Newman v. State, 700 S.W.2d 307 (Tex.App.—Houston [1st Dist.] 1985), *reversed on other grounds at* 743 S.W.2d 641 (Tex. Crim. App. 1988).

§14:54 Means of Making Inquiry

§14:54.1 Type of Question Permitted

It is permissible to use hypothetical questions to explain the application of the law. Atkins v. State, 951 S.W.2d 787 (Tex. Crim. App. 1997); Cuevas v. State, 742 S.W.2d 331 (Tex. Crim. App. 1987).

It is also allowable to use a hypothetical regarding the victim's status as a child because it seeks to ascertain potential bias. Maddux v. State, 862 S.W.2d 590 (Tex. Crim. App. 1993) (where defense counsel sought to inquire if the venire could consider 5 years probation where the victim of the offense was a child). It is not proper to inquire how a veniremember would respond to particular circumstances presented in a hypothetical question. Cuevas.

Questions that are not clearly improper on some other basis can be asked for the purpose of intelligently exercising peremptory challenges subject to reasonable time limits imposed by the trial court. Barajas v. State, 93 S.W.3d 36 (Tex. Crim. App. 2002).

Questions phrased, "Can you be fair and impartial..." are overly broad and the trial court does not abuse its discretion in disallowing them. Barajas.

§14:55 Mechanics of Voir Dire

A trial court does not have to permit the parties to videotape the voir dire process. Curry v. State, 910 S.W.2d 490 (Tex. Crim. App. 1995).

§14:56 Preserving Error

§14:56.1 Preserving Error Where Proper Inquiry Disallowed

A trial court generally will not abuse its discretion unless it limits the appellant from asking a proper question concerning a proper area of inquiry. Allridge v. State, 762 S.W.2d 146 (Tex. Crim. App. 1988).

Where the trial court disallows the defendant to ask a proper question on voir dire, it is a constitutional error. Jones v. State, 223 S.W.3d 379 (Tex. Crim. App. 2007); Samaripas v. State, 454 S.W.3d 1 (Tex. Crim. App. 2014).

Where the trial court excludes a proper question during voir dire, the effect is subject to harmless error analysis. Gonzales v. State, 994 S.W.2d 170 (Tex. Crim. App. 1999).

Defense counsel must state a proper question in precise form. Jones v. State, 843 S.W.2d 487 (Tex. Crim. App. 1992), *cert. denied,* 507 U.S. 1035, 113 S.Ct. 1858, 123 L. Ed. 2d 479 (1993).

The record must reflect the precise question that counsel sought to ask. Caldwell v. State, 818 S.W.2d 790 (Tex. Crim. App. 1991), *cert. denied,* 503 U.S. 990, 112 S.Ct. 1684, 118 L. Ed. 2d 399 (1992).

Where the court disallows a proper inquiry, a defendant preserves error only if:

- Defendant exhausts all of his peremptory challenges
- Defendant requests more challenges
- Request is denied, and
- Defendant identifies an objectionable person seated on the jury on whom he would have exercised a peremptory challenge

Anson v. State, 959 S.W.2d 203 (Tex. Crim. App. 1997); Narvaiz v. State, 840 S.W.2d 415 (Tex. Crim. App. 1992), *cert. denied,* 507 U.S. 975, 113 S.Ct. 1422, 122 L. Ed. 2d 791 (1993).

There is no distinction between capital and non-capital cases for the purpose of the steps required to preserve error for the erroneous denial of proper questioning. Anson.

Defense counsel should object that the intelligent use of the defendant's peremptory challenges has been affected and that he has been deprived of the effective assistance of counsel. Sawyers v. State, 724 S.W.2d 24 (Tex. Crim. App. 1986).

When an appellant challenges a trial judge's limitation on the voir dire process, the reviewing court must analyze the claim under an abuse of discretion standard, the focus of which is whether the appellant proffered a proper question concerning a proper area of inquiry. Caldwell, *supra.*

Voir dire error in a non-capital case cannot be error affecting the punishment phase only under CCP Art. 44.29 (b); therefore, the defendant has the right to begin a retrial in the guilt-innocence phase. Carson v. State, 6 S.W.3d 536 (Tex. Crim. App. 1999).

Defense counsel's reply in the negative to whether there is an objection to the jury as selected at the conclusion of jury selection does not constitute a waiver of a previously preserved claim of error during voir dire proceedings. Stairhime v. State, 463 S.W.3d 902, 908 (Tex. Crim. App. 2015).

§14:56.2 Preserving Error for Improper Inquiry by State

Where the defendant claims that the state has been allowed to ask an improper voir dire question, he must contemporaneously object in order to preserve error. Boyd v. State, 811 S.W.2d 105 (Tex. Crim. App. 1991), *cert. denied,* 502 U.S. 971, 112 S.Ct. 448, 116 L. Ed. 2d 466 (1991).

The objection must be made as soon as the basis for the objection becomes apparent. Johnson v. State, 803 S.W.2d 272 (Tex. Crim. App. 1990), *cert. denied,* 501 U.S. 1259, 111 S.Ct. 2914, 15 L. Ed. 2d 1078 (1991). Objections made after the questions are answered do not preserve error for appeal. Russell v. State, 904 S.W.2d 191 (Tex.App.—Amarillo 1995, *pet. ref'd*).

If the defendant's objection is sustained, he must go further and request an instruction to disregard, and, if granted, seek a motion for mistrial in order to preserve the complaint for appeal. Holloway v. State, 695 S.W.2d 112 (Tex.App.—Fort Worth 1985), *aff'd,* 751 S.W.2d 866 (Tex. Crim. App. 1988).

Voir dire error in a non-capital case cannot be error affecting the punishment phase only under CCP Art. 44.29 (b); therefore, the defendant has the right to begin a retrial in the guilt-innocence phase. Carson v. State, 6 S.W.3d 536 (Tex. Crim. App. 1999).

§14:57 Restrictions on Voir Dire

§14:57.1 In General

The trial court has the power to impose reasonable restrictions on the conduct of voir dire. Marquez v. State, 725 S.W.2d 217 (Tex. Crim. App. 1987), *cert. denied*, 484 U.S. 872, 108 S.Ct. 201, 98 L. Ed. 2d 152 (1987). The exercise of that power is within the sound discretion of the trial judge. Marquez.

In making the determination whether the trial court abused its discretion in restricting the appellant's voir dire examination, the test is whether the question or questions sought to be asked are proper. If the question was proper and the defendant was prevented from asking it, then the defendant has been prevented from intelligently exercising his peremptory challenges. Faulder v. State, 745 S.W.2d 327 (Tex. Crim. App. 1987).

A proper question is one that seeks to discover a veniremember's views on an issue applicable to the case. Caldwell v. State, 818 S.W.2d 790 (Tex. Crim. App. 1991), *cert. denied*, 503 U.S. 990, 112 S.Ct. 1684, 118 L. Ed. 2d 399 (1992).

To be constitutionally compelled, it is not enough that a defendant's voir dire questions might be helpful. Rather, the trial court's failure to allow these questions must render the defendant's trial fundamentally unfair. Jacobs v. State, 560 S.W.3d 205, 211 (Tex. Crim. App. 2019) (where the defendant's question of the venire regarding whether they could be fair if they knew that the defendant had been previously convicted of a sexual offense was found to only marginally enhance his ability to explore their biases over the question he was permitted, which was in terms of conviction of an assaultive offense).

Not all rulings limiting defense counsel's voir dire presentation are constitutional error as a *per se* violation of the right to counsel. Most rulings limiting voir dire will be subject to Rule 44.2(b) harm analysis. However, there may be instances when a judge's limitation on voir dire is so substantial as to warrant labeling the error as constitutional error subject to a Rule 44.2(a) harm analysis. Easley v. State, 424 S.W.3d 535, 538-9 (Tex. Crim. App. 2014).

TIP TO THE BENCH: Although the trial judge is vested with the authority to impose reasonable restrictions on the accused's exercise of his voir dire examination, conducting a criminal trial promptly and expeditiously must never be undertaken at the risk of denying a party on trial a substantial right. Faulder.

It is improper for a trial judge to impose restrictions based on the mere possibility that an otherwise proper question might lengthen the voir dire process. The trial judge must first allow the question, and may later curtail similar questions if the voir dire process proves to be unduly lengthy. Battie v. State, 551 S.W.2d 401 (Tex. Crim. App. 1977), *cert. denied,* 434 U.S. 1041, 98 S.Ct. 782, 54 L. Ed. 2d 790 (1978).

TIP TO THE BENCH: Should a trial judge determine that either party is prolonging the voir dire process, the simple and effective remedy is to call the attorneys to the bench and instruct them to expedite the process. McCarter v. State, 837 S.W.2d 117 (Tex. Crim. App. 1992).

§14:57.2 Permissible Voir Dire Restrictions

§14:57.2.1 Questions and Areas Subject to Restriction

§14:57.2.1.1 Confusing, Repetitious, Broad and Vague Questions

A trial court may restrict confusing or misleading voir dire questions. Jones v. State, 850 S.W.2d 223 (Tex.App.—Fort Worth 1993, *pet. ref'd*). The trial court also has the power to terminate needlessly duplicitous or repetitious questioning. Allridge v. State, 762 S.W.2d 146 (Tex. Crim. App. 1988), *cert. denied*, 489 U.S. 1040, 109 S.Ct. 1176, 103 L. Ed. 2d 238 (1989).

If a prospective juror states his or her position clearly, unequivocally, and without reservation, the trial court does not err in refusing to permit further questioning. Guerra v. State, 771 S.W.2d 453 (Tex. Crim. App. 1988), *cert. denied,* 492 U.S. 925, 109 S.Ct. 3260, 106 L. Ed. 2d 606 (1989). However, defense counsel may not be precluded from the traditional voir dire examination simply because the questions asked are repetitious of those asked by the court and prosecutor. Mathis v. State, 576 S.W.2d 835 (Tex. Crim. App. 1979). This is because each party has the right to question potential jurors in their own individual manner to emphasize a point or uncover a hidden bias, and may not be forced to rely on other parties to ask similar questions. Williams v. State, 804 S.W.2d 95 (Tex. Crim. App. 1991).

A trial court does not abuse its discretion in restricting counsel from asking potential jurors broad, global question such as those about "their thoughts" on issues. Boyd v. State, 811 S.W.2d 105 (Tex. Crim. App. 1991), *cert. denied,* 502 U.S. 971, 112 S.Ct. 448, 116 L. Ed. 2d 466 (1991).

Likewise, vague questions about "what things" the veniremember thinks are important in determining the issues in the case, wholly unrelated to the case's procedural scheme, amount to a fishing expedition going beyond the scope of proper voir dire. Boyd.

Questions phrased, "Can you be fair and impartial..." are overly broad and the trial court does not abuse its discretion in disallowing them. Barajas v. State, 93 S.W.3d 36 (Tex. Crim. App. 2002).

TIP TO THE BENCH: The trial court can prevent global fishing expeditions on voir dire by requiring the parties to phrase their questions with the precision necessary to glean relevant information.

Examples of impermissibly global questions include questions asking a prospective capital juror's understanding of a life sentence or of parole, Boyd; and asking what cases would be appropriate for the death penalty, Allridge, *supra.*

Whether a veniremember would want to know certain legal parameters before deciding the issues in a case is not strictly relevant to a juror's duties or any issue in the case. What the jurors want to know is immaterial; the trial court will give jurors the proper information about the application of the law. Sells v. State, 121 S.W.3d 748 (Tex. Crim. App. 2003).

§14:57.2.1.2 Commitment Questions

A trial judge may limit a defendant's voir dire where a question commits a veniremember to a particular course of action based on a specific set of facts. Atkins v. State, 951 S.W.2d 787 (Tex. Crim. App. 1997); Dinkins v. State, 894 S.W.2d 330 (Tex. Crim. App. 1995); Standefer v. State, 59 S.W.3d 177 (Tex. Crim. App. 2001).

A commitment question is one that asks veniremen to resolve or refrain from resolving an issue a certain way after learning a particular fact. Standefer, *supra.*

A question may be proper if it seeks to discover a juror's views on an issue applicable to the case or to commit the juror to follow the law; however, if the question adds a fact in addition to those necessary to establish a challenge for cause it is an improper "commitment" question. Standefer, *supra.*

PRACTICE TIP: The Court of Criminal Appeals hints that questions that ask veniremen "could there be circumstances" are permissible because this wording does not ask a venireman to take or refrain from taking any action based on the parameters of the question. See Standefer.

A trial court may properly disallow hypothetical questions that are based on the facts of the case on trial. Coleman v. State, 881 S.W.2d 344 (Tex. Crim. App. 1994); Cuevas v. State, 742 S.W.2d 331 (Tex. Crim. App. 1987), *cert. denied,* 485 U.S. 1015, 108 S.Ct. 1488, 99 L. Ed. 2d 716 (1988).

A trial court may disallow the asking of open-ended questions designed so the veniremember could set the hypothetical parameters for their decision-making. Allridge v. State, 850 S.W.2d 471 (Tex. Crim. App. 1991), *cert. denied,* 510 U.S. 831, 114 S.Ct. 101, 126 L. Ed. 2d 68 (1993); Standefer v. State, 59 S.W.3d 177 (Tex. Crim. App. 2001).

TIP TO THE BENCH: This is a very tricky area for trial judges. The judge may have to call counsel to the bench to inquire as to the facts of the case before ruling on an objection of this nature.

The appropriate test for harm when the state is allowed to improperly commit jurors to a set of facts is the harmless error analysis set out in Rule 44.2(b) of the Texas Rules of Appellate Procedure. Sanchez v. State, 165 S.W.3d 707 (Tex. Crim. App. 2005).

§14:57.2.1.2.1 Improper Commitment Questions

The question "[d]o each of you feel as though you could evaluate a witness and his testimony and decide if he's being truthful without automatically dismissing his testimony because of some criminal history?" is a commitment question because it asks jurors to resolve witness credibility (issue) based on the knowledge that the witness has a criminal history (particular fact). Lydia v. State, 109 S.W.3d 495 (Tex. Crim. App. 2003).

For example, the inquiry "what ... is a case that is proper for the death penalty to be imposed is improper because it committed the veniremember to a particular fact situation. Boyd v. State, 811 S.W.2d 105 (Tex. Crim. App.1991), *cert. denied,* 502 U.S. 971, 112 S.Ct. 448, 116 L. Ed. 2d 466 (1991).

Asking veniremembers if they agree that a particular fact situation would be self-defense is improperly contracting with the venire. Montes v. State, 870 S.W.2d 643 (Tex.App.—El Paso 1994, *no pet.*).

A question asking if the venireman would be more likely, or less likely, generally, to view a defendant as a continuing threat to society, if the venireman knew he could not be paroled for a minimum of 40 years, is improper because it seeks to determine whether the venireman will give the parole instruction mitigating or aggravating effect in the context of the future dangerousness punishment issue. Sells v. State, 121 S.W.3d 748 (Tex. Crim. App. 2003).

The question "[w]ould you presume someone guilty if he or she refused a breath test on their refusal alone?" is an improper commitment question because the question asks whether the prospective juror would resolve the issue of guilt against the defendant if the juror learns a particular fact—that the defendant has refused a breath test. The facts in this question would not lead to a valid challenge for cause because a juror may permissibly presume guilt from such evidence. Standefer.

The question "[w]hat circumstances in your opinion warrant the imposition of the death penalty?" is an improper commitment question. It is an example of an open-ended question that is a commitment question because it asks the prospective juror to set the hypothetical parameters for his decision-making. An example of a proper question that seeks to uncover the same type of information is "Do you think there might be circumstances that would mitigate against the death penalty?" Allridge.

The question "[w]ould the minimum length of time a defendant could serve in prison before he could be paroled be something you would want to know in answering the special issues?" is an improper commitment question. The question relates to how a particular fact (in this case, the minimum amount of time a capital life defendant must be incarcerated before becoming eligible for parole) might influence jury deliberations. These types of questions implicate the strictures imposed by Standefer against commitment questions and by Barajas against ambiguous questions. The question is an attempt to commit the venire members to giving mitigating or aggravating effect to the minimum parole eligibility requirement. Additionally, this matter is not strictly relevant to a juror's duties or any issue in the case. What the jurors want to know is immaterial; the trial court will give jurors the proper information about the application of the law. Sells.

The question, "Do you believe disabilities of a defendant would be something you'd take into consideration?" is an improper commitment question in a capital case because jurors are not required to consider any particular type of evidence as mitigating. Davis v. State, 313 S.W.3d 317 (Tex. Crim. App. 2010).

§14:57.2.1.2.2 Proper Commitment Questions

The prosecutor used the "mercy killing" scenario as an example of an intentional murder in the context of questioning on the range of punishment for first degree murder. Then the prosecutor asked: If taking into consideration the wide range of circumstances which could constitute murder, could the venire member keep an open mind as to the full range of punishment? The questions were not improper commitment questions. They did not attempt to bind the venire member to resolve or refrain from resolving an issue (*i.e.*, the minimum punishment for an intentional murder) on the basis of one or more facts contained in the questions. Halprin v. State, 170 S.W.3d 111 (Tex. Crim. App. 2005).

The prosecutor gave a hypothetical example of a case involving the testimony of a single witness in an indecency with a child case and then asked: ". . . [a]ssume for me that you do believe her testimony beyond a reasonable doubt, you do believe her, and it meets all the elements of the charge, indecency with a child by contact, assume for me that you do believe her beyond a reasonable doubt, could you convict that hypothetical defendant of that charge or would you require some other witness or some other evidence?" The question is a proper commitment question because it leads to a valid challenge for cause without adding extra facts. Lee v. State, 206 S.W.3d 620 (Tex. Crim. App. 2006).

The state's inquiry whether the venire was more likely to believe a self-defense claim made by a woman was a proper commitment question because it could lead to a challenge for cause to a veniremember for having a bias for the female defendant. Braxton v. State, 226 S.W.3d 602 (Tex. App.—Houston [1st Dist.] 2007, *pet. dism'd*).

In a prosecution for tampering with a government record, the question "Do you believe there is anything wrong with putting false information in a police report?" was found to not be a commitment question. Jurors are required to follow the law that makes it a crime for a police officer to tamper with a government record. Asking the venire if they agree that the commission of such a crime is wrong does not commit them to resolve an issue of fact in a certain way. Rather, it inquires into the jurors' general beliefs as to the wrongness of conduct that the law has made a crime. Wingo v. State, 189 S.W.3d 270 (Tex. Crim. App. 2006).

§14:57.2.1.2.3 Non-Commitment Questions

In a case where defense counsel asked the venire if they could be fair and impartial in a case in which the victim was nine years old, it was found to not be a commitment question because it does not ask the prospective juror to resolve or refrain from resolving any issue. However, this "fair and impartial" question is a license to go fishing without providing any concrete information for the intelligent use of peremptory challenges or challenges for cause. Vague questions about "what things" the venire member thought were important in determining the issues in a case amounts to a fishing expedition going beyond the scope of proper voir dire. Barajas v. State, 93 S.W.3d 36 (Tex. Crim. App. 2002), *overruling* Nunfio v. State, 808 S.W.2d 482 (Tex. Crim. App. 1991), which had held that the question "If the victim is a nun, could you be fair and impartial?" was not a commitment question.

The prosecutor, in the context of the mitigation special issue, used mental retardation as an example of a factor that jurors might consider mitigating. The prosecutor then asked the prospective juror: Would he be able to keep an open mind to a life sentence with regard to the mitigation special issue even if he had already found that the defendant was guilty of killing a police officer and will be a future danger? The prosecutor did not improperly commit the prospective juror. The prosecution's question did not attempt to bind the juror to resolve or refrain from resolving an issue (*i.e.*, the mitigating evidence special issue) on the basis of one or more facts contained in the question. The totality of the voir dire record reflected that the prosecution explained that the prospective juror should have an open mind and consider all of the evidence in answering the mitigating evidence special issue. Halprin v. State, 170 S.W.3d 111 (Tex. Crim. App. 2005).

In a non-capital case the question, "what factors ... are important" in sentencing is not an improper commitment question because it does not ask the jurors how particular facts would influence their deliberations. It is an inquiry into the jurors' general philosophies and does not ask jurors to define situations in which they would impose a specific sentence. This question may be an improper commitment question in a death penalty case because sentencing for a capital felony has only two possible outcomes, life in prison without parole or the death penalty. Davis v. State, 349 S.W.3d 517 (Tex. Crim. App. 2010).

§14:57.2.1.3 Questions in Improper Form

A trial court also may restrict questions asked in improper form. Guerra v. State, 771 S.W.2d 453 (Tex. Crim. App. 1988), *cert. denied,* 492 U.S. 925, 109 S.Ct. 3260, 106 L. Ed. 2d 606 (1989).

More specifically, the trial court must be given some discretion to regulate the form of the question (*i.e.,* its semantics or phrasing), and to preclude those questions that the court might, in its discretion, find improperly constructed. Howard v. State, 941 S.W.2d 102 (Tex. Crim. App. 1996).

PRACTICE TIP: Where opposing counsel attempts to question without adequately explaining the law to the venire or individual veniremember, an objection to the "form of the question" is a valid objection and can be sustained by the trial court under Howard and Guerra.

Where the trial court sustains the state's objection to an improper commitment question but the defense is later allowed to question on the substance of the question, the trial court, by not placing an absolute limitation on the subject matter of the inquiry, has merely sustained an objection to the form of the original question. Hernandez v. State, 390 S.W.3d 310, 316 (Tex. Crim. App. 2012) (where the defense had sought to ask if the venireperson could consider a particular piece of evidence as mitigating).

Where the trial court in sustaining an objection based on the form of the question places no absolute limitation on the underlying substance of a defendant's voir dire question, it is incumbent upon defense counsel to rephrase an improperly phrased query or else waive the voir dire restriction. Trevino v. State, 815 S.W.2d 592 (Tex. Crim. App. 1991).

PRACTICE TIP: Where the trial court has sustained an objection that a question was in improper form, it is incumbent on the proffering party to rephrase the question or the point will be waived on appeal. Guerra.

§14:57.2.1.4 Preserving Error of Improper Question by State

In order to preserve error where the state has asked an improper question, the defendant must timely object and take all of the steps required to preserve error where the court has improperly denied one of his challenges for cause. Morrow v. State, 753 S.W.2d 372 (Tex. Crim. App. 1988).

§14:57.2.2 Subject Matter That May Be Restricted

§14:57.2.2.1 Inquiries on Incorrect Law

The trial court can prevent inquiries into subject matters that are no longer the law. Spence v. State, 795 S.W.2d 743 (Tex. Crim. App. 1990), *cert. denied*, 499 U.S. 932, 111 S.Ct. 1339, 113 L. Ed. 2d 271 (1991).

It is not error for the trial court to refuse to allow questioning on an area of law as to which there is to be no evidence presented at the trial. Jones v. State, 843 S.W.2d 487 (Tex. Crim. App. 1992), *cert. denied*, 507 U.S. 1035, 113 S.Ct. 1858, 123 L. Ed. 2d 479 (1993).

A trial court may restrict counsel's questioning on irrelevant matters. Boyd v. State, 811 S.W.2d 105 (Tex. Crim. App.1991), *cert. denied*, 502 U.S. 971, 112 S.Ct. 448, 116 L. Ed. 2d 466 (1991).

A trial court does not abuse its discretion in disallowing inquiries of veniremembers as to whether they could follow jury charge language that is not required by law. Hankins v. State, 132 S.W.3d 380 (Tex. Crim. App. 2004) (where the defendant sought to voir dire on the disapproved Geesa instruction).

§14:57.2.2.2 Definitions of Legally Undefined Terms

The trial court has the discretion to prohibit voir dire on a veniremember's understanding of the definition of undefined terms in the charge. Wheatfall v. State, 882 S.W.2d 829 (Tex. Crim. App. 1994). For example, a trial court does not abuse its discretion by refusing to permit counsel to question a veniremember regarding his definition of the term "probability." Lagrone v. State, 942 S.W.2d 602 (Tex. Crim. App. 1997), *cert.*

denied, 522 U.S. 917, 118 S.Ct. 305, 139 L. Ed. 2d 235 (1997).

A trial court does not abuse its discretion by refusing voir dire questioning on definitions of the terms "deliberately, probability, and criminal acts of violence" in a capital case. Milton v. State, 599 S.W.2d 824 (Tex. Crim. App. 1980), *cert. denied*, 451 U.S. 1031, 101 S.Ct. 3022, 69 L. Ed. 2d 400 (1981).

A trial court is, however, required under certain circumstances to allow the parties to question the venire as to their comprehension of certain terms as those terms are relevant to the issues in the case. Teague v. State, 864 S.W.2d 505 (Tex. Crim. App. 1993), *overruled on other grounds by* Robertson v. State, 871 S.W.2d 701 (Tex. Crim. App. 1993) (where the issue was the difference between the terms "intentionally" and "deliberately").

A trial court is on firmer ground limiting voir dire on the understanding of undefined terms in capital murder cases because in those cases voir dire examination can take an unreasonable length of time. Woolridge v. State, 827 S.W.2d 900 (Tex. Crim. App. 1992).

TIP TO THE BENCH: In situations such as a death penalty case, it can save a great deal of time on voir dire to encourage the parties to agree on a definition for terms not defined by law which are crucial to a veniremember's function as a juror.

It is error for the trial court to allow a party to limit the venire to its definition of a phrase not defined by law. Jackson v. State, 822 S.W.2d 18 (Tex. Crim. App. 1990), *cert. denied*, 509 U.S. 921, 113 S.Ct. 3034, 125 L. Ed. 2d 722 (1993).

§14:57.2.2.3 Factual Matters

It is permissible for the trial court to disallow questioning on whether members of the venire remembered specific aspects of a newspaper article. Faulder v. State, 745 S.W.2d 327 (Tex. Crim. App. 1987).

Questions about parole are properly disallowed in capital cases. Jones v. State, 982 S.W.2d 346 (Tex. Crim. App. 1998).

§14:57.3 Time Restrictions

§14:57.3.1 In General

The control of the voir dire examination is within the sound discretion of the trial judge, and it is well settled that the trial judge's discretionary authority extends to imposing reasonable limitations on the time for which counsel may question the jury panel. Caldwell v. State, 818 S.W.2d 790 (Tex. Crim. App. 1991), *cert. denied*, 503 U.S. 990, 112 S.Ct. 1684, 118 L. Ed. 2d 399 (1992).

Accordingly, a trial court's decision to limit questioning during voir dire may be reviewed only to determine whether such a restraint amounts to an abuse of discretion. Smith v. State, 703 S.W.2d 641 (Tex. Crim. App. 1985).

§14:57.3.2 Relevant Factors

Where the defendant is restricted while questioning the venire *collectively*, two factors are relevant in determining whether a trial court abused its discretion in limiting voir dire time: (1) whether the defendant's voir dire examination reveals an attempt to prolong the voir dire; for example, whether the questions were irrelevant, immaterial or unnecessarily repetitious; and (2) whether the questions that the defendant was not permitted to ask members of the venire were proper voir dire questions. De la Rosa v. State, 414 S.W.2d 668 (Tex. Crim. App. 1967).

Where the defendant is restricted while attempting to question veniremembers *individually,* three factors are relevant in determining whether a trial court abused its discretion in limiting voir dire time:

- Did defendant's voir dire examination reveal an attempt to prolong the voir dire; for example, were the questions irrelevant, immaterial or unnecessarily repetitious
- Were the questions that the defendant was not permitted to ask members of the venire proper voir dire questions
- Can defendant show that he was not permitted to examine jurors who served on the jury. Ratliff v. State, 690 S.W.2d 597 (Tex. Crim. App. 1985)

This test remains the same in the capital context, except that the third prong is whether the party was not permitted to ask proper questions of veniremembers who actually served on the jury. Etheridge v. State, 903 S.W.2d 1 (Tex. Crim. App. 1994).

Defense counsel does not unnecessarily prolong voir dire merely by addressing matters previously covered by the trial court or the state. Tobar v. State, 850 S.W.2d 182 (Tex. Crim. App. 1993); McCarter v. State, 837 S.W.2d 117 (Tex. Crim. App. 1992).

If the parties have made an agreement to be bound to a certain voir dire time limit, a reviewing court may consider the agreement as a factor in deciding if the trial judge impermissibly limited voir dire examination. Cantu v. State, 842 S.W.2d 667 (Tex. Crim. App. 1992), *cert. denied,* 509 U.S. 926, 113 S.Ct. 3046, 125 L. Ed. 2d 731 (1993).

§14:57.3.3 Preserving Error for Improper Time Restriction

Where the defendant has been restricted from asking additional voir dire questions, he must do several things to preserve error:

- Request additional time
- Inform the judge that counsel has additional subject matter on which to question the veniremember
- Present to the trial court a specific question counsel wishes to proffer
- Obtain an adverse ruling as to that request
- Exercise peremptory challenges on those veniremembers not questioned (and exhaust peremptory challenges); and
- Identify an objectionable juror who sat on the jury

Caldwell v. State, 818 S.W.2d 790 (Tex. Crim. App. 1991), *cert. denied,* 503 U.S. 990, 112 S.Ct. 1684, 118 L. Ed. 2d 399 (1992).

PRACTICE TIP: When conducting voir dire, have several proper questions written down and ready to proffer to the trial court in the event that the trial court attempts to terminate counsel's questioning prematurely.

TIP TO THE BENCH: Because of the unpredictability of veniremember's attitudes and the possible issues involved in a case, the imposition of an absolute time limit is a slippery slope. The trial court is advised to impose a flexible goal rather than an absolute time limit. Any proper question has the potential to lengthen the voir dire portion of a trial. Woolridge v. State, 827 S.W.2d 900 (Tex. Crim. App. 1992).

§14:57.3.4 Form for Additional Time

See our companion book *Texas Criminal Forms* (James Publishing) for the following sample form:

- Form 14-11 Motion for Additional Time to Conduct Voir Dire.

§14:60 Challenges for Cause

§14:61 Statutory Law

CCP Art. 35.16. Reasons for Challenge for Cause

(a) A challenge for cause is an objection made to a particular juror, alleging some fact which renders the juror incapable or unfit to serve on the jury. A challenge for cause may be made by either the state or the defense for any one of the following reasons:

(1) That the juror is not a qualified voter in the state and county under the Constitution and laws of the state; provided, however, the failure to register to vote shall not be a disqualification;

(2) That the juror has been convicted of misdemeanor theft or a felony;

(3) That the juror is under indictment or other legal accusation for misdemeanor theft or a felony;

(4) That the juror is insane;

(5) That the juror has such defect in the organs of feeling or hearing, or such bodily or mental defect or disease as to render the juror unfit for jury service, or that the juror is legally blind and the court in its discretion is not satisfied that the juror is fit for jury service in that particular case;

(6) That the juror is a witness in the case;

(7) That the juror served on the grand jury which found the indictment;

(8) That the juror served on a petit jury in a former trial of the same case;

(9) That the juror has a bias or prejudice in favor of or against the defendant;

(10) That from hearsay, or otherwise, there is established in the mind of the juror such a conclusion as to the guilt or innocence of the defendant as would influence the juror in finding a verdict. To ascertain whether this cause of challenge exists, the juror shall first be asked whether, in the juror's opinion, the conclusion so established will influence the juror's verdict. If the juror answers in the affirmative, the juror shall be discharged without further interrogation by either party or the court. If the juror answers in the negative, the juror shall be further examined as to how the juror's conclusion was formed, and the extent to which it will affect the juror's action; and, if it appears to have been formed from reading newspaper accounts, communications, statements or reports, or mere rumor or hearsay, and if the juror states that the juror feels able, notwithstanding such opinion, to render an impartial verdict upon the law and the evidence, the court, if satisfied that the juror is impartial and will render such verdict, may, in its discretion, admit the juror as competent to serve in such case. If the court, in its discretion, is not satisfied that the juror is impartial, the juror shall be discharged;

(11) That the juror cannot read or write.

No juror shall be impaneled when it appears that the juror is subject to the second, third or fourth grounds of challenge for cause set forth above, although both parties may consent. All other grounds for challenge may be waived by the party or parties in whose favor such grounds of challenge exist.

In this subsection, "legally blind" shall mean having not more than 20/200 of visual acuity in the better eye with correcting lenses, or visual acuity greater than 20/200 but with a limitation in the field of vision such that the widest diameter of the visual field subtends an angle no greater than 20 degrees.

(b) A challenge for cause may be made by the State for any of the following reasons:

(1) That the juror has conscientious scruples in regard to the infliction of the punishment of death for crime, in a capital case, where the State is seeking the death penalty;

(2) That he is related within the third degree of consanguinity or affinity, as determined under Chapter 573, Government Code, to the defendant; and

(3) That he has a bias or prejudice against any phase of the law upon which the State is entitled to rely for conviction or punishment.

(c) A challenge for cause may be made by the defense for any of the following reasons:

(1) That he is related within the third degree of consanguinity or affinity, as determined under Chapter 573, Government Code, to the person injured by the commission of the offense, or to any prosecutor in the case; and

(2) That he has a bias or prejudice against any of the law applicable to the case upon which the defense is entitled to rely, either as a defense to some phase of the offense for which the defendant is being prosecuted or as a mitigation thereof or of the punishment therefor.

CCP Art. 35.18. Other evidence on challenge

Upon a challenge for cause, the examination is not confined to the answers of the juror, but other evidence may be heard for or against the challenge.

CCP Art. 35.21. Judge to decide qualifications

The court is the judge, after proper examination, of the qualifications of a juror, and shall decide all challenges without delay and without argument thereupon.

§14:62 In General

A challenge for cause is an objection made to a juror for some legal reason that renders him unfit or incapable to serve on the jury. Castillo v. State, 913 S.W.2d 529 (Tex. Crim. App. 1995).

A party may properly challenge a veniremember for cause when he reveals that he will not be able to follow the applicable law in the case. The burden of demonstrating that a veniremember should be excused for cause rests with the party urging the excusal. Castillo. This must be done by a preponderance of the evidence. Fuller v. State, 829 S.W.2d 191 (Tex. Crim. App. 1992), *cert. denied,* 508 U.S. 941, 113 S.Ct. 2418, 124 L. Ed. 2d 640 (1993); Hernandez v. State, 757 S.W.2d 744 (Tex. Crim. App. 1988), *cert. denied,* 504 U.S. 974, 112 S.Ct. 2944 (1992), *overruled on other grounds.*

Before a veniremember can be challenged for cause, he must have the law explained to him and then state unequivocally that he cannot follow it. Chambers v. State, 903 S.W.2d 21 (Tex. Crim. App. 1995). A trial court will not be shown to have abused its discretion in failing to grant a challenge for cause where the law has not been fully explained to the veniremember. Chambers.

Similarly, a trial court does not abuse its discretion in overruling a challenge for cause where the defendant misinforms the veniremember of the applicable law. Garcia v. State, 887 S.W.2d 846 (Tex. Crim. App. 1994), *cert. denied*, 514 U.S. 1005, 115 S.Ct. 1317, 131 L. Ed. 2d 198 (1995).

PRACTICE TIP: Where a party has failed to inform the venire of the law before challenging a veniremember for cause for failure to follow the law, the opposing party should object that the question is not proper as to form. Howard v. State, 941 S.W.2d 102 (Tex. Crim. App. 1996).

Grounds for the exclusion of a veniremember may be shown by evidence other than the voir dire examination of the veniremember under Art. 35.18. Jernigan v. State, 661 S.W.2d 936 (Tex. Crim. App. 1983).

§14:63 Basis for Challenges for Cause

There are two opposing lines of cases pertaining to whether Art. 35.16 contains a comprehensive list of challenges for cause. Maldonado v. State, 998 S.W.2d 239 (Tex. Crim. App. 1999, fn.14).

One line of cases holds that Art. 35.16 which enumerates the reasons for challenges for cause is a complete list of challenges. Butler v. State, 830 S.W.2d 125 (Tex. Crim. App. 1992). The other line of cases holds that a challenge for cause may be predicated on facts that show that the prospective juror would be incapable or unfit to serve on the jury whether or not those facts fit within the enumerated factors of Art. 35.16. Mason v. State, 905 S.W.2d 570 (Tex. Crim. App. 1995), *cert.*

denied, 516 U.S. 1051, 116 S.Ct. 717, 133 L. Ed. 2d 670 (1996) (where the trial court excused a college freshman who stated he would be an unfit juror due to his preoccupation with school).

Regardless of which line of cases is now prevailing, and even under Mason, a trial court is not required to grant a challenge for cause not enumerated in CCP Art. 35.16. Rather, challenges not based upon a ground specifically enumerated in Art. 35.16 are addressed to the sound discretion of the trial judge. Mason; Maldonado. For example, where jury service could adversely affect the health of the veniremember the trial court has the discretion to excuse the veniremember. Goodwin v. State, 799 S.W.2d 719 (Tex. Crim. App. 1990), *cert. denied,* 501 U.S. 1259111 S.Ct. 2913, 115 L. Ed. 2d 1076 (1991).

A trial court will not abuse its discretion in failing to grant a challenge for cause based on a reason not expressly enumerated in CCP Art. 35.16. Maldonado.

There is also a doctrine of implied bias which states that since the Sixth Amendment promise of an impartial jury applies to the states, regardless of what Art. 35.16 provides, a state court would be bound to grant a challenge for cause if the failure to do so would result in a biased jury. Only extreme situations will justify a finding of implied bias, such as, perhaps, a situation where an assistant district attorney is on the venire of a case prosecuted by his office. State v. Morales, 253 S.W.3d 686 (Tex. Crim. App. 2008).

§14:64 Bias or Prejudice as Matter of Law

Where a party wishes to exclude a juror because of bias, it is the party seeking exclusion who must demonstrate, through questioning, that the potential juror lacks impartiality. Buntion, 482 S.W.3d 58, 84 (Tex. Crim. App. 2016).

If bias or prejudice is established as a matter of law, a challenge for cause must be granted by the trial court. Williams v. State, 773 S.W.2d 525 (Tex. Crim. App. 1988), *cert. denied,* 493 U.S. 900, 110 S.Ct. 257, 107 L. Ed. 2d 207 (1989).

Once a juror expressly admits his bias against a phase of law upon which both the State and defense are entitled to rely, a sufficient foundation has been laid to support a challenge for cause. Cardenas v. State, 325 S.W.3d 179 (Tex. Crim. App. 2010).

The opposing party or trial judge may then examine the juror further to ensure that he fully understands and appreciates the position that he is taking, but unless there is further clarification or vacillation by the juror, the trial judge must grant a challenge for cause if the juror persists in his position. Cardenas.

§14:64.1 Bias and Prejudice Defined

Bias is defined as an inclination toward one side of an issue rather than to the other, which leads to the natural inference that the veniremember will not act with impartiality. Anderson v. State, 633 S.W.2d 851 (Tex. Crim. App. 1982).

Bias against the law is the refusal to consider or apply the relevant law. It exists when a veniremember's beliefs or opinions would prevent or substantially impair the performance of his duties as a juror in accordance with his instructions and oath. Sadler v. State, 977 S.W.2d 140 (Tex. Crim. App. 1998).

Prejudice is defined as a pre-judgment. Anderson.

§14:64.2 Bias and Prejudice Established

The proper question to determine bias against the law is whether, in the proper case, where the facts justify it and the law allows it, the veniremember can fully and fairly apply the law as properly explained to him. Sadler v. State, 977 S.W.2d 140 (Tex. Crim. App. 1998).

Bias is established as a matter of law when a veniremember:

- Admits that he is biased for or against the defendant
- Admits or demonstrates prejudice toward a racial or ethnic class of which the defendant is a member
- Admits resentment towards the defendant because of some prior contact with the defendant
- Is related to the state's primary witness

Anderson v. State, 633 S.W.2d 851 (Tex. Crim. App. 1982).

A veniremember can also be found biased as a matter of law when he admits prejudice against persons who use intoxicating beverages in a case where the defendant is charged with an offense involving liquor. Morales v. State, 875 S.W.2d 724 (Tex.App.—Fort Worth 1994, *no pet.*).

A veniremember is biased as a matter of law when his answers conclusively establish bias. Banda v. State, 890 S.W.2d 42 (Tex. Crim. App. 1994), *cert. denied,* 515 U.S. 1105, 115 S.Ct. 2253, 132 L. Ed. 2d 260 (1995).

The standard for bias does not require that a juror's bias be proved with unmistakable clarity because many venire members simply cannot be asked enough questions to reach the point where their bias has been made unmistakably clear. Buntion v. State, 482 S.W.3d 58, 84 (Tex. Crim. App. 2016).

To establish that a challenge for cause is proper, the proponent of the challenge must show that the prospective juror understood the requirements of the law and could not overcome his prejudice well enough to follow the law. Buntion, 482 S.W.3d at 84.

Before a prospective juror may be excused for cause on this basis, the law must be explained to him, and he must be asked whether he can follow that law, regardless of his personal views. Buntion, 482 S.W.3d at 84.

When a veniremember is conclusively shown to be biased as a matter of law, he must be excused when challenged, even if he states that he can set aside his bias and provide a fair trial. Anderson.

§14:64.3 Bias and Prejudice Not Established

A veniremember is not biased as a matter of law when, despite an earlier *non-conclusive* adverse position, he states that he can follow the court's instructions and render a verdict according to the evidence. Adanandus v. State, 866 S.W.2d 210 (Tex. Crim. App. 1993), *cert. denied,* 510 U.S. 1215, 114 S.Ct. 1338, 127 L. Ed. 2d 686 (1994).

Under CCP Art. 35.16(a)(9) (challenge on the basis of bias or prejudice in favor of or against the defendant), it is not necessary to show a particular bias or prejudice in favor of or against the defendant. It is sufficient that the juror states that he cannot be a fair and impartial juror. Sosa v. State, 769 S.W.2d 909 (Tex. Crim. App.1989).

Bias is not conclusively established where the veniremember answers "I don't think so" when asked if he could follow the law. Moore v. State, 542 S.W.2d 664 (Tex. Crim. App. 1976), *cert. denied,* 431 U.S. 949, 97 S.Ct. 2666, 53 L. Ed. 2d 266 (1977).

Where the record contains conflicting statements by the veniremember, bias is not conclusively established. Banda v. State, 890 S.W.2d 42 (Tex. Crim. App. 1994), *cert. denied,* 515 U.S. 1105, 115 S.Ct. 2253, 132 L. Ed. 2d 260 (1995).

Showing that a potential juror is unable to articulate a set of facts upon which he could follow the law is not sufficient to establish bias as a matter of law. Hogue v. State, 711 S.W.2d 9 (Tex. Crim. App. 1986), *cert. denied*, 479 U.S. 922, 107 S.Ct. 329, 93 L. Ed. 2d 301 (1986).

A venire member's saying that he wants to hear evidence from the defense, and expects to hear such evidence, does not automatically disqualify him for having a bias or prejudice against the law placing the burden of proof on the State, or being unable to acquit the defendant unless the defense produced some evidence. Banda v. State, 890 S.W.2d 42, 57 (Tex. Crim. App. 1994).

§14:65 Substantial Impairment Test

When bias or prejudice is not established as a matter of law, the trial court has discretion to determine whether bias or prejudice actually exists to such a degree that the veniremember is disqualified. Little v. State, 758 S.W.2d 551 (Tex. Crim. App. 1988), *cert. denied,* 488 U.S. 934, 109 S.Ct. 328, 102 L. Ed. 2d 346 (1988).

The proper analysis to be used by the trial court to decide if a veniremember is challengeable for cause is whether his views would prevent or substantially impair his duty to act in accordance with the judge's instructions and the oath taken by the jury. Kemp v. State, 846 S.W.2d 289 (Tex. Crim. App. 1992), *cert. denied*, 508 U.S. 918, 113 S.Ct. 2361, 124 L. Ed. 2d 268 (1993).

The decision of a trial court regarding a challenge for cause will not be disturbed absent an abuse of discretion. Banda v. State, 890 S.W.2d 42 (Tex. Crim. App. 1994), *cert. denied,* 515 U.S. 1105,

115 S.Ct. 2253, 132 L. Ed. 2d 260 (1995). In other words, the judge's decision will be affirmed unless it is so clearly wrong as to lie outside the zone within which reasonable persons might disagree. Kemp.

In determining whether abuse of discretion exists in a situation where the trial court has denied a challenge for cause, the appellate court reviews all of the veniremember's answers to determine if they support the trial court's implied finding that the veniremember's views would not substantially impair the performance of his or her duties as a juror in accordance with the juror's oath. Post v. State, 936 S.W.2d 343 (Tex.App.—Fort Worth 1996, *pet. ref'd*). The trial court does not abuse its discretion in overruling a challenge for cause where the record shows that, although a potential juror initially expressed reservations about an area of the law, he ultimately testifies that he would follow the law. Cooks v. State, 844 S.W.2d 697 (Tex. Crim. App. 1992), *cert. denied,* 509 U.S. 927, 113 S.Ct. 3048, 125 L. Ed. 2d 732 (1993).

An isolated statement by a veniremember will not require that he be struck for cause when his testimony as a whole indicates that he was not biased. Cooks.

§14:66 Vacillating Veniremember

A vacillating veniremember is one who gives contradictory responses to those voir dire questions that test the veniremember's ability to follow the legal mandates set by the Legislature and by the holdings of the appellate courts. Howard v. State, 941 S.W.2d 102 (Tex. Crim. App. 1996).

Where the veniremember has vacillated between two positions or equivocated as to his views, appellate courts will give great deference to the decision of the trial court. This is because the trial court was in the best position to judge the veniremember's demeanor, tone and inflection of voice, hesitation or lack thereof, and body language. Banda v. State, 890 S.W.2d 42 (Tex. Crim. App. 1994), *cert. denied,* 515 U.S. 1105, 115 S.Ct. 2253, 132 L. Ed. 2d 260 (1995).

If a veniremember vacillates so that a trial court's ruling that the veniremember is subject to a challenge for cause can be supported by the record, then the trial court's ruling on the challenge will not be disturbed on appeal. Trevino v. State, 991 S.W.2d 849 (Tex. Crim. App. 1999); Green v. State, 840 S.W.2d 394 (Tex. Crim. App. 1992), *cert. denied,* 507 U.S. 1020, 113 S.Ct. 1819, 123 L. Ed. 2d 449 (1993), *disavowed on other grounds*.

Reviewing courts also give great deference to the trial court's ruling when the prospective juror's answers are unclear or contradictory. Maldonado v. State, 998 S.W.2d 239 (Tex. Crim. App. 1999); Colburn v. State, 966 S.W.2d 511 (Tex. Crim. App. 1998).

§14:67 Predicate for Challenge for Cause

Before a veniremember can be properly challenged for cause, the law must be fully explained to him. Jones v. State, 982 S.W.2d 386 (Tex. Crim. App. 1998), *cert. denied,* 120 S.Ct. 444, 145 L. Ed. 2d 362 (1999); Chambers v. State, 903 S.W.2d 21 (Tex. Crim. App. 1995). Once the law has been explained to the veniremember, he must be asked whether he can follow that law regardless of his personal views. Jones; Chambers.

If the veniremember states that he can follow the law regardless of his personal views, he is not challengeable for cause. Riley v. State, 889 S.W.2d 290 (Tex. Crim. App. 1993).

If the veniremember's views would substantially impair his ability to follow the law, then bias or prejudice exists to such a degree that the veniremember is disqualified. Little v. State, 758 S.W.2d 551 (Tex. Crim. App. 1988), *cert. denied,* 488 U.S. 934, 109 S.Ct. 328, 102 L. Ed. 2d 346 (1988).

§14:68 Areas of Proper Challenge for Cause

§14:68.1 Challenges for Cause by State or Defendant

The following are challenges for cause that can be made by either the state or the defendant. Unless otherwise noted, the defense would cite CCP Art. 35.16(c)(2) and the state would cite CCP Art. 35.16(b)(3) as statutory authority for their respective challenges for cause where veniremember:

- Unable to consider the minimum of the range of punishment—Nethery v.

State, 692 S.W.2d 686 (Tex. Crim. App. 1985); Cardenas v. State, 325 S.W.3d 179 (Tex. Crim. App. 2010).

- Unable to consider the full range of punishment—Banda v. State, 890 S.W.2d 42 (Tex. Crim. App. 1994), *cert. denied*, 515 U.S. 1105, 115 S.Ct. 2253, 132 L. Ed. 2d 260 (1995); Cardenas v. State, 325 S.W.3d 179 (Tex. Crim. App. 2010).
- Unable to consider the full range of punishment for a defendant who might be found guilty as a principle or as a party—Johnson v. State, 982 S.W.2d 403 (Tex. Crim. App. 1998); Cardenas v. State, 325 S.W.3d 179 (Tex. Crim. App. 2010).
- Unable to consider probation for a defendant who is eligible—Pierce v. State, 696 S.W.2d 899 (Tex. Crim. App. 1985), *cert. denied*, 496 U.S. 912, 110 S.Ct. 2603, 110 L. Ed. 2d 283 (1990); Cardenas v. State, 325 S.W.3d 179 (Tex. Crim. App. 2010).
- Unable to disregard an illegally obtained confession—Phillips v. State, 701 S.W.2d 875 (Tex. Crim. App. 1985)
- Holds the defendant's invocation of his Fifth Amendment rights against him—Guerra v. State, 771 S.W.2d 453 (Tex. Crim. App. 1988), *cert. denied*, 492 U.S. 925, 109 S.Ct. 3260, 106 L. Ed. 2d 606 (1989)
- Unable to hold the state to its burden of proof—Wheatfall v. State, 882 S.W.2d 829 (Tex. Crim. App. 1994)
- Served as a witness in the case at any phase of the trial—Wyle v. State, 777 S.W.2d 709 (Tex. Crim. App. 1989); CCP Art. 35.16(a)(6).
- Deafness would impair his service as a juror—Nobles v. State, 843 S.W.2d 503 (Tex. Crim. App. 1992); CCP Art. 35.16(a)(5)
- Low intelligence would affect his ability to serve as a juror—Burton v. State, 805 S.W.2d 564 (Tex.App.—Dallas 1991, *pet. ref'd*); CCP Art. 35.16(a)(5)
- Inability to read and write would affect his service as a juror—Allridge v. State, 850 S.W.2d 471 (Tex. Crim. App. 1991), *cert. denied*, 510 U.S. 831, 114 S.Ct. 101, 126 L. Ed. 2d 68 (1993); Flores v. State, 871 S.W.2d 714 (Tex. Crim. App. 1993); Stillwell v. State, 466 S.W.3d 908, 912 (Tex. App.—Fort Worth 2015, no pet.). CCP Art. 35.16(a)(11)
- Family situation would prevent her from making a decision in the case rendering her less than a fair and impartial juror—Sosa v. State, 769 S.W.2d 909 (Tex. Crim. App. 1989)
- Biased against a class of witnesses—Perry v. State, 864 S.W.2d 794 (Tex. App.—Fort Worth 1993, *no pet.*)
- Biased as a matter of law in favor of a class of witnesses, e.g., believes unequivocally that a police officer would never lie—Lane v. State, 822 S.W.2d 35 (Tex. Crim. App. 1991), *cert. denied*, 504 U.S. 920, 112 S.Ct. 1968, 118 L. Ed. 2d 568 (1992)
- Unable to consider circumstantial evidence—Narvaiz v. State, 840 S.W.2d 415 (Tex. Crim. App. 1992), *cert. denied*, 507 U.S. 975, 113 S.Ct. 1422, 122 L. Ed. 2d 791 (1993)
- Unable to presume the defendant innocent; challengeable for cause under Article 35.16(c)(2) for having a bias or prejudice against the law—Banda v. State, 890 S.W.2d 42 (Tex. Crim. App. 1994), *cert. denied*, 515 U.S. 1105, 115 S.Ct. 2253, 132 L. Ed. 2d 260 (1995)
- Under CCP Art. 35.16(a)(10), has established in mind a conclusion as to the guilt or innocence of the defendant as would influence him in his action in finding a verdict. However, a prospective juror is not challengeable for cause merely because he has heard news reports about the crime or the suspect. Heiselbetz v. State, 906 S.W.2d 500 (Tex. Crim. App. 1995); Macias v. State, 733 S.W.2d 192 (Tex. Crim. App. 1987), *cert. denied*, 484 U.S. 1079, 108 S.Ct. 1059, 98 L. Ed. 2d 1021 (1988)
- Unable to avoid consideration of a factor not properly before the jury (such as parole considerations in a death penalty case)—Felder v. State, 758 S.W.2d 760 (Tex. Crim. App. 1988), *cert. denied*, 440 U.S. 950, 99 S.Ct. 1433, 59 L. Ed. 2d 640 (1979).

- Under CCP Art. 35.16(a)(8), Cannot impartially judge the credibility of witnesses—Hernandez v. State, 563 S.W.2d 947 (Tex. Crim. App. 1978)
- Would consider the indictment as some evidence of guilt—Homan v. State, 662 S.W.2d 372 (Tex. Crim. App. 1984)
- Unable to return a not guilty verdict based upon a "technicality"—White v. State, 779 S.W.2d 809 (Tex. Crim. App. 1989) (where the veniremember stated that he would find the defendant guilty even if the state had not proven that the offense had occurred on or about the date alleged in indictment)
- Incapable of understanding the difference in the burdens of proof extant in the justice system—Staley v. State, 887 S.W.2d 885 (Tex. Crim. App. 1994), *cert. denied*, 514 U.S. 1020, 115 S.Ct. 1366, 131 L. Ed. 2d 222 (1995)
- Not willing or not able to state whether or not he could follow the law—Soria v. State, 933 S.W.2d 46 (Tex. Crim. App. 1996)
- Not a qualified voter in the county of the trial—Mayo v. State, 4 S.W.3d 9 (Tex. Crim. App. 1999); Gaona v. State, 733 S.W.2d 611 (Tex.App.—Corpus Christi 1987, *pet. ref'd*); CCP Art. 35.16(a)(1)
- Has been convicted of theft—Holland v. State, 761 S.W.2d 307 (Tex. Crim. App. 1988), *cert. denied,* 489 U.S. 1091, 109 S.Ct. 1560, 103 L. Ed. 2d 863 (1989); CCP Art. 35.16(a)(2)
- Has been convicted of a felony—Chambers v. State, 903 S.W.2d 21 (Tex. Crim. App. 1995); CCP Art. 35.16(a)(2)
- Under indictment or other legal accusation for theft—Brown v. State, 716 S.W.2d 939 (Tex. Crim. App. 1986); CCP Art. 35.16(a)(3).
- Under indictment or other legal accusation for a felony—State v. Holloway, 886 S.W.2d 482 (Tex. App.—Houston [1st Dist.] 1984, *pet. ref'd*); CCP Art.35.16(a)(3)
- Is insane—Von Byrd v. State, 569 S.W.2d 883 (Tex. Crim. App. 1978); CCP Art. 35.16(a)(4)
- Served on the grand jury that found the indictment—CCP Art. 35.16(a)(7)
- Served on the petit jury in a former trial of the same case—CCP Art. 35.16(A)(8)

The following are based on grounds not specifically enumerated in CCP Art. 35.16, where veniremember:

- Would suffer adverse health affect—Goodwin v. State, 799 S.W.2d 719 (Tex. Crim. App. 1990), *cert. denied*, 501 U.S. 1259111 S.Ct. 2913, 115 L. Ed. 2d 1076 (1991)
- Not able to keep his mind on the trial such that his ability to serve effectively is impaired—Johnson v. State, 773 S.W.2d 322 (Tex. Crim. App. 1989), affirmed at 509 U.S. 350, 113 S.Ct. 2658, 125 L. Ed. 2d 290 (1993)
- Distracted by business or economic reasons to the extent that ability to serve would be impaired—Allridge v. State, 850 S.W.2d 471 (Tex. Crim. App. 1991), *cert. denied*, 510 U.S. 831, 114 S.Ct. 101, 126 L. Ed. 2d 68 (1993)
- Unfit for jury service because such service would have conflicted with plans to get married rendering veniremember preoccupied with personal problems—Nichols v. State, 754 S.W.2d 185 (Tex. Crim. App. 1988)
- Commitments to children represented an insurmountable burden to serving on the jury. Rogers v. State, 774 S.W.2d 247 (Tex. Crim. App. 1989), *cert. denied*, 493 U.S. 984, 110 S.Ct. 519, 107 L. Ed. 2d 520 (1989)
- So traumatized by the experience of being called to jury duty that physically shakes, may be properly struck for cause as unfit despite her assurances that she will survive the "crisis." Powell v. State, 897 S.W.2d 307 (Tex. Crim. App. 1994)

The state is allowed to avail itself of each of the challenges available to the defendant because the courts recognize that the state has a duty to see that justice is done in individual cases. Morrow v. State, 910 S.W.2d 471 (Tex. Crim. App. 1995), *cert. denied,* 517 U.S. 1192, 116 S.Ct. 1683, 134 L. Ed. 2d 784 (1996). However, a defendant is not entitled to excuse for cause a veniremember who has a bias against some phase of the law that the state is entitled to rely on because defense counsel has no corresponding duty to see that justice is done. Morrow, (where the defendant challenged a veniremember who wanted to hold the state to a higher burden than beyond a reasonable doubt).

§14:68.1.1 Bias Against Class of Witnesses

Although a veniremember is challengeable for cause if he cannot impartially judge the credibility of witnesses (*see* Hernandez, §14:68.1), he is not challengeable for cause where he would merely be skeptical of the testimony of a certain class of witnesses. The veniremember's testimony overall must show that he has a bias toward that class of witnesses. Jones v. State, 982 S.W.2d 386 (Tex. Crim. App. 1998), *cert. denied,* 120 S.Ct. 444, 145 L. Ed. 2d 362 (1999).

Furthermore, a veniremember is not challengeable for cause simply because he would give certain classes of witnesses a slight edge in terms of credibility; again, the edge in credibility must amount to an extreme or absolute position before the veniremember is challengeable. Ladd v. State, 3 S.W.3d 547 (Tex. Crim. App. 1999).

Where a veniremember's deference to the testimony of a law enforcement officer was not so much to any unfailing honesty he perceived on the part of all policemen so much as his own personal knowledge of their investigatory training and experience, he is not challengeable for cause for failure to fairly evaluate the testimony of a police officer. Gardner v. State, 730 S.W.2d 675 (Tex. Crim. App. 1987), *cert. denied,* 484 U.S. 905, 108 S.Ct. 248, 98 L. Ed. 2d 206 (1987).

§14:68.1.2 Failure to Consider Punishment Range

Both the State and the defense are entitled to jurors who can consider the entire range of punishment for the particular statutory offense—*i.e.,* from the maximum to the minimum and all points in between. Jurors must be able to consider both a situation in which the minimum penalty would be appropriate and a situation in which the maximum penalty would be appropriate. Therefore, both sides may question the panel on the range of punishment and may commit jurors to consider the entire range of punishment for the statutory offense. Cardenas v. State, 325 S.W.3d 179 (Tex. Crim. App. 2010).

A juror who states that he cannot consider the minimum or maximum punishment for a particular statutory offense is subject to a challenge for cause. Cardenas.

The requirement that the juror be able to consider the full punishment range for the offense includes consideration of each statutory variation of the offense, such as the age element in an aggravated sexual assault of a child case. Cardenas (where the offense was aggravated sexual assault of a child under 14).

Where a veniremember would only consider the minimum of the punishment range in a certain narrow circumstance, he is challengeable for cause. This is because the veniremember, in effect, would be creating his own statute concerning the punishment range. Sunday v. State, 745 S.W.2d 436 (Tex.App.—Beaumont 1988, *pet. ref'd*).

A trial court does not commit error in overruling a challenge for cause against a veniremember who is unable to articulate a set of facts meriting a minimum punishment because the inability to articulate such a set of facts does not equate to an inability to consider the full range of punishment. Penry v. State, 903 S.W.2d 715 (Tex. Crim. App. 1995), *cert. denied,* 516 U.S. 977, 116 S.Ct. 480, 133 L. Ed. 2d 408 (1995).

A veniremember is not challengeable as a matter of law if she states that she could consider probation in a proper case. Faulder v. State, 745 S.W.2d 327 (Tex. Crim. App. 1987). A veniremember is not required to state that he could consider probation in a given fact situation. Faulder.

A veniremember is not challengeable as a matter of law if he states that he could consider probation after he had heard all of the evidence. Rische v. State, 834 S.W.2d 942 (Tex.App.—Houston [14th Dist.] 1992, *pet. ref'd*).

A veniremember is not challengeable for cause merely because she will use the facts of the case to determine punishment. Sadler v. State, 977 S.W.2d 140 (Tex. Crim. App. 1998). This is even though a party is allowed to make inquiry of the venire as to whether they would decide the case (especially the punishment issues) any differently based on the status of the victim (where the victim was a nun);

and Maddux v. State, 862 S.W.2d 590 (Tex. Crim. App. 1993)(where the victim was a child).

A veniremember is not challengeable for cause where he is unable to consider the full range of punishment for a particular fact situation where he states that he can consider the full range for the offense as defined by law. Sadler.

A trial court does not err in refusing a challenge for cause for grounds that do not become relevant at trial. Soria v. State, 933 S.W.2d 46 (Tex. Crim. App. 1996) (where the veniremember could not consider the punishment range for the offense of murder but where murder was not raised as a lesser included offense).

§14:68.1.3 Relationship With Party or Witness

A casual acquaintance with a crime victim or with the facts of the case will not disqualify a juror who answers on voir dire that those circumstances will not affect his ability to render a verdict according to the evidence. Little v. State, 758 S.W.2d 551 (Tex. Crim. App. 1988), *cert. denied,* 488 U.S. 934, 109 S.Ct. 328, 102 L. Ed. 2d 346 (1988).

Similarly, a mere acquaintance with the defendant or a state's witness is not a sufficient basis for disqualification. Vaughn v. State, 833 S.W.2d 180 (Tex.App.—Dallas 1992, *pet. ref'd*).

The mere fact that a veniremember had been employed by the District Attorney's office that was prosecuting the case, without more, was not sufficient to disqualify the veniremember. Eubanks v. State, 635 S.W.2d 568 (Tex.App.—Houston [1st Dist.] 1982, *no pet.*).

§14:68.2 Challenges for Cause by Defendant

CCP Art. 35.16(c)(2) permits the defendant to challenge for cause a veniremember on the ground "[t]hat he has a bias or prejudice against any of the law applicable to the case upon which the defendant is entitled to rely, either as a defense to some phase of the offense for which the defendant is being prosecuted or as a mitigation thereof or of the punishment therefor." Cordova v. State, 733 S.W.2d 175 (Tex. Crim. App. 1987), *cert. denied,* 487 U.S. 1240, 108 S.Ct. 2915, 101 L. Ed. 2d 946 (1988).

In addition to the grounds listed in section 14:68.1 above, the following have been recognized as proper areas for defense challenges for cause:

- Inability to afford the defendant his right against self-incrimination—Montoya v. State, 810 S.W.2d 160 (Tex. Crim. App. 1989)
- Inability to consider the law of self-defense—Harris v. State, 784 S.W.2d 5 (Tex. Crim. App. 1989), *cert. denied,* 494 U.S. 1090, 110 S.Ct. 1837, 108 L. Ed. 2d 966 (1990)
- Biased against the defendant—Garcia v. State, 919 S.W.2d 370 (Tex. Crim. App. 1996); Curry v. State, 910 S.W.2d 490 (Tex. Crim. App. 1995)
- Inability to be objective in the case because of knowledge and biases gained from her employment—Sykora v. State, 872 S.W.2d 323 (Tex.App.—Beaumont 1994, *no pet.*)
- Would automatically believe a police officer—Hernandez v. State, 563 S.W.2d 947 (Tex. Crim. App. 1978)
- Related within the third degree of consanguinity or affinity to the injured party or prosecutor—CCP Art. 35.16(c)(1)

The state is allowed to avail itself of each of the challenges available to the defendant because the courts recognize that the state has a duty to see that justice is done in individual cases. Morrow v. State, 910 S.W.2d 471 (Tex. Crim. App. 1995), *cert. denied,* 517 U.S. 1192, 116 S.Ct. 1683, 134 L. Ed. 2d 784 (1996). However, a defendant is not entitled to excuse for cause a veniremember who has a bias against some phase of the law that the state is entitled to rely on because defense counsel has no corresponding duty to see that justice is done. Morrow, (where the defendant challenged a veniremember who wanted to hold the state to a higher burden than beyond a reasonable doubt).

Under Art. 35.16(a)(9) (challenge on the basis of bias or prejudice in favor of or against the defendant), it is not necessary to show a particular bias or prejudice in favor of or against the defendant. It is sufficient that the juror states that he cannot be a fair and impartial juror. Sosa v. State, 769 S.W.2d 909 (Tex. Crim. App.1989).

§14:68.2.1 Preservation of Error for Denial of Defense Challenges for Cause

In order to preserve error in a trial court's erroneous denial of a defense challenge for cause, the defendant must follow a rigorous predicate including:

- Voir dire of the individual veniremember was recorded and transcribed—Long v. State, 823 S.W.2d 259 (Tex. Crim. App. 1991), *cert. denied*, 505 U.S. 1224, 112 S.Ct. 3042, 120 L. Ed. 2d 910 (1992)
- Counsel asserted a clear and specific challenge for cause clearly articulating the grounds therefore—Miller v. State, 741 S.W.2d 382 (Tex. Crim. App. 1987), *cert. denied,* 486 U.S. 1061, 108 S.Ct. 2835, 100 L. Ed. 2d 935 (1988)
- Counsel used a peremptory challenge on that particular veniremember—Jones v. State, 833 S.W.2d 118 (Tex. Crim. App. 1992), *cert. denied,* 507 U.S. 921, 113 S.Ct. 1285, 122 L. Ed. 2d 678 (1993)
- Defense exhausted all peremptory challenges—Adanandus v. State, 866 S.W.2d 210 (Tex. Crim. App. 1993), *cert. denied,* 510 U.S. 1215, 114 S.Ct. 1338, 127 L. Ed. 2d 686 (1994)
- Trial judge denied request for additional peremptory challenges—Jacobs v. State, 787 S.W.2d 397 (Tex. Crim. App. 1990), *cert. denied,* 496 U.S. 882, 111 S.Ct. 231, 112 L. Ed. 2d 185 (1990)
- Defense was forced to accept an objectionable veniremember on the jury because of the trial court's erroneous actions, *i.e.*, that she would have exercised a peremptory challenge against that veniremember if she had one remaining—Trevino v. State, 991 S.W.2d 849 (Tex. Crim. App. 1999); Green v. State, 840 S.W.2d 394 (Tex. Crim. App. 1992), *cert. denied,* 507 U.S. 1020, 113 S.Ct. 1819, 123 L. Ed. 2d 449 (1993), *disavowed on other grounds.*

PRACTICE TIP: In order to preserve error, recite into the record prior to the jury being seated and sworn, that:

- Defense was forced to use a peremptory challenge on a juror against whom a challenge for cause had been made
- Defense exhausted all allotted peremptory challenges
- Defense requested additional peremptory challenges and that request has been denied (or if additional peremptory challenges were granted, that those also have been exhausted and an additional request has been made and denied); and
- An objectionable juror has been seated and that defense would exercise a peremptory challenge on that juror if afforded an additional peremptory.

Then request an additional peremptory challenge and (presumably) have the request denied.

This harm analysis is the same under Rule 44.2(b) of the Texas Rules of Appellate Procedure. Johnson v. State, 43 S.W.3d 1 (Tex. Crim. App. 2001).

The defendant must demonstrate that he exercised a peremptory challenge on the veniremember upon which he bases his claim. Brosky v. State, 915 S.W.2d 120 (Tex.App.—Fort Worth, *pet. ref'd*), *cert. denied*, 519 U.S. 1020, 117 S.Ct. 537, 136 L. Ed. 2d 422 (1996).

In order to do this, the defendant must generally make the jury strike lists part of the record on appeal. Hernandez v. State, 938 S.W.2d 503 (Tex. App.—Waco 1997, *pet. ref'd*).

The defendant cannot show that he was harmed by the trial court's erroneous denial of his challenge for cause where he uses one of his peremptory challenges on a veniremember out of the "strike zone" and leaves the objectionable juror unchallenged. Comeaux v. State, 445 S.W.3d 745, 751-2 (Tex. Crim. App. 2014).

In order to make a certain claim on appeal, the defendant must have objected on the same ground in the trial court. Trevino v. State, 991 S.W.2d 849 (Tex. Crim. App. 1999); Green v. State, 840 S.W.2d 394 (Tex. Crim. App. 1992), *cert. denied,* 507 U.S.

1020, 113 S.Ct. 1819, 123 L. Ed. 2d 449 (1993), *disavowed on other grounds*.

It is not necessary that the defendant go so far as to explain why the objectionable juror that he has named is, in fact, objectionable. Garcia v. State, 887 S.W.2d 846 (Tex. Crim. App. 1994), *cert. denied,* 514 U.S. 1005, 115 S.Ct. 1317, 131 L. Ed. 2d 198 (1995).

Where the trial court grants two additional peremptory challenges, the defendant must show that the trial court erred in denying his challenge for cause to three prospective jurors. Penry v. State, 903 S.W.2d 715 (Tex. Crim. App. 1995), *cert. denied,* 516 U.S. 977, 116 S.Ct. 480, 133 L. Ed. 2d 408 (1995). The defendant also must have exhausted any additional peremptory challenges granted in order to preserve error for appeal. Penry.

Voir dire error in a non-capital case cannot be error affecting the punishment phase only under CCP Art. 44.29(b); therefore, the defendant has the right to begin a retrial in the guilt-innocence phase. Carson v. State, 6 S.W.3d 536 (Tex. Crim. App. 1999).

§14:68.3 Challenges for Cause by State

CCP Art. 35.16(b)(3) allows the state to challenge for cause veniremembers who have a "bias or prejudice against any phase of the law upon which the state is entitled to rely for conviction or punishment." Unless otherwise noted, the grounds listed below are based on Art. 35.16(b)(3).

In addition to the grounds listed in section §14:68.1, the following have been recognized as valid reasons for a state to challenge a veniremember for cause:

- Holding the state to a higher burden than proof beyond a reasonable doubt—Crane v. State, 786 S.W.2d 338 (Tex. Crim. App. 1990)
- Requiring the state to prove premeditation—Crane.
- Requiring the state to prove motive—Esquivel v. State, 595 S.W.2d 516 (Tex. Crim. App. 1980), *cert. denied,* 449 U.S. 986, 101 S.Ct. 408, 66 L. Ed. 2d 251 (1980)
- Being related to the defendant within the third degree of consanguinity or affinity—Moss v. State, 850 S.W.2d 788 (Tex.App.—Houston [14th Dist.] 1993, *pet. ref'd*). CCP Art. 35.16(b)(2)
- Being unable to consider accomplice testimony. May v. State, 738 S.W.2d 261 (Tex. Crim. App. 1987), *cert. denied,* 484 U.S. 872, 108 S.Ct. 206, 98 L. Ed. 2d 158 (1987)
- Being incapable of convicting in a case without eyewitness testimony—White v. State, 779 S.W.2d 809 (Tex. Crim. App. 1989), *cert. denied,* 495 U.S. 962, 110 S.Ct. 2575, 109 L. Ed. 2d 757 (1990); Marquez v. State, 725 S.W.2d 217 (Tex. Crim. App.1987), *cert. denied*, 484 U.S. 872, 108 S.Ct. 201, 98 L. Ed. 2d 152 (1987)
- Being incapable of convicting in a case on the basis of a confession only. Butler v. State, 872 S.W.2d 227 (Tex. Crim. App. 1994), *cert. denied,* 513 U.S. 1157, 115 S.Ct. 1157, 130 L. Ed. 2d 1079 (1995)
- Requiring the state to prove a specific fact in addition to the elements of the crime before considering the maximum punishment is challengeable for cause—Fuller v. State, 829 S.W.2d 191 (Tex. Crim. App. 1992), *cert. denied,* 508 U.S. 941, 113 S.Ct. 2418, 124 L. Ed. 2d 640 (1993) (where the veniremember stated he would require the state to prove another murder before he could answer the future dangerousness special issue in the affirmative)
- Being unable to consider a defendant's guilt under the law of parties—Nichols v. State, 754 S.W.2d 185 (Tex. Crim. App. 1988), *cert. denied,* 488 U.S. 1019, 109 S.Ct. 819, 102 L. Ed. 2d 808 (1989)
- Being incapable of convicting even if believes one eyewitness beyond a reasonable doubt—Caldwell v. State, 818 S.W.2d 790 (Tex. Crim. App. 1991), *cert. denied*, 503 U.S. 990, 112 S.Ct. 1684, 118 L. Ed. 2d 399 (1992)
- States that without medical testimony to corroborate the complainant, he would not vote to convict even if the complainant's testimony established guilt beyond a reasonable doubt—

Decker v. State, 894 S.W.2d 475 (Tex. App.—Austin 1995, *pet. ref'd*)

- Clearly and unequivocally affirms that she could not clearly and fairly evaluate the testimony of a police officer—Sawyers v. State, 724 S.W.2d 24 (Tex. Crim. App. 1986)
- Being unable to give fair consideration to the maximum of the range of punishment. Cardenas v. State, 325 S.W.3d 179 (Tex. Crim. App. 2010).

TIP TO THE PROSECUTION: A challenge for cause has a greater likelihood of being granted by the trial court and upheld by an appellate court if it is premised upon a veniremember's inability to convict even where he believes the witness, evidence or other scenario beyond a reasonable doubt. Caldwell, supra.

The state is allowed to avail itself of each of the challenges available to the defendant because the courts recognize that the state has a duty to see that justice is done in individual cases. Morrow v. State, 910 S.W.2d 471 (Tex. Crim. App. 1995), *cert. denied,* 517 U.S. 1192, 116 S.Ct. 1683, 134 L. Ed. 2d 784 (1996). However, a defendant is not entitled to excuse for cause a veniremember who has a bias against some phase of the law that the state is entitled to rely on because defense counsel has no corresponding duty to see that justice is done. Morrow (where the defendant challenged a veniremember who wanted to hold the state to a higher burden than beyond a reasonable doubt).

§14:68.3.1 Challenges Where Beyond Reasonable Doubt Not Enough

Where the courts allow exclusion of veniremembers because they would require eyewitness testimony, medical testimony, etc., it is because the veniremember would be, in effect, requiring additional elements in the state's burden of proof. Howard v. State, 941 S.W.2d 102 (Tex. Crim. App. 1996).

However, a venireman who requires a particular type of evidence has not demonstrated an inability to abide by the law if his requirement is predicated upon his personal threshold of reasonable doubt. The state has the burden to show that the venireman's insistence on evidence of a particular type will prevent him from following the law regardless of whether he was otherwise convinced beyond a reasonable doubt, before it can be said it has met its burden to demonstrate the venireman cannot follow the law. Howard; Murphy v. State, 112 S.W.3d 192 (Tex. Crim. App. 2003).

The rationale behind allowing challenges for cause against veniremembers who would require a specific type of evidence before they could convict is that they are raising the state's burden of proof beyond that required by law. As such, they demonstrate a bias against some phase of the law on which the state is entitled to rely (that being the juror's obligation to return a verdict of guilty where the state proves its case beyond a reasonable doubt). Caldwell v. State, 818 S.W.2d 790 (Tex. Crim. App. 1991), *cert. denied*, 503 U.S. 990, 112 S.Ct. 1684, 118 L. Ed. 2d 399 (1992). In other words, factors other than those prescribed by law must not be made absolute prerequisites for imposition of any punishment, including the death penalty. Drew v. State, 743 S.W.2d 207 (Tex. Crim. App. 1987).

Contrast this situation with the classic Garrett situation where a veniremember who needs more than the minimum quantum of evidence (such as more than one witness) before he can find guilt beyond a reasonable doubt is not challengeable for cause. *See,* Garrett v. State, 851 S.W.2d 853 (Tex. Crim. App. 1993).

The difference in the above two situations is that in one situation (Caldwell) the veniremember would not convict even where the case were proven beyond a reasonable doubt, whereas in the second (Garrett) situation the veniremember would not find proof beyond a reasonable doubt.

A juror who says that he will always have a reasonable doubt in a one-witness case is not challengeable for cause on that ground. Zinger v. State, 932 S.W.2d 511 (Tex. Crim. App. 1996). This is because he needs more than the minimum quantum of evidence. *See*, Garrett.

A trial court commits no error in overruling a challenge for cause lodged against a potential juror who believes that a minimum or lesser punishment should be assessed only in a proper case. Fuller v. State, 829 S.W.2d 191 (Tex. Crim. App. 1992), *cert. denied,* 508 U.S. 941, 113 S.Ct. 2418, 124 L. Ed. 2d 640 (1993).

Conversely, a juror who says that even if he believes the one witness beyond a reasonable doubt he cannot convict can be successfully challenged for cause. Coleman v. State, 881 S.W.2d 344 (Tex. Crim. App. 1994), *cert. denied,* 513 U.S. 1096, 115 S.Ct. 763, 130 L. Ed. 2d 660 (1995). This is because the veniremember would be holding the state to a higher burden than proof beyond a reasonable doubt.

A venireman whose definition of a particular term requires a certain type of evidence is not subject to a challenge for cause. Murphy (the venireman's definition of the phrase "criminal acts of violence" required evidence that the defendant committed or attempted to commit other murders before the venireman could answer the future dangerousness special issue in the affirmative in a death penalty case).

The rule that a venire man who cannot convict upon the testimony of a single witness even where he believes that witness beyond a reasonable doubt is challengeable for cause remains in effect after the decision in the Standefer case. Lee v. State, 206 S.W.3d 620 (Tex. Crim. App. 2006).

§14:68.3.2 Preserving Error for Erroneous Granting of State's Challenge for Cause

To show error in the trial court's grant of a State's challenge for cause, a defendant must lodge a timely objection in the trial court. Kemp v. State, 846 S.W.2d 289 (Tex. Crim. App. 1992), *cert. denied*, 508 U.S. 918, 113 S.Ct. 2361, 124 L. Ed. 2d 268 (1993).

The objection is timely where it is lodged at the time of the trial court's ruling and is not abandoned. Zimmerman v. State, 860 S.W.2d 89 (Tex. Crim. App. 1993).

The defendant must next show either that the trial court applied the wrong legal standard in sustaining the challenge or that the trial court abused its discretion in applying the correct legal standard. Vuong v. State, 830 S.W.2d 929, 943 (Tex. Crim. App. 1992), *cert. denied*, 506 U.S. 997, 113 S.Ct. 595, 121 L. Ed. 2d 533 (1992).

Moreover, to obtain a reversal because of the error, a defendant must show that the error actually deprived him of a fair and impartial jury. Jones v. State, 982 S.W.2d 386 (Tex. Crim. App. 1998), *cert. denied,* 120 S.Ct. 444, 145 L. Ed. 2d 362 (1999).

Voir dire error in a non-capital case cannot be error affecting the punishment phase only under CCP Art. 44.29(b); therefore, the defendant has the right to begin a retrial in the guilt-innocence phase. Carson v. State, 6 S.W.3d 536 (Tex. Crim. App. 1999).

§14:70 *Sua Sponte* Excusals

§14:71 Applicable Statute

CCP Art. 35.03. Excuses

> *Sec. 1. Except as provided by Sections 2 and 3 of this article, the court shall then hear and determine excuses offered for not serving as a juror, including any claim of an exemption or a lack of qualification, and if the court considers the excuse sufficient, the court shall discharge the prospective juror or postpone the prospective juror's service to a date specified by the court, as appropriate.*
>
> *Sec. 2. Under a plan approved by the commissioners court of the county in the same manner as a plan is approved for jury selection under Section 62.011, Government Code, in a case other than a capital felony case, the court's designee may hear and determine an excuse offered for not serving as a juror, including any claim of an exemption or a lack of qualification. The court's designee may discharge the prospective juror or postpone the prospective juror's service to a date specified by the court's designee, as appropriate, if:*
>
> > *(1) the court's designee considers the excuse sufficient; and*
> >
> > *(2) the juror submits to the court's designee a statement of the ground of the exemption or lack of qualification or other excuse.*

§14:72 In General

Art. 35.03 affords the trial court wide latitude in determining sufficiency of excuses and discharging veniremembers. Ellason v. State, 815 S.W.2d 656 (Tex. Crim. App. 1991). Pursuant to Art. 35.03, a trial judge has broad discretion in excusing prospective jurors on any proper basis, either with or without the prompting of counsel. Kemp v. State, 846 S.W.2d 289 (Tex. Crim. App. 1992), *cert. denied*, 508 U.S. 918, 113 S.Ct. 2361, 124 L. Ed. 2d 268 (1993).

Article 35.03 provides no enumerated bases for excusing a potential juror; but rather recognizes that there are varying circumstances attendant in each new venire. Butler v. State, 830 S.W.2d 125 (Tex. Crim. App. 1992).

PRACTICE TIP: Where defense counsel opposes the trial court granting an excuse under Art. 35.03, it should be requested that the trial court make findings of fact and conclusions of law on the excuse to preserve the issue for appeal. This also will help prevent the trial court from having unfettered discretion in these matters.

The trial judge is not permitted to excuse and replace a juror after the jury has been sworn (except where there is a qualified alternate available). Gentry v. State, 881 S.W.2d 35 (Tex.App.—Dallas 1994, *pet. ref'd*).

A trial judge has no authority to use Art. 35.03 to excuse a prospective juror who is only subject to a challenge for cause under Art. 35.16. Warren v. State, 768 S.W.2d 300 (Tex. Crim. App.1989); Green v. State, 764 S.W.2d 242 (Tex. Crim. App. 1989).

However, the trial court may on its own motion excuse a prospective juror who is absolutely disqualified pursuant to CCP Art. 35.19. Warren.

Thus, if a prospective juror has been convicted or indicted for theft or any felony or is insane, the trial court has the authority to excuse the juror *sua sponte*. Green.

Also, if the trial court had a proper basis to excuse a veniremember under Art. 35.03, it is immaterial that the court cited Art. 35.16 for its actions. Coleman v. State, 881 S.W.2d 344 (Tex. Crim. App. 1994), *cert. denied,* 513 U.S. 1096, 115 S.Ct. 763, 130 L. Ed. 2d 660 (1995).

The erroneous excusal of a veniremember will call for reversal only if the record shows that the error deprived the defendant of a lawfully constituted jury. The question is whether or not the jurors who actually sat were impartial. Gamboa v. State, 296 S.W.3d 574 (Tex. Crim. App. 2009).

§14:73 Permissible Reasons for Sua Sponte Excusal

The following veniremember excuses have been recognized as valid under Art. 35.03:

- Concern over missing work—Butler v. State, 830 S.W.2d 125 (Tex. Crim. App. 1992)
- Upcoming wedding of the daughter—McFarland v. State, 845 S.W.2d 824 (Tex. Crim. App. 1992), *cert. denied,* 508 U.S. 963, 113 S.Ct. 2937, 124 L. Ed. 2d 686 (1992)
- Starting new job in a different county—Harris v. State, 784 S.W.2d 5 (Tex. Crim. App. 1989), *cert. denied,* 494 U.S. 1090, 110 S.Ct. 1837, 108 L. Ed. 2d 966 (1990)
- Health problems—Ellason v. State, 815 S.W.2d 656 (Tex. Crim. App. 1991)
- Mother's health problems—Leberta v. State, 770 S.W.2d 828 (Tex.App.—Fort Worth 1988, *pet. ref'd*)
- Wife's upcoming surgery—Kemp v. State, 846 S.W.2d 289 (Tex. Crim. App. 1992), *cert. denied*, 508 U.S. 918, 113 S.Ct. 2361, 124 L. Ed. 2d 268 (1993)
- Child care problems—Rousseau v. State, 855 S.W.2d 666 (Tex. Crim. App. 1993), *cert. denied,* 510 U.S. 919, 114 S.Ct. 313, 126 L. Ed. 2d 260 (1993)
- Responsibility to care for grandchildren—Johnson v. State, 773 S.W.2d 322 (Tex. Crim. App. 1989)
- Failure to return timely from a break—Brossette v. State, 885 S.W.2d 841 (Tex.App.—Dallas 1994, *pet. ref'd*)
- Upcoming job performance review. Brossette.
- Discomfort in hearing a murder case coupled with expectation of pressure from neighbors and residents. Murray

v. State, 861 S.W.2d 47 (Tex.App.—Texarkana 1993, *pet. ref'd*)

- Scheduled out of town vacation—Moody v. State, 827 S.W.2d 875 (Tex. Crim. App. 1992), *cert. denied,* 506 U.S. 839, 113 S.Ct. 119, 121 L. Ed. 2d 75 (1992); Crutsinger v. State, 206 S.W.3d 607 (Tex. Crim. App.)
- Where absolutely disqualified pursuant to CCP Art. 35.19. Warren v. State, 768 S.W.2d 300 (Tex. Crim. App. 1989)
- Where veniremember was the only doctor available to treat children at a clinic—Clark v. State, 881 S.W.2d 682 (Tex. Crim. App. 1994)
- Concerned about finances to the point that service would be impaired—Teague v. State, 864 S.W.2d 505 (Tex. Crim. App. 1993), *overruled on other grounds*, Robertson v. State, 871 S.W.2d 701 (Tex. Crim. App. 1993)
- Distracted because of family responsibilities—Saldivar v. State, 980 S.W.2d 475 (Tex.App.—Houston [14th Dist.] 1998, *pet. ref'd*)
- Cannot read and write—Liveoak v. State, 717 S.W.2d 691 (Tex.App.—San Antonio 1986, *pet. ref'd*)
- (Under Art. 35.03) because of ongoing campaign for public office—Narvaiz v. State, 840 S.W.2d 415 (Tex. Crim. App. 1992), *cert. denied,* 507 U.S. 975, 113 S.Ct. 1422, 122 L. Ed. 2d 791 (1993)
- That the juror is hearing impaired—Black v. State, 26 S.W.3d 895 (Tex. Crim. App. 2000)

§14:74 Timing of Sua Sponte Excusal

The power to grant an excusal from jury service pursuant to Art. 35.03 inheres to the trial judge from the first assemblage of the array until the juror is seated and sworn. Butler v. State, 830 S.W.2d 125 (Tex. Crim. App. 1992).

A trial judge may *sua sponte* excuse a veniremember who is absolutely disqualified. Martinez v. State, 621 S.W.2d 797 (Tex. Crim. App. 1981).

The trial judge has the authority to use Art. 35.03 to allow a veniremember to claim an exemption from jury service even after the veniremember has been questioned on voir dire. Rousseau v. State, 855 S.W.2d 666 (Tex. Crim. App. 1993), *cert. denied,* 510 U.S. 919, 114 S.Ct. 313, 126 L. Ed. 2d 260 (1993).

Since the trial court does not need consent of the parties to *sua sponte* excuse a venireman, before the panel is sworn the judge may *sua sponte* excuse a venireman out of the presence of the state and defendant. Wright v. State, 28 S.W.3d 526 (Tex. Crim. App. 2000).

§14:75 Preserving Error for *Sua Sponte* Excusal

Where the trial court erroneously excuses an otherwise qualified veniremember under Art. 35.03, the defendant must make a specific and proper objection and then call the attention of the trial court to the specific complaint to be raised on appeal. Little v. State, 758 S.W.2d 551 (Tex. Crim. App. 1988), *cert. denied,* 488 U.S. 934, 109 S.Ct. 328, 102 L. Ed. 2d 346 (1988); Alvarado v. State, 822 S.W.2d 236 (Tex.App.—Houston [14th Dist.], 1991, *pet. ref'd*).

The defendant also must assert that he has a legitimate objection to the jury chosen. Murray v. State, 861 S.W.2d 47 (Tex.App.—Texarkana 1993, *pet. ref'd*).

Any error in the excusal of a qualified veniremember will not result in reversal unless the defendant shows that he has been injured. Cooks v. State, 844 S.W.2d 697 (Tex. Crim. App. 1992), *cert. denied,* 509 U.S. 927, 113 S.Ct. 3048, 125 L. Ed. 2d 732 (1993).

If a *disqualified* prospective juror (*i.e.*, one who is subject to a challenge for cause, but otherwise qualified) is *sua sponte* excused by the trial court, the defendant must take the following steps to preserve error and establish harm:

- Object to the excusal of the juror
- At the conclusion of the voir dire, claim that he is to be tried by a jury to which he has a legitimate objection (and have a sufficient basis in the record for the complaint)
- Specifically identify the juror or jurors of whom he is complaining
- Exhaust all of peremptory challenges and request additional peremptory challenges (and show that the trial

court denied the request for additional peremptory challenges)

- Green v. State, 764 S.W.2d 242 (Tex. Crim. App. 1989).

PRACTICE TIP: In order to preserve error in the case of veniremembers who were summoned for jury duty but did not appear, the defendant must request writs of attachment for those veniremembers. Coleman v. State, 881 S.W.2d 344 (Tex. Crim. App. 1994), cert. denied, 513 U.S. 1096, 115 S.Ct. 763, 130 L. Ed. 2d 660 (1995). See §14:14.1, §14:14.2 for sample writs of attachment.

If a *qualified* juror is improperly excused *sua sponte*, the defendant can show harm and thus reversible error by establishing that the state exhausted its peremptory challenges. Green, *supra*.

Voir dire error in a non-capital case cannot be error affecting the punishment phase only under CCP Art. 44.29(b); therefore, the defendant has the right to begin a retrial in the guilt-innocence phase. Carson v. State, 6 S.W.3d 536 (Tex. Crim. App. 1999).

§14:80 Trial Court Questioning of Venire

Because the ultimate decision to dismiss potential jurors for cause is a heavy responsibility, the trial judge must be allowed to ask questions to clarify a veniremember's answers and make intelligent and informed decisions. Heiselbetz v. State, 906 S.W.2d 500 (Tex. Crim. App. 1995).

The court may ask questions in order to clarify a juror's position or instruct a juror. Gardner v. State, 730 S.W.2d 675 (Tex. Crim. App. 1987), *cert. denied,* 484 U.S. 905, 108 S.Ct. 248, 98 L. Ed. 2d 206 (1987). Usually such intervention is warranted for purposes of clarification and expedition. Gardner.

If a clear, potential source of bias is evident from the record in the case, the trial court bears the burden of making inquiry to discern whether actual bias exists and whether it exists to such a degree that it requires the excusal of the veniremember. Turner v. State, 671 S.W.2d 679 (Tex.App.—Dallas 1984, *pet. ref'd*).

§14:81 Comments by Trial Court in Front of Venire

§14:81.1 In General

CCP Art. 38.05. Judge Shall Not Discuss Evidence

> *In ruling upon the admissibility of evidence, the judge shall not discuss or comment upon the weight of the same or its bearing in the case, but shall simply decide whether or not it is admissible; nor shall he, at any stage of the proceeding previous to the return of the verdict, make any remark calculated to convey to the jury his opinion of the case.*

As a general rule, the trial court must refrain from commenting on the weight of the evidence in the jury's presence during any stage of the proceeding. CCP Art. 38.05; Livingston v. State, 782 S.W.2d 112 (Tex.App.—Dallas 1989).

A trial court's comments constitute reversible error only where they are reasonably calculated to benefit the state or prejudice the defendant. Livingston; Sharpe v. State, 648 S.W.2d 705 (Tex. Crim. App. 1983).

To determine whether the comment is either reasonably calculated to benefit the State or to prejudice the defendant, an appellate court must first examine whether the trial court's statement was material to the case. Clark v. State, 878 S.W.2d 224 (Tex.App.—Dallas 1994, *no pet.*); Burge v. State, 443 S.W.2d 720 (Tex. Crim. App.1969), *cert. denied*, 396 U.S. 934, 90 S.Ct. 277, 24 L. Ed. 2d 233 (1969).

A statement is material if it:

- Implies approval of the State's argument
- Indicates any disbelief in the defense's position; or
- Diminishes the credibility of the defense's approach to its case

Clark.

Generally, an instruction by the judge to disregard any comment made by him is sufficient to cure any error. Fletcher v. State, 960 S.W.2d 694 (Tex.App.—Tyler 1997); Marks v. State, 617 S.W.2d 250 (Tex. Crim. App. 1981).

TIP TO THE BENCH: Trial judges are advised to be very careful when making any superfluous statement during trial as jurors are prone to seize with alacrity upon any conduct or language of the trial judge which they may interpret as shedding light upon his view of the weight of the evidence, or the merits of the issues involved. Jones v. State, 788 S.W.2d 834 (Tex.App.—Dallas 1990, no pet.).

§14:81.2 Comments That Are Not Error

A trial judge's inadvertent reference to the complainant in the case as "the victim" was not erroneous in a case where the defendant has pled guilty before the jury. Cooper v. State, 910 S.W.2d 605 (Tex.App.—Tyler 1995, *no pet.*).

Where the trial court overruled a defense objection to testimony as hearsay and gave an instruction that, "[t]his testimony is being admitted by the court not for the truth of the statement but for the fact he was so told," the judge was properly explaining the law under Texas Rule of Evidence 103(B) regarding the admission of the statement. Rosales v. Texas, 932 S.W.2d 530 (Tex.App.—Tyler 1995, *pet. ref'd*).

Although it is ill-advised in the presence of the jury, numerous cases have held that when a judge chastises attorneys during a trial for unacceptable conduct it does not violate the rule against court comments. Rosales; Liveoak v. State, 717 S.W.2d 691 (Tex.App.—San Antonio 1986), *pet. ref'd* at 741 S.W.2d 451 (Tex. Crim. App. 1987) ("get on with the case"); Betancourt v. State, 657 S.W.2d 451 (Tex.App.—Corpus Christi 1983, *pet. ref'd*) (arguing with court).

A trial court's comment that, "I don't know if you need to do that anymore for appellate purposes" was not erroneous because it was not calculated to benefit State or prejudice the defendant. Rodela v. State, 829 S.W.2d 845 (Tex.App.—Houston [1st Dist.] 1992, *pet. ref'd*).

It is proper for a trial court to state for the record what he observed in the courtroom if it is done without editorial comment. Rankin v. State, 872 S.W.2d 279 (Tex.App.—Houston [14th Dist.] 1994), *vacated and remanded on other grounds,* 974 S.W.2d 707 (Tex. Crim. App. 1996).

Where a trial judge made comments expressing his dislike of vigorous cross examination of child victims he did not err because, although ill-advised, his remarks were directed at the defense attorney rather than the defendant. Latson v. State, 807 S.W.2d 372 (Tex.App.—Houston [14th Dist.] 1991).

§14:81.3 Comments That Are Error

A trial judge's comments before voir dire of the jury that, "I think that the evidence will show that he [defendant] went to her [victim's] house looking for her daughter and the State's evidence will probably show that Sara Whitehead [victim] was killed somewhere during that time..." were calculated to, and probably did, convey to the jury the court's opinion of the case on a pivotal issue. Graham v. State, 624 S.W.2d 785 (Tex.App.—Fort Worth 1981, *no pet.*).

A trial judge's comments in front of the jury, "[t]hat's fine. You can try that on appeal," were found to be erroneous, although cured by a strong instruction to disregard. Hernandez v. State, 507 S.W.2d 209 (Tex. Crim. App. 1974).

Where the trial judge informed the venire that the indictment alleged that the defendant had two prior convictions, she violated the defendant's right to a fair and impartial trial. Clark v. State, 878 S.W.2d 224 (Tex.App.—Dallas 1994, *no pet.*).

A trial judge committed reversible error at the punishment phase of a trial by instructing the jury to find as a matter of law that the enhancement paragraphs in the indictment were true, where the defendant had entered a plea of not true. Lewis v. State, 814 S.W.2d 513 (Tex.App.—Houston [14th Dist.] 1991, *pet. ref'd*).

A trial court committed error by referring, in the presence of the jury, to the State's expert witness as "a credible witness," but the comment was harmless in light of the court's instruction to disregard. Tennison v. State, 814 S.W.2d 484 (Tex. App.—Waco 1991, *no pet.*).

A trial judge's comment upon admitting evidence that, "I now find that there is evidence of a conspiracy," was reversible error where the defendant was charged with a drug conspiracy offense. Bachus

v. State, 803 S.W.2d 402, (Tex.App.—Dallas 1991, *pet. ref'd*). A trial judge's comment at the time a photo was admitted into evidence that, "I find that the probative value of State's Exhibit 4 greatly outweighs any potential prejudice. I admit 4," is an improper comment. Fletcher v. State, 960 S.W.2d 694 (Tex. App.—Tyler 1997); Marks v. State, 617 S.W.2d 250 (Tex. Crim. App. 1981).

A trial judge's comments to the jury making it clear that he believed that the case would require a punishment hearing was a reversible comment on the weight of the evidence. Jones v. State, 788 S.W.2d 834 (Tex.App.—Dallas 1990, *no pet.*). A trial court's statement that the state's argument was a "reasonable inference from the evidence" constituted reversible error. Dixon v. State, 775 S.W.2d 795 (Tex.App.—El Paso 1989, *pet. ref'd*).

A trial court's comments to the venire to the effect that the defendant had been contemplating the state's plea bargain offer and that defendants may not testify if they are guilty were found to be erroneous even though there was no objection to any of the comments. Blue v. State, 41 S.W.3d 129 (Tex. Crim. App. 2000) (plurality opinion).

A trial court's comment that he would further consider a matter "... as soon as I hear from the defendant ..." is reversible as a comment on the defendant's failure to testify. Bustamante v. State, 48 S.W.3d 761 (Tex. Crim. App. 200).

TIP TO THE BENCH: In carrying out its duties in ruling upon a defendant's objection, a trial court must not embellish the ruling with an unwarranted comment. Smith v. State, 595 S.W.2d 120 (Tex. Crim. App. 1980).

§14:81.4 Complaints About Court's Questions or Comments

In order to preserve error for appeal, the defendant must lodge a timely and specific objection at trial. Ladd v. State, 3 S.W.3d 547 (Tex. Crim. App. 1999); Sattiewhite v. State, 786 S.W.2d 271 (Tex. Crim. App. 1989). The objection must be made at the earliest opportunity. Wheatfall v. State, 882 S.W.2d 829 (Tex. Crim. App. 1994).

As in cases where the defendant is complaining of an improper argument, he should ask that the venire be instructed to disregard the remark. Livingston v. State, 782 S.W.2d 112 (Tex.App.—Dallas 1989). The defendant must show that members of the jury heard the remark. Nelson v. State, 881 S.W.2d 97 (Tex.App.—Houston [1st Dist.] 1994, *pet. ref'd*).

PRACTICE TIP: Where an improper comment is made during individual voir dire, exercise a peremptory challenge on the veniremember who heard the remark. Where the improper comment was made to the entire venire, immediately move to quash the venire. Nelson.

Improper comments made by the trial court to the venire are subject to harm analysis. Unkart v. State, 400 S.W.3d 94, 99 (Tex. Crim. App. 2013). The reviewing court will view the voir dire in its entirety in determining if the defendant has been harmed. Unkart.

Voir dire error in a non-capital case cannot be error affecting the punishment phase only under CCP Art. 44.29(b); therefore, the defendant has the right to begin a retrial in the guilt-innocence phase. Carson v. State, 6 S.W.3d 536 (Tex. Crim. App. 1999).

§14:90 Withholding Of Information By Veniremember

When a juror withholds material information during voir dire that the defense, using due diligence, could not uncover, the parties are denied the opportunity to exercise their challenges, which hinders their selection of an impartial jury. Franklin v. State, 138 S.W.3d 351, 355–56 (Tex. Crim. App. 2004); State v. Gutierrez, 541 S.W.3d 91, 99-100 (Tex. Crim. App. 2017).

A violation, by itself, is insufficient for reversal. The defendant must also have been harmed by that constitutional violation. Franklin, 138 S.W.3d at 358; State v. Gutierrez, 541 S.W.3d at 100;Tex. R. App. P. 44.2(a).

Error occurs where a prejudiced or biased juror is selected without fault or lack of diligence on the part of defense counsel, where defense counsel has acted in good faith on the juror's responses and has no knowledge of their accuracy. Gonzales v. State, 3 S.W.3d 915 (Tex. Crim. App. 1999); Brandon v. State, 599 S.W.2d 567 (Tex. Crim. App. 1979).

A defendant is entitled to a reversal where a juror has withheld information if two criteria are met: (1) the omission is material; and, (2) the defendant has exercised due diligence in eliciting that information. Jones v. State, 596 S.W.2d 134 (Tex. Crim. App. 1980).

Under these circumstances, a reversal is merited because withholding material information denies the defendant the opportunity to knowingly and intelligently exercise peremptory challenges, hence jeopardizing his right to a fair trial before an impartial jury. Salazar v. State, 562 S.W.2d 480 (Tex. Crim. App. 1978).

Where the juror's information is provided by way of a written questionnaire, counsel has a duty to follow up with additional questions to test the accuracy of the written response. Gonzales, *supra*.

§14:91 Material Information

It is not dispositive of the materiality issue that the juror did not intentionally mislead the defendant during voir dire because the good faith of a juror is largely irrelevant when considering the materiality of information withheld. Franklin v. State, 12 S.W.3d 473 (Tex. Crim. App. 2000).

The information need not prove that the juror is actually biased, but it must have a tendency to show bias. Franklin v. State, 138 S.W.3d 351, 355–56 (Tex. Crim. App. 2004).

If the withheld information has a tendency to show bias, an appropriate procedure is to hold a hearing at which evidence should be adduced regarding whether the juror is actually biased. Uranga v. State, 330 S.W.3d 301, 306 (Tex. Crim. App. 2010).

If a trial judge finds that the juror is not actually biased, and that finding is supported by the record, then the defendant has not been harmed by the violation of his constitutional right to an impartial jury. State v. Gutierrez, 541 S.W.3d 91, 100 (Tex. Crim. App. 2017).

If, however, the judge finds that the juror is actually biased, the only remedy is a mistrial. Franklin, 138 S.W.3d at 354.

Appellate courts review the trial court's determination of historical fact for an abuse of discretion (i.e., whether the juror was actually biased), and appellate courts should give almost total deference to the trial court's determination if it is supported by the record. State v. Gutierrez, 541 S.W.3d at 100.

Material information includes information regarding a juror's relationship with any party in a criminal proceeding. Von January v. State, 576 S.W.2d 43 (Tex. Crim. App. 1978). However, mere familiarity with a witness is not necessarily material information. Decker v. State, 717 S.W.2d 903 (Tex. Crim. App. 1986).

Material information also includes the fact that the veniremember has an extensive criminal history. State v. Read, 965 S.W.2d 74 (Tex.App.—Austin 1988, *no pet.*). The defendant should be allowed to pursue questioning of the juror to determine if withheld information is material. Franklin.

§14:92 Defendant's Burden

Defense counsel has an obligation to ask questions calculated to bring out information which might be said to indicate a juror's inability to be impartial and truthful. Jones v. State, 596 S.W.2d 134 (Tex. Crim. App. 1980).

Defense counsel has a duty to follow up on information contained in jury questionnaires; and if he does not, he cannot later complain that a veniremember withheld information or lied on the questionnaire. Gonzales v. State, 3 S.W.3d 915 (Tex. Crim. App. 1999).

Where defense counsel fails to fulfill his obligation to ask questions which will reveal that answers on a questionnaire are incorrect or to otherwise discover material information, there has been a lack of diligence. In this situation, appellate courts will find that the juror did not withhold any information and will not reverse the case. Gonzalez.

Counsel must ask specific questions and not rely on broad ones to satisfy this obligation. Brandon v. State, 599 S.W.2d 567 (Tex. Crim. App. 1979).

The defendant will not be found to have carried his burden where it appears from the record that the veniremember did not understand the question as propounded. Hernandez v. State, 952 S.W.2d 59 (Tex.App.—Austin 1997), *vacated on other grounds*, 957 S.W.2d 851 (Tex. Crim. App.1998). In order for the defendant to carry his burden to show that a veniremember withheld information about a criminal history, he must show that the veniremember knew of the situation. Hernandez.

PRACTICE TIP: Where the failure of a prospective juror to reveal information during voir dire comes to light during trial, ask for an immediate hearing, ask to question the juror on the matter, develop as extensive a record as possible and request a mistrial. See, Salazar v. State, 562 S.W.2d 480 (Tex. Crim. App. 1978).

Where a juror brings new information to light during trial and the trial court does not allow additional questioning of the juror, a defendant preserves error by propounding to the court specific areas of questions that he would ask the juror. Franklin v. State, 12 S.W.3d 473 (Tex. Crim. App. 2000).

TIP TO THE BENCH: If a juror brings new material information to light during a trial, the trial court should allow the defendant to conduct further voir dire examination of the juror to determine if bias exists. See, Franklin.

§14:93 Personal Information About Jurors and Juror Privacy Considerations

Art. 35.29. PERSONAL INFORMATION ABOUT JURORS. (a) Except as provided by Subsections (b) and (c), information collected by the court or by a prosecuting attorney during the jury selection process about a person who serves as a juror, including the juror's home address, home telephone number, social security number, driver's license number, and other personal information, is confidential and may not be disclosed by the court, the prosecuting attorney, the defense counsel, or any court personnel.

(b) On application by a party in the trial, or on application by a bona fide member of the news media acting in such capacity, to the court for the disclosure of information described by Subsection (a), the court shall, on a showing of good cause, permit disclosure of the information sought.

(c) The defense counsel may disclose information described by Subsection (a) to successor counsel representing the same defendant in a proceeding under Article 11.071 without application to the court or a showing of good cause.

§14:100 Peremptory Challenges

A defendant's constitutional rights are not violated when he chooses to use a peremptory challenge against a venireman that the trial court should have excused for cause because peremptory challenges are not of federal constitutional dimension. U.S. v. Martinez-Salazar, 528 U.S. 304, 120 S.Ct. 774, 145 L.Ed.2d 792 (2000).

§14:101 Statutory Law

CCP Art. 35.14. A Peremptory Challenge

A peremptory challenge is made to a juror without assigning any reason therefor.

CCP Art. 35.15. Number of Challenges

(a) In capital cases in which the State seeks the death penalty both the State and defendant shall be entitled to fifteen peremptory challenges. Where two or more defendants are tried together, the State shall be entitled to eight peremptory challenges for each defendant; and each defendant shall be entitled to eight peremptory challenges.

(b) In non-capital felony cases and in capital cases in which the State does not seek the death penalty, the State and defendant shall each be entitled to ten peremptory challenges. If two or more defendants are tried together each defendant shall be entitled to six

peremptory challenges and the State to six for each defendant.

(c) The State and the defendant shall each be entitled to five peremptory challenges in a misdemeanor tried in the district court and to three in the county court, or county court at law. If two or more defendants are tried together, each defendant shall be entitled to three such challenges and the State to three for each defendant in either court.

(d) The State and the defendant shall each be entitled to one peremptory challenge in addition to those otherwise allowed by law if one or two alternate jurors are to be impaneled and two peremptory challenges if three or four alternate jurors are to be impaneled. The additional peremptory challenges provided by this subsection may be used against an alternate juror only, and the other peremptory challenges allowed by law may not be used against an alternate juror.

CCP Art. 35.25. Making Peremptory Challenge

In non-capital cases and in capital cases in which the State's attorney has announced that he will not qualify the jury for, or seek the death penalty, the party desiring to challenge any juror peremptorily shall strike the name of such juror from the list furnished him by the clerk.

§14:102 When Peremptory Challenges Are Made

In non-capital cases, peremptory challenges are made after all the veniremembers have been examined. CCP Art. 35.25. The trial court shall allow adequate time for the making of peremptory challenges. Thomas v. State, 658 S.W.2d 175 (Tex. Crim. App. 1983).

Where an error has been made in the exercise of peremptory challenge or the trial court makes an error in impaneling the jury, it is incumbent on the aggrieved party to object before the jury is sworn or the error is waived. Truong v. State, 782 S.W.2d 904 (Tex.App.—Houston [14th Dist.] 1989, *pet. ref'd*).

A trial court's erroneous denial of a defendant's peremptory challenge to the seating of a juror where all seated jurors are qualified and unbiased, does not require automatic reversal of the defendant's conviction under the Due Process clause. Rivera v. Illinois, 556 U.S. ___, 129 S.Ct. 1446, 173 L.Ed.2d 320 (2009) (trial court had disallowed the defendant's peremptory because of a Batson violation).

§14:103 Forms for Peremptory Challenges

See our companion book *Texas Criminal Forms* (James Publishing) for the following sample forms:

- Form 14-13, Motion for Additional Peremptory Challenges.
- Form 14-15, Peremptory Challenge Tally Form.

§14:110 Batson v. Kentucky Hearings

CCP Art. 35.261. Peremptory Challenges Based on Race Prohibited

(a) After the parties have delivered their lists to the clerk under Article 35.26 of this code and before the court has impaneled the jury, the defendant may request the court to dismiss the array and call a new array in the case. The court shall grant the motion of a defendant for dismissal of the array if the court determines that the defendant is a member of an identifiable racial group, that the attorney representing the state exercised peremptory challenges for the purpose of excluding persons from the jury on the basis of their race, and that the defendant has offered evidence of relevant facts that tend to show that challenges made by the attorney representing the state were made for reasons based on race. If the defendant establishes a prima facie case, the burden then shifts to the attorney

representing the state to give a racially neutral explanation for the challenges. The burden of persuasion remains with the defendant to establish purposeful discrimination.

(b) If the court determines that the attorney representing the state challenged prospective jurors on the basis of race, the court shall call a new array in the case.

United States Constitution Fourteenth Amendment. "... nor shall any State deprive any person of life, liberty, or property, without due process of law, nor deny to any person within its jurisdiction the equal protection of the laws."

§14:111 Introduction and History

More than 115 years ago, the Supreme Court decided in Strauder v. West Virginia, 100 U.S. 303, 25 L.Ed. 664 (1880), that the exclusion of blacks from grand and petit juries in criminal proceedings violated the Equal Protection Clause of the Fourteenth Amendment, but the fact that a particular jury or a series of juries does not statistically reflect the racial composition of the community does not in itself make out an invidious discrimination forbidden by the Clause. Instead, a defendant in a criminal case is entitled to require that the State not deliberately and systematically deny to members of his race the right to participate as jurors in the administration of justice. Alexander v. Louisiana, 405 U.S. 625, 92 S.Ct. 1221, 31 L.Ed.2d 536 (1972).

§14:111.1 Systematic Exclusion Standard

In Swain v. Alabama, 380 U.S. 202, 85 S.Ct. 824, 13 L.Ed.2d 759 (1965), the Supreme Court recognized that a state's purposeful or deliberate denial to blacks of their right to participate as jurors in the administration of justice on the basis of race violates the Equal Protection Clause. Swain stood for the proposition that a black defendant could make out a *prima facie* case of purposeful discrimination by proving that the peremptory challenge system was being perverted by denying blacks the same right and opportunity to participate in the administration of justice enjoyed by the white population. However, at that time the Supreme Court further stated that peremptory challenges remained available to the prosecution in a given case to remove veniremen of a particular race, religion, or nationality for tactical reasons.

In order for a defendant to overcome the Swain presumption that peremptory challenges had not been used in a manner offensive to the Equal Protection Clause, he had to prove that the prosecution systematically used peremptory challenges to remove minority veniremen from service in repeated cases. A number of lower appellate courts interpreted Swain to require a repeated use of peremptory challenges over a number of cases as necessary to establish a violation of the Equal Protection Clause. Hence a "systematic exclusion" was the burden defendants shouldered in order to assert their rights. This burden required an emphasis on the history of jury selection practices by prosecutors in the jurisdiction, rather than a focus on the jury selection factors in the individual case.

§14:111.2 Transitional Cases

Since Swain, the Supreme Court has recognized that a defendant may make a *prima facie* showing of purposeful racial discrimination in the selection of his jury by relying solely on the jury selection practices in his particular case. Alexander v. Louisiana, 405 U.S. 625, 92 S.Ct. 1221, 31 L.Ed.2d 536 (1972). Subsequent cases then began to define the framework for what eventually became the court inquiry of Batson v. Kentucky, *infra*. Especially important to the formulation of the holding in Batson is the presumption that peremptory challenges can constitute a jury selection practice that permits those who are of a mind to discriminate to do so. Avery v. Georgia, 345 U.S. 559, 73 S.Ct. 891, 97 L.Ed.2d 1244 (1953).

The courts have specifically held that racially motivated peremptory challenges do not violate the Sixth Amendment fair cross-section requirements. Holland v. Illinois, 493 U.S. 474, 110 S.Ct. 803, 107 L.Ed.2d 905 (1990); Seubert v. State, 787 S.W.2d 68 (Tex. Crim. App.1990). The fair cross-section requirements of the Sixth Amendment govern only the assembly of a fair venire, not the petit jury itself. Holland and Seubert, *supra*.

§14:112 Batson and Beyond

§14:112.1 Batson Case

On April 30, 1986, the Supreme Court handed down its opinion in Batson v. Kentucky, 476 U.S. 79, 106 S.Ct. 1712, 90 L.Ed.2d 69 (1986). Batson requires trial courts to be sensitive to the racially discriminatory use of peremptory challenges and forces the mandate of equal protection while furthering the ends of justice. The Court reasoned that, due to the heterogeneous population of our nation, public respect for our criminal justice system and the rule of law will be strengthened by ensuring that no citizen is disqualified from jury service because of his race. The significance of the Batson case is that a defendant can now establish a *prima facie* case of purposeful discrimination based solely upon the prosecutor's exercise of peremptory challenges at the defendant's trial. No longer is it necessary to demonstrate any type of pattern or system of challenges outside of the case on trial.

The Batson case does not go so far as to hold that the Equal Protection Clause is violated when racially neutral criteria in the peremptory challenge process results in a disproportionate impact on minority jurors.

Batson further does not limit a party's right to challenge a venireman for cause. Tompkins v. State, 774 S.W.2d 195 (Tex. Crim. App. 1987), affirmed at 490 U.S. 754, 109 S.Ct. 2180, 104 L.Ed.2d 834 (1989).

Batson laid out a framework of inquiry which will be discussed below. The Batson Court did not go beyond this skeletal framework. It did not formulate particular procedures to be followed after a proper objection because of the variety of jury selection practices followed in the state and federal courts. This has, of course, spawned recurrent litigation to establish the practices to be followed in the individual jurisdictions.

§14:112.2 Non-Minority Defendants

In Powers v. Ohio, 499 U.S. 400, 111 S.Ct. 1364, 113 L.Ed.2d 411 (1991), the Supreme Court held that a defendant has standing to assert the equal protection rights of veniremen who are excused as jurors through the discriminatory use of peremptory challenges. Because this rationale does not require racial identity between the defendant and the illegally excused venireman, Powers has the effect of allowing non-minority defendants to object to the prosecution's racially motivated peremptory challenges. This rule was applied to Texas cases through Tex.Code Crim.Proc.Ann. Art. 35.261 (Vernon 1989) in State v. Oliver, 808 S.W.2d 492 (Tex. Crim. App. 1991). *See also*, Mead v. State, 819 S.W.2d 869 (Tex. Crim. App. 1991).

§14:112.3 Any Party

The Supreme Court has further expanded the scope of the Batson decision to prohibit a criminal defendant from engaging in purposeful racial discrimination in the exercise of peremptory challenges. Georgia McCollum, 505 U.S. 42, 112 S.Ct. 2348, 120 L.Ed.2d 33 (1992), on remand 422 S.E.2d 866. The Court found that the harms discussed in Batson are also inflicted by a defendant's racially discriminatory use of peremptory challenges. They found that, because a petit jury is a governmental body, the racially discriminatory use of peremptory challenges by any party invokes the Equal Protection Clause. Thus, the prosecution has third-party standing to challenge a defendant's discriminatory use of peremptory challenges. A prohibition against the discriminatory exercise of peremptory challenges does not violate a criminal defendant's constitutional rights because it is an affront to justice to argue that the right to a fair trial includes the right to discriminate against a group of citizens based upon their race. The only procedural mandate set out in McCollum was a Batson-type inquiry.

§14:112.4 Juvenile Proceedings

The principles of Batson derived through the Equal Protection Clause apply to civil proceedings which include juvenile adjudication proceedings. R.X.F. v. State, 921 S.W.2d 888 (Tex. App.—Waco 1996).

§14:112.5 Gender-Based Discrimination

The Supreme Court has also decided that Batson principles should apply to gender-based, as well as racial, discrimination. *See* J.E.B. v. Alabama ex rel. T.B., 5 11 U.S. 127, 114 S.Ct. 1419, 128 L.Ed.2d 89 (1994), on remand 641 So.2d 821, in which the State of Alabama used 9 of its 10 peremptory challenges to strike men in petitioner's paternity suit.

The Court of Criminal Appeals applied J.E.B., *supra,* to a Texas case in Fritz v. State, 946 S.W.2d 844 (Tex. Crim. App. 1997). The Court found that the prosecution violated the equal protection clause by using seven of its peremptory challenges to remove males.

§14:112.6 Religious Beliefs

Reasoning that "[t]o hold... that a veniremember may not be excluded on account of his religious preference is tantamount to a holding that he may not be struck on account of his beliefs[,]" the Texas Court of Criminal Appeals has declined to extend Batson principles to the use of peremptory challenges based on religious preference. Casarez v. State, 913 S.W.2d 468 (Tex. Crim. App. 1994) (*op. on rehearing*).

Thus, the trial court does not err in failing to grant a new trial where the State singles out Catholics for exclusion from the jury panel. Devoe v. State, 354 S.W.3d 457 (Tex. Crim. App. 2011).

§14:113 Texas Procedure

To codify and implement Batson in Texas, the legislature enacted Tex. Code Crim. Proc. Art. 35.261, which is intended to create uniform procedures and remedies to address claimed constitutional violations during jury selection. State v. Oliver, 808 S.W.2d 492 (Tex. Crim. App. 1991).

A trial court's Batson rulings are reviewed on appeal under the clearly erroneous standard of review. Gibson v. State, 144 S.W.3d 530 (Tex. Crim. App. 2004).

§14:113.1 Timing of Objections

Since the codification of Batson principles in Art. 35.261, the Texas courts have held that an objection to the jury composition must be made prior to the jury being impaneled. Rousseau v. State, 824 S.W.2d 579 (Tex. Crim. App. 1992). A jury is considered "impaneled" when the members have been both selected and sworn. Price v. State, 782 S.W.2d 266 (Tex.App. Beaumont 1989). This principle controls over prior cases which required the objection to also precede discharge of the remaining panel members. Hill v. State, 827 S.W.2d 860 (Tex. Crim. App. 1992). In a capital murder case where individual *voir dire is* being conducted, the objection will be timely when made immediately after the peremptory challenge in question is exercised. Rousseau v. State, *supra.*

§14:113.2 Specificity of Objections

A Batson objection must be reasonably specific in order to merit review. Invoking Batson, Swain, Article 35.261, or mentioning the practice of using peremptory challenges to exclude minorities will usually be sufficient to call for review. Merely objecting that a defendant has a federal and state constitutional right to a cross section of his peers is not sufficient. Rosales v. State, 841 S.W.2d 368 (Tex. Crim. App. 1992).

An objection was held insufficient in Bacy v. State, 827 S.W.2d 91 (Tex.App. Fort Worth 1992 *pet. ref'd*), where the movant stated his objection to exclusion of African-Americans and Hispanics and then on appeal opposed the peremptory challenge of an Asian not mentioned in his trial objection.

PRACTICE TIP: The movant should make sure the following is reflected by the record to properly preserve its Batson challenge:

(1) The racial makeup of the venire, as well as the race of those veniremen peremptorily challenged
(2) The race or ethnicity of the movant
(3) If the crime is interracial this should be placed in the record; see, Gerber v. State, 845 S.W.2d 460 (Tex. App. — Houston [1st Dist.] 1993), and Dewberg v. State, 776 S.W2d 589 (Tex. Cr App. 1989)
(4) Any inconsistent treatment by the challenging party of minority and non-minority veniremembers
(5) If possible, the racial composition of the population eligible for jury service

Keep in mind when making the objection that the race or ethnic origin of a movant or venireman cannot be conclusively proven for purposes of a Batson hearing based solely on the person's surname. Sanchez v. State, 797 S.W.2d 951 (Tex. App. —Dallas 1990). Furthermore, the fact that a veniremember has married a person of a particular ethnicity, and therefore has an ethnic surname, will not automatically categorize that individual as a member of their spouse's ethnicity. Aguilar v. State, 826 S.W.2d 760 (Tex.App. — Fort Worth 1992).

PRACTICE TIP: The easiest and most conclusive method of establishing the race or ethnicity of a person is to have the trial judge take judicial notice of the fact sought to be established. Most trial judges will take judicial notice of the obvious race or ethnicity. If the judge will not take judicial notice as requested, the fact must be proven by an alternate method such as by calling a witness or by stipulation between the parties.

§14:113.2.1 Batson Inapplicable

Appellate courts have refused to extend Batson principles to trial procedures other than peremptory challenges. Ladd v. State, 3 S.W.3d 547 (Tex. Crim. App. 1999) (jury shuffles); Wearren v. State, 877 S.W.2d 545 (Tex.App. — Beaumont 1994, *no pet.*) (jury shuffles).

Also, Batson principles do not extend to the state's non-use of peremptory challenges. Hamel v. State, 803 S.W.2d 878 (Tex.App. — Fort Worth 1991, *pet. ref'd*) (where the defendant argued that the state had acted with racial motivation in not using several peremptory challenges, which would have allowed the process to mathematically reach black veniremembers).

§14:113.3 Burdens of Proof/ Production/ Persuasion

In an opinion clarifying the procedure to be followed in a Batson inquiry, the Texas Court of Criminal Appeals has held that Batson contemplates a scheme of shifting burdens. In Williams v. State, 804 S.W.2d 95 (Tex. Crim. App. 1991), *cert. denied*, 501 U.S. 1239, 111 S.Ct. 2875, 115 L.Ed.2d 1038 (1991), the Court held:

- Movant has the initial burden of proof to establish a *prima facie* case of purposeful discrimination
- Burden (termed a burden of production) then shifts to the challenging party to supply a race-neutral explanation for the peremptory challenge(s)
- Burden (termed a burden of persuasion) next shifts back to the movant to rebut the challenging party's explanations so that it can be rationally inferred that the challenging party engaged in purposeful racial discrimination

§14:113.3.1 Burden of Proof

§14:113.3.1.1 Prima Facie Case Necessary

In order to be entitled to have a Batson or purposeful discrimination claim reviewed by a trial or appellate court, the movant must make out a *prima facie* case of racial discrimination in the challenging party's use of peremptory challenges. A *prima facie* case represents the minimum quantum of evidence necessary to support a rational inference that the allegation of fact is true. The party with the burden of proof must produce at least this much evidence to avoid a finding that the allegation is not true as a matter of law. Tompkins v. State, 774 S.W.2d 195 (Tex. Crim. App. 1987), affirmed at 490 U.S. 754, 109 S.Ct. 2180, 104 L.Ed.2d 834 (1989).

The minimum quantum of evidence necessary to support a rational inference of purposeful racial discrimination (*i.e.*, a *prima facie* case) is any relevant evidence with more than a modicum of probative value. Linscomb v. State, 829 S.W.2d 164 (Tex. Crim. App. 1992). A *prima facie* case is what raises the issue, not what eventually disposes of it. In determining, therefore, whether a *prima facie* case is reflected on the record, courts are not to resolve the question of deliberate racial discrimination on its merits by assessing the probative weight of competing inferences or conflicting evidence. They are simply to decide whether the issue has been raised. Linscomb, *supra*.

A defendant satisfies the requirements of *Batson's* first step by producing evidence sufficient to permit the trial judge to draw an inference that discrimination has occurred; and is not required to make a showing that discrimination occurred more likely than not. Johnson v. California, 545 U.S. 162, 125 S.Ct. 2410, 162 L.Ed.2d 129 (2005).

The following are illustrative of the types of evidence that can be used to raise the inference of discrimination:

- Veniremen peremptorily challenged have no characteristic in common other than race
- Pattern of peremptory challenges against minority jurors on the particular venire
- Past conduct of the challenging party in the use of peremptory challenges
- Type and manner of the challenging party's questions and statements during voir dire
- Type and manner of questions directed to the challenged juror, including a lack of questions, or a lack of meaningful questions
- Disparate treatment of members of the jury venire with the same characteristics, or who answer a question in the same or similar manner
- Disparate examination of members of the venire; *e.g.*, questioning some veniremen in a manner likely to provoke a certain response
- Use of all or most of the peremptory challenges to eliminate minorities from the venire

Keeton v. State, 749 S.W.2d 861 (Tex. Crim. App. 1988).

The mere fact that a party exercised a peremptory against a venireman of a cognizable minority group is not sufficient to meet a movant's *prima facie* burden for purposes of Batson. Staley v. State, 887 S.W.2d 885 (Tex. Crim. App. 1994). The movant must show from the circumstances that the challenging party used its peremptory challenge to exclude the venireman for racial reasons. Chambers v. State, 866 S.W.2d 9 (Tex. Crim. App. 1993).

§14:113.3.1.2 Elements of Prima Facie Case

A movant must demonstrate the following in order to establish a *prima facie* case:

- Challenged venireman was a member of a cognizable minority group
- Challenging party exercised a peremptory challenge to remove that venireman
- Those facts and any other relevant circumstances raise an inference that the challenging party excluded the venireman from the venire on the basis of race

Taylor v. State, 911 S.W.2d 906 (Tex.App.-Fort Worth 1995 *pet. ref'd*).

If the movant is successful in making out a *prima facie* case, it creates a rebuttable presumption that the challenging party engaged in purposeful racial discrimination in the use of its peremptory challenges. Lewis v. State, 779 S.W.2d 449 (Tex. App.Tyler 1989).

Once produced, however, the allegation must be found true unless it is contradicted, impeached, or rebutted by other evidence .... (footnote omitted). Tompkins v. State, 774 S.W.2d 195 (Tex. Crim. App. 1987), affirmed at 490 U.S. 754, 109 S.Ct. 2180, 104 L.Ed.2d 834 (1989).

TIP TO THE BENCH: Batson held that any discrimination, no matter how slight, constituted the prima facie showing necessary to shift the burden of production to the challenging party. In fact, the Texas Court of Criminal Appeals has stated that "the United States Constitution is offended by so much as a single strike exercised on the basis of race." Linscomb v. State, 829 S.W. 2d 164 (Tex. Crim. App. 1992).

§14:113.3.1.3 Cognizable Minorities

The test for whether a particular group is a cognizable racial group is whether it is a recognizable, distinct class singled out for different treatment under laws as written or as applied. U.S. v. Dennis, 804 F.2d 1208 (11th Cir. 1986). Because of the wording of Batson, most courts have not extended its principles to groups outside those which have historically experienced discrimination such as African-Americans, Hispanics, and Native Americans. This limitation draws the line short of threatening the existence and practice of allowing peremptory challenges.

It is axiomatic that African-Americans and Hispanics are cognizable minorities for Batson purposes. Some other groups which have been recognized are:

- An Indian or Pakistani. Bacy v. State, 827 S.W.2d 91 (Tex.App.—Fort Worth 1992, *pet. ref'd*) (where the venireman was described as "an individual of dark complexion, Indian or Pakistani"), and Minard v. State, 836 S.W.2d 287 (Tex. App. — Dallas 1992) ("an Indian who appeared to be from India or of Indian extraction").
- A person of Oriental extraction. Kline v. State, 737 S.W.2d 895 (Tex. App. — Houston [1st Dist.] 1987).
- Native Americans. U.S. v. Roan Eagle, 867 F.2d 436 (8th Cir. 1989), *cert. denied*, 490 U.S. 1028, 109 S.Ct. 1764, 104 L.Ed.2d 199 (1989).

"Race," for purposes of *Batson*, encompasses notions of ancestral line and ethnicity. Discrimination based on such considerations is racial discrimination under *Batson*. Wamget v. State, 67 S.W.3d 851 (Tex. Crim. App. 2001).

The party alleging discrimination based on nationality or ethnicity under *Batson* will not adequately establish the venireperson's ethnicity and cognizable racial group by showing only the country of their birth, and such party will likewise fail to meet its burden of persuasion of race discrimination by showing that the peremptory strike was based only on the country of the venireperson's birth. Wamget, *supra*.

§14:113.3.1.4 Texas Cases

The following Texas cases illustrate where a pattern of peremptory challenges rises to the level of a *prima facie* case:

- Miller-El v. State, 790 S.W.2d 351 (Tex. App.—Dallas 1990), where the prosecutor used five of 10 peremptory challenges to challenge all five remaining African-American veniremen
- Vann v. State, 788 S.W.2d 899 (Tex. App. — Dallas 1990), where the prosecution used six of 10 peremptory challenges to challenge six of nine African-American veniremen
- Salazar v. State, 745 S.W.2d 385 (Tex. App. — Fort Worth 1987), where the prosecutor peremptorily challenged the only Hispanic veniremember, *but see* Staley v. State, 887 S.W.2d 885 (Tex. Crim. App. 1994), where this was found to be insufficient
- Whitsey v. State, 796 S.W.2d 707 (Tex. Crim. App. 1989), where the prosecution peremptorily challenged all six African-American veniremen
- Chambers v. State, 784 S.W.2d 29 (Tex. Crim. App. 1989), where the prosecution peremptorily challenged all three remaining African-American veniremen
- Dewberry v. State, 776 S.W.2d 589 (Tex. Crim. App. 1989), where the prosecution peremptorily challenged five out of six black veniremen (*i.e.*, a disproportionate number of peremptory challenges used against African-Americans)
- Miller-El v. State, 748 S.W.2d 459 (Tex. Crim. App. 1988), where the prosecution used 10 out of 14 peremptory challenges to exclude all but one African-American venireman
- Linscomb v. State, 829 S.W.2d 164 (Tex. Crim. App. 1992), where the defendant established that the prosecutor exercised peremptory challenges against African-American veniremen at more than twice the rate one would expect from random selection

However, *see* Flores v. State, 920 S.W.2d 347 (Tex.App.— San Antonio 1996), where the trial court found that the defendant had not made out a *prima facie* case where the prosecution used five of 12 peremptory challenges (there were two defendants on trial) against Hispanics. The court reasoned that, because there were 14 Hispanics on the venire, four were peremptorily challenged by the defendants, and four were seated on the jury, there was no showing that the peremptory challenges were racially motivated.

Keep in mind that the movant may be able to raise a *prima facie* case merely by their assertions into the record where their assertions are not disputed by the challenging party. For example, in Emerson v. State, 820 S.W.2d 802 (Tex. Crim. App. 1991), the defense made certain assertions from their observations of the venire. The prosecution did not challenge those assertions, so the appellate courts had no choice but to treat the assertions as undisputed fact.

§14:113.3.1.5 Statistical Analysis

The courts also have recognized that a statistical analysis may be used to make out a *prima facie* case. *See* Linscomb v. State, 829 S.W.2d 164 (Tex. Crim. App. 1992), where the court found purposeful discrimination where the prosecution used 40% of its peremptory challenges to challenge members of an identifiable race which only made up 19% of the population. The Court reasoned that the prosecution used strikes against black veniremen at twice the rate one would expect if done by random selection and that, since the U.S. Constitution is offended by so much as a single strike exercised on the basis of race, there was purposeful discrimination.

TIP TO THE CHALLENGING PARTY: In order to argue on appeal that the movant did not make out a prima facie showing requiring the challenging party to go forward with race-neutral explanations, the challenging party must lodge an objection at trial to the trial court's finding of a prima facie case. Chambers v. State 866 S.W.2d 9 (Tex. Crim. App. 1993). Going forward with explanations without any prompting or inquiry by the trial court also will make the preliminary issue of a prima facie case a moot issue. Hill v. State, 827 S.W. 2d 860 (Tex. Crim. App. 1992).

It is difficult for the movant to make out a prima facie case based on statistical analysis alone where there is a small sample size, such as in a misdemeanor case. Hassan v. State, 369 S.W.3d 872, 877 (Tex. Crim. App. 2012).

Once the movant has established a *prima facie* case of purposeful racial discrimination, he is entitled to an adversarial hearing to pursue his claim. Mata v. State, 867 S.W.2d 798 (Tex. App. — El Paso 1993).

If the trial court does not find a *prima facie* case of purposeful discrimination, by implication, Batson ends the inquiry.

§14:113.3.2 Burden of Production

If the trial court finds that the movant has established a *prima facie* case, the challenging party will then be called upon to produce race-neutral explanations for the peremptory challenge(s).

§14:113.3.2.1 Race-Neutral Explanation

A race-neutral explanation is one based on something other than the race of the venireman. Francis v. State, 909 S.W.2d 158 (Tex. App. — Houston [14th Dist.] 1995). An explanation must apply equally to minority and non-minority veniremen in order to be truly race-neutral.

At this step of the inquiry, the issue is the facial validity of the challenging party's explanation. Unless a discriminatory intent is inherent in the explanation, the reason offered will be deemed race-neutral. Hernandez v. New York, 500 U.S. 352,

111 S.Ct. 1859, 114 L.Ed.2d 395 (1991) (plurality opinion).

The emphasis at this phase of the inquiry is whether the explanations are non-discriminatory rather than whether they are persuasive or even plausible. Purkett v. Elem, 115 S.Ct. 1769, 131 L.Ed.2d 834 (1995). What is meant by a "legitimate reason" is not a reason that makes sense, but a reason that does not deny equal protection. Purkett, *supra*.

§14:113.3.2.2 Quantum of Production

The showing required by the challenging party need not rise to the level of proof required for a challenge for cause. Keeton v. State, 749 S.W.2d 861 (Tex. Crim. App. 1988). However, the challenging party must present a clear, specific, and legitimate race-neutral explanation for each peremptory challenge. McGee v. State, 909 S.W.2d 516 (Tex.App. — Tyler 1995).

The challenging party must produce more than a mere assertion that it peremptorily challenged minority veniremen on the assumption, or through intuitive judgment, that they would be partial to the movant because of their shared race. Nor can a challenging party rebut the movant's case merely by denying that it had a discriminatory motive or by affirming good faith in the individual selections. The challenging party must, therefore, articulate a race-neutral explanation related to the particular case on trial. Williams v. State, 804 S.W.2d 95 (Tex. Crim. App. 1991), *cert. denied*, 501 U.S. 1239, 111 S.Ct. 2875, 115 L.Ed. 1038 (1991).

§14:113.3.2.3 Basis of Explanation

The challenging party is not required to ask a specific rubric of questions in order to substantiate its race-neutral explanations. It is sufficient that the *voir dire* establish the veracity of its apprehensions through any means. Chambers v. State, 866 S.W.2d 9 (Tex. Crim. App. 1993).

The race-neutral explanation can be based on the subjective belief of the challenging party. An otherwise legitimate race-neutral explanation will not fail merely because there is conflicting evidence in the record. Varizas v. State, 838 S.W.2d 552 (Tex. Crim. App. 1992) (where the venireman testified that she could not follow the law, then later testified that she could).

The challenging party is allowed to exercise peremptory challenges in the face of conflicting answers where it does not feel the venireman will be a favorable juror for its side or where it considers the venireman disingenuous because of the change of opinion. Mines v. State, 852 S.W.2d 941 (Tex. Crim. App. 1992).

The challenging party may exercise its peremptory challenges on the basis of its legitimate hunches and past experience, so long as racial discrimination is not the motive behind the challenge. Keeton v. State, 749 S.W.2d 861 (Tex. Crim. App. 1988).

The courts have found that intuitive assumptions on the part of the challenging party can be valid race-neutral explanations. However, these types of explanations are more strictly scrutinized on appeal than those corroborated by the record. Branch v. State, 774 S.W.2d 781 (Tex.App.—El Paso 1989, *pet. ref'd*).

The fact that a litigant mentions race in his explanation for a peremptory challenge is indicative, but not conclusive, of purposeful discrimination in the case. However, when the motives behind a challenged peremptory strike are "mixed," *i.e.*, both impermissible (race or gender-based) and permissible (race and gender-neutral), if the striking party shows that he would have struck the juror based solely on the neutral reasons, then the strike does not violate the juror's Fourteenth Amendment right to equal protection of the law. Guzman v. State, 85 S.W.3d 242 (Tex. Crim. App. 2002).

Under this "dual motivation" or "mixed motives" doctrine, if the opponent of a peremptory strike makes a *prima facie* showing of discriminatory purpose, the strike's proponent must demonstrate that he would have exercised the peremptory strike even if the improper factor had not existed or contributed to the decision to strike the prospective juror. The trial court then determines whether the opponent has carried his burden in proving purposeful discrimination. Guzman.

The challenging party must be sure that the explanations are supported by the record and also must be sure that the reason for the challenge does not apply to some non-minority venireman who was not peremptorily challenged. The trial judge has the duty to evaluate the challenging party's explanations in light of the circumstances of the case to determine whether the facially neutral explanations have been invented to mask racial discrimination. Whitsey v. State, 796 S.W.2d 707 (Tex. Crim. App. 1989).

However, the mere fact that the prosecutor struck a minority venireman while not striking a white venireman does not require a finding of discriminatory intent. Gibson v. State, 144 S.W.3d 530 (Tex. Crim. App. 2004).

It is not the correctness, but rather the genuineness of the explanation that is relevant in determining whether the challenge was racially motivated. Jasper v. State, 61 S.W.3d 413 (Tex. Crim. App. 2001).

The Supreme Court has recently said that they have no rule that requires the trial judge to have observed the demeanor cited in the prosecution's race-neutral explanation. Thaler v. Haynes, 175 L.Ed.2d 1003 (2010).

§14:113.3.2.4 Recognized Race-Neutral Explanations

The Texas courts have recognized a lengthy and non-exclusive list of race-neutral explanations for peremptory challenges:

- Relatives or friends in trouble with the law; Earhart v. State, 823 S.W.2d 607 (Tex. Crim. App. 1991), vacated and remanded on other grounds, 113 S.Ct. 3026, 125 L.Ed.2d 715 (1993)
- Tendency to assess minimum punishment; Earhart, *supra*
- Tendency to overspeculate on facts; Earhart, *supra*
- Family member convicted of crime; Wilson v. State, 769 S.W.2d 682 (Tex. App.—Beaumont 1989)
- Lack of steady employment; Johnson v. State, 740 S.W.2d 868 (Tex. App.— Houston [14th Dist.] 1987)
- Lack of ties to the community; Johnson, *supra*
- Rehabilitation a priority at punishment; Salazar v. State, 745 S.W.2d 385 (Tex. App.—Fort Worth 1987)
- Religious affiliation (Seventh Day Adventist); Salazar, *supra*
- Membership in a certain profession (quality control inspector); Salazar, *supra*
- Distrust of prosecutor; Rodgers v. State, 725 S.W.2d 477 (Tex.App.—Houston [1st Dist.] 1987)
- Age and youth; Yarbrough v. State, 732 S.W.2d 86 (Tex.App.—Beaumont 1987)
- No religion; Yarbrough, *supra*
- Prior jury service on a mistried case; Yarbrough, *supra*
- Venireman smiled at defendant; Yarbrough, *supra*
- Body English; Chambers v. State, 724 S.W.2d 440 (Tex. App.—Houston [14th Dist.] 1987)
- Membership in a "fringe" religion; Chambers, *supra*
- Unable to consider full punishment range; Rijo v. State, 721 S.W.2d 562 (Tex.App.—Amarillo 1986)
- Bad attitude toward police credibility; Rasco v. State, 739 S.W.2d 437 (Tex. App.—Houston [14th Dist.] 1987)
- Baptist minister where defense attorney is a minister; Rasco, *supra*
- Opposed to death penalty (in a capital case); Tennard v. State, 802 S.W.2d 678 (Tex. Crim. App. 1990), and Staley v. State, 887 S.W.2d 885 (Tex. Crim. App. 1994)
- Venireman had trouble understanding questions; Tennard, *supra*
- Venireman had reservations about relying on accomplice testimony; Tennard, *supra*
- Venireman had gone to school with defendant and was unable to state her closeness to him; Sims v. State, 768 S.W.2d 863 (Tex.App.—Texarkana 1989)
- Failure to properly fill out jury information form; Catley v. State, 763 S.W.2d 465 (Tex. App.—Houston [14th Dist.] 1988)
- Venireman seemed asleep during *voir dire;* Holman v. State, 772 S.W.2d 530 (Tex.App.—Beaumont 1989)
- Venireman worked at the same place as the defendant; Holton, *supra*
- Poor education; Rice v. State, 746 S.W.2d 356 (Tex.App.—Fort Worth 1988)

- Venireman seemed disinterested and arrogant; Ybarra v. State, 775 S.W.2d 409 (Tex.App.—Waco 1989)
- Marital status of venireman; Crawford v. State, 770 S.W.2d 51 (Tex.App.—Texarkana 1989)
- Venireman would hold prosecution to higher burden of proof than beyond a reasonable doubt; Keeton, *supra*
- An amicable association between the venireman and the defendant's family; Chambers v. State, *supra*
- Where the venireman lists punishment as a low priority in criminal justice system; Yarborough v. State, 868 S.W2d 913 (Tex.App.—Fort Worth 1994)
- Age, lack of employment and religious affiliation; Jack v. State, 867 S.W.2d 942 (Tex.App.—Beaumont 1993)
- Difficulty in being able to consider circumstantial evidence; Tompkins v. State, 774 S.W.2d 195 (Tex. Crim. App. 1987)
- Opposition to law as enacted; Spears v. State, 902 S.W.2d 512 (Tex.App.—Houston [1st Dist.] 1994)
- Venireman's incorrect completion of a blank on a juror information form; Jones v. State, 818 S.W.2d 532 (Tex. App.—Houston [1st Dist.] 1991)
- Prior testimony as a character witness; Gerber v. State, 845 S.W.2d 460 (Tex. App.—Houston [1st Dist.] 1993)
- Where the venireman wore sunglasses during voir dire preventing the prosecutor from being able to tell if his eyes were open; Gerber, *supra*
- Where the venireman was a minister and the prosecutor was concerned about his "ability to forgive"; Gerber, *supra*
- Venireman holds himself out to be a "liberal"; Henderson v. State, 816 S.W.2d 845 (Tex.App.—Fort Worth 1991, *no pet.*)
- A strong religious background; Dixon v. State, 828 S.W.2d 42 (Tex.App.—Tyler, *pet. ref'd*)
- Venireman's employment as a TDC guard; Prosper v. State, 788 S.W.2d 625 (Tex.App.—Houston [14th Dist.] 1990)
- Where the venireman stated that he knew people were sometimes convicted of crimes they did not commit (and prosecutor explained that her "instincts" told her that this venireman would not make a good juror); Chamberlain v. State, 998 S.W.2d 230 (Tex. Crim. App. No. 72,850, June 16, 1999)
- Where the venireman stated that he agreed with the verdict in the O.J. Simpson murder case. Shelling v. State, 52 S.W.3d 213 (Tex.App.—Houston [1st Dist.] 2001, *pet. ref'd*)
- The fact that the venireman "glared at" the prosecutor during voir dire. Nieto v. State, 365 S.W.3d 673 (Tex. Crim. App. 2012).
- The fact that the venireman had the same last name as a known criminal family. Nieto.

In addition to this non-exclusive list, it should be noted that the courts are sympathetic to the challenging party's reasoning that its trial strategy led to the exercise of peremptory challenges in a certain manner. This will not obviate the need for a race-neutral explanation, but will complement such an explanation.

It would also be important for the challenging party to demonstrate that it challenged non-minority veniremen with the same characteristics as the challenged minority veniremen. The challenging party should also demonstrate that it did not use a "pattern" to peremptorily challenge minority jurors. Keeton v. State, 749 S.W.2d 861 (Tex. Crim. App. 1988).

§14:113.3.2.5 Invalid Race-Neutral Explanations

The following are explanations for peremptory challenges of minority veniremen which have been found by the courts to not be race-neutral (*i.e.*, are racially motivated):

- Black veniremen associated with one another and appeared to be "very friendly;" Hill v. State, 775 S.W.2d 754 (Tex.App.—Dallas 1989), affirmed at 827 S.W.2d 860 (Tex. Crim. App. 1992)
- Venireman thought prone to identify with defendant; Hill, *supra*
- No eye contact with prosecutor (where prosecutor did not individually question

veniremen); Daniels v. State, 768 S.W.2d 314 (Tex.App.—Tyler 1988)

- Venireman paid more attention to defense attorney; Daniels, *supra*
- Venireman thought to be of low intelligence without asking individual questions; Chivers v. State, 796 S.W.2d 539 (Tex.App.—Dallas 1990)
- Venireman was a member of an organization concerned with the advancement of black person's rights (because the membership of the organization is predominantly black); Somerville v. State, 792 S.W.2d 265 (Tex.App.—Dallas 1990)
- Where prosecution exercised a "sight strike" and felt venireman would not fit with the category of jurors prosecution wanted; Davis v. State, 796 S.W.2d 813 (Tex.App.—Dallas 1990, *pet. ref'd*) (striking party must be able to articulate why he felt the venireman's appearance was unacceptable)
- Where the prosecutor could not recall the reasons that he peremptorily challenged two African-American veniremen; Brooks v. State, 802 S.W.2d 692 (Tex. Crim. App. 1991)
- Where the prosecutor explained that his peremptory challenge was based on a "visceral reaction"; Smith v. State, 790 S.W.2d 794 (Tex.App.—Beaumont 1990)
- Any instance where the challenging party exercises peremptories based upon a shared membership in a minority group; Miller-El v. State, 790 S.W.2d 351 (Tex.App.—Dallas 1990)
- Minority venirewoman indicated that she had been involved in a case where one of the state's medical witnesses had appeared as a witness and further details of her involvement were not pursued; Lewis v. State, 779 S.W.2d 449 (Tex. App.—Tyler 1989)
- Prosecution removed "all males under the age of 30 because of their potential bias or shared identity with appellant"; Fritz v. State, 946 S.W.2d 844 (Tex. Crim. App. 1997)

§14:113.3.3 Burden of Persuasion

After the movant has made a *prima facie* showing of racial discrimination in the use of peremptory challenges and the challenging party has produced explanations for its challenges, the movant has the ultimate burden of proving by a preponderance of the evidence that the peremptory challenges were actually used for a racially discriminatory purpose. Lewis v. State, 815 S.W.2d 560 (Tex. Crim. App. 1991); Marquez v. State, 921 S.W.2d 217 (Tex. Crim. App. 1996).

This burden can be sustained by impeaching or rebutting the challenging party's explanation, for example, by showing that the explanation was merely a sham or a pretext for discrimination. Bryant v. State, 923 S.W.2d 199 (Tex.App.—Waco 1996).

On a Batson challenge the inquiry is limited only to the facial validity of the challenging party's explanation for exercising a peremptory challenge on a minority venireman. Bridges v. State, 909 S.W.2d 151 (Tex.App.—Houston [14th Dist.] 1995).

Where the defendant has proven only that the reason given for the peremptory challenge was incorrect he has not met his burden of persuasion to successfully challenge the state's peremptory strikes at trial. Ford v. State, 1 S.W.3d 691 (Tex. Crim. App. 1999).

The defendant must demonstrate in the trial court that the state's reason for exercising the strike was racially motivated; it is not enough to make this argument on appeal. Ford.

A party's failure to offer any real rebuttal to a proffered race-neutral explanation can be fatal to his claim. Chamberlain v. State, 998 S.W.2d 230 (Tex. Crim. App. 1999).

§14:113.3.3.1 Trial Court's Duty

The trial court has the duty of determining whether the movant has carried its burden of showing purposeful racial discrimination. Francis v. State, 909 S.W.2d 158 (Tex.App.—Houston [14th Dist.] 1995).

The trial judge must assess the entirety of the voir dire on both an objective and subjective basis. Keeton v. State, 749 S.W.2d 861 (Tex. Crim. App. 1988). The court should not accept at face value the race-neutral explanations of the challenging party, but shall review the testimony as a whole to determine if the articulated reasons are merely given to conceal a truly racial motive. Keeton, *supra*.

In determining whether the exercise of a peremptory challenge was racially motivated, the trial judge may reflect upon his own personal experiences with voir dire examinations and compare his observations and assessments of veniremembers with those articulated by the challenging party. The judge also may consider his personal experiences with particular attorneys and any evidence offered by the movant to show a pattern or practice of purposeful discrimination. Ramirez v. State, 862 S.W.2d 648 (Tex.App.—Dallas 1993).

Often the best evidence of whether the movant has carried its burden of persuasion will be the demeanor of the attorney who exercises the challenge. Francis, *supra*.

In evaluating the credibility of the challenging party, the trial court may look to several factors:

- Whether the challenging party failed to question any of the minority veniremen, yet peremptorily challenged them anyway
- If the challenging party peremptorily challenged minority veniremen who gave answers similar to non-minority veniremen who were not challenged
- Whether the challenging party challenged minority veniremen who had the same characteristics professionally, socially, religiously, *etc.*, as non-minority veniremen who were not challenged

Young v. State, 848 S.W.2d 203 (Tex.App.—Dallas 1992).

§14:113.3.3.2 Trial Court Considerations

As a general rule, a trial court's resolution of a Batson challenge often involves the weighing of three broad circumstances:

- Strength of the movant's *prima facie* case
- Reason for each peremptory challenge and whether it was contrived to mask a discriminatory purpose
- Overview of the totality of the challenging party's explanations

Woods v. State, 801 S.W.2d 932 (Tex.App.—Austin 1990, *pet. ref'd*).

The trial court can look to the following factors in deciding whether the defendant has met his burden of persuasion: (1) the race-neutral explanations; (2) whether the state's strikes disproportionately impacted minorities; (3) whether the state engaged in disparate questioning; and (4) comparative juror analysis. These factors should be analyzed as a group. No single factor is deserving of more weight than another factor. Watkins v. State, 245 S.W.3d 444 (Tex. Crim. App. 2008).

All the relevant facts and circumstances must be considered by the courts in deciding whether a Batson violation has occurred. These factors can include the State's pattern of striking black prospective jurors in other trials, as well as the disparate questioning of black and white prospective jurors. The inquiry is not whether any one fact alone would require reversal, but whether the peremptory strikes were motivated in substantial part by discriminatory intent. Flowers v. Mississippi, 588 U.S. ___, ___ S.Ct. ___, ___ L.Ed.2d ___ (No. 17-9572, 2019 WL 2552489, June 21, 2019).

Strength of *Prima Facie* Case

In deciding if the movant has sustained its burden of persuasion, the trial (and appellate) court will look to several factors which may already have been examined in deciding if the movant made out a *prima facie* case. These include:

- Whether there were any minority jurors seated on the trial jury
- Percentage of minorities on the jury as contrasted with the percentage on the venire
- Whether a disproportionate number of peremptory challenges were exercised against minority veniremen

Where other members of the challenged venireman's race actually served on the jury, a trial judge may, in appropriate cases, infer that race was not a motive for the peremptory challenge. Jones v. State, 845 S.W.2d 419 (Tex.App.—Houston [1st Dist.] 1992). For example, in Salinas v. State, 888 S.W.2d 93 (Tex.App.—Corpus Christi

JURY SELECTION AND VOIR DIRE

1994), the Court found that, even though the defendant had made out a *prima facie* case of racial discrimination, there was not a Batson violation where the prosecution failed to proffer explanations for two of the peremptorily challenged Hispanics on the venire. The appellate court accepted the trial court's conclusion that the prosecution's use of its peremptory challenges was not race-based was supported by other circumstances, *i.e.,* the overwhelming number of Hispanics on the venire and the petit jury.

Strength of Race-Neutral Explanation

Factors to be considered by the trial court during a burden of persuasion analysis to determine whether a peremptory challenge was race-neutral are whether:

- Challenging party's explanation was related to the facts of the case
- There was a lack of meaningful questioning of the challenged venireman
- There was a disparate selection of veniremen, *i.e.*, were veniremen impaneled on the jury with the same or similar characteristics as the challenged veniremen
- There was a disparate examination of a venireman, *i.e.,* questioning of a venireman to evoke a particular response without asking the same questions of other veniremen
- Challenging party's explanation was based on a group bias where the group trait was not shown to apply to the challenged venireman specifically

Whitsey v. State, 796 S.W.2d 707 (Tex. Crim. App. 1989).

Disparate Questioning

The proffering of questions to a challenged venireman designed to provoke a certain response without the proffering of the same or similar question to non-minority veniremen may tend to show racial motivation in the use of peremptory challenges. Another example of disparate questioning recognized in Whitsey, *supra*, is the failure to question a challenged venireman.

Group Bias

If a challenging party makes an assumption about a venireman, but does not question him to verify the assumption, it is some indication that the challenge was not race-neutral. Woods v. State, 801 S.W.2d 932 (Tex. App. Austin 1990, *pet. ref'd*). In this situation, the challenging party is acting upon a group bias which has not been shown to apply to the individual. For example, in Woods the prosecutor exercised a peremptory challenge against a venireman because he wore a punk rock haircut which the prosecutor associated with liberals, radicals and the drug culture. The Court found the challenge racial in motivation because the prosecutor did not question the venireman to verify his assumptions.

Likewise, in Emerson v. State, 851 S.W.2d 269 (Tex. Crim. App. 1993), the state peremptorily challenged a black unemployed college student on grounds that his unemployment and the collegiate environment foment liberalism. The prosecutor asked no individual voir dire questions of the venireman to confirm any of the liberal characteristics he associated with this type of person. Thus, the prosecutor was relying on an improper group bias theory because he did not connect it to the particular venireman through voir dire questioning.

The trial court should not merely accept specific reasoning at face value, but must consider whether the challenging party contrived racially neutral explanations to avoid admitting acts of group discrimination. Miller-El v. State, 790 S.W.2d 351 (Tex.App. Dallas 1990).

While failure to show how a group bias applies directly to a venireman will weigh heavily against legitimacy of race-neutral explanations, this failure does not establish conclusively that the challenging party's reasons were a pretext for racial discrimination. The challenging party may base peremptory challenges on its legitimate hunches and past experiences as long as racial discrimination is not a motive. Frazier v. State, 909 S.W.2d 255 (Tex. App. Houston [14th Dist.] 1995).

Reliance on group bias factors have been upheld in several cases even where not shown to apply directly to the challenged venireman. In Frazier, *supra*, the court upheld the peremptory challenge of a state school employee where, in the opinion of the prosecution, she was the "stereotypical" equivalent of a social worker and possibly too lenient on defendants. In Reese v. State, 905 S.W.2d 631 (Tex.App. Texarkana 1995), the court upheld the prosecution's peremptory challenge of a woman married to a paralegal who worked with a lawyer in a criminal defense practice. In distinguishing the situation from one involving pure group bias, the court reasoned that the prosecution could have feared that the venireman would have been unfavorable to the state because of her presumed familiarity with the system. In Emerson v. State, 851 S.W.2d 269 (Tex. Crim. App. 1993), the court upheld the peremptory challenge of a nurse which was based on her professional status where the prosecution reasoned that she may expect medical testimony in a case where they had none to introduce.

Nonquantifiable Explanations

Race-neutral explanations which are based on nonverbal factors or subjective reasons are not necessarily insufficient to rebut presumption of discrimination, but they merit closer scrutiny than explanations based on objective reasons. Goodwin v. State, 898 S.W.2d 380 (Tex.App. San Antonio 1995). The protection of the constitutional guarantees that Batson recognizes requires the court to scrutinize such elusive, intangible, and easily contrived explanations with a healthy skepticism. Daniels v. State, 768 S.W.2d 314 (Tex.App. Tyler 1988, *pet. ref'd*).

Even if the challenging party was mistaken in some of his subjective beliefs, the trial court could still conclude that the strike was race-neutral if the court finds that the challenging party based the peremptory challenge on an honest belief. Jack v. State, 867 S.W.2d 942 (Tex.App. Beaumont 1993).

Corroboration Requirement

A counsel's statement about an occurrence in the courtroom, which was made for the purposes of the record, recorded by the court reporter, undisputed by the opposing counsel, and unquestioned and unqualified by the judge in whose presence the statement was made, established the occurrence for purposes of the appellate record. Yarborough v. State, 947 S.W.2d 892 (Tex. Crim. App. 1997)

Some courts have previously required that peremptory challenges based on nonquantifiable factors, such as hostility or lack of attentiveness, must have the nonquantifiable factor substantiated in the record by corroboration other than the challenging party's assertion. Roberson v. State, 866 S.W.2d 259 (Tex.App. Fort Worth 1993) (where the prosecutor made a sufficient showing through questioning of the venireman that it had appeared that he had been sleeping during voir dire). *See also*, Hill v. State, 827 S.W.2d 860 (Tex. Crim. App. 1992), *cert. denied*, 113 S.Ct. 297 (1992) (where the prosecutor stated that one of the reasons for a peremptory challenge of a minority venireman was a dislike of the way he answered the prosecutor's questions and the record was silent on specific body language and other nonverbal actions which would have substantiated the prosecutor's belief).

The record will be deemed substantiated where the trial court confirms the challenging party's recitations of courtroom observations. Muhammad v. State, 911 S.W.2d 823 (Tex.App. Texarkana 1995).

Prosecution assumptions which are not verified by questioning of the venireman are some evidence of racial discrimination. Cuestas v. State, 933 S.W.2d 731 (Tex.App. Fort Worth 1996).

PRACTICE TIP: If an assertion made during an objection or hearing is controverted by the opposing party, be sure to present additional evidence to bolster the assertion. For example, in Troff v. State 882 S.W.2d 905 (Tex.App. Houston [1st Dist.] 1994), the defendant made the assertion that two veniremen were Hispanic. The prosecutor challenged this assertion, and the appeals court found that since the assertion was challenged it could not be accepted at face value.

No Corroboration Requirement

Other courts have held that undisputed statements of the challenging party in a Batson hearing constitute "valid proof." Salinas v. State, 888 S.W.2d 93 (Tex.App. Corpus Christi 1994) (where the prosecutor feared that, since the challenged venireman had a business near the offense location, he had information which would affect his opinion), and Moss v. State, 877 S.W.2d

895 (Tex.App. Waco 1995) (where the state asserted that the challenged venireman was not attentive and generally disinterested; an assertion which was not challenged by the defense).

PRACTICE TIP: If the opposing party relies on subjective explanations or courtroom observations which can be refuted by testimony, make a record challenging those subjective beliefs or courtroom observations through the testimony of witnesses who did not observe the asserted factors.

The lack of an explanation by the challenging party, even if based on oversight, represents a failure in the challenging party's burden and constitutes reversible error. Wright v. State, 832 S.W.2d 601 (Tex. Crim. App. 1992).

Race as a Factor

The fact that a litigant mentions race in his explanation for a peremptory challenge is indicative, but not conclusive, of purposeful discrimination in the case. However, when the motives behind a challenged peremptory strike are "mixed," *i.e.*, both impermissible (race or gender-based) and permissible (race and gender-neutral), if the striking party shows that he would have struck the juror based solely on the neutral reasons, then the strike does not violate the juror's Fourteenth Amendment right to equal protection of the law. Guzman v. State, 85 S.W.3d 242 (Tex. Crim. App. 2002).

Under this "dual motivation" or "mixed motives" doctrine, if the opponent of a peremptory strike makes a prima facie showing of discriminatory purpose, the strike's proponent must demonstrate that he would have exercised the peremptory strike even if the improper factor had not existed or contributed to the decision to strike the prospective juror. The trial court then determines whether the opponent has carried his burden in proving purposeful discrimination. Guzman.

Disparate Treatment

If the challenging party is shown by the movant to have peremptorily challenged minority veniremembers with a particular characteristic while not challenging non-minority veniremen with the same or similar characteristic, the court may find that the challenging party has engaged in "disparate" treatment. Whitsey v. State, 796 S.W.2d 707 (Tex. Crim. App. 1989) (opinion on rehearing). A finding of disparate treatment voids the challenging party's race-neutral explanations and results in a finding that there was purposeful discrimination. Whitsey, *supra*.

Evidence of disparate treatment of veniremen weighs heavily against legitimacy of proffered raceneutral explanations given by the challenging party. Miller-El v. State, 790 S.W.2d 351 (Tex.App. Dallas 1990).

Disparate treatment can lead to an imputation of racial bias where the disparate treatment exists as the sole reason or primary reasons stated for the exercise of the peremptory challenge. Esteves v. State, 849 S.W.2d 822 (Tex. Crim. App. 1993).

Examples of Disparate Treatment

Examples of cases in which the challenging party has not peremptorily challenged persons with same or similar characteristics to a challenged minority venireman are as follows:

- Prosecution peremptorily challenged a minority venireman because her brother was in the penitentiary and did not challenge a similarly situated white venireman; Esteves v. State, 849 S.W.2d 822 (Tex. Crim. App. 1993)
- Prosecution peremptorily challenged three minority veniremen for the reason that they had relatives who had been in trouble with the law, while failing to peremptorily challenge seven similarly situated whites; Ramirez v. State, 862 S.W2d 648 (Tex.App. Dallas 1993)
- State peremptorily challenged a minority concluding that he was "more liberal minded" since he had a mustache while not peremptorily challenging a white venireman with a beard; Lewis v. State, 815 S.W.2d 560 (Tex. Crim. App. 1991)
- Black woman was peremptorily challenged for not having kids, but a white juror without kids was not peremptorily challenged; C.E.J. v. State, 788 S.W.2d 849 (Tex. App. — Dallas 1990)

Non-Disparate Treatment

Disparate treatment as such cannot automatically be imputed in every situation where one of the challenging party's reasons for challenging a venireperson would technically apply to another venireperson whom the challenging party found acceptable. McHenry v. State, 823 S.W2d 667 (Tex.App. Dallas 1991). The decision to peremptorily challenge a particular venireperson is not susceptible to rigid quantification, rather, it is a fluid process, often hinging on the interaction of a number of variables and permutations. *See* U.S. v. Lance, 853 E2d 1177 (5th Cir. 1988).

It is unlikely that two venirepersons on one panel will possess the same objectionable attribute or character trait in precisely the same degree. Such qualitative distinctions may cause a challenging party to challenge one venireman and not the other. Roy v. State, 813 S.W.2d 532 (Tex.App. — Dallas 199 1, *pet. ref'd*). For example, in Spears v. State, 902 S.W.2d 512 (Tex.App. Houston [1st Dist.] 1994), the state peremptorily challenged a minority venireman who had a relative who had been convicted of a crime. The state did not peremptorily challenge nonminority veniremen whose relatives had been convicted of a crime. The Court found that the state had not engaged in disparate treatment because the nonminority veniremen had unequivocally stated that their relatives had been treated fairly by the system. *See also*, Copeland v. State, 848 S.W.2d 199 (Tex.App. Corpus Christi 1992, *pet. ref'd*) (where the state peremptorily challenged a black teacher but did not challenge several white teachers; the court found that, since the black teacher had made statements during voir dire which the state took as negative, the challenge was not racially motivated).

In situations where the challenging party has offered more than one plausible reason for striking a venireperson, it is proper for the trial and appellate courts to review these reasons in their entirety in order to assess whether the challenging party's explanation was valid or merely pretextual. When the challenging party has offered numerous race-neutral reasons for its peremptory challenge, the mere fact that there were other acceptable jurors possessing one or more of these objectionable attributes is not sufficient to establish "disparate treatment" by the challenging party in use of its peremptory challenges. Cantu v. State, 842 S.W.2d 667 (Tex. Crim. App. 1992); Adanandus v. State, 866 S.W.2d 210 (Tex. Crim. App. 1993).

Use of Notes to Designate Minorities

One form of disparate treatment that supports an inference that facially valid race-neutral explanations are a sham is the designation in the notes of the challenging party as to which veniremen are minorities. Whitsey v. State, 796 S.W.2d 707 (Tex. Crim. App. 1989).

The use of notes to designate minority members on the panel can suggest that race was a factor in the challenging party's use of peremptory challenges. Francis v. State, 909 S.W.2d 158 (Tex. App. Houston [14th Dist.] 1995) (where the prosecutor's notes contained a circled "B" next to the names of black panel members); Whitsey, *supra*, (where the prosecutor identified black veniremen with the letter "B" written above their names).

§14:120 Death Penalty Case Voir Dire

§14:121 Formation of Venire

§14:121.1 Statutory Law

CCP Art. 34.01. Special Venire

A "special venire" is a writ issued in a capital case by order of the district court, commanding the sheriff to summon either verbally or by mail such a number of persons, not less than 50, as the court may order, to appear before the court on a day named in the writ from whom the jury for the trial of such case is to be selected. Where as many as one hundred jurors have been summoned in such county for regular service for the week in which such capital case is set for trial, the judge of the court having jurisdiction of a capital case in which a motion for a special venire has been made, shall grant or refuse such motion for a special venire, and upon such refusal require the case to be tried by regular jurors summoned for service in such county for the week in which such capital case is set for trial and such additional talesmen as may be summoned by the sheriff upon order of the court as provided in Article 34.02 of this Code, but the clerk of such court shall furnish the defendant or his counsel a list of the persons summoned as provided in Article 34.04.

CCP Art. 34.05. Mechanical or Electronic Selection Method

A mechanical or electronic method of jury selection as provided by Chapter 62, Government Code, may be used under this chapter.

§14:121.2 Assembling Venire

In a capital case, a venire may be assembled in two alternate manners: (1) as a special venire ordered by the court; or (2) as a "jury panel for the week" if as many as 100 jurors have been summoned in the county for regular service for the week. CCP Art. 34.01.

The district court has the discretion to grant or refuse a defendant's motion for a special venire and to order trial before a jury selected from the regular jury panel for the week where at least 100 jurors have been summoned. CCP Art. 34.01.

A special venire is an order of the court to the sheriff to summon verbally or by mail any number of persons not less than 50 to appear in court on a certain day as prospective jurors. CCP Art. 34.01.

§14:121.3 Summoning Additional Veniremembers

In obtaining additional veniremembers in a capital case, the jury wheel must be resorted to in jury wheel counties; in counties not using jury wheels, the sheriff summons them. CCP Art. 34.02.

Additionally, *see* §14:16.

§14:121.4 Furnishing Defendant With List of Veniremembers

CCP Art. 34.04. Notice of List

No defendant in a capital case in which the state seeks the death penalty shall be brought to trial until he shall have had at least two days (including holidays) a copy of the names of the persons summoned as veniremembers, for the week for which his case is set for trial except where he waives the right or is on bail. When such defendant is on bail, the clerk of the court in which the case is pending shall furnish such a list to the defendant or his counsel at least two days prior to the trial (including holidays) upon timely motion by the defendant or his counsel therefor at the office of such clerk, and the defendant shall not be brought to trial until such list has been furnished defendant or his counsel for at least two days (including holidays). Where the venire is exhausted, by challenges or otherwise, and additional names are drawn, the defendant shall not be entitled to two days service of the names additionally drawn, but the clerk shall compile a list of such names promptly after they are drawn and if the defendant is not on bail, the sheriff shall serve a copy of such list promptly upon the defendant, and if on bail, the clerk shall furnish a copy of such list to the defendant or his counsel upon request, but the proceedings shall not be delayed thereby.

A defendant who is tried for the death penalty and who is in custody is entitled to be furnished a list of the persons summoned at least two days before trial. CCP Art. 34.04. Where the defendant is on bail, he is entitled to the list two days before trial where he timely files a motion therefor. He must pick it up at the clerk's office. CCP Art. 34.04. The two-day requirement in Art. 34.04 is mandatory. Smith v. State, 744 S.W.2d 86 (Tex. Crim. App. 1987).

Service of the jury list does not have to be on the defendant personally. Delivery to his attorney is sufficient. Wyle v. State, 777 S.W.2d 709 (Tex. Crim. App. 1989), *affirmed on remand, at* 836 S.W.2d 796.

It is sufficient to serve the list on the defendant two days before the actual voir dire begins, even though the trial judge gave his introductory remarks to the venire on the same day the list was served. May v. State, 738 S.W.2d 261 (Tex. Crim. App.

1987), *cert. denied*, 484 U.S. 872, 108 S.Ct. 206, 98 L. Ed. 2d 158 (1987).

The defendant is entitled to be served two days before trial with the initial list of veniremembers but not with any list of supplemental veniremembers summoned. Wyle.

§14:121.5 Excusing Veniremembers From Special Venire

In the case of a special venire, the trial court cannot designate others to hear and determine veniremembers' excuses for not serving. Chambers v. State, 903 S.W.2d 21 (Tex. Crim. App. 1995); CCP Art. 35.03. However, this does not prevent others from hearing and determining excuses from veniremembers before they are assigned to the special venire. Chambers.

PRACTICE TIP: In order to preserve error in the case of veniremembers who were summoned for jury duty but did not appear, request writs of attachment for those veniremembers. Coleman v. State, 881 S.W.2d 344 (Tex. Crim. App. 1994), cert. denied, 513 U.S. 1096, 115 S.Ct. 763, 130 L. Ed. 2d 660 (1995).

Additionally, *see* §14:70.

TIP TO THE BENCH: The trial judge is able to conserve trial time and avoid possible error by calling the special venire from a group of veniremembers who have already been summoned and had their excuses heard by the jury room judge. Chambers, supra, allows others to hear excuses from veniremembers before they are assigned to a special venire. See Chambers.

§14:121.6 Forms for Venire Formation

See our companion book *Texas Criminal Forms* (James Publishing) for the following sample forms:

- Form 14-16 Defendant's Motion for Special Venire.
- Form 14-16.2 Defendant's Motion for the Court to Decide Jurors' Exemptions, Excuses and Disqualifications Pursuant to Art. 35.03.
- Form 14-17 Writ to Summon Special Venire.
- Form 14-18 Writ to Serve Defendant Jury List.

§14:130 Mechanics of Capital Voir Dire

CCP Art. 35.17. Voir Dire Examination

> *1. When the court in its discretion so directs, except as provided in Section 2, the state and defendant shall conduct the voir dire examination of prospective jurors in the presence of the entire panel.*
>
> *2. In a capital felony case in which the State seeks the death penalty, the court shall propound to the entire panel of prospective jurors questions concerning the principles, as applicable to the case on trial, of reasonable doubt, burden of proof, return of indictment by grand jury, presumption of innocence, and opinion. Then, on demand of the State or defendant, either is entitled to examine each juror on voir dire individually and apart from the entire panel, and may further question the juror on the principles propounded by the court.*

The trial court may restrict the extent of individual questioning in a death penalty case by requiring counsel to propound general voir dire questions to the venire or a mini-panel as long as questions pertaining specifically to the death penalty are propounded to the veniremembers individually. This procedure satisfies Art. 35.17(2). Martinez v. State, 867 S.W.2d 30 (Tex. Crim. App. 1993), *cert. denied,* 512 U.S. 1246, 114 S.Ct. 2765, 129 L. Ed. 2d 879 (1994).

The use of "mini-panels" in capital cases has been approved by the Court of Criminal Appeals. Butler v. State, 872 S.W.2d 227 (Tex. Crim. App. 1994), *cert. denied,* 513 U.S. 1157, 115 S.Ct. 1157,

130 L. Ed. 2d 1079 (1995); Hall v. State, 661 S.W.2d 113 (Tex. Crim. App. 1983).

In capital cases it is customary for peremptory challenges to be made after each veniremember has been individually questioned. CCP Art. 35.13.

The jury selection system set by the Code of Criminal Procedure requiring death penalty defendants to exercise peremptory challenges following examination of each individual juror, as compared to the non-capital case system in which peremptory challenges are exercised after all prospective jurors have been examined, does not violate state or federal due process or equal protection provisions. Janecka v. State, 739 S.W.2d 813 (Tex. Crim. App. 1987), *reversed on other grounds,* 823 S.W.2d 232 (Tex. Crim. App. 1990).

The practice of withholding peremptory challenges until a qualified pool of veniremembers is established also has been approved. Rousseau v. State, 824 S.W.2d 579 (Tex. Crim. App. 1992), *cert. denied,* 510 U.S. 919, 114 S.Ct. 313, 126 L. Ed. 2d 260 (1993).

However, the trial court does not violate a defendant's Sixth Amendment right to an impartial jury and his Fourteenth Amendment rights to due process and equal protection by denying his motion to postpone the exercise of peremptory challenges until the completion of voir dire. Ladd v. State, 3 S.W.3d 547 (Tex. Crim. App. 1999); Dowthitt v. State, 931 S.W.2d 244 (Tex. Crim. App. 1996).

§14:131 Trial Court Instructions to Capital Veniremembers

As quoted above, CCP Art. 35.17 requires that the trial court propound to the entire venire questions concerning certain aspects of capital murder law. For a sample set of instructions to be propounded by the trial court, *see* §14:131.1.

It is often helpful in death penalty case voir dire to distribute an outline of the law at the first meeting of the venire. The purpose of the outline is to give the veniremembers a reference to study prior to the voir dire interview. The outline should contain explanations of the laws, which will be explained to the veniremembers during the voir dire interview. For a sample voir dire outline for distribution to veniremembers, *see* §14:131.2.

§14:131.1 Form: Sample Instructions for Death Penalty Venire Panel

See our companion book *Texas Criminal Forms* (James Publishing) Form 14-21

§14:131.2 Form: Sample Law Outline for Death Penalty Venire Panel

See our companion book *Texas Criminal Forms* (James Publishing) Form 14-22

§14:132 Refusal to Allow Questioning

The trial court errs where it sustains the state's challenge for cause against a venire member without allowing defense counsel an opportunity to question. This error is subject to harmless error analysis under Rule 44.2(b) and will be found harmless where it is highly unlikely that the defendant would have been able to convince the venireman to agree with the law or that the trial court would have abused its discretion in dismissing the venireman for cause. Simpson v. State, 119 S.W.3d 262 (Tex. Crim. App. 2003).

There must be a contemporaneous objection to preserve error where the trial court has not allowed defense counsel to question a venireman. Ransom v. State, 789 S.W.2d 572 (Tex. Crim. App. 1989).

§14:140 Subject Matter for Capital Voir Dire

Texas Penal Code § 12.31. Capital Felony

> *(a) An individual adjudged guilty of a capital felony in a case in which the state seeks the death penalty shall be punished by imprisonment in the institutional division for life without parole or by death. An individual adjudged guilty of a capital felony in a case in which the state does not seek the death penalty shall be punished by imprisonment in the institutional division for life without parole.*
>
> *(b) In a capital felony trial in which the state seeks the death penalty, prospective jurors shall be informed that a sentence of*

life imprisonment without parole or death is mandatory on conviction of a capital felony. In a capital felony trial in which the state does not seek the death penalty, prospective jurors shall be informed that the state is not seeking the death penalty and that a sentence of life imprisonment without parole is mandatory on conviction of the capital felony.

The right to representation as afforded under Art. I, §10 of the Texas Constitution includes the right to properly question prospective jurors during voir dire in order to effectively exercise peremptory challenges or to establish a predicate for a challenge for cause. Ex parte McKay, 819 S.W.2d 478 (Tex. Crim. App. 1990).

Not all rulings limiting defense counsel's voir dire presentation are constitutional error as a *per se* violation of the right to counsel. Most rulings limiting voir dire will be subject to Rule 44.2(b) harm analysis. However, there may be instances when a judge's limitation on voir dire is so substantial as to warrant labeling the error as constitutional error subject to a Rule 44.2(a) harm analysis. Easley v. State, 424 S.W.3d 535, 538-9 (Tex. Crim. App. 2014).

See our companion book, *Texas Criminal Forms* (James Publishing) Form 14-9.1, for a motion to submit a jury questionnaire.

§14:141 Form: Sample Death Penalty Case Juror Questionnaire Instructions for Juror

See our companion book *Texas Criminal Forms* (James Publishing) Form 14-8 and 14-9.1.

§14:142 Informing Venire of Effect of Answers

The parties are allowed to inform the venire of the effect of their answers to the special issues. Jones v. State, 843 S.W.2d 487 (Tex. Crim. App. 1992), *cert. denied,* 507 U.S. 1035, 113 S.Ct. 1858, 123 L. Ed. 2d 479 (1993). However, CCP Art. 37.071 §2(a) provides that veniremembers not be informed "of the effect of failure of the jury to agree on an issue submitted under th[e] article." Thus, the parties may not inform the venire of the effect of their failure to answer a special issue. Clark v. State, 881 S.W.2d 682 (Tex. Crim. App. 1994); Sattiewhite v. State, 786 S.W.2d 271 (Tex. Crim. App. 1989), *cert. denied*, 498 U.S. 881, 111 S.Ct. 226, 112 L. Ed. 2d 181 (1990).

The error caused by so informing the venire is in large part cured by an appropriate jury charge informing the jurors of their duty under the law to vote according to the evidence on the special issues. Clark.

§14:143 Questioning on Definition of Terms

The trial court has the discretion to prohibit voir dire on a veniremember's understanding of the definition of undefined terms in the charge. Wheatfall v. State, 882 S.W.2d 829 (Tex. Crim. App. 1994) ("deliberate").

A trial court is on firmer ground limiting voir dire on the understanding of undefined terms in capital murder cases than in non-capital cases because in those cases voir dire examination can take an unreasonable length of time. Woolridge v. State, 827 S.W.2d 900 (Tex. Crim. App. 1992). For example, a trial court does not abuse its discretion by refusing to permit counsel to question a veniremember regarding his definition of the term "probability." Lagrone v. State, 942 S.W.2d 602 (Tex. Crim. App. 1997), *cert. denied,* 522 U.S. 917, 118 S.Ct. 305, 139 L. Ed. 2d 235 (1997).

A trial court does not abuse its discretion by refusing voir dire questioning on definitions of the terms "deliberately, probability, and criminal acts of violence" in a capital case. Milton v. State, 599 S.W.2d 824 (Tex. Crim. App. 1980), *cert. denied*, 451 U.S. 1031, 101 S.Ct. 3022, 69 L. Ed. 2d 400 (1981).

A trial court is, however, required under certain circumstances to allow the parties to question the venire as to their comprehension of certain terms as those terms are relevant to the issues in the case. Teague v. State, 864 S.W.2d 505 (Tex. Crim. App. 1993), *overruled on other grounds,* Robertson v. State, 871 S.W.2d 701 (Tex. Crim. App. 1993) (where the issue was the difference between the terms "intentionally" and "deliberately").

This is especially true where a party is making inquiry of the veniremember in an attempt to lay the proper predicate for a challenge for cause. Gardner v. State, 730 S.W.2d 675 (Tex. Crim. App. 1987), *cert. denied,* 484 U.S. 905, 108 S.Ct. 248, 98 L. Ed. 2d 206 (1987).

TIP TO THE BENCH: In situations such as a death penalty case, you can save a great deal of time on voir dire to encourage the parties to agree on a definition for terms not defined by law, which are crucial to a veniremember's function as a juror.

Where terms such as "criminal acts of violence" are not defined for the jury, error in the voir dire examination occurs when the state attempts to limit the venire to a specific definition. Jackson v. State, 822 S.W.2d 18 (Tex. Crim. App. 1990), *cert. denied*, 509 U.S. 921, 113 S.Ct. 3034, 125 L. Ed. 2d 722 (1993). However, a party may suggest a definition to a veniremember, where the party makes it clear that the veniremember will be required to determine for himself what the definition of the term should be. Coble v. State, 871 S.W.2d 192 (Tex. Crim. App. 1993), *cert. denied,* 513 U.S. 829, 115 S.Ct. 101, 130 L. Ed. 2d 50 (1994).

Although the word "society" in the future-dangerousness special issue is not statutorily defined and thus ordinarily should not be defined, the word must be limited to its technical meaning in this particular context in order for the Texas capital murder statute to comply with federal constitutional requirements. Curry v. State, 541 S.W.3d 751 (Tex. Crim. App. 2017), (dissent by Judge Alcala).

§14:144 Questioning About Facts of Case

A party is allowed to attempt to determine if a veniremember could consider, or conceive, that a crime against property can be a crime of violence; but must do so without committing each veniremember to a particular position. Burks v. State, 876 S.W.2d 877 (Tex. Crim. App. 1994), *cert. denied*, 513 U.S. 1114, 115 S.Ct. 909, 130 L. Ed. 2d 791 (1995). The trial court may restrict questions in which counsel attempts to commit a veniremember to a particular resolution based upon facts peculiar to the trial. Coleman v. State, 881 S.W.2d 344 (Tex. Crim. App.1994), *cert. denied*, 513 U.S. 1096, 115 S.Ct. 763, 130 L. Ed. 2d 660 (1995).

§14:145 Questions Regarding Mitigation of Punishment

Whether the veniremember could consider certain types of evidence in mitigation of punishment is a proper area of inquiry. Rhoades v. State, 934 S.W.2d 113 (Tex. Crim. App. 1996).

The trial court may restrict questions in which counsel attempts to commit a veniremember to a particular resolution based upon facts peculiar to the trial. Coleman v. State, 881 S.W.2d 344 (Tex. Crim. App.1994), *cert. denied*, 513 U.S. 1096, 115 S.Ct. 763, 130 L. Ed. 2d 660 (1995).

Counsel improperly committing a veniremember to a particular set of facts occurs where counsel asks a veniremember whether he *believes* a particular type of evidence to be mitigating, rather than asking whether he could *consider* the evidence as a possible mitigating circumstance. Rhoades.

Since a jury will now be instructed in the punishment phase that a defendant must serve a minimum of 40 years before becoming eligible for parole on a capital murder conviction, the trial court will probably not abuse its discretion in permitting questioning of the veniremen about their attitudes on parole. Sells v. State, 121 S.W.3d 748 (Tex. Crim. App. 2003).

§14:146 Questions About Race of Defendant and/or Victim

A capital murder defendant is entitled to have prospective jurors informed of the race of the victim and questioned on the issue of racial bias. Turner v. Murray, 476 U.S. 28, 106 S.Ct. 1683, 90 L. Ed. 2d 27 (1986).

§14:147 Improper Areas

The trial court does not err in preventing counsel from asking questions based on an improper statement of the law. Morrow v. State, 910 S.W.2d 471 (Tex. Crim. App. 1995), *cert. denied,* 517 U.S. 1192, 116 S.Ct. 1683, 134 L. Ed. 2d 784 (1996).

For offenses occurring before September 1, 1999 refusing to allow questioning about parole eligibility is not error in a capital case because this is not a proper area of inquiry. Eldridge v. State, 940 S.W.2d 646 (Tex. Crim. App. 1996).

Effective for offenses committed on or after September 1, 1999 juries are informed in the punishment charge as to the law of parole on a capital life sentence. It follows that counsel will be allowed to question on capital life sentence parole procedures the same as allowed on parole provisions applicable for other offenses.

However, precedent maintaining that parole is not a proper issue for jury consideration remains in effect except to the extent explicitly provided for in Article 37.071 §2(e)(2)(B). Hankins v. State, 132 S.W.3d 380 (Tex. Crim. App. 2004) (defendant sought to voir dire on whether the veniremembers could consider and give full deliberation to the parole board requirements for release.).

Questions which ask veniremembers whether or not he would consider a particular type of evidence as a mitigating factor are objectionable based on the form of the question as an attempt to bind the juror to a particular verdict. Soria v. State, 933 S.W.2d 46 (Tex. Crim. App. 1996).

§14:148 Preserving Error in Denial of Proper Inquiry

In a death penalty prosecution where jurors are examined individually, counsel must exercise a peremptory challenge on the specific veniremember in order to demonstrate harm where the trial court does not permit a proper question and thus denies the intelligent exercise of peremptory challenges. Janecka v. State, 739 S.W.2d 813 (Tex. Crim. App. 1987), *reversed on other grounds,* 823 S.W.2d 232 (Tex. Crim. App. 1990).

Voir dire error in a capital case can be error affecting the punishment phase only under CCP Art. 44.29(c); therefore, depending on the error the retrial could be for a punishment hearing only. Ransom v. State, 920 S.W.2d 288 (Tex. Crim. App. 1994).

§14:150 Challenges for Cause

CCP Art. 35.13. Passing juror for challenge

> *A juror in a capital case in which the state has made it known it will seek the death penalty, held to be qualified, shall be passed for acceptance or challenge first to the state and then to the defendant. Challenges to jurors are either peremptory or for cause.*

For a full discussion of the grounds for challenges for cause *see* §14:60.

§14:151 Excusing Veniremembers for Cause in Capital Cases

Potential jurors must be able to set aside their personal preferences and biases to consider death eligible all those defined as death eligible by §19.03 Penal Code and CCP Art. 37.071. Rachal v. State, 917 S.W.2d 799 (Tex. Crim. App. 1996) (plurality opinion).

A veniremember is challengeable for cause where his views against the death penalty are such that they would prevent or substantially impair the performance of his duties as a juror in accordance with his instructions and oath. Vuong v. State, 830 S.W.2d 929, 943 (Tex. Crim. App. 1992), *cert. denied*, 506 U.S. 997, 113 S.Ct. 595, 121 L. Ed. 2d 533 (1992); Wainwright v. Witt, 469 U.S. 412, 105 S.Ct. 844, 83 L. Ed. 2d 841 (1985). In other words, unless a veniremember is so irrevocably opposed to capital punishment that he could not follow the law or obey the instructions of the trial court, he is not disqualified from capital murder jury service. Riley v. State, 889 S.W.2d 290 (Tex. Crim. App. 1993).

The prosecution has the burden of demonstrating that because of the veniremember's views he will answer the special issues in a way that a life sentence rather than a death penalty will be imposed. Clark v. State, 929 S.W.2d 5 (Tex. Crim. App. 1996).

Veniremembers may not displace the broad legal categories of death eligibility with their personal agendas. Rachal, *supra*. Veniremembers who have stated an opposition to some phase of the law are nonetheless competent to serve as jurors where they state clearly that they are willing to temporarily set aside their own beliefs in deference to the rule of law. Ellis v. State, 726 S.W.2d 39 (Tex. Crim. App. 1986); Lockhart v. McCree, 476 U.S. 162, 106 S.Ct. 1758, 90 L. Ed. 2d 137 (1986). A veniremember may not be excluded from jury service based solely on his opinion of the death penalty when the record clearly demonstrates he is capable of following the law. Riley, *supra*.

If the veniremember equivocates or vacillates in his answers, it is within the sound discretion of the trial court to excuse or not depending on the observations of the court. Brown v. State, 913 S.W.2d 577 (Tex. Crim. App. 1996).

So long as a veniremember consistently affirms that he could in fact answer the special issues in accordance with the evidence, neither the difficulty he may have in doing so, nor the fact that it might violate his conscience, renders him a "vacillating" veniremember in any material sense. Therefore, in this situation a reviewing court will not defer to the findings of the trial court absent an abuse of discretion. Riley.

Where the veniremember states that he will distort his answers to reach a certain result he must be excused because of his inability to follow the law. Clark, *supra.* If a veniremember states that he will not consciously distort his answer to obtain a specific result (or to avoid one) he has not shown an inability to follow the law. Clark.

It is not enough that the veniremember's views on the death penalty might influence his decision making. Vuong, *supra.* Religious scruples against the death penalty in and of themselves are not sufficient to sustain a challenge for cause under the Vuong standard. Clark.

Deliberate failure to report a true verdict is a substantial impairment in any juror, and when supported by the record will authorize sustaining a challenge for cause. Rogers v. State, 774 S.W.2d 247 (Tex. Crim. App. 1989). If a juror is to obey his oath and follow the law, he must be willing not only to accept that in certain circumstances death is an acceptable penalty, but also to be able to answer the statutory questions without conscious distortion or bias. Gunter v. State, 858 S.W.2d 430 (Tex. Crim. App. 1993).

A refusal to take the juror's oath demonstrates that the veniremember's views prevent him from carrying out the duties of a juror. Gunter.

TIP TO THE PROSECUTION: In order to meet the burden under Vuong and Clark, the Court of Criminal Appeals has suggested that the state directly ask the veniremember whether his opposition to the death penalty is such as to cause him to answer one of the special issues in such a way as to ensure a life sentence, regardless of what the facts of the case may be.

A juror who would require more than the minimum quantum of evidence before answering the future dangerous special issue in the affirmative is not subject to a challenge for cause on this ground. Garrett v. State, 851 S.W.2d 853 (Tex. Crim. App. 1993).

§14:152 When to Make Challenge for Cause

A contemporaneous challenge for cause must be asserted against an objectionable veniremember in order to preserve error for appeal. Purtell v. State, 761 S.W.2d 360 (Tex. Crim. App. 1988). The challenge for cause generally must be made before the objectionable veniremember is dismissed and before the questioning of the next veniremember. Barefield v. State, 784 S.W.2d 38 (Tex. Crim. App. 1989).

CCP Art. 35.13 provides that in capital cases, the veniremembers shall be called individually, and upon completion of voir dire by both parties the veniremember is to be passed first to the state for acceptance, challenge for cause or peremptory challenge; and next to the defendant for acceptance, challenge for cause or peremptory challenge. Bigby v. State, 892 S.W.2d 864 (Tex. Crim. App. 1994), *cert. denied,* 515 U.S. 1102, 115 S.Ct. 2617, 132 L. Ed. 2d 860 (1995).

However, requiring the defendant to assert his challenge for cause before the state is required to accept the veniremember or exercise a peremptory challenge is not reversible error because the reversal of the order of challenges has no effect on the selection of jurors. Bigby. Where the challenge system employed ultimately requires the state to exercise its peremptory challenge on a veniremember before the defendant is called upon to exercise his, the defendant has not been harmed. Hughes v. State, 24 S.W.3d 833 (Tex. Crim. App. 2000).

Special circumstances may allow the use of a challenge for cause after the veniremember has been chosen as a juror. Jones v. State, 843 S.W.2d 487 (Tex. Crim. App. 1992), *cert. denied,* 507 U.S. 1035, 113 S.Ct. 1858, 123 L. Ed. 2d 479 (1993). These special circumstances may include that the trial court discovers that the juror may be challengeable for cause, excusable or otherwise disqualified. Draughon v. State, 831 S.W.2d 331 (Tex. Crim. App. 1992), *cert. denied,* 509 U.S. 926, 113 S.Ct. 3045, 125 L. Ed. 2d 730 (1993).

Where the entire jury has not yet been selected and no evidence received in trial of the cause, the judge is permitted general discretion to allow further examination and to entertain additional challenges when it comes to his attention that a previously selected juror may be objectionable for cause, excusable, or otherwise disqualified from jury service. Draughon.

§14:153 Challenges for Cause by Either Party

For an extensive list of grounds for challenges for cause in all types of criminal cases, *see* §14:68.1.

§14:153.1 Grounds Where Challenges Granted

The following have been recognized as proper areas for a challenge for cause by either party in a death penalty case. The defendant would cite CCP Art. 35.16(c)(2), and the state would cite CCP Art. 35.16(b)(1).

- Veniremember would not be able to reconsider the guilt phase evidence independently for purposes of resolving the special issues; Gardner v. State, 730 S.W.2d 675 (Tex. Crim. App. 1987), *cert. denied,* 484 U.S. 905, 108 S.Ct. 248, 98 L. Ed. 2d 206 (1987)
- Veniremember's inability to comprehend the limited function of a juror at the punishment phase of a capital case constitutes such a mental defect as to render him unfit for jury service; Matamoros v. State, 901 S.W.2d 470 (Tex. Crim. App. 1995); Gardner, *supra*.
- Veniremember could never find a murder committed during a robbery to be reasonable as a response to provocation, thus indicating a bias against the law with respect to special issue three; Garcia v. State, 919 S.W.2d 370 (Tex. Crim. App. 1994)
- Veniremember cannot differentiate between the terms "probability" and "possibility," because this renders the veniremember impaired in evaluating the evidence offered to prove future dangerousness; Patrick v. State, 906 S.W.2d 481 (Tex. Crim. App. 1995), *cert. denied,* 517 U.S. 406, 116 S.Ct. 1323, 134 L. Ed. 2d 475 (1996)
- Veniremember would automatically answer the first special issue in the affirmative; Caldwell v. State, 818 S.W.2d 790 (Tex. Crim. App. 1991), *cert. denied,* 503 U.S. 990, 112 S.Ct. 1684, 118 L. Ed. 2d 399 (1992)
- Veniremember would automatically answer the first two special issues in the affirmative, or who would place the burden of proof on the defense; Banda v. State, 890 S.W.2d 42 (Tex. Crim. App.1994), *cert. denied,* 515 U.S. 1105, 115 S.Ct. 2253, 132 L. Ed. 2d 260 (1995)
- Veniremember has a bias or prejudice against one of the special issues; Cumbo v. State, 760 S.W.2d 251 (Tex. Crim. App. 1988)
- A venireperson who says that she cannot answer honestly answer the special issues so that a death penalty results could not judge another person and did not think that she could follow the law in a capital case. Hernandez v. State, 390 S.W.3d 310, 321 (Tex. Crim. App. 2012).

§14:153.2 Grounds Where Challenges Not Granted

A veniremember is not challengeable for cause on the ground that he will not give mitigating effect to particular evidence. Lagrone v. State, 942 S.W.2d 602 (Tex. Crim. App. 1997), *cert. denied,* 522 U.S. 917, 118 S.Ct. 305, 139 L. Ed. 2d 235 (1997); Massey v. State, 933 S.W.2d 141 (Tex. Crim. App. 1996); Morrow v. State, 910 S.W.2d 471 (Tex. Crim. App. 1995), *cert. denied,* 517 U.S. 1192, 116 S.Ct. 1683, 134 L. Ed. 2d 784 (1996).

§14:154 Challenges for Cause by Defendant

CCP Art. 35.16(c)(2) permits the defendant to challenge for cause a veniremember on the ground "[t]hat he has a bias or prejudice against any of the law applicable to the case upon which the defendant is entitled to rely, either as a defense to some phase of the offense for which the defendant is being prosecuted or as a mitigation thereof or of the punishment therefor."

For an extensive list of grounds for challenges for cause by the defendant in all types of criminal cases, *see* §14:68.2.

§14:154.1 Grounds Where Challenges Will Be Granted

The following have been recognized as proper areas for defense challenges for cause in capital cases:

- Veniremember inability to consider the minimum punishment for a lesser included offense; Pierce v. State, 696 S.W.2d 899 (Tex. Crim. App. 1985)
- Veniremember inability to consider a life sentence as a possible punishment for capital murder; Cumbo v. State, 760 S.W.2d 251 (Tex. Crim. App. 1988)
- Veniremember inability to distinguish between deliberate and intentional conduct; Martinez v. State, 763 S.W.2d 413 (Tex. Crim. App. 1988)
- Veniremember inability to disregard parole as a factor in answering the special issues; Felder v. State, 758 S.W.2d 760 (Tex. Crim. App. 1988)
- Veniremember always answers the mitigation special issue in such a way that a death penalty would result; Staley v. State, 887 S.W.2d 885 (Tex. Crim. App. 1994), *cert. denied,* 514 U.S. 1020, 115 S.Ct. 1366, 131 L. Ed. 2d 222 (1995)

§14:154.2 Areas Where Challenges Will Not Be Granted

A veniremember is not challengeable for cause on the ground that he will not give mitigating effect to particular evidence. Lagrone v. State, 942 S.W.2d 602 (Tex. Crim. App. 1997), *cert. denied,* 522 U.S. 917, 118 S.Ct. 305, 139 L. Ed. 2d 235 (1997); Massey v. State, 933 S.W.2d 141 (Tex. Crim. App. 1996); Morrow v. State, 910 S.W.2d 471 (Tex. Crim. App. 1995), *cert. denied,* 517 U.S. 1192, 116 S.Ct. 1683, 134 L. Ed. 2d 784 (1996).

There is no law placing the burden of proof on the state as to the mitigation special issue, so a veniremember is not challengeable for cause simply because he would place the burden of proof on mitigation on the defense. Ladd v. State, 3 S.W.3d 547 (Tex. Crim. App. 1999); Barnes v. State, 876 S.W.2d 316 (Tex. Crim. App. 1994), *cert. denied*, 513 U.S. 861, 115 S.Ct. 174, 130 L. Ed. 2d 110 (1994).

A veniremember is not subject to a challenge for cause merely because he is unable to think of a situation in which he would answer the future dangerousness special issue in the negative. Pierce v. State, 777 S.W.2d 399 (Tex. Crim. App. 1989).

Similarly, a veniremember is not challengeable for cause simply because he cannot immediately envision a scenario in which the minimum punishment would be appropriate. The mere fact that a veniremember cannot envision such a scenario during voir dire does not indicate that he could not consider the minimum punishment if an appropriate scenario presented itself at trial. Ladd.

§14:154.3 Preserving Error for Denial of Defense Challenge

In order to preserve error in a trial court's erroneous denial of a defense challenge for cause, the defendant must follow a rigorous predicate showing that:

- Voir dire of the individual veniremember was recorded and transcribed; Long v. State, 823 S.W.2d 259 (Tex. Crim. App. 1991), *cert. denied,* 505 U.S. 1224, 112 S.Ct. 3042, 120 L. Ed. 2d 910 (1992)
- Defense asserted a clear and specific challenge for cause clearly articulating the grounds therefore; Miller v. State, 741 S.W.2d 382 (Tex. Crim. App. 1987), *cert. denied,* 486 U.S. 1061, 108 S.Ct. 2835, 100 L. Ed. 2d 935 (1988)
- Defense used a peremptory challenge on that particular veniremember; Jones v. State, 833 S.W.2d 118 (Tex. Crim. App. 1992), *cert. denied,* 507 U.S. 921, 113 S.Ct. 1285, 122 L. Ed. 2d 678 (1993)
- Defense exhausted all of his peremptory challenges; Adanandus v. State, 866 S.W.2d 210 (Tex. Crim. App. 1993), *cert. denied,* 510 U.S. 1215, 114 S.Ct. 1338, 127 L. Ed. 2d 686 (1994); Jacobs

v. State, 787 S.W.2d 397 (Tex. Crim. App. 1990), *cert. denied,* 496 U.S. 882, 111 S.Ct. 231, 112 L. Ed. 2d 185 (1990)

- Defense was forced to accept an objectionable veniremember on the jury because of the trial court's erroneous actions, *i.e*., that he would have exercised a peremptory challenge against that veniremember if he had one remaining; Trevino v. State, 991 S.W.2d 849 (Tex. Crim. App. 1999); Green v. State, 840 S.W.2d 394 (Tex. Crim. App. 1992), *cert. denied,* 507 U.S. 1020, 113 S.Ct. 1819, 123 L. Ed. 2d 449 (1993), *disavowed on other grounds.*

PRACTICE TIP: In order to preserve error, recite into the record prior to the jury being seated and sworn, that:

- Defense was forced to use a peremptory challenge on a juror against whom a challenge for cause had been made
- Defense exhausted all of his peremptory challenges
- Defense has requested additional peremptory challenges and that request has been denied (or if additional peremptory challenges were granted, that those also have been exhausted and an additional request has been made and denied)
- Objectionable juror has been seated and that he would exercise a peremptory challenge on that juror if afforded an additional peremptory. Counsel should then request an additional peremptory challenge and have the request denied

The defendant must demonstrate that he exercised a peremptory challenge on the veniremember upon which he bases his claim. Brosky v. State, 915 S.W.2d 120 (Tex.App.—Fort Worth , *pet. ref'd*), *cert. denied*, 117 S.Ct. 537 (1996).

In order to do this, the defendant bears the burden of including the jury strike lists as part of the record on appeal. Hernandez v. State, 938 S.W.2d 503 (Tex.App.—Waco 1997, *pet. ref'd*).

In order to make a certain claim on appeal, the defendant must have objected on the same ground in the trial court. Trevino v. State, 991 S.W.2d 849 (Tex. Crim. App. 1999); Green v. State, 840 S.W.2d 394 (Tex. Crim. App. 1992), *cert. denied,* 507 U.S. 1020, 113 S.Ct. 1819, 123 L. Ed. 2d 449 (1993), *disavowed on other grounds*. It is not necessary that the defendant go so far as to explain why the objectionable juror that he has named is, in fact, objectionable. Garcia v. State, 887 S.W.2d 846 (Tex. Crim. App. 1994), *cert. denied,* 514 U.S. 1005, 115 S.Ct. 1317, 131 L. Ed. 2d 198 (1995).

Where the trial court grants two additional peremptory challenges, the defendant must show that the trial court erred in denying his challenge for cause to three prospective jurors. Penry v. State, 903 S.W.2d 715 (Tex. Crim. App. 1995), *cert. denied,* 516 U.S. 977, 116 S.Ct. 480, 133 L. Ed. 2d 408 (1995).

Harm from the erroneous denial of a defense challenge occurs: (1) when a defendant exercises a peremptory challenge on a veniremember whom the trial court should have excused for cause at the defendant's request, (2) the defendant uses all of his statutorily allotted peremptory challenges, and (3) the defendant unsuccessfully requests an additional peremptory challenge which he claims he would use on another veniremember whom the defendant identifies as "objectionable" and who actually sits on the jury. When these conditions are met, this harms a defendant because he had to use a peremptory challenge to remove a veniremember who should have been removed for cause which, in effect, wrongfully deprives the defendant of one of his statutory peremptory challenges. Newbury v. State, 135 S.W.3d 22 (Tex. Crim. App. 2004)

Voir dire error in a capital case can be error affecting the punishment phase only under CCP Art. 44.29(c); therefore, depending on the error the retrial could be for a punishment hearing only. Ransom v. State, 920 S.W.2d 288 (Tex. Crim. App. 1994).

§14:155 Challenges for Causes by State in Capital Cases

CCP Art. 35.16(b)(3) allows the state to challenge for cause veniremembers who have a "bias or prejudice against any phase of the law upon which the state is entitled to rely for conviction or punishment."

In addition to the grounds listed here, *see* §14:68.3.

§14:155.1 Grounds Where Challenges Will Be Granted

The following have been recognized as valid reasons for the state to challenge a veniremember for cause in a capital case:

- Veniremember interprets the phrase "criminal acts of violence" to require that the state prove that the defendant has murdered before; Fuller v. State, 829 S.W.2d 191 (Tex. Crim. App. 1992), *cert. denied,* 508 U.S. 941, 113 S.Ct. 2418, 124 L. Ed. 2d 640 (1993); Drew v. State, 743 S.W.2d 207 (Tex. Crim. App. 1987).
- Veniremember would not be able to fairly consider the death penalty as a possible punishment; Long v. State, 823 S.W.2d 259 (Tex. Crim. App. 1991)
- Veniremember would require more than proof beyond a reasonable doubt because the state is seeking the death penalty; Chambers v. State, 866 S.W.2d 9 (Tex. Crim. App. 1993), *cert. denied,* 511 U.S. 1100, 114 S.Ct. 1871, 128 L. Ed. 2d 491 (1994)
- Veniremember is unable to consider the full range of punishment because of the youth of the defendant; Garcia v. State, 919 S.W.2d 370 (Tex. Crim. App. 1994)
- Veniremember is challengeable as having a bias or prejudice against a law upon which the state is entitled to rely if he would always answer the mitigating circumstances punishment issue, Art. 37.071, §2(e), in favor of the defendant; Wolfe v. State, 917 S.W.2d 270 (Tex. Crim. App. 1996)
- Likewise, a veniremember is challengeable under CCP Art. 35.16(b)(3) if he would never answer the future dangerousness issue in favor of the state. Wolfe.
- Veniremember acknowledges that violent property crimes could constitute a continuing threat to society, yet she required proof of violence against people. Such a position indicates that the veniremember would supplant the legal categories of death eligibility with her personal preferences and biases. Howard v. State, 941 S.W.2d 102 (Tex. Crim. App. 1996)
- Veniremember would require that the murder be premeditated before he could vote in such a way that the death penalty be assessed; Patrick v. State, 906 S.W.2d 481 (Tex. Crim. App. 1995), *cert. denied,* 517 U.S. 406, 116 S.Ct. 1323, 134 L. Ed. 2d 475 (1996)
- Veniremember would never vote in such a way that the death penalty would be assessed against a defendant who was a party to a capital murder; Selvage v. State, 680 S.W.2d 17 (Tex. Crim. App. 1984)
- Veniremember who would only consider the death penalty where one of his family members was the victim is not qualified; Fearance v. State, 771 S.W.2d 486 (Tex. Crim. App. 1988), *cert. denied,* 492 U.S. 927, 109 S.Ct. 3266, 106 L. Ed. 2d 611 (1989)
- Veniremember who will only consider the death penalty for a defendant who confessed to the crime is challengeable for cause; Monroe v. Blackburn, 748 F.2d 958 (5th Circ. 1984), *cert. denied,* 476 U.S. 1145, 106 S.Ct. 2261, 90 L. Ed. 2d 706 (1986)
- Veniremember who could only consider the death penalty for a person convicted of a murder for hire is challengeable for cause; Crane v. State, 786 S.W.2d 338 (Tex. Crim. App. 1990)
- Veniremember was properly removed for cause where he acknowledged that he could not resolve his religious conflicts with respect to the death penalty and as a consequence his performance as a juror would be substantially impaired; Cooks v. State, 844 S.W.2d 697 (Tex. Crim. App. 1992), *cert. denied,* 509 U.S. 927, 113 S.Ct. 3048, 125 L. Ed. 2d 732 (1993); Kinnamon v. State, 791

S.W.2d 84 (Tex. Crim. App. 1990), *overruled on other grounds*

- Veniremember ultimately guided by personal beliefs rather than the law and evidence is not qualified to sit on a jury; Castillo v. State, 739 S.W.2d 280 (Tex. Crim. App. 1987), *cert. denied,* 487 U.S. 1228, 108 S.Ct. 2889, 101 L. Ed. 2d 924 (1988)

§14:155.2 Grounds Where Challenges Will Not Be Granted

A veniremember is not subject to a challenge for cause by the state merely because he would require more evidence than the legal minimum in order to answer the future dangerousness special issue in the affirmative. Garrett v. State, 851 S.W.2d 853 (Tex. Crim. App. 1993).

A venireman who requires a particular type of evidence has not demonstrated an inability to abide by the law if his requirement is predicated upon his personal threshold of reasonable doubt. The state has the burden to show that the venireman's insistence on evidence of a particular type will prevent him from honestly answering the special issue regardless of whether he was otherwise convinced beyond a reasonable doubt of future dangerousness, before it can be said it has met its burden to demonstrate the venireman cannot follow the law. Howard v. State, 941 S.W.2d 102 (Tex. Crim. App. 1996); Murphy v. State, 112 S.W.3d 192 (Tex. Crim. App. 2003).

Similarly, a venireman whose definition of a particular term requires a certain type of evidence is not subject to a challenge for cause. Murphy.

It is not permissible for the prosecution to use the language of the Penal Code §12.31(b) oath to disqualify potential jurors; White v. State, 610 S.W.2d 504 (Tex. Crim. App. 1981); Adams v. Texas, 448 U.S. 38, 100 S.Ct. 2521, 65 L. Ed. 2d 581 (1980).

TIP TO THE PROSECUTION: A veniremember may state that he requires specific evidence above and beyond the facts of the case in order to return a death penalty sentence. Before you can excuse the veniremember for cause, it is incumbent on the state to show that the veniremember would refuse to answer the special circumstances questions in any manner in which the death penalty would be imposed, even if the other evidence is sufficient to prove beyond a reasonable doubt that the questions should be answered that way. In other words, the state must show that the veniremember's position is not merely a prediction of his attitude upon hearing the evidence, but is in fact, a categorical refusal to answer "yes" even if other evidence could convince him of the defendant's future dangerousness beyond a reasonable doubt. Howard v. State, 941 S.W.2d 102 (Tex. Crim. App. 1996).

It is the state's burden, as the challenging party, to show that the veniremember's refusal to return a death penalty is predicated upon something other than his understanding of proof beyond a reasonable doubt. Castillo v. State, 913 S.W.2d 529 (Tex. Crim. App. 1995).

§14:155.3 Preserving Error for Improperly Granted State's Challenge

The defendant preserves error in the trial court's improper granting of a state's challenge for cause where he lodges a specific objection before the veniremember is dismissed and before the questioning of the next veniremember begins. Barefield v. State, 784 S.W.2d 38 (Tex. Crim. App. 1989), *cert. denied*, 497 U.S. 1011, 110 S.Ct. 3256, 111 L. Ed. 2d 766 (1990).

An objection is untimely if it is made after that veniremember was excused and questioning begins on the next veniremember. Fuller v. State, 827 S.W.2d 919 (Tex. Crim. App. 1992), *cert. denied,* 509 U.S. 922, 113 S.Ct. 3035, 125 L. Ed. 2d 722 (1993).

In noncapital cases, for a reversal, the defendant must show that he was deprived of a lawfully constituted jury. However, in death penalty cases if a veniremember is wrongfully excused on a state's challenge for cause because of a general opposition to the death penalty, the exclusion of a juror by an unintentional mistake amounts to a constitutional violation requiring reversal. Jones v. State, 982

S.W.2d 386 (Tex. Crim. App. 1998), *cert. denied,* 120 S.Ct. 444, 145 L. Ed. 2d 362 (1999).

However, where the reason for the improper excusal of the venireman does not implicate *Witherspoon* or *Wainwright* error, the error is harmless. Murphy v. State, 112 S.W.3d 192 (Tex. Crim. App. 2003). (the venireman was improperly excused because her own definition of the phrase "criminal acts of violence" required evidence that the defendant committed or attempted to commit other murders).

Voir dire error in a capital case can be error affecting the punishment phase only under CCP Art. 44.29(c); therefore, depending on the error, the retrial could be for a punishment hearing only. Ransom v. State, 920 S.W.2d 288 (Tex. Crim. App. 1994).

§14:160 Veniremembers Excused by Agreement

CCP Art. 35.05. Excused by Consent

> *One summoned upon a special venire may by consent of both parties be excused from attendance by the court at any time before he is impaneled.*

The court may excuse a special veniremember from attendance by consent of both parties at any time before the veniremember is empaneled. CCP Art. 35.05.

§14:170 Peremptory Challenges in Death Penalty Cases

In capital cases, it is customary for peremptory challenges to be made after each veniremember has been individually questioned. CCP Art. 35.13.

The jury selection system set by the Code of Criminal Procedure requiring death penalty defendants to exercise peremptory challenges following examination of each individual juror, as compared to the noncapital case system in which peremptory challenges are exercised after all prospective jurors have been examined, does not violate state or federal due process or equal protection provisions. Janecka v. State, 739 S.W.2d 813 (Tex. Crim. App. 1987), *reversed on other grounds,* 823 S.W.2d 232 (Tex. Crim. App. 1990).

The procedure controlling the order and timing of the exercise of peremptory challenges is not an absolute requirement. Busby v. State, 990 S.W.2d 263 (Tex. Crim. App. 1999), *cert. denied*, 528 U.S. 1081, 120 S.Ct. 803, 145 L.Ed.2d 676 (2000).

Where the challenge system employed ultimately requires the state to exercise its peremptory challenge on a veniremember before the defendant is called upon to exercise his, the defendant has not been harmed. Hughes v. State, 24 S.W.3d 833 (Tex. Crim. App. 2000).

The practice of withholding peremptory challenges until a qualified pool of veniremembers is established has also been approved. Rousseau v. State, 824 S.W.2d 579 (Tex. Crim. App. 1992), *cert. denied,* 510 U.S. 919, 114 S.Ct. 313, 126 L. Ed. 2d 260 (1993).

However, the trial court does not violate a defendant's Sixth Amendment right to an impartial jury and his Fourteenth Amendment rights to due process and equal protection by denying his motion to postpone the exercise of peremptory challenges until the completion of voir dire. Ladd v. State, 3 S.W.3d 547 (Tex. Crim. App. 1999); Dowthitt v. State, 931 S.W.2d 244 (Tex. Crim. App.1996).

The trial court does not err in denying the defendant's request to make an out-of-time peremptory challenge. Beavers v. State, 856 S.W.2d 429 (Tex. Crim. App. 1993), *cert. denied,* 510 U.S. 951, 114 S.Ct. 399, 126 L. Ed. 2d 346 (1993).

The trial court does not err in allowing the state to use a peremptory challenge against the preceding juror after it has completed its voir dire examination of the succeeding juror, but before the defense has commenced. Barnard v. State, 730 S.W.2d 703 (Tex. Crim. App. 1987), *cert. denied,* 485 U.S. 929, 108 S.Ct. 1098, 99 L. Ed. 2d 261 (1988); *see also*, Franklin v. State, 693 S.W.2d 420 (Tex. Crim. App. 1985), *cert. denied,* 475 U.S. 1031, 106 S.Ct. 1238, 89 L. Ed. 2d 346 (1986).

The defendant must object in a timely fashion in order to preserve error where the state has been allowed to retroactively exercise a peremptory challenge. Fearance v. State, 771 S.W.2d 486 (Tex. Crim. App.1988), *cert. denied*, 492 U.S. 927, 109 S.Ct. 3266, 106 L. Ed. 2d 611 (1989).

A defendant's constitutional rights are not violated when he chooses to use a peremptory challenge against a venireman that the trial court should have excused for cause because peremptory challenges are not of federal constitutional dimension. U.S. v. Martinez-Salazar, 528 U.S. 304, 120 S.Ct. 774, 145 L.Ed.2d 792 (2000).

(This page intentionally left blank.)